Principles of Anatomy and Physiology

Eleventh Edition

Gerard J. Tortora
Bergen Community College

Bryan Derrickson
Valencia Community College

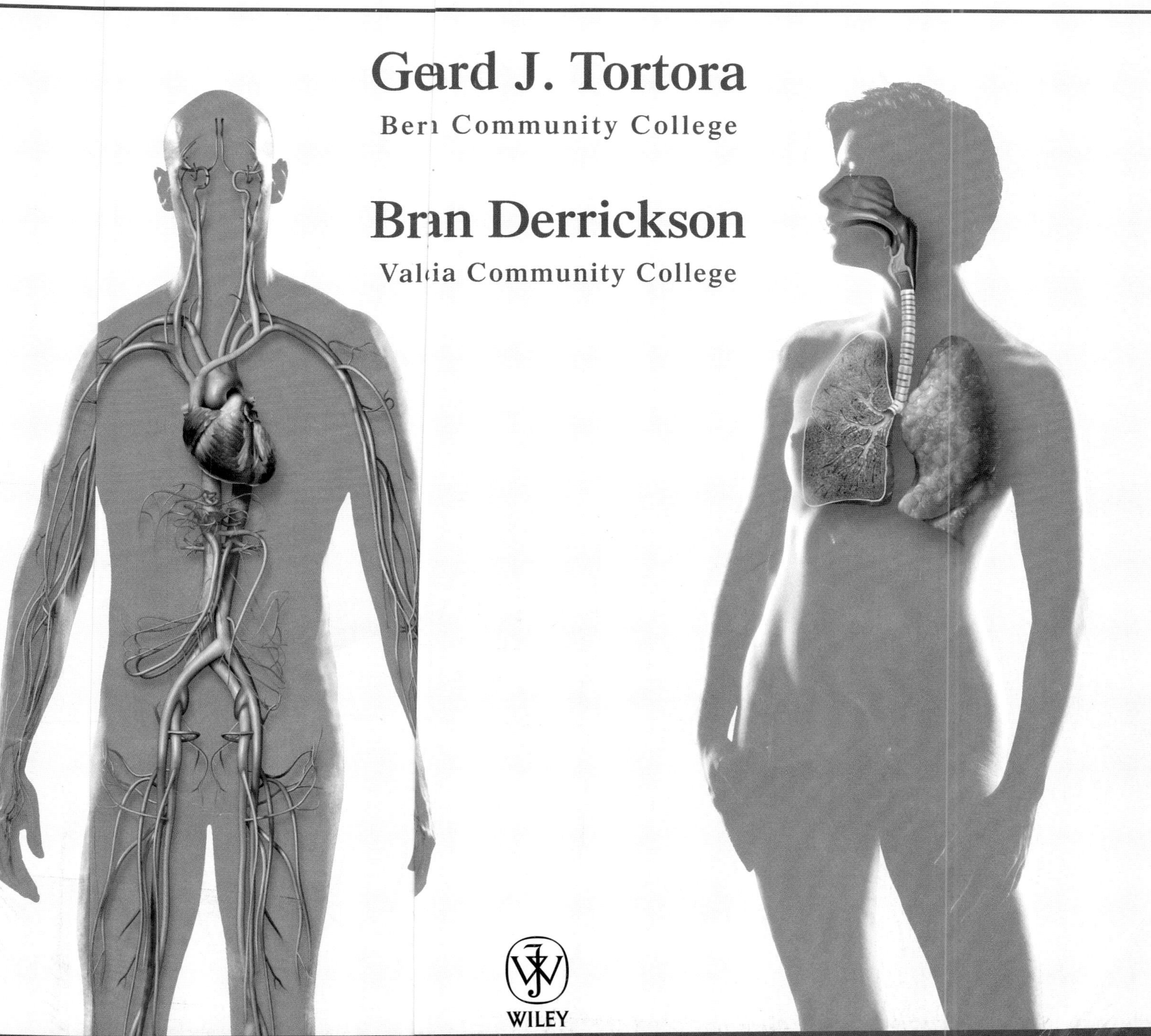

WILEY

John Wiley & Sons, Inc.

Executive Editor	Bonnie Roesch
Executive Marketing Manager	Clay Stone
Developmental Editor	Karen Trost
Associate Production Manager	Kelly Tavares
Text and Cover Designer	Karin Gerdes Kincheloe
Art Coordinator	Claudia Durrell
Photo Editor	Hilary Newman
Chapter Opener Illustrations	Keith Kasnot
Cover photos:	© Science Photo Library/Photo Researchers

Illustration and photo credits follow the Glossary.

This book was typeset by Progressive Information Technologies. It was printed and bound by Von Hoffmann Press, Inc. The cover was also printed by Von Hoffman Press, Inc.

The paper in this book was manufactured by a mill whose forest management programs include sustained yield harvesting of its timberlands. Sustained yield harvesting principles ensure that the number of trees cut each year does not exceed the amount of new growth.

This book is printed on acid-free paper.

This title published by John Wiley & Sons, Inc.

To order books or for customer service please, call 1(800)-CALL-WILEY (225-5945).

USA ISBN-10: 0-471-68934-3
WIE ISBN-10: 0471-71871-8
ISBN-13: 978-0-471-68934-3

Printed in the United States of America.

10 9 8 7 6 5 4 3

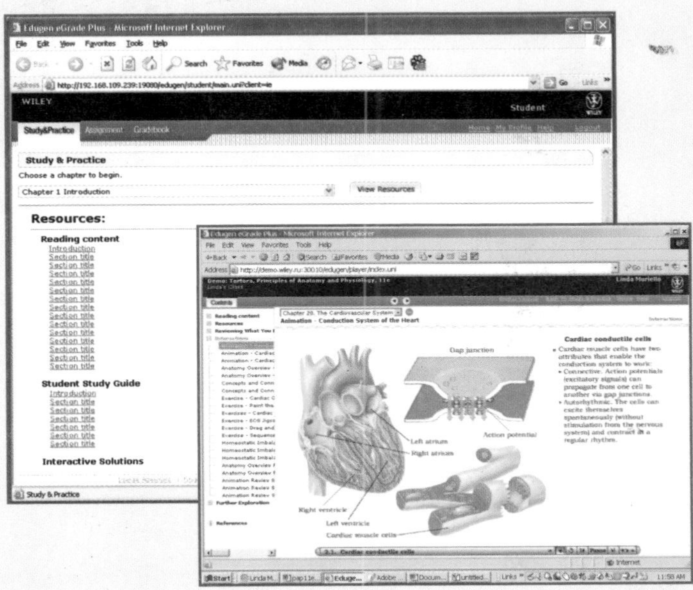

Study More Effectively

Get Immediate Feedback When You Practice on Your Own

eGrade Plus integrates everything you need to succeed with the relevant sections of the **electronic book content,** so that you can review the text while you study online. Resources include **animations**, **interactive exercises**, quizzes, flash cards, Web links and much more.

Complete Assignments / Get Help with Problem Solving

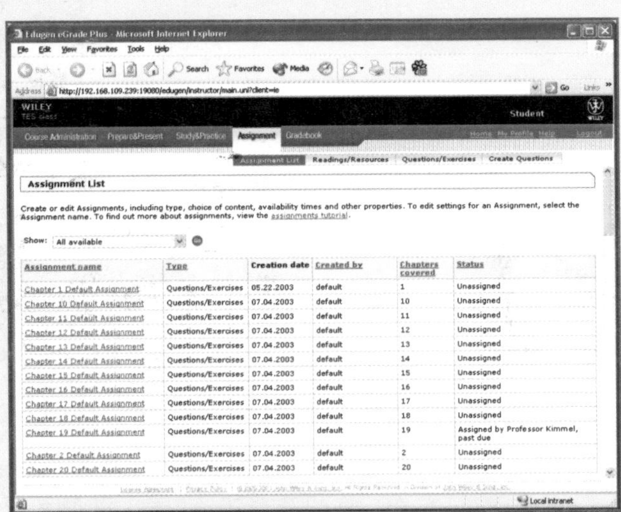

Assignment name	Type	Creation date	Created by	Chapters covered	Status
Chapter 1 Default Assignment	Questions/Exercises	05.22.2003	default	1	Unassigned
Chapter 10 Default Assignment	Questions/Exercises	07.04.2003	default	10	Unassigned
Chapter 11 Default Assignment	Questions/Exercises	07.04.2003	default	11	Unassigned
Chapter 12 Default Assignment	Questions/Exercises	07.04.2003	default	12	Unassigned
Chapter 13 Default Assignment	Questions/Exercises	07.04.2003	default	13	Unassigned
Chapter 14 Default Assignment	Questions/Exercises	07.04.2003	default	14	Unassigned
Chapter 15 Default Assignment	Questions/Exercises	07.04.2003	default	15	Unassigned
Chapter 16 Default Assignment	Questions/Exercises	07.04.2003	default	16	Unassigned
Chapter 17 Default Assignment	Questions/Exercises	07.04.2003	default	17	Unassigned
Chapter 18 Default Assignment	Questions/Exercises	07.04.2003	default	18	Unassigned
Chapter 19 Default Assignment	Questions/Exercises	07.04.2003	default	19	Assigned by Professor Kimmel, past due
Chapter 2 Default Assignment	Questions/Exercises	07.04.2003	default	2	Unassigned
Chapter 20 Default Assignment	Questions/Exercises	07.04.2003	default	20	Unassigned

An **Assignment** area keeps all of your assigned work in one location, making it easy for you to stay "on task." In addition, all pre-lecture quizzes contain a **link** to the relevant section of the **electronic book,** providing you with a text explanation to help you conquer problem-solving obstacles as they arise.

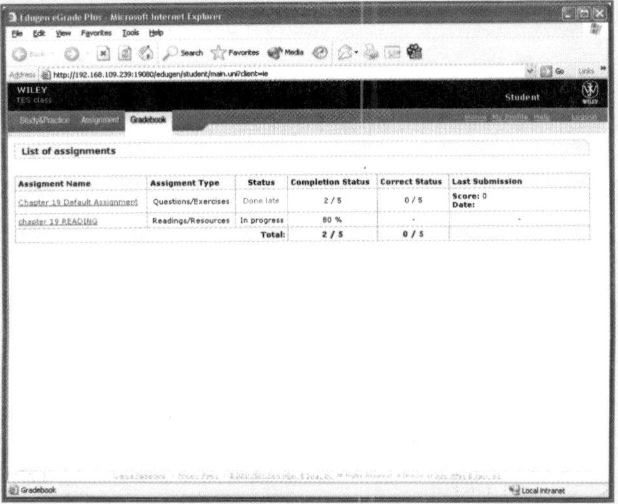

Assigment Name	Assigment Type	Status	Completion Status	Correct Status	Last Submission
Chapter 19 Default Assignment	Questions/Exercises	Done late	2 / 5	0 / 5	Score: 0 Date:
chapter 19 READING	Readings/Resources	In progress	80 %	-	
		Total:	2 / 5	0 / 5	

Keep Track of How You're Doing

A **Personal Gradebook** allows you to view your results from past assignments at any time.

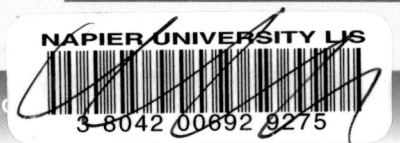

About the Authors

Gerard J. Tortora is Professor of Biology and former Coordinator at Bergen Community College in Paramus, New Jersey, where he teaches human anatomy and physiology as well as microbiology. He received his bachelor's degree in biology from Fairleigh Dickinson University and his master's degree in science education from Montclair State College. He is a member of many professional organizations, such as the Human Anatomy and Physiology Society (HAPS), the American Society of Microbiology (ASM), American Association for the Advancement of Science (AAAS), National Education Association (NEA), and the Metropolitan Association of College and University Biologists (MACUB).

Above all, Jerry is devoted to his students and their aspirations. In recognition of this commitment, Jerry was the recipient of MACUB's 1992 President's Memorial Award. In 1996, he received a National Institute for Staff and Organizational Development (NISOD) excellence award from the University of Texas and was selected to represent Bergen Community College in a campaign to increase awareness of the contributions of community colleges to higher education.

Jerry is the author of several best-selling science textbooks and laboratory manuals, a calling that often requires an additional 40 hours per week beyond his teaching responsibilities. Nevertheless, he still makes time for four or five weekly aerobic workouts that include biking and running. He also enjoys attending college basketball and professional hockey games and performances at the Metropolitan Opera House.

To my children, Lynne, Gerard, Kenneth, Anthony, and Andrew, who make it all worthwhile.
— G.J.T.

Bryan Derrickson is Professor of Biology at Valencia Community College in Orlando, Florida, where he teaches human anatomy and physiology as well as general biology and human sexuality. He received his bachelor's degree in biology from Morehouse College and his Ph.D. in Cell Biology from Duke University. Bryan's study at Duke was in the Physiology Division within the Department of Cell Biology, so while his degree is in Cell Biology his training focused on physiology. At Valencia, he frequently serves on faculty hiring committees. He has served as a member of the Faculty Senate, which is the governing body of the college, and as a member of the Faculty Academy Committee (now called the Teaching and Learning Academy), which sets the standards for the acquisition of tenure by faculty members. Nationally, he is a member of the Human Anatomy and Physiology Society (HAPS) and the National Association of Biology Teachers (NABT).

Bryan has always wanted to teach. Inspired by several biology professors while in college, he decided to pursue physiology with an eye to teaching at the college level. He is completely dedicated to the success of his students. He particularly enjoys the challenges of his diverse student population, in terms of their age, ethnicity, and academic ability, and finds being able to reach all of them, despite their differences, a rewarding experience. His students continually recognize Bryan's efforts and care by nominating him for a campus award known as the "Valencia Professor Who Makes Valencia A Better Place To Start." Bryan has received this award three times in the past 5 years.

Preface

An anatomy and physiology course can be the gateway to a gratifying career in a host of health-related professions. As active teachers of the course, we recognize both the rewards and challenges in providing a strong foundation for understanding the complexities of the human body to an increasingly diverse population of students. The eleventh edition of *Principles of Anatomy and Physiology* continues to offer a balanced presentation of content under the umbrella of our primary and unifying theme of homeostasis, supported by relevant discussions of disruptions to homeostasis. In addition, years of student feedback have convinced us that readers learn anatomy and physiology more readily when they remain mindful of the relationship between structure and function. As a writing team—an anatomist and a physiologist—our very different specializations offer practical advantages in fine-tuning the balance between anatomy and physiology.

Most importantly, our students continue to remind us of their needs for—and of the power of—simplicity, directness, and clarity. To meet these needs each chapter has been written and revised to include:

- clear, compelling, and up-to-date discussions of anatomy and physiology
- expertly executed and generously sized art
- classroom-tested pedagogy
- outstanding student study support.

As we revised the content for this edition we kept our focus on these important criteria for success in the anatomy and physiology classroom and have refined or added new elements to enhance the teaching and learning process.

HOMEOSTASIS: A UNIFYING THEME

The dynamic physiological constancy known as homeostasis is the prime theme in *Principles of Anatomy and Physiology*. We immediately introduce this unifying concept in Chapter 1 and describe how various feedback mechanisms work to maintain physiological processes within the narrow range that is compati-

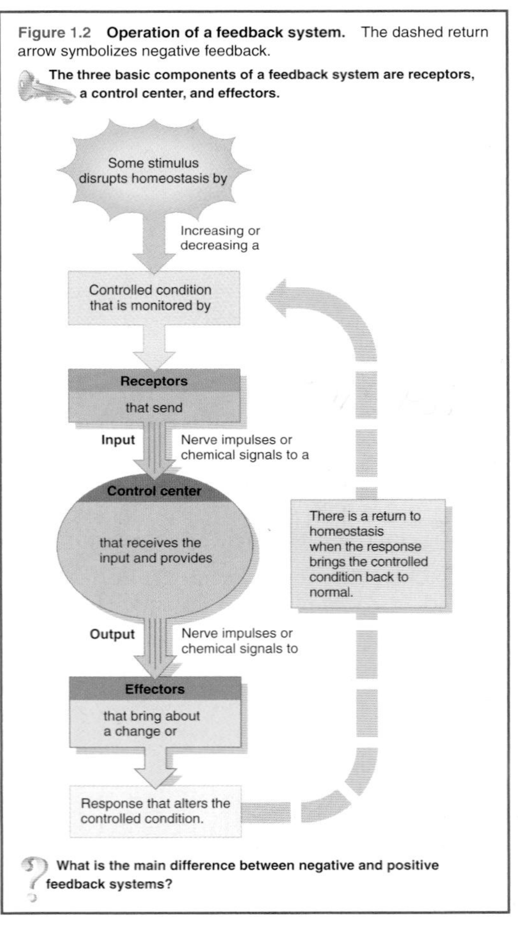

Some stimulus disrupts homeostasis by
Increasing or decreasing a
Controlled condition that is monitored by
Receptors
that send
Input
Nerve impulses or chemical signals to a
Control center
that receives the input and provides
There is a return to homeostasis when the response brings the controlled condition back to normal.
Output
Nerve impulses or chemical signals to
Effectors
that bring about a change or
Response that alters the controlled condition.

Figure 1.2 Operation of a feedback system. The dashed return arrow symbolizes negative feedback.

The three basic components of a feedback system are receptors, a control center, and effectors.

What is the main difference between negative and positive feedback systems?

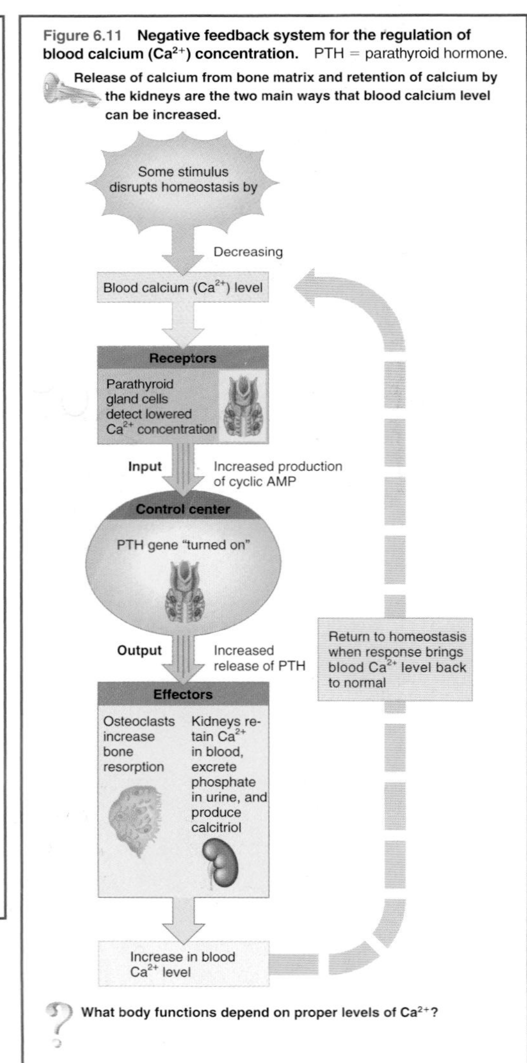

Some stimulus disrupts homeostasis by
Decreasing
Blood calcium (Ca^{2+}) level
Receptors
Parathyroid gland cells detect lowered Ca^{2+} concentration
Input
Increased production of cyclic AMP
Control center
PTH gene "turned on"
Output
Increased release of PTH
Return to homeostasis when response brings blood Ca^{2+} level back to normal
Effectors
Osteoclasts increase bone resorption
Kidneys retain Ca^{2+} in blood, excrete phosphate in urine, and produce calcitriol
Increase in blood Ca^{2+} level

Figure 6.11 Negative feedback system for the regulation of blood calcium (Ca^{2+}) concentration. PTH = parathyroid hormone.

Release of calcium from bone matrix and retention of calcium by the kidneys are the two main ways that blood calcium level can be increased.

What body functions depend on proper levels of Ca^{2+}?

ble with life. Homeostatic mechanisms are discussed throughout the book, and homeostatic processes are clarified through our well-received series of homeostasis feedback illustrations. Most chapters are bookended by content to reinforce this important concept. New for the eleventh edition, each chapter-opening page is designed with a brief statement linking the content of the chapter with its main contributions to overall body homeostasis. This is further reinforced by the ten *Focus on Homeostasis* features (one each for the integumentary, skeletal, muscular, nervous, endocrine, cardiovascular, lymphatic and immune, respiratory, digestive, and urinary systems) that were so well-received when we introduced them in the tenth edition. Incorporating both graphic and narrative elements, these pages explain, clearly and succinctly, how the system under consideration contributes to the homeostasis of each of the other body systems. Use of this feature will enhance student understanding of the links between body systems and how interactions among systems contribute to the homeostasis of the body as a whole.

In addition, we believe students can better understand normal physiological processes by examining situations in which diseases or disorders impair those processes. We offer three features that highlight these disruptions to homeostasis.

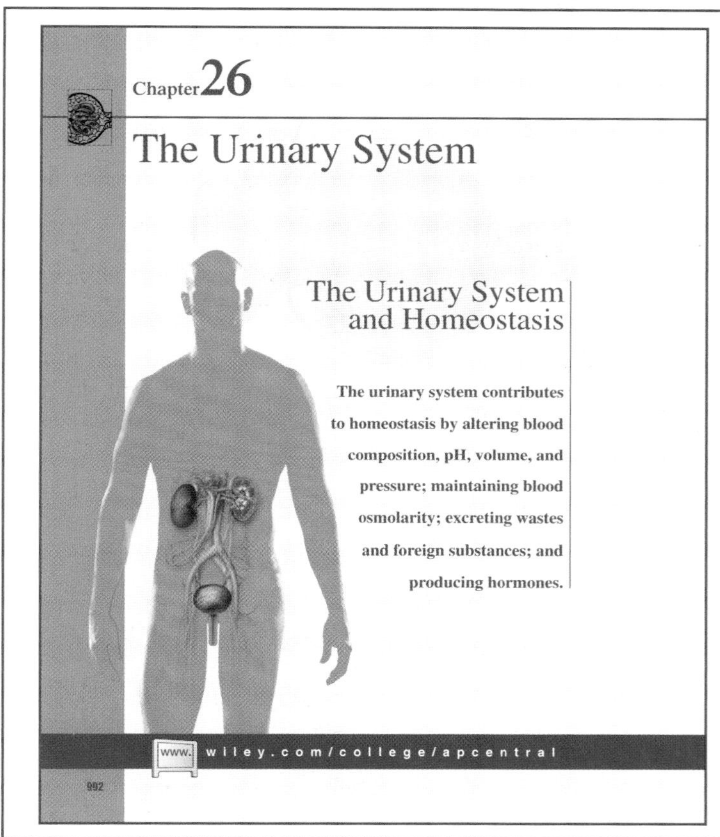
Chapter 26

The Urinary System

The Urinary System and Homeostasis

The urinary system contributes to homeostasis by altering blood composition, pH, volume, and pressure; maintaining blood osmolarity; excreting wastes and foreign substances; and producing hormones.

www. wiley.com/college/apcentral

992

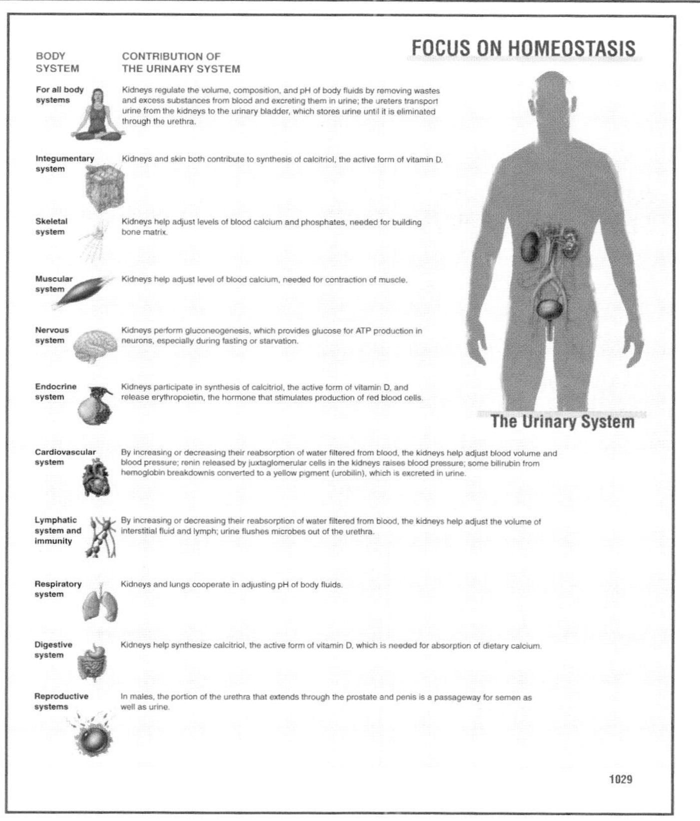
FOCUS ON HOMEOSTASIS

BODY SYSTEM	CONTRIBUTION OF THE URINARY SYSTEM
For all body systems	Kidneys regulate the volume, composition, and pH of body fluids by removing wastes and excess substances from blood and excreting them in urine; the ureters transport urine from the kidneys to the urinary bladder, which stores urine until it is eliminated through the urethra.
Integumentary system	Kidneys and skin both contribute to synthesis of calcitriol, the active form of vitamin D.
Skeletal system	Kidneys help adjust levels of blood calcium and phosphates, needed for building bone matrix.
Muscular system	Kidneys help adjust level of blood calcium, needed for contraction of muscle.
Nervous system	Kidneys perform gluconeogenesis, which provides glucose for ATP production in neurons, especially during fasting or starvation.
Endocrine system	Kidneys participate in synthesis of calcitriol, the active form of vitamin D, and release erythropoietin, the hormone that stimulates production of red blood cells.
Cardiovascular system	By increasing or decreasing their reabsorption of water filtered from blood, the kidneys help adjust blood volume and blood pressure; renin released by juxtaglomerular cells in the kidneys raises blood pressure; some bilirubin from hemoglobin breakdownis converted to a yellow pigment (urobilin), which is excreted in urine.
Lymphatic system and immunity	By increasing or decreasing their reabsorption of water filtered from blood, the kidneys help adjust the volume of interstitial fluid and lymph; urine flushes microbes out of the urethra.
Respiratory system	Kidneys and lungs cooperate in adjusting pH of body fluids.
Digestive system	Kidneys help synthesize calcitriol, the active form of vitamin D, which is needed for absorption of dietary calcium.
Reproductive systems	In males, the portion of the urethra that extends through the prostate and penis is a passageway for semen as well as urine.

The Urinary System

1029

A perennial favorite among students, the intriguing **Clinical Applications** in every chapter explore the clinical, professional, or everyday relevance of a particular anatomical structure or its related function. Many are new to this edition, and all have been reviewed for accuracy and relevance. Each application directly follows the discussion to which it relates.

Regeneration of Heart Cells

As noted earlier in the chapter, the heart of a heart attack survivor often has regions of infarcted (dead) cardiac muscle tissue that typically are replaced with noncontractile fibrous scar tissue over time. Our inability to repair damage from a heart attack has been attributed to a lack of stem cells in cardiac muscle and to the absence of mitosis in mature cardiac muscle fibers. A recent study of heart transplant recipients by American and Italian scientists, however, provides evidence for significant replacement of heart cells. The researchers studied men who had received a heart from a female, and then looked for the presence of a Y chromosome in heart cells. (All female cells except gametes have two X chromosomes and lack the Y chromosome.) Several years after the transplant surgery, between 7% and 16% of the heart cells in the transplanted tissue, including cardiac muscle fibers and endothelial cells in coronary arterioles and capillaries, had been replaced by the recipient's own cells, as evidenced by the presence of a Y chromosome. The study also revealed cells with some of the characteristics of stem cells in both transplanted hearts and control hearts. Evidently, stem cells can migrate from the blood into the heart and differentiate into functional muscle and endothelial cells. The hope is that researchers can learn how to "turn on" such regeneration of heart cells to treat people with heart failure or cardiomyopathy (diseased heart). ■

The **Disorders: Homeostatic Imbalances** sections at the end of most chapters include concise discussions of major diseases and disorders that illustrate departures from normal homeostasis. They provide answers to many questions that students ask about medical problems. Vocabulary-building glossaries of selected **Medical Terminology** and conditions also appear at the end of appropriate chapters. This feature has been expanded and updated for this edition.

542 CHAPTER 15 • THE AUTONOMIC NERVOUS SYSTEM

DISORDERS: HOMEOSTATIC IMBALANCES

Raynaud's Phenomenon

In **Raynaud's phenomenon** (rā-NŌZ) the digits (fingers and toes) become ischemic (lack blood) after exposure to cold or with emotional stress. The condition is due to excessive sympathetic stimulation of smooth muscle in the arterioles of the digits and a heightened response to stimuli that cause vasoconstriction. When arterioles in the digits vasoconstrict in response to sympathetic stimulation, blood flow is greatly diminished. As a result, the digits may blanc… cyanotic (look b… cases, the digits… With rewarming… fingers and toes… non have low b… adrenergic recep… occurs more ofte… should avoid ex… and feet warm. D… channel blocker… relaxes smooth r… of alcohol or illic…

Autonomic D…

Autonomic dy… response of the…

85% of individuals with spinal cord injury at or above the level of T6. The condition is seen after recovery from spinal shock (see page 467) and occurs due to interruption of the control of ANS neurons by higher centers. When certain sensory impulses, such as those resulting from stretching of a full urinary bladder, are unable to ascend the spinal cord, mass stimulation of the sympathetic nerves inferior to the level of injury occurs. Other triggers include stimulation of pain…

MEDICAL TERMINOLOGY

Autonomic nerve neuropathy (noo-ROP-a-thē) If a neuropathy (specifically a disorder of a cranial or spinal nerve) affects one or more autonomic nerves, there can be multiple effects on the autonomic nervous system that interfere with reflexes. These include fainting and low blood pressure when standing (orthostatic hypotension) due to decreased sympathetic control of the cardiovascular system, constipation, urinary incontinence, and impotence. This type of neuropathy is often caused by long-term diabetes mellitus and is known as ***diabetic retinopathy.***

Biofeedback A technique in which an individual is provided with information regarding an autonomic response such as heart rate, blood pressure, or skin temperature. Various electronic monitoring devices provide visual or auditory signals about the autonomic responses. By concentrating on positive thoughts, individuals learn to alter autonomic responses. For example, biofeedback has been used to decrease heart rate and blood pressure and increase skin temperature in order to decrease the severity of migraine headaches.

Dysautonomia (dis-aw-tō-NŌ-mē-a; *dys-* = difficult; *autonomia* = …

a visceral organ (such as the urinary bladder or colon) below the level of the injury results in intense activation of autonomic and somatic output from the spinal cord as reflex activity returns. The exaggerated response occurs because there is no inhibitory input from the brain. The mass reflex consists of flexor spasms of the lower limbs, evacuation of the urinary bladder and colon, and profuse sweating below the level of the lesion.

Megacolon (*mega-* = big) An abnormally large colon. In congenital megacolon, parasympathetic nerves to the distal segment of the colon do not develop properly. Loss of motor function in the segment causes massive dilation of the normal proximal colon. The condition results in extreme constipation, abdominal distension, and occasionally, vomiting. Surgical removal of the affected segment of the colon corrects the disorder.

Reflex sympathetic dystrophy (RSD) A syndrome that includes spontaneous pain, painful hypersensitivity to stimuli such as light touch, and excessive coldness and sweating in the involved body part. The disorder frequently involves the forearms, hands, knees, …

ORGANIZATION, SPECIAL TOPICS, AND CONTENT IMPROVEMENTS

The book follows the same unit sequence as its ten earlier editions, but the order of presentation of topics has changed slightly. It is divided into five principal sections: Unit 1, "Organization of the Human Body," provides an understanding of the structural and functional levels of the body, from molecules to organ systems. Unit 2, "Principles of Support and Movement," analyzes the anatomy and physiology of bones, joints, and muscles. Unit 3, "Control Systems of the Human Body," emphasizes the importance of neural communication in the immediate maintenance of homeostasis, the role of sensory receptors in providing information about the internal and external environments, and the significance of hormones in maintaining long-term homeostasis. Unit 4, "Maintenance of the Human Body," explains how body systems function to maintain homeostasis on a moment-to-moment basis through the processes of circulation, respiration, digestion, cellular metabolism, urinary functions, and buffer systems. Unit 5, "Continuity," covers the anatomy and physiology of the reproductive systems, development, and the basic concepts of genetics and inheritance.

Development We often tell our students that they can better appreciate the "logic" of human anatomy by becoming aware of how various structures developed in the first place. As in previous editions, illustrated discussions of development are found near the conclusion of most body system chapters. Placing this coverage at the end of chapters enables students to master the anatomical terminology they need before attempting to learn about embryonic and fetal structures. The fetus icon designates the start of each development discussion.

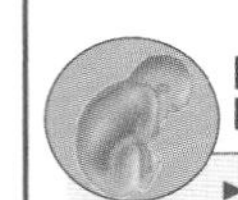

DEVELOPMENT OF THE NERVOUS SYSTEM

▶ OBJECTIVE

Describe how the parts of the brain develop.

Development of the nervous system begins in the third week of gestation with a thickening of the **ectoderm** called the **neural plate** (Figure 14.28). The plate folds inward and forms a longitudinal groove, the **neural groove.** The raised edges of the neural plate are called **neural folds.** As development continues, the neural folds increase in height and meet to form a tube called …

AGING AND BONE TISSUE

▶ OBJECTIVE

Describe the effects of aging on bone tissue.

From birth through adolescence, more bone tissue is produced than is lost during bone remodeling. In young adults the rates of bone deposition and resorption are about the same. As the level of sex hormones diminishes during middle age, especially in women after menopause, a decrease in bone mass occurs because bone resorption by osteoblasts outpaces bone deposition by osteoblasts. …

Aging Students need to be reminded from time to time that anatomy and physiology is not static. As the body ages, its structures and related functions subtly change. Moreover, aging is a professionally relevant topic for the majority of this book's readers, who will go on to careers in *health-related* fields in which the average age of the client population is steadily advancing. For these reasons, age-related changes in anatomy and physiology are discussed at the end of fifteen chapters.

EXERCISE AND THE HEART

▶ OBJECTIVE

Explain the relationship between exercise and the heart.

Regardless of the current level, a person's cardiovascular fitness can be improved at any age with regular exercise. Some types of exercise are more effective than others for improving the health of the cardiovascular system. **Aerobics,** any activity that works large body muscles for at least 20 minutes, elevates cardiac output and accelerates metabolic rate. Three to five such sessions a week are usually recommended for improving the health of the cardiovascular system. Brisk walking, running, bicycling, cross-country skiing, and swimming are examples of aerobic activities.

Exercise Physical exercise can produce favorable changes in some anatomical structures and enhance many physiological functions, most notably those associated with the muscular, skeletal, and cardiovascular systems. This information is especially relevant to readers embarking on careers in physical education, sports training, and dance. Hence, key chapters include brief discussions of exercise-related considerations, which are signaled by a distinctive running shoe icon.

Every chapter in the eleventh edition of *Principles of Anatomy and Physiology* incorporates a host of improvements to both the text and the art, many suggested by reviewers, educators, and students. In addition, most chapters offer new Clinical Applications. Here are some of the more noteworthy changes.

Chapter 1 An Introduction to the Human Body

Blood clotting is now included as a new example of positive feedback. Figures 1.1, 1.5, 1.7, and 1.9 have been redrawn, and Figures 1.9, 1.10, and 1.12 have been expanded. Table 1.3 (Common Medical Imaging Procedures) has also been expanded to include mammography, bone densitometry, contrast x rays, radionuclide scanning, and endoscopy

Chapter 2 The Chemical Level of Organization

The sections on saturated, monounsaturated, and polyunsaturated fats have undergone extensive revision. Figure 2.22 has been redrawn, and there is a new Clinical Application on fatty acids in health and disease.

Chapter 3 The Cellular Level of Organization

In the section on control of cell destiny there is new information on cyclin-dependent protein kinases and cyclins. The discussion of plasma membrane transport has been reorganized into two sections, one on kinetic energy transport and one on transport by transporter proteins. Reproductive cell division has been moved from Chapter 28 to Chapter 3 to bring it closer to the discussion of mitosis. Figures 3.7, 3.14, 3.15, and 3.30 have undergone changes in this edition, and two new Clinical Applications appear: smooth ER and drug tolerance, and mitotic spindle and cancer.

Chapter 4 The Tissue Level of Organization

There has been some reclassification of epithelial tissues in this chapter, and there is a new section on excitable cells. Figures 4.3 and 4.7 are new. Tables 4.1 and 4.3 have been expanded, and all table art has been redrawn. The chapter also features a new Clinical Application on basement membranes and disease, and a revised Medical Terminology listing.

Chapter 5 The Integumentary System

The skin chapter includes a new section on tattooing and body piercing and new coverage of cosmetic anti-aging treatments in the section on aging. In addition, coverage of the development of the integumentary system has been increased. Figures 5.6 and 5.7 are new to this chapter, Figures 5.1 and 5.9 have been expanded, and Figure 5.3 has undergone extensive revision. New Clinical Applications include lines of cleavage and surgery; chemotherapy and hair loss; and sun damage, sunscreens and sunblocks.

Chapter 6 The Skeletal System: Bone Tissue

Figures 6.2 and 6.9 have been expanded to include photographs illustrating bone tissue cell types and fractures, respectively. The chapter also includes a new Clinical Application on remodeling and orthodontics.

Chapter 7 The Skeletal System: The Axial Skeleton

Figures 7.13, 7.25, and 7.26 are new to the chapter on the axial skeleton, and Figure 7.15 has been expanded to better illustrate the position of the hyoid bone. Abnormal curves of the vertebral column and fractures of the vertebral column have been added to the Disorders section.

Chapter 8 The Skeletal System: The Appendicular Skeleton

The section on development of the skeletal system has been moved to this chapter from Chapter 6, and the figure accompanying it, Figure 8.18, has been expanded to illustrate the development of the skull along with the rest of the skeletal system. There are three new Clinical Applications, on pelvimetry, bone grafting, and fractures of the metatarsals.

Chapter 9 Joints

The joint chapter has been reorganized somewhat, so that types of synovial joints (planar, hinge, pivot, condyloid, saddle, and ball and socket) are now described after the types of movements at synovial joints (gliding, flexion, extension, etc.). There is a new figure (Figure 9.16) to illustrate the new section on arthroplasty, and a new exhibit on the temporomandibular joint, along with a new description of ankylosing spondylitis in the Disorders section.

Chapter 10 Muscular Tissue

This chapter has been retitled "Muscular Tissue" to reflect the latest terminology. Figures 10.2, 10.6, and 10.18 have been redrawn for this edition. The eleventh edition also features new Clinical Applications on electromyography and hypotonia and hypertonia.

Chapter 11 The Muscular System

In this chapter, Exhibit 11.8 has undergone extensive revision, and its accompanying figure (Figure 11.11) has been expanded to include two additional views of the muscles used in breathing. A new Clinical Application on intramuscular injections also appears. In addition, the tables in each Exhibit in this chapter now include an innervation column describing the nerve or nerves that cause contraction of each muscle.

Chapter 12 Nervous Tissue

The section on neuroglia has been expanded and reorganized into two sections, "Neuroglia of the CNS" and "Neuroglia of the PNS." The section entitled "Electrical Signals in Neurons" has been modified to include practical examples of nervous system functions. Specifically, this new information describes how the nervous system allows a person to feel the surface of a pen and then causes the muscles of the hand to contract so the person can write with it. A new illustration (Figure 12.10) has been rendered to illustrate this concept. Figures 12.6 and 12.7 are also new to this edition, Figures 12.17 and 12.20 have been redrawn, and much of the rest of the art in this chapter has undergone extensive revision to provide a more consistent color scheme.

Chapter 13 The Spinal Cord and Spinal Nerves

The art program for this chapter has undergone a number of changes. Figure 13.1 has been expanded, Figure 13.13 has been redrawn, and Figures 13.2, 13.3, 13.5, 13.7, and 13.12 have been revised extensively. The chapter has been reorganized so that the section on spinal nerves appears before the section on spinal cord physiology. In addition, there are new Clinical Applications on spinal nerve root damage, and reflexes and diagnosis, and a new Disorder section on spinal cord injuries.

Chapter 14 The Brain and Cranial Nerves

The art program for this chapter has undergone the most extensive revision. Figures 14.18 and 14.19 are new, and almost all of the other figures in the chapter have been redrawn and/or revised extensively. See, for example, Figure 14.29, which has been redrawn and completely revised for this edition. Another example is Figure 14.2, which has been expanded to illustrate the falx cerebri, falx cerebelli, and tentorium cerebelli in addition to the other protective coverings of the brain. New Clinical Applications appear on ataxia and damage to the basal ganglia, and there are new entries in the Disorders section on brain tumors and attention deficit hyperactivity disorder.

Chapter 15 The Autonomic Nervous System

This chapter has been moved forward in this unit to give readers a basic understanding of the ANS before they proceed to sensory, motor, and integrative systems, and to the chapter on the special senses. The section on anatomy of autonomic pathways has been reorganized, and there is a new section on autonomic tone in the part of the chapter dealing with physiological effects of the ANS. In addition, Figure 15.1 is new, Figure 15.4 has been enlarged to clarify the elements of the autonomic plexuses, and the chapter has a completely new Medical Terminology section.

Chapter 16 Sensory, Motor, and Integrative Systems

Figure 16.8 has been completely reworked to better illustrate direct motor pathways. Other changes include a new entry in the Disorders section on Parkinson disease, and a new Clinical Application on amnesia. In addition, the section entitled somatic sensory and motor maps in the cerebral cortex in the tenth edition has now been distributed so that the information on mapping appears in the separate sections on somatic sensory pathways and somatic motor pathways.

Chapter 17 The Special Senses

A new section on Aging and the Special Senses has been added to this chapter, along with new Clinical Applications on detached retina and LASIK.

Chapter 18 The Endocrine System

Figures 18.5 and 18.10 have undergone extensive revision, and Figures 18.13, 18.15, and 18.18 have been expanded to include cadaver views of important endocrine system organs. In addition, a new Clinical Application on Congenital Adrenal Hyperplasia is included.

Chapter 19 The Cardiovascular System: The Blood

Figure 19.14 is new to this edition, Figures 19.7 and 19.8 have been revised extensively, and Figures 19.2 and 19.10 have been expanded. There is also a new section on stem cell transplants from bone marrow and cord blood, and a new Clinical Application on bone marrow examination.

Chapter 20 The Cardiovascular System: The Heart

Figures 20.19 and 20.22 are new to this edition, and Figures 20.3, 20.8, 20.14, 20.18, and 20.21 have undergone extensive revision. There is a new section on development of the heart. The sections on atherosclerotic plaques, diagnosis of coronary artery disease, and arrhythmias have been revised extensively, and there are new Clinical Applications on myocarditis and endocardicis, myocardial ischemia and infarction, and sick sinus syndrome.

Chapter 21 The Cardiovascular System: Blood Vessels and Hemodynamics

An illustrated table on distinguishing features of blood vessels (Table 21.2) is new to this edition, along with a new Clinical Application on angiogenesis and disease and an expanded Medical Terminology listing.

Chapter 22 The Lymphatic System and Immunity

The title of this chapter has been changed to emphasize immunity as the primary function of the lymphatic system. Other changes include revised and updated sections on antigen processing, T cells, elimination of invaders, and the complement system. In addition, the art program for this chapter was reworked extensively: Figures 22.9, 22.11, 22.13, 22.14, and 22.16 have been redrawn, and Figures 22.12, 22,15, and 22.18 are new to this edition.

Chapter 23 The Respiratory System

Figure 23.25 has been redrawn, and Figure 23.12 has been expanded. There are new Clinical Applications on nasal polyps and respiratory distress syndrome, and expansion of the one on pneumothorax to include hemothorax. New entries in the Disorders section include asbestos-related diseases, sudden infant death syndrome, and severe acute respiratory syndrome.

Chapter 24 The Digestive System

A new section entitled "neural innervation of the GI tract" has been added to this chapter, which includes information about the enteric nervous system, autonomic nervous system, and GI reflex pathways. The information on deglutition (swallowing) and phases of digestion (cephalic, gastric, and intestinal) has been reorganized to appear in separate sections; the latter section also includes information on the various hormones involved in digestion. Figure 24.2 has been redrawn, and Figure 24.3 is new to this edition. There are also new Clinical Applications on Zollinger-Ellison syndrome, liver biopsy, and polyps in the colon, and the one on pancreatitis has been expanded to include pancreatic cancer.

Chapter 25 Metabolism and Nutrition
This chapter includes a new section on hypervitaminosis and a completely reworked illustration of the food pyramid for Figure 25.20.

Chapter 26 The Urinary System
Figure 26.22 is new to this edition, and Figure 26.5 has undergone extensive revision. New Clinical Applications on kidney transplant and cystoscopy also appear in this chapter.

Chapter 28 The Reproductive Systems
The coverage of ejaculation and erection has been revised extensively, and there are new Clinical Applications on testicular injuries, vasectomy, premature ejaculation, and ovarian cysts. In addition, an entry on erectile dysfunction has been added to the Disorders section.

Art has been added to Table 28.1, Figures 28.17 and 28.18 have been expanded, and Figures 28.23, 28.24, 28.25, and 28.26 have been redrawn to look more anatomical.

THE ILLUSTRATION PROGRAM

New Design A textbook with beautiful illustrations or photographs on most pages requires a carefully crafted and functional design. The design for the eleventh edition continues to assist students in making the most of the text's many features and outstanding art. Each page is carefully laid out to place related text, figures, and tables near one another, minimizing the need for page turning while reading a topic. New to this edition is the expanded use of red print used to indicate all callouts of figures and tables. Not only is the reader alerted to refer to the figure or table, but the color print also serves as a place locator for easy return to the narrative.

Distinctive icons incorporated throughout the chapters signal special features and make them easy to find during review. These include the **key** with Key Concept Statements; the **question mark** with the applicable questions that enhance every figure; the **stethoscope** indicating a clinical application within the chapter narrative; the **fetus icon** announcing the developmental anatomy section; the **running shoe** highlighting content relevant to exercise; and the icons that indicate the **study outline** and distinctive types of **chapter-ending questions.**

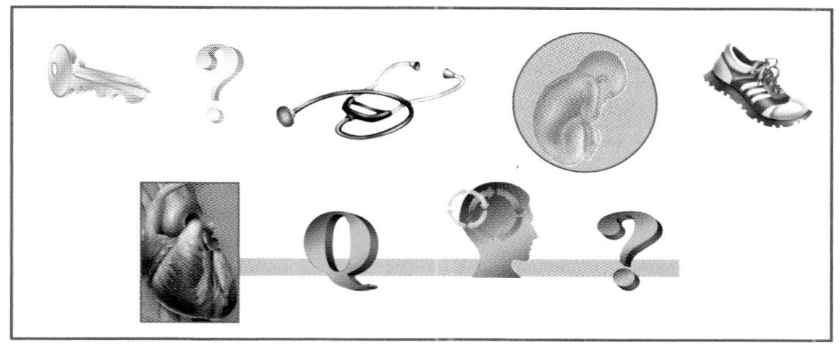

Helpful Orientation Diagrams Students sometimes need help figuring out the plane of view of anatomy illustrations —descriptions alone do not always suffice. An orientation diagram that depicts and explains the perspective of the view represented in the figure accompanies every major anatomy illustration. All have been redrawn for this edition using color and often a 3-D perspective making them more apparent as well as useful. There are three types of diagrams: (1) planes used to indicate where certain sections are made when a part of the body is cut; (2) diagrams containing a directional arrow and the word "View" to indicate the direction from which the body part is viewed, and (3) diagrams with arrows leading from or to them that direct attention to enlarged and detailed parts of illustrations.

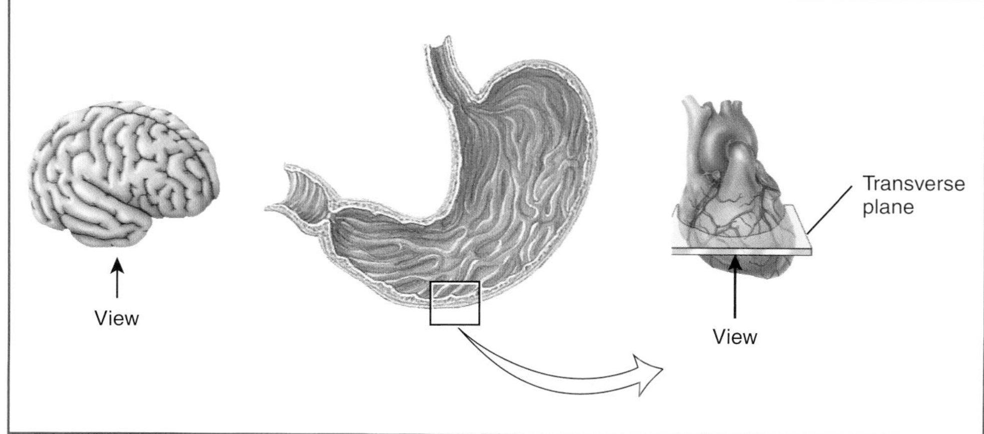

New Art Studying human anatomy and physiology is both a visual and a descriptive enterprise. Today, as more and more students identify themselves as visual learners, ensuring that the illustrations in the text are as helpful to them as possible is a high priority for us. Many of the figures depicting the toughest topics for students to grasp have been enhanced or re-developed for more clarity and ease of understanding. These include numerous illustrations depicting the structures and functions of the nervous system and special attention to tough topics like muscle contraction, capillary exchange, or the countercurrent mechanism. In all, approximately 100 figures are newly rendered for this edition. Most figures have been improved in some way.

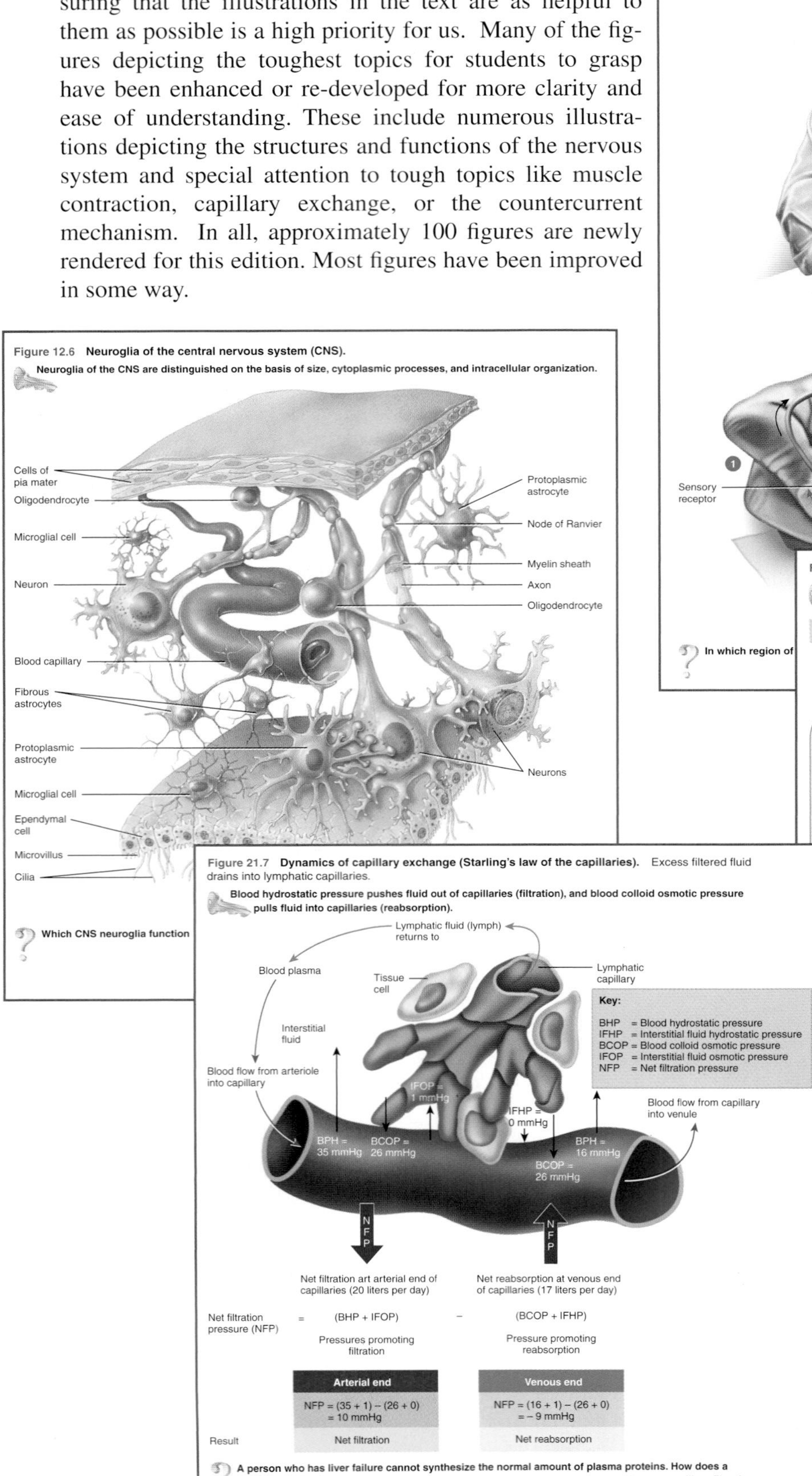

Figure 12.10 Overview of nervous system functions.

Graded potentials and nerve and muscle action potentials are involved in the relay of sensory stimuli, integrative functions such as perception, and motor activities.

Right side of brain
Left side of brain
Brain
Cerebral cortex
Interneuron
Upper motor neuron
Thalamus
Interneuron
Sensory neuron
Spinal cord
Lower motor neuron
Key:
Graded potential
Nerve action potential
Muscle action potential
Sensory receptor
Neuromuscular junction

In which region of

Figure 10.11 Summary of the events of contraction and relaxation in a skeletal muscle fiber.

Acetylcholine released at the neuromuscular junction triggers a muscle action potential, which leads to muscle contraction.

1 Nerve impulse arrives at axon terminal of motor neuron and triggers release of acetylcholine (ACh).
Nerve impulse
Muscle action potential
2 ACh diffuses across synaptic cleft, binds to its receptors in the motor end plate, and triggers a muscle action potential (AP).
Transverse tubule
4 Muscle AP travelling along transverse tubule opens Ca^{2+} release channels in the sarcoplasmic reticulum (SR) membrane, which allows calcium ions to flood into the sarcoplasm.
ACh receptor
Synaptic vesicle filled with ACh
3 Acetylcholinesterase in synaptic cleft destroys ACh so another muscle action potential does not arise unless more ACh is released from motor neuron.
Ca^{2+}
SR
9 Muscle relaxes.
8 Troponin–tropomyosin complex slides back into position where it blocks the myosin binding sites on actin.
5 Ca^{2+} binds to troponin on the thin filament, exposing the binding sites for myosin.
Elevated Ca^{2+}
Ca^{2+} active transport pumps
7 Ca^{2+} release channels in SR close and Ca^{2+} active transport pumps use ATP to restore low level of Ca^{2+} in sarcoplasm.
6 Contraction: power strokes use ATP; myosin heads bind to actin, swivel, and release; thin filaments are pulled toward center of sarcomere.

Which numbered steps in this figure are part of excitation–contraction coupling?

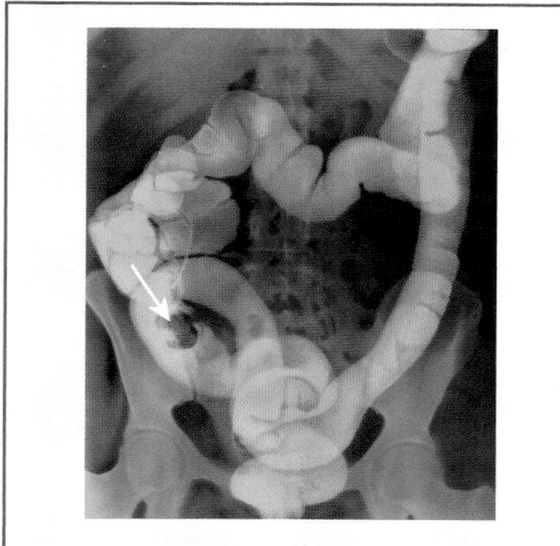

Barium contrast x-ray showing a cancer of the ascending colon (arrow)

New or Enhanced Photographs Approximately 40 new photos and micrographs have been added to this edition. These are found throughout the text and include medical images as well as histological micrographs. As before, we also provide an assortment of large, clear cadaver photos at strategic points in many chapters. The cadaver photos have been redesigned to be even more effective with the illustrations they accompany.

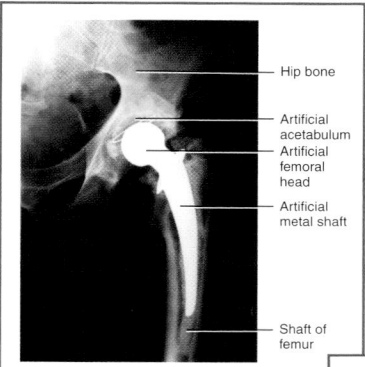

Hip bone
Artificial acetabulum
Artificial femoral head
Artificial metal shaft
Shaft of femur

(c) Radiograph of an artificial hip joint

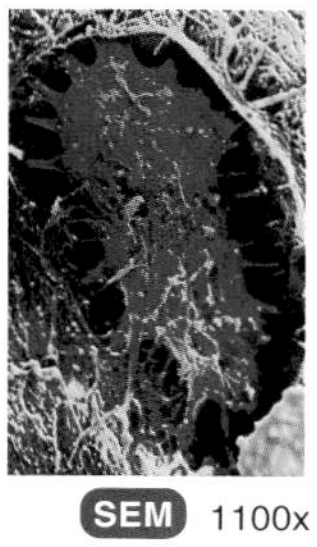

SEM 1100x

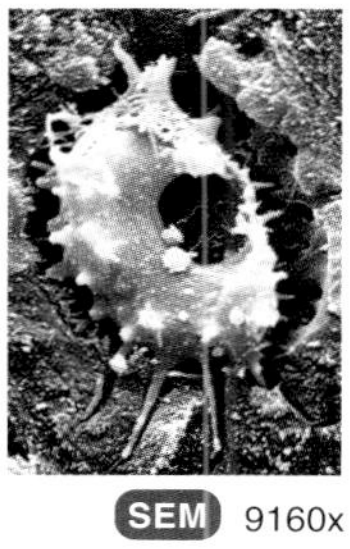

SEM 9160x

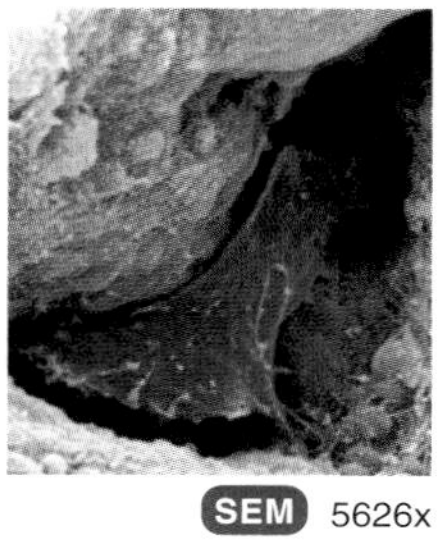

SEM 5626x

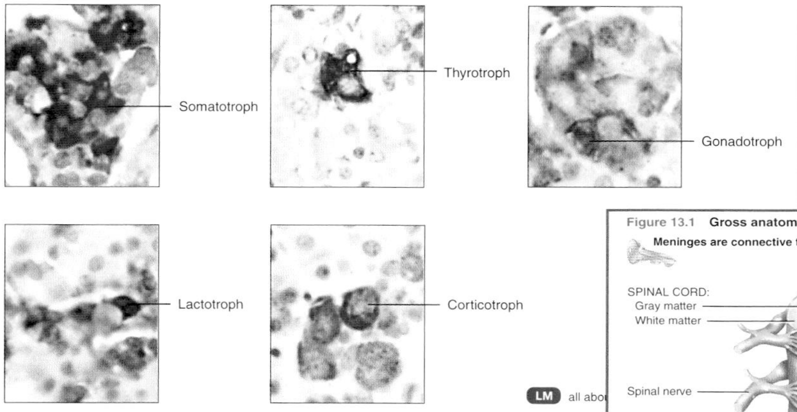

Somatotroph
Thyrotroph
Gonadotroph
Lactotroph
Corticotroph
LM all abo

(c) Histology of anterior pituitary

Figure 13.1 **Gross anatomy of the spinal cord.** The spinal meninges are evident in parts (a) and (c).

Meninges are connective tissue coverings that surround the spinal cord and brain.

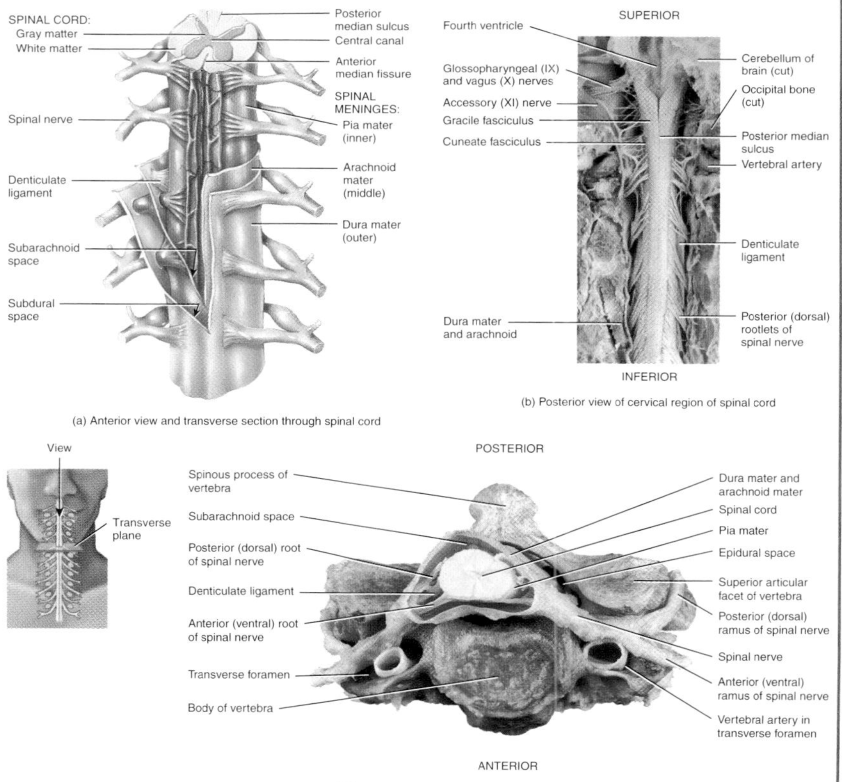

SPINAL CORD:
Gray matter
White matter
Posterior median sulcus
Central canal
Anterior median fissure
SPINAL MENINGES:
Pia mater (inner)
Arachnoid mater (middle)
Dura mater (outer)
Spinal nerve
Denticulate ligament
Subarachnoid space
Subdural space

(a) Anterior view and transverse section through spinal cord

SUPERIOR
Fourth ventricle
Glossopharyngeal (IX) and vagus (X) nerves
Accessory (XI) nerve
Gracile fasciculus
Cuneate fasciculus
Dura mater and arachnoid
Cerebellum of brain (cut)
Occipital bone (cut)
Posterior median sulcus
Vertebral artery
Denticulate ligament
Posterior (dorsal) rootlets of spinal nerve
INFERIOR

(b) Posterior view of cervical region of spinal cord

View
Transverse plane
POSTERIOR
Spinous process of vertebra
Subarachnoid space
Posterior (dorsal) root of spinal nerve
Denticulate ligament
Anterior (ventral) root of spinal nerve
Transverse foramen
Body of vertebra
Dura mater and arachnoid mater
Spinal cord
Pia mater
Epidural space
Superior articular facet of vertebra
Posterior (dorsal) ramus of spinal nerve
Spinal nerve
Anterior (ventral) ramus of spinal nerve
Vertebral artery in transverse foramen
ANTERIOR

(c) Transverse section of the spinal cord within a cervical vertebra

What are the superior and inferior boundaries of the spinal dura mater?

of synaptic end bulbs. Because calcium ions are more concentrated in the extracellular fluid, Ca^{2+} flows inward through the opened channels.

3. An increase in the concentration of Ca^{2+} inside the presynaptic neuron serves as a signal that triggers exocytosis of the synaptic vesicles. As vesicle membranes merge with the plasma membrane, neurotransmitter molecules within the vesicles are released into the synaptic cleft. Each synaptic vesicle contains several thousand molecules of neurotransmitter.
4. The neurotransmitter molecules diffuse across the synaptic cleft and bind to **neurotransmitter receptors** in the postsynaptic neuron's plasma membrane. The receptor shown in Figure 12.17 is part of a ligand-gated channel (see Figure 12.11b); in other cases the receptor may be a separate protein in the membrane.
5. Binding of neurotransmitter molecules to their receptors on ligand-gated channels opens the channels and allows particular ions to flow across the membrane.
6. As ions flow through the opened channels, the voltage across the membrane changes. This change in membrane voltage is a **postsynaptic potential.** Depending on which ions the channels admit, the postsynaptic potential may be a depolarization or hyperpolarization. For example, opening of Na^+ channels allows inflow of Na^+, which causes depolarization. However, opening of Cl^- or K^+ channels causes hyperpolarization. Opening Cl^- channels permits Cl^- to move into the cell, while opening the K^+ channels allows K^+ to move out—in either event, the inside of the cell becomes more negative.
7. When a depolarizing postsynaptic potential reaches threshold, it triggers an action potential.

Figure 12.17 **Signal transmission at a chemical synapse.** Through exocytosis of synaptic vesicles, a presynaptic neuron releases neurotransmitter molecules. After diffusing across the synaptic cleft, the neurotransmitter binds to receptors in the plasma membrane of the postsynaptic neuron and produces a postsynaptic potential.

At a chemical synapse, a presynaptic neuron converts an electrical signal (nerve impulse) into a chemical signal (neurotransmitter release). The postsynaptic neuron then converts the chemical signal back into an electrical signal (postsynaptic potential).

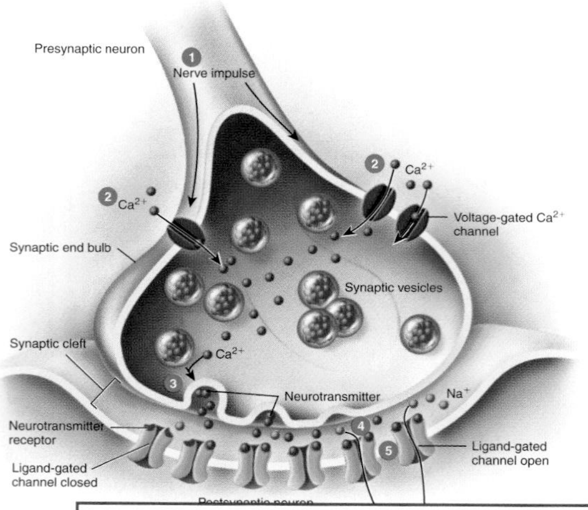

Why may electr... in only one direct...

Correlation of Sequential Processes Correlation of sequential processes in text and art is achieved through the use of special numbered lists in the narrative that correspond to numbered segments in the accompanying figure. This approach is used extensively throughout the book to lend clarity to the flow of complex processes.

Figure 10.1 **Organization of skeletal muscle and its connective tissue coverings.**

A skeletal muscle consists of individual muscle fibers (cells) bundled into fascicles and surrounded by three connective tissue layers that are extensions of the deep fascia.

Functions of Muscle Tissues

1. **Produce body movements.**
2. **Stabilize body positions.**
3. **Store and move substances within the body.**
4. **Generate heat (thermogenesis).**

Which connective tissue coat surrounds groups of muscle fibers, separating them into fascicles?

Functions Overview This feature, which appears as boxed text within selected figures, summarizes the functions of the anatomical structure or body system depicted. The juxtaposition of text and art further reinforces the connection between structure and function.

Key Concept Statements This art-related feature summarizes an idea that is discussed in the text and demonstrated in a figure. Each Key Concept Statement is positioned adjacent to its figure and is denoted by a distinctive key icon.

Figure Questions This highly applauded feature asks readers to synthesize verbal and visual information, think critically, or draw conclusions about what they see in a figure. Each Figure Question appears adjacent to its illustration and is highlighted in this edition by the blue question mark icon. Answers are located at the end of each chapter.

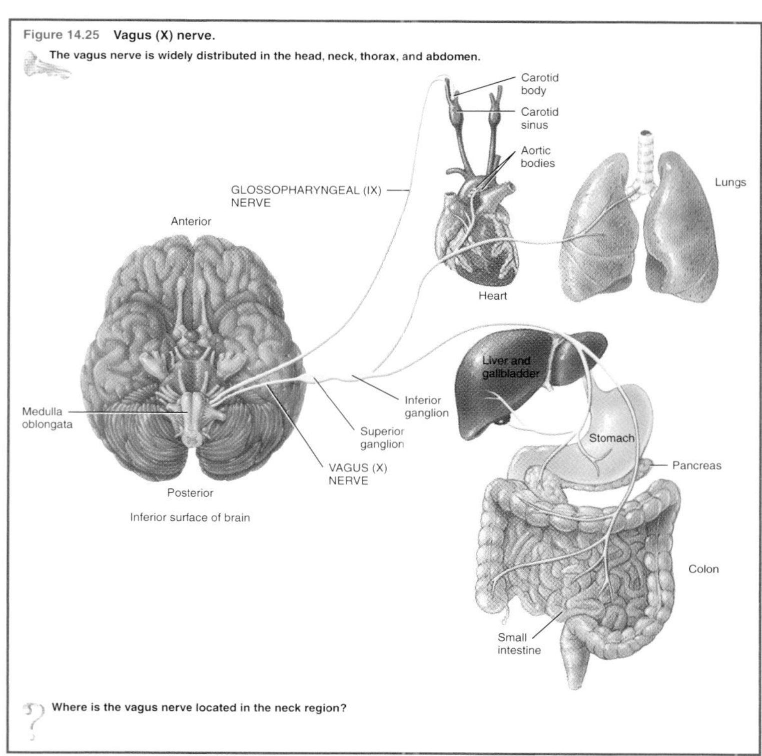

HALLMARK FEATURES

The eleventh edition of *Principles of Anatomy and Physiology* builds on the legacy of thoughtfully designed and class-tested pedagogical features that provide a complete learning system for students as they navigate their way through the text and course. All have been revised to reflect the enhancements to the text.

Chapter-opening Pages Each chapter begins with a chapter-opening page that includes a beautiful new piece of art related to the body system under consideration. Also included is a brief opening statement of how the topic under consideration contributes to the homeostasis of the human body.

Helpful Exhibits Students of anatomy and physiology need extra help learning the many structures that constitute certain body systems—most notably skeletal muscles, articulations, blood vessels, and nerves. As in previous editions, the chapters that present these topics are organized around **Exhibits,** each of which consists of an overview, a tabular summary of the relevant anatomy, and an associated suite of illustrations or photographs. Each Exhibit is prefaced by an Objective and closes with a Checkpoint activity. Many also incorporate a related clinical application. We trust you will agree that our spaciously designed Exhibits are ideal study vehicles for learning anatomically complex body systems.

Student Objectives and Checkpoints **Objectives** are found at the beginning of major sections throughout each chapter. Complementing this format, **Checkpoint** questions appear at strategic intervals within chapters to give students the chance to validate their understanding of the material they have just read.

Tools for Mastering Vocabulary Students—even the best ones—generally find it difficult at first to read and pronounce anatomical and physiological terms. Moreover, as teachers we are sympathetic to the needs of the growing ranks of college students who speak English as a second language. For these reasons, we have endeavored to ensure that this book has a strong and helpful vocabulary component. The key terms in every chapter are emphasized by use of **boldface type.** We include **pronunciation guides** when major, or especially hard-to-pronounce, structures and functions are introduced in the discussions, Tables, or Exhibits, and have increased the number in this edition. **Word roots** citing the Greek or Latin derivations of anatomical terms are offered as an additional aid. For additional help, we provide a list of **Medical Terminology** at the conclusion of most chapters, and a comprehensive **Glossary** at the back of the book. Also included at the end of the book is a list of the basic building blocks of medical terminology—**Combining Forms, Word Roots, Prefixes, and Suffixes.**

Study Outline As always, readers will benefit from the popular end-of-chapter **Study Outline** that is page referenced to the chapter discussions.

End-of-chapter Questions The **Self-Quiz Questions** are written in a variety of styles (true/false, multiple choice, matching) that are calculated to appeal to readers' different testing preferences. **Critical Thinking Questions** challenge readers to apply concepts to real-life situations. The style of these questions ought to make students smile on occasion as well as think! Answers to the Self-Quiz and Critical Thinking Questions are located in Appendix E.

COMPLETE TEACHING AND LEARNING PACKAGE

Continuing the tradition of providing a complete teaching and learning package, the eleventh edition of *Principles of Anatomy and Physiology* is available with a host of carefully planned supplementary materials that will help you and your students attain the maximal benefit from our textbook. Please contact your Wiley sales representative for additional information about any of these resources.

EgradePlus Helping Teachers Teach and Students Learn
www.wiley.com/college/tortora

This title is available with eGrade Plus, a powerful online tool that provides instructors and students with an integrated suite of teaching and learning resources in one easy to use Website. eGrade Plus is organized around the essential activities you and your students perform in class.

For Instructors

Prepare & Present: Create class presentations using a wealth of Wiley-provided resources – such as an online version of the textbook, PowerPoint slides, animations, overviews, and interactive case studies from *Interactions*, and images from the Wiley A&P Visual Library – making your preparation time more efficient. You may easily adapt, customize, and add to this content to meet the needs of your course.

Create Assignments: Automate the assigning and grading of homework or quizzes by using Wiley-provided question banks, or by writing your own. Student results will be automatically graded and recorded in your gradebook. eGrade Plus can link the pre-lecture quizzes and test bank questions to the relevant section of the online text, providing students with context-sensitive help.

Track Student Progress: Keep track of your students' progress via and instructor's gradebook, which allow you to analyze individual and overall class results to determine their progress and level of understanding.

Administer Your Course: eGrade Plus can easily be integrated with another course management system, gradebook, or other resources you are using in your class, providing you with the flexibility to build your course, your way.

For Students

Wiley's eGrade Plus provides immediate feedback on student assignments and a wealth of support materials. This powerful study tool will help your students develop their conceptual understanding of the class material and increase their ability to answer questions.

A **"Study and Practice"** area links directly to text content, allowing students to review the text while they study and answer. Resources include all of the *Interactions* content, inclusive of animations, interactive exercises and concept maps, and animated case studies. Also included are practice quizzes, anatomy drill and practice, a flash card tool, pronunciation dictionary, web explorations, pre-lecture quizzes, and other resources for study.

An **"Assignment"** area keeps all the work you want your students to complete in one location, making it easy for them to stay "on task". Students will have access to a variety of interactive self-assessment tools, as well as other resources for building their confidence and understanding. In addition, all of the pre-lecture quizzes contain a link to the relevant section of the multimedia book, providing students with context-sensitive help that allows them to conquer problem-solving obstacles as they arise.

A **Personal Gradebook** for each student will allow students to view their results from past assignments at any time.

Please view our online demo at **www.wiley.com/college/egradeplus**. Here you will find additional information about the features and benefits of eGrade Plus, how to request a "test drive" of eGrade Plus for this title, and how to adopt it for class use.

Materials Available For Students

New! A Brief Atlas of the Skeleton, Surface Anatomy, and Selected Medical Images (0-471-22377-8) Packaged for free with every new copy of the text, this all new atlas of stunning photographs provides a visual reference for both lecture and lab.

Dedicated Book Companion Website, A dynamic website rich with many activities for review and exploration includes: Chapter Overview and Objectives, Self-Quizzes for each chapter, Anatomy Drag and Drops, Cadaver Practicals, Pronunciation Dictionary with Flash Card Option and Terminology Quiz, Insights and Explorations – Web-based Activities, Crossword Puzzles, Disorder Search linked to Chapter Content, Weblinks linked to Chapter Content, Medical Tests and Procedures linked to Chapter Content, Essays on Wellness. In addition there are sections on study tips, determining your learning style, and correlations of what assets will work best with a specific learning style

Interactions: Exploring the Functions of the Human Body 2.0. Lancraft et. al. - Covering all body systems, this dynamic and highly acclaimed program includes anatomical overviews linking form and function; rich animations of complex physiological processes, a variety of creative interactive exercises, concept maps to help students make the connections, and animated clinical case studies. The 2.0 release boasts improvements to better facilitate navigation around the program and enhanced ease of use. Interactions is available in one **DVD (0-471-65419-1)**, or as a set of **9 individual CDs (0-471-20781-0)**, each focusing on one or two systems. If desired, the CDs may be purchased or adopted individually, as well.

Learning Guide (0-471- 68935-1) by Kathleen Schmidt Prezbindowski, College of Mount St. Joseph. Designed specifically to fit the needs of students with different learning styles, this well-received guide helps students to more closely examine important concepts through a variety of activities and exercises. The 29 chapters in the *Learning Guide* parallel those of the textbook and include many activities, quizzes and tests for review and study.

Illustrated Notebook (0471-68935-1) A true companion to the text, this unique notebook is a tool for organized note taking in class and for review during study. Following the sequence in the textbook, each left-handed page displays an unlabeled black and white copy of every text figure. Students can fill in the labels during lecture or lab at the instructor's directions and take additional notes on the lined right-handed pages.

Materials Available For Instructors

Wiley's Visual Library for Anatomy & Physiology 2.0 (0471-70001-0) This all-new cross-platform DVD includes all of the illustrations from the textbook in labelled, unlabeled, and unlabeled with leader lines format. In addition, many illustrations and photographs not included in the text, but which could easily be added to enhance lecture or lab, are included. Search for images by chapter, or by using key words. The Visual Library can also be accessed online at the Instructor's Companion Website, or through eGrade Plus. View a demo at
http://www.wiley.com/college/apvislibrarydemo/

Full-color Overhead Transparencies (0471-70002-9) A set of full-color overheads includes all of the figures in the text, including histology micrographs. All transparencies have been color-enhanced and carefully reviewed to maximize the labels for clear projection in the classroom.

Test Bank (0471-70000-2) by Janice Smith of Tarrant County CC - A testbank of nearly 3,000 questions, many new to the eleventh edition, is available. A variety of formats—multiple choice, short answer, matching and essay—are provided to accommodate different testing preferences. Available as a cross-platform CD-ROM, on the web at the Instructor's Companion Website, or integrated into eGradePlus, users can easily view, edit and add questions. Users can create questions in six different formats, import graphics and create graphs.

Instructor's Companion Web Site A dedicated companion website for instructors provides many resources for preparing and presenting lectures. Prepared by Lee Famiano of Cuyahoga CC, the site includes a brief Chapter Synopsis', Chapter Outlines and Objectives, Suggested Lecture Outlines, Teaching Tips, and both print and multimedia service. Also included are two or three Essay Questions for each chapter and "What's New and Different in this Chapter" to aid instructors who are transitioning to the new edition. Also available are three sets of **PowerPoint Slides** to ease in lecture presentation – Illustrated Lecture Slides prepared by Dieterich Steinmetz of Portland CC, slides of all illustrations in the text by chapter, and slides of all text tables.

Interactions: Exploring the Functions of the Human Body A copy of the *Interactions* DVD (or set of nine CDs, if preferred), is available to each professor who adopts the text for use in preparing dynamic lectures.

WebCT or **Blackboard** Course Management Systems with content prepared by Juville Dario-Becker of Central Virginia CC are available.

Personal Response System A full set of questions to use with Personal Response Systems are available. For more information, see your Wiley representative, or go to
www.wiley.com/college/prs

Faculty Resource Network Wiley's support structure to help instructors implement the dynamic new media that supports this text into their classrooms, laboratories, or online courses. Consult with your Wiley representative for details about this program or visit **www.wherefacultyconnect.com**.

For the Laboratory

Laboratory Manual for Anatomy and Physiology 2e (0-471-69122-4) by Connie Allen and Valerie Harper, Edison Community College. This newly revised laboratory manual presents material covered in the 2-semester undergraduate anatomy & physiology laboratory course in a clear and concise way, while maintaining a student-friendly tone. The manual is very interactive and contains activities and experiments that enhance students' ability to both visualize anatomical structures and understand physiological topics. ***New for the second edition***, each copy of the lab manual will also include our new ***PowerPhys*** simulation software for the laboratory. EGrade Plus, with a wealth of integrated resources include cat and fetal pig dissection video, is also available for adoption with this laboratory. The manual can be packaged with either the Cat Dissection Manual or the Fetal Pig Dissection Manual [depending on your needs] at no extra expense to the student.

Cat Dissection Manual 2e (0-471-70141-6) by Connie Allen and Valerie Harper, Edison Community College This manual includes photographs and illustrations of the cat along with guidelines for dissection. All photographs of the cat dissection, provided by Dennis Strete of McClennan CC, are new to this edition. It is available independently as well as bundled with the main manual depending upon your adoption needs.

Fetal Pig Dissection Manual 2e (0-471-701386) by Connie Allen and Valerie Harper, Edison Community College This manual includes photographs and illustrations of the fetal pig along with guidelines for dissection. All photographs of the fetal pig dissection, provided by Dennis Strete of McClennan CC, are new to this edition. It is available independently as well as bundled with the main manual depending upon your adoption needs.

New! PowerPhys (0-471-66289-5) by Allen, Harper, Ivlev, and Lancraft – 10 Self Contained Lab Modules for exploring physiological principles. Each module contains objectives with illustrated and animated review material, pre-lab quizzes, pre-lab reporting, data collection and analysis, and a full lab report with discussion and application questions. Experiments contain randomly generated data, allowing users to experiment multiple times, but still arrive at the same conclusions. Available as a stand-alone product, PowerPhys is also bundled with every new copy of the Allen and Harper Laboratory Manual and integrated into eGrade Plus. View a demo at **http://www.wiley.com/college/powerphysdemo/**

PowerAnatomy (0-471-44558-4) by Allen, Harper and Baxley - Developed in conjunctions with Primal Pictures, U.K., this is an on-line human anatomy laboratory manual, combining beautiful 3-D images of the human body alongside text, exercises and review questions focused on the undergraduate anatomy and anatomy & physiology student. Users can rotate the images, click on linked terms to see structures, and then answer self-assessing questions to test their knowledge. Included is a free 6-month subscription to Primal's acclaimed Anatomy.tv website. To view a demo of this product, go to **www.wiley.com/college/apcentral**.

A Photographic Atlas of the Human Body with Selected Cat, Sheep, and Cow Dissections, 2nd edition by Gerard Tortora **(0-471-42064-6)** This four-colored atlas is designed to support both study and laboratory experiences. Organized by body systems, the clearly labeled photographs provide a stunning visual reference to gross anatomy. Histological micrographs are also included. Many of the illustrations within *Principles of Anatomy and Physiology,* 11th edition are cross-referenced to this atlas.

Like each of our students, this book has a life of its own. The structure, content, and production values of *Principles of Anatomy and Physiology* are shaped as much by its relationship with educators and readers as by the vision that gave birth to the book eleven editions ago. Today you, our readers, are the "heart" of this book. We invite you to continue the tradition of sending your suggestions to us so that we can include them in the twelfth edition.

Gerard J. Tortora
Department of Science and Technology, S229
Bergen Community College
400 Paramus Road
Paramus, NJ 07652

Bryan Derrickson
Department of Science
Valencia Community College
PO Box 3028
Orlando, FL 32802
bderrickson@valenciacc.edu

Acknowledgements

For the eleventh edition of ***Principles of Anatomy and Physiology***, we have once again enjoyed the opportunity of collaborating with a group of dedicated and talented professionals. Accordingly, we would like to recognize and thank the members of our book team, who often worked evenings and weekends, as well as days, to bring this book to you. At John Wiley & Sons, Inc., our longtime Executive Editor, Bonnie Roesch, again illuminates the path toward ever better books with her creative ideas and dedication. There is no replacement for Bonnie. She is truly the heart and soul of the Tortora books. There is no way to thank her adequately. All we can say in print is thank you, thank you, thank you so very much! Karen Trost, Developmental Editor, shepherded the manuscript and electronic files during the revision process, as well as managing reviewer feedback. Her insights and editorial skills helped tremendously in polishing the eleventh edition to best meet the needs of students and faculty alike. Karen is now an indispensable member of our team. Thanks, Karen. Karin Kincheloe is the Wiley designer whose vision results in the beautiful presentation of this text. Karin laid out each page of the book to achieve the best possible placement of text, figures, and other elements. Both instructors and students will appreciate and benefit from the pedagogically effective and visually pleasing design elements that augment the content changes made to this edition. Karin's imprint is there for all of us to see. Thank you, Karin. Claudia Durrell, our Art Coordinator, has collaborated on this text for many editions. Her artistic ability, organizational skills, attention to detail, and understanding of our illustration preferences greatly enhance the visual appeal and style of the figures. She remains a cornerstone of our projects—thank you, Claudia, for all your contributions. We really appreciate all that you have done for us. Kelly Tavares, Associate Production Manager, demonstrated her untiring expertise during each step of the production process. She coordinated all aspects of actually making and manufacturing the book. Thank you, Kelly, for all the extra hours you spent to implement book-improving changes! We really value your talent and motivation. Hillary Newman, Photo Editor, provided us with all of the photos we requested and did it with efficiency, accuracy, and professionalism. Her efforts are so readily apparent. Mary O'Sullivan, Project Editor, coordinated the development of the many supplements that support this text. In her quiet way, Mary makes a tremendous impact. We are most appreciative! Thanks to Clay Stone, Executive Marketing Manager, for his best efforts in promoting our work to you and in returning your feedback to us for future revisions. Clay is the consummate professional whose efforts we truly appreciate. Wiley Editorial Assistants Maureen Powers, Alicia Romano, and Shannon Knoppel helped with various aspects of the project and took care of many details. Their behind-the-scenes activities make our job much easier. Thanks to all of you! There is no better team in all of publishing!

We wish to especially thank several of our academic colleagues for their helpful contributions to this edition. Thanks to Caryl Tickner of Stark State College, who revised the end-of-chapter Self-Quiz Questions and Critical Thinking Questions, for writing questions that students will appreciate. The high quality of her study activities ensures student success. We are grateful to the dedicated educators who have contributed to the high quality of the diverse supplementary material that accompanies this text: Connie Allen, Juville Dario-Becker, Christine Earls, Lee Famiano, Frances Frierson, Valerie Harper, Thomas Lancraft, Mark Nielsen, Janice Smith, Dietrich Steinmetz, Dennis Strete, and Charles Wert. And a special thank you to Kathleen Prezbindowski, who has authored the *Learning Guide* for so many editions.

Outstanding illustrations and photographs have always been a signature feature of *Principles of Anatomy and Physiology*. Respected scientific and medical illustrators on our team of exceptional artists include Mollie Borman, Leonard Dank, Sharon Ellis, Keith Kasnot, Steve Oh, Lynn O'Kelley, Hilda Muinos, Tomo Narashima, Nadine Sokol, and Kevin Somerville. Artists at Imagineering created the amazing computer graphic images and provided all labeling of figures. Mark Nielsen of the University of Utah provided many of the cadaver photos that appear in this edition, and Dr. Michael Ross of the University of Florida prepared and photographed many of the histology micrographs.

Reviewers

We are extremely grateful to our colleagues who reviewed the manuscript and offered insightful suggestions for improvement. The contributions of all these people, who generously provided their time and expertise to help us maintain the book's accuracy and clarity, are acknowledged in the list that follows.

Lynne Anderson	Meridian Community College
Theresa M. Arburn	Palo Alto College
Gordon Atkins	Andrews University
Tim Ballard	UNC-Wilmington
Ronald Beumer	Armstrong Atlantic State University
Charles Biggers	University of Memphis
Cheryl Black	Aiken Technical College
Nishi Bryska	UNC Charlotte
Stephen C.Burnett	Clayton College & State University
Paul Buttenhoff	College of St Catherine
James A.Carson	University of South Carolina
Redding Corbett	Midlands Technical College
Bruce Craig	Ball State University
James Tim Daniels	Southern Arkansas University
Rosemary Davenport	Gulf Coast Community College
Mary Dettman	Seminole Community College
Joseph Flanagin	Tarrant County Junior College South
Edward Fliss	St Louis Community College
Paul Florence	Jefferson Community College Louisville
Durwood Foote	Tarrant County Junior College NE
Marc Franco	South Seattle Community College
Purti Gadkari	Wharton County Junior College
Eric Genz-Mould	Shoreline Community College
Carol Gerding	Cuyahoga Community College
Louis Giacinti	Milwaukee Area Technical College
Tejendra Gill	University of Houston
Chaya Gopalan	St. Louis Community College
D. Bruce Gray	Simmons College
Richard Griner	Augusta State University
Pramila Gurrala	Wharton County Junior College
Clare Hays	Metropolitan State College
Randall Howell	Southern Union State Community College
Barbara Hunnicutt	Seminole Community College
Mandy Itiat	Florida Community College
Amy Jetton	Middle Tennessee State University
Sally Johnston	Community College of Southern Nevada
Marie Kotter	Weber State University
Susan Landesman	Mt. Hood Community College
J. Ellen Lathrop-Davis	The Community College of Baltimore County
Ronald Markle	Northern Arizona University
Kenneth Moore	Seattle Pacific University
Judi Nath	Lourdes College
Kerry L.Openshaw	Bemidji State University
Betsy Ott	Tyler Junior College
Amy Ouchley	University of Louisiana-Monroe
Betsy Peitz	California State University- Los Angeles
John Pellegrini	College of St Catherine
Davonya Person	Auburn University
Danny Pincivero	University of Toledo
Terrence Ravine	University of South Alabama
J. Orion Rogers	Radford University
Marilyn Shannon	Indiana Purdue University- Fort Wayne
John Simmons	Barton County Community College
Lori Smith	American River College
Dianne Snyder	Augusta State University
Claudia Stanescu	University of Arizona
Maura Stevenson	Community College of Allegheny County
Cynthia Surmacz	Bloomsburg University of PA
Yong Tang	Front Range Community College
Kent R.Thomas	Wichita State University
Caryl Tickner	Stark State College
Terry Tijerina	South Texas Community College
Teresa Trendler	Pasadena City College
Sarah Tringle	Mississippi Gulf Coast Community College
Richard Tsou	Gordon College
Vicki Veigl	University of Louisville
Leticia Vosotros	Ozarks Technical Community College
Judy Wallace	Middlesex Community College
Kathy Warren	Daytona Beach Community College
DeLoris Wenzel	University of Georgia
Judy Williams	Southeastern Oklahoma State University
Ruth Williams	Oakton Community College
Brian Witz	Nazareth College
Jeanne M.Workman	Duquesne University

To the Student

Your book has a variety of special features that will make your time studying anatomy and physiology a more rewarding experience. These have been developed based on feedback from students – like you – who have used previous editions of the text. A review of the preface will give you insight, both visually and in narrative, into all of the text's distinctive features.

Our experience in the classroom has taught us that students appreciate a hint – both visually and verbally – at the beginning of each chapter about what to expect from its contents. Each chapter of your book begins with a stunning illustration depicting the system or main content being covered in the chapter. In addition, a short introduction is included, explaining how the topic under consideration contributes to the homeostasis of the human body.

Chapter 12

Nervous Tissue

Nervous Tissue and Homeostasis

The exitable characteristic of nervous tissue allows for the generation of nerve impulses (action potentials) that provide communication and regulation of most body tissues.

www. wiley.com/college/apcentral

403

As you begin each narrative section of the chapter, be sure to take note of the **objectives** at the beginning of the section to help you focus on what is important as you read it.

SIGNAL TRANSMISSION AT SYNAPSES

▶ OBJECTIVES

Explain the events of signal transmission at a chemical synapse.

Distinguish between spatial and temporal summation.

Give examples of excitatory and inhibitory neurotransmitters, and describe how they act.

In Chapter 10 we described the events occurring at one type of synapse, the neuromuscular junction. Our focus in this chapter ... billions of neurons ... ial for homeostasis ... red and integrated. ... tion of particular ... some signals to be ... For example, the ... will determine how ... gy tests! Synapses ... s and neurological ... tic communication, and many therapeutic and addictive chemicals affect the body at these junctions.

At the end of the section, take time to try and answer the **Checkpoint** questions placed there. If you can, then you are ready to move on to the next section. If you experience difficulty answering the questions, you may want to re-read the section before continuing.

▶ CHECKPOINT

13. How is neurotransmitter removed from the synaptic cleft?
14. How are excitatory and inhibitory postsynaptic potentials similar and different?
15. Why are action potentials said to be "all-or-none," and EPSPs and IPSPs are described as "graded"?

Studying the figures (illustrations that include artwork and photographs) in this book is as important as reading the text. To get the most out of the visual parts of this book, use the tools we have added to the figures to help you understand the concepts being presented. Start by reading the **legend**, which explains what the figure is about. Next, study the **key concept statement**, which reveals a basic idea portrayed in the figure. Added to many figures you will also find an **orientation diagram** to help you understand the perspective from which you are viewing a particular piece of anatomical art. Finally, at the bottom of each figure you will find a **figure question**. If you try to answer these questions as you go along, they will serve as self-checks to help you understand the material. Often it will be possible to answer a question by examining the figure itself. Other questions will encourage you to integrate the knowledge you've gained by carefully reading the text associated with the figure. Still other questions may prompt you to think critically about the topic at hand or predict a consequence in advance of its description in the text.

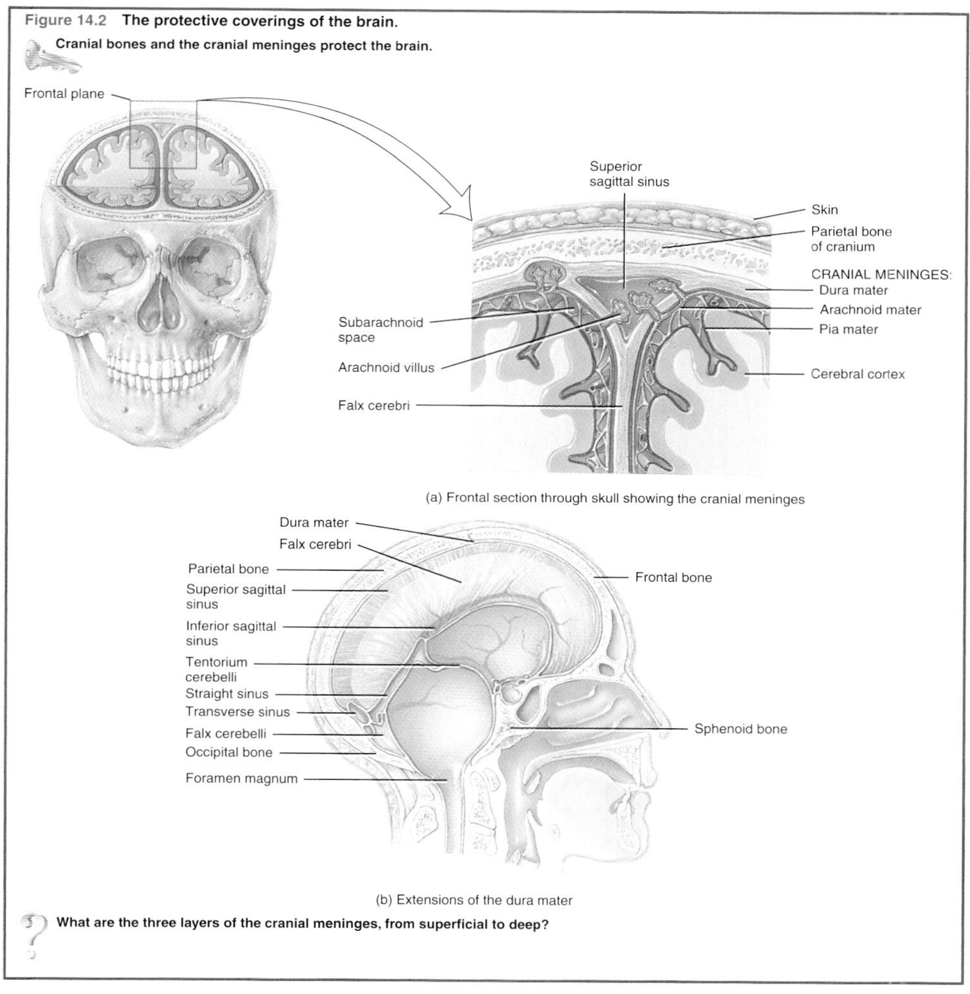

Figure 14.2 **The protective coverings of the brain.**

Cranial bones and the cranial meninges protect the brain.

What are the three layers of the cranial meninges, from superficial to deep?

At the end of each chapter are other resources that you will find useful. **The Study Outline** is a concise statement of important topics discussed in the chapter. Page numbers are listed next to key concepts so you can easily refer to the specific passages in the text for clarification or amplification.

STUDY OUTLINE

BRAIN ORGANIZATION, PROTECTION, AND BLOOD SUPPLY (p. 474)

1. The major parts of the brain are the brain stem, cerebellum, diencephalon, and cerebrum.
2. The brain is protected by cranial bones and the cranial meninges.
3. The cranial meninges are continuous with the spinal meninges. From superficial to deep they are the dura mater, arachnoid mater, and pia mater.
4. Blood flow to the brain is mainly via the internal carotid and vertebral arteries.
5. Any interruption of the oxygen or glucose supply to the brain can result in weakening of, permanent damage to, or death of brain cells.
6. The blood–brain barrier (BBB) causes different substances to move between the blood and the brain tissue at different rates and prevents the movement of some substances from blood into the brain.

helps maintain consciousness, causes awakening from slee contributes to regulating muscle tone.

THE CEREBELLUM (p. 486)

1. The cerebellum occupies the inferior and posterior aspects cranial cavity. It consists of two lateral hemispheres and a r constricted vermis.
2. It connects to the brain stem by three pairs of cerebellar ped
3. The cerebellum coordinates contractions of skeletal muscl maintains normal muscle tone, posture, and balance.

THE DIENCEPHALON (p. 488)

1. The diencephalon surrounds the third ventricle and consists thalamus, hypothalamus, and epithalamus.
2. The thalamus is superior to the midbrain and contains nucl

The **Self-quiz Questions** are designed to help you evaluate your understanding of the chapter contents. **Critical Thinking Questions** are word problems that allow you to apply the concepts you have studied in the chapter to specific situations.

SELF-QUIZ QUESTIONS

Fill in the blanks in the following statements.

1. The cerebral hemispheres are connected internally by a broad band of white matter known as the ___ .
2. List the five lobes of the cerebrum: , , , , .
3. The ___ separates the cerebrum into right and left halves.

Indicate whether the following statements are true or false.

4. The brain stem consists of the medulla oblongata, pons, and diencephalon.
5. You are the greatest student of anatomy and physiology and you

primary motor areas of the cerebral cortex; (b) helping maintain consciousness; (c) involved in nonlocalized perception of pain, pressure, and thermal sensations; (d) regulation of body temperature; (e) relaying sensory impulses to the cerebral cortex.

7. Which of the following statements is *false*? (a) The blood supply to the brain is provided mainly by the internal carotid and vertebral arteries. (b) Neurons in the brain rely almost exclusively on aerobic respiration to produce ATP. (c) An interruption of blood flow to the brain for even 20 seconds may impair brain function. (d) Glucose supply to the brain must be continuous. (e) Low levels result in unconsciousness.

CRITICAL THINKING QUESTIONS

1. An elderly relative suffered a CVA (stroke) and now has difficulty moving her right arm, and she also has speech problems. What areas of the brain were damaged by the stroke?
2. Nicky has recently had a viral infection and now she cannot move the muscles on the right side of her face. In addition, she is experiencing a loss of taste and a dry mouth, and she cannot close her right eye. What cranial nerve has been affected by the viral infection?
3. You have been hired by a pharmaceutical company to develop a drug to regulate a specific brain disorder. What is a major physiological roadblock to developing such a drug and how can you design a drug to bypass that roadblock so that the drug is delivered to the brain where it is needed?

Learning the language of anatomy and physiology can be one the more challenging aspects of taking this course. Throughout the text we have included pronunciations, and sometimes, word roots, for many terms that may be new to you. These appear in parentheses immediately following the new words, and the pronunciations are repeated in the glossary at the back of the book. The Companion Website offers you a review of these terms by chapter, pronounces them for you, and allows you the opportunity to create flash cards or quiz yourself on the many new terms.

Look at the words carefully and say them out loud several times. Learning to pronounce a new word will help you remember it and make it a useful part of your medical vocabulary. Take a few minutes to review the following pronunciation key, so it will be familiar to you when you encounter new words. The key is repeated at the beginning of the Glossary on page G-1 at the back of the book.

Pronunciation Key

1. The most strongly accented syllable appears in capital letters, for example, bilateral (bī-LAT-er-al) and diagnosis (dī-ag-NŌ-sis).
2. If there is a secondary accent, it is noted by a prime (′), for example, constitution (kon′-sti-TOO-shun) and physiology (fiz′-ē-OL-ō-jē). Any additional secondary accents are also noted by a prime, for example, decarboxylation (dē-kar-bok′-si-LĀ-shun).
3. Vowels marked by a line above the letter are pronounced with the long sound, as in the following common words:

ā as in *make*	*ō* as in *pole*
ē as in *be*	*ū* as in *cute*
ī as in *ivy*	

4. Vowels not marked by a line above the letter are pronounced with the short sound, as in the following words:

a as in *above* or *at*	o as in *not*
e as in *bet*	*u* as in *bud*
i as in *sip*	

5. Other vowel sounds are indicated as follows:
   oy as in *oil*
   oo as in *root*
6. Consonant sounds are pronounced as in the following words:

b as in *bat*	*m* as in *mother*
ch as in *chair*	*n* as in *no*
d as in *dog*	*p* as in *pick*
f as in *father*	*r* as in *rib*
g as in *get*	*s* as in *so*
h as in *hat*	*t* as in *tea*
j as in *jump*	*v* as in *very*
k as in *can*	*w* as in *welcome*
ks as in *tax*	*z* as in *zero*
kw as in *quit*	*zh* as in *lesion*
l as in *let*	

Brief Table of Contents

Contents

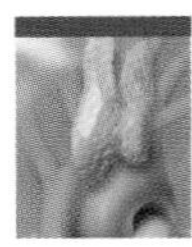

FOCUS ON HOMEOSTASIS: THE ENDOCRINE SYSTEM 657

UNIT 4 MAINTENANCE OF THE HUMAN BODY

Chapter 19 The Cardiovascular System: The Blood 666

Chapter 20 The Cardiovascular System: The Heart 695

Chapter 21 The Cardiovascular System: Blood Vessels and Hemodynamics 736

Chapter 22 The Lymphatic System and Immunity 804

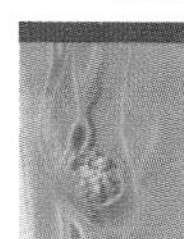

Chapter 23 The Respiratory System 846

Chapter 1

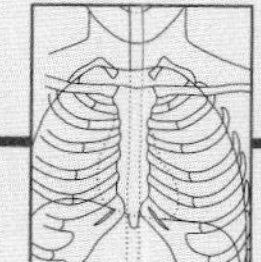

An Introduction to the Human Body

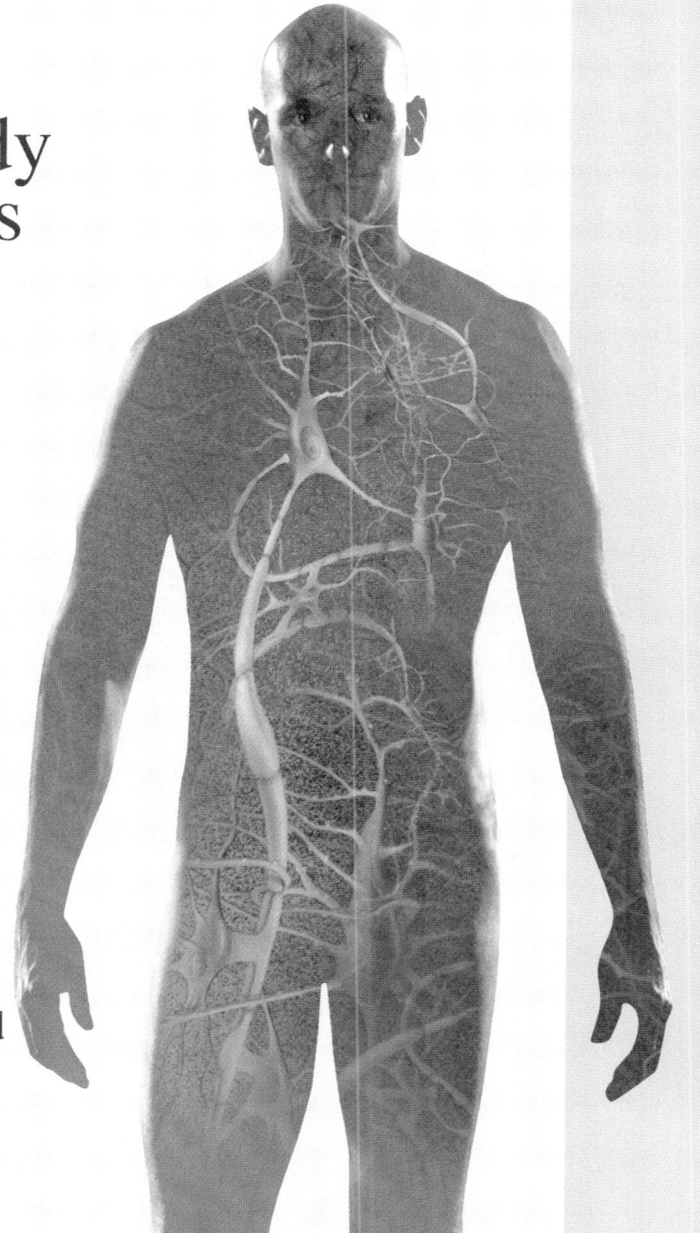

The Human Body and Homeostasis

Humans have many ways to maintain homeostasis, the state of relative stability of the body's internal environment. Disruptions to homeostasis often set in motion corrective cycles, called feedback systems, that help restore the conditions needed for health and life.

www. wiley.com/college/apcentral

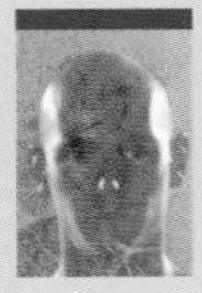

Our fascinating journey through the human body begins with an overview of the disciplines of anatomy and physiology, followed by a discussion of the organization of the human body and the properties that it shares with all living things. Next, you will discover how the body regulates its own internal environment; this unceasing process, called homeostasis, is a major theme in every chapter of this book. Finally, we introduce the basic vocabulary that will help you speak about the body in a way that is understood by scientists and health-care professionals alike.

ANATOMY AND PHYSIOLOGY DEFINED

▶ **OBJECTIVE**

Define anatomy and physiology, and name several subdisciplines of these sciences.

Two branches of science—anatomy and physiology—provide the foundation for understanding the body's parts and functions. **Anatomy** (a-NAT-ō-mē; *ana-* = up; *-tomy* = process of cutting) is the science of body *structures* and the relationships among them. It was first studied by **dissection** (dis-SEK-shun; *dis-* = apart; *-section* = act of cutting), the careful cutting apart of body structures to study their relationships. Today, a variety of imaging techniques (see Table 1.3) also contribute to the advancement of anatomical knowledge. Whereas anatomy deals with structures of the body, **physiology** (fiz′-ē-OL-o-jē; *physio-* = nature; *-logy* = study of) is the science of body *functions*—how the body parts work. Table 1.1 describes several subdisciplines of anatomy and physiology.

Because structure and function are so closely related, you will learn about the human body by studying its anatomy and physiology together. The structure of a part of the body allows performance of certain functions. For example, the bones of the skull join tightly to form a rigid case that protects the brain. The bones of the fingers are more loosely joined to allow a variety of movements. The walls of the air sacs in the lungs are very thin, permitting rapid movement of inhaled oxygen into the blood. The lining of the urinary bladder is much thicker to prevent the escape of urine into the pelvic cavity, yet its construction allows for considerable stretching as the urinary bladder fills with urine.

▶ **CHECKPOINT**

1. What body function might a respiratory therapist strive to improve? What structures are involved?
2. Give your own example of how the structure of a part of the body is related to its function.

TABLE 1.1 Selected Subdisciplines of Anatomy and Physiology

Subdisciplines of Anatomy	Study of
Embryology (em′-brē-OL-ō-jē; *embry-* = embryo; *-logy* = study of)	Structures that emerge from the time of the fertilized egg through the eighth week in utero.
Developmental biology	Structures that emerge from the time of the fertilized egg to the adult form.
Histology (his′-TOL-ō -jē; *hist-* = tissue)	Microscopic structure of tissues.
Surface anatomy	Anatomical landmarks on the surface of the body through visualization and palpation.
Gross anatomy	Structures that can be examined without using a microscope.
Systemic anatomy	Structure of specific systems of the body such as the nervous or respiratory systems.
Regional anatomy	Specific regions of the body such as the head or chest.
Radiographic anatomy (rā-dē-ō-GRAF-ik; *radio-* = ray; *-graphic* = to write)	Body structures that can be visualized with x rays.
Pathological anatomy (path′-ō-LOJ-i-kal; *path-* = disease)	Structural changes (from gross to microscopic) associated with disease.

Subdisciplines of Physiology	Study of
Neurophysiology (NOOR-ō-fiz-ē-ol′-ō-jē; *neuro-* = nerve)	Functional properties of nerve cells.
Endocrinology (en′-dō-kri-NOL-ō-jē; *endo-* = within; *-crin* = secretion)	Hormones (chemical regulators in the blood) and how they control body functions.
Cardiovascular physiology (kar-dē-ō-VAS-kū-lar; *cardi-* = heart; *-vascular* = blood vessels)	Functions of the heart and blood vessels.
Immunology (im′-ū-NOL-ō-jē; *immun-* = not susceptible)	How the body defends itself against disease-causing agents.
Respiratory physiology (RES-pir-a-to′-rē; *respira-* = to breathe)	Functions of the air passageways and lungs.
Renal physiology (RĒ-nal; *ren-* = kidney)	Functions of the kidneys.
Exercise physiology	Changes in cell and organ functions as a result of muscular activity.
Pathophysiology (PATH-ō-fiz-ē-ol′-ō-jē)	Functional changes associated with disease and aging.

LEVELS OF STRUCTURAL ORGANIZATION

▶ **OBJECTIVES**

Describe the levels of structural organization that make up the human body.

List the 11 systems of the human body, representative organs present in each, and their general functions.

The levels of organization of a language—letters, words, sentences, paragraphs, and so on—can be compared to the levels of organization of the human body. Your exploration of the human body will extend from elements and molecules to the whole person. From the smallest to the largest, six levels of organization are relevant to understanding anatomy and physiology: the chemical, cellular, tissue, organ, system, and organismal levels of organization (Figure 1.1).

1. The **chemical level,** which can be compared to the letters of the alphabet, includes **atoms,** the smallest units of matter that participate in chemical reactions, and **molecules,** two or more atoms joined together. Certain atoms, such as

Figure 1.1 Levels of structural organization in the human body.

The levels of structural organization are chemical, cellular, tissue, organ, system, and organismal.

1 CHEMICAL LEVEL

Atoms (C, H, O, N, P)

Molecule (DNA)

2 CELLULAR LEVEL

3 TISSUE LEVEL

4 ORGAN LEVEL

Serous membrane

Smooth muscle tissue layers

Stomach

Epithelial tissue

5 SYSTEM LEVEL

Esophagus

Liver

Stomach

Pancreas

Gallbladder

Small intestine

Large intestine

Digestive system

6 ORGANISMAL LEVEL

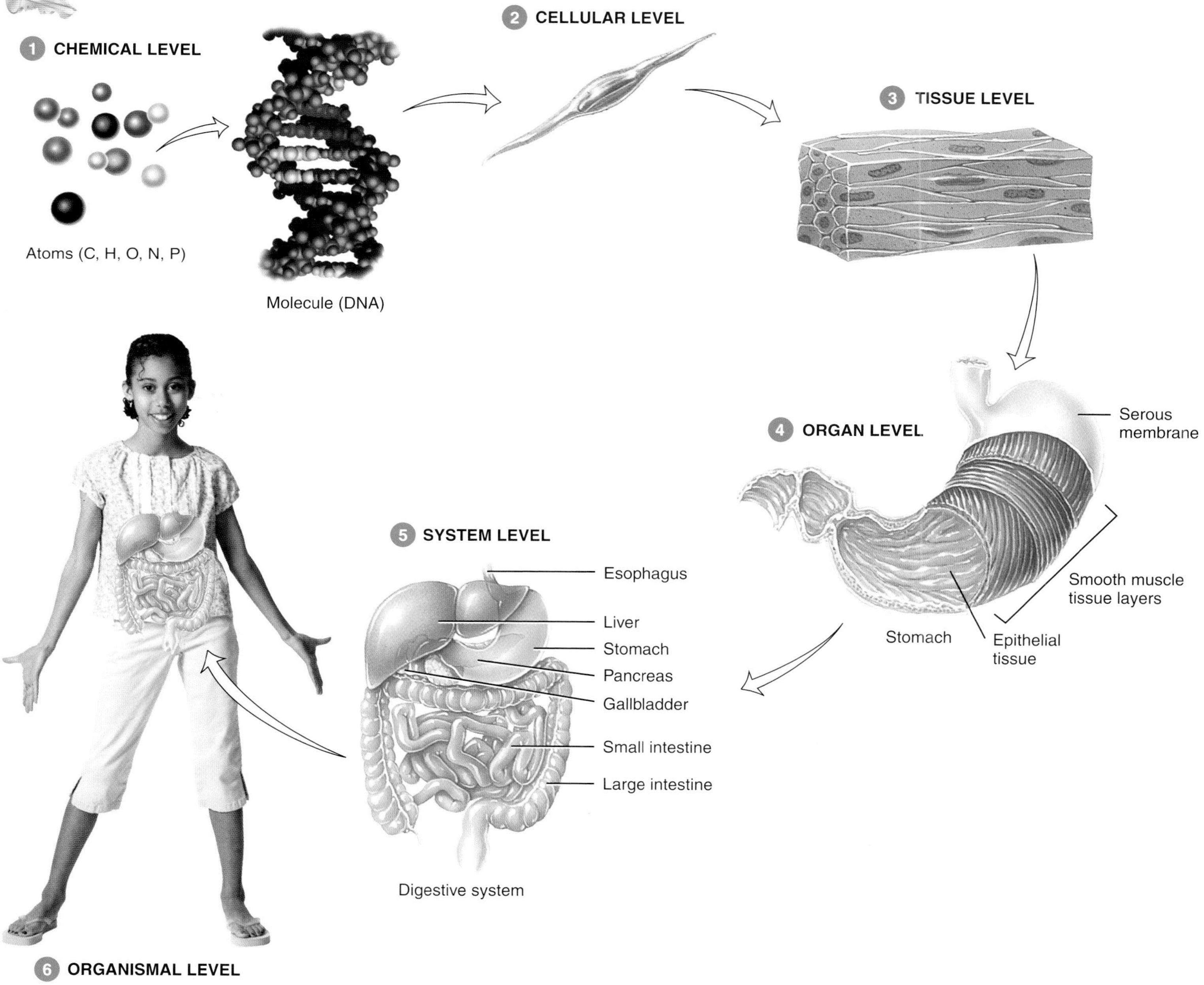

Which level of structural organization is composed of two or more different types of tissues that work together to perform a specific function?

carbon (C), hydrogen (H), oxygen (O), nitrogen (N), phosphorus (P), calcium (Ca), and sulfur (S), are essential for maintaining life. Two familiar molecules found in the body are deoxyribonucleic acid (DNA), the genetic material passed from one generation to the next, and glucose, commonly known as blood sugar. Chapters 2 and 25 focus on the chemical level of organization.

2 At the **cellular level,** molecules combine to form **cells,** the basic structural and functional units of an organism. Just as words are the smallest elements of language that make sense, cells are the smallest living units in the human body. Among the many kinds of cells in your body are muscle cells, nerve cells, and epithelial cells. Figure 1.1 shows a smooth muscle cell, one of the three types of muscle cells in the body. The cellular level of organization is the focus of Chapter 3.

3 The next level of structural organization is the **tissue level. Tissues** are groups of cells and the materials surrounding them that work together to perform a particular function, similar to the way words are put together to form sentences. There are just four basic types of tissue in your body: *epithelial tissue, connective tissue, muscular tissue,* and *nervous tissue.* Chapter 4 describes the tissue level of organization. Shown in Figure 1.1 is smooth muscle tissue, which consists of tightly packed smooth muscle cells.

4 At the **organ level,** different types of tissues are joined together. Similar to the relationship between sentences and paragraphs, **organs** are structures that are composed of two or more different types of tissues; they have specific functions and usually have recognizable shapes. Examples of organs are the skin, bones, stomach, heart, liver, lungs, and brain. Figure 1.1 shows how several tissues make up the stomach. The stomach's outer covering is a *serous membrane,* a layer of epithelial tissue and connective tissue that reduces friction when the stomach moves and rubs against other organs. Underneath are the *smooth muscle tissue layers,* which contract to churn and mix food and then push it into the next digestive organ, the small intestine. The innermost lining is an *epithelial tissue layer* that produces fluid and chemicals responsible for digestion in the stomach.

5 The next level of structural organization in the body is the **system level,** also called the **organ-system level.** A **system** (or chapter in our analogy) consists of related organs (paragraphs) with a common function. An example is the digestive system, which breaks down and absorbs food. Its organs include the mouth, salivary glands, pharynx (throat), esophagus, stomach, small intestine, large intestine, liver, gallbladder, and pancreas. Sometimes an organ is part of more than one system. The pancreas, for example, is part of both the digestive system and the hormone-producing endocrine system.

6 The largest organizational level is the **organismal level.** An **organism,** any living individual, can be compared to a book in our analogy. All the parts of the human body functioning together constitute the total organism.

TABLE 1.2 The Eleven Systems of the Human Body

Integumentary System (Chapter 5)

Components: Skin, and structures derived from it, such as hair, nails, sweat glands, and oil glands.

Functions: Protects the body; helps regulate body temperature; eliminates some wastes; helps make vitamin D; and detects sensations such as touch, pain, warmth, and cold.

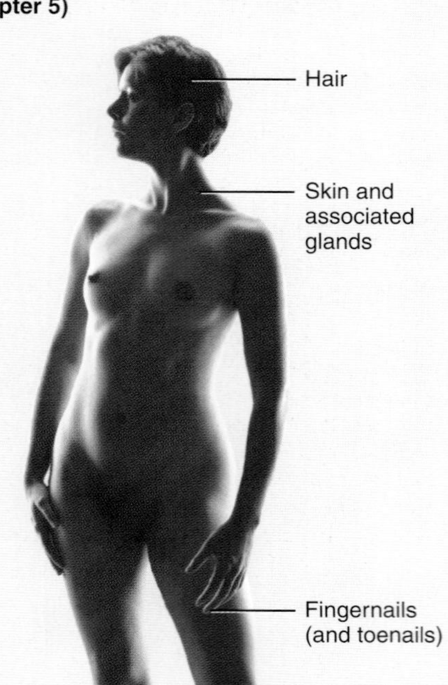

Skeletal System (Chapters 6–9)

Components: Bones and joints of the body and their associated cartilages.

Functions: Supports and protects the body; provides a surface area for muscle attachments; aids body movements; houses cells that produce blood cells; stores minerals and lipids (fats).

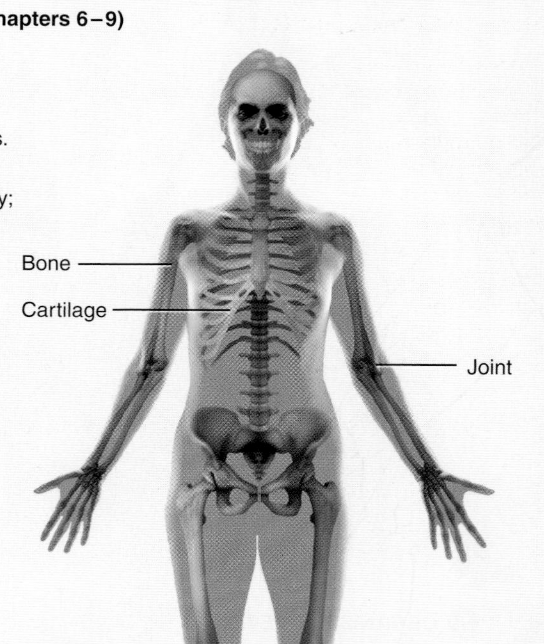

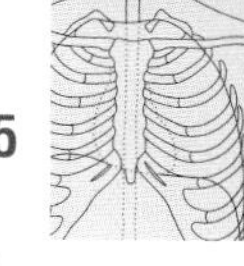

In the chapters that follow, you will study the anatomy and physiology of the body systems. Table 1.2 starting on page 4 lists the components and introduces the functions of these systems. You will also discover that all body systems influence one another. As you study each of the body systems in more detail, you will discover how they work together to maintain health, provide protection from disease, and allow for reproduction of the human species.

Noninvasive Diagnostic Techniques

Health-care professionals and students of anatomy and physiology commonly use several **noninvasive diagnostic techniques** to assess certain aspects of body structure and function. In **inspection,** the examiner observes the body for any changes that deviate from normal. Following this, one or more additional techniques may be employed. In **palpation** (pal-PĀ-shun; *palp-* = gently touching) the examiner feels body surfaces with the hands. An example is palpating the abdomen to detect enlarged or tender internal organs or abnormal masses. In **auscultation** (aws-kul-TĀ-shun; *auscult-* = listening) the examiner listens to body sounds to evaluate the functioning of certain organs, often using a stethoscope to amplify the sounds. An example is auscultation of the lungs during breathing to check for crackling sounds associated with abnormal fluid accumulation. In **percussion** (pur-KUSH-un; *percus-* = beat through) the examiner taps on the body surface with the fingertips and listens to the resulting echo. For example, percussion may reveal the abnormal presence of fluid in the lungs or air in the intestines. It may also provide information about the size, consistency, and position of an underlying structure. ■

▶ CHECKPOINT

3. Define the following terms: atom, molecule, cell, tissue, organ, system, and organism.
4. At what levels of organization would an exercise physiologist study the human body? *(Hint: Refer to Table 1.1.)*
5. Referring to Table 1.2, which body systems help eliminate wastes?

CHARACTERISTICS OF THE LIVING HUMAN ORGANISM

▶ OBJECTIVES

Define the important life processes of the human body.

Define homeostasis and explain its relationship to interstitial fluid.

Basic Life Processes

Certain processes distinguish organisms, or living things, from nonliving things. Following are the six most important life processes of the human body:

1. **Metabolism** (me-TAB-ō-lizm) is the sum of all the chemical processes that occur in the body. One phase of metabolism is

Muscular System (Chapters 10, 11)

Components: Muscles composed of skeletal muscle tissue, so-named because it is usually attached to bones.

Functions: Produces body movements, such as walking; stabilizes body position (posture); generates heat.

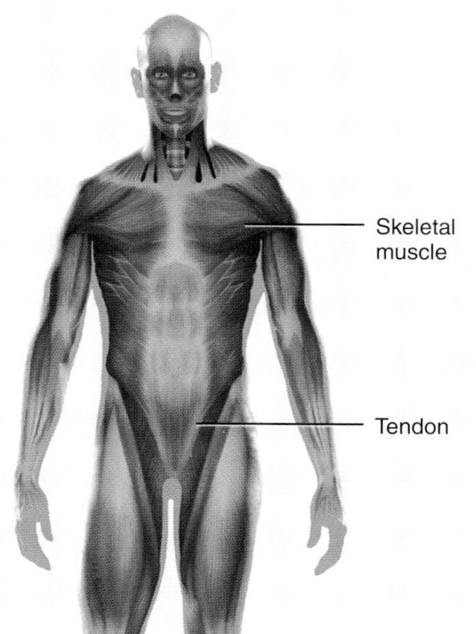

Nervous System (Chapters 12–17)

Components: Brain, spinal cord, nerves, and special sense organs, such as the eyes and ears.

Functions: Generates action potentials (nerve impulses) to regulate body activities; detects changes in the body's internal and external environment, interprets the changes, and responds by causing muscular contractions or glandular secretions.

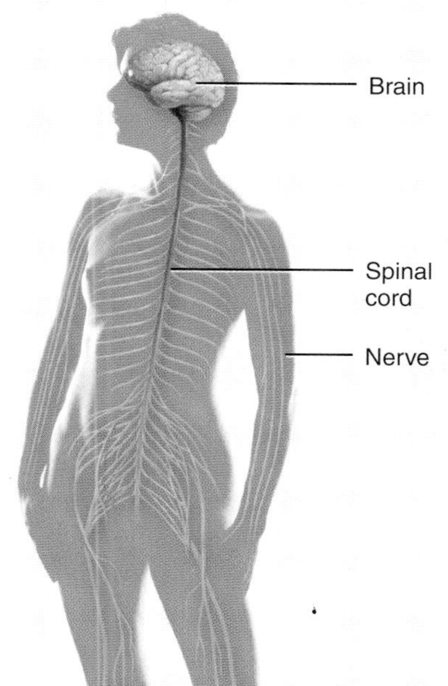

continues

TABLE 1.2 The Eleven Systems of the Human Body (continued)

Endocrine System (Chapter 18)

Components: Hormone-producing glands (pineal gland, hypothalamus, pituitary gland, thymus, thyroid gland, parathyroid glands, adrenal glands, pancreas, ovaries, and testes) and hormone-producing cells in several other organs.

Functions: Regulates body activities by releasing hormones, which are chemical messengers transported in blood from an endocrine gland to a target organ.

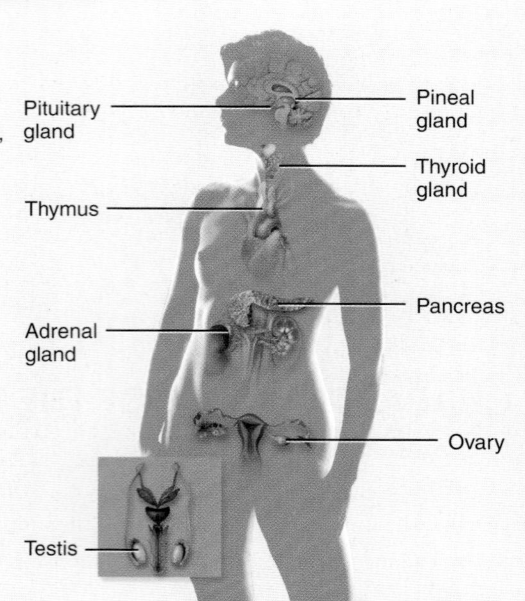

Lymphatic System and Immunity (Chapter 22)

Components: Lymphatic fluid and vessels; also includes spleen, thymus, lymph nodes, and tonsils.

Functions: Returns proteins and fluid to blood; carries lipids from gastrointestinal tract to blood; includes structures where lymphocytes that protect against disease-causing microbes mature and proliferate.

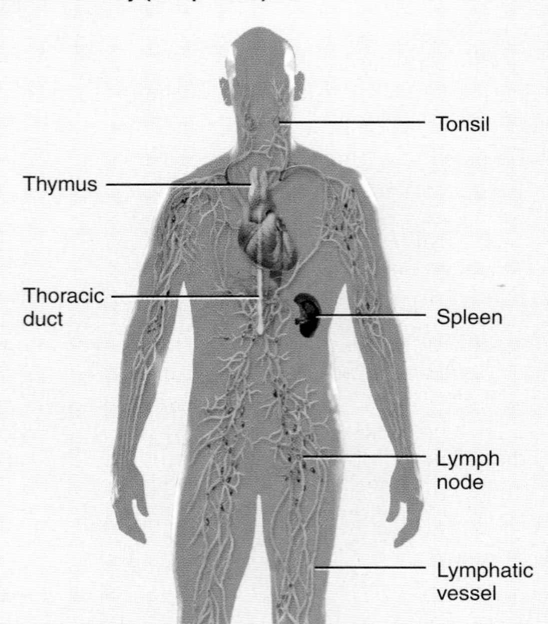

Cardiovascular System (Chapters 19–21)

Components: Blood, heart, and blood vessels.

Functions: Heart pumps blood through blood vessels; blood carries oxygen and nutrients to cells and carbon dioxide and wastes away from cells and helps regulate acid–base balance, temperature, and water content of body fluids; blood components help defend against disease and mend damaged blood vessels.

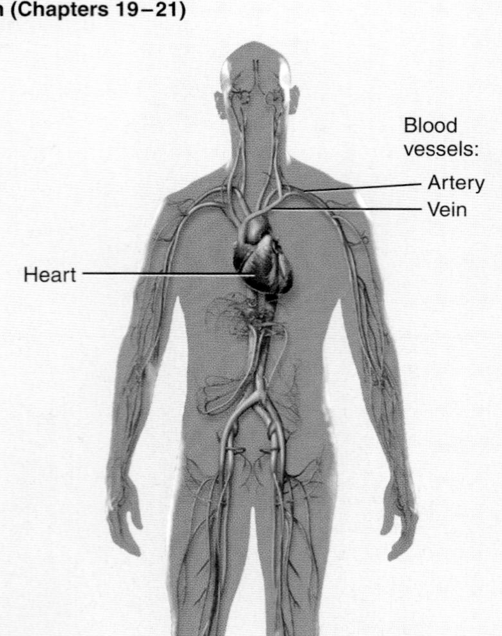

Respiratory System (Chapter 23)

Components: Lungs and air passageways such as the pharynx (throat), larynx (voice box), trachea (windpipe), and bronchial tubes leading into and out of them.

Functions: Transfers oxygen from inhaled air to blood and carbon dioxide from blood to exhaled air; helps regulate acid–base balance of body fluids; air flowing out of lungs through vocal cords produces sounds.

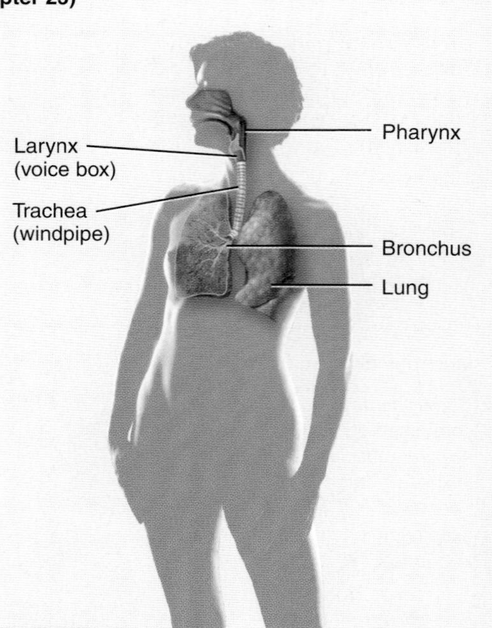

catabolism (ka-TAB-ō-lizm; *catabol-* = throwing down; *-ism* = a condition), the breakdown of complex chemical substances into simpler components. The other phase of metabolism is **anabolism** (a-NAB-ō-lizm; *anabol-* = a raising up), the building up of complex chemical substances from smaller, simpler components. For example, digestive processes catabolize (split) proteins in food into amino acids. These amino acids are then used to anabolize (build) new proteins that make up body structures such as muscles and bones.

2. **Responsiveness** is the body's ability to detect and respond to changes. For example, a decrease in body temperature represents a change in the internal environment, and turning your head

Digestive System (Chapter 24)

Components: Organs of gastrointestinal tract,a long tube that includes the mouth, pharynx (throat), esophagus, stomach, small and large intestines, and anus; also includes accessory organs that assist in digestive processes, such as the salivary glands, liver, gallbladder, and pancreas.

Functions: Achieves physical and chemical breakdown of food; absorbs nutrients; eliminates solid wastes.

Urinary System (Chapter 26)

Components: Kidneys, ureters, urinary bladder, and urethra.

Functions: Produces, stores, and eliminates urine; eliminates wastes and regulates volume and chemical composition of blood; helps maintain the acid–base balance of body fluids; maintains body's mineral balance; helps regulate production of red blood cells.

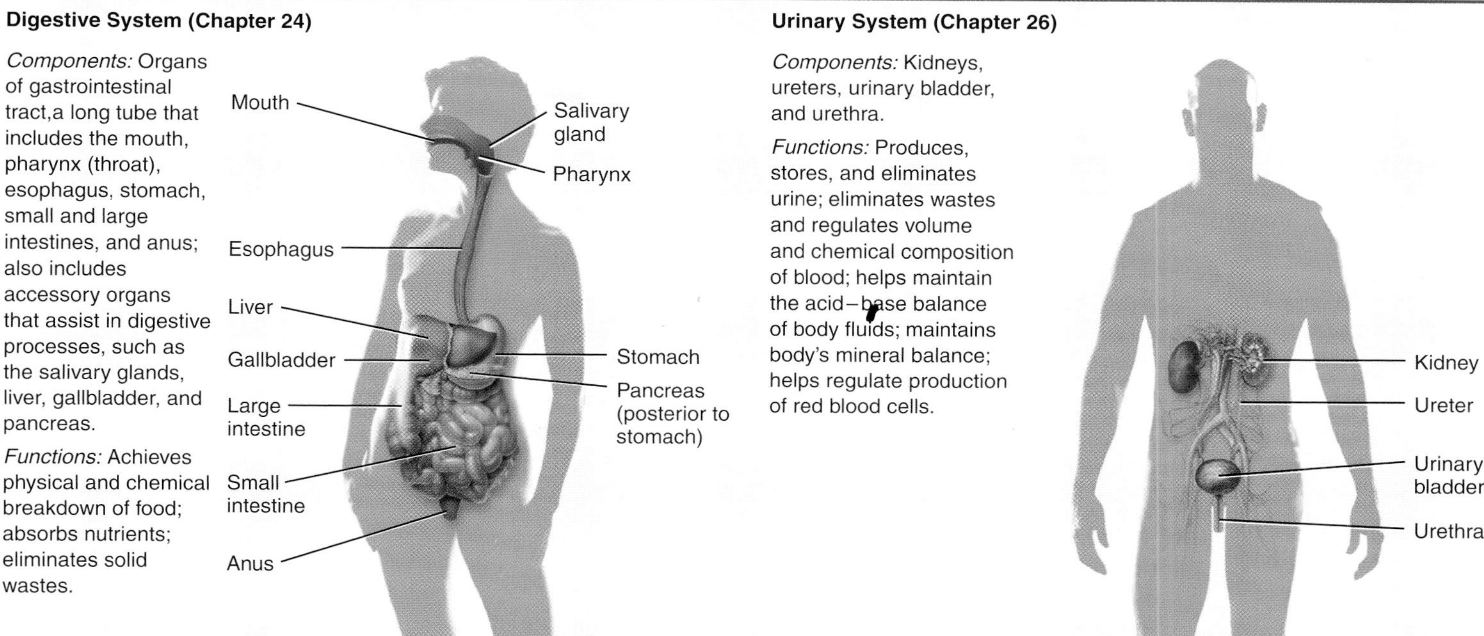

Reproductive Systems (Chapter 28)

Components: Gonads (testes in males and ovaries in females) and associated organs (uterine tubes, uterus, and vagina in females and epididymis, ductus deferens, and penis in males).

Functions: Gonads produce gametes (sperm or oocytes) that unite to form a new organism; gonads also release hormones that regulate reproduction and other body processes; associated organs transport and store gametes.

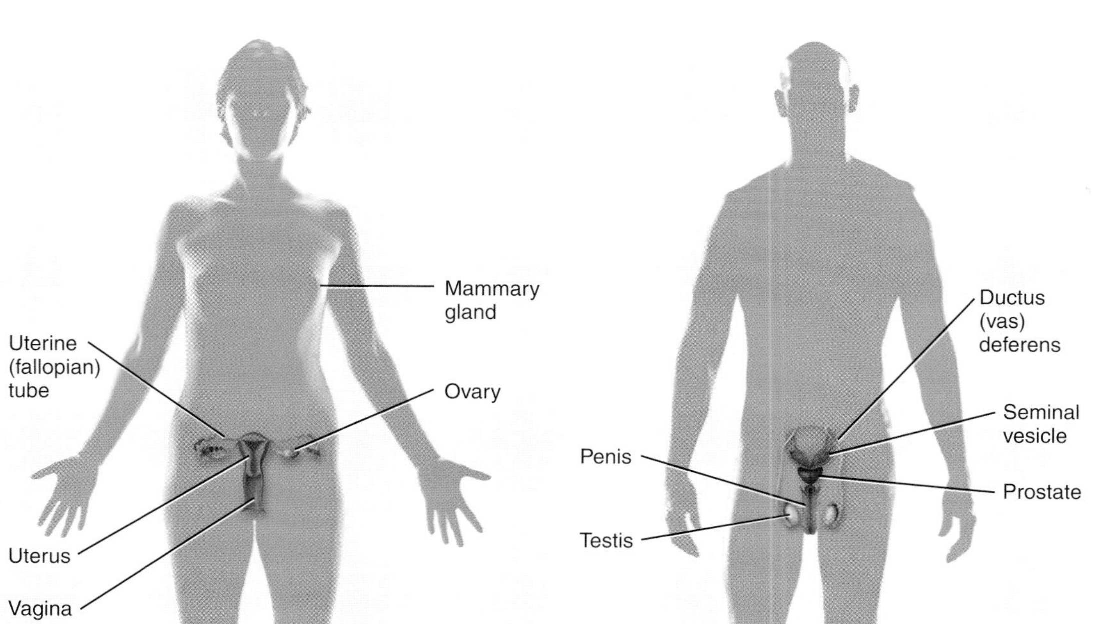

toward the sound of squealing brakes is a response to change in the external environment. Different cells in the body respond to environmental changes in characteristic ways. Nerve cells respond by generating electrical signals known as nerve impulses (action potentials). Muscle cells respond by contracting, which generates force to move body parts.

3. Movement includes motion of the whole body, individual organs, single cells, and even tiny structures inside cells. For example, the coordinated action of leg muscles moves your whole body from one place to another when you walk or run. After you eat a meal that contains fats, your gallbladder contracts and squirts bile into the gastrointestinal tract to aid in the digestion of fats. When a body tissue is damaged or infected, certain white blood cells move from the blood into the affected tissue to help clean up and repair the area. Inside the cell, various parts move from one position to another to carry out their functions.

4. **Growth** is an increase in body size that results from an increase in the size of existing cells, the number of cells, or both. In addition, a tissue sometimes increases in size because the amount of material between cells increases. In a growing bone, for example, mineral deposits accumulate between bone cells, causing the bone to enlarge in length and width.

5. **Differentiation** (dif′-er-en-shē-Ā-shun) is the development of a cell from an unspecialized to a specialized state. As you will see later in the text, each type of cell in the body has a specialized structure and function that differs from that of its precursor cells. For example, red blood cells and several types of white blood cells all arise from the same unspecialized precursor (ancestor) cells in red bone marrow. Such precursor cells, which can divide and give rise to cells that undergo differentiation, are known as **stem cells.** Also through differentiation, a fertilized egg (ovum) develops into an embryo, and then into a fetus, an infant, a child, and finally an adult.

6. **Reproduction** refers either to the formation of new cells for tissue growth, repair, or replacement, or to the production of a new individual. In humans, the former process occurs continuously throughout life, which continues from one generation to the next through the latter process, the fertilization of an ovum by a sperm cell.

When the life processes cease to occur properly, the result is death of cells and tissues, which may lead to death of the organism. Clinically, loss of the heartbeat, absence of spontaneous breathing, and loss of brain functions indicate death in the human body.

Homeostasis

Homeostasis (hō′mē-ō-STĀ-sis; *homeo-* = sameness; *-stasis* = standing still) is the condition of equilibrium (balance) in the body's internal environment due to the ceaseless interplay of the body's many regulatory processes. Homeostasis is a dynamic condition. In response to changing conditions, the body's equilibrium can shift among points in a narrow range that is compatible with maintaining life. For example, the level of glucose in blood normally stays between 70 and 110 milligrams of glucose per 100 milliliters of blood.* Each structure, from the cellular level to the systemic level, contributes in some way to keeping the internal environment of the body within normal limits.

*Appendix A describes metric measurements.

Body Fluids

An important aspect of homeostasis is maintaining the volume and composition of **body fluids,** dilute, watery solutions containing dissolved chemicals that are found inside cells as well as surrounding them. The fluid within cells is **intracellular fluid** (*intra-* = inside), abbreviated **ICF.** The fluid outside body cells is **extracellular fluid** (*extra-* = outside), abbreviated **ECF.** The ECF that fills the narrow spaces between cells of tissues is known as **interstitial fluid** (in′-ter-STISH-al; *inter-* = between). As you progress with your studies, you will learn that the ECF differs depending on where it occurs in the body: ECF within blood vessels is termed **blood plasma,** within lymphatic vessels it is called **lymph,** in and around the brain and spinal cord it is known as **cerebrospinal fluid,** in joints it is referred to as **synovial fluid,** and the ECF of the eyes is called **aqueous humor** and **vitreous body.**

The proper functioning of body cells depends on precise regulation of the composition of the fluid surrounding them. Because interstitial fluid surrounds all body cells, it is often called the body's *internal environment.* The composition of interstitial fluid changes as substances move back and forth between it and blood plasma. Such exchange of materials occurs across the thin walls of the smallest blood vessels in the body, the *blood capillaries.* This movement in both directions across capillary walls provides needed materials, such as glucose, oxygen, ions, and so on, to tissue cells. It also removes wastes, such as carbon dioxide, from interstitial fluid.

▶ **CHECKPOINT**

6. Which life process in the human body sustains all the others?
7. Describe the locations of intracellular fluid, extracellular fluid, interstitial fluid, and blood plasma.
8. Why is interstitial fluid called the internal environment of the body?

CONTROL OF HOMEOSTASIS

▶ **OBJECTIVES**

Describe the components of a feedback system.

Contrast the operation of negative and positive feedback systems.

Explain how homeostatic imbalances are related to disorders.

Homeostasis in the human body is continually being disturbed. Some disruptions come from the external environment (outside the body) in the form of physical insults such as the intense heat of a Texas summer or a lack of enough oxygen for that two-mile run. Other disruptions originate in the internal environment (within the body), such as a blood glucose level that falls too low when you skip breakfast. Homeostatic imbalances may also occur due to psychological stresses in our social environment—the demands of work and school, for example. In most cases the disruption of homeostasis is mild and temporary, and the responses of body cells quickly restore balance in the internal environment. However, in some cases the disruption of homeostasis may be intense and prolonged, as in poisoning, overexposure to temperature extremes, or severe infection.

Fortunately, the body has many regulating systems that can usually bring the internal environment back into balance. Most often, the nervous system and the endocrine system, working together or independently, provide the needed corrective measures. The nervous system regulates homeostasis by sending electrical signals known as *nerve impulses (action potentials)* to organs that can counteract deviations from the balanced state. The endocrine system includes many glands that secrete messenger molecules called *hormones* into the blood. Nerve impulses typically cause rapid changes, but hormones usually work more slowly. Both means of regulation, however, work toward the same end, usually through negative feedback systems.

Feedback Systems

The body can regulate its internal environment through a multitude of feedback systems. A **feedback system** or *feedback loop* is a cycle of events in which the status of a body condition is monitored, evaluated, changed, remonitored, reevaluated, and so on. Each monitored variable, such as body temperature, blood pressure, or blood glucose level, is termed a *controlled condition.* Any disruption that changes a controlled condition is called a *stimulus.* A feedback system includes three basic components—a receptor, a control center, and an effector (Figure 1.2).

1. A **receptor** is a body structure that monitors changes in a controlled condition and sends input to a control center. Typically, the input is in the form of nerve impulses or chemical signals. For example, certain nerve endings in the skin sense temperature and can detect changes, such as a dramatic drop in temperature.

2. A **control center** in the body, for example, the brain, sets the range of values within which a controlled condition should be maintained, evaluates the input it receives from receptors, and generates output commands when they are needed. Output from the control center typically occurs as nerve impulses, or hormones or other chemical signals. In our skin temperature example, the brain acts as the control center, receiving nerve impulses from the skin receptors and generating nerve impulses as output.

3. An **effector** is a body structure that receives output from the control center and produces a *response* or effect that changes the controlled condition. Nearly every organ or tissue in the body can behave as an effector. When your body temperature drops sharply, your brain (control center) sends nerve impulses (output) to your skeletal muscles (effectors).

Figure 1.2 **Operation of a feedback system.** The dashed return arrow symbolizes negative feedback.

The three basic components of a feedback system are receptors, a control center, and effectors.

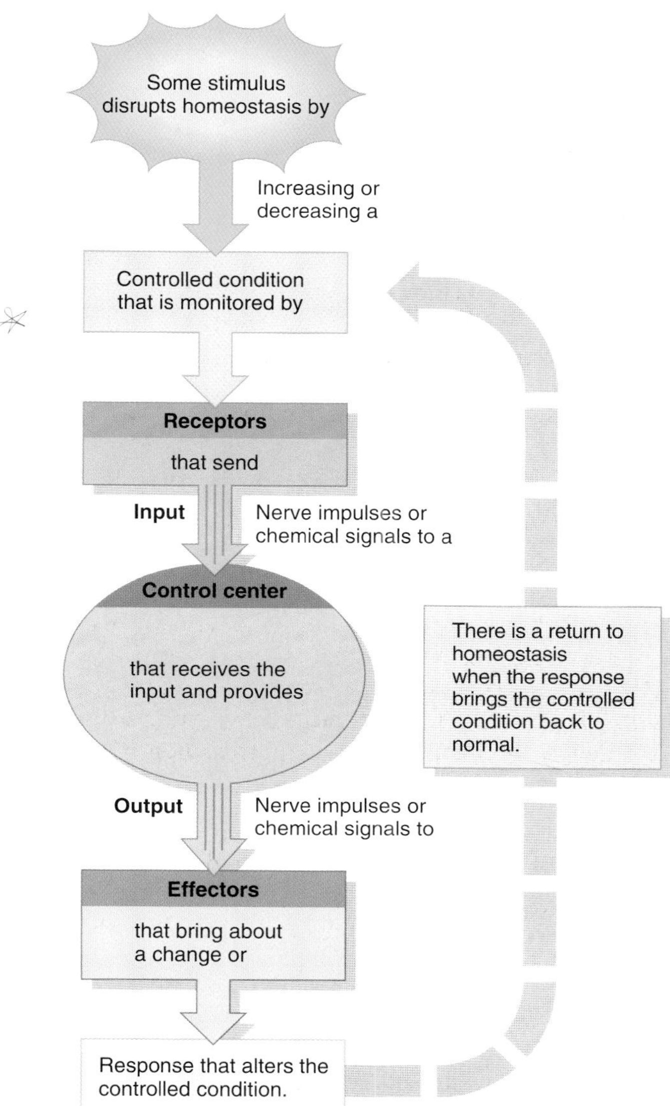

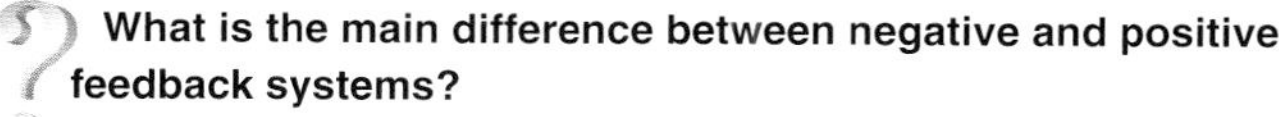

What is the main difference between negative and positive feedback systems?

The result is shivering, which generates heat and raises your body temperature.

A group of receptors and effectors communicating with their control center forms a feedback system that can regulate a controlled condition in the body's internal environment. In a feedback system, the response of the system "feeds back" information to change the controlled condition in some way, either negating it (negative feedback) or enhancing it (positive feedback).

Negative Feedback Systems

A **negative feedback system** *reverses* a change in a controlled condition. Consider the regulation of blood pressure. Blood pressure (BP) is the force exerted by blood as it presses against the walls of blood vessels. When the heart beats faster or harder, BP increases. If some internal or external stimulus causes blood pressure (controlled condition) to rise, the following sequence of events occurs (Figure 1.3). *Baroreceptors* (the receptors), pressure-sensitive nerve cells located in the walls of certain blood vessels, detect the higher pressure. The baroreceptors send nerve impulses (input) to the brain (control center), which interprets the impulses and responds by sending nerve impulses (output) to the heart (the effector). Heart rate decreases, which causes BP to decrease (response). This sequence of events quickly returns the controlled condition—blood pressure—to normal, and homeostasis is restored. Notice that the activity of the effector causes BP to drop, a result that negates the original stimulus (an increase in BP). This is why is it called a negative feedback system.

Positive Feedback Systems

A **positive feedback system** tends to *strengthen* or *reinforce* a change in one of the body's controlled conditions. A positive feedback system operates similarly to a negative feedback system, except for the way the response affects the controlled condition. The control center still provides commands to an effector, but this time the effector produces a physiological response that adds to or *reinforces* the initial change in the controlled condition. The action of a positive feedback system continues until it is interrupted by some mechanism.

Normal childbirth provides a good example of a positive feedback system (Figure 1.4). The first contractions of labor (stimulus) push part of the fetus into the cervix, the lowest part of the uterus, which opens into the vagina. Stretch-sensitive nerve cells (receptors) monitor the amount of stretching of the cervix (controlled condition). As stretching increases, they send more nerve impulses (input) to the brain (control center), which in turn releases the hormone oxytocin (output) into the blood. Oxytocin causes muscles in the wall of the uterus (effector) to contract even more forcefully. The contractions push the fetus farther down the uterus, which stretches the cervix even more. The cycle of stretching, hormone release, and ever-stronger contractions is

Figure 1.3 Homeostatic regulation of blood pressure by a negative feedback system. Note that the response is fed back into the system, and the system continues to lower blood pressure until there is a return to normal blood pressure (homeostasis).

If the response reverses the stimulus, a system is operating by negative feedback.

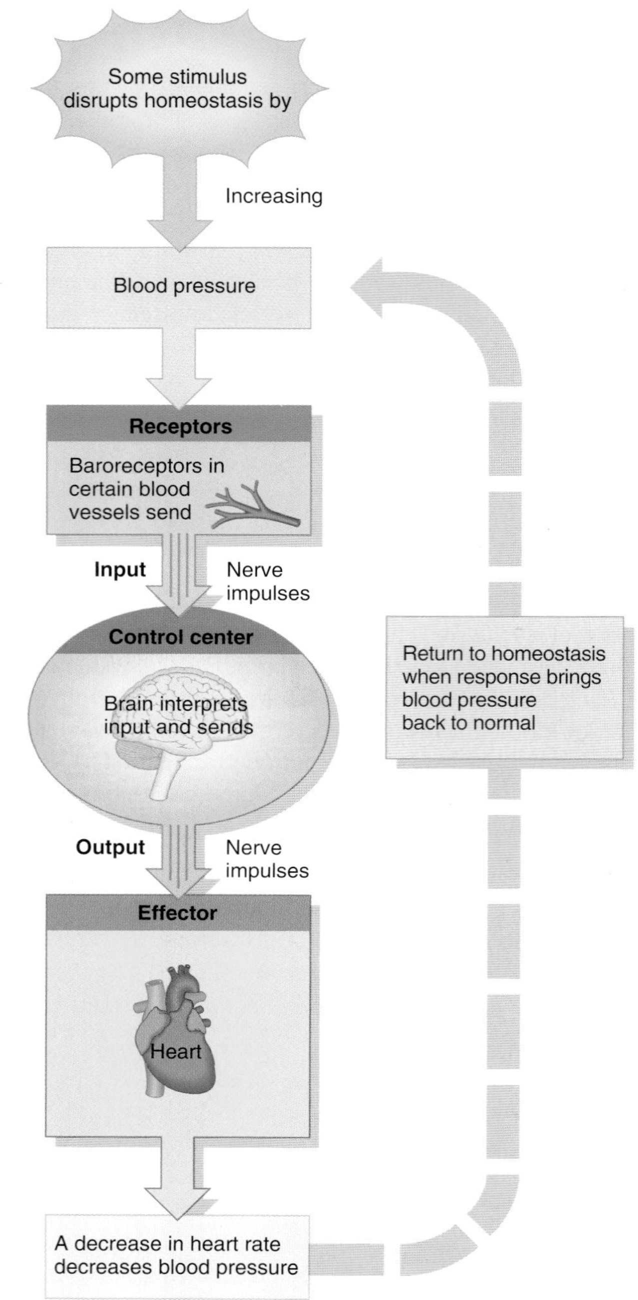
Some stimulus disrupts homeostasis by
Increasing
Blood pressure
Receptors
Baroreceptors in certain blood vessels send
Input
Nerve impulses
Control center
Brain interprets input and sends
Output
Nerve impulses
Effector
Heart
A decrease in heart rate decreases blood pressure
Return to homeostasis when response brings blood pressure back to normal

What would happen to heart rate if some stimulus caused blood pressure to decrease? Would this occur by way of positive or negative feedback?

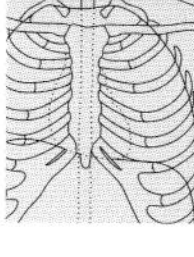

Figure 1.4 Positive feedback control of labor contractions during birth of a baby. The solid return arrow symbolizes positive feedback.

If the response enhances or intensifies the stimulus, a system is operating by positive feedback.

Contractions of wall of uterus force baby's head or body into the cervix, thus
Increasing
Stretching of cervix
Receptors
Stretch-sensitive nerve cells in cervix send
Input
Nerve impulses
Control center
Brain interprets input and releases
Positive feedback: Increased stretching of cervix causes release of more oxytocin, which results in more stretching of the cervix
Output
Oxytocin
Effectors
Muscles in wall of uterus contract more forcefully
Baby's body stretches cervix more
Interruption of cycle: Birth of baby decreases stretching of cervix, thus breaking the positive feedback cycle

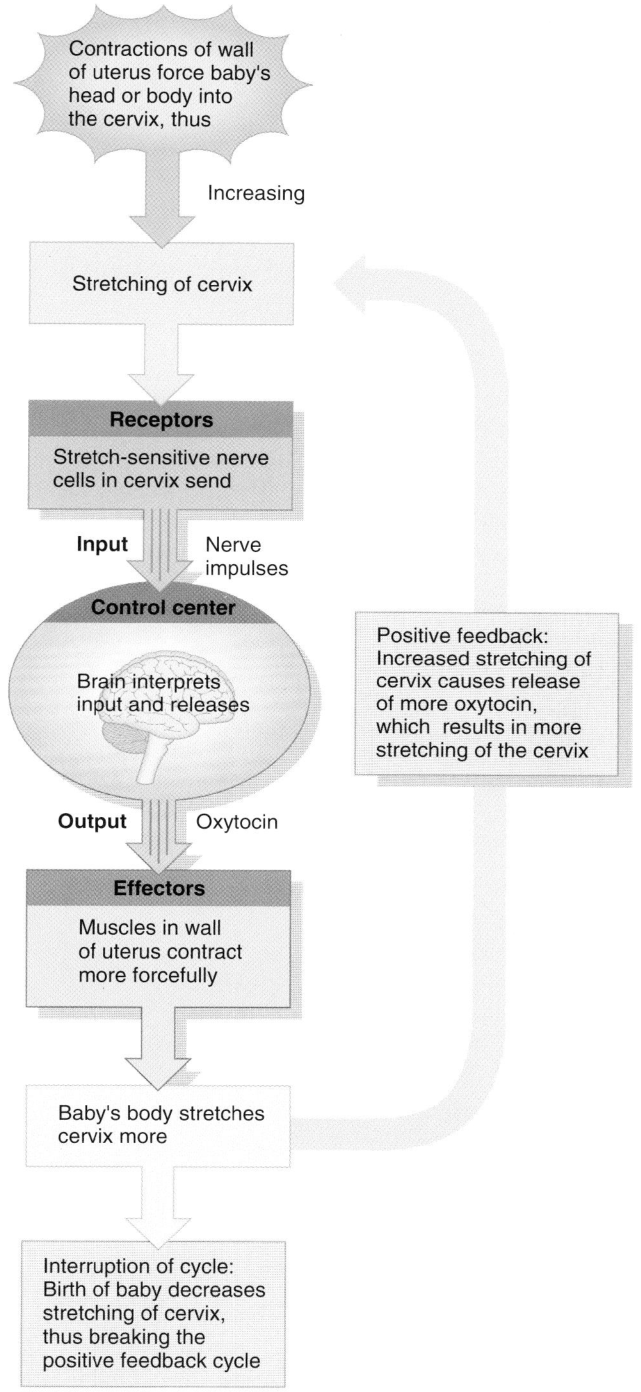

Why do positive feedback systems that are part of a normal physiological response include some mechanism that terminates the system?

interrupted only by the birth of the baby. Then, stretching of the cervix ceases and oxytocin is no longer released.

Another example of positive feedback is what happens to your body when you lose a great deal of blood. Under normal conditions, the heart pumps blood under sufficient pressure to body cells to provide them with oxygen and nutrients to maintain homeostasis. Upon severe blood loss, blood pressure drops and blood cells (including heart cells) receive less oxygen and function less efficiently. If the blood loss continues, heart cells become weaker, the pumping action of the heart decreases further, and blood pressure continues to fall. This is an example of a positive feedback cycle that has serious consequences and may even lead to death if there is no medical intervention. As you will see in Chapter 19, blood clotting is also an example of a positive feedback system.

These examples suggest some important differences between positive and negative feedback systems. Because a positive feedback system continually reinforces a change in a controlled condition, some event outside the system must shut it off. If the action of a positive feedback system is not stopped, it can "run away" and may even produce life-threatening conditions in the body. The action of a negative feedback system, by contrast, slows and then stops as the controlled condition returns to its normal state. Usually, positive feedback systems reinforce conditions that do not happen very often, and negative feedback systems regulate conditions in the body that remain fairly stable over long periods.

Homeostatic Imbalances

As long as all the body's controlled conditions remain within certain narrow limits, body cells function efficiently, negative feedback systems maintain homeostasis, and the body stays healthy. Should one or more components of the body lose their ability to contribute to homeostasis, however, the normal equilibrium among body processes may be disturbed. If the homeostatic imbalance is moderate, a disorder or disease may occur; if it is severe, death may result.

A **disorder** is any abnormality of structure or function. **Disease** is a more specific term for an illness characterized by a recognizable set of signs and symptoms. A *local disease* affects one part or a limited region of the body; a *systemic disease* affects either the entire body or several parts of it. Diseases alter body structures and functions in characteristic ways. A person with a disease may experience **symptoms,** *subjective* changes in body functions that are not apparent to an observer. Examples of symptoms are headache, nausea, and anxiety. *Objective* changes that a clinician can observe and measure are called **signs.** Signs of disease can be either anatomical, such as swelling or a rash, or physiological, such as fever, high blood pressure, or paralysis.

The science that deals with why, when, and where diseases occur and how they are transmitted among individuals in a community is known as **epidemiology** (ep′-i-dē-mē-OL-ō-jē; *epi-* =

upon; *-demi* = people). **Pharmacology** (far′-ma-KOL-ō-jē; *pharmac-* = drug) is the science that deals with the effects and uses of drugs in the treatment of disease.

Diagnosis of Disease

Diagnosis (dī′-ag-NŌ-sis; *dia-* = through; *-gnosis* = knowledge) is the science and skill of distinguishing one disorder or disease from another. The patient's symptoms and signs, his or her medical history, a physical exam, and laboratory tests provide the basis for making a diagnosis. Taking a *medical history* consists of collecting information about events that might be related to a patient's illness. These include the chief complaint (primary reason for seeking medical attention), history of present illness, past medical problems, family medical problems, social history, and review of symptoms. A *physical examination* is an orderly evaluation of the body and its functions. This process includes the noninvasive techniques of inspection, palpation, auscultation, and percussion that you learned about earlier in the chapter, along with measurement of vital signs (temperature, pulse, respiratory rate, and blood pressure), and sometimes laboratory tests. ■

▶ CHECKPOINT

9. What types of disturbances can act as stimuli that initiate a feedback system?
10. How are negative and positive feedback systems similar? How are they different?
11. What is the difference between symptoms and signs of a disease? Give examples of each.

ANATOMICAL TERMINOLOGY

▶ OBJECTIVES

Describe the orientation of the body in the anatomical position.

Relate the common names to the corresponding anatomical descriptive terms for various regions of the human body.

Define the anatomical planes, sections, and directional terms used to describe the human body.

Outline the major body cavities, the organs they contain, and their associated linings.

Scientists and health-care professionals use a common language of special terms when referring to body structures and their functions. The language of anatomy they use has precisely defined meanings that allow us to communicate clearly and precisely. For example, is it correct to say, "The wrist is above the fingers"? This might be true if your arms are at your sides. But if you hold your hands up above your head, your fingers would be above your wrists. To prevent this kind of confusion, anatomists developed a standard anatomical position and a special vocabulary for relating body parts to one another.

Body Positions

Descriptions of any region or part of the human body assume that it is in a specific stance called the **anatomical position.** In the anatomical position, the subject stands erect facing the observer, with the head level and the eyes facing directly forward. The feet are flat on the floor and directed forward, and the arms are at the sides with the palms turned forward (Figure 1.5). In the anatomical position, the body is upright. Two terms describe a reclining body. If the body is lying face down, it is in the **prone** position. If the body is lying face up, it is in the **supine** position.

Regional Names

The human body is divided into several major regions that can be identified externally. The principal regions are the head, neck, trunk, upper limbs, and lower limbs (Figure 1.5). The **head** consists of the skull and face. The *skull* encloses and protects the brain; the *face* is the front portion of the head that includes the eyes, nose, mouth, forehead, cheeks, and chin. The **neck** supports the head and attaches it to the trunk. The **trunk** consists of the chest, abdomen, and pelvis. Each **upper limb** attaches to the trunk and consists of the shoulder, armpit, arm (portion of the limb from the shoulder to the elbow), forearm (portion of the limb from the elbow to the wrist), wrist, and hand. Each **lower limb** also attaches to the trunk and consists of the buttock, thigh (portion of the limb from the buttock to the knee), leg (portion of the limb from the knee to the ankle), ankle, and foot. The *groin* is the area on the front surface of the body marked by a crease on each side, where the trunk attaches to the thighs.

Figure 1.5 shows the common names of major parts of the body. The corresponding anatomical descriptive form (adjective) for each part appears in parentheses next to the common name. For example, if you receive a tetanus shot in your *buttock,* it is a *gluteal* injection. Because the descriptive form of a body part usually is based on a Greek or Latin word, it may look different from the common name for the same part or area. For example, the Latin word for armpit is *axilla* (ak-SIL-a). Thus, one of the nerves passing within the armpit is named the axillary nerve. You will learn more about the Greek and Latin word roots of anatomical and physiological terms as you read this book.

Directional Terms

To locate various body structures, anatomists use specific **directional terms,** words that describe the position of one body part relative to another. Several directional terms are grouped in pairs that have opposite meanings, such as anterior (front) and posterior (back). Exhibit 1.1 on page 14 and Figure 1.6 on page 15 present the main directional terms.

Figure 1.5 The anatomical position. The common names and corresponding anatomical terms (in parentheses) are indicated for specific body regions. For example, the head is the cephalic region.

In the anatomical position, the subject stands erect facing the observer with the head level and the eyes facing forward. The feet are flat on the floor and directed forward, and the arms are at the sides with the palms facing forward.

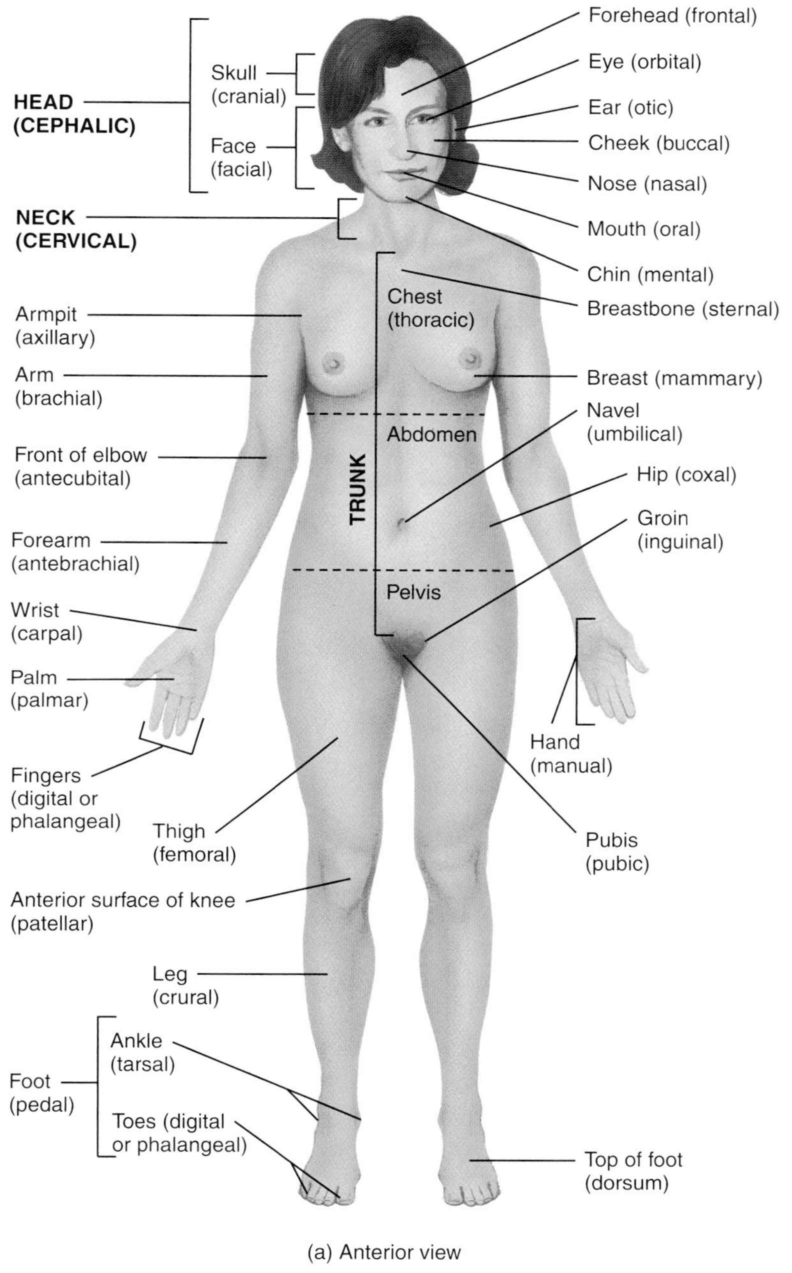

(a) Anterior view

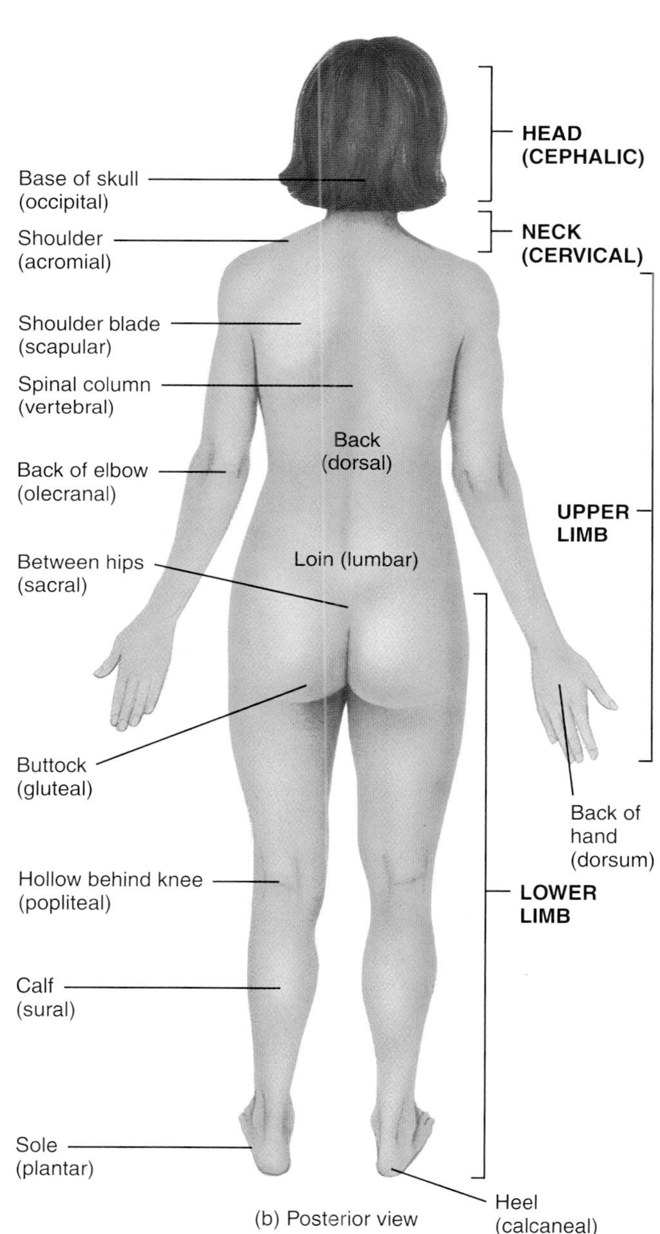

(b) Posterior view

What is the usefulness of defining one standard anatomical position?

EXHIBIT 1.1 DIRECTIONAL TERMS (FIGURE 1.6)

▶ **OBJECTIVE**

Define each directional term used to describe the human body.

Overview

Most of the directional terms used to describe the human body can be grouped into pairs that have opposite meanings. For example, **superior** means toward the upper part of the body, and **inferior** means toward the lower part of the body. It is important to understand that directional terms have relative meanings; they make sense only when used to describe the position of one structure relative to another. For example, your knee is superior to your ankle, even though both are located in the inferior half of the body. Study the directional terms below and the example of how each is used. As you read the examples, look at Figure 1.6 to see the location of each structure.

▶ **CHECKPOINT**

Which directional terms can be used to specify the relationships between (1) the elbow and the shoulder, (2) the left and right shoulders, (3) the sternum and the humerus, and (4) the heart and the diaphragm?

DIRECTIONAL TERM	DEFINITION	EXAMPLE OF USE
Superior (soo′-PĒR-ē-or) **(cephalic or cranial)**	Toward the head, or the upper part of a structure.	The heart is superior to the liver.
Inferior (in′-FĒR-ē-or) **(caudal)**	Away from the head, or the lower part of a structure.	The stomach is inferior to the lungs.
Anterior (an-TĒR-ē-or) **(ventral)***	Nearer to or at the front of the body.	The sternum (breastbone) is anterior to the heart.
Posterior (pos-TĒR-ē-or) **(dorsal)**	Nearer to or at the back of the body.	The esophagus is posterior to the trachea (windpipe).
Medial (MĒ-dē-al)	Nearer to the midline.†	The ulna is medial to the radius.
Lateral (LAT-er-al)	Farther from the midline.	The lungs are lateral to the heart.
Intermediate (in′-ter-MĒ-dē-at)	Between two structures.	The transverse colon is intermediate between the ascending and descending colons.
Ipsilateral (ip-si-LAT-er-al)	On the same side of the body as another structure.	The gallbladder and ascending colon are ipsilateral.
Contralateral (CON-tra-lat-er-al)	On the opposite side of the body from another structure.	The ascending and descending colons are contralateral.
Proximal (PROK-si-mal)	Nearer to the attachment of a limb to the trunk; nearer to the origination of a structure.	The humerus is proximal to the radius.
Distal (DIS-tal)	Farther from the attachment of a limb to the trunk; farther from the origination of a structure.	The phalanges are distal to the carpals.
Superficial (soo′-per-FISH-al)	Toward or on the surface of the body.	The ribs are superficial to the lungs.
Deep	Away from the surface of the body.	The ribs are deep to the skin of the chest and back.

*Ventral refers to the belly side, whereas dorsal refers to the back side. In four-legged animals, anterior = cephalic (toward the head), ventral = inferior, posterior = caudal (toward the tail), and dorsal = superior.

†The midline is an imaginary vertical line that divides the body into equal right and left sides.

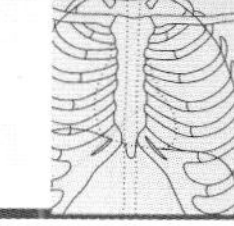

Figure 1.6 Directional terms.

Directional terms precisely locate various parts of the body relative to one another.

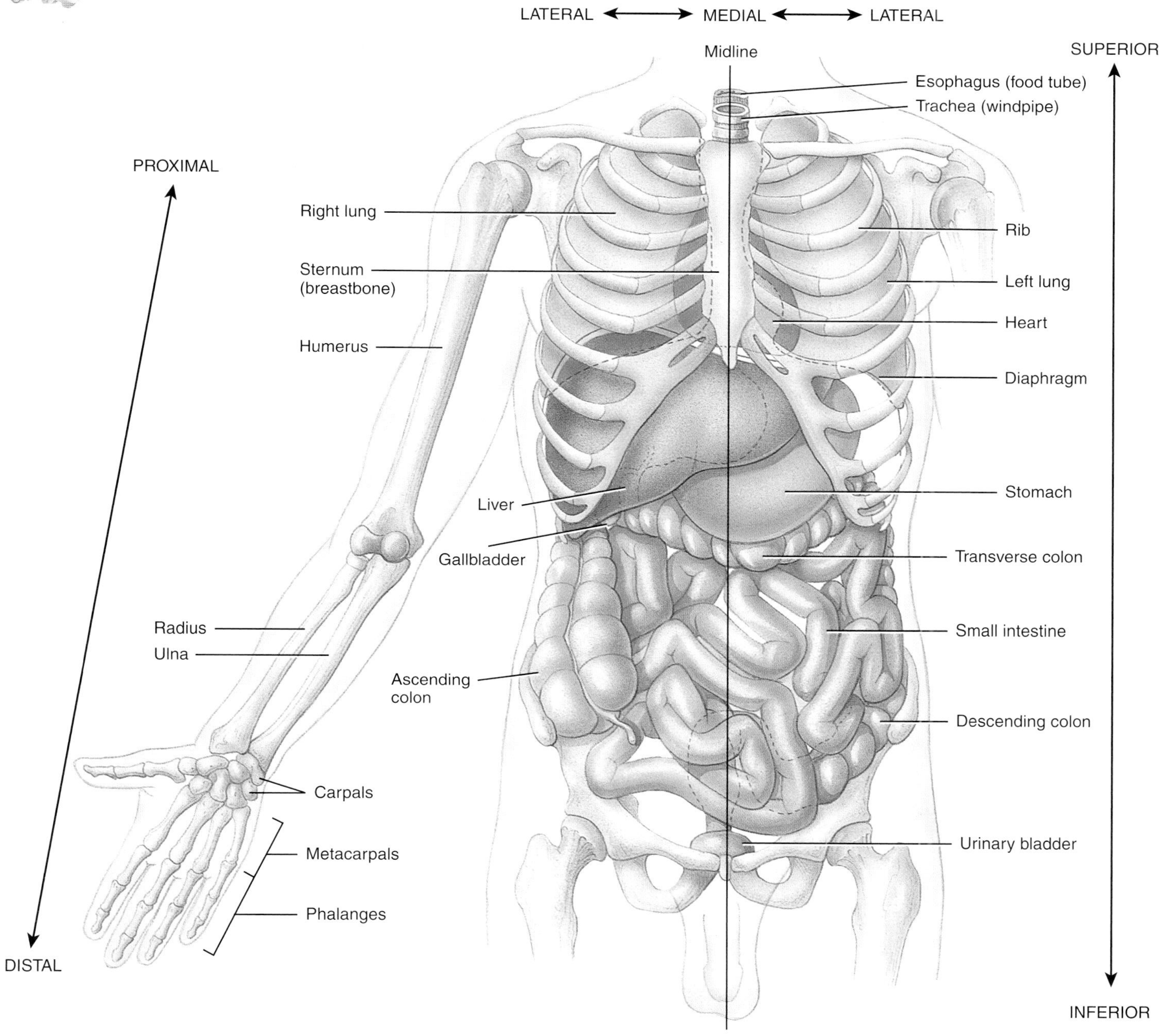

LATERAL ⟷ MEDIAL ⟷ LATERAL

Midline

SUPERIOR

INFERIOR

PROXIMAL

DISTAL

Esophagus (food tube)

Trachea (windpipe)

Right lung

Sternum (breastbone)

Humerus

Rib

Left lung

Heart

Diaphragm

Liver

Gallbladder

Stomach

Transverse colon

Small intestine

Radius

Ulna

Ascending colon

Descending colon

Carpals

Metacarpals

Phalanges

Urinary bladder

Anterior view of trunk and right upper limb

Is the radius proximal to the humerus? Is the esophagus anterior to the trachea? Are the ribs superficial to the lungs? Is the urinary bladder medial to the ascending colon? Is the sternum lateral to the descending colon?

Planes and Sections

You will also study parts of the body relative to **planes,** imaginary flat surfaces that pass through the body parts (Figure 1.7). A **sagittal plane** (SAJ-i-tal; *sagitt-* = arrow) is a vertical plane that divides the body or an organ into right and left sides. More specifically, when such a plane passes through the midline of the body or an organ and divides it into *equal* right and left sides, it is called a **midsagittal plane** or a **median plane.** If the sagittal plane does not pass through the midline but instead divides the body or an organ into *unequal* right and left sides, it is called a **parasagittal plane** (*para-* = near). A **frontal** or **coronal plane** (kō-RŌ-nal; *corona* = crown) divides the body or an organ into anterior (front) and posterior (back) portions. A **transverse plane** divides the body or an organ into superior (upper) and inferior (lower) portions. Other names for a transverse plane are a **cross-sectional** or **horizontal plane.** Sagittal, frontal, and transverse planes are all at right angles to one another. An **oblique plane,** by contrast, passes through the body or an organ at an angle between the transverse plane and either a sagittal or frontal plane.

When you study a body region, you often view it in section. A **section** is one flat surface of a three-dimensional structure or a cut along a plane. It is important to know the plane of the section so you can understand the anatomical relationship of one part to another. Figure 1.8 indicates how three different sections—*transverse, frontal,* and *midsagittal*—provide different views of the brain.

Figure 1.7 Planes through the human body.

Frontal, transverse, sagittal, and oblique planes divide the body in specific ways.

Frontal plane
Transverse plane
Parasagittal plane
Midsagittal plane
Oblique plane

Right anterolateral view

Which plane divides the heart into anterior and posterior portions?

Figure 1.8 Planes and sections through different parts of the brain. The diagrams (left) show the planes, and the photographs (right) show the resulting sections. Note: The arrows in the diagrams indicate the direction from which each section is viewed. This aid is used throughout the book to indicate viewing perspectives.

Planes divide the body in various ways to produce sections.

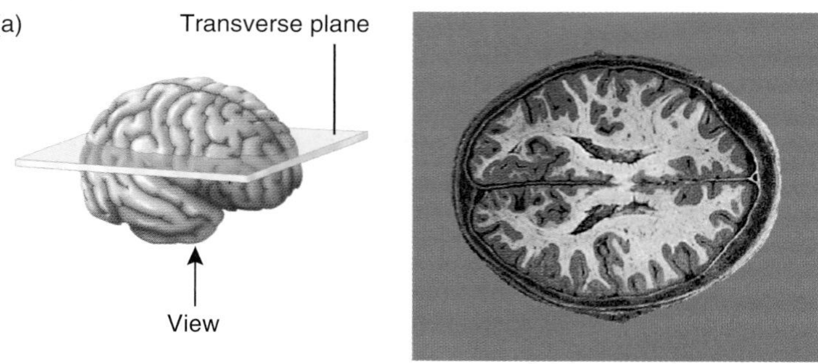

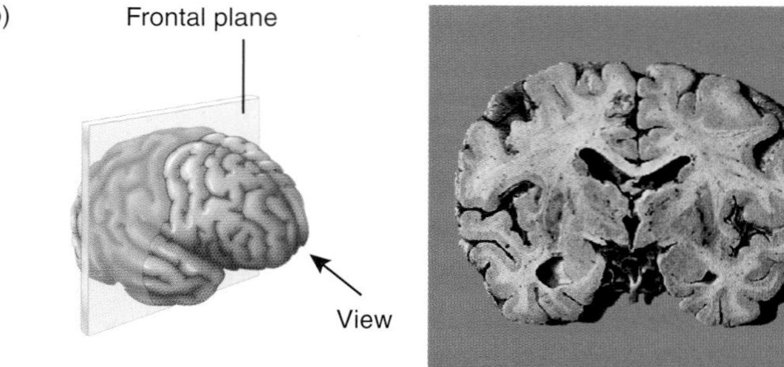

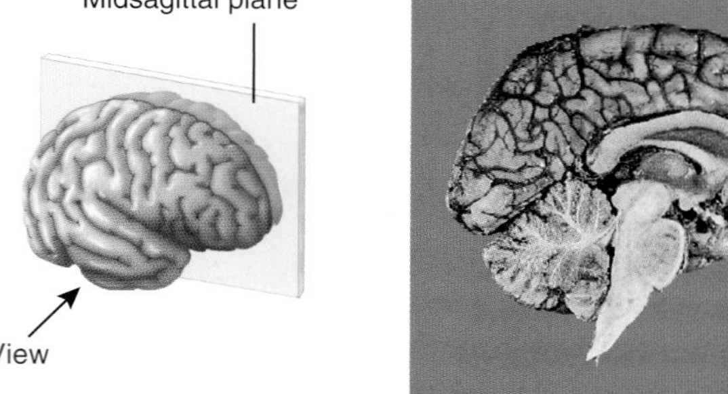

Which plane divides the brain into unequal right and left portions?

Body Cavities

Body cavities are spaces within the body that help protect, separate, and support internal organs. Bones, muscles, ligaments, and other structures separate the various body cavities from one another. Here we discuss several of the larger body cavities (Figure 1.9).

The cranial bones form the **cranial cavity,** which contains the brain. The bones of the vertebral column (backbone) form the **vertebral (spinal) cavity,** which contains the spinal cord. Three layers of protective tissue, the **meninges** (me-NIN-jēz), line the cranial cavity and the vertebral cavity.

The major body cavities of the trunk are the thoracic and abdominopelvic cavities. The **thoracic cavity** (thor-AS-ik; *thorac-* = chest) or chest cavity (Figure 1.10) is formed by the ribs, the muscles of the chest, the sternum (breastbone), and the thoracic portion of the vertebral column (backbone). Within the thoracic cavity are the **pericardial cavity** (per′-i-KAR-dē-al; *peri-* = around; *-cardial* = heart), a fluid-filled space that surrounds the heart, and two **pleural cavities** (PLOOR-al; *pleur-* = rib or side). Each pleural cavity surrounds one lung and contains a small amount of fluid. The central part of the thoracic cavity is called the **mediastinum** (mē′-dē-as-TĪ-num; *media-* = middle; *-stinum* = partition). It is between the lungs, extending from the sternum to the vertebral column and from the neck to the diaphragm (Figure 1.10a). The mediastinum contains all thoracic organs except the lungs themselves. Among the structures in the mediastinum are the heart, esophagus, trachea, thymus, and several large blood vessels. The **diaphragm** (DĪ-a-fram = partition or wall) is a dome-shaped muscle that separates the thoracic cavity from the abdominopelvic cavity.

The **abdominopelvic cavity** (ab-dom′-i-nō-PEL-vik; see Figure 1.9) extends from the diaphragm to the groin and is encircled by the abdominal wall and the bones and muscles of the pelvis. As the name suggests, the abdominopelvic cavity is

Figure 1.9 Body cavities. The dashed lines in (a) and (b) indicate the border between the abdominal and pelvic cavities. (See Tortora, *A Photographic Atlas of the Human Body, Second Edition,* Figures 6.5, 6.6, and 11.11.)

The major cavities of the trunk are the thoracic and abdominopelvic cavities.

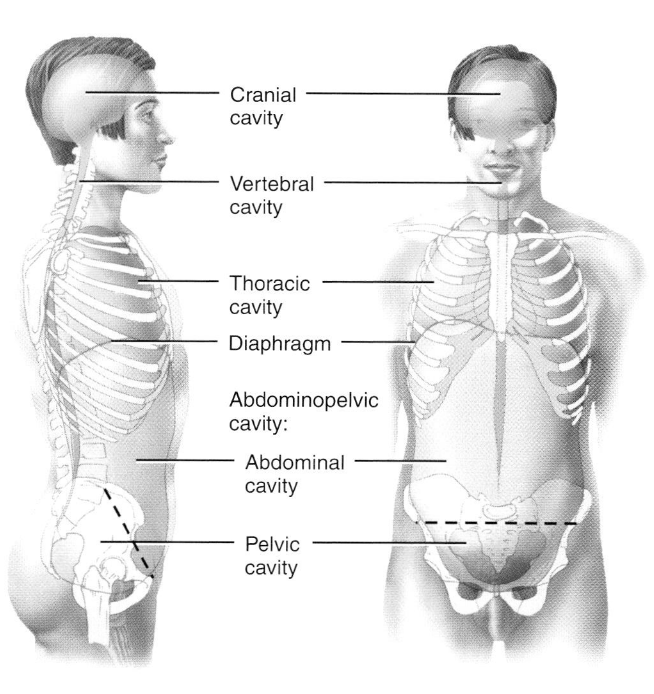

CAVITY	COMMENTS
Cranial cavity	Formed by cranial bones and contains brain.
Vertebral cavity	Formed by vertebral column and contains spinal cord and the beginnings of spinal nerves.
Thoracic cavity*	Chest cavity; contains pleural and pericardial cavities and mediastinum.
Pleural cavity	Each surrounds a lung; the serous membrane of the pleural cavities is the pleura.
Pericardial cavity	Surrounds the heart; the serous membrane of the pericardial cavity is the pericardium.
Mediastinum	Central portion of thoracic cavity between the lungs; extends from sternum to vertebral column and from neck to diaphragm; contains heart, thymus, esophagus, trachea, and several large blood vessels.
Abdominopelvic cavity	Subdivided into abdominal and pelvic cavities.
Abdominal cavity	Contains stomach, spleen, liver, gallbladder, small intestine, and most of large intestine; the serous membrane of the abdominal cavity is the peritoneum.
Pelvic cavity	Contains urinary bladder, portions of large intestine, and internal organs of reproduction.

* See figure 1.10 for details of the thoracic cavity

In which cavities are the following organs located: urinary bladder, stomach, heart, small intestine, lungs, internal female reproductive organs, thymus, spleen, liver? Use the following symbols for your response: T = thoracic cavity, A = abdominal cavity, or P = pelvic cavity.

Figure 1.10 The thoracic cavity. The dashed lines indicate the borders of the mediastinum. Note: When transverse sections are viewed inferiorly (from below), the anterior aspect of the body appears on top and the left side of the body appears on the right side of the illustration. (See Tortora, *A Photographic Atlas of the Human Body, Second Edition,* Figure 6.6.)

The thoracic cavity contains three smaller cavities and the mediastinum.

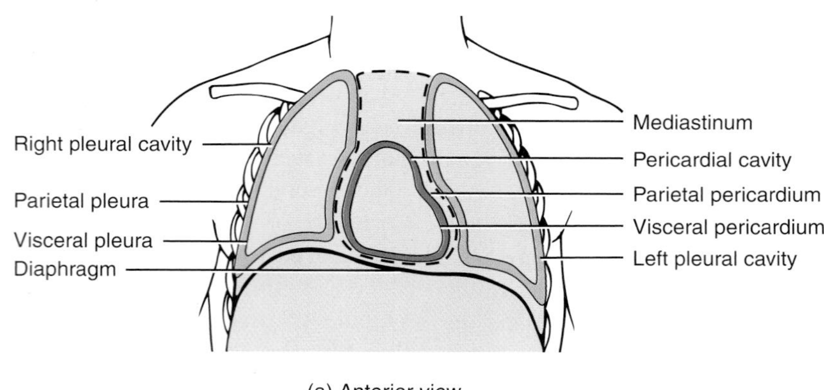

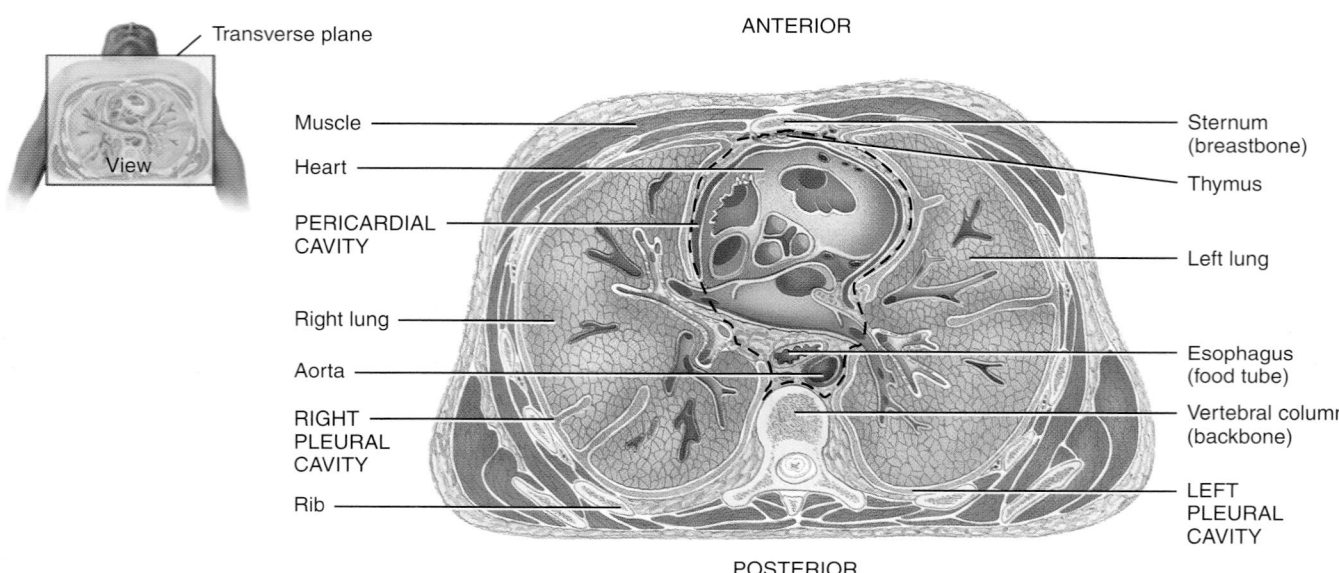

(a) Anterior view

Transverse plane
View
ANTERIOR
Muscle
Heart
PERICARDIAL CAVITY
Right lung
Aorta
RIGHT PLEURAL CAVITY
Rib
Sternum (breastbone)
Thymus
Left lung
Esophagus (food tube)
Vertebral column (backbone)
LEFT PLEURAL CAVITY
POSTERIOR

(b) Inferior view of transverse section of thoracic cavity

divided into two portions, even though no wall separates them (Figure 1.11). The superior portion, the **abdominal cavity** (*abdomin-* = belly), contains the stomach, spleen, liver, gallbladder, small intestine, and most of the large intestine. The inferior portion, the **pelvic cavity** (*pelv-* = basin), contains the urinary bladder, portions of the large intestine, and internal organs of the reproductive system. Organs inside the thoracic and abdominopelvic cavities are called **viscera** (VIS-er-a).

Thoracic and Abdominal Cavity Membranes

A thin, slippery, double layered **serous membrane** covers the viscera within the thoracic and abdominal cavities and also lines the walls of the thorax and abdomen. The parts of a serous membrane are (1) the *parietal layer* (pa-RĪ-e-tal), which lines the walls of the cavities, and (2) the *visceral layer,* which covers and adheres to the viscera within the cavities. Serous fluid between the two layers reduces friction, allowing the viscera to slide somewhat during movements, such as when the lungs inflate and deflate during breathing.

The serous membrane of the pleural cavities is called the **pleura** (PLOO-ra). The *visceral pleura* clings to the surface of the lungs, whereas the anterior part of the *parietal pleura* lines the chest wall, covering the superior surface of the diaphragm. In between is the pleural cavity, filled with a small volume of serous fluid. The serous membrane of the pericardial cavity is the **pericardium.** The *visceral pericardium* covers the surface of the heart, whereas the *parietal pericardium* lines the chest wall. Between them is the pericardial cavity. The **peritoneum** (per-i-tō-NĒum) is the serous membrane of the abdominal cavity. The *visceral peritoneum* covers the abdominal viscera, whereas the

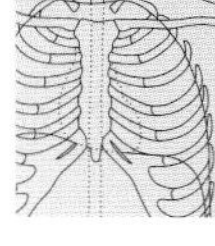

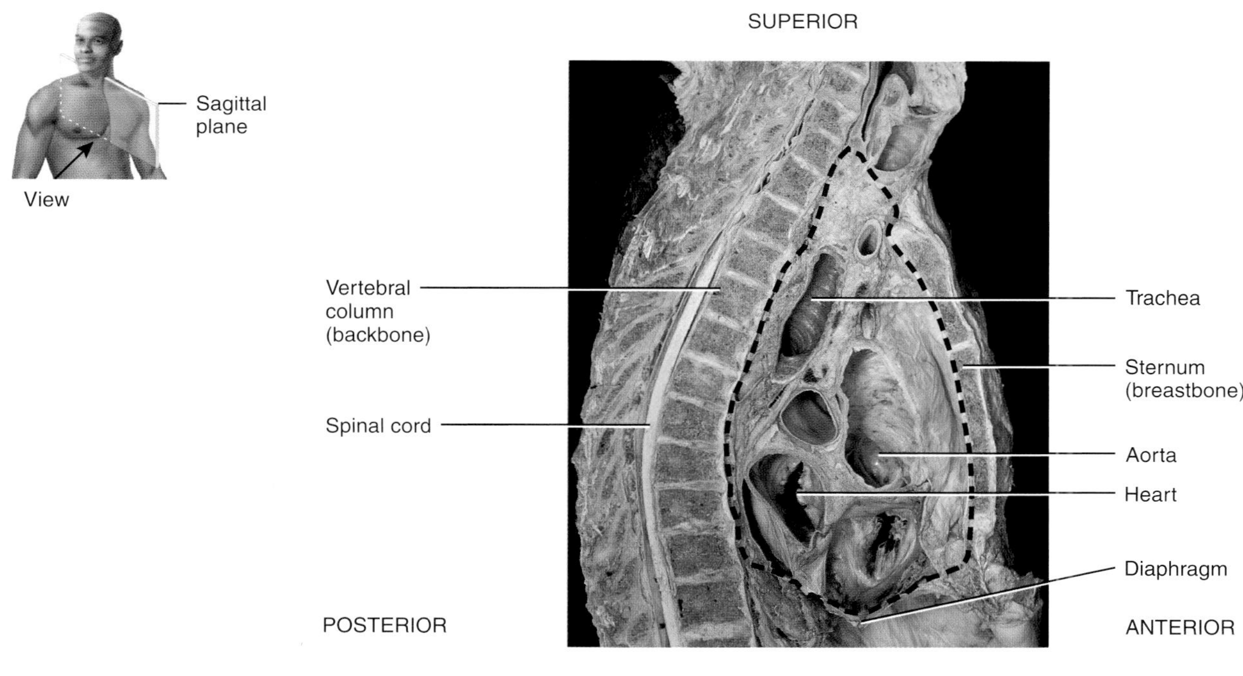

(c) Sagittal section of thoracic cavity

What is the name of the cavity that surrounds the heart? Which cavities surround the lungs?

parietal peritoneum lines the abdominal wall, covering the inferior surface of the diaphragm. Between them is the peritoneal cavity. Most abdominal organs are located in the peritoneal cavity. Some are located between the parietal peritoneum and the posterior abdominal wall. Such organs are said to be *retroperitoneal* (re′-trō-per-i-tō-NĒ-al; *retro-* = behind). The kidneys, adrenal glands, pancreas, duodenum of the small intestine, ascending and descending colons of the large intestine, and portions of the abdominal aorta and inferior vena cava are retroperitoneal.

In addition to the body cavities just described, you will also learn about other body cavities in later chapters. These include the *oral (mouth) cavity,* which contains the tongue and teeth; the *nasal cavity* in the nose; the *orbital cavities,* which contain the eyeballs; the *middle ear cavities,* which contain small bones in the middle ear; and *synovial cavities,* which are found in freely movable joints and contain synovial fluid. A summary of body cavities and their membranes is presented in the table included in Figure 1.9.

Abdominopelvic Regions and Quadrants

To describe the location of the many abdominal and pelvic organs more easily, anatomists and clinicians use two methods of dividing the abdominopelvic cavity into smaller areas. In the first method, two horizontal and two vertical lines, aligned like a tic-tac-toe grid, partition this cavity into nine **abdominopelvic regions** (Figure 1.12a, b). The top horizontal line, the *subcostal line,* is drawn just inferior to the rib cage, across the inferior portion of the stomach; the bottom horizontal line, the *transtubercular line,* is

Figure 1.11 **The abdominopelvic cavity.** The dashed line shows the approximate boundary between the abdominal and pelvic cavities. (See Tortora, *A Photographic Atlas of the Human Body, Second Edition,* Figure 12.2.)

The abdominopelvic cavity extends from the diaphragm to the groin.

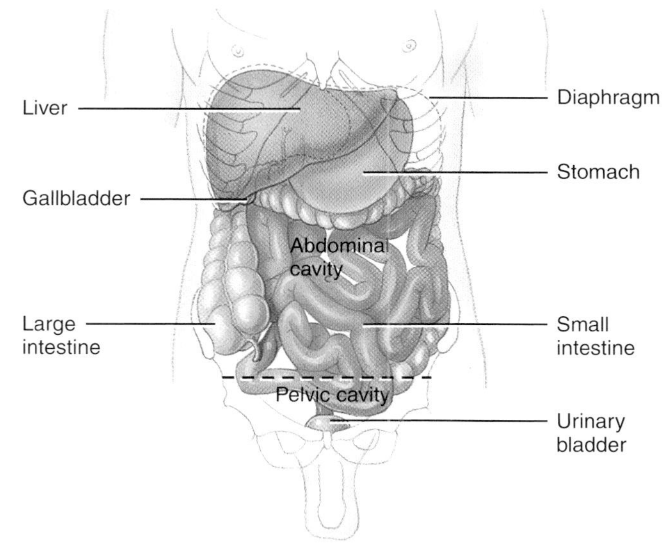

To which body systems do the organs shown here within the abdominal and pelvic cavities belong? *(Hint: Refer to Table 1.2.)*

Figure 1.12 Regions and quadrants of the abdominopelvic cavity.

The nine-region designation is used for anatomical studies; the quadrant designation is used to locate the site of pain, tumor, or some other abnormality.

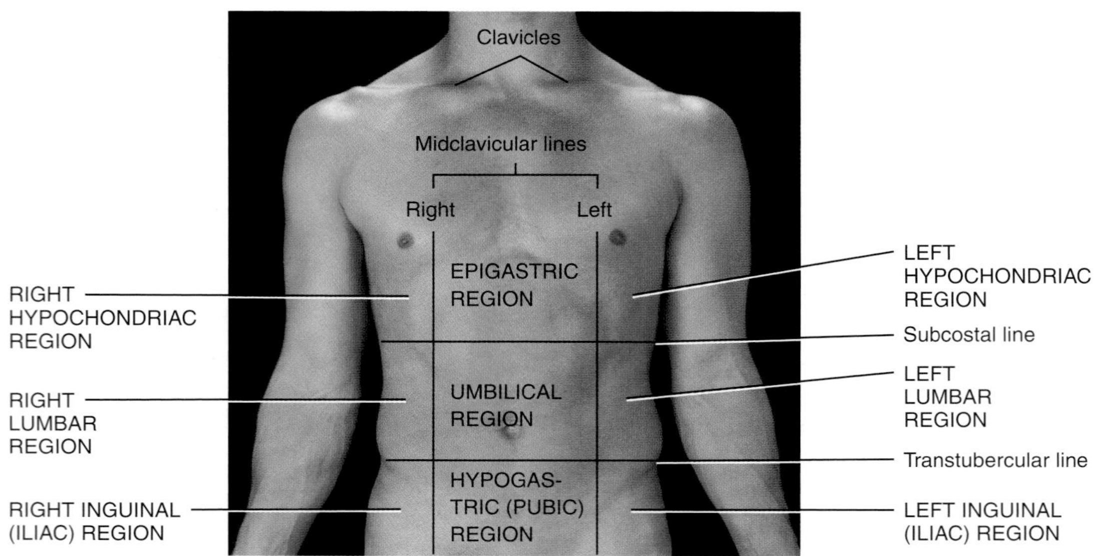

(a) Anterior view showing abdominopelvic regions

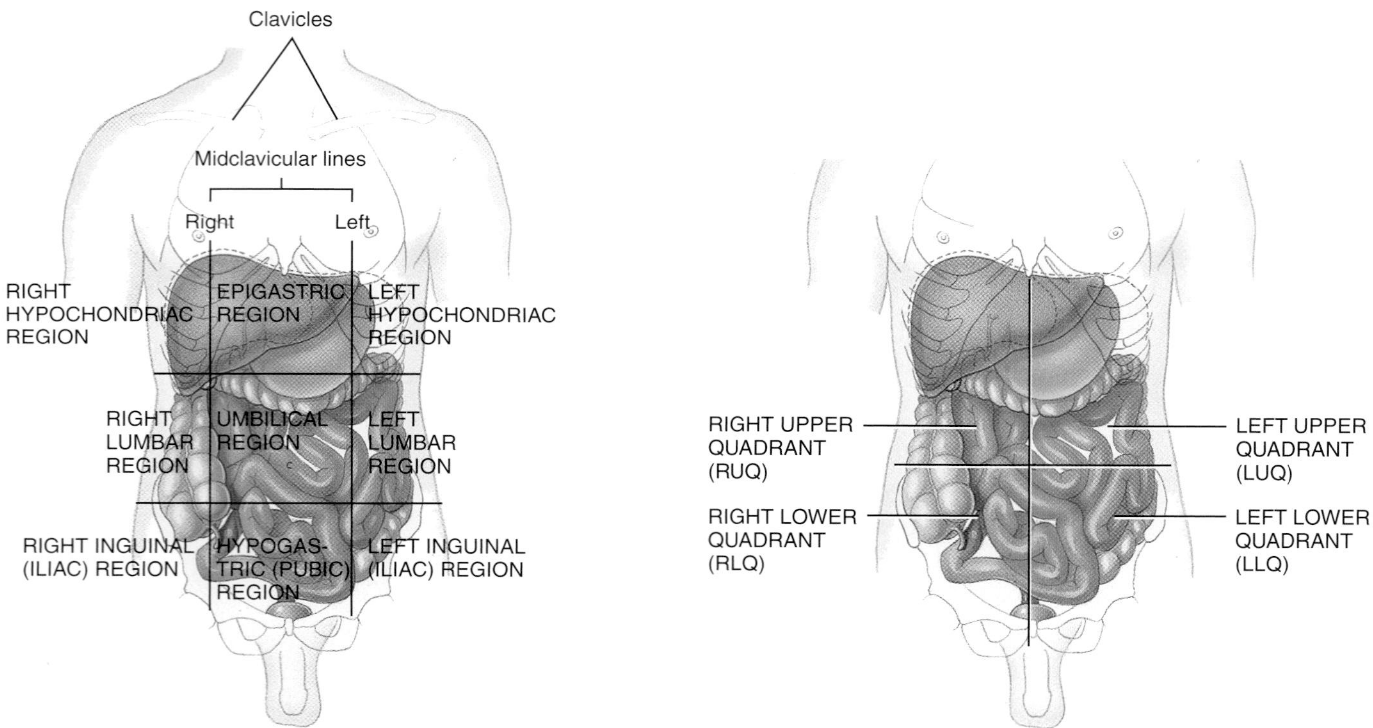

(b) Anterior view showing location of abdominopelvic regions

(c) Anterior view showing location of abdominopelvic quadrants

In which abdominopelvic region is each of the following found: most of the liver, transverse colon, urinary bladder, spleen? In which abdominopelvic quadrant would pain from appendicitis (inflammation of the appendix) be felt?

drawn just inferior to the tops of the hip bones. Two vertical lines, the left and right *midclavicular lines*, are drawn through the midpoints of the clavicles (collar bones), just medial to the nipples. The four lines divide the abdominopelvic cavity into a larger middle section and smaller left and right sections. The names of the nine abdominopelvic regions are right hypochondriac, epigastric, left hypochondriac, right lumbar, umbilical, left lumbar, right inguinal (iliac), hypogastric (pubic), and left inguinal (iliac).

The second method is simpler and divides the abdominopelvic cavity into **quadrants** (KWOD-rantz; *quad-* = one-fourth), as shown in Figure 1.12c. In this method, a vertical line and a horizontal line are passed through the **umbilicus** (um-bi-LĪ-kus; *umbilic-* = navel) or *belly button*. The names of the abdominopelvic quadrants are right upper quadrant (RUQ), left upper quadrant (LUQ), right lower quadrant (RLQ), and left lower quadrant (LLQ). The nine-region division is more widely used for anatomical studies, and quadrants are more commonly used by clinicians for describing the site of abdominopelvic pain, tumor, or other abnormality.

▶ CHECKPOINT

12. Locate each region shown in Figure 1.5 on your own body, and then identify it by its common name and the corresponding anatomical descriptive form.

13. What structures separate the various body cavities from one another?

14. Locate the nine abdominopelvic regions and the four abdominopelvic quadrants on yourself, and list some of the organs found in each.

MEDICAL IMAGING

▶ OBJECTIVE

Describe the principles and importance of medical imaging procedures in the evaluation of organ functions and the diagnosis of disease.

Various types of **medical imaging** procedures allow visualization of structures inside our bodies and are increasingly helpful for precise diagnosis of a wide range of anatomical and physiological disorders. The grandparent of all medical imaging techniques is conventional radiography (x rays), in medical use since the late 1940s. The newer imaging technologies not only contribute to diagnosis of disease, but they also are advancing our understanding of normal physiology. Table 1.3 describes some commonly used medical imaging techniques. Other imaging methods, such as cardiac catheterization, will be discussed in later chapters.

TABLE 1.3 Common Medical Imaging Procedures

Radiography

Procedure: A single barrage of x rays passes through the body, producing an image of interior structures on x ray-sensitive film. The resulting two-dimensional image is a *radiograph* (RĀ-dē-ō-graf′), commonly called an *x ray.*

Comments: Radiographs are relatively inexpensive, quick, and simple to perform, and usually provide sufficient information for diagnosis. X rays do not easily pass through dense structures so bones appear white. Hollow structures, such as the lungs, appear black. Structures of intermediate density, such as skin, fat, and muscle, appear as varying shades of gray. At low doses, x rays are useful for examining soft tissues such as the breast **(mammography)** and bone density **(bone densitometry).**

It is necessary to use a substance called a contrast medium to make hollow or fluid-filled structures visible in radiographs. X rays make structures that contain contrast media appear white. The medium may be introduced by injection, orally, or rectally depending on the structure to be imaged.

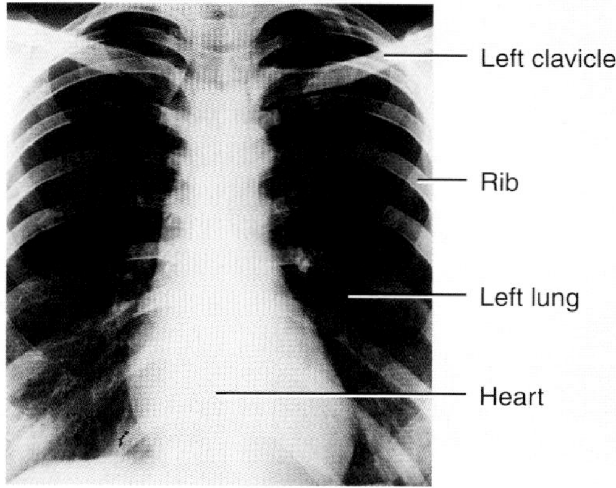

Radiograph of the thorax in anterior view

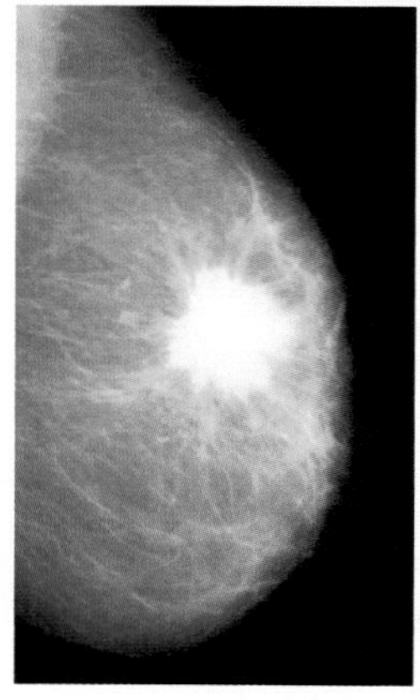

Mammogram of a female breast showing a cancerous tumor (white mass with uneven border)

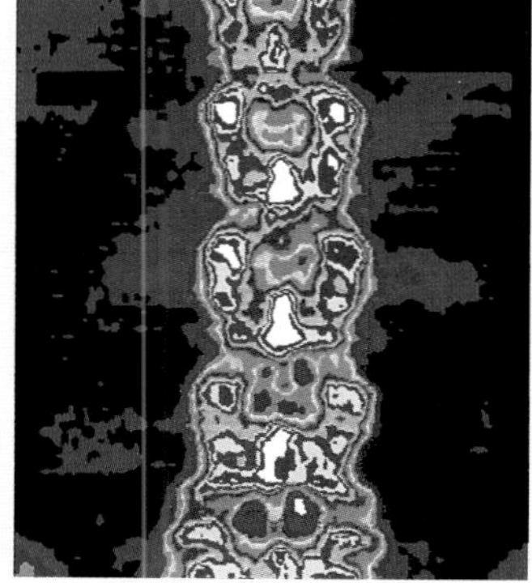

Bone densiometry scan of the lumbar spine in anterior view

continues

TABLE 1.3 Common Medical Imaging Procedures (continued)

Radiography (continued)

Contrast x rays are used to image blood vessels **(angiography),** the urinary system **(intravenous urography),** and the gastrointestinal tract **(barium contrast x ray).**

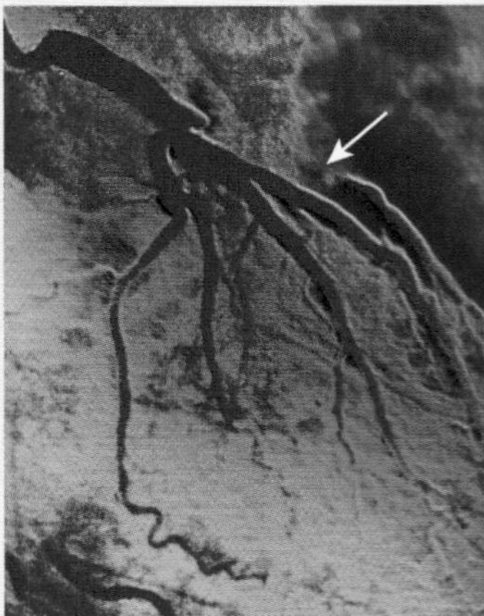

Angiogram of an adult human heart showing a blockage in a coronary artery (arrow)

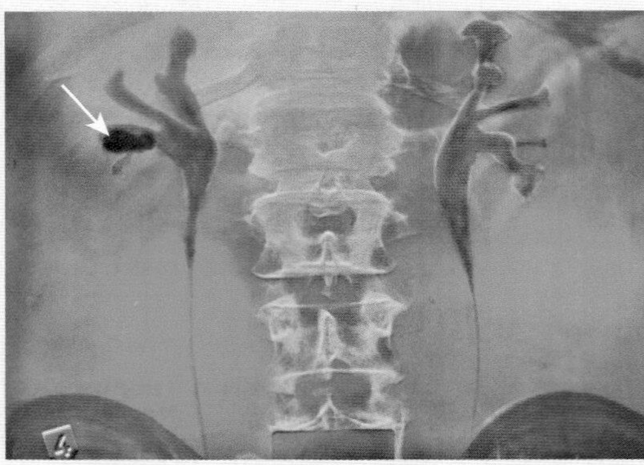

Intravenous urogram showing a kidney stone (arrow) in the right kidney

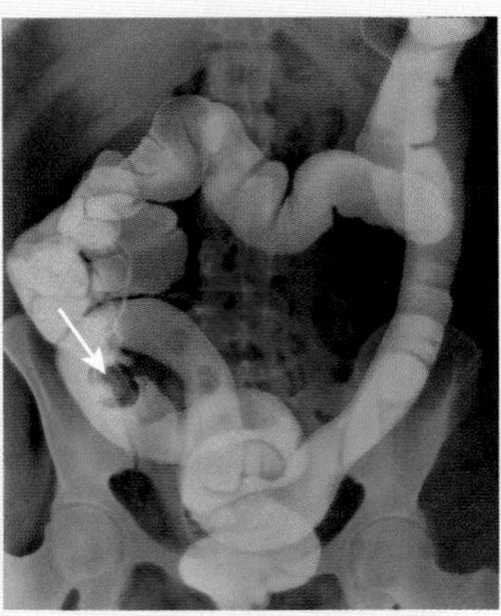

Barium contrast x-ray showing a cancer of the ascending colon (arrow)

Magnetic resonance imaging (MRI)

Procedure: The body is exposed to a high-energy magnetic field, which causes protons (small positive particles within atoms, such as hydrogen) in body fluids and tissues to arrange themselves in relation to the field. Then a pulse of radio waves "reads" these ion patterns, and a color-coded image is assembled on a video monitor. The result is a two- or three-dimensional blueprint of cellular chemistry.

Comments: Relatively safe, but can't be used on patients with metal in their bodies. Shows fine details for soft tissues but not for bones. Most useful for differentiating between normal and abnormal tissues. Used to detect tumors and artery-clogging fatty plaques, reveal brain abnormalities, measure blood flow, and detect a variety of musculoskeletal, liver, and kidney disorders.

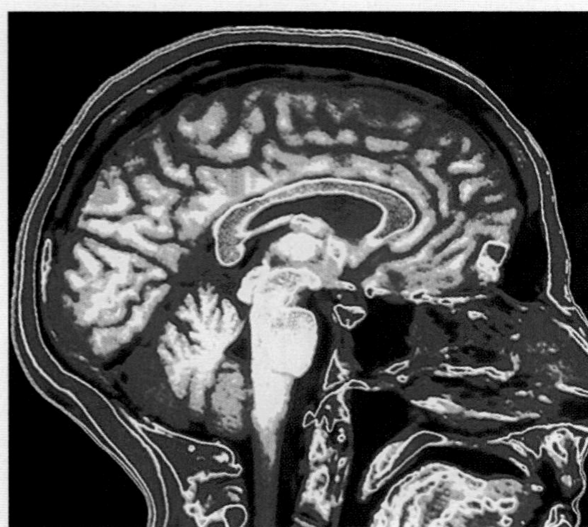

Magnetic resonance image of the brain in sagittal section

Computed tomography (CT) [formerly called computerized axial tomography (CAT) scanning]

Procedure: Computer-assisted radiography in which an x ray beam traces an arc at multiple angles around a section of the body. The resulting transverse section of the body, called a *CT scan,* is reproduced on a video monitor.

Comments: Visualizes soft tissues and organs with much more detail than conventional radiographs. Differing tissue densities show up as various shades of gray. Multiple scans can be assembled to build three-dimensional views of structures. In recent years, whole-body CT scanning has emerged. Typically, such scans actually target the torso. Whole-body CT scanning appears to provide the most benefit in screening for lung cancers, coronary artery disease, and kidney cancers.

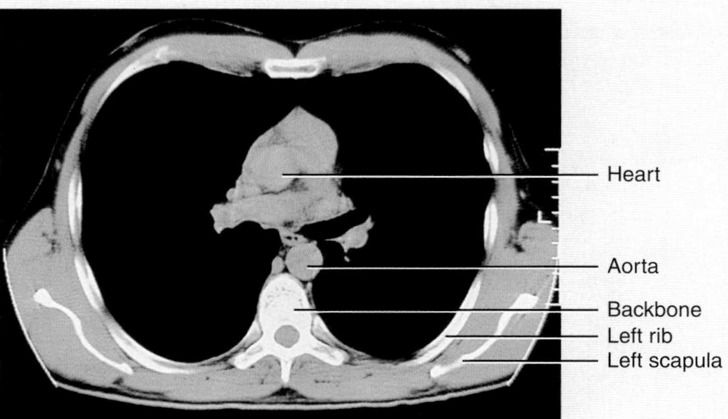

Computed tomography scan of the thorax in inferior view

Ultrasound scanning

Procedure: High-frequency sound waves produced by a handheld wand reflect off body tissues and are detected by the same instrument. The image, which may be still or moving, is called a *sonogram* (SON-ō-gram) and is reproduced on a video monitor.

Comments: Safe, noninvasive, painless, and uses no dyes. Most commonly used to visualize the fetus during pregnancy. Also used to observe the size, location, and actions of organs and blood flow through blood vessels **(doppler ultrasound).**

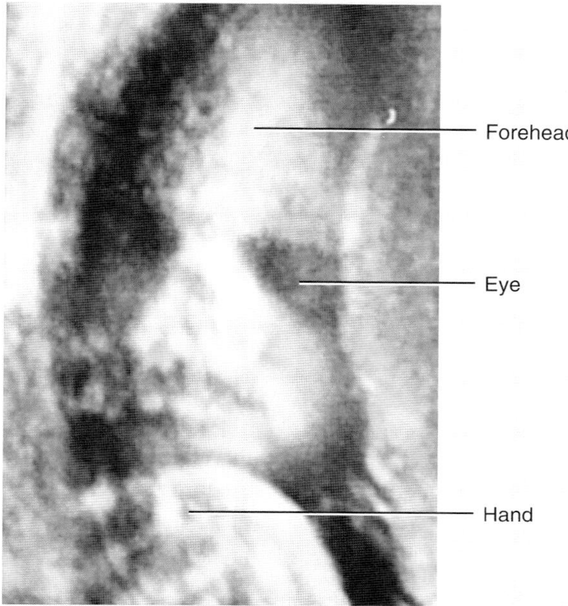

Sonogram of a fetus (Courtesy of Andrew Joseph Tortora and Damaris Soler)

Positron emission tomography (PET)

Procedure: A substance that emits positrons (positively charged particles) is injected into the body,where it is taken up by tissues. The collision of positrons with negatively charged electrons in body tissues produces gamma rays (similar to x rays) that are detected by gamma cameras positioned around the subject. A computer receives signals from the gamma cameras and constructs a *PET* scan image, displayed in color on a video monitor. The PET scan shows where the injected substance is being used in the body. In the PET scan image shown here, the black and blue colors indicate minimal activity, whereas the red, orange, yellow, and white colors indicate areas of increasingly greater activity.

Comments: Used to study the physiology of body structures, such as metabolism in the brain or heart.

ANTERIOR

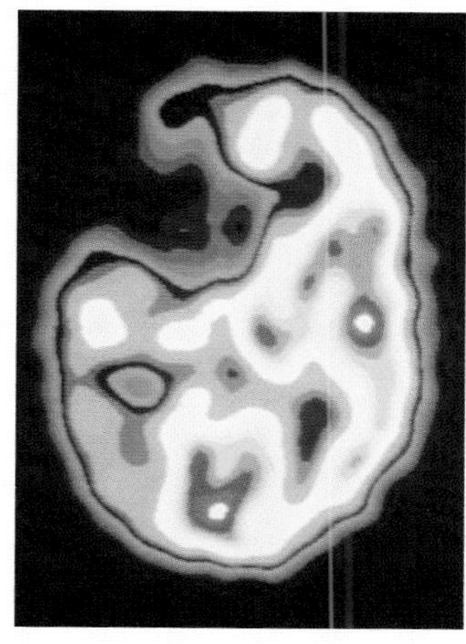

POSTERIOR

Positron emission tomography scan of a transverse section of the brain (darkened area at upper left indicates where a stroke has occurred)

Radionuclide scanning

Procedure: A *radionuclide* (radioactive substance) is introduced intravenously into the body and carried by the blood to the tissue to be imaged. Gamma rays emitted by the radionuclide are detected by a gamma camera outside the subject and fed into a computer. The computer constructs a *radionuclide image* and displays it in color on a video monitor. Areas of intense color take up a lot of the radionuclide and represent high tissue activity; areas of less intense color take up smaller amounts of the radionuclide and represent low tissue activity. **Single-photo-emission computerized tomography (SPECT) scanning** is a specialized type of radionuclide scanning that is especially useful for studying the brain, heart, lungs, and liver.

Comments: Used to study activity of a tissue or organ, such as the heart, thyroid gland, and kidneys.

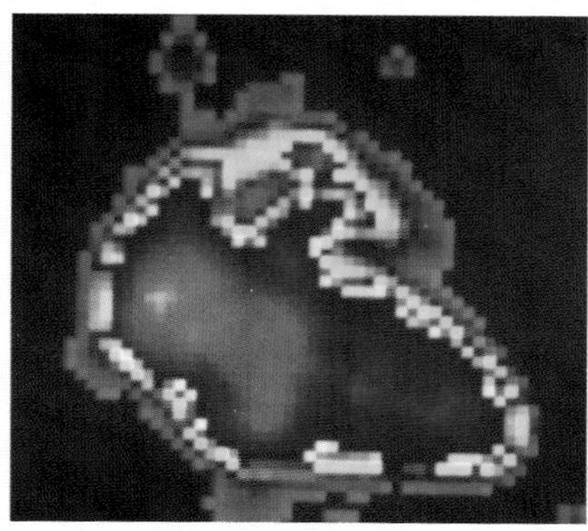

Radionuclide (nuclear) scan of a normal human heart

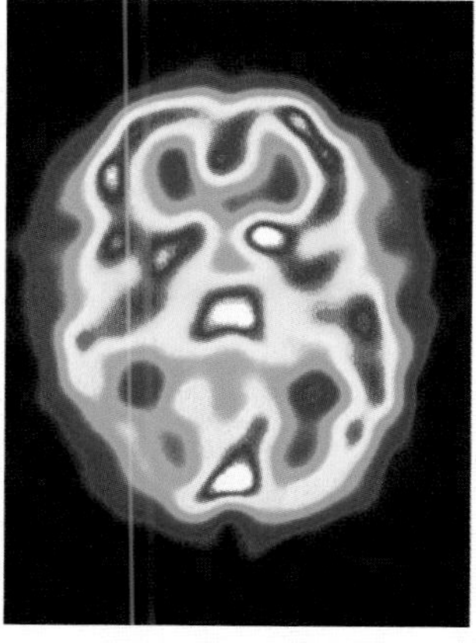

Single-photon-emission computerized tomography (SPECT) scan of a transverse section of the brain (green area at lower left indicates a migraine attack)

continues

TABLE 1.3 Common Medical Imaging Procedures (continued)

Endoscopy

Procedure: The visual examination of the inside of body organs or cavities using a lighted instrument with lenses called an *endoscope.* The image is viewed through an eyepiece on the endoscope or projected onto a monitor.

Comments: Examples of endoscopy include colonoscopy, laparoscopy, and arthroscopy. *Colonoscopy* is used to examine the interior of the colon, which is part of the large intestine. *Laparoscopy* is used to examine the organs within the abdominopelvic cavity. *Arthroscopy* is used to examine the interior of a joint, usually the knee.

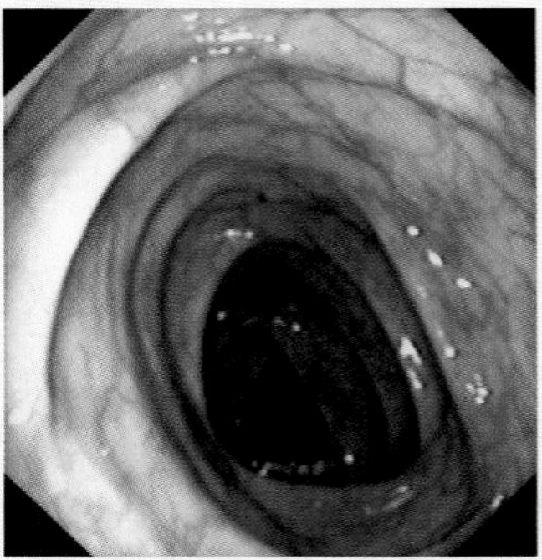

Interior view of the colon as shown by colonoscopy

STUDY OUTLINE

ANATOMY AND PHYSIOLOGY DEFINED (p. 2)

1. Anatomy is the science of body structures and the relationships among structures; physiology is the science of body functions.
2. Dissection is the careful cutting apart of body structures to study their relationships.
3. Some subdisciplines of anatomy are embryology, developmental biology, histology, surface anatomy, gross anatomy, systemic anatomy, regional anatomy, radiographic anatomy, and pathological anatomy (see Table 1.1 on page 2).
4. Some subdisciplines of physiology are neurophysiology, endocrinology, cardiovascular physiology, immunology, respiratory physiology, renal physiology, exercise physiology, and pathophysiology (see Table 1.1 on page 2).

LEVELS OF STRUCTURAL ORGANIZATION (p. 3)

1. The human body consists of six levels of structural organization: chemical, cellular, tissue, organ, system, and organismal.
2. Cells are the basic structural and functional living units of an organism and the smallest living units in the human body.
3. Tissues are groups of cells and the materials surrounding them that work together to perform a particular function.
4. Organs are composed of two or more different types of tissues; they have specific functions and usually have recognizable shapes.
5. Systems consist of related organs that have a common function.
6. An organism is any living individual.
7. Table 1.2 on pages 4–7 introduces the 11 systems of the human organism: the integumentary, skeletal, muscular, nervous, endocrine, cardiovascular, lymphatic, respiratory, digestive, urinary, and reproductive systems.

CHARACTERISTICS OF THE LIVING HUMAN ORGANISM (p. 5)

1. All organisms carry on certain processes that distinguish them from nonliving things.
2. The most important life processes of the human body are metabolism, responsiveness, movement, growth, differentiation, and reproduction.
3. Homeostasis is a condition of equilibrium in the body's internal environment produced by the interplay of all the body's regulatory processes.
4. Body fluids are dilute, watery solutions. Intracellular fluid (ICF) is inside cells, and extracellular fluid (ECF) is outside cells. Interstitial fluid is the ECF that fills spaces between tissue cells; plasma is the ECF within blood vessels.
5. Because it surrounds all body cells, interstitial fluid is called the body's internal environment.

CONTROL OF HOMEOSTASIS (p. 8)

1. Disruptions of homeostasis come from external and internal stimuli and psychological stresses.
2. When disruption of homeostasis is mild and temporary, responses of body cells quickly restore balance in the internal environment. If disruption is extreme, regulation of homeostasis may fail.
3. Most often, the nervous and endocrine systems acting together or separately regulate homeostasis. The nervous system detects body changes and sends nerve impulses to counteract changes in con-

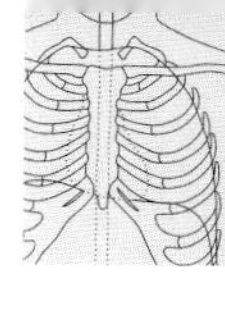

trolled conditions. The endocrine system regulates homeostasis by secreting hormones.

4. Feedback systems include three components. (1) Receptors monitor changes in a controlled condition and send input to a control center. (2) The control center sets the value at which a controlled condition should be maintained, evaluates the input it receives from receptors, and generates output commands when they are needed. (3) Effectors receive output from the control center and produce a response (effect) that alters the controlled condition.
5. If a response reverses the original stimulus, the system is operating by negative feedback. If a response enhances the original stimulus, the system is operating by positive feedback.
6. One example of negative feedback is the regulation of blood pressure. If a stimulus causes blood pressure (controlled condition) to rise, baroreceptors (pressure-sensitive nerve cells, the receptors) in blood vessels send impulses (input) to the brain (control center). The brain sends impulses (output) to the heart (effector). As a result, heart rate decreases (response) and blood pressure decreases to normal (restoration of homeostasis).
7. One example of positive feedback occurs during the birth of a baby. When labor begins, the cervix of the uterus is stretched (stimulus), and stretch-sensitive nerve cells in the cervix (receptors) send nerve impulses (input) to the brain (control center). The brain responds by releasing oxytocin (output), which stimulates the uterus (effector) to contract more forcefully (response). Movement of the fetus further stretches the cervix, more oxytocin is released, and even more forceful contractions occur. The cycle is broken with the birth of the baby.
8. Disruptions of homeostasis—homeostatic imbalances—can lead to disorders, diseases, and even death.
9. Disorder is a general term for any abnormality of structure or function. A disease is an illness with a definite set of signs and symptoms.
10. Symptoms are subjective changes in body functions that are not apparent to an observer, whereas signs are objective changes that can be observed and measured.

ANATOMICAL TERMINOLOGY (p. 12)

1. Descriptions of any region of the body assume the body is in the anatomical position, in which the subject stands erect facing the observer, with the head level and the eyes facing directly forward. The feet are flat on the floor and directed forward, and the arms are at the sides, with the palms turned forward.
2. A body lying face down is prone; a body lying face up is supine.
3. Regional names are terms given to specific regions of the body. The principal regions are the head, neck, trunk, upper limbs, and lower limbs.
4. Within the regions, specific body parts have common names and are specified by corresponding anatomical terms. Examples are chest (thoracic), nose (nasal), and wrist (carpal).
5. Directional terms indicate the relationship of one part of the body to another. Exhibit 1.1 on page 14 summarizes commonly used directional terms.
6. Planes are imaginary flat surfaces that are used to divide the body or organs to visualize interior structures. A midsagittal plane divides the body or an organ into equal right and left sides. A parasagittal plane divides the body or an organ into unequal right and left sides. A frontal plane divides the body or an organ into anterior and posterior portions. A transverse plane divides the body or an organ into superior and inferior portions. An oblique plane passes through the body or an organ at an angle between a transverse plane and either a midsagittal, parasagittal, or frontal plane.
7. Sections are flat surfaces of three-dimensional structures or cuts along a plane. They are named according to the plane along which the cut is made and include transverse, frontal, and sagittal sections.
8. Body cavities are spaces in the body that help protect, separate, and support internal organs.
9. The cranial cavity contains the brain and the vertebral cavity contains the spinal cord. The meninges are protective tissues that line the cranial cavity and vertebral cavity.
10. The diaphragm separates the thoracic cavity from the abdominopelvic cavity. The viscera are organs within the thoracic and abdominopelvic cavities. A serous membrane lines the wall of the cavity and adheres to the viscera.
11. The thoracic cavity is subdivided into three smaller cavities: a pericardial cavity, which contains the heart, and two pleural cavities, each of which contains a lung.
12. The central part of the thoracic cavity is the mediastinum. It is located between the pleural cavities, extending from the sternum to the vertebral column and from the neck to the diaphragm. It contains all thoracic viscera except the lungs.
13. The abdominopelvic cavity is divided into a superior abdominal and an inferior pelvic cavity.
14. Viscera of the abdominal cavity include the stomach, spleen, liver, gallbladder, small intestine, and most of the large intestine.
15. Viscera of the pelvic cavity include the urinary bladder, portions of the large intestine, and internal organs of the reproductive system.
16. Serous membranes line the walls of the thoracic and abdominal cavities and cover the organs within them. They include the pleura, associated with the lungs; the pericardium, associated with the heart; and the peritoneum, associated with the abdominal cavity.
17. Figure 1.9 on page 17 summarizes body cavities and their membranes.
18. To describe the location of organs more easily, the abdominopelvic cavity is divided into nine regions: right hypochondriac, epigastric, left hypochondriac, right lumbar, umbilical, left lumbar, right inguinal (iliac), hypogastric (pubic), and left inguinal (iliac).
19. To locate the site of an abdominopelvic abnormality in clinical studies, the abdominopelvic cavity is divided into quadrants: right upper quadrant (RUQ), left upper quadrant (LUQ), right lower quadrant (RLQ), and left lower quadrant (LLQ).

MEDICAL IMAGING (p. 21)

1. Medical imaging techniques allow visualization of internal structures to diagnose abnormal anatomy and deviations from normal physiology.
2. Table 1.3 on pages 21–24 summarizes and illustrates several medical imaging techniques.

SELF-QUIZ QUESTIONS

Fill in the blanks in the following statements.

1. A(n) ____ is a group of similar cells and their surrounding materials performing specific functions.
2. The sum of all of the body's chemical processes is ____. It consists of two parts: the phase that builds up new substances is ____, and the phase that breaks down substances is ____.
3. The fluid located within cells is the ____, whereas the fluid located outside of the cells is ____.

Indicate whether the following statements are true or false.

4. In a positive feedback system, the response enhances or intensifies the original stimulus.
5. A person lying face down would be in the supine position.
6. The highest level of structural organization is the system level.

Choose the one best answer to the following questions.

7. A plane that separates the body into unequal right and left sides is a (a) transverse plane, (b) frontal plane, (c) midsagittal plane, (d) coronal plane, (e) parasagittal plane.
8. Midway through a 5-mile workout, a runner begins to sweat profusely. The sweat glands producing the sweat would be considered which part of a feedback loop? (a) controlled condition, (b) receptors, (c) stimulus, (d) effectors, (e) control center.
9. An unspecialized stem cell becomes a brain cell during fetal development. This is an example of (a) differentiation, (b) growth, (c) organization, (d) responsiveness, (e) homeostasis.
10. An radiography technician needs to x-ray a growth on the urinary bladder. To accomplish this, the camera must be positioned on the ____ region. (a) left inguinal, (b) epigastric, (c) hypogastric, (d) right inguinal, (e) umbilical.
11. Which of the following would not be associated with the thoracic cavity? (1) pericardium, (2) mediastinum, (3) peritoneum, (4) pleura, (a) 2 and 3, (b) 2, (c) 3, (d) 1 and 4, (e) 3 and 4.

12. Choose the term that best fits the blank in each statement. Some answers may be used more than once.

____(a) Your eyes are ____ to your chin.	(1) superior
____(b) Your skin is ____ to your heart.	(2) inferior
____(c) Your right shoulder is ____ and ____ from your umbilicus (belly button).	(3) anterior
____(d) In the anatomical position, your thumb is ____.	(4) posterior
____(e) Your buttocks are ____.	(5) medial
____(f) Your right foot and right hand are ____.	(6) lateral
____(g) Your knee is ____ between your thigh and toes.	(7) intermediate
____(h) Your lungs are ____ to your spinal column.	(8) ipsilateral
____ (i) Your breastbone is ____ to your chin.	(9) contralateral
____(j) Your calf is ____ to your heel.	(10) proximal
	(11) distal
	(12) superficial
	(13) deep

13. Match the following cavities to their definitions:

____(a) a fluid-filled space that surrounds the heart	(1) cranial cavity
____(b) the cavity which contains the brain	(2) vertebral cavity
____(c) a cavity formed by the ribs, muscles of the chest, sternum and part of the vertebral column	(3) thoracic cavity
____(d) a cavity that contains the stomach, spleen, liver, gallbladder, small intestine and most of the large intestine	(4) pericardial cavity
____(e) fluid-filled space that surrounds a lung	(5) pleural cavity
____(f) the cavity which contains the urinary bladder, part of the large intestine and the organs of the reproductive system	(6) abdominal cavity
____(g) the cavity that contains the spinal cord	(7) pelvic cavity

14. Match the following systems with their functions:

____(a) nervous system	(1) regulates body activities through hormones (chemicals) transported in the blood to various target organs of the body
____(b) endocrine system	(2) produces gametes; releases hormones from gonads
____(c) urinary system	(3) protects against disease; returns fluids to blood
____(d) cardiovascular system	(4) protects body by forming a barrier to the outside environment; helps regulate body temperature
____(e) muscular system	(5) transports oxygen and nutrients to cells; protects against disease; carries wastes away from cells
____(f) respiratory system	(6) regulates body activities through action potentials (nerve impulses); receives sensory information; interprets and responds to the information
____(g) digestive system	(7) carries out the physical and chemical breakdown of food and absorption of nutrients
____(h) skeletal system	(8) transfers oxygen and carbon dioxide between air and blood
____ (i) integumentary system	(9) supports and protects the body; provides internal framework; provides a place for muscle attachment
____ (j) lymphatic system and immunity	(10) powers movements of the body and stabilizes body position
____(k) reproductive system	(11) eliminates wastes; regulates the volume and chemical composition of blood

15. Match the following common names and anatomical descriptive adjectives:

____(a) axillary	(1) skull
____(b) inguinal	(2) eye
____(c) cervical	(3) cheek
____(d) cranial	(4) armpit
____(e) oral	(5) arm
____(f) brachial	(6) groin
____(g) orbital	(7) buttock
____(h) gluteal	(8) neck
____(i) buccal	(9) mouth
____(j) coxal	(10) hip

CRITICAL THINKING QUESTIONS

1. You are studying for your first anatomy and physiology exam and want to know which areas of your brain are working hardest as you study. Your classmate suggests that you could have a computed tomography (CT) scan done to assess your brain activity. Would this be the best way to determine brain activity levels?
2. There is much interest in using stem cells to help in the treatment of diseases such as type I diabetes, which is due to a malfunction of some of the normal cells in the pancreas. What would make stem cells useful in disease treatment?
3. On her first anatomy and physiology exam, Heather defined homeostasis as "the condition in which the body approaches room temperature and stays there." Do you agree with Heather's definition?

ANSWERS TO FIGURE QUESTIONS

1.1 Organs are composed of two or more different types of tissues that work together to perform a specific function.

1.2 The difference between negative and positive feedback systems is that in negative feedback systems, the response reverses the original stimulus, whereas in positive feedback systems, the response enhances the original stimulus.

1.3 When something causes blood pressure to decrease, then heart rate increases due to operation of this negative feedback system.

1.4 Because positive feedback systems continually intensify or reinforce the original stimulus, some mechanism is needed to end the response.

1.5 Having one standard anatomical position allows directional terms to be clearly defined so that any body part can be described in relation to any other part.

1.6 No, the radius is distal to the humerus; No, the esophagus is posterior to the trachea; Yes, the ribs are superficial to the lungs; Yes, the urinary bladder is medial to the ascending colon; No, the sternum is medial to the descending colon.

1.7 The frontal plane divides the heart into anterior and posterior portions.

1.8 The parasagittal plane (not shown in the figure) divides the brain into unequal right and left portions.

1.9 Urinary bladder = P, stomach = A, heart = T, small intestine = A, lungs = T, internal female reproductive organs = P, thymus = T, spleen = A, liver = A.

1.10 The pericardial cavity surrounds the heart, and the pleural cavities surround the lungs.

1.11 The illustrated abdominal cavity organs all belong to the digestive system (liver, gallbladder, stomach, appendix, small intestine, and most of the large intestine). Illustrated pelvic cavity organs belong to the urinary system (the urinary bladder) and the digestive system (part of the large intestine).

1.12 The liver is mostly in the epigastric region; the transverse colon is in the umbilical region; the urinary bladder is in the hypogastric region; the spleen is in the left hypochondriac region. The pain associated with appendicitis would be felt in the right lower quadrant (RLQ).

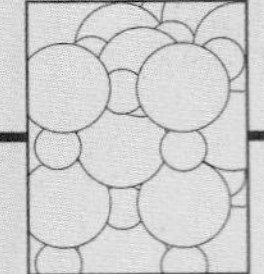

Chapter 2

The Chemical Level of Organization

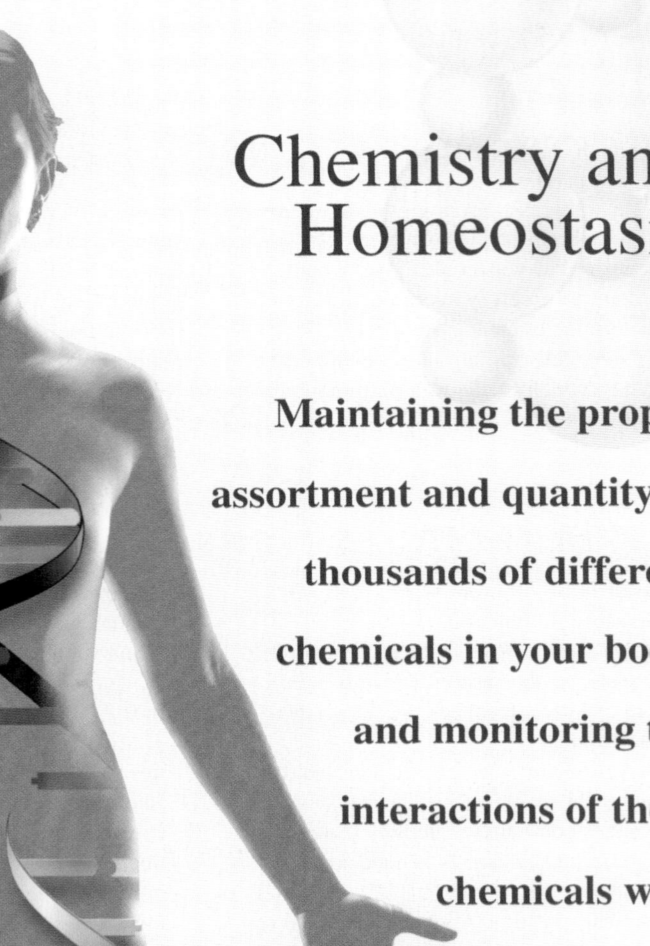

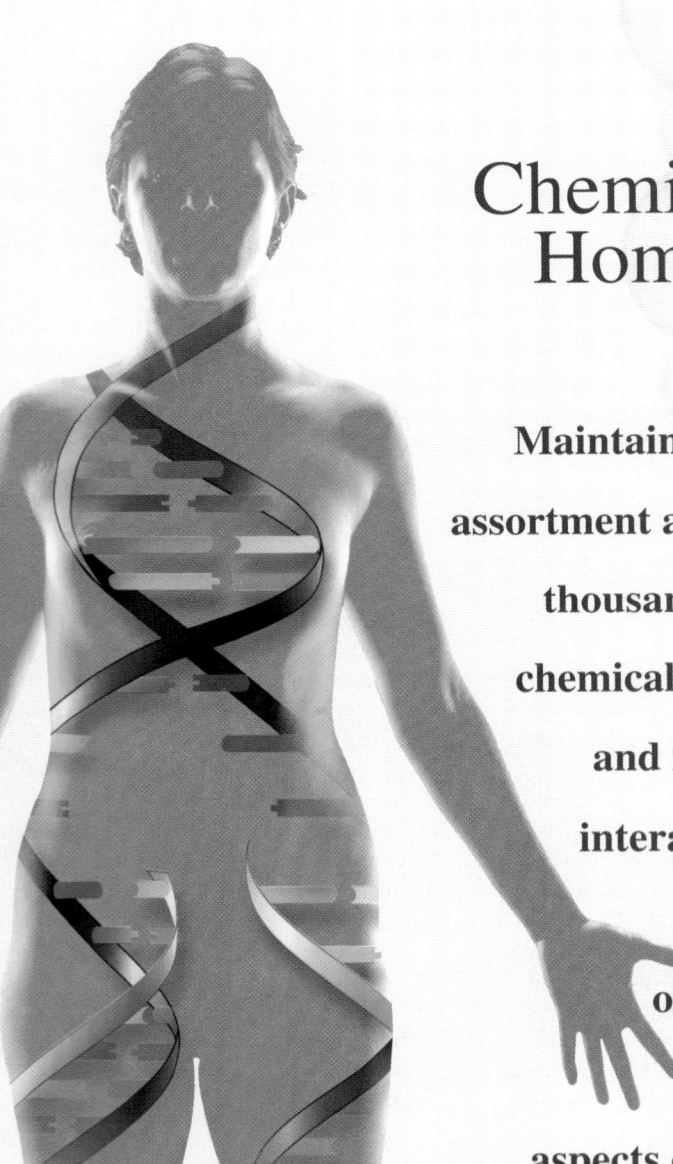

Chemistry and Homeostasis

Maintaining the proper assortment and quantity of thousands of different chemicals in your body, and monitoring the interactions of these chemicals with one another, are two important aspects of homeostasis.

wiley.com/college/apcentral

You learned in Chapter 1 that the chemical level of organization, the lowest level of structural organization, consists of atoms and molecules. These letters of the anatomical alphabet combine to form body structures and systems of astonishing size and complexity. In this chapter, we consider how atoms bond together to form molecules, and how atoms and molecules release or store energy in processes known as chemical reactions. You will also learn about the vital importance of water, which accounts for nearly two-thirds of your body weight, in chemical reactions and the maintenance of homeostasis. Finally, we present five families of molecules whose unique properties contribute to assembly of your body's structures and to powering the processes that enable you to live.

Chemistry (KEM-is-trē) is the science of the structure and interactions of matter. All living and nonliving things consist of **matter,** which is anything that occupies space and has **mass.** Mass is the amount of matter in any object, which does not change. *Weight,* the force of gravity acting on matter, does change. When objects are farther from Earth, the pull of gravity is weaker; this is why the weight of an astronaut is close to zero in outer space.

HOW MATTER IS ORGANIZED

▶ **OBJECTIVES**

Identify the main chemical elements of the human body.

Describe the structures of atoms, ions, molecules, free radicals, and compounds.

Chemical Elements

Matter exists in three states: solid, liquid, and gas. *Solids,* such as bones and teeth, are compact and have a definite shape and volume. *Liquids,* such as blood plasma, have a definite volume and assume the shape of their container. *Gases,* like oxygen and carbon dioxide, have neither a definite shape nor volume. *All* forms of matter—both living and nonliving—are made up of a limited number of building blocks called **chemical elements.** Each element is a substance that cannot be split into a simpler substance by ordinary chemical means. Scientists now recognize 112 elements. Of these, 92 occur naturally on Earth. The rest have been produced from the natural elements using particle accelerators or nuclear reactors. Each element is designated by a **chemical symbol,** one or two letters of the element's name

TABLE 2.1 Main Chemical Elements in the Body

Chemical Element (Symbol)	% Of Total Body Mass	Significance
MAJOR ELEMENTS		
Oxygen (O)	65.0	Part of water and many organic (carbon-containing) molecules; used to generate ATP, a molecule used by cells to temporarily store chemical energy.
Carbon (C)	18.5	Forms backbone chains and rings of all organic molecules: carbohydrates, lipids (fats), proteins, and nucleic acids (DNA and RNA).
Hydrogen (H)	9.5	Constituent of water and most organic molecules; ionized form (H^+) makes body fluids more acidic.
Nitrogen (N)	3.2	Component of all proteins and nucleic acids.
LESSER ELEMENTS		
Calcium (Ca)	1.5	Contributes to hardness of bones and teeth; ionized form (Ca^{2+}) needed for blood clotting, release of some hormones, contraction of muscle, and many other processes.
Phosphorus (P)	1.0	Component of nucleic acids and ATP; required for normal bone and tooth structure.
Potassium (K)	0.35	Ionized form (K^+) is the most plentiful cation (positively charged particle) in intracellular fluid; needed to generate action potentials.
Sulfur (S)	0.25	Component of some vitamins and many proteins.
Sodium (Na)	0.2	Ionized form (Na^+) is the most plentiful cation in extracellular fluid; essential for maintaining water balance; needed to generate action potentials.
Chlorine (Cl)	0.2	Ionized form (Cl^-) is the most plentiful anion (negatively charged particle) in extracellular fluid; essential for maintaining water balance.
Magnesium (Mg)	0.1	Ionized form (Mg^{2+}) needed for action of many enzymes, molecules that increase the rate of chemical reactions in organisms.
Iron (Fe)	0.005	Ionized forms (Fe^{2+} and Fe^{3+}) are part of hemoglobin (oxygen-carrying protein in red blood cells) and some enzymes (proteins that catalyze chemical reactions in living cells).
TRACE ELEMENTS	0.2	Aluminum (Al), Boron (B), Chromium (Cr), Cobalt (Co), Copper (Cu), Fluorine (F), Iodine (I), Manganese (Mn), Molybdenum (Mo), Selenium (Se), Silicon (Si), Tin (Sn), Vanadium (V), and Zinc (Zn).

in English, Latin, or another language. Examples of chemical symbols are H for hydrogen, C for carbon, O for oxygen, N for nitrogen, Ca for calcium, and Na for sodium (*natrium* = sodium).*

Twenty-six different elements normally are present in your body. Just four elements, called the *major elements,* constitute about 96% of the body's mass: oxygen, carbon, hydrogen, and nitrogen. Eight others, the *lesser elements,* contribute 3.8% to the body's mass: calcium, phosphorus (P), potassium (K), sulfur (S), sodium, chlorine (Cl), magnesium (Mg), and iron (Fe). An additional 14 elements—the *trace elements*—are present in tiny amounts. Together, they account for the remaining 0.2% of the body's mass. Several trace elements have important functions in the body. For example, iodine is needed to make thyroid hormones. The functions of some trace elements are unknown. Table 2.1 on page 29 lists the main chemical elements of the human body.

Structure of Atoms

Each element is made up of **atoms,** the smallest units of matter that retain the properties and characteristics of the element. Atoms are extremely small. Two hundred thousand of the largest atoms would fit on the period at the end of this sentence. Hydrogen atoms, the smallest atoms, have a diameter less than 0.1 nanometer (0.1×10^{-9} m = 0.0000000001 m), and the largest atoms are only five times larger.

Dozens of different **subatomic particles** compose individual atoms. However, only three types of subatomic particles are important for understanding the chemical reactions in the human body: protons, neutrons, and electrons (Figure 2.1). The dense central core of an atom is its **nucleus.** Within the nucleus are positively charged **protons (p^+)** and uncharged (neutral) **neutrons (n^0)**. The tiny, negatively charged **electrons (e^-)** move about in a large space surrounding the nucleus. They do not follow a fixed path or orbit but instead form a negatively charged "cloud" that envelops the nucleus (Figure 2.1a).

Even though their exact positions cannot be predicted, specific groups of electrons are most likely to move about within certain regions around the nucleus. These regions, called **electron shells,** are depicted as simple circles around the nucleus. Because each electron shell can hold a specific number of electrons, the electron shell model best conveys this aspect of atomic structure (Figure 2.1b). The first electron shell (nearest the nucleus) never holds more than 2 electrons. The second shell holds a maximum of 8 electrons, and the third can hold up to 18 electrons. The electron shells fill with electrons in a specific order, beginning with the first shell. For example, notice in Figure 2.2 that sodium (Na), which has 11 electrons total, contains 2 electrons in the first shell, 8 electrons in the second shell, and 1 electron in the third shell. The most massive element present in the human body is iodine, which has a total of 53 electrons: 2 in the first shell, 8 in the second shell, 18 in the third shell, 18 in the fourth shell, and 7 in the fifth shell.

The number of electrons in an atom of an element always equals the number of protons. Because each electron and proton carries one charge, the negatively charged electrons and the positively charged protons balance each other. Thus, each atom is electrically neutral; its total charge is zero.

Atomic Number and Mass Number

The *number of protons* in the nucleus of an atom is an atom's **atomic number.** Figure 2.2 shows that atoms of different elements have different atomic numbers because they have different numbers of protons. For example, oxygen has an atomic number of 8 because its nucleus has 8 protons, and sodium has an atomic number of 11 because its nucleus has 11 protons.

The **mass number** of an atom is the sum of its protons and neutrons. Because sodium has 11 protons and 12 neutrons, its mass number is 23 (Figure 2.2). Although all atoms of one element have the same number of protons, they may have different numbers of neutrons and thus different mass numbers. **Isotopes** are atoms of an element that have different numbers of neutrons and therefore different mass numbers. In a sample of oxygen, for

*The periodic table of elements, which lists all of the known chemical elements, can be found in Appendix B.

Figure 2.1 Two representations of the structure of an atom. Electrons move about the nucleus, which contains neutrons and protons. (a) In the electron cloud model of an atom, the shading represents the chance of finding an electron in regions outside the nucleus. (b) In the electron shell model, filled circles represent individual electrons, which are grouped into concentric circles according to the shells they occupy. Both models depict a carbon atom, with six protons, six neutrons, and six electrons.

An atom is the smallest unit of matter that retains the properties and characteristics of its element.

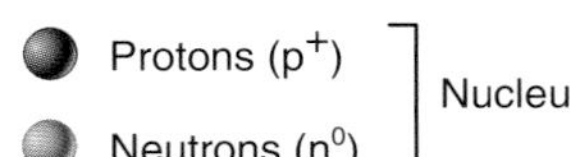

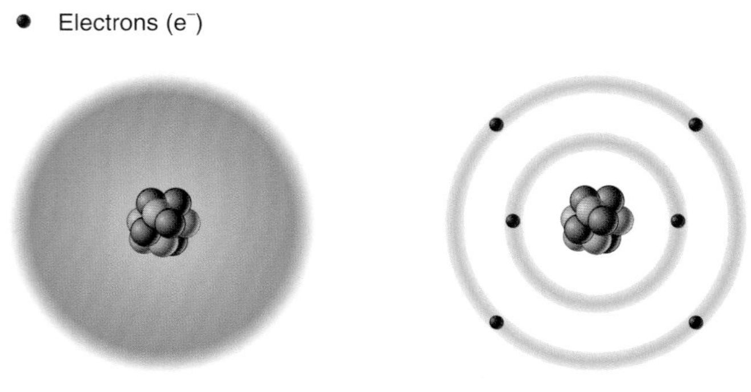

How are the electrons of carbon distributed between the first and second electron shells?

Figure 2.2 **Atomic structures of several stable atoms.**

The atoms of different elements have different atomic numbers because they have different numbers of protons.

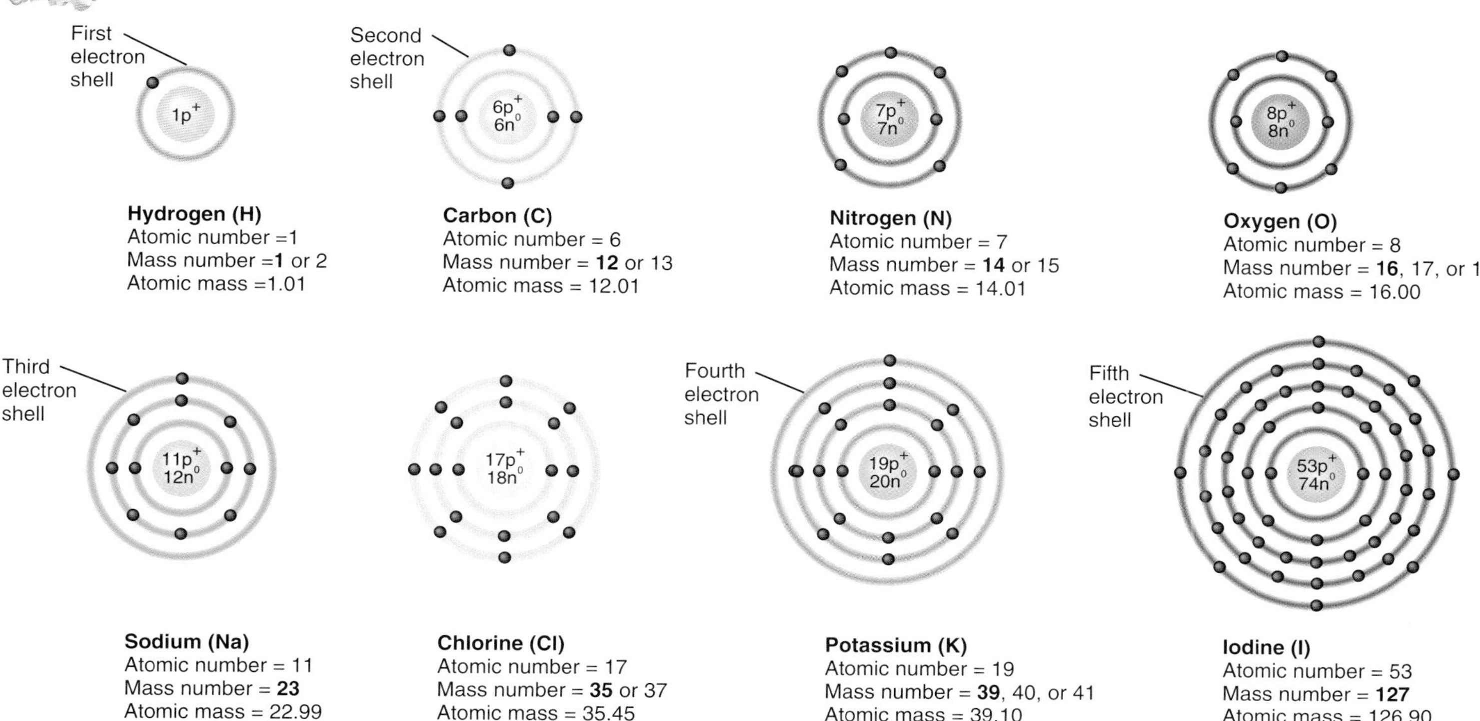

Atomic number = number of protons in an atom
Mass number = number of protons and neutrons in an atom (boldface indicates most common isotope)
Atomic mass = average mass of all stable atoms of a given element in daltons

Which four of these elements are present most abundantly in living organisms?

example, most atoms have 8 neutrons, and a few have 9 or 10, but all have 8 protons and 8 electrons. Most isotopes are stable, which means that their nuclear structure does not change over time. The stable isotopes of oxygen are designated ^{16}O, ^{17}O, and ^{18}O (or O-16, O-17, and O-18). As you may already have determined, the numbers indicate the mass number of each isotope. As you will discover shortly, the number of electrons of an atom determines its chemical properties. Although the isotopes of an element have different numbers of neutrons, they have identical chemical properties because they have the same number of electrons.

Certain isotopes called **radioactive isotopes** are unstable; their nuclei decay (spontaneously change) into a stable configuration. Examples are H-3, C-14, O-15, and O-19. As they decay, these atoms emit radiation—either subatomic particles or packets of energy—and in the process often transform into a different element. For example, the radioactive isotope of carbon, C-14, decays to N-14. The decay of a radioisotope may be as fast as a fraction of a second or as slow as millions of years. The **half-life** of an isotope is the time required for half of the radioactive atoms in a sample of that isotope to decay into a more stable form. The half-life of C-14, which is used to determine the age of organic samples, is 5600 years, whereas the half-life of I-131, an important clinical tool, is 8 days.

Harmful and Beneficial Effects of Radiation

Radioactive isotopes may have either harmful or helpful effects. Their radiations can break apart molecules, posing a serious threat to the human body by producing tissue damage and/or causing various types of cancer. Although the decay of naturally occurring radioactive isotopes typically releases just a small amount of radiation into the environment, localized accumulations can occur. Radon-222, a colorless and odorless gas that is a naturally occurring radioactive breakdown product of uranium, may seep out of the soil and accumulate in buildings. It is not only associated with many cases of lung cancer in smokers but has also been implicated in many cases of lung cancer in nonsmokers. Beneficial effects of certain radioisotopes include their use in medical imaging procedures to diagnose and treat certain disorders. Some radioisotopes can be used as **tracers** to follow the movement of certain substances through the body. Thallium-201 is used to monitor blood flow through the heart during an exercise stress test. Iodine-131 is used to detect cancer of the thyroid gland and to assess its size and activity, and may also be used to destroy part of an overactive thyroid gland. Cesium-137 is used to treat advanced cervical cancer and iridium is used to treat prostate cancer. ■

Atomic Mass

The standard unit for measuring the mass of atoms and their subatomic particles is a **dalton,** also known as an *atomic mass unit (amu)*. A neutron has a mass of 1.008 daltons, and a proton has a mass of 1.007 daltons. The mass of an electron, at 0.0005 dalton, is almost 2000 times smaller than the mass of a neutron or proton. The **atomic mass** (also called the *atomic weight*) of an element is the average mass of all its naturally occurring isotopes. Typically, the atomic mass of an element is close to the mass number of its most abundant isotope.

Ions, Molecules, and Compounds

As we discussed, atoms of the same element have the same number of protons. The atoms of each element have a characteristic way of losing, gaining, or sharing their electrons when interacting with other atoms to achieve stability. The way that electrons behave enables atoms in the body to exist in electrically charged forms called ions, or to join with each other into complex combinations called molecules. If an atom either *gives up* or *gains* electrons, it becomes an ion. An **ion** is an atom that has a positive or negative charge because it has unequal numbers of protons and electrons. *Ionization* is the process of giving up or gaining electrons. An ion of an atom is symbolized by writing its chemical symbol followed by the number of its positive (+) or negative (−) charges. Thus, Ca^{2+} stands for a calcium ion that has two positive charges because it has lost two electrons.

When two or more atoms *share* electrons, the resulting combination is called a **molecule** (MOL-e-kūl). A *molecular formula* indicates the elements and the number of atoms of each element that make up a molecule. A molecule may consist of two atoms of the same kind, such as an oxygen molecule (Figure 2.3a). The molecular formula for a molecule of oxygen is O_2. The subscript 2 indicates that the molecule contains two atoms of oxygen. Two or more different kinds of atoms may also form a molecule, as in a water molecule (H_2O). In H_2O one atom of oxygen shares electrons with two atoms of hydrogen.

Figure 2.3 Atomic structures of an oxygen molecule and a superoxide free radical.

A free radical has an unpaired electron in its outermost electron shell.

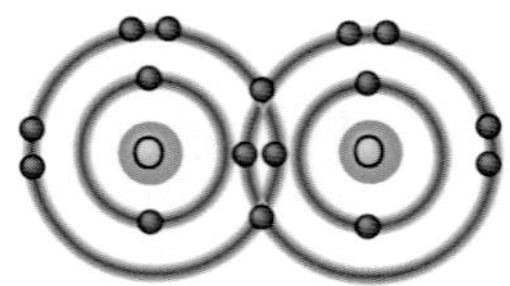

(a) Oxygen molecule (O_2)

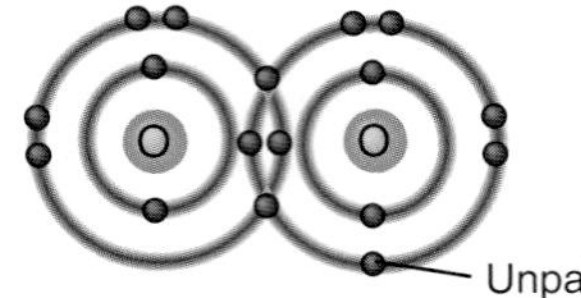

(b) Superoxide free radical (O_2^-)

What substances in the body can inactivate oxygen-derived free radicals?

A **compound** is a substance that contains atoms of two or more different elements. Most of the atoms in the body are joined into compounds. Water (H_2O) and sodium chloride (NaCl), common table salt, are compounds. However, a molecule of oxygen (O_2) is not a compound because it consists of atoms of only one element.

A **free radical** is an electrically charged atom or group of atoms with an unpaired electron in the outermost shell. A common example is superoxide, which is formed by the addition of an electron to an oxygen molecule (Figure 2.3b). Having an unpaired electron makes a free radical unstable, highly reactive, and destructive to nearby molecules. Free radicals become stable by either giving up their unpaired electron to, or taking on an electron from, another molecule. In so doing, free radicals may break apart important body molecules.

Free Radicals and Their Effects on Health

In our bodies, several processes can generate free radicals, including exposure to ultraviolet radiation in sunlight, exposure to x rays, and some reactions that occur during normal metabolic processes. Certain harmful substances, such as carbon tetrachloride (a solvent used in dry cleaning), also give rise to free radicals when they participate in metabolic reactions in the body. Among the many disorders, diseases, and conditions linked to oxygen-derived free radicals are cancer, atherosclerosis, Alzheimer's disease, emphysema, diabetes mellitus, cataracts, macular degeneration, rheumatoid arthritis, and deterioration associated with aging. Consuming more *antioxidants*—substances that inactivate oxygen-derived free radicals—is thought to slow the pace of damage caused by free radicals. Important dietary antioxidants include selenium, zinc, beta-carotene, and vitamins C and E. ■

▶ CHECKPOINT

1. List the names and chemical symbols of the 12 most abundant chemical elements in the human body.
2. What are the atomic number, mass number, and atomic mass of carbon? How are they related?
3. Define isotopes and free radicals.

CHEMICAL BONDS

▶ OBJECTIVES

Describe how valence electrons form chemical bonds.

Distinguish among ionic, covalent, and hydrogen bonds.

The forces that hold together the atoms of a molecule or a compound are **chemical bonds.** The likelihood that an atom will form a chemical bond with another atom depends on the number of electrons in its outermost shell, also called the **valence shell.** An atom with a valence shell holding eight electrons is *chemically stable,* which means it is unlikely to form chemical bonds with other atoms. Neon, for example, has eight electrons in its valence shell, and for this reason it does not bond easily with other atoms. The valence shell of hydrogen and helium is the first electron shell, which holds a maximum of two electrons. Because helium has two valence electrons, it too is stable and seldom bonds with other atoms. Hydrogen, on the other hand, has only one valence electron (see Figure 2.2), so it binds readily with other atoms.

The atoms of most biologically important elements do not have eight electrons in their valence shells. Under the right conditions, two or more atoms can interact in ways that produce a chemically stable arrangement of eight valence electrons for each atom. This chemical principle, called the **octet rule** (*octet* = set of eight), helps explain why atoms interact in predictable ways. One atom is more likely to interact with another atom if doing so will leave both with eight valence electrons. For this to happen, an atom either empties its partially filled valence shell, fills it with donated electrons, or shares electrons with other atoms. The way that valence electrons are distributed determines what kind of chemical bond results. We will consider three types of chemical bonds: ionic bonds, covalent bonds, and hydrogen bonds.

Ionic Bonds

As you have already learned, when atoms lose or gain one or more valence electrons, ions are formed. Positively and negatively charged ions are attracted to one another—opposites attract. The force of attraction that holds together ions with opposite charges is an **ionic bond.** Consider sodium and chlorine atoms, the components of common table salt. Sodium has one valence electron (Figure 2.4a). If sodium *loses* this electron, it is left with the eight electrons in its second shell, which becomes the valence shell. As a result, however, the total number of protons (11) exceeds the number of electrons (10). Thus, the sodium atom has become a **cation** (KAT-ī-on), or positively charged ion. A sodium ion has a charge of 1+ and is written Na^+. By contrast, chlorine has seven valence electrons (Figure 2.4b). If chlorine *gains* an electron from a neighboring atom, it will have a complete octet in its third electron shell. After gaining an electron, the total number of electrons (18) exceeds the number of protons (17), and the chlorine atom has become an **anion** (AN-ī-on), a negatively charged ion. The ionic form of chlorine is called a *chloride* ion. It has a charge of 1− and is written Cl^-. When an atom of sodium donates its sole valence electron to an atom of chlorine, the resulting positive and negative charges pull both ions tightly together, forming an ionic bond (Figure 2.4c). The resulting compound is sodium chloride, written NaCl.

In general, ionic compounds exist as solids, with an orderly, repeating arrangement of the ions, as in a crystal of NaCl (Figure 2.4d). A crystal of NaCl may be large or small—the total number of ions can vary—but the ratio of Na^+ to Cl^- is always 1:1. In the body, ionic bonds are found mainly in teeth and bones, where they give great strength to these important structural tissues. An ionic compound that breaks apart into positive and negative ions in solution is called an **electrolyte** (e-LEK-trō-līt). Most ions in the body are dissolved

Figure 2.4 Ions and ionic bond formation. (a) A sodium atom can have a complete octet of electrons in its outermost shell by losing one electron. (b) A chlorine atom can have a complete octet by gaining one electron. (c) An ionic bond may form between oppositely charged ions. (d) In a crystal of NaCl, each Na^+ is surrounded by six Cl^-. In (a), (b), and (c), the electron that is lost or accepted is colored red.

An ionic bond is the force of attraction that holds together oppositely charged ions.

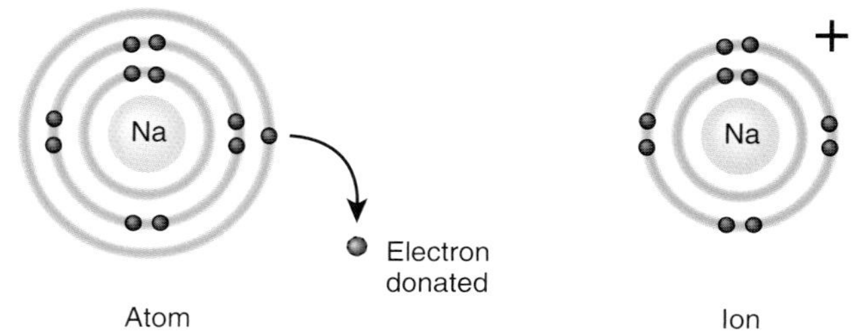

(a) Sodium: 1 valence electron

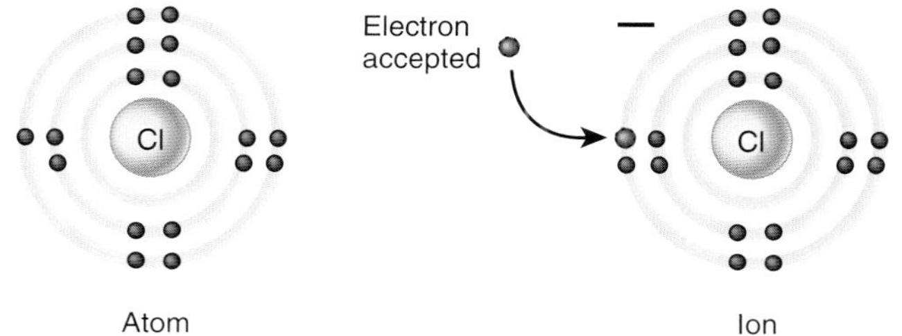

(b) Chlorine: 7 valence electrons

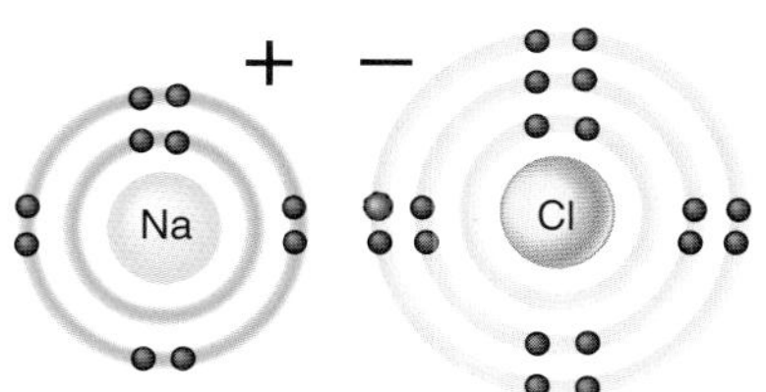

(c) Ionic bond in sodium chloride (NaCl)

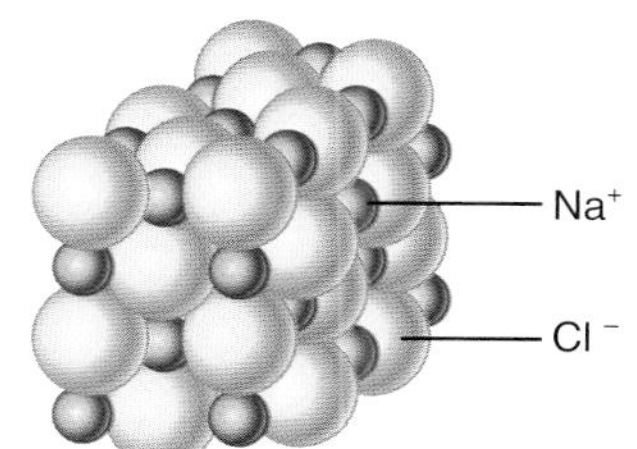

(d) Packing of ions in a crystal of sodium chloride

What are cations and anions?

in body fluids as electrolytes, so-named because their solutions can conduct an electric current. (In Chapter 27 we will discuss the chemistry and importance of electrolytes.) Table 2.2 lists the names and symbols of the most common ions in the body.

Covalent Bonds

When a **covalent bond** forms, two or more atoms *share* electrons rather than gaining or losing them. Atoms form a covalently bonded molecule by sharing one, two, or three pairs of valence electrons. The larger the number of electron pairs shared between two atoms, the stronger the covalent bond. Covalent bonds may form between atoms of the same element or between atoms of different elements. They are the most common chemical bonds in the body, and the compounds that result from them form most of the body's structures.

A **single covalent bond** results when two atoms share one electron pair. For example, a molecule of hydrogen forms when two hydrogen atoms share their single valence electrons (Figure 2.5a), which allows both atoms to have a full valence

Figure 2.5 Covalent bond formation. The red electrons are shared equally. In writing the structural formula of a covalently bonded molecule, each straight line between the chemical symbols for two atoms denotes a pair of shared electrons. In molecular formulas, the number of atoms in each molecule is noted by subscripts.

In a covalent bond, two atoms share one, two, or three pairs of valence electrons.

	DIAGRAMS OF ATOMIC AND MOLECULAR STRUCTURE		STRUCTURAL FORMULA	MOLECULAR FORMULA
(a)	Hydrogen atoms	Hydrogen molecule	H — H	H_2
(b)	Oxygen atoms	Oxygen molecule	O = O	O_2
(c)	Nitrogen atoms	Nitrogen molecule	N ≡ N	N_2
(d)	Carbon atom; Hydrogen atoms	Methane molecule	H — C — H (with H above and below C)	CH_4

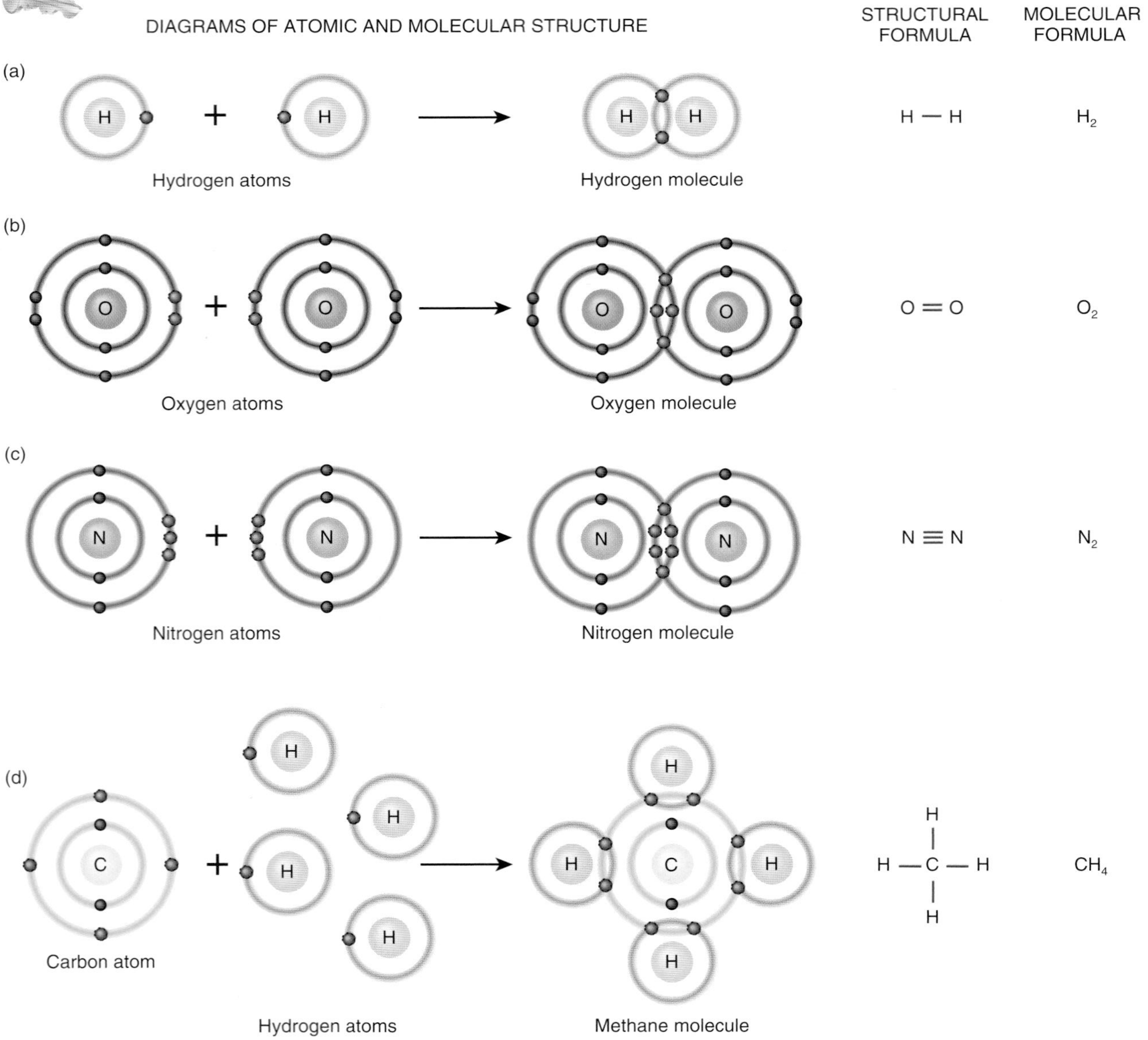

What is the principal difference between an ionic bond and a covalent bond?

TABLE 2.2 Common Ions and Ionic Compounds in the Body

Cations		Anions	
Name	**Symbol**	**Name**	**Symbol**
Hydrogen ion	H^+	Fluoride ion	F^-
Sodium ion	Na^+	Chloride ion	Cl^-
Potassium ion	K^+	Iodide ion	I^-
Ammonium ion	NH_4^+	Hydroxide ion	OH^-
Hydronium ion	H_3O^+	Nitrate ion	NO_3^-
Magnesium ion	Mg^{2+}	Bicarbonate ion	HCO_3^-
Calcium ion	Ca^{2+}	Oxide ion	O^{2-}
Iron (II) ion	Fe^{2+}	Sulfate ion	SO_4^{2-}
Iron (III) ion	Fe^{3+}	Phosphate ion	PO_4^{3-}

shell at least part of the time. A **double covalent bond** results when two atoms share two pairs of electrons, as happens in an oxygen molecule (Figure 2.5b). A **triple covalent bond** occurs when two atoms share three pairs of electrons, as in a molecule of nitrogen (Figure 2.5c). Notice in the *structural formulas* for covalently bonded molecules in Figure 2.5 that the number of lines between the chemical symbols for two atoms indicates whether the bond is a single (—), double (=), or triple (≡) covalent bond.

The same principles of covalent bonding that apply to atoms of the same element also apply to covalent bonds between atoms of different elements. The gas methane (CH_4) contains covalent bonds formed between the atoms of two different elements, one carbon and four hydrogens (Figure 2.5d). The valence shell of the carbon atom can hold eight electrons but has only four of its own. The single electron shell of a hydrogen atom can hold two electrons, but each hydrogen atom has only one of its own. A methane molecule contains four separate single covalent bonds. Each hydrogen atom shares one pair of electrons with the carbon atom.

In some covalent bonds, two atoms share the electrons equally—one atom does not attract the shared electrons more strongly than the other atom. This type of bond is a **nonpolar covalent bond.** The bonds between two identical atoms are always nonpolar covalent bonds (Figure 2.5a–c). The bonds between carbon and hydrogen atoms are also nonpolar, such as the four C—H bonds in a methane molecule (Figure 2.5d).

In a **polar covalent bond,** the sharing of electrons between two atoms is unequal—the nucleus of one atom attracts the shared electrons more strongly than the nucleus of the other atom. When polar covalent bonds form, the resulting molecule has a partial negative charge near the atom that attracts electrons more strongly. This atom has greater **electronegativity,** the power to attract electrons to itself. At least one other atom in the molecule then will have a partial positive charge. The partial charges are indicated by a lowercase Greek delta with a minus or plus sign: δ^- or δ^+. A very important example of a polar covalent bond in living systems is the bond between oxygen and hydrogen in a molecule of water (Figure 2.6); in this molecule, the nucleus of the oxygen atom attracts the electrons more strongly than the nuclei of the hydrogen atoms, so the oxygen atom is said to have greater electronegativity. Later in the chapter, we will see how polar covalent bonds allow water to dissolve many molecules that are important to life. Bonds between nitrogen and hydrogen and those between oxygen and carbon are also polar bonds.

Figure 2.6 **Polar covalent bonds between oxygen and hydrogen atoms in a water molecule.** The red electrons are shared unequally. Because the oxygen nucleus attracts the shared electrons more strongly, the oxygen end of a water molecule has a partial negative charge, written δ^-, and the hydrogen ends have partial positive charges, written δ^+.

A polar covalent bond occurs when one atomic nucleus attracts the shared electrons more strongly than does the nucleus of another atom in the molecule.

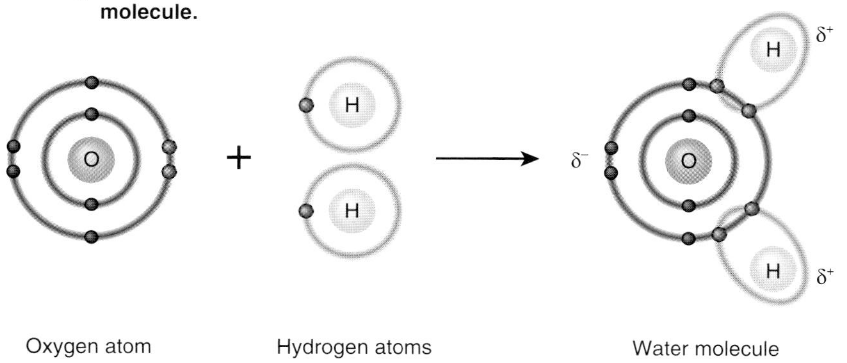

Which atom in a water molecule has greater electronegativity?

Hydrogen Bonds

The polar covalent bonds that form between hydrogen atoms and other atoms can give rise to a third type of chemical bond, a hydrogen bond (Figure 2.7). A **hydrogen bond** forms when a hydrogen atom with a partial positive charge (δ^+) attracts the partial negative charge (δ^-) of neighboring electronegative atoms, most often oxygen or nitrogen. Thus, hydrogen bonds result from attraction of oppositely charged parts of molecules rather than from sharing of electrons as in covalent bonds, or the loss or gain of electrons as in ionic bonds. Hydrogen bonds are weak compared to ionic and covalent bonds. Thus, they cannot bind atoms into molecules. However, hydrogen bonds do establish important links between molecules or between different parts of a large molecule, such as a protein or nucleic acid (both discussed later in this chapter).

The hydrogen bonds that link neighboring water molecules give water considerable *cohesion,* the tendency of like particles to stay together. The cohesion of water molecules creates a very high **surface tension,** a measure of the difficulty of stretching or breaking the surface of a liquid. At the boundary between water and air, water's surface tension is very high because the water molecules are much more attracted to one another than they are attracted to molecules in the air. This is readily seen when a spider walks on water or a leaf floats on water. The influence of water's surface tension on the body can be seen in the way it increases the work required for breathing. A thin film of watery fluid coats the air sacs of the lungs. So, each inhalation must have enough force to overcome the opposing effect of surface tension as the air sacs stretch and enlarge when taking in air.

Even though single hydrogen bonds are weak, very large molecules may contain thousands of these bonds. Acting collectively, hydrogen bonds provide considerable strength and stability and help determine the three-dimensional shape of large molecules. As you will see later in this chapter, a large molecule's shape determines how it functions.

Figure 2.7 Hydrogen bonding among water molecules. Each water molecule forms hydrogen bonds, indicated by dotted lines, with three to four neighboring water molecules.

Hydrogen bonds occur because hydrogen atoms in one water molecule are attracted to the partial negative charge of the oxygen atom in another water molecule.

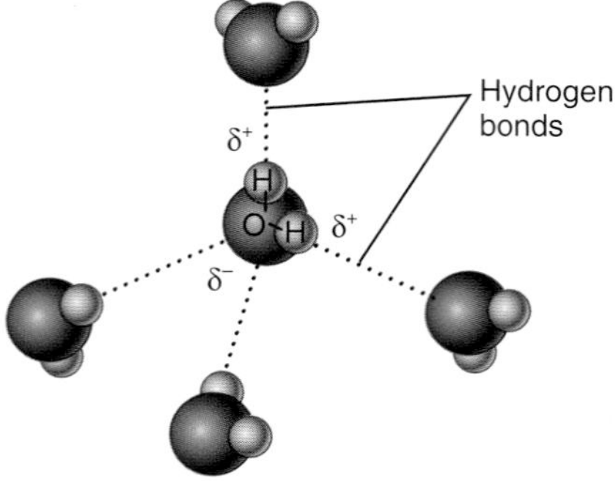

Why would you expect ammonia (NH_3) to form hydrogen bonds with water molecules?

▶ CHECKPOINT

4. Which electron shell is the valence shell of an atom, and what is its significance?
5. Compare the properties of ionic, covalent, and hydrogen bonds.
6. What information is conveyed when you write the molecular or structural formula for a molecule?

CHEMICAL REACTIONS

▶ OBJECTIVES

Define a chemical reaction.

Describe the various forms of energy.

Compare exergonic and endergonic chemical reactions.

Describe the role of activation energy and catalysts in chemical reactions.

Describe synthesis, decomposition, exchange, and reversible reactions.

A **chemical reaction** occurs when new bonds form or old bonds break between atoms. Chemical reactions are the foundation of all life processes, and as we have seen, the interactions of valence electrons are the basis of all chemical reactions. Consider how hydrogen and oxygen molecules react to form water molecules (Figure 2.8). The starting substances—two H_2 and one O_2—are known as the **reactants.** The ending substances—two molecules of H_2O—are the **products.** The arrow in the figure indicates the direction in which the reaction proceeds. In a chemical reaction, the total mass of the reactants equals the total mass of the products. Thus, the number of atoms of each element is the same before and after the reaction. However, because the atoms are rearranged, the reactants and products have different chemical properties. Through thousands of different chemical reactions, body structures are built and body functions are carried out. The term **metabolism** refers to all the chemical reactions occurring in the body.

Figure 2.8 The chemical reaction between two hydrogen molecules (H_2) and one oxygen molecule (O_2) to form two molecules of water (H_2O). Note that the reaction occurs by breaking old bonds and making new bonds.

The number of atoms of each element is the same before and after a chemical reaction.

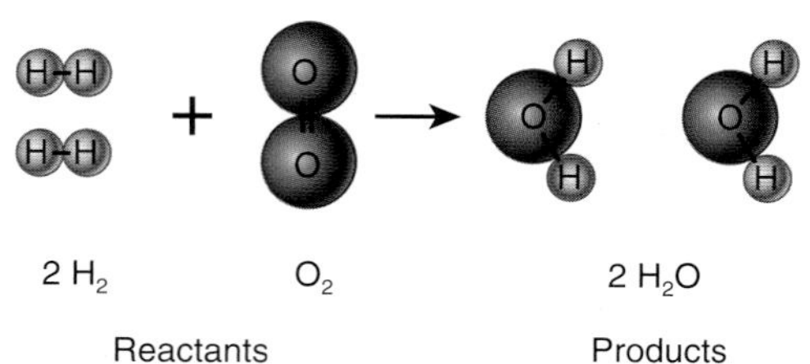

Why does this reaction require two molecules of H_2?

Forms of Energy and Chemical Reactions

Each chemical reaction involves energy changes. **Energy** (*en-* = in; *-ergy* = work) is the capacity to do work. Two principal forms of energy are **potential energy,** energy stored by matter due to its position, and **kinetic energy,** the energy associated with matter in motion. For example, the energy stored in water behind a dam or in a person poised to jump down some steps is potential energy. When the gates of the dam are opened or the person jumps, potential energy is converted into kinetic energy. **Chemical energy** is a form of potential energy that is stored in the bonds of compounds and molecules. The total amount of energy present at the beginning and end of a chemical reaction is the same. Although energy can be neither created nor destroyed, it may be converted from one form to another. This principle is known as the **law of conservation of energy.** For example, some of the chemical energy in the foods we eat is eventually converted into various forms of kinetic energy, such as mechanical energy used to walk and talk. Conversion of energy from one form to another generally releases heat, some of which is used to maintain normal body temperature.

Energy Transfer in Chemical Reactions

Chemical bonds represent stored chemical energy and chemical reactions occur when new bonds are formed or old bonds are broken between atoms. The *overall reaction* may either release energy or absorb energy. **Exergonic reactions** (*ex-* = out) release more energy than they absorb. By contrast, **endergonic reactions** (*end-* = within) absorb more energy than they release.

A key feature of the body's metabolism is the coupling of exergonic reactions and endergonic reactions. Energy released from an exergonic reaction often is used to drive an endergonic one. In general, exergonic reactions occur as nutrients, such as glucose, are broken down. Some of the energy released may be trapped in the covalent bonds of adenosine triphosphate (ATP), which we describe more fully later in this chapter. If a molecule of glucose is completely broken down, the chemical energy in its bonds can be used to produce as many as 38 molecules of ATP. The energy transferred to the ATP molecules is then used to drive endergonic reactions needed to build body structures, such as muscles and bones. The energy in ATP is also used to do the mechanical work involved in the contraction of muscle or the movement of substances into or out of cells.

Activation Energy

Because particles of matter such as atoms, ions, and molecules have kinetic energy, they are continuously moving and colliding with one another. A sufficiently forceful collision can disrupt the movement of valence electrons, causing an existing chemical bond to break or a new one to form. The collision energy needed to break the chemical bonds of the reactants is called the **activation energy** of the reaction (Figure 2.9). This initial energy "investment" is needed to start a reaction. The reactants must absorb enough energy for their chemical bonds to become unstable and their valence electrons to form new combinations. Then, as new bonds form, energy is released to the surroundings.

Figure 2.9 **Activation energy.**

Activation energy is the energy needed to break chemical bonds in the reactant molecules so a reaction can start.

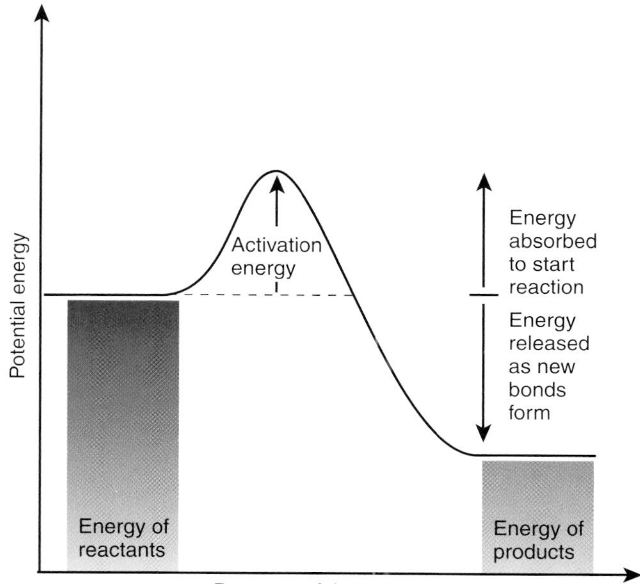

Why is the reaction illustrated here exergonic?

Both the concentration of particles and the temperature influence the chance that a collision will occur and cause a chemical reaction.

- ***Concentration.*** The more particles of matter present in a confined space, the greater the chance that they will collide (think of people crowding into a subway car at rush hour). The concentration of particles increases when more are added to a given space or when the pressure on the space increases, which forces the particles closer together so that they collide more often.
- ***Temperature.*** As temperature rises, particles of matter move about more rapidly. Thus, the higher the temperature of matter, the more forcefully particles will collide, and the greater the chance that a collision will produce a reaction.

Catalysts

As we have seen, chemical reactions occur when chemical bonds break or form after atoms, ions, or molecules collide with one another. Body temperature and the concentrations of molecules in body fluids, however, are far too low for most chemical reactions to occur rapidly enough to maintain life. Raising the temperature and the number of reacting particles of matter in the body could increase the frequency of collisions and thus increase the rate of chemical reactions, but doing so could also damage or kill the body's cells.

Substances called catalysts solve this problem. **Catalysts** are chemical compounds that speed up chemical reactions by lowering the activation energy needed for a reaction to occur (Figure 2.10). The most important catalysts in the body are enzymes, which we will discuss later in this chapter.

A catalyst does not alter the difference in potential energy between the reactants and the products. Rather, it lowers the amount of energy needed to start the reaction.

For chemical reactions to occur, some particles of matter—especially large molecules—not only must collide with sufficient force, but they must "hit" one another at precise spots. A catalyst helps to properly orient the colliding particles. Thus, they interact at the spots that make the reaction happen. Although the action of a catalyst helps to speed up a chemical reaction, the catalyst itself is unchanged at the end of the reaction. A single catalyst molecule can assist one chemical reaction after another.

Types of Chemical Reactions

After a chemical reaction takes place, the atoms of the reactants are rearranged to yield products with new chemical properties. In this section we will look at the types of chemical reactions common to all living cells. Once you have learned them, you will be able to understand the chemical reactions so important to the operation of the human body that are discussed throughout the book.

Figure 2.10 Comparison of energy needed for a chemical reaction to proceed with a catalyst (green curve) and without a catalyst (red curve).

Catalysts speed up chemical reactions by lowering the activation energy.

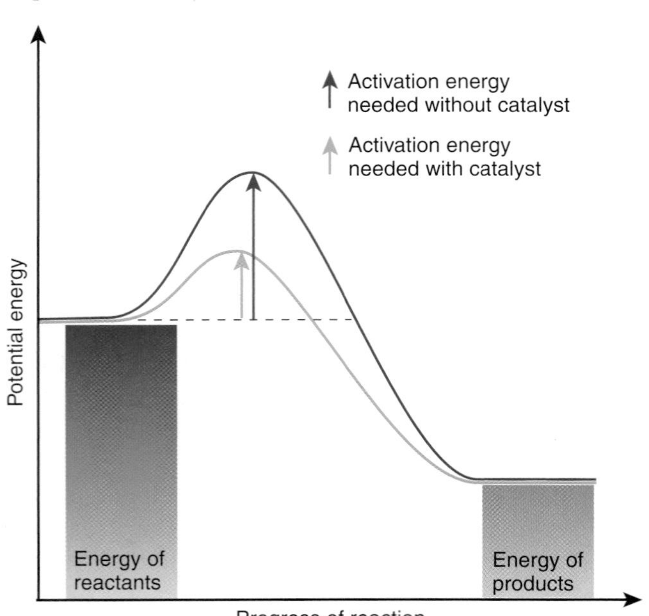

Does a catalyst change the potential energies of the products and reactants?

Synthesis Reactions—Anabolism

When two or more atoms, ions, or molecules combine to form new and larger molecules, the processes are called **synthesis reactions.** The word *synthesis* means "to put together." A synthesis reaction can be expressed as follows:

$$\underset{\text{Atom, ion, or molecule A}}{A} + \underset{\text{Atom, ion, or molecule B}}{B} \xrightarrow{\text{Combine to form}} \underset{\text{New molecule AB}}{AB}$$

One example of a synthesis reaction is the reaction between two hydrogen molecules and one oxygen molecule to form two molecules of water (see Figure 2.8). Another example of a synthesis reaction is the formation of ammonia from nitrogen and hydrogen:

$$\underset{\text{One nitrogen molecule}}{N_2} + \underset{\text{Three hydrogen molecules}}{3H_2} \xrightarrow{\text{Combine to form}} \underset{\text{Two ammonia molecules}}{2NH_3}$$

All the synthesis reactions that occur in your body are collectively referred to as **anabolism** (a-NAB-ō-lizm). Overall, anabolic reactions are usually endergonic because they absorb more energy than they release. Combining simple molecules like amino acids (discussed shortly) to form large molecules such as proteins is an example of anabolism.

Decomposition Reactions—Catabolism

Decomposition reactions split up large molecules into smaller atoms, ions, or molecules. A decomposition reaction is expressed as follows:

$$\underset{\text{Molecule AB}}{AB} \xrightarrow{\text{Breaks down into}} \underset{\text{Atom, ion, or molecule A}}{A} + \underset{\text{Atom, ion, or molecule B}}{B}$$

The decomposition reactions that occur in your body are collectively referred to as **catabolism** (ka-TAB-ō-lizm). Overall, catabolic reactions are usually exergonic because they release more energy than they absorb. For instance, the series of reactions that break down glucose to pyruvic acid, with the net production of two molecules of ATP, are important catabolic reactions in the body. These reactions will be discussed in Chapter 25.

Exchange Reactions

Many reactions in the body are **exchange reactions;** they consist of both synthesis and decomposition reactions. One type of exchange reaction works like this:

$$AB + CD \longrightarrow AD + BC$$

The bonds between A and B and between C and D break (decomposition), and new bonds then form (synthesis) between A and D and between B and C. An example of an exchange reaction is:

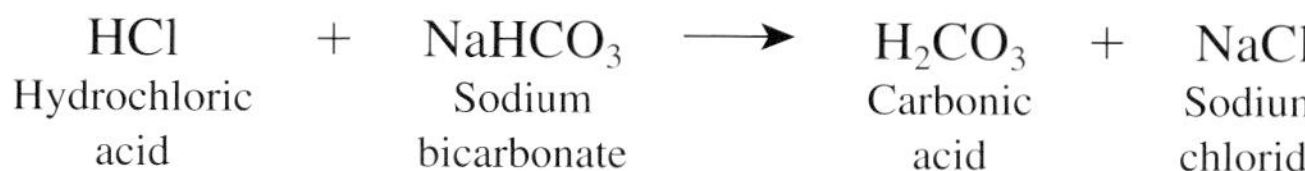

Notice that the ions in both compounds have "switched partners": The hydrogen ion (H^+) from HCl has combined with the bicarbonate ion (HCO_3^-) from $NaHCO_3$, and the sodium ion (Na^+) from $NaHCO_3$ has combined with the chloride ion (Cl^-) from HCl.

Reversible Reactions

Some chemical reactions proceed in only one direction, from reactants to products, as previously indicated by the single arrows. Other chemical reactions may be reversible. In a **reversible reaction,** the products can revert to the original reactants. A reversible reaction is indicated by two half arrows pointing in opposite directions:

$$AB \underset{\text{Combines to form}}{\overset{\text{Breaks down into}}{\rightleftharpoons}} A + B$$

Some reactions are reversible only under special conditions:

$$AB \underset{\text{Heat}}{\overset{\text{Water}}{\rightleftharpoons}} A + B$$

In that case, whatever is written above or below the arrows indicates the condition needed for the reaction to occur. In these reactions, AB breaks down into A and B only when water is added, and A and B react to produce AB only when heat is applied. Many reversible reactions in the body require catalysts called enzymes. Often, different enzymes guide the reactions in opposite directions.

▶ CHECKPOINT

7. What is the relationship between reactants and products in a chemical reaction?
8. Compare potential energy and kinetic energy.
9. How do catalysts affect activation energy?
10. How are anabolism and catabolism related to synthesis and decomposition reactions, respectively?

INORGANIC COMPOUNDS AND SOLUTIONS

▶ OBJECTIVES

Describe the properties of water and those of inorganic acids, bases, and salts.

Distinguish among solutions, colloids, and suspensions.

Define pH and explain the role of buffer systems in homeostasis.

Most of the chemicals in your body exist in the form of compounds. Biologists and chemists divide these compounds into two principal classes: inorganic compounds and organic compounds. **Inorganic compounds** usually lack carbon and are structurally simple. They include water and many salts, acids, and bases. Inorganic compounds may have either ionic or covalent bonds. Water makes up 55–60% of a lean adult's total body mass; all other inorganic compounds combined add 1–2%. Examples of inorganic compounds that contain carbon are carbon dioxide (CO_2), bicarbonate ion (HCO_3^-), and carbonic acid (H_2CO_3). **Organic compounds** always contain carbon, usually contain hydrogen, and always have covalent bonds. Most are large molecules and many are made up of long chains of carbon atoms. Organic compounds make up the remaining 38–43% of the human body.

Water

Water is the most important and abundant inorganic compound in all living systems. Although you might be able to survive for weeks without food, without water you would die in a matter of days. Nearly all the body's chemical reactions occur in a watery medium. Water has many properties that make it such an indispensable compound for life. We have already mentioned the most important property of water, its polarity—the uneven sharing of valence electrons that confers a partial negative charge near the one oxygen atom and two partial positive charges near the two hydrogen atoms in a water molecule (see Figure 2.6). This property alone makes water an excellent solvent for other ionic or polar substances, gives water molecules cohesion (the tendency to stick together), and allows water to resist temperature changes.

Water as a Solvent

In medieval times people searched in vain for a "universal solvent," a substance that would dissolve all other materials. They found nothing that worked as well as water. Although it is the most versatile solvent known, water is not the universal solvent sought by medieval alchemists. If it were, no container could hold it because it would dissolve all potential containers! What exactly is a solvent? In a **solution,** a substance called the **solvent** dissolves another substance called the **solute.** Usually there is more solvent than solute in a solution. For example, your sweat is a dilute solution of water (the solvent) plus small amounts of salts (the solutes).

The versatility of water as a solvent for ionized or polar substances is due to its polar covalent bonds and its bent shape, which allows each water molecule to interact with several neighboring ions or molecules. Solutes that are charged or contain polar covalent bonds are **hydrophilic** (*hydro-* = water; *-philic* = loving), which means they dissolve easily in water. Common examples of hydrophilic solutes are sugar and salt. Molecules that contain mainly nonpolar covalent bonds, by contrast, are **hydrophobic** (*-phobic* = fearing). They are not very water soluble. Examples of hydrophobic compounds include animal fats and vegetable oils.

To understand the dissolving power of water, consider what happens when a crystal of a salt such as sodium chloride (NaCl) is placed in water (Figure 2.11). The electronegative oxygen atom in water molecules attracts the sodium ions (Na^+), and the electropositive hydrogen atoms in water molecules attract the chloride ions (Cl^-). Soon, water molecules surround and separate Na^+ and Cl^- ions from each other at the surface of the crystal, breaking the ionic bonds that held NaCl together. The water molecules surrounding the ions also lessen the chance that Na^+ and Cl^- will come together and reform an ionic bond.

The ability of water to form solutions is essential to health and survival. Because water can dissolve so many different substances, it is an ideal medium for metabolic reactions. Water enables dissolved reactants to collide and form products. Water also dissolves waste products, which allows them to be flushed out of the body in the urine.

Figure 2.11 How polar water molecules dissolve salts and polar substances. When a crystal of sodium chloride is placed in water, the slightly negative oxygen end (red) of water molecules is attracted to the positive sodium ions (Na^+), and the slightly positive hydrogen portions (gray) of water molecules are attracted to the negative chloride ions (Cl^-). In addition to dissolving sodium chloride, water also causes it to dissociate, or separate into charged particles, which is discussed shortly.

Water is a versatile solvent because its polar covalent bonds, in which electrons are shared unequally, create positive and negative regions.

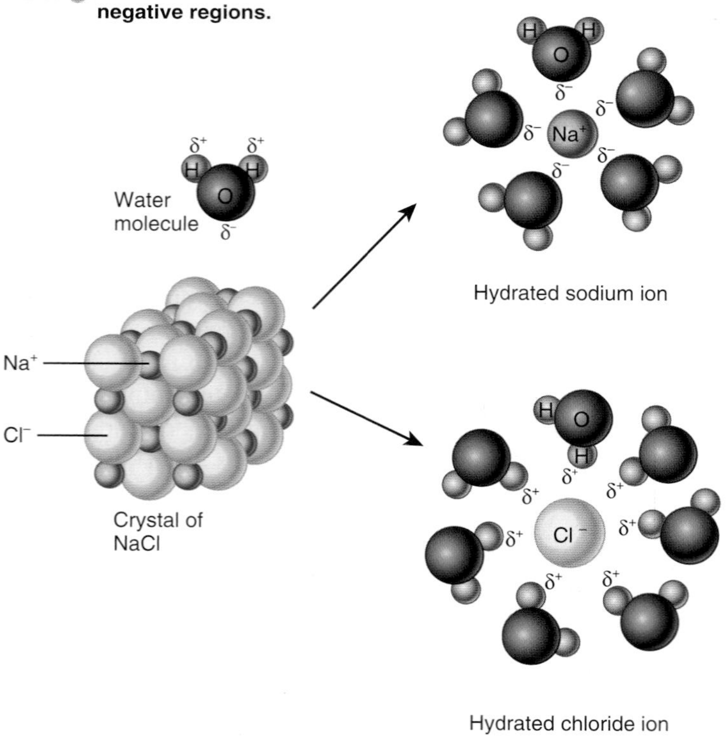

Table sugar (sucrose) easily dissolves in water but is not an electrolyte. Is it likely that all the covalent bonds between atoms in table sugar are nonpolar bonds? Why or why not?

Water in Chemical Reactions

Water serves as the medium for most chemical reactions in the body, and participates as a reactant or product in certain reactions. During digestion, for example, decomposition reactions break down large nutrient molecules into smaller molecules by the addition of water molecules. This type of reaction is called **hydrolysis** (hī-DROL-i-sis; *-lysis* = to loosen or break apart). Hydrolysis reactions enable dietary nutrients to be absorbed into the body. By contrast, when two smaller molecules join to form a larger molecule in a **dehydration synthesis reaction** (*de-* = from, down, or out; *hydra-* = water), a water molecule is one of the products formed. As you will see later in the chapter, such reactions occur during synthesis of proteins and other large molecules (for example, see Figure 2.22).

Thermal Properties of Water

In comparison to most substances, water can absorb or release a relatively large amount of heat with only a modest change in its own temperature. For this reason, water is said to have a high *heat capacity.* The reason for this property is the large number of hydrogen bonds in water. As water absorbs heat energy, some of the energy is used to break hydrogen bonds. Less energy is then left over to increase the motion of water molecules, which would increase the water's temperature. The high heat capacity of water is the reason it is used in automobile radiators; it cools the engine by absorbing heat without its own temperature rising to an unacceptably high level. The large amount of water in the body has a similar effect: It lessens the impact of environmental temperature changes, helping to maintain the homeostasis of body temperature.

Water also requires a large amount of heat to change from a liquid to a gas. Its *heat of vaporization* is high. As water evaporates from the surface of the skin, it removes a large quantity of heat, providing an important cooling mechanism.

Water as a Lubricant

Water is a major component of mucus and other lubricating fluids throughout the body. Lubrication is especially necessary in the chest (pleural and pericardial cavities) and abdomen (peritoneal cavity), where internal organs touch and slide over one another. It is also needed at joints, where bones, ligaments, and tendons rub against one another. Inside the gastrointestinal tract, mucus and other watery secretions moisten foods, which aids their smooth passage through the digestive system.

Solutions, Colloids, and Suspensions

A **mixture** is a combination of elements or compounds that are physically blended together but not bound by chemical bonds. For example, the air you are breathing is a mixture of gases that includes nitrogen, oxygen, argon, and carbon dioxide. Three common liquid mixtures are solutions, colloids, and suspensions.

Once mixed together, solutes in a solution remain evenly dispersed among the solvent molecules. Because the solute particles in a solution are very small, a solution looks clear and transparent.

A **colloid** differs from a solution mainly because of the size of its particles. The solute particles in a colloid are large enough to scatter light, just as water droplets in fog scatter light from a car's headlight beams. For this reason, colloids usually appear translucent or opaque. Milk is an example of a liquid that is both a colloid and a solution: The large milk proteins make it a colloid, whereas calcium salts, milk sugar (lactose), ions, and other small particles are in solution.

The solutes in both solutions and colloids do not settle out and accumulate on the bottom of the container. In a **suspension,** by contrast, the suspended material may mix with the liquid or suspending medium for some time, but eventually it will settle out. Blood is an example of a suspension. When freshly drawn from the body, blood has an even, reddish color. After blood sits for a while in a test tube, red blood cells settle out of the suspension and drift to the bottom of the tube (see Figure 19.1a on page 688). The upper layer, the liquid portion of blood, appears pale yellow and is called blood plasma. Blood plasma is both a solution of ions and other small solutes and a colloid due to the presence of larger plasma proteins.

The **concentration** of a solution may be expressed in several ways. One common way is by a mass per volume **percentage,** which gives the relative mass of a solute found in a given volume of solution. For example, you may have seen the following on the label of a bottle of wine: "Alcohol 14.1% by volume." Another way expresses concentration in units of **moles per liter (mol/L),** which relate to the total number of molecules in a given volume of solution. A **mole** is the amount of any substance that has a mass in grams equal to the sum of the atomic masses of all its atoms. For example, one mole of the element chlorine (atomic mass = 35.45) is 35.45 grams and one mole of the salt sodium chloride (NaCl) is 58.44 grams (22.99 for Na + 35.45 for Cl). Just as a dozen always means 12 of something, a mole of anything has the same number of particles: 6.023×10^{23}. This huge number is called *Avogadro's number.* Thus, measurements of substances that are stated in moles tell us about the numbers of atoms, ions, or molecules present. This is important when chemical reactions are occurring because each reaction requires a set number of atoms of specific elements. Table 2.3 describes these ways of expressing concentration.

TABLE 2.3 Percentage and Molarity

Definition	Example
Percentage (mass per volume)	
Number of grams of a substance per 100 milliliters (mL) of solution.	To make a 10% NaCl solution,take 10 gm of NaCl and add enough water to make a total of 100 mL of solution.
Molarity = moles (mol) per liter	
A 1 molar (1 M) solution = 1 mole of a solute in 1 liter of solution.	To make a 1 molar (1 M) solution of NaCl, dissolve1 mole of NaCl (58.44 gm) in enough water to make a total of 1 liter of solution.

Inorganic Acids, Bases, and Salts

When inorganic acids, bases, or salts dissolve in water, they **dissociate** (dis′-sō-sē-ĀT), that is, they separate into ions and become surrounded by water molecules. An **acid** (Figure 2.12a) is a substance that dissociates into one or more **hydrogen ions (H^+)** and one or more anions. Because H^+ is a single proton with one positive charge, an acid is also referred to as a **proton donor.** A **base,** by contrast (Figure 2.12b), removes H^+ from a solution and is therefore a **proton acceptor.** Many bases dissociate into one or more **hydroxide ions (OH^-)** and one or more cations.

A **salt,** when dissolved in water, dissociates into cations and anions, neither of which is H^+ or OH^- (Figure 2.12c). In the body, salts such as potassium chloride are electrolytes that are important for carrying electrical currents (ions flowing from one place to another), especially in nerve and muscular tissues. The ions of salts also provide many essential chemical elements in intracellular and extracellular fluids such as blood, lymph, and the interstitial fluid of tissues.

Acids and bases react with one another to form salts. For example, the reaction of hydrochloric acid (HCl) and potassium hydroxide (KOH), a base, produces the salt potassium chloride (KCl) and water (H_2O). This exchange reaction can be written as follows:

$$\underset{\text{Acid}}{HCl} + \underset{\text{Base}}{KOH} \longrightarrow \underbrace{H^+ + Cl^- + K^+ + OH^-}_{\text{Dissociated ions}} \longrightarrow \underset{\text{Salt}}{KCl} + \underset{\text{Water}}{H_2O}$$

Figure 2.12 Dissociation of inorganic acids, bases, and salts.

Dissociation is the separation of inorganic acids, bases, and salts into ions in a solution.

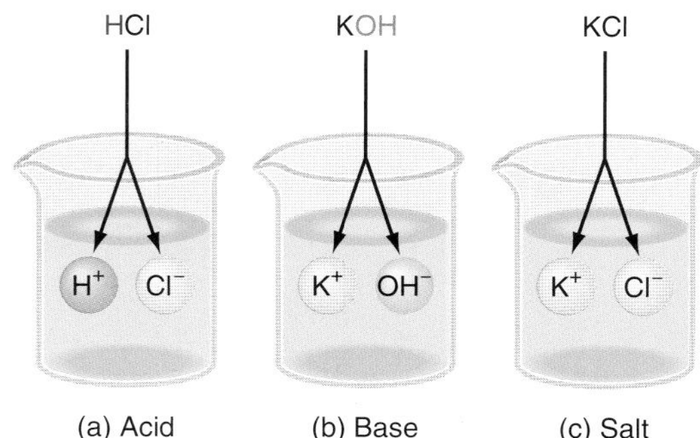

The compound $CaCO_3$ (calcium carbonate) dissociates into a calcium ion Ca^{2+} and a carbonate ion CO_3^{2-}. Is it an acid, a base, or a salt? What about H_2SO_4, which dissociates into two H^+ and one SO_4^{2-}?

Acid–Base Balance: The Concept of pH

To ensure homeostasis, intracellular and extracellular fluids must contain almost balanced quantities of acids and bases. The more hydrogen ions (H^+) dissolved in a solution, the more acidic the solution; the more hydroxide ions (OH^-), the more basic (alkaline) the solution. The chemical reactions that take place in the body are very sensitive to even small changes in the acidity or alkalinity of the body fluids in which they occur. Any departure from the narrow limits of normal H^+ and OH^- concentrations greatly disrupts body functions.

A solution's acidity or alkalinity is expressed on the **pH scale,** which extends from 0 to 14 (Figure 2.13). This scale is based on the concentration of H^+ in moles per liter. A pH of 7 means that a solution contains one ten-millionth (0.0000001) of a mole of hydrogen ions per liter. The number 0.0000001 is written as 1×10^{-7} in scientific notation, which indicates that the number is 1 with the decimal point moved seven places to the left. To convert this value to pH, the negative exponent (-7) is changed to a positive number (7). A solution with a H^+ concentration of 0.0001 (10^{-4}) moles per liter has a pH of 4; a solution with a H^+ concentration of 0.000000001 (10^{-9}) moles per liter has a pH of 9; and so on. It is important to realize that a change of one whole number on the pH scale represents a *tenfold* change in the number of H^+. A pH of 6 denotes 10 times more H^+ than a pH of 7, and a pH of 8 indicates 10 times fewer H^+ than a pH of 7 and 100 times fewer H^+ than a pH of 6.

The midpoint of the pH scale is 7, where the concentrations of H^+ and OH^- are equal. A substance with a pH of 7, such as pure water, is neutral. A solution that has more H^+ than OH^- is an **acidic solution** and has a pH below 7. A solution that has more OH^- than H^+ is a **basic (alkaline) solution** and has a pH above 7.

Maintaining pH: Buffer Systems

Although the pH of body fluids may differ, as we have discussed, the normal limits for each fluid are quite narrow. Table 2.4 shows the pH values for certain body fluids along with those of some common substances outside the body. Homeostatic mechanisms maintain the pH of blood between 7.35 and 7.45, which is slightly more basic than pure water. You will learn in Chapter 27 that if the pH of blood falls below 7.35, a condition called acidosis occurs, and if the pH rises above 7.45, it results in a condition called alkalosis; both conditions can seriously compromise homeostasis. Saliva is slightly acidic, and semen is slightly basic. Because the kidneys help remove excess acid from the body, urine can be quite acidic.

Even though strong acids and bases are continually taken into and formed by the body, the pH of fluids inside and outside cells remains almost constant. One important reason is the presence of **buffer systems,** which function to convert strong acids or bases into weak acids or bases. Strong acids (or bases) ionize easily and

Figure 2.13 The pH scale. A pH below 7 indicates an acidic solution—more H^+ than OH^-. $[H^+]$ = hydrogen ion concentration; $[OH^-]$ = hydroxide ion concentration.

The lower the numerical value of the pH, the more acidic is the solution because the H^+ concentration becomes progressively greater. A pH above 7 indicates a basic (alkaline) solution; that is, there are more OH^- than H^+. The higher the pH, the more basic the solution.

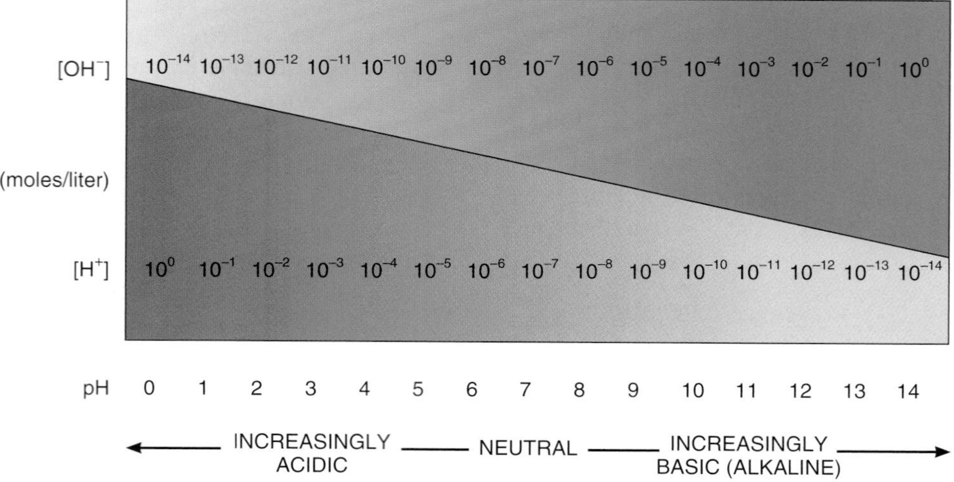

At pH 7 (neutrality), the concentrations of H^+ and OH^- are equal (10^{-7} mol/liter). What are the concentrations of H^+ and OH^- at pH 6? Which pH is more acidic, 6.82 or 6.91? Which pH is closer to neutral, 8.41 or 5.59?

TABLE 2.4 pH Values of Selected Substances

Substance	pH Value
• Gastric juice (found in the stomach)	1.2–3.0
Lemon juice	2.3
Vinegar	3.0
Carbonated soft drink	3.0–3.5
Orange juice	3.5
• Vaginal fluid	3.5–4.5
Tomato juice	4.2
Coffee	5.0
• Urine	4.6–8.0
• Saliva	6.35–6.85
Milk	6.8
Distilled (pure) water	7.0
• Blood	7.35–7.45
• Semen (fluid containing sperm)	7.20–7.60
• Cerebrospinal fluid (fluid associated with nervous system)	7.4
• Pancreatic juice (digestive juice of the pancreas)	7.1–8.2
• Bile (liver secretion that aids fat digestion)	7.6–8.6
Milk of magnesia	10.5
Lye (sodium hydroxide)	14.0

• Denotes substances in the human body.

contribute many H^+ or (OH^-) to a solution. Therefore, they can change pH drastically, which can disrupt the body's metabolism. Weak acids (or bases) do not ionize as much and contribute fewer H^+ (or OH^-). Hence, they have less effect on the pH. The chemical compounds that can convert strong acids or bases into weak ones are called **buffers.** They do so by removing or adding protons (H^+).

One important buffer system in the body is the **carbonic acid–bicarbonate buffer system.** Carbonic acid (H_2CO_3) can act as a weak acid, and the bicarbonate ion (HCO_3^-) can act as a weak base. Hence, this buffer system can compensate for either an excess or a shortage of H^+. For example, if there is an excess of H^+ (an acidic condition), HCO_3^- can function as a weak base and remove the excess H^+, as follows:

$$\underset{\text{Hydrogen}}{H^+} + \underset{\text{Bicarbonate ion (weak base)}}{HCO_3^-} \longrightarrow \underset{\text{Carbonic acid}}{H_2CO_3}$$

If there is a shortage of H^+ (an alkaline condition), by contrast, H_2CO_3 can function as a weak acid and provide needed H^+ as follows:

$$\underset{\text{Carbonic acid (weak acid)}}{H_2CO_3} \longrightarrow \underset{\text{Hydrogen}}{H^+} + \underset{\text{Bicarbonate ion}}{HCO_3^-}$$

Chapter 27 describes buffers and their roles in maintaining acid–base balance in more detail.

▶ CHECKPOINT

11. How do inorganic compounds differ from organic compounds?

12. Describe two ways to express the concentration of a solution.

13. What functions does water perform in the body?

14. How do bicarbonate ions prevent buildup of excess H^+?

ORGANIC COMPOUNDS

▶ OBJECTIVES

Describe the functional groups of organic molecules.

Identify the building blocks and functions of carbohydrates, lipids, proteins, and enzymes.

Describe the structure and functions of deoxyribonucleic acid (DNA), ribonucleic acid (RNA), and adenosine triphosphate (ATP).

Inorganic compounds are relatively simple. Their molecules have only a few atoms and cannot be used by cells to perform complicated biological functions. Many organic molecules, by contrast, are relatively large and have unique characteristics that allow them to carry out complex functions. Important categories of organic compounds include carbohydrates, lipids, proteins, nucleic acids, and adenosine triphosphate (ATP).

Carbon and Its Functional Groups

Carbon has several properties that make it particularly useful to living organisms. For one thing, it can form bonds with one to thousands of other carbon atoms to produce large molecules that can have many different shapes. Due to this property of carbon, the body can build many different organic compounds, each of which has a unique structure and function. Moreover, the large size of most carbon-containing molecules and the fact that some do not dissolve easily in water make them useful materials for building body structures.

Organic compounds are usually held together by covalent bonds. Carbon has four electrons in its outermost (valence) shell. It can bond covalently with a variety of atoms, including other carbon atoms, to form rings and straight or branched chains. Other elements that most often bond with carbon in organic compounds are hydrogen, oxygen, and nitrogen. Sulfur and phosphorus are also present in organic compounds. The other elements listed in Table 2.1 are present in a smaller number of organic compounds.

The chain of carbon atoms in an organic molecule is called the **carbon skeleton.** Many of the carbons are bonded to hydrogen atoms, yielding a hydrocarbon. Also attached to the carbon skeleton are distinctive **functional groups,** other atoms or molecules bound to the hydrocarbon skeleton. Each type of functional group has a specific arrangement of atoms that confers characteristic

TABLE 2.5 Major Functional Groups

Name and Structural Formula*	Occurrence and Significance
Hydroxyl R—O—H	Alcohols contain an —OH group, which is polar and hydrophilic due to its electronegative O atom. Molecules with many —OH groups dissolve easily in water.
Sulfhydryl R—S—H	*Thiols* have an —SH group, which is polar and and hydrophilic due to its electronegative S atom. Certain amino acids, the building blocks of proteins, contain —SH groups, which help stabilize the shape of proteins.An example is the amino acid cysteine.
Carbonyl R—C(=O)—R or R—C(=O)—H	*Ketones* contain a carbonyl group within the carbon skeleton. The carbonyl group is polar and hydrophilic due to its electronegative O atom. *Aldehydes* have a carbonyl group at the end of the carbon skeleton.
Carboxyl R—C(=O)—OH or R—C(=O)—O⁻	*Carboxylic acids* contain a carboxyl group at the end of the carbon skeleton. All amino acids have a —COOH group at one end. The negatively charged form predominates at the pH of body cells and is hydrophilic.
Ester R—C(=O)—O—R	*Esters* predominate in dietary fats and oils and also occur in our body triglycerides. Aspirin is an ester of salicylic acid, a pain-relieving molecule found in the bark of the willow tree.
Phosphate R—O—P(=O)(—O⁻)—O⁻	*Phosphates* contain a phosphate group(—PO_4^{2-}), which is very hydrophilic due to the dual negative charges. An important example is adenosine triphosphate (ATP), which transfers chemical energy between organic molecules during chemical reactions.
Amino R—N(H)(H) or R—N^{+}(H)(H)—H	*Amines* have an —NH_2 group, which can act as a base and pick up a hydrogen ion, giving the amino group a positive charge. At the pH of body fluids, most amino groups have a charge of 1+. All amino acids have an amino group at one end.

*R = variable group.

chemical properties on the organic molecule attached to it. Table 2.5 lists the most common functional groups of organic molecules and describes some of their properties. Because organic molecules often are big, there are shorthand methods for representing their structural formulas. Figure 2.14 shows two ways to indicate the structure of the sugar glucose, a molecule with a ring-shaped carbon skeleton that has several hydroxyl groups attached.

Figure 2.14 Alternative ways to write the structural formula for glucose.

In standard shorthand, carbon atoms are understood to be at locations where two bond lines intersect, and single hydrogen atoms are not indicated.

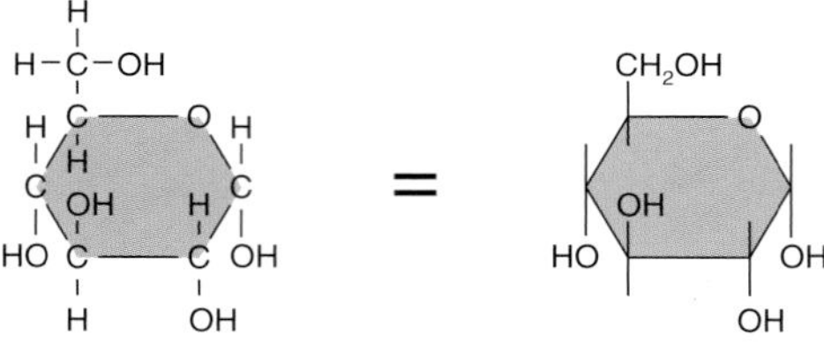

How many hydroxyl groups does a molecule of glucose have? How many carbon atoms are part of glucose's carbon skeleton?

Small organic molecules can combine into very large molecules that are called **macromolecules** (*macro-* = large). Macromolecules are usually **polymers** (*poly-* = many; *-mers* = parts). A polymer is a large molecule formed by the covalent bonding of many identical or similar small building-block molecules called **monomers** (*mono-* = one). Usually, the reaction that joins two monomers is a dehydration synthesis. In this type of reaction, a hydrogen atom is removed from one monomer and a hydroxyl group is removed from the other to form a molecule of water (see Figure 2.15). Macromolecules such as carbohydrates, lipids, proteins, and nucleic acids are assembled in cells via dehydration synthesis reactions.

Molecules that have the same molecular formula but different structures are called **isomers** (Ī-so-merz; *iso-* = equal or the same). For example, the molecular formulas for the sugars glucose and fructose are both $C_6H_{12}O_6$. The individual atoms, however, are positioned differently along the carbon skeleton (see Figure 2.15), giving the sugars different chemical properties.

Carbohydrates

Carbohydrates include sugars, glycogen, starches, and cellulose. Even though they are a large and diverse group of organic compounds and have several functions, carbohydrates represent only 2–3% of your total body mass. In humans and animals, carbohydrates function mainly as a source of chemical energy for generating ATP needed to drive metabolic reactions. Only a few carbohydrates are used for building structural units. One example is deoxyribose, a type of sugar that is a building block of deoxyribonucleic acid (DNA), the molecule that carries inherited genetic information.

Carbon, hydrogen, and oxygen are the elements found in carbohydrates. The ratio of hydrogen to oxygen atoms is usually 2:1, the same as in water. Although there are exceptions, carbohydrates generally contain one water molecule for each carbon

Figure 2.15 Structural and molecular formulas for the monosaccharides glucose and fructose and the disaccharide sucrose. In dehydration synthesis (read from left to right), two smaller molecules, glucose and fructose, are joined to form a larger molecule of sucrose. Note the loss of a water molecule. In hydrolysis (read from right to left), the addition of a water molecule to the larger sucrose molecule breaks the disaccharide into two smaller molecules, glucose and fructose.

Monosaccharides are the monomers used to build carbohydrates.

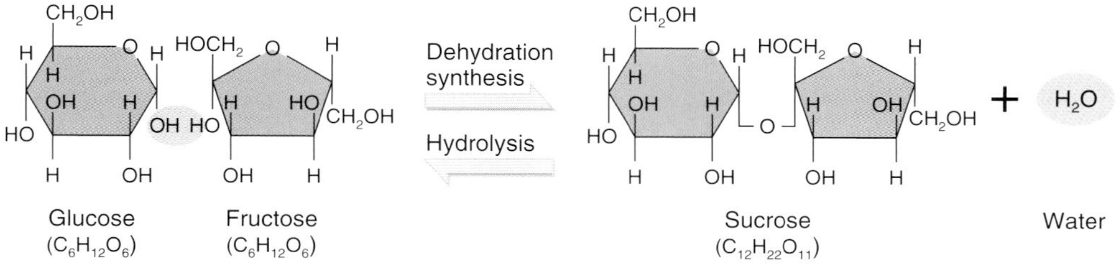

Dehydration synthesis
Hydrolysis
Glucose ($C_6H_{12}O_6$)
Fructose ($C_6H_{12}O_6$)
Sucrose ($C_{12}H_{22}O_{11}$)
+ H_2O
Water

How many carbon atoms can you count in fructose? In sucrose?

atom. This is the reason they are called carbohydrates, which means "watered carbon." The three major groups of carbohydrates, based on their sizes, are monosaccharides, disaccharides, and polysaccharides (Table 2.6).

Monosaccharides and Disaccharides: The Simple Sugars

Monosaccharides and disaccharides are known as **simple sugars.** The monomers of carbohydrates, **monosaccharides** (mon′-ō-SAK-a-rīds; *sacchar-* = sugar), contain from three to seven carbon atoms. They are designated by names ending in "-ose" with a prefix that indicates the number of carbon atoms. For example, monosaccharides with three carbons are called *trioses* (*tri-* = three). There are also *tetroses* (four-carbon sugars), *pentoses* (five-carbon sugars), *hexoses* (six-carbon sugars), and *heptoses* (seven-carbon sugars). Cells throughout the body break down the hexose glucose to produce ATP.

Two monosaccharide molecules can combine by dehydration synthesis to form one **disaccharide** (dī-SAK-a-rīd; *di-* = two) molecule and one molecule of water. For example, molecules of the monosaccharides glucose and fructose combine to form a molecule of the disaccharide sucrose (table sugar), as shown in Figure 2.15. Glucose and fructose are isomers. As you learned earlier in the chapter, isomers have the same molecular formula, but the relative positions of the oxygen and carbon atoms are different, causing the compounds to have different chemical properties. Notice that the formula for sucrose is $C_{12}H_{22}O_{11}$, not $C_{12}H_{24}O_{12}$, because a molecule of water is removed as the two monosaccharides are joined.

Disaccharides can also be split into smaller, simpler molecules by hydrolysis. A molecule of sucrose, for example, may be hydrolyzed into its components, glucose and fructose, by the addition of water. Figure 2.15 also illustrates this reaction. Some individuals use **artificial sweeteners** to limit their sugar consumption for medical reasons, while others do so to avoid calories that might result in weight gain. Artificial sweeteners are much sweeter than sucrose, have fewer calories, and do not cause tooth decay.

TABLE 2.6 Major Carbohydrate Groups

Type of Carbohydrate	Examples
Monosaccharides (Simple sugars that contain from 3 to 7 carbon atoms.)	Glucose (the main blood sugar). Fructose (found in fruits). Galactose (in milk sugar). Deoxyribose (in DNA). Ribose (in RNA).
Disaccharides (Simple sugars formed from the combination of two monosaccharides by dehydration synthesis.)	Sucrose (table sugar) = glucose + fructose. Lactose (milk sugar) = glucose + galactose. Maltose = glucose + glucose.
Polysaccharides (From tens to hundreds of monosaccharides joined by dehydration synthesis.)	Glycogen (the stored form of carbohydrates in animals). Starch (the stored form of carbohydrate in plants and main carbohydrate in food). Cellulose (part of cell walls in plants that cannot be digested by humans but aids movement of food through intestines).

Polysaccharides

The third major group of carbohydrates is the **polysaccharides** (pol′-ē-SAK-a-rīds). Each polysaccharide molecule contains tens or hundreds of monosaccharides joined through dehydration synthesis reactions. Unlike simple sugars, polysaccharides usually are insoluble in water and do not taste sweet. The main polysaccharide in the human body is **glycogen,** which is made entirely of glucose monomers linked to one another in branching chains (Figure 2.16). A limited amount of carbohydrates is stored as glycogen in the liver and skeletal muscles. **Starches** are polysaccharides formed from glucose by plants. They are found in foods such as pasta and potatoes and are the major carbohydrates in the diet. Like disaccharides, polysaccharides such as glycogen and starches can be broken down into monosaccharides through hydrolysis reactions. For example, when the blood glucose level falls, liver cells break down glycogen into glucose and release it into the blood, making it available to body cells, which break it down to synthesize ATP. **Cellulose** is a polysaccharide found in plants that cannot be digested by humans but does provide bulk to help eliminate feces.

Lipids

A second important group of organic compounds is **lipids** (*lip-* = fat). Lipids make up 18–25% of body mass in lean adults. Like carbohydrates, lipids contain carbon, hydrogen, and oxygen. Unlike carbohydrates, they do not have a 2:1 ratio of hydrogen to oxygen. The proportion of electronegative oxygen atoms in lipids is usually smaller than in carbohydrates, so there are fewer polar covalent bonds. As a result, most lipids are insoluble in polar solvents such as water; they are *hydrophobic*. Because they are hydrophobic, only the smallest lipids (some fatty acids) can dissolve in watery blood plasma. To become more soluble in blood plasma, other lipid molecules join with hydrophilic protein molecules. The resulting lipid/protein complexes are termed **lipoproteins.** Lipoproteins are soluble because the proteins are on the outside and the lipids are on the inside.

Figure 2.16 **Part of a glycogen molecule, the main polysaccharide in the human body.**

Glycogen is made up of glucose monomers and is the stored form of carbohydrate in the human body.

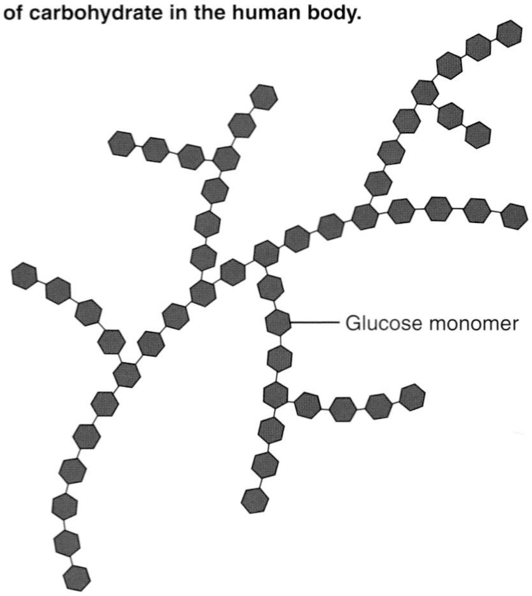

Which body cells store glycogen?

The diverse lipid family includes triglycerides (fats and oils), phospholipids (lipids that contain phosphorus), steroids (lipids that contain rings of carbon atoms), eicosanoids (20-carbon lipids), and a variety of other lipids, including fatty acids, fat-soluble vitamins (vitamins A, D, E, and K), and lipoproteins. Table 2.7 introduces the various types of lipids and highlights their roles in the human body.

Triglycerides

The most plentiful lipids in your body and in your diet are the **triglycerides** (trī-GLI-cer-īdes; *tri-* = three), also known as **triacylglycerols,** which may be either solids (fats) or liquids (oils) at room temperature. They are the body's most highly concentrated form of chemical energy. Triglycerides provide more than twice as much energy per gram as do carbohydrates and proteins. Our capacity to store triglycerides in adipose (fat) tissue is unlimited for all practical purposes. Excess dietary carbohydrates, proteins, fats, and oils all have the same fate: They are deposited in adipose tissue as triglycerides.

TABLE 2.7 Types of Lipids in the Body

Type of Lipid	Functions
Triglycerides *(fats and oils)*	Protection, insulation, energy storage.
Phospholipids	Major lipid component of cell membranes.
Steroids	
Cholesterol	Minor component of all animal cell membranes; precursor of bile salts,vitamin D, and steroid hormones.
Bile salts	Needed for digestion and absorption of dietary lipids.
Vitamin D	Helps regulate calcium level in the body; needed for bone growth and repair.
Adrenocortical hormones	Help regulate metabolism, resistance to stress, and salt and water balance.
Sex hormones	Stimulate reproductive functions and sexual characteristics.
Eicosanoids *(Prostaglandins and leukotrienes)*	Have diverse effects on modifying responses to hormones, blood clotting, inflammation, immunity, stomach acid secretion, airway diameter, lipid breakdown, and smooth muscle contraction.
Other lipids	
Fatty acids	Catabolized to generate adenosine triphosphate (ATP) or used to synthesize triglycerides and phospholipids.
Carotenes	Needed for synthesis of vitamin A, which is used to make visual pigments in the eyes. Also function as antioxidants.
Vitamin E	Promotes wound healing, prevents tissue scarring, contributes to the normal structure and function of the nervous system, and functions as an antioxidant.
Vitamin K	Required for synthesis of blood-clotting proteins.
Lipoproteins	Transport lipids in the blood, carry triglycerides and cholesterol to tissues, and remove excess choles terol from the blood.

A triglyceride consists of two types of building blocks, a single glycerol molecule and three fatty acid molecules. A three-carbon **glycerol** molecule forms the backbone of a triglyceride (Figure 2.17). Three **fatty acids** are attached by dehydration synthesis reactions, one to each carbon of the glycerol backbone. The chemical bond formed where each water molecule is removed is an *ester linkage* (see Table 2.5). The reverse reaction, hydrolysis, breaks down a single molecule of a triglyceride into three fatty acids and glycerol.

Saturated fats are triglycerides that contain only *single covalent bonds* between fatty acid carbon atoms. Because they lack double bonds, each carbon atom is *saturated with hydrogen atoms* (see, for example, palmitic acid and stearic acid in Figure 2.17c). Triglycerides with mainly saturated fatty acids usually are solid at room temperature. Although saturated fats occur mostly in meats (especially red meats) and nonskim dairy products (whole milk, cheese, and butter), they are also found in a few plant products, such as cocoa butter, palm oil, and coconut oil. Diets that contain large amounts of saturated fats are associated with disorders such as heart disease and colorectal cancer.

Monounsaturated fats contain fatty acids with *one double covalent bond* between two fatty acid carbon atoms. Thus, they are not completely saturated with hydrogen atoms (see, for example, oleic acid in Figure 2.17c). The double bonds in monounsaturated fatty acids (and polyunsaturated fatty acids) form kinks in the fatty acids. Olive oil, peanut oil, canola oil, most nuts, and avocados are rich in triglycerides with monounsaturated fatty acids. Monounsaturated fats are thought to decrease the risk of heart disease.

Polyunsaturated fats contain *more than one double covalent bond* between fatty acid carbon atoms. An example is linoleic acid. Corn oil, safflower oil, sunflower oil, soybean oil,

Figure 2.17 **The formation of a triglyceride (triacylglycerol) from a glycerol and three fatty acid molecules.** Each time a glycerol (a) and a fatty acid (b) are joined in dehydration synthesis, a molecule of water is removed. An ester linkage joins the glycerol to each of the three molecules of fatty acids, which vary in length and in the number and location of double bonds between carbon atoms (C=C). Shown here (c) is a triglyceride molecule that contains two saturated fatty acids and a monounsaturated fatty acid. The kink (bend) in the oleic acid occurs at the double bond.

One glycerol and three fatty acids are the building blocks of triglycerides.

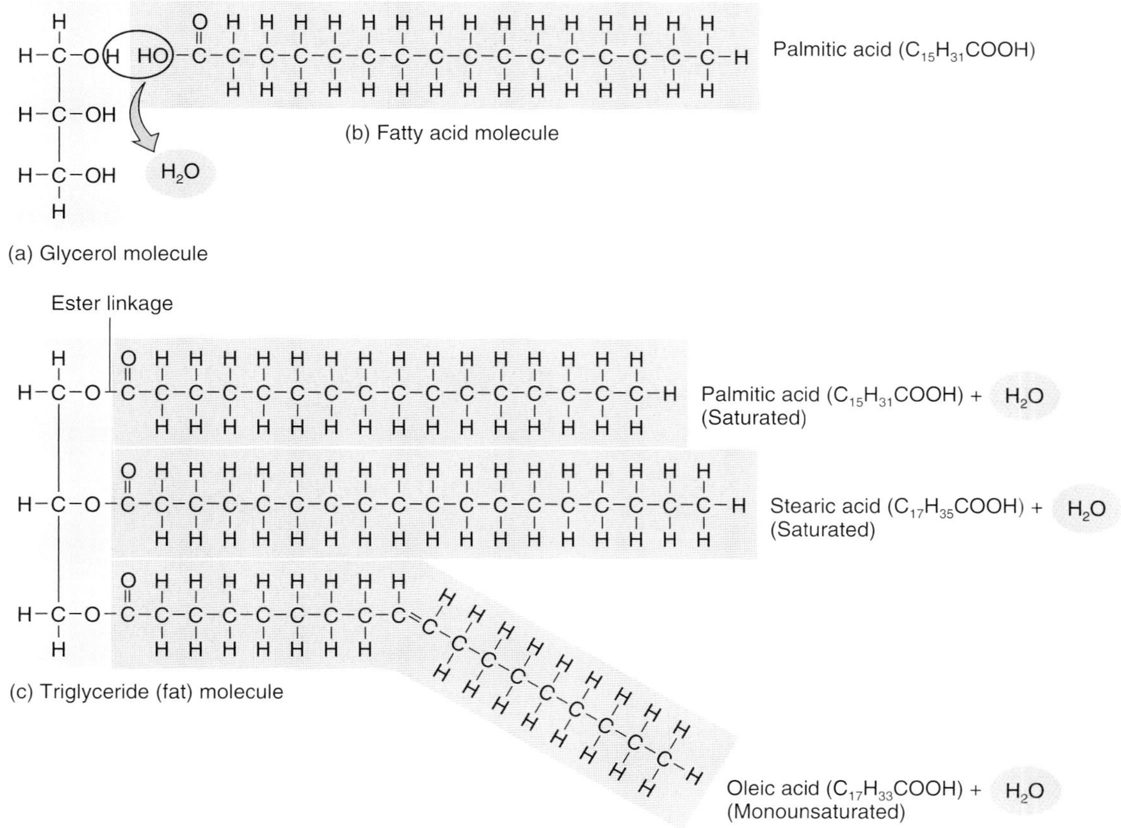

Palmitic acid ($C_{15}H_{31}COOH$)

(b) Fatty acid molecule

H_2O

(a) Glycerol molecule

Ester linkage

Palmitic acid ($C_{15}H_{31}COOH$) + H_2O
(Saturated)

Stearic acid ($C_{17}H_{35}COOH$) + H_2O
(Saturated)

(c) Triglyceride (fat) molecule

Oleic acid ($C_{17}H_{33}COOH$) + H_2O
(Monounsaturated)

Does the oxygen in the water molecule removed during dehydration synthesis come from the glycerol or from a fatty acid?

and fatty fish (salmon, tuna, and mackerel) contain a high percentage of polyunsaturated fatty acids. Polyunsaturated fats are also believed to decrease the risk of heart disease.

Fatty Acids in Health and Disease

As its name implies, a group of fatty acids called **essential fatty acids (EFAs)** is essential to human health. However, they cannot be made by the human body and must be obtained from foods or supplements. Among the more important EFAs are *omega-3 fatty acids, omega-6 fatty acids,* and cis-*fatty acids.*

Omega-3 and omega-6 fatty acids are polyunsaturated fatty acids that are believed to work together to promote health. They may have a protective effect against heart disease and stroke by lowering total cholesterol, raising HDL (high-density lipoproteins or "good cholesterol") and lowering LDL (low-density lipoproteins or "bad cholesterol"). In addition, omega-3 and omega-6 fatty acids decrease bone loss by increasing calcium utilization by the body; reduce symptoms of arthritis due to inflammation; promote wound healing; improve certain skin disorders (psoriasis, eczema, and acne); and improve mental functions. Primary sources of omega-3 fatty acids include flaxseed, fatty fish, oils that have large amounts of polyunsaturated fats, fish oils, and walnuts. Primary sources of omega-6 fatty acids include most processed foods (cereals, breads, white rice), eggs, baked goods, oils with large amounts of polyunsaturated fats, and meats (especially organ meats, such as liver).

Cis-fatty acids are nutritionally beneficial monounsaturated fatty acids that are used by the body to produce hormone-like regulators and cell membranes. However, when *cis*-fatty acids are heated, pressurized, and combined with a catalyst (usually nickel) in a process called *hydrogenation,* they are changed to unhealthy *trans*-fatty acids. Hydrogenation is used by manufacturers to make vegetable oils solid at room temperature and less likely to turn rancid. Hydrogenated or *trans*-fatty acids are common in commercially baked goods (crackers, cakes, and

Figure 2.18 Phospholipids. (a) In the synthesis of phospholipids, two fatty acids attach to the first two carbons of the glycerol backbone. A phosphate group links a small charged group to the third carbon in glycerol. In (b), the circle represents the polar head region, and the two wavy lines represent the two nonpolar tails. Double bonds in the fatty acid hydrocarbon chain often form kinks in the tail.

Phospholipids are amphipathic molecules, having both polar and nonpolar regions.

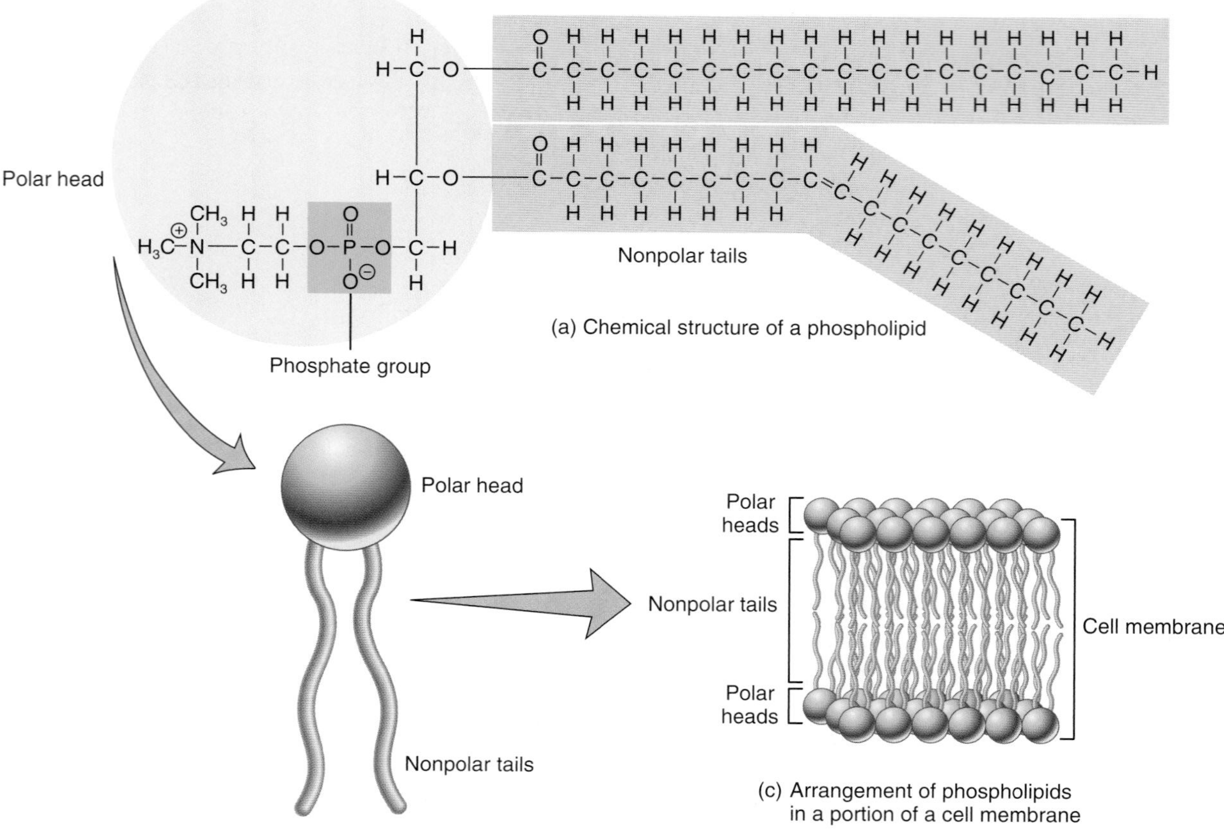

Which portion of a phospholipid is hydrophilic, and which portion is hydrophobic?

cookies), salty snack foods, some margarines, and fried foods (donuts and french fries). If a product label contains the words hydrogenated or partially hydrogenated, then the product contains *trans*-fatty acids. Among the adverse effects of *trans*-fatty acids are an increase in total cholesterol, a decrease in HDL, an increase in LDL, and an increase in triglycerides. These effects, which can increase the risk of heart disease and other cardiovascular diseases, are similar to those caused by saturated fats. ■

Phospholipids

Like triglycerides, **phospholipids** have a glycerol backbone and two fatty acid chains attached to the first two carbons. In the third position, however, a phosphate group (PO_4^{3-}) links a small charged group that usually contains nitrogen (N) to the backbone (Figure 2.18 on page 48). This portion of the molecule (the "head") is polar and can form hydrogen bonds with water molecules. The two fatty acids (the "tails"), by contrast, are nonpolar and can interact only with other lipids. Molecules that have both polar and nonpolar parts are said to be **amphipathic** (am-fi-PATH-ic; *amphi-* = on both sides; *-pathic* = feeling). Amphipathic phospholipids line up tail-to-tail in a double row to make up much of the membrane that surrounds each cell (Figure 2.18c).

Steroids

The structure of **steroids** differs considerably from that of the triglycerides. Steroids have four rings of carbon atoms (colored gold in Figure 2.19). Body cells synthesize other steroids from cholesterol (Figure 2.19a), which has a large nonpolar region consisting of the four rings and a hydrocarbon tail. In the body, the commonly encountered steroids, such as cholesterol, estrogens, testosterone, cortisol, bile salts, and vitamin D, are known as **sterols** because they also have at least one hydroxyl (alcohol) group (—OH). The polar hydroxyl groups make sterols weakly amphipathic. Cholesterol is needed for cell membrane structure; estrogens and testosterone are required for regulating sexual functions; cortisol is necessary for maintaining normal blood sugar levels; bile salts are needed for lipid digestion and absorption; and vitamin D is related to bone growth. In Chapter 10, we will discuss the use of anabolic steroids by athletes to increase muscle size, strength, and endurance.

Other Lipids

Eicosanoids (ī-KŌ-sa-noids; *eicosan-* = twenty) are lipids derived from a 20-carbon fatty acid called arachidonic acid. The two principal subclasses of eicosanoids are the **prostaglandins** (pros′-ta-GLAN-dins) and the **leukotrienes** (loo′-kō-TRĪ-ēnz). Prostaglandins have a wide variety of functions. They modify responses to hormones, contribute to the inflammatory response (Chapter 22), prevent stomach ulcers, dilate (enlarge) airways to the lungs, regulate body temperature, and influence formation of blood clots, to name just a few. Leukotrienes participate in allergic and inflammatory responses.

Other lipids also include fatty acids (which can undergo either hydrolysis to provide ATP or dehydration synthesis to build triglycerides and phospholipids); fat-soluble vitamins such as beta-carotenes (the yellow-orange pigments in egg yolk, carrots, and tomatoes that are converted to vitamin A); vitamins D, E, and K; and lipoproteins.

▶ **CHECKPOINT**

15. How are carbohydrates classified?
16. How are dehydration synthesis and hydrolysis reactions related?
17. What is the importance to the body of triglycerides, phospholipids, steroids, lipoproteins, and eicosanoids?
18. Distinguish among saturated, monounsaturated, and polyunsaturated fats.

Proteins

Proteins are large molecules that contain carbon, hydrogen, oxygen, and nitrogen. Some proteins also contain sulfur. A normal, lean adult body is 12–18% protein. Much more complex in structure than carbohydrates or lipids, proteins have many roles in the body and are largely responsible for the structure of body tissues. Enzymes are proteins that speed up most biochemical reactions. Other proteins work as "motors" to drive muscle contraction. Antibodies are proteins that defend against invading

Figure 2.19 **Steroids.** All steroids have four rings of carbon atoms.

Cholesterol, which is synthesized in the liver, is the starting material for synthesis of other steroids in the body.

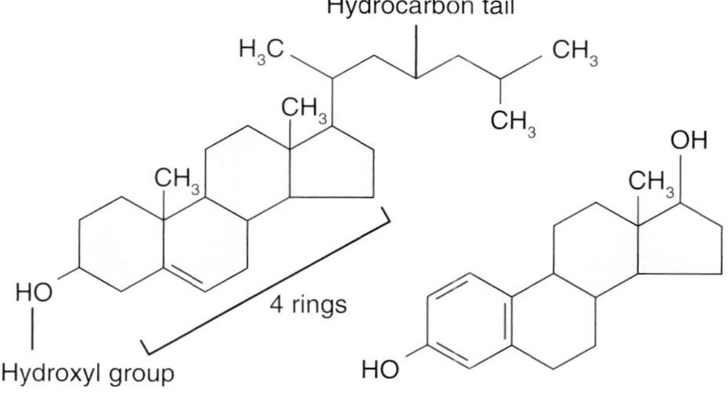

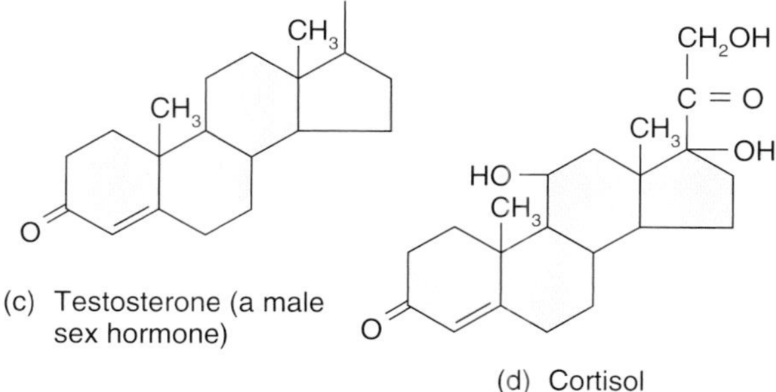

How is the structure of estradiol different from that of testosterone?

TABLE 2.8 Functions of Proteins

Type of Protein	Functions
Structural	Form structural framework of various parts of the body. *Examples:* collagen in bone and other connective tissues, and keratin in skin, hair, and fingernails.
Regulatory	Function as hormones that regulate various physiological processes; control growthand development; as neurotransmitters,mediate responses of the nervous system. *Examples:* the hormone insulin, which regulates blood glucose level, and a neurotransmitter known as substance P, which mediates sensation of pain in the nervous system.
Contractile	Allow shortening of muscle cells, which produces movement. *Examples:* myosin and actin.
Immunological	Aid responses that protect body against foreign substances and invading pathogens. *Examples:* antibodies and interleukins.
Transport	Carry vital substances throughout body. *Example:* hemoglobin, which transports most oxygen and some carbon dioxide in the blood.
Catalytic	Act as enzymes that regulate biochemical reactions. *Examples:* salivary amylase, sucrase, and ATPase.

microbes. Some hormones that regulate homeostasis also are proteins. Table 2.8 describes several important functions of proteins.

Amino Acids and Polypeptides

The monomers of proteins are **amino acids** (a-MĒ-nō). Each of the 20 different amino acids has three important functional groups attached to a central carbon atom (Figure 2.20a): (1) an amino group (—NH_2), (2) an acidic carboxyl group (—COOH), and (3) a side chain (R group). At the normal pH of body fluids, both the amino group and the carboxyl group are ionized (Figure 2.20b). The different side chains give each amino acid its distinctive chemical identity (Figure 2.20c).

A protein is synthesized in stepwise fashion—one amino acid is joined to a second, a third is then added to the first two, and so on. The covalent bond joining each pair of amino acids is a **peptide bond.** It always forms between the carbon of the carboxyl group (—COOH) of one amino acid and the nitrogen of the amino group (—NH_2) of another. As the peptide bond is formed, a molecule of water is removed (Figure 2.21), making this a dehydration synthesis reaction. Breaking a peptide bond, as occurs during digestion of dietary proteins, is a hydrolysis reaction (Figure 2.21).

When two amino acids combine, a **dipeptide** results. Adding another amino acid to a dipeptide produces a **tripeptide.** Further additions of amino acids result in the formation of a chainlike **peptide** (4–9 amino acids) or **polypeptide** (10–2000 or more amino acids). Small proteins may consist of a single polypeptide chain with as few as 50 amino acids. Larger proteins have hundreds or thousands of amino acids and may consist of two or more polypeptide chains folded together.

Figure 2.20 **Amino acids.** (a) In keeping with their name, amino acids have an amino group (shaded blue) and a carboxyl (acid) group (shaded red). The side chain (R group) is different in each amino acid. (b) At pH close to 7, both the amino group and the carboxyl group are ionized. (c) Glycine is the simplest amino acid; the side chain is a single H atom. Cysteine is one of two amino acids that contain sulfur (S). The side chain in tyrosine contains a six-carbon ring. Lysine has a second amino group at the end of its side chain.

Body proteins contain 20 different amino acids, each of which has a unique side chain.

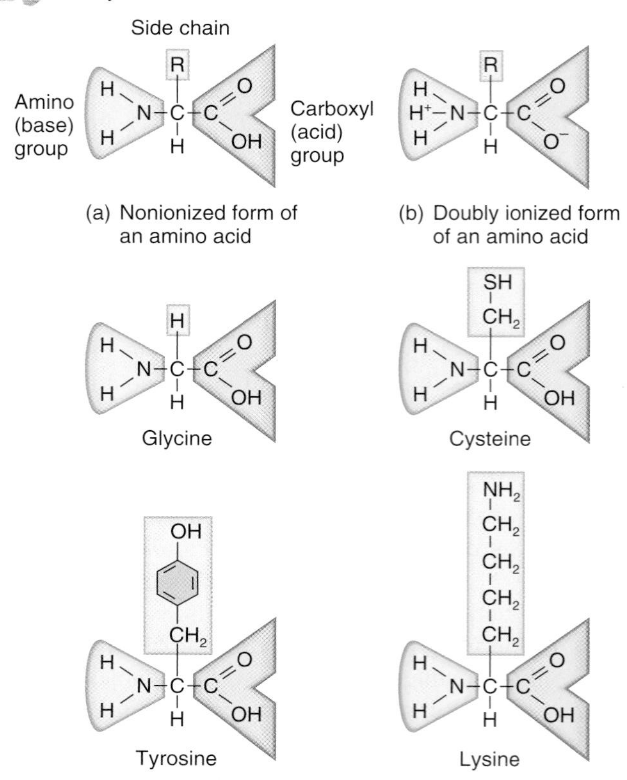

In an amino acid, what is the minimum number of carbon atoms? Of nitrogen atoms?

Because each variation in the number or sequence of amino acids can produce a different protein, a great variety of proteins is possible. The situation is similar to using an alphabet of 20 letters to form words. Each different amino acid is like a letter, and their various combinations give rise to a seemingly endless diversity of words (peptides, polypeptides, and proteins).

Levels of Structural Organization in Proteins

Proteins exhibit four levels of structural organization. The **primary structure** is the unique sequence of amino acids that are linked by covalent peptide bonds to form a polypeptide chain (Figure 2.22a on page 52). A protein's primary structure is genetically determined, and any changes in a protein's amino acid sequence can have serious consequences for body cells. In **sickle-cell disease,** for example, a nonpolar amino acid

Figure 2.21 Formation of a peptide bond between two amino acids during dehydration synthesis. In this example, glycine is joined to alanine, forming a dipeptide (read from left to right). Breaking a peptide bond occurs via hydrolysis (read from right to left).

Amino acids are the monomers used to build proteins.

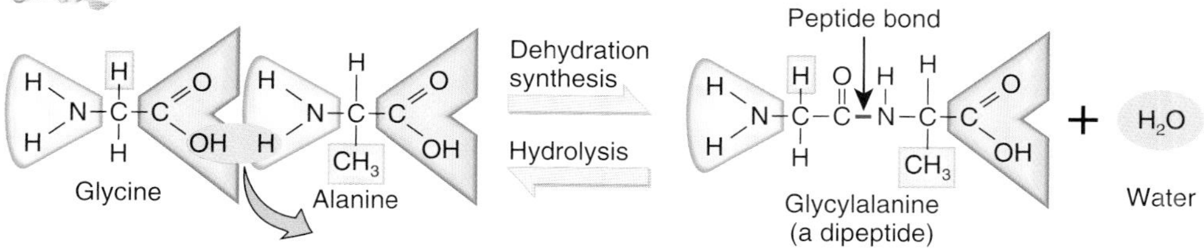

What type of reaction takes place during catabolism of proteins?

(valine) replaces a polar amino acid (glutamate) through two mutations in the oxygen-carrying protein hemoglobin. This change of amino acids diminishes hemoglobin's water solubility. As a result, the altered hemoglobin tends to form crystals inside red blood cells, producing deformed, sickle-shaped cells that cannot properly squeeze through narrow blood vessels. The symptoms and treatment of sickle-cell disease are discussed on page 689.

The **secondary structure** of a protein is the repeated twisting or folding of neighboring amino acids in the polypeptide chain (Figure 2.22b). Two common secondary structures are *alpha helixes* (clockwise spirals) and *beta pleated sheets.* The secondary structure of a protein is stabilized by hydrogen bonds, which form at regular intervals along the polypeptide backbone.

The **tertiary structure** (TUR-shē-er′-ē) refers to the three-dimensional shape of a polypeptide chain. Each protein has a unique tertiary structure that determines how it will function. The tertiary folding pattern may allow amino acids at opposite ends of the chain to be close neighbors (Figure 2.22c). Several types of bonds can contribute to a protein's tertiary structure. The strongest but least common bonds, S—S covalent bonds called *disulfide bridges,* form between the sulfhydryl groups of two monomers of the amino acid cysteine. Many weak bonds—hydrogen bonds, ionic bonds, and hydrophobic interactions—also help determine the folding pattern. Some parts of a polypeptide are attracted to water (hydrophilic), and other parts are repelled by it (hydrophobic). Because most proteins in our body exist in watery surroundings, the folding process places most amino acids with hydrophobic side chains in the central core, away from the protein's surface. Often, helper molecules known as *chaperones* aid the folding process.

In those proteins that contain more than one polypeptide chain (not all of them do), the arrangement of the individual polypeptide chains relative to one another is the **quaternary structure** (KWA-ter-ner′-ē; Figure 2.22d). The bonds that hold polypeptide chains together are similar to those that maintain the tertiary structure.

Proteins vary tremendously in structure. Different proteins have different architectures and different three-dimensional shapes. This variation in structure and shape is directly related to their diverse functions. In practically every case, the function of a protein depends on its ability to recognize and bind to some other molecule. Thus, a hormone binds to a specific protein on a cell in order to alter its function, and an antibody protein binds to a foreign substance (antigen) that has invaded the body. A protein's unique shape permits it to interact with other molecules to carry out a specific function.

Homeostatic mechanisms maintain the temperature and chemical composition of body fluids, which allow body proteins to keep their proper three-dimensional shapes. If a protein encounters an altered environment, it may unravel and lose its characteristic shape (secondary, tertiary, and quaternary structure). This process is called **denaturation.** Denatured proteins are no longer functional. Although in some cases denaturation can be reversed, a frying egg is a common example of permanent denaturation. In a raw egg the soluble egg-white protein (albumin) is a clear, viscous fluid. When heat is applied to the egg, the protein denatures, becomes insoluble, and turns white.

Enzymes

In living cells, most catalysts are protein molecules called **enzymes** (EN-zīms). Some enzymes consist of two parts—a protein portion, called the **apoenzyme** (ā′-pō-EN-zīm), and a nonprotein portion, called a **cofactor.** The cofactor may be a metal ion (such as iron, magnesium, zinc, or calcium) or an organic molecule called a *coenzyme.* Coenzymes often are derived from vitamins. The names of enzymes usually end in the suffix *-ase.* All enzymes can be grouped according to the types of chemical reactions they catalyze. For example, *oxidases* add oxygen, *kinases* add phosphate, *dehydrogenases* remove hydrogen, *ATPases* split ATP, *anhydrases* remove water, *proteases* break down proteins, and *lipases* break down triglycerides.

Enzymes catalyze specific reactions. They do so with great efficiency and with many built-in controls. Three important properties of enzymes are as follows:

1. ***Enzymes are highly specific.*** Each particular enzyme binds only to specific **substrates**—the reactant molecules on which the enzyme acts. Of the more than 1000 known enzymes in your body, each has a characteristic three-dimensional shape with a

Figure 2.22 Levels of structural organization in proteins. (a) The primary structure is the sequence of amino acids in the polypeptide. (b) Common secondary structures include alpha helixes and beta pleated sheets. For simplicity, the amino acid side groups are not shown here. (c) The tertiary structure is the overall folding pattern that produces a distinctive, three-dimensional shape. (d) The quaternary structure in a protein is the arrangement of two or more polypeptide chains relative to one another.

The unique shape of each protein permits it to carry out specific functions.

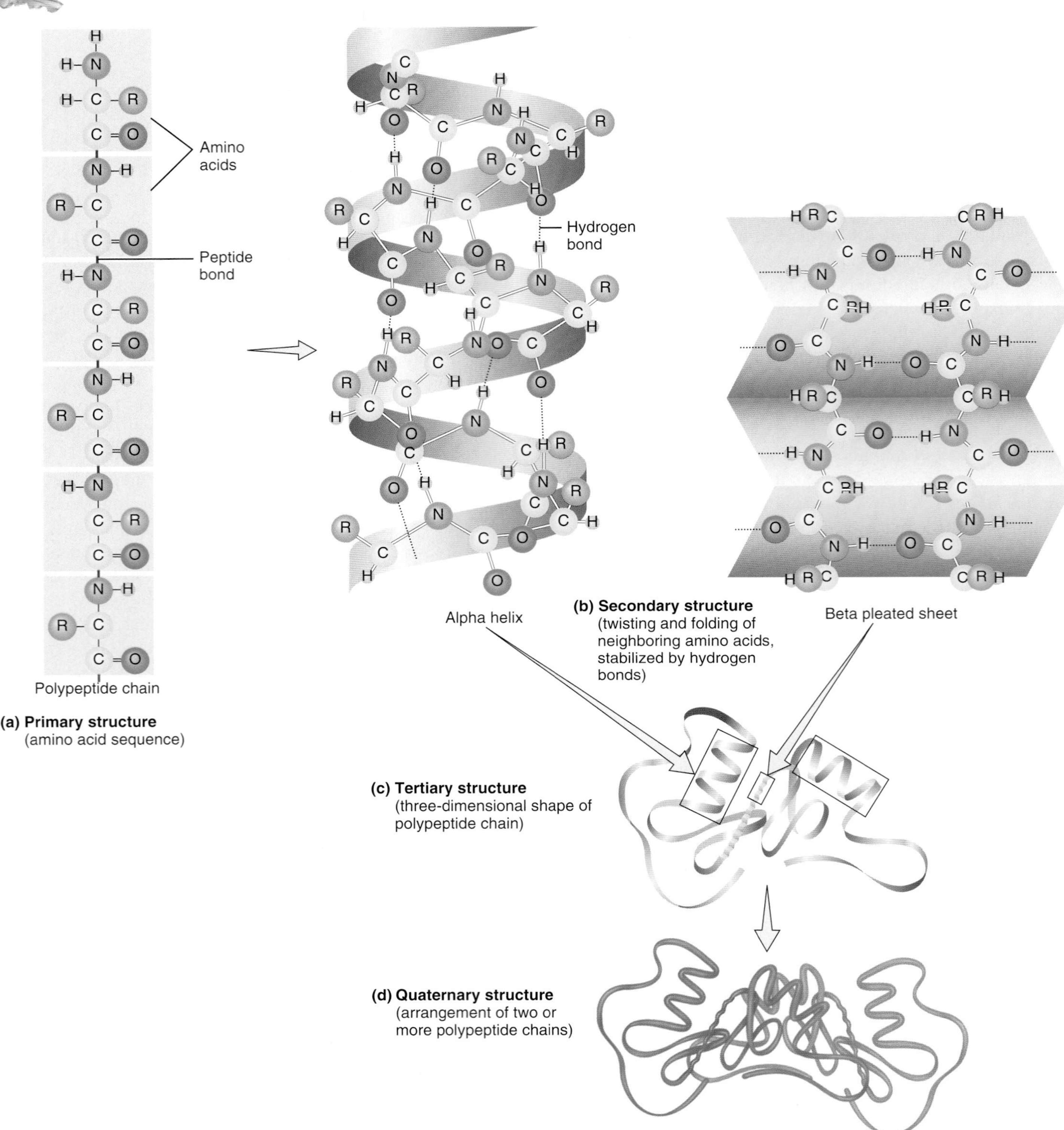

Do all proteins have a quaternary structure?

specific surface configuration, which allows it to recognize and bind to certain substrates. In some cases, the part of the enzyme that catalyzes the reaction, called the **active site,** is thought to "fit" the substrate like a key fits in a lock. In other cases the active site changes its shape to fit snugly around the substrate once the substrate enters the active site. This change in shape is known as an *induced fit.*

Not only is an enzyme matched to a particular substrate, it also catalyzes a specific reaction. From among the large number of diverse molecules in a cell, an enzyme must recognize the correct substrate and then take it apart or merge it with another substrate to form one or more specific products.

2. ***Enzymes are very efficient.*** Under optimal conditions, enzymes can catalyze reactions at rates that are from 100 million to 10 billion times more rapid than those of similar reactions occurring without enzymes. The number of substrate molecules that a single enzyme molecule can convert to product molecules in one second is generally between 1 and 10,000 and can be as high as 600,000.

3. ***Enzymes are subject to a variety of cellular controls.*** Their rate of synthesis and their concentration at any given time are under the control of a cell's genes. Substances within the cell may either enhance or inhibit the activity of a given enzyme. Many enzymes have both active and inactive forms in cells. The rate at which the inactive form becomes active or vice versa is determined by the chemical environment inside the cell.

Enzymes lower the activation energy of a chemical reaction by decreasing the "randomness" of the collisions between molecules. They also help bring the substrates together in the proper orientation so that the reaction can occur. Figure 2.23 depicts how an enzyme works:

1. The substrates make contact with the active site on the surface of the enzyme molecule, forming a temporary intermediate compound called the **enzyme-substrate complex.** In this reaction the two substrate molecules are sucrose (a disaccharide) and water.
2. The substrate molecules are transformed by the rearrangement of existing atoms, the breakdown of the substrate molecule, or the combination of several substrate molecules into the products of the reaction. Here the products are two monosaccharides: glucose and fructose.
3. After the reaction is completed and the reaction products move away from the enzyme, the unchanged enzyme is free to attach to other substrate molecules.

Sometimes a single enzyme may catalyze a reversible reaction in either direction, depending on the relative amounts of the substrates and products. For example, the enzyme *carbonic anhydrase* catalyzes the following reversible reaction:

$$\underset{\text{Carbon dioxide}}{CO_2} + \underset{\text{Water}}{H_2O} \xrightleftharpoons{\textit{Carbonic anhydrase}} \underset{\text{Carbonic acid}}{H_2CO_3}$$

During exercise, when more CO_2 is produced and released into the blood, the reaction flows to the right, increasing the amount of carbonic acid in the blood. Then, as you exhale CO_2, its level in the blood falls and the reaction flows to the left, converting carbonic acid to CO_2 and H_2O.

Nucleic Acids: Deoxyribonucleic Acid (DNA) and Ribonucleic Acid (RNA)

Nucleic acids (nū-KLĒ-ic), so named because they were first discovered in the nuclei of cells, are huge organic molecules that contain carbon, hydrogen, oxygen, nitrogen, and phosphorus. Nucleic acids are of two varieties. The first, **deoxyribonucleic acid (DNA)** (dē-ok′-sē-rī-bō-nū-KLĒ-īk), forms the inherited genetic material inside each human cell. In humans, each **gene** is a segment of a DNA molecule. Our genes determine the traits we inherit, and by controlling protein synthesis they regulate most of the activities that take place in body cells throughout a our lives. When a cell divides, its hereditary information passes on to the next generation of cells. **Ribonucleic acid (RNA),** the second type of nucleic acid, relays instructions from the genes to guide each cell's synthesis of proteins from amino acids.

Figure 2.23 How an enzyme works.

An enzyme speeds up a chemical reaction without being altered or consumed.

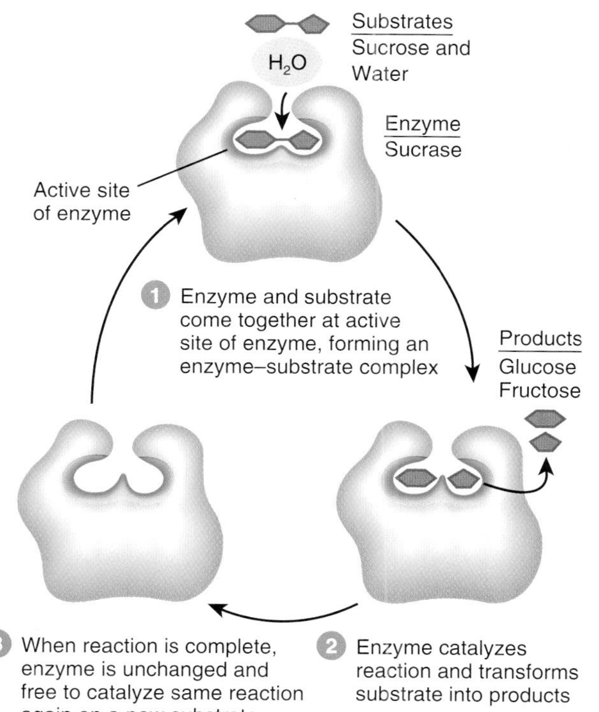
Substrates
Sucrose and Water
H2O
Enzyme
Sucrase
Active site of enzyme
1 Enzyme and substrate come together at active site of enzyme, forming an enzyme–substrate complex
Products
Glucose
Fructose
3 When reaction is complete, enzyme is unchanged and free to catalyze same reaction again on a new substrate
2 Enzyme catalyzes reaction and transforms substrate into products

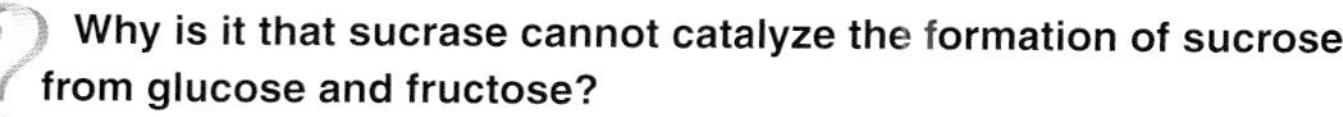
Why is it that sucrase cannot catalyze the formation of sucrose from glucose and fructose?

A nucleic acid is a chain of repeating monomers called **nucleotides.** Each nucleotide of DNA consists of three parts (Figure 2.24a):

1. ***Nitrogenous base.*** DNA contains four different nitrogenous bases, which contain atoms of C, H, O, and N. In DNA the four **nitrogenous bases** are adenine (A), thymine (T), cytosine (C), and guanine (G). Adenine and guanine are larger, double-ring bases called **purines** (PŪR-ēnz); thymine and cytosine are smaller, single-ring bases called **pyrimidines** (pī-RIM-i-dēnz). The nucleotides are named according to the base that is present. For instance, a nucleotide containing thymine is called a

Figure 2.24 DNA molecule. (a) A nucleotide consists of a base, a pentose sugar, and a phosphate group. (b) The paired bases project toward the center of the double helix. The structure is stabilized by hydrogen bonds (dotted lines) between each base pair. There are two hydrogen bonds between adenine and thymine and three between cytosine and guanine.

Nucleotides are the monomers of nucleic acids.

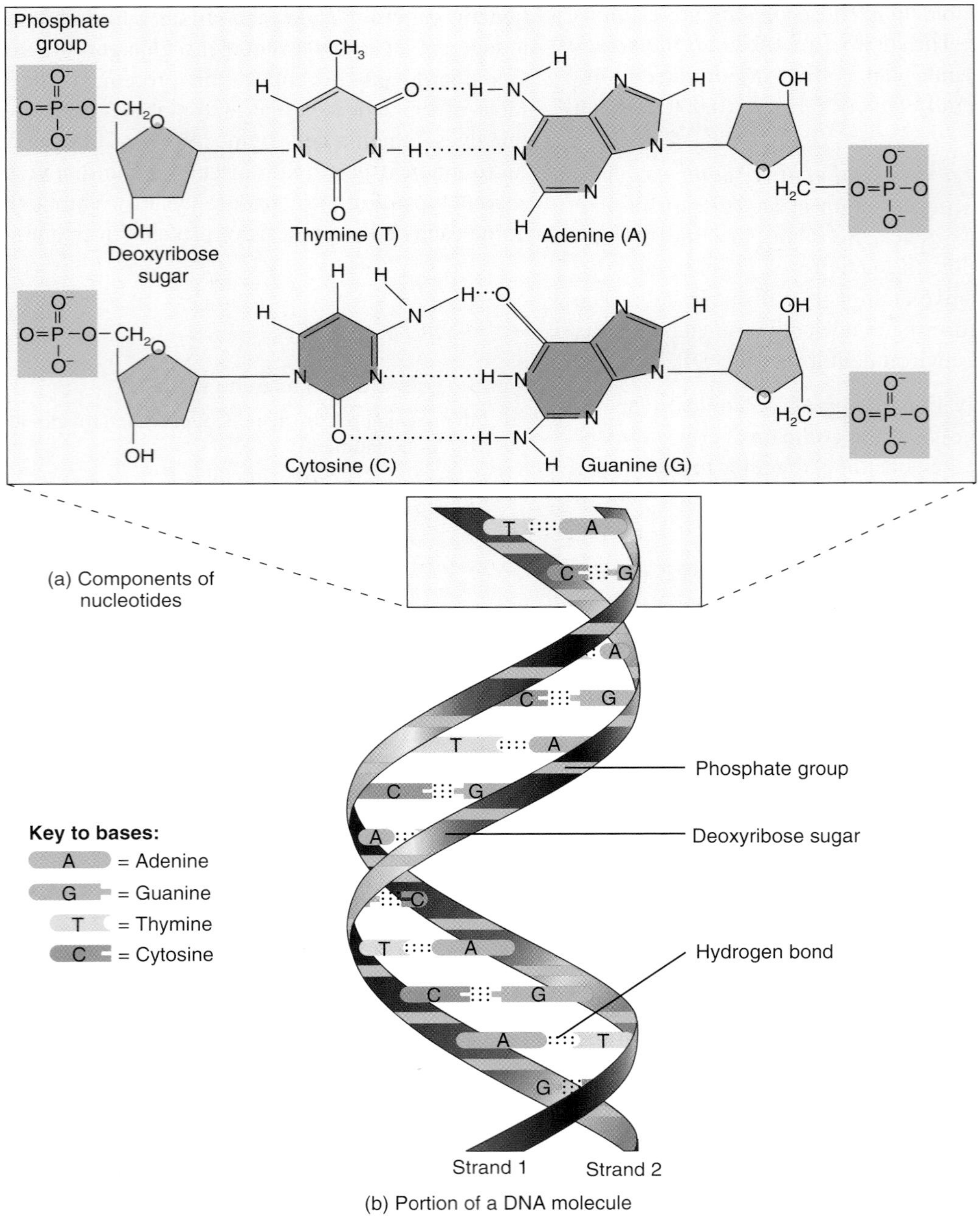

Phosphate group

Deoxyribose sugar

Thymine (T)

Adenine (A)

Cytosine (C)

Guanine (G)

(a) Components of nucleotides

Key to bases:

A = Adenine

G = Guanine

T = Thymine

C = Cytosine

Phosphate group

Deoxyribose sugar

Hydrogen bond

Strand 1

Strand 2

(b) Portion of a DNA molecule

Which bases always pair with one another?

thymine nucleotide, one containing adenine is called an adenine nucleotide, and so on.

2. ***Pentose sugar.*** A five-carbon sugar called **deoxyribose** attaches to each base in DNA.

3. ***Phosphate group.*** Phosphate groups (PO_4^{3-}) alternate with pentose sugars to form the "backbone" of a DNA strand; the bases project inward from the backbone chain (Figure 2.24b).

In 1953, F.H.C. Crick of Great Britain and J.D. Watson, a young American scientist, published a brief paper describing how these three components might be arranged in DNA. Their insights into data gathered by others led them to construct a model so elegant and simple that the scientific world immediately knew it was correct! In the Watson–Crick **double helix** model, DNA resembles a spiral ladder (Figure 2.24b). Two strands of alternating phosphate groups and deoxyribose sugars form the uprights of the ladder. Paired bases, held together by hydrogen bonds, form the rungs. Because adenine always pairs with thymine, and cytosine always pairs with guanine, if you know the sequence of bases in one strand of DNA, you can predict the sequence on the complementary (second) strand. Each time DNA is copied, as when living cells divide to increase their number, the two strands unwind. Each strand serves as the template or mold on which to construct a new second strand. Any change that occurs in the base sequence of a DNA strand is called a *mutation*. Some mutations can result in the death of a cell, cause cancer, or produce genetic defects in future generations.

RNA, the second variety of nucleic acid, differs from DNA in several respects. In humans, RNA is single-stranded. The sugar in the RNA nucleotide is the pentose **ribose,** and RNA contains the pyrimidine base uracil (U) instead of thymine. Cells contain three different kinds of RNA: messenger RNA, ribosomal RNA, and transfer RNA. Each has a specific role to perform in carrying out the instructions coded in DNA (described on page 88).

DNA Fingerprinting

A technique called **DNA fingerprinting** is used in research and in courts of law to ascertain whether a person's DNA matches the DNA obtained from samples or pieces of legal evidence such as blood stains or hairs. In each person, certain DNA segments contain base sequences that are repeated several times. Both the number of repeat copies in one region and the number of regions subject to repeat are different from one person to another. DNA fingerprinting can be done with minute quantities of DNA—for example, from a single strand of hair, a drop of semen, or a spot of blood. It also can be used to identify a crime victim or a child's biological parents and even to determine whether two people have a common ancestor. ■

Adenosine Triphosphate

Adenosine triphosphate (a-DEN-ō-sēn) or **ATP** is the "energy currency" of living systems (Figure 2.25). ATP transfers the

Figure 2.25 **Structures of ATP and ADP.** "Squiggles" (~) indicate the two phosphate bonds that can be used to transfer energy. Energy transfer typically involves hydrolysis of the last phosphate bond of ATP.

ATP transfers chemical energy to power cellular activities.

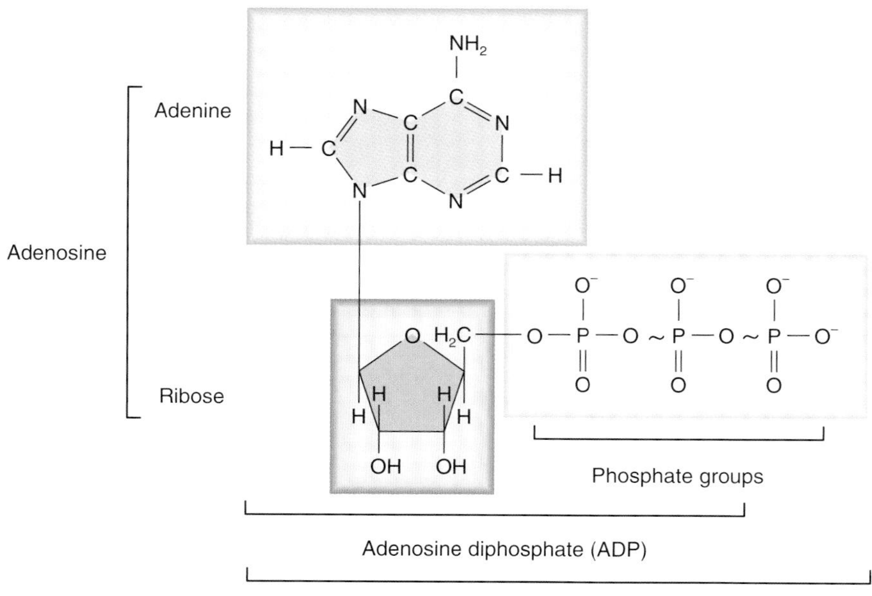

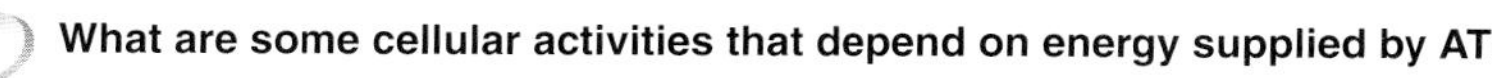
What are some cellular activities that depend on energy supplied by ATP?

energy liberated in exergonic catabolic reactions to power cellular activities that require energy (endergonic reactions). Among these cellular activities are muscular contractions, movement of chromosomes during cell division, movement of structures within cells, transport of substances across cell membranes, and synthesis of larger molecules from smaller ones. As its name implies, ATP consists of three phosphate groups attached to adenosine, a unit composed of adenine and the five-carbon sugar ribose.

When a water molecule is added to ATP, the third phosphate group (PO_4^{3-}), symbolized by Ⓟ in the following discussion, is removed, and the overall reaction liberates energy. The enzyme that catalyzes the hydrolysis of ATP is called *ATPase.* Removal of the third phosphate group produces a molecule called **adenosine diphosphate (ADP)** in the following reaction:

$$\underset{\text{Adenosine triphosphate}}{\text{ATP}} + \underset{\text{Water}}{H_2O} \xrightarrow{\textit{ATPase}} \underset{\text{Adenosine diphosphate}}{\text{ADP}} + \underset{\text{Phosphate group}}{Ⓟ} + \underset{\text{Energy}}{E}$$

As noted previously, the energy supplied by the catabolism of ATP into ADP is constantly being used by the cell. As the supply of ATP at any given time is limited, a mechanism exists to replenish it: The enzyme *ATP synthase* catalyzes the addition of a phosphate group to ADP in the following reaction:

$$\underset{\text{Adenosine diphosphate}}{\text{ADP}} + \underset{\text{Phosphate group}}{Ⓟ} + \underset{\text{Energy}}{E} \xrightarrow{\textit{ATP synthase}} \underset{\text{Adenosine triphosphate}}{\text{ATP}} + \underset{\text{Water}}{H_2O}$$

Where does the cell get the energy required to produce ATP? The energy needed to attach a phosphate group to ADP is supplied mainly by the catabolism of glucose in a process called cellular respiration. Cellular respiration has two phases, anaerobic and aerobic:

1. ***Anaerobic phase.*** In a series of reactions that do not require oxygen, glucose is partially broken down by a series of catabolic reactions into pyruvic acid. Each glucose molecule that is converted into a pyruvic acid molecule yields two molecules of ATP.
2. ***Aerobic phase.*** In the presence of oxygen, glucose is completely broken down into carbon dioxide and water. These reactions generate heat and 36 or 38 ATP molecules.

Chapters 10 and 25 cover the details of cellular respiration.

In Chapter 1, you learned that the human body is comprised of various levels of organization; this chapter has just showed you the alphabet of atoms and molecules that is the basis for the language of the body. Now that you have an understanding of the chemistry of the human body, you are ready to form words; in Chapter 3 you will see how atoms and molecules are organized to form structures of cells and perform the activities of cells that contribute to homeostasis.

▶ CHECKPOINT

19. Define a protein. What is a peptide bond?
20. Outline the levels of structural organization in proteins.
21. How do DNA and RNA differ?
22. In the reaction catalyzed by ATP synthase, what are the substrates and products? Is this an exergonic or endergonic reaction?

STUDY OUTLINE

HOW MATTER IS ORGANIZED (p. 29)

1. All forms of matter are composed of chemical elements.
2. Oxygen, carbon, hydrogen, and nitrogen make up about 96% of body mass.
3. Each element is made up of small units called atoms.
4. Atoms consist of a nucleus, which contains protons and neutrons, plus electrons that move about the nucleus in regions called electron shells.
5. The number of protons (the atomic number) distinguishes the atoms of one element from those of another element.
6. The mass number of an atom is the sum of its protons and neutrons.
7. Different atoms of an element that have the same number of protons but different numbers of neutrons are called isotopes. Radioactive isotopes are unstable and decay.
8. The atomic mass of an element is the average mass of all naturally occurring isotopes of that element.
9. An atom that *gives up* or *gains* electrons becomes an ion—an atom that has a positive or negative charge because it has unequal numbers of protons and electrons. Positively charged ions are cations; negatively charged ions are anions.
10. If two atoms share electrons, a molecule is formed. Compounds contain atoms of two or more elements.
11. A free radical is an electrically charged atom or group of atoms with an unpaired electron in its outermost shell. A common example is superoxide, which is formed by the addition of an electron to an oxygen molecule.

CHEMICAL BONDS (p. 32)

1. Forces of attraction called chemical bonds hold atoms together. These bonds result from gaining, losing, or sharing electrons in the valence shell.
2. Most atoms become stable when they have an octet of eight electrons in their valence (outermost) electron shell.

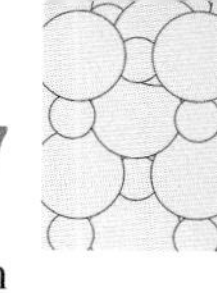

3. When the force of attraction between ions of opposite charge holds them together, an ionic bond has formed.
4. In a covalent bond, atoms share pairs of valence electrons. Covalent bonds may be single, double, or triple and either nonpolar or polar.
5. An atom of hydrogen that forms a polar covalent bond with an oxygen atom or a nitrogen atom may also form a weaker bond, called a hydrogen bond, with an electronegative atom. The polar covalent bond causes the hydrogen atom to have a partial positive charge (δ^+) that attracts the partial negative charge (δ^-) of neighboring electronegative atoms, often oxygen or nitrogen.

CHEMICAL REACTIONS (p. 36)

1. When atoms combine with or break apart from other atoms, a chemical reaction occurs. The starting substances are the reactants, and the ending ones are the products.
2. Energy, the capacity to do work, is of two principal kinds: potential (stored) energy and kinetic energy (energy of motion).
3. Endergonic reactions require energy; exergonic reactions release energy. ATP couples endergonic and exergonic reactions.
4. The initial energy investment needed to start a reaction is the activation energy. Reactions are more likely when the concentrations and the temperatures of the reacting particles are higher.
5. Catalysts accelerate chemical reactions by lowering the activation energy. Most catalysts in living organisms are protein molecules called enzymes.
6. Synthesis reactions involve the combination of reactants to produce larger molecules. The reactions are anabolic and usually endergonic.
7. In decomposition reactions, a substance is broken down into smaller molecules. The reactions are catabolic and usually exergonic.
8. Exchange reactions involve the replacement of one atom or atoms by another atom or atoms.
9. In reversible reactions, end products can revert to the original reactants.

INORGANIC COMPOUNDS AND SOLUTIONS (p. 39)

1. Inorganic compounds usually are small and usually lack carbon. Organic substances always contain carbon, usually contain hydrogen, and always have covalent bonds.
2. Water is the most abundant substance in the body. It is an excellent solvent and suspending medium, participates in hydrolysis and dehydration synthesis reactions, and serves as a lubricant. Because of its many hydrogen bonds, water molecules are cohesive, which causes a high surface tension. Water also has a high capacity for absorbing heat and a high heat of vaporization.
3. Inorganic acids, bases, and salts dissociate into ions in water. An acid ionizes into hydrogen ions (H^+) and anions and is a proton donor; many bases ionize into cations and hydroxide ions (OH^-) and all are proton acceptors. A salt ionizes into neither H^+ nor OH^-.
4. Mixtures are combinations of elements or compounds that are physically blended together but are not bound by chemical bonds. Solutions, colloids, and suspensions are mixtures with different properties.
5. Two ways to express the concentration of a solution are percentage (mass per volume), expressed in grams per 100 mL of a solution, and moles per liter. A mole (abbreviated mol) is the amount in grams of any substance that has a mass equal to the combined atomic mass of all its atoms.
6. The pH of body fluids must remain fairly constant for the body to maintain homeostasis. On the pH scale, 7 represents neutrality. Values below 7 indicate acidic solutions, and values above 7 indicate alkaline solutions. Normal blood pH is 7.35–7.45.
7. Buffer systems remove or add protons (H^+) to help maintain pH homeostasis.
8. One important buffer system is the carbonic acid–bicarbonate buffer system. The bicarbonate ion (HCO_3^-) acts as a weak base and removes excess H^+, and carbonic acid (H_2CO_3) acts as a weak acid and adds H^+.

ORGANIC COMPOUNDS (p. 43)

1. Carbon, with its four valence electrons, bonds covalently with other carbon atoms to form large molecules of many different shapes. Attached to the carbon skeletons of organic molecules are functional groups that confer distinctive chemical properties.
2. Small organic molecules are joined together to form larger molecules by dehydration synthesis reactions in which a molecule of water is removed. In the reverse process, called hydrolysis, large molecules are broken down into smaller ones by the addition of water.
3. Carbohydrates provide most of the chemical energy needed to generate ATP. They may be monosaccharides, disaccharides, or polysaccharides.
4. Lipids are a diverse group of compounds that include triglycerides (fats and oils), phospholipids, steroids, and eicosanoids. Triglycerides protect, insulate, provide energy, and are stored. Phospholipids are important cell membrane components. Steroids are important in cell membrane structure, regulating sexual functions, maintaining normal blood sugar level, aiding lipid digestion and absorption, and helping bone growth. Eicosanoids (prostaglandins and leukotrienes) modify hormone responses, contribute to inflammation, dilate airways, and regulate body temperature.
5. Proteins are constructed from amino acids. They give structure to the body, regulate processes, provide protection, help muscles contract, transport substances, and serve as enzymes. Levels of structural organization among proteins include primary, secondary, tertiary, and (sometimes) quaternary. Variations in protein structure and shape are related to their diverse functions.
6. Deoxyribonucleic acid (DNA) and ribonucleic acid (RNA) are nucleic acids consisting of nitrogenous bases, five-carbon (pentose) sugars, and phosphate groups. DNA is a double helix and is the primary chemical in genes. RNA takes part in protein synthesis.
7. Adenosine triphosphate (ATP) is the principal energy-transferring molecule in living systems. When it transfers energy to an endergonic reaction, it is decomposed to adenosine diphosphate (ADP) and a phosphate group. ATP is synthesized from ADP and a phosphate group using the energy supplied by various decomposition reactions, particularly those of glucose.

SELF-QUIZ QUESTIONS

Fill in the blanks in the following statements.

1. An atom with a mass number of 18 that contains 10 neutrons would have an atomic number of ____.
2. Matter exists in three forms: ____, ____, and ____.
3. The building blocks of carbohydrates are the monomers ____ while the building blocks of proteins are the monomers ____.

Indicate whether the following statements are true or false.

4. The elements that compose most of the body's mass are carbon, hydrogen, oxygen, and nitrogen.
5. Ionic bonds are created when atoms lose, gain, or share electrons in the valence shell.
6. Human blood has a normal pH between 7.35 and 7.45 and is considered slightly alkaline.

Choose the one best answer to the following questions.

7. Which of the following would be considered a compound? (1) $C_6H_{12}O_6$, (2) O_2, (3) Fe, (4) H_2, (5) CH_4. (a) all are compounds, (b) 1, 2, 4, and 5, (c) 1 and 5, (d) 2 and 4, (e) 3.
8. The monosaccharides glucose and fructose combine to form the disaccharide sucrose by a process known as (a) dehydration synthesis, (b) hydrolysis, (c) decomposition, (d) hydrogen bonding, (e) ionization.
9. Which of the following is *not* a function of proteins? (a) provide structural framework, (b) bring about contraction, (c) transport materials throughout the body, (d) store energy, (e) regulate many physiological processes.
10. Which of the following organic compounds are classified as lipids? (1) polysaccharides, (2) triglycerides, (3) steroids, (4) enzymes, (5) eicosanoids. (a) 1, 2, and 4, (b) 2, 3, and 5, (c) 2 and 5, (d) 2, 3, 4, and 5, (e) 2 and 3.
11. A compound dissociates in water and forms a cation other than H^+ and an anion other than OH^-. This substance most likely is a(n) (a) acid, (b) base, (c) enzyme, (d) buffer, (e) salt.
12. Which of the following statements regarding ATP are *true*? (1) ATP is the energy currency for the cell. (2) The energy supplied by the hydrolysis of ATP is constantly being used by cells. (3) Energy is required to produce ATP. (4) The production of ATP involves both aerobic and anaerobic phases. (5) The process of producing energy in the form of ATP is termed the law of conservation of energy. (a) 1, 2, 3, and 4, (b) 1, 2, 3, and 5, (c) 2, 4, and 5, (d) 1, 2, and 4, (e) 3, 4, and 5.
13. During the course of analyzing an unknown chemical, a chemist determines that the chemical is composed of carbon, hydrogen, and oxygen in the proportion of 1 carbon to 2 hydrogens to 1 oxygen. The chemical is probably (a) an amino acid, (b) DNA, (c) a triglyceride, (d) a protein, (e) a monosaccharide.
14. Match the following reactions with the term that describes them:

____(a) $H_2 + Cl_2 \longrightarrow 2\,HCl$
____(b) $3\,NaOH + H_3PO_4 \longrightarrow Na_3PO_4 + 3\,H_2O$
____(c) $CaCO_3 + CO_2 + H_2O \longrightarrow Ca(HCO_3)_2$
____(d) $NH_3 + H_2O \rightleftharpoons NH_4^+ + OH^-$
____(e) $C_{12}H_{22}O_{11} + H_2O \longrightarrow C_6H_{12}O_6 + C_6H_{12}O_6$

(1) synthesis reaction
(2) exchange reaction
(3) decomposition reaction
(4) reversible reaction

15. Match the following:

____(a) an abundant polar covalent molecule that serves as a solvent, has a high heat capacity, creates a high surface tension, and serves as a lubricant
____(b) a substance that dissociates into one or more hydrogen ions and one or more anions
____(c) a substance that dissociates into cations and anions, neither of which is a hydrogen ion or a hydroxyl ion
____(d) a proton acceptor
____(e) a measure of hydrogen ion concentration
____(f) a chemical compound that can convert strong acids and bases into weak ones
____(g) a catalyst for chemical reactions that is specific, efficient, and under cellular control
____(h) a single-stranded compound that contains a five-carbon sugar, and the bases adenine, cytosine, guanine, and uracil
____(i) a compound that functions to temporarily store and then transfer energy liberated in exergonic reactions to cellular activities that require energy
____(j) a double-stranded compound that contains a five-carbon sugar, the bases adenine, thymine, cytosine, and guanine, and the body's genetic material
____(k) a charged atom
____(l) a charged atom with an unpaired electron in its outermost shell

(1) acid
(2) free radical
(3) base
(4) buffer
(5) enzyme
(6) ion
(7) pH
(8) salt
(9) RNA
(10) ATP
(11) water
(12) DNA

CRITICAL THINKING QUESTIONS

1. Your best friend has decided to begin frying his breakfast eggs in margarine instead of butter because he has heard that eating butter is bad for his heart. Has he made a wise choice? Are there other alternatives?
2. A 4-month old baby is admitted to the hospital with a fever of 102°F (39°C). Why is it critical to treat the fever as quickly as possible?
3. During chemistry lab, Maria places sucrose (table sugar) in a glass beaker, adds water, and stirs. As the table sugar disappears, she loudly proclaims that she has chemically broken down the sucrose into fructose and glucose. Is Maria's chemical analysis correct?

ANSWERS TO FIGURE QUESTIONS

2.1 In carbon, the first shell contains two electrons and the second shell contains four electrons.

2.2 The four most plentiful elements in living organisms are oxygen, carbon, hydrogen, and nitrogen.

2.3 Antioxidants such as selenium, zinc, beta-carotene, vitamin C, and vitamin D can inactivate free radicals derived from oxygen.

2.4 A cation is a positively charged ion; an anion is a negatively charged ion.

2.5 An ionic bond involves the *loss* and *gain* of electrons; a covalent bond involves the *sharing* of pairs of electrons.

2.6 The oxygen atom in a water molecule has greater electronegativity than the hydrogen atoms.

2.7 The N atom in ammonia is electronegative. Because it attracts electrons more strongly than do the H atoms, the nitrogen end of ammonia acquires a slight negative charge allowing H atoms in water molecules (or in other ammonia molecules) to form hydrogen bonds with it. Likewise, O atoms in water molecules can form hydrogen bonds with H atoms in ammonia molecules.

2.8 The number of hydrogen atoms in the reactants must equal the number in the products—in this case, four hydrogen atoms total. Put another way, two molecules of H_2 are needed to react with each molecule of O_2 so that the number of H atoms and O atoms in the reactants is the same as the number of H atoms and O atoms in the products.

2.9 This reaction is exergonic because the reactants have more potential energy than the products.

2.10 No. A catalyst does not change the potential energies of the products and reactants; it only lowers the activation energy needed to get the reaction going.

2.11 Because sugar easily dissolves in a polar solvent (water), you can correctly predict that it has several polar covalent bonds.

2.12 $CaCO_3$ is a salt, and H_2SO_4 is an acid.

2.13 At pH = 6, $[H^+] = 10^{-6}$ mol/liter and $[OH^-] = 10^{-8}$ mol/liter. A pH of 6.82 is more acidic than a pH of 6.91. Both pH = 8.41 and pH = 5.59 are 1.41 pH units from neutral (pH = 7).

2.14 Glucose has five —OH groups and 6 carbon atoms.

2.15 There are 6 carbons in fructose and 12 in sucrose.

2.16 Cells in the liver and in skeletal muscle store glycogen.

2.17 The oxygen in the water molecule comes from a fatty acid.

2.18 The polar head is hydrophilic, and the nonpolar tails are hydrophobic.

2.19 The only differences between estradiol and testosterone are the number of double bonds in and the types of functional groups attached to ring A.

2.20 An amino acid has a minimum of two carbon atoms and one nitrogen atom.

2.21 Hydrolysis occurs during catabolism of proteins.

2.22 Proteins consisting of a single polypeptide chain do not have a quaternary structure.

2.23 Sucrase has specificity for the sucrose molecule and thus would not "recognize" glucose and fructose.

2.24 Thymine always pairs with adenine, and cytosine always pairs with guanine.

2.25 A few cellular activities that depend on energy supplied by ATP are muscular contractions, movement of chromosomes, transport of substances across cell membranes, and synthesis (anabolic) reactions.

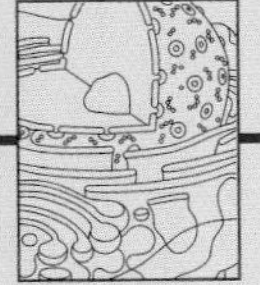

Chapter 3

The Cellular Level of Organization

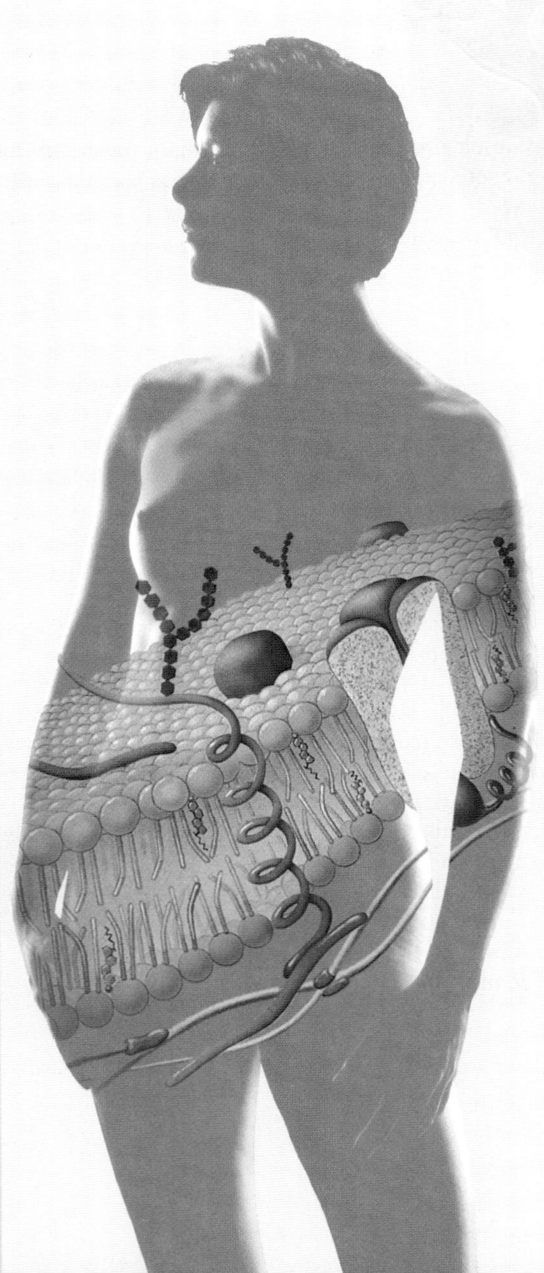

Cells and Homeostasis

About 200 different types of specialized cells carry out a multitude of functions that help each system contribute to the homeostasis of the entire body. At the same time, all cells share key structures and functions that support their intense activity.

www. wiley.com/college/apcentral

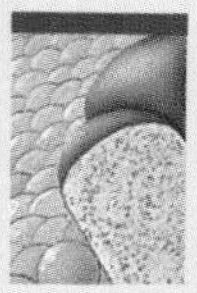

In the last chapter you learned about the atoms and molecules that comprise the alphabet of the language of the human body. These are combined into about 200 different types of words called **cells**—living structural and functional units enclosed by a membrane. All cells arise from existing cells by the process of **cell division,** in which one cell divides into two identical cells. Different types of cells fulfill unique roles that support homeostasis and contribute to the many functional capabilities of the human organism. **Cell biology** is the study of cellular structure and function. As you study the various parts of a cell and their relationships to one another, you will learn that cell structure and function are intimately related. In this chapter, you will learn that cells carry out a dazzling array of chemical reactions to create and maintain life processes—in part, by isolating specific types of chemical reactions within specialized cellular structures.

PARTS OF A CELL

▶ **OBJECTIVE**

Name and describe the three main parts of a cell.

Figure 3.1 provides an overview of the typical structures found in body cells. Most cells have many of the structures shown in this diagram, but no one cell has all of them. For ease of study, we divide the cell into three main parts: plasma membrane, cytoplasm, and nucleus.

- The **plasma membrane** forms the cell's flexible outer surface, separating the cell's internal environment from the external environment. It is a selective barrier that regulates the flow of materials into and out of a cell. This selectivity helps establish and maintain the appropriate environment for normal cellular activities. The plasma membrane also plays a key role in communication among cells and between cells and their external environment.

Figure 3.1 Typical structures found in body cells.

The cell is the basic living, structural and functional unit of the body.

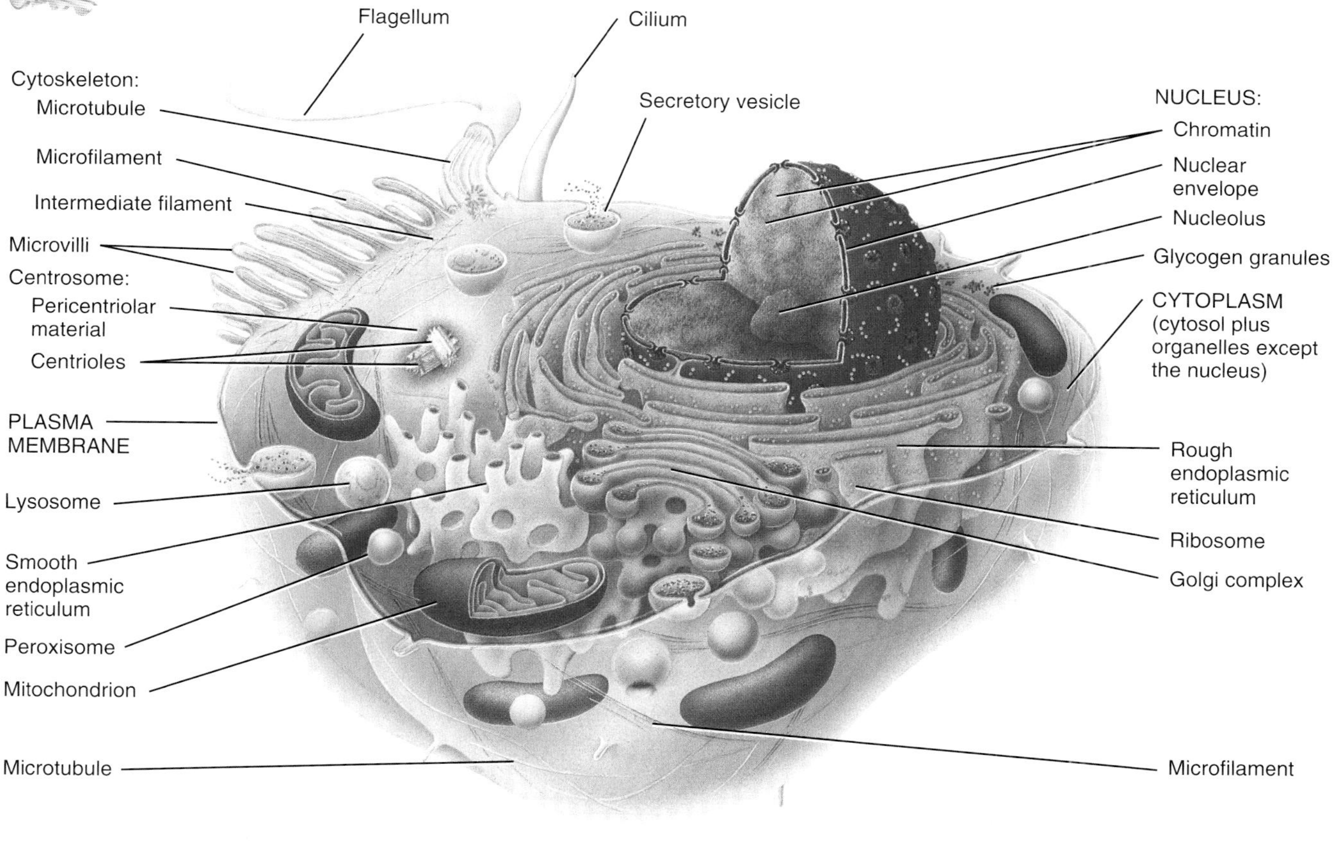

What are the three principal parts of a cell?

- The **cytoplasm** (SĪ-tō-plasm; *-plasm* = formed or molded) consists of all the cellular contents between the plasma membrane and the nucleus. This compartment has two components: cytosol and organelles. **Cytosol** (SĪ-tō-sol), the fluid portion of cytoplasm, contains water, dissolved solutes, and suspended particles. Surrounded by cytosol are several different types of **organelles** (or-ga-NELZ = little organs). Each type of organelle has a characteristic shape and specific functions. Examples include the cytoskeleton, ribosomes, endoplasmic reticulum, Golgi complex, lysosomes, peroxisomes, and mitochondria.
- The **nucleus** (NOO-klē-us = nut kernel) is a large organelle that houses most of a cell's DNA. Within the nucleus, each **chromosome** (*chromo-* = colored), a single molecule of DNA associated with several proteins, contains thousands of hereditary units called **genes** that control most aspects of cellular structure and function.

▶ CHECKPOINT

1. List the three main parts of a cell and explain their functions.

THE PLASMA MEMBRANE

▶ OBJECTIVES

Describe the structure and functions of the plasma membrane.

Explain the concept of selective permeability.

Define the electrochemical gradient and describe its components.

The **plasma membrane,** a flexible yet sturdy barrier that surrounds and contains the cytoplasm of a cell, is best described by using a structural model called the *fluid mosaic model.* According to this model, the molecular arrangement of the plasma membrane resembles an ever-moving sea of fluid lipids that contains a mosaic of many different proteins (Figure 3.2). Some proteins float freely like icebergs in the lipid sea, whereas others are anchored at specific locations like boats at a dock. The membrane lipids allow passage of several types of lipid-soluble molecules but act as a barrier to the entry or exit of charged or polar substances. Some of the proteins in the plasma membrane allow movement of polar molecules and ions into and out of the cell. Other proteins can act as signal receptors or adhesion molecules.

Figure 3.2 The fluid mosaic arrangement of lipids and proteins in the plasma membrane.

Membranes are fluid structures because the lipids and many of the proteins are free to rotate and move sideways in their own half of the bilayer.

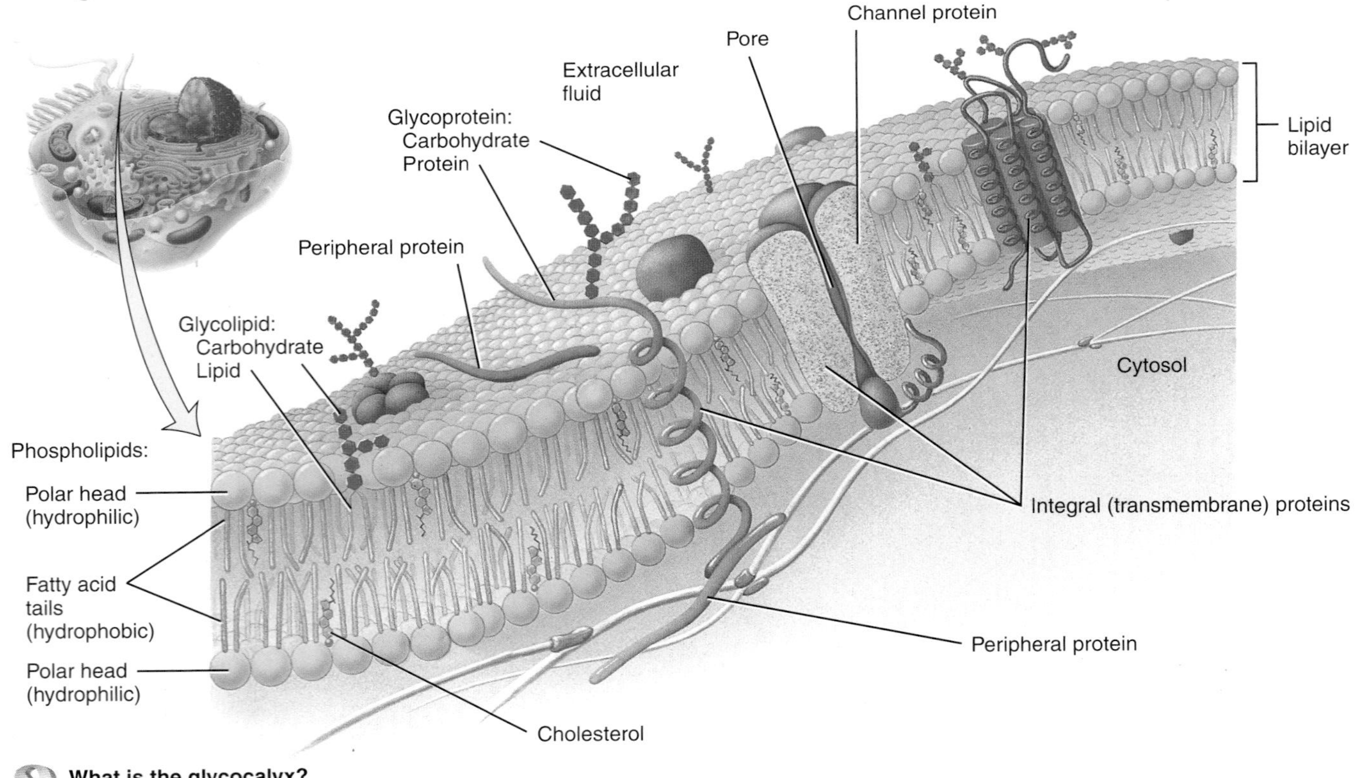

What is the glycocalyx?

The Lipid Bilayer

The basic structural framework of the plasma membrane is the **lipid bilayer,** two back-to-back layers made up of three types of lipid molecules—phospholipids, cholesterol, and glycolipids (Figure 3.2). About 75% of the membrane lipids are **phospholipids,** lipids that contain phosphate groups. Present in smaller amounts are **cholesterol** (about 20%), a steroid with an attached —OH (hydroxyl) group, and various **glycolipids** (about 5%), lipids with attached carbohydrate groups.

The bilayer arrangement occurs because the lipids are **amphipathic** (am-fē-PATH-ik) molecules, which means that they have both polar and nonpolar parts. In phospholipids (see Figure 2.18 on page 48), the polar part is the phosphate-containing "head," which is *hydrophilic* (*hydro-* = water; *-philic* = loving). The nonpolar parts are the two long fatty acid "tails," which are *hydrophobic* (*-phobic* = fearing) hydrocarbon chains. Because "like seeks like," the phospholipid molecules orient themselves in the bilayer with their hydrophilic heads facing outward. In this way, the heads face a watery fluid on either side—cytosol on the inside and extracellular fluid on the outside. The hydrophobic fatty acid tails in each half of the bilayer point toward one another, forming a nonpolar, hydrophobic region in the membrane's interior.

Cholesterol molecules are weakly amphipathic (see Figure 2.19a on page 49) and are interspersed among the other lipids in both layers of the membrane. The tiny —OH group is the only polar region of cholesterol, and it forms hydrogen bonds with the polar heads of phospholipids and glycolipids. The stiff steroid rings and hydrocarbon tail of cholesterol are nonpolar; they fit among the fatty acid tails of the phospholipids and glycolipids. The carbohydrate groups of glycolipids form a polar "head"; their fatty acid "tails" are nonpolar. Glycolipids appear only in the membrane layer that faces the extracellular fluid, which is one reason the two sides of the bilayer are asymmetric, or different.

Arrangement of Membrane Proteins

Membrane proteins are categorized as integral or peripheral according to whether they are firmly embedded in the membrane (Figure 3.2). **Integral proteins** extend into or through the lipid bilayer among the fatty acid tails and are firmly embedded in it. Most integral proteins are **transmembrane proteins,** which means that they span the entire lipid bilayer and protrude into both the cytosol and extracellular fluid. A few integral proteins are tightly attached to one side of the bilayer by covalent bonding to fatty acids. Like membrane lipids, integral membrane proteins are amphipathic. Their hydrophilic regions protrude into either the watery extracellular fluid or the cytosol, and their hydrophobic regions extend among the fatty acid tails.

As their name implies, **peripheral proteins** are not as firmly embedded in the membrane. They associate more loosely with the polar heads of membrane lipids or with integral proteins at the inner or outer surface of the membrane.

Many membrane proteins are **glycoproteins,** proteins with carbohydrate groups attached to the ends that protrude into the extracellular fluid. The carbohydrates are *oligosaccharides* (*oligo-* = few; *saccharides* = sugars), chains of 2 to 60 monosaccharides that may be straight or branched. The carbohydrate portions of glycolipids and glycoproteins form an extensive sugary coat called the **glycocalyx** (glī-kō-KĀL-iks). The glycocalyx acts like a molecular "signature" that enables cells to recognize one another. For example, a white blood cell's ability to detect a "foreign" glycocalyx is one basis of the immune response that helps us destroy invading organisms. In addition, the glycocalyx enables cells to adhere to one another in some tissues and protects cells from being digested by enzymes in the extracellular fluid. The hydrophilic properties of the glycocalyx attract a film of fluid to the surface of many cells. This action makes red blood cells slippery as they flow through narrow blood vessels and protects cells that line the airways and the gastrointestinal tract from drying out.

Functions of Membrane Proteins

Generally, the types of lipids in cellular membranes vary only slightly. In contrast, the membranes of different cells and various intracellular organelles have remarkably different assortments of proteins that determine many of the membrane's functions (Figure 3.3).

- Some integral membrane proteins form **ion channels,** *pores* or holes through which specific ions, such as potassium ions (K^+), can flow to get into or out of the cell. Most ion channels are *selective;* they allow only a single type of ion to pass through.
- Other integral proteins act as **transporters,** selectively moving a polar substance or ion from one side of the membrane to the other.
- Integral proteins called **receptors** serve as cellular recognition sites. Each type of receptor recognizes and binds a specific type of molecule. For instance, insulin receptors bind the hormone insulin. A specific molecule that binds to a receptor is called a **ligand** (LĪ-gand; *liga* = tied) of that receptor.
- Some integral proteins are **enzymes** that catalyze specific chemical reactions at the inside or outside surface of the cell.
- Integral proteins may also serve as **linkers,** which anchor proteins in the plasma membranes of neighboring cells to one another or to protein filaments inside and outside the cell. Peripheral proteins also serve as enzymes and linkers.
- Membrane glycoproteins and glycolipids often serve as **cell-identity markers.** They may enable a cell to recognize other cells of the same kind during tissue formation or to recognize and respond to potentially dangerous foreign cells. The ABO blood type markers are one example of cell-identity markers. When you receive a blood transfusion, the blood type must be compatible with your own.

Figure 3.3 **Functions of membrane proteins.**

Membrane proteins largely reflect the functions a cell can perform.

Extracellular fluid | Plasma membrane | Cytosol

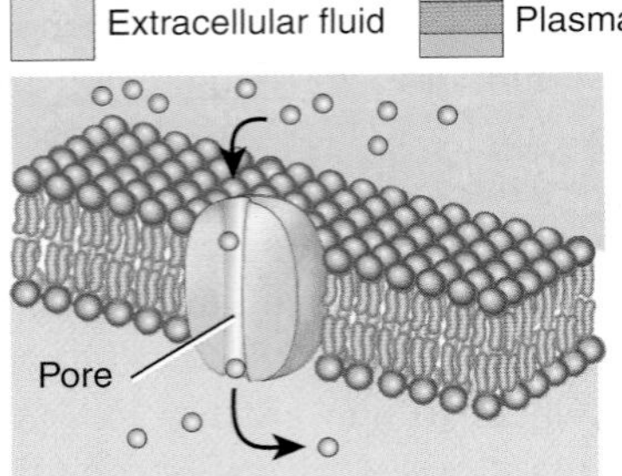

Ion channel (integral)
Allows specific ion (○) to move through water-filled pore. Most plasma membranes include specific channels for several common ions.

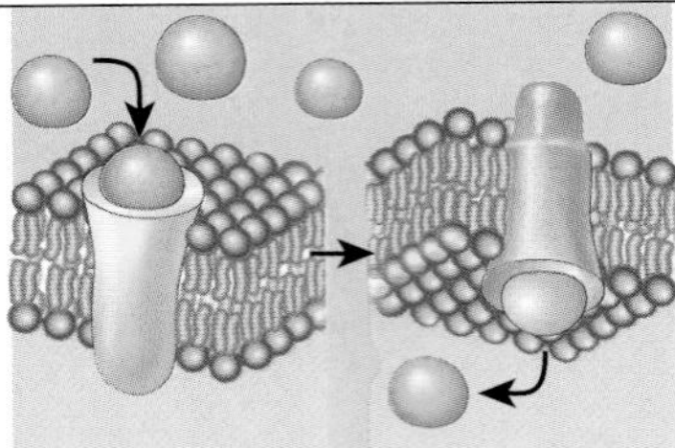

Transporter (integral)
Transports specific substances (○) across membrane by changing shape. For example, amino acids, needed to synthesize new proteins, enter body cells via transporters.

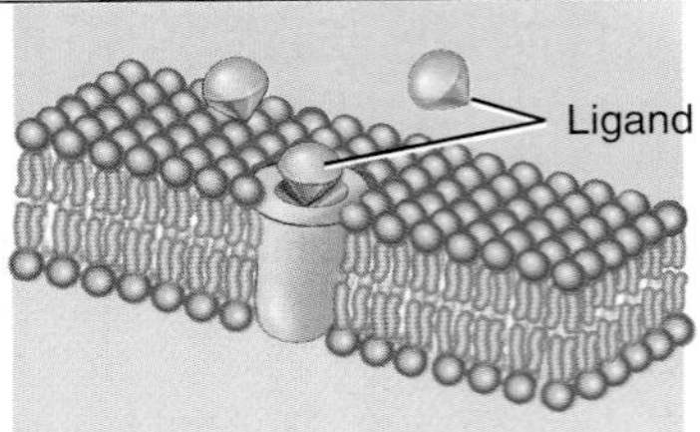

Receptor (integral)
Recognizes specific ligand (▽) and alters cell's function in some way. For example, antidiuretic hormone binds to receptors in the kidneys and changes the water permeability of certain plasma membranes.

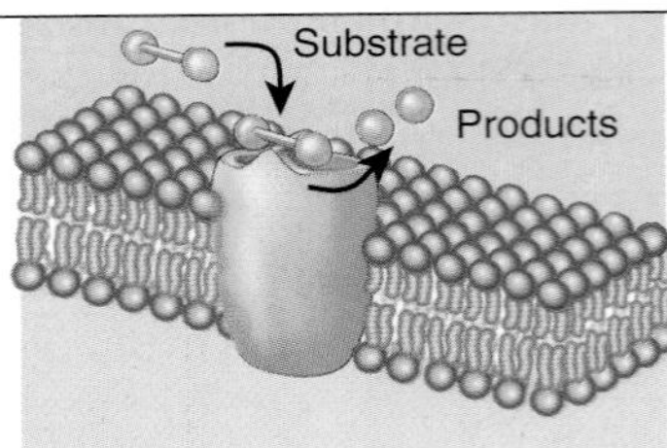

Enzyme (integral and peripheral)
Catalyzes reaction inside or outside cell (depending on which direction the active site faces). For example, lactase protruding from epithelial cells lining your small intestine splits the disaccharide lactose in the milk you drink.

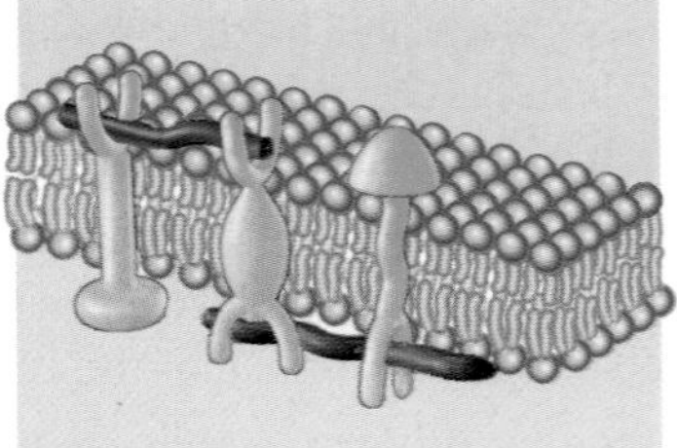

Linker (integral and peripheral)
Anchors filaments inside and outside the plasma membrane, providing structural stability and shape for the cell. May also participate in movement of the cell or link two cells together.

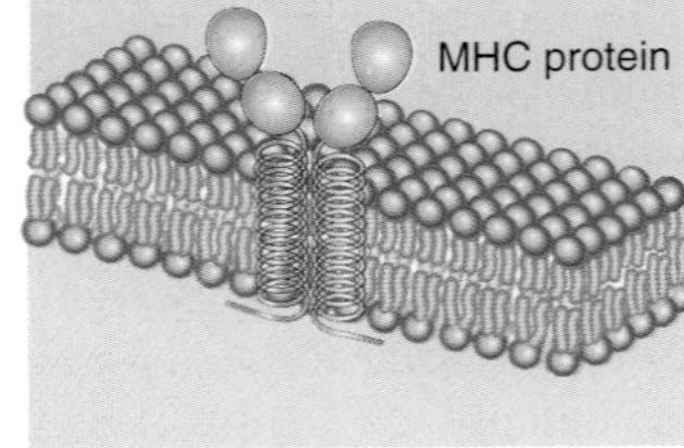

Cell identity marker (glycoprotein)
Distinguishes your cells from anyone else's (unless you are an identical twin). An important class of such markers are the major histocompatibility (MHC) proteins.

When stimulating a cell, the hormone insulin first binds to a protein in the plasma membrane. This action best represents which membrane protein function?

In addition, peripheral proteins help support the plasma membrane, anchor integral proteins, and participate in mechanical activities such as moving materials and organelles within cells, changing cell shape in dividing and muscle cells, and attaching cells to one another.

Membrane Fluidity

Membranes are fluid structures; that is, most of the membrane lipids and many of the membrane proteins easily rotate and move sideways in their own half of the bilayer. Neighboring lipid molecules exchange places about 10 million times per second and may wander completely around a cell in only a few minutes! Membrane fluidity depends both on the number of double bonds in the fatty acid tails of the lipids that make up the bilayer, and on the amount of cholesterol present. Each double bond puts a "kink" in the fatty acid tail (see Figure 2.18 on page 48), which increases membrane fluidity by preventing lipid molecules from packing tightly in the membrane. Membrane fluidity is an excellent compromise for the cell; a rigid membrane would lack mobility, and a completely fluid membrane would lack the structural organization and mechanical support required by the cell. Membrane fluidity allows interactions to occur within the plasma membrane, such as the assembly of membrane proteins. It also enables the movement of the membrane components responsible for cellular processes such as cell movement, growth, division, and secretion, and the formation of cellular junctions. Fluidity allows the lipid bilayer to self-seal if torn or punctured. When a needle is pushed through a plasma membrane and pulled out, the puncture site seals spontaneously, and the cell does not burst. This property of the lipid bilayer allows a procedure called intracytoplasmic sperm injection to help infertile couples conceive a child; scientists can fertilize an oocyte by injecting a sperm cell through a tiny syringe. It also permits removal and replacement of a cell's nucleus in cloning experiments, such as the one that created Dolly, the famous cloned sheep.

Despite the great mobility of membrane lipids and proteins in their own half of the bilayer, they seldom flip-flop from one half of the bilayer to the other, because it is difficult for hydrophilic parts of membrane molecules to pass through the hydrophobic core of the membrane. This difficulty contributes to the asymmetry of the membrane bilayer.

Because of the way it forms hydrogen bonds with neighboring phospholipid and glycolipid heads and fills the space between bent fatty acid tails, cholesterol makes the lipid bilayer stronger but less fluid at normal body temperature. At low temperatures, cholesterol has the opposite effect—it increases membrane fluidity.

Membrane Permeability

A membrane is said to be *permeable* to substances that can pass through it and *impermeable* to those that cannot. Although plasma membranes are not completely permeable to any substance, they do permit some substances to pass more readily than others. This property of membranes is termed **selective permeability.**

The lipid bilayer portion of the membrane is permeable to nonpolar, uncharged molecules, such as oxygen, carbon dioxide, and steroids, but is impermeable to ions and large, uncharged polar molecules such as glucose. It is also *slightly* permeable to small, uncharged polar molecules such as water and urea, a waste product from the breakdown of amino acids. The slight permeability to water and urea is an unexpected property since they are polar molecules. These two small molecules are thought to pass through the lipid bilayer in the following way. As the fatty acid tails of membrane phospholipids and glycolipids randomly move about, small gaps briefly appear in the hydrophobic environment of the membrane's interior. Water and urea molecules are small enough to move from one gap to another until they have crossed the membrane.

Transmembrane proteins that act as channels and transporters increase the plasma membrane's permeability to a variety of ions and uncharged polar molecules that, unlike water and urea molecules, cannot cross the lipid bilayer unassisted. Channels and transporters are very selective. Each one helps a specific molecule or ion to cross the membrane. Macromolecules, such as proteins, are so large that they are unable to pass across the plasma membrane except by endocytosis and exocytosis (discussed later in this chapter).

Gradients Across the Plasma Membrane

The selective permeability of the plasma membrane allows a living cell to maintain different concentrations of certain substances on either side of the plasma membrane. A **concentration gradient** is a difference in the concentration of a chemical from one place to another, such as from the inside to the outside of the plasma membrane. Many ions and molecules are more concentrated in either the cytosol or the extracellular fluid. For instance, oxygen molecules and sodium ions (Na^+) are more concentrated in the extracellular fluid than in the cytosol; the opposite is true of carbon dioxide molecules and potassium ions (K^+).

The plasma membrane also creates a difference in the distribution of positively and negatively charged ions between the two sides of the plasma membrane. Typically, the inner surface of the plasma membrane is more negatively charged and the outer surface is more positively charged. A difference in electrical charges between two regions constitutes an **electrical gradient.** Because it occurs across the plasma membrane, this charge difference is termed the **membrane potential.**

As you will see shortly, the concentration gradient and electrical gradient are important because they help move substances across the plasma membrane. In many cases a substance will move across a plasma membrane *down its concentration gradient.* That is to say, a substance will move "downhill," from where it is more concentrated to where it is less concentrated to reach equilibrium. Similarly, a positively charged substance will tend to move toward a negatively charged area, and a negatively charged substance will tend to move toward a positively charged area. The combined influence of the concentration gradient and the membrane potential on movement of a particular ion is referred to as its **electrochemical gradient.**

▶ CHECKPOINT

2. How do hydrophobic and hydrophilic regions govern the arrangement of membrane lipids in a bilayer?
3. What substances can and cannot diffuse through the lipid bilayer?
4. "The proteins present in a plasma membrane determine the functions that a membrane can perform." Is this statement true or false? Explain your answer.
5. How does cholesterol affect membrane fluidity?
6. Why are membranes said to have selective permeability?
7. What factors contribute to an electrochemical gradient?

TRANSPORT ACROSS THE PLASMA MEMBRANE

▶ OBJECTIVE

Describe the processes that transport substances across the plasma membrane.

Transport of materials across the plasma membrane is essential to the life of a cell. Certain substances must move into the cell to support metabolic reactions. Other substances that have been produced by the cell for export or as cellular waste products must move out of the cell.

Substances generally move across cellular membranes via transport processes that can be classified as active or passive, depending on whether they require cellular energy. In *passive processes,* a substance moves down its concentration or electrical gradient to cross the membrane using only its own kinetic energy. There is no input of energy from the cell. In *active processes,* cellular energy is used to drive the substance "uphill" against its concentration or electrical gradient. The cellular energy used is usually in the form of ATP.

Some materials cross cellular membranes simply by moving through the lipid bilayer or membrane channels using their own kinetic energy (energy of motion). Kinetic energy is intrinsic to the particles that are moving. Examples of processes that rely on kinetic energy include diffusion and osmosis. Other substances must bind to specific transporter proteins to cross a cellular membrane, as in facilitated diffusion and active transport. Still other substances pass through cellular membranes within small, spherical sacs called vesicles that bud off from an existing membrane. Examples include endocytosis, in which vesicles detach from the plasma membrane while bringing materials into a cell, and exocytosis, the merging of vesicles with the plasma membrane to release materials from a cell. Thus, materials may cross plasma membranes by using kinetic energy, transporter proteins, or vesicles.

Kinetic Energy Transport

Diffusion

Learning why materials diffuse across membranes requires an understanding of how diffusion occurs in a solution. **Diffusion** (di-FŪ-zhun; *diffus-* = spreading) is a passive process in which the random mixing of particles in a solution occurs because of the particles' kinetic energy. Both the *solutes,* the dissolved substances, and the *solvent,* the liquid that does the dissolving, undergo diffusion. If a particular solute is present in high concentration in one area of a solution and in low concentration in another area, solute molecules will diffuse toward the area of lower concentration—they move *down their concentration gradient.* After some time, the particles become evenly distributed throughout the solution and the solution is said to be at equilibrium. The particles continue to move about randomly due to their kinetic energy, but their concentrations do not change.

For example, when you place a crystal of dye in a water-filled container (Figure 3.4), the color is most intense in the area closest to the dye because its concentration is higher there. At increasing distances, the color is lighter and lighter because the dye concentration is lower. Some time later, the solution of water and dye will have a uniform color, because the dye molecules and water molecules have diffused down their concentration gradients until they are evenly mixed in solution—they are at equilibrium.

Figure 3.4 **Principle of diffusion.** At the beginning of our experiment, a crystal of dye placed in a cylinder of water dissolves (a) and then diffuses from the region of higher dye concentration to regions of lower dye concentration (b). At equilibrium (c), the dye concentration is uniform throughout, although random movement continues.

In diffusion, a substance moves down its concentration gradient.

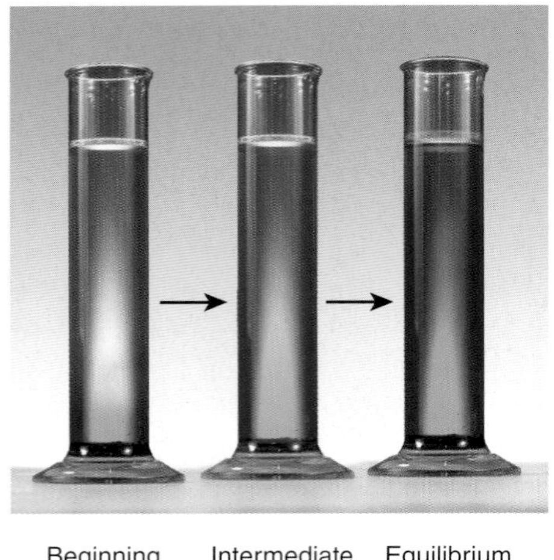

Beginning (a) — Intermediate (b) — Equilibrium (c)

How would having a fever affect body processes that involve diffusion?

In this simple example, no membrane was involved. Substances may also diffuse through a membrane, if the membrane is permeable to them. Several factors influence the diffusion rate of substances across plasma membranes:

1. ***Steepness of the concentration gradient.*** The greater the difference in concentration between the two sides of the membrane, the higher the rate of diffusion. When charged particles are diffusing, it is the steepness of the electrochemical gradient that determines diffusion rate across the membrane.
2. ***Temperature.*** The higher the temperature, the faster the rate of diffusion. All of the body's diffusion processes occur more rapidly in a person with a fever.
3. ***Mass of the diffusing substance.*** The larger the mass of the diffusing particle, the slower its diffusion rate. Smaller molecules diffuse more rapidly than larger ones.
4. ***Surface area.*** The larger the membrane surface area available for diffusion, the faster the diffusion rate. For example, the air sacs of the lungs have a large surface area available for diffusion of oxygen from the air into the blood. Some lung diseases, such as emphysema, reduce the surface area. This slows the rate of oxygen diffusion and makes breathing more difficult.
5. ***Diffusion distance.*** The greater the distance over which diffusion must occur, the longer it takes. Diffusion across a plasma membrane takes only a fraction of a second because the membrane is so thin. In pneumonia, fluid collects in the lungs; the additional fluid increases the diffusion distance because oxygen must move through both the built-up fluid and the membrane to reach the bloodstream.

DIFFUSION THROUGH THE LIPID BILAYER Nonpolar, hydrophobic molecules diffuse freely through the lipid bilayer of the plasma membranes of cells without the help of membrane transport proteins (Figure 3.5a). Such molecules include oxygen, carbon dioxide, and nitrogen gases; fatty acids, steroids, and fat-soluble vitamins (A, E, D, and K); small alcohols; and ammonia. As you have already learned, two small, uncharged polar molecules—water and urea—can diffuse through the lipid bilayer. Diffusion through the lipid bilayer is important in the movement of oxygen and carbon dioxide between blood and body cells, and between blood and air within the lungs during breathing. It also is the route for absorption of some nutrients and excretion of some wastes by body cells.

DIFFUSION THROUGH MEMBRANE ION CHANNELS Most membrane channels are *ion channels,* integral transmembrane proteins that allow passage of small, inorganic ions that are too hydrophilic to penetrate the nonpolar interior of the lipid bilayer (Figure 3.5a). Each ion can diffuse across the membrane only at certain sites. In typical plasma membranes, the most numerous ion channels are selective for K^+ (potassium ions) or Cl^- (chloride ions); fewer channels are available for Na^+ (sodium ions) or Ca^{2+} (calcium ions). Diffusion of ions through channels is generally slower than free diffusion through the lipid bilayer

Figure 3.5 Types of diffusion.

Nonpolar, hydrophobic molecules diffuse through the lipid bilayer; small, inorganic ions pass through channel proteins.

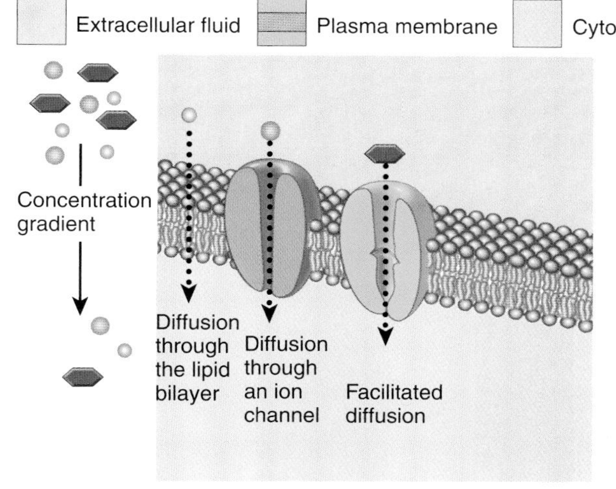

(a) Comparison of types of diffusion

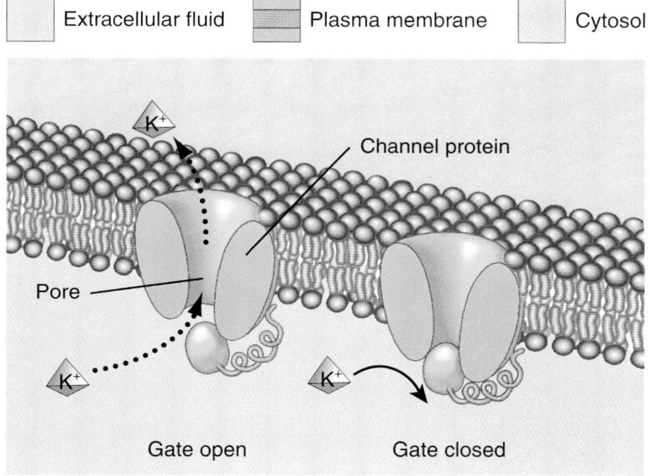

(b) Details of the K^+ channel

Why is diffusion through channel proteins slower than diffusion through the lipid bilayer?

because channels occupy a smaller fraction of the membrane's total surface area than lipids. Still, diffusion through channels is a very fast process: More than a million potassium ions can flow through a K^+ channel in one second!

A channel is said to be "gated" when part of the channel protein acts as a "plug" or "gate," changing shape in one way to open the pore and in another way to close it (Figure 3.5b). Some gated channels randomly alternate between the open and closed positions; others are regulated by chemical or electrical changes inside and outside the cell. When the gates of a channel are open, ions diffuse into or out of cells, down their electrochemical gradients. The plasma membranes of different types of cells may have different numbers of ion channels and thus display different permeabilities to various ions.

Osmosis

Osmosis (oz-MŌ-sis) is the net movement of a solvent through a selectively permeable membrane. Like diffusion, it is a passive process. In living systems, the solvent is water, which moves by osmosis across plasma membranes from an area of *higher water concentration* to an area of *lower water concentration.* Another way to understand this idea is to consider the solute concentration: In osmosis, water moves through a selectively permeable membrane from an area of *lower solute concentration* to an area of *higher solute concentration.* During osmosis, water molecules pass through a plasma membrane in two ways: (1) by moving through the lipid bilayer, as previously described, and (2) by moving through **aquaporins** (*aqua-* = water), integral membrane proteins that function as water channels.

Osmosis occurs only when a membrane is permeable to water but is not permeable to certain solutes. A simple experiment can demonstrate osmosis. Consider a U-shaped tube in which a selectively permeable membrane separates the left and right arms of the tube. A volume of pure water is poured into the left arm, and the same volume of a solution containing a solute that cannot pass through the membrane is poured into the right arm (Figure 3.6a). Because the *water* concentration is higher on the left and lower on the right, net movement of water molecules—osmosis—occurs from left to right, so that the water is moving down its concentration gradient. At the same time, the membrane prevents diffusion of the solute from the right arm into the left arm. As a result, the volume of water in the left arm decreases, and the volume of solution in the right arm increases (Figure 3.6b).

You might think that osmosis would continue until no water remained on the left side, but this is *not* what happens. In this experiment, the higher the column of solution in the right arm becomes, the more pressure it exerts on its side of the membrane. Pressure exerted in this way by a liquid, known as **hydrostatic pressure,** forces water molecules to move back into the left arm. Equilibrium is reached when just as many water molecules move from right to left due to the hydrostatic pressure as move from left to right due to osmosis (Figure 3.6b).

To further complicate matters, the solution with the impermeable solute also exerts a force, called the **osmotic pressure.** The osmotic pressure of a solution is proportional to the concentration of the solute particles that cannot cross the membrane—the higher the solute concentration, the higher the solution's osmotic pressure. Consider what would happen if a piston were used to apply more pressure to the fluid in the right arm of the tube in Figure 3.6. With enough pressure, the volume of fluid in each arm could be restored to the starting volume, and the concentration of solute in the right arm would be the same as it was at the beginning of the experiment (Figure 3.6c). The

Figure 3.6 Principle of Osmosis. Water molecules move through the selectively permeable membrane; the solute molecules in the right arm cannot pass through the membrane. (a) As the experiment starts, water molecules move from the left arm into the right arm, down the water concentration gradient. (b) After some time, the volume of water in the left arm has decreased and the volume of solution in the right arm has increased. At equilibrium, there is no net osmosis: Hydrostatic pressure forces just as many water molecules to move from right to left as osmosis forces water molecules to move from left to right. (c) If pressure is applied to the solution in the right arm, the starting conditions can be restored. This pressure, which stops osmosis, is equal to the osmotic pressure.

Osmosis is the movement of water molecules through a selectively permeable membrane.

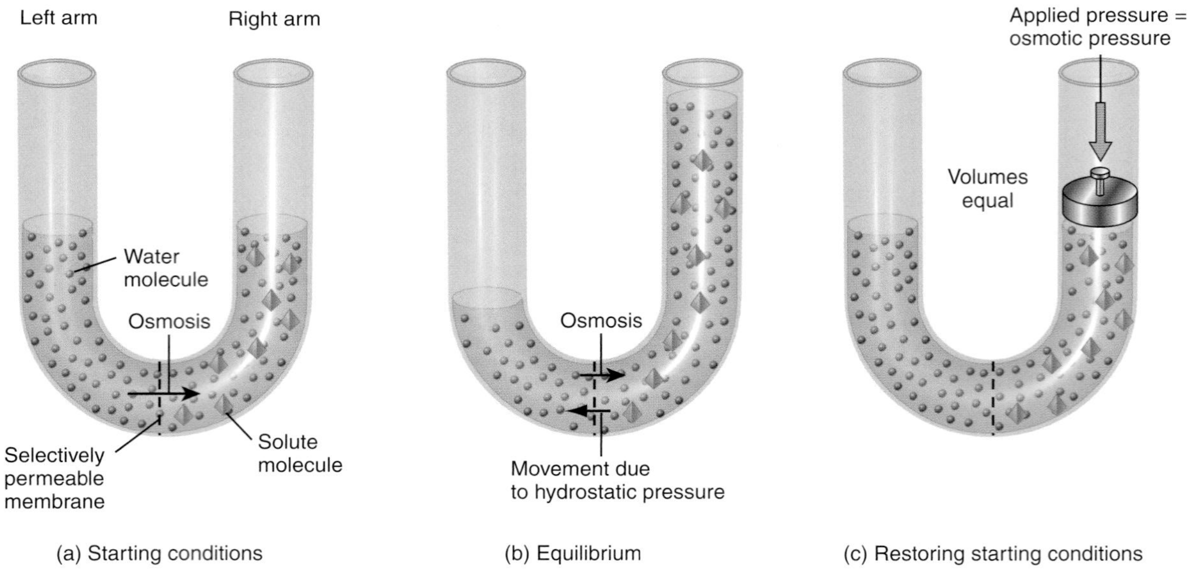

Will the fluid level in the right arm rise until the water concentrations are the same in both arms?

amount of pressure needed to restore the starting condition equals the osmotic pressure. So, in our experiment osmotic pressure is the pressure needed to stop the movement of water from the left tube into the right tube. Notice that the osmotic pressure of a solution does not produce the movement of water during osmosis. Rather it is the pressure that would *prevent* such water movement.

Normally, the osmotic pressure of the cytosol is the same as the osmotic pressure of the interstitial fluid outside cells. Because the osmotic pressure on both sides of the plasma membrane (which is selectively permeable) is the same, cell volume remains relatively constant. When body cells are placed in a solution having a different osmotic pressure than cytosol, however, the shape and volume of the cells changes. As water moves by osmosis into or out of the cells, their volume increases or decreases. A solution's **tonicity** (*tonic* = tension) is a measure of the solution's ability to change the volume of cells by altering their water content.

Any solution in which a cell—for example, a red blood cell (RBC)—maintains its normal shape and volume is an **isotonic solution** (*iso-* = same) (Figure 3.7a). The concentrations of solutes that cannot cross the plasma membrane are the same on both sides of the membrane in this solution. For instance, a 0.9% NaCl solution (0.9 grams of sodium chloride in 100 mL of solution), called a *normal (physiological) saline solution,* is isotonic for RBCs. The RBC plasma membrane permits the water to move back and forth, but it behaves as though it is impermeable to Na^+ and Cl^-, the solutes. (Any Na^+ or Cl^- ions that enter the cell through channels or transporters are immediately moved back out by active transport or other means.) When RBCs are bathed in 0.9% NaCl, water molecules enter and exit at the same rate, allowing the RBCs to keep their normal shape and volume.

A different situation results if RBCs are placed in a **hypotonic solution** (*hypo-* = less than), a solution that has a *lower* concentration of solutes than the cytosol inside the RBCs (Figure 3.7b). In this case, water molecules enter the cells faster than they leave, causing the RBCs to swell and eventually to burst. The rupture of RBCs in this manner is called **hemolysis** (hē-MOL-i-sis; *hemo-* = blood; *-lysis* = to loosen or split apart); the rupture of other types of cells due to placement in a hypotonic solution is referred to simply as **lysis.** Pure water is very hypotonic and causes rapid hemolysis.

A **hypertonic solution** (*hyper-* = greater than) has a *higher* concentration of solutes than does the cytosol inside RBCs (Figure 3.7c). One example of a hypertonic solution is a

Figure 3.7 Tonicity and its effects on red blood cells (RBCs). The arrows indicate the direction and degree of water movement into and out of the cells. One example of an isotonic solution for RBCs is 0.9% NaCl.

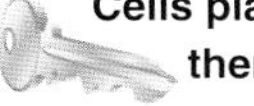

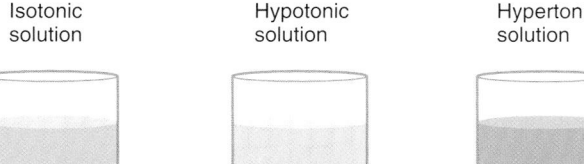

Cells placed in an isotonic solution maintain their shape because there is no net water movement into or out of the cell.

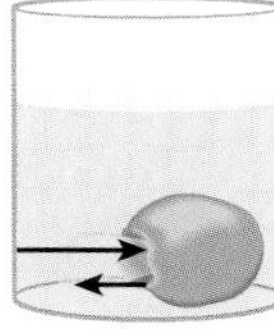

(a) Illustrations showing direction of water movement

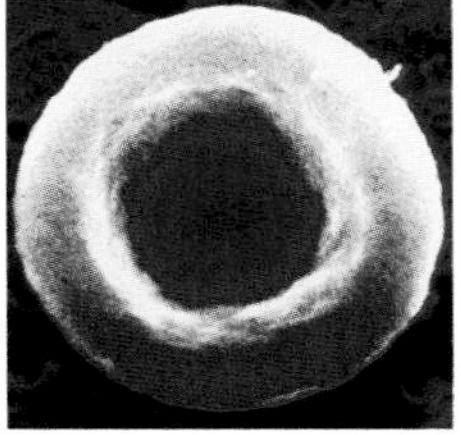

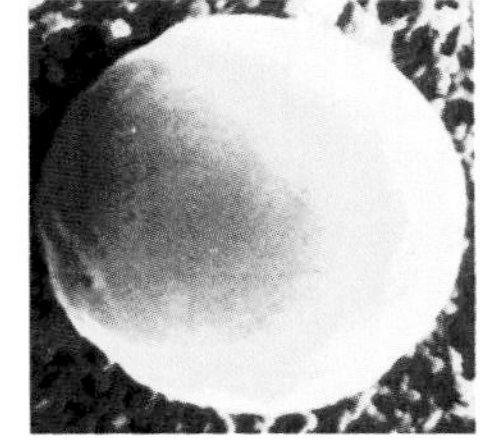

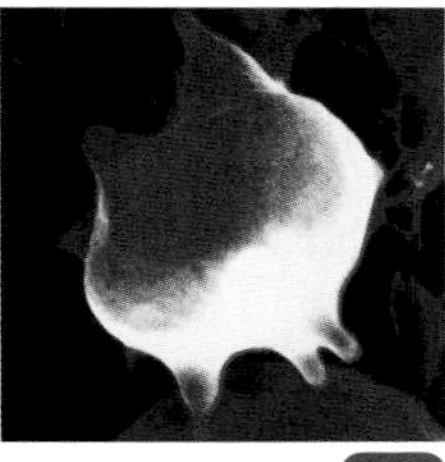

Normal RBC shape | RBC undergoes hemolysis | RBC undergoes crenation

(b) Scanning electron micrographs (all 800x)

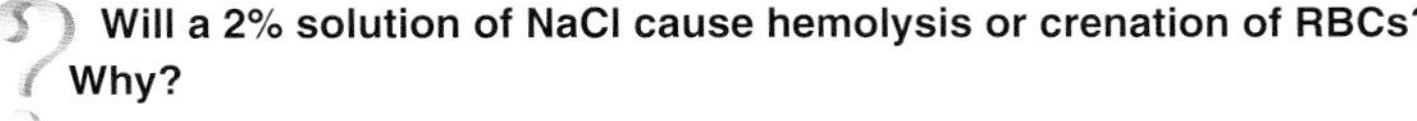

Will a 2% solution of NaCl cause hemolysis or crenation of RBCs? Why?

2% NaCl solution. In such a solution, water molecules move out of the cells faster than they enter, causing the cells to shrink. Such shrinkage of cells is called **crenation** (kre-NĀ-shun).

Medical Uses of Isotonic, Hypertonic, and Hypotonic Solutions

RBCs and other body cells may be damaged or destroyed if exposed to hypertonic or hypotonic solutions. For this reason, most **intravenous (IV) solutions,** liquids infused into the blood of a vein, are isotonic. Examples are isotonic saline (0.9% NaCl) and D5W, which stands for dextrose 5% in water. Sometimes infusion of a hypertonic solution such as mannitol is useful to treat patients who have *cerebral edema,* excess interstitial fluid in the brain. Infusion of such a solution relieves fluid overload by causing osmosis of water from interstitial fluid into the blood. The kidneys then excrete the excess water from the blood into the urine. Hypotonic solutions, given either orally or through an IV, can be used to treat people who are dehydrated. The water in the hypotonic solution moves from the blood into interstitial fluid and then into body cells to rehydrate them. Water and most sports drinks that you consume to "rehydrate" after a workout are hypotonic relative to your body cells. ■

▶ CHECKPOINT

8. What factors can increase the rate of diffusion?
9. What is osmotic pressure?
10. Which substances can diffuse directly through the lipid bilayer?
11. How does diffusion through membrane channels compare with facilitated diffusion?

Transport by Transporter Proteins

Facilitated Diffusion

Solutes that are too polar or highly charged to diffuse through the lipid bilayer and are too big to diffuse through membrane channels can cross the plasma membrane by **facilitated diffusion.** In this process, a solute binds to a specific transporter on one side of the membrane and is released on the other side after the transporter undergoes a change in shape (see Figure 3.5a).

Like diffusion, facilitated diffusion is a passive process. The net result of facilitated diffusion is movement down a concentration gradient. The solute binds more often to the transporter on the side of the membrane with a higher concentration of solute. Once the concentration is the same on both sides of the membrane, solute molecules bind to the transporter on the cytosolic side and move out to the extracellular fluid as rapidly as they bind to the transporter on the extracellular side and move into the cytosol. The rate of facilitated diffusion (how quickly it occurs) is determined by the steepness of the concentration gradient across the membrane.

The number of transporters available in a plasma membrane places an upper limit, called the *transport maximum,* on the rate at which facilitated diffusion can occur. Once all the transporters are occupied, the transport maximum is reached, and a further increase in the concentration gradient does not increase the rate of facilitated diffusion. Thus, much like a completely saturated sponge can absorb no more water, the process of facilitated diffusion exhibits *saturation.*

The selective permeability of the plasma membrane is often regulated to achieve homeostasis. For instance, the hormone insulin, via the action of the insulin receptor, promotes the insertion of many copies of a specific type of glucose transporter into the plasma membranes of certain cells. Thus, the effect of insulin is to elevate the transport maximum for facilitated diffusion of glucose into cells. With more transporters available, body cells can pick up glucose from the blood more rapidly. An inability to produce or utilize insulin is called diabetes mellitus (Chapter 18).

Active Transport

Some polar or charged solutes that must enter or leave body cells cannot cross the plasma membrane through any form of passive transport because they would need to move "uphill," *against* their concentration gradients. Such solutes may be able to cross the membrane by a process called **active transport.** Active transport is considered an active process because energy is required for transporter proteins to move solutes across the membrane against a concentration gradient. Two sources of cellular energy can be used to drive active transport: (1) Energy obtained from hydrolysis of ATP is the source in *primary active transport;* (2) energy stored in an ionic concentration gradient is the source in *secondary active transport.* Like facilitated diffusion, active transport processes exhibit a transport maximum and saturation. Solutes actively transported across the plasma membrane include several ions, such as Na^+, K^+, H^+, Ca^{2+}, I^- (iodide ions), and Cl^-; amino acids; and monosaccharides. (Note that some of these substances also cross the membrane via channels or facilitated diffusion when the proper channel proteins or transporters are present.)

PRIMARY ACTIVE TRANSPORT In **primary active transport,** energy derived from hydrolysis of ATP changes the shape of a transporter protein, which "pumps" a substance across a plasma membrane against its concentration gradient. Indeed, transporter proteins that carry out primary active transport are often called **pumps.** A typical body cell expends about 40% of the ATP it generates on primary active transport. Chemicals that turn off ATP production—for example, the poison cyanide—are lethal because they shut down active transport in cells throughout the body.

The most prevalent primary active transport mechanism expels sodium ions (Na^+) from cells and brings potassium ions (K^+) in. Because of the specific ions it moves, this transporter is called the **sodium-potassium pump.** Because a part of the sodium-potassium pump acts as an *ATPase,* an enzyme that hydrolyzes ATP, another name for this pump is **Na^+/K^+ ATPase.** All cells have thousands of sodium-potassium pumps in their plasma membranes. These sodium-potassium pumps maintain a low concentration of Na^+ in the cytosol by pumping them into the extracellular fluid against the Na^+ concentration gradient. At the same time, the pumps move K^+ into cells against the K^+ concentration gradient. Because K^+ and Na^+ slowly leak back across the plasma membrane down their electrochemical gradients—through passive transport or secondary active transport—the sodium-potassium pumps must work nonstop to maintain a low concentration of Na^+ and a high concentration of K^+ in the cytosol.

Figure 3.8 depicts the operation of the sodium-potassium pump:

1. Three Na^+ in the cytosol bind to the pump protein.
2. Binding of Na^+ triggers the hydrolysis of ATP into ADP, a reaction that also attaches a phosphate group (P) to the pump protein. This chemical reaction changes the shape of the pump protein, expelling the three Na^+ into the extracellular fluid. Now the shape of the pump protein favors binding of two K^+ in the extracellular fluid to the pump protein.
3. The binding of K^+ triggers release of the phosphate group from the pump protein. This reaction again causes the shape of the pump protein to change.
4. As the pump protein reverts to its original shape, it releases K^+ into the cytosol. At this point, the pump is again ready to bind three Na^+, and the cycle repeats.

The different concentrations of Na^+ and K^+ in cytosol and extracellular fluid are crucial for maintaining normal cell volume and for the ability of some cells to generate electrical signals such as action potentials. Recall that the tonicity of a solution is

Figure 3.8 The sodium-potassium pump (Na^+/K^+ ATPase) expels sodium ions (Na^+) and brings potassium ions (K^+) into the cell.

Sodium-potassium pumps maintain a low intracellular concentration of sodium ions.

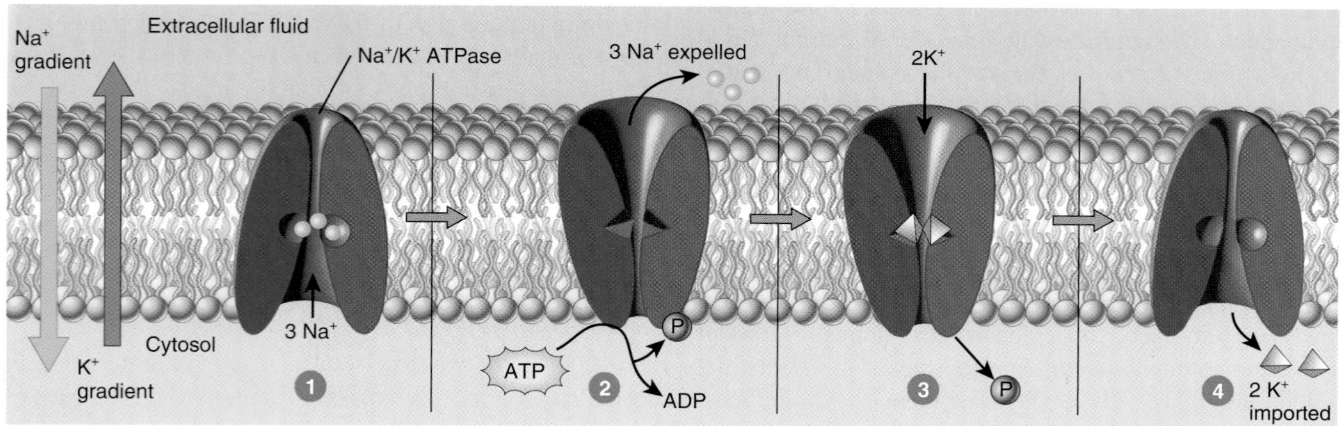

What is the role of ATP in the operation of this pump?

proportional to the concentration of its solute particles that cannot penetrate the membrane. Because sodium ions that diffuse into a cell or enter through secondary active transport are immediately pumped out, it is as if they never entered. In effect, sodium ions behave as if they cannot penetrate the membrane. Thus, sodium ions are an important contributor to the tonicity of the extracellular fluid. A similar condition holds for K^+ in the cytosol. By helping to maintain normal tonicity on each side of the plasma membrane, the sodium-potassium pumps ensure that cells neither shrink nor swell due to the movement of water by osmosis out of or into cells.

Secondary Active Transport In **secondary active transport,** the energy stored in a Na^+ or H^+ concentration gradient is used to drive other substances across the membrane against their own concentration gradients. Because a Na^+ or H^+ gradient is established by primary active transport, secondary active transport *indirectly* uses energy obtained from the hydrolysis of ATP.

The sodium-potassium pump maintains a steep concentration gradient of Na^+ across the plasma membrane. As a result, the sodium ions have stored or potential energy, just like water behind a dam. Accordingly, if there is a route for Na^+ to leak back in, some of the stored energy can be converted to kinetic energy (energy of motion) and used to transport other substances *against their concentration gradients.* In essence, secondary active transport proteins harness the energy in the Na^+ concentration gradient by providing routes for Na^+ to leak into cells. In secondary active transport, a transporter protein simultaneously binds to Na^+ and another substance and then changes its shape so that both substances cross the membrane at the same time. If these transporters move two substances in the same direction they are called **symporters** (*sym-* = same); **antiporters,** in contrast, move two substances in opposite directions across the membrane (*anti-* = against).

Plasma membranes contain several antiporters and symporters that are powered by the Na^+ gradient (Figure 3.9a). For instance, the concentration of calcium ions (Ca^{2+}) is low in the cytosol because Na^+/Ca^{2+} antiporters eject calcium ions. Likewise, Na^+/H^+ antiporters help regulate the cytosol's pH (H^+ concentration) by expelling excess H^+. By contrast, dietary glucose and amino acids are absorbed into cells that line the small intestine by Na^+/glucose and Na^+/amino acid symporters (Figure 3.9b). In each case, sodium ions are moving down their concentration gradient while the other solutes move "uphill," against their concentration gradients. Keep in mind that all these symporters and antiporters can do their job because the sodium-potassium pumps maintain a low concentration of Na^+ in the cytosol.

Digitalis Increases Ca^{2+} in Heart Muscle Cells

Digitalis often is given to patients with *heart failure,* a condition of weakened pumping action by the heart. Digitalis exerts its effect by slowing the action of the sodium-potassium pumps, which lets more Na^+ accumulate inside heart muscle cells. The

Figure 3.9 **Secondary active transport mechanisms.** (a) Antiporters carry two substances across the membrane in opposite directions. (b) Symporters carry two substances across the membrane in the same direction.

Secondary active transport mechanisms use the energy stored in an ionic concentration gradient (here, for Na^+). Because primary active transport pumps that hydrolyze ATP maintain the gradient, secondary active transport mechanisms consume ATP indirectly.

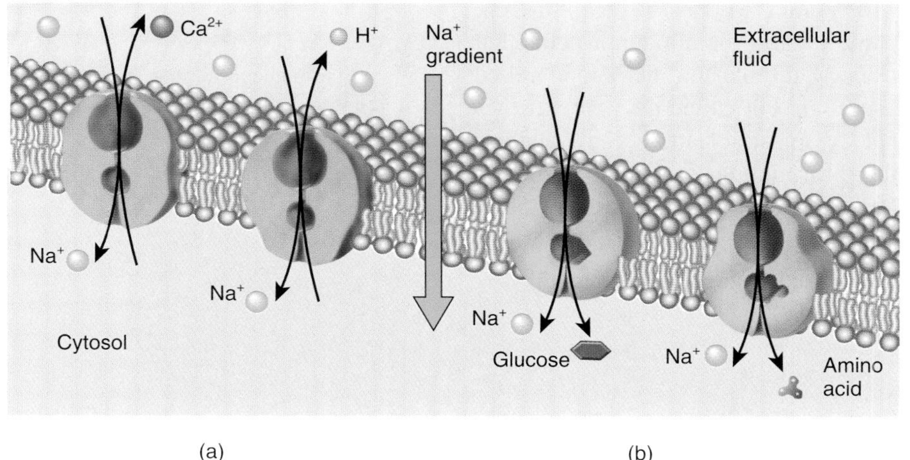

What is the main difference between primary and secondary active transport mechanisms?

result is a decreased Na^+ concentration gradient across the plasma membrane, which causes the Na^+/Ca^{2+} antiporters to slow down. As a result, more Ca^{2+} remains inside heart muscle cells. The slight increase in the level of Ca^{2+} in the cytosol of heart muscle cells increases the force of their contractions and thus strengthens the force of the heartbeat. ■

Transport in Vesicles

A **vesicle,** as noted earlier, is a small, spherical sac. As you will learn later in this chapter, a variety of substances are transported in vesicles from one structure to another within cells. Vesicles also import materials from and release materials into extracellular fluid. During **endocytosis** (*endo-* = within), materials move into a cell in a vesicle formed from the plasma membrane. In **exocytosis** (*exo-* = out), materials move out of a cell by the fusion with the plasma membrane of vesicles formed inside the cell. Both endocytosis and exocytosis require energy supplied by ATP. Thus, transport in vesicles is an active process.

Endocytosis

Here we consider three types of endocytosis: receptor-mediated endocytosis, phagocytosis, and bulk-phase endocytosis. **Receptor-mediated endocytosis** is a highly selective type of endocytosis by which cells take up specific ligands. (Recall that ligands are molecules that bind to specific receptors.) A vesicle forms after a receptor protein in the plasma membrane recognizes and binds to a particular particle in the extracellular fluid. For instance, cells take up cholesterol-containing low-density lipoproteins (LDLs), transferrin (an iron-transporting protein in the blood), some vitamins, antibodies, and certain hormones by receptor-mediated endocytosis. Receptor-mediated endocytosis of LDLs (and other ligands) occurs as follows (Figure 3.10):

Figure 3.10 Receptor-mediated endocytosis of a low-density lipoprotein (LDL) particle.

Receptor-mediated endocytosis imports materials that are needed by cells.

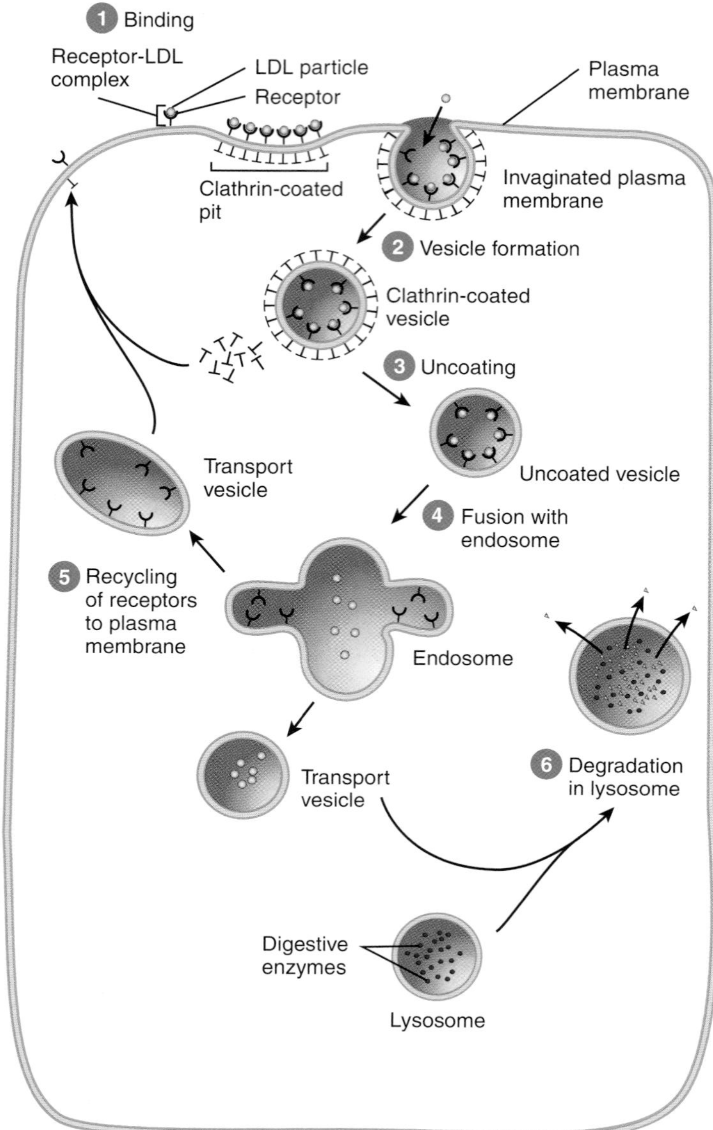

What are several other examples of ligands that can undergo receptor-mediated endocytosis?

1. ***Binding.*** On the extracellular side of the plasma membrane, an LDL particle that contains cholesterol binds to a specific receptor in the plasma membrane to form a receptor-LDL complex. The receptors are integral membrane proteins that are concentrated in regions of the plasma membrane called *clathrin-coated pits.* Here, a protein called *clathrin* attaches to the membrane on its cytoplasmic side. Many clathrin molecules come together, forming a basketlike structure around the receptor-LDL complexes that causes the membrane to invaginate (fold inward).
2. ***Vesicle formation.*** The invaginated edges of the membrane around the clathrin-coated pit fuse and a small piece of the membrane pinches off. The resulting vesicle, known as a *clathrin-coated vesicle,* contains the receptor-LDL complexes.
3. ***Uncoating.*** Almost immediately after it is formed, the clathrin-coated vesicle loses its clathrin coat to become an *uncoated vesicle.* Clathrin molecules either return to the inner surface of the plasma membrane or help form coats on other vesicles inside the cell.
4. ***Fusion with endosome.*** The uncoated vesicle quickly fuses with a vesicle known as an *endosome.* Within an endosome, the LDL particles separate from their receptors.
5. ***Recycling of receptors to plasma membrane.*** Most of the receptors accumulate in elongated protrusions of the endosome. These pinch off, forming transport vesicles that return

the receptors to the plasma membrane. An LDL receptor is returned to the plasma membrane about 10 minutes after it enters a cell.

6. ***Degradation in lysosomes.*** Other transport vesicles, which contain the LDL particles, bud off the endosome and soon fuse with a *lysosome.* Lysosomes contain many digestive enzymes. Certain enzymes break down the large protein and lipid molecules of the LDL particle into amino acids, fatty acids, and cholesterol. These smaller molecules then leave the lysosome. The cell uses cholesterol for rebuilding its membranes and for synthesis of steroids, such as estrogen. Fatty acids and amino acids can be used for ATP production or to build other molecules needed by the cell.

Viruses and Receptor-mediated Endocytosis

Although receptor-mediated endocytosis normally imports needed materials, some viruses are able to use this mechanism to enter and infect body cells. For example, the human immunodeficiency virus (HIV), which causes acquired immunodeficiency syndrome (AIDS), can attach to a receptor called CD4. This receptor is present in the plasma membrane of white blood cells called helper T cells. After binding to CD4, HIV enters the helper T cell via receptor-mediated endocytosis. ■

Phagocytosis (fag′-ō-sī-TŌ-sis; *phago-* = to eat) is a form of endocytosis in which the cell engulfs large solid particles, such as worn-out cells, whole bacteria, or viruses (Figure 3.11). Only a few body cells, termed **phagocytes,** are able to carry out phagocytosis. Two main types of phagocytes are *macrophages,* located in many body tissues, and *neutrophils,* a type of white blood cell. Phagocytosis begins when the particle binds to a plasma membrane receptor on the phagocyte, causing it to extend **pseudopods** (SOO-dō-pods; *pseudo-* = false; *-pods* = feet), projections of its plasma membrane and cytoplasm. Pseudopods surround the particle outside the cell, and the membranes fuse to form a vesicle called a *phagosome,* which enters the cytoplasm. The phagosome fuses with one or more lysosomes, and lysosomal enzymes break down the ingested material. In most cases, any undigested materials in the phagosome remain indefinitely in a vesicle called a *residual body.* The process of phagocytosis is a vital defense mechanism that helps protect the body from disease. Through phagocytosis, macrophages dispose of invading microbes and billions of aged, worn-out red blood cells every day; neutrophils also help rid the body of invading microbes. Pus is a mixture of dead neutrophils, macrophages, and tissue cells and fluid in an infected wound.

Most body cells carry out **bulk-phase endocytosis,** also called **pinocytosis** (pi-nō-sī-TŌ-sis; *pino-* = to drink), a form of endocytosis in which tiny droplets of extracellular fluid are taken

Figure 3.11 Phagocytosis. Pseudopods surround a particle and the membranes fuse to form a phagosome.

Phagocytosis is a vital defense mechanism that helps protect the body from disease.

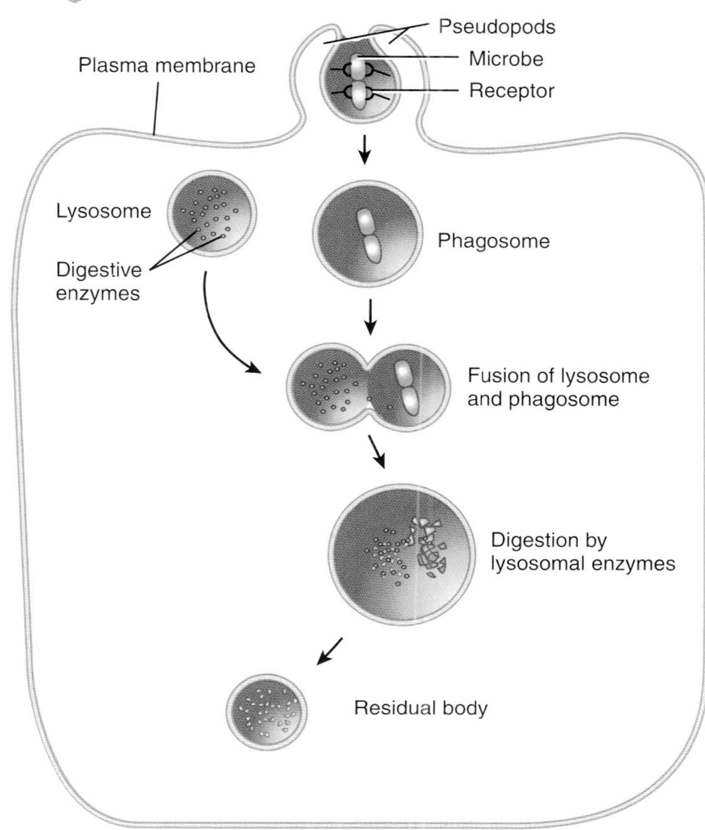

(a) Diagram of the process

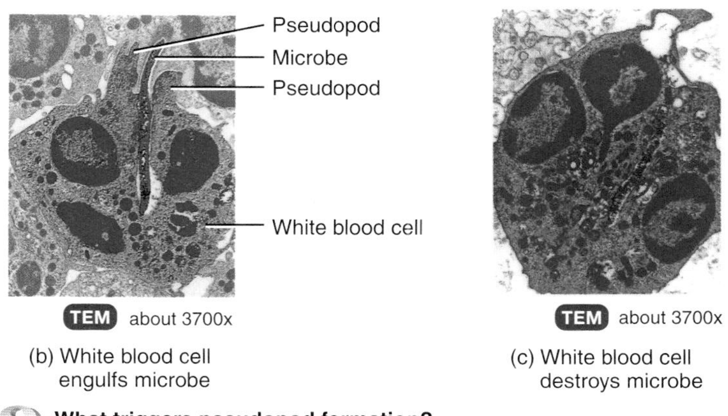

(b) White blood cell engulfs microbe

(c) White blood cell destroys microbe

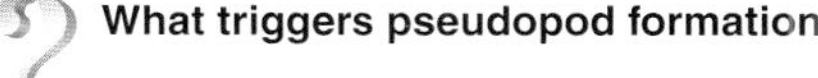

What triggers pseudopod formation?

up (Figure 3.12). No receptor proteins are involved; all solutes dissolved in the extracellular fluid are brought into the cell. During bulk-phase endocytosis, the plasma membrane folds inward and forms a vesicle containing a droplet of extracellular fluid. The vesicle detaches or "pinches off" from the plasma membrane and enters the cytosol. Within the cell, the vesicle fuses with a lysosome, where enzymes degrade the engulfed solutes. The resulting smaller molecules, such as amino acids and fatty acids, leave the lysosome to be used elsewhere in the cell. Bulk-phase endocytosis occurs in most cells, especially absorptive cells in the intestines and kidneys.

Exocytosis

In contrast with endocytosis, which brings materials into a cell, **exocytosis** releases materials from a cell. All cells carry out exocytosis, but it is especially important in two types of cells: (1) secretory cells that liberate digestive enzymes, hormones, mucus, or other secretions; (2) nerve cells that release substances called *neurotransmitters* (see Figure 12.17 on page 425). In some cases, wastes are also released by exocytosis. During exocytosis, membrane-enclosed vesicles called *secretory vesicles* form inside the cell, fuse with the plasma membrane, and release their contents into the extracellular fluid.

Figure 3.12 Bulk-phase endocytosis. The plasma membrane folds inward, forming a vesicle.

Most body cells carry out bulk-phase endocytosis, the nonselective uptake of tiny droplets of extracellular fluid.

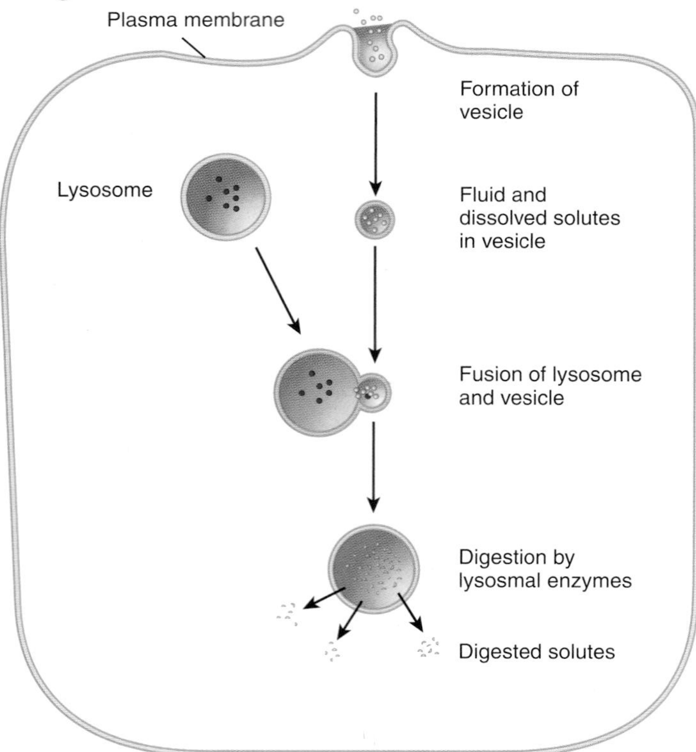

How do receptor-mediated endocytosis and phagocytosis differ from bulk-phase endocytosis?

Segments of the plasma membrane lost through endocytosis are recovered or recycled by exocytosis. The balance between endocytosis and exocytosis keeps the surface area of a cell's plasma membrane relatively constant. Membrane exchange is quite extensive in certain cells. In your pancreas, for example, the cells that secrete digestive enzymes can recycle an amount of plasma membrane equal to the cell's entire surface area in 90 minutes.

Transcytosis

Transport in vesicles may also be used to successively move a substance into, across, and out of a cell. In this active process, called **transcytosis,** vesicles undergo endocytosis on one side of a cell, move across the cell, and then undergo exocytosis on the opposite side. As the vesicles fuse with the plasma membrane, the vesicular contents are released into the extracellular fluid. Transcytosis occurs most often across the endothelial cells that line blood vessels and is a means for materials to move between blood plasma and interstitial fluid. For instance, when a woman is pregnant, some of her antibodies cross the placenta into the fetal circulation via transcytosis.

Table. 3.1 summarizes the processes by which materials move into and out of cells.

► CHECKPOINT

12. What is the key difference between passive and active transport?
13. How do symporters and antiporters carry out their functions?
14. What is the difference between primary and secondary active transport?
15. In what ways are endocytosis and exocytosis similar and different?

CYTOPLASM

► OBJECTIVE

Describe the structure and function of cytoplasm, cytosol, and organelles.

Cytoplasm consists of all the cellular contents within the plasma membrane except for the nucleus, and has two components: (1) the cytosol and (2) organelles, tiny structures that perform different functions in the cell.

Cytosol

The **cytosol (intracellular fluid)** is the fluid portion of the cytoplasm that surrounds organelles (see Figure 3.1) and constitutes about 55% of total cell volume. Although it varies in composition and consistency from one part of a cell to another, cytosol is 75–90% water plus various dissolved and suspended components. Among these are different types of ions, glucose, amino acids, fatty acids, proteins, lipids, ATP, and waste prod-

TABLE 3.1 Transport of Materials Into and Out of Cells

Transport Process	Description	Substances Transported
Kinetic Energy Transport		
Diffusion	Random mixing of molecules or ions due to their kinetic energy. A substance diffuses down a concentration gradient until it reaches equilibrium.	
Diffusion through the lipid bilayer	Passive diffusion of a substance through the lipid bilayer of the plasma membrane.	Nonpolar, hydrophobic solutes: oxygen, carbon dioxide, and nitrogen; fatty acids, steroids, and fat-soluble vitamins; glycerol, small alcohols; ammonia. Polar molecules such as water and urea.
Diffusion through membrane channels	Passive diffusion of a substance down its electrochemical gradient through channels that span a lipid bilayer; some channels are gated.	Small inorganic solutes, mainly ions: K^+, Cl^-, Na^+, and Ca^{2+}. Water.
Osmosis	Movement of water molecules across a selectively permeable membrane from an area of higher water concentration to an area of lower water concentration.	Solvent: water in living systems.
Transport by Transporter Proteins		
Facilitated Diffusion	Passive movement of a substance down its concentration gradient via transmembrane proteins that act as transporters; maximum diffusion rate is limited by number of available transporters.	Polar or charged solutes: glucose, fructose, galactose, and some vitamins.
Active Transport	Transport in which cell expends energy to move a substance across the membrane against its concentration gradient through transmembrane proteins that act as transporters; maximum transport rate is limited by number of available transporters.	Polar or charged solutes.
Primary active transport	Transport of a substance across the membrane against its concentration gradient by pumps; transmembrane proteins that use energy supplied by hydrolysis of ATP.	Na^+, K^+, Ca^{2+}, H^+, I^-, Cl^-, and other ions.
Secondary active transport	Coupled transport of two substances across the membrane using energy supplied by a Na^+ or H^+ concentration gradient maintained by primary active transport pumps. Antiporters move Na^+ (or H^+) and another substance in opposite directions across the membrane; symporters move Na^+ (or H^+) and another substance in the same direction across the membrane.	Antiport: Ca^{2+}, H^+ out of cells. Symport: glucose, amino acids into cells.
Transport in Vesicles	Movement of substances into or out of a cell in vesicles that bud from the plasma membrane; requires energy supplied by ATP.	
Endocytosis	Movement of substances into a cell in vesicles.	
Receptor-mediated endocytosis	Ligand–receptor complexes trigger infolding of a clathrin-coated pit that forms a vesicle containing ligands.	Ligands: transferrin, low-density lipoproteins (LDLs), some vitamins, certain hormones, and antibodies.
Phagocytosis	"Cell eating"; movement of a solid particle into a cell after pseudopods engulf it to form a phagosome.	Bacteria, viruses, and aged or dead cells.
Bulk-phase endocytosis	"Cell drinking"; movement of extracellular fluid into a cell by infolding of plasma membrane to form a vesicle.	Solutes in extracellular fluid.
Exocytosis	Movement of substances out of a cell in secretory vesicles that fuse with the plasma membrane and release their contents into the extracellular fluid.	Neurotransmitters, hormones, and digestive enzymes.
Transcytosis	Movement of a substance through a cell as a result of endocytosis on one side and exocytosis on the opposite side.	Substances, such as antibodies, across endothelial cells. This is a common route for substances to pass between blood plasma and interstitial fluid.

ucts, some of which we have already discussed. Also present in some cells are various organic molecules that aggregate into masses for storage. These aggregations may appear and disappear at different times in the life of a cell. Examples include *lipid droplets* that contain triglycerides, and clusters of glycogen molecules called *glycogen granules* (see Figure 3.1).

The cytosol is the site of many chemical reactions required for a cell's existence. For example, enzymes in cytosol catalyze *glycolysis,* a series of 10 chemical reactions that produces two molecules of ATP from one molecule of glucose (see Figure 25.4 on page 956). Other types of cytosolic reactions provide the building blocks for maintenance of cell structures and for cell growth.

Organelles

As noted earlier, **organelles** are specialized structures within the cell that have characteristic shapes; they perform specific functions in cellular growth, maintenance, and reproduction. Despite the many chemical reactions going on in a cell at any given time, there is little interference among reactions because they are confined to different organelles. Each type of organelle has its own set of enzymes that carry out specific reactions, and serves as a functional compartment for specific biochemical processes. The numbers and types of organelles vary in different cells, depending on the cell's function. Although they have different functions, organelles often cooperate to maintain homeostasis. Even though the nucleus is a large organelle, it is discussed in a separate section because of its special importance in directing the life of a cell.

The Cytoskeleton

The **cytoskeleton** is a network of protein filaments that extends throughout the cytosol (see Figure 3.1). Three types of filamentous proteins contribute to the cytoskeleton's structure, as well as the structure of other organelles. In the order of their increasing diameter, these structures are microfilaments, intermediate filaments, and microtubules.

Microfilaments, the thinnest elements of the cytoskeleton, are composed of the protein *actin* and are most prevalent at the periphery of a cell (Figure 3.13a). Microfilaments have two general functions: They help generate movement and provide mechanical support. With respect to movement, microfilaments are involved in muscle contraction, cell division, and cell locomotion, such as occurs during the migration of embryonic cells during development, the invasion of tissues by white blood cells to fight infection, or the migration of skin cells during wound healing.

Microfilaments provide much of the mechanical support that is responsible for the basic strength and shapes of cells. They anchor the cytoskeleton to integral proteins in the plasma membrane. Microfilaments also provide mechanical support for cell extensions called **microvilli** (*micro-* = small; *-villi* = tufts of hair), nonmotile, microscopic fingerlike projections of the plasma membrane. Within each microvillus is a core of parallel microfilaments. Because they greatly increase the surface area of

Figure 3.13 Cytoskeleton.

The cytoskeleton is a network of three types of protein filaments that extend throughout the cytoplasm: microfilaments, intermediate filaments, and microtubules.

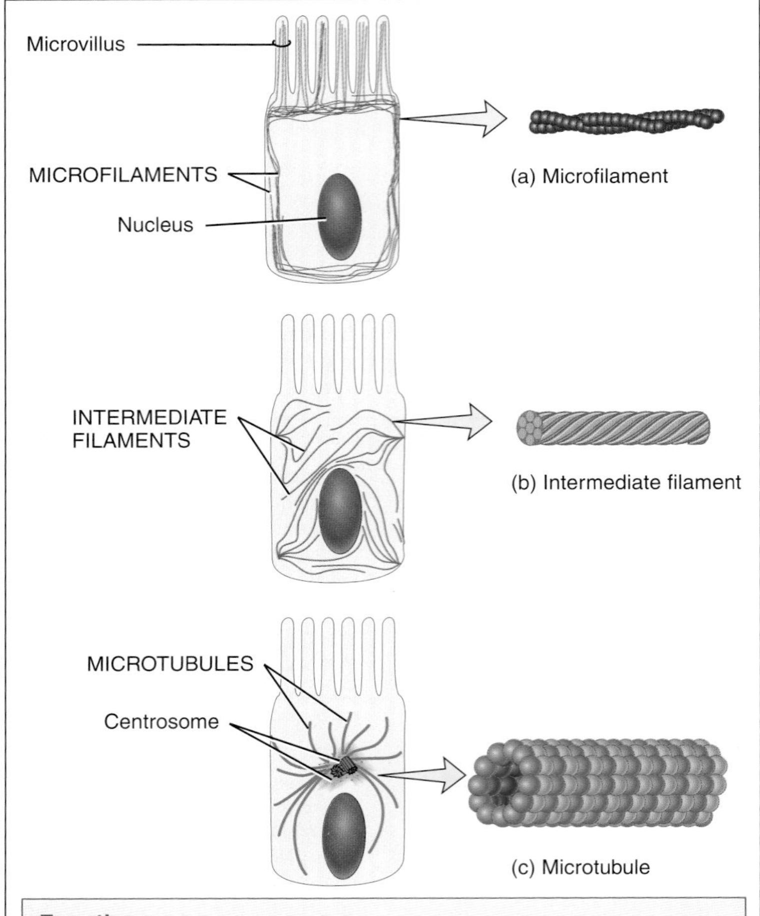

Functions

1. **Serves as a scaffold that helps to determine a cell's shape and to organize the cellular contents.**
2. **Aids movement of organelles within the cell, of chromosomes during cell division, and of whole cells such as phagocytes.**

Which cytoskeletal component helps form the structure of centrioles, cilia, and flagella?

the cell, microvilli are abundant on cells involved in absorption, such as the epithelial cells that line the small intestine.

As their name suggests, **intermediate filaments** are thicker than microfilaments but thinner than microtubules (Figure 3.13b). Several different proteins can compose intermediate filaments, which are exceptionally strong. They are found in parts of cells subject to mechanical stress, help stabilize the position of organelles such as the nucleus, and help attach cells to one another.

The largest of the cytoskeletal components, **microtubules** are long, unbranched hollow tubes composed mainly of the protein *tubulin.* The assembly of microtubules begins in an organelle called the centrosome (discussed shortly). The microtubules grow outward from the centrosome toward the periphery of the cell (Figure 3.13c). Microtubules help determine cell shape. They also function in the movement of organelles such as secretory vesicles, of chromosomes during cell division, and of specialized cell projections, such as cilia and flagella.

Centrosome

The **centrosome,** located near the nucleus, consists of two components: a pair of centrioles and pericentriolar material (Figure 3.14a). The two **centrioles** are cylindrical structures, each composed of nine clusters of three microtubules (triplets) arranged in a circular pattern (Figure 3.14b). The long axis of one centriole is at a right angle to the long axis of the other (Figure 3.14c). Surrounding the centrioles is **pericentriolar material** (per′-ē-sen′-trē-Ō-lar), which contains hundreds of ring-shaped complexes composed of the protein *tubulin.* These tubulin complexes are the organizing centers for growth of the mitotic spindle, which plays a critical role in cell division, and for microtubule formation in nondividing cells. During cell division, centrosomes replicate so that succeeding generations of cells have the capacity for cell division.

Cilia and Flagella

Microtubules are the dominant components of cilia and flagella, both of which are motile projections of the cell surface (Figure 3.15). **Cilia** (SIL-ē-a = eyelashes; singular is *cilium*) are numerous, short, hairlike projections that extend from the surface of the cell (see Figure 3.1). Each cilium contains a core of 20 microtubules surrounded by plasma membrane (Figure 3.15a). The microtubules are arranged such that one pair in the center is surrounded by nine clusters of two fused microtubules (doublets). Each cilium is anchored to a *basal body* just below the surface of the plasma membrane. A basal body is similar in structure to a centriole and functions in initiating the assembly of cilia and flagella.

A cilium displays an oarlike pattern of beating; it is relatively stiff during the power stroke (oar digging into the water), but more flexible during the recovery stroke (oar moving above the water preparing for a new stroke) (Figure 3.15b). The coordinated movement of many cilia on the surface of a cell causes the steady movement of fluid along the cell's surface. Many cells of the respiratory tract, for example, have hundreds of cilia that help sweep foreign particles trapped in mucus away from the lungs. In cystic fibrosis, the extremely thick mucus secretions that are produced interfere with ciliary action and the normal functions of the respiratory tract. The movement of cilia is also paralyzed by nicotine in cigarette smoke. For this reason, smokers cough often to remove foreign particles from their airways. Cells that line the uterine (fallopian) tubes also have cilia that sweep oocytes (egg cells) toward the uterus, and females who smoke have an increased risk of ectopic (outside the uterus) pregnancy.

Flagella (fla-JEL-a = whip; singular is *flagellum*) are similar in structure to cilia but are typically much longer. Flagella usually move an entire cell. A flagellum generates forward motion along its axis by rapidly wiggling in a wavelike pattern (Figure 3.15c). The only example of a flagellum in the human body is a sperm cell's tail, which propels the sperm toward its rendezvous with an oocyte.

Figure 3.14 Centrosome.

Located near the nucleus, the centrosome consists of a pair of centrioles and pericentriolar material.

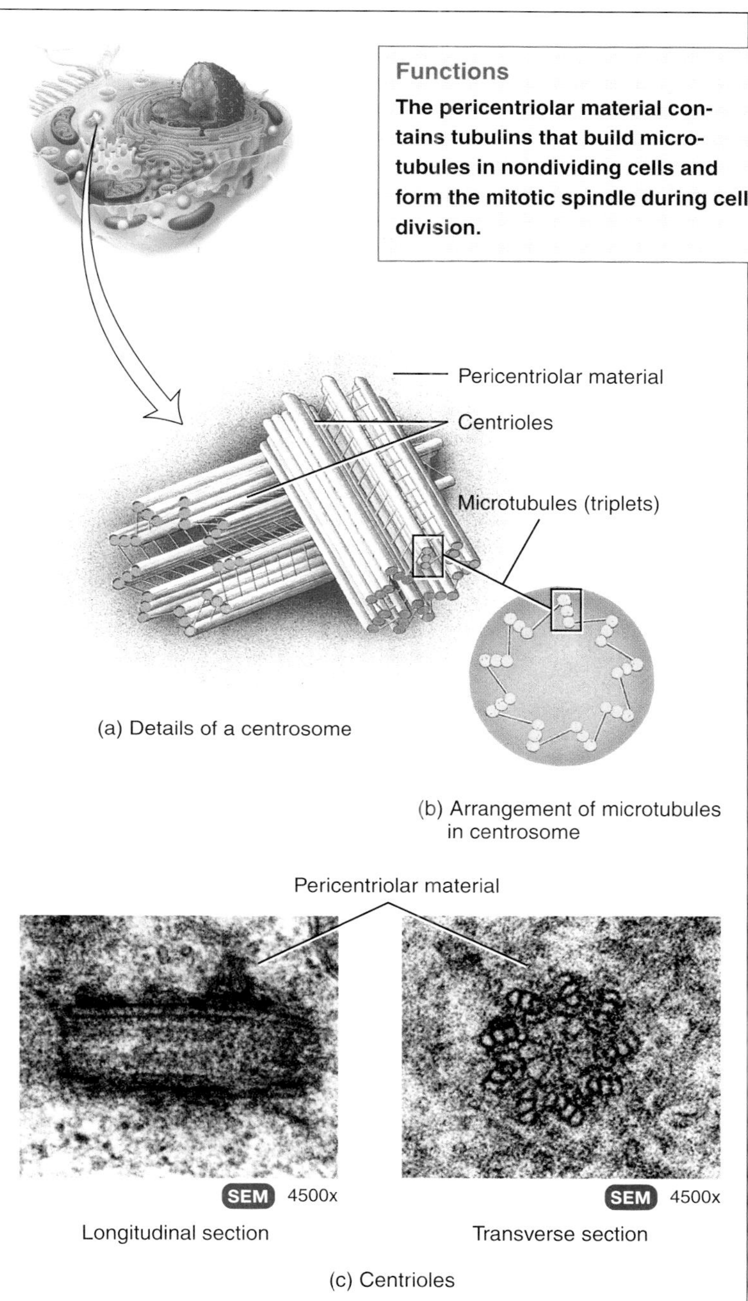

If you observed that a cell did not have a centrosome, what could you predict about its capacity for cell division?

Ribosomes

Ribosomes (RĪ-bō-sōms; *-somes* = bodies) are the sites of protein synthesis. The name of these tiny organelles reflects their high content of one type of ribonucleic acid, **ribosomal RNA (rRNA),**

but each one also includes more than 50 proteins. Structurally, a ribosome consists of two subunits, one about half the size of the other (Figure 3.16). The large and small subunits are made separately in the nucleolus, a spherical body inside the nucleus. Once produced, the large and small subunits exit the nucleus separately, then come together in the cytoplasm.

Figure 3.15 Cilia and flagella.

A cilium contains a core of microtubules with one pair in the center surrounded by nine clusters of doublet microtubules.

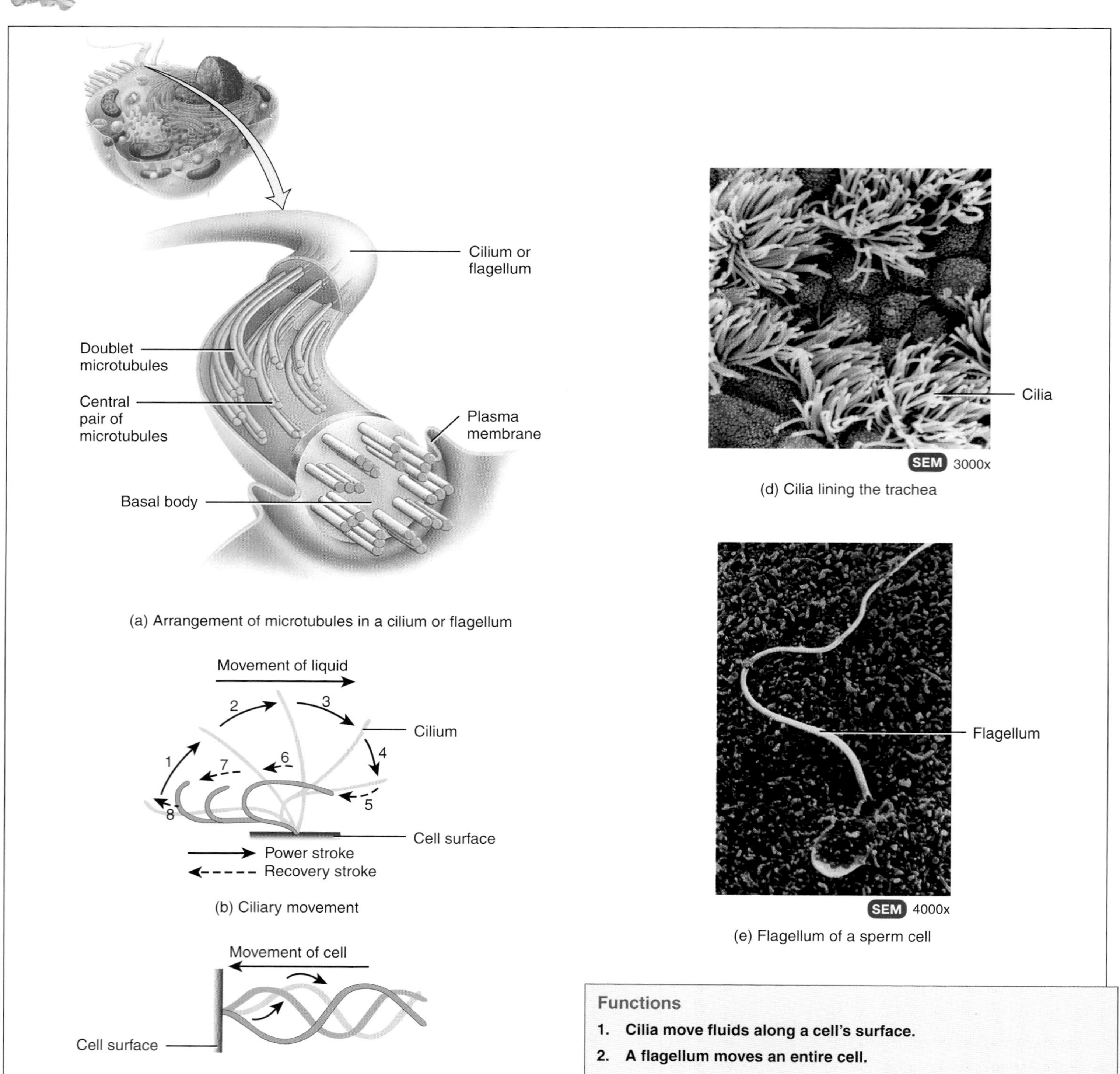

Functions

1. **Cilia move fluids along a cell's surface.**
2. **A flagellum moves an entire cell.**

What is the functional difference between cilia and flagella?

Some ribosomes are attached to the outer surface of the nuclear membrane and to an extensively folded membrane called the endoplasmic reticulum. These ribosomes synthesize proteins destined for specific organelles, for insertion in the plasma membrane, or for export from the cell. Other ribosomes are "free" or unattached to other cytoplasmic structures. Free ribosomes synthesize proteins used in the cytosol. Ribosomes are also located within mitochondria, where they synthesize mitochondrial proteins.

Endoplasmic Reticulum

The **endoplasmic reticulum** (en′-dō-PLAS-mik re-TIK-ū-lum; *-plasmic* = cytoplasm; *reticulum* = network) or **ER** is a network of membranes in the form of flattened sacs or tubules (Figure 3.17). The ER extends from the nuclear envelope (membrane around the nucleus), to which it is connected, throughout the cytoplasm. The ER is so extensive that it constitutes more than half of the membranous surfaces within the cytoplasm of most cells.

Cells contain two distinct forms of ER, which differ in structure and function. **Rough ER** is continuous with the nuclear membrane and usually is folded into a series of flattened sacs. The outer surface of rough ER is studded with ribosomes, the sites of protein synthesis. Proteins synthesized by ribosomes attached to rough ER enter spaces within the ER for processing

Figure 3.16 **Ribosomes.**

Ribosomes are the sites of protein synthesis.

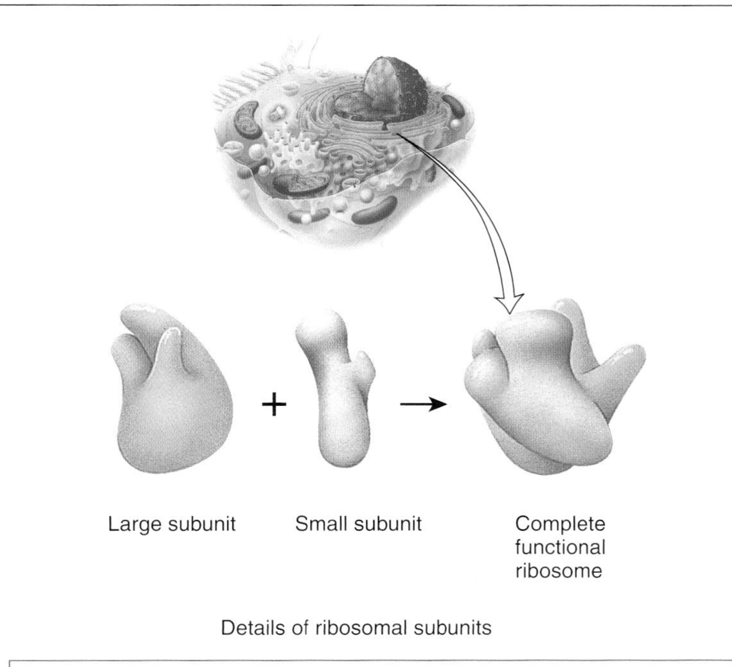

Functions

1. **Ribosomes associated with endoplasmic reticulum synthesize proteins destined for insertion in the plasma membrane or secretion from the cell.**
2. **Free ribosomes synthesize proteins used in the cytosol.**

Where are subunits of ribosomes synthesized and assembled?

Figure 3.17 **Endoplasmic reticulum.**

The endoplasmic reticulum is a network of membrane-enclosed sacs or tubules that extend throughout the cytoplasm and connect to the nuclear envelope.

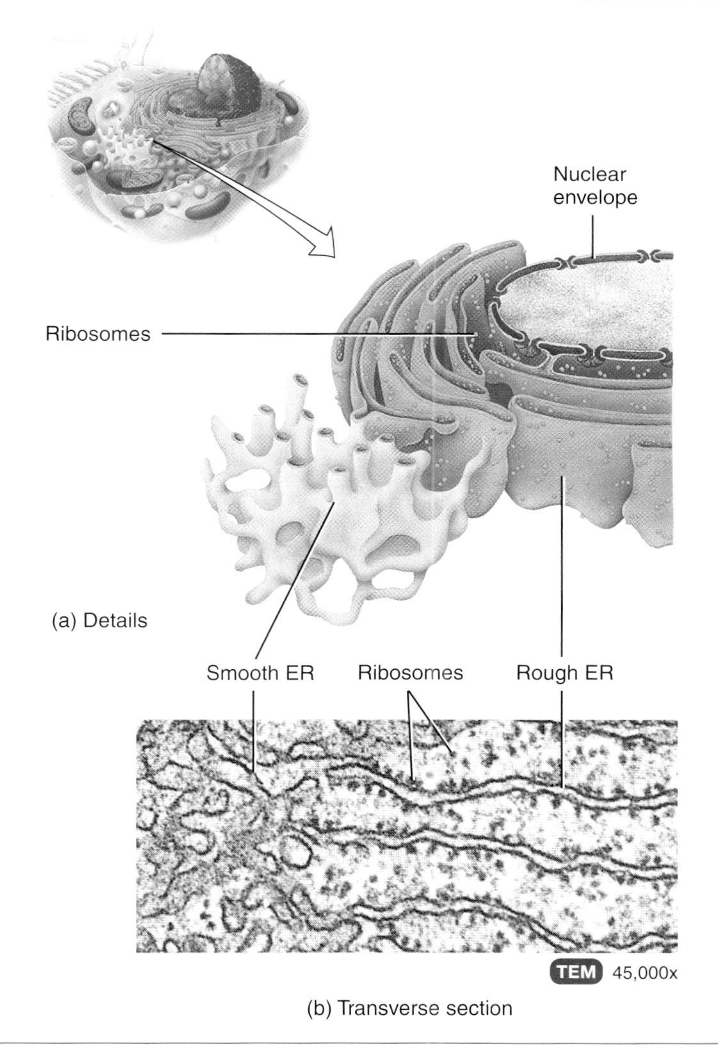

Functions

1. **Rough ER synthesizes glycoproteins and phospholipids that are transferred into cellular organelles, inserted into the plasma membrane, or secreted during exocytosis.**
2. **Smooth ER synthesizes fatty acids and steroids, such as estrogens and testosterone; inactivates or detoxifies drugs and other potentially harmful substances; removes the phosphate group from glucose-6-phosphate; and stores and releases calcium ions that trigger contraction in muscle cells.**

What are the structural and functional differences between rough and smooth ER?

and sorting. In some cases, enzymes attach the proteins to carbohydrates to form glycoproteins. In other cases, enzymes attach the proteins to phospholipids, also synthesized by rough ER. These molecules may be incorporated into the membranes of organelles, inserted into the plasma membrane or secreted via exocytosis. Thus rough ER produces secretory proteins, membrane proteins, and many organellar proteins.

Smooth ER extends from the rough ER to form a network of membrane tubules (Figure 3.17). Unlike rough ER, smooth ER does not have ribosomes on the outer surfaces of its membrane. However, smooth ER contains unique enzymes that make it functionally more diverse than rough ER. Because it lacks ribosomes, smooth ER does not synthesize proteins, but it does synthesize fatty acids and steroids, such as estrogens and testosterone. In liver cells, enzymes of the smooth ER help release glucose into the bloodstream and inactivate or detoxify lipid-soluble drugs or potentially harmful substances, such as alcohol, pesticides, and *carcinogens* (cancer-causing agents). In liver, kidney, and intestinal cells a smooth ER enzyme removes the phosphate group from glucose-6-phosphate, which allows the "free" glucose to enter the bloodstream. In muscle cells, the calcium ions that trigger contraction are released from the sarcoplasmic reticulum, a form of smooth ER.

Smooth ER and Drug Tolerance

One of the functions of smooth ER, as noted earlier, is to detoxify certain drugs. Individuals who repeatedly take such drugs, such as the sedative phenobarbital, develop changes in the smooth ER in their liver cells. Prolonged administration of phenobarbital results in increased tolerance to the drug; the same dose no longer produces the same degree of sedation. With repeated exposure to the drug, the amount of smooth ER and its enzymes increases to protect the cell from its toxic effects. As the amount of smooth ER increases, higher and higher dosages of the drug are needed to achieve the original effect. ■

Golgi Complex

Most of the proteins synthesized by ribosomes attached to rough ER are ultimately transported to other regions of the cell. The first step in the transport pathway is through an organelle called the **Golgi complex** (GOL-jē). It consists of 3 to 20 **cisternae** (sis-TER-nē = cavities; singular is *cisterna*), small, flattened membranous sacs with bulging edges that resemble a stack of pita bread (Figure 3.18). The cisternae are often curved, giving the Golgi complex a cuplike shape. Most cells have several Golgi complexes, and Golgi complexes are more extensive in cells that secrete proteins, a clue to the organelle's role in the cell.

Figure 3.18 Golgi complex.

The opposite faces of a Golgi complex differ in size, shape, content, and enzymatic activities.

Functions

1. **Modifies, sorts, packages, and transports proteins received from the rough ER.**
2. **Forms secretory vesicles that discharge processed proteins via exocytosis into extracellular fluid; forms membrane vesicles that ferry new molecules to the plasma membrane; forms transport vesicles that carry molecules to other organelles, such as lysosomes.**

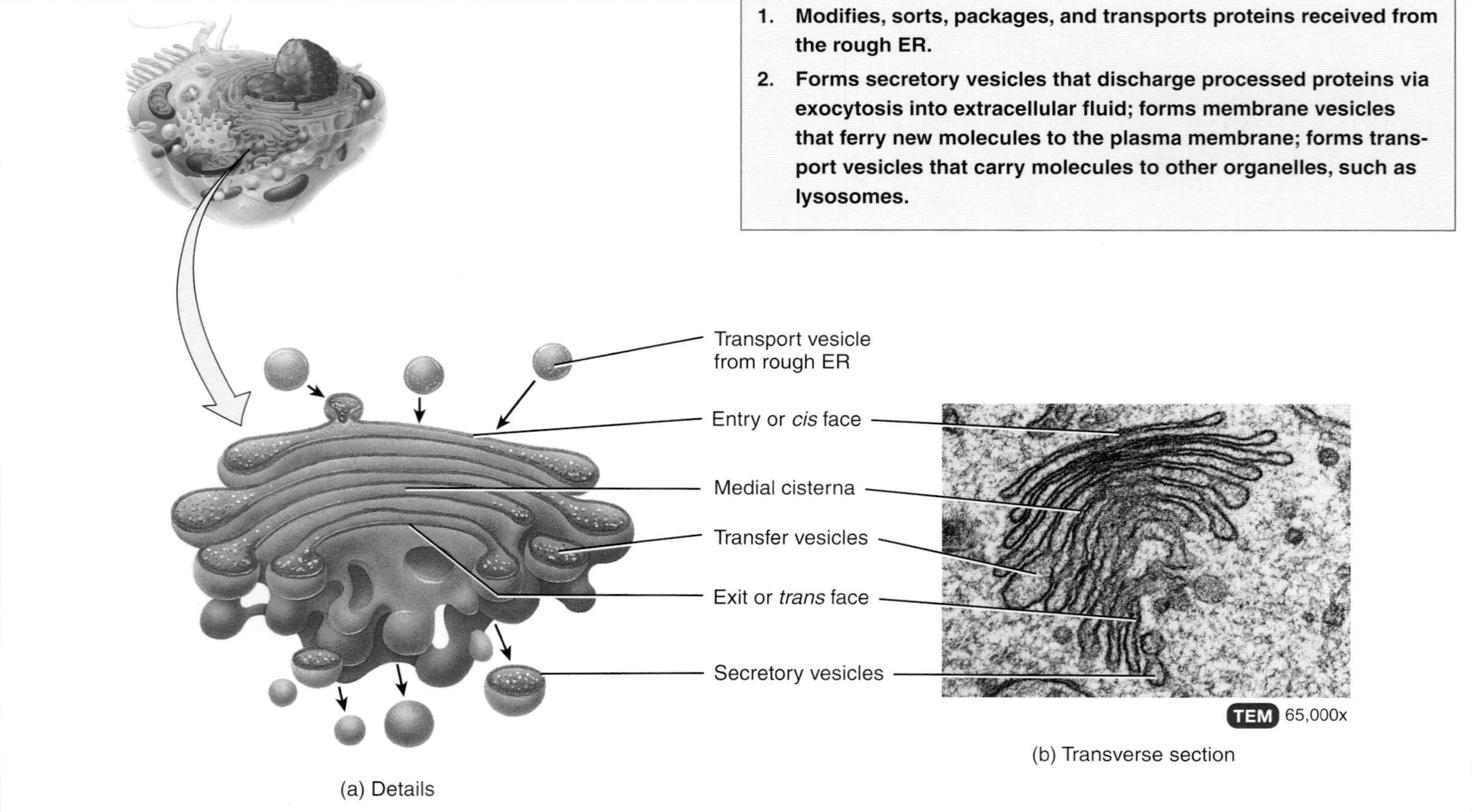

(a) Details

(b) Transverse section

How do the entry and exit faces differ in function?

The cisternae at the opposite ends of a Golgi complex differ from each other in size, shape, and enzymatic activity. The convex **entry** or ***cis* face** is a cisterna that faces the rough ER. The concave **exit** or ***trans* face** is a cisterna that faces the plasma membrane. Sacs between the entry and exit faces are called **medial cisternae.** Transport vesicles (described shortly) from the ER merge to form the entry face. From the entry face, the cisternae are thought to mature, in turn becoming medial and then exit cisternae.

Different enzymes in the entry, medial, and exit regions of the Golgi complex permit each of these areas to modify, sort, and package proteins for transport to different destinations. The entry face receives and modifies proteins produced by the rough ER. The medial cisternae add carbohydrates to proteins to form glycoproteins and lipids to proteins to form lipoproteins. The exit face modifies the molecules further and then sorts and packages them for transport to their destinations.

Proteins arriving at, passing through, and exiting the Golgi complex do so through maturation of the cisternae and exchanges that occur via transfer vesicles (Figure 3.19):

1. Proteins synthesized by ribosomes on the rough ER are surrounded by a piece of the ER membrane, which eventually buds from the membrane surface to form **transport vesicles.**
2. Transport vesicles move toward the entry face of the Golgi complex.
3. Fusion of several transport vesicles creates the entry face of the Golgi complex and releases proteins into its lumen (space).
4. The proteins move from the entry face into one or more medial cisternae. Enzymes in the medial cisternae modify the proteins to form glycoproteins, glycolipids, and lipoproteins. **Transfer vesicles** that bud from the edges of the cisternae move specific enzymes back toward the entry face and move some partially modified proteins toward the exit face.
5. The products of the medial cisternae move into the lumen of the exit face.
6. Within the exit face cisterna, the products are further modified and are sorted and packaged.
7. Some of the processed proteins leave the exit face and are stored in **secretory vesicles.** These vesicles deliver the proteins to the plasma membrane, where they are discharged by exocytosis into the extracellular fluid. For example, certain pancreatic cells release the hormone insulin in this way.
8. Other processed proteins leave the exit face in **membrane vesicles** that deliver their contents to the plasma membrane for incorporation into the membrane. In doing so, the Golgi complex adds new segments of plasma membrane as existing segments are lost and modifies the number and distribution of membrane molecules.

Figure 3.19 Processing and packaging of proteins by the Golgi complex.

All proteins exported from the cell are processed in the Golgi complex.

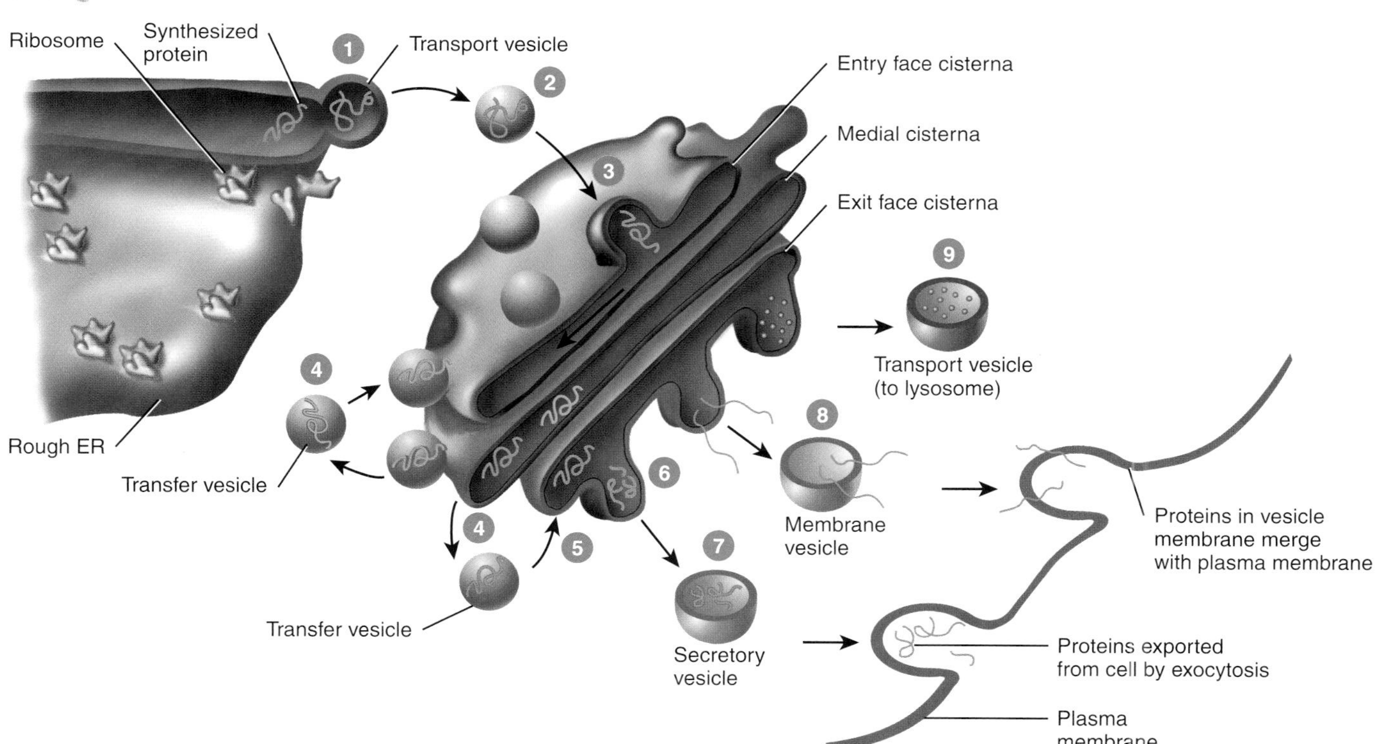

What are the three general destinations for proteins that leave the Golgi complex?

9 Finally, some processed proteins leave the exit face in transport vesicles that will carry the proteins to another cellular destination. For instance, transport vesicles ferry digestive enzymes to lysosomes; the structure and functions of these important organelles are discussed next.

Lysosomes

Lysosomes (LĪ-sō-sōms; *lyso-* = dissolving; *-somes* = bodies) are membrane-enclosed vesicles that form from the Golgi complex (Figure 3.20). Inside, as many as 60 kinds of powerful digestive and hydrolytic enzymes can break down a wide variety of molecules once lysosomes fuse with vesicles formed during endocytosis. Because lysosomal enzymes work best at an acidic pH, the lysosomal membrane includes active transport pumps that import hydrogen ions (H^+). Thus, the lysosomal interior has a pH of 5, which is 100 times more acidic than the pH of the cytosol (pH 7). The lysosomal membrane also includes transporters that move the final products of digestion, such as glucose, fatty acids, and amino acids, into the cytosol.

Lysosomal enzymes also help recycle worn-out cell structures. A lysosome can engulf another organelle, digest it, and return the digested components to the cytosol for reuse, a process known as **autophagy** (aw-TOF-a-jē; *auto-* = self; *-phagy* = eating). During autophagy, the organelle to be digested is enclosed by a membrane derived from the ER to create a vesicle called an *autophagosome,* which then fuses with a lysosome. In this way, a human liver cell recycles about half of its cytoplasmic contents every week. Lysosomal enzymes may also destroy the entire cell, a process known as **autolysis** (aw-TOL-i-sis). Autolysis occurs in some pathological conditions and is also responsible for the tissue deterioration that occurs just after death.

As we just discussed, most lysosomal enzymes act within a cell. However, some operate in extracellular digestion. One example occurs during fertilization. The head of a sperm cell releases lysosomal enzymes that aid its penetration of the oocyte by dissolving its protective coating in a process called the acrosomal reaction (see page 1127).

Tay-Sachs Disease

Some disorders are caused by faulty or absent lysosomal enzymes. For instance, **Tay-Sachs disease,** which most often affects children of Ashkenazi (eastern European Jewish) descent, is an inherited condition characterized by the absence of a single lysosomal enzyme called Hex A. This enzyme normally breaks down a membrane glycolipid called ganglioside G_{M2} that is especially prevalent in nerve cells. As the excess ganglioside G_{M2} accumulates, the nerve cells function less efficiently. Children with Tay-Sachs disease typically experience seizures and muscle rigidity. They gradually become blind, demented, and uncoordinated and usually die before the age of 5. Tests can now reveal whether an adult is a carrier of the defective gene. ■

Figure 3.20 Lysosomes.

Lysosomes contain several types of powerful digestive enzymes.

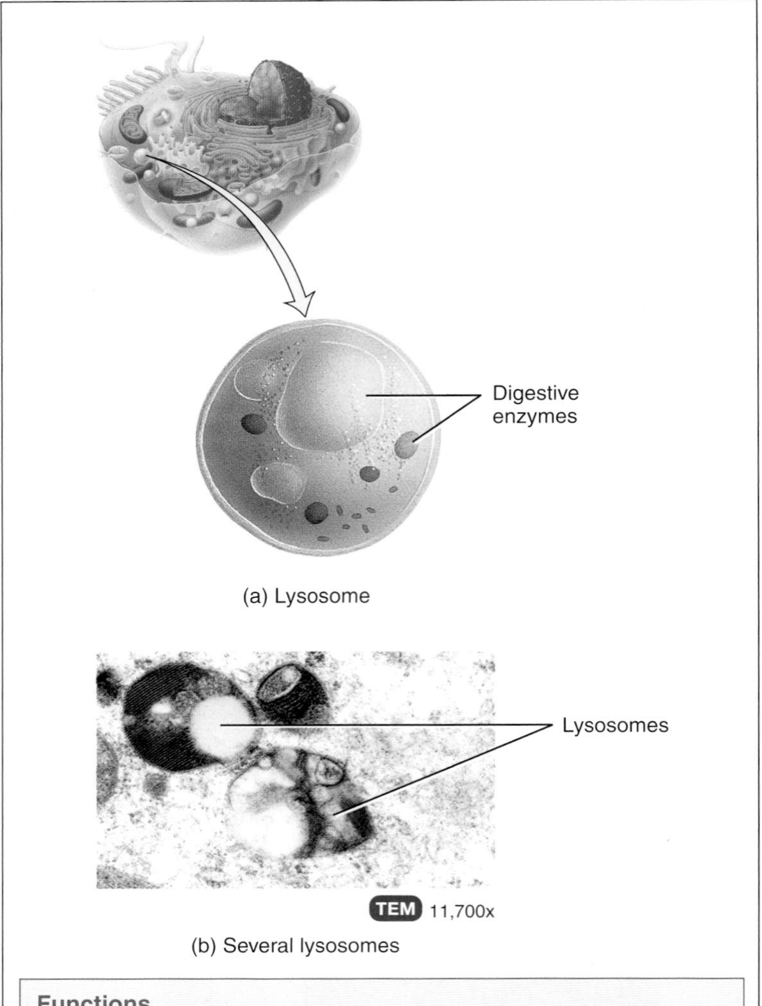

Functions

1. **Digest substances that enter a cell via endocytosis and transport final products of digestion into cytosol.**
2. **Carry out autophagy, the digestion of worn-out organelles.**
3. **Carry out autolysis, the digestion of entire cell.**
4. **Carry out extracellular digestion.**

What is the name of the process by which worn-out organelles are digested by lysosomes?

Peroxisomes

Another group of organelles similar in structure to lysosomes, but smaller, are the **peroxisomes** (pe-ROKS-i-sōms; *peroxi-* = peroxide; *somes* = bodies; see Figure 3.1). Peroxisomes contain several *oxidases,* enzymes that can oxidize (remove hydrogen atoms from) various organic substances. For instance, amino acids and fatty acids are oxidized in peroxisomes as part of normal metabolism. In addition, enzymes in peroxisomes oxidize toxic substances, such as alcohol. Thus, peroxisomes are very abundant in the liver, where detoxification of alcohol and other damaging substances occurs. A byproduct of the oxidation reactions is hydrogen peroxide (H_2O_2), a potentially toxic compound. However, peroxisomes also contain the enzyme *catalase,* which decomposes H_2O_2. Because production and degradation of H_2O_2 occurs within the same organelle, peroxisomes protect other parts of the cell from the toxic effects of H_2O_2. New peroxisomes form from preexisting ones.

Proteasomes

As you have just learned, lysosomes degrade proteins delivered to them in vesicles. Cytosolic proteins also require disposal at certain times in the life of a cell. Continuous destruction of unneeded, damaged, or faulty proteins is the function of tiny barrel-shaped structures called **proteasomes** (PRŌ-tē-a-sōmes = bodies). For example, proteins that are part of metabolic pathways are degraded after they have accomplished their function. Such protein destruction plays a part in negative feedback by halting a pathway once the appropriate response has been achieved. A typical body cell contains many thousands of proteasomes, in both the cytosol and the nucleus. Discovered only recently because they are far too small to discern under the light microscope and do not show up well in electron micrographs, proteasomes were so-named because they contain myriad *proteases,* enzymes that cut proteins into small peptides. Once the enzymes of a proteasome have chopped up a protein into smaller chunks, other enzymes then break down the peptides into amino acids, which can be recycled into new proteins.

Some diseases could result from failure of proteasomes to degrade abnormal proteins. For example, clumps of misfolded proteins accumulate in brain cells of people with Parkinson disease and Alzheimer disease. Discovering why the proteasomes fail to clear these abnormal proteins is a goal of ongoing research.

Mitochondria

Because they generate most of the ATP through aerobic (oxygen-requiring) respiration, **mitochondria** (mī-tō-KON-drē-a; *mito-* = thread; *-chondria* = granules; singular is *mitochondrion*) are referred to as the "powerhouses" of the cell. A cell may have as few as a hundred or as many as several thousand mitochondria, depending on how active the cell is. Active cells, such as those found in the muscles, liver, and kidneys, which use ATP at a high rate, have a large number of mitochondria. Mitochondria are usually located within the cell where oxygen enters the cell or where the ATP is used, for example, among the contractile proteins in muscle cells.

A mitochondrion consists of an **outer mitochondrial membrane** and an **inner mitochondrial membrane** with a small fluid-filled space between them (Figure 3.21). Both membranes are similar in structure to the plasma membrane. The inner mito-

Figure 3.21 Mitochondria.

Within mitochondria, chemical reactions of aerobic cellular respiration generate ATP.

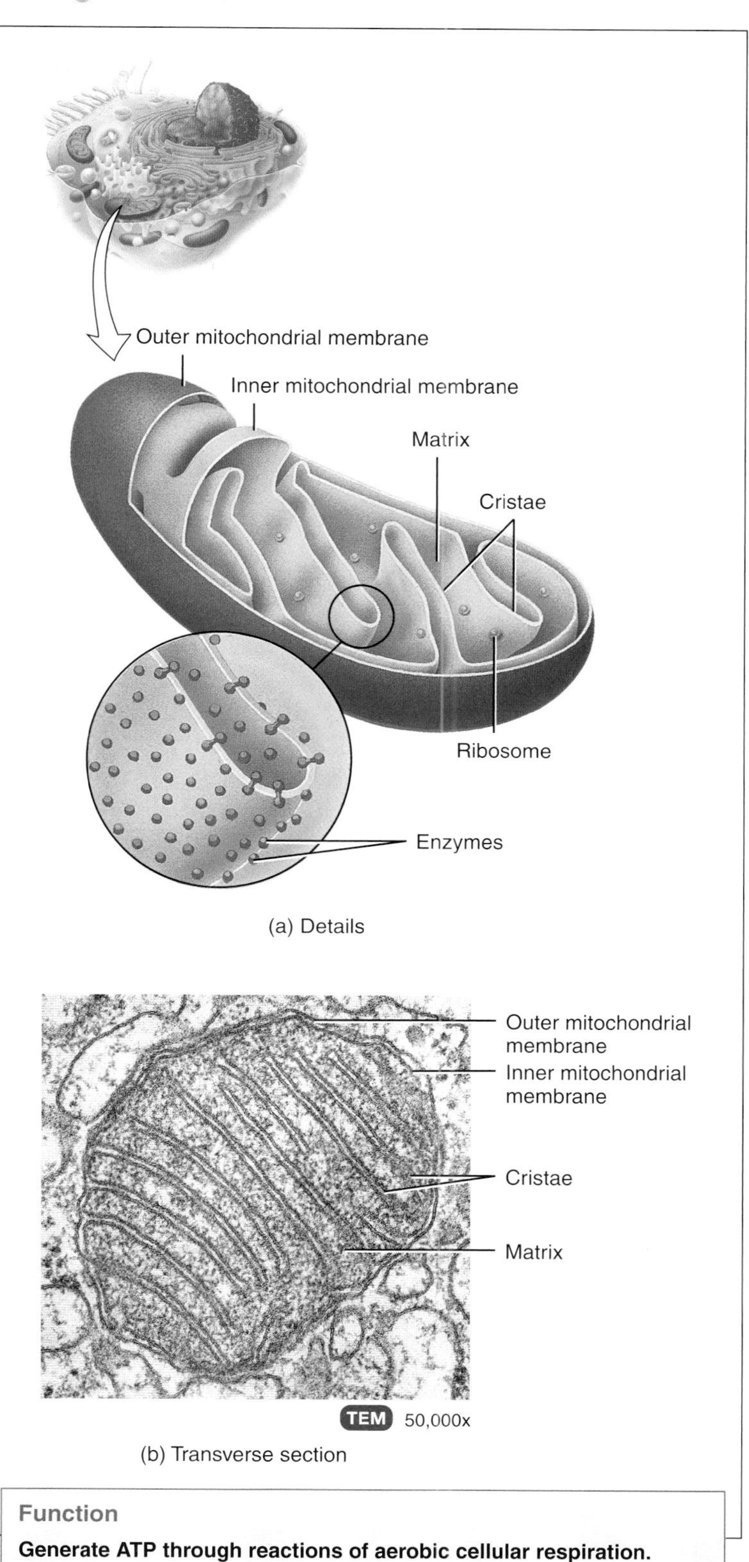

Function

Generate ATP through reactions of aerobic cellular respiration.

How do the cristae of a mitochondrion contribute to its ATP-producing function?

chondrial membrane contains a series of folds called **cristae** (KRIS-tē = ridges). The large central fluid-filled cavity of a mitochondrion, enclosed by the inner mitochondrial membrane, is the **matrix.** The elaborate folds of the cristae provide an enormous surface area for the chemical reactions that are part of the aerobic phase of *cellular respiration,* the reactions that produce most of a cell's ATP (see Chapter 25). The enzymes that catalyze these reactions are located on the cristae and in the matrix of the mitochondria.

Like peroxisomes, mitochondria self-replicate, a process that occurs during times of increased cellular energy demand or before cell division. Synthesis of some of the proteins needed for mitochondrial functions occurs on the ribosomes that are present in the mitochondria matrix. Mitochondria even have their own DNA, in the form of multiple copies of a circular DNA molecule that contains 37 genes. These mitochondrial genes control the synthesis of 2 ribosomal RNAs, 22 transfer RNAs, and 13 proteins that build mitochondrial components.

Although the nucleus of each somatic cell contains genes from both your mother and father, mitochondrial genes are inherited only from your mother. The head of a sperm (the part that penetrates and fertilizes an oocyte) normally lacks most

Figure 3.22 Nucleus.

The nucleus contains most of the cell's genes, which are located on chromosomes.

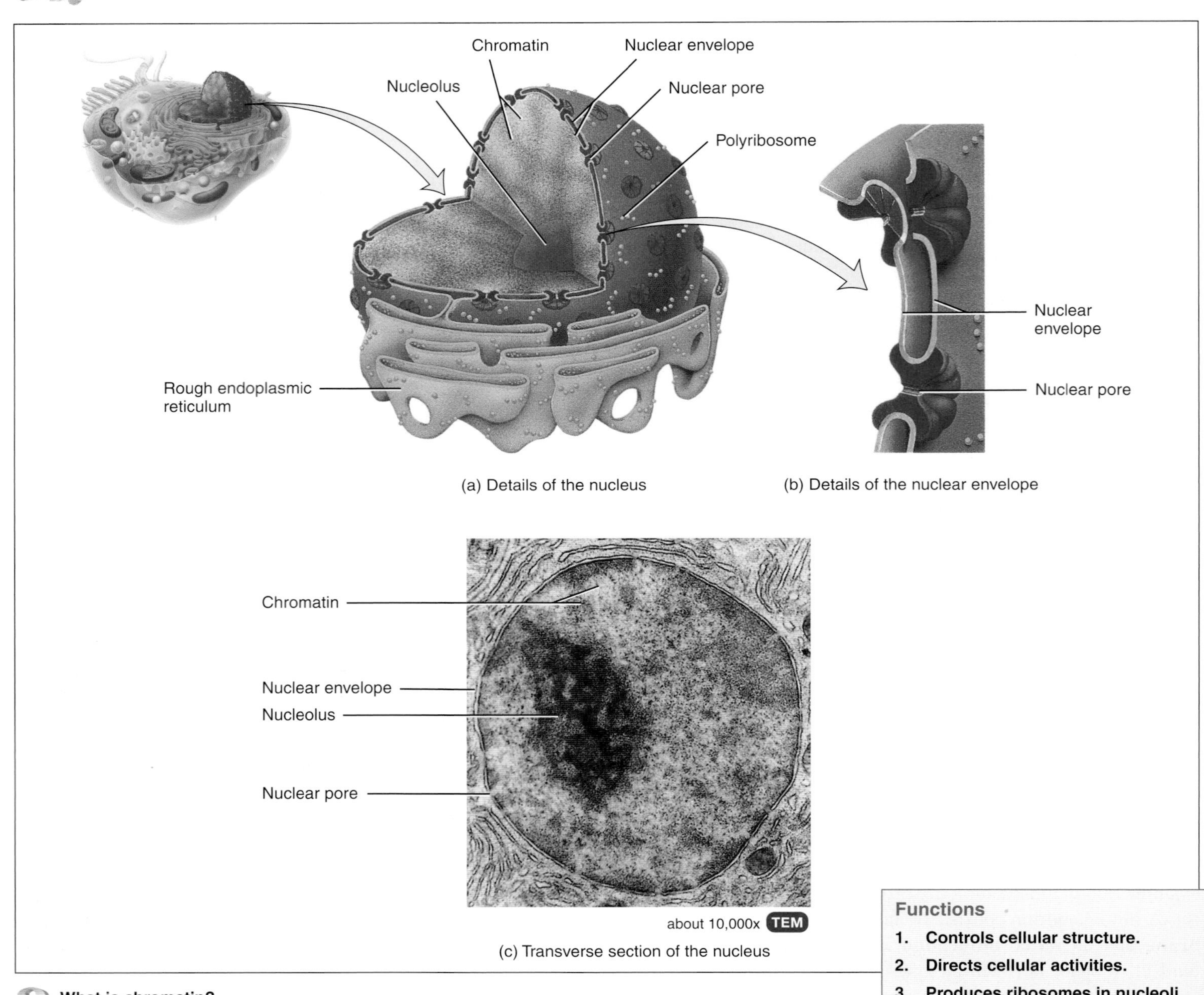

Functions

1. **Controls cellular structure.**
2. **Directs cellular activities.**
3. **Produces ribosomes in nucleoli.**

What is chromatin?

organelles, such as mitochondria, ribosomes, endoplasmic reticulum, and the Golgi complex, and any sperm mitochondria that do enter the oocyte are soon destroyed.

► CHECKPOINT

16. What does cytoplasm have that cytosol lacks?
17. Which organelles are surrounded by a membrane and which are not?
18. Which organelles contribute to synthesizing protein hormones and packaging them into secretory vesicles?
19. What happens on the cristae and in the matrix of mitochondria?

NUCLEUS

► OBJECTIVE

Describe the structure and function of the nucleus.

The **nucleus** is a spherical or oval-shaped structure that usually is the most prominent feature of a cell (Figure 3.22 on page 84). Most cells have a single nucleus, although some, such as mature red blood cells, have none. In contrast, skeletal muscle cells and a few other types of cells have multiple nuclei. A double membrane called the **nuclear envelope** separates the nucleus from the cytoplasm. Both layers of the nuclear envelope are lipid bilayers similar to the plasma membrane. The outer membrane of the nuclear envelope is continuous with rough ER and resembles it in structure. Many openings called **nuclear pores** extend through the nuclear envelope. Each nuclear pore consists of a circular arrangement of proteins surrounding a large central opening that is about 10 times wider than the pore of a channel protein in the plasma membrane.

Nuclear pores control the movement of substances between the nucleus and the cytoplasm. Small molecules and ions move through the pores passively by diffusion. Most large molecules, such as RNAs and proteins, cannot pass through the nuclear pores by diffusion. Instead, their passage involves an active transport process in which the molecules are recognized and selectively transported through the nuclear pore into or out of the nucleus. For example, proteins needed for nuclear functions move from the cytosol into the nucleus; newly formed RNA molecules move from the nucleus into the cytosol in this manner.

Inside the nucleus are one or more spherical bodies called **nucleoli** (noo′-KLĒ-ō-lī; singular is *nucleolus*) that function in producing ribosomes. Each nucleolus is simply a cluster of protein, DNA, and RNA; it is not enclosed by a membrane. Nucleoli are the sites of synthesis of rRNA and assembly of rRNA and proteins into ribosomal subunits. Nucleoli are quite prominent in cells that synthesize large amounts of protein, such as muscle and liver cells. Nucleoli disperse and disappear during cell division and reorganize once new cells are formed.

Figure 3.23 Packing of DNA into a chromosome in a dividing cell. When packing is complete, two identical DNA molecules and their histones form a pair of chromatids, which are held together by a centromere.

A chromosome is a highly coiled and folded DNA molecule that is combined with protein molecules.

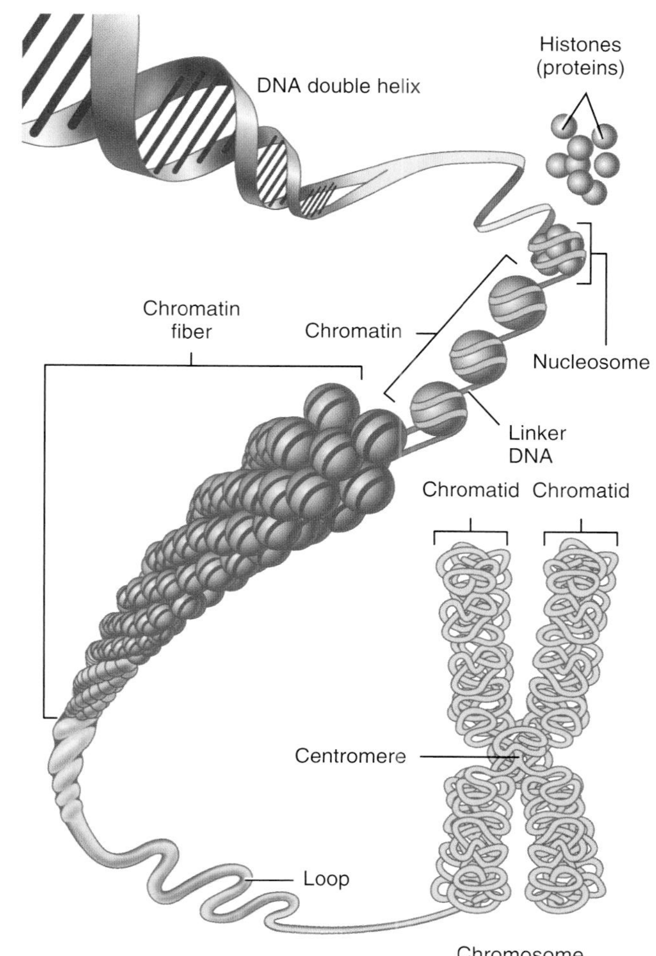

What are the components of a nucleosome?

Within the nucleus are most of the cell's hereditary units, called **genes,** which control cellular structure and direct cellular activities. Genes are arranged along **chromosomes** (*chromo-* = colored). Human somatic (body) cells have 46 chromosomes, 23 inherited from each parent. Each chromosome is a long molecule of DNA that is coiled together with several proteins (Figure 3.23). This complex of DNA, proteins, and some RNA is called **chromatin.** The total genetic information carried in a cell or an organism is its **genome.**

In cells that are not dividing, the chromatin appears as a diffuse, granular mass. Electron micrographs reveal that chromatin has a beads-on-a-string structure. Each bead is a **nucleosome** and consists of double-stranded DNA wrapped twice around a core of eight proteins called **histones,** which help organize the coiling and folding of DNA. The string between the beads is **linker DNA,** which holds adjacent nucleosomes together. In cells that are not dividing, another histone promotes coiling of nucleosomes into a larger-diameter **chromatin fiber,** which then folds into large loops. Just before cell division takes place, however, the DNA replicates (duplicates) and the loops condense even more, forming a pair of **chromatids.** As you will see shortly, during cell division a pair of chromatids constitutes a chromosome.

The main parts of a cell and their functions are summarized in Table 3.2.

Genomics

In the last decade of the twentieth century, the genomes of humans, mice, fruit flies, and more than 50 microbes were sequenced. As a result, research in the field of **genomics,** the study of the relationships between the genome and the biological functions of an organism, has flourished. The Human Genome Project began in June 1990 as an effort to sequence all of the nearly 3.2 billion nucleotides of our genome, and was completed in April 2003. More than 99.9% of the nucleotide bases are identical in everyone. Less than 0.1% of our DNA (1 in each 1000 bases) accounts for inherited differences among humans. Surprisingly, at least half of the human genome consists of repeated sequences that do not code for proteins, so-called "junk" DNA. The average gene consists of 3000 nucleotides, but sizes vary greatly. The largest known human gene, with 2.4 million nucleotides, codes for the protein dystrophin. Scientists now know that the total number of genes in the human genome is about 30,000, far fewer than the 100,000 previously predicted to exist. Information regarding the human genome and how it is affected by the environment seeks to identify and discover the functions of the specific genes that play a role in genetic diseases. Genomic medicine also aims to design new drugs and to provide screening tests to enable physicians to provide more effective counseling and treatment for disorders with significant genetic components such as hypertension (high blood pressure), obesity, diabetes, and cancer. ■

▶ CHECKPOINT

20. How do large particles enter and exit the nucleus?
21. Where is RNA produced?
22. How is DNA packed in the nucleus?

PROTEIN SYNTHESIS

▶ OBJECTIVE

Describe the sequence of events in protein synthesis.

Although cells synthesize many chemicals to maintain homeostasis, much of the cellular machinery is devoted to synthesizing large numbers of diverse proteins. The proteins, in turn, determine the physical and chemical characteristics of cells and, therefore, of the organisms formed from them. Some proteins help assemble cellular structures such as the plasma membrane, the cytoskeleton, and other organelles. Others serve as hormones, antibodies, and contractile elements in muscular tissue. Still others act as enzymes, regulating the rates of the numerous chemical reactions that occur in cells, or transporters, carrying various materials in the blood. Just as genome means all of the genes in an organism, **proteome** (PRŌ-te-ōm) refers to all of an organism's proteins.

In the process called **gene expression,** a gene's DNA is used as a template for synthesis of a specific protein. First, in a process aptly named transcription, the information encoded in a specific region of DNA is *transcribed* (copied) to produce a specific molecule of RNA (ribonucleic acid). In a second process, referred to as translation, the RNA attaches to a ribosome, where the information contained in RNA is *translated* into a corresponding sequence of amino acids to form a new protein molecule (Figure 3.24).

DNA and RNA store genetic information as sets of three nucleotides. A sequence of three such nucleotides in DNA is called a **base triplet.** Each DNA base triplet is transcribed as a complementary sequence of three nucleotides, called a **codon.** A given codon specifies a particular amino acid. The **genetic code** is the set of rules that relate the base triplet sequence of DNA to the corresponding codons of RNA and the amino acids they specify.

Figure 3.24 Overview of gene expression. Synthesis of a specific protein requires transcription of a gene's DNA into RNA and translation of RNA into a corresponding sequence of amino acids.

Transcription occurs in the nucleus; translation occurs in the cytoplasm.

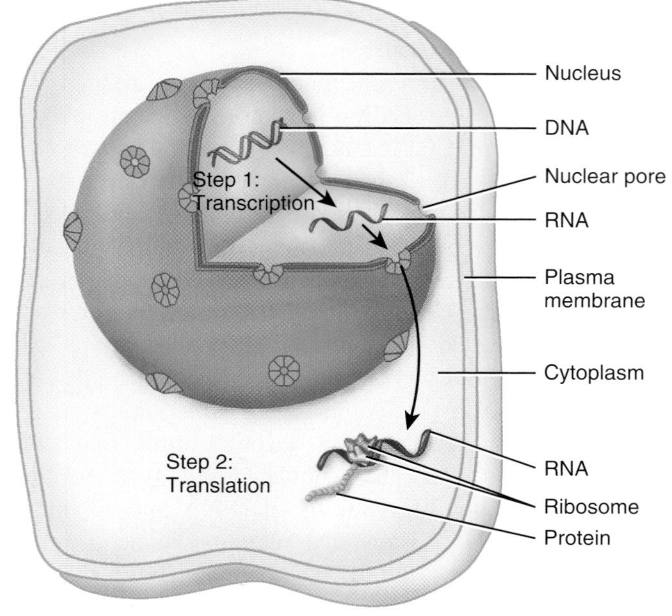

Why are proteins important in the life of a cell?

TABLE 3.2 Cell Parts and Their Functions

Part	Structure	Functions
Plasma Membrane	Fluid-mosaic lipid bilayer (phospholipids, cholesterol, and glycolipids) studded with proteins; surrounds cytoplasm.	Protects cellular contents; makes contact with other cells; contains channels, transporters, receptors, enzymes, cell-identity markers, and linker proteins; mediates the entry and exit of substances.
Cytoplasm	Cellular contents between the plasma membrane and nucleus—cytosol and organelles.	Site of all intracellular activities except those occurring in the nucleus.
Cytosol	Composed of water, solutes, suspended particles, lipid droplets, and glycogen granules.	Medium in which many of cell's metabolic reactions occur.
Organelles	Specialized structures with characteristic shapes.	Each organelle has specific functions.
Cytoskeleton	Network of three types of protein filaments: microfilaments, intermediate filaments, and microtubules.	Maintains shape and general organization of cellular contents; responsible for cellular movements.
Centrosome	A pair of centrioles plus pericentriolar material.	The pericentriolar material contains tubulins, which are used for growth of the mitotic spindle and microtubule formation.
Cilia and flagella	Motile cell surface projections that contain 20 microtubules and a basal body.	Cilia move fluids over a cell's surface; flagella move an entire cell.
Ribosome	Composed of two subunits containing ribosomal RNA and proteins; may be free in cytosol or attached to rough ER.	Protein synthesis.
Endoplasmic reticulum (ER)	Membranous network of flattened sacs or tubules. Rough ER is covered by ribosomes and is attached to nuclear envelope; smooth ER lacks ribosomes.	Rough ER synthesizes glycoproteins and phospholipids that are transferred to cellular organelles, inserted into the plasma membrane, or secreted during exocytosis. Smooth ER synthesizes fatty acids and steroids; inactivates or detoxifies drugs; removes phosphate group from glucose-6-phosphate; and stores and releases calcium ions in muscle cells.
Golgi complex	Consists of 3–20 flattened membranous sacs called cisternae; structurally and functionally divided into entry (*cis*) face, medial cisternae, and exit (*trans*) face.	Entry *(cis)* face accepts proteins from rough ER; medial cisternae form glycoproteins, glycolipids, and lipoproteins; exit *(trans)* face modifies the molecules further, then sorts and packages them for transport to their destinations.
Lysosome	Vesicle formed from Golgi complex; contains digestive enzymes.	Fuses with and digests contents of endosomes, pinocytic vesicles, and phagosomes and transports final products of digestion into cytosol; digests worn-out organelles (autophagy), entire cells (autolysis), and extracellular materials.
Peroxisome	Vesicle containing oxidases (oxidative enzymes) and catalase (decomposes hydrogen peroxide); new peroxisomes bud from preexisting ones.	Oxidizes amino acids and fatty acids; detoxifies harmful substances, such as alcohol; produces hydrogen peroxide.
Proteasome	Tiny structure that contains proteases (proteolytic enzymes).	Degrades unneeded, damaged, or faulty proteins by cutting them into small peptides.
Mitochondrion	Consists of outer and inner mitochondrial membranes, cristae, and matrix; new mitochondria form from preexisting ones.	Site of aerobic cellular respiration reactions that produce most of a cell's ATP.
Nucleus	Consists of nuclear envelope with pores, nucleoli, and chromosomes, which exist as a tangled mass of chromatin in interphase cells.	Nuclear pores control the movement of substances between the nucleus and cytoplasm, nucleoli produce ribosomes, and chromosomes consist of genes that control cellular structure and direct cellular functions.

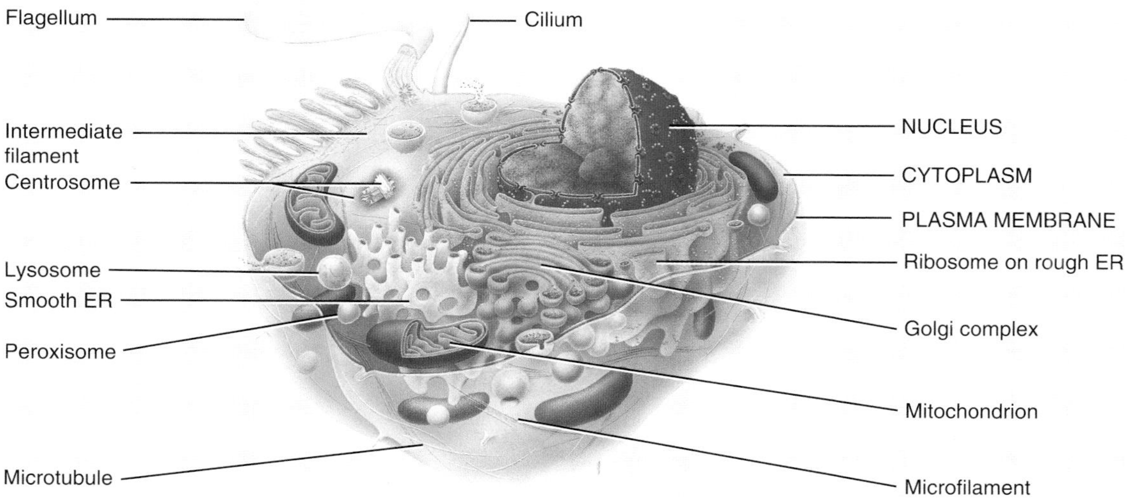

Transcription

During **transcription,** which occurs in the nucleus, the genetic information represented by the sequence of base triplets in DNA serves as a template for copying the information into a complementary sequence of codons. Three types of RNA are made from the DNA template:

1. **Messenger RNA (mRNA)** directs the synthesis of a protein.
2. **Ribosomal RNA (rRNA)** joins with ribosomal proteins to make ribosomes.
3. **Transfer RNA (tRNA)** binds to an amino acid and holds it in place on a ribosome until it is incorporated into a protein during translation. One end of the tRNA carries a specific amino acid, and the opposite end consists of a triplet of nucleotides called an **anticodon.** By pairing between complementary bases, the tRNA anticodon attaches to the mRNA codon. Each of the more than 20 different types of tRNA binds to only one of the 20 different amino acids.

The enzyme **RNA polymerase** catalyzes transcription of DNA. However, the enzyme must be instructed where to start the transcription process and where to end it. Only one of the two DNA strands serves as a template for RNA synthesis. The segment of DNA where transcription begins, a special nucleotide sequence called a **promoter,** is located near the beginning of a gene (Figure 3.25a). This is where RNA polymerase attaches to the DNA. During transcription, bases pair in a complementary manner: The bases cytosine (C), guanine (G), and thymine (T) in the DNA template pair with guanine, cytosine, and adenine (A), respectively, in the RNA strand (Figure 3.25b). However, adenine in the DNA template pairs with uracil (U), not thymine, in RNA:

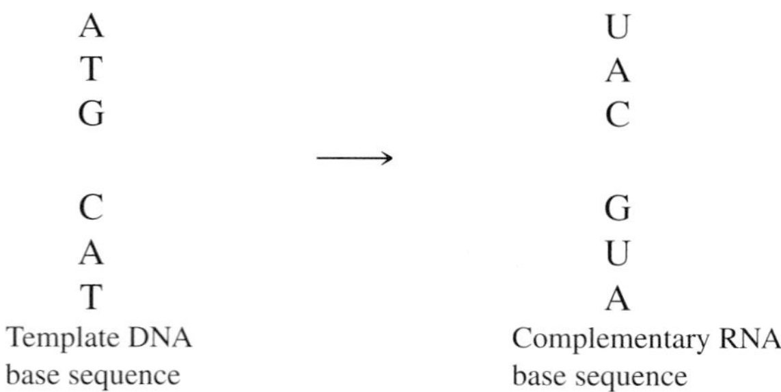

Transcription of the DNA strand ends at another special nucleotide sequence called a **terminator,** which specifies the end of the gene (Figure 3.25a). When RNA polymerase reaches the terminator, the enzyme detaches from the transcribed RNA molecule and the DNA strand.

Not all parts of a gene actually code for parts of a protein. Regions within a gene called **introns** *do not* code for parts of proteins. They are located between regions called **exons** that *do* code for segments of a protein. Immediately after transcription, the transcript includes information from both introns and exons and is called **pre-mRNA.** The introns are removed from pre-mRNA by **small nuclear ribonucleoproteins** (snRNPs,

Figure 3.25 **Transcription.** DNA transcription begins at a promoter and ends at a terminator.

During transcription, the genetic information in DNA is copied to RNA.

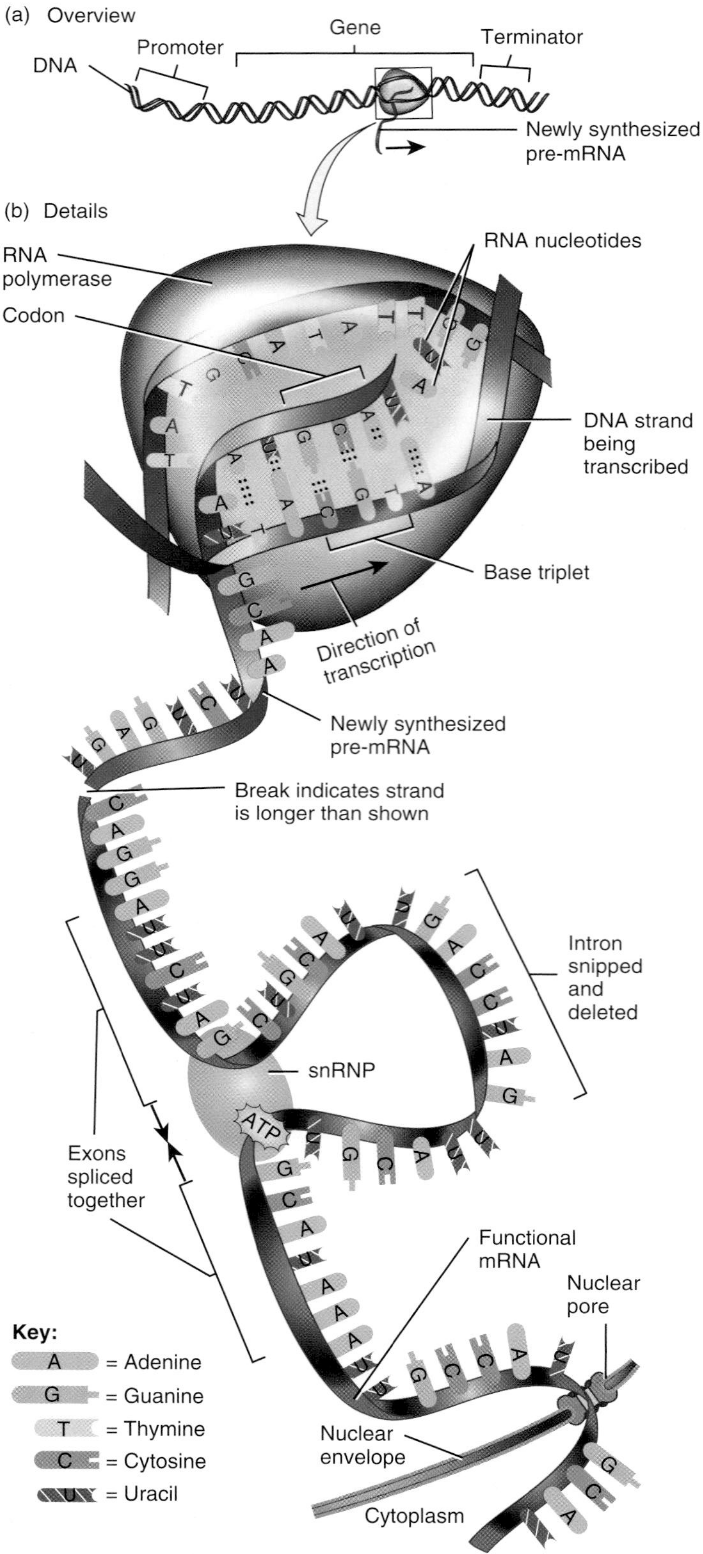

If the DNA template had the base sequence AGCT, what would be the mRNA base sequence, and what enzyme would catalyze DNA transcription?

pronounced "snurps"; Figure 3.25b). The snRNPs are enzymes that cut out the introns and splice together the exons. The resulting product is a functional mRNA molecule that passes through a pore in the nuclear envelope to reach the cytoplasm, where translation takes place.

Although the human genome contains around 30,000 genes, there are probably 500,000 to 1 million human proteins. How can so many proteins be coded for by so few genes? Part of the answer lies in **alternative splicing** of mRNA, a process in which the pre-mRNA transcribed from a gene is spliced in different ways to produce several different mRNAs. The different mRNAs are then translated into different proteins. In this way, one gene may code for 10 or more different proteins. In addition, chemical modifications are made to proteins after translation, for example, as proteins pass through the Golgi complex. Such chemical alterations can produce two or more different proteins from a single translation.

Translation

In the process of **translation,** the nucleotide sequence in an mRNA molecule specifies the amino acid sequence of a protein. Ribosomes in the cytoplasm carry out translation. The small subunit of a ribosome has a *binding site* for mRNA; the large subunit has two binding sites for tRNA molecules, a *P site* and an *A site* (Figure 3.26). The first tRNA molecule bearing its specific amino acid attaches to mRNA at the P site. The A site holds the next tRNA molecule bearing its amino acid. Translation occurs in the following way (Figure 3.27):

1. An mRNA molecule binds to the small ribosomal subunit at the mRNA binding site. A special tRNA, called *initiator tRNA,* binds to the start codon (AUG) on mRNA, where translation begins. The tRNA anticodon (UAC) attaches to the mRNA codon (AUG) by pairing between the complementary bases. Besides being the start codon, AUG is also the codon for the amino acid methionine. Thus, methionine is always the first amino acid in a growing polypeptide.
2. Next, the large ribosomal subunit attaches to the small ribosomal subunit–mRNA complex, creating a functional ribosome. The initiator tRNA, with its amino acid (methionine), fits into the P site of the ribosome.
3. The anticodon of another tRNA with its attached amino acid pairs with the second mRNA codon at the A site of the ribosome.
4. A component of the large ribosomal subunit catalyzes the formation of a peptide bond between methionine, which separates from its tRNA at the P site, and the amino acid carried by the tRNA at the A site.
5. After peptide bond formation, the tRNA at the P site detaches from the ribosome, and the ribosome shifts the mRNA strand by one codon. The tRNA in the A site bearing the two-peptide protein shifts into the P site, allowing another tRNA with its amino acid to bind to a newly exposed codon at the A site. Steps 3 through 5 occur repeatedly, and the protein lengthens progressively.
6. Protein synthesis ends when the ribosome reaches a stop codon at the A site, which causes the completed protein to detach from the final tRNA. When the tRNA vacates the A site, the ribosome splits into its large and small subunits.

Protein synthesis progresses at a rate of about 15 peptide bonds per second. As the ribosome moves along the mRNA and before it completes synthesis of the whole protein, another ribosome may attach behind it and begin translation of the same mRNA strand. Several ribosomes attached to the same mRNA constitute a **polyribosome.** The simultaneous movement of several ribosomes along the same mRNA molecule permits the translation of one mRNA into several identical proteins at the same time.

Scientists have developed techniques for inserting genes from other organisms into a variety of host cells. Manipulating the cell in this way can cause the host organism to produce proteins it normally does not synthesize. Organisms so altered are called **recombinants,** and their DNA—a combination of DNA from different sources—is called **recombinant DNA.** When recombinant DNA functions properly, the host will synthesize the protein specified by the new gene it has acquired. The technology that has arisen from the manipulation of genetic material is referred to as **genetic engineering.**

The practical applications of recombinant DNA technology are enormous. Strains of recombinant bacteria now produce large quantities of many important therapeutic substances, including *human growth hormone (hGH),* required for normal growth and

Figure 3.26 Translation. During translation, an mRNA molecule binds to a ribosome. Then, the mRNA nucleotide sequence specifies the amino acid sequence of a protein.

Ribosomes have a binding site for mRNA and a P site and an A site for attachment of tRNA.

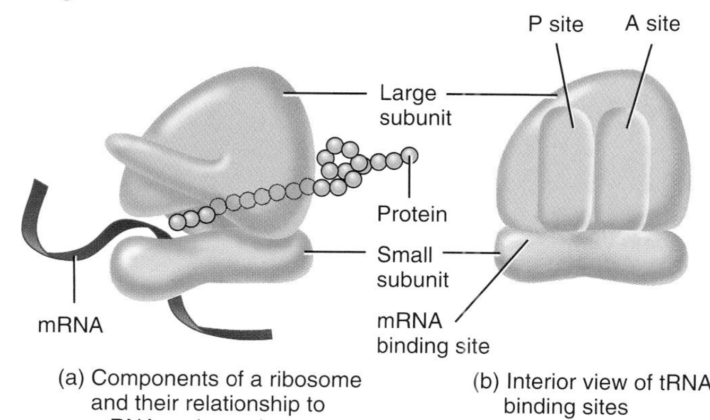

(a) Components of a ribosome and their relationship to mRNA and protein during translation

(b) Interior view of tRNA binding sites

What roles do the P and A sites serve?

Figure 3.27 **Protein elongation and termination of protein synthesis during translation.**

During protein synthesis the small and large ribosomal subunits join to form a functional ribosome. When the process is complete, they separate.

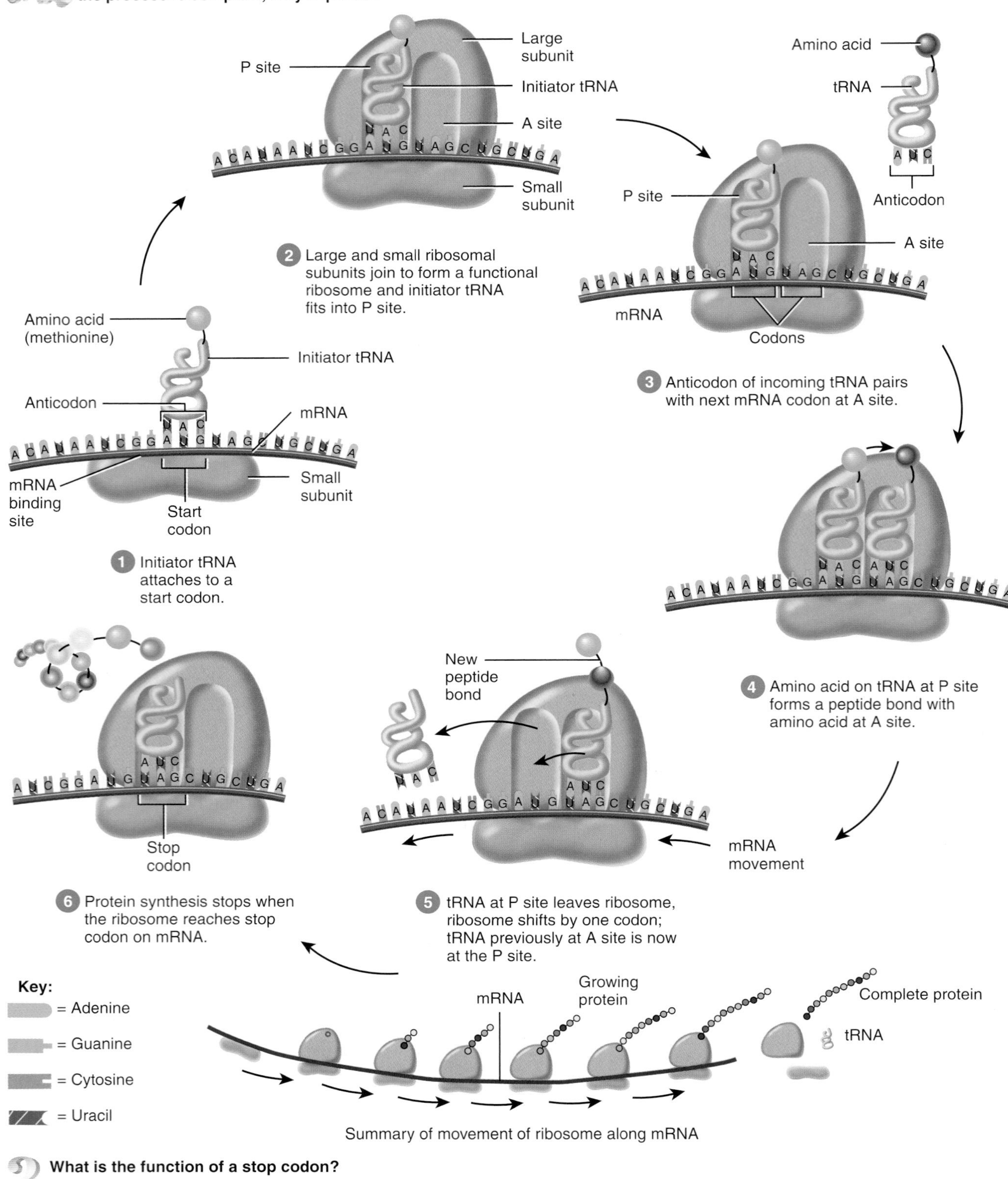
Amino acid (methionine)
Initiator tRNA
Anticodon
mRNA
mRNA binding site
Start codon
Small subunit
1 Initiator tRNA attaches to a start codon.
P site
Large subunit
Initiator tRNA
A site
Small subunit
2 Large and small ribosomal subunits join to form a functional ribosome and initiator tRNA fits into P site.
Amino acid
tRNA
Anticodon
P site
A site
mRNA
Codons
3 Anticodon of incoming tRNA pairs with next mRNA codon at A site.
4 Amino acid on tRNA at P site forms a peptide bond with amino acid at A site.
New peptide bond
mRNA movement
5 tRNA at P site leaves ribosome, ribosome shifts by one codon; tRNA previously at A site is now at the P site.
Stop codon
6 Protein synthesis stops when the ribosome reaches stop codon on mRNA.
Key:
= Adenine
= Guanine
= Cytosine
= Uracil
mRNA
Growing protein
Complete protein
tRNA
Summary of movement of ribosome along mRNA

What is the function of a stop codon?

metabolism; *insulin,* a hormone that helps regulate blood glucose level and is used by diabetics; *interferon (IFN),* an antiviral (and possibly anticancer) substance; and *erythropoietin (EPO),* a hormone that stimulates production of red blood cells. ■

▶ CHECKPOINT

23. What is the difference between transcription and translation?

CELL DIVISION

▶ OBJECTIVES

Discuss the stages, events, and significance of somatic and reproductive cell division.

Describe the signals that induce somatic cell division.

Most cells of the human body undergo **cell division,** the process by which cells reproduce themselves. The two types of cell division—somatic cell division and reproductive cell division—accomplish different goals for the organism.

A **somatic cell** (*soma* = body) is any cell of the body other than a germ cell, that is, a gamete (sperm or oocyte) or any precursor cell destined to become a gamete. In **somatic cell division,** a cell undergoes a nuclear division called **mitosis** and a cytoplasmic division called **cytokinesis** to produce two identical cells, each with the same number and kind of chromosomes as the original cell. Somatic cell division replaces dead or injured cells and adds new ones during tissue growth.

Reproductive cell division is the mechanism that produces gametes, the cells needed to form the next generation of sexually reproducing organisms. This process consists of a special two-step division called **meiosis,** in which the number of chromosomes in the nucleus is reduced by half.

Somatic Cell Division

The **cell cycle** is an orderly sequence of events by which a somatic cell duplicates its contents and divides in two. Human cells, such as those in the brain, stomach, and kidneys, contain 23 pairs of chromosomes, for a total of 46. One member of each pair is inherited from each parent. The two chromosomes that make up each pair are called **homologous chromosomes** (hō-MOL-ō-gus; *homo-* = same) or **homologs;** they contain similar genes arranged in the same (or almost the same) order. When examined under a light microscope, homologous chromosomes generally look very similar. The exception to this rule is one pair of chromosomes called the **sex chromosomes,** designated X and Y. In females the homologous pair of sex chromosome consists of two large X chromosomes; in males the pair consists of an X and a much smaller Y chromosome. Because somatic cells contain two sets of chromosomes, they are called **diploid cells** (DIP-loid; *dipl-* = double; *-oid* = form), symbolized **2n.**

When a cell reproduces, it must replicate (duplicate) all its chromosomes to pass its genes to the next generation of cells. The cell cycle consists of two major periods: interphase, when a cell is not dividing, and the mitotic (M) phase, when a cell is dividing (Figure 3.28).

Interphase

During **interphase** the cell replicates its DNA through a process that will be described shortly. It also produces additional organelles and cytosolic components in anticipation of cell division. Interphase is a state of high metabolic activity; it is during this time that the cell does most of its growing. Interphase consists of three phases: G_1, S, and G_2 (Figure 3.28). The S stands for *synthesis* of DNA. Because the G-phases are periods when there is no activity related to DNA duplication, they are thought of as *gaps* or interruptions in DNA duplication.

The **G_1 phase** is the interval between the mitotic phase and the S phase. During G_1, the cell is metabolically active; it replicates most of its organelles and cytosolic components but not its DNA. Replication of centrosomes also begins in the G_1 phase. Virtually all the cellular activities described in this chapter happen during G_1. For a cell with a total cell cycle time of 24 hours, G_1 lasts 8 to 10 hours. However, the duration of this phase is quite variable. It is very short in many embryonic cells or cancer cells. Cells that remain in G_1 for a very long time, perhaps destined never to divide again, are said to be in the **G_0 state.** Most nerve cells are in the G_0 state. Once a cell enters the S phase, however, it is committed to go through cell division.

Figure 3.28 The cell cycle. Not illustrated is cytokinesis, division of the cytoplasm, which occurs during late anaphase or early telophase of the mitotic phase.

In a complete cell cycle, a starting cell duplicates its contents and divides into two identical cells.

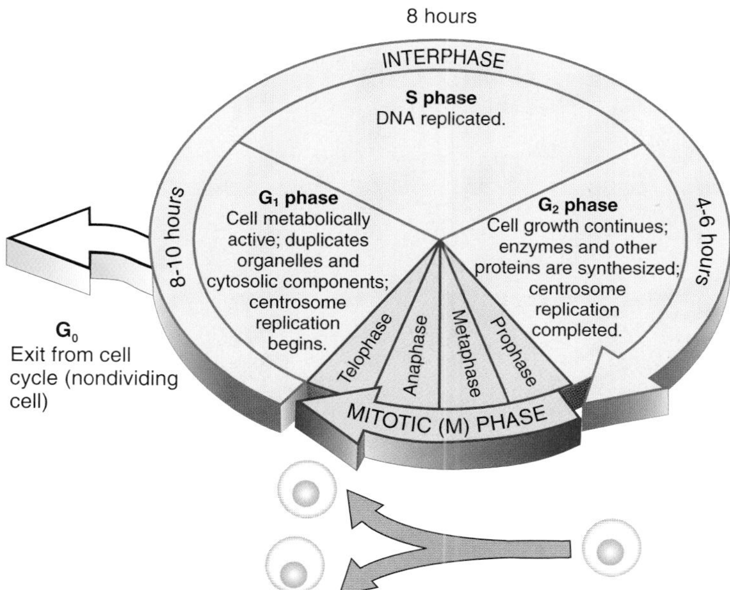

In which phase of the cell cycle does DNA replication occur?

The **S phase,** the interval between G_1 and G_2, lasts about 8 hours. During the S phase, DNA replication occurs. As a result, the two identical cells formed during cell division will have the same genetic material. The **G_2 phase** is the interval between the S phase and the mitotic phase. It lasts 4 to 6 hours. During G_2, cell growth continues, enzymes and other proteins are synthesized in preparation for cell division, and replication of centrosomes is completed. When DNA replicates during the S phase, its helical structure partially uncoils, and the two strands separate at the points where hydrogen bonds connect base pairs (Figure 3.29). Each exposed base of the old DNA strand then pairs with the complementary base of a newly synthesized nucleotide. A new DNA strand takes shape as chemical bonds form between neighboring nucleotides. The uncoiling and complementary base pairing continues until each of the two original DNA strands is joined with a newly formed complementary DNA strand. The original DNA molecule has become two identical DNA molecules.

A microscopic view of a cell during interphase shows a clearly defined nuclear envelope, a nucleolus, and a tangled mass of chromatin (Figure 3.30a). Once a cell completes its activi-ties during the G_1, S, and G_2 phases of interphase, the mitotic phase begins.

Figure 3.29 **Replication of DNA.** The two strands of the double helix separate by breaking the hydrogen bonds (shown as dotted lines) between nucleotides. New, complementary nucleotides attach at the proper sites, and a new strand of DNA is synthesized alongside each of the original strands. Arrows indicate hydrogen bonds forming again between pairs of bases.

Replication doubles the amount of DNA.

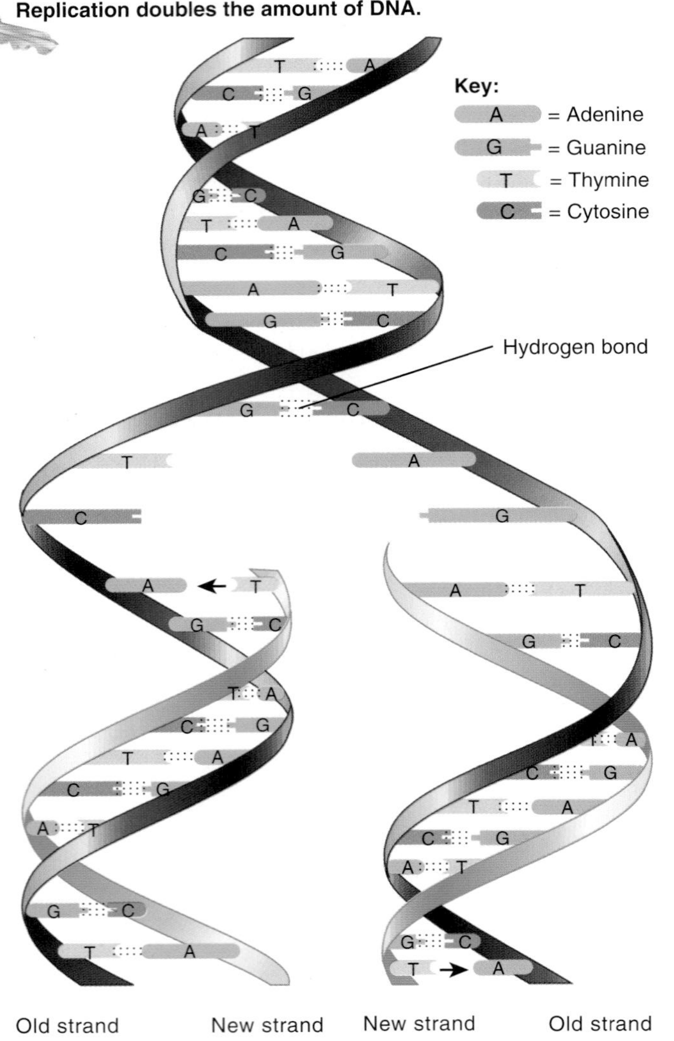

Why is it crucial that DNA replication occurs before cytokinesis in somatic cell division?

Mitotic Phase

The **mitotic (M) phase** of the cell cycle consists of a nuclear division, or mitosis, and a cytoplasmic division, or cytokinesis, to form two identical cells. The events that occur during mitosis and cytokinesis are plainly visible under a microscope because chromatin condenses into discrete chromosomes.

NUCLEAR DIVISION: MITOSIS **Mitosis** (mī-TŌ-sis; *mitos* = thread) is the distribution of two sets of chromosomes into two separate nuclei. The process results in the *exact* partitioning of genetic information. For convenience, biologists divide the process into four stages: prophase, metaphase, anaphase, and telophase. However, mitosis is a continuous process; one stage merges imperceptibly into the next.

1. **Prophase.** During early prophase, the chromatin fibers condense and shorten into chromosomes that are visible under the light microscope (Figure 3.30b). The condensation process may prevent entangling of the long DNA strands as they move during mitosis. Because DNA replication took place during the S phase of interphase, each prophase chromosome consists of a pair of identical, double-stranded *chromatids.* A constricted region called a **centromere** holds the chromatid pair together. At the outside of each centromere is a protein complex known as the **kinetochore** (ki-NET-ō-kor). Later in prophase, tubulins in the pericentriolar material of the centrosomes start to form the **mitotic spindle,** a football-shaped assembly of microtubules that attach to the kinetochore (Figure 3.30b). As the microtubules lengthen, they push the centrosomes to the poles (ends) of the cell so that the spindle extends from pole to pole. The mitotic spindle is responsible for the separation of chromatids to opposite poles of the cell. Then, the nucleolus disappears and the nuclear envelope breaks down.
2. **Metaphase.** During metaphase, the microtubules align the centromeres of the chromatid pairs at the exact center of the mitotic spindle (Figure 3.30c). This midpoint region is called the **metaphase plate.**
3. **Anaphase.** During anaphase, the centromeres split, separating the two members of each chromatid pair, which move toward opposite poles of the cell (Figure 3.30d). Once separated, the chromatids are termed chromosomes. As the chromosomes are pulled by the microtubules during anaphase, they appear V-shaped because the centromeres lead the way, dragging the trailing arms of the chromosomes toward the pole.

Figure 3.30 Cell division: mitosis and cytokinesis. Begin the sequence at 1 at the top of the figure and read clockwise to complete the process.

In somatic cell division, a single starting cell divides to produce two identical diploid cells.

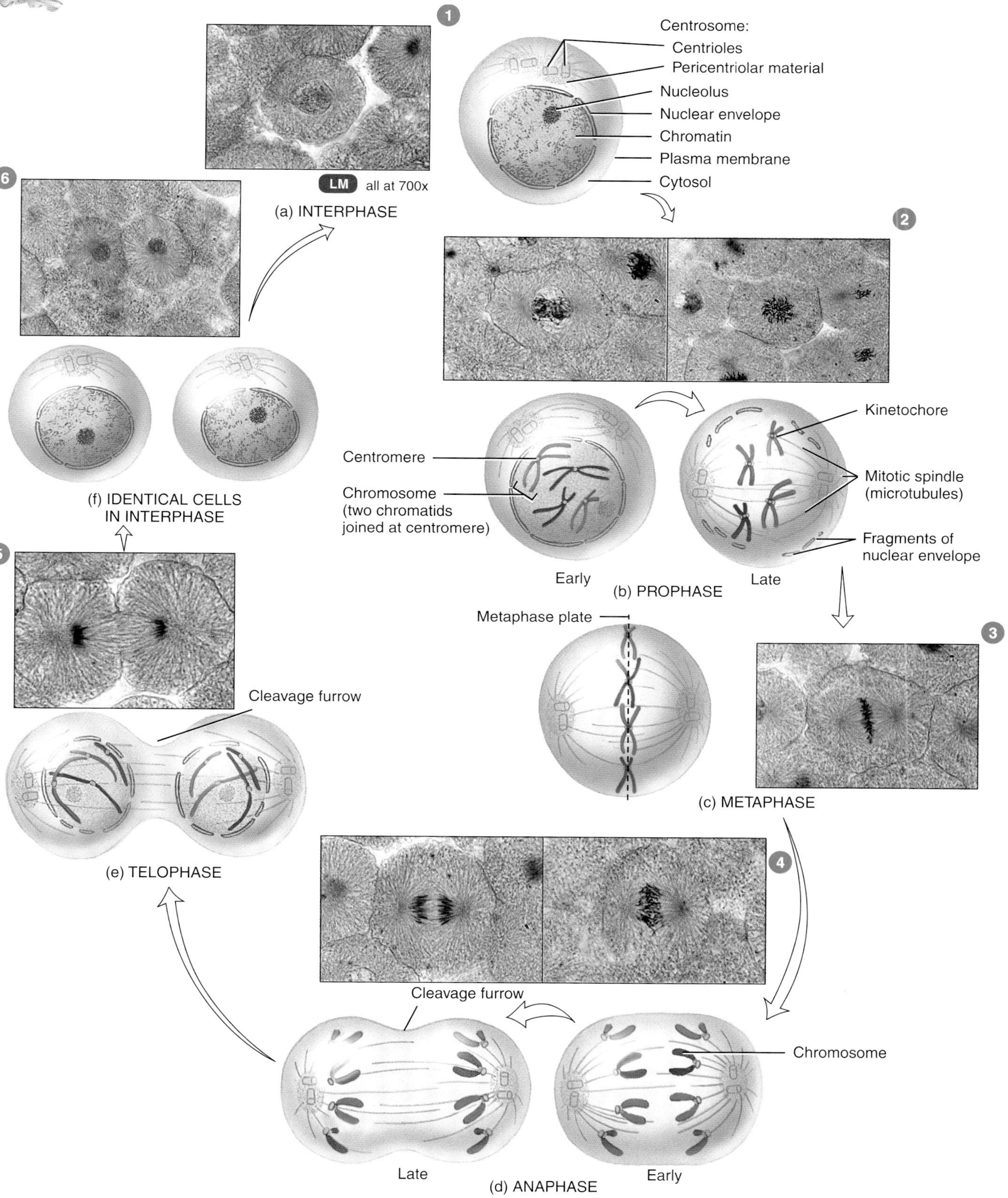

When does cytokinesis begin?

4. Telophase. The final stage of mitosis, telophase, begins after chromosomal movement stops (Figure 3.30e). The identical sets of chromosomes, now at opposite poles of the cell, uncoil and revert to the threadlike chromatin form. A nuclear envelope forms around each chromatin mass, nucleoli reappear in the identical nuclei, and the mitotic spindle breaks up.

CYTOPLASMIC DIVISION: CYTOKINESIS Division of a cell's cytoplasm and organelles into two identical cells is called **cytokinesis** (sī′-tō-ki-NĒ-sis; *-kinesis* = motion). This process begins in late anaphase with the formation of a **cleavage furrow,** a slight indentation of the plasma membrane, and is completed after telophase. The cleavage furrow usually appears midway between the centrosomes and extends around the periphery of the cell (Figure 3.30d and e). Actin microfilaments that lie just inside the plasma membrane form a *contractile ring* that pulls the plasma membrane progressively inward. The ring constricts the center of the cell, like tightening a belt around the waist, and ultimately pinches it in two. Because the plane of the cleavage furrow is always perpendicular to the mitotic spindle, the two sets of chromosomes end up in separate cells. When cytokinesis is complete, interphase begins (Figure 3.30f).

The sequence of events can be summarized as

$$G_1 \longrightarrow \text{S phase} \longrightarrow G_2 \text{ phase} \longrightarrow \text{mitosis} \longrightarrow \text{cytokinesis}$$

Table 3.3 summarizes the events of the cell cycle in somatic cells.

TABLE 3.3 Events of the Somatic Cell Cycle

Phase	Activity
Interphase	Period between cell divisions; chromosomes not visible under light microscope.
G_1 phase	Metabolically active cell duplicates organelles and cytosolic components; replication of chromosomes begins. (Cells that remain in the G_1 phase for a very long time, and possibly never divide again, are said to be in the G_0 state.)
S phase	Replication of DNA and centrosomes.
G_2 phase	Cell growth, enzyme and protein synthesis continues; replication of centrosomes complete.
Mitotic Phase	Parent cell produces identical cells with identical chromosomes; chromosomes visible under light microscope.
Mitosis	Nuclear division; distribution of two sets of chromosomes into separate nuclei.
Prophase	Chromatin fibers condense into paired chromatids; nucleolus and nuclear envelope disappear; each centrosome moves to an opposite pole of the cell.
Metaphase	Centromeres of chromatid pairs line up at metaphase plate.
Anaphase	Centromeres split; identical sets of chromosomes move to opposite poles of cell.
Telophase	Nuclear envelopes and nucleoli reappear; chromosomes resume chromatin form; mitotic spindle disappears.
Cytokinesis	Cytoplasmic division; contractile ring forms cleavage furrow around center of cell, dividing cytoplasm into separate and equal portions.

Mitotic Spindle and Cancer

One of the distinguishing features of cancer cells is uncontrolled division. The mass of cells resulting from this division is called a neoplasm or tumor. One of the ways to treat cancer is by chemotherapy, the use of anticancer drugs. Some of these drugs stop cell division by inhibiting the formation of the mitotic spindle. Unfortunately, these types of anticancer drugs also kill all types of rapidly dividing cells in the body, causing side effects such as nausea, diarrhea, hair loss, fatigue, and decreased resistance to disease. ■

Control of Cell Destiny

A cell has three possible destinies—to remain alive and functioning without dividing, to grow and divide, or to die. Homeostasis is maintained when there is a balance between cell proliferation and cell death. The signals that tell a cell when to exist in the G_0 phase, when to divide, and when to die have been the subjects of intense and fruitful research in recent years.

Within a cell, there are enzymes called **cyclin-dependent protein kinases (Cdks)** that can transfer a phosphate group from ATP to a protein to activate the protein; other enzymes can remove the phosphate group from the protein to deactivate it. The activation and deactivation of Cdks at the appropriate time is crucial in the initiation and regulation of DNA replication, mitosis, and cytokinesis.

Switching the Cdks on and off is the responsibility of cellular proteins called **cyclins** (SĪK-lins), so named because their levels rise and fall during the cell cycle. The joining of a specific cyclin and Cdk molecule triggers various events that control cell division.

The activation of specific cyclin-Cdk complexes is responsible for progression of a cell from G_1 to S to G_2 to mitosis in a specific order. If any step in the sequence is delayed, all subsequent steps are delayed in order to maintain the normal sequence. The levels of cyclins in the cell are very important in determining the timing and sequence of events in cell division. For example, the level of the cyclin that helps drive a cell from G_2 to mitosis rises throughout the G_1, S, and G_2 phases and into mitosis. The high level triggers mitosis, but toward the end of mitosis, the level declines rapidly and mitosis ends. Destruction of this cyclin, as well as others in the cell, is by proteasomes.

Cellular death is also regulated. Throughout the lifetime of an organism, certain cells undergo **apoptosis** (ap-ō-TŌ-sis = a falling off), an orderly, genetically programmed death. In apoptosis, a triggering agent from either outside or inside the cell causes "cell-suicide" genes to produce enzymes that damage the cell in several ways, including disruption of its cytoskeleton and nucleus. As a result, the cell shrinks and pulls away from neighboring cells. Although the plasma membrane remains intact, the DNA within the nucleus fragments and the cytoplasm shrinks. Phagocytes in the vicinity then ingest the dying cell. This function of phagocytes involves a receptor protein in the plasma membrane of the phagocyte that binds to a lipid in the plasma membrane of the suicidal cell. Apoptosis removes unneeded cells during fetal development, such as the webbing between digits. It continues to occur after birth to regulate the number of cells in a tissue and eliminate potentially dangerous cells such as cancer cells.

Apoptosis is a normal type of cell death; in contrast, **necrosis** (ne-KRŌ-sis = death) is a pathological type of cell death that results from tissue injury. In necrosis, many adjacent cells swell, burst, and spill their cytoplasm into the interstitial fluid. The cellular debris usually stimulates an inflammatory response by the immune system, a process that does not occur in apoptosis.

Tumor-suppressor Genes

Abnormalities in genes that regulate the cell cycle or apoptosis are associated with many diseases. For example, damage to genes called **tumor-suppressor genes,** which produce proteins that normally inhibit cell division, causes some types of cancer. Loss or alteration of a tumor-suppressor gene called *p53* on chromosome 17 is the most common genetic change leading to a wide variety of tumors, including breast and colon cancers. The normal p53 protein arrests cells in the G_1 phase, which prevents cell division. Normal p53 protein also assists in repair of damaged DNA and induces apoptosis in the cells where DNA repair was not successful. For this reason, the p53 gene is nicknamed "the guardian angel of the genome." ■

Reproductive Cell Division

In sexual reproduction, each new organism is the result of the union of two different gametes (fertilization), one produced by each parent. If gametes had the same number of chromosomes as somatic cells, the number of chromosomes would double at fertilization. **Meiosis** (mī-Ō-sis; *mei-* = lessening; *-osis* = condition of), the reproductive cell division that occurs in the gonads (ovaries and testes), produces gametes in which the number of chromosomes is reduced by half. As a result, gametes contain a single set of 23 chromosomes and thus are **haploid (n) cells** (HAP-loyd; *hapl-* = single). Fertilization restores the diploid number of chromosomes.

Meiosis

Unlike mitosis, which is complete after a single round, meiosis occurs in two successive stages: **meiosis I** and **meiosis II.** During the interphase that precedes meiosis I, the chromosomes of the diploid starting cell replicate. As a result of replication, each chromosome consists of two sister (genetically identical) chromatids, which are attached at their centromeres. This replication of chromosomes is similar to the one that precedes mitosis in somatic cell division.

MEIOSIS I Meiosis I, which begins once chromosomal replication is complete, consists of four phases: prophase I, metaphase I, anaphase I, and telophase I (Figure 3.31a). Prophase I is an extended phase in which the chromosomes shorten and thicken, the nuclear envelope and nucleoli disappear, and the mitotic spindle forms. Two events that are not seen in mitotic prophase occur during prophase I of meiosis. First, the two sister chromatids of each pair of homologous chromosomes pair off, an event called **synapsis** (Figure 3.31b). The resulting four chromatids form a structure called a **tetrad.** Second, parts of the chromatids of two homologous chromosomes may be exchanged with one another. Such an exchange between parts of nonsister (genetically different) chromatids is termed **crossing-over** (Figure 3.31b). This process, among others, permits an exchange of genes between chromatids of homologous chromosomes. Due to crossing-over, the resulting cells are genetically unlike each other and genetically unlike the starting cell that produced them. Crossing-over results in *genetic recombination*—that is, the formation of new combinations of genes—and accounts for part of the great genetic variation among humans and other organisms that form gametes via meiosis.

In metaphase I, the tetrads formed by the homologous pairs of chromosomes line up along the metaphase plate of the cell, with homologous chromosomes side by side (Figure 3.31a). During anaphase I, the members of each homologous pair of chromosomes separate as they are pulled to opposite poles of the cell by the microtubules attached to the centromeres. The paired chromatids, held by a centromere, remain together. (Recall that during mitotic anaphase, the centromeres split and the sister chromatids separate.) Telophase I and cytokinesis of meiosis are similar to telophase and cytokinesis of mitosis. The net effect of meiosis I is that each resulting cell contains the haploid number of chromosomes because it contains only one member of each pair of the homologous chromosomes present in the starting cell.

MEIOSIS II The second stage of meiosis, meiosis II, also consists of four phases: prophase II, metaphase II, anaphase II, and telophase II (Figure 3.31d). These phases are similar to those that occur during mitosis; the centromeres split, and the sister chromatids separate and move toward opposite poles of the cell.

In summary, meiosis I begins with a diploid starting cell and ends with two cells, each with the haploid number of chromosomes. During meiosis II, each of the two haploid cells formed during meiosis I divides; the net result is four haploid gametes that are genetically different from the original diploid starting cell.

Figure 3.32 on page 97 compares the events of meiosis and mitosis.

Figure 3.31 **Meiosis, reproductive cell division.** Details of events are discussed in the text.

In reproductive cell division, a single diploid starting cell undergoes meiosis I and meiosis II to produce four haploid gametes that are genetically different from the starting cell that produced them.

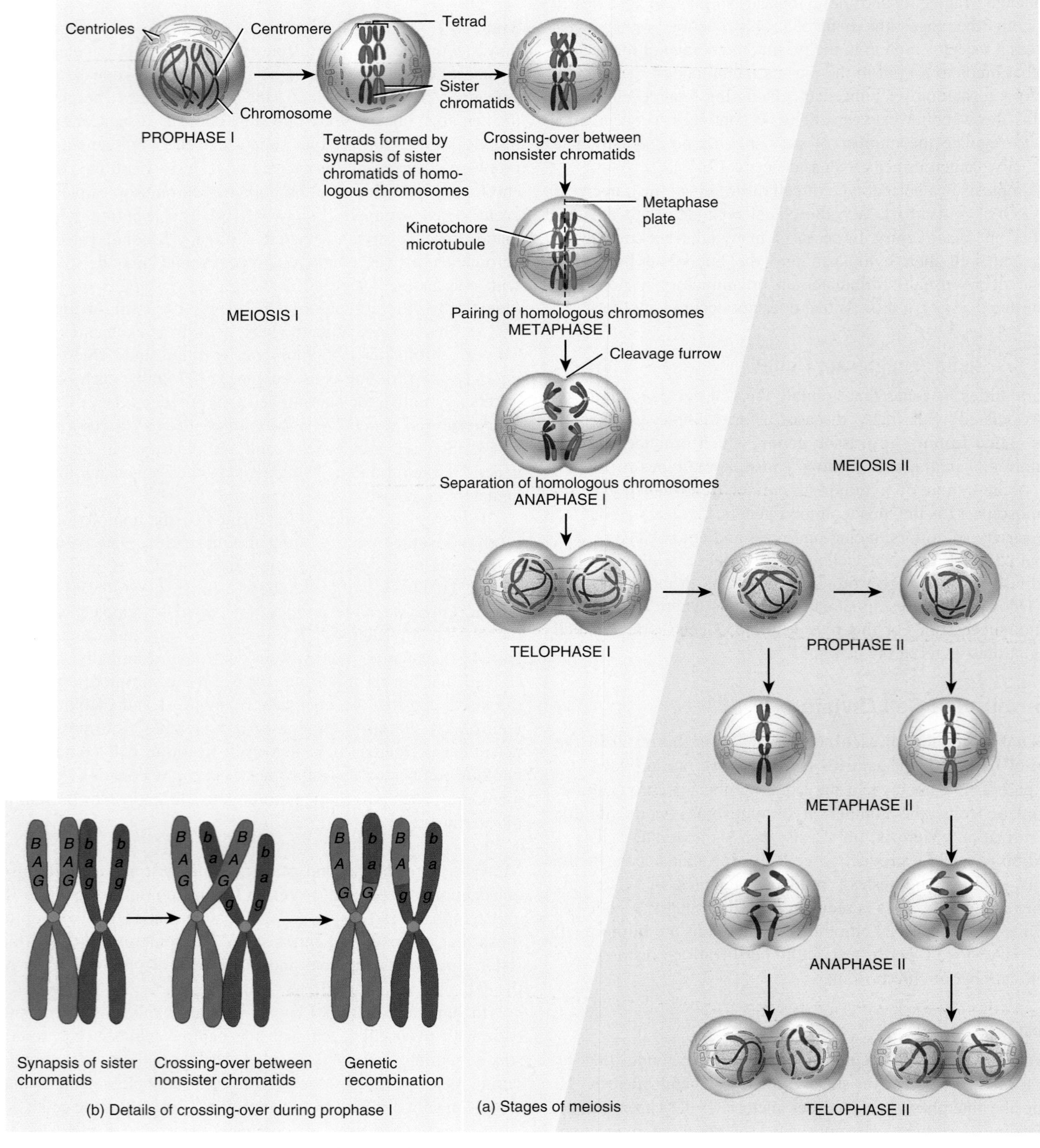

(b) Details of crossing-over during prophase I

(a) Stages of meiosis

How does crossing-over affect the genetic content of the haploid gametes?

Figure 3.32 Comparison between mitosis (left) and meiosis (right) in which the starting cell has two pairs of homologous chromosomes.

The phases of meiosis II and mitosis are similar.

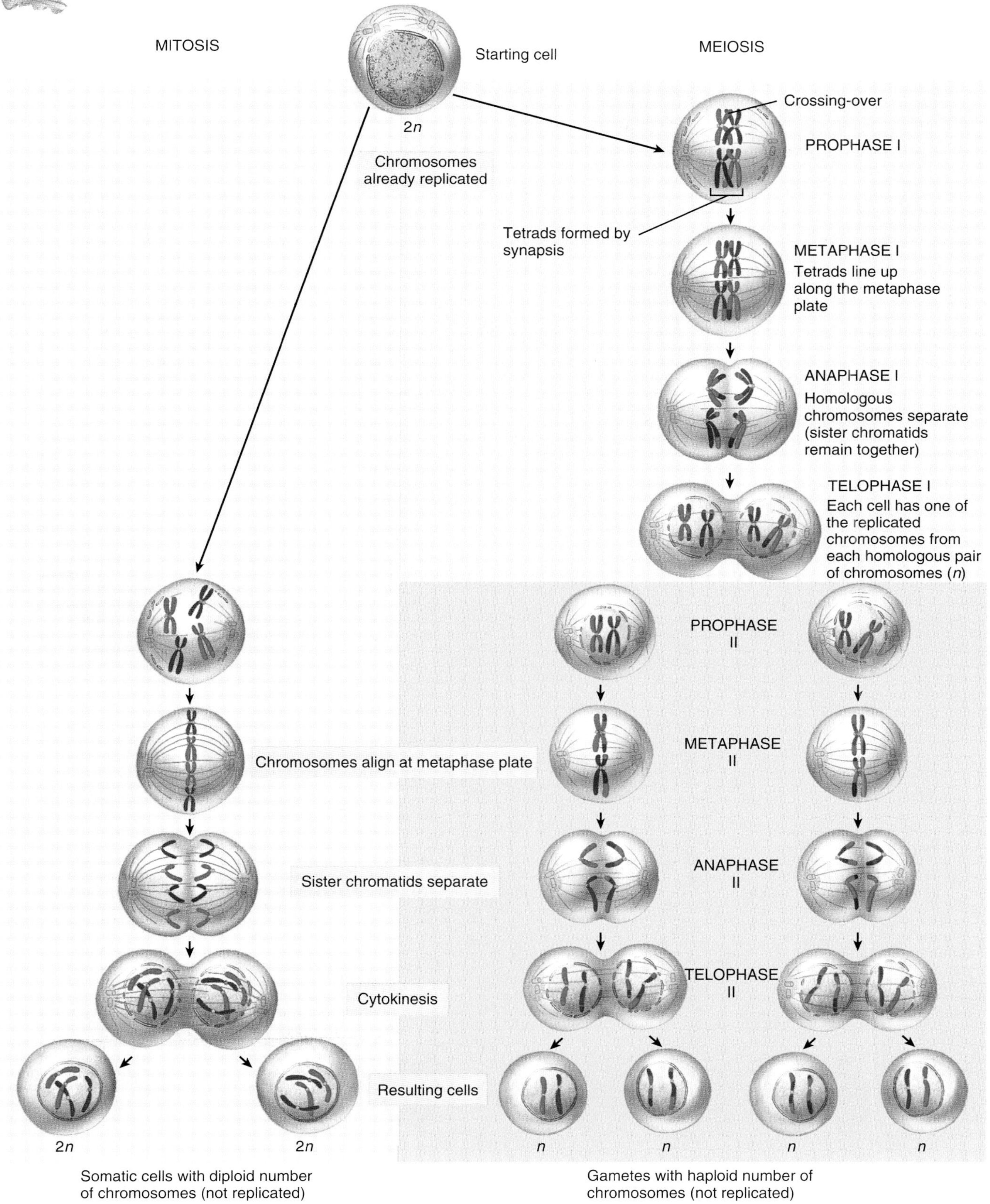

MITOSIS

Starting cell

2n

Chromosomes already replicated

MEIOSIS

Crossing-over

PROPHASE I

Tetrads formed by synapsis

METAPHASE I
Tetrads line up along the metaphase plate

ANAPHASE I
Homologous chromosomes separate (sister chromatids remain together)

TELOPHASE I
Each cell has one of the replicated chromosomes from each homologous pair of chromosomes (n)

PROPHASE II

Chromosomes align at metaphase plate

METAPHASE II

Sister chromatids separate

ANAPHASE II

Cytokinesis

TELOPHASE II

Resulting cells

2n 2n

n n n n

Somatic cells with diploid number of chromosomes (not replicated)

Gametes with haploid number of chromosomes (not replicated)

How does anaphase I of meiosis differ from anaphase of mitosis and anaphase II of meiosis?

▶ CHECKPOINT

24. Distinguish between the somatic and reproductive types of cell division. What is the importance of each?
25. Define interphase. When does DNA replicate?
26. What are the major events of each stage of the mitotic phase of the cell cycle?
27. How are apoptosis and necrosis similar and different?
28. How are haploid and diploid cells different?
29. What are homologous chromosomes?

CELLULAR DIVERSITY

▶ OBJECTIVE

Describe how cells differ in size and shape.

The body of an average human adult is composed of nearly 100 trillion cells. All of these cells can be classified into about 200 different cell types. Cells vary considerably in size. High-powered microscopes are needed to see the smallest cells of the body. The largest cell, a single oocyte, is barely visible to the unaided eye. The sizes of cells are measured in units called *micrometers*. One micrometer (μm) is equal to 1 one-millionth of a meter, or 10^{-6} m (1/25,000 of an inch). Whereas a red blood cell has a diameter of 8 μm, an oocyte has a diameter of about 140 μm.

The shapes of cells also vary considerably (Figure 3.33). They may be round, oval, flat, cuboidal, columnar, elongated, star-shaped, cylindrical, or disc-shaped. A cell's shape is related to its function in the body. For example, a sperm cell has a long whiplike tail (flagellum) that it uses for locomotion. The disc shape of a red blood cell gives it a large surface area that enhances its ability to pass oxygen to other cells. The long, spindle shape of a relaxed smooth muscle cell shortens as it contracts. This change in shape allows groups of smooth muscle cells to narrow or widen the passage for blood flowing through blood vessels. In this way, they regulate blood flow through various tissues. Some cells contain microvilli, which greatly increase their surface area. Microvilli are common in the epithelial cells that line the small intestine, where the large surface area speeds the absorption of digested food. Nerve cells have long extensions that permit them to conduct nerve impulses over great distances. As you will see in the following chapters, cellular diversity also permits organization of cells into more complex tissues and organs.

▶ CHECKPOINT

30. How is cell shape related to function? Give several of your own examples.

AGING AND CELLS

▶ OBJECTIVE

Describe the cellular changes that occur with aging.

Aging is a normal process accompanied by a progressive alteration of the body's homeostatic adaptive responses. It produces observable changes in structure and function and increases vulnerability to environmental stress and disease. The specialized branch of medicine that deals with the medical problems and care of elderly persons is **geriatrics** (jer′-ē-AT-riks; *ger-* = old age; *-iatrics* = medicine). **Gerontology** (jer′-on-TOL-ō-jē) is the scientific study of the process and problems associated with aging.

Although many millions of new cells normally are produced each minute, several kinds of cells in the body—skeletal muscle cells and nerve cells—do not divide because they are arrested permanently in the G_0 phase (see page 91). Experiments have

Figure 3.33 **Diverse shapes and sizes of human cells.** The relative difference in size between the smallest and largest cells is actually much greater than shown here.

The nearly 100 trillion cells in an average adult can be classified into about 200 different cell types.

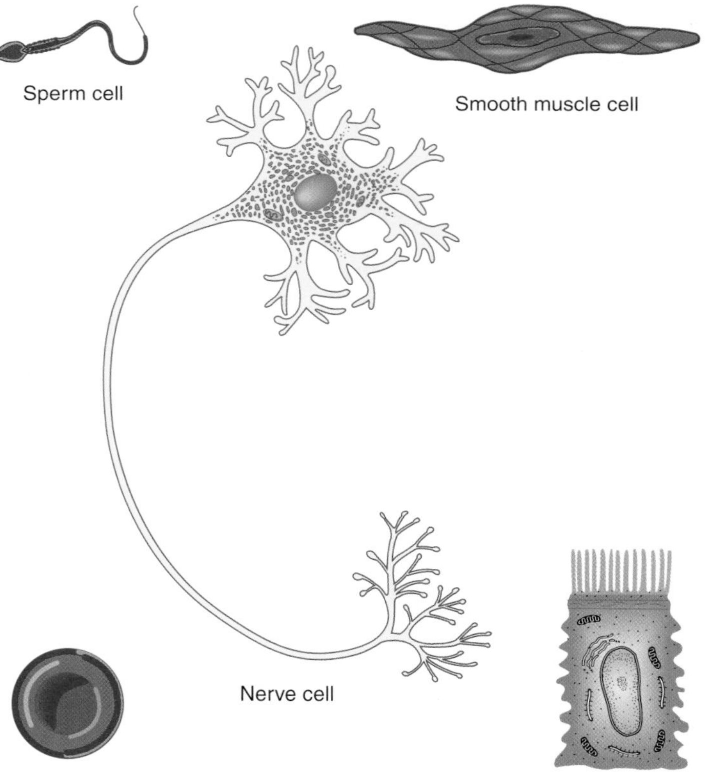

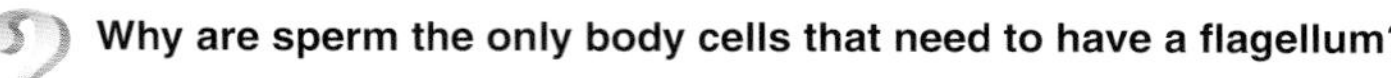
Why are sperm the only body cells that need to have a flagellum?

shown that many other cell types have only a limited capability to divide. Normal cells grown outside the body divide only a certain number of times and then stop. These observations suggest that cessation of mitosis is a normal, genetically programmed event. According to this view, "aging genes" are part of the genetic blueprint at birth. These genes have an important function in normal cells but their activities slow over time. They bring about aging by slowing down or halting processes vital to life.

Another aspect of aging involves **telomeres** (TĒ-lō-mērz), specific DNA sequences found only at the tips of each chromosome. These pieces of DNA protect the tips of chromosomes from erosion and from sticking to one another. However, in most normal body cells each cycle of cell division shortens the telomeres. Eventually, after many cycles of cell division, the telomeres can be completely gone and even some of the functional chromosomal material may be lost. These observations suggest that erosion of DNA from the tips of our chromosomes contributes greatly to aging and death of cells.

Glucose, the most abundant sugar in the body, plays a role in the aging process. It is haphazardly added to proteins inside and outside cells, forming irreversible cross-links between adjacent protein molecules. With advancing age, more cross-links form, which contributes to the stiffening and loss of elasticity that occur in aging tissues.

Free radicals produce oxidative damage in lipids, proteins, or nucleic acids by "stealing" an electron to accompany their unpaired electrons. Some effects are wrinkled skin, stiff joints, and hardened arteries. Normal metabolism—for example, aerobic cellular respiration in mitochondria—produces some free radicals. Others are present in air pollution, radiation, and certain foods we eat. Naturally occurring enzymes in peroxisomes and in the cytosol normally dispose of free radicals. Certain dietary substances, such as vitamin E, vitamin C, beta-carotene, zinc, and selenium, are antioxidants that inhibit free radical formation.

Whereas some theories of aging explain the process at the cellular level, others concentrate on regulatory mechanisms operating within the entire organism. For example, the immune system may start to attack the body's own cells. This *autoimmune response* might be caused by changes in cell-identity markers at the surface of cells that cause antibodies to attach to and mark the cell for destruction. As changes in the proteins on the plasma membrane of cells increase, the autoimmune response intensifies, producing the well-known signs of aging. In the chapters that follow, we will discuss the effects of aging on each body system in sections similar to this one.

Progeria and Werner Syndrome

Progeria (prō-JER-ē-a) is a disease characterized by normal development in the first year of life followed by rapid aging. The condition is expressed by dry and wrinkled skin, total baldness, and birdlike facial features. Death usually occurs around age 13. Although caused by a genetic defect in which telomeres are considerably shorter than usual, progeria is not an inherited disorder but a congenital (present at birth) abnormality in the genes.

Werner syndrome is a rare, inherited disease that causes a rapid acceleration of aging, usually while the person is only in his or her twenties. It is characterized by wrinkling of the skin, graying of the hair and baldness, cataracts, muscular atrophy, and a tendency to develop diabetes mellitus, cancer, and cardiovascular disease. Most afflicted individuals die before age 50. Recently, the gene that causes Werner syndrome has been identified. Researchers hope to use the information to gain insight into the mechanisms of aging, as well as to help those suffering from the disorder. ■

▶ CHECKPOINT

31. What is one reason that some tissues become stiffer as they age?

DISORDERS: HOMEOSTATIC IMBALANCES

Most chapters in the text are followed by concise discussions of major diseases and disorders that illustrate departures from normal homeostasis. They provide answers to many questions that you might ask about medical problems.

Cancer

Cancer is a group of diseases characterized by uncontrolled or abnormal cell proliferation. When cells in a part of the body divide without control, the excess tissue that develops is called a **tumor** or **neoplasm** (NĒ-ō-plazm; *neo-* = new). The study of tumors is called **oncology** (on-KOL-ō-jē; *onco-* = swelling or mass). Tumors may be cancerous and often fatal, or they may be harmless. A cancerous neoplasm is called a **malignant tumor** or **malignancy.** One property of most malignant tumors is their ability to undergo **metastasis** (me-TAS-ta-sis), the spread of cancerous cells to other parts of the body. A **benign tumor** is a neoplasm that does not metastasize. An example is a wart. Most benign tumors may be removed surgically if they interfere with normal body function or become disfiguring. Some benign tumors can be inoperable and perhaps fatal.

Growth and Spread of Cancer

Cells of malignant tumors duplicate rapidly and continuously. As malignant cells invade surrounding tissues, they often trigger **angiogenesis,** the growth of new networks of blood vessels. Proteins that stimulate angiogenesis in tumors are called **tumor angiogenesis factors (TAFs).** The formation of new blood vessels can occur either by overproduction of TAFs or by the lack of naturally occurring angiogenesis inhibitors. As the cancer grows, it begins to compete with normal tissues for space and nutrients. Eventually, the normal tissue decreases in size and dies. Some malignant cells may detach from the initial (primary) tumor and invade a body cavity or enter the blood or lymph, then circulate to and invade other body tissues, establishing secondary tumors. Malignant cells resist the antitumor defenses of the body. The pain associated with cancer develops when the tumor presses on nerves or blocks a passageway in an organ so that secretions build up pressure, or as a result of dying tissue or organs.

Causes of Cancer

Several factors may trigger a normal cell to lose control and become cancerous. One cause is environmental agents: substances in the air we breathe, the water we drink, and the food we eat. A chemical agent or radiation that produces cancer is called a **carcinogen** (car-SIN-ō-jen). Carcinogens induce **mutations,** permanent changes in the DNA base sequence of a gene. The World Health Organization estimates that carcinogens are associated with 60–90% of all human cancers. Examples of carcinogens are hydrocarbons found in cigarette tar, radon gas from the earth, and ultraviolet (UV) radiation in sunlight.

Intensive research efforts are now directed toward studying cancer-causing genes, or **oncogenes** (ON-kō-jēnz). When inappropriately activated, these genes have the ability to transform a normal cell into a cancerous cell. Most oncogenes derive from normal genes called **proto-oncogenes** that regulate growth and development. The proto-oncogene undergoes some change that causes it either to be expressed inappropriately or to make its products in excessive amounts or at the wrong time. Some oncogenes cause excessive production of growth factors, chemicals that stimulate cell growth. Others may trigger changes in a cell-surface receptor, causing it to send signals as though it were being activated by a growth factor. As a result, the growth pattern of the cell becomes abnormal.

Proto-oncogenes in every cell carry out normal cellular functions until a malignant change occurs. It appears that some proto-oncogenes are activated to oncogenes by mutations in which the DNA of the proto-oncogene is altered. Other proto-oncogenes are activated by a rearrangement of the chromosomes so that segments of DNA are exchanged. Rearrangement activates proto-oncogenes by placing them near genes that enhance their activity.

Some cancers have a viral origin. Viruses are tiny packages of nucleic acids, either RNA or DNA, that can reproduce only while inside the cells they infect. Some viruses, termed **oncogenic viruses,** cause cancer by stimulating abnormal proliferation of cells. For instance, the *human papillomavirus (HPV)* causes virtually all cervical cancers in women. The virus produces a protein that causes proteasomes to destroy p53, a protein that normally suppresses unregulated cell division. In the absence of this suppressor protein, cells proliferate without control.

Recent studies suggest that certain cancers may be linked to a cell having abnormal numbers of chromosomes. As a result, the cell could potentially have extra copies of oncogenes or too few copies of tumor-suppressor genes, which in either case could lead to uncontrolled cell proliferation. There is also some evidence suggesting that cancer may be caused by normal stem cells that develop into cancerous stem cells capable of forming malignant tumors.

Carcinogenesis: A Multistep Process

Carcinogenesis (kar′-si-nō-JEN-e-sis), the process by which cancer develops, is a multistep process in which as many as 10 distinct mutations may have to accumulate in a cell before it becomes cancerous. The progression of genetic changes leading to cancer is best understood for colon (colorectal) cancer. Such cancers, as well as lung and breast cancer, take years or decades to develop. In colon cancer, the tumor begins as an area of increased cell proliferation that results from one mutation. This growth then progresses to abnormal, but noncancerous, growths called adenomas. After two or three additional mutations, a mutation of the tumor-suppressor gene p53 occurs and a carcinoma develops. The fact that so many mutations are needed for a cancer to develop indicates that cell growth is normally controlled with many sets of checks and balances. A compromised immune system is also a significant component in carcinogenesis.

Treatment of Cancer

Many cancers are removed surgically. However, when cancer is widely distributed throughout the body or exists in organs such as the brain whose functioning would be greatly harmed by surgery, chemotherapy and radiation therapy may be used instead. Sometimes surgery, chemotherapy, and radiation therapy are used in combination. Chemotherapy involves administering drugs that cause death of cancerous cells. Radiation therapy breaks chromosomes, thus blocking cell division. Because cancerous cells divide rapidly, they are more vulnerable to the destructive effects of chemotherapy and radiation therapy than are normal cells. Unfortunately for the patients, hair follicle cells, red bone marrow cells, and cells lining the gastrointestinal tract also are rapidly dividing. Hence, the side effects of chemotherapy and radiation therapy include hair loss due to death of hair follicle cells, vomiting and nausea due to death of cells lining the stomach and intestines, and susceptibility to infection due to slowed production of white blood cells in red bone marrow.

Treating cancer is difficult because it is not a single disease and because the cells in a single tumor population rarely behave all in the same way. Although most cancers are thought to derive from a single abnormal cell, by the time a tumor reaches a clinically detectable size, it may contain a diverse population of abnormal cells. For example, some cancerous cells metastasize readily, and others do not. Some are sensitive to chemotherapy drugs and some are drug-resistant. Because of differences in drug resistance, a single chemotherapeutic agent may destroy susceptible cells but permit resistant cells to proliferate.

Another potential treatment for cancer that is currently under development is *virotherapy,* the use of viruses to kill cancer cells. The viruses employed in this strategy are designed so that they specifically target cancer cells without affecting the healthy cells of the body. For example, proteins (such as antibodies) that specifically bind to receptors found only in cancer cells are attached to viruses. Once inside the body, the viruses bind to cancer cells and then infect them. The cancer cells are eventually killed once the viruses cause cellular lysis.

Researchers are also investigating the role of *metastasis regulatory genes* that control the ability of cancer cells to undergo metastasis. Scientists hope to develop therapeutic drugs that can manipulate these genes and, therefore, block metastasis of cancer cells.

MEDICAL TERMINOLOGY

Most chapters in this text are followed by a glossary of key medical terms that include both normal and pathological conditions. You should familiarize yourself with these terms because they will play an essential role in your medical vocabulary.

Some of these conditions, as well as ones discussed in the text, are referred to as local or systemic. A *local disease* is one that affects one part or a limited area of the body. A *systemic disease* affects the entire body or several parts.

The science that deals with why, when, and where diseases occur and how they are transmitted in a human community is known as **epidemiology** (ep′-i-dē-mē-OL-ō-jē; *epidemios-* = prevalent; *-logos* = study of). The science that deals with the effects and uses of drugs in the treatment of disease is called **pharmacology** (far′-ma-KOL-ō-jē; *pharmakon-* = medicine).

Anaplasia (an′-a-PLĀ-zē-a; *an-* = not; *-plasia* = to shape) The loss of tissue differentiation and function that is characteristic of most malignancies.

Atrophy (AT-rō-fē; *a-* = without; *-trophy* = nourishment) A decrease in the size of cells, with a subsequent decrease in the size of the affected tissue or organ; wasting away.

Dysplasia (dis-PLĀ-zē-a; *dys-* = abnormal) Alteration in the size, shape, and organization of cells due to chronic irritation or inflammation; may progress to neoplasia (tumor formation, usually malignant) or revert to normal if the irritation is removed.

Hyperplasia (hī-per-PLĀ-zē-a; *hyper-* = over) Increase in the number of cells of a tissue due to an increase in the frequency of cell division.

Hypertrophy (hī-PER-trō-fē) Increase in the size of cells without cell division.

Metaplasia (met′-a-PLĀ-zē-a; *meta-* = change) The transformation of one type of cell into another.

Progeny (PROJ-e-nē; *pro-* = forward; *-geny* = production) Offspring or descendants.

Proteomics (prō′-tē-Ō-miks; *proteo-* = protein) The study of the proteome (all of an organism's proteins) in order to identify all the proteins produced; it involves determining how the proteins interact and ascertaining the three-dimensional structure of proteins so that drugs can be designed to alter protein activity to help in the treatment and diagnosis of disease.

Tumor marker A substance introduced into circulation by tumor cells that indicates the presence of a tumor, as well as the specific type. Tumor markers may be used to screen, diagnose, make a prognosis, evaluate a response to treatment, and monitor for recurrence of cancer.

STUDY OUTLINE

INTRODUCTION (p. 61)

1. A cell is the basic, living, structural and functional unit of the body.
2. Cell biology is the scientific study of cellular structure and function.

PARTS OF A CELL (p. 61)

1. Figure 3.1 provides an overview of the typical structures in body cells.
2. The principal parts of a cell are the plasma membrane; the cytoplasm, the cellular contents between the plasma membrane and nucleus; and the nucleus.

THE PLASMA MEMBRANE (p. 62)

1. The plasma membrane surrounds and contains the cytoplasm of a cell.
2. The membrane is composed of proteins and lipids that are held together by noncovalent forces.
3. According to the fluid mosaic model, the membrane is a mosaic of proteins floating like icebergs in a lipid bilayer sea.
4. The lipid bilayer consists of two back-to-back layers of phospholipids, cholesterol, and glycolipids. The bilayer arrangement occurs because the lipids are amphipathic, having both polar and nonpolar parts.
5. Integral proteins extend into or through the lipid bilayer; peripheral proteins associate with membrane lipids or integral proteins at the inner or outer surface of the membrane.
6. Many integral proteins are glycoproteins, with sugar groups attached to the ends that face the extracellular fluid. Together with glycolipids, the glycoproteins form a glycocalyx on the extracellular surface of cells.
7. Membrane proteins have a variety of functions. Integral proteins are channels and transporters that help specific solutes cross the membrane; receptors that serve as cellular recognition sites; enzymes that catalyze specific chemical reactions; and linkers that anchor proteins in the plasma membranes to protein filaments inside and outside the cell. Peripheral proteins serve as enzymes and linkers; support the plasma membrane; anchor integral proteins; and participate in mechanical activities. Membrane glycoproteins function as cell-identity markers.
8. Membrane fluidity is greater when there are more double bonds in the fatty acid tails of the lipids that make up the bilayer. Cholesterol makes the lipid bilayer stronger but less fluid at normal body temperature. Its fluidity allows interactions to occur within the plasma membrane, enables the movement of membrane components, and permits the lipid bilayer to self-seal when torn or punctured.
9. The membrane's selective permeability permits some substances to pass more readily than others. The lipid bilayer is permeable to

most nonpolar, uncharged molecules. It is impermeable to ions and charged or polar molecules other than water and urea. Channels and transporters increase the plasma membrane's permeability to small- and medium-sized polar and charged substances, including ions, that cannot cross the lipid bilayer.

10. The selective permeability of the plasma membrane supports the existence of concentration gradients, differences in the concentrations of chemicals between one side of the membrane and the other.

TRANSPORT ACROSS THE PLASMA MEMBRANE (p. 65)

1. In passive processes, a substance moves down its concentration gradient across the membrane using its own kinetic energy of motion. In active processes, cellular energy is used to drive the substance "uphill" against its concentration gradient.
2. Substances cross plasma membranes by using kinetic energy, by binding to specific transporter proteins, and by utilizing vesicles.
3. In diffusion, molecules or ions move from an area of higher concentration to an area of lower concentration until an equilibrium is reached.
4. The rate of diffusion across a plasma membrane is affected by the steepness of the concentration gradient, temperature, mass of the diffusing substance, surface area available for diffusion, and the distance over which diffusion must occur.
5. Nonpolar, hydrophobic molecules, such as oxygen, carbon dioxide, nitrogen, steroids, fat-soluble vitamins (A, E, D, and K), small alcohols, and ammonia, plus polar, uncharged water and urea diffuse through the lipid bilayer of the plasma membrane.
6. Ion channels selective for K^+, Cl^-, Na^+, and Ca^{2+} allow these small, inorganic ions (which are too hydrophilic to penetrate the membrane's nonpolar interior) to diffuse across the plasma membrane.
7. Osmosis is the net movement of water through a selectively permeable membrane from an area of higher water concentration to an area of lower water concentration.
8. In an isotonic solution, red blood cells maintain their normal shape; in a hypotonic solution, they undergo hemolysis; in a hypertonic solution, they undergo crenation.
9. In facilitated diffusion, a solute such as glucose binds to a specific transporter on one side of the membrane and is released on the other side after the transporter undergoes a change in shape.
10. Substances can cross the membrane against their concentration gradient by active transport. Actively transported substances include ions such as Na^+, K^+, H^+, Ca^{2+}, I^-, and Cl^-; amino acids; and monosaccharides.
11. Two sources of energy are used to drive active transport: Energy obtained from hydrolysis of ATP is the source in primary active transport, and energy stored in a Na^+ or H^+ concentration gradient is the source in secondary active transport.
12. The most prevalent primary active transport pump is the sodium-potassium pump, also known as Na^+/K^+ ATPase.
13. Secondary active transport mechanisms include both symporters and antiporters that are powered by either a Na^+ or H^+ concentration gradient. Symporters move two substances in the same direction across the membrane; antiporters move two substances in opposite directions.
14. In endocytosis, tiny vesicles detach from the plasma membrane to move materials across the membrane into a cell; in exocytosis, vesicles merge with the plasma membrane to move materials out of a cell.
15. Receptor-mediated endocytosis is the selective uptake of large molecules and particles (ligands) that bind to specific receptors in membrane areas called clathrin-coated pits.
16. Phagocytosis is the ingestion of solid particles. Some white blood cells destroy microbes that enter the body in this way.
17. In bulk-phase endocytosis (pinocytosis), the ingestion of extracellular fluid, a vesicle surrounds the fluid to take it into the cell.
18. In transcytosis, vesicles undergo endocytosis on one side of a cell, move across the cell, and undergo exocytosis on the opposite side.

CYTOPLASM (p. 74)

1. Cytoplasm is all the cellular contents within the plasma membrane except for the nucleus. It consists of cytosol and organelles.
2. Cytosol is the fluid portion of cytoplasm, containing water, ions, glucose, amino acids, fatty acids, proteins, lipids, ATP, and waste products. It is the site of many chemical reactions required for a cell's existence.
3. Organelles are specialized structures with characteristic shapes that have specific functions.
4. Components of the cytoskeleton, a network of several kinds of protein filaments that extend throughout the cytoplasm, include microfilaments, intermediate filaments, and microtubules. The cytoskeleton provides a structural framework for the cell and is responsible for cell movements.
5. The centrosome consists of a pair of centrioles and pericentriolar material. The pericentriolar material organizes microtubules in nondividing cells and the mitotic spindle in dividing cells.
6. Cilia and flagella, motile projections of the cell surface, are formed by basal bodies. Cilia move fluid along the cell surface; flagella move an entire cell.
7. Ribosomes consist of two subunits made in the nucleus that are composed of ribosomal RNA and ribosomal proteins. They serve as sites of protein synthesis.
8. Endoplasmic reticulum (ER) is a network of membranes that form flattened sacs or tubules; it extends from the nuclear envelope throughout the cytoplasm.
9. Rough ER is studded with ribosomes that synthesize proteins; the proteins then enter the space within the ER for processing and sorting. Rough ER produces secretory proteins, membrane proteins, and organelle proteins; forms glycoproteins; synthesizes phospholipids; and attaches proteins to phospholipids.
10. Smooth ER lacks ribosomes. It synthesizes fatty acids and steroids; inactivates or detoxifies drugs and other potentially harmful substances; removes phosphate from glucose-6-phosphate; and releases calcium ions that trigger contraction in muscle cells.
11. The Golgi complex consists of flattened sacs called cisternae. The entry, medial, and exit regions of the Golgi complex contain different enzymes that permit each to modify, sort, and package proteins for transport in secretory vesicles, membrane vesicles, or transport vesicles to different cellular destinations.
12. Lysosomes are membrane-enclosed vesicles that contain digestive enzymes. Endosomes, phagosomes, and pinocytic vesicles deliver materials to lysosomes for degradation. Lysosomes function in digestion of worn-out organelles (autophagy), digestion of a host cell (autolysis), and extracellular digestion.
13. Peroxisomes contain oxidases that oxidize amino acids, fatty acids, and toxic substances; the hydrogen peroxide produced in the process is destroyed by catalase.

14. The proteases contained in proteasomes continually degrade unneeded, damaged, or faulty proteins by cutting them into small peptides.
15. Mitochondria consist of a smooth outer membrane, an inner membrane containing cristae, and a fluid-filled cavity called the matrix. These so-called "powerhouses" of the cell produce most of a cell's ATP.

NUCLEUS (p. 85)

1. The nucleus consists of a double nuclear envelope; nuclear pores, which control the movement of substances between the nucleus and cytoplasm; nucleoli, which produce ribosomes; and genes arranged on chromosomes, which control cellular structure and direct cellular activities.
2. Human somatic cells have 46 chromosomes, 23 inherited from each parent. The total genetic information carried in a cell or an organism is its genome.

PROTEIN SYNTHESIS (p. 86)

1. Cells make proteins by transcribing and translating the genetic information contained in DNA.
2. The genetic code is the set of rules that relates the base triplet sequences of DNA to the corresponding codons of RNA and the amino acids they specify.
3. In transcription, the genetic information in the sequence of base triplets in DNA serves as a template for copying the information into a complementary sequence of codons in messenger RNA. Transcription begins on DNA in a region called a promoter. Regions of DNA that code for protein synthesis are called exons; those that do not are called introns.
4. Newly synthesized pre-mRNA is modified before leaving the nucleus.
5. In the process of translation, the nucleotide sequence of mRNA specifies the amino acid sequence of a protein. The mRNA binds to a ribosome, specific amino acids attach to tRNA, and anticodons of tRNA bind to codons of mRNA, bringing specific amino acids into position on a growing polypeptide. Translation begins at the start codon and ends at the stop codon.

CELL DIVISION (p. 91)

1. Cell division is the process by which cells reproduce themselves. It consists of nuclear division (mitosis or meiosis) and cytoplasmic division (cytokinesis).
2. Cell division that replaces cells or adds new ones is called somatic cell division and involves mitosis and cytokinesis.
3. Cell division that results in the production of gametes (sperm and ova) is called reproductive cell division and consists of meiosis and cytokinesis.

Somatic Cell Division (p. 91)

1. The cell cycle, an orderly sequence of events in which a somatic cell duplicates its contents and divides in two, consists of interphase and a mitotic phase.
2. Human somatic cells contain 23 pairs of homologous chromosomes and are thus diploid (2n).
3. Before the mitotic phase, the DNA molecules, or chromosomes, replicate themselves so that identical sets of chromosomes can be passed on to the next generation of cells.
4. A cell between divisions that is carrying on every life process except division is said to be in interphase, which consists of three phases: G_1, S, and G_2.
5. During the G_1 phase, the cell replicates its organelles and cytosolic components, and centrosome replication begins; during the S phase, DNA replication occurs; during the G_2 phase, enzymes and other proteins are synthesized and centrosome replication is completed.
6. Mitosis is the splitting of the chromosomes and the distribution of two identical sets of chromosomes into separate and equal nuclei; it consists of prophase, metaphase, anaphase, and telophase.
7. In cytokinesis, which usually begins in late anaphase and ends once mitosis is complete, a cleavage furrow forms at the cell's metaphase plate and progresses inward, pinching in through the cell to form two separate portions of cytoplasm.

Control of Cell Destiny (p. 94)

1. A cell can either remain alive and functioning without dividing, grow and divide, or die. The control of cell division depends on specific cyclin-dependent protein kinases and cyclins.
2. Apoptosis is normal, programmed cell death. It first occurs during embryological development and continues throughout the lifetime of an organism.
3. Certain genes regulate both cell division and apoptosis. Abnormalities in these genes are associated with a wide variety of diseases and disorders.

Reproductive Cell Division (p. 95)

1. In sexual reproduction, each new organism is the result of the union of two different gametes, one from each parent.
2. Gametes contain a single set of chromosomes (23) and thus are haploid (n).
3. Meiosis is the process that produces haploid gametes; it consists of two successive nuclear divisions called meiosis I and meiosis II.
4. During meiosis I, homologous chromosomes undergo synapsis (pairing) and crossing-over; the net result is two haploid cells that are genetically unlike each other and unlike the starting diploid parent cell that produced them.
5. During meiosis II, two haploid cells divide to form four haploid cells.

CELLULAR DIVERSITY (p. 98)

1. The almost 200 different types of cells in the body vary considerably in size and shape.
2. The sizes of cells are measured in micrometers. One micrometer (μm) equals 10^{-6} m (1/25,000 of an inch). Cells in the body range from 8 μm to 140 μm in size.
3. A cell's shape is related to its function.

AGING AND CELLS (p. 98)

1. Aging is a normal process accompanied by progressive alteration of the body's homeostatic adaptive responses.
2. Many theories of aging have been proposed, including genetically programmed cessation of cell division, buildup of free radicals, and an intensified autoimmune response.

SELF-QUIZ QUESTIONS

Fill in the blanks in the following statements.

1. The three major parts of the cell are the ____, ____, and ____.
2. Cell death that is genetically programmed is known as ____, while cell death which is due to tissue injury is known as ____.
3. ____ are special DNA sequences located at the ends of chromosomes and whose erosion contributes to cellular aging and death.
4. The mRNA base sequence that is complementary to the DNA base sequence ATC would be ____.

Indicate whether the following statements are true or false.

5. A small membrane surface area will increase the rate of diffusion across the cell membrane.
6. The cells created during meiosis are genetically different from the original cell.
7. An important and abundant active mechanism that helps maintain cellular tonicity is the Na^+/K^+ ATPase pump.

Choose the one best answer to the following questions.

8. If the concentration of solutes in the ECF and ICF are equal, the cell is in a(n) ____ solution. (a) hypertonic, (b) hydrophobic, (c) saturated, (d) hypotonic, (e) isotonic.
9. Which membrane protein is *incorrectly* matched with its function? (a) receptor: allows recognition of specific molecules, (b) ion channel: allows passage of specific ions through the membrane, (c) transporter: allows cells to recognize each other and foreign cells, (d) linker: allows binding of one cell to another and provides stability and shape to a cell, (e) enzyme: catalyzes cellular reactions.
10. Place the following steps in protein synthesis in the correct order. (a) anticodons of tRNA bind to codons of mRNA, (b) modification of newly synthesized pre-mRNA by snRNPs before leaving the nucleus and entering the cytoplasm, (c) attachment of RNA polymerase at promoter, (d) binding of mRNA to a ribosome's small subunit, (e) amino acids joined by peptide bonds, (f) large and small ribosomal subunits join to create a functional ribosome, (g) transcription of a segment of DNA onto mRNA, (h) protein detaches from ribosome when ribosome reaches stop codon on mRNA, (i) detachment of RNA polymerase after reaching terminator, (j) specific amino acids attach to tRNA, (k) initiator tRNA binds to start codon on mRNA.
11. Which of the following organelles function primarily in decomposition reactions? (1) ribosomes, (2) proteasomes, (3) lysosomes, (4) centrosomes, (5) peroxisomes (a) 2, 3, and 5, (b) 3 and 5, (c) 2, 4, and 5, (d) 1 and 4, (e) 2 and 5.
12. Which of the following statements regarding the nucleus are *true*? (1) Nucleoli within the nucleus are the sites of ribosome synthesis. (2) The nucleus contains the cell's hereditary units. (3) The nuclear membrane is a solid, impermeable membrane. (4) Protein synthesis occurs within the nucleus. (5) In nondividing cells, DNA is found in the nucleus in the form of chromatin. (a) 1, 2, and 3, (b) 1, 2, and 4, (c) 1, 2, and 5, (d) 2, 4, and 5, (e) 2, 3, and 4
13. Match the following:

____(a) mitosis
____(b) meiosis
____(c) prophase
____(d) metaphase
____(e) anaphase
____(f) telophase
____(g) cytokinesis
____(h) interphase

(1) cytoplasmic division
(2) somatic cell division resulting in the formation of two identical cells
(3) reproductive cell division that reduces the number of chromosomes by half
(4) stage of cell division when replication of DNA occurs
(5) stage when chromatin fibers condense and shorten to form chromosomes
(6) stage when centromeres split and sister chromatids move to opposite poles of the cell
(7) stage when centromeres of chromatid pairs line up at the center of the mitotic spindle
(8) stage when chromosomes uncoil and revert to chromatin

14. Match the following:

____(a) cytoskeleton
____(b) centrosome
____(c) ribosomes
____(d) rough ER
____(e) smooth ER
____(f) Golgi complex
____(g) lysosomes
____(h) peroxisomes
____(i) mitochondria
____(j) cilia
____(k) flagellum
____(l) proteasomes
____(m) vesicles

(1) membrane-enclosed vesicles formed in the Golgi complex that contain strong hydrolytic and digestive enzymes
(2) network of protein filaments that extend throughout the cytoplasm, providing cellular shape, organization, and movement
(3) sites of protein synthesis
(4) contain enzymes that break apart unneeded, damaged, or faulty proteins into small peptides
(5) site where secretory proteins and membrane molecules are synthesized
(6) membrane-enclosed vesicles that contain enzymes that oxidize various organic substances
(7) short microtubular structures extending from the plasma membrane and involved in movement of materials along the cell's surface
(8) modifies, sorts, packages, and transports molecules synthesized in the rough ER
(9) an organizing center for growth of the mitotic spindle
(10) function in ATP generation
(11) functions in synthesizing fatty acids and steroids, helping liver cells release glucose into the bloodstream, and detoxification
(12) membrane-bound sacs that transport, transfer, or secrete proteins
(13) long microtubular structure extending from the plasma membrane and involved in movement of a cell

15. Match the following:

____(a) diffusion
____(b) osmosis
____(c) facilitated diffusion
____(d) primary active transport
____(e) secondary active transport
____(f) vesicular transport
____(g) phagocytosis
____(h) pinocytosis
____(i) exocytosis
____(j) receptor-mediated endocytosis
____(k) transcytosis

(1) passive transport in which a solute binds to a specific transporter on one side of the membrane and is released on the other side
(2) movement of materials out of the cell by fusing of secretory vesicles with the plasma membrane
(3) the random mixing of particles in a solution due to the kinetic energy of the particles; substances move from high to low concentrations until equilibrium is reached
(4) transport of substances either into or out of the cell by means of small, spherical membranous sac formed by budding off from existing membranes
(5) uses energy derived from hydrolysis of ATP to change the shape of a transporter protein, which "pumps" a substance across a cellular membrane against its concentration gradient
(6) vesicular movement involving endocytosis on one side of a cell and subsequent exocytosis on the opposite side of the cell
(7) type of endocytosis that involves the nonselective uptake of tiny droplets of extracellular fluid
(8) type of endocytosis in which large solid particles are taken in
(9) movement of water from an area of higher to an area of lower water concentration through a selectively permeable membrane
(10) process that allows a cell to take specific ligands from the ECF by forming vesicles
(11) indirectly uses energy obtained from the breakdown of ATP; involves symporters and antiporters

CRITICAL THINKING QUESTIONS

1. Mucin is a protein present in saliva and other secretions. When mixed with water, it becomes the slippery substance known as mucus. Trace the route taken by mucin through the cell, from its synthesis to its secretion, listing all the organelles and processes involved.
2. Jason has decided to pierce his nose. He knows that the body is primarily water and wonders why he doesn't leak water after his nose is pierced. What could you tell him?
3. In order to lose weight, some individuals undergo surgical removal or bypass of a large part of the small intestine. Knowing what you do concerning cellular function, how does this contribute to weight loss?
4. Marathon runners can become dehydrated due to the extreme physical activity. What types of fluids should they consume in order to rehydrate their cells?

ANSWERS TO FIGURE QUESTIONS

3.1 The three main parts of a cell are the plasma membrane, cytoplasm, and nucleus.

3.2 The glycocalyx is the sugary coat on the extracellular surface of the plasma membrane. It is composed of the carbohydrate portions of membrane glycolipids and glycoproteins.

3.3 The membrane protein that binds to insulin acts as a receptor.

3.4 Because fever involves an increase in body temperature, the rates of all diffusion processes would increase.

3.5 Diffusion through channel proteins is slower than diffusion through the lipid bilayer because channel proteins occupy a smaller surface area of the total membrane than lipids.

3.6 The water concentrations can never be the same in the two arms because the left arm contains pure water and the right arm contains a solution that is less than 100% water.

3.7 A 2% solution of NaCl will cause crenation of RBCs because it is hypertonic.

3.8 ATP adds a phosphate group to the pump protein, which changes the pump's three-dimensional shape. ATP transfers energy to power the pump.

3.9 In secondary active transport, hydrolysis of ATP is used indirectly to drive the activity of symporter or antiporter proteins; this reaction directly powers the pump protein in primary active transport.

3.10 Iron, vitamins, and hormones are other examples of ligands that can undergo receptor-mediated endocytosis.

3.11 The binding of particles to a plasma membrane receptor triggers pseudopod formation.

3.12 Receptor-mediated endocytosis and phagocytosis involve receptor proteins; bulk-phase endocytosis does not.

3.13 Microtobules help to form centrioles, cilia, and flagella.

3.14 A cell without a centrosome probably would not be able to undergo cell division.

3.15 Cilia move fluids across cell surfaces; flagella move an entire cell.

3.16 Large and small ribosomal subunits are synthesized separately in the nucleolus in the nucleus and then come together in the cytoplasm.

3.17 Rough ER has attached ribosomes; smooth ER does not. Rough ER synthesizes proteins that will be exported from the cell; smooth ER is associated with lipid synthesis and other metabolic reactions.

3.18 The entry face receives and modifies proteins from rough ER; the exit face modifies, sorts, and packages molecules for transport to other destinations.

3.19 Some proteins are secreted from the cell by exocytosis, some are incorporated into the plasma membrane, and some occupy storage vesicles that become lysosomes.

3.20 Digestion of worn-out organelles by lysosomes is called autophagy.

3.21 Mitochondrial cristae increase the surface area available for chemical reactions and contain some of the enzymes needed for ATP production.

3.22 Chromatin is a complex of DNA, proteins, and some RNA.

3.23 A nucleosome is a double-stranded molecule of DNA wrapped twice around a core of eight histones (proteins).

3.24 Proteins determine the physical and chemical characteristics of cells.

3.25 The DNA base sequence AGCT would be transcribed into the RNA base sequence UCGA by RNA polymerase.

3.26 The P site holds the tRNA attached to the growing polypeptide. The A site holds the tRNA carrying the next amino acid to be added to the growing polypeptide.

3.27 When a ribosome encounters a stop codon at the A site, it releases the completed protein from the final tRNA.

3.28 DNA replicates during the S phase.

3.29 DNA replication occurs before cytokinesis so that each of the new cells will have a complete genome.

3.30 Cytokinesis usually starts in late anaphase.

3.31 The result of crossing-over is that cells that follow are genetically unlike each other and genetically unlike the starting cell that produced them.

3.32 During anaphase I of meiosis, the paired chromatids are held together by a centromere and do not separate. During anaphase II of meiosis and during mitosis, the centromeres split and the paired chromatids separate.

3.33 Sperm, which use the flagella for locomotion, are the only body cells required to move considerable distances.

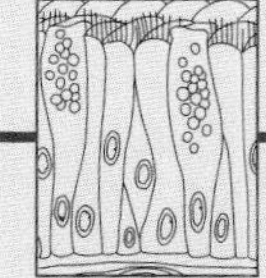

Chapter 4

The Tissue Level of Organization

Tissues and Homeostasis

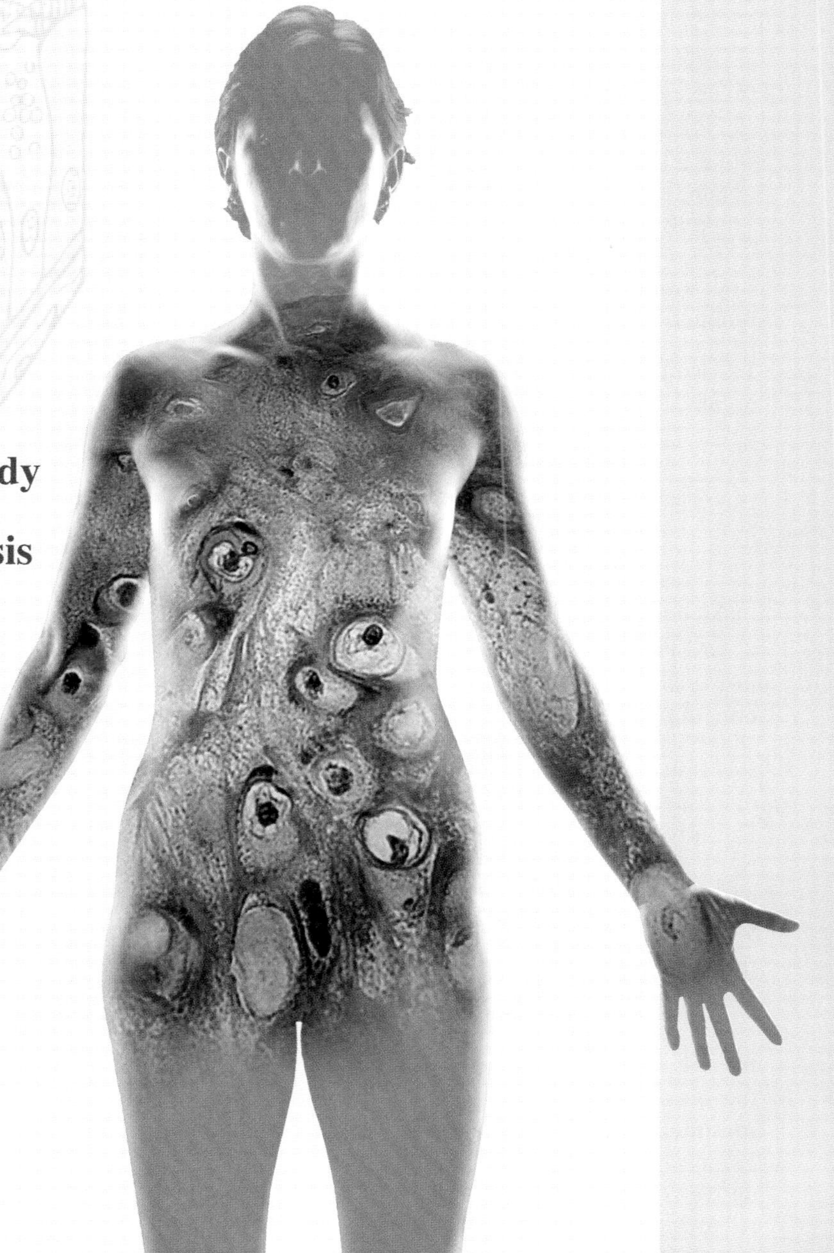

The four basic types of tissues in the human body contribute to homeostasis by providing diverse functions including protection, support, communication among cells, and resistance to disease, to name just a few.

wiley.com/college/apcentral

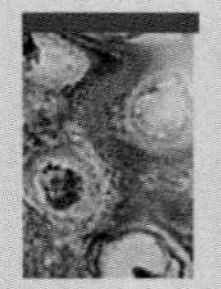

As you learned in Chapter 3, a cell is a complex collection of compartments, each of which carries out a host of biochemical reactions that make life possible. However, like words, cells seldom function as isolated units in the body. Instead, cells usually work together in groups called tissues, similar to the way words are linked into sentences. A **tissue** is a group of similar cells that usually have a common embryonic origin and functions together to carry out specialized activities. As you will learn, the structure and properties of a specific tissue are influenced by factors such as the nature of the extracellular material that surrounds the tissue cells and the connections between the cells that compose the tissue. Tissues may be hard (bone), semisolid (fat), or even liquid (blood) in their consistency. In addition, tissues vary tremendously with respect to the types of cells present, their arrangement, and whether fibers are present. **Histology** (hiss′-TOL-ō-jē; *histo-* = tissue; *-logy* = study of) is the science that deals with the study of tissues. A **pathologist** (pa-THOL-ō-gist; *patho-* = disease) is a physician who specializes in laboratory studies of cells and tissues to help other physicians make accurate diagnoses. One of the principal functions of a pathologist is to examine tissues for any changes that might indicate disease.

TYPES OF TISSUES AND THEIR ORIGINS

▶ **OBJECTIVE**

Name the four basic types of tissues that make up the human body and state the characteristics of each.

Body tissues can be classified into four basic types according to function and structure:

1. **Epithelial tissue** covers body surfaces and lines hollow organs, body cavities, and ducts. It also forms glands.
2. **Connective tissue** protects and supports the body and its organs. Various types of connective tissue bind organs together, store energy reserves as fat, and help provide immunity to disease-causing organisms.
3. **Muscular tissue** generates the physical force needed to make body structures move.
4. **Nervous tissue** detects changes in a variety of conditions inside and outside the body and responds by generating action potentials (nerve impulses) that help maintain homeostasis.

Epithelial tissue and most types of connective tissue are discussed in detail in this chapter. However, only the general features of bone tissue and blood (connective tissues) will be introduced here; detailed discussions are presented in Chapters 6 and 19, respectively. Similarly, we acquaint you with the structure and function of muscular tissue and nervous tissue here, but cover them in more detail in Chapters 10 and 12, respectively.

As you will see later in the text, tissues of the body develop from three **primary germ layers,** the first tissues formed in the human embryo called the **ectoderm, endoderm,** and **mesoderm.** Epithelial tissues develop from all three primary germ layers. All connective tissue and most muscle tissues derive from mesoderm. Nervous tissue develops from ectoderm. (Figure 29.7b on page 1113 illustrates the primary germ layers and Table 29.1 on page 1112 provides descriptions of the structures derived from the primary germ layers.)

Normally, most cells within a tissue remain anchored to other cells or structures. Only a few cells, such as phagocytes, move freely through the body, searching for invaders to destroy. However, many cells migrate extensively during the growth and development process before birth.

▶ **CHECKPOINT**

1. Define a tissue.
2. What are the four basic types of human tissues?

CELL JUNCTIONS

▶ **OBJECTIVE**

Describe the structure and functions of the five main types of cell junctions.

Most epithelial cells and some muscle and nerve cells are tightly joined into functional units. **Cell junctions** are contact points between the plasma membranes of tissue cells. Here we consider the five most important types of cell junctions: tight junctions, adherens junctions, desmosomes, hemidesmosomes, and gap junctions (Figure 4.1).

Tight Junctions

Tight junctions consist of weblike strands of transmembrane proteins that fuse the outer surfaces of adjacent plasma membranes together (Figure 4.1a). Cells of epithelial tissues that line the stomach, intestines, and urinary bladder have many tight junctions to retard the passage of substances between cells and prevent the contents of these organs from leaking into the blood or surrounding tissues.

Adherens Junctions

Adherens junctions (ad-HER-ens) contain **plaque,** a dense layer of proteins on the inside of the plasma membrane that attaches to both membrane proteins and to microfilaments of the cytoskeleton (Figure 4.1b). It is actually transmembrane glycoproteins called **cadherins** that join the cells. Each cadherin inserts into the plaque from the opposite side of the plasma membrane, partially crosses the intercellular space (the space between the cells), and connects to cadherins of an adjacent cell.

In epithelial cells, adherens junctions often form extensive zones called **adhesion belts** because they encircle the cell similar to the way a belt encircles your waist. Adherens junctions help epithelial surfaces resist separation during various contractile activities, as when food moves through the intestines.

Desmosomes

Like adherens junctions, **desmosomes** (DEZ-mō-sōms; *desmo-* = band) contain plaque and have transmembrane glycoproteins (cadherins) that extend into the intercellular space between adjacent cell membranes and attach cells to one another (Figure 4.1c). However, unlike adherens junctions, the plaque of desmosomes does not attach to microfilaments. Instead, a desmosome plaque attaches to other elements of the cytoskeleton known as intermediate filaments that consist of the protein keratin. The intermediate filaments extend from desmosomes on one side of the cell across the cytosol to desmosomes on the opposite side of the cell. This structural arrangement contributes to the stability of the cells and tissue. These spot-weld-like junctions are common among the cells that make up the epidermis (the outermost layer of the skin) and among cardiac muscle cells in

Figure 4.1 Cell junctions.

Most epithelial cells and some muscle and nerve cells contain cell junctions.

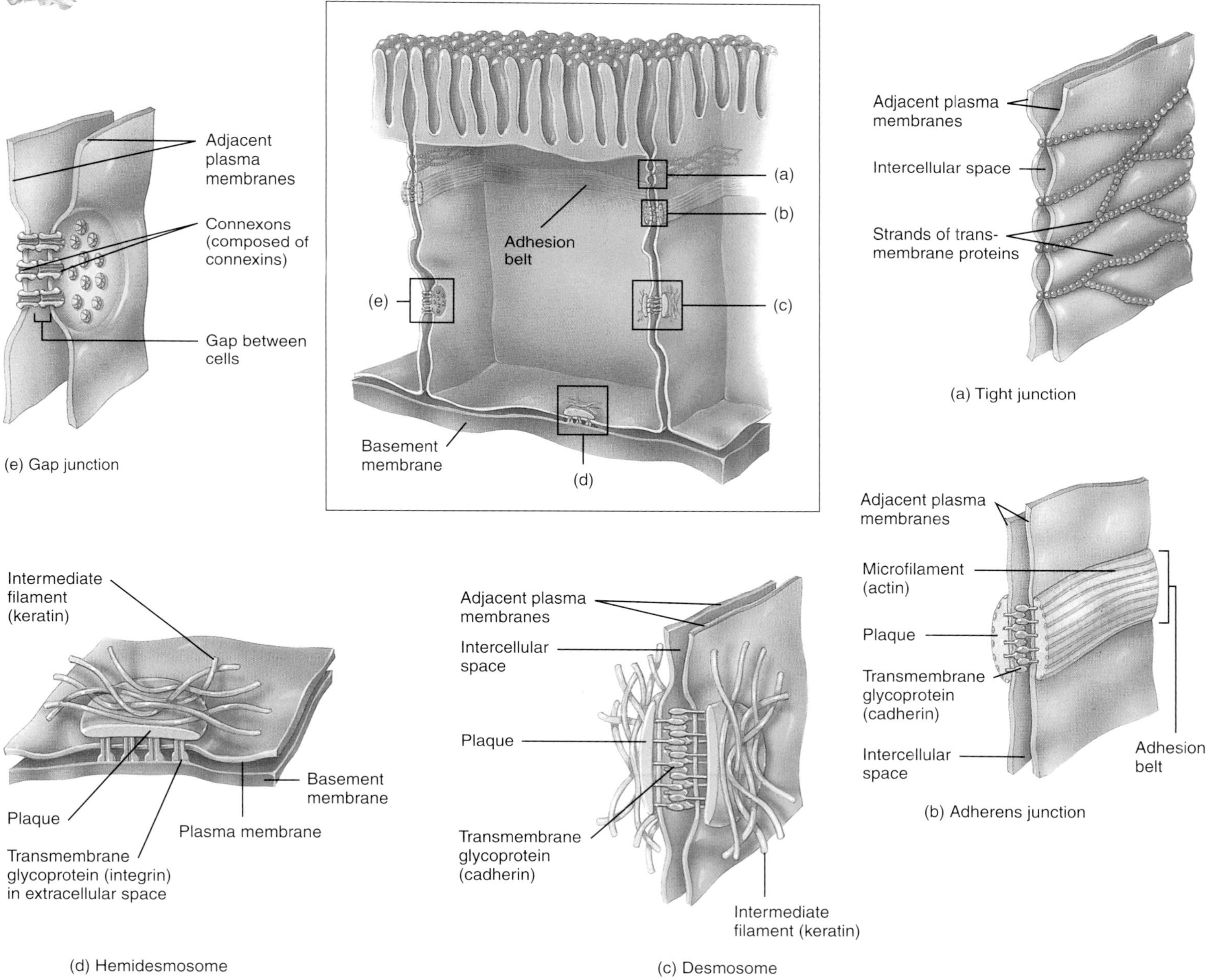

Which type of cell junction functions in communication between adjacent cells?

the heart. Desmosomes prevent epidermal cells from separating under tension and cardiac muscle cells from pulling apart during contraction.

Hemidesmosomes

Hemidesmosomes (*hemi-* = half) resemble desmosomes but they do not link adjacent cells. The name arises from the fact that they look like half of a desmosome (Figure 4.1d). However, the transmembrane glycoproteins in hemidesmosomes are **integrins** rather than cadherins. On the inside of the plasma membrane, integrins attach to intermediate filaments made of the protein keratin. On the outside of the plasma membrane, the integrins attach to the protein **laminin,** which is present in the basement membrane (discussed shortly). Thus, hemidesmosomes anchor cells not to each other but to the basement membrane.

Gap Junctions

At **gap junctions,** membrane proteins called **connexins** form tiny fluid-filled tunnels called **connexons** that connect neighboring cells (Figure 4.1e). The plasma membranes of gap junctions are not fused together as in tight junctions but are separated by a very narrow intercellular gap (space). Through the connexons, ions and small molecules can diffuse from the cytosol of one cell to another. The transfer of nutrients, and perhaps wastes, takes place through gap junctions in avascular tissues such as the lens and cornea of the eye. Gap junctions allow the cells in a tissue to communicate with one another. In a developing embryo, some of the chemical and electrical signals that regulate growth and cell differentiation travel via gap junctions. Gap junctions also enable nerve or muscle impulses to spread rapidly among cells, a process that is crucial for the normal operation of some parts of the nervous system and for the contraction of muscle in the heart, gastrointestinal tract, and uterus.

▶ CHECKPOINT

3. Which type of cell junctions allow cellular communication?
4. Which types of cell junctions are found in epithelial tissues?

EPITHELIAL TISSUE

▶ OBJECTIVES

Describe the general features of epithelial tissues.

List the location, structure, and function of each different type of epithelium.

An **epithelial tissue** (ep-i-THĒ-lē-al) or **epithelium** (plural is *epithelia*), consists of cells arranged in continuous sheets, in either single or multiple layers. Because the cells are closely packed and are held tightly together by many cell junctions, there is little intercellular space between adjacent plasma membranes.

The various surfaces of epithelial cells often differ in structure and have specialized functions. The **apical (free) surface** of an epithelial cell faces the body surface, a body cavity, the lumen (interior space) of an internal organ, or a tubular duct that receives cell secretions (Figure 4.2). Apical surfaces may contain cilia or microvilli. The **lateral surfaces** of an epithelial cell face the adjacent cells on either side. As you just learned, and saw in Figure 4.1, lateral surfaces may contain tight junctions, adherens junctions, desmosomes, and/or gap junctions. The **basal surface** of an epithelial cell is opposite the apical surface and the basal surfaces of the deepest layer of cells adhere to extracellular materials such as the basement membrane. Hemidesmosomes in the basal surfaces of the deepest layer of epithelial cells anchor the epithelium to the basement membrane. In discussing epithelia with multiple layers the term *apical layer* refers to the most superficial layer of cells, and the *basal layer* is the deepest layer of cells.

The **basement membrane** is a thin extracellular layer that commonly consists of two layers, the basal lamina and reticular lamina. The *basal lamina* (*lamina* = thin layer) is closer to—and secreted by—the epithelial cells. It contains proteins such as collagen (described shortly) and laminin, as well as glycoproteins

Figure 4.2 Surfaces of epithelial cells and the structure and location of the basement membrane.

The basement membrane is found between epithelium and connective tissue.

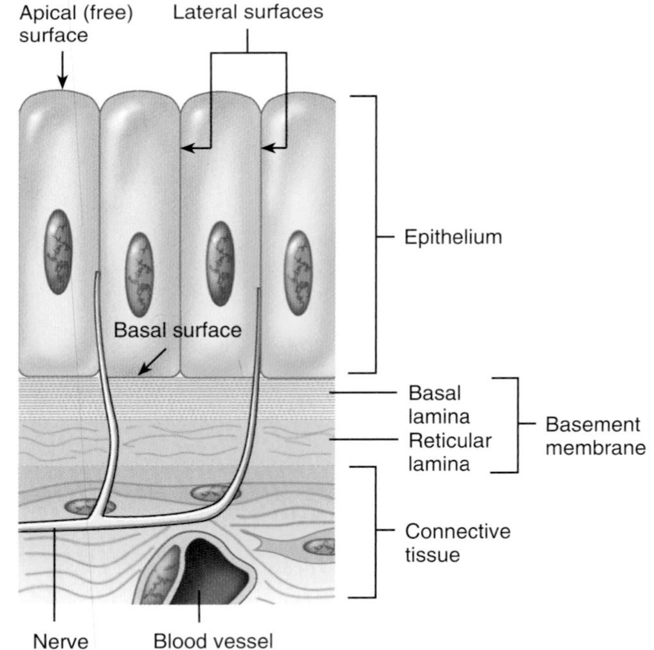

What is the function of the basement membrane?

and proteoglycans (also described shortly). As you have already learned, the laminin molecules in the basal lamina adhere to integrins in hemidesmosomes and thus attach epithelial cells to the basement membrane (see Figure 4.1d). The *reticular lamina* is closer to the underlying connective tissue and contains fibrous proteins produced by connective tissue cells called fibroblasts. The basement membrane functions as a point of attachment and support for the overlying epithelial tissue.

Basement Membranes and Disease

Under certain conditions, basement membranes become markedly thickened, due to increased production of collagen and laminin. In untreated cases of diabetes mellitus, the basement membrane of small blood vessels (capillaries) thickens, especially in the eyes and kidneys. Because of this the blood vessels cannot function properly and blindness and kidney failure may result. ■

Epithelial tissue has its own nerve supply, but is **avascular** (*a-* = without; *vascular* = vessel); that is, it lacks its own blood supply. The blood vessels that bring in nutrients and remove wastes are located in the adjacent connective tissue. Exchange of substances between epithelium and connective tissue occurs by diffusion.

Because epithelial tissue forms boundaries between the body's organs, or between the body and the external environment, it is repeatedly subjected to physical stress and injury. A high rate of cell division allows epithelial tissue to constantly renew and repair itself by sloughing off dead or injured cells and replacing them with new ones. Epithelial tissue plays many different roles in the body; the most important are protection, filtration, secretion, absorption, and excretion. In addition, epithelial tissue combines with nervous tissue to form special organs for smell, hearing, vision, and touch.

Epithelial tissue may be divided into two types. (1) **Covering and lining epithelium** forms the outer covering of the skin and some internal organs. It also forms the inner lining of blood vessels, ducts, and body cavities, and the interior of the respiratory, digestive, urinary, and reproductive systems. (2) **Glandular epithelium** makes up the secreting portion of glands such as the thyroid gland, adrenal glands, and sweat glands.

Covering and Lining Epithelium

The types of covering and lining epithelial tissue are classified according to two characteristics: the arrangement of cells into layers and the shapes of the cells (Figure 4.3).

1. ***Arrangement of cells in layers.*** The cells of covering and lining epithelia are arranged in one or more layers depending on the functions the epithelium performs:

a. *Simple epithelium* is a single layer of cells that functions in diffusion, osmosis, filtration, secretion, and absorption. **Secretion** is the production and release of substances such as mucus, sweat, or enzymes. **Absorption** is the intake of fluids or other substances such as digested food from the intestinal tract.

b. *Pseudostratified epithelium* (*pseudo-* = false) appears to

Figure 4.3 Cell shapes and arrangement of layers for covering and lining epithelium.

Cell shapes and arrangement of layers are the bases for classifying covering and lining epithelium.

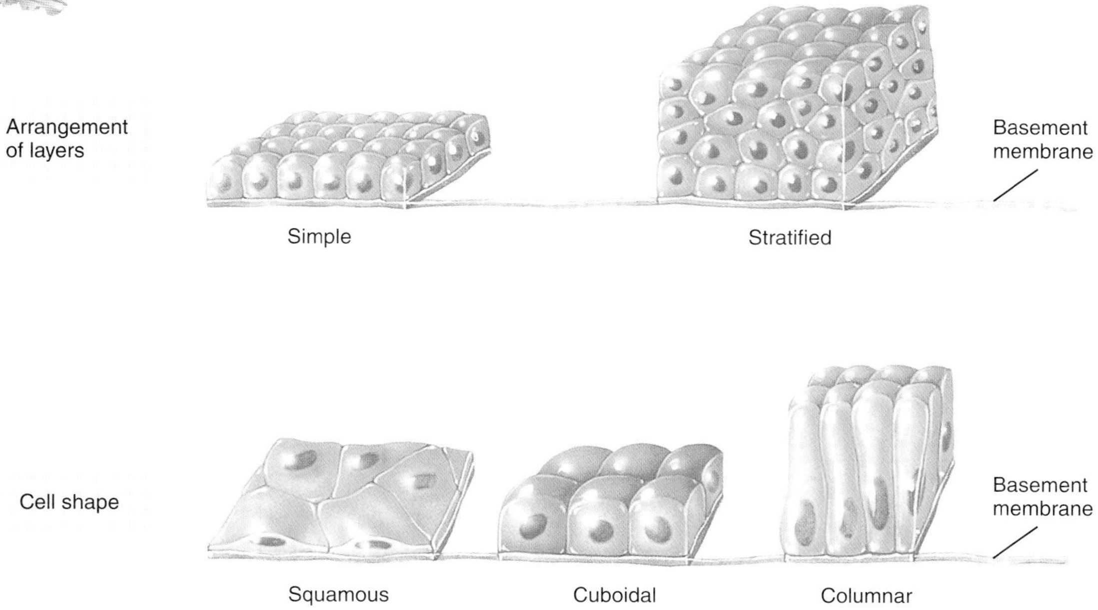

Which cell shape is best adapted for the rapid movement of substances from one cell to another?

have multiple layers of cells because the cell nuclei lie at different levels and not all cells reach the apical surface. Cells that do extend to the apical surface may contain cilia; others (goblet cells) secrete mucus. Pseudostratified epithelium is actually a simple epithelium because all its cells rest on the basement membrane.

   c. *Stratified epithelium* (*stratum* = layer) consists of two or more layers of cells that protect underlying tissues in locations where there is considerable wear and tear.

2. ***Cell shapes.***

   a. *Squamous* cells (SKWĀ-mus = flat) are arranged like floor tiles and are thin, which allows for the rapid passage of substances.

   b. *Cuboidal* cells are as tall as they are wide and are shaped like cubes or hexagons. They may have microvilli at their apical surface and function in either secretion or absorption.

   c. *Columnar* cells are much taller than they are wide, like columns, and protect underlying tissues. Their apical surfaces may have cilia or microvilli, and they often are specialized for secretion and absorption.

   d. *Transitional* cells change shape, from flat to cuboidal, as organs such as the urinary bladder stretch (distend) to a larger size and then collapse to a smaller size.

Combining the two characteristics (arrangements of layers and cell shapes), the types of covering and lining epithelia are as follows:

I. Simple epithelium
   A. Simple squamous epithelium
   B. Simple cuboidal epithelium
   C. Simple columnar epithelium (nonciliated and ciliated)
   D. Pseudostratified columnar epithelium (nonciliated and ciliated)

II. Stratified epithelium
   A. Stratified squamous epithelium (keratinized and nonkeratinized)*
   B. Stratified cuboidal epithelium*
   C. Stratified columnar epithelium*
   D. Transitional epithelium

Each of these covering and lining epithelia is described in the following sections and illustrated in Table 4.1. The illustration of each type consists of a photomicrograph, a corresponding diagram, and an inset that identifies a major location of the tissue in the body. Descriptions, locations, and functions of the tissues accompany each illustration.

*This classification is based on the shape of the cells at the *apical* surface.

Simple Epithelium

SIMPLE SQUAMOUS EPITHELIUM This tissue consists of a single layer of flat cells that resembles a tiled floor when viewed from the apical surface (Table 4.1A). The nucleus of each cell is a flattened oval or sphere and is centrally located. Simple squamous epithelium is present at sites where the processes of filtration (such as blood filtration in the kidneys) or diffusion (such as diffusion of oxygen into blood vessels of the lungs) occur. It is not found in body areas that are subject to mechanical stress (wear and tear).

The simple squamous epithelium that lines the heart, blood vessels, and lymphatic vessels is known as **endothelium** (*endo-* = within; *-thelium* = covering); the type that forms the epithelial layer of serous membranes such as the peritoneum is called **mesothelium** (*meso-* = middle). Unlike other epithelial tissues, which arise from embryonic ectoderm or endoderm, endothelium and mesothelium both are derived from embryonic mesoderm.

SIMPLE CUBOIDAL EPITHELIUM The cuboidal shape of the cells in this tissue (Table 4.1B) is obvious when the tissue is sectioned and viewed from the side. Cell nuclei are usually round and centrally located. Simple cuboidal epithelium is found in organs such as the thyroid gland and kidneys and performs the functions of secretion and absorption.

SIMPLE COLUMNAR EPITHELIUM When viewed from the side, the cells of simple columnar epithelium appear like columns, with oval nuclei near the base. Simple columnar epithelium exists in two forms: nonciliated simple columnar epithelium and ciliated simple columnar epithelium.

Nonciliated simple columnar epithelium contains two types of cells—columnar epithelial cells with microvilli at their apical surface, and goblet cells (Table 4.1C). **Microvilli,** fingerlike cytoplasmic projections, increase the surface area of the plasma membrane (see Figure 3.1 on page 61), thus increasing the rate of absorption by the cell. **Goblet cells** are modified columnar epithelial cells that secrete mucus, a slightly sticky fluid, at their apical surfaces. Before it is released, mucus accumulates in the upper portion of the cell, causing it to bulge out and making the whole cell resemble a goblet or wine glass. Secreted mucus serves as a lubricant for the linings of the digestive, respiratory, and reproductive tracts, and most of the urinary tract. Mucus also helps prevent destruction of the stomach lining by acidic gastric juice secreted by the stomach.

Ciliated simple columnar epithelium contains columnar epithelial cells with cilia at the apical surface (Table 4.1D). In certain parts of the airways of the upper respiratory tract, goblet cells are interspersed among ciliated columnar epithelia. Mucus secreted by the goblet cells forms a film over the airway surface that traps inhaled foreign particles. The cilia beat in unison, moving the mucus and any foreign particles toward the throat, where it can be coughed up and swallowed or spit out. Coughing and sneezing speed up the movement of cilia and mucus. Cilia also help move oocytes expelled from the ovaries through the uterine (fallopian) tubes into the uterus.

TABLE 4.1 Epithelial Tissues: Covering and Lining Epithelia

Simple Epithelium

A. Simple squamous epithelium

Description: Single layer of flat cells; centrally located nucleus.

Location: Lines heart, blood vessels, lymphatic vessels, air sacs of lungs, glomerular (Bowman's) capsule of kidneys, and inner surface of the tympanic membrane (eardrum); forms epithelial layer of serous membranes, such as the peritoneum.

Function: Filtration, diffusion, osmosis, and secretion in serous membranes.

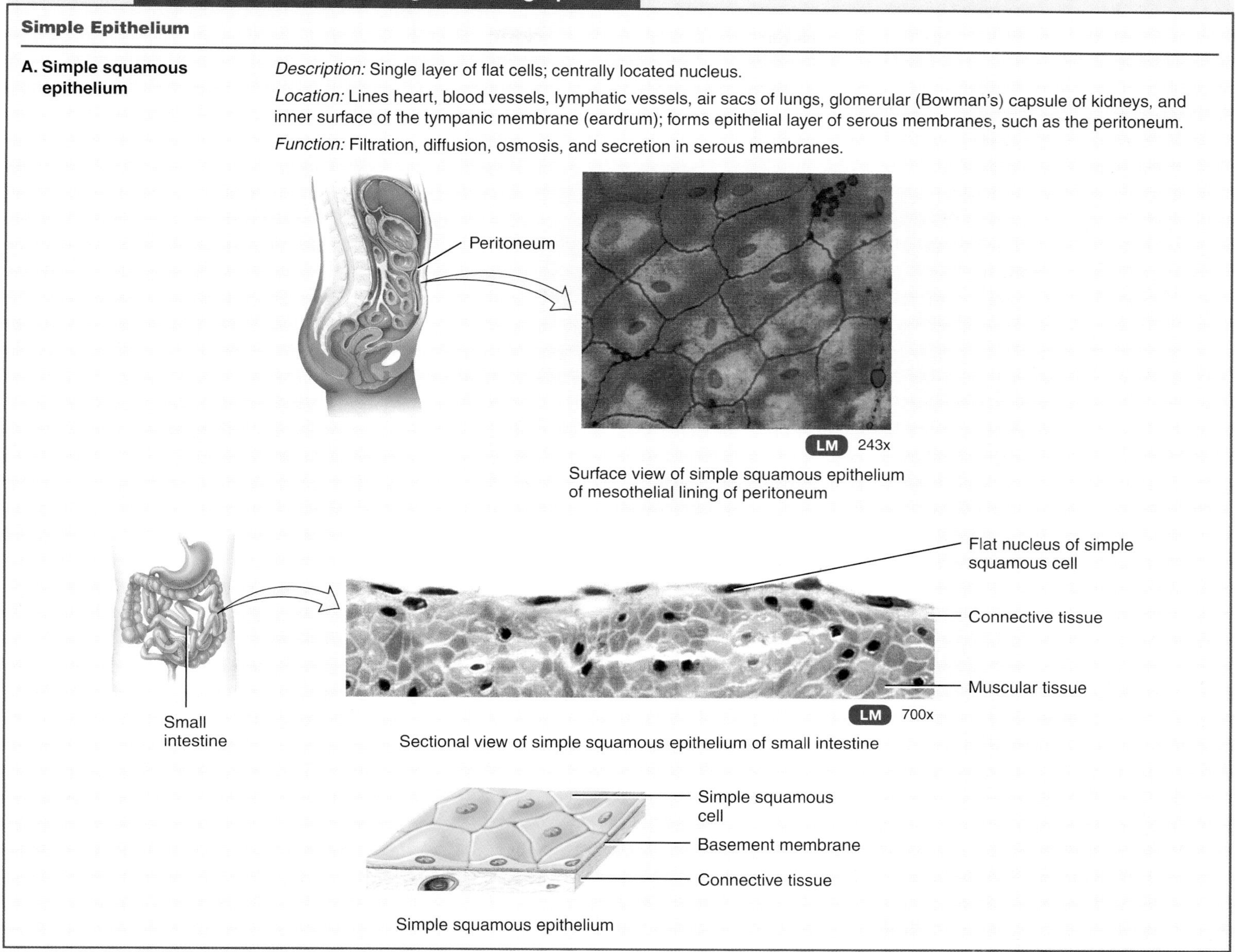

Surface view of simple squamous epithelium of mesothelial lining of peritoneum

Sectional view of simple squamous epithelium of small intestine

Simple squamous epithelium

continues

Pseudostratified Columnar Epithelium

As noted earlier, pseudostratified columnar epithelium appears to have several layers because the nuclei of the cells are at various depths (Table 4.1E). Even though all the cells are attached to the basement membrane in a single layer, some cells do not extend to the apical surface. When viewed from the side, these features give the false impression of a multilayered tissue—thus the name *pseudo*stratified epithelium (*pseudo-* = false). In *pseudostratified ciliated columnar epithelium,* the cells that extend to the surface either secrete mucus (goblet cells) or bear cilia. The secreted mucus traps foreign particles and the cilia sweep away mucus for eventual elimination from the body. *Pseudostratified nonciliated columnar epithelium* contains cells without cilia and lacks goblet cells.

Stratified Epithelium

In contrast to simple epithelium, stratified epithelium has two or more layers of cells. Because of this, it is more durable and can better protect underlying tissues. Some cells of stratified epithelia also produce secretions. The name of the specific kind of stratified epithelium depends on the shape of the cells in the apical layer.

Stratified Squamous Epithelium Cells in the apical layer of this type of epithelium are flat; those of the deep layers vary in shape from cuboidal to columnar (Table 4.1F). The basal (deepest) cells continually undergo cell division. As new cells grow, the cells of the basal layer are pushed upward toward the apical layer. As they move farther from the deeper layers and from their blood supply in the underlying connective tissue, they

TABLE 4.1 Epithelial Tissues: Covering and Lining Epithelia (continued)

Simple Epithelium

B. Simple cuboidal epithelium

Description: Single layer of cube-shaped cells; centrally located nucleus.

Location: Covers surface of ovary, lines anterior surface of capsule of the lens of the eye, forms the pigmented epithelium at the posterior surface of the eye, lines kidney tubules and smaller ducts of many glands, and makes up the secreting portion of some glands such as the thyroid gland and the ducts of some glands such as the pancreas.

Function: Secretion and absorption.

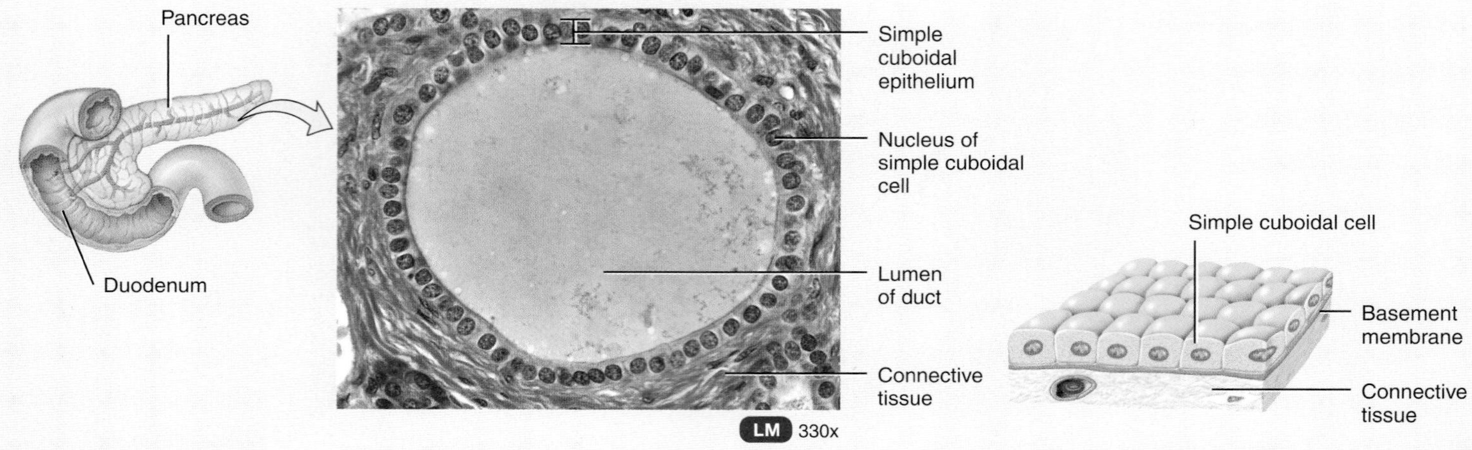

Sectional view of simple cuboidal epithelium of intralobular duct of pancreas

Simple cuboidal epithelium

C. Nonciliated simple columnar epithelium

Description: Single layer of nonciliated column-like cells with nuclei near base of cells; contains goblet cells and cells with microvilli in some locations.

Location: Lines the gastrointestinal tract (from the stomach to the anus), ducts of many glands, and gallbladder.

Function: Secretion and absorption.

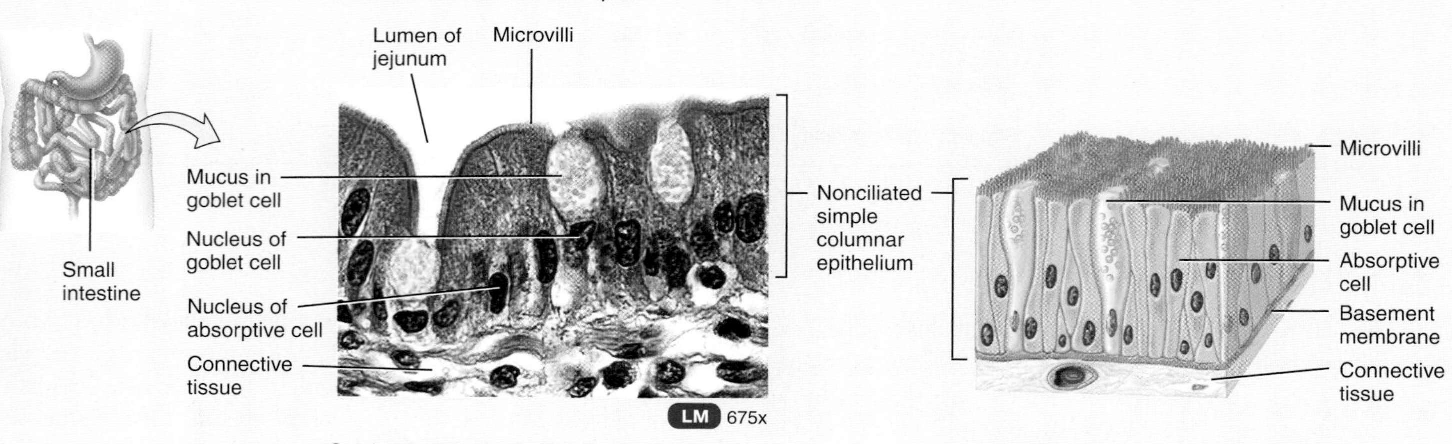

Sectional view of nonciliated simple columnar epithelium of lining of jejunum of small intestine

Nonciliated simple columnar epithelium

Simple Epithelium

D. Ciliated simple columnar epithelium

Description: Single layer of ciliated column-like cells with nuclei near base; contains goblet cells in some locations.

Location: Lines a few portions of upper respiratory tract, uterine (fallopian) tubes, uterus, some paranasal sinuses, central canal of spinal cord, and ventricles of the brain.

Function: Moves mucus and other substances by ciliary action.

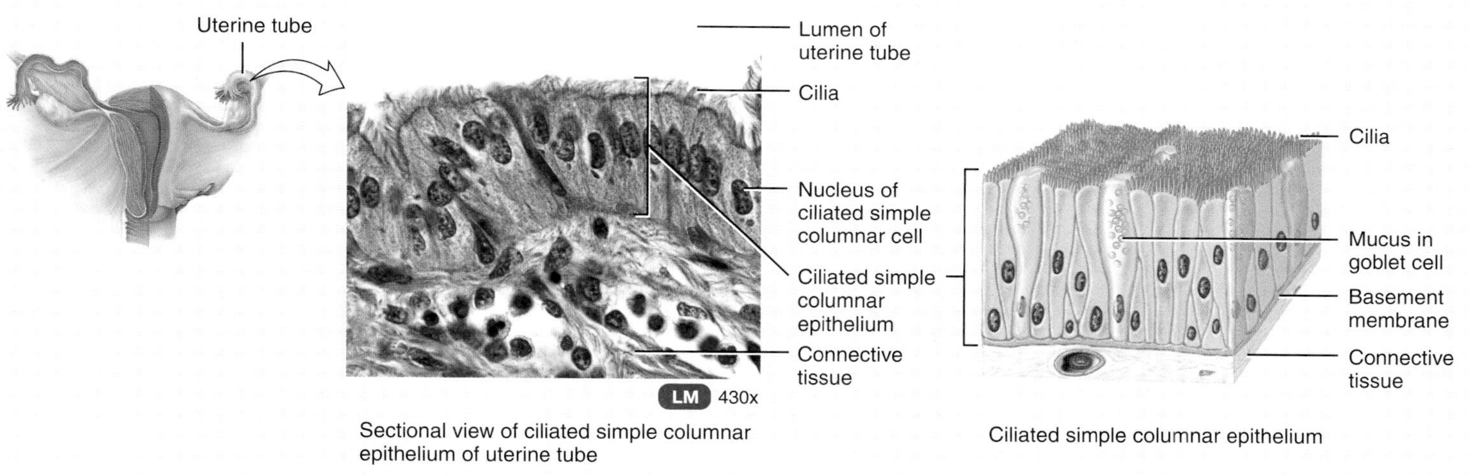

Sectional view of ciliated simple columnar epithelium of uterine tube

Ciliated simple columnar epithelium

E. Pseudostratified columnar epithelium

Description: Not a true stratified tissue; nuclei of cells are at different levels; all cells are attached to basement membrane, but not all reach the apical surface.

Location: Pseudostratified ciliated columnar epithelium lines the airways of most of upper respiratory tract; pseudostratified nonciliated columnar epithelium lines larger ducts of many glands, epididymis, and part of male urethra.

Function: Secretion and movement of mucus by ciliary action.

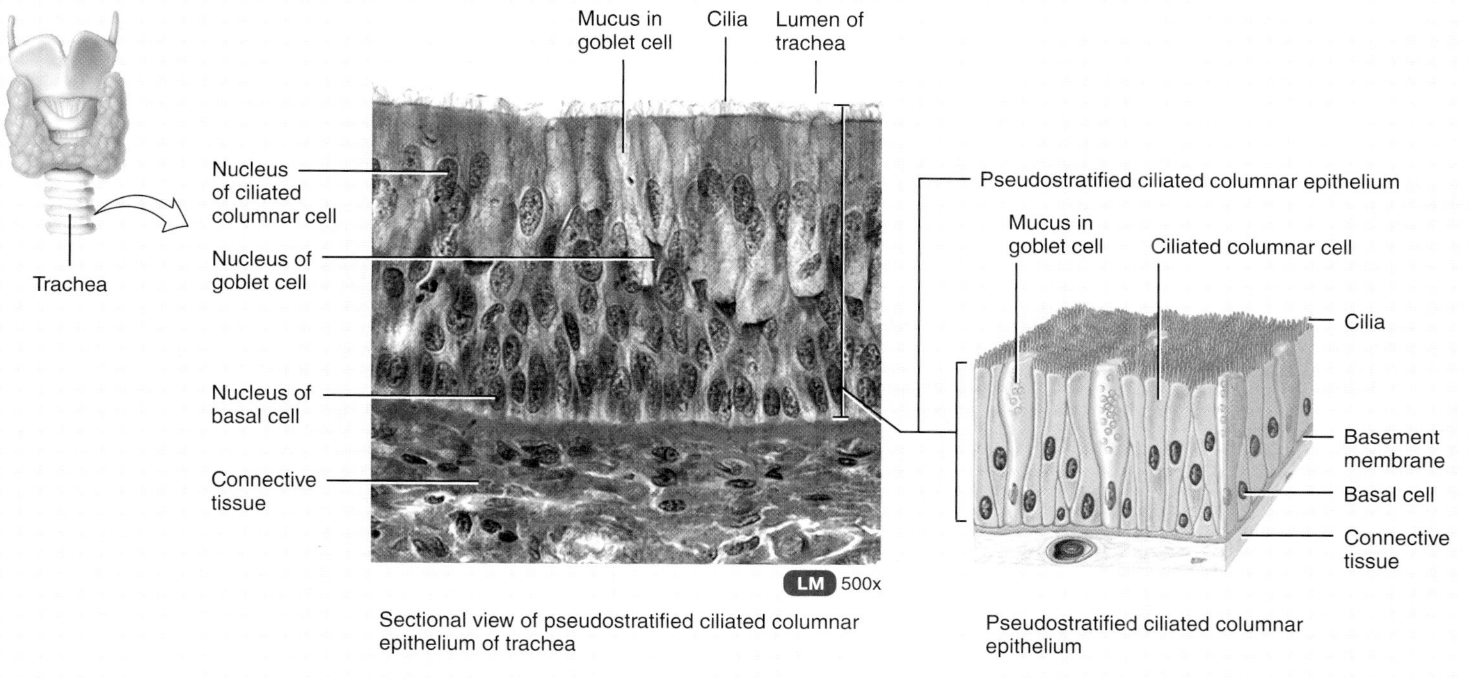

Sectional view of pseudostratified ciliated columnar epithelium of trachea

Pseudostratified ciliated columnar epithelium

continues

TABLE 4.1 Epithelial Tissues: Covering and Lining Epithelia (continued)

Stratified Epithelium

F. Stratified squamous epithelium

Description: Several layers of cells; cuboidal to columnar shape in deep layers; squamous cells form the apical layer and several layers deep to it; cells from the basal layer replace surface cells as they are lost.

Location: Keratinized variety forms superficial layer of skin; nonkeratinized variety lines wet surfaces, such as lining of the mouth, esophagus, part of epiglottis, part of pharynx, and vagina, and covers the tongue.

Function: Protection.

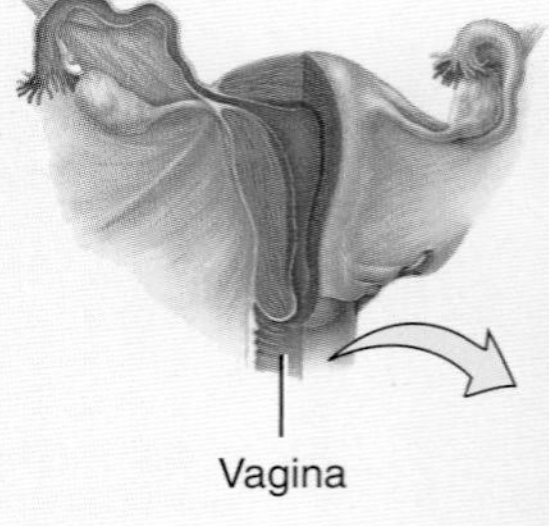

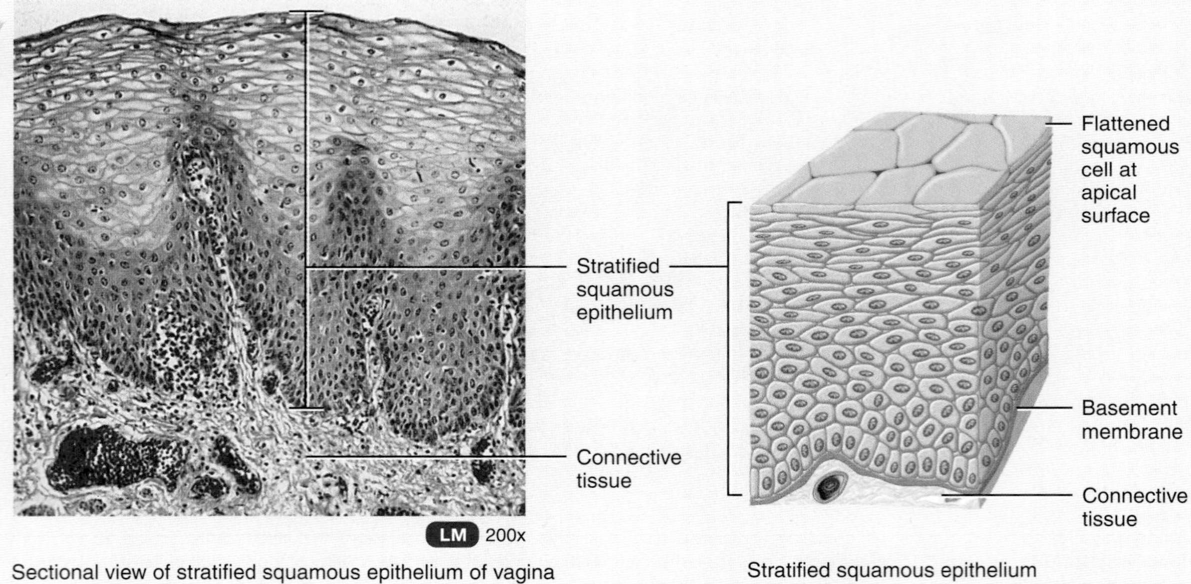

Sectional view of stratified squamous epithelium of vagina

Stratified squamous epithelium

G. Stratified cuboidal epithelium

Description: Two or more layers of cells in which the cells in the apical layer are cube-shaped.

Location: Ducts of adult sweat glands and esophageal glands and part of male urethra.

Function: Protection and limited secretion and absorption.

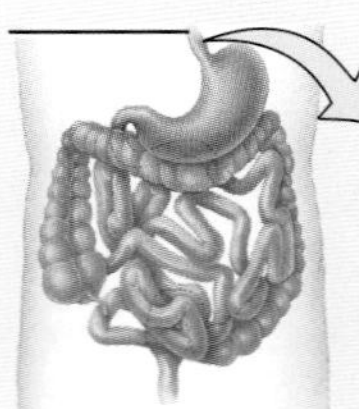

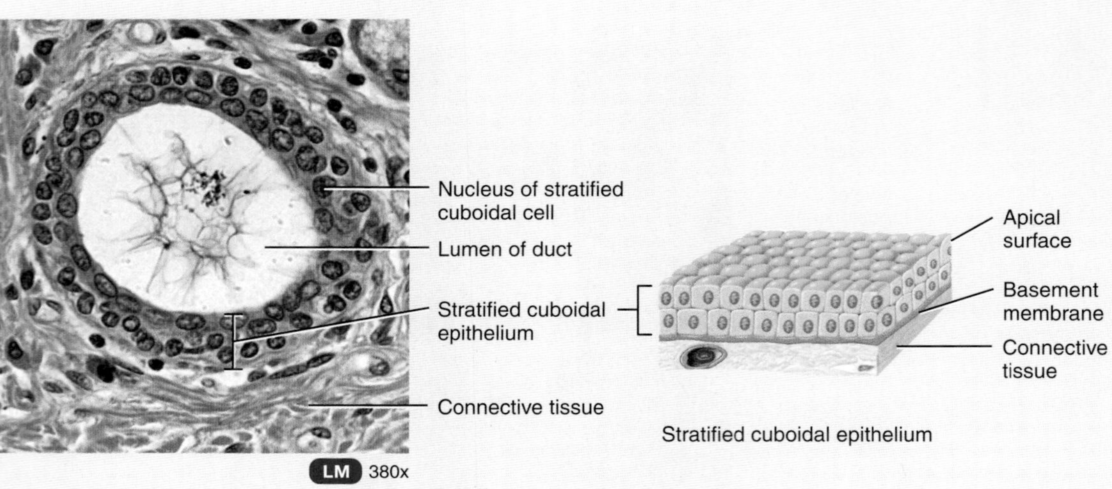

Sectional view of stratified cuboidal epithelium of the duct of an esophageal gland

Stratified cuboidal epithelium

Stratified Epithelium

H. Stratified columnar epithelium

Description: Several layers of irregularly shaped cells; only the apical layer has columnar cells.

Location: Lines part of urethra, large excretory ducts of some glands, such as esophageal glands, small areas in anal mucous membrane, and part of the conjunctiva of the eye.

Function: Protection and secretion.

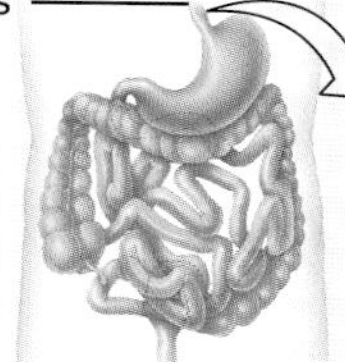

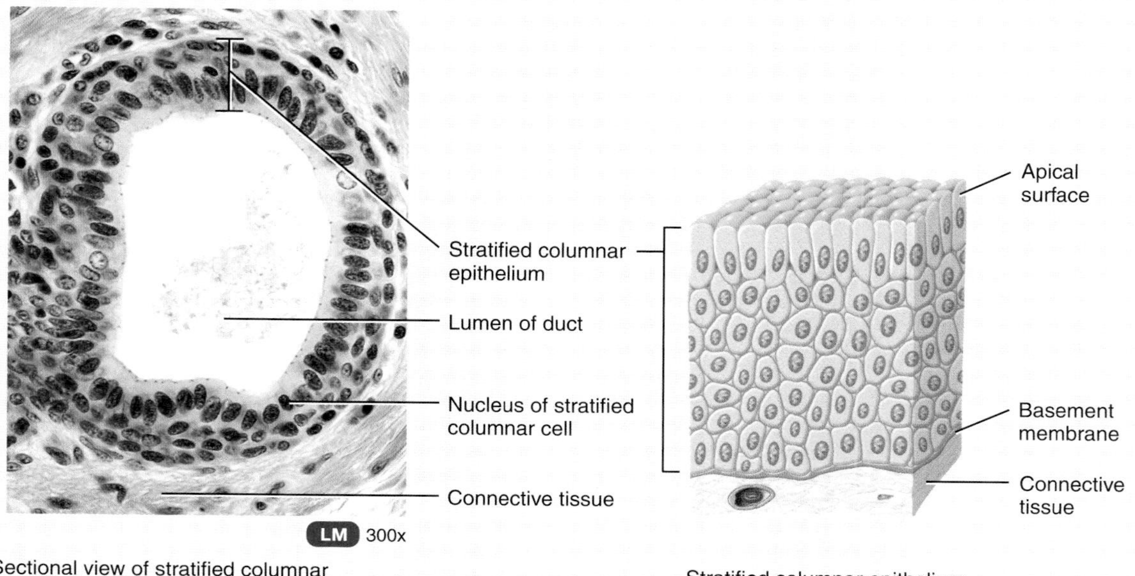

Sectional view of stratified columnar epithelium of the duct of an esophageal gland

Stratified columnar epithelium

I. Transitional epithelium

Description: Appearance is variable (transitional); shape of cells in apical layer ranges from squamous (when stretched) to cuboidal (when relaxed).

Location: Lines urinary bladder and portions of ureters and urethra.

Function: Permits distension.

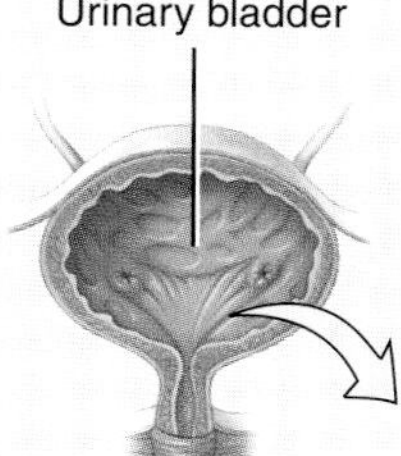

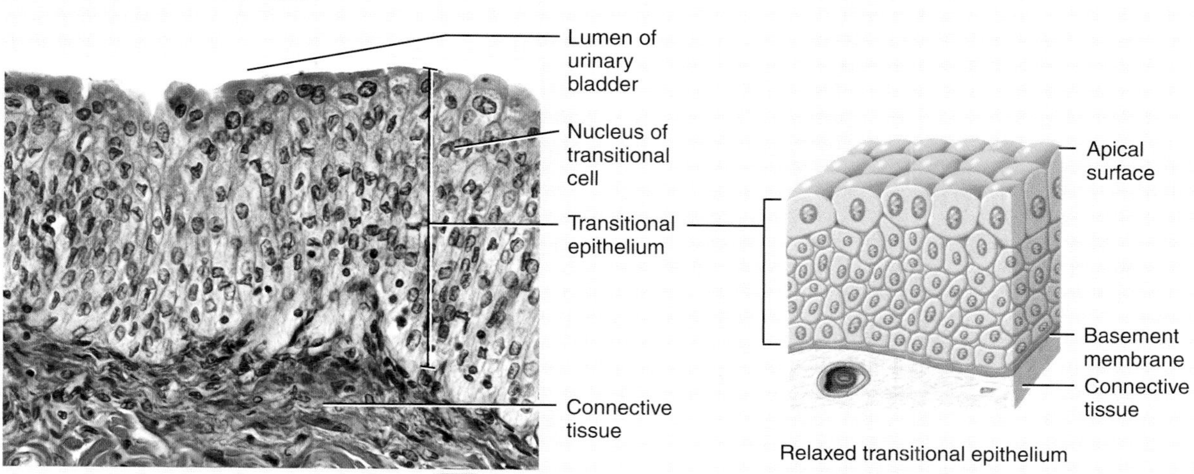

Sectional view of transitional epithelium of urinary bladder in relaxed state

Relaxed transitional epithelium

become dehydrated, shrunken, and harder, and then die. At the apical layer, after the dead cells lose their cell junctions they are sloughed off, but they are replaced continuously as new cells emerge from the basal layer.

Stratified squamous epithelium exists in both keratinized and nonkeratinized forms. In *keratinized stratified squamous epithelium,* the apical layer and several layers deep to it are partially dehydrated and contain a layer of **keratin,** a tough, fibrous protein that helps protect the skin and underlying tissues from heat, microbes, and chemicals. Keratinized stratified squamous epithelium forms the superficial layer of the skin. *Nonkeratinized stratified squamous epithelium,* which is found, for example, lining the mouth and esophagus, does not contain keratin in the apical layer and several layers deep to it and remains moist. Both types form the first line of defense against microbes.

Papanicolaou Test

A **Papanicolaou test** (pa-pa-NI-kō-lō), also called a **Pap test** or **Pap smear,** involves collection and microscopic examination of epithelial cells that have been scraped off the apical layer of a tissue. A very common type of Pap test involves examining the cells from the nonkeratinized stratified squamous epithelium of the vagina and cervix (inferior portion) of the uterus. This type of Pap test is performed mainly to detect early changes in the cells of the female reproductive system that may indicate cancer or a precancerous condition. An annual Pap test is recommended for all women as part of a routine pelvic exam.

Stratified Cuboidal Epithelium This is a fairly rare type of epithelium in which cells in the apical layer are cuboidal (Table 4.1G). Stratified cuboidal epithelium mainly serves a protective function, but it also has a limited role in secretion and absorption.

Stratified Columnar Epithelium Like stratified cuboidal epithelium, stratified columnar epithelium also is uncommon. Usually the basal layers consist of shortened, irregularly shaped cells; only the apical layer has cells that are columnar in shape (Table 4.1H). This type of epithelium functions in protection and secretion.

Transitional Epithelium Transitional epithelium, a type of stratified epithelium, is present only in the urinary system and has a variable appearance. In its relaxed or unstretched state (Table 4.1I), transitional epithelium looks like stratified cuboidal epithelium, except that the cells in the apical layer tend to be large and rounded. As the tissue is stretched, its cells become flatter, giving the appearance of stratified squamous epithelium (Table 4.1I). Because of its elasticity, transitional epithelium is ideal for lining hollow structures that are subjected to expansion from within, such as the urinary bladder. It allows the urinary bladder to stretch to hold a variable amount of fluid without rupturing.

Glandular Epithelium

The function of glandular epithelium, secretion, is accomplished by glandular cells that often lie in clusters deep to the covering and lining epithelium. A **gland** may consist of a single cell or a group of cells that secrete substances into ducts (tubes), onto a surface, or into the blood. All glands of the body are classified as either endocrine or exocrine.

The secretions of **endocrine glands** (Table 4.2A) enter the interstitial fluid and then diffuse directly into the bloodstream without flowing through a duct. These secretions, called *hormones,* regulate many metabolic and physiological activities to maintain homeostasis. The pituitary, thyroid, and adrenal glands are examples of endocrine glands. Endocrine glands will be described in detail in Chapter 18.

Exocrine glands (*exo-* = outside; *-crine* = secretion; Table 4.2B) secrete their products into ducts that empty onto the surface of a covering and lining epithelium such as the skin surface or the lumen of a hollow organ. The secretions of exocrine glands include mucus, sweat, oil, earwax, saliva, and digestive enzymes. Examples of exocrine glands include sudoriferous (sweat) glands, which produce sweat to help lower body temperature, and salivary glands, which secrete saliva. Saliva contains mucus and digestive enzymes among other substances. As you will learn later in the text, some glands of the body, such as the pancreas, ovaries, and testes, are mixed glands that contain both endocrine and exocrine tissue.

Structural Classification of Exocrine Glands

Exocrine glands are classified as unicellular or multicellular. As the name implies, **unicellular glands** are single-celled. Goblet cells are important unicellular exocrine glands that secrete mucus directly onto the apical surface of a lining epithelium. Most glands are **multicellular glands,** composed of many cells that form a distinctive microscopic structure or macroscopic organ. Examples include sudoriferous, sebaceous (oil), and salivary glands.

Multicellular glands are categorized according to two criteria: (1) whether their ducts are branched or unbranched and (2) the shape of the secretory portions of the gland (Figure 4.4 on page 120). If the duct of the gland does not branch, it is a **simple gland.** If the duct branches, it is a **compound gland.** Glands with tubular secretory parts are **tubular glands;** those with more rounded secretory portions are **acinar glands** (AS-i-nar; *acin-* = berry), also called *alveolar glands.* **Tubuloacinar glands** have both tubular and more rounded secretory parts.

Combinations of these features are the criteria for the following structural classification scheme for multicellular exocrine glands:

I. Simple glands
   A. **Simple tubular.** Tubular secretory part is straight and attaches to a single unbranched duct. Example: glands in the large intestine.

TABLE 4.2 Epithelial Tissue: Glandular Epithelium

A. Endocrine glands

Description: Secretory products (hormones) diffuse into blood after passing through interstitial fluid.

Location: Examples include pituitary gland at base of brain, pineal gland in brain, thyroid and parathyroid glands near larynx (voice box), adrenal glands superior to kidneys, pancreas near stomach, ovaries in pelvic cavity, testes in scrotum, and thymus in thoracic cavity.

Function: Produce hormones that regulate various body activities.

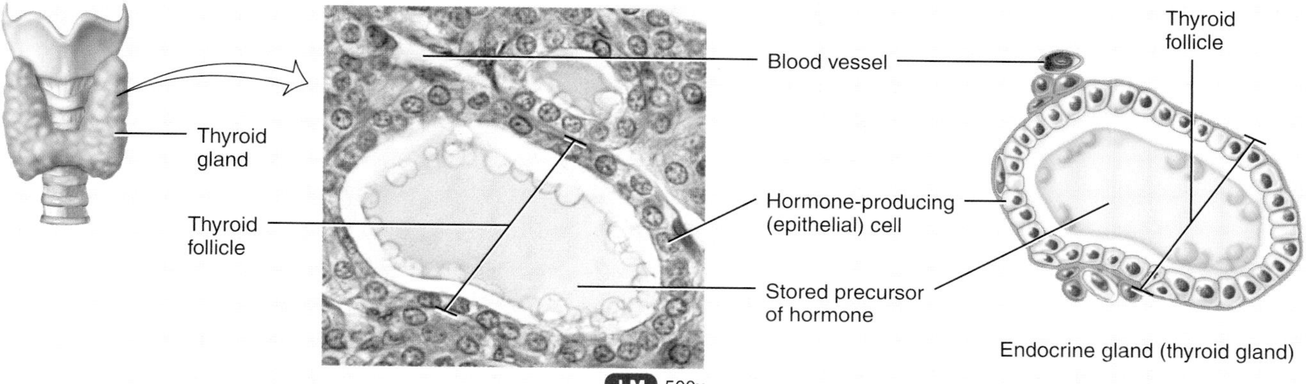

Sectional view of endocrine gland (thyroid gland)

Endocrine gland (thyroid gland)

B. Exocrine glands

Description: Secretory products released into ducts.

Location: Sweat, oil, and earwax glands of the skin; digestive glands such as salivary glands, which secrete into mouth cavity, and pancreas, which secretes into the small intestine.

Function: Produce substances such as sweat, oil, earwax, saliva, or digestive enzymes.

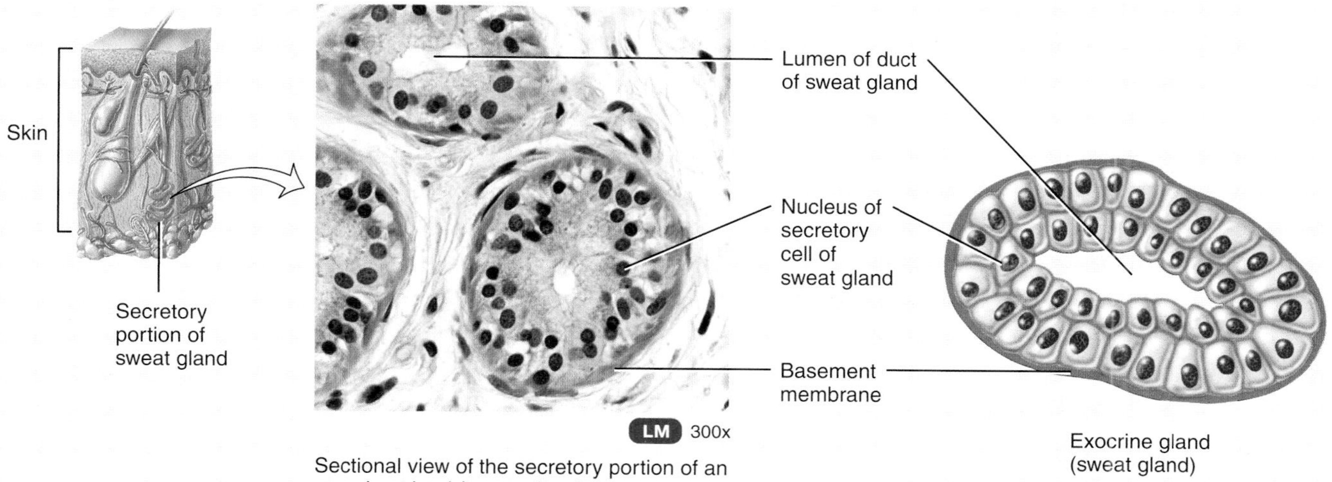

Sectional view of the secretory portion of an exocrine gland (sweat gland)

Exocrine gland (sweat gland)

Figure 4.4 Multicellular exocrine glands. Pink represents the secretory portion; lavender represents the duct.

Structural classification of multicellular exocrine glands is based on the branching pattern of the duct and the shape of the secreting portion.

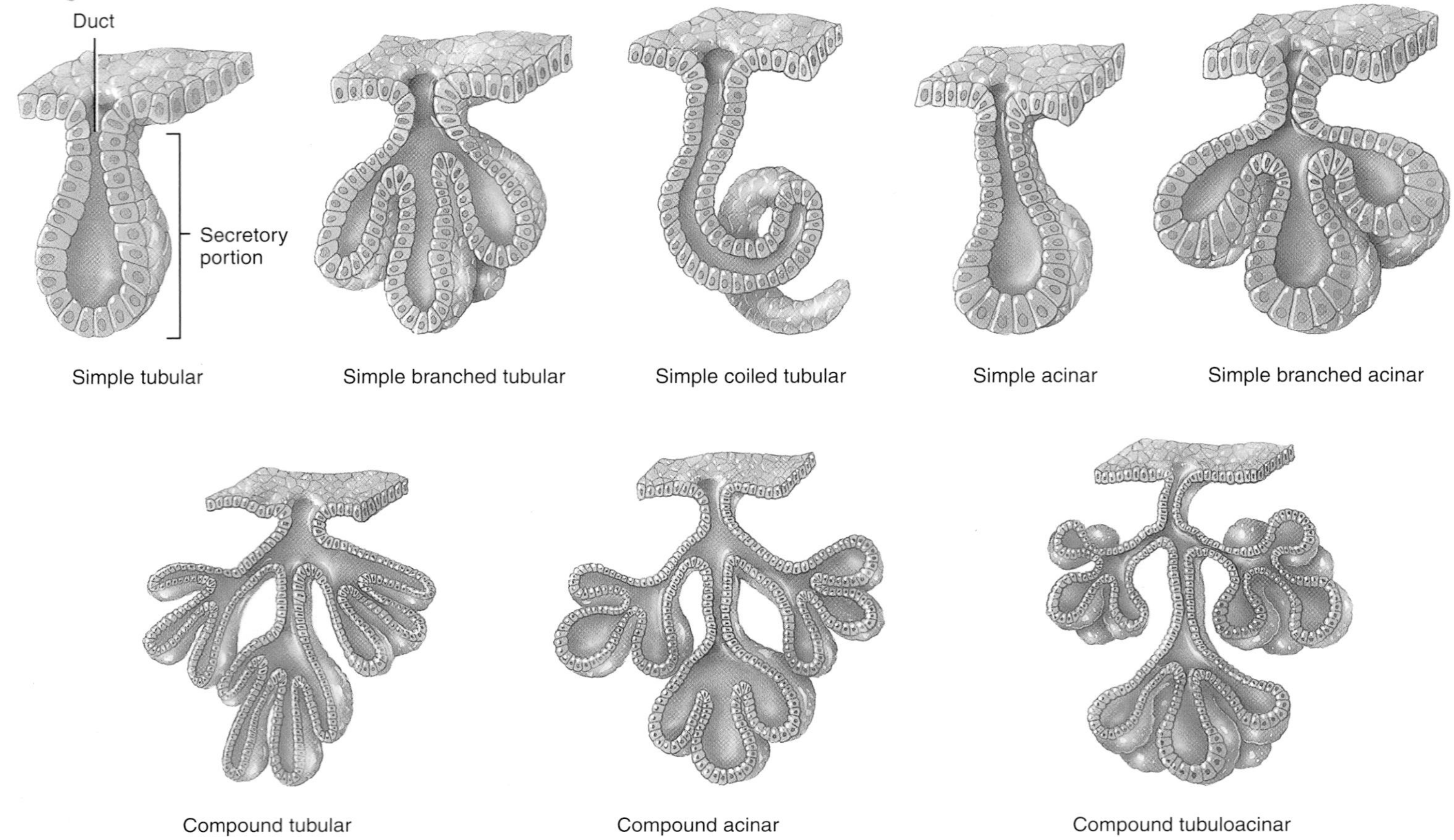

Duct

Secretory portion

Simple tubular

Simple branched tubular

Simple coiled tubular

Simple acinar

Simple branched acinar

Compound tubular

Compound acinar

Compound tubuloacinar

How do simple multicellular glands differ from compound ones?

B. **Simple branched tubular.** Tubular secretory part is branched and attaches to a single unbranched duct. Example: gastric glands.

C. **Simple coiled tubular.** Tubular secretory part is coiled and attaches to a single unbranched duct. Example: sweat glands.

D. **Simple acinar.** Secretory portion is rounded and attaches to a single unbranched duct. Example: glands of the penile urethra.

E. **Simple branched acinar.** Rounded secretory part is branched and attaches to a single unbranched duct. Example: sebaceous glands.

II. Compound glands

A. **Compound tubular.** Secretory portion is tubular and attaches to a branched duct. Example: bulbourethral (Cowper's) glands.

B. **Compound acinar.** Secretory portion is rounded and attaches to a branched duct. Example: mammary glands.

C. **Compound tubuloacinar.** Secretory portion is both tubular and rounded and attaches to a branched duct. Example: acinar glands of the pancreas.

Functional Classification of Exocrine Glands

The functional classification of exocrine glands is based on how their secretions are released. Secretions of **merocrine glands** (MER-ō-krin; *mero-* = a part) are synthesized on ribosomes attached to rough ER; processed, sorted, and packaged by the Golgi complex; and released from the cell in secretory vesicles via exocytosis (Figure 4.5a). Most exocrine glands of the body are merocrine glands. Examples include the salivary glands and pancreas. **Apocrine glands** (AP-ō-krin; *apo-* = from) accumulate their secretory product at the apical surface of the secreting cell. Then, that portion of the cell pinches off from the rest of the cell to release the secretion (Figure 4.5b). The remaining part of the cell repairs itself and repeats the process. Electron micrographic studies have called into question whether humans have apocrine glands. What were once thought to be apocrine glands— for example, mammary glands that secrete milk—are probably merocrine

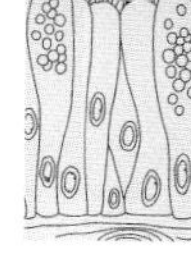

glands. The cells of **holocrine glands** (HŌ-lō-krin; *holo-* = entire) accumulate a secretory product in their cytosol. As the secretory cell matures, it ruptures and becomes the secretory product (Figure 4.5c). The sloughed off cell is replaced by a new cell. One example of a holocrine gland is a sebaceous gland of the skin.

Figure 4.5 Functional classification of multicellular exocrine glands.

The functional classification of exocrine glands is based on whether a secretion is a product of a cell or consists of an entire or a partial glandular cell.

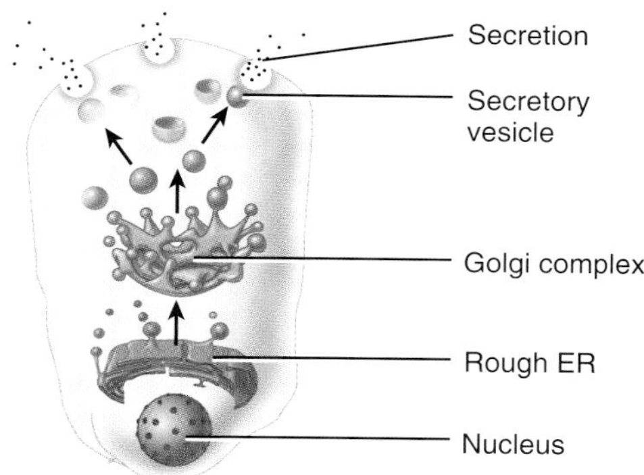

(a) Merocrine secretion

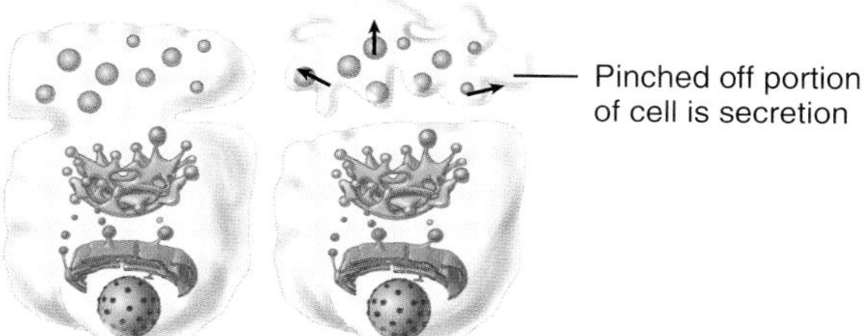

(b) Apocrine secretion

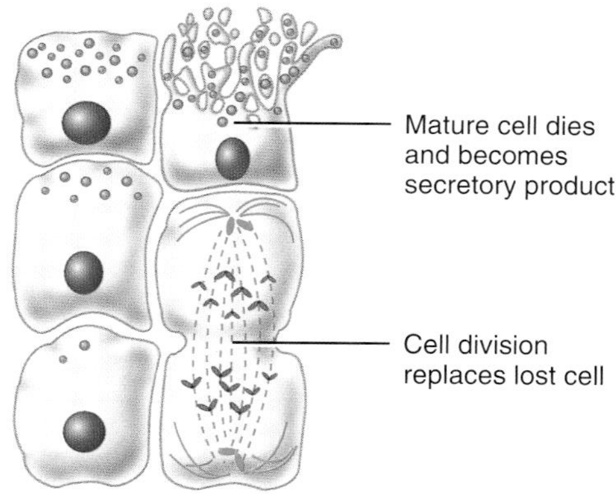

(c) Holocrine secretion

What class of glands are sebaceous (oil) glands? Salivary glands?

▶ CHECKPOINT

5. Describe the various layering arrangements and cell shapes of epithelium.
6. What characteristics are common to all epithelial tissues?
7. How is the structure of the following kinds of epithelium related to their functions: simple squamous, simple cuboidal, simple columnar (nonciliated and ciliated), pseudostratified columnar (ciliated and nonciliated), stratified squamous (keratinized and nonkeratinized), stratified cuboidal, stratified columnar, and transitional?
8. Where are endothelium and mesothelium located?
9. What distinguishes endocrine glands and exocrine glands? Name and give examples of the three functional classes of exocrine glands.

CONNECTIVE TISSUE

▶ OBJECTIVES

Describe the general features of connective tissue.

Describe the structure, location, and function of the various types of connective tissue.

Connective tissue is one of the most abundant and widely distributed tissues in the body. In its various forms, connective tissue has a variety of functions. It binds together, supports, and strengthens other body tissues; protects and insulates internal organs; compartmentalizes structures such as skeletal muscles; serves as the major transport system within the body (blood, a fluid connective tissue); is the primary location of stored energy reserves (adipose, or fat, tissue); and is the main source of immune responses.

General Features of Connective Tissue

Connective tissue consists of two basic elements: cells and extracellular matrix. A connective tissue's **extracellular matrix** is the material located between its widely spaced cells. The extracellular matrix consists of protein fibers and ground substance, the material between the cells and the fibers. The extracellular matrix is usually secreted by the connective tissue cells and determines the tissue's qualities. For instance, in cartilage, the extracellular matrix is firm but pliable. The extracellular matrix of bone, by contrast, is hard and inflexible.

In contrast to epithelia, connective tissues do not usually occur on body surfaces. Also unlike epithelia, connective tissues usually are highly vascular; that is, they have a rich blood supply. Exceptions include cartilage, which is avascular, and tendons, with a scanty blood supply. Except for cartilage, connective tissues, like epithelia, are supplied with nerves.

Connective Tissue Cells

Mesodermal embryonic cells called mesenchymal cells give rise to the cells of connective tissue. Each major type of connective tissue contains an immature class of cells with a name ending in *-blast,* which means "to bud or sprout." These immature cells are called *fibroblasts* in loose and dense connective tissue, *chondroblasts* in cartilage, and *osteoblasts* in bone. Blast cells retain the capacity for cell division and secrete the matrix that is characteristic of the tissue. In cartilage and bone, once the matrix is produced, the immature cells differentiate into mature cells with names ending in *-cyte,* namely chondrocytes and osteocytes. Mature cells have reduced capacities for cell division and matrix formation and are mostly involved in maintaining the matrix.

The types of connective tissue cells vary according to the type of tissue and include the following (Figure 4.6):

1. **Fibroblasts** (FĪ-brō-blasts; *fibro-* = fibers) are large, flat cells with branching processes. They are present in several connective tissues, and usually are the most numerous. Fibroblasts migrate through the connective tissue, secreting the fibers and ground substance of the extracellular matrix.
2. **Macrophages** (MAK-rō-fā-jez; *macro-* = large; *-phages* = eaters) develop from monocytes, a type of white blood cell. Macrophages have an irregular shape with short branching projections and are capable of engulfing bacteria and cellular debris by phagocytosis. *Fixed macrophages* reside in a particular tissue; examples include alveolar macrophages in the lungs or spleen macrophages in the spleen. *Wandering macrophages* have the ability to move throughout the tissue and gather at sites of infection or inflammation to carry on phagocytosis.
3. **Plasma cells** are small cells that develop from a type of white blood cell called a B lymphocyte. Plasma cells secrete antibodies, proteins that attack or neutralize foreign substances in the body. Thus, plasma cells are an important part of the body's immune response. Although they are found in many places in the body, most plasma cells reside in connective tissues, especially in the gastrointestinal and respiratory tracts. They are also abundant in the salivary glands, lymph nodes, spleen, and red bone marrow.
4. **Mast cells** are abundant alongside the blood vessels that supply connective tissue. They produce histamine, a chemical that dilates small blood vessels as part of the inflammatory response, the body's reaction to injury or infection. In addition, researchers have recently discovered that mast cells can bind to, ingest, and kill bacteria.
5. **Adipocytes,** also called fat cells or adipose cells, are connective tissue cells that store triglycerides (fats). They are found deep to the skin and around organs such as the heart and kidneys.
6. **White blood cells** are not found in significant numbers in normal connective tissue. However, in response to certain conditions they migrate from blood into connective tissues. For example, *neutrophils* gather at sites of infection, and *eosinophils* migrate to sites of parasitic invasions and allergic responses.

Connective Tissue Extracellular Matrix

Each type of connective tissue has unique properties, based on the specific extracellular materials between the cells. The extracellular matrix consists of two major components: (1) ground substance and (2) fibers.

Ground Substance

As noted earlier, the **ground substance** is the component of a connective tissue between the cells and fibers. The ground

Figure 4.6 Representative cells and fibers present in connective tissues.

Fibroblasts are usually the most numerous connective tissue cells.

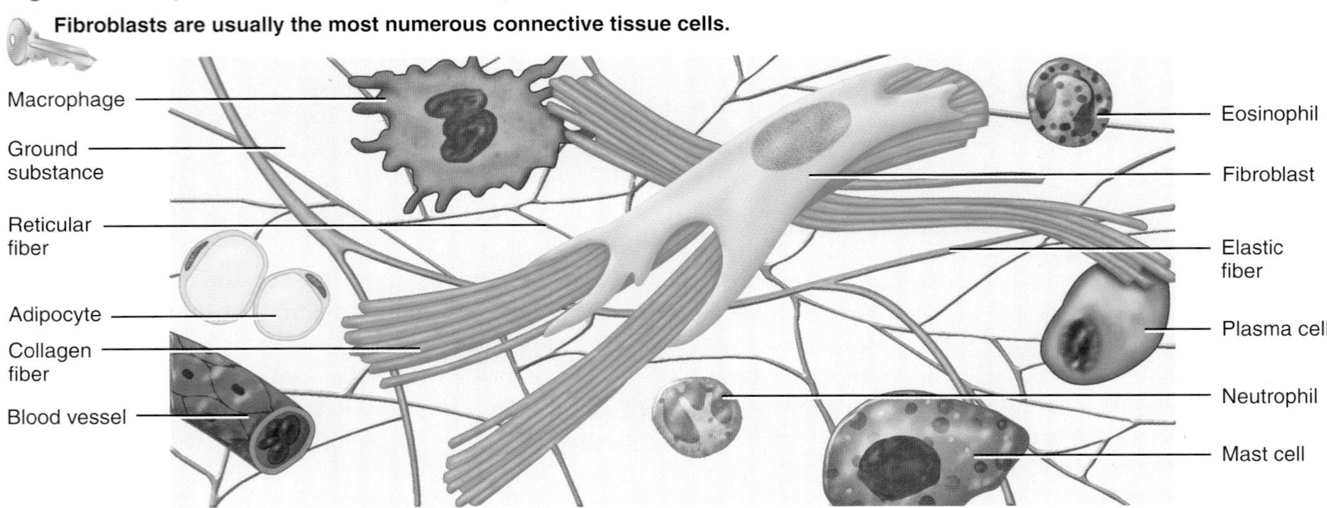

What is the function of fibroblasts?

substance may be fluid, semifluid, gelatinous, or calcified. The ground substance supports cells, binds them together, stores water, and provides a medium through which substances are exchanged between the blood and cells. It plays an active role in how tissues develop, migrate, proliferate, and change shape, and in how they carry out their metabolic functions.

Ground substance contains water and an assortment of large organic molecules, many of which are complex combinations of polysaccharides and proteins. The polysaccharides include hyaluronic acid, chondroitin sulfate, dermatan sulfate, and keratan sulfate. Collectively, they are referred to as **glycosaminoglycans** (glī-kos-a-mē′-nō-GLĪ-kans) or **GAGs.** Except for hyaluronic acid, the GAGs are associated with proteins called **proteoglycans** (prō-tē-ō-GLĪ-kans). The proteoglycans form a core protein and the GAGs project from the protein like the bristles of a brush. One of the most important properties of GAGs is that they trap water, making the ground substance more jellylike.

Hyaluronic acid (hī′-a-loo-RON-ik) is a viscous, slippery substance that binds cells together, lubricates joints, and helps maintain the shape of the eyeballs. White blood cells, sperm cells, and some bacteria produce *hyaluronidase,* an enzyme that breaks apart hyaluronic acid, thus causing the ground substance of connective tissue to become more liquid. The ability to produce hyaluronidase helps white blood cells move more easily through connective tissues to reach sites of infection and aids penetration of an oocyte by a sperm cell during fertilization. It also accounts for the rapid spread of bacteria through connective tissues. **Chondroitin sulfate** (kon-DROY-tin) provides support and adhesiveness in cartilage, bone, skin, and blood vessels. The skin, tendons, blood vessels, and heart valves contain **dermatan sulfate;** bone, cartilage, and the cornea of the eye contain **keratan sulfate.** Also present in the ground substance are **adhesion proteins,** which are responsible for linking components of the ground substance to one another and to the surfaces of cells. The main adhesion protein of connective tissue is **fibronectin,** which binds to both collagen fibers (discussed shortly) and ground substance, linking them together. It also attaches cells to the ground substance.

Fibers

Three types of **fibers** are embedded in the extracellular matrix between the cells: collagen fibers, elastic fibers, and reticular fibers. They function to strengthen and support connective tissues.

Collagen fibers (*colla* = glue) are very strong and resist pulling forces, but they are not stiff, which allows tissue flexibility. The properties of different types of collagen fibers vary from tissue to tissue. For example, the collagen fibers found in cartilage attract more water molecules than do the collagen fibers in bone, which gives cartilage a more cushioning consistency. Collagen fibers often occur in parallel bundles (Figure 4.6). The bundle arrangement adds great strength to the tissue. Chemically, collagen fibers consist of the protein *collagen,* which is the most abundant protein in your body, representing about 25% of the total. Collagen fibers are found in most types of connective tissues, especially bone, cartilage, tendons, and ligaments.

Elastic fibers, which are smaller in diameter than collagen fibers, branch and join together to form a network within a tissue. An elastic fiber consists of molecules of the protein *elastin* surrounded by a glycoprotein named *fibrillin,* which adds strength and stability. Because of their unique molecular structure, elastic fibers are strong but can be stretched up to 150% of their relaxed length without breaking. Equally important, elastic fibers have the ability to return to their original shape after being stretched, a property called *elasticity.* Elastic fibers are plentiful in skin, blood vessel walls, and lung tissue.

Reticular fibers (*reticul-* = net), consisting of *collagen* arranged in fine bundles with a coating of glycoprotein, provide support in the walls of blood vessels and form a network around the cells in some tissues, such as areolar connective tissue, adipose tissue, and smooth muscle tissue. Produced by fibroblasts, reticular fibers are much thinner than collagen fibers and form branching networks. Like collagen fibers, reticular fibers provide support and strength. Reticular fibers are plentiful in reticular connective tissue, which forms the **stroma** (= bed or covering) or supporting framework of many soft organs, such as the spleen and lymph nodes. These fibers also help form the basement membrane.

Classification of Connective Tissues

Because of the diversity of cells and extracellular matrix and the differences in their relative proportions, the classification of connective tissues is not always clear-cut. We offer the following scheme:

I. Embryonic connective tissue
   - A. Mesenchyme
   - B. Mucous connective tissue

II. Mature connective tissue
   - A. Loose connective tissue
     1. Areolar connective tissue
     2. Adipose tissue
     3. Reticular connective tissue
   - B. Dense connective tissue
     1. Dense regular connective tissue
     2. Dense irregular connective tissue
     3. Elastic connective tissue
   - C. Cartilage
     1. Hyaline cartilage
     2. Fibrocartilage
     3. Elastic cartilage
   - D. Bone tissue
   - E. Liquid connective tissue
     1. Blood tissue
     2. Lymph

Note that our classification scheme has two major subclasses of connective tissue: embryonic and mature. **Embryonic connective tissue** is present primarily in the *embryo,* the developing human from fertilization through the first two months of pregnancy, and in the *fetus,* the developing human from the third month of pregnancy to birth.

One example of embryonic connective tissue found almost exclusively in the embryo is **mesenchyme** (MEZ-en-kīm), the tissue from which all other connective tissues eventually arise (Table 4.3A). Mesenchyme is composed of irregularly shaped cells, a semifluid ground substance, and delicate reticular fibers. Another kind of embryonic tissue is **mucous connective tissue (Wharton's jelly),** found mainly in the umbilical cord of the fetus. Mucous connective tissue is a form of mesenchyme that contains widely scattered fibroblasts, a more viscous jellylike ground substance, and collagen fibers (Table 4.3B).

TABLE 4.3 Embryonic Connective Tissues

A. Mesenchyme

Description: Consists of irregularly shaped mesenchymal cells embedded in a semifluid ground substance that contains reticular fibers.

Location: Under skin and along developing bones of embryo; some mesenchymal cells are found in adult connective tissue, especially along blood vessels.

Function: Forms all other types of connective tissue.

B. Mucous connective tissue

Description: Consists of widely scattered fibroblasts embedded in a viscous, jellylike ground substance that contains fine collagen fibers.

Location: Umbilical cord of fetus.

Function: Support.

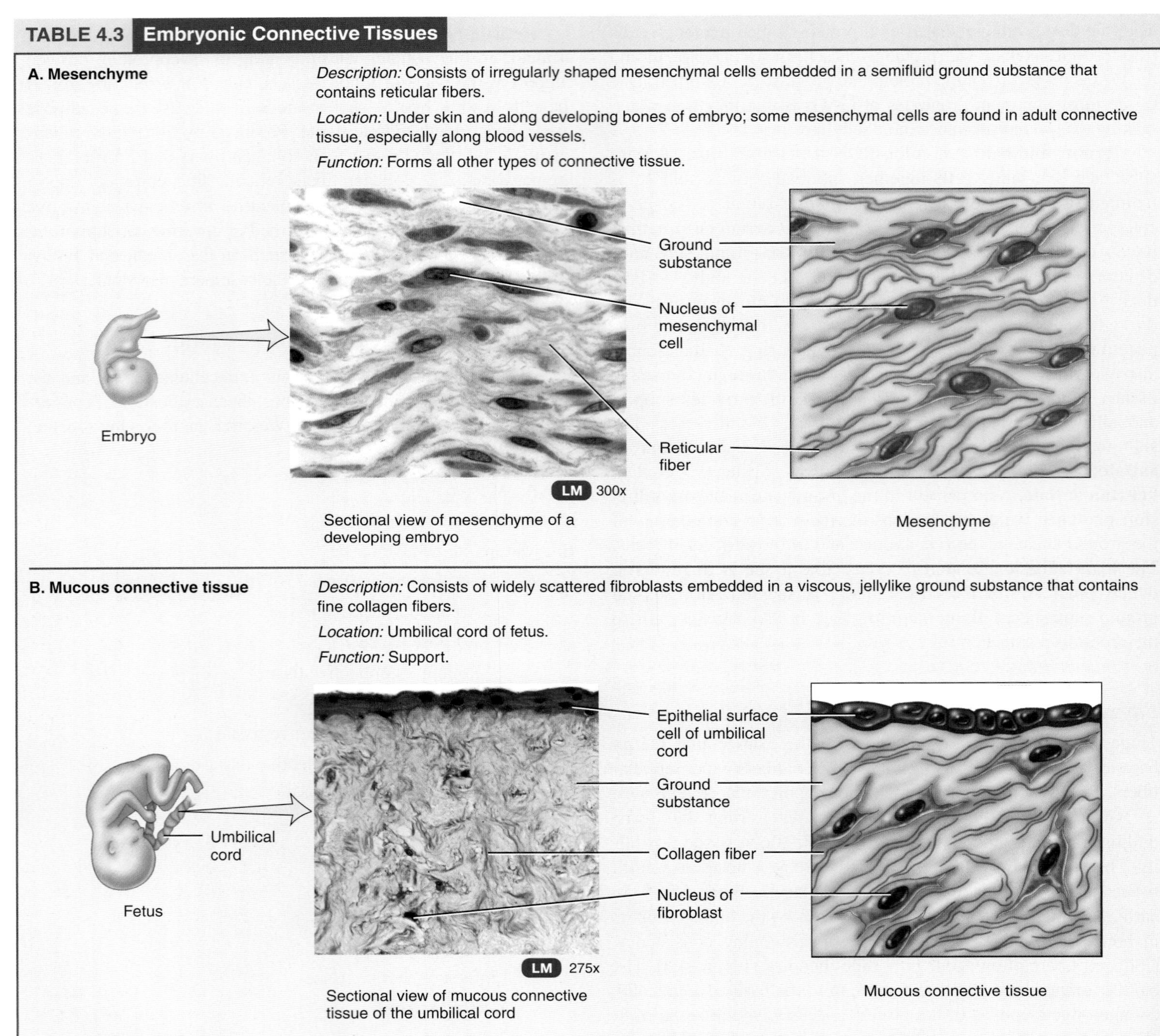

Sectional view of mesenchyme of a developing embryo

Mesenchyme

Sectional view of mucous connective tissue of the umbilical cord

Mucous connective tissue

The second major subclass of connective tissue, **mature connective tissue,** is present in the newborn. Its cells arise from mesenchyme. In the next section we explore the numerous types of mature connective tissue.

Types of Mature Connective Tissue

The five types of mature connective tissue are (1) loose connective tissue, (2) dense connective tissue, (3) cartilage, (4) bone tissue, and (5) liquid connective tissue (blood tissue and lymph). We now examine each in detail.

Loose Connective Tissue

The fibers of **loose connective tissue** are loosely intertwined between cells. The types of loose connective tissue are areolar connective tissue, adipose tissue, and reticular connective tissue.

AREOLAR CONNECTIVE TISSUE One of the most widely distributed connective tissues in the body is areolar connective tissue (a-RĒ-ō-lar; *areol-* = a small space). It contains several types of cells, including fibroblasts, macrophages, plasma cells, mast cells, adipocytes, and a few white blood cells (Table 4.4A). All three types of fibers—collagen, elastic, and reticular—are arranged randomly throughout the tissue. The ground substance contains hyaluronic acid, chondroitin sulfate, dermatan sulfate, and keratan sulfate. Combined with adipose tissue, areolar connective tissue forms the *subcutaneous layer,* the layer of tissue that attaches the skin to underlying tissues and organs.

ADIPOSE TISSUE Adipose tissue is a loose connective tissue in which the cells, called **adipocytes** (*adipo-* = fat), are specialized for storage of triglycerides (fats) (Table 4.4B). Adipocytes are derived from fibroblasts. Because the cell fills up with a single, large triglyceride droplet, the cytoplasm and nucleus are pushed to the periphery of the cell. Adipose tissue is found wherever areolar connective tissue is located. Adipose tissue is a good insulator and can therefore reduce heat loss through the skin. It is a major energy reserve and generally supports and protects various organs. As a person gains weight, the amount of adipose tissue increases and new blood vessels form. Thus, an obese person has many more blood vessels than does a lean person, a situation that can cause high blood pressure, since the heart has to work harder.

Most adipose tissue in adults is *white adipose tissue,* the type just described. Another type, called *brown adipose tissue (BAT),* obtains its darker color from a very rich blood supply, along with numerous pigmented mitochondria that participate in aerobic cellular respiration. Although BAT is widespread in the fetus and infant, in adults only small amounts are present. BAT generates considerable heat and probably helps to maintain body temperature in the newborn. The heat generated by the many mitochondria is carried away to other body tissues by the extensive blood supply.

Liposuction

A surgical procedure called **liposuction** (*lip-* = fat) or **suction lipectomy** (*-ectomy* = to cut out) involves suctioning small amounts of adipose tissue from various areas of the body. The technique can be used as a body-contouring procedure in regions such as the thighs, buttocks, arms, breasts, and abdomen. Postsurgical complications that may develop include fat that may obstruct blood flow in blood vessels, infection, fluid depletion, injury to internal structures, and severe postoperative pain. ■

RETICULAR CONNECTIVE TISSUE Reticular connective tissue consists of fine interlacing reticular fibers and reticular cells (Table 4.4C). Reticular connective tissue forms the stroma (supporting framework) of the liver, spleen, and lymph nodes and helps bind together smooth muscle cells. Additionally, reticular fibers in the spleen filter blood and remove worn-out blood cells, and reticular fibers in lymph nodes filter lymph and remove bacteria.

Dense Connective Tissue

Dense connective tissue contains more numerous, thicker, and denser fibers but considerably fewer cells than loose connective tissue. There are three types: dense regular connective tissue, dense irregular connective tissue, and elastic connective tissue.

DENSE REGULAR CONNECTIVE TISSUE In dense regular connective tissue, bundles of collagen fibers are *regularly* arranged in parallel patterns that provide the tissue with great strength (Table 4.4D). The tissue withstands pulling along the axis of the fibers. Fibroblasts, which produce the fibers and ground substance, appear in rows between the fibers. The tissue is silvery white and tough, yet somewhat pliable. Examples include tendons and most ligaments.

DENSE IRREGULAR CONNECTIVE TISSUE Dense irregular connective tissue contains collagen fibers that are packed more closely together than those of loose connective tissue and are usually *irregularly* arranged (Table 4.4E). Found in parts of the body where pulling forces are exerted in various directions, this tissue often occurs in sheets, such as in the dermis of the skin, which is deep to the epidermis, or the pericardium around the heart. Heart valves, the perichondrium (the membrane surrounding cartilage), and the periosteum (the membrane surrounding bone) are dense irregular connective tissues, although they have a fairly orderly arrangement of collagen fibers.

ELASTIC CONNECTIVE TISSUE Branching elastic fibers predominate in elastic connective tissue (Table 4.4F), giving the unstained tissue a yellowish color. Fibroblasts are present in the spaces between the fibers. Elastic connective tissue is quite strong and can recoil to its original shape after being stretched. Elasticity is important to the normal functioning of lung tissue, which recoils as you exhale, and elastic arteries, which recoil between heartbeats to maintain blood flow.

TABLE 4.4 Mature Connective Tissues

Loose Connective Tissue

A. Areolar connective tissue

Description: Consists of fibers (collagen, elastic, and reticular) and several kinds of cells (fibroblasts, macrophages, plasma cells, adipocytes, and mast cells) embedded in a semifluid ground substance.

Location: Subcutaneous layer deep to skin; papillary (superficial) region of dermis of skin; lamina propria of mucous membranes; and around blood vessels, nerves, and body organs.

Function: Strength, elasticity, and support.

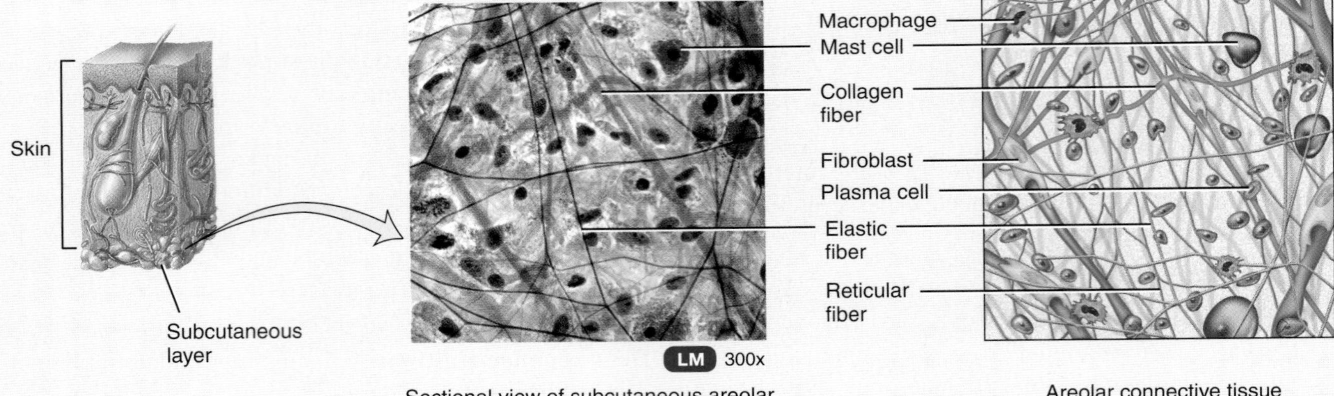

Sectional view of subcutaneous areolar connective tissue

Areolar connective tissue

B. Adipose tissue

Description: Consists of adipocytes, cells specialized to store triglycerides (fats) as a large centrally located droplet; nucleus and cytoplasm are peripherally located.

Location: Subcutaneous layer deep to skin, around heart and kidneys, yellow bone marrow, and padding around joints and behind eyeball in eye socket.

Function: Reduces heat loss through skin, serves as an energy reserve, supports, and protects. In newborns, brown adipose tissue generates considerable heat that helps maintain proper body temperature.

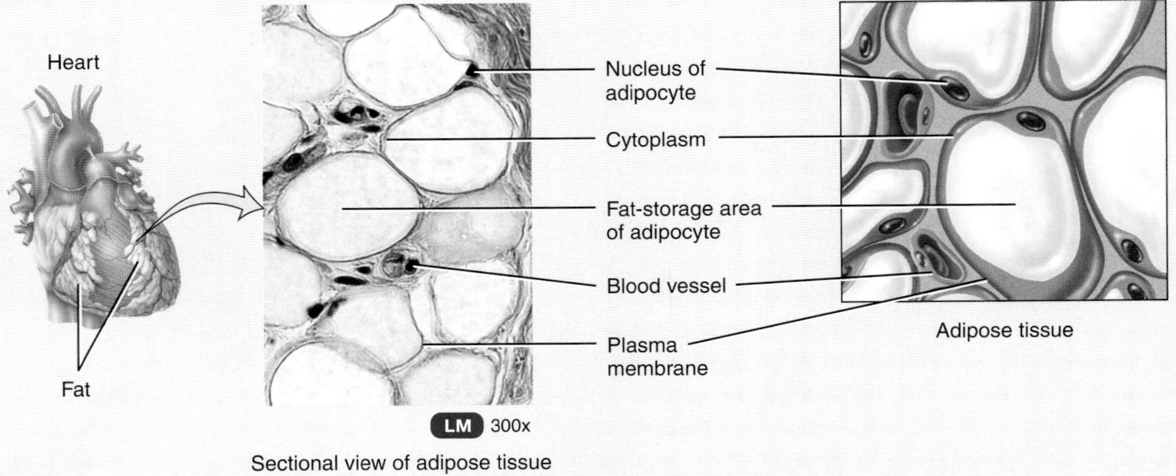

Sectional view of adipose tissue showing adipocytes of white fat

Adipose tissue

Cartilage

Cartilage consists of a dense network of collagen fibers and elastic fibers firmly embedded in chondroitin sulfate, a gel-like component of the ground substance. Cartilage can endure considerably more stress than loose and dense connective tissues. The strength of cartilage is due to its collagen fibers, and its resilience (ability to assume its original shape after deformation) is due to chondroitin sulfate.

The cells of mature cartilage, called **chondrocytes** (KON-drō-sīts; *chondro-* = cartilage), occur singly or in groups within spaces called **lacunae** (la-KOO-nē = little lakes; singular is *lacuna*) in the extracellular matrix. A membrane of dense irregular connective tissue called the **perichondrium** (per′-i-KON-drē-um *peri-* = around) covers the surface of most cartilage. Unlike other connective tissues, cartilage has no blood vessels

Loose Connective Tissue

C. Reticular connective tissue

Description: A network of interlacing reticular fibers and reticular cells.

Location: Stroma (supporting framework) of liver, spleen, lymph nodes; red bone marrow, which gives rise to blood cells; reticular lamina of the basement membrane; and around blood vessels and muscles.

Function: Forms stroma of organs; binds together smooth muscle tissue cells; filters and removes worn-out blood cells in the spleen and microbes in lymph nodes.

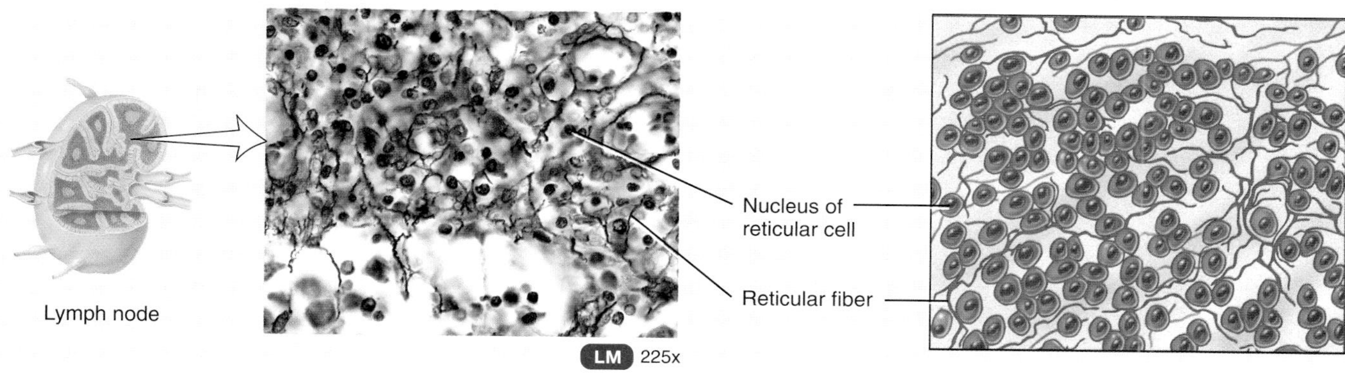

Sectional view of reticular connective tissue of a lymph node

Reticular connective tissue

Dense Connective Tissue

D. Dense regular connective tissue

Description: Extracellular matrix looks shiny white; consists mainly of collagen fibers arranged in bundles; fibroblasts present in rows between bundles.

Location: Forms tendons (attach muscle to bone), most ligaments (attach bone to bone), and aponeuroses (sheetlike tendons that attach muscle to muscle or muscle to bone).

Function: Provides strong attachment between various structures.

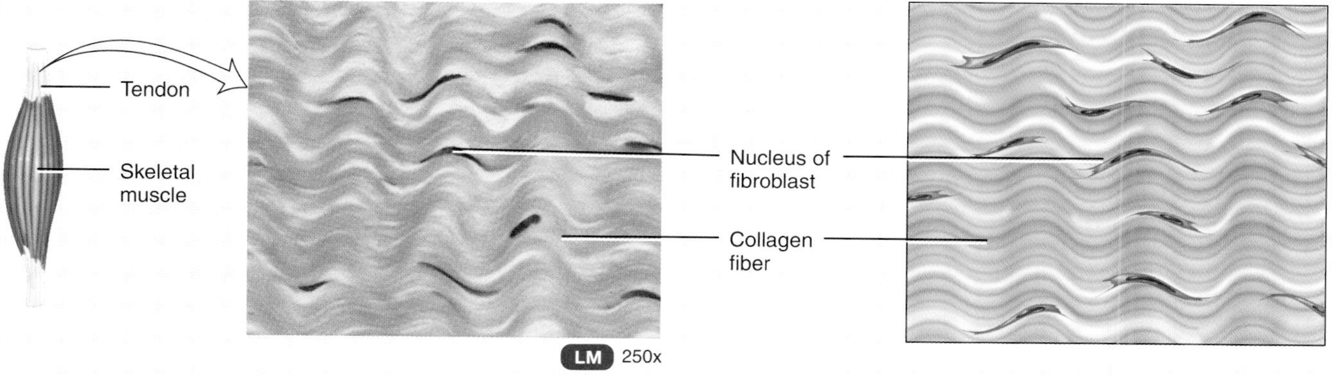

Sectional view of dense regular connective tissue of a tendon

Dense regular connective tissue

continues

or nerves, except in the perichondrium. Since cartilage has no blood supply, it heals poorly following an injury. There are three types of cartilage: hyaline cartilage, fibrocartilage, and elastic cartilage.

Hyaline Cartilage Hyaline cartilage contains a resilient gel as its ground substance and appears in the body as a bluish-white, shiny substance. The fine collagen fibers are not visible with ordinary staining techniques, and prominent chondrocytes are found in lacunae (Table 4.4G). Most hyaline cartilage is surrounded by a perichondrium. The exceptions are the articular cartilage in joints and at the epiphyseal plates, the regions where bones lengthen as a person grows. Hyaline cartilage is the most abundant cartilage in the body. It provides

TABLE 4.4 Mature Connective Tissues (continued)

Dense Connective Tissue

E. Dense irregular connective tissue

Description: Consists predominantly of randomly arranged collagen fibers and a few fibroblasts.

Location: Fasciae (tissue beneath skin and around muscles and other organs), reticular (deeper) region of dermis of skin, periosteum of bone, perichondrium of cartilage, joint capsules, membrane capsules around various organs (kidneys, liver, testes, lymph nodes), pericardium of the heart, and heart valves.

Function: Provides strength.

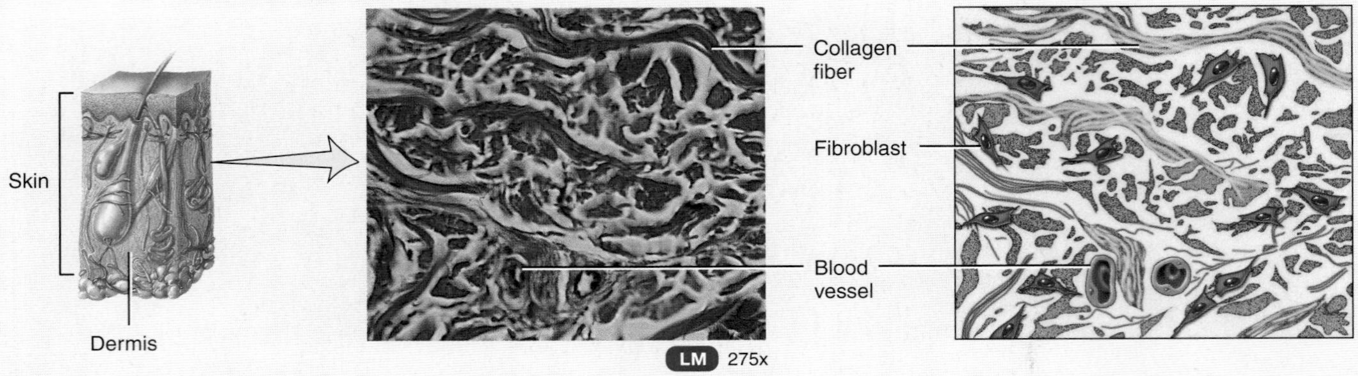

Sectional view of dense irregular connective tissue of reticular region of dermis

Dense irregular connective tissue

F. Elastic connective tissue

Description: Consists predominantly of freely branching elastic fibers; fibroblasts are present in spaces between fibers.

Location: Lung tissue, walls of elastic arteries, trachea, bronchial tubes, true vocal cords, suspensory ligament of penis, and ligaments between vertebrae.

Function: Allows stretching of various organs.

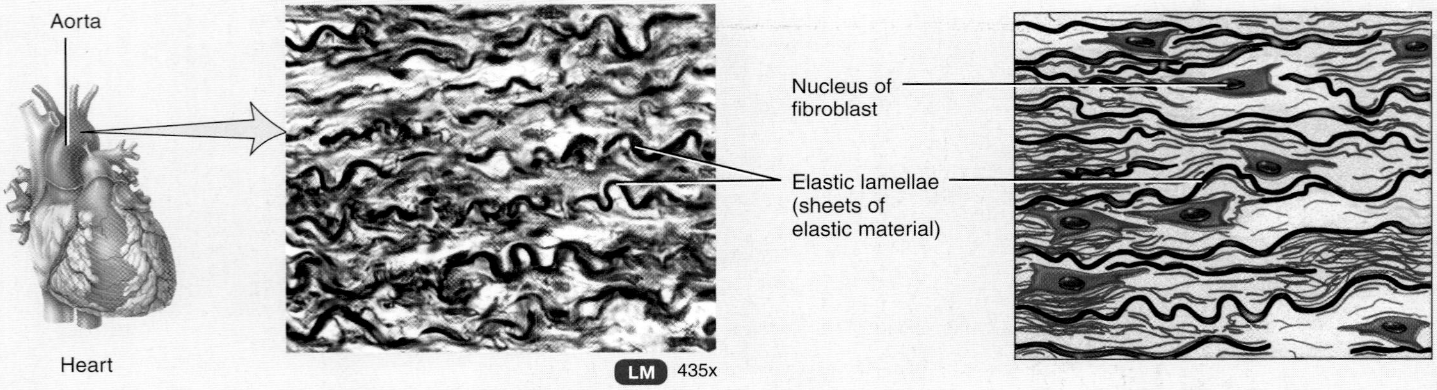

Sectional view of elastic connective tissue of aorta

Elastic connective tissue

flexibility and support and, at joints, reduces friction and absorbs shock. Hyaline cartilage is the weakest of the three types of cartilage.

Fibrocartilage Chondrocytes are scattered among clearly visible bundles of collagen fibers within the extracellular matrix of fibrocartilage (Table 4.4H). Fibrocartilage lacks a perichondrium. With a combination of strength and rigidity, this tissue is the strongest of the three types of cartilage. One location of fibrocartilage is the intervertebral discs, the disk-shaped material between the vertebrae (backbones).

Elastic Cartilage The chondrocytes of elastic cartilage are located within a threadlike network of elastic fibers within the extracellular matrix (Table 4.4I). A perichondrium is present.

Cartilage

G. Hyaline cartilage

Description: Consists of a bluish-white, shiny ground substance with fine collagen fibers and many chondrocytes; most abundant type of cartilage.

Location: Ends of long bones, anterior ends of ribs, nose, parts of larynx, trachea, bronchi, bronchial tubes, and embryonic and fetal skeleton.

Function: Provides smooth surfaces for movement at joints, as well as flexibility and support.

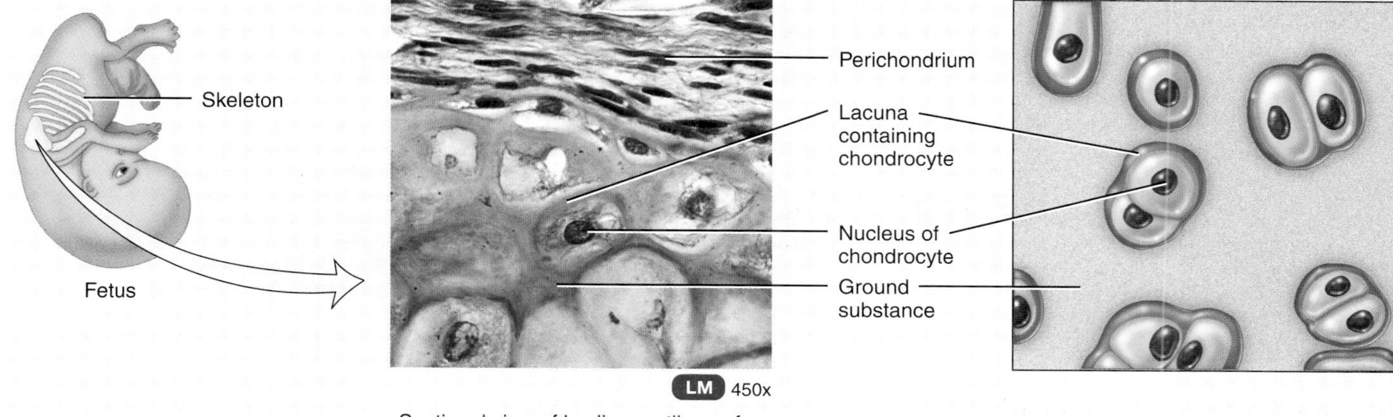

Sectional view of hyaline cartilage of a developing fetal bone

Hyaline cartilage

H. Fibrocartilage

Description: Consists of chondrocytes scattered among bundles of collagen fibers within the extracellular matrix.

Location: Pubic symphysis (point where hip bones join anteriorly), intervertebral discs (discs between vertebrae), menisci (cartilage pads) of knee, and portions of tendons that insert into cartilage.

Function: Support and fusion.

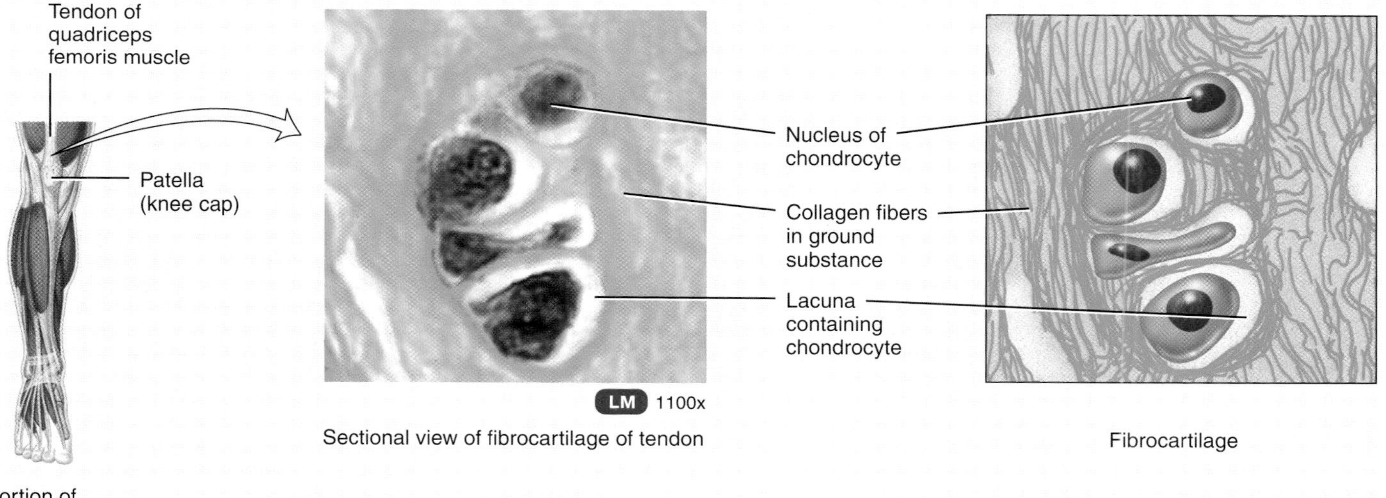

Sectional view of fibrocartilage of tendon

Fibrocartilage

Portion of right lower limb

continues

Elastic cartilage provides strength and elasticity and maintains the shape of certain structures, such as the external ear.

Repair and Growth of Cartilage Metabolically, cartilage is a relatively inactive tissue that grows slowly. When injured or inflamed, cartilage repair proceeds slowly, in large part because cartilage is avascular. Substances needed for repair and blood cells that participate in tissue repair must diffuse or migrate into the cartilage. The growth of cartilage follows two basic patterns: interstitial growth and appositional growth.

In **interstitial growth,** the cartilage increases rapidly in size due to the division of existing chondrocytes and the continuous deposition of increasing amounts of extracellular matrix by the chondrocytes. As the chondrocytes synthesize new matrix, they are pushed away from each other. These events cause the cartilage to expand from within like bread rising, which is the reason for the

TABLE 4.4 Mature Connective Tissues (continued)

Cartilage

I. Elastic cartilage

Description: Consists of chondrocytes located in a threadlike network of elastic fibers within the extracellular matrix.

Location: Lid on top of larynx (epiglottis), part of external ear (auricle), and auditory (eustachian) tubes.

Function: Gives support and maintains shape.

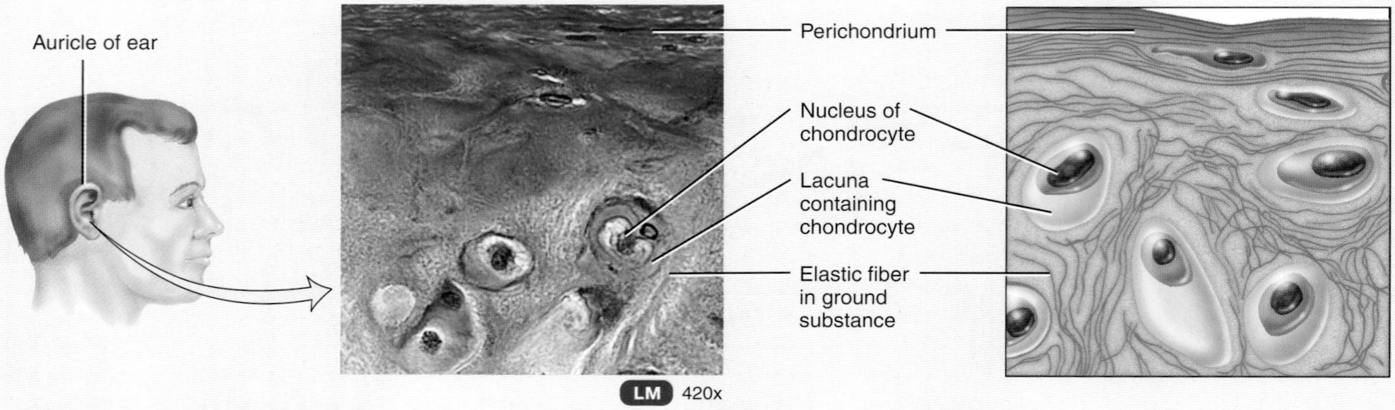

Sectional view of elastic cartilage of auricle of ear

Elastic cartilage

Bone Tissue

J. Compact bone

Description: Compact bone tissue consists of osteons (haversian systems) that contain lamellae, lacunae, osteocytes, canaliculi, and central (haversian) canals. By contrast, spongy bone tissue (see Figure 6.3 on page 176) consists of thin columns called trabeculae; spaces between trabeculae are filled with red bone marrow.

Location: Both compact and spongy bone tissue make up the various parts of bones of the body.

Function: Support, protection, storage; houses blood-forming tissue; serves as levers that act with muscle tissue to enable movement.

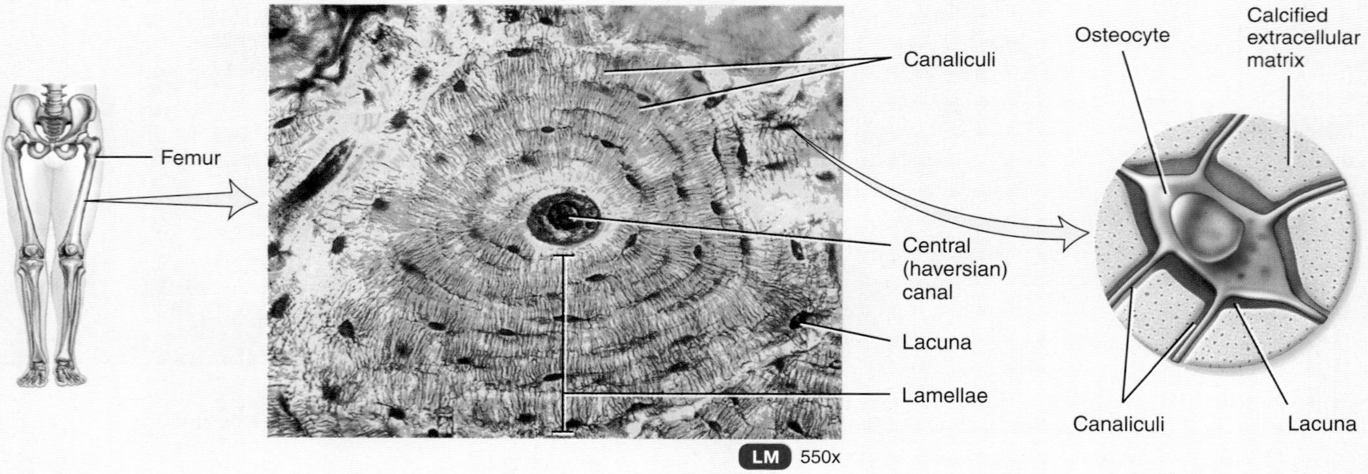

Sectional view of an osteon (haversian system) of femur (thigh bone)

Details of an osteocyte

term *inter*stitial. This growth pattern occurs while the cartilage is young and pliable, during childhood and adolescence.

In **appositional growth,** activity of cells in the inner chondrogenic layer of the perichondrium leads to growth. The deeper cells of the perichondrium, the fibroblasts, divide; some differentiate into chondroblasts. As differentiation continues, the chondroblasts surround themselves with extracellular matrix and become chondrocytes. As a result, matrix accumulates beneath the perichondrium on the outer surface of the cartilage, causing it to grow in width. Appositional growth starts later than interstitial growth and continues through adolescence.

Bone Tissue

Cartilage, joints, and bones make up the skeletal system. The skeletal system supports soft tissues, protects delicate structures, and works with skeletal muscles to generate movement. Bones

Liquid Connective Tissue

K. Blood

Description: Consists of blood plasma and formed elements: red blood cells (erythrocytes), white blood cells (leukocytes), and platelets (thrombocytes).

Location: Within blood vessels (arteries, arterioles, capillaries, venules, and veins) and within the chambers of the heart.

Function: Red blood cells transport oxygen and some carbon dioxide; white blood cells carry on phagocytosis and are involved in allergic reactions and immune system responses; platelets are essential for the clotting of blood.

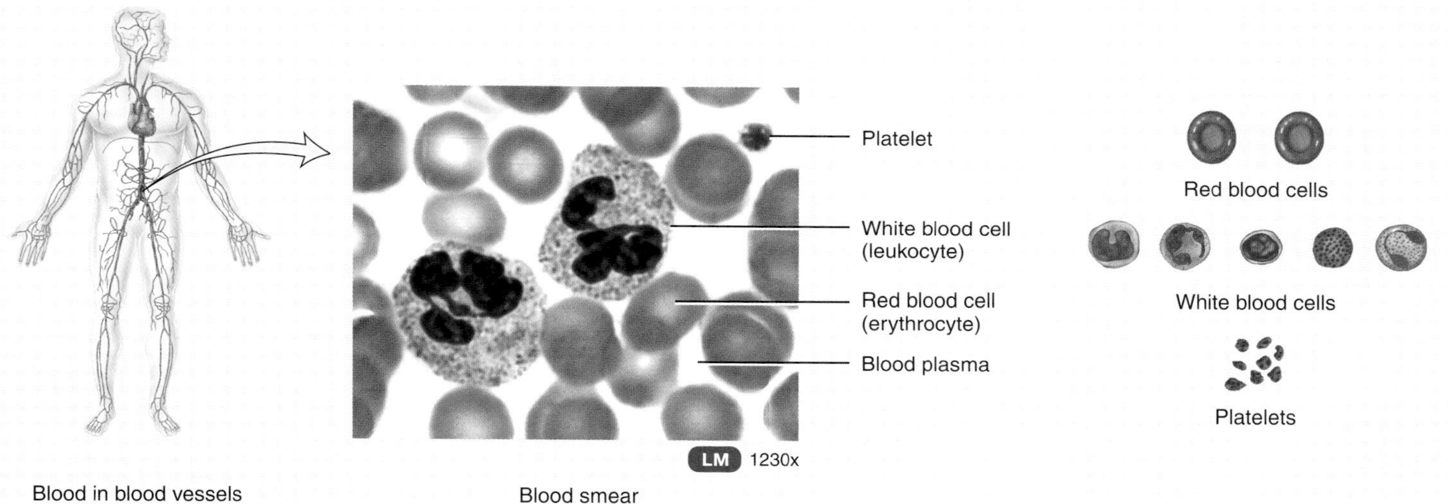

Blood in blood vessels

Blood smear

store calcium and phosphorus; house red bone marrow, which produces blood cells; and contain yellow bone marrow, a storage site for triglycerides. Bones are organs composed of several different connective tissues, including **bone** or **osseous tissue** (OS-ē-us), the periosteum, red and yellow bone marrow, and the endosteum (a membrane that lines a space within bone that stores yellow bone marrow). Bone tissue is classified as either compact or spongy, depending on how its extracellular matrix and cells are organized.

The basic unit of **compact bone** is an **osteon** or **haversian system** (Table 4.4J). Each osteon has four parts:

1. The **lamellae** (la-MEL-lē = little plates) are concentric rings of extracellular matrix that consist of mineral salts (mostly calcium and phosphates), which give bone its hardness, and collagen fibers, which give bone its strength. The lamellae are responsible for the compact nature of this type of bone tissue.

2. **Lacunae** are small spaces between lamellae that contain mature bone cells called **osteocytes.**

3. Projecting from the lacunae are **canaliculi** (kan-a-LIK-ū-lī = little canals), networks of minute canals containing the processes of osteocytes. Canaliculi provide routes for nutrients to reach osteocytes and for wastes to leave them.

4. A **central (haversian) canal** contains blood vessels and nerves.

Spongy bone lacks osteons. Rather, it consists of columns of bone called **trabeculae** (tra-BEK-ū-lē = little beams), which contain lamellae, osteocytes, lacunae, and canaliculi. Spaces between lamellae are filled with red bone marrow. Chapter 6 presents bone tissue histology in more detail.

Tissue Engineering

The technology of **tissue engineering** has allowed scientists to grow new tissues in the laboratory to replace damaged tissues in the body. Tissue engineers have already developed laboratory-grown versions of skin and cartilage using scaffolding beds of biodegradable synthetic materials or collagen as substrates that permit body cells to be cultured. As the cells divide and assemble, the scaffolding degrades; the new, permanent tissue is then implanted in the patient. Other structures currently under development include bones, tendons, heart valves, bone marrow, and intestines. Work is also under way to develop insulin-producing cells for diabetics, dopamine-producing cells for Parkinson disease patients, and even entire livers and kidneys. ■

Liquid Connective Tissue

BLOOD TISSUE **Blood tissue** (or simply blood) is a connective tissue with a liquid extracellular matrix called **blood plasma,** a pale yellow fluid that consists mostly of water with a wide variety of dissolved substances—nutrients, wastes, enzymes, plasma proteins, hormones, respiratory gases, and ions (Table 4.4K). Suspended in the blood plasma are formed elements—red blood cells (erythrocytes), white blood cells (leukocytes), and platelets (thrombocytes). **Red blood cells** transport oxygen to body cells and remove some carbon dioxide from them. **White blood cells** are involved in phagocytosis, immunity, and allergic reactions. **Platelets** participate in blood clotting. The details of blood are considered in Chapter 19.

LYMPH **Lymph** is the extracellular fluid that flows in lymphatic vessels. It is a connective tissue that consists of several types of cells in a clear liquid extracellular matrix that is similar to blood plasma but with much less protein. The composition of lymph varies from one part of the body to another. For example, lymph leaving lymph nodes includes many lymphocytes, a type of white blood cell, in contrast to lymph from the small intestine, which has a high content of newly absorbed dietary lipids. The details of lymph are considered in Chapter 22.

► **CHECKPOINT**

10. In what ways do connective tissues differ from epithelia?
11. What are the features of the cells, ground substance, and fibers that make up connective tissue?
12. How are connective tissues classified? List the various types.
13. Describe how the structures of the following connective tissues are related to their functions: areolar connective tissue, adipose tissue, reticular connective tissue, dense regular connective tissue, dense irregular connective tissue, elastic connective tissue, hyaline cartilage, fibrocartilage, elastic cartilage, bone tissue, blood tissue, and lymph.
14. What is the difference between interstitial and appositional growth of cartilage?

MEMBRANES

► **OBJECTIVES**

Define a membrane.

Describe the classification of membranes.

Membranes are flat sheets of pliable tissue that cover or line a part of the body. The combination of an epithelial layer and an underlying connective tissue layer constitutes an **epithelial membrane.** The principal epithelial membranes of the body are mucous membranes, serous membranes, and the cutaneous membrane, or skin. Another type of membrane, a **synovial membrane,** lines joints and contains connective tissue but no epithelium.

Epithelial Membranes

Mucous Membranes

A **mucous membrane** or **mucosa** lines a body cavity that opens directly to the exterior. Mucous membranes line the entire digestive, respiratory, and reproductive tracts, and much of the urinary tract. They consist of a lining layer of epithelium and an underlying layer of connective tissue (Figure 4.7a).

The epithelial layer of a mucous membrane is an important feature of the body's defense mechanisms because it is a barrier that microbes and other pathogens have difficulty penetrating. Usually, tight junctions connect the cells, so materials cannot leak in between them. Goblet cells and other cells of the epithelial layer of a mucous membrane secrete mucus, and this slippery fluid prevents the cavities from drying out. It also traps particles in the respiratory passageways and lubricates food as it moves through the gastrointestinal tract. In addition, the epithelial layer secretes some of the enzymes needed for digestion and is the site of food and fluid absorption in the gastrointestinal tract. The epithelia of mucous membranes vary greatly in differ-

Figure 4.7 **Membranes.**

A membrane is a flat sheet of pliable tissues that covers or lines a part of the body.

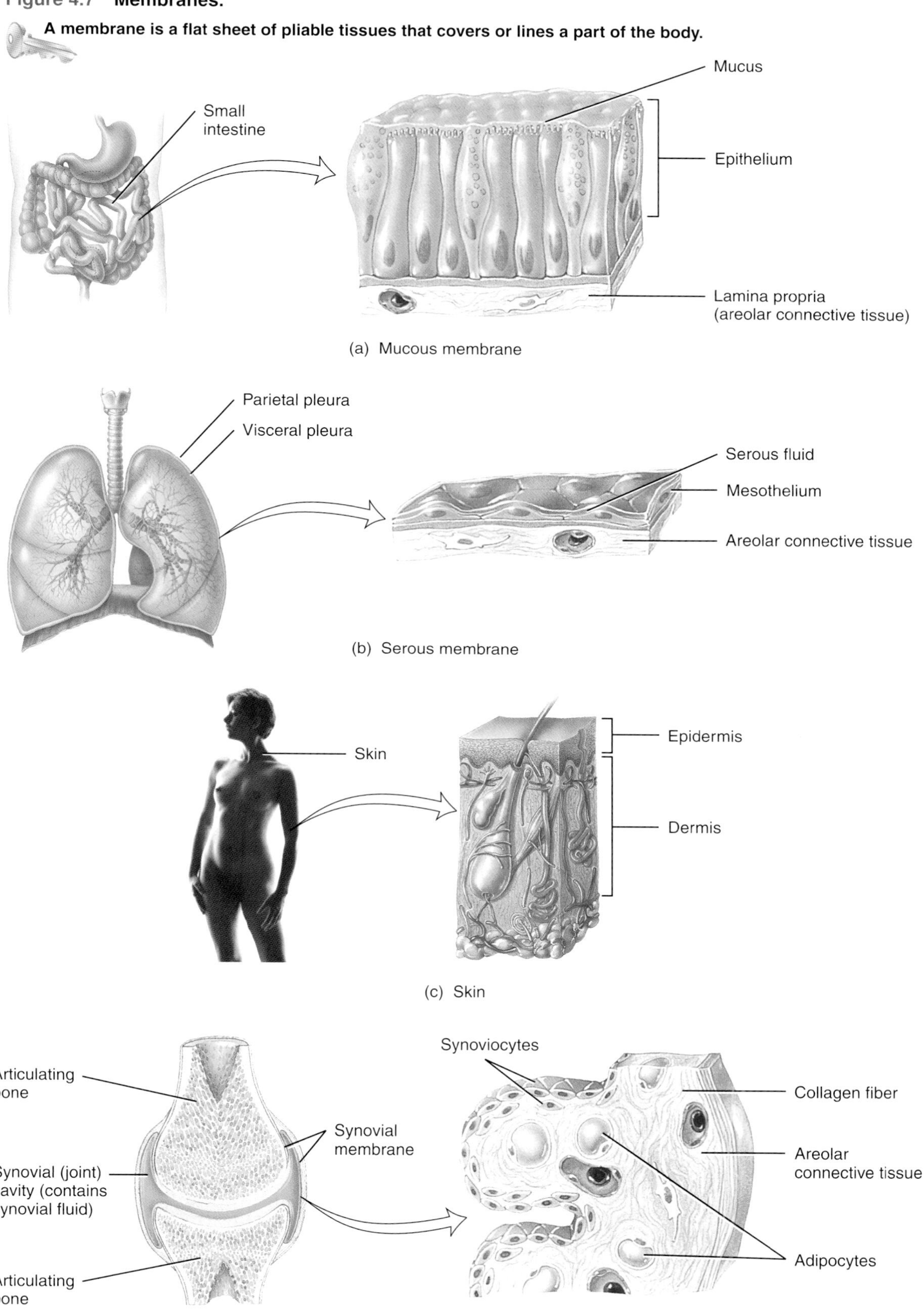

(a) Mucous membrane

(b) Serous membrane

(c) Skin

(d) Synovial membrane

What is an epithelial membrane?

ent parts of the body. For example, the epithelium of the mucous membrane of the small intestine is nonciliated simple columnar (see Table 4.1C), and the epithelium of the large airways to the lungs is pseudostratified ciliated columnar (see Table 4.1E).

The connective tissue layer of a mucous membrane is areolar connective tissue and is called the **lamina propria** (LAM-ī-na PRŌ-prē-a). The lamina propria is so named because it belongs to the mucous membrane (*propria* = one's own). The lamina propria supports the epithelium, binds it to the underlying structures, and allows some flexibility of the membrane. It also holds blood vessels in place and protects underlying muscles from abrasion or puncture. Oxygen and nutrients diffuse from the lamina propria to the epithelium covering it; carbon dioxide and wastes diffuse in the opposite direction.

Serous Membranes

A **serous membrane** (*serous* = watery) or **serosa** lines a body cavity that does not open directly to the exterior, and it covers the organs that lie within the cavity. Serous membranes consist of areolar connective tissue covered by mesothelium (simple squamous epithelium) (Figure 4.7b). Serous membranes have two layers: The layer attached to the cavity wall is called the **parietal layer** (pa-RĪ-e-tal; *pariet-* = wall); the layer that covers and attaches to the organs inside the cavity is the **visceral layer** (*viscer-* = body organ) (see Figure 1.10a on page 18). The mesothelium of a serous membrane secretes **serous fluid,** a watery lubricant that allows organs to glide easily over one another or to slide against the walls of cavities.

The serous membrane lining the thoracic cavity and covering the lungs is the **pleura.** The serous membrane lining the heart cavity and covering the heart is the **pericardium.** The serous membrane lining the abdominal cavity and covering the abdominal organs is the **peritoneum.**

Cutaneous Membrane

The **cutaneous membrane** or **skin** covers the surface of the body and consists of a superficial portion called the *epidermis* and a deeper portion called the *dermis* (Figure 4.7c). The epidermis consists of keratinized stratified squamous epithelium, which protects underlying tissues. The dermis consists of connective tissue (areolar connective tissue and dense irregular connective tissue). Details of the cutaneous membrane are presented in Chapter 5.

Synovial Membranes

Synovial membranes (sin-Ō-vē-al; *syn-* = together, referring here to a place where bones come together) line the cavities of freely movable joints. Like serous membranes, synovial membranes line structures that do not open to the exterior. Unlike mucous, serous, and cutaneous membranes, they lack an epithelium and are therefore not epithelial membranes. Synovial membranes are composed of a discontinuous layer of cells called **synoviocytes** (si-NŌ-vē-ō-sīts), which are closer to the synovial cavity (space between the bones) and a layer of connective tissue (areolar and adipose) deep to the synoviocytes (Figure 4.7d). Synoviocytes secrete some of the components of synovial fluid. **Synovial fluid** lubricates and nourishes the cartilage covering the bones at movable joints and contains macrophages that remove microbes and debris from the joint cavity.

▶ **CHECKPOINT**

15. Define the following kinds of membranes: mucous, serous, cutaneous, and synovial. How do they differ from one another?

16. Where is each type of membrane located in the body? What are their functions?

MUSCULAR TISSUE

▶ **OBJECTIVES**

Describe the general features of muscular tissue.

Contrast the structure, location, and mode of control of skeletal, cardiac, and smooth muscle tissue.

Muscular tissue consists of elongated cells called *muscle fibers* that can use ATP to generate force. As a result, muscular tissue produces body movements, maintains posture, and generates heat. It also provides protection. Based on its location and certain structural and functional features, muscular tissue is classified into three types: skeletal, cardiac, and smooth (Table 4.5).

Skeletal muscle tissue is named for its location—it is usually attached to the bones of the skeleton (Table 4.5A). Another feature is its *striations,* alternating light and dark bands within the fibers that are visible under a light microscope. Skeletal muscle is considered *voluntary* because it can be made to contract or relax by conscious control. A single skeletal muscle fiber is very long (up to 30–40 cm or about 12–16 in. in your longest muscles). A muscle fiber is roughly cylindrical in shape, and has many nuclei located at the periphery. Within a whole muscle, the individual muscle fibers are parallel to one another.

Cardiac muscle tissue forms most of the wall of the heart (Table 4.5B). Like skeletal muscle, it is striated. However, unlike skeletal muscle tissue, it is *involuntary;* its contraction is not consciously controlled. Cardiac muscle fibers are branched and usually have only one centrally located nucleus; an occasional cell has two nuclei. They attach end to end by transverse thickenings of the plasma membrane called **intercalated discs** (*intercalat-* = to insert between), which contain both desmosomes and gap junctions. Intercalated discs are unique to cardiac muscle. The desmosomes strengthen the tissue and hold the fibers together during their vigorous contractions. The gap junctions provide a route for quick conduction of muscle action potentials throughout the heart.

TABLE 4.5 Muscular Tissues

A. Skeletal muscle tissue

Description: Long, cylindrical, striated fibers with many peripherally located nuclei; voluntary control.

Location: Usually attached to bones by tendons.

Function: Motion, posture, heat production, and protection.

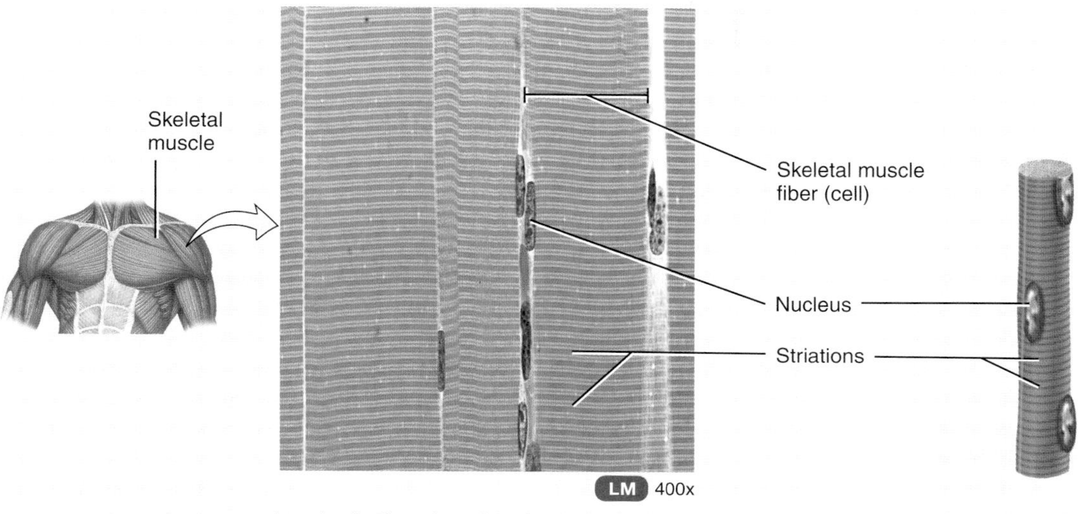

Longitudinal section of skeletal muscle tissue

Skeletal muscle fiber

B. Cardiac muscle tissue

Description: Branched striated fibers with one or two centrally located nuclei; contains intercalated discs; involuntary control.

Location: Heart wall.

Function: Pumps blood to all parts of the body.

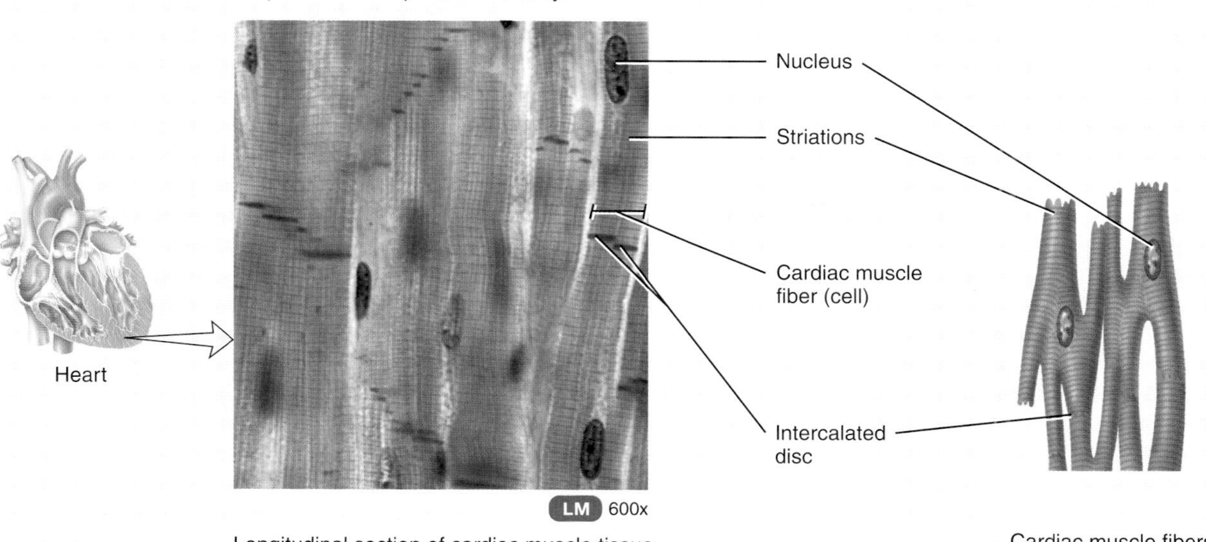

Longitudinal section of cardiac muscle tissue

Cardiac muscle fibers

continues

TABLE 4.5 Muscular Tissues (continued)

C. Smooth muscle tissue

Description: Spindle-shaped (thickest in middle and tapering at both ends), nonstriated fibers with one centrally located nucleus; involuntary control.

Location: Iris of the eyes, walls of hollow internal structures such as blood vessels, airways to the lungs, stomach, intestines, gallbladder, urinary bladder, and uterus.

Function: Motion (constriction of blood vessels and airways, propulsion of foods through gastrointestinal tract, contraction of urinary bladder and gallbladder).

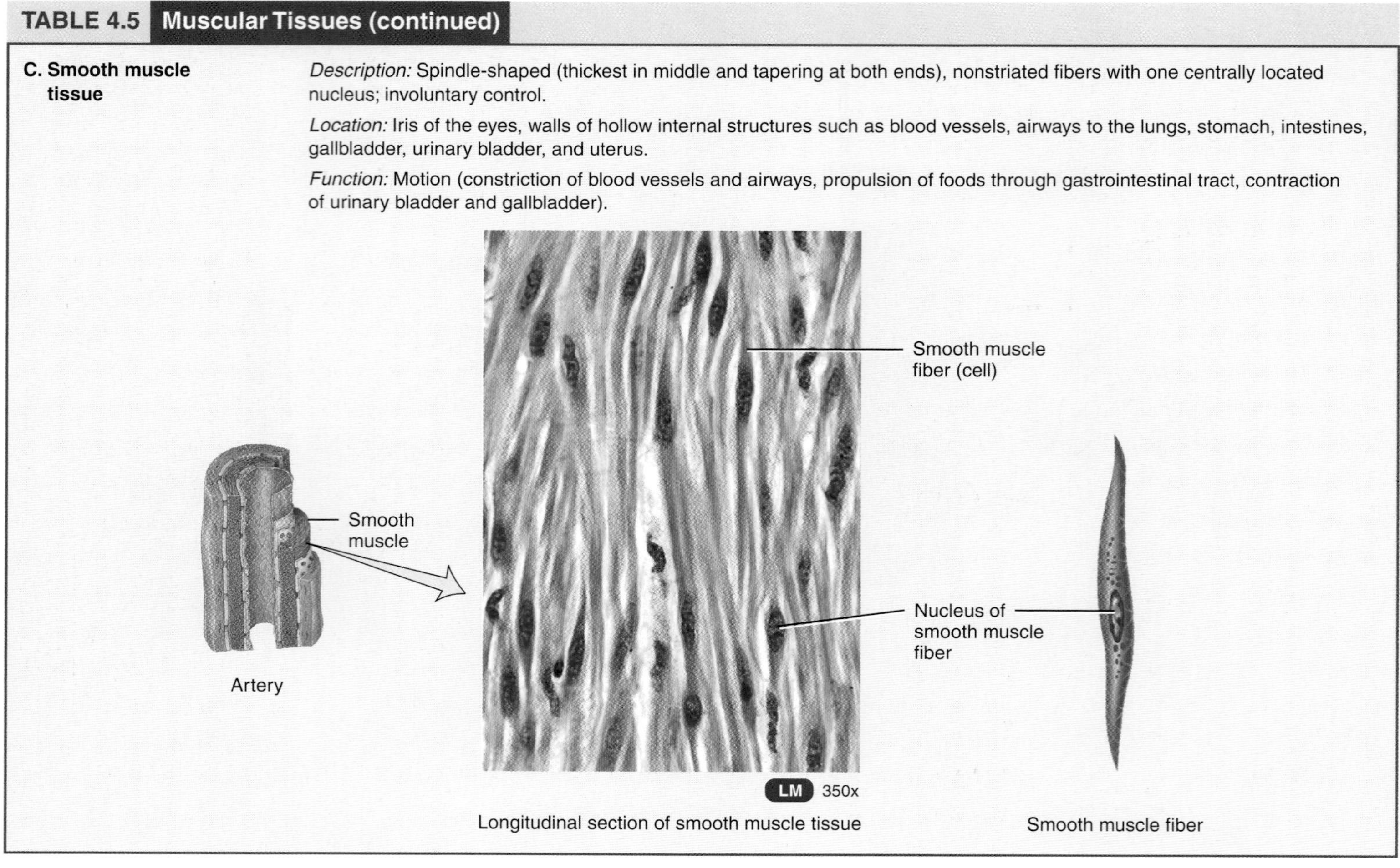

Longitudinal section of smooth muscle tissue

Smooth muscle fiber

Smooth muscle tissue is located in the walls of hollow internal structures such as blood vessels, airways to the lungs, the stomach, intestines, gallbladder, and urinary bladder (Table 4.5C). Its contraction helps constrict or narrow the lumen of blood vessels, physically break down and move food along the gastrointestinal tract, move fluids through the body, and eliminate wastes. Smooth muscle fibers are usually *involuntary,* and they are nonstriated (lack striations), hence the term *smooth.* A smooth muscle fiber is small, thickest in the middle, and tapering at each end. It contains a single, centrally located nucleus. Gap junctions connect many individual fibers in some smooth muscle tissues, for example, in the wall of the intestines. Such muscle tissues can produce powerful contractions as many muscle fibers contract in unison. In other locations, such as the iris of the eye, smooth muscle fibers contract individually, like skeletal muscle fibers, because gap junctions are absent. Chapter 10 provides a detailed discussion of muscular tissue.

► **CHECKPOINT**

17. Which muscles are striated and which are smooth?

18. Which types of muscular tissue have gap junctions?

NERVOUS TISSUE

► **OBJECTIVE**

Describe the structural features and functions of nervous tissue.

Despite the awesome complexity of the nervous system, it consists of only two principal types of cells: neurons and neuroglia. **Neurons,** or nerve cells, are sensitive to various stimuli. They convert stimuli into electrical signals called **action potentials (nerve impulses)** and conduct these action potentials to other neurons, to muscle tissue, or to glands. Most neurons consist of three basic parts: a cell body and two kinds of cell processes—dendrites and axons (Table 4.6). The **cell body** contains the nucleus and other organelles. **Dendrites** (*dendr-* = tree) are tapering, highly branched, and usually short cell processes. They are the major receiving or input portion of a neuron. The **axon** (*axo-* = axis) of a neuron is a single, thin, cylindrical process that may be very long. It is the output portion of a neuron, conducting nerve impulses toward another neuron or to some other tissue.

Even though **neuroglia** (noo-RŌG-lē-a; *-glia* = glue) do not generate or conduct nerve impulses, these cells do have many important supportive functions. The detailed structure

TABLE 4.6 Nervous Tissue

Description: Consists of neurons (nerve cells) and neuroglia. Neurons consist of a cell body and processes extending from the cell body (multiple dendrites and a single axon). Neuroglia do not generate or conduct nerve impulses but have other important supporting functions.

Location: Nervous system.

Function: Exhibits sensitivity to various types of stimuli, converts stimuli into nerve impulses (action potentials), and conducts nerve impulses to other neurons, muscle fibers, or glands.

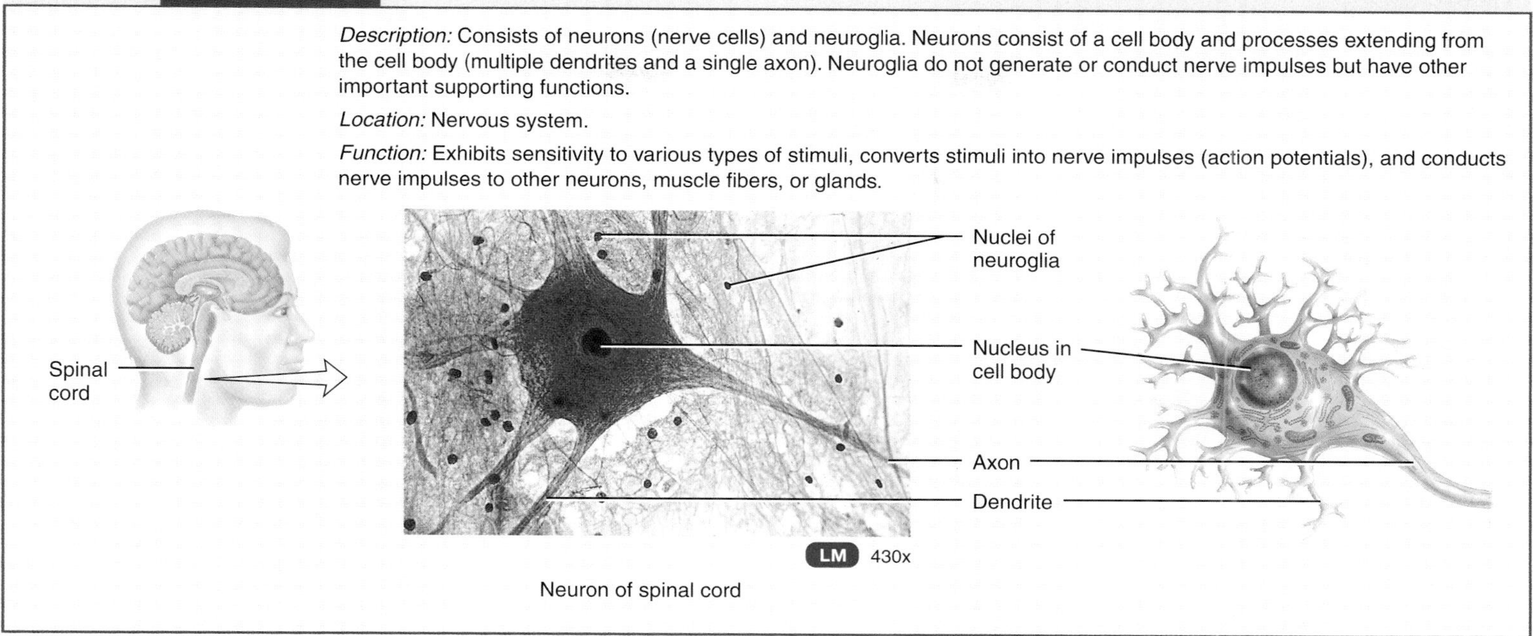

Neuron of spinal cord

and function of neurons and neuroglia are considered in Chapter 12.

► CHECKPOINT

19. What are the functions of the dendrites, cell body, and axon of a neuron?

EXCITABLE CELLS

► OBJECTIVE

Explain the concept of electrical excitability.

Neurons and muscle fibers are considered **excitable cells** because they exhibit **electrical excitability,** the ability to respond to certain stimuli by producing electrical signals such as *action potentials.* Action potentials can propagate (travel) along the plasma membrane of a neuron or muscle fiber due to the presence of specific voltage-gated ion channels. When an action potential forms in a neuron, the neuron releases chemicals called *neurotransmitters,* which allow neurons to communicate with other neurons, muscle fibers, or glands. When an action potential occurs in a muscle fiber, the muscle fiber contracts, resulting in activities such as the movement of the limbs, the propulsion of food through the small intestine, and the movement of blood out of the heart and into the blood vessels of the body. The muscle action potential and the nerve action potential are discussed in detail in Chapters 10 and 12, respectively.

► CHECKPOINT

20. Why is electrical excitability important to neurons and muscle fibers?

TISSUE REPAIR: RESTORING HOMEOSTASIS

► OBJECTIVE

Describe the role of tissue repair in restoring homeostasis.

Tissue repair is the replacement of worn-out, damaged, or dead cells. New cells originate by cell division from the **stroma,** the supporting connective tissue, or from the **parenchyma,** cells that constitute the functioning part of the tissue or organ. In adults, each of the four basic tissue types (epithelial, connective, muscle, and nervous) has a different capacity for replenishing parenchymal cells lost by damage, disease, or other processes.

Epithelial cells, which endure considerable wear and tear (and even injury) in some locations, have a continuous capacity for renewal. In some cases, immature, undifferentiated cells called **stem cells** divide to replace lost or damaged cells. For example, stem cells reside in protected locations in the epithelia of the skin and gastrointestinal tract to replenish cells sloughed from the apical layer, and stem cells in red bone marrow continually provide new red and white blood cells and platelets. In other cases, mature, differentiated cells can undergo cell division;

examples include hepatocytes (liver cells) and endothelial cells in blood vessels.

Some connective tissues also have a continuous capacity for renewal. One example is bone, which has an ample blood supply. Connective tissues such as cartilage can replenish cells much less readily, in part because of a smaller blood supply.

Muscular tissue has a relatively poor capacity for renewal of lost cells. Even though skeletal muscle tissue contains stem cells called *satellite cells,* they do not divide rapidly enough to replace extensively damaged muscle fibers. Cardiac muscle tissue lacks satellite cells, and existing cardiac muscle fibers do not undergo mitosis to form new cells. Recent evidence suggests that stem cells do migrate into the heart from the blood. There, they can differentiate and replace a limited number of cardiac muscle fibers and endothelial cells in heart blood vessels. Smooth muscle fibers can proliferate to some extent, but they do so much more slowly than the cells of epithelial or connective tissues.

Nervous tissue has the poorest capacity for renewal. Although experiments have revealed the presence of some stem cells in the brain, they normally do not undergo mitosis to replace damaged neurons. Discovering why this is so is a major goal of researchers who seek ways to repair nervous tissue damaged by injury or disease.

The restoration of an injured tissue or organ to normal structure and function depends entirely on whether parenchymal cells are active in the repair process. If parenchymal cells accomplish the repair, **tissue regeneration** is possible, and a near-perfect reconstruction of the injured tissue may occur. However, if fibroblasts of the stroma are active in the repair, the replacement tissue will be a new connective tissue. The fibroblasts synthesize collagen and other matrix materials that aggregate to form scar tissue, a process known as **fibrosis.** Because scar tissue is not specialized to perform the functions of the parenchymal tissue, the original function of the tissue or organ is impaired.

When tissue damage is extensive, as in large, open wounds, both the connective tissue stroma and the parenchymal cells are active in repair; fibroblasts divide rapidly, and new collagen fibers are manufactured to provide structural strength. Blood capillaries also sprout new buds to supply the healing tissue with the materials it needs. All these processes create an actively growing connective tissue called **granulation tissue.** This new tissue forms across a wound or surgical incision to provide a framework (stroma) that supports the epithelial cells that migrate into the open area and fill it. The newly formed granulation tissue also secretes a fluid that kills bacteria.

Three factors affect tissue repair: nutrition, blood circulation, and age. Nutrition is vital because the healing process places a great demand on the body's store of nutrients. Adequate protein in the diet is important because most of the structural components of a tissue are proteins. Several vitamins also play a direct role in wound healing and tissue repair. For example, vitamin C directly affects the normal production and maintenance of matrix materials, especially collagen, and strengthens and promotes the formation of new blood vessels. In a person with vitamin C deficiency, even superficial wounds fail to heal, and the walls of the blood vessels become fragile and are easily ruptured.

Proper blood circulation is essential to transport oxygen, nutrients, antibodies, and many defensive cells to the injured site. The blood also plays an important role in the removal of tissue fluid, bacteria, foreign bodies, and debris, elements that would otherwise interfere with healing. The third factor in tissue repair, age, is the topic of the next section.

Adhesions

Scar tissue can form **adhesions,** abnormal joining of tissues. Adhesions commonly form in the abdomen around a site of previous inflammation such as an inflamed appendix, and they can develop after surgery. Although adhesions do not always cause problems, they can decrease tissue flexibility, cause obstruction (such as in the intestine), and make a subsequent operation more difficult. An *adhesiotomy,* the surgical release of adhesions, may be required. ■

▶ CHECKPOINT

21. How are stromal and parenchymal repair of a tissue different?

22. What is the importance of granulation tissue?

AGING AND TISSUES

▶ OBJECTIVE

Describe the effects of aging on tissues.

Generally, tissues heal faster and leave less obvious scars in the young than in the aged. In fact, surgery performed on fetuses leaves no scars. The younger body is generally in a better nutritional state, its tissues have a better blood supply, and its cells have a higher metabolic rate. Thus, cells can synthesize needed materials and divide more quickly. The extracellular components of tissues also change with age. Glucose, the most abundant sugar in the body, plays a role in the aging process. As the body ages, glucose is haphazardly added to proteins inside and outside cells, forming irreversible cross-links between adjacent protein molecules. With advancing age, more cross-links form, which contributes to the stiffening and loss of elasticity that occur in aging tissues. Collagen fibers, responsible for the strength of tendons, increase in number and change in quality with aging. These changes in the collagen of arterial

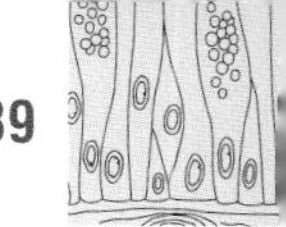

walls affect the flexibility of arteries as much as the fatty deposits associated with atherosclerosis (see page 756). Elastin, another extracellular component, is responsible for the elasticity of blood vessels and skin. It thickens, fragments, and acquires a greater affinity for calcium with age—changes that may also be associated with the development of atherosclerosis.

▶ CHECKPOINT

23. What common changes occur in epithelial and connective tissues with aging?

DISORDERS: HOMEOSTATIC IMBALANCES

Disorders of epithelial tissues are mainly specific to individual organs, such as peptic ulcer disease (PUD), which erodes the epithelial lining of the stomach or small intestine. For this reason, epithelial disorders are described along with the relevant body system throughout the text. The most prevalent disorders of connective tissues are **autoimmune diseases**—diseases in which antibodies produced by the immune system fail to distinguish what is foreign from what is self and attack the body's own tissues. One of the most common autoimmune disorders is rheumatoid arthritis, which attacks the synovial membranes of joints. Because connective tissue is one of the most abundant and widely distributed of the four main types of tissues, its disorders often affect multiple body systems. Common disorders of muscular tissue and nervous tissue are described at the ends of Chapters 10 and 12, respectively.

Sjögren's Syndrome

Sjögren's syndrome (SHŌ-grenz) is a common autoimmune disorder that causes inflammation and destruction of exocrine glands, especially the lacrimal (tear) glands and salivary glands. Signs include dryness of the eyes, mouth, nose, ears, skin, and vagina, and salivary gland enlargement. Systemic effects include fatigue, arthritis, difficulty in swallowing, pancreatitis (inflammation of the pancreas), pleuritis (inflammation of the pleurae of the lungs), and muscle and joint pain. The disorder affects females more than males by a ratio of 9 to 1. About 20% of older adults experience some signs of Sjögren's. Treatment is supportive, including using artificial tears to moisten the eyes, sipping fluids, chewing sugarless gum, using a saliva substitute to moisten the mouth, and using moisturizing creams for the skin.

Systemic Lupus Erythematosus

Systemic lupus erythematosus (er-i-thē-ma-TŌ-sus), **SLE,** or simply lupus, is a chronic inflammatory disease of connective tissue occurring mostly in nonwhite women during their childbearing years. It is an autoimmune disease that can cause tissue damage in every body system. The disease, which can range from a mild condition in most patients to a rapidly fatal disease, is marked by periods of exacerbation and remission. The prevalence of SLE is about 1 in 2000, with females more likely to be afflicted than males by a ratio of 8 or 9 to 1.

Although the cause of SLE is unknown, genetic, environmental, and hormonal factors all have been implicated. The genetic component is suggested by studies of twins and family history. Environmental factors include viruses, bacteria, chemicals, drugs, exposure to excessive sunlight, and emotional stress. Sex hormones, such as estrogens, may also trigger SLE.

Signs and symptoms of SLE include painful joints, low-grade fever, fatigue, mouth ulcers, weight loss, enlarged lymph nodes and spleen, sensitivity to sunlight, rapid loss of large amounts of scalp hair, and anorexia. A distinguishing feature of lupus is an eruption across the bridge of the nose and cheeks called a "butterfly rash." Other skin lesions may occur, including blistering and ulceration. The erosive nature of some SLE skin lesions was thought to resemble the damage inflicted by the bite of a wolf—thus, the name *lupus* (= wolf). The most serious complications of the disease involve inflammation of the kidneys, liver, spleen, lungs, heart, brain, and gastrointestinal tract. Because there is no cure for SLE, treatment is supportive, including anti-inflammatory drugs, such as aspirin, and immunosuppressive drugs.

MEDICAL TERMINOLOGY

Atrophy (AT-rō-fē; *a-* = without; *-trophy* = nourishment) A decrease in the size of cells, with a subsequent decrease in the size of the affected tissue or organ.

Biopsy (BĪ-op-sē; *bio-* = life; *-opsy* = to view) The removal of a sample of living tissue for microscopic examination to help diagnose disease.

Hypertrophy (hī-PER-trō-fē; *hyper-* = above or excessive) Increase in the size of a tissue because its cells enlarge without undergoing cell division.

Tissue rejection An immune response of the body directed at foreign proteins in a transplanted tissue or organ; immunosuppressive drugs, such as cyclosporine, have largely overcome tissue rejection in heart-, kidney-, and liver-transplant patients.

Tissue transplantation The replacement of a diseased or injured tissue or organ. The most successful transplants involve use of a person's own tissues or those from an identical twin.

Xenotransplantation (zen′-ō-trans-plan-TĀ-shun; *xeno-* = strange, foreign) The replacement of a diseased or injured tissue or organ with cells or tissues from an animal. Porcine (from pigs) and bovine (from cows) heart valves are used for some heart-valve replacement surgeries.

STUDY OUTLINE

TYPES OF TISSUES AND THEIR ORIGINS (p. 108)

1. A tissue is a group of similar cells, usually with a similar embryological origin, that is specialized for a particular function.
2. The various tissues of the body are classified into four basic types: epithelial, connective, muscular, and nervous.
3. All tissues of the body develop from three primary germ layers, the first tissues that form in a human embryo: ectoderm, mesoderm, and endoderm.

CELL JUNCTIONS (P. 108)

1. Cell junctions are points of contact between adjacent plasma membranes.
2. Tight junctions form fluid-tight seals between cells; adherens junctions, desmosomes, and hemidesmosomes anchor cells to one another or to the basement membrane; and gap junctions permit electrical and chemical signals to pass between cells.

EPITHELIAL TISSUE (P. 110)

1. The subtypes of epithelia include covering and lining epithelia and glandular epithelia.
2. An epithelium consists mostly of cells with little extracellular material between adjacent plasma membranes. The apical, lateral, and basal surfaces of epithelial cells are modified in various ways to carry out specific functions. Epithelium is arranged in sheets and attached to a basement membrane. Although it is avascular, it has a nerve supply. Epithelia are derived from all three primary germ layers and have a high capacity for renewal.
3. Epithelial layers can be simple (one layer) or stratified (several layers). The cell shapes may be squamous (flat), cuboidal (cube-like), columnar (rectangular), or transitional (variable).
4. Simple squamous epithelium consists of a single layer of flat cells (Table 4.1A). It is found in parts of the body where filtration or diffusion are priority processes. One type, endothelium, lines the heart and blood vessels. Another type, mesothelium, forms the serous membranes that line the thoracic and abdominopelvic cavities and cover the organs within them.
5. Simple cuboidal epithelium consists of a single layer of cube-shaped cells that function in secretion and absorption (Table 4.1B). It is found covering the ovaries, in the kidneys and eyes, and lining some glandular ducts.
6. Nonciliated simple columnar epithelium, a single layer of nonciliated rectangular cells (Table 4.1C), lines most of the gastrointestinal tract. Specialized cells containing microvilli perform absorption. Goblet cells secrete mucus.
7. Ciliated simple columnar epithelium consists of a single layer of ciliated rectangular cells (Table 4.1D). It is found in a few portions of the upper respiratory tract, where it moves foreign particles trapped in mucus out of the respiratory tract.
8. Pseudostratified columnar epithelium has only one layer but gives the appearance of many (Table 4.1E). A ciliated variety contains goblet cells and lines most of the upper respiratory tract; a nonciliated variety has no goblet cells and lines ducts of many glands, the epididymis, and part of the male urethra.
9. Stratified squamous epithelium consists of several layers of cells: Cells of the apical layer and several layers deep to it are flat (Table 4.1F). A nonkeratinized variety lines the mouth. A keratinized variety forms the epidermis, the most superficial layer of the skin.
10. Stratified cuboidal epithelium consists of several layers of cells: Cells at the apical layer are cube-shaped (Table 4.1G). It is found in adult sweat glands and a portion of the male urethra.
11. Stratified columnar epithelium consists of several layers of cells: Cells of the apical layer have a columnar shape (Table 4.1H). It is found in a portion of the male urethra and large excretory ducts of some glands.
12. Transitional epithelium consists of several layers of cells whose appearance varies with the degree of stretching (Table 4.1I). It lines the urinary bladder.
13. A gland is a single cell or a group of epithelial cells adapted for secretion.
14. Endocrine glands secrete hormones into interstitial fluid and then into the blood (Table 4.2A).
15. Exocrine glands (mucous, sweat, oil, and digestive glands) secrete into ducts or directly onto a free surface (Table 4.2B).
16. The structural classification of exocrine glands includes unicellular and multicellular glands.
17. The functional classification of exocrine glands includes holocrine, apocrine, and merocrine glands.

CONNECTIVE TISSUE (P. 121)

1. Connective tissue is one of the most abundant body tissues.
2. Connective tissue consists of relatively few cells and an abundant extracellular matrix of ground substance and fibers. It does not usually occur on free surfaces, has a nerve supply (except for cartilage), and is highly vascular (except for cartilage, tendons, and ligaments).
3. Cells in connective tissue are derived from mesenchymal cells.
4. Cell types include fibroblasts (secrete matrix), macrophages (perform phagocytosis), plasma cells (secrete antibodies), mast cells (produce histamine), adipocytes (store fat), and white blood cells (migrate from blood in response to infections).
5. The ground substance and fibers make up the extracellular matrix.
6. The ground substance supports and binds cells together, provides a medium for the exchange of materials, stores water, and is active in influencing cell functions.
7. Substances found in the ground substance include water and polysaccharides such as hyaluronic acid, chondroitin sulfate, dermatan sulfate, and keratan sulfate (glycosaminoglycans). Also present are proteoglycans and adhesion proteins.
8. The fibers in the extracellular matrix provide strength and support and are of three types: (a) Collagen fibers (composed of collagen) are found in large amounts in bone, tendons, and ligaments. (b) Elastic fibers (composed of elastin, fibrillin, and other glycoproteins) are found in skin, blood vessel walls, and lungs. (c) Reticular fibers (composed of collagen and glycoprotein) are found around fat cells, nerve fibers, and skeletal and smooth muscle cells.
9. The two major subclasses of connective tissue are embryonic connective tissue (found in the embryo and fetus) and mature connective tissue (present in the newborn).
10. The embryonic connective tissues are mesenchyme, which forms all other connective tissues (Table 4.3A), and mucous connective

tissue, found in the umbilical cord of the fetus, where it gives support (Table 4.3B).

11. Mature connective tissue differentiates from mesenchyme. It is subdivided into several types: loose or dense connective tissue, cartilage, bone tissue, and liquid connective tissue.
12. Loose connective tissue includes areolar connective tissue, adipose tissue, and reticular connective tissue.
13. Areolar connective tissue consists of the three types of fibers, several types of cells, and a semifluid ground substance (Table 4.4A). It is found in the subcutaneous layer, in mucous membranes, and around blood vessels, nerves, and body organs.
14. Adipose tissue consists of adipocytes, which store triglycerides (Table 4.4B). It is found in the subcutaneous layer, around organs, and in yellow bone marrow. Brown adipose tissue (BAT) generates heat.
15. Reticular connective tissue consists of reticular fibers and reticular cells and is found in the liver, spleen, and lymph nodes (Table 4.4C).
16. Dense connective tissue includes dense regular connective tissue, dense irregular connective tissue, and elastic connective tissue.
17. Dense regular connective tissue consists of parallel bundles of collagen fibers and fibroblasts (Table 4.4D). It forms tendons, most ligaments, and aponeuroses.
18. Dense irregular connective tissue usually consists of randomly arranged collagen fibers and a few fibroblasts (Table 4.4E). It is found in fasciae, the dermis of skin, and membrane capsules around organs.
19. Elastic connective tissue consists of branching elastic fibers and fibroblasts (Table 4.4F). It is found in the walls of large arteries, lungs, trachea, and bronchial tubes.
20. Cartilage contains chondrocytes and has a rubbery matrix (chondroitin sulfate) containing collagen and elastic fibers.
21. Hyaline cartilage, which consists of a gel-like ground substance and appears bluish white in the body, is found in the embryonic skeleton, at the ends of bones, in the nose, and in respiratory structures (Table 4.4G). It is flexible, allows movement, and provides support, and is usually surrounded by a perichondrium.
22. Fibrocartilage is found in the pubic symphysis, intervertebral discs, and menisci (cartilage pads) of the knee joint (Table 4.4H). It contains chondrocytes scattered among clearly visible bundles of collagen fibers.
23. Elastic cartilage, which maintains the shape of organs such as the epiglottis of the larynx, auditory (eustachian) tubes, and external ear (Table 4.4I), contains chondrocytes located within a threadlike network of elastic fibers, and has a perichondrium.
24. Cartilage enlarges by interstitial growth (from within) and appositional growth (from without).
25. Bone or osseous tissue consists of a matrix of mineral salts and collagen fibers that contribute to the hardness of bone, and osteocytes that are located in lacunae (Table 4.4J). It supports, protects, provides a surface area for muscle attachment, helps provide movement, stores minerals, and houses blood-forming tissue.
26. Blood tissue is liquid connective tissue that consists of blood plasma and formed elements—red blood cells, white blood cells, and platelets (Table 4.4K). Its cells transport oxygen and carbon dioxide, carry on phagocytosis, participate in allergic reactions, provide immunity, and bring about blood clotting.
27. Lymph, the extracellular fluid that flows in lymphatic vessels, is also a liquid connective tissue. It is a clear fluid similar to blood plasma but with less protein.

MEMBRANES (p. 132)

1. An epithelial membrane consists of an epithelial layer overlying a connective tissue layer. Examples are mucous, serous, and cutaneous membranes.
2. Mucous membranes line cavities that open to the exterior, such as the gastrointestinal tract.
3. Serous membranes line closed cavities (pleura, pericardium, peritoneum) and cover the organs in the cavities. These membranes consist of parietal and visceral layers.
4. Synovial membranes line joint cavities, bursae, and tendon sheaths and consist of areolar connective tissue instead of epithelium.

MUSCULAR TISSUE (p. 134)

1. Muscular tissue consists of fibers that are specialized for contraction. It provides motion, maintenance of posture, heat production, and protection.
2. Skeletal muscle tissue is attached to bones and is striated and voluntary (Table 4.5A).
3. The action of cardiac muscle tissue, which forms most of the heart wall and is striated, is involuntary (Table 4.5B).
4. Smooth muscle tissue is found in the walls of hollow internal structures (blood vessels and viscera) and is nonstriated and involuntary (Table 4.5C).

NERVOUS TISSUE (p. 136)

1. The nervous system is composed of neurons (nerve cells) and neuroglia (protective and supporting cells) (Table 4.6).
2. Neurons are sensitive to stimuli, convert stimuli into electrical signals called action potentials (nerve impulses), and conduct nerve impulses.
3. Most neurons consist of a cell body and two types of processes, dendrites and axons.

EXCITABLE CELLS (p. 137)

1. Electrical excitability is the ability to respond to certain stimuli by producing electrical signals such as action potentials.
2. Because neurons and muscle fibers exhibit electrical excitability, they are considered excitable cells.

TISSUE REPAIR: RESTORING HOMEOSTASIS (p. 137)

1. Tissue repair is the replacement of worn-out, damaged, or dead cells by healthy ones.
2. Stem cells may divide to replace lost or damaged cells.
3. If the injury is superficial, tissue repair involves parenchymal regeneration; if damage is extensive, granulation tissue is involved.
4. Good nutrition and blood circulation are vital to tissue repair.

AGING AND TISSUES (p. 138)

1. Tissues heal faster and leave less obvious scars in the young than in the aged; surgery performed on fetuses leaves no scars.
2. The extracellular components of tissues, such as collagen and elastic fibers, also change with age.

SELF-QUIZ QUESTIONS

Fill in the blanks in the following statements.

1. The four types of connective tissues are ____, ____, ____, and ____.
2. Epithelial tissue tends to be classified according to two criteria: ____ and ____.

Indicate whether the following statements are true or false.

3. Epithelial tissue cells have an apical surface at the top and are attached to a basement membrane at the bottom.
4. Connective tissue fibers that are arranged in bundles and lend strength and flexibility to a tissue are collagen fibers.

Choose the one best answer to the following questions.

5. Which of the following muscle tissues can be voluntarily controlled? (1) cardiac, (2), smooth, (3) skeletal. (a) 1, 2, and 3, (b) 2, (c) 1, (d) 1 and 3, (e) 3.
6. Which of the following tissues is avascular? (a) cardiac muscle, (b) stratified squamous epithelial, (c) compact bone, (d) skeletal muscle, (e) adipose.
7. If the lining of an organ produces and releases mucus, which of the following cells would likely be found in the tissue lining the organ? (a) goblet cells, (b) mast cells, (c) macrophages, (d) osteoblasts, (e) fibroblasts.
8. Why does damaged cartilage heal slowly? (a) Damaged cartilage undergoes fibrosis, which interferes with the movement of materials needed for repair. (b) Cartilage does not contain fibroblasts, which are needed to produce the fibers in cartilage tissue. (c) Cartilage is avascular, so materials needed for repair must diffuse from surrounding tissue. (d) Chondrocytes cannot be replaced once they are damaged. (e) Chondrocytes undergo mitosis slowly, which delays healing.
9. Which of the following is *true* concerning serous membranes? (a) A serous membrane lines a body part that opens directly to the body's exterior. (b) The parietal portion of a serous membrane attaches to the organ. (c) The visceral portion of a serous membrane attaches to a body cavity wall. (d) The serous membrane covering the heart is known as the peritoneum. (e) The serous membrane covering the lungs is known as the pleura.
10. The type of exocrine gland that forms its secretory product and simply releases it from the cell by exocytosis is the (a) apocrine gland, (b) merocrine gland, (c) holocrine gland, (d) endocrine gland, (e) tubular gland.
11. Tissue changes that occur with aging can be due to (1) cross-links forming between glucose and proteins, (2) a decrease in the amount of collagen fibers, (3) a decreased blood supply, (4) improper nutrition, (5) a higher cellular metabolic rate. (a) 1, 2, 3, 4 and 5, (b) 1, 2, 3, and 4, (c) 1 and 4, (d) 1, 3, and 4, (e) 1, 2, and 3.
12. What type of cell junction would be required for cells to communicate with one another? (a) adherens junction, (b) desmosome, (c) gap junction, (d) tight junction, (e) hemidesmosome.
13. For each of the following items, indicate the tissue type with which they are associated. Use **E** for epithelial tissue, **C** for connective tissue, **M** for muscle tissue, and **N** for nervous tissue.

____(a) binds, supports
____(b) contains elongated cells that generate force
____(c) neuroglia
____(d) avascular
____(e) may contain fibroblasts
____(f) tightly packed cells
____(g) intercalated discs
____(h) goblet cells
____(i) contains extracellular matrix
____(j) striated
____(k) generate action potentials
____(l) cilia
____(m) ground substance
____(n) apical surface
____(o) excitable

14. Match the following epithelial tissues to their descriptions:

____(a) contains a single layer of flat cells; found in the body where filtration (kidney) or diffusion (lungs) are priority processes
____(b) found in the superficial part of skin; provides protection from heat, microbes, and chemicals
____(c) contains cube-shaped cells functioning in secretion and absorption
____(d) lines the upper respiratory tract and uterine tubes; wavelike motion of cilia propels materials through the lumen
____(e) contains cells with microvilli and goblet cells; found in linings of the digestive, reproductive, and urinary tracts
____(f) found in the urinary bladder; contains cells that can change shape (stretch or relax)
____(g) contains cells that are all attached to the basement membrane, although some do not reach the surface; those cells that do extend to the surface secrete mucus or contain cilia
____(h) a fairly rare type of epithelium that has a mainly protective function

(1) pseudostratified ciliated columnar epithelium
(2) ciliated simple columnar epithelium
(3) transitional epithelium
(4) simple squamous epithelium
(5) simple cuboidal epithelium
(6) nonciliated simple columnar epithelium
(7) stratified cuboidal epithelium
(8) keratinized stratified squamous epithelium

15. Matching the following connective tissues to their descriptions:

____(a) the tissue from which all other connective tissues eventually arise
____(b) connective tissue with a clear, liquid matrix that flows in lymphatic vessels
____(c) connective tissue consisting of several kinds of cells, containing all three fiber types randomly arranged, and found in the subcutaneous layer of the skin
____(d) a loose connective tissue specialized for triglyceride storage
____(e) tissue that contains reticular fibers and reticular cells and forms the stroma of certain organs such as the spleen
____(f) tissue with irregularly arranged collagen fibers found in the dermis of the skin
____(g) tissue found in the lungs that is strong and can recoil back to its original shape after being stretched
____(h) tissue that affords flexibility at joints and reduces joint friction
____(i) tissue that provides strength and rigidity and is the strongest of the three types of cartilage
____(j) bundles of collagen arranged in parallel patterns; compose tendons and ligaments
____(k) tissue that forms the internal framework of the body and works with skeletal muscle to generate movement
____(l) tissue that contains a network of elastic fibers, providing strength, elasticity, and maintenance of shape; located in the external ear
____(m) connective tissue with formed elements suspended in a liquid matrix called plasma

(1) blood
(2) fibrocartilage
(3) mesenchyme
(4) dense regular connective tissue
(5) lymph
(6) hyaline cartilage
(7) dense irregular connective tissue
(8) areolar connective tissue
(9) reticular connective tissue
(10) bone (osseous tissue)
(11) elastic connective tissue
(12) elastic cartilage
(13) adipose tissue

CRITICAL THINKING QUESTIONS

1. Imagine that you live 50 years in the future, and you can custom-design a human to suit the environment. Your assignment is to customize the human's tissue so that the individual can survive on a large planet with gravity, a cold, dry climate, and a thin atmosphere. What adaptations would you incorporate into the structure and/or amount of tissues, and why?
2. You are entering a "Cutest Baby Contest" and have asked your colleagues to help you choose the most adorable picture of yourself as a baby. One of your colleagues rudely points out that you were quite chubby as an infant. You, however, are not offended and proceed to explain to your colleague the benefit of that "baby fat."
3. You've been on a "bread-and-water" diet for three weeks and have noticed that a cut on your shin won't heal and bleeds easily. Why?

ANSWERS TO FIGURE QUESTIONS

4.1 Gap junctions allow cellular communication via passage of electrical and chemical signals between adjacent cells.

4.2 The basement membrane provides a physical support for an epithelium.

4.3 Substances would move most rapidly through squamous cells because they are so thin.

4.4 Simple multicellular exocrine glands have a nonbranched duct; compound multicellular exocrine glands have a branched duct.

4.5 Sebaceous (oil) glands are holocrine glands, and salivary glands are merocrine glands.

4.6 Fibroblasts secrete the fibers and ground substance of the extracellular matrix.

4.7 An epithelial membrane is a membrane that consists of an epithelial layer and an underlying layer of connective tissue.

Chapter 5

The Integumentary System

The Integumentary System and Homeostasis

The integumentary system contributes to homeostasis by protecting the body and helping regulate body temperature. It also allows you to sense pleasurable, painful, and other stimuli in your external environment.

wiley.com/college/apcentral

Skin and its accessory structures—hair and nails, along with various glands, muscles, and nerves—make up the **integumentary system** (in-teg-ū-MEN-tar-ē; *inte-* = whole; *-gument* = body covering). The integumentary system protects the body, helps maintain a constant body temperature, and provides sensory information about the surrounding environment. Of all the body's organs, none is more easily inspected or more exposed to infection, disease, and injury than the skin. Although its location makes it vulnerable to damage from trauma, sunlight, microbes, and pollutants in the environment, the skin's protective features ward off such damage. Because of its visibility, skin reflects our emotions (frowning, blushing) and some aspects of normal physiology (such as sweating). Changes in skin color may also indicate homeostatic imbalances in the body. For example, the bluish skin color associated with hypoxia (oxygen deficiency at the tissue level) is one sign of heart failure as well as other disorders. Abnormal skin eruptions or rashes such as chickenpox, cold sores, or measles may reveal systemic infections or diseases of internal organs, while other conditions, such as warts, age spots, or pimples, may involve the skin alone. So important is the skin to self-image that many people spend a great deal of time and money to restore it to a more normal or youthful appearance. **Dermatology** (der′-ma-TOL-ō-jē; *dermato-* = skin; *-logy* = study of) is the medical specialty that deals with the diagnosis and treatment of integumentary system disorders.

STRUCTURE OF THE SKIN

▶ **OBJECTIVES**

Describe the layers of the epidermis and the cells that compose them.

Compare the composition of the papillary and reticular regions of the dermis.

Explain the basis for different skin colors.

The **skin** or **cutaneous membrane,** which covers the external surface of the body, is the largest organ of the body in both surface area and weight. In adults, the skin covers an area of about 2 square meters (22 square feet) and weighs 4.5–5 kg (10–11 lb), about 16% of total body weight. It ranges in thickness from 0.5 mm (0.02 in.) on the eyelids to 4.0 mm (0.16 in.) on the heels. However, over most of the body it is 1–2 mm (0.04–0.08 in.) thick. Structurally, the skin consists of two main parts (Figure 5.1). The superficial, thinner portion, which is composed of *epithelial tissue,* is the **epidermis** (ep′-i-DERM-is; *epi-* = above). The deeper, thicker *connective tissue* part is the **dermis.**

Deep to the dermis, but not part of the skin, is the **subcutaneous (subQ) layer.** Also called the **hypodermis** (*hypo-* = below), this layer consists of areolar and adipose tissues. Fibers that extend from the dermis anchor the skin to the subcutaneous layer, which, in turn, attaches to underlying tissues and organs. The subcutaneous layer serves as a storage depot for fat and contains large blood vessels that supply the skin. This region (and sometimes the dermis) also contains nerve endings called **lamellated (pacinian) corpuscles** (pa-SIN-ē-an) that are sensitive to pressure (Figure 5.1).

Epidermis

The **epidermis** is composed of keratinized stratified squamous epithelium. It contains four principal types of cells: keratinocytes, melanocytes, Langerhans cells, and Merkel cells (Figure 5.2 on page 148). About 90% of epidermal cells are **keratinocytes** (ker-a-TIN-ō-sīts; *keratino-* = hornlike; *-cytes* = cells), which are arranged in four or five layers and produce the protein **keratin** (Figure 5.2a). Recall from Chapter 4 that keratin is a tough, fibrous protein that helps protect the skin and underlying tissues from heat, microbes, and chemicals. Keratinocytes also produce lamellar granules, which release a water-repellent sealant that decreases water entry and loss and inhibits the entry of foreign materials.

About 8% of the epidermal cells are **melanocytes** (MEL-a-nō-sīts; *melano-* = black), which develop from the ectoderm of a developing embryo and produce the pigment melanin (Figure 5.2b). Their long, slender projections extend between the keratinocytes and transfer melanin granules to them. **Melanin** is a yellow-red or brown-black pigment that contributes to skin color and absorbs damaging ultraviolet (UV) light. Once inside keratinocytes, the melanin granules cluster to form a protective veil over the nucleus, on the side toward the skin surface. In this way, they shield the nuclear DNA from damage by UV light. Although their melanin granules effectively protect keratinocytes, melanocytes themselves are particularly susceptible to damage by UV light.

Langerhans cells (LANG-er-hans) arise from red bone marrow and migrate to the epidermis (Figure 5.2c), where they constitute a small fraction of the epidermal cells. They participate in immune responses mounted against microbes that invade the skin, and are easily damaged by UV light.

Merkel cells are the least numerous of the epidermal cells. They are located in the deepest layer of the epidermis, where they contact the flattened process of a sensory neuron (nerve cell), a structure called a **tactile (Merkel) disc** (Figure 5.2d). Merkel cells and tactile discs detect different aspects of touch sensations.

Several distinct layers of keratinocytes in various stages of development form the epidermis (Figure 5.3 on page 149). In most regions of the body the epidermis has four strata or layers—stratum basale, stratum spinosum, stratum granulosum, and a thin stratum corneum. This is called **thin skin.** Where exposure to friction is greatest, such as in the fingertips, palms, and soles, the epidermis has five layers—stratum basale, stratum spinosum, stratum granulosum, stratum lucidum, and a thick stratum corneum. This is called **thick skin.** The details of thin and thick skin are discussed later in the chapter.

Figure 5.1 Components of the integumentary system. The skin consists of a superficial, thin epidermis and a deep, thicker dermis. Deep to the skin is the subcutaneous layer, which attaches the dermis to underlying organs and tissues.

The integumentary system includes the skin and its accessory structures—hair, nails, and skin glands—along with associated smooth muscles and nerves.

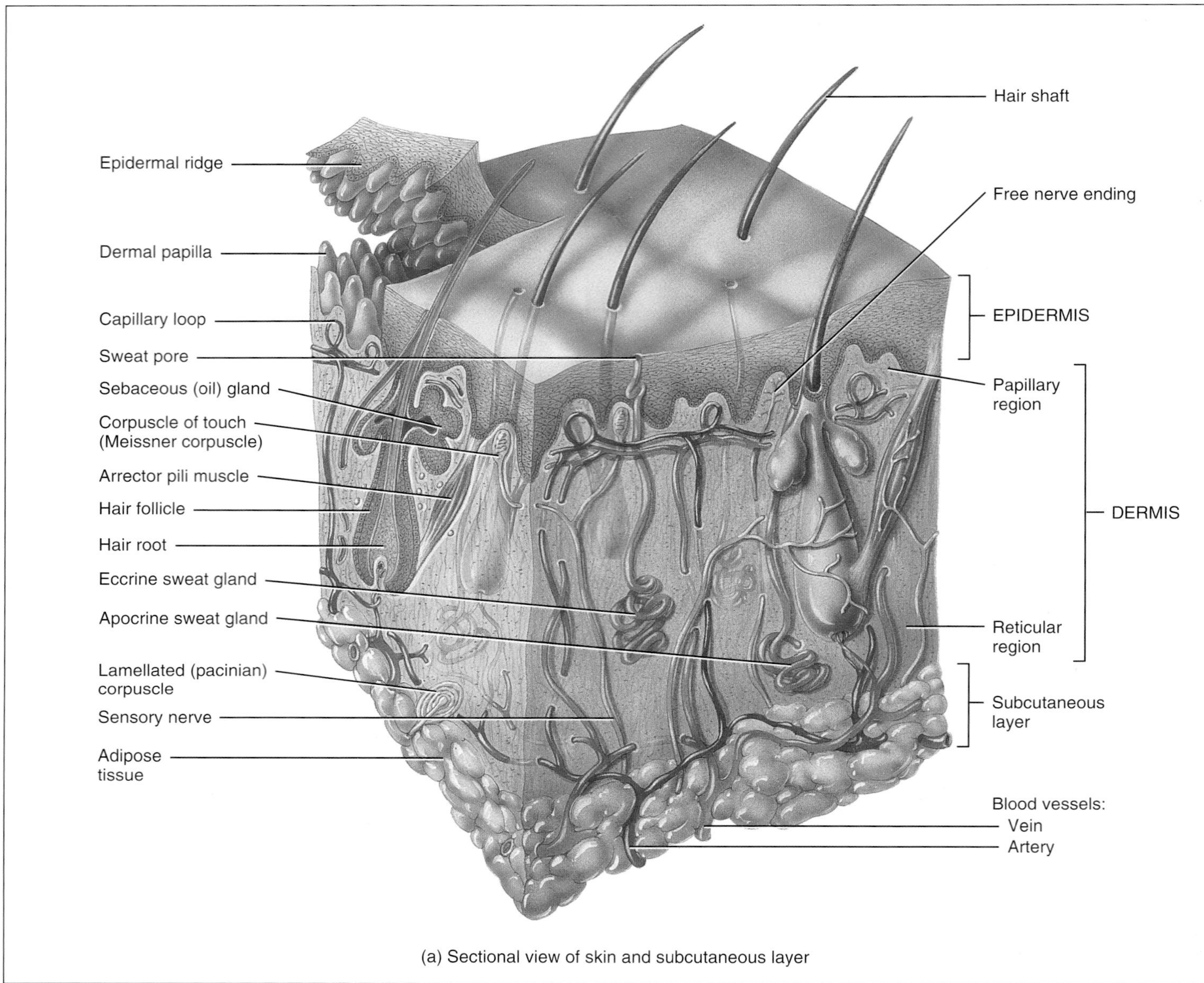

(a) Sectional view of skin and subcutaneous layer

Stratum Basale

The deepest layer of the epidermis is the **stratum basale** (ba-SA-lē; *basal-* = base), composed of a single row of cuboidal or columnar keratinocytes. Some cells in this layer are *stem cells* that undergo cell division to continually produce new keratinocytes. The nuclei of keratinocytes in the stratum basale are large, and their cytoplasm contains many ribosomes, a small Golgi complex, a few mitochondria, and some rough endoplasmic reticulum. The cytoskeleton within keratinocytes of the stratum basale includes scattered intermediate filaments, called *tonofilaments.* The tonofilaments are composed of a protein that will form keratin in more superficial epidermal layers. Tonofilaments attach to desmosomes, which bind cells of the stratum basale to each other and to the cells of the adjacent stratum spinosum, and to hemidesmosomes, which bind the keratinocytes to the basement membrane between the epidermis and the dermis. Melanocytes, Langerhans cells, and Merkel cells with their associated tactile discs are scattered among the keratinocytes of the basal layer. The stratum basale is also known as the **stratum germinativum** (jer′-mi-na-TĒ-vum; *germ-* = sprout) to indicate its role in forming new cells.

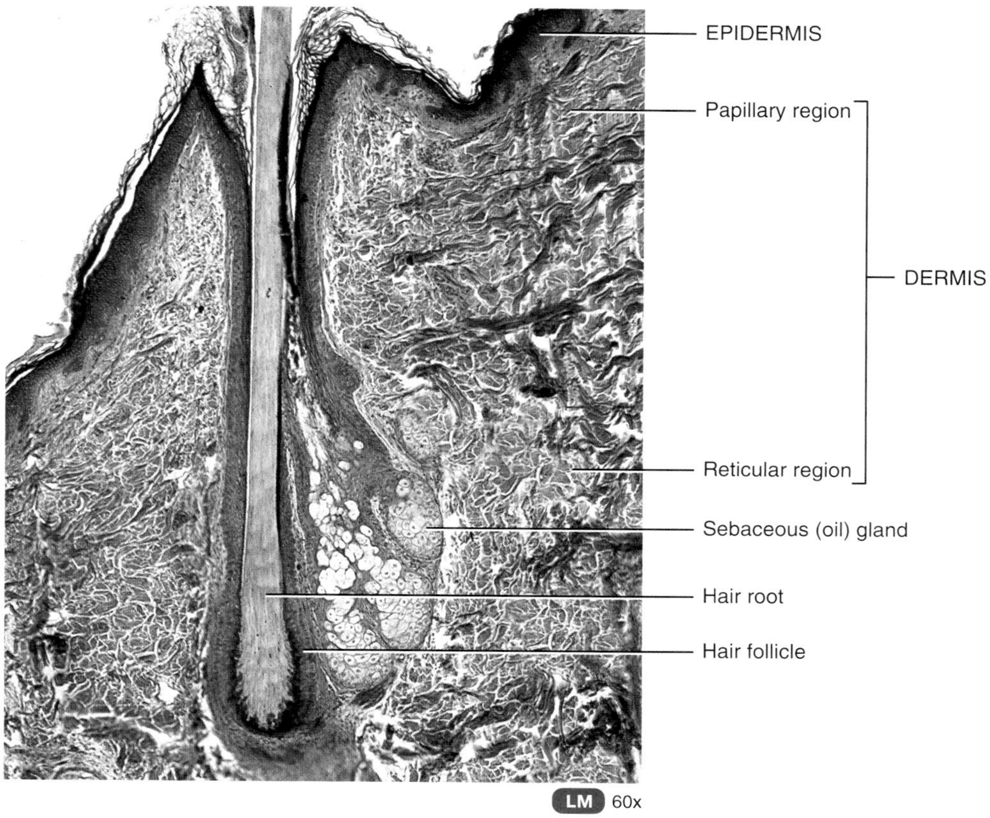

EPIDERMIS

Papillary region

DERMIS

Reticular region

Sebaceous (oil) gland

Hair root

Hair follicle

LM 60x

(b) Sectional view of skin

Functions

1. **Regulates body temperature.**
2. **Stores blood.**
3. **Protects body from external environment.**
4. **Detects cutaneous sensations.**
5. **Excretes and absorbs substances.**
6. **Synthesizes vitamin D.**

What types of tissues make up the epidermis and the dermis?

Skin Grafts

New skin cannot regenerate if an injury destroys a large area of the stratum basale and its stem cells. Skin wounds of this magnitude require skin grafts in order to heal. A **skin graft** involves covering the wound with a patch of healthy skin taken from a donor site. To avoid tissue rejection, the transplanted skin is usually taken from the same individual (*autograft*) or an identical twin (*isograft*). If skin damage is so extensive that an autograft would cause harm, a self-donation procedure called *autologous skin transplantation* (aw-TOL-ō-gus) may be used. In this procedure, performed most often for severely burned patients, small amounts of an individual's epidermis are removed, and the keratinocytes are cultured in the laboratory to produce thin sheets of skin. The new skin is transplanted back to the patient so that it covers the burn wound and generates a permanent skin. Also available as skin grafts for wound coverage are products (Apligraft and Transite) grown in the laboratory from the foreskins of circumcised infants. ■

Stratum Spinosum

Superficial to the stratum basale is the **stratum spinosum** (spi-NŌ-sum; *spinos-* = thornlike), where 8 to 10 layers of

Figure 5.2 Types of cells in the epidermis. Besides keratinocytes, the epidermis contains melanocytes, which produce the pigment melanin; Langerhans cells, which participate in immune responses; and Merkel cells, which function in the sensation of touch.

Most of the epidermis consists of keratinocytes, which produce the protein keratin (protects underlying tissues) and lamellar granules (contain a waterproof sealant).

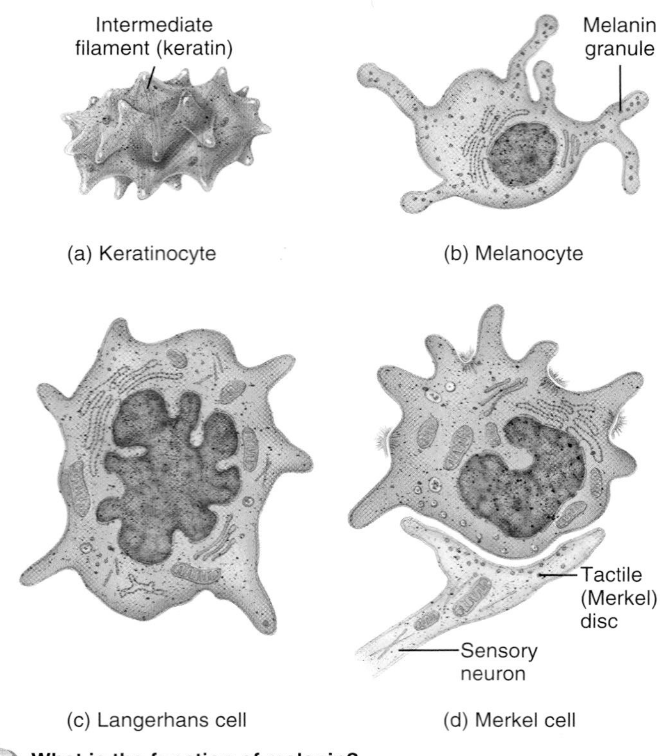

What is the function of melanin?

many-sided keratinocytes fit closely together. These keratinocytes have the same organelles as cells of the stratum basale. When cells of the stratum spinosum are prepared for microscopic examination, they shrink and pull apart so that they seem to be covered with thornlike spines (see Figure 5.2a), although they appear rounded and larger in living tissue. Each spiny projection in a prepared tissue section is a point where bundles of tonofilaments are inserting into a desmosome, tightly joining the cells to one another. This arrangement provides both strength and flexibility to the skin. Projections of both Langerhans cells and melanocytes also appear in this layer.

Stratum Granulosum

At about the middle of the epidermis, the **stratum granulosum** (gran-ū-LŌ-sum; *granulos-* = little grains) consists of three to five layers of flattened keratinocytes that are undergoing apoptosis. (Recall from Chapter 3 that apoptosis is an orderly, genetically programmed cell death in which the nucleus fragments before the cells die.) The nuclei and other organelles of these cells begin to degenerate, and tonofilaments become more apparent. A distinctive feature of cells in this layer is the presence of darkly staining granules of a protein called **keratohyalin** (ker′-a-tō-HĪ-a-lin), which converts the tonofilaments into keratin. Also present in the keratinocytes are membrane-enclosed **lamellar granules,** which release a lipid-rich secretion. This secretion fills the spaces between cells of the stratum granulosum, stratum lucidum, and stratum corneum. The lipid-rich secretion acts as a water-repellent sealant, retarding loss and entry of water and entry of foreign materials. As their nuclei break down during apoptosis, the keratinocytes of the stratum granulosum can no longer carry on vital metabolic reactions, and they die. Thus, the stratum granulosum marks the transition between the deeper, metabolically active strata and the dead cells of the more superficial strata.

Stratum Lucidum

The **stratum lucidum** (LOO-si-dum; *lucid-* = clear) is present only in the thick skin of the fingertips, palms, and soles. It consists of three to five layers of flattened clear, dead keratinocytes that contain large amounts of keratin and thickened plasma membranes.

Stratum Corneum

The **stratum corneum** (COR-nē-um; *corne-* = horn or horny) consists of 25 to 30 layers of flattened dead keratinocytes. These cells are continuously shed and replaced by cells from the deeper strata. The interior of the cells contains mostly keratin. Between the cells are lipids from lamellar granules that help make this layer an effective water-repellent barrier. Its multiple layers of dead cells also help to protect deeper layers from injury and microbial invasion. Constant exposure of skin to friction stimulates the formation of a *callus,* an abnormal thickening of the stratum corneum.

Keratinization and Growth of the Epidermis

Newly formed cells in the stratum basale are slowly pushed to the surface. As the cells move from one epidermal layer to the next, they accumulate more and more keratin, a process called **keratinization** (ker′-a-tin-i-ZĀ-shun). Then they undergo apoptosis. Eventually the keratinized cells slough off and are replaced by underlying cells that, in turn, become keratinized. The whole process by which cells form in the stratum basale, rise to the surface, become keratinized, and slough off takes about four weeks in an average epidermis of 0.1 mm (0.004 in.) thickness. The rate of cell division in the stratum basale increases when the outer layers of the epidermis are stripped away, as occurs in abrasions and burns. The mechanisms that regulate this remarkable growth are not well understood, but hormone-like proteins such as **epidermal growth factor (EGF)** play a role. An excessive amount of keratinized cells shed from the skin of the scalp is called **dandruff.**

Table 5.1 summarizes the distinctive features of the epidermal strata.

Figure 5.3 **Layers of the epidermis.**

The epidermis consists of keratinized stratified squamous epithelium.

Stratum corneum
Stratum lucidum
Stratum granulosum
Stratum spinosum
Stratum basale
Dead keratinocytes
Lamellar granules
Keratinocyte
Langerhans cell
Merkel cell
Tactile disc
Sensory neuron
Melanocyte
Dermis
Superficial
Deep

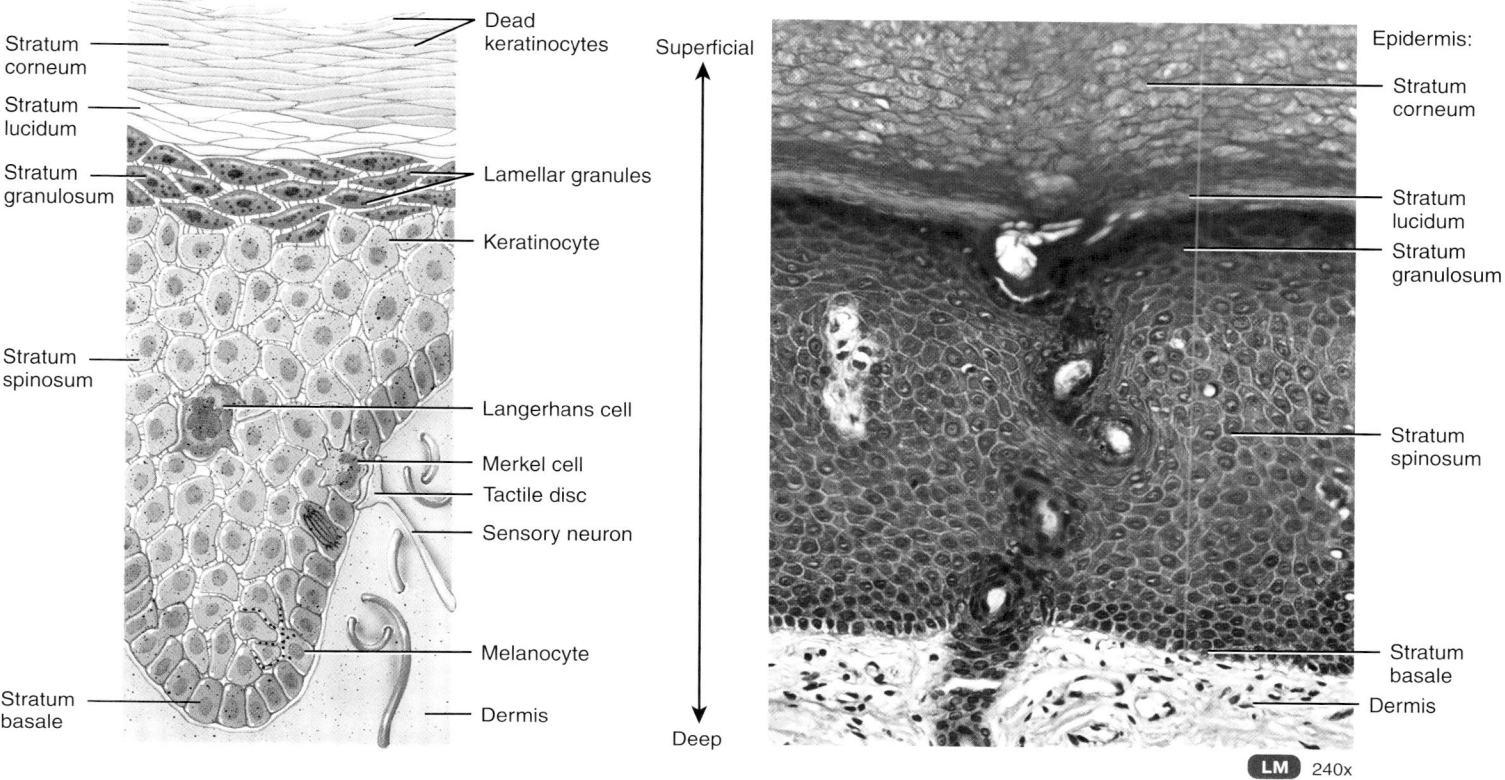

(a) Four principal cell types in epidermis

Epidermis:
Stratum corneum
Stratum lucidum
Stratum granulosum
Stratum spinosum
Stratum basale
Dermis
LM 240x

(b) Photomicrograph of a portion of the skin

Which epidermal layer includes stem cells that continually undergo cell division?

TABLE 5.1 Summary of Epidermal Strata

Stratum	Description
Basale	Deepest layer, composed of a single row of cuboidal or columnar keratinocytes that contain scattered tonofilaments (intermediate filaments); stem cells undergo cell division to produce new keratinocytes; melanocytes, Langerhans cells, and Merkel cells associated with tactile discs are scattered among the keratinocytes.
Spinosum	Eight to 10 rows of many-sided keratinocytes with bundles of tonofilaments; includes projections of melanocytes and Langerhans cells.
Granulosum	Three to five rows of flattened keratinocytes, in which organelles are beginning to degenerate; cells contain the protein keratohyalin, which converts tonofilaments into keratin, and lamellar granules, which release a lipid-rich, water-repellent secretion.
Lucidum	Present only in skin of fingertips, palms, and soles; consists of three to five rows of clear, flat, dead keratinocytes with large amounts of keratin.
Corneum	Twenty-five to 30 rows of dead, flat keratinocytes that contain mostly keratin.

Psoriasis

Psoriasis is a common and chronic skin disorder in which keratinocytes divide and move more quickly than normal from the stratum basale to the stratum corneum. They are shed prematurely in as little as 7 to 10 days. The immature keratinocytes make an abnormal keratin, which forms flaky, silvery scales at the skin surface, most often on the knees, elbows, and scalp. Effective treatments—various topical ointments and UV phototherapy—suppress cell division, decrease the rate of cell growth, or inhibit keratinization. ■

Dermis

The second, deeper part of the skin, the **dermis,** is composed mainly of connective tissue. Blood vessels, nerves, glands, and hair follicles are embedded in dermal tissue. Based on its tissue structure, the dermis can be divided into a papillary region and a reticular region.

The **papillary region** makes up about one-fifth of the thickness of the total layer (see Figure 5.1). It consists of areolar connective tissue containing fine elastic fibers. Its surface area is greatly increased by small, fingerlike structures called **dermal**

papillae (pa-PIL-ē = nipples). These nipple-shaped structures project into the epidermis and some contain **capillary loops** (blood capillaries). Some dermal papillae also contain tactile receptors called **corpuscles of touch** or **Meissner corpuscles,** nerve endings that are sensitive to touch, and **free nerve endings,** dendrites that lack any apparent structural specialization. Different free nerve endings initiate signals that give rise to sensations of warmth, coolness, pain, tickling, and itching.

The **reticular region** *(reticul-* = netlike), which is attached to the subcutaneous layer, consists of dense irregular connective tissue containing fibroblasts, bundles of collagen, and some coarse elastic fibers. The collagen fibers in the reticular region interlace in a netlike manner. A few adipose cells, hair follicles, nerves, sebaceous (oil) glands, and sudoriferous (sweat) glands occupy the spaces between fibers.

The combination of collagen and elastic fibers in the reticular region provides the skin with strength, **extensibility** (ability to stretch), and **elasticity** (ability to return to original shape after stretching). The extensibility of skin can be readily seen around joints and in pregnancy and obesity. Extreme stretching may produce small tears in the dermis, causing **striae** (STRĪ-ē = streaks), or stretch marks, visible as red or silvery white streaks on the skin surface.

Lines of Cleavage and Surgery

In certain regions of the body, collagen fibers tend to orient more in one direction than another. **Lines of cleavage (tension lines)** in the skin indicate the predominant direction of underlying collagen fibers. The lines are especially evident on the palmar surfaces of the fingers, where they are aligned with the long axis of the digits. Knowledge of lines of cleavage is especially important to plastic surgeons. For example, a surgical incision running parallel to the collagen fibers will heal with only a fine scar. A surgical incision made across the rows of fibers disrupts the collagen, and the wound tends to gape open and heal in a broad, thick scar. ■

The surfaces of the palms, fingers, soles, and toes have a series of ridges and grooves. They appear either as straight lines or as a pattern of loops and whorls, as on the tips of the digits. These **epidermal ridges** develop during the third month of fetal development as downward projections of the epidermis into the dermis between the dermal papillae of the papillary region (see Figure 5.1). The ridges increase the surface area of the epidermis and thus increase the grip of the hand or foot by increasing friction. Because the ducts of sweat glands open on the tops of the epidermal ridges as sweat pores, the sweat and ridges form **fingerprints** (or **footprints**) upon touching a smooth object. The epidermal ridge pattern is genetically determined and is unique for each individual. Normally, the ridge pattern does not change during life, except to enlarge, and thus can serve as the basis for identification. The study of the pattern of epidermal ridges is called **dermatoglyphics** (der′-ma-tō-GLIF-iks; *glyphe* = carved work).

TABLE 5.2 Summary of Papillary and Reticular Regions of the Dermis

Region	Description
Papillary	The superficial portion of the dermis (about one-fifth); consists of areolar connective tissue with elastic fibers; contains dermal papillae that house capillaries, corpuscles of touch, and free nerve endings.
Reticular	The deeper portion of the dermis (about four-fifths); consists of dense irregular connective tissue with bundles of collagen and some coarse elastic fibers. Spaces between fibers contain some adipose cells, hair follicles, nerves, sebaceous glands, and sudoriferous glands.

Table 5.2 summarizes the structural features of the papillary and reticular regions of the dermis.

The Structural Basis of Skin Color

Melanin, hemoglobin, and carotene are three pigments that impart a wide variety of colors to skin. The amount of **melanin** causes the skin's color to vary from pale yellow to red to tan to black. The difference between the two forms of melanin, *pheomelanin* (yellow to red) and *eumelanin* (brown to black), is most apparent in the hair. Melanocytes, the melanin-producing cells, are most plentiful in the epidermis of the penis, nipples of the breasts, area just around the nipples (areolae), face, and limbs. They are also present in mucous membranes. Because the *number* of melanocytes is about the same in all people, differences in skin color are due mainly to the *amount of pigment* the melanocytes produce and transfer to keratinocytes. In some people, melanin accumulates in patches called *freckles.* With age, *age (liver) spots* may develop. These flat blemishes look like freckles and range in color from light brown to black. Like freckles, age spots are accumulations of melanin. A round, flat, or raised area that represents a benign localized overgrowth of melanocytes and usually develops in childhood or adolescence is called a **nevus** (NĒ-vus), or a **mole.**

Melanocytes synthesize melanin from the amino acid *tyrosine* in the presence of an enzyme called *tyrosinase.* Synthesis occurs in an organelle called a **melanosome.** Exposure to UV light increases the enzymatic activity within melanosomes and thus increases melanin production. Both the amount and darkness of melanin increase upon UV exposure, which gives the skin a tanned appearance and helps protect the body against further UV radiation. Melanin absorbs UV radiation, prevents damage to DNA in epidermal cells, and neutralizes free radicals that form in the skin following damage by UV radiation. Thus, within limits, melanin serves a protective function. As you will see later, however, repeatedly exposing the skin to UV light may cause skin cancer. A tan is lost when the melanin-containing keratinocytes are shed from the stratum corneum.

Dark-skinned individuals have large amounts of melanin in the epidermis. Consequently, the epidermis has a dark pigmentation and skin color ranges from yellow to red to tan to black. Light-skinned individuals have little melanin in the epidermis. Thus, the epidermis appears translucent and skin color ranges from pink to red depending on the amount and oxygen content of the blood moving through capillaries in the dermis. The red color is due to **hemoglobin,** the oxygen-carrying pigment in red blood cells.

Carotene (KAR-ō-tēn; *carot* = carrot) is a yellow-orange pigment that gives egg yolk and carrots their color. This precursor of vitamin A, which is used to synthesize pigments needed for vision, accumulates in the stratum corneum and fatty areas of the dermis and subcutaneous layer in response to excessive dietary intake. In fact, so much carotene may be deposited in the skin after eating large amounts of carotene-rich foods that the skin color actually turns orange, which is especially apparent in light-skinned individuals.

Albinism (AL-bin-izm; *albin-* = white) is the inherited inability of an individual to produce melanin. Most **albinos** (al-BĪ-nōs), people affected by albinism, have melanocytes that are unable to synthesize tyrosinase. Melanin is missing from their hair, eyes, and skin.

In another condition, called **vitiligo** (vit-i-LĪ-gō), the partial or complete loss of melanocytes from patches of skin produces irregular white spots. The loss of melanocytes may be related to an immune system malfunction in which antibodies attack the melanocytes.

Skin Color as a Diagnostic Clue

The color of skin and mucous membranes can provide clues for diagnosing certain conditions. When blood is not picking up an adequate amount of oxygen from the lungs, as in someone who has stopped breathing, the mucous membranes, nail beds, and skin appear bluish or **cyanotic** (sī-a-NOT-ik; *cyan-* = blue). **Jaundice** (JON-dis; *jaund-* = yellow) is due to a buildup of the yellow pigment bilirubin in the skin. This condition gives a yellowish appearance to the skin and the whites of the eyes, and usually indicates liver disease. **Erythema** (er-e-THĒ-ma; *eryth-* = red), redness of the skin, is caused by engorgement of capillaries in the dermis with blood due to skin injury, exposure to heat, infection, inflammation, or allergic reactions. **Pallor** (PAL-or), or paleness of the skin, may occur in conditions such as shock and anemia. All skin color changes are observed most readily in people with lighter-colored skin and may be more difficult to discern in people with darker skin. However, examination of the nail beds and gums can provide some information about circulation in individuals with darker skin. ■

Tattooing and Body Piercing

Tattooing is a permanent coloration of the skin in which a foreign pigment is deposited with a needle into the dermis. It is believed that the practice originated in Ancient Egypt between 4000 and 2000 B.C. Today, tattooing is performed in one form or another by nearly all peoples of the world, and it is estimated that about one in five U.S. college students has one or more. Tattoos can be removed by lasers, which use concentrated beams of light. In the procedure, which requires a series of treatments, the tattoo inks and pigments selectively absorb the high-intensity laser light without destroying normal surrounding skin tissue. The laser causes the tattoo to dissolve into small ink particles that are eventually removed by the immune system. Laser removal of tattoos involves a considerable investment in time and money and can be quite painful.

Body piercing, the insertion of jewelry through an artificial opening, is also an ancient practice employed by Egyptian pharaohs and Roman soldiers, and a current tradition among many Americans. Today it is estimated that about one in three U.S. college students has had a body piercing. For most piercing locations, the piercer cleans the skin with an antiseptic, retracts the skin with forceps, and pushes a needle through the skin. Then the jewelry is connected to the needle and pushed through the skin. Total healing can take up to a year. Among the sites that are pierced are the ears, nose, eyebrows, lips, tongue, nipples, navel, and genitals. Potential complications of body piercing are infections, allergic reactions, and anatomical damage (such as nerve damage or cartilage deformation). In addition, body piercing jewelry may interfere with certain medical procedures such as masks used for resuscitation, airway management procedures, urinary catheterization, radiographs, and delivery of a baby.

▶ CHECKPOINT

1. What structures are included in the integumentary system?
2. How does the process of keratinization occur?
3. What are the structural and functional differences between the epidermis and dermis?
4. How are epidermal ridges formed?
5. What are the three pigments in the skin and how do they contribute to skin color?
6. What is a tattoo? What are some potential problems associated with body piercing?

ACCESSORY STRUCTURES OF THE SKIN

▶ OBJECTIVE

Contrast the structure, distribution, and functions of hair, skin glands, and nails.

Accessory structures of the skin—hair, skin glands, and nails—develop from the embryonic epidermis. They have a host of important functions. For example, hair and nails protect the body, and sweat glands help regulate body temperature.

Hair

Hairs, or *pili* (PI-lē), are present on most skin surfaces except the palms, palmar surfaces of the fingers, the soles, and plantar surfaces of the feet. In adults, hair usually is most heavily distributed across the scalp, in the eyebrows, in the axillae (armpits), and around the external genitalia. Genetic and hormonal influences largely determine the thickness and the pattern of distribution of hairs.

Although the protection it offers is limited, hair on the head guards the scalp from injury and the sun's rays. It also decreases heat loss from the scalp. Eyebrows and eyelashes protect the eyes from foreign particles, as does hair in the nostrils and in the external ear canal. Touch receptors (hair root plexuses) associated with hair follicles are activated whenever a hair is moved even slightly. Thus, hairs also function in sensing light touch.

Anatomy of a Hair

Each hair is composed of columns of dead, keratinized cells bonded together by extracellular proteins. The **shaft** is the superficial portion of the hair, which projects above the surface of the skin (Figure 5.4a). The **root** is the portion of the hair deep to the shaft that penetrates into the dermis, and sometimes into the subcutaneous layer. The shaft and root of the hair both consist of three concentric layers of cells: medulla, cortex, and cuticle of the hair (Figure 5.4c, d). The inner *medulla,* which may be lacking in thinner hair, is composed of two or three rows of irregularly shaped cells. The middle *cortex* forms the major part of the shaft and consists of elongated cells. The *cuticle of the hair,* the outermost layer, consists of a single layer of thin, flat cells that are the most heavily keratinized. Cuticle cells on the shaft are arranged like shingles on the side of a house, with their free edges pointing toward the end of the hair (Figure 5.4b).

Surrounding the root of the hair is the **hair follicle,** which is made up of an external root sheath and an internal root sheath, together referred to as an **epithelial root sheath** (Figure 5.4c, d). The *external root sheath* is a downward continuation of the epidermis. The *internal root sheath* is produced by the matrix (described shortly) and forms a cellular tubular sheath of epithelium between the external root sheath and the hair. The dense dermis surrounding the hair follicle is called the **dermal root sheath.**

The base of each hair follicle is an onion-shaped structure, the **bulb** (Figure 5.4c). This structure houses a nipple-shaped indentation, the **papilla of the hair,** which contains areolar connective tissue and many blood vessels that nourish the growing hair follicle. The bulb also contains a germinal layer of cells called the **matrix.** The matrix cells arise from the stratum basale, the site of cell division. Hence, matrix cells are responsible for the growth of existing hairs, and they produce new hairs when old hairs are shed. This replacement process occurs within the same follicle. Matrix cells also give rise to the cells of the internal root sheath.

Hair Removal

A substance that removes hair is called a **depilatory.** It dissolves the protein in the hair shaft, turning it into a gelatinous mass that can be wiped away. Because the hair root is not affected, regrowth of the hair occurs. In **electrolysis,** an electric current is used to destroy the hair matrix so the hair cannot regrow. **Laser treatments** may also be used to remove hair. ■

Sebaceous (oil) glands (discussed shortly) and a bundle of smooth muscle cells are also associated with hairs (Figure 5.4a). The smooth muscle is the **arrector pili** (a-REK-tor PI-lē; *arrect-* = to raise). It extends from the superficial dermis of the skin to the dermal root sheath around the side of the hair follicle. In its normal position, hair emerges at an angle to the surface of the skin. Under physiologic or emotional stress, such as cold or fright, autonomic nerve endings stimulate the arrector pili muscles to contract, which pulls the hair shafts perpendicular to the skin surface. This action causes "goose bumps" or "gooseflesh" because the skin around the shaft forms slight elevations.

Surrounding each hair follicle are dendrites of neurons, called **hair root plexuses,** that are sensitive to touch (Figure 5.4a). The hair root plexuses generate nerve impulses if their hair shafts are moved.

Hair Growth

Each hair follicle goes through a growth cycle, which consists of a growth stage and a resting stage. During the **growth stage,** cells of the matrix differentiate, keratinize, and die. As new cells are added at the base of the hair root, the hair grows longer. In time, the growth of the hair stops and the **resting stage** begins. After the resting stage, a new growth cycle begins. The old hair root falls out or is pushed out of the hair follicle, and a new hair begins to grow in its place. Scalp hair grows for 2 to 6 years and rests for about 3 months. At any time, about 85% of scalp hairs are in the growth stage. Visible hair is dead, but until the hair is pushed out of its follicle by a new hair, portions of its root within the scalp are alive.

Normal hair loss in the adult scalp is about 70–100 hairs per day. Both the rate of growth and the replacement cycle may be altered by illness, radiation therapy, chemotherapy, age, genetics, gender, and severe emotional stress. Rapid weight-loss diets that severely restrict calories or protein increase hair loss. The rate of shedding also increases for three to four months after childbirth. **Alopecia** (al′-o-PĒ-shē-a), the partial or complete lack of hair, may result from genetic factors, aging, endocrine disorders, chemotherapy, or skin disease.

Chemotherapy and Hair Loss

Chemotherapy is the treatment of disease, usually cancer, by means of chemical substances or drugs. Chemotherapeutic agents interrupt the life cycle of rapidly dividing cancer cells. Unfortunately, the drugs also affect other rapidly dividing cells

Figure 5.4 Hair.

Hairs are growths of epidermis composed of dead, keratinized cells.

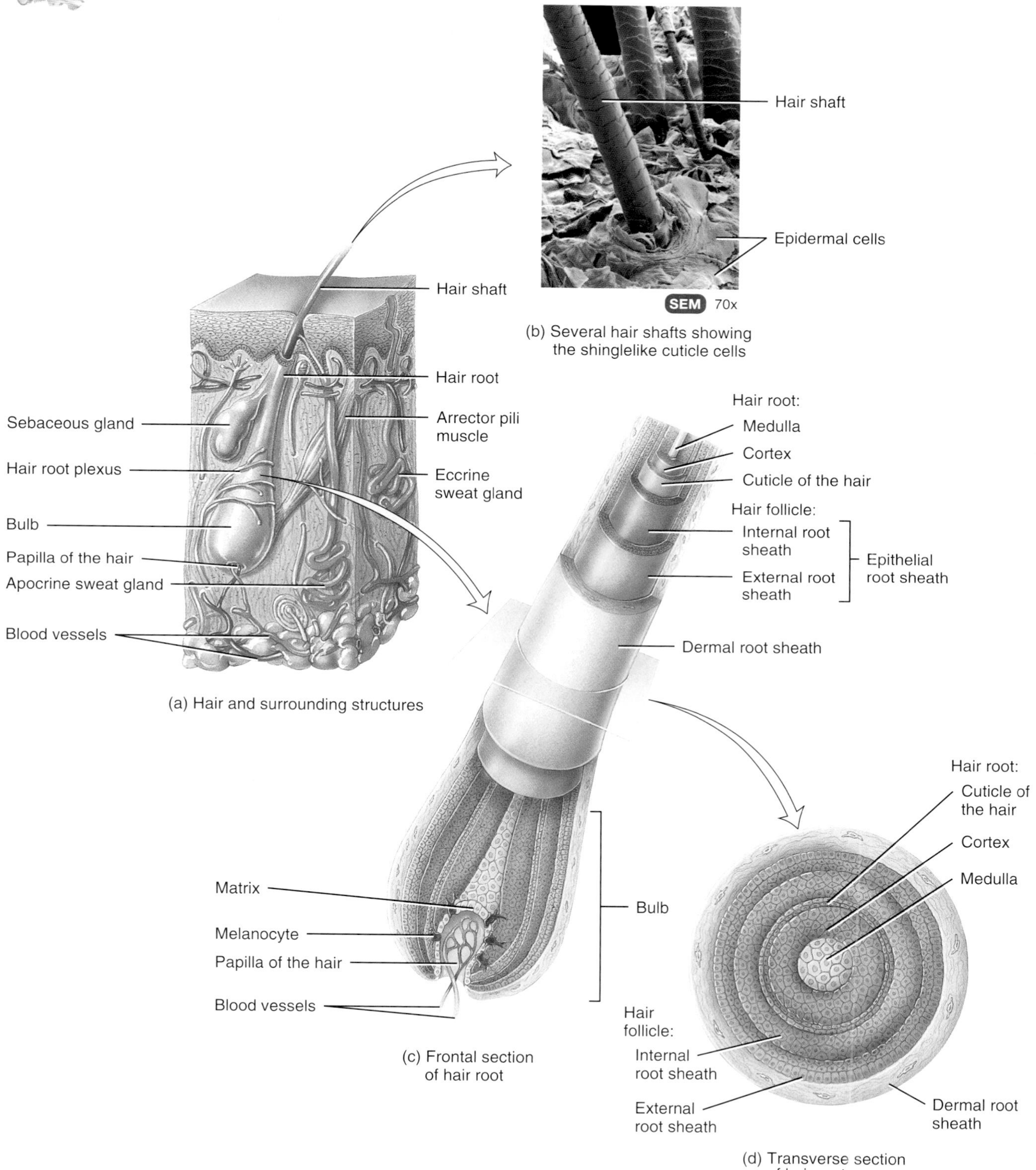

Hair shaft
Epidermal cells
SEM 70x

(b) Several hair shafts showing the shinglelike cuticle cells

Hair shaft
Hair root
Sebaceous gland
Arrector pili muscle
Hair root plexus
Eccrine sweat gland
Bulb
Papilla of the hair
Apocrine sweat gland
Blood vessels

(a) Hair and surrounding structures

Hair root:
Medulla
Cortex
Cuticle of the hair
Hair follicle:
Internal root sheath
External root sheath
Epithelial root sheath
Dermal root sheath
Matrix
Melanocyte
Papilla of the hair
Blood vessels
Bulb

(c) Frontal section of hair root

Hair root:
Cuticle of the hair
Cortex
Medulla
Hair follicle:
Internal root sheath
External root sheath
Dermal root sheath

(d) Transverse section of hair root

Why does it hurt when you pluck a hair out but not when you have a haircut?

in the body, such as the matrix cells of a hair. It is for this reason that individuals undergoing chemotherapy experience hair loss. Since about 15% of the matrix cells of scalp hairs are in the resting stage, these cells are not affected by chemotherapy. Once chemotherapy is stopped, the matrix cells replace lost hair follicles and hair growth resumes. ■

Types of Hairs

Hair follicles develop between the ninth and twelfth weeks after fertilization. Usually by the fifth month of development, the follicles produce very fine, nonpigmented hairs called **lanugo** (la-NOO-gō = wool or down) that cover the body of the fetus. This hair is shed before birth, except in the scalp, eyebrows, and eyelashes. A few months after birth, slightly thicker hairs replace these downy hairs. Over the remainder of the body of an infant, a new growth of short, fine hair occurs. These hairs are known as **vellus hairs** (VEL-us = fleece), commonly called "peach fuzz." In response to hormones (androgens) secreted at puberty, coarse pigmented and frequently curly hair develops in the axillae (armpits) and pubic region. In males, these hairs also appear on the face and other parts of the body. The coarse hairs that develop at puberty, together with those of the head, eyebrows, and eyelashes, are heavily pigmented and are called **terminal hairs.** About 95% of body hair on males is terminal hair (5% vellus hair), whereas about 35% of body hair on females is terminal (65% vellus hair).

Hair Color

The color of hair is due primarily to the amount and type of melanin in its keratinized cells. Melanin is synthesized by melanocytes scattered in the matrix of the bulb and passes into cells of the cortex and medulla of the hair (Figure 5.4c). Dark-colored hair contains mostly eumelanin; blond and red hair contain variants of pheomelanin. Hair becomes gray because of a progressive decline in melanin production. White hair results from the lack of melanin and the accumulation of air bubbles in the shaft.

Hair and Hormones

At puberty, when the testes begin secreting significant quantities of androgens (masculinizing sex hormones), males develop the typical male pattern of hair growth, including a beard and a hairy chest. In females at puberty, the ovaries and the adrenal glands produce small quantities of androgens, which promote hair growth in the axillae and pubic region. Occasionally, a tumor of the adrenal glands, testes, or ovaries produces an excessive amount of androgens. The result in females or prepubertal males is **hirsutism** (HER-soo-tizm; *hirsut-* = shaggy), a condition of excessive body hair.

Surprisingly, androgens also must be present for occurrence of the most common form of baldness, **androgenic alopecia** or **male-pattern baldness.** In genetically predisposed adults, androgens inhibit hair growth. In men, hair loss usually begins with a receding hairline followed by hair loss in the temples and crown. Women are more likely to have thinning of hair on top of the head. The first drug approved for enhancing scalp hair growth was minoxidil (Rogaine®). It causes vasodilation (widening of blood vessels), thus increasing circulation. In about a third of the people who try it, minoxidil improves hair growth, causing scalp follicles to enlarge and lengthening the growth cycle. For many, however, the hair growth is meager. Minoxidil does not help people who already are bald. ■

Skin Glands

Recall from Chapter 4 that glands are epithelial cells that secrete a substance. Several kinds of exocrine glands are associated with the skin: sebaceous (oil) glands, sudoriferous (sweat) glands, and ceruminous glands. Mammary glands, which are specialized sudoriferous glands that secrete milk, are discussed in Chapter 28 along with the female reproductive system.

Sebaceous Glands

Sebaceous glands (se-BĀ-shus; *sebace-* = greasy) or **oil glands** are simple, branched acinar glands. With few exceptions, they are connected to hair follicles (see Figures 5.1 and 5.4a). The secreting portion of a sebaceous gland lies in the dermis and usually opens into the neck of a hair follicle. In some locations, such as the lips, glans penis, labia minora, and tarsal glands of the eyelids, sebaceous glands open directly onto the surface of the skin. Absent in the palms and soles, sebaceous glands are small in most areas of the trunk and limbs, but large in the skin of the breasts, face, neck, and superior chest.

Sebaceous glands secrete an oily substance called **sebum** (SĒ-bum), a mixture of triglycerides, cholesterol, proteins, and inorganic salts. Sebum coats the surface of hairs and helps keep them from drying and becoming brittle. Sebum also prevents excessive evaporation of water from the skin, keeps the skin soft and pliable, and inhibits the growth of certain bacteria.

Acne

Acne is an inflammation of sebaceous glands that usually begins at puberty, when the sebaceous glands grow in size and increase their production of sebum. Androgens from the testes, ovaries, and adrenal glands play the greatest role in stimulating sebaceous glands. Acne occurs predominantly in sebaceous follicles that have been colonized by bacteria, some of which thrive in the lipid-rich sebum. The infection may cause a cyst or sac of connective tissue cells to form, which can destroy and displace epidermal cells. This condition, called **cystic acne,** can permanently scar the epidermis. Treatment consists of gently washing the affected areas once or twice daily with a mild soap, topical antibiotics (such as clindamycin, and erythromycin), topical drugs such as benzoyl peroxide or tretinoin, and oral antibiotics (such as tetracycline, minocycline, erythromycin, and isotretinoin). Contrary to popular belief, foods such as chocolate or fried foods do not cause or worsen acne. ■

Sudoriferous Glands

There are three to four million **sweat glands,** or **sudoriferous glands** (soo′-dor-IF-er-us; *sudori-* = sweat; *-ferous* = bearing). The cells of these glands release sweat, or perspiration, into hair follicles or onto the skin surface through pores. Sweat glands are divided into two main types, eccrine and apocrine, based on their structure, location, and type of secretion.

Eccrine sweat glands (*eccrine* = secreting outwardly), also known as **merocrine sweat glands,** are simple, coiled tubular glands that are much more common than apocrine sweat glands (see Figures 5.1 and 5.4a). They are distributed throughout the skin of most regions of the body, especially in the skin of the forehead, palms, and soles. Eccrine sweat glands are not present, however, in the margins of the lips, nail beds of the fingers and toes, glans penis, glans clitoris, labia minora, and eardrums. The secretory portion of eccrine sweat glands is located mostly in the deep dermis (sometimes in the upper subcutaneous layer). The excretory duct projects through the dermis and epidermis and ends as a pore at the surface of the epidermis (see Figure 5.1).

The sweat produced by eccrine sweat glands (about 600 mL per day) consists of water, ions (mostly Na^+ and Cl^-), urea, uric acid, ammonia, amino acids, glucose, and lactic acid. The main function of eccrine sweat glands is to help regulate body temperature through evaporation. As sweat evaporates, large quantities of heat energy leave the body surface. Eccrine sweat also plays a small role in eliminating wastes such as urea, uric acid, and ammonia from the body. Sweat that evaporates from the skin before it is perceived as moisture is termed **insensible perspiration.** Sweat that is excreted in larger amounts and is seen as moisture on the skin is called **sensible perspiration.**

Apocrine sweat glands are also simple, coiled tubular glands (see Figures 5.1 and 5.4a). They are found mainly in the skin of the axilla (armpit), groin, areolae (pigmented areas around the nipples) of the breasts, and bearded regions of the face in adult males. These glands were once thought to release their secretions in an apocrine manner (see page 120 and Figure 4.5b on page 121)—by pinching off a portion of the cell. We now know, however, that their secretion is via exocytosis, which is characteristic of merocrine glands (see Figure 4.5a on page 121). Nevertheless, the term *apocrine* is still used. The secretory portion of these sweat glands is located mostly in the subcutaneous layer, and the excretory duct opens into hair follicles (see Figure 5.1). Their secretory product is slightly viscous compared to eccrine secretions and contains the same components as eccrine sweat plus lipids and proteins. Eccrine sweat glands start to function soon after birth, but apocrine sweat glands do not begin to function until puberty. Apocrine sweat glands are stimulated during emotional stress and sexual excitement; these secretions are commonly known as a "cold sweat."

Table 5.3 presents a comparison of eccrine and apocrine sweat glands.

TABLE 5.3 Comparison of Eccrine and Apocrine Sweat Glands

Feature	Eccrine Sweat Glands	Apocrine Sweat Glands
Distribution	Throughout skin of most regions of the body, especially in skin of forehead, palms, and soles.	Skin of the axilla, groin, areolae, bearded regions of the face, clitoris, and labia minora.
Location of secretory portion	Mostly in deep dermis.	Mostly in subcutaneous layer.
Termination of excretory duct	Surface of epidermis.	Hair follicle.
Secretion	Less viscous; consists of water, ions (Na^+, Cl^-), urea, uric acid, ammonia, amino acids, glucose, and lactic acid.	More viscous; consists of the same components as eccrine sweat glands plus lipids and proteins.
Functions	Regulation of body temperature and waste removal.	Stimulated during emotional stress and sexual excitement.
Onset of function	Soon after birth.	Puberty.

Ceruminous Glands

Modified sweat glands in the external ear, called **ceruminous glands** (se-RŪ-mi-nus; *cer-* = wax), produce a waxy secretion. The secretory portions of ceruminous glands lie in the subcutaneous layer, deep to sebaceous glands. Their excretory ducts open either directly onto the surface of the external auditory canal (ear canal) or into ducts of sebaceous glands. The combined secretion of the ceruminous and sebaceous glands is called **cerumen,** or earwax. Cerumen, together with hairs in the external auditory canal, provides a sticky barrier that impedes the entrance of foreign bodies.

Impacted Cerumen

Some people produce an abnormally large amount of cerumen in the external auditory canal. If it accumulates until it becomes impacted (firmly wedged), sound waves may be prevented from reaching the eardrum. Treatments for **impacted cerumen** include periodic ear irrigation with enzymes to dissolve the wax and removal of wax with a blunt instrument by trained medical personnel. The use of cotton-tipped swabs or sharp objects is not recommended for this purpose because they may push the cerumen further into the external auditory canal and damage the eardrum. ■

Nails

Nails are plates of tightly packed, hard, dead, keratinized epidermal cells that form a clear, solid covering over the dorsal surfaces of the distal portions of the digits. Each nail consists of a nail body, a free edge, and a nail root (Figure 5.5). The **nail body** is the visible portion of the nail, the **free edge** is the part that may extend past the distal end of the digit, and the **nail root** is the portion that is buried in a fold of skin. Below the nail body is a region of epithelium and a deeper layer of dermis. Most of the nail body appears pink because of blood flowing through the capillaries in the underlying dermis. The free edge is white because there are no underlying capillaries. The whitish, crescent-shaped area of the proximal end of the nail body is called the **lunula** (LOO-noo-la = little moon). It appears whitish because the vascular tissue underneath does not show through due to a thickened region of epithelium in the area. Beneath the free edge is a thickened region of stratum corneum called the **hyponychium** (hī′-pō-NIK-ē-um; *hypo-* = below; *-onych* = nail), which secures the nail to the fingertip. The **eponychium** (ep′-ō-NIK-ē-um; *ep-* = above) or **cuticle** is a narrow band of epidermis that extends from and adheres to the margin (lateral border) of the nail wall. It occupies the proximal border of the nail and consists of stratum corneum.

The proximal portion of the epithelium deep to the nail root is the **nail matrix,** where cells divide by mitosis to produce growth. Nail growth occurs by the transformation of superficial cells of the matrix into nail cells. The growth rate of nails is determined by the rate of mitosis in matrix cells, which is influenced by factors such as a person's age, health, and nutritional status. Nail growth also varies according to the season, the time of day, and environmental temperature. The average growth in the length of fingernails is about 1 mm (0.04 in.) per week. The growth rate is somewhat slower in toenails.

Functionally, nails help us grasp and manipulate small objects in various ways, provide protection against trauma to the ends of the digits, and allow us to scratch various parts of the body.

▶ CHECKPOINT

7. Describe the structure of a hair. What causes "goose bumps"?
8. Contrast the locations and functions of sebaceous (oil) glands, sudoriferous (sweat) glands, and ceruminous glands.
9. Describe the parts of a nail.

TYPES OF SKIN

▶ OBJECTIVE

Compare structural and functional differences in thin and thick skin.

Although the skin over the entire body is similar in structure, there are quite a few local variations related to thickness of the epidermis, strength, flexibility, degree of keratinization, distribution and type of hair, density and types of glands, pigmentation, vascularity (blood supply), and innervation (nerve supply). Two major types of skin are recognized on the basis of certain structural and functional properties: **thin (hairy) skin** and **thick (hairless) skin.**

Table 5.4 presents a comparison of the features of thin and thick skin.

▶ CHECKPOINT

10. What criteria are used to distinguish thin and thick skin?

Figure 5.5 Nails. Shown is a fingernail.

Nail cells arise by transformation of superficial cells of the nail matrix.

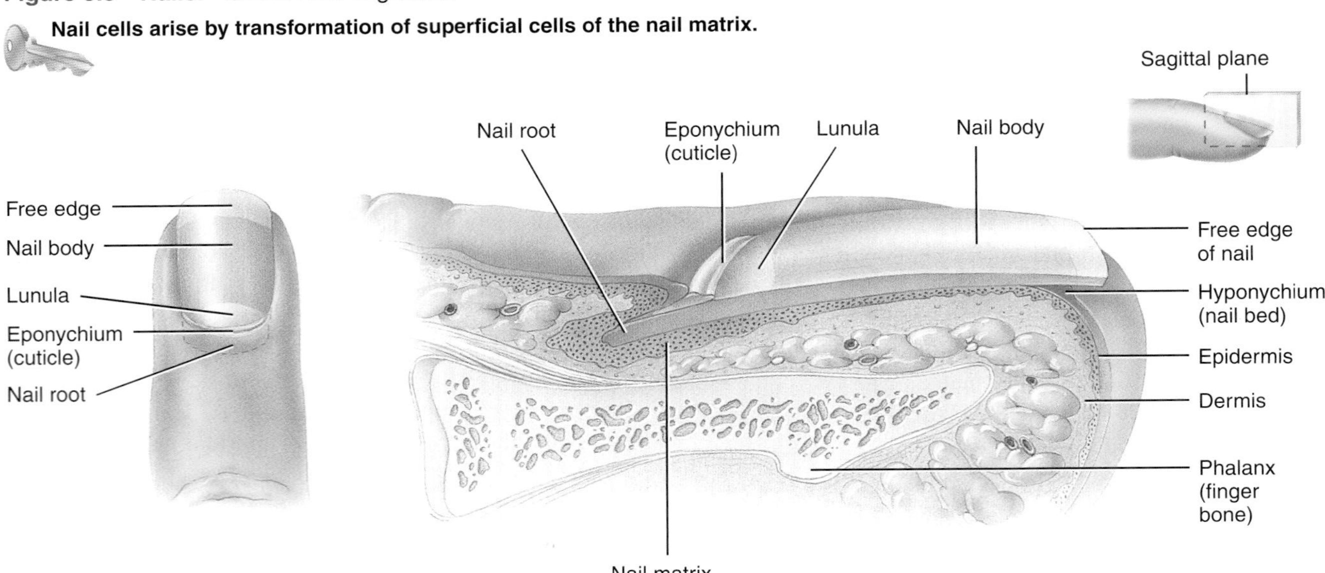

Why are nails so hard?

TABLE 5.4 Comparison of Thin and Thick Skin

Feature	Thin Skin	Thick Skin
Distribution	All parts of the body except palms and palmar surface of digits, and soles.	Palms, palmar surface of digits, and soles.
Epidermal thickness	0.10–0.15 mm (0.004–0.006 in.).	0.6–4.5 mm (0.024–0.18 in.).
Epidermal strata	Stratum lucidum essentially lacking; thinner strata spinosum and corneum.	Thick strata lucidum, spinosum, and corneum.
Epidermal ridges	Lacking due to poorly developed and fewer dermal papillae.	Present due to well-developed and more numerous dermal papillae.
Hair follicles and arrector pili muscles	Present.	Absent.
Sebaceous glands	Present.	Absent.
Sudoriferous glands	Fewer.	More numerous.
Sensory receptors	Sparser.	Denser.

FUNCTIONS OF THE SKIN

▶ **OBJECTIVE**

Describe how the skin contributes to regulation of body temperature, storage of blood, protection, sensation, excretion and absorption, and synthesis of vitamin D.

Now that you have a basic understanding of the structure of the skin, you can better appreciate its many functions, which were introduced at the beginning of this chapter. The numerous functions of the integumentary system (mainly the skin) include thermoregulation, storage of blood, protection, cutaneous sensations, excretion and absorption, and synthesis of vitamin D.

Thermoregulation

The skin contributes to **thermoregulation,** the homeostatic regulation of body temperature, in two ways: by liberating sweat at its surface and by adjusting the flow of blood in the dermis. In response to high environmental temperature or heat produced by exercise, sweat production increases; the evaporation of sweat from the skin surface helps lower body temperature. In addition, blood vessels in the dermis of the skin dilate (become wider); consequently, more blood flows through the dermis, which increases the amount of heat loss from the body. In response to low environmental temperature, production of sweat is decreased, which helps conserve heat. Also, the blood vessels in the dermis of the skin constrict (become narrow), which decreases blood flow through the skin and reduces heat loss from the body.

Blood Reservoir

The dermis houses an extensive network of blood vessels that carry 8–10% of the total blood flow in a resting adult. For this reason, the skin acts as a **blood reservoir.**

Protection

The skin provides **protection** to the body in various ways. Keratin protects underlying tissues from microbes, abrasion, heat, and chemicals and the tightly interlocked keratinocytes resist invasion by microbes. Lipids released by lamellar granules retard evaporation of water from the skin surface, thus guarding against dehydration; they also retard entry of water across the skin surface during showers and swims. The oily sebum from the sebaceous glands keeps skin and hairs from drying out and contains bactericidal chemicals that kill surface bacteria. The acidic pH of perspiration retards the growth of some microbes. The pigment melanin helps shield against the damaging effects of UV light. Two types of cells carry out protective functions that are immunological in nature. Epidermal Langerhans cells alert the immune system to the presence of potentially harmful microbial invaders by recognizing and processing them, and macrophages in the dermis phagocytize bacteria and viruses that manage to bypass the Langerhans cells of the epidermis.

Cutaneous Sensations

Cutaneous sensations are sensations that arise in the skin, including tactile sensations—touch, pressure, vibration, and tickling—as well as thermal sensations such as warmth and coolness. Another cutaneous sensation, pain, usually is an indication of impending or actual tissue damage. There is a wide variety of nerve endings and receptors distributed throughout the skin, including the tactile discs of the epidermis, the corpuscles of touch in the dermis, and hair root plexuses around each hair follicle. Chapter 16 provides more details on the topic of cutaneous sensations.

Excretion and Absorption

The skin normally has a small role in **excretion,** the elimination of substances from the body, and **absorption,** the passage of materials from the external environment into body cells. Despite the almost waterproof nature of the stratum corneum, about 400 mL of water evaporates through it daily. A sedentary person loses an additional 200 mL per day as sweat; a physically active person loses much more. Besides removing water and heat from the body, sweat also is the vehicle for excretion of small amounts of salts, carbon dioxide, and two organic molecules that result from the breakdown of proteins—ammonia and urea.

The absorption of water-soluble substances through the skin is negligible, but certain lipid-soluble materials do penetrate the skin. These include fat-soluble vitamins (A, D, E, and K), certain drugs, and the gases oxygen and carbon dioxide. Toxic materials that can be absorbed through the skin include organic solvents such as acetone (in some nail polish removers) and carbon tetrachloride (dry-cleaning fluid); salts of heavy metals such as lead, mercury, and arsenic; and the substances in poison ivy and poison oak. Since topical (applied to the skin) steroids, such as cortisone, are lipid-soluble, they move easily into the papillary region of the dermis. Here, they exert their anti-inflammatory properties by inhibiting histamine production by mast cells (recall that histamine contributes to inflammation).

Transdermal Drug Administration

Most drugs are either absorbed into the body through the digestive system or injected into subcutaneous tissue or muscle. An alternative route, **transdermal (transcutaneous) drug administration,** enables a drug contained within an adhesive skin patch to pass across the epidermis and into the blood vessels of the dermis. The drug is released continuously at a controlled rate over a period of one to several days. This method of administration is especially useful for drugs that are quickly eliminated from the body because such drugs, if taken in other forms, would have to be taken quite frequently. Because the major barrier to penetration of most drugs is the stratum corneum, transdermal absorption is most rapid in regions of the skin where this layer is thin, such as the scrotum, face, and scalp. A growing number of drugs are available for transdermal administration, including nitroglycerin, for prevention of angina pectoris (chest pain associated with heart disease); scopolamine, for motion sickness; estradiol, used for estrogen-replacement therapy during menopause; ethinyl estradiol and norelgestromin in contraceptive patches; nicotine, used to help people stop smoking; and fentanyl, used to relieve severe pain in cancer patients. ■

Synthesis of Vitamin D

Synthesis of vitamin D requires activation of a precursor molecule in the skin by UV rays in sunlight. Enzymes in the liver and kidneys then modify the activated molecule, finally producing *calcitriol,* the most active form of vitamin D. Calcitriol is a hormone that aids in the absorption of calcium in foods from the gastrointestinal tract into the blood.

▶ CHECKPOINT

11. In what two ways does the skin help regulate body temperature?

12. How does the skin serve as a protective barrier?

13. What sensations arise from stimulation of neurons in the skin?

14. What types of molecules can penetrate the stratum corneum?

MAINTAINING HOMEOSTASIS: SKIN WOUND HEALING

▶ OBJECTIVE

Explain how epidermal wounds and deep wounds heal.

Skin damage sets in motion a sequence of events that repairs the skin to its normal (or near-normal) structure and function. Two kinds of wound-healing processes can occur, depending on the depth of the injury. Epidermal wound healing occurs following wounds that affect only the epidermis; deep wound healing occurs following wounds that penetrate the dermis.

Epidermal Wound Healing

Even though the central portion of an epidermal wound may extend to the dermis, the edges of the wound usually involve only slight damage to superficial epidermal cells. Common types of epidermal wounds include abrasions, in which a portion of skin has been scraped away, and minor burns.

In response to an epidermal injury, basal cells of the epidermis surrounding the wound break contact with the basement membrane. The cells then enlarge and migrate across the wound (Figure 5.6a). The cells appear to migrate as a sheet until advancing cells from opposite sides of the wound meet. When epidermal cells encounter one another, they stop migrating due to a cellular response called **contact inhibition.** Migration of the epidermal cells stops completely when each is finally in contact with other epidermal cells on all sides.

As the basal epidermal cells migrate, a hormone called *epidermal growth factor* stimulates basal stem cells to divide and replace the ones that have moved into the wound. The relocated basal epidermal cells divide to build new strata, thus thickening the new epidermis (Figure 5.6b).

Deep Wound Healing

Deep wound healing occurs when an injury extends to the dermis and subcutaneous layer. Because multiple tissue layers must be repaired, the healing process is more complex than in epidermal wound healing. In addition, because scar tissue is formed, the healed tissue loses some of its normal function. Deep wound healing occurs in four phases: an inflammatory phase, a migratory phase, a proliferative phase, and a maturation phase.

During the **inflammatory phase,** a blood clot forms in the wound and loosely unites the wound edges (Figure 5.6c). As its name implies, this phase of deep wound healing involves **inflammation,** a vascular and cellular response that helps eliminate microbes, foreign material, and dying tissue in preparation for repair. The vasodilation and increased permeability of blood vessels associated with inflammation enhance delivery of helpful cells. These include phagocytic white blood cells called neutrophils; monocytes, which develop into macrophages that phagocytize microbes; and mesenchymal cells, which develop into fibroblasts.

Figure 5.6 **Skin wound healing.**

In an epidermal wound, the injury is restricted to the epidermis; in a deep wound, the injury extends deep into the dermis.

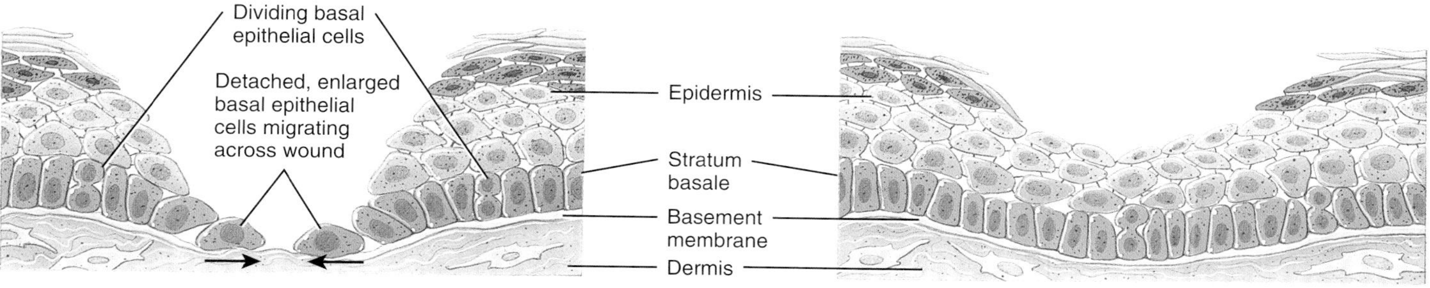

(a) Division of basal epithelial cells and migration across wound

(b) Thickening of epidermis

Epidermal wound healing

Blood clot in wound
Epithelium migrating across wound
Fibroblast
Collagen fibers
Monocyte (macrophage)
Neutrophil
Dilated blood vessel
Damaged blood vessel
End of clot

(c) Inflammatory phase

Scab
Resurfaced epithelium
Collagen fibers
Scar tissue
Fibroblast
Restored blood vessel

(d) Maturation phase

Deep wound healing

Would you expect an epidermal wound to bleed? Why or why not?

The three phases that follow do the work of repairing the wound. In the **migratory phase,** the clot becomes a scab, and epithelial cells migrate beneath the scab to bridge the wound. Fibroblasts migrate along fibrin threads and begin synthesizing scar tissue (collagen fibers and glycoproteins), and damaged blood vessels begin to regrow. During this phase, the tissue filling the wound is called **granulation tissue.** The **proliferative phase** is characterized by extensive growth of epithelial cells beneath the scab, deposition by fibroblasts of collagen fibers in random patterns, and continued growth of blood vessels. Finally, during the **maturation phase,** the scab sloughs off once the epidermis has been restored to normal thickness. Collagen fibers become more organized, fibroblasts decrease in number, and blood vessels are restored to normal (Figure 5.6d).

The process of scar tissue formation is called **fibrosis.** Sometimes, so much scar tissue is formed during deep wound healing that a raised scar—one that is elevated above the normal epidermal surface—results. If such a scar remains within the boundaries of the original wound, it is a **hypertrophic scar.** If it extends beyond the boundaries into normal surrounding tissues, it is a **keloid scar.** Scar tissue differs from normal skin in that its collagen fibers are more densely arranged, it has decreased elasticity, it has fewer blood vessels, and it may or may not contain the same number of hairs, skin glands, or sensory structures as undamaged skin. Because of the arrangement of collagen fibers and the scarcity of blood vessels, scars usually are lighter in color than normal skin.

▶ CHECKPOINT

15. Why doesn't epidermal wound healing result in scar formation?

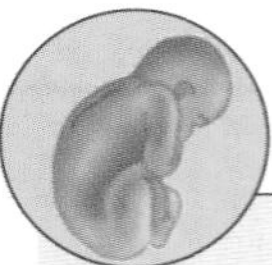

DEVELOPMENT OF THE INTEGUMENTARY SYSTEM

▶ OBJECTIVE

Describe the development of the epidermis, its accessory structures, and the dermis.

The *epidermis* is derived from the **ectoderm,** which covers the surface of the embryo. Initially, at about the fourth week after fertilization, the epidermis consists of only a single layer of ectodermal cells (Figure 5.7a). At the beginning of the seventh week the single layer, called the **basal layer,** divides and forms a superficial protected layer of flattened cells called the **periderm** (Figure 5.7b). The peridermal cells are continuously sloughed off and by the fifth month of development, secretions from sebaceous glands mix with them and hairs to form a fatty substance called **vernix caseosa** (VER-niks KĀ-sē-ō-sa; *vernix* = varnish, *caseosa* = cheese). This substance covers and protects the skin of the fetus from the constant exposure to the amniotic fluid in which it is bathed. In addition, the vernix caseosa facilitates the birth of the fetus because of its slippery nature and protects the skin from being damaged by the nails.

By about 11 weeks, the basal layer forms an intermediate layer of cells (Figure 5.7c). Proliferation of the basal cells eventually forms all layers of the epidermis, which are present at birth (Figure 5.7d). *Epidermal ridges* form along with the epidermal layers (Figure 5.7c). By about the eleventh week, cells from the ectoderm migrate into the dermis and differentiate into *melanoblasts* (Figure 5.7c). As you will see later, the neural crest develops into cranial and spinal nerves, among other nervous tissue structures. These cells soon enter the epidermis and differentiate into *melanocytes.* Later in the first trimester of pregnancy, *Langerhans cells,* which arise from red bone marrow, invade the epidermis. *Merkel cells* appear in the epidermis in the fourth to sixth months; their origin is unknown.

The *dermis* arises from **mesoderm** located deep to the surface ectoderm. The mesoderm gives rise to a loosely organized embryonic connective tissue called **mesenchyme** (MEZ-en-kīm; see Figure 5.7a). By 11 weeks, the mesenchymal cells differentiate into fibroblasts and begin to form collagen and elastic fibers. As the epidermal ridges form, parts of the superficial dermis project into the epidermis and develop into the *dermal papillae,* which contain capillary loops, corpuscles of touch, and free nerve endings (Figure 5.7c).

Hair follicles develop between the ninth and twelfth weeks as downgrowths of the basal layer of the epidermis into the deeper dermis. The downgrowths are called **hair buds** (Figure 5.7e). As the hair buds penetrate deeper into the dermis, their distal ends become club-shaped and are called **hair bulbs** (Figure 5.7f). Invaginations of the hair bulbs, called **papillae of the hair,** fill with mesoderm in which blood vessels and nerve endings develop (Figure 5.7g). Cells in the center of a hair bulb develop into the *matrix,* which forms the *hair,* and the peripheral cells of the hair bulb form the *epithelial root sheath* (Figure 5.7h). Mesenchyme in the surrounding dermis develops into the *dermal root sheath* and *arrector pili muscle* (Figure 5.7h). By the fifth month, the hair follicles produce lanugo (delicate fetal hair; see page 154). It is produced first on the head and then on other parts of the body, and is usually shed prior to birth.

Most *sebaceous (oil) glands* develop as outgrowths from the sides of hair follicles at about four months and remain connected to the follicles (Figure 5.7f). Most *sudoriferous (sweat) glands* are derived from downgrowths **(buds)** of the stratum basale of the epidermis into the dermis (Figure 5.7e). As the buds penetrate into the dermis, the proximal portion forms the duct of the sweat gland and the distal portion coils and forms the secretory portion of the gland (Figure 5.7h). Sweat glands appear at about five months on the palms and soles and a little later in other regions.

Nails are developed at about 10 weeks. Initially they consist of a thick layer of epithelium called the **primary nail field.** The nail itself is keratinized epithelium and grows distally from its base. It is not until the ninth month that the nails actually reach the tips of the digits.

▶ CHECKPOINT

16. Which structures develop as downgrowths of the stratum basale?

AGING AND THE INTEGUMENTARY SYSTEM

▶ OBJECTIVE

Describe the effects of aging on the integumentary system.

The pronounced effects of skin aging do not become noticeable until people reach their late forties. Most of the age-related changes occur in the dermis. Collagen fibers in the dermis begin to decrease in number, stiffen, break apart, and disorganize into a shapeless, matted tangle. Elastic fibers lose some of their elasticity, thicken into clumps, and fray, an effect that is greatly accelerated in the skin of smokers. Fibroblasts, which produce both collagen and elastic fibers, decrease in number. As a result, the skin forms the characteristic crevices and furrows known as *wrinkles.*

With further aging, Langerhans cells dwindle in number and macrophages become less-efficient phagocytes, thus decreasing the skin's immune responsiveness. Moreover, decreased size of sebaceous glands leads to dry and broken skin that is more susceptible to infection. Production of sweat diminishes, which probably contributes to the increased incidence of heat stroke in the elderly. There is a decrease in the number of functioning melanocytes, resulting in gray hair and atypical skin pigmentation. An increase in the size of some melanocytes produces pigmented blotching (age spots). Walls of blood vessels in the dermis become thicker and less permeable, and subcutaneous adipose tissue is lost. Aged skin (especially the dermis) is thinner than young skin, and the migration of cells from the basal layer to the epidermal surface slows considerably. With the onset

Figure 5.7 Development of the integumentary system.

The epidermis develops from ectoderm, and the dermis develops from mesoderm.

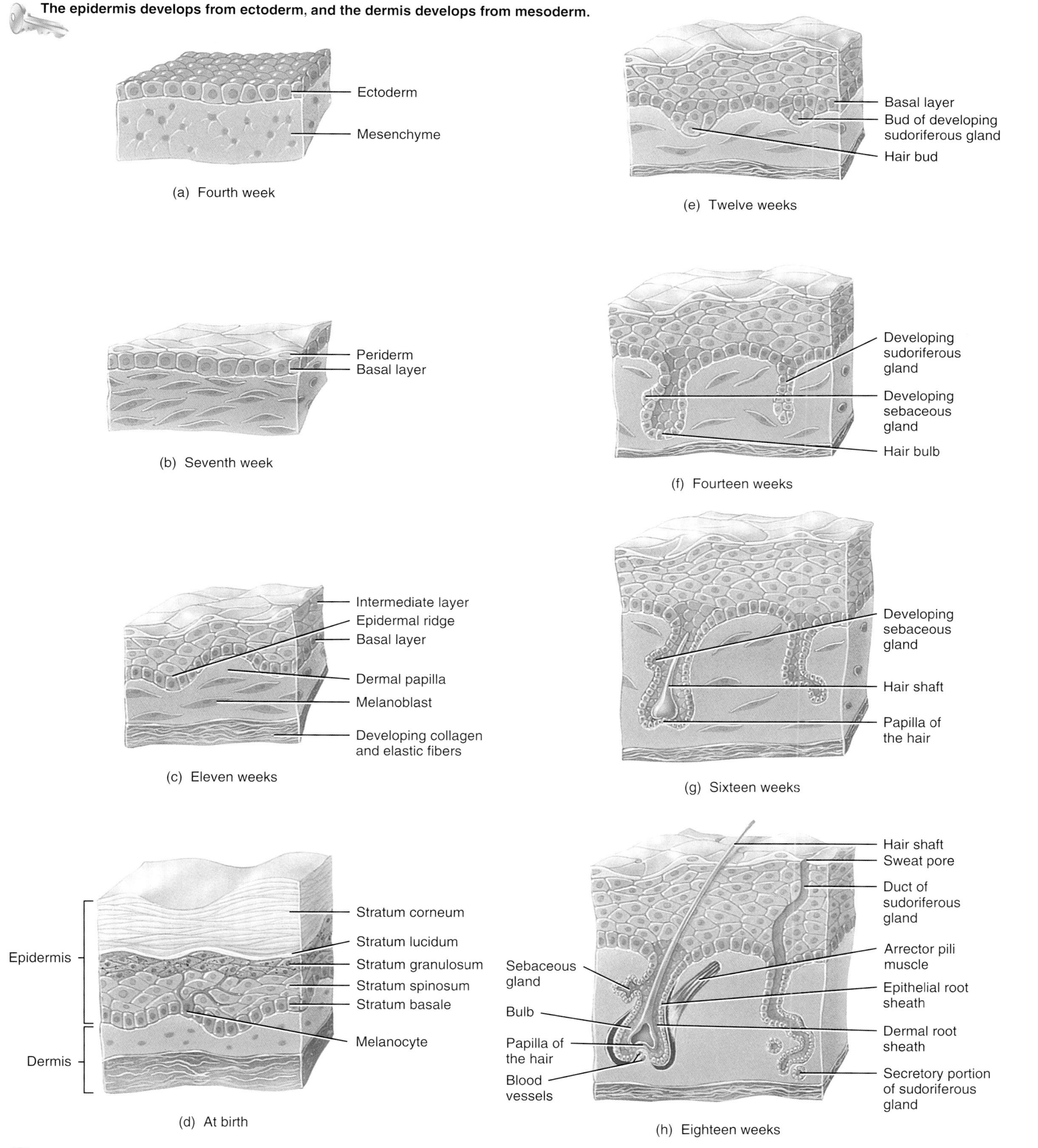

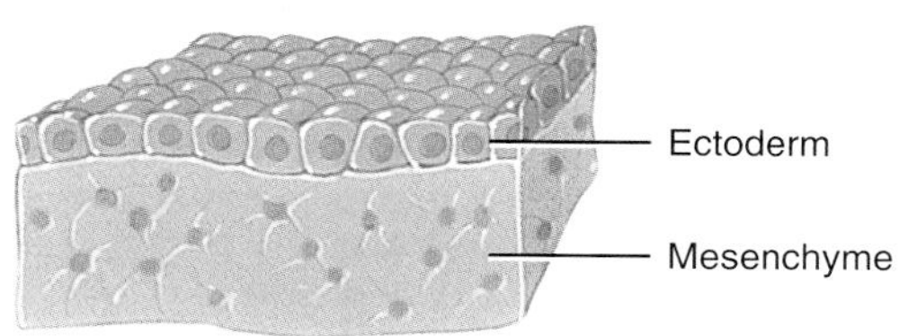

Ectoderm
Mesenchyme
(a) Fourth week
Periderm
Basal layer
(b) Seventh week
Intermediate layer
Epidermal ridge
Basal layer
Dermal papilla
Melanoblast
Developing collagen and elastic fibers
(c) Eleven weeks
Epidermis
Dermis
Stratum corneum
Stratum lucidum
Stratum granulosum
Stratum spinosum
Stratum basale
Melanocyte
(d) At birth
Basal layer
Bud of developing sudoriferous gland
Hair bud
(e) Twelve weeks
Developing sudoriferous gland
Developing sebaceous gland
Hair bulb
(f) Fourteen weeks
Developing sebaceous gland
Hair shaft
Papilla of the hair
(g) Sixteen weeks
Hair shaft
Sweat pore
Duct of sudoriferous gland
Arrector pili muscle
Epithelial root sheath
Dermal root sheath
Secretory portion of sudoriferous gland
Sebaceous gland
Bulb
Papilla of the hair
Blood vessels
(h) Eighteen weeks

What is the composition of vernix caseosa?

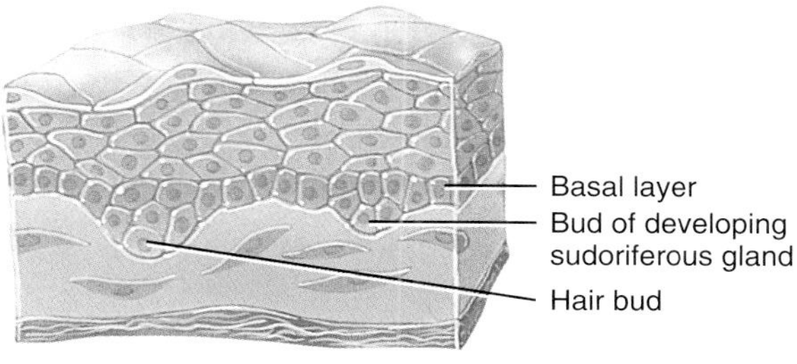

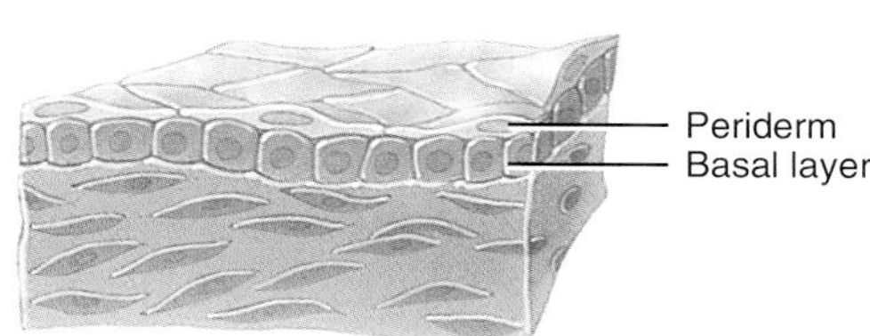

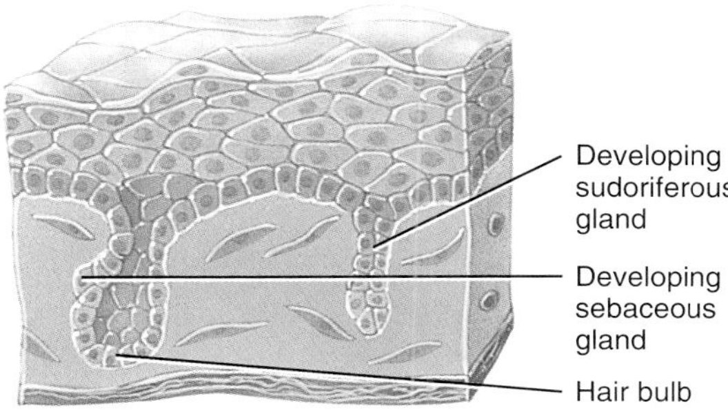

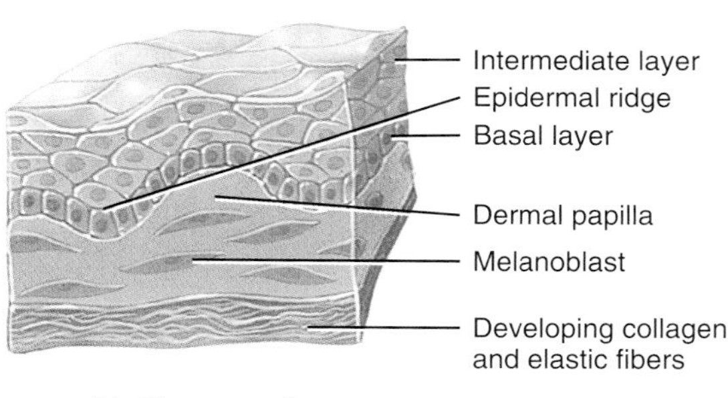

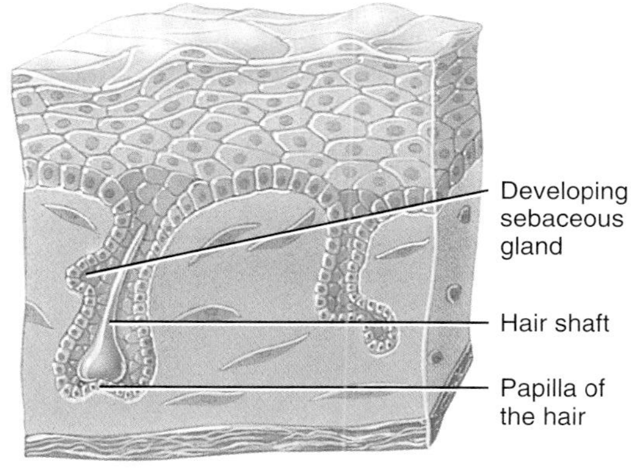

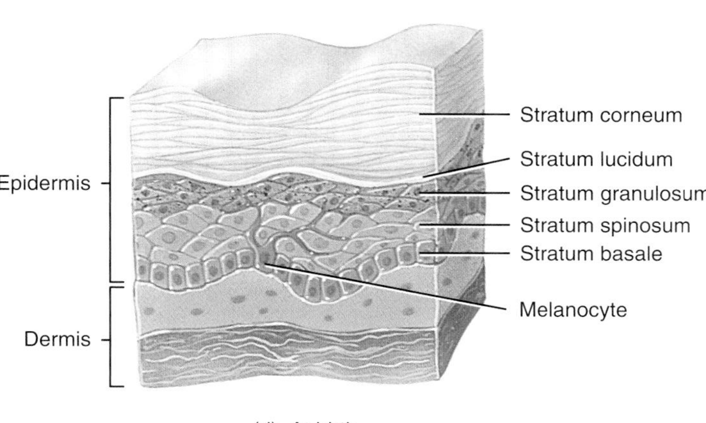

of old age, skin heals poorly and becomes more susceptible to pathological conditions such as skin cancer and pressure sores. **Rosacea** (ro-ZĀ-shē-a = rosy) is a skin condition that affects mostly light-skinned adults between the ages of 30 and 60. It is characterized by redness, tiny pimples, and noticeable blood vessels, usually in the central area of the face.

Growth of nails and hair slows during the second and third decades of life. The nails also may become more brittle with age, often due to dehydration or repeated use of cuticle remover or nail polish.

Several cosmetic anti-aging treatments are available to diminish the effects of aging or sun-damaged skin, including **topical products** that bleach the skin to tone down blotches and blemishes (hydroquinone) or decrease fine wrinkles and roughness (retinoic acid); **microdermabrasion** (mī-krō-DER-ma-brā′-zhun; *mikros-* = small; *derm* = skin; *-abrasio* = to wear away), the use of tiny crystals under pressure to remove and vacuum the skin's surface cells to improve skin texture and reduce blemishes; **chemical peel,** application of a mild acid (such as glycolic acid) to the skin to remove surface cells to improve skin texture and reduce blemishes; **laser resurfacing,** the use of a laser to clear up blood vessels near the skin surface, even out blotches and blemishes, and decrease fine wrinkles; **dermal fillers,** injections of collagen from cows, hyaluronic acid, or calcium hydroxylapatite that plumps up the skin to smooth out wrinkles and fill in furrows, such as those around the nose and mouth and between the eyebrows; **fat transplantation,** in which fat from one part of the body is injected into another location such as around the eyes; **botulinum toxin** or **Botox®,** a diluted version of the toxin that causes food poisoning that is injected into the skin to paralyze muscles that cause the skin to wrinkle; **radio frequency nonsurgical facelift,** the use of radio frequency emissions to tighten skin of the jowls, neck, and sagging eyebrows and eyelids; and **facelift, browlift,** or **necklift,** invasive surgery in which loose skin and fat are removed surgically and the underlying connective tissue and muscle are tightened.

Sun Damage, Sunscreens, and Sunblocks

Although basking in the warmth of the sun may feel good, it is not a healthy practice. There are two forms of ultraviolet radiation that affect the health of the skin. Longer-wavelength ultraviolet A (UVA) rays make up nearly 95% of the ultraviolet radiation that reaches the earth. UVA rays are not absorbed by the ozone layer. They penetrate the furthest into the skin, where they are absorbed by melanocytes and thus are involved in sun tanning. UVA rays also depress the immune system. Shorter-wavelength ultraviolet B (UVB) rays are partially absorbed by the ozone layer and do not penetrate the skin as deeply as UVA rays. UVB rays cause sunburn and are responsible for most of the tissue damage (production of oxygen free radicals that disrupt collagen and elastic fibers) that results in wrinkling and aging of the skin and cataract formation. Both UVA and UVB rays are thought to cause skin cancer. Long-term overexposure to sunlight results in dilated blood vessels, age spots, freckles, and changes in skin texture.

Exposure to ultraviolet radiation (either natural sunlight or the artificial light of a tanning booth) may also produce **photosensitivity,** a heightened reaction of the skin after consumption of certain medications or contact with certain substances. Photosensitivity is characterized by redness, itching, blistering, peeling, hives, and even shock. Among the medications or substances that may cause a photosensitivity reaction are certain antibiotics (tetracycline), nonsteroidal anti-inflammatory drugs (ibuprofen or naproxen), certain herbal supplements (St. John's Wort), some birth control pills, some high blood pressure medications, some antihistamines, and certain artificial sweeteners, perfumes, after shaves, lotions, detergents, and medicated cosmetics.

Self-tanning lotions (sunless tanners), topically applied substances, contain a color additive (dihydroxyacetone) that produces a tanned appearance by interacting with proteins in the skin.

Sunscreens are topically applied preparations that contain various chemical agents (such as benzophenone or one of its derivatives) that absorb UVB rays, but let most of the UVA rays pass through.

Sunblocks are topically applied preparations that contain substances such as zinc oxide that reflect and scatter both UVB and UVA rays.

Both sunscreens and sunblocks are graded according to a *sun protection factor (SPF)* rating, which measures the level of protection they supposedly provide against UV rays. The higher the rating, presumably the greater the degree of protection. As a precautionary measure, individuals who plan to spend a significant amount of time in the sun should use a sunscreen or a sunblock with an SPF of 15 or higher. Although sunscreens protect against sunburn, there is considerable debate as to whether they actually protect against skin cancer. In fact, some studies suggest that sunscreens increase the incidence of skin cancer because of the false sense of security they provide. ■

▶ CHECKPOINT

17. What factors contribute to the susceptibility of aging skin to infection?

• • •

To appreciate the many ways that skin contributes to homeostasis of other body systems, examine *Focus on Homeostasis: The Integumentary System.* This feature is the first of 10, found at the end of selected chapters, that explain how the body system under consideration contributes to the homeostasis of all other body systems. Next, in Chapter 6, we will explore how bone tissue is formed and how bones are assembled into the skeletal system, which, like the skin, protects many of our internal organs.

FOCUS ON HOMEOSTASIS

The Integumentary System

BODY SYSTEM	CONTRIBUTION OF THE INTEGUMENTARY SYSTEM
For all body systems	Skin and hair provide barriers that protect all internal organs from damaging agents in the external in the external environment; sweat glands and skin blood vessels regulate body temperature, needed for proper functioning of other body systems.
Skeletal system 	Skin helps activate vitamin D, needed for proper absorption of dietary calcium and phosphorus to build and maintain bones.
Muscular system 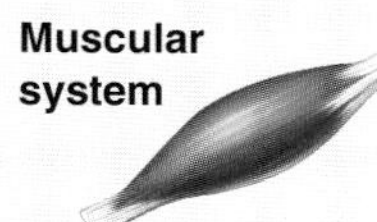	Skin helps provide calcium ions, needed for muscle contraction.
Nervous system 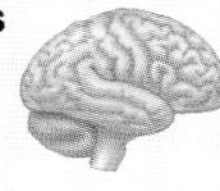	Nerve endings in skin and subcutaneous tissue provide input to the brain for touch, pressure, thermal, and pain sensations.
Endocrine system 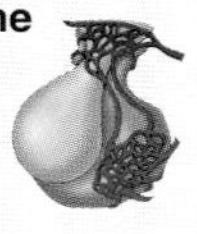	Keratinocytes in skin help activate vitamin D to calcitriol, a hormone that aids absorption of dietary calcium and phosphorus.
Cardiovascular system 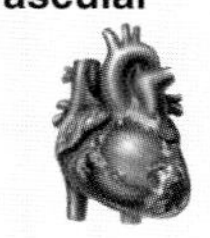	Local chemical changes in dermis cause widening and narrowing of skin blood vessels, which help adjust blood flow to the skin.
Lymphatic system and immunity 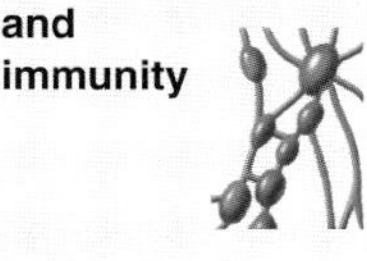	Skin is "first line of defense" in immunity, providing mechanical barriers and chemical secretions that discourage penetration and growth of microbes; Langerhans cells in epidermis participate in immune responses by recognizing and processing foreign antigens; macrophages in the dermis phagocytize microbes that penetrate the skin surface.
Respiratory system 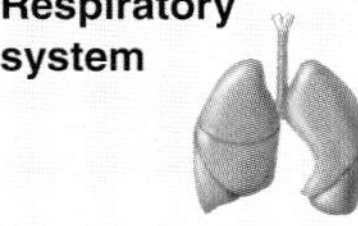	Hairs in nose filter dust particles from inhaled air; stimulation of pain nerve endings in skin may alter breathing rate.
Digestive system 	Skin helps activate vitamin D to the hormone calcitriol, which promotes absorption of dietary calcium and phosphorus in the small intestine.
Urinary system 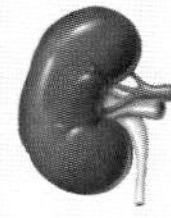	Kidney cells receive partially activated vitamin D hormone from skin and convert it to calcitriol; some waste products are excreted from body in sweat, contributing to excretion by urinary system.
Reproductive system 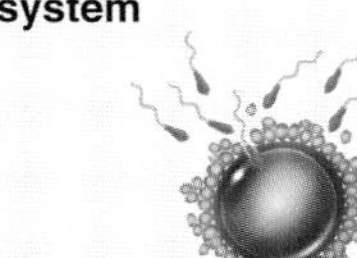	Nerve endings in skin and subcutaneous tissue respond to erotic stimuli, thereby contributing to sexual pleasure; suckling of a baby stimulates nerve endings in skin, leading to milk ejection; mammary glands (modified sweat glands) produce milk; skin stretches during pregnancy as fetus enlarges.

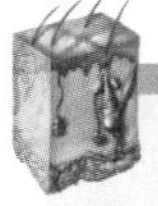

DISORDERS: HOMEOSTATIC IMBALANCES

Skin Cancer

Excessive exposure to the sun has caused virtually all of the one million cases of **skin cancer** diagnosed annually in the United States. There are three common forms of skin cancer. **Basal cell carcinomas** account for about 78% of all skin cancers. The tumors arise from cells in the stratum basale of the epidermis and rarely metastasize. **Squamous cell carcinomas,** which account for about 20% of all skin cancers, arise from squamous cells of the epidermis, and they have a variable tendency to metastasize. Most arise from preexisting lesions of damaged tissue on sun-exposed skin. Basal and squamous cell carcinomas are together known as *nonmelanoma skin cancer.* They are 50% more common in males than in females.

Malignant melanomas arise from melanocytes and account for about 2% of all skin cancers. The estimated lifetime risk of developing melanoma is now 1 in 75, double the risk only 20 years ago. In part, this increase is due to depletion of the ozone layer, which absorbs some UV light high in the atmosphere. But the main reason for the increase is that more people are spending more time in the sun and in tanning beds. Malignant melanomas metastasize rapidly and can kill a person within months of diagnosis.

The key to successful treatment of malignant melanoma is early detection. The early warning signs of malignant melanoma are identified by the acronym ABCD (Figure 5.8). *A* is for *asymmetry;* malignant melanomas tend to lack symmetry. *B* is for *border;* malignant melanomas have irregular—notched, indented, scalloped, or indistinct—borders. *C* is for *color;* malignant melanomas have uneven coloration and may contain several colors. *D* is for *diameter;* ordinary moles typically are smaller than 6 mm (0.25 in.), about the size of a pencil eraser. Once a malignant melanoma has the characteristics of A, B, and C, it is usually larger than 6 mm.

Among the risk factors for skin cancer are the following:

1. ***Skin type.*** Individuals with light-colored skin who never tan but always burn are at high risk.
2. ***Sun exposure.*** People who live in areas with many days of sunlight per year and at high altitudes (where ultraviolet light is more intense) have a higher risk of developing skin cancer. Likewise, people who engage in outdoor occupations and those who have suffered three or more severe sunburns have a higher risk.
3. ***Family history.*** Skin cancer rates are higher in some families than in others.
4. ***Age.*** Older people are more prone to skin cancer owing to longer total exposure to sunlight.
5. ***Immunological status.*** Immunosuppressed individuals have a higher incidence of skin cancer.

Figure 5.8 Comparison of a normal nevus (mole) and a malignant melanoma.

Excessive exposure to the sun accounts for almost all cases of skin cancer.

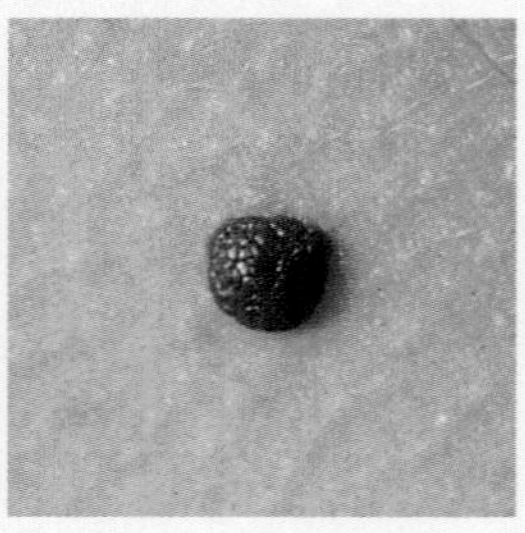

(a) Normal nevus (mole)

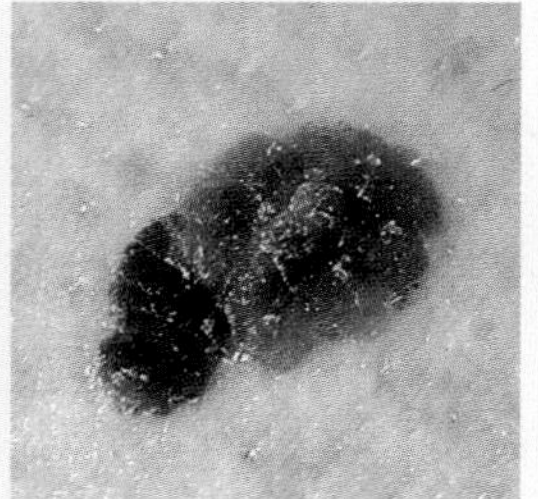

(b) Malignant melanoma

Which is the most common type of skin cancer?

Burns

A **burn** is tissue damage caused by excessive heat, electricity, radioactivity, or corrosive chemicals that denature the proteins in the skin cells. Burns destroy some of the skin's important contributions to homeostasis—protection against microbial invasion and desiccation, and thermoregulation.

Burns are graded according to their severity. A *first-degree burn* involves only the epidermis (Figure 5.9a). It is characterized by mild pain and erythema (redness) but no blisters. Skin functions remain intact. Immediate flushing with cold water may lessen the pain and damage caused by a first-degree burn. Generally, healing of a first-degree burn will occur in 3 to 6 days and may be accompanied by flaking or peeling. One example of a first-degree burn is mild sunburn.

A *second-degree burn* destroys the epidermis and part of the dermis (Figure 5.9b). Some skin functions are lost. In a second-degree burn, redness, blister formation, edema, and pain result. In a blister the epidermis separates from the dermis due to the accumulation of tissue fluid between them. Associated structures, such as hair follicles, sebaceous glands, and sweat glands, usually are not injured. If there is no infection, second-degree burns heal without skin grafting in about 3 to 4 weeks, but scarring may result. First- and second-degree burns are collectively referred to as *partial-thickness burns.*

A *third-degree burn* or *full-thickness burn* destroys the epidermis, dermis, and subcutaneous layer (Figure 5.9c). Most skin functions are lost. Such burns vary in appearance from marble-white to mahogany colored to charred, dry wounds. There is marked edema, and the burned region is numb because sensory nerve endings have been destroyed. Regeneration occurs slowly, and much granulation tissue forms before being covered by epithelium. Skin grafting may be required to promote healing and to minimize scarring.

The injury to the skin tissues directly in contact with the damaging agent is the *local effect* of a burn. Generally, however, the *systemic effects* of a major burn are a greater threat to life. The systemic effects of a burn may include (1) a large loss of water, plasma, and plasma proteins, which causes shock; (2) bacterial infection; (3) reduced circulation of blood; (4) decreased production of urine; and (5) diminished immune responses.

The seriousness of a burn is determined by its depth and extent of area involved, as well as the person's age and general health. According to the American Burn Association's classification of burn injury, a major burn includes third-degree burns over 10% of body surface area; or second-degree burns over 25% of body surface area; or any third-degree burns on the face, hands, feet, or *perineum* (per-i-NĒ-um, which includes the anal and urogenital regions). When the burn area exceeds

Figure 5.9 **Burns.**

A burn is tissue damage caused by agents that destroy the proteins in skin cells.

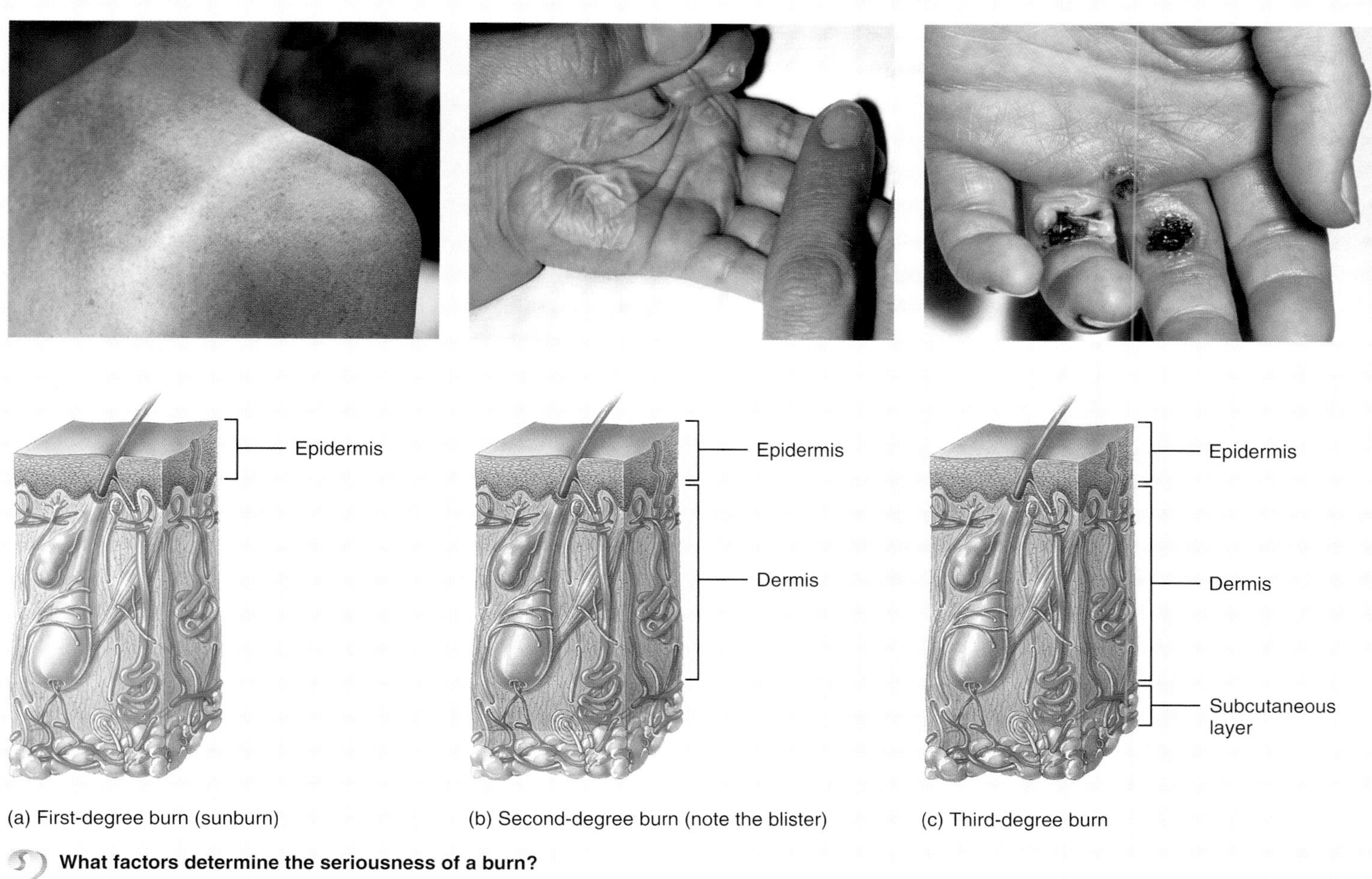

(a) First-degree burn (sunburn)

(b) Second-degree burn (note the blister)

(c) Third-degree burn

What factors determine the seriousness of a burn?

70%, more than half the victims die. A quick means for estimating the surface area affected by a burn in an adult is the **rule of nines** (Figure 5.10):

1. Count 9% if both the anterior and posterior surfaces of the head and neck are affected.
2. Count 9% for both the anterior and posterior surfaces of each upper limb (total of 18% for both upper limbs).
3. Count four times nine or 36% for both the anterior and posterior surfaces of the trunk, including the buttocks.
4. Count 9% for the anterior and 9% for the posterior surfaces of each lower limb as far up as the buttocks (total of 36% for both lower limbs).
5. Count 1% for the perineum.

Many people who have been burned in fires also inhale smoke. If the smoke is unusually hot or dense or if inhalation is prolonged, serious problems can develop. The hot smoke can damage the trachea (windpipe), causing its lining to swell. As the swelling narrows the trachea, airflow into the lungs is obstructed. Further, small airways inside the lungs can also narrow, producing wheezing or shortness of breath. A person who has inhaled smoke is given oxygen through a face mask and a tube may be inserted into the trachea to assist breathing.

Pressure Ulcers

Pressure ulcers, also known as *decubitus ulcers* (dē-KŪ-bi-tus) or *bedsores,* are caused by a constant deficiency of blood flow to tissues (Figure 5.11). Typically the affected tissue overlies a bony projection that has been subjected to prolonged pressure against an object such as a bed, cast, or splint. If the pressure is relieved in a few hours, redness occurs but no lasting tissue damage results. Blistering of the affected area may indicate superficial damage; a reddish-blue discoloration may indicate deep tissue damage. Prolonged pressure causes tissue ulceration. Small breaks in the epidermis become infected, and the sensitive subcutaneous layer and deeper tissues are damaged. Eventually, the tissue dies. Pressure ulcers occur most often in bedridden patients. With proper care, pressure ulcers are preventable, but they can develop very quickly in patients who are very old or very ill.

Figure 5.10 Rule-of-nines method for determining the extent of a burn. The percentages are the approximate proportions of the body surface area.

The rule of nines is a quick rule for estimating the surface area affected by a burn in an adult.

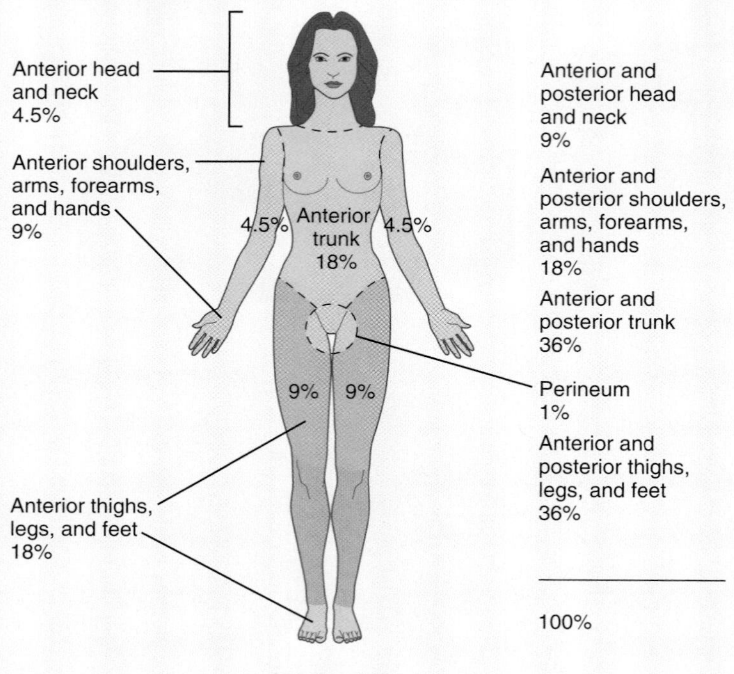

What percentage of the body would be burned if only the anterior trunk and anterior left upper limb were involved?

Figure 5.11 Pressure ulcers.

A pressure ulcer is a shedding of epithelium caused by a constant deficiency of blood flow to tissues.

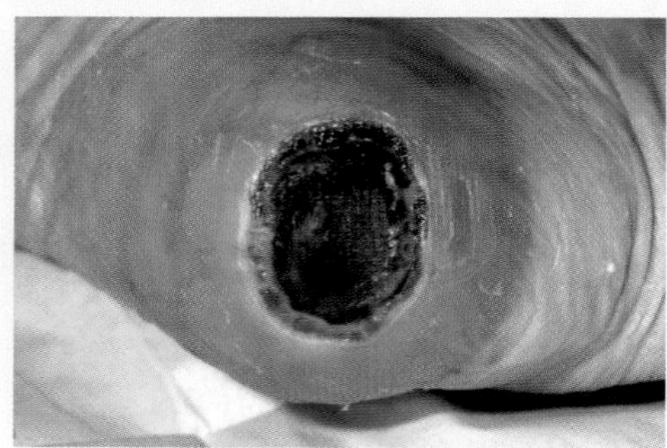

Pressure ulcer on heel

What parts of the body are usually affected by pressure ulcers?

MEDICAL TERMINOLOGY

Abrasion (a-BRĀ-shun; *ab-* = away; *-rasion* = scraped) An area where skin has been scraped away.

Athlete's foot A superficial fungal infection of the skin of the foot.

Blister A collection of serous fluid within the epidermis or between the epidermis and dermis, due to short-term but severe friction. The term **bulla** (BUL-a) refers to a large blister.

Callus (KAL-lus = hardskin) An area of hardened and thickened skin that is usually seen in palms and soles and is due to persistent pressure and friction.

Cold sore A lesion, usually in oral mucous membrane, caused by Type 1 herpes simplex virus (HSV) transmitted by oral or respiratory routes. The virus remains dormant until triggered by factors such as ultraviolet light, hormonal changes, and emotional stress. Also called a **fever blister.**

Comedo (KOM-ē-dō; *comedo* = to eat up) A collection of sebaceous material and dead cells in the hair follicle and excretory duct of the sebaceous (oil) gland. Usually found over the face, chest, and back, and more commonly during adolescence. Also called a **blackhead.**

Contact dermatitis (der-ma-TĪ-tis; *dermat-* = skin; *-itis* = inflammation of) Inflammation of the skin characterized by redness, itching, and swelling and caused by exposure of the skin to chemicals that bring about an allergic reaction, such as poison ivy toxin.

Corn A painful conical thickening of the stratum corneum of the epidermis found principally over toe joints and between the toes, often caused by friction or pressure. Corns may be hard or soft, depending on their location. Hard corns are usually found over toe joints, and soft corns are usually found between the fourth and fifth toes.

Cyst (SIST; *cyst* = sac containing fluid) A sac with a distinct connective tissue wall, containing a fluid or other material.

Eczema (EK-ze-ma; *ekzeo-* = to boil over) An inflammation of the skin characterized by patches of red, blistering, dry, extremely itchy skin. It occurs mostly in skin creases in the wrists, backs of the knees, and fronts of the elbows. It typically begins in infancy and many children outgrow the condition. The cause is unknown but is linked to genetics and allergies.

Frostbite Local destruction of skin and subcutaneous tissue on exposed surfaces as a result of extreme cold. In mild cases, the skin is blue and swollen and there is slight pain. In severe cases there is considerable swelling, some bleeding, no pain, and blistering. If untreated, gangrene may develop. Frostbite is treated by rapid rewarming.

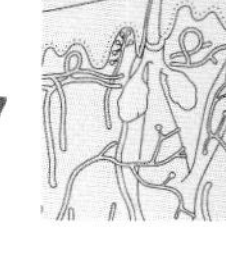

Hemangioma (he-man′-jē-Ō-ma; *hem-* = blood; *-angi-* = blood vessel; *-oma* = tumor) Localized tumor of the skin and subcutaneous layer that results from an abnormal increase in blood vessels. One type is a **portwine stain,** a flat, pink, red, or purple lesion present at birth, usually at the nape of the neck.

Hives Reddened elevated patches of skin that are often itchy. Most commonly caused by infections, physical trauma, medications, emotional stress, food additives, and certain food allergies. Also called **urticaria** (ūr-ti-KAR-ē-a).

Keratosis (ker′-a-TŌ-sis; *kera-* = horn) Formation of a hardened growth of epidermal tissue, such as *solar keratosis,* a premalignant lesion of the sun-exposed skin of the face and hands.

Laceration (las-er-Ā-shun; *lacer-* = torn) An irregular tear of the skin.

Papule (PAP-ūl; *papula* = pimple) A small, round skin elevation less than 1 cm in diameter. One example is a pimple.

Pruritus (proo-RĪ-tus; *pruri-* = to itch) Itching, one of the most common dermatological disorders. It may be caused by skin disorders (infections), systemic disorders (cancer, kidney failure), psychogenic factors (emotional stress), or allergic reactions.

Topical In reference to a medication, applied to the skin surface rather than ingested or injected.

Wart Mass produced by uncontrolled growth of epithelial skin cells; caused by a papillomavirus. Most warts are noncancerous.

STUDY OUTLINE

STRUCTURE OF THE SKIN (p. 145)

1. The integumentary system consists of the skin and its accessory structures—hair, nails, glands, muscles, and nerves.
2. The skin is the largest organ of the body in surface area and weight. The principal parts of the skin are the epidermis (superficial) and dermis (deep).
3. The subcutaneous layer (hypodermis) is deep to the dermis and not part of the skin. It anchors the dermis to underlying tissues and organs, and it contains lamellated (pacinian) corpuscles.
4. The types of cells in the epidermis are keratinocytes, melanocytes, Langerhans cells, and Merkel cells.
5. The epidermal layers, from deep to superficial, are the stratum basale, stratum spinosum, stratum granulosum, stratum lucidum (in thick skin only), and stratum corneum (see Table 5.1). Stem cells in the stratum basale undergo continuous cell division, producing keratinocytes for the other layers.
6. The dermis consists of papillary and reticular regions. The papillary region is composed of areolar connective tissue containing fine elastic fibers, dermal papillae, and Meissner corpuscles. The reticular region is composed of dense irregular connective tissue containing interlaced collagen and coarse elastic fibers, adipose tissue, hair follicles, nerves, sebaceous (oil) glands, and ducts of sudoriferous (sweat) glands.
7. Epidermal ridges provide the basis for fingerprints and footprints.
8. The color of skin is due to melanin, carotene, and hemoglobin.
9. In tattooing, a pigment is deposited with a needle in the dermis. Body piercing is the insertion of jewelry through an artificial opening.

ACCESSORY STRUCTURES OF THE SKIN (p. 151)

1. Accessory structures of the skin—hair, skin glands, and nails—develop from the embryonic epidermis.
2. A hair consists of a shaft, most of which is superficial to the surface, a root that penetrates the dermis and sometimes the subcutaneous layer, and a hair follicle.
3. Associated with each hair follicle is a sebaceous (oil) gland, an arrector pili muscle, and a hair root plexus.
4. New hairs develop from division of matrix cells in the bulb; hair replacement and growth occur in a cyclic pattern consisting of alternating growth and resting stages.
5. Hairs offer a limited amount of protection—from the sun, heat loss, and entry of foreign particles into the eyes, nose, and ears. They also function in sensing light touch.
6. Sebaceous (oil) glands are usually connected to hair follicles; they are absent from the palms and soles. Sebaceous glands produce sebum, which moistens hairs and waterproofs the skin. Clogged sebaceous glands may produce acne.
7. There are two types of sudoriferous (sweat) glands: eccrine and apocrine. Eccrine sweat glands have an extensive distribution; their ducts terminate at pores at the surface of the epidermis. Apocrine sweat glands are limited to the skin of the axillae, groin, and areolae; their ducts open into hair follicles. They begin functioning at puberty and are stimulated during emotional stress and sexual excitement. Mammary glands are specialized sudoriferous glands that secrete milk.
8. Ceruminous glands are modified sudoriferous glands that secrete cerumen. They are found in the external auditory canal (ear canal).
9. Lanugo of the fetus is shed before birth, except in the scalp, eyebrows, and eyelashes. Most body hair on males is terminal (coarse, pigmented); most body hair on females is vellus (fine).
10. Nails are hard, keratinized epidermal cells over the dorsal surfaces of the distal portions of the digits.
11. The principal parts of a nail are the nail body, free edge, nail root, lunula, eponychium, and matrix. Cell division of the matrix cells produces new nails.

TYPES OF SKIN (p. 156)

1. Thin skin covers all parts of the body except for the palms, palmar surfaces of the digits, and the soles.
2. Thick skin covers the palms, palmar surfaces of the digits, and soles.

FUNCTIONS OF THE SKIN (p. 157)

1. Skin functions include body temperature regulation, blood storage, protection, sensation, excretion and absorption, and synthesis of vitamin D.
2. The skin participates in thermoregulation by liberating sweat at its surface and by adjusting the flow of blood in the dermis.
3. The skin provides physical, chemical, and biological barriers that help protect the body.
4. Cutaneous sensations include tactile sensations, thermal sensations, and pain.

MAINTAINING HOMEOSTASIS: SKIN WOUND HEALING (p. 158)

1. In an epidermal wound, the central portion of the wound usually extends down to the dermis; the wound edges involve only superficial damage to the epidermal cells.
2. Epidermal wounds are repaired by enlargement and migration of basal cells, contact inhibition, and division of migrating and stationary basal cells.
3. During the inflammatory phase of deep wound healing, a blood clot unites the wound edges, epithelial cells migrate across the wound, vasodilation and increased permeability of blood vessels enhance delivery of phagocytes, and mesenchymal cells develop into fibroblasts.
4. During the migratory phase, fibroblasts migrate along fibrin threads and begin synthesizing collagen fibers and glycoproteins.
5. During the proliferative phase, epithelial cells grow extensively.
6. During the maturation phase, the scab sloughs off, the epidermis is restored to normal thickness, collagen fibers become more organized, fibroblasts begin to disappear, and blood vessels are restored to normal.

DEVELOPMENT OF THE INTEGUMENTARY SYSTEM (p. 160)

1. The epidermis develops from the embryonic ectoderm, and the accessory structures of the skin (hair, nails, and skin glands) are epidermal derivatives.
2. The dermis is derived from mesodermal cells.

AGING AND THE INTEGUMENTARY SYSTEM (p. 160)

1. Most effects of aging begin to occur when people reach their late forties.
2. Among the effects of aging are wrinkling, loss of subcutaneous adipose tissue, atrophy of sebaceous glands, and decrease in the number of melanocytes and Langerhans cells.

SELF-QUIZ QUESTIONS

Fill in the blanks in the following statements.

1. The epidermal layer that is found in thick skin but not in thin skin is the ____.
2. The most common sweat glands that release a watery secretion are ____ sweat glands; modified sweat glands in the ear are ____ glands; sweat glands located in the axillae, groin, areolae, and beards of males and that release a thick, lipid-rich secretion are ____ sweat glands.

Indicate whether the following statements are true or false.

3. An individual with a dark skin color has more melanocytes than a fair-skinned person.
4. In order to permanently prevent growth of an unwanted hair, you must destroy the hair matrix.

Choose the one best answer to the following questions.

5. The layer of the epidermis that contains stem cells undergoing mitosis is the (a) stratum corneum, (b) stratum lucidum, (c) stratum basale, (d) stratum spinosum, (e) stratum granulosum.
6. The substance that helps promote mitosis in epidermal skin cell is (a) keratohyalin, (b) melanin, (c) carotene, (d) collagen, (e) epidermal growth factor.
7. Which of the following is *not* a function of skin? (a) calcium production, (b) vitamin D synthesis, (c) protection, (d) excretion of wastes, (e) temperature regulation.
8. To expose underlying tissues in the bottom of the foot, a foot surgeon must first cut through the skin. Place the following layers in the order that the scalpel would cut. (1) stratum lucidum, (2) stratum corneum, (3) stratum basale, (4) stratum granulosum, (5) stratum spinosum. (a) 3, 5, 4, 1, 2, (b) 2, 1, 5, 4, 3, (c) 2, 1, 4, 5, 3, (d) 1, 3, 5, 4, 2, (e) 3, 4, 5, 1, 2.
9. Aging of the skin can result in (a) an increase in collagen and elastic fibers, (b) a decrease in the activity of sebaceous glands, (c) a thickening of the skin, (d) an increased blood flow to the skin, (e) an increase in toenail growth.
10. Which of the following is *not* true? (a) Albinism is an inherited inability of melanocytes to produce melanin. (b) Striae occurs when the dermis is overstretched to the point of tearing. (c) In order to prevent excessive scarring, surgeons should cut parallel to the lines of cleavage. (d) The papillary layer of the dermis is responsible for fingerprints. (e) Much of the body's fat is located in the dermis of the skin.
11. A patient is brought into the emergency room suffering from a burn. The patient does not feel any pain at the burn site. Using a gentle pull on a hair, the examining physician can remove entire hair follicles from the patient's arm. This patient is suffering from what type of burn? (a) third degree, (b) second degree, (c) first degree, (d) partial-thickness, (e) localized.
12. Which of the following statements are *true*? (1) Nails are composed of tightly packed, hard, keratinized cells of the epidermis that form a clear, solid covering over the dorsal surface of the terminal end of digits. (2) The free edge of the nail is white due to the absence of capillaries. (3) Nails help us grasp and manipulate small objects. (4) Nails protect the ends of digits from trauma. (5) Nail color is due to a combination of melanin and carotene. (1) 1, 2, and 3, (b) 1, 3, and 4, (c) 1, 2, 3, and 4, (d) 2, 3, and 4, (e) 1, 3, and 5.

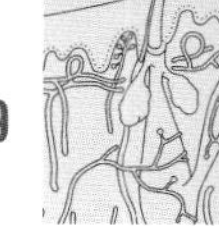

13. Match the following:

____(a) produce the protein that helps protect the skin and underlying tissues from light, heat, microbes, and many chemicals
____(b) produce a pigment that contributes to skin color and absorbs ultraviolet light
____(c) cells that arise from red bone marrow, migrate to the epidermis, and participate in immune responses
____(d) cells thought to function in the sensation of touch
____(e) located in the dermis, they function in the sensations of warmth, coolness, pain, itching, and tickling
____(f) smooth muscles associated with the hair follicles; when contracted, they pull the hair shafts perpendicular to the skin's surface
____(g) an abnormal thickening of the epidermis
____(h) release a lipid-rich secretion that functions as a water-repellent sealant in the stratum granulosum
____(i) pressure-sensitive cells found mostly in the subcutaneous layer
____(j) a fatty substance that covers and protects the skin of the fetus from the constant exposure to amniotic fluid
____(k) associated with hair follicles, these secrete an oily substance that helps prevent hair from becoming brittle, prevents evaporation of water from the skin's surface, and inhibits the growth of certain bacteria

(1) Merkel cells
(2) callus
(3) keratinocytes
(4) Langerhans cells
(5) melanocytes
(6) free nerve endings
(7) sebaceous glands
(8) lamellar granules
(9) lamellated (pacinian) corpuscles
(10) vernix caseosa
(11) arrector pili

14. Match the following:

____(a) deep region of the dermis composed primarily of dense irregular connective tissue
____(b) composed of keratinized stratified squamous epithelial tissue
____(c) not considered part of the skin, it contains areolar and adipose tissues and blood vessels; attaches skin to underlying tissues and organs
____(d) superficial region of the dermis; composed of areolar connective tissue

(1) subcutaneous layer (hypodermis)
(2) papillary region
(3) reticular region
(4) epidermis

15. Match the following and place the phases of deep wound healing in the correct order:

____(a) epithelial cells migrate under scab to bridge the wound; formation of granulation tissue
____(b) sloughing of scab; reorganization of collagen fibers; blood vessels return to normal
____(c) vasodilation and increased permeability of blood vessels to deliver cells involved in phagocytosis; clot formation
____(d) extensive growth of epithelial cells beneath scab; random deposition of collagen fibers; continued growth of blood vessels

(1) proliferative phase
(2) inflammatory phase
(3) maturation phase
(4) migratory phase

Correct order of phases:
1)____, 2) ____, 3) ____, 4) ____

CRITICAL THINKING QUESTIONS

1. The amount of dust that collects in a house with an assortment of dogs, cats, and people is truly amazing. A lot of these dust particles had a previous "life" as part of the home's living occupants. Where did the dust originate on the human body?
2. Janet has just returned from the hairdresser with a new, shorter haircut. She states that the new hairstyle has made her hair much thicker. Is this possible? Why or why not?
3. Six months ago, Chef Eduardo sliced through the end of his right thumbnail. Although the surrounding nail grows normally, this part of his nail remains split and doesn't seem to want to "heal." What has happened to cause this?

ANSWERS TO FIGURE QUESTIONS

5.1 The epidermis is composed of epithelial tissue; the dermis is made up of connective tissue.

5.2 Melanin protects DNA of keratinocytes from the damaging effects of UV light.

5.3 The stratum basale is the layer of the epidermis with stem cells that continually undergo cell division.

5.4 Plucking a hair stimulates hair root plexuses in the dermis, some of which are sensitive to pain. Because the cells of a hair shaft are already dead and the hair shaft lacks nerves, cutting hair is not painful.

5.5 Nails are hard because they are composed of tightly packed, hard, keratinized epidermal cells.

5.6 Since the epidermis is avascular, an epidermal wound would not produce any bleeding.

5.7 Vernix caseosa consists of secretions from sebaceous glands, sloughed off peridermal cells, and hairs.

5.8 Basal cell carcinoma is the most common type of skin cancer.

5.9 The seriousness of a burn is determined by the depth and extent of the area involved, the individual's age, and general health.

5.10 About 22.5% of the body would be involved (4.5% [arm] + 18% [anterior trunk]).

5.11 Pressure ulcers typically develop in tissues that overlie bony projections subjected to pressure, such as the shoulders, hips, buttocks, heels, and ankles.

Chapter 6

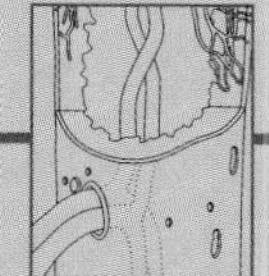

The Skeletal System: Bone Tissue

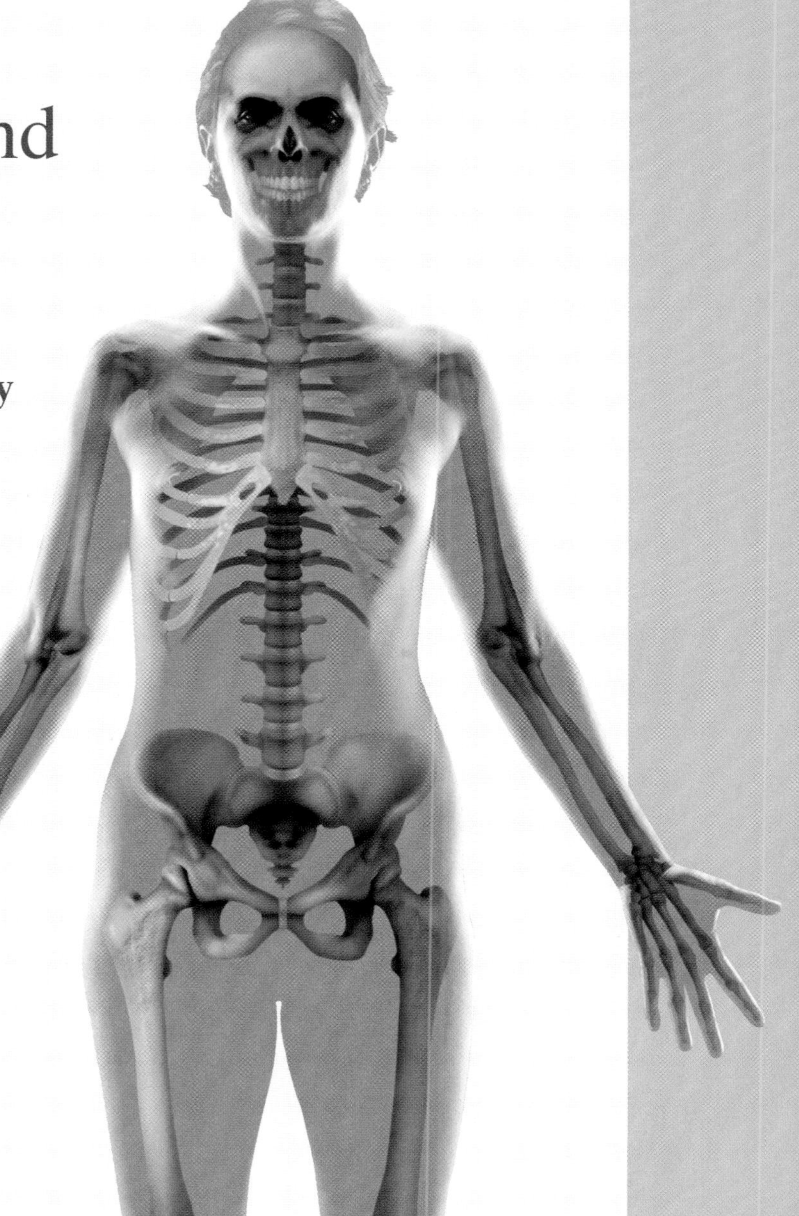

Bone Tissue and Homeostasis

Bone tissue is continuously growing, remodeling, and repairing itself. It contributes to homeostasis of the body by providing support, protection, the production of blood cells, and the storage of minerals and triglycerides.

wiley.com/college/apcentral

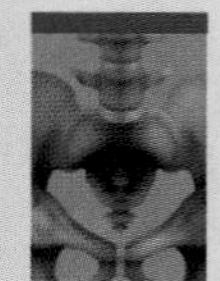

A bone is made up of several different tissues working together: bone or osseous tissue, cartilage, dense connective tissues, epithelium, adipose tissue, and nervous tissue. For this reason, each individual bone in your body is considered an organ. Bone tissue, a complex and dynamic living tissue, continually engages in a process called remodeling—the construction of new bone tissue and breaking down of old bone tissue. The entire framework of bones and their cartilages constitutes the **skeletal system.** In this chapter we will survey the various components of bones to help you understand how bones form and age, and how exercise affects their density and strength. The study of bone structure and the treatment of bone disorders is called **osteology** (os-tē-OL-ō-jē; *osteo-* = bone; *-logy* = study of).

FUNCTIONS OF BONE AND THE SKELETAL SYSTEM

► **OBJECTIVE**

Describe the six main functions of the skeletal system.

Bone tissue makes up about 18% of the weight of the human body. The skeletal system performs several basic functions:

1. ***Support.*** The skeleton serves as the structural framework for the body by supporting soft tissues and providing attachment points for the tendons of most skeletal muscles.

2. ***Protection.*** The skeleton protects the most important internal organs from injury. For example, cranial bones protect the brain, vertebrae (backbones) protect the spinal cord, and the rib cage protects the heart and lungs.

3. ***Assistance in movement.*** Most skeletal muscles attach to bones; when they contract, they pull on bones to produce movement. This function is discussed in detail in Chapter 10.

4. ***Mineral homeostasis.*** Bone tissue stores several minerals, especially calcium and phosphorus, which contribute to the strength of bone. On demand, bone releases minerals into the blood to maintain critical mineral balances (homeostasis) and to distribute the minerals to other parts of the body.

5. ***Blood cell production.*** Within certain bones, a connective tissue called **red bone marrow** produces red blood cells, white blood cells, and platelets, a process called **hemopoiesis** (hēm-ō-poy-Ē-sis; *hemo-* = blood; *poiesis-* = making). Red bone marrow consists of developing blood cells, adipocytes, fibroblasts, and macrophages within a network of reticular fibers. It is present in developing bones of the fetus and in some adult bones, such as the pelvis, ribs, breastbone, vertebrae (backbones), skull, and ends of the bones of the arm and thigh.

6. ***Triglyceride storage.*** **Yellow bone marrow** consists mainly of adipose cells, which store triglycerides. The stored triglycerides are a potential chemical energy reserve.

► **CHECKPOINT**

1. What types of tissues make up the skeletal system?
2. How do red and yellow bone marrow differ in composition, location, and function?

STRUCTURE OF BONE

► **OBJECTIVE**

Describe the functions of each part of a long bone.

We will now examine the structure of bone at the macroscopic level. Macroscopic bone structure may be analyzed by considering the parts of a long bone, such as the humerus (the arm bone) shown in Figure 6.1a. A *long bone* is one that has greater length than width. A typical long bone consists of the following parts:

1. The **diaphysis** (dī-AF-i-sis = growing between) is the bone's shaft or body—the long, cylindrical, main portion of the bone.

2. The **epiphyses** (e-PIF-i-sēz = growing over; singular is *epiphysis*) are the distal and proximal ends of the bone.

3. The **metaphyses** (me-TAF-i-sēz; *meta-* = between; singular is *metaphysis*) are the regions in a mature bone where the diaphysis joins the epiphyses. In a growing bone, each metaphysis includes an **epiphyseal plate** (ep′-i-FIZ-ē-al), a layer of hyaline cartilage that allows the diaphysis of the bone to grow in length (described later in the chapter). When a bone ceases to grow in length at about ages 18–21, the cartilage in the epiphyseal plate is replaced by bone; the resulting bony structure is known as the **epiphyseal line.**

4. The **articular cartilage** is a thin layer of hyaline cartilage covering the part of the epiphysis where the bone forms an articulation (joint) with another bone. Articular cartilage reduces friction and absorbs shock at freely movable joints.

5. The **periosteum** (per′-ē-OS-tē-um; *peri-* = round) is a tough sheath of dense irregular connective tissue that surrounds the bone surface wherever it is not covered by articular cartilage. The bone-forming cells of the periosteum enable bone to grow in thickness, but not in length. The periosteum also protects the bone, assists in fracture repair, helps nourish bone tissue, and serves as an attachment point for ligaments and tendons. It is attached to the underlying bone through **perforating (Sharpey's) fibers,** thick bundles of collagen fibers that extend from the periosteum into the extracellular bone matrix.

6. The **medullary cavity** (MED-ū-lar′-ē; *medulla-* = marrow, pith) or **marrow cavity** is the space within the diaphysis that contains fatty yellow bone marrow in adults.

Figure 6.1 Parts of a long bone. The spongy bone tissue of the epiphyses and metaphysis contains red bone marrow, whereas the medullary cavity of the diaphysis contains yellow bone marrow (in adults).

A long bone is covered by articular cartilage at its proximal and distal epiphyses and by periosteum around the diaphysis.

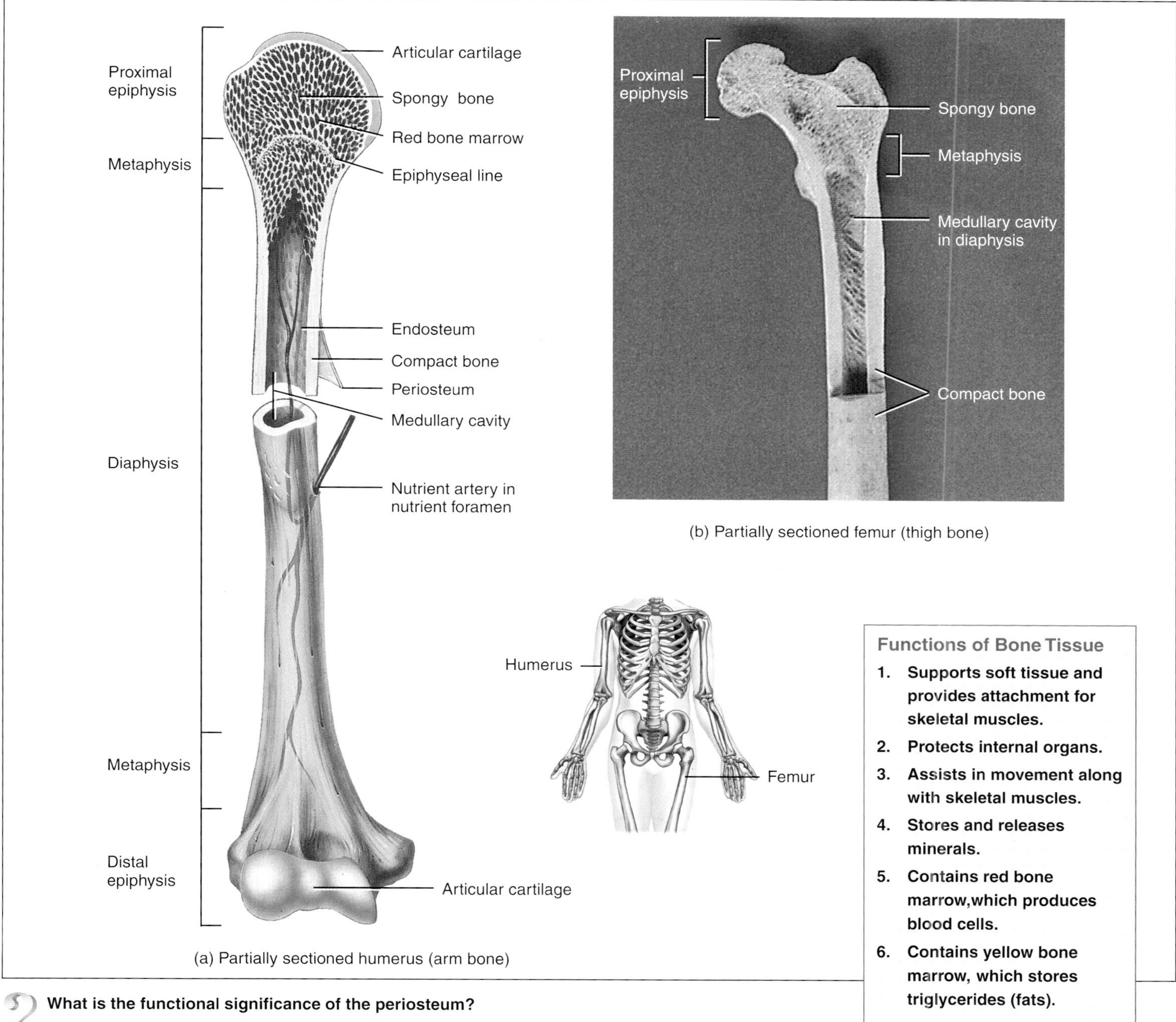

Functions of Bone Tissue

1. **Supports soft tissue and provides attachment for skeletal muscles.**
2. **Protects internal organs.**
3. **Assists in movement along with skeletal muscles.**
4. **Stores and releases minerals.**
5. **Contains red bone marrow,which produces blood cells.**
6. **Contains yellow bone marrow, which stores triglycerides (fats).**

What is the functional significance of the periosteum?

7. The **endosteum** (end-OS-tē-um; *endo-* = within) is a thin membrane that lines the medullary cavity. It contains a single layer of bone-forming cells and a small amount of connective tissue.

▶ CHECKPOINT

3. Diagram the parts of a long bone, and list the functions of each part.

HISTOLOGY OF BONE TISSUE

▶ OBJECTIVE

Describe the histological features of bone tissue.

We will now examine the structure of bone at the microscopic level. Like other connective tissues, **bone,** or **osseous tissue** (OS-ē-us), contains an abundant extracellular matrix that surrounds widely separated cells. The extracellular matrix is about 25% water, 25% collagen fibers, and 50% crystallized mineral salts. The most abundant mineral salt is calcium phosphate [$Ca_3(PO_4)_2$]. It combines with another mineral salt, calcium hydroxide [$Ca(OH)_2$], to form crystals of **hydroxyapatite.** As the crystals form, they combine with still other mineral salts, such as calcium carbonate ($CaCO_3$), and ions such as magnesium, fluoride, potassium, and sulfate. As these mineral salts are deposited in the framework formed by the collagen fibers of the extracellular matrix, they crystallize and the tissue hardens. This process of **calcification** (kal′-si-fi-KĀ-shun) is initiated by bone-building cells called osteoblasts.

It was once thought that calcification simply occurred when enough mineral salts were present to form crystals. We now know that the process requires the presence of collagen fibers. Mineral salts first begin to crystallize in the microscopic spaces between collagen fibers. After the spaces are filled, mineral crystals accumulate around the collagen fibers.

Although a bone's *hardness* depends on the crystallized inorganic mineral salts, a bone's *flexibility* depends on its collagen fibers. Like reinforcing metal rods in concrete, collagen fibers and other organic molecules provide *tensile strength,* resistance to being stretched or torn apart. Soaking a bone in an acidic solution, such as vinegar, dissolves its mineral salts, causing the bone to become rubbery and flexible. As you will see shortly, when the need for particular minerals arises or as part of bone formation or breakdown, bone cells called osteoclasts secrete enzymes and acids that break down both the mineral salts and the collagen fibers of bone extracellular matrix.

Four types of cells are present in bone tissue: osteogenic cells, osteoblasts, osteocytes, and osteoclasts (Figure 6.2).

1. **Osteogenic cells** (os′-tē-ō-JEN-ik; *-genic* = producing) are unspecialized stem cells derived from mesenchyme, the tissue from which all connective tissues are formed. They are the only bone cells to undergo cell division; the resulting cells develop into osteoblasts. Osteogenic cells are found along the inner portion of the periosteum, in the endosteum, and in the canals within bone that contain blood vessels.
2. **Osteoblasts** (OS-tē-ō-blasts′; *-blasts* = buds or sprouts) are bone-building cells. They synthesize and secrete collagen fibers and other organic components needed to build the extracellular matrix of bone tissue, and they initiate calcification (described shortly). As osteoblasts surround themselves with extracellular matrix, they become trapped in their secretions and become osteocytes. (Note: *Blasts* in bone or any other connective tissue secrete extracellular matrix.)
3. **Osteocytes** (OS-tē-ō-sīts′; *-cytes* = cells), mature bone cells, are the main cells in bone tissue and maintain its daily metabolism, such as the exchange of nutrients and wastes with the blood. Like osteoblasts, osteocytes do not undergo cell division. (Note: *Cytes* in bone or any other tissue maintain the tissue.)

Figure 6.2 Types of cells in bone tissue.

Osteogenic cells undergo cell division and develop into osteoblasts, which secrete bone extracellular matrix.

Ruffled border

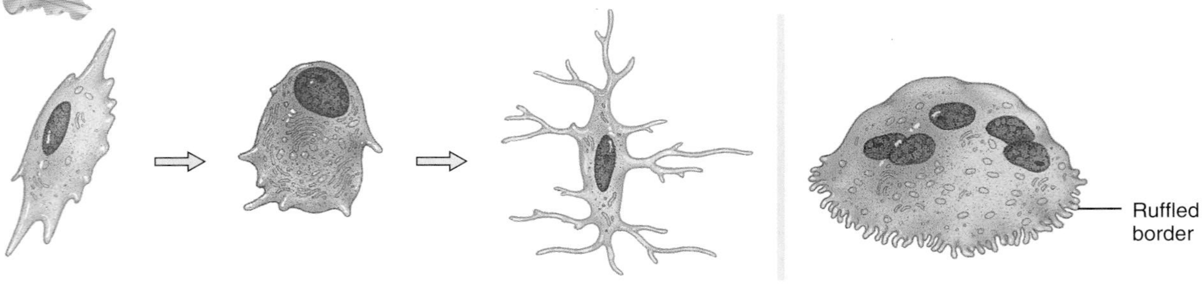

Osteogenic cell (develops into an osteoblast)

Osteoblast (forms bone matrix)

Osteocyte (maintains bone tissue)

Osteoclast (functions in resorption, the breakdown of bone matrix)

SEM 1100x

SEM 9160x

SEM 5626x

Why is bone resorption important?

4. Osteoclasts (OS-tē-ō-clasts′; *-clast* = break) are huge cells derived from the fusion of as many as 50 monocytes (a type of white blood cell) and are concentrated in the endosteum. On the side of the cell that faces the bone surface, the osteoclast's plasma membrane is deeply folded into a *ruffled border*. Here the cell releases powerful lysosomal enzymes and acids that digest the protein and mineral components of the underlying bone matrix. This breakdown of bone extracellular matrix, termed **resorption** (rē-SORP-shun), is part of the normal development, growth, maintenance, and repair of bone. (Note: *Clasts* in bone break down extracellular matrix.) As you will see later, in response to certain hormones, osteoclasts help regulate blood calcium level (see page 188). They are also target cells for drug therapy used to treat osteoporosis (see page 189).

Bone is not completely solid but has many small spaces between its cells and extracellular matrix components. Some spaces serve as channels for blood vessels that supply bone cells with nutrients. Other spaces act as storage areas for red bone marrow. Depending on the size and distribution of the spaces, the regions of a bone may be categorized as compact or spongy (see Figure 6.1). Overall, about 80% of the skeleton is compact bone and 20% is spongy bone.

Compact Bone Tissue

Compact bone tissue contains few spaces (Figure 6.3a) and is the strongest form of bone tissue. It is found beneath the periosteum of all bones and makes up the bulk of the diaphyses of long bones. Compact bone tissue provides protection and support and resists the stresses produced by weight and movement.

Blood vessels, lymphatic vessels, and nerves from the periosteum penetrate compact bone through transverse **perforating** or **Volkmann's** (FOLK-mans) **canals.** The vessels and nerves of the perforating canals connect with those of the medullary cavity, periosteum, and **central** or **haversian canals** (ha-VER-shun). The central canals run longitudinally through the bone. Around the central canals are **concentric lamellae** (la-MEL-ē)—rings of calcified extracellular matrix much like the rings of a tree trunk. Between the lamellae are small spaces called **lacunae** (la-KOO-nē = little lakes; singular is *lacuna*), which contain osteocytes. Radiating in all directions from the lacunae are tiny **canaliculi** (kan′-a-LIK-ū-lī = small channels) filled with extracellular fluid. Inside the canaliculi are slender fingerlike processes of osteocytes (see inset at the right in Figure 6.3a). Neighboring osteocytes communicate via gap junctions. The canaliculi connect lacunae with one another and with the central canals, forming an intricate, miniature system of interconnected canals throughout the bone. This system provides many routes for nutrients and oxygen to reach the osteocytes and for the removal of wastes.

The components of compact bone tissue are arranged into repeating units called **osteons** or **haversian systems** (Figure 6.3a). Each osteon consists of a central (haversian) canal with its concentrically arranged lamellae, lacunae, osteocytes, and canaliculi. Osteons in compact bone tissue are aligned in the same direction along lines of stress. In the shaft, for example, they are parallel to the long axis of the bone. As a result, the shaft of a long bone resists bending or fracturing even when considerable force is applied from either end. The osteons of a long bone can be compared to a stack of logs; each log is made up of rings of hard material, and together it requires considerable force to fracture them all. The lines of stress in a bone change as a baby learns to walk and in response to repeated strenuous physical activity, such as weight training. The lines of stress in a bone also can change in response to fractures or physical deformity. Thus, the organization of osteons is not static but changes over time in response to the physical demands placed on the skeleton.

The areas between osteons contain **interstitial lamellae** (in′-ter-STISH-al), which also have lacunae with osteocytes and canaliculi. Interstitial lamellae are fragments of older osteons that have been partially destroyed during bone rebuilding or growth. Lamellae that encircle the bone just beneath the periosteum or encircle the medullary cavity are called **circumferential lamellae.**

Spongy Bone Tissue

In contrast to compact bone tissue, **spongy bone tissue** does not contain osteons. Despite what the name seems to imply, the term "spongy" does not refer to the texture of the bone, only its appearance (Figure 6.3b). Spongy bone consists of lamellae arranged in an irregular lattice of thin columns called **trabeculae** (tra-BEK-ū-lē = little beams; singular is *trabecula*). The macroscopic spaces between the trabeculae help make bones lighter and can sometimes be filled with red bone marrow. Within each trabecula are lacunae that contain osteocytes. Canaliculi radiate outward from the lacunae. Because the osteocytes of spongy bone are located on the superficial surfaces of trabeculae, they receive nourishment directly from the blood circulating through the medullary cavities.

Spongy bone tissue makes up most of the bone tissue of short, flat, and irregularly shaped bones. It also forms most of the epiphyses of long bones and a narrow rim around the medullary cavity of the diaphysis of long bones.

At first glance, the structure of the osteons of compact bone tissue appears to be highly organized, and the trabeculae of spongy bone tissue appear to be randomly arranged. However, the trabeculae of spongy bone tissue are precisely oriented along lines of stress, a characteristic that helps bones resist stresses and transfer force without breaking. Spongy bone tissue tends to be located where bones are not heavily stressed or where stresses are applied from many directions.

Spongy bone tissue is different from compact bone tissue in two respects. First, spongy bone tissue is light, which reduces the overall weight of a bone so that it moves more readily when pulled by a skeletal muscle. Second, the trabeculae of spongy bone tissue support and protect the red bone marrow. The spongy bone tissue in the hip bones, ribs, breastbone, vertebrae, and the ends of long bones is where red bone marrow is stored and, thus, where hemopoiesis (blood cell production) occurs in adults.

Figure 6.3 Histology of compact and spongy bone. (a) Sections through the diaphysis of a long bone, from the surrounding periosteum on the right, to compact bone in the middle, to spongy bone and the medullary cavity on the left. The inset at the upper right shows an osteocyte in a lacuna. (b and c) Details of spongy bone. See Table 4.4J on page 130 for a photomicrograph of compact bone tissue and Figure 6.12a for a scanning electron micrograph of spongy bone tissue.

Osteocytes lie in lacunae arranged in concentric circles around a central (haversian) canal in compact bone and in irregularly arranged lacunae in the trabeculae of spongy bone.

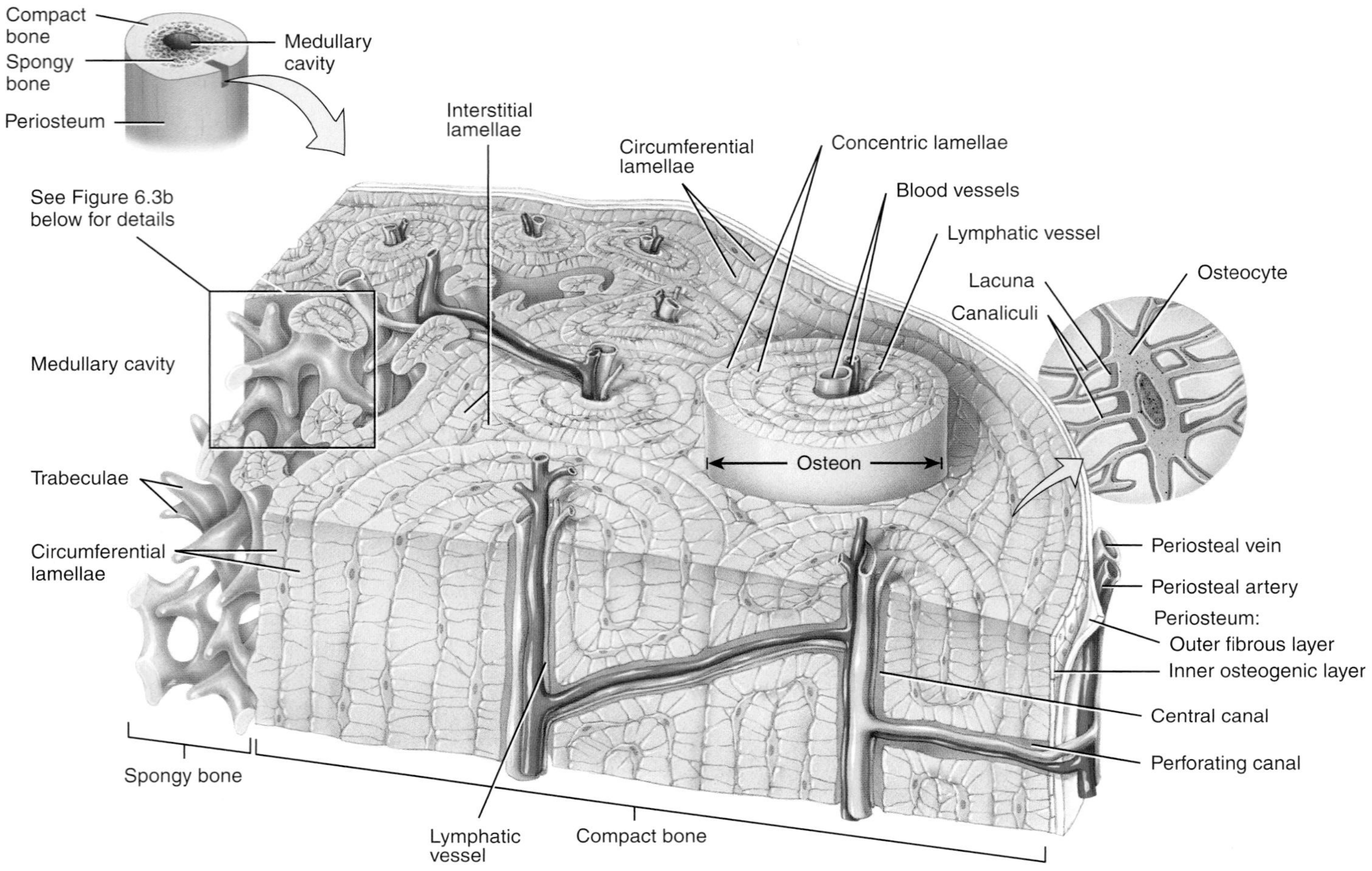

Compact bone
Spongy bone
Periosteum
Medullary cavity
Interstitial lamellae
Circumferential lamellae
Concentric lamellae
Blood vessels
Lymphatic vessel
Lacuna
Canaliculi
Osteocyte
See Figure 6.3b below for details
Medullary cavity
Trabeculae
Circumferential lamellae
Osteon
Periosteal vein
Periosteal artery
Periosteum:
Outer fibrous layer
Inner osteogenic layer
Central canal
Perforating canal
Spongy bone
Lymphatic vessel
Compact bone

(a) Osteons (haversian systems) in compact bone and trabeculae in spongy bone

Space for red bone marrow
Trabeculae
Lacuna
Interstitial lamellae
Canaliculi
Osteocyte
Osteoclast
Osteoblasts aligned along trabeculae of new bone

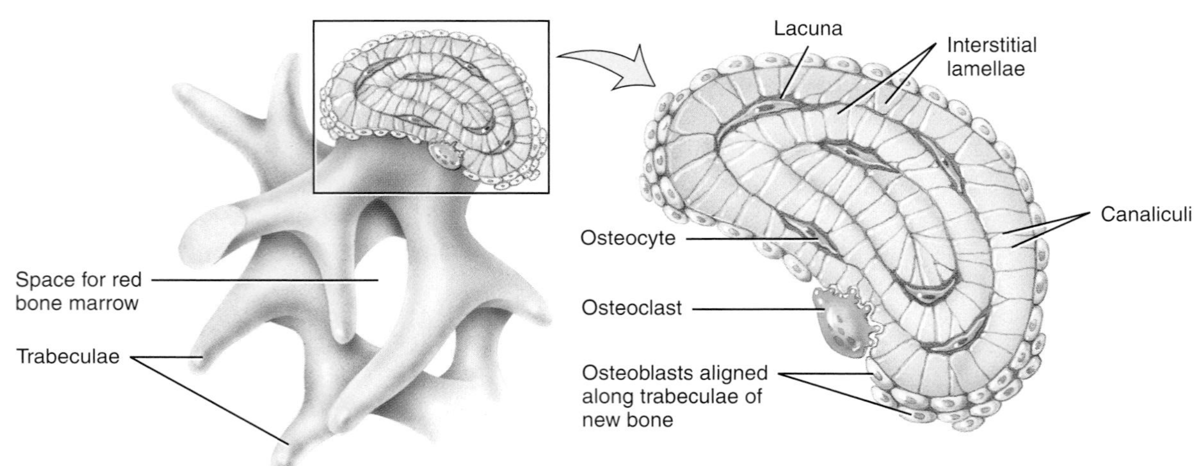

(b) Enlarged aspect of spongy bone trabeculae

(c) Details of a section of a trabecula

As people age, some central (haversian) canals may become blocked. What effect would this have on the surrounding osteocytes?

Bone Scan

A **bone scan** is a diagnostic procedure that takes advantage of the fact that bone is living tissue. A small amount of a radioactive tracer compound that is readily absorbed by bone is injected intravenously. The degree of uptake of the tracer is related to the amount of blood flow to the bone. A scanning device (gamma camera) measures the radiation emitted from the bones, and the information is translated into a photograph that can be read like an X ray on a monitor. Normal bone tissue is identified by a consistent gray color throughout because of its uniform uptake of the radioactive tracer. Darker or lighter areas may indicate bone abnormalities. Darker areas called "hot spots" are areas of increased metabolism that absorb more of the radioactive tracer due to increased blood flow. Hot spots may indicate bone cancer, abnormal healing of fractures, or abnormal bone growth. Lighter areas called "cold spots" are areas of decreased metabolism that absorb less of the radioactive tracer due to decreased blood flow. Cold spots may indicate problems such as degenerative bone disease, decalcified bone, fractures, bone infections, Paget's disease, and rheumatoid arthritis. A bone scan detects abnormalities 3 to 6 months sooner than standard x-ray procedures and exposes the patient to less radiation. A bone scan is the standard test for bone density screening, particularly important in screening females for osteoporosis. ■

▶ CHECKPOINT

4. Why is bone considered a connective tissue?
5. List the four types of cells in bone tissue and their functions.
6. What is the composition of the extracellular matrix of bone tissue?
7. How are spongy and compact bone tissues different in microscopic appearance, location, and function?

BLOOD AND NERVE SUPPLY OF BONE

▶ OBJECTIVE

Describe the blood and nerve supply of bone.

Bone is richly supplied with blood. Blood vessels, which are especially abundant in portions of bone containing red bone marrow, pass into bones from the periosteum. We will consider the blood supply of a long bone such as the mature tibia (shin bone) shown in Figure 6.4.

Periosteal arteries (per-ē-OS-tē-al) accompanied by nerves enter the diaphysis through many perforating (Volkmann's)

Figure 6.4 Blood supply of a mature long bone, the tibia (shin bone).

Bone is richly supplied with blood vessels.

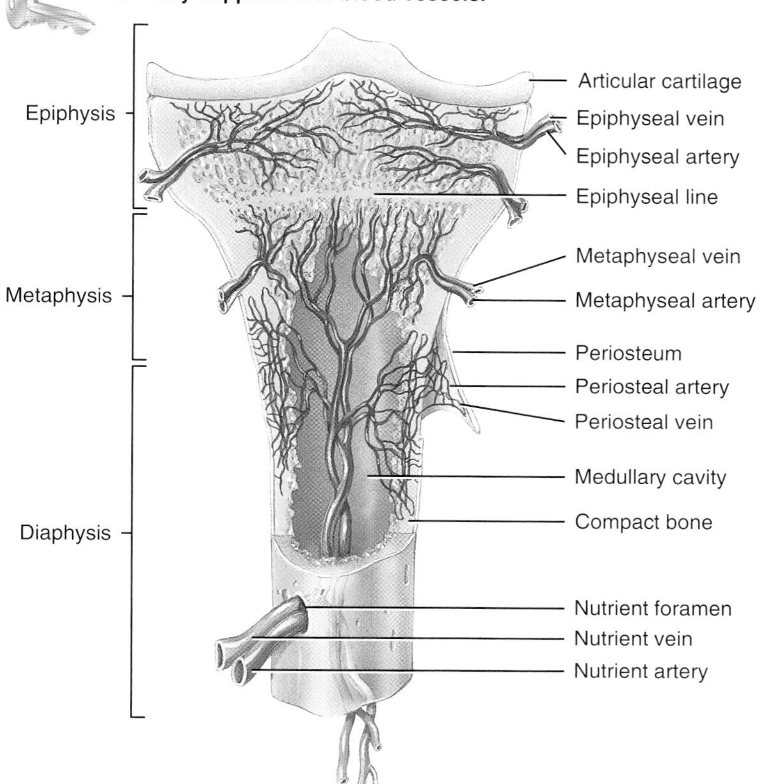

Where do periosteal arteries enter bone tissue?

canals and supply the periosteum and outer part of the compact bone (see Figure 6.3a). Near the center of the diaphysis, a large **nutrient artery** passes through a hole in compact bone called the **nutrient foramen.** On entering the medullary cavity, the nutrient artery divides into proximal and distal branches that supply both the inner part of compact bone tissue of the diaphysis and the spongy bone tissue and red marrow as far as the epiphyseal plates (or lines). Some bones, like the tibia, have only one nutrient artery; others like the femur (thigh bone) have several. The ends of long bones are supplied by the metaphyseal and epiphyseal arteries, which arise from arteries that supply the associated joint. The **metaphyseal arteries** (met-a-FIZ-ē-al) enter the metaphyses of a long bone and, together with the nutrient artery, supply the red bone marrow and bone tissue of the metaphyses. The **epiphyseal arteries** (ep′-i-FIZ-ē-al) enter the epiphyses of a long bone and supply the red bone marrow and bone tissue of the epiphyses.

Veins that carry blood away from long bones are evident in three places: (1) One or two **nutrient veins** accompany the nutrient artery in the diaphysis; (2) numerous **epiphyseal veins** and **metaphyseal veins** exit with their respective arteries in the epiphyses; and (3) many small **periosteal veins** exit with their respective arteries in the periosteum.

Nerves accompany the blood vessels that supply bones. The periosteum is rich in sensory nerves, some of which carry pain sensations. These nerves are especially sensitive to tearing or tension, which explains the severe pain resulting from a fracture or a bone tumor. For the same reason there is some pain associated with a bone marrow needle biopsy. In this procedure, a needle is inserted into the middle of the bone to withdraw a sample of red bone marrow to examine it for conditions such as leukemias, metastatic neoplasms, lymphoma, Hodgkin's disease, and aplastic anemia. As the needle penetrates the periosteum, pain is felt. Once it passes through, there is little pain.

▶ CHECKPOINT

8. Explain the location and roles of the nutrient arteries, nutrient foramina, epiphyseal arteries, and periosteal arteries.
9. Which part of a bone contains sensory nerves associated with pain? Describe one situation in which this is important.

BONE FORMATION

▶ OBJECTIVE

Describe the steps of intramembranous and endochondral ossification.

The process by which bone forms is called **ossification** (os′-i-fi-KĀ-shun; *ossi-* = bone; *-fication* = making) or **osteogenesis** (os′-tē-ō-JEN-e-sis). The "skeleton" of a human embryo is composed of loose mesenchymal cells, which are shaped like bones and are the sites where ossification occurs. These "bones" provide the template for subsequent ossification, which begins during the sixth week of embryonic development and follows one of two patterns.

The two methods of bone formation, which both involve the replacement of a preexisting connective tissue with bone, do not lead to differences in the structure of mature bones, but are simply different methods of bone development. In the first type of ossification, called **intramembranous ossification** (in′-tra-MEM-bra-nus; *intra-* = within; *membran-* = membrane), bone forms directly within mesenchyme arranged in sheetlike layers that resemble membranes. In the second type, **endochondral ossification** (en′-dō-KON-dral; *endo-* = within; *-chondral* = cartilage), bone forms within hyaline cartilage that develops from mesenchyme.

Intramembranous Ossification

Intramembranous ossification is the simpler of the two methods of bone formation. The flat bones of the skull and mandible (lower jawbone) are formed in this way. Also, the "soft spots" that help the fetal skull pass through the birth canal later harden as they undergo intramembranous ossification, which occurs as follows (Figure 6.5):

1. ***Development of the ossification center.*** At the site where the bone will develop, specific chemical messages cause the mesenchymal cells to cluster together and differentiate, first into osteogenic cells and then into osteoblasts. (Recall that *mesenchyme* is the tissue from which all other connective tissues arise.) The site of such a cluster is called an **ossification center.** Osteoblasts secrete the organic extracellular matrix of bone until they are surrounded by it.
2. ***Calcification.*** Next, the secretion of extracellular matrix stops and the cells, now called osteocytes, lie in lacunae and extend their narrow cytoplasmic processes into canaliculi that radiate in all directions. Within a few days, calcium and other mineral salts are deposited and the extracellular matrix hardens or calcifies (calcification).
3. ***Formation of trabeculae.*** As the bone extracellular matrix forms, it develops into trabeculae that fuse with one another to form spongy bone. Blood vessels grow into the spaces between the trabeculae. Connective tissue that is associated with the blood vessels in the trabeculae differentiates into red bone marrow.
4. ***Development of the periosteum.*** At the periphery of the bone, the mesenchyme condenses and develops into the periosteum. Eventually, a thin layer of compact bone replaces the surface layers of the spongy bone, but spongy bone remains in the center. Much of the newly formed bone is remodeled (destroyed and reformed) as the bone is transformed into its adult size and shape.

Figure 6.5 Intramembranous ossification. Illustrations ① and ② show a smaller field of vision at higher magnification than Illustrations ③ and ④. Refer to this figure as you read the corresponding numbered paragraphs in the text.

Intramembranous ossification involves the formation of bone within mesenchyme arranged in sheetlike layers that resemble membranes.

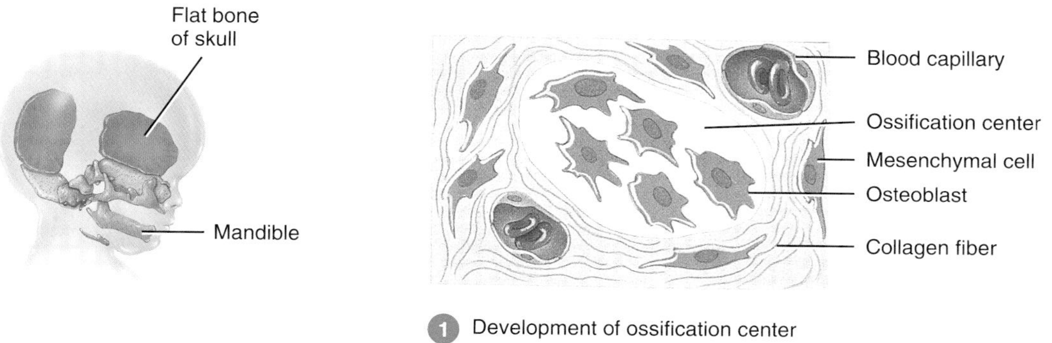

① Development of ossification center

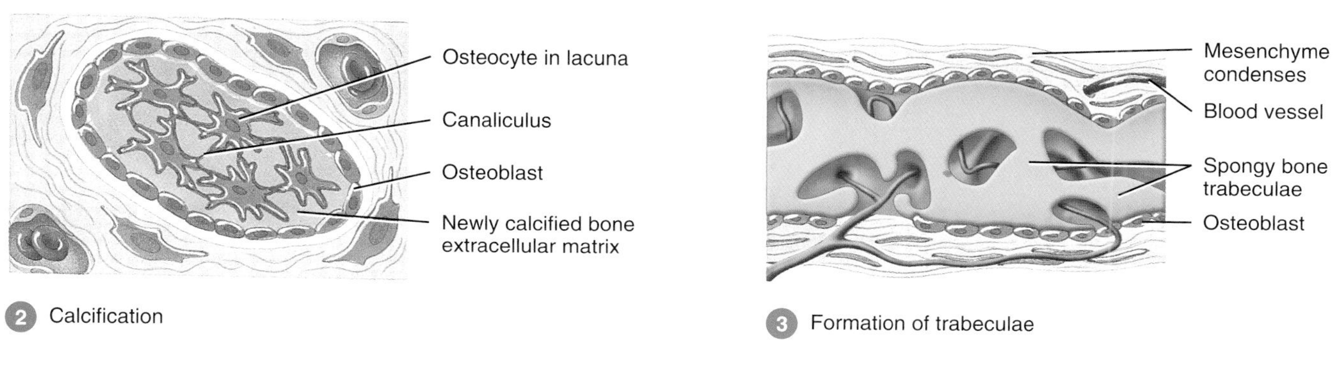

② Calcification

③ Formation of trabeculae

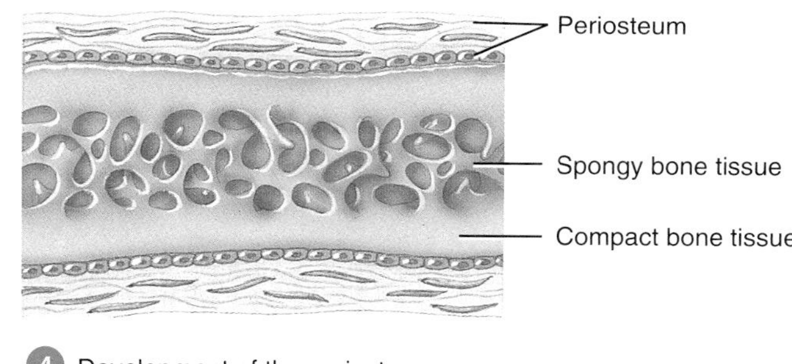

④ Development of the periosteum

Which bones of the body develop by intramembranous ossification?

Endochondral Ossification

The replacement of cartilage by bone is called **endochondral ossification.** Although most bones of the body are formed in this way, the process is best observed in a long bone. It proceeds as follows (Figure 6.6):

1 ***Development of the cartilage model.*** At the site where the bone is going to form, specific chemical messages cause the mesenchymal cells to crowd together in the shape of the future bone, and then develop into chondroblasts. The chondroblasts secrete cartilage extracellular matrix,

Figure 6.6 Endochondral ossification.

During endochondral ossification, bone gradually replaces a cartilage model.

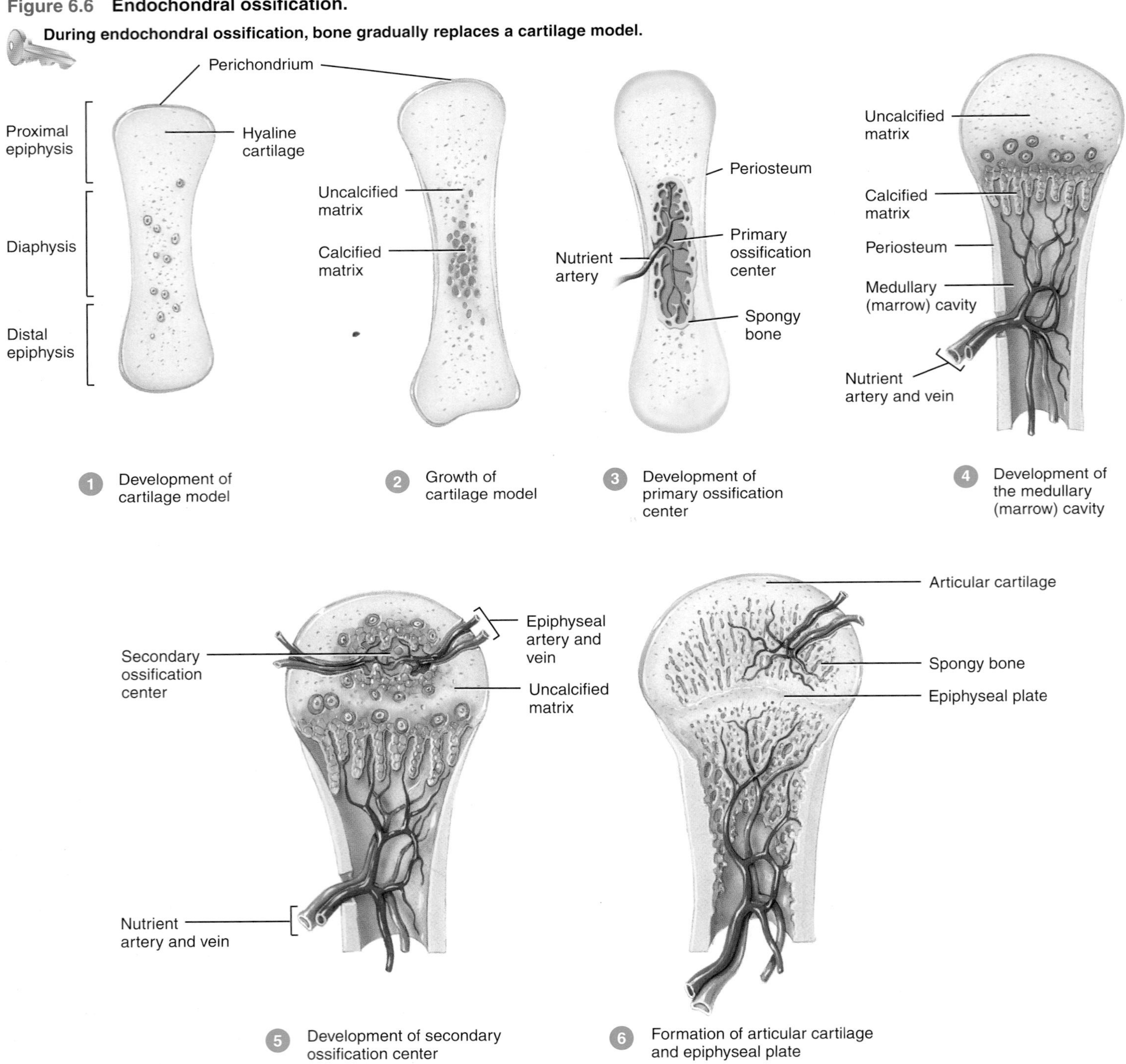

If radiographs of an 18-year-old basketball player show clear epiphyseal plates but no epiphyseal lines, is she likely to grow taller?

producing a **cartilage model** consisting of hyaline cartilage. A membrane called the **perichondrium** (per-i-KON-drē-um) develops around the cartilage model.

2 ***Growth of the cartilage model.*** Once chondroblasts become deeply buried in the cartilage extracellular matrix, they are called chondrocytes. The cartilage model grows in length by continual cell division of chondrocytes accompanied by further secretion of the cartilage extracellular matrix. This type of growth is termed **interstitial growth** and results in an increase in length. In contrast, growth of the cartilage in thickness is due mainly to the addition of more extracellular matrix material to the periphery of the model by new chondroblasts that develop from the perichondrium. This growth pattern, in which extracellular matrix is deposited on the cartilage surface, is called **appositional growth** (a-pō-ZISH-i-nal).

As the cartilage model continues to grow, chondrocytes in its mid-region hypertrophy (increase in size) and the surrounding cartilage extracellular matrix begins to calcify. Other chondrocytes within the calcifying cartilage die because nutrients can no longer diffuse quickly enough through the extracellular matrix. As these chondrocytes die, lacunae form and eventually merge into small cavities.

3 ***Development of the primary ossification center.*** Primary ossification proceeds *inward* from the external surface of the bone. A nutrient artery penetrates the perichondrium and the calcifying cartilage model through a nutrient foramen in the mid-region of the cartilage model, stimulating osteogenic cells in the perichondrium to differentiate into osteoblasts. Once the perichondrium starts to form bone, it is known as the **periosteum.** Near the middle of the model, periosteal capillaries grow into the disintegrating calcified cartilage, inducing growth of a **primary ossification center,** a region where bone tissue will replace most of the cartilage. Osteoblasts then begin to deposit bone extracellular matrix over the remnants of calcified cartilage, forming spongy bone trabeculae.

4 ***Development of the medullary (marrow) cavity.*** As the primary ossification center grows toward the ends of the bone, osteoclasts break down some of the newly formed spongy bone trabeculae. This activity leaves a cavity, the medullary (marrow) cavity, in the diaphysis (shaft). Eventually, most of the wall of the diaphysis is replaced by compact bone.

5 ***Development of the secondary ossification centers.*** When branches of the epiphyseal artery enter the epiphyses, **secondary ossification centers** develop, usually around the time of birth. Bone formation is similar to that in primary ossification centers. One difference, however, is that spongy bone remains in the interior of the epiphyses (no medullary cavities are formed there). In contrast to primary ossification, secondary ossification proceeds *outward* from the center of the epiphysis toward the outer surface of the bone.

6 ***Formation of articular cartilage and the epiphyseal plate.*** The hyaline cartilage that covers the epiphyses becomes the articular cartilage. Prior to adulthood, hyaline cartilage remains between the diaphysis and epiphysis as the **epiphyseal plate,** which is responsible for the lengthwise growth of long bones.

▶ CHECKPOINT

10. What are the major events of intramembranous ossification and endochondral ossification and how are they different?

BONE GROWTH

▶ OBJECTIVES

Describe how bone grows in length and thickness.

Explain the role of nutrients and hormones in regulating bone growth.

During childhood, bones throughout the body grow in thickness by appositional growth, and long bones lengthen by the addition of bone material on the diaphyseal side of the epiphyseal plate by interstitial growth.

Growth in Length

To understand how a bone grows in length, you need to know some of the details of the structure of the epiphyseal plate (Figure 6.7). The **epiphyseal plate** (ep-i-FIZ-ē-al) is a layer of hyaline cartilage in the metaphysis of a growing bone that consists of four zones (Figure 6.7b):

1. ***Zone of resting cartilage.*** This layer is nearest the epiphysis and consists of small, scattered chondrocytes. The term "resting" is used because the cells do not function in bone growth. Rather, they anchor the epiphyseal plate to the epiphysis of the bone.
2. ***Zone of proliferating cartilage.*** Slightly larger chondrocytes in this zone are arranged like stacks of coins. These chondrocytes divide to replace those that die at the diaphyseal side of the epiphyseal plate.
3. ***Zone of hypertrophic cartilage*** (hī-per-TRŌ-fik). This layer consists of large, maturing chondrocytes arranged in columns.
4. ***Zone of calcified cartilage.*** The final zone of the epiphyseal plate is only a few cells thick and consists mostly of chondrocytes that are dead because the extracellular matrix around them has calcified. Osteoclasts dissolve the calcified cartilage, and osteoblasts and capillaries from the diaphysis invade the area. The osteoblasts lay down bone extracellular matrix, replacing the calcified cartilage. As a result, the zone of calcified cartilage becomes "new diaphysis" that is firmly cemented to the rest of the diaphysis of the bone.

The activity of the epiphyseal plate is the only way that the diaphysis can increase in length. As a bone grows, new chondrocytes are formed on the epiphyseal side of the plate, while old chondrocytes on the diaphyseal side of the plate are replaced by bone. In this way the thickness of the epiphyseal plate remains relatively constant, but the bone on the diaphyseal side increases in length.

At about age 18 in females and 21 in males, the epiphyseal plates close; the epiphyseal cartilage cells stop dividing, and bone replaces all the cartilage. The epiphyseal plate fades, leaving a bony structure called the **epiphyseal line.** The appearance of the epiphyseal line signifies that the bone has stopped growing in length. The clavicle is the last bone to stop growing. If a bone fracture damages the epiphyseal plate, the fractured bone may be shorter than normal once adult stature is reached. This is because damage to cartilage accelerates closure of the epiphyseal plate, thus inhibiting lengthwise growth of the bone.

Figure 6.7 **The epiphyseal plate is a layer of hyaline cartilage in the metaphysis of a growing bone.** The epiphyseal plate appears as a dark band between whiter calcified areas in the radiograph shown in part (a).

The epiphyseal plate allows the diaphysis of a bone to increase in length.

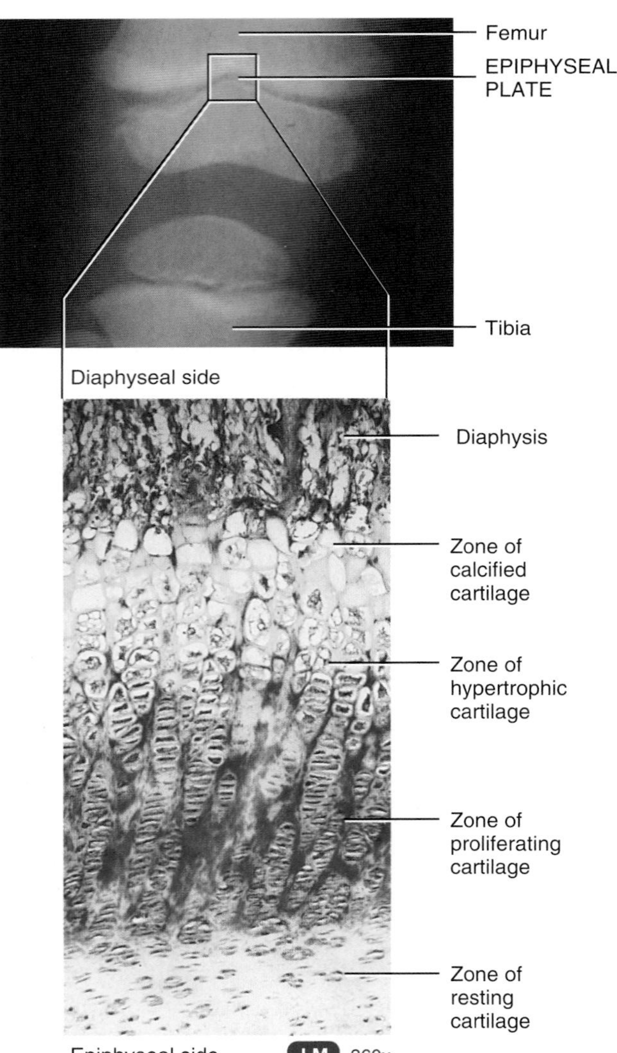

What activities of the epiphyseal plate account for the lengthwise growth of the diaphysis?

Growth in Thickness

Unlike cartilage, which can thicken by both interstitial and appositional growth, bone can grow in thickness (diameter) only by **appositional growth** (Figure 6.8):

1. At the bone surface, cells in the periosteum differentiate into osteoblasts, which secrete collagen fibers and other organic molecules that form bone extracellular matrix. The osteoblasts become surrounded by extracellular matrix and develop into osteocytes. This process forms bone ridges on either side of a periosteal blood vessel. The ridges slowly enlarge and create a groove for the periosteal blood vessel.
2. Eventually, the ridges fold together and fuse, and the groove becomes a tunnel that encloses the blood vessel. The former periosteum now becomes the endosteum that lines the tunnel.
3. Osteoblasts in the endosteum deposit bone extracellular matrix, forming new concentric lamellae. The formation of additional concentric lamellae proceeds inward toward the periosteal blood vessel. In this way, the tunnel fills in, and a new osteon is created.
4. As an osteon is forming, osteoblasts under the periosteum deposit new outer circumferential lamellae, further increasing the thickness of the bone. As additional periosteal blood vessels become enclosed as in step 2, the growth process continues.

As new bone tissue is deposited on the outer surface of bone, the bone tissue lining the medullary cavity is destroyed by osteoclasts in the endosteum. In this way, the medullary cavity enlarges as the bone increases in thickness.

▶ CHECKPOINT

11. Describe the zones of the epiphyseal plate and their functions, and the significance of the epiphyseal line.
12. How is bone growth in length different from bone growth in thickness?
13. How could the metaphyseal area of a bone help determine the age of a skeleton?

Figure 6.8 Bone growth in thickness: appositional growth.

Cartilage can grow by both interstitial and appositional growth, but bone can grow in diameter only by appositional growth.

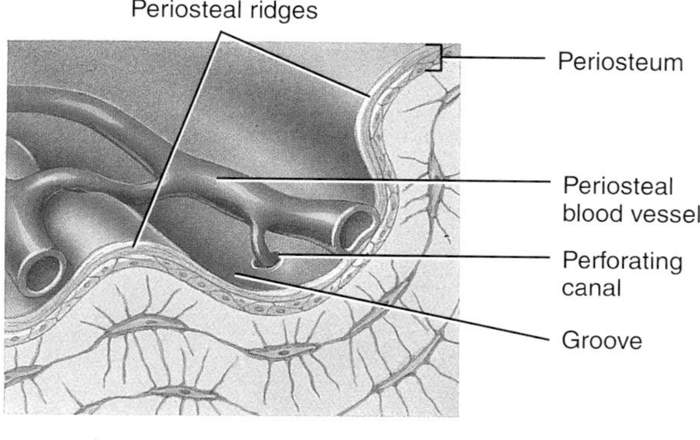

1 Ridges in periosteum create groove for periosteal blood vessel.

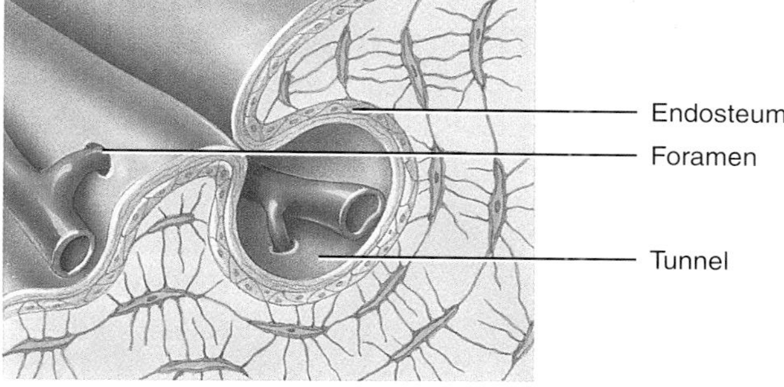

2 Periosteal ridges fuse, forming an endosteum-lined tunnel.

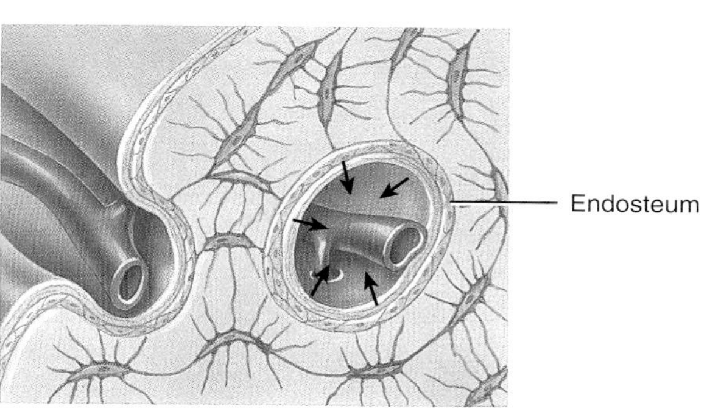

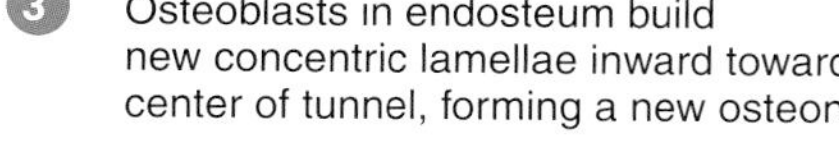

3 Osteoblasts in endosteum build new concentric lamellae inward toward center of tunnel, forming a new osteon.

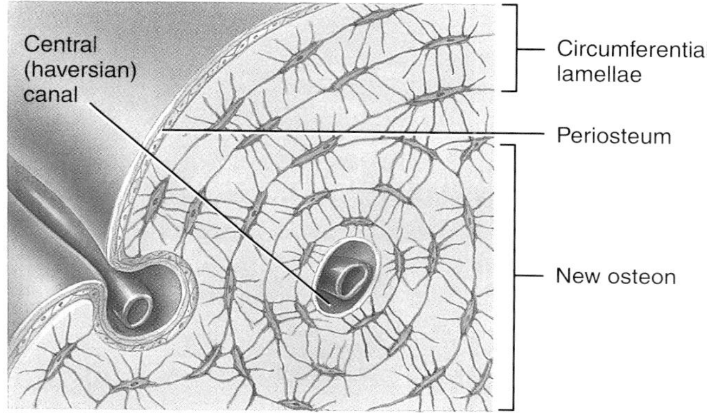

4 Bone grows outward as osteoblasts in periosteum build new circumferential lamellae. Osteon formation repeats as new periosteal ridges fold over blood vessels.

How does the medullary cavity enlarge during growth in thickness?

BONES AND HOMEOSTASIS

► **OBJECTIVES**

Describe the processes involved in bone remodeling.

Describe the sequence of events in repair of a fracture.

Describe the role of bone in calcium homeostasis.

Bone Remodeling

Like skin, bone forms before birth but continually renews itself thereafter. **Bone remodeling** is the ongoing replacement of old bone tissue by new bone tissue. It involves **bone resorption,** the removal of minerals and collagen fibers from bone by osteoclasts, and **bone deposition,** the addition of minerals and collagen fibers to bone by osteoblasts. Thus, bone resorption results in the destruction of bone extracellular matrix, while bone deposition results in the formation of bone extracellular matrix. At any given time, about 5% of the total bone mass in the body is remodeled. The renewal rate for compact bone tissue is about 4% per year and for spongy bone tissue it is about 20% per year. Remodeling also takes place at different rates in different regions of the body. The distal portion of the thighbone (femur) is replaced about every four months. By contrast, bone in certain areas of the shaft of the femur will not be replaced completely during an individual's life. Even after bones have reached their adult shapes and sizes, old bone is continually destroyed and new bone is formed in its place. Remodeling also removes injured bone, replacing it with new bone tissue. Remodeling may be triggered by factors such as exercise, sedentary lifestyle, and changes in diet.

Remodeling has several other benefits. Since the strength of bone is related to the degree to which it is stressed, if newly formed bone is subjected to heavy loads, it will grow thicker and therefore be stronger than the old bone. Also, the shape of a bone can be altered for proper support based on the stress patterns experienced during the remodeling process. Finally, new bone is more resistant to fracture than old bone.

Remodeling and Orthodontics

Orthodontics (or-thō-DON-tiks) is the branch of dentistry concerned with the prevention and correction of poorly aligned teeth. The movement of teeth by braces places a stress on the bone that forms the sockets that anchor the teeth. In response to this artificial stress, osteoclasts and osteoblasts remodel the sockets so that the teeth align properly. ■

During the process of bone resorption, an osteoclast attaches tightly to the bone surface at the endosteum or periosteum and forms a leakproof seal at the edges of its ruffled border (see Figure 6.2). Then it releases protein-digesting lysosomal enzymes and several acids into the sealed pocket. The enzymes digest collagen fibers and other organic substances while the acids dissolve the bone minerals. Working together, several osteoclasts carve out a small tunnel in the old bone. The degraded bone proteins and extracellular matrix minerals, mainly calcium and phosphorus, enter an osteoclast by endocytosis, cross the cell in vesicles, and undergo exocytosis on the side opposite the ruffled border. Now in the interstitial fluid, the products of bone resorption diffuse into nearby blood capillaries. Once a small area of bone has been resorbed, osteoclasts depart and osteoblasts move in to rebuild the bone in that area.

A delicate balance exists between the actions of osteoclasts and osteoblasts. Should too much new tissue be formed, the bones become abnormally thick and heavy. If too much mineral material is deposited in the bone, the surplus may form thick bumps, called *spurs,* on the bone that interfere with movement at joints. Excessive loss of calcium or tissue weakens the bones, and they may break, as occurs in osteoporosis, or they may become too flexible, as in rickets and osteomalacia. (For more on these disorders, see the Disorders: Homeostatic Imbalances section at the end of the chapter.) Abnormal acceleration of the remodeling process results in a condition called Paget's disease, in which the newly formed bone, especially that of the pelvis, limbs, lower vertebrae, and skull, becomes hard and brittle and fractures easily.

Factors Affecting Bone Growth and Bone Remodeling

Normal bone metabolism—growth in the young and bone remodeling in the adult—depends on several factors. These include adequate dietary intake of minerals and vitamins, as well as sufficient levels of several hormones.

1. *Minerals.* Large amounts of calcium and phosphorus are needed while bones are growing, as are smaller amounts of fluoride, magnesium, iron, and manganese. These minerals are also necessary during bone remodeling.

2. *Vitamins.* Vitamin C is needed for synthesis of collagen, the main bone protein, and also for differentiation of osteoblasts into osteocytes. Vitamins K and B_{12} also are needed for protein synthesis, whereas vitamin A stimulates activity of osteoblasts.

3. *Hormones.* During childhood, the hormones most important to bone growth are the insulinlike growth factors (IGFs), which are produced by the liver and bone tissue (see page 628). IGFs stimulate osteoblasts, promote cell division at the epiphyseal plate and in the periosteum, and enhance synthesis of the proteins needed to build new bone. IGFs are produced in response to the secretion of human growth hormone (hGH) from the anterior lobe of the pituitary gland (see page 629). Thyroid hormones (T_3 and T_4) from the thyroid gland also promote bone growth by stimulating osteoblasts.

At puberty, the secretion of hormones known as sex hormones causes a dramatic effect on bone growth. The **sex hormones** include estrogens (produced by the ovaries) and androgens such as testosterone (produced by the testes). Although females have much higher levels of estrogens and males have higher levels of androgens, females also have low levels of androgens, and males have low levels of estrogens. The adrenal glands of both sexes produce androgens, and other tissues, such as adipose tissue, can convert androgens to estrogens. These hormones are responsible for increased osteoblast activity and synthesis of bone extracellular matrix and the sudden "growth spurt" that occurs during the teenage years. Estrogens also promote changes in the skeleton that are typical of females, such as widening of the pelvis. Ultimately sex hormones, especially estrogens in both sexes, shut down growth at epiphyseal plates, causing elongation of the bones to cease. Lengthwise growth of bones typically ends earlier in females than in males due to their higher levels of estrogens.

During adulthood, sex hormones contribute to bone remodeling by slowing resorption of old bone and promoting deposition of new bone. One way that estrogens slow resorption is by promoting apoptosis (programmed death) of osteoclasts. As you will see shortly, parathyroid hormone, calcitriol (the active form of vitamin D), and calcitonin are other hormones that can affect bone remodeling.

Hormonal Abnormalities that Affect Height

Excessive or deficient secretion of hormones that normally govern bone growth can cause a person to be abnormally tall or short. Oversecretion of hGH during childhood produces giantism, in which a person becomes much taller and heavier than normal. Undersecretion of hGH produces pituitary dwarfism, in which a person has short stature. Because estrogens terminate growth at the epiphyseal plates, both men and women who lack estrogens or receptors for estrogens grow taller than normal. ■

Fracture and Repair of Bone

A **fracture** is any break in a bone. Fractures are named according to their severity, the shape or position of the fracture line, or even the physician who first described them. Among the common kinds of fractures are the following (Figure 6.9):

Figure 6.9 Types of bone fractures. Illustrations are shown on the left and radiographs are shown on the right.

A fracture is any break in a bone.

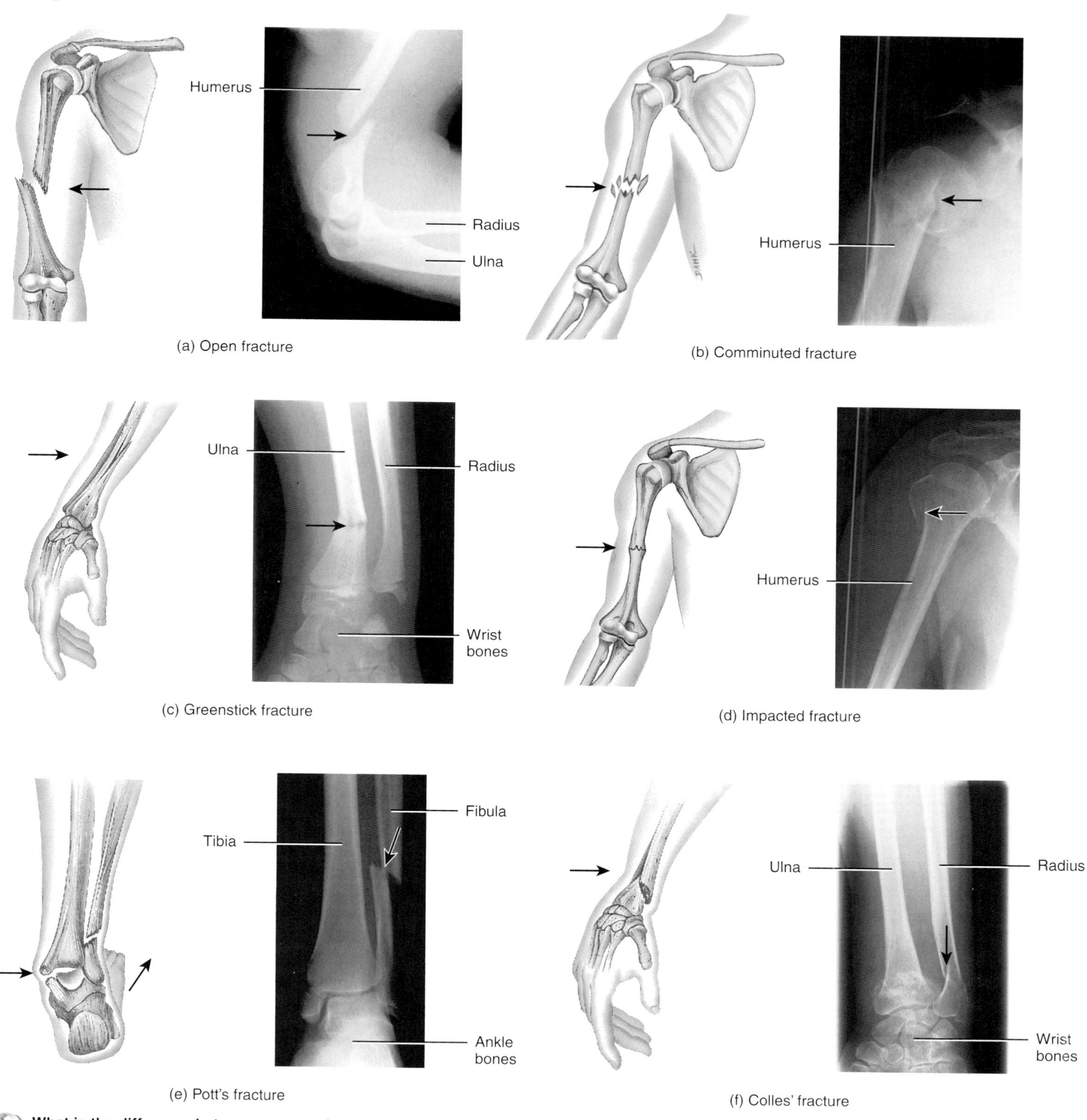

What is the difference between an open fracture and a closed fracture?

- **Open (compound) fracture:** The broken ends of the bone protrude through the skin (Figure 6.9a). Conversely, a **closed (simple) fracture** does not break the skin.
- **Comminuted fracture** (KOM-i-noo-ted; *com-* = together; *-minuted* = crumbled): The bone splinters at the site of impact, and smaller bone fragments lie between the two main fragments (Figure 6.9b).
- **Greenstick fracture:** A partial fracture in which one side of the bone is broken and the other side bends; occurs only in children, whose bones are not yet fully ossified and contain more organic material than inorganic material (Figure 6.9c).
- **Impacted fracture:** One end of the fractured bone is forcefully driven into the interior of the other (Figure 6.9d).
- **Pott's fracture:** A fracture of the distal end of the lateral leg bone (fibula), with serious injury of the distal tibial articulation (Figure 6.9e).
- **Colles' fracture** (KOL-ez): A fracture of the distal end of the lateral forearm bone (radius) in which the distal fragment is displaced posteriorly (Figure 6.9f).

In some cases, a bone may fracture without visibly breaking. A **stress fracture** is a series of microscopic fissures in bone that forms without any evidence of injury to other tissues. In healthy adults, stress fractures result from repeated, strenuous activities such as running, jumping, or aerobic dancing. Stress fractures also result from disease processes that disrupt normal bone calcification, such as osteoporosis (discussed on page 189). About 25% of stress fractures involve the tibia. Although standard x-ray images often fail to reveal the presence of stress fractures, they show up clearly in a bone scan.

The repair of a bone fracture involves the following steps (Figure 6.10):

1. *Formation of fracture hematoma.* Blood vessels crossing the fracture line are broken. As blood leaks from the torn ends of the vessels, it forms a clot around the site of the fracture. This clot, called a **fracture hematoma** (hē′-ma-TŌ-ma; *hemat-* = blood; *-oma* = tumor), usually forms 6 to 8 hours after the injury. Because the circulation of blood stops at the site where the fracture hematoma forms, nearby bone cells die. Swelling and inflammation occur in response to dead bone cells, producing additional cellular debris. Phagocytes (neutrophils and macrophages) and osteoclasts begin to remove the dead or damaged tissue in and around the fracture hematoma. This stage may last up to several weeks.
2. *Fibrocartilaginous callus formation.* Fibroblasts from the periosteum invade the fracture site and produce collagen fibers. In addition, cells from the periosteum develop into chondroblasts and begin to produce fibrocartilage in this region. These events lead to the development of a **fibrocartilaginous callus** (fī-brō-kar-ti-LAJ-i-nus), a mass of repair tissue consisting of collagen fibers and cartilage that bridges the broken ends of the bone. Formation of the fibrocartilaginous callus takes about 3 weeks.
3. *Bony callus formation.* In areas closer to well-vascularized healthy bone tissue, osteogenic cells develop into osteoblasts, which begin to produce spongy bone trabeculae. The trabeculae join living and dead portions of the original bone fragments. In time, the fibrocartilage is converted to spongy bone, and the callus is then referred to as a **bony callus.** The bony callus lasts about 3 to 4 months.
4. *Bone remodeling.* The final phase of fracture repair is **bone remodeling** of the callus. Dead portions of the original fragments of broken bone are gradually resorbed by osteoclasts. Compact bone replaces spongy bone around the periphery of the fracture. Sometimes, the repair process is so thorough that the fracture line is undetectable, even in a radiograph (X ray). However, a thickened area on the surface of the bone remains as evidence of a healed fracture, and eventually a healed bone may be stronger than it was before the break.

Although bone has a generous blood supply, healing sometimes takes months. The calcium and phosphorus needed to strengthen and harden new bone are deposited only gradually, and bone cells generally grow and reproduce slowly. The temporary disruption in their blood supply also helps explain the slowness of healing of severely fractured bones.

Treatments for Fractures

Treatments for fractures vary according to age, type of fracture, and the bone involved. The ultimate goals of fracture treatment are realignment of the bone fragments, immobilization to maintain realignment, and restoration of function. For bones to unite properly, the fractured ends must be brought into alignment, a process called **reduction.** In **closed reduction,** the fractured ends of a bone are brought into alignment by manual manipulation, and the skin remains intact. In **open reduction,** the fractured ends of a bone are brought into alignment by a surgical procedure in which internal fixation devices such as screws, plates, pins, rods, and wires are used. Following reduction, a fractured bone may be kept immobilized by a cast, sling, splint, elastic bandage, external fixation device, or a combination of these devices. ■

Bone's Role in Calcium Homeostasis

Bone is the body's major calcium reservoir, storing 99% of total body calcium. One way to maintain the level of calcium in the blood is to control the rates of calcium resorption from bone into blood and of calcium deposition from blood into bone. Both nerve and muscle cells depend on a stable level of calcium ions (Ca^{2+}) in extracellular fluid to function properly. Blood clotting also requires Ca^{2+}. Also, many enzymes require Ca^{2+} as a cofactor (an additional substance needed for an enzymatic reaction to occur). For this reason, the blood plasma level of Ca^{2+} is very

Figure 6.10 **Steps in repair of a bone fracture.**

Bone heals more rapidly than cartilage because its blood supply is more plentiful.

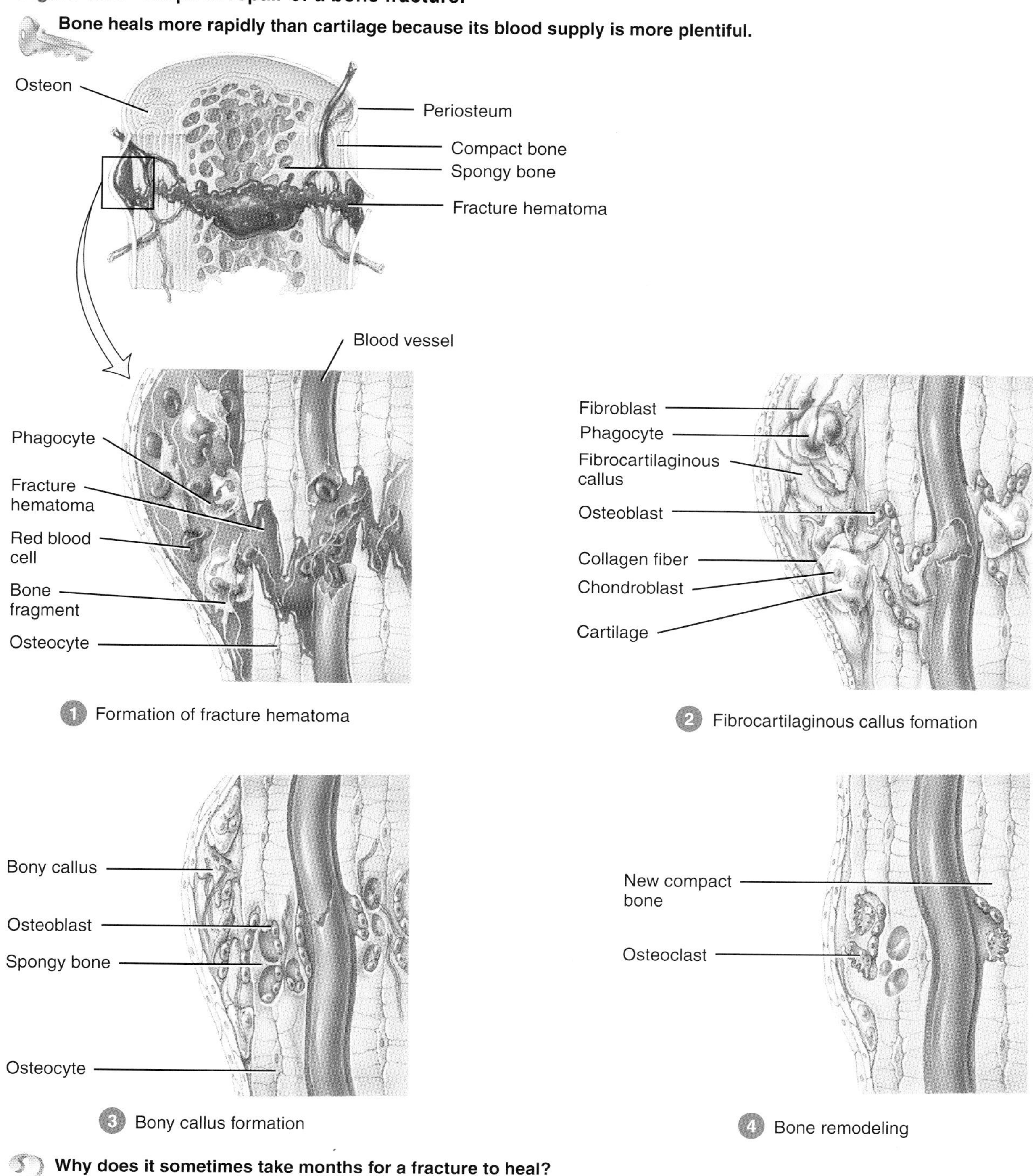

Why does it sometimes take months for a fracture to heal?

closely regulated between 9 and 11 mg/100 mL. Even small changes in Ca^{2+} concentration outside this range may prove fatal—the heart may stop (cardiac arrest) if the concentration goes too high, or breathing may cease (respiratory arrest) if the level falls too low. The role of bone in calcium homeostasis is to help "buffer" the blood Ca^{2+} level, releasing Ca^{2+} into blood plasma (using osteoclasts) when the level decreases, and absorbing Ca^{2+} (using osteoblasts) when the level rises.

Ca^{2+} exchange is regulated by hormones, the most important of which is **parathyroid hormone (PTH)** secreted by the parathyroid glands (see Figure 18.13 on page 622). This hormone increases blood Ca^{2+} level. PTH secretion operates via a negative feedback system (Figure 6.11). If some stimulus causes the blood Ca^{2+} level to decrease, parathyroid gland cells (receptors) detect this change and increase their production of a molecule known as cyclic adenosine monophosphate (cyclic AMP). The gene for PTH within the nucleus of a parathyroid gland cell (the control center) detects the intracellular increase in cyclic AMP (the input). As a result, PTH synthesis speeds up, and more PTH (the output) is released into the blood. The presence of higher levels of PTH increases the number and activity of osteoclasts (effectors), which step up the pace of bone resorption. The resulting release of Ca^{2+} from bone into blood returns the blood Ca^{2+} level to normal.

Figure 6.11 Negative feedback system for the regulation of blood calcium (Ca^{2+}) concentration. PTH = parathyroid hormone.

Release of calcium from bone matrix and retention of calcium by the kidneys are the two main ways that blood calcium level can be increased.

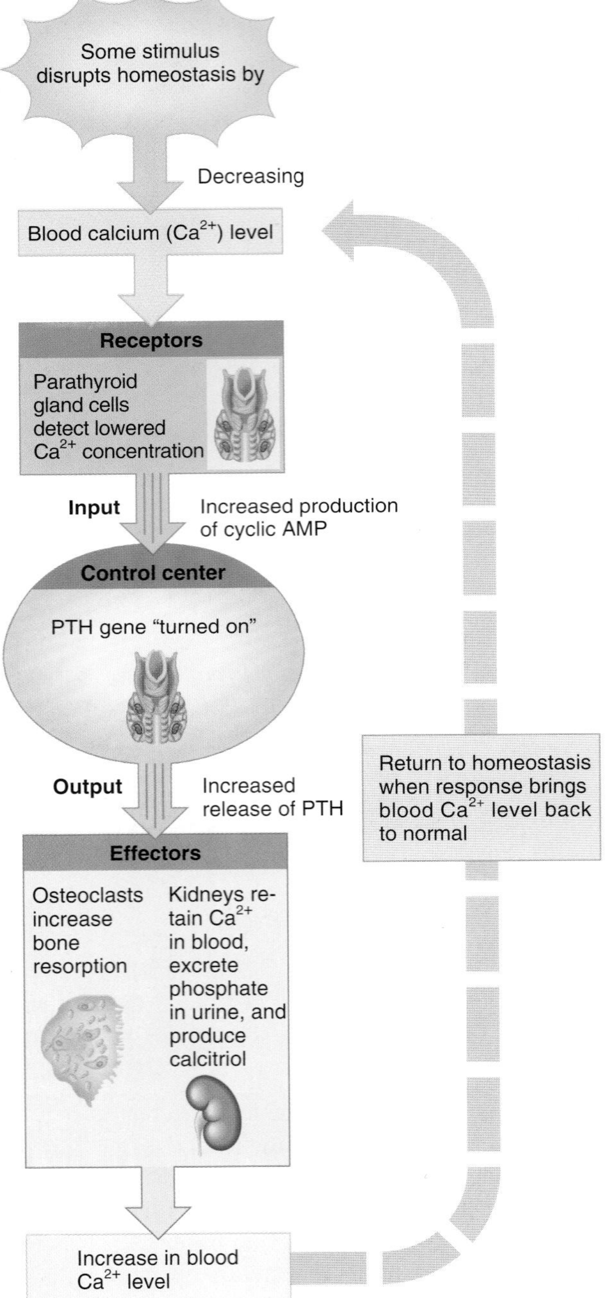
Some stimulus disrupts homeostasis by
Decreasing
Blood calcium (Ca^{2+}) level
Receptors
Parathyroid gland cells detect lowered Ca^{2+} concentration
Input
Increased production of cyclic AMP
Control center
PTH gene "turned on"
Output
Increased release of PTH
Effectors
Osteoclasts increase bone resorption
Kidneys retain Ca^{2+} in blood, excrete phosphate in urine, and produce calcitriol
Increase in blood Ca^{2+} level

Return to homeostasis when response brings blood Ca^{2+} level back to normal

What body functions depend on proper levels of Ca^{2+}?

PTH also acts on the kidneys (effectors) to decrease loss of Ca^{2+} in the urine, so more is retained in the blood. And PTH stimulates formation of **calcitriol** (the active form of vitamin D), a hormone that promotes absorption of calcium from foods in the gastrointestinal tract into the blood. Both of these actions also help elevate blood Ca^{2+} level.

Another hormone works to decrease blood Ca^{2+} level. When blood Ca^{2+} rises above normal, *parafollicular cells* in the thyroid gland secrete **calcitonin (CT).** CT inhibits activity of osteoclasts, speeds blood Ca^{2+} uptake by bone, and accelerates Ca^{2+} deposition into bones. The net result is that CT promotes bone formation and decreases blood Ca^{2+} level. Despite these effects, the role of CT in normal calcium homeostasis is uncertain because it can be completely absent without causing symptoms. Nevertheless, calcitonin harvested from salmon (Miacalcin®) is an effective drug for treating osteoporosis because it slows bone resorption.

Figure 18.14 on page 640 summarizes the roles of parathyroid hormone, calcitriol, and calcitonin in regulation of blood Ca^{2+} level.

▶ **CHECKPOINT**

14. Define remodeling, and describe the roles of osteoblasts and osteoclasts in the process.

15. What factors affect bone growth and bone remodeling?

16. List the types of fractures and outline the four steps involved in fracture repair.

17. How do hormones act on bone to regulate calcium homeostasis?

EXERCISE AND BONE TISSUE

▶ **OBJECTIVE**

Describe how exercise and mechanical stress affect bone tissue.

Within limits, bone tissue has the ability to alter its strength in response to changes in mechanical stress. When placed under stress, bone tissue becomes stronger through increased deposition of mineral salts and production of collagen fibers by osteoblasts. Without mechanical stress, bone does not remodel normally because bone resorption occurs more quickly than bone formation.

The main mechanical stresses on bone are those that result from the pull of skeletal muscles and the pull of gravity. If a person is bedridden or has a fractured bone in a cast, the strength of the unstressed bones diminishes because of the loss of bone minerals and decreased numbers of collagen fibers. Astronauts subjected to the microgravity of space also lose bone mass. In both cases, bone loss can be dramatic—as much as 1% per week. Bones of athletes, which are repetitively and highly stressed, become notably thicker and stronger than those of nonathletes. Weight-bearing activities, such as walking or moderate weight lifting, help build and retain bone mass. Adolescents and young adults should engage in regular weight-bearing exercise prior to the closure of the epiphyseal plates to help build total mass prior to its inevitable reduction with aging. Even elderly people can strengthen their bones by engaging in weight-bearing exercise.

▶ CHECKPOINT

18. What types of mechanical stresses may be used to strengthen bone tissue?

19. Would children raised in space ever be able to return to Earth?

AGING AND BONE TISSUE

▶ OBJECTIVE

Describe the effects of aging on bone tissue.

From birth through adolescence, more bone tissue is produced than is lost during bone remodeling. In young adults the rates of bone deposition and resorption are about the same. As the level of sex hormones diminishes during middle age, especially in women after menopause, a decrease in bone mass occurs because bone resorption by osteoblasts outpaces bone deposition by osteoblasts. In old age, loss of bone through resorption occurs more rapidly than bone gain. Because women's bones generally are smaller and less massive than men's bones to begin with, loss of bone mass in old age typically has a greater adverse effect in females. These factors contribute to the higher incidence of osteoporosis in females.

There are two principal effects of aging on bone tissue: loss of bone mass and brittleness. Loss of bone mass results from **demineralization** (dē-min′-er-al-i-ZĀ-shun), the loss of calcium and other minerals from bone extracellular matrix. This loss usually begins after age 30 in females, accelerates greatly around age 45 as levels of estrogens decrease, and continues until as much as 30% of the calcium in bones is lost by age 70. Once bone loss begins in females, about 8% of bone mass is lost every 10 years. In males, calcium loss typically does not begin until after age 60, and about 3% of bone mass is lost every 10 years. The loss of calcium from bones is one of the problems in osteoporosis (described shortly).

The second principal effect of aging on the skeletal system, brittleness, results from a decreased rate of protein synthesis. Recall that the organic part of bone extracellular matrix, mainly collagen fibers, gives bone its tensile strength. The loss of tensile strength causes the bones to become very brittle and susceptible to fracture. In some elderly people, collagen fiber synthesis slows, in part, due to diminished production of human growth hormone. In addition to increasing the susceptibility to fractures, loss of bone mass also leads to deformity, pain, loss of height, and loss of teeth.

▶ CHECKPOINT

20. What is demineralization, and how does it affect the functioning of bone?

21. What changes occur in the organic part of bone extracellular matrix with aging?

DISORDERS: HOMEOSTATIC IMBALANCES

Osteoporosis

Osteoporosis (os′-tē-ō-pō-RŌ-sis; *por-* = passageway; *-osis* = condition) is literally a condition of porous bones (Figure 6.12). The basic problem is that bone resorption outpaces bone deposition. In large part this is due to depletion of calcium from the body—more calcium is lost in urine, feces, and sweat than is absorbed from the diet. Bone mass becomes so depleted that bones fracture, often spontaneously, under the mechanical stresses of everyday living. For example, a hip fracture might result from simply sitting down too quickly. In the United States, osteoporosis results in more than a million fractures a year, mainly in the hip, wrist, and vertebrae. Osteoporosis afflicts the entire skeletal system. In addition to fractures, osteoporosis causes shrinkage of vertebrae, height loss, hunched backs, and bone pain.

Thirty million people in the United States suffer from osteoporosis. The disorder primarily affects middle-aged and elderly people, 80% of them women. Older women suffer from osteoporosis more often than men for two reasons: Women's bones are less massive than men's bones, and production of estrogens in women declines dramatically at menopause, but production of the main androgen, testosterone, wanes gradually and only slightly in older men. Estrogens and testosterone stimulate osteoblast activity and synthesis of bone extracellular matrix. Besides gender, risk factors for developing osteoporosis include a family history of the disease,

Figure 6.12 Comparison of spongy bone tissue from (a) a normal young adult and (b) a person with osteoporosis. Notice the weakened trabeculae in (b). Compact bone tissue is similarly affected by osteoporosis.

In osteoporosis, bone resorption outpaces bone formation, so bone mass decreases.

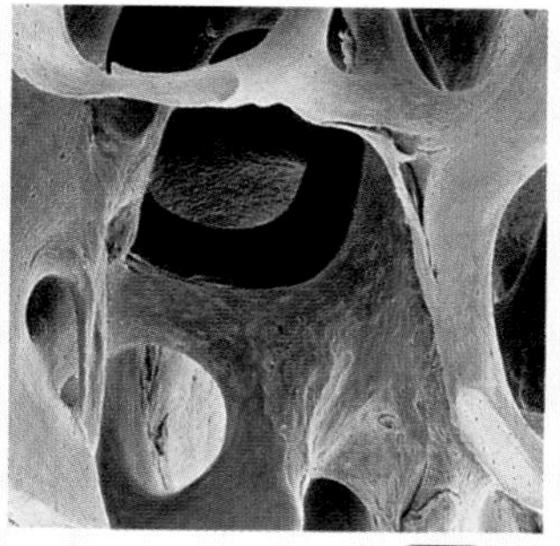

(a) Normal bone

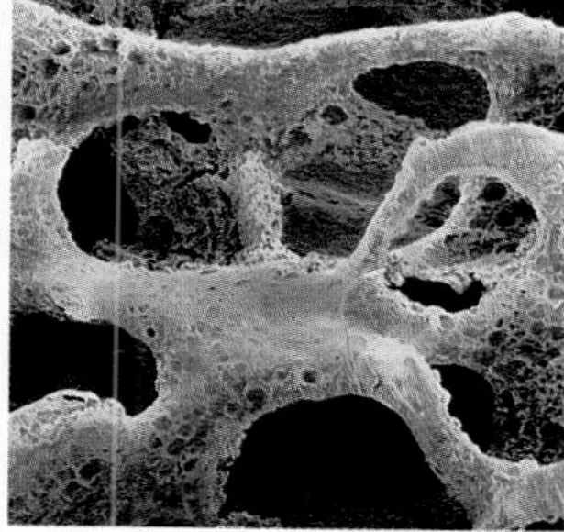

(b) Osteoporotic bone

If you wanted to develop a drug to lessen the effects of osteoporosis, would you look for a chemical that inhibits the activity of osteoblasts or that of osteoclasts?

European or Asian ancestry, thin or small body build, an inactive lifestyle, cigarette smoking, a diet low in calcium and vitamin D, more than two alcoholic drinks a day, and the use of certain medications.

In postmenopausal women, treatment of osteoporosis may include estrogen replacement therapy (ERT; low doses of estrogens) or hormone replacement therapy (HRT; a combination of estrogens and progesterone, another sex steroid). Although such treatments help combat osteoporosis, they increase cell metabolism in the entire body, which may increase a woman's risk of breast cancer. The drug raloxifene (Evista®) mimics the beneficial effects of estrogens on bone without increasing the risk of breast cancer. Another drug that may be used is the nonhormone drug alendronate (Fosamax®), which blocks resorption of bone by osteoclasts.

Perhaps more important than treatment is prevention. Adequate calcium intake and weight-bearing exercise, particularly when a woman is young, may be more beneficial to a woman than drugs and calcium supplements when she is older.

Rickets and Osteomalacia

Rickets and osteomalacia (os′-tē-ō-ma-LĀ-shē-a; *-malacia* = softness) are disorders in which bones fail to calcify. Although the organic matrix is still produced, calcium salts are not deposited, and the bones become "soft" or rubbery and easily deformed. Rickets affects the growing bones of children. Because new bone formed at the epiphyseal plates fails to ossify, bowed legs and deformities of the skull, rib cage, and pelvis are common. In osteomalacia, sometimes called "adult rickets," new bone formed during remodeling fails to calcify. This disorder causes varying degrees of pain and tenderness in bones, especially in the hip and leg. Bone fractures also result from minor trauma. Rickets and osteomalacia are typically caused by a deficiency of vitamin D, either due to insufficient sunlight or a lack of vitamin D in the diet. A recombinant human parathyroid hormone (rPTH), called Forteo, builds bone tissue by stimulating osteoblasts.

MEDICAL TERMINOLOGY

Osteoarthritis (os′-tē-ō-ar-THRĪ-tis; *arthr* = joint) The degeneration of articular cartilage such that the bony ends touch; the resulting friction of bone against bone worsens the condition. Usually associated with the elderly.

Osteogenic sarcoma (os′-tē-ō-JEN-ik sar-KŌ-ma; *sarcoma* = connective tissue tumor) Bone cancer that primarily affects osteoblasts and occurs most often in teenagers during their growth spurt; the most common sites are the metaphyses of the thigh bone (femur), shin bone (tibia), and arm bone (humerus). Metastases occur most often in lungs; treatment consists of multidrug chemotherapy and removal of the malignant growth, or amputation of the limb.

Osteomyelitis (os′-tē-ō-mī-e-LĪ-tis) An infection of bone characterized by high fever, sweating, chills, pain, and nausea, pus formation, edema, and warmth over the affected bone and rigid overlying muscles. It is often caused by bacteria, usually *Staphylococcus aureus.* The bacteria may reach the bone from outside the body (through open fractures, penetrating wounds, or orthopedic surgical procedures); from other sites of infection in the body (abscessed teeth, burn infections, urinary tract infections, or upper respiratory infections) via the blood; and from adjacent soft tissue infections (as occurs in diabetes mellitus).

Osteopenia (os′-tē-ō-PĒ-nē-a; *penia* = poverty) Reduced bone mass due to a decrease in the rate of bone synthesis to a level too low to compensate for normal bone resorption; any decrease in bone mass below normal. An example is osteoporosis.

STUDY OUTLINE

INTRODUCTION (p. 172)

1. A bone is made up of several different tissues: bone or osseous tissue, cartilage, dense connective tissues, epithelium, adipose tissue, and nervous tissue.
2. The entire framework of bones and their cartilages constitutes the skeletal system.

FUNCTIONS OF BONE AND THE SKELETAL SYSTEM (p. 172)

1. The skeletal system functions in support, protection, movement, mineral homeostasis, blood cell production, and triglyceride storage.

STRUCTURE OF BONE (p. 172)

1. Parts of a typical long bone are the diaphysis (shaft), proximal and distal epiphyses (ends), metaphyses, articular cartilage, periosteum, medullary (marrow) cavity, and endosteum.

HISTOLOGY OF BONE TISSUE (p. 173)

1. Bone tissue consists of widely separated cells surrounded by large amounts of extracellular matrix.
2. The four principal types of cells in bone tissue are osteogenic cells, osteoblasts, osteocytes, and osteoclasts.
3. The extracellular matrix of bone contains abundant mineral salts (mostly hydroxyapatite) and collagen fibers.
4. Compact bone tissue consists of osteons (haversian systems) with little space between them.
5. Compact bone tissue lies over spongy bone tissue in the epiphyses and makes up most of the bone tissue of the diaphysis. Functionally, compact bone tissue is the strongest form of bone and protects, supports, and resists stress.
6. Spongy bone tissue does not contain osteons. It consists of trabeculae surrounding many red bone marrow-filled spaces.
7. Spongy bone tissue forms most of the structure of short, flat, and irregular bones, and the interior of the epiphyses in long bones. Functionally, spongy bone tissue trabeculae offer resistance along

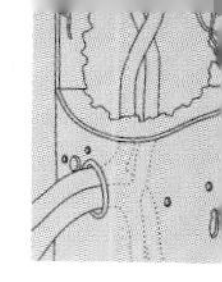

lines of stress, support and protect red bone marrow, and make bones lighter for easier movement.

BLOOD AND NERVE SUPPLY OF BONE (p. 177)

1. Long bones are supplied by periosteal, nutrient, and epiphyseal arteries; veins accompany the arteries.
2. Nerves accompany blood vessels in bone; the periosteum is rich in sensory neurons.

BONE FORMATION (p.178)

1. Bone forms by a process called ossification (osteogenesis), which begins when mesenchymal cells become transformed into osteogenic cells. These undergo cell division and give rise to cells that differentiate into osteoblasts, osteoclasts, and osteocytes.
2. Ossification begins during the sixth week of embryonic life. The two types of ossification, intramembranous and endochondral, involve the replacement of a preexisting connective tissue with bone.
3. Intramembranous ossification refers to bone formation directly within mesenchyme arranged in sheetlike layers that resemble membranes.
4. Endochondral ossification refers to bone formation within hyaline cartilage that develops from mesenchyme. The primary ossification center of a long bone is in the diaphysis. Cartilage degenerates, leaving cavities that merge to form the medullary cavity. Osteoblasts lay down bone. Next, ossification occurs in the epiphyses, where bone replaces cartilage, except for the epiphyseal plate.

BONE GROWTH (p. 181)

1. The epiphyseal plate consists of four zones: zone of resting cartilage, zone of proliferating cartilage, zone of hypertrophic cartilage, and zone of calcified cartilage.
2. Because of the cell division in the epiphyseal plate, the diaphysis of a bone increases in length.
3. Bone grows in thickness or diameter due to the addition of new bone tissue by periosteal osteoblasts around the outer surface of the bone (appositional growth).

BONES AND HOMEOSTASIS (p. 183)

1. Bone remodeling is an ongoing process in which osteoclasts carve out small tunnels in old bone tissue and then osteoblasts rebuild it.
2. In bone resorption, osteoclasts release enzymes and acids that degrade collagen fibers and dissolve mineral salts.
3. Dietary minerals (especially calcium and phosphorus) and vitamins (C, K, and B_{12}) are needed for bone growth and mainteance. Insulin-like growth factors (IGFs), human growth hormone, thyroid hormones, estrogens, and androgens stimulate bone growth.
4. Sex hormones slow resorption of old bone and promote new bone deposition.
5. A fracture is any break in a bone.
6. Fracture repair involves formation of a fracture hematoma, a fibrocartilaginous callus, and a bony callus, and bone remodeling.
7. Types of fractures include closed (simple), open (compound), comminuted, greenstick, impacted, stress, Pott's, and Colles'.
8. Bone is the major reservoir for calcium in the body.
9. Parathyroid hormone (PTH) secreted by the parathyroid gland increases blood Ca^{2+} level, whereas calcitonin (CT) from the thyroid gland has the potential to decrease blood Ca^{2+} level. Vitamin D enhances absorption of calcium and phosphate and thus raises the blood levels of these substances.

EXERCISE AND BONE TISSUE (p. 188)

1. Mechanical stress increases bone strength by increasing deposition of mineral salts and production of collagen fibers.
2. Removal of mechanical stress weakens bone through demineralization and collagen fiber reduction.

AGING AND BONE TISSUE (p. 189)

1. The principal effect of aging is demineralization, a loss of calcium from bones, which is due to reduced osteoblast activity.
2. Another effect is decreased production of extracellular matrix proteins (mostly collagen fibers), which makes bones more brittle and thus more susceptible to fracture.

SELF-QUIZ QUESTIONS

Fill in the blanks in the following statements.

1. Bone growth in length is called ____ growth, and bone growth in diameter (thickness) is called ____ growth.
2. The crystallized inorganic mineral salts in bone contribute to bone's ____, while the collagen fibers and other organic molecules provide bone with ____.

Indicate whether the following statements are true or false.

3. Bone resorption involves increased activity of osteoclasts.
4. The formation of bone from cartilage is known as endochondral ossification.
5. The growth of bone is controlled primarily by hormones.

Choose the one best answer to the following questions.

6. Place in order the steps involved in intramembranous ossification. (1) Bony matrices fuse to form trabeculae. (2) Clusters of osteoblasts form a center of ossification that secretes the organic extracellular matrix. (3) Spongy bone is replaced with compact bone on the bone's surface. (4) Periosteum develops on the bone's periphery. (5) The extracellular matrix hardens by deposition of calcium and mineral salts. (a) 2, 4, 5, 1, 3; (b) 4, 3, 5, 1, 2; (c) 1, 2, 5, 4, 3; (d) 2, 5, 1, 4, 3; (e) 5, 1, 3, 4, 2.
7. Place in order the steps involved in endochondral ossification. (1) Nutrient artery invades the perichondrium. (2) Osteoclasts create a marrow cavity. (3) Chondrocytes enlarge and calcify. (4) Secondary ossification centers appear at epiphyses. (5) Osteoblasts become active in the primary ossification center. (a) 3, 1, 5, 2, 4; (b) 3, 1, 5, 4, 2; (c) 1, 3, 5, 2, 4; (d) 1, 2, 3, 5, 4; (e) 2, 5, 4, 3, 1.
8. Spongy bone differs from compact bone because spongy bone (a) is composed of numerous osteons (haversian systems); (b) is found primarily in the diaphyses of long bones, and compact bone is found primarily in the epiphyses of long bones; (c) contains osteons all aligned in the same direction along lines of stress; (d) does not contain osteocytes contained in lacunae; (e) is composed of trabeculae that are oriented along lines of stress.

9. A primary effect that weight-bearing exercise has on bones is to (a) provide oxygen for bone development; (b) increase the demineralization of bone; (c) maintain and increase bone mass; (d) stimulate the release of sex hormones for bone growth; (e) utilize the stored triglycerides from the yellow bone marrow.

10. Place in order the steps involved in the repair of a bone fracture. (1) Osteoblast production of trabeculae and bony callus formation; (2) formation of a hematoma at the site of fracture; (3) resorption of remaining bone fragments and remodeling of bone; (4) migration of fibroblasts to the fracture site; (5) bridging of broken ends of bones by a fibrocartilagenous callus. (a) 2, 4, 5, 1, 3; (b) 2, 5, 4, 1, 3; (c) 1, 2, 5, 4, 3; (d) 2, 5, 1, 3, 4; (e) 5, 2, 4, 1, 3.

11. Match the following:

____(a) space within the shaft of the bone that contains yellow bone marrow
____(b) triglyceride storage tissue
____(c) hemopoietic tissue
____(d) thin layer of hyaline cartilage covering the ends of bones where they form a joint
____(e) distal and proximal ends of bones
____(f) the long, cylindrical main portion of the bone; the shaft
____(g) in a growing bone, the region that contains the epiphyseal plate
____(h) the tough membrane that surrounds the bone surface wherever cartilage is not present
____(i) a layer of hyaline cartilage in the area between the shaft and end of a growing bone
____(j) membrane lining the medullary cavity
____(k) a remnant of the active epiphyseal plate; a sign that the bone has stopped growing in length
____(l) bundles of collagen fibers that attach periosteum to bone

(1) articular cartilage
(2) endosteum
(3) medullary cavity
(4) diaphysis
(5) epiphyses
(6) metaphysis
(7) periosteum
(8) red bone marrow
(9) yellow bone marrow
(10) perforating (Sharpey's) fibers
(11) epiphyseal line
(12) epiphyseal plate

12. Match the following:

____(a) small spaces between lamellae that contain osteocytes
____(b) perforating canals that penetrate compact bone; carry blood vessels, lymphatic vessels, and nerves from the periosteum
____(c) areas between osteons; fragments of old osteons
____(d) cells that secrete the components required to build bone
____(e) microscopic unit of compact bone tissue
____(f) interconnected, tiny canals filled with extracellular fluid; connect lacunae to each other and to the central canal
____(g) canals that extend longitudinally through the bone and connect blood vessels and nerves to the osteocytes
____(h) large cells derived from monocytes and involved in bone resorption
____(i) irregular lattice of thin-columns of bone found in spongy bone tissue
____(j) rings of hard calcified matrix found just beneath the periosteum and in the medullary cavity
____(k) mature cells that maintain the daily metabolism of bone
____(l) an opening in the shaft of the bone allowing an artery to pass into the bone
____(m) unspecialized stem cells derived from mesenchyme

(1) osteogenic cells
(2) osteocytes
(3) osteon (haversian system)
(4) Volkmann's canals
(5) circumferential lamellae
(6) osteoblasts
(7) trabeculae
(8) interstitial lamellae
(9) canaliculi
(10) osteoclasts
(11) nutrient foramen
(12) lacunae
(13) haversian (central) canals

13. Match the following:

____(a) decreases blood calcium levels by accelerating calcium deposition in bones and inhibiting osteoclasts
____(b) required for collagen synthesis
____(c) during childhood, it promotes growth at epiphyseal plate; production stimulated by human growth hormone
____(d) involved in bone growth by increasing osteoblast activity; causes long bones to stop growing in length
____(e) required for protein synthesis
____(f) active form of vitamin D; raises blood calcium levels by increasing absorption of calcium from digestive tract
____(g) raises blood calcium levels by increasing bone resorption

(1) PTH
(2) CT
(3) calcitriol
(4) insulinlike growth factors
(5) sex hormones
(6) vitamin C
(7) vitamin K

14. Match the following:

____(a) column-like layer of maturing chondrocytes
____(b) layer of small, scattered chondrocytes anchoring the epiphyseal plate to the bone
____(c) layer of actively dividing chondrocytes
____(d) region of dead chondrocytes

(1) zone of hypertrophic cartilage
(2) zone of calcified cartilage
(3) zone of proliferating cartilage
(4) zone of resting cartilage

15. Match the following:

____(a) a broken bone in which one end of the fractured bone is driven into the other end
____(b) a condition of porous bones characterized by decreased bone mass and increased susceptibility to fractures
____(c) splintered bone, with smaller fragments lying between main fragments
____(d) a broken bone that does not break through the skin
____(e) a partial break in a bone in which one side of the bone is broken and the other side bends
____(f) a broken bone that protrudes through the skin
____(g) microscopic bone breaks resulting from inability to withstand repeated stressful impact
____(h) a degeneration of articular cartilage allowing the bony ends to touch; worsens due to friction between the bones
____(i) condition characterized by failure of new bone formed by remodeling to calcify in adults
____(j) an infection of bone

(1) closed (simple) fracture
(2) open (compound) fracture
(3) impacted fracture
(4) greenstick fracture
(5) stress fracture
(6) comminuted fracture
(7) osteoporosis
(8) osteomalacia
(9) osteoarthritis
(10) osteomyelitis

CRITICAL THINKING QUESTIONS

1. Taryn is a high school senior who is undergoing a strenuous running regimen for several hours a day in order to qualify for her state high school track meet. Lately she has experienced intense pain in her right leg that is hindering her workouts. Her physician performs an examination of her right leg. The doctor doesn't notice any outward evidence of injury; he then orders a bone scan. What does her doctor suspect the problem is?
2. While playing basketball, nine-year-old Marcus fell and broke his left arm. The arm was placed in a cast and appeared to heal normally. As an adult, Marcus is puzzled because it seems that his right arm is longer than his left arm. He measured both arms and he was correct—his right arm *is* longer! How would you explain to Marcus what happened?
3. Astronauts in space exercise as part of their daily routine, yet they still have problems with bone weakness after prolonged stays in space. Why does this happen?

ANSWERS TO FIGURE QUESTIONS

6.1 The periosteum is essential for growth in bone thickness, bone repair, and bone nutrition. It also serves as a point of attachment for ligaments and tendons.

6.2 Bone resorption is necessary for the development, growth, maintenance, and repair of bone.

6.3 The central (haversian) canals are the main blood supply to the osteocytes of an osteon (haversian system), so their blockage would lead to death of the osteocytes.

6.4 Periosteal arteries enter bone tissue through perforations (Volkmann's canals).

6.5 Flat bones of the skull and mandible (lower jawbone) develop by intramembranous ossification.

6.6 Yes, she probably will grow taller. Epiphyseal lines are indications of growth zones that have ceased to function. The absence of epiphyseal lines indicates that the bone is still lengthening.

6.7 The lengthwise growth of the diaphysis is caused by cell divisions in the zone of proliferating cartilage and maturation of the cells in the zone of hypertrophic cartilage.

6.8 The medullary cavity enlarges by activity of the osteoclasts in the endosteum.

6.9 In an open fracture the ends of the bone break through the skin; in a closed fracture they do not.

6.10 Healing of bone fractures can take months because calcium and phosphorus deposition is a slow process, and bone cells generally grow and reproduce slowly.

6.11 Heartbeat, respiration, nerve cell functioning, enzyme functioning, and blood clotting all depend on proper levels of calcium.

6.12 A drug that inhibits the activity of osteoclasts might lessen the effects of osteoporosis.

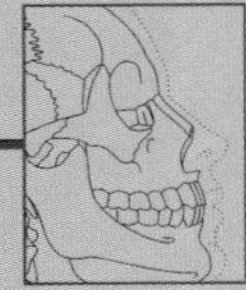

Chapter 7

The Skeletal System: The Axial Skeleton

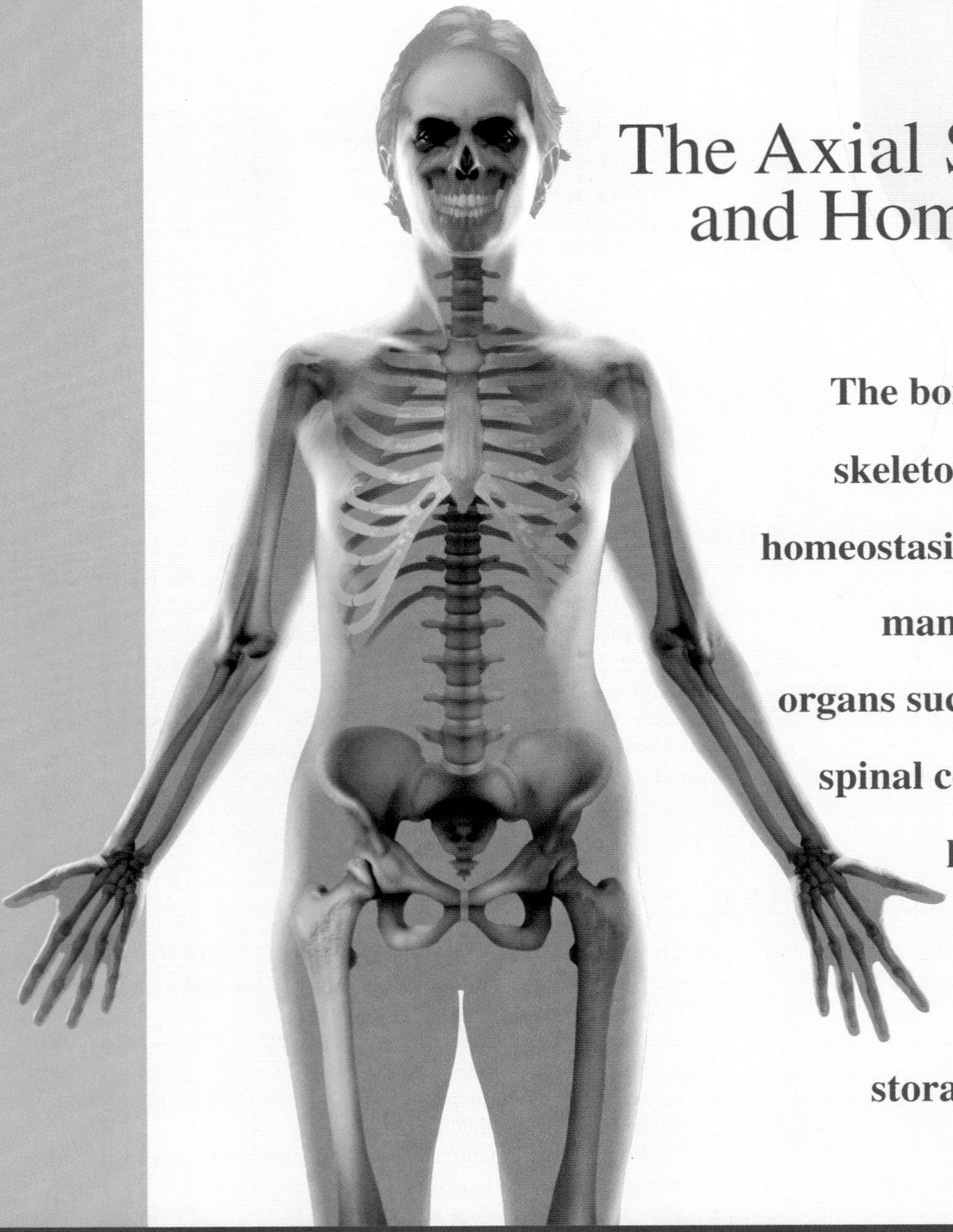

The Axial Skeleton and Homeostasis

The bones of the axial skeleton contribute to homeostasis by protecting many of the body's organs such as the brain, spinal cord, heart, and lungs. They are also important in calcium storage and release.

wiley.com/college/apcentral

Without bones, you could not survive. You would be unable to perform movements such as walking or grasping, and the slightest blow to your head or chest could damage your brain or heart. Because the skeletal system forms the framework of the body, a familiarity with the names, shapes, and positions of individual bones will help you locate and name many other anatomical features. For example, the radial artery, the site where the pulse is usually taken, is named for its closeness to the radius, the lateral bone of the forearm. The ulnar nerve is named for its proximity to the ulna, the medial bone of the forearm. The frontal lobe of the brain lies deep to the frontal (forehead) bone. The tibialis anterior muscle lies along the anterior surface of the tibia (shin bone). Parts of certain bones also serve to locate structures within the skull and to outline the lungs, heart, and abdominal and pelvic organs.

Movements such as throwing a ball, biking, and walking require interactions between bones and muscles. To understand how muscles produce different movements, you will learn where the muscles attach on individual bones and you will learn the types of joints acted on by the contracting muscles. The bones, muscles, and joints together form an integrated system called the **musculoskeletal system.** The branch of medical science concerned with the prevention or correction of disorders of the musculoskeletal system is called **orthopedics** (or′-thō-PĒ-diks; *ortho-* = correct; *pedi* = child).

DIVISIONS OF THE SKELETAL SYSTEM

▶ **OBJECTIVE**

Describe how the skeleton is divided into axial and appendicular divisions.

The adult human skeleton consists of 206 named bones, most of which are paired, with one member of each pair on the right and left sides of the body. The skeletons of infants and children have more than 206 bones because some of their bones fuse later in life. Examples are the hip bones and some bones of the vertebral column (backbone).

Bones of the adult skeleton are grouped into two principal divisions: the **axial skeleton** and the **appendicular skeleton** (*appendic-* = to hang onto). Table 7.1 presents the 80 bones of the axial skeleton and the 126 bones of the appendicular skeleton. Figure 7.1 shows how both divisions join to form the complete skeleton (the bones of the axial skeleton are shown in blue). The axial skeleton consists of the bones that lie around the longitudinal **axis** of the human body, an imaginary vertical line that runs through the body's center of gravity from the head to the space between the feet: skull bones, auditory ossicles (ear bones), hyoid bone (see Figure 7.4), ribs, sternum (breastbone), and bones of the vertebral column. The appendicular skeleton consists of the bones of the **upper** and **lower limbs (extremities),** plus the bones forming the **girdles** that connect the limbs to the axial skeleton. Functionally, the auditory ossicles in the middle ear, which vibrate in response to sound waves that strike the eardrum, are not part of either the axial or appendicular skeleton, but they are grouped with the axial skeleton for convenience (see Chapter 17).

We will organize our study of the skeletal system around the two divisions of the skeleton, with emphasis on how the many bones of the body are interrelated. In this chapter we focus on the axial skeleton, looking first at the skull and then at the bones of the vertebral column and the chest. In Chapter 8 we explore the appendicular skeleton, examining in turn the bones of the pectoral (shoulder) girdle and upper limbs, and then the pelvic (hip) girdle and the lower limbs. Before we examine the axial skeleton, we direct your attention to some general characteristics of bones.

TABLE 7.1 The Bones of the Adult Skeletal System

Division of the Skeleton	Structure	Number of Bones
Axial Skeleton	**Skull**	
	Cranium	8
	Face	14
	Hyoid	1
	Auditory ossicles	6
	Vertebral column	26
	Thorax	
	Sternum	1
	Ribs	24
	Subtotal =	80
Appendicular Skeleton	**Pectoral (shoulder) girdles**	
	Clavicle	2
	Scapula	2
	Upper limbs (extremities)	
	Humerus	2
	Ulna	2
	Radius	2
	Carpals	16
	Metacarpals	10
	Phalanges	28
	Pelvic (hip) girdle	
	Hip, pelvic, or coxal bone	2
	Lower limbs (extremities)	
	Femur	2
	Patella	2
	Fibula	2
	Tibia	2
	Tarsals	14
	Metatarsals	10
	Phalanges	28
	Subtotal =	126
	Total =	206

Figure 7.1 **Divisions of the skeletal system.** The axial skeleton is indicated in blue, the appendicular skeleton in yellow. (Note the position of the hyoid bone in Figure 7.4.) (See Tortora, *A Photographic Atlas of the Human Body, Second Edition,* Figure 3.1.)

The adult human skeleton consists of 206 bones grouped into axial and appendicular divisions.

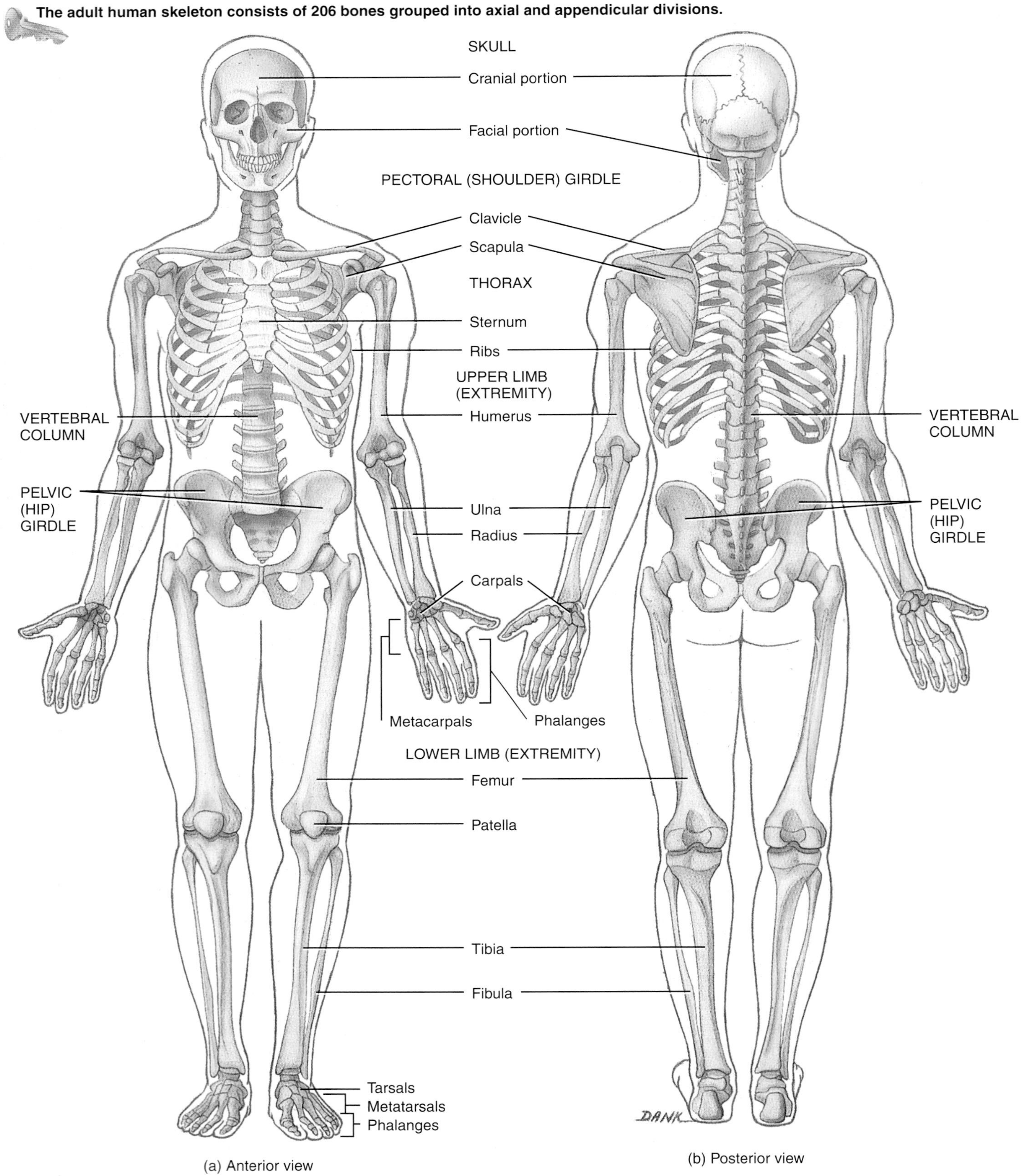

Which of the following structures are part of the axial skeleton, and which are part of the appendicular skeleton? Skull, clavicle, vertebral column, shoulder girdle, humerus, pelvic girdle, and femur.

▶ CHECKPOINT

1. Which bones make up the axial and appendicular divisions of the skeleton?

TYPES OF BONES

▶ OBJECTIVE

Classify bones based on their shape or location.

Almost all bones of the body can be classified into five main types based on shape: long, short, flat, irregular, and sesamoid (Figure 7.2). As you learned in Chapter 6, **long bones** have greater length than width, consist of a shaft and a variable number of extremities (ends), and are slightly curved for strength. A curved bone absorbs the stress of the body's weight at several different points, so that it is evenly distributed. If bones were straight, the weight of the body would be unevenly distributed, and the bone would fracture more easily. Long bones consist mostly of *compact bone tissue* in their diaphyses but have considerable amounts of *spongy bone tissue* in their epiphyses. Long bones vary tremendously in size and include those in the thigh (femur), leg (tibia and fibula), arm (humerus), forearm (ulna and radius), and fingers and toes (phalanges).

Short bones are somewhat cube-shaped and are nearly equal in length and width. They consist of spongy bone tissue except at the surface, which has a thin layer of compact bone tissue.

Figure 7.2 **Types of bones based on shape.** The bones are not drawn to scale.

The shapes of bones largely determine their functions.

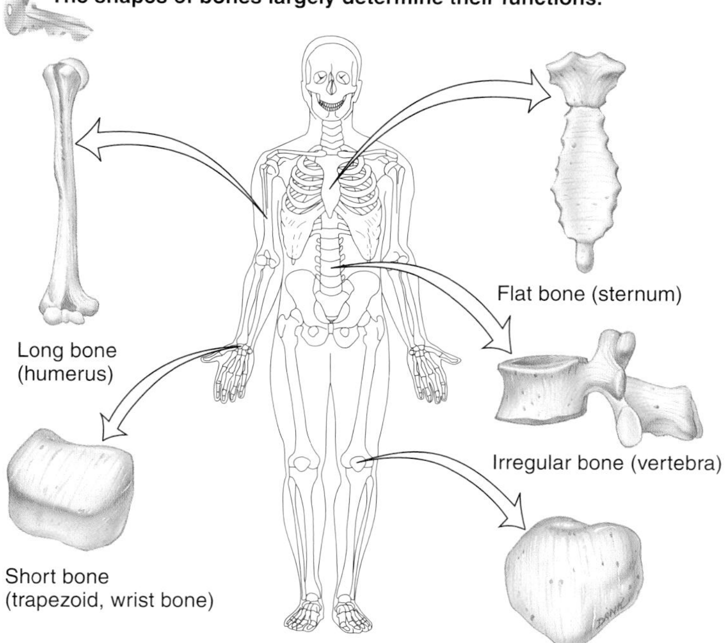

Which type of bone primarily provides protection and a large surface area for muscle attachment?

Examples of short bones are the carpal (wrist) bones (except for the pisiform, which is a sesamoid bone) and the tarsal (ankle) bones (except for the calcaneus or heel bone, which is an irregular bone).

Flat bones are generally thin and composed of two nearly parallel plates of compact bone tissue enclosing a layer of spongy bone tissue. Flat bones afford considerable protection and provide extensive areas for muscle attachment. Flat bones include the cranial bones, which protect the brain; the sternum (breastbone) and ribs, which protect organs in the thorax; and the scapulae (shoulder blades).

Irregular bones have complex shapes and cannot be grouped into any of the previous categories. They vary in the amount of spongy and compact bone present. Such bones include the vertebrae (backbones), hip bones, certain facial bones, and the calcaneus.

Sesamoid bones (SES-a-moyd = shaped like a sesame seed) develop in certain tendons where there is considerable friction, tension, and physical stress, such as the palms and soles. They may vary in number from person to person, are not always completely ossified, and typically measure only a few millimeters in diameter. Notable exceptions are the two patellae (kneecaps), large sesamoid bones located in the quadriceps femoris tendon (see Figure 11.20a on page 385) that are normally present in everyone. Functionally, sesamoid bones protect tendons from excessive wear and tear, and they often change the direction of pull of a tendon, which improves the mechanical advantage at a joint.

An additional type of bone is classified by location rather than shape. **Sutural bones** (SOO-chur-al; *sutur-* = seam) are small bones located in sutures (immovable joints) between certain cranial bones (see Figure 7.6). Their number varies greatly from person to person.

Recall from Chapter 6 that in adults, red bone marrow is restricted to flat bones such as the ribs, sternum (breastbone), and skull; irregular bones such as vertebrae (backbones) and hip bones; long bones such as the proximal epiphyses of the femur (thigh bone) and humerus (arm bone); and some short bones.

▶ CHECKPOINT

2. Give examples of long, short, flat, and irregular bones.

BONE SURFACE MARKINGS

▶ OBJECTIVE

Describe the principal surface markings on bones and the functions of each.

Bones have characteristic **surface markings,** structural features adapted for specific functions. Most are not present at birth but develop in response to certain forces and are most prominent in the adult skeleton. In response to tension on a bone surface from tendons, ligaments, aponeuroses, and fasciae, new bone is deposited, resulting in raised or roughened areas. Conversely, compression on a bone surface results in a depression.

There are two major types of surface markings: (1) *depressions and openings,* which form joints or allow the passage of soft tissues (such as blood vessels and nerves), and (2) *processes,* projections or outgrowths that either help form joints or serve as attachment points for connective tissue (such as ligaments and tendons). Table 7.2 describes the various surface markings and provides examples of each.

▶ CHECKPOINT

3. List and describe several bone surface markings, and give an example of each. Check your list against Table 7.2.

SKULL

▶ OBJECTIVES

Name the cranial and facial bones and indicate whether they are paired or single.

Describe the following special features of the skull: sutures, paranasal sinuses, and fontanels.

The **skull,** with its 22 bones, rests on the superior end of the vertebral column (backbone). The bones of the skull are grouped into two categories: cranial bones and facial bones. The **cranial bones** (*crani-* = brain case) form the cranial cavity, which encloses and protects the brain. The eight cranial bones are the frontal bone, two parietal bones, two temporal bones, the occipital bone, the sphenoid bone, and the ethmoid bone. Fourteen **facial bones** form the face: two nasal bones, two maxillae (or maxillas), two zygomatic bones, the mandible, two lacrimal bones, two palatine bones, two inferior nasal conchae, and the vomer. Figures 7.3 through 7.8 illustrate these bones from different viewing directions.

General Features and Functions

Besides forming the large cranial cavity, the skull also forms several smaller cavities, including the nasal cavity and orbits (eye sockets), which open to the exterior. Certain skull bones also contain cavities called paranasal sinuses that are lined with mucous membranes and open into the nasal cavity. Other small cavities within the skull house the structures involved in hearing and equilibrium.

TABLE 7.2 Bone Surface Markings

Marking	Description	Example
Depressions and openings: Sites allowing the passage of soft tissue (nerves, blood vessels, ligaments, tendons) or formation of joints		
Fissure (FISH-ur)	Narrow slit between adjacent parts of bones through which blood vessels or nerves pass.	Superior orbital fissure of the sphenoid bone (Figure 7.12).
Foramen (fō-RĀ-men = hole; plural is *foramina*)	Opening through which blood vessels, nerves, or ligaments pass.	Optic foramen of the sphenoid bone (Figure 7.12).
Fossa (FOS-a = trench)	Shallow depression.	Coronoid fossa of the humerus (Figure 8.5a on page 236).
Sulcus (SUL-kus = groove)	Furrow along a bone surface that accommodates a blood vessel, nerve, or tendon.	Intertubercular sulcus of the humerus (Figure 8.5a on page 236).
Meatus (mē-Ā-tus = passageway)	Tubelike opening.	External auditory meatus of the temporal bone (Figure 7.4a).
Processes: Projections or outgrowths on bone that form joints or attachment points for connective tissue, such as ligaments and tendons.		
Processes that form joints		
Condyle (KON-dīl = knuckle)	Large, round protuberance at the end of a bone.	Lateral condyle of the femur (Figure 8.13a on page 246).
Facet	Smooth flat articular surface.	Superior articular facet of a vertebra (Figure 7.18d).
Head	Rounded articular projection supported on the neck (constricted portion) of a bone.	Head of the femur (Figure 8.13a on page 246).
Processes that form attachment points for connective tissue		
Crest	Prominent ridge or elongated projection.	Iliac crest of the hip bone (Figure 8.10b on page 241).
Epicondyle (*epi-* = above)	Projection above a condyle.	Medial epicondyle of the femur (Figure 8.13a on page 246).
Line (linea)	Long, narrow ridge or border (less prominent than a crest).	Linea aspera of the femur (Figure 8.13b on page 246).
Spinous process	Sharp, slender projection.	Spinous process of a vertebra (Figure 7.17).
Trochanter (trō-KAN-ter)	Very large projection.	Greater trochanter of the femur (Figure 8.13b on page 246).
Tubercle (TOO-ber-kul; *tuber-* = knob)	Small, rounded projection.	Greater tubercle of the humerus (Figure 8.5a on page 236).
Tuberosity	Large, rounded, usually roughened projection.	Ischial tuberosity of the hip bone (Figure 8.10b on page 241).

Other than the auditory ossicles, which are involved in hearing and are located within the temporal bones, the mandible is the only movable bone of the skull. Immovable joints called sutures fuse most of the skull bones together and are especially noticeable on the outer surface of the skull.

The skull has many surface markings, such as foramina and fissures through which blood vessels and nerves pass. You will learn the names of important skull bone surface markings as we describe each bone.

In addition to protecting the brain, the cranial bones stabilize the positions of the brain, blood vessels, lymphatic vessels, and nerves through the attachment of their inner surfaces to meninges (membranes). The outer surfaces of cranial bones provide large areas of attachment for muscles that move various parts of the head. The bones also provide attachment for some muscles that produce facial expressions such as the frown of concentration you wear when studying this book. The facial bones form the framework of the face and provide support for the entrances to the digestive and respiratory systems. Together, the cranial and facial bones protect and support the delicate special sense organs for vision, taste, smell, hearing, and equilibrium (balance).

Cranial Bones

Frontal Bone

The **frontal bone** forms the forehead (the anterior part of the cranium), the roofs of the orbits (eye sockets), and most of the anterior part of the cranial floor (Figure 7.3). Soon after birth, the left and right sides of the frontal bone are united by the *metopic suture,* which usually disappears between the ages of six and eight.

Note the *frontal squama,* a scalelike plate of bone that forms the forehead in the anterior view of the skull shown in Figure 7.3. It gradually slopes inferiorly from the coronal suture, on the top of the skull, then angles abruptly and becomes almost

Figure 7.3 Anterior view of skull. (See Tortora, *A Photographic Atlas of the Human Body, Second Edition,* Figure 3.2.)

The skull consists of cranial bones and facial bones.

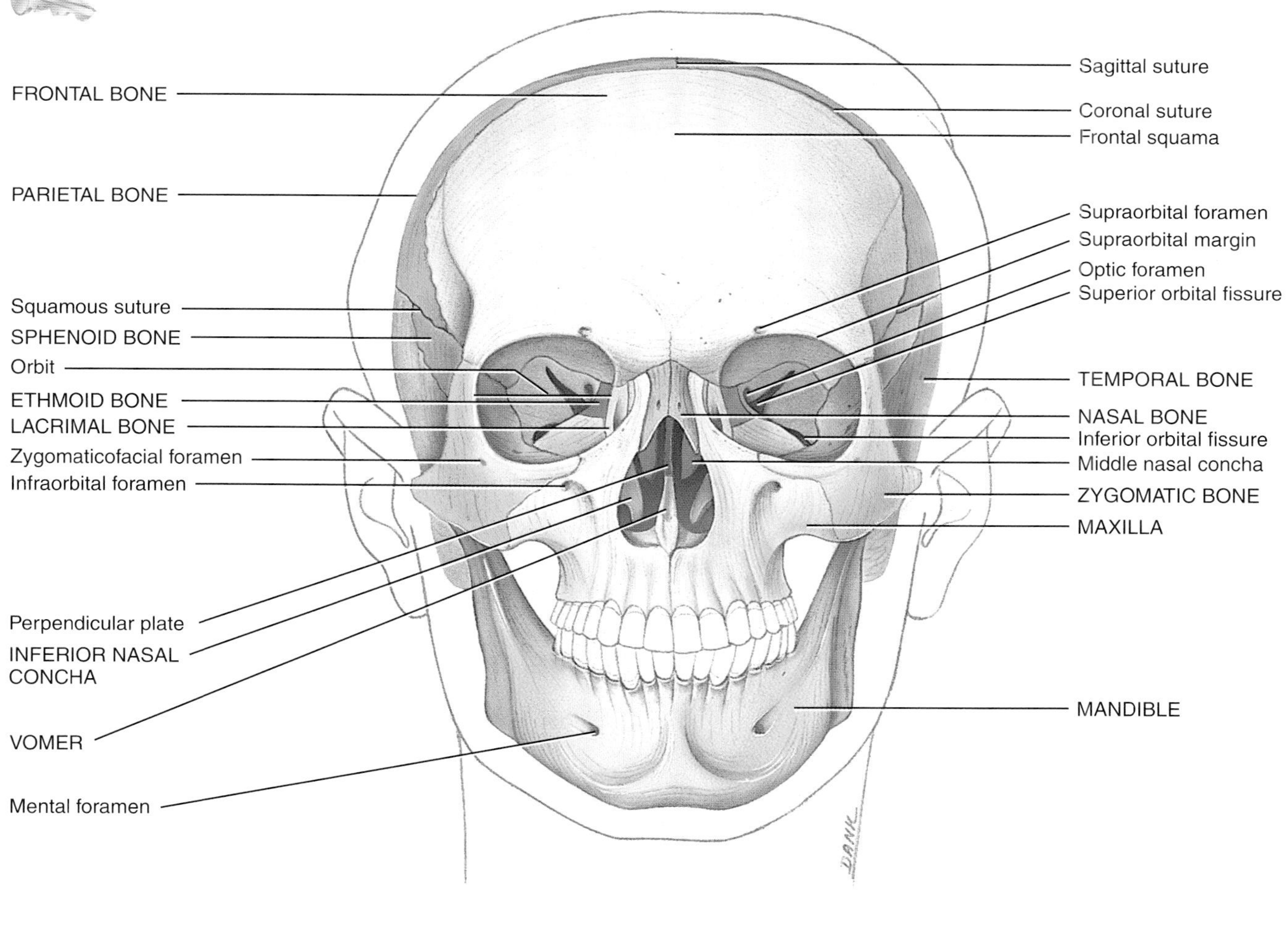

Which of the bones shown here are cranial bones?

vertical. Superior to the orbits the frontal bone thickens, forming the *supraorbital margin* (*supra-* = above; *-orbi* = circle). From this margin, the frontal bone extends posteriorly to form the roof of the orbit, which is part of the floor of the cranial cavity. Within the supraorbital margin, slightly medial to its midpoint, is a hole called the *supraorbital foramen.* Sometimes the foramen is incomplete and is called the *supraorbital notch.* As you read about each foramen associated with a cranial bone, refer to Table 7.3 on page 210 to note which structures pass through it. The *frontal sinuses* lie deep to the frontal squama. Sinuses, or more technically paranasal sinuses, are mucous membrane-lined cavities in certain skull bones that will be discussed later.

Black Eye

Just superior to the supraorbital margin is a sharp ridge. A blow to the ridge often fractures the bone or lacerates the skin over it, resulting in bleeding. Bruising of the skin over the ridge causes tissue fluid and blood to accumulate in the surrounding connective tissue. The resulting swelling and discoloration is called a **black eye.** ■

Parietal Bones

The two **parietal bones** (pa-RĪ-e-tal; *pariet-* = wall) form the greater portion of the sides and roof of the cranial cavity (Figure 7.4). The internal surfaces of the parietal bones contain many protrusions and depressions that accommodate the blood vessels supplying the dura mater, the superficial connective tissue covering of the brain.

Temporal Bones

The paired **temporal bones** (*tempor-* = temple) form the inferior lateral aspects of the cranium and part of the cranial floor. In

Figure 7.4 Right lateral view of skull. Although the hyoid bone is not part of the skull, it is included in the illustration for reference. (See Tortora, *A Photographic Atlas of the Human Body, Second Edition,* Figure 3.3.)

The zygomatic arch is formed by the zygomatic process of the temporal bone and the temporal process of the zygomatic bone.

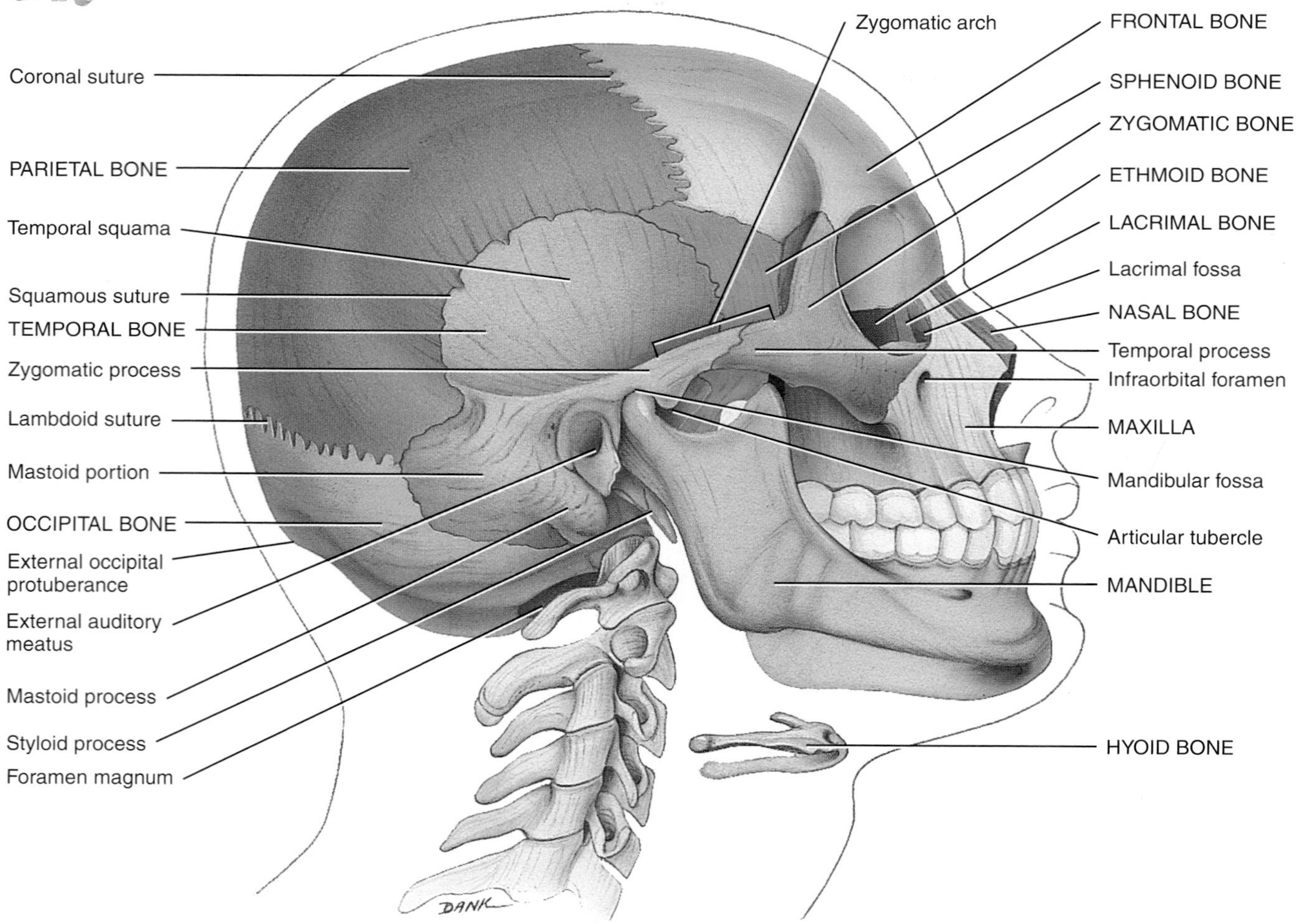

Right lateral view

What major bones are joined by the squamous suture, the lambdoid suture, and the coronal suture?

Figure 7.4, note the *temporal squama* (= scale), the thin, flat part of the temporal bone that forms the anterior and superior part of the temple. Projecting from the inferior portion of the temporal squama is the *zygomatic process,* which articulates (forms a joint) with the temporal process of the zygomatic (cheek) bone. Together, the zygomatic process of the temporal bone and the temporal process of the zygomatic bone form the *zygomatic arch.*

A socket called the *mandibular fossa* is located on the inferior posterior surface of the zygomatic process of each temporal bone. Anterior to the mandibular fossa is a rounded elevation, the *articular tubercle* (Figure 7.4). The mandibular fossa and articular tubercle articulate with the mandible (lower jawbone) to form the *temporomandibular joint (TMJ).*

The *mastoid portion* (*mastoid* = breast-shaped; see Figure 7.4) of the temporal bone is located posterior and inferior to the *external auditory meatus* (*meatus* = passageway), or ear canal, which directs sound waves into the ear. In an adult, this portion of the bone contains several *mastoid "air cells."* These tiny air-filled compartments are separated from the brain by thin bony partitions. In cases of **mastoiditis** (inflammation of the mastoid air cells caused, for example, by a middle-ear infection), the infection may spread to the brain.

The *mastoid process* is a rounded projection of the mastoid portion of the temporal bone posterior to the external auditory meatus. It is the point of attachment for several neck muscles. The *internal auditory meatus* (Figure 7.5) is the opening through which the facial (VII) nerve and vestibulocochlear (VIII) nerve pass. The *styloid process* (*styl-* = stake or pole) projects inferiorly from the inferior surface of the temporal bone and serves as a point of attachment for muscles and ligaments of the tongue and neck (see Figure 7.4). Between the styloid process and the mastoid process is the *stylomastoid foramen,* through which the facial (VII) nerve and stylomastoid artery pass (see Figure 7.7).

Figure 7.5 Medial view of sagittal section of skull. (See Tortora, *A Photographic Atlas of the Human Body, Second Edition,* Figure 3.4.)

The cranial bones are the frontal, parietal, temporal, occipital, sphenoid, and ethmoid bones. The facial bones are the nasal bones, maxillae, zygomatic bones, lacrimal bones, palatine bones, mandible, and vomer.

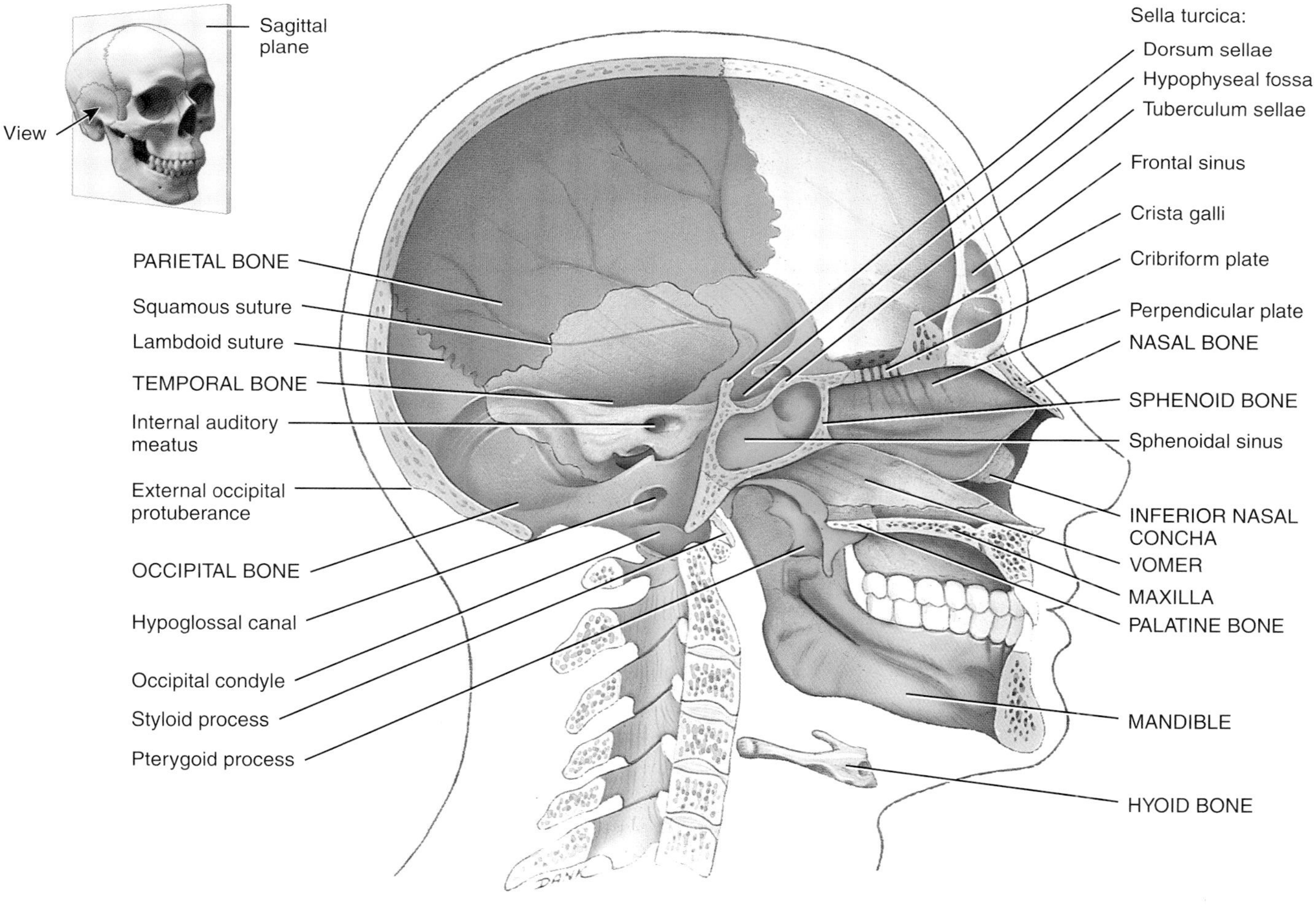

Medial view of sagittal section

With which bones does the temporal bone articulate?

At the floor of the cranial cavity (see Figure 7.8a) is the *petrous portion* (*petrous* = rock) of the temporal bone. This triangular part, located at the base of the skull between the sphenoid and occipital bones, houses the internal ear and the middle ear, structures involved in hearing and equilibrium (balance). It also contains the *carotid foramen,* through which the carotid artery passes (see Figure 7.7). Posterior to the carotid foramen and anterior to the occipital bone is the *jugular foramen,* a passageway for the jugular vein.

Occipital Bone

The **occipital bone** (ok-SIP-i-tal; *occipit-* = back of head) forms the posterior part and most of the base of the cranium (Figure 7.6; also see Figure 7.4). Also view the occipital bone and surrounding structures in the inferior view of the skull in Figure 7.7. The *foramen magnum* (= large hole) is in the inferior part of the bone. The medulla oblongata (inferior part of the brain) connects with the spinal cord within this foramen, and the vertebral and spinal arteries also pass through it. The *occipital condyles,* oval processes with convex surfaces on either side of the foramen magnum (Figure 7.7), articulate with depressions on the first cervical vertebra (atlas) to form the *atlanto-occipital joint,* which allows you to nod your head "yes." Superior to each occipital condyle on the inferior surface of the skull is the *hypoglossal canal* (*hypo-* = under; *-glossal* = tongue). (See Figure 7.5.)

The *external occipital protuberance* is a prominent midline projection on the posterior surface of the bone just above the foramen magnum. You may be able to feel this structure as a bump on the back of your head, just above your neck. (See Figure 7.4.) A large fibrous, elastic ligament, the *ligamentum nuchae* (*nucha-* = nape of neck), extends from the external occipital protuberance to the seventh cervical vertebra to help support the head. Extending laterally from the protuberance

Figure 7.6 Posterior view of skull. The sutures are exaggerated for emphasis. (See Tortora, *A Photographic Atlas of the Human Body, Second Edition,* Figure 3.5.)

The occipital bone forms most of the posterior and inferior portions of the cranium.

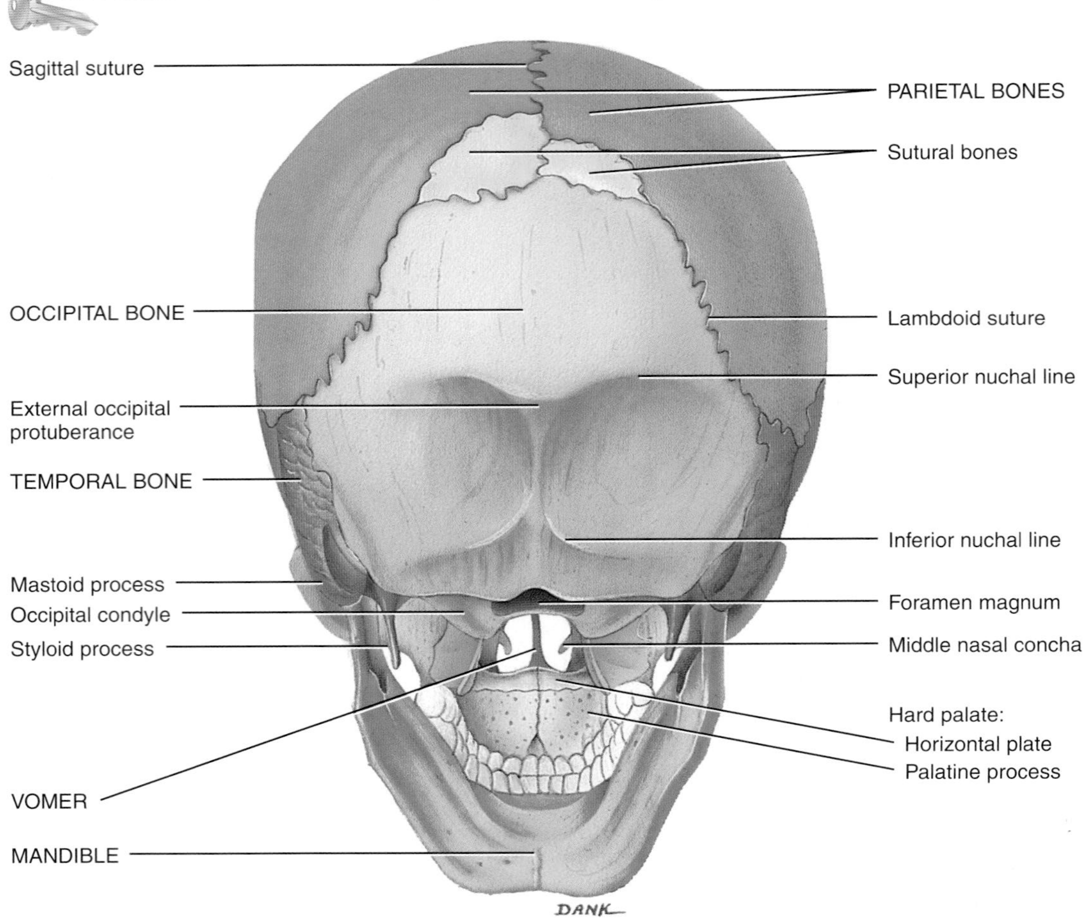

Which bones form the posterior, lateral portion of the cranium?

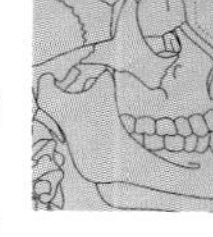

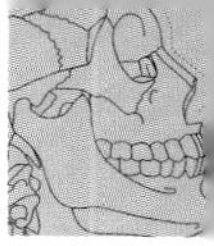

are two curved ridges, the *superior nuchal lines,* and below these are two *inferior nuchal lines,* which are areas of muscle attachment (Figure 7.7).

Sphenoid Bone

The **sphenoid bone** (SFĒ-noyd = wedge-shaped) lies at the middle part of the base of the skull (Figures 7.7 and 7.8). This bone is the keystone of the cranial floor because it articulates with all the other cranial bones, holding them together. View the floor of the cranium superiorly (Figure 7.8a) and note the sphenoid articulations. It joins anteriorly with the frontal bone, laterally with the temporal bones, and posteriorly with the occipital bone. The sphenoid lies posterior and slightly superior to the nasal cavity and forms part of the floor, side walls, and rear wall of the orbit (see Figure 7.12).

The shape of the sphenoid resembles a bat with outstretched wings (see Figure 7.8b). The *body* of the sphenoid is the cube-like medial portion between the ethmoid and occipital bones. It contains the *sphenoidal sinuses,* which drain into the nasal cavity (see Figure 7.13). The *sella turcica* (SEL-a TUR-si-ka; *sella* = saddle; *turcica* = Turkish) is a bony saddle-shaped structure on the superior surface of the body of the sphenoid (Figure 7.8a). The anterior part of the sella turcica, which forms

Figure 7.7 **Inferior view of skull.** The mandible (lower jawbone) has been removed. (See Tortora, *A Photographic Atlas of the Human Body, Second Edition,* Figure 3.7.)

The occipital condyles of the occipital bone articulate with the first cervical vertebra to form the atlanto-occipital joints.

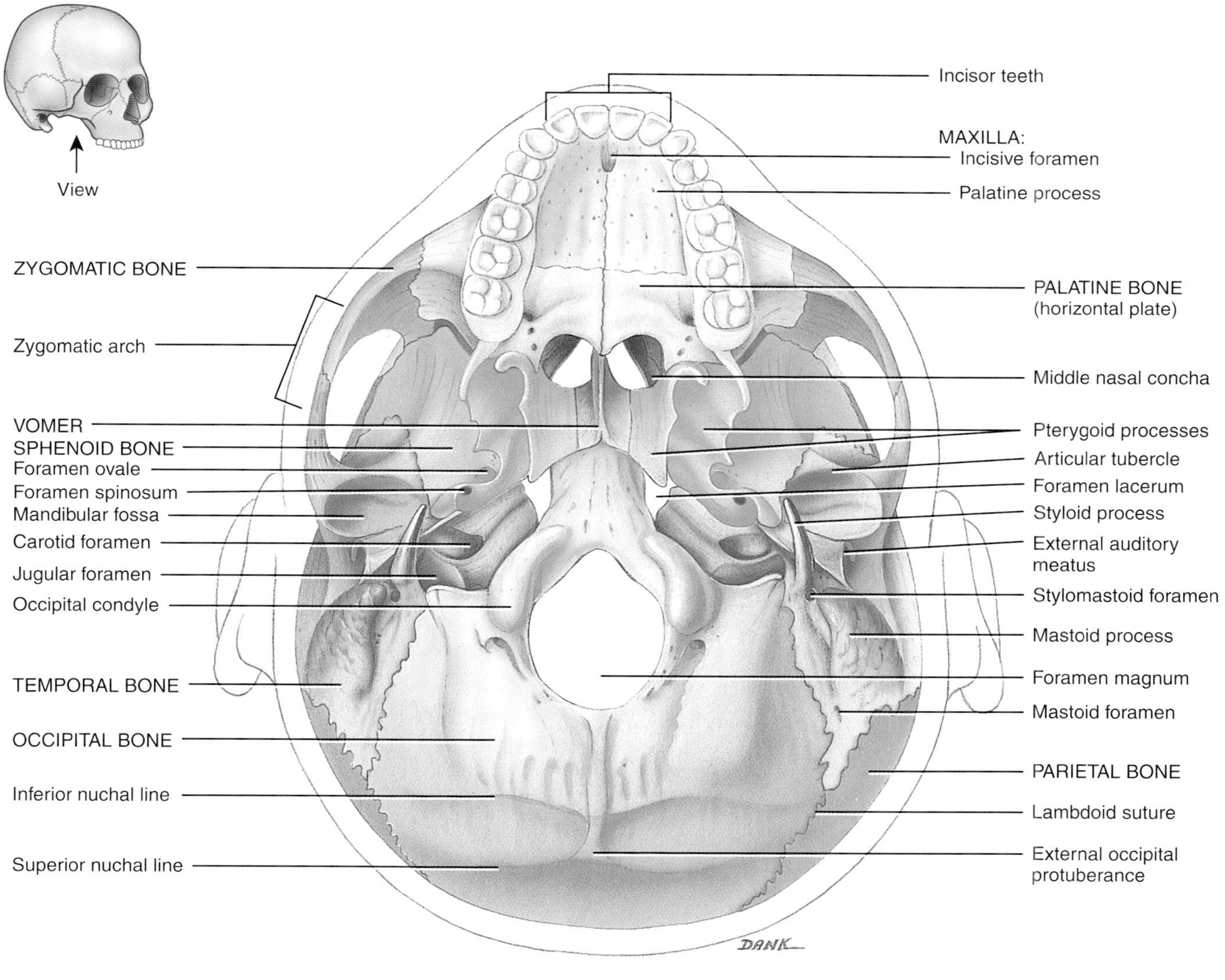

View

Incisor teeth

MAXILLA:

Incisive foramen

Palatine process

ZYGOMATIC BONE

PALATINE BONE (horizontal plate)

Zygomatic arch

Middle nasal concha

VOMER

SPHENOID BONE

Foramen ovale

Foramen spinosum

Mandibular fossa

Carotid foramen

Jugular foramen

Occipital condyle

Pterygoid processes

Articular tubercle

Foramen lacerum

Styloid process

External auditory meatus

Stylomastoid foramen

Mastoid process

TEMPORAL BONE

Foramen magnum

Mastoid foramen

OCCIPITAL BONE

PARIETAL BONE

Inferior nuchal line

Lambdoid suture

Superior nuchal line

External occipital protuberance

DANK

Inferior view

What parts of the nervous system join together within the foramen magnum?

Figure 7.8 Sphenoid bone. (See Tortora, *A Photographic Atlas of the Human Body, Second Edition,* Figures 3.8 and 3.9.)

The sphenoid bone is called the keystone of the cranial floor because it articulates with all other cranial bones, holding them together.

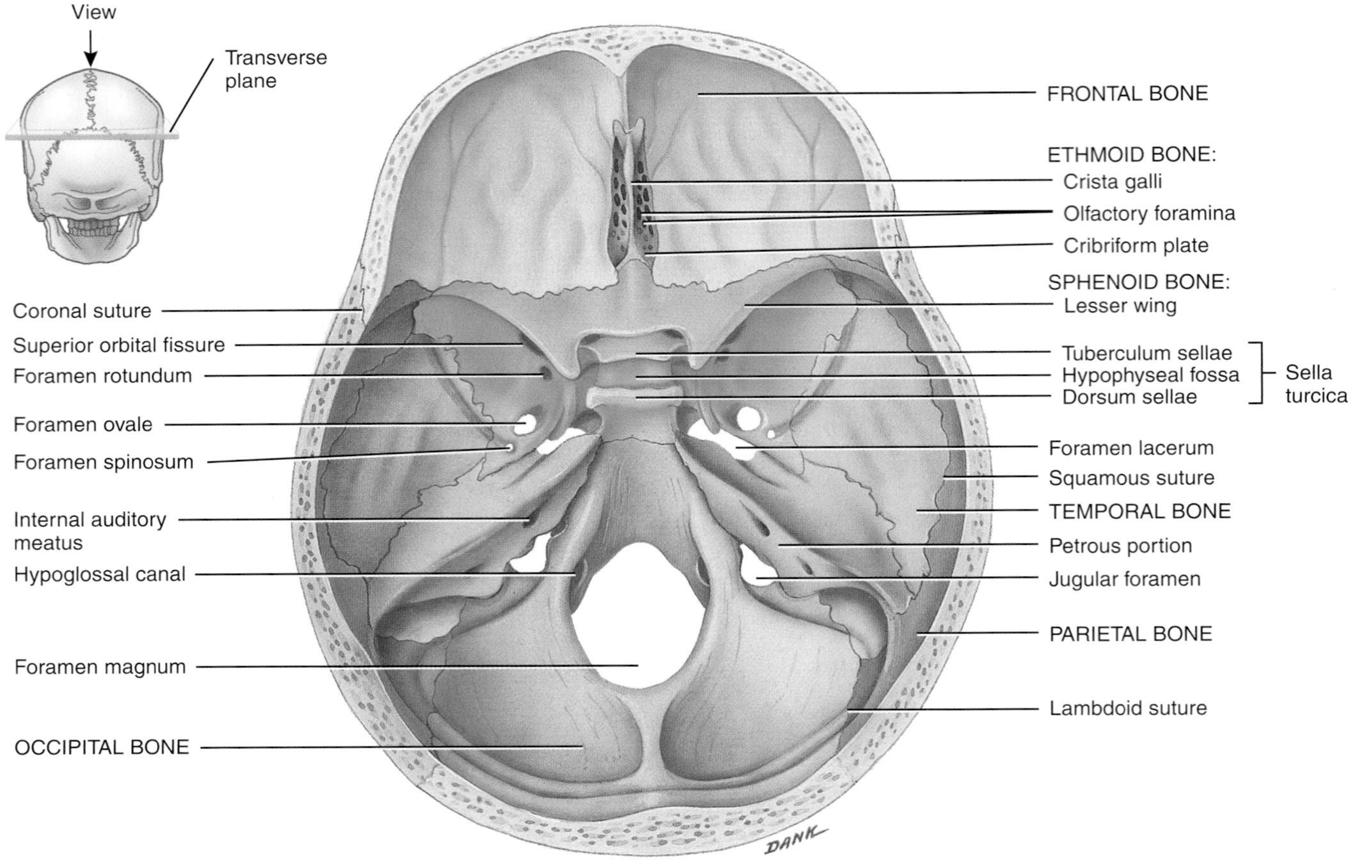

View
Transverse plane
Coronal suture
Superior orbital fissure
Foramen rotundum
Foramen ovale
Foramen spinosum
Internal auditory meatus
Hypoglossal canal
Foramen magnum
OCCIPITAL BONE
FRONTAL BONE
ETHMOID BONE:
Crista galli
Olfactory foramina
Cribriform plate
SPHENOID BONE:
Lesser wing
Tuberculum sellae
Hypophyseal fossa
Dorsum sellae
Sella turcica
Foramen lacerum
Squamous suture
TEMPORAL BONE
Petrous portion
Jugular foramen
PARIETAL BONE
Lambdoid suture
DANK

(a) Superior view of sphenoid bone in floor of cranium

Frontal plane
View
Greater wings
Lesser wings
Optic foramen
Sphenoidal sinus
Foramen rotundum
Superior orbital fissure
Body
Pterygoid processes
DANK

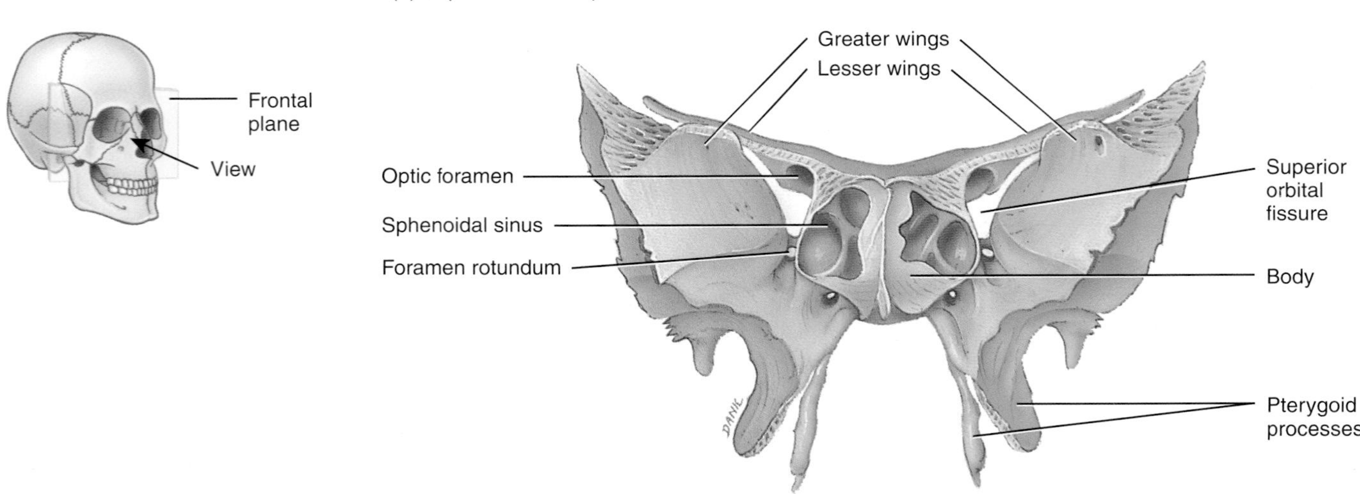

(b) Anterior view of sphenoid bone

Name the bones that articulate with the sphenoid bone, starting at the crista galli of the ethmoid bone and going in a clockwise direction.

the horn of the saddle, is a ridge called the *tuberculum sellae.* The seat of the saddle is a depression, the *hypophyseal fossa* (hī-pō-FIZ-ē-al), which contains the pituitary gland. The posterior part of the sella turcica, which forms the back of the saddle, is another ridge called the *dorsum sellae.*

The *greater wings* of the sphenoid project laterally from the body and form the anterolateral floor of the cranium. The greater wings also form part of the lateral wall of the skull just anterior to the temporal bone and can be viewed externally. The *lesser wings,* which are smaller, form a ridge of bone anterior and superior to the greater wings. They form part of the floor of the cranium and the posterior part of the orbit of the eye.

Between the body and lesser wing just anterior to the sella turcica is the *optic foramen* (*optic* = eye), through which the optic (II) nerve and ophthalmic artery pass. Lateral to the body between the greater and lesser wings is a triangular slit called the *superior orbital fissure.* This fissure may also be seen in the anterior view of the orbit in Figure 7.12.

The *pterygoid processes* (TER-i-goyd = winglike) project inferiorly from the points where the body and greater wings of the sphenoid bone unite; they form the lateral posterior region of the nasal cavity (see Figures 7.7 and 7.8b). Some of the muscles that move the mandible attach to the pterygoid processes. At the base of the lateral pterygoid process in the greater wing is the *foramen ovale* (= oval hole). The *foramen lacerum* (= lacerated), covered in part by a layer of fibrocartilage in living subjects, is bounded anteriorly by the sphenoid bone and medially by the sphenoid and occipital bones. It transmits a branch of the ascending pharyngeal artery. Another foramen associated with the sphenoid bone is the *foramen rotundum* (= round hole) located at the junction of the anterior and medial parts of the sphenoid bone. The maxillary branch of the trigeminal (V) nerve passes through the foramen rotundum.

Ethmoid Bone

The **ethmoid bone** (ETH-moyd = like a sieve) is spongelike in appearance and is located on the midline in the anterior part of the cranial floor medial to the orbits (Figure 7.9). It is anterior to the sphenoid and posterior to the nasal bones. The ethmoid bone forms (1) part of the anterior portion of the cranial floor; (2) the medial wall of the orbits; (3) the superior portion of the nasal septum, a partition that divides the nasal cavity into right and left sides; and (4) most of the superior sidewalls of the nasal cavity. The ethmoid bone is a major superior supporting structure of the nasal cavity.

The *cribriform plate* (*cribri-* = sieve) of the ethmoid bone lies in the anterior floor of the cranium and forms the roof of the nasal cavity. The cribriform plate contains the *olfactory foramina* (*olfact-* = to smell) through which the olfactory nerves pass. Projecting superiorly from the cribriform plate is a triangular process called the *crista galli* (*crista* = crest; *galli* = cock), which serves as a point of attachment for the membranes that cover the brain. Projecting inferiorly from the cribriform plate is the *perpendicular plate,* which forms the superior portion of the nasal septum (see Figure 7.11).

The *lateral masses* of the ethmoid bone compose most of the wall between the nasal cavity and the orbits. They contain 3 to 18 air spaces, or cells. The ethmoidal cells together form the *ethmoidal sinuses* (see Figure 7.13). The lateral masses contain two thin, scroll-shaped projections lateral to the nasal septum. These are called the *superior nasal concha* (KONG-ka = shell) or *turbinate* and the *middle nasal concha (turbinate).* The plural form is *conchae* (KONG-kē). A third pair of conchae, the inferior nasal conchae, are separate bones (discussed shortly). The conchae increase the vascular and mucous membrane surface area in the nasal cavities, which aids in the sense of smell and warms, moistens, and filters inhaled air before it passes into the lungs. The conchae filter air by causing the inhaled air to swirl; as a result, many inhaled particles strike and become trapped in the mucus that lines the nasal passageways. In this way the conchae help cleanse inhaled air before it passes into the rest of the respiratory tract. The superior nasal conchae also participate in the sense of smell.

Facial Bones

The shape of the face changes dramatically during the first two years after birth. The brain and cranial bones expand, the first set of teeth form and erupt (emerge), and the paranasal sinuses increase in size. Growth of the face ceases at about 16 years of age. The 14 facial bones include two nasal bones, two maxillae (or maxillas), two zygomatic bones, the mandible, two lacrimal bones, two palatine bones, two inferior nasal conchae, and the vomer.

Nasal Bones

The paired **nasal bones** meet at the midline (see Figure 7.3) and form part of the bridge of the nose. The rest of the supporting tissue of the nose consists of cartilage.

Maxillae

The paired **maxillae** (mak-SIL-ē = jawbones; singular is *maxilla*) unite to form the upper jawbone. They articulate with every bone of the face except the mandible (lower jawbone) (see Figures 7.4 and 7.7). The maxillae form part of the floors of the orbits, part of the lateral walls and floor of the nasal cavity, and most of the hard palate. The hard palate is the bony roof of the mouth, and is formed by the palatine processes of the maxillae and horizontal plates of the palatine bones. The hard palate separates the nasal cavity from the oral cavity.

Each maxilla contains a large *maxillary sinus* that empties into the nasal cavity (see Figure 7.13). The *alveolar process* (al-VĒ-ō-lar; *alveol-* = small cavity) of the maxilla is an arch that contains the *alveoli* (sockets) for the maxillary (upper) teeth. The *palatine process* is a horizontal projection of the maxilla

Figure 7.9 Ethmoid bone. (See Tortora, *A Photographic Atlas of the Human Body, Second Edition,* Figure 3.10.)

The ethmoid bone forms part of the anterior portion of the cranial floor, the medial wall of the orbits, the superior portions of the nasal septum, and most of the side walls of the nasal cavity.

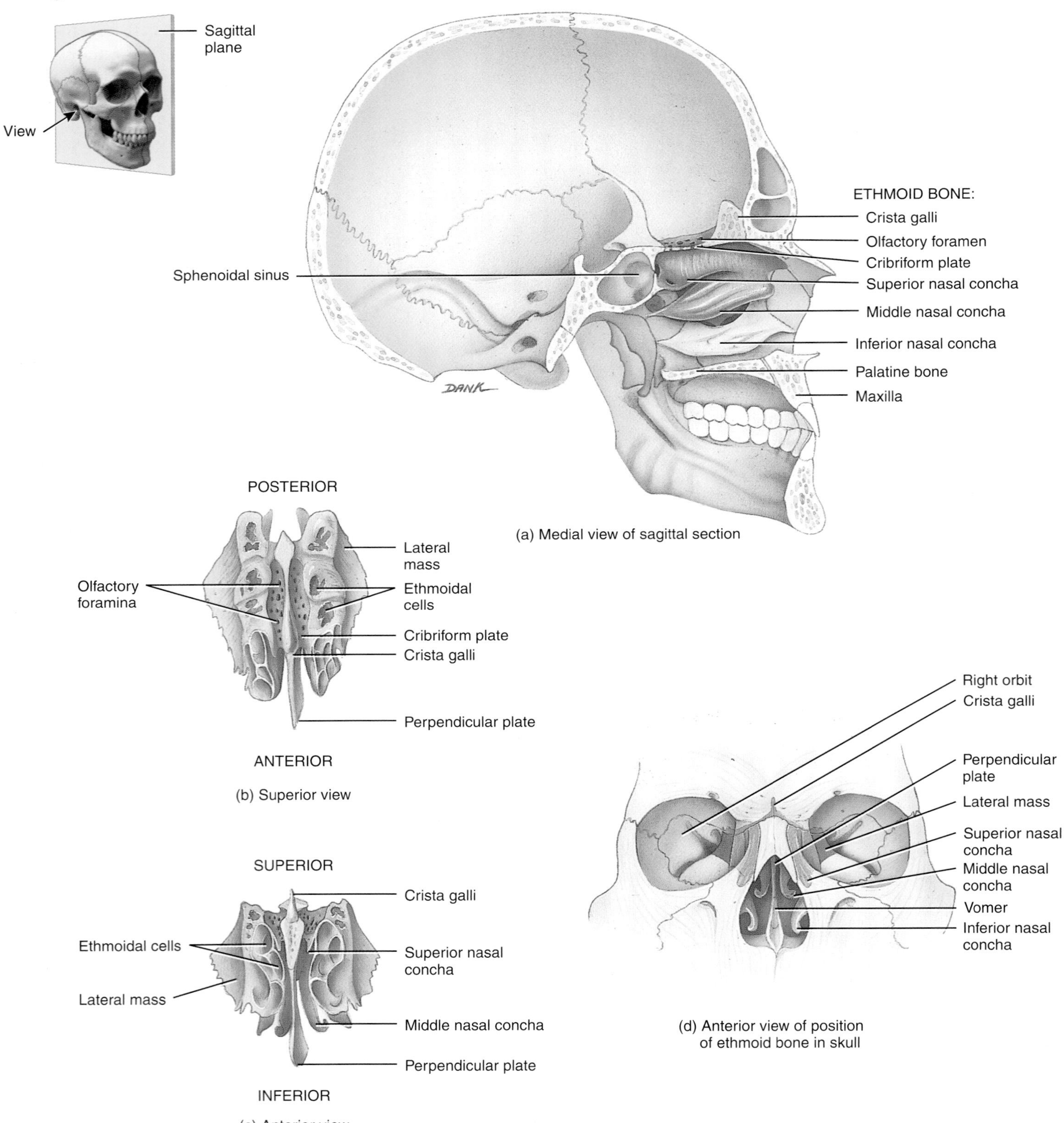

(a) Medial view of sagittal section

(b) Superior view

(c) Anterior view

(d) Anterior view of position of ethmoid bone in skull

What part of the ethmoid bone forms the superior part of the nasal septum? The medial walls of the orbits?

that forms the anterior three-quarters of the hard palate. The union and fusion of the maxillary bones normally is completed before birth. If this fusion fails, this condition is referred to as a cleft palate (described later).

The *infraorbital foramen* (*infra-* = below; *orbital* = orbit; see Figure 7.3), an opening in the maxilla inferior to the orbit, allows passage of the infraorbital nerve and blood vessels and a branch of the maxillary division of the trigeminal (V) nerve. Another prominent foramen in the maxilla is the *incisive foramen* (= incisor teeth) just posterior to the incisor teeth (see Figure 7.7). It transmits branches of the greater palatine blood vessels and nasopalatine nerve. A final structure associated with the maxilla and sphenoid bone is the *inferior orbital fissure,* located between the greater wing of the sphenoid and the maxilla (see Figure 7.12).

Cleft Palate and Cleft Lip

Usually the palatine processes of the maxillary bones unite during weeks 10 to 12 of embryonic development. Failure to do so can result in one type of **cleft palate.** The condition may also involve incomplete fusion of the horizontal plates of the palatine bones (see Figure 7.7). Another form of this condition, called **cleft lip,** involves a split in the upper lip. Cleft lip and cleft palate often occur together. Depending on the extent and position of the cleft, speech and swallowing may be affected. In addition, children with cleft palate tend to have many ear infections, which can lead to hearing loss. Facial and oral surgeons recommend closure of cleft lip during the first few weeks following birth, and surgical results are excellent. Repair of cleft palate typically is completed between 12 and 18 months of age, ideally before the child begins to talk. Because the palate is important for pronouncing consonants, speech therapy may be required, and orthodontic therapy may be needed to align the teeth. Again, results are usually excellent. Supplementation with folic acid (one of the B vitamins) during pregnancy decreases the incidence of cleft palate and cleft lip. ■

Zygomatic Bones

The two **zygomatic bones** (*zygo-* = yokelike), commonly called cheekbones, form the prominences of the cheeks and part of the lateral wall and floor of each orbit (see Figure 7.12). They articulate with the frontal, maxilla, sphenoid, and temporal bones.

The *temporal process* of the zygomatic bone projects posteriorly and articulates with the zygomatic process of the temporal bone to form the *zygomatic arch* (see Figure 7.4).

Lacrimal Bones

The paired **lacrimal bones** (LAK-ri-mal; *lacrim-* = teardrops) are thin and roughly resemble a fingernail in size and shape (see Figures 7.3, 7.4, and 7.12). These bones, the smallest bones of the face, are posterior and lateral to the nasal bones and form a part of the medial wall of each orbit. The lacrimal bones each contain a *lacrimal fossa,* a vertical groove formed with the maxilla, that houses the lacrimal sac, a structure that gathers tears and passes them into the nasal cavity (see Figure 7.12).

Palatine Bones

The two L-shaped **palatine bones** (PAL-a-tīn) form the posterior portion of the hard palate, part of the floor and lateral wall of the nasal cavity, and a small portion of the floors of the orbits (Figures 7.7 and 7.12). The posterior portion of the hard palate is formed by the *horizontal plates* of the palatine bones (Figures 7.6 and 7.7).

Inferior Nasal Conchae

The two **inferior nasal conchae,** which are inferior to the middle nasal conchae of the ethmoid bone, are separate bones, not part of the ethmoid bone (see Figures 7.3 and 7.9a). These scroll-like bones form a part of the inferior lateral wall of the nasal cavity and project into the nasal cavity. All three pairs of nasal conchae (superior, middle, and inferior) help swirl and filter air before it passes into the lungs. However, only the superior nasal conchae of the ethmoid bone are involved in the sense of smell.

Vomer

The **vomer** (VŌ-mer = plowshare) is a roughly triangular bone on the floor of the nasal cavity that articulates superiorly with the perpendicular plate of the ethmoid bone and inferiorly with both the maxillae and palatine bones along the midline (see Figures 7.3, 7.7, and 7.11). It forms the inferior portion of the nasal septum.

Mandible

The **mandible** (*mand-* = to chew), or lower jawbone, is the largest, strongest facial bone (Figure 7.10). It is the only

Figure 7.10 Mandible.

The mandible is the largest and strongest facial bone.

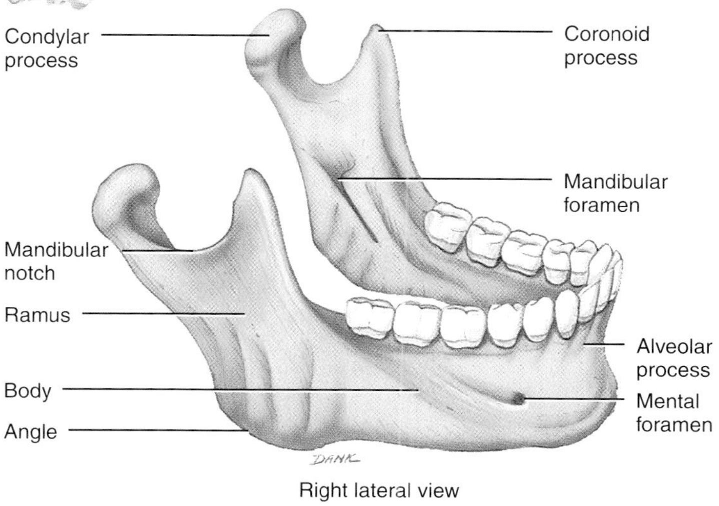

What is the distinctive functional feature of the mandible among all the skull bones?

movable skull bone (other than the auditory ossicles). In the lateral view, you can see that the mandible consists of a curved, horizontal portion, the *body,* and two perpendicular portions, the *rami* (RĀ-mī = branches; singular is *ramus*). The *angle* of the mandible is the area where each *ramus* meets the body. Each ramus has a posterior *condylar process* (KON-di-lar) that articulates with the mandibular fossa and articular tubercle of the temporal bone (see Figure 7.4) to form the **temporomandibular joint (TMJ),** and an anterior *coronoid process* (KOR-ō-noyd) to which the temporalis muscle attaches. The depression between the coronoid and condylar processes is called the *mandibular notch.* The *alveolar process* is an arch containing the *alveoli* (sockets) for the mandibular (lower) teeth.

The *mental foramen* (*ment-* = chin) is approximately inferior to the second premolar tooth. It is near this foramen that dentists reach the mental nerve when injecting anesthetics. Another foramen associated with the mandible is the *mandibular foramen* on the medial surface of each ramus, another site often used by dentists to inject anesthetics. The mandibular foramen is the beginning of the *mandibular canal,* which runs obliquely in the ramus and anteriorly to the body. Through the canal pass the inferior alveolar nerves and blood vessels, which are distributed to the mandibular teeth.

Temporomandibular Joint Syndrome

One problem associated with the temporomandibular joint is **temporomandibular joint (TMJ) syndrome.** It is characterized by dull pain around the ear, tenderness of the jaw muscles, a clicking or popping noise when opening or closing the mouth, limited or abnormal opening of the mouth, headache, tooth sensitivity, and abnormal wearing of the teeth. TMJ syndrome can be caused by improperly aligned teeth, grinding or clenching the teeth, trauma to the head and neck, or arthritis. Treatments include application of moist heat or ice, limiting the diet to soft foods, administration of pain relievers such as aspirin, muscle retraining, adjustment or reshaping of the teeth (orthodontic treatment), and surgery. ■

Nasal Septum

The inside of the nose, called the nasal cavity, is divided into right and left sides by a vertical partition called the **nasal septum,** which consists of bone and cartilage. The three components of the nasal septum are the vomer, septal cartilage, and the perpendicular plate of the ethmoid bone (Figure 7.11). The anterior border of the vomer articulates with the septal cartilage, which is hyaline cartilage, to form the anterior portion of the septum. The superior border of the vomer articulates with the perpendicular plate of the ethmoid bone to form the remainder of the nasal septum. The term "broken nose," in most cases, refers to damage to the septal cartilage rather than the nasal bones themselves.

Deviated Nasal Septum

A **deviated nasal septum** is one that is deflected laterally from the midline of the nose. The deviation usually occurs at the junction of the vomer with the septal cartilage. Septal deviations may occur due to a developmental abnormality or trauma. If the deviation is severe, it may block the nasal passageway entirely. Even a partial blockage may lead to infection. If inflammation occurs, it may cause nasal congestion, blockage of the paranasal sinus openings, chronic sinusitis, headache, and nosebleeds. The condition usually can be corrected, or at least improved, surgically. ■

Figure 7.11 Nasal septum. (See Tortora, *A Photographic Atlas of the Human Body, Second Edition,* Figure 3.4.)

The structures that form the nasal septum are the perpendicular plate of the ethmoid bone, the vomer, and septal cartilage.

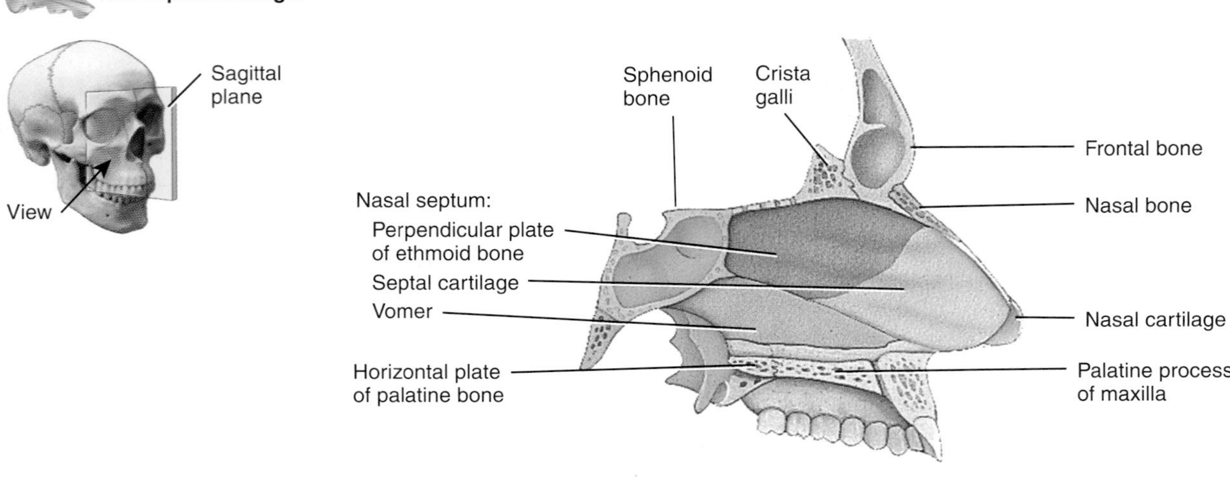

What is the function of the nasal septum?

Orbits

Seven bones of the skull join to form each **orbit** (eye socket), which contains the eyeball and associated structures (Figure 7.12). The three cranial bones of the orbit are the frontal, sphenoid, and ethmoid; the four facial bones are the palatine, zygomatic, lacrimal, and maxilla. Each pyramid-shaped orbit has four regions that converge posteriorly:

1. Parts of the frontal and sphenoid bones comprise the *roof* of the orbit.
2. Parts of the zygomatic and sphenoid bones form the *lateral wall* of the orbit.
3. Parts of the maxilla, zygomatic, and palatine bones make up the *floor* of the orbit.
4. Parts of the maxilla, lacrimal, ethmoid, and sphenoid bones form the *medial wall* of the orbit.

Associated with each orbit are five openings:

1. The *optic foramen* is at the junction of the roof and medial wall.
2. The *superior orbital fissure* is at the superior lateral angle of the apex.
3. The *inferior orbital fissure* is at the junction of the lateral wall and floor.
4. The *supraorbital foramen* is on the medial side of the supraorbital margin of the frontal bone.
5. The *lacrimal fossa* is in the lacrimal bone.

Foramina

We mentioned most of the **foramina** (openings for blood vessels, nerves, or ligaments) of the skull in the descriptions of the cranial and facial bones that they penetrate. As preparation for studying other systems of the body, especially the nervous and cardiovascular systems, these foramina and the structures passing through them are listed in Table 7.3. For your convenience and for future reference, the foramina are listed alphabetically.

Unique Features of the Skull

The skull exhibits several unique features not seen in other bones of the body. These include sutures, paranasal sinuses, and fontanels.

Sutures

A **suture** (SOO-chur = seam) is an immovable joint in an adult that is found only between skull bones and that holds most skull bones together. Sutures in the skulls of infants and children often are movable. The names of many sutures reflect the bones they unite. For example, the frontozygomatic suture is between the frontal bone and the zygomatic bone. Similarly, the sphenoparietal suture is between the sphenoid bone and the parietal bone. In other cases, however, the names of sutures are not so obvious. Of the many sutures found in the skull, we will identify only four prominent ones:

Figure 7.12 Details of the orbit (eye socket). (See Tortora, *A Photographic Atlas of the Human Body, Second Edition,* Figure 3.11.)

The orbit is a pyramid-shaped structure that contains the eyeball and associated structures.

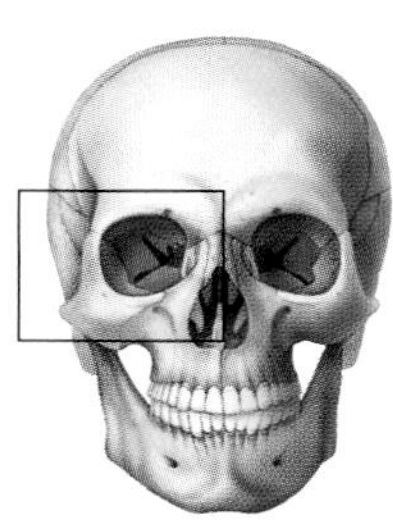

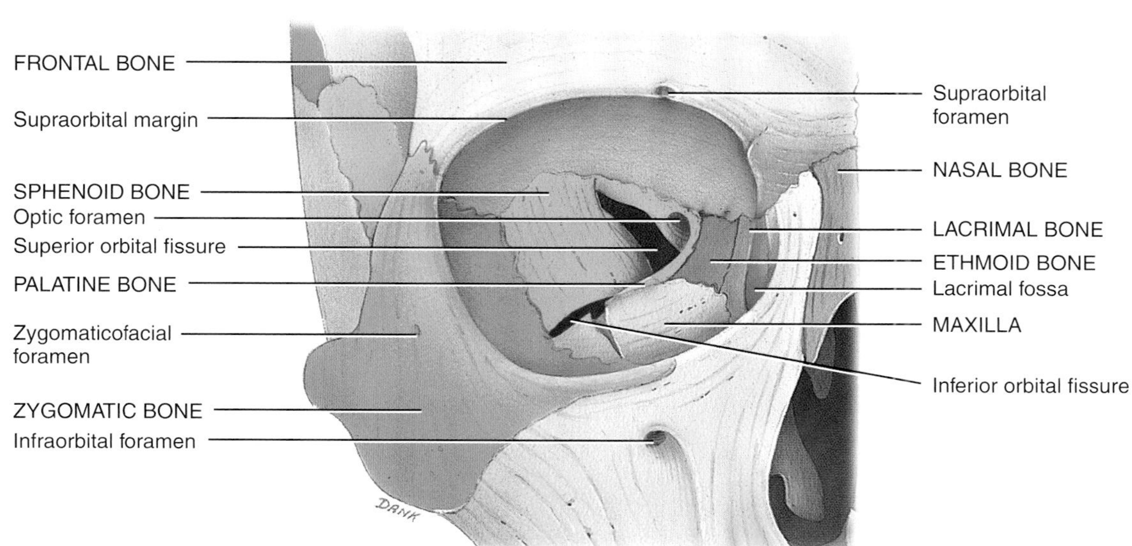

Anterior view showing the bones of the right orbit

Which seven bones form the orbit?

TABLE 7.3 **Principal Foramina of the Skull**

Foramen	Location	Structures Passing Through*
Carotid (relating to carotid artery in neck)	Petrous portion of temporal bone (Figure 7.7).	Internal carotid artery and sympathetic nerves for eyes.
Hypoglossal (*hypo-* = under; *glossus* = tongue)	Superior to base of occipital condyles (Figure 7.8a).	Cranial nerve XII (hypoglossal) and branch of ascending pharyngeal artery.
Infraorbital (*infra-* = below)	Inferior to orbit in maxilla (Figure 7.12).	Infraorbital nerve and blood vessels and a branch of the maxillary division of cranial nerve V (trigeminal).
Jugular (*jugul-* = the throat)	Posterior to carotid canal between petrous portion of temporal bone and occipital bone (Figure 7.8a).	Internal jugular vein, cranial nerves IX (glossopharyngeal), X (vagus), and XI (accessory).
Lacerum (*lacerum* = lacerated)	Bounded anteriorly by sphenoid bone, posteriorly by petrous portion of temporal bone, and medially by sphenoid and occipital bones (Figure 7.8a).	Branch of ascending pharyngeal artery.
Magnum (= large)	Occipital bone (Figure 7.7).	Medulla oblongata and its membranes (meninges), cranial nerve XI (accessory), and vertebral and spinal arteries.
Mandibular (*mand-* = to chew)	Medial surface of ramus of mandible (Figure 7.10).	Inferior alveolar nerve and blood vessels.
Mastoid (= breast-shaped)	Posterior border of mastoid process of temporal bone (Figure 7.7).	Emissary vein to transverse sinus and branch of occipital artery to dura mater.
Mental (*ment-* = chin)	Inferior to second premolar tooth in mandible (Figure 7.10).	Mental nerve and vessels.
Olfactory (*olfact-* = to smell)	Cribriform plate of ethmoid bone (Figure 7.8a).	Cranial nerve I (olfactory).
Optic (= eye)	Between superior and inferior portions of small wing of sphenoid bone (Figure 7.12).	Cranial nerve II (optic) and ophthalmic artery.
Ovale (= oval)	Greater wing of sphenoid bone (Figure 7.8a).	Mandibular branch of cranial nerve V (trigeminal).
Rotundum (= round)	Junction of anterior and medial parts of sphenoid bone (Figure 7.8a and b).	Maxillary branch of cranial nerve V (trigeminal).
Stylomastoid (*stylo-* = stake or pole)	Between styloid and mastoid processes of temporal bone (Figure 7.7).	Cranial nerve VII (facial) and stylomastoid artery.
Supraorbital (*supra-* = above)	Supraorbital margin of orbit in frontal bone (Figure 7.12).	Supraorbital nerve and artery.

*The cranial nerves listed here are described in Table 14.3 on page 500.

1. The **coronal suture** (KŌ-rō-nal; *coron-* = crown) unites the frontal bone and both parietal bones (see Figure 7.4).

2. The **sagittal suture** (SAJ-i-tal; *sagitt-* = arrow) unites the two parietal bones on the superior midline of the skull (see Figure 7.6). The sagittal suture is so named because in the infant, before the bones of the skull are firmly united, the suture and the fontanels (soft spots) associated with it resemble an arrow.

3. The **lambdoid suture** (LAM-doyd) unites the two parietal bones to the occipital bone. This suture is so named because of its resemblance to the Greek letter lambda (Λ), as can be seen in Figure 7.6. Sutural bones may occur within the sagittal and lambdoid sutures.

4. The **squamous sutures** (SKWĀ-mus; *squam-* = flat) unite the parietal and temporal bones on the lateral aspects of the skull (see Figure 7.4).

Paranasal Sinuses

The **paranasal sinuses** (*para-* = beside) are cavities in certain cranial and facial bones near the nasal cavity (Figure 7.13). The paranasal sinuses are lined with mucous membranes that are continuous with the lining of the nasal cavity. The frontal, sphenoid, ethmoid, and maxillary bones all contain paranasal sinuses. Besides producing mucus, the paranasal sinuses serve as resonating chambers for sound as we speak or sing.

Sinusitis

Secretions produced by the mucous membranes of the paranasal sinuses drain into the nasal cavity. An inflammation of the membranes due to an allergic reaction or infection is called **sinusitis** (sī-nū-SĪ-tis). If the membranes swell enough to block drainage into the nasal cavity, fluid pressure builds up in the paranasal sinuses, and a sinus headache results. A severely deviated nasal septum or nasal polyps, growths that can be removed surgically, may also cause chronic sinusitis. ■

Fontanels

The skeleton of a newly formed embryo consists of cartilage or mesenchyme arranged in sheetlike layers that resemble membranes shaped like the bones they will become. Gradually, ossification occurs—bone replaces most of the cartilage and mesenchyme. At birth, mesenchyme-filled spaces called

Figure 7.13 **Paranasal sinuses.** (See Tortora, *A Photographic Atlas of the Human Body, Second Edition,* Figure 3.4.)

Paranasal sinuses are mucous membrane-lined spaces in the frontal, sphenoid, ethmoid, and maxillary bones that connect to the nasal cavity.

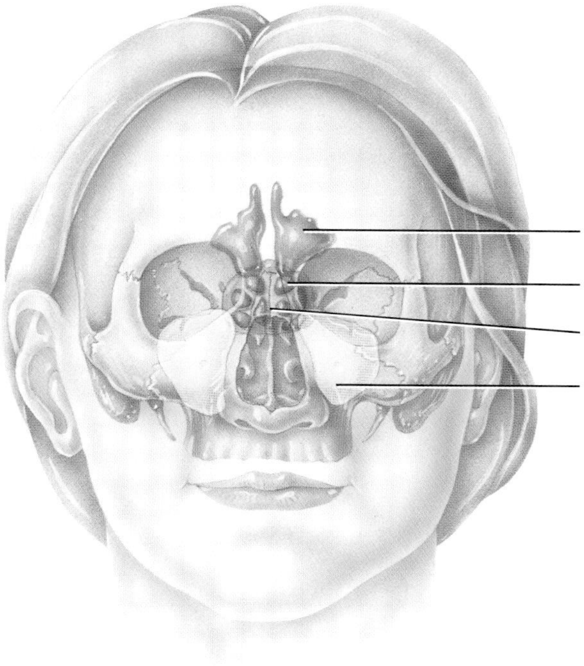

Frontal sinus
Ethmoidal cells
Sphenoidal sinus
Maxillary sinus

(a) Anterior view

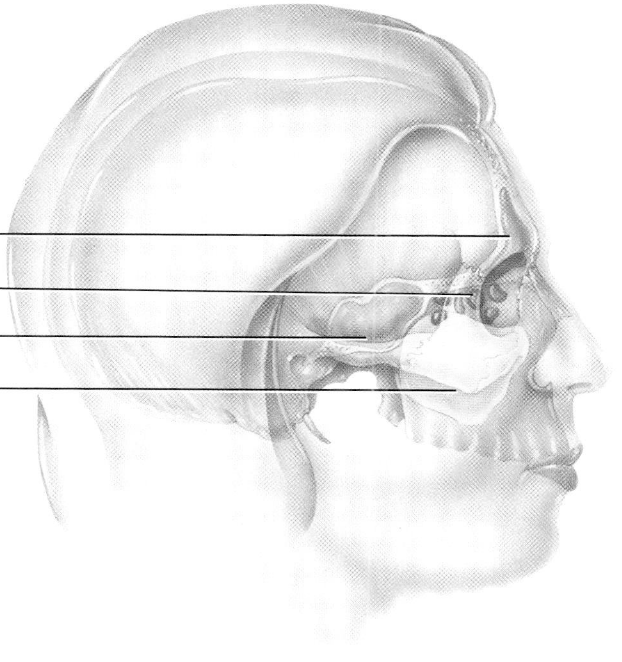

(b) Right lateral view

What are the functions of the paranasal sinuses?

fontanels (fon-ta-NELZ = little fountains), commonly called "soft spots," are present between the cranial bones (Figure 7.14). Fontanels are areas of unossified mesenchyme. Eventually, they will be replaced with bone by intramembranous ossification to become sutures. Functionally, the fontanels provide some flexibility to the fetal skull, allowing the skull to change shape as it passes through the birth canal and later permitting rapid growth of the brain during infancy. Although an infant may have many fontanels at birth, the form and location of six are fairly constant:

- The unpaired **anterior fontanel,** the largest fontanel, is located at the midline between the two parietal bones and the frontal bone, and is roughly diamond-shaped. It usually closes 18 to 24 months after birth.
- The unpaired **posterior fontanel** is located at the midline between the two parietal bones and the occipital bone. Because it is much smaller than the anterior fontanel, it generally closes about 2 months after birth.
- The paired **anterolateral fontanels,** located laterally between the frontal, parietal, temporal, and sphenoid bones, are small and irregular in shape. Normally, they close about 3 months after birth.

Figure 7.14 **Fontanels at birth.** (See Tortora, *A Photographic Atlas of the Human Body, Second Edition,* Figure 3.12.)

Fontanels are mesenchyme-filled spaces between cranial bones that are present at birth.

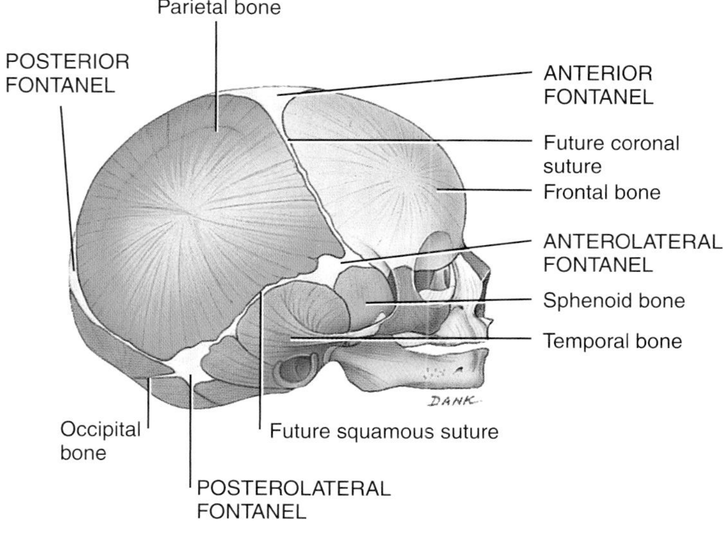

Right lateral view

Which fontanel is bordered by four different skull bones?

- The paired **posterolateral fontanels,** located laterally between the parietal, occipital, and temporal bones, are irregularly shaped. They begin to close 1 to 2 months after birth, but closure is generally not complete until 12 months.

The amount of closure in fontanels helps a physician gauge the degree of brain development. In addition, the anterior fontanel serves as a landmark for withdrawal of blood for analysis from the superior sagittal sinus (a large vein on the midline surface of the brain).

HYOID BONE

▶ **OBJECTIVE**

Describe the relationship of the hyoid bone to the skull.

The single **hyoid bone** (= U-shaped) is a unique component of the axial skeleton because it does not articulate with any other bone. Rather, it is suspended from the styloid processes of the temporal bones by ligaments and muscles. Located in the anterior neck between the mandible and larynx (Figure 7.15a), the hyoid bone supports the tongue, providing attachment sites for some tongue muscles and for muscles of the neck and pharynx. The hyoid bone consists of a horizontal *body* and paired projections called the *lesser horns* and the *greater horns* (Figure 7.15b and c). Muscles and ligaments attach to these paired projections.

The hyoid bone and the cartilages of the larynx and trachea are often fractured during strangulation. As a result, they are carefully examined at autopsy when strangulation is suspected.

▶ **CHECKPOINT**

4. Describe the general features of the skull.
5. What bones constitute the orbit?
6. What structures make up the nasal septum?
7. Define the following: foramen, suture, paranasal sinus, and fontanel.
8. What are the functions of the hyoid bone?

Figure 7.15 **Hyoid bone.** (See Tortora, *A Photographic Atlas of the Human Body, Second Edition,* Figure 3.13.)

The hyoid bone supports the tongue, providing attachment sites for muscles of the tongue, neck, and pharynx.

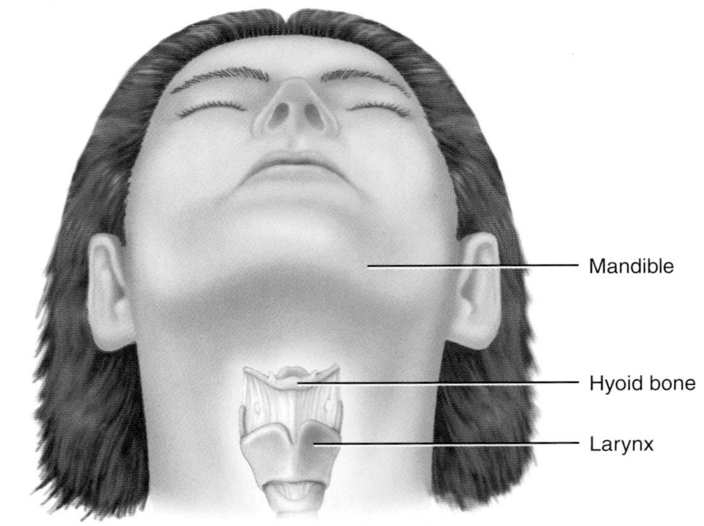

(a) Position of hyoid

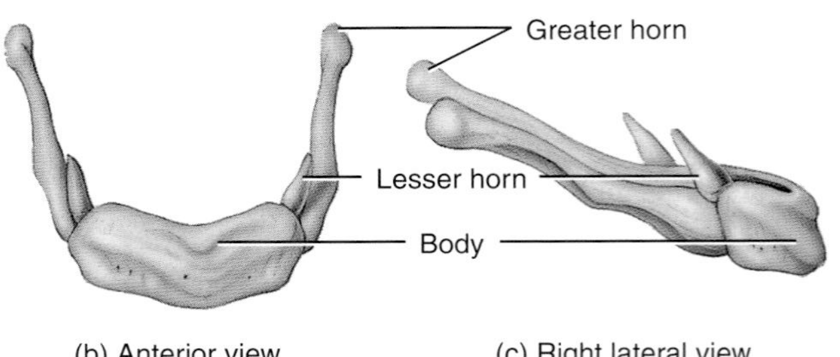

(b) Anterior view (c) Right lateral view

In what way is the hyoid bone different from all the other bones of the axial skeleton?

VERTEBRAL COLUMN

▶ **OBJECTIVE**

Identify the regions and normal curves of the vertebral column and describe its structural and functional features.

The **vertebral column,** also called the *spine* or *backbone,* makes up about two-fifths of your total height and is composed of a series of bones called **vertebrae** (VER-te-brē; singular is *vertebra*). The vertebral column, the sternum, and the ribs form the skeleton of the trunk of the body. The vertebral column consists of bone and connective tissue; the spinal cord that it surrounds and protects consists of nervous and connective tissues. At about 71 cm (28 in.) in an average adult male and about 61 cm (24 in.) in an average adult female, the vertebral column functions as a strong, flexible rod with elements that can move forward, backward, and sideways, and rotate. In addition to enclosing and protecting the spinal cord, it supports the head, and serves as a point of attachment for the ribs, pelvic girdle, and muscles of the back.

The total number of vertebrae during early development is 33. As a child grows, several vertebrae in the sacral and coccygeal regions fuse. As a result, the adult vertebral column, also called the spinal column, typically contains 26 vertebrae (Figure 7.16a). These are distributed as follows:

Figure 7.16 Vertebral column. The numbers in parentheses in (a) indicate the number of vertebrae in each region. In (d), the relative size of the disc has been enlarged for emphasis. A "window" has been cut in the annulus fibrosus to view the nucleus pulposus. (See Tortora, *A Photographic Atlas of the Human Body, Second Edition,* Figure 3.15.)

The adult vertebral column typically contains 26 vertebrae.

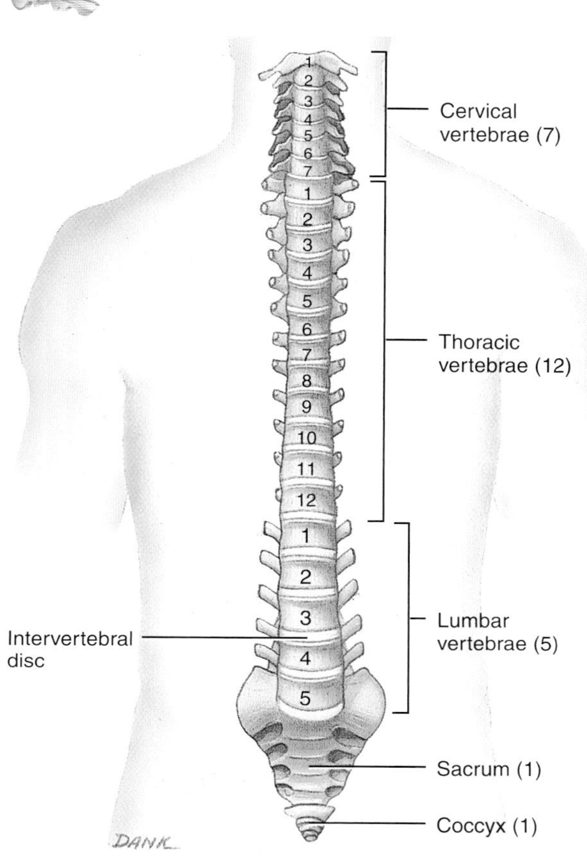

(a) Anterior view showing regions of the vertebral column

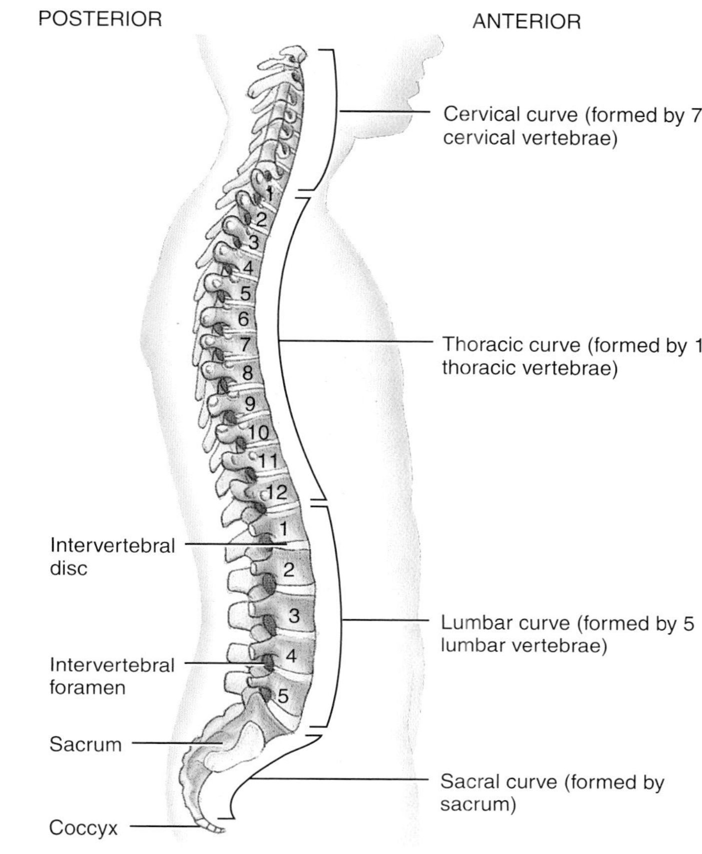

(b) Right lateral view showing four normal curves

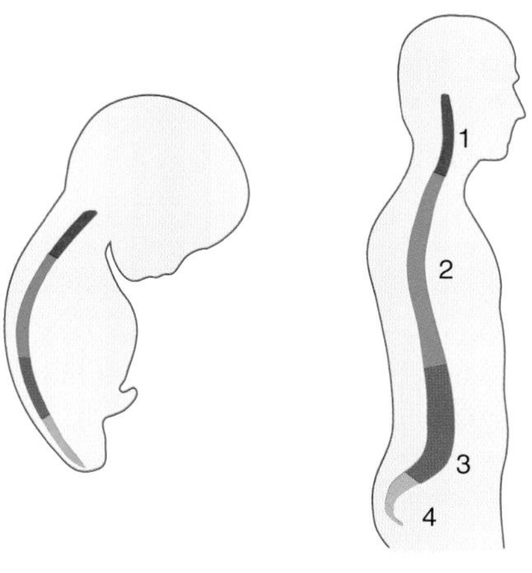

(c) Fetal and adult curves

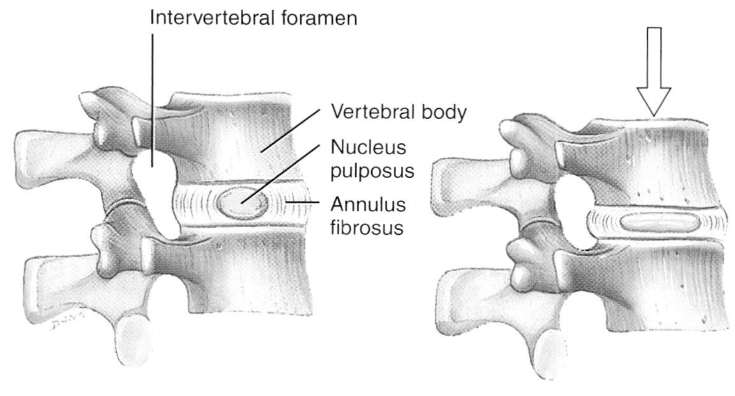

(d) Intervertebral disc

Which curves of the adult vertebral column are concave (relative to the anterior side of the body)?

- 7 **cervical vertebrae** (*cervic-* = neck) are in the neck region.
- 12 **thoracic vertebrae** (*thorax* = chest) are posterior to the thoracic cavity.
- 5 **lumbar vertebrae** (*lumb-* = loin) support the lower back.
- 1 **sacrum** (SĀ-krum = sacred bone) consists of five fused **sacral vertebrae.**
- 1 **coccyx** (KOK-siks = cuckoo, because the shape resembles the bill of a cuckoo bird) usually consists of four fused **coccygeal vertebrae** (kok-SIJ-ē-al).

The cervical, thoracic, and lumbar vertebrae are movable, but the sacrum and coccyx are not. We will discuss each of these regions in detail shortly.

Normal Curves of the Vertebral Column

When viewed from the side, the adult vertebral column shows four slight bends called **normal curves** (Figure 7.16b). Relative to the front of the body, the *cervical* and *lumbar curves* are convex (bulging out); the *thoracic* and *sacral curves* are concave (cupping in). The curves of the vertebral column increase its strength, help maintain balance in the upright position, absorb shocks during walking, and help protect the vertebrae from fracture.

The fetus has a single anteriorly concave curve (Figure 7.16c). At about the third month after birth, when an infant begins to hold its head erect, the cervical curve develops. Later, when the child sits up, stands, and walks, the lumbar curve develops. The thoracic and sacral curves are called *primary curves* because they form first during fetal development. The cervical and lumbar curves are known as *secondary curves* because they begin to form later, several months after birth. All curves are fully developed by age 10. However, secondary curves may be progressively lost in old age.

Various conditions may exaggerate the normal curves of the vertebral column, or the column may acquire a lateral bend, resulting in **abnormal curves** of the vertebral column. Three such abnormal curves—kyphosis, lordosis, and scoliosis—are described in the Disorders: Homeostatic Imbalances section on page 225.

Intervertebral Discs

Intervertebral discs are found between the bodies of adjacent vertebrae from the second cervical vertebra to the sacrum (Figure 7.16d). Each disc has an outer fibrous ring consisting of fibrocartilage called the *annulus fibrosus* (*annulus* = ringlike) and an inner soft, pulpy, highly elastic substance called the *nucleus pulposus* (*pulposus* = pulplike). The discs form strong joints, permit various movements of the vertebral column, and absorb vertical shock. Under compression, they flatten and broaden; with age, the nucleus pulposus hardens and becomes less elastic. Narrowing of the discs and compression of vertebrae results in a decrease in height with age.

Parts of a Typical Vertebra

Vertebrae in different regions of the spinal column vary in size, shape, and detail, but they are similar enough that we can discuss the structures (and the functions) of a typical vertebra (Figure 7.17). Vertebrae typically consist of a body, a vertebral arch, and several processes.

Body

The **body,** the thick, disc-shaped anterior portion, is the weight-bearing part of a vertebra. Its superior and inferior surfaces are roughened for the attachment of cartilaginous intervertebral discs. The anterior and lateral surfaces contain nutrient foramina, openings through which blood vessels deliver nutrients and oxygen and remove carbon dioxide and wastes from bone tissue.

Vertebral Arch

Two short, thick processes, the *pedicles* (PED-i-kuls = little feet), project posteriorly from the vertebral body to unite with the flat *laminae* (LAM-i-nē = thin layers), to form the **vertebral arch.** The vertebral arch extends posteriorly from the body of the vertebra; together, the body of the vertebra and the vertebral arch surround the spinal cord by forming the *vertebral foramen.* The vertebral foramen contains the spinal cord, adipose tissue, areolar connective tissue, and blood vessels. Collectively, the vertebral foramina of all vertebrae form the **vertebral (spinal) cavity.** The pedicles exhibit superior and inferior indentations called *vertebral notches.* When the vertebral notches are stacked on top of one another, they form an opening between adjoining vertebrae on both sides of the column. Each opening, called an *intervertebral foramen,* permits the passage of a single spinal nerve that passes to a specific region of the body.

Processes

Seven **processes** arise from the vertebral arch. At the point where a lamina and pedicle join, a *transverse process* extends laterally on each side. A single *spinous process (spine)* projects posteriorly from the junction of the laminae. These three processes serve as points of attachment for muscles. The remaining four processes form joints with other vertebrae above or below. The two *superior articular processes* of a vertebra articulate (form joints) with the two inferior articular processes of the vertebra immediately above them. In turn, the two *inferior articular processes* of that vertebra articulate with the two superior articular processes of the vertebra immediately below them, and so on. The articulating surfaces of the articular processes,

Figure 7.17 **Structure of a typical vertebra, as illustrated by a thoracic vertebra.** In (b), only one spinal nerve has been included, and it has been extended beyond the intervertebral foramen for clarity. The sympathetic chain is part of the autonomic nervous system (see Figure 15.2 on page 528). (See Tortora, *A Photographic Atlas of the Human Body, Second Edition,* Figure 3.16.)

A vertebra consists of a body, a vertebral arch, and several processes.

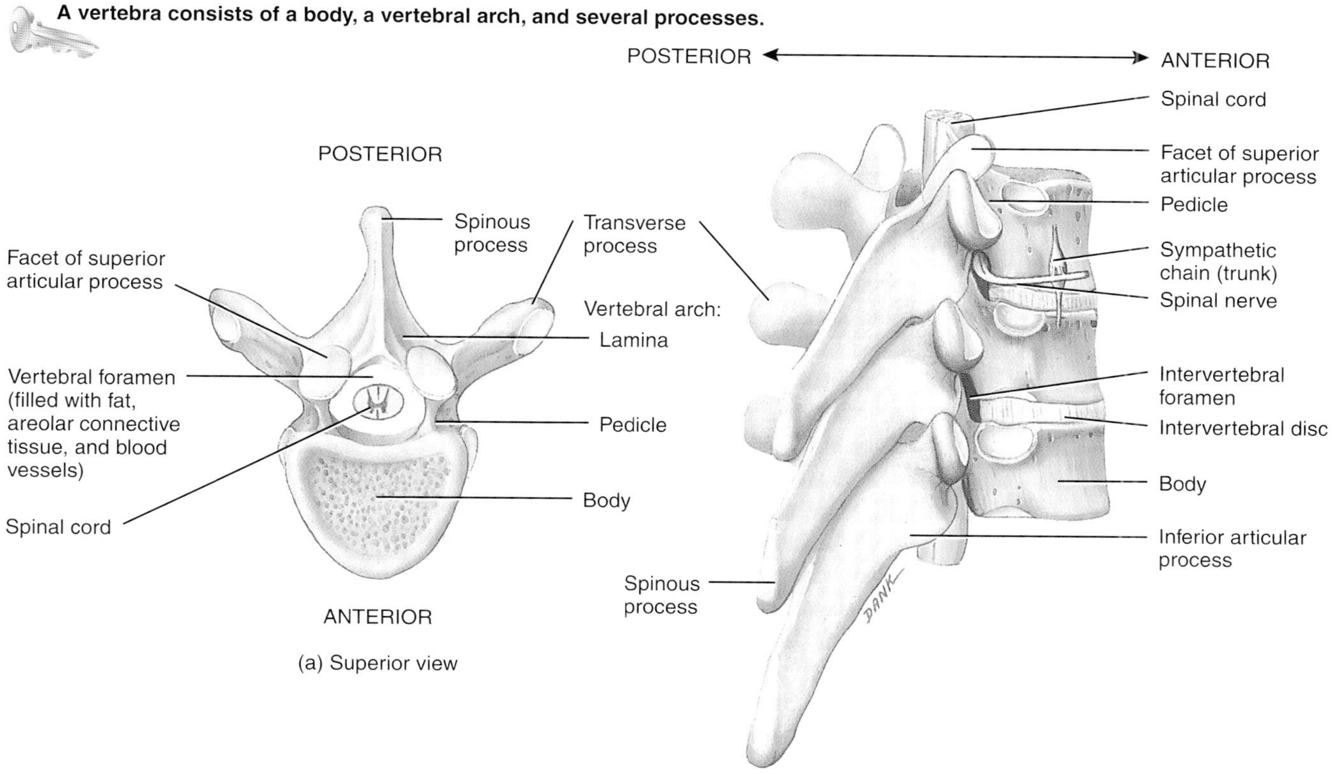

What are the functions of the vertebral and intervertebral foramina?

which are referred to as *facets* (= little faces), are covered with hyaline cartilage. The articulations formed between the bodies and articular facets of successive vertebrae are termed *intervertebral joints.*

Regions of the Vertebral Column

We turn now to the five regions of the vertebral column, beginning superiorly and moving inferiorly. Note that vertebrae in each region are numbered in sequence, from superior to inferior. When you actually view the bones of the vertebral column, you will notice that the transition from one region to the next is not abrupt but gradual, a feature that helps the vertebrae fit together.

Cervical Region

The bodies of **cervical vertebrae** (C1–C7) are smaller than all other vertebrae except those that form the coccyx (Figure 7.18a). Their vertebral arches, however, are larger. All cervical vertebrae have three foramina: one vertebral foramen and two transverse foramina (Figure 7.18d). The vertebral foramina of cervical vertebrae are the largest in the spinal column because they house the cervical enlargement of the spinal cord. Each cervical transverse process contains a *transverse foramen* through which the vertebral artery and its accompanying vein and nerve fibers pass. The spinous processes of C2 through C6 are often *bifid*—that is, split into two parts (Figure 7.18a, d).

The first two cervical vertebrae differ considerably from the others. The **atlas** (C1), named after the mythological Atlas who supported the world on his shoulders, is the first cervical vertebra inferior to the skull (Figure 7.18a, b). The atlas is a ring of bone with *anterior* and *posterior arches* and large *lateral masses.* It lacks a body and a spinous process. The superior surfaces of the lateral masses, called *superior articular facets,* are concave. They articulate with the occipital condyles

Figure 7.18 Cervical vertebrae. (See Tortora, *A Photographic Atlas of the Human Body, Second Edition,* Figure 3.17.)

The cervical vertebrae are found in the neck region.

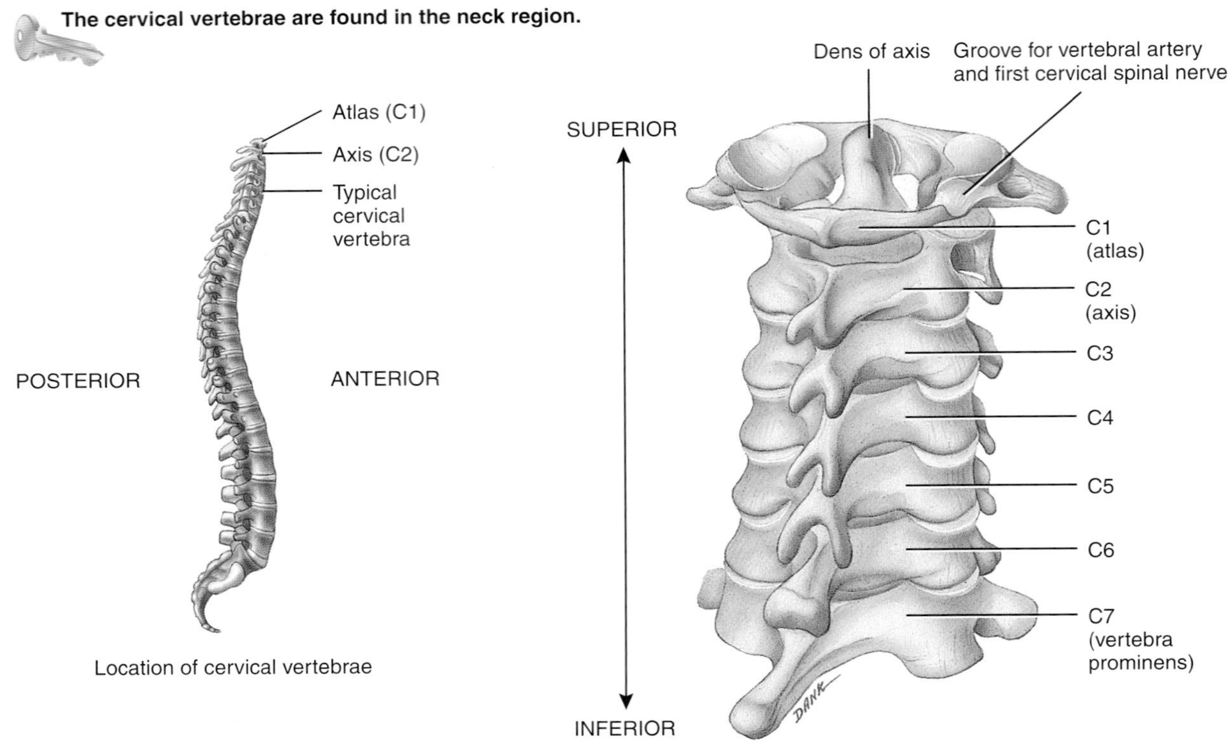

(a) Posterior view of articulated cervical vertebrae

of the occipital bone to form the paired *atlanto-occipital joints.* These articulations permit you to move your head to signify "yes." The inferior surfaces of the lateral masses, the *inferior articular facets,* articulate with the second cervical vertebra. The transverse processes and transverse foramina of the atlas are quite large.

The second cervical vertebra (C2), the **axis** (see Figure 7.18a, c), does have a body. A peglike process called the *dens* (= tooth) or *odontoid process* projects superiorly through the anterior portion of the vertebral foramen of the atlas. The dens makes a pivot on which the atlas and head rotate. This arrangement permits side-to-side movement of the head, as when you move your head to signify "no." The articulation formed between the anterior arch of the atlas and dens of the axis, and between their articular facets, is called the *atlanto-axial joint.* In some instances of trauma, the dens of the axis may be driven into the medulla oblongata of the brain. This type of injury is the usual cause of death from whiplash injuries.

The third through sixth cervical vertebrae (C3–C6), represented by the vertebra in Figure 7.18d, correspond to the structural pattern of the typical cervical vertebra previously described. The seventh cervical vertebra (C7), called the *vertebra prominens,* is somewhat different (see Figure 7.18a). Its single large spinous process may be seen and felt at the base of the neck.

Thoracic Region

Thoracic vertebrae (T1–T12; Figure 7.19 on page 218) are considerably larger and stronger than cervical vertebrae. In addition, the spinous processes on T1 and T2 are long, laterally flattened, and directed inferiorly. In contrast, the spinous processes on T11 and T12 are shorter, broader, and directed more posteriorly. Compared to cervical vertebrae, thoracic vertebrae also have longer and larger transverse processes.

The feature of the thoracic vertebrae that distinguishes them from other vertebrae is that they articulate with the ribs. Except for T11 and T12, the transverse processes have facets for articulating with the *tubercles* of the ribs. The bodies of thoracic vertebrae also have either facets or demifacets (half facets) for

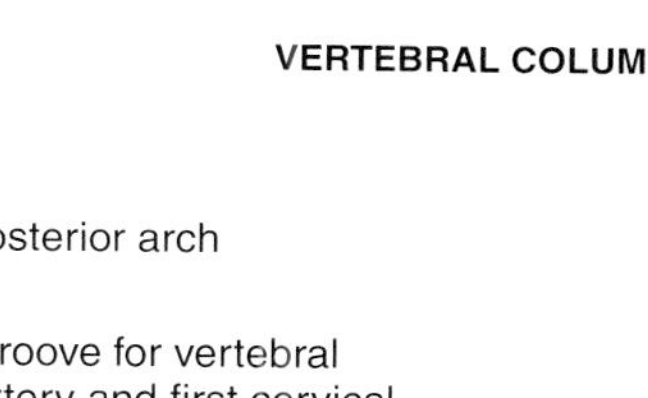

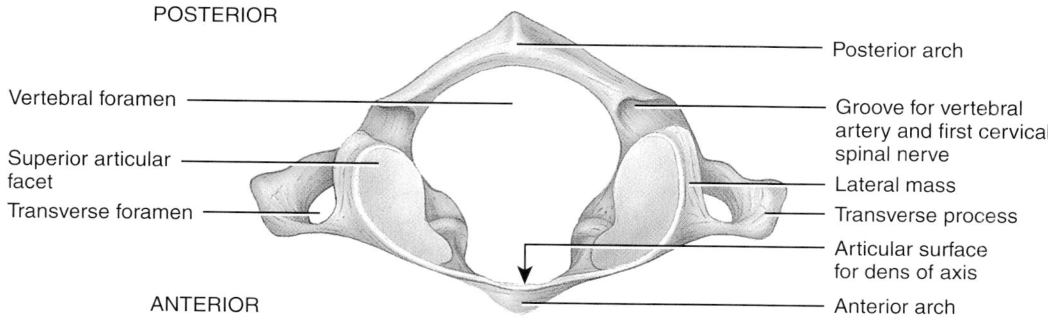

(b) Superior view of the atlas (C1)

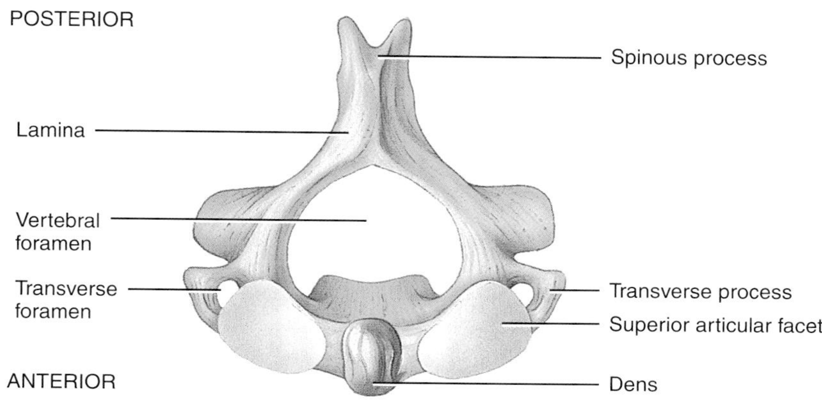

(c) Superior view of the axis (C2)

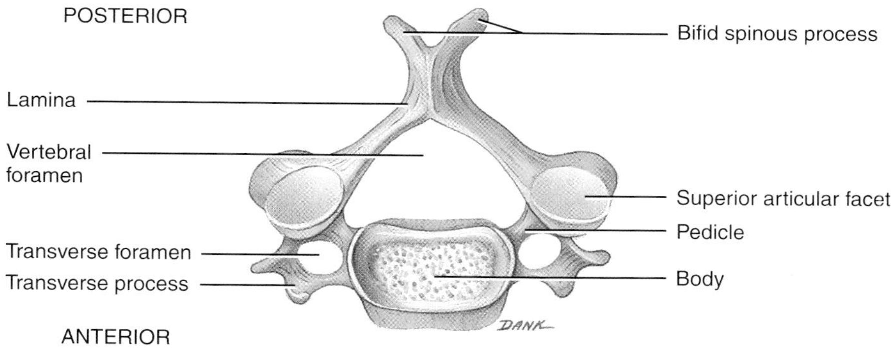

(d) Superior view of a typical cervical vertebra

Which bones permit you to move your head to signify "no"?

Figure 7.19 Thoracic vertebrae. (See Tortora, *A Photographic Atlas of the Human Body, Second Edition,* Figure 3.16.)

The thoracic vertebrae are found in the chest region and articulate with the ribs.

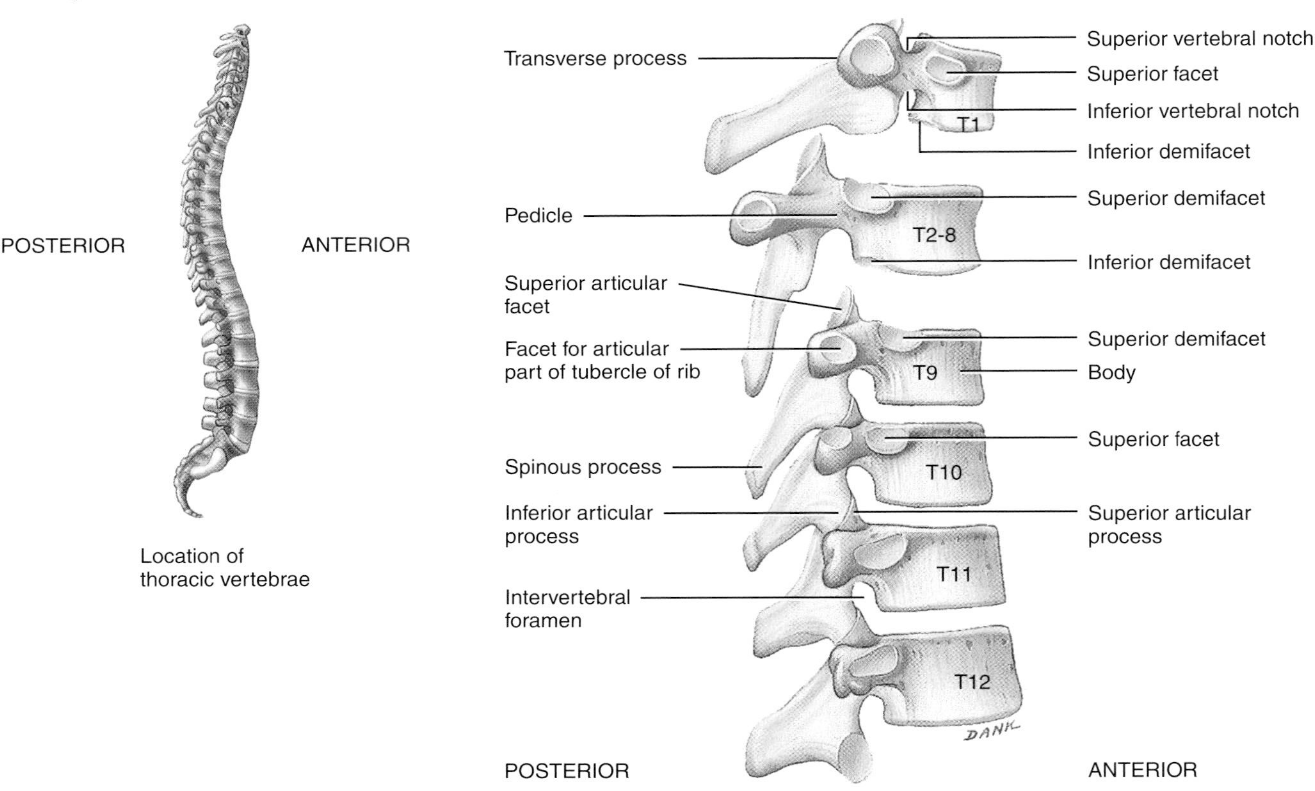

(a) Right lateral view of several articulated thoracic vertebrae

articulation with the *heads* of the ribs (see Figure 7.23). The articulations between the thoracic vertebrae and ribs, called *vertebrocostal joints,* occur on both sides of the vertebral body. As you can see in Figure 7.19, T1 has a superior facet and an inferior demifacet. T2–T8 have superior and inferior demifacets. T9 has a superior demifacet, and T10–T12 have a superior facet. Movements of the thoracic region are limited by the attachment of the ribs to the sternum.

Lumbar Region

The **lumbar vertebrae** (L1–L5) are the largest and strongest in the vertebral column (Figure 7.20 on page 220) because the amount of body weight supported by the vertebrae increases toward the inferior end of the backbone. Their various projections are short and thick. The superior articular processes are directed medially instead of superiorly, and the inferior articular processes are directed laterally instead of inferiorly. The spinous processes are quadrilateral in shape, thick and broad, and project nearly straight posteriorly. The spinous processes are well adapted for the attachment of the large back muscles.

A summary of the major structural differences among cervical, thoracic, and lumbar vertebrae is presented in Table 7.4 on page 221.

Sacrum

The **sacrum** is a triangular bone formed by the union of five sacral vertebrae (S1–S5) (Figure 7.21a on page 221). The sacral vertebrae begin to fuse in individuals between 16 and 18 years of age, a process usually completed by age 30. Positioned at the posterior portion of the pelvic cavity medial to the two hip bones, the sacrum serves as a strong foundation for the pelvic girdle. The female sacrum is shorter, wider, and more curved between S2 and S3 than the male sacrum (see Table 8.1 on page 244).

The concave anterior side of the sacrum faces the pelvic cavity. It is smooth and contains four *transverse lines (ridges)* that mark the joining of the sacral vertebral bodies

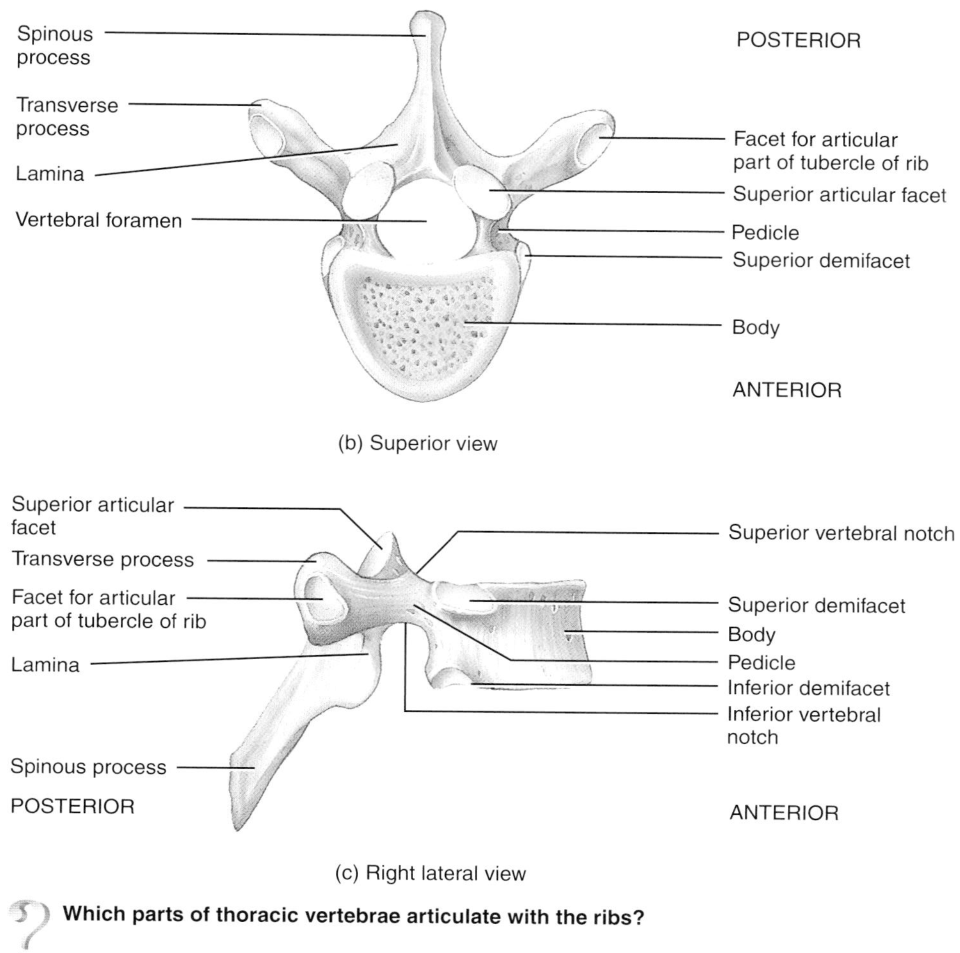

Which parts of thoracic vertebrae articulate with the ribs?

(Figure 7.21a on page 221). At the ends of these lines are four pairs of *anterior sacral foramina.* The lateral portion of the superior surface of the sacrum contains a smooth surface called the *sacral ala* (= wing), which is formed by the fused transverse processes of the first sacral vertebra (S1).

The convex, posterior surface of the sacrum contains a *median sacral crest,* the fused spinous processes of the upper sacral vertebrae; a *lateral sacral crest,* the fused transverse processes of the sacral vertebrae; and four pairs of *posterior sacral foramina* (Figure 7.21b). These foramina connect with anterior sacral foramina to allow passage of nerves and blood vessels. The *sacral canal* is a continuation of the vertebral cavity. The laminae of the fifth sacral vertebra, and sometimes the fourth, fail to meet. This leaves an inferior entrance to the vertebral canal called the *sacral hiatus* (hī-Ā-tus = opening). On either side of the sacral hiatus are the *sacral cornua* (KOR-noo-a; *cornu* = horn), the inferior articular processes of the fifth sacral vertebra. They are connected by ligaments to the coccyx.

The narrow inferior portion of the sacrum is known as the *apex.* The broad superior portion of the sacrum is called the *base.* The anteriorly projecting border of the base, called the *sacral promontory* (PROM-on-tō-rē), is one of the points used for measurements of the pelvis. On both lateral surfaces the sacrum has a large ear-shaped *auricular surface* that articulates with the ilium of each hipbone to form the *sacroiliac joint.* Posterior to the auricular surface is a roughened surface, the *sacral tuberosity,* which contains depressions for the attachment of ligaments. The sacral tuberosity unites with the hip bones to form the sacroiliac joints. The *superior articular processes* of the sacrum articulate with the inferior articular processes of the fifth lumbar vertebra, and the base of the sacrum articulates with the body of the fifth lumbar vertebra to form the *lumbosacral joint.*

Coccyx

The **coccyx,** like the sacrum, is triangular in shape. It is formed by the fusion of usually four coccygeal vertebrae, indicated in

Figure 7.20 Lumbar vertebrae. (See Tortora, *A Photographic Atlas of the Human Body, Second Edition,* Figure 3.18.)

Lumbar vertebrae are found in the lower back.

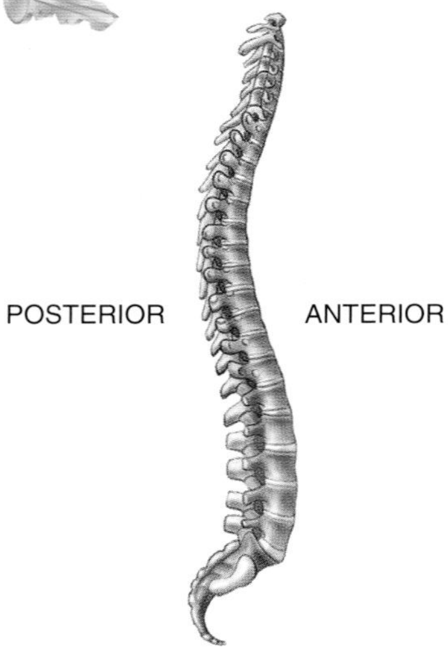

POSTERIOR
ANTERIOR

Location of lumbar vertebrae

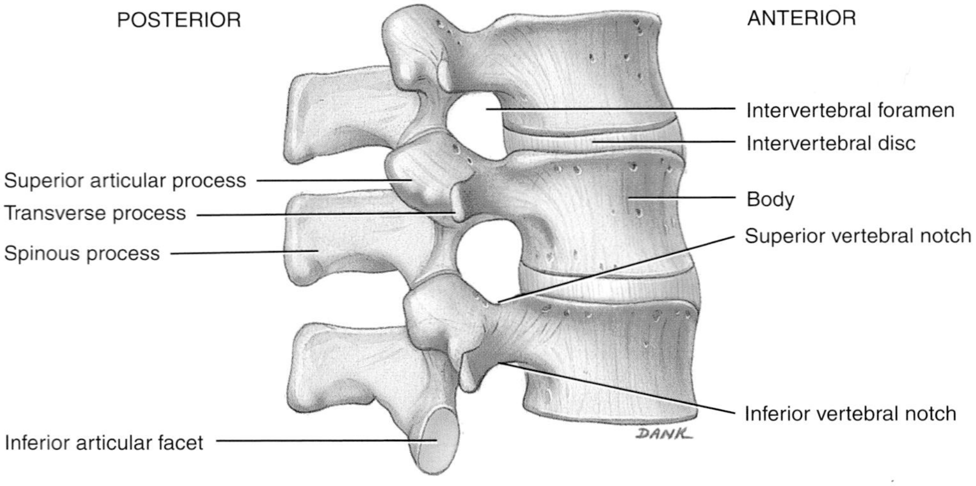

POSTERIOR
ANTERIOR
Intervertebral foramen
Intervertebral disc
Superior articular process
Transverse process
Spinous process
Body
Superior vertebral notch
Inferior vertebral notch
Inferior articular facet
DANK

(a) Right lateral view of articulated lumbar vertebrae

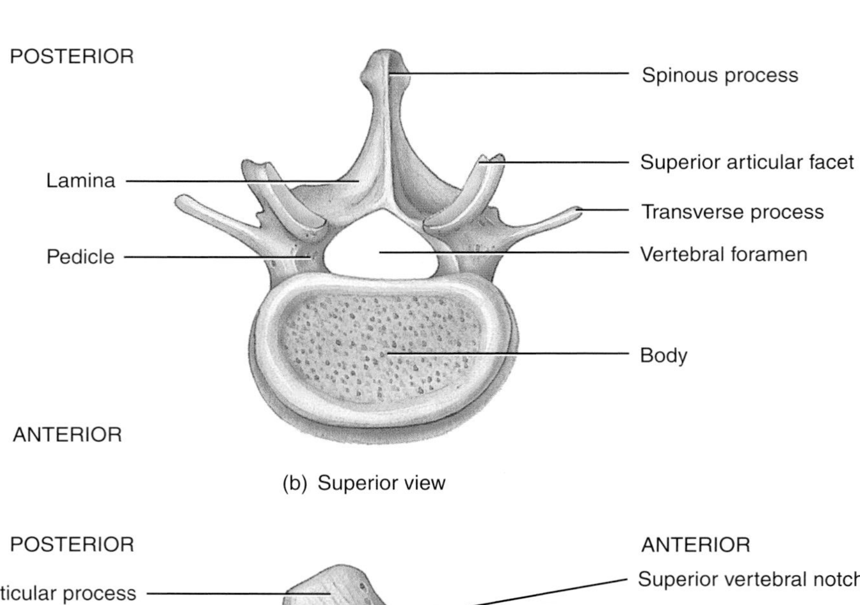

(b) Superior view

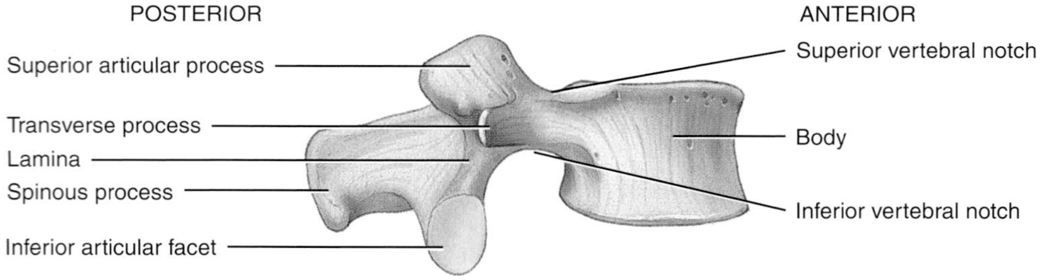

POSTERIOR
ANTERIOR
Superior articular process
Superior vertebral notch
Transverse process
Lamina
Body
Spinous process
Inferior vertebral notch
Inferior articular facet

(c) Right lateral view

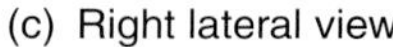

Why are the lumbar vertebrae the largest and strongest in the vertebral column?

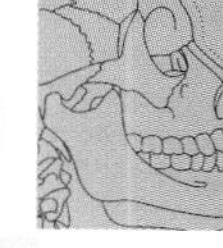

TABLE 7.4 Comparison of Major Structural Features of Cervical, Thoracic, and Lumbar Vetebrae

Characteristic	Cervical	Thoracic	Lumbar
Overall structure	See Figure 7.18d	See Figure 7.19b	See Figure 7.20b
Body	Small	Larger	Largest
Foramina	One vertebral and two transverse	One vertebral	One vertebral
Spinous processes	Slender and often bifid (C2–C6)	Long and fairly thick (most project inferiorly)	Short and blunt (project posteriorly rather than inferiorly)
Transverse processes	Small	Fairly large	Large and blunt
Articular facets for ribs	Absent	Present	Absent
Direction of articular facets			
Superior	Posterosuperior	Posterolateral	Medial
Inferior	Anteroinferior	Anteromedial	Lateral
Size of intervertebral discs	Thick relative to size of vertebral bodies	Thin relative to size of vertebral bodies	Massive

Figure 7.21 as Co1–Co4. The coccygeal vertebrae fuse somewhat later than the sacral vertebrae, between the ages of 20 and 30. The dorsal surface of the body of the coccyx contains two long *coccygeal cornua* that are connected by ligaments to the sacral cornua. The coccygeal cornua are the pedicles and superior articular processes of the first coccygeal vertebra. On the lateral surfaces of the coccyx are a series of *transverse processes;* the first pair are the largest. The coccyx articulates superiorly with the apex of the sacrum. In females, the coccyx points inferiorly to allow the passage of a baby during birth; in males, it points anteriorly (see Table 8.1 on page 244).

Caudal Anesthesia

Anesthetic agents that act on the sacral and coccygeal nerves are sometimes injected through the sacral hiatus, a procedure called

Figure 7.21 Sacrum and coccyx. (See Tortora, *A Photographic Atlas of the Human Body, Second Edition,* Figure 3.19.)

The sacrum is formed by the union of five sacral vertebrae, and the coccyx is formed by the union of usually four coccygeal vertebrae.

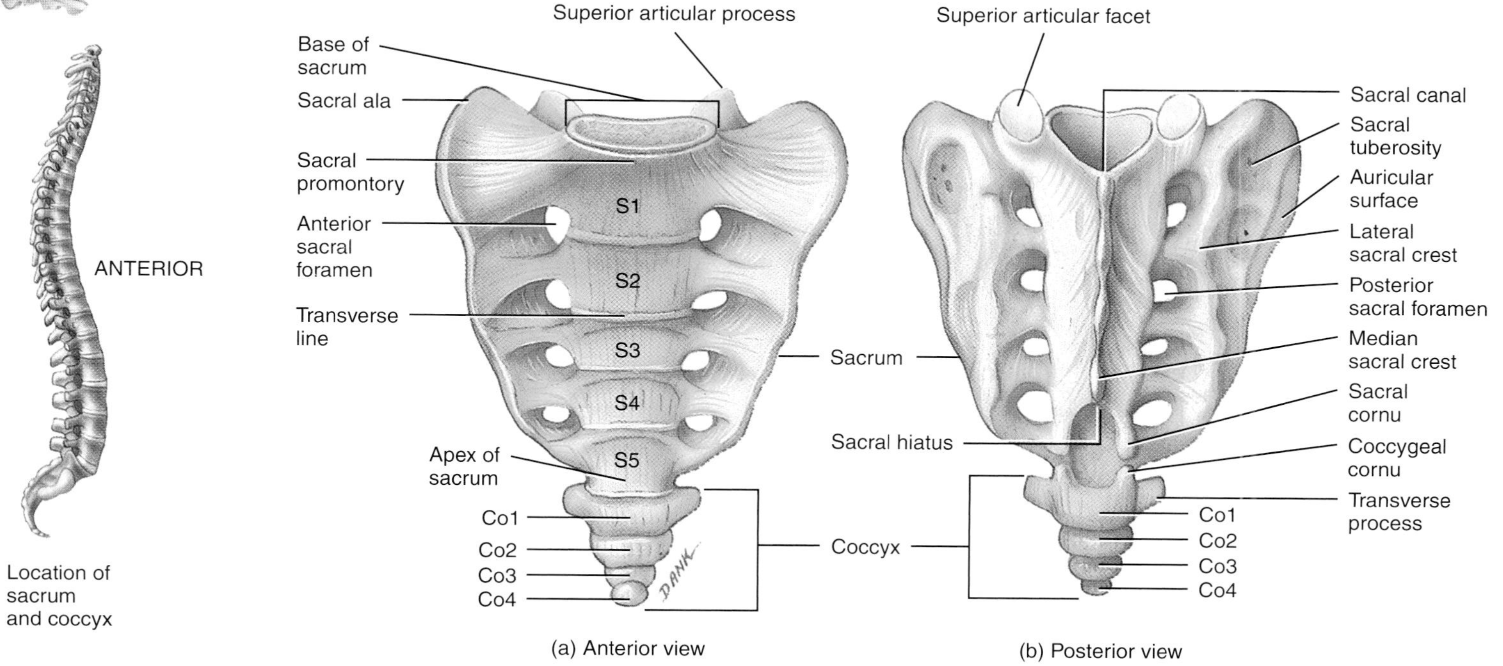

Location of sacrum and coccyx

ANTERIOR

Superior articular process
Base of sacrum
Sacral ala
Sacral promontory
Anterior sacral foramen
Transverse line
Apex of sacrum
S1
S2
S3
S4
S5
Co1
Co2
Co3
Co4
Sacrum
Coccyx
Sacral hiatus
Superior articular facet
Sacral canal
Sacral tuberosity
Auricular surface
Lateral sacral crest
Posterior sacral foramen
Median sacral crest
Sacral cornu
Coccygeal cornu
Transverse process

(a) Anterior view

(b) Posterior view

How many foramina pierce the sacrum, and what is their function?

caudal anesthesia or **epidural block.** The procedure is used most often to relieve pain during labor and to provide anesthesia to the perineal area. Because the sacral hiatus is between the sacral cornua, the cornua are important bony landmarks for locating the hiatus. Anesthetic agents may also be injected through the posterior sacral foramina. ■

▶ CHECKPOINT

9. What are the functions of the vertebral column?

10. When do the secondary vertebral curves develop?

11. What are the principal distinguishing characteristics of the bones of the various regions of the vertebral column?

THORAX

▶ OBJECTIVE

Identify the bones of the thorax.

The term **thorax** refers to the entire chest. The skeletal part of the thorax, the **thoracic cage,** is a bony enclosure formed by the sternum, costal cartilages, ribs, and the bodies of the thoracic vertebrae (Figure 7.22). The thoracic cage is narrower at its superior end and broader at its inferior end and is flattened from front to back. It encloses and protects the organs in the thoracic and superior abdominal cavities and provides support for the bones of the shoulder girdle and upper limbs.

Sternum

The **sternum,** or breastbone, is a flat, narrow bone located in the center of the anterior thoracic wall that measures about 15 cm (6 in.) in length and consists of three parts (Figure 7.22). The superior part is the **manubrium** (ma-NOO-brē-um = handle-like); the middle and largest part is the **body;** and the inferior, smallest part is the **xiphoid process** (ZĪ-foyd = sword-shaped). The segments of the sternum typically fuse by age 25 and the points of fusion are marked by transverse ridges.

The junction of the manubrium and body forms the *sternal angle.* The manubrium has a depression on its superior surface, the *suprasternal notch.* Lateral to the suprasternal notch are *clavicular notches* that articulate with the medial ends of the clavicles to form the *sternoclavicular joints.* The manubrium also articulates with the costal cartilages of the first and second ribs. The body of the sternum articulates directly or indirectly with the costal cartilages of the second through tenth ribs. The xiphoid process consists of hyaline cartilage during infancy and childhood and does not completely ossify until about age 40. No ribs are attached to it, but the xiphoid process provides attachment for some abdominal muscles. Incorrect positioning of the hands of a rescuer during cardiopulmonary resuscitation (CPR) may fracture the xiphoid process, driving it into internal organs. During thoracic surgery, the sternum may be split along the midline and the halves spread apart to allow surgeons access to structures in the thoracic cavity such as the thymus, heart, and great vessels of the heart. After surgery, the halves of the sternum are held together with wire sutures.

Ribs

Twelve pairs of **ribs** give structural support to the sides of the thoracic cavity (see Figure 7.22b). The ribs increase in length from the first through seventh, and then decrease in length to the twelfth rib. Each rib articulates posteriorly with its corresponding thoracic vertebra.

The first through seventh pairs of ribs have a direct anterior attachment to the sternum by a strip of hyaline cartilage called *costal cartilage* (*cost-* = rib). The costal cartilages contribute to the elasticity of the thoracic cage and prevent various blows to the chest from fracturing the sternum and/or ribs. The ribs that have costal cartilages and attach directly to the sternum are called *true (vertebrosternal) ribs.* The articulations formed between the true ribs and the sternum are called *sternocostal joints.* The remaining five pairs of ribs are termed *false ribs* because their costal cartilages either attach indirectly to the sternum or do not attach to the sternum at all. The cartilages of the eighth, ninth, and tenth pairs of ribs attach to one another and then to the cartilages of the seventh pair of ribs. These false ribs are called *vertebrochondral ribs.* The eleventh and twelfth pairs of ribs are false ribs designated as *floating (vertebral) ribs* because the costal cartilage at their anterior ends does not attach to the sternum at all. These ribs attach only posteriorly to the thoracic vertebrae. Inflammation of one or more costal cartilages, called *costochondritis,* is characterized by local tenderness and pain in the anterior chest wall that may radiate. The symptoms mimic the chest pain associated with a heart attack (angina pectoris).

Figure 7.23a on page 224 shows the parts of a typical (third through ninth) rib. The *head* is a projection at the posterior end of the rib. The facet of the head fits into a facet on the body of a single vertebra or into the demifacets of two adjoining vertebrae to form a *vertebrocostal joint.* The *neck* is a constricted portion of a rib just lateral to the head. A knoblike structure on the posterior surface, where the neck joins the body, is called a *tubercle* (TOO-ber-kul). The *nonarticular part* of the tubercle attaches to the transverse process of a vertebra by a ligament (lateral costotransverse ligament). The *articular part* of the tubercle articulates with the facet of a transverse process of the inferior of the two vertebrae (Figure 7.23c). These articulations also form vertebrocostal joints. The *body (shaft)* is the main part of the rib. A short distance beyond the tubercle, an abrupt change in the curvature of the shaft occurs. This point is called the *costal angle.* The inner surface of the rib has a *costal groove* that protects blood vessels and a small nerve.

In summary, the posterior portion of the rib connects to a thoracic vertebra by its head and the articular part of a tubercle. The facet of the head fits into a facet on the body of one vertebra

Figure 7.22 Skeleton of the thorax. (See Tortora, *A Photographic Atlas of the Human Body, Second Edition,* Figure 3.20.)

The bones of the thorax enclose and protect organs in the thoracic cavity and in the superior abdominal cavity.

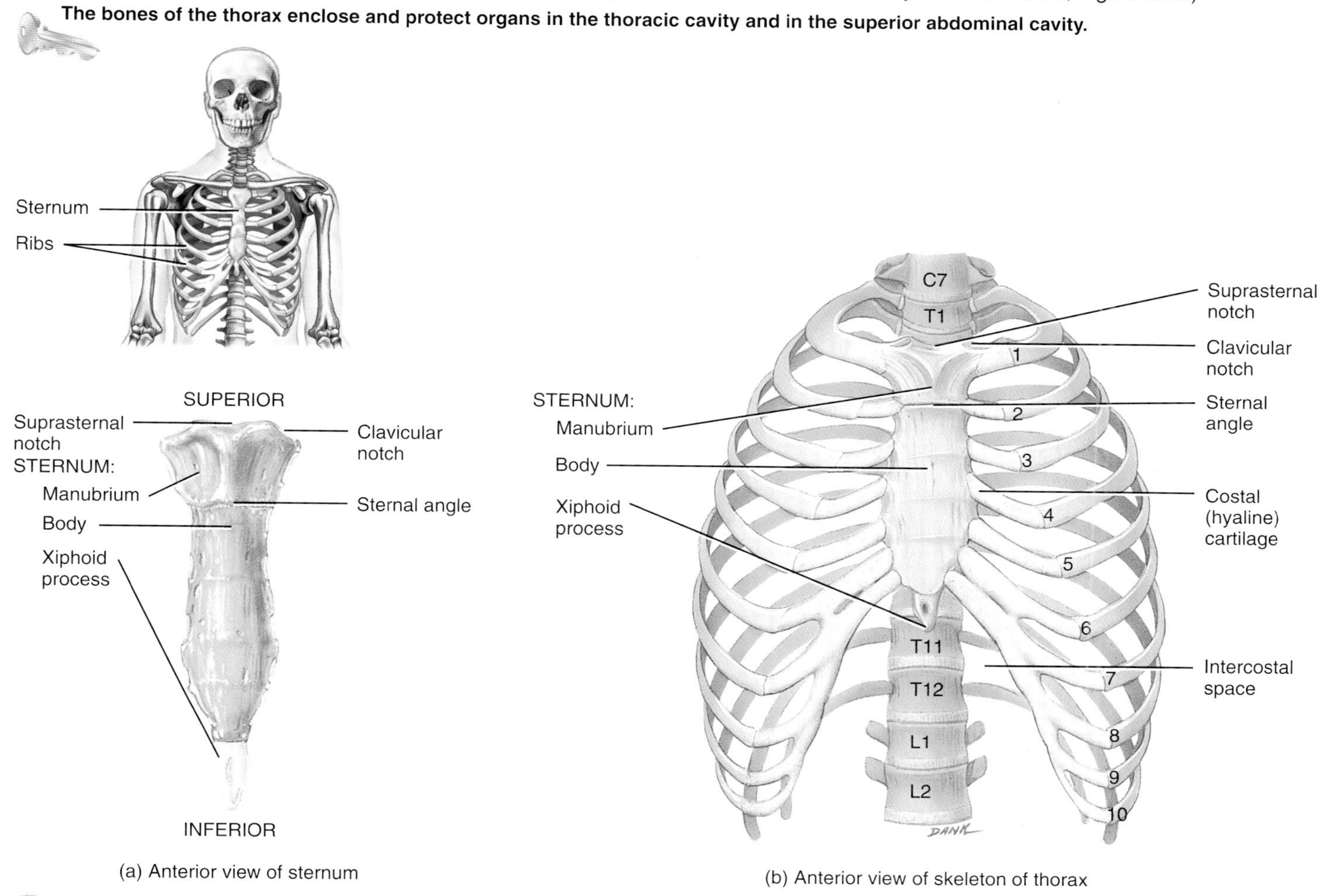

(a) Anterior view of sternum

(b) Anterior view of skeleton of thorax

Which ribs are true ribs, false ribs, and floating ribs?

or into the demifacets of two adjoining vertebrae. The articular part of the tubercle articulates with the facet of the transverse process of the vertebra.

Spaces between ribs, called *intercostal spaces,* are occupied by intercostal muscles, blood vessels, and nerves. The lungs or other structures in the thoracic cavity are commonly accessed surgically through an intercostal space. Special rib retractors are used to create a wide separation between ribs. The costal cartilages are sufficiently elastic in younger individuals to permit *considerable* bending without breaking.

Rib Fractures, Dislocations, and Separations

Rib fractures are the most common chest injuries. They usually result from direct blows, most often from impact with a steering wheel, falls, or crushing injuries to the chest. Ribs tend to break at the point where the greatest force is applied, but they may also break at their weakest point—the site of greatest curvature, just anterior to the costal angle. The middle ribs are the most commonly fractured. In some cases, fractured ribs may puncture the heart, great vessels of the heart, lungs, trachea, bronchi, esophagus, spleen, liver, and kidneys. Rib fractures are usually quite painful. Rib fractures are no longer bound with bandages because of the pneumonia that would result from lack of proper lung ventilation.

Dislocated ribs, which are common in body contact sports, involve displacement of a costal cartilage from the sternum, with resulting pain, especially during deep inhalations.

Separated ribs involve displacement of a rib and its costal cartilage; as a result, a rib may move superiorly, overriding the rib above and causing severe pain. ■

CHECKPOINT

12. What bones form the skeleton of the thorax?

13. What are the functions of the bones of the thorax?

14. How are ribs classified?

Figure 7.23 The structure of ribs. Each rib has a head, a neck, and a body. The facets and the articular part of the tubercle are where the rib articulates with a vertebra. (See Tortora, *A Photographic Atlas of the Human Body, Second Edition,* Figure 3.21.)

Each rib articulates posteriorly with its corresponding thoracic vertebra.

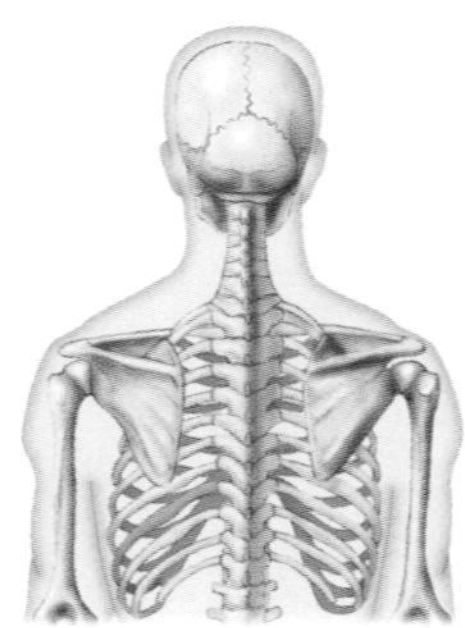

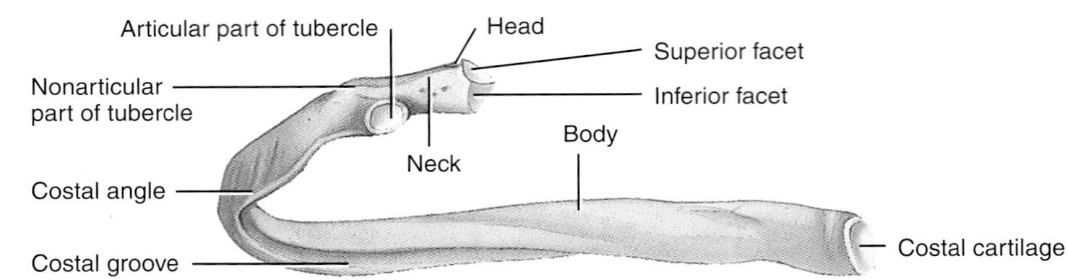

(a) Posterior view of left rib

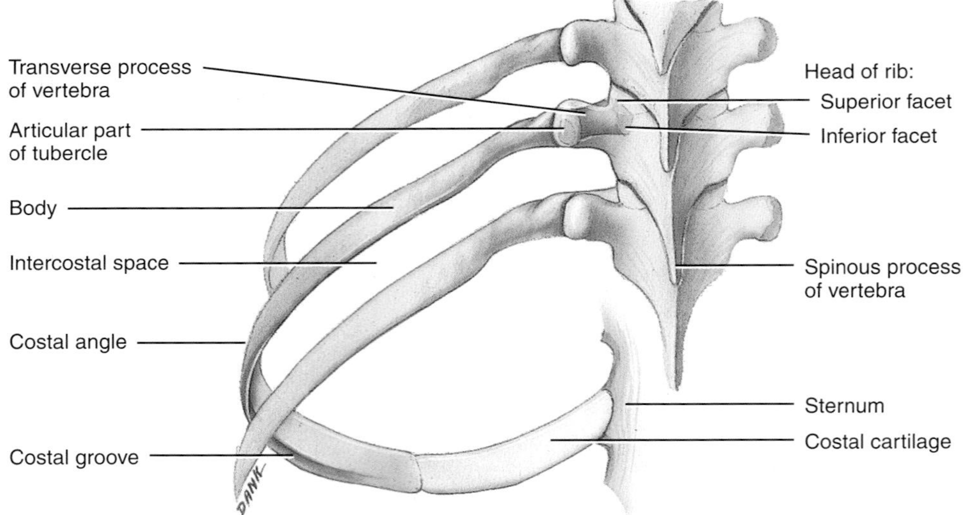

(b) Posterior view of left ribs articulated with thoracic vertebrae and sternum

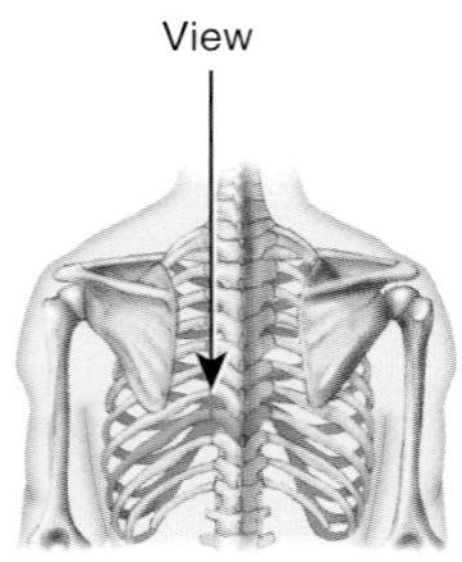

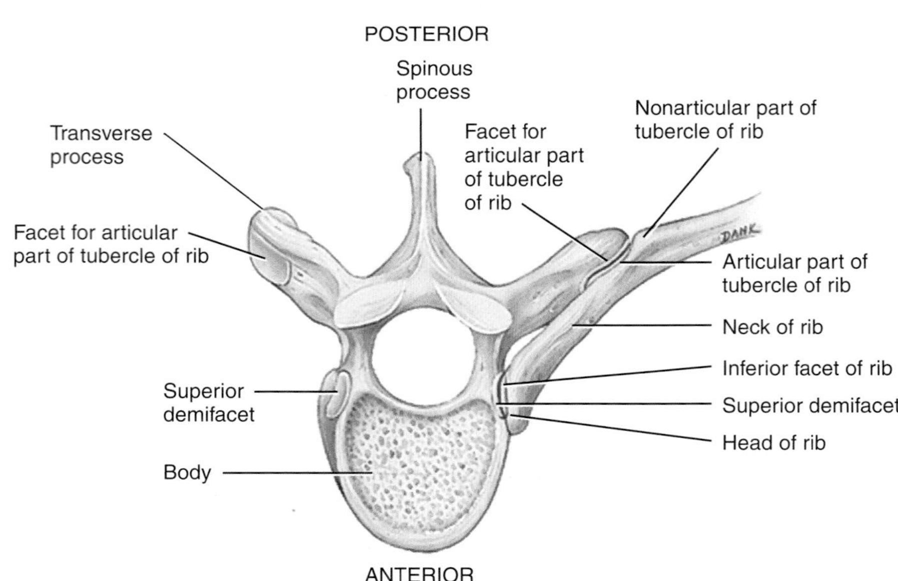

(c) Superior view of left rib articulated with thoracic vertebra

How does a rib articulate with a thoracic vertebra?

DISORDERS: HOMEOSTATIC IMBALANCES

Herniated (Slipped) Disc

In their function as shock absorbers, intervertebral discs are constantly being compressed. If the anterior and posterior ligaments of the discs become injured or weakened, the pressure developed in the nucleus pulposus may be great enough to rupture the surrounding fibrocartilage (annulus fibrosus). If this occurs, the nucleus pulposus may herniate (protrude) posteriorly or into one of the adjacent vertebral bodies (Figure 7.24). This condition is called a **herniated (slipped) disc.** Because the lumbar region bears much of the weight of the body, and is the region of the most flexing and bending, herniated discs most often occur in the lumbar area.

Frequently, the nucleus pulposus slips posteriorly toward the spinal cord and spinal nerves. This movement exerts pressure on the spinal nerves, causing local weakness and acute pain. If the roots of the sciatic nerve, which passes from the spinal cord to the foot, are compressed, the pain radiates down the posterior thigh, through the calf, and occasionally into the foot. If pressure is exerted on the spinal cord itself, some of its neurons may be destroyed. Treatment options include bed rest, medications for pain, physical therapy and exercises, and traction. A person with a herniated disc may also undergo a *laminectomy,* a procedure in which parts of the laminae of the vertebra and intervertebral disc are removed to relieve pressure on the nerves.

Abnormal Curves of the Vertebral Column

Various conditions may exaggerate the normal curves of the vertebral column, or the column may acquire a lateral bend, resulting in **abnormal curves** of the vertebral column.

Scoliosis (skō-lē-Ō-sis; *scolio* = crooked), the most common of the abnormal curves, is a lateral bending of the vertebral column, usually in the thoracic region (Figure 7.25a). It may result from congenitally (present at birth) malformed vertebrae, chronic sciatica, paralysis of muscles on one side of the vertebral column, poor posture, or one leg being shorter than the other.

Kyphosis (kī-FŌ-sis; *kyphos-* = hump; *-osis* = condition) is an exaggeration of the thoracic curve of the vertebral column (Figure 7.25b). In tuberculosis of the spine, vertebral bodies may partially collapse, causing an acute angular bending of the vertebral column. In the elderly, degeneration of the intervertebral discs leads to kyphosis. Kyphosis may also be caused by rickets and poor posture. It is also common in females with advanced osteoporosis. The term *round-shouldered* is an expression for mild kyphosis.

Lordosis (lor-DŌ-sis; *lord-* = bent backward), sometimes called *hollow back,* is an exaggeration of the lumbar curve of the vertebral column (Figure 7.25c). It may result from increased weight of the abdomen as in pregnancy, or extreme obesity, poor posture, rickets, osteoporosis, or tuberculosis of the spine.

Spina Bifida

Spina bifida (SPĪ-na BIF-i-da) is a congenital defect of the vertebral column in which laminae of L5 and/or S1 fail to develop normally and unite at the midline. The least serious form is called *spina bifida occulta.* It occurs in L5 or S1 and produces no symptoms. The only evidence of its presence is a small dimple with a tuft of hair in the overlying skin. Several types of spina bifida involve protrusion of meninges (membranes) and/or spinal cord through the defect in the laminae and are collectively termed *spina bifida cystica* because of the presence of a cystlike sac protruding from the backbone

Figure 7.24 Herniated (slipped) disc.

Most often the nucleus pulposus herniates posteriorly.

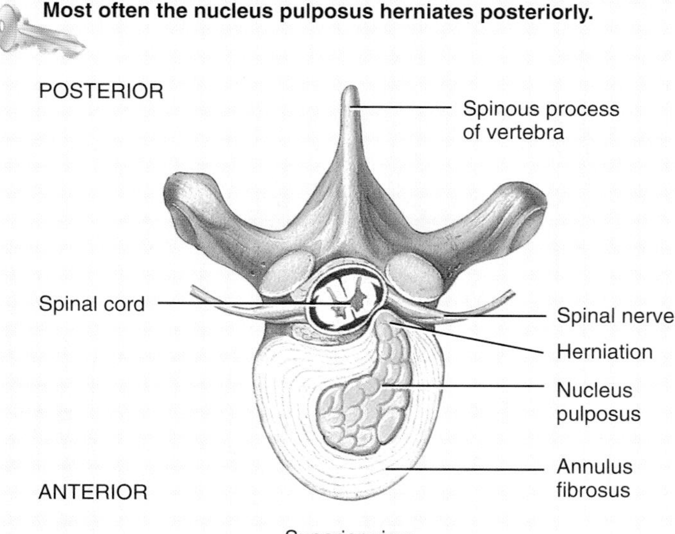

Why do most herniated discs occur in the lumbar region?

Figure 7.25 Abnormal curves of the vertebral column.

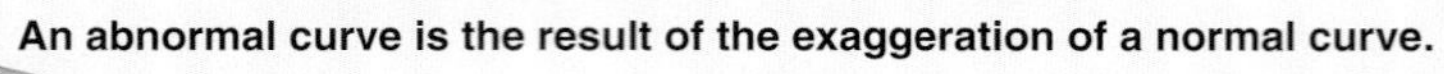

An abnormal curve is the result of the exaggeration of a normal curve.

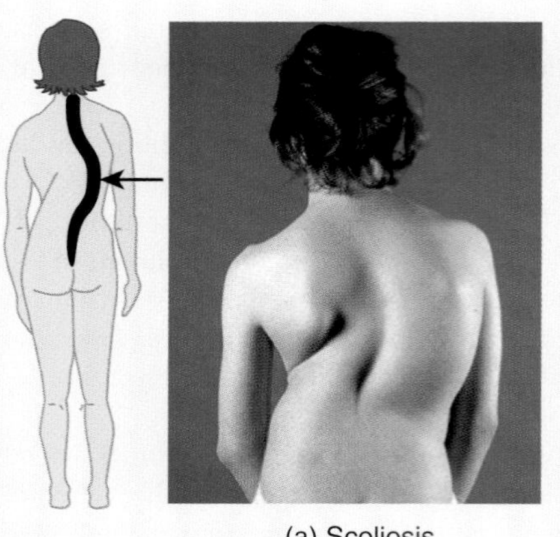

(a) Scoliosis

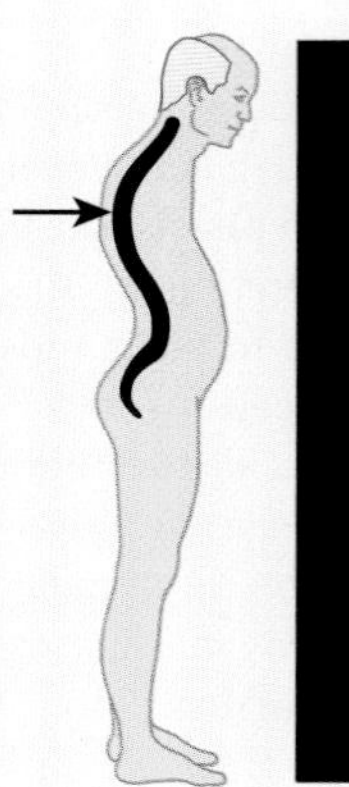

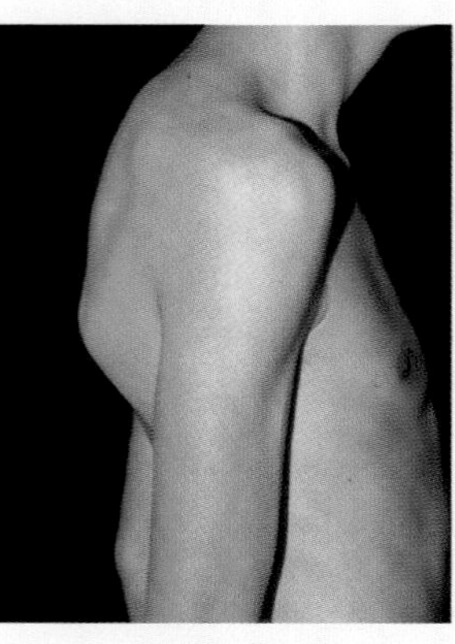

(b) Kyphosis

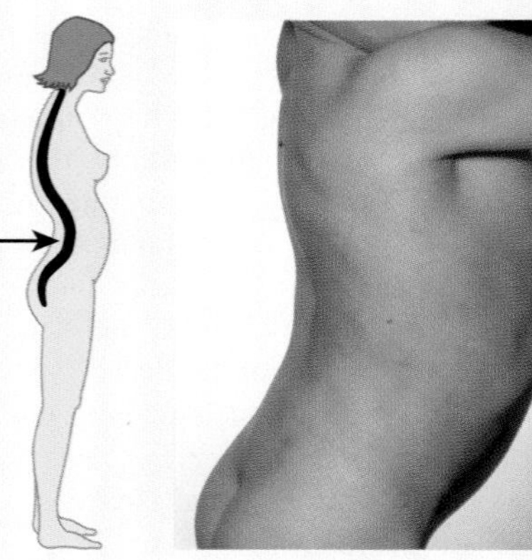

(c) Lordosis

Which abnormal curve is common in women with advanced osteoporosis?

(Figure 7.26). If the sac contains the meninges from the spinal cord and cerebrospinal fluid, the condition is called *spina bifida with meningocele* (me-NING-gō-sēl). If the spinal cord and/or its nerve roots are in the sac, the condition is called *spina bifida with meningomyelocele* (me-ning-gō-MĪ-ē-lō-sēl). The larger the cyst and the number of neural structures it contains, the more serious the neurological problems. In severe cases, there may be partial or complete paralysis, partial or complete loss of urinary bladder and bowel control, and the absence of reflexes. An increased risk of spina bifida is associated with low levels of a B vitamin called folic acid during pregnancy. Spina bifida may be diagnosed prenatally by a test of the mother's blood for a substance produced by the fetus called alphafetoprotein, by sonography, or by amniocentesis (withdrawal of amniotic fluid for analysis).

Figure 7.26 Spina bifida. Shown here is spina bifida with meningomyelocele.

Spina bifida is caused by a failure of laminae to unite at the midline.

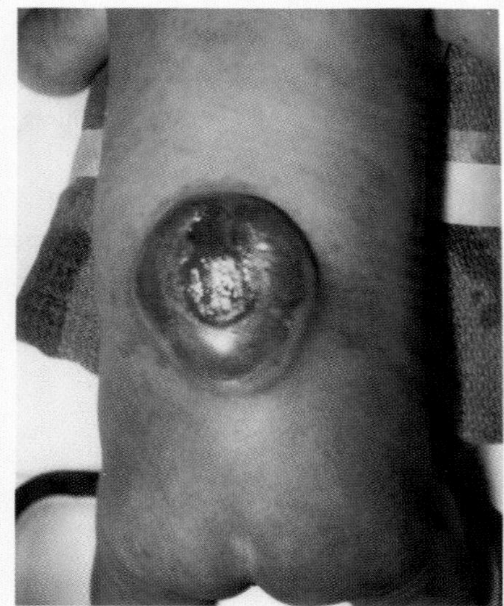

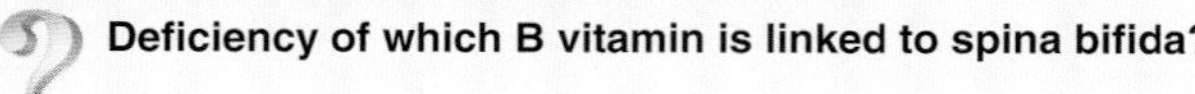

Deficiency of which B vitamin is linked to spina bifida?

Fractures of the Vertebral Column

Fractures of the vertebral column often involve C1, C2, C4–T7, and T12–L2. Cervical or lumbar fractures usually result from a flexion-compression type of injury such as might be sustained in landing on the feet or buttocks after a fall or having a weight fall on the shoulders. Cervical vertebrae may be fractured or dislodged by a fall on the head with acute flexion of the neck, as might happen on diving into shallow water, or being thrown from a horse. Dislocation may result from the sudden forward-then-backward jerk ("whiplash") that may occur in an automobile crash. Spinal cord or spinal nerve damage may occur as a result of fractures of the vertebral column.

MEDICAL TERMINOLOGY

Craniostenosis (krā-nē-ō-sten-Ō-sis; *cranio-* = skull; *-stenosis* = narrowing). Premature closure of one or more cranial sutures during the first 18 to 20 months of life, resulting in a distorted skull. Premature closure of the sagittal suture produces a long narrow skull; premature closure of the coronal suture results in a broad skull. Premature closure of all sutures restricts brain growth and development; surgery is necessary to prevent brain damage.

Craniotomy (krā-nē-OT-ō-mē; *cranio-* = skull; *-tone* = cutting) Surgical procedure in which part of the cranium is removed. It may be performed to remove a blood clot, a brain tumor, or a sample of brain tissue for biopsy.

Laminectomy (lam′-i-NEK-tō-mē; *lamina-* = layer) Surgical procedure to remove a vertebral lamina. It may be performed to access the vertebral cavity and relieve the symptoms of a herniated disc.

Lumbar spine stenosis (*sten-* = narrowed) Narrowing of the spinal cavity in the lumbar part of the vertebral column, due to hypertrophy of surrounding bone or soft tissues. It may be caused by arthritic changes in the intervertebral discs and is a common cause of back and leg pain.

Spinal fusion (FŪ-zhun) Surgical procedure in which two or more vertebrae of the vertebral column are stabilized with a bone graft or synthetic device. It may be performed to treat a fracture of a vertebra or following removal of a herniated disc.

Whiplash injury Injury to the neck region due to severe hyperextension (backward tilting) of the head followed by severe hyperflexion (forward tilting) of the head, usually associated with a rear-end automobile collision. Symptoms are related to stretching and tearing of ligaments and muscles, vertebral fractures, and herniated vertebral discs.

STUDY OUTLINE

INTRODUCTION (p. 195)

1. Bones protect soft body parts and make movement possible; they also serve as landmarks for locating parts of other body systems.
2. The musculoskeletal system is composed of the bones, joints, and muscles working together.

DIVISIONS OF THE SKELETAL SYSTEM (p. 195)

1. The axial skeleton consists of bones arranged along the longitudinal axis. The parts of the axial skeleton are the skull, auditory ossicles (ear bones), hyoid bone, vertebral column, sternum, and ribs. See Table 7.1 on page 195.
2. The appendicular skeleton consists of the bones of the girdles and the upper and lower limbs (extremities). The parts of the appendicular skeleton are the pectoral (shoulder) girdles, bones of the upper limbs, pelvic (hip) girdles, and bones of the lower limbs. See Table 7.1 on page 195.

TYPES OF BONES (p. 197)

1. On the basis of shape, bones are classified as long, short, flat, irregular, or sesamoid. Sesamoid bones develop in tendons or ligaments.
2. Sutural bones are found within the sutures of some cranial bones.

BONE SURFACE MARKINGS (p. 197)

1. Surface markings are structural features visible on the surfaces of bones.
2. Each marking—whether a depression, an opening, or a process—is structured for a specific function, such as joint formation, muscle attachment, or passage of nerves and blood vessels (see Table 7.2 on page 198).

SKULL (p. 198)

1. The 22 bones of the skull include cranial bones and facial bones.
2. The eight cranial bones are the frontal, parietal (2), temporal (2), occipital, sphenoid, and ethmoid.
3. The 14 facial bones are the nasal (2), maxillae (2), zygomatic (2), lacrimal (2), palatine (2), inferior nasal conchae (2), vomer, and mandible.
4. The nasal septum consists of the vomer, perpendicular plate of the ethmoid, and septal cartilage. The nasal septum divides the nasal cavity into left and right sides.
5. Seven skull bones form each of the orbits (eye sockets).
6. The foramina of the skull bones provide passages for nerves and blood vessels (See Table 7.3 on page 210).
7. Sutures are immovable joints that connect most bones of the skull. Examples are the coronal, sagittal, lambdoid, and squamous sutures.
8. Paranasal sinuses are cavities in bones of the skull that are connected to the nasal cavity. The frontal, sphenoid, and ethmoid bones and the maxillae contain paranasal sinuses.
9. Fontanels are mesenchyme-filled spaces between the cranial bones of fetuses and infants. The major fontanels are the anterior, posterior, anterolaterals (2), and posterolaterals (2). After birth, the fontanels fill in with bone and become sutures.

HYOID BONE (p. 212)

1. The hyoid bone is a U-shaped bone that does not articulate with any other bone.
2. It supports the tongue and provides attachment for some tongue muscles and for some muscles of the pharynx and neck.

VERTEBRAL COLUMN (p. 212)

1. The vertebral column, sternum, and ribs constitute the skeleton of the body's trunk.

2. The 26 bones of the adult vertebral column are the cervical vertebrae (7), the thoracic vertebrae (12), the lumbar vertebrae (5), the sacrum (5 fused vertebrae), and the coccyx (usually 4 fused vertebrae).
3. The adult vertebral column contains four normal curves (cervical, thoracic, lumbar, and sacral) that provide strength, support, and balance.
4. Each vertebra usually consists of a body, vertebral arch, and seven processes. Vertebrae in the different regions of the column vary in size, shape, and detail.

THORAX (p. 222)

1. The thoracic skeleton consists of the sternum, ribs, costal cartilages, and thoracic vertebrae.
2. The thoracic cage protects vital organs in the chest area and upper abdomen.

Q SELF-QUIZ QUESTIONS

Fill in the blanks in the following statements.

1. Membrane-filled spaces between cranial bones that enable the fetal skull to modify its size and shape for passage through the birth canal are called ____.
2. The hypophyseal fossa of the sella turcica of the sphenoid bone contains the____.
3. The regions of the vertebral column that consist of fused vertebrae are the ____ and the ____.

Indicate whether the following statements are true or false.

4. The atlanto-occipital joints allow you to rotate the head, as in signifying "no."
5. Ribs that are not attached to the sternum are known as the true ribs.

Choose the one best answer to the following questions.

6. In which of the following bones are paranasal sinuses *not* found? (a) frontal bone, (b) sphenoid bone, (c) lacrimal bones, (d) ethmoid bone, (e) maxillae.
7. Which of the following pairs are mismatched? (a) mandible: only movable bone in the skull, (b) hyoid: bone that does not articulate with any other bone, (c) sacrum: supports lower back, (d) thoracic vertebrae: articulate with thoracic ribs posteriorly, (e) inferior nasal conchae: classified as facial bones.
8. Which of the following bones are *not* paired? (a) vomer, (b) palatine, (c) lacrimal, (d) maxilla, (e) nasal.
9. The suture located between a parietal and temporal bone is the (a) lambdoid, (b) sagittal, (c) coronal, (d) anterolateral, (e) squamous.
10. The primary vertebral curves that appear during fetal development are the (1) cervical curve, (2) thoracic curve, (3) lumbar curve, (4) coccyx curve, (5) sacral curve. (a) 2 and 3, (b) 1 and 2, (c) 2 and 4, (d) 2 and 5, (e) 1 and 3.
11. Which of the following are functions of the cranial bones? (1) Protection of the brain; (2) attachment of muscles that move the head; (3) protection of the special sense organs; (4) attachment to the meninges; (5) attachment of muscles that produce facial expressions. (a) 1, 2, and 5; (b) 1, 2, 4, and 5; (c) 2 and 5; (d) 1, 2, 3, and 5; (e) 1, 2, 3, 4, and 5.

12. Match the following:

____ (a) prominent ridge or elongated projection	(1) foramen
____ (b) tubelike opening	(2) tuberosity
____ (c) large round protuberance at the end of a bone	(3) spinous process
____ (d) smooth, flat articular surface	(4) crest
____ (e) sharp, slender projection	(5) facet
____ (f) opening for passage of blood vessels, nerves, or ligaments	(6) fissure
____ (g) large, rounded, rough projection	(7) condyle
____ (h) shallow depression	(8) fossa
____ (i) narrow slit between adjacent parts of bones for passage of blood vessels or nerves	(9) meatus

13. Match the following:

____ (a) supraorbital foramen	(1) temporal bone
____ (b) temporomandibular joint	(2) sphenoid bone
____ (c) external auditory meatus	(3) cervical vertebrae
____ (d) foramen magnum	(4) ethmoid bone
____ (e) optic foramen	(5) articulation of mandibular fossa and articular tubercle of the temporal bone to the mandible
____ (f) cribriform plate	(6) occipital bone
____ (g) palatine process	(7) frontal bone
____ (h) ramus, body, and condylar process	(8) maxillae
____ (i) transverse foramen, bifid spinous processes	(9) mandible
____ (j) dens	(10) axis
____ (k) promontory	(11) sacrum
____ (l) costal cartilages	(12) sternum
____ (m) xiphoid process	(13) ribs

14. Match the following (the same answer may be used more than once):

____ (a) bones that have greater length than width and consist of a shaft and a variable number of extremities
____ (b) cube-shaped bones that are nearly equal in length and width
____ (c) bones that develop in certain tendons where there is considerable friction, tension, and physical stress
____ (d) small bones located within joints between certain cranial bones
____ (e) thin bones composed of two nearly parallel plates of compact bone enclosing a layer of spongy bone
____ (f) bones with complex shapes, including the vertebrae and some facial bones
____ (g) patella is an example
____ (h) bones that provide considerable protection and extensive areas for muscle attachment
____ (i) include femur, tibia, fibula, humerus, ulna, and radius
____ (j) include cranial bones, sternum, and ribs
____ (k) include almost all of the carpal (wrist) and tarsal (ankle) bones

(1) irregular bones
(2) long bones
(3) short bones
(4) flat bones
(5) sesamoid bones
(6) sutural bones

15. Match the following:

____ (a) forms the forehead
____ (b) form the inferior lateral aspects of the cranium and part of the cranial floor; contain zygomatic process and mastoid process
____ (c) forms part of the anterior portion of the cranial floor, medial wall of the orbits, superior portions of nasal septum, most of the side walls of the nasal cavity; is a major supporting structure of the nasal cavity
____ (d) form the prominence of the cheek and part of the lateral wall and floor of each orbit
____ (e) the largest, strongest facial bone; is the only movable skull bone
____ (f) a roughly triangular bone on the floor of the nasal cavity; one of the components of the nasal septum
____ (g) form greater portion of the sides and roof of the cranial cavity
____ (h) forms the posterior part and most of the base of the cranium; contains the foramen magnum
____ (i) called the keystone of the cranial floor; contains the sella turcica, optic foramen, and pterygoid processes
____ (j) form the bridge of the nose
____ (k) the smallest bones of the face; contain a vertical groove that houses a structure that gathers tears and passes them into the nasal cavity
____ (l) does not articulate with any other bone
____ (m) unite to form the upper jawbone and articulate with every bone of the face except the lower jawbone
____ (n) form the posterior part of the hard palate, part of the floor and lateral wall of the nasal cavity, and a small portion of the floors of the orbits
____ (o) scroll-like bones that form a part of the lateral walls of the nasal cavity; functions in the turbulent circulation and filtration of air

(1) temporal bones
(2) parietal bones
(3) frontal bone
(4) occipital bone
(5) sphenoid bone
(6) ethmoid bone
(7) nasal bones
(8) maxillae
(9) zygomatic bones
(10) lacrimal bones
(11) palatine bones
(12) vomer
(13) mandible
(14) inferior nasal conchae
(15) hyoid bone

CRITICAL THINKING QUESTIONS

1. Jimmy is in a car accident. He can't open his mouth and has been told that he suffers from the following: black eye, broken nose, broken cheek, broken upper jaw, damaged eye socket, and punctured lung. Describe *exactly* what structures have been affected by his car accident.
2. Bubba is a tug-of-war expert. He practices day and night by pulling on a rope attached to an 800-lb anchor. What kinds of changes would you expect that he develops in his bone structure?
3. A new mother brings her newborn infant home and has been told by her well-meaning friend not to wash the baby's hair for several months because the water and soap could "get through that soft area in the top of the head and cause brain damage." Explain to her why this is not true.

ANSWERS TO FIGURE QUESTIONS

7.1 The skull and vertebral column are part of the axial skeleton. The clavicle, shoulder girdle, humerus, pelvic girdle, and femur are part of the appendicular skeleton.

7.2 Flat bones protect underlying organs and provide a large surface area for muscle attachment.

7.3 The frontal, parietal, sphenoid, ethmoid, and temporal bones are cranial bones.

7.4 The parietal and temporal bones are joined by the squamous suture, the parietal and occipital bones are joined by the lambdoid suture, and the parietal and frontal bones are joined by the coronal suture.

7.5 The temporal bone articulates with the parietal, sphenoid, zygomatic, and occipital bones.

7.6 The parietal bones form the posterior, lateral portion of the cranium.

7.7 The medulla oblongata of the brain connects with the spinal cord in the foramen magnum.

7.8 From the crista galli of the ethmoid bone, the sphenoid articulates with the frontal, parietal, temporal, occipital, temporal, parietal, and frontal bones, ending again at the crista galli of the ethmoid bone.

7.9 The perpendicular plate of the ethmoid bone forms the superior part of the nasal septum, and the lateral masses compose most of the medial walls of the orbits.

7.10 The mandible is the only movable skull bone, other than the auditory ossicles.

7.11 The nasal septum divides the nasal cavity into right and left sides.

7.12 Bones forming the orbit are the frontal, sphenoid, zygomatic, maxilla, lacrimal, ethmoid, and palatine.

7.13 The paranasal sinuses produce mucus and serve as resonating chambers for vocalization.

7.14 The anterolateral fontanel is bordered by four different skull bones, the frontal, parietal, temporal, and sphenoid.

7.15 The hyoid bone does not articulate with any other bone.

7.16 The thoracic and sacral curves of the vertebral column are concave relative to the anterior of the body.

7.17 The vertebral foramina enclose the spinal cord; the intervertebral foramina provide spaces through which spinal nerves exit the vertebral column.

7.18 The atlas moving on the axis permits movement of the head to signify "no."

7.19 The facets and demifacets on the bodies of the thoracic vertebrae articulate with the heads of the ribs, and the facets on the transverse processes of these vertebrae articulate with the tubercles of the ribs.

7.20 The lumbar vertebrae are the largest and strongest in the body because the amount of weight supported by vertebrae increases toward the inferior end of the vertebral column.

7.21 There are four pairs of sacral foramina, for a total of eight. Each anterior sacral foramen joins a posterior sacral foramen at the intervertebral foramen. Nerves and blood vessels pass through these tunnels in the bone.

7.22 Pairs 1–7 are the true ribs, pairs 8–12 are the false ribs, and pairs 11 and 12 are also known as the floating ribs.

7.23 The facet on the head of a rib fits into a facet on the body of a vertebra, and the articular part of the tubercle of a rib articulates with the facet of the transverse process of a vertebra.

7.24 Most herniated discs occur in the lumbar region because it bears most of the body weight and most flexing and bending occurs there.

7.25 Kyphosis is common in individuals with advanced osteoporosis.

7.26 Deficiency of folic acid is associated with spina bifida.

Chapter 8

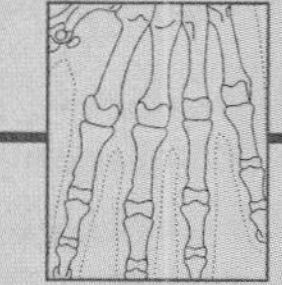

The Skeletal System: The Appendicular Skeleton

The Appendicular Skeleton and Homeostasis

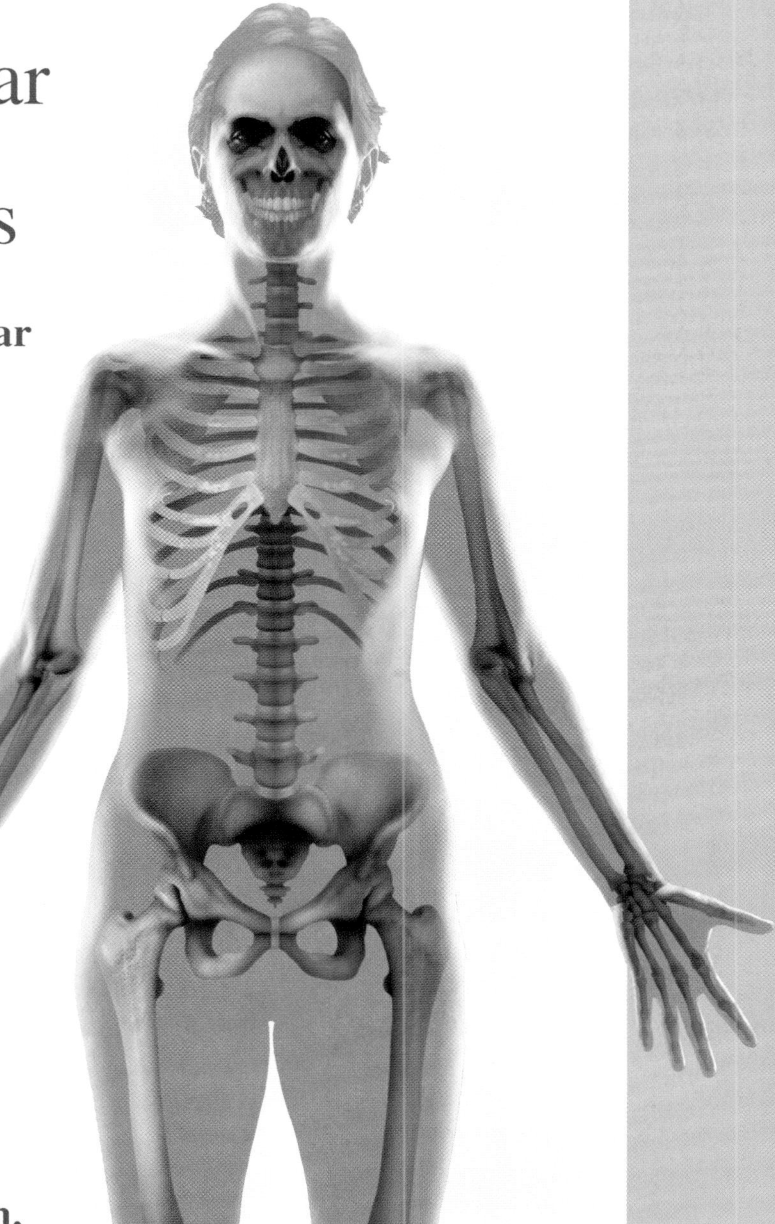

The bones of the appendicular skeleton contribute to homeostasis by providing attachment points and leverage for muscles, which aids body movements; providing support and protection of internal organs, such as the reproductive organs; and by storing and releasing calcium.

wiley.com/college/apcentral

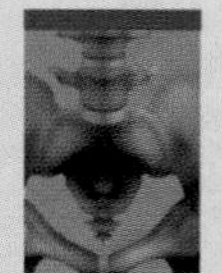

As noted in Chapter 7, the two main divisions of the skeletal system are the axial skeleton and the appendicular skeleton. As you learned in that chapter, the general function of the axial skeleton is the protection of internal organs; the primary function of the appendicular skeleton, the focus of this chapter, is movement. The appendicular skeleton includes the bones that make up the upper and lower limbs as well as the bones of the two girdles that attach the limbs to the axial skeleton. The bones of the appendicular skeleton are connected with one another and with skeletal muscles, permitting us to walk, write, use a computer, dance, swim, and play a musical instrument.

PECTORAL (SHOULDER) GIRDLE

▶ **OBJECTIVE**

Identify the bones of the pectoral (shoulder) girdle and their principal markings.

The human body has two **pectoral** (PEK-tō-ral) or **shoulder girdles** that attach the bones of the upper limbs to the axial skeleton (Figure 8.1). Each of the two pectoral girdles consists of a clavicle and a scapula. The *clavicle* is the anterior bone and articulates with the manubrium of the sternum at the *sternoclavicular joint.* The scapula articulates with the clavicle at the *acromioclavicular joint* and with the humerus at the *glenohumeral (shoulder) joint.* The pectoral girdles do not articulate with the vertebral column and are held in position by muscle attachments.

Clavicle

Each slender, S-shaped **clavicle** (KLAV-i-kul = key), or *collarbone,* lies horizontally across the anterior part of the thorax superior to the first rib (Figure 8.2). The bone is S-shaped because the medial half is convex anteriorly, and the lateral half is concave anteriorly. The medial end, called the *sternal end,* is rounded and articulates with the manubrium of the sternum to form the *sternoclavicular joint.* The broad, flat, lateral end, the *acromial end* (a-KRŌ-mē-al), articulates with the acromion of the scapula to form the *acromioclavicular joint* (see Figure 8.1). The *conoid tubercle* (KŌ-noyd = conelike) on the inferior surface of the lateral end of the bone is a point of attachment for the conoid ligament, which attaches the clavicle and scapula. As its name implies, the *impression for the costoclavicular ligament* on the inferior surface of the sternal end is a point of attachment for

Figure 8.1 Right pectoral (shoulder) girdle. (See Tortora, *A Photographic Atlas of the Human Body, Second Edition,* Figure 3.1.)

The clavicle is the anterior bone of the pectoral girdle and the scapula is the posterior bone.

What is the function of the pectoral girdles?

Figure 8.2 **Right clavicle.**

The clavicle articulates medially with the manubrium of the sternum and laterally with the acromion of the scapula.

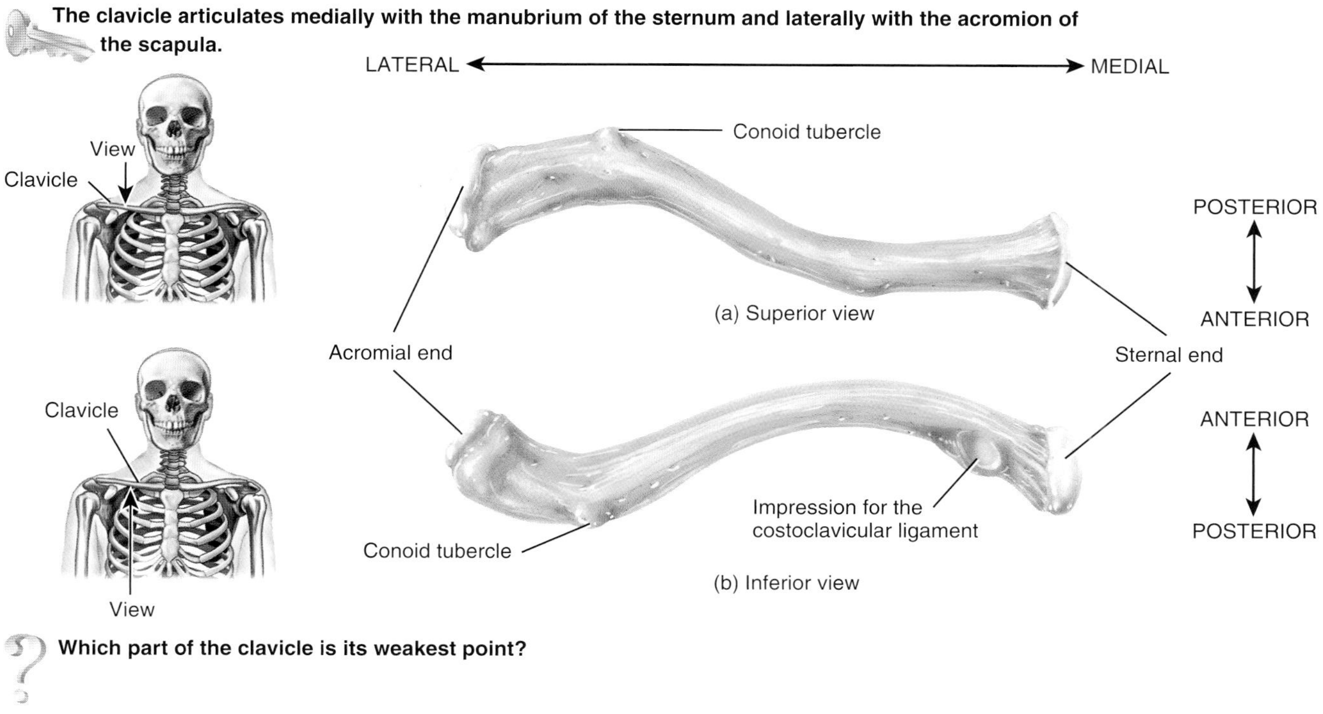

Which part of the clavicle is its weakest point?

the costoclavicular ligament (see Figure 8.2b). The costoclavicular ligament attaches the clavicle and first rib.

Fractured Clavicle

The clavicle transmits mechanical force from the upper limb to the trunk. If the force transmitted to the clavicle is excessive, as when you fall on your outstretched arm, a **fractured clavicle** may result. The clavicle is one of the most frequently broken bones in the body. Because the junction of the two curves of the clavicle is its weakest point, the clavicular midregion is the most frequent fracture site. Even in the absence of fracture, compression of the clavicle as a result of automobile accidents involving the use of shoulder harness seatbelts often causes damage to the median nerve, which lies between the clavicle and the second rib. A fractured clavicle is usually treated with a regular sling to keep the arm from moving outward. ■

Scapula

Each **scapula** (SCAP-ū-la; plural is *scapulae*), or *shoulder blade,* is a large, triangular, flat bone situated in the superior part of the posterior thorax between the levels of the second and seventh ribs (Figure 8.3). A prominent ridge called the *spine* runs diagonally across the posterior surface of the flattened, triangular *body* of the scapula (Figure 8.3b). The lateral end of the spine projects as a flattened, expanded process called the *acromion* (a-KRŌ-mē-on; *acrom-* = topmost), easily felt as the high point of the shoulder. Tailors measure the length of the upper limb from the acromion. Inferior to the acromion is a shallow depression, the *glenoid cavity,* that accepts the head of the humerus (arm bone) to form the *glenohumeral joint* (see Figure 8.1).

The thin edge of the scapula closer to the vertebral column is called the *medial (vertebral) border.* The thick edge of the scapula closer to the arm is called the *lateral (axillary) border.* The medial and lateral borders join at the *inferior angle.* The superior edge of the scapula, called the *superior border,* joins the medial border at the *superior angle.* The *scapular notch* is a prominent indentation along the superior border through which the suprascapular nerve passes.

At the lateral end of the superior border of the scapula is a projection of the anterior surface called the *coracoid process* (KOR-a-koyd = like a crow's beak), to which the tendons of muscles (pectoralis minor, coracobrachialis, and biceps brachii) and ligaments (coracoacromial, conoid, and trapezoid) attach. Superior and inferior to the spine on the posterior surface of the scapula are two fossae: the *supraspinous fossa* (sū-pra-SPĪ-nus) and the *infraspinous fossa* (in-fra-SPĪ-nus), respectively. Both serve as surfaces of attachment for the tendons of the supraspinatus and infraspinatus muscles of the shoulder. On the anterior surface of the scapula is a slightly hollowed-out area called the *subscapular fossa,* also a surface of attachment for the tendons of shoulder muscles.

▶ CHECKPOINT

1. Which bones or parts of bones of the pectoral girdle form the sternoclavicular, acromioclavicular, and glenohumeral joints?

Figure 8.3 Right scapula (shoulder blade). (See Tortora, *A Photographic Atlas of the Human Body, Second Edition,* Figure 3.22.)

The glenoid cavity of the scapula articulates with the head of the humerus to form the glenohumeral (shoulder) joint.

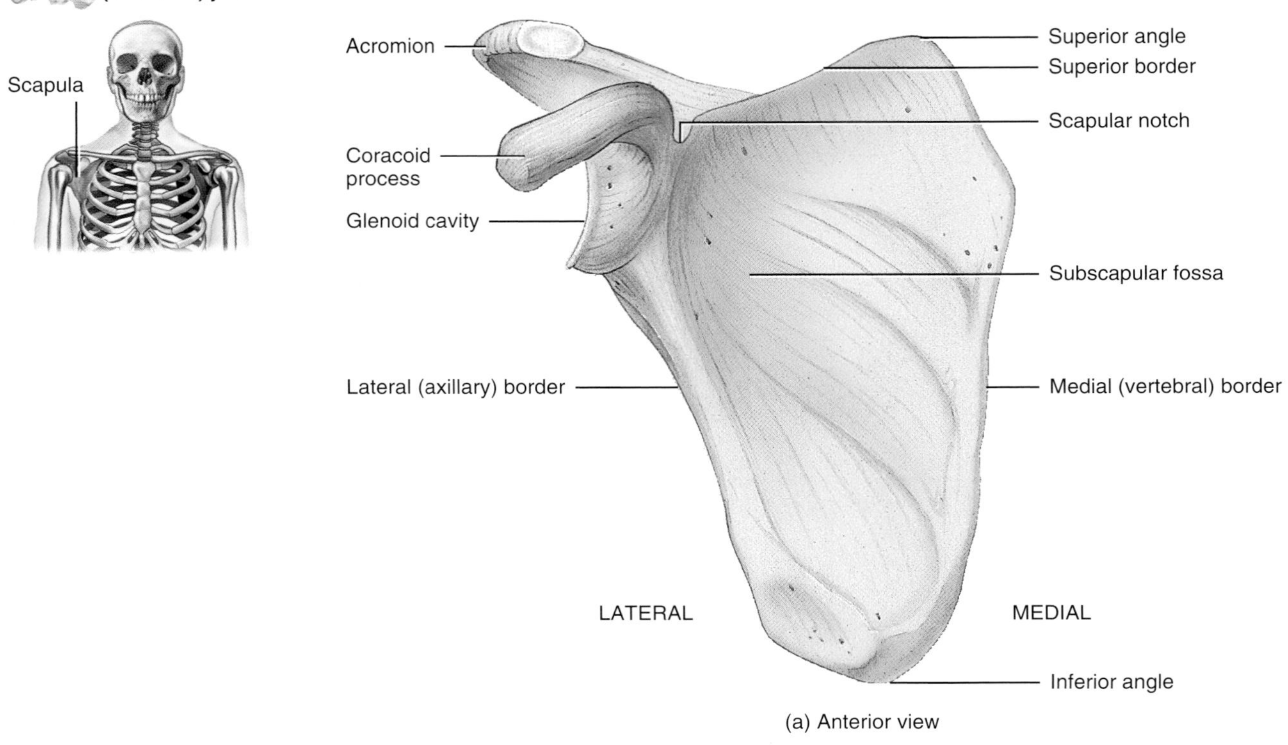

(a) Anterior view

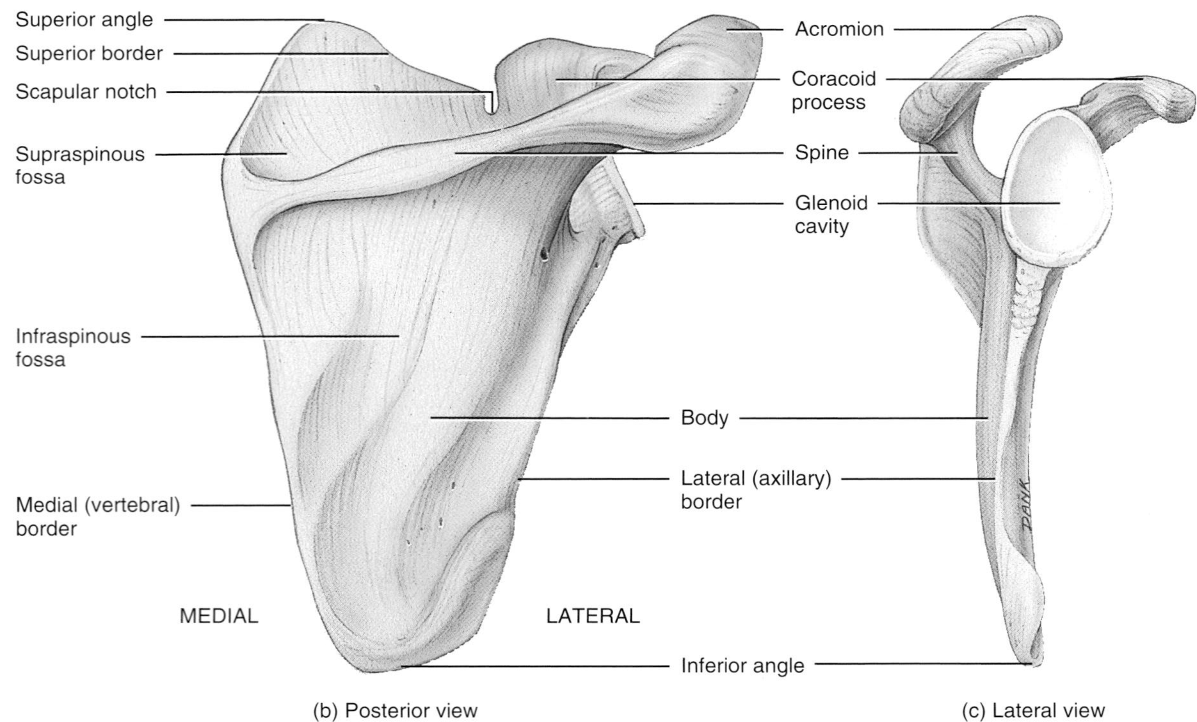

(b) Posterior view

(c) Lateral view

Which part of the scapula forms the high point of the shoulder?

UPPER LIMB (EXTREMITY)

▶ OBJECTIVES

Identify the bones of the upper limb and their principal markings.

Describe the joints between the upper limb bones.

Each **upper limb (upper extremity)** has 30 bones in three locations—(1) the humerus in the arm; (2) the ulna and radius in the forearm; and (3) the 8 carpals in the carpus (wrist), the 5 metacarpals in the metacarpus (palm), and the 14 phalanges (bones of the digits) in the hand (Figure 8.4).

Figure 8.4 **Right upper limb.**

Each upper limb includes a humerus, ulna, radius, carpals, metacarpals, and phalanges.

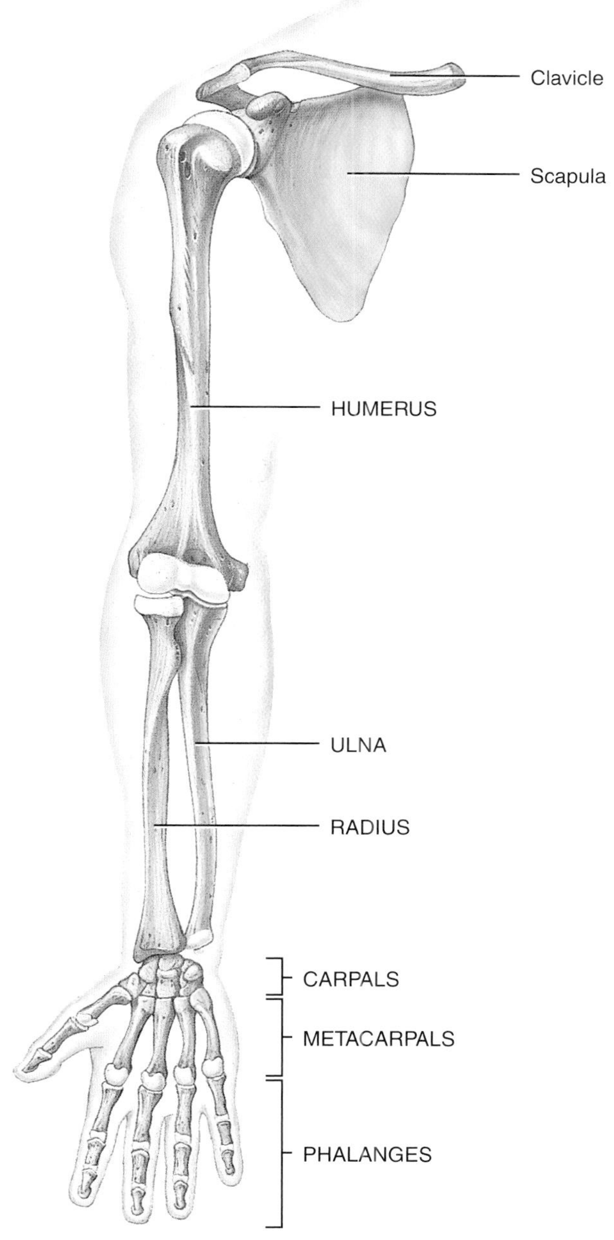

How many bones make up each upper limb?

Humerus

The **humerus** (HŪ-mer-us), or arm bone, is the longest and largest bone of the upper limb (Figure 8.5). It articulates proximally with the scapula and distally at the elbow with two bones, the ulna and the radius.

The proximal end of the humerus features a rounded *head* that articulates with the glenoid cavity of the scapula to form the *glenohumeral joint.* Distal to the head is the *anatomical neck,* which is visible as an oblique groove. The *greater tubercle* is a lateral projection distal to the anatomical neck. It is the most laterally palpable bony landmark of the shoulder region. The *lesser tubercle* projects anteriorly. Between the two tubercles there is a groove named the *intertubercular sulcus.* The *surgical neck* is a constriction in the humerus just distal to the tubercles, where the head tapers to the shaft; it is so named because fractures often occur here.

The *body (shaft)* of the humerus is roughly cylindrical at its proximal end, but it gradually becomes triangular until it is flattened and broad at its distal end. Laterally, at the middle portion of the shaft, there is a roughened, V-shaped area called the *deltoid tuberosity.* This area serves as a point of attachment for the tendons of the deltoid muscle.

Several prominent features are evident at the distal end of the humerus. The *capitulum* (ka-PIT-ū-lum; *capit-* = head) is a rounded knob on the lateral aspect of the bone that articulates with the head of the radius. The *radial fossa* is an anterior depression that receives the head of the radius when the forearm is flexed (bent). The *trochlea* (TRŌK-lē-a), located medial to the capitulum, is a spool-shaped surface that articulates with the ulna. The *coronoid fossa* (KŌR-o-noyd = crown-shaped) is an anterior depression that receives the coronoid process of the ulna when the forearm is flexed. The *olecranon fossa* (ō-LEK-ra-non = elbow) is a posterior depression that receives the olecranon of the ulna when the forearm is extended (straightened). The *medial epicondyle* and *lateral epicondyle* are rough projections on either side of the distal end of the humerus to which the tendons of most muscles of the forearm are attached. The ulnar nerve, the one that makes you see stars when you hit your elbow, may be easily palpated by rolling a finger over the skin surface above the posterior surface of the medial epicondyle.

Figure 8.5 Right humerus in relation to the scapula, ulna, and radius. (See Tortora, *A Photographic Atlas of the Human Body, Second Edition,* Figure 3.23.)

The humerus is the longest and largest bone of the upper limb.

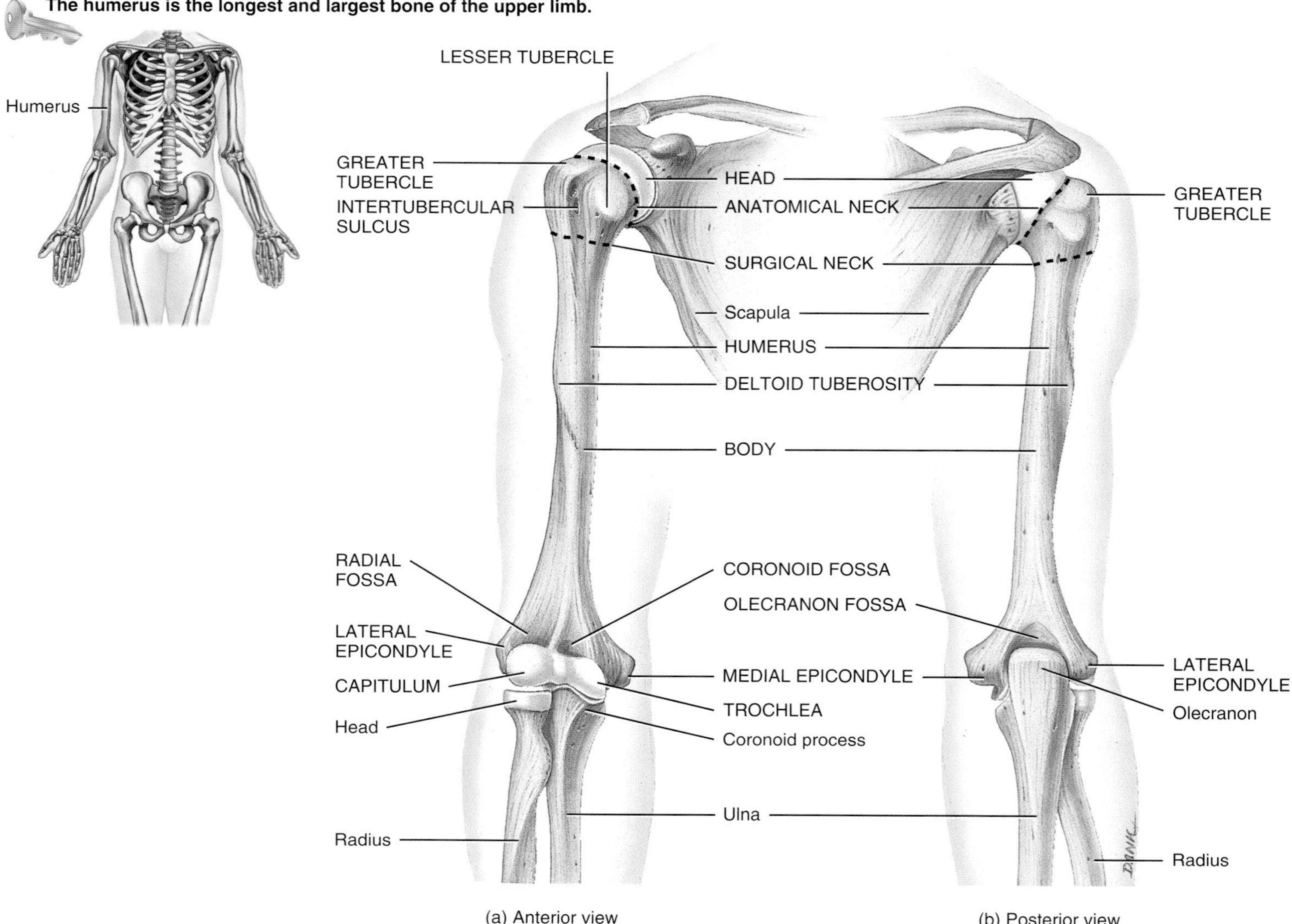

Which parts of the humerus articulate with the radius at the elbow? With the ulna at the elbow?

Ulna and Radius

The **ulna** is located on the medial aspect (the little-finger side) of the forearm and is longer than the radius (Figure 8.6). You may find it convenient to use an aid called a *mnemonic device* (nē-MON-ik = memory) to learn new or unfamiliar information. One such mnemonic to help you remember the location of the ulna in relation to the hand is "p.u." (the **p**inky is on the **u**lna side). At the proximal end of the ulna (Figure 8.6b) is the *olecranon,* which forms the prominence of the elbow. With the olecranon, an anterior projection called the *coronoid process* (Figure 8.6a) receives the trochlea of the humerus. The *trochlear notch* is a large curved area between the olecranon and coronoid process that forms part of the elbow joint (see Figure 8.7b). On the lateral side of the coronoid process is a depression, the *radial notch,* which receives the head of the radius. Just inferior to the coronoid process is the *ulnar tuberosity,* to which the biceps brachii muscle attaches. The distal end of the ulna consists of a *head* that is separated from the wrist by a disc of fibrocartilage. A *styloid process* is located on the posterior side of the ulna's distal end. It provides attachment for the ulnar collateral ligament to the wrist.

The **radius** is located on the lateral aspect (thumb side) of the forearm (see Figure 8.6). The proximal end of the radius has a disc-shaped *head* that articulates with the capitulum of the humerus and the radial notch of the ulna. Inferior to the head is the constricted *neck.* A roughened area inferior to the neck on the medial side, called the *radial tuberosity,* is a point of attachment for the tendons of the biceps brachii muscle. The shaft of the radius widens distally to form a *styloid process* on the lateral side, which can be felt proximal to the thumb. The styloid

Figure 8.6 Right ulna and radius in relation to the humerus and carpals. (See Tortora, *A Photographic Atlas of the Human Body, Second Edition,* Figure 3.24.)

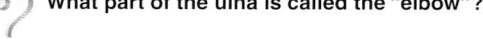

In the forearm, the longer ulna is on the medial side, and the shorter radius is on the lateral side.

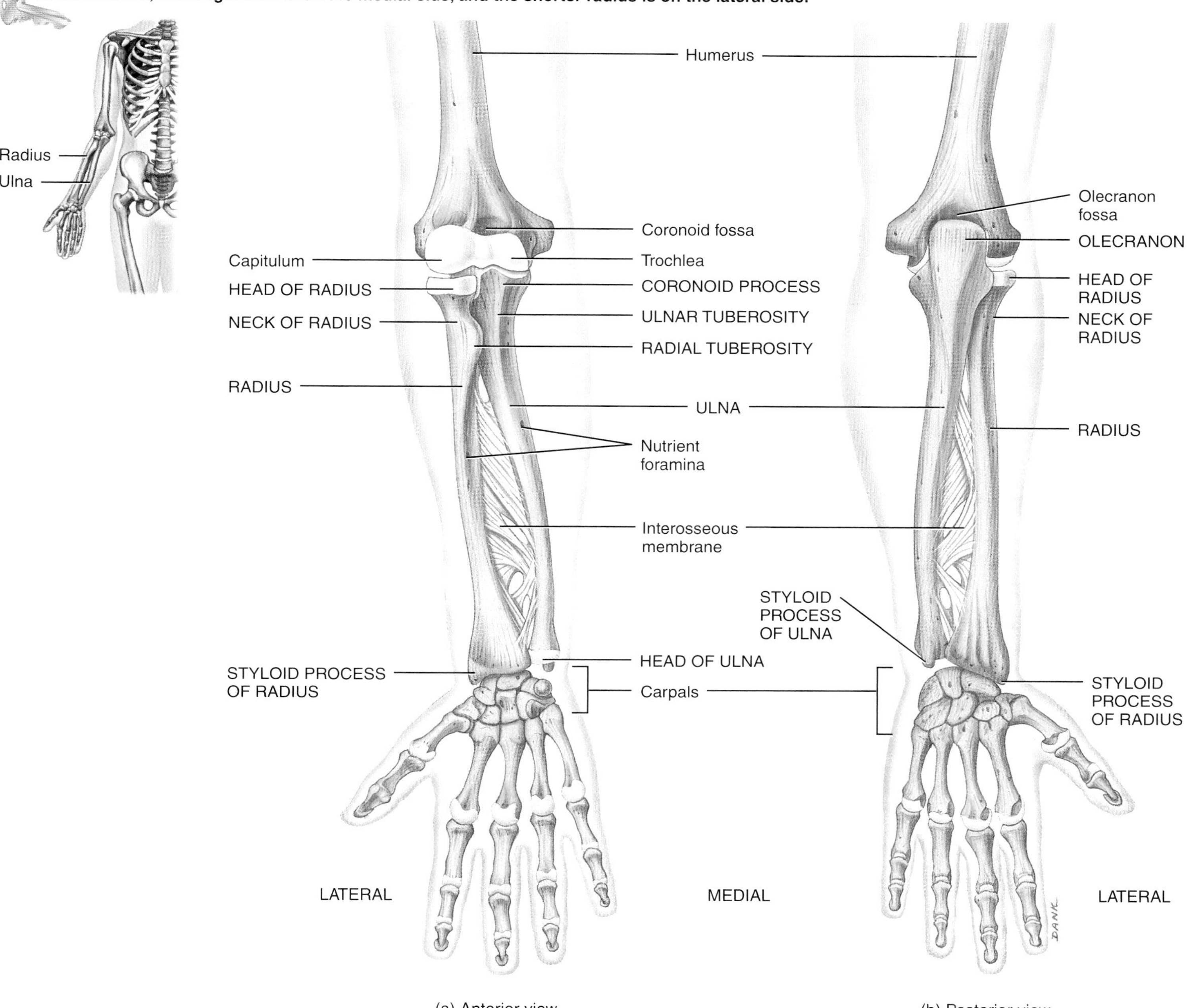

(a) Anterior view

(b) Posterior view

What part of the ulna is called the "elbow"?

process provides attachment for the brachioradialis muscle and for attachment of the radial collateral ligament to the wrist. Fracture of the distal end of the radius is the most common fracture in adults older than 50 years.

The ulna and radius articulate with the humerus at the *elbow joint.* The articulation occurs in two places: where the head of the radius articulates with the capitulum of the humerus (Figure 8.7a), and where the trochlear notch of the ulna receives the trochlea of the humerus (Figure 8.7b).

The ulna and the radius connect with one another at three sites. First, a broad, flat, fibrous connective tissue called the *interosseous membrane* (in-ter-OS-ē-us; *inter-* = between, *osse-* = bone) joins the shafts of the two bones. This membrane also provides a site of attachment for some tendons of deep skeletal muscles of the forearm. The ulna and radius articulate directly at their proximal and distal ends. Proximally, the head of the radius articulates with the ulna's *radial notch,* a depression that is lateral and inferior to the trochlear notch (Figure 8.7b). This articulation is the *proximal radioulnar joint.* Distally, the head of the ulna articulates with the *ulnar notch* of the radius (Figure 8.7c). This articulation is the *distal radioulnar joint.* Finally, the distal end of the radius articulates with three bones of the wrist—the lunate, the scaphoid, and the triquetrum—to form the *radiocarpal (wrist) joint.*

Carpals, Metacarpals, and Phalanges

The **carpus** (wrist) is the proximal region of the hand and consists of eight small bones, the **carpals,** joined to one another by ligaments (Figure 8.8). Articulations among carpal bones are called *intercarpal joints.* The carpals are arranged in two transverse rows of four bones each. Their names reflect their shapes. The carpals in the proximal row, from lateral to medial, are the **scaphoid** (SKAF-oyd = boatlike), **lunate** (LOO-nāt = moon-shaped), **triquetrum** (trī-KWĒ-trum = three-cornered), and **pisiform** (PIS-i-form = pea-shaped). The carpals in the distal row, from lateral to medial, are the **trapezium** (tra-PĒ-zē-um = four-sided figure with no two sides parallel), **trapezoid** (TRAP-e-zoyd = four-sided figure with two sides parallel), **capitate** (KAP-i-tāt = head-shaped), and **hamate** (HAM-āt = hooked).

Figure 8.7 Articulations formed by the ulna and radius. (a) Elbow joint. (b) Joint surfaces at proximal end of the ulna. (c) Joint surfaces at distal ends of radius and ulna. The ulna and radius are also attached by the interosseous membrane.

The elbow joint is formed by two articulations: (1) the trochlear notch of the ulna with the trochlea of the humerus and (2) the head of the radius with the capitulum of the humerus.

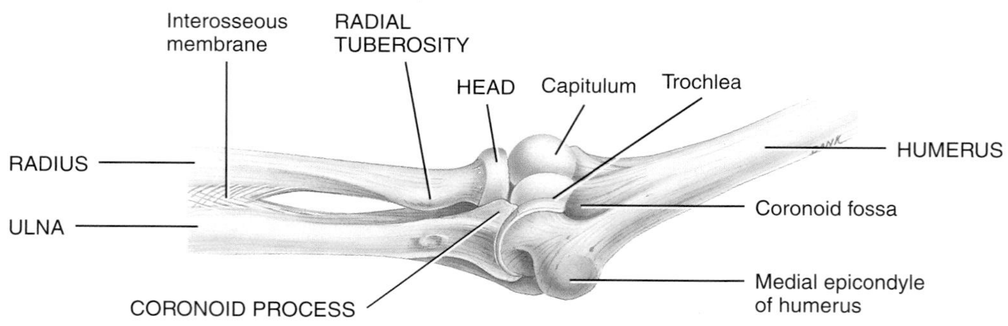

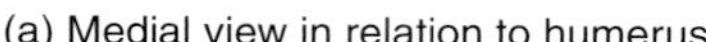
(a) Medial view in relation to humerus

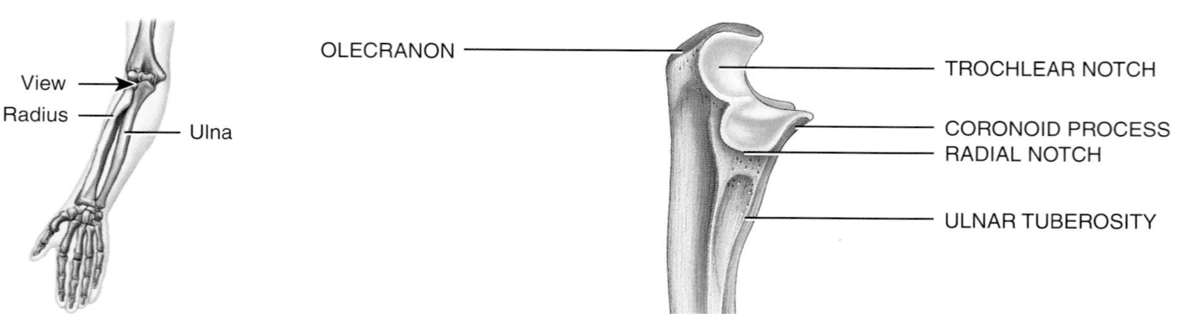

(b) Lateral view of proximal end of ulna

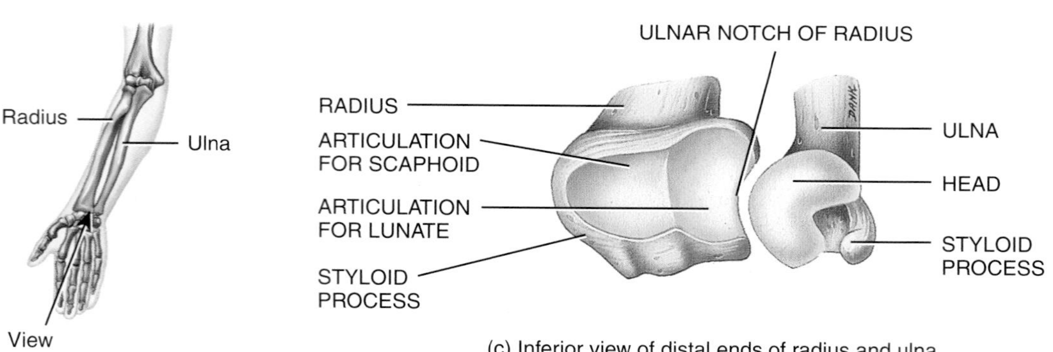

(c) Inferior view of distal ends of radius and ulna

How many points of attachment are there between the radius and ulna?

The capitate is the largest carpal bone; its rounded projection, the head, articulates with the lunate. The hamate is named for a large hook-shaped projection on its anterior surface. In about 70% of carpal fractures, only the scaphoid is broken. This is because the force of a fall on an outstretched hand is transmitted from the capitate through the scaphoid to the radius.

The concave space formed by the pisiform and hamate (on the ulnar side), and the scaphoid and trapezium (on the radial side), plus the *flexor retinaculum* (fibrous bands of deep fascia) is the **carpal tunnel.** The long flexor tendons of the digits and thumb and the median nerve pass through the carpal tunnel. Narrowing of the carpal tunnel, due to such factors as inflammation, may give rise to a condition called carpal tunnel syndrome (described on page 374).

A mnemonic for learning the names of the carpal bones is shown in Figure 8.8. The first letter of the carpal bones from lateral to medial (proximal row, then distal row) corresponds to the first letter of each word in the mnemonic.

The **metacarpus** (*meta-* = beyond), or palm, is the intermediate region of the hand and consists of five bones called **metacarpals.** Each metacarpal bone consists of a proximal *base,* an intermediate *shaft,* and a distal *head* (Figure 8.8b).

Figure 8.8 Right wrist and hand in relation to the ulna and radius.

The skeleton of the hand consists of the proximal carpals, the intermediate metacarpals, and the distal phalanges.

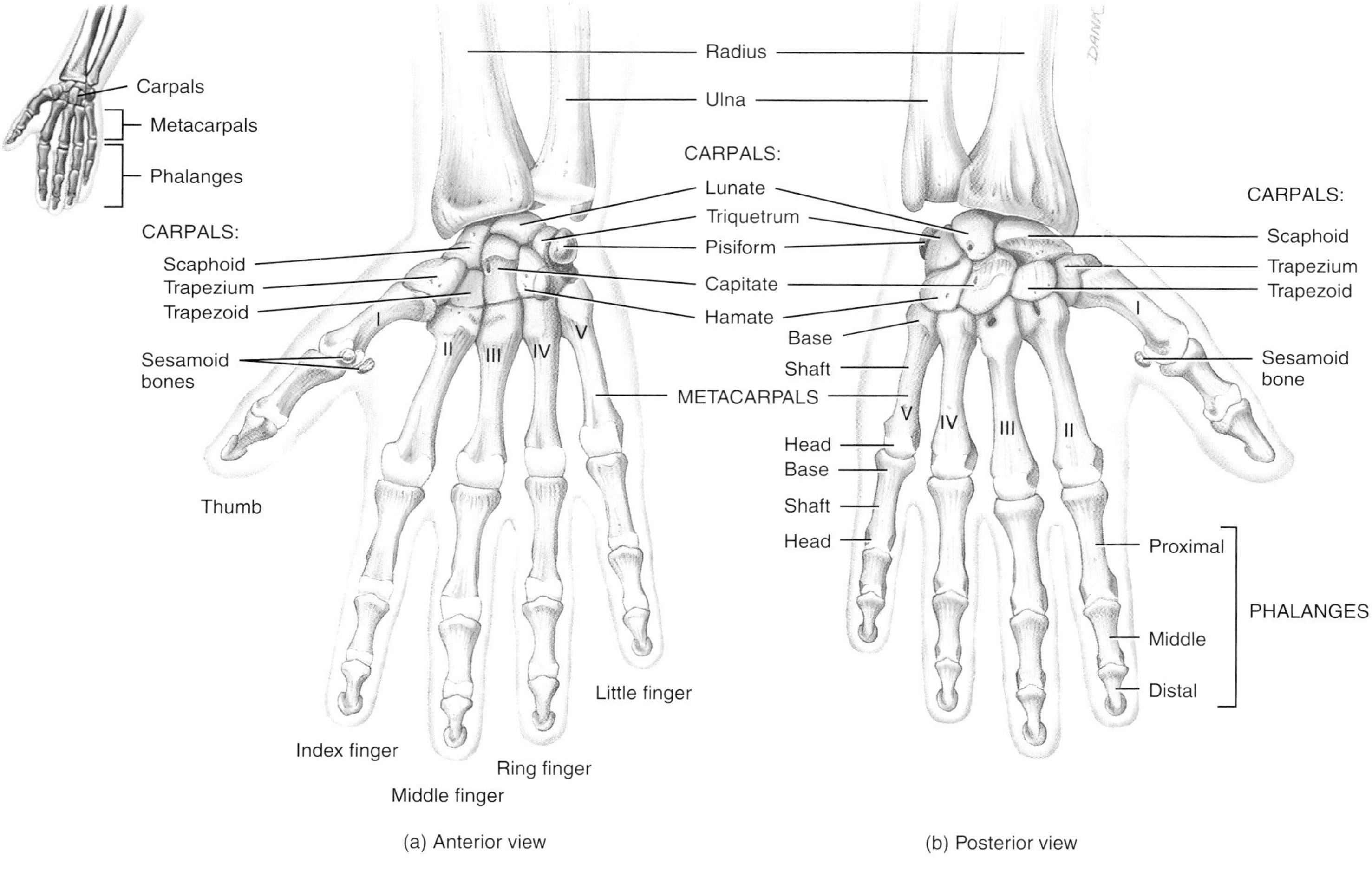

(a) Anterior view

(b) Posterior view

MNEMONIC for carpal bones*:

Stop **L**etting **T**hose **P**eople **T**ouch **T**he **C**adaver's **H**and.

Scaphoid **L**unate **T**riquetrum **P**isiform | **T**rapezium **T**rapezoid **C**apitate **H**amate

Proximal row: Lateral → Medial | Distal row: Lateral → Medial

* *Edward Tanner, University of Alabama, SOM*

Which is the most frequently fractured wrist bone?

The metacarpal bones are numbered I to V (or 1–5), starting with the thumb, from lateral to medial. The bases articulate with the distal row of carpal bones to form the *carpometacarpal joints.* The heads articulate with the proximal phalanges to form the *metacarpophalangeal joints.* The heads of the metacarpals, commonly called "knuckles," are readily visible in a clenched fist.

The **phalanges** (fa-LAN-jēz; *phalan-* = a battle line), or bones of the digits, make up the distal part of the hand. There are 14 phalanges in the five digits of each hand and, like the metacarpals, the digits are numbered I to V (or 1–5), beginning with the thumb, from lateral to medial. A single bone of a digit is referred to as a **phalanx** (FĀ-lanks). Each phalanx consists of a proximal *base,* an intermediate *shaft,* and a distal *head.* The thumb (*pollex*) has two phalanges, and there are three phalanges in each of the other four digits. In order from the thumb, these other four digits are commonly referred to as the index finger, middle finger, ring finger, and little finger. The first row of phalanges, the *proximal row,* articulates with the metacarpal bones and second row of phalanges. The second row of phalanges, the *middle row,* articulates with the proximal row and the third row, called the *distal row.* The thumb has no middle phalanx. Joints between phalanges are called *interphalangeal joints.*

► CHECKPOINT

2. Name the bones that form the upper limb, from proximal to distal.
3. Describe the joints of the upper limb.

PELVIC (HIP) GIRDLE

► OBJECTIVES

Identify the bones of the pelvic girdle and their principal markings.

Describe the division of the pelvic girdle into false and true pelves.

The **pelvic (hip) girdle** consists of the two **hip bones,** also called **coxal bones** or **os coxa** (KOK-sal; *cox-* = hip) (Figure 8.9). The hip bones unite anteriorly at a joint called the **pubic symphysis** (PŪ-bik SIM-fi-sis). They unite posteriorly with the sacrum at the *sacroiliac joints.* The complete ring composed of the hip bones, pubic symphysis, and sacrum forms a deep, basinlike structure called the **bony pelvis** (*pelv-* = basin). The plural is *pelves* (PEL-vēz) or *pelvises.* Functionally, the bony pelvis provides a strong and stable support for the vertebral column and pelvic organs. The pelvic girdle of the bony pelvis also connects the bones of the lower limbs to the axial skeleton.

Each of the two hip bones of a newborn consists of three bones separated by cartilage: a superior *ilium,* an inferior and anterior *pubis,* and an inferior and posterior *ischium* (IS-kē-um). By age 23, the three separate bones fuse together (Figure 8.10a). Although the hip bones function as single bones, anatomists commonly discuss them as three separate bones.

Ilium

The **ilium** (IL-ē-um = flank), the largest of the three components of the hip bone (Figure 8.10b, c), is composed of a

Figure 8.9 Bony pelvis. Shown here is the female bony pelvis. (See Tortora, *A Photographic Atlas of the Human Body, Second Edition,* Figure 3.27.)

The hip bones unite anteriorly at the pubic symphysis and posteriorly at the sacrum to form the bony pelvis.

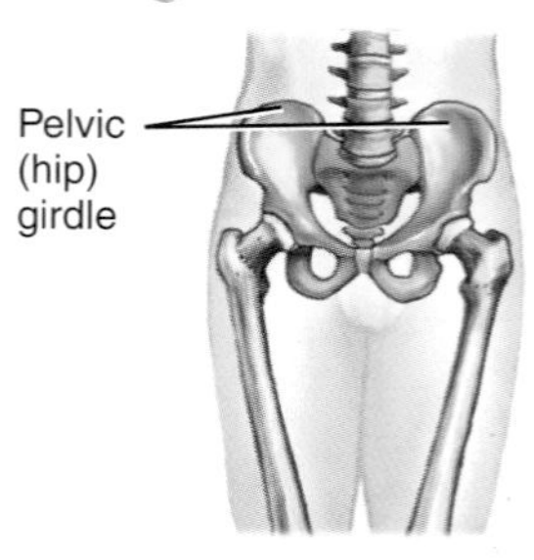

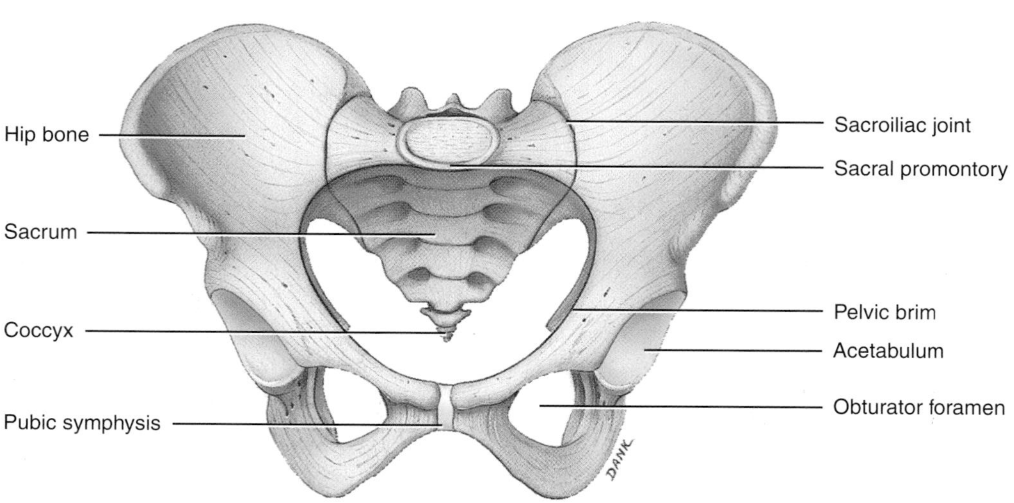

What are the functions of the bony pelvis?

Figure 8.10 Right hip bone. The lines of fusion of the ilium, ischium, and pubis depicted in (a) are not always visible in an adult. (See Tortora, *A Photographic Atlas of the Human Body, Second Edition,* Figure 3.26.)

The acetabulum is the socket formed by the convergence of the three parts of the hip bone.

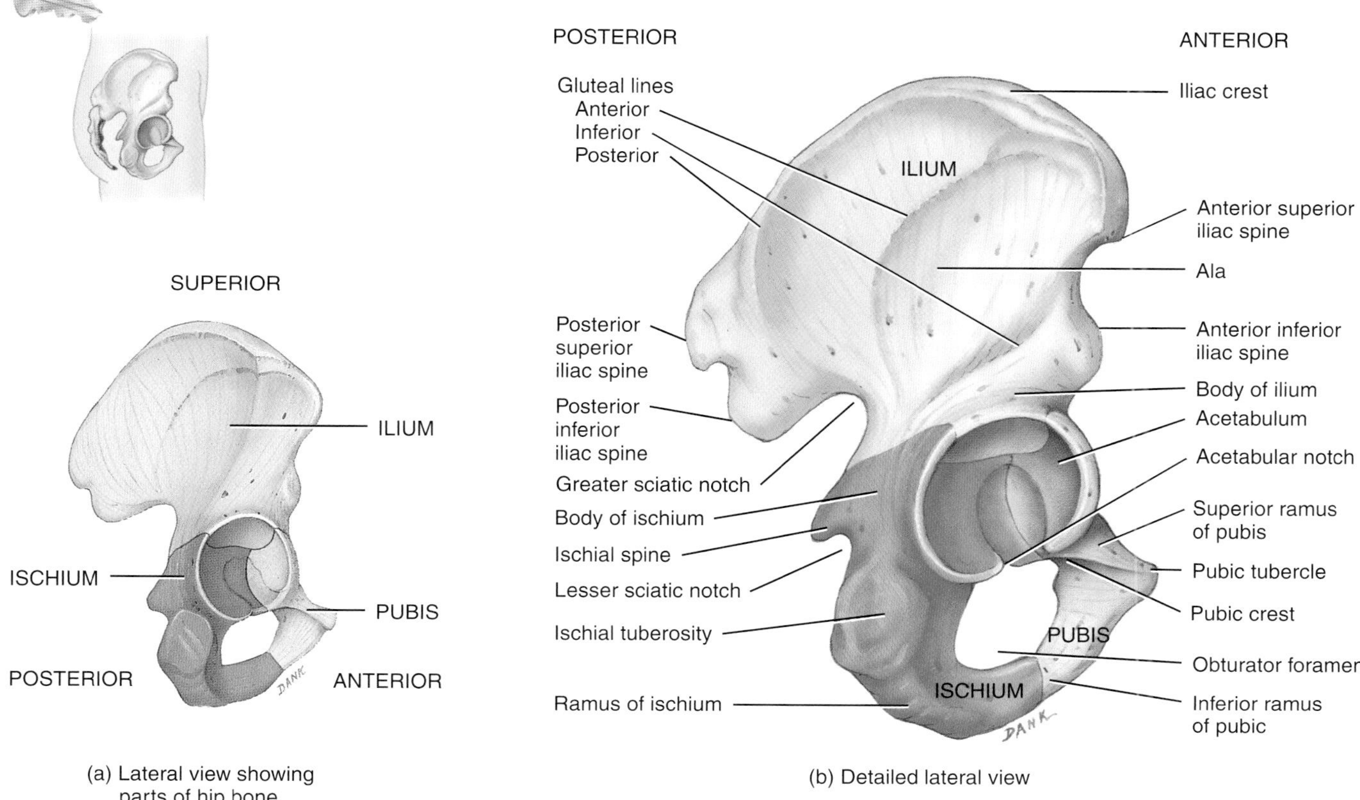

(a) Lateral view showing parts of hip bone

(b) Detailed lateral view

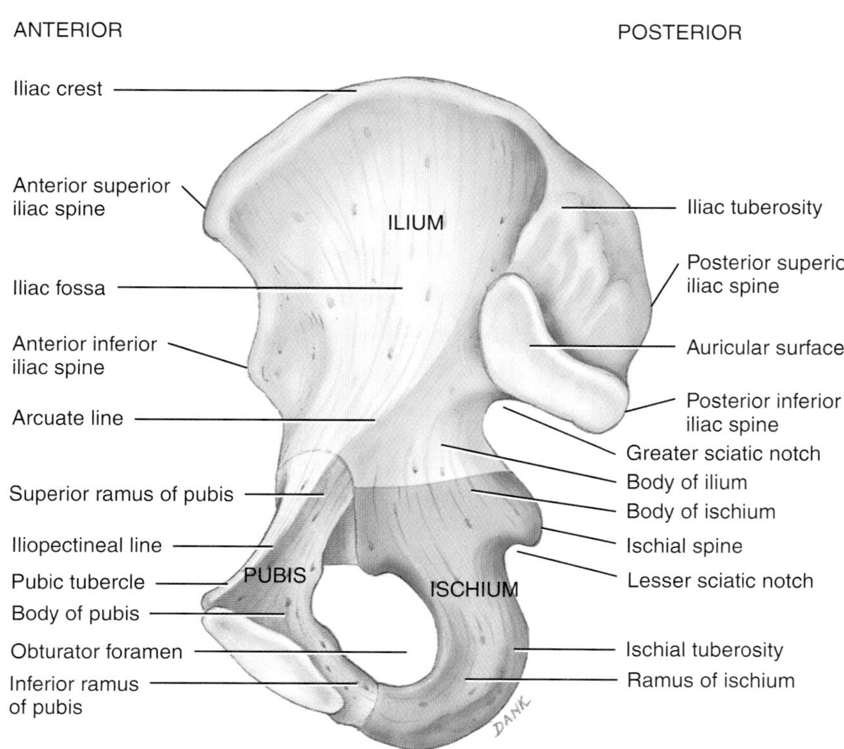

(c) Detailed medial view

Which part of the hip bone articulates with the femur? With the sacrum?

superior *ala* (= wing) and an inferior *body*. The body helps form the *acetabulum,* the socket for the head of the femur. The superior border of the ilium, the *iliac crest,* ends anteriorly in a blunt *anterior superior iliac spine.* Bruising of the anterior superior iliac spine and associated soft tissues, such as occurs in body contact sports, is called a **hip pointer.** Below this spine is the *anterior inferior iliac spine.* Posteriorly, the iliac crest ends in a sharp *posterior superior iliac spine.* Below this spine is the *posterior inferior iliac spine.* The spines serve as points of attachment for the tendons of the muscles of the trunk, hip, and thighs. Below the posterior inferior iliac spine is the *greater sciatic notch* (sī-AT-ik), which allows passage of the sciatic nerve, the longest nerve in the body.

The medial surface of the ilium contains the *iliac fossa,* a concavity where the tendon of the iliacus muscle attaches. Posterior to this fossa are the *iliac tuberosity,* a point of attachment for the sacroiliac ligament, and the *auricular surface* (*auric-* = ear-shaped), which articulates with the sacrum to form the *sacroiliac joint* (see Figure 8.9). Projecting anteriorly and inferiorly from the auricular surface is a ridge called the *arcuate line* (AR-kū–āt; *arc-* = bow).

The other conspicuous markings of the ilium are three arched lines on its lateral surface called the *posterior gluteal line* (*glut-* = buttock), the *anterior gluteal line,* and the *inferior gluteal line.* The tendons of the gluteal muscles attach to the ilium between these lines.

Ischium

The **ischium** (IS-kē-um = hip), the inferior, posterior portion of the hip bone (Figure 8.10b, c), is comprised of a superior *body* and an inferior *ramus* (*ram-* = branch; plural is *rami*). The ramus is the portion of the ischium that fuses with the pubis. Features of the ischium include the prominent *ischial spine,* a *lesser sciatic notch* below the spine, and a rough and thickened *ischial tuberosity.* This prominent tuberosity may hurt a person's thigh when you sit on his or her lap. Together, the ramus and the pubis surround the *obturator foramen* (OB-too-rā-tōr; *obtur-* = closed up), the largest foramen in the skeleton. The foramen is so named because, even though blood vessels and nerves pass through it, it is nearly completely closed by the fibrous *obturator membrane.*

Pubis

The **pubis** or **os pubis,** meaning pubic bone, is the anterior and inferior part of the hip bone (Figure 8.10b, c). A *superior ramus,* an *inferior ramus,* and a *body* between the rami comprise the pubis. The anterior border of the body is the *pubic crest,* and at its lateral end is a projection called the *pubic tubercle.* This tubercle is the beginning of a raised line, the *iliopectineal line* (il-ē-ō-pek-TIN-ē-al), which extends superiorly and laterally along the superior ramus to merge with the arcuate line of the ilium. These lines, as you will see shortly, are important landmarks for distinguishing the superior and inferior portions of the bony pelvis.

The *pubic symphysis* is the joint between the two hip bones (see Figure 8.9). It consists of a disc of fibrocartilage. Inferior to this joint, the inferior rami of the two pubic bones converge to form the *pubic arch.* In the later stages of pregnancy, the hormone relaxin (produced by the ovaries and placenta) increases the flexibility of the pubic symphysis to ease delivery of the baby. Weakening of the joint, together with an already altered center of gravity due to an enlarged uterus, also changes the gait during pregnancy.

The *acetabulum* (as-e-TAB-ū-lum = vinegar cup) is a deep fossa formed by the ilium, ischium, and pubis. It functions as the socket that accepts the rounded head of the femur. Together, the acetabulum and the femoral head form the *hip (coxal) joint.* On the inferior side of the acetabulum is a deep indentation, the *acetabular notch,* that forms a foramen through which blood vessels and nerves pass and serves as a point of attachment for ligaments of the femur (for example, the ligament of the head of the femur).

False and True Pelves

The bony pelvis is divided into superior and inferior portions by a boundary called the *pelvic brim* (Figure 8.11a). You can trace the pelvic brim by following the landmarks around parts of the hip bones to form the outline of an oblique plane. Beginning posteriorly at the *sacral promontory* of the sacrum, trace laterally and inferiorly along the *arcuate lines* of the ilium. Continue inferiorly along the *iliopectineal lines* of the pubis. Finally, trace anteriorly to the superior portion of the pubic symphysis. Together, these points form an oblique plane that is higher in the back than in the front. The circumference of this plane is the pelvic brim.

The portion of the bony pelvis superior to the pelvic brim is referred to as the **false (greater) pelvis** (Figure 8.11b). It is bordered by the lumbar vertebrae posteriorly, the upper portions of the hip bones laterally, and the abdominal wall anteriorly. The space enclosed by the false pelvis is part of the abdomen; it does not contain pelvic organs, except for the urinary bladder (when it is full) and the uterus during pregnancy.

The portion of the bony pelvis inferior to the pelvic brim is the **true (lesser) pelvis** (Figure 8.11b). It has an inlet, an outlet, and a cavity. It is bounded by the sacrum and coccyx posteriorly, inferior portions of the ilium and ischium laterally, and the pubic bones anteriorly. The true pelvis surrounds the pelvic cavity (see Figure 1.9 on page 17). The superior opening of the true pelvis, bordered by the pelvic brim, is called the *pelvic inlet;* the inferior opening of the true pelvis is the *pelvic outlet.* The *pelvic axis* is an imaginary line that curves through the true pelvis from the central point of the plane of the pelvic inlet to the central point of the plane of the pelvic outlet. During childbirth the pelvic axis is the route taken by the baby's head as it descends through the pelvis.

Figure 8.11 **True and false pelves.** Shown here is the female pelvis. For simplicity, in part (a) the landmarks of the pelvic brim are shown only on the left side of the body, and the outline of the pelvic brim is shown only on the right side. The entire pelvic brim is shown in Figure 8.9. (See Tortora, *A Photographic Atlas of the Human Body, Second Edition,* Figure 3.27.)

The true and false pelves are separated by the pelvic brim.

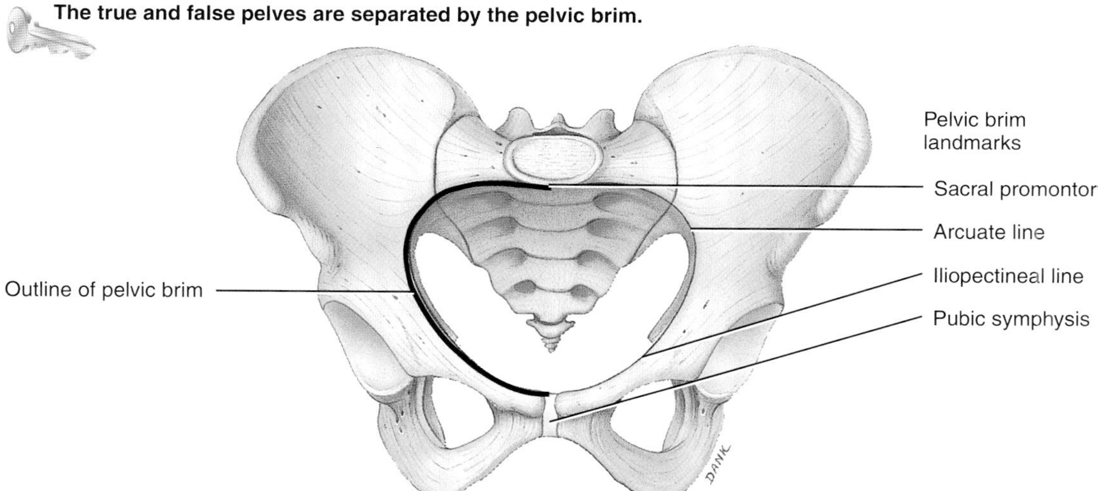

(a) Anterior view of borders of pelvic brim

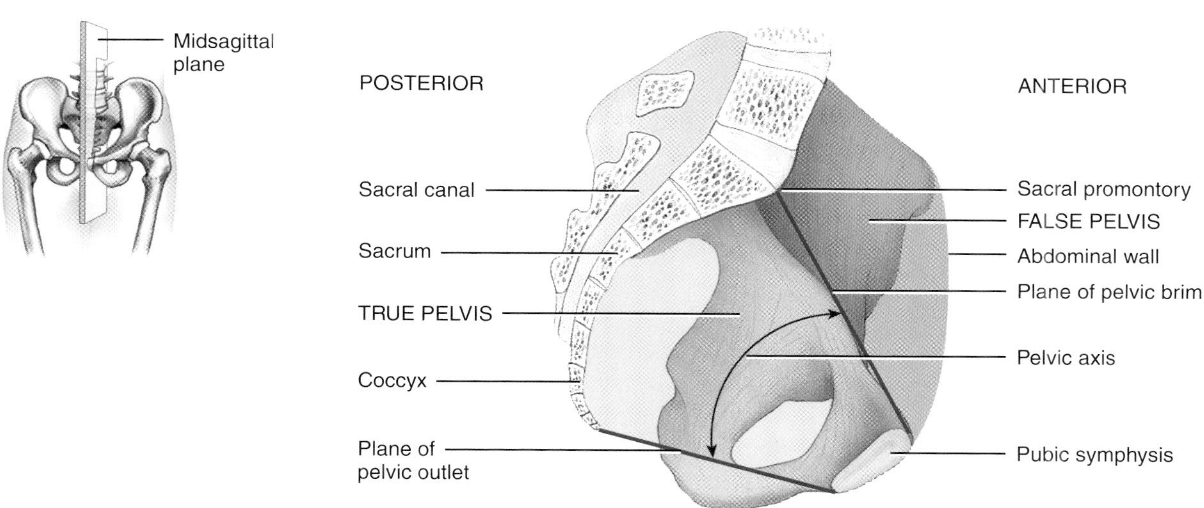

(b) Midsagittal section indicating locations of true and false pelves

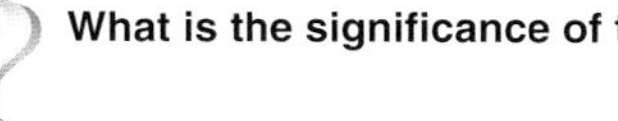

What is the significance of the pelvic axis?

TABLE 8.1 Comparison of Female and Male Pelves

Point of Comparison	Female	Male
General structure	Light and thin.	Heavy and thick.
False (greater) pelvis	Shallow.	Deep.
Pelvic brim (inlet)	Larger and more oval.	Smaller and heart-shaped.
Acetabulum	Small and faces anteriorly.	Large and faces laterally.
Obturator foramen	Oval.	Round.
Pubic arch	Greater than 90° angle.	Less than 90° angle.

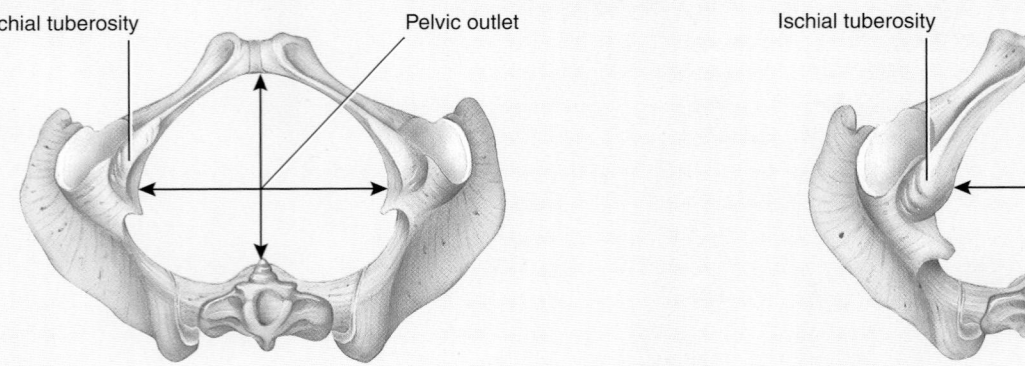

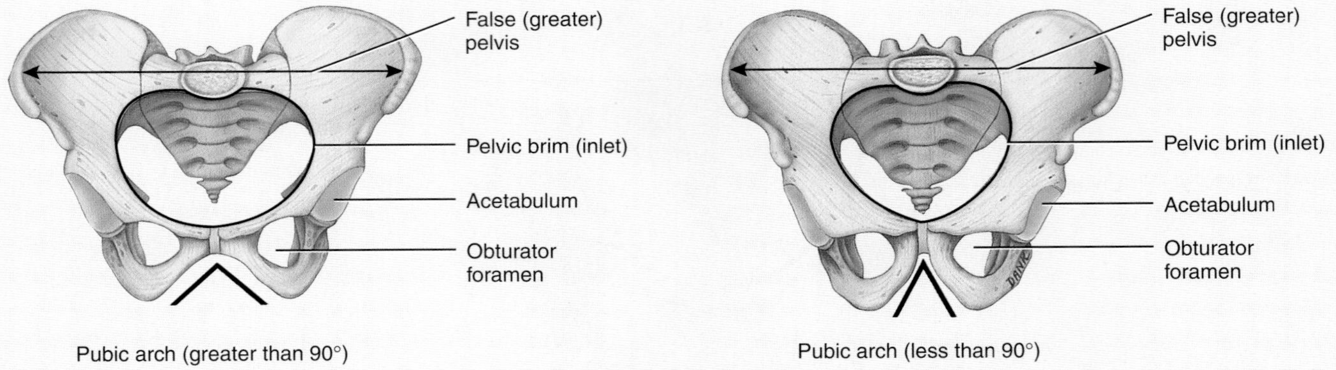

False (greater) pelvis
Pelvic brim (inlet)
Acetabulum
Obturator foramen

Pubic arch (greater than 90°)

False (greater) pelvis
Pelvic brim (inlet)
Acetabulum
Obturator foramen

Pubic arch (less than 90°)

Anterior views

Point of Comparison	Female	Male
Iliac crest	Less curved.	More curved.
Ilium	Less vertical.	More vertical.
Greater sciatic notch	Wide.	Narrow.
Coccyx	More movable and more curved anteriorly.	Less movable and less curved anteriorly.
Sacrum	Shorter, wider (see anterior views), and less curved anteriorly.	Longer, narrower (see anterior views), and more curved anteriorly.

Iliac crest
Ilium
Greater sciatic notch
Sacrum
Coccyx

Iliac crest
Ilium
Greater sciatic notch
Sacrum
Coccyx

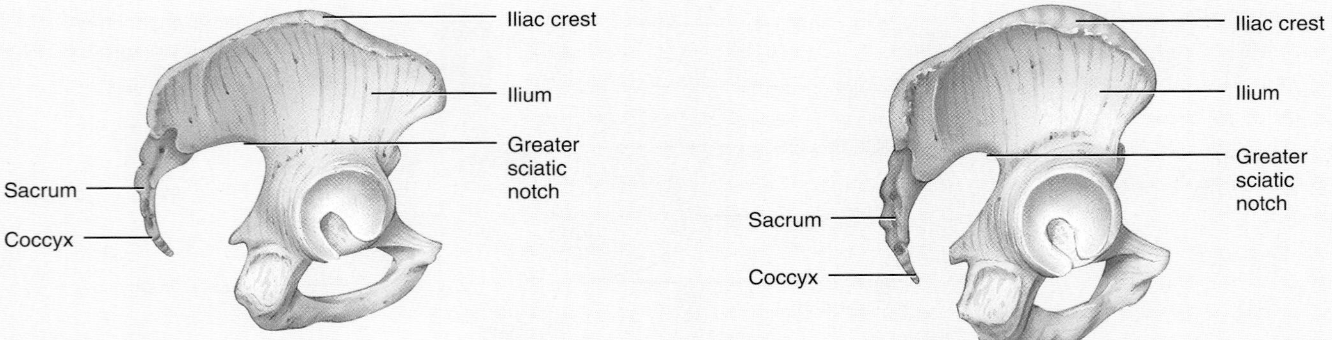

Right lateral views

Point of Comparison	Female	Male
Pelvic outlet	Wider.	Narrower.
ischial tuberosity	Shorter, farther apart, and more medially projecting.	Longer, closer together, and more laterally projecting

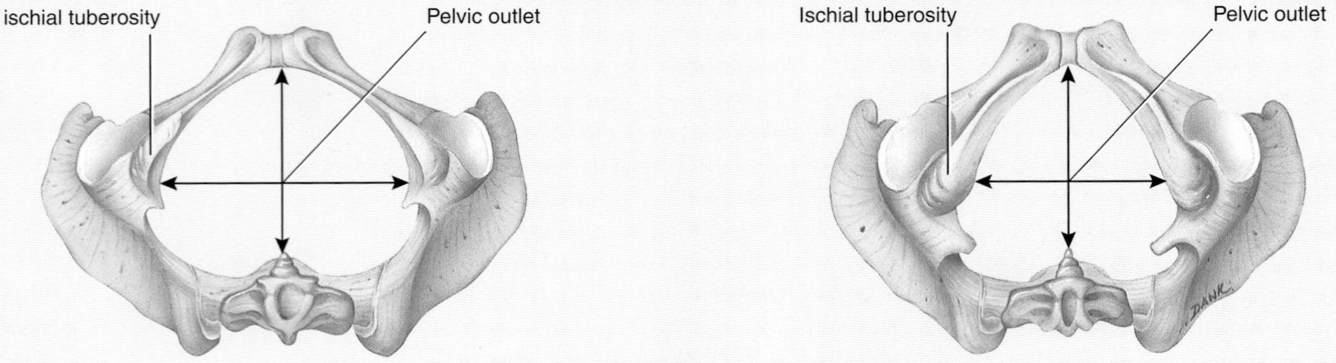

ischial tuberosity
Pelvic outlet

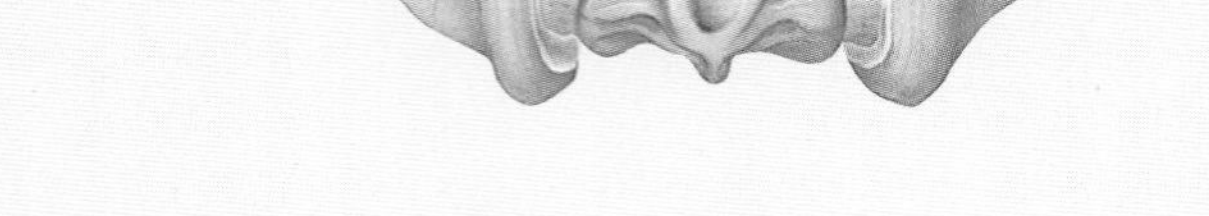

Ischial tuberosity
Pelvic outlet

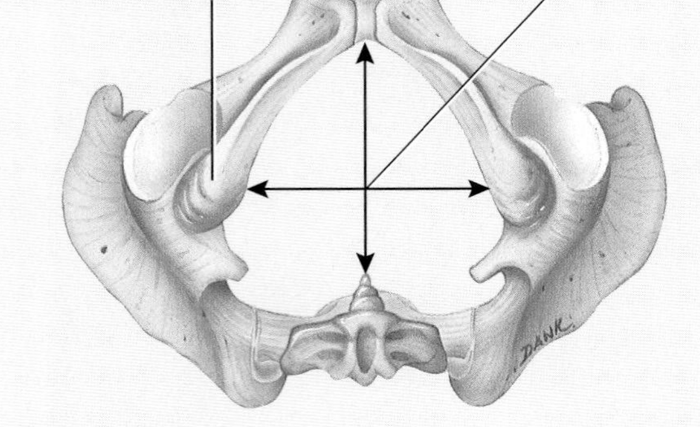

Inferior views

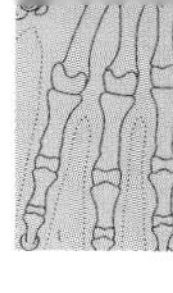

Pelvimetry

Pelvimetry is the measurement of the size of the inlet and outlet of the birth canal, which may be done by ultrasonography or physical examination. Measurement of the pelvic cavity in pregnant females is important because the fetus must pass through the narrower opening of the pelvis at birth. A cesarean section is usually planned if it is determined that the pelvic cavity is too small to permit passage of the baby. ■

▶ CHECKPOINT

4. Describe the distinguishing characteristics of the individual bones of the pelvic girdle.
5. Distinguish between the false and true pelves.

COMPARISON OF FEMALE AND MALE PELVES

▶ OBJECTIVE

Compare the principal structural differences between female and male pelves.

Generally, the bones of males are larger and heavier and possess larger surface markings than those of females of comparable age and physical stature. Sex-related differences in the features of bones are readily apparent when comparing the female and male pelves. Most of the structural differences in the pelves are adaptations to the requirements of pregnancy and childbirth. The female's pelvis is wider and shallower than the male's. Consequently, there is more space in the true pelvis of the female, especially in the pelvic inlet and pelvic outlet, to accommodate the passage of the infant's head at birth. Other significant structural differences between the pelves of females and males are listed and illustrated in Table 8.1.

▶ CHECKPOINT

6. Why are structural differences between female and male pelves important?

LOWER LIMB (EXTREMITY)

▶ OBJECTIVE

Identify the bones of the lower limb and their principal markings.

Each **lower limb (lower extremity)** has 30 bones in four locations—(1) the femur in the thigh; (2) the patella (kneecap); (3) the tibia and fibula in the leg; (4) and the 7 tarsals in the tarsus (ankle), the 5 metatarsals in the metatarsus, and the 14 phalanges (bones of the digits) in the foot (Figure 8.12).

Figure 8.12 **Right lower limb.**

Each lower limb includes a femur, patella (kneecap), tibia, fibula, tarsals (ankle bones), metatarsals, and phalanges (bones of the digits).

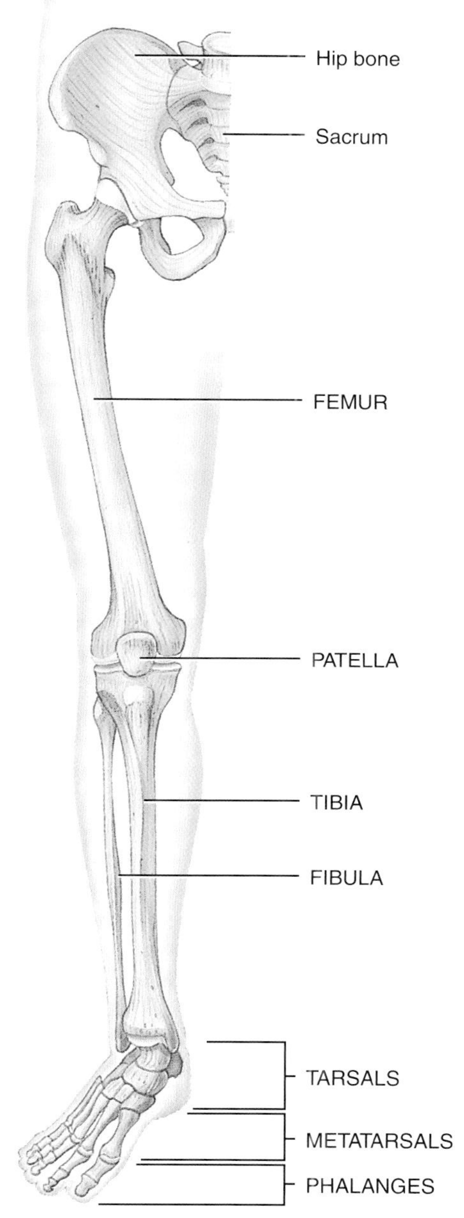

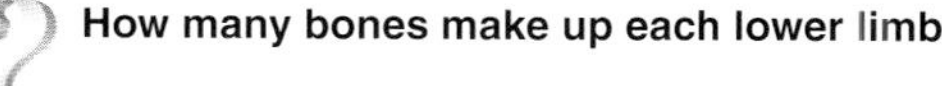

How many bones make up each lower limb?

Figure 8.13 **Right femur in relation to the hip bone, patella, tibia, and fibula.** (See Tortora, *A Photographic Atlas of the Human Body, Second Edition,* Figure 3.28.)

The acetabulum of the hip bone and head of the femur articulate to form the hip joint.

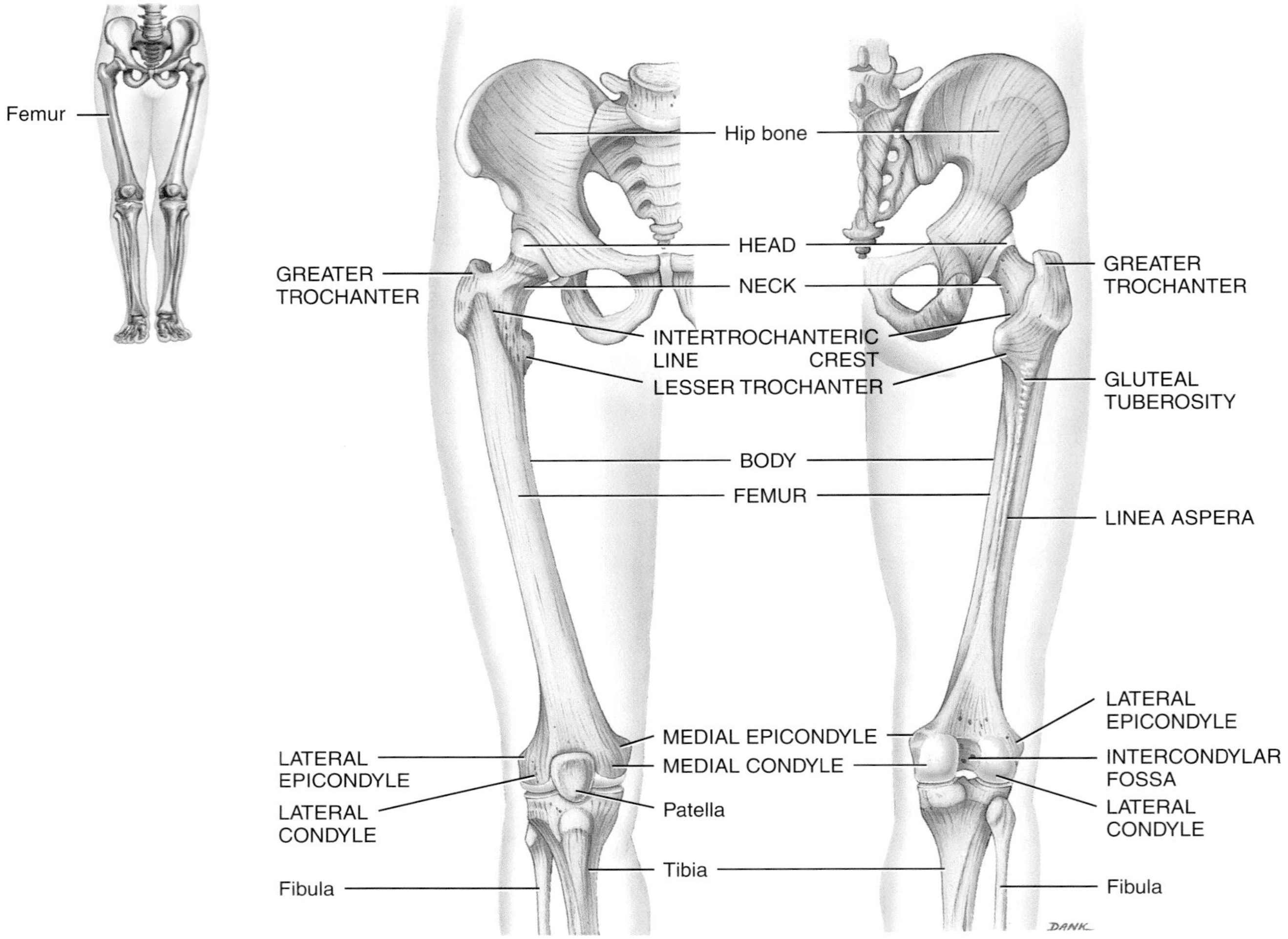

(a) Anterior view

(b) Posterior view

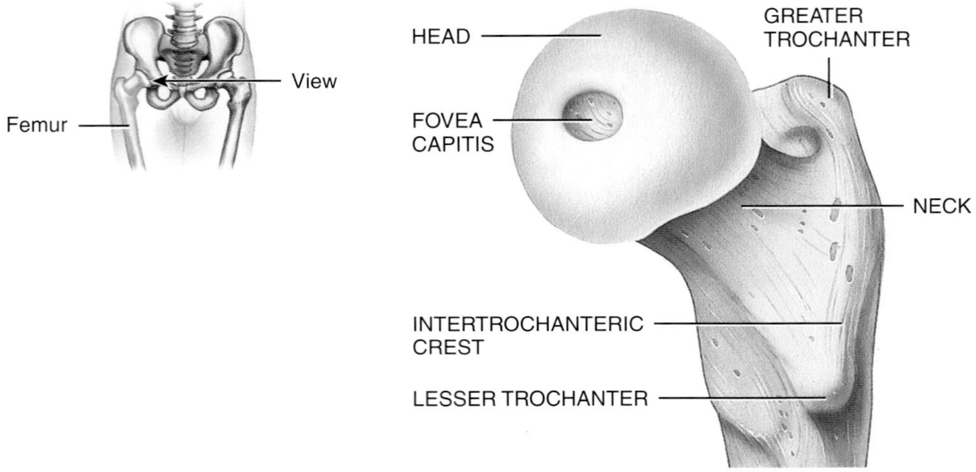

(c) Medial view of proximal end of femur

Why is the angle of convergence of the femurs greater in females than males?

Femur

The **femur,** or thigh bone, is the longest, heaviest, and strongest bone in the body (Figure 8.13). Its proximal end articulates with the acetabulum of the hip bone. Its distal end articulates with the tibia and patella. The *body (shaft)* of the femur angles medially and, as a result, the knee joints are closer to the midline. The angle is greater in females because the female pelvis is broader.

The proximal end of the femur consists of a rounded *head* that articulates with the acetabulum of the hip bone to form the *hip (coxal) joint.* The head contains a small centered depression (pit) called the *fovea capitis* (FŌ-vē-a CAP-i-tis; *fovea* = pit; *capitis* = of the head). The ligament of the head of the femur connects the fovea capitis of the femur to the acetabulum of the hip bone. The *neck* of the femur is a constricted region distal to the head. A "broken hip" is more often associated with a fracture in the neck of the femur than fractures of the hip bones. The *greater trochanter* (trō-KAN-ter) and *lesser trochanter* are projections from the junction of the neck and shaft that serve as points of attachment for the tendons of some of the thigh and buttock muscles. The greater trochanter is the prominence felt and seen anterior to the hollow on the side of the hip. It is a landmark commonly used to locate the site for intramuscular injections into the lateral surface of the thigh. The lesser trochanter is inferior and medial to the greater trochanter. Between the anterior surfaces of the trochanters is a narrow *intertrochanteric line* (Figure 8.13a). A ridge called the *intertrochanteric crest* appears between the posterior surfaces of the trochanters (Figure 8.13b).

Inferior to the intertrochanteric crest on the posterior surface of the body of the femur is a vertical ridge called the *gluteal tuberosity.* It blends into another vertical ridge called the *linea aspera* (LIN-ē-a AS-per-a; *asper* = rough). Both ridges serve as attachment points for the tendons of several thigh muscles.

The expanded distal end of the femur includes the *medial condyle* and the *lateral condyle.* These articulate with the medial and lateral condyles of the tibia. Superior to the condyles are the *medial epicondyle* and the *lateral epicondyle,* to which ligaments of the knee joint attach. A depressed area between the condyles on the posterior surface is called the *intercondylar fossa* (in-ter-KON-di-lar). The *patellar surface* is located between the condyles on the anterior surface.

Patella

The **patella** (= little dish), or kneecap, is a small, triangular bone located anterior to the knee joint (Figure 8.14). The broad superior end of this sesamoid bone, which develops in the tendon of the quadriceps femoris muscle, is called the *base;* the pointed inferior end is referred to as the *apex.* The posterior surface contains two *articular facets,* one for the medial condyle of the femur and another for the lateral condyle of the femur. The patellar ligament attaches the patella to the tibial tuberosity. The *patellofemoral joint,* between the posterior surface of the patella and the patellar surface of the femur, is the intermediate component of the *tibiofemoral (knee) joint.* The patella increases the leverage of the tendon of the quadriceps femoris muscle, maintains the position of the tendon when the knee is bent (flexed), and protects the knee joint.

Patellofemoral Stress Syndrome

Patellofemoral stress syndrome ("runner's knee") is one of the most common problems runners experience. During normal flexion and extension of the knee, the patella tracks (glides) superiorly and inferiorly in the groove between the femoral condyles. In patellofemoral stress syndrome, normal tracking does not occur; instead, the patella tracks laterally as well as

Figure 8.14 Right patella.

The patella articulates with the lateral and medial condyles of the femur.

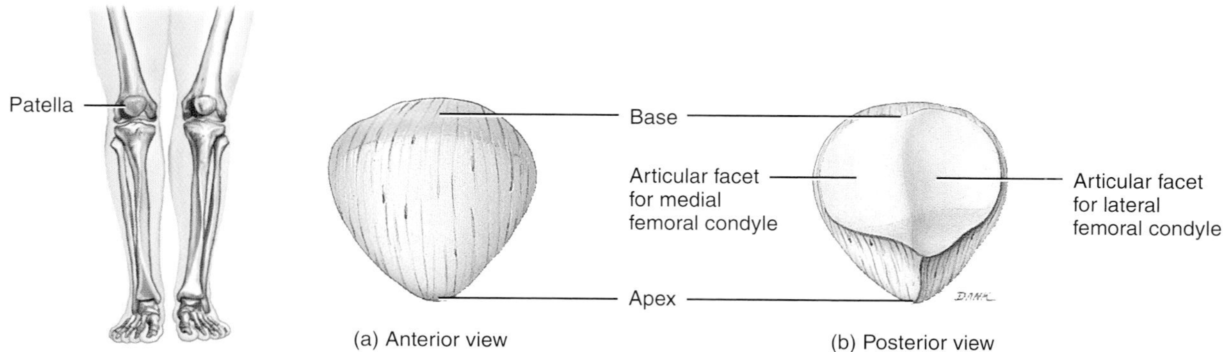

The patella is classified as which type of bone? Why?

superiorly and inferiorly, and the increased pressure on the joint causes aching or tenderness around or under the patella. The pain typically occurs after a person has been sitting for awhile, especially after exercise. It is worsened by squatting or walking down stairs. One cause of runner's knee is constantly walking, running, or jogging on the same side of the road. Because roads slope down on the sides, the knee that is closer to the center of the road endures greater mechanical stress because it does not fully extend during a stride. Other predisposing factors include running on hills, running long distances, and an anatomical deformity called knock-knee (see page 255). ■

Tibia and Fibula

The **tibia,** or shin bone, is the larger, medial, weight-bearing bone of the leg (Figure 8.15). The tibia articulates at its proximal end with the femur and fibula, and at its distal end with the fibula and the talus bone of the ankle. The tibia and fibula, like the ulna and radius, are connected by an interosseous membrane.

The proximal end of the tibia is expanded into a *lateral condyle* and a *medial condyle.* These articulate with the condyles of the femur to form the lateral and medial *tibiofemoral (knee) joints.* The inferior surface of the lateral condyle articulates with the head of the fibula. The slightly concave condyles are separated by an upward projection called the *intercondylar eminence* (Figure 8.15b). The *tibial tuberosity* on the anterior surface is a point of attachment for the patellar ligament. Inferior to and continuous with the tibial tuberosity is a sharp ridge that can be felt below the skin known as the *anterior border (crest)* or *shin.*

The medial surface of the distal end of the tibia forms the *medial malleolus* (mal-LĒ-ō-lus = hammer). This structure articulates with the talus of the ankle and forms the prominence that can be felt on the medial surface of the ankle. The *fibular notch* (Figure 8.15c) articulates with the distal end of the fibula to form the *distal tibiofibular joint.* Of all the long bones of the body, the tibia is the most frequently fractured and is also the most frequent site of an open (compound) fracture.

The **fibula** is parallel and lateral to the tibia, but it is considerably smaller. (See Figure 8.15 for a mnemonic describing the relative positions of the tibia and fibula.) The *head* of the fibula, the proximal end, articulates with the inferior surface of the lateral condyle of the tibia below the level of the knee joint to form the *proximal tibiofibular joint.* The distal end is more arrowhead-shaped and has a projection called the *lateral malleolus* that articulates with the talus of the ankle. This forms the prominence on the lateral surface of the ankle. As noted previously, the fibula also articulates with the tibia at the fibular notch to form the distal tibiofibular joint.

Bone Grafting

Bone grafting generally consists of taking a piece of bone, along with its periosteum and nutrient artery, from one part of the body to replace missing bone in another part of the body. The transplanted bone restores the blood supply to the transplanted site and healing occurs as in a fracture. The fibula is a common source of bone for grafting because even after a piece of the fibula has been removed, walking, running, and jumping can be normal. Recall that the tibia is the weight-bearing bone of the leg. ■

Tarsals, Metatarsals, and Phalanges

The **tarsus** (ankle) is the proximal region of the foot and consists of seven **tarsal bones** (Figure 8.16 on page 250). They include the **talus** (TĀ-lus = ankle bone) and **calcaneus** (kal-KĀ-nē-us = heel), located in the posterior part of the foot. The calcaneus is the largest and strongest tarsal bone. The anterior tarsal bones are the **navicular** (= like a little boat), three **cuneiform bones** (= wedge-shaped) called the **third (lateral), second (intermediate),** and **first (medial) cuneiforms,** and the **cuboid** (= cube-shaped). (A mnemonic to help you remember the names of the tarsal bones is included in Figure 8.16.) Joints between tarsal bones are called *intertarsal joints.* The talus, the most superior tarsal bone, is the only bone of the foot that articulates with the fibula and tibia. It articulates on one side with the medial malleolus of the tibia and on the other side with the lateral malleolus of the fibula. These articulations form the *talocrural (ankle) joint.* During walking, the talus transmits about half the weight of the body to the calcaneus. The remainder is transmitted to the other tarsal bones.

The **metatarsus,** the intermediate region of the foot, consists of five **metatarsal bones** numbered I to V (or 1–5) from the medial to lateral position (Figure 8.16 on page 250). Like the metacarpals of the palm of the hand, each metatarsal consists of a proximal *base,* an intermediate *shaft,* and a distal *head.* The metatarsals articulate proximally with the first, second, and third cuneiform bones and with the cuboid to form the *tarsometatarsal joints.* Distally, they articulate with the proximal row of phalanges to form the *metatarsophalangeal joints.* The first metatarsal is thicker than the others because it bears more weight.

Fractures of the Metatarsals

Fractures of the metatarsals occur when a heavy object falls on the foot or when a heavy object rolls over the foot. Such fractures are also common among dancers, especially female ballet dancers. If a ballet dancer is on the tip of her toes and loses her balance, the full body weight is placed on the metatarsals, causing one or more of them to fracture. ■

The **phalanges** comprise the distal component of the foot and resemble those of the hand both in number and arrangement. The toes are numbered I to V (or 1–5) beginning with the great toe, from medial to lateral. Each *phalanx* (singular) consists of a proximal *base,* an intermediate *shaft,* and a distal *head.* The great or big toe (*hallux*) has two large, heavy phalanges called proximal and distal phalanges. The other four toes each have three phalanges—

Figure 8.15 Right tibia and fibula in relation to the femur, patella, and talus. (See Tortora, *A Photographic Atlas of the Human Body, Second Edition,* Figure 3.30.)

The tibia articulates with the femur and fibula proximally, and with the fibula and talus distally.

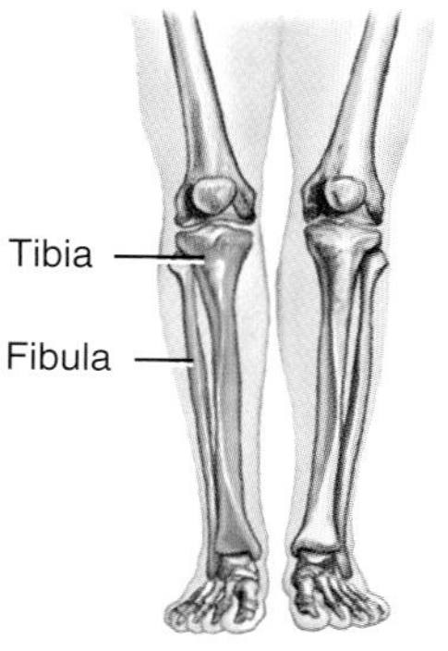

Tibia
Fibula

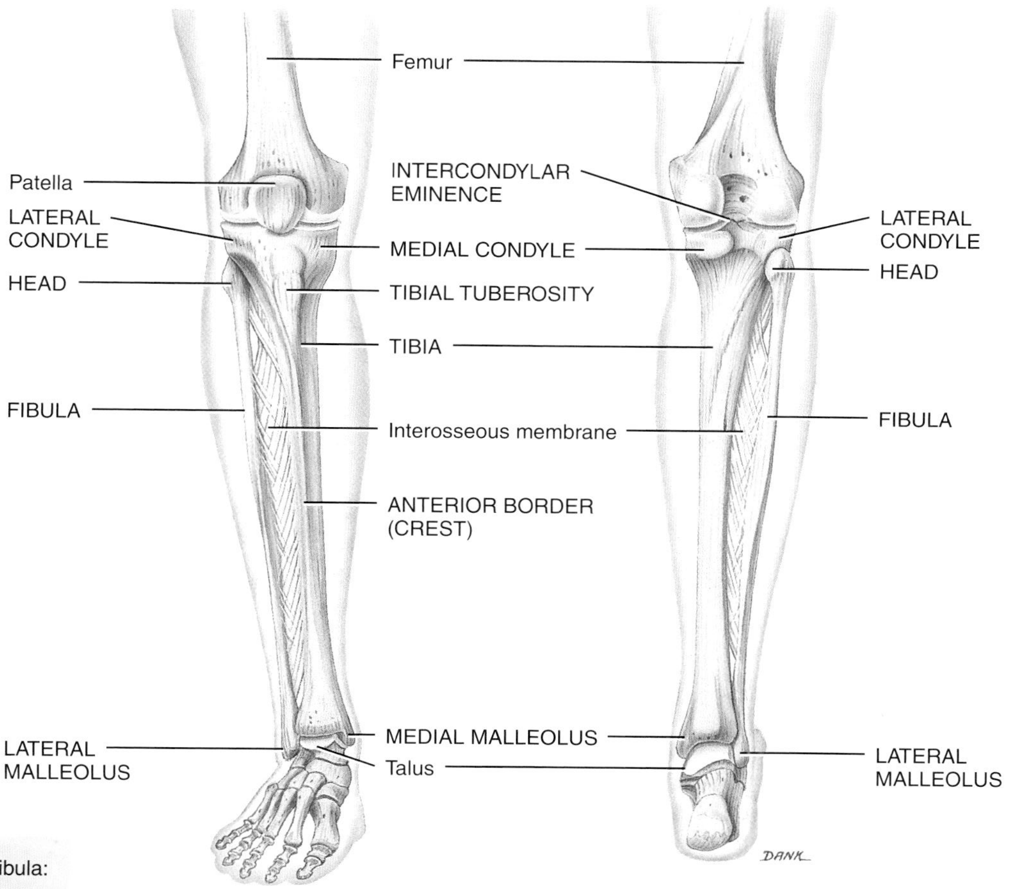

Femur
Patella
LATERAL CONDYLE
HEAD
FIBULA
LATERAL MALLEOLUS
INTERCONDYLAR EMINENCE
MEDIAL CONDYLE
TIBIAL TUBEROSITY
TIBIA
Interosseous membrane
ANTERIOR BORDER (CREST)
MEDIAL MALLEOLUS
Talus
LATERAL CONDYLE
HEAD
FIBULA
LATERAL MALLEOLUS
DANK

(a) Anterior view

(b) Posterior view

MNEMONIC for location of tibia and fibula:
The fibu**LA** is **LA**teral.

Tibia
View

POSTERIOR
ANTERIOR
FIBULAR NOTCH
MEDIAL MALLEOLUS
DANK

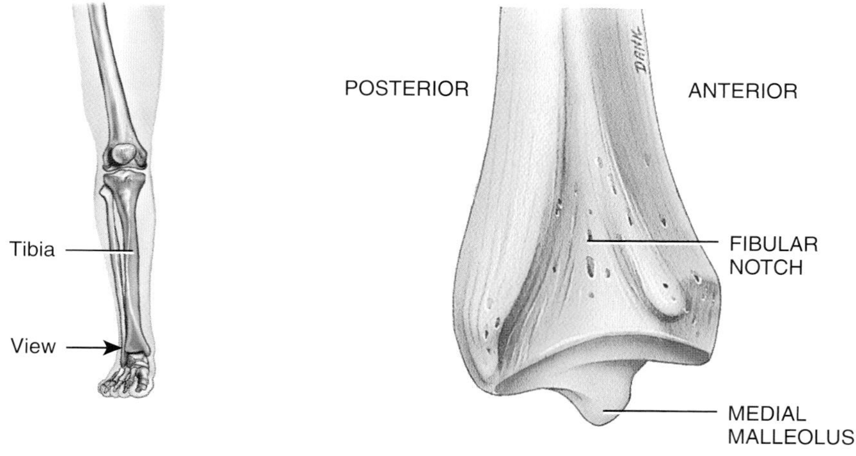

(c) Lateral view of distal end of tibia

Which leg bone bears the weight of the body?

Figure 8.16 Right foot. (See Tortora, *A Photographic Atlas of the Human Body, Second Edition,* Figure 3.31.)

The skeleton of the foot consists of the proximal tarsals, the intermediate metatarsals, and the distal phalanges.

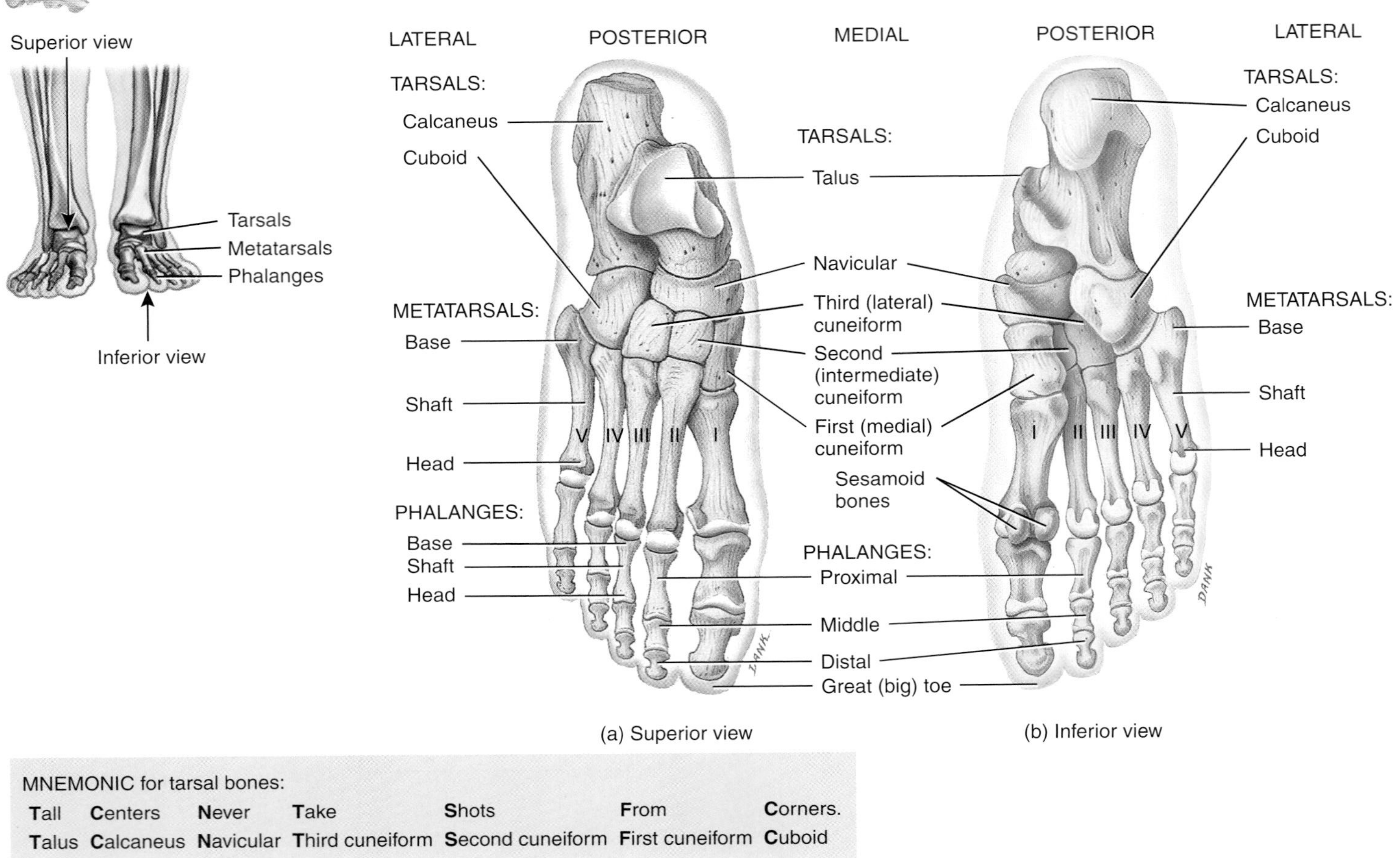

Which tarsal bone articulates with the tibia and fibula?

proximal, middle, and distal. Joints between phalanges of the foot, like those of the hand, are called *interphalangeal joints.*

Arches of the Foot

The bones of the foot are arranged in two **arches** that are held in position by ligaments and tendons (Figure 8.17). The arches enable the foot to support the weight of the body, provide an ideal distribution of body weight over the soft and hard tissues of the foot, and provide leverage while walking. The arches are not rigid; they yield as weight is applied and spring back when the weight is lifted, thus storing energy for the next step and helping to absorb shocks. Usually, the arches are fully developed by age 12 or 13.

The **longitudinal arch** has two parts, both of which consist of tarsal and metatarsal bones arranged to form an arch from the anterior to the posterior part of the foot. The *medial part* of the longitudinal arch, which originates at the calcaneus, rises to the talus and descends through the navicular, the three cuneiforms, and the heads of the three medial metatarsals. The *lateral part* of the longitudinal arch also begins at the calcaneus. It rises at the cuboid and descends to the heads of the two lateral metatarsals. The medial portion of the longitudinal arch is so high that the medial portion of the foot between the ball and heel does not touch the ground when you walk on a hard surface.

The **transverse arch** is found between the medial and lateral aspects of the foot and is formed by the navicular, three cuneiforms, and the bases of the five metatarsals.

As noted earlier, one function of the arches is to distribute body weight over the soft and hard tissues of the body. Normally, the ball of the foot carries about 40 percent of the weight and the heel carries about 60 percent. The ball of the foot is the padded portion of the sole superficial to the heads of the metatarsals. When a person wears high-heeled shoes, however, the distribution of weight changes so that the ball of the foot

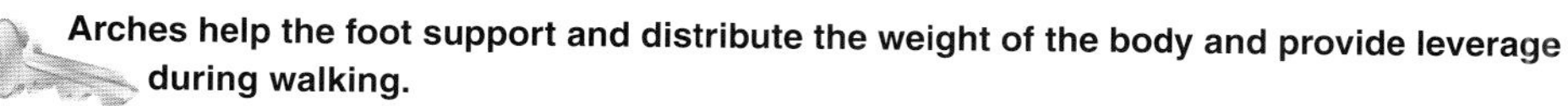

Figure 8.17 Arches of the right foot.

Arches help the foot support and distribute the weight of the body and provide leverage during walking.

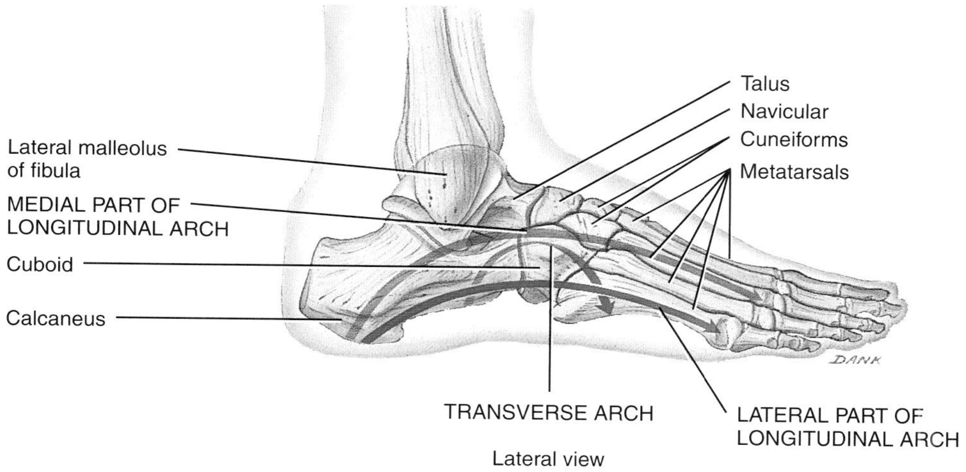

What structural feature of the arches allows them to absorb shocks?

may carry up to 80 percent and the heel 20 percent. As a result, the fat pads at the ball of the foot are damaged, joint pain develops, and structural changes in bones may occur.

Flatfoot and Clawfoot

The bones composing the arches of the foot are held in position by ligaments and tendons. If these ligaments and tendons are weakened, the height of the medial longitudinal arch may decrease or "fall." The result is **flatfoot,** the causes of which include excessive weight, postural abnormalities, weakened supporting tissues, and genetic predisposition. Fallen arches may lead to inflammation of the deep fascia of the sole (plantar fasciitis), Achilles tendinitis, shinsplints, stress fractures, bunions, and calluses. A custom-designed arch support often is prescribed to treat flatfoot.

Clawfoot is a condition in which the medial longitudinal arch is abnormally elevated. It is often caused by muscle deformities, such as may occur in diabetics whose neurological lesions lead to atrophy of muscles of the foot. ■

▶ CHECKPOINT

7. Name the bones that form the lower limb, from proximal to distal.
8. Describe the joints of the lower limbs.
9. What are the functions of the arches of the foot?

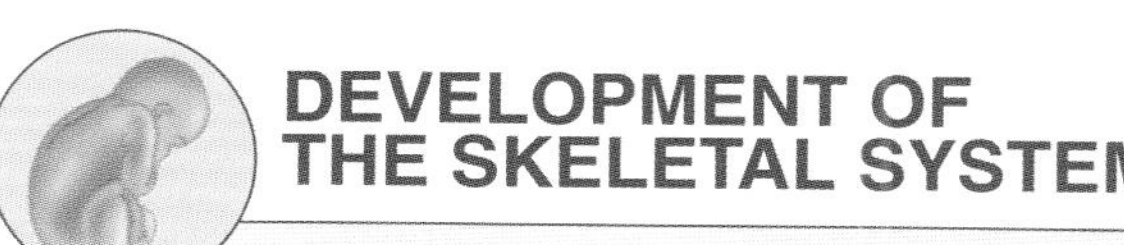

DEVELOPMENT OF THE SKELETAL SYSTEM

▶ OBJECTIVE

Describe the development of the skeletal system.

All skeletal tissue arises from *mesenchymal cells,* connective tissue cells derived from **mesoderm.** The mesenchymal cells condense and form models of bones in areas where the bones themselves will ultimately form. In some cases, the bones form directly within the mesenchyme (intramembranous ossification; see Figure 6.5 on page 179). In other cases, the bones form within hyaline cartilage that develops from mesenchyme (endochondral ossification; see Figure 6.6 on page 180).

The *skull* begins development during the fourth week after fertilization. It develops from mesenchyme around the developing brain and consists of two major portions: **neurocranium,** which forms the bones of the skull, and **viscerocranium,** which forms the bones of the face (Figure 8.18a). The neurocranium is divided into two parts called the **cartilaginous neurocranium** and **membranous neurocranium.** The cartilaginous neurocranium consists of hyaline cartilage developed from mesenchyme at the base of the developing skull. It later undergoes endochondral ossification to form the *flat bones at the base of the skull.* The membranous neurocranium consists of mesenchyme and later undergoes intramembranous ossification to form the *flat bones that make up the roof and sides of the skull.* During fetal life and infancy the flat bones are separated by membrane-filled spaces called fontanels (see Figure 7.14 on page 211). The viscerocranium, like the

Figure 8.18 **Development of the skeletal system.** Bones that develop from the cartilaginous neurocranium are indicated in light blue; from the cartilaginous viscerocranium in dark blue; from the membranous neurocranium in dark red; and from the membranous viscerocranium in light red.

After the limb buds develop, endochondral ossification of the limb bones begins by the end of the eighth embryonic week.

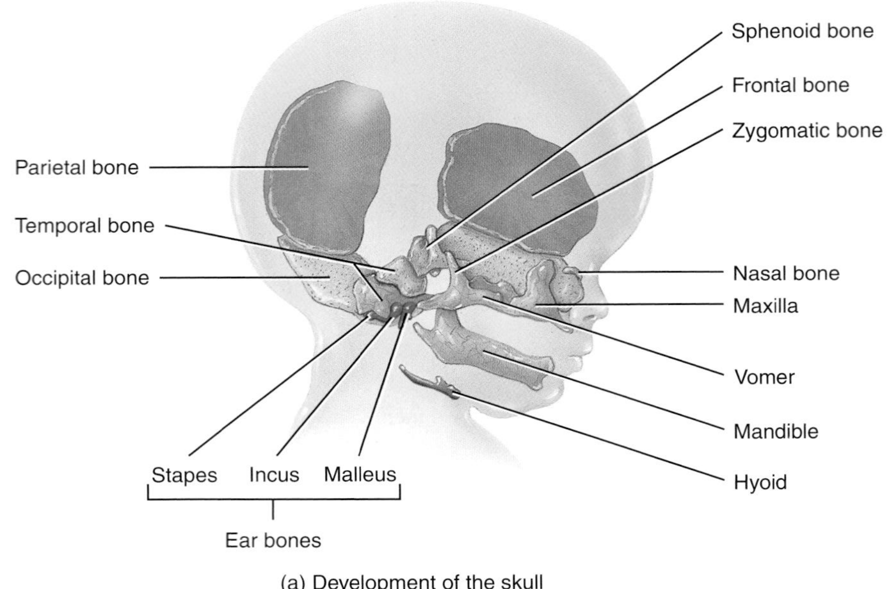

(a) Development of the skull

neurocranium, is divided into two parts: **cartilaginous viscerocranium** and **membranous viscerocranium.** The cartilaginous viscerocranium is derived from the cartilage of the first two pharyngeal (branchial) arches (see Figure 29.13 on page 1120). Endochondral ossification of these cartilages forms the *ear bones* and *hyoid bone.* The membranous viscerocranium is derived from mesenchyme in the first pharyngeal arch and, following intramembranous ossification, forms the facial bones.

Vertebrae are derived from portions of cube-shaped masses of mesoderm called somites (see Figure 10.19 on page 318). Mesenchymal cells from these regions surround the notochord (see Figure 10.19) at about four weeks after fertilization. The **notochord** is a solid cylinder of mesodermal cells that induces (stimulates) the mesenchymal cells to form the *vertebral bodies.* Between the vertebral bodies, the notochord induces mesenchymal cells to form the *nucleus pulposus* of an intervertebral disc and surrounding mesenchymal cells form the *annulus fibrosus* of an intervertebral disc. As development continues, other parts of a vertebra form and the *vertebral arch* surrounds the spinal cord (failure of the vertebral arch to develop properly results in a condition called spina bifida; see page 225 in Chapter 7). In the thoracic region, processes from the vertebrae develop into the *ribs.* The *sternum* develops from mesoderm in the ventral body wall.

The *skeleton of the limbs* is derived from mesoderm. During the middle of the fourth week after fertilization, the upper limbs appear as small elevations at the sides of the trunk called **upper limb buds** (Figure 8.18b). About 2 days later, the **lower limb buds** appear. The limb buds consist of **mesenchyme** covered by **ectoderm.** At this point, a mesenchymal skeleton exists in the limbs; some of the masses of mesoderm surrounding the developing bones will become the skeletal muscles of the limbs.

By the sixth week, the limb buds develop a constriction around the middle portion. The constriction produces flattened distal segments of the upper buds called **hand plates** and distal segments of the lower buds called **foot plates** (Figure 8.18c). These plates represent the beginnings of the hands and feet, respectively. At this stage of limb development, a cartilaginous skeleton formed from mesenchyme is present. By the seventh week (Figure 8.18d), the

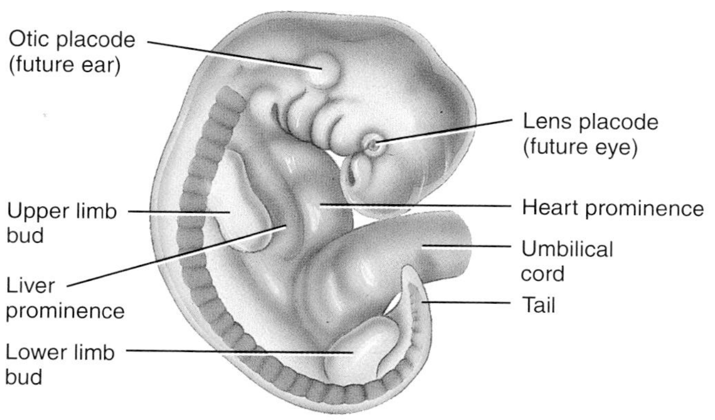

(b) Four-week embryo showing development of limb buds

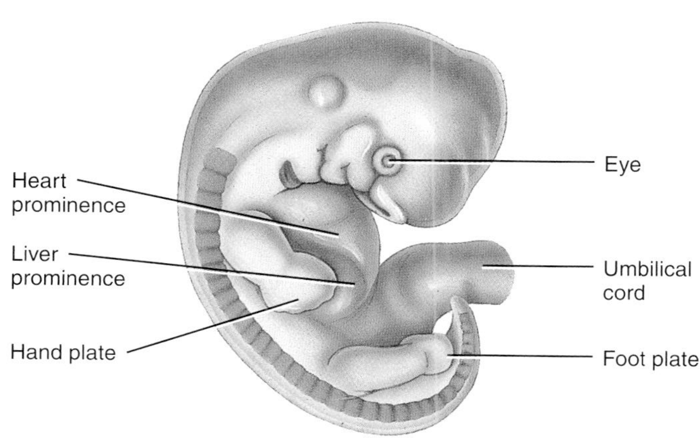

(c) Six-week embryo showing development of hand and foot plates

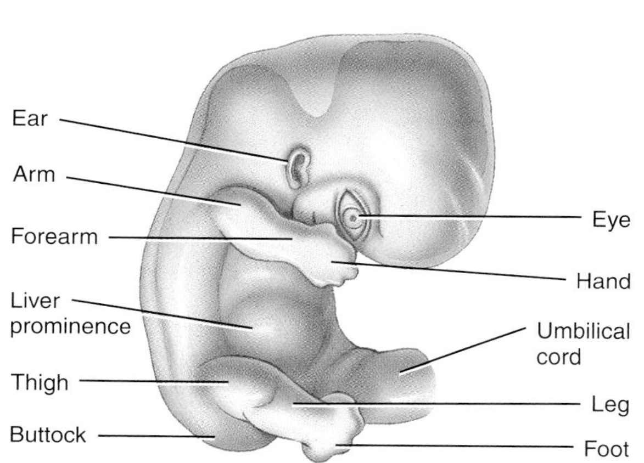

(d) Seven-week embryo showing development of arm, forearm, and hand in upper limb bud and thigh, leg, and foot in lower limb bud

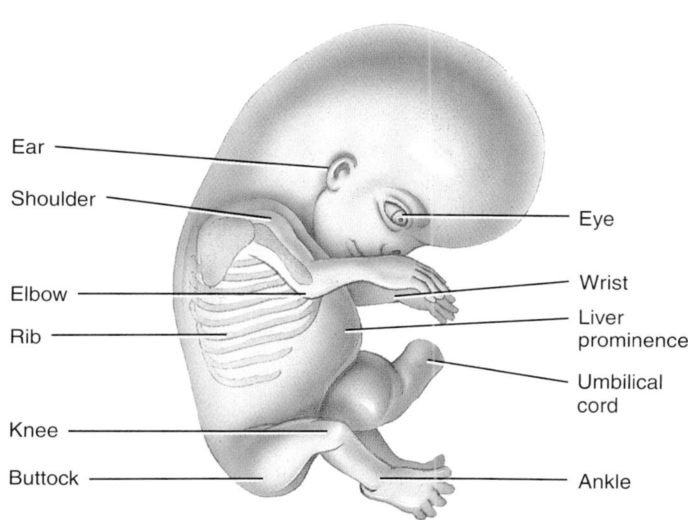

(e) Eight-week embryo in which limb buds have developed into upper and lower limbs

Which of the three basic embryonic tissues—ectoderm, mesoderm, and endoderm—gives rise to the skeletal system?

arm, forearm, and *hand* are evident in the upper limb bud, and the *thigh, leg,* and *foot* appear in the lower limb bud. By the eight week (Figure 8.18e), as the shoulder, elbow, and wrist areas become apparent, the upper limb bud is appropriately called the upper limb, and the lower limb bud is now the lower limb.

Endochondral ossification of the limb bones begins by the end of the eighth week after fertilization. By the twelfth week, primary ossification centers are present in most of the limb bones. Most secondary ossification centers appear after birth.

► CHECKPOINT

10. When and how do the limbs develop?

• • •

To appreciate the skeletal system's contributions to homeostasis of other body systems, examine *Focus on Homeostasis: The Skeletal System.* Next, in Chapter 9, we will see how joints both hold the skeleton together and permit it to participate in movements.

FOCUS ON HOMEOSTASIS

The Skeletal System

BODY SYSTEM	CONTRIBUTION OF THE SKELETAL SYSTEM
For all body systems	Bones provide support and protection for internal organs; bones store and release calcium, which is needed for proper functioning of most body tissues.
Integumentary system	Bones provide strong support for overlying muscles and skin while joints provide flexibility that allows skin to bend.
Muscular system 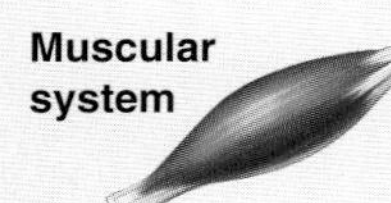	Bones provide attachment points for muscles and leverage for muscles to bring about body movements; contraction of skeletal muscle requires calcium ions.
Nervous system 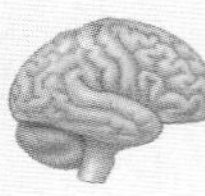	Skull and vertebrae protect brain and spinal cord; normal blood level of calcium is needed for normal functioning of neurons and neuroglia.
Endocrine system 	Bones store and release calcium, needed during exocytosis of hormone-filled vesicles and for normal actions of many hormones.
Cardiovascular system	Red bone marrow carries out hemopoiesis (blood cell formation); rhythmic beating of the heart requires calcium ions.
Lymphatic system and immunity	Red bone marrow produces lymphocytes, white blood cells that are . involved in immune responses
Respiratory system 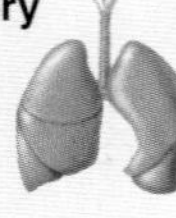	Axial skeleton of thorax protects lungs; rib movements assist in breathing; some muscles used for breathing attach to bones via tendons.
Digestive system 	Teeth masticate (chew) food; rib cage protects esophagus, stomach, and liver; pelvis protects portions of the intestines.
Urinary system 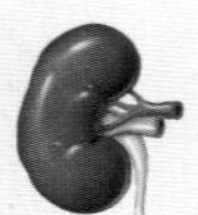	Ribs partially protect kidneys; pelvis protects urinary bladder and urethra.
Reproductive systems	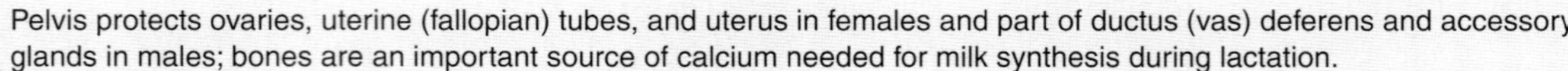 Pelvis protects ovaries, uterine (fallopian) tubes, and uterus in females and part of ductus (vas) deferens and accessory glands in males; bones are an important source of calcium needed for milk synthesis during lactation.

DISORDERS: HOMEOSTATIC IMBALANCES

Hip Fracture

Although any region of the hip girdle may fracture, the term **hip fracture** most commonly applies to a break in the bones associated with the hip joint—the head, neck, or trochanteric regions of the femur, or the bones that form the acetabulum. In the United States, 300,000 to 500,000 people sustain hip fractures each year. The incidence of hip fractures is increasing, due in part to longer life spans. Decreases in bone mass due to osteoporosis (which occurs more often in females), along with an increased tendency to fall, predispose elderly people to hip fractures.

Hip fractures often require surgical treatment, the goal of which is to repair and stabilize the fracture, increase mobility, and decrease pain. Sometimes the repair is accomplished by using surgical pins, screws, nails, and plates to secure the head of the femur. In severe hip fractures, the femoral head or the acetabulum of the hip bone may be replaced by prostheses (artificial devices). The procedure of replacing either the femoral head or the acetabulum is *hemiarthroplasty* (hem-ē-AR-thrō-plas-tē; *hemi-* = one half; *arthro-* = joint; *-plasty* = molding). Replacement of both the femoral head and acetabulum is *total hip arthroplasty.* The acetabular prosthesis is made of plastic, and the femoral prosthesis is metal; both are designed to withstand a high degree of stress. The prostheses are attached to healthy portions of bone with acrylic cement and screws (see Figure 9.16 on page 284).

MEDICAL TERMINOLOGY

Clubfoot or ***talipes equinovarus*** (*-pes* = foot; *equino-* = horse) An inherited deformity in which the foot is twisted inferiorly and medially, and the angle of the arch is increased; occurs in 1 of every 1000 births. Treatment consists of manipulating the arch to a normal curvature by casts or adhesive tape, usually soon after birth. Corrective shoes or surgery may also be required.

Genu valgum (JĒ-noo VAL-gum; *genu-* = knee; *valgum* = bent outward) A deformity in which the knees are abnormally close together and the space between the ankles is increased due to a lateral angulation of the tibia. Also called **knock-knee.**

Genu varum (JĒ-noo VAR-um; *varum* = bent toward the midline) A deformity in which the knees are abnormally separated and the lower limbs are bowed medially. Also called **bowleg.**

Hallux valgus (HAL-uks VAL-gus; *hallux* = great toe) Angulation of the great toe away from the midline of the body, typically caused by wearing tightly fitting shoes. When the great toe angles toward the next toe, there is a bony protrusion at the base of the great toe. Also called a **bunion.**

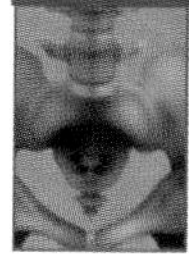

STUDY OUTLINE

PECTORAL (SHOULDER) GIRDLE (p. 232)

1. Each of the body's two pectoral (shoulder) girdles consists of a clavicle and scapula.
2. Each pectoral girdle attaches an upper limb to the axial skeleton.

UPPER LIMB (EXTREMITY) (p. 235)

1. Each of the two upper limbs (extremities) contains 30 bones.
2. The bones of each upper limb include the humerus, the ulna, the radius, the carpals, the metacarpals, and the phalanges.

PELVIC (HIP) GIRDLE (p.240)

1. The pelvic (hip) girdle consists of two hip bones.
2. Each hip bone consists of three fused bones: the ilium, pubis, and ischium.
3. The hip bones, sacrum, and pubic symphysis form the bony pelvis. It supports the vertebral column and pelvic viscera and attaches the lower limbs to the axial skeleton.
4. The true pelvis is separated from the false pelvis by the pelvic brim.

COMPARISON OF FEMALE AND MALE PELVES (p. 245)

1. Bones of males are generally larger and heavier than bones of females. They also have more prominent markings for muscle attachments.
2. The female pelvis is adapted for pregnancy and childbirth. Sex-related differences in pelvic structure are listed and illustrated in Table 8.1.

LOWER LIMB (EXTREMITY) (p. 245)

1. Each of the two lower limbs (extremities) contains 30 bones.
2. The bones of each lower limb include the femur, the patella, the tibia, the fibula, the tarsals, the metatarsals, and the phalanges.
3. The bones of the foot are arranged in two arches, the longitudinal arch and the transverse arch, to provide support and leverage.

DEVELOPMENT OF THE SKELETAL SYSTEM (p. 251)

1. Bone forms from mesoderm by intramembranous or endochondral ossification.
2. Limbs develop from limb buds, which consist of mesoderm and ectoderm.

SELF-QUIZ QUESTIONS

Fill in the blanks in the following statements.

1. The bones that comprise the palm are the ____.
2. List the three bones that fuse to form a coxal (hip) bone: ____, ____, ____.
3. The portion of the bony pelvis that is inferior to the pelvic brim is the ____ pelvis; the portion superior to the pelvic brim is the ____ pelvis.

Indicate whether the following statements are true or false.

4. The largest carpal bone is the lunate.
5. The anterior joint formed by the two coxal (hip) bones is the pubic symphysis.

Choose the one best answer to the following questions.

6. Which of the following statements are *true?* (1) The pectoral girdle consists of the scapula, the clavicle, and the sternum. (2) Although the joints of the pectoral girdle are not very stable, they allow free movement in many directions. (3) The anterior component of the pectoral girdle is the scapula. (4) The pectoral girdle articulates directly with the vertebral column. (5) The posterior component of the pectoral girdle is the sternum. (a) 1, 2, and 3; (b) 2 only; (c) 4 only; (d) 2, 3, and 5; (e) 3, 4, and 5.
7. Which of the following are *true* concerning the elbow joint? (1) When the forearm is extended, the olecranon fossa receives the olecranon. (2) When the forearm is flexed, the radial fossa receives the coronoid process. (3) The head of the radius articulates with the capitulum. (4) The trochlea articulates with the trochlear notch. (5) The head of the ulna articulates with the ulnar notch of the radius. (a) 1, 2, 3, 4, and 5; (b) 1, 3, and 4; (c) 1, 3, 4, and 5; (d) 1, 2, 3, and 4; (e) 2, 3, and 4.
8. Which of the following is the most superior of the tarsals and articulates with the distal end of the tibia? (a) calcaneus, (b) navicular, (c) cuboid, (d) cuneiform, (e) talus.
9. Which is(are) *not* true concerning the scapula? (1) The lateral border is also known as the axillary border. (2) The scapular notch accommodates the head of the humerus. (3) The scapula is also known as the collarbone. (4) The acromion process articulates with the clavicle. (5) The coracoid process is utilized for muscle attachment. (a) 1, 2, and 3; (b) 3 only; (c) 2 and 3; (d) 3 and 4; (e) 2, 3, and 5.
10. Which of the following is *false*? (a) A decrease in the height of the medial longitudinal arch creates a condition known as clawfoot. (b) The transverse arch is formed by the navicular, cuneiforms, and bases of the five metatarsals. (c) The longitudinal arch has medial and lateral parts, both of which originate at the calcaneus. (d) Arches help to absorb shocks. (e) Arches enable the foot to support the body's weight.
11. The bones that form the pectoral girdle are the (a) vertebrae, clavicle, and scapula; (b) sternum and clavicle; (c) clavicle, scapula, and sternum; (d) clavicle and scapula; (e) clavicle, scapula, and humerus.
12. The greater sciatic notch is located on the (a) ilium, (b) ischium, (c) femur, (d) pubis, (e) sacrum.
13. Match the following:

____(a) a large, triangular, flat bone found in the posterior part of the thorax
____(b) an S-shaped bone lying horizontally in the superior and anterior part of the thorax
____(c) articulates proximally with the scapula and distally with the radius and ulna
____(d) located on the medial aspect of the forearm
____(e) located on the lateral aspect of the forearm
____(f) the longest, heaviest, and strongest bone of the body
____(g) the larger, medial bone of the leg
____(h) the smaller, lateral bone of the leg
____(i) heel bone
____(j) sesamoid bone that articulates with the femur and tibia

(1) calcaneus
(2) scapula
(3) patella
(4) radius
(5) femur
(6) clavicle
(7) ulna
(8) tibia
(9) humerus
(10) fibula

14. Match the following:

____(a) largest and strongest tarsal bone
____(b) most medial bone in the distal row of carpals; has a hook-shaped projection on anterior surface
____(c) most medial, pea-shaped bone located in the proximal row of carpals
____(d) articulate with metatarsals I–III and cuboid
____(e) located in the proximal row of carpals; its name means "moon-shaped"
____(f) most lateral bone in the distal row of carpals
____(g) largest carpal bone
____(h) generally classified as proximal, middle, and distal
____(i) most lateral bone in the proximal row of carpals
____(j) articulates with the tibia and fibula
____(k) located in the proximal row of carpals; its name indicates that it is "three-cornered"
____(l) lateral bone that articulates with the calcaneus and metatarsals; IV–V
____(m) articulates with metacarpal II
____(n) boat-shaped bone that articulates with the talus

(1) cuboid
(2) triquetrum
(3) calcaneus
(4) pisiform
(5) capitate
(6) phalanges
(7) trapezoid
(8) hamate
(9) lunate
(10) scaphoid
(11) cuneiforms
(12) navicular
(13) trapezium
(14) talus

15. Match the following (some answers will be used more than once):

____(a) olecranon
____(b) olecranon fossa
____(c) trochlea
____(d) greater trochanter
____(e) medial malleolus
____(f) acromial extremity
____(g) capitulum
____(h) acromion
____(i) radial tuberosity
____(j) acetabulum
____(k) lateral malleolus
____(l) glenoid cavity
____(m) coronoid process
____(n) linea aspera
____(o) anterior border
____(p) anterior-superior iliac spine
____(q) fovea capitis
____(r) greater tubercle
____(s) trochlear notch
____(t) obturator foramen
____(u) styloid process

(1) clavicle
(2) scapula
(3) humerus
(4) ulna
(5) radius
(6) femur
(7) tibia
(8) fibula
(9) coxal bone

CRITICAL THINKING QUESTIONS

1. Mr. Smith's dog Rover dug up a complete set of human bones in the woods near his house. After examining the scene, the local police collected the bones and transported them to coroner's office for identification. Later, Mr. Smith read in the newspaper that the bones belonged to an elderly female. How was this determined?
2. A proud dad holds his 5-month old baby girl upright on her feet while supporting her under her arms. He states that she can never be a dancer because her feet are too flat. Is this true? Why or why not?
3. The local newspaper reported that Farmer White caught his hand in a piece of machinery last Tuesday. He lost the lateral two fingers of his left hand. His daughter, who is taking high school science, reports that Farmer White has three remaining phalanges. Is she correct, or does she need a refresher course in anatomy? Support your answer.

ANSWERS TO FIGURE QUESTIONS

8.1 The pectoral girdles attach the upper limbs to the axial skeleton.
8.2 The weakest part of the clavicle is its midregion at the junction of the two curves.
8.3 The acromion of the scapula forms the high point of the shoulder.
8.4 Each upper limb has 30 bones.
8.5 The radius articulates at the elbow with the capitulum and radial fossa of the humerus. The ulna articulates at the elbow with the trochlea, coronoid fossa, and olecranon fossa of the humerus.
8.6 The olecranon is the "elbow" part of the ulna.
8.7 The radius and ulna form the proximal and distal radioulnar joints. Their shafts are also connected by the interosseous membrane.
8.8 The scaphoid is the most frequently fractured wrist bone.
8.9 The bony pelvis attaches the lower limbs to the axial skeleton and supports the backbone and pelvic viscera.
8.10 The femur articulates with the acetabulum of the hip bone; the sacrum articulates with the auricular surface of the hip bone.
8.11 The pelvic axis is the course taken by a baby's head as it descends through the pelvis during childbirth.
8.12 Each lower limb has 30 bones.
8.13 The angle of convergence of the femurs is greater in females than males because the female pelvis is broader.
8.14 The patella is classified as a sesamoid bone because it develops in a tendon (the tendon of the quadriceps femoris muscle of the thigh).
8.15 The tibia is the weight-bearing bone of the leg.
8.16 The talus is the only tarsal bone that articulates with the tibia and the fibula.
8.17 Because the arches are not rigid, they yield when weight is applied and spring back when weight is lifted, allowing them to absorb the shock of walking.
8.18 The skeletal system arises from embryonic mesoderm.

Chapter 9

Joints

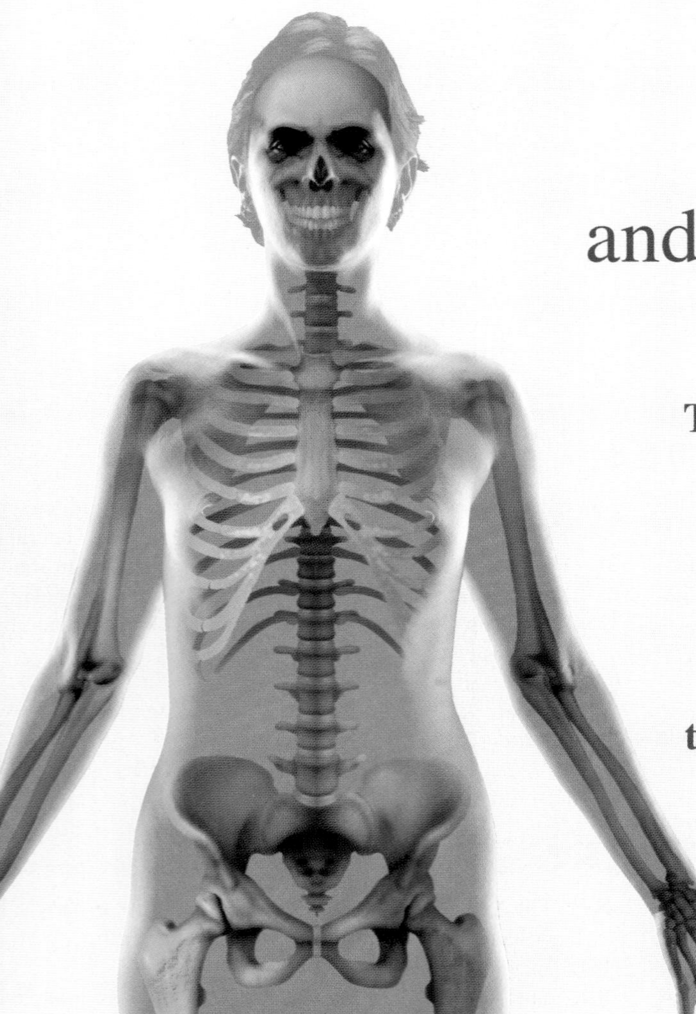

Joints and Homeostasis

The joints of the skeletal system contribute to homeostasis by holding bones together in ways that allow for movement and flexibility.

wiley.com/college/apcentral

Bones are too rigid to bend without being damaged. Fortunately, flexible connective tissues form joints that hold bones together while still permitting, in most cases, some degree of movement. A **joint,** also called an **articulation** (ar-tik-ū-LĀ-shun) or **arthrosis** (ar-THRŌ-sis), is a point of contact between two bones, between bone and cartilage, or between bone and teeth. When we say one bone *articulates* with another bone, we mean that the bones form a joint. You can appreciate the importance of joints if you have ever had a cast over your knee joint, which makes walking difficult, or a splint on your finger, which limits your ability to manipulate small objects. The scientific study of joints is termed **arthrology** (ar-THROL-ō-jē; *arthr-* = joint; *-logy* = study of). The study of motion of the human body is called **kinesiology** (ki-nē-sē′-OL-ō-jē; *kinesi* = movement).

JOINT CLASSIFICATIONS

▶ OBJECTIVE

Describe the structural and functional classifications of joints.

Joints are classified structurally, based on their anatomical characteristics, and functionally, based on the type of movement they permit.

The structural classification of joints is based on two criteria: (1) the presence or absence of a space between the articulating bones, called a synovial cavity, and (2) the type of connective tissue that binds the bones together. Structurally, joints are classified as one of the following types:

- **Fibrous joints** (FĪ-brus): There is no synovial cavity and the bones are held together by fibrous connective tissue that is rich in collagen fibers.
- **Cartilaginous joints** (kar-ti-LAJ-i-nus): There is no synovial cavity and the bones are held together by cartilage.
- **Synovial joints** (sī-NŌ-vē-al): The bones forming the joint have a synovial cavity and are united by the dense irregular connective tissue of an articular capsule, and often by accessory ligaments.

The functional classification of joints relates to the degree of movement they permit. Functionally, joints are classified as one of the following types:

- **Synarthrosis** (sin′-ar-THRŌ-sis; *syn-* = together): An immovable joint. The plural is *synarthroses.*
- **Amphiarthrosis** (am′-fē-ar-THRŌ-sis; *amphi-* = on both sides): A slightly movable joint. The plural is *amphiarthroses.*
- **Diarthrosis** (dī-ar-THRŌ-sis = movable joint): A freely movable joint. The plural is *diarthroses.* All diarthroses are synovial joints. They have a variety of shapes and permit several different types of movements.

The following sections present the joints of the body according to their structural classifications. As we examine the structure of each type of joint, we will also outline its functions.

▶ CHECKPOINT

1. On what basis are joints classified?

FIBROUS JOINTS

▶ OBJECTIVE

Describe the structure and functions of the three types of fibrous joints.

As previously noted, **fibrous joints** lack a synovial cavity, and the articulating bones are held very closely together by fibrous connective tissue. Fibrous joints permit little or no movement. The three types of fibrous joints are sutures, syndesmoses, and gomphoses.

Sutures

A **suture** (SOO-chur; *sutur-* = seam) is a fibrous joint composed of a thin layer of dense fibrous connective tissue; sutures occur only between bones of the skull. An example is the coronal suture between the parietal and frontal bones (Figure 9.1a). The irregular, interlocking edges of sutures give them added strength and decrease their chance of fracturing. Because a suture is immovable, it is classified functionally as a synarthrosis.

Some sutures that are present during childhood are replaced by bone in the adult. Such a suture is an example of a **synostosis** (sin′-os-TŌ-sis; *os-* = bone), or bony joint—a joint in which there is a complete fusion of two separate bones into one bone. For example, the frontal bone grows in halves that join together across a suture line. Usually they are completely fused by age 6 and the suture becomes obscure. If the suture persists beyond age 6, it is called a *metopic suture* (me-TŌ-pik; *metopon* = forehead). A synostosis is also classified functionally as a synarthrosis.

Syndesmoses

A **syndesmosis** (sin′-dez-MŌ-sis; *syndesmo-* = band or ligament) is a fibrous joint in which there is a greater distance between the articulating bones and more fibrous connective tissue than in a suture. The fibrous connective tissue is arranged either as a bundle (ligament) or as a sheet (interosseous membrane). One example of a syndesmosis is the distal tibiofibular joint, where the anterior tibiofibular ligament connects the tibia and fibula. Another example is the interosseous membrane between the parallel borders of the tibia and fibula (Figure 9.1b). Because it permits slight movement, a syndesmosis is classified functionally as an amphiarthrosis.

Figure 9.1 **Fibrous joints.**

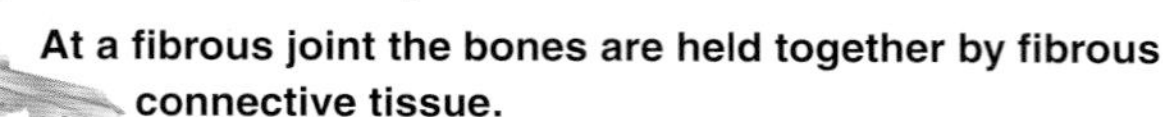

At a fibrous joint the bones are held together by fibrous connective tissue.

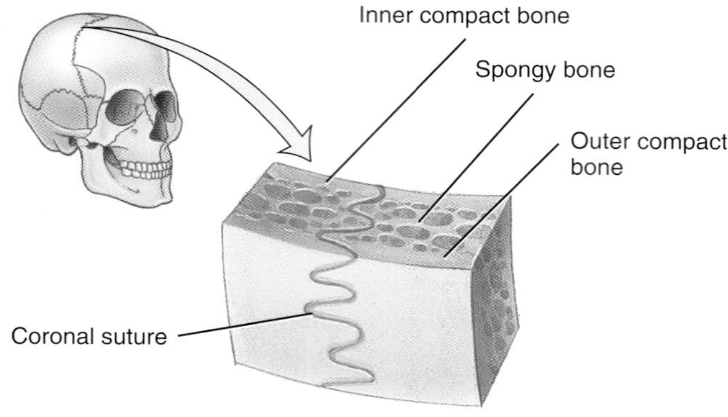

(a) Suture between skull bones

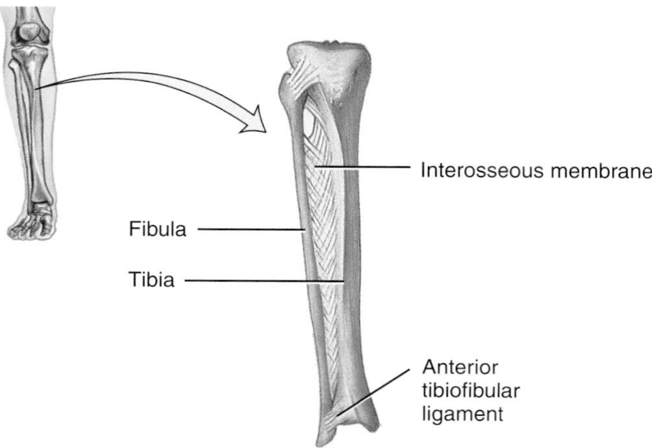

(b) Syndesmoses between tibia and fibula

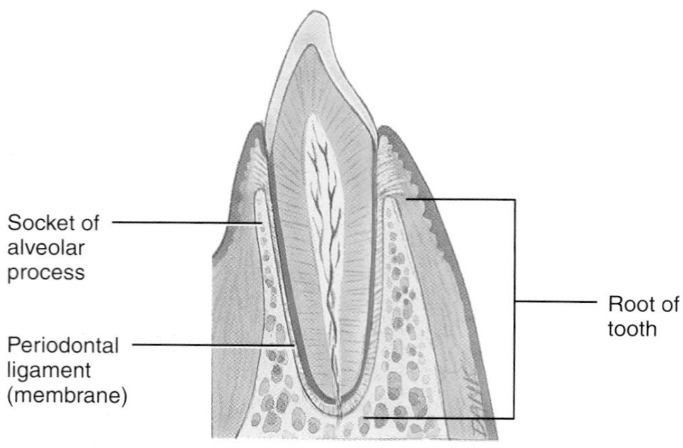

(c) Gomphosis between tooth and socket of alveolar process

Functionally, why are sutures classified as synarthroses, and syndesmoses as amphiarthroses?

Gomphoses

A **gomphosis** (gom-FŌ-sis; *gompho-* = a bolt or nail) or *dentoalveolar joint* is a type of fibrous joint in which a cone-shaped peg fits into a socket. The only examples of gomphoses in the human body are the articulations of the roots of the teeth with the sockets (alveoli) of the alveolar processes of the maxillae and mandible (Figure 9.1c). The dense fibrous connective tissue between a tooth and its socket is the periodontal ligament (membrane). A gomphosis is classified functionally as a synarthrosis, an immovable joint. Inflammation and degeneration of the gums, periodontal ligament, and bone is called *periodontal disease.*

▶ CHECKPOINT

2. Which fibrous joints are synarthroses? Which are amphiarthroses?

CARTILAGINOUS JOINTS

▶ OBJECTIVE

Describe the structure and functions of the two types of cartilaginous joints.

Like a fibrous joint, a **cartilaginous joint** lacks a synovial cavity and allows little or no movement. Here the articulating bones are tightly connected by either hyaline cartilage or fibrocartilage (see Table 4.4G, H). The two types of cartilaginous joints are synchondroses and symphyses.

Synchondroses

A **synchondrosis** (sin′-kon-DRŌ-sis; *chondro-* = cartilage) is a cartilaginous joint in which the connecting material is hyaline cartilage. An example of a synchondrosis is the epiphyseal plate that connects the epiphysis and diaphysis of a growing bone (Figure 9.2a). A photomicrograph of the epiphyseal plate is shown in Figure 6.7a on page 182. Functionally, a synchondrosis is a synarthrosis. When bone elongation ceases, bone replaces the hyaline cartilage, and the synchondrosis becomes a synostosis, a bony joint. Another example of a synchondrosis is the joint between the first rib and the manubrium of the sternum, which also ossifies during adult life and becomes an immovable synostosis (see Figure 7.22b on page 223).

Symphyses

A **symphysis** (SIM-fi-sis = growing together) is a cartilaginous joint in which the ends of the articulating bones are covered with hyaline cartilage, but a broad, flat disc of fibrocartilage connects the bones. All symphyses occur in the midline of the body. The pubic symphysis between the anterior surfaces of the hip bones is one example of a symphysis (Figure 9.2b). This type of joint is also found at the junction of the manubrium and body of the sternum (see Figure 7.22 on page 223) and at the intervertebral

Figure 9.2 **Cartilaginous joints.**

At a cartilaginous joint the bones are held together by cartilage.

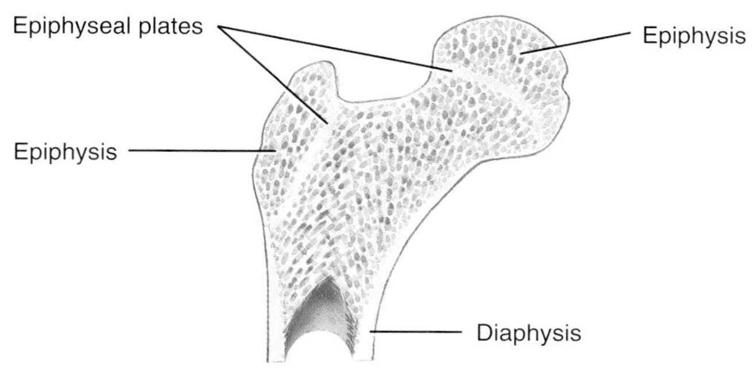

(a) Synchondrosis

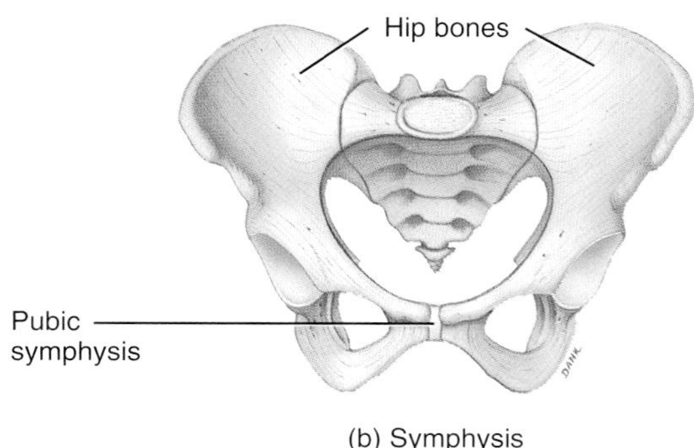

(b) Symphysis

What is the structural difference between a synchondrosis and a symphysis?

Figure 9.3 **Structure of a typical synovial joint.** Note the two layers of the articular capsule—the fibrous capsule and the synovial membrane. Synovial fluid fills the joint cavity between the synovial membrane and the articular cartilage.

The distinguishing feature of a synovial joint is the synovial cavity between the articulating bones.

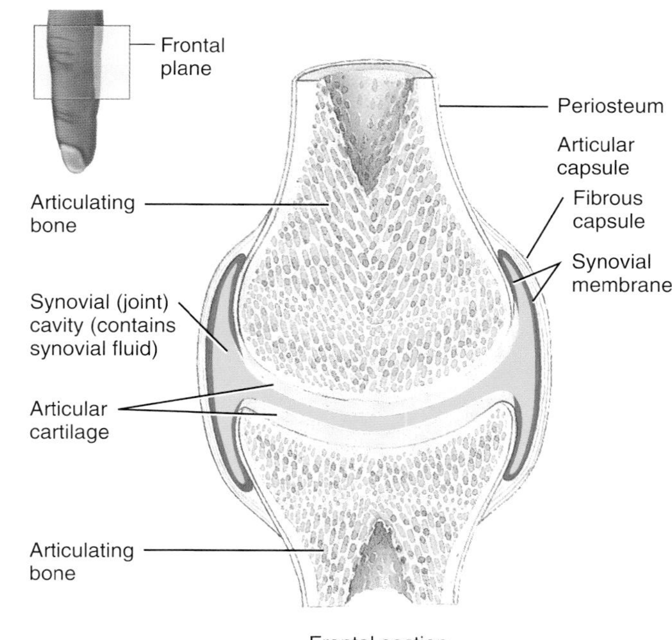

Frontal section

What is the functional classification of synovial joints?

joints between the bodies of vertebrae (see Figure 7.20a on page 220). A portion of the intervertebral disc is composed of fibrocartilage. A symphysis is an amphiarthrosis, a slightly movable joint.

▶ CHECKPOINT

3. Which cartilaginous joints are synarthroses? Which are amphiarthroses?

SYNOVIAL JOINTS

▶ OBJECTIVES

Describe the structure of synovial joints.

Describe the structure and function of bursae and tendon sheaths.

Structure of Synovial Joints

Synovial joints (si-NŌ-vē-al) have certain characteristics that distinguish them from other joints. The unique characteristic of a synovial joint is the presence of a space called a **synovial (joint) cavity** between the articulating bones (Figure 9.3). Because the synovial cavity allows a joint to be freely movable, all synovial joints are classified functionally as diarthroses. The bones at a synovial joint are covered by a layer of hyaline cartilage called **articular cartilage.** The cartilage covers the articulating surface of the bones with a smooth, slippery surface but does not bind them together. Articular cartilage reduces friction between bones in the joint during movement and helps to absorb shock.

Articular Capsule

A sleevelike **articular capsule** surrounds a synovial joint, encloses the synovial cavity, and unites the articulating bones. The articular capsule is composed of two layers, an outer fibrous capsule and an inner synovial membrane (Figure 9.3). The **fibrous capsule** usually consists of dense irregular connective tissue (mostly collagen fibers) that attaches to the periosteum of the articulating bones. The flexibility of the fibrous capsule permits considerable movement at a joint while its great tensile strength (resistance to stretching) helps prevent the bones from dislocating. The fibers of some fibrous capsules are arranged as parallel bundles of dense regular connective tissue that are highly adapted for resisting strains. The strength of these fiber bundles, called **ligaments** (*liga-* = bound or tied), is one of the principal mechanical factors that hold bones close together in a synovial joint. The inner layer of the articular capsule, the **synovial membrane,** is composed of areolar connective tissue with elastic fibers. At many synovial joints the synovial membrane includes accumulations of adipose tissue, called **articular fat pads.** An example is the infrapatellar fat pad in the knee (see Figure 9.15c).

A "double-jointed" person does not really have extra joints. Individuals who are "double-jointed" have greater flexibility in their articular capsules and ligaments; the resulting increase in range of motion allows them to entertain fellow partygoers with activities such as touching their thumbs to their wrists and putting their ankles or elbows behind their necks. Unfortunately, such flexible joints are less structurally stable and are more easily dislocated.

Synovial Fluid

The synovial membrane secretes **synovial fluid** (*ov-* = egg), a viscous, clear or pale yellow fluid named for its similarity in appearance and consistency to uncooked egg white. Synovial fluid consists of hyaluronic acid secreted by fibroblast-like cells in the synovial membrane and interstitial fluid filtered from blood plasma. It forms a thin film over the surfaces within the articular capsule. Its functions include reducing friction by lubricating the joint, absorbing shocks, and supplying oxygen and nutrients to and removing carbon dioxide and metabolic wastes from the chondrocytes within articular cartilage. (Recall that cartilage is an avascular tissue, so it does not have blood vessels to perform the latter function.) Synovial fluid also contains phagocytic cells that remove microbes and the debris that results from normal wear and tear in the joint. When a synovial joint is immobile for a time, the fluid becomes quite viscous (gel-like), but as joint movement increases, the fluid becomes less viscous. One of the benefits of warming up before exercise is that it stimulates the production and secretion of synovial fluid; more fluid means less stress on the joints during exercise.

We are all familiar with the cracking sounds heard as certain joints move, or the popping sounds that arise when people crack their knuckles. According to one theory, when the synovial cavity expands, the pressure of the synovial fluid decreases, creating a partial vacuum. The suction draws carbon dioxide and oxygen out of blood vessels in the synovial membrane, forming bubbles in the fluid. When the bubbles are forced to burst, as when fingers are hyperflexed, the cracking or popping sound is heard.

Accessory Ligaments and Articular Discs

Many synovial joints also contain **accessory ligaments** called extracapsular ligaments and intracapsular ligaments. *Extracapsular ligaments* lie outside the articular capsule. Examples are the fibular and tibial collateral ligaments of the knee joint (see Figure 9.15d). *Intracapsular ligaments* occur within the articular capsule but are excluded from the synovial cavity by folds of the synovial membrane. Examples are the anterior and posterior cruciate ligaments of the knee joint (see Figure 9.15d).

Inside some synovial joints, such as the knee, pads of fibrocartilage lie between the articular surfaces of the bones and are attached to the fibrous capsule. These pads are called **articular discs** or **menisci** (me-NIS-sī or me-NIS-kī; singular is *meniscus*). Figure 9.15d on page 283 depicts the lateral and medial menisci in the knee joint. The discs usually subdivide the synovial cavity into two spaces, allowing separate movements to occur in each space. As you will see later, separate movements also occur in the respective compartments of the temporomandibular joint (TMJ) (see page 275). By modifying the shape of the joint surfaces of the articulating bones, articular discs allow two bones of different shapes to fit together more tightly. Articular discs also help to maintain the stability of the joint and direct the flow of synovial fluid to the areas of greatest friction.

Torn Cartilage and Arthroscopy

The tearing of articular discs (menisci) in the knee, commonly called **torn cartilage,** occurs often among athletes. Such damaged cartilage will begin to wear and may precipitate arthritis unless it is surgically removed (meniscectomy). Surgical repair of the torn cartilage is required because of the avascular nature of cartilage and may be assisted by **arthroscopy** (ar-THROS-kō-pē; *-scopy* = observation), the visual examination of the interior of a joint, usually the knee, with an *arthroscope,* a lighted, pencil-thin instrument. Arthroscopy is used to determine the nature and extent of damage following knee injury and to monitor the progression of disease and the effects of therapy. In addition, the insertion of surgical instruments through the arthroscope or other incisions enables a physician to remove torn cartilage and repair damaged cruciate ligaments in the knee; to remodel poorly formed cartilage; to obtain tissue samples for analysis; and to perform surgery on other joints, such as the shoulder, elbow, ankle, and wrist. ■

Nerve and Blood Supply

The nerves that supply a joint are the same as those that supply the skeletal muscles that move the joint. Synovial joints contain many nerve endings that are distributed to the articular capsule and associated ligaments. Some of the nerve endings convey information about pain from the joint to the spinal cord and brain for processing. Other nerve endings respond to the degree of movement and stretch at a joint. The spinal cord and brain may respond by sending impulses through different nerves to the muscles to adjust body movements.

Although many of the components of synovial joints are avascular, arteries in the vicinity send out numerous branches that penetrate the ligaments and articular capsule to deliver oxygen and nutrients. Veins remove carbon dioxide and wastes from the joints. The arterial branches from several different arteries typically merge around a joint before penetrating the articular capsule. The chondrocytes of articular cartilage of a synovial joint receive oxygen and nutrients from synovial fluid derived from blood; all other joint tissues are supplied directly by arteries. Carbon dioxide and wastes pass from chondrocytes of articular cartilage into synovial fluid and then into veins; carbon dioxide and wastes from all other joint structures pass directly into veins.

Sprain and Strain

A **sprain** is the forcible wrenching or twisting of a joint that stretches or tears its ligaments but does not dislocate the bones. It occurs when the ligaments are stressed beyond their normal capacity. Sprains also may damage surrounding blood vessels, muscles, tendons, or nerves. Severe sprains may be so painful that the joint cannot be moved. There is considerable swelling, which results from chemicals released by the damaged cells and hemorrhage of ruptured blood vessels. The ankle joint is most often sprained; the lower back is another frequent location. A **strain** is a stretched or partially torn muscle. It often occurs when a muscle contracts suddenly and powerfully—such as the leg muscles of sprinters when they spring from the blocks. ■

Bursae and Tendon Sheaths

The various movements of the body create friction between moving parts. Saclike structures called **bursae** (BER-sē = purses; singular is *bursa*) are strategically situated to alleviate friction in some joints, such as the shoulder and knee joints (see Figures 9.12 and 9.15c). Bursae are not strictly part of synovial joints, but they do resemble joint capsules because their walls consist of connective tissue lined by a synovial membrane. They are filled with a small amount of fluid that is similar to synovial fluid. Bursae can be located between the skin and bones, tendons and bones, muscles and bones, or ligaments and bones. The fluid-filled bursal sacs cushion the movement of these body parts against one another.

Structures called tendon sheaths also reduce friction at joints. **Tendon sheaths** are tubelike bursae that wrap around certain tendons that experience considerable friction. This occurs where tendons pass through synovial cavities, such as the tendon of the biceps brachii muscle at the shoulder joint (see Figure 9.12c). Tendon sheaths are also found at the wrist and ankle, where many tendons come together in a confined space (see Figure 11.23 on page 397), and in the fingers and toes, where there is a great deal of movement (see Figure 11.18 on page 376).

Bursitis

An acute or chronic inflammation of a bursa, called **bursitis,** is usually caused by irritation from repeated, excessive exertion of a joint. The condition may also be caused by trauma, an acute or chronic infection (including syphilis and tuberculosis), or by rheumatoid arthritis (described on page 285). Symptoms include pain, swelling, tenderness, and limited movement. Treatment may include oral anti-inflammatory agents and injections of cortisol-like steroids. ■

▶ **CHECKPOINT**

4. How does the structure of synovial joints classify them as diarthroses?
5. What are the functions of articular cartilage, synovial fluid, and articular discs?
6. What types of sensations are perceived at joints, and from what sources do joints receive nourishment?
7. In what ways are bursae similar to joint capsules? How do they differ?

TYPES OF MOVEMENTS AT SYNOVIAL JOINTS

▶ **OBJECTIVE**

Describe the types of movements that can occur at synovial joints.

Anatomists, physical therapists, and kinesiologists (professionals who treat disease by movements of various kinds) use specific terminology to designate the movements that can occur at synovial joints. These precise terms may indicate the form of motion, the direction of movement, or the relationship of one body part to another during movement. Movements at synovial joints are grouped into four main categories: (1) gliding, (2) angular movements, (3) rotation, and (4) special movements.

Gliding

Gliding is a simple movement in which relatively flat bone surfaces move back-and-forth and from side-to-side with respect to one another (Figure 9.4). There is no significant alteration of

Figure 9.4 Gliding movements at synovial joints.

Gliding movements consist of side-to-side and back-and-forth motions.

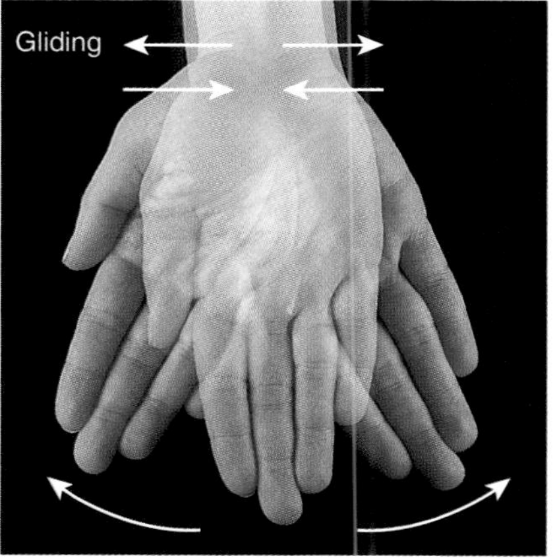

Intercarpal joints

What are two examples of joints that permit gliding movements?

the angle between the bones. Gliding movements are limited in range due to the structure of the articular capsule and associated ligaments and bones. The intercarpal and intertarsal joints are examples of articulations where gliding movements occur.

Angular Movements

In **angular movements,** there is an increase or a decrease in the angle between articulating bones. The major angular movements are flexion, extension, lateral extension, hyperextension, abduction, adduction, and circumduction. These movements are discussed with respect to the body in the anatomical position (see Figure 1.5 on page 13).

Flexion, Extension, Lateral Flexion, and Hyperextension

Flexion and extension are opposite movements. In **flexion** (FLEK-shun; *flex-* = to bend) there is a decrease in the angle between articulating bones; in **extension** (eks-TEN-shun; *exten-* = to stretch out) there is an increase in the angle between articulating bones, often to restore a part of the body to the anatomical position after it has been flexed (Figure 9.5). Both movements usually occur along the sagittal plane. All of the following are examples of flexion (as you have probably already guessed, extension is simply the reverse of these movements):

- Bending the head toward the chest at the atlanto-occipital joint between the atlas (the first vertebra) and the occipital bone of the skull, and at the cervical intervertebral joints between the cervical vertebrae (Figure 9.5a)
- Bending the trunk forward at the intervertebral joints
- Moving the humerus forward at the shoulder joint, as in swinging the arms forward while walking (Figure 9.5b)
- Moving the forearm toward the arm at the elbow joint between the humerus, ulna, and radius (Figure 9.5c)
- Moving the palm toward the forearm at the wrist or radiocarpal joint between the radius and carpals (Figure 9.5d)
- Bending the digits of the hand or feet at the interphalangeal joints between phalanges
- Moving the femur forward at the hip joint between the femur and hip bone, as in walking (Figure 9.5e)
- Moving the leg toward the thigh at the tibiofemoral joint between the tibia, femur, and patella, as occurs when bending the knee (Figure 9.5f)

Although flexion and extension usually occur along the sagittal plane, there are a few exceptions. For example, flexion of the thumb involves movement of the thumb medially across the palm at the carpometacarpal joint between the trapezium and metacarpal of the thumb, as when you touch your thumb to the opposite side of your palm (see Figure 11.18d on page 277). Another example is movement of the trunk sideways to the right or left at the waist. This movement, which occurs along the frontal plane and involves the intervertebral joints, is called **lateral flexion** (Figure 9.5g).

Continuation of extension beyond the anatomical position is called **hyperextension** (*hyper-* = beyond or excessive). Examples of hyperextension include:

- Bending the head backward at the atlanto-occipital and cervical intervertebral joints (Figure 9.5a)
- Bending the trunk backward at the intervertebral joints
- Moving the humerus backward at the shoulder joint, as in swinging the arms backward while walking (Figure 9.5b)
- Moving the palm backward at the wrist joint (Figure 9.5d)
- Moving the femur backward at the hip joint, as in walking (Figure 9.5e)

Hyperextension of hinge joints, such as the elbow, interphalangeal, and knee joints, is usually prevented by the arrangement of ligaments and the anatomical alignment of the bones.

Abduction, Adduction, and Circumduction

Abduction (ab-DUK-shun; *ab-* = away; *-duct* = to lead) is the movement of a bone away from the midline; **adduction** (ad-DUK-shun; *ad-* = toward) is the movement of a bone toward the midline. Both movements usually occur along the frontal plane. Examples of abduction include moving the humerus laterally at the shoulder joint, moving the palm laterally at the wrist joint, and moving the femur laterally at the hip joint (Figure 9.6a–c on page 266). The movement that returns each of these body parts to the anatomical position is adduction (Figure 9.6a–c).

The midline of the body is not used as a point of reference for abduction and adduction of the digits. In abduction of the fingers (but not the thumb), an imaginary line is drawn through the longitudinal axis of the middle (longest) finger, and the fingers move away (spread out) from the middle finger (Figure 9.6d). In abduction of the thumb, the thumb moves away from the palm in the sagittal plane (see Figure 11.18d on page 377). Abduction of the toes is relative to an imaginary line drawn through the second toe. Adduction of the fingers and toes returns them to the anatomical position. Adduction of the thumb moves the thumb toward the palm in the sagittal plane (see Figure 11.18d).

Circumduction (ser-kum-DUK-shun; *circ-* = circle) is movement of the distal end of a body part in a circle (Figure 9.7 on page 266). Circumduction is not an isolated movement by itself but rather a continuous sequence of flexion, abduction, extension, and adduction. Therefore, circumduction does not occur along a separate axis or plane of movement. Examples of circumduction are moving the humerus in a circle at the shoulder joint (Figure 9.7a), moving the hand in a circle at the wrist joint, moving the thumb in a circle at the carpometacarpal joint, moving the fingers in a circle at the metacarpophalangeal joints (between the metacarpals and phalanges), and moving the femur in a circle at the hip joint (Figure 9.7b). Both the shoulder and hip joints permit circumduction. Flexion, abduction, extension, and adduction

Figure 9.5 Angular movements at synovial joints—flexion, extension, hyperextension, and lateral flexion.

In angular movements, there is an increase or decrease in the angle between articulating bones.

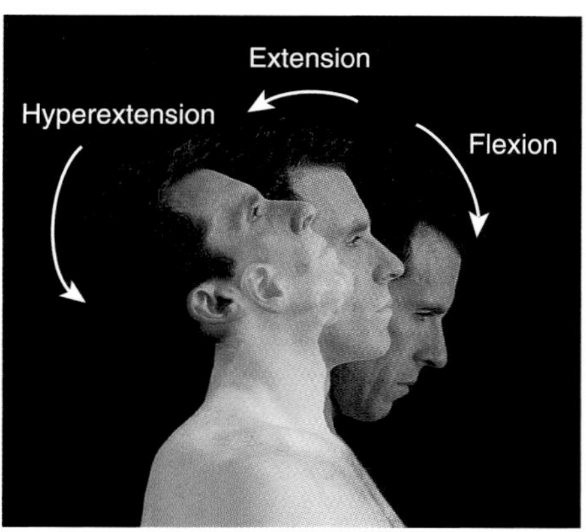

(a) Atlanto-occipital and cervical intervertebral joints

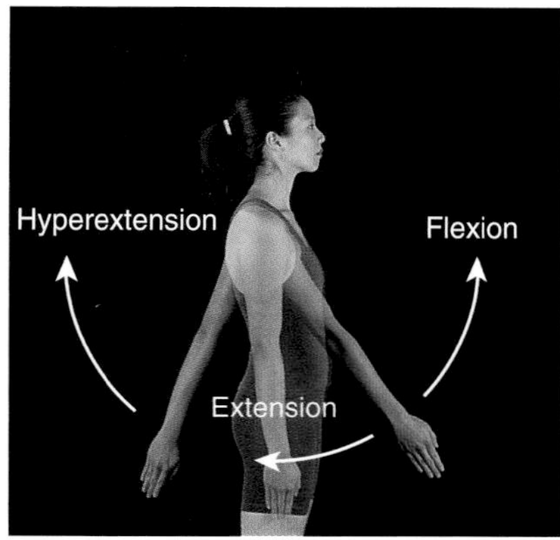

(b) Shoulder joint

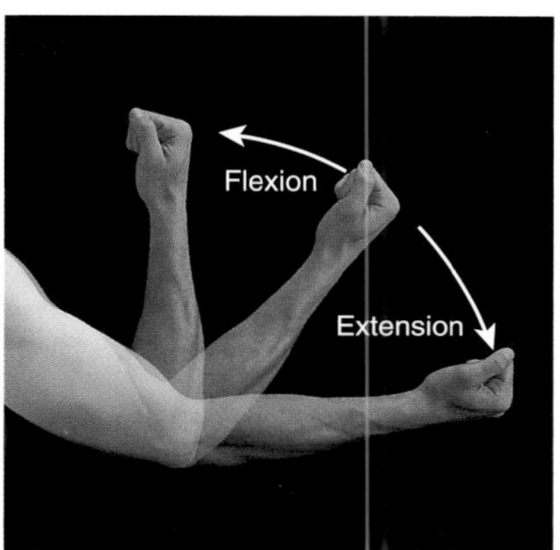

(c) Elbow joint

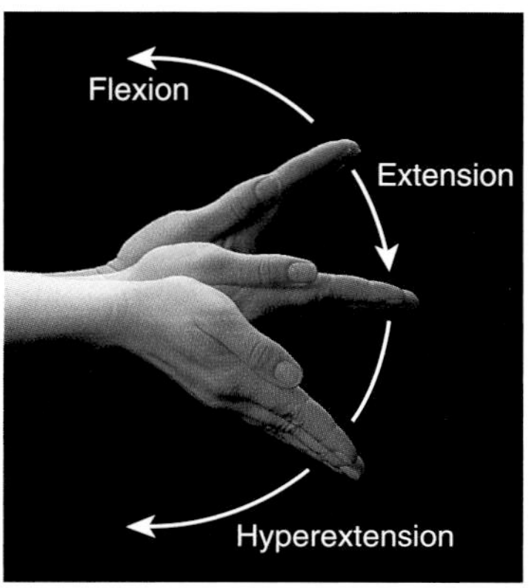

(d) Wrist joint

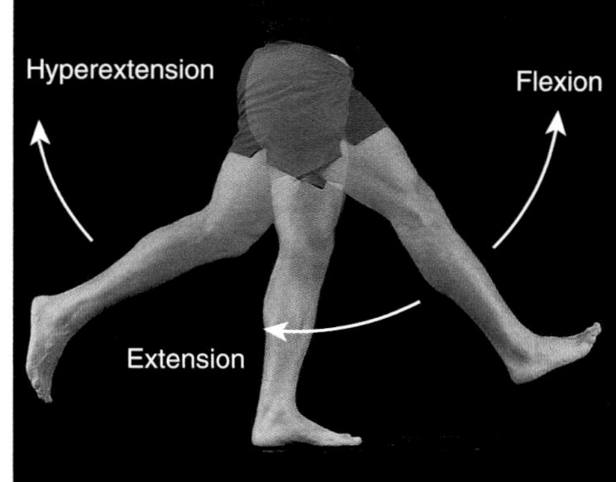

(e) Hip joint

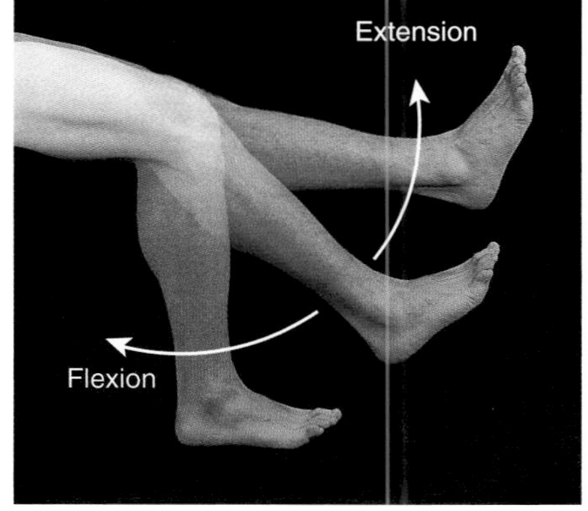

(f) Knee joint

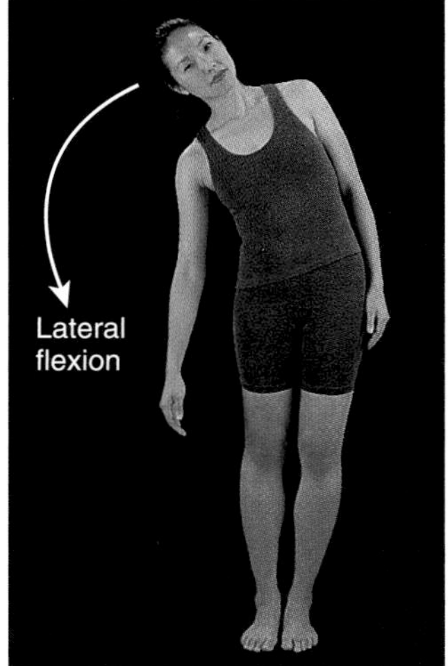

(g) Intervertebral joints

What are two examples of flexion that do not occur along the sagittal plane?

are more limited in the hip joints than in the shoulder joints due to the tension on certain ligaments and muscles (see Exhibits 9.2 and 9.4).

Rotation

In **rotation** (rō-TĀ-shun; *rota-* = revolve), a bone revolves around its own longitudinal axis. One example is turning the head from side to side at the atlanto-axial joint (between the atlas and axis), as when you shake your head "no" (Figure 9.8a on page 267). Another is turning the trunk from side to side at the intervertebral joints while keeping the hips and lower limbs in the anatomical position. In the limbs, rotation is defined relative to the midline, and specific qualifying terms are used. If the anterior surface of a bone of the limb is turned toward the midline, the movement is called *medial (internal) rotation.* You can medially rotate the humerus at the shoulder joint as follows: Starting in the anatomical position, flex your elbow and then draw your palm across the chest (Figure 9.8b). Medial rotation of the forearm at the radioulnar joints (between the radius and ulna) involves turning the palm medially from the anatomical position (see Figure 9.9h). You can medially rotate

Figure 9.6 Angular movements at synovial joints—abduction and adduction.

Abduction and adduction usually occur along the frontal plane.

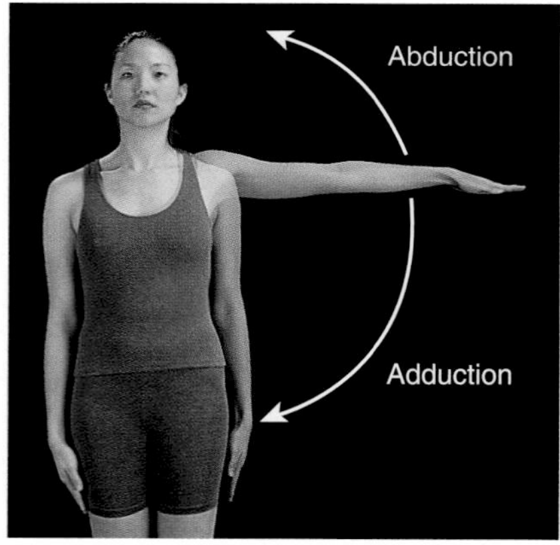

(a) Shoulder joint

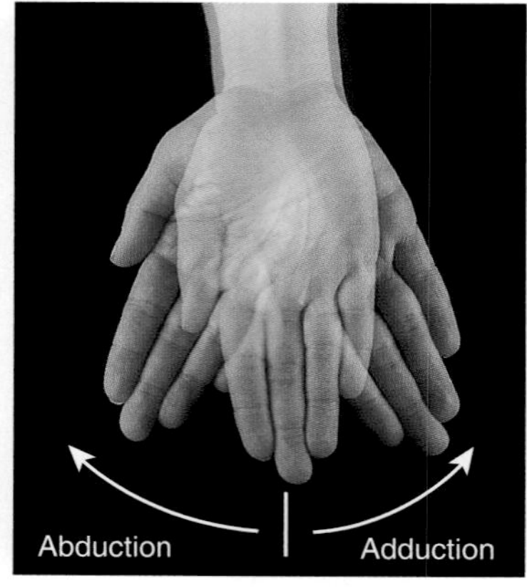

(b) Wrist joint

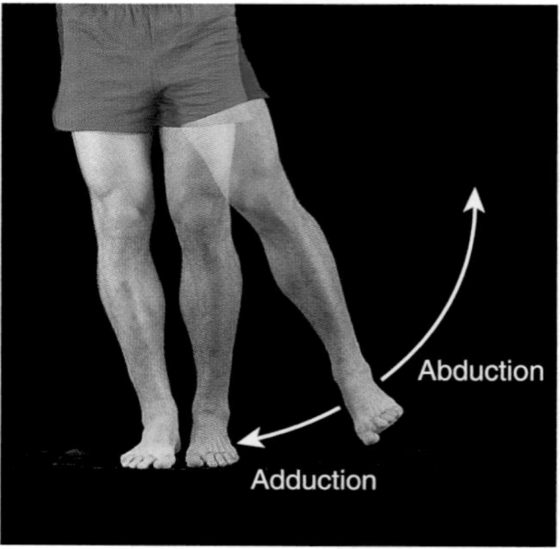

(c) Hip joint

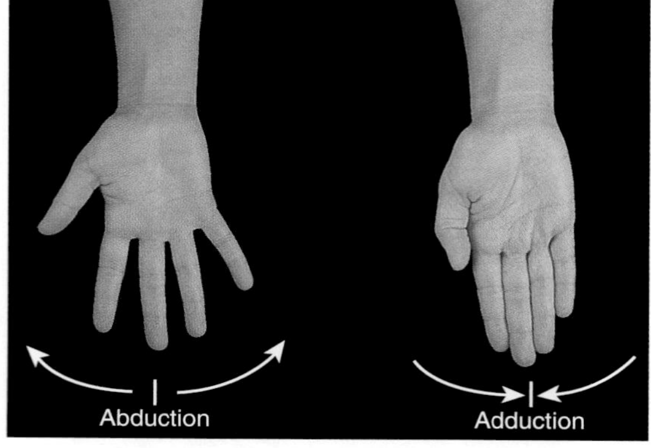

(d) Metacarpophalangeal joints of the fingers (not the thumb)

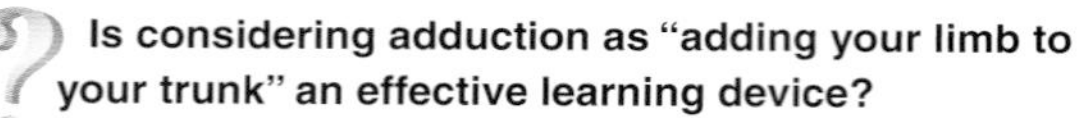

Is considering adduction as "adding your limb to your trunk" an effective learning device?

the femur at the hip joint as follows: Lie on your back, bend your knee, and then move your leg and foot laterally from the midline. Although you are moving your leg and foot laterally, the femur is rotating medially (Figure 9.8c). Medial rotation of the leg at the knee joint can be produced by sitting on a chair, bending your knee, raising your lower limb off the floor, and turning your toes medially. If the anterior surface of the bone of a limb is turned away from the midline, the movement is called *lateral (external) rotation* (see Figure 9.8b, c).

Special Movements

Special movements occur only at certain joints. They include elevation, depression, protraction, retraction, inversion, eversion, dorsiflexion, plantar flexion, supination, pronation, and opposition (Figure 9.9):

Figure 9.7 Angular movements at synovial joints—circumduction.

Circumduction is the movement of the distal end of a body part in a circle.

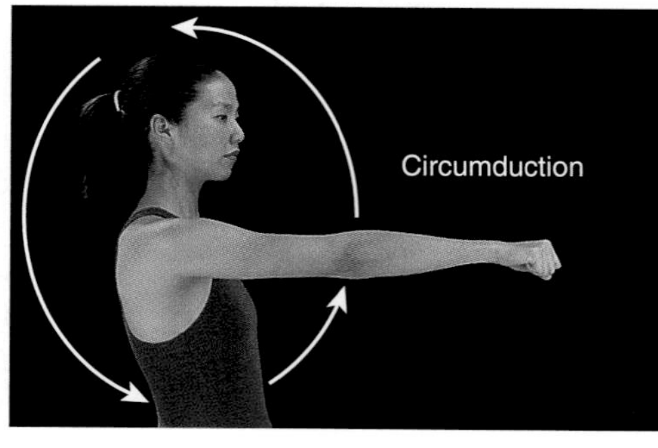

(a) Shoulder joint

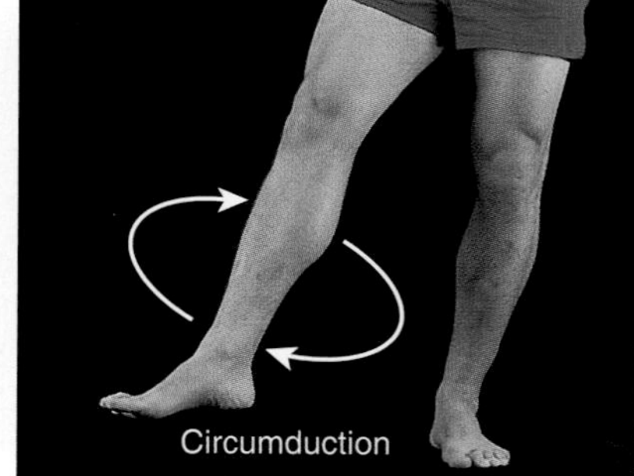

(b) Hip joint

Which movements in continuous sequence produce circumduction?

Figure 9.8 **Rotation at synovial joints.**

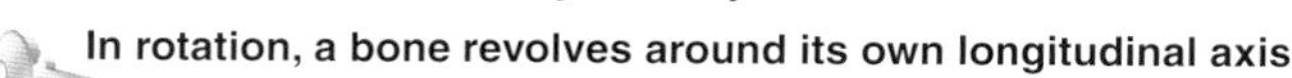

In rotation, a bone revolves around its own longitudinal axis.

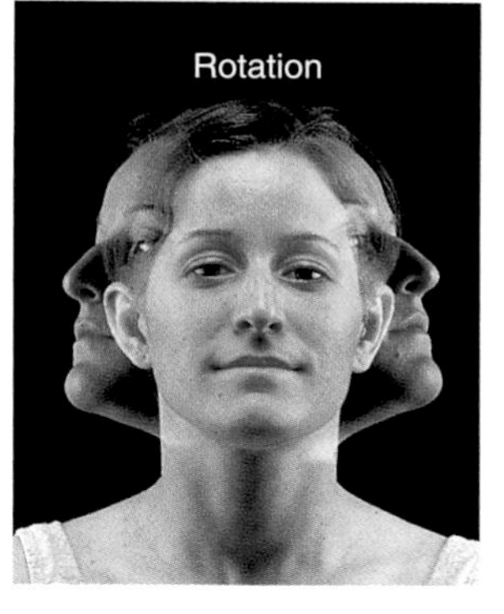

(a) Atlanto-axial joint

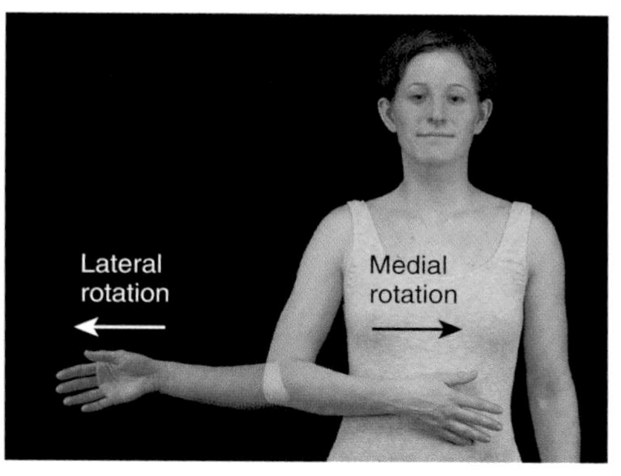

(b) Shoulder joint

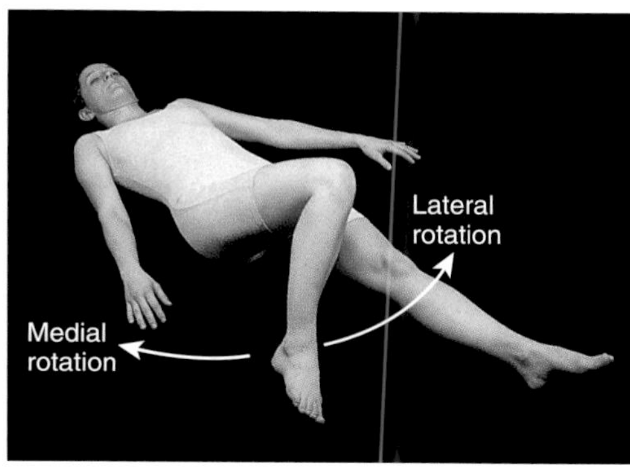

(c) Hip joint

How do medial and lateral rotation differ?

- **Elevation** (el-e-VĀ-shun = to lift up) is an upward movement of a part of the body, such as closing the mouth at the temporomandibular joint (between the mandible and temporal bone) to elevate the mandible (Figure 9.9a) or shrugging the shoulders at the acromioclavicular joint to elevate the scapula. Its opposing movement is depression. Other bones that may be elevated (or depressed) include the hyoid, clavicle, and ribs.
- **Depression** (de-PRESH-un = to press down) is a downward movement of a part of the body, such as opening the mouth to depress the mandible (Figure 9.9b) or returning shrugged shoulders to the anatomical position to depress the scapula.
- **Protraction** (prō-TRAK-shun = to draw forth) is a movement of a part of the body anteriorly in the transverse plane. Its opposing movement is retraction. You can protract your

Figure 9.9 **Special movements at synovial joints.**

Special movements occur only at certain synovial joints.

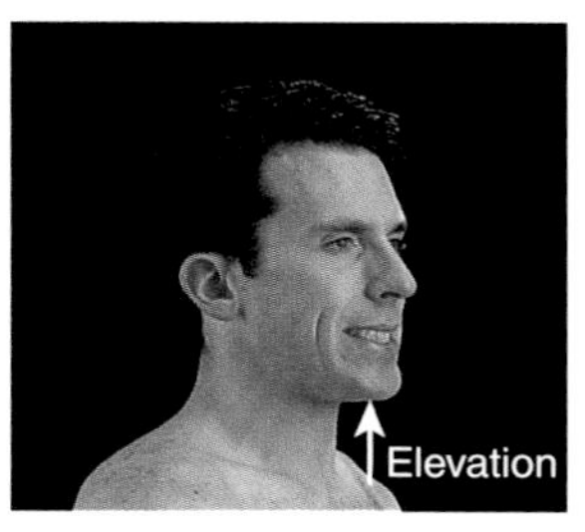

(a) Temporomandibular joint (b)

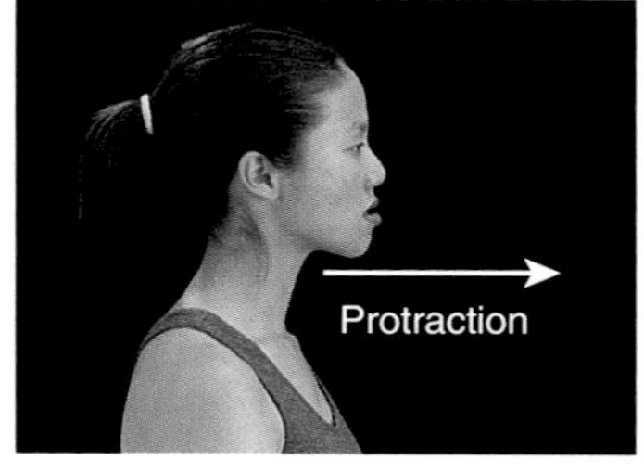

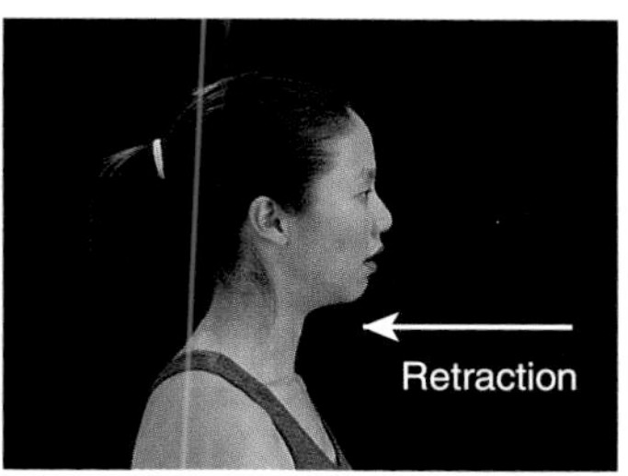

(c) Temporomandibular joint (d)

Inversion

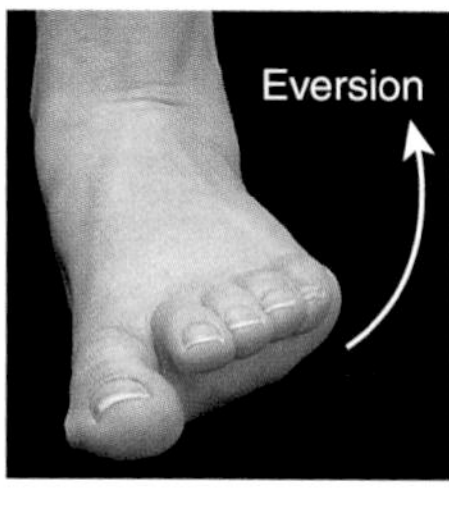

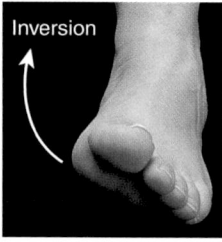

(e) Intertarsal joints (f)

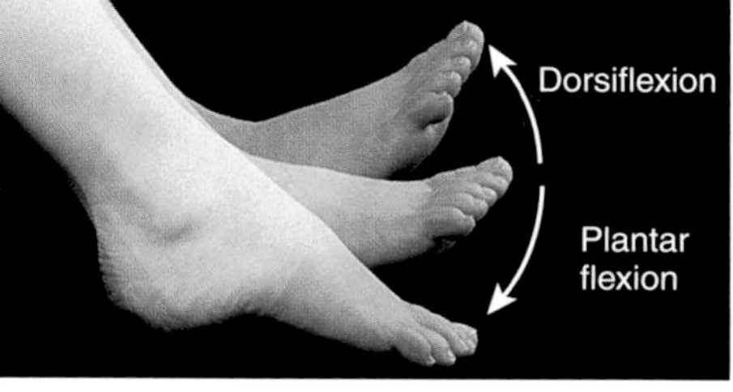

(g) Ankle joint

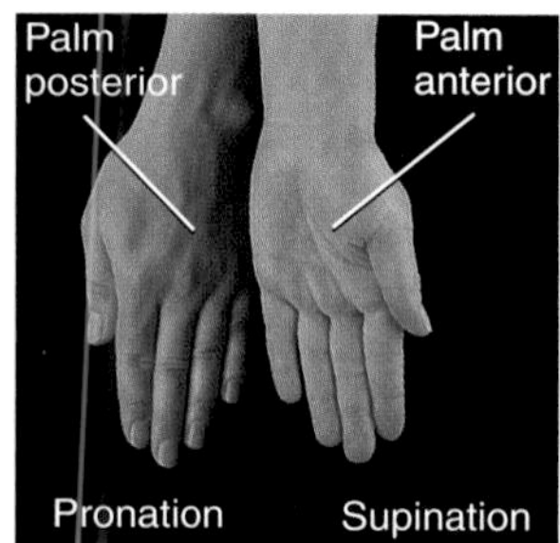

(h) Radioulnar joint

What movement of the shoulder girdle occurs when you bring your arms forward until the elbows touch?

mandible at the temporomandibular joint by thrusting it outward (Figure 9.9c) or protract your clavicles at the acromioclavicular and sternoclavicular joints by crossing your arms.

- **Retraction** (rē-TRAK-shun = to draw back) is a movement of a protracted part of the body back to the anatomical position (Figure 9.9d).
- **Inversion** (in-VER-zhun = to turn inward) is movement of the soles medially at the intertarsal joints (between the tarsals) so that the soles face each other (Figure 9.9e). Its opposing movement is eversion. Physical therapists also refer to inversion of the feet as *supination.*
- **Eversion** (ē-VER-zhun = to turn outward) is a movement of the soles laterally at the intertarsal joints so that the soles face away from each other (Figure 9.9f). Physical therapists also refer to eversion of the feet as *pronation.*
- **Dorsiflexion** (dor-si-FLEK-shun) refers to bending of the foot at the ankle or talocrural joint (between the tibia, fibula, and talus) in the direction of the dorsum (superior surface) (Figure 9.9g). Dorsiflexion occurs when you stand on your heels. Its opposing movement is plantar flexion.
- **Plantar flexion** involves bending of the foot at the ankle joint in the direction of the plantar or inferior surface (see Figure 9.9g), as when you elevate your body by standing on your toes.
- **Supination** (soo-pi-NĀ-shun) is a movement of the forearm at the proximal and distal radioulnar joints in which the palm is turned anteriorly (Figure 9.9h). This position of the palms is one of the defining features of the anatomical position. Its opposing movement is pronation.
- **Pronation** (prō-NĀ-shun) is a movement of the forearm at the proximal and distal radioulnar joints in which the distal end of the radius crosses over the distal end of the ulna and the palm is turned posteriorly (Figure 9.9h).
- **Opposition** (op-ō-ZISH-un) is the movement of the thumb at the carpometacarpal joint (between the trapezium and metacarpal of the thumb) in which the thumb moves across the palm to touch the tips of the fingers on the same hand (see Figure 11.18d on page 377). This is the distinctive digital movement that gives humans and other primates the ability to grasp and manipulate objects very precisely.

A summary of the movements that occur at synovial joints is presented in Table 9.1.

▶ **CHECKPOINT**

8. What are the four major categories of movements that occur at synovial joints?

9. On yourself or with a partner, demonstrate each movement listed in Table 9.1.

TABLE 9.1 Summary of Movements at Synovial Joints

Movement	Description	Movement	Description
Gliding	Movement of relatively flat bone surfaces back-and-forth and side-to-side over one another; little change in the angle between bones.	**Rotation**	Movement of a bone around its longitudinal axis; in the limbs, it may be medial (toward midline) or lateral (away from midline).
Angular	Increase or decrease in the angle between bones.	**Special**	Occurs at specific joints.
Flexion	Decrease in the angle between articulating bones, usually in the sagittal plane.	**Elevation**	Superior movement of a body part.
Lateral flexion	Movement of the trunk in the frontal plane.	**Depression**	Inferior movement of a body part.
Extension	Increase in the angle between articulating bones, usually in the sagittal plane.	**Protraction**	Anterior movement of a body part in the transverse plane.
Hyperextension	Extension beyond the anatomical position.	**Retraction**	Posterior movement of a body part in the transverse plane.
Abduction	Movement of a bone away from the midline, usually in the frontal plane.	**Inversion**	Medial movement of the soles so that they face each other.
Adduction	Movement of a bone toward the midline, usually in the frontal plane.	**Eversion**	Lateral movement of the soles so that they face away from each other.
Circumduction	Flexion, abduction, extension, and adduction in succession, in which the distal end of a body part moves in a circle.	**Dorsiflexion**	Bending the foot in the direction of the dorsum (superior surface).
		Plantar flexion	Bending the foot in the direction of the plantar surface (sole).
		Supination	Movement of the forearm that turns the palm anteriorly.
		Pronation	Movement of the forearm that turns the palm posteriorly.
		Opposition	Movement of the thumb across the palm to touch fingertips on the same hand.

TYPES OF SYNOVIAL JOINTS

▶ **OBJECTIVE**

Describe the six subtypes of synovial joints.

Although all synovial joints are similar in structure, the shapes of the articulating surfaces vary; thus, many types of movements are possible. Synovial joints are divided into six categories based on type of movement: planar, hinge, pivot, condyloid, saddle, and ball-and-socket.

Planar Joints

The articulating surfaces of bones in a **planar joint** are flat or slightly curved (Figure 9.10a). Planar joints primarily permit gliding movements. These joints are said to be *nonaxial* because the motion they allow does not occur around an axis or along a plane. Examples of planar joints are the intercarpal joints (between carpal bones at the wrist), intertarsal joints (between tarsal bones at the ankle), sternoclavicular joints (between the manubrium of the sternum and the clavicle), acromioclavicular joints (between the acromion of the scapula and the clavicle), sternocostal joints (between the sternum and ends of the costal cartilages at the tips of the second through seventh pairs of ribs), and vertebrocostal joints (between the heads and tubercles of ribs and transverse processes of thoracic vertebrae). X-ray films made during wrist and ankle movements reveal some rotation of the small carpal and tarsal bones in addition to their gliding movements.

Hinge Joints

In a **hinge joint,** the convex surface of one bone fits into the concave surface of another bone (Figure 9.10b). As the name implies, hinge joints produce an angular, opening-and-closing motion like that of a hinged door. In most joint movements, one bone remains in a fixed position while the other moves around an axis. Hinge joints are *monaxial (uniaxial)* because they typically allow motion around a single axis. Hinge joints permit only flexion and extension. Examples of hinge joints are the knee, elbow, ankle, and interphalangeal joints.

Pivot Joints

In a **pivot joint,** the rounded or pointed surface of one bone articulates with a ring formed partly by another bone and partly by a ligament (Figure 9.10c). A pivot joint is monaxial because it allows rotation only around its own longitudinal axis. Examples of pivot joints are the atlanto-axial joint, in which the atlas rotates around the axis and permits the head to turn from side to side as when you shake your head "no" (see Figure 9.8a), and the radioulnar joints that enable the palms to turn anteriorly and posteriorly (see Figure 9.9h).

Figure 9.10 **Subtypes of synovial joints.** For each subtype, a drawing of the actual joint and a simplified diagram are shown.

Synovial joints are classified into subtypes based on the shapes of the articulating bone surfaces.

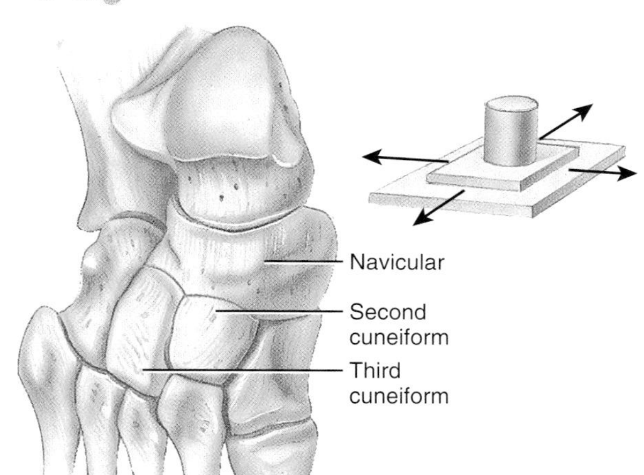

(a) Planar joint between the navicular and second and third cuneiforms of the tarsus in the foot

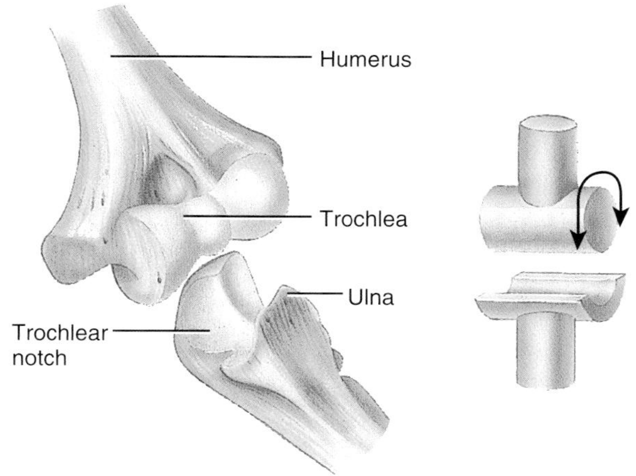

(b) Hinge joint between trochlea of humerus and trochlear notch of ulna at the elbow

continues

Figure 9.10 **(Continued)**

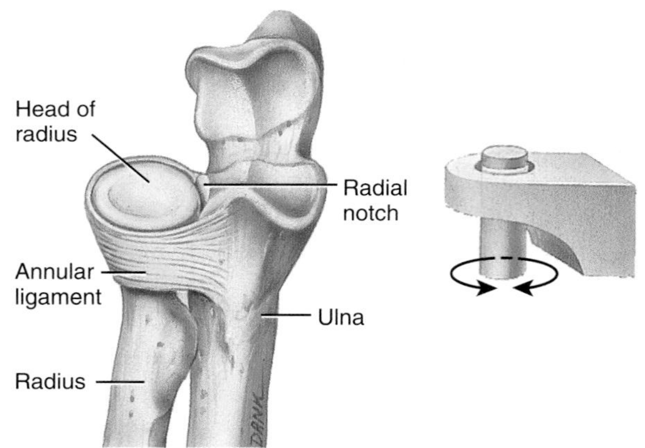

(c) Pivot joint between head of radius and radial notch of ulna

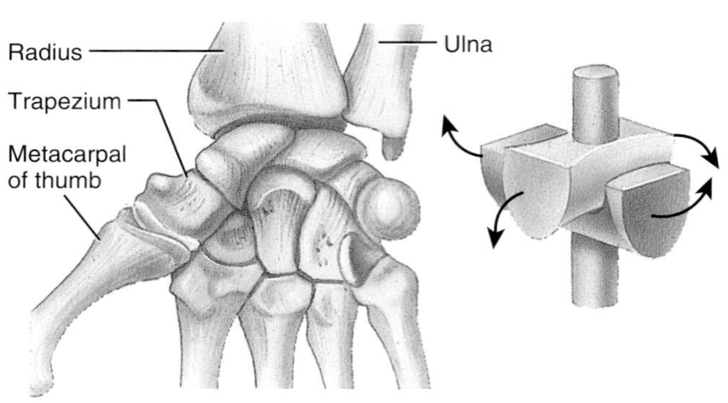

(e) Saddle joint between trapezium of carpus (wrist) and metacarpal of thumb

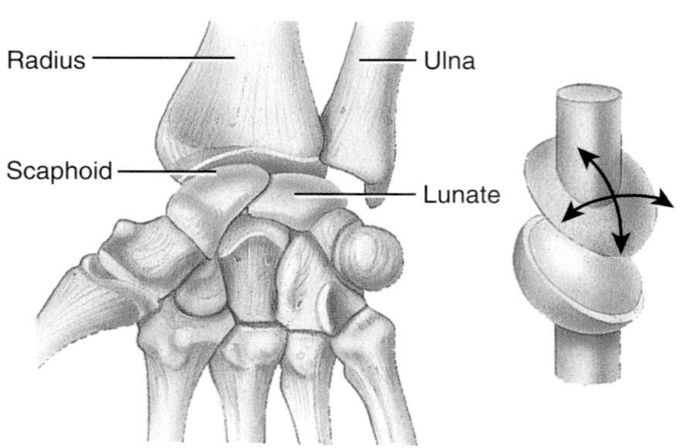

(d) Condyloid joint between radius and scaphoid and lunate bones of the carpus (wrist)

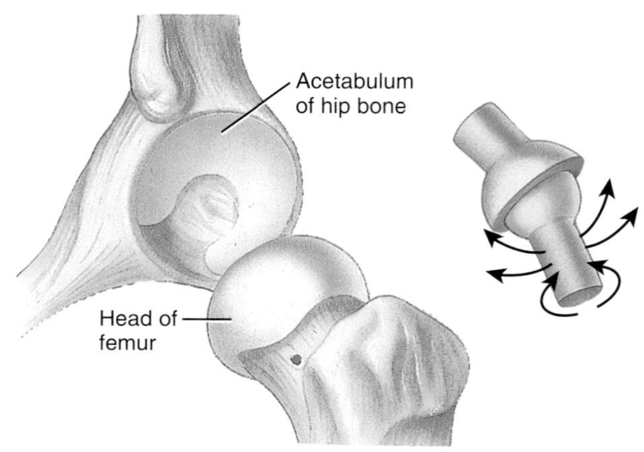

(f) Ball-and-socket joint between head of the femur and acetabulum of the hip bone

Which of the joints shown here are biaxial?

Condyloid Joints

In a **condyloid joint** (KON-di-loyd; *condyl-* = knuckle) or *ellipsoidal* joint, the convex oval-shaped projection of one bone fits into the oval-shaped depression of another bone (Figure 9.10d). A condyloid joint is *biaxial* because the movement it permits is around two axes. Flexion, extension, abduction, adduction, and circumduction are movements that can occur at condyloid joints. Examples of condyloid joints are the wrist and metacarpophalangeal joints for the second through fifth digits.

Saddle Joints

In a **saddle joint,** the articular surface of one bone is saddle-shaped, and the articular surface of the other bone fits into the "saddle" as a sitting rider would sit (Figure 9.10e). A saddle joint is a modified condyloid joint in which the movement is somewhat freer. Saddle joints are *biaxial,* allowing flexion, extension, abduction, adduction, and circumduction to occur. An example of a saddle joint is the carpometacarpal joint between the trapezium of the carpus and metacarpal of the thumb.

Ball-and-Socket Joints

A **ball-and-socket joint** consists of the ball-like surface of one bone fitting into a cuplike depression of another bone (Figure 9.10f). Such joints are *multiaxial (polyaxial)* because they permit movement around three axes plus all directions in between. Therefore, flexion, extension, abduction, adduction, circumduction, and rotation can occur at ball-and-socket

TABLE 9.2 Summary of Structural and Functional Classifications of Joints

Structural Classification	Description	Functional Classification	Example
Fibrous	**No synovial cavity; articulating bones held together by fibrous connective tissue.**		
Suture	Articulating bones united by a thin layer of dense fibrous connective tissue, found between bones of the skull. With age, some sutures are replaced by a synostosis, in which separate cranial bones fuse into a single bone.	Synarthrosis (immovable).	Coronal suture.
Syndesmosis	Articulating bones united by dense fibrous connective tissue, either a ligament or an interosseous membrane.	Amphiarthrosis (slightly movable).	Distal tibiofibular joint.
Gomphosis	Articulating bones united by periodontal ligament; cone-shaped peg fits into a socket.	Synarthrosis.	At roots of teeth in alveoli (sockets) of maxillae and mandible.
Cartilaginous	**No synovial cavity; articulating bones united by cartilage.**		
Synchondrosis	Connecting material is hyaline cartilage; becomes a synostosis when bone elongation ceases.	Synarthrosis.	Epiphyseal plate between the diaphysis and epiphysis of a long bone.
Symphysis	Connecting material is a broad, flat disc of . fibrocartilage	Amphiarthrosis.	Intervertebral joints and pubic symphysis.
Synovial	**Characterized by a synovial cavity, articular cartilage, and an articular capsule; may contain accessory ligaments, articular discs, and bursae.**		
Planar	Articulated surfaces are flat or slightly curved.	Nonaxial diarthrosis (freely movable); gliding motion.	Intercarpal, intertarsal, sternocostal (between sternum and the second–seventh pairs of ribs), and vertebrocostal joints.
Hinge	Convex surface fits into a concave surface.	Monaxial diarthrosis; flexion and extension.	Elbow, ankle, and interphalangeal joints.
Pivot	Rounded or pointed surface fits into a ring formed partly by bone and partly by a ligament.	Monaxial diarthrosis; rotation.	Atlanto-axial and radioulnar joints.
Condyloid	Oval-shaped projection fits into an oval-shaped depression.	Biaxial diarthrosis; flexion, extension,abduction, adduction, and circumduction.	Radiocarpal and metacarpophalangeal joints.
Saddle	Articular surface of one bone is saddle-shaped, and the articular surface of the other bone "sits" in the saddle.	Biaxial diarthrosis; flexion, extension, abduction, adduction, and circumduction.	Carpometarcarpal joint between trapezium and thumb.
Ball-and-socket	Ball-like surface fits into a cuplike depression.	Multiaxial diarthrosis; flexion, extension, abduction, adduction, circumduction, and rotation.	Shoulder and hip joints.

joints. Examples of functional ball-and-socket joints are the shoulder and hip joints. At the shoulder joint, the head of the humerus fits into the glenoid cavity of the scapula. At the hip joint, the head of the femur fits into the acetabulum of the hip bone.

Table 9.2 summarizes the structural and functional categories of joints.

▶ **CHECKPOINT**

10. Which types of joints are nonaxial, monaxial, biaxial, and multiaxial?

FACTORS AFFECTING CONTACT AND RANGE OF MOTION AT SYNOVIAL JOINTS

▶ **OBJECTIVE**

Describe six factors that influence the type of movement and range of motion possible at a synovial joint.

The articular surfaces of synovial joints contact one another and determine the type and range of motion that is possible. **Range of motion (ROM)** refers to the range, measured in degrees of a circle, through which the bones of a joint can be moved. The following factors contribute to keeping the articular surfaces in contact and affect range of motion:

1. ***Structure or shape of the articulating bones.*** The structure or shape of the articulating bones determines how closely they can fit together. The articular surfaces of some bones have a complementary relationship. This spatial relationship is very obvious at the hip joint, where the head of the femur articulates with the acetabulum of the hip bone. An interlocking fit allows rotational movement.

2. ***Strength and tension (tautness) of the joint ligaments.*** The different components of a fibrous capsule are tense or taut only when the joint is in certain positions. Tense ligaments not only restrict the range of motion but also direct the movement of the articulating bones with respect to each other. In the knee joint, for example, the anterior cruciate ligament is taut and the posterior cruciate ligament is loose when the knee is straightened, and the reverse occurs when the knee is bent.

3. ***Arrangement and tension of the muscles.*** Muscle tension reinforces the restraint placed on a joint by its ligaments, and thus restricts movement. A good example of the effect of muscle tension on a joint is seen at the hip joint. When the thigh is flexed with the knee extended, the movement is restricted by the tension of the hamstring muscles on the posterior surface of the thigh. But if the knee is flexed, the tension on the hamstring muscles is lessened, and the thigh can be raised farther.

4. ***Contact of soft parts.*** The point at which one body surface contacts another may limit mobility. For example, if you bend your arm at the elbow, it can move no farther after the anterior surface of the forearm meets with and presses against the biceps brachii muscle of the arm. Joint movement may also be restricted by the presence of adipose tissue.

5. ***Hormones.*** Joint flexibility may also be affected by hormones. For example, relaxin, a hormone produced by the placenta and ovaries, increases the flexibility of the fibrocartilage of the pubic symphysis and loosens the ligaments between the sacrum, hip bone, and coccyx toward the end of pregnancy. These changes permit expansion of the pelvic outlet, which assists in delivery of the baby.

6. ***Disuse.*** Movement at a joint may be restricted if a joint has not been used for an extended period. For example, if an elbow joint is immobilized by a cast, range of motion at the joint may be limited for a time after the cast is removed. Disuse may also result in decreased amounts of synovial fluid, diminished flexibility of ligaments and tendons, and *muscular atrophy,* a reduction in size or wasting of a muscle.

▶ **CHECKPOINT**

11. How do the strength and tension of ligaments determine range of motion?

SELECTED JOINTS OF THE BODY

In Chapters 7 and 8, we discussed the major bones and their markings. In this chapter we have examined how joints are classified according to their structure and function, and we have introduced the movements that occur at joints. Table 9.3 (selected joints of the axial skeleton) and Table 9.4 on page 274 (selected joints of the appendicular skeleton) will help you integrate the information you have learned in all three chapters. These tables list some of the major joints of the body according to their articular components (the bones that enter into their formation), their structural and functional classification, and the type(s) of movement that occurs at each joint.

Next we examine in detail several selected joints of the body in a series of exhibits. Each exhibit considers a specific synovial joint and contains (1) a definition—a description of the type of joint and the bones that form the joint; (2) the anatomical components—a description of the major connecting ligaments, articular disc, articular capsule, and other distinguishing features of the joint; and (3) the joint's possible movements. Each exhibit also refers you to a figure that illustrates the joint. The joints described are the temporomandibular joint (TMJ), shoulder (humeroscapular or glenohumeral) joint, elbow joint, hip (coxal) joint, and knee (tibiofemoral) joint. Because these joints are described in Exhibits 9.1 through 9.5, they are not included in Tables 9.3 and 9.4.

TABLE 9.3 Selected Joints of the Axial Skeleton

Joint	Articular Components	Classification	Movements
Suture	Between skull bones.	*Structural:* fibrous. *Functional:* synarthrosis.	None.
Atlanto-occipital	Between superior articular facets of atlas and occipital condyles of occipital bone.	*Structural:* synovial (condyloid). *Functional:* diarthrosis.	Flexion and extension of head and slight lateral flexion of head to either side.
Atlanto-axial	(1) Between dens of axis and anterior arch of atlas and (2) between lateral masses of atlas and axis.	*Structural:* synovial (pivot) between dens and anterior arch, and synovial (planar) between lateral masses. *Functional:* diarthrosis.	Rotation of head.
Intervertebral	(1) Between vertebral bodies and (2) between vertebral arches.	*Structural:* cartilaginous (symphysis) between vertebral bodies, and synovial (planar) between vertebral arches. *Functional:* amphiarthrosis between vertebral bodies, and diarthrosis between vertebral arches.	Flexion, extension, lateral flexion, and rotation of vertebral column.
Vertebrocostal	(1) Between facets of heads of ribs and facets of bodies of adjacent thoracic vertebrae and intervertebral discs between them and (2) between articular part of tubercles of ribs and facets of transverse processes of thoracic vertebrae.	*Structural:* synovial (planar). *Functional:* diarthrosis.	Slight gliding.
Sternocostal	Between sternum and first seven pairs of ribs.	*Structural:* cartilaginous (synchondrosis) between sternum and first pair of ribs, and synovial (planar) between sternum and second through seventh pairs of ribs. *Functional:* synarthrosis between sternum and first pair of ribs, and diarthrosis between sternum and second through seventh pairs of ribs.	None between sternum and first pair of ribs; slight gliding between sternum and second through seventh pairs of ribs.
Lumbosacral	(1) Between body of fifth lumbar vertebra and base of sacrum and (2) between inferior articular facets of fifth lumbar vertebra and superior articular facets of first vertebra of sacrum.	*Structural:* cartilaginous (symphysis) between body and base, and synovial (planar) between articular facets. *Functional:* amphiarthrosis between body and base, and diarthrosis between articular facets.	Flexion, extension, lateral flexion, and rotation of vertebral column.

TABLE 9.4 Selected Joints of the Appendicular Skeleton

Joint	Articular Components	Classification	Movements
Sternoclavicular	Between sternal end of clavicle, manubrium of sternum, and first costal cartilage.	*Structural:* synovial (planar and pivot). *Functional:* diarthrosis.	Gliding, with limited movements in nearly every direction.
Acromioclavicular	Between acromion of scapula and acromial end of clavicle.	*Structural:* synovial (planar). *Functional:* diarthrosis.	Gliding and rotation of scapula on clavicle.
Radioulnar	Proximal radioulnar joint between head of radius and radial notch of ulna; distal radioulnar joint between ulnar notch of radius and head of ulna.	*Structural:* synovial (pivot). *Functional:* diarthrosis.	Rotation of forearm.
Wrist (radiocarpal)	Between distal end of radius and scaphoid, lunate, and triquetrum of carpus.	*Structural:* synovial (condyloid). *Functional:* diarthrosis.	Flexion, extension, abduction, adduction, circumduction, and slight hyperextension of wrist.
Intercarpal	Between proximal row of carpal bones, distal row of carpal bones, and between both rows of carpal bones (midcarpal joints).	*Structural:* synovial (planar), except for hamate, scaphoid, and lunate (midcarpal) joint, which is synovial (saddle). *Functional:* diarthrosis.	Gliding plus flexion, extension, abduction, adduction, and slight rotation at midcarpal joints.
Carpometacarpal	Carpometacarpal joint of thumb between trapezium of carpus and first metacarpal; carpometacarpal joints of remaining digits formed between carpus and second through fifth metacarpals.	*Structural:* synovial (saddle) at thumb and synovial (planar) at remaining digits. *Functional:* diarthrosis.	Flexion, extension, abduction, adduction, and circumduction at thumb, and gliding at remaining digits.
Metacarpophalangeal and **metatarsophalangeal**	Between heads of metacarpals (or metatarsals) and bases of proximal phalanges.	*Structural:* synovial (condyloid). *Functional:* diarthrosis.	Flexion, extension, abduction, adduction, and circumduction of phalanges.
Interphalangeal	Between heads of phalanges and bases of more distal phalanges.	*Structural:* synovial (hinge). *Functional:* diarthrosis.	Flexion and extension of phalanges.
Sacroiliac	Between auricular surfaces of sacrum and ilia of hip bones.	*Structural:* synovial (planar). *Functional:* diarthrosis.	Slight gliding (even more so during pregnancy).
Pubic symphysis	Between anterior surfaces of hip bones.	*Structural:* cartilaginous (symphysis). *Functional:* amphiarthrosis.	Slight movements (even more so during pregnancy).
Tibiofibular	Proximal tibiofibular joint between lateral condyle of tibia and head of fibula; distal tibiofibular joint between distal end of fibula and fibular notch of tibia.	*Structural:* synovial (planar) at proximal joint, and fibrous (syndesmosis) at distal joint. *Functional:* diarthrosis at proximal joint, and amphiarthrosis at distal joint.	Slight gliding at proximal joint, and slight rotation of fibula during dorsiflexion of foot.
Ankle (talocrural)	(1) Between distal end of tibia and its medial malleolus and talus and (2) between lateral malleolus of fibula and talus.	*Structural:* synovial (hinge). *Functional:* diarthrosis.	Dorsiflexion and plantar flexion of foot.
Intertarsal	Subtalar joint between talus and calcaneus of tarsus; talocalcaneonavicular joint between talus and calcaneus and navicular of tarsus; calcaneocuboid joint between calcaneus and cuboid of tarsus.	*Structural:* synovial (planar) at subtalar and calcaneocuboid joints, and synovial at talocalcaneonavicular joint. *Functional:* diarthrosis.	Inversion and eversion of foot.
Tarsometatarsal	Between three cuneiforms of tarsus and bases of five metatarsal bones.	*Structural:* synovial (planar). *Functional:* diarthrosis.	Slight gliding.

► OBJECTIVE

Describe the anatomical components of the temporomandibular joint and explain the movements that can occur at this joint.

Definition

The **temporomandibular joint (TMJ)** is a combined hinge and planar joint formed by the condylar process of the mandible and the mandibular fossa and articular tubercle of the temporal bone. The temporomandibular joint is the only movable joint between skull bones; all other skull joints are sutures and therefore immovable.

Anatomical Components

1. ***Articular disc (meniscus).*** Fibrocartilage disc that separates the joint cavity into superior and inferior compartments, each with a synovial membrane (Figure 9.11c).
2. ***Articular capsule.*** Thin, fairly loose envelope around the circumference of the joint (Figure 9.11a, b).
3. ***Lateral ligament.*** Two short bands on the lateral surface of the articular capsule that extend inferiorly and posteriorly from the inferior border and tubercle of the zygomatic process of the temporal bone to the lateral and posterior aspect of the neck of the mandible. The lateral ligament is covered by the parotid gland and helps prevent displacement of the mandible (Figure 9.11a).
4. ***Sphenomandibular ligament.*** Thin band that extends inferiorly and anteriorly from the spine of the sphenoid bone to the ramus of the mandible (Figure 9.11b).
5. ***Stylomandibular ligament.*** Thickened band of deep cervical fascia that extends from the styloid process of the temporal bone to the inferior and posterior border of the ramus of the mandible. This ligament separates the parotid gland from the submandibular gland (Figure 9.11a, b).

Movements

In the temporomandibular joint, only the mandible moves because the temporal bone is firmly anchored to other bones of the skull by sutures. Accordingly, the mandible may function in depression (jaw opening) and elevation (jaw closing), which occurs in the inferior compartment, and protraction, retraction, lateral displacement, and slight rotation, which occur in the superior compartment (see Figure 9.9a–d).

Dislocated Mandible

A **dislocation** (dis′-lō-KĀ-shun; *dis-* = apart) or *luxation* (luks-Ā-shun; *luxatio* = dislocation) is the displacement of a bone from a joint with tearing of ligaments, tendons, and articular capsules. It is usually caused by a blow or fall, although unusual physical effort may be a factor. For example, if the condylar processes of the mandible pass anterior to the articular tubercles when you yawn or take a large bite, a dislocated mandible (anterior displacement) may occur. When the mandible is displaced in this manner, the mouth remains wide open and the person is unable to close it. This may be corrected by pressing the thumbs downward on the lower molar teeth and pushing the mandible backward. Other causes of a dislocated mandible include a lateral blow to the chin when the mouth is open and a fracture of the mandible. ■

► CHECKPOINT

What distinguishes the temporomandibular joint from the other joints of the skull?

Figure 9.11 **Right temporomandibular joint (TMJ).**

The TMJ is the only movable joint between skull bones.

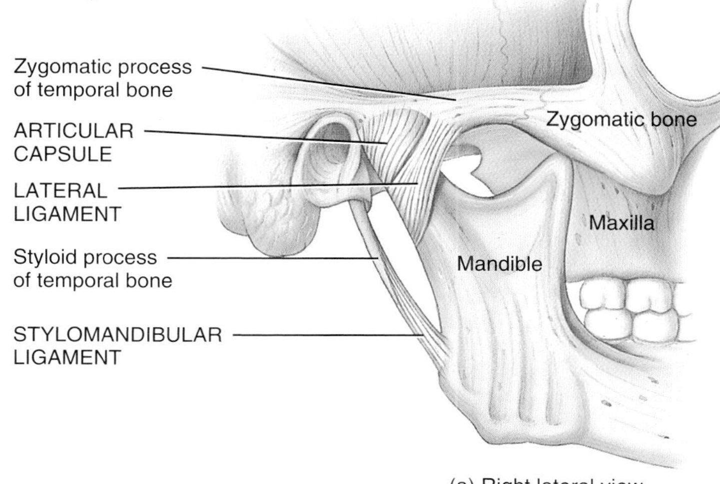

(a) Right lateral view

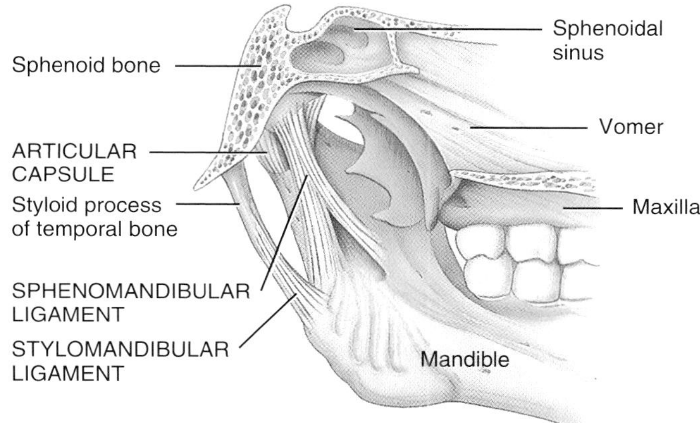

(b) Medial view

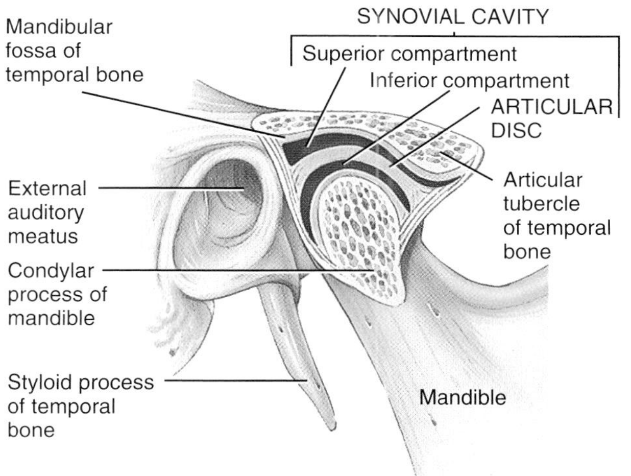

(c) Sagittal section

Which ligament prevents displacement of the mandible?

▶ **OBJECTIVE**

Describe the anatomical components of the shoulder joint and the movements that can occur at this joint.

Definition

The **shoulder joint** is a ball-and-socket joint formed by the head of the humerus and the glenoid cavity of the scapula. It also is referred to as the *humeroscapular* or *glenohumeral joint.*

Anatomical Components

1. ***Articular capsule.*** Thin, loose sac that completely envelops the joint and extends from the glenoid cavity to the anatomical neck of the humerus. The inferior part of the capsule is its weakest area (Figure 9.12).
2. ***Coracohumeral ligament.*** Strong, broad ligament that strengthens the superior part of the articular capsule and extends from the coracoid process of the scapula to the greater tubercle of the humerus (Figure 9.12a, b).
3. ***Glenohumeral ligaments.*** Three thickenings of the articular capsule over the anterior surface of the joint that extend from the glenoid cavity to the lesser tubercle and anatomical neck of the humerus. These ligaments are often indistinct or absent and provide only minimal strength (Figure 9.12a, b).
4. ***Transverse humeral ligament.*** Narrow sheet extending from the greater tubercle to the lesser tubercle of the humerus (Figure 9.12a).
5. ***Glenoid labrum.*** Narrow rim of fibrocartilage around the edge of the glenoid cavity that slightly deepens and enlarges the glenoid cavity (Figure 9.12b, c).
6. ***Bursae.*** Four *bursae* (see page 263) are associated with the shoulder joint. They are the *subscapular bursa* (Figure 9.12a), *subdeltoid bursa, subacromial bursa* (Figure 9.12a–c), and *subcoracoid bursa.*

Movements

The shoulder joint allows flexion, extension, abduction, adduction, medial rotation, lateral rotation, and circumduction of the arm (see Figures 9.5–9.8). It has more freedom of movement than any other joint of the body. This freedom results from the looseness of the articular capsule and shallowness of the glenoid cavity in relation to the large size of the head of the humerus.

Although the ligaments of the shoulder joint strengthen it to some extent, most of the strength results from the muscles that surround the joint, especially the *rotator cuff muscles.* These muscles (supraspinatus, infraspinatus, teres minor, and subscapularis) join the scapula to the humerus (see also Figure 11.15 on pages 364–365). The tendons of the rotator cuff muscles encircle the joint (except for the inferior portion) and fuse with the articular capsule. The rotator cuff muscles work as a group to hold the head of the humerus in the glenoid cavity.

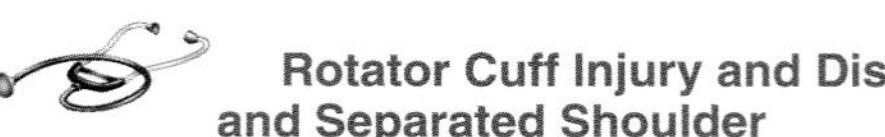

Rotator Cuff Injury and Dislocated and Separated Shoulder

Rotator cuff injury is a strain or tear in the rotator cuff muscles and is a common injury among baseball pitchers, volleyball players, racket sports players, swimmers and violinists, due to shoulder movements that involve vigorous circumduction. It also occurs as a result of wear and tear, aging, trauma, poor posture, improper lifting, and repetitive motions in certain jobs, such as placing items on a shelf above your head. Most often, there is tearing of the supraspinatus muscle tendon of the rotator cuff. This tendon is especially predisposed to wear-and-tear because of its location between the head of the humerus and acromion of the scapula, which compresses the tendon during shoulder movements.

Figure 9.12 Right shoulder (humeroscapular or glenohumeral) joint. (See Tortora, *A Photographic Atlas of the Human Body, Second Edition,* Figures 4.1 and 4.2.)

Most of the stability of the shoulder joint results from the arrangement of the rotator cuff muscles.

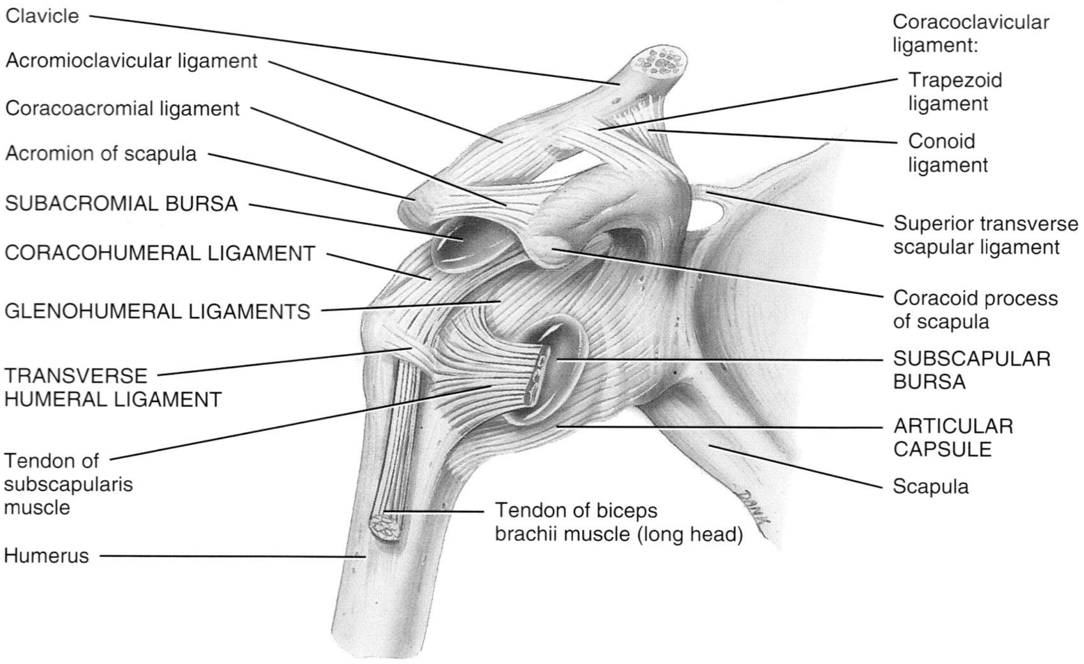

(a) Anterior view

The joint most commonly dislocated in adults is the shoulder joint because its socket is quite shallow and the bones are held together by supporting muscles. Usually in a **dislocated shoulder,** the head of the humerus becomes displaced inferiorly, where the articular capsule is least protected. Dislocations of the mandible, elbow, fingers, knee, or hip are less common.

A **separated shoulder** refers to an injury of the acromioclavicular joint, a joint formed by the acromion of the scapula and the acromial end of the clavicle. This condition is usually the result of forceful trauma to the joint, as when the shoulder strikes the ground in a fall. ■

▶ CHECKPOINT

Which tendons at the shoulder joint of a baseball pitcher are most likely to be torn due to excessive circumduction?

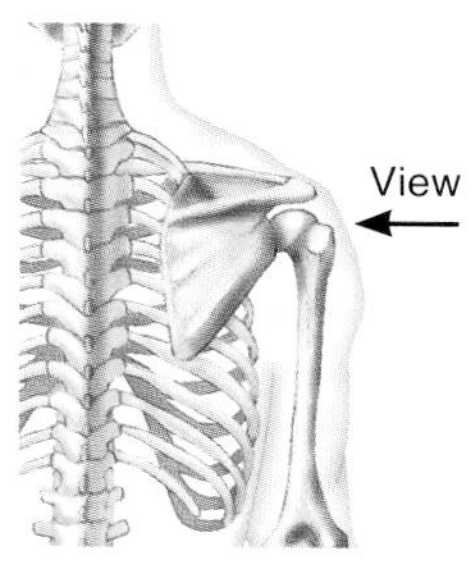

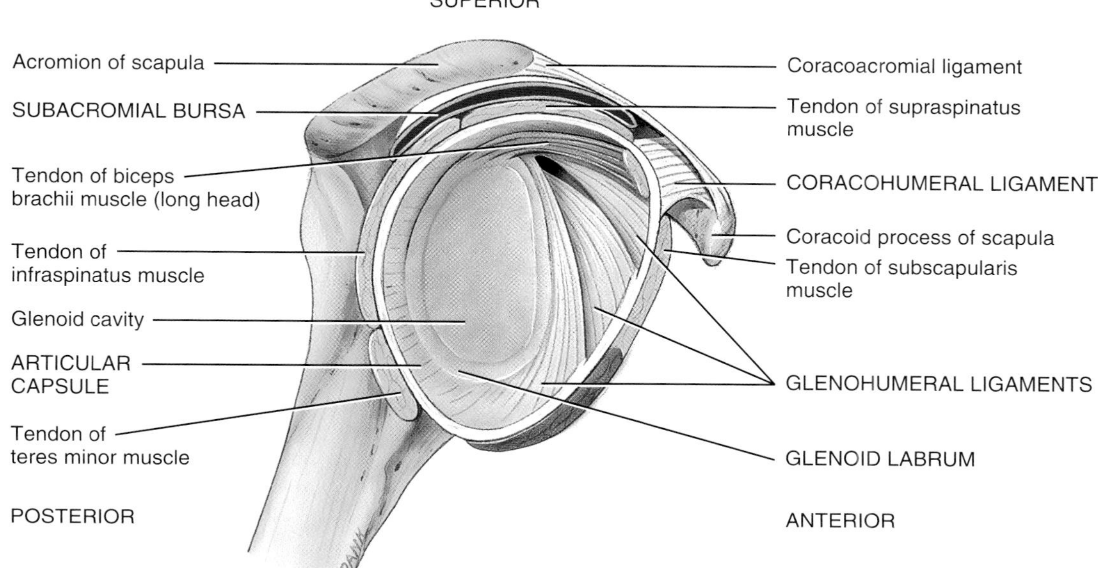

(b) Lateral view (opened)

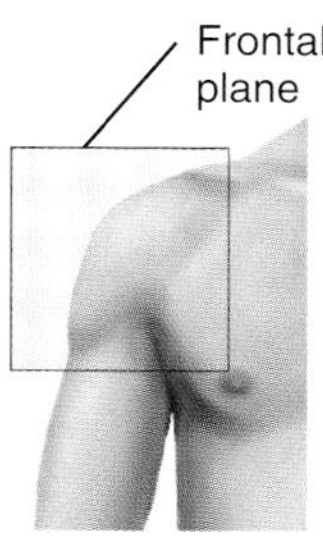

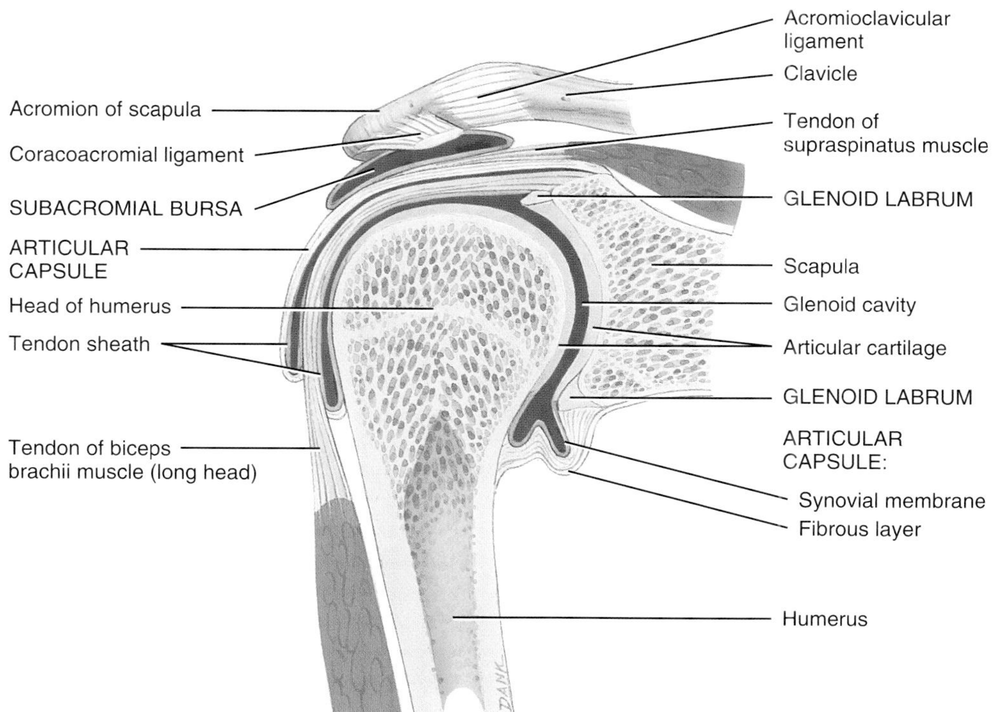

(c) Frontal section

Why does the shoulder joint have more freedom of movement than any other joint of the body?

▶ **OBJECTIVE**

Describe the anatomical components of the elbow joint and the movements that can occur at this joint.

Definition

The **elbow joint** is a hinge joint formed by the trochlea of the humerus, the trochlear notch of the ulna, and the head of the radius.

Anatomical Components

1. ***Articular capsule.*** The anterior part of the articular capsule covers the anterior part of the elbow joint, from the radial and coronoid fossae of the humerus to the coronoid process of the ulna and the annular ligament of the radius. The posterior part extends from the capitulum, olecranon fossa, and lateral epicondyle of the humerus to the annular ligament of the radius, the olecranon of the ulna, and the ulna posterior to the radial notch (Figure 9.13a, b).
2. ***Ulnar collateral ligament.*** Thick, triangular ligament that extends from the medial epicondyle of the humerus to the coronoid process and olecranon of the ulna (Figure 9.13a).
3. ***Radial collateral ligament.*** Strong, triangular ligament that extends from the lateral epicondyle of the humerus to the annular ligament of the radius and the radial notch of the ulna (Figure 9.13b).

Figure 9.13 Right elbow joint. (See Tortora, *A Photographic Atlas of the Human Body, Second Edition,* Figures 4.3 and 4.4.)

The elbow joint is formed by parts of three bones: humerus, ulna, and radius.

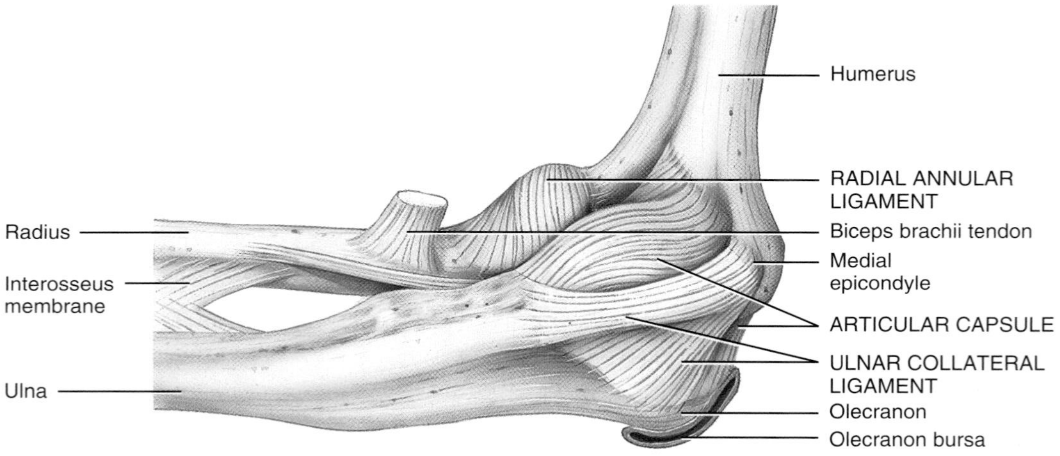

(a) Medial aspect

Movements

The elbow joint allows flexion and extension of the forearm (see Figure 9.5c).

Tennis Elbow, Little-League Elbow, and Dislocation of the Radial Head

Tennis elbow most commonly refers to pain at or near the lateral epicondyle of the humerus, usually caused by an improperly executed backhand. The extensor muscles strain or sprain, resulting in pain. **Little-league elbow** typically develops as a result of a heavy pitching schedule and/or a schedule that involves throwing curve balls, especially among youngsters. In this disorder, the elbow may enlarge, fragment, or separate.

A **dislocation of the radial head (nursemaid's elbow)** is the most common upper limb dislocation in children. In this injury, the head of the slides past or ruptures the radial annular ligament, a ligament that forms a collar around the head of the radius at the proximal radioulnar joint. Dislocation is most apt to occur when a strong pull is applied to the forearm while it is extended and supinated, for instance while swinging a child around with outstretched arms. ■

▶ CHECKPOINT

At the elbow joint, which ligaments connect (a) the humerus and the ulna, and (b) the humerus and the radius?

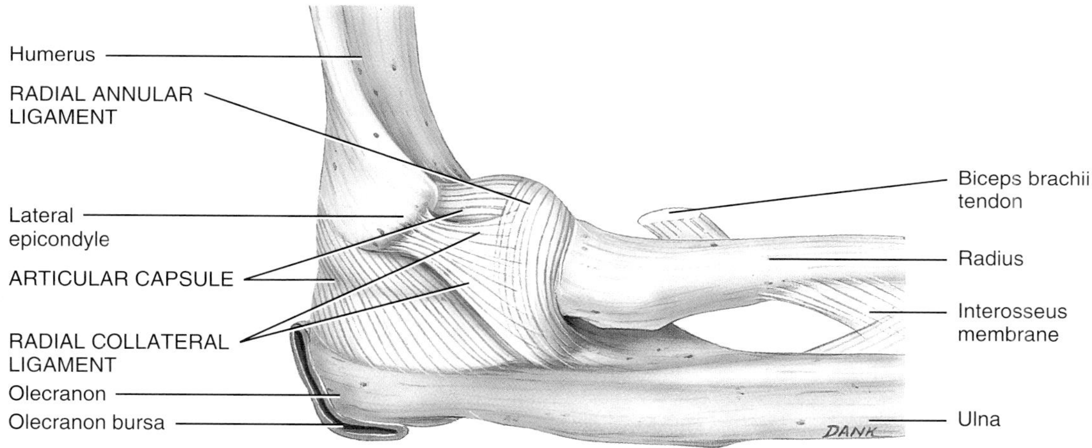

(b) Lateral aspect

Which movements are possible at a hinge joint?

▶ OBJECTIVE

Describe the anatomical components of the hip joint and the movements that can occur at this joint.

Definition

The **hip joint** *(coxal joint)* is a ball-and-socket joint formed by the head of the femur and the acetabulum of the hip bone.

Anatomical Components

1. ***Articular capsule.*** Very dense and strong capsule that extends from the rim of the acetabulum to the neck of the femur (Figure 9.14b). One of the strongest structures of the body, the capsule consists of circular and longitudinal fibers. The circular fibers, called the *zona orbicularis,* form a collar around the neck of the femur. Accessory ligaments known as the iliofemoral ligament, pubofemoral ligament, and ischiofemoral ligament reinforce the longitudinal fibers of the articular capsule.
2. ***Iliofemoral ligament.*** Thickened portion of the articular capsule that extends from the anterior inferior iliac spine of the hip bone to the intertrochanteric line of the femur (Figure 9.14a, c).
3. ***Pubofemoral ligament.*** Thickened portion of the articular capsule that extends from the pubic part of the rim of the acetabulum to the neck of the femur (Figure 9.14a).
4. ***Ischiofemoral ligament.*** Thickened portion of the articular capsule that extends from the ischial wall of the acetabulum to the neck of the femur (Figure 9.14c).
5. ***Ligament of the head of the femur.*** Flat, triangular band that extends from the fossa of the acetabulum to the fovea capitis of the head of the femur (Figure 9.14b).
6. ***Acetabular labrum.*** Fibrocartilage rim attached to the margin of the acetabulum that enhances the depth of the acetabulum. Because the diameter of the acetabular rim is smaller than that of the head of the femur, dislocation of the femur is rare (Figure 9.14b).
7. ***Transverse ligament of the acetabulum.*** Strong ligament that crosses over the acetabular notch. It supports part of the acetabular labrum and is connected with the ligament of the head of the femur and the articular capsule (Figure 9.14b).

Movements

The hip joint allows flexion, extension, abduction, adduction, circumduction, medial rotation, and lateral rotation of thigh (see Figures 9.5–9.8). The extreme stability of the hip joint is related to the very strong articular capsule and its accessory ligaments, the manner in which the femur fits into the acetabulum, and the muscles surrounding the joint. Although the shoulder and hip joints are both ball-and-socket joints, the movements at the hip joints do not have as wide a range of motion. Flexion is limited by the anterior surface of the thigh coming into contact with the anterior abdominal wall when

Figure 9.14 Right hip (coxal) joint. (See Tortora, *A Photographic Atlas of the Human Body, Second Edition,* Figure 4.5.)

The articular capsule of the hip joint is one of the strongest structures in the body.

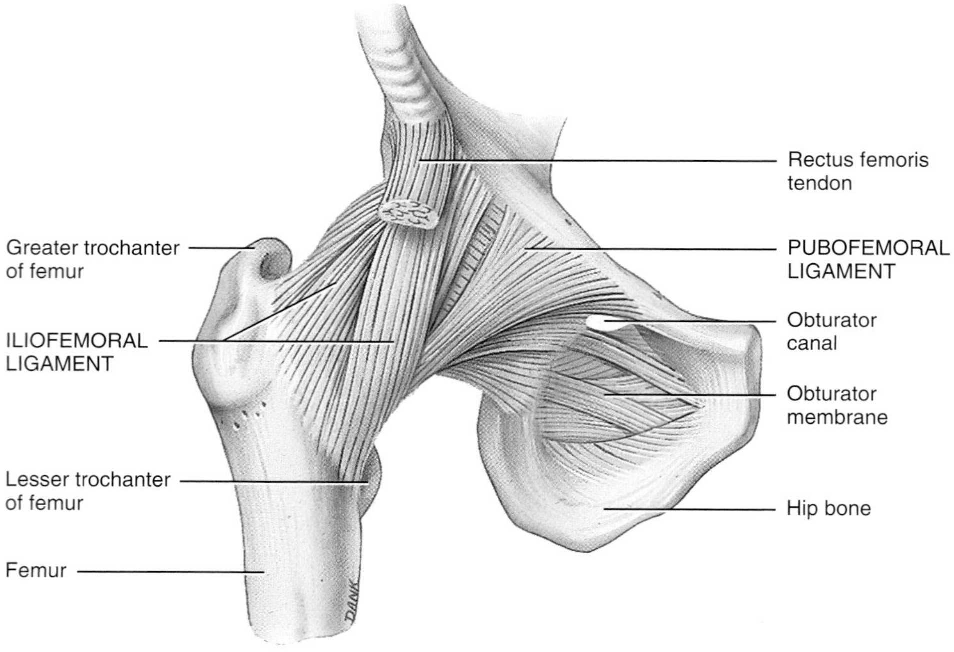

(a) Anterior view

the knee is flexed and by tension of the hamstring muscles when the knee is extended. Extension is limited by tension of the iliofemoral, pubofemoral, and ischiofemoral ligaments. Abduction is limited by the tension of the pubofemoral ligament, and adduction is limited by contact with the opposite limb and tension in the ligament of the head of the femur. Medial rotation is limited by the tension in the ischiofemoral ligament, and lateral rotation is limited by tension in the iliofemoral and pubofemoral ligaments.

▶ CHECKPOINT

What factors limit the degree of flexion and abduction at the hip joint?

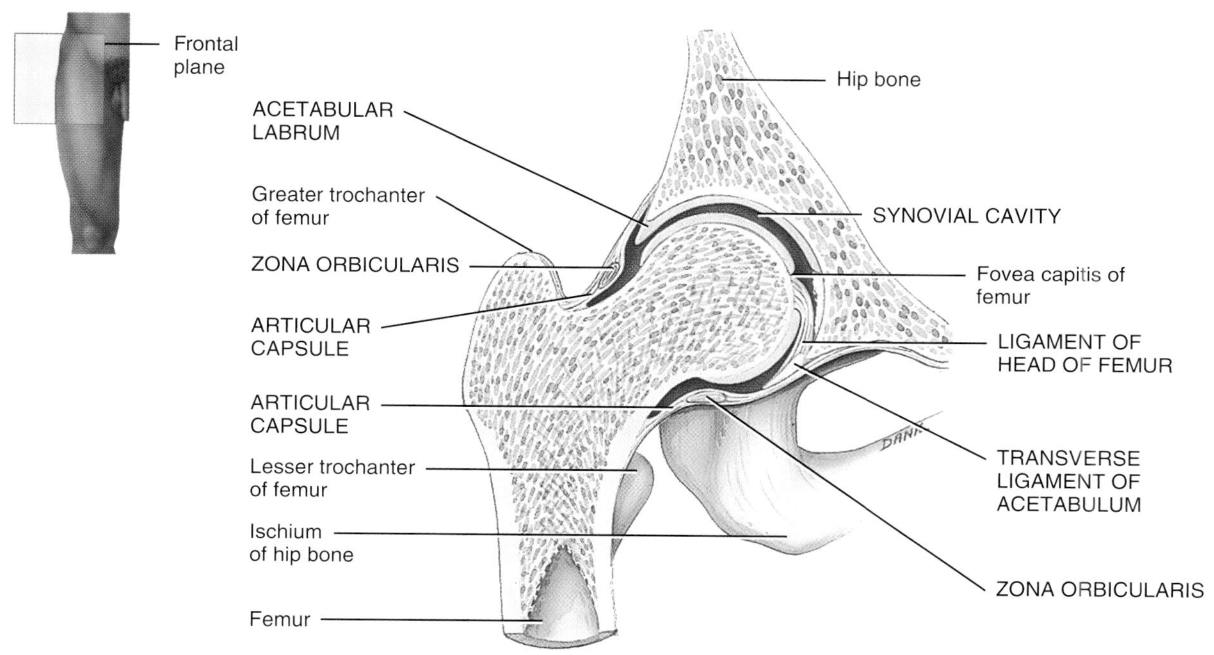

(b) Frontal section

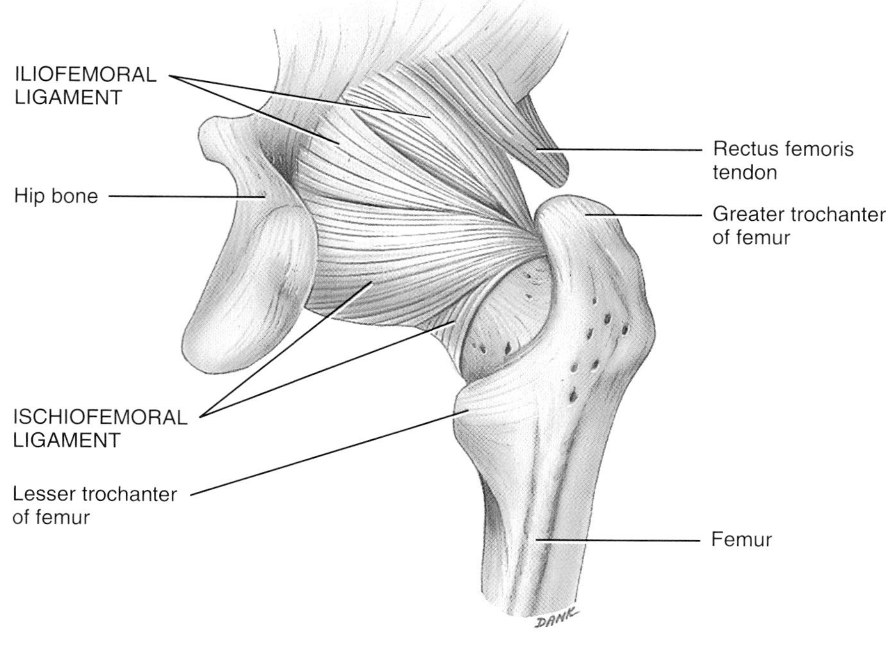

(c) Posterior view

Which ligaments limit the degree of extension that is possible at the hip joint?

EXHIBIT 9.5 KNEE JOINT (FIGURE 9.15)

▶ **OBJECTIVE**

Describe the main anatomical components of the knee joint and explain the movements that can occur at this joint.

Definition

The **knee joint** (*tibiofemoral joint*) is the largest and most complex joint of the body, actually consisting of three joints within a single synovial cavity:

1. Laterally is a tibiofemoral joint, between the lateral condyle of the femur, lateral meniscus, and lateral condyle of the tibia. It is a modified hinge joint.
2. Medially is a second tibiofemoral joint, between the medial condyle of the femur, medial meniscus, and medial condyle of the tibia. It is also a modified hinge joint.
3. An intermediate patellofemoral joint, between the patella and the patellar surface of the femur, is a planar joint.

Anatomical Components

1. ***Articular capsule.*** No complete, independent capsule unites the bones of the knee joint. The ligamentous sheath surrounding the joint consists mostly of muscle tendons or their expansions (Figure 9.15a, b). There are, however, some capsular fibers connecting the articulating bones.
2. ***Medial and lateral patellar retinacula.*** Fused tendons of insertion of the quadriceps femoris muscle and the fascia lata (deep fascia of thigh) that strengthen the anterior surface of the joint (Figure 9.15a).
3. ***Patellar ligament.*** Continuation of the common tendon of insertion of the quadriceps femoris muscle that extends from the patella to the tibial tuberosity. This ligament also strengthens the anterior surface of the joint. The posterior surface of the ligament is separated from the synovial membrane of the joint by an infrapatellar fat pad (Figure 9.15a, c).
4. ***Oblique popliteal ligament.*** Broad, flat ligament that extends from the intercondylar fossa of the femur to the head of the tibia (Figure 9.15b). The tendon of the semimembranosus muscle is superficial to the ligament and passes from the medial condyle of the tibia to the lateral condyle of the femur. The ligament and tendon strengthen the posterior surface of the joint.
5. ***Arcuate popliteal ligament.*** Extends from the lateral condyle of the femur to the styloid process of the head of the fibula. It strengthens the lower lateral part of the posterior surface of the joint (Figure 9.15b).
6. ***Tibial collateral ligament.*** Broad, flat ligament on the medial surface of the joint that extends from the medial condyle of the femur to the medial condyle of the tibia (Figure 9.15a, b, d). Tendons of the sartorius, gracilis, and semitendinosus muscles, all of which strengthen the medial aspect of the joint, cross the ligament. Because the tibial collateral ligament is firmly attached to the medial meniscus, tearing of the ligament frequently results in tearing of the meniscus and damage to the anterior cruciate ligament, described under 8a.
7. ***Fibular collateral ligament.*** Strong, rounded ligament on the lateral surface of the joint that extends from the lateral condyle of the femur to the lateral side of the head of the fibula (Figure 9.15a, b, d). It strengthens the lateral aspect of the joint. The ligament is covered by the tendon of the biceps femoris muscle. The tendon of the popliteal muscle is deep to the ligament.
8. ***Intracapsular ligaments.*** Ligaments within the capsule that connect the tibia and femur. The anterior and posterior cruciate ligaments (KROO-shē-āt = like a cross) are named based on their origins relative to the intercondylar area of the tibia. From their origins, they cross on their way to their destinations on the femur.
    a. ***Anterior cruciate ligament (ACL).*** Extends posteriorly and laterally from a point *anterior* to the intercondylar area of the tibia to the posterior part of the medial surface of the lateral condyle of the femur (Figure 9.15d). The ACL limits hyperextension of the knee and prevents the anterior sliding of the tibia on the femur. This ligament is stretched or torn in about 70% of all serious knee injuries.
    b. ***Posterior cruciate ligament (PCL).*** Extends anteriorly and medially from a depression on the *posterior* intercondylar area of the tibia and lateral meniscus to the anterior part of the lateral surface of the medial condyle of the femur (Figure 9.15d). The PCL prevents the posterior sliding of the tibia (and anterior sliding of the femur) when the knee is flexed. This is very important when walking down stairs or a steep incline.
9. ***Articular discs (menisci).*** Two fibrocartilage discs between the tibial and femoral condyles help compensate for the irregular shapes of the bones and circulate synovial fluid.
    a. ***Medial meniscus.*** Semicircular piece of fibrocartilage (C-shaped). Its anterior end is attached to the anterior intercondylar fossa of the tibia, anterior to the anterior cruciate ligament. Its posterior end is attached to the posterior intercondylar fossa of the tibia between the attachments of the posterior cruciate ligament and lateral meniscus (Figure 9.15d).
    b. ***Lateral meniscus.*** Nearly circular piece of fibrocartilage (approaches an incomplete O in shape) (Figure 9.15c, d). Its anterior end is attached anteriorly to the intercondylar eminence of the tibia, and laterally and posteriorly to the anterior cruciate ligament. Its posterior end is attached posteriorly to the intercondylar eminence of the tibia, and anteriorly to the posterior end of the medial meniscus. The medial and lateral menisci are connected to each other by the *transverse ligament* (Figure 9.15d) and to the margins of the head of the tibia by the *coronary ligaments* (not illustrated).
10. The more important *bursae* of the knee include the following:
    a. *Prepatellar bursa* between the patella and skin (Figure 9.15c).
    b. *Infrapatellar bursa* between superior part of tibia and patellar ligament (Figure 9.15a, c).
    c. *Suprapatellar bursa* between inferior part of femur and deep surface of quadriceps femoris muscle (Figure 9.15a, c).

Movements

The knee joint allows flexion, extension, slight medial rotation, and lateral rotation of leg in the flexed position (see Figures 9.5f and 9.8c).

Knee Injuries

The knee joint is the joint most vulnerable to damage because it is a mobile, weight-bearing joint and its stability depends almost entirely on its associated ligaments and muscles. Further, there is no correspondence of the articulating bones. A **swollen knee** may occur immediately or hours after an injury. Immediate swelling is due to escape of blood from damaged blood vessels adjacent to areas involving rupture of the anterior cruciate ligament, damage to synovial membranes, torn menisci, fractures, or collateral ligament sprains. Delayed swelling is due to excessive production of synovial fluid, a condition commonly referred to as "water on the knee." A common type of knee injury in football is **rupture of the tibial collateral ligaments,** often associated with tearing of the anterior cruciate ligament and medial meniscus (torn cartilage). Usually, a hard blow to the lateral side of the knee while the foot is fixed on the ground causes the damage. A **dislocated knee** refers to the displacement of the tibia relative to the femur. The most common type is dislocation anteriorly, resulting from hyperextension of the knee. A frequent consequence of a dislocated knee is damage to the popliteal artery. ■

▶ **CHECKPOINT**

What are the opposing functions of the anterior and posterior cruciate ligaments?

Figure 9.15 Right knee (tibiofemoral) joint. (See Tortora, *A Photographic Atlas of the Human Body, Second Edition,* Figures 4.6 through 4.8.)

The knee joint is the largest and most complex joint in the body.

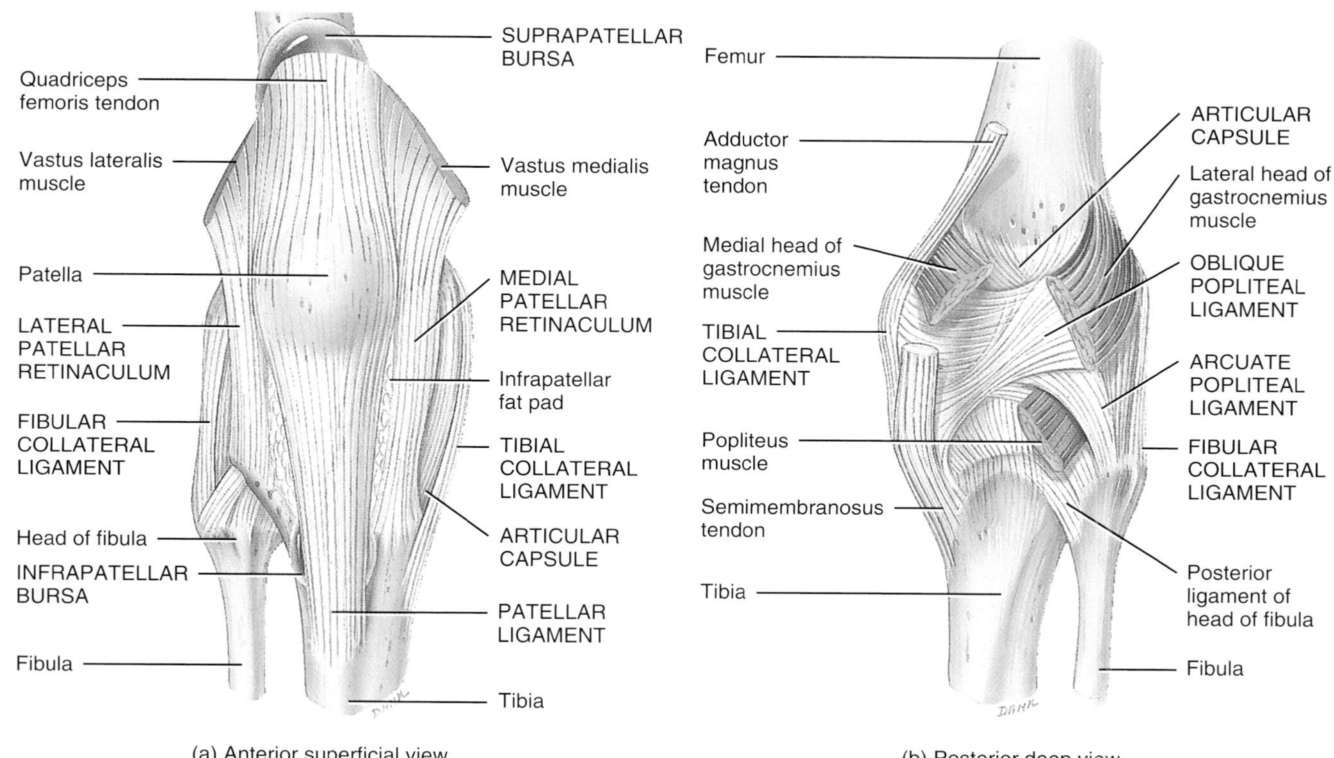

(a) Anterior superficial view

(b) Posterior deep view

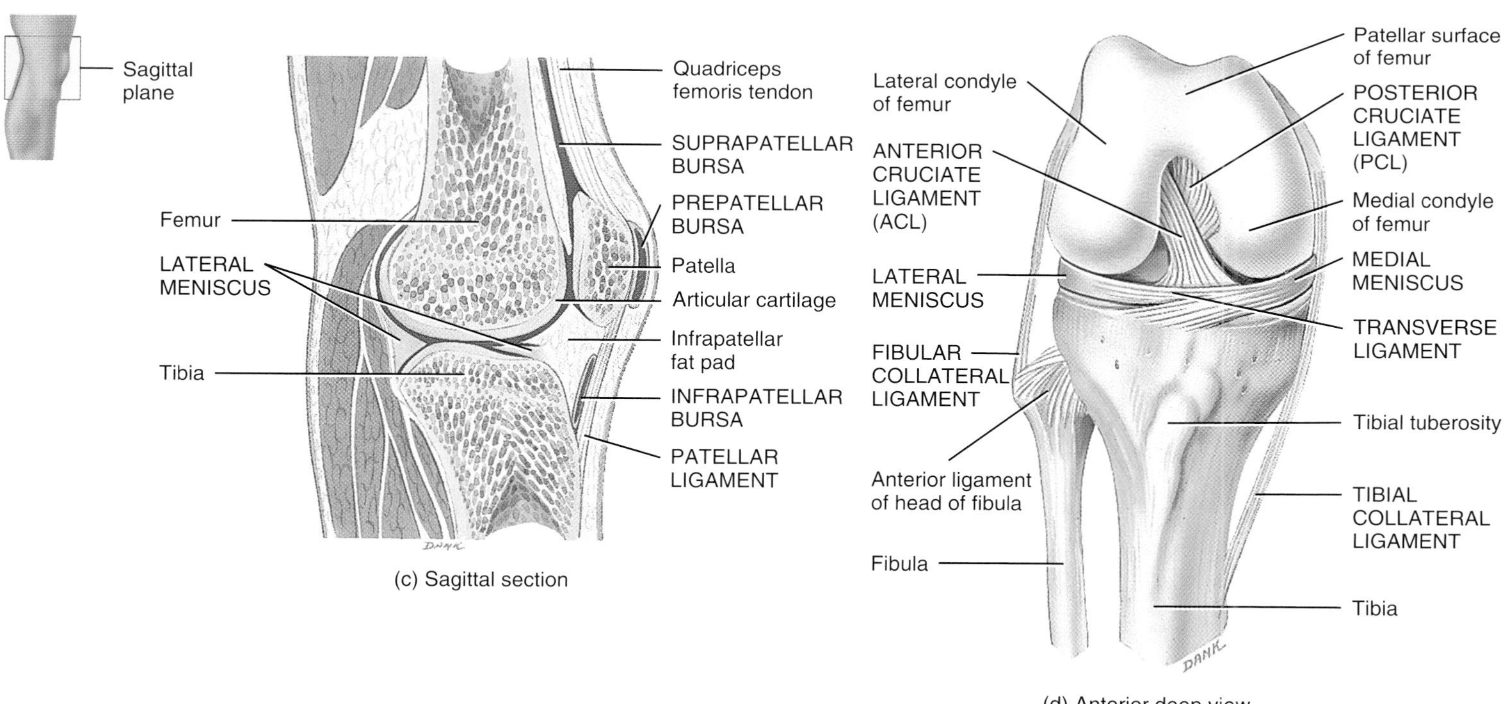

(c) Sagittal section

(d) Anterior deep view

What movement occurs at the knee joint when the quadriceps femoris (anterior thigh) muscles contract?

AGING AND JOINTS

▶ **OBJECTIVE**

Explain the effects of aging on joints.

Aging usually results in decreased production of synovial fluid in joints. In addition, the articular cartilage becomes thinner with age, and ligaments shorten and lose some of their flexibility. The effects of aging on joints are influenced by genetic factors and by wear and tear, and vary considerably from one person to another. Although degenerative changes in joints may begin as early as age 20, most changes do not occur until much later. By age 80, almost everyone develops some type of degeneration in the knees, elbows, hips, and shoulders. It is also common for elderly individuals to develop degenerative changes in the vertebral column, resulting in a hunched-over posture and pressure on nerve roots. One type of arthritis, called osteoarthritis (see the Disorders: Homeostatic Imbalances section on page 285), is at least partially age-related. Nearly everyone over age 70 has evidence of some osteoarthritic changes. Stretching and aerobic exercises that attempt to maintain full range of motion are helpful in minimizing the effects of aging. They help to maintain the effective functioning of ligaments, tendons, muscles, synovial fluid, and articular cartilage.

▶ **CHECKPOINT**

12. Which joints show evidence of degeneration in nearly all individuals as aging progresses?

ARTHROPLASTY

▶ **OBJECTIVE**

Explain the procedures involved in arthroplasty, and describe how a total hip replacement is performed.

Joints that have been severely damaged by diseases such as arthritis, or by injury, may be replaced surgically with artificial joints in a procedure referred to as **arthroplasty** (AR-thrō-plas′-tē; *arthr-* = joint; *-plasty* = plastic repair of). Although most joints in the body can undergo arthroplasty, the ones most commonly replaced are the hips, knees, and shoulders. During the procedure, the ends of the damaged bones are removed and the metal, ceramic, or plastic components are fixed in place. The goals of arthroplasty are to relieve pain and increase range of motion.

Thousands of *partial hip replacements,* involving only the femur, are performed annually. A *total hip replacement* involves both the acetabulum and head of the femur (Figure 9.16). The damaged portions of the acetabulum and the head of the femur are replaced by prefabricated prostheses (artificial devices). The acetabulum is shaped to accept the new socket, the head of the femur is removed, and the center of the femur is shaped to fit the femoral component. The acetabular component consists of polyethylene, and the femoral component is composed of cobalt-chrome, titanium alloys, or stainless steel. These materials are designed to withstand a high degree of stress as well as not provoke a response by the immune system. Once the appropriate acetabular and femoral components are selected, they are attached to the healthy portion of bone with

Figure 9.16 Total hip replacement.

In a total hip replacement, damaged portions of the acetabulum and the head of the femur are replaced by prostheses.

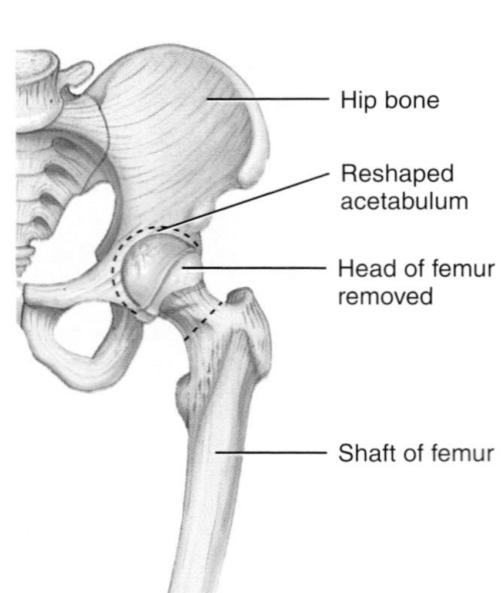

(a) Preparation for total hip replacement

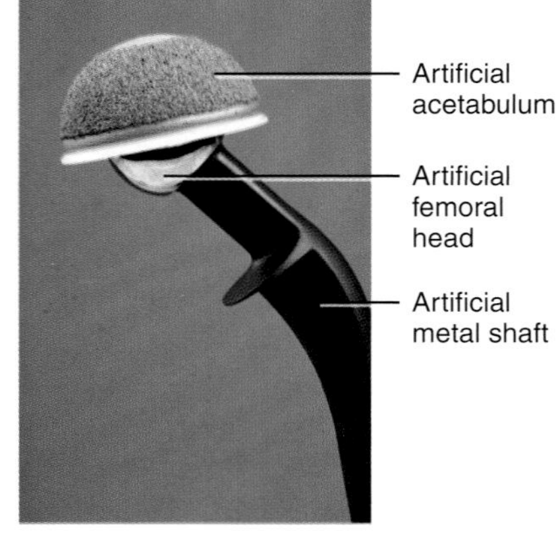

(b) Artificial hip joint

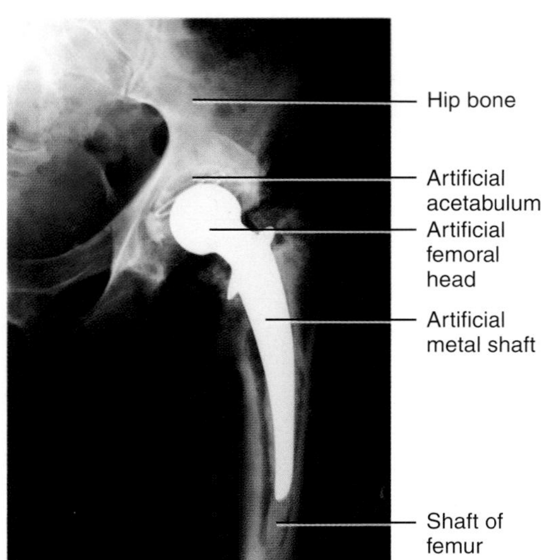

(c) Radiograph of an artificial hip joint

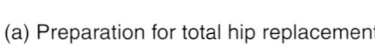

What is the purpose of arthroplasty?

acrylic cement, which forms an interlocking mechanical bond. Researchers are continually seeking to improve the strength of the cement and devise ways to stimulate bone growth around the implanted area. Potential complications of arthroplasty include infection, blood clots, loosening or dislocation of the replacement components, and nerve injury.

▶ CHECKPOINT

13. Which joints of the body most commonly undergo arthroplasty?

DISORDERS: HOMEOSTATIC IMBALANCES

Rheumatism and Arthritis

Rheumatism (ROO-ma-tizm) is any painful disorder of the supporting structures of the body—bones, ligaments, tendons, or muscles— that is not caused by infection or injury. **Arthritis** is a form of rheumatism in which the joints are swollen, stiff, and painful. It afflicts about 45 million people in the United States, and is the leading cause of physical disability among adults over age 65.

Osteoarthritis

Osteoarthritis (OA) (os′-tē-ō-ar-THRĪ-tis) is a degenerative joint disease in which joint cartilage is gradually lost. It results from a combination of aging, obesity, irritation of the joints, muscle weakness, and wear and abrasion. Commonly known as "wear-and-tear" arthritis, osteoarthritis is the most common type of arthritis.

Osteoarthritis is a progressive disorder of synovial joints, particularly weight-bearing joints. Articular cartilage deteriorates and new bone forms in the subchondral areas and at the margins of the joint. The cartilage slowly degenerates, and as the bone ends become exposed, spurs (small bumps) of new osseous tissue are deposited on them in a misguided effort by the body to protect against the friction. These spurs decrease the space of the joint cavity and restrict joint movement. Unlike rheumatoid arthritis (described next), osteoarthritis affects mainly the articular cartilage, although the synovial membrane often becomes inflamed late in the disease. Two major distinctions between osteoarthritis and rheumatoid arthritis are that osteoarthritis first afflicts the larger joints (knees, hips) and is due to wear and tear, whereas rheumatoid arthritis first strikes smaller joints and is an active attack of the cartilage. Osteoarthritis is the most common reason for hip- and knee-replacement surgery.

Rheumatoid Arthritis

Rheumatoid arthritis (RA) is an autoimmune disease in which the immune system of the body attacks its own tissues—in this case, its own cartilage and joint linings. RA is characterized by inflammation of the joint, which causes swelling, pain, and loss of function. Usually, this form of arthritis occurs bilaterally: If one wrist is affected, the other is also likely to be affected, although often not to the same degree.

The primary symptom of rheumatoid arthritis is inflammation of the synovial membrane. If untreated, the membrane thickens, and synovial fluid accumulates. The resulting pressure causes pain and tenderness. The membrane then produces an abnormal granulation tissue, called pannus, that adheres to the surface of the articular cartilage and sometimes erodes the cartilage completely. When the cartilage is destroyed, fibrous tissue joins the exposed bone ends. The fibrous tissue ossifies and fuses the joint so that it becomes immovable—the ultimate crippling effect of rheumatoid arthritis. The growth of the granulation tissue causes the distortion of the fingers that characterizes hands of RA sufferers.

Gouty Arthritis

Uric acid (a substance that gives urine its name) is a waste product produced during the metabolism of nucleic acid (DNA and RNA) subunits. A person who suffers from **gout** (GOWT) either produces excessive amounts of uric acid or is not able to excrete as much as normal. The result is a buildup of uric acid in the blood. This excess acid then reacts with sodium to form a salt called sodium urate. Crystals of this salt accumulate in soft tissues such as the kidneys and in the cartilage of the ears and joints.

In **gouty arthritis,** sodium urate crystals are deposited in the soft tissues of the joints. Gout most often affects the joints of the feet, especially at the base of the big toe. The crystals irritate and erode the cartilage, causing inflammation, swelling, and acute pain. Eventually, the crystals destroy all joint tissues. If the disorder is untreated, the ends of the articulating bones fuse, and the joint becomes immovable. Treatment consists of pain relief (ibuprofen, naproxen, colchicine, and cortisone) followed by administration of allopurinol to keep uric acid levels low so that crystals do not form.

Lyme Disease

A spiral-shaped bacterium called *Borrelia burgdorferi* causes **Lyme disease,** named for the town of Lyme, Connecticut, where it was first reported in 1975. The bacteria are transmitted to humans mainly by deer ticks (*Ixodes dammini*). These ticks are so small that their bites often go unnoticed. Within a few weeks of the tick bite, a rash may appear at the site. Although the rash often resembles a bull's-eye target, there are many variations, and some people never develop a rash. Other symptoms include joint stiffness, fever and chills, headache, stiff neck, nausea, and low back pain. In advanced stages of the disease, arthritis is the main complication. It usually afflicts the larger joints such as the knee, ankle, hip, elbow, or wrist. Antibiotics are generally effective against Lyme disease, especially if they are given promptly. However, some symptoms may linger for years.

Ankylosing Spondylitis

Ankylosing spondylitis (ang′-ki-LŌ-sing spon′-di-LĪ-tis; *ankyle* = stiff; *spondyl* = vertebra) is an inflammatory disease of unknown origin that affects joints between vertebrae (intervertebral) and between the sacrum and hip bone (sacroiliac joint). The disease, which is more common in males, sets in between ages 20 and 40. It is characterized by pain and stiffness in the hips and lower back that progress upward along the backbone. Inflammation can lead to *ankylosis* (severe or complete loss of movement at a joint) and *kyphosis* (hunchback). Treatment consists of anti-inflammatory drugs, heat, massage, and supervised exercise.

MEDICAL TERMINOLOGY

Arthralgia (ar-THRAL-jē-a; *arthr-* = joint; *-algia* = pain) Pain in a joint.

Bursectomy (bur-SEK-tō-mē; *-ectomy* = removal of) Removal of a bursa.

Chondritis (kon-DRĪ-tis; *chondr-* = cartilage) Inflammation of cartilage.

Subluxation (sub-luks-Ā-shun) A partial or incomplete dislocation.

Synovitis (sin′-ō-VĪ-tis) Inflammation of a synovial membrane in a joint.

STUDY OUTLINE

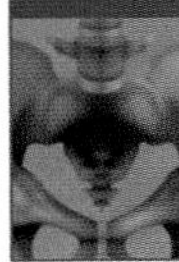

INTRODUCTION (p. 259)

1. A joint (articulation or arthrosis) is a point of contact between two bones, between bone and cartilage, or between bone and teeth.
2. A joint's structure may permit no movement, slight movement, or free movement.

JOINT CLASSIFICATIONS (p. 259)

1. Structural classification is based on the presence or absence of a synovial cavity and the type of connective tissue. Structurally, joints are classified as fibrous, cartilaginous, or synovial.
2. Functional classification of joints is based on the degree of movement permitted. Joints may be synarthroses (immovable), amphiarthroses (slightly movable), or diarthroses (freely movable).

FIBROUS JOINTS (p. 259)

1. The bones of fibrous joints are held together by fibrous connective tissue.
2. These joints include immovable sutures (found between skull bones), slightly movable syndesmoses (such as the distal tibiofibular joint), and immovable gomphoses (roots of teeth in the sockets in the mandible and maxilla).

CARTILAGINOUS JOINTS (p. 260)

1. The bones of cartilaginous joints are held together by cartilage.
2. These joints include immovable synchondroses united by hyaline cartilage (epiphyseal plates between diaphyses and epiphyses) and slightly movable symphyses united by fibrocartilage (pubic symphysis).

SYNOVIAL JOINTS (p. 261)

1. Synovial joints contain a space between bones called the synovial cavity. All synovial joints are diarthroses.
2. Other characteristics of synovial joints are the presence of articular cartilage and an articular capsule, made up of a fibrous capsule and a synovial membrane.
3. The synovial membrane secretes synovial fluid, which forms a thin, viscous film over the surfaces within the articular capsule.
4. Many synovial joints also contain accessory ligaments (extracapsular and intracapsular) and articular discs (menisci).
5. Synovial joints contain an extensive nerve and blood supply. The nerves convey information about pain, joint movements, and the degree of stretch at a joint. Blood vessels penetrate the articular capsule and ligaments.
6. Bursae are saclike structures, similar in structure to joint capsules, that alleviate friction in joints such as the shoulder and knee joints.
7. Tendon sheaths are tubelike bursae that wrap around tendons where there is considerable friction.

TYPES OF MOVEMENTS AT SYNOVIAL JOINTS (p. 263)

1. In a gliding movement, the nearly flat surfaces of bones move back-and-forth and side-to-side.
2. In angular movements, a change in the angle between bones occurs. Examples are flexion–extension, lateral flexion, hyperextension, and abduction–adduction. Circumduction refers to flexion, abduction, extension, and adduction in succession.
3. In rotation, a bone moves around its own longitudinal axis.
4. Special movements occur at specific synovial joints. Examples are elevation–depression, protraction–retraction, inversion–-eversion, dorsiflexion–plantar flexion, supination–pronation, and opposition.
5. Table 9.1 summarizes the various types of movements at synovial joints.

TYPES OF SYNOVIAL JOINTS (p. 269)

1. Subtypes of synovial joints are planar, hinge, pivot, condyloid, saddle, and ball-and-socket.
2. In a planar joint the articulating surfaces are flat, and the bones glide back-and-forth and side-to-side (nonaxial); examples are joints between carpals and tarsals.
3. In a hinge joint, the convex surface of one bone fits into the concave surface of another, and the motion is angular around one axis (monaxial); examples are the elbow, knee, and ankle joints.
4. In a pivot joint, a round or pointed surface of one bone fits into a ring formed by another bone and a ligament, and movement is rotational (monaxial); examples are the atlanto-axial and radioulnar joints.
5. In a condyloid joint, an oval projection of one bone fits into an oval cavity of another, and motion is angular around two axes (biaxial); examples include the wrist joint and metacarpophalangeal joints of the second through fifth digits.

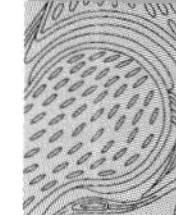

6. In a saddle joint, the articular surface of one bone is shaped like a saddle and the other bone fits into the "saddle" like a sitting rider; motion is angular around two axes (biaxial). An example is the carpometacarpal joint between the trapezium and the metacarpal of the thumb.
7. In a ball-and-socket joint, the ball-shaped surface of one bone fits into the cuplike depression of another; motion is angular and rotational around three axes and all directions in between (multiaxial). Examples include the shoulder and hip joints.
8. Table 9.2 summarizes the structural and functional categories of joints.

FACTORS AFFECTING CONTACT AND RANGE OF MOTION AT SYNOVIAL JOINTS (p. 272)

1. The ways that articular surfaces of synovial joints contact one another determines the type of movement that is possible.
2. Factors that contribute to keeping the surfaces in contact and affect range of motion are structure or shape of the articulating bones, strength and tension of the joint ligaments, arrangement and tension of the muscles, apposition of soft parts, hormones, and disuse.

SELECTED JOINTS OF THE BODY (p. 272)

1. A summary of selected joints of the body, including articular components, structural and functional classifications, and movements, is presented in Tables 9.3 and 9.4.
2. The temporomandibular joint (TMJ) is between the condyle of the mandible and mandibular fossa and articular tubercle of the temporal bone (Exhibit 9.1).
3. The shoulder (humeroscapular or glenohumeral) joint is between the head of the humerus and glenoid cavity of the scapula (Exhibit 9.2).
4. The elbow joint is between the trochlea of the humerus, the trochlear notch of the ulna, and the head of the radius (Exhibit 9.3).
5. The hip (coxal) joint is between the head of the femur and acetabulum of the hip bone (Exhibit 9.4).
6. The knee (tibiofemoral) joint is between the patella and patellar surface of the femur; the lateral condyle of the femur, the lateral meniscus, and the lateral condyle of the tibia; and the medial condyle of the femur, the medial meniscus, and the medial condyle of the tibia (Exhibit 9.5).

AGING AND JOINTS (p. 284)

1. With aging, a decrease in synovial fluid, thinning of articular cartilage, and decreased flexibility of ligaments occur.
2. Most individuals experience some degeneration in the knees, elbows, hips, and shoulders due to the aging process.

ARTHROPLASTY (p. 284)

1. Arthroplasty refers to the surgical replacement of joints.
2. The most commonly replaced joints are the hips, knees, and shoulders.

SELF-QUIZ QUESTIONS

Fill in the blanks in the following statements.

1. A point of contact between two bones, between bone and cartilage, or between bone and teeth is called a(n) ____.
2. The surgical procedure in which a severely damaged joint is replaced with an artificial joint is known as ____.

Indicate whether the following statements are true or false.

3. Menisci are fluid-filled sacs located outside of the joint cavity to ease friction between bones and softer tissue.
4. Shrugging your shoulders involves flexion and extension.
5. Synovial fluid becomes more viscous (thicker) as movement at the joint increases.

Choose the one best answer to the following questions.

6. Which of the following are structural classifications of joints? (1) amphiarthrosis, (2) cartilaginous, (3) synovial, (4) synarthrosis, (5) fibrous. (a) 1, 2, 3, 4, and 5; (b) 2 and 5; (c) 1 and 4; (d) 1, 2, 4, and 5; (e) 2, 3, and 5.
7. Which of the following joints could be classified functionally as synarthroses? (1) syndesmosis, (2) symphysis, (3) synovial, (4) gomphosis, (5) suture. (a) 1 and 2; (b) 3 and 5; (c) 1, 2, and 3; (d) 4 and 5; (e) 5 only.
8. The most common degenerative joint disease in the elderly, often caused by wear-and-tear, is (a) rheumatoid arthritis, (b) osteoarthritis, (c) rheumatism, (d) gouty arthritis, (e) ankylosing spondylitis.
9. Chewing your food involves (1) flexion, (2) extension, (3) hyperextension, (4) elevation, (5) depression. (a) 1 and 2; (b) 1 and 3; (c) 4 and 5; (d) 3 and 5; (e) 1 and 4.
10. Synovial fluid functions to (1) absorb shocks at joints, (2) lubricate joints, (3) form a blood clot in a joint injury, (4) supply oxygen and nutrients to chondrocytes, (5) provide phagocytes to remove debris from joints. (a) 1, 2, 4 and 5; (b) 1, 2, 3, 4, and 5; (c) 1, 2, and 4; (d) 3 and 4; (e) 2, 4, and 5.
11. Which of the following statements are *true* concerning a synovial joint? (1) The bones at a synovial joint are covered by a mucous membrane. (2) The articular capsule surrounds a synovial joint, encloses the synovial cavity, and unites the articulating bones. (3) The fibrous portion of the articular capsule permits considerable movement at a joint. (4) The tensile strength of the fibrous capsule helps prevent bones from disarticulating. (5) All joints contain a fibrous capsule. (a) 1, 2, 3, and 4; (b) 2, 3, 4, and 5; (c) 2, 3, and 4; (d) 1, 2, and 3; (e) 2, 4, and 5.
12. Which of the following keep the articular surfaces of synovial joints in contact and affect range of motion? (1) structure or shape of the articulating bones, (2) strength and tension of the joint ligaments, (3) arrangement and tension of muscles, (4) lack of use, (5) contact of soft parts. (a) 1, 2, 3, and 5; (b) 2, 3, 4, and 5; (c) 1, 3, 4, and 5; (d) 1, 3, and 5; (e) 1, 2, 3, 4, and 5.

13. Match the following:

____(a) a fibrous joint that unites the bones of the skull; a synarthrosis
____(b) a fibrous joint between the tibia and fibula; an amphiarthrosis
____(c) the articulation between bone and teeth
____(d) the epiphyseal plate
____(e) joint between the two pubic bones
____(f) joint with a cavity between the bones; diarthrosis
____(g) a bony joint

(1) hinge joint
(2) saddle joint
(3) ball-and-socket joint
(4) planar joint
(5) condyloid joint
(6) pivot joint

14. Match the following:

____(a) rounded or pointed surface of one bone articulates with a ring formed by another bone and a ligament; allows rotation around its own axis
____(b) articulating bone surfaces are flat or slightly curved; permit gliding movement
____(c) convex, oval projection of one bone fits into oval depression of another bone; permits movement in two axes
____(d) convex surface of one bone articulates with concave surface of another bone; permits flexion and extension
____(e) ball-shaped surface of one bone articulates with cuplike depression of another bone; permits largest degree of movement in three axes
____(f) modified condyloid joint where articulating bones resemble a rider sitting in a saddle

(1) synostosis
(2) synchondrosis
(3) syndesmosis
(4) synovial
(5) suture
(6) symphysis
(7) gomphosis

15. Match the following:

____(a) upward movement of a body part
____(b) downward movement of a body part
____(c) movement of bone toward midline
____(d) movement in which relatively flat bone surfaces move back-and-forth and side-to-side with respect to one another
____(e) movement of a body part anteriorly in the transverse plane
____(f) decrease in angle between bones
____(g) movement of an anteriorly projected body part back to the anatomical position
____(h) movement of the soles medially
____(i) movement of the soles laterally
____(j) movement of bone away from midline
____(k) action that occurs when you stand on your heels
____(l) action that occurs when you stand on your toes
____(m) movement of the forearm to turn the palm anteriorly
____(n) movement of the forearm to turn the palm posteriorly
____(o) movement of thumb across the palm to touch the tips of the fingers of the same hand
____(p) increase in angle between bones
____(q) movement of distal end of a part of the body in a circle
____(r) bone revolves around its own longitudinal axis

(1) pronation
(2) plantar flexion
(3) eversion
(4) abduction
(5) rotation
(6) retraction
(7) opposition
(8) elevation
(9) flexion
(10) adduction
(11) depression
(12) inversion
(13) gliding
(14) extension
(15) protraction
(16) dorsiflexion
(17) circumduction
(18) supination

CRITICAL THINKING QUESTIONS

1. Katie loves pretending that she's a human cannonball. As she jumps off the diving board, she assumes the proper position before she pounds into the water: head and thighs tucked against her chest; back rounded; arms pressed against her sides while her forearms, crossed in front of her shins, hold her legs tightly folded against her chest. Use the proper anatomical terms to describe the position of Katie's back, head, and limbs.
2. During football practice, Jeremiah was tackled and twisted his lower leg. There was a sharp pain, followed immediately by swelling of the knee joint. The pain and swelling worsened throughout the remainder of the afternoon until Jeremiah could barely walk. The coach told Jeremiah to see a doctor who might want to "drain the water off his knee." What was the coach referring to and what specifically do you think happened to Jeremiah's knee joint to cause these symptoms?
3. Since her stay in the hospital, elderly Aunt Agnes is now bragging to the other residents in the nursing home that she has become the "bionic woman" and is making bets that soon she will be able to "swing" her legs behind her head because she has some "new bones!" What do you suppose Agnes had done in the hospital and why?

ANSWERS TO FIGURE QUESTIONS

9.1 Functionally, sutures are classified as synarthroses because they are immovable; syndesmoses are classified as amphiarthroses because they are slightly movable.

9.2 The structural difference between a synchondrosis and a symphysis is the type of cartilage that holds the joint together: hyaline cartilage in a synchondrosis and fibrocartilage in a symphysis.

9.3 Functionally, synovial joints are diarthroses, freely movable joints.

9.4 Gliding movements occur at intercarpal joints and at intertarsal joints.

9.5 Two examples of flexion that do not occur along the sagittal plane are flexion of the thumb and lateral flexion of the trunk.

9.6 When you adduct your arm or leg, you bring it closer to the midline of the body, thus "adding" it to the trunk.

9.7 Circumduction involves flexion, abduction, extension, and adduction in continuous sequence.

9.8 The anterior surface of a bone or limb rotates toward the midline in medial rotation, and away from the midline in lateral rotation.

9.9 Bringing your arms forward until the elbows touch is an example of protraction.

9.10 Condyloid and saddle joints are biaxial joints.

9.11 The lateral ligament prevents displacement of the mandible.

9.12 The shoulder joint is the most freely movable joint in the body because of the looseness of its articular capsule and the shallowness of the glenoid cavity in relation to the size of the head of the humerus.

9.13 A hinge joint permits flexion and extension.

9.14 Tension in three ligaments—iliofemoral, pubofemoral, and ischiofemoral—limits the degree of extension at the hip joint.

9.15 Contraction of the quadriceps femoris muscle causes extension at the knee joint.

9.16 The purpose of arthroplasty is to relieve joint pain and permit greater range of motion.

Chapter 10

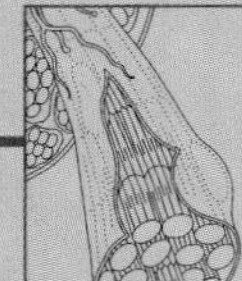

Muscular Tissue

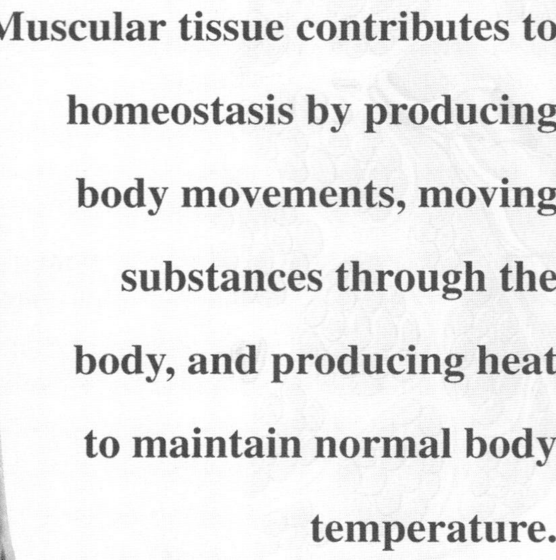

Muscular Tissue and Homeostasis

Muscular tissue contributes to homeostasis by producing body movements, moving substances through the body, and producing heat to maintain normal body temperature.

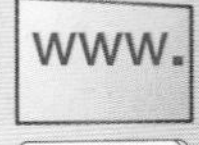

wiley.com/college/apcentral

Although bones provide leverage and form the framework of the body, they cannot move body parts by themselves. Motion results from the alternating contraction and relaxation of muscles, which make up 40–50% of total adult body weight. Your muscular strength reflects the primary function of muscle—the transformation of chemical energy into mechanical energy to generate force, perform work, and produce movement. In addition, muscle tissues stabilize body position, regulate organ volume, generate heat, and propel fluids and food matter through various body systems. The scientific study of muscles is known as **myology** (mī-OL-ō-jē; *myo-* = muscle; *-logy* = study of).

OVERVIEW OF MUSCULAR TISSUE

▶ **OBJECTIVES**

Explain the structural differences between the three types of muscular tissue.

Compare the functions and special properties of the three types of muscular tissue.

Types of Muscular Tissue

The three types of muscular tissue, skeletal, cardiac, and smooth, were introduced in Chapter 4 (see Table 4.5 on pages 135–136). Although the different types of muscular tissue share some properties, they differ from one another in their microscopic anatomy, location, and how they are controlled by the nervous and endocrine systems.

Skeletal muscle tissue is so named because most skeletal muscles move bones of the skeleton. (A few skeletal muscles attach to and move the skin or other skeletal muscles.) Skeletal muscle tissue is *striated*: Alternating light and dark bands (*striations*) are seen when the tissue is examined with a microscope (see Figure 10.4). Skeletal muscle tissue works mainly in a *voluntary* manner. Its activity can be consciously controlled by neurons (nerve cells) that are part of the somatic (voluntary) division of the nervous system. (Figure 12.1 on page 405 depicts the divisions of the nervous system.) Most skeletal muscles also are controlled subconsciously to some extent. For example, your diaphragm continues to alternately contract and relax without conscious control so that you don't stop breathing. Also, you do not need to consciously think about contracting the skeletal muscles that maintain your posture or stabilize body positions.

Only the heart contains **cardiac muscle tissue,** which forms most of the heart wall. Cardiac muscle is also *striated,* but its action is *involuntary.* The alternating contraction and relaxation of the heart is not consciously controlled. Rather, the heart beats because it has a pacemaker that initiates each contraction. This built-in rhythm is termed **autorhythmicity.** Several hormones and neurotransmitters can adjust heart rate by speeding or slowing the pacemaker.

Smooth muscle tissue is located in the walls of hollow internal structures, such as blood vessels, airways, and most organs in the abdominopelvic cavity. It is also found in the skin, attached to hair follicles. Under a microscope, this tissue lacks the striations of skeletal and cardiac muscle tissue. For this reason, it looks *nonstriated,* which is why it is referred to as *smooth.* The action of smooth muscle is usually *involuntary,* and some smooth muscle tissue, such as the muscles that propel food through your gastrointestinal tract, has autorhythmicity. Both cardiac muscle and smooth muscle are regulated by neurons that are part of the autonomic (involuntary) division of the nervous system and by hormones released by endocrine glands.

Functions of Muscular Tissue

Through sustained contraction or alternating contraction and relaxation, muscular tissue has four key functions: producing body movements, stabilizing body positions, storing and moving substances within the body, and generating heat.

1. ***Producing body movements.*** Movements of the whole body such as walking and running, and localized movements such as grasping a pencil or nodding the head, rely on the integrated functioning of bones, joints, and skeletal muscles.
2. ***Stabilizing body positions.*** Skeletal muscle contractions stabilize joints and help maintain body positions, such as standing or sitting. Postural muscles contract continuously when you are awake; for example, sustained contractions of your neck muscles hold your head upright.
3. ***Storing and moving substances within the body.*** Storage is accomplished by sustained contractions of ringlike bands of smooth muscle called *sphincters,* which prevent outflow of the contents of a hollow organ. Temporary storage of food in the stomach or urine in the urinary bladder is possible because smooth muscle sphincters close off the outlets of these organs. Cardiac muscle contractions of the heart pump blood through the blood vessels of the body. Contraction and relaxation of smooth muscle in the walls of blood vessels help adjust blood vessel diameter and thus regulate the rate of blood flow. Smooth muscle contractions also move food and substances such as bile and enzymes through the gastrointestinal tract, push gametes (sperm and oocytes) through the passageways of the reproductive systems, and propel urine through the urinary system. Skeletal muscle contractions promote the flow of lymph and aid the return of blood to the heart.
4. ***Generating heat.*** As muscular tissue contracts, it produces heat, a process known as **thermogenesis.** Much of the heat generated by muscle is used to maintain normal body temperature. Involuntary contractions of skeletal muscle, known as *shivering,* can increase the rate of heat production.

Properties of Muscular Tissue

Muscular tissue has four special properties that enable it to function and contribute to homeostasis:

1. **Electrical excitability,** a property of both muscle and nerve cells that was introduced in Chapter 4, is the ability to respond to certain stimuli by producing electrical signals called *action potentials.* Chapter 12 provides more detail about how action potentials arise; see page 418. Action potentials can travel along a cell's plasma membrane due to the presence of specific voltage-gated channels. For muscle cells, two main types of stimuli trigger action potentials. One is autorhythmic electrical signals arising in the muscular tissue itself, as in the heart's pacemaker. The other is chemical stimuli, such as neurotransmitters released by neurons, hormones distributed by the blood, or even local changes in pH.
2. **Contractility** is the ability of muscular tissue to contract forcefully when stimulated by an action potential. When a muscle contracts, it generates tension (force of contraction) while pulling on its attachment points. If the tension generated is great enough to overcome the resistance of the object to be moved, the muscle shortens and movement occurs.
3. **Extensibility** is the ability of muscular tissue to stretch without being damaged. Extensibility allows a muscle to contract forcefully even if it is already stretched. Normally, smooth muscle is subject to the greatest amount of stretching. For example, each time your stomach fills with food, the muscle in its wall is stretched. Cardiac muscle also is stretched each time the heart fills with blood.
4. **Elasticity** is the ability of muscular tissue to return to its original length and shape after contraction or extension.

This chapter focuses mainly on the structure and function of skeletal muscle tissue. Cardiac muscle and smooth muscle are examined in detail in later chapters.

▶ CHECKPOINT

1. What features distinguish the three types of muscular tissue?
2. List the general functions of muscular tissue.
3. Describe the properties of muscular tissue.

SKELETAL MUSCLE TISSUE

▶ OBJECTIVES

Explain the importance of connective tissue components, blood vessels, and nerves to skeletal muscles.

Describe the microscopic anatomy of a skeletal muscle fiber.

Distinguish thick filaments from thin filaments.

Each of your skeletal muscles is a separate organ composed of hundreds to thousands of cells, which are called **muscle fibers** because of their elongated shapes. Thus, *muscle cell* and *muscle fiber* are two terms for the same structure. Skeletal muscle also contains connective tissues surrounding muscle fibers and whole muscles, and blood vessels and nerves (Figure 10.1). To understand how contraction of skeletal muscle can generate tension, you must first understand its gross and microscopic anatomy.

Connective Tissue Components

Connective tissue surrounds and protects muscular tissue. A **fascia** (FASH-ē-a = bandage) is a sheet or broad band of fibrous connective tissue that supports and surrounds muscles and other organs of the body. The **superficial fascia (subcutaneous layer** or **hypodermis)**, which separates muscle from skin (see Figure 11.21 on page 390), is composed of areolar connective tissue and adipose tissue. It provides a pathway for nerves, blood vessels, and lymphatic vessels to enter and exit muscles. The adipose tissue of superficial fascia stores most of the body's triglycerides, serves as an insulating layer that reduces heat loss, and protects muscles from physical trauma. **Deep fascia** is dense irregular connective tissue that lines the body wall and limbs and holds muscles with similar functions together (see Figure 11.21 on page 390). Deep fascia allows free movement of muscles, carries nerves, blood vessels, and lymphatic vessels, and fills spaces between muscles.

Three layers of connective tissue extend from the deep fascia to protect and strengthen skeletal muscle (Figure 10.1). The outermost layer, encircling the entire muscle, is the **epimysium** (ep-i-MĪZ-ē-um; *epi-* = upon). **Perimysium** (per-i-MĪZ-ē-um; *peri-* = around) surrounds groups of 10 to 100 or more muscle fibers, separating them into bundles called **fascicles** (FAS-i-kuls = little bundles). Many fascicles are large enough to be seen with the naked eye. They give a cut of meat its characteristic "grain"; if you tear a piece of meat, it rips apart along the fascicles. Both epimysium and perimysium are dense irregular connective tissue. Penetrating the interior of each fascicle and separating individual muscle fibers from one another is **endomysium** (en′-dō-MĪZ-ē-um; *endo-* = within), a thin sheath of areolar connective tissue.

The epimysium, perimysium, and endomysium all are continuous with the connective tissue that attaches skeletal muscle to other structures, such as bone or another muscle. All three connective tissue layers may extend beyond the muscle fibers to form a **tendon**—a cord of dense regular connective tissue composed of parallel bundles of collagen fibers that attach a muscle to the periosteum of a bone. An example is the calcaneal (Achilles) tendon of the gastrocnemius (calf) muscle, which attaches the muscle to the calcaneus (shown in Figure 11.22c on page 394). When the connective tissue elements extend as a broad, flat layer, the tendon is called an **aponeurosis** (*apo-* = from; *neur-* = a sinew). An example is the epicranial aponeurosis on top of the skull between the frontal and occipital bellies of the occipitofrontalis muscle (shown in parts a and c of Figure 11.4 on pages 338–339).

Figure 10.1 Organization of skeletal muscle and its connective tissue coverings.

A skeletal muscle consists of individual muscle fibers (cells) bundled into fascicles and surrounded by three connective tissue layers that are extensions of the deep fascia.

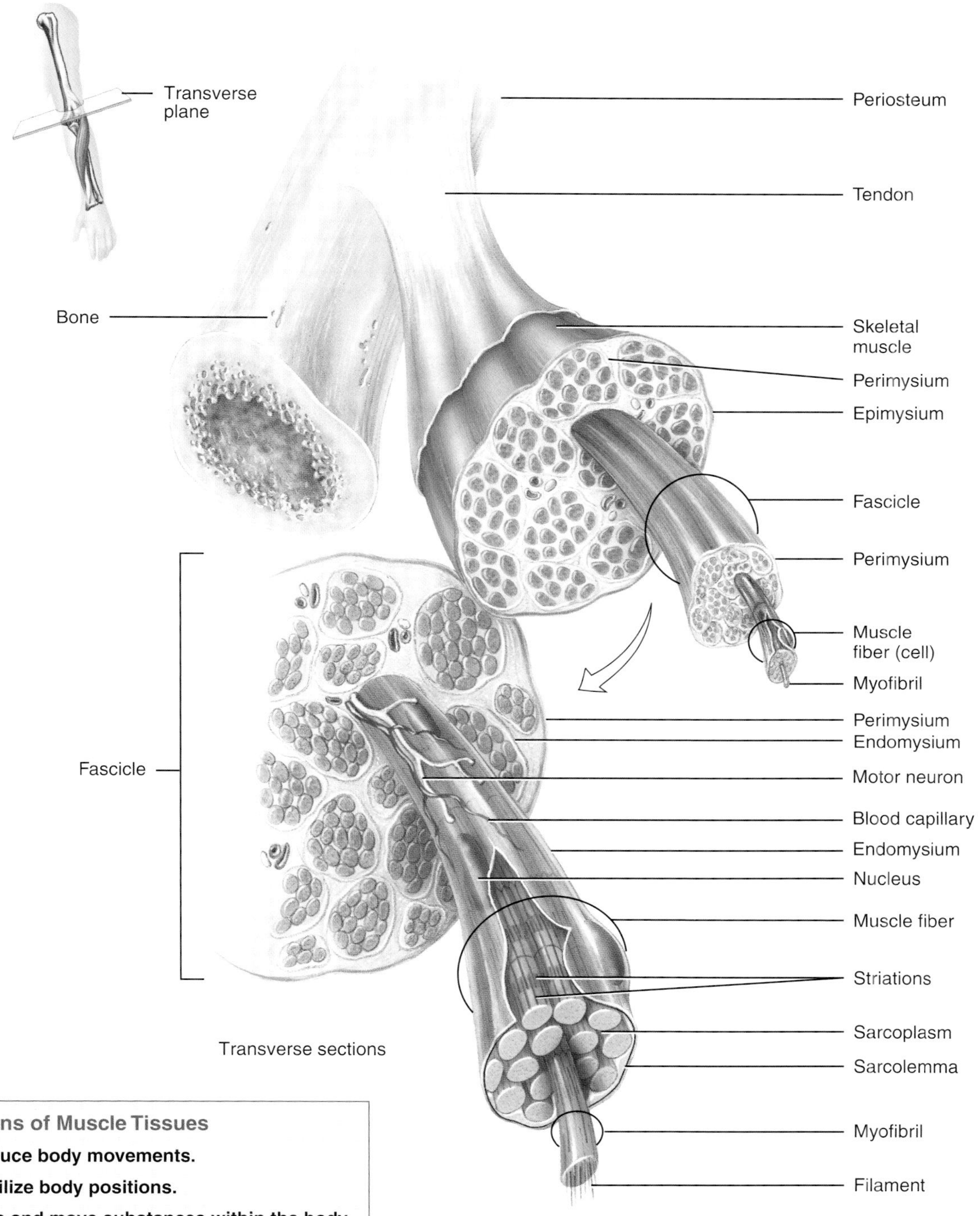

Transverse plane

Periosteum

Tendon

Bone

Skeletal muscle

Perimysium

Epimysium

Fascicle

Perimysium

Muscle fiber (cell)

Myofibril

Perimysium

Endomysium

Fascicle

Motor neuron

Blood capillary

Endomysium

Nucleus

Muscle fiber

Striations

Transverse sections

Sarcoplasm

Sarcolemma

Myofibril

Filament

Functions of Muscle Tissues

1. **Produce body movements.**
2. **Stabilize body positions.**
3. **Store and move substances within the body.**
4. **Generate heat (thermogenesis).**

Which connective tissue coat surrounds groups of muscle fibers, separating them into fascicles?

Certain tendons, especially those of the wrist and ankle, are enclosed by tubes of fibrous connective tissue called **tendon (synovial) sheaths,** which are similar in structure to bursae. The inner layer of a tendon sheath, the *visceral layer,* is attached to the surface of the tendon. The outer layer, known as the *parietal layer,* is attached to bone (see Figure 11.18a on pages 376–377). Between the layers is a cavity that contains a film of synovial fluid. Tendon sheaths reduce friction as tendons slide back and forth.

Nerve and Blood Supply

Skeletal muscles are well supplied with nerves and blood vessels. Generally, an artery and one or two veins accompany each nerve that penetrates a skeletal muscle. The neurons that stimulate skeletal muscle to contract are *somatic motor neurons.* Each somatic motor neuron has a threadlike axon that extends from the brain or spinal cord to a group of skeletal muscle fibers (see Figure 10.10d). The axon of a somatic motor neuron typically branches many times, each branch extending to a different skeletal muscle fiber.

Microscopic blood vessels called capillaries are plentiful in muscular tissue; each muscle fiber is in close contact with one or more capillaries (Figure 10.10d). The blood capillaries bring in oxygen and nutrients and remove heat and the waste products of muscle metabolism. Especially during contraction, a muscle fiber synthesizes and uses considerable ATP (adenosine triphosphate). These reactions, which you will learn more about later on, require oxygen, glucose, fatty acids, and other substances that are delivered to the muscle fiber in the blood.

Microscopic Anatomy of a Skeletal Muscle Fiber

The most important components of a skeletal muscle are the muscle fibers themselves. The diameter of a mature skeletal muscle fiber ranges from 10 to 100 μm.* The typical length of a mature skeletal muscle fiber is about 10 cm (4 in.), although some are as long as 30 cm (12 in.). Because each skeletal muscle fiber arises during embryonic development from the fusion of a hundred or more small mesodermal cells called *myoblasts* (Figure 10.2a), each mature skeletal muscle fiber has a hundred or more nuclei. Once fusion has occurred, the muscle fiber loses its ability to undergo mitosis. Thus, the number of skeletal muscle fibers is set before you are born, and most of these cells last a lifetime.

The dramatic muscle growth that occurs after birth occurs mainly by **hypertrophy** (hī-PER-trō-fē; *hyper-* = above or excessive), an enlargement of existing muscle fibers, rather than by **hyperplasia** (hi-per-PLĀ-zē-a; *-plasis* = molding), an increase in the number of fibers. During childhood, human growth hormone and other hormones stimulate an increase in the size of skeletal muscle fibers. The hormone testosterone (from the testes in males and in small amounts from other tissues in females) promotes further enlargement of muscle fibers. A few myoblasts do persist in mature skeletal muscle as *satellite cells* (Figure 10.2a). These cells retain the capacity to fuse with one another or with damaged muscle fibers to regenerate functional muscle fibers. However, the number of new skeletal muscle fibers formed is not enough to compensate for significant skeletal muscle damage or degeneration. In such cases, skeletal muscle tissue undergoes **fibrosis,** the replacement of muscle fibers by fibrous scar tissue. For this reason, skeletal muscle can regenerate only to a limited extent.

Sarcolemma, Transverse Tubules, and Sarcoplasm

The multiple nuclei of a skeletal muscle fiber are located just beneath the **sarcolemma** (*sarc-* = flesh; *-lemma* = sheath), the plasma membrane of a muscle cell (Figure 10.2b, c). Thousands of tiny invaginations of the sarcolemma, called **transverse (T) tubules,** tunnel in from the surface toward the center of each muscle fiber. T tubules are open to the outside of the fiber and thus are filled with interstitial fluid. Muscle action potentials travel along the sarcolemma and through the T tubules, quickly spreading throughout the muscle fiber. This arrangement ensures that an action potential excites all parts of the muscle fiber at essentially the same instant.

Within the sarcolemma is the **sarcoplasm,** the cytoplasm of a muscle fiber. Sarcoplasm includes a substantial amount of glycogen, which is a large molecule composed of many glucose molecules. Glycogen can be used for synthesis of ATP. In addition, the sarcoplasm contains a red-colored protein called **myoglobin** (mī-ō-GLŌB-in). This protein, found only in muscle, binds oxygen molecules that diffuse into muscle fibers from interstitial fluid. Myoglobin releases oxygen when it is needed by the mitochondria for ATP production. The mitochondria lie in rows throughout the muscle fiber, strategically close to the muscle proteins that use ATP during contraction (Figure 10.2c).

Myofibrils and Sarcoplasmic Reticulum

At high magnification, the sarcoplasm appears stuffed with little threads. These small structures are the **myofibrils** (mī-o-FĪ-brils; *myo-* = muscle; *-fibrilla* = little fiber), the contractile organelles of skeletal muscle (Figure 10.2c). Myofibrils are about 2 μm in diameter and extend the entire length of a muscle fiber. Their prominent striations make the entire skeletal muscle fiber appear striated.

A fluid-filled system of membranous sacs called the **sarcoplasmic reticulum** (sar′-kō-PLAZ-mik re-TIK-ū-lum) or **SR** encircles each myofibril (Figure 10.2c). This elaborate system is similar to smooth endoplasmic reticulum in nonmuscular cells. Dilated end sacs of the sarcoplasmic reticulum called **terminal cisterns** (= reservoirs) butt against the T tubule from both sides. A transverse tubule and the two terminal cisterns on either side of it form a **triad** (*tri-* = three). In a relaxed muscle fiber, the sarcoplasmic reticulum stores calcium ions (Ca^{2+}). Release of Ca^{2+} from the terminal cisterns of the sarcoplasmic reticulum triggers muscle contraction.

*One micrometer (μm) is 10^{-6} meter (1/25,000 in.).

Figure 10.2 Microscopic organization of skeletal muscle. (a) During embryonic development, many myoblasts fuse to form one skeletal muscle fiber. Once fusion has occurred, a skeletal muscle fiber loses the ability to undergo cell division, but satellite cells retain this ability. (b and c) The sarcolemma of the fiber encloses sarcoplasm and myofibrils, which are striated. Sarcoplasmic reticulum wraps around each myofibril. Thousands of tranverse tubules, filled with interstitial fluid, invaginate from the sarcolemma toward the center of the muscle fiber. A triad is a transverse tubule and the two terminal cisterns of the sarcoplasmic reticulum on either side of it. A photomicrograph of skeletal muscle tissue is shown in Table 4.5a on page 135.

The contractile elements of muscle fibers, the myofibrils, contain overlapping thick and thin filaments.

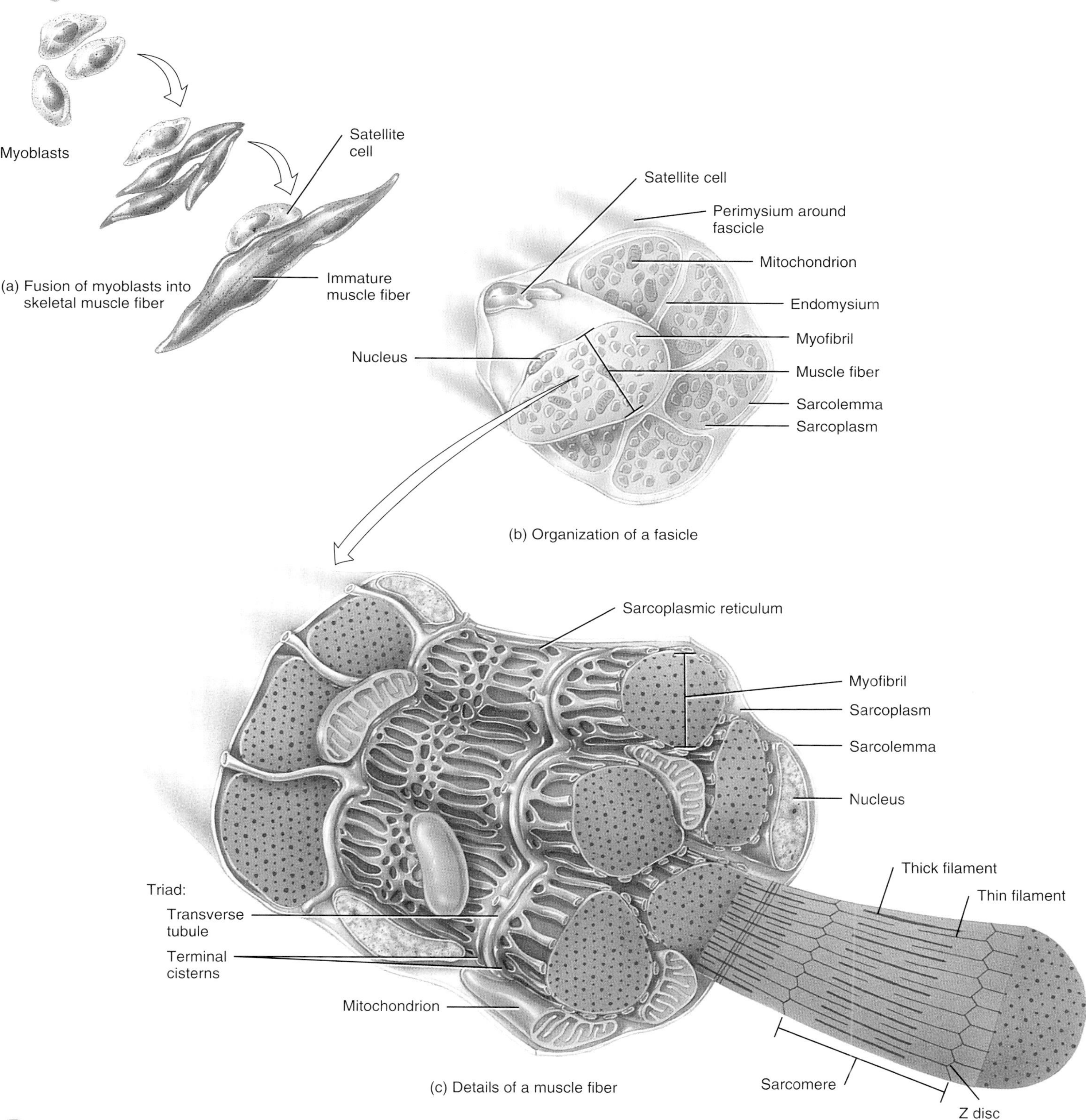

Which structure shown here releases calcium ions to trigger muscle contraction?

Muscular Atrophy and Hypertrophy

Muscular atrophy (A-trō-fē; *a-* = without, *-trophy* = nourishment) is a wasting away of muscles. Individual muscle fibers decrease in size because of progressive loss of myofibrils. Atrophy that occurs because muscles are not used is termed *disuse atrophy*. Bedridden individuals and people with casts experience disuse atrophy because the flow of nerve impulses (nerve action potentials) to inactive skeletal muscle is greatly reduced. The condition is reversible. If instead the nerve supply to a muscle is disrupted or cut, the muscle undergoes *denervation atrophy*. Over a period of 6 months to 2 years, the muscle shrinks to about one-fourth its original size, and the muscle fibers are irreversibly replaced by fibrous connective tissue.

As noted previously, **muscular hypertrophy** is an increase in the diameter of muscle fibers due to increased production of myofibrils, mitochondria, sarcoplasmic reticulum, and other organelles. It results from very forceful, repetitive muscular activity, such as strength training. Because hypertrophied muscles contain more myofibrils, they are capable of more forceful contractions. ■

Filaments and the Sarcomere

Within myofibrils are smaller structures called **filaments** (Figure 10.2c). *Thin filaments* are 8 nm in diameter and 1–2 μm long,* while *thick filaments* are 16 nm in diameter and 1–2 μm long. Both thin and thick filaments are directly involved in the contractile process. Overall, there are two thin filaments for every thick filament in the regions of filament overlap. The filaments inside a myofibril do not extend the entire length of a muscle fiber. Instead, they are arranged in compartments called **sarcomeres** (*-mere* = part), the basic functional units of a myofibril (Figure 10.3a). Narrow, plate-shaped regions of dense material called **Z discs** separate one sarcomere from the next.

The thick and thin filaments overlap one another to a greater or lesser extent, depending on whether the muscle is contracted, relaxed, or stretched. The pattern of their overlap, consisting of a

*One nanometer (nm) is 10^{-9} meter (0.001 μm); one micrometer (μ) = 1/25,000 of an inch.

Figure 10.3 The arrangement of filaments within a sarcomere. A sarcomere extends from one Z disc to the next.

Myofibrils contain two types of filaments: thick filaments and thin filaments.

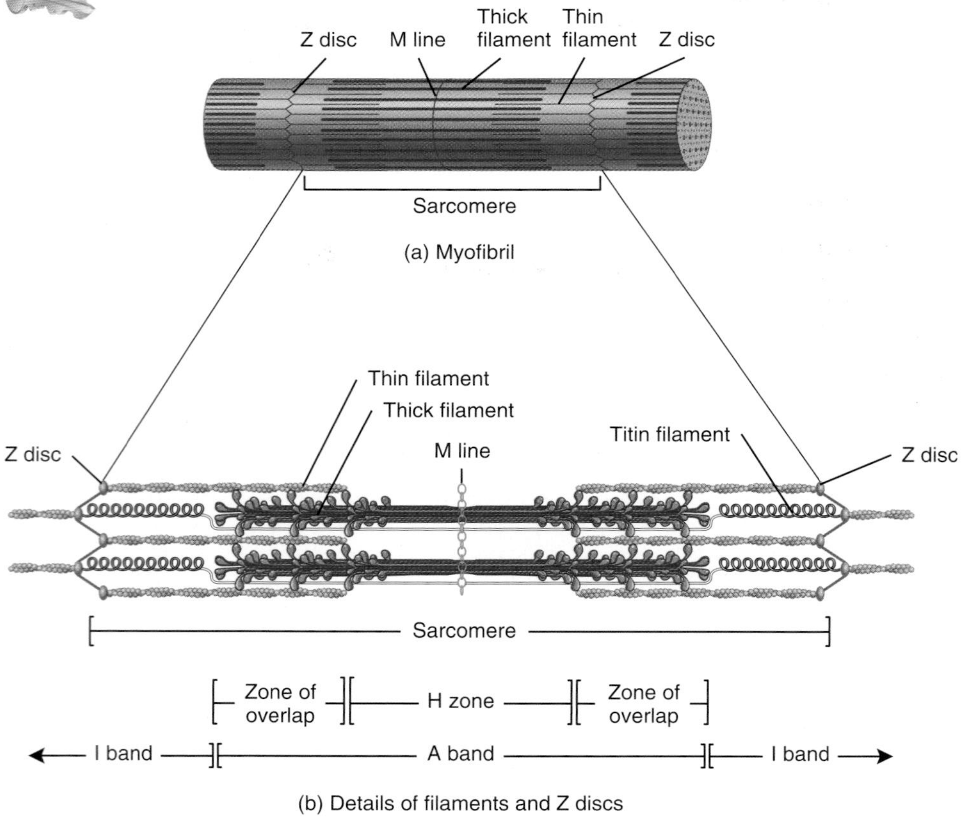

Which of the following is the smallest: muscle fiber, thick filament, or myofibril? Which is largest?

variety of zones and bands (Figure 10.3b), creates the striations that can be seen both in single myofibrils and in whole muscle fibers. The darker middle part of the sarcomere is the **A band,** which extends the entire length of the thick filaments (Figure 10.3b). Toward each end of the A band is a *zone of overlap,* where the thick and thin filaments lie side by side. The **I band** is a lighter, less dense area that contains the rest of the thin filaments but no thick filaments (Figure 10.3b). A Z disc passes through the center of each I band. A narrow **H zone** in the center of each A band contains thick but not thin filaments. Supporting proteins that hold the thick filaments together at the center of the H zone form the **M line,** so named because it is at the *middle* of the sarcomere. Figure 10.4 shows the relations of the zones, bands, and lines as seen in a transmission electron micrograph.

Exercise-Induced Muscle Damage

Comparison of electron micrographs of muscle tissue taken from athletes before and after intense exercise reveal considerable exercise-induced muscle damage, including torn sarcolemmas in some muscle fibers, damaged myofibrils, and disrupted Z discs. Microscopic muscle damage after exercise also is indicated by increases in blood levels of proteins, such as myoglobin and the enzyme creatine kinase, that are normally confined within muscle fibers. From 12 to 48 hours after a period of strenuous exercise, skeletal muscles often become sore. Such **delayed onset muscle soreness (DOMS)** is accompanied by stiffness, tenderness, and swelling. Although the causes of DOMS are not completely understood, microscopic muscle damage appears to be a major factor. ■

Muscle Proteins

Myofibrils are built from three kinds of proteins: (1) contractile proteins, which generate force during contraction; (2) regulatory proteins, which help switch the contraction process on and off; and (3) structural proteins, which keep the thick and thin filaments in the proper alignment, give the myofibril elasticity and extensibility, and link the myofibrils to the sarcolemma and extracellular matrix.

The two *contractile proteins* in muscle are myosin and actin, which are the main components of thick and thin filaments, respectively. **Myosin** functions as a *motor protein* in all three types of muscle tissue. Motor proteins push or pull various cellular structures to achieve movement by converting the chemical energy in ATP to the mechanical energy of motion or the produc-

Figure 10.4 Characteristic zones and bands of a sarcomere.

The striations of skeletal muscle are alternating darker A bands and lighter I bands.

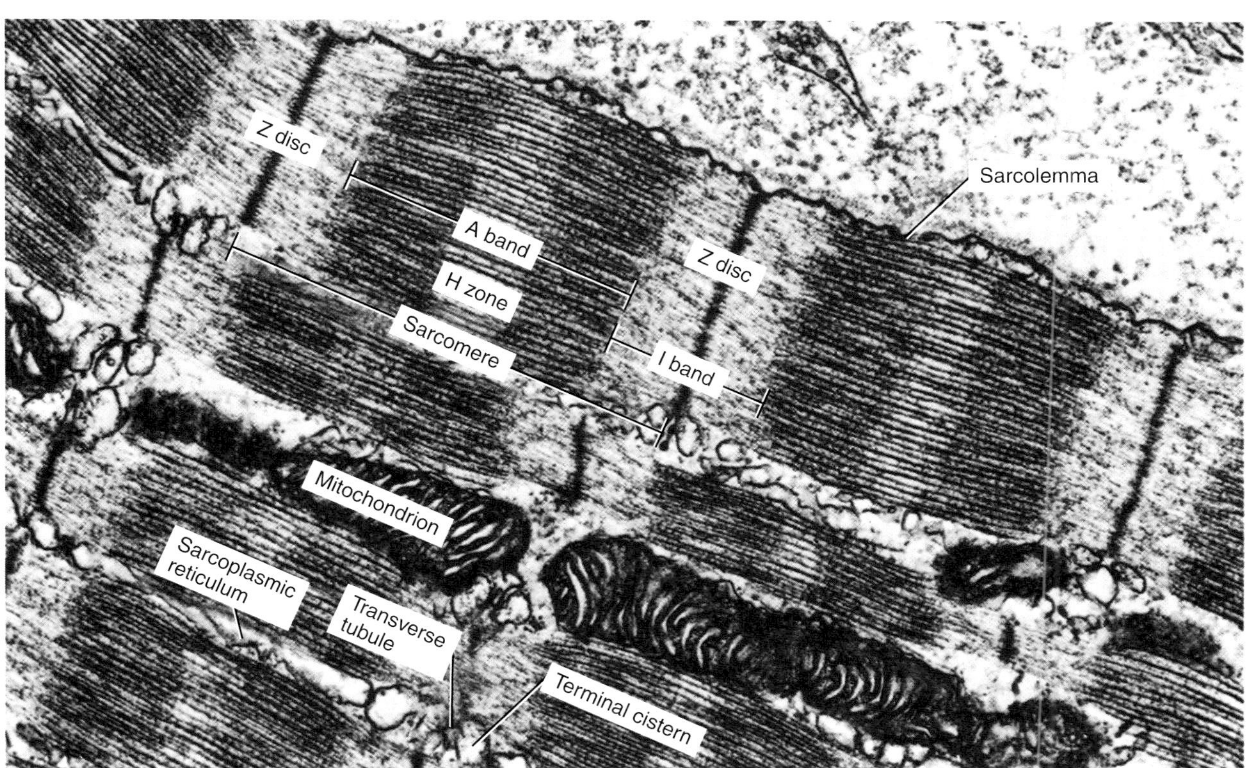

How are sarcomeres separated from one another?

tion of force. In skeletal muscle, about 300 molecules of myosin form a single thick filament. Each myosin molecule is shaped like two golf clubs twisted together (Figure 10.5a). The *myosin tail* (twisted golf club handles) points toward the M line in the center of the sarcomere. Tails of neighboring myosin molecules lie parallel to one another, forming the shaft of the thick filament. The two projections of each myosin molecule (golf club heads) are called *myosin heads.* The heads project outward from the shaft in a spiraling fashion, each extending toward one of the six thin filaments that surround each thick filament.

Thin filaments are anchored to Z discs (see Figure 10.3b). Their main component is the protein **actin.** Individual actin molecules join to form an actin filament that is twisted into a helix (Figure 10.5b). On each actin molecule is a *myosin-binding site,* where a myosin head can attach. Smaller amounts of two *regulatory proteins*—**tropomyosin** and **troponin**—are also part of the thin filament. In relaxed muscle, myosin is blocked from binding to actin because strands of tropomyosin cover the *myosin-binding sites* on actin. The tropomyosin strands, in turn, are held in place by troponin molecules.

Besides contractile and regulatory proteins, muscle contains about a dozen *structural proteins,* which contribute to the alignment, stability, elasticity, and extensibility of myofibrils. Several key structural proteins are titin, myomesin, nebulin, and dystrophin. *Titin* (*titan* = gigantic) is the third most plentiful protein in skeletal muscle (after actin and myosin). This molecule's name reflects its huge size. With a molecular weight of about 3 million daltons, titin is 50 times larger than an average-sized protein. Each titin molecule spans half a sarcomere, from a Z disc to an M line (see Figure 10.3b), a distance of 1 to 1.2 μm in relaxed muscle. Titin anchors a thick filament to both a Z disc and the M line, thereby helping stabilize the position of the thick filament. The part of the titin molecule that extends from the Z disc to the beginning of the thick filament is very elastic. Because it can stretch to at least four times its resting length and then spring back unharmed, titin accounts for much of the elasticity and extensibility of myofibrils. Titin probably helps the sarcomere return to its resting length after a muscle has contracted or been stretched, may help prevent overextension of sarcomeres, and maintains the central location of the A bands.

Molecules of the protein *myomesin* form the M line. The M line proteins bind to titin and connect adjacent thick filaments to one another. *Nebulin* is a long, nonelastic protein wrapped around the entire length of each thin filament. It helps anchor the thin filaments to the Z discs and regulates the length of thin filaments during development. *Dystrophin* is a cytoskeletal protein that links thin filaments of the sarcomere to integral membrane proteins of the sarcolemma, which are attached in turn to proteins in the connective tissue extracellular matrix that surrounds muscle fibers. Dystrophin and its associated proteins are thought to reinforce the sarcolemma and help transmit the tension generated by the sarcomeres to the tendons. The relationship of dystrophin to muscular dystrophy is discussed on page 319.

Figure 10.5 Structure of thick and thin filaments. (a) A thick filament contains about 300 myosin molecules, one of which is shown enlarged. The myosin tails form the shaft of the thick filament, and the myosin heads project outward toward the surrounding thin filaments. (b) Thin filaments contain actin, troponin, and tropomyosin.

Contractile proteins (myosin and actin) generate force during contraction; regulatory proteins (troponin and tropomyosin) help switch contraction on and off.

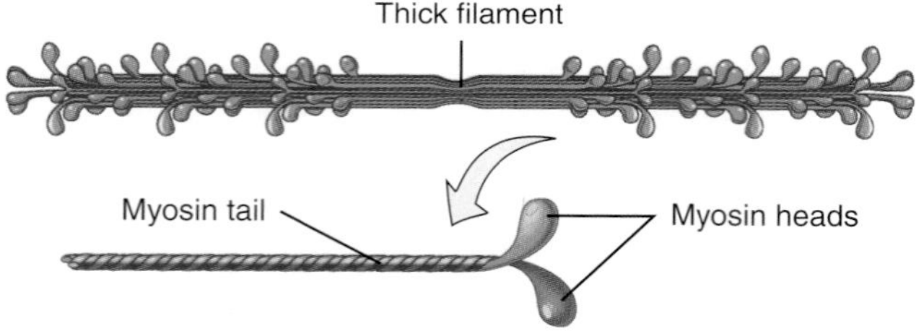

(a) One thick filament (above) and a myosin molecule (below)

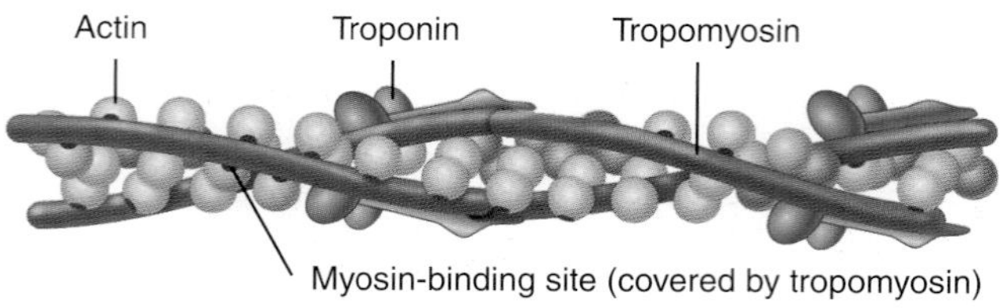

(b) Portion of a thin filament

Which proteins connect into the Z disc? Which proteins are present in the A band? In the I band?

▶ **CHECKPOINT**

4. What types of fascia cover skeletal muscles?
5. Why is a rich blood supply important for muscle contraction?
6. How are the structures of thin and thick filaments different?

CONTRACTION AND RELAXATION OF SKELETAL MUSCLE FIBERS

▶ **OBJECTIVES**

Outline the steps involved in the sliding filament mechanism of muscle contraction.

Describe how muscle action potentials arise at the neuromuscular junction.

When scientists examined the first electron micrographs of skeletal muscle in the mid-1950s, they were surprised to see that the lengths of the thick and thin filaments were the same in both relaxed and contracted muscle. It had been thought that muscle contraction must be a folding process, somewhat like closing an accordion. Instead, researchers discovered that skeletal muscle shortens during contraction because the thick and thin filaments slide past one another. The model describing this process is known as the **sliding filament mechanism.**

The Sliding Filament Mechanism

Muscle contraction occurs because myosin heads attach to and "walk" along the thin filaments at both ends of a sarcomere, progressively pulling the thin filaments toward the M line (Figure 10.6). As a result, the thin filaments slide inward and meet at the center of a sarcomere. They may even move so far inward that their ends overlap (Figure 10.6c). As the thin filaments slide inward, the Z discs come closer together, and the sarcomere shortens. However, the lengths of the individual thick and thin filaments do not change. Shortening of the sarcomeres causes shortening of the whole muscle fiber, which in turn leads to shortening of the entire muscle.

The Contraction Cycle

At the onset of contraction, the sarcoplasmic reticulum releases calcium ions (Ca^{2+}) into the cytosol. There, they bind to troponin and cause the troponin–tropomyosin complexes to move away from the myosin-binding sites on actin. Once the binding sites are "free," the **contraction cycle**—the repeating sequence of events that causes the filaments to slide—begins. The contraction cycle consists of four steps (Figure 10.7):

1. ***ATP hydrolysis.*** The myosin head includes an ATP-binding site and an ATPase, an enzyme that hydrolyzes ATP into ADP (adenosine diphosphate) and a phosphate group. This hydrolysis reaction reorients and energizes the myosin head. Notice that the products of ATP hydrolysis—ADP and a phosphate group—are still attached to the myosin head.
2. ***Attachment of myosin to actin to form crossbridges.*** The energized myosin head attaches to the myosin-binding site on actin and releases the previously hydrolyzed phosphate group. When the myosin heads attach to actin during contraction, they are referred to as **crossbridges.**

Figure 10.6 Sliding filament mechanism of muscle contraction, as it occurs in two adjacent sarcomeres.

During muscle contractions, thin filaments move toward the M line of each sarcomere.

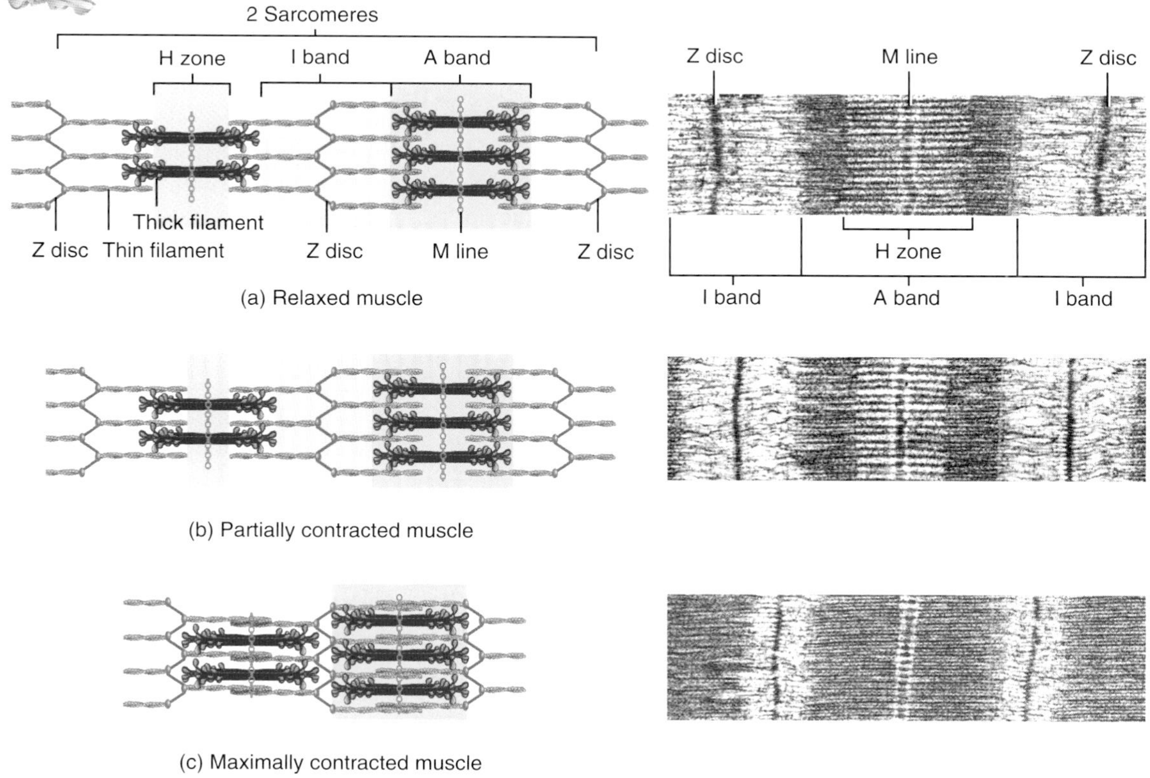

2 Sarcomeres
H zone
I band
A band
Thick filament
Z disc
Thin filament
Z disc
M line
Z disc
M line
H zone
I band
A band
I band

(a) Relaxed muscle

(b) Partially contracted muscle

(c) Maximally contracted muscle

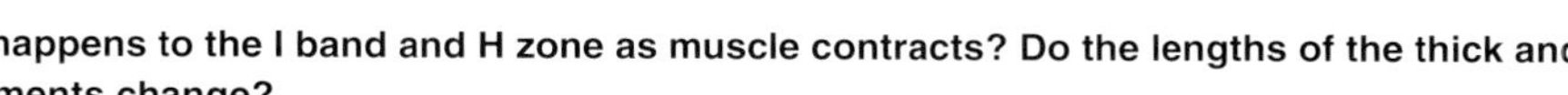
What happens to the I band and H zone as muscle contracts? Do the lengths of the thick and thin filaments change?

Figure 10.7 The contraction cycle. Sarcomeres exert force and shorten through repeated cycles during which the myosin heads attach to actin (crossbridges), rotate, and detach.

During the power stroke of contraction, crossbridges rotate and move the thin filaments past the thick filaments toward the center of the sarcomere.

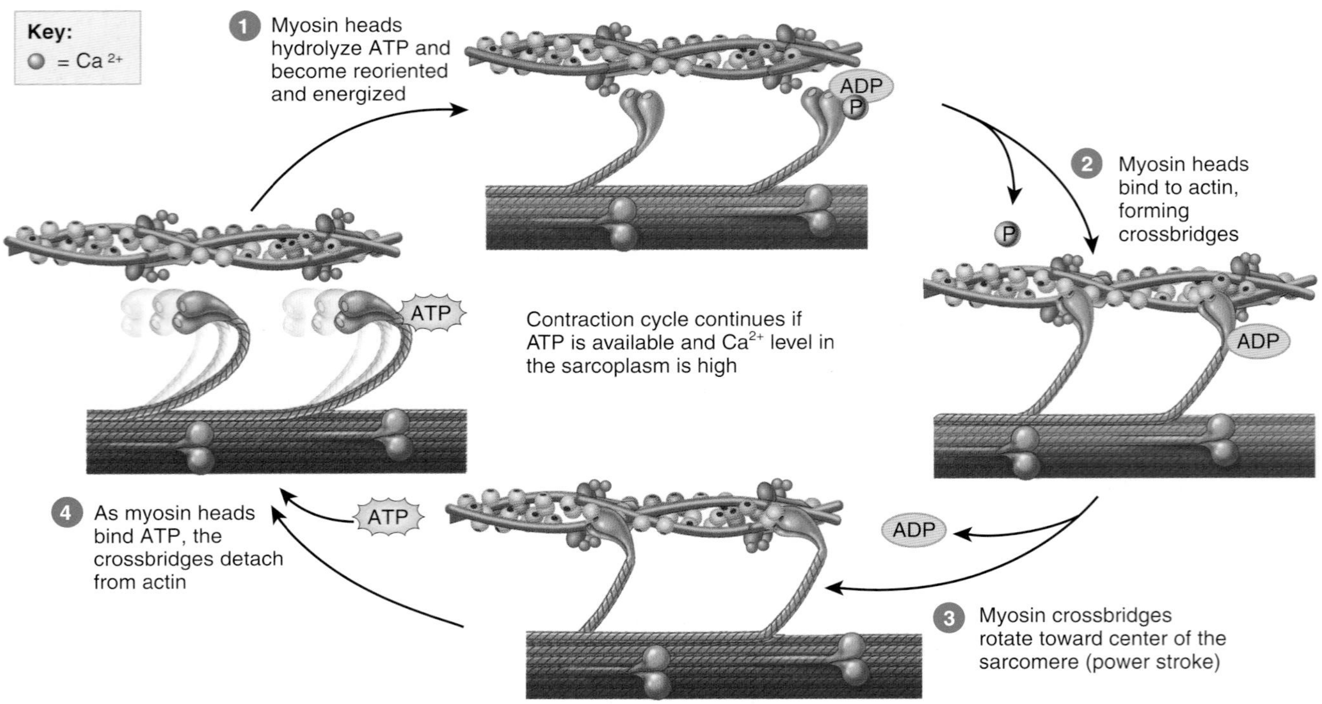

Key:
= Ca 2+
1 Myosin heads hydrolyze ATP and become reoriented and energized
ADP
P
2 Myosin heads bind to actin, forming crossbridges
P
ATP
Contraction cycle continues if ATP is available and Ca2+ level in the sarcoplasm is high
ADP
4 As myosin heads bind ATP, the crossbridges detach from actin
ATP
ADP
3 Myosin crossbridges rotate toward center of the sarcomere (power stroke)

What would happen if ATP suddenly were not available after the sarcomere had started to shorten?

3 ***Power stroke.*** After the crossbridges form, the power stroke occurs. During the power stroke, the site on the crossbridge where ADP is still bound opens. As a result, the crossbridge rotates and releases the ADP. The crossbridge generates force as it rotates toward the center of the sarcomere, sliding the thin filament past the thick filament toward the M line.

4 ***Detachment of myosin from actin.*** At the end of the power stroke, the crossbridge remains firmly attached to actin until it binds another molecule of ATP. As ATP binds to the ATP-binding site on the myosin head, the myosin head detaches from actin.

The contraction cycle repeats as the myosin ATPase hydrolyzes the newly bound molecule of ATP, and continues as long as ATP is available and the Ca^{2+} level near the thin filament is sufficiently high. The crossbridges keep rotating back and forth with each power stroke, pulling the thin filaments toward the M line. Each of the 600 crossbridges in one thick filament attaches and detaches about five times per second. At any one instant, some of the myosin heads are attached to actin, forming crossbridges and generating force, and other myosin heads are detached from actin and getting ready to bind again.

Contraction is analogous to running on a foot-powered treadmill. One foot (crossbridge) strikes the belt (thin filament) and pushes it backward (toward the M line). Then the other foot comes down and imparts a second push. The belt (thin filament) moves smoothly while the runner (thick filament) remains stationary. Each crossbridge progressively "walks" along a thin filament, coming closer to the Z disc with each "step," while the thin filament moves toward the M line. And like the legs of a runner, the crossbridge needs a constant supply of energy to keep going—one molecule of ATP for each contraction cycle!

As the contraction cycle continues, movement of crossbridges applies the force that draws the Z discs toward each other, and the sarcomere shortens. During a maximal muscle contraction, the distance between two Z discs can decrease to half the resting length. The Z discs, in turn, pull on neighboring sarcomeres, and the whole muscle fiber shortens. Some of the components of a muscle are elastic: They stretch slightly before they transfer the tension generated by the sliding filaments. The elastic components include titin molecules, connective tissue around the muscle fibers (endomysium, perimysium, and epimysium), and tendons that attach muscle to bone. As the cells of a skeletal muscle start to shorten, they first pull on their connective tissue coverings and tendons. The coverings and tendons stretch and then become taut, and the tension passed through the tendons pulls on the bones to which they are attached. The result is movement of a part of the body. You will soon learn, however, that the contraction cycle does not always result in shortening of the muscle fibers and the whole muscle. In some contractions, the crossbridges rotate and generate tension, but the thin filaments cannot slide inward because the tension they generate is not large enough to move the load on the muscle.

Excitation–Contraction Coupling

An increase in Ca^{2+} concentration in the cytosol starts muscle contraction, and a decrease stops it. When a muscle fiber is relaxed, the concentration of Ca^{2+} in its cytosol is very low, only about 0.1 micromole per liter (0.1 μm/L). However, a huge amount of Ca^{2+} is stored inside the sarcoplasmic reticulum (Figure 10.8a). As a muscle action potential propagates along the sarcolemma and into the T tubules, it causes **Ca^{2+} release channels** in the SR membrane to open (Figure 10.8b). When these channels open, Ca^{2+} flows out of the SR into the cytosol around the thick and thin filaments. As a result, the Ca^{2+} concentration in the cytosol rises tenfold or more. The released calcium ions combine with troponin, causing it to change shape. This conformational change moves the troponin–tropomyosin complex away from the myosin-binding sites on actin. Once these binding sites are free, myosin heads bind to them to form crossbridges, and the contraction cycle begins. The events just described constitute **excitation–contraction coupling,** the steps that connect excitation (a muscle action potential propagating along the sarcolemma and into the T tubules) to contraction (sliding of the filaments).

The sarcoplasmic reticulum membrane also contains **Ca^{2+} active transport pumps** that use ATP to move Ca^{2+} constantly from the cytosol into the SR (Figure 10.8). While muscle action potentials continue to propagate through the T tubules, the Ca^{2+} release channels are open. Calcium ions flow into the cytosol more rapidly than they are transported back by the pumps. After the last action potential has propagated throughout the T tubules, the Ca^{2+} release channels close. As the pumps move Ca^{2+} back into the SR, the concentration of calcium ions in the cytosol quickly decreases. Inside the SR, molecules of a calcium-binding protein, appropriately called **calsequestrin,** bind to the Ca^{2+}, enabling even more Ca^{2+} to be sequestered or stored within the SR. As a result, the concentration of Ca^{2+} is 10,000 times higher in the SR than in the cytosol in a relaxed muscle fiber. As the Ca^{2+} level in the cytosol drops, the troponin–tropomyosin complexes cover the myosin-binding sites, and the muscle fiber relaxes.

Figure 10.8 The role of Ca^{2+} in the regulation of contraction by troponin and tropomyosin. (a) During relaxation, the level of Ca^{2+} in the sarcoplasm is low, only 0.1 μM (0.001 mM), because calcium ions are pumped into the sarcoplasmic reticulum by Ca^{2+} active transport pumps. (b) A muscle action potential propagating along a transverse tubule opens Ca^{2+} release channels in the sarcoplasmic reticulum, calcium ions flow into the cytosol, and contraction begins.

An increase in the Ca^{2+} level in the sarcoplasm starts the sliding of thin filaments. When the level of Ca^{2+} in the sarcoplasm declines, sliding stops.

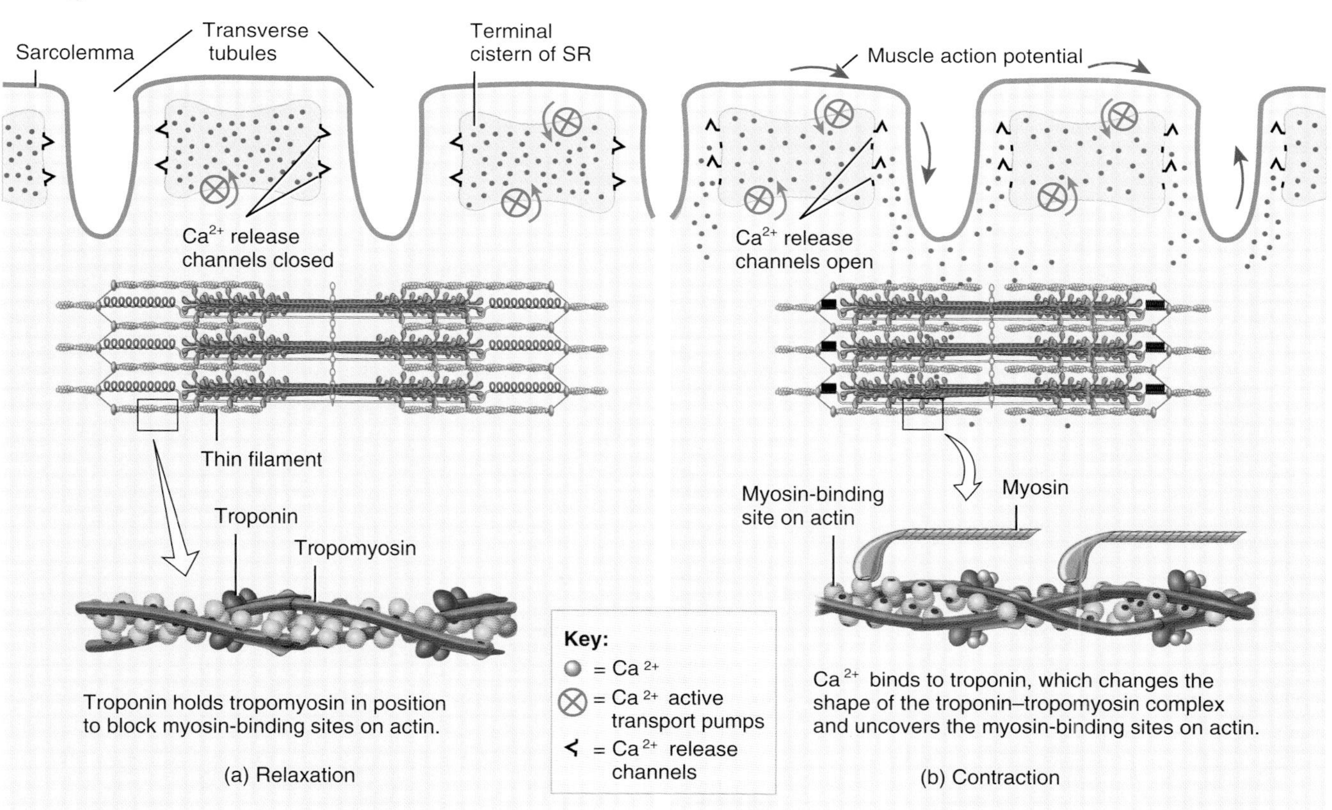

What are three functions of ATP in muscle contraction?

Rigor Mortis

After death, cellular membranes become leaky. Calcium ions leak out of the sarcoplasmic reticulum into the cytosol and allow myosin heads to bind to actin. ATP synthesis ceases shortly after breathing stops, however, so the crossbridges cannot detach from actin. The resulting condition, in which muscles are in a state of rigidity (cannot contract or stretch), is called **rigor mortis** (rigidity of death). Rigor mortis begins 3–4 hours after death and lasts about 24 hours; then it disappears as proteolytic enzymes from lysosomes digest the crossbridges. ■

Length–Tension Relationship

Figure 10.9 shows the **length–tension relationship** for skeletal muscle, which indicates how the forcefulness of muscle contraction depends on the length of the sarcomeres within a muscle *before contraction begins.* At a sarcomere length of about 2.0–2.4 μm (which is very close to the resting length in most muscles), the zone of overlap in each sarcomere is optimal, and the muscle fiber can develop maximum tension. Notice in Figure 10.9 that maximum tension (100%) occurs when the zone of overlap between a thick and thin filament extends from the edge of the H zone to one end of a thick filament.

As the sarcomeres of a muscle fiber are stretched to a longer length, the zone of overlap shortens, and fewer myosin heads can make contact with thin filaments. Therefore, the tension the fiber can produce decreases. When a skeletal muscle fiber is stretched to 170% of its optimal length, there is no overlap between the thick and thin filaments. Because none of the myosin heads can bind to thin filaments, the muscle fiber cannot contract, and tension is zero. As sarcomere lengths become increasingly shorter than the optimum, the tension that can develop again decreases. This is because thick filaments crumple as they are compressed by the Z discs, resulting in fewer myosin heads making contact with thin filaments. Normally, resting muscle fiber length is held very close to the optimum by firm attachments of skeletal muscle to bones (via their tendons) and to other inelastic tissues.

Figure 10.9 Length–tension relationship in a skeletal muscle fiber. Maximum tension during contraction occurs when the resting sarcomere length is 2.0–2.4 μm.

A muscle fiber develops its greatest tension when there is an optimal zone of overlap between thick and thin filaments.

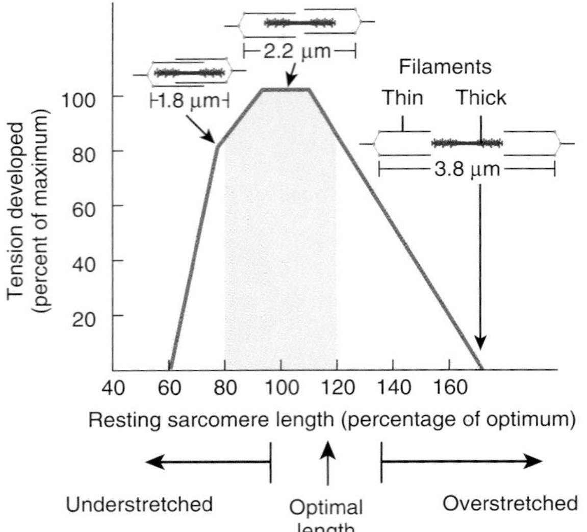

Why is tension maximal at a sarcomere length of 2.2 μm?

The Neuromuscular Junction

As noted earlier in the chapter, the neurons that stimulate skeletal muscle fibers to contract are called **somatic motor neurons.** Each somatic motor neuron has a threadlike axon that extends from the brain or spinal cord to a group of skeletal muscle fibers. A muscle fiber contracts in response to one or more action potentials propagating along its sarcolemma and through its system of T tubules. Muscle action potentials arise at the **neuromuscular junction (NMJ),** the synapse between a somatic motor neuron and a skeletal muscle fiber (Figure 10.10a). A **synapse** is a region where communication occurs between two neurons, or between a neuron and a target cell—in this case, between a somatic motor neuron and a muscle fiber. At most synapses a small gap, called the **synaptic cleft,** separates the two cells. Because the cells do not physically touch, the action potential cannot "jump the gap" from one cell to another. Instead, the first cell communicates with the second by releasing a chemical called a **neurotransmitter.**

At the NMJ, the end of the motor neuron, called the axon terminal, divides into a cluster of synaptic end bulbs (Figure 10.10a, b). Suspended in the cytosol within each synaptic end bulb are hundreds of membrane-enclosed sacs called **synaptic vesicles.** Inside each synaptic vesicle are thousands of molecules of **acetylcholine** (as′-ē-til-KŌ-lēn), abbreviated **ACh,** the neurotransmitter released at the NMJ.

The region of the sarcolemma opposite the synaptic end bulbs, called the **motor end plate** (Figure 10.10b, c), is the muscle fiber part of the NMJ. Within each motor end plate are 30 to 40 million **acetylcholine receptors,** integral transmembrane proteins that bind specifically to ACh. As you will see, the ACh receptors are ligand-gated ion channels. A neuromuscular junction thus includes all the synaptic end bulbs on one side of the synaptic cleft, plus the motor end plate of the muscle fiber on the other side.

A nerve impulse (nerve action potential) elicits a muscle action potential in the following way (Figure 10.10c):

1. ***Release of acetylcholine.*** Arrival of the nerve impulse at the synaptic end bulbs causes many synaptic vesicles to undergo exocytosis. During exocytosis, the synaptic vesicles fuse with the motor neuron's plasma membrane, liberating ACh into the synaptic cleft. The ACh then diffuses across the synaptic cleft between the motor neuron and the motor end plate.

2. ***Activation of ACh receptors.*** Binding of two molecules of ACh to the receptor on the motor end plate opens an ion channel in the ACh receptor. Once the channel is open, small cations, most importantly Na^+, can flow across the membrane.
3. ***Production of muscle action potential.*** The inflow of Na^+ (down its electrochemical gradient) makes the inside of the muscle fiber more positively charged. This change in the membrane potential triggers a muscle action potential. Each nerve impulse normally elicits one muscle action potential. The muscle action potential then propagates along the sarcolemma into the T tubule system. This causes the sarcoplasmic reticulum to release its stored Ca^{2+} into the sarcoplasm and the muscle fiber subsequently contracts.
4. ***Termination of ACh activity.*** The effect of ACh binding lasts only briefly because ACh is rapidly broken down by an

Figure 10.10 Structure of the neuromuscular junction (NMJ), the synapse between a somatic motor neuron and a skeletal muscle fiber.

Synaptic end bulbs at the tips of axon terminals contain synaptic vesicles filled with acetylcholine (ACh).

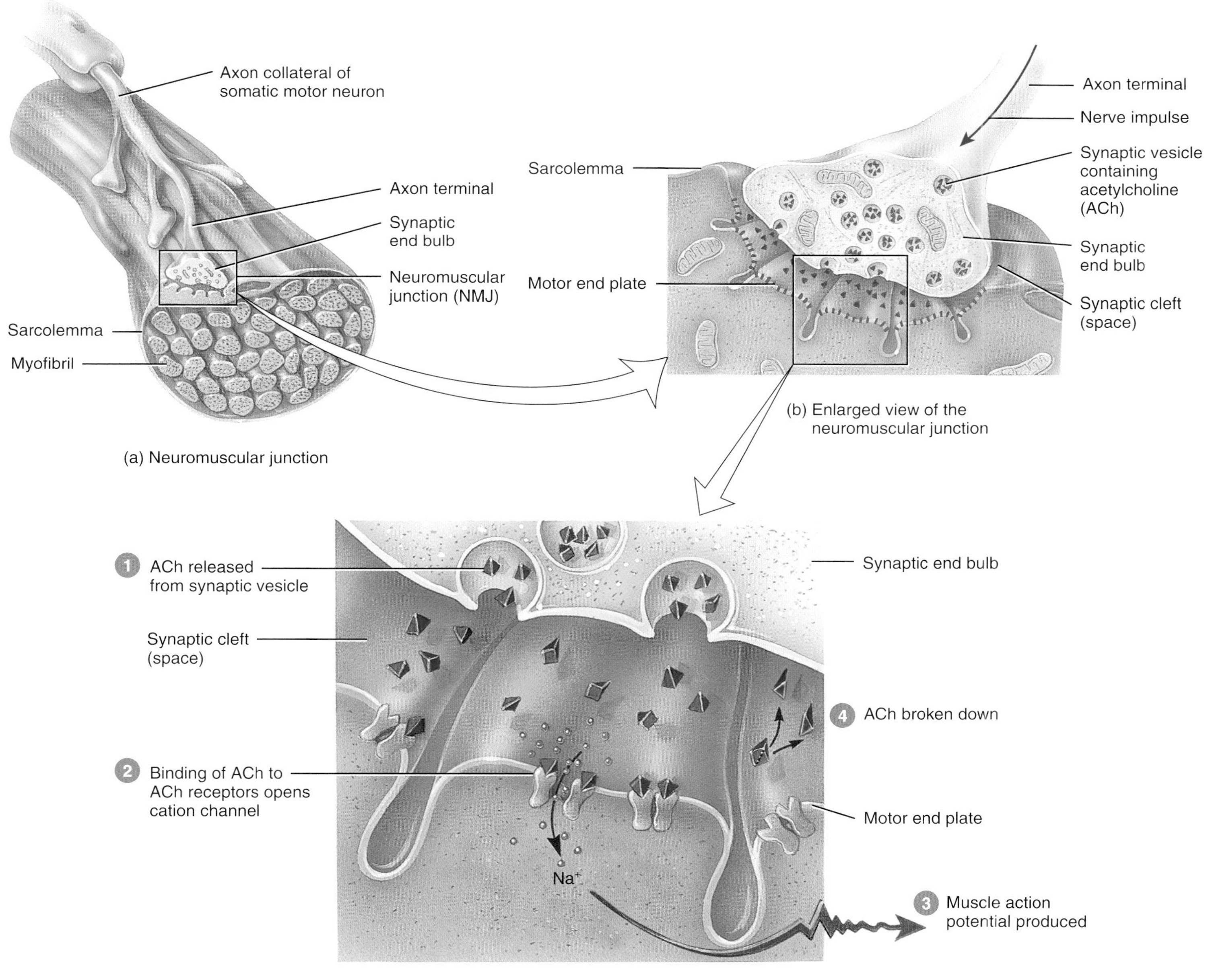

(a) Neuromuscular junction

(b) Enlarged view of the neuromuscular junction

(c) Binding of acetylcholine to ACh receptors in the motor end plate

Figure 10.10 **(continued)**

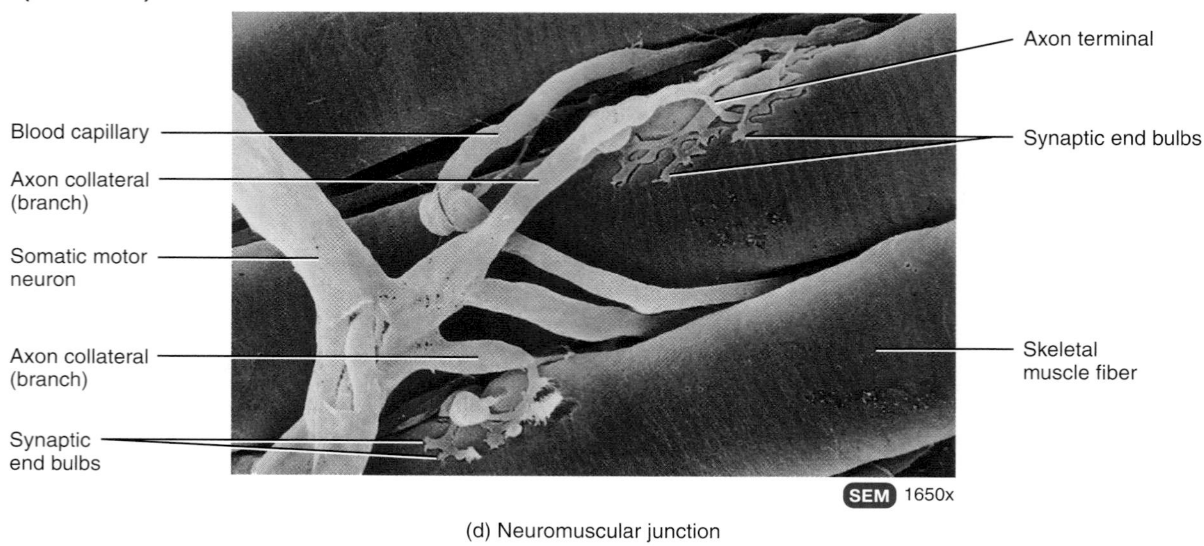

(d) Neuromuscular junction

What part of the sarcolemma contains acetylcholine receptors?

enzyme called **acetylcholinesterase (AChE).** This enzyme is attached to collagen fibers in the extracellular matrix of the synaptic cleft. AChE breaks down ACh into acetyl and choline, products that cannot activate the ACh receptor.

If another nerve impulse releases more acetylcholine, steps 2 and 3 repeat. When action potentials in the motor neuron cease, ACh is no longer released, and AChE rapidly breaks down the ACh already present in the synaptic cleft. This ends the production of muscle action potentials, and the Ca^{2+} release channels in the sarcoplasmic reticulum membrane close.

The NMJ usually is near the midpoint of a skeletal muscle fiber. Muscle action potentials that arise at the NMJ propagate toward both ends of the fiber. This arrangement permits nearly simultaneous activation (and thus contraction) of all parts of the muscle fiber.

Figure 10.11 summarizes the events that occur during contraction and relaxation of a skeletal muscle fiber.

Several plant products and drugs selectively block certain events at the NMJ. *Botulinum toxin,* produced by the bacterium *Clostridium botulinum,* blocks exocytosis of synaptic vesicles at the NMJ. As a result, ACh is not released, and muscle contraction does not occur. The bacteria proliferate in improperly canned foods, and their toxin is one of the most lethal chemicals known. A tiny amount can cause death by paralyzing skeletal muscles. Breathing stops due to paralysis of respiratory muscles, including the diaphragm. Yet it is also the first bacterial toxin to be used as a medicine (Botox®). Injections of Botox into the affected muscles can help patients who have strabismus (crossed eyes), blepharospasm (uncontrollable blinking), or spasms of the vocal cords that interfere with speech. It is also used as a cosmetic treatment to relax muscles that cause facial wrinkles and to alleviate chronic back pain due to muscle spasms in the lumbar region.

The plant derivative *curare,* a poison used by South American Indians on arrows and blowgun darts, causes muscle paralysis by binding to and blocking ACh receptors. In the presence of curare, the ion channels do not open. Curare-like drugs are often used during surgery to relax skeletal muscles.

A family of chemicals called *anticholinesterase agents* have the property of slowing the enzymatic activity of acetylcholinesterase, thus slowing removal of ACh from the synaptic cleft. At low doses, these agents can strengthen weak muscle contractions. One example is neostigmine, which is used to treat patients with myasthenia gravis (see page 319). Neostigmine is also used as an antidote for curare poisoning and to terminate the effects of curare-like drugs after surgery.

Electromyography

Electromyography (e-lek′-trō-mī-OG-ra-fē; *electro-* = electricity; *myo-* = muscle; *-graph* = to write) or **EMG** is a test that measures the electrical activity (muscle action potentials) in resting and contracting muscles. Normally, resting muscle produces no electrical activity; a slight contraction produces some electrical activity; and a more forceful contraction produces increased electrical activity. In the procedure, a ground electrode is placed over the muscle to be tested to eliminate background electrical activity. Then, a fine needle attached by wires to a recording instrument is inserted into the muscle. The electrical activity of the muscle is displayed as waves on an oscilloscope and heard through a loudspeaker.

EMG helps to determine if muscle weakness or paralysis is due to a malfunction of the muscle itself or the nerves supplying the muscle. EMG is also used to diagnose certain muscle disorders, such as muscular dystrophy. ■

Figure 10.11 **Summary of the events of contraction and relaxation in a skeletal muscle fiber.**

Acetylcholine released at the neuromuscular junction triggers a muscle action potential, which leads to muscle contraction.

Nerve impulse
1 Nerve impulse arrives at axon terminal of motor neuron and triggers release of acetylcholine (ACh).
Muscle action potential
Transverse tubule
2 ACh diffuses across synaptic cleft, binds to its receptors in the motor end plate, and triggers a muscle action potential (AP).
ACh receptor
Synaptic vesicle filled with ACh
3 Acetylcholinesterase in synaptic cleft destroys ACh so another muscle action potential does not arise unless more ACh is released from motor neuron.
4 Muscle AP travelling along transverse tubule opens Ca^{2+} release channels in the sarcoplasmic reticulum (SR) membrane, which allows calcium ions to flood into the sarcoplasm.
SR
Ca^{2+}
9 Muscle relaxes.
8 Troponin–tropomyosin complex slides back into position where it blocks the myosin binding sites on actin.
5 Ca^{2+} binds to troponin on the thin filament, exposing the binding sites for myosin.
Elevated Ca^{2+}
Ca^{2+} active transport pumps
7 Ca^{2+} release channels in SR close and Ca^{2+} active transport pumps use ATP to restore low level of Ca^{2+} in sarcoplasm.
6 Contraction: power strokes use ATP; myosin heads bind to actin, swivel, and release; thin filaments are pulled toward center of sarcomere.

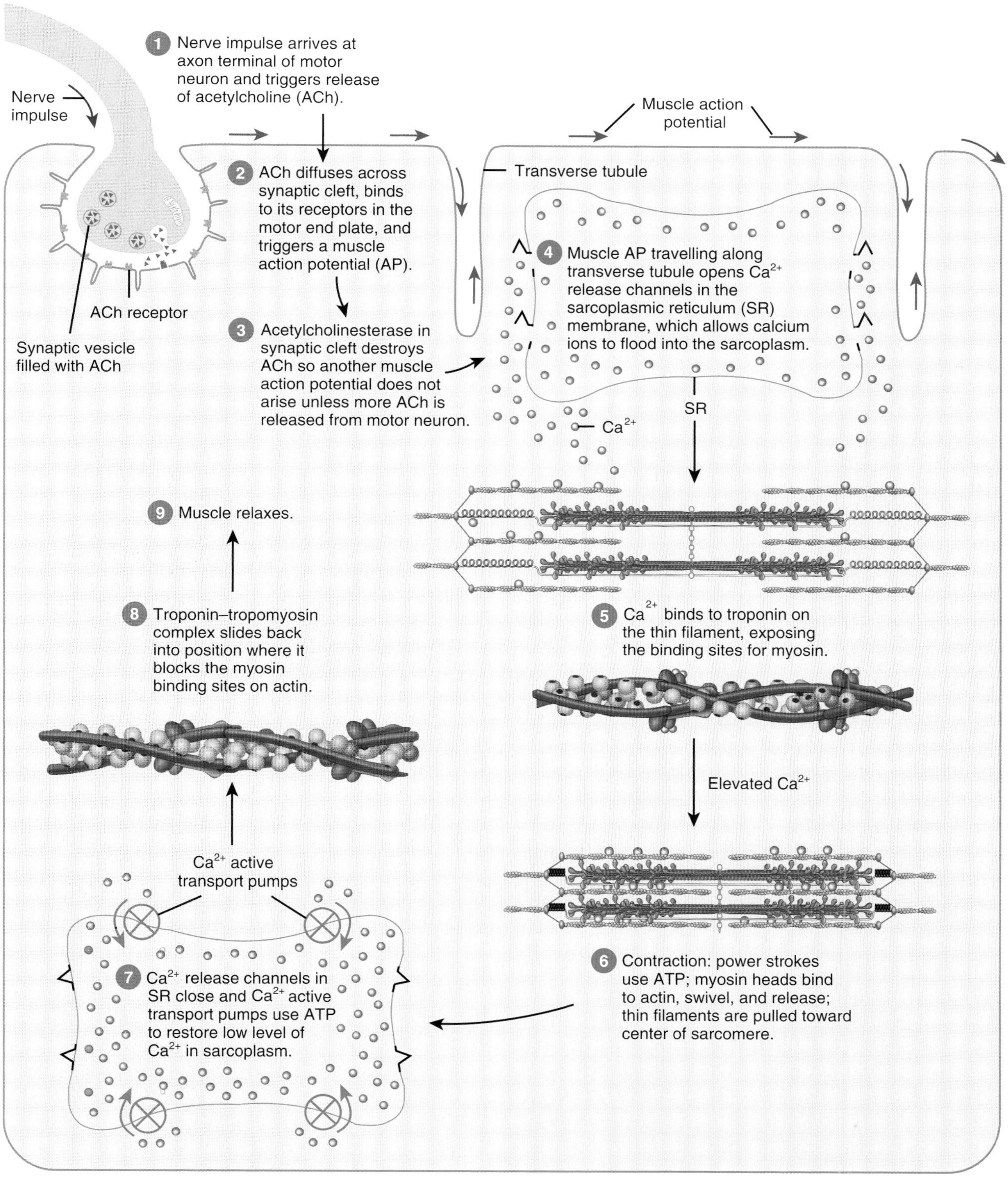

Which numbered steps in this figure are part of excitation–contraction coupling?

► CHECKPOINT

7. What roles do contractile, regulatory, and structural proteins play in muscle contraction and relaxation?
8. How do calcium ions and ATP contribute to muscle contraction and relaxation?
9. How does sarcomere length influence the maximum tension that is possible during muscle contraction?
10. How is the motor end plate different from other parts of the sarcolemma?

MUSCLE METABOLISM

► OBJECTIVES

Describe the reactions by which muscle fibers produce ATP.

Distinguish between anaerobic and aerobic cellular respiration.

Describe the factors that contribute to muscle fatigue.

Production of ATP in Muscle Fibers

Unlike most cells of the body, skeletal muscle fibers often switch between a low level of activity, when they are relaxed and using only a modest amount of ATP, and a high level of activity, when they are contracting and using ATP at a rapid pace. A huge amount of ATP is needed to power the contraction cycle, to pump Ca^{2+} into the sarcoplasmic reticulum, and for other metabolic reactions involved in muscle contraction. However, the ATP present inside muscle fibers is enough to power contraction for only a few seconds. If strenuous exercise continues past that time, the muscle fibers must make more ATP. Muscle fibers have three ways to produce ATP: (1) from creatine phosphate, (2) by anaerobic cellular respiration, and (3) by aerobic cellular respiration (Figure 10.12). The use of creatine phosphate for ATP production is unique to muscle fibers, but all body cells make ATP by the reactions of anaerobic and aerobic cellular respiration. We consider the events of cellular respiration briefly here and then in detail in Chapter 25.

Creatine Phosphate

While muscle fibers are relaxed, they produce more ATP than they need for resting metabolism. The excess ATP is used to synthesize **creatine phosphate,** an energy-rich molecule that is found only in muscle fibers (Figure 10.12a). The enzyme *creatine kinase (CK)* catalyzes the transfer of one of the high-energy phosphate groups from ATP to creatine, forming creatine phosphate and ADP. **Creatine** is a small, amino acid-like molecule that is synthesized in the liver, kidneys, and pancreas and then transported to muscle fibers. Creatine phosphate is three to six times more plentiful than ATP in the sarcoplasm of a relaxed muscle fiber. When contraction begins and the ADP level starts to rise, CK catalyzes the transfer of a high-energy phosphate group from creatine phosphate back to ADP. This direct phosphorylation reaction quickly regenerates new ATP molecules. Together, creatine phosphate and ATP provide enough energy for muscles to contract maximally for about 15 seconds. This amount of energy is sufficient for maximal short bursts of activity—for example, to run a 100-meter dash.

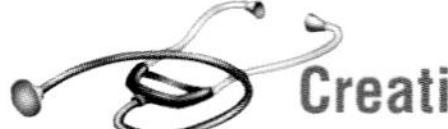

Creatine Supplementation

Creatine is both synthesized in the body (in the liver, kidneys, and pancreas) and derived from foods such as milk, red meat, and some fish. Adults need to synthesize and ingest a total of about 2 grams of creatine daily to make up for the urinary loss of creatinine, the breakdown product of creatine. Some studies have demonstrated improved performance during explosive movements, such as sprinting. Other studies, however, have failed to find a performance-enhancing effect of creatine supplementation. Moreover, ingesting extra creatine decreases the body's own synthesis of creatine, and it is not known whether natural synthesis recovers after long-term creatine supplementation. In addition, creatine supplementation can cause dehydration and may cause kidney dysfunction. Further research is needed to determine both the long-term safety and the value of creatine supplementation. ■

Anaerobic Cellular Respiration

Anaerobic cellular respiration is a series of ATP-producing reactions that do not require oxygen. When muscle activity continues and the supply of creatine phosphate within the muscle fiber is depleted, glucose is catabolized to generate ATP. Glucose easily passes from the blood into contracting muscle fibers via facilitated diffusion, and it is also produced by the breakdown of glycogen within muscle fibers (Figure 10.12b). Then, a series of 10 reactions known as *glycolysis* quickly breaks down each glucose molecule into two molecules of pyruvic acid. (Figure 25.4 on page 956 shows the reactions of glycolysis.) These reactions use two molecules of ATP but produce four, for a net gain of two molecules of ATP.

Ordinarily, the pyruvic acid formed by glycolysis in the cytosol enters mitochondria, where it undergoes a series of oxygen-requiring reactions called aerobic cellular respiration (described next) that produce a large amount of ATP. During some activities, however, not enough oxygen is available. In such cases, anaerobic reactions convert most of the pyruvic acid to lactic acid in the cytosol. About 80% of the lactic acid produced in this way diffuses out of the skeletal muscle fibers into the blood. Liver cells can convert some of the lactic acid back to glucose. In addition to providing new glucose molecules, this conversion reduces acidity of the blood. Anaerobic cellular respiration can provide enough energy for about 30 to 40 seconds of maximal muscle activity. Together, conversion of creatine phosphate and glycolysis can provide enough ATP to run a 400-meter race.

Aerobic Cellular Respiration

Muscle activity that lasts longer than half a minute depends increasingly on **aerobic cellular respiration,** a series of oxygen-requiring reactions that produce ATP in mitochondria. If sufficient oxygen is present, pyruvic acid enters the mitochondria, where it is completely oxidized in reactions that generate ATP, carbon dioxide, water, and heat (Figure 10.12c). Although aerobic cellular respiration is slower than glycolysis, it yields much more ATP. Each molecule of glucose yields about 36 molecules of ATP; a typical fatty acid molecule yields more than 100 molecules of ATP via aerobic cellular respiration.

Muscle tissue has two sources of oxygen: (1) oxygen that diffuses into muscle fibers from the blood and (2) oxygen released by myoglobin within muscle fibers. Both myoglobin (found only in muscle cells) and hemoglobin (found only in red blood cells) are oxygen-binding proteins. They bind oxygen when it is plentiful and release oxygen when it is scarce.

Aerobic cellular respiration supplies enough ATP for prolonged activity provided sufficient oxygen and nutrients are available. These nutrients include the pyruvic acid obtained from the glycolysis of glucose, fatty acids from the breakdown of triglycerides in adipose cells, and amino acids from the breakdown of proteins. In activities that last more than 10 minutes, the aerobic system provides more than 90% of the needed ATP. At the end of an endurance event such as a marathon race, nearly 100% of the ATP is being produced by aerobic cellular respiration.

Muscle Fatigue

The inability of a muscle to maintain force of contraction after prolonged activity is called **muscle fatigue.** Fatigue results mainly from changes within muscle fibers. Even before actual muscle fatigue occurs, a person may have feelings of tiredness and the desire to cease activity; this response, called *central*

Figure 10.12 **Production of ATP for muscle contraction.** (a) Creatine phosphate, formed from ATP while the muscle is relaxed, transfers a high-energy phosphate group to ADP, forming ATP, during muscle contraction. (b) Breakdown of muscle glycogen into glucose and production of pyruvic acid from glucose via glycolysis produce both ATP and lactic acid. Because no oxygen is needed, this is an anaerobic pathway. (c) Within mitochondria, pyruvic acid, fatty acids, and amino acids are used to produce ATP via aerobic cellular respiration, an oxygen-requiring set of reactions.

During a long-term event such as a marathon race, most ATP is produced aerobically.

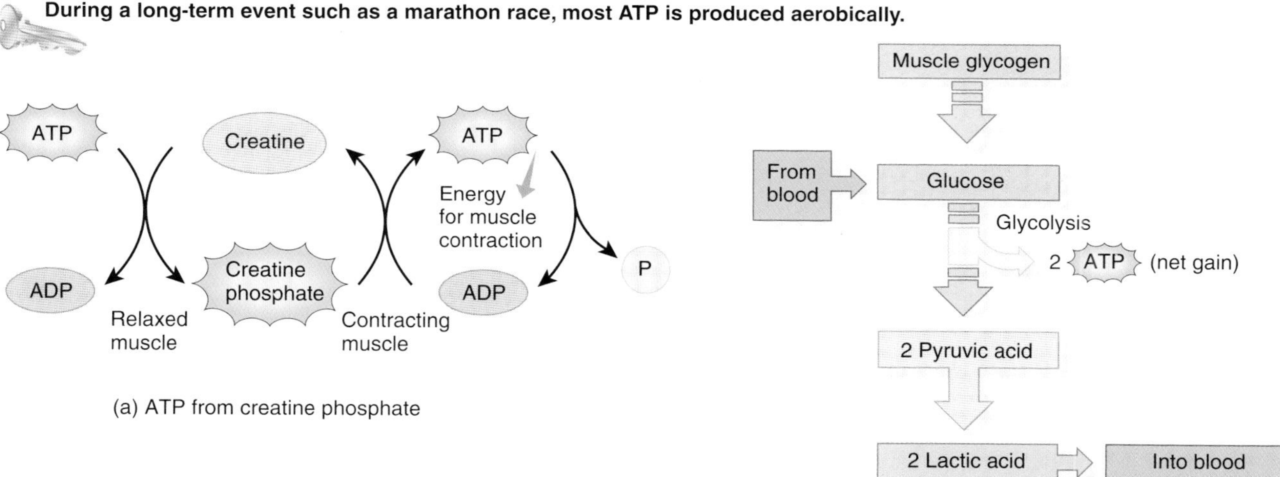

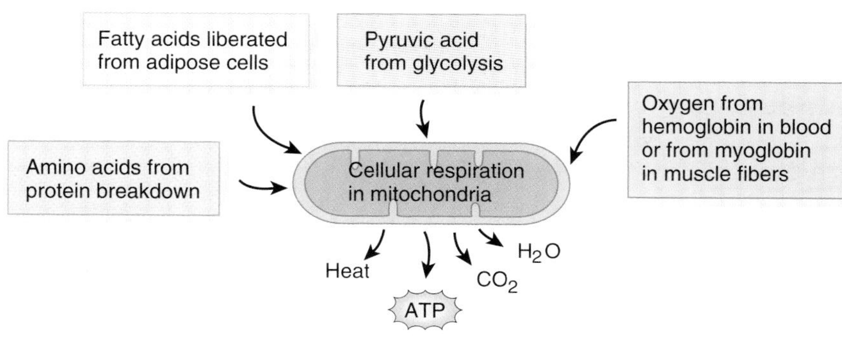

Where inside a skeletal muscle fiber are the events shown here occurring?

fatigue, is caused by changes in the central nervous system (brain and spinal cord). Although its exact mechanism is unknown, it may be a protective mechanism to stop a person from exercising before muscles become damaged. As you will see, certain types of skeletal muscle fibers fatigue more quickly than others.

Although the precise mechanisms that cause muscle fatigue are still not clear, several factors are thought to contribute. One is inadequate release of calcium ions from the SR, resulting in a decline of Ca^{2+} concentration in the sarcoplasm. Depletion of creatine phosphate also is associated with fatigue, but surprisingly, the ATP levels in fatigued muscle often are not much lower than those in resting muscle. Other factors that contribute to muscle fatigue include insufficient oxygen, depletion of glycogen and other nutrients, buildup of lactic acid and ADP, and failure of action potentials in the motor neuron to release enough acetylcholine.

Oxygen Consumption After Exercise

During prolonged periods of muscle contraction, increases in breathing rate and blood flow enhance oxygen delivery to muscle tissue. After muscle contraction has stopped, heavy breathing continues for a while, and oxygen consumption remains above the resting level. Depending on the intensity of the exercise, the recovery period may be just a few minutes, or it may last as long as several hours. The term **oxygen debt** refers to the added oxygen, over and above the resting oxygen consumption, that is taken into the body after exercise. This extra oxygen is used to "pay back" or restore metabolic conditions to the resting level in three ways: (1) to convert lactic acid back into glycogen stores in the liver, (2) to resynthesize creatine phosphate and ATP in muscle fibers, and (3) to replace the oxygen removed from myoglobin.

The metabolic changes that occur *during exercise* can account for only some of the extra oxygen used *after exercise.* Only a small amount of glycogen resynthesis occurs from lactic acid. Instead, most glycogen is made much later from dietary carbohydrates. Much of the lactic acid that remains after exercise is converted back to pyruvic acid and used for ATP production via aerobic cellular respiration in the heart, liver, kidneys, and skeletal muscle. Oxygen use after exercise also is boosted by ongoing changes. First, the elevated body temperature after strenuous exercise increases the rate of chemical reactions throughout the body. Faster reactions use ATP more rapidly, and more oxygen is needed to produce the ATP. Second, the heart and the muscles used in breathing are still working harder than they were at rest, and thus they consume more ATP. Third, tissue repair processes are occurring at an increased pace. For these reasons, **recovery oxygen uptake** is a better term than oxygen debt for the elevated use of oxygen after exercise.

▶ CHECKPOINT

11. Which ATP-producing reactions are aerobic and which are anaerobic?
12. Which sources provide ATP during a 1000-meter run?
13. What factors contribute to muscle fatigue?
14. Why is the term *recovery oxygen uptake* more accurate than *oxygen debt*?

CONTROL OF MUSCLE TENSION

▶ OBJECTIVES

Describe the structure and function of a motor unit, and define motor unit recruitment.

Explain the phases of a twitch contraction.

Describe how frequency of stimulation affects muscle tension, and how muscle tone is produced.

Distinguish between isotonic and isometric contractions.

A single nerve impulse in a somatic motor neuron elicits a single muscle action potential in all the skeletal muscle fibers with which it forms synapses. Action potentials always have the same size in a given neuron or muscle fiber. In contrast, the force of muscle fiber contraction does vary; a muscle fiber is capable of producing a much greater force than the one that results from a single action potential. The total force or tension that a single muscle fiber can produce depends mainly on the rate at which nerve impulses arrive at the neuromuscular junction. The number of impulses per second is the *frequency of stimulation.* Maximum tension is also affected by the amount of stretch before contraction (see Figure 10.9) and by nutrient and oxygen availability. The total tension a whole muscle can produce depends on the number of muscle fibers that are contracting in unison.

Motor Units

Even though each skeletal muscle fiber has only a single neuromuscular junction, the axon of a somatic motor neuron branches out and forms neuromuscular junctions with many different muscle fibers. A **motor unit** consists of a somatic motor neuron plus all the skeletal muscle fibers it stimulates (Figure 10.13). A single somatic motor neuron makes contact with an average of 150 skeletal muscle fibers, and all of the muscle fibers in one motor unit contract in unison. Typically, the muscle fibers of a motor unit are dispersed throughout a muscle rather than clustered together.

Whole muscles that control precise movements consist of many small motor units. For instance, muscles of the larynx (voice box) that control voice production have as few as two or three muscle fibers per motor unit, and muscles controlling eye movements may have 10 to 20 muscle fibers per motor unit. In contrast, skeletal muscles responsible for large-scale and powerful movements, such as the biceps brachii muscle in the arm and the gastrocnemius muscle in the calf of the leg, have as many as 2000 to 3000 muscle fibers in some motor units. Because all

Figure 10.13 **Motor units.** Two somatic motor neurons (one purple and one green) are shown, each supplying the muscle fibers of its motor unit.

A motor unit consists of a somatic motor neuron plus all the muscle fibers it stimulates.

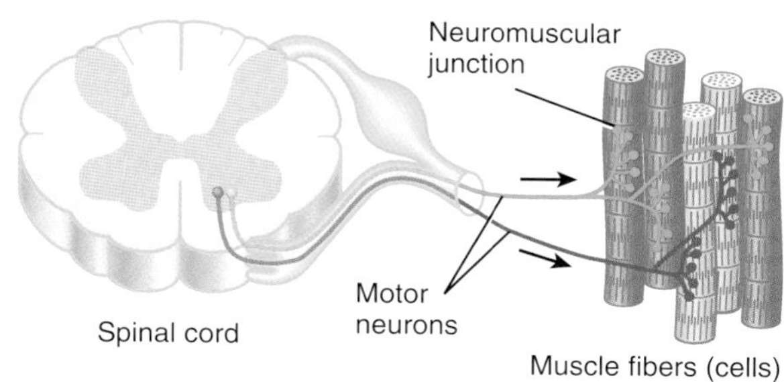

What is the effect of the size of a motor unit on its strength of contraction? (Assume that each muscle fiber can generate about the same amount of tension.)

Figure 10.14 **Myogram of a twitch contraction.** The arrow indicates the time at which the stimulus occurred.

A myogram is a record of a muscle contraction.

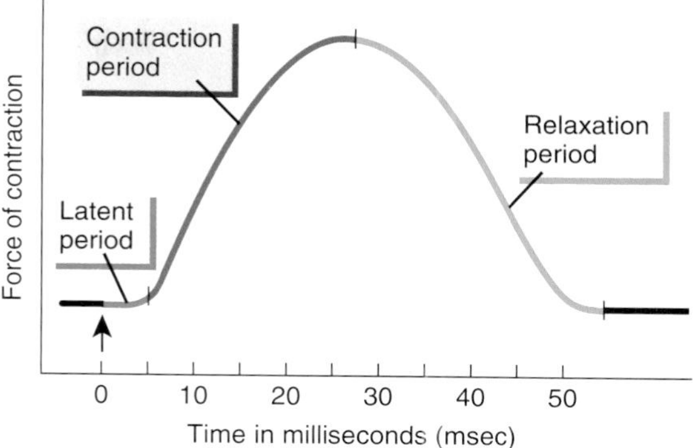

What events occur during the latent period?

the muscle fibers of a motor unit contract and relax together, the total strength of a contraction depends, in part, on the size of the motor units and the number that are activated at a given time.

Twitch Contraction

A **twitch contraction** is the brief contraction of all the muscle fibers in a motor unit in response to a single action potential in its motor neuron. In the laboratory, a twitch can be produced by direct electrical stimulation of a motor neuron or its muscle fibers. The record of a muscle contraction, called a **myogram,** is shown in Figure 10.14. Twitches of skeletal muscle fibers last anywhere from 20 to 200 msec. This is very long compared to the brief 1–2 msec* that a muscle action potential lasts.

Note that a brief delay occurs between application of the stimulus (time zero on the graph) and the beginning of contraction. The delay, which lasts about two milliseconds, is termed the **latent period.** During the latent period, the muscle action potential sweeps over the sarcolemma and calcium ions are released from the sarcoplasmic reticulum. The second phase, the **contraction period,** lasts 10–100 msec. During this time, Ca^{2+} binds to troponin, myosin-binding sites on actin are exposed, and crossbridges form. Peak tension develops in the muscle fiber. During the third phase, the **relaxation period,** also lasting 10–100 msec, Ca^{2+} is actively transported back into the sarcoplasmic reticulum, myosin-binding sites are covered by tropomyosin, myosin heads detach from actin, and tension in the muscle fiber decreases. The actual duration of these periods depends on the type of skeletal muscle fiber. Some fibers, such as the fast-twitch fibers that move the eyes (described shortly), have contraction periods as brief as 10 msec and equally brief relaxation periods. Others, such as the slow-twitch fibers that move the legs, have contraction and relaxation periods of about 100 msec each.

If two stimuli are applied, one immediately after the other, the muscle will respond to the first stimulus but not to the second. When a muscle fiber receives enough stimulation to contract, it temporarily loses its excitability and cannot respond for a time. The period of lost excitability, called the **refractory period,** is a characteristic of all muscle and nerve cells. The duration of the refractory period varies with the muscle involved. Skeletal muscle has a short refractory period of about five milliseconds; cardiac muscle has a longer refractory period of about 300 milliseconds.

Frequency of Stimulation

When a second stimulus occurs after the refractory period of the first stimulus is over, but before the skeletal muscle fiber has relaxed, the second contraction will actually be stronger than the first (Figure 10.15b). This phenomenon, in which stimuli arriving at different times cause larger contractions, is called **wave summation.** When a skeletal muscle fiber is stimulated at a rate of 20 to 30 times per second, it can only partially relax between stimuli. The result is a sustained but wavering contraction called **unfused (incomplete) tetanus** (*tetan-* = rigid, tense; Figure 10.15c). When a skeletal muscle fiber is stimulated at a higher rate of 80 to 100 times per second, it does not relax at all. The result is **fused (complete) tetanus,** a sustained contraction in which individual twitches cannot be detected (Figure 10.15d).

Wave summation and both kinds of tetanus occur when additional Ca^{2+} is released from the sarcoplasmic reticulum by subsequent stimuli while the levels of Ca^{2+} in the sarcoplasm are still elevated from the first stimulus. Because of the buildup in the Ca^{2+} level, the peak tension generated during fused tetanus is 5 to 10 times larger than the peak tension produced during a single twitch. Even so, smooth, sustained voluntary

*One millisecond (msec) is 10^{-3} seconds (0.001 sec).

Figure 10.15 Myograms showing the effects of different frequencies of stimulation. (a) Single twitch. (b) When a second stimulus occurs before the muscle fiber has relaxed, the second contraction is stronger than the first, a phenomenon called wave summation. (The dashed line indicates the force of contraction expected in a single twitch.) (c) Unfused tetanus produces a jagged curve due to partial relaxation of the muscle fiber between stimuli. (d) In fused tetanus, which occurs when there are 80–100 stimuli per second, the myogram line, like the contraction force, is steady and sustained.

Due to wave summation, the tension produced during a sustained contraction is greater than that produced by a single twitch.

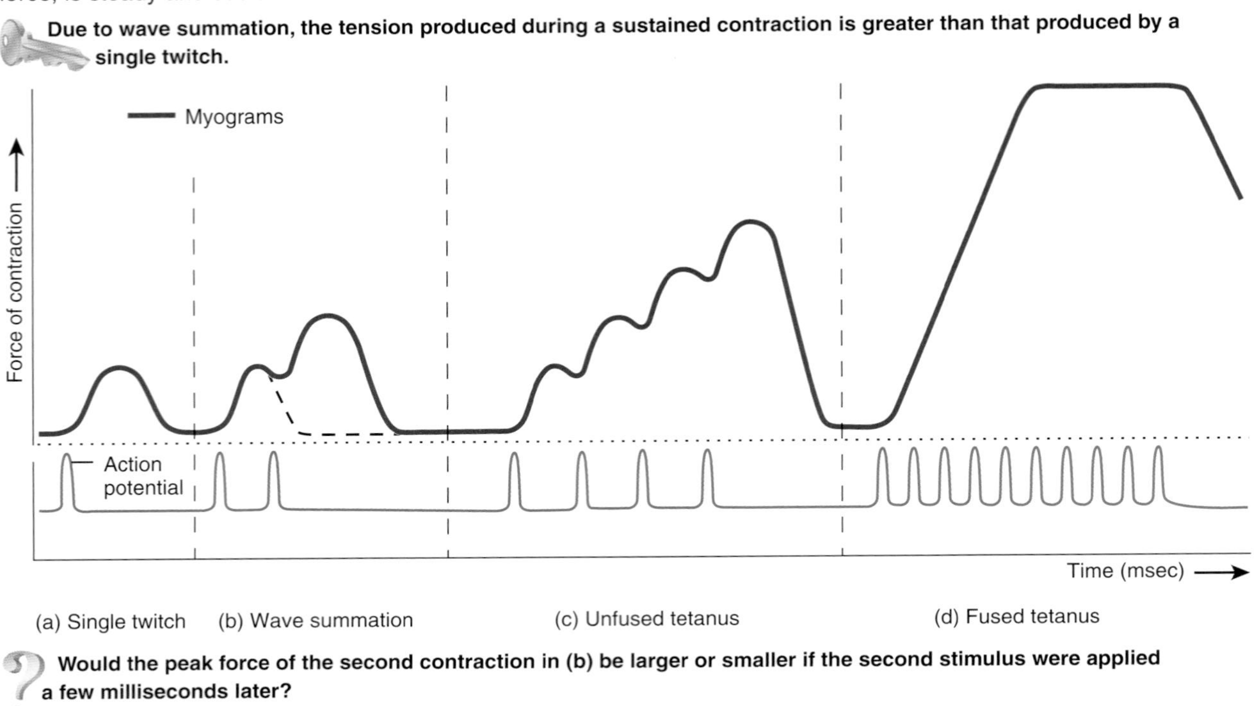

Would the peak force of the second contraction in (b) be larger or smaller if the second stimulus were applied a few milliseconds later?

muscle contractions are achieved mainly by out-of-synchrony unfused tetanus in different motor units.

The stretch of elastic components, such as tendons and connective tissues around muscle fibers, also affects wave summation. During wave summation, elastic components are not given much time to spring back between contractions, and thus remain taut. While in this state, the elastic components do not require very much stretching before the beginning of the next muscular contraction. The combination of the tautness of the elastic components and the partially contracted state of the filaments enables the force of another contraction to be greater than the one before.

Motor Unit Recruitment

The process in which the number of active motor units increases is called **motor unit recruitment.** Typically, the different motor units of an entire muscle are not stimulated to contract in unison. While some motor units are contracting, others are relaxed. This pattern of motor unit activity delays muscle fatigue and allows contraction of a whole muscle to be sustained for long periods. The weakest motor units are recruited first, with progressively stronger motor units added if the task requires more force.

Recruitment is one factor responsible for producing smooth movements rather than a series of jerks. As mentioned, the number of muscle fibers innervated by one motor neuron varies greatly. Precise movements are brought about by small changes in muscle contraction. Therefore, the small muscles that produce precise movements are made up of small motor units. For this reason, when a motor unit is recruited or turned off, only slight changes occur in muscle tension. By contrast, large motor units are active where large tension is needed and precision is less important.

Aerobic Training versus Strength Training

Regular, repeated activities such as jogging or aerobic dancing increase the supply of oxygen-rich blood available to skeletal muscles for aerobic cellular respiration. By contrast, activities such as weight lifting rely more on anaerobic production of ATP through glycolysis. Such anaerobic activities stimulate synthesis of muscle proteins and result, over time, in increased muscle size (muscle hypertrophy). As a result, aerobic training builds endurance for prolonged activities; in contrast, anaerobic training builds muscle strength for short-term feats. **Interval training** is a workout regimen that incorporates both types of training—for example, alternating sprints with jogging. ■

Muscle Tone

Even at rest, a skeletal muscle exhibits **muscle tone** (*tonos* = tension), a small amount of tautness or tension in the muscle due to weak, involuntary contractions of its motor units. Recall that skeletal muscle contracts only after it is activated by acetylcholine released by nerve impulses in its motor neurons. Hence, muscle tone is established by neurons in the brain and spinal cord that excite the muscle's motor neurons. When the motor neurons serving a skeletal muscle are damaged or cut, the muscle becomes **flaccid** (FLAK-sid or FLAS-sid = flabby), a state of limpness in which muscle tone is lost. To sustain muscle tone, small groups of motor units are alternately active and inactive in a constantly shifting pattern. Muscle tone keeps skeletal muscles firm, but it does not result in a force strong enough to produce movement. For example, when the muscles in the back of the neck are in normal tonic contraction, they keep the head upright and prevent it from slumping forward on the chest. Muscle tone also is important in smooth muscle tissues, such as those found in the gastrointestinal tract, where the walls of the digestive organs maintain a steady pressure on their contents. The tone of smooth muscle fibers in the walls of blood vessels plays a crucial role in maintaining blood pressure.

Hypotonia and Hypertonia

Hypotonia (*hypo-* = below) refers to decreased or lost muscle tone. Such muscles are said to be flaccid. Flaccid muscles are loose and appear flattened rather than rounded; the affected limbs are hyperextended. Certain disorders of the nervous system and disruptions in the balance of electrolytes (especially sodium, calcium, and, to a lesser extent, magnesium) may result in **flaccid paralysis,** which is characterized by loss of muscle tone, loss or reduction of tendon reflexes, and atrophy (wasting away) and degeneration of muscles.

Hypertonia (*hyper-* = above) refers to increased muscle tone and is expressed in two ways: spasticity or rigidity. **Spasticity** (spas-TIS-i-tē) is characterized by increased muscle tone (stiffness) associated with an increase in tendon reflexes and pathological reflexes (such as the Babinski sign, in which the great toe extends with or without fanning of the other toes in response to stroking the outer margin of the sole). Certain disorders of the nervous system and electrolyte disturbances such as those previously noted may result in **spastic paralysis,** partial paralysis in which the muscles exhibit spasticity. **Rigidity** refers to increased muscle tone in which reflexes are not affected, as occurs in tetanus. ■

Isotonic and Isometric Contractions

Muscle contractions are classified as either isotonic or isometric. In an **isotonic contraction** (*iso-* = equal; *-tonic* = tension), the tension (force of contraction) developed by the muscle remains almost constant while the muscle changes its length. Isotonic contractions are used for body movements and for moving objects. The two types of isotonic contractions are concentric and eccentric. In a **concentric isotonic contraction,** if the tension generated is great enough to overcome the resistance of the object to be moved, the muscle shortens and pulls on another structure, such as a tendon, to produce movement and to reduce the angle at a joint. Picking a book up off a table involves concentric isotonic contractions of the biceps brachii muscle in the arm (Figure 10.16a). By contrast, as you lower the book to place it back on the table, the previously shortened biceps lengthens in a controlled manner while it continues to contract. When the length of a muscle increases during a contraction, the contraction is an **eccentric isotonic contraction** (Figure 10.16b).

Figure 10.16 Comparison between isotonic (concentric and eccentric) and isometric contractions. Parts (a) and (b) show isotonic contractions of the biceps brachii muscle in the arm; part (c) shows isometric contraction of shoulder and arm muscles.

In an isotonic contraction, tension remains constant as muscle length decreases or increases; in an isometric contraction, tension increases greatly without a change in muscle length.

(a) Concentric contraction while picking up a book

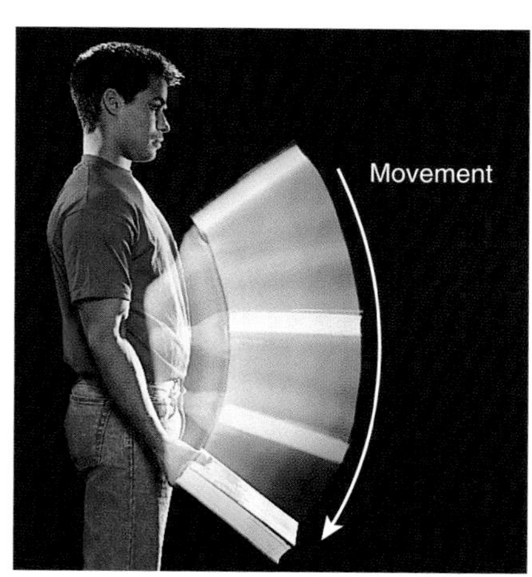

(b) Eccentric contraction while lowering a book

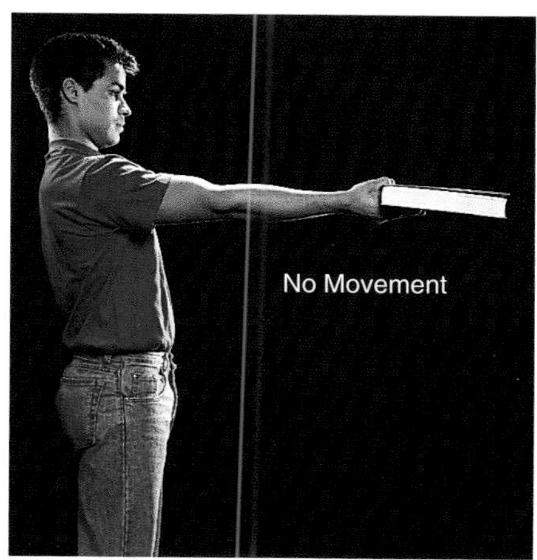

(c) Isometric contraction while holding a book steady

What type of contraction occurs in your neck muscles while you are walking?

During an eccentric contraction, the tension exerted by the myosin crossbridges resists movement of a load (the book, in this case) and slows the lengthening process. For reasons that are not well understood, repeated eccentric isotonic contractions (for example, walking downhill) produce more muscle damage and more delayed-onset muscle soreness than do concentric isotonic contractions.

In an **isometric contraction** (*-metro* = measure or length), the tension generated is not enough to exceed the resistance of the object to be moved and the muscle does not change its length. An example would be holding a book steady using an outstretched arm (Figure 10.16c). These contractions are important for maintaining posture and for supporting objects in a fixed position. Although isometric contractions do not result in body movement, energy is still expended. The book pulls the arm downward, stretching the shoulder and arm muscles. The isometric contraction of the shoulder and arm muscles counteracts the stretch. Isometric contractions are important because they stabilize some joints as others are moved. Most activities include both isotonic and isometric contractions.

▶ CHECKPOINT

15. How are the sizes of motor units related to the degree of muscular control they allow?

16. What is motor unit recruitment?

17. Why is muscle tone important?

18. Define each of following terms: concentric isotonic contraction, eccentric isotonic contraction, and isometric contraction.

19. Demonstrate an isotonic contraction. How does it feel? What do you think causes the physical discomfort you are experiencing?

TYPES OF SKELETAL MUSCLE FIBERS

▶ OBJECTIVE

Compare the structure and function of the three types of skeletal muscle fibers.

Skeletal muscle fibers are not all alike in composition and function. For example, muscle fibers vary in their content of myoglobin, the red-colored protein that binds oxygen in muscle fibers. Skeletal muscle fibers that have a high myoglobin content are termed *red muscle fibers* and appear darker (the dark meat in chicken legs and thighs); those that have a low content of myoglobin are called *white muscle fibers* and appear lighter (the white meat in chicken breasts). Red muscle fibers also contain more mitochondria and are supplied by more blood capillaries.

Skeletal muscle fibers also contract and relax at different speeds, and vary in which metabolic reactions they use to generate ATP and in how quickly they fatigue. For example, a fiber is categorized as either slow or fast depending on how rapidly the ATPase in its myosin heads hydrolyzes ATP. Based on all these structural and functional characteristics, skeletal muscle fibers are classified into three main types: (1) slow oxidative fibers, (2) fast oxidative-glycolytic fibers, and (3) fast glycolytic fibers.

Slow Oxidative Fibers

Slow oxidative (SO) fibers are smallest in diameter and thus are the least powerful type of muscle fibers. They appear dark red because they contain large amounts of myoglobin and many blood capillaries. Because they have many large mitochondria, SO fibers generate ATP mainly by aerobic cellular respiration, which is why they are called oxidative fibers. These fibers are said to be "slow" because the ATPase in the myosin heads hydrolyzes ATP relatively slowly and the contraction cycle proceeds at a slower pace than in "fast" fibers. As a result, SO fibers have a slow speed of contraction. Their twitch contractions last from 100 to 200 msec, and they take longer to reach peak tension. However, slow fibers are very resistant to fatigue and are capable of prolonged, sustained contractions for many hours. These slow-twitch, fatigue-resistant fibers are adapted for maintaining posture and for aerobic, endurance-type activities such as running a marathon.

Fast Oxidative-Glycolytic Fibers

Fast oxidative-glycolytic (FOG) fibers are intermediate in diameter between the other two types of fibers. Like slow oxidative fibers, they contain large amounts of myoglobin and many blood capillaries. Thus, they also have a dark red appearance. FOG fibers can generate considerable ATP by aerobic cellular respiration, which gives them a moderately high resistance to fatigue. Because their intracellular glycogen level is high, they also generate ATP by anaerobic glycolysis. FOG fibers are "fast" because the ATPase in their myosin heads hydrolyzes ATP three to five times faster than the myosin ATPase in SO fibers, which makes their speed of contraction faster. Thus, twitches of FOG fibers reach peak tension more quickly than those of SO fibers but are briefer in duration—less than 100 msec. FOG fibers contribute to activities such as walking and sprinting.

Fast Glycolytic Fibers

Fast glycolytic (FG) fibers are largest in diameter and contain the most myofibrils. Hence, they can generate the most powerful contractions. FG fibers have low myoglobin content, relatively few blood capillaries, few mitochondria, and appear white in color. They contain large amounts of glycogen and generate ATP mainly by glycolysis. Due to their large size and their ability to hydrolyze ATP rapidly, FG fibers contract strongly and quickly. These fast-twitch fibers are adapted for intense anaerobic movements of short duration, such as weight lifting or throwing a ball, but they fatigue quickly. Strength training programs that engage a person in activities requiring great strength for short times increase the size, strength, and glycogen content of fast glycolytic fibers. The FG fibers of a weight lifter may be 50% larger than those of a sedentary person or endurance athlete. The increase in size is due to increased synthesis of muscle proteins. The overall result is muscle enlargement due to hypertrophy of the FG fibers.

Distribution and Recruitment of Different Types of Fibers

Most skeletal muscles are a mixture of all three types of skeletal muscle fibers; about half the fibers in a typical skeletal muscle are SO fibers. However, the proportions vary somewhat, depending on the action of the muscle, the person's training regimen, and genetic factors. For example, the continually active postural muscles of the neck, back, and legs have a high proportion of SO fibers. Muscles of the shoulders and arms, in contrast, are not constantly active but are used briefly now and then to produce large amounts of tension, such as in lifting and throwing. These muscles have a high proportion of FG fibers. Leg muscles, which not only support the body but are also used for walking and running, have large numbers of both SO and FOG fibers.

Within a particular motor unit, all of the skeletal muscle fibers are of the same type. The different motor units in a muscle are recruited in a specific order, depending on need. For example, if weak contractions suffice to perform a task, only SO motor units are activated. If more force is needed, the motor units of FOG fibers are also recruited. Finally, if maximal force is required, motor units of FG fibers are also called into action. Activation of various motor units is controlled by the brain and spinal cord.

Table 10.1 summarizes the characteristics of the three types of skeletal muscle fibers.

TABLE 10.1 Characteristics of the Three Types of Skeletal Muscle Fibers

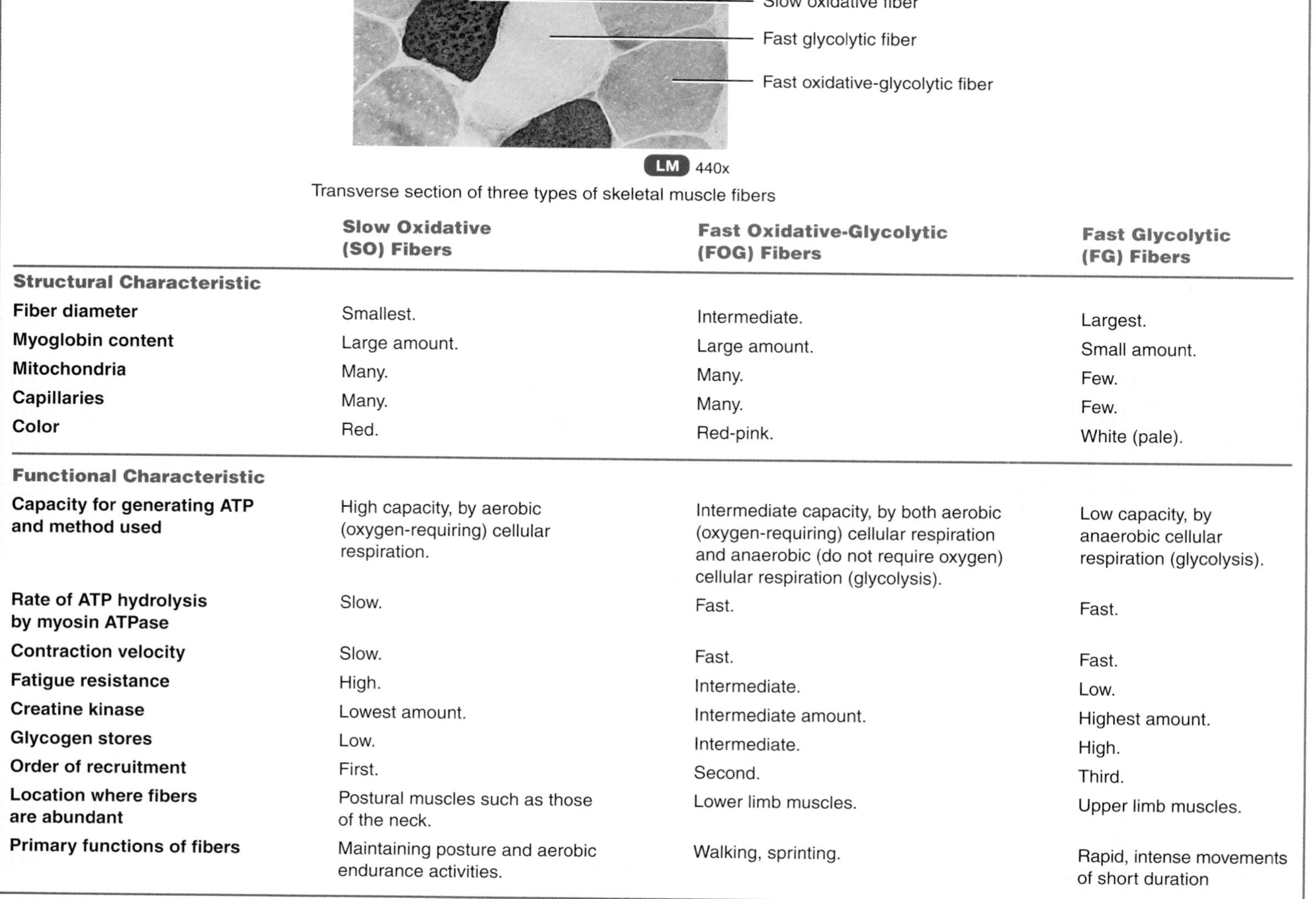

Transverse section of three types of skeletal muscle fibers

	Slow Oxidative (SO) Fibers	Fast Oxidative-Glycolytic (FOG) Fibers	Fast Glycolytic (FG) Fibers
Structural Characteristic			
Fiber diameter	Smallest.	Intermediate.	Largest.
Myoglobin content	Large amount.	Large amount.	Small amount.
Mitochondria	Many.	Many.	Few.
Capillaries	Many.	Many.	Few.
Color	Red.	Red-pink.	White (pale).
Functional Characteristic			
Capacity for generating ATP and method used	High capacity, by aerobic (oxygen-requiring) cellular respiration.	Intermediate capacity, by both aerobic (oxygen-requiring) cellular respiration and anaerobic (do not require oxygen) cellular respiration (glycolysis).	Low capacity, by anaerobic cellular respiration (glycolysis).
Rate of ATP hydrolysis by myosin ATPase	Slow.	Fast.	Fast.
Contraction velocity	Slow.	Fast.	Fast.
Fatigue resistance	High.	Intermediate.	Low.
Creatine kinase	Lowest amount.	Intermediate amount.	Highest amount.
Glycogen stores	Low.	Intermediate.	High.
Order of recruitment	First.	Second.	Third.
Location where fibers are abundant	Postural muscles such as those of the neck.	Lower limb muscles.	Upper limb muscles.
Primary functions of fibers	Maintaining posture and aerobic endurance activities.	Walking, sprinting.	Rapid, intense movements of short duration

▶ CHECKPOINT

20. Why are some skeletal muscle fibers classified as "fast" and others are said to be "slow"?

21. In what order are the various types of skeletal muscle fibers recruited when you sprint to make it to the bus stop?

EXERCISE AND SKELETAL MUSCLE TISSUE

▶ OBJECTIVE

Describe the effects of exercise on different types of skeletal muscle fibers.

The relative ratio of fast glycolytic (FG) and slow oxidative (SO) fibers in each muscle is genetically determined and helps account for individual differences in physical performance. For example, people with a higher proportion of FG fibers often excel in activities that require periods of intense activity, such as weight lifting or sprinting. People with higher percentages of SO fibers are better at activities that require endurance, such as long-distance running.

Although the total number of skeletal muscle fibers usually does not increase, the characteristics of those present can change to some extent. Various types of exercises can induce changes in the fibers in a skeletal muscle. Endurance-type (aerobic) exercises, such as running or swimming, cause a gradual transformation of some FG fibers into fast oxidative-glycolytic (FOG) fibers. The transformed muscle fibers show slight increases in diameter, number of mitochondria, blood supply, and strength. Endurance exercises also result in cardiovascular and respiratory changes that cause skeletal muscles to receive better supplies of oxygen and nutrients but do not increase muscle mass. By contrast, exercises that require great strength for short periods produce an increase in the size and strength of FG fibers. The increase in size is due to increased synthesis of thick and thin filaments. The overall result is muscle enlargement (hypertrophy), as evidenced by the bulging muscles of body builders.

Anabolic Steroids

The use of **anabolic steroids** by athletes has received widespread attention. These steroid hormones, similar to testosterone, are taken to increase muscle size and thus strength during athletic contests. However, the large doses needed to produce an effect have damaging, sometimes even devastating side effects, including liver cancer, kidney damage, increased risk of heart disease, stunted growth, wide mood swings, increased acne, and increased irritability and aggression. Additionally, females who take anabolic steroids may experience atrophy of the breasts and uterus, menstrual irregularities, sterility, facial hair growth, and deepening of the voice. Males may experience diminished testosterone secretion, atrophy of the testes, sterility, and baldness. ■

▶ CHECKPOINT

22. On a cellular level, what causes muscle hypertrophy?

CARDIAC MUSCLE TISSUE

▶ OBJECTIVE

Describe the main structural and functional characteristics of cardiac muscle tissue.

The principal tissue in the heart wall is **cardiac muscle tissue** (described in more detail in Chapter 20 and illustrated in Figure 20.9 on page 709). Between the layers of **cardiac muscle fibers,** the contractile cells of the heart, are sheets of connective tissue that contain blood vessels, nerves, and the conduction system of the heart. Cardiac muscle fibers have the same arrangement of actin and myosin and the same bands, zones, and Z discs as skeletal muscle fibers. However, *intercalated discs* (in-TER-ka-lāt-ed; *intercal-* = to insert between) are unique to cardiac muscle fibers. These microscopic structures are irregular transverse thickenings of the sarcolemma that connect the ends of cardiac muscle fibers to one another. The discs contain *desmosomes,* which hold the fibers together, and *gap junctions,* which allow muscle action potentials to spread from one cardiac muscle fiber to another (see Figure 4.1e on page 109).

In response to a single action potential, cardiac muscle tissue remains contracted 10 to 15 times longer than skeletal muscle tissue (see Figure 20.11 on page 712). The long contraction is due to prolonged delivery of Ca^{2+} into the sarcoplasm. In cardiac muscle fibers, Ca^{2+} enters the sarcoplasm both from the sarcoplasmic reticulum (as in skeletal muscle fibers) and from the interstitial fluid that bathes the fibers. Because the channels that allow inflow of Ca^{2+} from interstitial fluid stay open for a relatively long time, a cardiac muscle contraction lasts much longer than a skeletal muscle twitch.

We have seen that skeletal muscle tissue contracts only when stimulated by acetylcholine released by a nerve impulse in a motor neuron. In contrast, cardiac muscle tissue contracts when stimulated by its own autorhythmic muscle fibers. Under normal resting conditions, cardiac muscle tissue contracts and relaxes about 75 times a minute. This continuous, rhythmic activity is a major physiological difference between cardiac and skeletal muscle tissue. The mitochondria in cardiac muscle fibers are larger and more numerous than in skeletal muscle fibers. This structural feature correctly suggests that cardiac muscle depends largely on aerobic cellular respiration to generate ATP, and thus requires a constant supply of oxygen. Cardiac muscle fibers can also use lactic acid produced by skeletal muscle fibers to make ATP, a benefit during exercise.

▶ CHECKPOINT

23. What are the similarities among and differences between skeletal and cardiac muscle?

SMOOTH MUSCLE TISSUE

▶ OBJECTIVE

Describe the main structural and functional characteristics of smooth muscle tissue.

Like cardiac muscle tissue, **smooth muscle tissue** is usually activated involuntarily. Of the two types of smooth muscle tissue, the more common type is **visceral (single-unit) smooth muscle tissue** (Figure 10.17a). It is found in tubular arrangements that form part of the walls of small arteries and veins and of hollow organs such as the stomach, intestines, uterus, and urinary bladder. Like cardiac muscle, visceral smooth muscle is autorhythmic. The fibers connect to one another by gap junctions, forming a network through which muscle action potentials can spread. When a neurotransmitter, hormone, or autorhythmic signal stimulates one fiber, the muscle action potential is transmitted to neighboring fibers, which then contract in unison, as a single unit.

The second type of smooth muscle tissue, **multiunit smooth muscle tissue** (Figure 10.17b), consists of individual fibers, each with its own motor neuron terminals and with few gap junctions between neighboring fibers. Stimulation of one visceral muscle fiber causes contraction of many adjacent fibers, but stimulation of one multiunit fiber causes contraction of that fiber only. Multiunit smooth muscle tissue is found in the walls of large arteries, in airways to the lungs, in the arrector pili muscles that attach to hair follicles, in the muscles of the iris that adjust pupil diameter, and in the ciliary body that adjusts focus of the lens in the eye.

Microscopic Anatomy of Smooth Muscle

A single relaxed smooth muscle fiber is 30–200 μm long. It is thickest in the middle (3–8 μm) and tapers at each end (Figure 10.18). Within each fiber is a single, oval, centrally located nucleus. The sarcoplasm of smooth muscle fibers contains both *thick filaments* and *thin filaments,* in ratios between 1:10 and 1:15, but they are not arranged in orderly sarcomeres as in striated muscle. Smooth muscle fibers also contain *intermediate filaments.* Because the various filaments have no regular pattern of overlap, smooth muscle fibers do not exhibit striations (see Table 4.5C on page 136), causing a smooth appearance. Smooth muscle fibers also lack transverse tubules and have only a small

Figure 10.17 **Two types of smooth muscle tissue.** In (a), one autonomic motor neuron synapses with several visceral smooth muscle fibers, and action potentials spread to neighboring fibers through gap junctions. In (b), three autonomic motor neurons synapse with individual multiunit smooth muscle fibers. Stimulation of one multiunit fiber causes contraction of that fiber only.

Visceral smooth muscle fibers connect to one another by gap junctions and contract as a single unit. Multiunit smooth muscle fibers lack gap junctions and contract independently.

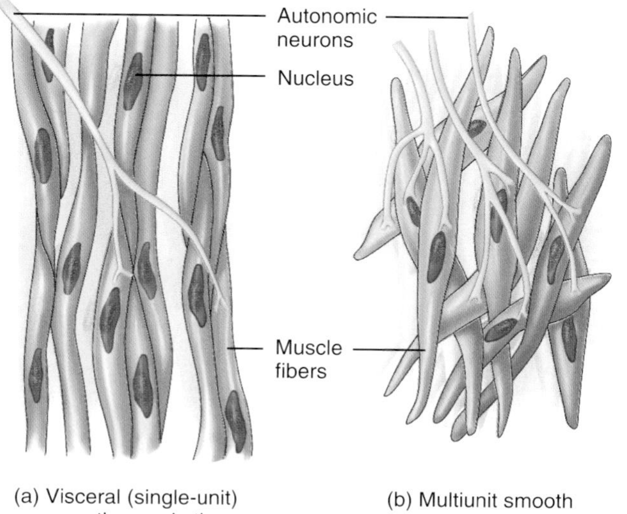

Which type of smooth muscle is more like cardiac muscle than skeletal muscle, with respect to both its structure and function?

Figure 10.18 **Microscopic anatomy of a smooth muscle fiber.** A photomicrograph of smooth muscle is shown in Table 4.5C on page 136.

Smooth muscle fibers have thick and thin filaments but no transverse tubules and scanty sarcoplasmic reticulum.

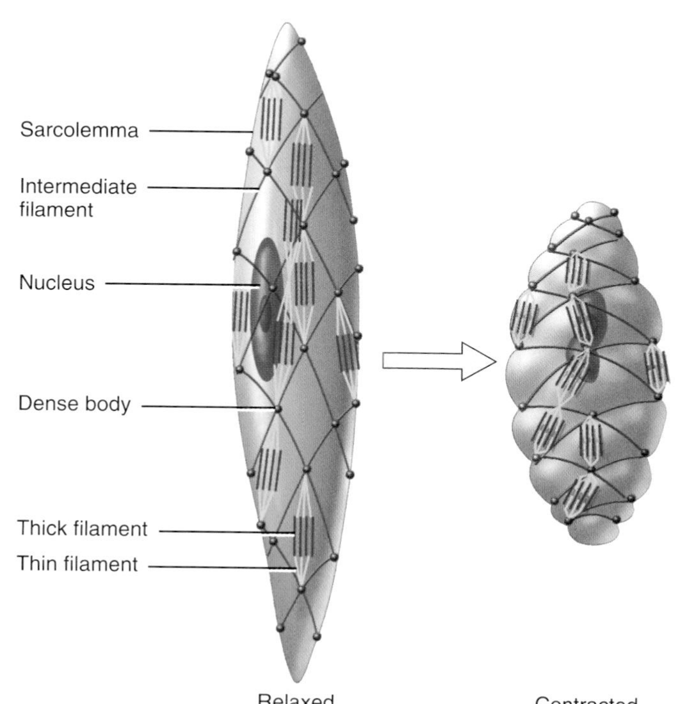

How does the speed of onset and duration of contraction in a smooth muscle fiber compare with that in a skeletal muscle fiber?

amount of sarcoplasmic reticulum for storage of Ca^{2+}. Although there are no transverse tubules in smooth muscle tissue, there are small pouchlike invaginations of the plasma membrane called **caveolae** (kav′-ē-Ō-lē; *cavus* = space) that contain extracellular Ca^{2+} that can be used for muscular contraction.

In smooth muscle fibers, the thin filaments attach to structures called **dense bodies,** which are functionally similar to Z discs in striated muscle fibers. Some dense bodies are dispersed throughout the sarcoplasm; others are attached to the sarcolemma. Bundles of intermediate filaments also attach to dense bodies and stretch from one dense body to another (Figure 10.18). During contraction, the sliding filament mechanism involving thick and thin filaments generates tension that is transmitted to intermediate filaments. These, in turn, pull on the dense bodies attached to the sarcolemma, causing a lengthwise shortening of the muscle fiber. As a smooth muscle fiber contracts, it rotates as a corkscrew turns. The fiber twists in a helix as it contracts, and rotates in the opposite direction as it relaxes.

Physiology of Smooth Muscle

Although the principles of contraction are similar, smooth muscle tissue exhibits some important physiological differences from cardiac and skeletal muscle tissue. Contraction in a smooth muscle fiber starts more slowly and lasts much longer than skeletal muscle fiber contraction. Another difference is that smooth muscle can both shorten and stretch to a greater extent than the other muscle types.

An increase in the concentration of Ca^{2+} in the cytosol of a smooth muscle fiber initiates contraction, just as in striated muscle. Sarcoplasmic reticulum (the reservoir for Ca^{2+} in striated muscle) is found in small amounts in smooth muscle. Calcium ions flow into smooth muscle cytosol from both the interstitial fluid and sarcoplasmic reticulum. Because there are no transverse tubules in smooth muscle fibers (there are caveolae instead), it takes longer for Ca^{2+} to reach the filaments in the center of the fiber and trigger the contractile process. This accounts, in part, for the slow onset of contraction of smooth muscle.

Several mechanisms regulate contraction and relaxation of smooth muscle cells. In one such mechanism, a regulatory protein called **calmodulin** (cal-MOD-ū-lin) binds to Ca^{2+} in the cytosol. (Recall that troponin takes this role in striated muscle fibers.) After binding to Ca^{2+}, calmodulin activates an enzyme called *myosin light chain kinase.* This enzyme uses ATP to add a phosphate group to a portion of the myosin head. Once the phosphate group is attached, the myosin head can bind to actin, and contraction can occur. Because myosin light chain kinase works rather slowly, it contributes to the slowness of smooth muscle contraction.

Not only do calcium ions enter smooth muscle fibers slowly, they also move slowly out of the muscle fiber, which delays relaxation. The prolonged presence of Ca^{2+} in the cytosol provides for **smooth muscle tone,** a state of continued partial contraction. Smooth muscle tissue can thus sustain long-term tone, which is important in the gastrointestinal tract, where the walls maintain a steady pressure on the contents of the tract, and in the walls of blood vessels called arterioles, which maintain a steady pressure on blood.

Most smooth muscle fibers contract or relax in response to action potentials from the autonomic nervous system. In addition, many smooth muscle fibers contract or relax in response to stretching, hormones, or local factors such as changes in pH, oxygen and carbon dioxide levels, temperature, and ion concentrations. For example, the hormone epinephrine, released by the adrenal medulla, causes relaxation of smooth muscle in the airways and in some blood vessel walls (those that have so-called β_2 receptors; see Table 15.2 on page 536).

Unlike striated muscle fibers, smooth muscle fibers can stretch considerably and still maintain their contractile function. When smooth muscle fibers are stretched, they initially contract, developing increased tension. Within a minute or so, the tension decreases. This phenomenon, which is called the **stress–relaxation response,** allows smooth muscle to undergo great changes in length while retaining the ability to contract effectively. Thus, even though smooth muscle in the walls of blood vessels and hollow organs such as the stomach, intestines, and urinary bladder can stretch, the pressure on the contents within them changes very little. After the organ empties, the smooth muscle in the wall rebounds, and the wall retains its firmness.

▶ CHECKPOINT

24. What are the differences between visceral and multiunit smooth muscle?

25. How are skeletal and smooth muscle similar? How do they differ?

REGENERATION OF MUSCULAR TISSUE

▶ OBJECTIVE

Explain how muscle fibers regenerate.

Because mature skeletal muscle fibers have lost the ability to undergo cell division, growth of skeletal muscle after birth is due mainly to **hypertrophy,** the enlargement of existing cells, rather than to **hyperplasia,** an increase in the number of fibers. Satellite cells divide slowly and fuse with existing fibers to assist both in muscle growth and in repair of damaged fibers. Thus, skeletal muscle tissue can regenerate only to a limited extent.

Until recently it was believed that damaged cardiac muscle fibers could not be replaced and that healing took place exclusively by fibrosis, the formation of scar tissue. New research described in Chapter 20 indicates that, under certain circumstances, cardiac muscle tissue can regenerate. In addition, cardiac muscle fibers can undergo hypertrophy in response to increased workload. Hence, many athletes have enlarged hearts.

Smooth muscle tissue, like skeletal and cardiac muscle tissue, can undergo hypertrophy. In addition, certain smooth muscle fibers, such as those in the uterus, retain their capacity for division and thus can grow by hyperplasia. Also, new smooth muscle fibers can arise from cells called *pericytes,* stem cells found in association with blood capillaries and small veins. Smooth muscle fibers can also proliferate in certain pathological conditions, such as occur in the development of atherosclerosis (see page 798). Compared with the other two types of muscle tissue, smooth muscle tissue has considerably greater powers of regeneration. Such powers are still limited when compared with other tissues, such as epithelium.

Table 10.2 summarizes the major characteristics of the three types of muscular tissue.

TABLE 10.2 Summary of the Major Features of the Three Types of Muscular Tissue

Characteristic	Skeletal Muscle	Cardiac Muscle	Smooth Muscle
Microscopic appearance and features	Long cylindrical fiber with many peripherally located nuclei; striated.	Branched cylindrical fiber with one centrally located nucleus; intercalated discs join neighboring fibers; striated.	Fiber is thickest in middle, tapered at each end, and has one centrally positioned nucleus; not striated.
Location	Most commonly attached by tendons to bones.	Heart.	Walls of hollow viscera, airways, blood vessels, iris and ciliary body of eye, arrector pili muscles of hair follicles.
Fiber diameter	Very large (10–100 μm).	Large (10–20 μm).	Small (3–8 μm).
Connective tissue components	Endomysium, perimysium, and epimysium.	Endomysium.	Endomysium.
Fiber length	100 μm–30 cm.	50–100 μm.	30–200 μm.
Contractile proteins organized into sarcomeres	Yes.	Yes.	No.
Sarcoplasmic reticulum	Abundant.	Some.	Very little.
Transverse tubules present	Yes, aligned with each A–I band junction.	Yes, aligned with each Z disc.	No.
Junctions between fibers	None.	Intercalated discs contain gap junctions and desmosomes.	Gap junctions in visceral smooth muscle; none in multiunit smooth muscle.
Autorhythmicity	No.	Yes.	Yes, in visceral smooth muscle.
Source of Ca^{2+} for contraction	Sarcoplasmic reticulum.	Sarcoplasmic reticulum and interstitial fluid.	Sarcoplasmic reticulum and interstitial fluid.
Regulator proteins for contraction	Troponin and tropomyosin.	Troponin and tropomyosin.	Calmodulin and myosin light chain kinase.
Speed of contraction	Fast.	Moderate.	Slow.
Nervous control	Voluntary (somatic nervous system).	Involuntary (autonomic nervous system).	Involuntary (autonomic nervous system).
Contraction regulated by:	Acetylcholine released by somatic motor neurons.	Acetylcholine and norepinephrine released by autonomic motor neurons; several hormones.	Acetylcholine and norepinephrine released by autonomic motor neurons; several hormones; local chemical changes; stretching.
Capacity for regeneration	Limited, via satellite cells.	Limited, under certain conditions.	Considerable, via pericytes (compared with other muscle tissues, but limited compared with epithelium).

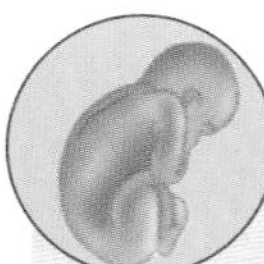

DEVELOPMENT OF MUSCLE

▶ **OBJECTIVE**

Describe the development of muscles.

Except for the muscles such as those of the iris of the eyes and the arrector pili muscles attached to hairs, all muscles of the body are derived from **mesoderm.** As the mesoderm develops, part of it becomes arranged in dense columns on either side of the developing nervous system. These columns of mesoderm undergo segmentation into a series of cube-shaped structures called **somites** (SŌ-mīts) (Figure 10.19a). The first pair of somites appears on the 20th day of embryonic development. Eventually, 42 to 44 pairs of somites are formed by the end of the fifth week. The number of somites can be correlated to the approximate age of the embryo.

With the exception of the skeletal muscles of the head and limbs, *skeletal muscles* develop from the **mesoderm of somites.** Because there are very few somites in the head region of the embryo, most of the skeletal muscles there develop from the **general mesoderm** in the head region. The skeletal muscles of the limbs develop from masses of general mesoderm around developing bones in embryonic limb buds (origins of future limbs; see Figure 8.18b on page 253).

The cells of a somite differentiate into three regions: (1) a **myotome,** which forms the skeletal muscles of the head, neck, and limbs; (2) a **dermatome,** which forms the connective tissues, including the dermis of the skin; and (3) a **sclerotome,** which gives rise to the vertebrae (Figure 10.19b).

Cardiac muscle develops from **mesodermal cells** that migrate to and envelop the developing heart while it is still in the form of primitive heart tubes (see Figure 20.18 on page 725).

Smooth muscle develops from **mesodermal cells** that migrate to and envelop the developing gastrointestinal tract and viscera.

▶ **CHECKPOINT**

26. From which embryonic tissues do the three types of muscles develop?

Figure 10.19 Location and structure of somites, key structures in the development of the muscular system.

Most muscles are derived from mesoderm.

HEAD END

Developing nervous system:

Neural plate

Neural folds

Neural groove

Somite

Transverse plane through somite

TAIL END

(a) Dorsal view of an embryo showing somites, about 22 days

Somite:

Sclerotome

Myotome

Dermatome

Developing nervous system

Notochord

Blood vessel (future aorta)

(b) Transverse section through a somite

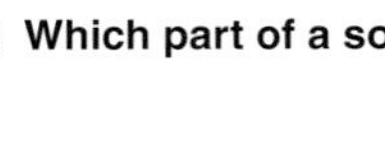

Which part of a somite differentiates into skeletal muscle?

AGING AND MUSCULAR TISSUE

▶ **OBJECTIVE**

Explain how aging affects skeletal muscle.

With aging, humans undergo a slow, progressive loss of skeletal muscle mass that is replaced largely by fibrous connective tissue and adipose tissue. In part, this decline is due to decreased levels of physical activity. Accompanying the loss of muscle mass is a decrease in maximal strength, a slowing of muscle reflexes, and a loss of flexibility. Muscle strength at age 85 is about half that at age 25. In some muscles, a selective loss of muscle fibers of a given type may occur. With aging, the relative number of slow oxidative fibers appears to increase. This could be due to either atrophy of the other fiber types or their conversion into slow oxidative fibers. Whether these are effects of aging itself or mainly reflections of the more limited physical activity of older people is unclear. Nevertheless, aerobic activities and strength training programs are effective in older people and can slow or even reverse the age-associated decline in muscular performance.

▶ **CHECKPOINT**

27. Why does muscle strength decrease with aging?

28. Why do you think a healthy 30-year-old can lift a 25-lb load much more comfortably than an 80-year-old?

DISORDERS: HOMEOSTATIC IMBALANCES

Abnormalities of skeletal muscle function may be due to disease or damage of any of the components of a motor unit: somatic motor neuron, neuromuscular junctions, or muscle fibers. The term **neuromuscular disease** encompasses problems at all three sites; the term **myopathy** (mī-OP-a-thē; *-pathy* = disease) signifies a disease or disorder of the skeletal muscle tissue itself.

Myasthenia Gravis

Myasthenia gravis (mī-as-THĒ-nē-a GRAV-is; *mys-* = muscle; *aisthesis* = sensation) is an autoimmune disease that causes chronic, progressive damage of the neuromuscular junction. The immune system inappropriately produces antibodies that bind to and block some ACh receptors, thereby decreasing the number of functional ACh receptors at the motor end plates of skeletal muscles (see Figure 10.10). Because 75% of patients with myasthenia gravis have hyperplasia or tumors of the thymus, it is thought that thymic abnormalities cause the disorder. As the disease progresses, more ACh receptors are lost. Thus, muscles become increasingly weaker, fatigue more easily, and may eventually cease to function.

Myasthenia gravis occurs in about 1 in 10,000 people and is more common in women, typically ages 20 to 40 at onset; men usually are ages 50 to 60 at onset. The muscles of the face and neck are most often affected. Initial symptoms include weakness of the eye muscles, which may produce double vision, and weakness of the throat muscles that may produce difficulty in swallowing. Later, the person has difficulty chewing and talking. Eventually the muscles of the limbs may become involved. Death may result from paralysis of the respiratory muscles, but often the disorder does not progress to this stage.

Anticholinesterase drugs such as pyridostigmine (Mestinon®) or neostigmine, the first line of treatment, act as inhibitors of acetylcholinesterase, the enzyme that breaks down ACh. Thus, the inhibitors raise the level of ACh that is available to bind with still-functional receptors. More recently, steroid drugs such as prednisone have been used with success to reduce antibody levels. Another treatment is plasmapheresis, a procedure that removes the antibodies from the blood. Often, surgical removal of the thymus (thymectomy) is helpful.

Muscular Dystrophy

The term **muscular dystrophy** (DIS-trō-fē; *dys-* = difficult; *-trophy* = nourishment) refers to a group of inherited muscle-destroying diseases that cause progressive degeneration of skeletal muscle fibers. The most common form of muscular dystrophy is *Duchenne muscular dystrophy* (doo-SHĀN) or *DMD.* Because the mutated gene is on the X chromosome, and males have only one, DMD strikes boys almost exclusively. (Sex-linked inheritance is described in Chapter 29.) Worldwide, about 1 in every 3500 male babies—21,000 in all—are born with DMD each year. The disorder usually becomes apparent between the ages of 2 and 5, when parents notice the child falls often and has difficulty running, jumping, and hopping. By age 12 most boys with DMD are unable to walk. Respiratory or cardiac failure usually causes death by age 20.

In DMD, the gene that codes for the protein dystrophin is mutated, so little or no dystrophin is present in the sarcolemma. Without the reinforcing effect of dystrophin, the sarcolemma tears easily during muscle contraction, causing muscle fibers to rupture and die. The dystrophin gene was discovered in 1987, and by 1990 the first attempts were made to treat DMD patients with gene therapy. The muscles of three boys with DMD were injected with myoblasts bearing functional dystrophin genes, but only a few muscle fibers gained the ability to produce dystrophin. Similar clinical trials with additional patients have also failed. An alternative approach to the problem is to find a way to induce muscle fibers to produce the protein *utrophin,* which is similar to dystrophin. Experiments with dystrophin-deficient mice suggest this approach may work.

Fibromyalgia

Fibromyalgia (fī-brō-mī-AL-jē-a; *algia* = painful condition) is a painful, nonarticular rheumatic disorder that usually appears between the ages of 25 and 50. An estimated 3 million people in the United States suffer from fibromyalgia, which is 15 times more common in women than in men. The disorder affects the fibrous connective tissue components of muscles, tendons, and ligaments. A striking sign is pain that results from gentle pressure at specific "tender points." Even without pressure, there is pain, tenderness, and stiffness of muscles, tendons, and surrounding soft tissues. Besides muscle pain, those with fibromyalgia report severe fatigue, poor sleep, headaches, depression, and inability to carry out their daily activities. Treatment consists of stress reduction, regular exercise, application of heat, gentle massage, physical therapy, medication for pain, and a low-dose antidepressant to help improve sleep.

Abnormal Contractions of Skeletal Muscle

One kind of abnormal muscular contraction is a **spasm,** a sudden involuntary contraction of a single muscle in a large group of muscles. A painful spasmodic contraction is known as a **cramp.** Cramps may be caused by inadequate blood flow to muscles, overuse of a muscle, dehydration, injury, holding a position for prolonged periods, and low blood levels of electrolytes, such as potassium. A **tic** is a spasmodic twitching made involuntarily by muscles that are ordinarily under voluntary control. Twitching of the eyelid and facial muscles are examples of tics. A **tremor** is a rhythmic, involuntary, purposeless contraction that produces a quivering or shaking movement. A **fasciculation** (fa-sik-ū-LĀ-shun) is an involuntary, brief twitch of an entire motor unit that is visible under the skin; it occurs irregularly and is not associated with movement of the affected muscle. Fasciculations may be seen in multiple sclerosis (see page 433) or in amyotrophic lateral sclerosis (Lou Gehrig's disease, see page 562). A **fibrillation** (fi-bri-LĀ-shun) is a spontaneous contraction of a single muscle fiber that is not visible under the skin but can be recorded by electromyography. Fibrillations may signal destruction of motor neurons.

MEDICAL TERMINOLOGY

Muscle strain Tearing of a muscle because of forceful impact, accompanied by bleeding and severe pain. Also known as a *charley horse* or pulled muscle. It often occurs in contact sports and typically affects the quadriceps femoris muscle on the anterior surface of the thigh. The condition is treated by RICE therapy: rest (R), ice immediately after the injury (I), compression via a supportive wrap (C), and elevation of the limb (E).

Myalgia (mī-AL-jē-a; *-algia* = painful condition) Pain in or associated with muscles.

Myoma (mī-Ō-ma; *-oma* = tumor) A tumor consisting of muscle tissue.

Myomalacia (mī′-ō-ma-LĀ-shē-a; *-malacia* = soft) Pathological softening of muscle tissue.

Myositis (mī′-ō-SĪ-tis; *-itis* = inflammation of) Inflammation of muscle fibers (cells).

Myotonia (mī′-ō-TŌ-nē-a; *-tonia* = tension) Increased muscular excitability and contractility, with decreased power of relaxation; tonic spasm of the muscle.

Volkmann's contracture (FŌLK-manz kon-TRAK-tur; *contra-*= against) Permanent shortening (contracture) of a muscle due to replacement of destroyed muscle fibers by fibrous connective tissue, which lacks extensibility. Destruction of muscle fibers may occur from interference with circulation caused by a tight bandage, a piece of elastic, or a cast.

STUDY OUTLINE

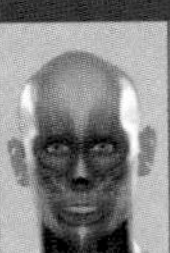

INTRODUCTION (p. 291)

1. Motion results from alternating contraction and relaxation of muscles, which constitute 40–50% of total body weight.
2. The prime function of muscle is changing chemical energy into mechanical energy to perform work.

OVERVIEW OF MUSCULAR TISSUE (p. 291)

1. The three types of muscular tissue are skeletal, cardiac, and smooth. Skeletal muscle tissue is primarily attached to bones; it is striated and voluntary. Cardiac muscle tissue forms the wall of the heart; it is striated and involuntary. Smooth muscle tissue is located primarily in internal organs; it is nonstriated (smooth) and involuntary.
2. Through contraction and relaxation, muscular tissue performs four important functions: producing body movements; stabilizing body positions; moving substances within the body and regulating organ volume; and producing heat.
3. Four special properties of muscular tissues are (1) electrical excitability, the property of responding to stimuli by producing action potentials; (2) contractility, the ability to generate tension to do work; (3) extensibility, the ability to be extended (stretched); and (4) elasticity, the ability to return to original shape after contraction or extension.

SKELETAL MUSCLE TISSUE (p. 292)

1. Connective tissues surrounding muscle are the epimysium, covering the entire muscle; perimysium, covering fascicles; and the endomysium, covering muscle fibers. Superficial fascia separates muscle from skin.
2. Tendons and aponeuroses are extensions of connective tissue beyond muscle fibers that attach the muscle to bone or to other muscle. A tendon is generally ropelike in shape; an aponeurosis is wide and flat.
3. Skeletal muscles are well supplied with nerves and blood vessels. Generally, an artery and one or two veins accompany each nerve that penetrates a skeletal muscle.
4. Somatic motor neurons provide the nerve impulses that stimulate skeletal muscle to contract.
5. Blood capillaries bring in oxygen and nutrients and remove heat and waste products of muscle metabolism.
6. The major cells of skeletal muscle tissue are termed skeletal muscle fibers. Each muscle fiber has 100 or more nuclei because it arises from the fusion of many myoblasts. Satellite cells are myoblasts that persist after birth. The sarcolemma is a muscle fiber's plasma membrane; it surrounds the sarcoplasm. Transverse tubules are invaginations of the sarcolemma.
7. Each muscle fiber (cell) contains hundreds of myofibrils, the contractile elements of skeletal muscle. Sarcoplasmic reticulum surrounds each myofibril. Within a myofibril are thin and thick filaments, arranged in compartments called sarcomeres.
8. The overlapping of thick and thin filaments produces striations. Darker A bands alternate with lighter I bands.
9. Myofibrils are composed of three types of proteins: contractile, regulatory, and structural. The contractile proteins are myosin (thick filament) and actin (thin filament). Regulatory proteins are tropomyosin and troponin, both of which are part of the thin filament. Structural proteins include titin (links Z disc to M line and stabilizes thick filament), myomesin (forms M line), nebulin (anchors thin filaments to Z discs and regulates length of thin filaments during development), and dystrophin (links thin filaments to sarcolemma).
10. Projecting myosin heads contain actin-binding and ATP-binding sites and are the motor proteins that power muscle contraction.

CONTRACTION AND RELAXATION OF SKELETAL MUSCLE FIBERS (p. 298)

1. Muscle contraction occurs because crossbridges attach to and "walk" along the thin filaments at both ends of a sarcomere,

progressively pulling the thin filaments toward the center of a sarcomere. As the thin filaments slide inward, the Z discs come closer together, and the sarcomere shortens.

2. The contraction cycle is the repeating sequence of events that causes sliding of the filaments: (1) myosin ATPase hydrolyzes ATP and becomes energized; (2) the myosin head attaches to actin, forming a crossbridge; (3) the crossbridge generates force as it rotates toward the center of the sarcomere (power stroke); and (4) binding of ATP to the myosin head detaches it from actin. The myosin head again hydrolyzes the ATP, returns to its original position, and binds to a new site on actin as the cycle continues.
3. An increase in Ca^{2+} concentration in the cytosol starts filament sliding; a decrease turns off the sliding process.
4. The muscle action potential propagating into the T tubule system causes opening of Ca^{2+} release channels in the SR membrane. Calcium ions diffuse from the SR into the cytosol and combine with troponin. This binding causes the troponin–tropomyosin complex to move away from the myosin-binding sites on actin.
5. Ca^{2+} active transport pumps continually remove Ca^{2+} from the sarcoplasm into the SR. When the concentration of calcium ions in the cytosol decreases, the troponin–tropomyosin complexes slide back over and block the myosin-binding sites, and the muscle fiber relaxes.
6. A muscle fiber develops its greatest tension when there is an optimal zone of overlap between thick and thin filaments. This dependency is the length–tension relationship.
7. The neuromuscular junction (NMJ) is the synapse between a somatic motor neuron and a skeletal muscle fiber. The NMJ includes the axon terminals and synaptic end bulbs of a motor neuron, plus the adjacent motor end plate of the muscle fiber sarcolemma.
8. When a nerve impulse reaches the synaptic end bulbs of a somatic motor neuron, it triggers exocytosis of the synaptic vesicles, which releases acetylcholine (ACh). ACh diffuses across the synaptic cleft and binds to ACh receptors, initiating a muscle action potential. Acetylcholinesterase then quickly breaks down ACh into its component parts.

MUSCLE METABOLISM (p. 306)

1. Muscle fibers have three sources for ATP production: creatine, anaerobic cellular respiration, and aerobic cellular respiration.
2. Creatine kinase catalyzes the transfer of a high-energy phosphate group from creatine phosphate to ADP to form new ATP molecules. Together, creatine phosphate and ATP provide enough energy for muscles to contract maximally for about 15 seconds.
3. Glucose is converted to pyruvic acid in the reactions of glycolysis, which yield two ATPs without using oxygen. Such anaerobic cellular respiration can provide enough energy for 30–40 seconds of maximal muscle activity.
4. Muscular activity that lasts longer than half a minute depends on aerobic cellular respiration, mitochondrial reactions that require oxygen to produce ATP.
5. The inability of a muscle to contract forcefully after prolonged activity is muscle fatigue.
6. Elevated oxygen use after exercise is called recovery oxygen uptake.

CONTROL OF MUSCLE TENSION (p. 308)

1. A motor neuron and the muscle fibers it stimulates form a motor unit. A single motor unit may contain as few as two or as many as 3000 muscle fibers.
2. Recruitment is the process of increasing the number of active motor units.
3. A twitch contraction is a brief contraction of all the muscle fibers in a motor unit in response to a single action potential.
4. A record of a contraction is called a myogram. It consists of a latent period, a contraction period, and a relaxation period.
5. Wave summation is the increased strength of a contraction that occurs when a second stimulus arrives before the muscle fiber has completely relaxed after a previous stimulus.
6. Repeated stimuli can produce unfused (incomplete) tetanus, a sustained muscle contraction with partial relaxation between stimuli. More rapidly repeating stimuli produce fused (complete) tetanus, a sustained contraction without partial relaxation between stimuli.
7. Continuous involuntary activation of a small number of motor units produces muscle tone, which is essential for maintaining posture.
8. In a concentric isotonic contraction, the muscle shortens to produce movement and to reduce the angle at a joint. During an eccentric isotonic contraction, the muscle lengthens.
9. Isometric contractions, in which tension is generated without muscle changing its length, are important because they stabilize some joints as others are moved.

TYPES OF SKELETAL MUSCLE FIBERS (p. 312)

1. On the basis of their structure and function, skeletal muscle fibers are classified as slow oxidative (SO), fast oxidative-glycolytic (FOG), and fast glycolytic (FG) fibers.
2. Most skeletal muscles contain a mixture of all three fiber types. Their proportions vary with the typical action of the muscle.
3. The motor units of a muscle are recruited in the following order: first SO fibers, then FOG fibers, and finally FG fibers.
4. Table 10.1 summarizes the three types of skeletal muscle fibers.

EXERCISE AND SKELETAL MUSCLE TISSUE (p. 314)

1. Various types of exercises can induce changes in the fibers in a skeletal muscle. Endurance-type (aerobic) exercises cause a gradual transformation of some fast glycolytic (FG) fibers into fast oxidative-glycolytic (FOG) fibers.
2. Exercises that require great strength for short periods produce an increase in the size and strength of fast-glycolytic (FG) fibers. The increase in size is due to increased synthesis of thick and thin filaments.

CARDIAC MUSCLE TISSUE (p. 314)

1. Cardiac muscle is found only in the heart. Cardiac muscle fibers have the same arrangement of actin and myosin and the same bands, zones, and Z discs as skeletal muscle fibers. The fibers connect to one another through intercalated discs, which contain both desmosomes and gap junctions.
2. Cardiac muscle tissue remains contracted 10 to 15 times longer than skeletal muscle tissue due to prolonged delivery of Ca^{2+} into the sarcoplasm.
3. Cardiac muscle tissue contracts when stimulated by its own

autorhythmic fibers. Due to its continuous, rhythmic activity, cardiac muscle depends greatly on aerobic cellular respiration to generate ATP.

SMOOTH MUSCLE TISSUE (p. 315)

1. Smooth muscle is nonstriated and involuntary.
2. Smooth muscle fibers contain intermediate filaments and dense bodies; the function of dense bodies is similar to that of the Z discs in striated muscle.
3. Visceral (single-unit) smooth muscle is found in the walls of hollow viscera and of small blood vessels. Many fibers form a network that contracts in unison.
4. Multiunit smooth muscle is found in large blood vessels, large airways to the lungs, arrector pili muscles, and the eye, where it adjusts pupil diameter and lens focus. The fibers operate independently rather than in unison.
5. The duration of contraction and relaxation of smooth muscle is longer than in skeletal muscle since it takes longer for Ca^{2+} to reach the filaments.
6. Smooth muscle fibers contract in response to nerve impulses, hormones, and local factors.
7. Smooth muscle fibers can stretch considerably and still maintain their contractile function.

REGENERATION OF MUSCULAR TISSUE (p. 316)

1. Skeletal muscle fibers cannot divide and have limited powers of regeneration; cardiac muscle fibers can regenerate under limited circumstances; and smooth muscle fibers have the best capacity for division and regeneration.
2. Table 10.2 summarizes the major characteristics of the three types of muscular tissue.

DEVELOPMENT OF MUSCLE (p. 318)

1. With few exceptions, muscles develop from mesoderm.
2. Skeletal muscles of the head and limbs develop from general mesoderm. Other skeletal muscles develop from the mesoderm of somites.

AGING AND MUSCULAR TISSUE (p. 318)

1. Beginning at about 30 years of age, there is a slow, progressive loss of skeletal muscle, which is replaced by fibrous connective tissue and fat.
2. Aging also results in a decrease in muscle strength, slower muscle reflexes, and loss of flexibility.

SELF-QUIZ QUESTIONS

Fill in the blanks in the following statements.

1. A single somatic motor neuron and all of the muscle fibers it stimulates is known as a ____.
2. The wasting away of muscle due to lack of use is known as____ while the replacement of skeletal muscle fibers with scar tissue is known as ____.
3. The synaptic end bulbs of somatic motor neurons contain synaptic vesicles filled with the neurotransmitter____.

Indicate whether the following statements are true or false.

4. The ability of muscle cells to respond to stimuli and produce electrical signals is known as excitability.
5. The sequence of events resulting in skeletal muscle contraction are (a) generation of a nerve impulse, (b) release of the neurotransmitter acetylcholine, (c) generation of a muscle action potential, (d) release of calcium ions from the sarcoplasmic reticulum, (e) calcium ion binding to the troponin-tropomyosin complex, (f) power stroke with actin and myosin binding and release.

Choose the one best answer to the following questions.

6. In muscle physiology, the latent period refers to (a) the period of lost excitability that occurs when two stimuli are applied immediately one after the other, (b) the brief contraction of a motor unit, (c) the period of elevated oxygen use after exercise, (d) an inability of a muscle to contract forcefully after prolonged activity, (e) a brief delay that occurs between application of a stimulus and the beginning of contraction.
7. Which of the following muscle proteins and their descriptions are mismatched? (a) titin: regulatory protein that holds troponin in place; (b) myosin: contractile motor protein; (c) tropomyosin: regulatory protein that blocks myosin-binding sites; (d) actin: contractile anchoring protein that contains myosin-binding sites; (e) calsequestrin: calcium-binding protein
8. During muscle contraction all of the following occur *except* (a) crossbridges are formed when the energized myosin head attaches to actin's myosin-binding site, (b) ATP undergoes hydrolysis, (c) the thick filaments slide inward toward the M line, (d) calcium concentration in the cytosol increases, (e) the Z discs are drawn toward each other.
9. Which of the following is *not* true concerning muscle fiber length–tension relationships? (a) If sarcomeres are stretched, the tension in the fiber decreases. (b) If a muscle cell is stretched so that there is no overlap of the filaments, no tension is generated. (c) Extremely compressed sarcomeres result in less muscle tension. (d) Maximum tension occurs when the zone of overlap between a thick and thin filament extends from the edge of the H zone to one end of a thick filament. (e) If sarcomeres shorten, the tension in them increases.
10. Which of the following are sources of ATP for muscle contraction? (1) creatine phosphate, (2) glycolysis, (3) anaerobic cellular respiration, (4) aerobic cellular respiration, (5) acetylcholine (a) 1, 2, and 3; (b) 2, 3, and 4; (c) 2, 3, and 5; (d) 1, 2, 3, and 4; (e) 2, 3, 4, and 5
11. What would happen if ATP were suddenly unavailable after the sarcomere had begun to shorten? (a) Nothing. The contraction would proceed normally. (b) The myosin heads would be unable to detach from actin. (c) Troponin would bind with the myosin heads. (d) Actin and myosin filaments would separate completely and be unable to recombine. (e) The myosin heads would detach completely from actin and bind to the troponin-tropomyosin complex.

12. Match the following:

____(a) a sheath of areolar connective tissue that wraps around individual skeletal muscle fibers
____(b) dense irregular connective tissue that separates a muscle into groups of individual muscle fibers
____(c) bundles of muscle fibers
____(d) the outermost connective tissue layer that encircles an entire skeletal muscle
____(e) dense irregular connective tissue that lines the body wall and limbs and holds functional muscle units together
____(f) a cord of dense regular connective tissue that attaches muscle to the periosteum of bone
____(g) elongated muscle cell
____(h) areolar and adipose connective tissue that separates muscle from skin
____(i) connective tissue elements extended as a broad, flat layer
____(j) a two-layer tube of fibrous connective tissue enclosing certain tendons

(1) aponeurosis
(2) deep fascia
(3) superficial fascia
(4) tendon
(5) endomysium
(6) perimysium
(7) epimysium
(8) tendon (synovial) sheath
(9) fascicles
(10) muscle fiber

13. Match the following:

____(a) synapse between a motor neuron and a muscle fiber
____(b) invaginations of the sarcolemma from the surface toward the center of the muscle fiber
____(c) myoblasts that persist in mature skeletal muscle
____(d) plasma membrane of a muscle fiber
____(e) oxygen-binding protein found only in muscle fibers
____(f) Ca^{2+}-storing tubular system similar to smooth endoplasmic reticulum
____(g) the contracting unit of a skeletal muscle fiber
____(h) middle area in the sarcomere where thick and thin filaments are found
____(i) area in the sarcomere where only thin filaments are present but thick filaments are not
____(j) separates the sarcomeres from each other
____(k) area of only thick filaments
____(l) cytoplasm of a muscle fiber
____(m) composed of supporting proteins holding thick filaments together at the H zone

(1) A band
(2) I band
(3) Z disc
(4) H zone
(5) M line
(6) sarcomere
(7) neuromuscular junction
(8) myoglobin
(9) satellite cells
(10) transverse tubules
(11) sarcoplasmic reticulum
(12) sarcolemma
(13) sarcoplasm

14. Match the following:

____(a) the smooth muscle action that allows the fibers to maintain their contractile function even when stretched
____(b) a brief contraction of all the muscle fibers in a motor unit of a muscle in response to a single action potential in its motor neuron
____(c) sustained contraction of a muscle, with no relaxation between stimuli
____(d) larger contractions resulting from stimuli arriving at different times
____(e) process of increasing the number of activated motor units
____(f) contraction in which the muscle shortens
____(g) inability of a muscle to maintain its strength of contraction or tension during prolonged activity
____(h) sustained, but wavering contraction with partial relaxation between stimuli
____(i) produced by the continual involuntary activation of a small number of skeletal muscle motor units; results in firmness in skeletal muscle
____(j) contraction in which muscle tension is generated without shortening of the muscle
____(k) amount of oxygen needed to restore the body's metabolic conditions back to resting levels after exercise
____(l) contraction in which a muscle lengthens

(1) muscle fatigue
(2) twitch contraction
(3) wave summation
(4) fused (complete) tetanus
(5) concentric isotonic contraction
(6) motor unit recruitment
(7) muscle tone
(8) eccentric isotonic contraction
(9) isometric contraction
(10) stress-relaxation response
(11) recovery oxygen uptake
(12) unfused (incomplete) tetanus

15. Match the following (some questions will have more than one answer):

____(a) has fibers joined by intercalated discs
____(b) thick and thin filaments are not arranged as orderly sarcomeres
____(c) uses satellite cells to repair damaged muscle fibers
____(d) striated
____(e) contraction begins slowlybut lasts for long periods
____(f) has an extended contraction due to prolonged calcium delivery from both the sarcoplasmic reticulum and the interstitial fluid
____(g) does not exhibit autorhythmicity
____(h) uses pericytes to repair damaged muscle fibers
____(i) uses troponin as a regulatory protein
____(j) can be classified as single-unit or multiunit
____(k) can be autorhythmic
____(l) uses calmodulin as a regulatory protein

(1) skeletal muscle
(2) cardiac muscle
(3) smooth muscle

CRITICAL THINKING QUESTIONS

1. Weight-lifter Jamal has been practicing many hours a day and his muscles have gotten noticeably bigger. He tells you that his muscle cells are "multiplying like crazy and making him get stronger and stronger." Do you believe his explanation? Why or why not?
2. Chicken breasts are composed of "white meat" while chicken legs are composed of "dark meat." The breasts and legs of migrating ducks are dark meat. The breasts of both chickens and ducks are used in flying. How can you explain the differences in the color of the meat (muscles)? How are they adapted for their particular functions?
3. Polio is a disease caused by a virus that can attack the somatic motor neurons in the central nervous system. Individuals who suffer from polio can develop muscle weakness and atrophy. In a certain percentage of cases, the individuals may die due to respiratory paralysis. Relate your knowledge of how muscle fibers function to the symptoms exhibited by infected individuals.

ANSWERS TO FIGURE QUESTIONS

10.1 Perimysium bundles groups of muscle fibers into fascicles.

10.2 The sarcoplasmic reticulum releases calcium ions to trigger muscle contraction.

10.3 The following are arranged from smallest to largest: thick filament, myofibril, muscle fiber.

10.4 Sarcomeres are separated from one another by Z discs.

10.5 Actin and titin anchor into the Z disc. A bands contain myosin, actin, troponin, tropomyosin, and titin; I bands contain actin, troponin, tropomyosin, and titin.

10.6 The I bands and H zones disappear during muscle contraction; the lengths of the thin and thick filaments do not change.

10.7 If ATP were not available, the crossbridges would not be able to detach from actin. The muscles would remain in a state of rigidity, as occurs in rigor mortis.

10.8 Three functions of ATP in muscle contraction include the following: (1) its hydrolysis by an ATPase activates the myosin head so it can bind to actin and rotate; (2) its binding to myosin causes detachment from actin after the power stroke; and (3) it powers the pumps that transport Ca^{2+} from the cytosol back into the sarcoplasmic reticulum.

10.9 A sarcomere length of 2.2 μm gives a generous zone of overlap between the parts of the thick filaments that have myosin heads and the thin filaments without the overlap being so extensive that sarcomere shortening is limited.

10.10 The portion of the sarcolemma that contains acetylcholine receptors is the motor end plate.

10.11 Steps 4 through 6 are part of excitation–contraction coupling (muscle action potential through binding of myosin heads to actin).

10.12 Glycolysis, exchange of phosphate between creatine phosphate and ADP, and glycogen breakdown occur in the cytosol. Oxidation of pyruvic acid, amino acids, and fatty acids (aerobic cellular respiration) occurs in mitochondria.

10.13 Motor units having many muscle fibers are capable of more forceful contractions than those having only a few fibers.

10.14 During the latent period the events of excitation–contraction coupling occur: Release of Ca^{2+} from SR and binding to troponin result in attachment of myosin heads to actin and the onset of rotation.

10.15 If the second stimulus were applied a little later, the second contraction would be smaller than the one illustrated in (b).

10.16 Holding your head upright without movement involves mainly isometric contractions.

10.17 Visceral smooth muscle is more like cardiac muscle; both contain gap junctions, which allow action potentials to spread from each cell to its neighbors.

10.18 Contraction in a smooth muscle fiber starts more slowly and lasts much longer than contraction in a skeletal muscle fiber.

10.19 The myotome of a somite differentiates into skeletal muscle.

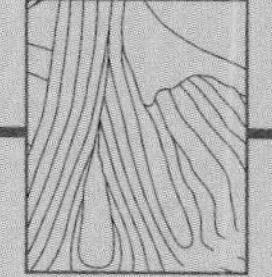

The Muscular System

The Muscular System and Homeostasis

The muscular system and muscular tissue of your body contribute to homeostasis by stabilizing body position, producing movements, regulating organ volume, moving substances within the body, and producing heat.

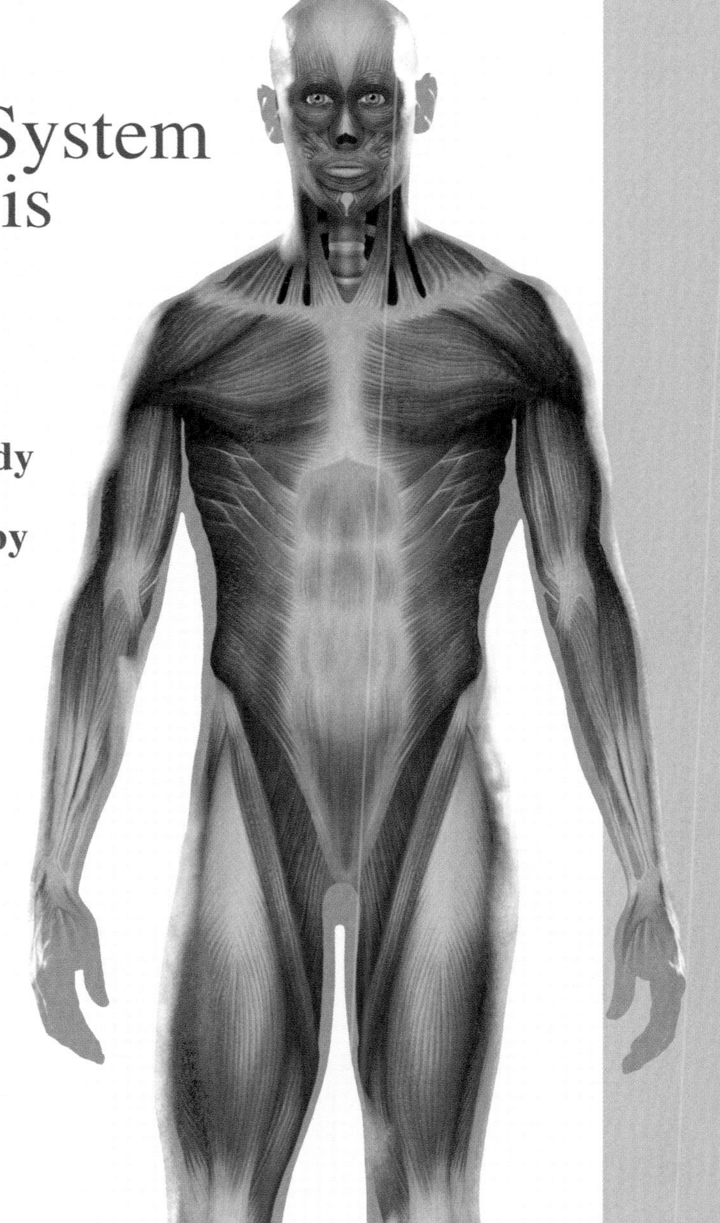

www. wiley.com/college/apcentral

Together, the voluntarily controlled muscles of your body comprise the **muscular system.** Almost all of the 700 individual muscles that make up the muscular system, for instance, the biceps brachii muscle, include both skeletal muscle tissue and connective tissue. The function of most muscles is to produce movements of body parts. A few muscles function mainly to stabilize bones so that other skeletal muscles can execute a movement more effectively. This chapter presents many of the major skeletal muscles in the body, most of which are found on both the right and left sides. We will identify the attachment sites and innervation (the nerve or nerves that stimulate contraction) of each muscle described. Developing a working knowledge of these key aspects of skeletal muscle anatomy will enable you to understand how normal movements occur. This knowledge is especially crucial for professionals, such as those in the allied health and physical rehabilitation fields, who work with patients whose normal patterns of movement and physical mobility have been disrupted by physical trauma, surgery, or muscular paralysis.

HOW SKELETAL MUSCLES PRODUCE MOVEMENTS

▶ **OBJECTIVES**

Describe the relationship between bones and skeletal muscles in producing body movements.

Define lever and fulcrum, and compare the three types of levers based on location of the fulcrum, effort, and load.

Identify the types of fascicle arrangements in a skeletal muscle, and relate the arrangements to strength of contraction and range of motion.

Explain how the prime mover, antagonist, synergist, and fixator in a muscle group work together to produce movements.

Muscle Attachment Sites: Origin and Insertion

Skeletal muscles that produce movements do so by exerting force on tendons, which in turn pull on bones or other structures (such as skin). Most muscles cross at least one joint and are usually attached to the articulating bones that form the joint (Figure 11.1a).

When a skeletal muscle contracts, it pulls one of the articulating bones toward the other. The two articulating bones usually do not move equally in response to contraction. One bone remains stationary or near its original position, either because other muscles stabilize that bone by pulling it in the opposite direction or because its structure makes it less movable. Ordinarily, the attachment of a muscle's tendon to the stationary bone is called the **origin;** the attachment of the muscle's other tendon to the movable bone is called the **insertion.** A good analogy is a spring on a door. In this example, the part of the spring attached to the frame is the origin; the part attached to the door represents the insertion. A useful rule of thumb is that the origin is usually proximal and the insertion distal, especially in the limbs; the insertion is usually pulled toward the origin. The fleshy portion of the muscle between the tendons is called the **belly** (*gaster*), the coiled middle portion of the spring in our example. The **actions** of a muscle are the main movements that occur when the muscle contracts. In our spring example, this would be the closing of the door.

Muscles that move a body part often do not cover the moving part. Figure 11.1b shows that although one of the functions of the biceps brachii muscle is to move the forearm, the belly of the muscle lies over the humerus, not over the forearm. You will also see that muscles that cross two joints, such as the rectus femoris and sartorius of the thigh, have more complex actions than muscles that cross only one joint.

Tenosynovitis

Tenosynovitis (ten′-ō-sin-ō-VĪ-tis) is an inflammation of the tendons, tendon sheaths, and synovial membranes surrounding certain joints. The tendons most often affected are at the wrists, shoulders, elbows (resulting in *tennis elbow*), finger joints (resulting in *trigger finger*), ankles, and feet. The affected sheaths sometimes become visibly swollen because of fluid accumulation. Tenderness and pain are frequently associated with movement of the body part. The condition often follows trauma, strain, or excessive exercise. Tenosynovitis of the dorsum of the foot may be caused by tying shoelaces too tightly. Gymnasts are prone to developing the condition as a result of chronic, repetitive, and maximum hyperextension at the wrists. Other repetitive movements involving activities such as typing, haircutting, carpentry, and assembly line work can also result in tenosynovitis. ■

Lever Systems and Leverage

In producing movement, bones act as levers, and joints function as the fulcrums of these levers. A **lever** is a rigid structure that can move around a fixed point called a **fulcrum,** symbolized by ▲. A lever is acted on at two different points by two different forces: the **effort** (E), which causes movement, and the **load** L or **resistance,** which opposes movement. The effort is the force exerted by muscular contraction; the load is typically the weight of the body part that is moved. Motion occurs when the effort applied to the bone at the insertion exceeds the load. Consider the biceps brachii flexing the forearm at the elbow as an object is lifted (Figure 11.1b). When the forearm is raised, the elbow is the fulcrum. The weight of the forearm plus the weight of the object in the hand is the load. The force of contraction of the biceps brachii pulling the forearm up is the effort.

Levers produce trade-offs between effort and the speed and range of motion. A lever operates at a *mechanical advantage*—has **leverage**—when a smaller effort can move a heavier load.

Figure 11.1 Relationship of skeletal muscles to bones. (a) Muscles are attached to bones by tendons known as the origins and insertions. (b) Skeletal muscles produce movements by pulling on bones. Bones serve as levers, and joints act as fulcrums for the levers. Here the lever–fulcrum principle is illustrated by the movement of the forearm. Note where the load (resistance) and effort are applied in this example.

In the limbs, the origin of a muscle is usually proximal and the insertion is usually distal.

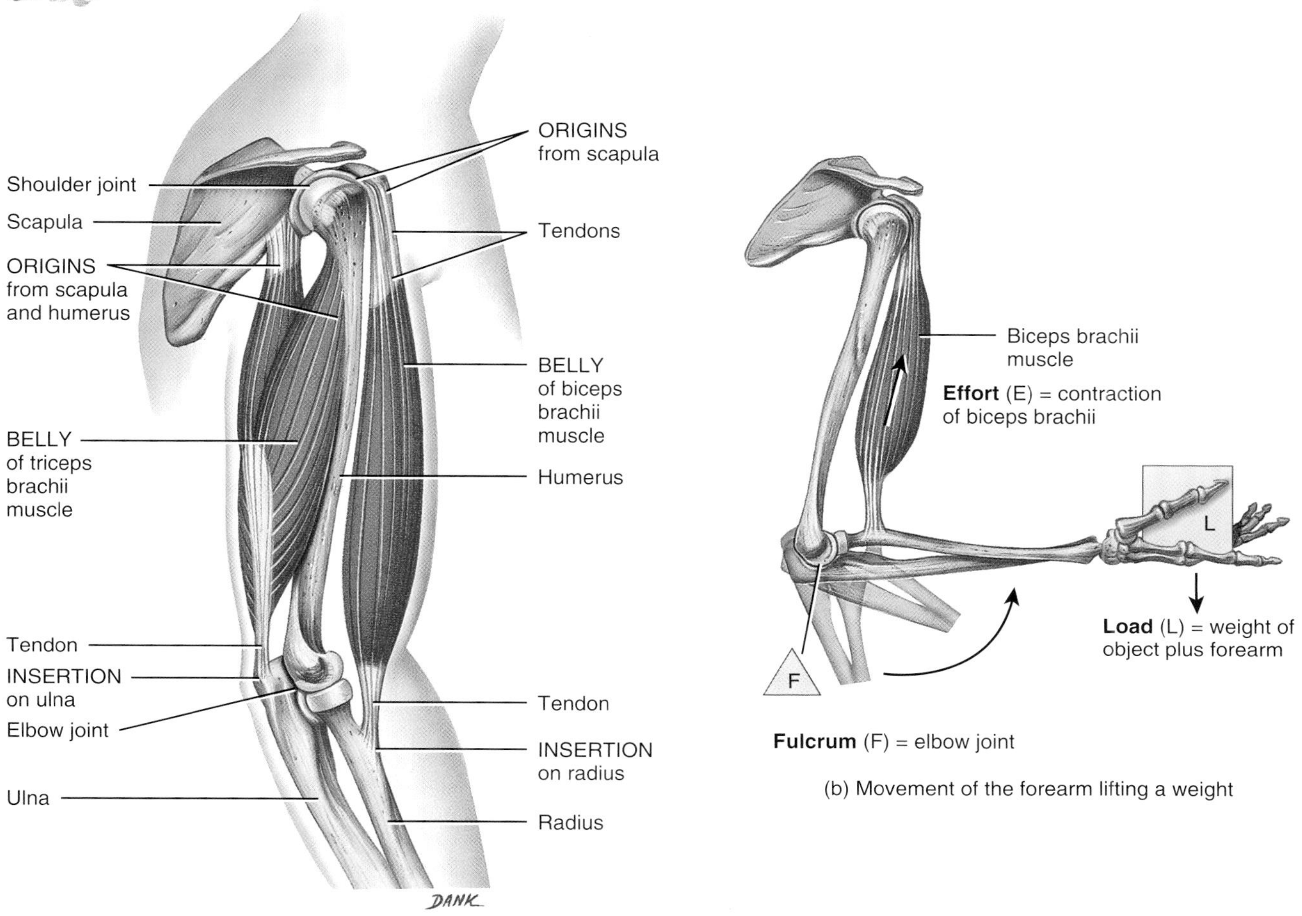

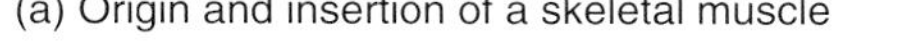

(a) Origin and insertion of a skeletal muscle

(b) Movement of the forearm lifting a weight

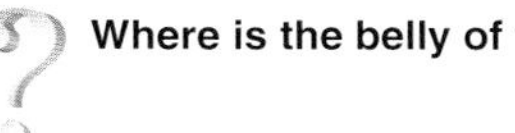

Where is the belly of the muscle that extends the forearm located?

Here the trade-off is that the effort must move a greater distance (must have a longer range of motion) and must be faster than the load. Recall from Chapter 9 that range of motion refers to the range, measured in degrees of a circle, through which the bones of a joint can be moved. The lever formed by the mandible at the temporomandibular joints (fulcrums) and the effort provided by contraction of the jaw muscles produce a high mechanical advantage that crushes food. In contrast, a lever operates at a *mechanical disadvantage* when a larger effort moves a lighter load. In this case the trade-off is that the effort must move more slowly and for a shorter distance than the load. The lever formed by the humerus at the shoulder joint (fulcrum) and the effort provided by the back and shoulder muscles produces a mechanical "disadvantage" that enables a major-league pitcher to hurl a baseball at nearly 100 miles per hour!

The positions of the effort, load, and fulcrum on the lever determine whether the lever operates at a mechanical advantage or disadvantage. When the load is close to the fulcrum and the effort is applied farther away, the lever operates at a mechanical advantage. When you chew food, the load (the food) is positioned close to the fulcrums (your temporomandibular joints) while your jaw muscles exert effort farther out from the joints. By contrast, when the effort is applied close to the fulcrum and the load is farther away, the lever operates at a mechanical disadvantage. When a pitcher throws a baseball, the back and shoulder muscles apply intense effort very close to the fulcrum (the shoulder joint) while the lighter load (the ball) is propelled at the far end of the lever (the arm bone).

Levers are categorized into three types according to the positions of the fulcrum, the effort, and the load:

1. The fulcrum is between the effort and the load in **first-class levers** (Figure 11.2a). (Think E**F**L.) Scissors and seesaws are examples of first-class levers. A first-class lever can produce either a mechanical advantage or disadvantage depending on whether the effort or the load is closer to the fulcrum. (Think of an adult and a child on a seesaw.) As we have seen in the preceding examples, if the effort (child) is farther from the fulcrum than the load (adult), a heavy load can be moved, but not very far or fast. If the effort is closer to the fulcrum than the load, only a lighter load can be moved, but it moves far and fast.

There are few first-class levers in the body. One example is the lever formed by the head resting on the vertebral column (Figure 11.2a). When the head is raised, the contraction of the posterior neck muscles provides the effort (E), the joint between the atlas and the occipital bone (atlanto-occipital joint) forms the fulcrum (F), and the weight of the anterior portion of the skull is the load.

2. The load is between the fulcrum and the effort in **second-class levers** (Figure 11.2b). (Think F**L**E.) They operate like a wheelbarrow. Second-class levers always produce a mechanical advantage because the load is always closer to the fulcrum than the effort. This arrangement sacrifices speed and range of motion for force; this type of lever produces the most force. Most authorities believe that there are no second-class levers in the body.

3. The effort is between the fulcrum and the load in **third-class levers** (Figure 11.2c). (Think F**E**L.) These levers operate like a pair of forceps and are the most common levers in the body. Third-class levers always produce a mechanical disadvantage because the effort is always closer to the fulcrum than the load. In the body, this arrangement favors speed and range of motion over force. The elbow joint, the biceps brachii muscle, and the bones of the arm and forearm provide an example of a third-class lever (Figure 11.2c). As we have

Figure 11.2 **Types of levers.**

Levers are divided into three types based on the placement of the fulcrum, effort, and load (resistance).

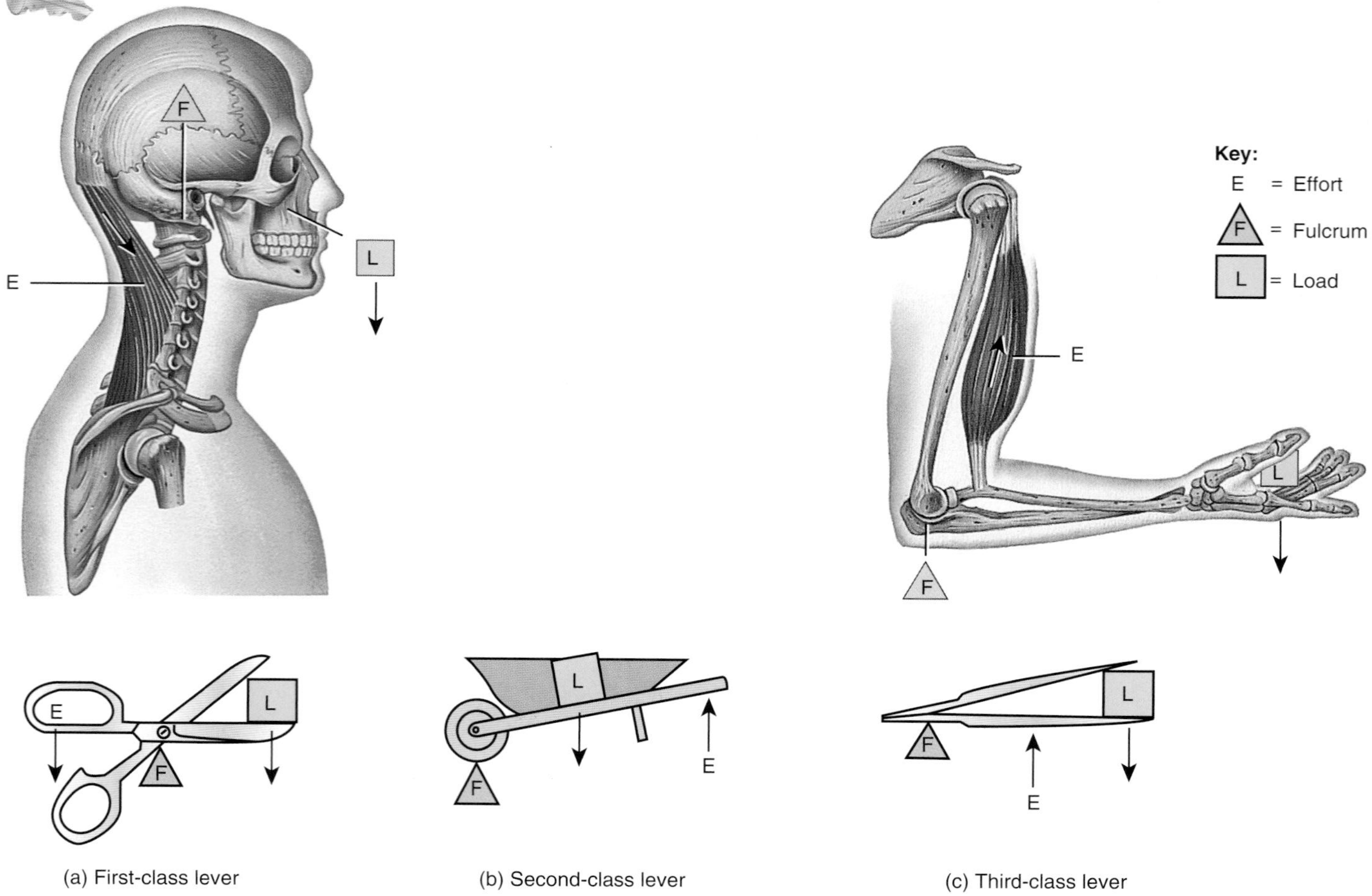

Key:
E = Effort
F = Fulcrum
L = Load

(a) First-class lever

(b) Second-class lever

(c) Third-class lever

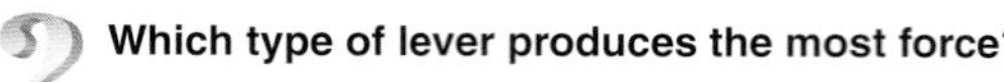

Which type of lever produces the most force?

seen, in flexing the forearm at the elbow, the elbow joint is the fulcrum, the contraction of the biceps brachii muscle provides the effort (E), and the weight of the hand and forearm is the load. Another example of the action of a third-class lever is adduction of the thigh, in which the hip joint is the fulcrum, the contraction of the adductor muscles is the effort, and the thigh is the load.

Effects of Fascicle Arrangement

Recall from Chapter 10 that the skeletal muscle fibers (cells) within a muscle are arranged in bundles known as **fascicles.** Within a fascicle, all muscle fibers are parallel to one another. The fascicles, however, may form one of five patterns with respect to the tendons: parallel, fusiform (shaped like a cigar), circular, triangular, or pennate (shaped like a feather) (Table 11.1).

Fascicular arrangement affects a muscle's power and range of motion. As a muscle fiber contracts, it shortens to about 70% of its resting length. The longer the fibers in a muscle, the greater the range of motion it can produce. However, the power of a muscle depends not on length but on its total cross-sectional area; a short fiber can contract as forcefully as a long one. Fascicular arrangement often represents a compromise between power and range of motion. Pennate muscles, for instance, have a large number of fascicles distributed over their tendons, giving them greater power but a smaller range of motion. Parallel muscles, in contrast, have comparatively few fascicles that extend the length of the muscle, so they have a greater range of motion but less power.

TABLE 11.1 Arrangement of Fascicles

Parallel

Fascicles parallel to longitudinal axis of muscle; terminate at either end in flat tendons.

Example: Stylohyoid muscle (see Figure 11.8)

Fusiform

Fascicles nearly parallel to longitudinal axis of muscle; terminate in flat tendons; muscle tapers toward tendons, where diameter is less than at belly.

Example: Digastric muscle (see Figure 11.8)

Circular

Fascicles in concentric circular arrangements form sphincter muscles that enclose an orifice (opening).

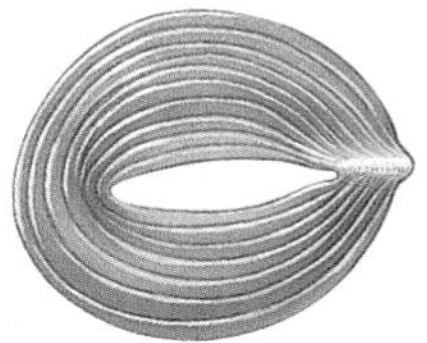

Example: Orbicularis oculi muscle (see Figure 11.4)

Triangular

Fascicles spread over broad area converge at thick central tendon; gives muscle a triangular appearance.

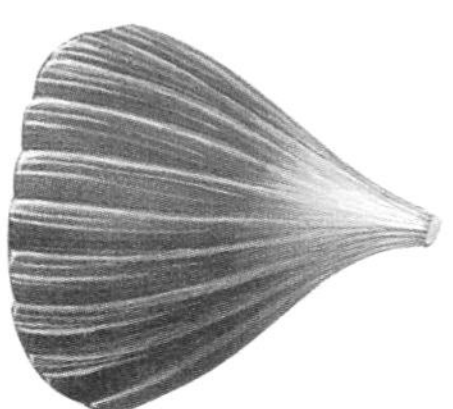

Example: Pectoralis major muscle (see Figure 11.3a)

Pennate

Short fascicles in relation to total muscle length; tendon extends nearly entire length of muscle.

Unipennate

Fascicles are arranged on only one side of tendon.

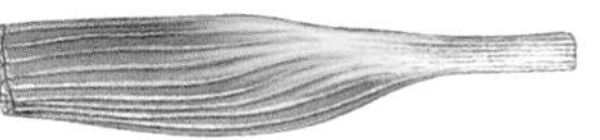

Example: Extensor digitorum longus muscle (see Figure 11.22b)

Bipennate

Fascicles are arranged on both sides of centrally positioned tendons.

Example: Rectus femoris muscle (see Figure 11.20a)

Multipennate

Fascicles attach obliquely from many directions to several tendons.

Example: Deltoid muscle (see Figure 11.10b)

Intramuscular Injections

An **intramuscular (IM) injection** penetrates the skin and subcutaneous tissue to enter the muscle itself. Intramuscular injections are preferred when prompt absorption is desired, when larger doses than can be given subcutaneously are indicated, or when the drug is too irritating to give subcutaneously. The common sites for intramuscular injections include the gluteus medius muscle of the buttock (see Figure 11.3b), lateral side of the thigh in the midportion of the vastus lateralis muscle (see Figure 11.3a), and the deltoid muscle of the shoulder (see Figure 11.3b). Muscles in these areas, especially the gluteal muscles in the buttock, are fairly thick, and absorption is promoted by their extensive blood supply. To avoid injury, intramuscular injections are given deep within the muscle, away from major nerves and blood vessels. Intramuscular injections have a faster speed of delivery than oral medications, but are slower than intravenous infusions. ■

Coordination Within Muscle Groups

Movements often are the result of several skeletal muscles acting as a group. Most skeletal muscles are arranged in opposing (antagonistic) pairs at joints—that is, flexors—extensors, abductors–adductors, and so on. Within opposing pairs, one muscle, called the **prime mover** or **agonist** (= leader), contracts to cause an action while the other muscle, the **antagonist** (*anti-* = against), stretches and yields to the effects of the prime mover. In the process of flexing the forearm at the elbow, for instance, the biceps brachii is the prime mover, and the triceps brachii is the antagonist (see Figure 11.1). The antagonist and prime mover are usually located on opposite sides of the bone or joint, as is the case in this example.

Within an opposing pair of muscles, the roles of the prime mover and antagonist can switch for different movements. For example, while extending the forearm at the elbow (i.e., lowering the load shown in Figure 11.1), the triceps brachii becomes the prime mover, and the biceps brachii is the antagonist. The roles of the two muscles reverse during flexion of the elbow. If a prime mover and its antagonist contract at the same time with equal force, there will be no movement.

Sometimes a prime mover crosses other joints before it reaches the joint at which its primary action occurs. The biceps brachii, for example, spans both the shoulder and elbow joints, with primary action on the forearm. To prevent unwanted movements at intermediate joints or to otherwise aid the movement of the prime mover, muscles called **synergists** (SIN-er-gists; *syn-* = together; *-ergon* = work) contract and stabilize the intermediate joints. As an example, muscles that flex the fingers (prime movers) cross the intercarpal and radiocarpal joints (intermediate joints). If movement at these intermediate joints was unrestrained, you would not be able to flex your fingers without flexing the wrist at the same time. Synergistic contraction of the wrist extensor muscles stabilizes the wrist joint and prevents unwanted movement, while the flexor muscles of the fingers contract to bring about the primary action, which is flexion of the fingers. Synergists are usually located close to the prime mover.

Some muscles in a group also act as **fixators,** stabilizing the origin of the prime mover so that it can act more efficiently. Fixators steady the proximal end of a limb while movements occur at the distal end. For example, the scapula in the pectoral (shoulder) girdle is a freely movable bone that serves as the origin for several muscles that move the arm. When the arm muscles contract, the scapula must be held steady. In abduction of the arm, the deltoid muscle serves as the prime mover, whereas fixators (pectoralis minor, trapezius, subclavius, serratus anterior muscles, and others) hold the scapula firmly against the back of the thorax (see Figure 11.14). The insertion of the deltoid muscle pulls on the humerus to abduct the arm. For different movements and at different times, muscles may act as prime movers, antagonists, synergists, or fixators.

In the limbs, a **compartment** is a group of skeletal muscles, their associated blood vessels and associated nerves, all of which have a common function. In the upper limbs, for example, flexor compartment muscles are anterior, and extensor compartment muscles are posterior.

Benefits of Stretching

The overall goal of **stretching** is to achieve normal range of motion of joints and mobility of soft tissues surrounding the joints. For most individuals, the best stretching routine involves *static stretching,* that is, slow sustained stretching that holds a muscle in a lengthened position. The muscles should be stretched to the point of slight discomfort (not pain) and held for about 15–30 seconds. Stretching should be done after warming up to increase the range of motion most effectively. Among the benefits of stretching are the following:

1. ***Improved physical performance.*** A flexible joint has the ability to move through a greater range of motion, which improves performance.
2. ***Decreased risk of injury.*** Stretching decreases resistance in various soft tissues so there is less likelihood of exceeding maximum tissue extensibility during an activity (i.e., injuring the soft tissues).
3. ***Reduced muscle soreness.*** Stretching can reduce some of the muscle soreness that results after exercise.

4. ***Improved posture.*** Poor posture results from improper position of various parts of the body and the effects of gravity over a number of years. Stretching can help realign soft tissues to improve and maintain good posture. ■

▶ CHECKPOINT

1. Using the terms origin, insertion, and belly in your discussion, describe how skeletal muscles produce body movements by pulling on bones.
2. Describe the three types of levers, and give an example of a first- and third- class lever found in the body.
3. Describe the various arrangements of fascicles.
4. Why can a parallel muscle have a greater range of motion than a pennate muscle?
5. Define the roles of the prime mover (agonist), antagonist, synergist, and fixator in producing various movements of the upper limb.

HOW SKELETAL MUSCLES ARE NAMED

▶ OBJECTIVE

Explain seven features used in naming skeletal muscles.

Several features of skeletal muscles provide descriptive ways to name muscles. The names of most of the nearly 700 skeletal muscles contain combinations of word roots for their distinctive features. Learning the terms that refer to these features will help you remember the names of muscles. Such muscle features include the pattern of the muscle's fascicles; the size, shape, action, number of origins, and location of the muscle; and the sites of origin and insertion of the muscle. Study Table 11.2 to become familiar with the terms used in muscle names.

▶ CHECKPOINT

6. Select 10 muscles in Figure 11.3 and identify the features on which their names are based. (*Hint: Use the prefix, suffix, and root of each muscle's name as a guide.*)

PRINCIPAL SKELETAL MUSCLES

Exhibits 11.1 through 11.20 will assist you in learning the names of the principal skeletal muscles in various regions of the body. The muscles in the exhibits are divided into groups according to the part of the body on which they act. As you study groups of muscles in the exhibits, refer to Figure 11.3 on pages 334–335 to see how each group is related to the others.

The exhibits contain the following elements:

- ***Objective.*** This statement describes what you should learn from the exhibit.
- ***Overview.*** These paragraphs provide a general introduction to the muscles under consideration and emphasize how the muscles are organized within various regions. The discussion also highlights any distinguishing features of the muscles.
- ***Muscle names.*** The muscles to be discussed are included in the form of a table. The word roots indicate how the muscles are named. Once you have mastered the naming of the muscles, you can more easily understand their actions.
- ***Origins, insertions, actions, and innervations.*** In these tables you are also given the origin, insertion, actions, and innervation of each muscle. The innervation section lists the nerve or nerves that cause contraction of each muscle. In general, cranial nerves, which arise from the lower parts of the brain, serve muscles in the head region. Spinal nerves, which arise from the spinal cord within the vertebral column, innervate muscles in the rest of the body. Cranial nerves are designated by both a name and a Roman numeral—for example, the facial (VII) nerve. Spinal nerves are numbered in groups according to the part of the spinal cord from which they arise: C = cervical (neck region), T = thoracic (chest region), L = lumbar (lower back region), and S = sacral (buttocks region). An example is T1, the first thoracic spinal nerve.
- ***Relating muscles to movements.*** These exercises will help you organize the muscles according to the actions they produce.
- ***Checkpoint questions.*** These knowledge checkpoints relate specifically to information in each exhibit, and take the form of review, critical thinking, and/or application questions.
- ***Clinical applications.*** Selected exhibits include clinical applications, which, like those in the text, explore the clinical, professional, or everyday relevance of a particular muscle or its function through descriptions of disorders or clinical procedures.
- ***Figures.*** The figures in the exhibits may present superficial and deep, anterior and posterior, or medial and lateral views to show each muscle's position as clearly as possible. The muscle names in all capital letters are specifically referred to in the table part of the exhibit.

The following is a list of the exhibits and accompanying figures that describe the principal skeletal muscles:

TABLE 11.2 Characteristics Used to Name Muscles

Name	Meaning	Example	Figure
DIRECTION: Orientation of muscle fascicles relative to the body's midline.			
Rectus	Parallel to midline	Rectus abdominis	11.10c
Transverse	Perpendicular to midline	Transversus abdominis	11.10c
Oblique	Diagonal to midline	External oblique	11.10a
SIZE: Relative size of the muscle.			
Maximus	Largest	Gluteus maximus	11.3b
Medius	Intermediate	Gluteus medius	11.20c
Minimus	Smallest	Gluteus minimus	11.20c
Longus	Long	Adductor longus	11.20a
Brevis	Short	Adductor brevis	11.20b
Latissimus	Widest	Latissimus dorsi	11.15b
Longissimus	Longest	Longissimus capitis	11.19a
Magnus	Large	Adductor magnus	11.20a
Major	Larger	Pectoralis major	11.10b
Minor	Smaller	Pectoralis minor	11.14a
Vastus	Huge	Vastus lateralis	11.20a
SHAPE: Relative shape of the muscle.			
Deltoid	Triangular	Deltoid	11.10b
Trapezius	Trapezoid	Trapezius	11.3b
Serratus	Saw-toothed	Serratus anterior	11.14b
Rhomboid	Diamond-shaped	Rhomboid major	11.15c
Orbicularis	Circular	Orbicularis oculi	11.4a
Pectinate	Comblike	Pectineus	11.20a
Piriformis	Pear-shaped	Piriformis	11.20c
Platys	Flat	Platysma	11.4c
Quadratus	Square, four-sided	Quadratus femoris	11.20c
Gracilis	Slender	Gracilis	11.20a

- ► *Exhibit 11.3* Muscles that Move the Mandible (Lower Jaw Bone) (Figure 11.6), p. 342.
- ► *Exhibit 11.4* Muscles that Move the Tongue—Extrinsic Tongue Muscles (Figure 11.7), p. 344.
- ► *Exhibit 11.5* Muscles of the Anterior Neck (Figure 11.8), p. 346.
- ► *Exhibit 11.6* Muscles that Move the Head (Figure 11.9), p. 348.
- ► *Exhibit 11.7* Muscles that Act on the Abdominal Wall (Figure 11.10), p. 350.
- ► *Exhibit 11.8* Muscles Used in Breathing (Figure 11.11), p. 353.
- ► *Exhibit 11.9* Muscles of the Pelvic Floor (Figure 11.12), p. 356.
- ► *Exhibit 11.10* Muscles of the Perineum (Figures 11.12 and 11.13), p. 358.
- ► *Exhibit 11.11* Muscles that Move the Pectoral (Shoulder) Girdle (Figure 11.14), p. 360.
- ► *Exhibit 11.12* Muscles that Move the Humerus (Arm Bone) (Figure 11.15), p. 362.
- ► *Exhibit 11.13* Muscles that Move the Radius and Ulna (Forearm Bones) (Figure 11.16), p. 366.
- ► *Exhibit 11.14* Muscles that Move the Wrist, Hand, and Digits (Figure 11.17), p. 370.

Name	Meaning	Example	Figure
ACTION: Principal action of the muscle.			
Flexor	Decreases a joint angle	Flexor carpi radialis	11.17a
Extensor	Increases a joint angle	Extensor carpi ulnaris	11.17c
Abductor	Moves a bone away from the midline	Abductor pollicis longus	11.17c
Adductor	Moves a bone closer to the midline	Adductor longus	11.20a
Levator	Raises or elevates a body part	Levator scapulae	11.14a
Depressor	Lowers or depresses a body part	Depressor labii inferioris	11.4b
Supinator	Turns palm anteriorly	Supinator	11.17b
Pronator	Turns palm posteriorly	Pronator teres	11.17a
Sphincter	Decreases the size of an opening	External anal sphincter	11.12
Tensor	Makes a body part rigid	Tensor fasciae latae	11.20a
Rotator	Rotates a bone around its longitudinal axis	Rotatores	11.19a
NUMBER OF ORIGINS: Number of tendons of origin.			
Biceps	Two origins	Biceps brachii	11.16a
Triceps	Three origins	Triceps brachii	11.16b
Quadriceps	Four origins	Quadriceps femoris	11.20a
LOCATION: Structure near which a muscle is found.			
Example: Temporalis, a muscle near the temporal bone.			11.4c
ORIGIN AND INSERTION: Sites where muscle originates and inserts.			
Example: Sternocleidomastoid, originating on the sternum and clavicle and inserting onto mastoid process of temporal bone.			11.3a

• • •

To appreciate the many ways the muscular system contributes to homeostasis of other body systems, examine *Focus on Homeostasis: The Muscular System* on page 398. In the next chapter (Chapter 12), you will see how the nervous system is organized, how neurons generate nerve impulses that activate muscle tissues as well as other neurons, and how synapses function.

Figure 11.3 **Principal superficial skeletal muscles.**

Most movements require several skeletal muscles acting in groups rather than individually.

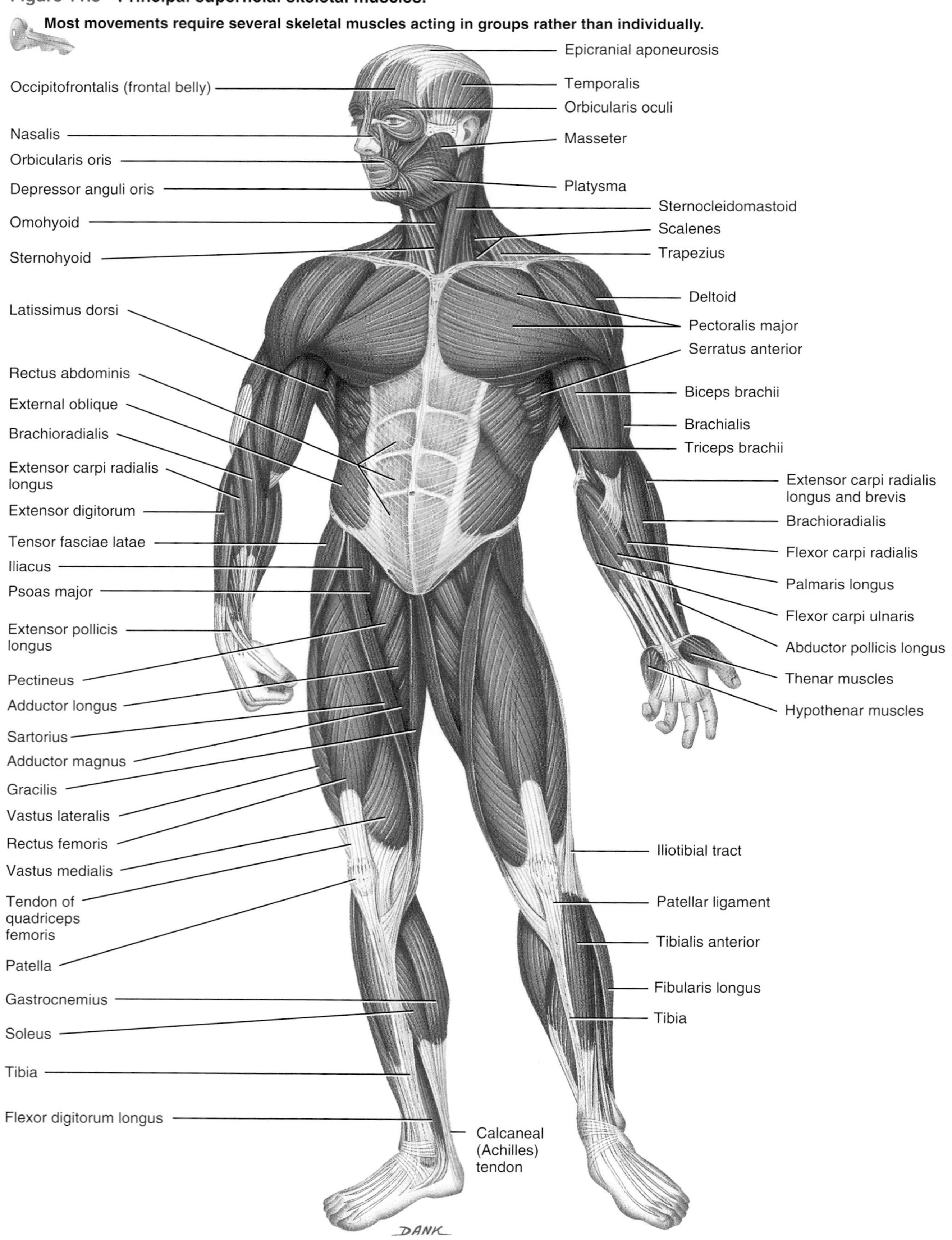

(a) Anterior view

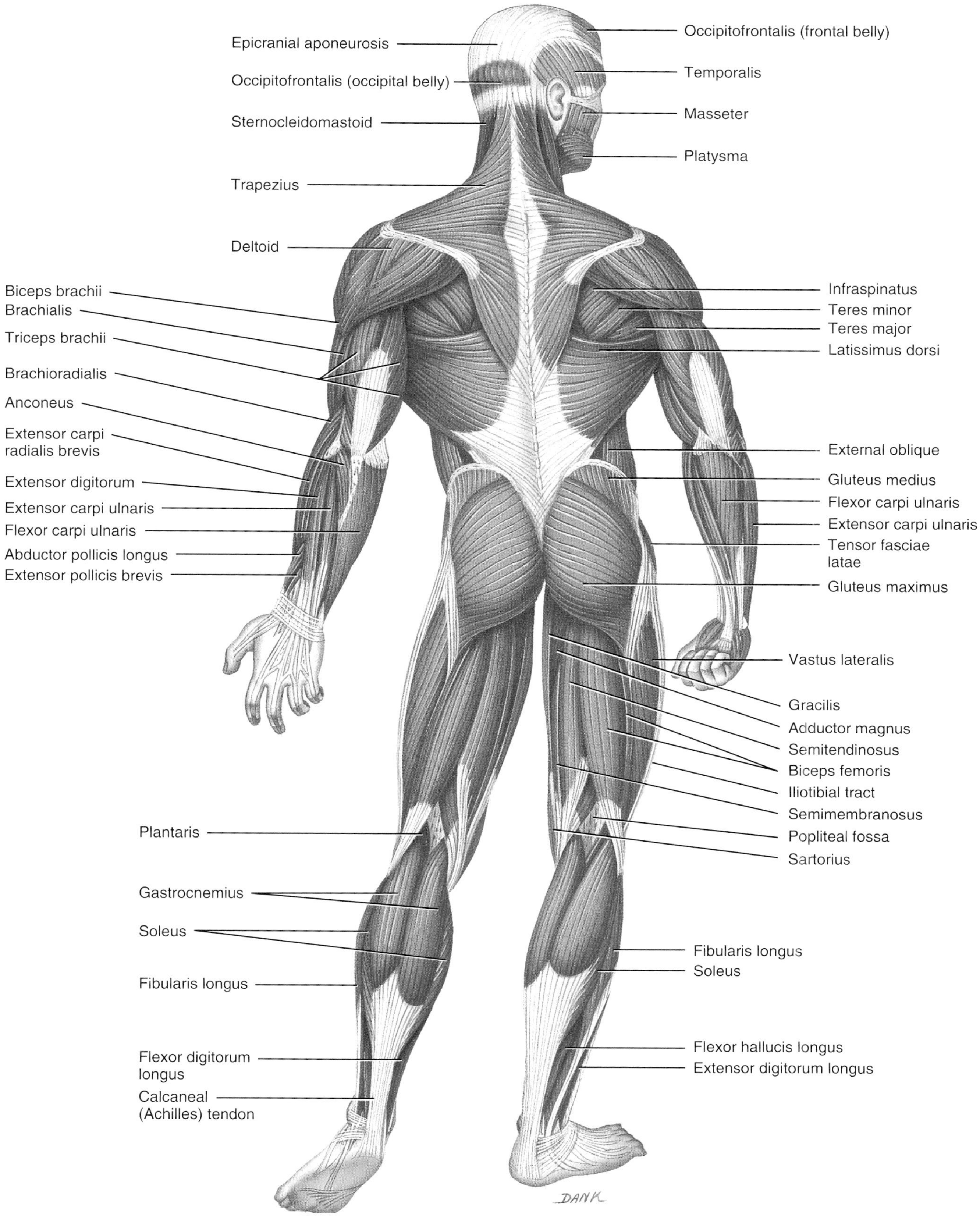

Epicranial aponeurosis
Occipitofrontalis (occipital belly)
Sternocleidomastoid
Trapezius
Deltoid
Occipitofrontalis (frontal belly)
Temporalis
Masseter
Platysma
Biceps brachii
Brachialis
Triceps brachii
Brachioradialis
Anconeus
Extensor carpi radialis brevis
Extensor digitorum
Extensor carpi ulnaris
Flexor carpi ulnaris
Abductor pollicis longus
Extensor pollicis brevis
Infraspinatus
Teres minor
Teres major
Latissimus dorsi
External oblique
Gluteus medius
Flexor carpi ulnaris
Extensor carpi ulnaris
Tensor fasciae latae
Gluteus maximus
Vastus lateralis
Gracilis
Adductor magnus
Semitendinosus
Biceps femoris
Iliotibial tract
Semimembranosus
Popliteal fossa
Sartorius
Plantaris
Gastrocnemius
Soleus
Fibularis longus
Flexor digitorum longus
Calcaneal (Achilles) tendon
Fibularis longus
Soleus
Flexor hallucis longus
Extensor digitorum longus

(b) Posterior view

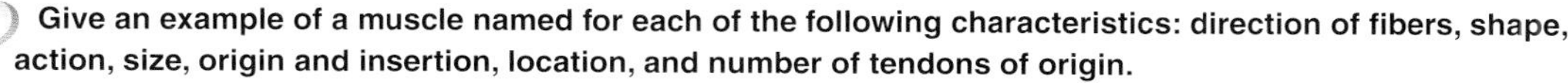

Give an example of a muscle named for each of the following characteristics: direction of fibers, shape, action, size, origin and insertion, location, and number of tendons of origin.

EXHIBIT 11.1 MUSCLES OF FACIAL EXPRESSION (FIGURE 11.4)

▶ **OBJECTIVE**

Describe the origin, insertion, action, and innervation of the muscles of facial expression.

The muscles of facial expression, which provide us with the ability to express a wide variety of emotions, lie within the layers of superficial fascia. They usually originate in the fascia or bones of the skull and insert into the skin. Because of their insertions, the muscles of facial expression move the skin rather than a joint when they contract.

Among the noteworthy muscles in this group are those surrounding the orifices (openings) of the head such as the eyes, nose, and mouth. These muscles function as *sphincters* (SFINGK-ters), which close the orifices, and *dilators,* which dilate or open the orifices. For example, the **orbicularis oculi** muscle closes the eye, and the **levator palpebrae superioris** muscle opens it. The **occipitofrontalis** is an unusual muscle in this group because it is made up of two parts: an anterior part called the **frontal belly,** which is superficial to the frontal bone, and a posterior part called the **occipital belly,** which is superficial to the occipital bone. The two muscular portions are held together by a strong aponeurosis (sheetlike tendon) that covers the superior and lateral surfaces of the skull, the **epicranial aponeurosis** (ep-i-KRĀ-nē-al ap-ō-noo-RŌ-sis), also called the **galea aponeurotica** (GĀ-lē-a ap-ō-noo′-RŌ-ti-ka). The **buccinator** muscle forms the major muscular portion of the cheek. The duct of the parotid gland (a salivary gland) passes through the buccinator muscle to reach the oral cavity. The buccinator muscle is so named because it compresses the cheeks (*bucc-* = cheek) during blowing—for example, when a musician plays a wind instrument such as a trumpet. It functions in whistling, blowing, and sucking and assists in chewing.

MUSCLE	ORIGIN	INSERTION	ACTION	INNERVATION
Scalp Muscles				
Occipitofrontalis (ok-sip′-i-tō-frun-TĀ-lis)				
Frontal belly	Epicranial aponeurosis.	Skin superior to supraorbital margin.	Draws scalp anteriorly, raises eyebrows, and wrinkles skin of forehead horizontally as in a look of surprise.	Facial (VII) nerve.
Occipital belly (*occipit-* = back of the head)	Occipital bone and mastoid process of temporal bone.	Epicranial aponeurosis.	Draws scalp posteriorly.	Facial (VII) nerve.
Mouth Muscles				
Orbicularis oris (or-bi′-kū-LAR-is OR-is; *orb-* = circular; *oris* = of the mouth)	Muscle fibers surrounding opening of mouth.	Skin at corner of mouth.	Closes and protrudes lips, as in kissing; compresses lips against teeth; and shapes lips during speech.	Facial (VII) nerve.
Zygomaticus major (zī-gō-MA-ti-kus; *zygomatic* = cheek bone; *major* = greater)	Zygomatic bone.	Skin at angle of mouth and orbicularis oris.	Draws angle of mouth superiorly and laterally, as in smiling.	Facial (VII) nerve.
Zygomaticus minor (*minor* = lesser)	Zygomatic bone.	Upper lip.	Raises (elevates) upper lip, exposing maxillary teeth.	Facial (VII) nerve.
Levator labii superioris (le-VĀ-tor LĀ-bē-ī soo-per′-ē-OR- is; *levator* = raises or elevates; *labii* = lip; *superioris* = upper)	Superior to infraorbital foramen of maxilla.	Skin at angle of mouth and orbicularis oris.	Raises upper lip.	Facial (VII) nerve.
Depressor labii inferioris (de-PRE-sor LĀ-bē-ī *depressor* = depresses or lowers; *inferioris* = lower)	Mandible.	Skin of lower lip.	Depresses (lowers) lower lip.	Facial (VII) nerve.
Depressor anguli oris (*angul* = angle or corner)	Mandible.	Angle of mouth.	Draws angle of mouth laterally and inferiorly, as in opening mouth.	Facial (VII) nerve.
Levator anguli oris	Inferior to infraorbital foramen.	Skin of lower lip and orbicularis oris.	Draws angle of mouth laterally and superiorly.	Facial (VII) nerve.

Bell's Palsy

Bell's palsy, also known as **facial paralysis,** is a unilateral paralysis of the muscles of facial expression. It is due to damage or disease of the facial (VII) nerve. Possible causes include inflammation of the facial nerve due to an ear infection, ear surgery that damages the facial nerve, or infection by the herpes simplex virus. The paralysis causes the entire side of the face to droop in severe cases. The person cannot wrinkle the forehead, close the eye, or pucker the lips on the affected side. Drooling and difficulty in swallowing also occur. Eighty percent of patients recover completely within a few weeks to a few months. For others, paralysis is permanent. The symptoms of Bell's palsy mimic those of a stroke. ■

Relating Muscles to Movements

Arrange the muscles in this exhibit into two groups: (1) those that act on the mouth and (2) those that act on the eyes.

▶ CHECKPOINT

Why do the muscles of facial expression move the skin rather than a joint?

MUSCLE	ORIGIN	INSERTION	ACTION	INNERVATION
Mouth Muscles (continued)				
Buccinator (BUK-si-nā′-tor; *bucc-* = cheek)	Alveolar processes of maxilla and mandible and pterygomandibular raphe (fibrous band extending from the pterygoid process of the sphenoid bone to the mandible).	Orbicularis oris.	Presses cheeks against teeth and lips, as in whistling, blowing, and sucking; draws corner of mouth laterally; and assists in mastication (chewing) by keeping food between the teeth (and not between teeth and cheeks).	Facial (VII) nerve.
Risorius (ri-ZOR-ē-us; *risor* = laughter)	Fascia over parotid (salivary) gland.	Skin at angle of mouth.	Draws angle of mouth laterally, as in grimacing.	Facial (VII) nerve.
Mentalis (men-TĀ-lis; *ment-* = the chin)	Mandible.	Skin of chin.	Elevates and protrudes lower lip and pulls skin of chin up, as in pouting.	Facial (VII) nerve.
Neck Muscle				
Platysma (pla-TIZ-ma; *platy* = flat, broad)	Fascia over deltoid and pectoralis major muscles.	Mandible, muscle around angle of mouth, and skin of lower face.	Draws outer part of lower lip inferiorly and posteriorly as in pouting; depresses mandible.	Facial (VII) nerve.
Orbit and Eyebrow Muscles				
Orbicularis oculi (or-bi′-kū-LAR-is OK-ū-lī; *oculi* = of the eye)	Medial wall of orbit.	Circular path around orbit.	Closes eye.	Facial (VII) nerve.
Corrugator supercilii (KOR-a-gā′-tor soo-per-SI-lē-ī; *corrugat* = wrinkle; *supercilii* = of the eyebrow)	Medial end of superciliary arch of frontal bone.	Skin of eyebrow.	Draws eyebrow inferiorly and wrinkles skin of forehead vertically as in frowning.	Facial (VII) nerve.
Levator palpebrae superioris (le-VĀ-tor PAL-pe-brē soo-per′-ē-OR-is; *palpebrae* = eyelids) (see also Figure 11.5a)	Roof of orbit (lesser wing of sphenoid bone).	Skin of upper eyelid.	Elevates upper eyelid (opens eye).	Oculomotor (III) nerve.

continues

Figure 11.4 Muscles of facial expression. (See Tortora, *A Photographic Atlas of the Human Body, Second Edition,* Figures 5.2 through 5.4.)

When they contract, muscles of facial expression move the skin rather than a joint.

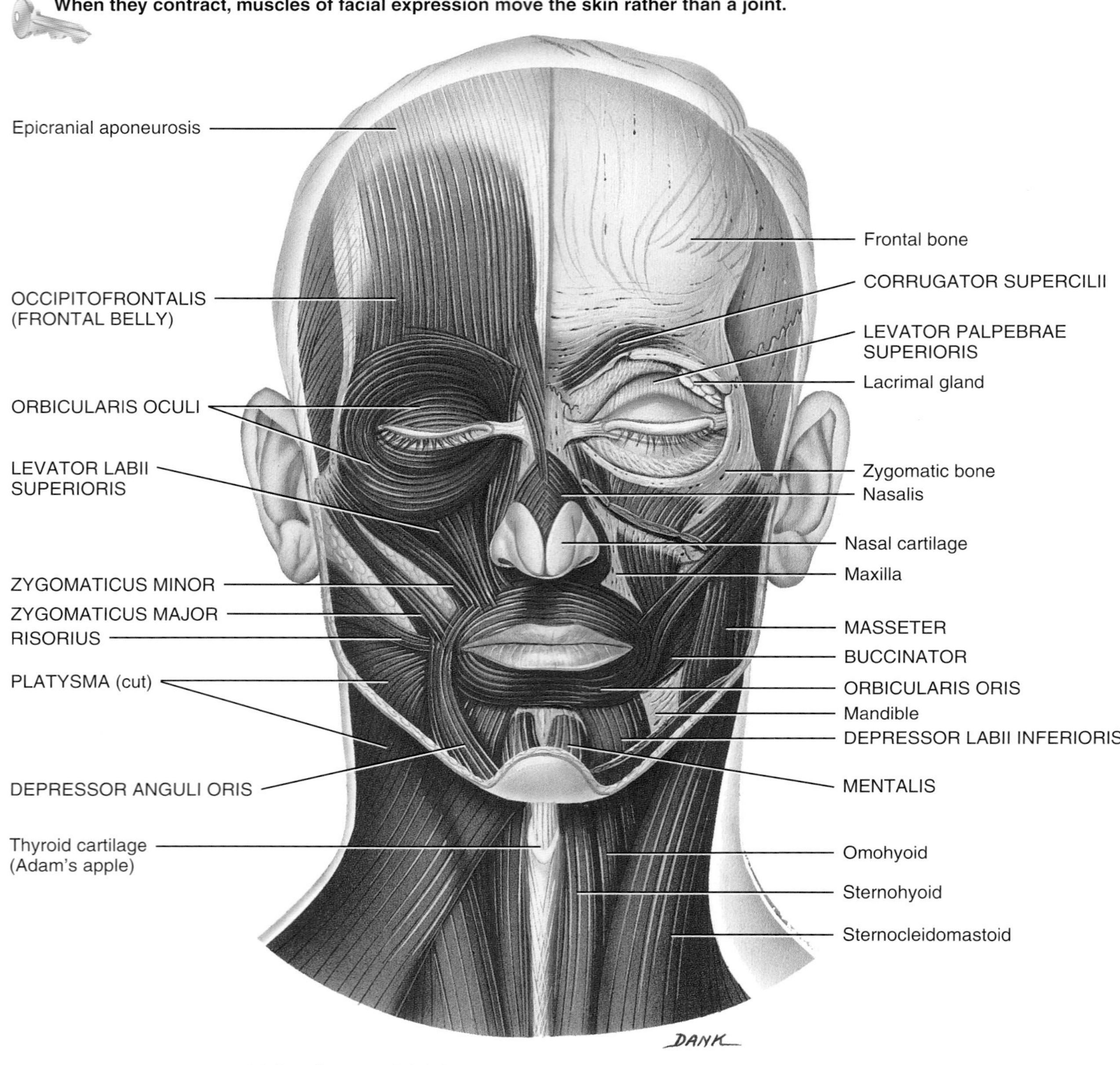

(a) Anterior superficial view (b) Anterior deep view

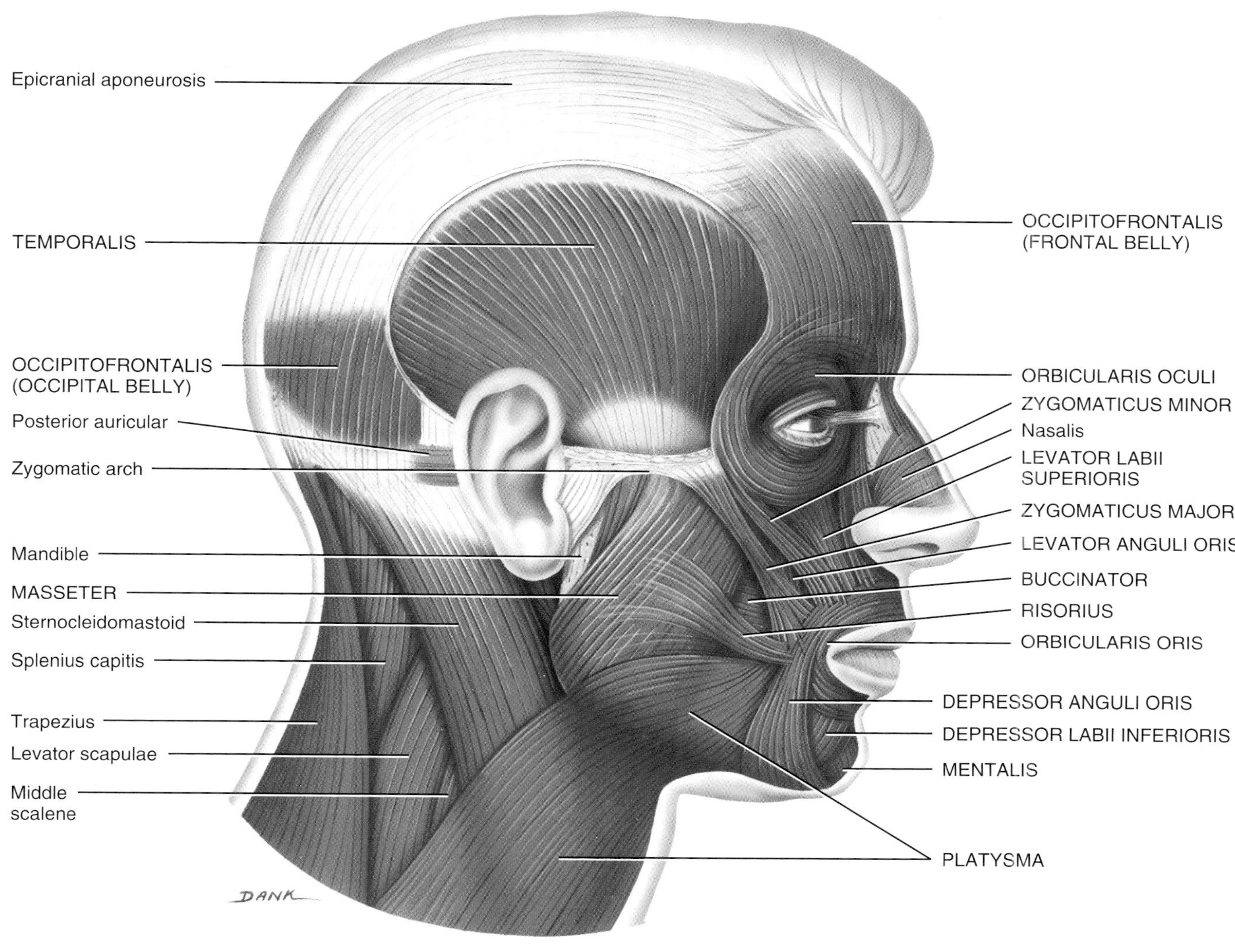

(c) Right lateral superficial view

Which muscles of facial expression cause frowning, smiling, pouting, and squinting?

EXHIBIT 11.2 MUSCLES THAT MOVE THE EYEBALLS—EXTRINSIC EYE MUSCLES (FIGURE 11.5)

▶ **OBJECTIVE**

Describe the origin, insertion, action, and innervation of the extrinsic eye muscles.

Muscles that move the eyeballs are called **extrinsic eye muscles** because they originate outside the eyeballs (in the orbit) and insert on the outer surface of the sclera ("white of the eye"). The extrinsic eye muscles are some of the fastest contracting and most precisely controlled skeletal muscles in the body.

Three pairs of extrinsic eye muscles control movements of the eyeballs: (1) superior and inferior recti, (2) lateral and medial recti, and (3) superior and inferior oblique. The four recti muscles (superior, inferior, lateral, and medial) arise from a tendinous ring in the orbit and insert into the sclera of the eye. As their names imply, the **superior** and **inferior recti** move the eyeballs superiorly and inferiorly; the **lateral** and **medial recti** move the eyeballs laterally and medially.

The actions of the oblique muscles cannot be deduced from their names. The **superior oblique** muscle originates posteriorly near the tendinous ring, then passes anteriorly, and ends in a round tendon. The tendon extends through a pulleylike loop called the *trochlea* (= pulley) in the anterior and medial part of the roof of the orbit. Finally, the tendon turns and inserts on the posterolateral aspect of the eyeballs. Accordingly, the superior oblique muscle moves the eyeballs inferiorly and laterally. The **inferior oblique** muscle originates on the maxilla at the anteromedial aspect of the floor of the orbit. It then passes posteriorly and laterally and inserts on the posterolateral aspect of the eyeballs. Because of this arrangement, the inferior oblique muscle moves the eyeballs superiorly and laterally.

Strabismus

Strabismus (stra-BIZ-mus; *strabismos* = squinting) is a condition in which the two eyes are not properly aligned. This can be hereditary or it can be due to birth injuries, poor attachments of the muscles, problems with the brain's control center, or localized disease. Strabismus can be constant or intermittent. In strabismus, each eye sends an image to a different area of the brain and because the brain usually ignores the messages sent by one of the eyes, the ignored eye becomes weaker, hence "lazy eye" or *amblyopia,* develops. *External strabismus* results when a lesion in the oculomotor (III) nerve causes the eyeball to move laterally when at rest, and results in an inability to move the eyeball medially and inferiorly. A lesion in the abducens (VI) nerve results in *internal strabismus,* a condition in which the eyeball moves medially when at rest and cannot move laterally.

Treatment options for strabismus depend on the specific type of problem and include surgery, visual therapy (retraining the brain's control center), and orthoptics (eye muscle training to straighten the eyes). ■

Relating Muscles to Movements

Arrange the muscles in this exhibit according to their actions on the eyeballs: (1) elevation, (2) depression, (3) abduction, (4) adduction, (5) medial rotation, and (6) lateral rotation. The same muscle may be mentioned more than once.

▶ **CHECKPOINT**

Which muscles contract and relax in each eye as you gaze to your left without moving your head?

MUSCLE	ORIGIN	INSERTION	ACTION	INNERVATION
Superior rectus (*rectus* = fascicles parallel to midline)	Common tendinous ring (attached to orbit around optic foramen).	Superior and central part of eyeball.	Moves eyeball superiorly (elevation) and medially (adduction), and rotates it medially.	Oculomotor (III) nerve.
Inferior rectus	Same as above.	Inferior and central part of eyeball.	Moves eyeball inferiorly (depression) and medially (adduction), and rotates it medially.	Oculomotor (III) nerve.
Lateral rectus	Same as above.	Lateral side of eyeball.	Moves eyeball laterally (abduction).	Abducens (VI) nerve.
Medial rectus	Same as above.	Medial side of eyeball.	Moves eyeball medially (adduction).	Oculomotor (III) nerve.
Superior oblique (*oblique* = fascicles diagonal to midline)	Sphenoid bone, superior and medial to the tendinous ring in the orbit.	Eyeball between superior and lateral recti. The muscle inserts into the superior and lateral surfaces of the eyeball via a tendon that passes through the trochlea.	Moves eyeball inferiorly (depression) and laterally (abduction), and rotates it medially.	Trochlear (IV) nerve.
Inferior oblique	Maxilla in floor of orbit.	Eyeball between inferior and lateral recti.	Moves eyeball superiorly (elevation) and laterally (abduction) and rotates it laterally.	Oculomotor (III) nerve.

Figure 11.5 **Extrinsic muscles of the eyeball.** (See Tortora, *A Photographic Atlas of the Human Body, Second Edition,* Figure 5.5).

The extrinsic muscles of the eyeball are among the fastest contracting and most precisely controlled skeletal muscles in the body.

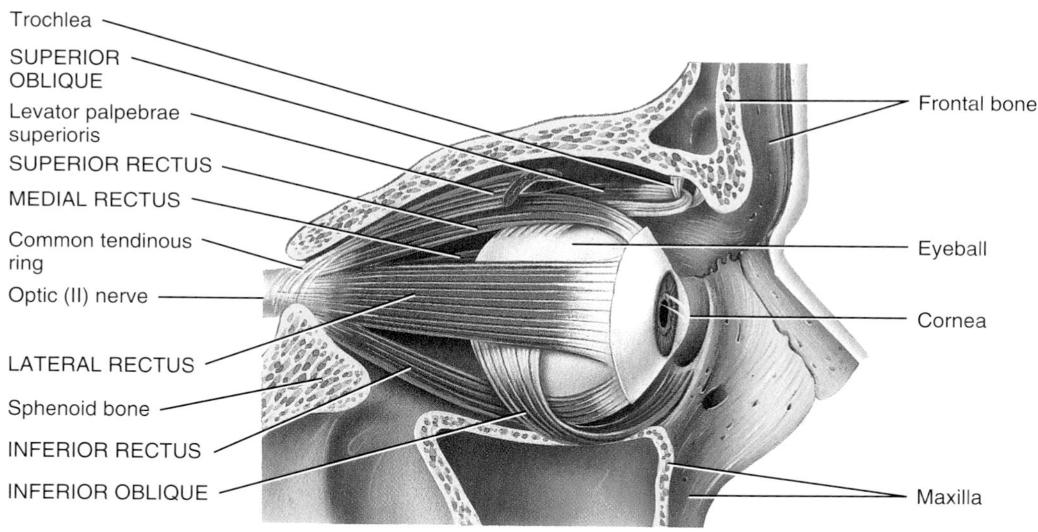

(a) Lateral view of right eyeball

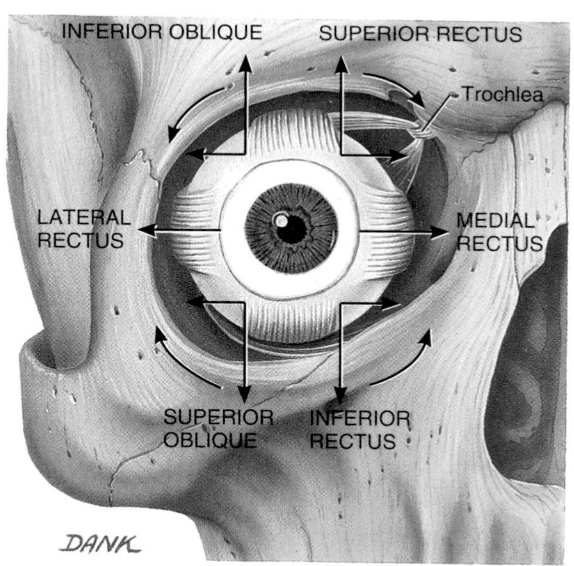

(b) Movements of right eyeball in response to contraction of extrinsic muscles

How does the inferior oblique muscle move the eyeball superiorly and laterally?

EXHIBIT 11.3 MUSCLES THAT MOVE THE MANDIBLE (LOWER JAW BONE) (FIGURE 11.6)

▶ **OBJECTIVE**

Describe the origin, insertion, action, and innervation of the muscles that move the mandible.

The muscles that move the mandible (lower jaw bone) at the temporomandibular joint (TMJ) are known as the muscles of mastication (chewing). Of the four pairs of muscles involved in mastication, three are powerful closers of the jaw and account for the strength of the bite: **masseter, temporalis,** and **medial pterygoid.** Of these, the masseter is the strongest muscle of mastication. The medial and **lateral pterygoid** muscles assist in mastication by moving the mandible from side to side to help grind food. Additionally, these muscles protract (protrude) the mandible.

Relating Muscles to Movements

Arrange the muscles in this exhibit according to their actions on the mandible: (1) elevation, (2) depression, (3) retraction, (4) protraction, and (5) side-to-side movement. The same muscle may be mentioned more than once.

▶ **CHECKPOINT**

What would happen if you lost tone in the masseter and temporalis muscles?

MUSCLE	ORIGIN	INSERTION	ACTION	INNERVATION
Masseter (MA-se-ter = a chewer) (see Figure 11.4c)	Maxilla and zygomatic arch.	Angle and ramus of mandible.	Elevates mandible, as in closing mouth.	Mandibular division of trigeminal (V) nerve.
Temporalis (tem′-pō-RĀ-lis; *tempor-* = time or temples)	Temporal bone.	Coronoid process and ramus of mandible.	Elevates and retracts mandible.	Mandibular division of trigeminal (V) nerve.
Medial pterygoid (TER-i-goyd; *medial* = closer to midline; *pterygoid* = like a wing)	Medial surface of lateral portion of pterygoid process of sphenoid bone; maxilla.	Angle and ramus of mandible.	Elevates and protracts (protrudes) mandible and moves mandible from side to side.	Mandibular division of trigeminal (V) nerve.
Lateral pterygoid (TER-i-goyd; *lateral* = farther from midline)	Greater wing and lateral surface of lateral portion of pterygoid process of sphenoid bone.	Condyle of mandible; temporomandibular joint (TMJ).	Protracts mandible, depresses mandible as in opening mouth, and moves mandible from side to side.	Mandibular division of trigeminal (V) nerve.

Figure 11.6 **Muscles that move the mandible (lower jaw bone).** (See Tortora, *A Photographic Atlas of the Human Body, Second Edition,* Figure 5.4).

The muscles that move the mandible are also known as muscles of mastication.

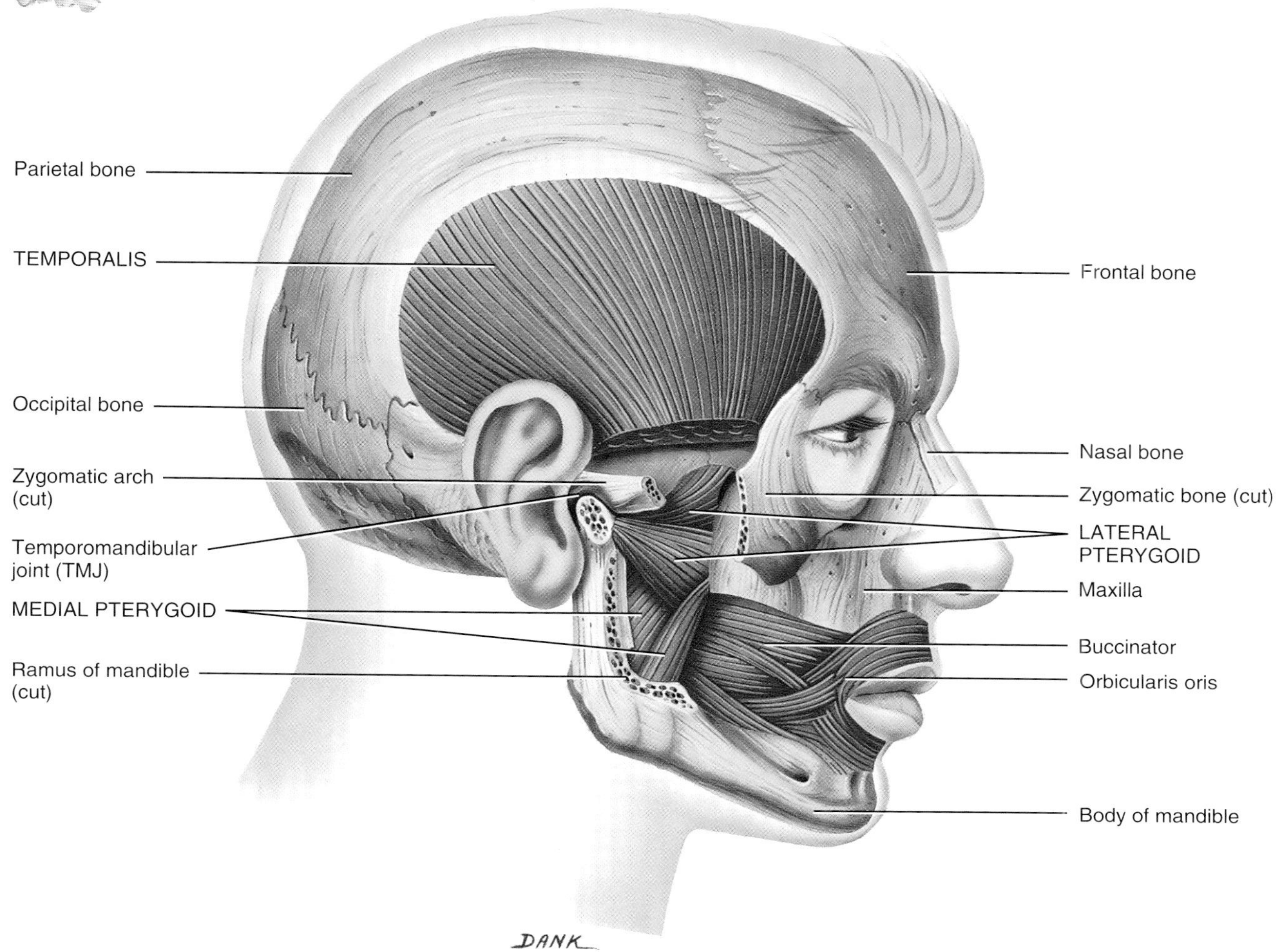

Right lateral superficial view

Which is the strongest muscle of mastication?

EXHIBIT 11.4 MUSCLES THAT MOVE THE TONGUE—EXTRINSIC TONGUE MUSCLES (FIGURE 11.7)

▶ **OBJECTIVE**

Describe the origin, insertion, action, and innervation of the extrinsic muscles of the tongue.

The tongue is a highly mobile structure that is vital to digestive functions such as mastication, detection of taste, and deglutition (swallowing). It is also important in speech. The tongue's mobility is greatly aided by its attachment to the mandible, styloid process of the temporal bone, and hyoid bone.

The tongue is divided into lateral halves by a median fibrous septum. The septum extends throughout the length of the tongue. Inferiorly, the septum attaches to the hyoid bone. Muscles of the tongue are of two principal types: extrinsic and intrinsic. **Extrinsic tongue muscles** originate outside the tongue and insert into it. They move the entire tongue in various directions, such as anteriorly, posteriorly, and laterally. **Intrinsic tongue muscles** originate and insert within the tongue. These muscles alter the shape of the tongue rather than moving the entire tongue. The extrinsic and intrinsic muscles of the tongue insert into both lateral halves of the tongue.

When you study the extrinsic tongue muscles, you will notice that all of their names end in *glossus,* meaning tongue. You will also notice that the actions of the muscles are obvious, considering the positions of the mandible, styloid process, hyoid bone, and soft palate, which serve as origins for these muscles. For example, the **genioglossus** (origin: the mandible) pulls the tongue downward and forward, the **styloglossus** (origin: the styloid process) pulls the tongue upward and backward, the **hyoglossus** (origin: the hyoid bone) pulls the tongue downward and flattens it, and the **palatoglossus** (origin: the soft palate) raises the back portion of the tongue.

Intubation During Anesthesia

When general anesthesia is administered during surgery, a total relaxation of the muscles results. Once the various types of drugs for anesthesia have been given (especially the paralytic agents), the patient's airway must be protected and the lungs ventilated because the muscles involved with respiration are among those paralyzed. Paralysis of the genioglossus muscle causes the tongue to fall posteriorly, which may obstruct the airway to the lungs. To avoid this, the mandible is either manually thrust forward and held in place (known as the "sniffing position"), or a tube is inserted from the lips through the laryngopharynx (inferior portion of the throat) into the trachea (endotracheal intubation). People can also be intubated nasally (through the nose). ■

Relating Muscles to Movements

Arrange the muscles in this exhibit according to the following actions on the tongue: (1) depression, (2) elevation, (3) protraction, and (4) retraction. The same muscle may be mentioned more than once.

▶ **CHECKPOINT**

When your physician says, "Open your mouth, stick out your tongue, and say *ahh,*" to examine the inside of your mouth for possible signs of infection, which muscles do you contract?

MUSCLE	ORIGIN	INSERTION	ACTION	INNERVATION
Genioglossus (jē′-nē-ō-GLOS-us; *genio-* = the chin; *glossus* = tongue)	Mandible.	Undersurface of tongue and hyoid bone.	Depresses tongue and thrusts it anteriorly (protraction).	Hypoglossal (XII) nerve.
Styloglossus (stī′-lō-GLOS-us; *stylo* = stake or pole; styloid process of temporal bone)	Styloid process of temporal bone.	Side and undersurface of tongue.	Elevates tongue and draws it posteriorly (retraction).	Hypoglossal (XII) nerve.
Palatoglossus (pal′-a-tō-GLOS- us; *palato-* = the roof of the mouth or palate)	Anterior surface of soft palate.	Side of tongue.	Elevates posterior portion of tongue and draws soft palate down on tongue.	Pharyngeal plexus, which contains axons from both the vagus (X) and accessory (XI) nerves.
Hyoglossus (hī′-ō-GLOS-us)	Greater horn and body of hyoid bone.	Side of tongue.	Depresses tongue and draws down its sides.	Hypoglossal (XII) nerve.

Figure 11.7 Muscles that move the tongue.

The extrinsic and intrinsic tongue muscles are arranged in both halves of the tongue.

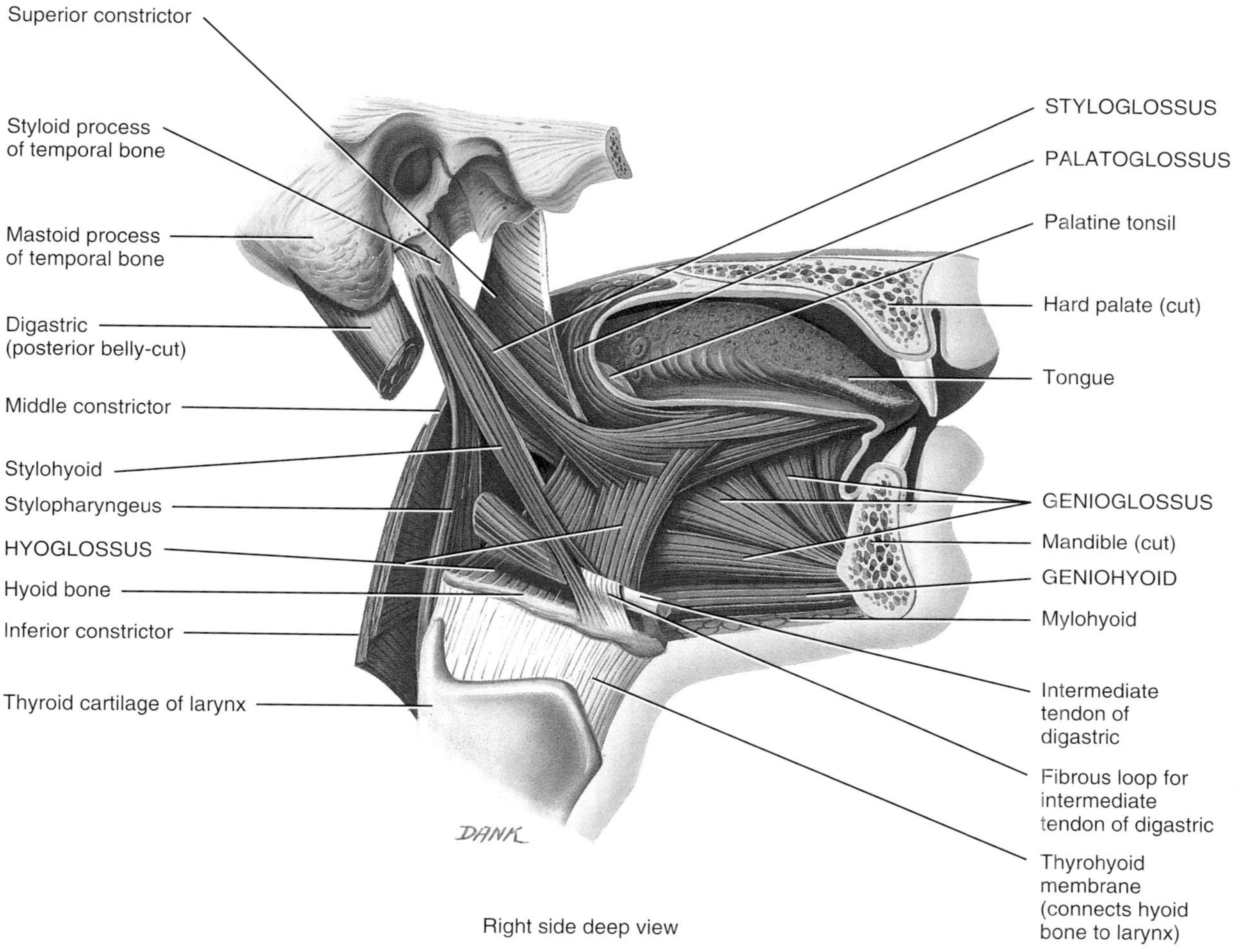

Right side deep view

What are the functions of the tongue?

MUSCLES OF THE ANTERIOR NECK (FIGURE 11.8)

▶ OBJECTIVE

Describe the origin, insertion, action, and innervation of the muscles of the anterior neck.

Two groups of muscles are associated with the anterior aspect of the neck: (1) the **suprahyoid muscles,** so called because they are located superior to the hyoid bone, and (2) the **infrahyoid muscles,** named for their position inferior to the hyoid bone. Both groups of muscles stabilize the hyoid bone, allowing it to serve as a firm base on which the tongue can move.

As a group, the suprahyoid muscles elevate the hyoid bone, floor of the oral cavity, and tongue during swallowing. As its name suggests, the **digastric** muscle has two bellies, anterior and posterior, united by an intermediate tendon that is held in position by a fibrous loop (see Figure 11.7). This muscle elevates the hyoid bone and larynx (voice box) during swallowing and speech and depresses the mandible. The **stylohyoid** muscle elevates and draws the hyoid bone posteriorly, thus elongating the floor of the oral cavity during swallowing. The **mylohyoid** muscle elevates the hyoid bone and helps press the tongue against the roof of the oral cavity during swallowing to move food from the oral cavity into the throat. The **geniohyoid** muscle (see Figure 11.7) elevates and draws the hyoid bone anteriorly to shorten the floor of the oral cavity and to widen the throat to receive food that is being swallowed. It also depresses the mandible.

The infrahyoid muscles are sometimes called "strap" muscles because of their ribbonlike appearance. Most of the infrahyoid muscles depress the hyoid bone and some move the larynx during swallowing and speech. The **omohyoid** muscle, like the digastric muscle, is composed of two bellies connected by an intermediate tendon. In this case, however, the two bellies are referred to as *superior* and *inferior,* rather than anterior and posterior. Together, the omohyoid, **sternohyoid,** and **thyrohyoid** muscles depress the hyoid bone. In addition, the **sternothyroid** muscle depresses the thyroid cartilage (Adam's apple) of the larynx, and the thyrohyoid muscle elevates the thyroid cartilage. The actions are necessary during phonation to produce low and high tones, respectively.

Relating Muscles to Movements

Arrange the muscles in this exhibit according to the following actions on the hyoid bone: (1) elevating it, (2) drawing it anteriorly, (3) drawing it posteriorly, and (4) depressing it; and on the thyroid cartilage: (1) elevating it and (2) depressing it. The same muscle may be mentioned more than once.

▶ CHECKPOINT

Which tongue, facial, and mandibular muscles do you use for chewing?

MUSCLE	ORIGIN	INSERTION	ACTION	INNERVATION
Suprahyoid Muscles				
Digastric (dī′-GAS-trik; *di-* = two; *gastr-* = belly)	Anterior belly from inner side of inferior border of mandible; posterior belly from temporal bone.	Body of hyoid bone via an intermediate tendon.	Elevates hyoid bone and depresses mandible, as in opening the mouth.	Anterior belly: mandibular division of trigeminal (V) nerve. Posterior belly: facial (VII) nerve.
Stylohyoid (stī′-lō-HĪ-oid; *stylo-* = stake or pole, styloid process of temporal bone; *hyo-* = U-shaped, pertaining to hyoid bone)	Styloid process of temporal bone.	Body of hyoid bone.	Elevates hyoid bone and draws it posteriorly.	Facial (VII) nerve.
Mylohyoid (mī′-lō-HĪ-oid) (*mylo-* = mill)	Inner surface of mandible.	Body of hyoid bone.	Elevates hyoid bone and floor of mouth and depresses mandible.	Mandibular division of trigeminal (V) nerve.
Geniohyoid (jē′-nē-ō-HĪ-oid; *genio-* = chin) (see Figure 11.7)	Inner surface of mandible.	Body of hyoid bone.	Elevates hyoid bone, draws hyoid bone and tongue anteriorly, and depresses mandible.	First cervical spinal nerve.
Infrahyoid Muscles				
Omohyoid (ō-mō-HĪ-oid; *omo-* = relationship to the shoulder)	Superior border of scapula and superior transverse ligament.	Body of hyoid bone.	Depresses hyoid bone.	Branches of spinal nerves C1–C3.
Sternohyoid (ster′-nō-HĪ-oid; *sterno-* = sternum)	Medial end of clavicle and manubrium of sternum.	Body of hyoid bone.	Depresses hyoid bone.	Branches of spinal nerves C1–C3.
Sternothyroid (ster′-nō-THĪ-roid; *thyro-* = thyroid gland)	Manubrium of sternum.	Thyroid cartilage of larynx.	Depresses thyroid cartilage of larynx.	Branches of spinal nerves C1–C3.
Thyrohyoid (thī′-rō-HĪ-oid)	Thyroid cartilage of larynx.	Greater horn of hyoid bone.	Elevates thyroid cartilage and depresses hyoid bone.	Branches of spinal nerves C1–C2 and descending hypoglossal (XII) nerve.

Figure 11.8 **Muscles of the floor of the oral cavity and front of the neck.**

The suprahyoid muscles elevate the hyoid bone, the floor of the oral cavity, and the tongue during swallowing.

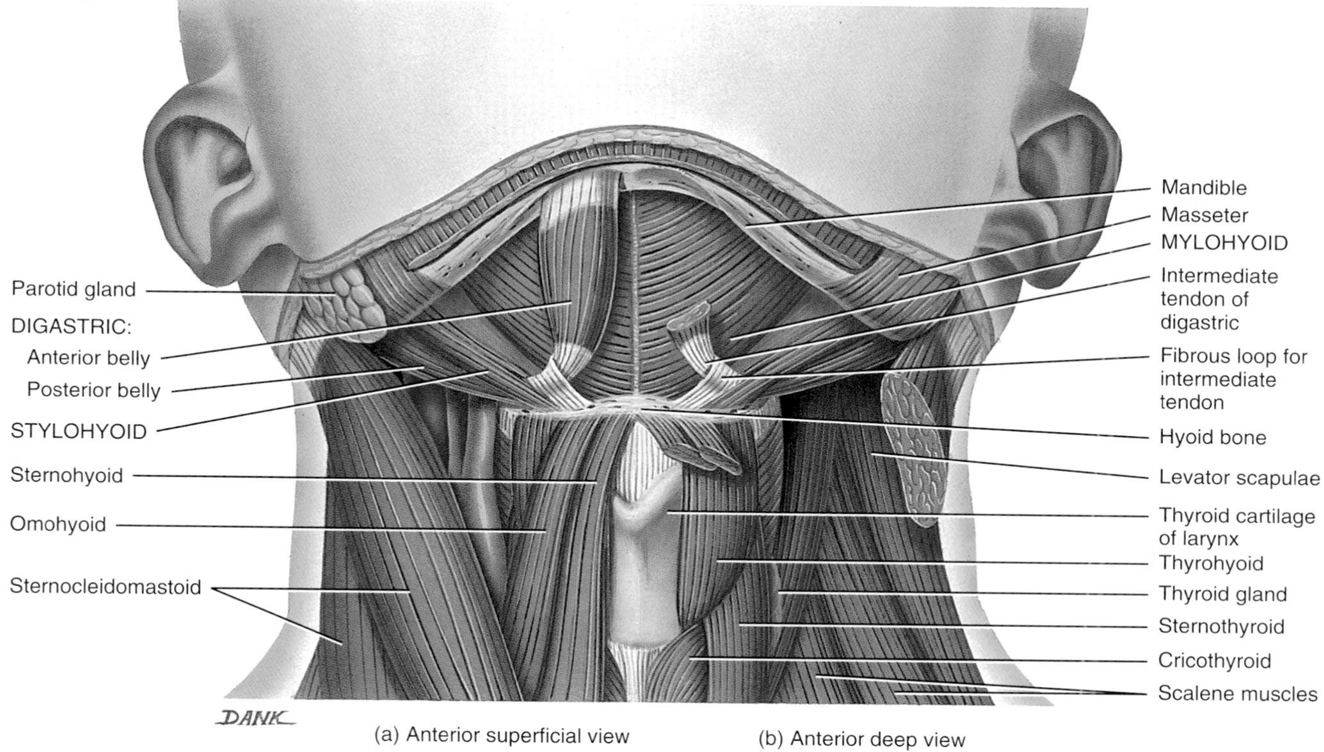

(a) Anterior superficial view (b) Anterior deep view

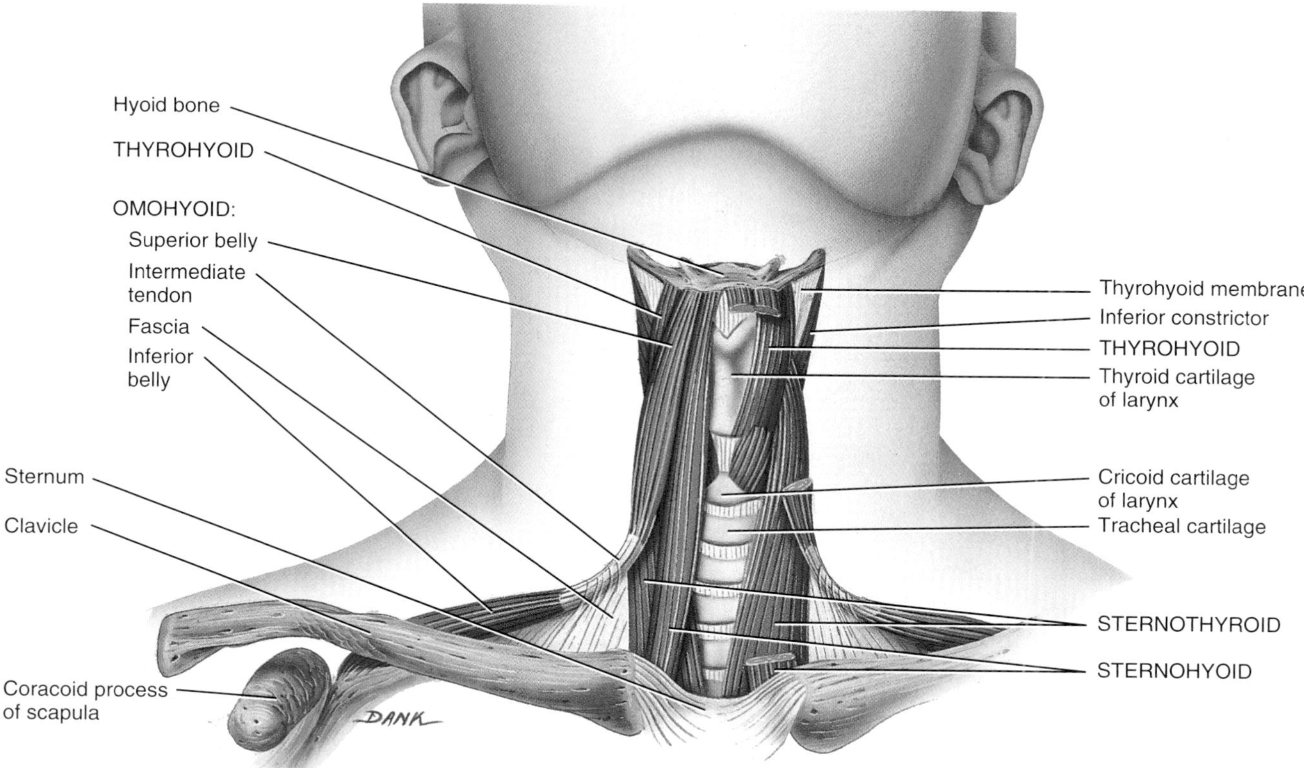

Anterior superficial view (c) Anterior deep view

What is the combined action of the suprahyoid and infrahyoid muscles?

MUSCLES THAT MOVE THE HEAD (FIGURE 11.9)

▶ **OBJECTIVE**

Describe the origin, insertion, action, and innervation of the muscles that move the head.

The head is attached to the vertebral column at the atlanto-occipital joint formed by the atlas and occipital bone. Balance and movement of the head on the vertebral column involves the action of several neck muscles. For example, acting together (bilaterally), contraction of the two **sternocleidomastoid** muscles flexes the cervical portion of the vertebral column and flexes the head. Acting singly (unilaterally), each sternocleidomastoid muscle laterally extends and rotates the head. Bilateral contraction of the **semispinalis capitis, splenius capitis,** and **longissimus capitis** muscles extends the head. However, when these same muscles contract unilaterally, their actions are quite different, involving primarily rotation of the head.

The sternocleidomastoid muscle is an important landmark that divides the neck into two major triangles: anterior and posterior. The triangles are important because of the structures that lie within their boundaries.

The **anterior triangle** is bordered superiorly by the mandible, inferiorly by the sternum, medially by the cervical midline, and laterally by the anterior border of the sternocleidomastoid muscle. The anterior triangle is subdivided into an unpaired submental triangle and three paired triangles: submandibular, carotid, and muscular. The anterior triangle contains submental, submandibular, and deep cervical lymph nodes; the submandibular salivary gland and a portion of the parotid salivary gland; the facial artery and vein; common carotid arteries and internal jugular vein; and the following cranial nerves: glossopharyngeal (IX), vagus (X), accessory (XI), and hypoglossal (XII).

The **posterior triangle** is bordered inferiorly by the clavicle, anteriorly by the posterior border of the sternocleidomastoid muscle, and posteriorly by the anterior border of the trapezius muscle. The posterior triangle is subdivided into two triangles, occipital and supraclavicular (omoclavicular), by the inferior belly of the omohyoid muscle. The posterior triangle contains part of the subclavian artery, external jugular vein, cervical lymph nodes, brachial plexus, and the accessory (XI) nerve.

Relating Muscles to Movements

Arrange the muscles in this exhibit according to the following actions on the head: (1) flexion, (2) lateral flexion, (3) extension, (4) rotation to side opposite contracting muscle, and (5) rotation to same side as contracting muscle. The same muscle may be mentioned more than once.

▶ **CHECKPOINT**

What muscles do you contract to signify "yes" and "no"?

MUSCLE	ORIGIN	INSERTION	ACTION	INNERVATION
Sternocleidomastoid (ster′-nō-klī′-dō-MAS-toid; *sterno-* = breastbone; *cleido-* = clavicle; *mastoid* = mastoid process of temporal bone)	Sternum and clavicle.	Mastoid process of temporal bone.	Acting together (bilaterally), flex cervical portion of vertebral column, flex head, and elevate sternum during forced inhalation; acting singly (unilaterally), laterally extend and rotate head to side opposite contracting muscle.	Accessory (XI) nerve.
Semispinalis capitis (se′-mē-spi-NĀ-lis KAP-i-tis; *semi-* = half; *spine* = spinous process; *capit-* = head) (see Figure 11.19a)	Transverse processes of first six or seven thoracic vertebrae and seventh cervical vertebra, and articular processes of fourth, fifth, and sixth cervical vertebrae.	Occipital bone between superior and inferior nuchal lines.	Acting together, extend head; acting singly, rotate head to side opposite contracting muscle.	Cervical spinal nerves.
Splenius capitis (SPLĒ-nē-us KAP-i-tis; *splenion-* = bandage) (see Figure 11.19a)	Ligamentum nuchae and spinous processes of seventh cervical vertebra and first three or four thoracic vertebrae.	Occipital bone and mastoid process of temporal bone.	Acting together, extend head; acting singly, laterally flex and rotate head to same side as contracting muscle.	Cervical spinal nerves.
Longissimus capitis (lon-JIS-i-mus KAP-i-tis; *longissimus* = longest) (see Figure 11.19a)	Transverse processes of upper four thoracic vertebrae and articular processes of last four cervical vertebrae.	Mastoid process of temporal bone.	Acting together, extend head; acting singly, laterally flex and rotate head to same side as contracting muscle.	Cervical spinal nerves.

Figure 11.9 Triangles of the neck. (See Tortora, *A Photographic Atlas of the Human Body, Second Edition,* Figure 5.3.)

The sternocleidomastoid muscle divides the neck into two principal triangles: anterior and posterior.

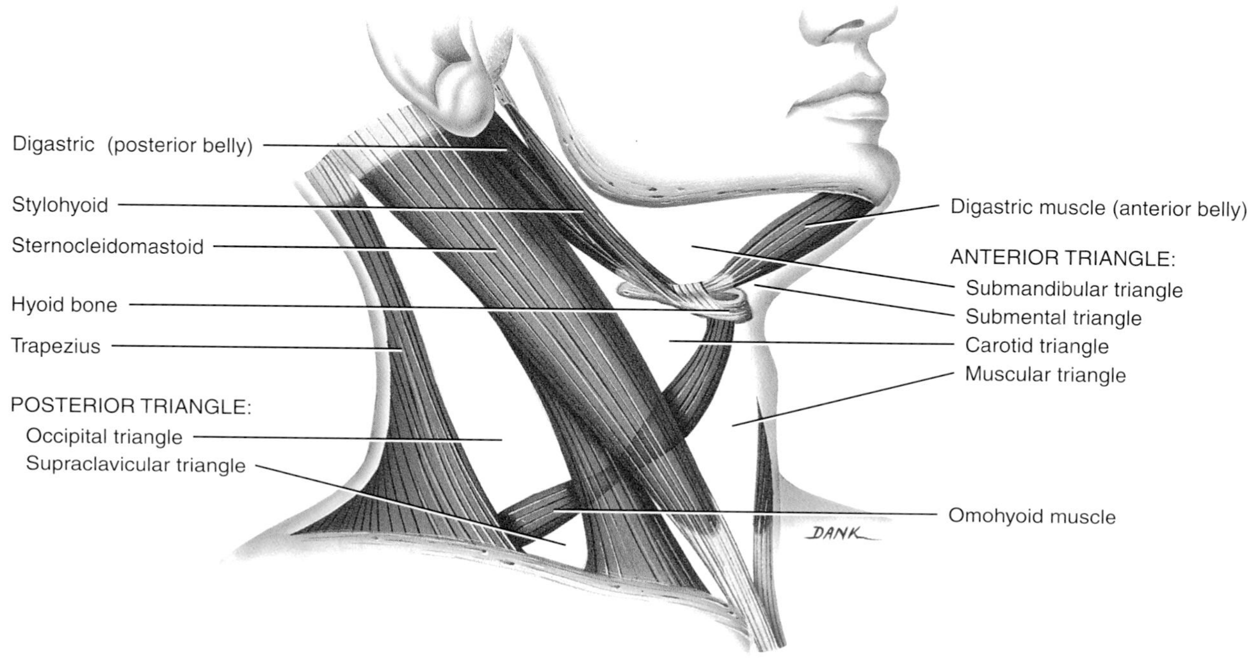

Right lateral view

Why are triangles important?

EXHIBIT 11.7 MUSCLES THAT ACT ON THE ABDOMINAL WALL (FIGURE 11.10)

▶ OBJECTIVE

Describe the origin, insertion, action, and innervation of the muscles that act on the abdominal wall.

The anterolateral abdominal wall is composed of skin, fascia, and four pairs of muscles: the external oblique, internal oblique, transversus abdominis, and rectus abdominis. The first three muscles named are arranged from superficial to deep. The **external oblique** is the superficial muscle. Its fascicles extend inferiorly and medially. The **internal oblique** is the intermediate flat muscle. Its fascicles extend at right angles to those of the external oblique. The **transversus abdominis** is the deep muscle, with most of its fascicles directed transversely around the abdominal wall. Together, the external oblique, internal oblique, and transversus abdominis form three layers of muscle around the abdomen. In each layer, the muscle fascicles extend in a different direction. This is a structural arrangement that affords considerable protection to the abdominal viscera, especially when the muscles have good tone.

The **rectus abdominis** muscle is a long muscle that extends the entire length of the anterior abdominal wall, originating at the pubic crest and pubic symphysis and inserting on the cartilages of ribs 5–7 and the xiphoid process of the sternum. The anterior surface of the muscle is interrupted by three transverse fibrous bands of tissue called **tendinous intersections,** believed to be remnants of septa that separated myotomes during embryological development (see Figure 10.19 on page 318).

As a group, the muscles of the anterolateral abdominal wall help contain and protect the abdominal viscera; flex, laterally flex, and rotate the vertebral column at the intervertebral joints; compress the abdomen during forced exhalation; and produce the force required for defecation, urination, and childbirth.

The aponeuroses (sheathlike tendons) of the external oblique, internal oblique, and transversus abdominis muscles form the **rectus sheaths,** which enclose the rectus abdominis muscles. The sheaths meet at the midline to form the **linea alba** (= white line), a tough, fibrous band that extends from the xiphoid process of the sternum to the pubic symphysis. In the latter stages of pregnancy, the linea alba stretches to increase the distance be-

MUSCLE	ORIGIN	INSERTION	ACTION	INNERVATION
Rectus abdominis (REK-tus ab-DOM-in-is; *rectus-* = fascicles parallel to midline; *abdomin* = abdomen)	Pubic crest and pubic symphysis.	Cartilage of fifth to seventh ribs and xiphoid process.	Flexes vertebral column, especially lumbar portion, and compresses abdomen to aid in defecation, urination, forced exhalation, and childbirth.	Thoracic spinal nerves T7–T12.
External oblique (ō-BLĒK; *external* = closer to surface; *oblique* = fascicles diagonal to midline)	Inferior eight ribs.	Iliac crest and linea alba.	Acting together (bilaterally), compress abdomen and flex vertebral column; acting singly (unilaterally), laterally flex vertebral column, especially lumbar portion, and rotate vertebral column.	Thoracic spinal nerves T7–T12 and the iliohypogastric nerve.
Internal oblique (ō-BLĒK; *internal* = farther from surface)	Iliac crest, inguinal ligament, and thoracolumbar fascia.	Cartilage of last three or four ribs and linea alba.	Acting together, compress abdomen and flex vertebral column; acting singly, laterally flex vertebral column, especially lumbar portion, and rotate vertebral column.	Thoracic spinal nerves T8–T12, iliohypogastric nerve, and ilioinguinal nerve.
Transversus abdominis (tranz-VER-sus ab-DOM-in-is; *transverse* = fascicles perpendicular to midline)	Iliac crest, inguinal ligament, lumbar fascia, and cartilages of inferior six ribs.	Xiphoid process, linea alba, and pubis.	Compresses abdomen.	Thoracic spinal nerves T8–T12, iliohypogastric nerve, and ilioinguinal nerve.
Quadratus lumborum (kwod-RĀ-tus lum-BOR-um; *quad-* = four; *lumbo-* = lumbar region) (see Figure 11.11)	Iliac crest and iliolumbar ligament.	Inferior border of twelfth rib and first four lumbar vertebrae.	Acting together, pull twelfth ribs inferiorly during forced exhalation, fix twelfth ribs to prevent their elevation during deep inhalation, and help extend lumbar portion of vertebral column; acting singly, laterally flex vertebral column, especially lumbar portion.	Thoracic spinal nerve T12 and lumbar spinal nerves L1–L3 or L1–L4.

tween the rectus abdominis muscles. The inferior free border of the external oblique aponeurosis forms the **inguinal ligament,** which runs from the anterior superior iliac spine to the pubic tubercle (see Figure 11.20a). Just superior to the medial end of the inguinal ligament is a triangular slit in the aponeurosis referred to as the **superficial inguinal ring,** the outer opening of the **inguinal canal** (see Figure 28.2 on page 1059). The inguinal canal contains the spermatic cord and ilioinguinal nerve in males, and the round ligament of the uterus and ilioinguinal nerve in females.

The posterior abdominal wall is formed by the lumbar vertebrae, parts of the ilia of the hip bones, psoas major and iliacus muscles (described in Exhibit 11.17), and quadratus lumborum muscle. The anterolateral abdominal wall can contract and distend; the posterior abdominal wall is bulky and stable by comparison.

Inguinal Hernia

A **hernia** is a protrusion of an organ through a structure that normally contains it, which creates a lump that can be seen or felt through the skin's surface. The inguinal region is a weak area in the abdominal wall. It is often the site of an **inguinal hernia,** a rupture or separation of a portion of the inguinal area of the abdominal wall resulting in the protrusion of a part of the small intestine. Hernia is much more common in males than in females because the inguinal canals in males are larger to accommodate the spermatic cord and ilioinguinal nerve. Treatment of hernias most often involves surgery. The organ that protrudes is "tucked" back into the abdominal cavity and the defect in the abdominal muscles is repaired. In addition, a mesh is often applied to reinforce the area of weakness. ■

Relating Muscles to Movements

Arrange the muscles in this exhibit according to the following actions on the vertebral column: (1) flexion, (2) lateral flexion, (3) extension, and (4) rotation. The same muscle may be mentioned more than once.

▶ CHECKPOINT

Which muscles do you contract when you "suck in your gut," thereby compressing the anterior abdominal wall?

Figure 11.10 Muscles of the male anterolateral abdominal wall. (See Tortora, *A Photographic Atlas of the Human Body, Second Edition,* Figure 5.7.)

The anterolateral abdominal muscles protect the abdominal viscera, move the vertebral column, and assist in forced exhalation, defecation, urination, and childbirth (in the female).

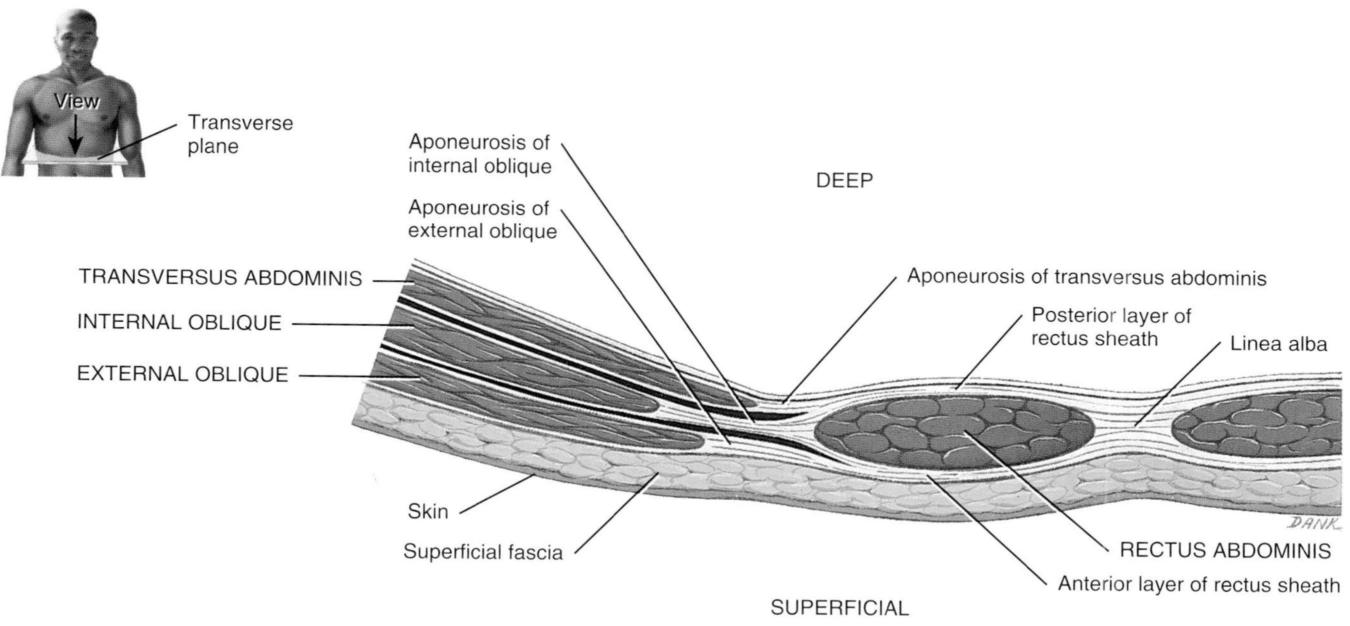

(a) Transverse section of anterior abdominal wall superior to umbilicus (navel)

continues

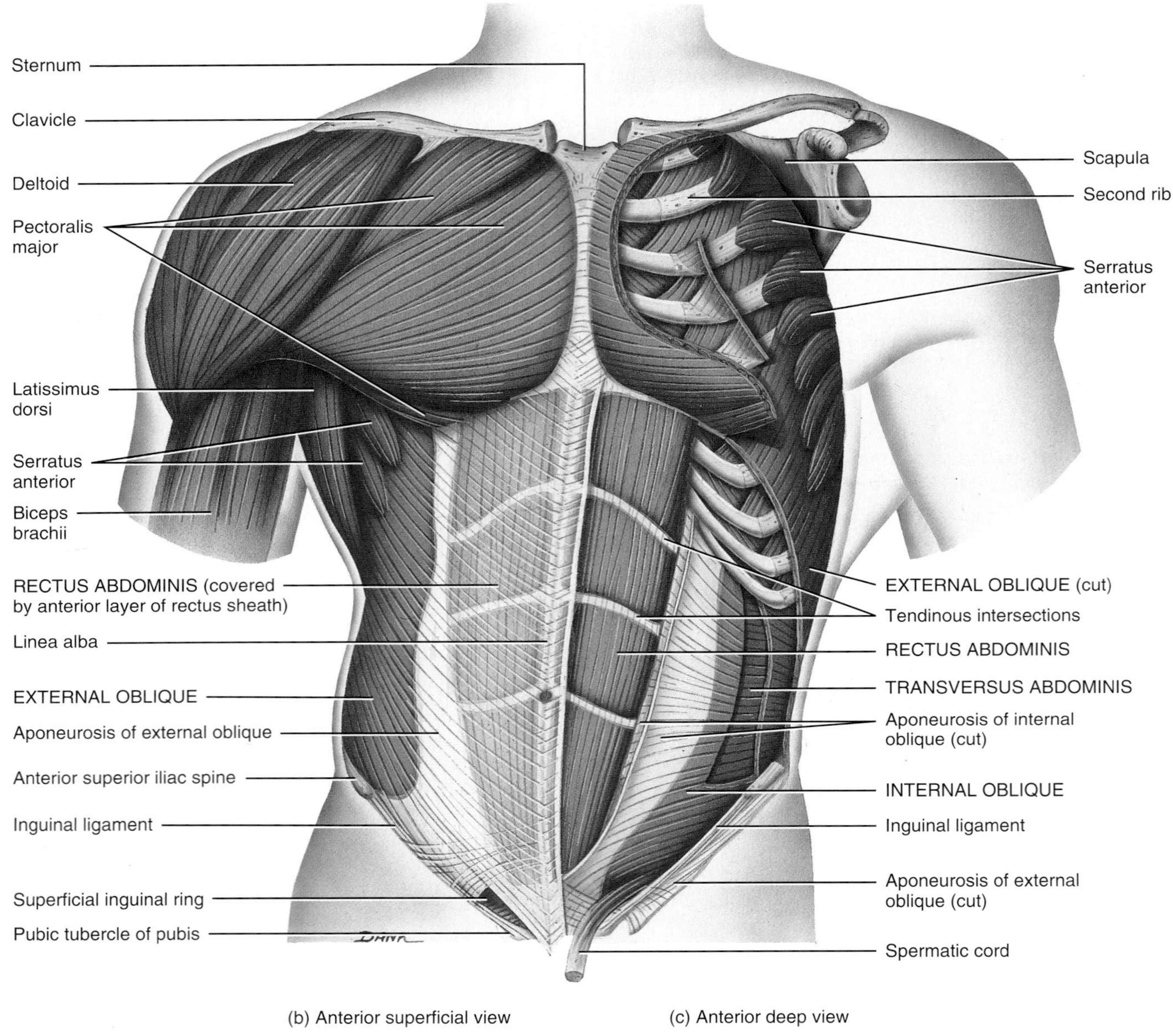

(b) Anterior superficial view (c) Anterior deep view

Which abdominal muscle aids in urination?

MUSCLES USED IN BREATHING (FIGURE 11.11)

▶ **OBJECTIVE**

Describe the origin, insertion, action, and innervation of the muscles used in breathing.

The muscles described here alter the size of the thoracic cavity so that breathing can occur. Inhalation (breathing in) occurs when the thoracic cavity increases in size, and exhalation (breathing out) occurs when the thoracic cavity decreases in size.

The dome-shaped **diaphragm** is the most important muscle that powers breathing. It also separates the thoracic and abdominal cavities. The diaphragm has a convex, superior surface that forms the floor of the thoracic cavity (Figure 11.11c) and a concave, inferior surface that forms the roof of the abdominal cavity (Figure 11.11d). The **peripheral muscular portion** of the diaphragm originates on the xiphoid process of the sternum, the inferior six ribs and their costal cartilages, and the lumbar vertebrae and their intervertebral discs and the twelfth rib (Figure 11.11d). From their various origins, the fibers of the muscular portion converge and insert into the **central tendon,** a strong aponeurosis located near the center of the muscle (Figure 11.11c, d). The central tendon fuses with the inferior surface of the pericardium (covering of the heart) and the pleurae (coverings of the lungs).

The diaphragm has three major openings through which various structures pass between the thorax and abdomen. These structures include the aorta, along with the thoracic duct and azygous vein, which pass through the **aortic hiatus;** the esophagus with accompanying vagus (X) nerves, which pass through the **esophageal hiatus;** and the inferior vena cava, which passes through the **caval opening (foramen for the vena cava).** In a condition called a hiatus hernia, the stomach protrudes superiorly through the esophageal hiatus.

Movements of the diaphragm also help return venous blood passing through abdominal veins to the heart. Together with the anterolateral abdominal muscles, the diaphragm helps to increase intra-abdominal pressure to evacuate the pelvic contents during defecation, urination, and childbirth. This mechanism is further assisted when you take a deep breath and close the rima glottidis (the space between vocal folds). The trapped air in the respiratory system prevents the diaphragm from elevating. The increase in intra-abdominal pressure also helps support the vertebral column and prevents flexion during weight lifting. This greatly assists the back muscles in lifting a heavy weight.

Other muscles involved in breathing, called **intercostal muscles,** span the intercostal spaces, the spaces between ribs. These muscles are arranged in three layers. The 11 pairs of **external intercostal muscles** occupy the superficial layer, and their fibers run in an oblique direction inferiorly and anteriorly from the rib above to the rib below. They elevate the ribs during inhalation to help expand the thoracic cavity. The 11 pairs of **internal intercostal muscles** occupy the intermediate layer of the intercostal spaces. The fibers of these muscles run at right angles to the external intercostals, in an oblique direction inferiorly and posteriorly from the inferior border of the rib above to the superior border of the rib below. They draw adjacent ribs together during forced exhalation to help decrease the size of the thoracic cavity. The deepest muscle layer is made up of the paired **innermost intercostal muscles.** These poorly developed muscles (not illustrated) extend in the same direction as the internal intercostals and may have the same role.

As you will see in Chapter 23, the diaphragm and external intercostal muscles are used during quiet inhalation and exhalation. However, during deep, forceful inhalation (during exercise or playing a wind instrument), the sternocleidomastoid, scalene, and pectoralis minor muscles are also used; during deep, forceful exhalation, the external oblique, internal oblique, transversus abdominis, rectus abdominis, and internal intercostals are also used.

Relating Muscles to Movements

Arrange the muscles in this exhibit according to the following actions: (1) increase in vertical length, (2) increase in lateral and anteroposterior dimensions, and (3) decrease in lateral and anteroposterior dimensions of the thorax.

▶ **CHECKPOINT**

What are the names of the three openings in the diaphragm, and which structures pass through each?

MUSCLE	ORIGIN	INSERTION	ACTION	INNERVATION
Diaphragm (DĪ-a-fram; *dia-* = across; *-phragm* = wall)	Xiphoid process of the sternum, costal cartilages and adjacent portions of the inferior six ribs, lumbar vertebrae and their intervertebral discs, and the twelfth rib.	Central tendon.	Contraction of the diaphragm causes it to flatten and increases the vertical dimension of the thoracic cavity, resulting in inhalation; relaxation of the diaphragm causes it to move superiorly and decreases the vertical dimension of the thoracic cavity, resulting in exhalation.	Phrenic nerve, which contains axons from cervical spinal nerves (C3–C5).
External intercostals (in′-ter-KOS-tals; *external* = closer to surface; *inter-* = between; *costa* = rib)	Inferior border of rib above.	Superior border of rib below.	Contraction elevates the ribs and increases the anteroposterior and lateral dimensions of the thoracic cavity, resulting in inhalation; relaxation depresses the ribs and decreases the anteroposterior and lateral dimensions of the thoracic cavity, resulting in exhalation.	Thoracic spinal nerves T2–T12.
Internal intercostals (in′-ter-KOS-tals; *internal* = further from surface)	Superior border of rib below.	Inferior border of rib above.	Contraction draws adjacent ribs together to further decrease the anteroposterior and lateral dimensions of the thoracic cavity during forced exhalation.	Thoracic spinal nerves T2–T12.

continues

Figure 11.11 **Muscles used in breathing, as seen in a male.**

Openings in the diaphragm permit the passage of the aorta, esophagus, and inferior vena cava.

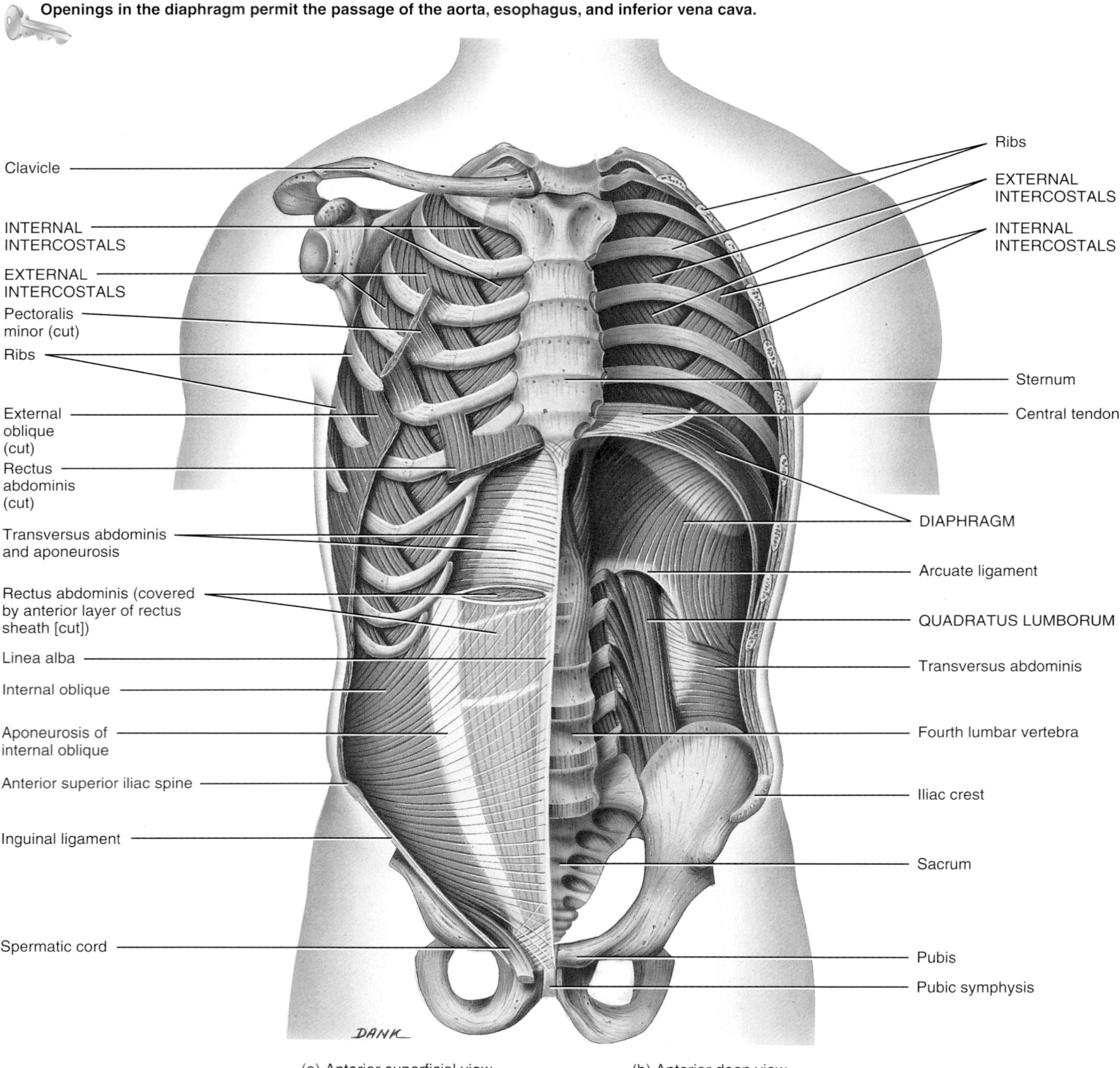

(a) Anterior superficial view

(b) Anterior deep view

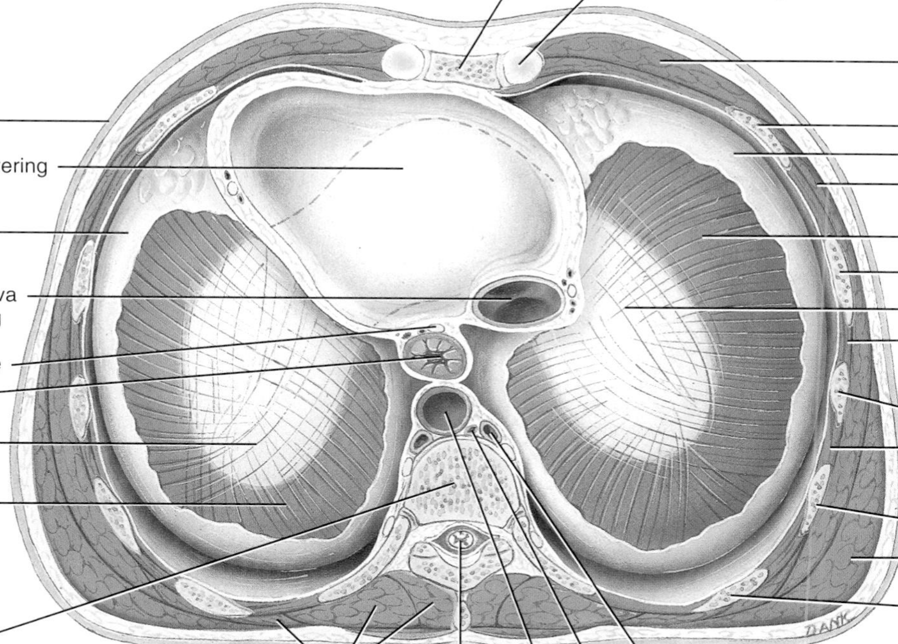

Sternum
Fifth costal cartilage
Pectoralis major muscle
Skin
Fifth rib
Pericardium covering central tendon
Pleura (cut)
Serratus anterior muscle
Pleura (cut)
DIAPHRAGM
Sixth rib
Inferior vena cava in caval opening
Central tendon
EXTERNAL INTERCOSTAL MUSCLE
In esophageal hiatus
Vagus (X) nerve
Esophagus
Seventh rib
Central tendon
INTERNAL INTERCOSTAL MUSCLE
DIAPHRAGM
Eighth rib
Latissimus dorsi muscle
Ninth rib
Body of T10
Azygos vein
Thoracic duct
In aortic hiatus
Erector spinae muscle
Aorta
Spinal cord

(c) Superior view of diaphragm

Xiphoid process of sternum
Costal cartilages
DIAPHRAGM
Inferior vena cava
Central tendon
Vagus (X) nerve
Esophagus
Tenth rib
Central tendon
Aorta
Azygos vein
Twelfth rib
Quadratus lumborum muscle
Thoracic duct
Second lumbar vertebra
Psoas major muscle
Third lumbar vertebra

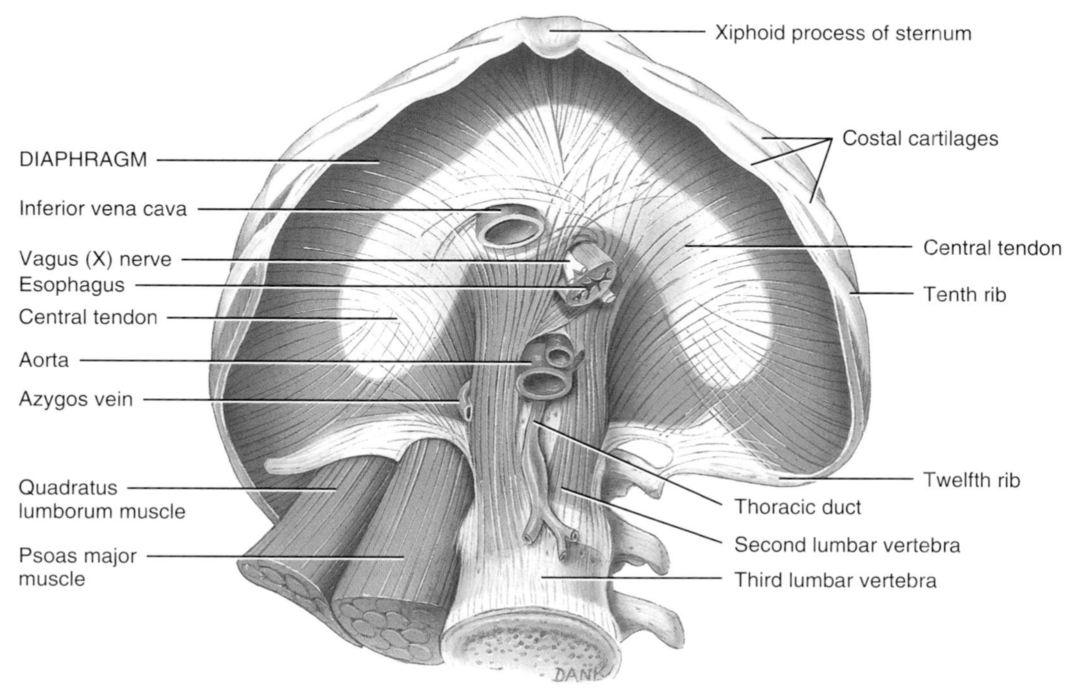

(d) Inferior view of diaphragm

Which muscle associated with breathing is innervated by the phrenic nerve?

EXHIBIT 11.9 MUSCLES OF THE PELVIC FLOOR (FIGURE 11.12)

▶ **OBJECTIVE**

Describe the origin, insertion, action, and innervation of the muscles of the pelvic floor.

The muscles of the pelvic floor are the levator ani and coccygeus. Together with the fascia covering their internal and external surfaces, these muscles are referred to as the **pelvic diaphragm,** which stretches from the pubis anteriorly to the coccyx posteriorly, and from one lateral wall of the pelvis to the other. This arrangement gives the pelvic diaphragm the appearance of a funnel suspended from its attachments. The anal canal and urethra pierce the pelvic diaphragm in both sexes, and the vagina also goes through it in females.

The two components of the **levator ani** muscle are the **pubococcygeus** and **iliococcygeus.** Figure 11.12 shows these muscles in the female and Figure 11.13 illustrates them in the male. The levator ani is the largest and most important muscle of the pelvic floor. It supports the pelvic viscera and resists the inferior thrust that accompanies increases in intra-abdominal pressure during functions such as forced exhalation, coughing, vomiting, urination, and defecation. The muscle also functions as a sphincter at the anorectal junction, urethra, and vagina. In addition to assisting the levator ani, the **coccygeus** pulls the coccyx anteriorly after it has been pushed posteriorly during defecation or childbirth.

Injury of Levator Ani and Urinary Stress Incontinence

During childbirth, the levator ani muscle supports the head of the fetus, and the muscle may be injured during a difficult childbirth or traumatized during an *episiotomy* (a cut made with surgical scissors to prevent or direct tearing of the perineum during the birth of a baby). The consequence of such injury may be **urinary stress incontinence,** that is, the leakage of urine whenever intra-abdominal pressure is increased—for example, during coughing. One way to treat urinary stress incontinence is to strengthen and tighten the muscles that support the pelvic viscera. This is accomplished by *Kegel exercises,* the alternate contraction and relaxation of muscles of the pelvic floor. To find the correct muscles, the person imagines that she is urinating and then contracts the muscles as if stopping in midstream. The muscles should be held for a count of three, then relaxed for a count of three. This should be done 5–10 times each hour—sitting, standing, and lying down. Kegel exercises are also encouraged during pregnancy to strengthen the muscles for delivery. ■

Relating Muscles to Movements

Arrange the muscles in this exhibit according to the following actions: (1) supporting and maintaining the position of the pelvic viscera; (2) resisting an increase in intra-abdominal pressure; (3) constriction of the anus, urethra, and vagina. The same muscle may be mentioned more than once.

▶ **CHECKPOINT**

Which muscles are strengthened by Kegel exercises?

MUSCLE	ORIGIN	INSERTION	ACTION	INNERVATION
Levator ani (le-VĀ-tor Ā-nē; *levator* = raises; *ani* = anus)	This muscle is divisible into two parts, the pubococcygeus muscle and the iliococcygeus muscle.			
Pubococcygeus (pū′-bō-kok-SIJ-ē-us; *pubo-* = pubis; *coccygeus* = coccyx)	Pubis.	Coccyx, urethra, anal canal, perineal body of the perineum (a wedge-shaped mass of fibrous tissue in the center of the perineum), and anococcygeal raphe (narrow fibrous band that extends from anus to coccyx).	Supports and maintains position of pelvic viscera; resists increase in intra-abdominal pressure during forced exhalation, coughing, vomiting, urination, and defecation; constricts anus, urethra, and vagina.	Sacral spinal nerves S2–S4.
Iliococcygeus (il′-ē-ō-kok-SIJ-ē-us; *ilio-* = ilium)	Ischial spine.	Coccyx.	As above.	Sacral spinal nerves S2–S4.
Coccygeus (kok-SIJ-ē-us)	Ischial spine.	Lower sacrum and upper coccyx.	Supports and maintains position of pelvic viscera; resists increase in intra-abdominal pressure during forced exhalation, coughing, vomiting, urination, and defecation; and pulls coccyx anteriorly following defecation or childbirth.	Sacral spinal nerves S4–S5.

Figure 11.12 Muscles of the pelvic floor, as seen in the female perineum.

The pelvic diaphragm supports the pelvic viscera.

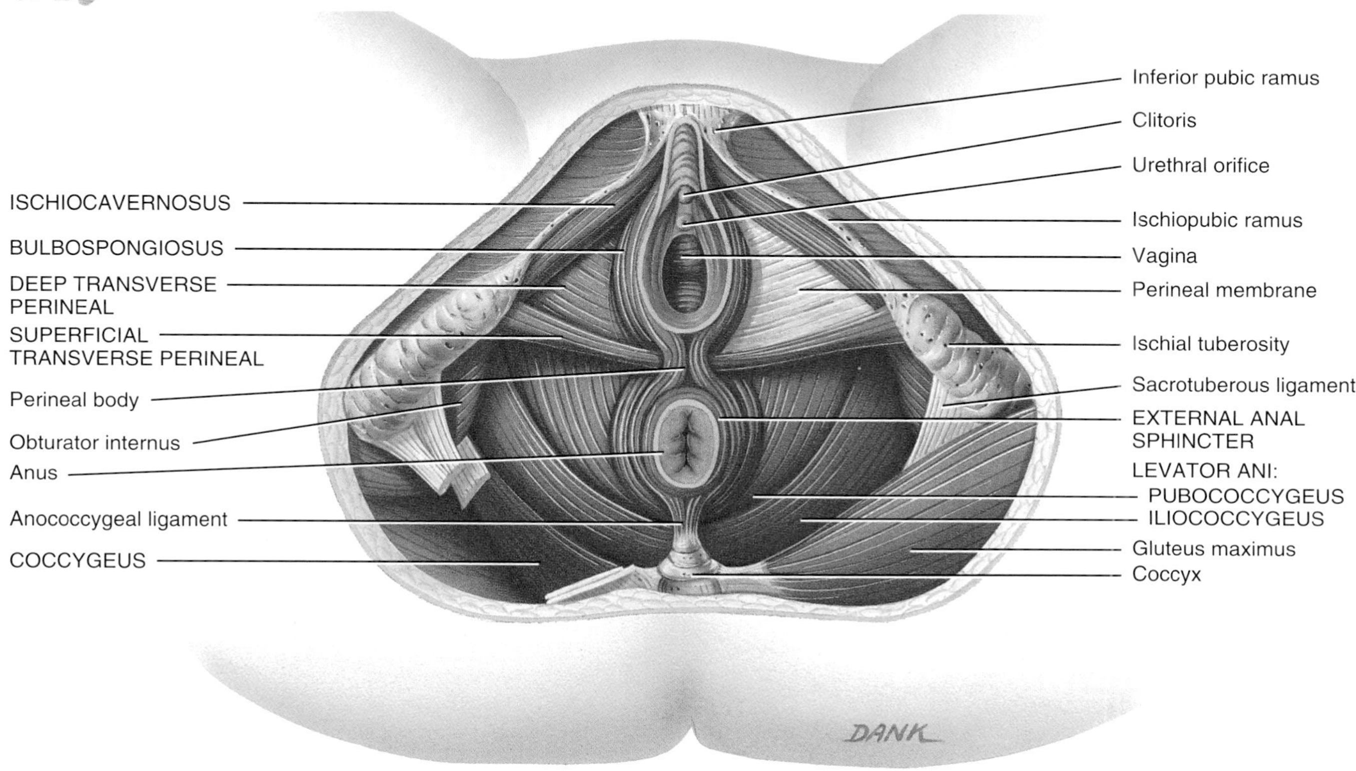

Inferior superficial view

What are the borders of the pelvic diaphragm?

MUSCLES OF THE PERINEUM (FIGURES 11.12 AND 11.13)

▶ **OBJECTIVE**

Describe the origin, insertion, action, and innervation of the muscles of the perineum.

The **perineum** is the region of the trunk inferior to the pelvic diaphragm. It is a diamond-shaped area that extends from the pubic symphysis anteriorly, to the coccyx posteriorly, and to the ischial tuberosities laterally. The female and the male perineums may be compared in Figures 11.12 and 11.13, respectively. A transverse line drawn between the ischial tuberosities divides the perineum into an anterior **urogenital triangle** that contains the external genitals and a posterior **anal triangle** that contains the anus (see Figure 28.21 on page 1082). Several perineal muscles insert into the perineal body of the perineum (described on page 1082). Clinically, the perineum is very important to physicians who care for women during pregnancy and treat disorders related to the female genital tract, urogenital organs, and the anorectal region.

The muscles of the perineum are arranged in two layers: **superficial** and **deep.** The muscles of the superficial layer are the **superficial transverse perineal muscle,** the **bulbospongiosus**, and the **ischiocavernosus**. The deep muscles of the perineum are the **deep transverse perineal muscle** and the **external urethral sphincter.** The deep muscles of the perineum assist in urination and ejaculation in males and urination and orgasm in females. The **external anal sphincter** closely adheres to the skin around the margin of the anus and keeps the anal canal and anus closed except during defecation.

Relating Muscles to Movements

Arrange the muscles in this exhibit according to the following actions: (1) expulsion of urine and semen, (2) erection of the clitoris and penis, (3) closure of the anal orifice, and (4) constriction of the vaginal orifice. The same muscle may be mentioned more than once.

▶ **CHECKPOINT**

What are the borders and contents of the urogenital triangle and the anal triangle?

MUSCLE	ORIGIN	INSERTION	ACTION	INNERVATION
Superficial Perineal Muscles				
Superficial transverse perineal (per-i-NĒ-al; *superficial* = closer to surface; *transverse* = across; *perineus* = perineum)	Ischial tuberosity.	Perineal body of perineum.	Stabilizes perineal body of perineum.	Perineal branch of the pudendal nerve of the sacral plexus.
Bulbospongiosus (bul′-bō-spon′-jē-Ō-sus; *bulb* = a bulb; *spongio-* = sponge)	Perineal body of perineum.	Perineal membrane of deep muscles of perineum, corpus spongiosum of penis, and deep fascia on dorsum of penis in male; pubic arch and root and dorsum of clitoris in female.	Helps expel urine during urination, helps propel semen along urethra, assists in erection of the penis in male; constricts vaginal orifice and assists in erection of clitoris in female.	Perineal branch of the pudendal nerve of the sacral plexus.
Ischiocavernosus (is′-kē-ō-ka′-ver-NŌ-sus; *ischio-* = the hip)	Ischial tuberosity and ischial and pubic rami.	Corpus cavernosum of penis in male and clitoris in female.	Maintains erection of penis in male and clitoris in female.	Perineal branch of the pudendal nerve of the sacral plexus.
Deep Perineal Muscles				
Deep transverse perineal (per-i-NĒ-al; *deep* = farther from surface)	Ischial rami.	Perineal body of perineum.	Helps expel last drops of urine and semen in male and urine in female.	Perineal branch of the pudendal nerve of the sacral plexus.
External urethral sphincter (ū-RĒ-thral SFINGK-ter) (see Figure 26.21)	Ischial and pubic rami.	Median raphe in male and vaginal wall in female.	Helps expel last drops of urine and semen in male and urine in female.	Sacral spinal nerve S4 and the inferior rectal branch of the pudendal nerve.
External anal sphincter (Ā-nal)	Anococcygeal ligament.	Perineal body of perineum.	Keeps anal canal and anus closed.	Sacral spinal nerve S4 and the inferior rectal branch of the pudendal nerve.

Figure 11.13 **Muscles of the male perineum.**

The deep muscles of the perineum assist in urination in females and males and ejaculation in males, and help strengthen the pelvic floor.

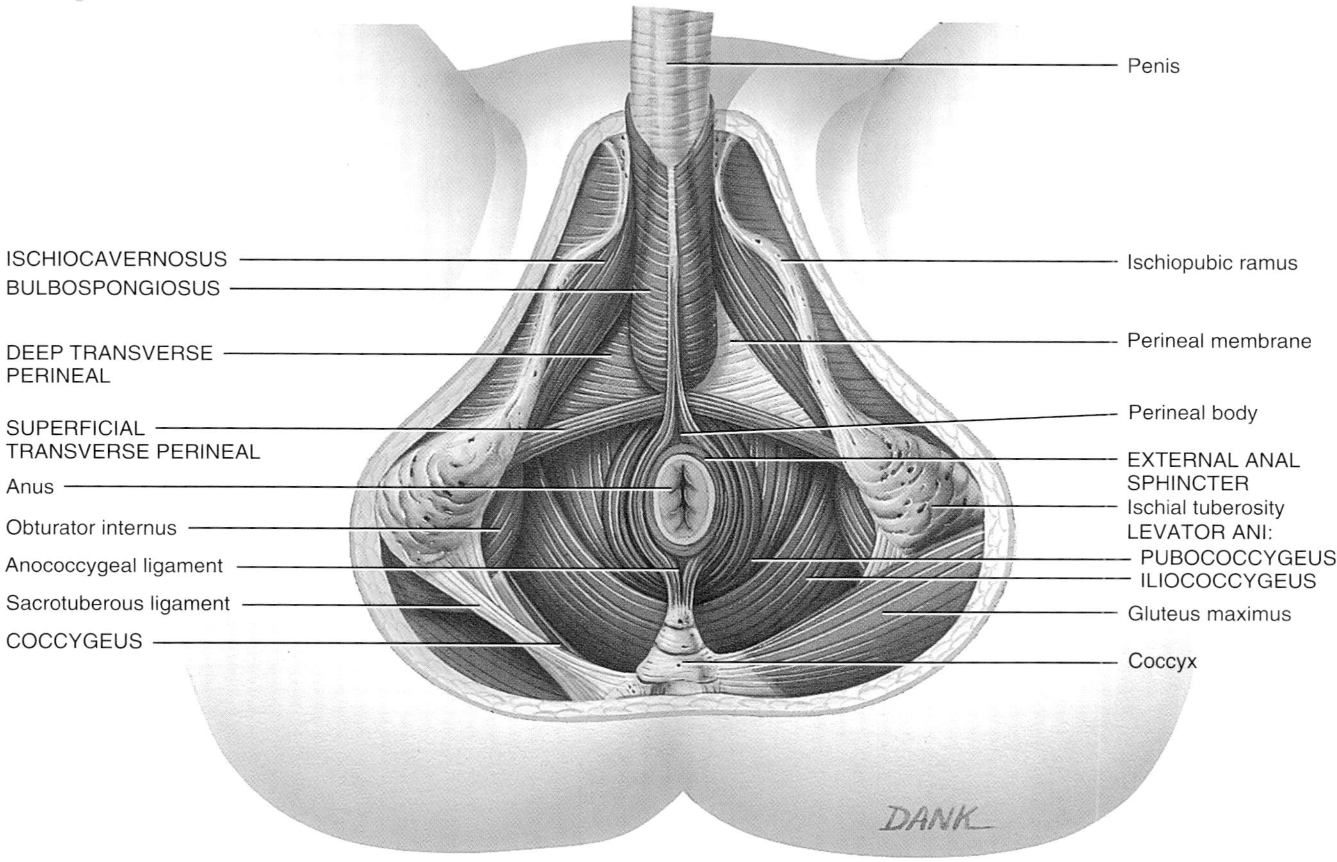

Inferior superficial view

What are the borders of the perineum?

EXHIBIT 11.11 MUSCLES THAT MOVE THE PECTORAL (SHOULDER) GIRDLE (FIGURE 11.14)

▶ **OBJECTIVE**

Describe the origin, insertion, action, and innervation of the muscles that move the pectoral girdle.

The main action of the muscles that move the pectoral girdle is to stabilize the scapula so it can function as a steady origin for most of the muscles that move the humerus. Because scapular movements usually accompany humeral movements in the same direction, the muscles also move the scapula to increase the range of motion of the humerus. For example, it would not be possible to abduct the humerus past the horizontal (raise your hand in class) if the scapula did not move with the humerus. During abduction, the scapula follows the humerus by rotating upward.

Muscles that move the pectoral girdle can be classified into two groups based on their location in the thorax: **anterior** and **posterior thoracic muscles.** The anterior thoracic muscles are the subclavius, pectoralis minor, and serratus anterior. The **subclavius** is a small, cylindrical muscle under the clavicle that extends from the clavicle to the first rib. It steadies the clavicle during movements of the pectoral girdle. The **pectoralis minor** is a thin, flat, triangular muscle that is deep to the pectoralis major. Besides its role in movements of the scapula, the pectoralis minor muscle assists in forced inhalation. The **serratus anterior** is a large, flat, fan-shaped muscle between the ribs and scapula. It is so named because of the saw-toothed appearance of its origins on the ribs.

The posterior thoracic muscles are the trapezius, levator scapulae, rhomboid major, and rhomboid minor. The **trapezius** is a large, flat, triangular sheet of muscle extending from the skull and vertebral column medially to the pectoral girdle laterally. It is the most superficial back muscle and covers the posterior neck region and superior portion of the trunk. The two trapezius muscles form a trapezoid (diamond-shaped quadrangle)—hence its name. The **levator scapulae** is a narrow, elongated muscle in the posterior portion of the neck. It is deep to the sternocleidomastoid and trapezoid muscles. As its name suggests, one of its actions is to elevate the scapula. The **rhomboid major** and **rhomboid minor** lie deep to the trapezius and are not always distinct from each other. They appear as parallel bands that pass inferiorly and laterally from the vertebrae to the scapula. Their names are based on their shape—that is, a rhomboid (an oblique parallelogram). The rhomboid major is about two times wider than the rhomboid minor. Both muscles are used when forcibly lowering the raised upper limbs, as in driving a stake with a sledgehammer.

MUSCLE	ORIGIN	INSERTION	ACTION	INNERVATION
Anterior Thoracic Muscles				
Subclavius (sub-KLĀ-vē-us; *sub-* = under; *clavius* = clavicle	First rib.	Clavicle.	Depresses and moves clavicle anteriorly and helps stabilize pectoral girdle.	Subclavian nerve.
Pectoralis minor (pek′-tō-RĀ-lis; *pector-* = the breast, chest, thorax; *minor* = lesser)	Second through fifth, third through fifth, or second through fourth ribs.	Coracoid process of scapula.	Abducts scapula and rotates it downward; elevates third through fifth ribs during forced inhalation when scapula is fixed.	Medial pectoral nerve.
Serratus anterior (ser-Ā-tus; *serratus* = saw-toothed; *anterior* = front)	Superior eight or nine ribs.	Vertebral border and inferior angle of scapula.	Abducts scapula and rotates it upward; elevates ribs when scapula is stabilized; known as "boxer's muscle" because it is important in horizontal arm movements such as punching and pushing.	Long thoracic nerve.
Posterior Thoracic Muscles				
Trapezius (tra-PĒ-zē-us; *trapezi-* = trapezoid-shaped)	Superior nuchal line of occipital bone, ligamentum nuchae, and spines of seventh cervical and all thoracic vertebrae.	Clavicle and acromion and spine of scapula.	Superior fibers elevate scapula and can help extend head; middle fibers adduct scapula; inferior fibers depress scapula; superior and inferior fibers together rotate scapula upward; stabilizes scapula.	Accessory (XI) nerve and cervical spinal nerves C3–C5.
Levator scapulae (le-VĀ-tor SKA-pū-lē; *levator* = raises; *scapulae* = of the scapula)	Superior four or five cervical vertebrae.	Superior vertebral border of scapula.	Elevates scapula and rotates it downward.	Dorsal scapular nerve and cervical spinal nerves C3–C5.
Rhomboid major (rom-BOYD; *rhomboid* = rhomboid or diamond-shaped) (see Figure 11.15c)	Spines of second to fifth thoracic vertebrae.	Vertebral border of scapula inferior to spine.	Elevates and adducts scapula and rotates it downward; stabilizes scapula.	Dorsal scapular nerve.
Rhomboid minor (rom-BOYD) (see Figure 11.15c)	Spines of seventh cervical and first thoracic vertebrae.	Vertebral border of scapula superior to spine.	Elevates and adducts scapula and rotates it downward; stabilizes scapula.	Dorsal scapular nerve.

To understand the actions of muscles that move the scapula, it is first helpful to review the various movements of the scapula:

- **Elevation:** superior movement of the scapula, such as shrugging the shoulders or lifting a weight over the head.
- **Depression:** inferior movement of the scapula, as in pulling down on a rope attached to a pulley.
- **Abduction (protraction):** movement of the scapula laterally and anteriorly, as in doing a "push-up" or punching.
- **Adduction (retraction):** movement of the scapula medially and posteriorly, as in pulling the oars in a rowboat.
- **Upward rotation:** movement of the inferior angle of the scapula laterally so that the glenoid cavity is moved upward. This movement is required to abduct the humerus past the horizontal as in raising the arms in a "jumping jack."
- **Downward rotation:** movement of the inferior angle of the scapula medially so that the glenoid cavity is moved downward. This movement is seen when a gymnast on parallel bars supports the weight of the body on the hands.

Relating Muscles to Movements

Arrange the muscles in this exhibit according to the following actions on the scapula: (1) depression, (2) elevation, (3) abduction, (4) adduction, (5) upward rotation, and (6) downward rotation. The same muscle may be mentioned more than once.

► CHECKPOINT

What muscles in this exhibit are used to raise your shoulders, lower your shoulders, join your hands behind your back, and join your hands in front of your chest?

Figure 11.14 Muscles that move the pectoral (shoulder) girdle. (See Tortora, *A Photographic Atlas of the Human Body, Second Edition,* Figure 5.8.)

Muscles that move the pectoral girdle originate on the axial skeleton and insert on the clavicle or scapula.

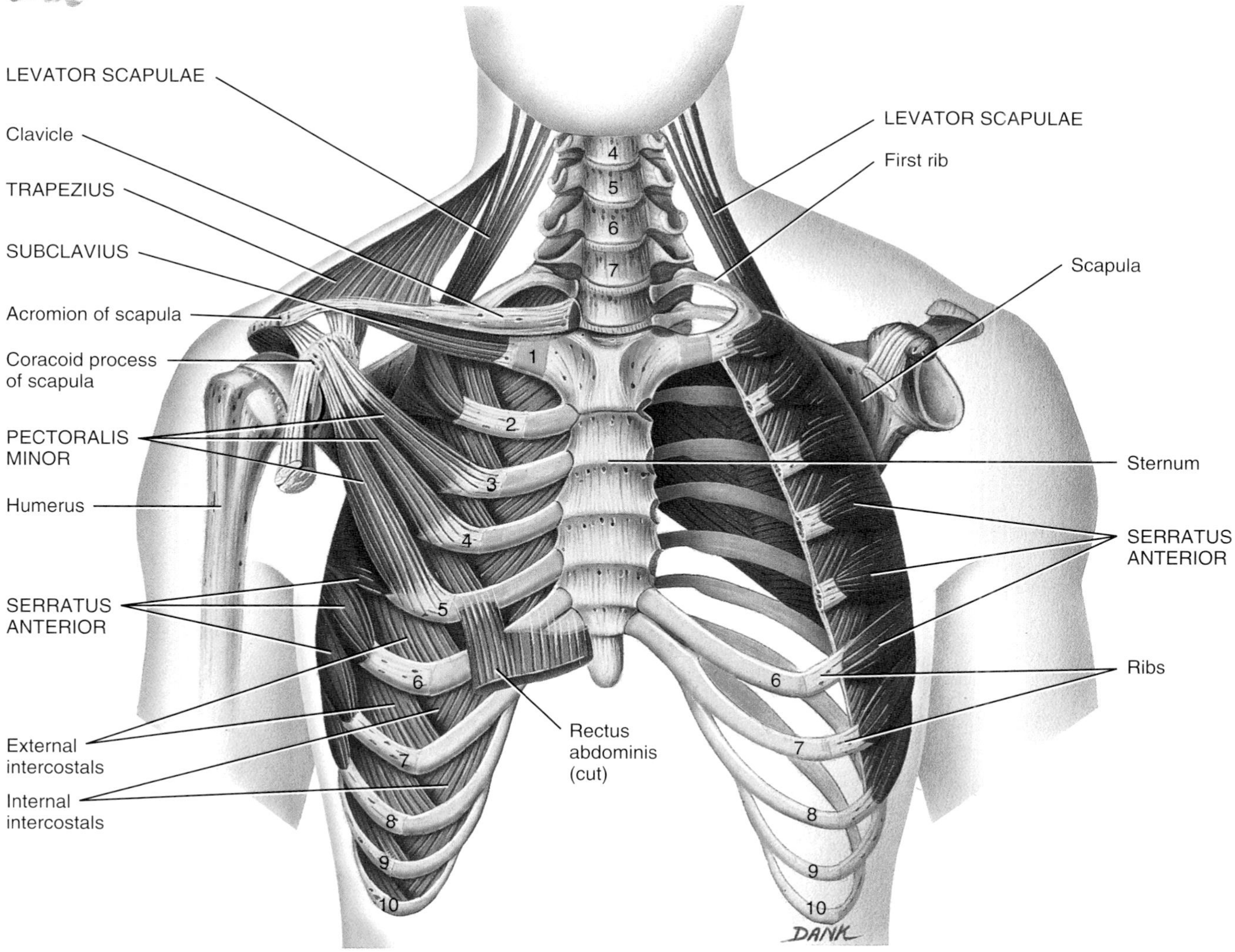

(a) Anterior deep view

(b) Anterior deeper view

What is the main action of the muscles that move the pectoral girdle?

▶ **OBJECTIVE**

Describe the origin, insertion, action, and innervation of the muscles that move the humerus.

Of the nine muscles that cross the shoulder joint, all except the pectoralis major and latissimus dorsi originate on the scapula. The pectoralis major and latissimus dorsi thus are called **axial muscles** because they originate on the axial skeleton. The remaining seven muscles, the **scapular muscles,** arise from the scapula.

Of the two axial muscles that move the humerus, the **pectoralis major** is a large, thick, fan-shaped muscle that covers the superior part of the thorax. It has two origins: a smaller clavicular head and a larger sternocostal head. The **latissimus dorsi** is a broad, triangular muscle located on the inferior part of the back. It is commonly called the "swimmer's muscle" because its many actions are used while swimming; consequently, many competitive swimmers have well-developed "lats."

Among the scapular muscles, the **deltoid** is a thick, powerful shoulder muscle that covers the shoulder joint and forms the rounded contour of the shoulder. This muscle is a frequent site of intramuscular injections. As you study the deltoid, note that its fascicles originate from three different points and that each group of fascicles moves the humerus differently. The **subscapularis** is a large triangular muscle that fills the subscapular fossa of the scapula and forms part of the posterior wall of the axilla. The **supraspinatus,** a rounded muscle named for its location in the supraspinous fossa of the scapula, lies deep to the trapezius. The **infraspinatus** is a triangular muscle, also named for its location in the infraspinous fossa of the scapula. The **teres major** is a thick, flattened muscle inferior to the teres minor that also helps form part of the posterior wall of the axilla. The **teres minor** is a cylindrical, elongated muscle, often inseparable from the infraspinatus, which lies along its superior border. The **coracobrachialis** is an elongated, narrow muscle in the arm.

Four deep muscles of the shoulder—subscapularis, supraspinatus, infraspinatus, and teres minor—strengthen and stabilize the shoulder joint. These muscles join the scapula to the humerus. Their flat tendons fuse together to form the **rotator (musculotendinous) cuff,** a nearly complete circle of tendons around the shoulder joint, like the cuff on a shirtsleeve. The supraspinatus muscle is especially subject to wear and tear because of its location between the head of the humerus and acromion of the scapula, which compress its tendon during shoulder movements, especially abduction of the arm.

Impingement Syndrome

One of the most common causes of shoulder pain and dysfunction in athletes is known as **impingement syndrome,** which is sometimes confused with another common complaint, compartment syndrome, discussed on page 399. The repetitive movement of the arm over the head that is common in baseball, overhead racquet sports, lifting weights over the head, spiking a volleyball, and swimming puts these athletes at risk. Impingement syndrome may also be caused by a direct blow or stretch injury. Continual pinching of the supraspinatus tendon as a result of overhead motions causes it to become inflamed and results in pain. If movement is continued despite the pain, the tendon may degenerate near the attachment to the humerus and ultimately may tear away from the bone (rotator cuff injury). Treatment consists of resting the injured tendons, strengthening the shoulder through exercise, and surgery if the injury is particularly severe. ■

Relating Muscles to Movements

Arrange the muscles in this exhibit according to the following actions on the humerus at the shoulder joint: (1) flexion, (2) extension, (3) abduction, (4) adduction, (5) medial rotation, and (6) lateral rotation. The same muscle may be mentioned more than once.

▶ **CHECKPOINT**

Why are the two muscles that cross the shoulder joint called axial muscles, and the seven others called scapular muscles?

MUSCLE	ORIGIN	INSERTION	ACTION	INNERVATION
Axial Muscles that Move the Humerus				
Pectoralis major (pek′-tō-RĀ-lis; *pector-* = chest; *major* = larger) (see also Figure 11.10c)	Clavicle (clavicular head), sternum, and costal cartilages of second to sixth ribs and sometimes first to seventh ribs (sternocostal head).	Greater tubercle and lateral lip of the intertubercular sulcus of humerus.	As a whole, adducts and medially rotates arm at shoulder joint; clavicular head flexes arm, and sternocostal head extends the flexed arm to side of trunk.	Medial and lateral pectoral nerves.
Latissimus dorsi (la-TIS-i-mus DOR-sī; *latissimus* = widest; *dorsi* = of the back)	Spines of inferior six thoracic vertebrae, lumbar vertebrae, crests of sacrum and ilium, inferior four ribs.	Intertubercular sulcus of humerus.	Extends, adducts, and medially rotates arm at shoulder joint; draws arm inferiorly and posteriorly.	Thoracodorsal nerve.
Scapular Muscles that Move the Humerus				
Deltoid (DEL-toyd = triangularly shaped)	Acromial extremity of clavicle (anterior fibers), acromion of scapula (lateral fibers), and spine of scapula (posterior fibers).	Deltoid tuberosity of humerus.	Lateral fibers abduct arm at shoulder joint; anterior fibers flex and medially rotate arm at shoulder joint; posterior fibers extend and laterally rotate arm at shoulder joint.	Axillary nerve.
Subscapularis (sub-scap′-ū-LĀ-ris; *sub-* = below; *scapularis* = scapula).	Subscapular fossa of scapula.	Lesser tubercle of humerus.	Medially rotates arm at shoulder joint.	Upper and lower subscapular nerve.
Supraspinatus (soo-pra-spī-NĀ-tus; *supra-* = above; *spina-* = spine [of the scapula])	Supraspinous fossa of scapula.	Greater tubercle of humerus.	Assists deltoid muscle in abducting arm at shoulder joint.	Suprascapular nerve.
Infraspinatus (in′-fra-spī-NĀ-tus; *infra-* = below)	Infraspinous fossa of scapula.	Greater tubercle of humerus.	Laterally rotates and adducts arm at shoulder joint.	Suprascapular nerve.
Teres major (TE-rēz; *teres* = long and round)	Inferior angle of scapula.	Medial lip of intertubercular sulcus of humerus.	Extends arm at shoulder joint and assists in adduction and medial rotation of arm at shoulder joint.	Lower subscapular nerve.
Teres minor (TE-rēz)	Inferior lateral border of scapula.	Greater tubercle of humerus.	Laterally rotates, extends, and adducts arm at shoulder joint.	Axillary nerve.
Coracobrachialis (kor′-a-kō-brā-kē-Ā-lis; *coraco-* = coracoid process [of the scapula]; *brachi-* = arm)	Coracoid process of scapula.	Middle of medial surface of shaft of humerus.	Flexes and adducts arm at shoulder joint.	Musculocutaneous nerve.

continues

Figure 11.15 Muscles that move the humerus (arm bone). (See Tortora, *A Photographic Atlas of the Human Body, Second Edition,* Figures 5.9 and 5.10.)

The strength and stability of the shoulder joint are provided by the tendons that form the rotator cuff.

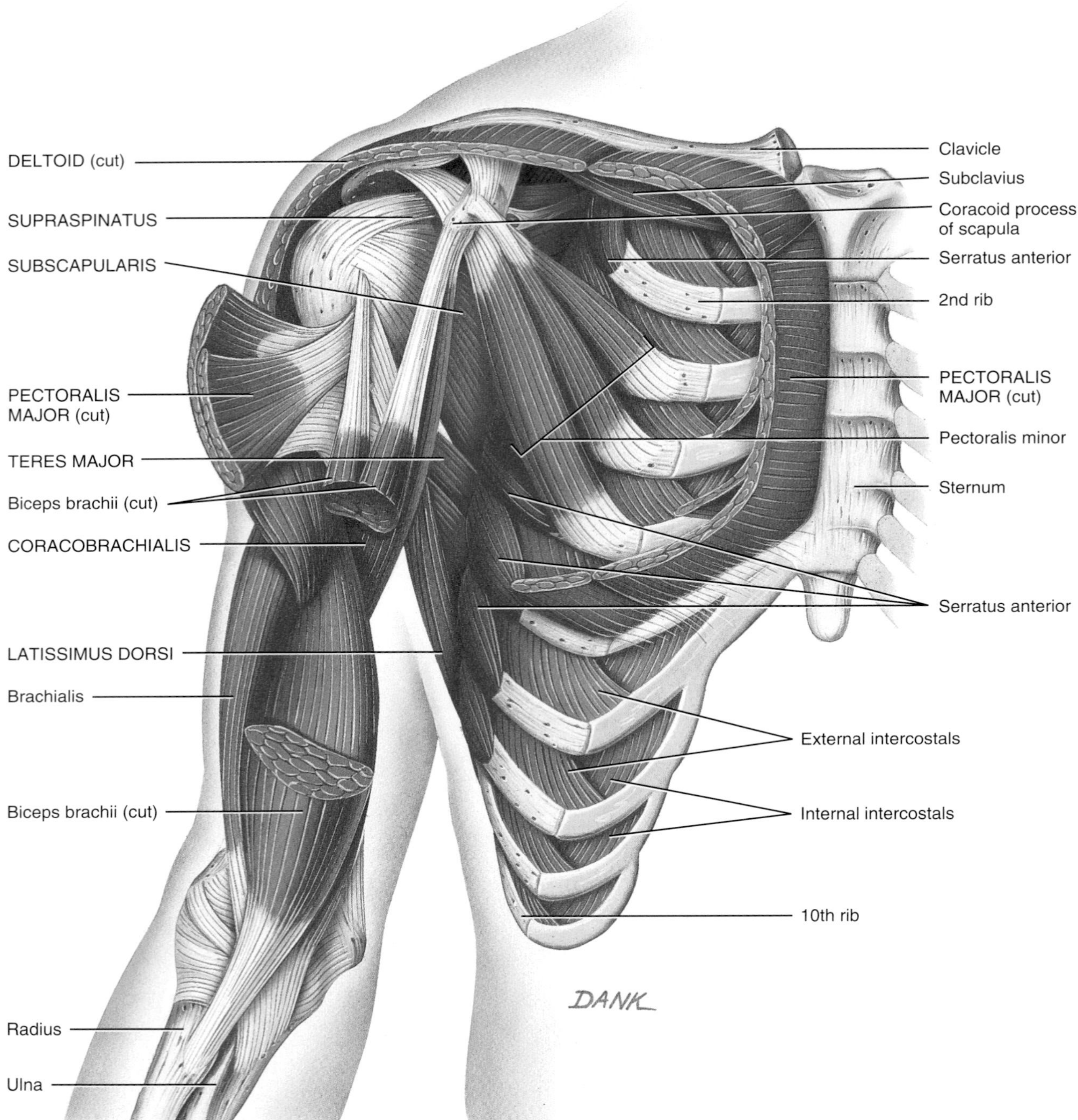

(a) Anterior deep view (the intact pectoralis major muscle is shown in figure 11.12a)

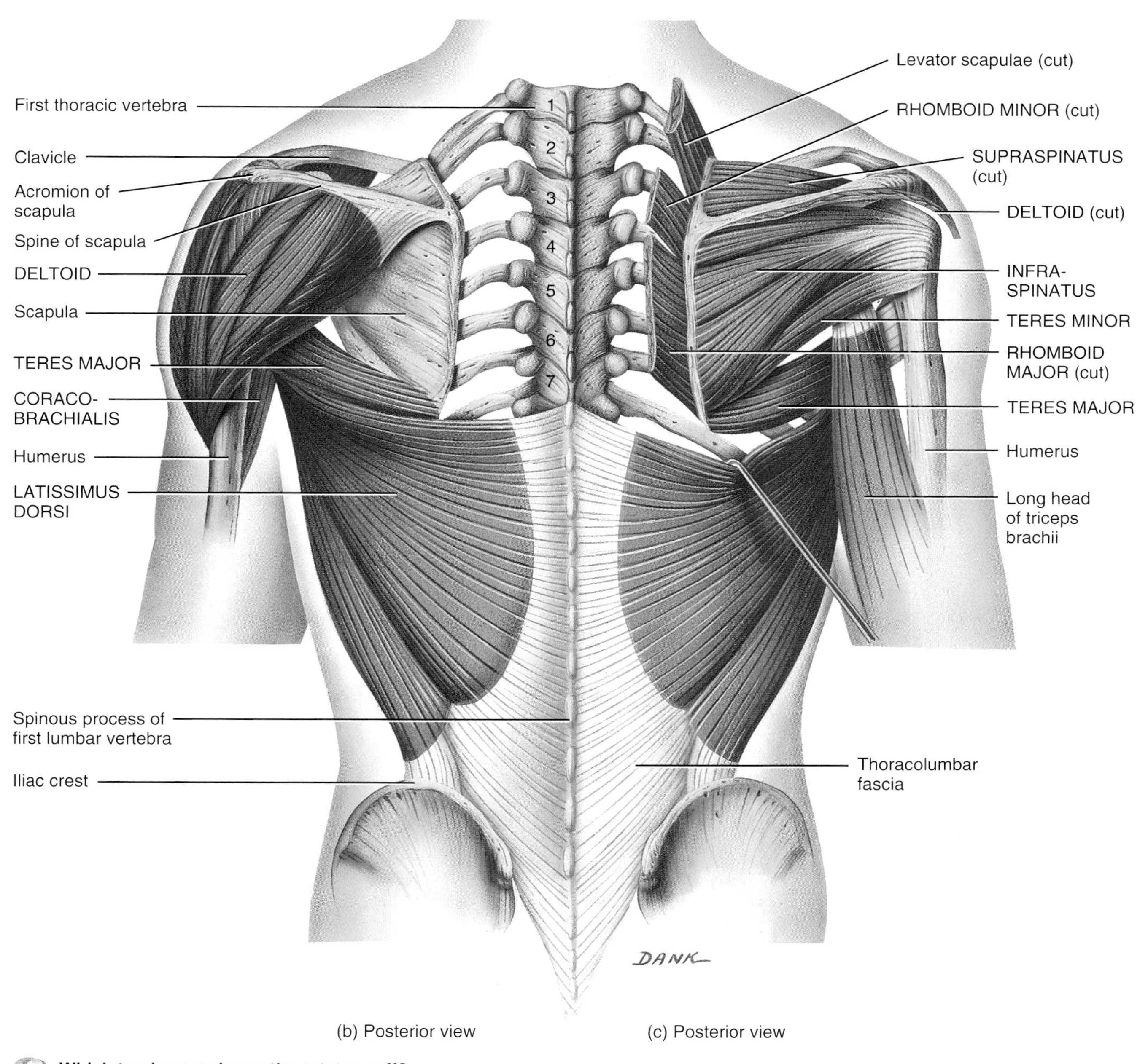

(b) Posterior view (c) Posterior view

Which tendons make up the rotator cuff?

EXHIBIT 11.13 MUSCLES THAT MOVE THE RADIUS AND ULNA (FOREARM BONES) (FIGURE 11.16)

▶ **OBJECTIVE**

Describe the origin, insertion, action, and innervation of the muscles that move the radius and ulna.

Most of the muscles that move the radius and ulna (forearm bones) cause flexion and extension at the elbow, which is a hinge joint. The biceps brachii, brachialis, and brachioradialis muscles are the flexor muscles. The extensor muscles are the triceps brachii and the anconeus.

The **biceps brachii** is the large muscle located on the anterior surface of the arm. As indicated by its name, it has two heads of origin (long and short), both from the scapula. The muscle spans both the shoulder and elbow joints. In addition to its role in flexing the forearm at the elbow joint, it also supinates the forearm at the radioulnar joints and flexes the arm at the shoulder joint. The **brachialis** is deep to the biceps brachii muscle. It is the most powerful flexor of the forearm at the elbow joint. For this reason, it is called the "workhorse" of the elbow flexors. The **brachioradialis** flexes the forearm at the elbow joint, especially when a quick movement is required or when a weight is lifted slowly during flexion of the forearm.

The **triceps brachii** is the large muscle located on the posterior surface of the arm. It is the more powerful of the extensors of the forearm at the elbow joint. As its name implies, it has three heads of origin, one from the scapula (long head) and two from the humerus (lateral and medial heads). The long head crosses the shoulder joint; the other heads do not. The **anconeus** is a small muscle located on the lateral part of the posterior aspect of the elbow that assists the triceps brachii in extending the forearm at the elbow joint.

MUSCLE	ORIGIN	INSERTION	ACTION	INNERVATION
Forearm Flexors				
Biceps brachii (BĪ-ceps BRĀ-kē-ī; *biceps* = two heads of origin; *brachii* = arm)	Long head originates from tubercle above glenoid cavity of scapula (supraglenoid tubercle); short head originates from coracoid process of scapula.	Radial tuberosity of radius and bicipital aponeurosis.*	Flexes forearm at elbow joint, supinates forearm at radioulnar joints, and flexes arm at shoulder joint.	Musculocutaneous nerve.
Brachialis (brā–kē-Ā-lis)	Distal, anterior surface of humerus.	Ulnar tuberosity and coronoid process of ulna.	Flexes forearm at elbow joint.	Musculocutaneous and radial nerves.
Brachioradialis (brā′-kē-ō-rā-dē-Ā-lis; *radi* = radius) (see Figure 11.17a)	Lateral border of distal end of humerus.	Superior to styloid process of radius.	Flexes forearm at elbow joint; supinates and pronates forearm at radioulnar joints to neutral position.	Radial nerve.
Forearm Extensors				
Triceps brachii (TRĪ-ceps BRĀ-kē-ī; *triceps* = three heads of origin)	Long head: infraglenoid tubercle, a projection inferior to glenoid cavity of scapula. Lateral head: lateral and posterior surface of humerus superior to radial groove. Medial head: entire posterior surface of humerus inferior to a groove for the radial nerve.	Olecranon of ulna.	Extends forearm at elbow joint and extends arm at shoulder joint.	Radial nerve.
Anconeus (an-KŌ-nē-us; *ancon* = the elbow) (see also Figure 11.17c)	Lateral epicondyle of humerus.	Olecranon and superior portion of shaft of ulna.	Extends forearm at elbow joint.	Radial nerve.

*The **bicipital aponeurosis** is a broad aponeurosis from the tendon of insertion of the biceps brachii muscle that descends medially across the brachial artery and fuses with deep fascia over the forearm flexor muscles.

Some muscles that move the radius and ulna are involved in pronation and supination at the radioulnar joints. The pronators, as suggested by their names, are the **pronator teres** and **pronator quadratus** muscles. The supinator of the forearm is aptly named the **supinator** muscle. You use the powerful action of the supinator when you twist a corkscrew or turn a screw with a screwdriver.

In the limbs, functionally related skeletal muscles and their associated blood vessels and nerves are grouped together by fascia into regions called **compartments.** In the arm, the biceps brachii, brachialis, and coracobrachialis muscles comprise the **anterior (flexor) compartment.** The triceps brachii muscle forms the **posterior (extensor) compartment.**

Relating Muscles to Movements

Arrange the muscles in this exhibit according to the following actions on the elbow joint: (1) flexion and (2) extension; the following actions on the forearm at the radioulnar joints: (1) supination and (2) pronation; and the following actions on the humerus at the shoulder joint: (1) flexion and (2) extension. The same muscle may be mentioned more than once.

▶ CHECKPOINT

Flex your arm. Which group of muscles is contracting? Which group of muscles must relax so that you can flex your arm?

MUSCLE	ORIGIN	INSERTION	ACTION	INNERVATION
Forearm Pronators				
Pronator teres (PRŌ-nā-tor TE-rēz; *pronator* = turns palm posteriorly) (see Figure 11.17a)	Medial epicondyle of humerus and coronoid process of ulna.	Midlateral surface of radius.	Pronates forearm at radioulnar joints and weakly flexes forearm at elbow joint.	Median nerve.
Pronator quadratus (PRŌ-nā-tor kwod-RĀ-tus; *quadratus* = square, four-sided) (see Figure 11.17a)	Distal portion of shaft of ulna.	Distal portion of shaft of radius.	Pronates forearm at radioulnar joints.	Median nerve.
Forearm Supinator				
Supinator (SOO-pi-nā-tor; *supinator* = turns palm anteriorly) (see Figure 11.17b)	Lateral epicondyle of humerus and ridge near radial notch of ulna (supinator crest).	Lateral surface of proximal one-third of radius.	Supinates forearm at radioulnar joints.	Deep radial nerve.

continues

Figure 11.16 Muscles that move the radius and ulna (forearm bones). (See Tortora, *A Photographic Atlas of the Human Body, Second Edition,* Figure 5.11.)

The anterior arm muscles flex the forearm; the posterior arm muscles extend it.

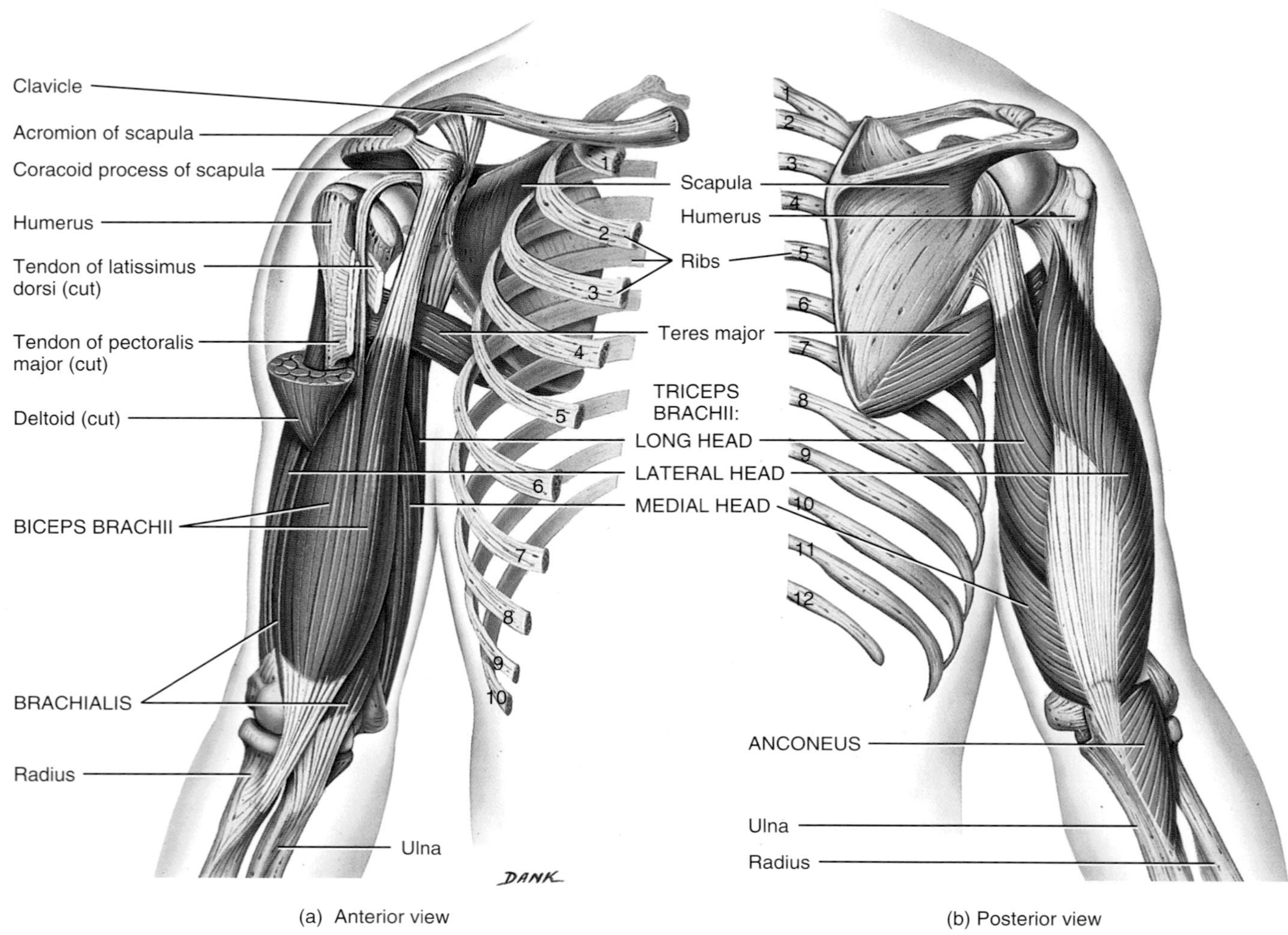

(a) Anterior view

(b) Posterior view

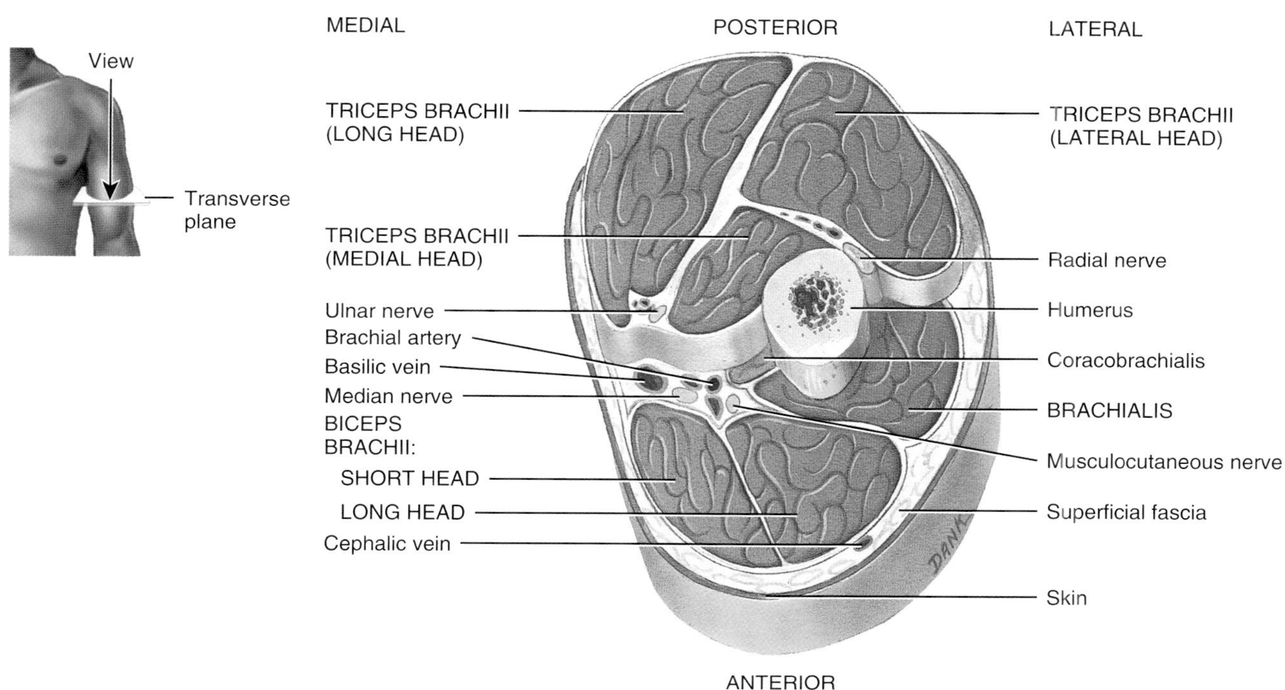

(c) Superior view of transverse section of arm

Which muscle is the most powerful flexor of the forearm? The most powerful extensor?

MUSCLES THAT MOVE THE WRIST, HAND, AND DIGITS (FIGURE 11.17)

▶ **OBJECTIVE**

Describe the origin, insertion, action, and innervation of the muscles that move the wrist, hand, thumb, and fingers.

Muscles of the forearm that move the wrist, hand, thumb, and fingers are many and varied. Those in this group that act on the digits are known as **extrinsic muscles of the hand** (*ex-* = outside) because they originate *outside* the hand and insert within it. As you will see, the names for the muscles that move the wrist, hand, and digits give some indication of their origin, insertion, or action. Based on location and function, the muscles of the forearm are divided into two groups: (1) anterior compartment muscles and (2) posterior compartment muscles. The **anterior (flexor) compartment** muscles of the forearm originate on the humerus, typically insert on the carpals, metacarpals, and phalanges, and function as flexors. The bellies of these muscles form the bulk of the forearm. One of the muscles in the superficial anterior compartment, the **palmaris longus** muscle, is missing in about 10% of individuals (usually in the left forearm) and is commonly used for tendon repair. The **posterior (extensor) compartment** muscles of the forearm originate on the humerus, insert on the metacarpals and phalanges, and function as extensors. Within each compartment, the muscles are grouped as superficial or deep.

The **superficial anterior compartment** muscles are arranged in the following order from lateral to medial: **flexor carpi radialis, palmaris longus** and **flexor carpi ulnaris** (the ulnar nerve and artery are just lateral to the tendon of this muscle at the wrist). The **flexor digitorum superficialis** muscle is deep to the other three muscles and is the largest superficial muscle in the forearm.

The **deep anterior compartment** muscles are arranged in the following order from lateral to medial: **flexor pollicis longus** (the only flexor of the distal phalanx of the thumb) and **flexor digitorum profundus** (ends in four tendons that insert into the distal phalanges of the fingers).

The **superficial posterior compartment** muscles are arranged in the following order from lateral to medial: **extensor carpi radialis longus, ex-**

MUSCLE	ORIGIN	INSERTION	ACTION	INNERVATION
Superficial Anterior (Flexor) Compartment of the Forearm				
Flexor carpi radialis (FLEK-sor KAR-pē rā′-dē-Ā-lis; *flexor* = decreases angle at joint; *carpi* = of the wrist; *radi-* = radius)	Medial epicondyle of humerus.	Second and third metacarpals.	Flexes and abducts hand (radial deviation) at wrist joint.	Median nerve.
Palmaris longus (pal-MA-ris LON-gus; *palma* = palm; *longus* = long)	Medial epicondyle of humerus.	Flexor retinaculum and palmar aponeurosis (deep fascia in center of palm).	Weakly flexes hand at wrist joint.	Median nerve.
Flexor carpi ulnaris (FLEK-sor KAR-pē ul-NAR-is; *ulnar-* = ulna)	Medial epicondyle of humerus and superior posterior border of ulna.	Pisiform, hamate, and base of fifth metacarpal.	Flexes and adducts hand (ulnar deviation) at wrist joint.	Ulnar nerve.
Flexor digitorum superficialis (FLEK-sor di-ji-TOR-um soo′- per-fish′-ē-Ā-lis; *digit* = finger or toe; *superficialis* = closer to surface)	Medial epicondyle of humerus, coronoid process of ulna, and a ridge along lateral margin of anterior surface (anterior oblique line) of radius.	Middle phalanx of each finger.*	Flexes middle phalanx of each finger at proximal interphalangeal joint, proximal phalanx of each finger at metacarpophalangeal joint, and hand at wrist joint.	Median nerve.
Deep Anterior (Flexor) Compartment of the Forearm				
Flexor pollicis longus (FLEK-sor POL-li-sis LON-gus; *pollic-* = thumb)	Anterior surface of radius and interosseous membrane (sheet of fibrous tissue that holds shafts of ulna and radius together).	Base of distal phalanx of thumb.	Flexes distal phalanx of thumb at interphalangeal joint.	Median nerve.
Flexor digitorum profundus (FLEK-sor di′-ji-TOR-um prō-FUN-dus; *profundus* = deep)	Anterior medial surface of body of ulna.	Base of distal phalanx of each finger.	Flexes distal and middle phalanges of each finger at interphalangeal joints, proximal phalanx of each finger at metacarpophalangeal joint, and hand at wrist joint.	Median and ulnar nerves.
Superficial Posterior (Extensor) Compartment of the Forearm				
Extensor carpi radialis longus (eks-TEN-sor KAR-pē rā′-dē-Ā-lis LON-gus; *extensor* = increases angle at joint)	Lateral supracondylar ridge of humerus.	Second metacarpal.	Extends and abducts hand at wrist joint.	Radial nerve.

*Reminder: The thumb or pollex is the first digit and has two phalanges: proximal and distal. The remaining digits, the fingers, are numbered II–V (2–5), and each has three phalanges: proximal, middle, and distal.

tensor carpi radialis brevis, extensor digitorum (occupies most of the posterior surface of the forearm and divides into four tendons that insert into the middle and distal phalanges of the fingers), **extensor digiti minimi** (a slender muscle usually connected to the extensor digitorum), and the **extensor carpi ulnaris.**

The **deep posterior compartment** muscles are arranged in the following order from lateral to medial: **abductor pollicis longus, extensor pollicis brevis, extensor pollicis longus,** and **extensor indicis.**

The tendons of the muscles of the forearm that attach to the wrist or continue into the hand, along with blood vessels and nerves, are held close to bones by strong fasciae. The tendons are also surrounded by tendon sheaths. At the wrist, the deep fascia is thickened into fibrous bands called **retinacula** (*retinacul* = a holdfast). The **flexor retinaculum** is located over the palmar surface of the carpal bones. The long flexor tendons of the digits and wrist and the median nerve pass deep to the flexor retinaculum. The **extensor retinaculum** is located over the dorsal surface of the carpal bones. The extensor tendons of the wrist and digits pass deep to it.

Relating Muscles to Movements

Arrange the muscles in this exhibit according to the following actions on the wrist joint: (1) flexion, (2) extension, (3) abduction, and (4) adduction; the following actions on the fingers at the metacarpophalangeal joints: (1) flexion and (2) extension; the following actions on the fingers at the interphalangeal joints: (1) flexion and (2) extension; the following actions on the thumb at the carpometacarpal, metacarpophalangeal, and interphalangeal joints: (1) extension and (2) abduction; and the following action on the thumb at the interphalangeal joint: flexion. The same muscle may be mentioned more than once.

► CHECKPOINT

Which muscles and actions of the wrist, hand, and digits are used when writing?

MUSCLE	ORIGIN	INSERTION	ACTION	INNERVATION
Superficial Posterior (Extensor) Compartment of the Forearm (continued)				
Extensor carpi radialis brevis (eks-TEN-sor KAR-pē rā′-dē-Ā-lis BREV-is; *brevis* = short)	Lateral epicondyle of humerus.	Third metacarpal.	Extends and abducts hand at wrist joint.	Radial nerve.
Extensor digitorum (eks-TEN-sor di′-ji-TOR-um)	Lateral epicondyle of humerus.	Distal and middle phalanges of each finger.	Extends distal and middle phalanges of each finger at interphalangeal joints, proximal phalanx of each finger at metacarpophalangeal joint, and hand at wrist joint.	Radial nerve.
Extensor digiti minimi (eks-TEN-sor DIJ-i-tē MIN-i-mē; *digit* = finger or toe; *minimi* = smallest)	Lateral epicondyle of humerus.	Tendon of extensor digitorum on fifth phalanx.	Extends proximal phalanx of little finger at metacarpophalangeal joint and hand at wrist joint.	Deep radial nerve.
Extensor carpi ulnaris (eks-TEN-sor KAR-pē ul-NAR-is)	Lateral epicondyle of humerus and posterior border of ulna.	Fifth metacarpal.	Extends and adducts hand at wrist joint.	Deep radial nerve.
Deep Posterior (Extensor) Compartment of the Forearm				
Abductor pollicis longus (ab-DUK-tor POL-li-sis LON-gus; *abductor* = moves part away from midline)	Posterior surface of middle of radius and ulna and interosseous membrane.	First metacarpal.	Abducts and extends thumb at carpometacarpal joint and abducts hand at wrist joint.	Deep radial nerve.
Extensor pollicis brevis (eks-TEN-sor POL-li-sis BREV-is)	Posterior surface of middle of radius and interosseous membrane.	Base of proximal phalanx of thumb.	Extends proximal phalanx of thumb at metacarpophalangeal joint, first metacarpal of thumb at carpometacarpal joint, and hand at wrist joint.	Deep radial nerve.
Extensor pollicis longus (eks-TEN-sor POL-li-sis LON-gus)	Posterior surface of middle of ulna and interosseous membrane.	Base of distal phalanx of thumb.	Extends distal phalanx of thumb at interphalangeal joint, first metacarpal of thumb at carpometacarpal joint, and abducts hand at wrist joint.	Deep radial nerve.
Extensor indicis (eks-TEN-sor IN-di-kis; *indicis* = index)	Posterior surface of ulna.	Tendon of extensor digitorum of index finger.	Extends distal and middle phalanges of index finger at interphalangeal joints, proximal phalanx of index finger at metacarpophalangeal joint, and hand at wrist joint.	Deep radial nerve.

continues

Figure 11.17 Muscles that move the wrist, hand, and digits. (See Tortora, *A Photographic Atlas of the Human Body, Second Edition,* Figures 5.12 and 5.13.)

The anterior compartment muscles function as flexors, and the posterior compartment muscles function as extensors.

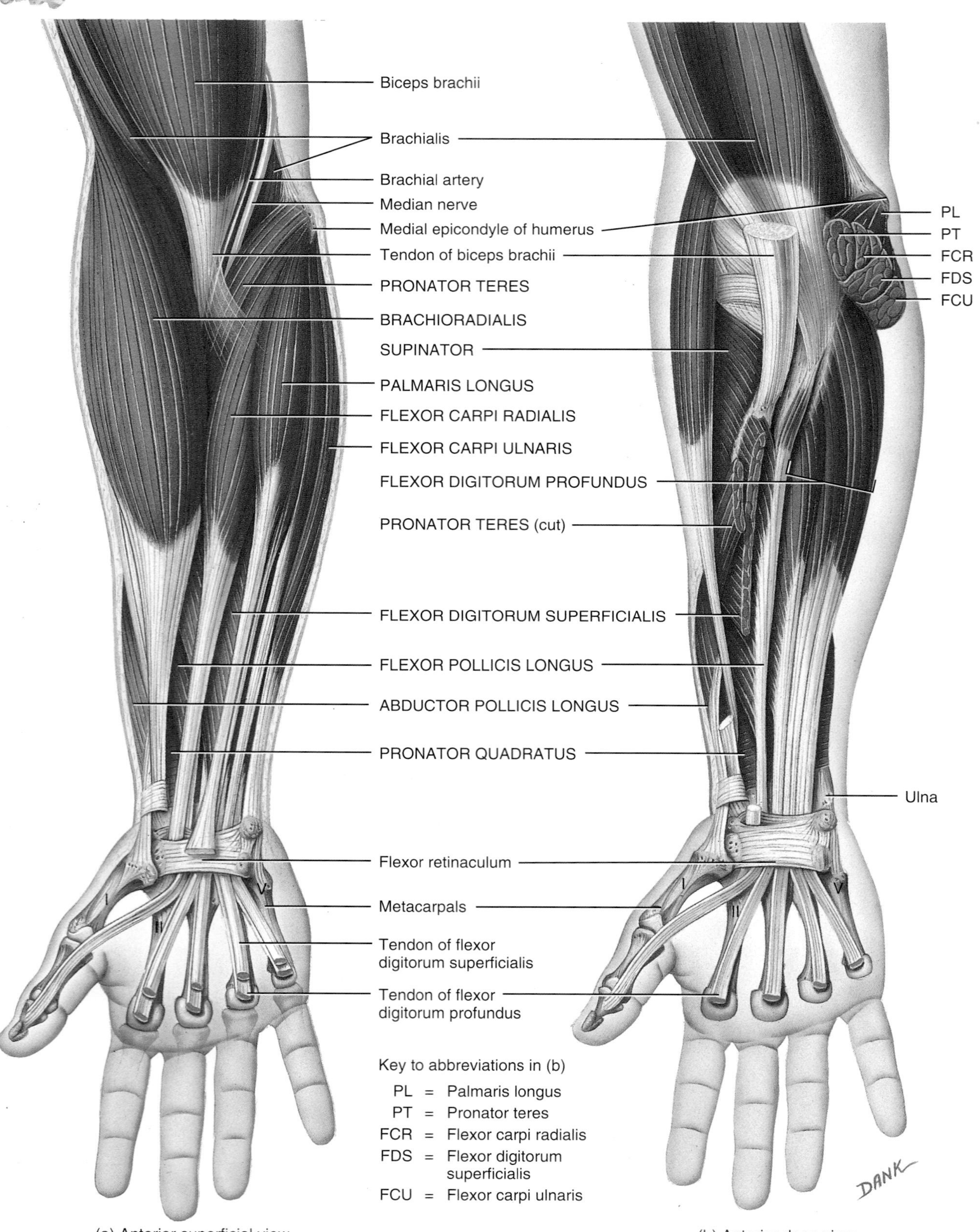

(a) Anterior superficial view

(b) Anterior deep view

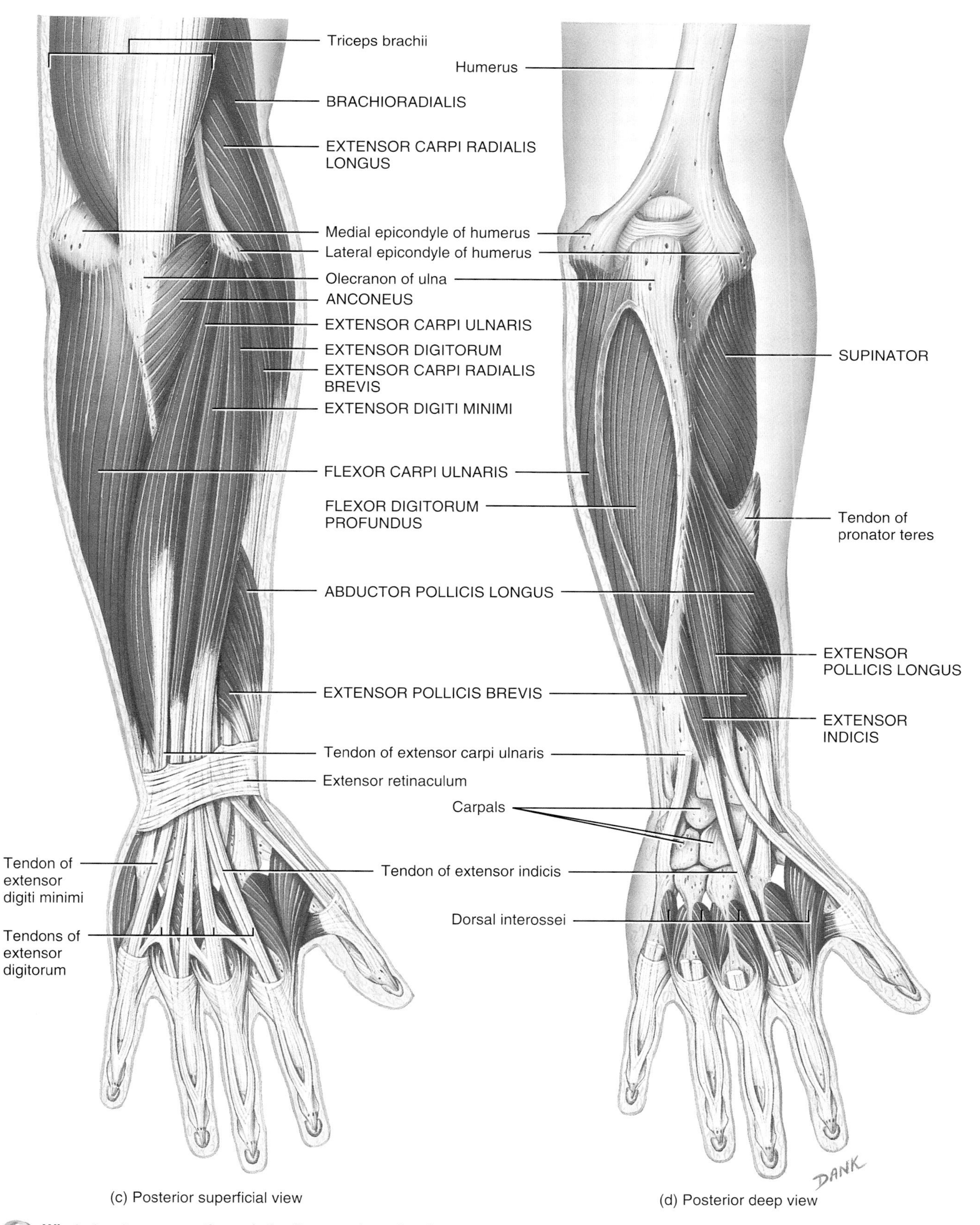

(c) Posterior superficial view

(d) Posterior deep view

What structures pass through the flexor retinaculum?

INTRINSIC MUSCLES OF THE HAND (FIGURE 11.18)

▶ **OBJECTIVE**

Describe the origin, insertion, action, and innervation of the intrinsic muscles of the hand.

Several of the muscles discussed in Exhibit 11.14 move the digits in various ways and are known as extrinsic muscles of the hand. They produce the powerful but crude movements of the digits. The **intrinsic muscles of the hand** in the palm produce the weak but intricate and precise movements of the digits that characterize the human hand. The muscles in this group are so named because their origins and insertions are *within* the hand.

The intrinsic muscles of the hand are divided into three groups: (1) **thenar,** (2) **hypothenar,** and (3) **intermediate.** The four thenar muscles act on the thumb and form the **thenar eminence,** the lateral rounded contour on the palm that is also called the ball of the thumb. The thenar muscles include the abductor pollicis brevis, opponens pollicis, flexor pollicis brevis, and adductor pollicis. The **abductor pollicis brevis** is a thin, short, relatively broad superficial muscle on the lateral side of the thenar eminence. The **opponens pollicis** is a small, triangular muscle that is deep to the abductor pollicis brevis muscle. The **flexor pollicis brevis** is a short, wide muscle that is medial to the abductor pollicis brevis muscle. The **adductor pollicis** is fan-shaped and has two heads (oblique and transverse) separated by a gap through which the radial artery passes.

The three hypothenar muscles act on the little finger and form the **hypothenar eminence,** the medial rounded contour on the palm that is also called the ball of the little finger. The hypothenar muscles are the abductor digiti minimi, flexor digiti minimi brevis, and opponens digiti minimi. The **abductor digiti minimi** is a short, wide muscle and is the most superficial of the hypothenar muscles. It is a powerful muscle that plays an important role in grasping an object with outspread fingers. The **flexor digiti minimi brevis** muscle is also short and wide and is lateral to the abductor digiti minimi muscle. The **opponens digiti minimi** muscle is triangular and deep to the other two hypothenar muscles.

The 11 intermediate (midpalmar) muscles act on all the digits except the thumb. The intermediate muscles include the lumbricals, palmar interossei, and dorsal interossei. The **lumbricals** (= worm-shaped)**,** as their name indicates, are worm-shaped. They originate from and insert into the tendons of other muscles (flexor digitorum profundus and extensor digitorum). The **palmar interossei** are the smaller and most superficial of the interossei muscles. The **dorsal interossei** are the deep interossei muscles. Both sets of interossei muscles are located between the metacarpals and are important in abduction, adduction, flexion, and extension of the fingers, and in movements in skilled activities such as writing, typing, and playing a piano.

The functional importance of the hand is readily apparent when you consider that certain hand injuries can result in permanent disability. Most of the dexterity of the hand depends on movements of the thumb. The general activities of the hand are free motion, power grip (forcible movement of the fingers and thumb against the palm, as in squeezing), precision handling (a change in position of a handled object that requires exact control of finger and thumb positions, as in winding a watch or threading a needle), and pinch (compression between the thumb and index finger or between the thumb and first two fingers).

Movements of the thumb are very important in the precise activities of the hand, and they are defined in different planes from comparable movements of other digits because the thumb is positioned at a right angle to the other digits. The five principal movements of the thumb are illustrated in Figure 11.18d and include *flexion* (movement of the thumb medially across the palm), *extension* (movement of the thumb laterally away from the palm), *abduction* (movement of the thumb in an anteroposterior plane away from the palm), *adduction* (movement of the thumb in an anteroposterior plane toward the palm), and *opposition* (movement of the thumb across the palm so that the tip of the thumb meets the tip of a finger). Opposition is the single most distinctive digital movement that gives humans and other primates the ability to grasp and manipulate objects precisely.

Carpal Tunnel Syndrome

The **carpal tunnel** is a narrow passageway formed anteriorly by the flexor retinaculum and posteriorly by the carpal bones. Through this tunnel pass the median nerve, the most superficial structure, and the long flexor tendons for the digits (Figure 11.18c). Structures within the carpal tunnel, especially the median nerve, are vulnerable to compression, and the resulting condition is called **carpal tunnel syndrome.** Compression of the median nerve leads to sensory changes over the lateral side of the hand and muscle weakness in the thenar eminence. This results in pain, numbness, and tingling of the fingers. The condition may be caused by inflammation of the digital tendon sheaths, fluid retention, excessive exercise, infection, trauma, and/or repetitive activities that involve flexion of the wrist, such as keyboarding, cutting hair, and playing a piano. Treatment may involve the use of nonsteroidal anti-inflammatory drugs (such as ibuprofen or aspirin), wearing a wrist splint, corticosteroid injections, or surgery to cut the flexor retinaculum and release presssure on the median nerve. ■

Relating Muscles to Movements

Arrange the muscles in this exhibit according to the following actions on the thumb at the carpometacarpal and metacarpophalangeal joints: (1) abduction, (2) adduction, (3) flexion, and (4) opposition; and the following actions on the fingers at the metacarpophalangeal and interphalangeal joints: (1) abduction, (2) adduction, (3) flexion, and (4) extension. The same muscle may be mentioned more than once.

▶ **CHECKPOINT**

How do the actions of the extrinsic and intrinsic muscles of the hand differ?

MUSCLE	ORIGIN	INSERTION	ACTION	INNERVATION
Thenar (Lateral Aspect of Palm)				
Abductor pollicis brevis (ab-DUK-tor POL-li-sis BREV-is; *abductor* = moves part away from middle; *pollic-* = the thumb; *brevis* = short)	Flexor retinaculum, scaphoid, and trapezium.	Lateral side of proximal phalanx of thumb.	Abducts thumb at carpometacarpal joint.	Median nerve.
Opponens pollicis (op-PŌ-nenz POL-li-sis; *opponens* = opposes)	Flexor retinaculum and trapezium.	Lateral side of first metacarpal (thumb).	Moves thumb across palm to meet little finger (opposition) at the carpometacarpal joint.	Median nerve.
Flexor pollicis brevis (FLEK-sor POL-li-sis BREV-is; *flexor* = decreases angle at joint)	Flexor retinaculum, trapezium, capitate, and trapezoid.	Lateral side of proximal phalanx of thumb.	Flexes thumb at carpometacarpal and metacarpophalangeal joints.	Median and ulnar nerves.
Adductor pollicis (ad-DUK-tor POL-li-sis; *adductor* = moves part toward midline)	Oblique head: capitate and second and third metacarpals; transverse head: third metacarpal.	Medial side of proximal phalanx of thumb by a tendon containing a sesamoid bone.	Adducts thumb at carpometacarpal and metacarpophalangeal joints.	Ulnar nerve.
Hypothenar (Medial Aspect of Palm)				
Abductor digiti minimi (ab-DUK-tor DIJ-i-tē MIN-i-mē; *digit* = finger or toe; *minimi* = little)	Pisiform and tendon of flexor carpi ulnaris.	Medial side of proximal phalanx of little finger.	Abducts and flexes little finger at metacarpophalangeal joint.	Ulnar nerve.
Flexor digiti minimi brevis (FLEK-sor DIJ-i-tē MIN-i-mē BREV-is)	Flexor retinaculum and hamate.	Medial side of proximal phalanx of little finger.	Flexes little finger at carpometacarpal and metacarpophalangeal joints.	Ulnar nerve.
Opponens digiti minimi (op-PŌ-nenz DIJ-i-tē MIN-i-mē)	Flexor retinaculum and hamate.	Medial side of fifth metacarpal (little finger).	Moves little finger across palm to meet thumb (opposition) at the carpometacarpal joint.	Ulnar nerve.
Intermediate (Midpalmar)				
Lumbricals (LUM-bri-kals; *lumbric-* = earthworm) (four muscles)	Lateral sides of tendons and flexor digitorum profundus of each finger.	Lateral sides of tendons of extensor digitorum on proximal phalanges of each finger.	Flex each finger at metacarpophalangeal joints and extend each finger at interphalangeal joints.	Median and ulnar nerves.
Palmar interossei (in′-ter-OS-ē-ī (*palmar* = palm; *inter-* = between; *ossei* = bones) (three muscles)	Sides of shafts of metacarpals of all digits (except the middle one).	Sides of bases of proximal phalanges of all digits (except the middle one).	Adduct each finger at metacarpophalangeal joints; flex each finger at metacarpophalangeal joints.	Ulnar nerve.
Dorsal interossei (in′-ter-OS-ē-ī; *dorsal* = back surface) (four muscles)	Adjacent sides of metacarpals.	Proximal phalanx of each finger.	Abduct fingers 2–4 at metacarpophalangeal joints; flex fingers 2–4 at metacarpophalangeal joints; and extend each finger at interphalangeal joints.	Ulnar nerve.

continues

Figure 11.18 Intrinsic muscles of the hand. (See Tortora, *A Photographic Atlas of the Human Body, Second Edition,* Figures 5.14 and 5.15.)

The intrinsic muscles of the hand produce the intricate and precise movements of the digits that characterize the human hand.

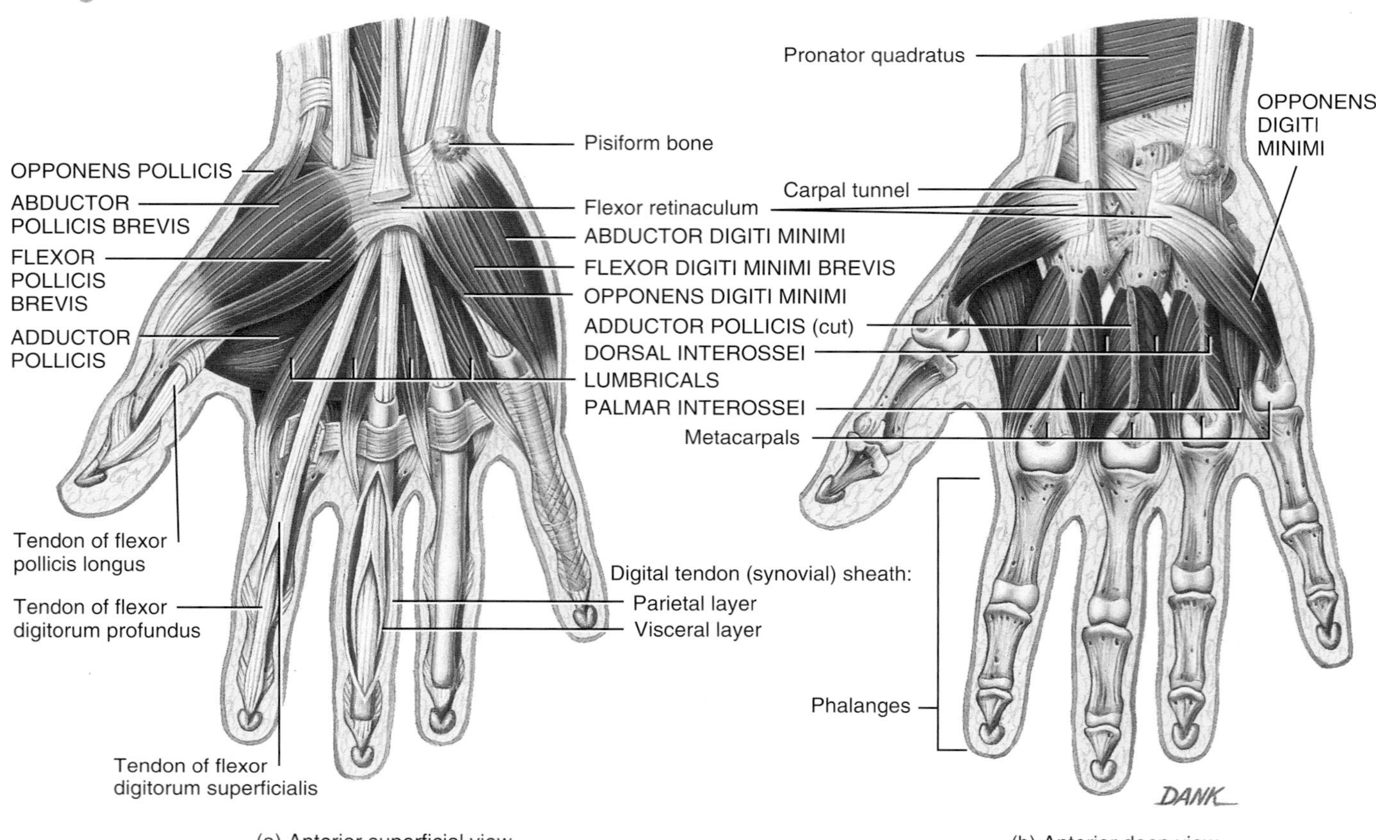

(a) Anterior superficial view

(b) Anterior deep view

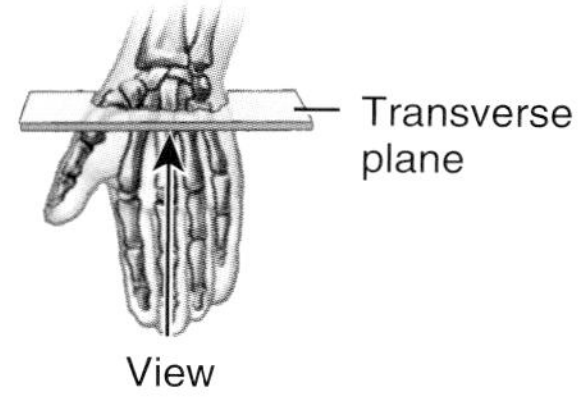

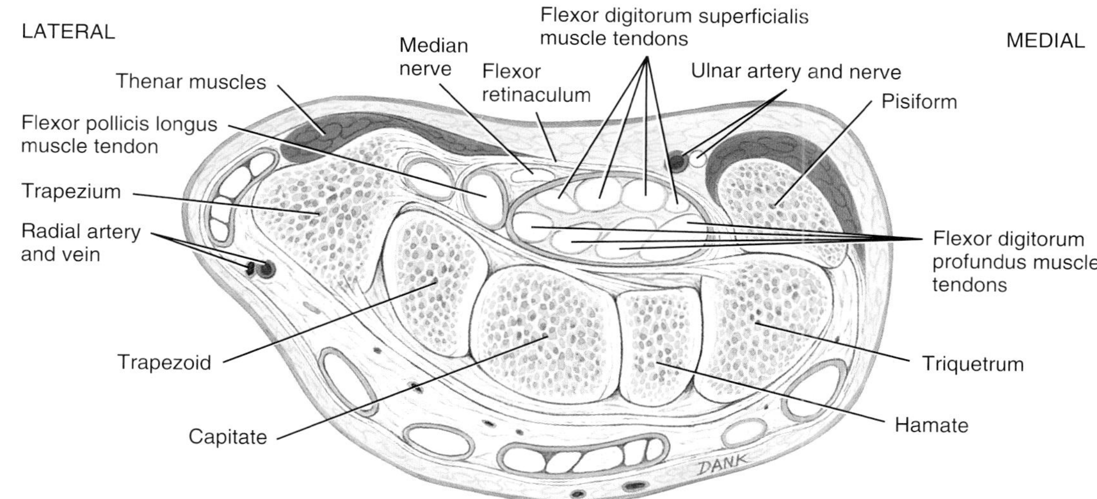

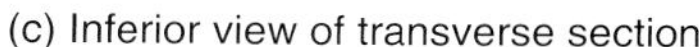

(c) Inferior view of transverse section

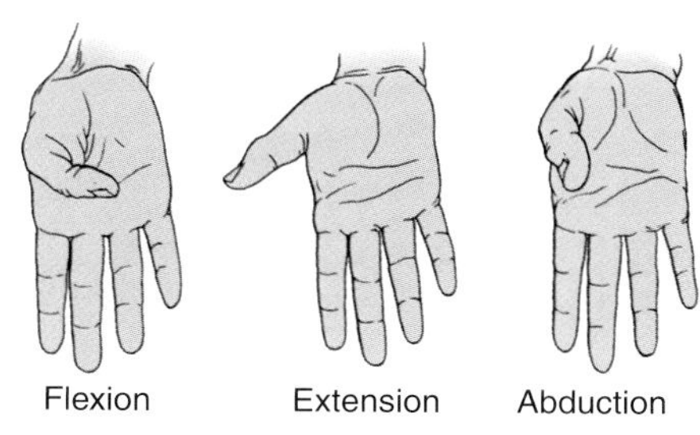

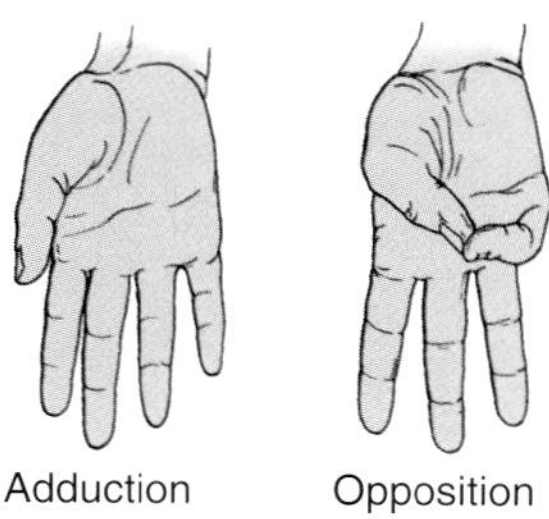

(d) Movements of the thumb

Muscles of the thenar eminence act on which digit?

MUSCLES THAT MOVE THE VERTEBRAL COLUMN (BACKBONE) (FIGURE 11.19)

▶ **OBJECTIVE**

Describe the origin, insertion, action, and innervation of the muscles that move the vertebral column.

The muscles that move the vertebral column (backbone) are quite complex because they have multiple origins and insertions and there is considerable overlap among them. One way to group the muscles is on the basis of the general direction of the muscle bundles and their approximate lengths. For example, the splenius muscles arise from the midline and extend laterally and superiorly to their insertions (Figure 11.19a). The erector spinae muscle group (consisting of the iliocostalis, longissimus, and spinalis muscles) arises from either the midline or more laterally but usually runs almost longitudinally, with neither a significant lateral nor medial direction as it is traced superiorly. The muscles of the transversospinalis group (semispinalis, rotatore, and multifidus) arise laterally but extend toward the midline as they are traced superiorly. Deep to these three muscle groups are small segmental muscles that extend between spinous processes or transverse processes of vertebrae.

MUSCLE	ORIGIN	INSERTION	ACTION	INNERVATION
Splenius (SPLĒ-nē-us)				
Splenius capitis (KAP-i-tis; *splenium* = bandage; *capit-* = head)	Ligamentum nuchae and spinous processes of seventh cervical vertebra and first three or four thoracic vertebrae.	Occipital bone and mastoid process of temporal bone.	Acting together (bilaterally), extend head; acting singly (unilaterally), laterally flex and rotate head to same side as contracting muscle.	Middle cervical spinal nerves.
Splenius cervicis (SER-vi-kis; *cervic-* = neck)	Spinous processes of third through sixth thoracic vertebrae.	Transverse processes of first two or four cervical vertebrae.	Acting together, extend head; acting singly, laterally flex and rotate head to same side as contracting muscle.	Inferior cervical spinal nerves.
Erector spinae (e-REK-tor SPI-nē) Consists of iliocostalis muscles (lateral), longissimus muscles (intermediate), and spinalis muscles (medial).				
Iliocostalis Group (Lateral)				
Iliocostalis cervicis (il′-ē-ō-kos-TĀL-is SER-vi-kis; *ilio-* = flank; *costa-* = rib)	Superior six ribs.	Transverse processes of fourth to sixth cervical vertebrae.	Acting together, muscles of each region (cervical, thoracic, and lumbar) extend and maintain erect posture of vertebral column of their respective regions; acting singly, laterally flex vertebral column of their respective regions.	Cervical and thoracic spinal nerves.
Iliocostalis thoracis (il′-ē-ō-kos-TĀL-is thō-RĀ-sis; *thorac-* = chest)	Inferior six ribs.	Superior six ribs.		Thoracic spinal nerves.
Iliocostalis lumborum (il′-ē-ō-kos-TĀL-is lum-BOR-um)	Iliac crest.	Inferior six ribs.		Lumbar spinal nerves.
Longissimus Group (Intermediate)				
Longissimus capitis (lon-JIS-i-mus KAP-i-tis; *longissimus* = longest)	Transverse processes of superior four thoracic vertebrae and articular processes of inferior four cervical vertebrae.	Mastoid process of temporal bone.	Acting together, both longissimus capitis muscles extend head; acting singly, rotate head to s ame side as contracting muscle.	Middle and inferior cervical spinal nerves.
Longissimus cervicis (lon-JIS-i-mus SER-vi-kis)	Transverse processes of fourth and fifth thoracic vertebrae.	Transverse processes of second to sixth cervical vertebrae.	Acting together, longissimus cervicis and both longissimus thoracis muscles extend vertebral column of their respective regions; acting singly, laterally flex vertebral column of their respective regions.	Cervical and superior thoracic, spinal nerves.
Longissimus thoracis (lon-JIS-i-mus thō-RĀ-sis)	Transverse processes of lumbar vertebrae.	Transverse processes of all thoracic and superior lumbar vertebrae and ninth and tenth ribs.		Thoracic and lumbar spinal nerves.
Spinalis Group (Medial)				
Spinalis capitis (spi-NĀ-lis KAP-i-tis; *spinal-* = vertebral column)	Arises with semispinalis capitis.	Occipital bone.	Acting together, muscles of each region (cervical, thoracic, and lumbar) extend vertebral column of their respective regions.	Cervical and superior thoracic spinal nerves.
Spinalis cervicis (spi-NĀ-lis SER-vi-kis)	Ligamentum nuchae and spinous process of seventh cervical vertebra.	Spinous process of axis.		Inferior cervical and thoracic spinal nerves.
Spinalis thoracis (spi-NĀ-lis thō-RĀ-sis)	Spinous processes of superior lumbar and inferior thoracic vertebrae.	Spinous processes of superior thoracic vertebrae.		Thoracic spinal nerves.

Because the scalene muscles assist in moving the vertebral column, they are also included in this exhibit. As noted in Exhibit 11.7, the rectus abdominis, external oblique, internal oblique, and quadratus lumborum muscles also play a role in moving the vertebral column.

The bandage-like **splenius** muscles are attached to the sides and back of the neck. The two muscles in this group are named on the basis of their superior attachments (insertions): **splenius capitis** (head region) and **splenius cervicis** (cervical region). They extend the head and laterally flex and rotate the head.

The **erector spinae** is the largest muscle mass of the back, forming a prominent bulge on either side of the vertebral column. It is the chief extensor of the vertebral column. It is also important in controlling flexion, lateral flexion, and rotation of the vertebral column and in maintaining the lumbar curve, because the main mass of the muscle is in the lumbar region. As just noted, it consists of three groups: iliocostalis (laterally placed), longissimus (intermediately placed), and spinalis (medially placed). These groups, in turn, consist of a series of overlapping muscles, and the muscles within the groups are named according to the regions of the body with which they are associated.

MUSCLE	ORIGIN	INSERTION	ACTION	INNERVATION
Transversospinales (trans-ver-sō-spi-NĀ-lēz)				
Semispinalis capitis (sem′-ē-spi-NĀ-lis KAP-i-tis; *semi-* = partially or one half)	Transverse processes of first six or seven thoracic vertebrae and seventh cervical vertebra, and articular processes of fourth, fifth, and sixth cervical vertebrae.	Occipital bone.	Acting together, extend head; acting singly, rotate head to side opposite contracting muscle.	Cervical and thoracic spinal nerves.
Semispinalis cervicis (sem′-ē-spi-NĀ-lis SER-vi-kis)	Transverse processes of superior five or six thoracic vertebrae.	Spinous processes of first to fifth cervical vertebrae.	Acting together, both semispinalis cervicis and both semispinalis thoracis muscles extend vertebral column of their respective regions; acting singly, rotate head to side opposite contracting muscle.	Cervical and thoracic spinal nerves.
Semispinalis thoracis (sem′-ē-spi-NĀ-lis thō-RĀ-sis)	Transverse processes of sixth to tenth thoracic vertebrae.	Spinous processes of superior four thoracic and last two cervical vertebrae.		Thoracic spinal nerves.
Multifidus (mul-TIF-i-dus; *multi* = many; *fid-* = segmented)	Sacrum, ilium, transverse processes of lumbar, thoracic, and inferior four cervical vertebrae.	Spinous process of a more superior vertebra.	Acting together, extend vertebral column; acting singly, laterally flex vertebral column and rotate head to side opposite contracting muscle.	Cervical, thoracic, and lumbar spinal nerves.
Rotatores (rō′-ta-TŌ-rēz; singular is **rotatore;** *rotatore* = to rotate)	Transverse processes of all vertebrae.	Spinous process of vertebra superior to the one of origin.	Acting together, extend vertebral column; acting singly, rotate vertebral column to side opposite contracting muscle.	Cervical, thoracic, and lumbar spinal nerves.
Segmental (seg-MEN-tal)				
Interspinales (in-ter-spī-NĀ-lēz; *inter-* = between)	Superior surface of all spinous processes.	Inferior surface of spinous process of vertebra superior to the one of origin.	Acting together, extend vertebral column; acting singly, stabilize vertebral column during movement.	Cervical, thoracic, and lumbar spinal nerves.
Intertransversarii (in′-ter- trans-vers-AR-ē-ī; singular is *intertransversarius*)	Transverse processes of all vertebrae.	Transverse process of vertebra superior to the one of origin.	Acting together, extend vertebral column; acting singly, laterally flex vertebral column and stabilize it during movements.	Cervical, thoracic, and lumbar spinal nerves.
Scalenes (SKĀ-lēnz)				
Anterior scalene (SKĀ-lēn; *anterior* = front; *scalene* = uneven)	Transverse processes of third through sixth cervical vertebrae.	First rib.	Acting together, right and left anterior scalene and middle scalene muscles flex head and elevate first ribs during deep inhalation; acting singly, laterally flex head and rotate head to side opposite contracting muscle.	Cervical spinal nerves C5–C6.
Middle scalene (SKĀ-lēn)	Transverse processes of inferior six cervical vertebrae.	First rib.		Cervical spinal nerves C3–C8.
Posterior scalene (SKĀ-lēn)	Transverse processes of fourth through sixth cervical vertebrae.	Second rib.	Acting together, flex head and elevate second ribs during deep inhalation; acting singly, laterally flex head and rotate head to side opposite contracting muscle.	Cervical spinal nerves C6–C8.

continues

The **iliocostalis group** consists of three muscles: the **iliocostalis cervicis** (cervical region), **iliocostalis thoracis** (thoracic region), and **iliocostalis lumborum** (lumbar region). The **longissimus group** resembles a herringbone and consists of three muscles: the **longissimus capitis** (head region), **longissimus cervicis** (cervical region), and **longissimus thoracis** (thoracic region). The **spinalis group** also consists of three muscles: the **spinalis capitis, spinalis cervicis,** and **spinalis thoracis.**

The **transversospinales** are so named because their fibers run from the transverse processes to the spinous processes of the vertebrae. The semispinalis muscles in this group are also named according to the region of the body with which they are associated: **semispinalis capitis** (head region), **semispinalis cervicis** (cervical region), and **semispinalis thoracis** (thoracic region). These muscles extend the vertebral column and rotate the head. The **multifidus** muscle in this group, as its name implies, is segmented into several bundles. It extends and laterally flexes the vertebral column and rotates the head. The **rotatores** muscles of this group are short and are found along the entire length of the vertebral column. They extend and rotate the vertebral column.

Within the **segmental** muscle group (Figure 11.19b), the **interspinales** and **intertransversarii** muscles unite the spinous and transverse processes of consecutive vertebrae. They function primarily in stabilizing the vertebral column during its movements.

Within the **scalene** group (Figure 11.19c), the **anterior scalene** muscle is anterior to the middle scalene muscle. The **middle scalene** muscle is intermediate in placement and is the longest and largest of the scalene muscles. The **posterior scalene** muscle is posterior to the middle scalene muscle and is the smallest of the scalene muscles. These muscles flex, laterally flex, and rotate the head and assist in deep inhalation.

Back Injuries and Heavy Lifting

Next to headaches, medical experts note that back problems are the most common medical complaint that lead people to seek treatment. Second only to the common cold as the greatest cause of lost workdays, back injuries cost U.S. industry \$10–14 billion in workers' compensation costs and about 100 million lost workdays annually.

The four factors associated with increased risk of back injury are amount of force, repetition, posture, and stress applied to the backbone. Poor physical condition, poor posture, lack of exercise, and excessive body weight contribute to the number and severity of sprains and strains. Back pain caused by a muscle strain or ligament sprain will normally heal within a short time and may never cause further problems. However, if ligaments and muscles are weak, discs in the lower back can become weakened and may herniate (rupture) with excessive lifting or a sudden fall. After years of back abuse, or with aging, the discs may simply wear out and cause chronic pain. Degeneration of the spine due to aging is often misdiagnosed as a sprain or strain.

Full flexion at the waist, as in touching your toes, overstretches the erector spinae muscles. Muscles that are overstretched cannot contract effectively since the zone of overlap in a sarcomere shortens and fewer cross-bridges make contact with thin filaments (see Figure 10.9 on page 302). Straightening up from such a position is therefore initiated by the hamstring muscles on the back of the thigh and the gluteus maximus muscles of the buttocks. The erector spinae muscles join in as the degree of flexion decreases. Improperly lifting a heavy weight, however, can strain the erector spinae muscles. The result can be painful muscle spasms, tearing of tendons and ligaments of the lower back, and herniating of inter-vertebral discs. The lumbar muscles are adapted for maintaining posture, not for lifting. This is why it is important to bend at the knees and use the powerful extensor muscles of the thighs and buttocks while lifting a heavy load. ■

Relating Muscles to Movements

Arrange the muscles in this exhibit according to the following actions on the head at the atlanto-occipital and intervertebral joints: (1) flexion, (2) extension, (3) lateral flexion, (4) rotation to same side as contracting muscle, and (5) rotation to opposite side as contracting muscle; the following actions on the vertebral column at the intervertebral joints: (1) flexion, (2) extension, (3) lateral flexion, (4) rotation, and (5) stabilization; and the following action on the ribs: elevation during deep inhalation. The same muscle may be mentioned more than once.

► CHECKPOINT

What is the largest muscle group of the back?

Figure 11.19 **Muscles that move the vertebral column (backbone).**

The erector spinae group (iliocostalis, longissimus, and spinalis muscles) is the largest muscular mass of the body and is the chief extensor of the vertebral column.

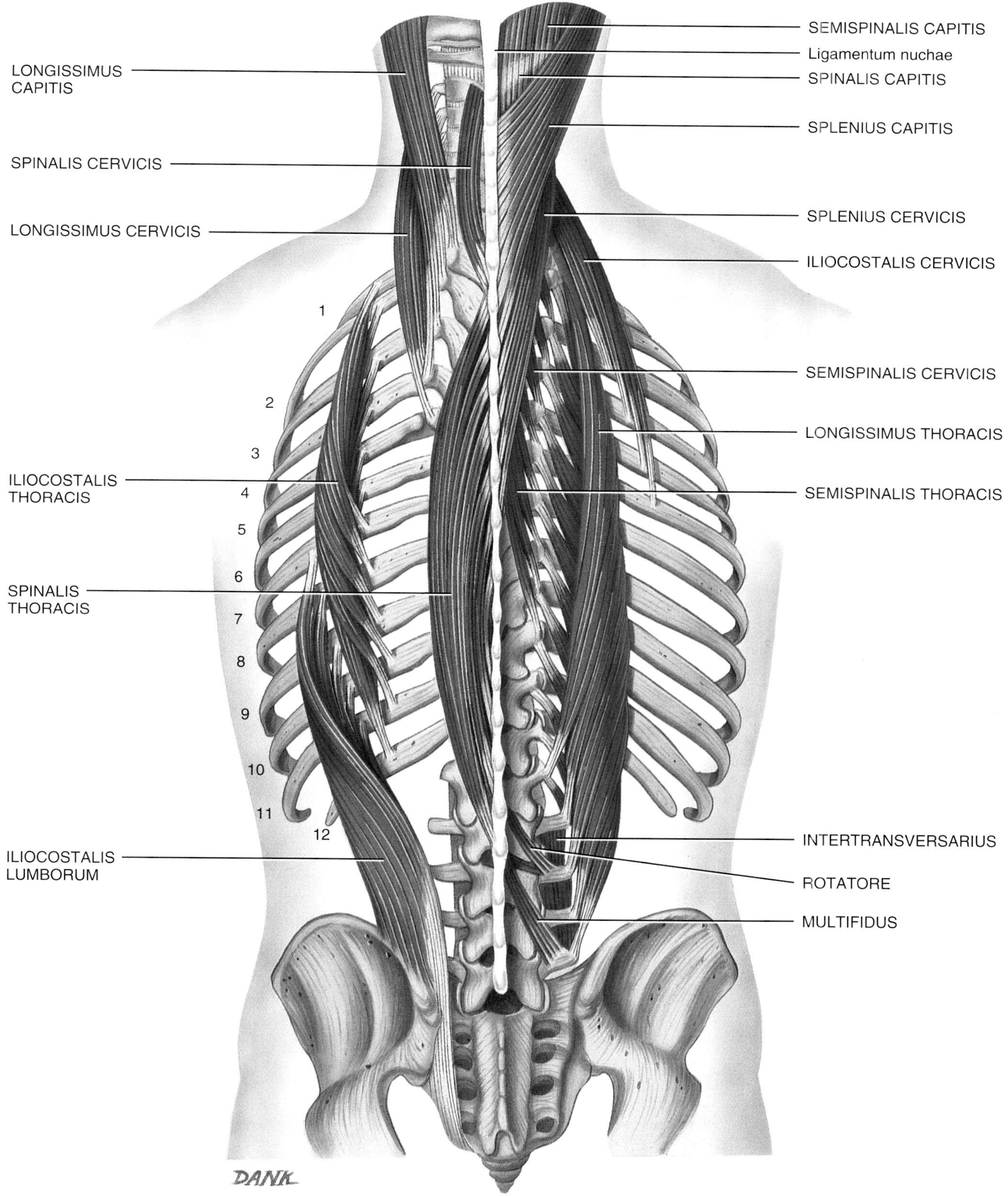

(a) Posterior view

continues

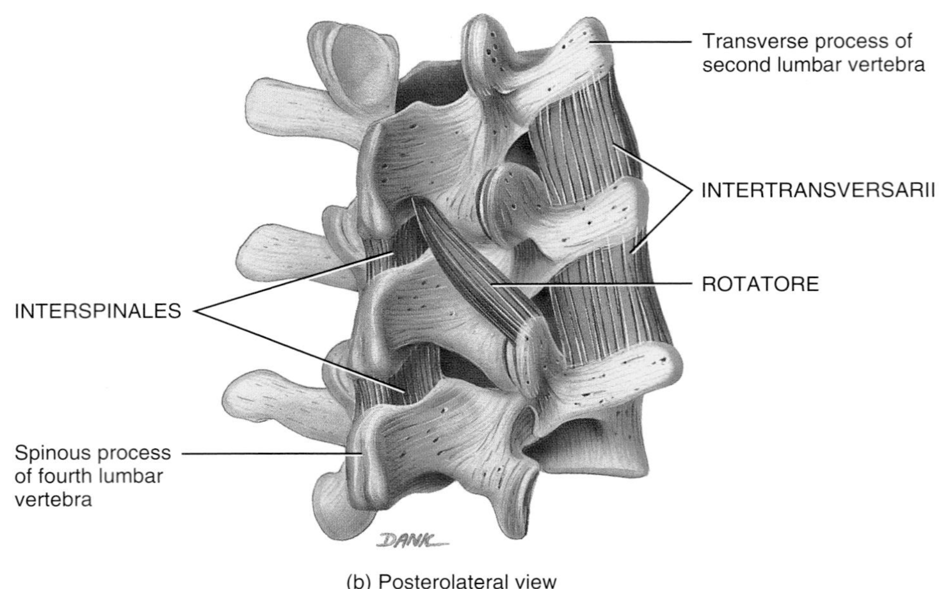

(b) Posterolateral view

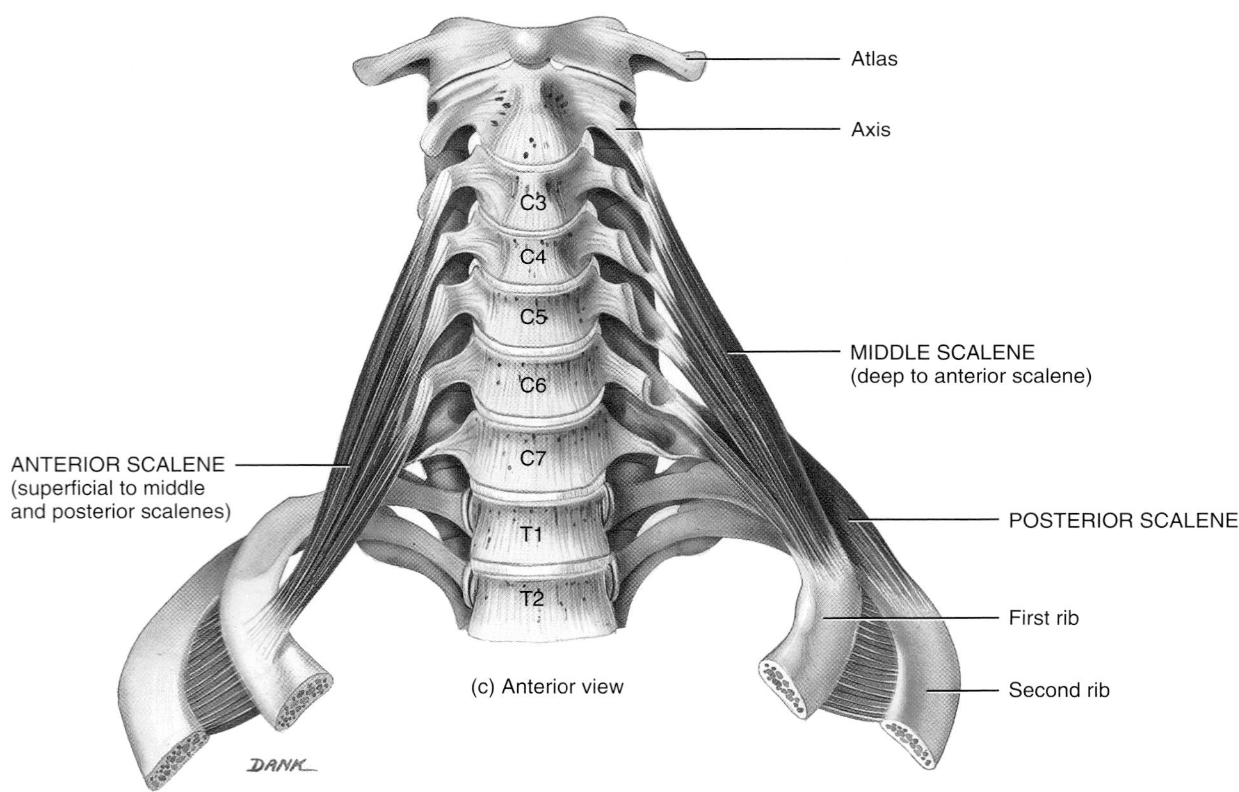

(c) Anterior view

What muscles originate at the midline and extend laterally and upward to their insertion?

EXHIBIT 11.17 MUSCLES THAT MOVE THE FEMUR (THIGH BONE) (FIGURE 11.20)

▶ **OBJECTIVE**

Describe the origin, insertion, action, and innervation of the muscles that move the femur.

As you will see, muscles of the lower limbs are larger and more powerful than those of the upper limbs because of differences in function. While upper limb muscles are characterized by versatility of movement, lower limb muscles function in stability, locomotion, and maintenance of posture. In addition, muscles of the lower limbs often cross two joints and act equally on both.

The majority of muscles that move the femur originate on the pelvic girdle and insert on the femur. The **psoas major** and **iliacus** muscles share a common insertion (lesser trochanter of femur) and are collectively known as the **iliopsoas** (il′-ē-ō-SŌ-as) muscle. There are three gluteal muscles: gluteus maximus, gluteus medius, and gluteus minimus. The **gluteus maximus** is the largest and heaviest of the three muscles and is one of the largest muscles in the body. It is the chief extensor of the femur. The **gluteus medius** is mostly deep to the gluteus maximus and is a powerful abductor of the femur at the hip joint. It is a common site for an intramuscular injection. The **gluteus minimus** is the smallest of the gluteal muscles and lies deep to the gluteus medius.

The **tensor fasciae latae** muscle is located on the lateral surface of the thigh. The *fascia lata* is a layer of deep fascia, composed of dense connective tissue, that encircles the entire thigh. It is well developed laterally where, together with the tendons of the tensor fasciae and gluteus maximus muscles, it forms a structure called the **iliotibial tract.** The tract inserts into the lateral condyle of the tibia.

The **piriformis, obturator internus, obturator externus, superior gemellus, inferior gemellus,** and **quadratus femoris** muscles are all deep to the gluteus maximus muscle and function as lateral rotators of the femur at the hip joint.

Three muscles on the medial aspect of the thigh are the **adductor longus, adductor brevis,** and **adductor magnus.** They originate on the pubic bone and insert on the femur. All three muscles adduct, flex, and medially rotate the femur at the hip joint. The **pectineus** muscle also adducts and flexes the femur at the hip joint.

Technically, the adductor muscles and pectineus muscles are components of the medial compartment of the thigh and could be included in Exhibit 11.18. However, they are included here because they act on the femur.

At the junction between the trunk and lower limb is a space called the **femoral triangle.** The base is formed superiorly by the inguinal ligament, medially by the lateral border of the adductor longus muscle, and laterally by the medial border of the sartorius muscle. The apex is formed by the crossing of the adductor longus by the sartorius muscle (Figure 11.20a). The contents of the femoral triangle, from lateral to medial, are the femoral nerve and its branches, the femoral artery and several of its branches, the femoral vein and its proximal tributaries, and the deep inguinal lymph nodes.

Groin Pull

The five major muscles of the inner thigh function to move the legs medially. This muscle group is important in activities such as sprinting, hurdling, and horseback riding. A rupture or tear of one or more of these muscles can cause a **groin pull.** Groin pulls most often occur during sprinting or twisting, or from kicking a solid, perhaps stationary object. Symptoms of a groin pull may be sudden, or may not surface until the day after the injury, and include sharp pain in the inguinal region, swelling, bruising, or inability to contract the muscles. As with most strain injuries, treatment involves RICE therapy, which stands for *R*est, *I*ce, *C*ompression, and *E*levation. Ice should be applied immediately, and the injured part should be elevated and rested. An elastic bandage should be applied, if possible, to compress the injured tissue. ■

Relating Muscles to Movements

Arrange the muscles in this exhibit according to the following actions on the thigh at the hip joint: (1) flexion, (2) extension, (3) abduction, (4) adduction, (5) medial rotation, and (6) lateral rotation. The same muscle may be mentioned more than once.

▶ **CHECKPOINT**

What is the origin of most muscles that move the femur?

continues

MUSCLE	ORIGIN	INSERTION	ACTION	INNERVATION
Iliopsoas				
Psoas major (SŌ-as; *psoa* = a muscle of the loin)	Transverse processes and bodies of lumbar vertebrae.	With iliacus into lesser trochanter of femur.	Psoas major and iliacus muscles acting together flex thigh at hip joint, rotate thigh laterally, and flex trunk on the hip as in sitting up from the supine position.	Lumbar spinal nerves L2–L3.
Iliacus (il′-ē-A-cus; *iliac-* = ilium)	Iliac fossa and sacrum.	With psoas major into lesser trochanter of femur.		Femoral nerve.
Gluteus maximus (GLOO-tē-us MAK-si-mus; *glute-* = rump or buttock; *-maximus* = largest)	Iliac crest, sacrum, coccyx, and aponeurosis of sacrospinalis.	Iliotibial tract of fascia lata and lateral part of linea aspera (gluteal tuberosity) under greater trochanter of femur.	Extends thigh at hip joint and laterally rotates thigh.	Inferior gluteal nerve.
Gluteus medius (GLOO-tē-us MĒ-de-us; *medi-* = middle)	Ilium.	Greater trochanter of femur.	Abducts thigh at hip joint and medially rotates thigh.	Superior gluteal nerve.
Gluteus minimus (GLOO-tē-us MIN-i-mus; *minim-* = smallest)	Ilium.	Greater trochanter of femur.	Abducts thigh at hip joint and medially rotates thigh.	Superior gluteal nerve.
Tensor fasciae latae (TEN-sor FA-shē-ē LĀ-tē; *tensor* = makes tense; *fasciae* = of the band; *lat-* = wide)	Iliac crest.	Tibia by way of the iliotibial tract.	Flexes and abducts thigh at hip joint.	Superior gluteal nerve.
Piriformis (pir-i-FOR-mis; *piri-* = pear; *form-* = shape)	Anterior sacrum.	Superior border of greater trochanter of femur.	Laterally rotates and abducts thigh at hip joint.	Sacral spinal nerves S1 or S2, mainly S1.
Obturator internus (OB-too-rā′-tor in-TER-nus; *obturator* = obturator foramen; *intern-* = inside)	Inner surface of obturator foramen, pubis, and ischium.	Medial surface of greater trochanter of femur.	Laterally rotates and abducts thigh at hip joint.	Nerve to obturator internus.
Obturator externus (OB-too-rā′-tor ex-TER-nus; *extern-* = outside)	Outer surface of obturator membrane.	Deep depression inferior to greater trochanter (trochanteric fossa) of femur.	Laterally rotates and abducts thigh at hip joint.	Obturator nerve.
Superior gemellus (jem-EL-lus; *superior* = above; *gemell-* = twins)	Ischial spine.	Medial surface of greater trochanter of femur.	Laterally rotates and abducts thigh at hip joint.	Nerve to obturator internus.
Inferior gemellus (jem-EL-lus; *inferior* = below)	Ischial tuberosity.	Medial surface of greater trochanter of femur.	Laterally rotates and abducts thigh at hip joint.	Nerve to quadratus femoris.
Quadratus femoris (kwod-RĀ-tus FEM-or-is; *quad-* = square, four-sided; *femoris* = femur)	Ischial tuberosity.	Elevation superior to mid-portion of intertrochanteric crest (quadrate tubercle) on posterior femur.	Laterally rotates and stabilizes hip joint.	Nerve to quadratus femoris.
Adductor longus (LONG-us; *adductor* = moves part closer to midline; *longus* = long)	Pubic crest and pubic symphysis.	Linea aspera of femur.	Adducts and flexes thigh at hip joint and laterally rotates thigh.	Obturator nerve.
Adductor brevis (BREV-is; *brevis* = short)	Inferior ramus of pubis.	Superior half of linea aspera of femur.	Adducts and flexes thigh at hip joint and medially rotates thigh.	Obturator nerve.
Adductor magnus (MAG-nus; *magnus* = large)	Inferior ramus of pubis and ischium to ischial tuberosity.	Linea aspera of femur.	Adducts thigh at hip joint and laterally rotates thigh; anterior part flexes thigh at hip joint, and posterior part extends thigh at hip joint.	Obturator and sciatic nerves.
Pectineus (pek-TIN-ē-us; *pectin-* = a comb)	Superior ramus of pubis.	Pectineal line of femur, between lesser trochanter and linea aspera.	Flexes and adducts thigh at hip joint.	Femoral nerve.

Figure 11.20 Muscles that move the femur (thigh bone). (See Tortora, *A Photographic Atlas of the Human Body, Second Edition,* Figures 5.16 and 5.17.)

Most muscles that move the femur originate on the pelvic (hip) girdle and insert on the femur.

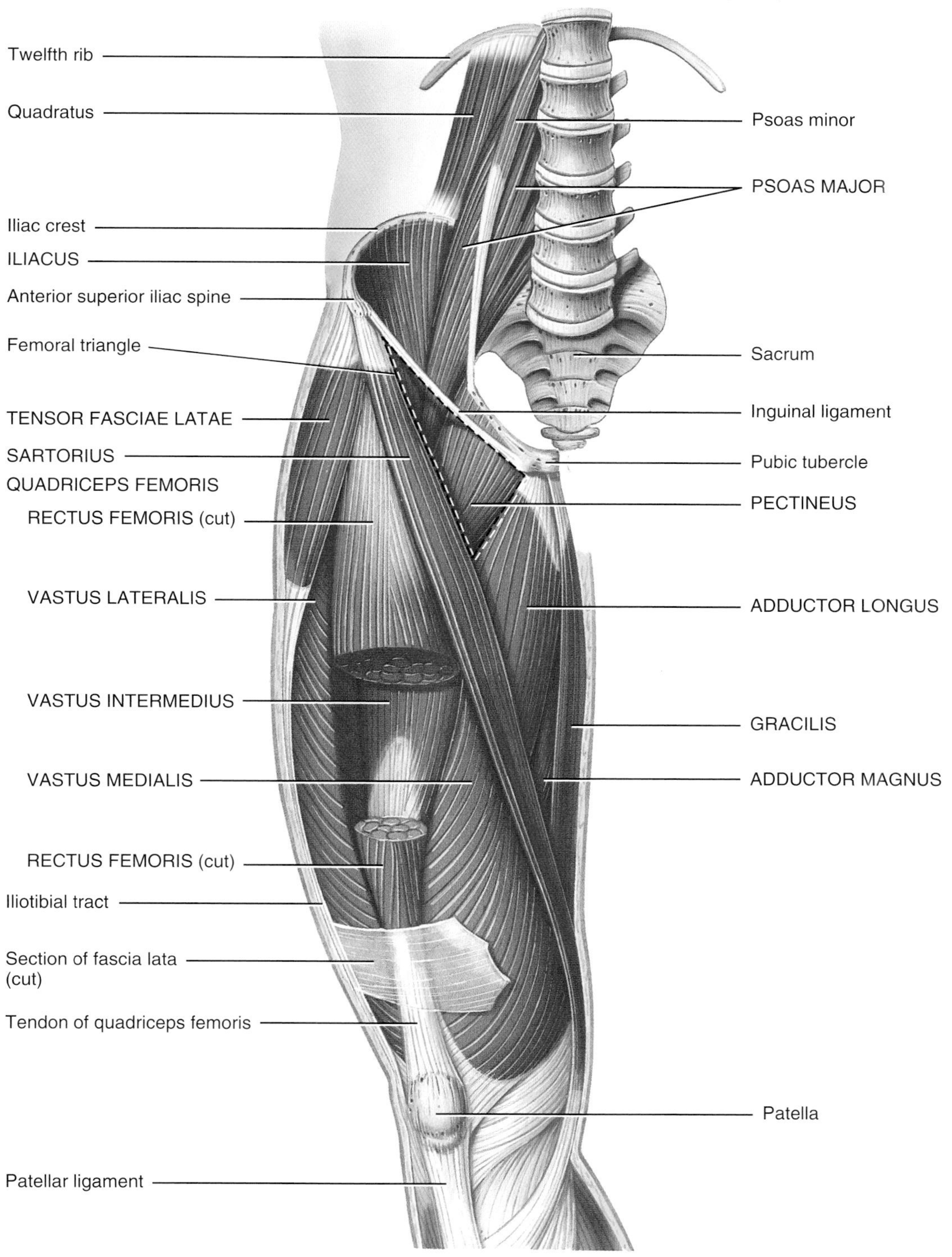

(a) Anterior superficial view (the femoral triangle is indicated by a dashed line)

continues

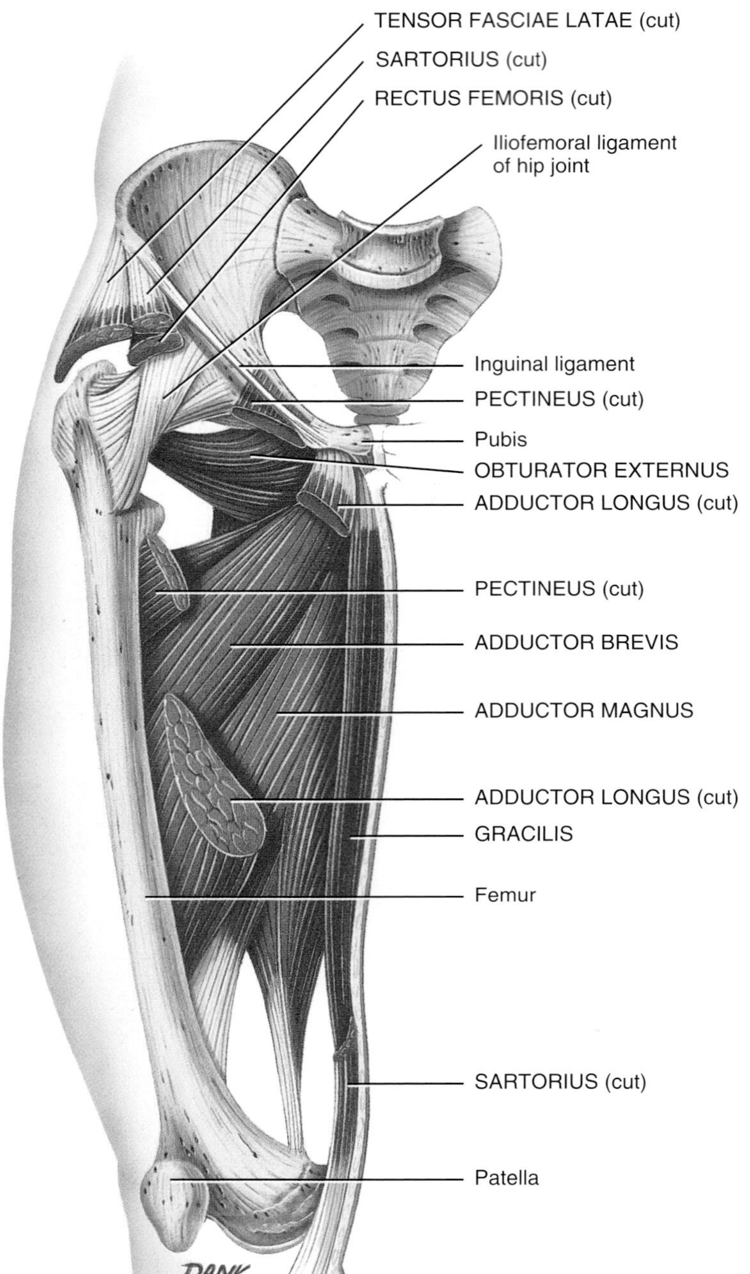

(b) Anterior deep view (femur rotated laterally)

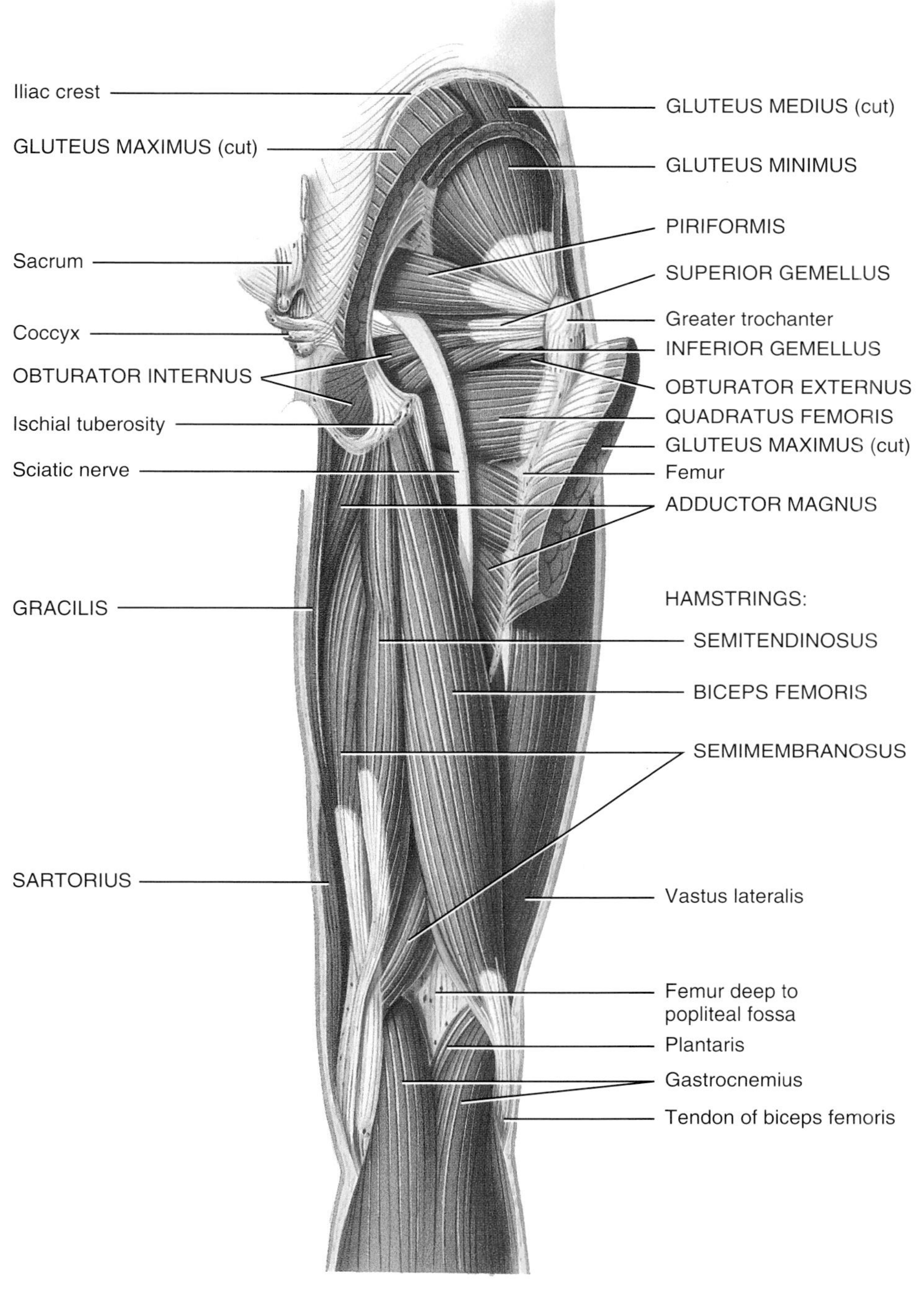

(c) Posterior superficial view

What are the principal differences between the muscles of the upper and lower limbs?

EXHIBIT 11.18 MUSCLES THAT ACT ON THE FEMUR (THIGH BONE) AND TIBIA AND FIBULA (LEG BONES) (FIGURES 11.20 AND 11.21)

▶ **OBJECTIVE**

Describe the origin, insertion, action, and innervation of the muscles that act on the femur, tibia, and fibula.

Deep fascia separate the muscles that act on the femur (thigh bone) and tibia and fibula (leg bones) into medial, anterior, and posterior compartments. The muscles of the **medial (adductor) compartment of the thigh** adduct the femur at the hip joint. (See the adductor magnus, adductor longus, adductor brevis, and pectineus, which are components of the medial compartment, in Exhibit 11.17.) The **gracilis,** the other muscle in the medial compartment, not only adducts the thigh, but also flexes the leg at the knee joint. For this reason, it is discussed here. The gracilis is a long, straplike muscle on the medial aspect of the thigh and knee.

The muscles of the **anterior (extensor) compartment of the thigh** extend the leg (and flex the thigh). This compartment contains the quadriceps femoris and sartorius muscles. The **quadriceps femoris** muscle is the largest muscle in the body, covering most of the anterior surface and sides of the thigh. The muscle is actually a composite muscle, usually described as four separate muscles: (1) **rectus femoris,** on the anterior aspect of the thigh; (2) **vastus lateralis,** on the lateral aspect of the thigh; (3) **vastus medialis,** on the medial aspect of the thigh; and (4) **vastus intermedius,** located deep to the rectus femoris between the vastus lateralis and vastus medialis. The common tendon for the four muscles is known as the **quadriceps tendon,** which inserts into the patella. The tendon continues below the patella as the **patellar ligament,** which attaches to the tibial tuberosity. The quadriceps femoris muscle is the great extensor muscle of the leg. The **sartorius** is a long, narrow muscle that forms a band across the thigh from the ilium of the hip bone to the medial side of the tibia. The various movements it produces (flexion of the leg at the knee joint and flexion, abduction, and lateral rotation at the hip joint) help effect the cross-legged sitting position in which the heel of one limb is placed on the knee of the opposite limb. It is known as the tailor's muscle because tailors often assume this cross-legged sitting position. (Because the major action of the sartorius muscle is to move the thigh rather than the leg, it could have been included in Exhibit 11.17.)

The muscles of the **posterior (flexor) compartment of the thigh** flex the leg (and extend the thigh). This compartment is composed of three muscles collectively called the **hamstrings:** (1) **biceps femoris**, (2) **semitendinosus,** and (3) **semimembranosus.** The hamstrings are so named because their tendons are long and stringlike in the popliteal area. Because the hamstrings span two joints (hip and knee), they are both extensors of the thigh and flexors of the leg. The **popliteal fossa** is a diamond-shaped space on the posterior aspect of the knee bordered laterally by the tendons of the biceps femoris muscle and medially by the tendons of the semitendinosus and semimembranosus muscles.

Pulled Hamstrings

A strain or partial tear of the proximal hamstring muscles is referred to as **pulled hamstrings** or **hamstring strains.** Like pulled groins (see Exhibit 11.17), they are common sports injuries in individuals who run very hard and/or are required to perform quick starts and stops. Sometimes the violent muscular exertion required to perform a feat tears away a part of the tendinous origins of the hamstrings, especially the biceps femoris, from the ischial tuberosity. This is usually accompanied by a contusion (bruising), tearing of some of the muscle fibers, and rupture of blood vessels, producing a hematoma (collection of blood) and sharp pain. Adequate training with good balance between the quadriceps femoris and hamstrings and stretching exercises before running or competing are important in preventing this injury. ■

Relating Muscles to Movements

Arrange the muscles in this exhibit according to the following actions on the thigh at the hip joint: (1) abduction, (2) adduction, (3) lateral rotation, (4) flexion, and (5) extension; and according to the following actions on the leg at the knee joint: (1) flexion and (2) extension. The same muscle may be mentioned more than once.

▶ **CHECKPOINT**

Which muscles are part of the medial, anterior, and posterior compartments of the thigh?

MUSCLE	ORIGIN	INSERTION	ACTION	INNERVATION
Medial (Adductor) Compartment of the Thigh				
Adductor magnus (MAG-nus)	See Exhibit 11.17			
Adductor longus (LONG-us)	See Exhibit 11.17			
Adductor brevis (BREV-is)	See Exhibit 11.17			
Pectineus (pek-TIN-ē-us)	See Exhibit 11.17			
Gracilis (gra-SIL-is; *gracilis* = slender)	Body and inferior ramus of pubis.	Medial surface of body of tibia.	Adducts thigh at hip joint, medially rotates thigh, and flexes leg at knee joint.	Obturator nerve.
Anterior (Extensor) Compartment of the Thigh				
Quadriceps femoris (KWOD-ri-ceps FEM-or-is; *quadriceps* = four heads [of origin]; *femoris* = femur)				
Rectus femoris (REK-tus FEM-or-is; *rectus* = fascicles parallel to midline)	Anterior inferior iliac spine.	Patella via quadriceps tendon and then tibial tuberosity via patellar ligament.	All four heads extend leg at knee joint; rectus femoris muscle acting alone also flexes thigh at hip joint.	Femoral nerve.
Vastus lateralis (VAS-tus lat′-e-RĀ-lis; *vast* = huge; *lateralis* = lateral)	Greater trochanter and linea aspera of femur.	Patella via quadriceps tendon and then tibial tuberosity via patellar ligament.	All four heads extend leg at knee joint; rectus femoris muscle acting alone also flexes thigh at hip joint.	Femoral nerve.
Vastus medialis (VAS-tus mē-dē-Ā-lis; *medialis* = medial)	Linea aspera of femur.	Patella via quadriceps tendon and then tibial tuberosity via patellar ligament.	All four heads extend leg at knee joint; rectus femoris muscle acting alone also flexes thigh at hip joint.	Femoral nerve.
Vastus intermedius (VAS-tus in′-ter-MĒ-dē-us; *intermedius* = middle)	Anterior and lateral surfaces of body of femur.	Patella via quadriceps tendon and then tibial tuberosity via patellar ligament.	All four heads extend leg at knee joint; rectus femoris muscle acting alone also flexes thigh at hip joint.	Femoral nerve.
Sartorius (sar-TOR-ē-us; *sartor* = tailor; longest muscle in body)	Anterior superior iliac spine.	Medial surface of body of tibia.	Flexes leg at knee joint; flexes, abducts, and laterally rotates thigh at hip joint.	Femoral nerve.
Posterior (Flexor) Compartment of the Thigh				
Hamstrings A collective designation for three separate muscles.				
Biceps femoris (BĪ-ceps FEM-or-is; *biceps* = two heads of origin)	Long head arises from ischial tuberosity; short head arises from linea aspera of femur.	Head of fibula and lateral condyle of tibia.	Flexes leg at knee joint and extends thigh at hip joint.	Tibial and common peroneal nerves from the sciatic nerve.
Semitendinosus (sem′-ē-ten-di-NŌ-sus; *semi-* = half; *tendo* = tendon)	Ischial tuberosity.	Proximal part of medial surface of shaft of tibia.	Flexes leg at knee joint and extends thigh at hip joint.	Tibial nerve from the sciatic nerve.
Semimembranosus (sem′-ē-mem-bra-NŌ-sus; *membran-* = membrane)	Ischial tuberosity.	Medial condyle of tibia.	Flexes leg at knee joint and extends thigh at hip joint.	Tibial nerve from the sciatic nerve.

continues

Figure 11.21 Muscles that act on the femur (thigh bone) and tibia and fibula (leg bones). (See Tortora, *A Photographic Atlas of the Human Body, Second Edition,* Figures 5.16 and 5.17.)

Muscles that act on the leg originate in the hip and thigh and are separated into compartments by deep fascia.

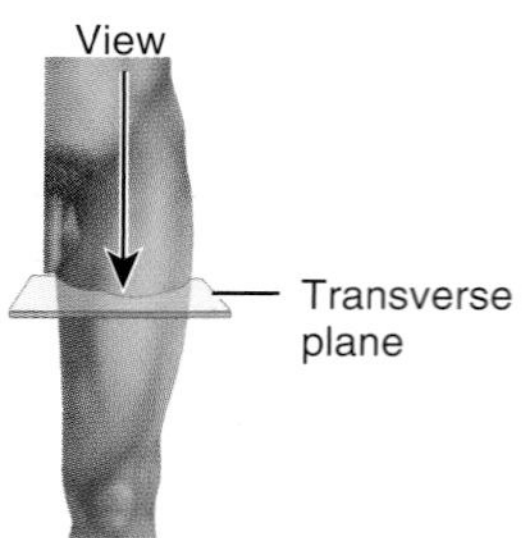

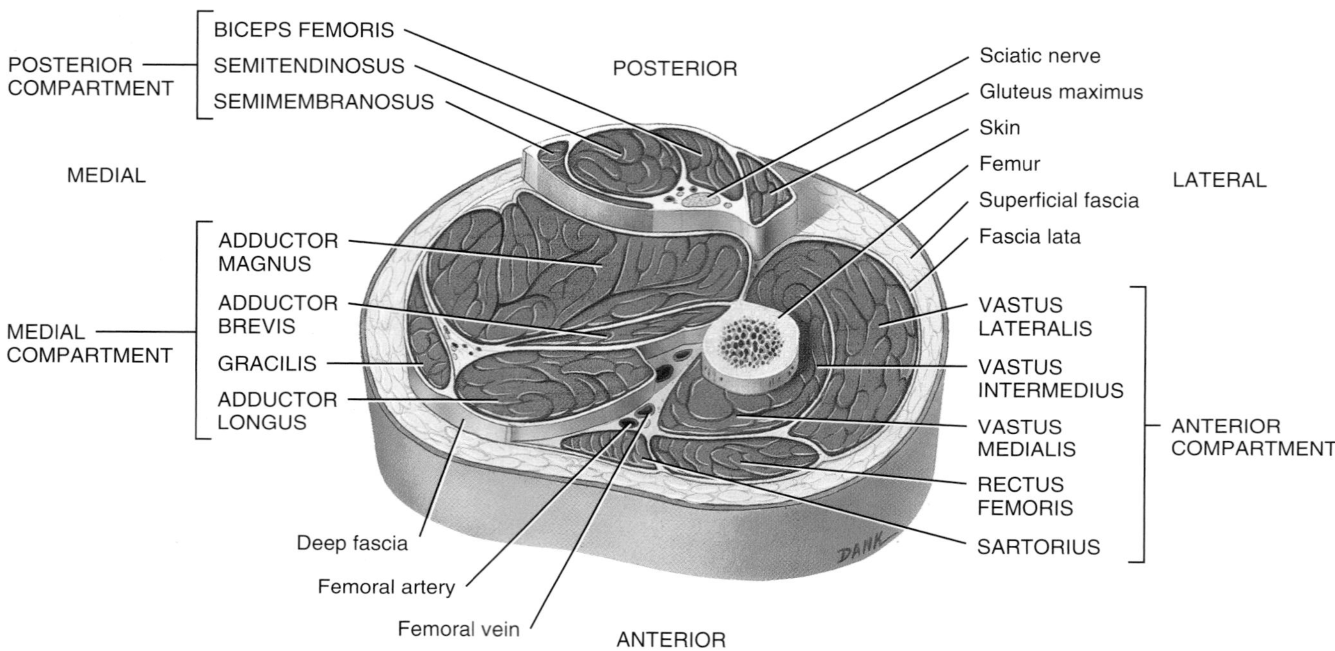

Superior view of transverse section of thigh

Which muscles constitute the quadriceps femoris and hamstring muscles?

▶ **OBJECTIVE**

Describe the origin, insertion, action, and innervation of the muscles that move the foot and toes.

Muscles that move the foot and toes are located in the leg. The muscles of the leg, like those of the thigh, are divided by deep fascia into three compartments: anterior, lateral, and posterior. The **anterior compartment of the leg** consists of muscles that dorsiflex the foot. In a situation analogous to the wrist, the tendons of the muscles of the anterior compartment are held firmly to the ankle by thickenings of deep fascia called the **superior extensor retinaculum** (*transverse ligament of the ankle*) and **inferior extensor retinaculum** (*cruciate ligament of the ankle*).

Within the anterior compartment, the **tibialis anterior** is a long, thick muscle against the lateral surface of the tibia, where it is easy to palpate (feel). The **extensor hallucis longus** is a thin muscle between and partly deep to the tibialis anterior and **extensor digitorum longus** muscles. This featherlike muscle is lateral to the tibialis anterior muscle, where it can also be palpated easily. The **fibularis (peroneus) tertius** muscle is part of the extensor digitorum longus, with which it shares a common origin.

The **lateral (fibular) compartment of the leg** contains two muscles that plantar flex and evert the foot: the **fibularis (peroneus) longus** and **fibularis (peroneus) brevis.**

The **posterior compartment of the leg** consists of muscles in superficial and deep groups. The superficial muscles share a common tendon of insertion, the **calcaneal (Achilles) tendon,** the strongest tendon of the body. It inserts into the calcaneal bone of the ankle. The superficial and most of the deep muscles plantar flex the foot at the ankle joint. The superficial muscles of the posterior compartment are the gastrocnemius, soleus, and plantaris—the so-called calf muscles. The large size of these muscles is directly related to the characteristic upright stance of humans. The **gastrocnemius** is the most superficial muscle and forms the prominence of the calf. The **soleus,** which lies deep to the gastrocnemius, is broad and flat. It derives its name from its resemblance to a flat fish (sole). The **plantaris** is a small muscle that may be absent; conversely, sometimes there are two of them in each leg. It runs obliquely between the gastrocnemius and soleus muscles.

The deep muscles of the posterior compartment are the popliteus, tibialis posterior, flexor digitorum longus, and flexor hallucis longus. The **popliteus** is a triangular muscle that forms the floor of the popliteal fossa. The **tibialis posterior** is the deepest muscle in the posterior compartment. It lies between the flexor digitorum longus and flexor hallucis longus muscles. The **flexor digitorum longus** is smaller than the **flexor hallucis longus,** even though the former flexes four toes, and the latter flexes only the great toe at the interphalangeal joint.

Shin Splint Syndrome

Shin splint syndrome, or simply **shin splints,** refers to pain or soreness along the tibia, specifically the medial, distal two-thirds. It may be caused by tendinitis of the anterior compartment muscles, especially the tibialis anterior muscle, inflammation of the periosteum (periostitis) around the tibia, or stress fractures of the tibia. The tendinitis usually occurs when poorly conditioned runners run on hard or banked surfaces with poorly supportive running shoes. The condition may also occur with vigorous activity of the legs following a period of relative inactivity or running in cold weather without proper warmup. The muscles in the anterior compartment (mainly the tibialis anterior) can be strengthened to balance the stronger posterior compartment muscles. ■

Relating Muscles to Movements

Arrange the muscles in this exhibit according to the following actions on the foot at the ankle joint: (1) dorsiflexion and (2) plantar flexion; according to the following actions on the foot at the intertarsal joints: (1) inversion and (2) eversion; and according to the following actions on the toes at the metatarsophalangeal and interphalangeal joints: (1) flexion and (2) extension. The same muscle may be mentioned more than once.

▶ **CHECKPOINT**

What are the superior extensor retinaculum and inferior extensor retinaculum?

continues

MUSCLE	ORIGIN	INSERTION	ACTION	INNERVATION
Anterior Compartment of the Leg				
Tibialis anterior (tib′-ē-Ā-lis = tibia; *anterior* = front)	Lateral condyle and body of tibia and interosseous membrane (sheet of fibrous tissue that holds shafts of tibia and fibula together).	First metatarsal and first (medial) cuneiform.	Dorsiflexes foot at ankle joint and inverts foot at intertarsal joints.	Deep fibular (peroneal) nerve.
Extensor hallucis longus (HAL-ū-sis LON-gus; extensor = increases angle at joint; *halluc-* = hallux or great toe; *longus* = long)	Anterior surface of fibula and interosseous membrane.	Distal phalanx of great toe.	Dorsiflexes foot at ankle joint and extends proximal phalanx of great toe at metatarsophalangeal joint.	Deep fibular (peroneal) nerve.
Extensor digitorum longus (di′-ji-TOR-um LON-gus)	Lateral condyle of tibia, anterior surface of fibula, and interosseous membrane.	Middle and distal phalanges of toes 2–5.*	Dorsiflexes foot at ankle joint and extends distal and middle phalanges of each toe at interphalangeal joints and proximal phalanx of each toe at metatarsophalangeal joint.	Deep fibular (peroneal) nerve.
Fibularis (Peroneus) tertius (fib-ū-LĀ-ris TER-shus; *peron-* = fibula; *tertius* = third)	Distal third of fibula and interosseous membrane.	Base of fifth metatarsal.	Dorsiflexes foot at ankle joint and everts foot at intertarsal joints.	Deep fibular (peroneal) nerve.
Lateral (Fibular) Compartment of the Leg				
Fibularis (Peroneus) longus (LON-gus)	Head and body of fibula and lateral condyle of tibia.	First metatarsal and first cuneiform.	Plantar flexes foot at ankle joint and everts foot at intertarsal joints.	Superficial fibular (peroneal) nerve.
Fibularis (Peroneus) brevis (BREV-is; *brevis* = short)	Body of fibula.	Base of fifth metatarsal.	Plantar flexes foot at ankle joint and everts foot at intertarsal joints.	Superficial fibular (peroneal) nerve.
Superficial Posterior Compartment of the Leg				
Gastrocnemius (gas′-trok-NĒ-mē-us; *gastro-* = belly; *cnem-* = leg)	Lateral and medial condyles of femur and capsule of knee.	Calcaneus by way of calcaneal (Achilles) tendon.	Plantar flexes foot at ankle joint and flexes leg at knee joint.	Tibial nerve.
Soleus (SŌ-lē-us; *sole* = a type of flat fish)	Head of fibula and medial border of tibia.	Calcaneus by way of calcaneal (Achilles) tendon.	Plantar flexes foot at ankle joint.	Tibial nerve.
Plantaris (plan-TĀR-is; *plantar-* = sole of foot)	Femur superior to lateral condyle.	Calcaneus by way of calcaneal (Achilles) tendon.	Plantar flexes foot at ankle joint and flexes leg at knee joint.	Tibial nerve.
Deep Posterior Compartment of the Leg				
Popliteus (pop-LIT-ē-us; *poplit-* = the back of the knee)	Lateral condyle of femur.	Proximal tibia.	Flexes leg at knee joint and medially rotates tibia to unlock the extended knee.	Tibial nerve.
Tibialis posterior (tib′-ē-Ā-lis; *posterior* = back)	Tibia, fibula, and interosseous membrane.	Second, third, and fourth metatarsals; navicular; all three cuneiforms; and cuboid.	Plantar flexes foot at ankle joint and inverts foot at intertarsal joints.	Tibial nerve.
Flexor digitorum longus (di′-ji-TOR-um LON-gus; *digit* = finger or toe)	Posterior surface of tibia.	Distal phalanges of toes 2–5.	Plantar flexes foot at ankle joint; flexes distal and middle phalanges of each toe at interphalangeal joints and proximal phalanx of each toe at metatarsophalangeal joint.	Tibial nerve.
Flexor hallucis longus (HAL-ū-sis; *flexor* = decreases angle at joint)	Inferior two-thirds of fibula.	Distal phalanx of great toe.	Plantar flexes foot at ankle joint; flexes distal phalanx of great toe at interphalangeal joint and proximal phalanx of great toe at metatarsophalangeal joint.	Tibial nerve.

*Reminder: The great toe or hallux is the first toe and has two phalanges: proximal and distal. The remaining toes are numbered II–V (2–5), and each has three phalanges: proximal, middle, and distal.

Figure 11.22 **Muscles that move the foot and toes.** (See Tortora, *A Photographic Atlas of the Human Body, Second Edition,* Figures 5.18 and 5.19.)

The superficial muscles of the posterior compartment share a common tendon of insertion, the calcaneal (Achilles) tendon, that inserts into the calcaneal bone of the ankle.

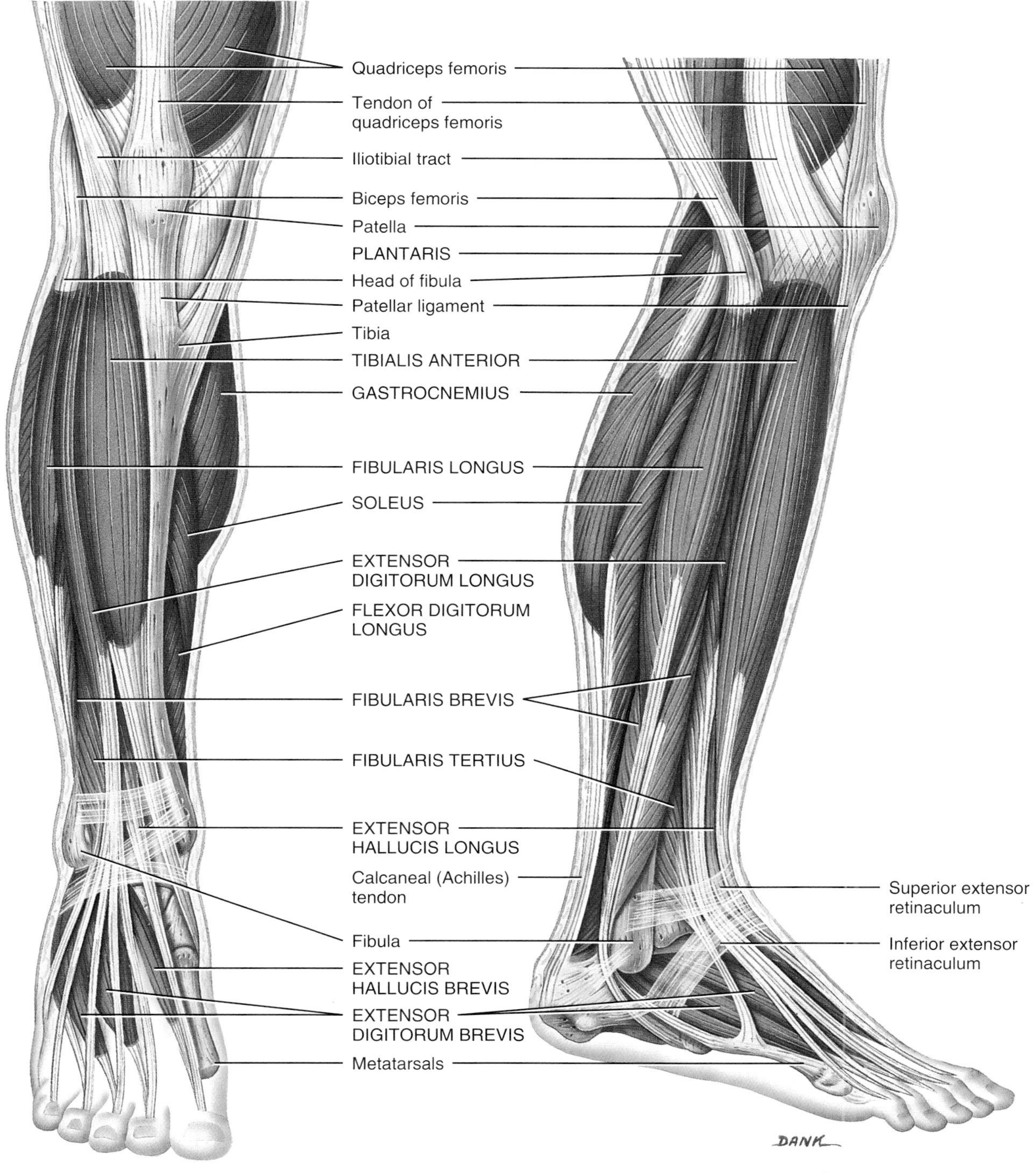

(a) Anterior superficial view

(b) Right lateral superficial view

continues

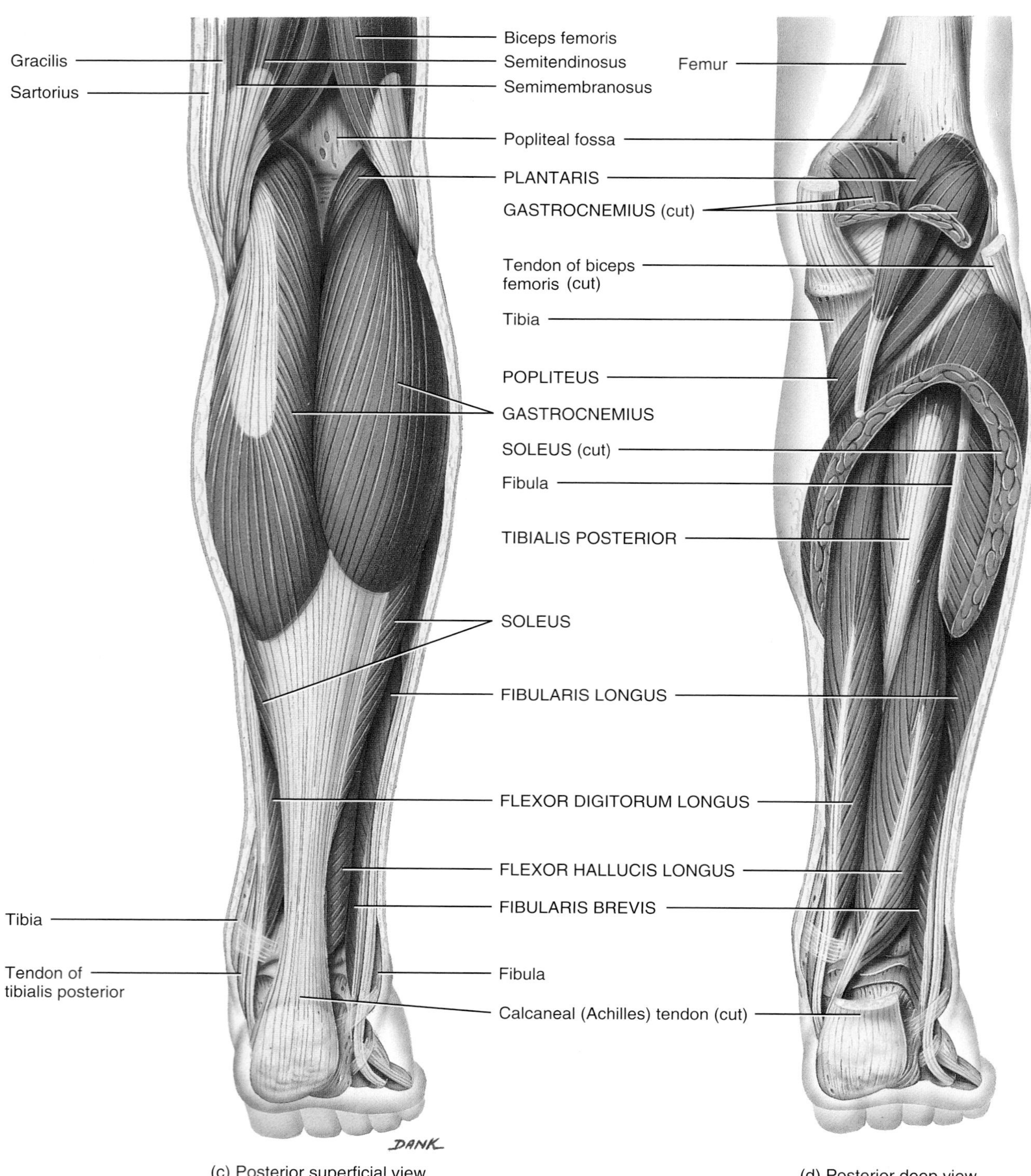

(c) Posterior superficial view

(d) Posterior deep view

What structures firmly hold the tendons of the anterior compartment muscles to the ankle?

INTRINSIC MUSCLES OF THE FOOT (FIGURE 11.23)

▶ **OBJECTIVE**

Describe the origin, insertion, action, and innervation of the intrinsic muscles of the foot.

The muscles in this exhibit are termed **intrinsic muscles of the foot** because they originate and insert *within* the foot. The muscles of the hand are specialized for precise and intricate movements, but those of the foot are limited to support and locomotion. The deep fascia of the foot forms the **plantar aponeurosis (fascia)** that extends from the calcaneus bone to the phalanges of the toes. The aponeurosis supports the longitudinal arch of the foot and encloses the flexor tendons of the foot.

The intrinsic muscles of the foot are divided into two groups: **dorsal** and **plantar.** There is only one dorsal muscle, the **extensor digitorum brevis,** a four-part muscle deep to the tendons of the extensor digitorum longus muscle, which extends toes 2–5 at the metatarsophalangeal joints.

The plantar muscles are arranged in four layers. The most superficial layer is called the first layer. Three muscles are in the first layer. The **abductor hallucis,** which lies along the medial border of the sole, is comparable to the abductor pollicis brevis in the hand, and abducts the great toe at the metatarsophalangeal joint. The **flexor digitorum brevis,** which lies in the middle of the sole, flexes toes 2–5 at the interphalangeal and metatarsophalangeal joints. The **abductor digiti minimi,** which lies along the lateral border of the sole, is comparable to the same muscle in the hand, and abducts the little toe.

The second layer consists of the **quadratus plantae,** a rectangular muscle that arises by two heads and flexes toes 2–5 at the metatarsophalangeal joints, and the **lumbricals,** four small muscles that are similar to the lumbricals in the hands. They flex the proximal phalanges and extend the distal phalanges of toes 2–5.

Three muscles comprise the third layer. The **flexor hallucis brevis,** which lies adjacent to the plantar surface of the metatarsal of the great toe, is comparable to the same muscle in the hand, and flexes the great toe. The **adductor hallucis,** which has an oblique and transverse head like the adductor pollicis in the hand, adducts the great toe. The **flexor digiti minimi brevis,** which lies superficial to the metatarsal of the little toe, is comparable to the same muscle in the hand, and flexes the little toe.

The fourth layer is the deepest and consists of two muscle groups. The **dorsal interossei** are four muscles that abduct toes 2–4, flex the proximal phalanges, and extend the distal phalanges. The three **plantar interossei** abduct toes 3–5, flex the proximal phalanges, and extend the distal phalanges. The interossei of the feet are similar to those of the hand. However, their actions are relative to the midline of the second digit rather than the third digit as in the hand.

Plantar Fasciitis

Plantar fasciitis (fas-ē-Ī-tis) or **painful heel syndrome** is an inflammatory reaction due to chronic irritation of the plantar aponeurosis (fascia) at its origin on the calcaneus (heel bone). The aponeurosis becomes less elastic with age. This condition is also related to weight-bearing activities (walking, jogging, lifting heavy objects), improperly constructed or fitting shoes, excess weight (puts pressure on the feet), and poor biomechanics (flat feet, high arches, and abnormalities in gait may cause uneven distribution of weight on the feet). Plantar fasciitis is the most common cause of heel pain in runners and arises in response to the repeated impact of running. Treatments include ice, deep heat, stretching exercises, weight loss, prosthetics (such as shoe inserts or heel lifts), steroid injections, and surgery. ■

Relating Muscles to Movements

Arrange the muscles in this exhibit according to the following actions on the great toe at the metatarsophalangeal joint: (1) flexion, (2) extension, (3) abduction, and (4) adduction; and according to the following actions on toes 2–5 at the metatarsophalangeal and interphalangeal joints: (1) flexion, (2) extension, (3) abduction, and (4) adduction. The same muscle may be mentioned more than once.

▶ **CHECKPOINT**

How do the intrinsic muscles of the hand and foot differ in function?

continues

MUSCLE	ORIGIN	INSERTION	ACTION	INNERVATION
Dorsal				
Extensor digitorum brevis (*extensor* = increases angle at joint; *digit* = finger or toe; *brevis* = short) (see Figure 11.22a, b)	Calcaneus and inferior extensor retinaculum.	Tendons of extensor digitorum longus on toes 2–4 and proximal phalanx of great toe.*	Extensor hallucis brevis extends great toe at metatarsophalangeal joint and extensor digitorum brevis extends toes 2–4 at interphalangeal joints.	Deep fibular (peroneal) nerve.
Plantar				
First Layer (most superficial)				
Abductor hallucis (*abductor* = moves part away from midline; *hallucis* = hallux or great toe)	Calcaneus, plantar aponeurosis, and flexor retinaculum.	Medial side of proximal phalanx of great toe with the tendon of the flexor hallucis brevis.	Abducts and flexes great toe at metatarsophalangeal joint.	Medial plantar nerve.
Flexor digitorum brevis (*flexor* = decreases angle at joint)	Calcaneus and plantar aponeurosis.	Sides of middle phalanx of toes 2–5.	Flexes toes 2–5 at proximal interphalangeal and metatarsophalangeal joints.	Medial plantar nerve.
Abductor digiti minimi (*minimi* = little)	Calcaneus and plantar aponeurosis.	Lateral side of proximal phalanx of little toe with the tendon of the flexor digiti minimi brevis.	Abducts and flexes little toe at metatarsophalangeal joint.	Lateral plantar nerve.
Second Layer				
Quadratus plantae (kwod-RĀ-tus; *quad-* = square, four-sided; *planta* = the sole)	Calcaneus.	Tendon of flexor digitorum longus.	Assists flexor digitorum longus to flex toes 2–5 at interphalangeal and metatarsophalangeal joints.	Lateral plantar nerve.
Lumbricals (LUM-bri-kals; *lumbric-* = earthworm)	Tendons of flexor digitorum longus.	Tendons of extensor digitorum longus on proximal phalanges of toes 2–5.	Extend toes 2–5 at interphalangeal joints and flex toes 2–5 at metatarsophalangeal joints.	Medial and lateral plantar nerves.
Third Layer				
Flexor hallucis brevis	Cuboid and third (lateral) cuneiform.	Medial and lateral sides of proximal phalanx of great toe via a tendon containing a sesamoid bone.	Flexes great toe at metatarsophalangeal joint.	Medial plantar nerve.
Adductor hallucis	Metatarsals 2–4, ligaments of 3–5 metatarsophalangeal joints, and tendon of peroneus longus.	Lateral side of proximal phalanx of great toe.	Adducts and flexes great toe at metatarsophalangeal joint.	Lateral plantar nerve.
Flexor digiti minimi brevis	Metatarsal 5 and tendon of peroneus longus.	Lateral side of proximal phalanx of little toe.	Flexes little toe at metatarsophalangeal joint.	Lateral plantar nerve.
Fourth Layer (deepest)				
Dorsal interossei (in-ter-OS-ē-ī) (not illustrated)	Adjacent side of metatarsals.	Proximal phalanges: both sides of toe 2 and lateral side of toes 3 and 4.	Abduct and flex toes 2–4 at metatarsophalangeal joints and extend toes at interphalangeal joints.	Lateral plantar nerve.
Plantar interossei	Metatarsals 3–5.	Medial side of proximal phalanges of toes 3–5.	Adduct and flex proximal metatarsophalangeal joints and extend toes at interphalangeal joints.	Lateral plantar nerve.

*The tendon that inserts into the proximal phalanx of the great toe, together with its belly, is often described as a separate muscle, the extensor hallucis brevis.

Figure 11.23 **Intrinsic muscles of the foot.**

The intrinsic muscles of the hand are specialized for precise and intricate movements; those of the foot are limited to support and movement.

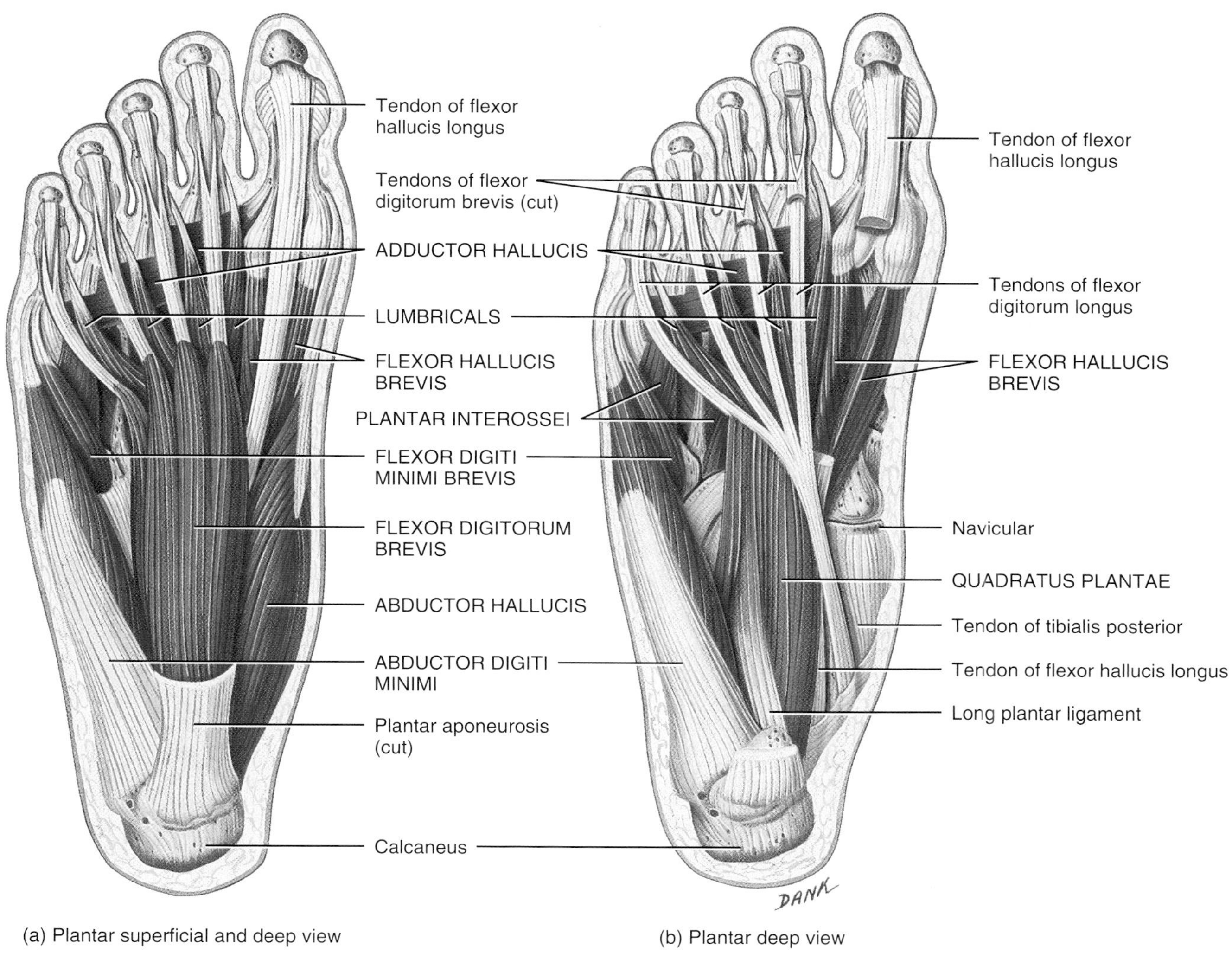

(a) Plantar superficial and deep view

(b) Plantar deep view

What structure supports the longitudinal arch and encloses the flexor tendons of the foot?

FOCUS ON HOMEOSTASIS

The Muscular System

BODY SYSTEM	CONTRIBUTION OF THE MUSCULAR SYSTEM
For all body systems 	The muscular system and muscle tissues produce body movements, stabilize body positions, move substances within the body, and produce heat that helps maintain normal body temperature.
Integumentary system 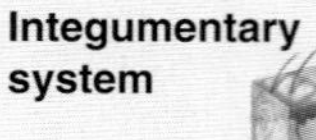	Pull of skeletal muscles on attachments to skin of face causes facial expressions; muscular exercise increases skin blood flow.
Skeletal system 	Skeletal muscle causes movement of body parts by pulling on attachments to bones; skeletal muscle provides stability for bones and joints.
Nervous system 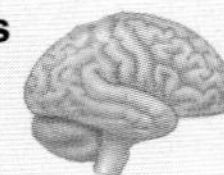	Smooth, cardiac, and skeletal muscles carry out commands for the nervous system; shivering—involuntary contraction of skeletal muscles that is regulated by the brain—generates heat to raise body temperature.
Endocrine system 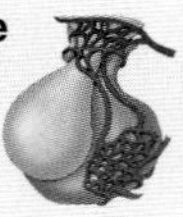	Regular activity of skeletal muscles (exercise) improves action and signaling mechanisms of some hormones, such as insulin; muscles protect some endocrine glands.
Cardiovascular system 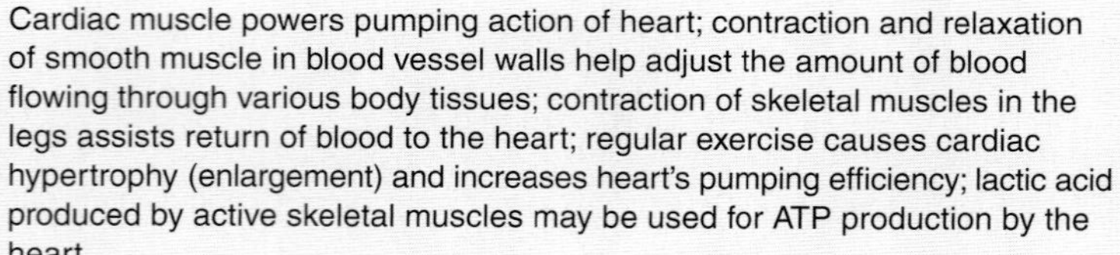	Cardiac muscle powers pumping action of heart; contraction and relaxation of smooth muscle in blood vessel walls help adjust the amount of blood flowing through various body tissues; contraction of skeletal muscles in the legs assists return of blood to the heart; regular exercise causes cardiac hypertrophy (enlargement) and increases heart's pumping efficiency; lactic acid produced by active skeletal muscles may be used for ATP production by the heart.
Lymphatic system and immunity 	Skeletal muscles protect some lymph nodes and lymphatic vessels and promote flow of lymph inside lymphatic vessels; exercise may increase or decrease some immune responses.
Respiratory system 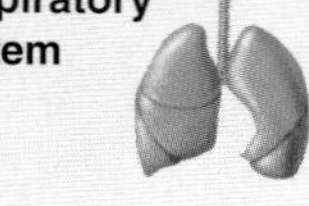	Skeletal muscles involved with breathing cause air to flow into and out of the lungs; smooth muscle fibers adjust size of airways; vibrations in skeletal muscles of larynx control air flowing past vocal cords, regulating voice production; coughing and sneezing, due to skeletal muscle contractions, help clear airways; regular exercise improves efficiency of breathing.
Digestive system 	Skeletal muscles protect and support organs in the abdominal cavity; alternating contraction and relaxation of skeletal muscles power chewing and initiate swallowing; smooth muscle sphincters control volume of organs of the gastrointestinal (GI) tract; smooth muscles in walls of GI tract mix and move its contents through the tract.
Urinary system 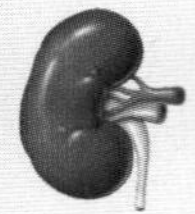	Skeletal and smooth muscle sphincters and smooth muscle in wall of urinary bladder control whether urine is stored in the urinary bladder or voided (urination).
Reproductive systems 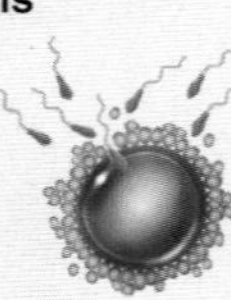	Skeletal and smooth muscle contractions eject semen from male; smooth muscle contractions propel oocyte along uterine tube, help regulate flow of menstrual blood from uterus, and force baby from uterus during childbirth; during intercourse, skeletal muscle contractions are associated with orgasm and pleasurable sensations in both sexes.

DISORDERS: HOMEOSTATIC IMBALANCES

Running Injuries

Many individuals who jog or run sustain some type of running-related injury. Although such injuries may be minor, some can be quite serious. Untreated or inappropriately treated minor injuries may become chronic. Among runners, common sites of injury include the ankle, knee, calcaneal (Achilles) tendon, hip, groin, foot, and back. Of these, the knee often is the most severely injured area.

Running injuries are frequently related to faulty training techniques. This may involve improper or lack of sufficient warm-up routines, running too much, or running too soon after an injury. Or it might involve extended running on hard and/or uneven surfaces. Poorly constructed or worn-out running shoes can also contribute to injury, as can any biomechanical problem (such as a fallen arch) aggravated by running.

Most sports injuries should be treated initially with RICE therapy, which stands for *R*est, *I*ce, *C*ompression, and *E*levation. Immediately apply ice, and rest and elevate the injured part. Then apply an elastic bandage, if possible, to compress the injured tissue. Continue using RICE for 2 to 3 days, and resist the temptation to apply heat, which may worsen the swelling. Follow-up treatment may include alternating moist heat and ice massage to enhance blood flow in the injured area. Sometimes it is helpful to take nonsteroidal anti-inflammatory drugs (NSAIDs) or to have local injections of corticosteroids. During the recovery period, it is important to keep active, using an alternative fitness program that does not worsen the original injury. This activity should be determined in consultation with a physician. Finally, careful exercise is needed to rehabilitate the injured area itself.

Compartment Syndrome

As noted earlier in this chapter, skeletal muscles in the limbs are organized into functional units called *compartments.* In a disorder called **compartment syndrome,** some external or internal pressure constricts the structures within a compartment, resulting in damaged blood vessels and subsequent reduction of the blood supply (ischemia) to the structures within the compartment. Symptoms include pain, burning, pressure, pale skin, and paralysis. Common causes of compartment syndrome include crushing and penetrating injuries, contusion (damage to subcutaneous tissues without the skin being broken), muscle strain (overstretching of a muscle), or an improperly fitted cast. The pressure increase in the compartment can have serious consequences, such as hemorrhage, tissue injury, and edema (buildup of interstitial fluid). Because deep fasciae (connective tissue coverings) that enclose the compartments are very strong, accumulated blood and interstitial fluid cannot escape, and the increased pressure can literally choke off the blood flow and deprive nearby muscles and nerves of oxygen. One treatment option is **fasciotomy** (fash-ē-OT-ō-mē), a surgical procedure in which muscle fascia is cut to relieve the pressure. Without intervention, nerves can suffer damage, and muscles can develop scar tissue that results in permanent shortening of the muscles, a condition called *contracture.* If left untreated, tissues may die and the limb may no longer be able to function. Once the syndrome has reached this stage, amputation may be the only treatment option.

STUDY OUTLINE

HOW SKELETAL MUSCLES PRODUCE MOVEMENTS (p. 326)

1. Skeletal muscles that produce movement do so by pulling on bones.
2. The attachment to the more stationary bone is the origin; the attachment to the more movable bone is the insertion.
3. Bones serve as levers, and joints serve as fulcrums. Two different forces act on the lever: load (resistance) and effort.
4. Levers are categorized into three types—first-class, second-class, and third-class (most common)—according to the positions of the fulcrum, the effort, and the load on the lever.
5. Fascicular arrangements include parallel, fusiform, circular, triangular, and pennate. Fascicular arrangement affects a muscle's power and range of motion.
6. A prime mover produces the desired action; an antagonist produces an opposite action. Synergists assist a prime mover by reducing unnecessary movement. Fixators stabilize the origin of a prime mover so that it can act more efficiently.

HOW SKELETAL MUSCLES ARE NAMED (p. 331)

1. Distinctive features of different skeletal muscles include direction of muscle fascicles; size, shape, action, number of origins (or heads), and location of the muscle; and sites of origin and insertion of the muscle.
2. Most skeletal muscles are named based on combinations of these features.

PRINCIPAL SKELETAL MUSCLES (p. 331)

1. Muscles of facial expression move the skin rather than a joint when they contract, and they permit us to express a wide variety of emotions.
2. The extrinsic muscles that move the eyeballs are among the fastest contracting and most precisely controlled skeletal muscles in the body. They permit us to elevate, depress, abduct, adduct, and medially and laterally rotate the eyeballs.
3. Muscles that move the mandible (lower jaw) are also known as the muscles of mastication because they are involved in chewing.
4. The extrinsic muscles that move the tongue are important in chewing, swallowing, and speech.
5. Muscles of the floor of the anterior neck, called suprahyoid muscles, are located above the hyoid bone. They elevate the hyoid bone, oral cavity, and tongue during swallowing.
6. Muscles that move the head alter its position and help balance the head on the vertebral column.

7. Muscles that act on the abdominal wall help contain and protect the abdominal viscera, move the vertebral column, compress the abdomen, and produce the force required for defecation, urination, vomiting, and childbirth.
8. Muscles used in breathing alter the size of the thoracic cavity so that ventilation can occur and assist in venous return of blood to the heart.
9. Muscles of the pelvic floor support the pelvic viscera, resist the thrust that accompanies increases in intra-abdominal pressure, and function as sphincters at the anorectal junction, urethra, and vagina.
10. Muscles of the perineum assist in urination, erection of the penis and clitoris, ejaculation, female orgasm, and defecation.
11. Muscles that move the pectoral (shoulder) girdle stabilize the scapula so it can function as a stable point of origin for most of the muscles that move the humerus.
12. Muscles that move the humerus (arm bone) originate for the most part on the scapula (scapular muscles); the remaining muscles originate on the axial skeleton (axial muscles).
13. Muscles that move the radius and ulna (forearm bones) are involved in flexion and extension at the elbow joint and are organized into flexor and extensor compartments.
14. Muscles that move the wrist, hand, thumb, and fingers are many and varied; those muscles that act on the digits are called extrinsic muscles.
15. The intrinsic muscles of the hand are important in skilled activities and provide humans with the ability to grasp and manipulate objects precisely.
16. Muscles that move the vertebral column are quite complex because they have multiple origins and insertions and because there is considerable overlap among them.
17. Muscles that move the femur (thigh bone) originate for the most part on the pelvic girdle and insert on the femur; these muscles are larger and more powerful than comparable muscles in the upper limb.
18. Muscles that move the femur (thigh bone) and tibia and fibula (leg bones) are separated into medial (adductor), anterior (extensor), and posterior (flexor) compartments.
19. Muscles that move the foot and toes are divided into anterior, lateral, and posterior compartments.
20. Intrinsic muscles of the foot, unlike those of the hand, are limited to the functions of support and locomotion.

SELF-QUIZ QUESTIONS

Fill in the blanks in the following statements.

1. The muscle that forms the major portion of the cheek is the ____.
2. The three superficial posterior plantar flexors of the leg are the ____, ____, and ____. All three of these muscles insert on the ____ by way of the Achilles tendon.

Indicate whether the following statements are true or false.

3. Longer fibers in a muscle result in a greater range of motion.
4. When flexing the forearm, the biceps brachii acts as the prime mover and the triceps brachii acts as the antagonist.

Choose the one best answer to the following questions.

5. Which of the following muscles does *not* flex the thigh? (a) rectus femoris, (b) gracilis, (c) sartorius, (d) iliacus, (e) tensor fascia latae.
6. The iliotibial tract is composed of the tendon of the gluteus maximus muscle, the deep fascia that encircles the thigh, and the tendon of which of the following muscles? (a) iliacus, (b) gluteus minimus, (c) tensor fascia latae, (d) adductor longus, (e) vastus lateralis.
7. In order for movement to occur, (1) muscles generally need to cross a joint, (2) contraction of the muscle will pull on the origin, (3) muscles that move a body part cannot cover the moving part, (4) muscles need to exert force on tendons that pull on bones, (5) the insertion must act to stabilize the joint. (a) 1, 2, 3, 4, and 5; (b) 1, 2, 3 and 4; (c) 1, 2 and 4; (d) 1, 3, and 4; (e) 3 and 4.
8. Because you did not do well on your recent anatomy and physiology exam, you leave the classroom pouting. Which one of these muscles are you using? (a) mentalis, (b) orbicularis oris, (c) risorius, (d) levator labii superioris, (e) zygomaticus minor.
9. The rectus femoris has fascicles arranged on both sides of a centrally positioned tendon. This pattern of fascicle arrangement is (a) unipennate, (b) fusiform, (c) multipennate, (d) parallel, (e) bipennate.
10. Which of the following muscle names and their naming descriptors are mismatched? (a) adductor brevis: short muscle that moves a bone closer to the midline, (b) rectus abdominis: muscle with fibers parallel to the midline of the abdomen, (c) levator scapula: muscle that raises the scapula, (d) sternohyoid: muscle attached to the sternum and hyoid, (e) serratus anterior: comblike muscle located on the body's anterior surface.
11. Match the following:

____(a) muscle that stabilizes the origin of the prime mover	(1) compartment
____(b) site of muscle attachment to a stationary bone	(2) origin
____(c) muscle that stretches to allow desired motion	(3) insertion
____(d) muscle that contracts to stabilize intermediate joints	(4) belly
____(e) site of muscle attachment to a movable bone	(5) synergist
____(f) group of muscles, along with their blood and nerves, that have a common function	(6) fixator
____(g) contracting muscle that produces the desired motion	(7) prime mover (agonist)
____(h) fleshy part of the muscle	(8) antagonist

12. Match the following:

____(a) compression of median nerve resulting in pain and numbness and tingling in the fingers
____(b) tendinitis of the anterior compartment muscles of the leg; inflammation of the tibial periosteum
____(c) improperly aligned eyeballs due to lesions in either the oculomotor or abducens nerves
____(d) stretching or tearing of distal attachments of adductor muscles
____(e) rupture of a portion of the inguinal area of the abdominal wall resulting in protrusion of part of the small intestine
____(f) caused by repetitive movement of the arm over the head that results in inflammation of the supraspinatus tendon
____(g) inflammation due to chronic irritation of the plantar aponeurosis at its origin on the calcaneus; most common cause of heel pain in runners
____(h) painful inflammation of tendons, tendon sheaths, and synovial membranes of joints
____(i) paralysis of facial muscles as a result of damage to the facial nerve
____(j) common in individuals who perform quick starts and stops; tearing away of part of the tendinous origins from the ischial tuberosity
____(k) permanent shortening of a muscle due to nerve damage and scar tissue development
____(l) may occur as a result of injury to levator ani muscle
____(m) external or internal pressure constricts structures in a compartment, causing a reduction of blood supply to the structures

(1) tenosynovitis
(2) Bell's palsy
(3) inguinal hernia
(4) urinary stress incontinence
(5) compartment syndrome
(6) groin strain
(7) pulled hamstrings
(8) strabismus
(9) shin splints
(10) plantar fasciitis
(11) impingement syndrome
(12) contracture
(13) carpal tunnel syndrome

13. Match the following:

____(a) rectus femoris, vastus lateralis, vastus medialis, vastus intermedius
____(b) biceps femoris, semitendinosus, semimembranosus
____(c) erector spinae; includes iliocostalis, longissimus, and spinalis groups
____(d) thenar, hypothenar, intermediate
____(e) biceps brachii, brachialis, coracobrachialis
____(f) latissimus dorsi
____(g) subscapularis, supraspinatus, infraspinatus, teres minor
____(h) diaphragm, external intercostals, internal intercostals
____(i) trapezius, levator scapulae, rhomboid major, rhomboid minor

(1) breathing muscles
(2) constitute flexor compartment of the arm
(3) hamstrings
(4) intrinsic muscle groups of the hand
(5) muscles that strengthen and stabilize the shoulder joint; the rotator cuff
(6) quadriceps femoris muscle
(7) largest muscle mass of the back
(8) posterior thoracic muscles
(9) swimmer's muscle

14. Match the following (some answers may be used more than once):

____(a) trapezius
____(b) orbicularis oculi
____(c) levator ani
____(d) rectus abdominis
____(e) triceps brachii
____(f) gastrocnemius
____(g) temporalis
____(h) external anal sphincter
____(i) external oblique
____(j) iliocostalis thoracis
____(k) digastric
____(l) styloglossus
____(m) masseter
____(n) adductor longus
____(o) zygomaticus major
____(p) latissimus dorsi
____(q) flexor carpi radialis
____(r) pronator teres
____(s) sternocleidomastoid
____(t) quadriceps femoris
____(u) deltoid
____(v) tibialis anterior
____(w) sartorius
____(x) gluteus maximus
____(y) superior rectus
____(z) trapezius

(1) muscle of facial expression
(2) muscle of mastication
(3) muscle that moves the eyeballs
(4) extrinsic muscle that moves the tongue
(5) suprahyoid muscle
(6) muscle of the perineum
(7) muscle that moves the head
(8) abdominal wall muscle
(9) pelvic floor muscle
(10) pectoral girdle muscle
(11) muscle that moves the humerus
(12) muscle that moves the radius and ulna
(13) muscle that moves the wrist, hand, and digits
(14) muscle that moves the vertebral column
(15) muscle that moves the femur
(16) muscle that acts on the femur, tibia, and fibula
(17) muscle that moves the foot and toes

15. Match the following (some answers may be used more than once):

____(a) most common lever in the body
____(b) lever formed by the head resting on the vertebral column
____(c) always produces a mechanical advantage
____(d) EFL
____(e) FLE
____(f) FEL
____(g) adduction of the thigh

(1) first-class lever
(2) second-class lever
(3) third-class lever

CRITICAL THINKING QUESTIONS

1. During a facelift, the cosmetic surgeon accidentally severs the facial nerve on the right side of the face. What are some of the effects this would have on the patient and what muscles are involved?
2. While taking the bus to the supermarket, eleven-year-old Desmond informs his mother that he has to "go to the bathroom" (urinate). His mother tells him he must "hold it" until they arrive at the store. What muscles must remain contracted in order for him to prevent urination?
3. Minor-league pitcher José has been throwing a hundred pitches a day in order to perfect his curve ball. Lately he has experienced pain in his pitching arm. The doctor diagnosed a torn rotator cuff. José was confused because he thought cuffs were only found on shirt sleeves, not inside his shoulder. Explain to José what the doctor means and how this injury could affect his arm movement.

ANSWERS TO FIGURE QUESTIONS

11.1 The belly of the muscle that extends the forearm, the triceps brachii, is located posterior to the humerus.
11.2 Second-class levers produce the most force.
11.3 For muscles named after their various characteristics, here are possible correct responses (for others, see Table 11.2): direction of fibers: external oblique; shape: deltoid; action: extensor digitorum; size: gluteus maximus; origin and insertion: sternocleidomastoid; location: tibialis anterior; number of tendons of origin: biceps brachii.
11.4 The corrugator supercilii is involved in frowning; the zygomaticus major muscle contracts when you smile; the mentalis and platysma muscles contribute to pouting; the orbicularis oculi muscle contributes to squinting.
11.5 The inferior oblique muscle moves the eyeball superiorly and laterally because it originates at the anteromedial aspect of the floor of the orbit and inserts on the posterolateral aspect of the eyeball.
11.6 The masseter is the strongest of the chewing muscles (muscles of mastication).
11.7 Functions of the tongue include chewing, detection of taste, swallowing, and speech.
11.8 The suprahyoid and infrahyoid muscles stabilize the hyoid bone to assist in tongue movements.
11.9 The triangles in the neck formed by the sternocleidomastoid muscles are important anatomically and surgically because of the structures that lie within their boundaries.
11.10 The rectus abdominis muscle aids in urination.
11.11 The diaphragm is innervated by the phrenic nerve.
11.12 The borders of the pelvic diaphragm are the pubic symphysis anteriorly, the coccyx posteriorly, and the walls of the pelvis laterally.
11.13 The borders of the perineum are the pubic symphysis anteriorly, the coccyx posteriorly, and the ischial tuberosities laterally.
11.14 The main action of the muscles that move the pectoral girdle is to stabilize the scapula to assist in movements of the humerus.
11.15 The rotator cuff consists of the flat tendons of the subscapularis, supraspinatus, infraspinatus, and teres minor muscles that form a nearly complete circle around the shoulder joint.
11.16 The brachialis is the most powerful forearm flexor; the triceps brachii is the most powerful forearm extensor.
11.17 Flexor tendons of the digits and wrist and the median nerve pass through the flexor retinaculum.
11.18 Muscles of the thenar eminence act on the thumb.
11.19 The splenius muscles arise from the midline and extend laterally and superiorly to their insertions.
11.20 Upper limb muscles exhibit diversity of movement; lower limb muscles function in stability, locomotion, and maintenance of posture. In addition, lower limb muscles usually cross two joints and act equally on both.
11.21 The quadriceps femoris consists of the rectus femoris, vastus lateralis, vastus medialis, and vastus intermedius; the hamstrings consist of the biceps femoris, semitendinosus, and semimembranosus.
11.22 The superior and inferior extensor retinacula firmly hold the tendons of the anterior compartment muscles to the ankle.
11.23 The plantar aponeurosis supports the longitudinal arch and encloses the flexor tendons of the foot.

Chapter 12

Nervous Tissue

Nervous Tissue and Homeostasis

The exitable characteristic of nervous tissue allows for the generation of nerve impulses (action potentials) that provide communication and regulation of most body tissues.

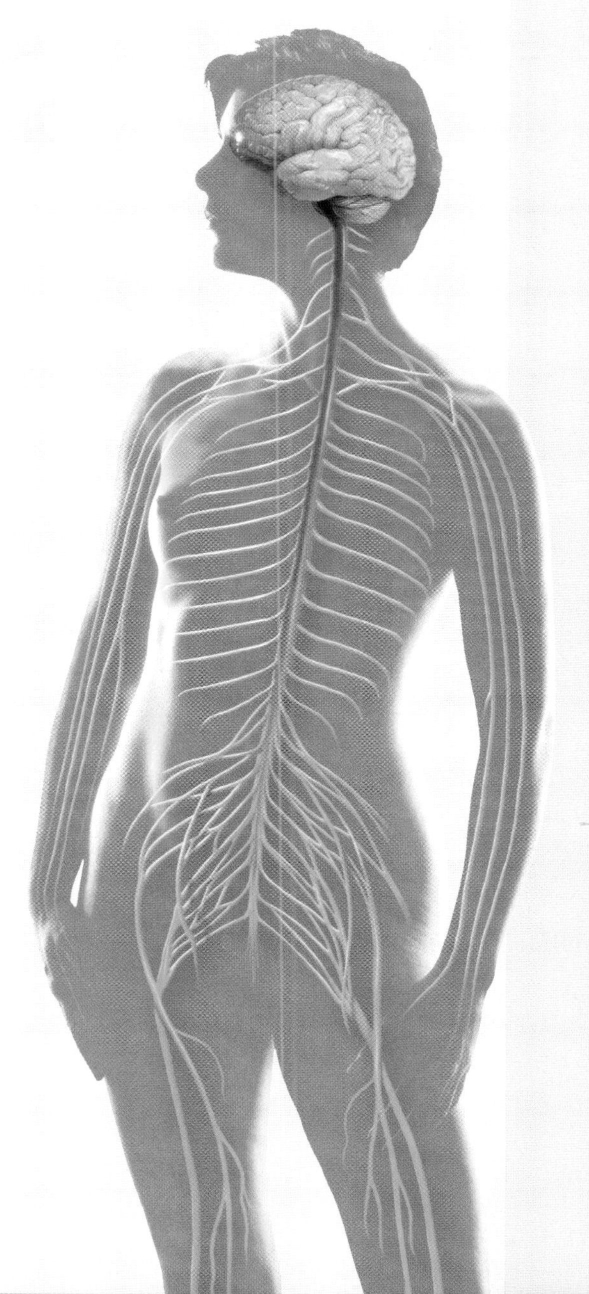

www. wiley.com/college/apcentral

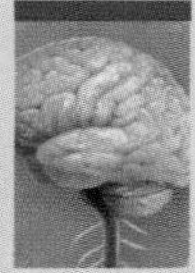

Together, the nervous system and the endocrine system share responsibility for maintaining homeostasis. Their objective is the same—to keep controlled conditions within limits that maintain life—but the two systems achieve that objective very differently. The nervous system regulates body activities by responding rapidly using nerve impulses (action potentials); the endocrine system responds more slowly, though no less effectively, by releasing hormones. The roles of the nervous and endocrine systems in maintaining homeostasis are compared in Chapter 18, on page 617.

Besides helping to maintain homeostasis, the nervous system is also responsible for our perceptions, behaviors, and memories, and initiates all voluntary movements. Because the nervous system is quite complex, we will consider different aspects of its structure and function in several related chapters. In this chapter we focus on the organization of the nervous system and the properties of the cells that make up nervous tissue—neurons (nerve cells) and neuroglia (cells that support the activities of neurons). In chapters that follow, we will examine the structure and functions of the spinal cord and spinal nerves (Chapter 13), and of the brain and cranial nerves (Chapter 14). The autonomic nervous system, the part of the nervous system that operates without voluntary control, will be covered in Chapter 15. Then we will discuss the somatic senses—touch, pressure, warmth, cold, pain, and others—and their sensory and motor pathways to understand how nerve impulses pass into the spinal cord and brain or from the spinal cord and brain to muscles and glands (Chapter 16). Our exploration of the nervous system concludes with a discussion of the special senses: smell, taste, vision, hearing, and equilibrium (Chapter 17).

The branch of medical science that deals with the normal functioning and disorders of the nervous system is **neurology** (noo-ROL-ō-jē; *neuro-* = nerve or nervous system; *-logy* = study of). A **neurologist** is a physician who specializes in the diagnosis and treatment of disorders of the neuromuscular system.

OVERVIEW OF THE NERVOUS SYSTEM

▸ **OBJECTIVES**

List the structures and basic functions of the nervous system.

Describe the organization of the nervous system.

Structures of the Nervous System

With a mass of only 2 kg (4.5 lb), about 3% of total body weight, the **nervous system** is one of the smallest and yet the most complex of the 11 body systems. The nervous system is an intricate, highly organized network of billions of neurons and even more neuroglia. The structures that make up the nervous system include the brain, cranial nerves and their branches, the spinal cord, spinal nerves and their branches, ganglia, enteric plexuses, and sensory receptors (Figure 12.1).

The skull encloses the **brain,** which contains about 100 billion (10^{11}) neurons. Twelve pairs (right and left) of **cranial nerves,** numbered I through XII, emerge from the base of the brain. A **nerve** is a bundle of hundreds to thousands of axons plus associated connective tissue and blood vessels that lies outside the brain and spinal cord. Each nerve follows a defined path and serves a specific region of the body. For example, cranial nerve I carries signals for the sense of smell from the nose to the brain.

The **spinal cord** connects to the brain through the foramen magnum of the skull and is encircled by the bones of the vertebral column. It contains about 100 million neurons. Thirty-one pairs of **spinal nerves** emerge from the spinal cord, each serving a specific region on the right or left side of the body. **Ganglia** (GANG-lē-a = swelling or knot) are small masses of nervous tissue, consisting primarily of neuron cell bodies, that are located outside the brain and spinal cord. Ganglia are closely associated with cranial and spinal nerves. In the walls of organs of the gastrointestinal tract, extensive networks of neurons, called **enteric plexuses,** help regulate the digestive system. The term **sensory receptor** is used to refer to the dendrites of sensory neurons (described shortly) as well as separate, specialized cells that monitor changes in the internal or external environment, such as photoreceptors in the retina of the eye (see Chapter 17).

Functions of the Nervous System

The nervous system carries out a complex array of tasks. It allows us to sense various smells, produce speech, and remember past events; in addition, it provides signals that control body movements, and regulates the operation of internal organs. These diverse activities can be grouped into three basic functions: sensory, integrative, and motor.

- ***Sensory function.*** Sensory receptors *detect* internal stimuli, such as an increase in blood acidity, and external stimuli, such as a raindrop landing on your arm. Neurons called **sensory** or **afferent neurons** (AF-er-ent NOO-ronz; *af-* = toward; *-ferrent* = carried) carry this sensory information into the brain and spinal cord through cranial and spinal nerves.
- ***Integrative function.*** The nervous system *integrates* (processes) sensory information by analyzing and storing some of it and by making decisions for appropriate responses. An important integrative function is **perception,** the conscious awareness of sensory stimuli. Perception occurs in the brain. Many of the neurons that participate in integration are **interneurons,** with axons that extend only for a short distance and contact nearby neurons in the brain or spinal cord. The vast majority of neurons in the body are interneurons.

Figure 12.1 **Major structures of the nervous system.**

The nervous system includes the brain, cranial nerves, spinal cord, spinal nerves, ganglia, enteric plexuses, and sensory receptors.

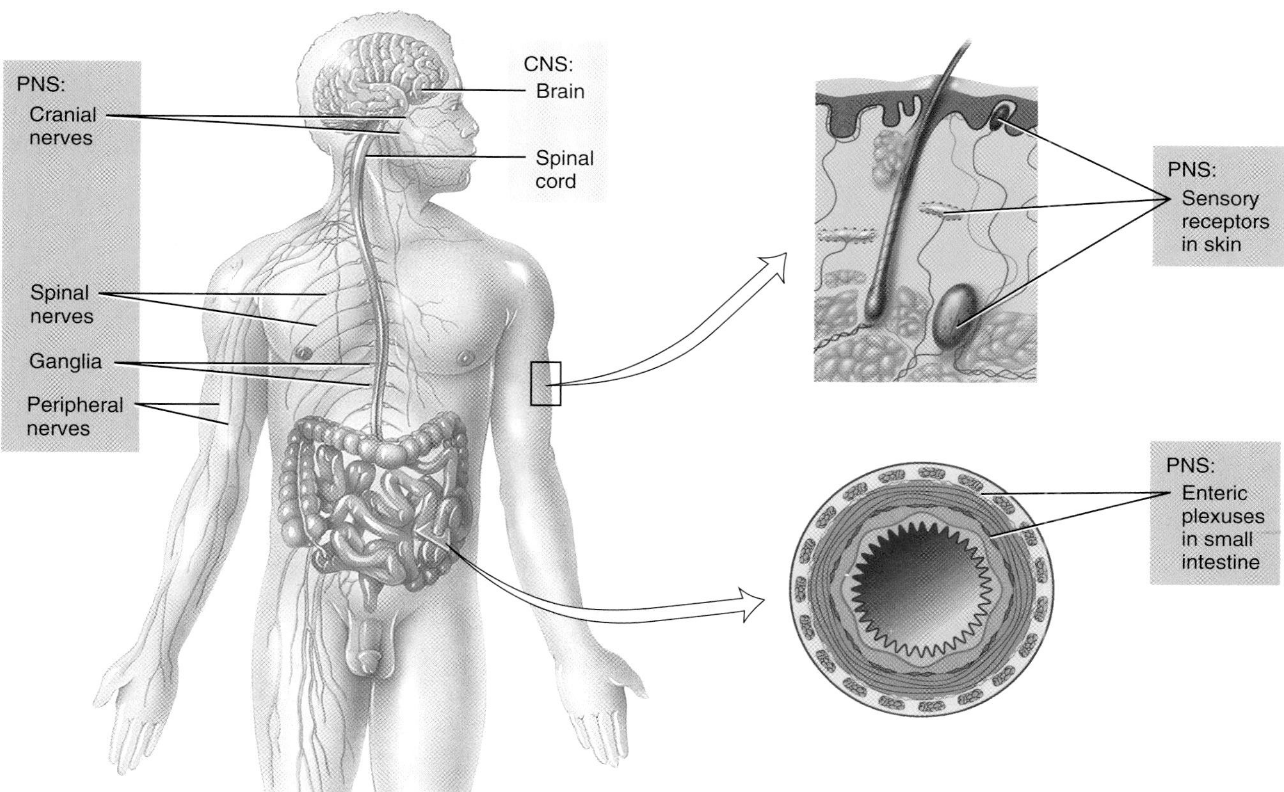

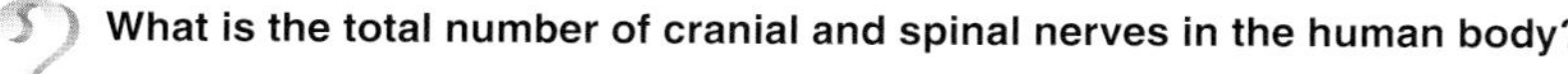

What is the total number of cranial and spinal nerves in the human body?

- ***Motor function.*** Once sensory information is integrated, the nervous system may elicit an appropriate motor response such as muscle contraction or gland secretion. The neurons that serve this function are called **motor** or **efferent neurons** (EF-er-ent; *ef-* = away from). Motor neurons carry information from the brain toward the spinal cord or out of the brain and spinal cord to **effectors** (muscles and glands) through cranial and spinal nerves. Stimulation of the effectors by motor neurons causes muscles to contract and glands to secrete.

Organization of the Nervous System

The two main subdivisions of the nervous system are the **central nervous system (CNS),** which consists of the brain and spinal cord, and the **peripheral** (pe-RIF-er-al) **nervous system (PNS),** which includes all nervous tissue outside the CNS. The CNS processes many different kinds of incoming sensory information. It is also the source of thoughts, emotions, and memories. Most nerve impulses that stimulate muscles to contract and glands to secrete originate in the CNS. Components of the PNS include cranial nerves and their branches, spinal nerves and their branches, ganglia, and sensory receptors. The PNS may be subdivided further into a **somatic nervous system (SNS)** (*somat-* = body), an **autonomic nervous system (ANS)** (*auto-* = self; *-nomic* = law), and an **enteric nervous system (ENS)** (*enter-* = intestines) (Figure 12.2). The SNS consists of (1) sensory neurons that convey information from somatic receptors in the head, body wall, and limbs and from receptors for the special senses of vision, hearing, taste, and smell to the CNS and (2) motor neurons that conduct impulses from the CNS to *skeletal muscles* only. Because these motor responses can be consciously controlled, the action of this part of the PNS is *voluntary.*

The ANS consists of (1) sensory neurons that convey information from autonomic sensory receptors, located primarily in visceral organs such as the stomach and lungs, to the CNS, and (2) motor neurons that conduct nerve impulses from the CNS to *smooth muscle, cardiac muscle,* and *glands.* Because its motor responses are not normally under conscious control, the action of the ANS is *involuntary.* The motor part of the ANS consists of two branches, the **sympathetic division** and the **parasympathetic division.** With a few exceptions, effectors receive nerves

Figure 12.2 **Organization of the nervous system.** Subdivisions of the PNS are the somatic nervous system (SNS), the autonomic nervous system (ANS), and the enteric nervous system (ENS).

The two main subsystems of the nervous system are (1) the central nervous system (CNS), consisting of the brain and spinal cord, and (2) the peripheral nervous system (PNS), consisting of all nervous tissue outside the CNS.

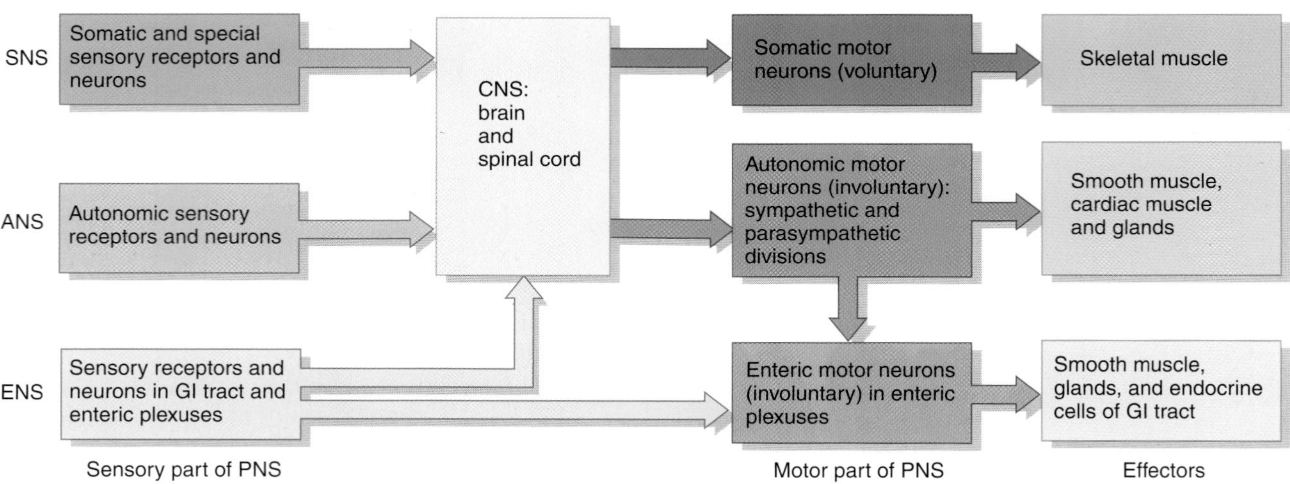

What terms are given to neurons that carry input to the CNS? That carry output from the CNS?

from both divisions, and usually the two divisions have opposing actions. For example, sympathetic neurons increase heart rate, and parasympathetic neurons slow it down. In general, the sympathetic division helps support exercise or emergency actions, so-called "fight-or-flight" responses, and the parasympathetic division takes care of "rest-and-digest" activities.

The operation of the ENS, the "brain of the gut," is involuntary. Once considered part of the ANS, the ENS consists of approximately 100 million neurons in enteric plexuses that extend most of the length of the gastrointestinal (GI) tract. Many of the neurons of the enteric plexuses function independently of the ANS and CNS to some extent, although they also communicate with the CNS via sympathetic and parasympathetic neurons. Sensory neurons of the ENS monitor chemical changes within the GI tract as well as the stretching of its walls. Enteric motor neurons govern contraction of GI tract smooth muscle to propel food through the GI tract, secretions of the GI tract organs such as acid from the stomach, and activity of GI tract endocrine cells, which secrete hormones.

► CHECKPOINT

1. What are the components of the CNS and PNS?
2. What kinds of problems would result from damage of sensory neurons, interneurons, and motor neurons?
3. What are the components and functions of the SNS, ANS, and ENS?
4. Which subdivisions of the PNS control voluntary actions? Involuntary actions?

HISTOLOGY OF NERVOUS TISSUE

► OBJECTIVES

Contrast the histological characteristics and the functions of neurons and neuroglia.

Distinguish between gray matter and white matter.

Nervous tissue consists of two types of cells: neurons and neuroglia. Neurons provide most of the unique functions of the nervous system, such as sensing, thinking, remembering, controlling muscle activity, and regulating glandular secretions. Neuroglia support, nourish, and protect the neurons and maintain homeostasis in the interstitial fluid that bathes them.

Neurons

Like muscle cells, **neurons (nerve cells)** possess **electrical excitability,** the ability to respond to a stimulus and convert it into an action potential. A **stimulus** is any change in the environment that is strong enough to initiate an action potential. An **action potential (nerve impulse)** is an electrical signal that propagates (travels) along the surface of the membrane of a neuron. It begins and travels due to the movement of ions (such as sodium and potassium) between interstitial fluid and the inside of a neuron through specific ion channels in its plasma membrane. Once begun, a nerve impulse travels rapidly and at a constant strength.

Some neurons are tiny and propagate impulses over a short distance (less than 1 mm) within the CNS. Others are the longest cells in the body. The motor neurons that enable you to wiggle your toes, for example, extend from the lumbar region of your spinal cord (just above waist level) to the muscles in your foot. Some sensory neurons are even longer. Those that allow you to feel the position of your wiggling toes stretch all the way from your foot to the lower portion of your brain. Nerve impulses travel these great distances at speeds ranging from 0.5 to 130 meters per second (1 to 280 mi/hr).

Parts of a Neuron

Most neurons have three parts: (1) a cell body, (2) dendrites, and (3) an axon (Figure 12.3). The **cell body (perikaryon)** contains a nucleus surrounded by cytoplasm that includes typical cellular organelles such as lysosomes, mitochondria, and a Golgi complex. Neuronal cell bodies also contain free ribosomes and prominent clusters of rough endoplasmic reticulum, termed *Nissl bodies.* The ribosomes are the sites of protein synthesis. Newly synthesized proteins produced by Nissl bodies are used to replace cellular components as material for growth of neurons and to regenerate damaged axons in the PNS. The cytokeleton includes both *neurofibrils,* composed of bundles of intermediate filaments that provide the cell shape and support, and *microtubules,* which assist in moving materials between the cell body and axon. Many neurons also contain *lipofuscin,* a pigment that occurs as clumps of yellowish brown granules in the cytoplasm. Lipofuscin is a product of neuronal lysosomes that accumulates as the neuron ages, but does not seem to harm the neuron.

A **nerve fiber** is a general term for any neuronal process or extension that emerges from the cell body of a neuron. Most neurons have two kinds of processes: multiple dendrites and a single axon. **Dendrites** (= little trees) are the receiving or input portions of a neuron. They usually are short, tapering, and highly branched. In many neurons the dendrites form a tree-shaped array of processes extending from the cell body. Their cytoplasm contains Nissl bodies, mitochondria, and other organelles.

The single **axon** (= axis) of a neuron propagates nerve impulses toward another neuron, a muscle fiber, or a gland cell. An axon is a long, thin, cylindrical projection that often joins the cell body at a cone-shaped elevation called the **axon hillock** (= small hill). The part of the axon closest to the axon hillock is the **initial segment.** In most neurons, nerve impulses arise at the junction of the axon hillock and the initial segment, an area called the **trigger zone,** from which they travel along the axon to their destination. An axon contains mitochondria, microtubules, and neurofibrils. Because rough endoplasmic reticulum is not present, protein synthesis does not occur in the axon. The cytoplasm of an axon, called **axoplasm,** is surrounded by a plasma membrane known as the **axolemma** (*lemma* = sheath or husk). Along the length of an axon, side branches called **axon collaterals** may branch off, typically at a right angle to the axon. The axon and its collaterals end by dividing into many fine processes called **axon terminals (telodendria).**

The site of communication between two neurons or between a neuron and an effector cell is called a **synapse** (SIN-aps). The tips of some axon terminals swell into bulb-shaped structures called **synaptic end bulbs;** others exhibit a string of swollen bumps called **varicosities.** Both synaptic end bulbs and varicosities contain many tiny membrane-enclosed sacs called **synaptic vesicles** that store a chemical **neurotransmitter.** Many neurons contain two or even three types of neurotransmitters, each with different effects on the postsynaptic cell. When neurotransmitter molecules are released from synaptic vesicles, they excite or inhibit other neurons, muscle fibers, or gland cells.

Because some substances synthesized or recycled in the neuron cell body are needed in the axon or at the axon terminals, two types of transport systems carry materials from the cell body to the axon terminals and back. The slower system, which moves materials about 1–5 mm per day, is called **slow axonal transport.** It conveys axoplasm in one direction only—from the cell body toward the axon terminals. Slow axonal transport supplies new axoplasm to developing or regenerating axons and replenishes axoplasm in growing and mature axons.

Fast axonal transport, which is capable of moving materials a distance of 200–400 mm per day, uses proteins that function as "motors" to move materials in both directions—away from and toward the cell body—along the surfaces of microtubules. Fast axonal transport moves various organelles and materials that form the membranes of the axolemma, synaptic end bulbs, and synaptic vesicles. Some materials transported back to the cell body are degraded or recycled; others influence neuronal growth.

Structural Diversity in Neurons

Neurons display great diversity in size and shape. For example, their cell bodies range in diameter from 5 micrometers (μm) (slightly smaller than a red blood cell) up to 135 μm (barely large enough to see with the unaided eye). The pattern of dendritic branching is varied and distinctive for neurons in different parts of the nervous system. A few small neurons lack an axon, and many others have very short axons. As we have already discussed, the longest axons are almost as long as a person is tall, extending from the toes to the lowest part of the brain.

Both functional and structural features are used to classify the various neurons in the body. Recall that neurons are classified as sensory neurons, interneurons, or motor neurons based on function. Structurally, neurons are classified according to the number of processes extending from the cell body (Figure 12.4 on page 409).

1. **Multipolar neurons** usually have several dendrites and one axon (see also Figure 12.3). Most neurons in the brain and spinal cord are of this type.

2. **Bipolar neurons** have one main dendrite and one axon. They are found in the retina of the eye, in the inner ear, and in the olfactory area of the brain.

3. **Unipolar neurons** are sensory neurons that begin in the embryo as bipolar neurons. During development, the axon and dendrite fuse into a single process that divides into two branches a short distance from the cell body. Both branches have the characteristic structure and function of an axon. They are long, cylindrical processes that propagate action potentials. However,

Figure 12.3 Structure of a multipolar neuron (a neuron with a large cell body, several short dendrites, and a single long axon). Arrows indicate the direction of information flow: dendrites → cell body → axon → axon terminals.

The basic parts of a neuron are dendrites, a cell body, and an axon.

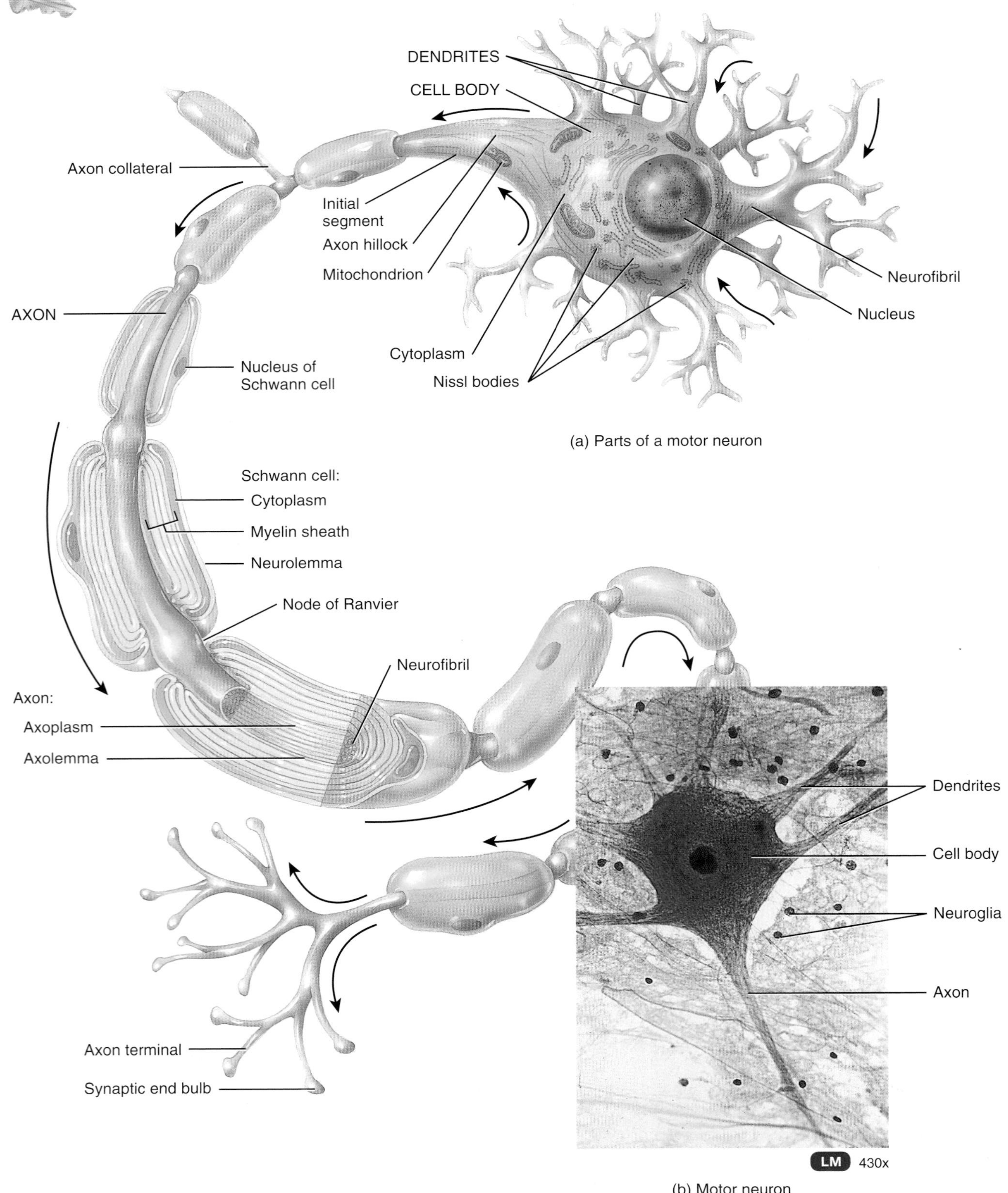

(b) Motor neuron

What roles do the dendrites, cell body, and axon play in communication of signals?

Figure 12.4 Structural classification of neurons. Breaks indicate that axons are longer than shown.

A multipolar neuron has many processes extending from the cell body, a bipolar neuron has two, and a unipolar neuron has one.

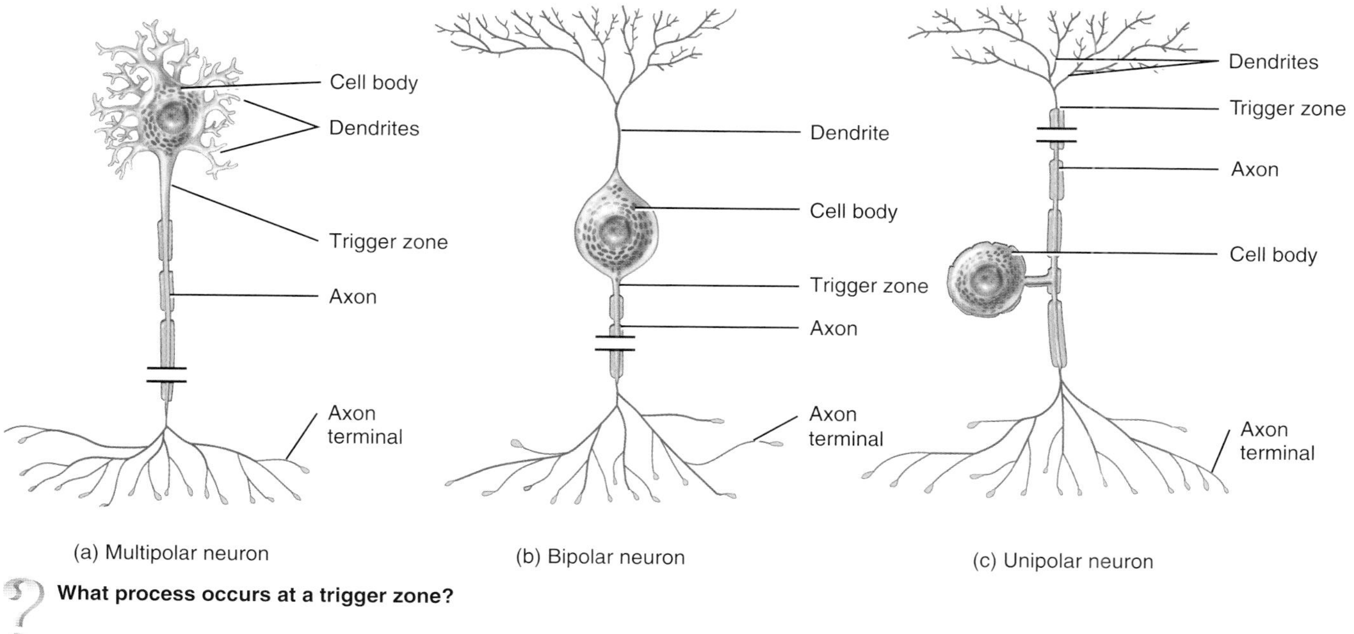

What process occurs at a trigger zone?

the axon branch that extends into the periphery has dendrites at its distal tip, whereas the axon branch that extends into the CNS ends in synaptic end bulbs. The dendrites monitor a sensory stimulus such as touch or stretching. The trigger zone for nerve impulses in a unipolar neuron is at the junction of the dendrites and axon (Figure 12.4c). The impulses then propagate toward the synaptic end bulbs. The cell bodies of most unipolar neurons are located in the ganglia of spinal and cranial nerves.

Some neurons are named for the histologist who first described them or for an aspect of their shape or appearance; examples include **Purkinje cells** (pur-KIN-jē) in the cerebellum (Figure 12.5a) and **pyramidal cells** (pi-RAM-i-dal), found in the cerebral cortex of the brain, which have pyramid-shaped cell bodies (Figure 12.5b). Often, a neuron can be identified by a distinctive pattern of dendritic branching.

Neuroglia

Neuroglia (noo-RŌG-lē-a; *-glia* = glue) or **glia** make up about half the volume of the CNS. Their name derives from the idea of early histologists that they were the "glue" that held nervous tissue together. We now know that neuroglia are not merely passive bystanders but rather actively participate in the activities of nervous tissue. Generally, neuroglia are smaller than neurons, and they are 5 to 50 times more numerous. In contrast to neurons, glia do not generate or propagate action potentials, and they can multiply and divide in the mature nervous system. In cases of injury or disease, neuroglia multiply to fill in the spaces formerly occupied by neurons. Brain tumors derived from glia, called **gliomas,** tend to be highly malignant and

Figure 12.5 Two examples of CNS neurons. Arrows indicate the direction of information flow.

The dendritic branching pattern often is distinctive for a particular type of neuron.

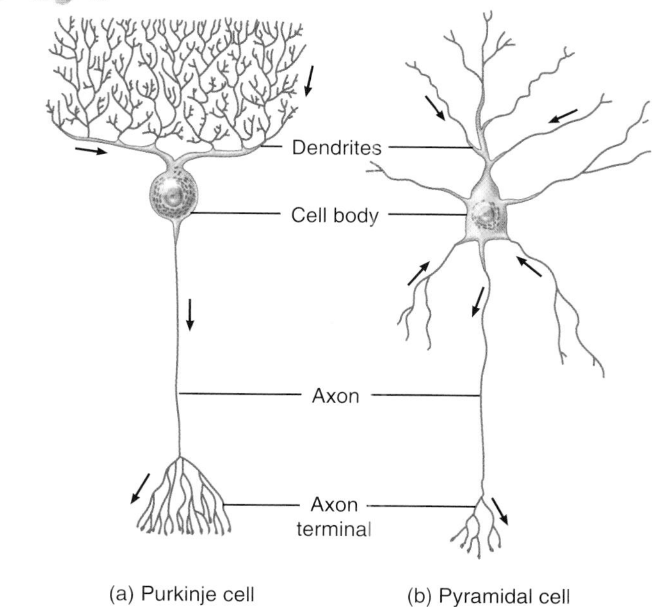

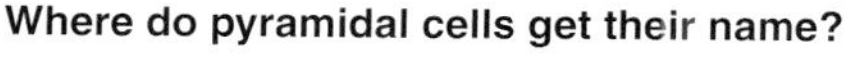

Where do pyramidal cells get their name?

to grow rapidly. Of the six types of neuroglia, four—astrocytes, oligodendrocytes, microglia, and ependymal cells—are found only in the CNS. The remaining two types—Schwann cells and satellite cells—are present in the PNS.

Neuroglia of the CNS

Neuroglia of the CNS can be classified on the basis of size, cytoplasmic processes, and intracellular organization into four types: astrocytes, oligodendrocytes, microglia, and ependymal cells (Figure 12.6).

ASTROCYTES (AS-trō-sīts; *astro-* = star; *-cyte* = cell) These star-shaped cells have many processes and are the largest and most numerous of the neuroglia. There are two types of **astrocytes.** *Protoplasmic astrocytes* have many short branching processes and are found in gray matter (described shortly). *Fibrous astrocytes* have many long unbranched processes and are located mainly in white matter (also described shortly). The processes of astrocytes make contact with blood capillaries, neurons, and the pia mater (a thin membrane around the brain and spinal cord).

The functions of astrocytes include the following: (1) Astrocytes contain microfilaments that give them considerable strength, which enables them to support neurons. (2) Processes of astrocytes wrapped around blood capillaries isolate neurons of the CNS from various potentially harmful substances in blood by secreting chemicals that maintain the unique selective permeability characteristics of the endothelial cells of the capillaries. In effect, the endothelial cells create a *blood–brain barrier,* which restricts the movement of substances between the blood and interstitial fluid of the CNS. Details of the blood–brain barrier are discussed in Chapter 14. (3)

Figure 12.6 Neuroglia of the central nervous system (CNS).

Neuroglia of the CNS are distinguished on the basis of size, cytoplasmic processes, and intracellular organization.

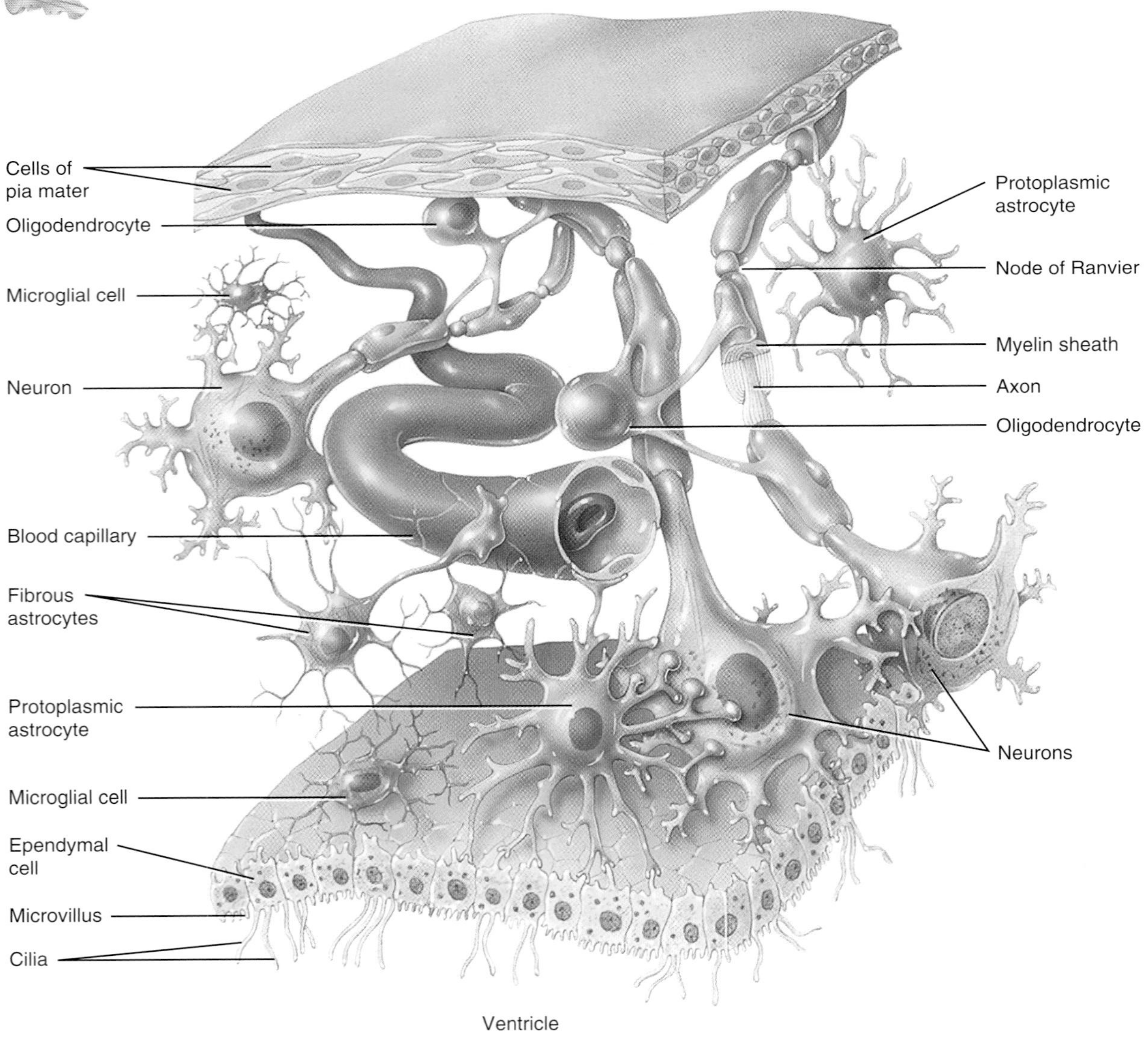

Which CNS neuroglia function as phagocytes?

In the embryo, astrocytes secrete chemicals that appear to regulate the growth, migration, and interconnection among neurons in the brain. (4) Astrocytes help to maintain the appropriate chemical environment for the generation of nerve impulses. For example, they regulate the concentration of important ions such as K^+; take up excess neurotransmitters; and serve as a conduit for the passage of nutrients and other substances between blood capillaries and neurons. (5) Astrocytes may also play a role in learning and memory by influencing the formation of neural synapses (see page 567).

OLIGODENDROCYTES (OL-i-gō-den′-drō-sīts; *oligo-* = few; *dendro-* = tree) These resemble astrocytes, but are smaller and contain fewer processes. **Oligodendrocyte** processes are responsible for forming and maintaining the myelin sheath around CNS axons. As you will see shortly, the **myelin sheath** is a multilayered lipid and protein covering around some axons that insulates them and increases the speed of nerve impulse conduction. Such axons are said to be **myelinated.**

MICROGLIA (mī-KROG-lē-a; *micro-* = small) These neuroglia are small cells with slender processes that give off numerous spinelike projections. **Microglia** function as phagocytes. Like tissue macrophages, they remove cellular debris formed during normal development of the nervous system and phagocytize microbes and damaged nervous tissue.

EPENDYMAL CELLS (ep-EN-de-mal; *epen-* = above; *dym-* = garment) **Ependymal cells** are cuboidal to columnar cells arranged in a single layer that possess microvilli and cilia. These cells line the ventricles of the brain and central canal of the spinal cord (spaces filled with cerebrospinal fluid, which protects and nourishes the brain and spinal cord). Functionally, ependymal cells produce, possibly monitor, and assist in the circulation of cerebrospinal fluid. They also form the blood–cerebrospinal fluid barrier, which is discussed in Chapter 14.

Neuroglia of the PNS

Neuroglia of the PNS completely surround axons and cell bodies. The two types of glial cells in the PNS are Schwann cells and satellite cells (Figure 12.7).

SCHWANN CELLS (SCHVON or SCHWON) These cells encircle PNS axons. Like oligodendrocytes, they form the myelin sheath around axons. However, a single oligodendrocyte myelinates several axons, but each **Schwann cell** myelinates a single axon (Figure 12.7a; see also Figure 12.8a, c). A single Schwann cell can also enclose as many as 20 or more unmyelinated axons (axons that lack a myelin sheath) (Figure 12.7b). Schwann cells participate in axon regeneration, which is more easily accomplished in the PNS than in the CNS.

SATELLITE CELLS (SAT-i-līt) These flat cells surround the cell bodies of neurons of PNS ganglia (Figure 12.7c). (Recall that ganglia are collections of neuronal cell bodies outside the CNS.) Besides providing structural support, **satellite cells** regulate the exchanges of materials between neuronal cell bodies and interstitial fluid.

Figure 12.7 Neuroglia of the peripheral nervous system (PNS).

Neuroglia of the PNS completely surround axons and cell bodies of neurons.

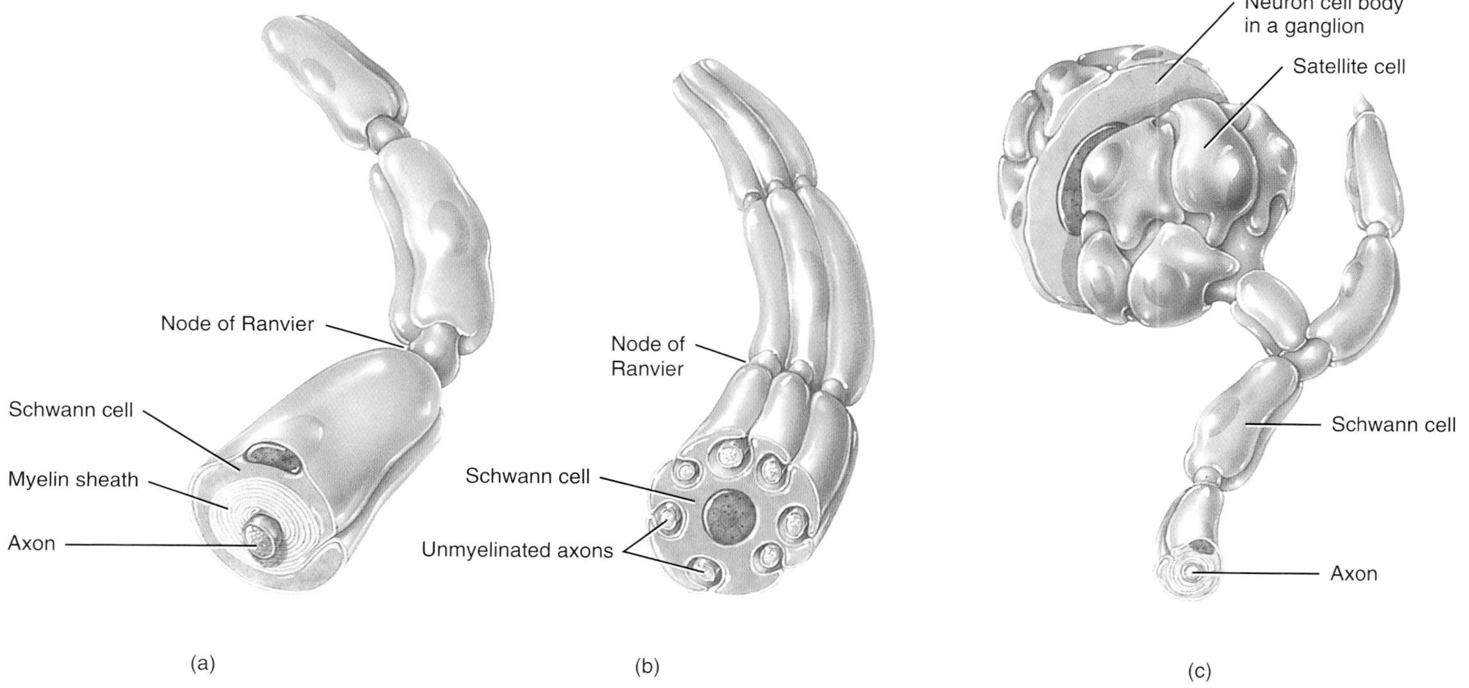

How do Schwann cells and oligodendrocytes differ with respect to the number of axons they myelinate?

Myelination

As you have already learned, axons surrounded by a multilayered lipid and protein covering, called the myelin sheath, are said to be myelinated (Figure 12.8a). The sheath electrically insulates the axon of a neuron and increases the speed of nerve impulse conduction. Axons without such a covering are said to be **unmyelinated** (Figure 12.8b).

Two types of neuroglia produce myelin sheaths: Schwann cells (in the PNS) and oligodendrocytes (in the CNS). Schwann cells begin to form myelin sheaths around axons during fetal

Figure 12.8 Myelinated and unmyelinated axons. Notice that one layer of Schwann cell plasma membrane surrounds unmyelinated axons.

Axons of neurons surrounded by a myelin sheath produced by Schwann cells in the PNS and by oligodendrocytes in the CNS are said to be myelinated.

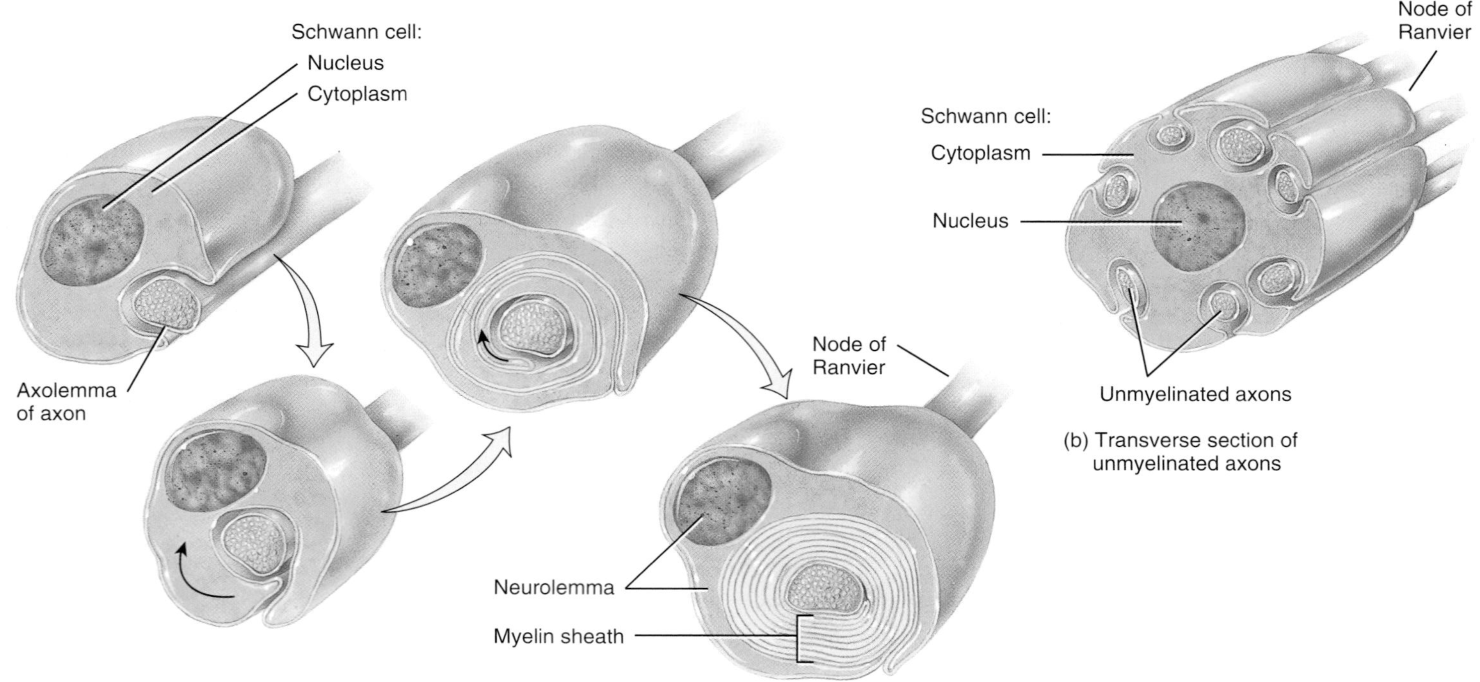

(a) Transverse sections of stages in the formation of a myelin sheath

(b) Transverse section of unmyelinated axons

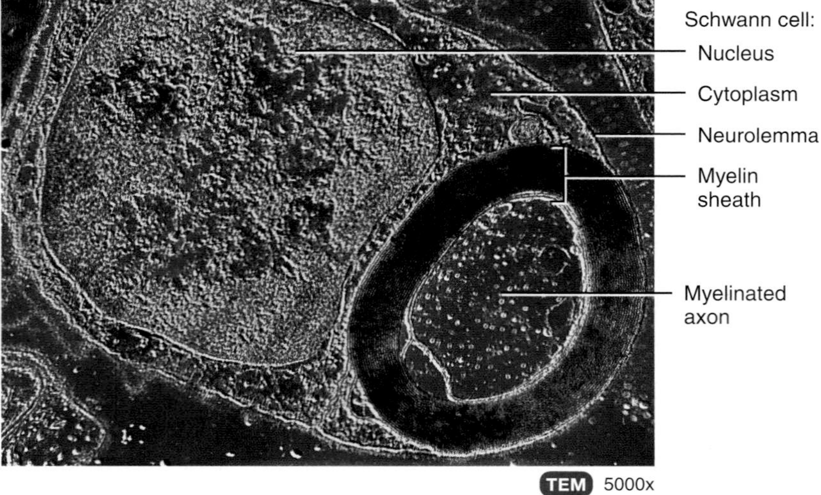

(c) Transverse section of myelinated axon

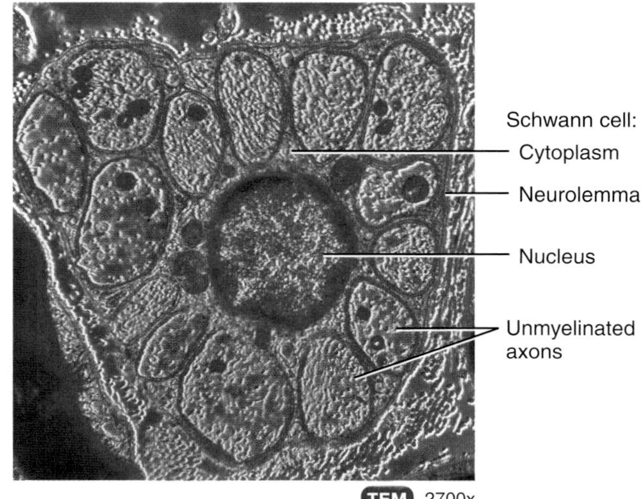

(d) Transverse section of unmyelinated axons

What is the functional advantage of myelination?

development. Each Schwann cell wraps about 1 millimeter (1 mm = 0.04 in.) of a single axon's length by spiraling many times around the axon (Figure 12.8a). Eventually, multiple layers of glial plasma membrane surround the axon, with the Schwann cell's cytoplasm and nucleus forming the outermost layer. The inner portion, consisting of up to 100 layers of Schwann cell membrane, is the myelin sheath. The outer nucleated cytoplasmic layer of the Schwann cell, which encloses the myelin sheath, is the **neurolemma (sheath of Schwann).** A neurolemma is found only around axons in the PNS. When an axon is injured, the neurolemma aids regeneration by forming a regeneration tube that guides and stimulates regrowth of the axon. Gaps in the myelin sheath, called **nodes of Ranvier** (RON-vē-ā), appear at intervals along the axon (see Figure 12.3). Each Schwann cell wraps one axon segment between two nodes.

In the CNS, an oligodendrocyte myelinates parts of several axons. Each oligodendrocyte puts forth about 15 broad, flat processes that spiral around CNS axons, forming a myelin sheath. A neurolemma is not present, however, because the oligodendrocyte cell body and nucleus do not envelop the axon. Nodes of Ranvier are present, but they are fewer in number. Axons in the CNS display little regrowth after injury. This is thought to be due, in part, to the absence of a neurolemma, and in part to an inhibitory influence exerted by the oligodendrocytes on axon regrowth.

The amount of myelin increases from birth to maturity, and its presence greatly increases the speed of nerve impulse conduction. An infant's responses to stimuli are neither as rapid nor as coordinated as those of an older child or an adult, in part because myelination is still in progress during infancy.

Demyelination

Demyelination (dē-mī-e-li-NĀ-shun) refers to the loss or destruction of myelin sheaths around axons. It may result from disorders such as multiple sclerosis (see page 433) or Tay-Sachs disease (see page 82), or from medical treatments such as radiation therapy and chemotherapy. Any single episode of demyelination may cause deterioration of affected nerves. ■

Gray and White Matter

In a freshly dissected section of the brain or spinal cord, some regions look white and glistening, and others appear gray (Figure 12.9). **White matter** is composed primarily of myelinated axons. The whitish color of myelin gives white matter its name. The **gray matter** of the nervous system contains neuronal cell bodies, dendrites, unmyelinated axons, axon terminals, and neuroglia. It appears grayish, rather than white, because the Nissl bodies impart a gray color and there is little or no myelin in these areas. Blood vessels are present in both white and gray matter.

In the spinal cord, the white matter surrounds an inner core of gray matter that, depending on how imaginative you are, is shaped like a butterfly or the letter H; in the brain, a thin shell of gray matter covers the surface of the largest portions of the brain, the cerebrum and cerebellum (Figure 12.9). When used to describe nervous tissue, a **nucleus** is a cluster of neuronal cell bodies within the CNS. (Recall that the term *ganglion* refers to a similar arrangement within the PNS.) Many nuclei of gray matter lie deep within the brain. The arrangement of gray matter and white matter in the spinal cord and brain is discussed more extensively in Chapters 13 and 14, respectively.

Figure 12.9 Distribution of gray and white matter in the spinal cord and brain.

White matter primarily consists of myelinated axons of many neurons. Gray matter consists of neuron cell bodies, dendrites, axon terminals, unmyelinated axons, and neuroglia.

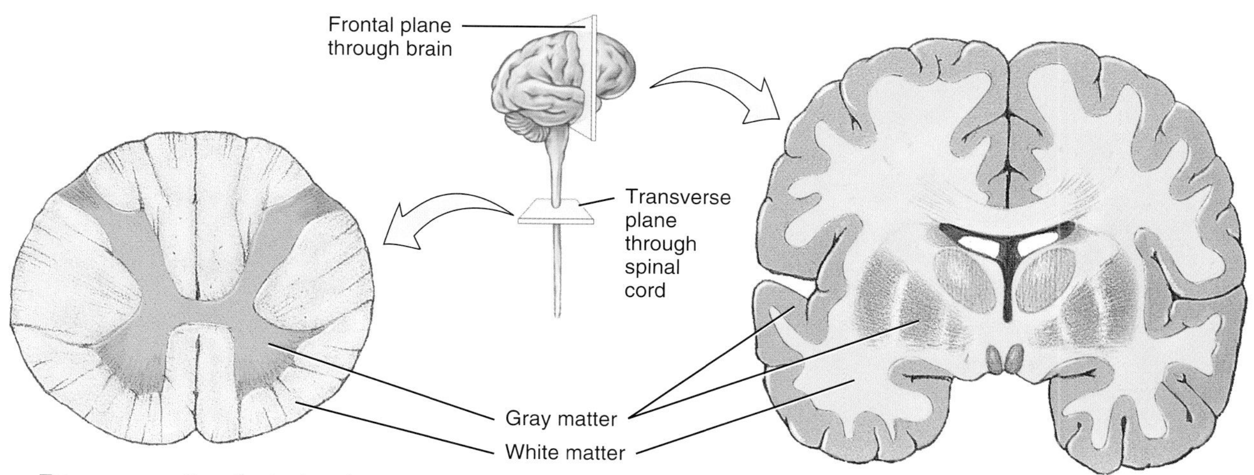

What is responsible for the white appearance of white matter?

▶ CHECKPOINT

5. Describe the parts of a neuron and the functions of each.
6. Give several examples of the structural diversity of neurons.
7. What is a neurolemma and why is it important?
8. With reference to the nervous system, what is a nucleus?

ELECTRICAL SIGNALS IN NEURONS

▶ OBJECTIVES

Describe the cellular properties that permit communication among neurons and effectors.

Compare the basic types of ion channels, and explain how they relate to action potentials and graded potentials.

Describe the factors that maintain a resting membrane potential.

List the sequence of events that generate an action potential.

Like muscle fibers, neurons are electrically excitable. They communicate with one another using two types of electrical signals: (1) *Graded potentials* are used for short-distance communication only. (2) *Action potentials* allow communication over both short and long distances within the body. Recall that an action potential in a muscle fiber is called a *muscle action potential.* When an action potential occurs in a neuron (nerve cell), it is called a *nerve action potential (nerve impulse).* To understand the functions of graded potentials and action potentials, consider how the nervous system allows you to feel the smooth surface of a pen that you have picked up off of a table (Figure 12.10):

1. As you touch the pen, a graded potential develops in a sensory receptor in the skin of the fingers.
2. The graded potential triggers the axon of the sensory neuron to form a nerve action potential, which travels along the axon into the CNS and ultimately causes the release of neurotransmitter at a synapse with an interneuron.
3. The neurotransmitter stimulates the interneuron to form a graded potential in its dendrites and cell body.
4. In response to the graded potential, the axon of the interneuron forms a nerve action potential. The nerve action potential travels along the axon, which results in neurotransmitter release at the next synapse with another interneuron.
5. This process of neurotransmitter release at a synapse followed by the formation of a graded potential and then a nerve action potential occurs over and over as interneurons in higher parts of the brain (such as the thalamus and cerebral cortex) are activated. Once interneurons in the **cerebral cortex,** the outer part of the brain, are activated, perception occurs and you are able to feel the smooth surface of the pen touch your fingers. As you will learn in Chapter 14, **perception,** the conscious awareness of a sensation, is primarily a function of the cerebral cortex.

Suppose that you want to use the pen to write a letter. The nervous system would respond in the following way (Figure 12.10):

6. A stimulus in the brain causes a graded potential to form in the dendrites and cell body of an **upper motor neuron,** a type of motor neuron that synapses with a lower motor neuron farther down in the CNS in order to contract a skeletal muscle. The graded potential subsequently causes a nerve action potential to occur in the axon of the upper motor neuron, followed by neurotransmitter release.
7. The neurotransmitter generates a graded potential in a **lower motor neuron,** a type of motor neuron that directly supplies skeletal muscle fibers. The graded potential triggers the formation of a nerve action potential and then release of neurotransmitter at neuromuscular junctions formed with skeletal muscle fibers that control movements of the fingers.
8. The neurotransmitter stimulates the formation of muscle action potentials in these muscle fibers. The muscle action potentials cause the muscle fibers of the fingers to contract, which allows you to write with the pen.

The production of graded potentials and action potentials depends on two basic features of the plasma membrane of excitable cells: the existence of a resting membrane potential and the presence of specific types of ion channels. Like most other cells in the body, the plasma membrane of excitable cells exhibits a **membrane potential,** an electrical voltage difference across the membrane. In excitable cells, this voltage is termed the **resting membrane potential.** The membrane potential is like voltage stored in a battery. If you connect the positive and negative terminals of a battery with a piece of wire, electrons will flow along the wire. This flow of charged particles is called **current.** In living cells, the flow of ions (rather than electrons) constitutes the electrical current.

Graded potentials and action potentials occur because the membranes of neurons contain many different kinds of ion channels that open or close in response to specific stimuli. Because the lipid bilayer of the plasma membrane is a good electrical insulator, the main paths for current to flow across the membrane are through the ion channels.

Ion Channels

When ion channels are open, they allow specific ions to move across the plasma membrane, down their **electrochemical gradient**—a concentration (chemical) difference plus an electrical difference. Recall that ions move from areas of higher concentration to areas of lower concentration (the chemical part of the gradient). Also, positively charged cations move toward a negatively charged area, and negatively charged anions move

Figure 12.10 Overview of nervous system functions.

Graded potentials and nerve and muscle action potentials are involved in the relay of sensory stimuli, integrative functions such as perception, and motor activities.

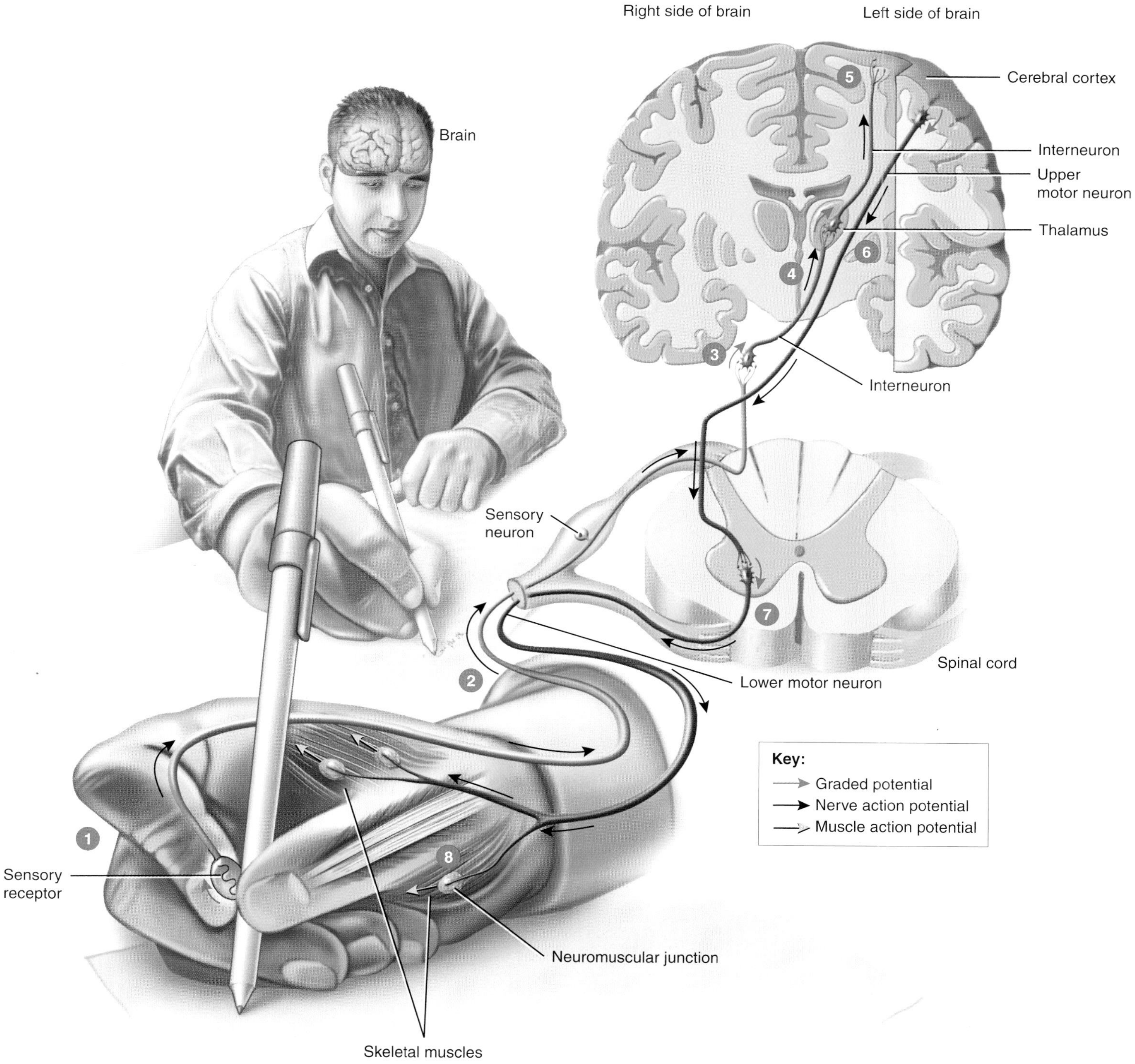

In which region of the brain does perception primarily occur?

toward a positively charged area (the electrical aspect of the gradient). As ions move, they create a flow of electrical current that can change the membrane potential.

Ion channels open and close due to the presence of "gates." The gate is a part of the channel protein that can seal the channel pore shut or move aside to open the pore (see Figure 3.5b on page 67). The electrical signals produced by neurons and muscle fibers rely on four types of ion channels: leakage channels, voltage-gated channels, ligand-gated channels, and mechanically gated channels.

1. The gates of **leakage channels** randomly alternate between open and closed positions. Typically, plasma membranes have many more potassium ion (K^+) leakage channels than sodium ion (Na^+) leakage channels, and the potassium ion leakage channels are leakier than the sodium ion leakage channels. Thus, the membrane's permeability to K^+ is much higher than its permeability to Na^+.
2. A **voltage-gated channel** opens in response to a change in membrane potential (voltage) (Figure 12.11a). Voltage-gated channels participate in the generation and conduction of action potentials.
3. A **ligand-gated channel** opens and closes in response to a specific chemical stimulus. A wide variety of chemical ligands—including neurotransmitters, hormones, and particular ions—can open or close ligand-gated channels. The neurotransmitter acetylcholine, for example, opens cation channels that allow Na^+ and Ca^{2+} to diffuse inward and K^+ to diffuse outward (Figure 12.11b). Ligand-gated channels operate in two basic ways. The ligand molecule itself may open or close the channel by binding to a portion of the channel protein, as in the case of acetylcholine. Or the ligand may act *indirectly* via a type of membrane protein called a G protein that activates another molecule, a "second messenger," in the cytosol, which in turn operates the channel's gate. Some hormones and neurotransmitters work by such second-messenger systems (see Figure 18.4 on page 623).
4. A **mechanically gated channel** opens or closes in response to mechanical stimulation in the form of vibration (such as sound waves), pressure (such as touch), or tissue stretching. The force distorts the channel from its resting position, opening the gate. Examples of mechanically gated channels are those found in auditory receptors in the ears, in receptors that monitor stretching of internal organs, and in touch receptors in the skin.

Figure 12.11 Voltage-gated and ligand-gated channels in the plasma membrane. (a) A change in membrane potential opens voltage-gated K^+ channels during an action potential. (b) A chemical stimulus—here, the neurotransmitter acetylcholine—opens a ligand-gated channel.

Gated channels open and close in response to a particular type of stimulus.

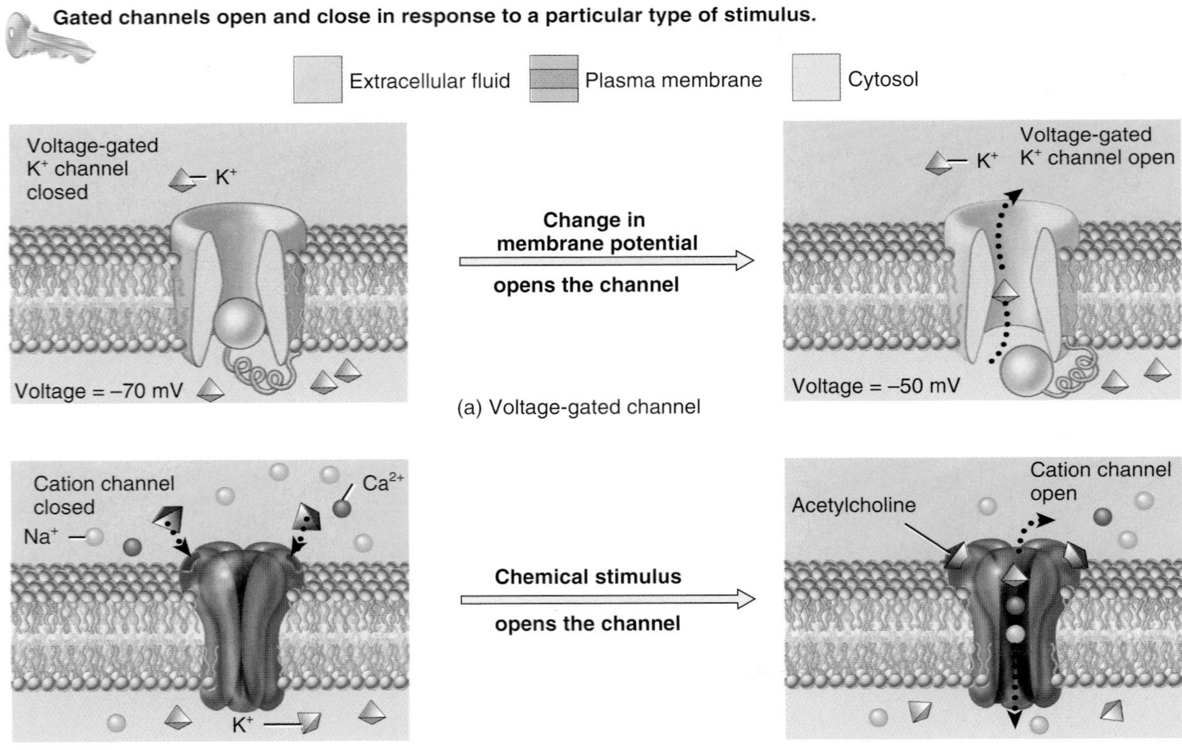

What type of gated channel (not shown here) is activated by a touch on the arm?

Resting Membrane Potential

The resting membrane potential exists because of a small buildup of negative ions in the cytosol along the inside of the membrane, and an equal buildup of positive ions in the extracellular fluid along the outside surface of the membrane (Figure 12.12a). Such a separation of positive and negative electrical charges is a form of potential energy, which is measured in volts or millivolts (1 mV = 0.001 V). The greater the difference in charge across the membrane, the larger the membrane potential (voltage). Notice in Figure 12.12a that the buildup of charge occurs only very close to the membrane. The cytosol or extracellular fluid elsewhere in the cell contains equal numbers of positive and negative charges and is electrically neutral.

In neurons, the resting membrane potential ranges from −40 to −90 mV. A typical value is −70 mV. The minus sign indicates that the inside of the cell is negative relative to the outside. A cell that exhibits a membrane potential is said to be **polarized.** Most body cells are polarized; the membrane potential varies from +5 mV to −100 mV in different types of cells.

The resting membrane potential arises from the unequal distributions of various ions in extracellular fluid and cytosol (Figure 12.12b). Extracellular fluid is rich in Na^+ and chloride ions (Cl^-). In cytosol, however, the main cation is K^+, and the two dominant anions are phosphates attached to molecules, such as the three phosphates in ATP, and amino acids in proteins. Because the concentration of K^+ is higher in cytosol and because the plasma membranes have many K^+ leakage channels, potassium ions diffuse down their concentration gradient—out of cells into the extracellular fluid. As more and more positive potassium ions exit, the inside of the membrane becomes increasingly negative, and the outside of the membrane becomes increasingly positive. Another factor contributes to the negativity inside: Most negatively charged ions inside the cell are not free to leave. They cannot follow the K^+ out of the cell because they are attached either to large proteins or to other large molecules. Because the negative charges inside attract K^+ back into the cell, eventually just as many K^+ are entering the cell due to the negativity inside as are exiting the cell because of the concentration difference.

Membrane permeability to Na^+ is very low because there are only a few sodium leakage channels. Nevertheless, sodium ions do slowly diffuse inward, down their concentration gradient. Left unchecked, such inward leakage of Na^+ would eventually destroy the resting membrane potential. The small inward Na^+ leak and outward K^+ leak are offset by the sodium-potassium pumps (Na^+/K^+ ATPase; see Figure 3.8 on page 70). These pumps help maintain the resting membrane potential by pumping out Na^+ as fast as it leaks in. At the same time, the sodium-potassium pumps bring in K^+. However, the potassium ions also redistribute according to electrical and chemical gradients, as previously described.

Because the sodium-potassium pumps expel three Na^+ for each two K^+ imported, they are *electrogenic,* which means they contribute to the negativity of the resting membrane potential. Their total contribution, however, is very small, only −3 mV of the total −70 mV resting membrane potential in a typical neuron.

Figure 12.12 Distributions of (a) charges and (b) ions that produce the resting membrane potential.

The resting membrane potential is due to a small buildup of anions, mainly phosphates (PO_4^{3-}) and proteins, in the cytosol just inside the membrane and an equal buildup of cations, mainly sodium ions (Na^+), in the extracellular fluid just outside the membrane.

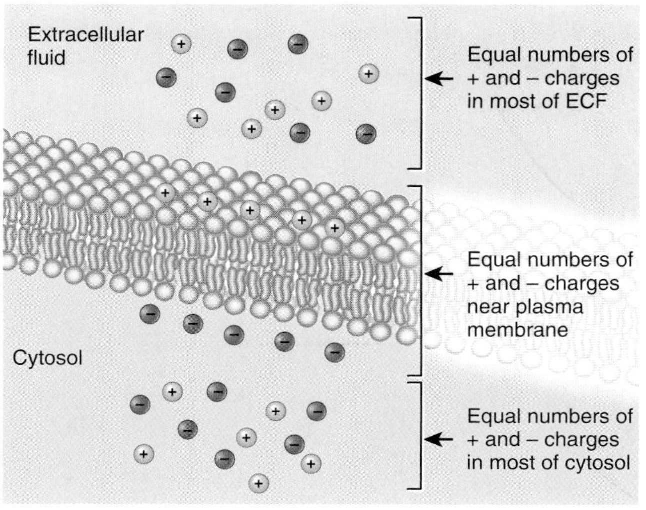

(a) Distribution of charges

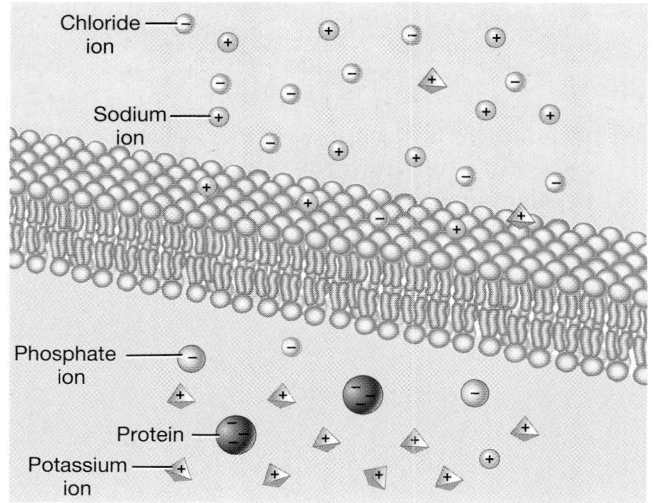

(b) Distribution of ions

What is a typical value for the resting membrane potential of a neuron?

Graded Potentials

When a stimulus causes ligand-gated or mechanically gated channels to open or close in an excitable cell's plasma membrane, a **graded potential** arises. A graded potential is a small deviation from the membrane potential that makes the membrane either more polarized (inside more negative) or less polarized (inside less negative). When the response makes the membrane even more polarized (inside more negative), it is termed a **hyperpolarizing graded potential** (Figure 12.13a). When the response makes the membrane less polarized (inside less negative), it is termed a **depolarizing graded potential** (Figure 12.13b).

To say that these electrical signals are *graded* means that they vary in amplitude (size), depending on the strength of the stimulus. They are larger or smaller depending on how many ion channels have opened (or closed) and how long each remains open. The opening or closing of ion channels alters the flow of specific ions across the membrane, producing a flow of current that is *localized,* which means that it spreads along the plasma membrane for a short distance and then dies out. For this reason, graded potentials are useful for communication only when the distance is less than a few hundred micrometers.

Typically, ligand-gated channels and mechanically gated channels can be present in the dendrites of sensory neurons, and ligand-gated channels are numerous in the dendrites and cell bodies of interneurons and motor neurons. Ligand-gated channels are occasionally present in axons. Hence, graded potentials occur most often in the dendrites and cell body of a neuron, and less often in the axon. Graded potentials have different names depending on which type of stimulus causes them and where they occur. For example, when a graded potential occurs in the dendrites or cell body of a neuron in response to a neurotransmitter, it is called a *postsynaptic potential* (explained shortly). On the other hand, the graded potentials that occur in sensory receptors and sensory neurons are termed *receptor potentials* and *generator potentials* (explained in Chapter 16).

Figure 12.13 **Graded potentials.** Most graded potentials occur in the dendrites and cell body (areas colored blue in the inset).

During a hyperpolarizing graded potential, the membrane polarization is more negative than the resting level. During a depolarizing graded potential, the membrane polarization is less negative than the resting level.

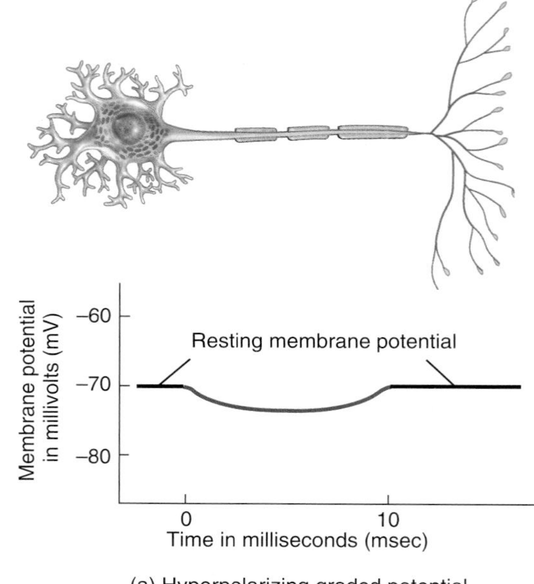

(a) Hyperpolarizing graded potential

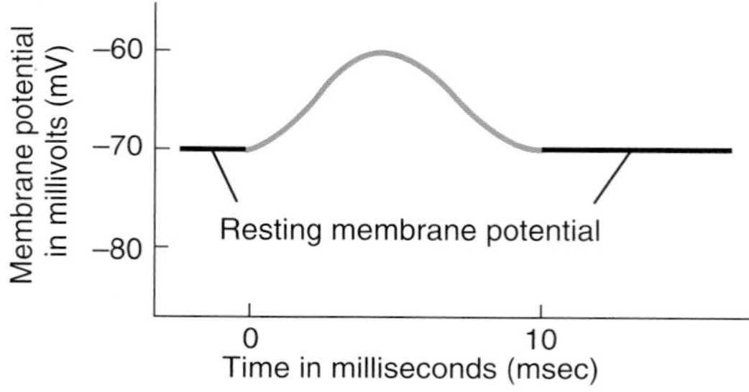

(b) Depolarizing graded potential

What types of ion channels produce graded potentials when they open or close?

Generation of Action Potentials

An **action potential (AP)** or **impulse** is a sequence of rapidly occurring events that take place in two phases (Figure 12.14). During the **depolarizing phase,** the negative membrane potential becomes less negative, reaches zero, and then becomes positive. During the **repolarizing phase,** the membrane potential is restored to the resting state of −70 mV. During an action potential, two types of voltage-gated channels open and then close. These channels are present mainly in the axon plasma membrane and axon terminals. The first channels that open, the Na^+ channels, allow Na^+ to rush into the cell, which causes the depolarizing phase. Then K^+ channels open, allowing K^+ to flow out, which produces the repolarizing phase. Together, the depolarization and repolarization phases last about 1 msec (0.001 sec) in a typical neuron.

Action potentials arise according to the **all-or-none principle.** When depolarization reaches a certain level termed the **threshold** (about −55 mV in many neurons), the voltage-gated Na^+ channels open, and an action potential occurs. The action potential is always the same size (amplitude). This situation is similar to pushing on the first domino in a long row of standing dominos. When the push on the first domino is strong enough (when depolarization reaches threshold), that domino falls against the second domino, and the *entire* row topples (an action potential occurs). Stronger pushes on the first domino produce the identical effect—toppling of the entire row. Thus,

Figure 12.14 Action potential (AP) or impulse. When a stimulus depolarizes the membrane to threshold (−55 mV), an AP is generated. The action potential arises at the trigger zone (here, at the junction of the axon hillock and the initial segment) and then propagates along the axon to the axon terminals. The green-colored regions of the neuron in the inset indicate the parts that typically have voltage-gated Na^+ and K^+ channels (axon plasma membrane and axon terminals).

An action potential consists of depolarizing and repolarizing phases.

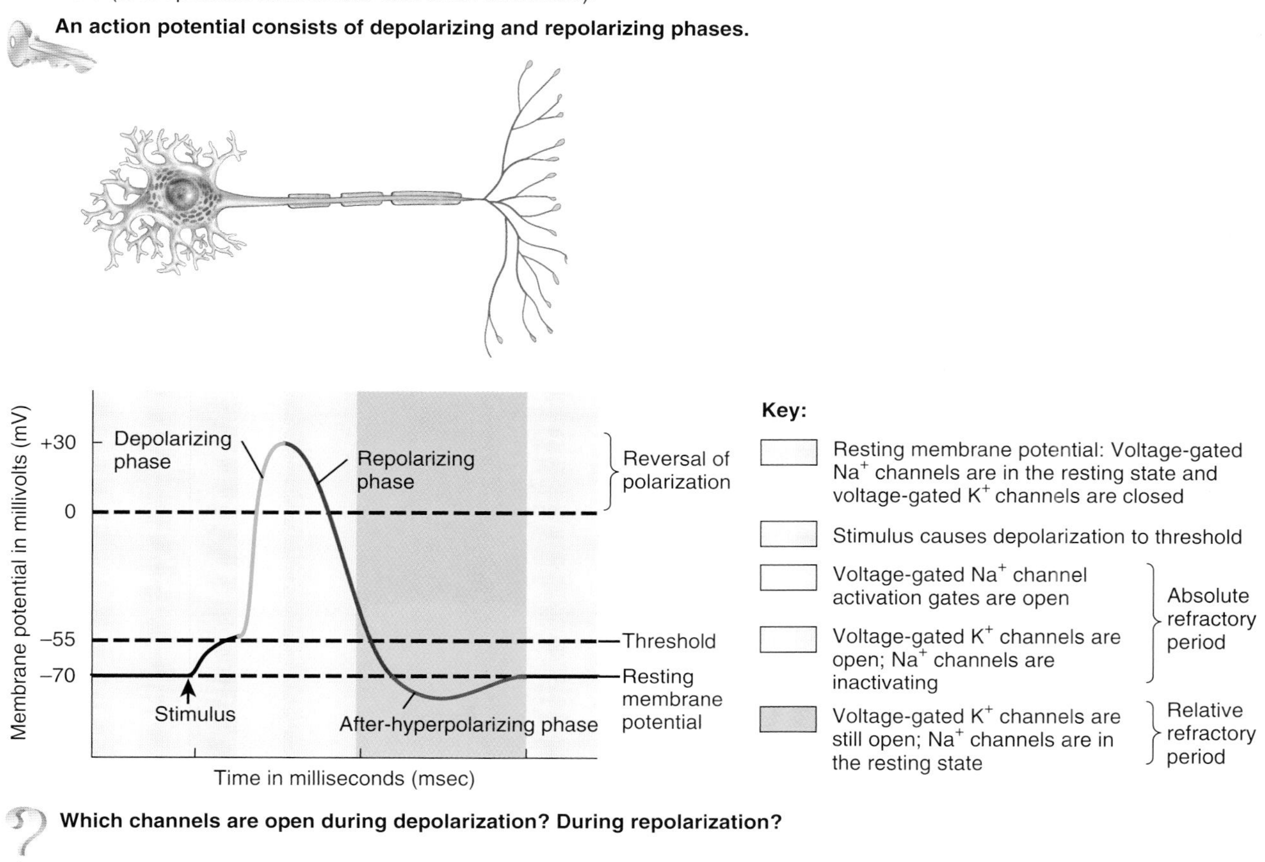

Which channels are open during depolarization? During repolarization?

pushing on the first domino produces an all-or-none event: The dominoes all fall or none fall. Because they can travel long distances without dying out, APs function in communication over both short and long distances. Different neurons may have different thresholds for generation of an action potential, but the threshold in a particular neuron usually is constant.

Depolarizing Phase

When a depolarizing graded potential or some other stimulus causes the membrane to depolarize to threshold, voltage-gated Na^+ channels open rapidly. Both the electrical and the chemical gradients favor inward movement of Na^+, and the resulting inrush of Na^+ causes the depolarizing phase of the action potential (Figure 12.14). The inflow of Na^+ changes the membrane potential from −55 mV to +30 mV. At the peak of the action potential, the inside of the membrane is 30 mV more positive than the outside.

Each voltage-gated Na^+ channel has two separate gates, an *activation gate* and an *inactivation gate.* In the *resting state* of a voltage-gated Na^+ channel, the inactivation gate is open, but the activation gate is closed (step 1 in Figure 12.15). As a result, Na^+ cannot move into the cell through these channels. At threshold, voltage-gated Na^+ channels are activated. In the *activated state* of a voltage-gated Na^+ channel, both the activation and inactivation gates in the channel are open and Na^+ inflow begins (step 2 in Figure 12.15). As more channels open, Na^+ inflow increases, the membrane depolarizes further, and more Na^+ channels open. This is an example of a positive feedback mechanism.

Shortly after the activation gates open the inactivation gates close (step 3 in Figure 12.15). Now the channel is in an *inactivated state.* During the few ten-thousandths of a second that the voltage-gated Na^+ channel is open, about 20,000 Na^+ flow across the membrane and change the membrane potential considerably. But the concentration of Na^+ hardly changes

Figure 12.15 Changes in ion flow through voltage-gated channels during the depolarizing and repolarizing phases of an action potential. Leakage channels and sodium-potassium pumps are not shown.

Inflow of sodium ions (Na^+) causes the depolarizing phase, and outflow of potassium ions (K^+) causes the repolarizing phase of an action potential.

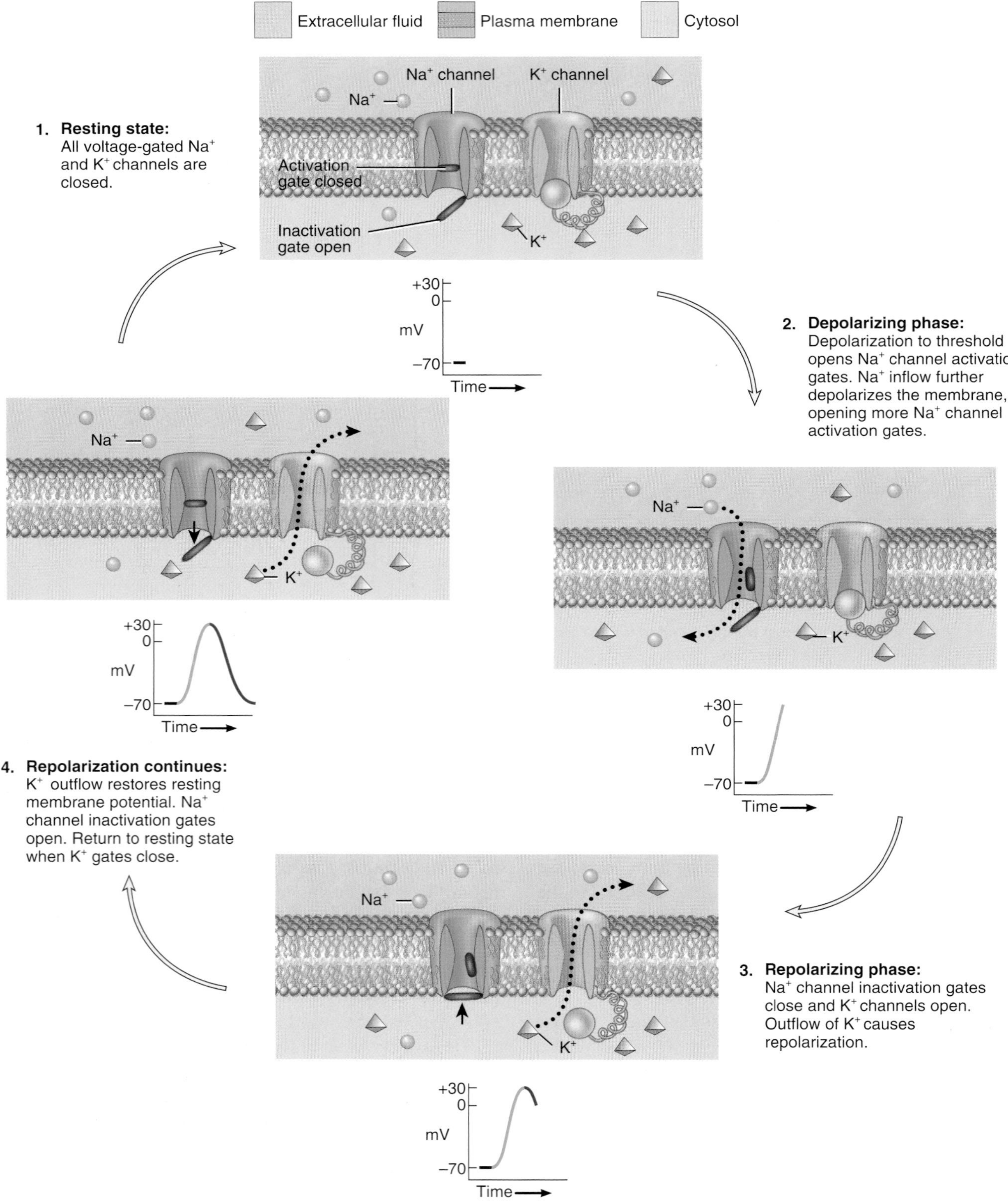

Extracellular fluid
Plasma membrane
Cytosol
Na⁺ channel
K⁺ channel
Na⁺
Activation gate closed
Inactivation gate open
K⁺
1. Resting state:
All voltage-gated Na⁺ and K⁺ channels are closed.
+30
0
mV
–70
Time
2. Depolarizing phase:
Depolarization to threshold opens Na⁺ channel activation gates. Na⁺ inflow further depolarizes the membrane, opening more Na⁺ channel activation gates.
3. Repolarizing phase:
Na⁺ channel inactivation gates close and K⁺ channels open. Outflow of K⁺ causes repolarization.
4. Repolarization continues:
K⁺ outflow restores resting membrane potential. Na⁺ channel inactivation gates open. Return to resting state when K⁺ gates close.

Given the existence of leakage channels for both K^+ and Na^+, could the membrane repolarize if the voltage-gated K^+ channels did not exist?

because of the millions of Na^+ present in the extracellular fluid. The sodium-potassium pumps easily bail out the 20,000 or so Na^+ that enter the cell during a single action potential and maintain the low concentration of Na^+ inside the cell.

Repolarizing Phase

In addition to opening voltage-gated Na^+ channels, a threshold-level depolarization also opens voltage-gated K^+ channels (steps 3 and 4 in Figure 12.15). Because the voltage-gated K^+ channels open more slowly, their opening occurs at about the same time the voltage-gated Na^+ channels are closing. The slower opening of voltage-gated K^+ channels and the closing of previously open Na^+channels produce the repolarizing phase of the action potential. As the Na^+ channels are inactivated, Na^+ inflow slows. At the same time, the K^+ channels are opening, accelerating K^+ outflow. Slowing of Na^+ inflow and acceleration of K^+ outflow causes the membrane potential to change from 30 mV to −70 mV. Repolarization also allows inactivated Na^+ channels to revert to the resting state.

While the voltage-gated K^+ channels are open, outflow of K^+ may be large enough to cause an **after-hyperpolarizing phase** of the action potential (see Figure 12.14). During this phase, the membrane is even more permeable to K^+ than in the resting state, and the membrane potential becomes even more negative (about −90 mV). As the voltage-gated K^+ channels close, activity of the sodium-potassium pump causes the membrane potential to return to the resting level of −70 mV. Unlike voltage-gated Na^+ channels, most voltage-gated K^+ channels do not exhibit an inactivated state. Instead, they alternate between closed (resting) and open (activated) states.

Refractory Period

The period of time after an action potential begins during which an excitable cell cannot generate another action potential is called the **refractory period** (see key in Figure 12.14). During the **absolute refractory period,** even a very strong stimulus cannot initiate a second action potential. This period coincides with the period of Na^+ channel activation and inactivation (steps 2–4 in Figure 12.15). Inactivated Na^+ channels cannot reopen; they first must return to the resting state (step 1 in Figure 12.15). In contrast to action potentials, graded potentials do not exhibit a refractory period.

Large-diameter axons have a larger surface area and have a brief absolute refractory period of about 0.4 msec. Because a second nerve impulse can arise very quickly, up to 1000 impulses per second are possible. Small-diameter axons have absolute refractory periods as long as 4 msec, enabling them to transmit a maximum of 250 impulses per second. Under normal body conditions, the maximum frequency of nerve impulses in different axons ranges between 10 and 1000 per second.

The **relative refractory period** is the period of time during which a second action potential can be initiated, but only by a larger-than-normal stimulus. It coincides with the period when the voltage-gated K^+ channels are still open after inactivated Na^+ channels have returned to their resting state (see Figure 12.14).

Propagation of Nerve Impulses

To communicate information from one part of the body to another, nerve impulses must travel from where they arise at a trigger zone, to the axon terminals. This mode of travel is called **propagation** or **conduction,** and it depends on positive feedback. As you have already learned, when sodium ions flow in, they cause voltage-gated Na^+ channels in adjacent segments of the membrane to open. Thus, the nerve impulse travels along the membrane rather like the activity of that long row of dominoes. A nerve impulse normally moves in one direction only—from the trigger zone toward the axon terminals.

Neurotoxins and Local Anesthetics

Certain shellfish and other organisms contain **neurotoxins,** substances that produce their poisonous effects by acting on the nervous system. One particularly lethal neurotoxin is tetrodotoxin (TTX), present in the viscera of Japanese pufferfish. TTX effectively blocks action potentials by inserting itself into voltage-gated Na^+ channels so they cannot open.

Local anesthetics are drugs that block pain and other somatic sensations. Examples include procaine (Novocaine®) and Lidocaine, which may be used to produce anesthesia in the skin during suturing of a gash, in the mouth during dental work, or in the lower body during childbirth. Like TTX, these drugs act by blocking the opening of voltage-gated Na^+ channels. Nerve impulses cannot propagate past the obstructed region, so pain signals do not reach the CNS.

Localized cooling of a nerve can also produce an anesthetic effect because axons propagate impulses at lower speeds when cooled. The application of ice to injured tissue can reduce pain because propagation of the pain sensations along axons is partially blocked. ■

Continuous and Saltatory Conduction

The type of impulse propagation described so far occurs in muscle fibers and unmyelinated axons. Such step-by-step depolarization and repolarization of each adjacent segment of the plasma membrane is called **continuous conduction** (Figure 12.16a). In continuous conduction, ions flow through their voltage-gated channels in each adjacent segment of the membrane. Note that the impulse propagates only a relatively short distance in a few milliseconds.

Nerve impulses propagate more rapidly along myelinated axons than along unmyelinated axons. If you compare parts a and b in Figure 12.16 you will see that the impulse propagates much farther along the myelinated axon in the same period of time. **Saltatory conduction** (SAL-ta-tō-rē; *saltat-* = leaping), the special mode of impulse propagation that occurs along myelinated axons, occurs because of the uneven distribution of voltage-gated channels. Few voltage-gated channels are present in regions where a myelin sheath covers the axolemma. By contrast, at the nodes of Ranvier (where there is no myelin sheath), the axolemma has many voltage-gated channels. Hence, current carried by Na^+ and K^+ flows across the membrane mainly at the nodes.

When a nerve impulse propagates along a myelinated axon, an electric current (carried by ions) flows through the extracellular fluid surrounding the myelin sheath and through the cytosol from one node to the next. The nerve impulse at the first node generates ionic currents in the cytosol and extracellular fluid that depolarize the membrane to threshold, opening voltage-gated Na^+ channels at the second node. The resulting ionic flow through the opened channels constitutes a nerve impulse at the second node. Then, the nerve impulse at the second node generates an ionic current that opens voltage-gated Na^+ channels at the third node, and so on. Each node repolarizes after it depolarizes.

The flow of current across the membrane only at the nodes of Ranvier has two consequences.

1. The impulse appears to "leap" from node to node as each nodal area depolarizes to threshold, thus the name "saltatory." Because an impulse leaps across long segments of the myelinated axolemma as current flows from one node to the next, it

Figure 12.16 Conduction (propagation) of a nerve impulse after it arises at the trigger zone. Dotted lines indicate ionic current flow. The insets show the path of current flow. (a) In continuous conduction along an unmyelinated axon, ionic currents flow across each adjacent segment of the membrane. (b) In saltatory conduction along a myelinated axon, the nerve impulse at the first node generates ionic currents in the cytosol and interstitial fluid that open voltage-gated Na^+ channels at the second node, and so on at each subsequent node.

Unmyelinated axons exhibit continuous conduction; myelinated axons exhibit saltatory conduction.

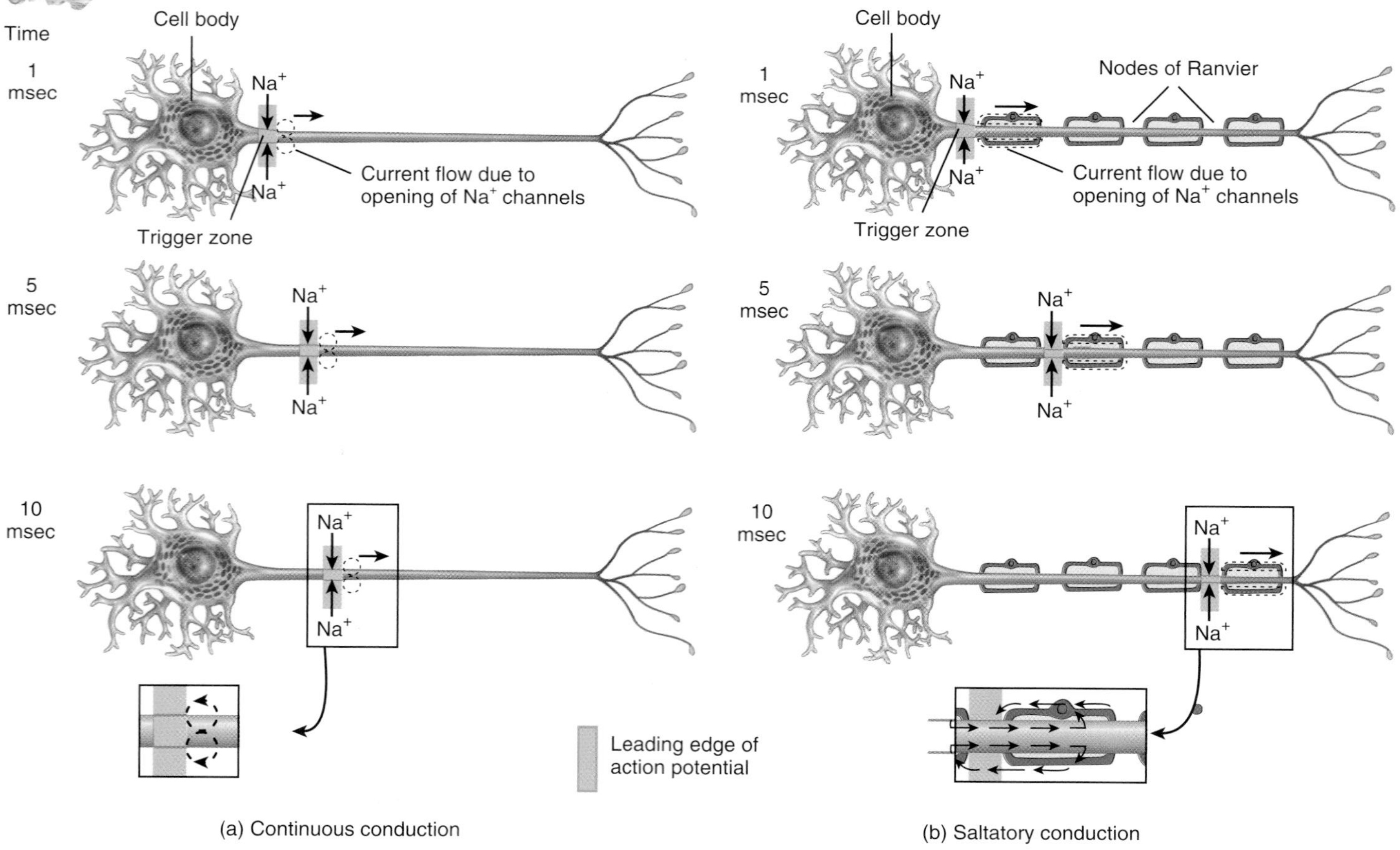

(a) Continuous conduction

(b) Saltatory conduction

What factors determine the speed of propagation of a nerve impulse?

travels much faster than it would in an unmyelinated axon of the same diameter.

2. Opening a smaller number of channels only at the nodes, rather than many channels in each adjacent segment of membrane, represents a more energy-efficient mode of conduction. Because only small regions of the membrane depolarize and repolarize, minimal inflow of Na^+ and outflow of K^+ occurs each time a nerve impulse passes by. Thus, less ATP is used by sodium-potassium pumps to maintain the low intracellular concentration of Na^+ and the low extracellular concentration of K^+.

Effect of Axon Diameter

Larger-diameter axons propagate impulses faster than smaller ones due to their larger surface areas. All of the largest-diameter axons (about 5–20 μm), called **A fibers,** are myelinated. A fibers have a brief absolute refractory period and conduct impulses at speeds of 12 to 130 m/sec (27–280 mi/hr). The axons of sensory neurons that propagate impulses associated with touch, pressure, position of joints, and some thermal sensations are A fibers, as are the axons of motor neurons that conduct impulses to skeletal muscles.

B fibers are axons with diameters of 2–3 μm. Like A fibers, B fibers are myelinated and exhibit saltatory conduction at speeds up to 15 m/sec (32 mi/hr). B fibers have a somewhat longer absolute refractory period than A fibers. B fibers conduct sensory nerve impulses from the viscera to the brain and spinal cord. They also constitute all the axons of the autonomic motor neurons that extend from the brain and spinal cord to the ANS relay stations called autonomic ganglia.

C fibers are the smallest diameter axons (0.5–1.5 μm) and all are unmyelinated. Nerve impulse propagation along a C fiber ranges from 0.5 to 2 m/sec (1–4 mi/hr). C fibers exhibit the longest absolute refractory periods. These unmyelinated axons conduct some sensory impulses for pain, touch, pressure, heat, and cold from the skin, and pain impulses from the viscera. Autonomic motor fibers that extend from autonomic ganglia to stimulate the heart, smooth muscle, and glands are C fibers. Examples of motor functions of B and C fibers are constricting and dilating the pupils, increasing and decreasing the heart rate, and contracting and relaxing the urinary bladder.

Encoding of Stimulus Intensity

How can your sensory systems detect stimuli of differing intensities if all nerve impulses are the same size? Why does a light touch feel different from firmer pressure? The main answer to this question is the *frequency of impulses*—how often they are generated at the trigger zone. A light touch generates a low frequency of nerve impulses. A firmer pressure elicits nerve impulses that pass down the axon at a higher frequency. In addition to this "frequency code," a second factor is the number of sensory neurons recruited (activated) by the stimulus. A firm pressure stimulates a larger number of pressure-sensitive neurons than does a light touch.

Comparison of Electrical Signals Produced by Excitable Cells

We have seen that excitable cells—neurons and muscle fibers—produce two types of electrical signals, graded potentials and action potentials (impulses). One obvious difference between them is that the propagation of action potentials permits communication over long distances, but graded potentials can function only in short-distance communication because they are not propagated. Table 12.1 presents a summary of the differences between graded potentials and action potentials.

As we discussed in Chapter 10, propagation of a muscle action potential along the sarcolemma and into the T tubule system initiates the events of muscle contraction. Although action potentials in muscle fibers and in neurons are similar, there are some notable differences. The typical resting membrane potential of a neuron is −70 mV, but it is closer to −90 mV in skeletal and cardiac muscle fibers. The duration of a nerve impulse is 0.5–2 msec, but a muscle action potential is considerably longer—about 1.0–5.0 msec for skeletal muscle fibers and 10–300 msec for cardiac and smooth muscle fibers. Finally, the conduction speed of action potentials along the

TABLE 12.1 Comparison of Graded Potentials and Action Potentials

Characteristic	Graded Potentials	Action Potentials
Origin	Arise mainly in dendrites and cell body (some arise in axons).	Arise at trigger zones and propagate along the axon.
Types of channels	Ligand-gated or mechanically gated ion channels.	Voltage-gated channels for Na^+ and K^+.
Conduction	Not propagated; localized and thus permit communication over a few micrometers.	Propagate and thus permit communication over longer distances.
Amplitude	Depending on strength of stimulus, varies from less than 1 mV to more than 50 mV.	All-or-none; typically about 100 mV.
Duration	Typically longer, ranging from several msec to several min.	Shorter, ranging from 0.5 to 2 msec.
Polarity	May be hyperpolarizing (inhibitory to generation of an action potential) or depolarizing (excitatory to generation of an action potential).	Always consist of depolarizing phase followed by repolarizing phase and return to resting membrane potential.
Refractory period	Not present, thus spatial and temporal summation can occur (described shortly).	Present, thus summation cannot occur.

largest-diameter, myelinated axons is about 18 times faster than the conduction speed along the sarcolemma of a skeletal muscle fiber.

▶ CHECKPOINT

9. Define the terms *resting membrane potential, depolarization, repolarization, nerve impulse,* and *refractory period* and identify the factors responsible for each.
10. How is saltatory conduction different from continuous conduction?
11. What factors determine the speed of propagation of nerve impulses?
12. How can you tell the difference between a stroke on the cheek and a slap across the face?

SIGNAL TRANSMISSION AT SYNAPSES

▶ OBJECTIVES

Explain the events of signal transmission at a chemical synapse.

Distinguish between spatial and temporal summation.

Give examples of excitatory and inhibitory neurotransmitters, and describe how they act.

In Chapter 10 we described the events occurring at one type of synapse, the neuromuscular junction. Our focus in this chapter is on synaptic communication among the billions of neurons in the nervous system. Synapses are essential for homeostasis because they allow information to be filtered and integrated. During learning, the structure and function of particular synapses change. The changes may allow some signals to be transmitted while others are blocked. For example, the changes in your synapses from studying will determine how well you do on your anatomy and physiology tests! Synapses are also important because some diseases and neurological disorders result from disruptions of synaptic communication, and many therapeutic and addictive chemicals affect the body at these junctions.

At a synapse between neurons, the neuron sending the signal is called the **presynaptic neuron,** and the neuron receiving the message is called the **postsynaptic neuron.** Most synapses are either **axodendritic** (from axon to dendrite), **axosomatic** (from axon to cell body), or **axoaxonic** (from axon to axon). The two types of synapses—electrical and chemical—differ both structurally and functionally.

Electrical Synapses

At an **electrical synapse,** action potentials (impulses) conduct directly between adjacent cells through structures called **gap junctions.** Each gap junction contains a hundred or so tubular *connexons,* which act like tunnels to connect the cytosol of the two cells directly (see Figure 4.1e on page 109). As ions flow from one cell to the next through the connexons, the action potential spreads from cell to cell. Gap junctions are common in visceral smooth muscle, cardiac muscle, and the developing embryo. They also occur in the CNS.

Electrical synapses have two main advantages:

1. ***Faster communication.*** Because action potentials conduct directly through gap junctions, electrical synapses are faster than chemical synapses. At an electrical synapse, the action potential passes directly from the presynaptic cell to the postsynaptic cell. The events that occur at a chemical synapse take some time and delay communication slightly.
2. ***Synchronization.*** Electrical synapses can synchronize the activity of a group of neurons or muscle fibers. In other words, a large number of neurons or muscle fibers can produce action potentials in unison if they are connected by gap junctions. The value of synchronized action potentials in the heart or in visceral smooth muscle is coordinated contraction of these fibers to produce a heartbeat or move food through the gastrointestinal tract.

Chemical Synapses

Although the plasma membranes of presynaptic and postsynaptic neurons in a **chemical synapse** are close, they do not touch. They are separated by the **synaptic cleft,** a space of 20–50 nm* that is filled with interstitial fluid. Nerve impulses cannot conduct across the synaptic cleft, so an alternate, indirect form of communication occurs. In response to a nerve impulse, the presynaptic neuron releases a neurotransmitter that diffuses through the fluid in the synaptic cleft and binds to receptors in the plasma membrane of the postsynaptic neuron. The postsynaptic neuron receives the chemical signal and, in turn, produces a **postsynaptic potential,** a type of graded potential. Thus, the presynaptic neuron converts an electrical signal (nerve impulse) into a chemical signal (released neurotransmitter). The postsynaptic neuron receives the chemical signal and, in turn, generates an electrical signal (postsynaptic potential). The time required for these processes at a chemical synapse, a **synaptic delay** of about 0.5 msec, is the reason that chemical synapses relay signals more slowly than electrical synapses.

A typical chemical synapse transmits a signal as follows (Figure 12.17):

1. A nerve impulse arrives at a synaptic end bulb (or at a varicosity) of a presynaptic axon.
2. The depolarizing phase of the nerve impulse opens **voltage-gated Ca^{2+} channels,** which are present in the membrane

*1 nanometer (nm) = 10^{-9} (0.000,000,001) meter.

of synaptic end bulbs. Because calcium ions are more concentrated in the extracellular fluid, Ca^{2+} flows inward through the opened channels.

3 An increase in the concentration of Ca^{2+} inside the presynaptic neuron serves as a signal that triggers exocytosis of the synaptic vesicles. As vesicle membranes merge with the plasma membrane, neurotransmitter molecules within the vesicles are released into the synaptic cleft. Each synaptic vesicle contains several thousand molecules of neurotransmitter.

4 The neurotransmitter molecules diffuse across the synaptic cleft and bind to **neurotransmitter receptors** in the postsynaptic neuron's plasma membrane. The receptor shown in Figure 12.17 is part of a ligand-gated channel (see Figure 12.11b); in other cases the receptor may be a separate protein in the membrane.

5 Binding of neurotransmitter molecules to their receptors on ligand-gated channels opens the channels and allows particular ions to flow across the membrane.

6 As ions flow through the opened channels, the voltage across the membrane changes. This change in membrane voltage is a **postsynaptic potential.** Depending on which ions the channels admit, the postsynaptic potential may be a depolarization or hyperpolarization. For example, opening of Na^+ channels allows inflow of Na^+, which causes depolarization. However, opening of Cl^- or K^+ channels causes hyperpolarization. Opening Cl^- channels permits Cl^- to move into the cell, while opening the K^+ channels allows K^+ to move out—in either event, the inside of the cell becomes more negative.

7 When a depolarizing postsynaptic potential reaches threshold, it triggers an action potential.

Figure 12.17 Signal transmission at a chemical synapse. Through exocytosis of synaptic vesicles, a presynaptic neuron releases neurotransmitter molecules. After diffusing across the synaptic cleft, the neurotransmitter binds to receptors in the plasma membrane of the postsynaptic neuron and produces a postsynaptic potential.

At a chemical synapse, a presynaptic neuron converts an electrical signal (nerve impulse) into a chemical signal (neurotransmitter release). The postsynaptic neuron then converts the chemical signal back into an electrical signal (postsynaptic potential).

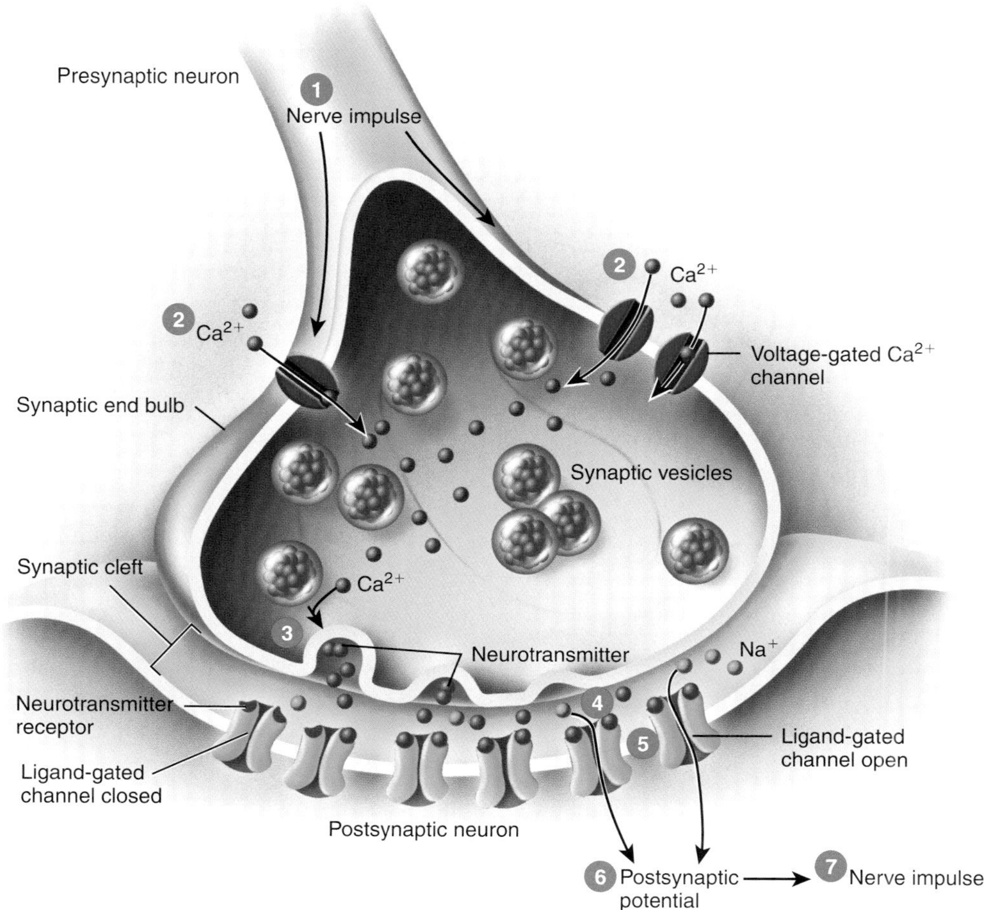

Why may electrical synapses work in two directions, but chemical synapses can transmit a signal in only one direction?

At most chemical synapses, only *one-way information transfer* can occur—from a presynaptic neuron to a postsynaptic neuron or an effector, such as a muscle fiber or a gland cell. For example, synaptic transmission at a neuromuscular junction (NMJ) proceeds from a somatic motor neuron to a skeletal muscle fiber (but not in the opposite direction). Only synaptic end bulbs of presynaptic neurons can release neurotransmitter, and only the postsynaptic neuron's membrane has the receptor proteins that can recognize and bind that neurotransmitter. As a result, action potentials move in one direction.

Excitatory and Inhibitory Postsynaptic Potentials

A neurotransmitter causes either an excitatory or an inhibitory graded potential. A neurotransmitter that *depolarizes* the postsynaptic membrane is excitatory because it brings the membrane closer to threshold (see Figure 12.13b). A depolarizing postsynaptic potential is called an **excitatory postsynaptic potential (EPSP).** Often, EPSPs result from opening of *cation* channels. These channels allow passage of the three most plentiful cations (Na^+, K^+, and Ca^{2+}) through the postsynaptic cell membrane, but Na^+ inflow is greater than either Ca^{2+} inflow or K^+ outflow. Although a single EPSP normally does not initiate a nerve impulse, the postsynaptic cell does become more excitable. Because it is partially depolarized, it is more likely to reach threshold when the next EPSP occurs.

A neurotransmitter that causes *hyperpolarization* of the postsynaptic membrane (see Figure 12.13a) is inhibitory. During hyperpolarization, generation of an action potential is more difficult than usual because the membrane potential becomes more negative and thus even farther from threshold than in its resting state. A hyperpolarizing postsynaptic potential is termed an **inhibitory postsynaptic potential (IPSP).** IPSPs often result from the opening of ligand-gated Cl^- or K^+ channels. When Cl^- channels open, a larger number of chloride ions diffuse inward. When K^+ channels open, a larger number of potassium ions diffuses outward. In both cases, the ionic flow causes the inside of the postsynaptic cell to become more negative (hyperpolarized).

Removal of Neurotransmitter

Removal of the neurotransmitter from the synaptic cleft is essential for normal synaptic function. If a neurotransmitter could linger in the synaptic cleft, it would influence the postsynaptic neuron, muscle fiber, or gland cell indefinitely. Neurotransmitter is removed in three ways:

1. ***Diffusion.*** Some of the released neurotransmitter molecules diffuse away from the synaptic cleft. Once a neurotransmitter molecule is out of reach of its receptors, it can no longer exert an effect.

2. ***Enzymatic degradation.*** Certain neurotransmitters are inactivated through enzymatic degradation. For example, the enzyme acetylcholinesterase breaks down acetylcholine in the synaptic cleft.

3. ***Uptake by cells.*** Many neurotransmitters are actively transported back into the neuron that released them (reuptake). Others are transported into neighboring neuroglia (uptake). The neurons that release norepinephrine, for example, rapidly take up the norepinephrine and recycle it into new synaptic vesicles. The membrane proteins that accomplish such uptake are called *neurotransmitter transporters.* Several therapeutically important drugs selectively block reuptake of specific neurotransmitters by interfering with these transporters. For example, the drug fluoxetine (Prozac®) is a **selective serotonin reuptake inhibitor (SSRI).** By inhibiting serotonin transporters, Prozac prolongs the activity of this neurotransmitter at synapses in the brain. SSRIs provide relief for those suffering from some forms of depression.

Spatial and Temporal Summation of Postsynaptic Potentials

A typical neuron in the CNS receives input from 1000 to 10,000 synapses. Integration of these inputs, which is known as **summation,** occurs at the trigger zone. The greater the summation of EPSPs, the greater the chance that threshold will be reached. At threshold, one or more nerve impulses (action potentials) arise.

When summation results from buildup of neurotransmitter released simultaneously by *several* presynaptic end bulbs, it is called **spatial summation** (Figure 12.18a). When summation results from buildup of neurotransmitter released by a *single* presynaptic end bulb two or more times in rapid succession, it is called **temporal summation** (Figure 12.18b). Because a typical EPSP lasts about 15 msec, the second (and subsequent) release of neurotransmitter must occur soon after the first one if temporal summation is to occur. Summation is rather like a vote on the Internet. Many people voting "yes" or "no" on an issue at the same time can be compared to spatial summation. One person voting repeatedly and rapidly is like temporal summation. Most of the time, spatial and temporal summations are acting together to influence the chance that a neuron fires an impulse.

A single postsynaptic neuron receives input from many presynaptic neurons, some of which release excitatory neurotransmitters and some of which release inhibitory neurotransmitters. The sum of all the excitatory and inhibitory effects at any given time determines the effect on the postsynaptic neuron, which may respond in the following ways:

1. ***EPSP.*** If the total excitatory effects are greater than the total inhibitory effects but less than the threshold level of stimulation, the result is an EPSP that does not reach threshold. Following an EPSP, subsequent stimuli can more easily generate a nerve impulse through summation because the neuron is partially depolarized.

Figure 12.18 Spatial and temporal summation. (a) When presynaptic neurons a and b separately cause EPSPs (arrows) in postsynaptic neuron c, the threshold level is not reached in neuron c. Spatial summation occurs only when neurons a and b act simultaneously on neuron c; their EPSPs sum to reach the threshold level and trigger a nerve impulse. (b) Temporal summation occurs when stimuli applied to the same axon in rapid succession (arrows) cause overlapping EPSPs that sum. When depolarization reaches the threshold level, a nerve impulse is triggered.

The sum of all excitatory and inhibitory postsynaptic potentials determines whether an action potential is generated.

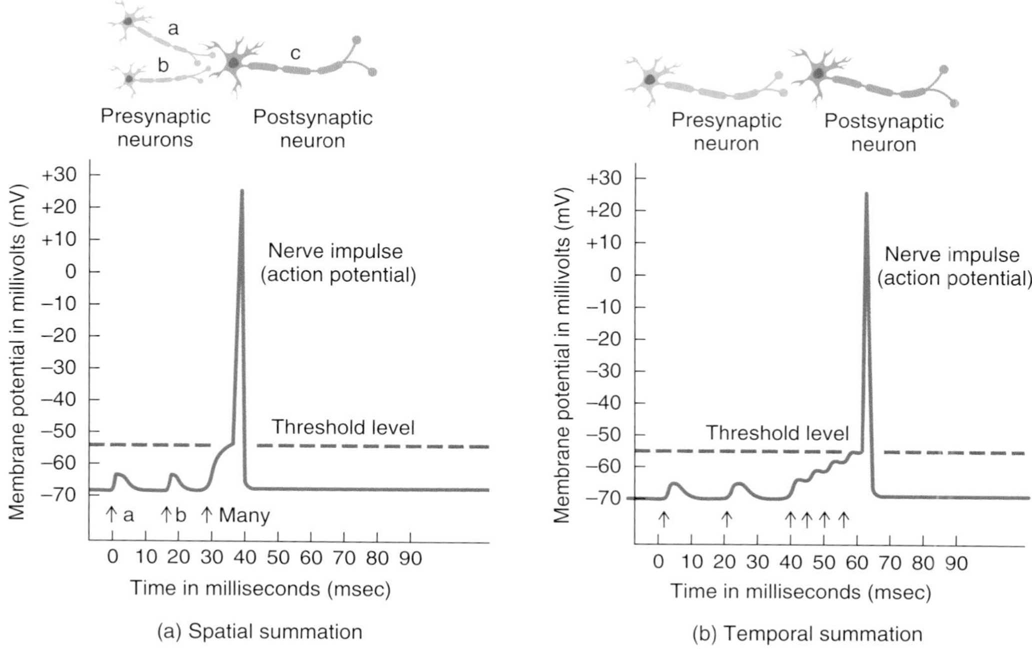

What would be the result if in addition to the four EPSPs indicated by arrows in part (b) an IPSP occurred at time 55 msec?

2. ***Nerve impulse(s).*** If the total excitatory effects are greater than the total inhibitory effects and threshold is reached, one or more nerve impulses will be triggered. Impulses continue to be generated as long as the EPSP is at or above the threshold level.
3. ***IPSP.*** If the total inhibitory effects are greater than the excitatory effects, the membrane hyperpolarizes (IPSP). The result is inhibition of the postsynaptic neuron and an inability to generate a nerve impulse.

Table 12.2 summarizes the structural and functional elements of a neuron.

Strychnine Poisoning

The importance of inhibitory neurons can be appreciated by observing what happens when their activity is blocked. Normally, inhibitory neurons in the spinal cord called *Renshaw cells* release the neurotransmitter glycine (described shortly) at inhibitory synapses with somatic motor neurons. This inhibitory input to their motor neurons prevents excessive contraction of skeletal muscles. Strychnine is a lethal poison that binds to and blocks glycine receptors. The normal, delicate balance between excitation and inhibition in the CNS is disturbed, and motor neurons generate nerve impulses without restraint. All skeletal muscles, including the diaphragm, contract fully and remain contracted. Because the diaphragm cannot relax, the victim cannot inhale, and suffocation results. ■

▶ CHECKPOINT

13. How is neurotransmitter removed from the synaptic cleft?
14. How are excitatory and inhibitory postsynaptic potentials similar and different?
15. Why are action potentials said to be "all-or-none," and EPSPs and IPSPs are described as "graded"?

TABLE 12.2 Summary of Neuronal Structure and Function

Structure	Functions
Dendrites	Receive stimuli through activation of ligand-gated or mechanically gated ion channels; in sensory neurons, produce generator or receptor potentials; in motor neurons and interneurons, produce excitatory and inhibitory postsynaptic potentials (EPSPs and IPSPs).
Cell body	Receives stimuli and produces EPSPs and IPSPs through activation of ligand-gated or mechanically gated ion channels.
Junction of axon hillock and initial segment of axon	Trigger zone in many neurons; integrates EPSPs and IPSPs and, if sum is a depolarization that reaches threshold, initiates action potential (nerve impulse).
Axon	Propagates (conducts) nerve impulses from initial segment (or from dendrites of sensory neurons) to axon terminals in a self-reinforcing manner; impulse amplitude does not change as it propagates along the axon.
Axon terminals and synaptic end bulbs (or varicosities)	Inflow of Ca^{2+} caused by depolarizing phase of nerve impulse triggers exocytosis of neurotransmitter from synaptic vesicles.

Plasma membrane includes chemically gated channels

Plasma membrane includes voltage-gated Na^+ and K^+ channels

Plasma membrane includes voltage-gated Ca^{2+} channels

NEUROTRANSMITTERS

▶ OBJECTIVE

Describe the classes and functions of neurotransmitters.

About 100 substances are either known or suspected neurotransmitters. Some neurotransmitters bind to their receptors and act quickly to open or close ion channels in the membrane. Others act more slowly via second-messenger systems to influence chemical reactions inside cells. The result of either process can be excitation or inhibition of postsynaptic neurons. Many neurotransmitters are also hormones released into the bloodstream by endocrine cells in organs throughout the body. Within the brain, certain neurons, called **neurosecretory cells,** also secrete hormones. Neurotransmitters can be divided into two classes based on size: small-molecule neurotransmitters and neuropeptides.

Small-Molecule Neurotransmitters

The small-molecule neurotransmitters include acetylcholine, amino acids, biogenic amines, ATP and other purines, and nitric oxide.

Acetylcholine

The best-studied neurotransmitter is **acetylcholine (ACh),** which is released by many PNS neurons and by some CNS neurons. ACh is an excitatory neurotransmitter at some synapses, such as the neuromuscular junction, where it acts directly to open ligand-gated cation channels. It is also an inhibitory neurotransmitter at other synapses, where its effects on ion channels occur indirectly via receptors that link to a G protein. For example, ACh slows heart rate at inhibitory synapses made by parasympathetic neurons of the vagus (X) nerve. The enzyme *acetylcholinesterase (AChE)* inactivates ACh by splitting it into acetate and choline fragments.

Amino Acids

Several amino acids are neurotransmitters in the CNS. **Glutamate** (glutamic acid) and **aspartate** (aspartic acid) have powerful excitatory effects. Most excitatory neurons in the CNS and perhaps half of the synapses in the brain communicate via glutamate. At some glutamate synapses, binding of the neurotransmitter to its receptors opens Ca^{2+} channels. The consequent inflow of calcium ions produces an EPSP. Inactivation of glutamate occurs via reuptake. Glutamate transporters actively transport glutamate back into the synaptic end bulbs and neighboring neuroglia.

Gamma aminobutyric (GAM-ma am-i-nō-bū-TIR-ik) **acid (GABA)** and **glycine** are important inhibitory neurotransmitters. Both cause IPSPs by opening Cl^- channels. GABA is found only in the CNS, where it is the most common inhibitory neurotransmitter. As many as one-third of all brain synapses use GABA. Antianxiety drugs such as diazepam (Valium®) enhance the action of GABA. About half of the inhibitory synapses in the spinal cord use the amino acid glycine; the rest use GABA.

Excitotoxicity

A high level of glutamate in the interstitial fluid of the CNS causes **excitotoxicity**—destruction of neurons through prolonged activation of excitatory synaptic transmission. The most common cause of excitotoxicity is oxygen deprivation of the brain due to ischemia (inadequate blood flow), as happens during a stroke. Lack of oxygen causes the glutamate transporters to fail, and glutamate accumulates in the interstitial spaces between neurons and glia, literally stimulating the neurons to death. Clinical trials are underway to see if antiglutamate drugs administered after a stroke can offer some protection from excitotoxicity. ■

Biogenic Amines

Certain amino acids are modified and decarboxylated (carboxyl group removed) to produce biogenic amines. Those that are prevalent in the nervous system include norepinephrine, epinephrine, dopamine, and serotonin. There are three or more different types of receptors for each biogenic amine; biogenic amines may cause either excitation or inhibition, depending on the type of receptor at the synapse.

Norepinephrine (NE) plays roles in arousal (awakening from deep sleep), dreaming, and regulating mood. A smaller number of neurons in the brain use **epinephrine** as a neurotransmitter. Both epinephrine and norepinephrine also serve as hormones. Cells of the adrenal medulla, the inner portion of the adrenal gland, release them into the blood.

Brain neurons containing the neurotransmitter **dopamine (DA)** are active during emotional responses, addictive behaviors, and pleasurable experiences. In addition, dopamine-releasing neurons help regulate skeletal muscle tone and some aspects of movement due to contraction of skeletal muscles. The muscular stiffness that occurs in Parkinson disease is due to degeneration of neurons that release dopamine (see page 568). One form of schizophrenia is due to accumulation of excess dopamine.

Norepinephrine, dopamine, and epinephrine are classified chemically as **catecholamines** (cat-e-KŌL-a-mēns). They all include an amino group ($—NH_2$) and a catechol ring composed of six carbons and two adjacent hydroxyl (—OH) groups. Catecholamines are synthesized from the amino acid tyrosine. Inactivation of catecholamines occurs via reuptake into synaptic end bulbs. Then they are either recycled back into the synaptic vesicles or destroyed by the enzymes. The two enzymes that break down catecholamines are **catechol-*O*-methyltransferase** (kat′-e-kōl-ō-meth-il-TRANS-fer-ās), or **COMT,** and **monoamine oxidase** (mon-ō-AM-īn OK-si-dās), or **MAO.**

Serotonin, also known as **5-hydroxytryptamine (5-HT),** is concentrated in the neurons in a part of the brain called the raphe nucleus. It is thought to be involved in sensory perception, temperature regulation, control of mood, appetite, and the induction of sleep.

ATP and Other Purines

The characteristic ring structure of the adenosine portion of ATP (shown in Figure 2.25 on page 55) is called a purine ring. Adenosine itself, as well as its triphosphate, diphosphate, and monophosphate derivatives (ATP, ADP, and AMP), is an excitatory neurotransmitter in both the CNS and the PNS. Most of the synaptic vesicles that contain ATP also contain another neurotransmitter. In the PNS, ATP and norepinephrine are released together from some sympathetic neurons; some parasympathetic neurons release ATP and acetylcholine in the same vesicles.

Nitric Oxide

The simple gas **nitric oxide (NO)** is an important neurotransmitter that has widespread effects throughout the body. NO includes a single nitrogen atom, in contrast to nitrous oxide (N_2O), or laughing gas, which has two nitrogen atoms. N_2O is sometimes used as an anesthetic during dental procedures.

The enzyme **nitric oxide synthase (NOS)** catalyzes formation of NO from the amino acid arginine. Based on the presence of NOS, it is estimated that more than 2% of the neurons in the brain produce NO. Unlike all previously known neurotransmitters, NO is not synthesized in advance and packaged into synaptic vesicles. Rather, it is formed on demand and acts immediately. Its action is brief because NO is a highly reactive free radical. It exists for less than 10 seconds before it combines with oxygen and water to form inactive nitrates and nitrites. Because NO is lipid soluble, it diffuses from cells that produce it into neighboring cells, where it activates an enzyme for production of a second messenger called cyclic GMP. Some research suggests that NO plays a role in memory and learning.

The first recognition of NO as a regulatory molecule was the discovery in 1987 that a chemical called EDRF (endothelium-derived relaxing factor) was actually NO. Endothelial cells in blood vessel walls release NO, which diffuses into neighboring smooth muscle cells and causes relaxation. The result is vasodilation, an increase in blood vessel diameter. The effects of such vasodilation range from a lowering of blood pressure to erection of the penis in males. Sildenafil (Viagra®) alleviates erectile dysfunction (impotence) by enhancing the effect of NO. In larger quantities, NO is highly toxic. Phagocytic cells, such as macrophages and certain white blood cells, produce NO to kill microbes and tumor cells.

Neuropeptides

Neurotransmitters consisting of 3 to 40 amino acids linked by peptide bonds are called **neuropeptides** (noor-ō-PEP-tīds). Numerous and widespread in both the CNS and the PNS, neuropeptides have both excitatory and inhibitory actions. Neuropeptides are formed in the neuron cell body, packaged into vesicles, and transported to axon terminals. Besides their role as neurotransmitters, many neuropeptides serve as hormones that regulate physiological responses elsewhere in the body.

Scientists discovered that certain brain neurons have plasma membrane receptors for opiate drugs such as morphine and heroin. The quest to find the naturally occurring substances that use these receptors brought to light the first neuropeptides: two molecules, each a chain of five amino acids, named **enkephalins** (en-KEF-a-lins). Their potent analgesic (pain-relieving) effect is 200 times stronger than morphine. Other so-called *opioid peptides* include the **endorphins** (en-DOR-fins) and **dynorphins** (dī-NOR-fins). It is thought that opioid peptides are the body's natural painkillers. Acupuncture may produce analgesia (loss of pain sensation) by increasing the release of opioids. These neuropeptides have also been linked to improved memory and learning; feelings of pleasure or euphoria; control of body temperature; regulation of hormones that affect the onset of puberty, sexual drive, and reproduction; and mental illnesses such as depression and schizophrenia.

Another neuropeptide, **substance P,** is released by neurons that transmit pain-related input from peripheral pain receptors into the central nervous system, enhancing the perception of pain. Enkephalin and endorphin suppress the release of substance P, thus decreasing the number of nerve impulses being relayed to the brain for pain sensations. Substance P has also been shown to counter the effects of certain nerve-damaging chemicals, prompting speculation that it might prove useful as a treatment for nerve degeneration.

Table 12.3 provides brief descriptions of these neuropeptides, as well as others that will be discussed in later chapters.

Modifying the Effects of Neurotransmitters

Substances naturally present in the body as well as drugs and toxins can ***modify the effects of neurotransmitters*** in several ways:

1. Neurotransmitter synthesis can be stimulated or inhibited. For instance, many patients with Parkinson disease (see page 568) receive benefit from the drug L-dopa because it is a precursor of dopamine. For a limited period of time, taking L-dopa boosts dopamine production in affected brain areas.

2. Neurotransmitter release can be enhanced or blocked. Amphetamines promote release of dopamine and norepinephrine. Botulinum toxin causes paralysis by blocking release of acetylcholine from somatic motor neurons.

3. The neurotransmitter receptors can be activated or blocked. An agent that binds to receptors and enhances or mimics the effect of a natural neurotransmitter is an **agonist.** Isoproterenol (Isuprel®) is a powerful agonist of epinephrine and norepinephrine. It can be used to dilate the airways during an asthma attack. An agent that binds to and blocks neurotransmitter receptors is an **antagonist.** Zyprexa®, a drug prescribed for schizophrenia, is an antagonist of serotonin and dopamine.

4. Neurotransmitter removal can be stimulated or inhibited. For example, cocaine produces euphoria—intensely pleasurable feelings—by blocking transporters for dopamine reuptake. This action allows dopamine to linger longer in synaptic clefts, producing excessive stimulation of certain brain regions. ■

► CHECKPOINT

16. Which neurotransmitters are excitatory and which are inhibitory? How do they exert their effects?

17. In what ways is nitric oxide different from all previously known neurotransmitters?

TABLE 12.3 Neuropeptides

Substance	Description
Substance P	Found in sensory neurons, spinal cord pathways, and parts of brain associated with pain; enhances perception of pain.
Enkephalins	Inhibit pain impulses by suppressing release of substance P; may have a role in memory and learning, control of body temperature, sexual activity, and mental illness.
Endorphins	Inhibit pain by blocking release of substance P; may have a role in memory and learning, sexual activity, control of body temperature, and mental illness.
Dynorphins	May be related to controlling pain and registering emotions.
Hypothalamic releasing and inhibiting hormones	Produced by the hypothalamus; regulate the release of hormones by the anterior pituitary.
Angiotensin II	Stimulates thirst; may regulate blood pressure in the brain. As a hormone causes vasoconstriction and promotes release of aldosterone, which increases the rate of salt and water reabsorption by the kidneys.
Cholecystokinin (CCK)	Found in the brain and small intestine; may regulate feeding as a "stop eating" signal. As a hormone, regulates pancreatic enzyme secretion during digestion, and contraction of smooth muscle in the gastrointestinal tract.

NEURAL CIRCUITS

▶ **OBJECTIVE**

Identify the various types of neural circuits in the nervous system.

The CNS contains billions of neurons organized into complicated networks called **neural circuits,** functional groups of neurons that process specific types of information. In a **simple series circuit,** a presynaptic neuron stimulates a single postsynaptic neuron. The second neuron then stimulates another, and so on. However, most neural circuits are more complex.

A single presynaptic neuron may synapse with several postsynaptic neurons. Such an arrangement, called **divergence,** permits one presynaptic neuron to influence several postsynaptic neurons (or several muscle fibers or gland cells) at the same time. In a **diverging circuit,** the nerve impulse from a single presynaptic neuron causes the stimulation of increasing numbers of cells along the circuit (Figure 12.19a). For example, a small number of neurons in the brain that govern a particular body movement stimulate a much larger number of neurons in the spinal cord. Sensory signals are also arranged in diverging circuits, allowing a sensory impulse to be relayed to several regions of the brain. This arrangement amplifies the signal.

In another arrangement, called **convergence,** several presynaptic neurons synapse with a single postsynaptic neuron. This arrangement permits more effective stimulation or inhibition of the postsynaptic neuron. In a **converging circuit** (Figure 12.19b), the postsynaptic neuron receives nerve impulses from several different sources. For example, a single motor neuron that synapses with skeletal muscle fibers at neuromuscular junctions receives input from several pathways that originate in different brain regions.

Some circuits are constructed so that once the presynaptic cell is stimulated, it will cause the postsynaptic cell to transmit a series of nerve impulses. One such circuit is called a **reverberating circuit** (Figure 12.19c). In this pattern, the incoming impulse stimulates the first neuron, which stimulates the second, which stimulates the third, and so on. Branches from later neurons synapse with earlier ones. This arrangement sends impulses back through the circuit again and again. The output signal may last from a few seconds to many hours, depending on the number of synapses and the arrangement of neurons in the circuit. Inhibitory neurons may turn off a reverberating circuit after a period of time. Among the body responses thought to be the result of output signals from reverberating circuits are breathing, coordinated muscular activities, waking up, and short-term memory.

A fourth type of circuit is the **parallel after-discharge circuit** (Figure 12.19d). In this circuit, a single presynaptic cell stimulates a group of neurons, each of which synapses with a common postsynaptic cell. A differing number of synapses between the first and last neurons imposes varying synaptic delays, so that the last neuron exhibits multiple EPSPs or IPSPs.

Figure 12.19 Examples of neural circuits.

A neural circuit is a functional group of neurons that processes a specific kind of information.

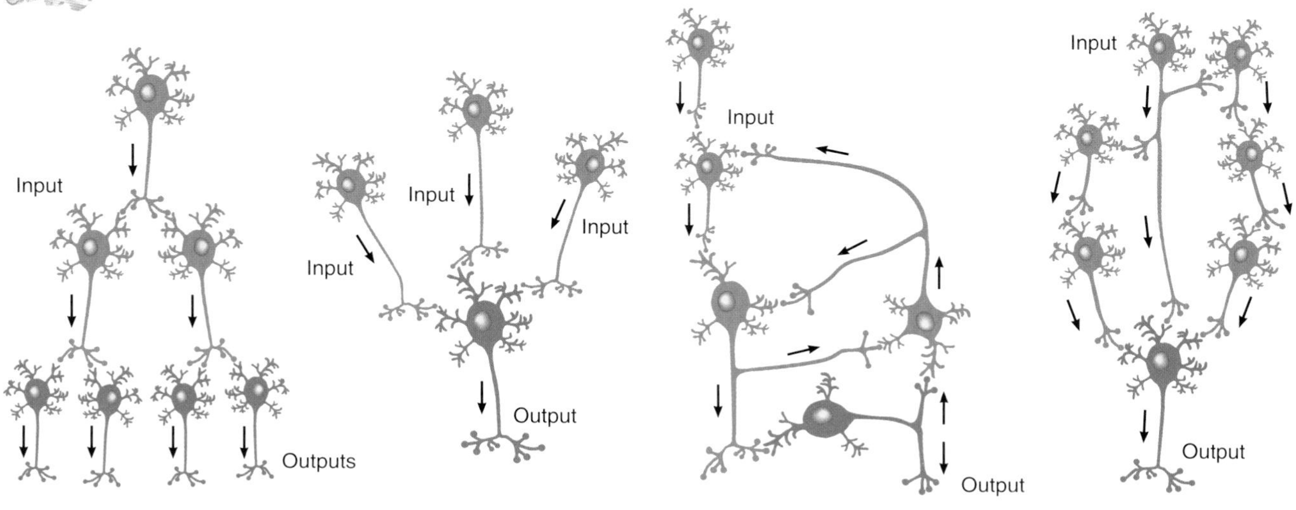

(a) Diverging circuit (b) Converging circuit (c) Reverberating circuit (d) Parallel after-discharge circuit

? A motor neuron in the spinal cord typically receives input from neurons that originate in several different regions of the brain. Is this an example of convergence or divergence?

If the input is excitatory, the postsynaptic neuron then can send out a stream of impulses in quick succession. Parallel after-discharge circuits may be involved in precise activities such as mathematical calculations.

▶ CHECKPOINT

18. What is a neural circuit?

19. What are the functions of diverging, converging, reverberating, and parallel after-discharge circuits?

REGENERATION AND REPAIR OF NERVOUS TISSUE

▶ OBJECTIVES

Define plasticity and neurogenesis.

Describe the events involved in damage and repair of peripheral nerves.

Throughout your life, your nervous system exhibits **plasticity,** the capability to change based on experience. At the level of individual neurons, the changes that can occur include the sprouting of new dendrites, synthesis of new proteins, and changes in synaptic contacts with other neurons. Undoubtedly, both chemical and electrical signals drive the changes that occur. Despite plasticity, however, mammalian neurons have very limited powers of **regeneration,** the capability to replicate or repair themselves. In the PNS, damage to dendrites and myelinated axons may be repaired if the cell body remains intact and if the Schwann cells that produce myelination remain active. In the CNS, little or no repair of damage to neurons occurs. Even when the cell body remains intact, a severed axon cannot be repaired or regrown.

Neurogenesis in the CNS

Neurogenesis—the birth of new neurons from undifferentiated stem cells—occurs regularly in some animals. For example, new neurons appear and disappear every year in some songbirds. Until recently, the dogma in humans and other primates was "no new neurons" in the adult brain. Then, in 1992, Canadian researchers published their unexpected finding that **epidermal growth factor (EGF)** stimulated cells taken from the brains of adult mice to proliferate into both neurons and astrocytes. Previously, EGF was known to trigger mitosis in a variety of nonneuronal cells and to promote wound healing and tissue regeneration. In 1998 scientists discovered that significant numbers of new neurons do arise in the adult human hippocampus, an area of the brain that is crucial for learning.

The nearly complete lack of neurogenesis in other regions of the brain and spinal cord seems to result from two factors: (1) inhibitory influences from neuroglia, particularly oligodendrocytes, and (2) absence of growth-stimulating cues that were present during fetal development. Axons in the CNS are myelinated by oligodendrocytes that do not form neurolemmas (sheaths of Schwann). In addition, CNS myelin is one of the factors inhibiting regeneration of neurons. Perhaps this same mechanism stops axonal growth once a target region has been reached during development. Also, after axonal damage, nearby astrocytes proliferate rapidly, forming a type of scar tissue that acts as a physical barrier to regeneration. Thus, injury of the brain or spinal cord usually is permanent. Ongoing research seeks ways to improve the environment for existing spinal cord axons to bridge the injury gap. Scientists also are trying to find ways to stimulate dormant stem cells to replace neurons lost through damage or disease and to develop tissue-cultured neurons that can be used for transplantation purposes.

Damage and Repair in the PNS

Axons and dendrites that are associated with a neurolemma may undergo repair if the cell body is intact, if the Schwann cells are functional, and if scar tissue formation does not occur too rapidly (Figure 12.20). Most nerves in the PNS consist of processes that are covered with a neurolemma. A person who injures axons of a nerve in an upper limb, for example, has a good chance of regaining nerve function.

When there is damage to an axon, changes usually occur both in the cell body of the affected neuron and in the portion of the axon distal to the site of injury. Changes also may occur in the portion of the axon proximal to the site of injury.

About 24 to 48 hours after injury to a process of a normal peripheral neuron (Figure 12.20a), the Nissl bodies break up into fine granular masses. This alteration is called **chromatolysis** (krō′-ma-TOL-i-sis; *chromato-* = color; *-lysis* = destruction). By the third to fifth day, the part of the axon distal to the damaged region becomes slightly swollen and then breaks up into fragments; the myelin sheath also deteriorates (Figure 12.20b). Even though the axon and myelin sheath degenerate, the neurolemma remains. Degeneration of the distal portion of the axon and myelin sheath is called **Wallerian degeneration.**

Following chromatolysis, signs of recovery in the cell body become evident. Macrophages phagocytize the debris. Synthesis of RNA and protein accelerates, which favors rebuilding or **regeneration** of the axon. The Schwann cells on either side of the injured site multiply by mitosis, grow toward each other, and may form a **regeneration tube** across the injured area (Figure 12.20c). The tube guides growth of a new axon from the proximal area across the injured area into the distal area previously occupied by the original axon. However, new axons cannot grow if the gap at the site of injury is too large or if the gap becomes filled with collagen fibers.

Figure 12.20 Damage and repair of a neuron in the PNS.

Myelinated axons in the peripheral nervous system may be repaired if the cell body remains intact and if Schwann cells remain active.

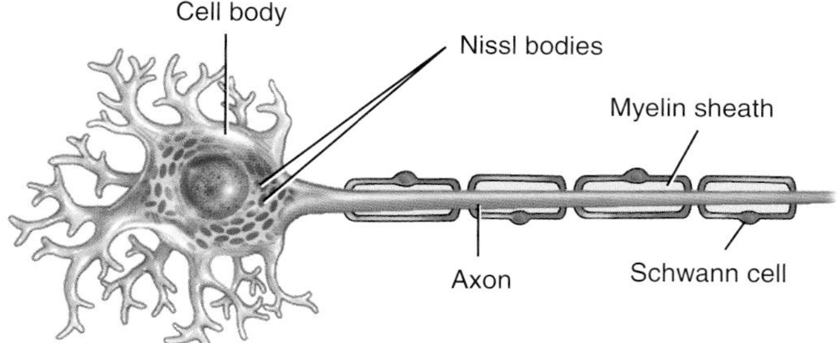

(a) Normal neuron

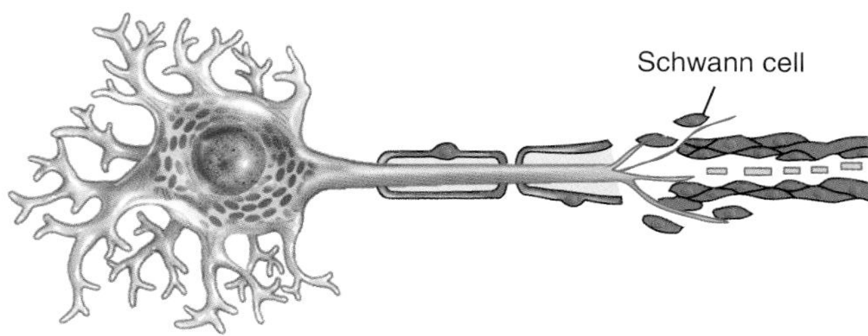

(b) Chromatolysis and Wallerian degeneration

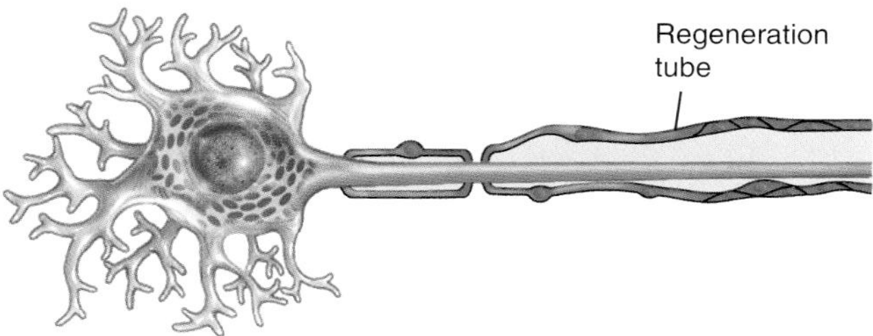

(c) Regeneration

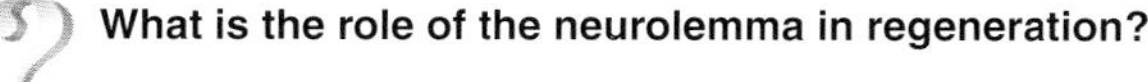

What is the role of the neurolemma in regeneration?

During the first few days following damage, buds of regenerating axons begin to invade the tube formed by the Schwann cells (Figure 12.20b). Axons from the proximal area grow at a rate of about 1.5 mm (0.06 in.) per day across the area of damage, find their way into the distal regeneration tubes, and grow toward the distally located receptors and effectors. Thus, some sensory and motor connections are reestablished and some functions restored. In time, the Schwann cells form a new myelin sheath.

Recent studies suggest that the formation of this new myelin sheath along the regenerated axon may be regulated by the blood-clotting protein fibrin (see page 682). As a PNS axon becomes injured, it may be exposed to fibrin and other blood components leaking from nearby blood vessels that are also damaged during the injury. As the damaged axon regenerates, the presence of fibrin prevents Schwann cells from producing myelin. Once the fibrin concentration decreases (due to tissue repair), the Schwann cells produce myelin along the regenerated axon. It is thought that this process slows down myelination long enough for the damaged axon to completely reform and reestablish synapses with the appropriate effectors.

► CHECKPOINT

20. What factors contribute to a lack of neurogenesis in most parts of the brain?

21. What is the function of the regeneration tube in repair of neurons?

DISORDERS: HOMEOSTATIC IMBALANCES

Multiple Sclerosis

Multiple sclerosis (MS) is a disease that causes a progressive destruction of myelin sheaths of neurons in the CNS. It afflicts about 350,000 people in the United States and 2 million people worldwide. It usually appears between the ages of 20 and 40, affecting females twice as often as males. MS is most common in whites, less common in blacks, and rare in Asians. MS is an autoimmune disease—the body's own immune system spearheads the attack. The condition's name describes the anatomical pathology: In *multiple* regions the myelin sheaths deteriorate to *scleroses,* which are hardened scars or plaques. Magnetic resonance imaging (MRI) studies reveal numerous plaques in the white matter of the brain and spinal cord. The destruction of myelin sheaths slows and then short-circuits propagation of nerve impulses.

The most common form of the condition is relapsing-remitting MS, which usually appears in early adulthood. The first symptoms may include a feeling of heaviness or weakness in the muscles, abnormal sensations, or double vision. An attack is followed by a period of remission during which the symptoms temporarily disappear. One attack follows another over the years, usually every year or two. The result is a progressive loss of function interspersed with remission periods, during which symptoms abate.

Although the cause of MS is unclear, both genetic susceptibility and exposure to some environmental factor (perhaps a herpes virus) appear to contribute. Since 1993 many patients with relapsing-remitting MS have been treated with injections of beta interferon. This treatment lengthens the time between relapses, decreases the severity of relapses, and slows formation of new lesions in some cases. Unfortunately, not all MS patients can tolerate beta interferon, and therapy becomes less effective as the disease progresses.

Epilepsy

Epilepsy is characterized by short, recurrent attacks of motor, sensory, or psychological malfunction, although it almost never affects intelligence. The attacks, called *epileptic seizures,* afflict about 1% of the world's population. They are initiated by abnormal, synchronous electrical discharges from millions of neurons in the brain, perhaps resulting from abnormal reverberating circuits. The discharges stimulate many of the neurons to send nerve impulses over their conduction pathways. As a result, lights, noise, or smells may be sensed when the eyes, ears, and nose have not been stimulated. Moreover, the skeletal muscles of a person having a seizure may contract involuntarily. *Partial seizures* begin in a small focus on one side of the brain and produce milder symptoms; *generalized seizures* involve larger areas on both sides of the brain and loss of consciousness.

Epilepsy has many causes, including brain damage at birth (the most common cause); metabolic disturbances (hypoglycemia, hypocalcemia, uremia, hypoxia); infections (encephalitis or meningitis); toxins (alcohol, tranquilizers, hallucinogens); vascular disturbances (hemorrhage, hypotension); head injuries; and tumors and abscesses of the brain. Seizures associated with fever are most common in children under the age of two. However, most epileptic seizures have no demonstrable cause.

Epileptic seizures often can be eliminated or alleviated by antiepileptic drugs, such as phenytoin, carbamazepine, and valproate sodium. An implantable device that stimulates the vagus (X) nerve has produced dramatic results in reducing seizures in some patients whose epilepsy was not well-controlled by drugs. In very severe cases, surgical intervention may be an option.

MEDICAL TERMINOLOGY

Guillain-Barré Syndrome (GBS) (GĒ-an ba-RĀ) An acute demyelinating disorder in which macrophages strip myelin from axons in the PNS. It is the most common cause of acute paralysis in North America and Europe and may result from the immune system's response to a bacterial infection. Most patients recover completely or partially, but about 15% remain paralyzed.

Neuroblastoma (noor-ō-blas-TŌ-ma) A malignant tumor that consists of immature nerve cells (neuroblasts); occurs most commonly in the abdomen and most frequently in the adrenal glands. Although rare, it is the most common tumor in infants.

Neuropathy (noo-ROP-a-thē; *neuro-* = a nerve; *-pathy* = disease) Any disorder that affects the nervous system but particularly a disorder of a cranial or spinal nerve. An example is *facial neuropathy* (Bell's palsy), a disorder of the facial (VII) nerve.

Rabies (RĀ-bēz; *rabi-* = mad, raving) A fatal disease caused by a virus that reaches the CNS via fast axonal transport. It is usually transmitted by the bite of an infected dog or other meat-eating animal. The symptoms are excitement, aggressiveness, and madness, followed by paralysis and death.

STUDY OUTLINE

OVERVIEW OF THE NERVOUS SYSTEM (p. 404)

1. The central nervous system (CNS) consists of the brain and spinal cord. The peripheral nervous system (PNS) consists of all nervous tissue outside the CNS.
2. Structures that make up the nervous system include the brain, 12 pairs of cranial nerves and their branches, the spinal cord, 31 pairs of spinal nerves and their branches, ganglia, enteric plexuses, and sensory receptors.
3. The nervous system helps maintain homeostasis and integrates all body activities by sensing changes (sensory function), interpreting them (integrative function), and reacting to them (motor function).
4. Sensory (afferent) neurons carry sensory information from cranial and spinal nerves into the brain and spinal cord or from a lower to a higher level in the spinal cord and brain. Interneurons have short axons that contact nearby neurons in the brain or spinal cord. Motor (efferent) neurons carry information from the brain toward the spinal cord or out of the brain and spinal cord into cranial or spinal nerves.
5. Components of the PNS include the somatic nervous system (SNS), autonomic nervous system (ANS), and enteric nervous system (ENS).
6. The SNS consists of neurons that conduct impulses from somatic and special sense receptors to the CNS and motor neurons from the CNS to skeletal muscles.
7. The ANS contains sensory neurons from visceral organs and motor neurons that convey impulses from the CNS to smooth muscle tissue, cardiac muscle tissue, and glands.
8. The ENS consists of neurons in enteric plexuses in the gastrointestinal (GI) tract that function somewhat independently of the ANS and CNS. The ENS monitors sensory changes in and controls operation of the GI tract.

HISTOLOGY OF NERVOUS TISSUE (p. 406)

1. Nervous tissue consists of neurons (nerve cells) and neuroglia. Neurons have the property of electrical excitability and are responsible for most unique functions of the nervous system: sensing, thinking, remembering, controlling muscle activity, and regulating glandular secretions.
2. Most neurons have three parts. The dendrites are the main receiving or input region. Integration occurs in the cell body, which includes typical cellular organelles. The output part typically is a single axon, which propagates nerve impulses toward another neuron, a muscle fiber, or a gland cell.
3. Synapses are the site of functional contact between two excitable cells. Axon terminals contain synaptic vesicles filled with neurotransmitter molecules.
4. Slow axonal transport and fast axonal transport are systems for conveying materials to and from the cell body and axon terminals.
5. On the basis of their structure, neurons are classified as multipolar, bipolar, or unipolar.
6. Neuroglia support, nurture, and protect neurons and maintain the interstitial fluid that bathes them. Neuroglia in the CNS include astrocytes, oligodendrocytes, microglia, and ependymal cells. Neuroglia in the PNS include Schwann cells and satellite cells.
7. Two types of neuroglia produce myelin sheaths: Oligodendrocytes myelinate axons in the CNS, and Schwann cells myelinate axons in the PNS.
8. White matter consists of aggregates of myelinated processes; gray matter contains cell bodies, dendrites, and axon terminals of neurons, unmyelinated axons, and neuroglia.
9. In the spinal cord, gray matter forms an H-shaped inner core that is surrounded by white matter. In the brain, a thin, superficial shell of gray matter covers the cerebral and cerebellar hemispheres.

ELECTRICAL SIGNALS IN NEURONS (p. 414)

1. Neurons communicate with one another using graded potentials, which are used for short-distance communication only, and action potentials, which allow communication over both short and long distances within the body.
2. The electrical signals produced by neurons and muscle fibers rely on four kinds of ion channels: leakage channels, voltage-gated channels, ligand-gated channels, and mechanically gated channels.
3. A typical value for the resting membrane potential is -70 mV. A cell that exhibits a membrane potential is polarized.
4. A graded potential is a small deviation from the resting membrane potential that occurs because ligand-gated or mechanically gated channels open or close. In hyperpolarization, a graded potential makes the membrane potential more negative (more polarized); in depolarization a graded potential makes the membrane potential less negative (less polarized).
5. The amplitude of a graded potential varies, depending on the strength of the stimulus.
6. According to the all-or-none principle, if a stimulus is strong enough to generate an action potential, the impulse generated is of a constant size. A stronger stimulus does not generate a larger action potential.
7. During an action potential, voltage-gated Na^+ and K^+ channels open and close in sequence. This results first in depolarization, the reversal of membrane polarization (from -70 mV to $+30$ mV). Then repolarization, the recovery of the resting membrane potential (from $+30$ mV to -70 mV), occurs.
8. During the first part of the refractory period (RP), another impulse cannot be generated at all (absolute RP); a little later, it can be triggered only by a larger-than-normal stimulus (relative RP).
9. Because an action potential travels from point to point along the membrane without getting smaller, it is useful for long-distance communication. Table 12.1 on page 423 compares graded potentials and action potentials.
10. Nerve impulse propagation in which the impulse "leaps" from one node of Ranvier to the next along a myelinated axon is saltatory conduction. Saltatory conduction is faster than continuous conduction.
11. Axons with larger diameters conduct impulses at higher speeds than do axons with smaller diameters.
12. The intensity of a stimulus is encoded in the frequency of action potentials and in the number of sensory neurons that are recruited.

SIGNAL TRANSMISSION AT SYNAPSES (p. 424)

1. A synapse is the functional junction between one neuron and another, or between a neuron and an effector such as a muscle or a gland. The two types of synapses are electrical and chemical.
2. A chemical synapse produces one-way information transfer—from a presynaptic neuron to a postsynaptic neuron.
3. An excitatory neurotransmitter is one that can depolarize the postsynaptic neuron's membrane, bringing the membrane potential closer to threshold. An inhibitory neurotransmitter hyperpolarizes the membrane of the postsynaptic neuron, moving it further from threshold.
4. Neurotransmitter is removed from the synaptic cleft in three ways: diffusion, enzymatic degradation, and uptake by cells (neurons and neuroglia).
5. If several presynaptic end bulbs release their neurotransmitter at about the same time, the combined effect may generate a nerve impulse, due to summation. Summation may be spatial or temporal.
6. The postsynaptic neuron is an integrator. It receives excitatory and inhibitory signals, integrates them, and then responds accordingly.
7. Table 12.2 on page 428 summarizes the structural and functional elements of a neuron.

NEUROTRANSMITTERS (p. 428)

1. Both excitatory and inhibitory neurotransmitters are present in the CNS and the PNS. A given neurotransmitter may be excitatory in some locations and inhibitory in others.
2. Neurotransmitters can be divided into two classes based on size: (1) small-molecule neurotransmitters (acetylcholine, amino acids, biogenic amines, ATP and other purines, and nitric oxide), and (2) neuropeptides, which are composed of 3 to 40 amino acids.
3. Chemical synaptic transmission may be modified by affecting synthesis, release, or removal of a neurotransmitter or by blocking or stimulating neurotransmitter receptors.
4. Table 12.3 on page 430 describes several important neuropeptides.

NEURAL CIRCUITS (p. 431)

1. Neurons in the central nervous system are organized into networks called neural circuits.
2. Neural circuits include simple series, diverging, converging, reverberating, and parallel after-discharge circuits.

REGENERATION AND REPAIR OF NERVOUS TISSUE (p. 432)

1. The nervous system exhibits plasticity (the capability to change based on experience), but it has very limited powers of regeneration (the capability to replicate or repair damaged neurons).
2. Neurogenesis, the birth of new neurons from undifferentiated stem cells, is normally very limited. Repair of damaged axons does not occur in most regions of the CNS.
3. Axons and dendrites that are associated with a neurolemma in the PNS may undergo repair if the cell body is intact, the Schwann cells are functional, and scar tissue formation does not occur too rapidly.

SELF-QUIZ QUESTIONS

Fill in the blanks in the following statements.

1. The subdivisions of the PNS are the ____, ____, and ____.
2. The two divisions of the autonomic nervous system are the ____ division and the ____ division.

Indicate whether the following statements are true or false.

3. At a chemical synapse between two neurons, the neuron receiving the signal is called the presynaptic neuron, and the neuron sending the signal is called the postsynaptic neuron.
4. Neurons in the PNS are always capable of repair while those in the CNS are not.

Choose the one best answer to the following questions.

5. Which of the following statements are *true*? (1) The sensory function of the nervous system involves sensory receptors sensing certain changes in the internal and external environments. (2) Sensory neurons receive electrical signals from sensory receptors. (3) The integrative function of the nervous system involves analyzing sensory information, storing some of it, and making decisions regarding appropriate responses. (4) Interneurons carry nerve impulses to effectors. (5) Motor function involves responding to integration decisions. (a) 1, 2, 3, and 4; (b) 2, 4, and 5; (c) 1, 2, 3, and 5; (d) 1, 2, and 4; (e) 2, 3, 4, and 5.
6. A neuron's resting membrane potential is established and maintained by (1) a high concentration of K^+ in the extracellular fluid and a high concentration of Na^+ in the cytosol, (2) the plasma membrane's higher permeability to Na^+ because of the presence of numerous Na^+ leakage channels, (3) differences in both ion concentrations and electrical gradients, (4) the fact that there are numerous large, nondiffusible anions in the cytosol, (5) sodium-potassium pumps that help to maintain the proper distribution of sodium and potassium. (a) 1, 2, and 5; (b) 1, 2, and 3; (c) 2, 3, and 4; (d) 3, 4, and 5; (e) 1, 2, 3, 4, and 5.
7. Place the following events in a chemical synapse in the correct order: (1) release of neurotransmitters into the synaptic cleft, (2) arrival of nerve impulse at the presynaptic neuron's synaptic end bulb (or varicosity), (3) either depolarization or hyperpolarization of postsynaptic membrane, (4) inward flow of Ca^{2+} through activated voltage-gated Ca^{2+} channels in the synaptic end bulb membrane, (5) exocytosis of synaptic vesicles, (6) opening of ligand-gated channels on the postsynaptic plasma membrane, (7) binding of neurotransmitters to receptors in the postsynaptic neuron's plasma membrane. (a) 2, 1, 5, 4, 7, 6, 3; (b) 1, 2, 4, 5, 7, 6, 3; (c) 2, 4, 5, 1, 7, 6, 3; (d) 4, 5, 1, 7, 6, 3, 2; (e) 2, 5, 1, 4, 6, 7, 3.
8. Several neurons in the brain sending impulses to a single motor neuron that terminates at a neuromuscular junction is an example of a ____ circuit. (a) reverberating, (b) simple series, (c) parallel after-discharge, (d) diverging, (e) converging.
9. Which of the following statements are *true*? (1) If the excitatory effect is greater than the inhibitory effect but less than the threshold of stimulation, the result is a subthreshold EPSP. (2) If the excitatory effect is greater than the inhibitory effect and reaches or surpasses the threshold level of stimulation, the result is a threshold or suprathreshold EPSP and one or more nerve impulses. (3) If the inhibitory effect is greater than the excitatory effect, the membrane hyperpolarizes, resulting in inhibition of the postsynaptic neuron and the inability of the neuron to generate a nerve impulse. (4) The greater the summation of hyperpolarizations, the more likely a nerve impulse will be initiated. (a) 1 and 4; (b) 2 and 4; (c) 1, 3, and 4; (d) 2, 3, and 4; (e) 1, 2, and 3.
10. Which of the following statements are *true*? (1) The basic types of ion channels are gated, leakage, and electrical. (2) Ion channels allow for the development of graded potentials and action potentials. (3) The major stimuli that operate gated ion channels are voltage changes, ligands (chemicals), and mechanical stimulation. (4) Ligand-gated channels may open either directly due to the presence of the ligand molecule itself or indirectly through the activation of a "second messenger" by a G protein. (5) A graded potential is useful only for communication over short distances. (a) 1, 2, and 3; (b) 2, 3, and 4; (c) 2, 3, and 5; (d) 2, 3, 4, and 5; (e) 1, 3, and 5.
11. Which of the following statements are *true*? (1) The frequency of impulses and number of activated sensory neurons encodes differences in stimuli intensity. (2) Larger-diameter axons conduct nerve impulses faster than smaller-diameter ones. (3) Continuous conduction is faster than saltatory conduction. (4) The diameter of an axon and the presence or absence of a myelin sheath are the most important factors that determine the speed of nerve impulse propagation. (5) Action potentials are localized, but graded potentials are propagated. (a) 1, 3, and 5; (b) 3 and 4; (c) 2, 4, and 5; (d) 2 and 4; (e) 1, 2, and 4.
12. Neurotransmitters are removed from the synaptic cleft by (1) axonal transport, (2) diffusion away from the cleft, (3) neurosecretory cells, (4) enzymatic breakdown, (5) cellular uptake. (a) 1, 2, 3, and 4; (b) 2, 4, and 5; (c) 2, 3, and 4; (d) 1, 4, and 5; (e) 1, 2, 3, 4, and 5.

13. Match the following:

____(a) neurons with just one process extending from the cell body; are always sensory neurons
____(b) small phagocytic neuroglia
____(c) help maintain an appropriate chemical environment for generation of action potentials by neurons; part of the blood–brain barrier
____(d) provide myelin sheath for CNS axons
____(e) contains neuronal cell bodies, dendrites, axon terminals, unmyelinated axons and neuroglia
____(f) a cluster of cell bodies within the CNS
____(g) form CSF and assist in its circulation; form blood–cerebrospinal barrier
____(h) neurons having several dendrites and one axon; most common neuronal type
____(i) neurons with one main dendrite and one axon; found in the retina of the eye
____(j) provide myelin sheath for PNS axons
____(k) support neurons in PNS ganglia
____(l) a cluster of neuronal cell bodies located outside the brain and spinal cord
____(m) aggregation of myelinated processes from many neurons
____(n) bundles of axons and associated connective tissue and blood vessels lying outside of the CNS
____(o) extensive neuronal networks that help regulate the digestive system

(1) astrocytes
(2) oligodendrocytes
(3) ganglia
(4) ependymal cells
(5) satellite cells
(6) unipolar neurons
(7) bipolar neurons
(8) multipolar neurons
(9) gray matter
(10) white matter
(11) enteric plexus
(12) microglia
(13) Schwann cells
(14) nucleus
(15) nerve

14. Match the following:

____(a) a sequence of rapidly occurring events that decreases and eventually reverses the membrane potential and then restores it to the resting state; a nerve impulse
____(b) a small deviation from the resting membrane potential that makes the membrane either more or less polarized
____(c) period of time when a second action potential can be initiated with a very strong stimulus
____(d) the minimum level of depolarization required for a nerve impulse to be generated
____(e) the recovery of the resting membrane potential
____(f) a neurotransmitter-caused depolarization of the postsynaptic membrane
____(g) a neurotransmitter-caused hyperpolarization of the postsynaptic membrane
____(h) time during which a neuron cannot produce an action potential even with a very strong stimulus
____(i) polarization that is less negative than the resting level
____(j) results from the buildup of neurotransmitter released simultaneously by several presynaptic end bulbs
____(k) the hyperpolarization that occurs after the repolarizing phase of an action potential
____(l) polarization that is more negative than the resting level
____(m) results from the buildup of neurotransmitter from the rapid, successive release by a single presynaptic end bulb

(1) graded potential
(2) action potential
(3) excitatory postsynaptic potential
(4) inhibitory postsynaptic potential
(5) absolute refractory period
(6) repolarization
(7) after-hyperpolarizing phase
(8) spatial summation
(9) threshold
(10) relative refractory period
(11) temporal summation
(12) depolarizing graded potential
(13) hyperpolarizing graded potential

15. Match the following:

____(a) the part of the neuron that contains the nucleus and organelles
____(b) rough endoplasmic reticulum in neurons; site of protein synthesis
____(c) store neurotransmitter
____(d) the process that propagates nerve impulses toward another neuron, muscle fiber, or gland cell
____(e) the highly branched receiving or input portions of a neuron
____(f) a multilayered lipid and protein covering for axons produced by neuroglia
____(g) the outer nucleated cytoplasmic layer of the Schwann cell
____(h) first portion of the axon, closest to the axon hillock
____(i) site of communication between two neurons or between a neuron and an effector cell
____(j) form the cytoskeleton of a neuron
____(k) gaps in the myelin sheath of an axon
____(l) general term for any neuronal process
____(m) area where the axon joins the cell body
____(n) area where nerve impulses arise
____(o) the numerous fine processes at the ends of an axon and its collaterals
____(p) interstitial fluid-filled space separating two neurons

(1) myelin sheath
(2) neurolemma
(3) nodes of Ranvier
(4) cell body
(5) Nissl bodies
(6) neurofibrils
(7) dendrites
(8) axon
(9) axon hillock
(10) initial segment
(11) trigger zone
(12) synaptic cleft
(13) nerve fiber
(14) axon terminals
(15) synapse
(16) synaptic vesicles

CRITICAL THINKING QUESTIONS

1. The buzzing of the alarm clock woke Carrie. She stretched, yawned, and started to salivate as she smelled the brewing coffee. She could feel her stomach rumble. List the divisions of the nervous system that are involved in each of these actions.
2. Baby Ming is learning to crawl. He also likes to pull himself onto window sills, gnawing on the painted wood of his century-old home as he looks out the windows. Lately his mother, an anatomy and physiology student, has noticed some odd behavior and took Ming to the pediatrician. Blood work determined that Ming had a high level of lead in his blood, ingested from the old leaded paint on the window sill. The doctor indicated that lead poisoning is a type of demyelination disorder. Why should Ming's mother be concerned?
3. As a torture procedure for his enemies, mad scientist Dr. Moro is trying to develop a drug that will enhance the effects of substance P. What cellular mechanisms could he enlist to design such a drug?

ANSWERS TO FIGURE QUESTIONS

12.1 The total number of cranial and spinal nerves in your body is $(12 \times 2) + (31 \times 2) = 86$.

12.2 Sensory or afferent neurons carry input to the CNS; motor or efferent neurons carry output from the CNS.

12.3 Dendrites receive (motor neurons or interneurons) input or generate (sensory neurons) signals; the cell body receives input; the axon conducts nerve impulses (action potentials) and transmits the message to another neuron or effector cell by releasing a neurotransmitter at its synaptic end bulbs.

12.4 Nerve impulses arise at the trigger zone.

12.5 The cell body of a pyramidal cell is shaped like a pyramid.

12.6 Microglia function as phagocytes in the central nervous system.

12.7 One Schwann cell myelinates a single axon; one oligodendrocyte myelinates several axons.

12.8 Myelination increases the speed of nerve impulse conduction.

12.9 Myelin makes white matter look shiny and white.

12.10 Perception primarily occurs in the cerebral cortex.

12.11 A touch on the arm activates mechanically gated channels.

12.12 A typical value for the resting membrane potential in a neuron is -70 mV.

12.13 Graded potentials occur when ligand- or mechanically gated channels open or close.

12.14 Voltage-gated Na^+ channels are open during the depolarizing phase, and voltage-gated K^+ channels are open during the repolarizing phase.

12.15 Yes, because the leakage channels would still allow K^+ to exit more rapidly than Na^+ could enter the axon. Some mammalian myelinated axons have only a few voltage-gated K^+ channels.

12.16 The diameter of an axon, presence or absence of a myelin sheath, and temperature determine the speed of propagation of a nerve impulse.

12.17 In some electrical synapses (gap junctions), ions may flow equally well in either direction, so either neuron may be the presynaptic one. At a chemical synapse, one neuron releases neurotransmitter and the other neuron has receptors that bind this chemical. Thus, the signal can proceed in only one direction.

12.18 If an IPSP occurred at 55 msec, threshold depolarization would likely not be reached, and a nerve impulse would not be generated.

12.19 A motor neuron receiving input from several other neurons is an example of convergence.

12.20 The neurolemma provides a regeneration tube that guides regrowth of a severed axon.

Chapter 13

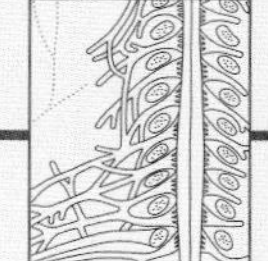

The Spinal Cord and Spinal Nerves

The Spinal Cord and Spinal Nerves and Homeostasis

The spinal cord and spinal nerves contribute to homeostasis by providing quick, reflexive responses to many stimuli. The spinal cord is the pathway for sensory input to the brain and motor output from the brain.

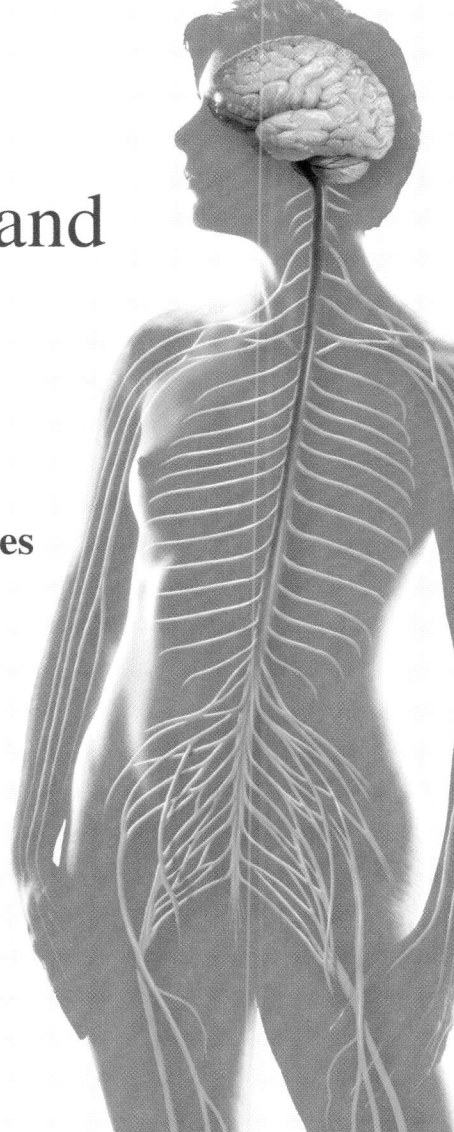

www. wiley.com/college/apcentral

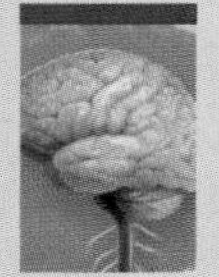

The spinal cord and spinal nerves contain neural circuits that mediate some of your most rapid reactions to environmental changes. If you pick up something hot, the grasping muscles may relax and you may drop the hot object even before you are consciously aware of the extreme heat or pain. This is an example of a spinal cord reflex—a quick, automatic response to certain kinds of stimuli that involves neurons only in the spinal nerves and spinal cord. Besides processing reflexes, the gray matter of the spinal cord also is a site for integration (summing) of excitatory postsynaptic potentials (EPSPs) and inhibitory postsynaptic potentials (IPSPs), which you learned about in Chapter 12. These graded potentials arise as neurotransmitter molecules interact with their receptors at synapses in the spinal cord. The white matter of the spinal cord contains a dozen major sensory and motor tracts, which function as the "highways" along which sensory input travels to the brain and motor output travels from the brain to effector tissues. Recall that the spinal cord is continuous with the brain and that together they comprise the central nervous system (CNS).

SPINAL CORD ANATOMY

► OBJECTIVES

Describe the protective structures and the gross anatomical features of the spinal cord.

Describe how spinal nerves are connected to the spinal cord.

Protective Structures

Two types of connective tissue coverings—bony vertebrae and tough, connective tissue meninges—plus a cushion of cerebrospinal fluid (produced in the brain) surround and protect the delicate nervous tissue of the spinal cord.

Vertebral Column

The spinal cord is located within the vertebral cavity of the vertebral column. As you learned in Chapter 7, the vertebral foramina of all the vertebrae, stacked one on top of the other, form the vertebral cavity. The surrounding vertebrae provide a sturdy shelter for the enclosed spinal cord (see Figure 13.1c). The vertebral ligaments, meninges, and cerebrospinal fluid provide additional protection.

Meninges

The **meninges** (me-NIN-jēz) are three connective tissue coverings that encircle the spinal cord and brain. The **spinal meninges** surround the spinal cord (Figure 13.1a) and are continuous with the **cranial meninges,** which encircle the brain (shown in Figure 14.4a on page 479). The most superficial of the three spinal meninges, the **dura mater** (DOO-ra MĀ-ter = tough mother), is composed of dense, irregular connective tissue. It forms a sac from the level of the foramen magnum in the occipital bone, where it is continuous with the dura mater of the brain, to the second sacral vertebra. The spinal cord is also protected by a cushion of fat and connective tissue located in the **epidural space,** a space between the dura mater and the wall of the vertebral cavity (Figure 13.1c).

The middle **meninx** (MĒ-ninks; singular form of *meninges*) is an avascular covering called the **arachnoid mater** (a-RAK-noyd; *arachn-* = spider; *-oid* = similar to) because of its spider's web arrangement of delicate collagen fibers and some elastic fibers. It is deep to the dura mater and is continuous with the arachnoid mater of the brain. Between the dura mater and the arachnoid mater is a thin **subdural space,** which contains interstitial fluid.

The innermost meninx is the **pia mater** (PĒ-a MĀ-ter; *pia* = delicate), a thin transparent connective tissue layer that adheres to the surface of the spinal cord and brain. It consists of interlacing bundles of collagen fibers and some fine elastic fibers. Within the pia mater are many blood vessels that supply oxygen and nutrients to the spinal cord. Between the arachnoid mater and the pia mater is the **subarachnoid space,** which contains cerebrospinal fluid.

Spinal Tap

In a **spinal tap (lumbar puncture),** a local anesthetic is given, and a long needle is inserted into the subarachnoid space. In adults, a spinal tap is normally performed between the third and fourth or fourth and fifth lumbar vertebrae. Because this region is inferior to the lowest portion of the spinal cord, it provides relatively safe access. (A line drawn across the highest points of the iliac crests, called the *supracristal line,* passes through the spinous process of the fourth lumbar vertebra.) The procedure is used to withdraw cerebrospinal fluid (CSF) for diagnostic purposes; to introduce antibiotics, contrast media for myelography, or anesthetics; to administer chemotherapy; to measure CSF pressure; and/or to evaluate the effects of treatment for diseases such as meningitis. ■

All three spinal meninges cover the spinal nerve roots up to the point where they exit the spinal column through the intervertebral foramina. As you will see later in the chapter, spinal nerve roots are structures that connect spinal nerves to the spinal cord. Triangular-shaped membranous extensions of the pia mater suspend the spinal cord in the middle of its dural sheath. These extensions, called **denticulate ligaments** (den-TIK-ū-lāt = small tooth), are thickenings of the pia mater. They project laterally and fuse with the arachnoid mater and inner surface of the dura mater between the anterior and posterior nerve roots of spinal nerves on either side (Figure 13.1a, b). Extending all along the length of the spinal cord, the denticulate ligaments protect the spinal cord against sudden displacement that could result in shock.

Figure 13.1 **Gross anatomy of the spinal cord.** The spinal meninges are evident in parts (a) and (c).

Meninges are connective tissue coverings that surround the spinal cord and brain.

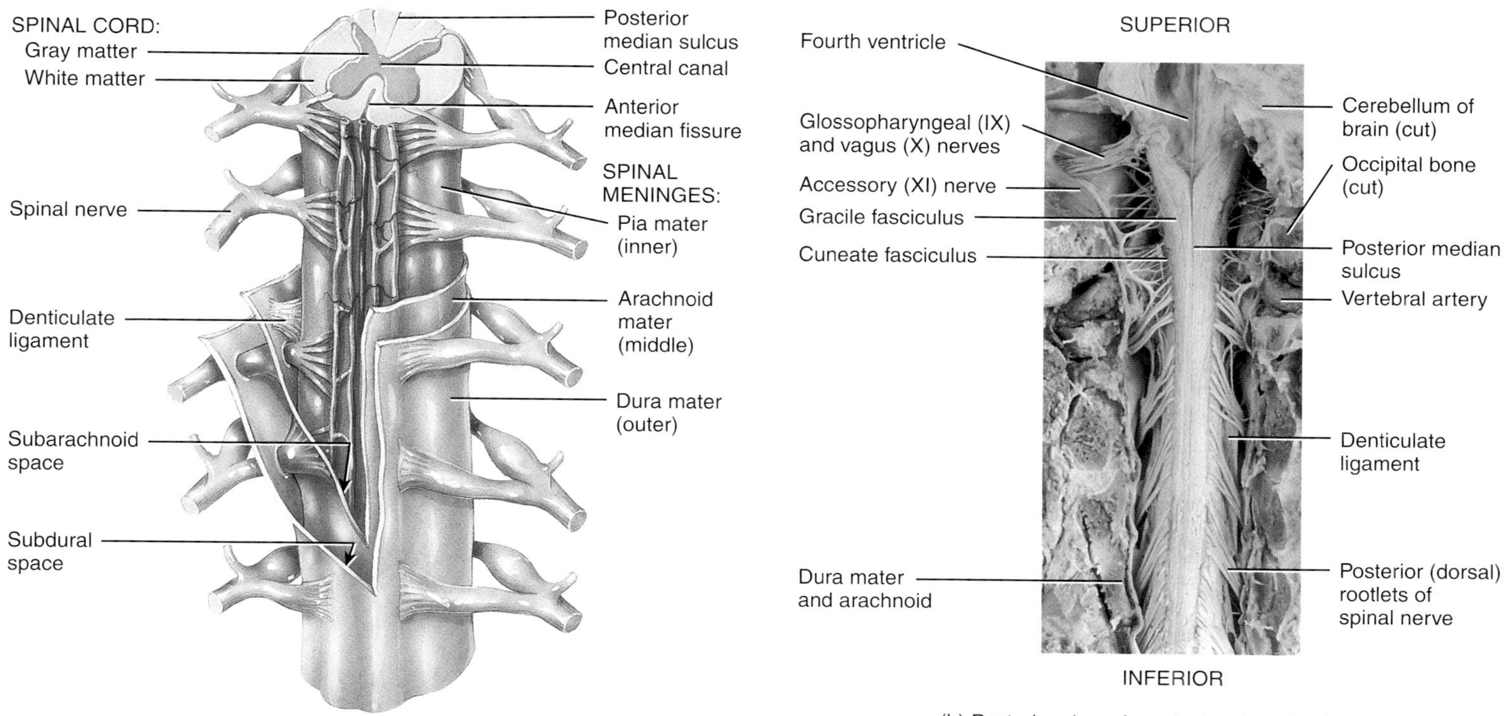

(a) Anterior view and transverse section through spinal cord

(b) Posterior view of cervical region of spinal cord

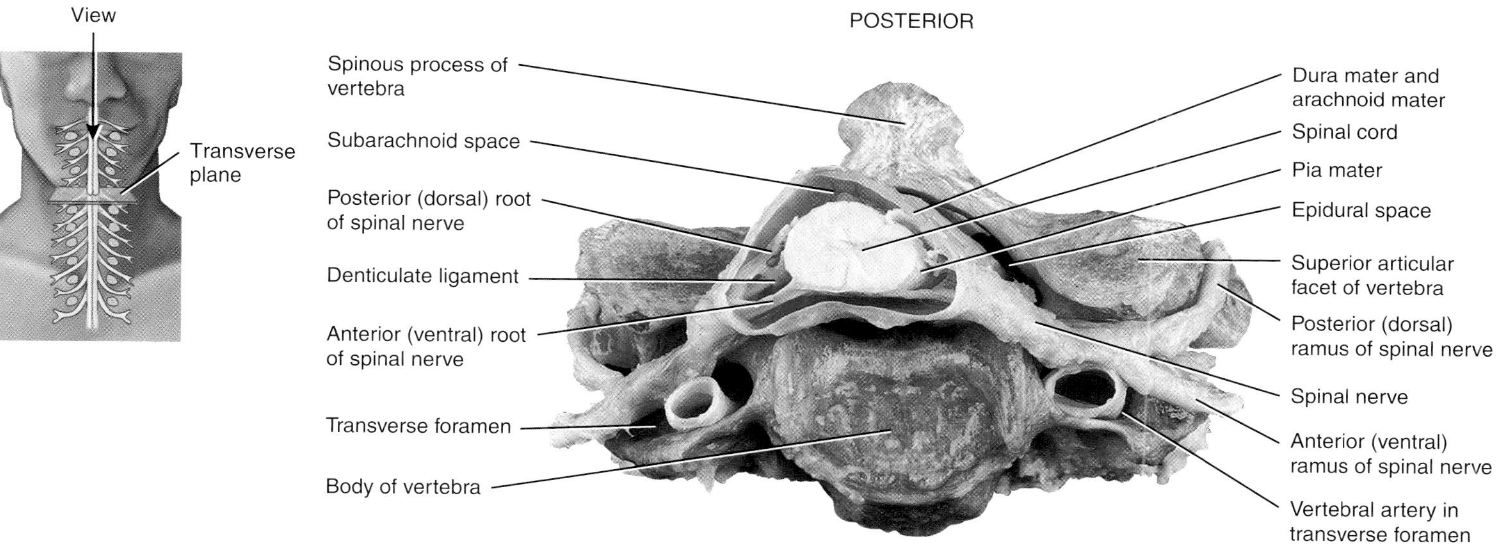

(c) Transverse section of the spinal cord within a cervical vertebra

What are the superior and inferior boundaries of the spinal dura mater?

External Anatomy of the Spinal Cord

The **spinal cord,** although roughly cylindrical, is flattened slightly in its anterior–posterior dimension. In adults, it extends from the medulla oblongata, the inferior part of the brain, to the superior border of the second lumbar vertebra (Figure 13.2). In newborn infants, it extends to the third or fourth lumbar vertebra. During early childhood, both the spinal cord and the vertebral column grow longer as part of overall body growth. Elongation of the spinal cord stops around age 4 or 5, but growth of the vertebral column continues. Thus, the spinal cord does not extend the entire length of the adult vertebral column. The length of the adult spinal cord ranges from 42 to 45 cm (16–18 in.). Its diameter is about 2 cm (0.75 in.) in the midthoracic region, somewhat larger in the lower cervical and midlumbar regions, and smallest at the inferior tip.

When the spinal cord is viewed externally, two conspicuous enlargements can be seen. The superior enlargement, the **cervical enlargement,** extends from the fourth cervical vertebra to the first thoracic vertebra. Nerves to and from the upper limbs arise from the cervical enlargement. The inferior enlargement, called the **lumbar enlargement,** extends from the ninth to the twelfth thoracic vertebra. Nerves to and from the lower limbs arise from the lumbar enlargement.

Inferior to the lumbar enlargement, the spinal cord terminates as a tapering, conical structure called the **conus medullaris** (KŌ-nus med-ū-LAR-is; *conus* = cone), which ends at the level of the intervertebral disc between the first and second lumbar vertebrae in adults. Arising from the conus medullaris is the **filum terminale** (FĪ-lum ter-mi-NAL-ē = terminal filament), an extension of the pia mater that extends inferiorly and anchors the spinal cord to the coccyx.

Because the spinal cord is shorter than the vertebral column, nerves that arise from the lumbar, sacral, and coccygeal regions of the spinal cord do not leave the vertebral column at the same level they exit the cord. The roots of these spinal nerves angle inferiorly in the vertebral cavity from the end of the spinal cord like wisps of hair. Appropriately, the roots of these nerves are collectively named the **cauda equina** (KAW-da ē-KWĪ-na), meaning "horse's tail" (Figure 13.2).

Spinal nerves are the paths of communication between the spinal cord and the nerves supplying specific regions of the body. Spinal cord organization appears to be segmented because the 31 pairs of spinal nerves emerge at regular intervals from intervertebral foramina (Figure 13.2). Indeed, each pair of spinal nerves is said to arise from a *spinal segment.* Within the spinal cord there is no obvious segmentation but, for convenience, the naming of spinal nerves is based on the segment in which they are located. There are 8 pairs of *cervical nerves* (represented in Figure 13.2 as C1–C8), 12 pairs of *thoracic nerves* (T1–T12), 5 pairs of *lumbar nerves* (L1–L5), 5 pairs of *sacral nerves* (S1–S5), and 1 pair of *coccygeal nerves* (Co1).

Two bundles of axons, called **roots,** connect each spinal nerve to a segment of the cord (see Figure 13.3a). The **posterior (dorsal) root** contains only sensory axons, which conduct nerve impulses from sensory receptors in the skin, muscles, and internal organs into the central nervous system. Each posterior root has a swelling, the **posterior (dorsal) root ganglion,** which contains the cell bodies of sensory neurons. The **anterior (ventral) root** contains axons of motor neurons, which conduct nerve impulses from the CNS to effector organs and cells.

Spinal Nerve Root Damage

As you have just learned, spinal nerve roots exit from the vertebral cavity through intervertebral foramina. The most common cause of **spinal nerve root damage** is a herniated intervertebral disc. Damage to vertebrae as result of osteoporosis, osteoarthritis, cancer, or injury can also damage spinal nerve roots. Symptoms of spinal nerve root damage include pain, muscle weakness, and loss of feeling. Rest, physical therapy, pain medications, and epidural injections are the most widely used conservative treatments. It is recommended that 6 to 12 weeks of conservative therapy be attempted first. If the pain continues, is intense, or is impairing normal functioning, surgery is often the next step. ■

Internal Anatomy of the Spinal Cord

Two grooves penetrate the white matter of the spinal cord and divide it into right and left sides (Figure 13.3 on page 444). The **anterior median fissure** is a deep, wide groove on the anterior (ventral) side. The **posterior median sulcus** is a shallower, narrow furrow on the posterior (dorsal) side. The gray matter of the spinal cord is shaped like the letter H or a butterfly and is surrounded by white matter. The gray matter consists of dendrites and cell bodies of neurons, unmyelinated axons, and neuroglia. The white matter consists primarily of bundles of myelinated axons of neurons. The **gray commissure** (KOM-mi-shur) forms the crossbar of the H. In the center of the gray commissure is a small space called the **central canal;** it extends the entire length of the spinal cord and is filled with cerebrospinal fluid. At its superior end, the central canal is continuous with the fourth ventricle (a space that contains cerebrospinal fluid) in the medulla oblongata of the brain. Anterior to the gray commissure is the **anterior (ventral) white commissure,** which connects the white matter of the right and left sides of the spinal cord.

In the gray matter of the spinal cord and brain, clusters of neuronal cell bodies form functional groups called **nuclei.** *Sensory nuclei* receive input from sensory receptors via sensory neurons, and *motor nuclei* provide output to effector tissues via motor neurons. The gray matter on each side of the spinal cord is subdivided into regions called **horns.** The **anterior (ventral) gray horns** contain somatic motor nuclei, which provide nerve impulses for contraction of skeletal muscles. The **posterior (dorsal) gray horns** contain somatic and autonomic sensory

Figure 13.2 External anatomy of the spinal cord and the spinal nerves. (See Tortora, *A Photographic Atlas of the Human Body, Second Edition,* Figure 8.3.)

The spinal cord extends from the medulla oblongata of the brain to the superior border of the second lumbar vertebra.

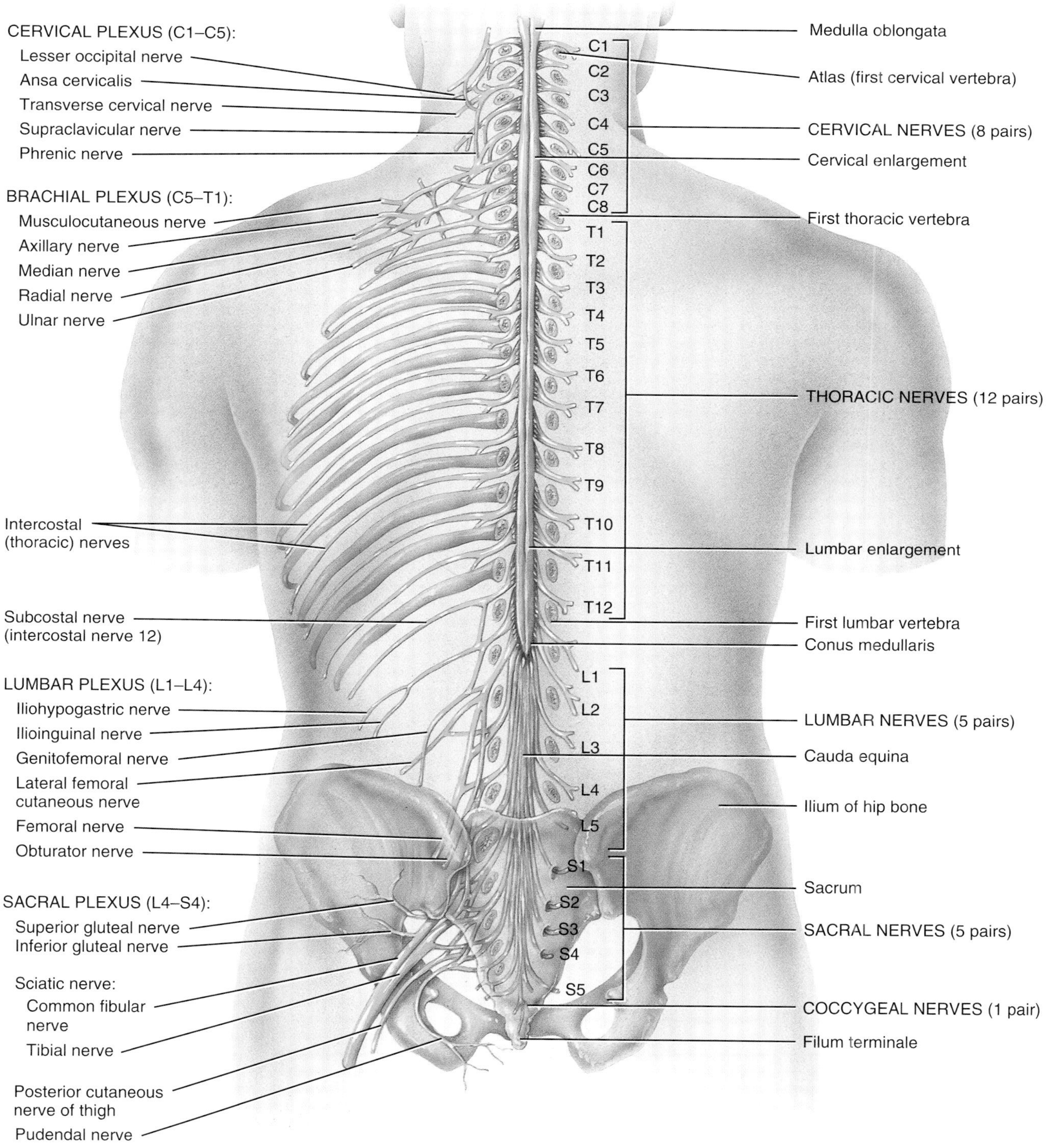

CERVICAL PLEXUS (C1–C5):
Lesser occipital nerve
Ansa cervicalis
Transverse cervical nerve
Supraclavicular nerve
Phrenic nerve

BRACHIAL PLEXUS (C5–T1):
Musculocutaneous nerve
Axillary nerve
Median nerve
Radial nerve
Ulnar nerve

Intercostal (thoracic) nerves

Subcostal nerve (intercostal nerve 12)

LUMBAR PLEXUS (L1–L4):
Iliohypogastric nerve
Ilioinguinal nerve
Genitofemoral nerve
Lateral femoral cutaneous nerve
Femoral nerve
Obturator nerve

SACRAL PLEXUS (L4–S4):
Superior gluteal nerve
Inferior gluteal nerve
Sciatic nerve:
Common fibular nerve
Tibial nerve
Posterior cutaneous nerve of thigh
Pudendal nerve

C1
C2
C3
C4
C5
C6
C7
C8
T1
T2
T3
T4
T5
T6
T7
T8
T9
T10
T11
T12
L1
L2
L3
L4
L5
S1
S2
S3
S4
S5

Medulla oblongata
Atlas (first cervical vertebra)
CERVICAL NERVES (8 pairs)
Cervical enlargement
First thoracic vertebra
THORACIC NERVES (12 pairs)
Lumbar enlargement
First lumbar vertebra
Conus medullaris
LUMBAR NERVES (5 pairs)
Cauda equina
Ilium of hip bone
Sacrum
SACRAL NERVES (5 pairs)
COCCYGEAL NERVES (1 pair)
Filum terminale

Posterior view of entire spinal cord and portions of spinal nerves

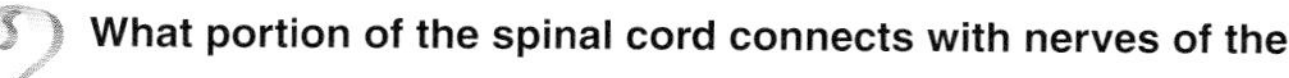

What portion of the spinal cord connects with nerves of the upper limbs?

Figure 13.3 Internal anatomy of the spinal cord: the organization of gray matter and white matter. For simplicity, dendrites are not shown in this and several other illustrations of transverse sections of the spinal cord. Blue and red arrows in (a) indicate the direction of nerve impulse propagation.

In the spinal cord, white matter surrounds the gray matter.

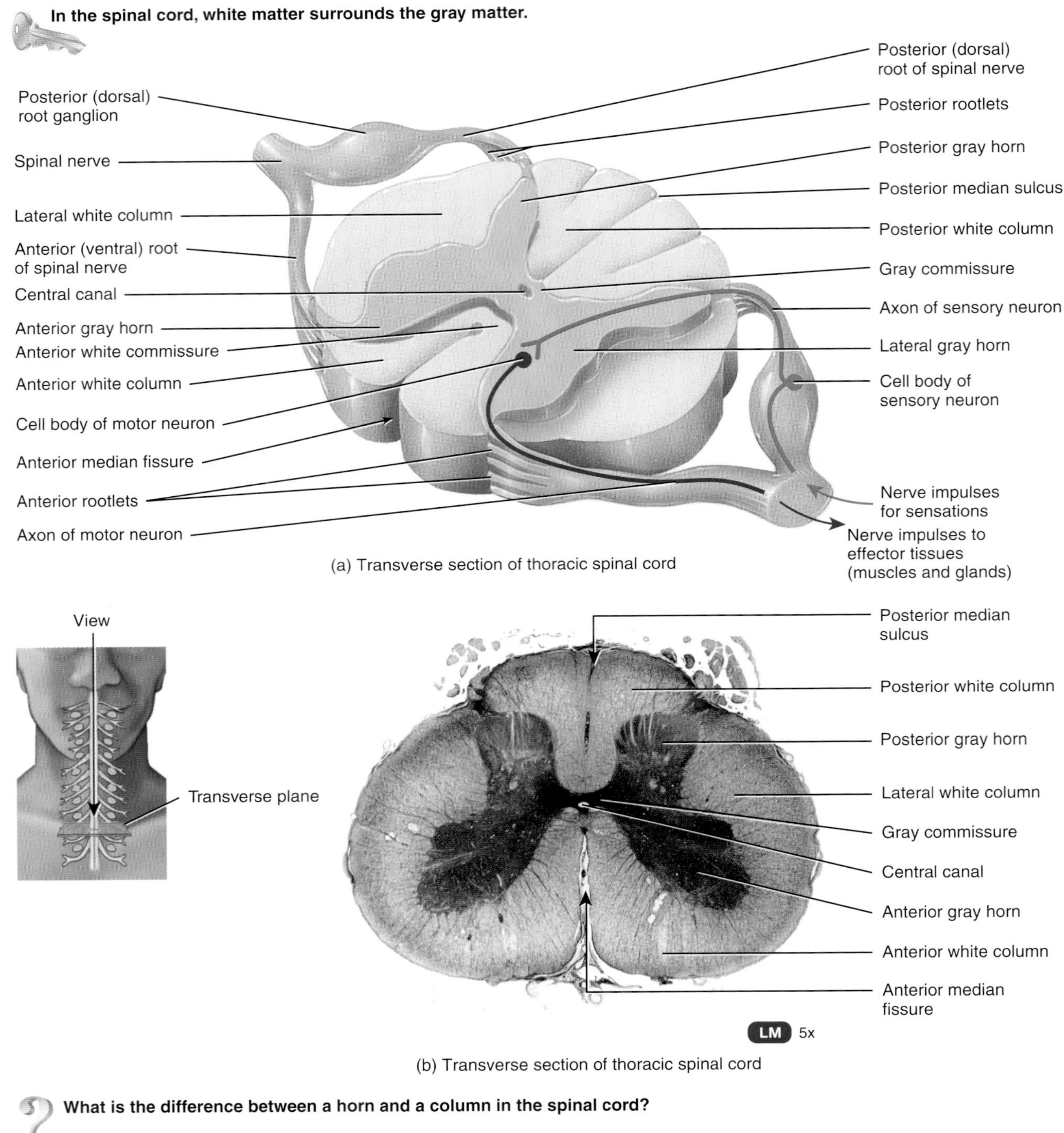

(a) Transverse section of thoracic spinal cord

(b) Transverse section of thoracic spinal cord

What is the difference between a horn and a column in the spinal cord?

nuclei. Between the anterior and posterior gray horns are the **lateral gray horns,** which are present only in the thoracic, upper lumbar, and sacral segments of the spinal cord. The lateral horns contain autonomic motor nuclei that regulate the activity of smooth muscle, cardiac muscle, and glands.

The white matter, like the gray matter, is organized into regions. The anterior and posterior gray horns divide the white matter on each side into three broad areas called **columns:** (1) **anterior (ventral) white columns,** (2) **posterior (dorsal) white columns,** and (3) **lateral white columns.** Each column, in turn, contains distinct bundles of axons having a common origin or destination and carrying similar information. These bundles, which may extend long distances up or down the spinal cord, are called **tracts.** Tracts are bundles of axons in the CNS; recall that nerves are bundles of axons in the PNS. **Sensory (ascending) tracts** consist of axons that conduct nerve impulses toward the brain. Tracts consisting of axons that carry nerve impulses from the brain are called **motor (descending) tracts.** Sensory and motor tracts of the spinal cord are continuous with sensory and motor tracts in the brain.

The various spinal cord segments vary in size, shape, relative amounts of gray and white matter, and distribution and shape of gray matter. These features are summarized in Table 13.1.

▶ **CHECKPOINT**

1. Where are the spinal meninges located? Where are the epidural, subdural, and subarachnoid spaces located?
2. What are the cervical and lumbar enlargements?
3. Define conus medullaris, filum terminale, and cauda equina. What is a spinal segment? How is the spinal cord partially divided into right and left sides?
4. What does each of the following terms mean? Gray commissure, central canal, anterior gray horn, lateral gray horn, posterior gray horn, anterior white column, lateral white column, posterior white column, ascending tract, and descending tract.

TABLE 13.1 Comparison of Various Spinal Cord Segments

Segment	Distinguishing Characteristics
Cervical (Segment C1) (Segment C8)	Relatively large diameter, relatively large amounts of white matter, oval in shape; in upper cervical segments (C1–C6), posterior gray horn is large, but anterior gray horn is relatively small; in lower cervical segments (C6 and below), posterior gray horns are enlarged and anterior gray horns are well-developed.
Thoracic (Segment T2)	Small diameter is due to relatively small amounts of gray matter; except for first thoracic segment, anterior and posterior gray horns are relatively small; a small lateral gray horn is present.
Lumbar (Segment L4)	Nearly circular; very large anterior and posterior gray horns; relatively less white matter than cervical segments.
Sacral (Segment S3)	Relatively small, but with relatively large amounts of gray matter; relatively small amounts of white matter; anterior and posterior gray horns are large and thick.
Coccygeal	Resemble lower sacral spinal segments, but much smaller.

SPINAL NERVES

▶ **OBJECTIVES**

Describe the components, connective tissue coverings, and branching of a spinal nerve.

Define plexus, and identify the distribution of nerves of the cervical, brachial, lumbar, and sacral plexuses.

Describe the clinical significance of dermatomes.

Spinal nerves and the nerves that branch from them are part of the peripheral nervous system (PNS). They connect the CNS to sensory receptors, muscles, and glands in all parts of the body. The 31 pairs of spinal nerves are named and numbered according to the region and level of the vertebral column from which they emerge (see Figure 13.2). The first cervical pair emerges between the atlas (first cervical vertebra) and the occipital bone. All other spinal nerves emerge from the vertebral column through the intervertebral foramina between adjoining vertebrae.

Not all spinal cord segments are aligned with their corresponding vertebrae. Recall that the spinal cord ends near the level of the superior border of the second lumbar vertebra, and that the roots of the lumbar, sacral, and coccygeal nerves descend at an angle to reach their respective foramina before emerging from the vertebral column. This arrangement constitutes the cauda equina (see Figure 13.2).

As noted earlier, a typical **spinal nerve** has two connections to the cord: a posterior root and an anterior root (see Figure 13.3a). The posterior and anterior roots unite to form a spinal nerve at the intervertebral foramen. Because the posterior root contains sensory axons and the anterior root contains motor axons, a spinal nerve is classified as a **mixed nerve.** The posterior root contains a posterior root ganglion in which cell bodies of sensory neurons are located.

Connective Tissue Coverings of Spinal Nerves

Each spinal nerve and cranial nerve consists of many individual axons and contains layers of protective connective tissue coverings (Figure 13.4). Individual axons within a nerve, whether myelinated or unmyelinated, are wrapped in **endoneurium** (en′-dō-NOO-rē-um; *endo-* = within or inner), the innermost layer. Groups of axons with their endoneurium are arranged in bundles called **fascicles,** each of which is wrapped in

Figure 13.4 **Organization and connective tissue coverings of a spinal nerve.** (Part (b): From Richard G. Kessel and Randy H. Kardon, *Tissues and Organs: A Text-Atlas of Scanning Electron Microscopy.* Copyright © 1979 by W. H. Freeman and Company. Reprinted by perimission.)

Three layers of connective tissue wrappings protect axons: Endoneurium surrounds individual axons, perineurium surrounds bundles of axons (fascicles), and epineurium surrounds an entire nerve.

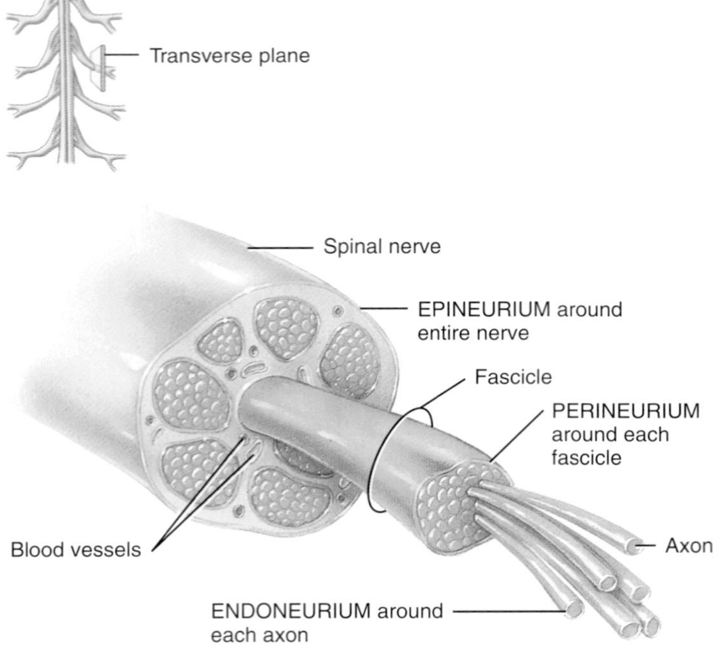

(a) Transverse section showing the coverings of a spinal nerve

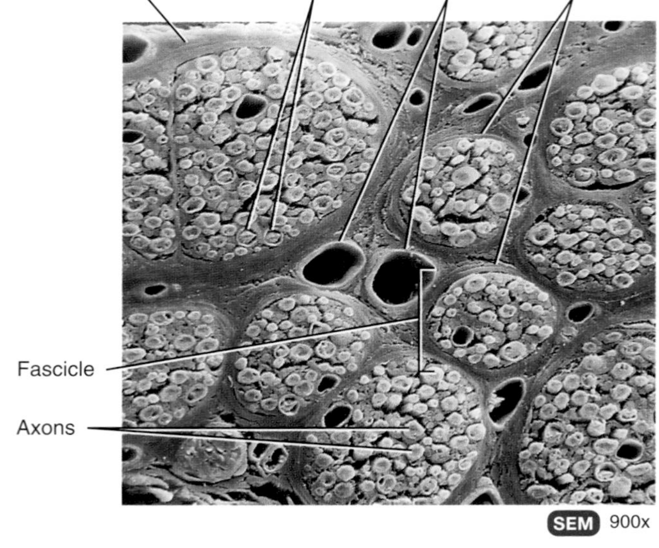

(b) Transverse section of 12 nerve fascicles

Why are all spinal nerves classified as mixed nerves?

perineurium (per′-i-NOO-rē-um; *peri-* = around), the middle layer. The outermost covering over the entire nerve is the **epineurium** (ep′-i-NOO-rē-um; *epi-* = over). Extensions of the epineurium also occur between fascicles. The dura mater of the spinal meninges fuses with the epineurium as the nerve passes through the intervertebral foramen. Note the presence of many blood vessels, which nourish nerves, within the perineurium and epineurium (Figure 13.4b). You may recall from Chapter 10 that the connective tissue coverings of skeletal muscles—endomysium, perimysium, and epimysium—are similar in organization to those of nerves.

Distribution of Spinal Nerves

Branches

A short distance after passing through its intervertebral foramen, a spinal nerve divides into several branches (Figure 13.5). These branches are known as **rami** (RĀ-mī = branches). The **posterior (dorsal) ramus** (RĀ-mus; singular form) serves the deep muscles and skin of the dorsal surface of the trunk. The **anterior (ventral) ramus** serves the muscles and structures of the upper and lower limbs and the skin of the lateral and ventral surfaces of the trunk. In addition to posterior and anterior rami, spinal nerves also give off a **meningeal branch.** This branch reenters the vertebral cavity through the intervertebral foramen and supplies the vertebrae, vertebral ligaments, blood vessels of the spinal cord, and meninges. Other branches of a spinal nerve are the **rami communicantes** (kō-mū-ni-KAN-tēz), components of the autonomic nervous system that will be discussed in Chapter 15.

Plexuses

Axons from the anterior rami of spinal nerves, except for thoracic nerves T2–T12, do not go directly to the body structures

Figure 13.5 **Branches of a typical spinal nerve, shown in transverse section through the thoracic portion of the spinal cord.** (See also Figure 13.1c.)

The branches of a spinal nerve are the posterior ramus, the anterior ramus, the meningeal branch, and the rami communicantes.

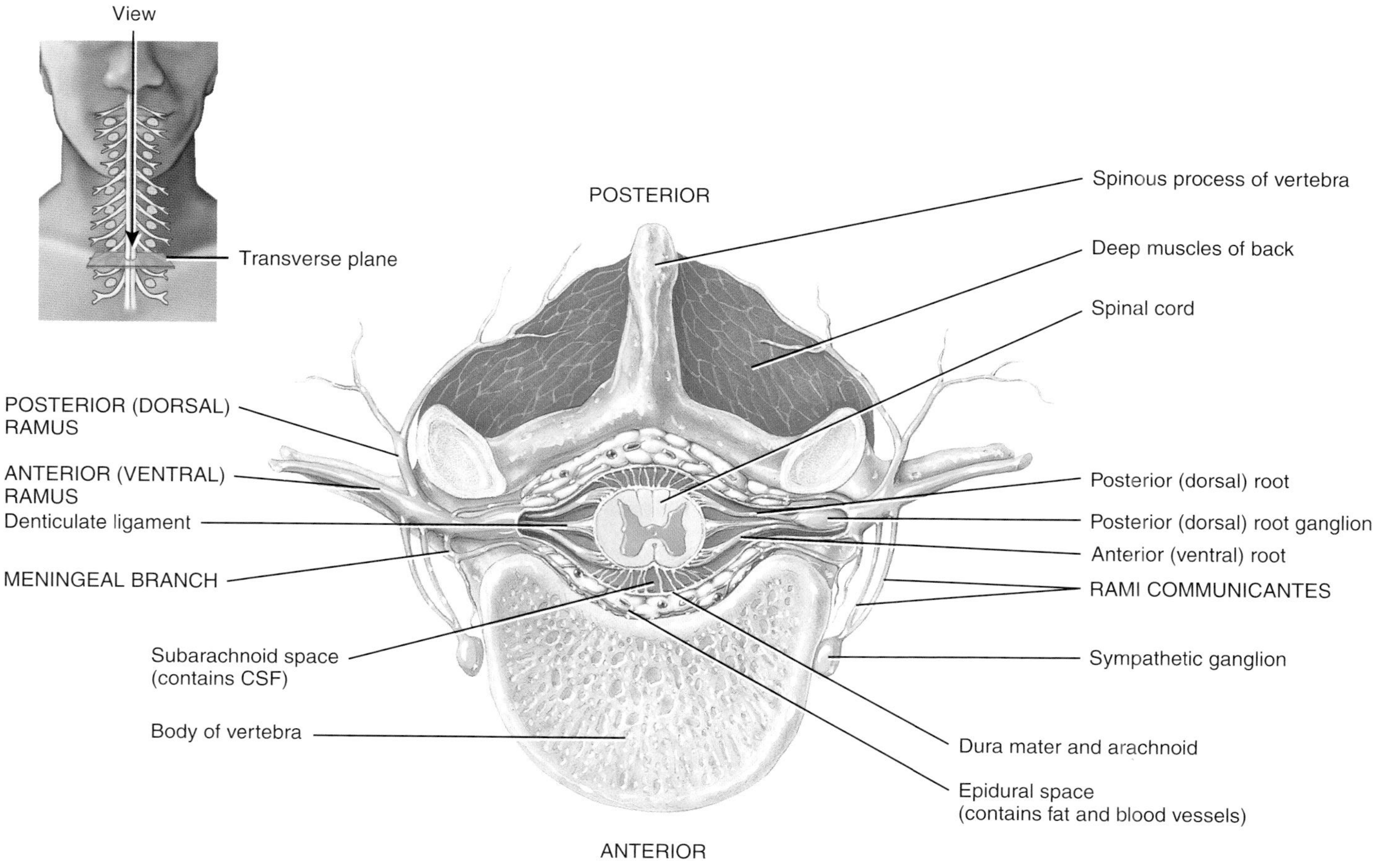

Which spinal nerve branches serve the upper and lower limbs?

CERVICAL PLEXUS (FIGURE 13.6)

▶ **OBJECTIVE**

Describe the origin and distribution of the cervical plexus.

The **cervical plexus** (SER-vi-kul) is formed by the roots (anterior rami) of the first four cervical nerves (C1–C4), with contributions from C5. There is one on each side of the neck alongside the first four cervical vertebrae.

The cervical plexus supplies the skin and muscles of the head, neck, and superior part of the shoulders and chest. The phrenic nerves arise from the cervical plexuses and supply motor fibers to the diaphragm. Branches of the cervical plexus also run parallel to two cranial nerves, the accessory (XI) nerve and hypoglossal (XII) nerve.

Injuries to the Phrenic Nerves

Complete severing of the spinal cord above the origin of the phrenic nerves (C3, C4, and C5) causes respiratory arrest. Breathing stops because the phrenic nerves no longer send nerve impulses to the diaphragm. ■

▶ **CHECKPOINT**

Which nerve that arises from the cervical plexus causes contraction of the diaphragm?

NERVE	ORIGIN	DISTRIBUTION
Superficial (Sensory) Branches		
Lesser occipital	C2	Skin of scalp posterior and superior to ear.
Great auricular (aw-RIK-ū-lar)	C2–C3	Skin anterior, inferior, and over ear, and over parotid glands.
Transverse cervical	C2–C3	Skin over anterior aspect of neck.
Supraclavicular	C3–C4	Skin over superior portion of chest and shoulder.
Deep (Largely Motor) Branches		
Ansa cervicalis (AN-sa ser-vi-KAL-is)		This nerve divides into superior and inferior roots.
Superior root	C1	Infrahyoid and geniohyoid muscles of neck.
Inferior root	C2–C3	Infrahyoid muscles of neck.
Phrenic (FREN-ik)	C3–C5	Diaphragm.
Segmental branches	C1–C5	Prevertebral (deep) muscles of neck, levator scapulae, and middle scalene muscles.

they supply. Instead, they form networks on both the left and right sides of the body by joining with various numbers of axons from anterior rami of adjacent nerves. Such a network of axons is called a **plexus** (= braid or network). The principal plexuses are the **cervical plexus, brachial plexus, lumbar plexus,** and **sacral plexus.** A smaller **coccygeal plexus** is also present. Refer to Figure 13.2 to see their relationships to one another. Emerging from the plexuses are nerves bearing names that are often descriptive of the general regions they serve or the course they take. Each of the nerves, in turn, may have several branches named for the specific structures they innervate.

Exhibits 13.1–13.4 summarize the principal plexuses. The anterior rami of spinal nerves T2–T12 are called intercostal nerves and will be discussed next.

Figure 13.6 Cervical plexus in anterior view. (See Tortora, *A Photographic Atlas of the Human Body, Second Edition,* Figure 8.7.)

The cervical plexus supplies the skin and muscles of the head, neck, superior portion of the shoulders and chest, and diaphragm.

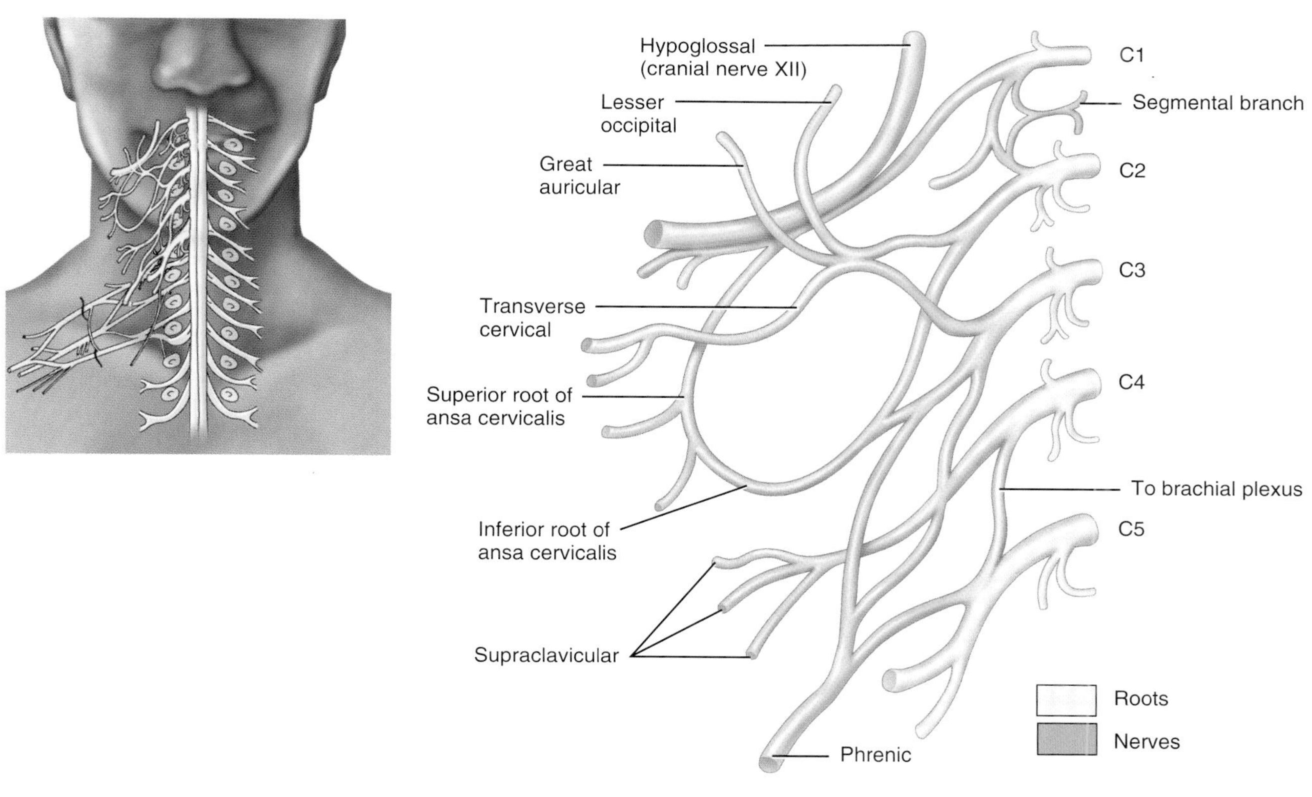

Origin of cervical plexus

Why does complete severing of the spinal cord at level C2 cause respiratory arrest?

Intercostal Nerves

The anterior rami of spinal nerves T2–T12 do not enter into the formation of plexuses and are known as **intercostal** or **thoracic nerves.** These nerves directly connect to the structures they supply in the intercostal spaces. After leaving its intervertebral foramen, the anterior ramus of nerve T2 innervates the intercostal muscles of the second intercostal space and supplies the skin of the axilla and posteromedial aspect of the arm. Nerves T3–T6 extend along the costal grooves of the ribs and then to the intercostal muscles and skin of the anterior and lateral chest wall. Nerves T7–T12 supply the intercostal muscles and abdominal muscles, and the overlying skin. The posterior rami of the intercostal nerves supply the deep back muscles and skin of the posterior aspect of the thorax.

EXHIBIT 13.2 BRACHIAL PLEXUS (FIGURES 13.7 AND 13.8)

▶ **OBJECTIVE**

Describe the origin, distribution, and effects of damage to the brachial plexus.

The roots (anterior rami) of spinal nerves C5–C8 and T1 form the **brachial plexus** (BRĀ-kē-al), which extends inferiorly and laterally on either side of the last four cervical and first thoracic vertebrae (Figure 13.7a). It passes above the first rib posterior to the clavicle and then enters the axilla.

Since the brachial plexus is so complex, an explanation of its various parts is helpful. As with the cervical and other plexuses, the **roots** are the anterior rami of the spinal nerves. The roots of several spinal nerves unite to form **trunks** in the inferior part of the neck. These are the *superior, middle,* and *inferior trunks.* Posterior to the clavicles, the trunks divide into **divisions,** called the *anterior* and *posterior divisions.* In the axillae, the divisions unite to form **cords** called the *lateral, medial,* and *posterior cords.* The cords are named for their relationship to the axillary artery, a large artery that supplies blood to the upper limb. The principal **nerves** of the brachial plexus branch from the cords.

The brachial plexus provides the entire nerve supply of the shoulders and upper limbs (Figure 13.7b). Five important nerves arise from the brachial plexus: (1) The **axillary nerve** supplies the deltoid and teres minor muscles. (2) The **musculocutaneous nerve** supplies the flexors of the arm. (3) The **radial nerve** supplies the muscles on the posterior aspect of the arm and forearm. (4) The **median nerve** supplies most of the muscles of the anterior forearm and some of the muscles of the hand. (5) The **ulnar nerve** supplies the anteromedial muscles of the forearm and most of the muscles of the hand.

Injuries to Nerves Emerging from the Brachial Plexus

Injury to the superior roots of the brachial plexus (C5–C6) may result from forceful pulling away of the head from the shoulder, as might occur from a heavy fall on the shoulder or excessive stretching of an infant's neck during childbirth. The presentation of this injury is characterized by an upper limb in

NERVE	ORIGIN	DISTRIBUTION
Dorsal scapular (SKAP-ū-lar)	C5	Levator scapulae, rhomboid major, and rhomboid minor muscles.
Long thoracic (thor-RAS-ik)	C5–C7	Serratus anterior muscle.
Nerve to subclavius (sub-KLĀ-vē-us)	C5–C6	Subclavius muscle.
Suprascapular	C5–C6	Supraspinatus and infraspinatus muscles.
Musculocutaneous (mus′-kū-lō-kū-TĀN-ē-us)	C5–C7	Coracobrachialis, biceps brachii, and brachialis muscles.
Lateral pectoral (PEK-to-ral)	C5–C7	Pectoralis major muscle.
Upper subscapular	C5–C6	Subscapularis muscle.
Thoracodorsal (tho-RĀ-kō-dor-sal)	C6–C8	Latissimus dorsi muscle.
Lower subscapular	C5–C6	Subscapularis and teres major muscles.
Axillary (AK-si-lar-ē)	C5–C6	Deltoid and teres minor muscles; skin over deltoid and superior posterior aspect of arm.
Median	C5–T1	Flexors of forearm, except flexor carpi ulnaris and some muscles of the hand (lateral palm); skin of lateral two-thirds of palm of hand and fingers.
Radial		Triceps brachii and other extensor muscles of arm and extensor muscles of forearm; skin of posterior arm and forearm, lateral two-thirds of dorsum of hand, and fingers over proximal and middle phalanges.
Medial pectoral	C8–T1	Pectoralis major and pectoralis minor muscles.
Medial cutaneous nerve of arm (kū′-TĀ-nē-us)	C8–T1	Skin of medial and posterior aspects of distal third of arm.
Medial cutaneous nerve of forearm	C8–T1	Skin of medial and posterior aspects of forearm.
Ulnar	C8–T1	Flexor carpi ulnaris, flexor digitorum profundus, and most muscles of the hand; skin of medial side of hand, little finger, and medial half of ring finger.

which the shoulder is adducted, the arm is medially rotated, the elbow is extended, the forearm is pronated, and the wrist is flexed (Figure 13.8a). This condition is called **Erb-Duchenne palsy** or **waiter's tip position.** There is loss of sensation along the lateral side of the arm.

Radial (and axillary) **nerve injury** can be caused by improperly administered intramuscular injections into the deltoid muscle. The radial nerve may also be injured when a cast is applied too tightly around the mid-humerus. Radial nerve injury is indicated by **wrist drop,** the inability to extend the wrist and fingers (Figure 13.8b). Sensory loss is minimal due to the overlap of sensory innervation by adjacent nerves.

Median nerve injury may result in **median nerve palsy,** which is indicated by numbness, tingling, and pain in the palm and fingers. There is also inability to pronate the forearm and flex the proximal interphalangeal joints of all digits and the distal interphalangeal joints of the second and third digits (Figure 13.8c). In addition, wrist flexion is weak and is accompanied by adduction, and thumb movements are weak.

Ulnar nerve injury may result in **ulnar nerve palsy,** which is indicated by an inability to abduct or adduct the fingers, atrophy of the interosseus muscles of the hand, hyperextension of the metacarpophalangeal joints, and flexion of the interphalangeal joints, a condition called **clawhand** (Figure 13.8d). There is also loss of sensation over the little finger.

Long thoracic nerve injury results in paralysis of the serratus anterior muscle. The medial border of the scapula protrudes, giving it the appearance of a wing. When the arm is raised, the vertebral border and inferior angle of the scapula pull away from the thoracic wall and protrude outward, causing the medial border of the scapula to protrude; because the scapula looks like a wing, this condition is called **winged scapula** (Figure 13.8e). The arm cannot be abducted beyond the horizontal position. ■

▶ CHECKPOINT

Injury of which nerve could cause paralysis of the serratus anterior muscle?

Figure 13.7 **Brachial plexus in anterior view.** (See Tortora, *A Photographic Atlas of the Human Body, Second Edition,* Figures 8.8 and 8.9.)

The brachial plexus supplies the shoulders and upper limbs.

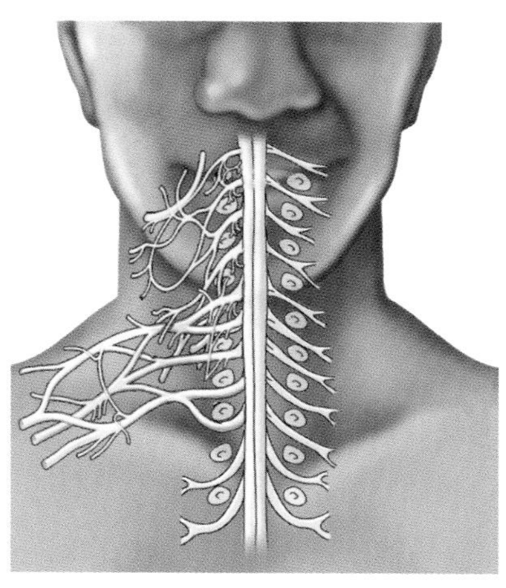

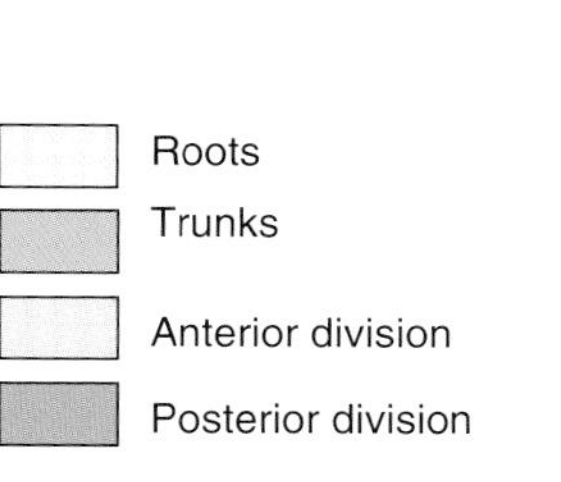

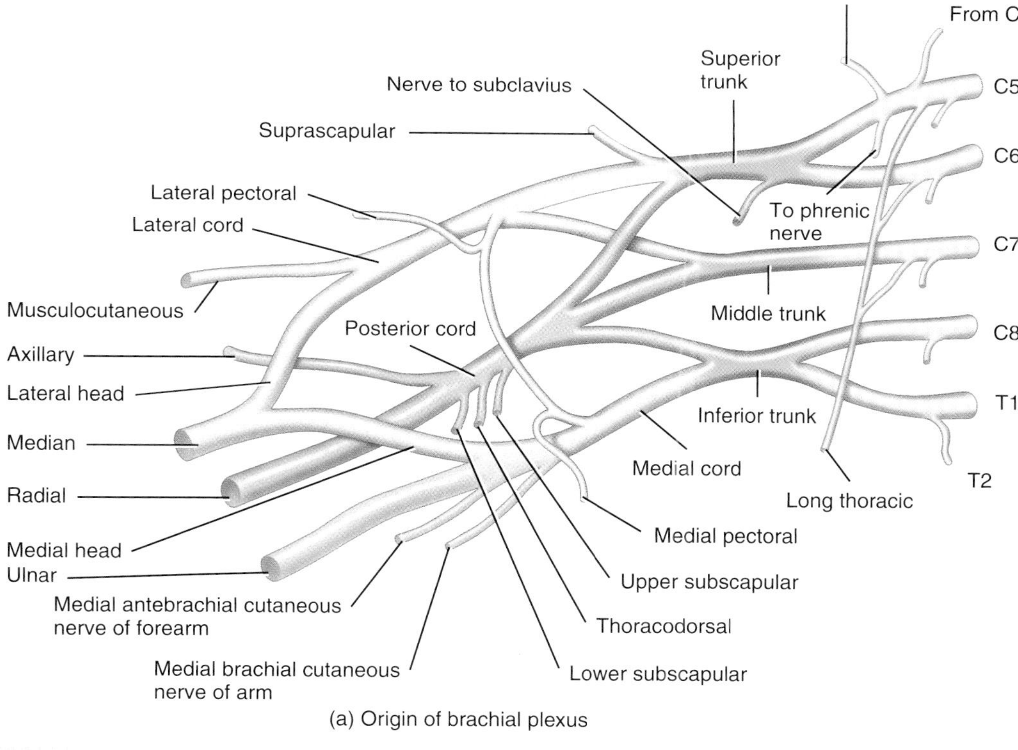

(a) Origin of brachial plexus

MNEMONIC for subunits of the brachial plexus:

Risk	**T**akers	**D**on't	**C**autiously	**B**ehave.
Roots	**T**runks	**D**ivisions	**C**ords	**B**ranches

continues

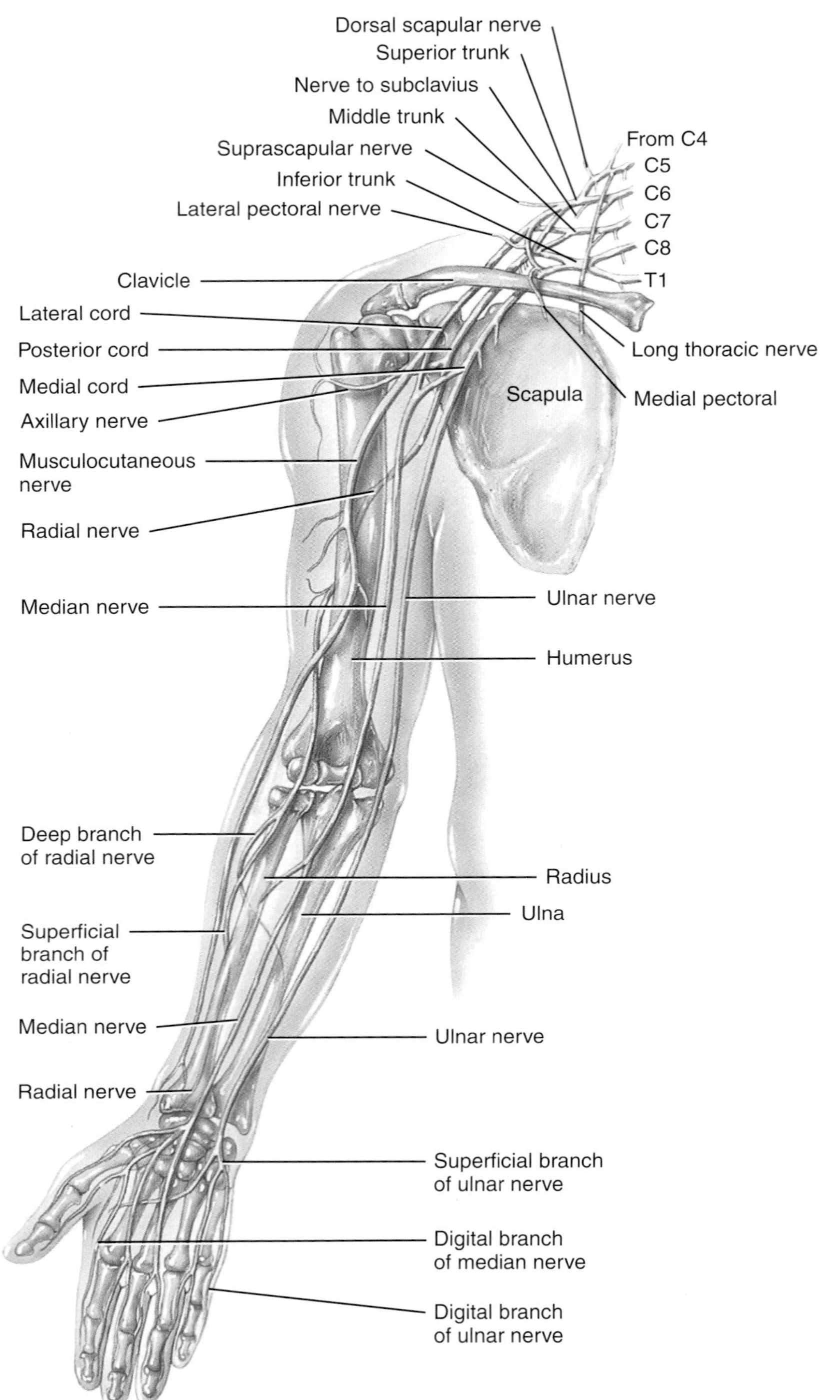

(b) Distribution of nerves from the brachial plexus

What five important nerves arise from the brachial plexus?

Figure 13.8 Injuries to the brachial plexus.

Injuries to the brachial plexus affect the sensations and movements of the upper limbs.

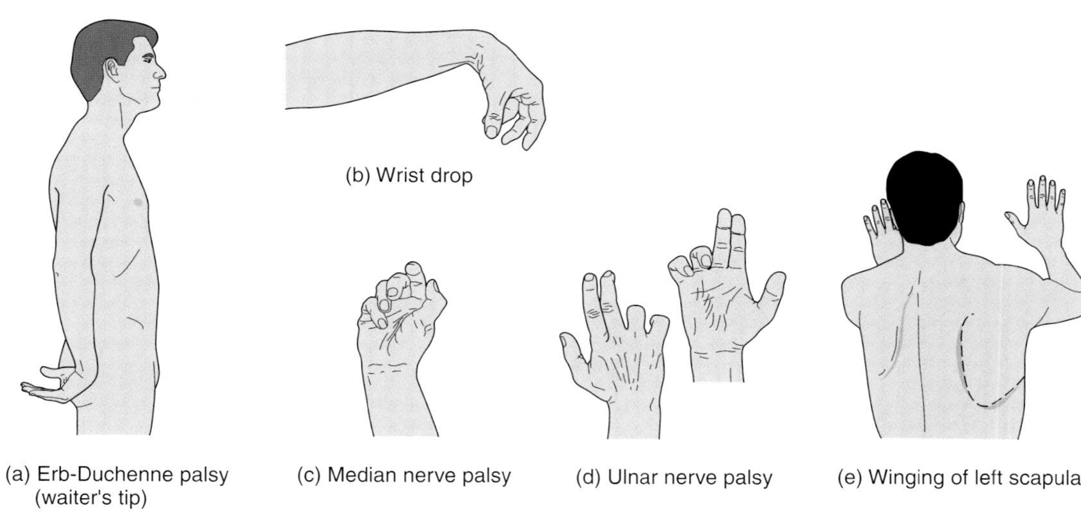

(a) Erb-Duchenne palsy (waiter's tip)

(b) Wrist drop

(c) Median nerve palsy

(d) Ulnar nerve palsy

(e) Winging of left scapula

Injury to which nerve of the brachial plexus affects sensations on the palm and fingers?

► OBJECTIVE

Describe the origin and distribution of the lumbar plexus.

The roots (anterior rami) of spinal nerves L1–L4 form the **lumbar plexus** (LUM-bar) (Figure 13.9). Unlike the brachial plexus, there is no intricate intermingling of fibers in the lumbar plexus. On either side of the first four lumbar vertebrae, the lumbar plexus passes obliquely outward, posterior to the psoas major muscle and anterior to the quadratus lumborum muscle. It then gives rise to its peripheral nerves.

The lumbar plexus supplies the anterolateral abdominal wall, external genitals, and part of the lower limbs.

Lumbar Plexus Injuries

The largest nerve arising from the lumbar plexus is the femoral nerve. **Femoral nerve injury,** which can occur in stab or gunshot wounds, is indicated by an inability to extend the leg and by loss of sensation in the skin over the anteromedial aspect of the thigh.

Obturator nerve injury results in paralysis of the adductor muscles of the leg and loss of sensation over the medial aspect of the thigh. It may result from pressure on the nerve by the fetal head during pregnancy. ■

► CHECKPOINT

What is the largest nerve arising from the lumbar flexes?

Figure 13.9 **Lumbar plexus in anterior view.** (See Tortora, *A Photographic Atlas of the Human Body, Second Edition,* Figure 8.10.)

The lumbar plexus supplies the anterolateral abdominal wall, external genitals, and part of the lower limbs.

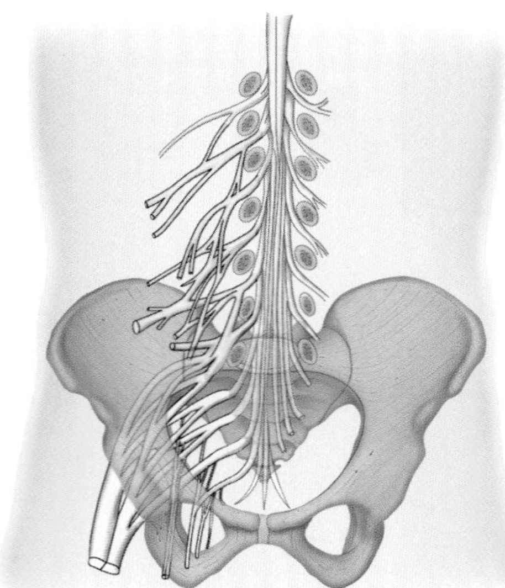

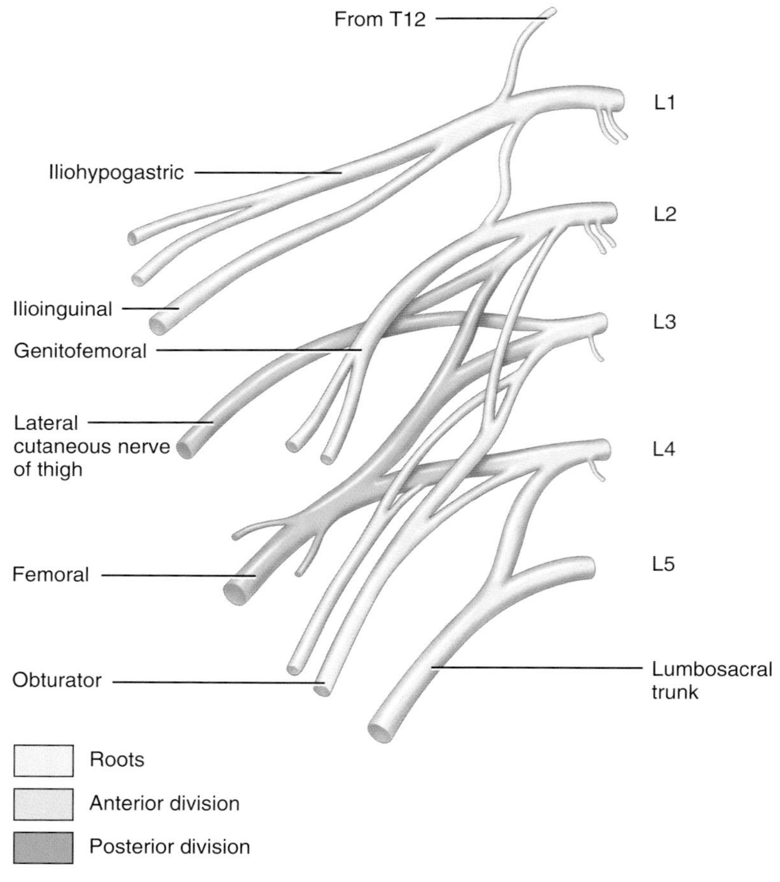

(a) Origin of lumbar plexus

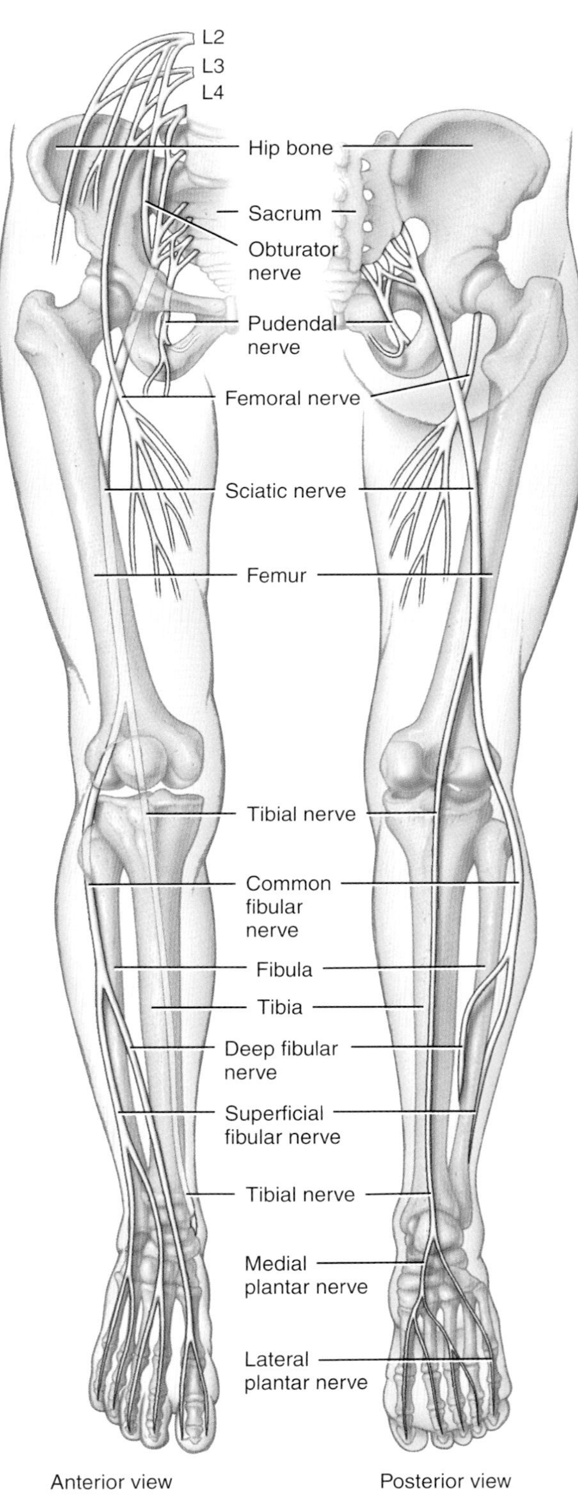

(b) Distribution of nerves from the lumbar and sacral plexuses

What are the signs of femoral nerve injury?

NERVE	ORIGIN	DISTRIBUTION
Iliohypogastric (il′-ē-ō-hī-pō-GAS-trik)	L1	Muscles of anterolateral abdominal wall; skin of inferior abdomen and buttock.
Ilioinguinal (il′-ē-ō-IN-gwi-nal)	L1	Muscles of anterolateral abdominal wall; skin of superior medial aspect of thigh, root of penis and scrotum in male, and labia majora and mons pubis in female.
Genitofemoral (jen′-i-tō-FEM-or-al)	L1–L2	Cremaster muscle; skin over middle anterior surface of thigh, scrotum in male, and labia majora in female.
Lateral cutaneous nerve of thigh	L2–L3	Skin over lateral, anterior, and posterior aspects of thigh.
Femoral	L2–L4	Flexor muscles of thigh and extensor muscles of leg; skin over anterior and medial aspect of thigh and medial side of leg and foot.
Obturator (OB-too-rā-tor)	L2–L4	Adductor muscles of leg; skin over medial aspect of thigh.

SACRAL AND COCCYGEAL PLEXUSES (FIGURE 13.10)

▶ **OBJECTIVE**

Describe the origin and distribution of the sacral plexus.

The roots (anterior rami) of spinal nerves L4–L5 and S1–S4 form the **sacral plexus** (SĀ-kral) (Figure 13.10). This plexus is situated largely anterior to the sacrum. The sacral plexus supplies the buttocks, perineum, and lower limbs. The largest nerve in the body—the sciatic nerve—arises from the sacral plexus.

The roots (anterior rami) of spinal nerves S4–S5 and the coccygeal nerves form a small **coccygeal plexus,** which supplies a small area of skin in the coccygeal region.

Sciatic Nerve Injury

The most common form of back pain is caused by compression or irritation of the sciatic nerve, the longest nerve in the human body. Injury to the sciatic nerve and its branches results in **sciatica,** pain that may extend from the buttock down the posterior and lateral aspect of the leg and the lateral aspect of the foot. The sciatic nerve may be injured because of a herniated (slipped) disc, dislocated hip, osteoarthritis of the lumbosacral spine, pressure from the uterus during pregnancy, inflammation, irritation, or an improperly administered gluteal intramuscular injection.

In the majority of sciatic nerve injuries, the common fibular portion is the most affected, frequently from fractures of the fibula or by pressure from casts or splints. Damage to the common fibular nerve causes the foot to be plantar flexed, a condition called **footdrop,** and inverted, a condition called **equinovarus.** There is also loss of function along the anterolateral aspects of the leg and dorsum of the foot and toes. Injury to the tibial portion of the sciatic nerve results in dorsiflexion of the foot plus eversion, a condition called **calcaneovalgus.** Loss of sensation on the sole also occurs. Treatments for sciatica are similar to those outlined earlier for a herniated (slipped) disc—rest, pain medications, exercises, ice or heat, and massage. ■

▶ **CHECKPOINT**

Injury of which nerve causes footdrop?

NERVE	ORIGIN	DISTRIBUTION
Superior gluteal (GLOO-tē-al)	L4–L5 and S1	Gluteus minimus and gluteus medius muscles and tensor fasciae latae.
Inferior gluteal	L5–S2	Gluteus maximus muscle.
Nerve to piriformis (pir-i-FORM-is)	S1–S2	Piriformis muscle.
Nerve to quadratus femoris (quad-RĀ-tus FEM-or-is) and **inferior gemellus** (jem-EL-us)	L4–L5 and S1	Quadratus femoris and inferior gemellus muscles.
Nerve to obturator internus (OB-too-rā′-tor in-TER-nus) and **superior gemellus**	L5–S2	Obturator internus and superior gemellus muscles.
Perforating cutaneous (kū′-TĀ-ne-us)	S2–S3	Skin over inferior medial aspect of buttock.
Posterior cutaneous nerve of thigh	S1–S3	Skin over anal region, inferior lateral aspect of buttock, superior posterior aspect of thigh, superior part of calf, scrotum in male, and labia majora in female.
Sciatic (sī-AT-ik)	L4–S3	Actually two nerves—tibial and common fibular—bound together by a common sheath of connective tissue. It splits into its two divisions, usually at the knee. (See below for distributions.) As the sciatic nerve descends through the thigh, it sends branches to hamstring muscles and the adductor magnus.
Tibial (TIB-ē-al)	L4–S3	Gastrocnemius, plantaris, soleus, popliteus, tibialis posterior, flexor digitorum longus, and flexor hallucis longus muscles. Branches of tibial nerve in foot are medial plantar nerve and lateral plantar nerve.
Medial plantar (PLAN-tar) (see Figure 13.9b)		Abductor hallucis, flexor digitorum brevis, and flexor hallucis brevis muscles; skin over medial two-thirds of plantar surface of foot.
Lateral plantar (see Figure 13.9b)		Remaining muscles of foot not supplied by medial plantar nerve; skin over lateral third of plantar surface of foot.
Common fibular (FIB-ū-lar)	L4–S2	Divides into a superficial fibular and a deep fibular branch.
Superficial fibular		Fibularis longus and fibularis brevis muscles; skin over distal third of anterior aspect of leg and dorsum of foot.
Deep fibular		Tibialis anterior, extensor hallucis longus, fibularis tertius, and extensor digitorum longus and extensor digitorum brevis muscles; skin on adjacent sides of great and second toes.
Pudendal (pū-DEN-dal)	S2–S4	Muscles of perineum; skin of penis and scrotum in male and clitoris, labia majora, labia minora, and vagina in female.

Figure 13.10 **Sacral and coccygeal plexuses in anterior view.** The distribution of the nerves of the sacral plexus is shown in Figure 13.9b. (See Tortora, *A Photographic Atlas of the Human Body, Second Edition,* Figure 8.11.)

The sacral plexus supplies the buttocks, perineum, and lower limbs.

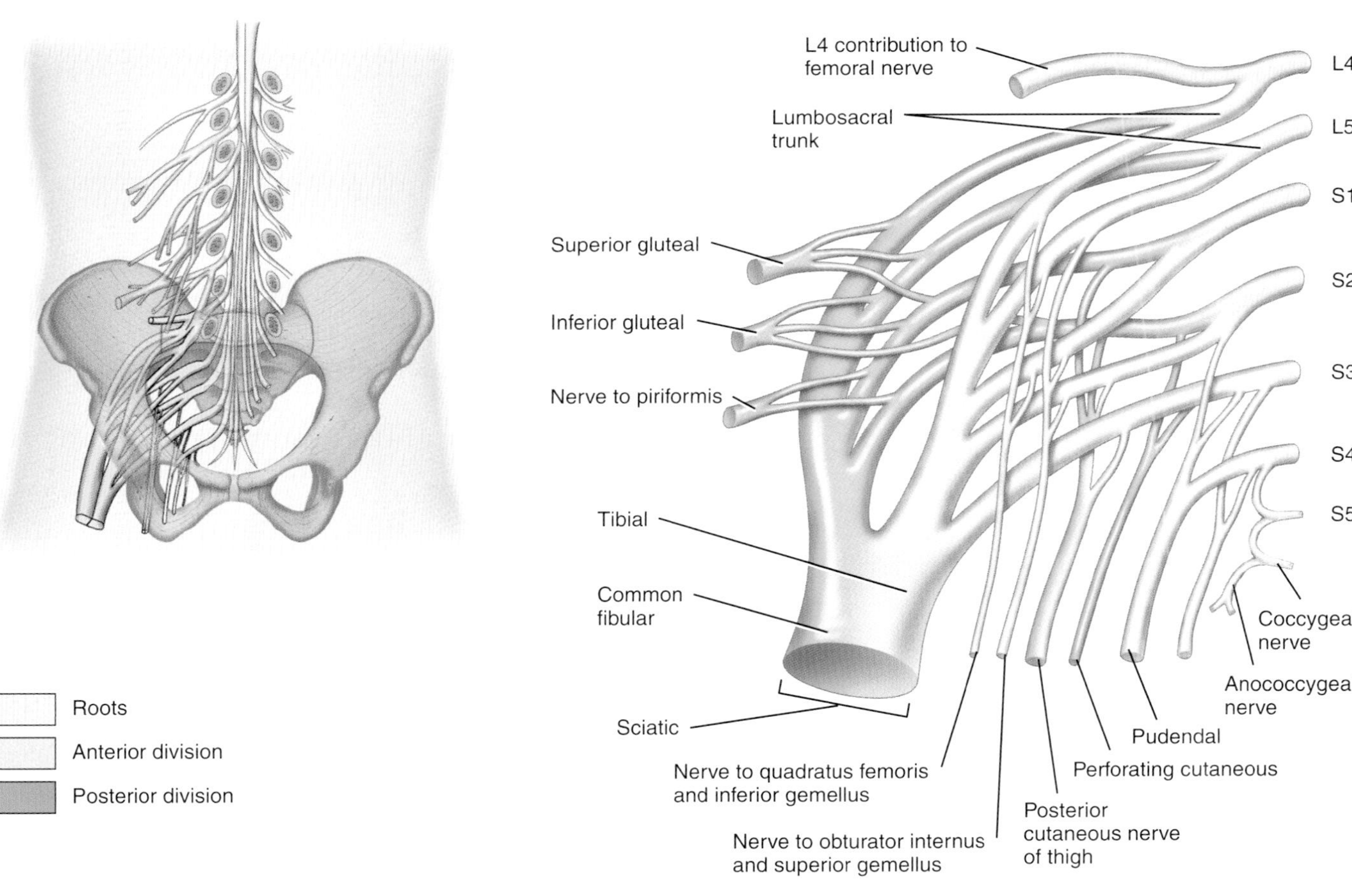

Origin of sacral plexus

What is the origin of the sacral plexus?

Dermatomes

The skin over the entire body is supplied by somatic sensory neurons that carry nerve impulses from the skin into the spinal cord and brain. Each spinal nerve contains sensory neurons that serve a specific, predictable segment of the body. One of the cranial nerves, the trigeminal (V) nerve, serves most of the skin of the face and scalp. The area of the skin that provides sensory input to the CNS via one pair of spinal nerves or the trigeminal (V) nerve is called a **dermatome** (*derma-* = skin; *-tome* = thin segment) (Figure 13.11). The nerve supply in adjacent dermatomes overlaps somewhat. Knowing which spinal cord segments supply each dermatome makes it possible to locate damaged regions of the spinal cord. If the skin in a particular region is stimulated but the sensation is not perceived, the nerves supplying that dermatome are probably damaged. In regions where the overlap is considerable, little loss of sensation may result if only one of the nerves supplying the dermatome is damaged. Information about the innervation patterns of spinal nerves can also be used therapeutically. Cutting posterior roots or infusing local anesthetics can block pain either permanently or transiently. Because dermatomes overlap, deliberate production of a region of complete anesthesia may require that at least three adjacent spinal nerves be cut or blocked by an anesthetic drug.

Figure 13.11 Distribution of dermatomes.

A dermatome is an area of skin that provides sensory input to the CNS via the posterior roots of one pair of spinal nerves or via the trigeminal (V) nerve.

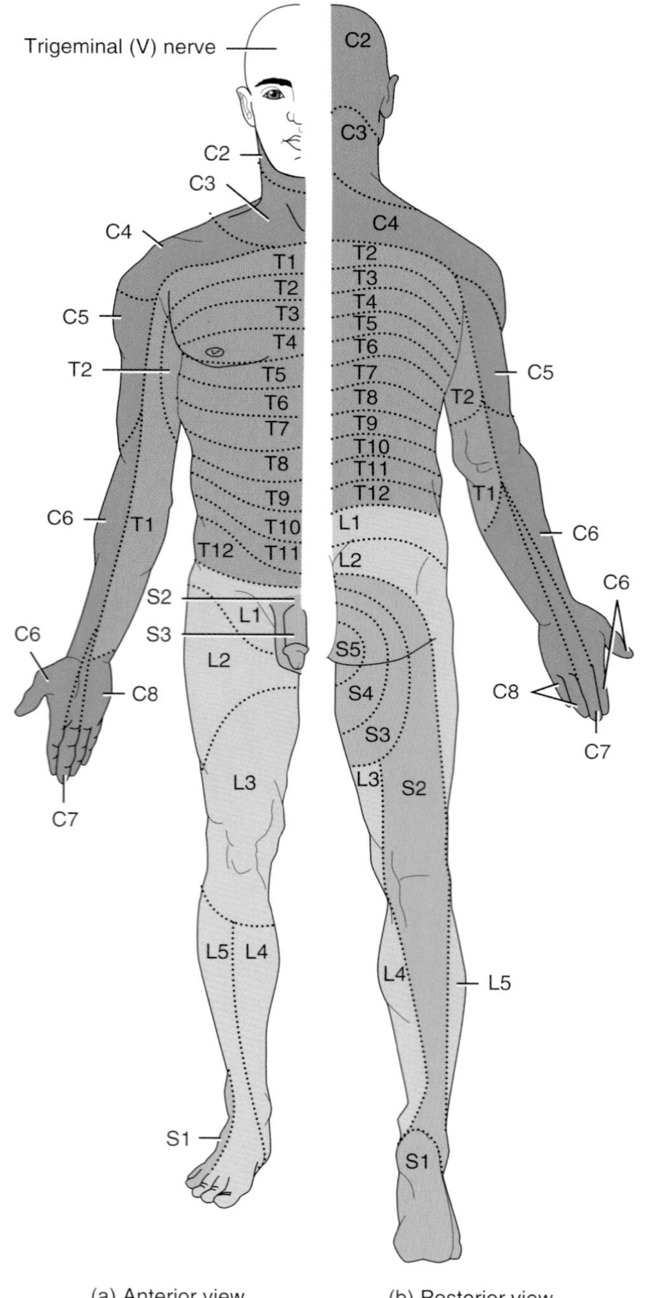

Which is the only spinal nerve that does not have a corresponding dermatome?

▶ CHECKPOINT

5. How are spinal nerves named and numbered? Why are all spinal nerves classified as mixed nerves?
6. How do spinal nerves connect to the spinal cord?
7. Which regions of the body are supplied by plexuses and by intercostal nerves?

SPINAL CORD PHYSIOLOGY

▶ OBJECTIVES

Describe the functions of the major sensory and motor tracts of the spinal cord.

Describe the functional components of a reflex arc and the ways reflexes maintain homeostasis.

The spinal cord has two principal functions in maintaining homeostasis: nerve impulse propagation and integration of information. The *white matter tracts* in the spinal cord are highways for nerve impulse propagation. Sensory input travels along these tracts toward the brain, and motor output travels from the brain along these tracts toward skeletal muscles and other effector tissues. The *gray matter* of the spinal cord receives and integrates incoming and outgoing information.

Sensory and Motor Tracts

As noted above, one of the ways the spinal cord promotes homeostasis is by conducting nerve impulses along tracts. Often, the name of a tract indicates its position in the white matter and where it begins and ends. For example, the anterior spinothalamic tract is located in the *anterior* white column; it begins in the *spinal cord* and ends in the *thalamus* (a region of the brain). Notice that the location of the axon terminals comes last in the name. This regularity in naming allows you to determine the direction of information flow along any tract named according to this convention. Because the anterior spinothalamic tract conveys nerve impulses from the spinal cord toward the brain, it is a sensory (ascending) tract. Figure 13.12 highlights the major sensory and motor tracts in the spinal cord. These tracts are described in detail in Chapter 16 and summarized in Tables 16.3 and 16.4 on pages 559 and 563.

Nerve impulses from sensory receptors propagate up the spinal cord to the brain along two main routes on each side: the spinothalamic tracts and the posterior columns. The **lateral** and **anterior spinothalamic tracts** convey nerve impulses for sensing pain, warmth, coolness, itching, tickling, deep pressure, and a crude, poorly localized sense of touch. The right and left **posterior columns** carry nerve impulses for several kinds of sensation. These include (1) proprioception, awareness of the positions and movements of muscles, tendons, and joints; (2) discriminative touch, the ability to feel exactly what part of the body is touched; (3) two-point discrimination, the ability to distinguish the touching of two different points on the skin, even though they are close together; and (4) vibration sensations.

The sensory systems keep the CNS informed of changes in the external and internal environments. The sensory information is integrated (processed) by interneurons in the spinal cord and brain. Responses to the integrative decisions are brought about by motor activities (muscular contractions and glandular secretions). The cerebral cortex, the outer part of the brain, plays a major role

Figure 13.12 **Locations of major sensory and motor tracts, shown in a transverse section of the spinal cord.** Sensory tracts are indicated on one half and motor tracts on the other half of the cord, but actually all tracts are present on both sides.

The name of a tract often indicates its location in the white matter and where it begins and ends.

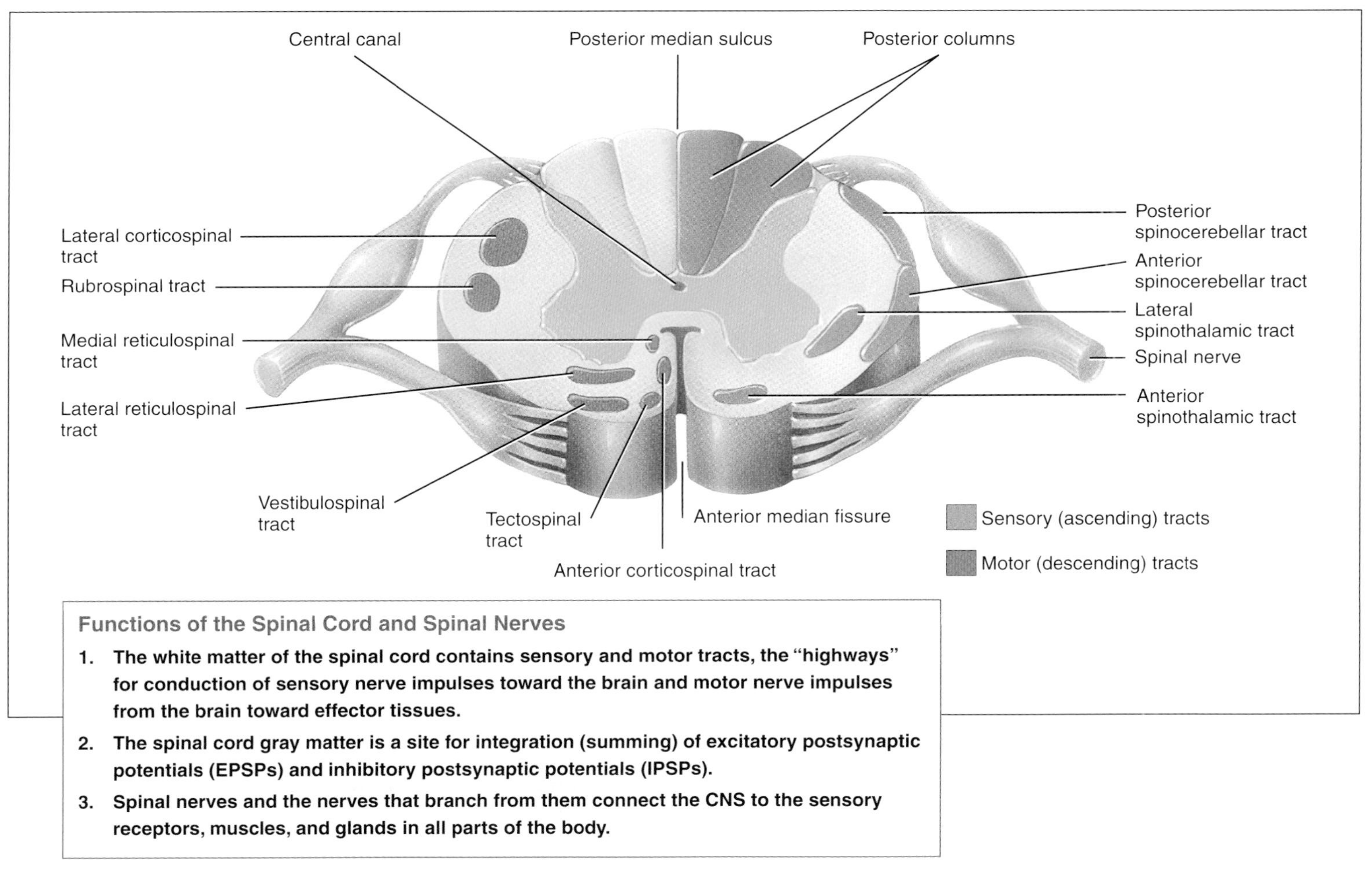

? Based on its name, what are the position in the spinal cord, origin, and destination of the anterior corticospinal tract? Is this a sensory or a motor tract?

in controlling precise voluntary muscular movements. Other brain regions provide important integration for regulation of automatic movements, such as arm swinging during walking. Motor output to skeletal muscles travels down the spinal cord in two types of descending pathways: direct and indirect. The **direct pathways** include the *lateral corticospinal, anterior corticospinal,* and *corticobulbar tracts.* They convey nerve impulses that originate in the cerebral cortex and are destined to cause precise, *voluntary* movements of skeletal muscles. **Indirect pathways** include the *rubrospinal, tectospinal,* and *vestibulospinal tracts.* They convey nerve impulses from the brain stem and other parts of the brain that govern *automatic movements* and help coordinate body movements with visual stimuli. Indirect pathways also maintain skeletal muscle tone, maintain contraction of postural muscles, and play a major role in equilibrium by regulating muscle tone in response to movements of the head.

Reflexes and Reflex Arcs

The second way the spinal cord promotes homeostasis is by serving as an integrating center for some reflexes. A **reflex** is a fast, automatic, unplanned sequence of actions that occurs in response to a particular stimulus. Some reflexes are inborn, such as pulling your hand away from a hot surface before you even feel that it is hot. Other reflexes are learned or acquired. For instance, you learn many reflexes while acquiring driving expertise. Slamming on the brakes in an emergency is one example. When integration takes place in the spinal cord gray matter, the reflex is a **spinal reflex.** An example is the familiar patellar reflex (knee jerk). If integration occurs in the brain stem rather than the spinal cord, the reflex is called a **cranial reflex.** An example is the tracking movements of your eyes as you read this sentence. You are probably most aware of **somatic reflexes,** which involve contraction of skeletal muscles. Equally important, however, are the **autonomic (visceral) reflexes,** which generally are not consciously perceived. They involve responses of smooth muscle, cardiac muscle, and glands. As you will see in Chapter 15, body functions such as heart rate, digestion, urination, and defecation are controlled by the autonomic nervous system through autonomic reflexes.

Nerve impulses propagating into, through, and out of the CNS follow specific pathways, depending on the kind of information, its origin, and its destination. The pathway followed by nerve impulses that produce a reflex is a **reflex arc (reflex circuit).** A reflex arc includes the following five functional components (Figure 13.13):

1. **Sensory receptor.** The distal end of a sensory neuron (dendrite) or an associated sensory structure serves as a sensory receptor. It responds to a specific **stimulus**—a change in the internal or external environment—by producing a graded potential called a generator (or receptor) potential (described on page 549). If a generator potential reaches the threshold level of depolarization, it will trigger one or more nerve impulses in the sensory neuron.
2. **Sensory neuron.** The nerve impulses propagate from the sensory receptor along the axon of the sensory neuron to the axon terminals, which are located in the gray matter of the spinal cord or brain stem.
3. **Integrating center.** One or more regions of gray matter within the CNS act as an integrating center. In the simplest type of reflex, the integrating center is a single synapse between a sensory neuron and a motor neuron. A reflex pathway having only one synapse in the CNS is termed a **monosynaptic reflex arc** (*mono-* = one). More often, the integrating center consists of one or more interneurons, which may relay impulses to other interneurons as well as to a motor neuron. A **polysynaptic reflex arc** (*poly-* = many) involves more than two types of neurons and more than one CNS synapse.
4. **Motor neuron.** Impulses triggered by the integrating center propagate out of the CNS along a motor neuron to the part of the body that will respond.
5. **Effector.** The part of the body that responds to the motor nerve impulse, such as a muscle or gland, is the effector. Its action is called a reflex. If the effector is skeletal muscle, the reflex is a **somatic reflex.** If the effector is smooth muscle, cardiac muscle, or a gland, the reflex is an **autonomic (visceral) reflex.**

Because reflexes are normally so predictable, they provide useful information about the health of the nervous system and can greatly aid diagnosis of disease. Damage or disease anywhere along its reflex arc can cause a reflex to be absent or abnormal. For example, tapping the patellar ligament normally causes reflex extension of the knee joint. Absence of the patellar reflex could indicate damage of the sensory or motor neurons, or a spinal cord injury in the lumbar region. Somatic reflexes generally can be tested simply by tapping or stroking the body surface.

Next, we examine four important somatic spinal reflexes: the stretch reflex, the tendon reflex, the flexor (withdrawal) reflex, and the crossed extensor reflex.

The Stretch Reflex

A **stretch reflex** causes contraction of a skeletal muscle (the effector) in response to stretching of the muscle. This type of reflex occurs via a monosynaptic reflex arc. The reflex can occur by activation of a single sensory neuron that forms one synapse in the CNS with a single motor neuron. Stretch reflexes can be elicited by tapping on tendons attached to muscles at the elbow, wrist, knee, and ankle joints.

A stretch reflex operates as follows (Figure 13.14):

1. Slight stretching of a muscle stimulates sensory receptors in the muscle called **muscle spindles** (shown in more detail in Figure 16.4 on page 554). The spindles monitor changes in the length of the muscle.

Figure 13.13 General components of a reflex arc. The arrows show the direction of nerve impulse propagation.

A reflex is a fast, predictable sequence of involuntary actions that occur in response to certain changes in the environment.

2 SENSORY NEURON (axon conducts impulses from receptor to integrating center)
1 SENSORY RECEPTOR (responds to a stimulus by producing a generator or receptor potential)
Interneuron
3 INTEGRATING CENTER (one or more regions within the CNS that relay impulses from sensory to motor neurons)
4 MOTOR NEURON (axon conducts impulses from integrating center to effector)
5 EFFECTOR (muscle or gland that responds to motor nerve impulses)

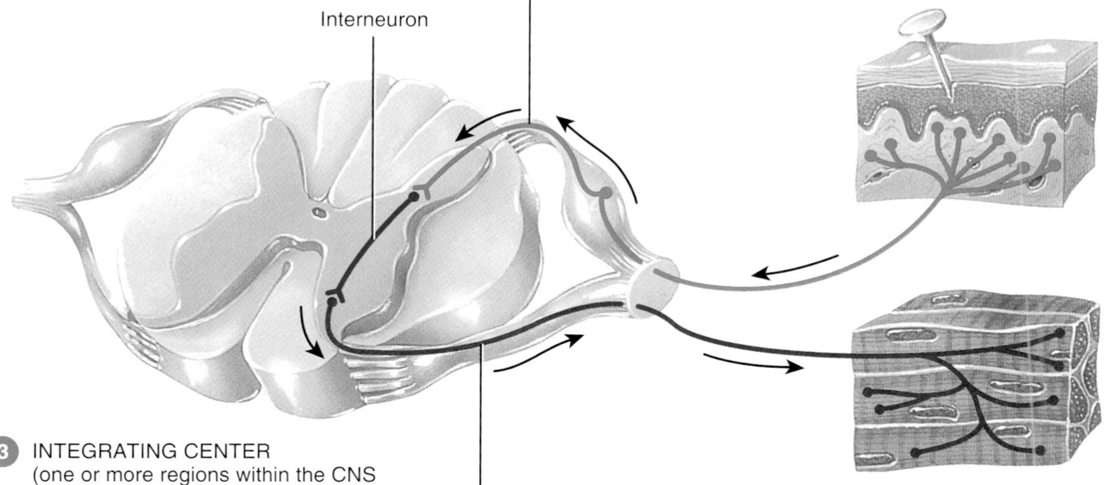

What initiates a nerve impulse in a sensory neuron? Which branch of the nervous system includes all integrating centers for reflexes?

2. In response to being stretched, a muscle spindle generates one or more nerve impulses that propagate along a somatic sensory neuron through the posterior root of the spinal nerve and into the spinal cord.
3. In the spinal cord (integrating center), the sensory neuron makes an excitatory synapse with and thereby activates a motor neuron in the anterior gray horn.
4. If the excitation is strong enough, one or more nerve impulses arise in the motor neuron and propagate along its axon, which extends from the spinal cord into the anterior root and through peripheral nerves to the stimulated muscle. The axon terminals of the motor neuron form neuromuscular junctions (NMJs) with skeletal muscle fibers of the stretched muscle.
5. Acetylcholine released by nerve impulses at the NMJs triggers one or more muscle action potentials in the stretched muscle (effector), and the muscle contracts. Thus, muscle stretch is followed by muscle contraction, which relieves the stretching.

In the reflex arc just described, sensory nerve impulses enter the spinal cord on the same side from which motor nerve impulses leave it. This arrangement is called an **ipsilateral reflex** (ip′-si-LAT-er-al = same side). All monosynaptic reflexes are ipsilateral.

In addition to the large-diameter motor neurons that innervate typical skeletal muscle fibers, smaller-diameter motor neurons innervate smaller, specialized muscle fibers within the muscle spindles themselves. The brain regulates muscle spindle sensitivity through pathways to these smaller motor neurons. This regulation ensures proper muscle spindle signaling over a wide range of muscle lengths during voluntary and reflex contractions. By adjusting how vigorously a muscle spindle responds to stretching, the brain sets an overall level of **muscle tone,** which is the small degree of contraction present while the muscle is at rest. Because the stimulus for the stretch reflex is stretching of muscle, this reflex helps avert injury by preventing overstretching of muscles.

Although the stretch reflex pathway itself is monosynaptic (just two neurons and one synapse), a polysynaptic reflex arc to the antagonistic muscles operates at the same time. This arc involves three neurons and two synapses. An axon collateral (branch) from the muscle spindle sensory neuron also synapses with an inhibitory interneuron in the integrating center. In turn, the interneuron synapses with and inhibits a motor neuron that normally excites the antagonistic muscles (Figure 13.14). Thus, when the stretched muscle contracts during a stretch reflex, antagonistic muscles that oppose the contraction relax. This type of arrangement, in which the components of a neural circuit simultaneously cause contraction of one muscle and relaxation of its antagonists, is termed **reciprocal innervation.** Reciprocal innervation prevents conflict between opposing muscles and is vital in coordinating body movements.

Axon collaterals of the muscle spindle sensory neuron also relay nerve impulses to the brain over specific ascending pathways. In this way, the brain receives input about the state of stretch or contraction of skeletal muscles, enabling it to coordinate muscular movements. The nerve impulses that pass to the brain also allow conscious awareness that the reflex has occurred.

The stretch reflex can also help maintain posture. For example, if a standing person begins to lean forward, the gastrocnemius and other calf muscles are stretched. Consequently, stretch reflexes are initiated in these muscles, which cause them to contract and reestablish the body's upright posture. Similar types of stretch reflexes occur in the muscles of the shin when a standing person begins to lean backward.

The Tendon Reflex

The stretch reflex operates as a feedback mechanism to control muscle *length* by causing muscle contraction. In contrast, the

Figure 13.14 Stretch reflex. This monosynaptic reflex arc has only one synapse in the CNS—between a single sensory neuron and a single motor neuron. A polysynaptic reflex arc to antagonistic muscles that includes two synapses in the CNS and one interneuron is also illustrated. Plus signs (+) indicate excitatory synapses; the minus sign (−) indicates an inhibitory synapse.

The stretch reflex causes contraction of a muscle that has been stretched.

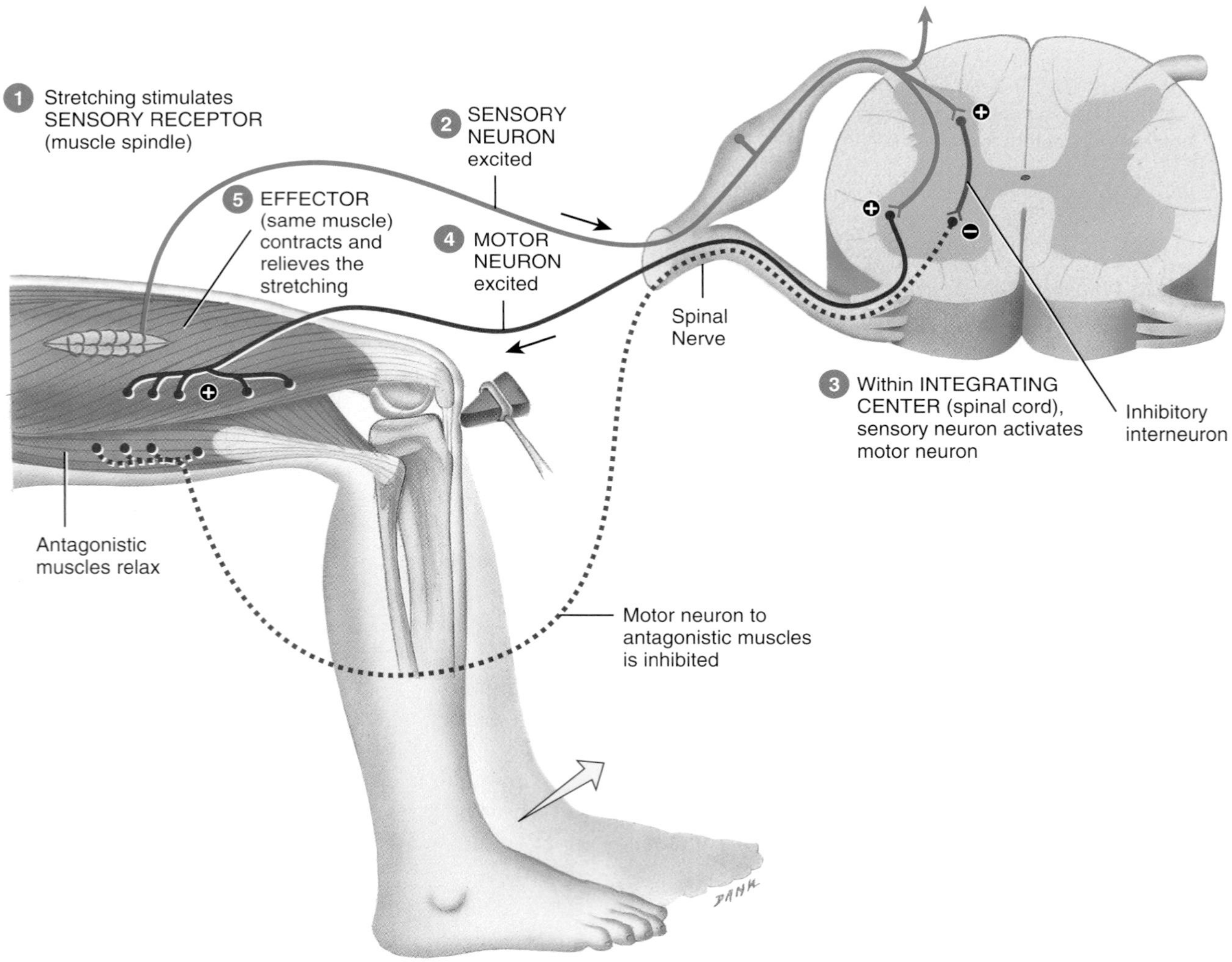

What makes this an ipsilateral reflex?

tendon reflex operates as a feedback mechanism to control muscle *tension* by causing muscle relaxation before muscle force becomes so great that tendons might be torn. Although the tendon reflex is less sensitive than the stretch reflex, it can override the stretch reflex when tension is great, making you drop a very heavy weight, for example. Like the stretch reflex, the tendon reflex is ipsilateral. The sensory receptors for this reflex are called **tendon (Golgi tendon) organs** (shown in more detail in Figure 16.4 on page 554), which lie within a tendon near its junction with a muscle. In contrast to muscle spindles, which are sensitive to changes in muscle length, tendon organs detect and respond to changes in muscle tension that are caused by passive stretch or muscular contraction.

A tendon reflex operates as follows (Figure 13.15):

1. As the tension applied to a tendon increases, the tendon organ (sensory receptor) is stimulated (depolarized to threshold).
2. Nerve impulses arise and propagate into the spinal cord along a sensory neuron.
3. Within the spinal cord (integrating center), the sensory neuron activates an inhibitory interneuron that synapses with a motor neuron.
4. The inhibitory neurotransmitter inhibits (hyperpolarizes) the motor neuron, which then generates fewer nerve impulses.
5. The muscle relaxes and relieves excess tension.

Figure 13.15 Tendon reflex. This reflex arc is polysynaptic—more than one CNS synapse and more than two different neurons are involved in the pathway. The sensory neuron synapses with two interneurons. An inhibitory interneuron causes relaxation of the effector, and a stimulatory interneuron causes contraction of the antagonistic muscle. Plus signs (+) indicate excitatory synapses; the minus sign (−) indicates an inhibitory synapse.

The tendon reflex causes relaxation of the muscle attached to the stimulated tendon organ.

To brain
Inhibitory interneuron
5 EFFECTOR (muscle attached to same tendon) relaxes and relieves excess tension
4 MOTOR NEURON inhibited
2 SENSORY NEURON excited
1 Increased tension stimulates SENSORY RECEPTOR (tendon organ)
Spinal nerve
3 Within INTEGRATING CENTER (spinal cord), sensory neuron activates inhibitory interneuron
Excitatory interneuron
Antagonistic muscles contract
Motor neuron to antagonistic muscles is excited
DANK

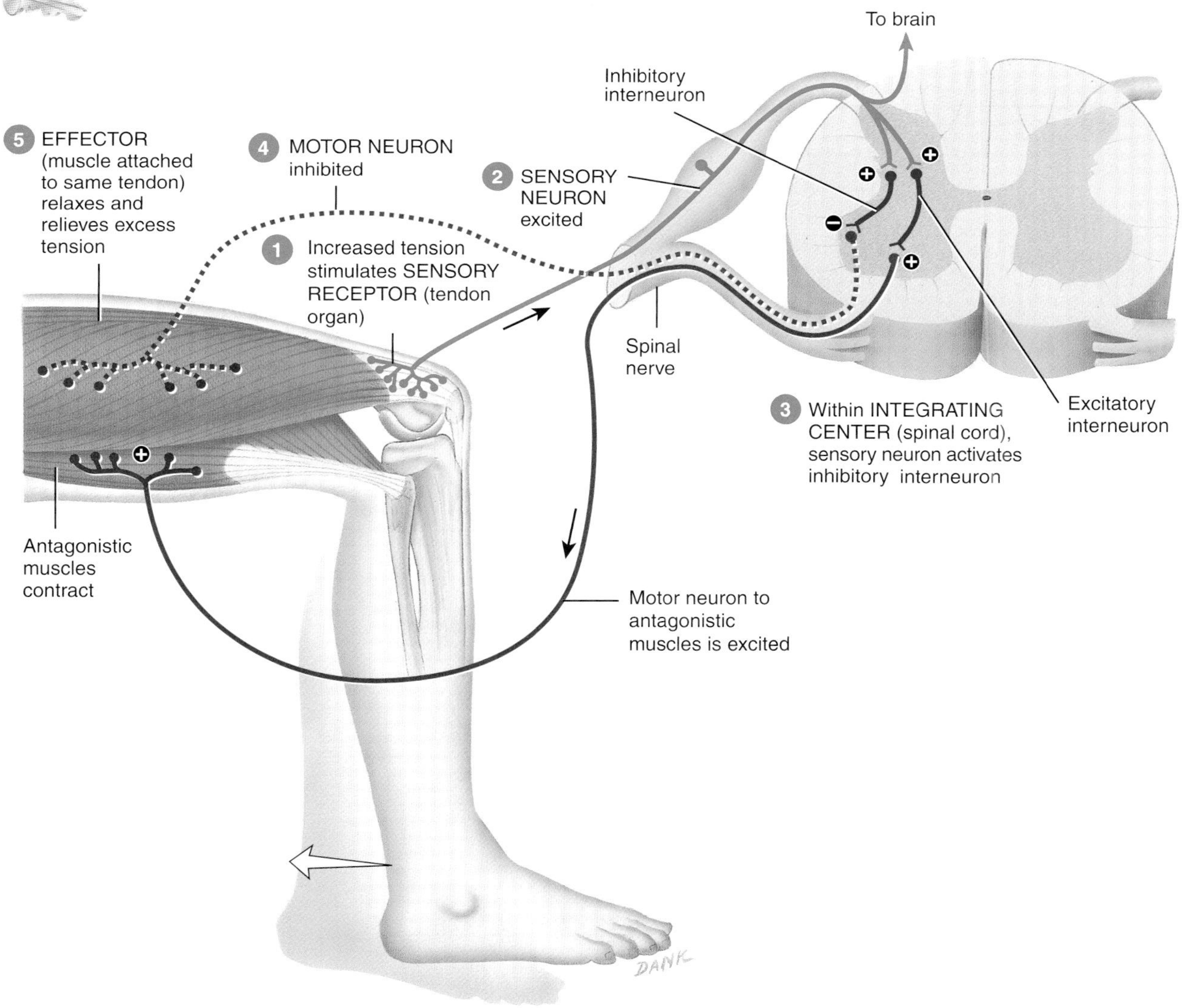

What is reciprocal innervation?

Thus, as tension on the tendon organ increases, the frequency of inhibitory impulses increases; inhibition of the motor neurons to the muscle developing excess tension (effector) causes relaxation of the muscle. In this way, the tendon reflex protects the tendon and muscle from damage due to excessive tension.

Note in Figure 13.15 that the sensory neuron from the tendon organ also synapses with an excitatory interneuron in the spinal cord. The excitatory interneuron, in turn, synapses with motor neurons controlling antagonistic muscles. Thus, while the tendon reflex brings about relaxation of the muscle attached to the tendon organ, it also triggers contraction of antagonists. Here we have another example of reciprocal innervation. The sensory neuron also relays nerve impulses to the brain by way of sensory tracts, thus informing the brain about the state of muscle tension throughout the body.

The Flexor and Crossed Extensor Reflexes

Another reflex involving a polysynaptic reflex arc results when, for instance, you step on a tack. In response to such a painful stimulus, you immediately withdraw your leg. This reflex, called the **flexor** or **withdrawal reflex,** operates as follows (Figure 13.16):

1. Stepping on a tack stimulates the dendrites (sensory receptor) of a pain-sensitive neuron.
2. This sensory neuron then generates nerve impulses, which propagate into the spinal cord.
3. Within the spinal cord (integrating center), the sensory neuron activates interneurons that extend to several spinal cord segments.
4. The interneurons activate motor neurons in several spinal cord segments. As a result, the motor neurons generate nerve impulses, which propagate toward the axon terminals.
5. Acetylcholine released by the motor neurons causes the flexor muscles in the thigh (effectors) to contract, producing withdrawal of the leg. This reflex is protective because contraction of flexor muscles moves a limb away from the source of a possibly damaging stimulus.

The flexor reflex, like the stretch reflex, is ipsilateral—the incoming and outgoing impulses propagate into and out of the same side of the spinal cord. The flexor reflex also illustrates another feature of polysynaptic reflex arcs. Moving your entire lower or upper limb away from a painful stimulus involves contraction of more than one muscle group. Hence, several motor neurons must simultaneously convey impulses to several limb muscles. Because nerve impulses from one sensory neuron ascend and descend in the spinal cord and activate interneurons in several segments of the spinal cord, this type of reflex is called an **intersegmental reflex arc** (*inter-* = between). Through intersegmental reflex arcs, a single sensory neuron can activate several motor neurons, thereby stimulating more than one effector. The monosynaptic stretch reflex, in contrast, involves muscles receiving nerve impulses from one spinal cord segment only.

Something else may happen when you step on a tack: You may start to lose your balance as your body weight shifts to the other foot. Besides initiating the flexor reflex that causes you to withdraw the limb, the pain impulses from stepping on the tack also initiate a **crossed extensor reflex** to help you maintain your balance; it operates as follows (Figure 13.17):

1. Stepping on a tack stimulates the sensory receptor of a pain-sensitive neuron in the right foot.
2. This sensory neuron then generates nerve impulses, which propagate into the spinal cord.
3. Within the spinal cord (integrating center), the sensory neuron activates several interneurons that synapse with motor neurons on the left side of the spinal cord in several spinal cord segments. Thus, incoming pain signals cross to the opposite side through interneurons at that level, and at several levels above and below the point of entry into the spinal cord.
4. The interneurons excite motor neurons in several spinal cord segments that innervate extensor muscles. The motor neurons, in turn, generate more nerve impulses, which propagate toward the axon terminals.
5. Acetylcholine released by the motor neurons causes extensor muscles in the thigh (effectors) of the unstimulated left limb to contract, producing extension of the left leg. In this way, weight can be placed on the foot that must now support the entire body. A comparable reflex occurs with painful stimulation of the left lower limb or either upper limb.

Unlike the flexor reflex, which is an ipsilateral reflex, the crossed extensor reflex involves a **contralateral reflex arc** (kon′-tra-LAT-er-al = opposite side): Sensory impulses enter one side of the spinal cord and motor impulses exit on the opposite side. Thus, a crossed extensor reflex synchronizes the extension of the contralateral limb with the withdrawal (flexion) of the stimulated limb. Reciprocal innervation also occurs in both the flexor reflex and the crossed extensor reflex. In the flexor reflex, when the flexor muscles of a painfully stimulated lower limb are contracting, the extensor muscles of the same limb are relaxing to some degree. If both sets of muscles contracted at the same time, the two sets of muscles would pull on the bones in opposite directions, which might immobilize the limb. Because of reciprocal innervation, one set of muscles contracts while the other relaxes.

Figure 13.16 Flexor (withdrawal) reflex. This reflex arc is polysynaptic and ipsilateral. Plus signs (+) indicate excitatory synapses.

The flexor reflex causes withdrawal of a part of the body in response to a painful stimulus.

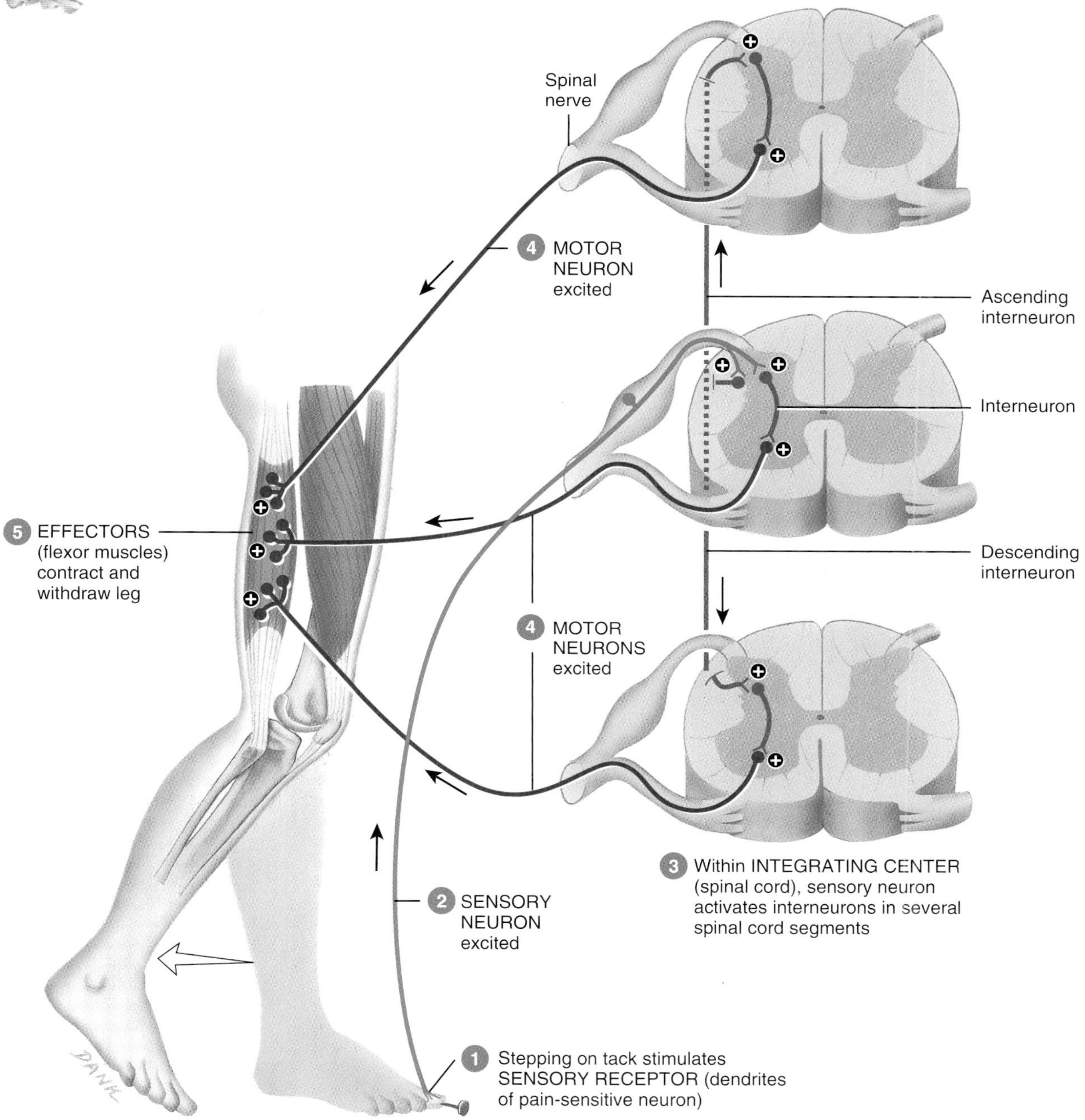

Spinal nerve

4 MOTOR NEURON excited

Ascending interneuron

Interneuron

5 EFFECTORS (flexor muscles) contract and withdraw leg

Descending interneuron

4 MOTOR NEURONS excited

3 Within INTEGRATING CENTER (spinal cord), sensory neuron activates interneurons in several spinal cord segments

2 SENSORY NEURON excited

1 Stepping on tack stimulates SENSORY RECEPTOR (dendrites of pain-sensitive neuron)

Why is the flexor reflex classified as an intersegmental reflex arc?

Figure 13.17 **Crossed extensor reflex.** The flexor reflex arc is shown (at left) to enable comparison with the crossed extensor reflex arc. Plus signs (+) indicate excitatory synapses.

A crossed extensor reflex causes contraction of muscles that extend joints in the limb opposite a painful stimulus.

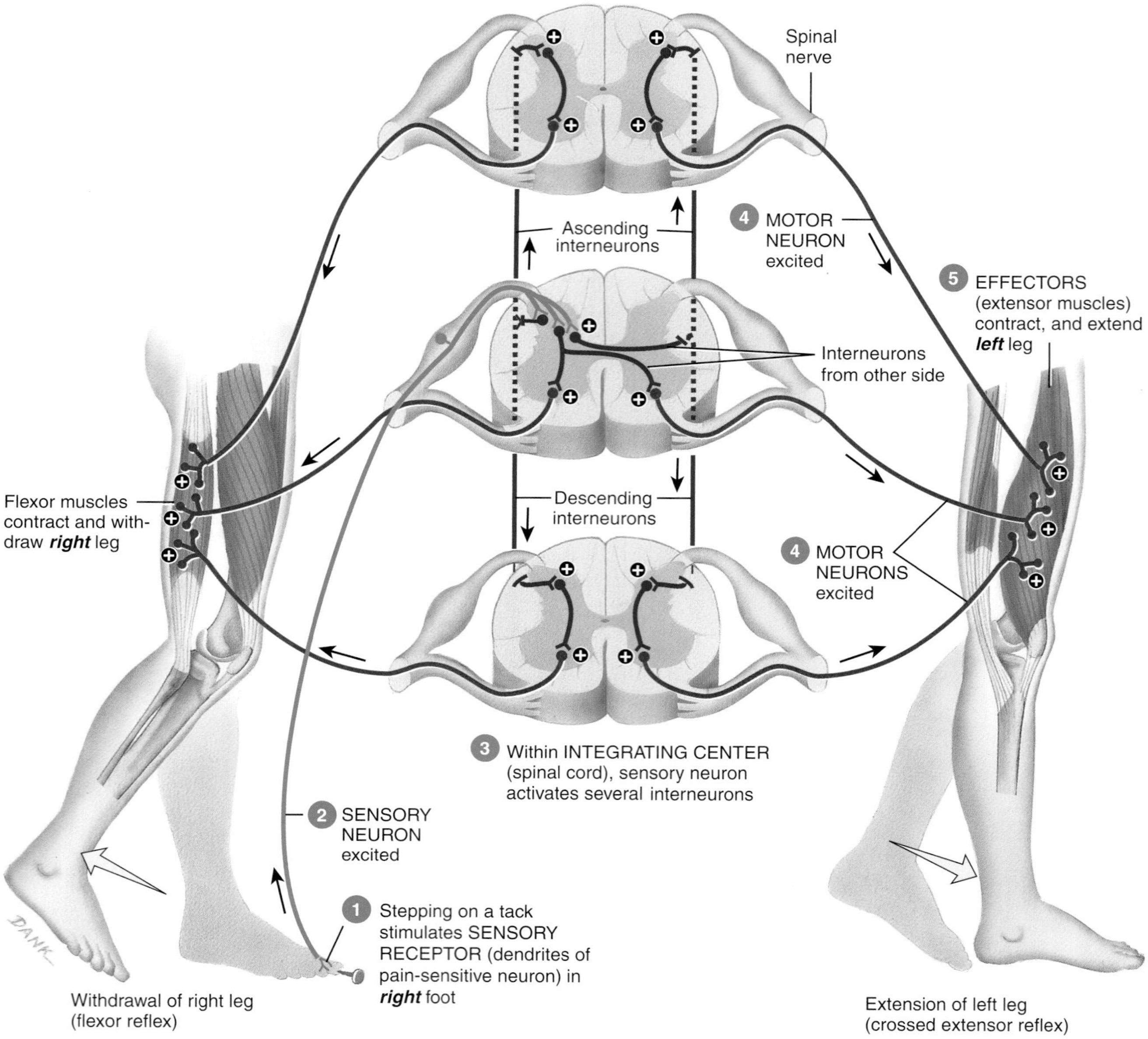

Spinal nerve

Ascending interneurons

4 MOTOR NEURON excited

5 EFFECTORS (extensor muscles) contract, and extend ***left*** leg

Interneurons from other side

Flexor muscles contract and withdraw ***right*** leg

Descending interneurons

4 MOTOR NEURONS excited

3 Within INTEGRATING CENTER (spinal cord), sensory neuron activates several interneurons

2 SENSORY NEURON excited

1 Stepping on a tack stimulates SENSORY RECEPTOR (dendrites of pain-sensitive neuron) in ***right*** foot

Withdrawal of right leg (flexor reflex)

Extension of left leg (crossed extensor reflex)

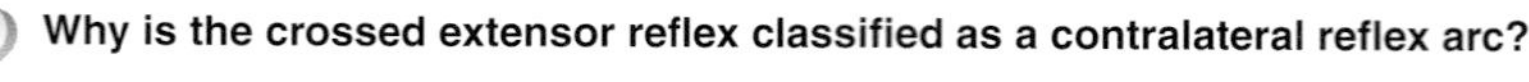

Why is the crossed extensor reflex classified as a contralateral reflex arc?

Reflexes and Diagnosis

Reflexes are often used for diagnosing disorders of the nervous system and locating injured tissue. If a reflex ceases to function or functions abnormally, the physician may suspect that the damage lies somewhere along a particular conduction pathway. Many somatic reflexes can be tested simply by tapping or stroking the body. Among the somatic reflexes of clinical significance are the following:

- **Patellar reflex (knee jerk).** This reflex involves extension of the leg at the knee joint by contraction of the quadriceps femoris muscle in response to tapping the patellar ligament (see Figure 13.14). This reflex is blocked by damage to the sensory or motor nerves supplying the muscle or to the integrating centers in the second, third, or fourth lumbar segments of the spinal cord. It is often absent in people with chronic diabetes mellitus or neurosyphilis, both of which cause degeneration of nerves. It is exaggerated in disease or injury involving certain motor tracts descending from the higher centers of the brain to the spinal cord.

- **Achilles reflex (ankle jerk).** This stretch reflex involves extension (plantar flexion) of the foot by contraction of the gastrocnemius and soleus muscles in response to tapping the calcaneal (Achilles) tendon. Absence of the Achilles reflex indicates damage to the nerves supplying the posterior leg muscles or to neurons in the lumbrosacral region of the spinal cord. This reflex may also disappear in people with chronic diabetes, neurosyphilis, alcoholism, and subarachnoid hemorrhages. An exaggerated Achilles reflex indicates cervical cord compression or a lesion of the motor tracts of the first or second sacral segments of the cord.
- **Babinski sign.** This reflex results from gentle stroking of the lateral outer margin of the sole. The great toe dorsiflexes, with or without a lateral fanning of the other toes. This phenomenon normally occurs in children under $1^1/_2$ years of age and is due to incomplete myelination of fibers in the corticospinal tract. A positive Babinski sign after age $1^1/_2$ is abnormal and indicates an interruption of the corticospinal tract as the result of a lesion of the tract, usually in the upper portion. The normal response after age $1^1/_2$ is the **plantar flexion reflex,** or **negative Babinski**—a curling under of all the toes.
- **Abdominal reflex.** This reflex involves contraction of the muscles that compress the abdominal wall in response to stroking the side of the abdomen. The response is an abdominal muscle contraction that causes the umbilicus to move in the direction of the stimulus. Absence of this reflex is associated with lesions of the corticospinal tracts. It may also be absent because of lesions of the peripheral nerves, lesions of integrating centers in the thoracic part of the cord, or multiple sclerosis.

Most autonomic reflexes are not practical diagnostic tools because it is difficult to stimulate visceral effectors, which are deep inside the body. An exception is the pupillary light reflex, in which the pupils of both eyes decrease in diameter when either eye is exposed to light. Because the reflex arc includes synapses in lower parts of the brain, the **absence of a normal pupillary light reflex** may indicate brain damage or injury. ■

▶ CHECKPOINT

8. Which spinal cord tracts are ascending tracts? Which are descending tracts?
9. How are somatic and autonomic reflexes similar and different?
10. Describe the mechanism and function of a stretch reflex, tendon reflex, flexor (withdrawal) reflex, and crossed extensor reflex.
11. What does each of the following terms mean in relation to reflex arcs? Monosynaptic, ipsilateral, polysynaptic, intersegmental, contralateral, and reciprocal innervation.

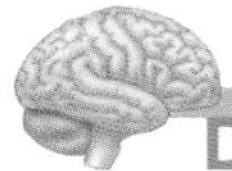

DISORDERS: HOMEOSTATIC IMBALANCES

The spinal cord can be damaged in several ways. Outcomes range from little or no long-term neurological deficits to severe deficits and even death.

Traumatic Injuries

Most spinal cord injuries are due to trauma as a result of factors such as automobile accidents, falls, contact sports, diving, or acts of violence. The effects of the injury depend on the extent of direct trauma to the spinal cord or compression of the cord by fractured or displaced vertebrae or blood clots. Although any segment of the spinal cord may be involved, most common sites of injury are in the cervical, lower thoracic, and upper lumbar regions. Depending on the location and extent of spinal cord damage, paralysis may occur. **Monoplegia** (*mono-* = one; *-plegia* = blow or strike) is paralysis of one limb only. **Diplegia** (*di-* = two) is paralysis of both upper limbs or both lower limbs. **Paraplegia** (*para-* = beyond) is paralysis of both lower limbs. **Hemiplegia** (*hemi-* = half) is paralysis of the upper limb, trunk, and lower limb on one side of the body, and **quadriplegia** (*quad-* = four) is paralysis of all four limbs.

Complete transection (tran-SEK-shun; *trans-* = across; *-section* = a cut) of the spinal cord means that the cord is severed from one side to the other, thus cutting all sensory and motor tracts. It results in a loss of all sensations and voluntary movement *below* the level of the transection. A person will have permanent loss of all sensations in dermatomes below the injury because ascending nerve impulses cannot propagate past the transection to reach the brain. At the same time, voluntary muscle contractions will be lost below the transection because nerve impulses descending from the brain also cannot pass. The extent of paralysis of skeletal muscles depends on the level of injury. The following list outlines which muscle functions may be *retained* at progressively lower levels of spinal cord transection.

- C1–C3: no function maintained from the neck down; ventilator needed for breathing
- C4–C5: diaphragm, which allows breathing

- C6–C7: some arm and chest muscles, which allows feeding, some dressing, and propelling wheelchair
- T1–T3: intact arm function
- T4–T9: control of trunk above the umbilicus
- T10–L1: most thigh muscles, which allows walking with long leg braces
- L1–L2: most leg muscles, which allows walking with short leg braces

Hemisection is a partial transection of the cord on either the right or left side. After hemisection, three main symptoms, known together as *Brown-Sequard syndrome* (se-KAR) occur below the level of injury: (1) Damage of the posterior column causes loss of proprioception and fine touch sensations on the ipsilateral (same) side as the injury. (2) Damage of the lateral corticospinal tract causes ipsilateral paralysis. (3) Damage of the spinothalamic tract causes loss of pain and temperature sensations on the contralateral (opposite) side.

Following complete transection, and to varying degrees after hemisection, spinal shock occurs. **Spinal shock** is an immediate response to spinal cord injury characterized by temporary **areflexia** (a′-rē-FLEX-sē-a), loss of reflex function. The areflexia occurs in parts of the body served by spinal nerves below the level of the injury. Signs of acute spinal shock include slow heart rate, low blood pressure, flaccid paralysis of skeletal muscles, loss of somatic sensations, and urinary bladder dysfunction. Spinal shock may begin within 1 hour after injury and may last from several minutes to several months, after which reflex activity gradually returns.

In many cases of traumatic injury of the spinal cord, the patient may have an improved outcome if an anti-inflammatory corticosteroid drug called methylprednisolone is given within 8 hours of the injury. This is because the degree of neurologic deficit is greatest immediately following traumatic injury as a result of edema (collection of fluid within tissues) as the immune system responds to injury.

Spinal Cord Compression

Although the spinal cord is normally protected by the vertebral column, certain disorders may put pressure on it and disrupt its normal functions. Spinal cord compression may result from fractured vertebrae, herniated intervertebral discs, tumors, osteoporosis, or infections. If the source of the compression is determined before neural tissue is destroyed, spinal cord function usually returns to normal. Depending on the location and degree of compression, symptoms include pain, weakness or paralysis, and either decreased or complete loss of sensation below the level of the injury.

Degenerative Diseases

A number of degenerative diseases affect the functions of the spinal cord. One of these is multiple sclerosis, the details of which were presented in Chapter 12 (see page 433). Another progressive degenerative disease is amyotrophic lateral sclerosis (Lou Gehrig's disease), which affects motor neurons of the brain and spinal cord and results in muscle weakness and atrophy. Details are presented in Chapter 16 (page 562).

Shingles

Shingles is an acute infection of the peripheral nervous system caused by herpes zoster (HER-pēz ZOS-ter), the virus that also causes chickenpox. After a person recovers from chickenpox, the virus retreats to a posterior root ganglion. If the virus is reactivated, the immune system usually prevents it from spreading. From time to time, however, the reactivated virus overcomes a weakened immune system, leaves the ganglion, and travels down sensory neurons of the skin by fast axonal transport (described on page 407). The result is pain, discoloration of the skin, and a characteristic line of skin blisters. The line of blisters marks the distribution (dermatome) of the particular cutaneous sensory nerve belonging to the infected posterior root ganglion.

Poliomyelitis

Poliomyelitis, or simply **polio,** is caused by a virus called poliovirus. The onset of the disease is marked by fever, severe headache, a stiff neck and back, deep muscle pain and weakness, and loss of certain somatic reflexes. In its most serious form, the virus produces paralysis by destroying cell bodies of motor neurons, specifically those in the anterior horns of the spinal cord and in the nuclei of the cranial nerves. Polio can cause death from respiratory or heart failure if the virus invades neurons in vital centers that control breathing and heart functions in the brain stem. Even though polio vaccines have virtually eradicated polio in the United States, outbreaks of polio continue throughout the world. Due to international travel, polio could be easily reintroduced into North America if individuals are not vaccinated appropriately.

Several decades after suffering a severe attack of polio and following their recovery from it, some individuals develop a condition called **post-polio syndrome.** This neurological disorder is characterized by progressive muscle weakness, extreme fatigue, loss of function, and pain, especially in muscles and joints. Post-polio syndrome seems to involve a slow degeneration of motor neurons that innervate muscle fibers. Triggering factors appear to be a fall, a minor accident, surgery, or prolonged bed rest. Possible causes include overuse of surviving motor neurons over time, smaller motor neurons because of the initial infection by the virus, reactivation of dormant polio viral particles, immune-mediated responses, hormone deficiencies, and environmental toxins. Treatment consists of muscle-strengthening exercises, administration of pyridostigmine to enhance the action of acetylcholine in stimulating muscle contraction, and administration of nerve growth factors to stimulate both nerve and muscle growth.

MEDICAL TERMINOLOGY

Epidural block Injection of an anesthetic drug into the epidural space, the space between the dura mater and the vertebral column, in order to cause a temporary loss of sensation. Such injections in the lower lumbar region are used to control pain during childbirth.

Meningitis (men-in-JĪ-tis; *-itis* = inflammation) Inflammation of the meninges due to an infection, usually caused by a bacterium or virus. Symptoms include fever, headache, stiff neck, vomiting, confusion, lethargy, and drowsiness. Bacterial meningitis is much more serious and is treated with antibiotics. Viral meningitis has no specific treatment. Bacterial meningitis may be fatal if not treated promptly; viral meningitis usually resolves on its own in 1–2 weeks. A vaccine is available to help protect against some types of bacterial meningitis.

Myelitis (mī-e-LĪ-tis; *myel-* = spinal cord) Inflammation of the spinal cord.

Myelography (mī-e-LOG-ra-fē; *myel-* = marrow, *-graph* = to write) A procedure in which a CT scan or x-ray image of the spinal cord is taken after injection of a radiopaque dye (contrast medium) to diagnose abnormalities such as tumors and herniated intervertebral discs. MRI has largely replaced myelography because the former shows greater detail, is safer, and is simpler.

Nerve block Loss of sensation in a region due to injection of a local anesthetic; an example is local dental anesthesia.

Neuralgia (noo-RAL-jē-a; *neur-* = nerve; *-algia* = pain) Attacks of pain along the entire course or a branch of a sensory nerve.

Neuritis (*neur-* = nerve; *-itis* = inflammation) Inflammation of one or several nerves that may result from irritation to the nerve produced by direct blows, bone fractures, contusions, or penetrating injuries. Additional causes include infections, vitamin deficiency (usually thiamine), and poisons such as carbon monoxide, carbon tetrachloride, heavy metals, and some drugs.

Paresthesia (par-es-THĒ-zē-a; *par-* = departure from normal; *-esthesia* = sensation) An abnormal sensation such as burning, pricking, tickling, or tingling resulting from a disorder of a sensory nerve.

STUDY OUTLINE

SPINAL CORD ANATOMY (p. 440)

1. The spinal cord is protected by the vertebral column, the meninges, cerebrospinal fluid, and denticulate ligaments.
2. The three meninges are coverings that run continuously around the spinal cord and brain. They are the dura mater, arachnoid mater, and pia mater.
3. The spinal cord begins as a continuation of the medulla oblongata and ends at about the second lumbar vertebra in an adult.
4. The spinal cord contains cervical and lumbar enlargements that serve as points of origin for nerves to the limbs.
5. The tapered inferior portion of the spinal cord is the conus medullaris, from which arise the filum terminale and cauda equina.
6. Spinal nerves connect to each segment of the spinal cord by two roots. The posterior or dorsal root contains sensory axons, and the anterior or ventral root contains motor neuron axons.
7. The anterior median fissure and the posterior median sulcus partially divide the spinal cord into right and left sides.
8. The gray matter in the spinal cord is divided into horns, and the white matter into columns. In the center of the spinal cord is the central canal, which runs the length of the spinal cord.
9. Parts of the spinal cord observed in transverse section are the gray commissure; central canal; anterior, posterior, and lateral gray horns; and anterior, posterior, and lateral white columns, which contain ascending and descending tracts. Each part has specific functions.
10. The spinal cord conveys sensory and motor information by way of ascending and descending tracts, respectively.

SPINAL NERVES (p. 446)

1. The 31 pairs of spinal nerves are named and numbered according to the region and level of the spinal cord from which they emerge.
2. There are 8 pairs of cervical, 12 pairs of thoracic, 5 pairs of lumbar, 5 pairs of sacral, and 1 pair of coccygeal nerves.
3. Spinal nerves typically are connected with the spinal cord by a posterior root and an anterior root. All spinal nerves contain both sensory and motor axons (are mixed nerves).
4. Three connective tissue coverings associated with spinal nerves are the endoneurium, perineurium, and epineurium.
5. Branches of a spinal nerve include the posterior ramus, anterior ramus, meningeal branch, and rami communicantes.
6. The anterior rami of spinal nerves, except for T2–T12, form networks of nerves called plexuses.
7. Emerging from the plexuses are nerves bearing names that typically describe the general regions they supply or the route they follow.
8. Nerves of the cervical plexus supply the skin and muscles of the head, neck, and upper part of the shoulders; they connect with some cranial nerves and innervate the diaphragm.
9. Nerves of the brachial plexus supply the upper limbs and several neck and shoulder muscles.
10. Nerves of the lumbar plexus supply the anterolateral abdominal wall, external genitals, and part of the lower limbs.
11. Nerves of the sacral plexus supply the buttocks, perineum, and part of the lower limbs.
12. Nerves of the coccygeal plexus supply the skin of the coccygeal region.
13. Anterior rami of nerves T2–T12 do not form plexuses and are called intercostal (thoracic) nerves. They are distributed directly to the structures they supply in intercostal spaces.
14. Sensory neurons within spinal nerves and the trigeminal (V) nerve serve specific, constant segments of the skin called dermatomes.
15. Knowledge of dermatomes helps a physician determine which segment of the spinal cord or which spinal nerve is damaged.

SPINAL CORD PHYSIOLOGY (p. 458)

1. The white matter tracts in the spinal cord are highways for nerve impulse propagation. Along these tracts, sensory input travels

toward the brain, and motor output travels from the brain toward skeletal muscles and other effector tissues.

2. Sensory input travels along two main routes in the white matter of the spinal cord: the posterior columns and the spinothalamic tracts.
3. Motor output travels along two main routes in the white matter of the spinal cord: direct pathways and indirect pathways.
4. A second major function of the spinal cord is to serve as an integrating center for spinal reflexes. This integration occurs in the gray matter.
5. A reflex is a fast, predictable sequence of involuntary actions, such as muscle contractions or glandular secretions, which occurs in response to certain changes in the environment.
6. Reflexes may be spinal or cranial and somatic or autonomic (visceral).
7. The components of a reflex arc are sensory receptor, sensory neuron, integrating center, motor neuron, and effector.
8. Somatic spinal reflexes include the stretch reflex, the tendon reflex, the flexor (withdrawal) reflex, and the crossed extensor reflex; all exhibit reciprocal innervation.
9. A two-neuron or monosynaptic reflex arc consists of one sensory neuron and one motor neuron. A stretch reflex, such as the patellar reflex, is an example.
10. The stretch reflex is ipsilateral and is important in maintaining muscle tone.
11. A polysynaptic reflex arc contains sensory neurons, interneurons, and motor neurons. The tendon reflex, flexor (withdrawal) reflex, and crossed extensor reflexes are examples.
12. The tendon reflex is ipsilateral and prevents damage to muscles and tendons when muscle force becomes too extreme. The flexor reflex is ipsilateral and moves a limb away from the source of a painful stimulus. The crossed extensor reflex extends the limb contralateral to a painfully stimulated limb, allowing the weight of the body to shift when a supporting limb is withdrawn.
13. Several important somatic reflexes are used to diagnose various disorders. These include the patellar reflex, Achilles reflex, Babinski sign, and abdominal reflex.

SELF-QUIZ QUESTIONS

Fill in the blanks in the following statements.

1. Because they contain both sensory and motor axons, spinal nerves are considered to be ____ nerves.
2. The five components of a reflex arc, in order from the beginning to the end, are (1) ____, (2)____, (3) ____, (4) ____, and (5) ____.

Indicate whether the following statements are true or false.

3. Gray matter of the spinal cord contains somatic motor and sensory nuclei, autonomic motor and sensory nuclei, and functions to receive and integrate both incoming and outgoing information.
4. The epidural space is located between the wall of the vertebral canal and the pia mater.

Choose the one best answer to the following questions.

5. Which of the following is *not* true? (1) Dermatomes are areas of the body that are stimulated by motor neurons exiting specific spinal nerves. (2) The stretch reflex helps to maintain muscle tone. (3) The Achilles reflex is an example of a stretch reflex. (4) The abdominal reflex is used to diagnose problems with autonomic reflexes. (5) Spinal nerves T2–T12 do not enter into the formation of a plexus. (a) 1, 2, and 4; (b) 2 and 5; (c) 1 and 4; (d) 1, 3, and 5; (e) 1, 3, and 4.
6. While identifying and labeling cadaver muscles, your lab partner accidentally pokes your finger with a pin. Place the following steps in the correct order from beginning to end of your body's response. (1) Impulses travel through anterior (ventral) root of spinal nerve(s). (2) Sensory neuron relays impulse to spinal cord. (3) Motor impulses reach muscles, causing withdrawal of the affected limb. (4) Integrating centers interpret sensory impulses, and then generate motor impulses. (5) Sensory receptor activated by stimulus. (6) Impulse travels through posterior (dorsal) root of spinal nerve. (a) 5, 3, 6, 4, 1, 2; (b) 5, 2, 1, 4, 6, 3; (c) 5, 2, 6, 4, 1, 3; (d) 3, 5, 1, 2, 4, 6; (e) 2, 1, 5, 4, 6, 3.
7. The connective tissue surrounding each individual axon is (a) endoneurium, (b) epineurium, (c) perineurium, (d) fascicle, (e) arachnoid mater.
8. The tracts of the posterior column are involved in (1) proprioception, (2) discriminative touch, (3) pain, (4) thermal sensations, (5) pressure, (6) vibration, (7) two-point discrimination. (a) 1, 2, 4, and 5; (b) 2, 4, 6, and 7; (c) 1, 2, 6, and 7; (d) 3, 4, 5, 6, and 7; (e) 1, 3, 5, 6, and 7.
9. Which of the following is a motor tract? (a) posterior spinocerebellar, (b) lateral spinothalamic, (c) anterior spinocerebellar, (d) lateral corticospinal, (e) posterior column.
10. Cutting the posterior root of a spinal nerve would (a) interfere with the circulation of cerebrospinal fluid, (b) impair motor control of skeletal muscles, (c) interfere with the ability of the brain to transmit motor impulses, (d) impair motor control of organs, (e) interfere with the flow of sensory impulses.
11. Which of the following statements is *false*? (a) The two main spinal cord sensory paths are the spinothalamic and anterior columns. (b) The spinothalamic tracts convey impulses for sensing pain, temperature, touch, and deep pressure. (c) Direct pathways convey nerve impulses destined to cause precise, voluntary movements of skeletal muscles. (d) Indirect pathways convey nerve impulses that program automatic movements, help coordinate body movements with visual stimuli, maintain skeletal muscle tone and posture, and contribute to equilibrium. (e) The direct pathways are motor pathways.
12. Which of the following are *true*? (1) The anterior (ventral) gray horns contain cell bodies of neurons that cause skeletal muscle contraction. (2) The gray commissure connects the white matter of the right and left sides of the spinal cord. (3) Cell bodies of autonomic motor neurons are located in the lateral gray horns. (4) Sensory (ascending) tracts conduct motor impulses down the spinal cord. (5) Gray matter in the spinal cord consists of cell bodies of neurons, neuroglia, unmyelinated axons and dendrites of interneurons and motor neurons. (a) 1, 2, 3, and 5; (b) 2 and 4; (c) 2, 3, 4; and 5; (d) 1, 3, and 5; (e) 1, 2, 3, and 4.

13. Match the following (some answers may be used more than once):

____(a) a reflex resulting in the contraction of a skeletal muscle when it is stretched
____(b) receptors that monitor changes in muscle length
____(c) a balance-maintaining reflex
____(d) operates as a feedback mechanism to control muscle tension by causing muscle relaxation when muscle force becomes too extreme
____(e) reflex arc that consists of one sensory and one motor neuron
____(f) acts as a feedback mechanism to control muscle length by causing muscle contraction
____(g) sensory impulses enter on one side of the spinal cord and motor impulses exit on the opposite side
____(h) occurs when sensory nerve impulse travels up and down the spinal cord, thereby activating several motor neurons and more than one effector
____(i) polysynaptic reflex initiated in response to a painful stimulus
____(j) receptors that monitor changes in muscle tension
____(k) maintains proper muscle tone
____(l) reflex pathway that contains sensory neurons, interneurons, and motor neurons
____(m) motor nerve impulses exit the spinal cord on the same side that sensory impulses entered the spinal cord
____(n) protects the tendon and muscle from damage due to excessive tension
____(o) a neural circuit that coordinates body movements by causing contraction of onemuscle and relaxation of antagonistic muscles or relaxation of a muscle and contraction of the antagonists

(1) stretch reflex
(2) tendon reflex
(3) flexor (withdrawal) reflex
(4) crossed extensor reflex
(5) intersegmental reflex arc
(6) contralateral reflex arc
(7) ipsilateral reflex arc
(8) muscle spindles
(9) tendon (Golgi tendon) organs
(10) reciprocal innervation
(11) monosynaptic reflex
(12) polysynaptic reflex

14. Match the following:

____(a) the joining together of the anterior rami of adjacent nerves
____(b) spinal nerve branches that serve the deep muscles and skin of the posterior surface of the trunk
____(c) spinal nerve branches that serve the muscles and structures of the upper and lower limbs and the lateral and ventral trunk
____(d) area of the spinal cord from which nerves to and from the upper limbs arise
____(e) area of the spinal cord from which nerves to and from the lower limbs arise
____(f) the roots form the nerves that arise from the inferior part of the spinal cord but do not leave the vertebral column at the same level as they exit the cord
____(g) contains motor neuron axons and conducts impulses from the spinal cord to the peripheral organs and cells
____(h) avascular covering of spinal cord composed of delicate collagen fibers and some elastic fibers
____(i) contains sensory neuron axons and conducts impulses from the peripheral receptors into the spinal cord
____(j) superficial spinal cord covering of dense, irregular connective tissue
____(k) an extension of the pia mater that anchors the spinal cord to the coccyx
____(l) extending the length of the spinal cord, these pia mater thickenings fuse with the arachnoid mater and dura mater and help to protect the spinal cord from shock and sudden displacement
____(m) thin transparent connective tissue composed of interlacing bundles of collagen fibers and some elastic fibers adhering to the spinal cord's surface
____(n) space within the spinal cord filled with cerebrospinal fluid
____(o) spinal nerve branch that supplies vertebrae, vertebral ligaments, blood vessels of the spinal cord, and meninges

(1) cervical enlargement
(2) lumbar enlargement
(3) central canal
(4) denticulate ligaments
(5) cauda equina
(6) meningeal branch
(7) pia mater
(8) arachnoid mater
(9) dura mater
(10) posterior (dorsal) root
(11) anterior (ventral) root
(12) posterior (dorsal) ramus
(13) anterior (ventral) ramus
(14) plexus
(15) filum terminale

15. Match the following:

____(a) provides the entire nerve supply of the shoulders and upper limbs		(1) cervical plexus
____(b) provides the nerve supply of the skin and muscles of the head, neck, and superior part of the shoulders and chest		(2) brachial plexus
____(c) provides the nerve supply of the anterolateral abdominal wall, external genitals, and part of the lower limbs		(3) lumbar plexus
____(d) supplies the buttocks, perineum, and lower limbs		(4) sacral plexus
____(e) formed by the anterior rami of C1–C4 with some contribution by C5		(5) coccygeal plexus
____(f) formed by anterior rami of S4–S5 and coccygeal nerves		
____(g) formed by the anterior rami of L1–L4		
____(h) formed by the anterior rami of C5–C8 and T1		
____(i) formed by the anterior rami of L4–L5 and S1–S4		
____(j) phrenic nerve arises from this plexus		
____(k) median nerve arises from this plexus		
____(l) sciatic nerve arises from this plexus		
____(m) femoral nerve arises from this plexus		
____(n) supplies a small area of skin in coccygeal region		
____(o) injury to this plexus can affect breathing		

CRITICAL THINKING QUESTIONS

1. Evalina's severe headaches and other symptoms were suggestive of meningitis, so her physician ordered a spinal tap. List the structures that the needle will pierce from the most superficial to the deepest. Why would the physician order a test in the spinal region to check a problem in Evalina's head?
2. Sunil has developed an infection that is destroying cells in the anterior gray horns in the lower cervical region of the spinal cord. What kinds of symptoms would you expect to occur?
3. Allyson is in a car accident and suffers spinal cord compression in the lower spinal cord. Although she is in pain, she cannot distinguish when the doctor is touching her calf or her toes and she is having trouble telling how her lower limbs are positioned. What part of the spinal cord has been affected by the accident?

ANSWERS TO FIGURE QUESTIONS

13.1 The superior boundary of the spinal dura mater is the foramen magnum of the occipital bone. The inferior boundary is the second sacral vertebra.

13.2 The cervical enlargement connects with sensory and motor nerves of the upper limbs.

13.3 A horn is an area of gray matter, and a column is a region of white matter in the spinal cord.

13.4 All spinal nerves are classified as mixed (with both sensory and motor components) because their posterior roots contain sensory axons and their anterior roots contain motor axons.

13.5 The anterior rami serve the upper and lower limbs.

13.6 Severing the spinal cord at level C2 causes respiratory arrest because it prevents descending nerve impulses from reaching the phrenic nerve, which stimulates contraction of the diaphragm, the main muscle needed for breathing.

13.7 The axillary, musculocutaneous, radial, median, and ulnar nerves are five important nerves that arise from the brachial plexus.

13.8 Injury to the median nerve affects sensations on the palm and fingers.

13.9 Signs of femoral nerve injury include inability to extend the leg and loss of sensation in the skin over the anterolateral aspect of the thigh.

13.10 The origin of the sacral plexus is the anterior rami of spinal nerves L4–L5 and S1–S4.

13.11 The only spinal nerve without a corresponding dermatome is C1.

13.12 The anterior corticospinal tract is located on the anterior side of the spinal cord, originates in the cortex of the cerebrum, and ends in the spinal cord. Because "spinal" comes last in the name, you know it contains descending axons and thus is a motor tract.

13.13 A sensory receptor produces a generator potential, which triggers a nerve impulse if the generator potential reaches threshold. Reflex integrating centers are in the CNS.

13.14 In an ipsilateral reflex, the sensory and motor neurons are on the same side of the spinal cord.

13.15 Reciprocal innervation is a type of arrangement of a neural circuit involving simultaneous contraction of one muscle and relaxation of its antagonist.

13.16 The flexor reflex is intersegmental because impulses go out over motor neurons located in several spinal nerves, each arising from a different segment of the spinal cord.

13.17 The crossed extensor reflex is a contralateral reflex arc because the motor impulses leave the spinal cord on the side opposite the entry of sensory impulses.

Chapter 14

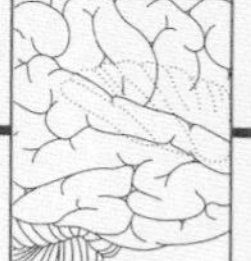

The Brain and Cranial Nerves

The Brain, Cranial Nerves and Homeostasis

Your brain contributes to homeostasis by receiving sensory input, integrating new and stored information, making decisions, and causing motor activities.

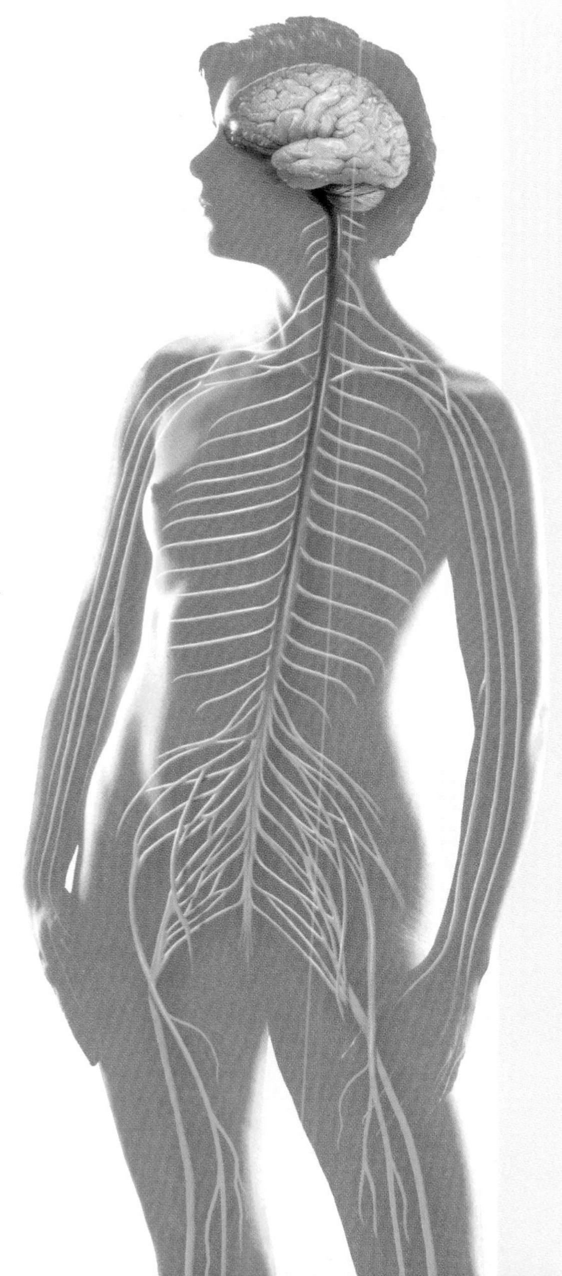

www. wiley.com/college/apcentral

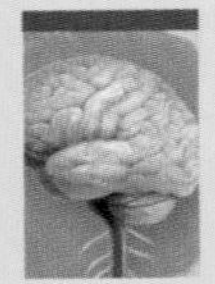

Solving an equation, feeling hungry, laughing—the neural processes needed for each of these activities occurs in different regions of the **brain,** that portion of the central nervous system contained within the cranium. About 100 billion neurons and 10–50 trillion neuroglia make up the brain, which has a mass of about 1300 g (almost 3 lb) in adults. On average, each neuron forms 1000 synapses with other neurons. Thus, the total number of synapses, about a thousand trillion or 10^{15}, is larger than the number of stars in the galaxy.

The brain is the center for registering sensations, correlating them with one another and with stored information, making decisions, and taking actions. It also is the center for the intellect, emotions, behavior, and memory. But the brain encompasses yet a larger domain: It directs our behavior toward others. With ideas that excite, artistry that dazzles, or rhetoric that mesmerizes, one person's thoughts and actions may influence and shape the lives of many others. As you will see shortly, different regions of the brain are specialized for different functions. Different parts of the brain also work together to accomplish certain shared functions. This chapter explores how the brain is protected and nourished, what functions occur in the major regions of the brain, and how the spinal cord and the 12 pairs of cranial nerves connect with the brain to form the control center of the human body.

BRAIN ORGANIZATION, PROTECTION, AND BLOOD SUPPLY

▶ **OBJECTIVES**

Identify the major parts of the brain.

Describe how the brain is protected.

Describe the blood supply of the brain.

Knowledge of the embryological development of the brain is necessary to understand the terminology used for the principal parts of the adult brain. The development of the brain is introduced briefly here, and discussed in more detail at the end of the chapter.

The brain and spinal cord develop from ectoderm arranged in a tubular structure called the neural tube (see Figure 14.28). The anterior part of the neural tube expands, and constrictions appear that create three regions called primary brain vesicles: prosencephalon (forebrain), mesencephalon (midbrain), and rhombencephalon (hindbrain) (see Figure 14.29). The mesencephalon gives rise to the midbrain and aqueduct of the midbrain (cerebral aqueduct). Both the prosencephalon and rhombencephalon subdivide further, forming secondary brain vesicles. The prosencephalon gives rise to the telencephalon and diencephalon, and the rhombencephalon develops into the metencephalon and myelencephalon. The telencephalon develops into the cerebrum and lateral ventricles. The diencephalon forms the thalamus, hypothalamus, and epithalamus. The metencephalon becomes the pons, cerebellum, and upper part of the fourth ventricle. Finally, the myelencephalon forms the medulla oblongata and lower part of the fourth ventricle. The various parts of the brain will be described shortly.

These relationships are summarized in Table 14.1.

Major Parts of the Brain

The adult brain consists of four major parts: brain stem, cerebellum, diencephalon, and cerebrum (Figure 14.1). The **brain stem**

TABLE 14.1 Development of the Brain

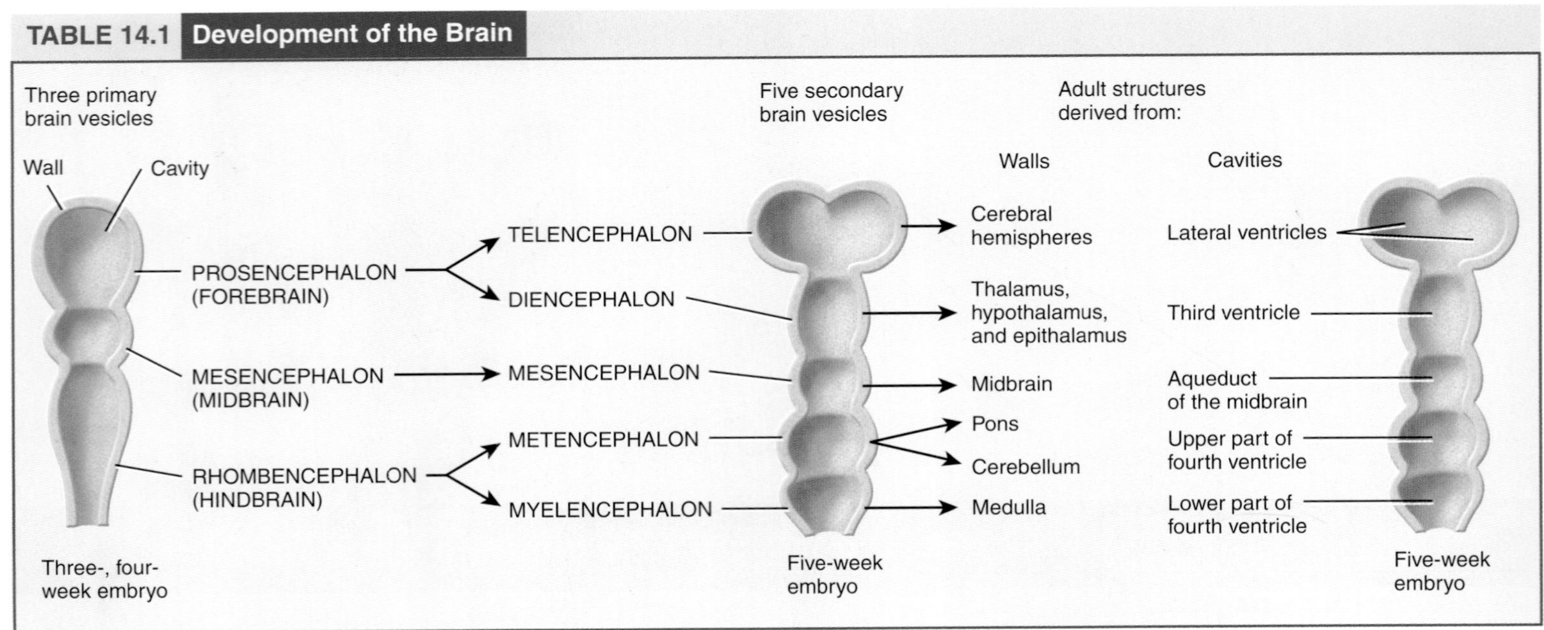

Figure 14.1 The brain. The pituitary gland is discussed with the endocrine system in Chapter 18. (See Tortora, *A Photographic Atlas of the Human Body, Second Edition,* Figures 8.12, 8.13, and 8.15.)

The four principal parts of the brain are the brain stem, cerebellum, diencephalon, and cerebrum.

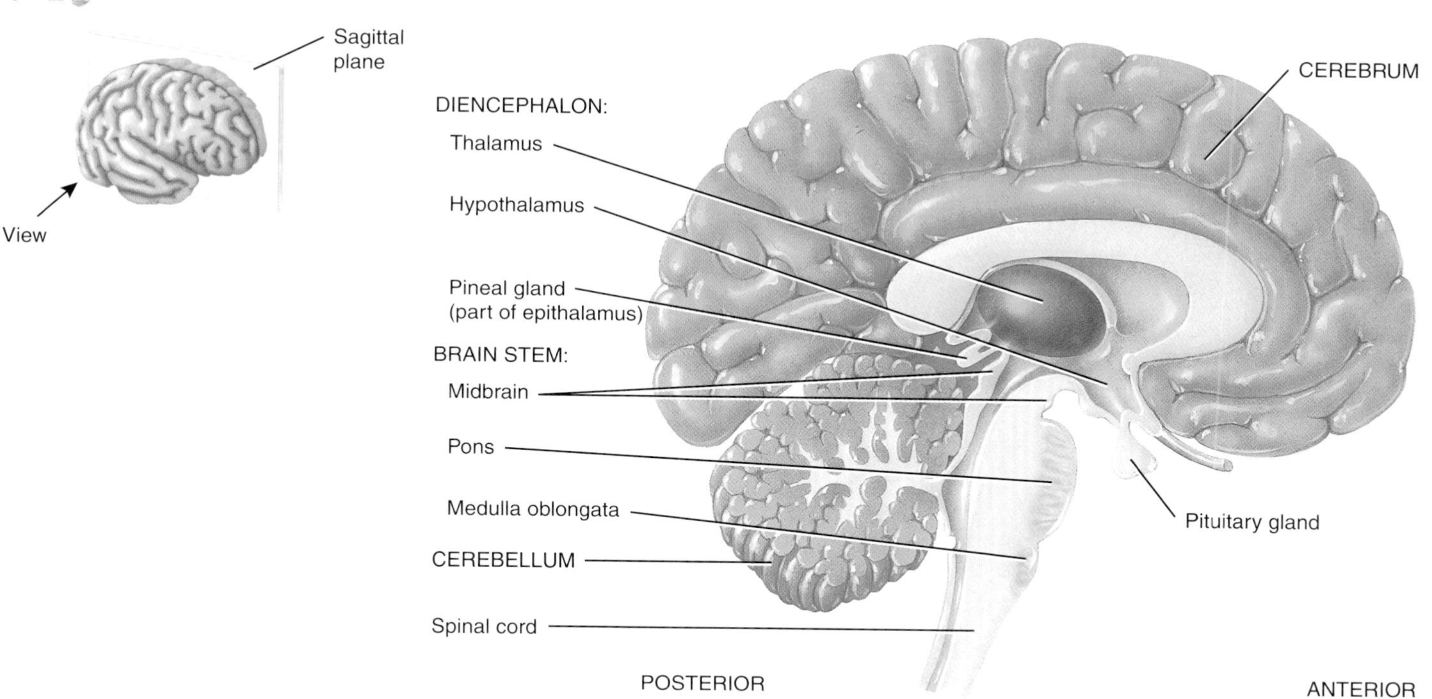

(a) Sagittal section, medial view

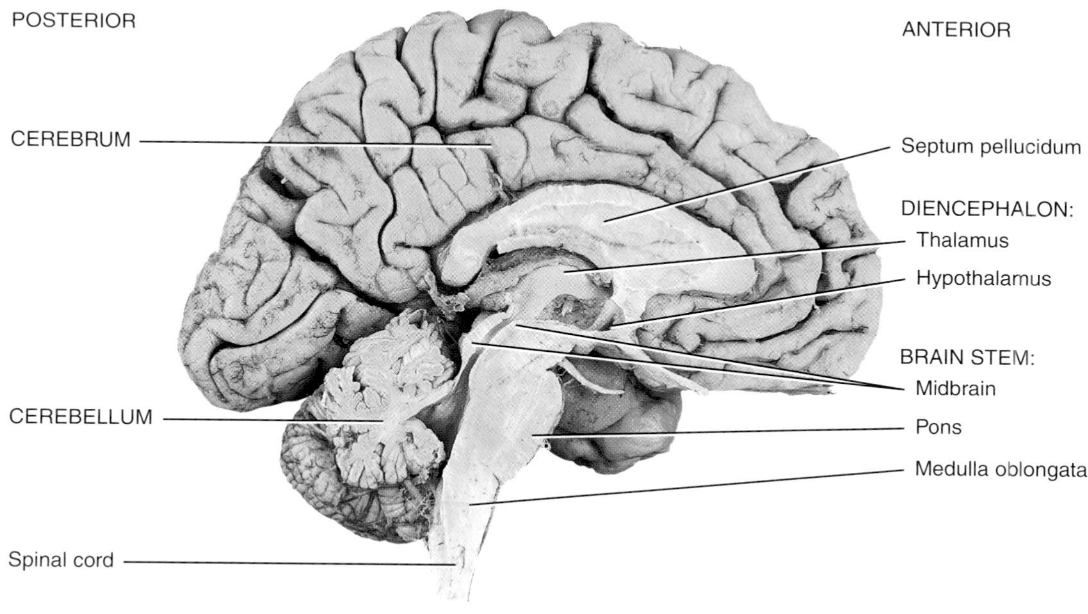

(b) Sagittal section, medial view

 **Which part of the brain is the largest?**

is continuous with the spinal cord and consists of the medulla oblongata, pons, and midbrain. Posterior to the brain stem is the **cerebellum** (ser′-e-BEL-um = little brain). Superior to the brain stem is the **diencephalon** (dī′-en-SEF-a-lon; *di-* = through; *-encephalon* = brain), which consists of the thalamus, hypothalamus, and epithalamus. Supported on the diencephalon and brain stem is the **cerebrum** (se-RĒ-brum = brain), the largest part of the brain.

Protective Coverings of the Brain

The cranium (see Figure 7.4 on page 200) and the cranial meninges surround and protect the brain. The **cranial meninges** (me-NIN-jēz) are continuous with the spinal meninges, have the same basic structure, and bear the same names: the outer **dura mater,** the middle **arachnoid mater,** and the inner **pia mater** (Figure 14.2). However, the cranial dura mater has two layers;

Figure 14.2 The protective coverings of the brain.

Cranial bones and the cranial meninges protect the brain.

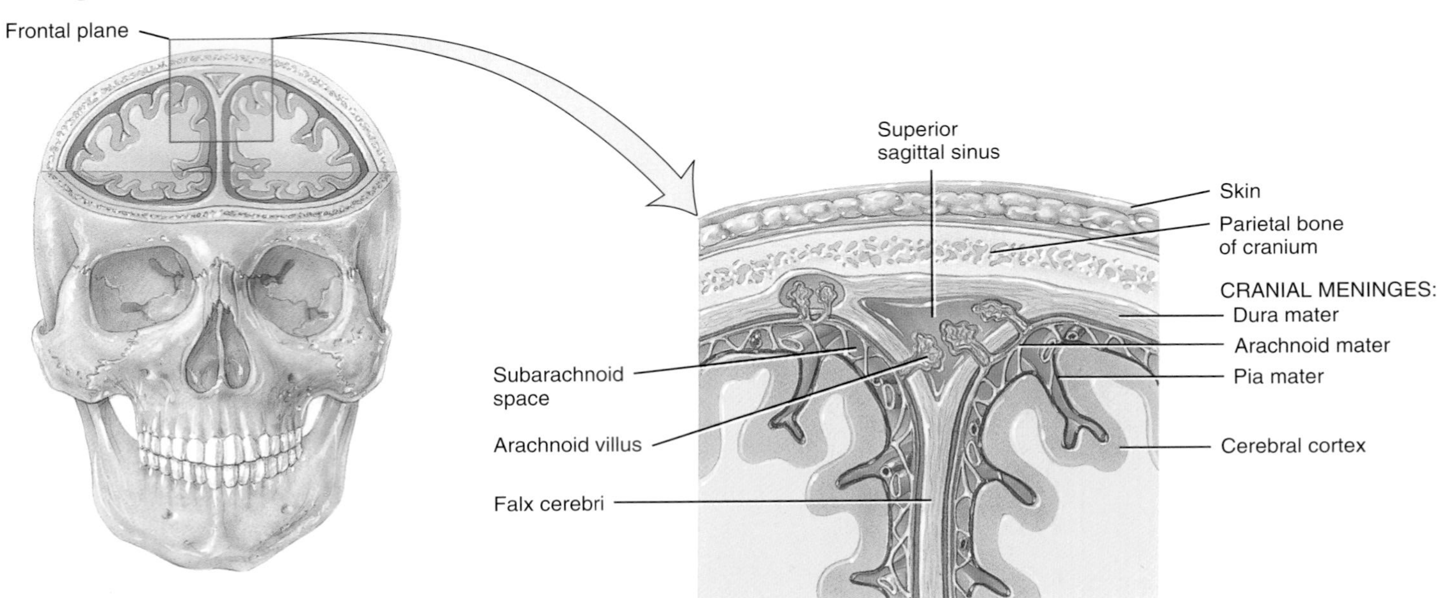

(a) Frontal section through skull showing the cranial meninges

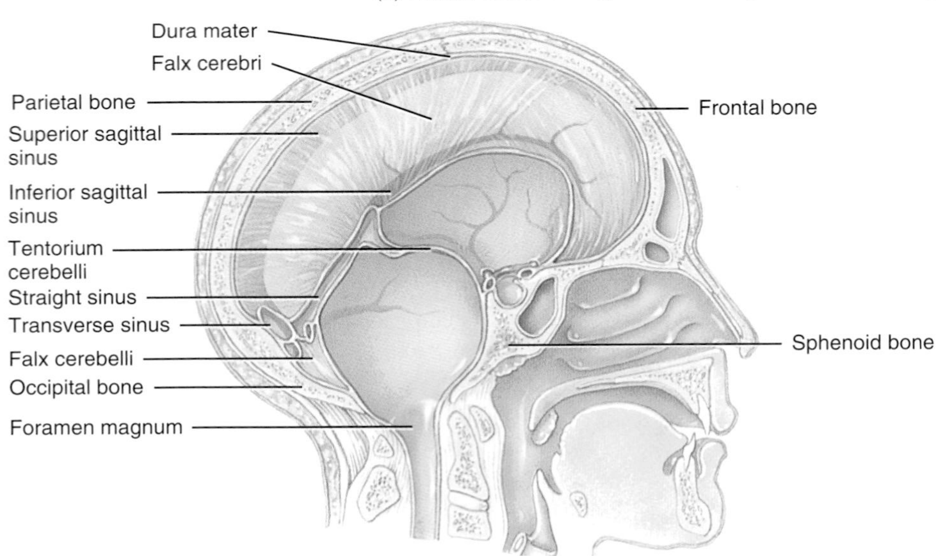

(b) Extensions of the dura mater

What are the three layers of the cranial meninges, from superficial to deep?

the spinal dura mater has only one. The two dural layers around the brain are fused together except where they separate to enclose the dural venous sinuses (endothelial-lined venous channels) that drain venous blood from the brain and deliver it into the internal jugular veins. Also, there is no epidural space around the brain. Blood vessels that enter brain tissue pass along the surface of the brain, and as they penetrate inward, they are sheathed by a loose-fitting sleeve of pia mater. Three extensions of the dura mater separate parts of the brain. (1) The **falx cerebri** (FALKS CER-e-brē; *falx* = sickle-shaped) separates the two hemispheres (sides) of the cerebrum. (2) The **falx cerebelli** (cer-e-BEL-ī) separates the two hemispheres of the cerebellum. (3) The **tentorium cerebelli** (ten-TŌ-rē-um = tent) separates the cerebrum from the cerebellum.

Brain Blood Flow and the Blood–Brain Barrier

Blood flows to the brain mainly via the internal carotid and vertebral arteries (see Figure 21.19 on page 767); the internal jugular veins return blood from the head to the heart see Figure 21.24 on page 780.

In an adult, the brain represents only 2% of total body weight, but consumes about 20% of the oxygen and glucose used even at rest. Neurons synthesize ATP almost exclusively from glucose via reactions that use oxygen. When activity of neurons and neuroglia increases in a region of the brain, blood flow to that area also increases. Even a brief slowing of brain blood flow may cause unconsciousness. Typically, an interruption in blood flow for 1 or 2 minutes impairs neuronal function, and total deprivation of oxygen for about 4 minutes causes permanent injury. Because virtually no glucose is stored in the brain, the supply of glucose also must be continuous. If blood entering the brain has a low level of glucose, mental confusion, dizziness, convulsions, and loss of consciousness may occur.

The existence of a **blood–brain barrier (BBB)** protects brain cells from harmful substances and pathogens by preventing passage of many substances from blood into brain tissue. The blood–brain barrier consists mainly of tight junctions (see Figure 4.1a on page 109) that seal together the endothelial cells of brain capillaries, along with a thick basement membrane around the capillaries. The processes of many astrocytes, which as you learned in Chapter 12 are one type of neuroglia, press up against the capillaries and secrete chemicals that maintain the permeability characteristics of the tight junctions. A few water-soluble substances, such as glucose, cross the BBB by active transport. Other substances, such as creatinine, urea, and most ions, cross the BBB very slowly. Still other substances—proteins and most antibiotic drugs—do not pass at all from the blood into brain tissue. However, lipid-soluble substances, such as oxygen, carbon dioxide, alcohol, and most anesthetic agents, easily cross the blood–brain barrier. Trauma, certain toxins, and inflammation can cause a breakdown of the blood–brain barrier.

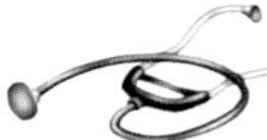

Breaching the Blood–Brain Barrier

We have seen how the BBB prevents the passage into brain tissue of potentially harmful substances. But another consequence of the BBB's efficient protection is that it also prevents the passage of certain drugs that could be therapeutic for brain cancer or other CNS disorders. Researchers are exploring ways to move drugs past the BBB. In one method, the drug is injected in a concentrated sugar solution. The high osmotic pressure of the sugar solution causes the endothelial cells of the capillaries to shrink, which opens gaps between their tight junctions and makes the BBB more leaky. As a result, the drug can enter the brain tissue. ■

▶ CHECKPOINT

1. Compare the sizes and locations of the cerebrum and cerebellum.
2. Describe the locations of the cranial meninges.
3. Explain the blood supply to the brain and the importance of the blood–brain barrier.

CEREBROSPINAL FLUID

▶ OBJECTIVE

Explain the formation and circulation of cerebrospinal fluid.

Cerebrospinal fluid (CSF) is a clear, colorless liquid that protects the brain and spinal cord from chemical and physical injuries. It also carries oxygen, glucose, and other needed chemicals from the blood to neurons and neuroglia. CSF continuously circulates through cavities in the brain and spinal cord and around the brain and spinal cord in the subarachnoid space (between the arachnoid mater and pia mater).

Figure 14.3 shows the four CSF-filled cavities within the brain, which are called **ventricles** (VEN-tri-kuls = little cavities). A **lateral ventricle** is located in each hemisphere of the cerebrum. Anteriorly, the lateral ventricles are separated by a thin membrane, the **septum pellucidum** (SEP-tum pe-LOO-si-dum; *pellucid* = transparent). The **third ventricle** is a narrow cavity along the midline superior to the hypothalamus and between the right and left halves of the thalamus. The **fourth ventricle** lies between the brain stem and the cerebellum.

The total volume of CSF is 80 to 150 mL (3 to 5 oz) in an adult. CSF contains glucose, proteins, lactic acid, urea, cations (Na^+, K^+, Ca^{2+}, Mg^{2+}), and anions (Cl^- and HCO_3^-); it also contains some white blood cells. The CSF contributes to homeostasis in three main ways:

Figure 14.3 Locations of ventricles within a "transparent" brain. One interventricular foramen on each side connects a lateral ventricle to the third ventricle, and the aqueduct of the midbrain connects the third ventricle to the fourth ventricle.

Ventricles are cavities within the brain that are filled with cerebrospinal fluid.

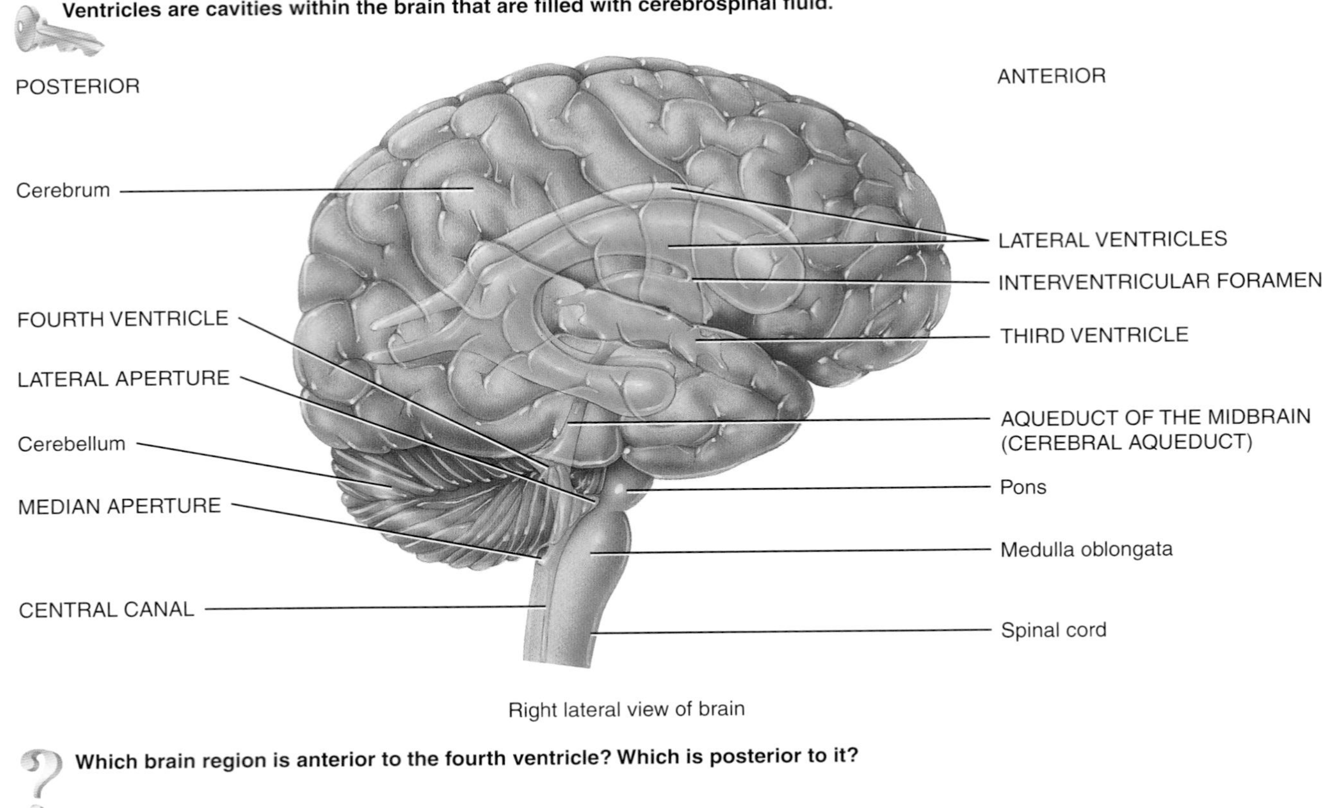

Right lateral view of brain

Which brain region is anterior to the fourth ventricle? Which is posterior to it?

1. ***Mechanical protection.*** CSF serves as a shock-absorbing medium that protects the delicate tissues of the brain and spinal cord from jolts that would otherwise cause them to hit the bony walls of the cranial and vertebral cavities. The fluid also buoys the brain so that it "floats" in the cranial cavity.

2. ***Chemical protection.*** CSF provides an optimal chemical environment for accurate neuronal signaling. Even slight changes in the ionic composition of CSF within the brain can seriously disrupt production of action potentials and postsynaptic potentials.

3. ***Circulation.*** CSF allows exchange of nutrients and waste products between the blood and nervous tissue.

Formation of CSF in the Ventricles

The sites of CSF production are the **choroid plexuses** (KŌ-royd = membrane like), networks of capillaries (microscopic blood vessels) in the walls of the ventricles. The capillaries are covered by ependymal cells that form cerebrospinal fluid from blood plasma by filtration and secretion. Because the ependymal cells are joined by tight junctions, materials entering CSF from choroid capillaries cannot leak between these cells; instead, they must pass through the ependymal cells. This **blood–cerebrospinal fluid barrier** permits certain substances to enter the CSF but excludes others, protecting the brain and spinal cord from potentially harmful blood-borne substances.

Circulation of CSF

The CSF formed in the choroid plexuses of each lateral ventricle flows into the third ventricle through two narrow, oval openings, the **interventricular foramina** (singular is *foramen;* Figure 14.4a). More CSF is added by the choroid plexus in the roof of the third ventricle. The fluid then flows through the **aqueduct of the midbrain (cerebral aqueduct),** which passes through the midbrain, into the fourth ventricle. The choroid plexus of the fourth ventricle contributes more fluid. CSF enters the subarachnoid space through three openings in the roof of the fourth ventricle: a **median aperture** and the paired **lateral apertures,** one on each side. CSF then circulates in the central canal of the spinal cord and in the subarachnoid space around the surface of the brain and spinal cord.

Figure 14.4 **Pathways of circulating cerebrospinal fluid.** (See Tortora, *A Photographic Atlas of the Human Body, Second Edition,* Figures 8.15 and 8.18.)

CSF is formed by ependymal cells that cover the choroid plexuses of the ventricles.

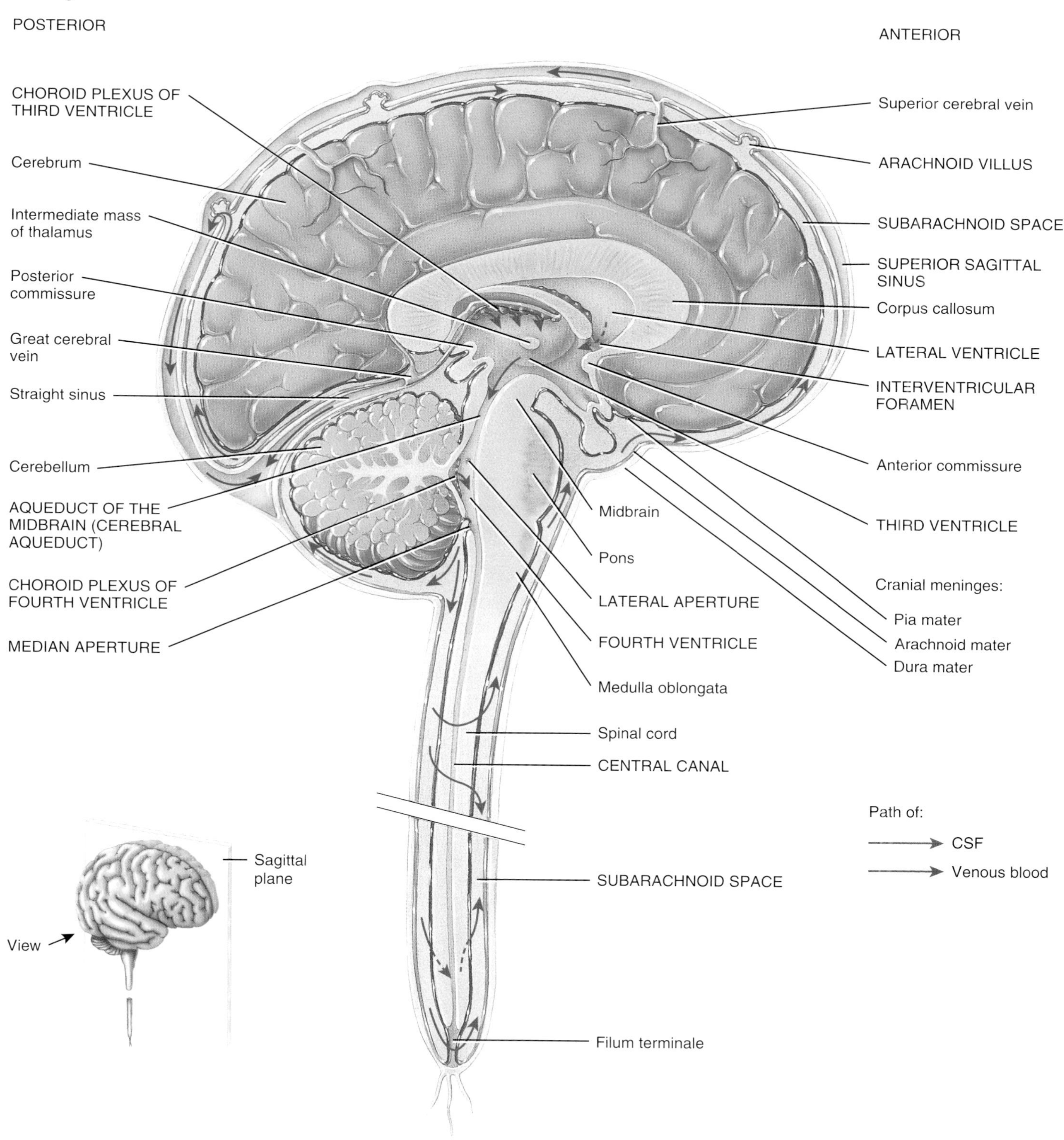

(a) Sagittal section of brain and spinal cord

continues

Figure 14.4 **(continued)**

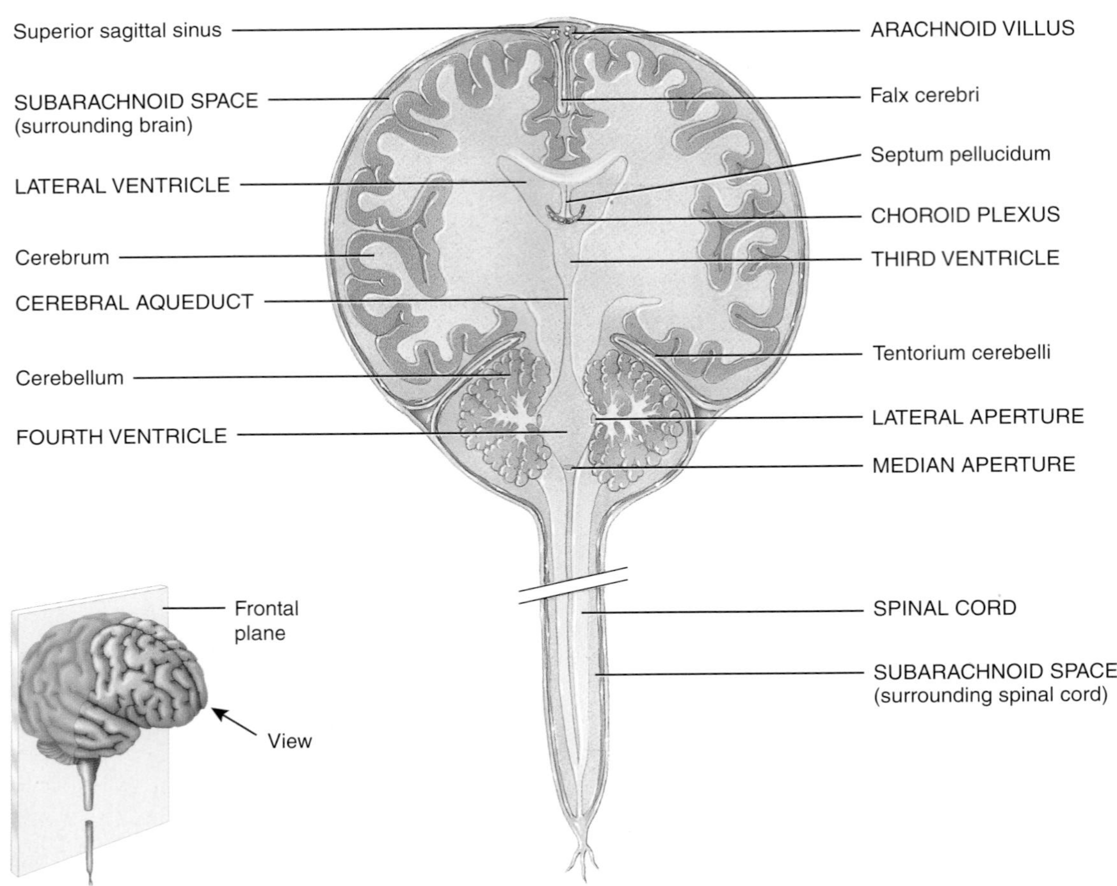

(b) Frontal section of brain and spinal cord

CSF is gradually reabsorbed into the blood through **arachnoid villi,** fingerlike extensions of the arachnoid that project into the dural venous sinuses, especially the **superior sagittal sinus** (see Figure 14.2). (A cluster of arachnoid villi is called an **arachnoid granulation.**) Normally, CSF is reabsorbed as rapidly as it is formed by the choroid plexuses, at a rate of about 20 mL/hr (480 mL/day). Because the rates of formation and reabsorption are the same, the pressure of CSF normally is constant. Figure 14.4c summarizes the production and flow of CSF.

Hydrocephalus

Abnormalities in the brain—tumors, inflammation, or developmental malformations—can interfere with the drainage of CSF from the ventricles into the subarachnoid space. When excess CSF accumulates in the ventricles, the CSF pressure rises. Elevated CSF pressure causes a condition called **hydrocephalus** (hī′-drō-SEF-a-lus; *hydro-* = water; *cephal-* = head). In a baby whose fontanels have not yet closed, the head bulges due to the increased pressure. If the condition persists, the fluid buildup compresses and damages the delicate nervous tissue. Hydrocephalus is relieved by draining the excess CSF. A neurosurgeon may implant a drain line, called a shunt, into the lateral ventricle to divert CSF into the superior vena cava or abdominal cavity, where it can be absorbed by the blood. In adults, hydrocephalus may occur after head injury, meningitis, or subarachnoid hemorrhage. This condition can quickly become life-threatening and requires immediate intervention; since the adult skull bones have already fused, nervous tissue damage occurs quickly. ■

▶ CHECKPOINT

4. What structures produce CSF, and where are they located?
5. What is the difference between the blood–brain barrier and the blood–cerebrospinal fluid barrier?

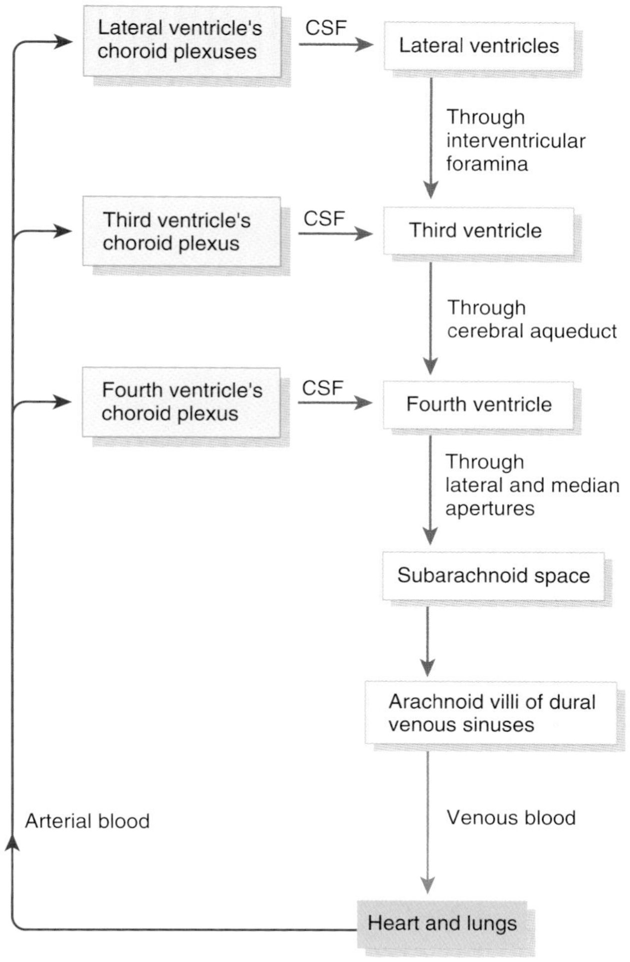

(c) Summary of the formation, circulation, and absorption of cerebrospinal fluid (CSF)

Where is cerebrospinal fluid reabsorbed?

THE BRAIN STEM

▶ OBJECTIVE

Describe the structures and functions of the brain stem.

The brain stem is the part of the brain between the spinal cord and the diencephalon; it consists of the (1) medulla oblongata, (2) pons, and (3) midbrain. Extending through the brain stem is the reticular formation, a netlike region of interspersed gray and white matter.

Medulla Oblongata

The **medulla oblongata** (me-DOOL-la ob′-long-GA-ta), or more simply the **medulla,** is continuous with the superior part of the spinal cord; it forms the inferior part of the brain stem (Figure 14.5; see also Figure 14.1). The medulla begins at the foramen magnum and extends to the inferior border of the pons, a distance of about 3 cm (1.2 in.).

The medulla's white matter contains all sensory (ascending) and motor (descending) tracts that extend between the spinal cord and other parts of the brain. Some of the white matter forms bulges on the anterior aspect of the medulla. These protrusions, called the **pyramids** (Figure 14.6 on page 483; see also Figure 14.5), are formed by the large corticospinal tracts that pass from the cerebrum to the spinal cord. Just superior to the junction of the medulla with the spinal cord, 90% of the axons in the left pyramid cross to the right side, and 90% of the axons in the right pyramid cross to the left side. This crossing is called the **decussation of pyramids** (dē′-ku-SĀ-shun; *decuss-* = crossing) and explains why each side of the brain controls movements on the opposite side of the body.

The medulla also contains several **nuclei,** masses of gray matter where neurons form synapses with one another. Several of these nuclei control vital body functions. The **cardiovascular center** regulates the rate and force of the heartbeat and the diameter of blood vessels. The **medullary rhythmicity area** of the respiratory center adjusts the basic rhythm of breathing. Other nuclei in the medulla control reflexes for vomiting, coughing, swallowing, hiccupping, and sneezing.

Just lateral to each pyramid is an oval-shaped swelling called an **olive** (see Figures 14.5 and 14.6). Within the olive is the **inferior olivary nucleus.** Neurons here relay impulses from proprioceptors (monitoring joint and muscle positions) to the cerebellum.

Nuclei associated with sensations of touch, conscious proprioception, and vibration are located in the posterior part of the medulla. These nuclei are the right and left **gracile nucleus** (GRAS-il = slender) and **cuneate nucleus** (KŪ-nē-āt = wedge). Many ascending sensory axons form synapses in these nuclei, and postsynaptic neurons then relay the sensory information to the thalamus on the opposite side of the brain (see Figure 16.5 on page 557). The axons ascend to the thalamus in a band of white matter called the **medial lemniscus** (lem-NIS-kus = ribbon), which extends through the medulla, pons, and midbrain (see Figure 14.7b).

Finally, the medulla contains nuclei associated with five pairs of cranial nerves (see Figure 14.5): vestibulocochlear (VIII) nerves, glossopharyngeal (IX) nerves, vagus (X) nerves, accessory (XI) nerves (cranial portion), and hypoglossal (XII) nerves. You will learn about these and other cranial nerves later in the chapter.

Injury of the Medulla

Given the many vital activities controlled by the medulla, it is not surprising that a hard blow to the back of the head or upper neck can be fatal. Damage to the medullary rhythmicity area is particularly serious and can rapidly lead to death. Symptoms of nonfatal injury to the medulla may include cranial nerve malfunctions on the same side of the body as the injury, paralysis and loss of sensation on the opposite side of the body, and irregularities in breathing or heart rhythm. ■

Figure 14.5 Medulla oblongata in relation to the rest of the brain stem. (See Tortora, *A Photographic Atlas of the Human Body, Second Edition,* Figure 8.19.)

The brain stem consists of the medulla oblongata, pons, and midbrain.

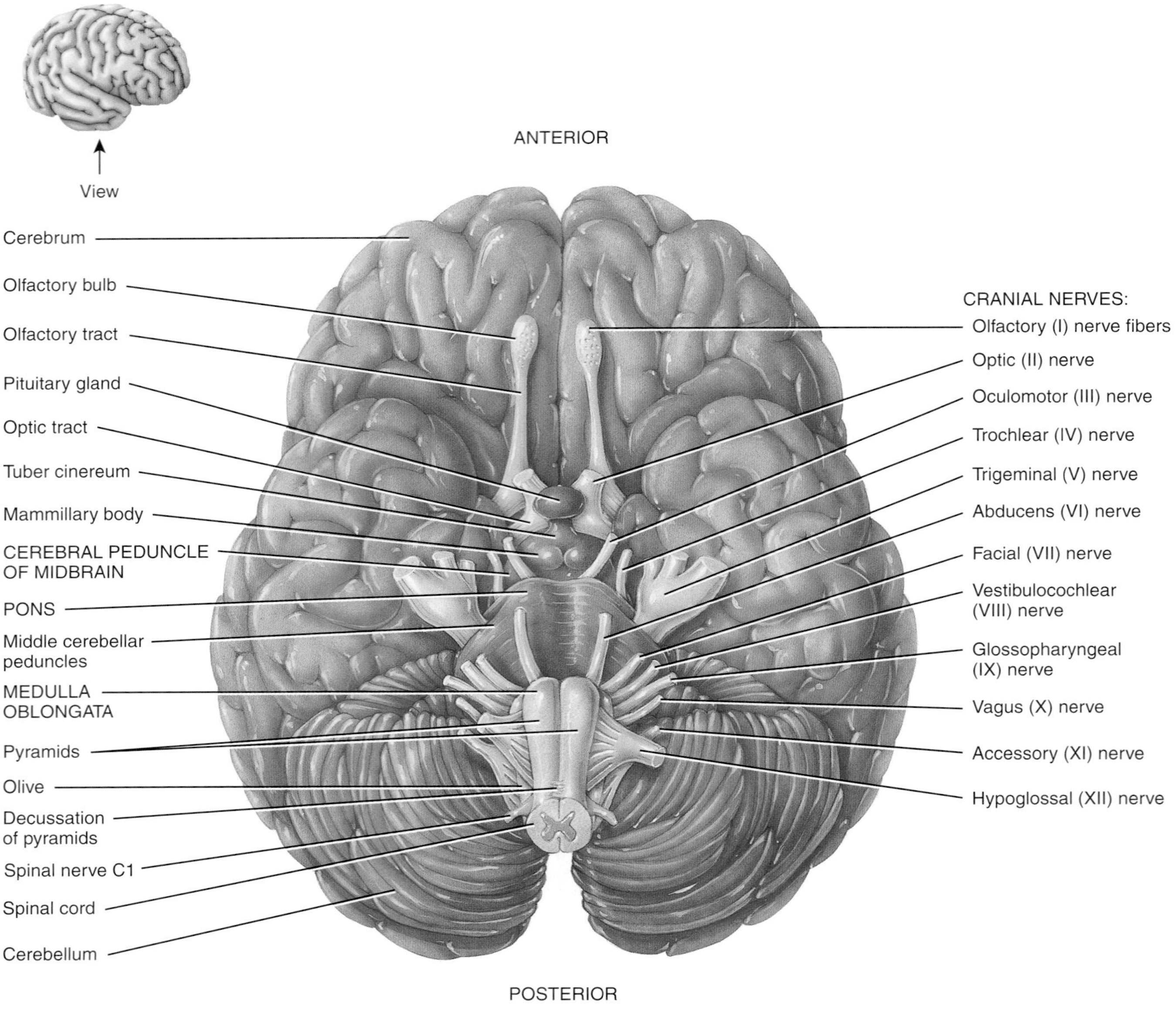

View

ANTERIOR

Cerebrum
Olfactory bulb
Olfactory tract
Pituitary gland
Optic tract
Tuber cinereum
Mammillary body
CEREBRAL PEDUNCLE OF MIDBRAIN
PONS
Middle cerebellar peduncles
MEDULLA OBLONGATA
Pyramids
Olive
Decussation of pyramids
Spinal nerve C1
Spinal cord
Cerebellum

CRANIAL NERVES:
Olfactory (I) nerve fibers
Optic (II) nerve
Oculomotor (III) nerve
Trochlear (IV) nerve
Trigeminal (V) nerve
Abducens (VI) nerve
Facial (VII) nerve
Vestibulocochlear (VIII) nerve
Glossopharyngeal (IX) nerve
Vagus (X) nerve
Accessory (XI) nerve
Hypoglossal (XII) nerve

POSTERIOR

Inferior aspect of brain

What part of the brain stem contains the pyramids? The cerebral peduncles? Literally means "bridge"?

Pons

The **pons** (= bridge) lies directly superior to the medulla and anterior to the cerebellum and is about 2.5 cm (1 in.) long (see Figures 14.1 and 14.5). Like the medulla, the pons consists of both nuclei and tracts. As its name implies, the pons is a bridge that connects parts of the brain with one another. These connections are provided by bundles of axons. Some axons of the pons connect the right and left sides of the cerebellum. Others are part of ascending sensory tracts and descending motor tracts.

Signals for voluntary movements originate in the cerebral cortex and are relayed through several **pontine nuclei** (PON-tīn) into the cerebellum. Other nuclei in the pons are the **pneumotaxic area** (noo-mō-TAK-sik) and the **apneustic area** (ap-NOO-stik), shown in Figure 23.25 on page 879. Together with the

Figure 14.6 **Internal anatomy of the medulla oblongata.**

The pyramids of the medulla contain the large corticospinal tracts that run from the cerebrum to the spinal cord.

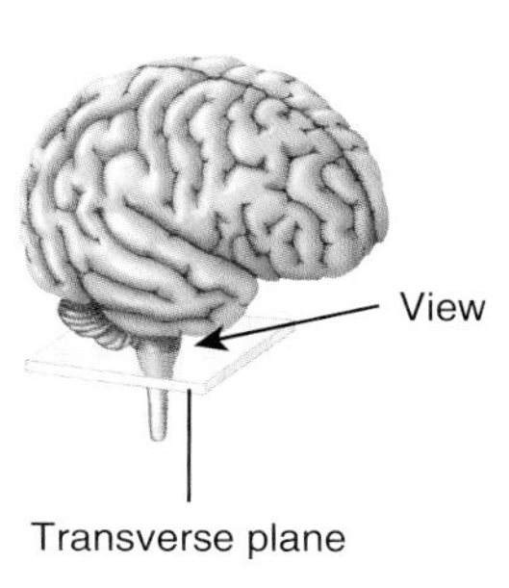

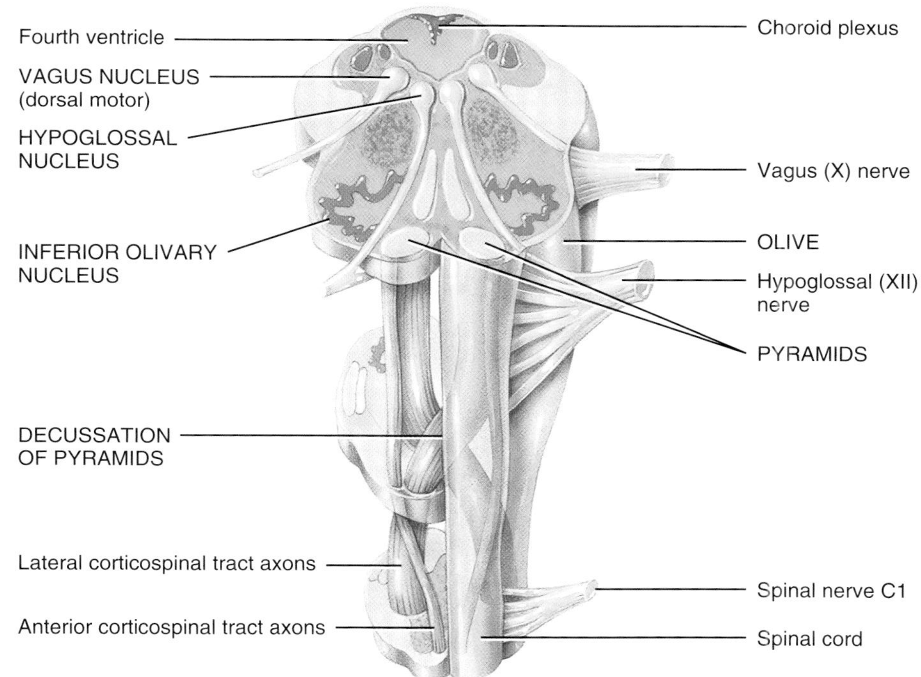

Transverse section and anterior surface of medulla oblongata

What does decussation mean? What is the functional consequence of decussation of the pyramids?

medullary rhythmicity area, the pneumotaxic and apneustic areas help control breathing.

The pons also contains nuclei associated with the following four pairs of cranial nerves (see Figure 14.5): trigeminal (V) nerves, abducens (VI) nerves, facial (VII) nerves, and vestibulocochlear (VIII) nerves.

Midbrain

The **midbrain** or **mesencephalon** extends from the pons to the diencephalon (see Figures 14.1 and 14.5) and is about 2.5 cm (1 in.) long. The cerebral aqueduct passes through the midbrain, connecting the third ventricle above with the fourth ventricle below. Like the medulla and the pons, the midbrain contains both tracts and nuclei.

The anterior part of the midbrain contains a pair of tracts called **cerebral peduncles** (pe-DUNK-kuls or PĒ-dung-kuls = little feet; see Figures 14.5 and 14.7b). These tracts contain axons of corticospinal, corticopontine, and corticobulbar motor neurons, which conduct nerve impulses from the cerebrum to the spinal cord, pons, and medulla, respectively. The cerebral peduncles also contain axons of sensory neurons that extend from the medulla to the thalamus.

The posterior part of the midbrain, called the **tectum** (TEK-tum = roof), contains four rounded elevations (Figure 14.7a). The two superior elevations, nuclei known as the **superior colliculi** (ko-LIK-ū-lī = little hills; singular is *colliculus*), serve as reflex centers for certain visual activities. Through neural circuits from the retina of the eye to the superior colliculi to the extrinsic eye muscles, visual stimuli elicit eye movements for tracking moving images (such as a moving car) and scanning stationary images (as you are doing to read this sentence). The superior colliculi are also responsible for reflexes that govern movements of the eyes, head, and neck in response to visual stimuli. The two inferior elevations, the **inferior colliculi,** are part of the auditory pathway, relaying impulses from the receptors for hearing in the ear to the thalamus. These two nuclei are also reflex centers for the *startle reflex,* sudden movements of the head and body that occur when you are surprised by a loud noise such as a gunshot.

The midbrain contains several other nuclei, including the left and right **substantia nigra** (sub-STAN-shē-a = substance; NĪ-gra = black), which are large and darkly pigmented (Figure 14.7b). Neurons that release dopamine, extending from the substantia nigra to the basal ganglia, help control subconscious muscle activities. Loss of these neurons is associated with

Parkinson disease (see page 568). Also present are the left and right **red nuclei** (Figure 14.7b), which look reddish due to their rich blood supply and an iron-containing pigment in their neuronal cell bodies. Axons from the cerebellum and cerebral cortex form synapses in the red nuclei, which function with the cerebellum to coordinate muscular movements.

Figure 14.7 Midbrain. (See Tortora, *A Photographic Atlas of the Human Body, Second Edition,* Figure 8.23.)

The midbrain connects the pons to the diencephalon.

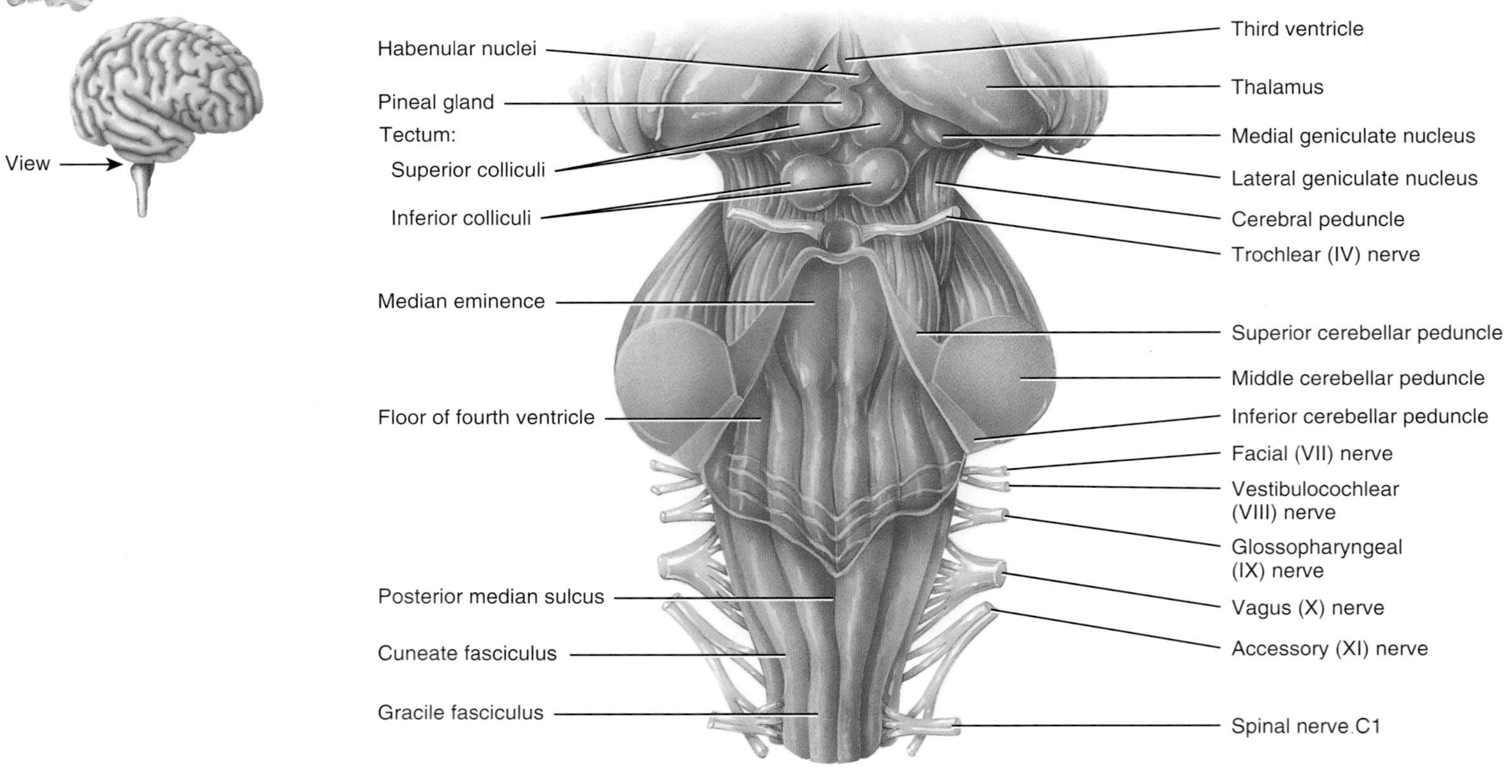

View
Habenular nuclei
Pineal gland
Tectum:
Superior colliculi
Inferior colliculi
Median eminence
Floor of fourth ventricle
Posterior median sulcus
Cuneate fasciculus
Gracile fasciculus
Third ventricle
Thalamus
Medial geniculate nucleus
Lateral geniculate nucleus
Cerebral peduncle
Trochlear (IV) nerve
Superior cerebellar peduncle
Middle cerebellar peduncle
Inferior cerebellar peduncle
Facial (VII) nerve
Vestibulocochlear (VIII) nerve
Glossopharyngeal (IX) nerve
Vagus (X) nerve
Accessory (XI) nerve
Spinal nerve C1

(a) Posterior view of midbrain in relation to brain stem

POSTERIOR
View
Transverse plane
Tectum
Reticular formation
Medial lemniscus
Cerebral peduncle
Superior colliculus
Gray matter
Aqueduct of the midbrain (cerebral aqueduct)
Medial geniculate nucleus
Oculomotor nucleus
Red nucleus
Substantia nigra
Corticospinal, corticopontine, and corticobulbar axons
Oculomotor (III) nerve
ANTERIOR

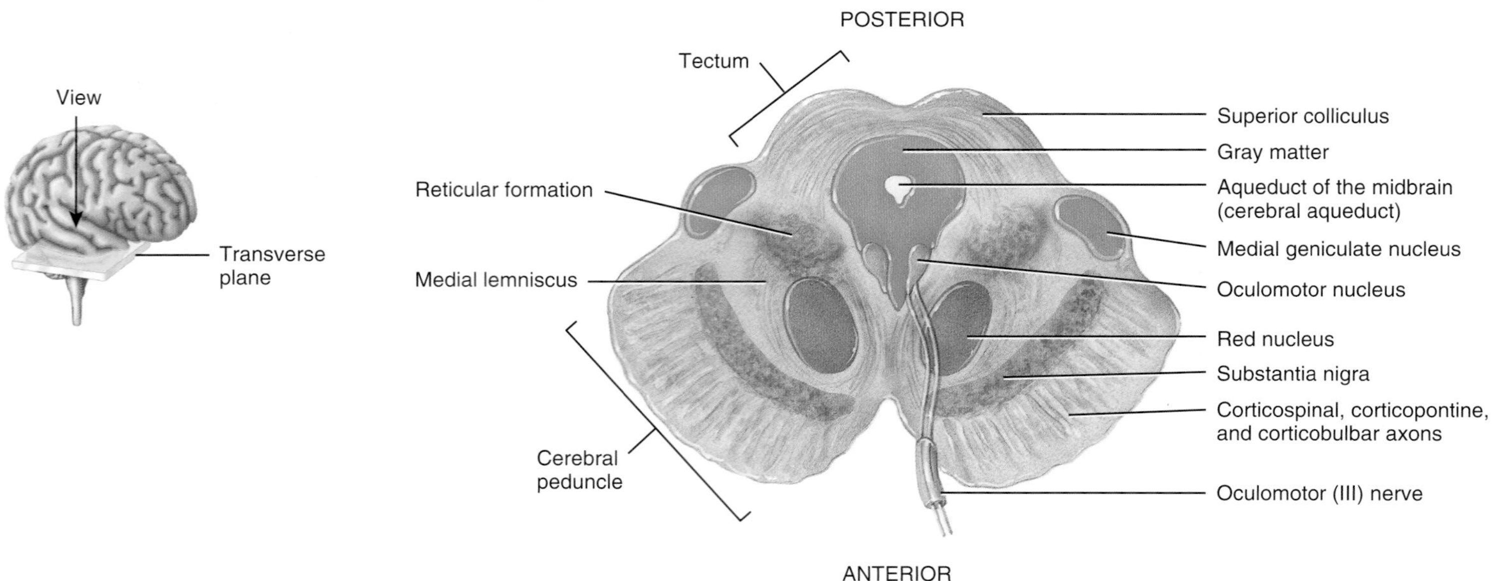

(b) Transverse section of midbrain

What is the importance of the cerebral peduncles?

Still other nuclei in the midbrain are associated with two pairs of cranial nerves (see Figure 14.5): oculomotor (III) nerves and trochlear (IV) nerves.

Reticular Formation

In addition to the well-defined nuclei already described, much of the brain stem consists of small clusters of neuronal cell bodies (gray matter) interspersed among small bundles of myelinated axons (white matter). The broad region where white matter and gray matter exhibit a netlike arrangement is known as the **reticular formation** (*ret-* = net Figure 14.7b). It extends from the upper part of the spinal cord, throughout the brain stem, and into the lower part of the diencephalon. Neurons within the reticular formation have both ascending (sensory) and descending (motor) functions. Part of the reticular formation, called the **reticular activating system (RAS),** consists of sensory axons that project to the cerebral cortex (see Figure 16.10 on page 566). The RAS helps maintain consciousness and is active during awakening from sleep. For example, we awaken to the sound of an alarm clock, to a flash of lightning, or to a painful pinch because of RAS activity that arouses the cerebral cortex. The reticular formation's main descending function is to help regulate *muscle tone,* the slight degree of contraction in normal resting muscles.

The functions of the brain stem are summarized in Table 14.2.

TABLE 14.2 Summary of Functions of Principal Parts of the Brain

Part	Function
Brain stem	
Medulla oblongata	*Medulla oblongata:* Relays sensory input and motor output between other parts of the brain and the spinal cord. Reticular formation (also in pons, midbrain, and diencephalon) functions in consciousness and arousal. Vital centers regulate heartbeat, blood vessel diameter, and breathing (together with pons). Other centers coordinate swallowing, vomiting, coughing, sneezing, and hiccupping. Contains nuclei of origin for cranial nerves VIII, IX, X, XI, and XII.
Pons	*Pons:* Relays impulses from one side of the cerebellum to the other and between the medulla and midbrain. Contains nuclei of origin for cranial nerves V, VI, VII, and VIII. Pneumotaxic area and apneustic area, together with the medulla, help control breathing.
Midbrain	*Midbrain:* Relays motor output from the cerebral cortex to the pons and sensory input from the spinal cord to the thalamus. Superior colliculi coordinate movements of the eyeballs in response to visual and other stimuli, and the inferior colliculi coordinate movements of the head and trunk in response to auditory stimuli. Most of substantia nigra and red nucleus contribute to control of movement. Contains nuclei of origin for cranial nerves III and IV.
Cerebellum 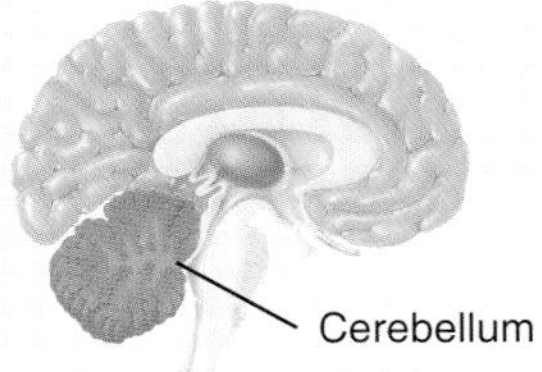	Compares intended movements with what is actually happening to smooth and coordinate complex, skilled movements. Regulates posture and balance. May have a role in cognition and language processing.
Diencephalon Epithalamus, Thalamus, Hypothalamus	*Thalamus:* Relays almost all sensory input to the cerebral cortex. Provides crude perception of touch, pressure, pain, and temperature. Includes nuclei involved in movement planning and control. *Hypothalamus:* Controls and integrates activities of the autonomic nervous system and pituitary gland. Regulates emotional and behavioral patterns and circadian rhythms. Controls body temperature and regulates eating and drinking behavior. Helps maintain the waking state and establishes patterns of sleep. Produces the hormones oxytocin and antidiuretic hormone (ADH). *Epithalamus:* Consists of pineal gland, which secretes melatonin, and the habenular nuclei.
Cerebrum	Sensory areas are involved in the perception of sensory information; motor areas control muscular movement; and association areas deal with more complex integrative functions such as memory, personality traits, and intelligence. Basal ganglia coordinate gross, automatic muscle movements and regulate muscle tone. Limbic system functions in emotional aspects of behavior related to survival.

▶ CHECKPOINT

6. Where are the medulla, pons, and midbrain located relative to one another?
7. Define decussation of pyramids. Why is it important?
8. What body functions are governed by nuclei in the brain stem?
9. What are two important functions of the reticular formation?

THE CEREBELLUM

▶ OBJECTIVE

Describe the structure and functions of the cerebellum.

The **cerebellum,** second only to the cerebrum in size, occupies the inferior and posterior aspects of the cranial cavity. It accounts for about a tenth of the brain mass yet contains nearly half of the neurons in the brain. The cerebellum is posterior to the medulla and pons and inferior to the posterior portion of the cerebrum (see Figure 14.1). A deep groove known as the **transverse fissure,** along with the **tentorium cerebelli,** which supports the posterior part of the cerebrum, separate the cerebellum from the cerebrum (see Figures 14.2b and 14.4b).

In superior or inferior views, the shape of the cerebellum resembles a butterfly. The central constricted area is the **vermis** (= worm), and the lateral "wings" or lobes are the **cerebellar hemispheres** (Figure 14.8a, b). Each hemisphere consists of lobes separated by deep and distinct fissures. The **anterior lobe** and **posterior lobe** govern subconscious aspects of skeletal muscle movements. The **flocculonodular lobe** (flok-ū-lō-NOD-ū-lar; *flocculo-* = wool-like tuft) on the inferior surface contributes to equilibrium and balance.

The superficial layer of the cerebellum, called the **cerebellar cortex,** consists of gray matter in a series of slender, parallel ridges called **folia** (= leaves). Deep to the gray matter are tracts of white matter called **arbor vitae** (= tree of life) that resemble branches of a tree. Even deeper, within the white matter, are the **cerebellar nuclei,** regions of gray matter that give rise to axons carrying impulses from the cerebellum to other brain centers and to the spinal cord.

Three paired **cerebellar peduncles** (pe-DUNG-kuls) attach the cerebellum to the brain stem (see Figures 14.7a and 14.8b). These bundles of white matter consist of axons that conduct impulses between the cerebellum and other parts of the brain. The **inferior cerebellar peduncles** carry sensory information from the vestibular apparatus of the inner ear and from proprioceptors throughout the body into the cerebellum; their axons extend from the inferior olivary nucleus of the medulla and from the spinocerebellar tracts of the spinal cord into the cerebellum. The **middle cerebellar peduncles** are the largest peduncles; their axons carry commands for voluntary movements (those that originate in motor areas of the cerebral cortex) from the pontine nuclei into the cerebellum. The **superior cerebellar peduncles** contain axons that extend from the cerebellum to the red nuclei of the midbrain and to several nuclei of the thalamus.

The primary function of the cerebellum is to evaluate how well movements initiated by motor areas in the cerebrum are actually being carried out. When movements initiated by the cerebral motor areas are not being carried out correctly, the cerebellum detects the discrepancies. It then sends feedback signals to motor areas of the cerebral cortex, via its connections to the red nucleus and thalamus. The feedback signals help correct the errors, smooth the movements, and coordinate complex sequences of skeletal muscle contractions. Aside from this coordination of skilled movements, the cerebellum is the main brain region that regulates posture and balance. These aspects of cerebellar function make possible all skilled muscular activities, from catching a baseball to dancing to speaking. The presence of reciprocal connections between the cerebellum and association areas of the cerebral cortex suggests that the cerebellum may also have nonmotor functions such as cognition (acquisition of knowledge) and language processing. This view is supported by imaging studies using MRI and PET. Studies also suggest that the cerebellum may play a role in processing sensory information.

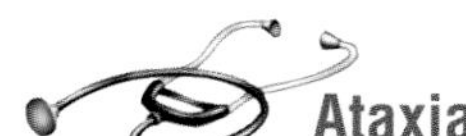

Ataxia

Damage to the cerebellum through trauma or disease disrupts muscle coordination, a condition called **ataxia** (*a-* = without; *-taxia* = order). Blindfolded people with ataxia cannot touch the tip of their nose with a finger because they cannot coordinate movement with their sense of where a body part is located. Another sign of ataxia is changed speech pattern due to uncoordinated speech muscles. Cerebellar damage may also result in staggering or abnormal walking movements. People who consume too much alcohol show signs of ataxia because alcohol inhibits activity of the cerebellum. Alcohol overdose also suppresses the medullary rhythmicity area and may result in death. ■

The functions of the cerebellum are summarized in Table 14.2.

▶ CHECKPOINT

10. Describe the location and principal parts of the cerebellum.
11. Where do the axons of each of the three pairs of cerebellar peduncles begin and end? What are their functions?

Figure 14.8 Cerebellum. (See Tortora, *A Photographic Atlas of the Human Body, Second Edition,* Figure 8.25.)

The cerebellum coordinates skilled movements and regulates posture and balance.

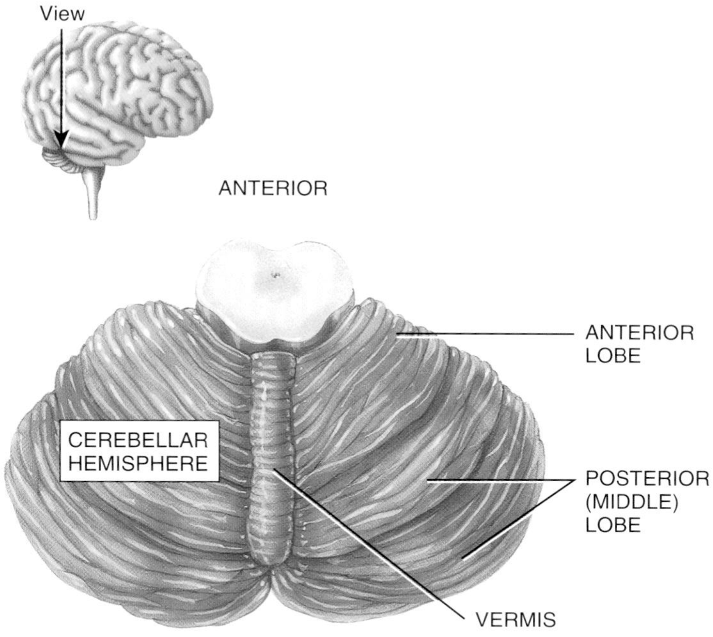

View

ANTERIOR

ANTERIOR LOBE

CEREBELLAR HEMISPHERE

POSTERIOR (MIDDLE) LOBE

VERMIS

POSTERIOR

(a) Superior view

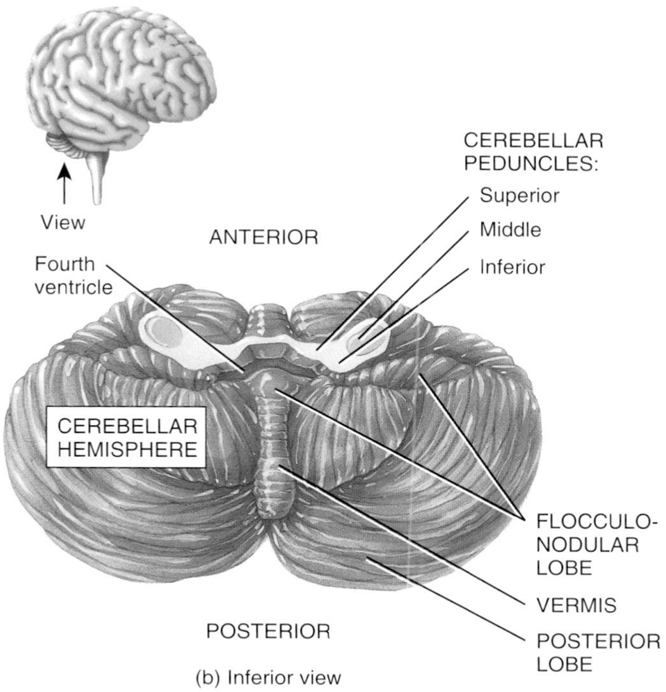

View

ANTERIOR

CEREBELLAR PEDUNCLES:

Superior

Middle

Inferior

Fourth ventricle

CEREBELLAR HEMISPHERE

FLOCCULO-NODULAR LOBE

VERMIS

POSTERIOR LOBE

POSTERIOR

(b) Inferior view

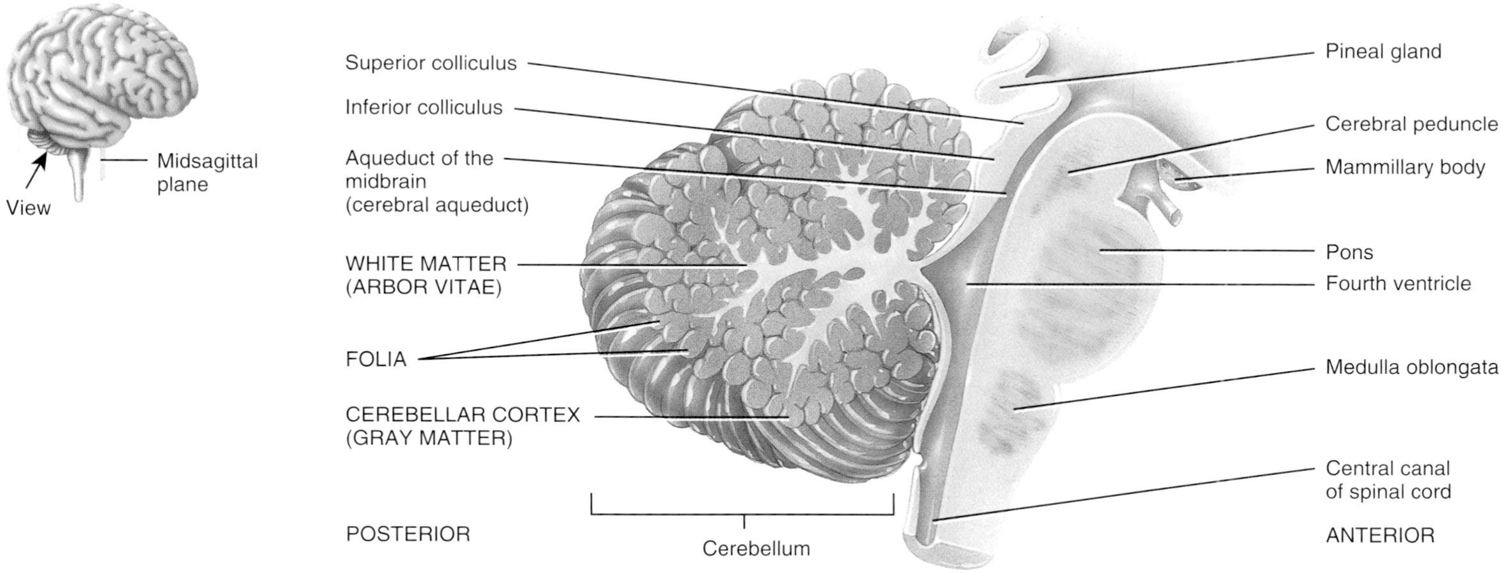

Midsagittal plane

View

Superior colliculus

Inferior colliculus

Aqueduct of the midbrain (cerebral aqueduct)

WHITE MATTER (ARBOR VITAE)

FOLIA

CEREBELLAR CORTEX (GRAY MATTER)

POSTERIOR

Cerebellum

Pineal gland

Cerebral peduncle

Mammillary body

Pons

Fourth ventricle

Medulla oblongata

Central canal of spinal cord

ANTERIOR

(c) Midsagittal section of cerebellum and brain stem

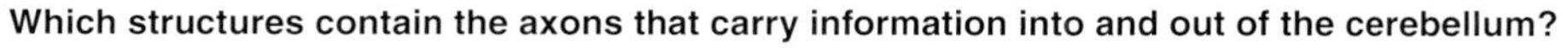

Which structures contain the axons that carry information into and out of the cerebellum?

THE DIENCEPHALON

▶ **OBJECTIVE**

Describe the components and functions of the diencephalon.

The **diencephalon** extends from the brain stem to the cerebrum and surrounds the third ventricle; it includes the thalamus, hypothalamus, and epithalamus.

Thalamus

The **thalamus** (THAL-a-mus = inner chamber), which measures about 3 cm (1.2 in.) in length and makes up 80% of the diencephalon, consists of paired oval masses of gray matter organized into nuclei with interspersed tracts of white matter (Figure 14.9). A bridge of gray matter called the **intermediate mass (interthalamic adhesion)** joins the right and left halves of the thalamus in about 70% of human brains.

The thalamus is the major relay station for most sensory impulses that reach the primary sensory areas of the cerebral cortex from the spinal cord and brain stem. Although crude perception of painful, thermal, and pressure sensations arises at the level of the thalamus, precise localization of these sensations depends on nerve impulses arriving at the cerebral cortex.

The thalamus contributes to motor functions by transmitting information from the cerebellum and basal ganglia to the primary motor area of the cerebral cortex. It also relays nerve impulses between different areas of the cerebrum, and plays a role in the regulation of autonomic activities and the maintenance of consciousness. Axons that connect the thalamus and cerebral cortex pass through the **internal capsule,** a thick band of white matter lateral to the thalamus (see Figure 14.13b).

Figure 14.9 Thalamus. Note the position of the thalamus in (a), the lateral view, and in (b), the medial view. The various thalamic nuclei shown in (c) and (d) are correlated by color to the cortical regions to which they project in (a) and (b). (See Tortora, *A Photographic Atlas of the Human Body, Second Edition,* Figures 8.18 and 8.22.)

The thalamus is the main relay station for sensory impulses that reach the cerebral cortex from other parts of the brain and the spinal cord.

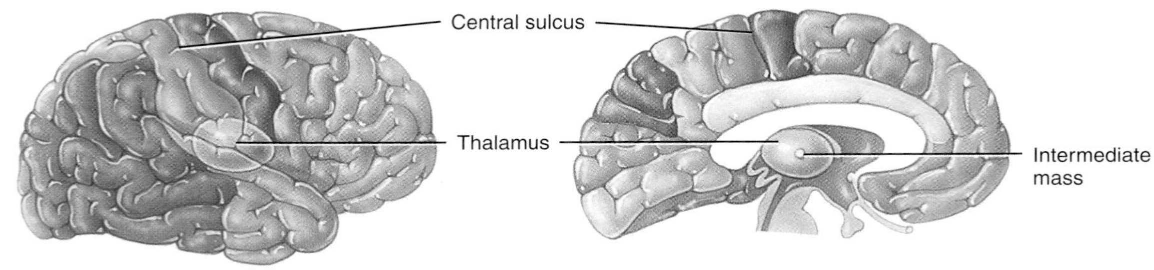

(a) Lateral view of right cerebral hemisphere

(b) Medial view of left cerebral hemisphere

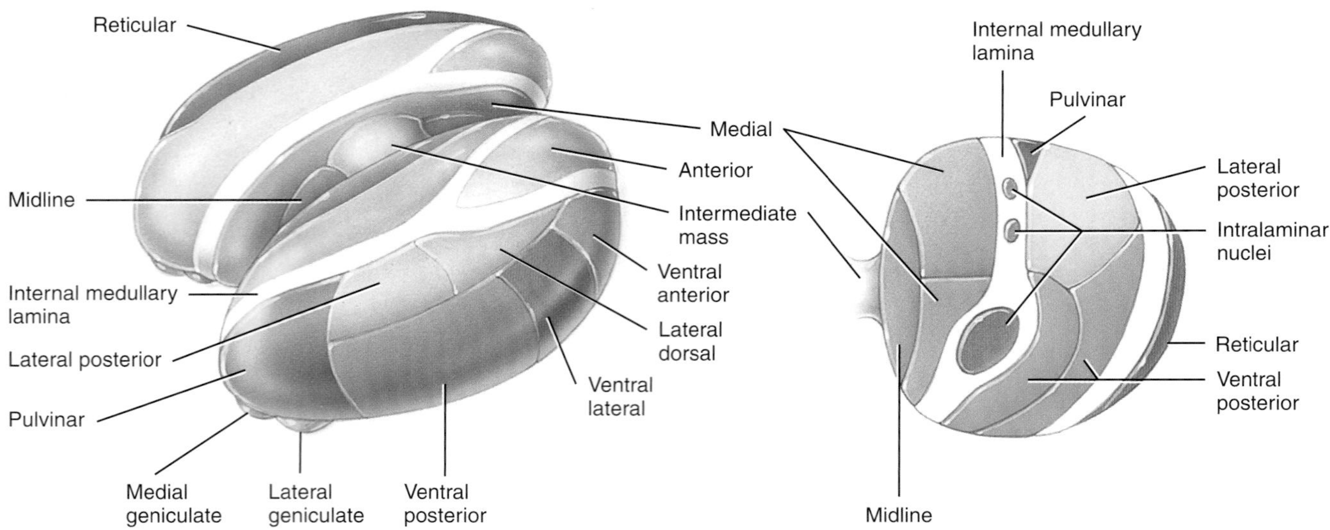

(c) Superolateral view of thalamus showing locations of thalamic nuclei (reticular nucleus is shown on the left side only; all other nuclei are shown on the right side)

(d) Transverse section of right side of thalamus showing locations of thalamic nuclei

A vertical Y-shaped sheet of white matter called the **internal medullary lamina** divides the gray matter of the right and left sides of the thalamus (Figure 14.9c). It consists of myelinated axons that enter and leave the various thalamic nuclei.

Based on their positions and functions, there are seven major groups of nuclei on each side of the thalamus (Figure 14.9c, d).

1. The **anterior nucleus** connects to the hypothalamus and limbic system (described on page 495). It functions in emotions, regulation of alertness, and memory.
2. The **medial nuclei** connect to the cerebral cortex, limbic system, and basal ganglia. They function in emotions, learning, memory, awareness, and cognition (thinking and knowing).
3. Nuclei in the **lateral group** connect to the superior colliculi, limbic system, and cortex in all lobes of the cerebrum. The **lateral dorsal nucleus** functions in the expression of emotions. The **lateral posterior nucleus** and **pulvinar nucleus** help integrate sensory information.
4. Five nuclei are part of the **ventral group.** The **ventral anterior nucleus** contributes to motor functions, possibly movement planning. The **ventral lateral nucleus** connects to the cerebellum and motor parts of the cerebral cortex. Its neurons are active during movements on the opposite side of the body. The **ventral posterior nucleus** relays impulses for somatic sensations such as touch, pressure, proprioception, vibration, heat, cold, and pain from the face and body to the cerebral cortex. The **lateral geniculate nucleus** (je-NIK-ū-lat = bent like a knee) relays visual impulses for sight from the retina to the primary visual area of the cerebral cortex. The **medial geniculate nucleus** relays auditory impulses for hearing from the ear to the primary auditory area of the cerebral cortex.
5. **Intralaminar nuclei** lie within the internal medullary lamina and make connections with the reticular formation, cerebellum, basal ganglia, and wide areas of the cerebral cortex. They function in pain perception, integration of sensory and motor information, and arousal (activation of the cerebral cortex from the brain stem reticular formation).
6. The **midline nucleus** forms a thin band adjacent to the third ventricle and has a presumed function in memory and olfaction.
7. The **reticular nucleus** surrounds the lateral aspect of the thalamus, next to the internal capsule. This nucleus monitors, filters, and integrates activities of other thalamic nuclei.

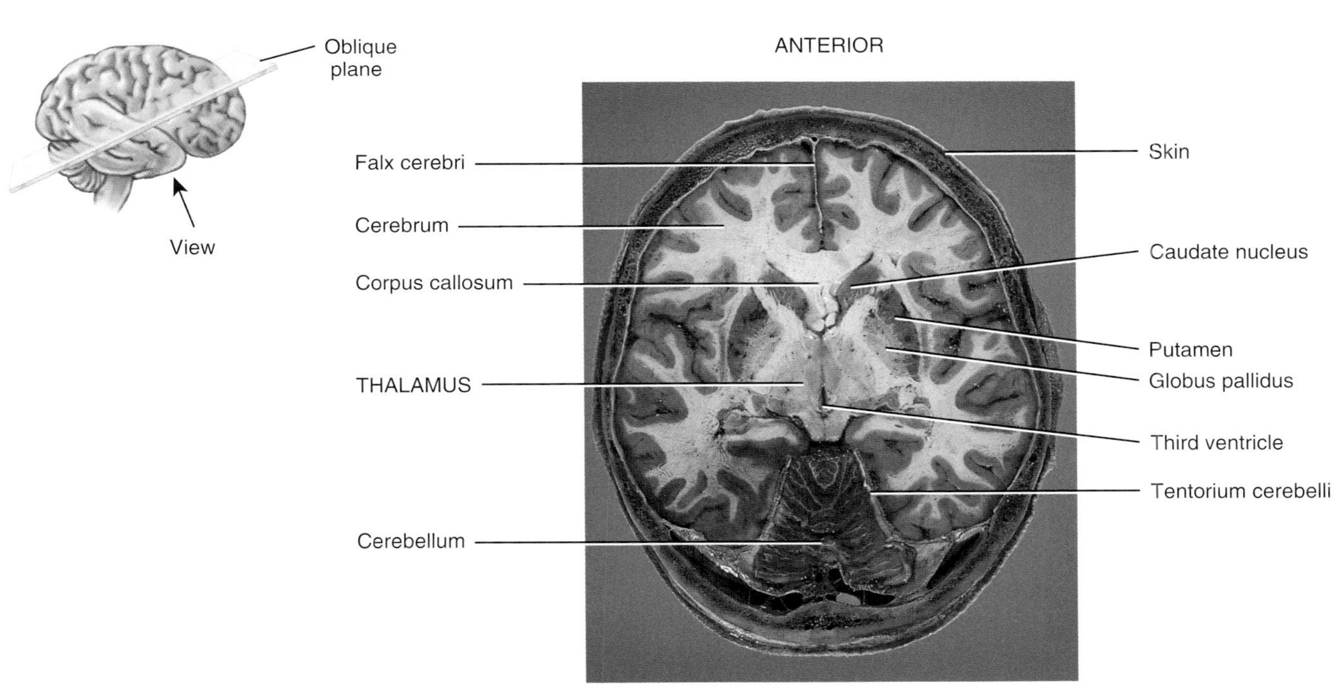

(e) Oblique section of brain

What structure usually connects the right and left halves of the thalamus?

Hypothalamus

The **hypothalamus** (*hypo-* = under) is a small part of the diencephalon located inferior to the thalamus. It is composed of a dozen or so nuclei in four major regions:

1. The **mammillary region** (*mammill-* = nipple-shaped), adjacent to the midbrain, is the most posterior part of the hypothalamus. It includes the mammillary bodies and posterior hypothalamic nuclei (Figure 14.10). The **mammillary bodies** are two, small, rounded projections that serve as relay stations for reflexes related to the sense of smell (see also Figure 14.5).

2. The **tuberal region,** the widest part of the hypothalamus, includes the *dorsomedial nucleus, ventromedial nucleus,* and *arcuate nucleus,* plus the stalklike **infundibulum** (in-fun-DIB-ū-lum = funnel), which connects the pituitary gland to the hypothalamus (Figure 14.10). The **median eminence** is a slightly raised region that encircles the infundibulum (see Figure 14.7a).

3. The **supraoptic region** (*supra-* = above; *-optic* = eye) lies superior to the optic chiasm (point of crossing of optic nerves) and contains the *paraventricular nucleus, supraoptic nucleus, anterior hypothalamic nucleus,* and *suprachiasmatic nucleus* (Figure 14.10). Axons from the paraventricular and supraoptic nuclei form the hypothalamohypophyseal tract, which extends through the infundibulum to the posterior lobe of the pituitary.

4. The **preoptic region** anterior to the supraoptic region is usually considered part of the hypothalamus because it participates with the hypothalamus in regulating certain autonomic activities. The preoptic region contains the *medial* and *lateral preoptic nuclei* (Figure 14.10).

The hypothalamus controls many body activities and is one of the major regulators of homeostasis. Sensory impulses related to both somatic and visceral senses arrive at the hypothalamus, as do impulses from receptors for vision, taste, and smell. Other receptors within the hypothalamus itself continually monitor osmotic pressure, glucose level, certain hormone concentrations, and the temperature of blood. The hypothalamus has several very important connections with the pituitary gland and produces a variety of hormones, which are described in more detail in Chapter 18. Some functions can be attributed to specific hypothalamic nuclei, but others are not so precisely localized. Important functions of the hypothalamus include the following:

- ***Control of the ANS.*** The hypothalamus controls and integrates activities of the autonomic nervous system, which regulates contraction of smooth and cardiac muscle and the secretions of many glands. Axons extend from the hypothalamus to sympathetic and parasympathetic nuclei in the brain

Figure 14.10 Hypothalamus. Selected portions of the hypothalamus and a three-dimensional representation of hypothalamic nuclei are shown (after Netter).

The hypothalamus controls many body activities and is an important regulator of homeostasis.

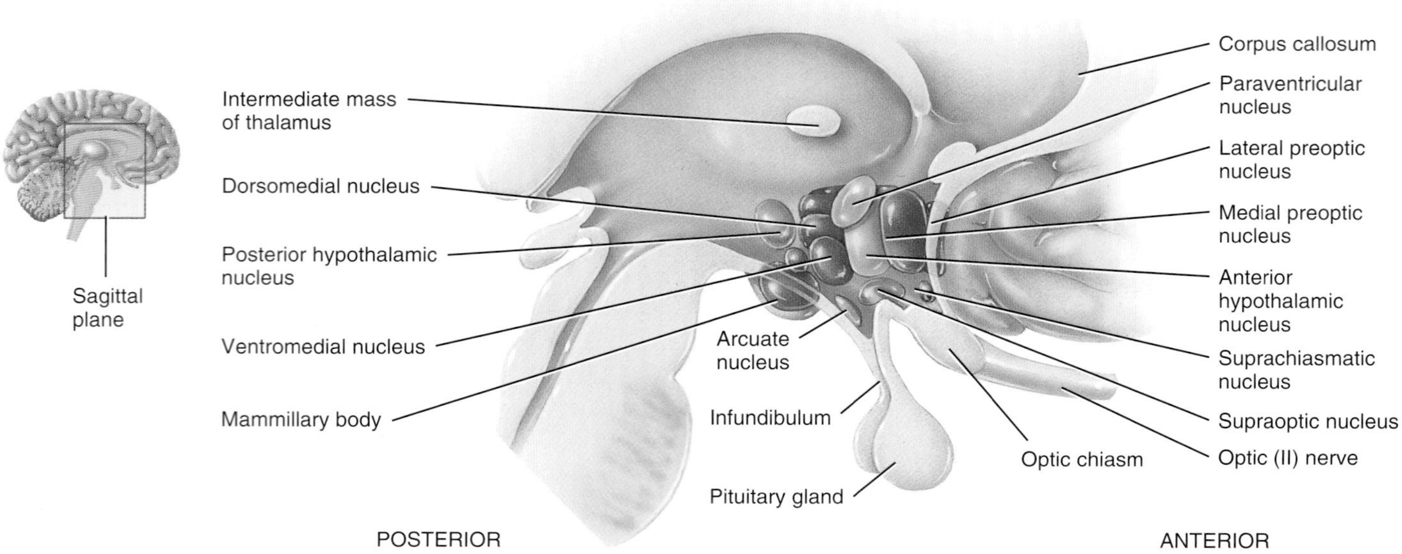

Sagittal section of brain showing hypothalamic nuclei

What are the four major regions of the hypothalamus, from posterior to anterior?

stem and spinal cord. Through the ANS, the hypothalamus is a major regulator of visceral activities, including regulation of heart rate, movement of food through the gastrointestinal tract, and contraction of the urinary bladder.

- ***Production of hormones.*** The hypothalamus produces several hormones and has two types of important connections with the pituitary gland, an endocrine gland located inferior to the hypothalamus (see Figure 14.1). First, hypothalamic hormones are released into capillary networks in the median eminence. The bloodstream carries these hormones directly to the anterior lobe of the pituitary, where they stimulate or inhibit secretion of anterior pituitary hormones. Second, axons extend from the paraventricular and supraoptic nuclei through the infundibulum into the posterior lobe of the pituitary. The cell bodies of these neurons make one of two hormones (oxytocin or antidiuretic hormone). Their axons transport the hormones to the posterior pituitary, where they are released.
- ***Regulation of emotional and behavioral patterns.*** Together with the limbic system (described shortly), the hypothalamus participates in expressions of rage, aggression, pain, and pleasure, and the behavioral patterns related to sexual arousal.
- ***Regulation of eating and drinking.*** The hypothalamus regulates food intake through the arcuate and paraventricular nuclei. It also contains a **thirst center.** When certain cells in the hypothalamus are stimulated by rising osmotic pressure of the extracellular fluid, they cause the sensation of thirst. The intake of water by drinking restores the osmotic pressure to normal, removing the stimulation and relieving the thirst.
- ***Control of body temperature.*** If the temperature of blood flowing through the hypothalamus is above normal, the hypothalamus directs the autonomic nervous system to stimulate activities that promote heat loss. When blood temperature is below normal, by contrast, the hypothalamus generates impulses that promote heat production and retention.
- ***Regulation of circadian rhythms and states of consciousness.*** The suprachiasmatic nucleus establishes patterns of awakening and sleep that occur on a circadian schedule (cycle of about 24 hours). This nucleus receives input from the eyes (retina) and sends output to other hypothalamic nuclei, the reticular formation, and the pineal gland.

Epithalamus

The **epithalamus** (*epi-* = above), a small region superior and posterior to the thalamus, consists of the pineal gland and habenular nuclei. The **pineal gland** (PĪN-ē-al = pinecone-like) is about the size of a small pea and protrudes from the posterior midline of the third ventricle (see Figure 14.1). The pineal gland is considered part of the endocrine system because it secretes the hormone **melatonin.** As more melatonin is liberated during darkness than in light, this hormone is thought to promote sleepiness. Melatonin also appears to contribute to the setting of the body's biological clock. The **habenular nuclei** (ha-BEN-ū-lar), shown in Figure 14.7a, are involved in olfaction, especially emotional responses to odors such as a loved one's cologne or Mom's chocolate chip cookies baking in the oven.

The functions of the three parts of the diencephalon are summarized in Table 14.2.

Circumventricular Organs

Parts of the diencephalon, called **circumventricular** (ser′-kum-ven-TRIK-ū-lar) **organs (CVOs)** because they lie in the walls of the third and fourth ventricles, can monitor chemical changes in the blood because they lack a blood–brain barrier. CVOs include part of the hypothalamus, the pineal gland, the pituitary gland, and a few other nearby structures. Functionally, these regions coordinate homeostatic activities of the endocrine and nervous systems, such as the regulation of blood pressure, fluid balance, hunger, and thirst. CVOs are also thought to be the sites of entry into the brain of HIV, the virus that causes AIDS. Once in the brain, HIV may cause dementia (irreversible deterioration of mental state) and other neurological disorders.

▶ CHECKPOINT

12. Why is the thalamus considered a "relay station" in the brain?

13. Why is the hypothalamus considered part of both the nervous system and the endocrine system?

THE CEREBRUM

▶ OBJECTIVES

Describe the cortex, convolutions, fissures, and sulci of the cerebrum.

List and locate the lobes of the cerebrum.

Describe the nuclei that comprise the basal ganglia.

List the structures and describe the functions of the limbic system.

The cerebrum is the "seat of intelligence." It provides us with the ability to read, write, and speak; to make calculations and compose music; and to remember the past, plan for the future, and imagine things that have never existed before.

The right and left halves of the cerebrum, called **cerebral hemispheres,** are separated by the falx cerebri. The hemispheres consist of an outer rim of gray matter, an internal region of cerebral white matter, and gray matter nuclei deep within the white matter. The outer rim of gray matter is the **cerebral cortex** (*cortex* = rind or bark) (Figure 14.11a). Although only 2–4 mm (0.08–0.16 in.) thick, the cerebral cortex contains billions of neurons. Deep to the cerebral cortex lies the cerebral white matter.

During embryonic development, when brain size increases rapidly, the gray matter of the cortex enlarges much faster than the deeper white matter. As a result, the cortical region rolls and folds upon itself. The folds are called **gyri** (JĪ-rī = circles; singular is *gyrus*) or **convolutions** (Figure 14.11a, b). The deepest grooves between folds are known as **fissures;** the shallower grooves between folds are termed **sulci** (SUL-sī = grooves; singular is *sulcus.*) The most prominent fissure, the **longitudinal fissure,** separates the cerebrum into right and left halves called **cerebral hemispheres.** The hemispheres are connected internally by the **corpus callosum** (kal-LŌ-sum; *corpus* = body; *callosum* = hard), a broad band of white matter containing axons that extend between the hemispheres (see Figure 14.12).

Lobes of the Cerebrum

Each cerebral hemisphere can be further subdivided into four lobes. The lobes are named after the bones that cover them: frontal, parietal, temporal, and occipital lobes (see Figure 14.11a, b). The **central sulcus** (SUL-kus) separates the **frontal lobe** from

Figure 14.11 Cerebrum. Because the insula cannot be seen externally, it has been projected to the surface in (b). (See Tortora, *A Photographic Atlas of the Human Body, Second Edition,* Figure 8.14.)

The cerebrum is the "seat of intelligence"; it provides us with the ability to read, write, and speak; to make calculations and compose music; to remember the past and plan for the future; and to create.

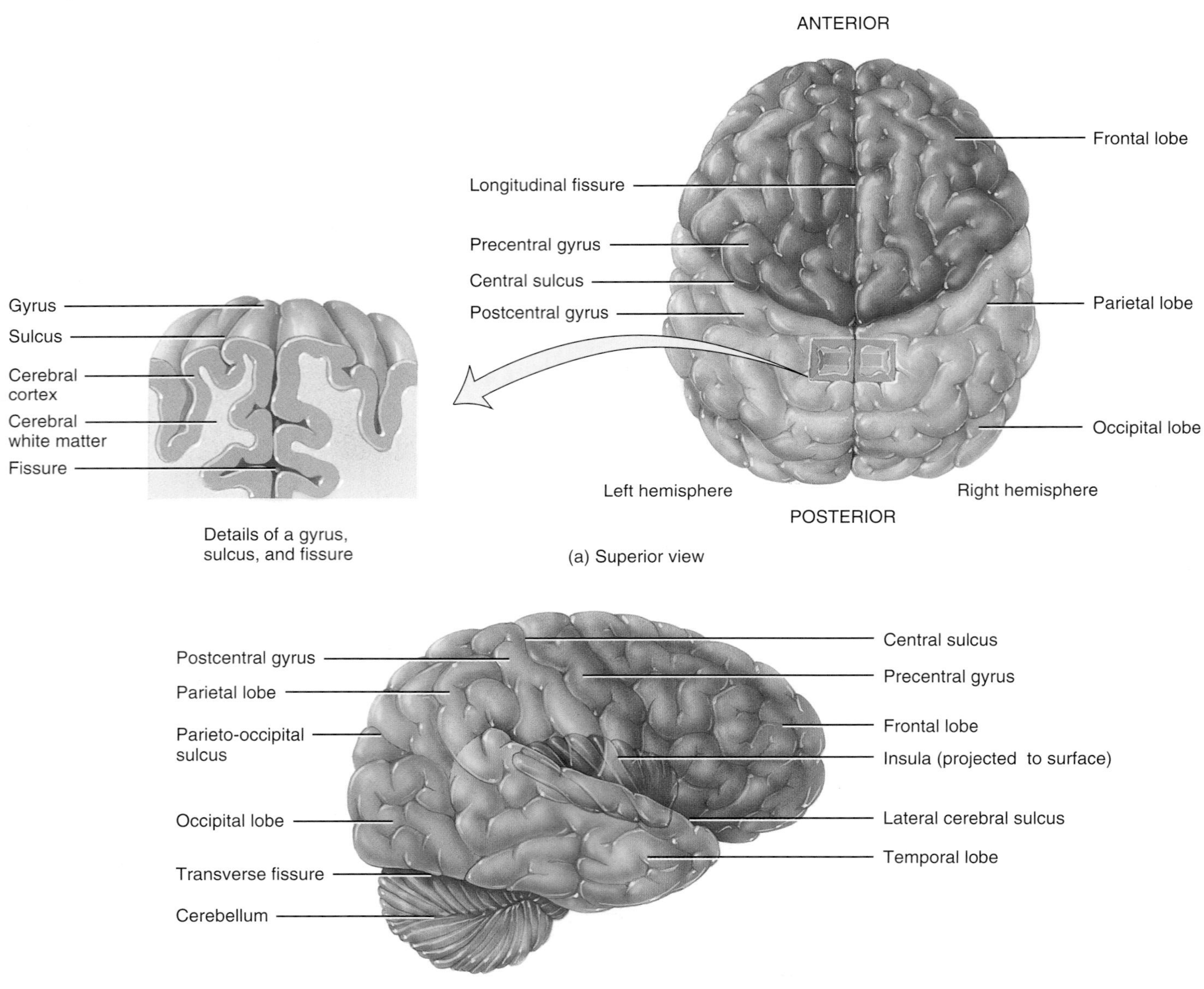

During development, which part of the brain—gray matter or white matter—enlarges more rapidly? What are the brain folds, shallow grooves, and deep grooves called?

the **parietal lobe.** A major gyrus, the **precentral gyrus**—located immediately anterior to the central sulcus—contains the primary motor area of the cerebral cortex. Another major gyrus, the **postcentral gyrus,** which is located immediately posterior to the central sulcus, contains the primary somatosensory area of the cerebral cortex. The **lateral cerebral sulcus (fissure)** separates the **frontal lobe** from the **temporal lobe.** The **parieto-occipital sulcus** separates the **parietal lobe** from the **occipital lobe.** A fifth part of the cerebrum, the **insula,** cannot be seen at the surface of the brain because it lies within the lateral cerebral sulcus, deep to the parietal, frontal, and temporal lobes (Figure 14.11b).

Cerebral White Matter

The **cerebral white matter** consists of myelinated and unmyelinated axons in three types of tracts (Figure 14.12 and see also Figure 14.4a):

1. **Association tracts** contain axons that conduct nerve impulses between gyri in the same hemisphere.
2. **Commissural tracts** contain axons that conduct nerve impulses from gyri in one cerebral hemisphere to corresponding gyri in the other cerebral hemisphere. Three important groups of commissural tracts are the **corpus callosum** (the largest fiber bundle in the brain, containing about 300 million fibers), **anterior commissure,** and **posterior commissure.**
3. **Projection tracts** contain axons that conduct nerve impulses from the cerebrum to lower parts of the CNS (thalamus, brainstem, or spinal cord) or from lower parts of the CNS to the cerebrum. An example is the **internal capsule,** a thick band of white matter that contains both ascending and descending axons (see Figure 14.13b).

Basal Ganglia

Deep within each cerebral hemisphere are three nuclei (masses of gray matter) that are collectively termed the **basal ganglia** (Figure 14.13). Recall that "ganglion" usually means a collection of neuronal cell bodies *outside* the CNS. The name here is the one exception to that general rule. An alternative term used in some textbooks—*basal nuclei*—is not used by most neuroscientists. It can be confused with the names of other brain regions, such as the nucleus basalis, which deteriorates in people who suffer from Alzheimer disease.

Two of the basal ganglia are side-by-side, just lateral to the thalamus. The **globus pallidus** (GLŌ-bus PAL-i-dus; *globus* = ball; *pallidus* = pale) is closer to the thalamus, and the **putamen** (pū-TĀ-men = shell) is closer to the cerebral cortex. Together, the globus pallidus and putamen are referred to as the **lentiform nucleus** (LEN-ti-fom = shaped like a lens). The third basal ganglion is the **caudate nucleus** (KAW-dāt; *caud-* = tail), which has a large "head" connected to a smaller "tail" by a long comma-shaped "body." Together, the lentiform and caudate nuclei are known as the **corpus striatum** (strī-Ā-tum; *corpus* = body; *striatum* = striated). The term corpus striatum refers to the striated (striped) appearance of the internal capsule as it passes among the basal ganglia. Nearby structures that are functionally linked to the basal ganglia are the *substantia nigra* of the midbrain (see Figure 14.7b) and the *subthalamic nuclei* (see Figure 14.13b). Axons from the substantia nigra terminate in the caudate nucleus and putamen. The subthalamic nuclei interconnect with the globus pallidus.

The basal ganglia receive input from the cerebral cortex and provide output back to motor parts of the cortex via medial and ventral group nuclei of the thalamus. In addition, the nuclei of the basal

Figure 14.12 **Organization of white matter tracts of the left cerebral hemisphere.**

Association tracts, commissural tracts, and projection tracts form white matter areas in the cerebral hemispheres.

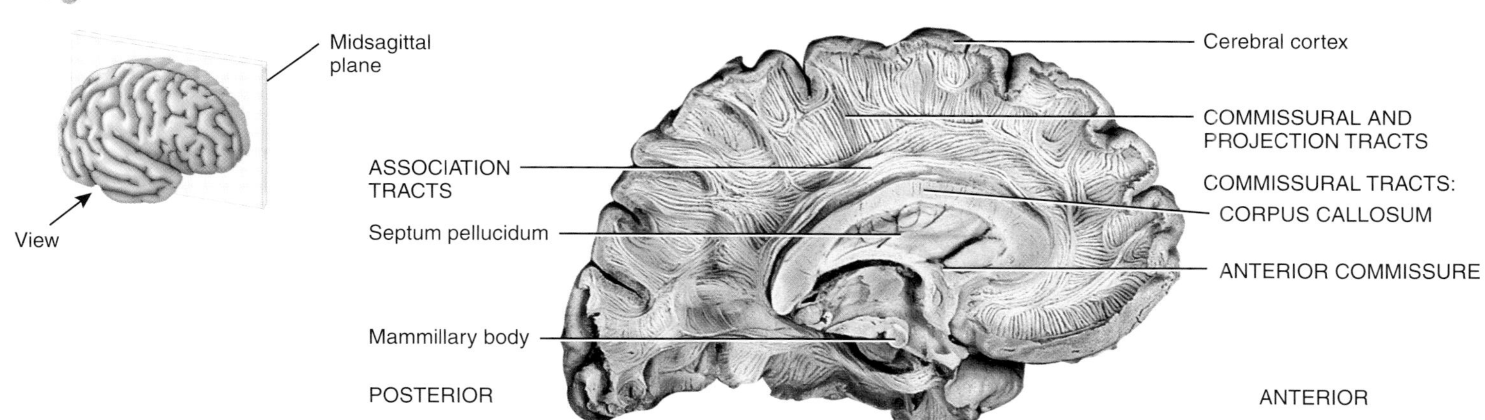

Medial view of tracts revealed by removing gray matter from a midsagittal section

Which tracts carry impulses between gyri of the same hemisphere? Between gyri in opposite hemispheres? From the cerebrum to the thalamus, brain stem, and spinal cord?

Figure 14.13 **Basal ganglia.** In (a) the basal ganglia have been projected to the surface and are shown in blue; in (b) they are also shown in blue. (See Tortora, *A Photographic Atlas of the Human Body, Second Edition,* Figures 8.17, 8.18, and 8.22.)

The basal ganglia control automatic movements of skeletal muscles and muscle tone.

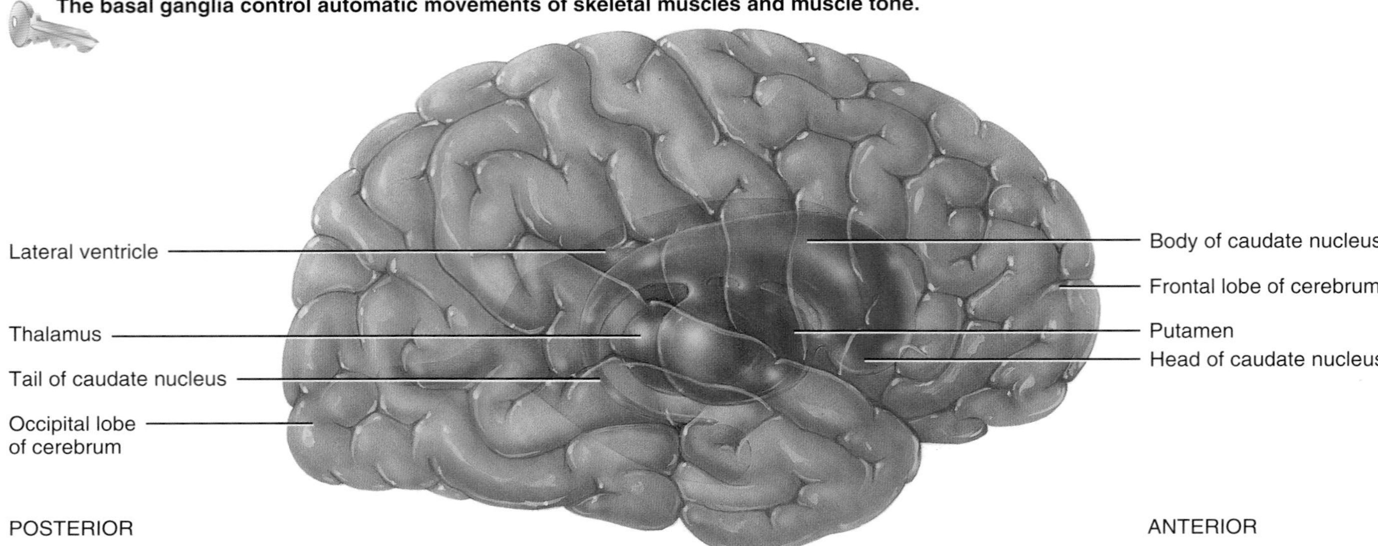

(a) Lateral view of right side of brain

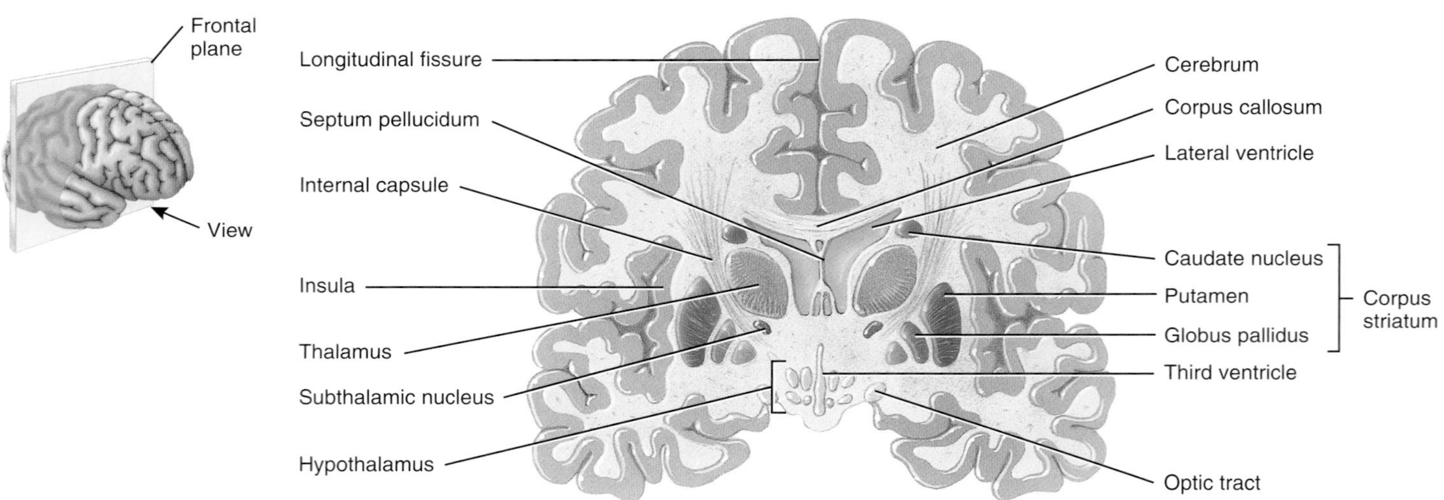

(b) Anterior view of frontal section

Where are the basal ganglia located relative to the thalamus?

ganglia have extensive connections with one another. A major function of the basal ganglia is to help regulate initiation and termination of movements. Activity of neurons in the putamen precedes or anticipates body movements, and activity of neurons in the caudate nucleus occurs prior to eye movements. The globus pallidus helps regulate the muscle tone required for specific body movements. The basal ganglia also control subconscious contractions of skeletal muscles. Examples include automatic arm swings while walking and true laughter in response to a joke (not the kind you consciously initiate to humor your anatomy and physiology instructor).

Damage to the Basal Ganglia

Damage to the basal ganglia results in uncontrollable shaking (tremor), muscular rigidity (stiffness), and involuntary muscle movements. Movement disruptions such as these are a hallmark of disorders like Parkinson disease (see page 568). In this disorder, neurons that extend from the substantia nigra to the putamen and caudate nucleus degenerate, causing the disruptions. ■

The basal ganglia have other roles in addition to influencing motor function. They help initiate and terminate some cognitive processes, such as attention, memory, and planning, and may act with the limbic system to regulate emotional behaviors. Some psychiatric disorders, such as obsessive-compulsive disorder, schizophrenia, and chronic anxiety, are thought to involve dysfunction of circuits between the basal ganglia and the limbic system.

The Limbic System

Encircling the upper part of the brain stem and the corpus callosum is a ring of structures on the inner border of the cerebrum and floor of the diencephalon that constitutes the **limbic system** (*limbic* = border). The main components of the limbic system are as follows (Figure 14.14):

- The so-called **limbic lobe** is a rim of cerebral cortex on the medial surface of each hemisphere. It includes the **cingulate gyrus** (*cingul-* = belt), which lies above the corpus callosum, and the **parahippocampal gyrus** (par′-a-hip-ō-KAM-pal), which is in the temporal lobe below. The **hippocampus** (= seahorse) is a portion of the parahippocampal gyrus that extends into the floor of the lateral ventricle.
- The **dentate gyrus** (*dentate* = toothed) lies between the hippocampus and parahippocampal gyrus.
- The **amygdala** (*amygda-* = almond-shaped) is composed of several groups of neurons located close to the tail of the caudate nucleus.
- The **septal nuclei** are located within the septal area formed by the regions under the corpus callosum and the paraterminal gyrus (a cerebral gyrus).
- The **mammillary bodies of the hypothalamus** are two round masses close to the midline near the cerebral peduncles.
- Two nuclei of the thalamus, the **anterior nucleus** and the **medial nucleus,** participate in limbic circuits (see Figure 14.9c, d).
- The **olfactory bulbs** are flattened bodies of the olfactory pathway that rest on the cribriform plate.
- The **fornix, stria terminalis, stria medullaris, medial forebrain bundle,** and **mammillothalamic tract** (mam-i-lō-tha-LAM-ik) are linked by bundles of interconnecting myelinated axons.

The limbic system is sometimes called the "emotional brain" because it plays a primary role in a range of emotions, including pain, pleasure, docility, affection, and anger. It also is involved in olfaction (smell) and memory. Experiments have shown that when different areas of animals' limbic systems are stimulated, the animals' reactions indicate intense pain or extreme pleasure. Stimulation of other limbic system areas in animals produces tameness and signs of affection. Stimulation of a cat's amygdala or certain nuclei of the hypothalamus produces a behavioral pattern called rage—the cat extends its claws, raises its tail, opens its eyes wide, hisses, and spits. By contrast, removal of the amygdala produces an animal that lacks both fear and aggression. A person whose amygdala is damaged fails to recognize fearful expressions in others or to express fear in appropriate situations.

The hippocampus, together with other parts of the cerebrum, functions in memory. People with damage to certain limbic

Figure 14.14 Components of the limbic system and surrounding structures.

The limbic system governs emotional aspects of behavior.

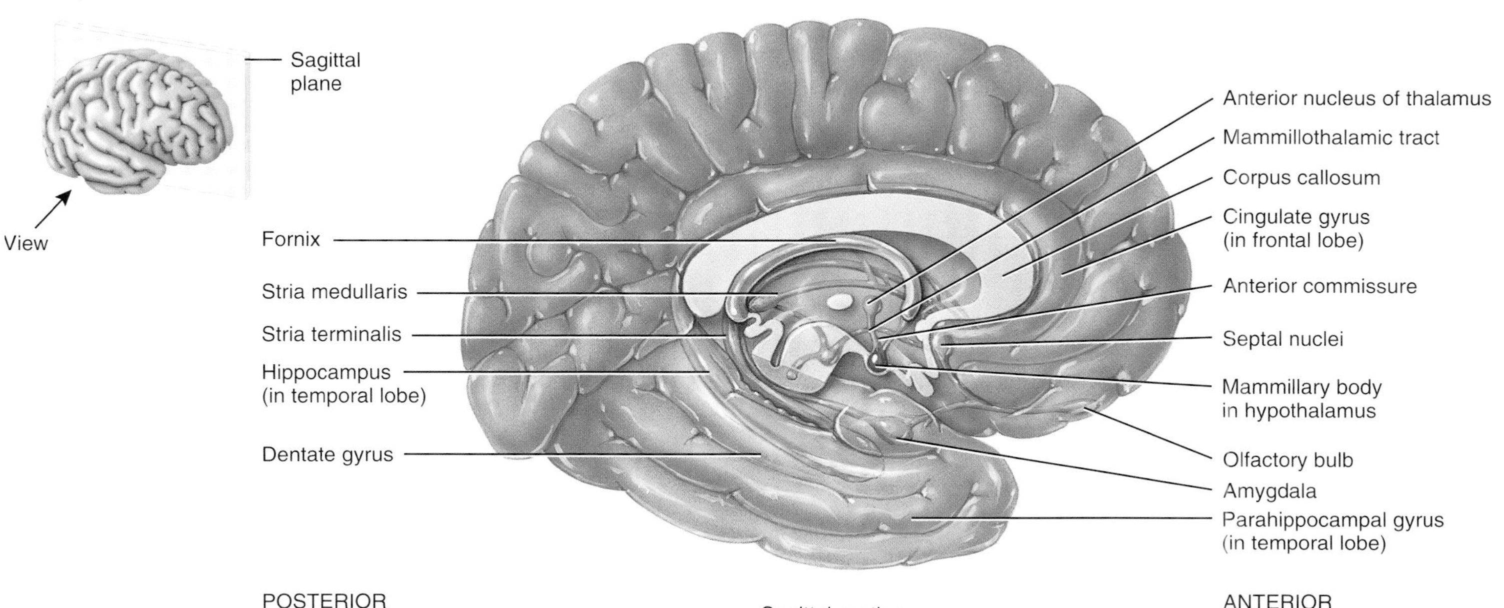

Which part of the limbic system functions with the cerebrum in memory?

system structures forget recent events and cannot commit anything to memory.

The functions of the cerebrum are summarized in Table 14.2.

Brain Injuries

Brain injuries are commonly associated with head trauma and result in part from displacement and distortion of neural tissue at the moment of impact. Additional tissue damage occurs when normal blood flow is restored after a period of ischemia (reduced blood flow). The sudden increase in oxygen level produces large numbers of oxygen free radicals (charged oxygen molecules with an unpaired electron). Brain cells recovering from the effects of a stroke or cardiac arrest also release free radicals. Free radicals cause damage by disrupting cellular DNA and enzymes and by altering plasma membrane permeability. Brain injuries can also result from hypoxia (oxygen deprivation).

Various degrees of brain injury are described by specific terms. A **concussion** is an injury characterized by an abrupt, but temporary, loss of consciousness (from seconds to hours), disturbances of vision, and problems with equilibrium. It is caused by a blow to the head or the sudden stopping of a moving head (as in an automobile accident) and is the most common brain injury. A concussion produces no obvious bruising of the brain. Signs of a concussion are headache, drowsiness, nausea and/or vomiting, lack of concentration, confusion, or post-traumatic amnesia (memory loss).

A **contusion** is bruising of the brain due to trauma and includes the leakage of blood from microscopic vessels. It is usually associated with a concussion. In a contusion, the pia mater may be torn, allowing blood to enter the subarachnoid space. The area most commonly affected is the frontal lobe. A contusion usually results in an immediate loss of consciousness (generally lasting no longer than 5 minutes), loss of reflexes, transient cessation of respiration, and decreased blood pressure. Vital signs typically stabilize in a few seconds.

A **laceration** is a tear of the brain, usually from a skull fracture or a gunshot wound. A laceration results in rupture of large blood vessels, with bleeding into the brain and subarachnoid space. Consequences include cerebral hematoma (localized pool of blood, usually clotted, that swells against the brain tissue), edema, and increased intracranial pressure. If the blood clot is small enough, it may pose no major threat and may be absorbed. If the blood clot is large it may require surgical evacuation. Swelling infringes on the limited space that the brain occupies in the cranial cavity. Swelling causes excruciating headaches. Brain tissue can become necrotic (die) due to the swelling; if the swelling is severe enough, the brain can herniate through the foramen magnum, resulting in death. ■

▶ **CHECKPOINT**

14. Describe the cortex, convolutions, fissures, and sulci of the cerebrum.

15. List and locate the lobes of the cerebrum. How are they separated from one another? What is the insula?

16. Describe the organization of cerebral white matter and indicate the function of each major group of fibers.

17. Give the name and function of each of the nuclei that form the basal ganglia, and describe the effects of damage to the basal ganglia.

18. Define the limbic system and list several of its functions.

FUNCTIONAL ORGANIZATION OF THE CEREBRAL CORTEX

▶ **OBJECTIVES**

Describe the locations and functions of the sensory, association, and motor areas of the cerebral cortex.

Explain the significance of hemispheric lateralization.

Define brain waves and indicate their significance.

Specific types of sensory, motor, and integrative signals are processed in certain regions of the cerebral cortex (Figure 14.15). Generally, **sensory areas** receive sensory information and are involved in **perception,** the conscious awareness of a sensation; **motor areas** initiate movements; and **association areas** deal with more complex integrative functions such as memory, emotions, reasoning, will, judgment, personality traits, and intelligence.

Sensory Areas

Sensory information arrives mainly in the posterior half of both cerebral hemispheres, in regions behind the central sulci. In the cortex, primary sensory areas have the most direct connections with peripheral sensory receptors.

Secondary sensory areas and sensory association areas often are adjacent to the primary areas. They usually receive input both from the primary areas and from other brain regions. Secondary sensory areas and sensory association areas integrate sensory experiences to generate meaningful patterns of recognition and awareness. Whereas a person with damage in the *primary* visual area would be blind in at least part of his visual field, a person with damage to a visual *association* area might see normally yet be unable to recognize her best friend.

The following are some important sensory areas (Figure 14.15; the significance of the numbers in parentheses is explained in the figure caption):

- The **primary somatosensory area** (areas 1, 2, and 3) is located directly posterior to the central sulcus of each cerebral hemisphere in the postcentral gyrus of each parietal lobe. It extends from the lateral cerebral sulcus, along the lateral surface of the parietal lobe to the longitudinal fissure, and then along the medial surface of the parietal lobe within the longitudinal fissure.
- The primary somatosensory area receives nerve impulses for touch, *proprioception* (joint and muscle position), pain,

itching, tickle, and temperature and is involved in the perception of these sensations. A "map" of the entire body is present in the primary somatosensory area: Each point within the area receives impulses from a specific part of the body (see Figure 16.6a on page 558). The size of the cortical area receiving impulses from a particular part of the body depends on the number of receptors present there rather than on the size of the body part. For example, a larger region of the somatosensory area receives impulses from the lips and fingertips than from the thorax or hip. The primary somatosensory area allows you to pinpoint where sensations originate, so that you know exactly where on your body to swat that mosquito.

- The **primary visual area** (area 17), located at the posterior tip of the occipital lobe mainly on the medial surface (next to the longitudinal fissure), receives visual information and is involved in visual perception.
- The **primary auditory area** (areas 41 and 42), located in the superior part of the temporal lobe near the lateral cerebral sulcus, receives information for sound and is involved in auditory perception.
- The **primary gustatory area** (area 43), located at the base of the postcentral gyrus superior to the lateral cerebral sulcus in the parietal cortex, receives impulses for taste and is involved in gustatory perception.
- The **primary olfactory area** (area 28), located in the temporal lobe on the medial aspect (and thus not visible in Figure 14.15), receives impulses for smell and is involved in olfactory perception.

Motor Areas

Motor output from the cerebral cortex flows mainly from the anterior part of each hemisphere. Among the most important motor areas are the following (Figure 14.15):

- The **primary motor area** (area 4) is located in the precentral gyrus of the frontal lobe. Each region in the primary motor area controls voluntary contractions of specific muscles or groups of muscles (see Figure 16.6b on page 558). Electrical stimulation of any point in the primary motor area causes contraction of specific skeletal muscle fibers on

Figure 14.15 Functional areas of the cerebrum. Broca's speech area and Wernicke's area are in the left cerebral hemisphere of most people; they are shown here to indicate their relative locations. The numbers, still used today, are from K. Brodmann's map of the cerebral cortex, first published in 1909.

Particular areas of the cerebral cortex process sensory, motor, and integrative signals.

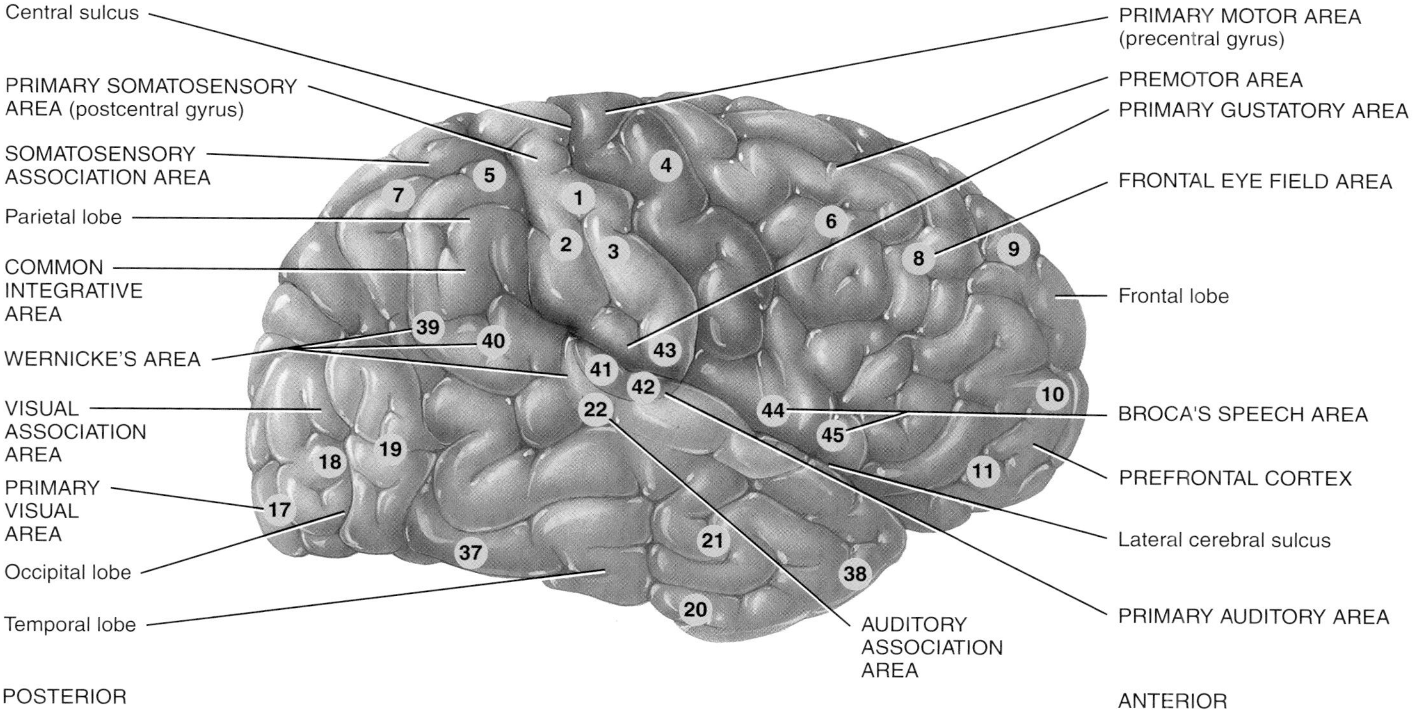

Lateral view of right cerebral hemisphere

What area(s) of the cerebrum integrate(s) interpretation of visual, auditory, and somatic sensations? Translates thoughts into speech? Controls skilled muscular movements? Interprets sensations related to taste? Interprets pitch and rhythm? Interprets shape, color, and movement of objects? Controls voluntary scanning movements of the eyes?

the opposite side of the body. As is true for the primary somatosensory area, body parts do not "map" to the primary motor area in proportion to their size. More cortical area is devoted to those muscles involved in skilled, complex, or delicate movement. For instance, the cortical region devoted to muscles that move the fingers is much larger than the region for muscles that move the toes.

- **Broca's** (BRŌ-kaz) **speech area** (areas 44 and 45), located in the frontal lobe close to the lateral cerebral sulcus, is involved in the articulation of speech. In most people, Broca's speech area is localized in the *left* cerebral hemisphere. Neural circuits established between Broca's speech area, the premotor area, and primary motor area activate muscles of the larynx, pharynx, and mouth and breathing muscles. The coordinated contractions of your speech and breathing muscles enable you to speak your thoughts. People who suffer a cerebrovascular accident (CVA) or stroke in this area can still have clear thoughts, but are unable to form words (nonfluent aphasia; see the clinical application on page 499).

Association Areas

The association areas of the cerebrum consist of some motor and sensory areas, plus large areas on the lateral surfaces of the occipital, parietal, and temporal lobes and on the frontal lobes anterior to the motor areas. Association areas are connected with one another by association tracts and include the following (Figure 14.15):

- The **somatosensory association area** (areas 5 and 7) is just posterior to and receives input from the primary somatosensory area, as well as from the thalamus and other parts of the brain. This area permits you to determine the exact shape and texture of an object without looking at it, to determine the orientation of one object with respect to another as they are felt, and to sense the relationship of one body part to another. Another role of the somatosensory association area is the storage of memories of past sensory experiences, enabling you to compare current sensations with previous experiences. For example, the somatosensory association area allows you to recognize objects such as a pencil and a paperclip simply by touching them.
- The **prefrontal cortex (frontal association area)** is an extensive area in the anterior portion of the frontal lobe that is well-developed in primates, especially humans (areas 9, 10, 11, and 12; area 12 is not illustrated since it can be seen only in a medial view). This area has numerous connections with other areas of the cerebral cortex, thalamus, hypothalamus, limbic system, and cerebellum. The prefrontal cortex is concerned with the makeup of a person's personality, intellect, complex learning abilities, recall of information, initiative, judgment, foresight, reasoning, conscience, intuition, mood, planning for the future, and development of abstract ideas. A person with bilateral damage to the prefrontal cortices typically becomes rude, inconsiderate, incapable of accepting advice, moody, inattentive, less creative, unable to plan for the future, and incapable of anticipating the consequences of rash or reckless words or behavior.
- The **visual association area** (areas 18 and 19), located in the occipital lobe, receives sensory impulses from the primary visual area and the thalamus. It relates present and past visual experiences and is essential for recognizing and evaluating what is seen. For example, the visual association area allows you to recognize an object such as a spoon simply by looking at it.
- The **auditory association area** (area 22), located inferior and posterior to the primary auditory area in the temporal cortex, allows you to recognize a particular sound as speech, music, or noise.
- **Wernicke's (posterior language) area** (VER-ni-kēz) (area 22, and possibly areas 39 and 40), a broad region in the *left* temporal and parietal lobes, interprets the meaning of speech by recognizing spoken words. It is active as you translate words into thoughts. The regions in the *right* hemisphere that correspond to Broca's and Wernicke's areas in the left hemisphere also contribute to verbal communication by adding emotional content, such as anger or joy, to spoken words. Unlike those who have CVAs in Broca's area, people who suffer strokes in Wernicke's area can still speak, but cannot arrange words in a coherent fashion (fluent aphasia, or "word salad"; see clinical application on page 499).
- The **common integrative area** (areas 5, 7, 39, and 40) is bordered by somatosensory, visual, and auditory association areas. It receives nerve impulses from these areas and from the primary gustatory area, primary olfactory area, the thalamus, and parts of the brain stem. This area integrates sensory interpretations from the association areas and impulses from other areas, allowing the formation of thoughts based on a variety of sensory inputs. It then transmits signals to other parts of the brain for the appropriate response to the sensory signals it has interpreted.
- The **premotor area** (area 6) is a motor association area that is immediately anterior to the primary motor area. Neurons in this area communicate with the primary motor cortex, the sensory association areas in the parietal lobe, the basal ganglia, and the thalamus. The premotor area deals with learned motor activities of a complex and sequential nature. It generates nerve impulses that cause specific groups of muscles to contract in a specific sequence, as when you write your name. The premotor area also serves as a memory bank for such movements.
- The **frontal eye field area** (area 8) in the frontal cortex is sometimes included in the premotor area. It controls voluntary scanning movements of the eyes—like those you just used in reading this sentence.

Aphasia

Much of what we know about language areas comes from studies of patients with language or speech disturbances that have resulted from brain damage. Broca's speech area, Wernicke's area, and other language areas are located in the left cerebral hemisphere of most people, regardless of whether they are left-handed or right-handed. Injury to language areas of the cerebral cortex results in **aphasia** (a-FĀ-zē-a; *a-* = without; *-phasia* = speech), an inability to use or comprehend words. Damage to Broca's speech area results in *nonfluent aphasia,* an inability to properly articulate or form words; people with nonfluent aphasia know what they wish to say but cannot speak. Damage to Wernicke's area, the common integrative area, or auditory association area results in *fluent aphasia,* characterized by faulty understanding of spoken or written words. A person experiencing this type of aphasia may fluently produce strings of words that have no meaning ("word salad"). For example, someone with fluent aphasia might say, "I rang car porch dinner light river pencil." The underlying deficit may be **word deafness** (an inability to understand spoken words), **word blindness** (an inability to understand written words), or both. ■

Hemispheric Lateralization

Although the brain is almost symmetrical on its right and left sides, subtle anatomical differences between the two hemispheres exist. For example, in about two-thirds of the population, the planum temporale, a region of the temporal lobe that includes Wernicke's area, is 50% larger on the left side than on the right side. This asymmetry appears in the human fetus at about 30 weeks of gestation. Physiological differences also exist; although the two hemispheres share performance of many functions, each hemisphere also specializes in performing certain unique functions. This functional asymmetry is termed **hemispheric lateralization.**

In the most obvious example of hemispheric lateralization, the left hemisphere receives somatic sensory signals from and controls muscles on the right side of the body, whereas the right hemisphere receives sensory signals from and controls the left side of the body. In most people the left hemisphere is more important for reasoning, numerical and scientific skills, spoken and written language, and the ability to use and understand sign language. Patients with damage in the left hemisphere, for example, often exhibit aphasia. Conversely, the right hemisphere is more specialized for musical and artistic awareness; spatial and pattern perception; recognition of faces and emotional content of language; discrimination of different smells; and generating mental images of sight, sound, touch, taste, and smell to compare relationships among them. Patients with damage in right hemisphere regions that correspond to Broca's and Wernicke's areas in the left hemisphere speak in a monotonous voice, having lost the ability to impart emotional inflection to what they say.

An example of hemispheric lateralization is shown in Figure 14.16. The photograph shows superimposed MRI and PET images, averaged from six men and five women, as several pleasant scents were presented to both nostrils. The MRI image depicts brain regions whereas the PET image demonstrates areas of increased blood flow, corresponding to areas of increased neuronal activity. In both hemispheres, an area termed the *piriform cortex,* near the junction of the frontal and temporal lobes, "lights up" in the PET images. This cortical area receives input from neurons in the olfactory bulb on the same side. In addition, activation occurs mainly on the right side in a region termed the *orbitofrontal cortex,* corresponding roughly to Brodmann's area 11 (see Figure 14.15). Indeed, people who suffer damage to the right orbitofrontal cortex have great difficulty identifying odors and discriminating among different odors.

Despite some dramatic differences in functions of the two hemispheres, there is considerable variation from one person to another. Also, lateralization seems less pronounced in females than in males, both for language (left hemisphere) and for visual and spatial skills (right hemisphere). For instance, females are less likely than males to suffer aphasia after damage to the left hemisphere. A possibly related observation is that the anterior commissure is 12% larger and the corpus callosum has a broader posterior portion in females. Recall that both the anterior commissure and the corpus callosum are commissural tracts that provide communication between the two hemispheres.

Figure 14.16 **Cortical areas activated by olfactory stimuli.** This is a composite picture in which multiple MRI (magnetic resonance image) and PET (positron emission tomography) scans were superimposed. The MRI scans reveal brain tissue and an outline of the skull whereas the PET scans "light up" regions of increased blood flow, an indicator of increased activity of neurons.

Hemispheric lateralization means that each cerebral hemisphere performs unique functions.

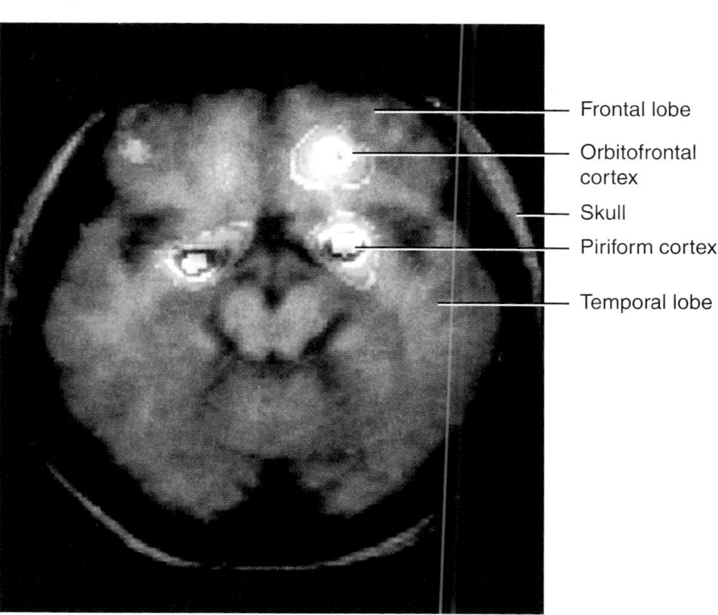

Transverse section through head

Which area of the cerebrum contains neurons that demonstrate hemispheric lateralization for olfaction?

TABLE 14.3 Functional Differences Between the Two Cerebral Hemispheres

Left Hemisphere Functions	Right Hemisphere Functions
Receives somatic sensory signals from and controls muscles on right side of body.	Receives somatic sensory signals from and controls muscles on left side of body.
Reasoning.	Musical and artistic awareness.
Numerical and scientific skills.	Space and pattern perception.
Ability to use and understand sign language.	Recognition of faces and emotional content of facial expressions.
Spoken and written language.	Generating emotional content of language.
	Generating mental images to
	Identifying and discriminating among odors.

Table 14.3 summarizes some of the distinctive functions that are more likely to exhibit hemispheric lateralization.

Brain Waves

At any instant, brain neurons are generating millions of nerve impulses (action potentials). Taken together, these electrical signals are called **brain waves.** Brain waves generated by neurons close to the brain surface, mainly neurons in the cerebral cortex, can be detected by sensors called electrodes placed on the forehead and scalp. A record of such waves is called an **electroencephalogram** (e-lek′-trō-en-SEF-a-lō-gram; *electro-* = electricity; *-gram* = recording) or **EEG.** Electroencephalograms are useful both in studying normal brain functions, such as changes that occur during sleep, and in diagnosing a variety of brain disorders, such as epilepsy, tumors, trauma, hematomas, metabolic abnormalities, sites of trauma, and degenerative diseases. The EEG is also utilized to determine if "life" is present, that is, to establish or confirm that brain death has occurred.

Figure 14.17 Types of brain waves recorded in an electroencephalogram (EEG).

Brain waves indicate electrical activity of the cerebral cortex.

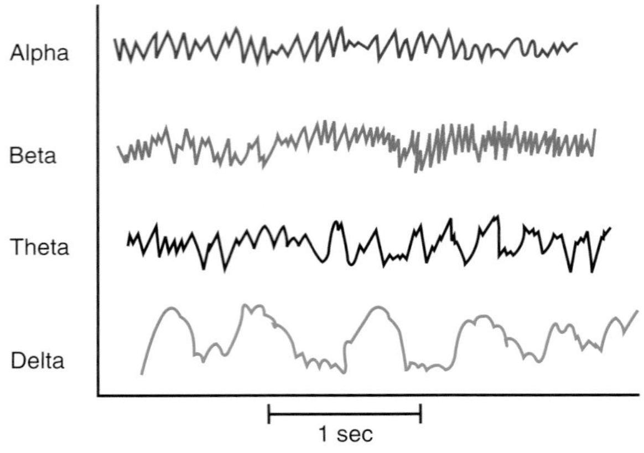

Which type of brain wave indicates emotional stress?

Patterns of activation of brain neurons produce four types of brain waves (Figure 14.17):

1. **Alpha waves.** These rhythmic waves occur at a frequency of about 8–13 cycles per second. (The unit commonly used to express frequency is the hertz [Hz]. One hertz is one cycle per second.) Alpha waves are present in the EEGs of nearly all normal individuals when they are awake and resting with their eyes closed. These waves disappear entirely during sleep.
2. **Beta waves.** The frequency of these waves is between 14 and 30 Hz. Beta waves generally appear when the nervous system is active—that is, during periods of sensory input and mental activity.
3. **Theta waves.** These waves have frequencies of 4–7 Hz. Theta waves normally occur in children and adults experiencing emotional stress. They also occur in many disorders of the brain.
4. **Delta waves.** The frequency of these waves is 1–5 Hz. Delta waves occur during deep sleep in adults, but they are normal in awake infants. When produced by an awake adult, they indicate brain damage.

▶ **CHECKPOINT**

19. Compare the functions of the sensory, motor, and association areas of the cerebral cortex.
20. What is hemispheric lateralization?
21. What is the diagnostic value of an EEG?

CRANIAL NERVES

▶ **OBJECTIVE**

Identify the cranial nerves by name, number, and type, and give the functions of each.

The 12 pairs of **cranial nerves** are so-named because they pass through various foramina in the bones of the cranium. Like the 31 pairs of spinal nerves, they are part of the peripheral nervous system (PNS). Each cranial nerve has both a number, designated by a roman numeral, and a name (see Figure 14.5). The numbers indicate the order, from anterior to posterior, in which the nerves arise from the brain. The names designate a nerve's distribution or function.

Cranial nerves emerge from the nose (cranial nerve I), the eyes (cranial nerve II), the inner ear (cranial nerve VIII), the brain stem (cranial nerves III–XII), and the spinal cord (part of cranial nerve XI). Two cranial nerves (cranial nerves I and II) contain only sensory axons and thus are called **sensory nerves.** The rest are classified as **mixed nerves** because they contain axons of both sensory and motor neurons. Cranial nerves III, IV, VI, XI, and XII are mainly motor. They contain a few sensory axons from muscle proprioceptors, but most of their axons are motor neurons that innervate skeletal muscles. Cranial nerves III, VII, IX, and X include both somatic and autonomic motor axons. The somatic axons innervate skeletal muscles; the autonomic axons, which are part of the parasympathetic division, innervate glands, smooth muscle, and cardiac muscle. Although the cranial nerves are mentioned singly in the following descriptions of their type, location, and function, remember that they are paired structures. The cell bodies of sensory neurons are located in ganglia outside the brain; the cell bodies of motor neurons lie in nuclei within the brain.

Olfactory (I) Nerve

The **olfactory (I) nerve** (ol-FAK-tō-rē; *olfact-* = to smell) is entirely sensory; it contains axons that conduct nerve impulses for olfaction, the sense of smell (Figure 14.18). The olfactory epithelium occupies the superior part of the nasal cavity, covering the inferior surface of the cribriform plate and extending down along the superior nasal concha. The olfactory receptors within the olfactory epithelium are bipolar neurons. Each has a single odor-sensitive dendrite projecting from one side of the cell body and an unmyelinated axon extending from the other side. Bundles of axons of olfactory receptors extend through about 20 olfactory foramina in the cribriform plate of the ethmoid bone on each side of the nose. These 40 or so bundles of axons collectively form the right and left olfactory nerves.

Olfactory nerves end in the brain in paired masses of gray matter called the **olfactory bulbs,** two extensions of the brain that rest on the cribriform plate. Within the olfactory bulbs, the axon terminals of olfactory receptors form synapses with the dendrites and cell bodies of the next neurons in the olfactory pathway. The axons of these neurons make up the **olfactory tracts,** which extend posteriorly from the olfactory bulbs (see

Figure 14.18 Olfactory (I) nerve.

The olfactory epithelium is located on the inferior surface of the cribriform plate and superior nasal conchae.

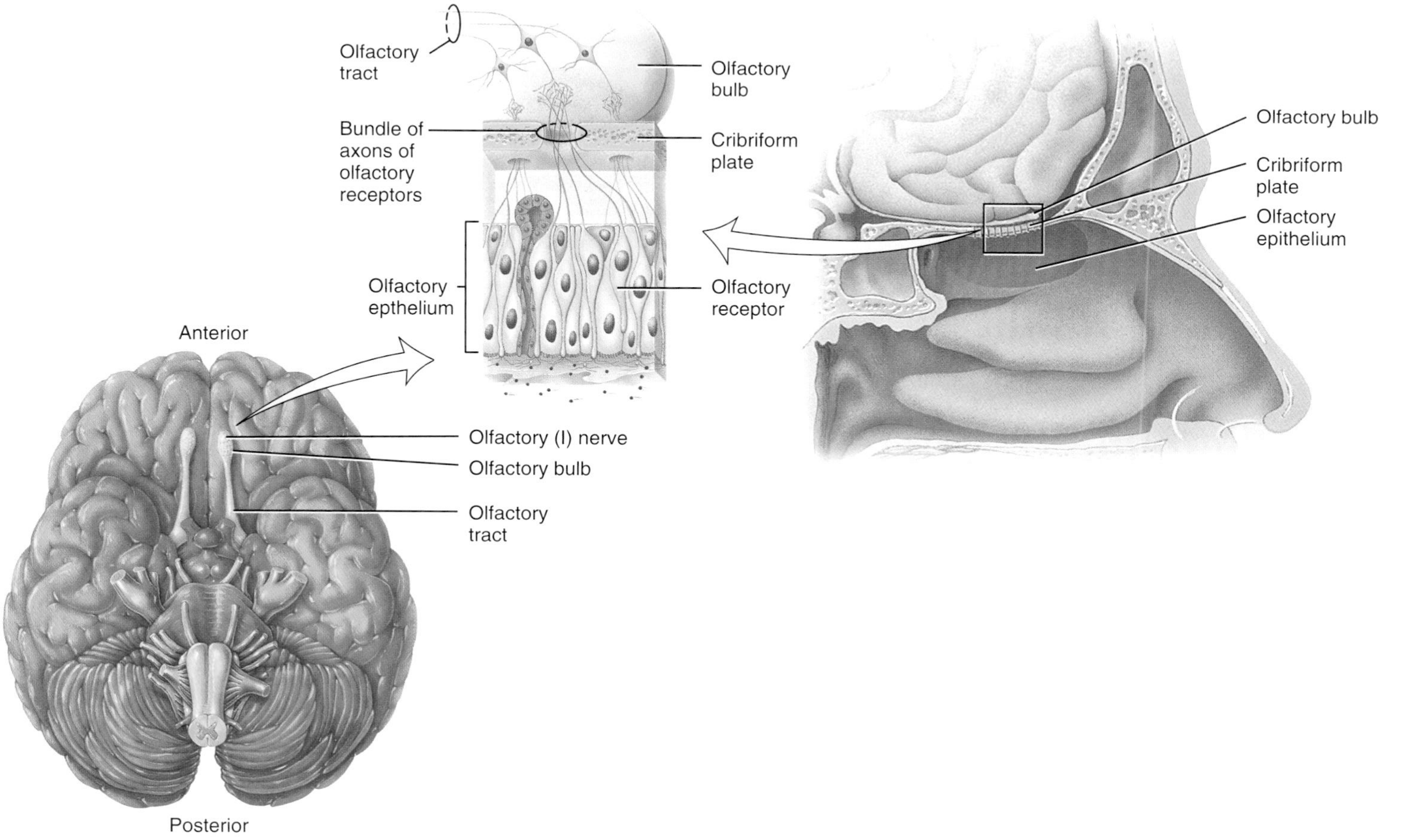

Where do axons in the olfactory tracts terminate?

Figure 14.5). Axons in the olfactory tracts end in the primary olfactory area in the temporal lobe of the cerebral cortex.

Optic (II) Nerve

The **optic (II) nerve** (OP-tik; *opti-* = the eye, vision) is entirely sensory; it contains axons that conduct nerve impulses for vision (Figure 14.19). In the retina, rods and cones initiate visual signals and relay them to bipolar cells, which transmit the signals to ganglion cells. Axons of all the ganglion cells in the retina of each eye join to form an optic nerve, which passes through the optic foramen. About 10 mm (0.4 in.) posterior to the eyeball, the two optic nerves merge to form the **optic chiasm** (KĪ-azm = a crossover, as in the letter X). Within the chiasm, axons from the medial half of each eye cross to the opposite side; axons from the lateral half remain on the same side. Posterior to the

Figure 14.19 Optic (II) nerve.

In sequence, visual signals are relayed from rods and cones to bipolar cells to ganglion cells.

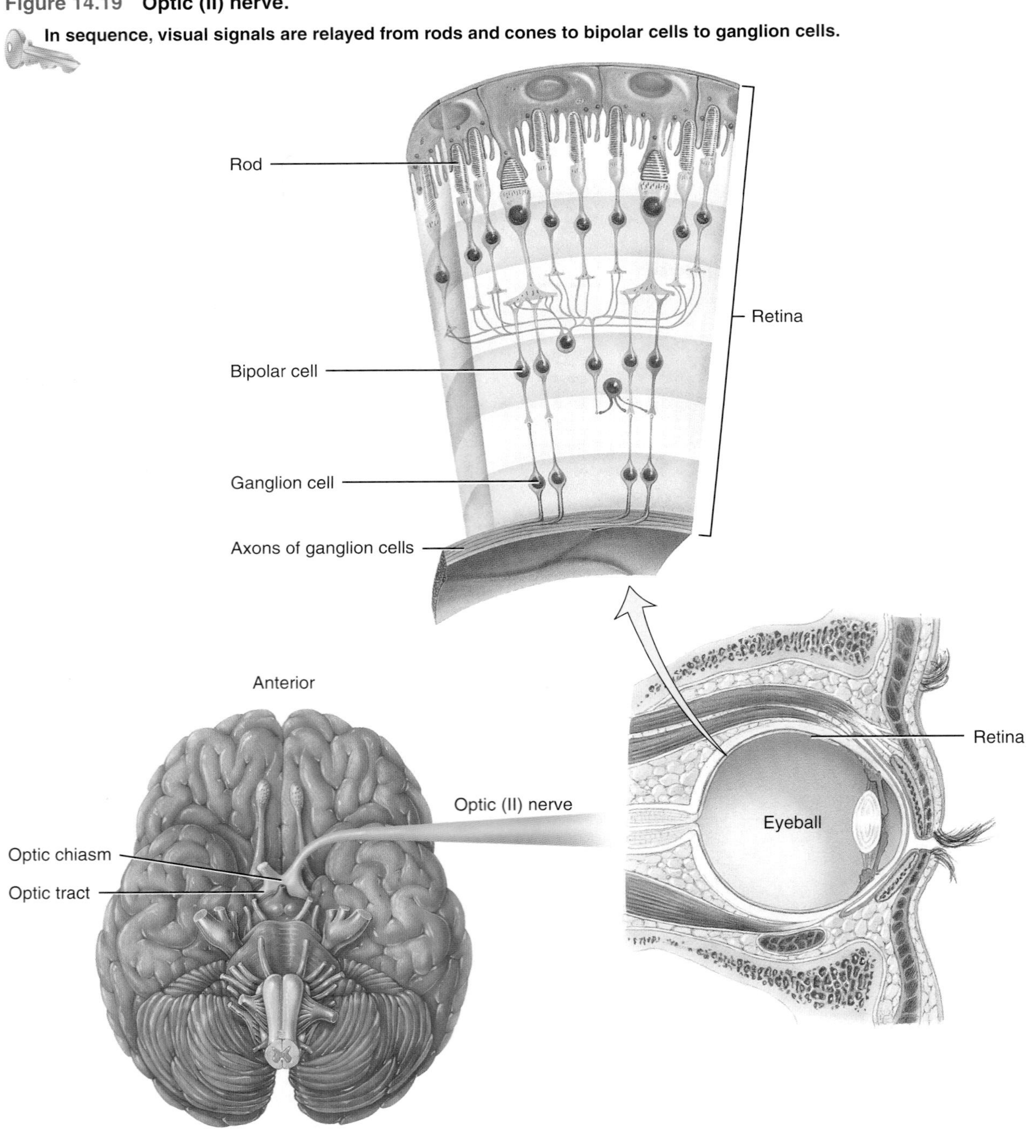

Where do most axons in the optic tracts terminate?

chiasm, the regrouped axons, some from each eye, form the **optic tracts.** Most axons in the optic tracts end in the lateral geniculate nucleus of the thalamus. There they synapse with neurons whose axons extend to the primary visual area in the occipital lobe of the cerebral cortex (area 17 in Figure 14.15). A few axons pass through the optic chiasm and then extend to the superior colliculi of the midbrain. They synapse with motor neurons that control the extrinsic and intrinsic eye muscles.

Oculomotor (III) Nerve

The **oculomotor (III) nerve** (ok′-ū-lō-MŌ-tor; *oculo-* = eye; *-motor* = a mover) is a mixed but mainly motor cranial nerve. Its motor nucleus is in the ventral part of the midbrain (Figure 14.20a). The oculomotor nerve extends anteriorly and divides into superior and inferior branches, both of which pass through the superior orbital fissure into the orbit. Axons in the superior branch innervate the superior rectus (an extrinsic

Figure 14.20 Oculomotor (III), trochlear (IV), and abducens (VI) nerves.

The oculomotor nerve has the widest distribution among extrinsic eye muscles.

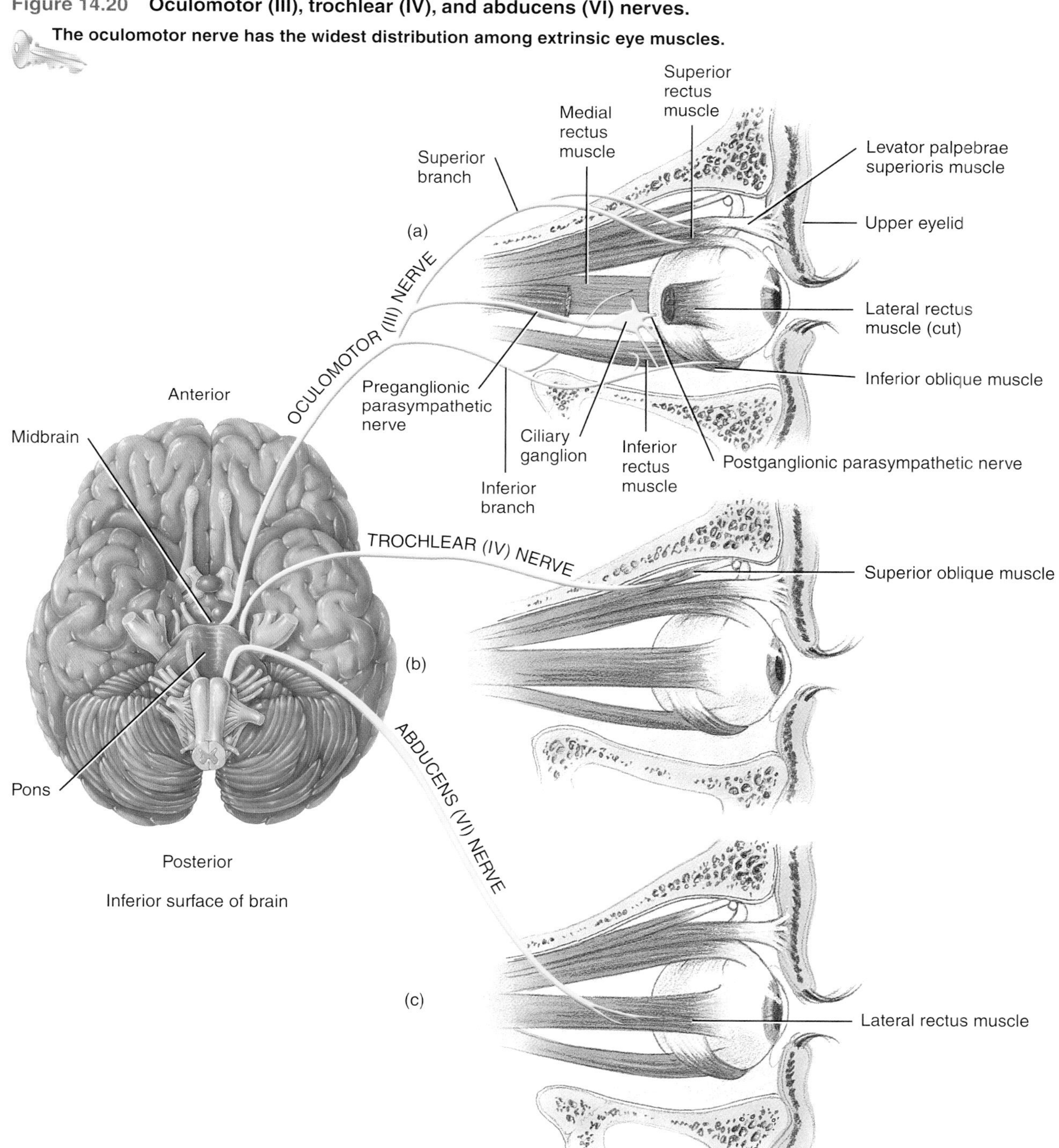

Which branch of the oculomotor nerve is distributed to the superior rectus muscle? Which is the smallest cranial nerve?

eyeball muscle) and the levator palpebrae superioris (the muscle of the upper eyelid). Axons in the inferior branch supply the medial rectus, inferior rectus, and inferior oblique muscles—all extrinsic eyeball muscles. These somatic motor neurons control movements of the eyeball and upper eyelid.

The inferior branch of the oculomotor nerve also provides parasympathetic innervation to intrinsic eyeball muscles, which are smooth muscle. They include the ciliary muscle of the eyeball and the circular muscles (sphincter pupillae) of the iris. Parasympathetic impulses propagate from the oculomotor nucleus in the midbrain to the **ciliary ganglion,** a relay center of the autonomic nervous system. From the ciliary ganglion, parasympathetic axons extend to the ciliary muscle, which adjusts the lens for near vision. Other parasympathetic axons stimulate the circular muscles of the iris to contract when bright light stimulates the eye, causing a decrease in the size of the pupil (constriction).

The sensory portion of the oculomotor nerve consists of afferent axons extending from proprioceptors in the extrinsic eyeball muscles supplied by the nerve to the midbrain. These axons convey nerve impulses for **proprioception,** the nonvisual perception of the movements and position of the body.

Trochlear (IV) Nerve

The **trochlear (IV) nerve** (TRŌK-lē-ar; *trochle-* = a pulley) is a mixed but mainly motor cranial nerve. It is the smallest of the 12 cranial nerves and is the only one that arises from the posterior aspect of the brain stem.

The motor portion originates in a nucleus in the midbrain, and axons from the nucleus pass through the superior orbital fissure of the orbit (Figure 14.20b). These somatic motor axons innervate the superior oblique muscle of the eyeball, another extrinsic eyeball muscle that controls movement of the eyeball.

The sensory portion of the trochlear nerve consists of axons that extend from proprioceptors in the superior oblique muscle to a nucleus of the nerve in the midbrain. Like those of the oculomotor nerve, these axons convey nerve impulses for proprioception.

Trigeminal (V) Nerve

The **trigeminal (V) nerve** (trī-JEM-i-nal = triple, for its three branches), the largest of the cranial nerves, is a mixed cranial nerve. The trigeminal nerve emerges from two roots on the ventrolateral surface of the pons. The large sensory root has a swelling called the **trigeminal ganglion,** which is located in a fossa on the inner surface of the petrous portion of the temporal bone. The ganglion contains cell bodies of most of the primary sensory neurons. The smaller motor root originates in a nucleus in the pons.

As indicated by its name, the trigeminal nerve has three branches: ophthalmic, maxillary, and mandibular (Figure 14.21). The **ophthalmic nerve** (of-THAL-mik; *ophthalm-* = the eye), the smallest branch, enters the orbit via the superior orbital fissure. The **maxillary nerve** (*maxilla* = upper jaw bone) is intermediate in size between the ophthalmic and mandibular nerves and enters the foramen rotundum. The **mandibular nerve** (*mandibula* = lower jaw bone), the largest branch, exits through the foramen ovale.

Sensory axons in the trigeminal nerve carry nerve impulses for touch, pain, and thermal sensations. The ophthalmic nerve contains sensory axons from the skin over the upper eyelid, eyeball, lacrimal glands, upper part of the nasal cavity, side of the nose, forehead, and anterior half of the scalp. The maxillary nerve includes sensory axons from the mucosa of the nose, palate, part of the pharynx, upper teeth, upper lip, and lower eyelid. The mandibular nerve contains sensory axons from the anterior two-thirds of the tongue (not taste), cheek and mucosa deep to it, lower teeth, skin over the mandible and side of the head anterior to the ear, and mucosa of the floor of the mouth. The sensory axons from the three branches enter the semilunar ganglion and terminate in nuclei in the pons. The trigeminal nerve also contains sensory fibers from proprioceptors located in the muscles of mastication.

Somatic motor axons of the trigeminal nerve are part of the mandibular nerve and supply muscles of mastication (masseter, temporalis, medial pterygoid, lateral pterygoid, anterior belly of digastric, and mylohyoid muscles). These motor neurons control chewing movements.

Dental Anesthesia

The inferior alveolar nerve, a branch of the mandibular nerve, supplies all the teeth in one-half of the mandible; it is often anesthetized in dental procedures. The same procedure will anesthetize the lower lip because the mental nerve is a branch of the inferior alveolar nerve. Because the lingual nerve runs very close to the inferior alveolar nerve near the mental foramen, it too is often anesthetized at the same time. For anesthesia to the upper teeth, the superior alveolar nerve endings, which are branches of the maxillary nerve, are blocked by inserting the needle beneath the mucous membrane. The anesthetic solution is then infiltrated slowly throughout the area of the roots of the teeth to be treated. ■

Abducens (VI) Nerve

The **abducens (VI) nerve** (ab-DOO-senz; *ab-* = away; *-ducens* = to lead) is a mixed but mainly motor cranial nerve that originates from a nucleus in the pons (see Figure 14.20c). Somatic motor axons extend from the nucleus to the lateral rectus muscle of the eyeball, an extrinsic eyeball muscle, through the superior orbital fissure of the orbit. The abducens nerve is so named because nerve impulses cause abduction of the eyeball (lateral rotation). The sensory axons extend from proprioceptors in the lateral rectus muscle to the pons.

Figure 14.21 Trigeminal (V) nerve.

The three branches of the trigeminal nerve leave the cranium through the superior orbital fissure, foramen rotundum, and foramen ovale.

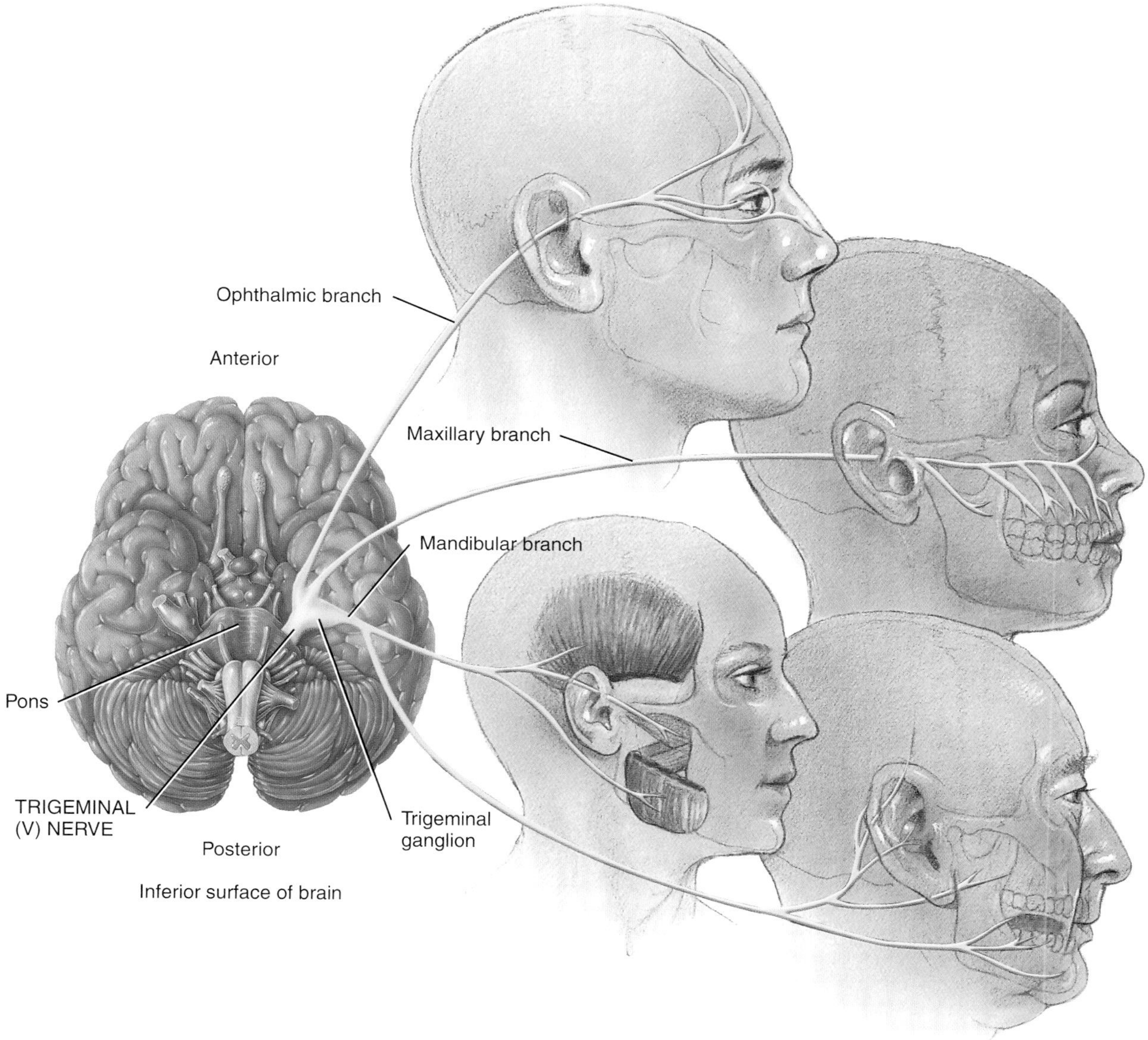

Ophthalmic branch
Anterior
Maxillary branch
Mandibular branch
Pons
TRIGEMINAL (V) NERVE
Posterior
Trigeminal ganglion
Inferior surface of brain

How does the trigeminal nerve compare in size with the other cranial nerves?

Facial (VII) Nerve

The **facial (VII) nerve** (FĀ-shal = face) is a mixed cranial nerve. Its sensory axons extend from the taste buds of the anterior two-thirds of the tongue through the **geniculate ganglion** (je-NIK-ū-lāt), a cluster of cell bodies of sensory neurons that lie beside the facial nerve, and end in the pons (Figure 14.22). The sensory portion of the facial nerve also contains axons from proprioceptors in muscles of the face and scalp.

Axons of somatic motor neurons arise from a nucleus in the pons, enter the petrous portion of the temporal bone, and innervate facial, scalp, and neck muscles. Nerve impulses propagating along these axons cause contraction of the muscles of facial expression plus the stylohyoid muscle and the posterior belly of the digastric muscle.

Axons of parasympathetic neurons that are part of the facial nerve end in two parasympathetic ganglia: the **pterygopalatine** (ter′-i-gō-PAL-a-tīn) **ganglion** and the **submandibular ganglion.** From the two ganglia, other parasympathetic axons extend to lacrimal glands (which secrete tears), nasal glands, palatine glands, and saliva-producing sublingual and submandibular glands.

Vestibulocochlear (VIII) Nerve

The **vestibulocochlear (VIII) nerve** (vest-tib-ū-lō-KOK-lē-ar; *vestibulo-* = small cavity; *-cochlear* = a spiral, snail-like) was formerly known as the **acoustic** or **auditory nerve.** It is a mixed but mainly sensory cranial nerve and has two branches, the vestibular branch and the cochlear branch (Figure 14.23). The **vestibular branch** carries impulses for equilibrium; the **cochlear branch** carries impulses for hearing.

Sensory axons in the vestibular branch arise from the semicircular canals, the saccule, and the utricle of the inner ear; extend to the **vestibular ganglion,** where their cell bodies are located (see Figure 17.19b on page 599); and end in vestibular

Figure 14.22 Facial (VII) nerve.

The facial nerve causes contraction of the muscles of facial expression.

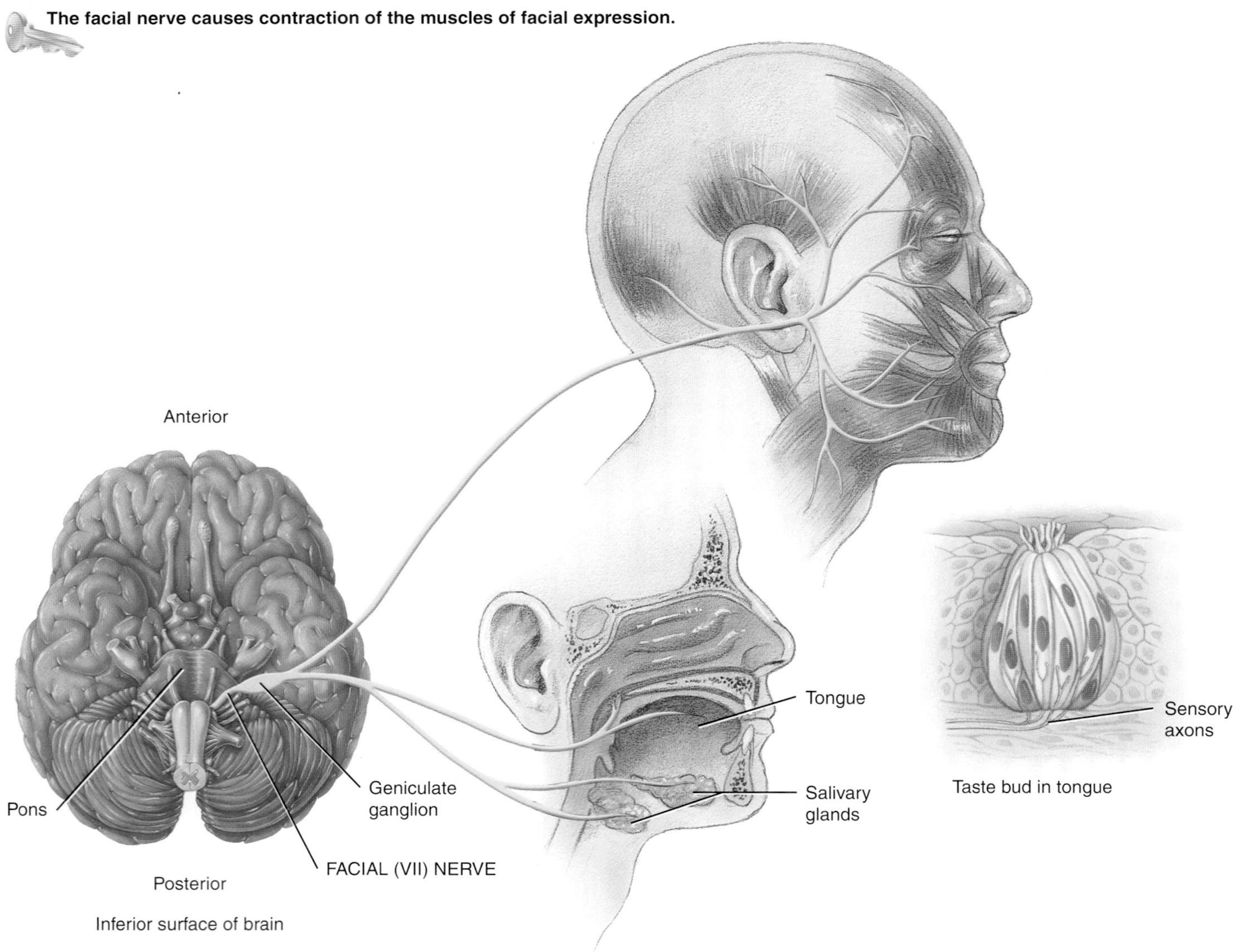

Where do the motor axons of the facial nerve originate?

nuclei in the pons. Some sensory axons also enter the cerebellum via the inferior cerebellar peduncle. Axons of motor neurons in the vestibular branch project from the pons to hair cells of the semicircular canals, saccule, and utricle.

Sensory axons in the cochlear branch arise in the spiral organ (organ of Corti) in the cochlea of the inner ear. The cell bodies of cochlear branch sensory neurons are located in the **spiral ganglion** of the cochlea (see Figure 17.19b). From there, axons extend to nuclei in the medulla oblongata. Axons of motor neurons in the cochlear branch project from the pons to hair cells of the spiral organ.

Glossopharyngeal (IX) Nerve

The **glossopharyngeal (IX) nerve** (glos′-ō-fa-RIN-jē-al; *glosso-* = tongue; *-pharyngeal* = throat) is a mixed cranial nerve. Sensory axons of the glossopharyngeal nerve arise from taste buds and somatic sensory receptors on the posterior one-third of the tongue, from proprioceptors in swallowing muscles supplied by the motor portion, from baroreceptors (stretch receptors) in the carotid sinus, and from chemoreceptors in the carotid body near the carotid arteries (Figure 14.24). The cell bodies of these sensory neurons are located in the superior and inferior ganglia. From the ganglia, sensory axons pass through the jugular foramen and end in the medulla.

Axons of motor neurons in the glossopharyngeal nerve arise in nuclei of the medulla and exit the skull through the jugular foramen. Somatic motor neurons innervate the stylopharyngeus muscle, which elevates the pharynx and larynx, and autonomic motor neurons (parasympathetic) stimulate the parotid gland to secrete saliva. Some of the cell bodies of parasympathetic motor neurons are located in the **otic ganglion.**

Figure 14.23 **Vestibulocochlear (VIII) nerve.**

The vestibular branch carries impulses for equilibrium, while the cochlear branch carries impulses for hearing.

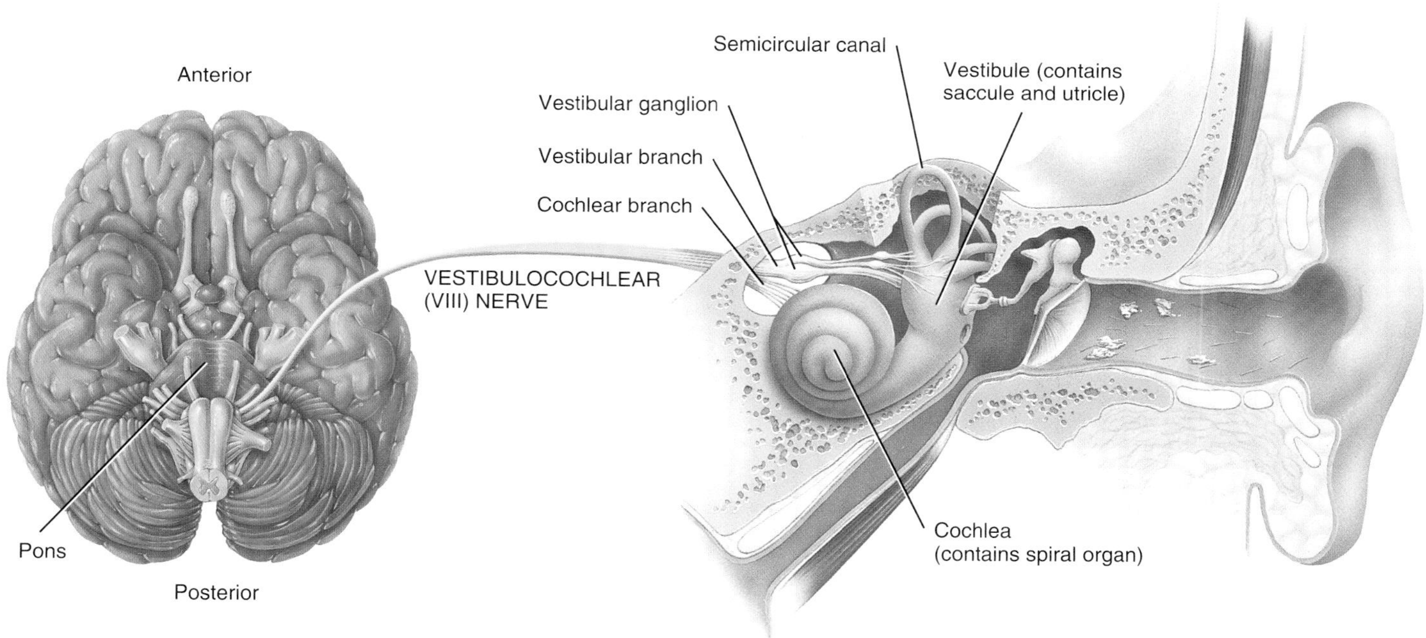

What structures are found in the vestibular and spiral ganglia?

Figure 14.24 Glossopharyngeal (IX) nerve.

Sensory axons in the glossopharyngeal nerve carry signals from the taste buds.

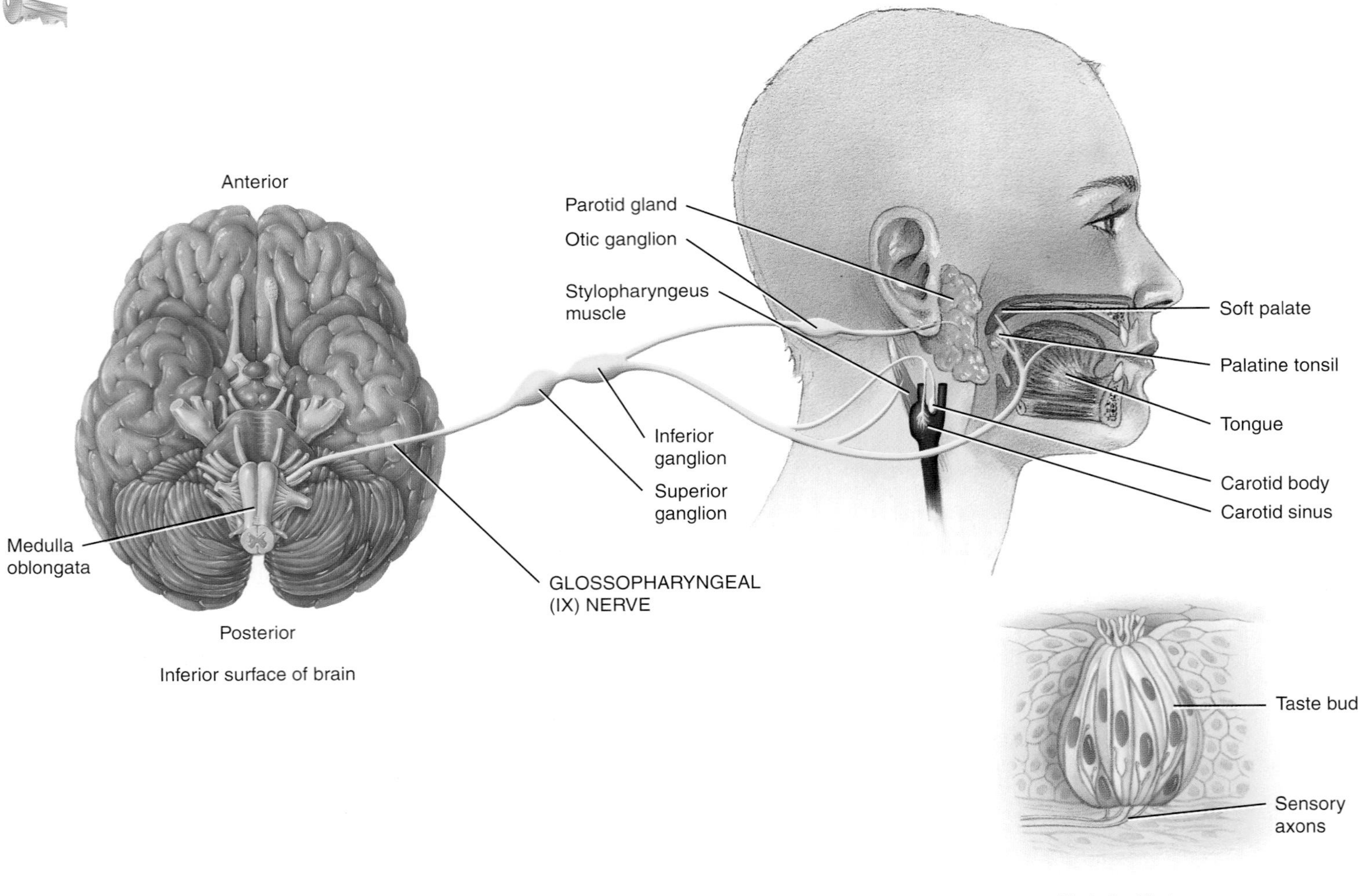

Anterior
Parotid gland
Otic ganglion
Stylopharyngeus muscle
Soft palate
Palatine tonsil
Tongue
Inferior ganglion
Superior ganglion
Carotid body
Carotid sinus
Medulla oblongata
GLOSSOPHARYNGEAL (IX) NERVE
Posterior

Inferior surface of brain

Taste bud
Sensory axons

Taste bud in tongue

Through which foramen does the glossopharyngeal nerve exit the skull?

Vagus (X) Nerve

The **vagus (X) nerve** (VĀ-gus = vagrant or wandering) is a mixed cranial nerve that is distributed from the head and neck into the thorax and abdomen (Figure 14.25). The nerve derives its name from its wide distribution. In the neck, it lies medial and posterior to the internal jugular vein and common carotid artery.

Sensory axons in the vagus nerve arise from the skin of the external ear, a few taste buds in the epiglottis and pharynx, and proprioceptors in muscles of the neck and throat. Also, sensory axons come from baroreceptors (stretch receptors) in the arch of the aorta; chemoreceptors in the aortic bodies near the arch of the aorta; and visceral sensory receptors in most organs of the thoracic and abdominal cavities. These axons pass through the jugular foramen and end in the medulla and pons.

Axons of autonomic motor neurons (parasympathetic) in the vagus nerve originate in nuclei of the medulla and end in the lungs and heart. Vagal parasympathetic axons also supply glands of the gastrointestinal (GI) tract and smooth muscle of the respiratory passageways, esophagus, stomach, gallbladder, small intestine, and most of the large intestine (see Figure 15.3 on page 529).

Accessory (XI) Nerve

The **accessory (XI) nerve** (ak-SES-ō-rē = assisting) is a mixed cranial nerve. It differs from all other cranial nerves

Figure 14.25 **Vagus (X) nerve.**

The vagus nerve is widely distributed in the head, neck, thorax, and abdomen.

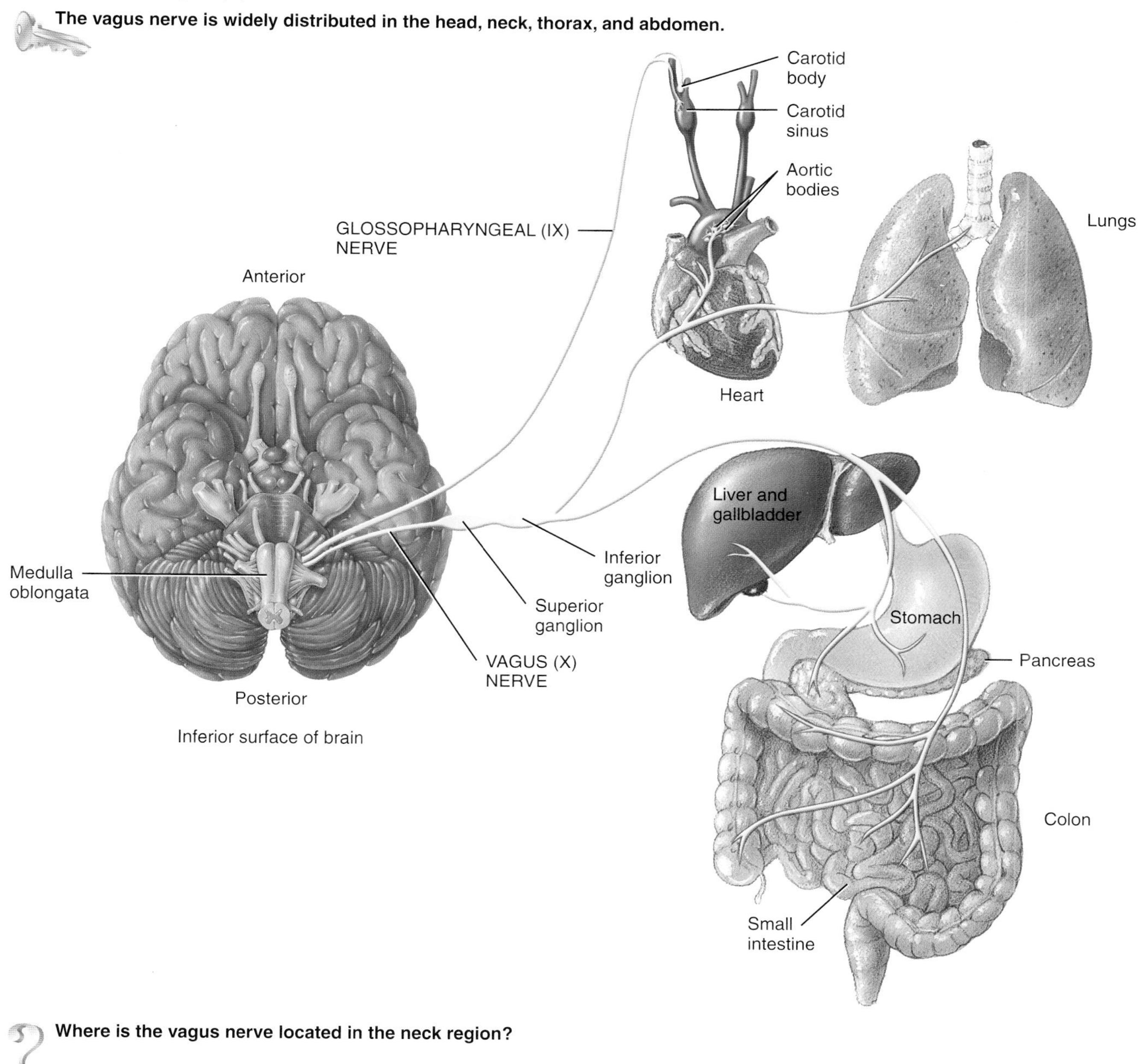

Where is the vagus nerve located in the neck region?

because it originates from *both* the brain stem and the spinal cord (Figure 14.26). The **cranial root** is motor and arises from nuclei in the medulla oblongata, passes through the jugular foramen, and supplies the voluntary muscles of the pharynx, larynx, and soft palate that are used in swallowing. The **spinal root** is mixed but mainly motor. Its motor axons arise in the anterior gray horn of the first five segments of the cervical portion of the spinal cord. The axons from the segments come together, pass through the foramen magnum, and then exit through the jugular foramen along with axons in the cranial root. The spinal root conveys motor impulses to the sternocleidomastoid and trapezius muscles to coordinate head movements. Sensory axons in the spinal root originate from proprioceptors in the muscles supplied by its motor neurons and end in the medulla oblongata.

Hypoglossal (XII) Nerve

The **hypoglossal (XII) nerve** (hī′-pō-GLOS-al; *hypo-* = below; *-glossal* = tongue) is a mixed cranial nerve. The sensory portion of the hypoglossal nerve consists of axons that originate from proprioceptors in the tongue muscles and end in

Figure 14.26 Accessory (XI) nerve.

The accessory nerve exits the cranium through the jugular foramen.

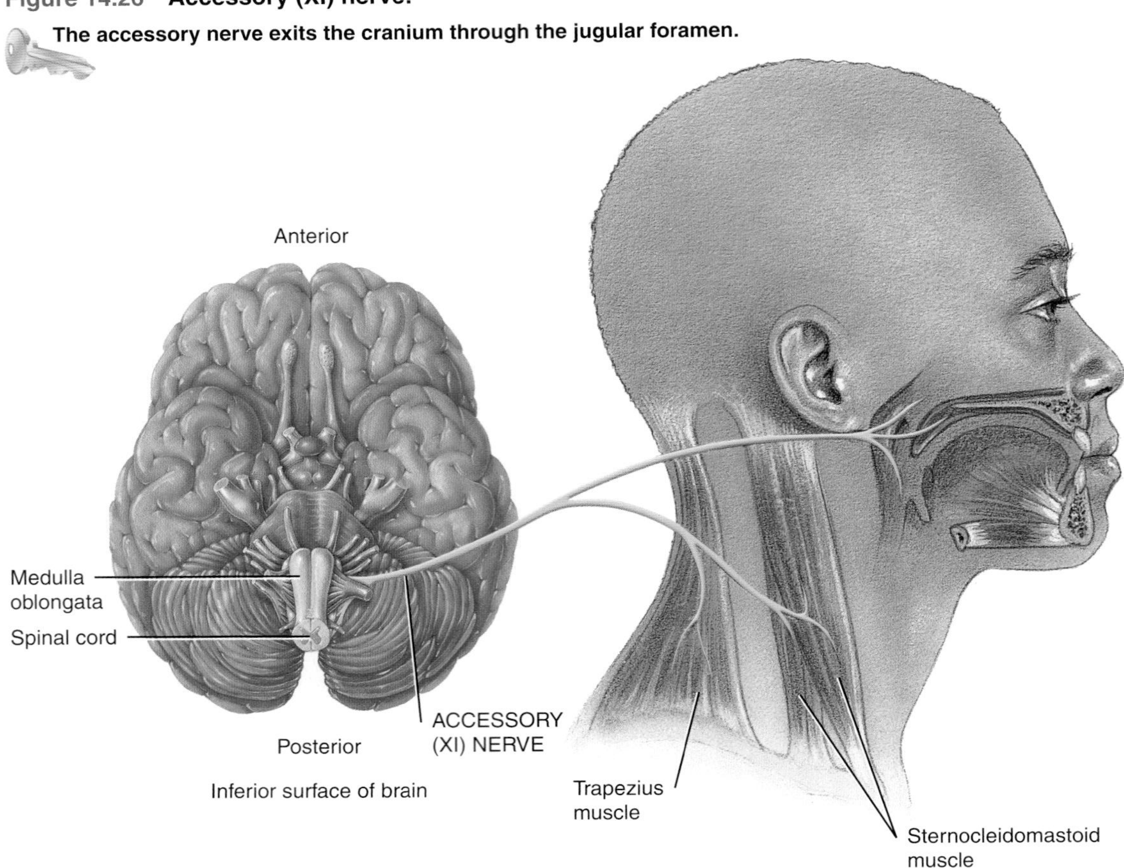

How does the accessory nerve differ from the other cranial nerves?

Figure 14.27 Hypoglossal (XII) nerve.

The hypoglossal nerve exits the cranium through the hypoglossal canal.

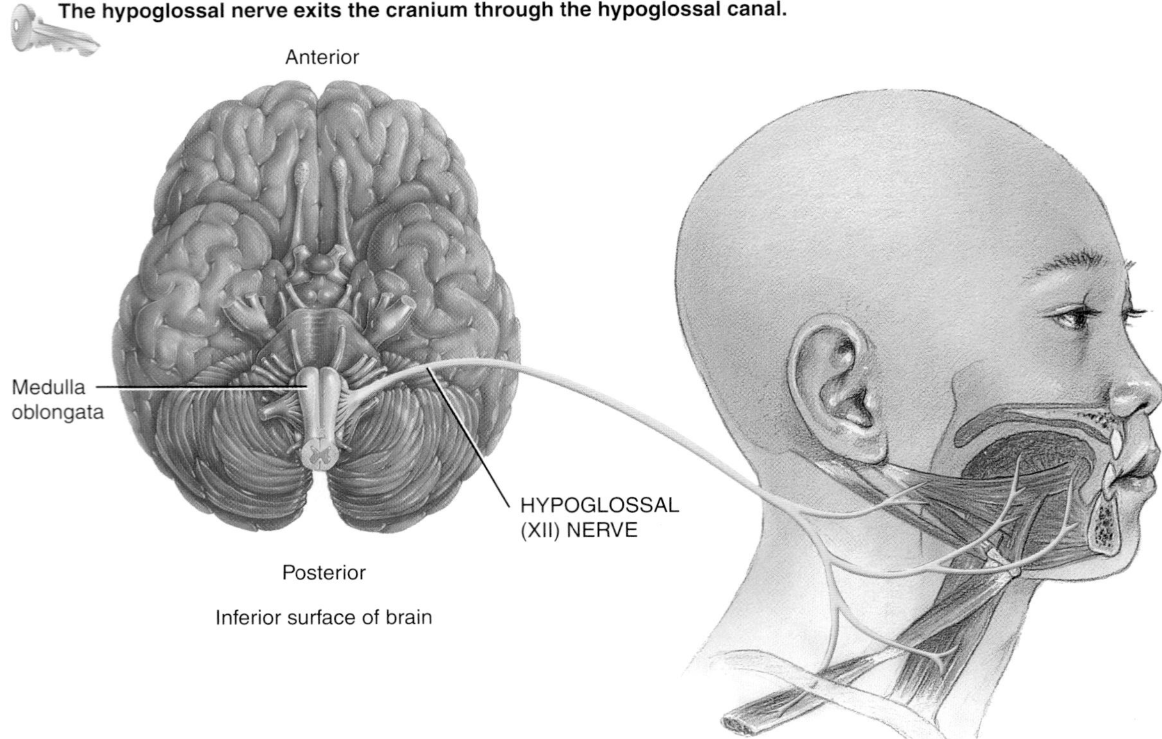

What important motor functions does the hypoglossal nerve mediate?

the medulla oblongata (Figure 14.27 on page 510). The sensory fibers conduct nerve impulses for proprioception. The somatic motor axons originate in a nucleus in the medulla oblongata, pass through the hypoglossal canal, and supply the muscles of the tongue. These axons conduct nerve impulses for speech and swallowing.

Table 14.4 presents a summary of cranial nerves, including clinical applications related to their dysfunctions.

▶ CHECKPOINT

22. How are cranial nerves named and numbered?
23. What is the difference between a mixed cranial nerve and a sensory cranial nerve?
24. What sort of test could reveal damage to each of the 12 cranial nerves?

TABLE 14.4 Summary of Cranial Nerves

Number and Name*	Type and Location	Function and Clinical Application
Olfactory (I) nerve Olfactory bulb Olfactory nerve Olfactory tract	**Sensory** Arises in olfactory mucosa, passes through foramina in the cribriform plate of the ethmoid bone, and ends in the olfactory bulb. The olfactory tract extends via two pathways to olfactory areas of cerebral cortex.	*Function:* Smell. *Clinical application:* Loss of the sense of smell, called *anosmia* (an-OZ-mē-a), may result from head injuries in which the cribriform plate of the ethmoid bone is fractured or from lesions along the olfactory pathway.
Optic (II) nerve 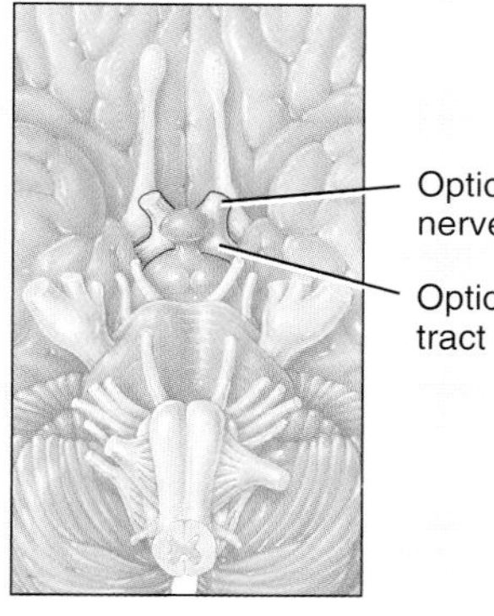	**Sensory** Arises in the retina of the eye, passes through the optic foramen, forms the optic chiasm and then the optic tracts, and terminates in the lateral geniculate nuclei of thalamus. From the thalamus, axons extend to the primary visual area (area 17) of the cerebral cortex.	*Function:* Vision *Clinical application:* Fractures in the orbit, damage along the visual pathway, and diseases of the nervous system may result in visual field defects and loss of visual acuity. Blindness due to a defect in or loss of one or both eyes is called *anopia.*
Oculomotor (III) nerve 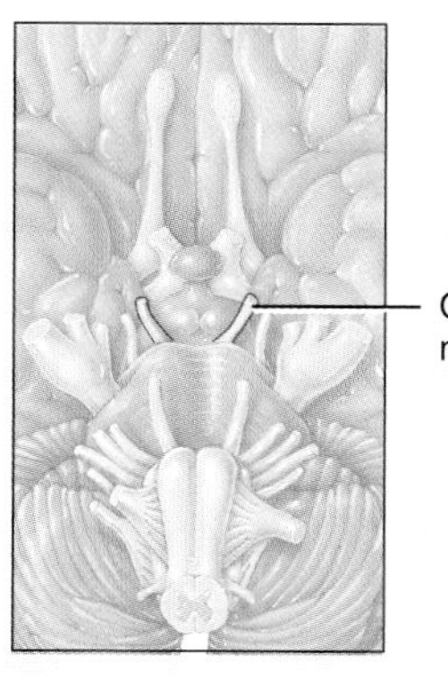	**Mixed (mainly motor)** *Sensory portion:* Consists of axons from proprioceptors in eyeball muscles that pass through the superior orbital fissure and terminate in the midbrain. *Motor portion:* Originates in the midbrain and passes through the superior orbital fissure. Axons of somatic motor neurons innervate the levator palpebrae superioris muscle of the upper eyelid and four extrinsic eyeball muscles (superior rectus, medial rectus, inferior rectus, and inferior oblique). Parasympathetic axons innervate the ciliary muscle of the eyeball and the circular muscles (sphincter pupillae) of the iris.	*Sensory function:* Proprioception. *Somatic motor function:* Movement of upper eyelid and eyeball. *Autonomic motor function (parasympathetic):* Accommodation of lens for near vision and constriction of pupil. *Clinical application:* Nerve damage causes *strabismus* (a deviation of the eye in which both eyes do not fix on the same object), *ptosis* (drooping) of the upper eyelid, dilation of the pupil, movement of the eyeball downward and outward on the damaged side, loss of accommodation for near vision, and *diplopia* (double vision).

continued

TABLE 14.4 Summary of Cranial Nerves (continued)

Nerve	Type and components	Functions and clinical application
Trochlear (IV) nerve 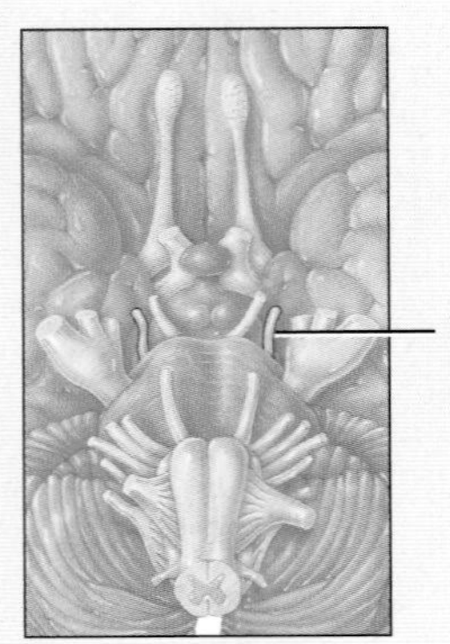 	**Mixed (mainly motor)** *Sensory portion:* Consists of axons from proprioceptors in the superior oblique muscles, which pass through the superior orbital fissure and terminate in the midbrain. *Motor portion:* Originates in the midbrain and passes through the superior orbital fissure. Innervates the superior oblique muscle, an extrinsic eyeball muscle.	*Sensory function:* Proprioception. *Somatic motor function:* Movement of the eyeball. *Clinical application:* In trochlear nerve paralysis, diplopia and strabismus occur.
Trigeminal (V) nerve 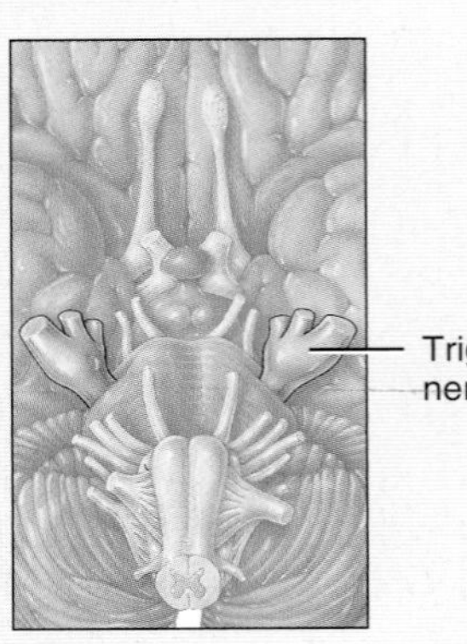 	**Mixed** *Sensory portion:* Consists of three branches, all of which end in the pons. (1) The **ophthalmic nerve** (*ophthalm-* = the eye) contains axons from the skin over the upper eyelid, eyeball, lacrimal glands, nasal cavity, side of nose, forehead, and anterior half of scalp that pass through superior orbital fissure. (2) The **maxillary nerve** (*maxilla* = upper jaw bone) contains axons from the mucosa of the nose,palate, parts of the pharynx, upper teeth, upper lip, and lower eyelid that pass through the foramen rotundum. (3) The **mandibular nerve** (*mandibula* = lower jaw bone) contains axons from the anterior two-thirds of the tongue (somatic sensory axons but not axons for the special sense of taste), the lower teeth, skin over mandible, cheek and mucosa deep to it, and side of head in front of ear that pass through the foramen ovale. *Motor portion:* Is part of the mandibular branch, which originates in the pons, passes through the foramen ovale, and innervates muscles of mastication (masseter, temporalis, medial pterygoid, lateral pterygoid, anterior belly of digastric, and mylohyoid muscles).	*Sensory function:* Conveys impulses for touch, pain,and temperature sensations and proprioception. *Somatic motor function:* Chewing. *Clinical application: Neuralgia* (pain) of one or more branches of the trigeminal nerve is called *trigeminal neuralgia (tic douloureux).* Injury of the mandibular nerve may cause paralysis of the chewing muscles and a loss of the sensations of touch, temperature, and proprioception in the lower part of the face. Dentists apply anesthetic drugs to branches of the maxillary nerve for anesthesia of upper teeth and to branches of the mandibular nerve for anesthesia of lower teeth.
Abducens (VI) nerve 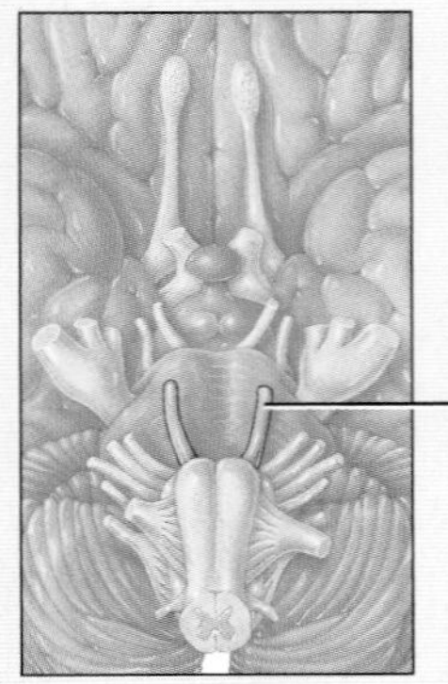 	**Mixed (mainly motor)** *Sensory portion:* Consists of axons from proprioceptors in the lateral rectus muscle, which pass through the superior orbital fissure and end in the pons. *Motor portion:* Originates in the pons, passes through the superior orbital fissure, and innervates the lateral rectus muscle, an extrinsic eyeball muscle.	*Sensory function:* Proprioception. *Somatic motor function:* Movement of the eyeball. *Clinical application:* With damage to this nerve, the affected eyeball cannot move laterally beyond the midpoint, and the eye usually is directed medially.

Nerve	Components	Function
Facial (VII) nerve 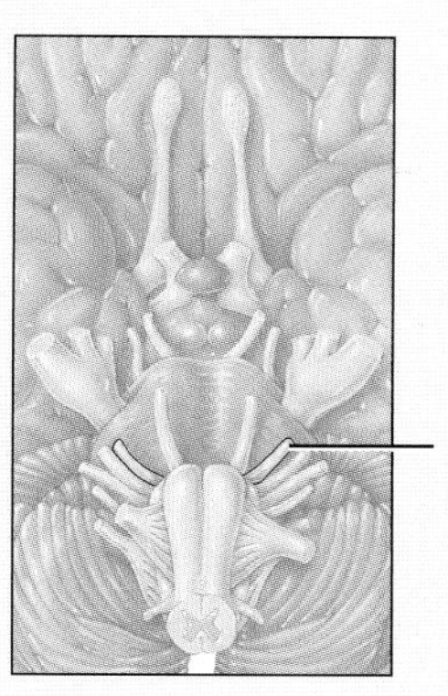	**Mixed** *Sensory portion:* Arises from taste buds on the anterior two-thirds of the tongue, passes through the stylomastoid foramen and geniculate ganglion (located beside the facial nerve), and ends in the pons. From there, axons extend to the thalamus, and then to the gustatory areas of the cerebral cortex. Also contains axons from proprioceptors in muscles of the face and scalp. *Motor portion:* Originates in the pons and passes through the stylomastoid foramen. Axons of somatic motor neurons innervate facial, scalp, and neck muscles. Parasympathetic axons innervate lacrimal, sublingual, submandibular, nasal, and palatine glands.	*Sensory function:* Proprioception and taste. *Somatic motor function:* Facial expression. *Autonomic motor function (parasympathetic):* Secretion of saliva and tears. *Clinical application:* Damage due to viral infection (shingles) or a bacterial infection (Lyme disease) produces *Bell's palsy* (paralysis of the facial muscles), loss of taste, decreased salivation, and loss of ability to close the eyes, even during sleep.
Vestibulocochlear (VIII) nerve 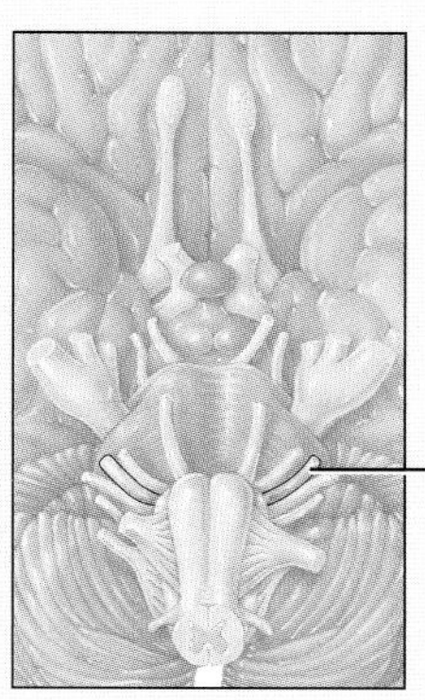	**Mixed (mainly sensory)** *Vestibular branch, sensory portion:* Arises in the semicircular canals, saccule, and utricle and forms the vestibular ganglion. Axons end in the pons and cerebellum. *Vestibular branch, motor portion:* Originates in the pons and terminates on hair cells of the semicircular canals, saccule, and utricle. *Cochlear branch, sensory portion:* Arises in the spiral organ (organ of Corti), forms the spiral ganglion, passes through nuclei in the medulla, and ends in the thalamus. Axons synapse with thalamic neurons that relay impulses to the primary auditory area (areas 41 and 42) of the cerebral cortex. *Cochlear branch, motor portion:* Originates in the pons and terminates on hair cells of the spiral organ.	*Vestibular branch, sensory function:* Conveys impulses related to equilibrium. *Vestibular branch, motor function:* Adjusts sensitivity of hair cells. *Cochlear branch, sensory function:* Conveys impulses for hearing. *Cochlear branch, motor function:* Modifies function of hair cells by altering their response to sound waves. *Clinical application:* Injury to the vestibular branch may cause *vertigo,* a subjective feeling that one's own body or the environment is rotating, *ataxia* (muscular incoordination), and *nystagmus* (involuntary rapid movement of the eyeball). Injury to the cochlear branch may cause *tinnitus* (ringing in the ears) or deafness.
Glossopharyngeal (IX) nerve 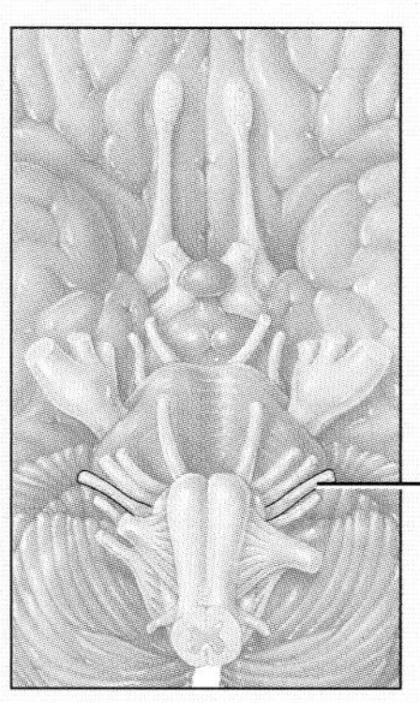	**Mixed** *Sensory portion:* Consists of axons from taste buds and somatic sensory receptors on posterior one-third of the tongue, from proprioceptors in swallowing muscles supplied by the motor portion, and from baroreceptors in carotid sinus and chemoreceptors in carotid body near the carotid arteries. Axons pass through the jugular foramen and end in the medulla. *Motor portion:* Originates in the medulla and passes through the jugular foramen. Axons of somatic motor neurons innervate the stylopharyngeus muscle, a muscle of the pharynx that elevates the larynx during swallowing. Parasympathetic axons innervate the parotid salivary) gland.	*Sensory function:* Taste and somatic sensations (touch, pain, temperature) from posterior third of tongue; proprioception in swallowing muscles; monitoring of blood pressure; monitoring of O_2 and CO_2 in blood for regulation of breathing rate and depth. *Somatic motor function:* Elevates the pharynx during swallowing and speech. *Autonomic motor function (parasympathetic):* Stimulates secretion of saliva. *Clinical application:* Injury causes difficulty in swallowing, reduced secretion of saliva, loss of sensation in the throat, and loss of taste sensation.

continued

TABLE 14.4 Summary of Cranial Nerves (continued)

Nerve	Type and portions	Functions
Vagus (X) nerve 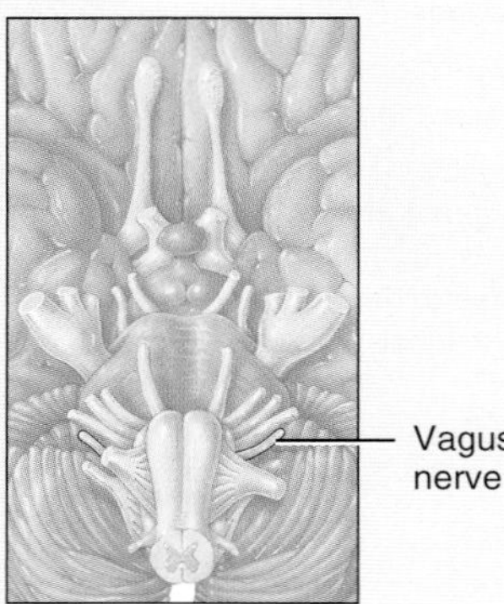 	**Mixed** *Sensory portion:* Consists of axons from small number of taste buds in the epiglottis and pharynx, proprioceptors in muscles of the neck and throat, baroreceptors in the arch of the aorta, chemoreceptors in the aortic bodies near the arch of the aorta, and visceral sensory receptors in most organs of the thoracic and abdominal cavities. Axons pass through the jugular foramen and end in the medulla and pons. *Motor portion:* Originates in medulla and passes through the jugular foramen. Axons of somatic motor neurons innervate skeletal muscles in the throat and neck. Parasympathetic axons innervate smooth muscle in the airways, esophagus, stomach, small intestine, most of large intestine, and gallbladder; cardiac muscle in the heart; and glands of the gastrointestinal (GI) tract.	*Sensory function:* Taste and somatic sensations (touch, pain, temperature, and proprioception) from epiglottis and pharynx; monitoring of blood pressure; monitoring of O_2 and CO_2 in blood for regulation of breathing rate and depth; sensations from visceral organs in thorax and abdomen. *Somatic motor function:* Swallowing, coughing, and voice production. *Autonomic motor function (parasympathetic):* Smooth muscle contraction and relaxation in organs of the GI tract; slowing of the heart rate; secretion of digestive fluids. *Clinical application:* Injury interrupts sensations from many organs in the thoracic and abdominal cavities, interferes with swallowing, paralyzes vocal cords, and causes heart rate to increase.
Accessory (XI) nerve 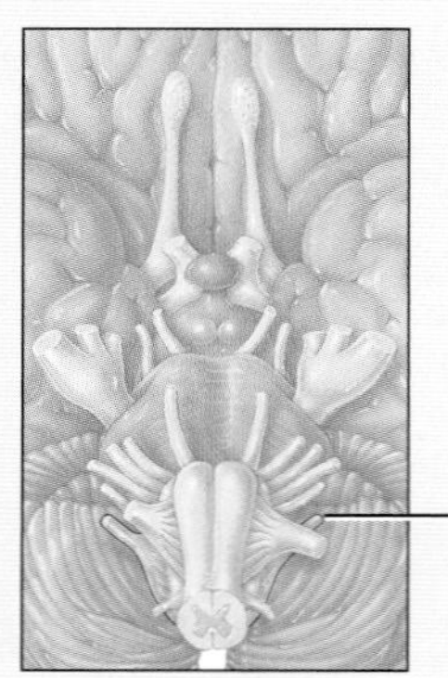 	**Mixed (mainly motor)** *Sensory portion:* Consists of axons from proprioceptors in muscles of the pharynx, larynx, and soft palate that pass through the jugular foramen and end in the medulla. *Motor portion:* Consists of a cranial root and a spinal root. *Cranial root* arises in the medulla, passes through the jugular foramen, and supplies muscles of the pharynx, larynx, and soft palate. *Spinal root* originates in the anterior gray horn of the first five cervical segments of the spinal cord, passes through the jugular foramen, and supplies the sternocleidomastoid and trapezius muscles.	*Sensory function:* Proprioception. *Somatic motor function:* Cranial root mediates swallowing movements; spinal root mediates movement of head and shoulders. *Clinical application:* If nerves are damaged, the sternocleidomastoid and trapezius muscles become paralyzed, with resulting inability to raise the shoulders and difficulty in turning the head.
Hypoglossal (XII) nerve 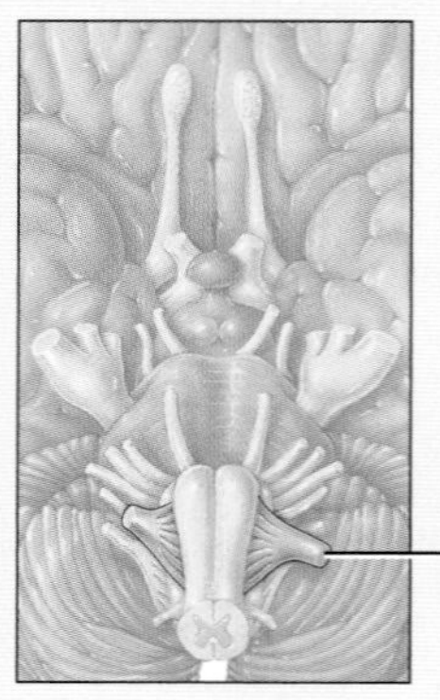 	**Mixed (mainly motor)** *Sensory portion:* Consists of axons from proprioceptors in tongue muscles that pass through the hypoglossal canal and end in the medulla. *Motor portion:* Originates in the medulla, passes through the hypoglossal canal, and supplies muscles of the tongue.	*Sensory function:* Proprioception. *Motor function:* Movement of tongue during speech and swallowing. *Clinical application:* Injury results in difficulty in chewing, speaking, and swallowing. The tongue, when protruded, curls toward the affected side, and the affected side atrophies.

MNEMONIC for cranial nerves:

Oh	**O**h	**O**h	**T**o	**T**ouch	**A**nd	**F**eel	**V**ery	**G**reen	**V**egetables	**AH!**	
Olfactory	**O**ptic	**O**culomotor	**T**rochlear	**T**rigeminal	**A**bducens	**F**acial	**V**estibulocochlear	**G**lossopharyngeal	**V**agus	**A**ccessory	**H**ypoglossal

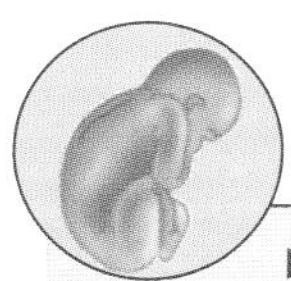

DEVELOPMENT OF THE NERVOUS SYSTEM

▶ OBJECTIVE

Describe how the parts of the brain develop.

Development of the nervous system begins in the third week of gestation with a thickening of the **ectoderm** called the **neural plate** (Figure 14.28). The plate folds inward and forms a longitudinal groove, the **neural groove.** The raised edges of the neural plate are called **neural folds.** As development continues, the neural folds increase in height and meet to form a tube called the **neural tube.**

Three layers of cells differentiate from the wall that encloses the neural tube. The outer or **marginal layer** cells develop into the *white matter* of the nervous system. The middle or **mantle layer** cells develop into the *gray matter.* The inner or **ependymal layer** cells eventually form the *lining of the central canal of the spinal cord* and *ventricles of the brain.*

The **neural crest** is a mass of tissue between the neural tube and the skin ectoderm (Figure 14.28b). It differentiates and eventually forms the *posterior (dorsal) root ganglia of spinal nerves, spinal nerves, ganglia of cranial nerves, cranial nerves, ganglia of the autonomic nervous system, adrenal medulla,* and *meninges.*

Figure 14.28 Origin of the nervous system. (a) Dorsal view of an embryo in which the neural folds have partially united, forming the early neural tube. (b) Transverse sections through the embryo showing the formation of the neural tube.

The nervous system begins developing in the third week from a thickening of ectoderm called the neural plate.

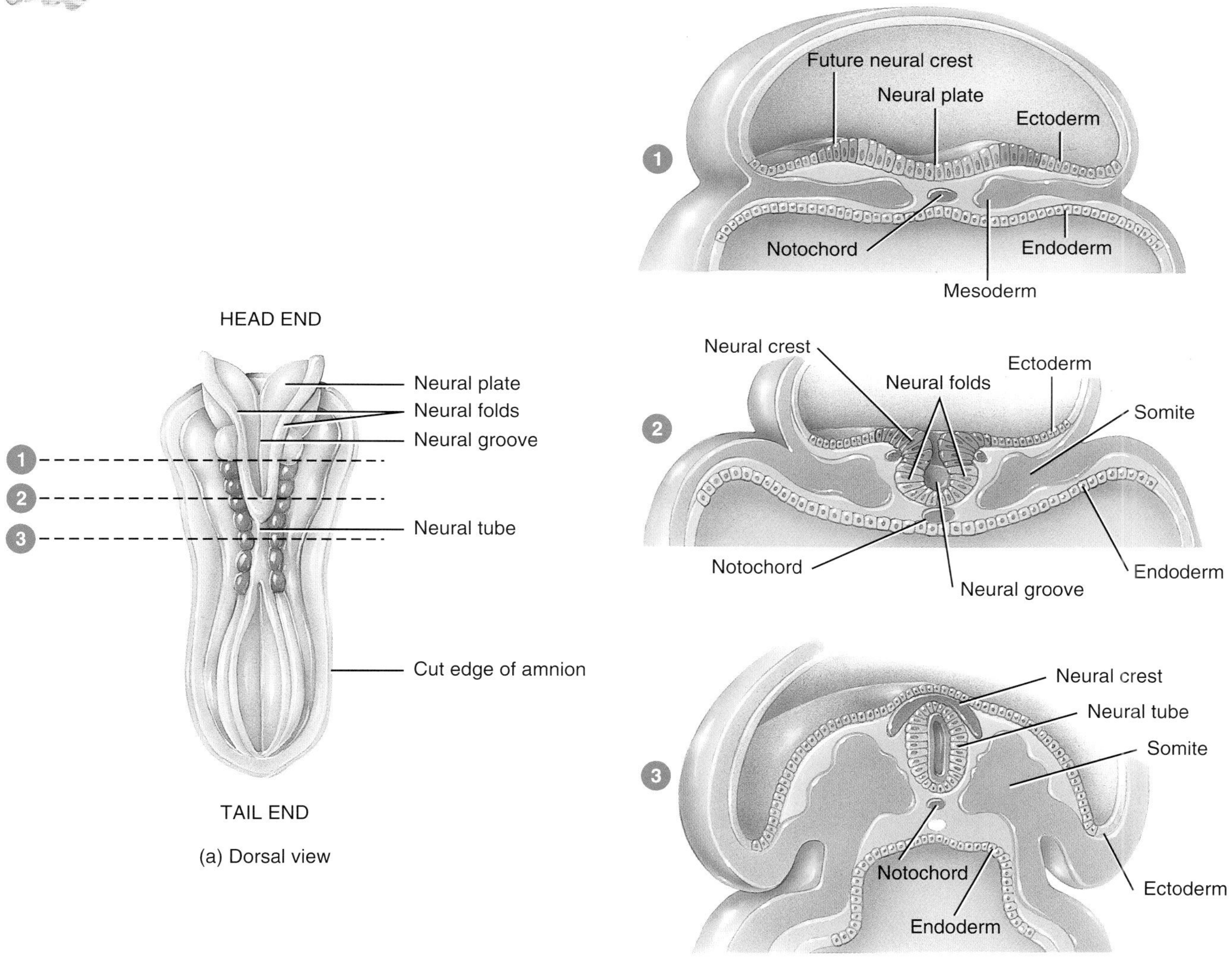

(a) Dorsal view

(b) Transverse sections

What is the origin of the gray matter of the nervous system?

As discussed at the beginning of this chapter, during the third to fourth week of embryonic development, the anterior part of the neural tube develops into three enlarged areas called **primary brain vesicles** that are named for their relative positions. These are the **prosencephalon** (prōs′-en-SEF-a-lon; *pros-* = before) or forebrain, **mesencephalon** (mes′-en-SEF-a- lon; *mes-* = middle) or midbrain, and **rhombencephalon** (rom′-ben-SEF-a-lon; *rhomb-* = behind) or hindbrain (Figure 14.29a; see also Table 14.1). During the fifth week of development, **secondary brain vesicles** begin to develop. The prosencephalon develops into two secondary brain vesicles called the **telencephalon** (tel′-en-SEF-a-lon; *tel-* = distant) and the **diencephalon** (dī-en-SEF-a-lon; *di-* = through) (Figure 14.29b). The rhombencephalon also develops into two secondary brain vesicles called the **metencephalon** (met′-en-SEF-a-lon; *met-* = after) and the **myelencephalon** (mī-el-en-SEF-a-lon; *myel-* = marrow). The area of the neural tube inferior to the myelencephalon gives rise to the *spinal cord.*

The brain vesicles continue to develop as follows (Figure 14.29c, d; see also Table 14.1):

- The telencephal on develops into the *cerebral hemispheres,* including the *basal ganglia,* and houses the paired *lateral ventricles.*
- The diencephalon develops into the *thalamus, hypothalamus,* and *epithalamus.*

Figure 14.29 Development of the brain and spinal cord.

The various parts of the brain develop from the primary brain vesicles.

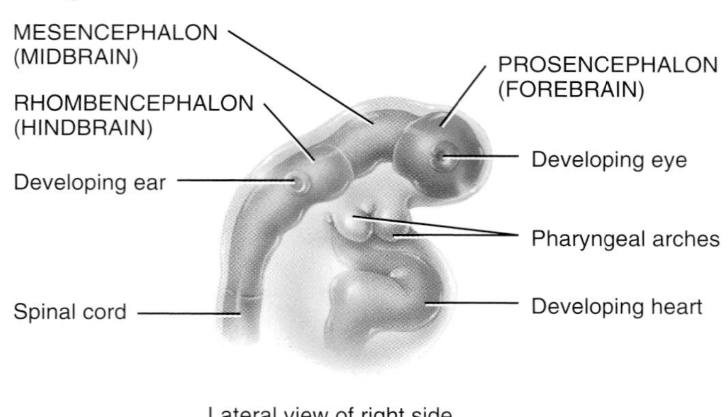

(a) Three-four week embryo showing primary brain vesicles

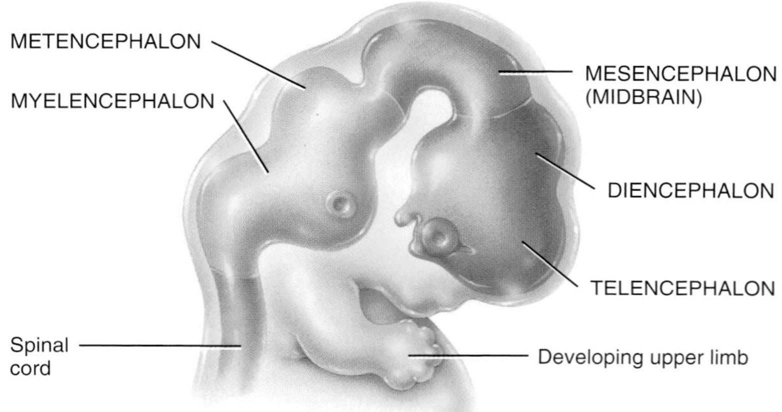

(b) Seven-week embryo showing secondary brain vesicles

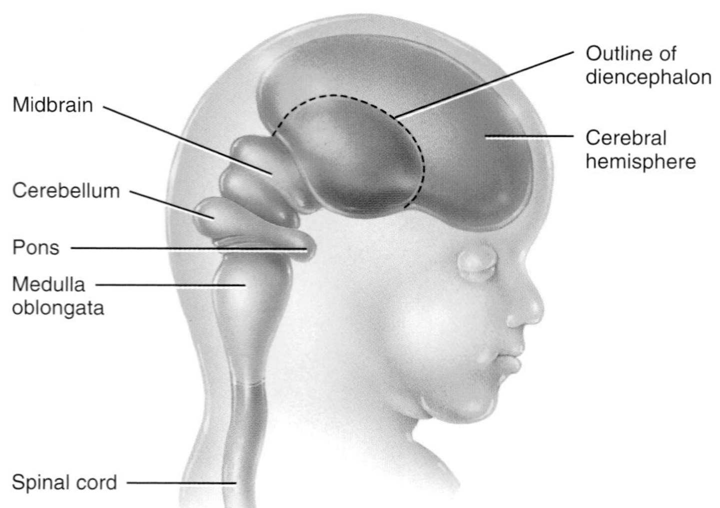

(c) Eleven-week fetus showing expanding cerebral hemispheres overgrowing the diencephalon

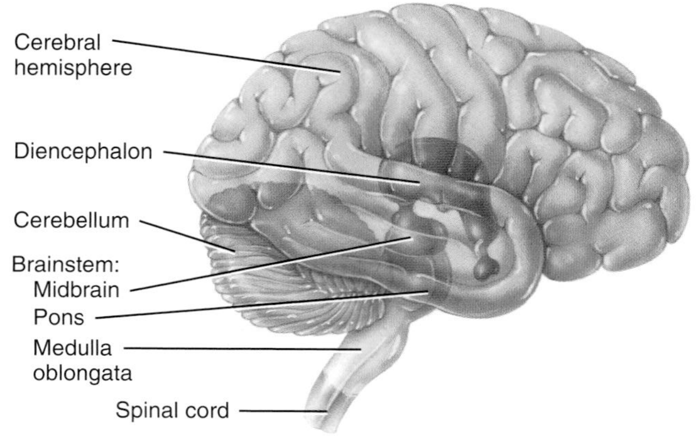

(d) Brain at birth (the diencephalon and superior portion of the brain stem have been projected to the surface)

Which primary brain vesicle does not develop into a secondary brain vesicle?

- The mesencephalon develops into the *midbrain,* which surrounds the *aqueduct of the midbrain (cerebral aqueduct).*
- The metencephalon becomes the *pons* and *cerebellum* and houses part of the *fourth ventricle.*
- The myelencephalon develops into the *medulla oblongata* and houses the remainder of the *fourth ventricle.*

Two neural tube defects—spina bifida (see page 225) and anencephaly (absence of the skull and cerebral hemispheres, discussed on page 1116)—are associated with low levels of folic acid (folate), one of the B vitamins, in the first few weeks of development. Many foods, especially grain products such as cereals and bread, are now fortified with folic acid; however, the incidence of both disorders is greatly reduced when women who are or may become pregnant take folic acid supplements.

▶ CHECKPOINT

25. What parts of the brain develop from each primary brain vesicle?

AGING AND THE NERVOUS SYSTEM

▶ OBJECTIVE

Describe the effects of aging on the nervous system.

The brain grows rapidly during the first few years of life. Growth is due mainly to an increase in the size of neurons already present, the proliferation and growth of neuroglia, the development of dendritic branches and synaptic contacts, and continuing myelination of axons. From early adulthood onward, brain mass declines. By the time a person reaches 80, the brain weighs about 7% less than it did in young adulthood. Although the number of neurons present does not decrease very much, the number of synaptic contacts declines. Associated with the decrease in brain mass is a decreased capacity for sending nerve impulses to and from the brain. As a result, processing of information diminishes. Conduction velocity decreases, voluntary motor movements slow down, and reflex times increase.

▶ CHECKPOINT

26. How is brain mass related to age?

DISORDERS: HOMEOSTATIC IMBALANCES

Cerebrovascular Accident

The most common brain disorder is a **cerebrovascular accident (CVA),** also called a **stroke** or **brain attack.** CVAs affect 500,000 people a year in the United States and represent the third leading cause of death, behind heart attacks and cancer. A CVA is characterized by abrupt onset of persisting neurological symptoms, such as paralysis or loss of sensation, that arise from destruction of brain tissue. Common causes of CVAs are intracerebral hemorrhage (from a blood vessel in the pia mater or brain), emboli (blood clots), and atherosclerosis (formation of cholesterol-containing plaques that block blood flow) of the cerebral arteries.

Among the risk factors implicated in CVAs are high blood pressure, high blood cholesterol, heart disease, narrowed carotid arteries, transient ischemic attacks (TIAs; discussed next), diabetes, smoking, obesity, and excessive alcohol intake.

A clot-dissolving drug called tissue plasminogen activator (t-PA) is now being used to open up blocked blood vessels in the brain. The drug is most effective when administered within three hours of the onset of the CVA, however, and is helpful only for CVAs due to a blood clot. Use of t-PA can decrease the permanent disability associated with these types of CVAs by 50%. New studies show that "cold therapy" might be successful in limiting the amount of residual damage from a CVA. These "cooling" therapies developed from knowledge obtained following examination of cold water drowning victims. States of hypothermia seem to trigger a survival response in which the body requires less oxygen. Some commercial companies now provide "CVA survival kits," which include cooling blankets that can be kept in the home.

Transient Ischemic Attacks

A **transient ischemic attack (TIA)** is an episode of temporary cerebral dysfunction caused by impaired blood flow to the brain. Symptoms include dizziness, weakness, numbness, or paralysis in a limb or in one side of the body; drooping of one side of the face; headache; slurred speech or difficulty understanding speech; and a partial loss of vision or double vision. Sometimes nausea or vomiting also occurs. The onset of symptoms is sudden and reaches maximum intensity almost immediately. A TIA usually persists for 5 to 10 minutes and only rarely lasts as long as 24 hours. It leaves no permanent neurological deficits. The causes of the impaired blood flow that lead to TIAs are blood clots, atherosclerosis, and certain blood disorders. About one-third of patients who experience a TIA will have a CVA eventually. Therapy for TIAs includes drugs such as aspirin, which blocks the aggregation of blood platelets, and anticoagu-lants; cerebral artery bypass grafting; and carotid endarterectomy (removal of the cholesterol-containing plaques and inner lining of an artery).

Alzheimer Disease

Alzheimer disease (ALTZ-hī-mer) or **AD** is a disabling senile dementia, the loss of reasoning and ability to care for oneself, that afflicts about 11% of the population over age 65. In the United States, about 4 million people suffer from AD. Claiming over 100,000 lives a year, AD is the fourth leading cause of death among the elderly, after heart disease, cancer, and stroke. The cause of most AD cases is still unknown, but evidence suggests it is due to a combination of genetic factors, environmental or lifestyle factors, and the aging process. Mutations in three different genes (coding for presenilin-1, presenilin-2, and amyloid precursor protein) lead to early-onset forms of AD in afflicted families but account for less than 1% of all cases. An environmental risk factor for developing AD is a history of head injury. A similar dementia occurs in boxers, probably caused by repeated blows to the head.

Individuals with AD initially have trouble remembering recent events. They then become confused and forgetful, often repeating questions or getting lost while traveling to familiar places. Disorientation grows and memories of past events disappear, and episodes of paranoia, hallucination, or violent changes in mood may occur. As their minds continue to deteriorate, they lose their ability to read, write, talk, eat, or walk. The disease culminates in dementia. A person with AD usually dies of some complication that afflicts bedridden patients, such as pneumonia.

At autopsy, brains of AD victims show three distinct structural abnormalities:

1. ***Loss of neurons that liberate acetylcholine.*** A major center of neurons that liberate ACh is the nucleus basalis, which is below the globus pallidus. Axons of these neurons project widely throughout the cerebral cortex and limbic system. Their destruction is a hallmark of Alzheimer's disease.
2. ***Beta-amyloid plaques,*** clusters of abnormal proteins deposited outside neurons.
3. ***Neurofibrillary tangles,*** abnormal bundles of filaments inside neurons in affected brain regions. These filaments consist of a protein called *tau* that has been hyperphosphorylated (meaning that too many phosphate groups have been added to it).

Drugs that inhibit acetylcholinesterase (AChE), the enzyme that inactivates ACh, improve alertness and behavior in about 5% of AD patients. Tacrine®, the first anticholinesterase inhibitor approved for treatment of AD in the United States, has significant side effects and requires dosing four times a day. Donepezil®, approved in 1998, is less toxic to the liver and has the advantage of once-a-day dosing. Some evidence suggests that vitamin E (an antioxidant), estrogen, ibuprofen, and ginkgo biloba extract may have slight beneficial effects in AD patients. In addition, researchers are currently exploring ways to develop drugs that will prevent beta-amyloid plaque formation by inhibiting the enzymes involved in beta-amyloid synthesis and by increasing the activity of the enzymes involved in beta-amyloid degradation. Researchers are also trying to develop drugs that will reduce the formation of neurofibrillary tangles by inhibiting the enzymes that hyperphosphorylate tau.

Brain Tumors

A **brain tumor** is an abnormal growth of tissue in the brain that may be malignant or benign. Unlike most other tumors in the body, malignant and benign tumors may be equally serious, compressing adjacent tissues and causing a buildup of pressure in the skull. The most common malignant tumors are secondary tumors that metastasize from other cancers in the body, such as those in the lungs, breasts, skin (malignant melanoma), blood (leukemia), and lymphatic organs (lymphoma). Most primary brain tumors (those that originate within the brain) are gliomas, which develop in neuroglia. The symptoms of a brain tumor depend on its size, location, and rate of growth. Among the symptoms are headache, poor balance and coordination, dizziness, double vision, slurred speech, nausea and vomiting, fever, abnormal pulse and breathing rates, personality changes, numbness and weakness of the limbs, and seizures. Treatment options for brain tumors vary with their size, location, and type and may include surgery, radiation therapy, and/or chemotherapy. Unfortunately, chemotherapeutic agents do not readily cross the blood–brain barrier.

Attention Deficit Hyperactivity Disorder

Attention deficit hyperactivity disorder (ADHD) is a learning disorder characterized by poor or short attention span, a consistent level of hyperactivity, and impulsiveness inappropriate for the child's age. ADHD is believed to affect about 5% of children and is diagnosed 10 times more in boys than girls. The condition typically begins in childhood and continues into adolescence and adulthood. Symptoms of ADHD develop in early childhood, often before age four, and include difficulty in organizing and finishing tasks, lack of attention to details, short attention span and inability to concentrate, difficulty following instructions, talking excessively and frequently interrupting others, frequent running or climbing excessively, inability to play quietly alone, and difficulty waiting or taking turns.

The causes of ADHD are not fully understood, but it does have a strong genetic component. Some evidence also suggests that ADHD is related to problems with neurotransmitters. In addition, recent imaging studies have demonstrated that people with ADHD have less nervous tissue in specific regions of the brain such as the frontal and temporal lobes, caudate nucleus, and cerebellum. Treatment may involve remedial education, behavioral modification techniques, restructuring routines, and drugs that calm the child and help focus attention.

MEDICAL TERMINOLOGY

Agnosia (ag-NŌ-zē-a; *a-* = without; *-gnosia* = knowledge) Inability to recognize the significance of sensory stimuli such as sounds, sights, smells, tastes, and touch.

Apraxia (a-PRAK-sē-a; *-praxia* = coordinated) Inability to carry out purposeful movements in the absence of paralysis.

Consciousness (KON-shus-nes) A state of wakefulness in which an individual is fully alert, aware, and oriented, partly as a result of feedback between the cerebral cortex and reticular activating system.

Delirium (dē-LIR-ē-um = off the track) A transient disorder of abnormal cognition and disordered attention accompanied by disturbances of the sleep–wake cycle and psychomotor behavior (hyperactivity or hypoactivity of movements and speech). Also called **acute confusional state (ACS).**

Dementia (de-MEN-shē-a; *de-* = away from; *-mentia* = mind) Permanent or progressive general loss of intellectual abilities, including impairment of memory, judgment, abstract thinking, and changes in personality.

Encephalitis (en′-sef-a-LĪ-tis) An acute inflammation of the brain caused by either a direct attack by any of several viruses or an allergic reaction to any of the many viruses that are normally harmless to the central nervous system. If the virus affects the spinal cord as well, the condition is called **encephalomyelitis.**

Encephalopathy (en-sef′-a-LOP-a-thē; *encephalo* = brain; *-pathos* = disease) Any disorder of the brain.

Lethargy (LETH-ar-jē) A condition of functional sluggishness.

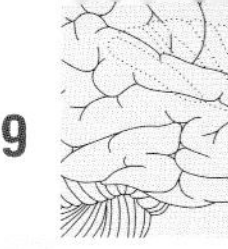

Microcephaly (mī-krō-SEF-a-lē; *micro-* = small; *-cephal* = head) A congenital condition that involves the development of a small brain and skull and frequently results in mental retardation.

Reye's (RĪZ) syndrome Occurs after a viral infection, particularly chickenpox or influenza, most often in children or teens who have taken aspirin; characterized by vomiting and brain dysfunction (disorientation, lethargy, and personality changes) that may progress to coma and death.

Stupor (STOO-por) Unresponsiveness from which a patient can be aroused only briefly and only by vigorous and repeated stimulation.

STUDY OUTLINE

BRAIN ORGANIZATION, PROTECTION, AND BLOOD SUPPLY (p. 474)

1. The major parts of the brain are the brain stem, cerebellum, diencephalon, and cerebrum.
2. The brain is protected by cranial bones and the cranial meninges.
3. The cranial meninges are continuous with the spinal meninges. From superficial to deep they are the dura mater, arachnoid mater, and pia mater.
4. Blood flow to the brain is mainly via the internal carotid and vertebral arteries.
5. Any interruption of the oxygen or glucose supply to the brain can result in weakening of, permanent damage to, or death of brain cells.
6. The blood–brain barrier (BBB) causes different substances to move between the blood and the brain tissue at different rates and prevents the movement of some substances from blood into the brain.

CEREBROSPINAL FLUID (p. 477)

1. Cerebrospinal fluid (CSF) is formed in the choroid plexuses and circulates through the lateral ventricles, third ventricle, fourth ventricle, subarachnoid space, and central canal. Most of the fluid is absorbed into the blood across the arachnoid villi of the superior sagittal sinus.
2. Cerebrospinal fluid provides mechanical protection, chemical protection, and circulation of nutrients.

THE BRAIN STEM (p. 481)

1. The medulla oblongata is continuous with the superior part of the spinal cord and contains both motor and sensory tracts. It contains nuclei that are reflex centers for regulation of heart rate, respiratory rate, vasoconstriction, swallowing, coughing, vomiting, hiccupping, and sneezing. It also contains nuclei associated with cranial nerves VIII through XII.
2. The pons is superior to the medulla. It connects parts of the brain with one another by way of tracts. Pontine nuclei relay nerve impulses related to voluntary skeletal movements from the cerebral cortex to the cerebellum. The pons contains the pneumotaxic and apneustic centers, which help control breathing. It contains nuclei associated with cranial nerves V–VII and the vestibular branch of cranial nerve VIII.
3. The midbrain connects the pons and diencephalon and surrounds the cerebral aqueduct. It conveys motor impulses from the cerebrum to the cerebellum and spinal cord, sends sensory impulses from the spinal cord to the thalamus, and regulates auditory and visual reflexes. It also contains nuclei associated with cranial nerves III and IV.
4. A large part of the brain stem consists of small areas of gray matter and white matter called the reticular formation, which helps maintain consciousness, causes awakening from sleep, and contributes to regulating muscle tone.

THE CEREBELLUM (p. 486)

1. The cerebellum occupies the inferior and posterior aspects of the cranial cavity. It consists of two lateral hemispheres and a medial, constricted vermis.
2. It connects to the brain stem by three pairs of cerebellar peduncles.
3. The cerebellum coordinates contractions of skeletal muscles and maintains normal muscle tone, posture, and balance.

THE DIENCEPHALON (p. 488)

1. The diencephalon surrounds the third ventricle and consists of the thalamus, hypothalamus, and epithalamus.
2. The thalamus is superior to the midbrain and contains nuclei that serve as relay stations for sensory impulses to the cerebral cortex. It also allows crude appreciation of pain, temperature, and pressure and mediates some motor activities.
3. The hypothalamus is inferior to the thalamus. It controls and integrates the autonomic nervous system, connects the nervous and endocrine systems, functions in rage and aggression, controls body temperature, regulates food and fluid intake, and establishes circadian rhythms.
4. The epithalamus consists of the pineal gland and the habenular nuclei. The pineal gland secretes melatonin, which is thought to promote sleep and to help set the body's biological clock.
5. Circumventricular organs (CVOs) can monitor chemical changes in the blood because they lack the blood–brain barrier.

THE CEREBRUM (p. 491)

1. The cerebrum is the largest part of the brain. Its cortex contains gyri (convolutions), fissures, and sulci.
2. The cerebral hemispheres are divided into four lobes: frontal, parietal, temporal, and occipital.
3. The white matter of the cerebrum is deep to the cortex and consists of myelinated and unmyelinated axons extending to other regions as association, commissural, and projection fibers.
4. The basal ganglia are several groups of nuclei in each cerebral hemisphere. They help control large, automatic movements of skeletal muscles and help regulate muscle tone.
5. The limbic system encircles the upper part of the brain stem and the corpus callosum. It functions in emotional aspects of behavior and memory.
6. Table 14.2 summarizes the functions of various parts of the brain.

FUNCTIONAL ORGANIZATION OF THE CEREBRAL CORTEX (p. 496)

1. The sensory areas of the cerebral cortex allow perception of sensory impulses. The motor areas are the regions that govern muscular movement. The association areas are concerned with more complex integrative functions.
2. The primary somatosensory area (areas 1, 2, and 3) receives nerve impulses from somatic sensory receptors for touch, proprioception, pain, and temperature. Each point within the area receives impulses from a specific part of the face or body.
3. The primary visual area (area 17) receives impulses that convey visual information. The primary auditory area (areas 41 and 42) interprets the basic characteristics of sound such as pitch and rhythm. The primary gustatory area (area 43) receives impulses for taste. The primary olfactory area (area 28) receives impulses for smell.
4. Motor areas include the primary motor area (area 4), which controls voluntary contractions of specific muscles or groups of muscles, and Broca's speech area (areas 44 and 45), which controls production of speech.
5. The prefrontal cortex (areas 9, 10, 11, and 12) is concerned with personality, intellect, complex learning abilities, judgment, reasoning, intuition, and development of abstract ideas.
6. The somatosensory association area (areas 5 and 7) permits you to determine the exact shape and texture of an object without looking at it and to sense the relationship of one body part to another. The visual association area (areas 18 and 19) relates present to past visual experiences and is essential for recognizing and evaluating what is seen. The auditory association area (area 22) deals with the meanings of sounds.
7. Wernicke's area (area 22 and possibly 39 and 40) interprets the meaning of speech by translating words into thoughts. The common integrative area (areas 5, 7, 39, and 40) integrates sensory interpretations from the association areas and impulses from other areas, allowing thoughts based on sensory inputs.
8. The premotor area (area 6) generates nerve impulses that cause specific groups of muscles to contract in specific sequences. The frontal eye field area (area 8) controls voluntary scanning movements of the eyes.
9. Subtle anatomical differences exist between the two hemispheres, and each has unique functions. Each hemisphere receives sensory signals from and controls the opposite side of the body. The left hemisphere is more important for language, numerical and scientific skills, and reasoning. The right hemisphere is more important for musical and artistic awareness, spatial and pattern perception, recognition of faces, emotional content of language, identifying odors, and generating mental images of sight, sound, touch, taste, and smell.
10. Brain waves generated by the cerebral cortex are recorded from the surface of the head in an electroencephalogram (EEG). The EEG may be used to diagnose epilepsy, infections, and tumors.

CRANIAL NERVES (p. 500)

1. Twelve pairs of cranial nerves originate from the nose, eyes, inner ear, brain stem, and spinal cord.
2. They are named primarily based on their distribution and are numbered I–XII in order of attachment to the brain. Table 14.4 summarizes the types, locations, functions, and disorders of the cranial nerves.

DEVELOPMENT OF THE NERVOUS SYSTEM (p. 515)

1. The development of the nervous system begins with a thickening of a region of the ectoderm called the neural plate.
2. During embryological development, primary brain vesicles form from the neural tube and serve as forerunners of various parts of the brain.
3. The telencephalon forms the cerebrum, the diencephalon develops into the thalamus and hypothalamus, the mesencephalon develops into the midbrain, the metencephalon develops into the pons and cerebellum, and the myelencephalon forms the medulla.

AGING AND THE NERVOUS SYSTEM (p. 517)

1. The brain grows rapidly during the first few years of life.
2. Age-related effects involve loss of brain mass and decreased capacity for sending nerve impulses.

SELF-QUIZ QUESTIONS

Fill in the blanks in the following statements.

1. The cerebral hemispheres are connected internally by a broad band of white matter known as the ____.
2. List the five lobes of the cerebrum: ____, ____, ____, ____, ____.
3. The ____ separates the cerebrum into right and left halves.

Indicate whether the following statements are true or false.

4. The brain stem consists of the medulla oblongata, pons, and diencephalon.
5. You are the greatest student of anatomy and physiology and you are well-prepared for your exam on the brain. As you confidently answer the questions, your brain is exhibiting beta waves.

Choose the one best answer to the following questions.

6. Which of the following is *not* a function of the thalamus? (a) relaying information from the cerebellum and basal ganglia to primary motor areas of the cerebral cortex; (b) helping maintain consciousness; (c) nonlocalized perception of pain, pressure, and thermal sensations; (d) regulating body temperature; (e) relaying sensory impulses to the cerebral cortex.
7. Which of the following statements is *false*? (a) The blood supply to the brain is provided mainly by the internal carotid and vertebral arteries. (b) Neurons in the brain rely almost exclusively on aerobic respiration to produce ATP. (c) An interruption of blood flow to the brain for even 20 seconds may impair brain function. (d) Glucose supply to the brain must be continuous. (e) Low levels of glucose in the blood to the brain may result in unconsciousness.
8. In which of the following ways does cerebrospinal fluid contribute to homeostasis? (1) mechanical protection, (2) chemical protection, (3) electrical protection, (4) circulation, (5) immunity. (a) 1, 2, and 3; (b) 2, 3, and 4; (c) 3, 4, and 5; (d) 1, 2, and 4; (e) 2, 4, and 5.

9. Which of the following are functions of the hypothalamus? (1) control of the ANS, (2) control of the pituitary gland, (3) regulation of emotional and behavioral patterns, (4) regulation of eating and drinking, (5) control of body temperature, (6) regulation of circadian rhythms and states of consciousness. (a) 1, 2, 4, and 6; (b) 2, 3, 5, and 6; (c) 1, 3, 5, and 6; (d) 1, 4, 5, and 6; (e) 1, 2, 3, 4, 5, and 6.

10. Which of the following statements is *false*? (a) Association tracts transmit nerve impulses between gyri in the same hemisphere. (b) Commissural tracts transmit impulses from the gyri in one cerebral hemisphere to the corresponding gyri in the other hemisphere. (c) Projection tracts form descending and ascending tracts that transmit impulses from the cerebrum and other parts of the brain to the spinal cord, or from the spinal cord to the brain. (d) The internal capsule is an example of commissural tracts. (e) The corpus callosum is an example of commissural tracts.

11. Which of the following statements is *true*? (a) The right and left hemispheres of the cerebrum are completely symmetrical. (b) The left hemisphere controls the left side of the body. (c) The right hemisphere is more important for spoken and written language. (d) The left hemisphere is more important for musical and artistic awareness. (e) Hemispheric lateralization is more pronounced in males than in females.

12. Match the following (some answers will be used more than once):

____(a) oculomotor
____(b) trigeminal
____(c) abducens
____(d) vestibulocochlear
____(e) accessory
____(f) vagus
____(g) facial
____(h) glossopharyngeal
____(i) olfactory
____(j) trochlear
____(k) optic
____(l) hypoglossal
____(m) functions in sense of smell
____(n) functions in hearing and equilibrium
____(o) functions in chewing
____(p) functions in facial expression and secretion of saliva and tears
____(q) functions in movement of tongue during speech and swallowing
____(r) functions in secretion of digestive fluids
____(s) functions in secretion of saliva, taste, regulation of blood pressure, and muscle sense
____(t) sensory only
____(u) functions in eye movement by controlling extrinsic eye muscles
____(v) functions in swallowing and head movements

(1) cranial nerve I
(2) cranial nerve II
(3) cranial nerve III
(4) cranial nerve IV
(5) cranial nerve V
(6) cranial nerve VI
(7) cranial nerve VII
(8) cranial nerve VIII
(9) cranial nerve IX
(10) cranial nerve X
(11) cranial nerve XI
(12) cranial nerve XII

13. Match the following (some answers may be used more than once):

____(a) emotional brain; involved in olfaction and memory
____(b) bridge connecting parts of the brain with each other
____(c) sensory relay area
____(d) alerts the cerebral cortex to incoming sensory signals and helps regulate muscle tone
____(e) the motor command center; regulates posture and balance
____(f) lacks a blood–brain barrier; can monitor chemical changes in the blood
____(g) site of decussation of pyramids
____(h) site of pneumotaxic and apneustic areas
____(i) secretes melatonin
____(j) contains sensory, motor, and association areas
____(k) responsible for maintaining consciousness and awakening from sleep
____(l) controls ANS
____(m) contains reflex centers for movements of the eyes, head, and neck in response to visual and other stimuli, and reflex center for movements of the head and trunk in response to auditory stimuli
____(n) plays an essential role in awareness and in the acquisition of knowledge; cognition
____(o) several groups of nuclei that control large autonomic movements of skeletal muscles and help regulate muscle tone required for specific body movements
____(p) produces hormones that regulate endocrine gland function
____(q) contains the vital cardiovascular center and medullary rhythmicity center

(1) medulla oblongata
(2) pons
(3) midbrain (mesencephalon)
(4) cerebellum
(5) pineal gland
(6) thalamus
(7) hypothalamus
(8) cerebrum
(9) limbic system
(10) reticular formation
(11) circumventricular organs
(12) reticular activating system
(13) basal ganglia

14. Match the following:

____(a) protrusions in the medulla formed by the large corticospinal tracts
____(b) dura mater extension that separates the two cerebral hemispheres
____(c) fingerlike extensions of arachnoid mater where CSF is reabsorbed
____(d) dura mater extension that separates the two cerebellar hemispheres
____(e) located in the hypothalamus; relay stations for reflexes related to smell
____(f) folds in the cerebral cortex
____(g) shallow grooves in the cerebral cortex
____(h) bundles of white matter that relay information between the cerebellum and other parts of the brain
____(i) a thick band of sensory and motor tracts that connect the cerebral cortex with the brain stem and spinal cord
____(j) dura mater extension that separates the cerebrum from the cerebellum
____(k) thin membranous partition between the lateral ventricles

(1) gyri
(2) internal capsule
(3) mammillary bodies
(4) tentorium cerebelli
(5) pyramids
(6) falx cerebelli
(7) septum pellucidum
(8) cerebellar peduncles
(9) falx cerebri
(10) sulci
(11) arachnoid villi

15. Match the following:

____(a) allows planning and production of speech
____(b) interprets pitch and rhythm
____(c) controls voluntary contraction of muscles
____(d) allows recognition and evaluation of visual experiences
____(e) integration and interpretation of somatic sensations; comparison of past to present sensations
____(f) receives impulses for touch, proprioception, pain, and temperature
____(g) receives impulses for taste
____(h) interpretation of sounds as speech, music, or noise
____(i) receives impulses from many sensory and association areas as well as the thalamus and brain stem; allows formation of thoughts so appropriate action can occur
____(j) translates words into thoughts
____(k) receives impulses for smell
____(l) allows interpretation of shape, color, and movement
____(m) coordinates muscle movement for complex, learned sequential motor activities
____(n) involved in scanning eye movements

(1) primary visual area
(2) primary auditory area
(3) primary gustatory area
(4) primary olfactory area
(5) primary somatosensory area
(6) primary motor area
(7) somatosensory association area
(8) visual association area
(9) frontal eye field
(10) Broca's area
(11) auditory association area
(12) premotor area
(13) Wernicke's area
(14) common integrative area

CRITICAL THINKING QUESTIONS

1. An elderly relative suffered a CVA (stroke) and now has difficulty moving her right arm, and she also has speech problems. What areas of the brain were damaged by the stroke?
2. Nicky has recently had a viral infection and now she cannot move the muscles on the right side of her face. In addition, she is experiencing a loss of taste and a dry mouth, and she cannot close her right eye. What cranial nerve has been affected by the viral infection?
3. You have been hired by a pharmaceutical company to develop a drug to regulate a specific brain disorder. What is a major physiological roadblock to developing such a drug and how can you design a drug to bypass that roadblock so that the drug is delivered to the brain where it is needed?

ANSWERS TO FIGURE QUESTIONS

14.1 The largest part of the brain is the cerebrum.

14.2 From superficial to deep, the three cranial meninges are the dura mater, arachnoid, and pia mater.

14.3 The brain stem is anterior to the fourth ventricle, and the cerebellum is posterior to it.

14.4 Cerebrospinal fluid is reabsorbed by the arachnoid villi that project into the dural venous sinuses.

14.5 The medulla oblongata contains the pyramids; the midbrain contains the cerebral peduncles; "pons" means "bridge."

14.6 Decussation means crossing to the opposite side. The functional consequence of decussation of the pyramids is that each side of the cerebrum controls muscles on the opposite side of the body.

14.7 The cerebral peduncles are the main sites through which tracts extend and nerve impulses are conducted between the superior parts of the brain and the inferior parts of the brain and the spinal cord.

14.8 The cerebellar peduncles contain the axons that carry information into and out of the cerebellum.

14.9 The intermediate mass connects the right and left halves of the thalamus.

14.10 From posterior to anterior, the four major regions of the hypothalamus are the mammillary, tuberal, supraoptic, and preoptic regions.

14.11 The gray matter enlarges more rapidly during development, in the process producing convolutions or gyri (folds), sulci (shallow grooves), and fissures (deep grooves).

14.12 Association tracts connect gyri of the same hemisphere; commissural tracts connect gyri in opposite hemispheres; projection tracts connect the cerebrum with the thalamus, brain stem, and spinal cord.

14.13 The basal ganglia are lateral, superior, and inferior to the thalamus.

14.14 The hippocampus is the component of the limbic system that functions with the cerebrum in memory.

14.15 The common integrative area integrates interpretation of visual, auditory, and somatic sensations; Broca's speech area translates thoughts into speech; the premotor area controls skilled muscular movements; the gustatory areas interpret sensations related to taste; the auditory areas interpret pitch and rhythm; the visual areas interpret shape, color, and movement of objects; the frontal eye field controls voluntary scanning movements of the eyes.

14.16 The orbitofrontal cortex in the right hemisphere exhibits hemispheric lateralization for olfaction.

14.17 In an EEG, theta waves indicate emotional stress.

14.18 Axons in the olfactory tracts terminate in the primary olfactory area in the temporal lobe of the cerebral cortex.

14.19 Most axons in the optic tracts terminate in the lateral geniculate nucleus of the thalamus.

14.20 The superior branch of the oculomotor nerve is distributed to the superior rectus muscle; the trochlear nerve is the smallest cranial nerve.

14.21 The trigeminal nerve is the largest cranial nerve.

14.22 Motor axons of the facial nerve originate in the pons.

14.23 The vestibular ganglion contains cell bodies from sensory axons that arise in the semicircular canals, saccule, and utricle; the spiral ganglion contains cell bodies from axons that arise in the spiral organ.

14.24 The glossopharyngeal nerve exits the skull through the jugular foramen.

14.25 The vagus nerve is located between and behind the internal jugular vein and common carotid artery in the neck.

14.26 The accessory nerve is the only cranial nerve that originates from both the brain and spinal cord.

14.27 Two important motor functions of the hypoglossal nerve are speech and swallowing.

14.28 The gray matter of the nervous system derives from the mantle layer cells of the neural tube.

14.29 The mesencephalon does not develop into a secondary brain vesicle.

Figure 15.1 **Motor neuron pathways in the (a) somatic nervous system and (b) autonomic nervous system (ANS).** Note that autonomic motor neurons release either acetylcholine (ACh) or norepinephrine (NE); somatic motor neurons release ACh.

Somatic nervous system stimulation always excites its effectors (skeletal muscle fibers); stimulation by the autonomic nervous system either excites or inhibits visceral effectors.

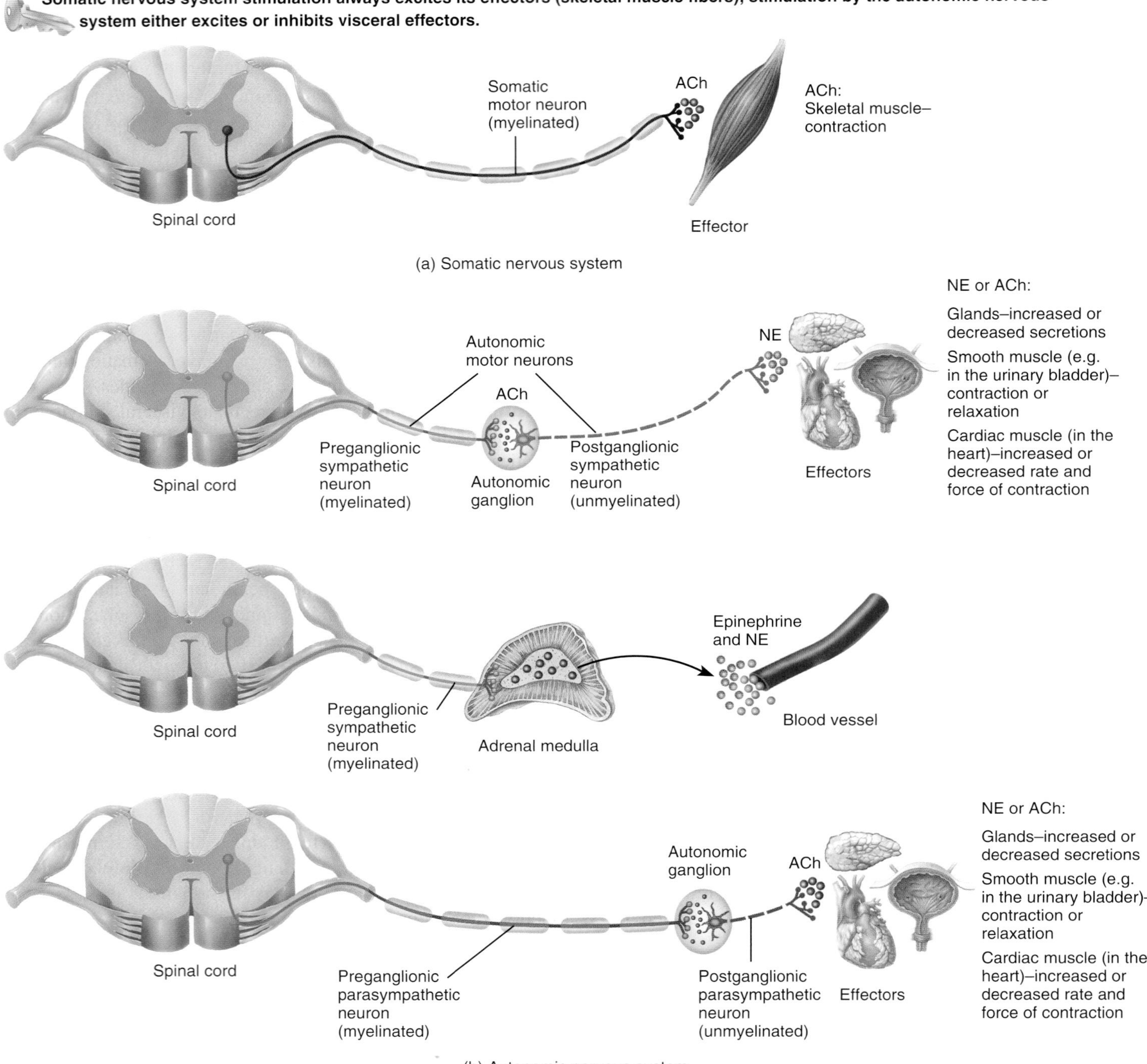

What does dual innervation mean?

nomic ganglion; its unmyelinated axon extends directly from the ganglion to the effector (smooth muscle, cardiac muscle, or a gland). Alternately, in some autonomic pathways, the first motor neuron extends to the adrenal medullae (inner portion of the adrenal glands) rather than an autonomic ganglion. In addition, all somatic motor neurons release only acetylcholine (ACh) as their neurotransmitter, but autonomic motor neurons release either ACh or norepinephrine (NE).

The output (motor) part of the ANS has two principal branches: the **sympathetic division** and the **parasympathetic division.** Most organs have **dual innervation:** They receive impulses from both sympathetic and parasympathetic neurons.

TABLE 15.1 Summary of Somatic and Autonomic Nervous Systems

	Somatic Nervous System	Autonomic Nervous System
Sensory input	Special senses and somatic senses.	Mainly from interoceptors; some from special senses and somatic senses.
Control of motor output	Voluntary control from cerebral cortex, with contributions from basal ganglia, cerebellum, brain stem, and spinal cord.	Involuntary control from limbic system, hypothalamus, brain stem, and spinal cord; limited control from cerebral cortex.
Motor neuron pathway	One-neuron pathway: Somatic motor neurons extending from CNS synapse directly with effector.	Usually two-neuron pathway: Preganglionic neurons extending from CNS synapse with postganglionic neurons in an autonomic ganglion, and postganglionic neurons extending from ganglion synapse with a visceral effector. Alternatively, preganglionic neurons may extend from CNS to synapse with cells of adrenal medullae.
Neurotransmitters and hormones	All somatic motor neurons release ACh.	All preganglionic axons release acetylcholine (ACh); most sympathetic postganglionic neurons release norepinephrine (NE); those to most sweat glands release ACh; all parasympathetic postganglionic neurons release ACh; adrenal medullae release epinephrine and norepinephrine.
Effectors	Skeletal muscle.	Smooth muscle, cardiac muscle, and glands.
Responses	Contraction of skeletal muscle.	Contraction or relaxation of smooth muscle; increased or decreased rate and force of contraction of cardiac muscle; increased or decreased secretions of glands.

In general, nerve impulses from one division of the ANS stimulate the organ to increase its activity (excitation), and impulses from the other division decrease the organ's activity (inhibition). For example, an increased rate of nerve impulses from the sympathetic division increases heart rate, and an increased rate of nerve impulses from the parasympathetic division decreases heart rate. Table 15.1 summarizes the similarities and differences between the somatic and autonomic nervous systems.

▶ **CHECKPOINT**

1. How do the autonomic nervous system and somatic nervous system compare in structure and function?
2. What are the main input and output components of the autonomic nervous system?

ANATOMY OF AUTONOMIC MOTOR PATHWAYS

▶ **OBJECTIVES**

Describe preganglionic and postganglionic neurons of the autonomic nervous system.

Compare the anatomical components of the sympathetic and parasympathetic divisions of the autonomic nervous system.

Anatomical Components

The first of the two motor neurons in any autonomic motor pathway is called a **preganglionic neuron** (Figure 15.1b). Its cell body is in the brain or spinal cord, and its axon exits the CNS as part of a cranial or spinal nerve. The axon of a preganglionic neuron is a small-diameter, myelinated type B fiber that usually extends to an autonomic ganglion, where it synapses with a **postganglionic neuron,** the second neuron in the autonomic motor pathway (Figure 15.1b). Notice that the postganglionic neuron lies entirely outside the CNS. Its cell body and dendrites are located in an autonomic ganglion, where it forms synapses with one or more preganglionic axons. The axon of a postganglionic neuron is a small-diameter, unmyelinated type C fiber that terminates in a visceral effector. Thus, preganglionic neurons convey nerve impulses from the CNS to autonomic ganglia, and postganglionic neurons relay the impulses from autonomic ganglia to visceral effectors.

Preganglionic Neurons

In the sympathetic division, the preganglionic neurons have their cell bodies in the lateral horns of the gray matter in the 12 thoracic segments and the first two (and sometimes three) lumbar segments of the spinal cord (Figure 15.2). For this reason, the sympathetic division is also called the **thoracolumbar division** (thōr′-a-kō-LUM-bar), and the axons of the sympathetic preganglionic neurons are known as the **thoracolumbar outflow.**

Cell bodies of preganglionic neurons of the parasympathetic division are located in the nuclei of four cranial nerves in the brain stem (III, VII, IX, and X) and in the lateral gray horns of the second through fourth sacral segments of the spinal cord (see Figure 15.3). Hence, the parasympathetic division is also known as the **craniosacral division** (krā′-nē-ō-SĀK-ral), and the axons of the parasympathetic preganglionic neurons are referred to as the **craniosacral outflow.**

Figure 15.2 Structure of the sympathetic division of the autonomic nervous system. Solid lines represent preganglionic axons; dashed lines represent postganglionic axons. Although the innervated structures are shown for only one side of the body for diagrammatic purposes, the sympathetic division actually innervates tissues and organs on both sides.

Cell bodies of sympathetic preganglionic neurons are located in the lateral horns of gray matter in the 12 thoracic and first two lumbar segments of the spinal cord.

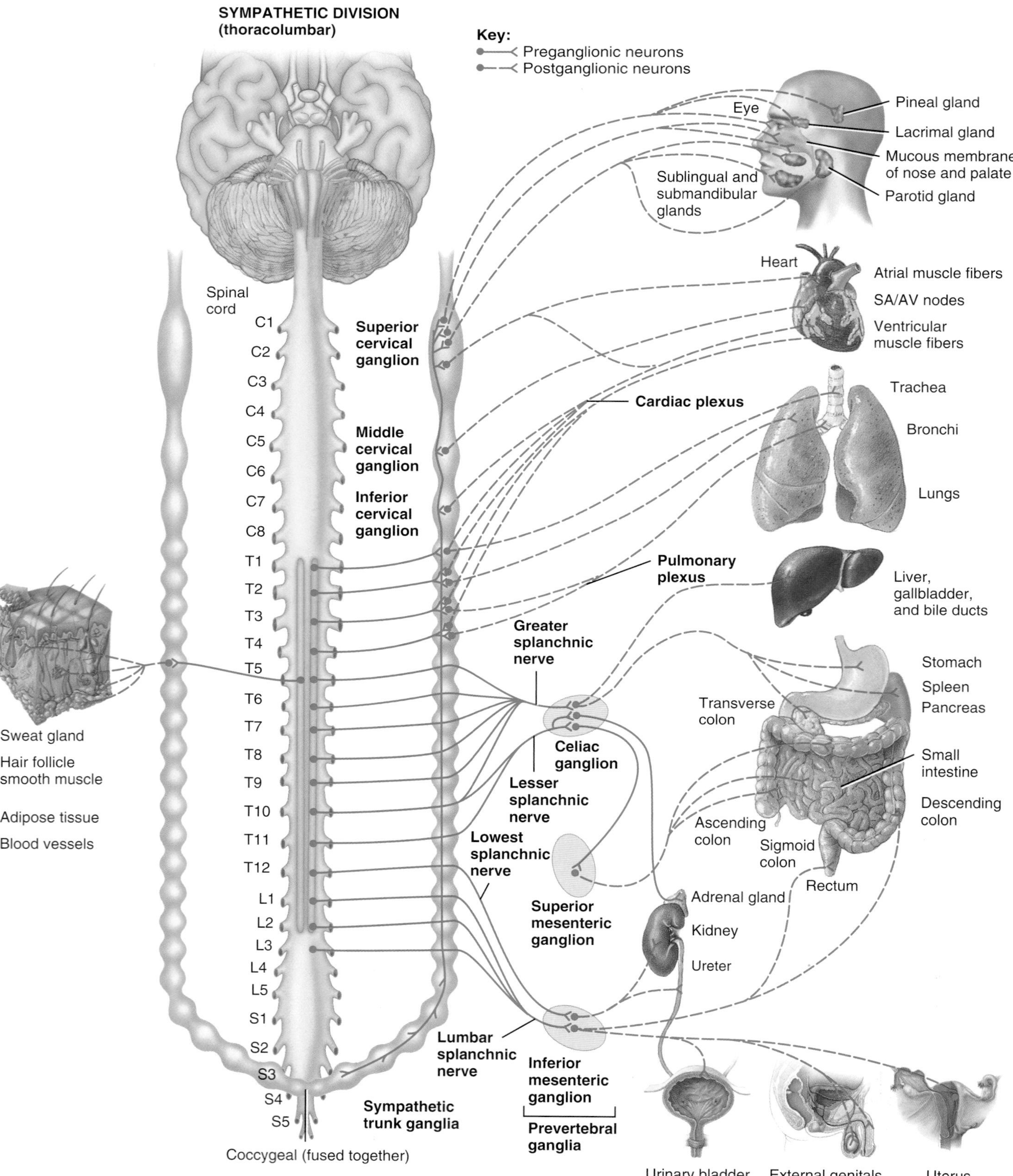

Which division, sympathetic or parasympathetic, has longer preganglionic axons? Why?

Figure 15.3 Structure of the parasympathetic division of the autonomic nervous system. Solid lines represent preganglionic axons; dashed lines represent postganglionic axons. Although the innervated structures are shown only for one side of the body for diagrammatic purposes, the parasympathetic division actually innervates tissues and organs on both sides.

Cell bodies of parasympathetic preganglionic neurons are located in brain stem nuclei and in the lateral horns of gray matter in the second through fourth sacral segments of the spinal cord.

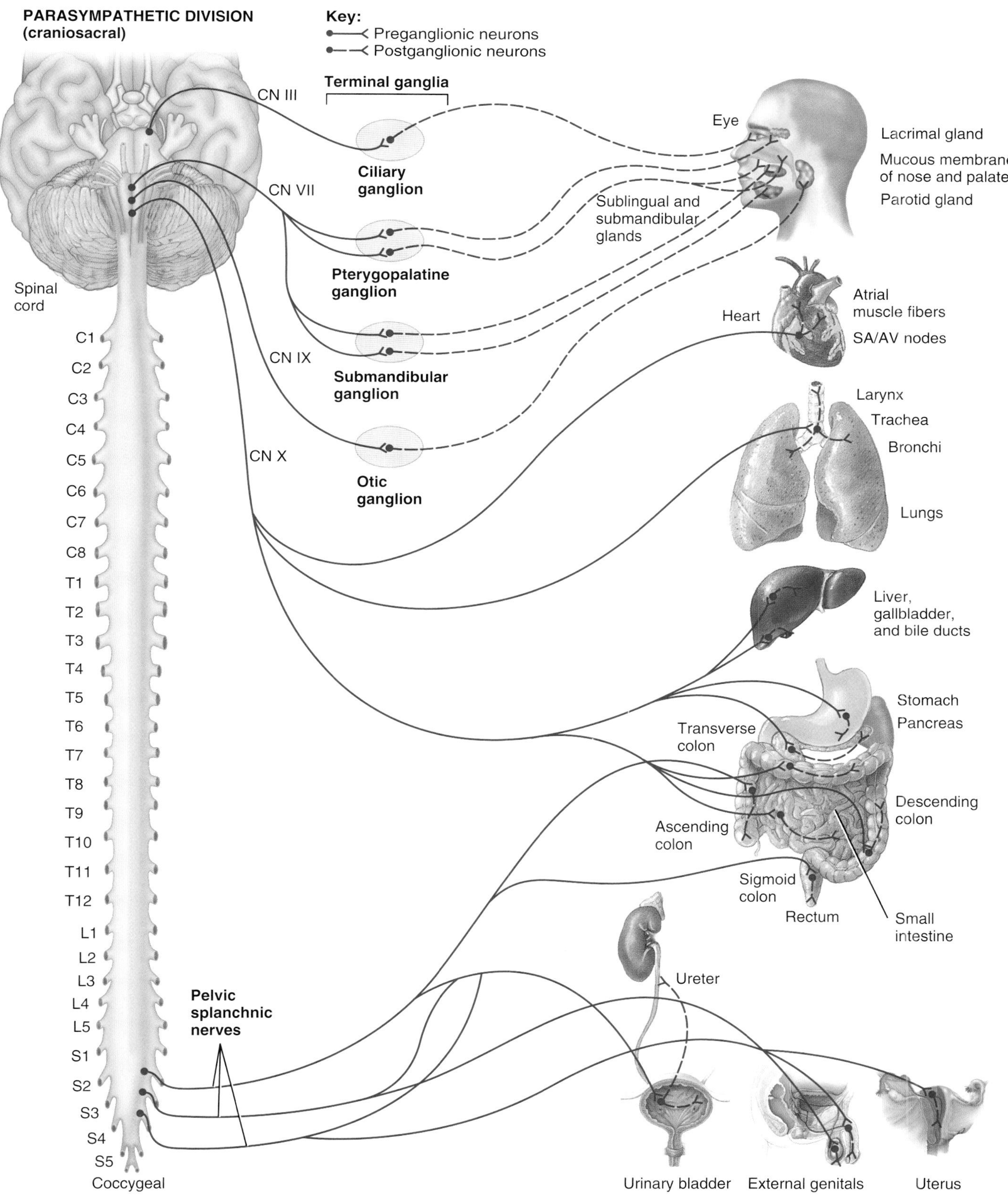

Which ganglia are associated with the parasympathetic division? Sympathetic division?

Autonomic Ganglia

The autonomic ganglia may be divided into three general groups: Two of the groups are components of the sympathetic division, and one group is a component of the parasympathetic division.

SYMPATHETIC GANGLIA The sympathetic ganglia are the sites of synapses between sympathetic preganglionic and postganglionic neurons. The two groups of sympathetic ganglia are sympathetic trunk ganglia and prevertebral ganglia. **Sympathetic trunk ganglia** (also called *vertebral chain ganglia* or *paravertebral ganglia*) lie in a vertical row on either side of the vertebral column. These ganglia extend from the base of the skull to the coccyx (Figure 15.2). Because the sympathetic trunk ganglia are near the spinal cord, most sympathetic preganglionic axons are short. Postganglionic axons from sympathetic trunk ganglia mostly innervate organs above the diaphragm. Examples of sympathetic trunk ganglia are the **superior, middle,** and **inferior cervical ganglia** (Figure 15.2).

The second group of sympathetic ganglia, the **prevertebral** (*collateral*) **ganglia,** lies anterior to the vertebral column and close to the large abdominal arteries. In general, postganglionic axons from prevertebral ganglia innervate organs below the diaphragm. There are three major prevertebral ganglia: (1) The **celiac ganglion** (SĒ-lē-ak) is on either side of the celiac artery just inferior to the diaphragm. (2) The **superior mesenteric ganglion** is near the beginning of the superior mesenteric artery in the upper abdomen. (3) The **inferior mesenteric ganglion** is near the beginning of the inferior mesenteric artery in the middle of the abdomen (Figure 15.2; see also Figure 15.5).

PARASYMPATHETIC GANGLIA Preganglionic axons of the parasympathetic division synapse with postganglionic neurons in **terminal** (*intramural*) **ganglia.** Most of these ganglia are located close to or actually within the wall of a visceral organ. Because the axons of parasympathetic preganglionic neurons extend from the CNS to a terminal ganglion in an innervated organ, they are longer than most of the axons of sympathetic preganglionic neurons. Examples of terminal ganglia include the **ciliary ganglion, pterygopalatine ganglion, submandibular ganglion,** and **otic ganglion** (Figure 15.3 on page 529).

Postganglionic Neurons

Once axons of sympathetic preganglionic neurons pass to sympathetic trunk ganglia, they may connect with postganglionic neurons in one of the following ways (Figure 15.4):

1. An axon may synapse with postganglionic neurons in the ganglion it first reaches.
2. An axon may ascend or descend to a higher or lower ganglion before synapsing with postganglionic neurons. The axons of incoming sympathetic preganglionic neurons that pass up or down the sympathetic trunk collectively form the **sympathetic chains,** the fibers on which the ganglia are strung.
3. An axon may continue, without synapsing, through the sympathetic trunk ganglion to end at a prevertebral ganglion and synapse with postganglionic neurons there.

In addition, some preganglionic sympathetic axons extend to and terminate in the adrenal medullae.

A single sympathetic preganglionic fiber has many axon collaterals (branches) and may synapse with 20 or more postganglionic neurons. This pattern of projection is an example of divergence and helps explain why many sympathetic responses affect almost the entire body simultaneously. After exiting their ganglia, the postganglionic axons typically terminate in several visceral effectors (see Figure 15.2).

Axons of preganglionic neurons of the parasympathetic division pass to terminal ganglia near or within a visceral effector (see Figure 15.3). In the ganglion, the presynaptic neuron usually synapses with only four or five postsynaptic neurons, all of which supply a single visceral effector, allowing parasympathetic responses to be localized to a single effector.

Autonomic Plexuses

In the thorax, abdomen, and pelvis, axons of both sympathetic and parasympathetic neurons form tangled networks called **autonomic plexuses,** many of which lie along major arteries. The autonomic plexuses also may contain sympathetic ganglia and axons of autonomic sensory neurons. The major plexuses in the thorax are the **cardiac plexus,** which supplies the heart, and the **pulmonary plexus,** which supplies the bronchial tree (Figure 15.5 on page 532; see also Figure 15.2).

The abdomen and pelvis also contain major autonomic plexuses (Figure 15.5) and often the plexuses are named after the artery along which they are distributed. The **celiac** (*solar*) **plexus** is the largest autonomic plexus and surrounds the celiac and superior mesenteric arteries. It contains two large celiac ganglia and a dense network of autonomic axons and is distributed to the liver, gallbladder, stomach, pancreas, spleen, kidneys, adrenal medullae, testes, and ovaries. The **superior mesenteric plexus** contains the superior mesenteric ganglion and supplies the small and large intestine. The **inferior mesenteric plexus** contains the inferior mesenteric ganglion, which innervates the large intestine. The **hypogastric plexus** is anterior to the fifth lumbar vertebra and supplies pelvic viscera. The **renal plexuses,** located near the kidneys, contain the renal ganglia and supply the renal arteries within the kidneys and the ureters.

With this background in mind, we can now examine some of the specific structural features of the sympathetic and parasympathetic divisions of the ANS in more detail.

Figure 15.4 Types of connections between ganglia and postganglionic neurons in the sympathetic division of the ANS. Numbers correspond to descriptions in the text. Also illustrated are the gray and white rami communicantes.

Sympathetic ganglia lie in two chains on either side of the vertebral column (sympathetic trunk ganglia) and near large abdominal arteries anterior to the vertebral column (prevertebral ganglia).

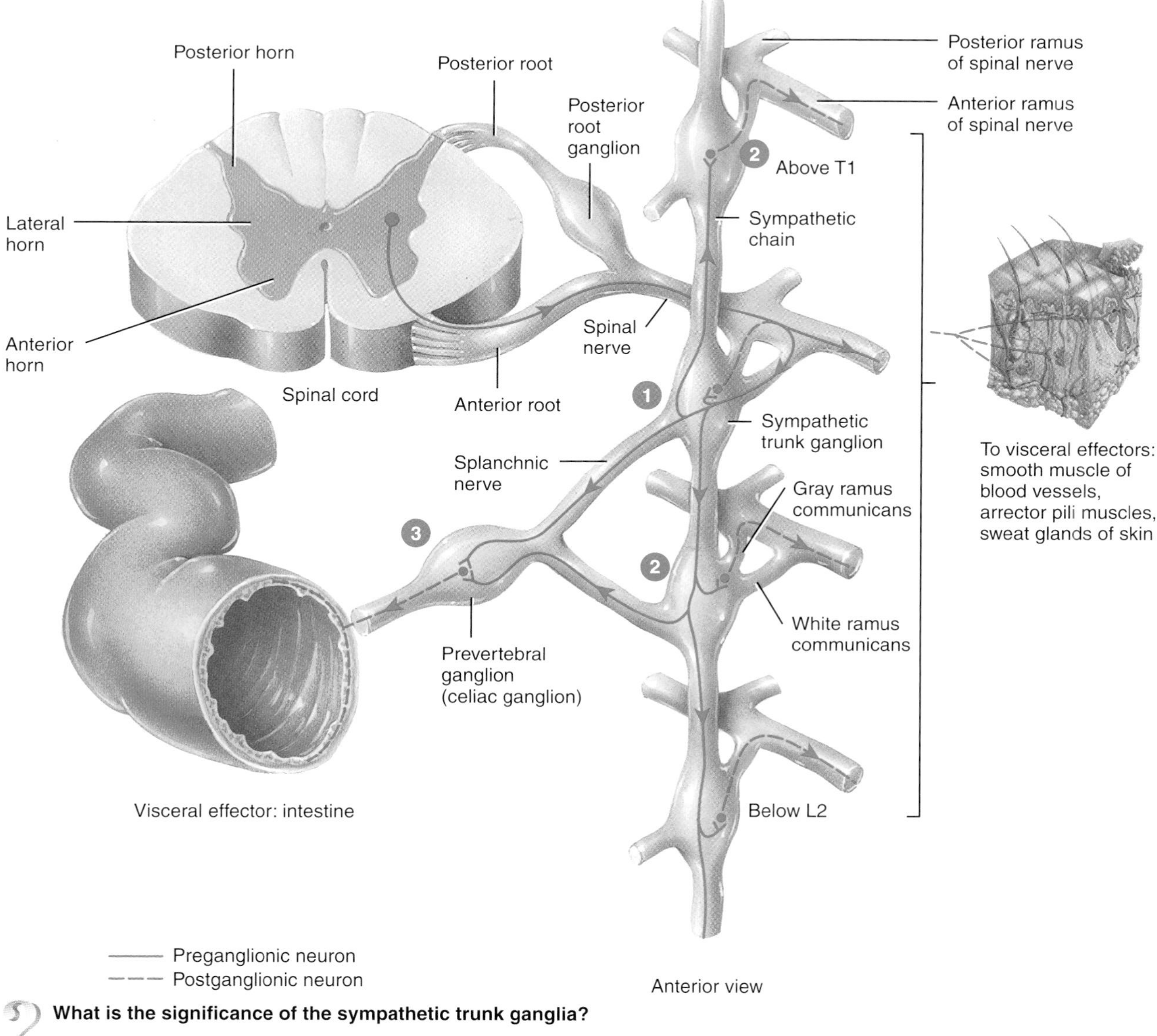

What is the significance of the sympathetic trunk ganglia?

Structure of the Sympathetic Division

Cell bodies of sympathetic preganglionic neurons are part of the lateral horns of all thoracic segments and of the first two lumbar segments of the spinal cord (see Figure 15.2). The preganglionic axons leave the spinal cord through the anterior root of a spinal nerve along with the somatic motor neurons at the same segmental level. After exiting through the intervertebral foramina, the myelinated preganglionic sympathetic axons enter a short pathway called a **white ramus** before passing to the nearest sympathetic trunk ganglion on the same side (see Figure 15.4). Collectively, the white rami are called the **white rami communicantes** (kō-mū-ni-KAN-tēz; singular is **ramus communicans**). Thus, white rami communicantes are structures containing sympathetic preganglionic axons that connect the anterior ramus of the spinal nerve with the ganglia of the sympathetic trunk. The "white" in their name indicates that they contain myelinated axons. Only the thoracic and first two or three lumbar nerves have white rami communicantes.

Recall that some of the incoming sympathetic preganglionic neurons synapse with postganglionic neurons in the sympathetic

Figure 15.5 Autonomic plexuses in the thorax, abdomen, and pelvis.

An autonomic plexus is a network of sympathetic and parasympathetic axons that sometimes also includes autonomic sensory axons and sympathetic ganglia.

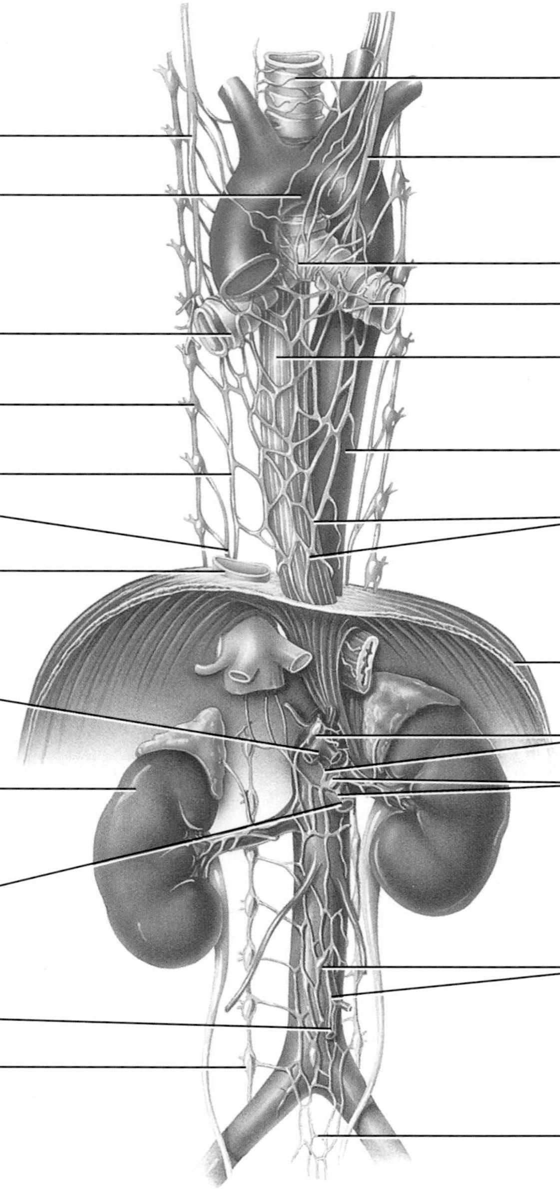

Which is the largest autonomic plexus?

trunk, either in the ganglion at the level of entry or in a ganglion farther up or down the sympathetic trunk. The axons of some of these postganglionic neurons leave the sympathetic trunk by entering a short pathway called a **gray ramus** and then merge with the anterior ramus of a spinal nerve to supply visceral effectors such as sweat glands, smooth muscle in blood vessels, and arrector pili muscles of hair follicles. Therefore, **gray rami communicantes** are structures containing sympathetic postganglionic axons that connect the ganglia of the sympathetic trunk to spinal nerves (see Figure 15.4). The "gray" in their name indicates that they contain unmyelinated axons. Gray rami communicantes outnumber the white rami because there is a gray ramus leading to each of the 31 pairs of spinal nerves.

The paired sympathetic trunk ganglia are arranged anterior and lateral to the vertebral column, one on either side. Typically, there are 3 cervical, 11 or 12 thoracic, 4 or 5 lumbar, 4 or 5 sacral sympathetic trunk ganglia, and 1 coccygeal ganglion. The right and left coccygeal ganglia are fused together and usually lie at the midline. Although the sympathetic trunk ganglia extend inferiorly from the neck, chest, and abdomen to the coccyx, they receive preganglionic axons only from the thoracic and lumbar segments of the spinal cord (see Figure 15.2).

The cervical portion of each sympathetic trunk is located in the neck and is subdivided into superior, middle, and inferior ganglia (see Figure 15.2). Postganglionic neurons leaving the **superior cervical ganglion** serve the head and heart. They are distributed to sweat glands, smooth muscle of the eye, blood vessels of the face, lacrimal glands, nasal mucosa, the heart, and the submandibular, sublingual, and parotid salivary glands. Postganglionic neurons leaving the **middle cervical ganglion** and the **inferior cervical ganglion** innervate the heart.

The thoracic portion of each sympathetic trunk lies anterior to the necks of the corresponding ribs. This region of the sympathetic trunk receives most of the sympathetic preganglionic axons. Postganglionic neurons from the thoracic sympathetic trunk innervate the heart, lungs, bronchi, and other thoracic viscera. In the skin, these neurons also innervate sweat glands, blood vessels, and arrector pili muscles of hair follicles. The lumbar portion of each sympathetic trunk lies lateral to the corresponding lumbar vertebrae. The sacral region of the sympathetic trunk lies in the pelvic cavity on the medial side of the sacral foramina.

Recall that some sympathetic preganglionic axons pass through the sympathetic trunk without terminating in it. Beyond the trunk, they form nerves known as **splanchnic nerves** (SPLANK-nik; see Figures 15.2 and 15.4), which extend to and terminate in the outlying prevertebral ganglia. Splanchnic nerves from the thoracic area terminate in the **celiac ganglion,** where the preganglionic neurons synapse with postganglionic cell bodies. Preganglionic axons from the fifth through ninth or tenth thoracic ganglia (T5–T9 or T10) form the **greater splanchnic nerve.** It pierces the diaphragm, and enters the celiac ganglion of the celiac plexus. From there, postganglionic neurons extend to the stomach, spleen, liver, kidney, and small intestine. Preganglionic axons from the tenth and eleventh thoracic ganglia (T10–T11) form the **lesser splanchnic nerve.** It pierces the diaphragm, and passes through the celiac plexus to enter the superior mesenteric ganglion of the superior mesenteric plexus. Postganglionic neurons from this ganglion innervate the small intestine and colon. The **lowest splanchnic nerve,** not always present, is formed by preganglionic axons from the twelfth thoracic ganglia (T12) or a branch of the lesser splanchnic nerve. It passes through the diaphragm, and enters the renal plexus near the kidney. Postganglionic neurons from the renal plexus supply kidney arterioles and the ureter. Preganglionic axons that form the **lumbar splanchnic nerve** from the first through third lumbar ganglia (L1–L3) enter the inferior mesenteric plexus and terminate in the inferior mesenteric ganglion, where they synapse with postganglionic neurons. Axons of postganglionic neurons extend through the hypogastric plexus and supply the distal colon and rectum, urinary bladder, and genital organs. Postganglionic axons leaving the prevertebral ganglia follow the course of various arteries to abdominal and pelvic visceral effectors.

Sympathetic preganglionic neurons also extend to the adrenal medullae. Developmentally, the adrenal medullae and sympathetic ganglia are derived from the same tissue, the neural crest (see Figure 14.28 on page 515). The adrenal medullae are modified sympathetic ganglia, and their cells are similar to sympathetic postganglionic neurons. Rather than extending to another organ, however, these cells release hormones into the blood. Upon stimulation by sympathetic preganglionic neurons, the adrenal medullae release a mixture of catecholamine hormones—about 80% **epinephrine,** 20% **norepinephrine,** and a trace amount of **dopamine.**

Horner's Syndrome

In **Horner's syndrome,** the sympathetic innervation to one side of the face is lost due to an inherited mutation, an injury, or a disease that affects sympathetic outflow through the superior cervical ganglion. Symptoms occur on the affected side and include ptosis (drooping of the upper eyelid), miosis (constricted pupil), and anhidrosis (lack of sweating). ■

Structure of the Parasympathetic Division

Cell bodies of parasympathetic preganglionic neurons are found in nuclei in the brain stem and in the lateral horns of the second through fourth sacral segments of the spinal cord (see Figure 15.3). Their axons emerge as part of a cranial nerve or as part of the anterior root of a spinal nerve. The **cranial parasympathetic outflow** consists of preganglionic axons that extend from the brain stem in four cranial nerves. The **sacral parasympathetic outflow** consists of preganglionic axons in anterior roots of the second through fourth sacral nerves. The preganglionic axons of both the cranial and sacral outflows end in terminal ganglia, where they synapse with postganglionic neurons.

The cranial outflow has four pairs of ganglia and the plexuses associated with the vagus (X) nerve. The four pairs of cranial parasympathetic ganglia innervate structures in the head and are located close to the organs they innervate (see Figure 15.3).

1. The **ciliary ganglia** lie lateral to each optic (II) nerve near the posterior aspect of the orbit. Preganglionic axons pass with the oculomotor (III) nerves to the ciliary ganglia. Postganglionic axons from the ganglia innervate smooth muscle fibers in the eyeball.
2. The **pterygopalatine ganglia** (ter′-i-gō-PAL-a-tīn) are located laterally to the sphenopalatine foramen, between the sphenoid and palatine bones. They receive preganglionic axons from the facial (VII) nerve and send postganglionic axons to the nasal mucosa, palate, pharynx, and lacrimal glands.
3. The **submandibular ganglia** are found near the ducts of the submandibular salivary glands. They receive preganglionic axons from the facial nerves and send postganglionic axons to the submandibular and sublingual salivary glands.
4. The **otic ganglia** are situated just inferior to each foramen ovale. They receive preganglionic axons from the glossopharyngeal (IX) nerves and send postganglionic axons to the parotid salivary glands.

Preganglionic axons that leave the brain as part of the vagus (X) nerves carry nearly 80% of the total craniosacral outflow. Vagal axons extend to many terminal ganglia in the thorax and abdomen. Because the terminal ganglia are close to or in the walls of their visceral effectors, postganglionic parasympathetic axons are very short. As the vagus nerve passes through the thorax, it sends axons to the heart and the airways of the lungs. In the abdomen, it supplies the liver, gallbladder, stomach, pancreas, small intestine, and part of the large intestine.

The sacral parasympathetic outflow consists of preganglionic axons from the anterior roots of the second through fourth sacral nerves (S2–S4) and they form the **pelvic splanchnic nerves** (see Figure 15.3). These nerves synapse with parasympathetic postganglionic neurons located in terminal ganglia in the walls of the innervated viscera. From the ganglia, parasympathetic postganglionic axons innervate smooth muscle and glands in the walls of the colon, ureters, urinary bladder, and reproductive organs.

▶ **CHECKPOINT**

3. Why is the sympathetic division called the thoracolumbar division even though its ganglia extend from the cervical to the sacral region?
4. List the organs served by each sympathetic and parasympathetic ganglion.
5. Describe the locations of sympathetic trunk ganglia, prevertebral ganglia, and terminal ganglia. Which types of autonomic neurons synapse in each type of ganglion?
6. Why may the sympathetic division produce simultaneous effects throughout the body, whereas parasympathetic effects typically are localized to specific organs?

ANS NEUROTRANSMITTERS AND RECEPTORS

▶ **OBJECTIVE**

Describe the neurotransmitters and receptors involved in autonomic responses.

Based on the neurotransmitter they produce and release, autonomic neurons are classified as either cholinergic or adrenergic. The receptors for the neurotransmitters are integral membrane proteins located in the plasma membrane of the postsynaptic neuron or effector cell.

Cholinergic Neurons and Receptors

Cholinergic neurons (kō′-lin-ER-jik) release the neurotransmitter **acetylcholine (ACh).** In the ANS, the cholinergic neurons include (1) all sympathetic and parasympathetic preganglionic neurons, (2) sympathetic postganglionic neurons that innervate most sweat glands, and (3) all parasympathetic postganglionic neurons (Figure 15.6).

ACh is stored in synaptic vesicles and released by exocytosis. It then diffuses across the synaptic cleft and binds with specific **cholinergic receptors,** integral membrane proteins in the *postsynaptic* plasma membrane. The two types of cholinergic receptors, both of which bind ACh, are nicotinic receptors and muscarinic receptors. **Nicotinic receptors** are present in the plasma membrane of dendrites and cell bodies of both sympathetic and parasympathetic postganglionic neurons (Figure 15.6a, b) and in the motor end plate at the neuromuscular junction. They are so named because nicotine mimics the action of ACh by binding to these receptors. (Nicotine, a natural substance in tobacco leaves, is not a naturally occurring substance in humans and is not normally present in nonsmokers.) **Muscarinic receptors** are present in the plasma membranes of all effectors (smooth muscle, cardiac muscle, and glands) innervated by parasympathetic postganglionic axons. In addition, most sweat glands receive their innervation from *cholinergic* sympathetic postganglionic neurons and possess muscarinic receptors (see Figure 15.6b). These receptors are so named because a mushroom poison called muscarine mimics the actions of ACh by binding to them. Nicotine does not activate muscarinic receptors, and muscarine does not activate nicotinic receptors, but ACh does activate both types of cholinergic receptors.

Activation of nicotinic receptors by ACh causes depolarization and thus excitation of the postsynaptic cell, which can be a postganglionic neuron, an autonomic effector, or a skeletal muscle fiber. Activation of muscarinic receptors by ACh sometimes causes depolarization (excitation) and sometimes causes hyperpolarization (inhibition), depending on which particular cell bears the muscarinic receptors. For example, binding of ACh to muscarinic receptors inhibits (relaxes) smooth muscle sphincters in the gastrointestinal tract. By contrast, ACh excites muscarinic receptors in smooth muscle fibers in the circular

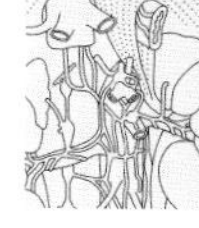

Figure 15.6 **Cholinergic neurons (blue) and adrenergic neurons (orange) in the sympathetic and parasympathetic divisions.** Cholinergic neurons release acetylcholine; adrenergic neurons release norepinephrine. Cholinergic and adrenergic receptors all are integral membrane proteins located in the plasma membrane of a postsynaptic neuron or an effector cell.

Most sympathetic postganglionic neurons are adrenergic; other autonomic neurons are cholinergic.

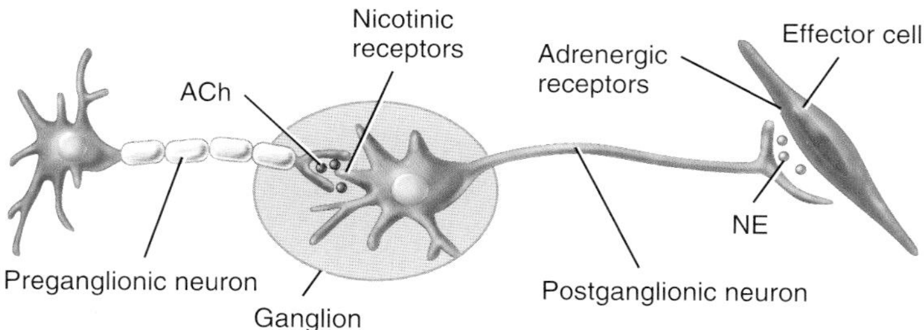

(a) Sympathetic division–innervation to most effector tissues

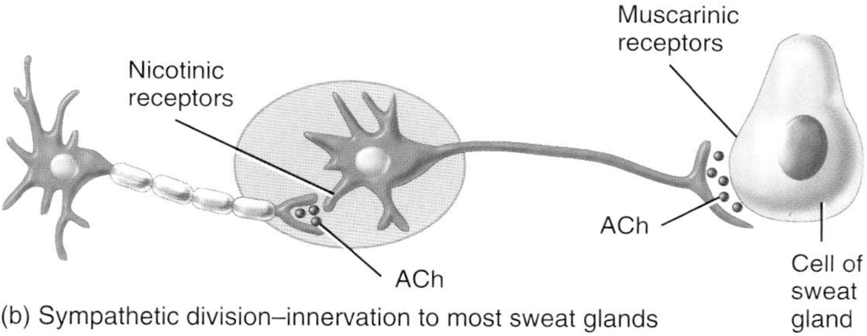

(b) Sympathetic division–innervation to most sweat glands

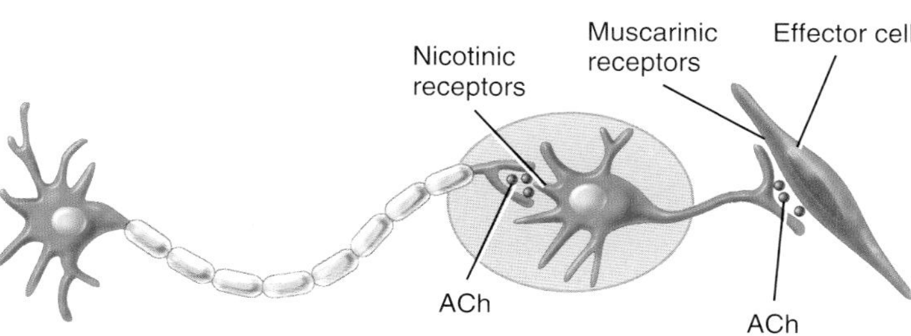

(c) Parasympathetic division

Which neurons are cholinergic and possess nicotinic ACh receptors? What type of receptors for ACh do the effector tissues innervated by these neurons possess?

muscles of the iris of the eye, causing them to contract. Because acetylcholine is quickly inactivated by the enzyme **acetylcholinesterase (AChE),** effects triggered by cholinergic neurons are brief.

Adrenergic Neurons and Receptors

In the ANS, **adrenergic neurons** (ad′-ren-ER-jik) release **norepinephrine (NE),** also known as **noradrenalin** (Figure 15.6a). Most sympathetic postganglionic neurons are adrenergic. Like ACh, NE is synthesized and stored in synaptic vesicles and released by exocytosis. Molecules of NE diffuse across the synaptic cleft and bind to specific adrenergic receptors on the postsynaptic membrane, causing either excitation or inhibition of the effector cell.

Adrenergic receptors bind both norepinephrine and epinephrine. The norepinephrine can be either released as a neurotransmitter by sympathetic postganglionic neurons or released as a hormone into the blood by the adrenal medullae; epinephrine is released as a hormone. The two main types of adrenergic receptors are **alpha (α) receptors** and **beta (β) receptors,** which are found on visceral effectors innervated by most sympathetic postganglionic axons. These receptors are further classified into subtypes—α_1, α_2, β_1, β_2, and β_3—based on the specific responses they elicit and by their selective binding of drugs that activate or block them. Although there are some exceptions, activation of α_1 and β_1 receptors generally produces excitation, and activation of α_2 and β_2 receptors causes inhibition of effector tissues. β_3 receptors are present only on cells of brown adipose tissue, where their activation causes thermogenesis (heat production). Cells of most effectors contain either alpha or beta receptors; some visceral effector cells contain both. Norepinephrine stimulates alpha receptors more strongly than beta receptors; epinephrine is a potent stimulator of both alpha and beta receptors.

The activity of norepinephrine at a synapse is terminated either when the NE is taken up by the axon that released it or when the NE is enzymatically inactivated by either **catechol-*O*-methyltransferase (COMT)** or **monoamine oxidase (MAO).** Compared to ACh, norepinephrine lingers in the synaptic cleft for a longer time. Thus, effects triggered by adrenergic neurons typically are longer lasting than those triggered by cholinergic neurons.

Table 15.2 describes the locations of cholinergic and adrenergic receptors and summarizes the responses that occur when each type of receptor is activated.

Receptor Agonists and Antagonists

A large variety of drugs and natural products can selectively activate or block specific cholinergic or adrenergic receptors. An **agonist** is a substance that binds to and activates a receptor, in the process mimicking the effect of a natural neurotransmitter or hormone. Phenylephrine, an adrenergic agonist at α_1 receptors, is a common ingredient in cold and sinus medications. Because it constricts blood vessels in the nasal mucosa, phenylephrine

TABLE 15.2 Locations and Responses of Adrenergic and Cholinergic Receptors

Type of Receptor	Major Locations	Effects of Receptor Activation
Cholinergic	Integral proteins in postsynaptic plasma membranes; activated by the neurotransmitter acetylcholine.	
Nicotinic	Plasma membrane of postganglionic sympathetic and parasympathetic neurons.	Excitation → impulses in postganglionic neurons.
	Cells of adrenal medullae.	Epinephrine and norepinephrine secretion.
	Sarcolemma of skeletal muscle fibers (motor end plate).	Excitation → contraction.
Muscarinic	Effectors innervated by parasympathetic postganglionic neurons.	In some receptors, excitation; in others, inhibition.
	Sweat glands innervated by cholinergic sympathetic postganglionic neurons.	Increased sweating.
	Skeletal muscle blood vessels innervated by cholinergic sympathetic postganglionic neurons.	Inhibition → relaxation → vasodilation.
Adrenergic	Integral proteins in postsynaptic plasma membranes; activated by the neurotransmitter norepinephrine, and by the hormones norepinephrine and epinephrine.	
α_1	Smooth muscle fibers in blood vessels that serve salivary glands, skin, mucosal membranes, kidneys, and abdominal viscera; radial muscle in iris of eye; sphincter muscles of stomach and urinary bladder.	Excitation → contraction, which causes vasoconstriction, dilation of pupil, and closing of sphincters.
	Salivary gland cells.	Secretion of K^+ and water.
	Sweat glands on palms and soles.	Increased sweating.
α_2	Smooth muscle fibers in some blood vessels.	Inhibition → relaxation → vasodilation.
	Cells of pancreatic islets that secrete the hormone insulin (beta cells).	Decreased insulin secretion.
	Pancreatic acinar cells.	Inhibition of digestive enzyme secretion.
	Platelets in blood.	Aggregation to form platelet plug.
β_1	Cardiac muscle fibers.	Excitation → increased force and rate of contraction.
	Juxtaglomerular cells of kidneys.	Renin secretion.
	Posterior pituitary.	Secretion of antidiuretic hormone.
	Adipose cells.	Breakdown of triglycerides → release of fatty acids into blood.
β_2	Smooth muscle in walls of airways; in blood vessels that serve the heart, skeletal muscle, adipose tissue, and liver; and in walls of visceral organs, such as the urinary bladder.	Inhibition → relaxation, which causes dilation of airways, vasodilation, and relaxation of organ walls.
	Ciliary muscle in eye.	Inhibition → relaxation.
	Hepatocytes in liver.	Glycogenolysis (breakdown of glycogen into glucose).
β_3	Brown adipose tissue.	Thermogenesis (heat production).

reduces production of mucus, thus relieving nasal congestion. An **antagonist** is a substance that binds to and blocks a receptor, thereby preventing a natural neurotransmitter or hormone from exerting its effect. For example, atropine blocks muscarinic ACh receptors, dilates the pupils, reduces glandular secretions, and relaxes smooth muscle in the gastrointestinal tract. As a result, it is used to dilate the pupils during eye examinations, in the treatment of smooth muscle disorders such as iritis and intestinal hypermotility, and as an antidote for chemical warfare agents that inactivate acetylcholinesterase.

Propranolol (Inderal®) often is prescribed for patients with hypertension (high blood pressure). It is a nonselective beta blocker, meaning it binds to all types of beta receptors and prevents their activation by epinephrine and norepinephrine. The desired effects of propranolol are due to its *blockade* of β_1 receptors—namely, decreased heart rate and force of contraction and a consequent decrease in blood pressure. Undesired effects due to blockade of β_2 receptors may include hypoglycemia (low blood glucose), resulting from decreased glycogen breakdown and decreased gluconeogenesis (the conversion of a noncarbohydrate into glucose in the liver), and mild bronchoconstriction (narrowing of the airways). If these side effects pose a threat to the patient, a selective β_1 blocker such as metoprolol (Lopressor®) can be prescribed instead of propranolol.

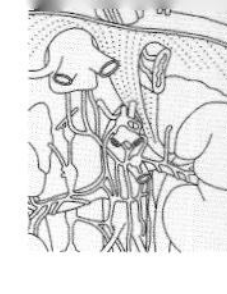

► CHECKPOINT

7. Why are cholinergic and adrenergic neurons so named?
8. What neurotransmitters and hormones bind to adrenergic receptors?
9. What do the terms *agonist* and *antagonist* mean?

PHYSIOLOGICAL EFFECTS OF THE ANS

► OBJECTIVE

Describe the major responses of the body to stimulation by the sympathetic and parasympathetic divisions of the ANS.

Autonomic Tone

As noted earlier, most body organs receive innervation from both divisions of the ANS, which typically work in opposition to one another. The balance between sympathetic and parasympathetic activity, called **autonomic tone,** is regulated by the hypothalamus. Typically, the hypothalamus turns up sympathetic tone at the same time it turns down parasympathetic tone, and vice versa. The two divisions can affect body organs differently because their postganglionic neurons release different neurotransmitters and because the effector organs possess different adrenergic and cholinergic receptors. A few structures receive only sympathetic innervation—sweat glands, arrector pili muscles attached to hair follicles in the skin, the kidneys, the spleen, most blood vessels, and the adrenal medullae (see Figure 15.2). In these structures there is no opposition from the parasympathetic division. Still, an increase in sympathetic tone has one effect, and a decrease in sympathetic tone produces the opposite effect.

Sympathetic Responses

During physical or emotional stress, the sympathetic division dominates the parasympathetic division. High sympathetic tone favors body functions that can support vigorous physical activity and rapid production of ATP. At the same time, the sympathetic division reduces body functions that favor the storage of energy. Besides physical exertion, a variety of emotions—such as fear, embarrassment, or rage—stimulate the sympathetic division. Visualizing body changes that occur during "E situations" such as exercise, emergency, excitement, and embarrassment will help you remember most of the sympathetic responses. Activation of the sympathetic division and release of hormones by the adrenal medullae set in motion a series of physiological responses collectively called the **fight-or-flight response,** which includes the following effects:

- The pupils of the eyes dilate.
- Heart rate, force of heart contraction, and blood pressure increase.
- The airways dilate, allowing faster movement of air into and out of the lungs.
- The blood vessels that supply the kidneys and gastrointestinal tract constrict, which decreases blood flow through these tissues. The result is a slowing of urine formation and digestive activities, which are not essential during exercise.
- Blood vessels that supply organs involved in exercise or fighting off danger—skeletal muscles, cardiac muscle, liver, and adipose tissue—dilate, allowing greater blood flow through these tissues.
- Liver cells perform glycogenolysis (breakdown of glycogen to glucose), and adipose tissue cells perform lipolysis (breakdown of triglycerides to fatty acids and glycerol).
- Release of glucose by the liver increases blood glucose level.
- Processes that are not essential for meeting the stressful situation are inhibited. For example, muscular movements of the gastrointestinal tract and digestive secretions slow down or even stop.

The effects of sympathetic stimulation are longer lasting and more widespread than the effects of parasympathetic stimulation for three reasons: (1) Sympathetic postganglionic axons diverge more extensively; as a result, many tissues are activated simultaneously. (2) Acetylcholinesterase quickly inactivates acetylcholine, whereas norepinephrine lingers in the synaptic cleft for a longer period. (3) Epinephrine and norepinephrine secreted into the blood from the adrenal medulla intensify and prolong the responses caused by NE liberated from sympathetic postganglionic axons. These blood-borne hormones circulate throughout the body, affecting all tissues that have alpha and beta receptors. In time, blood-borne NE and epinephrine are inactivated by enzymatic destruction in the liver.

Parasympathetic Responses

In contrast to the "fight-or-flight" activities of the sympathetic division, the parasympathetic division enhances "rest and digest" activities. Parasympathetic responses support body functions that conserve and restore body energy during times of rest and recovery. In the quiet intervals between periods of exercise, parasympathetic impulses to the digestive glands and the smooth muscle of the gastrointestinal tract predominate over sympathetic impulses. This allows energy-supplying food to be digested and absorbed. At the same time, parasympathetic responses reduce body functions that support physical activity.

The acronym *SLUDD* can be helpful in remembering five parasympathetic responses. It stands for salivation (S), lacrimation (L), urination (U), digestion (D), and defecation (D). All of these activities are stimulated mainly by the parasympathetic division. Besides the increasing SLUDD responses, other important parasympathetic responses are "three decreases": decreased heart rate, decreased diameter of airways (bronchoconstriction), and decreased diameter (constriction) of the pupils.

Table 15.3 compares the structural and functional features of the sympathetic and parasympathetic divisions of the ANS. Table 15.4 lists the responses of glands, cardiac muscle, and smooth muscle to stimulation by the sympathetic and parasympathetic divisions of the ANS.

TABLE 15.3 Comparison of Sympathetic and Parasympathetic Divisions of the ANS

	Sympathetic (thoracolumbar)	Parasympathetic (craniosacral)
Distribution	Wide regions of the body: skin, sweat glands, arrector pili muscles of hair follicles, adipose tissue, smooth muscle of blood vessels.	Limited mainly to head and to viscera of thorax, abdomen, abdomen, and pelvis; some blood vessels.
Location of preganglionic neuron cell bodies and site of outflow	Cell bodies of preganglionic neurons are located in lateral gray horns of spinal cord segments T1–L2. Axons of preganglionic neurons constitute thoracolumbar outflow.	Cell bodies of preganglionic neurons are located in the nuclei of cranial nerves III, VII, IX, and X and the lateral gray horns of spinal cord segments S2–S4. Axons of preganglionic neurons constitute craniosacral outflow.
Associated ganglia	Two types: sympathetic trunk ganglia and prevertebral ganglia.	One type: terminal ganglia.
Ganglia locations	Close to CNS and distant from visceral effectors.	Typically near or within wall of visceral effectors.
Axon length and divergence	Preganglionic neurons with short axons synapse with many postganglionic neurons with long axons that pass to many visceral effectors.	Preganglionic neurons with long axons usually synapse with four to five postganglionic neurons with short axons that pass to a single visceral effector.

TABLE 15.4 Effects of Sympathetic and Parasympathetic Divisions of the ANS

Visceral Effector	Effect of Sympathetic Stimulation (α or β adrenergic receptors, except as noted)*	Effect of Parasympathetic Stimulation (muscarinic ACh receptors)
Glands		
Adrenal medullae	Secretion of epinephrine and norepinephrine (nicotinic ACh receptors).	No known effect.
Lacrimal (tear)	Slight secretion of tears (α).	Secretion of tears.
Pancreas	Inhibits secretion of digestive enzymes and the hormone insulin (α_2); promotes secretion of the hormone glucagon (β_2).	Secretion of digestive enzymes and the hormone insulin.
Posterior pituitary	Secretion of antidiuretic hormone (ADH) (β_1).	No known effect.
Pineal	Increases synthesis and release of melatonin (β).	No known effect.
Sweat	Increases sweating in most body regions (muscarinic ACh receptors); sweating on palms and soles (α_1).	No known effect.
Adipose tissue[†]	Lipolysis (breakdown of triglycerides into fatty acids and glycerol) (β_1); release of fatty acids into blood (β_1 and β_3).	No known effect.
Liver[†]	Glycogenolysis (conversion of glycogen into glucose); gluconeogenesis (conversion of noncarbohydrates into glucose); decreased bile secretion (α and β_2).	Glycogen synthesis; increased bile secretion.
Kidney, juxtaglomerular cells[†]	Secretion of renin (β_1).	No known effect.
Cardiac (Heart) Muscle	Increased heart rate and force of atrial and ventricular contractions (β_1).	Decreased heart rate; decreased force of atrial contraction.
Smooth Muscle		
Iris, radial muscle	Contraction → dilation of pupil (α_1).	No known effect.
Iris, circular muscle	No known effect.	Contraction → constriction of pupil.
Ciliary muscle of eye	Relaxation for distant vision (β_2).	Contraction for close vision.
Lungs, bronchial muscle	Relaxation → airway dilation (β_2).	Contraction → airway constriction.
Gallbladder and ducts	Relaxation (β_2).	Contraction → increased release of bile intosmall intestine.
Stomach and intestines	Decreased motility and tone (α_1, α_2, β_2); contraction of sphincters (α_1).	Increased motility and tone; relaxation of sphincters.

	Sympathetic (thoracolumbar)	Parasympathetic (craniosacral)
Rami communicantes	Both present; white rami communicantes contain myelinated preganglionic axons, and gray rami communicantes contain unmyelinated postganglionic axons.	Neither present.
Neurotransmitters	Preganglionic neurons release acetylcholine (ACh), which is excitatory and stimulates postganglionic neurons; most postganglionic neurons release norepinephrine (NE); postganglionic neurons that innervate most sweat glands and some blood vessels in skeletal muscle release ACh.	Preganglionic neurons release acetylcholine (ACh), which is excitatory and stimulates postganglionic neurons; postganglionic neurons release ACh.
Physiological effects	Fight-or-flight responses.	Rest-and-digest activities.

Visceral Effector	Effect of Sympathetic Stimulation (α or β adrenergic receptors, except as noted)*	Effect of Parasympathetic Stimulation (muscarinic ACh receptors)
Smooth Muscle (continued)		
Spleen	Contraction and discharge of stored blood into general circulation (α_1).	No known effect.
Ureter	Increases motility (α_1).	Increases motility (?).
Urinary bladder	Relaxation of muscular wall (β_2); contraction of sphincter (α_1).	Contraction of muscular wall; relaxation of sphincter.
Uterus	Inhibits contraction in nonpregnant women (β_2); promotes contraction in pregnant women (α_1).	Minimal effect.
Sex organs	In males: contraction of smooth muscle of ductus (vas) deferens, seminal vesicle, prostate → ejaculation of semen (α_1).	Vasodilation; erection of clitoris (females) and penis (males).
Hair follicles, arrector pili muscle	Contraction → erection of hairs (α_1).	No known effect.
Vascular Smooth Muscle		
Salivary gland arterioles	Vasoconstriction, which decreases secretion (α_1).	Vasodilation, which increases K^+ and water secretion.
Gastric gland arterioles	Vasoconstriction, which inhibits secretion (α_1).	Secretion of gastric juice.
Intestinal gland arterioles	Vasoconstriction, which inhibits secretion (α_1).	Secretion of intestinal juice.
Coronary (heart) arterioles	Relaxation → vasodilation (β_2); contraction → vasoconstriction (α_1, α_2); contraction → vasoconstriction (muscarinic ACh receptors).	Contraction → vasoconstriction.
Skin and mucosal arterioles	Contraction: vasoconstriction (α_1).	Vasodilation, which may not be physiologically significant.
Skeletal muscle arterioles	Contraction → vasoconstriction (α_1); relaxation → vasodilation (β_2); relaxation → vasodilation (muscarinic ACh receptors).	No known effect.
Abdominal viscera arterioles	Contraction → vasoconstriction (α_1, β_2).	No known effect.
Brain arterioles	Slight contraction → vasoconstriction (α_1).	No known effect.
Kidney arterioles	Constriction of blood vessels → decreased urine volume (α_1).	No known effect.
Systemic veins	Contraction → constriction (α_1); relaxation → dilation (β_2).	No known effect.

*Subcategories of α and β receptors are listed if known.

†Grouped with glands because they release substances into the blood.

► CHECKPOINT

10. Define autonomic tone.

11. What are some examples of the antagonistic effects of the sympathetic and parasympathetic divisions of the autonomic nervous system?

12. What happens during the fight-or-flight response?

13. Why is the parasympathetic division of the ANS called an energy conservation/restoration system?

14. Describe the sympathetic response in a frightening situation for each of the following body parts: hair follicles, iris of eye, lungs, spleen, adrenal medullae, urinary bladder, stomach, intestines, gallbladder, liver, heart, arterioles of the abdominal viscera, and arterioles of skeletal muscles.

INTEGRATION AND CONTROL OF AUTONOMIC FUNCTIONS

► OBJECTIVES

Describe the components of an autonomic reflex.
Explain the relationship of the hypothalamus to the ANS.

Autonomic Reflexes

Autonomic reflexes are responses that occur when nerve impulses pass through an autonomic reflex arc. These reflexes play a key role in regulating controlled conditions in the body, such as *blood pressure,* by adjusting heart rate, force of ventricular contraction, and blood vessel diameter; *digestion,* by adjusting the motility (movement) and muscle tone of the gastrointestinal tract; and *defecation* and *urination,* by regulating the opening and closing of sphincters.

The components of an autonomic reflex arc are as follows:

- **Receptor.** Like the receptor in a somatic reflex arc (see Figure 13.13 on page 461), the receptor in an autonomic reflex arc is the distal end of a sensory neuron, which responds to a stimulus and produces a change that will ultimately trigger nerve impulses. Autonomic sensory receptors are mostly associated with interoceptors.
- **Sensory neuron.** Conducts nerve impulses from receptors to the CNS.
- **Integrating center.** Interneurons within the CNS relay signals from sensory neurons to motor neurons. The main integrating centers for most autonomic reflexes are located in the hypothalamus and brain stem. Some autonomic reflexes, such as those for urination and defecation, have integrating centers in the spinal cord.
- **Motor neurons.** Nerve impulses triggered by the integrating center propagate out of the CNS along motor neurons to an effector. In an autonomic reflex arc, two motor neurons connect the CNS to an effector: The preganglionic neuron conducts motor impulses from the CNS to an autonomic ganglion, and the postganglionic neuron conducts motor impulses from an autonomic ganglion to an effector (see Figure 15.1).
- **Effector.** In an autonomic reflex arc, the effectors are smooth muscle, cardiac muscle, and glands, and the reflex is called an autonomic reflex.

Autonomic Control by Higher Centers

Normally, we are not aware of muscular contractions of our digestive organs, our heartbeat, changes in the diameter of our blood vessels, and pupil dilation and constriction because the integrating centers for these autonomic responses are in the spinal cord or the lower regions of the brain. Somatic or autonomic sensory neurons deliver input to these centers, and autonomic motor neurons provide output that adjusts activity in the visceral effector, usually without our conscious perception.

The hypothalamus is the major control and integration center of the ANS. The hypothalamus receives sensory input related to visceral functions, olfaction (smell), and gustation (taste), as well as changes in temperature, osmolarity, and levels of various substances in blood. It also receives input relating to emotions from the limbic system. Output from the hypothalamus influences autonomic centers both in the brain stem (such as the cardiovascular, salivation, swallowing, and vomiting centers) and the spinal cord (such as the defecation and urination reflex centers in the sacral spinal cord).

Anatomically, the hypothalamus is connected to both the sympathetic and parasympathetic divisions of the ANS by axons of neurons with dendrites and cell bodies in various hypothalamic nuclei. The axons form tracts from the hypothalamus to sympathetic and parasympathetic nuclei in the brain stem and spinal cord through relays in the reticular formation. The posterior and lateral parts of the hypothalamus control the sympathetic division. Stimulation of these areas produces an increase in heart rate and force of contraction, a rise in blood pressure due to constriction of blood vessels, an increase in body temperature, dilation of the pupils, and inhibition of the gastrointestinal tract. In contrast, the anterior and medial parts of the hypothalamus control the parasympathetic division. Stimulation of these areas results in a decrease in heart rate, lowering of blood pressure, constriction of the pupils, and increased secretion and motility of the gastrointestinal tract.

► CHECKPOINT

15. Give three examples of controlled conditions in the body that are kept in homeostatic balance by autonomic reflexes.

16. How does an autonomic reflex arc differ from a somatic reflex arc?

• • •

Now that we have discussed the structure and function of the nervous system, you can appreciate the many ways that this system contributes to homeostasis of other body systems by examining *Focus on Homeostasis: The Nervous System.*

FOCUS ON HOMEOSTASIS

The Nervous System

BODY SYSTEM	CONTRIBUTION OF THE NERVOUS SYSTEM
For all body systems	Together with hormones from the endocrine system, nerve impulses provide communication and regulation of most body tissues.
Integumentary system 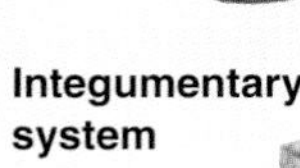	Sympathetic nerves of the autonomic nervous system (ANS) control contraction of smooth muscles attached to hair follicles and secretion of perspiration from sweat glands.
Skeletal system 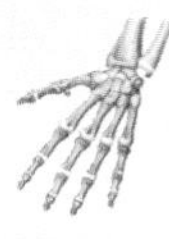	Pain receptors in bone tissue warn of bone trauma or damage.
Muscular system 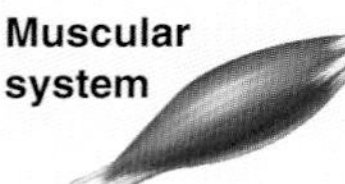	Somatic motor neurons receive instructions from motor areas of the brain and stimulate contraction of skeletal muscles to bring about body movements; basal ganglia and reticular formation set level of muscle tone; cerebellum coordinates skilled movements.
Endocrine system 	Hypothalamus regulates secretion of hormones from anterior and posterior pituitary; ANS regulates secretion of hormones from adrenal medulla and pancreas.
Cardiovascular system 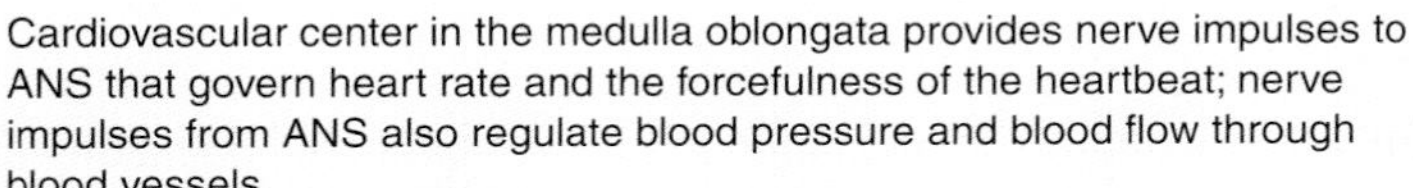	Cardiovascular center in the medulla oblongata provides nerve impulses to ANS that govern heart rate and the forcefulness of the heartbeat; nerve impulses from ANS also regulate blood pressure and blood flow through blood vessels.
Lymphatic system and immunity 	Certain neurotransmitters help regulate immune responses; activity in nervous system may increase or decrease immune responses.
Respiratory system 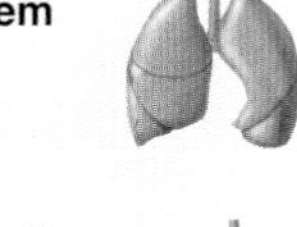	Respiratory areas in brain stem control breathing rate and depth; ANS helps regulate diameter of airways.
Digestive system 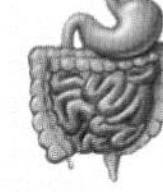	ANS and enteric nervous system (ENS) help regulate digestion; parasympathetic division of ANS stimulates many digestive processes.
Urinary system 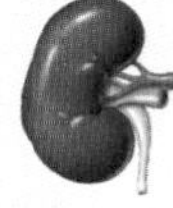	ANS helps regulate blood flow to kidneys, thereby influencing the rate of urine formation; brain and spinal cord centers govern emptying of the urinary bladder.
Reproductive systems 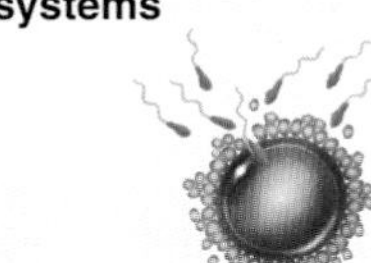	Hypothalamus and limbic system govern a variety of sexual behaviors; ANS brings about erection of penis in males and clitoris in females and ejaculation of semen in males; hypothalamus regulates release of anterior pituitary hormones that control gonads (ovaries and testes); nerve impulses elicited by touch stimuli from suckling infant cause release of oxytocin and milk ejection in nursing mothers.

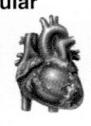

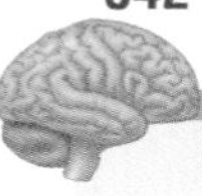

DISORDERS: HOMEOSTATIC IMBALANCES

Raynaud's Phenomenon

In **Raynaud's phenomenon** (rā-NŌZ) the digits (fingers and toes) become ischemic (lack blood) after exposure to cold or with emotional stress. The condition is due to excessive sympathetic stimulation of smooth muscle in the arterioles of the digits and a heightened response to stimuli that cause vasoconstriction. When arterioles in the digits vasoconstrict in response to sympathetic stimulation, blood flow is greatly diminished. As a result, the digits may blanch (look white due to blockage of blood flow) or become cyanotic (look blue due to deoxygenated blood in capillaries). In extreme cases, the digits may become necrotic from lack of oxygen and nutrients. With rewarming after cold exposure, the arterioles may dilate, causing the fingers and toes to look red. Many patients with Raynaud's phenomenon have low blood pressure. Some have increased numbers of alpha-adrenergic receptors. Raynaud's is most common in young women and occurs more often in cold climates. Patients with Raynaud's phenomenon should avoid exposure to cold, wear warm clothing, and keep the hands and feet warm. Drugs used to treat Raynaud's include nifedipine, a calcium channel blocker that relaxes vascular smooth muscle, and prazosin, which relaxes smooth muscle by blocking alpha receptors. Smoking and the use of alcohol or illicit drugs can exacerbate the symptoms of this condition.

Autonomic Dysreflexia

Autonomic dysreflexia (dis′-rē-FLEX-sē-a) is an exaggerated response of the sympathetic division of the ANS that occurs in about 85% of individuals with spinal cord injury at or above the level of T6. The condition is seen after recovery from spinal shock (see page 467) and occurs due to interruption of the control of ANS neurons by higher centers. When certain sensory impulses, such as those resulting from stretching of a full urinary bladder, are unable to ascend the spinal cord, mass stimulation of the sympathetic nerves inferior to the level of injury occurs. Other triggers include stimulation of pain receptors and the visceral contractions resulting from sexual stimulation, labor/delivery, and bowel stimulation. Among the effects of increased sympathetic activity is severe vasoconstriction, which elevates blood pressure. In response, the cardiovascular center in the medulla oblongata (1) increases parasympathetic output via the vagus (X) nerve, which decreases heart rate, and (2) decreases sympathetic output, which causes dilation of blood vessels superior to the level of the injury.

Autonomic dysreflexia is characterized by a pounding headache; hypertension; flushed, warm skin with profuse sweating above the injury level; pale, cold, and dry skin below the injury level; and anxiety. It is an emergency condition that requires immediate intervention. The first approach is to quickly identify the problematic stimulus and remove it. If this does not relieve the symptoms, an antihypertensive drug such as clonidine or nitroglycerine can be administered. If left untreated, autonomic dysreflexia can cause seizures, stroke, or heart attack.

MEDICAL TERMINOLOGY

Autonomic nerve neuropathy (noo-ROP-a-thē) If a neuropathy (specifically a disorder of a cranial or spinal nerve) affects one or more autonomic nerves, there can be multiple effects on the autonomic nervous system that interfere with reflexes. These include fainting and low blood pressure when standing (orthostatic hypotension) due to decreased sympathetic control of the cardiovascular system, constipation, urinary incontinence, and impotence. This type of neuropathy is often caused by long-term diabetes mellitus and is known as ***diabetic retinopathy.***

Biofeedback A technique in which an individual is provided with information regarding an autonomic response such as heart rate, blood pressure, or skin temperature. Various electronic monitoring devices provide visual or auditory signals about the autonomic responses. By concentrating on positive thoughts, individuals learn to alter autonomic responses. For example, biofeedback has been used to decrease heart rate and blood pressure and increase skin temperature in order to decrease the severity of migraine headaches.

Dysautonomia (dis-aw-tō-NŌ-mē-a; *dys-* = difficult; *autonomia* = self-governing) An inherited disorder in which the autonomic nervous system functions abnormally, resulting in reduced tear gland secretions, poor vasomotor control, motor incoordination, skin blotching, absence of pain sensation, difficulty in swallowing, hyporeflexia, excessive vomiting, and emotional instability.

Hyperhydrosis (hi′-per-hī-DRŌ-sis; *hyper-* = above or too much; *hidrosis* = sweat; *-osis* = condition) Excessive or profuse sweating due to intense stimulation of sweat glands.

Mass reflex In cases of severe spinal cord injury above the level of the sixth thoracic vertebra, stimulation of the skin or overfilling of a visceral organ (such as the urinary bladder or colon) below the level of the injury results in intense activation of autonomic and somatic output from the spinal cord as reflex activity returns. The exaggerated response occurs because there is no inhibitory input from the brain. The mass reflex consists of flexor spasms of the lower limbs, evacuation of the urinary bladder and colon, and profuse sweating below the level of the lesion.

Megacolon (*mega-* = big) An abnormally large colon. In congenital megacolon, parasympathetic nerves to the distal segment of the colon do not develop properly. Loss of motor function in the segment causes massive dilation of the normal proximal colon. The condition results in extreme constipation, abdominal distension, and occasionally, vomiting. Surgical removal of the affected segment of the colon corrects the disorder.

Reflex sympathetic dystrophy (RSD) A syndrome that includes spontaneous pain, painful hypersensitivity to stimuli such as light touch, and excessive coldness and sweating in the involved body part. The disorder frequently involves the forearms, hands, knees, and feet. It appears that activation of the sympathetic division of the autonomic nervous system due to traumatized nociceptors as a result of trauma or surgery on bones or joints is involved. Treatment consists of anesthetics and physical therapy. Recent clinical studies also suggest that the drug baclofen can be used to reduce pain and restore normal function to the affected body part. Also called ***complex regional pain syndrome type 1.***

Vagotomy (vā-GOT-ō-mē; *-tome* = incision) Cutting the vagus (X) nerve. It is frequently done to decrease the production of hydrochloric acid in persons with ulcers.

STUDY OUTLINE

COMPARISON OF SOMATIC AND AUTONOMIC NERVOUS SYSTEMS (p. 525)

1. The somatic nervous system operates under conscious control; the ANS usually operates without conscious control.
2. Sensory input for the somatic nervous system is mainly from the special senses and somatic senses; sensory input for the ANS is from interoceptors, in addition to special senses and somatic senses.
3. The axons of somatic motor neurons extend from the CNS and synapse directly with an effector. Autonomic motor pathways consist of two motor neurons in series. The axon of the first motor neuron extends from the CNS and synapses in a ganglion with the second motor neuron; the second neuron synapses with an effector.
4. The output (motor) portion of the ANS has two divisions: sympathetic and parasympathetic. Most body organs receive dual innervation; usually one ANS division causes excitation and the other causes inhibition.
5. Somatic nervous system effectors are skeletal muscles; ANS effectors include cardiac muscle, smooth muscle, and glands.
6. Table 15.1 compares the somatic and autonomic nervous systems.

ANATOMY OF AUTONOMIC MOTOR PATHWAYS (p. 527)

1. Preganglionic neurons are myelinated; postganglionic neurons are unmyelinated.
2. The cell bodies of sympathetic preganglionic neurons are in the lateral gray horns of the 12 thoracic and the first two or three lumbar segments of the spinal cord; the cell bodies of parasympathetic preganglionic neurons are in four cranial nerve nuclei (III, VII, IX, and X) in the brain stem and lateral gray horns of the second through fourth sacral segments of the spinal cord.
3. Autonomic ganglia are classified as sympathetic trunk ganglia (on both sides of vertebral column), prevertebral ganglia (anterior to vertebral column), and terminal ganglia (near or inside visceral effectors).
4. Sympathetic preganglionic neurons synapse with postganglionic neurons in ganglia of the sympathetic trunk or in prevertebral ganglia; parasympathetic preganglionic neurons synapse with postganglionic neurons in terminal ganglia.

ANS NEUROTRANSMITTERS AND RECEPTORS (p. 534)

1. Cholinergic neurons release acetylcholine, which binds to nicotinic or muscarinic cholinergic receptors.
2. In the ANS, the cholinergic neurons include all sympathetic and parasympathetic preganglionic neurons, all parasympathetic postganglionic neurons, and sympathetic postganglionic neurons that innervate most sweat glands.
3. In the ANS, adrenergic neurons release norepinephrine. Both epinephrine and norepinephrine bind to alpha and beta adrenergic receptors.
4. Most sympathetic postganglionic neurons are adrenergic.
5. Table 15.2 summarizes the types of cholinergic and adrenergic receptors.
6. An agonist is a substance that binds to and activates a receptor, mimicking the effect of a natural neurotransmitter or hormone. An antagonist is a substance that binds to and blocks a receptor, thereby preventing a natural neurotransmitter or hormone from exerting its effect.

PHYSIOLOGICAL EFFECTS OF THE ANS (p. 537)

1. The sympathetic division favors body functions that can support vigorous physical activity and rapid production of ATP (fight-or-flight response); the parasympathetic division regulates activities that conserve and restore body energy.
2. The effects of sympathetic stimulation are longer lasting and more widespread than the effects of parasympathetic stimulation.
3. Table 15.3 compares structural and functional features of the sympathetic and parasympathetic divisions.
4. Table 15.4 lists sympathetic and parasympathetic responses.

INTEGRATION AND CONTROL OF AUTONOMIC FUNCTIONS (p. 540)

1. An autonomic reflex adjusts the activities of smooth muscle, cardiac muscle, and glands.
2. An autonomic reflex arc consists of a receptor, a sensory neuron, an integrating center, two autonomic motor neurons, and a visceral effector.
3. The hypothalamus is the major control and integration center of the ANS. It is connected to both the sympathetic and the parasympathetic divisions.

SELF-QUIZ QUESTIONS

Fill in the blanks in the following statements.

1. Cholinergic neurons release ____ and adrenergic neurons release ____.
2. Because of the location of the preganglionic cell bodies, the sympathetic division of the ANS is also called the ____ division; the parasympathetic division is also called the ____ division.

Indicate whether the following statements are true or false.

3. The vagus nerves transmit 80% of the outflow of the parasympathetic preganglionic axons.
4. Organs that receive both sympathetic and parasympathetic motor impulses are said to have dual innervation.

Choose the one best answer to the following questions.

5. Which of the following statements is *false*? (a) A single sympathetic preganglionic fiber may synapse with 20 or more postganglionic fibers, which partly explains why sympathetic responses are widespread throughout the body. (b) Parasympathetic effects tend to be localized because parasympathetic neurons usually synapse in the terminal ganglia with only four or five postsynaptic

neurons (all of which supply a single effector). (c) Some sympathetic preganglionic neurons extend to and terminate in the adrenal medullae. (d) The parasympathetic preganglionic neurons synapse with the postganglionic axons in the prevertebral ganglia. (e) Parasympathetic preganglionic neurons emerge from the CNS as part of a cranial nerve or anterior root of a spinal nerve.

6. Which autonomic plexus supplies the large intestine? (1) renal, (2) inferior mesenteric, (3) hypogastric, (4) superior mesenteric, (5) celiac. (a) 2, 3, and 4; (b) 1, 2, 3, 4, and 5; (c) 3 and 4; (d) 4 and 5; (e) 2 and 4.

7. Which of the following statements are *true*? (1) The somatic nervous system and the ANS both include sensory and motor neurons. (2) Somatic motor neurons release the neurotransmitter norepinephrine. (3) The effect of an autonomic motor neuron is either excitation or inhibition, but that of a somatic motor neuron is always excitation. (4) Autonomic sensory neurons are mostly associated with interoceptors. (5) Autonomic motor pathways consist of two motor neurons in series. (6) Somatic motor pathways consist of two motor neurons in series. (a) 1, 2, 3, 4, and 5; (b) 1, 3, 4, and 5; (c) 2, 3, 5, and 6; (d) 1, 3, 5, and 6; (e) 2, 4, 5, and 6.

8. Which of the following statements is *false*? (a) The first neuron in an autonomic pathway is the preganglionic neuron. (b) The axons of preganglionic neurons are located in spinal or cranial nerves. (c) The postganglionic neuron's cell body is within the CNS. (d) Postganglionic neurons relay impulses from autonomic ganglia to visceral effectors. (e) All somatic motor neurons release acetylcholine.

9. Which of the following is *true*? (1) Monoamine oxidase enzymatically breaks down norepinephrine. (2) Activation of α_2 and β_2 receptors generally produces excitation in the effectors. (3) A beta blocker works by preventing activation of β receptors by epinephrine and norepinephrine. (4) An agonist is a substance that binds to a receptor and prevents the natural neurotransmitter from exerting its effect. (5) Activation of nicotinic receptors always causes excitation of the postsynaptic cell. (a) 2 and 3; (b) 1, 2, and 3; (c) 2, 4, and 5; (d) 1, 2, 3, 4, and 5; (e) 1, 3, and 5.

10. Which of the following are cholinergic neurons? (1) all sympathetic preganglionic neurons, (2) all parasympathetic preganglionic neurons, (3) all parasympathetic postganglionic neurons, (4) all sympathetic postganglionic neurons, (5) some sympathetic postganglionic neurons. (a) 1, 2, 3, and 5; (b) 1, 2, 3, and 4; (c) 2, 3, and 5; (d) 2 and 5; (e) 1, 3, and 5.

11. Which of the following statements are *true*? (1) Most sympathetic postganglionic axons are adrenergic. (2) Cholinergic receptors are classified as nicotinic and muscarinic. (3) Adrenergic receptors are classified as alpha and beta. (4) Muscarinic receptors are present on all effectors innervated by parasympathetic postganglionic axons. (5) In general, norepinephrine stimulates alpha receptors more vigorously than beta receptors; epinephrine is a potent stimulator of both alpha and beta receptors. (a) 1, 2, 3, 4, and 5; (b) 2, 3, 4 and 5; (c) 1, 3, 4, and 5; (d) 3, 4, and 5; (e) 1, 2, 3, and 4.

12. Which of the following are reasons why the effects of sympathetic stimulation are longer lasting and more widespread than those of parasympathetic stimulation? (1) There is greater divergence of sympathetic postganglionic fibers. (2) There is less divergence of sympathetic postganglionic fibers. (3) Acetylcholinesterase quickly inactivates ACh, whereas norepinephrine lingers in the synaptic cleft for a longer time. (4) Norepinephrine and epinephrine secreted into the blood by the adrenal medullae intensify the actions of the sympathetic division. (5) ACh remains in the synaptic cleft until norepinephrine is produced. (a) 1 and 3; (b) 1, 3, and 5; (c) 1, 3, and 4; (d) 2, 3, and 4; (e) 2, 3, and 5.

13. Place the following components of an autonomic reflex arc in the correct order from beginning to end. (a) postganglionic neuron, (b) sensory neuron, (c) effector, (d) autonomic ganglion, (e) receptor, (f) preganglionic neuron, (g) integrating center.

14. Match the following:

____(a) also known as intramural ganglia	(1) sympathetic trunk ganglia
____(b) includes the celiac, superior mesenteric, and inferior mesenteric ganglia	(2) prevertebral ganglia
____(c) also called vertebral chain or paravertebral ganglia	(3) terminal ganglia
____(d) lie in a vertical row on either side of the vertebral column	(4) white rami communicantes
____(e) postganglionic fibers, in general, innervate organs below the diaphragm	(5) gray rami communicantes
____(f) ganglia located at the end of an autonomic motor pathway close to or actually within the wall of a visceral organ	
____(g) includes ciliary, pterygopalatine, submandibular, and otic ganglia	
____(h) extend from base of the skull to the coccyx	
____(i) myelinated preganglionic fibers that connect the anterior rami of spinal nerves with the ganglia of the sympathetic trunk	
____(j) also known as collateral ganglia	
____(k) unmyelinated postganglionic axons that connect the ganglia of the sympathetic trunk to spinal nerves	

15. Match the following:

____(a) stimulates urination and defecation	(1) increased activity of the sympathetic division of the ANS
____(b) prepares the body for emergency situations	(2) increased activity of the parasympathetic division of the ANS
____(c) fight-or-flight response	
____(d) promotes digestion and absorption of food	
____(e) concerned primarily with processes involving the expenditure of energy	
____(f) controlled by the posterior and lateral portions of the hypothalamus	
____(g) controlled by the anterior and medial portions of the hypothalamus	
____(h) causes a decrease in heart rate	

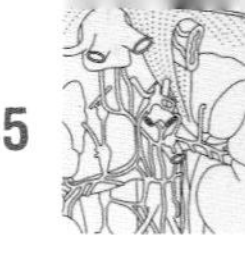

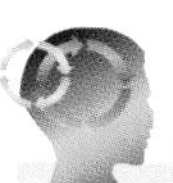

CRITICAL THINKING QUESTIONS

1. You've been to the "all-you-can-eat" buffet and have consumed large amounts of food. After returning home, you recline on the couch to watch television. Which division of the nervous system will be handling your body's after-dinner activities? List several organs involved, the major nerve supply to each organ, and the effects of the nervous system on their functions.
2. Ciara is driving home from school, listening to her favorite music, when a dog darts into the street in front of her car. She manages to swerve to avoid hitting the dog. As she continues on her way, she notices her heart is racing, she has "goose-bumps," and her hands are sweaty. Why is she experiencing these effects?
3. Mrs. Young is experiencing a bout of diarrhea that is keeping her house-bound. She would like to attend a birthday party for her brother but is afraid to attend because of her diarrhea. What type of drug, related to the autonomic nervous system function, could she take to help relieve her diarrhea?

ANSWERS TO FIGURE QUESTIONS

15.1 Dual innervation means that a body organ innervated by the ANS receives both sympathetic and parasympathetic neurons.

15.2 Most parasympathetic preganglionic axons are longer than most sympathetic preganglionic axons because most parasympathetic ganglia are in the walls of visceral organs, but most sympathetic ganglia are close to the spinal cord in the sympathetic trunk.

15.3 Terminal ganglia are associated with the parasympathetic division; sympathetic trunk and prevertebral ganglia are associated with the sympathetic division.

15.4 Sympathetic trunk ganglia contain sympathetic postganglionic neurons that lie in a vertical row on either side of the vertebral column.

15.5 The largest autonomic plexus is the celiac (solar) plexus.

15.6 Cholinergic neurons that have nicotinic ACh receptors include sympathetic postganglionic neurons innervating sweat glands and all parasympathetic postganglionic neurons. The effectors innervated by these cholinergic neurons possess muscarinic receptors.

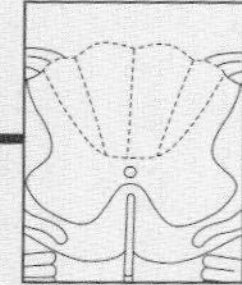

Chapter 16

Sensory, Motor, and Integrative Systems

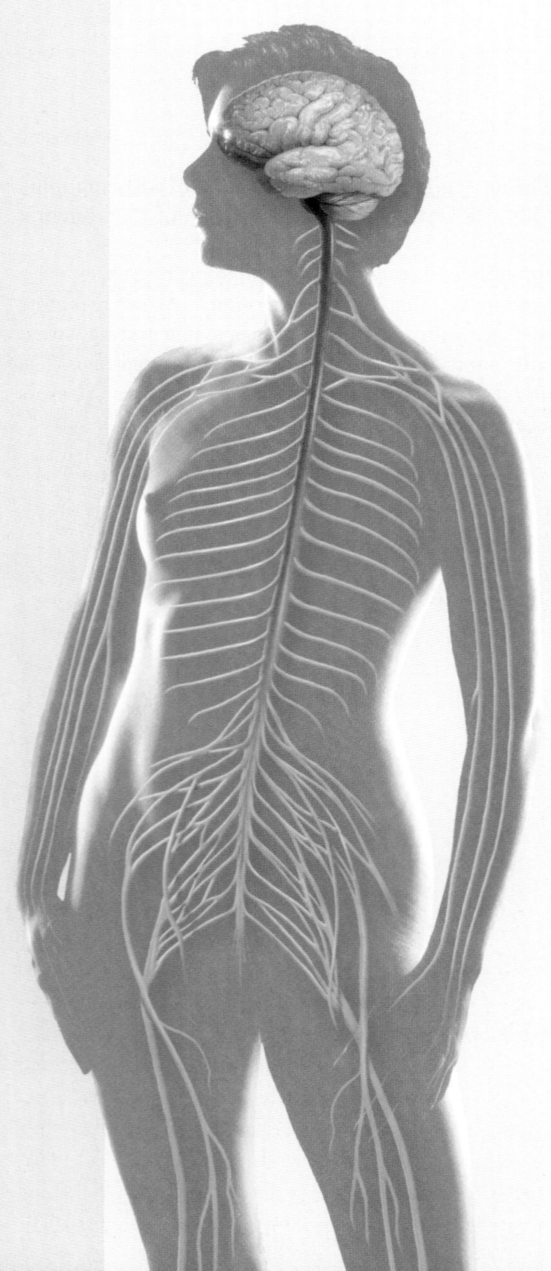

Sensory, Motor, and Integrative Systems and Homeostasis

The sensory and motor pathways of the body provide routes for input into the brain and spinal cord and output to targeted organs for responses, such as muscle contraction.

www. wiley.com/college/apcentral

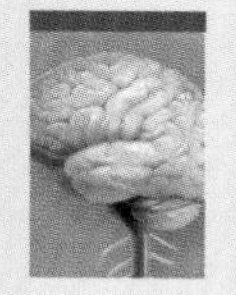

In the previous four chapters we described the organization of the nervous system. In this chapter, we explore the levels and components of sensation. We also examine the pathways that convey somatic sensory nerve impulses from the body to the brain, and the pathways that carry impulses from the brain to skeletal muscles to produce movements. As sensory impulses reach the CNS, they become part of a large pool of sensory input. However, not every nerve impulse transmitted to the CNS elicits a response. Rather, each piece of incoming information is combined with other arriving and previously stored information in a process called integration. Integration occurs at many places along pathways in the CNS, such as the spinal cord, brain stem, cerebellum, basal ganglia, and cerebral cortex. You will also learn how the motor responses that govern muscle contraction are modified at several of these levels. To conclude this chapter, we introduce two complex integrative functions of the brain: (1) wakefulness and sleep and (2) learning and memory.

SENSATION

▶ OBJECTIVES

Define sensation, and discuss the components of sensation.

Describe the different ways to classify sensory receptors.

In its broadest definition, **sensation** is the conscious or subconscious awareness of changes in the external or internal environment. The nature of the sensation and the type of reaction generated vary according to the ultimate destination of nerve impulses that convey sensory information to the CNS. In a spinal reflex such as the stretch reflex, sensory impulses serve as the input. Sensory impulses that reach the lower brain stem elicit more complex reflexes, such as changes in heart rate or breathing rate. When sensory impulses reach the cerebral cortex, we become consciously aware of the sensory stimuli and can precisely locate and identify specific sensations such as touch, pain, hearing, or taste. As you have learned in Chapter 14, **perception** is the conscious awareness and interpretation of sensations and is primarily a function of the cerebral cortex. We have no perception of some sensory information because it never reaches the cerebral cortex. For example, certain sensory receptors constantly monitor the pressure of blood in blood vessels. Because the nerve impulses conveying blood pressure information propagate to the cardiovascular center in the medulla oblongata rather than to the cerebral cortex, blood pressure is not consciously perceived.

Sensory Modalities

Each unique type of sensation—such as touch, pain, vision, or hearing—is called a **sensory modality.** A given sensory neuron carries information for only one sensory modality. Neurons relaying impulses for touch to the somatosensory area of the cerebral cortex do not transmit impulses for pain. Likewise, nerve impulses from the eyes are perceived as sight, and those from the ears are perceived as sounds.

The different sensory modalities can be grouped into two classes: general senses and special senses.

1. The **general senses** refer to both **somatic senses** (*somat-* = of the body) and **visceral senses.** Somatic sensory modalities include tactile sensations (touch, pressure, and vibration), thermal sensations (warm and cold), pain sensations, and proprioceptive sensations, which allow perception of both the static (nonmoving) positions of limbs and body parts (joint and muscle position sense) and movements of the limbs and head. Visceral sensations provide information about conditions within internal organs.

2. The **special senses** include the sensory modalities of smell, taste, vision, hearing, and equilibrium or balance.

In this chapter we discuss the somatic senses and visceral pain. The special senses are the focus of Chapter 17. Visceral senses were discussed in Chapter 15 and will be discussed in association with individual organs in later chapters.

The Process of Sensation

The process of sensation begins in a **sensory receptor,** which can either be a specialized cell or the dendrites of a sensory neuron. As previously noted, a given sensory receptor responds vigorously to one particular kind of **stimulus,** a change in the environment that can activate certain sensory receptors. A sensory receptor responds only weakly or not at all to other stimuli. This characteristic of sensory receptors is known as **selectivity.**

For a sensation to arise, the following four events typically occur:

1. ***Stimulation of the sensory receptor.*** An appropriate stimulus must occur within the sensory receptor's *receptive field,* that is, the body region where stimulation produces a response.

2. ***Transduction of the stimulus.*** A sensory receptor *transduces* (converts) energy in a stimulus into a graded potential. Recall that graded potentials vary in amplitude (size), depending on the strength of the stimulus that causes them, and are not propagated. (See page 423 to review the differences between action potentials and graded potentials.) Each type of sensory receptor exhibits selectivity: It can transduce only one kind of stimulus. For example, odorant molecules in the air stimulate olfactory (smell) receptors in the nose, which transduce the molecules' chemical energy into electrical energy in the form of a graded potential.

3. ***Generation of nerve impulses.*** When a graded potential in a sensory neuron reaches threshold, it triggers one or more nerve impulses, which then propagate toward the CNS. Sensory neurons that conduct impulses from the PNS into the CNS are called **first-order neurons.**

4. ***Integration of sensory input.*** A particular region of the CNS receives and integrates the sensory nerve impulses. Conscious sensations or perceptions are integrated in the cerebral cortex. You seem to see with your eyes, hear with your ears, and feel pain in an injured part of your body because sensory impulses from each part of the body arrive in a specific region of the cerebral cortex, which interprets the sensation as coming from the stimulated sensory receptors.

Sensory Receptors

Types of Sensory Receptors

Several structural and functional characteristics of sensory receptors can be used to group them into different classes. On a microscopic level, sensory receptors may be (1) free nerve endings of first-order sensory neurons, (2) encapsulated nerve endings of first-order sensory neurons, or (3) separate cells that synapse with first-order sensory neurons (Figure 16.1).

Free nerve endings are bare dendrites; they lack any structural specializations that can be seen under a light microscope

Figure 16.1 **Types of sensory receptors and their relationship to first-order sensory neurons.** (a) Free nerve endings, in this case, a cold-sensitive receptor. These endings are bare dendrites of first-order neurons with no apparent structural specialization. (b) An encapsulated nerve ending, in this case a pressure-sensitive receptor. Encapsulated nerve endings are dendrites of first-order neurons. (c) A separate receptor cell—here, a gustatory (taste) receptor—and its synapse with a first-order neuron.

Free nerve endings and encapsulated nerve endings produce generator potentials that trigger nerve impulses in first-order neurons. Separate sensory receptors produce a receptor potential that causes the release of the neurotransmitter. The neurotransmitter then triggers nerve impulses in a first-order neuron.

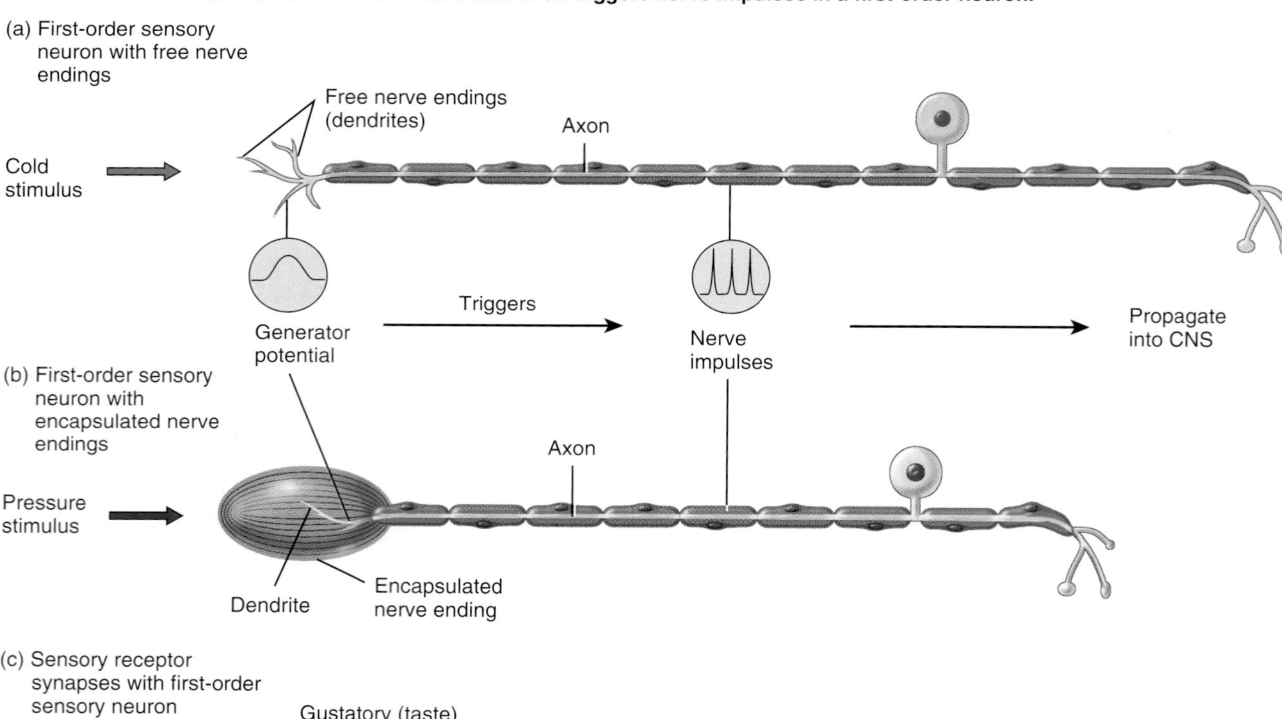

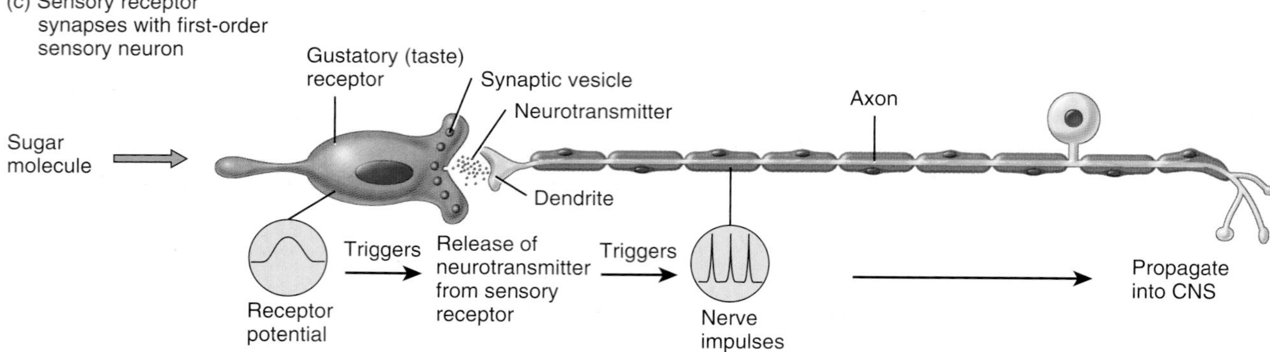

Which senses are served by receptors that are separate cells?

(Figure 16.1a). Receptors for pain, thermal, tickle, itch, and some touch sensations are free nerve endings. Receptors for other somatic and visceral sensations, such as touch, pressure, and vibration, are **encapsulated nerve endings.** Their dendrites are enclosed in a connective tissue capsule that has a distinctive microscopic structure—for example, lamellated (pacinian) corpuscles (Figure 16.1b). The different types of capsules enhance the sensitivity or specificity of the receptor. Sensory receptors for some special senses are specialized, **separate cells** that synapse with sensory neurons. These include *hair cells* for hearing and equilibrium in the inner ear, *gustatory receptor cells* in taste buds (Figure 16.1c), and *photoreceptors* in the retina of the eye for vision; you will learn more about separate cells in Chapter 17.

Sensory receptors produce two different kinds of graded potentials—generator potentials and receptor potentials—in response to a stimulus. When stimulated, the dendrites of free nerve endings, encapsulated nerve endings, and the receptive part of olfactory receptors produce a **generator potential** (Figure 16.1a, b). When a generator potential is large enough to reach threshold, it triggers one or more nerve impulses in the axon of a first-order sensory neuron. The resulting nerve impulse propagates along the axon into the CNS. Thus, generator potentials generate action potentials.

Hair cells of the inner ear, gustatory receptor cells, and photoreceptors, by contrast, do not generate action potentials. Their graded potentials, termed **receptor potentials,** trigger release of neurotransmitter through exocytosis of synaptic vesicles (Figure 16.1c). The neurotransmitter molecules liberated from synaptic vesicles diffuse across the synaptic cleft and produce a postsynaptic potential (PSP) in the first-order neuron. In turn, the PSPs may trigger one or more nerve impulses, which propagate along the axon into the CNS.

The amplitude of both generator potentials and receptor potentials varies with the intensity of the stimulus, with an intense stimulus producing a large potential and a weak stimulus eliciting a small one. Similarly, large generator potentials or receptor potentials trigger nerve impulses at high frequencies in the first-order neuron, whereas small generator potentials or receptor potentials trigger nerve impulses at lower frequencies.

Another way to group sensory receptors is based on the location of the receptors and the origin of the stimuli that activate them.

- **Exteroceptors** (EKS-ter-ō-sep′-tors) are located at or near the external surface of the body; they are sensitive to stimuli originating outside the body and provide information about the *external* environment. The sensations of hearing, vision, smell, taste, touch, pressure, vibration, temperature, and pain are conveyed by exteroceptors.
- **Interoceptors** (IN-ter-ō-sep′-tors) are located in blood vessels, visceral organs, muscles, and the nervous system and monitor conditions in the *internal* environment. The nerve impulses produced by interoceptors usually are not consciously perceived; occasionally, however, activation of interoceptors by strong stimuli may be felt as pain or pressure.
- **Proprioceptors** (PRŌ-prē-ō-sep′-tors; *proprio-* = one's own) are located in muscles, tendons, joints, and the inner ear. They provide information about body position, muscle length and tension, and the position and movement of our joints.

A third way to group sensory receptors is according to the type of stimulus they detect. Most stimuli are in the form of mechanical energy, such as sound waves or pressure changes; electromagnetic energy, such as light or heat; or chemical energy, such as in a molecule of glucose.

- **Mechanoreceptors** are sensitive to mechanical stimuli such as the deformation, stretching, or bending of cells. Mechanoreceptors provide sensations of touch, pressure, vibration, proprioception, and hearing and equilibrium. They also monitor the stretching of blood vessels and internal organs.
- **Thermoreceptors** detect changes in temperature.
- **Nociceptors** respond to painful stimuli resulting from physical or chemical damage to tissue.
- **Photoreceptors** detect light that strikes the retina of the eye.
- **Chemoreceptors** detect chemicals in mouth (taste), nose (smell), and body fluids.
- **Osmoreceptors** detect the osmotic pressure of body fluids.

Table 16.1 summarizes the classification of sensory receptors.

Adaptation in Sensory Receptors

A characteristic of most sensory receptors is **adaptation,** in which the generator potential or receptor potential decreases in amplitude during a maintained, constant stimulus. As you may already have guessed, this causes the frequency of nerve impulses in the first-order neuron to decrease. Because of adaptation, the perception of a sensation may fade or disappear even though the stimulus persists. For example, when you first step into a hot shower, the water may feel very hot, but soon the sensation decreases to one of comfortable warmth even though the stimulus (the high temperature of the water) does not change.

Receptors vary in how quickly they adapt. **Rapidly adapting receptors** adapt very quickly. They are specialized for signaling *changes* in a stimulus. Receptors associated with pressure, touch, and smell are rapidly adapting. **Slowly adapting receptors,** by contrast, adapt slowly and continue to trigger nerve impulses as long as the stimulus persists. Slowly adapting receptors monitor stimuli associated with pain, body position, and chemical composition of the blood.

TABLE 16.1 Classification of Sensory Receptors

Basis of Classification	Description
Microscopic Features	
Free nerve endings	Bare dendrites associated with pain, thermal, tickle, itch, and some touch sensations.
Encapsulated nerve endings	Dendrites enclosed in a connective tissue capsule, such as a corpuscle of touch.
Separate cells	Receptor cells synapse with first-order sensory neurons; located in the retina of the eye (photoreceptors), inner ear (hair cells), and taste buds of the tongue (gustatory receptor cells).
Receptor Location and Activating Stimuli	
Exteroceptors	Located at or near body surface; sensitive to stimuli originating outside body; provide information about external environment; convey visual, smell, taste, touch, pressure, vibration, thermal, and pain sensations.
Interoceptors	Located in blood vessels, visceral organs, and nervous system; provide information about internal environment; impulses produced usually are not consciously perceived but occasionally may be felt as pain or pressure.
Proprioceptors	Located in muscles, tendons, joints, and inner ear; provide information about body position, muscle length and tension, position and motion of joints, and equilibrium (balance).
Type of Stimulus Detected	
Mechanoreceptors	Detect mechanical pressure; provide sensations of touch, pressure, vibration, proprioception, and hearing and equilibrium; also monitor stretching of blood vessels and internal organs.
Thermoreceptors	Detect changes in temperature.
Nociceptors	Respond to painful stimuli resulting from physical or chemical damage to tissue.
Photoreceptors	Detect light that strikes the retina of the eye.
Chemoreceptors	Detect chemicals in mouth (taste), nose (smell), and body fluids.
Osmoreceptors	Sense the osmotic pressure of body fluids.

▶ **CHECKPOINT**

1. How is sensation different from perception?
2. What is a sensory modality?
3. How are generator potentials and receptor potentials similar? How are they different?
4. What is the difference between rapidly adapting and slowly adapting receptors?

SOMATIC SENSATIONS

▶ **OBJECTIVES**

Describe the location and function of the somatic sensory receptors for tactile, thermal, and pain sensations.

Identify the receptors for proprioception and describe their functions.

Somatic sensations arise from stimulation of sensory receptors embedded in the skin or subcutaneous layer; in mucous membranes of the mouth, vagina, and anus; in muscles, tendons, and joints; and in the inner ear. The sensory receptors for somatic sensations are distributed unevenly—some parts of the body surface are densely populated with receptors, and others contain only a few. The areas with the highest density of somatic sensory receptors are the tip of the tongue, the lips, and the fingertips. Somatic sensations that arise from stimulating the skin surface are **cutaneous sensations** (kū-TĀ-nē-us; *cutane-* = skin). There are four modalities of somatic sensation: tactile, thermal, pain, and proprioceptive.

Tactile Sensations

The **tactile sensations** (TAK-tīl; *tact-* = touch) include touch, pressure, vibration, itch, and tickle. Although we perceive differences among these sensations, they arise by activation of some of the same types of receptors. Several types of encapsulated mechanoreceptors attached to large-diameter myelinated A fibers mediate sensations of touch, pressure, and vibration. Other touch sensations, as well as itch and tickle sensations, are detected by free nerve endings attached to small-diameter, unmyelinated C fibers. Recall that larger diameter, myelinated axons propagate nerve impulses more rapidly than do smaller diameter, unmyelinated axons. Tactile receptors in the skin or subcutaneous layer include corpuscles of touch, hair root plexuses, type I and II cutaneous mechanoreceptors, lamellated corpuscles, and free nerve endings (Figure 16.2).

Touch

Sensations of **touch** generally result from stimulation of tactile receptors in the skin or subcutaneous layer. **Crude touch** is the ability to perceive that something has contacted the skin, even though its exact location, shape, size, or texture cannot be determined. **Fine touch** provides specific information about a touch

Figure 16.2 **Structure and location of sensory receptors in the skin and subcutaneous layer.**

The somatic sensations of touch, pressure, vibration, warmth, cold, and pain arise from sensory receptors in the skin, subcutaneous layer, and mucous membranes.

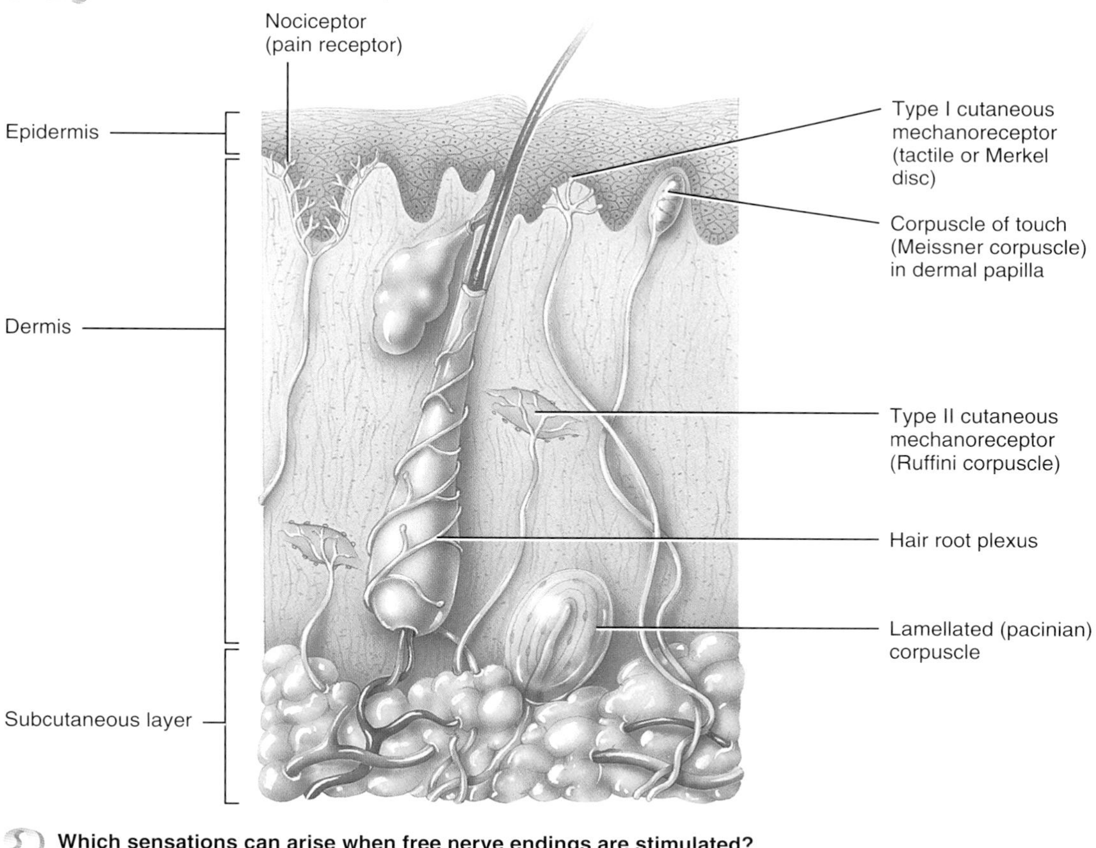

Which sensations can arise when free nerve endings are stimulated?

sensation, such as exactly what point on the body is touched plus the shape, size, and texture of the source of stimulation.

There are two types of rapidly adapting touch receptors. **Corpuscles of touch,** or **Meissner corpuscles** (MĪS-ner), are receptors for fine touch that are located in the dermal papillae of hairless skin. Each corpuscle is an egg-shaped mass of dendrites enclosed by a capsule of connective tissue. Because corpuscles of touch are rapidly adapting receptors, they generate nerve impulses mainly at the onset of a touch. They are abundant in the fingertips, hands, eyelids, tip of the tongue, lips, nipples, soles, clitoris, and tip of the penis. **Hair root plexuses** are rapidly adapting crude touch receptors found in hairy skin; they consist of free nerve endings wrapped around hair follicles. Hair root plexuses detect movements on the skin surface that disturb hairs. For example, an insect landing on a hair causes movement of the hair shaft that stimulates the free nerve endings.

There also are two types of slowly adapting touch receptors. **Type I cutaneous mechanoreceptors,** also known as **Merkel discs,** function in fine touch. Merkel discs are saucer-shaped, flattened free nerve endings that contact Merkel cells of the stratum basale (see Figure 5.2d on page 148). These mechanoreceptors are plentiful in the fingertips, hands, lips, and external genitalia. **Type II cutaneous mechanoreceptors,** or **Ruffini corpuscles,** are elongated, encapsulated receptors located deep in the dermis, and in ligaments and tendons. Present in the hands and abundant on the soles, they are most sensitive to stretching that occurs as digits or limbs are moved.

Pressure and Vibration

Pressure, a sustained sensation that is felt over a larger area than touch, occurs with deformation of deeper tissues. Receptors that contribute to sensations of pressure include corpuscles of touch, type I mechanoreceptors, and lamellated corpuscles. A **lamellated,** or **pacinian, corpuscle** (pa-SIN-ē-an) is a large oval structure composed of a multilayered connective tissue capsule that encloses a dendrite. Like corpuscles of touch, lamellated corpuscles adapt rapidly. They are widely distributed in the body: in the dermis and subcutaneous layer; in submucosal tissues that underlie mucous and serous membranes; around joints, tendons, and muscles; in the periosteum; and in the mammary glands, external genitalia, and certain viscera, such as the pancreas and urinary bladder.

Sensations of **vibration** result from rapidly repetitive sensory signals from tactile receptors. The receptors for vibration sensations are corpuscles of touch and lamellated corpuscles. Corpuscles of touch can detect lower-frequency vibrations, and lamellated corpuscles detect higher-frequency vibrations.

Itch and Tickle

The **itch** sensation results from stimulation of free nerve endings by certain chemicals, such as bradykinin, often because of a local inflammatory response (bradykinin is a kinin and is a potent vasodilator). Free nerve endings and lamellated corpuscles are thought to mediate the **tickle** sensation. This intriguing sensation typically arises only when someone else touches you, not when you touch yourself. The solution to this puzzle seems to lie in the impulses that conduct to and from the cerebellum when you are moving your fingers and touching yourself that don't occur when someone else is tickling you.

Phantom Limb Sensation

Patients who have had a limb amputated may still experience sensations such as itching, pressure, tingling, or pain as if the limb were still there. This phenomenon is called **phantom limb sensation.** One explanation for phantom limb sensations is that the cerebral cortex interprets impulses arising in the proximal portions of sensory neurons that previously carried impulses from the limb as coming from the nonexistent (phantom) limb. Another explanation for phantom limb sensations is that the brain itself contains networks of neurons that generate sensations of body awareness. In the latter view, neurons in the brain that previously received sensory impulses from the missing limb are still active, giving rise to false sensory perceptions. Phantom limb pain can be very distressing to an amputee. Many report that the pain is severe or extremely intense, and that it often does not respond to traditional pain medication therapy. In such cases, alternative treatments may include electrical nerve stimulation, acupuncture, and biofeedback. ■

Thermal Sensations

Thermoreceptors are free nerve endings that have receptive fields about 1 mm in diameter on the skin surface. Two distinct **thermal sensations**—coldness and warmth—are mediated by different receptors. **Cold receptors** are located in the stratum basale of the epidermis and are attached to medium-diameter, myelinated A fibers, although a few connect to small-diameter, unmyelinated C fibers. Temperatures between 10° and 40°C (50–105°F) activate cold receptors. **Warm receptors,** which are not as abundant as cold receptors, are located in the dermis and are attached to small-diameter, unmyelinated C fibers; they are activated by temperatures between 32° and 48°C (90–118°F). Cold and warm receptors both adapt rapidly at the onset of a stimulus, but as noted earlier in the chapter they continue to generate impulses at a lower frequency throughout a prolonged stimulus. Temperatures below 10°C and above 48°C primarily stimulate pain receptors, rather than thermoreceptors, producing painful sensations, which we discuss next.

Pain Sensations

Pain is indispensable for survival. It serves a protective function by signaling the presence of noxious, tissue-damaging conditions. From a medical standpoint, the subjective description and indication of the location of pain may help pinpoint the underlying cause of disease.

Nociceptors (*noci-* = harmful), the receptors for pain, are free nerve endings found in every tissue of the body except the brain (Figure 16.2). Intense thermal, mechanical, or chemical stimuli can activate nociceptors. Tissue irritation or injury releases chemicals such as prostaglandins, kinins, and potassium ions (K^+) that stimulate nociceptors. Pain may persist even after a pain-producing stimulus is removed because pain-mediating chemicals linger, and because nociceptors exhibit very little adaptation. Conditions that elicit pain include excessive distention (stretching) of a structure, prolonged muscular contractions, muscle spasms, or ischemia (inadequate blood flow to an organ).

Types of Pain

There are two types of pain: fast and slow. The perception of **fast pain** occurs very rapidly, usually within 0.1 second after a stimulus is applied, because the nerve impulses propagate along medium-diameter, myelinated A fibers. This type of pain is also known as acute, sharp, or pricking pain. The pain felt from a needle puncture or knife cut to the skin are examples of fast pain. Fast pain is not felt in deeper tissues of the body. The perception of **slow pain,** by contrast, begins a second or more after a stimulus is applied. It then gradually increases in intensity over a period of several seconds or minutes. Impulses for slow pain conduct along small-diameter, unmyelinated C fibers. This type of pain, which may be excruciating, is also referred to as chronic, burning, aching, or throbbing pain. Slow pain can occur both in the skin and in deeper tissues or internal organs. An example is the pain associated with a toothache. You can perceive the difference in onset of these two types of pain best when you injure a body part that is far from the brain because the conduction distance is long. When you stub your toe, for example, you first feel the sharp sensation of fast pain and then feel the slower, aching sensation of slow pain.

Pain that arises from stimulation of receptors in the skin is called **superficial somatic pain;** stimulation of receptors in skeletal muscles, joints, tendons, and fascia causes **deep somatic pain. Visceral pain** results from stimulation of nociceptors in visceral organs. If stimulation is *diffuse* (involves large areas), visceral pain can be severe. Diffuse stimulation of visceral nociceptors might result from distention or ischemia of an internal organ. For example, a kidney stone or a gallstone might cause severe pain by obstructing and distending a ureter or bile duct.

Localization of Pain

Fast pain is very precisely localized to the stimulated area. For example, if someone pricks you with a pin, you know exactly which part of your body was stimulated. Somatic slow pain also is well localized but more diffuse (involves large areas); it usually appears to come from a larger area of the skin. In some instances of visceral slow pain, the affected area is where the pain is felt. If the pleural membranes around the lungs are inflamed, for example, you experience chest pain.

However, in many instances of visceral pain, the pain is felt in or just deep to the skin that overlies the stimulated organ, or in a surface area far from the stimulated organ. This phenomenon is called **referred pain.** Figure 16.3 shows skin regions to which visceral pain may be referred. In general, the visceral organ involved and the area to which the pain is referred are served by the same segment of the spinal cord. For example, sensory fibers from the heart, the skin over the heart, and the skin along the medial aspect of the left arm enter spinal cord segments T1 to T5. Thus, the pain of a heart attack typically is felt in the skin over the heart and along the left arm.

Analgesia: Relief from Pain

Pain sensations sometimes occur out of proportion to minor damage, persist chronically due to an injury, or even appear for no obvious reason. In such cases, **analgesia** (*an-* = without; *-algesia* = pain) or pain relief is needed. Analgesic drugs such as aspirin and ibuprofen (for example, Advil® or Motrin®) block formation of prostaglandins, which stimulate nociceptors. Local anesthetics, such as Novocaine®, provide short-term pain relief by blocking conduction of nerve impulses along the axons of first-order pain neurons. Morphine and other opiate drugs alter the quality of pain perception in the brain; pain is still sensed, but it is no longer perceived as being so noxious. Many pain clinics use anticonvulsant and antidepressant medications to treat those suffering from chronic pain. ■

Proprioceptive Sensations

Proprioceptive sensations allow us to know where our head and limbs are located and how they are moving even if we are not looking at them, so that we can walk, type, or dress without using our eyes. **Kinesthesia** (kin′-es-THĒ-zē-a; *kin-* = motion; *-esthesia* = perception) is the perception of body movements. Proprioceptive sensations arise in receptors termed **proprioceptors.** Those proprioceptors embedded in muscles (especially postural muscles) and tendons inform us of the degree to which muscles are contracted, the amount of tension on tendons, and the positions of joints. Hair cells of the inner ear monitor the orientation of the head relative to the ground and head position during movements. The way they provide information for maintaining balance and equilibrium will be described in Chapter 17. Because proprioceptors adapt slowly and only slightly, the brain continually receives nerve impulses related to the position of different body parts and makes adjustments to ensure coordination.

Figure 16.3 Distribution of referred pain. The colored parts of the diagrams indicate skin areas to which visceral pain is referred.

Nociceptors are present in almost every tissue of the body.

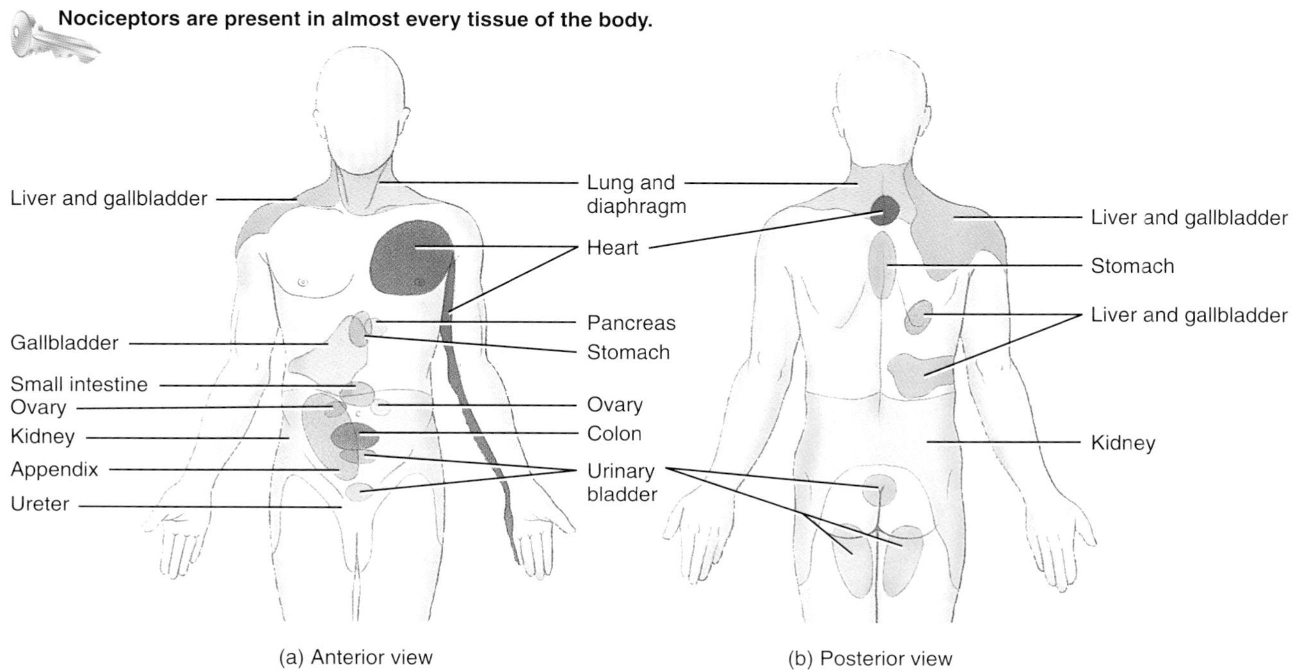

Which visceral organ has the broadest area for referred pain?

Proprioceptive sensations also allow us to estimate the weight of objects and determine the muscular effort necessary to perform a task. For example, as you pick up a bag you quickly realize whether it contains popcorn or books, and you then exert the correct amount of effort needed to lift it. Here we discuss three types of proprioceptors: muscle spindles within skeletal muscles, tendon organs within tendons, and joint kinesthetic receptors within synovial joint capsules.

Muscle Spindles

Muscle spindles are the proprioceptors in skeletal muscles that monitor changes in the length of skeletal muscles and participate in stretch reflexes (shown in Figure 13.14 on page 462). By adjusting how vigorously a muscle spindle responds to stretching of a skeletal muscle, the brain sets an overall level of **muscle tone,** the small degree of contraction that is present while the muscle is at rest.

Each **muscle spindle** consists of several slowly adapting sensory nerve endings that wrap around 3 to 10 specialized muscle fibers, called **intrafusal muscle fibers** (*intrafusal* = within a spindle). A connective tissue capsule encloses the sensory nerve endings and intrafusal fibers and anchors the spindle to the endomysium and perimysium (Figure 16.4). Muscle spindles are interspersed among most skeletal muscle fibers and aligned parallel to them. In muscles that produce finely controlled movements, such as those of the fingers or eyes, muscle spindles are plentiful. Muscles involved in coarser but more forceful movements, like the quadriceps femoris and hamstring muscles of the thigh, have fewer muscle spindles. The only skeletal muscles that lack spindles are the tiny muscles of the middle ear.

Figure 16.4 **Two types of proprioceptors: a muscle spindle and a tendon organ.** In muscle spindles, which monitor changes in skeletal muscle length, sensory nerve endings wrap around the central portion of intrafusal muscle fibers. In tendon organs, which monitor the force of muscle contraction, sensory nerve endings are activated by increasing tension on a tendon. If you examine Figure 13.14 on page 462, you can see the relationship of a muscle spindle to the spinal cord as a component of a stretch reflex. In Figure 13.15 on page 463, you can see the relationship of a tendon organ to the spinal cord as a component of a tendon reflex.

Proprioceptors provide information about body position and movement.

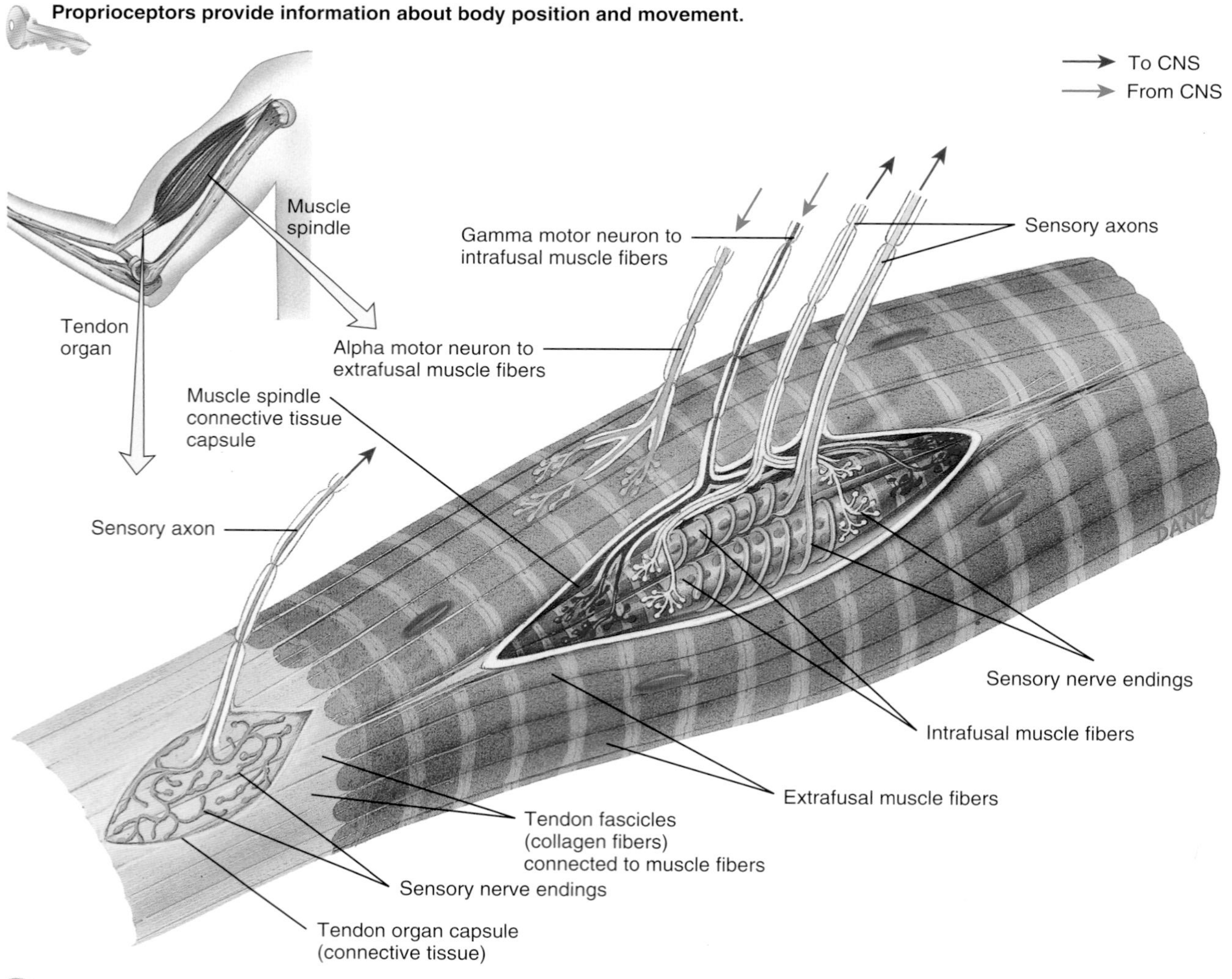

How is a muscle spindle activated?

The main function of muscle spindles is to measure *muscle length*—how much a muscle is being stretched. Either sudden or prolonged stretching of the central areas of the intrafusal muscle fibers stimulates the sensory nerve endings. The resulting nerve impulses propagate into the CNS. Information from muscle spindles arrives quickly at the somatic sensory areas of the cerebral cortex, which allows conscious perception of limb positions and movements. At the same time, impulses from muscle spindles pass to the cerebellum, where the input is used to coordinate muscle contractions.

In addition to their sensory nerve endings near the middle of intrafusal fibers, muscle spindles contain motor neurons called **gamma motor neurons.** These motor neurons terminate near both ends of the intrafusal fibers and adjust the tension in a muscle spindle to variations in the length of the muscle. For example, when a muscle shortens, gamma motor neurons stimulate the ends of the intrafusal fibers to contract slightly. This keeps the intrafusal fibers taut and maintains the sensitivity of the muscle spindle to stretching of the muscle. As the frequency of impulses in its gamma motor neuron increases, a muscle spindle becomes more sensitive to stretching of its midregion.

Surrounding muscle spindles are ordinary skeletal muscle fibers, called **extrafusal muscle fibers** (*extrafusal* = outside a spindle), which are supplied by large-diameter A fibers called **alpha motor neurons.** The cell bodies of both gamma and alpha motor neurons are located in the anterior gray horn of the spinal cord (or in the brain stem for muscles in the head). During the stretch reflex, impulses in muscle spindle sensory axons propagate into the spinal cord and brain stem and activate alpha motor neurons that connect to extrafusal muscle fibers in the same muscle. In this way, activation of its muscle spindles causes contraction of a skeletal muscle, which relieves the stretching.

Tendon Organs

Tendon organs are located at the junction of a tendon and a muscle. By initiating tendon reflexes (see Figure 13.15 on page 463), tendon organs protect tendons and their associated muscles from damage due to excessive tension. (When a muscle contracts, it exerts a force that pulls the points of attachment of the muscle at either end toward each other. This force is the muscle tension.) Each **tendon organ** consists of a thin capsule of connective tissue that encloses a few tendon fascicles (bundles of collagen fibers) (Figure 16.4).

TABLE 16.2 Summary of Receptors for Somatic Sensations

Receptor Type	Receptor Structure and Location	Sensations	Adaptation Rate
Tactile Receptors			
Corpuscles of touch (Meissner corpuscles)	Capsule surrounds mass of dendrites in dermal papillae of hairless skin.	Fine touch, pressure, and slow vibrations.	Rapid.
Hair root plexuses	Free nerve endings wrapped around hair follicles in skin.	Crude touch.	Rapid.
Type I cutaneous mechanoreceptors (tactile or Merkel discs)	Saucer-shaped free nerve endings make contact with Merkel cells in epidermis.	Fine touch and pressure.	Slow.
Type II cutaneous mechanoreceptors (Ruffini corpuscles)	Elongated capsule surrounds dendrites deep in dermis and in ligaments and tendons.	Stretching of skin.	Slow.
Lamellated (pacinian) corpuscles	Oval, layered capsule surrounds dendrites; present in dermis and subcutaneous layer, submucosal tissues, joints, periosteum, and some viscera.	Pressure, fast vibrations, and tickling.	Rapid.
Itch and **tickle receptors**	Free nerve endings and lamellated corpuscles in skin and mucous membranes.	Itching and tickling.	Both slow and rapid.
Thermoreceptors			
Warm receptors and **cold receptors**	Free nerve endings in skin and mucous membranes of mouth, vagina, and anus.	Warmth or cold.	Initially rapid, then slow.
Pain Receptors			
Nociceptors	Free nerve endings in every tissue of the body except the brain.	Pain.	Slow.
Proprioceptors			
Muscle spindles	Sensory nerve endings wrap around central area of encapsulated intrafusal muscle fibers within most skeletal muscles.	Muscle length.	Slow.
Tendon organs	Capsule encloses collagen fibers and sensory nerve endings at junction of tendon and muscle.	Muscle tension.	Slow.
Joint kinesthetic receptors	Lamellated corpuscles, Ruffini corpuscles, tendon organs, and free nerve endings.	Joint position and movement.	Rapid.

Penetrating the capsule are one or more sensory nerve endings that entwine among and around the collagen fibers of the tendon. When tension is applied to a muscle, the tendon organs generate nerve impulses that propagate into the CNS, providing information about changes in muscle tension. Tendon reflexes decrease muscle tension by causing muscle relaxation.

Joint Kinesthetic Receptors

Several types of **joint kinesthetic receptors** are present within and around the articular capsules of synovial joints. Free nerve endings and type II cutaneous mechanoreceptors (Ruffini corpuscles) in the capsules of joints respond to pressure. Small lamellated (pacinian) corpuscles in the connective tissue outside articular capsules respond to acceleration and deceleration of joints during movement. Articular ligaments contain receptors similar to tendon organs that adjust reflex inhibition of the adjacent muscles when excessive strain is placed on the joint.

Table 16.2 summarizes the types of somatic sensory receptors and the sensations they convey.

▶ CHECKPOINT

5. Which somatic sensory receptors are encapsulated?
6. Why do some receptors adapt slowly, and others adapt rapidly?
7. Which somatic sensory receptors mediate fine touch sensations?
8. How does fast pain differ from slow pain?
9. What is referred pain, and how is it useful in diagnosing internal disorders?
10. What aspects of muscle function are monitored by muscle spindles and tendon organs?

SOMATIC SENSORY PATHWAYS

▶ OBJECTIVE

Describe the neuronal components and functions of the posterior column–medial lemniscus pathway, the anterolateral pathway, and the spinocerebellar pathway.

Somatic sensory pathways relay information from the somatic sensory receptors just described to the primary somatosensory area in the cerebral cortex and to the cerebellum. The pathways to the cerebral cortex consist of thousands of sets of three neurons: a first-order neuron, a second-order neuron, and a third-order neuron.

1. **First-order neurons** conduct impulses from somatic receptors into the brain stem or spinal cord. From the face, mouth, teeth, and eyes, somatic sensory impulses propagate along *cranial nerves* into the brain stem. From the neck, trunk, limbs, and posterior aspect of the head, somatic sensory impulses propagate along *spinal nerves* into the spinal cord.
2. **Second-order neurons** conduct impulses from the brain stem and spinal cord to the thalamus. Axons of second-order neurons *decussate* (cross over to the opposite side) in the brain stem or spinal cord before ascending to the ventral posterior nucleus of the thalamus. Thus, all somatic sensory information from one side of the body reaches the thalamus on the opposite side.
3. **Third-order neurons** conduct impulses from the thalamus to the primary somatosensory area of the cortex on the same side.

Somatic sensory impulses entering the spinal cord ascend to the cerebral cortex via two general pathways: (1) the posterior column–medial lemniscus pathway and (2) the anterolateral (spinothalamic) pathways. Somatic sensory impulses entering the spinal cord reach the cerebellum via the spinocerebellar tracts.

Posterior Column–Medial Lemniscus Pathway to the Cortex

Nerve impulses for conscious proprioception and most tactile sensations ascend to the cerebral cortex along the **posterior column–medial lemniscus pathway** (Figure 16.5a). The name of the pathway comes from the names of two white-matter tracts that convey the impulses: the posterior column of the spinal cord and the medial lemniscus of the brain stem.

First-order neurons extend from sensory receptors in the trunk and limbs into the spinal cord and ascend to the medulla oblongata on the same side of the body. The cell bodies of these first-order neurons are in the posterior (dorsal) root ganglia of spinal nerves. In the spinal cord, their axons form the **posterior (dorsal) columns,** which consist of two parts: the **gracile fasciculus** (GRAS-īl fa-SIK-ū-lus) and the **cuneate** (KŪ-nē-Āt) **fasciculus.** (See Table 16.3.) The axon terminals synapse with second-order neurons whose cell bodies are located in the gracile nucleus or cuneate nucleus of the medulla. Impulses from the neck, upper limbs, and upper trunk propagate along axons in the cuneate fasciculus and arrive at the cuneate nucleus. Impulses from the lower trunk and lower limbs propagate along axons in the gracile fasciculus and arrive at the gracile nucleus. Most somatic sensory axons carrying impulses from the face are part of the trigeminal (V) nerve.

The axons of the second-order neurons cross to the opposite side of the medulla and enter the **medial lemniscus** (lem-NIS-kus = ribbon), a thin ribbonlike projection tract that extends from the medulla to the ventral posterior nucleus of the thalamus. In the thalamus, the axon terminals of second-order neurons synapse with third-order neurons, which project their axons to the primary somatosensory area of the cerebral cortex.

Figure 16.5 Somatic sensory pathways.

Nerve impulses propagate along sets of first-order, second-order, and third-order neurons to the primary somatosensory area (postcentral gyrus) of the cerebral cortex.

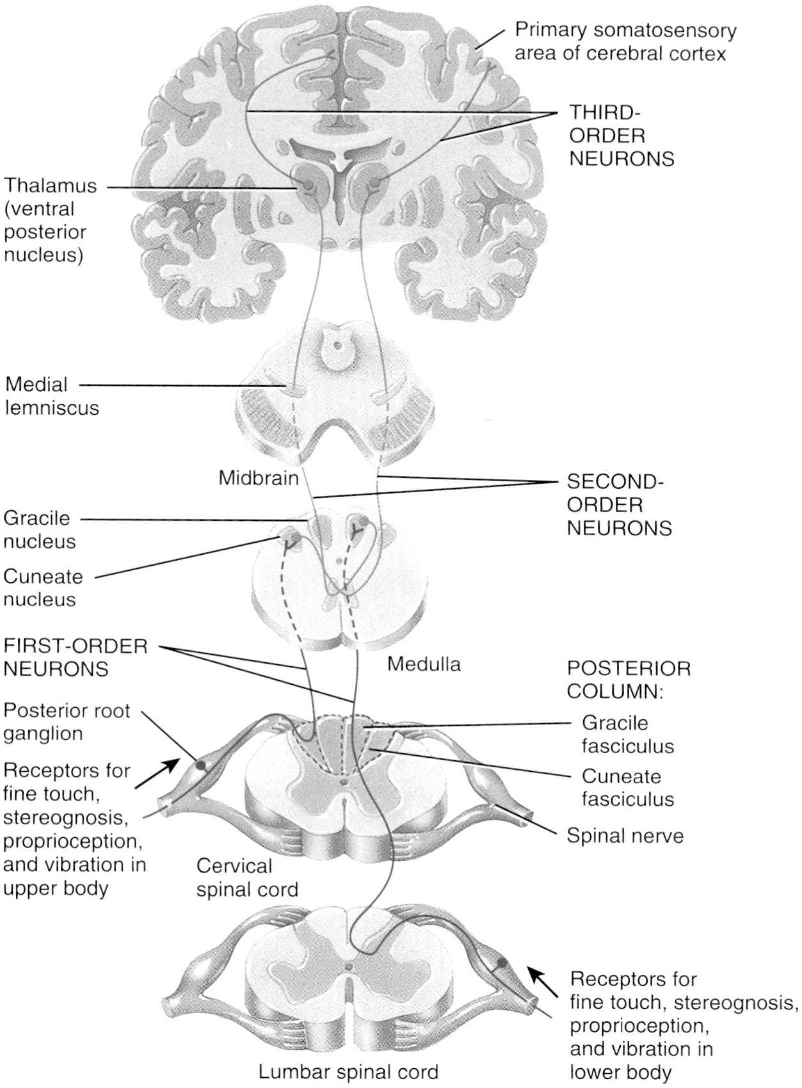

RIGHT SIDE OF BODY
LEFT SIDE OF BODY
Primary somatosensory area of cerebral cortex
THIRD-ORDER NEURONS
Thalamus (ventral posterior nucleus)
Medial lemniscus
Midbrain
SECOND-ORDER NEURONS
Gracile nucleus
Cuneate nucleus
FIRST-ORDER NEURONS
Medulla
POSTERIOR COLUMN:
Gracile fasciculus
Cuneate fasciculus
Posterior root ganglion
Receptors for fine touch, stereognosis, proprioception, and vibration in upper body
Spinal nerve
Cervical spinal cord
Lumbar spinal cord
Receptors for fine touch, stereognosis, proprioception, and vibration in lower body

(a) Posterior column-medial lemniscus pathway

RIGHT SIDE OF BODY
LEFT SIDE OF BODY
Primary somatosensory area of cerebral cortex
THIRD-ORDER NEURON
Thalamus (ventral posterior nucleus)
Midbrain
SECOND-ORDER NEURON
Medulla
Posterior gray horn
FIRST-ORDER NEURON
Posterior root ganglion
LATERAL SPINOTHALAMIC TRACT
Spinal nerve
ANTERIOR SPINOTHALAMIC TRACT
Receptors for pain, cold, warmth, crude touch, pressure, tickle, and itch
Spinal cord

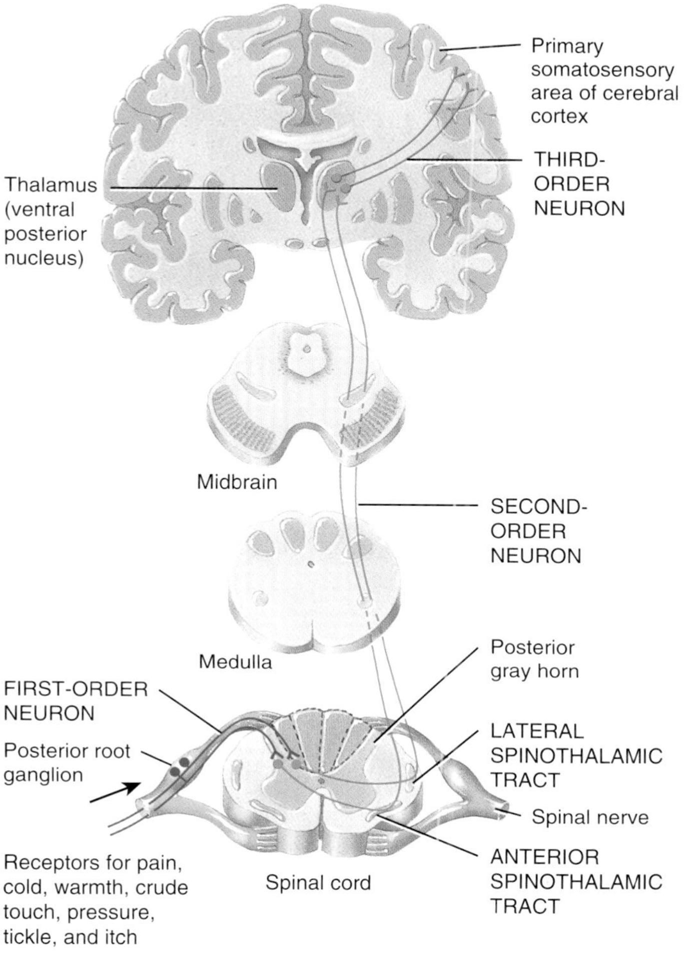

(b) Anterolateral (spinothalamic) pathways

What types of sensory deficits could be produced by damage to the right lateral spinothalamic tract?

Impulses conducted along the posterior column–medial lemniscus pathway give rise to several highly evolved and refined sensations:

- **Fine touch** is the ability to recognize specific information about a touch sensation, such as what point on the body is touched plus the shape, size, and texture of the source of stimulation.
- **Stereognosis** is the ability to recognize the size, shape, and texture of an object by feeling it. Examples are reading Braille or identifying a paperclip by feeling it.
- **Proprioception** is the awareness of the precise position of body parts, and **kinesthesia** is the awareness of directions of movement. Proprioceptors also allow **weight discrimination,** the ability to assess the weight of an object.
- **Vibratory sensations** arise when rapidly fluctuating touch stimuli are present.

Anterolateral Pathways to the Cortex

Like the posterior column-medial leminiscus pathway, the **anterolateral** or **spinothalamic pathways** (spī-nō-tha-LAM-ik) are composed of three-neuron sets (Figure 16.5b). The first-order neurons connect a receptor of the neck, trunk, or limbs with the spinal cord. The cell bodies of the first-order neurons are in the posterior root ganglion. The axon terminals of the first-order neurons synapse with second-order neurons, whose cell bodies are located in the posterior gray horn of the spinal cord.

The axons of the second-order neurons cross to the opposite side of the spinal cord. Then, they pass upward to the brain stem in either the **lateral spinothalamic tract** or the **anterior spinothalamic tract.** The lateral spinothalamic tract conveys sensory impulses for pain and temperature; the anterior spinothalamic tract conveys impulses for tickle, itch, crude touch, and pressure. The axons of the second-order neurons end in the ventral posterior nucleus of the thalamus, where they synapse with the third-order neurons. The axons of the third-order neurons project to the primary somatosensory area on the same side of the cerebral cortex as the thalamus.

Mapping the Primary Somatosensory Area

Specific areas of the cerebral cortex receive somatic sensory input from particular parts of the body and other areas of the cerebral cortex provide output instructions for movement of particular parts of the body. The *somatic sensory map* and the *somatic motor map* relate body parts to these cortical areas.

Precise localization of somatic sensations occurs when nerve impulses arrive at the **primary somatosensory area** (areas 1, 2, and 3 in Figure 14.15 on page 497), which occupies the postcentral gyri of the parietal lobes of the cerebral cortex.

Figure 16.6 Somatic sensory and somatic motor maps in the cerebral cortex. (a) Primary somatosensory area (postcentral gyrus) and (b) primary motor area (precentral gyrus) of the right cerebral hemisphere. The left hemisphere has similar representation. (After Penfield and Rasmussen.)

Each point on the body surface maps to a specific region in both the primary somatosensory area and the primary motor area.

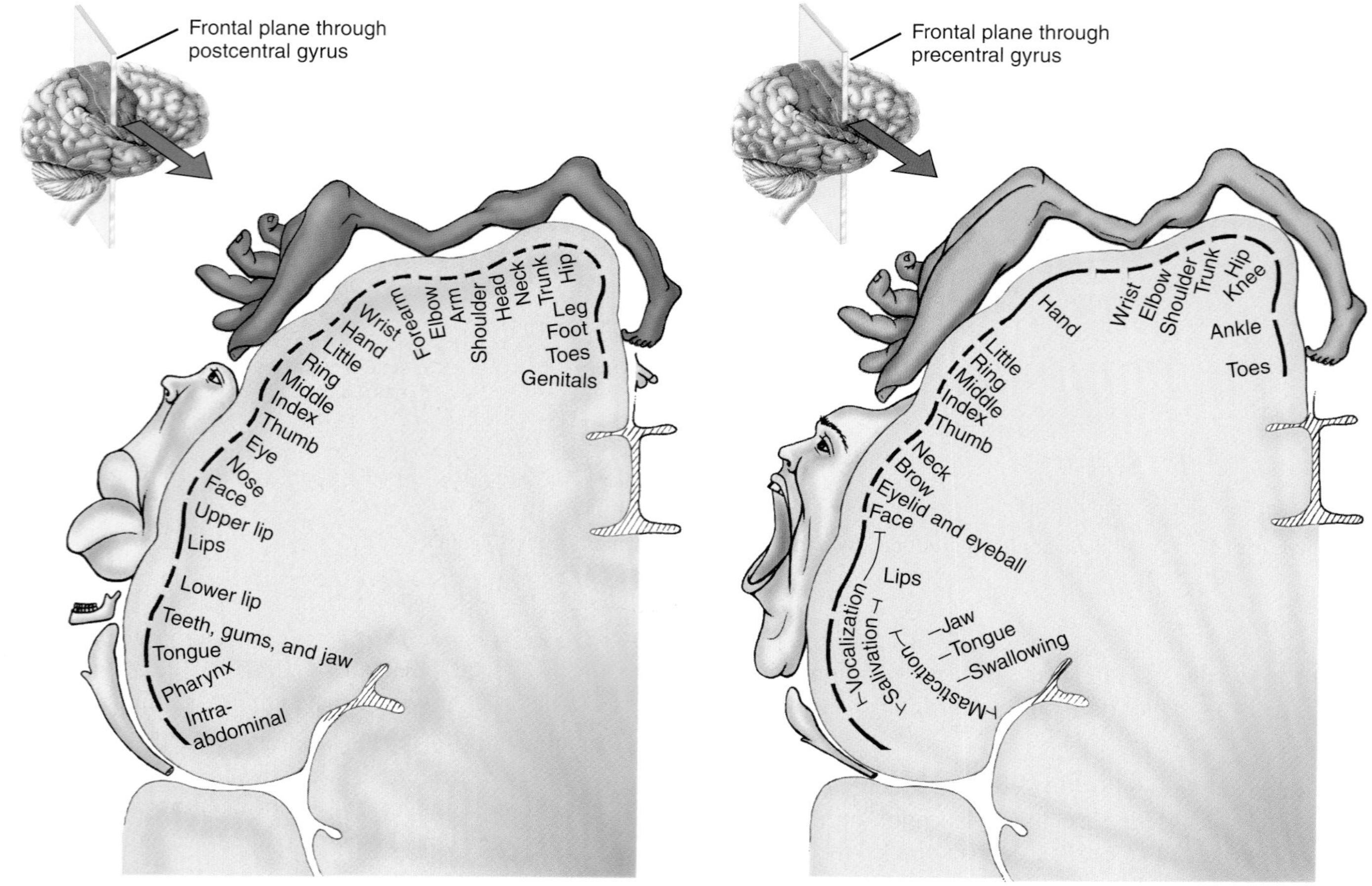

(a) Frontal section of primary somatosensory area in right cerebral hemisphere

(b) Frontal section of primary motor area in right cerebral hemisphere

How do the somatosensory and motor representations compare for the hand, and what does this difference imply?

Each region in this area receives sensory input from a different part of the body. Figure 16.6a maps the destination of somatic sensory signals from different parts of the left side of the body in the somatosensory area of the right cerebral hemisphere. The left cerebral hemisphere has a similar primary somatosensory area that receives sensory input from the right side of the body.

Note that some parts of the body—chiefly the lips, face, tongue, and thumb—provide input to large regions in the somatosensory area. Other parts of the body, such as the trunk and lower limbs, project to much smaller cortical regions. The relative sizes of these regions in the somatosensory area are proportional to the number of specialized sensory receptors within the corresponding part of the body. For example, there are many sensory receptors in the skin of the lips but few in the skin of the trunk. The size of the cortical region that represents a body part may expand or shrink somewhat, depending on the quantity of sensory impulses received from that body part. For example, people who learn to read Braille eventually have a larger cortical region in the somatosensory area to represent the fingertips.

Somatic Sensory Pathways to the Cerebellum

Two tracts in the spinal cord—the **posterior spinocerebellar tract** (spī-nō-ser-e-BEL-ar) and the **anterior spinocerebellar tract**—are the major routes proprioceptive impulses take to reach the cerebellum. Although they are not consciously perceived, sensory impulses conveyed to the cerebellum along these two pathways are critical for posture, balance, and coordination of skilled movements.

Table 16.3 summarizes the major somatic sensory tracts in the spinal cord and pathways in the brain.

TABLE 16.3 Major Somatic Sensory Tracts in the Spinal Cord and Pathways in the Brain

Tract and Location	Functions and Pathways
Posterior column: Gracile fasciculus, Cuneate fasciculus	**Posterior column:** Conveys nerve impulses for the sensations of fine touch, stereognosis, conscious proprioception, kinesthesia, weight discrimination, and vibration. Axons of first-order neurons from one side of the body form the posterior column on the same side and end in the medulla, where they synapse with dendrites and cell bodies of second-order neurons. Axons of second-order neurons decussate, enter the medial lemniscus on the opposite side, and extend to the thalamus. Third-order neurons transmit nerve impulses from the thalamus to the primary somatosensory cortex on the side opposite the site of stimulation.
Lateral spinothalamic tract; Anterior spinothalamic tract	**Lateral spinothalamic:** Conveys nerve impulses for pain and thermal sensations. Axons of first-order neurons from one side of the body synapse with dendrites and cell bodies of second-order neurons in the posterior gray horn on the same side of the body. Axons of second-order neurons decussate, enter the lateral spinothalamic tract on the opposite side, and extend to the thalamus. Third-order neurons transmit nerve impulses from the thalamus to the primary somatosensory cortex on the side opposite the site of stimulation. **Anterior spinothalamic:** Conveys nerve impulses for itch, tickle, pressure, and crude, poorly localized touch sensations. Axons of first-order neurons from one side of the body synapse with dendrites and cell bodies of second-order neurons in the posterior gray horn on the same side of the body. Axons of second-order neurons decussate, enter the anterior spinothalamic tract on the opposite side, and extend to the thalamus. Third-order neurons transmit nerve impulses from the thalamus to the primary somatosensory cortex on the side opposite the site of stimulation.
Posterior spinocerebellar tract; Anterior spinocerebellar tract	**Anterior and posterior spinocerebellar:** Convey nerve impulses from proprioceptors in the trunk and lower limb of one side of the body to the same side of the cerebellum. The proprioceptive input informs the cerebellum of actual movements, allowing it to coordinate, smooth, and refine skilled movements and maintain posture and balance.

Syphilis

Syphilis is a sexually transmitted disease caused by the bacterium *Treponema pallidum.* Because it is a bacterial infection, it can be treated with antibiotics. However, if the infection is not treated, the third stage of syphilis typically causes debilitating neurological symptoms. A common outcome is progressive degeneration of the posterior portions of the spinal cord, including the posterior columns, posterior spinocerebellar tracts, and posterior roots. Somatic sensations are lost, and the person's gait becomes uncoordinated and jerky because proprioceptive impulses fail to reach the cerebellum. ■

▶ CHECKPOINT

11. What are the functional differences between the posterior column–medial lemniscus pathway and the anterolateral pathways?

12. Which body parts have the largest representation in the primary somatosensory area?

13. What type of sensory information is carried in the spinocerebellar tracts, and what is its usefulness?

SOMATIC MOTOR PATHWAYS

▶ OBJECTIVES

Identify the locations and functions of the different types of neurons in the somatic motor pathways.

Compare the locations and functions of the direct and indirect motor pathways.

Explain how the basal ganglia and cerebellum contribute to movements.

Neural circuits in the brain and spinal cord orchestrate all voluntary and involuntary movements. Ultimately, all excitatory and inhibitory signals that control movement converge on the motor neurons that extend out of the brain stem and spinal cord to innervate skeletal muscles in the head and body. These neurons, also known as **lower motor neurons (LMNs),** have their cell bodies in the brain stem and spinal cord. Their axons extend from the motor nuclei of cranial nerves to skeletal muscles of the face and head and from all levels of the spinal cord to skeletal muscles of the limbs and trunk. Only lower motor neurons provide output from the CNS to skeletal muscle fibers. For this reason, they are also called the *final common pathway.*

Neurons in four distinct but highly interactive neural circuits, collectively termed the **somatic motor pathways,** participate in control of movement by providing input to lower motor neurons (Figure 16.7):

1. ***Local circuit neurons.*** Input arrives at lower motor neurons from nearby interneurons called **local circuit neurons.** These neurons are located close to the lower motor neuron cell bodies in the brain stem and spinal cord. Local circuit neurons receive input from somatic sensory receptors, such as nociceptors and muscle spindles, as well as from higher centers in the brain. They help coordinate rhythmic activity in specific muscle groups, such as alternating flexion and extension of the lower limbs during walking.
2. ***Upper motor neurons.*** Both local circuit neurons and lower motor neurons receive input from **upper motor neurons (UMNs).** Most upper motor neurons synapse with local circuit neurons, which in turn synapse with lower motor neurons. (A few upper motor neurons synapse directly with lower motor neurons.) UMNs from the cerebral cortex are essential for planning, initiating, and directing sequences of voluntary movements. Other UMNs originate in motor centers of the brain stem: the red nucleus, the vestibular nucleus, the superior colliculus, and the reticular formation. UMNs from the brain stem regulate muscle tone, control postural muscles, and help maintain balance and orientation of the head and body. Both the basal ganglia and cerebellum exert influence on upper motor neurons.
3. ***Basal ganglia neurons.*** Basal ganglia neurons assist movement by providing input to upper motor neurons. Neural circuits interconnect the basal ganglia with motor areas of the cerebral cortex, thalamus, subthalamic nucleus, and substantia nigra. These circuits help initiate and terminate movements, suppress unwanted movements, and establish a normal level of muscle tone.
4. ***Cerebellar neurons.*** Cerebellar neurons also aid movement by controlling the activity of upper motor neurons. Neural circuits interconnect the cerebellum with motor areas of the cerebral cortex (via the thalamus) and the brain stem. A prime function of the cerebellum is to monitor differences between intended movements and movements actually performed. Then, it issues commands to upper motor neurons to reduce errors in movement. The cerebellum thus coordinates body movements and helps maintain normal posture and balance.

Paralysis

Damage or disease of *lower* motor neurons produces **flaccid paralysis** of muscles on the same side of the body. There is neither voluntary nor reflex action of the innervated muscle fibers, muscle tone is decreased or lost, and the muscle remains limp or flaccid. Injury or disease of *upper* motor neurons in the cerebral cortex causes **spastic paralysis** of muscles on the opposite side of the body. In this condition muscle tone is increased, reflexes are exaggerated, and pathological reflexes such as the Babinski sign (see page 467) appear. ■

Organization of Upper Motor Neuron Pathways

The axons of upper motor neurons extend from the brain to lower motor neurons via two types of somatic motor pathways—direct

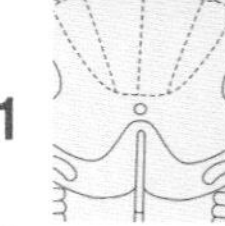

Figure 16.7 Somatic motor pathways for coordination and control of movement. Lower motor neurons receive input directly from 1 local circuit neurons (purple arrow) and 2 upper motor neurons in the cerebral cortex and brain stem (green arrows). Neural circuits involving basal ganglia neurons 3 and cerebellar neurons 4 regulate activity of upper motor neurons (red arrows).

Because lower motor neurons provide all output to skeletal muscles, they are called the final common pathway.

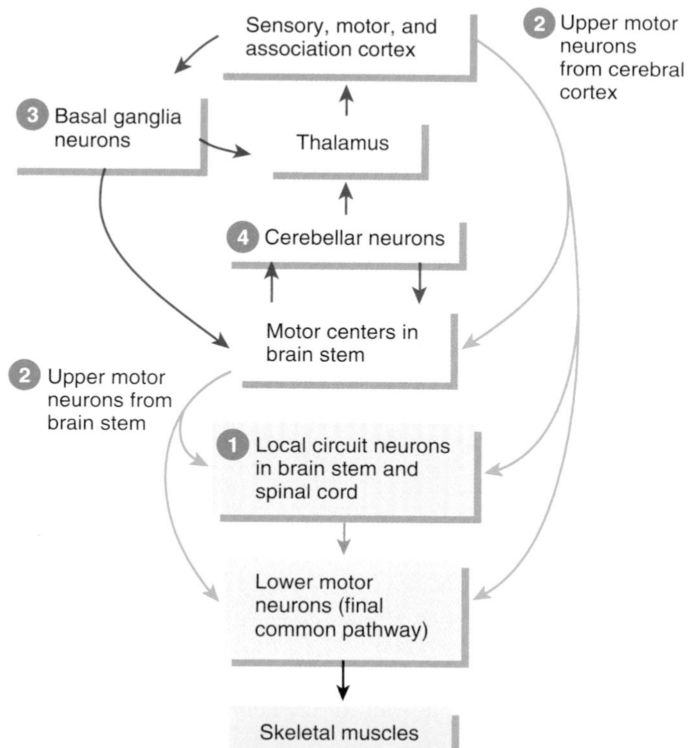

How do the functions of upper motor neurons from the cerebral cortex and from the brain stem differ?

and indirect. **Direct motor pathways** provide input to lower motor neurons via axons that extend directly from the cerebral cortex. **Indirect motor pathways** provide input to lower motor neurons from motor centers in the brain stem. These brain stem centers, in turn, receive signals from neurons in the basal ganglia, cerebellum, and cerebral cortex. Direct and indirect pathways both govern generation of nerve impulses in the lower motor neurons, the neurons that stimulate contraction of skeletal muscles.

Before we examine these pathways we consider the role of the motor cortex in voluntary movement.

Mapping the Motor Areas

Control of body movements occurs via neural circuits in several regions of the brain. The **primary motor area** (area 4 in Figure 14.15 on page 497), located in the precentral gyrus of the frontal lobe (see Figure 16.6b) of the cerebral cortex, is a major control region for planning and initiating voluntary movements. The adjacent **premotor area** (area 6) also contributes axons to the descending motor pathways. As is true for somatic sensory representation in the somatosensory area, different muscles are represented unequally in the primary motor area. The cortical area devoted to a muscle is proportional to the number of motor units in that muscle. Muscles in the thumb, fingers, lips, tongue, and vocal cords have large representations; the trunk has a much smaller representation. By comparing Figures 16.6a and b, you can see that somatosensory and somatic motor representations are similar but not identical for most parts of the body.

Direct Motor Pathways

Nerve impulses for voluntary movements propagate from the cerebral cortex to lower motor neurons via the direct motor pathways (Figure 16.8), also known as the *pyramidal pathways.* Areas of the cerebral cortex that contain large, pyramid-shaped cell bodies of upper motor neurons include not only the primary motor area in the precentral gyrus (area 4 in Figure 14.15 on page 497) but also the premotor area (area 6). Axons of these cortical UMNs descend through the internal capsule of the cerebrum. In the medulla oblongata, the axon bundles form the ventral bulges known as the pyramids.

About 90% of the axons of upper motor neurons *decussate* (cross over) to the *contralateral* (opposite) side in the medulla oblongata. The 10% that remain on the *ipsilateral* (same) side eventually decussate at the spinal cord levels where they synapse with an interneuron or lower motor neuron. Thus, the right cerebral cortex controls muscles on the left side of the body, and the left cerebral cortex controls muscles on the right side of the body.

Three tracts contain axons of upper motor neurons that are part of the direct motor pathways:

1. ***Lateral corticospinal tracts.*** Axons of UMNs that decussate in the medulla form the **lateral corticospinal tracts** (kor′-ti-kō-SPĪ-nal) in the right and left lateral white columns of the spinal cord (Figure 16.8 and Table 16.4). These motor neurons control muscles located in distal parts of the limbs. The distal muscles are responsible for precise, agile, and highly skilled movements of the limbs, hands, and feet. Examples include the movements needed to button a shirt or play the piano.
2. ***Anterior corticospinal tracts.*** Axons of cortical UMNs that do not decussate in the medulla form the **anterior corticospinal tracts** in the right and left anterior white columns (Figure 16.8 and Table 16.4). At each spinal cord level, some of these axons decussate via the anterior white commissure. Then, they synapse with interneurons or lower motor neurons in the anterior gray horn. Axons of these lower motor neurons exit the cervical and upper thoracic segments of the cord in the anterior roots of spinal nerves. They terminate in skeletal muscles that control movements of the neck and part of the trunk, thus coordinating movements of the axial skeleton.
3. ***Corticobulbar tracts.*** Some axons of upper motor neurons that conduct impulses for the control of skeletal muscles in the head form the **corticobulbar tracts** (kor′-ti-kō-BUL-bar),

Figure 16.8 Direct motor pathways in which signals initiated by the primary motor area in the right hemisphere control skeletal muscles on the left side of the body. Spinal cord tracts carrying impulses of direct motor pathways are the lateral corticospinal tract and anterior corticospinal tract.

Direct pathways convey impulses that result in precise, voluntary movements.

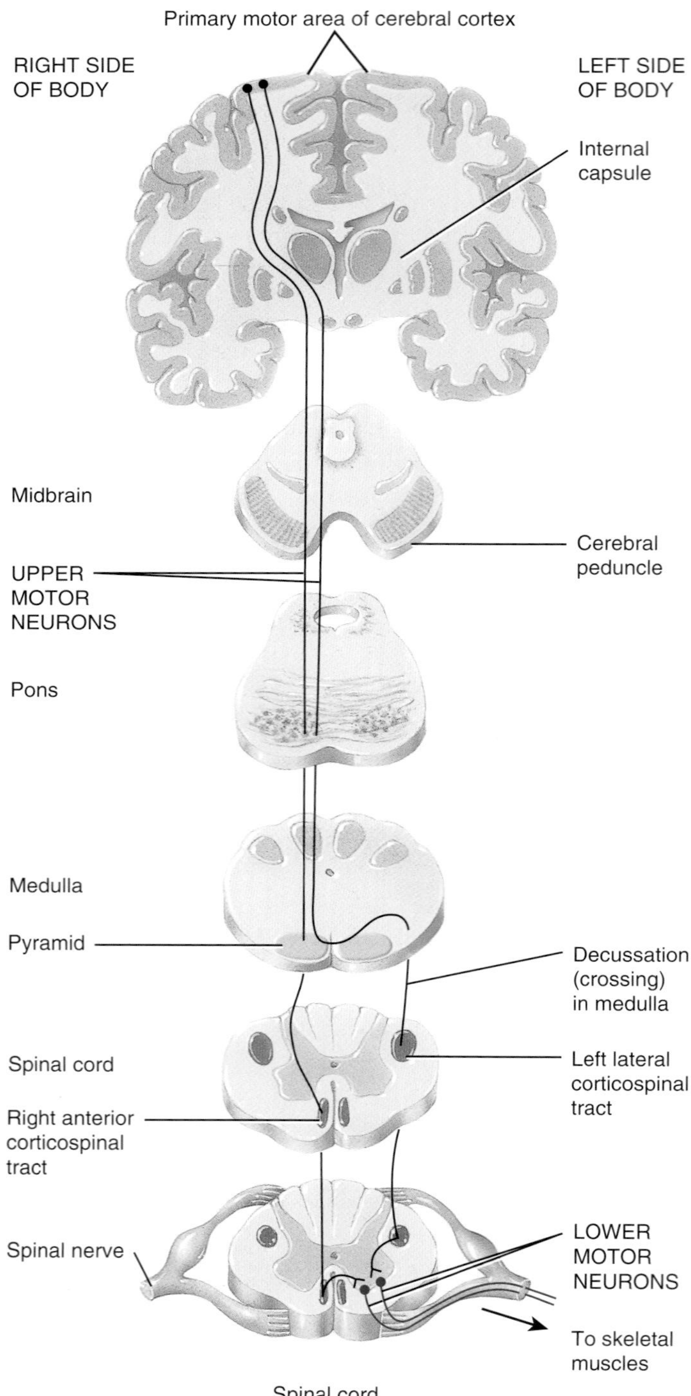

What two other tracts (not shown in the figure) convey impulses resulting in precise, voluntary movements?

which descend from the cerebral cortex to the brain stem (see Table 16.4). Some of the axons decussate, whereas others do not. The axons terminate in the motor nuclei of nine pairs of cranial nerves in the brain stem: the oculomotor (III), trochlear (IV), trigeminal (V), abducens (VI), facial (VII), glossopharyngeal (IX), vagus (X), accessory (XI), and hypoglossal (XII). The lower motor neurons of cranial nerves convey impulses that control precise, voluntary movements of the eyes, tongue, and neck, plus chewing, facial expression, and speech.

Table 16.4 summarizes the functions and pathways of the tracts in the direct motor pathways.

Amyotrophic Lateral Sclerosis

Amyotrophic lateral sclerosis (ALS) (ā′-mī-ō-TROF-ik; *a-* = without; *myo-* = muscle; *trophic* = nourishment) is a progressive degenerative disease that attacks motor areas of the cerebral cortex, axons of upper motor neurons in the lateral white columns (corticospinal and rubrospinal tracts), and lower motor neuron cell bodies. It causes progressive muscle weakness and atrophy. ALS often begins in sections of the spinal cord that serve the hands and arms but rapidly spreads to involve the whole body and face, without affecting intellect or sensations. Death typically occurs in 2 to 5 years. ALS is commonly known as *Lou Gehrig's disease* after the New York Yankees baseball player who died of it at age 37 in 1941.

Inherited mutations account for about 15% of all cases of ALS (familial ALS). Noninherited (sporadic) cases of ALS appear to have several implicating factors. According to one theory there is a buildup in the synaptic cleft of the neurotransmitter glutamate released by motor neurons due to a mutation of the protein that normally deactivates and recycles the neurotransmitter. The excess glutamate causes motor neurons to malfunction and eventually die. The drug riluzole, which is used to treat ALS, reduces damage to motor neurons by decreasing the release of glutamate. Other factors may include damage to motor neurons by free radicals, autoimmune responses, viral infections,, deficiency of nerve growth factor, apoptosis (programmed cell death), environmental toxins, and trauma.

In addition to riluzole, ALS is treated with drugs that relieve symptoms such as fatigue, muscle pain and spasticity, excessive saliva, and difficulty sleeping. The only other treatment is supportive care provided by physical, occupational, and speech therapists; nutritionists; social workers; and home care and hospice nurses. ■

Indirect Motor Pathways

The **indirect motor pathways** or **extrapyramidal pathways** include all somatic motor tracts other than the corticospinal and corticobulbar tracts. Nerve impulses conducted along the indirect pathways follow complex, polysynaptic circuits that involve the motor cortex, basal ganglia, thalamus, cerebellum, reticular formation, and nuclei in the brain stem. Axons of

TABLE 16.4 Major Somatic Motor Pathways in the Brain and Tracts in the Midbrain and Spinal Cord

Tract and Location	Functions and Pathways
Direct (pyramidal) tracts	
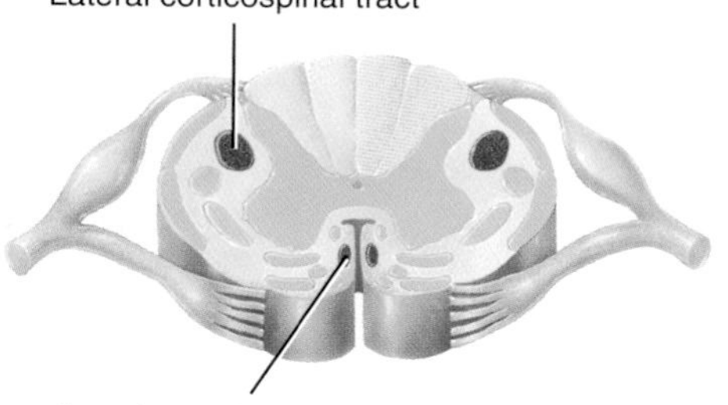 Spinal cord	**Lateral corticospinal:** Conveys nerve impulses from the motor cortex to skeletal muscles on opposite side of body for precise, voluntary movements of the limbs, hands, and feet. Axons of upper motor neurons (UMNs) descend from the precentral gyrus of the cortex into the medulla. Here 90% decussate (cross over to the opposite side) and then enter the contralateral side of the spinal cord to form this tract. At their level of termination, these UMNs end in the anterior gray horn on the same side. They provide input to lower motor neurons, which innervate skeletal muscles. **Anterior corticospinal:** Conveys nerve impulses from the motor cortex to skeletal muscles on opposite side of body for movements of the axial skeleton. Axons of UMNs descend from the cortex into the medulla. Here the 10% that do not decussate enter the spinal cord and form this tract. At their level of termination, these UMNs decussate and end in the anterior gray horn on the opposite side of the body. They provide input to lower motor neurons, which innervate skeletal muscles.
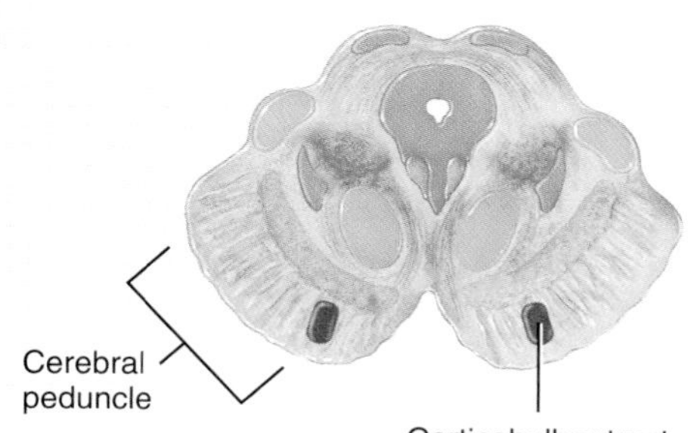 Midbrain of brain stem	**Corticobulbar:** Conveys nerve impulses from the motor cortex to skeletal muscles of the head and neck to coordinate precise, voluntary movements. Axons of UMNs descend from the cortex into the brain stem, where some decussate and others do not. They provide input to lower motor neurons in the nuclei of cranial nerves III, IV, V, VI, VII, IX, X, XI, and XII, which control voluntary movements of the eyes, tongue and neck; chewing; facial expression; and speech.
Indirect (extrapyramidal) tracts	
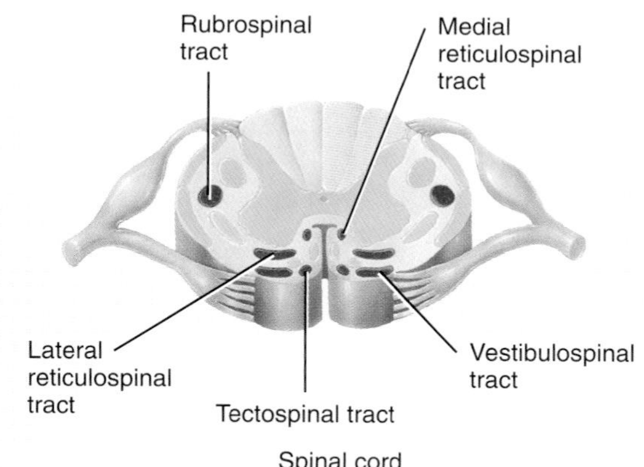 Spinal cord	**Rubrospinal:** Conveys nerve impulses from the red nucleus (which receives input from the cerebral cortex and cerebellum) to contralateral skeletal muscles that govern precise movements of the distal parts of the limbs. **Tectospinal:** Conveys nerve impulses from the superior colliculus to contralateral skeletal muscles that move the head and eyes in response to visual stimuli. **Vestibulospinal:** Conveys nerve impulses from the vestibular nucleus (which receives input about head movements from the inner ear) to regulate ipsilateral muscle tone for maintaining balance in response to head movements. **Lateral reticulospinal:** Conveys nerve impulses from the reticular formation to facilitate flexor reflexes, inhibit extensor reflexes, and decrease muscle tone in muscles of the axial skeleton and proximal parts of the limbs. **Medial reticulospinal:** Conveys nerve impulses from the reticular formation to facilitate extensor reflexes, inhibit flexor reflexes, and increase muscle tone in muscles of the axial skeleton and proximal parts of the limbs.

upper motor neurons that carry nerve impulses from the indirect pathways descend from various nuclei of the brain stem into five major tracts of the spinal cord and terminate on local circuit neurons or lower motor neurons. These tracts are the **rubrospinal** (ROO-brō-spī-nal), **tectospinal** (TEK-tō-spī-nal), **vestibulospinal** (ves-TIB-ū-lō-spī-nal), **lateral reticulospinal** (re-TIK-ū-lō-spī-nal), and **medial reticulospinal tracts.**

Table 16.4 summarizes the functions and pathways of the tracts in the indirect motor pathways.

Roles of the Basal Ganglia

As previously noted, the basal ganglia and cerebellum influence movement through their effects on upper motor neurons. Two parts of the basal ganglia, the caudate nucleus and the putamen, receive input from sensory, association, and motor areas of the cerebral cortex and from the substantia nigra. Output from the basal ganglia comes from the globus pallidus and substantia nigra, which send feedback signals to the motor cortex by way of the thalamus. (Figure 14.13b on page 494 shows these parts of the basal ganglia.) This circuit—from cortex to basal ganglia to thalamus to cortex—appears to function in initiating and terminating movements. Neurons in the putamen generate impulses just before body movements occur; and neurons in the caudate nucleus generate impulses just before eye movements occur.

The basal ganglia also suppress unwanted movements by their inhibitory effects on the thalamus and superior colliculus, and influence muscle tone. The globus pallidus sends impulses into the reticular formation that reduce muscle tone. Damage or destruction of some basal ganglia connections causes a generalized increase in muscle tone.

In addition to their motor functions, the basal ganglia also influence many aspects of cortical function, including sensory, limbic, cognitive, and linguistic functions.

Damage to the Basal Ganglia

Damage to the basal ganglia results in uncontrollable, abnormal body movements (such as "pill rolling" with the fingertips), often accompanied by muscle rigidity and tremors (shaking) while at rest. One example is **Parkinson disease (PD)** (see page 568).

Huntington disease (HD) is an inherited disorder in which the caudate nucleus and putamen degenerate, with loss of neurons that normally release GABA or acetylcholine. A key sign of HD is **chorea** (KŌ-rē-a = a dance) in which rapid, jerky movements occur involuntarily and without purpose. Progressive mental deterioration also occurs. Symptoms of HD often do not appear until age 30 or 40. Death occurs 10 to 20 years after symptoms first appear. ■

Modulation of Movement by the Cerebellum

In addition to maintaining proper posture and balance, the cerebellum is active in both learning and performing rapid, coordinated, highly skilled movements such as hitting a golf ball, speaking, and swimming. Cerebellar function involves four activities (Figure 16.9):

1. The cerebellum *monitors intentions for movement* by receiving impulses from the motor cortex and basal ganglia via the pontine nuclei in the pons regarding what movements are planned (red lines).
2. The cerebellum *monitors actual movement* by receiving input from proprioceptors in joints and muscles that reveals what actually is happening (blue lines). These nerve impulses travel in the anterior and posterior spinocerebellar tracts. Nerve impulses from the vestibular (equilibrium-sensing) apparatus in the inner ear and from the eyes also enter the cerebellum.
3. The cerebellum *compares the command signals* (intentions for movement) *with sensory information* (actual movement performed).
4. If there is a discrepancy between intended and actual movement, the cerebellum *sends out corrective feedback* to upper motor neurons. This information travels via the thalamus to UMNs in the cerebral cortex and goes directly to UMNs in brain stem motor centers (green lines). As movements occur, the cerebellum continuously provides error corrections to upper motor neurons, which decreases errors and smoothes the motion. It also contributes over longer periods to the learning of new motor skills.

Skilled activities such as tennis or volleyball provide good examples of the contribution of the cerebellum to movement. To make a good serve or to block a spike, you must bring your racket or arms forward just far enough to make solid contact. How do you stop at exactly the right point? Before you even hit the ball, the cerebellum has sent nerve impulses to the cerebral cortex and basal ganglia informing them where your swing must stop. In response to impulses from the cerebellum, the cortex and basal ganglia transmit motor impulses to opposing body muscles to stop the swing.

▶ CHECKPOINT

14. Trace the path of a motor impulse from the upper motor neurons through the final common pathway.
15. Which parts of the body have the largest representation in the motor cortex? Which have the smallest?
16. Explain why the two main somatic motor pathways are called "direct" and "indirect."
17. Explain the role of the cerebellum in performing rapid, coordinated, highly skilled movements.

INTEGRATIVE FUNCTIONS OF THE CEREBRUM

▶ OBJECTIVES

Compare the integrative cerebral functions of wakefulness and sleep, and learning and memory.

Describe the four stages of sleep.

Explain the factors that contribute to memory.

We turn now to a fascinating, though incompletely understood, function of the cerebrum: integration, the processing of sensory information by analyzing and storing it and making decisions for various responses. The **integrative functions** include cerebral activities such as sleep and wakefulness, learning and memory, and emotional responses. (The role of the limbic system in emotional behavior was discussed in Chapter 14.)

Figure 16.9 **Input to and output from the cerebellum.**

The cerebellum coordinates and smoothes contractions of skeletal muscles during skilled movements and helps maintain posture and balance.

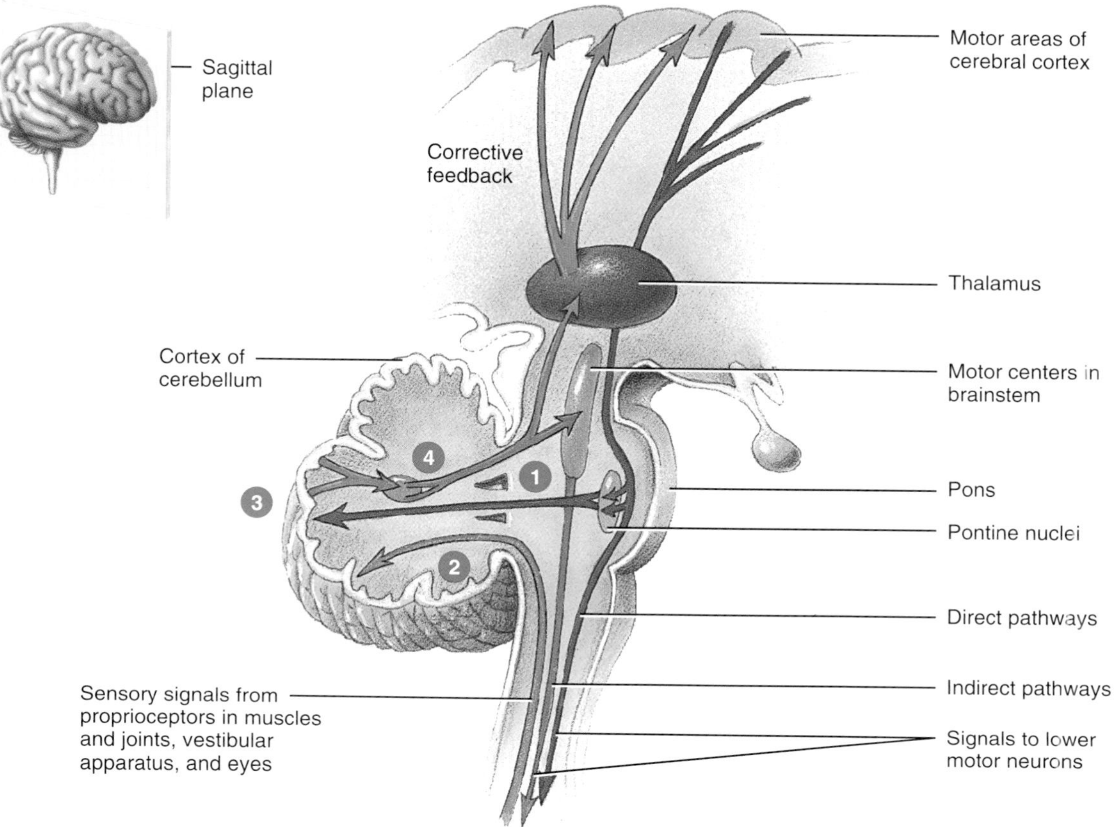

Sagittal section through brain and spinal cord

Which tracts carry information from proprioceptors in joints and muscles to the cerebellum?

Wakefulness and Sleep

Humans sleep and awaken in a 24-hour cycle called a **circadian rhythm** (ser-KĀ-dē-an; *circa-* = about; *-dia* = a day) that is established by the suprachiasmatic nucleus of the hypothalamus (see Figure 14.10 on page 490). A person who is awake is in a state of readiness and is able to react consciously to various stimuli. EEG recordings show that the cerebral cortex is very active during wakefulness; fewer impulses arise during most stages of sleep.

The Role of the Reticular Activating System in Awakening

How does your nervous system make the transition between these two states? Because stimulation of some of its parts increases activity of the cerebral cortex, a portion of the reticular formation is known as the **reticular activating system (RAS)** (Figure 16.10). When this area is active, many nerve impulses are transmitted to widespread areas of the cerebral cortex, both directly and via the thalamus. The effect is a generalized increase in cortical activity.

Arousal, or awakening from sleep, also involves increased activity in the RAS. For arousal to occur, the RAS must be stimulated. Many sensory stimuli can activate the RAS: painful stimuli detected by nociceptors, touch and pressure on the skin, movement of the limbs, bright light, or the buzz of an alarm clock. Once the RAS is activated, the cerebral cortex is also activated, and arousal occurs. The result is a state of wakefulness called **consciousness.** Notice in Figure 16.10 that even though the RAS receives input from somatic sensory receptors, the eyes, and the ears, there is no input from olfactory receptors; even strong odors may fail to cause arousal. People who die in house fires usually succumb to smoke inhalation without awakening. For this reason, all sleeping areas should have a nearby smoke detector that emits a loud alarm. A vibrating pillow or flashing light can serve the same purpose for those who are hearing impaired.

Figure 16.10 The reticular activating system (RAS) consists of neurons whose axons project from the reticular formation through the thalamus to the cerebral cortex.

Increased activity of the RAS causes awakening from sleep (arousal).

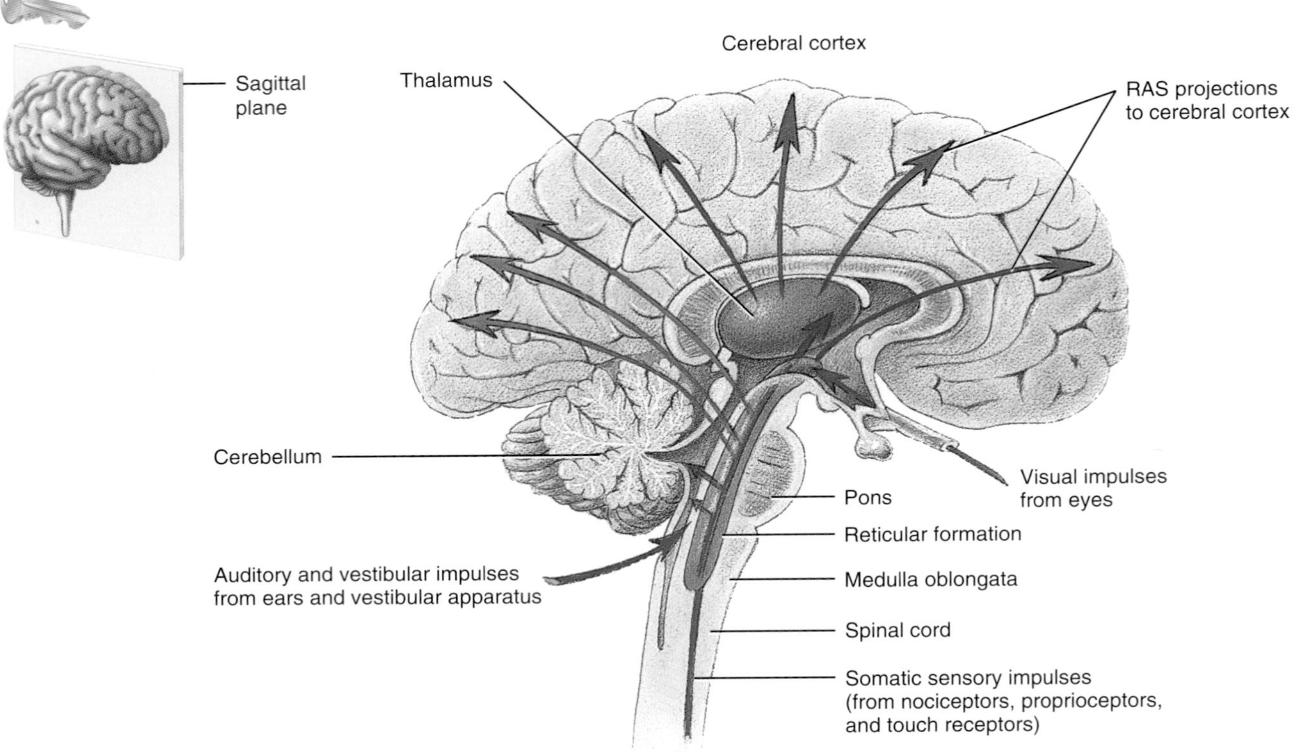

Sagittal section through brain and spinal cord

Why should every sleeping room have a smoke detector?

Sleep

Sleep is a state of altered consciousness or partial unconsciousness from which an individual can be aroused. Although it is essential, the exact functions of sleep are still unclear. Sleep deprivation impairs attention, learning, and performance. Normal sleep consists of two components: non-rapid eye movement (NREM) sleep and rapid eye movement (REM) sleep. **NREM sleep** consists of four gradually merging stages:

1. *Stage 1* is a transition stage between wakefulness and sleep that normally lasts 1–7 minutes. The person is relaxed with eyes closed and has fleeting thoughts. People awakened during this stage often say they have not been sleeping.
2. *Stage 2* or *light sleep* is the first stage of true sleep. In it, a person is a little more difficult to awaken. Fragments of dreams may be experienced, and the eyes may slowly roll from side to side.
3. *Stage 3* is a period of moderately deep sleep. Body temperature and blood pressure decrease and it is difficult to awaken the person. This stage occurs about 20 minutes after falling asleep.
4. *Stage 4* is the deepest level of sleep. Although brain metabolism decreases significantly and body temperature drops slightly at this time, most reflexes are intact, and muscle tone is decreased only slightly. When sleepwalking occurs, it does so during this stage.

Typically, a person goes from stage 1 to stage 4 of NREM sleep in less than an hour. During a typical 7- or 8-hour sleep period, there are three to five episodes of **REM sleep,** during which the eyes move rapidly back and forth under closed eyelids. The person may rapidly ascend through stages 3 and 2 before entering REM sleep. The first episode of REM sleep lasts 10–20 minutes. Then, another interval of NREM sleep follows.

REM and NREM sleep alternate throughout the night. REM periods, which occur approximately every 90 minutes, gradually lengthen, until the final one lasts about 50 minutes. In adults, REM sleep totals 90–120 minutes during a typical sleep period. As a person ages, the average total time spent sleeping decreases, and the percentage of REM sleep declines. As much as 50% of an infant's sleep is REM sleep, as opposed to 35% for 2-year-olds and 25% for adults. Although we do not yet understand the function of REM sleep, the high percentage of REM sleep in infants and children is thought to be important for the maturation of the brain. Neuronal activity is high during

REM sleep—brain blood flow and oxygen use is higher during REM sleep than during intense mental or physical activity while awake.

Different parts of the brain mediate NREM and REM sleep. Neurons in the preoptic area of the hypothalamus, the basal forebrain, and the medulla oblongata govern NREM sleep; neurons in the pons and midbrain turn REM sleep on and off. Several lines of evidence suggest the existence of sleep-inducing chemicals in the brain. One apparent sleep-inducer is adenosine, which accumulates during periods of high ATP (adenosine triphosphate) use by the nervous system. Adenosine binds to specific receptors, called A1 receptors, and inhibits certain cholinergic (acetylcholine-releasing) neurons of the RAS that participate in arousal. Thus, activity in the RAS during sleep is low due to the inhibitory effect of adenosine. Caffeine (in coffee) and theophylline (in tea)—substances known for their ability to maintain wakefulness—bind to and block the A1 receptors, preventing adenosine from binding and inducing sleep.

Several physiological changes occur during sleep. Most dreaming occurs during REM sleep, and the EEG readings are similar to those of a person who is awake. With the exception of motor neurons that govern breathing and eye movements, most somatic motor neurons are inhibited during REM sleep, which decreases muscle tone and even paralyzes the skeletal muscles. Many people experience a momentary feeling of paralysis if they are awakened during REM sleep. During sleep, activity in the parasympathetic division of the autonomic nervous system (ANS) increases while sympathetic activity decreases. Heart rate and blood pressure decrease during NREM sleep and decrease further during REM sleep. Increased parasympathetic activity during REM sleep sometimes causes erection of the penis, even when dream content is not sexual. The presence of penile erections during REM sleep in a man with erectile dysfunction (inability to attain an erection while awake) indicates that his problem has a psychological, rather than a physical, cause.

Learning and Memory

Without memory, we would repeat mistakes and be unable to learn. Similarly, we would not be able to repeat our successes or accomplishments, except by chance. Although both learning and memory have been extensively studied, we still have no completely satisfactory explanation for how we recall information or how we remember events. However, we do know something about how information is acquired and stored, and it is clear that there are different categories of memory.

Learning is the ability to acquire new information or skills through instruction or experience. **Memory** is the process by which information acquired through learning is stored and retrieved. For an experience to become part of memory, it must produce persistent structural and functional changes that represent the experience in the brain. This capability for change associated with learning is termed **plasticity.** Nervous system plasticity underlies our ability to change our behavior in response to stimuli from the external and internal environments. It involves changes in individual neurons—for example, synthesis of different proteins or sprouting of new dendrites—as well as changes in the strengths of synaptic connections among neurons. The parts of the brain known to be involved with memory include the association areas of the frontal, parietal, occipital, and temporal lobes; parts of the limbic system, especially the hippocampus and amygdala; and the diencephalon. The primary somatosensory and primary motor areas in the brain also exhibit plasticity. If a particular body part is used more intensively or in a newly learned activity, such as reading Braille, the cortical areas devoted to that body part gradually expand.

Memory occurs in stages over a period of time. **Immediate memory** is the ability to recall ongoing experiences for a few seconds. It provides a perspective to the present time that allows us to know where we are and what we are doing. **Short-term memory** is the temporary ability to recall a few pieces of information for seconds to minutes. One example is when you look up an unfamiliar telephone number, cross the room to the phone, and then dial the new number. If the number has no special significance, it is usually forgotten within a few seconds. Brain areas involved in immediate and short-term memory include the hippocampus, the mammillary bodies, and two nuclei of the thalamus (anterior and medial nuclei). Some evidence supports the notion that short-term memory depends more on electrical and chemical events in the brain than on structural changes, such as the formation of new synapses.

Information in short-term memory may later be transformed into a more permanent type of memory, called **long-term memory,** which lasts from days to years. If you use that new telephone number often enough, it becomes part of long-term memory. Information in long-term memory usually can be retrieved for use whenever needed. The reinforcement that results from the frequent retrieval of a piece of information is called **memory consolidation.** Long-term memories for information that can be expressed by language, such as a telephone number, apparently are stored in wide regions of the cerebral cortex. Memories for motor skills, such as how to serve a tennis ball, are stored in the basal ganglia and cerebellum as well as in the cerebral cortex.

Amnesia

Amnesia (am-NĒ-zē-a = forgetfulness) refers to the lack or loss of memory. It is a total or partial inability to remember past experiences. In *anterograde amnesia,* there is memory loss for events that occur *after* the trauma or disease that caused the condition. In other words, it is an inability to form new memories. In *retrograde amnesia,* there is a memory loss for events that occurred *before* the trauma or disease that caused the condition. In other words, it is an inability to recall past events. ■

Although the brain receives many stimuli, we pay attention to only a few of them at a time. It has been estimated that only 1% of all the information that comes to our consciousness is stored as long-term memory. Moreover, much of what goes into long-term memory is eventually forgotten. Memory does not record every detail as if it were magnetic tape. Even when details are lost, we can often explain the idea or concept using our own words and ways of viewing things.

Several conditions that inhibit the electrical activity of the brain, such as anesthesia, coma, electroconvulsive therapy (ECT), and ischemia of the brain, disrupt retention of recently acquired information without altering previously established long-term memories. People who suffer retrograde amnesia cannot remember anything that occurred during the 30 minutes or so before the amnesia developed. As a person recovers from amnesia, the most recent memories return last.

Anatomical changes occur in neurons when they are stimulated. For example, electron micrographs of neurons subjected to prolonged, intense activity reveal an increase in the number of presynaptic terminals and enlargement of synaptic end bulbs in presynaptic neurons, as well as an increase in the number of dendritic branches in postsynaptic neurons. Moreover, neurons grow new synaptic end bulbs with increasing age, presumably because of increased use. Opposite changes occur when neurons are inactive. For example, the cerebral cortex in the visual area of animals that have lost their eyesight becomes thinner.

A phenomenon called **long-term potentiation (LTP)** is believed to underlie some aspects of memory; transmission at some synapses within the hippocampus is enhanced (potentiated) for hours or weeks after a brief period of high-frequency stimulation. The neurotransmitter released is glutamate, which acts on NMDA* glutamate receptors on the postsynaptic neurons. In some cases, induction of LTP depends on the release of nitric oxide (NO) from the postsynaptic neurons after they have been activated by glutamate. The NO, in turn, diffuses into the presynaptic neurons and causes LTP.

▶ **CHECKPOINT**

18. Describe how sleep and wakefulness are related to the reticular activating system (RAS).

19. What are the four stages of non-rapid eye movement (NREM) sleep? How is NREM sleep distinguished from rapid eye movement (REM) sleep?

20. Define memory. What are the three kinds of memory? What is memory consolidation?

21. What is long-term potentiation?

*Named after the chemical N-methyl D-aspartate, which is used to detect this type of glutamate receptor.

DISORDERS: HOMEOSTATIC IMBALANCES

Parkinson Disease

Parkinson disease (PD) is a progressive disorder of the CNS that typically affects its victims around age 60. Neurons that extend from the substantia nigra to the putamen and caudate nucleus, where they release the neurotransmitter dopamine (DA), degenerate in PD. The caudate nucleus of the basal ganglia contains neurons that liberate the neurotransmitter acetylcholine (ACh). Although the level of ACh does not change as the level of DA declines, the imbalance of neurotransmitter activity—too little DA and too much ACh—is thought to cause most of the symptoms. The cause of PD is unknown, but toxic environmental chemicals, such as pesticides, herbicides, and carbon monoxide, are suspected contributing agents. Only 5% of PD patients have a family history of the disease.

In PD patients, involuntary skeletal muscle contractions often interfere with voluntary movement. For instance, the muscles of the upper limb may alternately contract and relax, causing the hand to shake. This shaking, called **tremor,** is the most common symptom of PD. Also, muscle tone may increase greatly, causing rigidity of the involved body part. Rigidity of the facial muscles gives the face a masklike appearance. The expression is characterized by a wide-eyed, unblinking stare and a slightly open mouth with uncontrolled drooling.

Motor performance is also impaired by **bradykinesia** (*brady-* = slow), slowness of movements. Activities such as shaving, cutting food, and buttoning a blouse take longer and become increasingly more difficult as the disease progresses. Muscular movements also exhibit **hypokinesia** (*hypo-* = under), decreasing range of motion. For example, words are written smaller, letters are poorly formed, and eventually handwriting becomes illegible. Often, walking is impaired; steps become shorter and shuffling, and arm swing diminishes. Even speech may be affected.

Treatment of PD is directed toward increasing levels of DA and decreasing levels of ACh. Although people with PD do not manufacture enough dopamine, taking it orally is useless because DA cannot cross the blood–brain barrier. Even though symptoms are partially relieved by a drug developed in the 1960s called levodopa (L-dopa), a precursor of DA, the drug does not slow the progression of the disease. As more and more affected brain cells die, the drug becomes useless. Another drug called selegiline (Deprenyl®) is used to inhibit monoamine oxidase, an enzyme that degrades catecholamine neurotransmitters such as dopamine. This drug slows progression of PD and may be used together with levodopa. Anticholinergic drugs such as benzotropine and trihexyphenidyl can also be used to block the effects of ACh at some of the synapses between basal ganglia neurons, which helps to restore the balance between ACh and DA. Anticholinergic drugs effectively reduce symptomatic tremor, rigidity, and drooling.

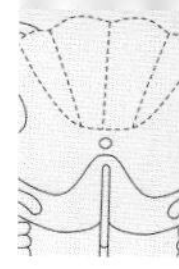

For more than a decade, surgeons have sought to reverse the effects of Parkinson disease by transplanting dopamine-rich fetal nervous tissue into the basal ganglia (usually the putamen) of patients with severe PD. Only a few postsurgical patients have shown any degree of improvement, such as less rigidity and improved quickness of motion. Another surgical technique that has produced improvement for some patients is *pallidotomy,* in which a part of the globus pallidus that generates tremors and produces muscle rigidity is destroyed. In addition, some patients are being treated with a surgical procedure called *deep-brain stimulation (DBS),* which involves the implantation of electrodes into the subthalamic nucleus. The electrical currents released by the implanted electrodes reduce many of the symptoms of PD.

MEDICAL TERMINOLOGY

Acupuncture (ak-ū-PUNK-chur) The use of fine needles (lasers, ultrasound, or electricity) inserted into specific exterior body locations (acupoints) and manipulated to relieve pain and provide therapy for various conditions. The placement of needles may cause the release of neurotransmitters such as endorphins, painkillers that may inhibit pain pathways.

Cerebral palsy (CP) A motor disorder that results in the loss of muscle control and coordination; caused by damage of the motor areas of the brain during fetal life, birth, or infancy. Radiation during fetal life, temporary lack of oxygen during birth, and hydrocephalus during infancy may also cause cerebral palsy.

Coma (KŌ-ma) A state of unconsciousness in which a person's responses to stimuli are reduced or absent. In a *light coma,* an individual may respond to certain stimuli, such as sound, touch, or light, and move his or her eyes, cough, and even murmur. In a *deep coma,* a person does not respond to any stimuli and does not make any movements. Causes of coma include head injuries, cardiac arrest, stroke, brain tumors, infections (encephalitis and meningitis), seizures, alcoholic intoxication, drug overdose, severe lung disorders (chronic obstructive pulmonary disease, pulmonary edema, pulmonary embolism), inhalation of large amounts of carbon monoxide, liver or kidney failure, low or high blood sugar or sodium levels, and low or high body temperature. If brain damage is minor or reversible, a person may come out of a coma and recover fully; if brain damage is severe and irreversible, recovery is unlikely.

Insomnia (in-SOM-nē-a; *in-* = not; *-somnia* = sleep) Difficulty in falling asleep and staying asleep.

Narcolepsy (NAR-kō-lep-sē; *narco-* = numbness; *-lepsy* = a seizure) A condition in which REM sleep cannot be inhibited during waking periods. As a result, involuntary periods of sleep that last about 15 minutes occur throughout the day.

Pain threshold The smallest intensity of a painful stimulus at which a person perceives pain. All individuals have the same pain threshold.

Pain tolerance The greatest intensity of painful stimulation that a person is able to tolerate. Individuals vary in their tolerance to pain.

Sleep apnea (AP-nē-a; *a-* = without; *-pnea* = breath) A disorder in which a person repeatedly stops breathing for 10 or more seconds while sleeping. Most often, it occurs because loss of muscle tone in pharyngeal muscles allows the airway to collapse.

Synesthesia (sin-es-THĒ-zē-a; *syn-* = together; *aisthesis* = sensation) A condition in which sensations of two or more modalities accompany one another. In some cases, a stimulus for one sensation is perceived as a stimulus for another; for example, a sound produces a sensation of color. In other cases, a stimulus from one part of the body is experienced as coming from a different part.

STUDY OUTLINE

SENSATION (p. 547)

1. Sensation is the awareness of changes in the external or internal environment.
2. The nature of a sensation and the type of reaction generated vary according to the destination of sensory impulses in the CNS.
3. Each different type of sensation is a sensory modality; usually, a given sensory neuron serves only one modality.
4. General senses include somatic senses (touch, pressure, vibration, warmth, cold, pain, itch, tickle, and proprioception) and visceral senses; special senses include the modalities of smell, taste, vision, hearing, and equilibrium.
5. For a sensation to arise, four events typically occur: stimulation, transduction, generation of impulses, and integration.
6. Simple receptors, consisting of free nerve endings and encapsulated nerve endings, are associated with the general senses; complex receptors are associated with the special senses.
7. Sensory receptors respond to stimuli by producing receptor or generator potentials.
8. Table 16.1 on page 550 summarizes the classification of sensory receptors.
9. Adaptation is a decrease in sensitivity during a long-lasting stimulus. Receptors are either rapidly adapting or slowly adapting.

SOMATIC SENSATIONS (p. 550)

1. Somatic sensations include tactile sensations (touch, pressure, vibration, itch, and tickle), thermal sensations (warmth and cold), pain, and proprioception.
2. Receptors for tactile, thermal, and pain sensations are located in the skin, subcutaneous layer, and mucous membranes of the mouth, vagina, and anus.
3. Receptors for proprioceptive sensations (position and movement of body parts) are located in muscles, tendons, joints, and the inner ear.

4. Receptors for touch are (a) hair root plexuses and corpuscles of touch (Meissner corpuscles), which are rapidly adapting, and (b) slowly adapting type I cutaneous mechanoreceptors (tactile or Merkel discs). Type II cutaneous mechanoreceptors (Ruffini corpuscles), which are slowly adapting, are sensitive to stretching. Receptors for pressure include corpuscles of touch, type I mechanoreceptors, and lamellated (pacinian) corpuscles. Receptors for vibration are corpuscles of touch and lamellated corpuscles. Itch receptors are free nerve endings; both free nerve endings and lamellated corpuscles mediate the tickle sensation.
5. Thermoreceptors are free nerve endings. Cold receptors are located in the stratum basale of the epidermis; warm receptors are located in the dermis.
6. Pain receptors (nociceptors) are free nerve endings that are located in nearly every body tissue.
7. Nerve impulses for fast pain propagate along medium-diameter, myelinated A fibers, whereas those for slow pain conduct along small-diameter, unmyelinated C fibers.
8. Proprioceptors include muscle spindles, tendon organs, joint kinesthetic receptors, and hair cells of the inner ear.
9. Table 16.2 on page 555 summarizes the somatic sensory receptors and the sensations they convey.

SOMATIC SENSORY PATHWAYS (p. 556)

1. Somatic sensory pathways from receptors to the cerebral cortex involve three-neuron sets: first-order, second-order, and third-order neurons.
2. Axon collaterals (branches) of somatic sensory neurons simultaneously carry signals into the cerebellum and the reticular formation of the brain stem.
3. Impulses propagating along the posterior column–medial lemniscus pathway relay fine touch, stereognosis, proprioception, and vibratory sensations.
4. The neural pathway for pain and thermal sensations is the lateral spinothalamic tract.
5. The neural pathway for tickle, itch, crude touch, and pressure sensations is the anterior spinothalamic pathway.
6. The pathways to the cerebellum are the anterior and posterior spinocerebellar tracts, which transmit impulses for subconscious muscle and joint position sense from the trunk and lower limbs.
7. Table 16.3 on page 559 summarizes the major somatic sensory pathways.
8. Specific regions of the primary somatosensory area (postcentral gyrus) of the cerebral cortex receive somatic sensory input from different parts of the body.
9. The primary motor area (precentral gyrus) of the cortex is the major control region for planning and initiating voluntary movements.

SOMATIC MOTOR PATHWAYS (p. 560)

1. All excitatory and inhibitory signals that control movement converge on the motor neurons, also known as lower motor neurons (LMNs) or the final common pathway.
2. Neurons in four neural circuits, collectively termed the somatic motor pathways, participate in control of movement by providing input to lower motor neurons: local circuit neurons, upper motor neurons, basal ganglia neurons, and cerebellar neurons.
3. The axons of upper motor neurons extend from the brain to lower motor neurons via direct and indirect motor pathways. The direct (pyramidal) pathways include the lateral and anterior corticospinal tracts and corticobulbar tracts. Indirect (extrapyramidal) pathways extend from several motor centers of the brain stem into the spinal cord.
4. Neurons of the basal ganglia assist movement by providing input to the upper motor neurons. They help initiate and terminate movements, suppress unwanted movements, and establish a normal level of muscle tone.
5. The cerebellum is active in learning and performing rapid, coordinated, highly skilled movements. It also contributes to maintaining balance and posture.
6. Table 16.4 on page 563 summarizes the major somatic motor pathways.

INTEGRATIVE FUNCTIONS OF THE CEREBRUM (p. 564)

1. Sleep and wakefulness are integrative functions that are controlled by the suprachiasmatic nucleus and the reticular activating system (RAS).
2. Non-rapid eye movement (NREM) sleep consists of four stages.
3. Most dreaming occurs during rapid eye movement (REM) sleep.
4. Memory, the ability to store and recall thoughts, involves persistent changes in the brain, a capability called plasticity. Three types are immediate, short-term, and long-term memory.

SELF-QUIZ QUESTIONS

Fill in the blanks in the following statements.

1. ____ is the conscious or subconscious awareness of external or internal stimuli; ____ is the conscious awareness and interpretation of sensory input.
2. The term used to describe the crossing over of axons from one side of the brain or spinal cord to the other side is ____.

Indicate whether the following statements are true or false.

3. Touch, pressure, and pain are all classified as tactile sensations.
4. Awakening from sleep involves increased activity in the reticular activating system.

Choose the one best answer to the following questions.

5. A nurse touches the lower back of a patient, but the patient does not feel the sensation. Which of the following could explain the lack of sensation? (1) The stimulus was not in the receptive field. (2) The generator potential has not reached threshold. (3) There is damage to the somatosensory region of the cerebral cortex. (4) The

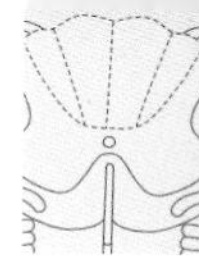

nurse was stimulating a proprioceptor. (5) A slowly adapting receptor has been stimulated. (a) 1, 3 and 5; (b) 3, 4, and 5; (c) 1, 2, and 3; (d) 2, 3, and 4; (e) 1 only.

6. Which of the following statements is *false*? (1) Upper motor neurons transmit impulses from the CNS to skeletal muscle fibers. (2) Lower motor neurons have their cell bodies in the brain stem and spinal cord. (3) Local circuit neurons receive input from somatic sensory receptors and help coordinate rhythmic activity in specific muscle groups. (4) The activity of upper motor neurons is influenced by both the basal ganglia and cerebellum. (5) The cerebellum helps to monitor differences between intended movements and actual movements for coordination, posture, and balance.

7. Which of the following statements are *true*? (1) Slow pain is a result of impulse propagation along myelinated A nerve fibers. (2) Visceral pain occurs when nociceptors in the skin are stimulated. (3) Referred pain is pain felt in an area far from the stimulated organ. (4) Nociceptors exhibit very little adaptation. (5) Nociceptors are located in every body tissue. (a) 1, 3, 4, and 5; (b) 2, 3, and 5; (c) 1 and 5; (d) 3 and 4; (e) 3, 4, and 5.

8. You cannot "hear" with your eyes because (a) hearing is a somatic sense and vision is a special sense, (b) the sensory neurons for sight carry information only for the modality of vision, (c) the impulses for hearing are transmitted to the somatosensory area of the cerebral cortex, (d) hearing receptors are selective and vision receptors are not, (e) hearing receptors produce a generator potential and vision receptors produce a receptor potential.

9. Which of the following statements is *false*? (a) First-order sensory neurons carry signals from the somatic receptors into either the brain stem or spinal cord. (b) Second-order neurons carry signals from the spinal cord and brain stem to the thalamus. (c) Third-order neurons project to the primary somatosensory area of the cortex, where conscious perception of the sensation results. (d) The somatic sensory pathways to the cerebellum are the posterior column–medial lemniscus pathway and the anterolateral pathway. (e) Axons of second-order neurons decussate (cross) in the spinal cord or brain stem before ascending to the thalamus.

10. Which of the following is *not* an aspect of cerebellar function? (a) monitoring movement intention, (b) monitoring actual movement, (c) comparing intent with actual performance, (d) sending out corrective signals, (e) directing sensory input to effectors.

11. During REM sleep (1) neuronal activity in the pons and midbrain is high, (2) most somatic motor neurons are inhibited, (3) most dreaming occurs, (4) sleepwalking can occur, (5) there is an increase in heart rate and blood pressure. (a) 1, 2, 4, and 5; (b) 2, 3, and 5; (c) 1, 2, 3, 4, and 5; (d) 2, 3, and 4; (e) 1, 2, and 3.

12. Which of the following statements is *incorrect*? (a) The graded potentials produced by receptors that serve the senses of touch, pressure, stretching, vibration, pain, proprioception, and smell are generator potentials. (b) The graded potentials produced by receptors that serve the special senses of vision, hearing, equilibrium, and taste are receptor potentials. (c) When a generator potential is large enough to reach threshold, it generates one or more nerve impulses in its first-order sensory neuron. (d) A receptor potential generates nerve impulses in a second-order neuron. (e) The amplitude of both generator and receptor potentials varies with the intensity of the stimulus.

13. Match the following:

____(a) located in the precentral gyrus, this is the major control region of the cerebral cortex for initiation of voluntary movements

____(b) direct pathways conveying impulses from the cerebral cortex to the spinal cord that result in precise, voluntary movements

____(c) contain motor neurons that control skilled movements of the hands and feet

____(d) tracts include rubrospinal, tectospinal, vestibulospinal, lateral reticulospinal, and medial reticulospinal

____(e) contain neurons that help initiate and terminate movements; can suppress unwanted movements; influence muscle tone

____(f) carry impulses for pain, temperature, tickle, itch, crude touch, and pressure

____(g) the major routes relaying proprioceptive input to the cerebellum; critical for posture, balance, and coordination of skilled movements

____(h) composed of axons of first-order neurons; include the gracile fasciculus and cuneate fasciculus

____(i) contain motor neurons that coordinate movements of the axial skeleton

____(j) contain axons that convey impulses for precise, voluntary movements of the eyes, tongue, and neck, plus chewing, facial expression, and speech

____(k) convey sensations of fine touch, stereognosis, proprioception, weight discrimination, and vibration to the cerebral cortex.

(1) posterior column
(2) anterolateral (spinothalamic) pathways
(3) spinocerebellar tracts
(4) lateral corticospinal tracts
(5) anterior corticospinal tracts
(6) corticobulbar tracts
(7) extrapyramidal pathways
(8) pyramidal pathways
(9) primary motor area
(10) basal ganglia
(11) posterior column–medial lemniscus pathways

14. Match the following (some answers may be used more than once):

____(a) receptors located in muscles, tendons, joints, and the inner ear
____(b) receptors located in blood vessels, visceral organs, muscles, and the nervous system
____(c) receptors that detect temperature changes
____(d) receptors that detect light that strikes the retina of the eye
____(e) receptors located at or near the external surface of the body
____(f) bare dendrites associated with pain, thermal, tickle, itch, and some touch sensations
____(g) receptors that provide information about body position, muscle tension, and position and activity of joints
____(h) receptors that sense osmotic pressures of body fluids
____(i) receptors that detect chemicals in the mouth, nose, and body fluids
____(j) receptors that detect mechanical pressure or stretching
____(k) receptors that respond to stimuli resulting from physical or chemical damage to tissues
____(l) dendrites enclosed in a connective tissue capsule

(1) exteroceptors
(2) interoceptors
(3) proprioceptors
(4) mechanoreceptors
(5) thermoreceptors
(6) nociceptors
(7) photoreceptors
(8) chemoreceptors
(9) free nerve endings
(10) encapsulated nerve endings
(11) osmoreceptors

15. Match the following:

____(a) specialized groupings of muscle fibers interspersed among regular skeletal muscle fibers and oriented parallel to them; monitor changes in the length of a skeletal muscle
____(b) inform the CNS about changes in muscle tension
____(c) widely distributed free nerve ending receptors for pain
____(d) encapsulated receptors for touch located in the dermal papillae; found in hairless skin, eyelids, tip of the tongue, and lips
____(e) lamellated corpuscles that detect pressure
____(f) type II cutaneous mechanoreceptors; most sensitive to stretching that occurs as digits or limbs are moved
____(g) located in the stratum basale and activated by low temperatures
____(h) located in the dermis and activated by high temperatures
____(i) found within and around the articular capsules of synovial joints; respond to pressure and acceleration and deceleration of joints
____(j) type I cutaneous mechanoreceptors that function in fine touch

(1) Meissner corpuscles
(2) Merkel discs
(3) Ruffini corpuscles
(4) pacinian corpuscles
(5) cold receptors
(6) warm receptors
(7) nociceptors
(8) tendon organs
(9) joint kinesthetic receptors
(10) muscle spindles

CRITICAL THINKING QUESTIONS

1. When Joni first stepped onto the sailboat, she smelled the tangy sea air and felt the motion of water beneath her feet. After a few minutes, she no longer noticed the smell, but unfortunately she was aware of the rolling motion for hours. What types of receptors are involved in smell and detection of motion? Why did her sensation of smell fade but the rolling sensation remain?

2. Monique sticks her left hand into a hot tub heated to about 43°C (110°F) in order to decide if she wants to enter the hot tub. Trace the pathway involved in transmitting the sensation of heat from her left hand to the somatosensory area in the cerebral cortex.

3. Marvin has had trouble sleeping. Last night his mother found him sleepwalking and gently led him back to his bed. When Marvin was awakened by his alarm clock the next day, he had no recollection of sleepwalking and, in fact, told his mother about the vivid dreams he had. What specific stages of sleep did Marvin undergo during the night? What neurological mechanism awakened Marvin in the morning?

ANSWERS TO FIGURE QUESTIONS

16.1 The special senses of vision, taste, hearing, and equilibrium are served by separate sensory cells.

16.2 Pain, thermal sensations, and tickle and itch arise with activation of different free nerve endings.

16.3 The kidneys have the broadest area for referred pain.

16.4 Muscle spindles are activated when the central areas of the intrafusal fibers are stretched.

16.5 Damage to the right lateral spinothalamic tract could result in loss of pain and thermal sensations on the left side of the body.

16.6 The hand has a larger representation in the motor area than in the somatosensory area, which implies greater precision in the hand's movement control than fine ability in its sensation.

16.7 Cerebral cortex UMNs are active in planning, initiating, and directing sequences of voluntary movements. Brain stem UMNs regulate muscle tone, control postural muscles, and help maintain balance and orientation of the head and body.

16.8 The corticobulbar and rubrospinal tracts convey impulses that result in precise, voluntary movements (see Table 16.4).

16.9 The anterior and posterior spinocerebellar tracts carry information from proprioceptors in joints and muscles to the cerebellum.

16.10 Olfactory input does not stimulate the RAS; a smoke detector responds to smoke by sounding a loud bell or buzzer, which awakens sleepers by providing auditory input that stimulates the RAS.

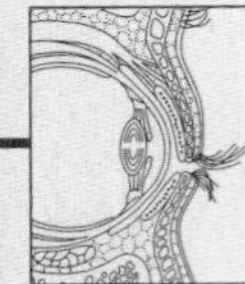

Chapter 17

The Special Senses

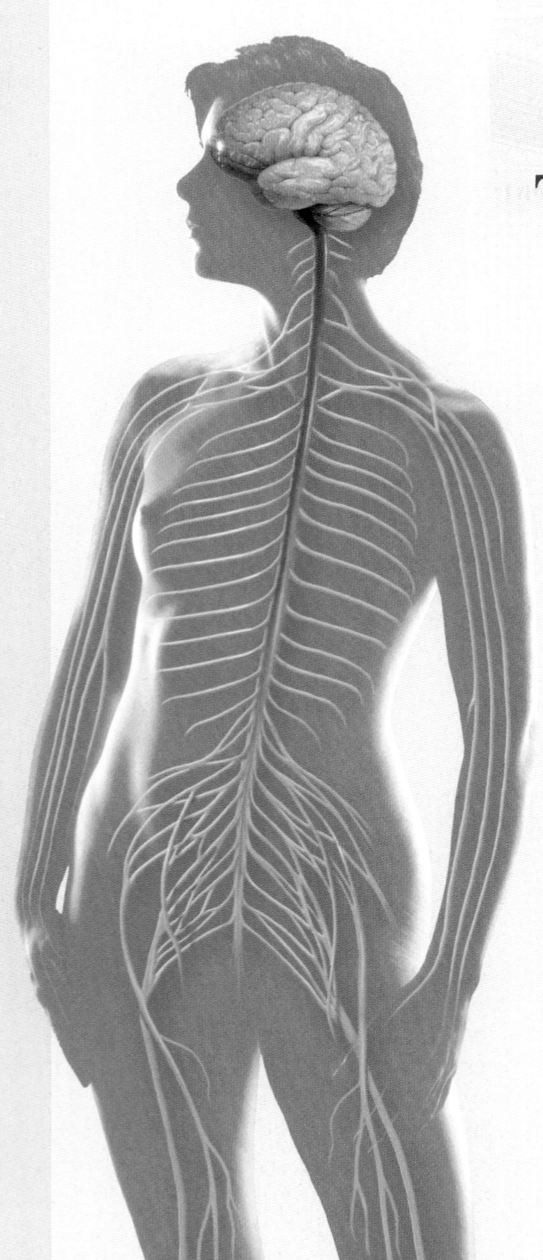

The Special Senses and Homeostasis

Sensory organs have special receptors that allow us to smell, taste, see, hear, and maintain equilibrium or balance. Information conveyed from these receptors to the central nervous system is used to help maintain homeostasis.

www. wiley.com/college/apcentral

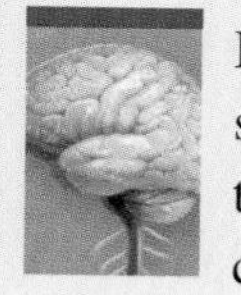

Recall from Chapter 16 that the general senses include somatic senses (tactile, thermal, pain, and proprioceptive) and visceral sensations. As you learned in that chapter, receptors for the general senses are scattered throughout the body and are relatively simple in structure. Receptors for the special senses—smell, taste, vision, hearing, and equilibrium—are anatomically distinct from one another and are concentrated in specific locations in the head. They are usually embedded in the epithelial tissue within complex sensory organs such as the eyes and ears. Neural pathways for the special senses are also more complex than those for the general senses.

In this chapter we examine the structure and function of the special sense organs, and the pathways involved in conveying information from them to the central nervous system. **Ophthalmology** (of-thal-MOL-ō-jē; *ophthalmo-* = eye; *-logy* = study of) is the science that deals with the eye and its disorders. The other special senses are, in large part, the concern of **otorhinolaryngology** (ō-tō-rī′-nō-lar-in-GOL-ō-jē; *oto-* = ear; *rhino-* = nose; *laryngo-* = larynx), the science that deals with the ears, nose, and throat and their disorders.

OLFACTION: SENSE OF SMELL

▶ **OBJECTIVE**

Describe the olfactory receptors and the neural pathway for olfaction.

Both smell and taste are chemical senses, because the sensations arise from the interaction of molecules with smell or taste receptors. Because impulses for smell and taste propagate to the limbic system (and to higher cortical areas as well), certain odors and tastes can evoke strong emotional responses or a flood of memories.

Anatomy of Olfactory Receptors

The nose contains 10–100 million receptors for the sense of smell or **olfaction** (ol-FAK-shun; *olfact-* = smell), contained within an area called the *olfactory epithelium.* With a total area of 5 cm^2 (a little less than 1 $in.^2$), the olfactory epithelium occupies the superior part of the nasal cavity, covering the inferior surface of the cribriform plate and extending along the superior nasal concha (Figure 17.1a). The olfactory epithelium consists of three kinds of cells: olfactory receptors, supporting cells, and basal cells (Figure 17.1b).

Olfactory receptors are the first-order neurons of the olfactory pathway. Each olfactory receptor is a bipolar neuron with an exposed knob-shaped dendrite and an axon projecting through the cribriform plate and ending in the olfactory bulb. The sites of olfactory transduction are the **olfactory hairs,** cilia that project from the dendrite. (Recall that *transduction* is the conversion of stimulus energy into a graded potential in a sensory receptor.) Chemicals that have an odor and can therefore stimulate the olfactory hairs are called **odorants.** Olfactory receptors respond to the chemical stimulation of an odorant molecule by producing a generator potential, thus initiating the olfactory response.

Supporting cells are columnar epithelial cells of the mucous membrane lining the nose. They provide physical support, nourishment, and electrical insulation for the olfactory receptors, and they help detoxify chemicals that come in contact with the olfactory epithelium. **Basal cells** are stem cells located between the bases of the supporting cells. They continually undergo cell division to produce new olfactory receptors, which live for only a month or so before being replaced. This process is remarkable considering that olfactory receptors are neurons, and as you have already learned, mature neurons are generally not replaced.

Within the connective tissue that supports the olfactory epithelium are **olfactory (Bowman's) glands,** which produce mucus that is carried to the surface of the epithelium by ducts. The secretion moistens the surface of the olfactory epithelium and dissolves odorants so that transduction can occur. Both supporting cells of the nasal epithelium and olfactory glands are innervated by branches of the facial (VII) nerve, which can be stimulated by certain chemicals. Impulses in these nerves, in turn, stimulate the lacrimal glands in the eyes and nasal mucous glands. The result is tears and a runny nose after inhaling substances such as pepper or the vapors of household ammonia.

Physiology of Olfaction

Many attempts have been made to distinguish among and classify "primary" sensations of smell. Genetic evidence now suggests the existence of hundreds of primary odors. Our ability to recognize about 10,000 different odors probably depends on patterns of activity in the brain that arise from activation of many different combinations of olfactory receptors.

Olfactory receptors react to odorant molecules in the same way that most sensory receptors react to their specific stimuli: A generator potential (depolarization) develops and triggers one or more nerve impulses. In some cases, an odorant binds to a receptor linked to proteins in the plasma membrane called G proteins and activates the enzyme adenylate cyclase (see page 623). The result is the following chain of events: production of cyclic adenosine monophosphate (cAMP) → opening of sodium ion (Na^+) channels → inflow of Na^+ → depolarizing generator potential → generation of nerve impulse and propagation along axon of olfactory receptor.

Odor Thresholds and Adaptation

Olfaction, like all the special senses, has a low threshold. Only a few molecules of certain substances need be present in air to be perceived as an odor. A good example is the chemical methyl mercaptan, which smells like rotten cabbage and can be detected in concentrations as low as 1/25 billionth of a milligram per milliliter of air. Because the natural gas used for cooking and heating is odorless but lethal and potentially explosive if it accumulates, a small amount of methyl mercaptan is added to natural gas to provide olfactory warning of gas leaks.

Adaptation (decreasing sensitivity) to odors occurs rapidly. Olfactory receptors adapt by about 50% in the first second or so after stimulation but adapt very slowly thereafter. Still, complete insensitivity to certain strong odors occurs about a minute after exposure. Apparently, reduced sensitivity involves an adaptation process in the central nervous system as well.

The Olfactory Pathway

On each side of the nose, bundles of the slender, unmyelinated axons of olfactory receptors extend through about 20 olfactory foramina in the cribriform plate of the ethmoid bone (Figure 17.1b). These 40 or so bundles of axons collectively form the right and left **olfactory (I) nerves.** The olfactory nerves terminate in the brain in paired masses of gray matter called the **olfactory bulbs,** which are located below the frontal lobes of the cerebrum and lateral to the crista galli of the ethmoid bone. Within the olfactory bulbs, the axon terminals of olfactory receptors—the first-order neurons—form synapses with the dendrites and cell bodies of second-order neurons in the olfactory pathway.

Axons of olfactory bulb neurons extend posteriorly and form the **olfactory tract** (Figure 17.1a). Some of the axons of the olfactory tract project to the primary olfactory area; located

Figure 17.1 Olfactory epithelium and olfactory receptors. (a) Location of olfactory epithelium in nasal cavity. (b) Anatomy of olfactory receptors, consisting of first-order neurons whose axons extend through the cribriform plate and terminate in the olfactory bulb. (See Tortora, *A Photographic Atlas of the Human Body, Second Edition,* Figures 9.1b and 9.1c.)

The olfactory epithelium consists of olfactory receptors, supporting cells, and basal cells.

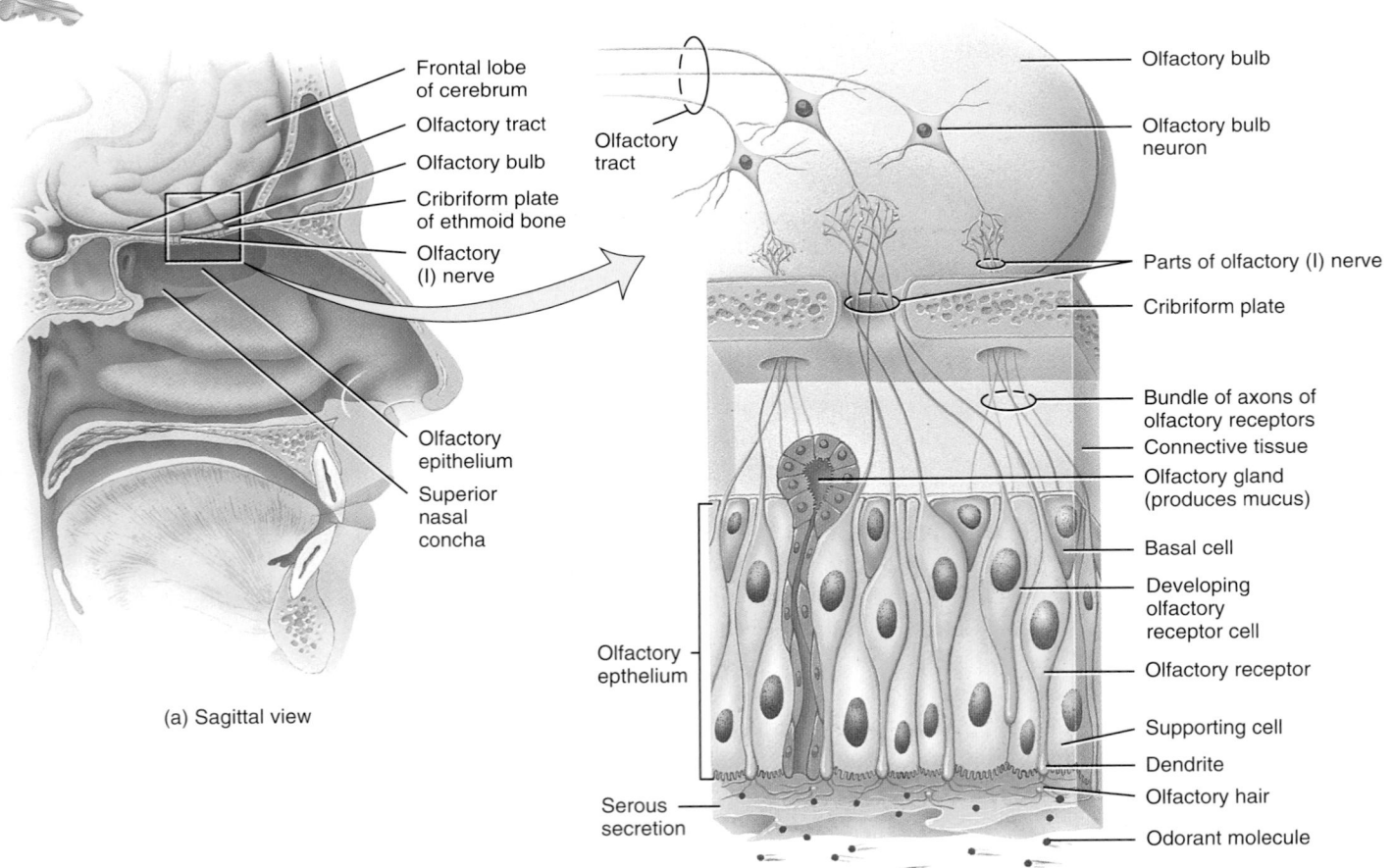

(a) Sagittal view

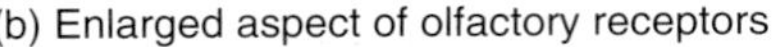

(b) Enlarged aspect of olfactory receptors

Which part of an olfactory receptor detects an odorant molecule?

at the inferior and medial surface of the temporal lobe, the primary olfactory area is where conscious awareness of smell begins. Other axons of the olfactory tract project to the limbic system and hypothalamus; these connections account for our emotional and memory-evoked responses to odors. Examples include sexual excitement upon smelling a certain perfume, nausea upon smelling a food that once made you violently ill, or an odor-evoked memory of a childhood experience.

From the primary olfactory area, pathways also extend to the frontal lobe, both directly and indirectly via the thalamus. An important region for odor identification and discrimination is the orbitofrontal area (area 11 in Figure 14.15 on page 497). People who suffer damage in this area have difficulty identifying different odors. Positron emission tomography (PET) studies suggest some degree of hemispheric lateralization: The orbitofrontal area of the *right* hemisphere exhibits greater activity during olfactory processing.

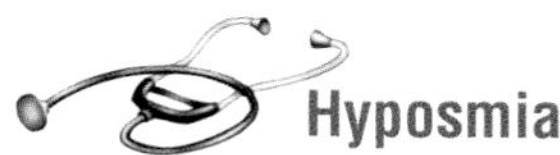

Hyposmia

Women often have a keener sense of smell than men do, especially at the time of ovulation. Smoking seriously impairs the sense of smell in the short term and may cause long-term damage to olfactory receptors. With aging the sense of smell deteriorates. **Hyposmia** (hī-POZ-mē-a; *osmi* = smell, odor), a reduced ability to smell, affects half of those over age 65 and 75% of those over age 80. Hyposmia also can be caused by neurological changes, such as a head injury, Alzheimer disease, or Parkinson disease; certain drugs, such as antihistamines, analgesics, or steroids; and the damaging effects of smoking. ■

► CHECKPOINT

1. How do basal cells contribute to olfaction?
2. What is the sequence of events from the binding of an odorant molecule to an olfactory hair to the arrival of a nerve impulse in the orbitofrontal area?

GUSTATION: SENSE OF TASTE

► OBJECTIVE

Describe the gustatory receptors and the neural pathway for gustation.

Taste or **gustation** (gus-TĀ-shun; *gust-* = taste), like olfaction, is a chemical sense. However, it is much simpler than olfaction in that only five primary tastes can be distinguished: *sour, sweet, bitter, salty,* and *umami* (ū-MAM-ē). The umami taste, recently reported by Japanese scientists, is described as "meaty" or "savory." Umami is believed to arise from taste receptors that are stimulated by monosodium glutamate (MSG), a substance naturally present in many foods and added to others as a flavor enhancer. All other flavors, such as chocolate, pepper, and coffee, are combinations of the five primary tastes, plus accompanying olfactory and tactile (touch) sensations. Odors from food can pass upward from the mouth into the nasal cavity, where they stimulate olfactory receptors. Because olfaction is much more sensitive than taste, a given concentration of a food substance may stimulate the olfactory system thousands of times more strongly than it stimulates the gustatory system. When you have a cold or are suffering from allergies and cannot taste your food, it is actually olfaction that is blocked, not taste.

Anatomy of Taste Buds and Papillae

The receptors for sensations of taste are located in the taste buds (Figure 17.2). Most of the nearly 10,000 taste buds of a young adult are on the tongue, but some are found on the soft palate (posterior portion of the roof of the mouth), pharynx (throat), and epiglottis (cartilage lid over voice box). The number of taste buds declines with age. Each **taste bud** is an oval body consisting of three kinds of epithelial cells: supporting cells, gustatory receptor cells, and basal cells (see Figure 17.2c). The **supporting cells** surround about 50 **gustatory receptor cells** in each taste bud. A single, long microvillus, called a **gustatory hair,** projects from each gustatory receptor cell to the external surface through the **taste pore,** an opening in the taste bud. **Basal cells,** stem cells found at the periphery of the taste bud near the connective tissue layer, produce supporting cells, which then develop into gustatory receptor cells, each of which has a life span of about 10 days. At their base, the gustatory receptor cells synapse with dendrites of the first-order neurons that form the first part of the gustatory pathway. The dendrites of each first-order neuron branch profusely and contact many gustatory receptor cells in several taste buds.

Taste buds are found in elevations on the tongue called **papillae** (pa-PIL-ē; singular is *papilla*), which provide a rough texture to the upper surface of the tongue (Figure 17.2a, b). Three types of papillae contain taste buds.

1. About 12 very large, circular **vallate (circumvallate) papillae** (VAL-āt = wall-like) form an inverted V-shaped row at the back of the tongue. Each of these papillae houses 100–300 taste buds.
2. **Fungiform papillae** (FUN-ji-form = mushroomlike) are mushroom-shaped elevations scattered over the entire surface of the tongue that contain about five taste buds each.
3. **Foliate papillae** (FŌ-lē-āt = leaflike) are located in small trenches on the lateral margins of the tongue but most of their taste buds degenerate in early childhood.

In addition, the entire surface of the tongue has **filiform papillae** (FIL-i-form = threadlike). These pointed, threadlike structures contain tactile receptors but no taste buds. They increase friction between the tongue and food, making it easier for the tongue to move food in the oral cavity.

Figure 17.2 The relationship of gustatory receptor cells in taste buds to tongue papillae. (See Tortora, *A Photographic Atlas of the Human Body, Second Edition,* Figure 9.2.)

Gustatory receptor cells are located in taste buds.

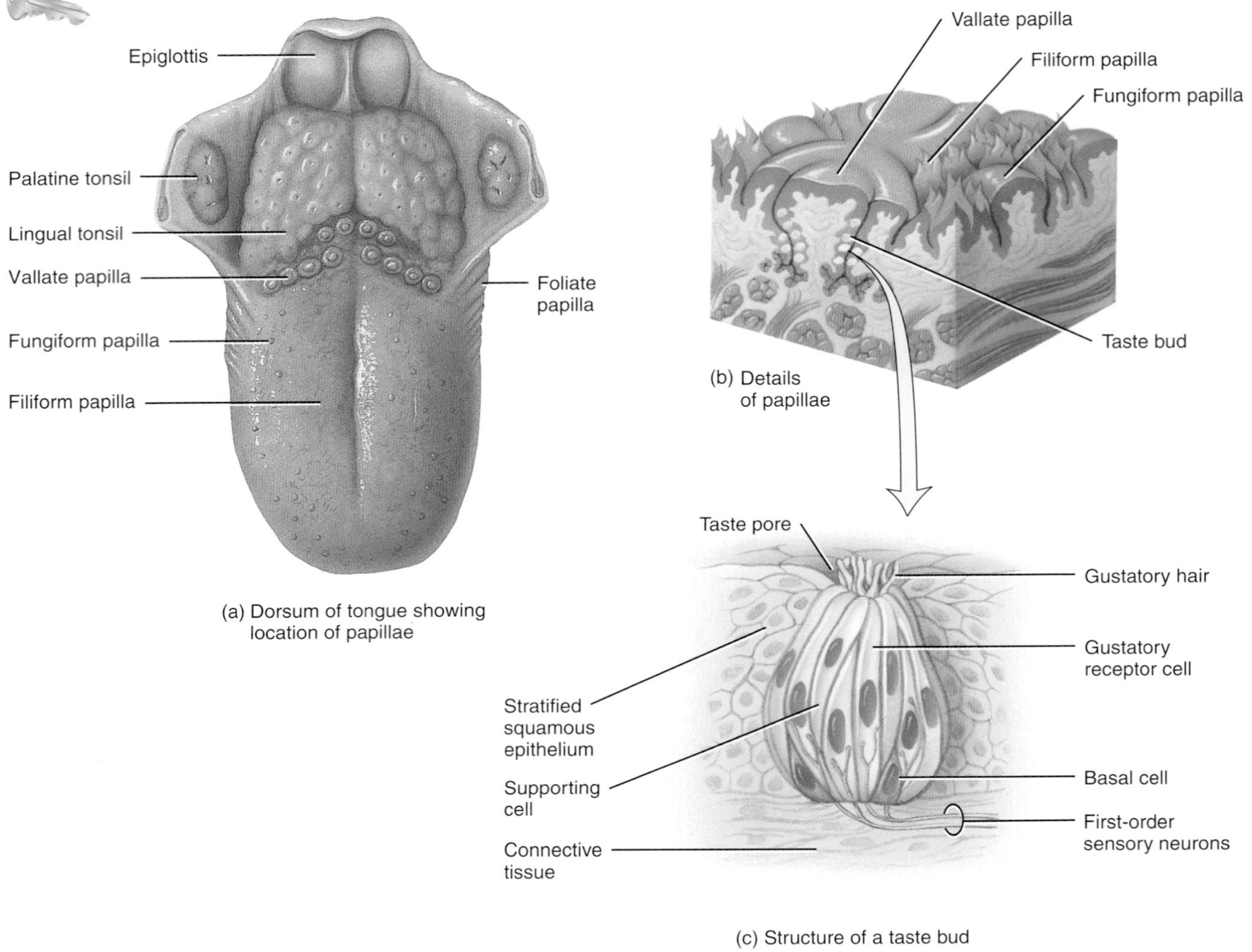

Epiglottis
Palatine tonsil
Lingual tonsil
Vallate papilla
Fungiform papilla
Filiform papilla
Foliate papilla

(a) Dorsum of tongue showing location of papillae

Vallate papilla
Filiform papilla
Fungiform papilla
Taste bud

(b) Details of papillae

Taste pore
Gustatory hair
Gustatory receptor cell
Stratified squamous epithelium
Supporting cell
Basal cell
First-order sensory neurons
Connective tissue

(c) Structure of a taste bud

What role do supporting cells in taste buds play?

Physiology of Gustation

Chemicals that stimulate gustatory receptor cells are known as **tastants.** Once a tastant is dissolved in saliva, it can make contact with the plasma membrane of the gustatory hairs, which are the sites of taste transduction. The result is a receptor potential that stimulates exocytosis of synaptic vesicles from the gustatory receptor cell. In turn, the liberated neurotransmitter molecules trigger nerve impulses in the first-order sensory neurons that synapse with gustatory receptor cells.

The receptor potential arises differently for different tastants. The sodium ions (Na^+) in a salty food enter gustatory receptor cells via Na^+ channels in the plasma membrane. The accumulation of Na^+ inside causes depolarization, which opens Ca^{2+} channels. In turn, inflow of Ca^{2+} triggers exocytosis of synaptic vesicles and liberation of neurotransmitter. The hydrogen ions (H^+) in sour tastants may flow into gustatory receptor cells via H^+ channels. They also influence opening and closing of other types of ion channels. Again, the result is a depolarization that leads to release of neurotransmitter.

Other tastants, responsible for stimulating sweet, bitter, and umami tastes, do not themselves enter gustatory receptor cells. Rather, they bind to receptors on the plasma membrane that are linked to G proteins. The G-proteins then activate several different chemicals known as second messengers inside the gustatory receptor cell. Different second messengers cause depolarization in different ways, but the result is the same—release of neurotransmitter.

If all tastants cause release of neurotransmitter from many gustatory receptor cells, why do foods taste different? The answer to this question is thought to lie in the patterns of nerve impulses in groups of first-order taste neurons that synapse with the gustatory receptor cells. Different tastes arise from activation of different groups of taste neurons. In addition, although each individual gustatory receptor cell responds to more than one of the five primary tastes, it may respond more strongly to some tastants than to others.

Taste Thresholds and Adaptation

The threshold for taste varies for each of the primary tastes. The threshold for bitter substances, such as quinine, is lowest. Because poisonous substances often are bitter, the low threshold (or high sensitivity) may have a protective function. The threshold for sour substances, such as lemon, as measured by using hydrochloric acid, is somewhat higher. The thresholds for salty substances, represented by sodium chloride, and for sweet substances, as measured by using sucrose, are similar, and are higher than those for bitter or sour substances.

Complete adaptation to a specific taste can occur in 1–5 minutes of continuous stimulation. Taste adaptation is due to changes that occur in the taste receptors, in olfactory receptors, and in neurons of the gustatory pathway in the CNS.

The Gustatory Pathway

Three cranial nerves contain axons of the first-order gustatory neurons that innervate the taste buds. The facial (VII) nerve serves taste buds in the anterior two-thirds of the tongue; the glossopharyngeal (IX) nerve serves taste buds in the posterior one-third of the tongue; and the vagus (X) nerve serves taste buds in the throat and epiglottis. From the taste buds, nerve impulses propagate along these cranial nerves to the medulla oblongata. From the medulla, some axons carrying taste signals project to the limbic system and the hypothalamus; others project to the thalamus. Taste signals that project from the thalamus to the primary gustatory area in the parietal lobe of the cerebral cortex (see area 43 in Figure 14.15 on page 497) give rise to the conscious perception of taste.

Taste Aversion

Probably because of taste projections to the hypothalamus and limbic system, there is a strong link between taste and pleasant or unpleasant emotions. Sweet foods evoke reactions of pleasure while bitter ones cause expressions of disgust even in newborn babies. This phenomenon is the basis for **taste aversion,** in which people and animals quickly learn to avoid a food if it upsets the digestive system. The advantage of avoiding foods that cause such illness is longer survival. However, the drugs and radiation treatments used to combat cancer often cause nausea and gastrointestinal upset regardless of what foods are consumed. Thus, cancer patients may lose their appetite because they develop taste aversions for most foods. ■

▶ CHECKPOINT

3. How do olfactory receptors and gustatory receptor cells differ in structure and function?
4. Trace the path of a gustatory stimulus from contact of a tastant with saliva to the primary gustatory area in the cerebral cortex.
5. Compare the olfactory and gustatory pathways.

VISION

▶ OBJECTIVES

List and describe the accessory structures of the eye and the structural components of the eyeball.

Discuss image formation by describing refraction, accommodation, and constriction of the pupil.

Describe the processing of visual signals in the retina and the neural pathway for vision.

Vision is extremely important to human survival. More than half the sensory receptors in the human body are located in the eyes, and a large part of the cerebral cortex is devoted to processing visual information. In this section of the chapter, we examine the accessory structures of the eye, the eyeball itself, the formation of visual images, the physiology of vision, and the visual pathway from the eye to the brain.

Accessory Structures of the Eye

The **accessory structures** of the eye include the eyelids, eyelashes, eyebrows, the lacrimal (tearing) apparatus, and extrinsic eye muscles.

Eyelids

The upper and lower **eyelids,** or **palpebrae** (PAL-pe-brē; singular is *palpebra*), shade the eyes during sleep, protect the eyes from excessive light and foreign objects, and spread lubricating secretions over the eyeballs (Figure 17.3). The upper eyelid is more movable than the lower and contains in its superior region the **levator palpebrae superioris muscle.** Sometimes a person may experience an annoying *twitch* in an eyelid, an involuntary quivering similar to muscle twitches in the hand, forearm, leg, or foot. Twitches are almost always harmless and usually last for only a few seconds. They are often associated with stress and fatigue. The space between the upper and lower eyelids that exposes the eyeball is the **palpebral fissure.** Its angles are

Figure 17.3 Surface anatomy of the right eye.

The palpebral fissure is the space between the upper and lower eyelids that exposes the eyeball.

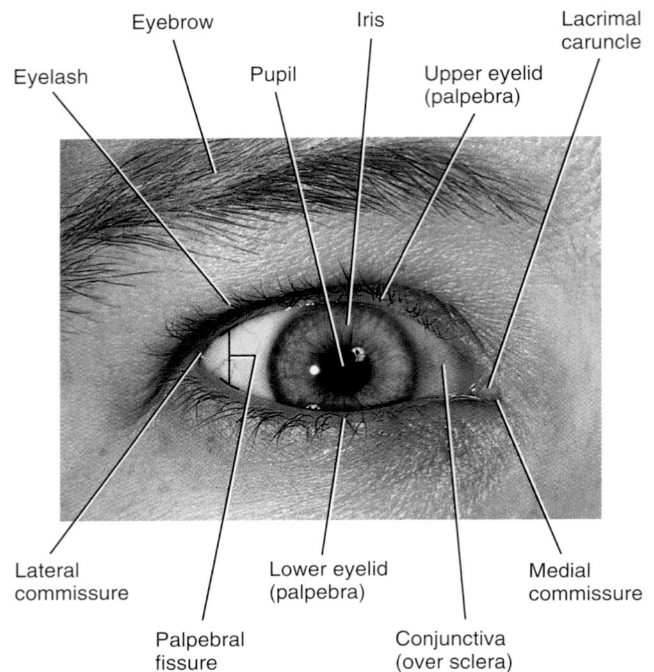

Which structure shown here is continuous with the inner lining of the eyelids?

known as the **lateral commissure** (KOM-i-shur), which is narrower and closer to the temporal bone, and the **medial commissure,** which is broader and nearer the nasal bone. In the medial commissure is a small, reddish elevation, the **lacrimal caruncle** (KAR-ung-kul), which contains sebaceous (oil) glands and sudoriferous (sweat) glands. The whitish material that sometimes collects in the medial commissure comes from these glands.

From superficial to deep, each eyelid consists of epidermis, dermis, subcutaneous tissue, fibers of the orbicularis oculi muscle, a tarsal plate, tarsal glands, and conjunctiva (Figure 17.4a). The **tarsal plate** is a thick fold of connective tissue that gives form and support to the eyelids. Embedded in each tarsal plate is a row of elongated modified sebaceous glands, known as **tarsal** or **Meibomian glands** (mī-BŌ-mē-an), that secrete a fluid that helps keep the eyelids from adhering to each other. Infection of the tarsal glands produces a tumor or cyst on the eyelid called a **chalazion** (ka-LĀ-zē-on = small bump). The **conjunctiva** (kon′-junk-TĪ-va) is a thin, protective mucous membrane composed of stratified columnar epithelium with numerous goblet cells that is supported by areolar connective tissue. The **palpebral conjunctiva** lines the inner aspect of the eyelids, and the **bulbar conjunctiva** passes from the eyelids onto the surface of the eyeball, where it covers the sclera (the "white" of the eye) but not the cornea, which is a transparent region that forms the outer anterior surface of the eyeball. Both the sclera and the cornea will be discussed in more detail shortly. Dilation and congestion of the blood vessels of the bulbar conjunctiva due to local irritation or infection are the cause of **bloodshot eyes.**

Eyelashes and Eyebrows

The **eyelashes,** which project from the border of each eyelid, and the **eyebrows,** which arch transversely above the upper eyelids, help protect the eyeballs from foreign objects, perspiration, and the direct rays of the sun. Sebaceous glands at the base of the hair follicles of the eyelashes, called **sebaceous ciliary glands,** release a lubricating fluid into the follicles. Infection of these glands is called a **sty.**

The Lacrimal Apparatus

The **lacrimal apparatus** (*lacrim-* = tears) is a group of structures that produces and drains **lacrimal fluid** or **tears.** The **lacrimal glands,** each about the size and shape of an almond, secrete lacrimal fluid, which drains into 6–12 **excretory lacrimal ducts** that empty tears onto the surface of the conjunctiva of the upper lid (Figure 17.4b). From here the tears pass medially over the anterior surface of the eyeball to enter two small openings called **lacrimal puncta** (singular is *punctum*). Tears then pass into two ducts, the **lacrimal canals,** which lead into the **lacrimal sac** and then into the **nasolacrimal duct.** This duct carries the lacrimal fluid into the nasal cavity just inferior to the inferior nasal concha. An infection of the lacrimal sacs is called **dacryocystitis** (*dacryo-* = lacrimal sac; *-itis* = inflammation of). It is usually caused by a bacterial infection and results in blockage of the nasolacrimal ducts.

The lacrimal glands are supplied by parasympathetic fibers of the facial (VII) nerves. The lacrimal fluid produced by these glands is a watery solution containing salts, some mucus, and **lysozyme,** a protective bactericidal enzyme. The fluid protects, cleans, lubricates, and moistens the eyeball. After being secreted from the lacrimal gland, lacrimal fluid is spread medially over the surface of the eyeball by the blinking of the eyelids. Each gland produces about 1 mL of lacrimal fluid per day.

Normally, tears are cleared away as fast as they are produced, either by evaporation or by passing into the lacrimal canals and then into the nasal cavity. If an irritating substance makes contact with the conjunctiva, however, the lacrimal glands are stimulated to oversecrete, and tears accumulate (watery eyes). Lacrimation is a protective mechanism, as the tears dilute and wash away the irritating substance. Watery eyes also occur when an inflammation of the nasal mucosa, such as occurs with a cold, obstructs the nasolacrimal ducts and blocks drainage of tears. Only humans express emotions, both happiness and sadness, by **crying.** In response to parasympathetic stimulation, the lacrimal glands produce excessive lacrimal fluid that may spill over the edges of the eyelids and even fill the nasal cavity with fluid. This is how crying produces a runny nose.

Figure 17.4 **Accessory structures of the eye.**

Accessory structures of the eye include the eyelids, eyelashes, eyebrows, the lacrimal apparatus, and extrinsic eye muscles.

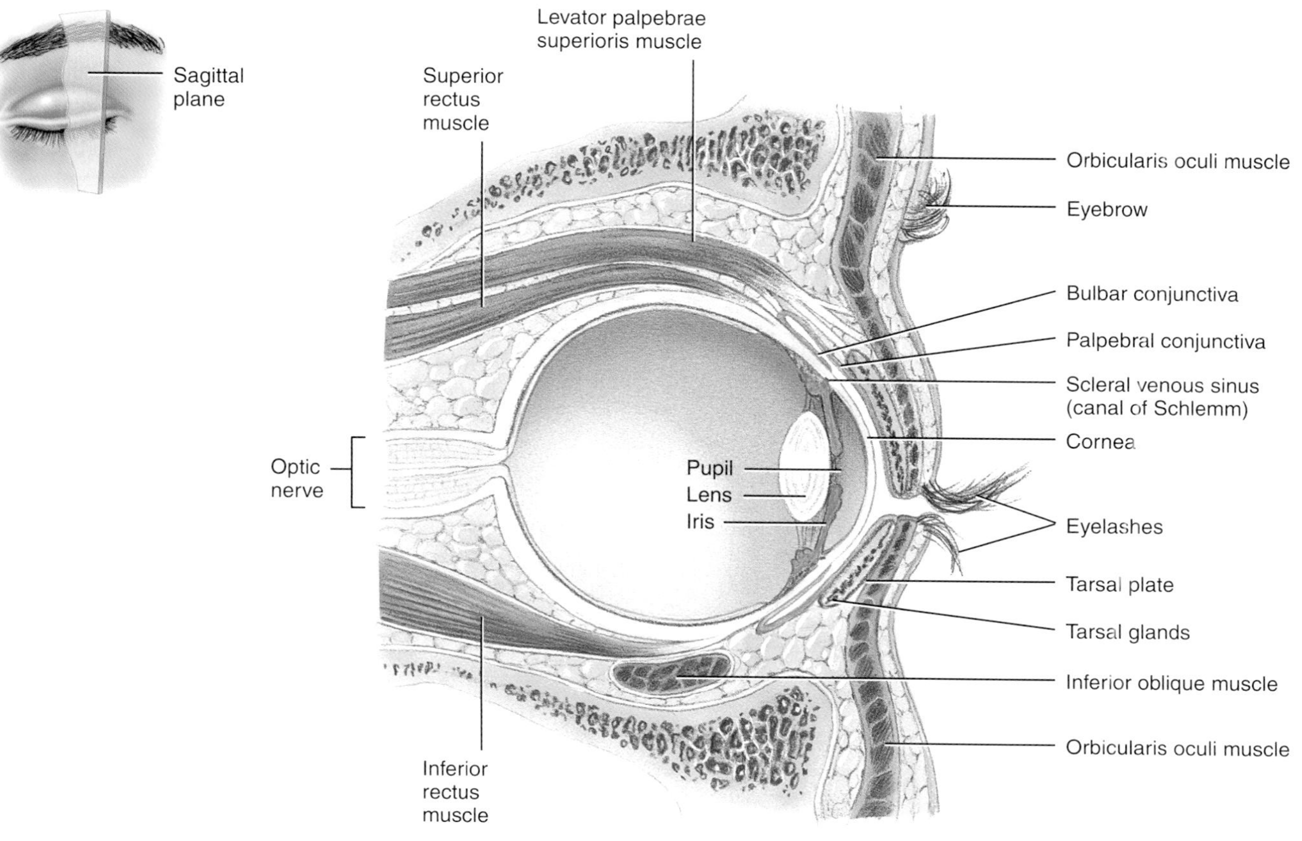

(a) Sagittal section of eye and its accessory structures

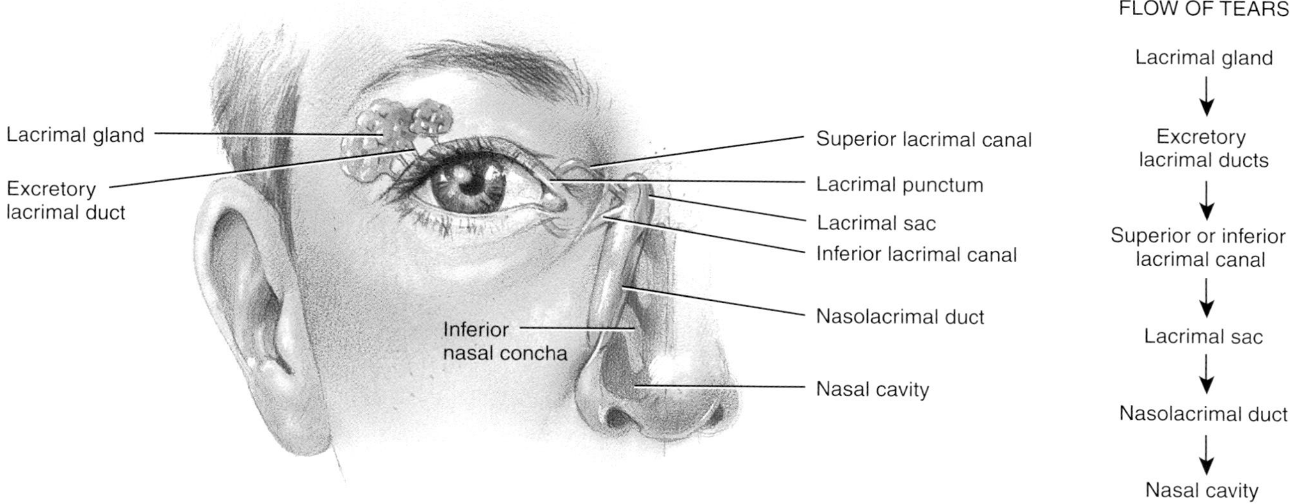

(b) Anterior view of the lacrimal apparatus

What is lacrimal fluid, and what are its functions?

Extrinsic Eye Muscles

Six extrinsic eye muscles move each eye: the **superior rectus, inferior rectus, lateral rectus, medial rectus, superior oblique,** and **inferior oblique** (Figures 17.4a and 17.5). They are supplied by cranial nerves III, IV, or VI. In general, the motor units in these muscles are small. Some motor neurons serve only two or three muscle fibers—fewer than in any other part of the body except the larynx (voice box). Such small motor units permit smooth, precise, and rapid movement of the eyes. As indicated in Exhibit 11.2 on page 340, the extrinsic eye muscles move the eyeball laterally, medially, superiorly, and inferiorly. For example, looking to the right requires simultaneous contraction of the right lateral rectus and left medial rectus muscles and relaxation of the left lateral rectus and right medial rectus. The oblique muscles preserve rotational stability of the eyeball. Neural circuits in the brain stem and cerebellum coordinate and synchronize the movements of the eyes.

Anatomy of the Eyeball

The adult **eyeball** measures about 2.5 cm (1 in.) in diameter. Of its total surface area, only the anterior one-sixth is exposed; the remainder is recessed and protected by the orbit, into which

Figure 17.5 Anatomy of the eyeball. (See Tortora, *A Photographic Atlas of the Human Body, Second Edition,* Figure 9.3c.)

The wall of the eyeball consists of three layers: the fibrous tunic, the vascular tunic, and the retina.

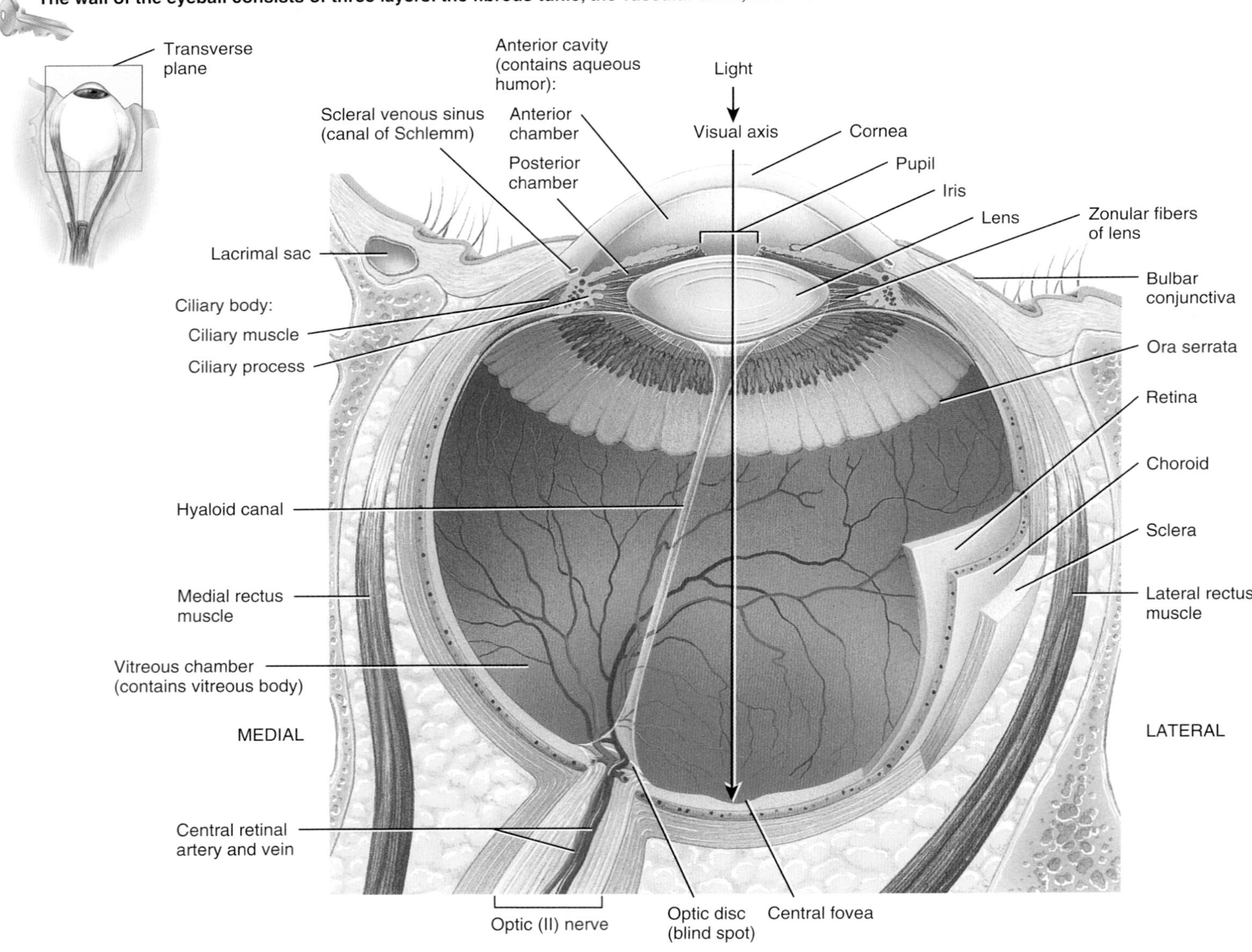

Superior view of transverse section of right eyeball

What are the components of the fibrous tunic and vascular tunic?

it fits. Anatomically, the wall of the eyeball consists of three layers: fibrous tunic, vascular tunic, and retina.

Fibrous Tunic

The **fibrous tunic,** is the superficial coat of the eyeball and consists of the anterior cornea and posterior sclera (Figure 17.5). The **cornea** (KOR-nē-a) is a transparent coat that covers the colored iris. Because it is curved, the cornea helps focus light onto the retina. Its outer surface consists of nonkeratinized stratified squamous epithelium. The middle coat of the cornea consists of collagen fibers and fibroblasts, and the inner surface is simple squamous epithelium. Since the central part of the cornea receives oxygen from the outside air, contact lenses that are worn for long periods of time must be permeable to permit oxygen to pass through them. The **sclera** (SKLE-ra; *scler-* = hard), the "white" of the eye, is a layer of dense connective tissue made up mostly of collagen fibers and fibroblasts. The sclera covers the entire eyeball except the cornea; it gives shape to the eyeball, makes it more rigid, and protects its inner parts. At the junction of the sclera and cornea is an opening known as the **scleral venous sinus (canal of Schlemm).** A fluid called aqueous humor drains into this sinus (Figure 17.5).

Vascular Tunic

The **vascular tunic** or **uvea** (Ū-vē-a) is the middle layer of the eyeball. It is composed of three parts: choroid, ciliary body, and iris (Figure 17.5). The highly vascularized **choroid** (KŌ-royd), which is the posterior portion of the vascular tunic, lines most of the internal surface of the sclera. Its numerous blood vessels provide nutrients to the posterior surface of the retina. The choroid also contains melanocytes that produce the pigment melanin, which causes this layer to appear dark brown in color. Melanin in the choroid absorbs stray light rays, which prevents reflection and scattering of light within the eyeball. As a result, the image cast on the retina by the cornea and lens remains sharp and clear. Albinos lack melanin in all parts of the body, including the eye. They often need to wear sunglasses, even indoors, because even moderately bright light is perceived as bright glare due to light scattering.

In the anterior portion of the vascular tunic, the choroid becomes the **ciliary body** (SIL-ē-ar′-ē). It extends from the **ora serrata** (Ō-ra ser-RĀ-ta), the jagged anterior margin of the retina, to a point just posterior to the junction of the sclera and cornea. Like the choroid, the ciliary body appears dark brown in color because it contains melanin-producing melanocytes. In addition, the ciliary body consists of ciliary processes and ciliary muscle. The **ciliary processes** are protrusions or folds on the internal surface of the ciliary body. They contain blood capillaries that secrete aqueous humor. Extending from the ciliary process are **zonular fibers (suspensory ligaments)** that attach to the lens. The **ciliary muscle** is a circular band of smooth muscle. Contraction or relaxation of the ciliary muscle changes the tightness of the zonular fibers, which alters the shape of the lens, adapting it for near or far vision.

The **iris,** the colored portion of the eyeball, is shaped like a flattened donut. It is suspended between the cornea and the lens and is attached at its outer margin to the ciliary processes. It consists of melanocytes and circular and radial smooth muscle fibers. The amount of melanin in the iris determines the eye color. The eyes appear brown to black when the iris contains a large amount of melanin, blue when its melanin concentration is very low, and green when its melanin concentration is moderate.

A principal function of the iris is to regulate the amount of light entering the eyeball through the **pupil,** the hole in the center of the iris. The pupil appears black because, as you look through the lens, you see the heavily pigmented back of the eye (choroid and retina). However, if bright light is directed into the pupil, the reflected light is red because of the blood vessels on the surface of the retina. It is for this reason that a person's eyes appear red in a photograph ("red eye") when a bright light is directed into the pupil. Autonomic reflexes regulate pupil diameter in response to light levels (Figure 17.6). When bright light stimulates the eye, parasympathetic fibers of the oculomotor (III) nerve stimulate the **circular muscles (sphincter pupillae)** of the iris to contract, causing a decrease in the size of the pupil (constriction). In dim light, sympathetic neurons stimulate the **radial muscles (dilator pupillae)** of the iris to contract, causing an increase in the pupil's size (dilation).

Retina

The third and inner coat of the eyeball, the **retina,** lines the posterior three-quarters of the eyeball and is the beginning of the visual pathway (see Figure 17.5). An *ophthalmoscope* (of-THAL-mō-skōp; *ophthalmos-* = eye; *-skopeo* = to examine) is an instrument that shines light into the eye and allows an observer to peer through the pupil, providing a magnified image

Figure 17.6 Responses of the pupil to light of varying brightness.

Contraction of the circular muscles causes constriction of the pupil; contraction of the radial muscles causes dilation of the pupil.

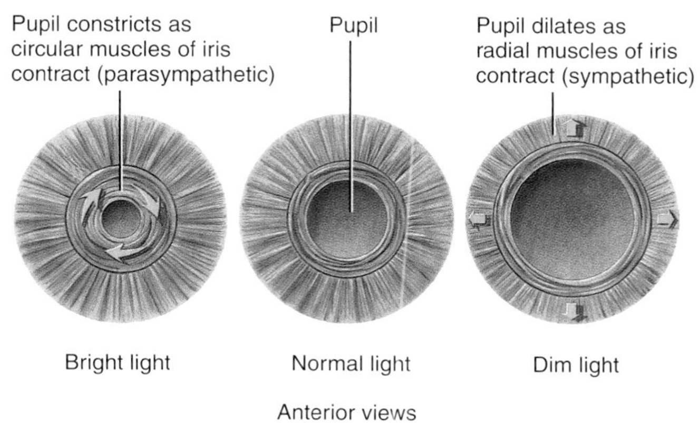

Which division of the autonomic nervous system causes pupillary constriction? Which causes pupillary dilation?

of the retina and its blood vessels as well as the optic (II) nerve (Figure 17.7). The surface of the retina is the only place in the body where blood vessels can be viewed directly and examined for pathological changes, such as those that occur with hypertension, diabetes mellitus, cataracts, and age-related macular disease. Several landmarks are visible through an ophthalmoscope. The **optic disc** is the site where the optic (II) nerve exits the eyeball. Bundled together with the optic nerve are the **central retinal artery,** a branch of the ophthalmic artery, and the **central retinal vein** (see Figure 17.5). Branches of the central retinal artery fan out to nourish the anterior surface of the retina; the central retinal vein drains blood from the retina through the optic disc. Also visible are the macula lutea and central fovea, which are described shortly.

The retina consists of a pigmented layer and a neural layer. The **pigmented layer** is a sheet of melanin-containing epithelial cells located between the choroid and the neural part of the retina. The melanin in the pigmented layer of the retina, like in the choroid, also helps to absorb stray light rays. The **neural layer** of the retina is a multilayered outgrowth of the brain that processes visual data extensively before sending nerve impulses into axons that form the optic nerve. Three distinct layers of retinal neurons—the **photoreceptor layer,** the **bipolar cell layer,** and the **ganglion cell layer**—are separated by two zones, the *outer* and *inner synaptic layers,* where synaptic contacts are made (Figure 17.8). Note that light passes through the ganglion and bipolar cell layers and both synaptic layers before it reaches the photoreceptor layer. Two other types of cells present in the bipolar cell layer of the retina are called **horizontal cells** and **amacrine cells.** These cells form laterally directed neural circuits that modify the signals being transmitted along the pathway from photoreceptors to bipolar cells to ganglion cells.

Figure 17.7 A normal retina, as seen through an ophthalmoscope. Blood vessels in the retina can be viewed directly and examined for pathological changes.

The optic disc is the site where the optic nerve exits the eyeball. The central fovea is the area of highest visual acuity.

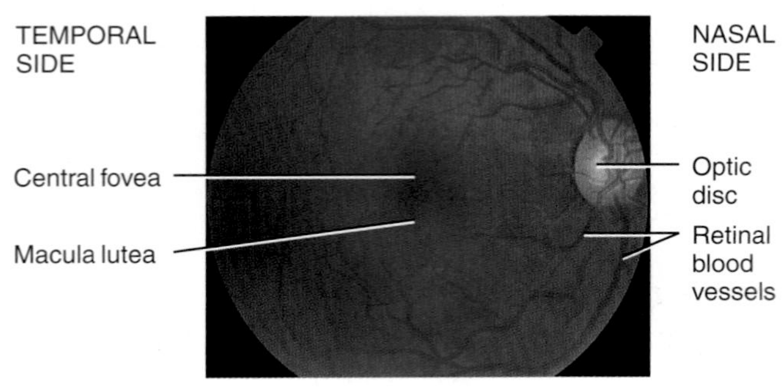

Evidence of what diseases may be seen through an ophthalmoscope?

Detached Retina

A **detached retina** may occur due to trauma, such as a blow to the head, in various eye disorders, or as a result of age-related degeneration. The detachment occurs between the neural portion of the retina and the pigment epithelium. Fluid accumulates between these layers, forcing the thin, pliable retina to billow outward. The result is distorted vision and blindness in the corresponding field of vision. The retina may be reattached by laser surgery or cryosurgery (localized application of extreme cold) and reattachment must be accomplished quickly to avoid permanent damage to the retina. ■

Photoreceptors are specialized cells that begin the process by which light rays are ultimately converted to nerve impulses. There are two types of photoreceptors: rods and cones. Each retina has about 6 million cones and 120 million rods. **Rods** allow us to see in dim light, such as moonlight. Because rods do not provide color vision, in dim light we can see only shades of gray. Brighter lights stimulate **cones,** which produce color vision. Three types of cones are present in the retina: (1) *blue cones,* which are sensitive to blue light, (2) *green cones,* which are sensitive to green light, and (3) *red cones,* which are sensitive to red light. Color vision results from the stimulation of various combinations of these three types of cones. Most of our experiences are mediated by the cone system, the loss of which produces legal blindness. A person who loses rod vision mainly has difficulty seeing in dim light and thus should not drive at night.

From photoreceptors, information flows through the outer synaptic layer to bipolar cells and then from bipolar cells through the inner synaptic layer to ganglion cells. The axons of ganglion cells extend posteriorly to the optic disc and exit the eyeball as the optic (II) nerve. The optic disc is also called the **blind spot.** Because it contains no rods or cones, we cannot see an image that strikes the blind spot. Normally, you are not aware of having a blind spot, but you can easily demonstrate its presence. Cover your left eye and gaze directly at the cross below. Then increase or decrease the distance between the book and your eye. At some point the square will disappear as its image falls on the blind spot.

+ ■

The **macula lutea** (MAK-ū-la LOO-tē-a; *macula* = a small, flat spot; *lute-* = yellowish) is in the exact center of the posterior portion of the retina, at the visual axis of the eye. The **central fovea** (see Figures 17.5 and 17.7), a small depression in the center of the macula lutea, contains only cones. In addition, the layers of bipolar and ganglion cells, which scatter light to some extent, do not cover the cones here; these layers are displaced to the periphery of the central fovea. As a result, the central fovea is the area of highest **visual acuity** or **resolution** (sharpness of vision). A main reason that you move your head and eyes while looking at something is to place images of interest on your central fovea—as you do to read the words in this sentence!

Figure 17.8 **Microscopic structure of the retina.** The downward blue arrow at left indicates the direction of the signals passing through the neural layer of the retina. Eventually, nerve impulses arise in ganglion cells and propagate along their axons, which make up the optic (II) nerve. (See Tortora, *A Photographic Atlas of the Human Body, Second Edition,* Figure 9.3b.)

In the retina, visual signals pass from photoreceptors to bipolar cells to ganglion cells.

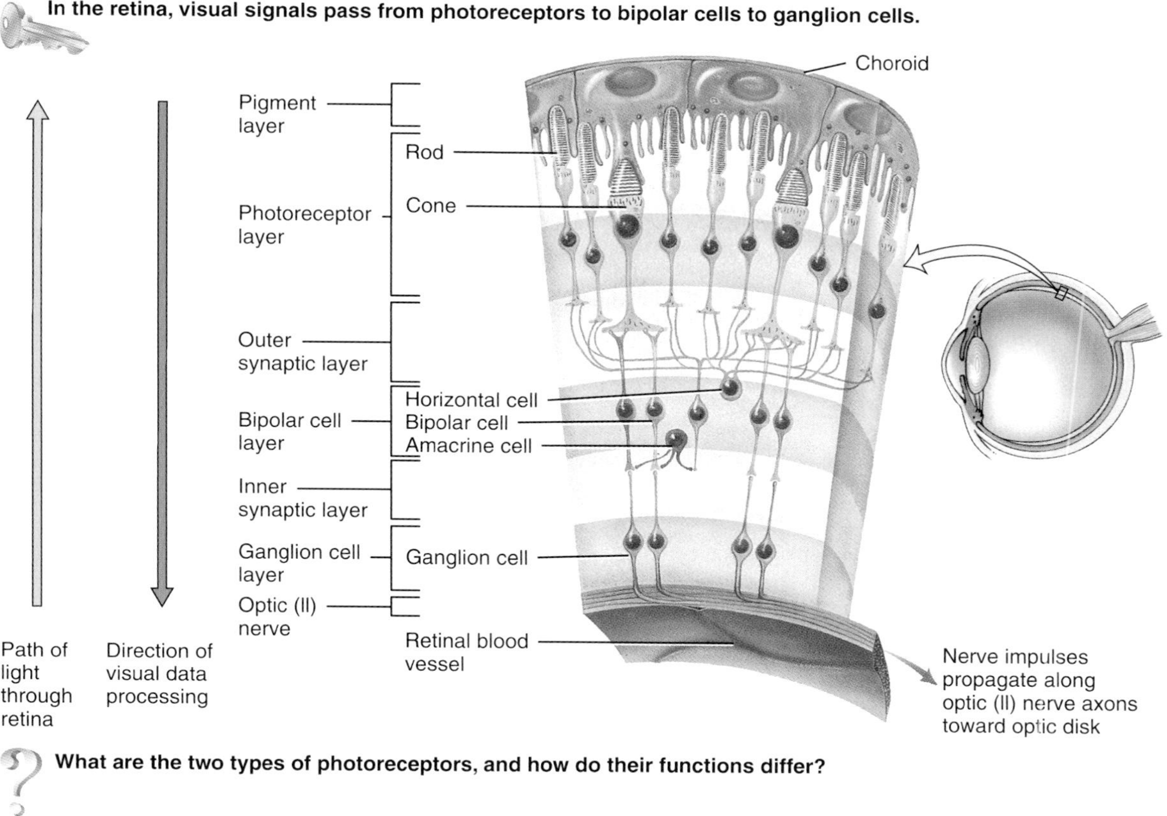

What are the two types of photoreceptors, and how do their functions differ?

Rods are absent from the central fovea and are more plentiful toward the periphery of the retina. Because rod vision is more sensitive than cone vision, you can see a faint object (such as a dim star) better if you gaze slightly to one side rather than looking directly at it.

Age-related Macular Disease

Age-related macular disease (AMD), also known as *macular degeneration,* is a degenerative disorder of the retina in persons 50 years of age and older. In AMD, abnormalities occur in the region of the macula lutea, which is ordinarily the area of most acute vision. Victims of advanced AMD retain their peripheral vision but lose the ability to see straight ahead. For instance, they cannot see facial features to identify a person in front of them. AMD is the leading cause of blindness in those over age 75, afflicting 13 million Americans, and is 2.5 times more common in pack-a-day smokers than in nonsmokers. Initially, a person may experience blurring and distortion at the center of the visual field. In "dry" AMD, central vision gradually diminishes because the pigmented layer atrophies and degenerates. There is no effective treatment. In about 10% of cases, dry AMD progresses to "wet" AMD, in which new blood vessels form in the choroid and leak plasma or blood under the retina. Vision loss can be slowed by using laser surgery to destroy the leaking blood vessels. ■

Lens

Behind the pupil and iris, within the cavity of the eyeball, is the **lens** (see Figure 17.5). Proteins called **crystallins,** arranged like the layers of an onion, make up the lens, which normally is perfectly transparent and lacks blood vessels. It is enclosed by a clear connective tissue capsule and held in position by encircling zonular fibers, which attach to the ciliary processes. The lens helps focus images on the retina to facilitate clear vision.

Interior of the Eyeball

The lens divides the interior of the eyeball into two cavities: the anterior cavity and vitreous chamber. The **anterior cavity**—the space anterior to the lens—consists of two chambers. The **anterior chamber** lies between the cornea and the iris. The **posterior chamber** lies behind the iris and in front of the zonular fibers and lens (Figure 17.9). Both chambers of the anterior cavity are filled with **aqueous humor** (*aqua* = water), a watery fluid that nourishes the lens and cornea. Aqueous humor continually filters out of blood capillaries in the ciliary processes and enters the posterior chamber. It then flows forward between the iris and the lens, through the pupil, and into the anterior chamber. From the anterior chamber, aqueous humor drains into the scleral venous sinus (canal of Schlemm) and then into the blood. Normally, aqueous humor is completely replaced about every 90 minutes.

The second, and larger, cavity of the eyeball is the **vitreous chamber,** which lies between the lens and the retina. Within the vitreous chamber is the **vitreous body,** a jellylike substance that holds the retina flush against the choroid, giving the retina an even surface for the reception of clear images. Unlike the aqueous humor, the vitreous body does not undergo constant replacement. It is formed during embryonic life and is not replaced thereafter. The vitreous body also contains phagocytic cells that remove debris, keeping this part of the eye clear for unobstructed vision. Occasionally, collections of debris may cast a shadow on the retina and create the appearance of specks that dart in and out of the field of vision. These *vitreal floaters,* which are more common in older individuals, are usually harmless and do not require treatment. The **hyaloid canal** is a narrow channel that runs through the vitreous body from the optic disc to the posterior aspect of the lens. In the fetus, it is occupied by the hyaloid artery (see Figure 17.23d).

The pressure in the eye, called **intraocular pressure,** is produced mainly by the aqueous humor and partly by the vitreous body; normally it is about 16 mmHg (millimeters of mercury). The intraocular pressure maintains the shape of the eyeball and prevents it from collapsing. Puncture wounds to the eyeball may cause the loss of aqueous humor and the vitreous body. This in turn causes a decrease in intraocular pressure, a detached retina, and in some cases blindness.

Table 17.1 summarizes the structures associated with the eyeball.

Figure 17.9 The iris separates the anterior and posterior chambers of the eye. The section is through the anterior portion of the eyeball at the junction of the cornea and sclera. Arrows indicate the flow of aqueous humor.

The lens separates the posterior chamber of the anterior cavity from the vitreous chamber.

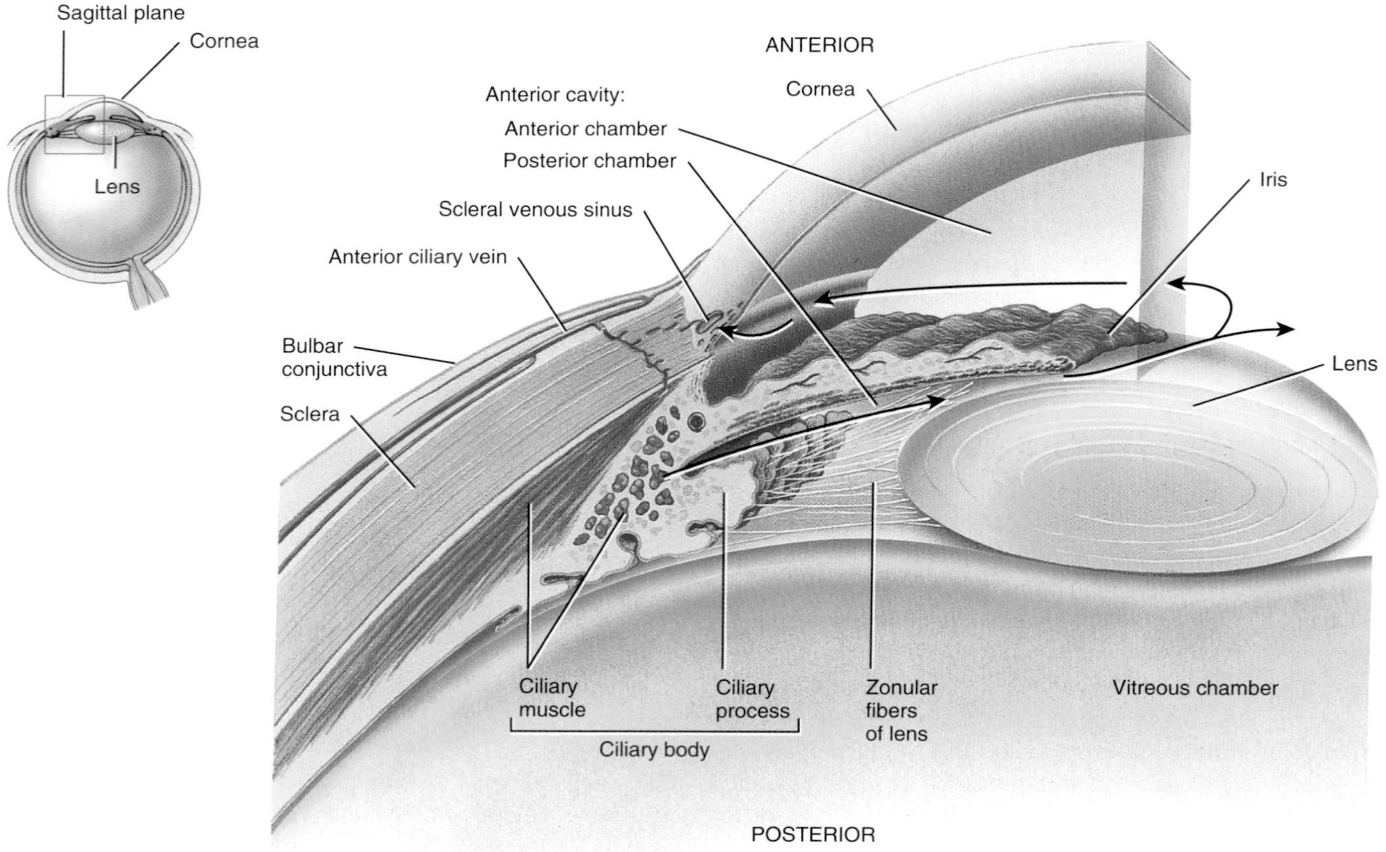

Where is aqueous humor produced, what is its circulation path, and where does it drain from the eyeball?

TABLE 17.1 Summary of the Structures of the Eyeball

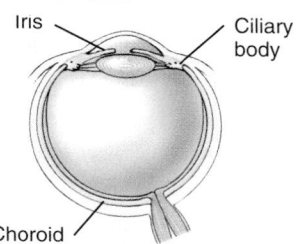

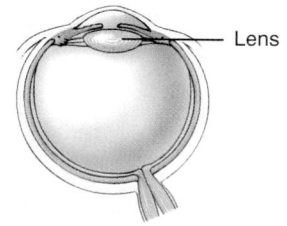

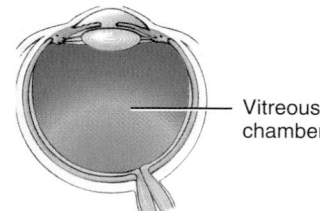

Structure	Function
Fibrous tunic (Cornea; Sclera)	*Cornea:* Admits and refracts (bends) light. *Sclera:* Provides shape and protects inner parts.
Vascular tunic (Iris; Ciliary body; Choroid)	*Iris:* Regulates amount of light that enters eyeball. *Ciliary body:* Secretes aqueous humor and alters shape of lens for near or far vision (accommodation). *Choroid:* Provides blood supply and absorbs scattered light.
Retina	Receives light and converts it into receptor potentials and nerve impulses. Output to brain is via axons of ganglion cells, which form the optic (II) nerve.
Lens	Refracts light.
Anterior cavity	Contains aqueous humor that helps maintain shape of eyeball and supplies oxygen and nutrients to lens and cornea.
Vitreous chamber	Contains vitreous body that helps maintain shape of eyeball and keeps the retina attached to the choroid.

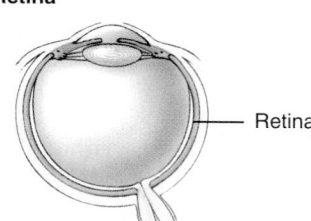

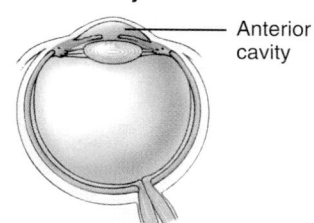

Image Formation

In some ways the eye is like a camera: Its optical elements focus an image of some object on a light-sensitive "film"—the retina—while ensuring the correct amount of light to make the proper "exposure." To understand how the eye forms clear images of objects on the retina, we must examine three processes: (1) the refraction or bending of light by the lens and cornea; (2) accommodation, the change in shape of the lens; and (3) constriction or narrowing of the pupil.

Refraction of Light Rays

When light rays traveling through a transparent substance (such as air) pass into a second transparent substance with a different density (such as water), they bend at the junction between the two substances. This bending is called **refraction** (Figure 17.10a). As light rays enter the eye, they are refracted at the anterior and posterior surfaces of the cornea. Both surfaces of the lens of the eye further refract the light rays so they come into exact focus on the retina.

Images focused on the retina are inverted (upside down) (Figure 17.10b, c). They also undergo right-to-left reversal; that is, light from the right side of an object strikes the left side of the retina, and vice versa. The reason the world does not look inverted and reversed is that the brain "learns" early in life to coordinate visual images with the orientations of objects. The brain stores the inverted and reversed images we acquired when we first reached for and touched objects and interprets those visual images as being correctly oriented in space.

About 75% of the total refraction of light occurs at the cornea. The lens provides the remaining 25% of focusing power and also changes the focus to view near or distant objects. When an object is 6 m (20 ft) or more away from the viewer, the light rays reflected from the object are nearly parallel to one another (Figure 17.10b). The lens must bend these parallel rays just enough so that they fall exactly focused on the central fovea, where vision is sharpest. Because light rays that are reflected from objects closer than 6 m (20 ft) are divergent rather than parallel (Figure 17.10c), the rays must be refracted more if they are to be focused on the retina. This additional refraction is accomplished through a process called accommodation.

Accommodation and the Near Point of Vision

A surface that curves outward, like the surface of a ball, is said to be *convex*. When the surface of a lens is convex, that lens will refract incoming light rays toward each other, so that they eventually intersect. If the surface of a lens curves inward, like the inside of a hollow ball, the lens is said to be *concave* and causes light rays to refract away from each other. The lens of the eye is convex on both its anterior and posterior surfaces, and its focusing power increases as its curvature becomes greater. When the

Figure 17.10 Refraction of light rays. (a) Refraction is the bending of light rays at the junction of two transparent substances with different densities. (b) The cornea and lens refract light rays from distant objects so the image is focused on the retina. (c) In accommodation, the lens becomes more spherical, which increases the refraction of light.

Images focused on the retina are inverted and left-to-right reversed.

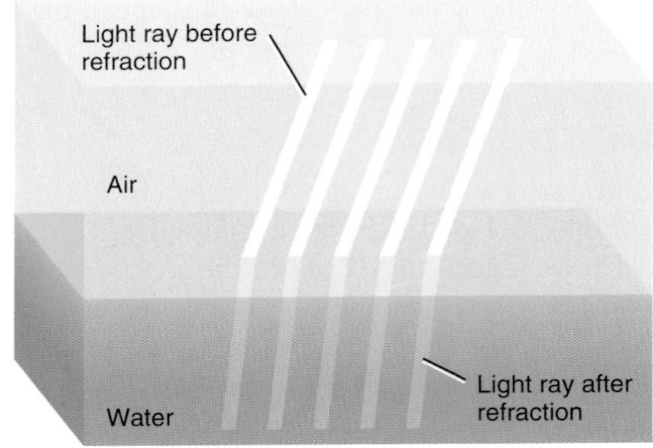

(a) Refraction of light rays

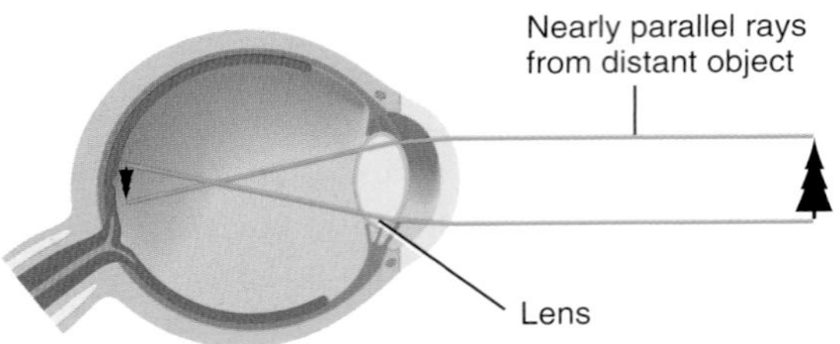

(b) Viewing distant object

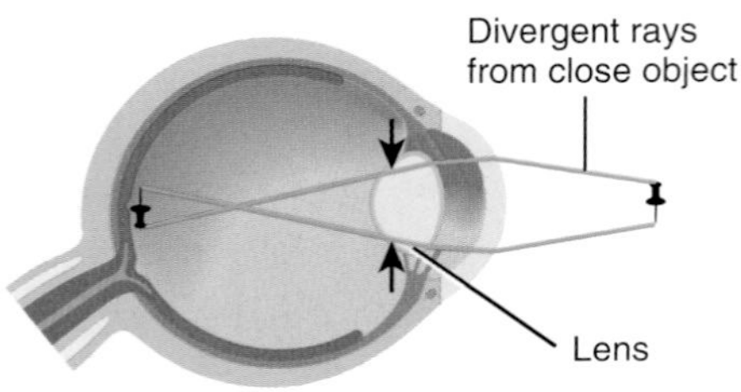

(c) Accommodation

What sequence of events occurs during accommodation?

eye is focusing on a close object, the lens becomes more curved, causing greater refraction of the light rays. This increase in the curvature of the lens for near vision is called **accommodation** (Figure 17.10c). The **near point of vision** is the minimum distance from the eye that an object can be clearly focused with maximum accommodation. This distance is about 10 cm (4 in.) in a young adult.

How does accommodation occur? When you are viewing distant objects, the ciliary muscle of the ciliary body is relaxed and the lens is flatter because it is stretched in all directions by taut zonular fibers. When you view a close object, the ciliary muscle contracts, which pulls the ciliary process and choroid forward toward the lens. This action releases tension on the lens and zonular fibers. Because it is elastic, the lens becomes more spherical (more convex), which increases its focusing power and causes greater convergence of the light rays. Parasympathetic fibers of the oculomotor (III) nerve innervate the ciliary muscle of the ciliary body and, therefore, mediate the process of accommodation.

Presbyopia

With aging, the lens loses elasticity and thus its ability to curve to focus on objects that are close. Therefore, older people cannot read print at the same close range as can youngsters. This condition is called **presbyopia** (prez-bē-Ō-pē-a; *presby-* = old; *-opia* = pertaining to the eye or vision). By age 40 the near point of vision may have increased to 20 cm (8 in.), and at age 60 it may be as much as 80 cm (31 in.). Presbyopia usually begins in the mid-forties. At about that age, people who have not previously worn glasses begin to need them for reading. Those who already wear glasses typically start to need bifocals, lenses that can focus for both distant and close vision. ■

Refraction Abnormalities

The normal eye, known as an **emmetropic eye** (em′-e- TROP-ik), can sufficiently refract light rays from an object 6 m (20 ft) away so that a clear image is focused on the retina. Many people, however, lack this ability because of refraction abnormalities. Among these abnormalities are **myopia** (mī-Ō-pē-a), or nearsightedness, which occurs when the eyeball is too long relative to the focusing power of the cornea and lens, or when the lens is thicker than normal, so an image converges in front of the retina. Myopic individuals can see close objects clearly, but not distant objects. In **hyperopia** (hī-per-Ō-pē-a) or farsightedness, also known as **hypermetropia** (hī′-per-me-TRŌ-pē-a), the eyeball length is short relative to the focusing power of the cornea and lens, or the lens is thinner than normal, so an image converges behind the retina. Hyperopic individuals can see distant objects clearly, but not close ones. Figure 17.11 illustrates these conditions and explains how they are corrected. Another refraction abnormality is **astigmatism** (a-STIG-ma-tizm), in which either the cornea or the lens has an irregular curvature. As a result, parts

Figure 17.11 **Refraction abnormalities in the eyeball and their correction.** (a) Normal (emmetropic) eye. (b) In the nearsighted or myopic eye, the image is focused in front of the retina. The condition may result from an elongated eyeball or thickened lens. (c) Correction of myopia is by use of a concave lens that diverges entering light rays so that they come into focus directly on the retina. (d) In the farsighted or hyperopic eye, the image is focused behind the retina. The condition results from a shortened eyeball or a thin lens. (e) Correction of hyperopia is by a convex lens that converges entering light rays so that they focus directly on the retina.

In myopia (nearsightedness), only close objects can be seen clearly; in hyperopia (farsightedness), only distant objects can be seen clearly.

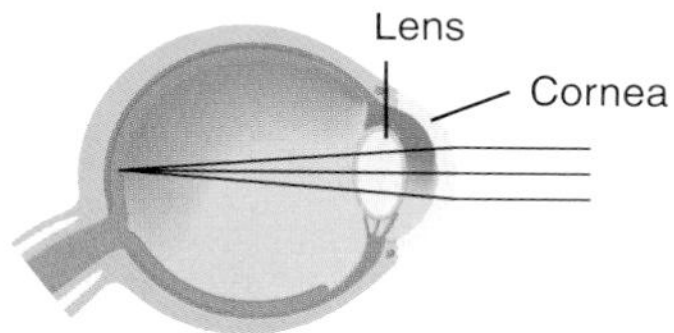

(a) Normal (emmetropic) eye

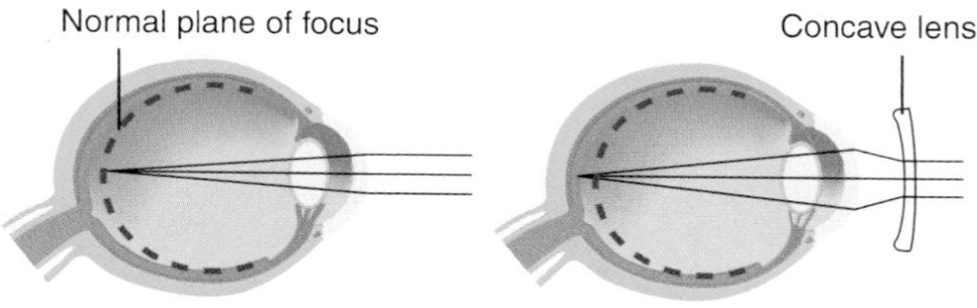

(b) Nearsighted (myopic) eye, uncorrected

(c) Nearsighted (myopic) eye, corrected

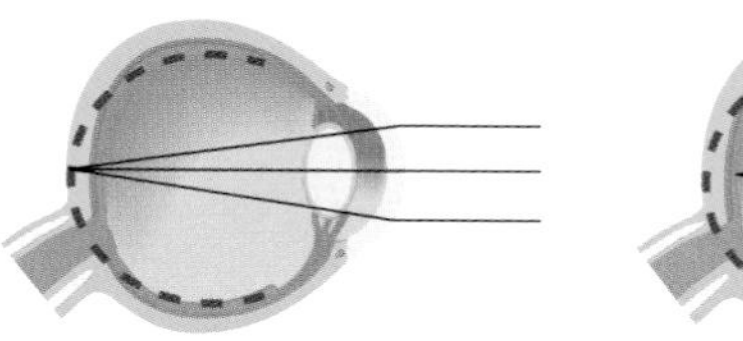

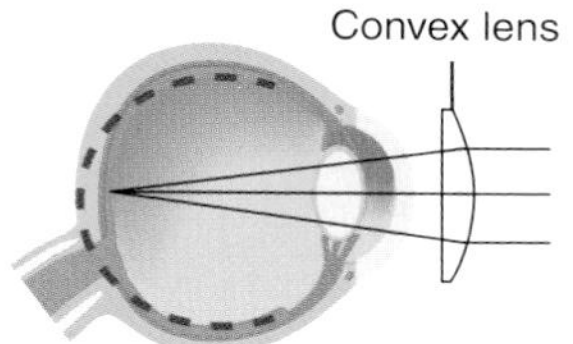

(d) Farsighted (hypermetropic) eye, uncorrected

(e) Farsighted (hypermetropic) eye, corrected

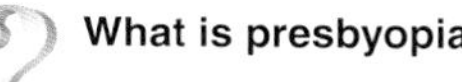

What is presbyopia?

of the image are out of focus, and thus vision is blurred or distorted.

Most errors of vision can be corrected by eyeglasses, contact lenses, or surgical procedures. A contact lens floats on a film of tears over the cornea. The anterior outer surface of the contact lens corrects the visual defect, and its posterior surface matches the curvature of the cornea. LASIK involves reshaping the cornea to permanently correct refraction abnormalities.

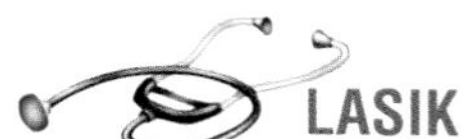

LASIK

An increasingly popular alternative to wearing glasses or contact lenses is refractive surgery to correct the curvature of the cornea for conditions such as farsightedness, nearsightedness, and astigmatism. The most common type of refractive surgery is **LASIK** (laser-assisted in-situ keratomileusis). After anesthetic drops are placed in the eye, a circular flap of tissue from the center of the cornea is cut. The flap is folded out of the way and the underlying layer of cornea is reshaped with a laser, one microscopic layer at a time. A computer assists the physician in removing very precise layers of the cornea. After the sculpting is complete, the corneal flap is repositioned over the treated area. A patch is placed over the eye overnight and the flap quickly reattaches to the rest of the cornea. ■

Constriction of the Pupil

The circular muscle fibers of the iris also have a role in the formation of clear retinal images. As you have already learned, **constriction of the pupil** is a narrowing of the diameter of the hole through which light enters the eye due to the contraction of the circular muscles of the iris. This autonomic reflex occurs simultaneously with accommodation and prevents light rays from entering the eye through the periphery of the lens. Light rays entering at the periphery would not be brought to focus on the retina and would result in blurred vision. The pupil, as noted earlier, also constricts in bright light.

Convergence

Because of the position of their eyes in their heads, many animals, such as horses and goats, see one set of objects off to the left through one eye, and an entirely different set of objects off to the right through the other. In humans, both eyes focus on only one set of objects—a characteristic called **binocular vision.** This feature of our visual system allows the perception of depth and an appreciation of the three-dimensional nature of objects.

Binocular vision occurs when light rays from an object strike corresponding points on the two retinas. When we stare straight ahead at a distant object, the incoming light rays are aimed directly at both pupils and are refracted to comparable spots on the retinas of both eyes. As we move closer to an object, however, the eyes must rotate medially if the light rays from the object are to strike the same points on both retinas. The term **convergence** refers to this medial movement of the two eyeballs so that both are directed toward the object being viewed, for example, tracking a pencil moving toward your eyes. The nearer the object, the greater the degree of convergence needed to maintain binocular vision. The coordinated action of the extrinsic eye muscles brings about convergence.

Physiology of Vision

Photoreceptors and Photopigments

Rods and cones were named for the different appearance of the *outer segment*—the distal end next to the pigmented layer—of each of these types of photoreceptors. The outer segments of rods are cylindrical or rod-shaped; those of cones are tapered or cone-shaped (Figure 17.12). Transduction of light energy into a receptor potential occurs in the outer segment of both rods and cones. The photopigments are integral proteins in the plasma membrane of the outer segment. In cones the plasma membrane is folded back and forth in a pleated fashion; in rods the pleats pinch off from the plasma membrane to form discs. The outer segment of each rod contains a stack of about 1000 discs, piled up like coins inside a wrapper.

Photoreceptor outer segments are renewed at an astonishingly rapid pace. In rods, one to three new discs are added to the base of the outer segment every hour while old discs slough off at the tip and are phagocytized by pigment epithelial cells. The *inner segment* contains the cell nucleus, Golgi complex, and many mitochondria. At its proximal end, the photoreceptor expands into bulblike synaptic terminals filled with synaptic vesicles.

The first step in visual transduction is absorption of light by a **photopigment,** a colored protein that undergoes structural changes when it absorbs light, in the outer segment of a photoreceptor. Light absorption initiates the events that lead to the production of a receptor potential. The single type of photopigment in rods is **rhodopsin** (*rhod-* = rose; *opsin-* = related to vision). Three different **cone photopigments** are present in the retina, one in each of the three types of cones. Color vision results from different colors of light selectively activating the different cone photopigments.

All photopigments associated with vision contain two parts: a glycoprotein known as **opsin** and a derivative of vitamin A called **retinal.** Vitamin A derivatives are formed from carotene, the plant pigment that gives carrots their orange color. Good vision depends on adequate dietary intake of carotene-rich vegetables such as carrots, spinach, broccoli, and yellow squash, or foods that contain vitamin A, such as liver.

Retinal is the light-absorbing part of all visual photopigments. In the human retina, there are four different opsins, three in the cones and one in the rods (rhodopsin). Small variations

Figure 17.12 Structure of rod and cone photoreceptors. The inner segments contain the metabolic machinery for synthesis of photopigments and production of ATP. The photopigments are embedded in the membrane discs or folds of the outer segments. New discs in rods and new folds in cones form at the base of the outer segment. Pigment epithelial cells phagocytize the old discs and folds that slough off the distal tip of the outer segments.

Transduction of light energy into a receptor potential occurs in the outer segments of rods and cones.

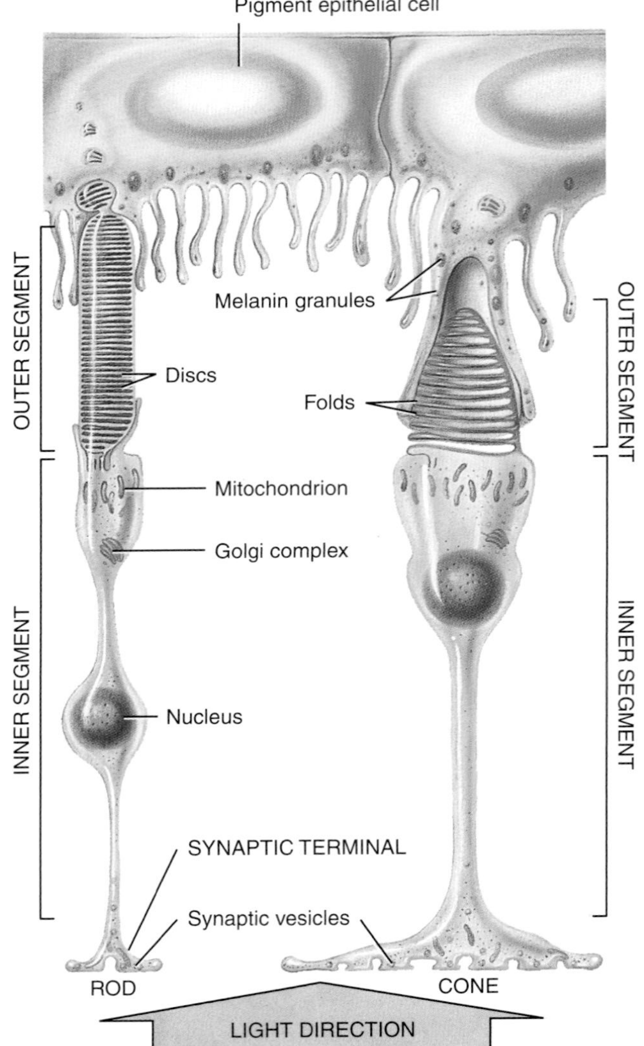

What are the functional similarities between rods and cones?

in the amino acid sequences of the different opsins permit the rods and cones to absorb different colors (wavelengths) of incoming light.

Photopigments respond to light in the following cyclical process (Figure 17.13):

1. In darkness, retinal has a bent shape, called *cis*-retinal, which fits snugly into the opsin portion of the photopigment. When *cis*-retinal absorbs a photon of light, it straightens out to a shape called *trans*-retinal. This *cis*-to-*trans* conversion is called **isomerization** and is the first step in visual transduction. After retinal isomerizes, several unstable chemical intermediates form and disappear. These chemical changes lead to production of a receptor potential (see Figure 17.14).
2. In about a minute, *trans*-retinal completely separates from opsin. The final products look colorless, so this part of the cycle is termed **bleaching** of photopigment.
3. An enzyme called **retinal isomerase** converts *trans*-retinal back to *cis*-retinal.
4. The *cis*-retinal then can bind to opsin, reforming a functional photopigment. This part of the cycle—resynthesis of a photopigment—is called **regeneration.**

The pigmented layer of the retina adjacent to the photoreceptors stores a large quantity of vitamin A and contributes to the regeneration process in rods. The extent of rhodopsin regeneration decreases drastically if the retina detaches from the pigmented layer. Cone photopigments regenerate much more quickly than the rhodopsin in rods and are less dependent on the pigmented layer. After complete bleaching, regeneration of half of the rhodopsin takes 5 minutes; half of the cone photopigments regenerate in only 90 seconds. Full regeneration of bleached rhodopsin takes 30 to 40 minutes.

Figure 17.13 The cyclical bleaching and regeneration of photopigment. Blue arrows indicate bleaching steps; black arrows indicate regeneration steps.

Retinal, a derivative of vitamin A, is the light-absorbing part of all visual photopigments.

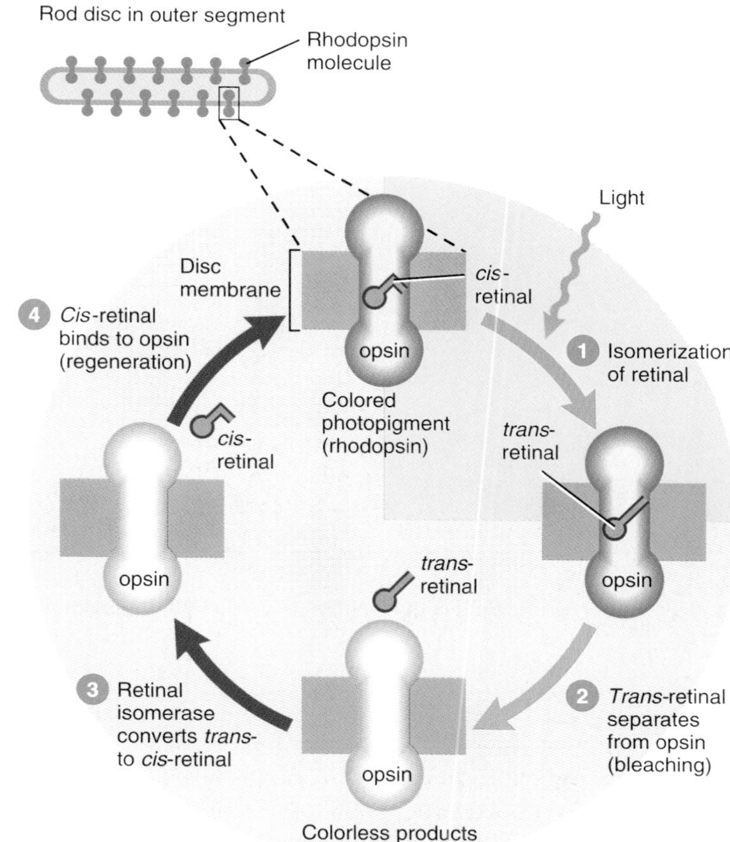

What is the conversion of *cis*-retinal to *trans*-retinal called?

Light and Dark Adaptation

When you emerge from dark surroundings (say, a tunnel) into the sunshine, **light adaptation** occurs—your visual system adjusts in seconds to the brighter environment by decreasing its sensitivity. On the other hand, when you enter a darkened room such as a theater, your visual system undergoes **dark adaptation**—its sensitivity increases slowly over many minutes. The difference in the rates of bleaching and regeneration of the photopigments in the rods and cones accounts for some (but not all) of the sensitivity changes during light and dark adaptation.

As the light level increases, more and more photopigment is bleached. While light is bleaching some photopigment molecules, however, others are being regenerated. In daylight, regeneration of rhodopsin cannot keep up with the bleaching process, so rods contribute little to daylight vision. In contrast, cone photopigments regenerate rapidly enough that some of the *cis* form is always present, even in very bright light.

If the light level decreases abruptly, sensitivity increases rapidly at first and then more slowly. In complete darkness, full regeneration of cone photopigments occurs during the first 8 minutes of dark adaptation. During this time, a threshold (barely perceptible) light flash is seen as having color. Rhodopsin regenerates more slowly, and our visual sensitivity increases until even a single photon (the smallest unit of light) can be detected. In that situation, although much dimmer light can be detected, threshold flashes appear gray-white, regardless of their color. At very low light levels, such as starlight, objects appear as shades of gray because only the rods are functioning.

Release of Neurotransmitter by Photoreceptors

As mentioned previously, the absorption of light and isomerization of retinal initiates chemical changes in the photoreceptor outer segment that lead to production of a receptor potential. To understand how the receptor potential arises, however, we

Figure 17.14 Operation of rod photoreceptors.

Light causes a hyperpolarizing receptor potential in photoreceptors, which decreases release of an inhibitory neurotransmitter (glutamate).

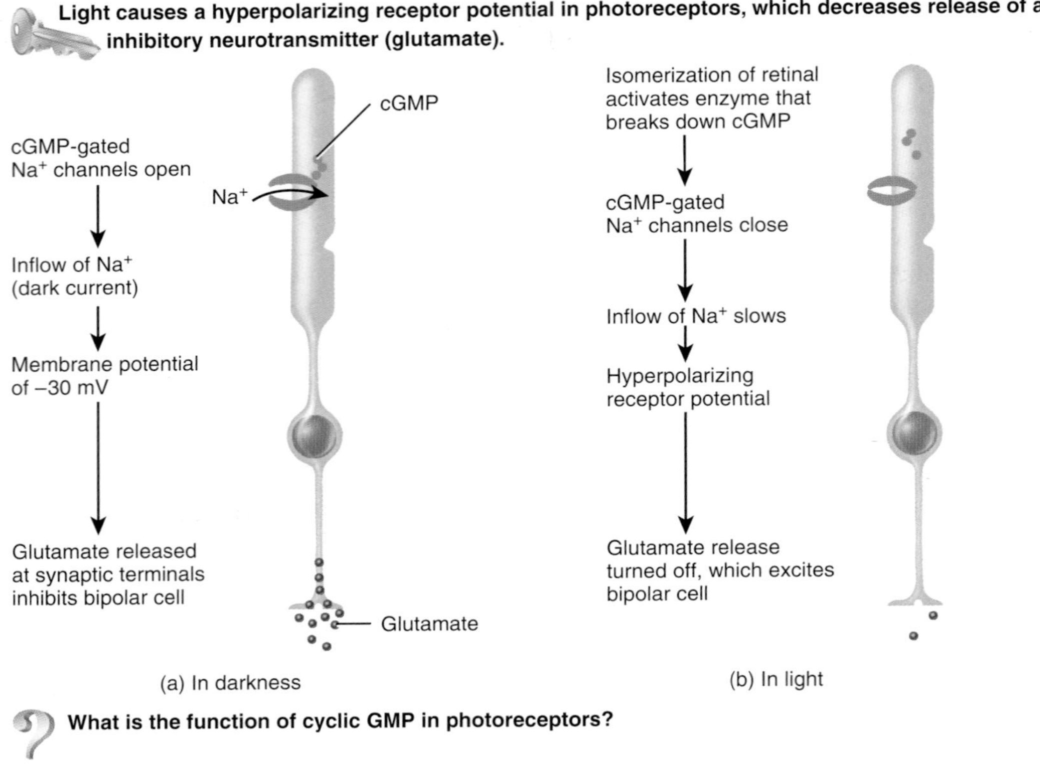

(a) In darkness (b) In light

What is the function of cyclic GMP in photoreceptors?

first need to examine the operation of photoreceptors in the absence of light. In darkness, sodium ions (Na^+) flow into photoreceptor outer segments through ligand-gated Na^+ channels (Figure 17.14a). The ligand that holds these channels open is **cyclic GMP (guanosine monophosphate)** or **cGMP.** The inflow of Na^+, called the "dark current," partially depolarizes the photoreceptor. As a result, in darkness the membrane potential of a photoreceptor is about −30 mV. This is much closer to zero than a typical neuron's resting membrane potential of −70 mV. The partial depolarization during darkness triggers continual release of neurotransmitter at the synaptic terminals. The neurotransmitter in rods, and perhaps in cones, is the amino acid glutamate (glutamic acid). At synapses between rods and some bipolar cells, glutamate is an inhibitory neurotransmitter: It triggers inhibitory postsynaptic potentials (IPSPs) that hyperpolarize the bipolar cells and prevent them from sending signals on to the ganglion cells.

When light strikes the retina and *cis*-retinal undergoes isomerization, enzymes are activated that break down cGMP. As a result, some cGMP-gated Na^+ channels close, Na^+ inflow decreases, and the membrane potential becomes more negative, approaching −70 mV (Figure 17.14b). This sequence of events produces a hyperpolarizing receptor potential that decreases the release of glutamate. Dim lights cause small and brief receptor potentials that partially turn off glutamate release; brighter lights elicit larger and longer receptor potentials that more completely shut down neurotransmitter release. Thus, light excites the bipolar cells that synapse with rods by turning off the release of an inhibitory neurotransmitter! The excited bipolar cells subsequently stimulate the ganglion cells to form action potentials in their axons.

Color Blindness and Night Blindness

Most forms of **color blindness,** an inherited inability to distinguish between certain colors, result from the absence or deficiency of one of the three types of cones. The most common type is *red-green color blindness,* in which red cones or green cones are missing. As a result, the person cannot distinguish between red and green. Prolonged vitamin A deficiency and the resulting below-normal amount of rhodopsin may cause **night blindness** or **nyctalopia** (nik′-ta-LŌ-pē-a), an inability to see well at low light levels. ■

The Visual Pathway

Visual signals in the retina undergo considerable processing at synapses among the various types of neurons (horizontal cells, bipolar cells, and amacrine cells; see Figure 17.8). Then, the axons of retinal ganglion cells provide output from the retina to the brain, exiting the eyeball as the **optic (II) nerve.**

Processing of Visual Input in the Retina

Within the retina, certain features of visual input are enhanced while other features may be discarded. Input from several cells may either converge upon a smaller number of postsynaptic neurons or diverge to a large number. Overall, convergence predominates: There are only 1 million ganglion cells, but 126 million photoreceptors in the human eye.

Once receptor potentials arise in the outer segments of rods and cones, they spread through the inner segments to the synaptic terminals. Neurotransmitter molecules released by rods and cones induce local graded potentials in both bipolar cells and horizontal cells. Between 6 and 600 rods synapse with a single bipolar cell in the outer synaptic layer of the retina; a cone more often synapses with a single bipolar cell. The convergence of many rods onto a single bipolar cell increases the light sensitivity of rod vision but slightly blurs the image that is perceived. Cone vision, although less sensitive, is sharper because of the one-to-one synapses between cones and their bipolar cells. Stimulation of rods by light excites bipolar cells; cone bipolar cells may be either excited or inhibited when a light is turned on.

Horizontal cells transmit inhibitory signals to bipolar cells in the areas lateral to excited rods and cones. This lateral inhibition enhances contrasts in the visual scene between areas of the retina that are strongly stimulated and adjacent areas that are more weakly stimulated. Horizontal cells also assist in the differentiation of various colors. Amacrine cells, which are excited by bipolar cells, synapse with ganglion cells and transmit information to them that signals a change in the level of illumination of the retina. When bipolar or amacrine cells transmit excitatory signals to ganglion cells, the ganglion cells become depolarized and initiate nerve impulses.

Brain Pathway and Visual Fields

The axons within the optic nerve pass through the **optic chiasm** (kī-AZ-m = a crossover, as in the letter X), a crossing point of the optic nerves (Figure 17.15a, b on page 594). Some axons cross to the opposite side, but others remain uncrossed. After passing through the optic chiasm, the axons, now part of the **optic tract,** enter the brain and terminate in the lateral geniculate nucleus of the thalamus. Here they synapse with neurons whose axons form the **optic radiations,** which project to the primary visual areas in the occipital lobes of the cerebral cortex (area 17 in Figure 14.15 on page 497) and visual perception begins.

Everything that can be seen by one eye is that eye's **visual field.** As noted earlier, because our eyes are located anteriorly in our heads, the visual fields overlap considerably (Figure 17.15b). We have binocular vision due to the large region where the visual fields of the two eyes overlap—the **binocular visual field.** The visual field of each eye is divided into two regions: the **nasal** or **central half** and the **temporal** or **peripheral half** (Figure 17.15c, d). For each eye, light rays from an object in the nasal half of the visual field fall on the temporal half of the retina, and light rays from an object in the temporal half of the visual field fall on the nasal half of the retina. Visual information from the *right* half of each visual field is conveyed to the *left* side of the brain, and visual information from the *left* half of each visual field is conveyed to the *right* side of the brain, as follows (Figure 17.15c, d):

1. The axons of all retinal ganglion cells in one eye exit the eyeball at the optic disc and form the optic nerve on that side.
2. At the optic chiasm, axons from the temporal half of each retina do not cross but continue directly to the lateral geniculate nucleus of the thalamus on the same side.
3. In contrast, axons from the nasal half of each retina cross the optic chiasm and continue to the opposite thalamus.
4. Each optic tract consists of crossed and uncrossed axons that project from the optic chiasm to the thalamus on one side.
5. Axon collaterals (branches) of the retinal ganglion cells project to the midbrain, where they participate in neural circuits that govern constriction of the pupils in response to light and coordination of head and eye movements. Collaterals also extend to the suprachiasmatic nucleus of the hypothalamus, which establishes patterns of sleep and other activities that occur on a circadian or daily schedule in response to intervals of light and darkness.
6. The axons of thalamic neurons form the optic radiations as they project from the thalamus to the primary visual area of the cortex on the same side.

Although we have just described the visual pathway as a single pathway, visual signals are thought to be processed by at least three separate systems in the cerebral cortex, each with its own function. One system processes information related to the shape of objects, another system processes information regarding color of objects, and a third system processes information about movement, location, and spatial organization.

CHECKPOINT

6. What is the function of the lacrimal apparatus?
7. What types of cells make up the neural layer and the pigmented layer of the retina?
8. How do photopigments respond to light and recover in darkness?
9. How do receptor potentials arise in photoreceptors?
10. By what pathway do nerve impulses triggered by an object in the nasal half of the visual field of the left eye reach the primary visual area of the cortex?

View

Visual pathway:

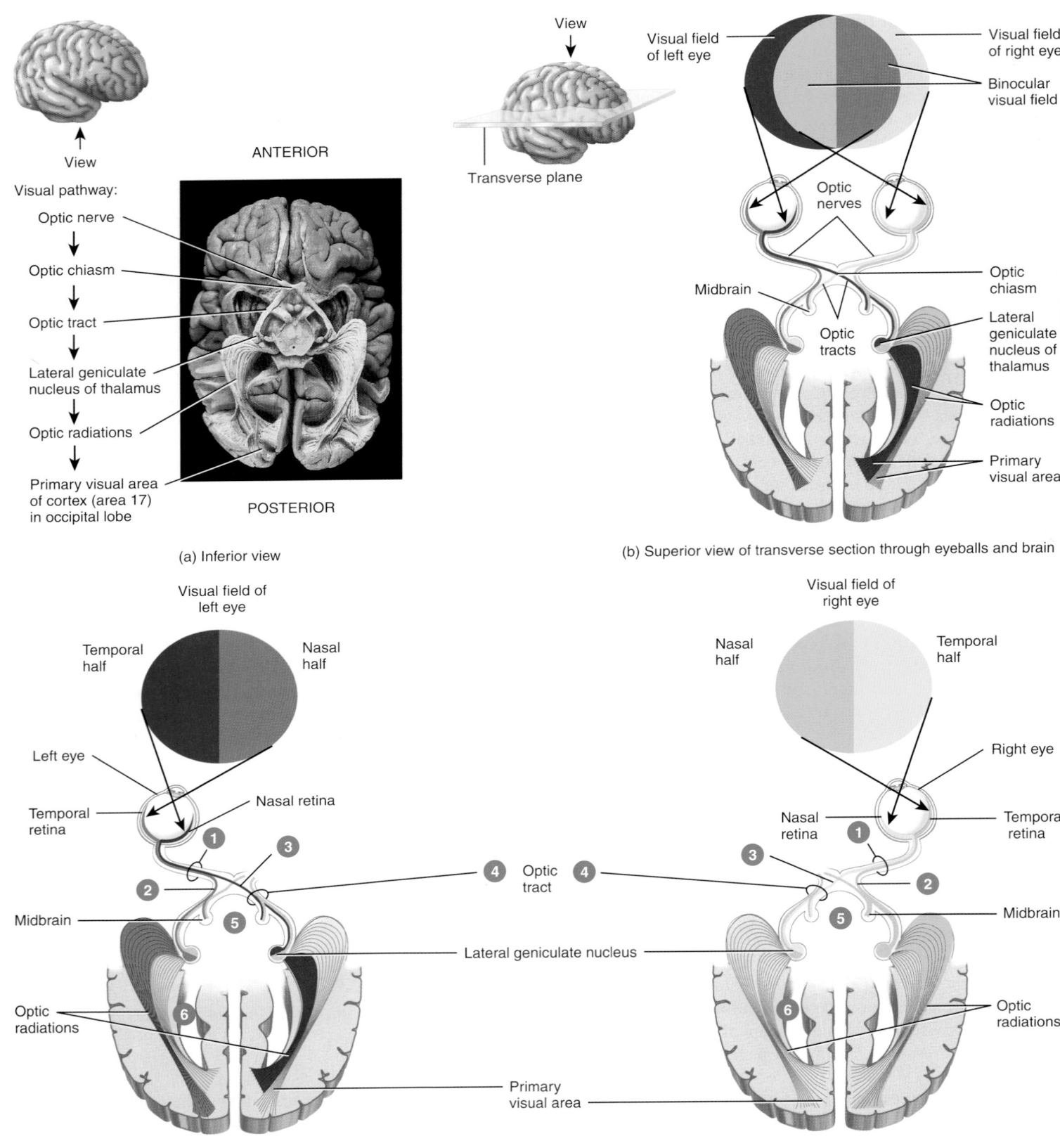

Optic nerve
↓
Optic chiasm
↓
Optic tract
↓
Lateral geniculate nucleus of thalamus
↓
Optic radiations
↓
Primary visual area of cortex (area 17) in occipital lobe

ANTERIOR

POSTERIOR

(a) Inferior view

View

Transverse plane

Visual field of left eye

Visual field of right eye

Binocular visual field

Optic nerves

Optic chiasm

Midbrain

Optic tracts

Lateral geniculate nucleus of thalamus

Optic radiations

Primary visual area

(b) Superior view of transverse section through eyeballs and brain

Visual field of left eye

Temporal half

Nasal half

Left eye

Temporal retina

Nasal retina

1

2

3

4 Optic tract 4

5

Midbrain

Lateral geniculate nucleus

6

Optic radiations

Primary visual area

Visual field of right eye

Nasal half

Temporal half

Right eye

Nasal retina

Temporal retina

1

2

3

5

Midbrain

6

Optic radiations

(c) Left eye and its pathways

(d) Right eye and its pathways

Light rays from an object in the temporal half of the visual field strike which half of the retina?

Figure 17.15 The visual pathway. (a) Partial dissection of the brain reveals the optic radiations (axons extending from the thalamus to the occipital lobe). (b) An object in the binocular visual field can be seen with both eyes. In (c) and (d), note that information from the right side of the visual field of each eye projects to the left side of the brain, and information from the left side of the visual field of each eye projects to the right side of the brain.

The axons of ganglion cells in the temporal half of each retina extend to the thalamus on the same side; the axons of ganglion cells in the nasal half of each retina extend to the thalamus on the opposite side.

HEARING AND EQUILIBRIUM

► OBJECTIVES

- **Describe the anatomy of the structures in the three main regions of the ear.**
- **List the major events in the physiology of hearing.**
- **Identify the receptor organs for equilibrium, and describe how they function.**
- **Describe the auditory and equilibrium pathways.**

The ear is an engineering marvel because its sensory receptors can transduce sound vibrations with amplitudes as small as the diameter of an atom of gold (0.3 nm) into electrical signals 1000 times faster than photoreceptors can respond to light. Besides receptors for sound waves, the ear also contains receptors for equilibrium.

Anatomy of the Ear

The ear is divided into three main regions: the external ear, which collects sound waves and channels them inward; the middle ear, which conveys sound vibrations to the oval window; and the internal ear, which houses the receptors for hearing and equilibrium.

External (Outer) Ear

The **external (outer) ear** consists of the auricle, external auditory canal, and eardrum (Figure 17.16). The **auricle (pinna)** is a flap of elastic cartilage shaped like the flared end of a trumpet and covered by skin. The rim of the auricle is the **helix;** the inferior portion is the **lobule.** Ligaments and muscles attach the auricle to the head. The **external auditory canal** (*audit-* = hearing) is a curved tube about 2.5 cm (1 in.) long that lies in the temporal bone and leads from the auricle to the eardrum. The **eardrum** or **tympanic membrane** (tim-PAN-ik; *tympan-* = a drum) is a thin, semitransparent partition between the external auditory canal and middle ear. The eardrum is covered by epidermis and lined by simple cuboidal epithelium. Between the epithelial layers is connective tissue composed of collagen, elastic fibers, and fibroblasts. Tearing of the tympanic membrane is called a **perforated eardrum.** It may be due to pressure from a cotton swab, trauma, or a middle ear infection, and usually heals within a month. The eardrum may be examined directly by an **otoscope** (*oto-* = ear; *-skopeo* = to view), a viewing instrument that illuminates and magnifies the external auditory canal and eardrum.

Near the exterior opening, the external auditory canal contains a few hairs and specialized sweat glands called **ceruminous glands** (se-ROO-mi-nus) that secrete earwax or **cerumen** (se-ROO-men). The combination of hairs and cerumen helps prevent dust and foreign objects from entering the ear. Cerumen usually dries up and falls out of the ear canal. However, some people produce a large amount of cerumen, which can become impacted and can muffle incoming sounds. The treatment for **impacted cerumen** is usually periodic ear irrigation or removal of wax with a blunt instrument by trained medical personnel.

Middle Ear

The **middle ear** is a small, air-filled cavity in the temporal bone that is lined by epithelium (Figure 17.17 on page 597). It is separated from the external ear by the eardrum and from the internal ear by a thin bony partition that contains two small membrane-covered openings: the oval window and the round window. Extending across the middle ear and attached to it by ligaments are the three smallest bones in the body, the **auditory ossicles** (OS-si-kuls), which are connected by synovial joints. The bones, named for their shapes, are the malleus, incus, and stapes—commonly called the hammer, anvil, and stirrup, respectively. The "handle" of the **malleus** (MAL-ē-us) attaches to the internal surface of the eardrum. The head of the malleus articulates with the body of the incus. The **incus** (ING-kus), the middle bone in the series, articulates with the head of the stapes. The base or footplate of the **stapes** (STĀ-pēz) fits into the **oval window.** Directly below the oval window is another opening, the **round window,** which is enclosed by a membrane called the **secondary tympanic membrane.**

Besides the ligaments, two tiny skeletal muscles also attach to the ossicles (Figure 17.17). The **tensor tympani muscle,** which is supplied by the mandibular branch of the trigeminal (V) nerve, limits movement and increases tension on the eardrum to prevent damage to the inner ear from loud noises. The **stapedius muscle,** which is supplied by the facial (VII) nerve, is the smallest skeletal muscle in the human body. By dampening large vibrations of the stapes due to loud noises, it protects the oval window, but it also decreases the sensitivity of hearing. For this reason, paralysis of the stapedius muscle is associated with **hyperacusia** (abnormally sensitive hearing). Because it takes a fraction of a second for the tensor tympani and stapedius muscles to

Figure 17.16 Anatomy of the ear. (See Tortora, *A Photographic Atlas of the Human Body, Second Edition,* Figure 9.4a.)

The ear has three principal regions: the external (outer) ear, the middle ear, and the internal (inner) ear.

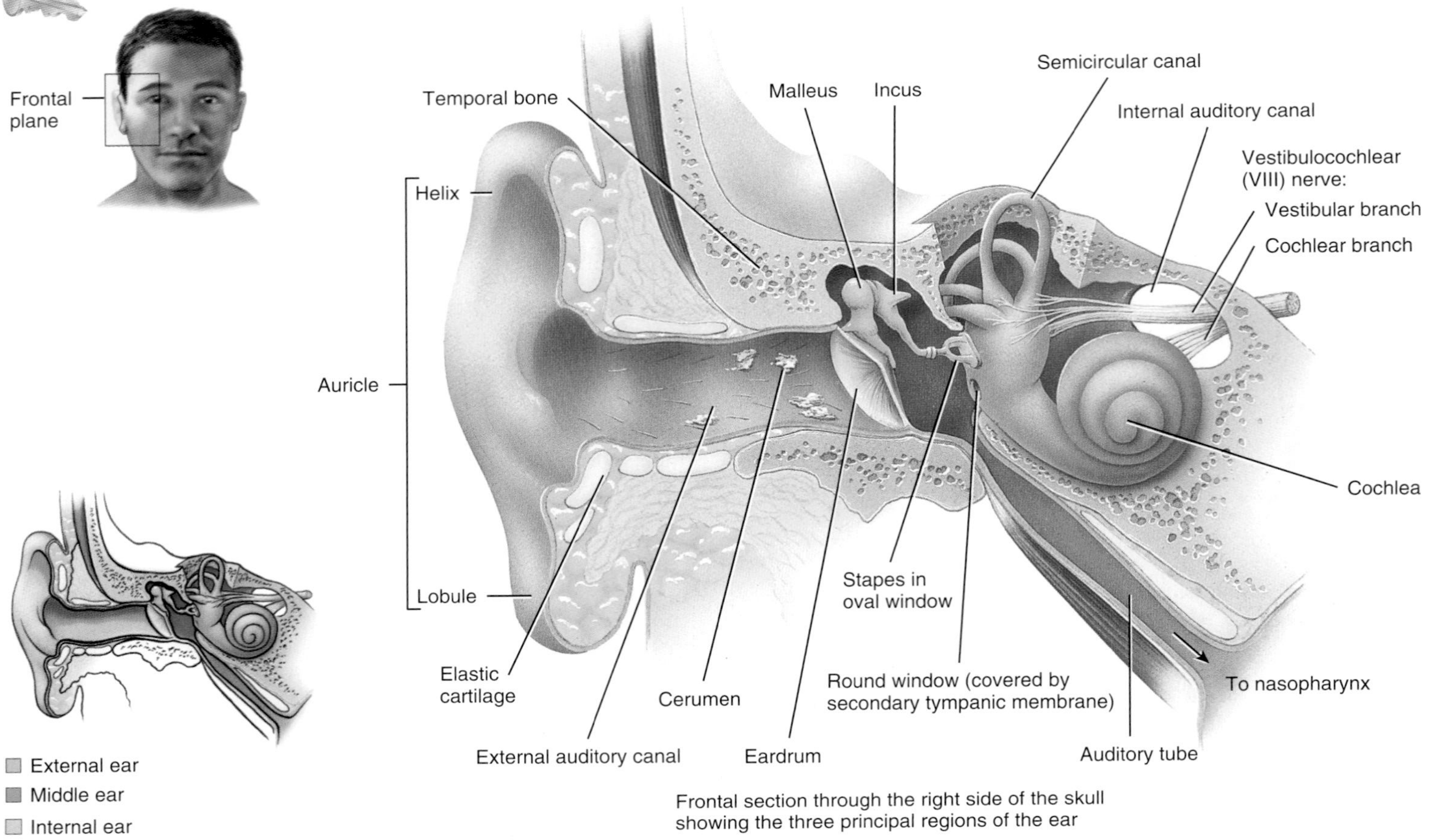

External ear
Middle ear
Internal ear

Frontal section through the right side of the skull showing the three principal regions of the ear

To which structure of the external ear does the malleus of the middle ear attach?

contract, they can protect the inner ear from prolonged loud noises, but not from brief ones such as a gunshot.

The anterior wall of the middle ear contains an opening that leads directly into the **auditory (pharyngotympanic) tube,** commonly known as the **eustachian tube.** The auditory tube, which consists of both bone and hyaline cartilage, connects the middle ear with the nasopharynx (upper portion of the throat). It is normally closed at its medial (pharyngeal) end. During swallowing and yawning, it opens, allowing air to enter or leave the middle ear until the pressure in the middle ear equals the atmospheric pressure. Most of us have experienced our ears popping as the pressures equalize. When the pressures are balanced, the eardrum vibrates freely as sound waves strike it. If the pressure is not equalized, intense pain, hearing impairment, ringing in the ears, and vertigo could develop. The auditory tube also is a route for pathogens to travel from the nose and throat to the middle ear, causing the most common type of ear infection (see Otitis Media in the Disorders section at the end of this chapter).

Internal (Inner) Ear

The **internal (inner) ear** is also called the **labyrinth** (LAB-i-rinth) because of its complicated series of canals (Figure 17.18 on page 598). Structurally, it consists of two main divisions: an outer bony labyrinth that encloses an inner membranous labyrinth. The **bony labyrinth** is a series of cavities in the temporal bone divided into three areas: (1) the semicircular canals and (2) the vestibule, both of which contain receptors for equilibrium, and (3) the cochlea, which contains receptors for hearing. The bony labyrinth is lined with periosteum and contains **perilymph.** This fluid, which is chemically similar to cerebrospinal fluid, surrounds the **membranous labyrinth,** a series of sacs and tubes inside the bony labyrinth with the same gen-

Figure 17.17 The right middle ear containing the auditory ossicles. (See Tortora, *A Photographic Atlas of the Human Body, Second Edition,* Figure 3.14.)

Common names for the malleus, incus, and stapes are the hammer, anvil, and stirrup, respectively.

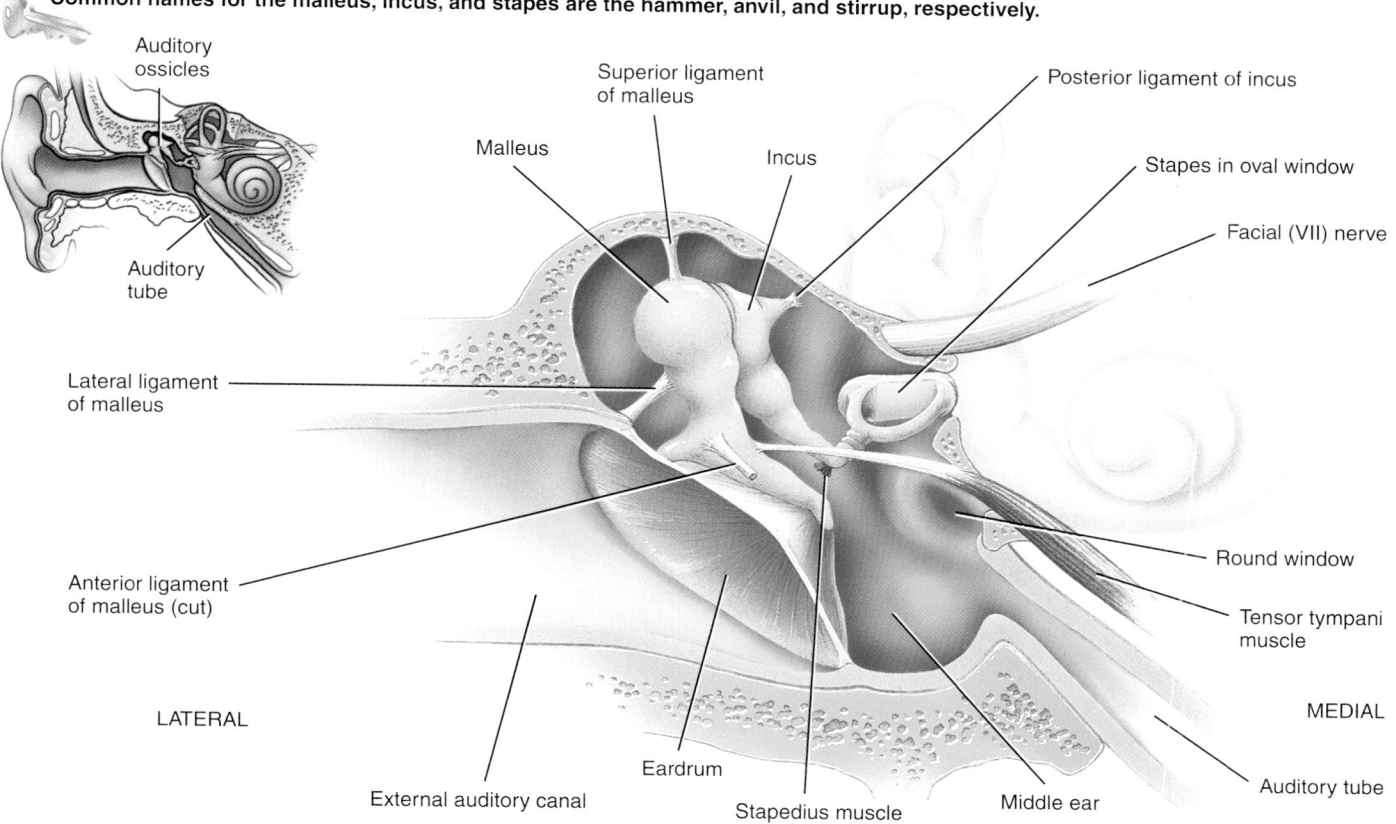

Frontal section showing location of auditory ossicles

What structures separate the middle ear from the internal ear?

eral form. The membranous labyrinth is lined by epithelium and contains **endolymph.** The level of potassium ions (K^+) in endolymph is unusually high for an extracellular fluid, and potassium ions play a role in the generation of auditory signals (described shortly).

The **vestibule** (VES-ti-būl) is the oval central portion of the bony labyrinth. The membranous labyrinth in the vestibule consists of two sacs called the **utricle** (Ū-tri-kl = little bag) and the **saccule** (SAK-ūl = little sac), which are connected by a small duct. Projecting superiorly and posteriorly from the vestibule are the three bony **semicircular canals,** each of which lies at approximately right angles to the other two. Based on their positions, they are named the anterior, posterior, and lateral semicircular canals. The anterior and posterior semicircular canals are vertically oriented; the lateral one is horizontally oriented. At one end of each canal is a swollen enlargement called the **ampulla** (am-PUL-la = saclike duct). The portions of the membranous labyrinth that lie inside the bony semicircular canals are called the **semicircular ducts.** These structures connect with the utricle of the vestibule.

The vestibular branch of the vestibulocochlear (VIII) nerve consists of *ampullary, utricular,* and *saccular nerves.* These nerves contain both first-order sensory neurons and motor neurons that synapse with receptors for equilibrium. The first-order sensory neurons carry sensory information from the receptors, and the motor neurons carry feedback signals to the receptors, apparently to modify their sensitivity. Cell bodies of the sensory neurons are located in the **vestibular ganglia** (see Figure 17.19b).

Anterior to the vestibule is the **cochlea** (KOK-lē-a = snail-shaped), a bony spiral canal (Figure 17.19a on page 599) that resembles a snail's shell and makes almost three turns around a central bony core called the **modiolus** (mō-DĪ-ō-lus; Figure 17.19b). Sections through the cochlea reveal that it is divided into three channels: cochlear duct, scala vestibuli, and scala

Figure 17.18 **The right internal ear.** The outer, cream-colored area is part of the bony labyrinth; the inner, pink-colored area is the membranous labyrinth.

The bony labyrinth contains perilymph, and the membranous labyrinth contains endolymph.

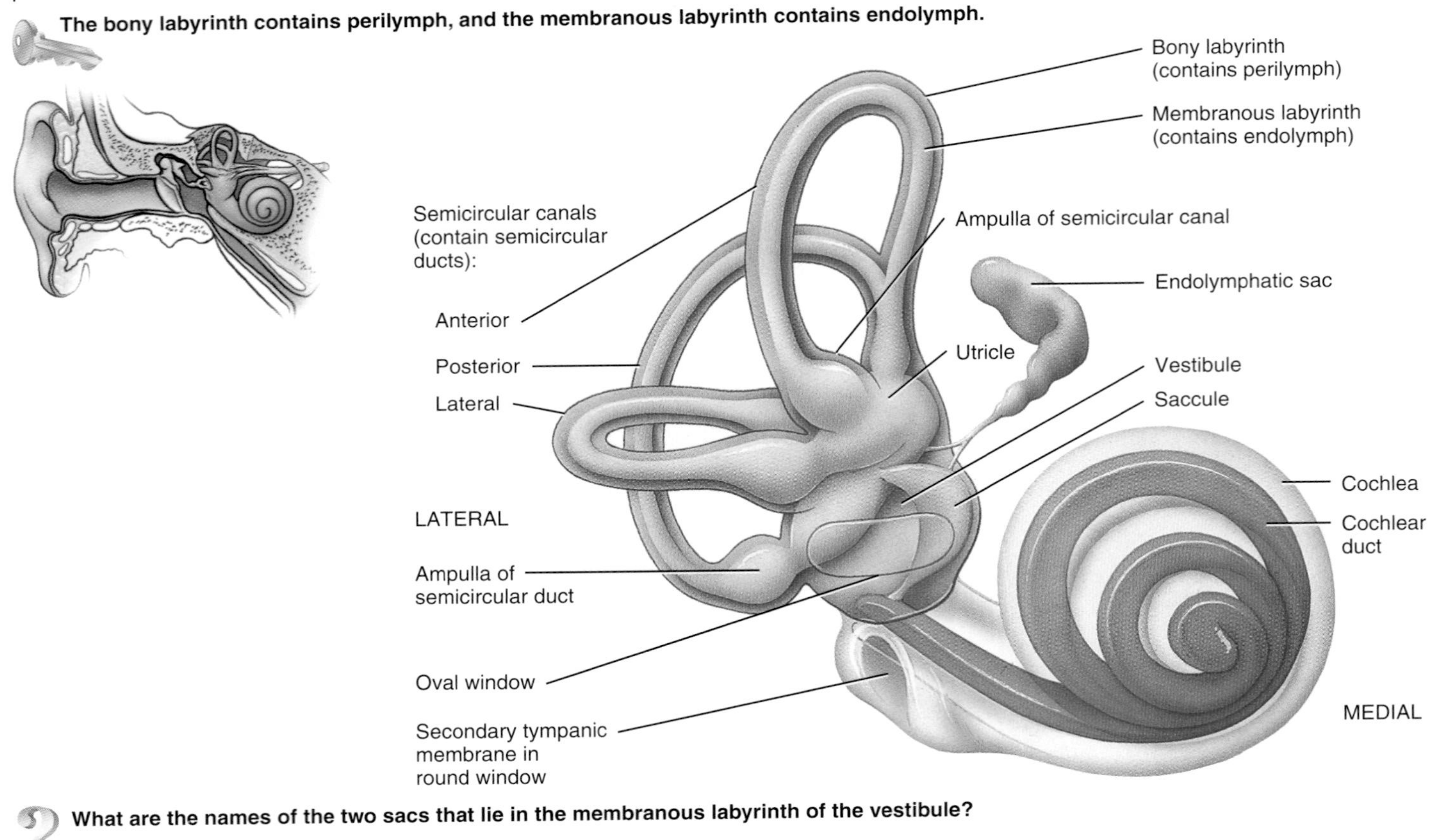

What are the names of the two sacs that lie in the membranous labyrinth of the vestibule?

tympani (Figure 17.19 a–c). The **cochlear duct** or **scala media** is a continuation of the membranous labyrinth into the cochlea; it is filled with endolymph. The channel above the cochlear duct is the **scala vestibuli,** which ends at the oval window. The channel below is the **scala tympani,** which ends at the round window. Both the scala vestibuli and scala tympani are part of the bony labyrinth of the cochlea; therefore, these chambers are filled with perilymph. The scala vestibuli and scala tympani are completely separated, except for an opening at the apex of the cochlea, the **helicotrema** (hel-i-kō-TRĒ-ma; Figure 17.19b). The cochlea adjoins the wall of the vestibule, into which the scala vestibuli opens. The perilymph in the vestibule is continuous with that of the scala vestibuli.

The **vestibular membrane** separates the cochlear duct from the scala vestibuli, and the **basilar membrane** separates the cochlear duct from the scala tympani. Resting on the basilar membrane is the **spiral organ** or **organ of Corti** (Figure 17.19c, d). The spiral organ is a coiled sheet of epithelial cells, including supporting cells and about 16,000 **hair cells,** which are the receptors for hearing. There are two groups of hair cells: The *inner hair cells* are arranged in a single row whereas the *outer hair cells* are arranged in three rows. At the apical tip of each hair cell is a **hair bundle,** consisting of 30–100 stereocilia that extend into the endolymph of the cochlear duct. Despite their name, stereocilia are actually long, hairlike microvilli arranged in several rows of graded height.

At their basal ends, inner and outer hair cells synapse both with first-order sensory neurons and with motor neurons from the cochlear branch of the vestibulocochlear (VIII) nerve. Cell bodies of the sensory neurons are located in the **spiral ganglion** (Figure 17.19b, c). Although outer hair cells outnumber them by 3 to 1, the inner hair cells synapse with 90–95% of the first-order sensory neurons in the cochlear nerve that relay auditory information to the brain. By contrast, 90% of the motor neurons in the cochlear nerve synapse with outer hair cells. The **tectorial membrane** (*tector-* = covering), a flexible gelatinous membrane, covers the hair cells of the spiral organ (Figure 17.19d).

The Nature of Sound Waves

In order to understand the physiology of hearing, it is necessary to learn something about its input, which occurs in the form of sound waves. **Sound waves** are alternating high- and low-pressure regions traveling in the same direction through some medium (such as air). They originate from a vibrating object in much the same way that ripples arise and travel over the surface of a pond when you toss a stone into it. The *frequency* of a sound vibration is its *pitch.* The higher the frequency of vibration, the higher is the pitch.

Figure 17.19 Semicircular canals, vestibule, and cochlea of the right ear. Note that the cochlea makes nearly three complete turns.

The three channels in the cochlea are the scala vestibuli, the scala tympani, and the cochlear duct.

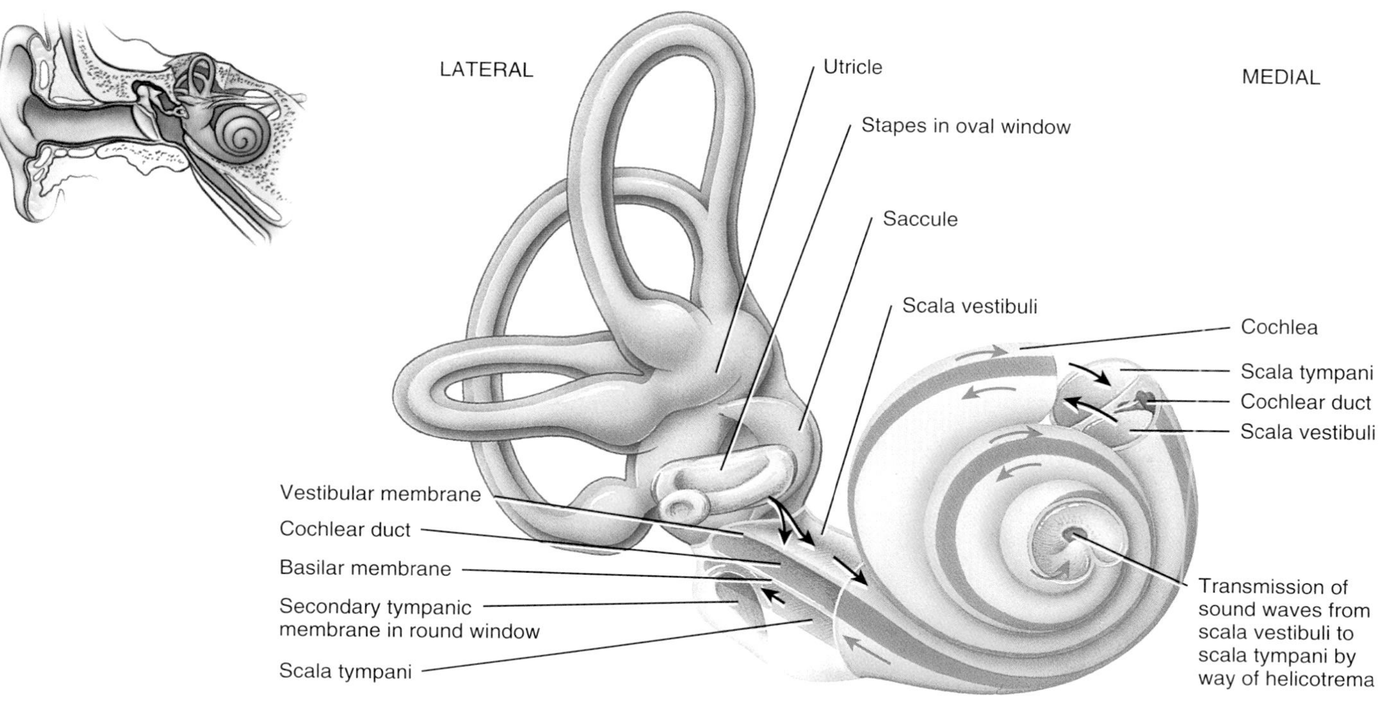

(a) Sections through the cochlea

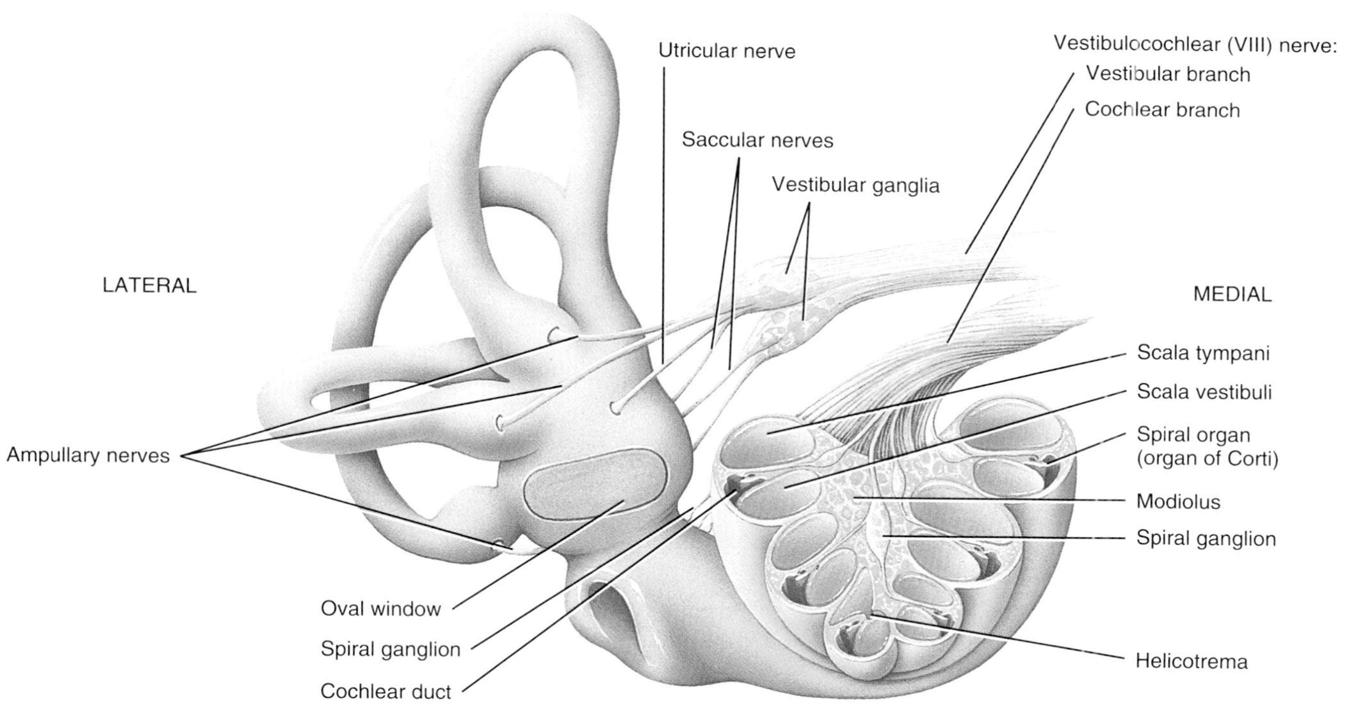

(b) Components of the vestibulocochlear nerve (cranial nerve VIII)

continues

Figure 17.19 **(continued)**

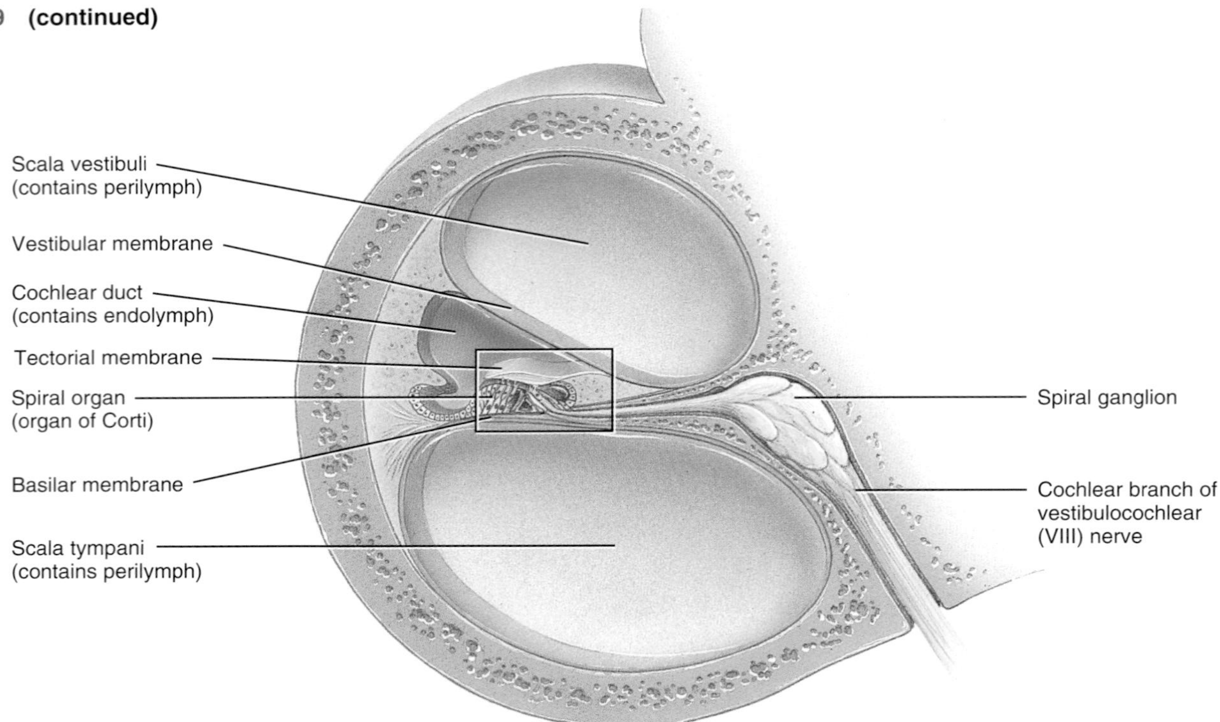

(c) Section through one turn of the cochlea

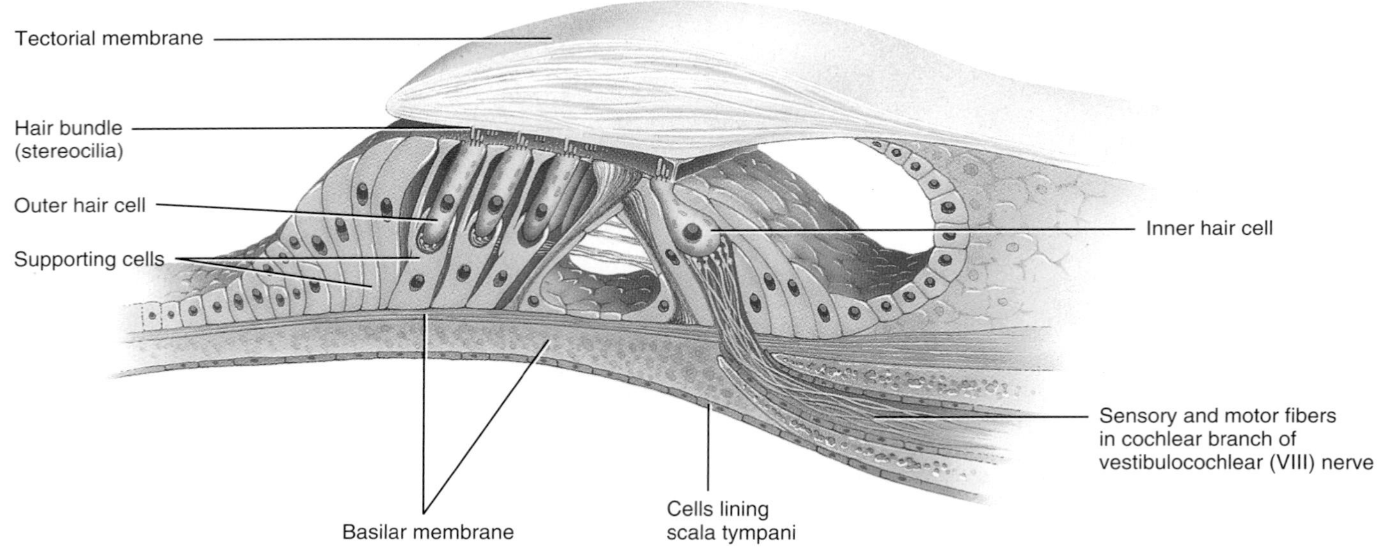

(d) Enlargement of spiral organ (organ of Corti)

What are the three subdivisions of the bony labyrinth?

The sounds heard most acutely by the human ear are those from sources that vibrate at frequencies between 500 and 5000 hertz (Hz; 1 Hz = 1 cycle per second). The entire audible range extends from 20 to 20,000 Hz. Sounds of speech primarily contain frequencies between 100 and 3000 Hz, and the "high C" sung by a soprano has a dominant frequency at 1048 Hz. The sounds from a jet plane several miles away range from 20 to 100 Hz.

The larger the *intensity* (size or amplitude) of the vibration, the *louder* is the sound. Sound intensity is measured in units called **decibels (dB).** An increase of one decibel represents a tenfold increase in sound intensity. The hearing threshold—the point at which an average young adult can just distinguish sound from silence—is defined as 0 dB at 1000 Hz. Rustling leaves have a decibel level of 15; whispered speech, 30; normal conver-

sation, 60; a vacuum cleaner, 75; shouting, 80; and a nearby motorcycle or jackhammer, 90. Sound becomes uncomfortable to a normal ear at about 120 dB, and painful above 140 dB.

Loud Sounds and Hair Cell Damage

Exposure to loud music and the engine roar of jet planes, revved-up motorcycles, lawn mowers, and vacuum cleaners damages hair cells of the cochlea. Because prolonged noise exposure causes hearing loss, employers in the United States must require workers to use hearing protectors when occupational noise levels exceed 90 dB. Rock concerts and even inexpensive headphones can easily produce sounds over 110 dB. Continued exposure to high-intensity sounds is one cause of **deafness,** a significant or total hearing loss. The louder the sounds, the more rapid is the hearing loss. Deafness usually begins with loss of sensitivity for high-pitched sounds. If you are listening to music through headphones and bystanders can hear it, the dB level is in the damaging range. Most people fail to notice their progressive hearing loss until destruction is extensive and they begin having difficulty understanding speech. Wearing earplugs with a noise-reduction rating of 30 dB while engaging in noisy activities can protect the sensitivity of your ears. ■

Physiology of Hearing

The following events are involved in hearing (Figure 17.20):

1. The auricle directs sound waves into the external auditory canal.
2. When sound waves strike the eardrum, the alternating high- and low-pressure of the air causes the eardrum to vibrate back and forth. The distance it moves, which is very small, depends on the intensity and frequency of the sound waves. The eardrum vibrates slowly in response to low-frequency (low-pitched) sounds and rapidly in response to high-frequency (high-pitched) sounds.
3. The central area of the eardrum connects to the malleus, which also starts to vibrate. The vibration is transmitted from the malleus to the incus and then to the stapes.
4. As the stapes moves back and forth, it pushes the membrane of the oval window in and out. The oval window vibrates about 20 times more vigorously than the eardrum because the ossicles efficiently transmit small vibrations spread over a large surface area (eardrum) into larger vibrations of a smaller surface (oval window).
5. The movement of the oval window sets up fluid pressure waves

Figure 17.20 Events in the stimulation of auditory receptors in the right ear. The numbers correspond to the events listed in the text. The cochlea has been uncoiled to more easily visualize the transmission of sound waves and their distortion of the vestibular and basilar membranes of the cochlear duct.

Hair cells of the spiral organ (organ of Corti) convert a mechanical vibration (stimulus) into an electrical signal (receptor potential).

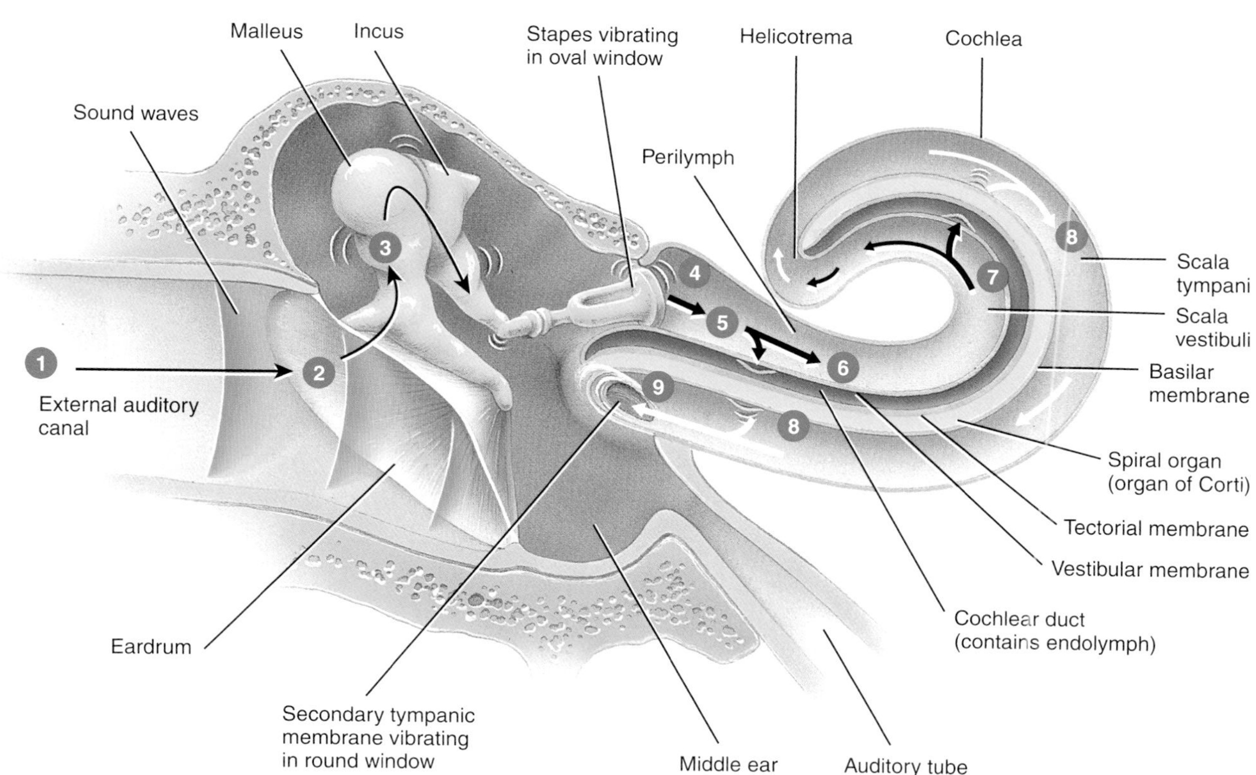

Which part of the basilar membrane vibrates most vigorously in response to high-frequency (high-pitched) sounds?

in the perilymph of the cochlea. As the oval window bulges inward, it pushes on the perilymph of the scala vestibuli.

6 Pressure waves are transmitted from the scala vestibuli to the scala tympani and eventually to the round window, causing it to bulge outward into the middle ear. (See 9 in the figure.)

7 As the pressure waves deform the walls of the scala vestibuli and scala tympani, they also push the vestibular membrane back and forth, creating pressure waves in the endolymph inside the cochlear duct.

8 The pressure waves in the endolymph cause the basilar membrane to vibrate, which moves the hair cells of the spiral organ against the tectorial membrane. Bending of the hair cell stereocilia produces receptor potentials that ultimately lead to the generation of nerve impulses.

Sound waves of various frequencies cause certain regions of the basilar membrane to vibrate more intensely than other regions. Each segment of the basilar membrane is "tuned" for a particular pitch. Because the membrane is narrower and stiffer at the base of the cochlea (portion closer to the oval window), high-frequency (high-pitched) sounds near 20,000 Hz induce maximal vibrations in this region. Toward the apex of the cochlea near the helicotrema, the basilar membrane is wider and more flexible; low-frequency (low-pitched) sounds near 20 Hz cause maximal vibration of the basilar membrane there. As noted previously, loudness is determined by the intensity of sound waves. High-intensity sound waves cause larger vibrations of the basilar membrane, which leads to a higher frequency of nerve impulses reaching the brain. Louder sounds also may stimulate a larger number of hair cells.

The hair cells transduce mechanical vibrations into electrical signals. As the basilar membrane vibrates, the hair bundles at the apex of the hair cell bend back and forth and slide against one another. A *tip link* protein connects the tip of each stereocilium to a mechanically gated ion channel called the **transduction channel** in its taller stereocilium neighbor. As the stereocilia bend in the direction of the taller stereocilia, the tip links tug on the transduction channels and open them. These channels allow cations in the endolymph, primarily K^+, to enter the hair cell cytosol. As cations enter, they produce a depolarizing receptor potential. Depolarization quickly spreads along the plasma membrane and opens voltage-gated Ca^{2+} channels in the base of the hair cell. The resulting inflow of Ca^{2+} triggers exocytosis of synaptic vesicles containing a neurotransmitter, which is probably glutamate. As more neurotransmitter is released, the frequency of nerve impulses in the first-order sensory neurons that synapse with the base of the hair cell increases. Bending of the stereocilia in the opposite direction closes the transduction channels, allows repolarization or even hyperpolarization to occur, and reduces neurotransmitter release from the hair cells. This decreases the frequency of nerve impulses in the sensory neurons.

Besides its role in detecting sounds, the cochlea has the surprising ability to produce sounds. These usually inaudible sounds, called **otoacoustic emissions,** can be picked up by placing a sensitive microphone next to the eardrum. They are caused by vibrations of the outer hair cells that occur in response to sound waves and to signals from motor neurons. As they depolarize and repolarize, the outer hair cells rapidly shorten and lengthen. This vibratory behavior appears to change the stiffness of the tectorial membrane and is thought to enhance the movement of the basilar membrane, which amplifies the responses of the inner hair cells. At the same time, the outer hair cell vibrations set up a traveling wave that goes back toward the stapes and leaves the ear as an otoacoustic emission. Detection of these inner ear–produced sounds is a fast, inexpensive, and noninvasive way to screen newborns for hearing defects. In deaf babies, otoacoustic emissions are not produced or are greatly reduced in size.

The Auditory Pathway

First-order sensory neurons in the cochlear branch of each vestibulocochlear (VIII) nerve terminate in the cochlear nuclei of the medulla oblongata on the same side. From there, axons carrying auditory signals project to the superior olivary nuclei in the pons on both sides. Slight differences in the timing of impulses arriving from the two ears at the olivary nuclei allow us to locate the source of a sound. From both the cochlear nuclei and the olivary nuclei, axons ascend to the inferior colliculus in the midbrain, and then to the medial geniculate nucleus of the thalamus. From the thalamus, auditory signals project to the primary auditory area in the superior temporal gyrus of the cerebral cortex (Brodmann's areas 41 and 42 in Figure 14.15 on page 497) and perception of sound occurs. Because many auditory axons decussate (cross over) in the medulla while others remain on the same side, the right and left primary auditory areas receive nerve impulses from both ears.

Cochlear Implants

A **cochlear implant** is a device that translates sounds into electrical signals that can be interpreted by the brain. Such a device is useful for people with deafness that is caused by damage to hair cells in the cochlea. The external parts of a cochlear implant consist of (1) a *microphone* worn around the ear that picks up sound waves, (2) a *sound processor,* which may be placed in a shirt pocket, that converts sound waves into electrical signals, and (3) a *transmitter,* worn behind the ear, which receives signals from the sound processor and passes them to an internal receiver. The internal parts of a cochlear implant are the (1) *internal receiver,* which relays signals to (2) *electrodes* implanted in the cochlea, where they trigger nerve impulses in sensory neurons in the cochlear branch of the vestibulocochlear (VIII) nerve. These artificially induced nerve impulses propagate over their normal pathways to the brain. The perceived sounds are crude compared to normal hearing, but they provide a sense of rhythm and loudness; infor-mation about certain noises, such as those made by telephones and automobiles; and the pitch and cadence of speech. Some patients hear well enough with a cochlear implant to use the telephone. ■

Physiology of Equilibrium

There are two types of **equilibrium** (balance). **Static equilibrium** refers to the maintenance of the position of the body (mainly the head) relative to the force of gravity. **Dynamic equilibrium** is the maintenance of body position (mainly the head) in response to sudden movements such as rotation, acceleration, and deceleration. Collectively, the receptor organs for equilibrium are called the **vestibular apparatus** (ves-TIB-ū-lar); these include the saccule, utricle, and semicircular ducts.

Otolithic Organs: Saccule and Utricle

The walls of both the utricle and the saccule contain a small, thickened region called a **macula** (MAK-ū-la; Figure 17.21). The two maculae (plural), which are perpendicular to one another, are the receptors for static equilibrium. They provide sensory information on the position of the head in space and are essential for maintaining appropriate posture and balance. The maculae also contribute to some aspects of dynamic equilibrium; they detect linear acceleration and deceleration—the sensations you feel while in an elevator or a car that is speeding up or slowing down.

The two maculae consist of two kinds of cells: **hair cells,** which are the sensory receptors, and **supporting cells.** Hair cells feature **hair bundles** that consist of 70 or more *stereocilia* (which are actually microvilli), plus one *kinocilium,* a conventional cilium anchored firmly to its basal body and extending beyond the longest stereocilia. As in the cochlea, the stereocilia are connected by tip links. Scattered among the hair cells are columnar supporting cells that probably secrete the thick, gelatinous, glycoprotein layer, called the **otolithic membrane,** that rests on the hair cells. A layer of dense calcium carbonate crystals, called **otoliths** (*oto-* = ear; *-liths* = stones) extends over the entire surface of the otolithic membrane.

Because the otolithic membrane sits on top of the macula, if you tilt your head forward, the otolithic membrane (and the otoliths as well) is pulled by gravity. It slides "downhill" over the hair cells in the direction of the tilt, bending the hair bundles. However, if you are sitting upright in a car that suddenly jerks forward, the otolithic membrane lags behind the head movement, pulls on the hair bundles, and makes them bend in the other direction. Bending of the hair bundles in one direction stretches the tip links, which pull open transduction channels thereby producing depolarizing receptor potentials; bending in the opposite direction closes the transduction channels and produces repolarization.

As the hair cells depolarize and repolarize, they release neurotransmitter at a faster or slower rate. The hair cells synapse with first-order sensory neurons in the vestibular branch of the vestibulocochlear (VIII) nerve (see Figure 17.19b). These neurons fire impulses at a slow or rapid pace depending on the amount of neurotransmitter present. Motor neurons also synapse with the hair cells and sensory neurons. Evidently, the motor neurons regulate the sensitivity of the hair cells and sensory neurons.

Semicircular Ducts

The three semicircular ducts, together with the saccule and the utricle, function in dynamic equilibrium. The ducts lie at right angles to one another in three planes (Figure 17.22 on page 605): The two vertical ducts are the anterior and posterior semicircular ducts, and the horizontal one is the lateral semicircular duct (see also Figure 17.18). This positioning permits detection of rotational acceleration or deceleration. In the **ampulla,** the dilated portion of each duct, is a small elevation called the **crista.** Each crista contains a group of **hair cells** and **supporting cells.** Covering the crista is a mass of gelatinous material called the **cupula** (KŪ-pū-la). When you move your head, the attached semicircular ducts and hair cells move with it. The endolymph within the ampulla, however, is not attached and lags behind. As the moving hair cells drag along the stationary endolymph, the hair bundles bend. Bending of the hair bundles produces receptor potentials. In turn, the receptor potentials lead to nerve impulses that pass along the vestibular branch of the vestibulocochlear (VIII) nerve.

Equilibrium Pathways

Most of the vestibular branch axons of the vestibulocochlear (VIII) nerve enter the brain stem and terminate in several vestibular nuclei in the medulla and pons. The remaining axons enter the cerebellum through the inferior cerebellar peduncle (see Figure 14.7a). Bidirectional pathways connect the vestibular nuclei and cerebellum.

Axons from all the vestibular nuclei extend to the nuclei of cranial nerves that control eye movements—oculomotor (III), trochlear (IV), and abducens (VI). Other axons from vestibular nuclei extend to the nucleus of the accessory (XI) nerve, which helps control head and neck movements. In addition, axons from the lateral vestibular nucleus form the vestibulospinal tract, which conveys impulses to skeletal muscles that regulate muscle tone in response to head movements.

Various pathways between the vestibular nuclei, cerebellum, and cerebrum enable the cerebellum to play a key role in maintaining equilibrium. The cerebellum continuously receives updated sensory information from the utricle and saccule. It monitors this information and makes corrective adjustments. Essentially, in response to input from the utricle, saccule, and semicircular ducts, the cerebellum continuously sends nerve impulses to the motor areas of the cerebrum. This feedback allows correction of signals from the motor cortex to specific skeletal muscles to maintain equilibrium.

Figure 17.21 **Location and structure of receptors in the maculae of the right ear.** Both first-order sensory neurons (blue) and motor neurons (red) synapse with the hair cells.

The movement of stereocilia initiates depolarizing receptor potentials.

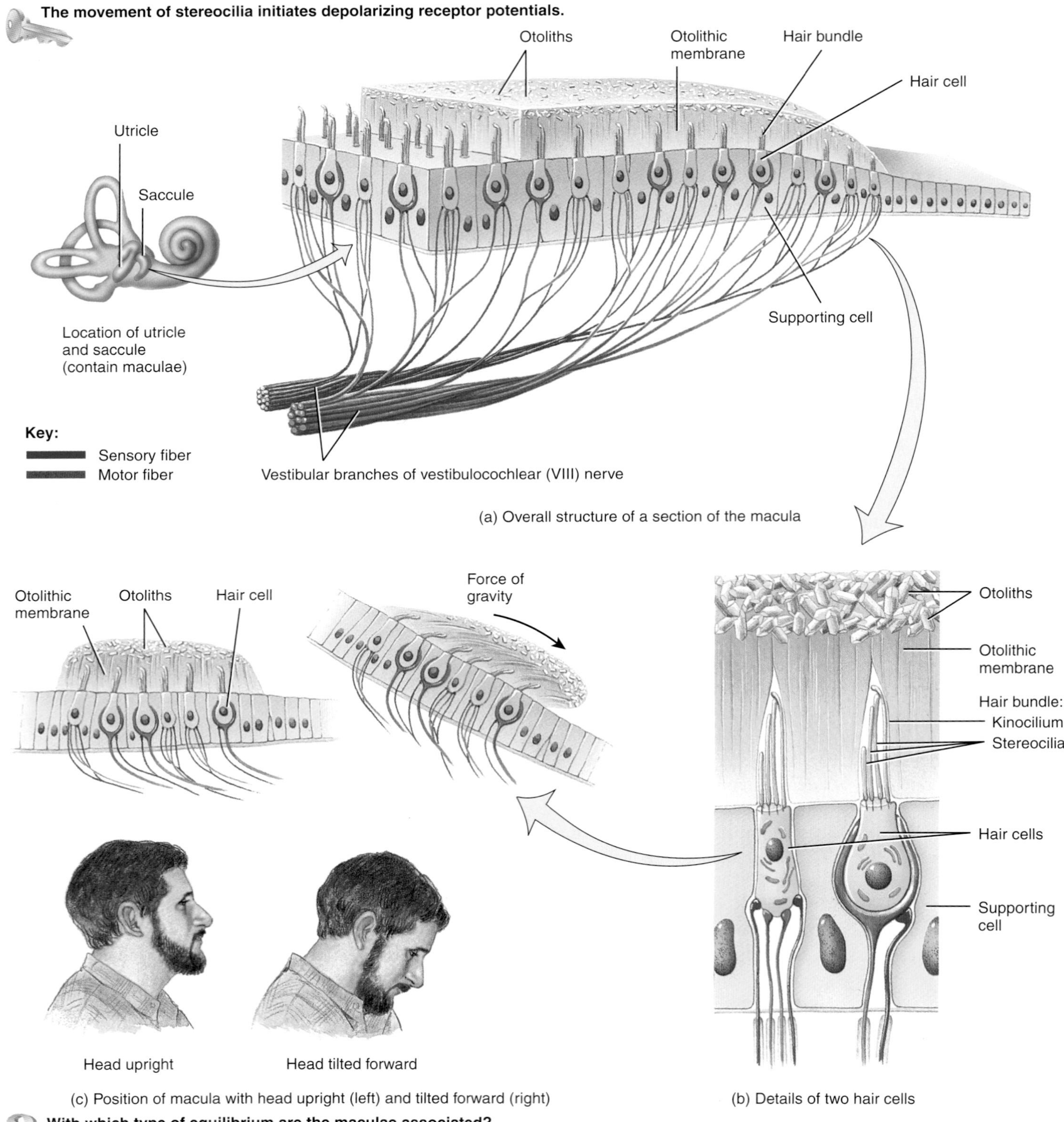

(a) Overall structure of a section of the macula

(b) Details of two hair cells

(c) Position of macula with head upright (left) and tilted forward (right)

With which type of equilibrium are the maculae associated?

Figure 17.22 **Location and structure of the semicircular ducts of the right ear.** Both first-order sensory neurons (blue) and motor neurons (red) synapse with the hair cells. The ampullary nerves are branches of the vestibular division of the vestibulocochlear (VIII) nerve.

The positions of the semicircular ducts permit detection of rotational movements.

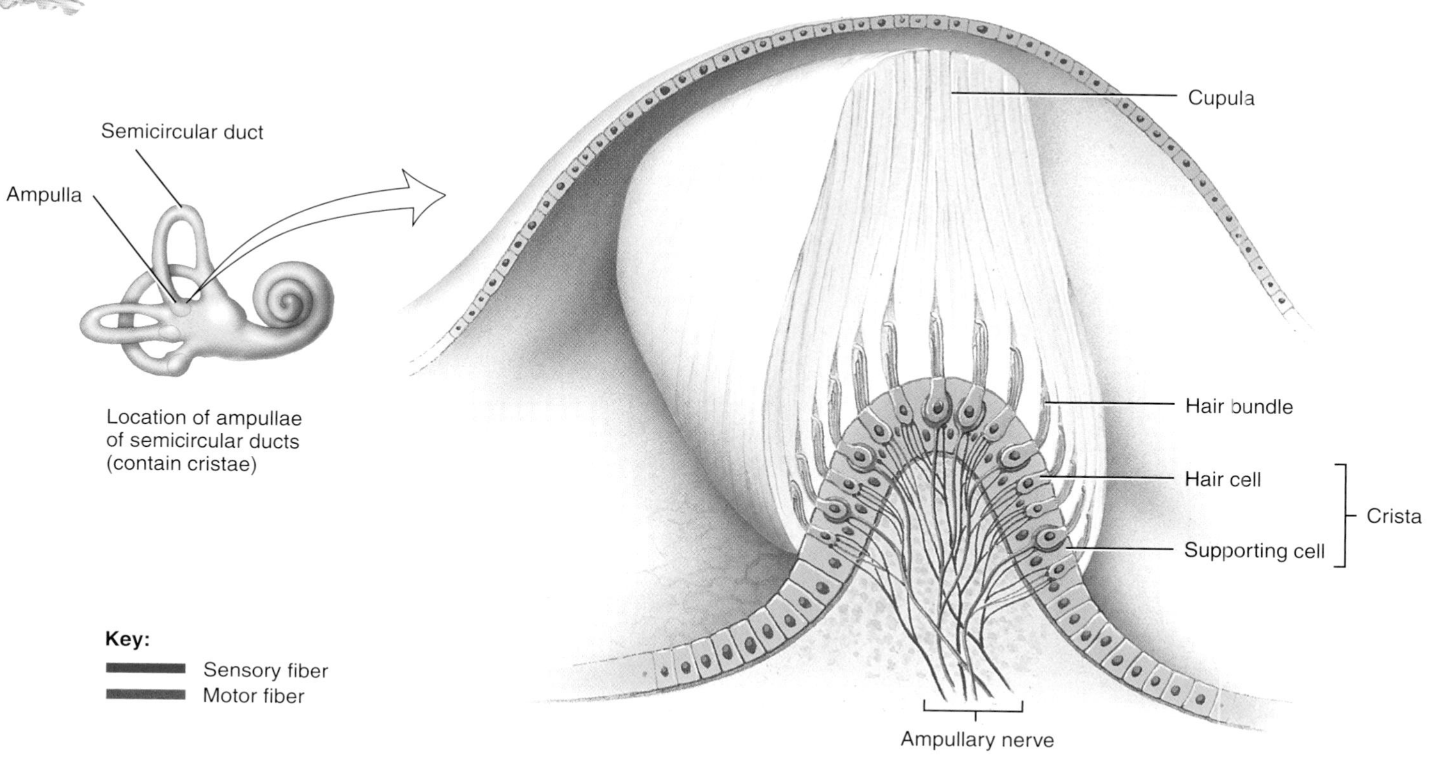

(a) Details of a crista

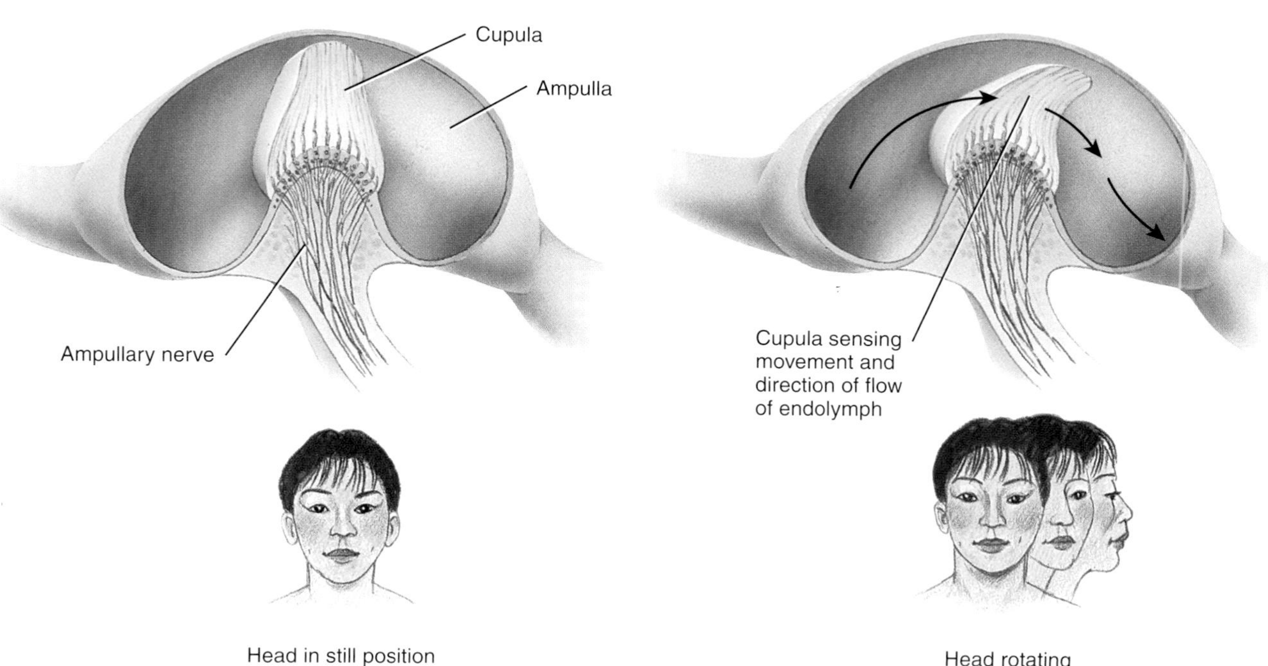

(b) Position of a cupula with the head in the still position (left) and when the head rotates (right)

With which type of equilibrium are the semicircular ducts associated?

Table 17.2 summarizes the structures of the ear related to hearing and equilibrium.

▶ **CHECKPOINT**

11. How are sound waves transmitted from the auricle to the spiral organ of Corti?

12. How do hair cells in the cochlea and vestibular apparatus transduce mechanical vibrations into electrical signals?

13. What is the pathway for auditory impulses from the cochlea to the cerebral cortex?

14. Compare the function of the maculae in maintaining static equilibrium with the role of the cristae in maintaining dynamic equilibrium.

15. What is the role of vestibular input to the cerebellum?

16. Describe the equilibrium pathways.

TABLE 17.2 Summary of Structures of the Ear

Regions of the Ear and Key Structures	Function
External (outer) ear External auditory canal Auricle Eardrum	*Auricle (pinna):* Collects sound waves. *External auditory canal (meatus):* Directs sound waves to the eardrum. *Eardrum (tympanic membrane):* Sound waves cause it to vibrate, which, in turn, causes the malleus to vibrate.
Middle ear 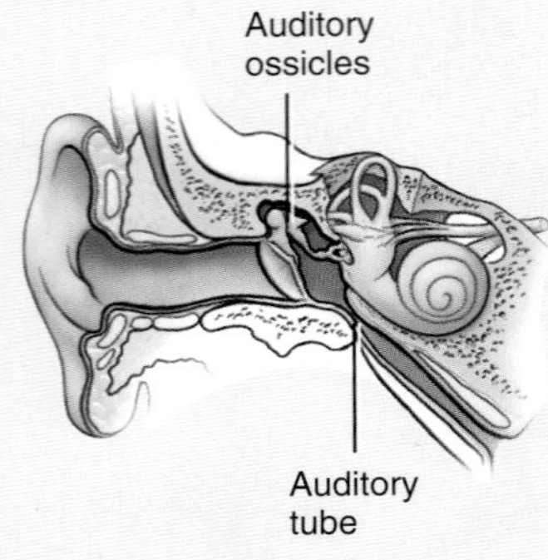 	*Auditory ossicles:* Transmit and amplify vibrations from tympanic membrane to oval window. *Auditory tube:* Equalizes air pressure on both sides of the tympanic membrane.
Internal (inner) ear 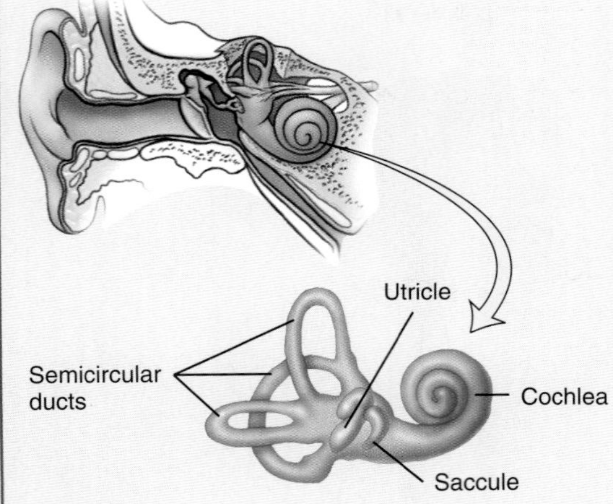 	*Cochlea:* Contains a series of fluids, channels, and membranes that transmit vibrations to the spiral organ (organ of Corti), the organ of hearing; hair cells in the spiral organ produce receptor potentials, which elicit nerve impulses in the cochlear branch of the vestibulocochlear (VIII) nerve. *Vestibular apparatus:* Includes semicircular ducts, utricle, and saccule, which generate nerve impulses that propagate along the vestibular branch of the vestibulocochlear (VIII) nerve. *Semicircular ducts:* Contain cristae, site of hair cells for dynamic equilibrium. *Utricle:* Contains macula, site of hair cells for static and dynamic equilibrium. *Saccule:* Contains macula, site of hair cells for static and dynamic equilibrium.

DEVELOPMENT OF THE EYES AND EARS

▶ OBJECTIVE

Describe the development of the eyes and the ears.

Eyes

The *eyes* begin to develop about 22 days after fertilization when the **ectoderm** of the lateral walls of the prosencephalon (forebrain) bulges out to form a pair of shallow grooves called the **optic grooves** (Figure 17.23a). Within a few days, as the neural tube is closing, the optic grooves enlarge and grow toward the surface ectoderm and become known as the **optic vesicles** (Figure 17.23b). When the optic vesicles reach the surface ectoderm, the surface ectoderm thickens to form the **lens placodes.** In addition, the distal portions of the optic vesicles invaginate (Figure 17.23c), forming the **optic cups;** they remain attached to the prosencephalon by narrow, hollow proximal structures called **optic stalks** (Figure 17.23d).

Figure 17.23 **Development of the eyes.**

The eyes begin to develop about 22 days after fertilization from ectoderm of the prosencephalon.

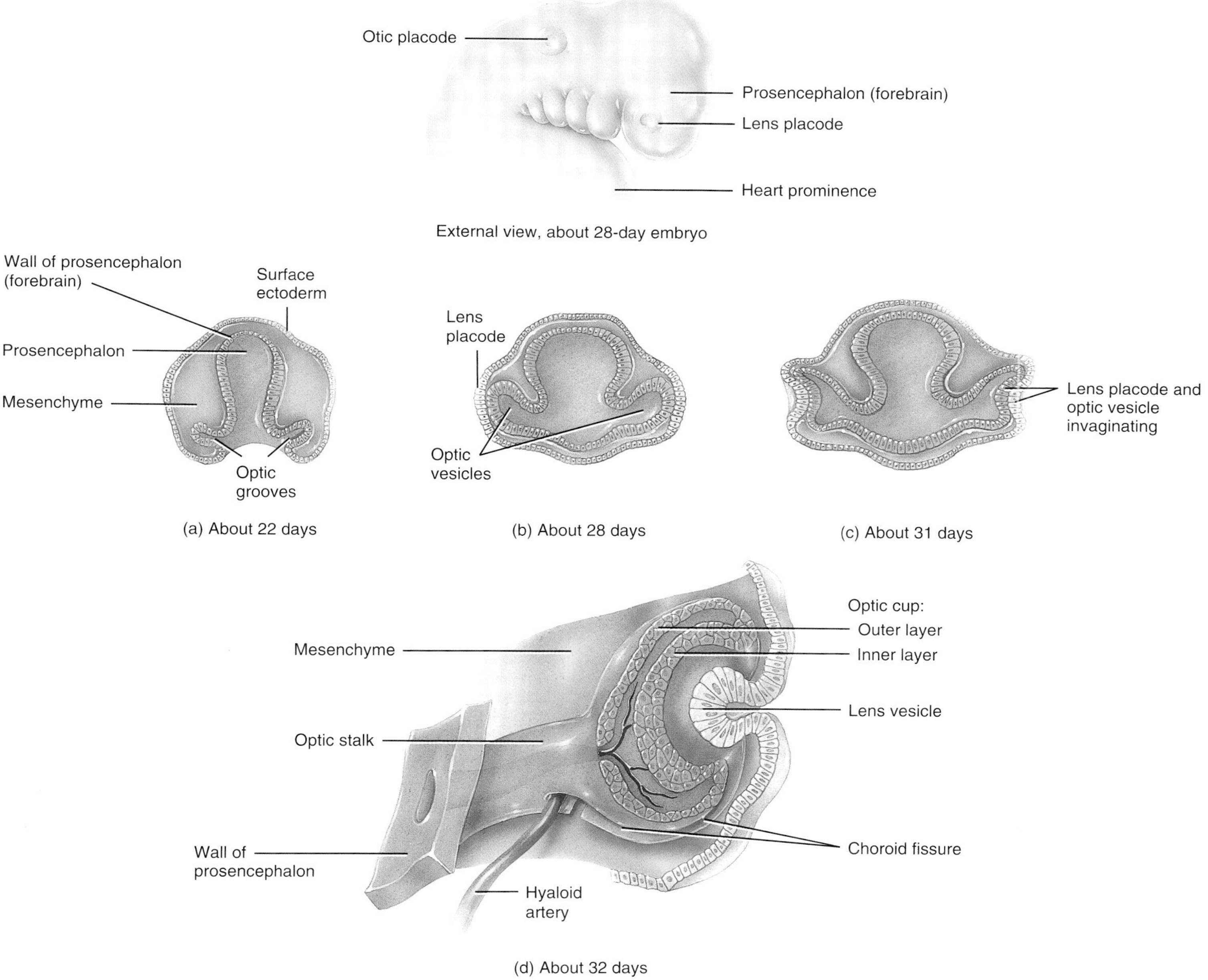

Otic placode
Prosencephalon (forebrain)
Lens placode
Heart prominence

External view, about 28-day embryo

Wall of prosencephalon (forebrain)
Surface ectoderm
Prosencephalon
Mesenchyme
Optic grooves

(a) About 22 days

Lens placode
Optic vesicles

(b) About 28 days

Lens placode and optic vesicle invaginating

(c) About 31 days

Optic cup:
Outer layer
Inner layer
Mesenchyme
Lens vesicle
Optic stalk
Choroid fissure
Wall of prosencephalon
Hyaloid artery

(d) About 32 days

Which structure gives rise to the neural and pigmented layers of the optic part of the retina?

The lens placodes also invaginate and develop into lens vesicles that sit in the optic cups. The lens vesicles eventually develop into the *lenses.* Blood is supplied to the developing lenses (and retina) by the hyaloid arteries. These arteries gain access to the developing eyes through a groove on the inferior surface of the optic cup and optic stalk called the **choroid fissure.** As the lenses mature, part of the hyaloid arteries that pass through the vitreous chamber degenerate; the remaining portions of the hyaloid arteries become the *central retinal arteries.*

The inner wall of the optic cup forms the *neural layer* of the optic part of the retina, while the outer layer forms the *pigmented layer* of the optic part of the retina. Axons from the neural layer grow through the optic stalk to the brain, converting the optic stalk to the *optic (II) nerve.* Although myelination of the optic nerves begins late in fetal life, it is not completed until the tenth week after birth.

The anterior portion of the optic cup forms the epithelium of the *ciliary body, iris,* and *circular and radial muscles* of the iris. The connective tissue of the ciliary body, *ciliary muscle,* and *zonular fibers* of the lens develop from **mesenchyme** around the anterior portion of the optic cup.

Mesenchyme surrounding the optic cup and optic stalk differentiates into an inner layer that gives rise to the *choroid* and an outer layer that develops into the *sclera* and part of the *cornea.* The remainder of the cornea is derived from surface ectoderm.

The *anterior chamber* develops from a cavity that forms in the mesenchyme between the iris and cornea; the *posterior chamber* develops from a cavity that forms in the mesenchyme between the iris and lens.

Some mesenchyme around the developing eye enters the optic cup through the choroid fissure. This mesenchyme occupies the space between the lens and retina and differentiates into a delicate network of fibers. Later the spaces between the fibers fill with a jellylike substance, thus forming the *vitreous body* in the vitreous chamber.

The *eyelids* form from surface ectoderm and mesenchyme. The upper and lower eyelids meet and fuse at about eight weeks of development and remain closed until about 26 weeks of development.

Ears

The first portion of the ear to develop is the *internal ear.* It begins to form about 22 days after fertilization as a thickening of the surface ectoderm, called **otic placodes** (Figure 17.24a), that appear on either side of the rhombencephalon (hindbrain). The otic placodes invaginate quickly (Figure 17.24b) to form the **otic pits** (Figure 17.24c). Next, the otic pits pinch off from the surface ectoderm to form the **otic vesicles** within the mesenchyme of the head (Figure 17.24d). During later development, the otic vesicles will form the structures associated with the *membranous labyrinth* of the internal ear. Mesenchyme around the otic vesicles produces cartilage that later ossifies to form the bone associated with the *bony labyrinth* of the internal ear.

The *middle ear* develops from a structure called the first **pharyngeal (branchial) pouch,** an **endoderm**-lined outgrowth of the primitive pharynx (see the inset in Figure 17.24). The pharyngeal pouches are discussed in detail in Chapter 29 on page 1120. The *auditory ossicles* develop from the first and second pharyngeal pouches.

The *external ear* develops from the first **pharyngeal cleft,** an endoderm-lined groove between the first and second pharyngeal pouches (see the inset in Figure 17.24). The pharyngeal clefts are discussed in detail in Chapter 29 on page 1120.

▶ CHECKPOINT

17. How do the origins of the eyes and ears differ?

AGING AND THE SPECIAL SENSES

▶ OBJECTIVE

Describe the age-related changes that occur in the eyes and ears.

Most people do not experience any problems with the senses of smell and taste until about age 50. This is due to a gradual loss of olfactory receptors and gustatory receptor cells coupled with their slower rate of replacement as we age.

Several age-related changes occur in the eyes. As noted earlier, the lens loses some of its elasticity and thus cannot change shape as easily, resulting in presbyopia (see page 588). Cataracts (loss of transparency of the lenses) also occur with aging (see page 610). In old age, the sclera ("white" of the eye) becomes thick and rigid and develops a yellowish or brownish coloration due to many years of exposure to ultraviolet light, wind, and dust. The sclera may also develop random splotches of pigment, especially in people with dark complexions. The iris fades or develops irregular pigment. The muscles that regulate the size of the pupil weaken with age and the pupils become smaller, react more slowly to light, and dilate more slowly in the dark. For these reasons, elderly people find that objects are not as bright, their eyes may adjust more slowly when going outdoors, and they have problems going from brightly lit to darkly lit places. Some diseases of the retina are more likely to occur in old age, including age-related macular disease (see page 585) and detached retina (see page 584). A disorder called glaucoma (see page 610) develops in the eyes of aging people as a result of the buildup of aqueous humor. Tear production and the number of mucous cells in the conjunctiva may decrease with age, resulting in dry eyes. The eyelids

Figure 17.24 **Development of the ears.**

The first parts of the ears to develop are the internal ears, which begin to form about 22 days after fertilization as thickenings of surface ectoderm.

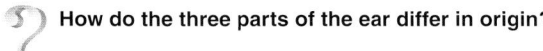

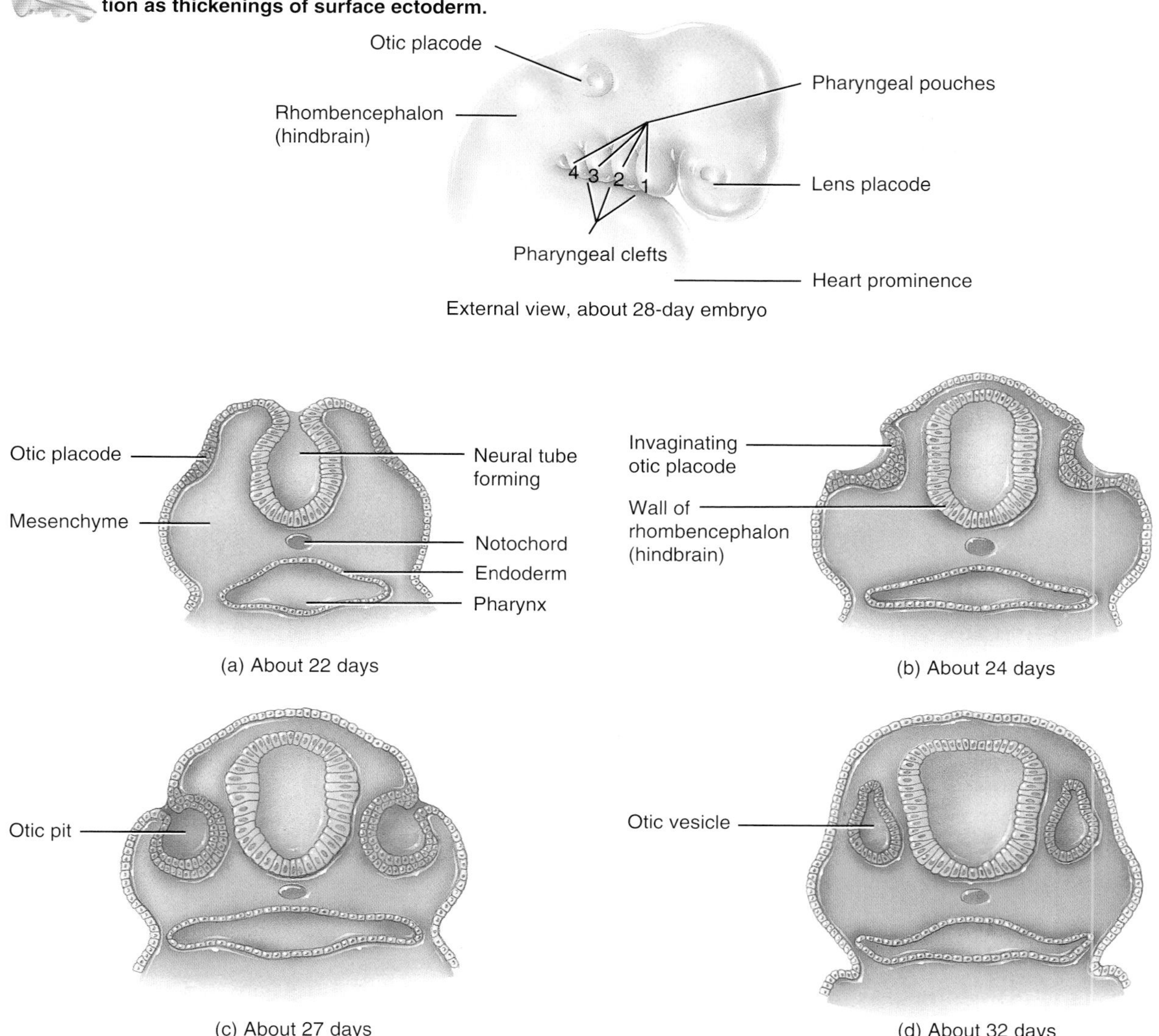

How do the three parts of the ear differ in origin?

lose their elasticity, becoming baggy and wrinkled. The amount of fat around the orbits may decrease, causing the eyeballs to sink into the orbits. Finally, as we age the sharpness of vision decreases, color and depth perception are reduced, and "vitreal floaters" increase.

By about age 60, around 25 percent of individuals experience a noticeable hearing loss, especially for higher-pitched sounds. The age-associated progressive loss of hearing in both ears is called **presbycusis** (pres′-bī-KŪ-sis; *presby-* = old; *-acou* = hearing; *-sis* = condition). It may be related to damaged and lost hair cells in the spiral organ or degeneration of the nerve pathway for hearing. Tinnitus (ringing in the ears) and vestibular imbalance also occur more frequently in the elderly.

▶ CHECKPOINT

18. What changes in the eyes and ears are related to the aging process, and how do they take place?

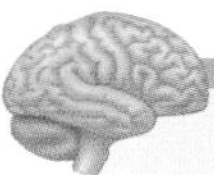

DISORDERS: HOMEOSTATIC IMBALANCES

Cataracts

A common cause of blindness is a loss of transparency of the lens known as a **cataract** (CAT-a-rakt = waterfall). The lens becomes cloudy (less transparent) due to changes in the structure of the lens proteins. Cataracts often occur with aging but may also be caused by injury, excessive exposure to ultraviolet rays, certain medications (such as long-term use of steroids), or complications of other diseases (for example, diabetes). People who smoke also have increased risk of developing cataracts. Fortunately, sight can usually be restored by surgical removal of the old lens and implantation of a new artificial one.

Glaucoma

Glaucoma (glaw-KŌ-ma) is the most common cause of blindness in the United States, afflicting about 2% of the population over age 40. Glaucoma is an abnormally high intraocular pressure due to a buildup of aqueous humor within the anterior cavity. The fluid compresses the lens into the vitreous body and puts pressure on the neurons of the retina. Persistent pressure results in a progression from mild visual impairment to irreversible destruction of neurons of the retina, damage to the optic nerve, and blindness. Glaucoma is painless, and the other eye compensates largely, so a person may experience considerable retinal damage and loss of vision before the condition is diagnosed. Because glaucoma occurs more often with advancing age, regular measurement of intraocular pressure is an increasingly important part of an eye exam as people grow older. Risk factors include race (blacks are more susceptible), increasing age, family history, and past eye injuries and disorders.

Deafness

Deafness is significant or total hearing loss. **Sensorineural deafness** is caused by either impairment of hair cells in the cochlea or damage of the cochlear branch of the vestibulocochlear (VIII) nerve. This type of deafness may be caused by atherosclerosis, which reduces blood supply to the ears; by repeated exposure to loud noise, which destroys hair cells of the spiral organ; and/or by certain drugs such as aspirin and streptomycin. **Conduction deafness** is caused by impairment of the external and middle ear mechanisms for transmitting sounds to the cochlea. Causes of conduction deafness include otosclerosis, the deposition of new bone around the oval window; impacted cerumen; injury to the eardrum; and aging, which often results in thickening of the eardrum and stiffening of the joints of the auditory ossicles. A hearing test called *Weber's test* is used to distinguish between sensorineural and conduction deafness. In the test, the stem of a vibrating fork is held to the forehead. In people with normal hearing, the sound is heard equally in both ears. If the sound is heard best in the affected ear, the deafness is probably of the conduction type; if the sound is heard best in the normal ear, it is probably of the sensorineural type.

Ménière's Disease

Ménière's disease (men'- ē-ĀRZ) results from an increased amount of endolymph that enlarges the membranous labyrinth. Among the symptoms are fluctuating hearing loss (caused by distortion of the basilar membrane of the cochlea) and roaring tinnitus (ringing). Spinning or whirling vertigo (dizziness) is characteristic of Ménière's disease. Almost total destruction of hearing may occur over a period of years.

Otitis Media

Otitis media is an acute infection of the middle ear caused mainly by bacteria and associated with infections of the nose and throat. Symptoms include pain, malaise, fever, and a reddening and outward bulging of the eardrum, which may rupture unless prompt treatment is received. (This may involve draining pus from the middle ear.) Bacteria passing into the auditory tube from the nasopharynx are the primary cause of middle ear infections. Children are more susceptible than adults to middle ear infections because their auditory tubes are almost horizontal, which decreases drainage. If otitis media occurs frequently, a surgical procedure called **tympanotomy** (tim'-pa-NOT-ō-mē; *tympano-* = drum; *-tome* = incision) is often employed. This consists of the insertion of a small tube into the eardrum to provide a pathway for the drainage of fluid from the middle ear.

MEDICAL TERMINOLOGY

Ageusia (a-GŪ-sē-ā; *a-* = without; *-geusis* = taste) Loss of the sense of taste.

Amblyopia (am'-blē-Ō-pē-a; *ambly-* = dull or dim) Term used to describe the loss of vision in an otherwise normal eye that, because of muscle imbalance, cannot focus in synchrony with the other eye. Sometimes called "wandering eyeball" or a "lazy eye."

Anosmia (an-OZ-mē-a; *a-* = without; *osmi* = smell, odor) Total lack of the sense of smell.

Barotrauma (bar'-ō-TRAW-ma; *baros-* = weight) Damage or pain, mainly affecting the middle ear, as a result of pressure changes. It occurs when pressure on the outer side of the tympanic membrane is higher than on the inner side, for example, when flying in an airplane or diving. Swallowing or holding your nose and exhaling with your mouth closed usually opens the auditory tubes, allowing air into the middle ear to equalize the pressure.

Blepharitis (blef-a-RĪ-tis; *blephar-* = eyelid; *-itis* = inflammation of) An inflammation of the eyelid.

Conjunctivitis (pinkeye) An inflammation of the conjunctiva; when caused by bacteria such as pneumococci, staphylococci, or *Hemophilus influenzae,* it is very contagious and more common in children. Conjunctivitis may also be caused by irritants, such as dust, smoke, or pollutants in the air, in which case it is not contagious.

Corneal abrasion (KOR-nē-al a-BRĀ-zhun) A scratch on the surface of the cornea, for example, from a speck of dirt or damaged contact lenses. Symptoms include pain, redness, watering, blurry vision, sensitivity to bright light, and frequent blinking.

Corneal transplant A procedure in which a defective cornea is removed and a donor cornea of similar diameter is sewn in. It is the most common and most successful transplant operation.

Since the cornea is avascular, antibodies in the blood that might cause rejection do not enter the transplanted tissue, and rejection rarely occurs. The shortage of donor corneas has been partially overcome by the development of artificial corneas made of plastic.

Diabetic retinopathy (ret-i-NOP-a-thē; *retino-* = retina; *-pathos* = suffering) Degenerative disease of the retina due to diabetes mellitus, in which blood vessels in the retina are damaged or new ones grow and interfere with vision.

Exotropia (ek′-sō-TRŌ-pē-a; *ex-* = out; *-tropia* = turning) Turning outward of the eyes.

Keratitis (ker′-a-TĪ-tis; *kerat-* = cornea) An inflammation or infection of the cornea.

Miosis (mī-Ō-sis) Constriction of the pupil.

Motion sickness Paleness, restlessness, nausea, weakness, dizziness, and malaise that may progress to vomiting caused by increased activity of the semicircular canals. It occurs during motion, for example, in a car, on a boat, on a train, or in an airplane. Usually, when motion stops, symptoms improve. Over-the-counter medications such as meclizine (Bonine®) or dimenhydrinate (Dramamine®) can be taken before embarking on a trip. A prescription skin patch that contains scopolamine (Transderm Scop®) can also be taken before symptoms occur.

Mydriasis (mi-DRĪ-a-sis) Dilation of the pupil.

Nystagmus (nis-TAG-mus; *nystagm-* = nodding or drowsy) A rapid involuntary movement of the eyeballs, possibly caused by a disease of the central nervous system. It is associated with conditions that cause vertigo.

Otalgia (ō-TAL-jē-a; *oto-* = ear; *-algia* = pain) Earache.

Photophobia (fō′-tō-FŌ-bē-a; *photo-* = light; *-phobia* = fear) Abnormal visual intolerance to light.

Ptosis (TŌ-sis = fall) Falling or drooping of the eyelid (or slippage of any organ below its normal position).

Retinoblastoma (ret-i-nō-blas-TŌ-ma; *-oma* = tumor) A tumor arising from immature retinal cells; it accounts for 2% of childhood cancers.

Scotoma (skō-TŌ-ma = darkness) An area of reduced or lost vision in the visual field.

Strabismus (stra-BIZ-mus; *strabismos* = squinting) Misalignment of the eyeballs so that the eyes do not move in unison when viewing an object; the affected eye turns either medially or laterally with respect to the normal eye and the result is double vision (diplopia). It may be caused by physical trauma, vascular injuries, or tumors of the extrinsic eye muscle or the oculomotor (III), trochlear (IV), or abducens (VI) cranial nerves.

Tinnitus (ti-NĪ-tus) A ringing, roaring, or clicking in the ears.

Tonometer (tō-NOM-ē-ter; *tono-* = tension or pressure; *-metron* = measure) An instrument for measuring pressure, especially intraocular pressure.

Trachoma (tra-KŌ-ma) A serious form of conjunctivitis and the greatest single cause of blindness in the world. It is caused by the bacterium *Chlamydia trachomatis*. The disease produces an excessive growth of subconjunctival tissue and invasion of blood vessels into the cornea, which progresses until the entire cornea is opaque.

Vertigo (VER-ti-gō = dizziness) A sensation of spinning or movement in which the world seems to revolve or the person seems to revolve in space, often associated with nausea and, in some cases, vomiting. It may be caused by arthritis of the neck or an infection of the vestibular apparatus.

STUDY OUTLINE

OLFACTION: SENSE OF SMELL (p. 575)

1. The receptors for olfaction, which are bipolar neurons, are in the nasal epithelium along with olfactory glands, which produce mucus that dissolves odorants.
2. In olfactory reception, a generator potential develops and triggers one or more nerve impulses.
3. The threshold of smell is low, and adaptation to odors occurs quickly.
4. Axons of olfactory receptors form the olfactory (I) nerves, which convey nerve impulses to the olfactory bulbs, olfactory tracts, limbic system, and cerebral cortex (temporal and frontal lobes).

GUSTATION: SENSATION OF TASTE (p. 577)

1. The receptors for gustation, the gustatory receptor cells, are located in taste buds.
2. Dissolved chemicals, called tastants, stimulate gustatory receptor cells by flowing through ion channels in the plasma membrane or by binding to receptors attached to G-proteins in the membrane.
3. Receptor potentials developed in gustatory receptor cells cause the release of neurotransmitter, which can generate nerve impulses in first-order sensory neurons.
4. The threshold varies with the taste involved, and adaptation to taste occurs quickly.
5. Gustatory receptor cells trigger nerve impulses in cranial nerves VII, IX, and X. Taste signals then pass to the medulla oblongata, thalamus, and cerebral cortex (parietal lobe).

VISION (p. 579)

1. Accessory structures of the eyes include the eyebrows, eyelids, eyelashes, lacrimal apparatus, and extrinsic eye muscles.
2. The lacrimal apparatus consists of structures that produce and drain tears.
3. The eye is constructed of three layers: (a) fibrous tunic (sclera and cornea), (b) vascular tunic (choroid, ciliary body, and iris), and (c) retina.
4. The retina consists of a pigmented layer and a neural layer that includes a photoreceptor layer, bipolar cell layer, ganglion cell layer, horizontal cells, and amacrine cells.

5. The anterior cavity contains aqueous humor; the vitreous chamber contains the vitreous body.
6. Image formation on the retina involves refraction of light rays by the cornea and lens, which focus an inverted image on the central fovea of the retina.
7. For viewing close objects, the lens increases its curvature (accommodation) and the pupil constricts to prevent light rays from entering the eye through the periphery of the lens.
8. The near point of vision is the minimum distance from the eye at which an object can be clearly focused with maximum accommodation.
9. In convergence, the eyeballs move medially so they are both directed toward an object being viewed.
10. The first step in vision is the absorption of light by photopigments in rods and cones and isomerization of *cis*-retinal. Receptor potentials in rods and cones decrease the release of inhibitory neurotransmitter, which induces graded potentials in bipolar cells and horizontal cells.
11. Horizontal cells transmit inhibitory signals to bipolar cells; bipolar or amacrine cells transmit excitatory signals to ganglion cells, which depolarize and initiate nerve impulses.
12. Impulses from ganglion cells are conveyed into the optic (II) nerve, through the optic chiasm and optic tract, to the thalamus. From the thalamus, impulses for vision propagate to the cerebral cortex (occipital lobe). Axon collaterals of retinal ganglion cells extend to the midbrain and hypothalamus.

HEARING AND EQUILIBRIUM (p. 595)

1. The external (outer) ear consists of the auricle, external auditory canal, and eardrum (tympanic membrane).
2. The middle ear consists of the auditory tube, ossicles, oval window, and round window.
3. The internal (inner) ear consists of the bony labyrinth and membranous labyrinth. The internal ear contains the spiral organ (organ of Corti), the organ of hearing.
4. Sound waves enter the external auditory canal, strike the eardrum, pass through the ossicles, strike the oval window, set up waves in the perilymph, strike the vestibular membrane and scala tympani, increase pressure in the endolymph, vibrate the basilar membrane, and stimulate hair bundles on the spiral organ (organ of Corti).
5. Hair cells convert mechanical vibrations into a receptor potential, which releases neurotransmitter that can initiate nerve impulses in first-order sensory neurons.
6. Sensory axons in the cochlear branch of the vestibulocochlear (VIII) nerve terminate in the medulla oblongata. Auditory signals then pass to the inferior colliculus, thalamus, and temporal lobes of the cerebral cortex.
7. Static equilibrium is the orientation of the body relative to the pull of gravity. The maculae of the utricle and saccule are the sense organs of static equilibrium.
8. Dynamic equilibrium is the maintenance of body position in response to movement. The cristae in the semicircular ducts are the main sense organs of dynamic equilibrium.
9. Most vestibular branch axons of the vestibulocochlear nerve enter the brain stem and terminate in the medulla and pons; other axons enter the cerebellum.

DEVELOPMENT OF THE EYES AND EARS (p. 607)

1. The eyes begin their development about 22 days after fertilization from ectoderm of the lateral walls of the prosencephalon (forebrain).
2. The ears begin their development about 22 days after fertilization from a thickening of ectoderm on either side of the rhombencephalon (hindbrain). The sequence of development of the ear is internal ear, middle ear, and external ear.

AGING AND THE SPECIAL SENSES (p. 608)

1. Most people do not experience problems with the senses of smell and taste until about age 50.
2. Among the age-related changes to the eyes are presbyopia, cataracts, difficulty adjusting to light, macular disease, glaucoma, dry eyes, and decreased sharpness of vision.
3. With age there is a progressive loss of hearing and tinnitus occurs more frequently.

SELF-QUIZ QUESTIONS

Fill in the blanks in the following statements.

1. The five primary taste sensations are ____, ____, ____, ____, and ____.
2. ____ equilibrium refers to the maintenance of the position of the body relative to the force of gravity; ____ equilibrium refers to the maintenance of body position in response to sudden movements such as rotation, acceleration, and deceleration.

Indicate whether the following statements are true or false.

3. Of all of the special senses, only smell and taste sensations project both to higher cortical areas and to the limbic system.
4. The ability to change the curvature of the lens for near vision is convergence.

Choose the one best answer to the following questions.

5. Which of the following are *true*? (1) The sites of olfactory transduction are the olfactory hairs. (2) The olfactory bulbs transmit impulses to the temporal lobe of the brain. (3) The axons of olfactory receptors pass through the olfactory foramina in the cribriform plate of the ethmoid bone. (4) The olfactory nerves are bundles of axons that terminate in the olfactory tracts. (5) Within the olfactory bulbs, the first-order neurons synapse with the second-order neurons. (a) 1, 2, and 4; (b) 2, 3, 4, and 5; (c) 1, 2, 3, 4, and 5; (d) 1, 3, and 5; (e) 1, 2, 3, and 5.
6. Which of the following statements is *incorrect*? (a) Olfactory receptors respond to the chemical stimulation of an odorant molecule by producing a receptor potential. (b) Basal stem cells

continually produce new olfactory receptors. (c) Adaptation to odors is rapid and occurs in both olfactory receptors and the CNS. (d) Production of nasal mucus by olfactory glands serves to moisten the olfactory epithelium and dissolve odorants. (e) The orbitofrontal area is an important region for odor identification and discrimination.

7. Which of the following statements is *incorrect*? (a) Taste is a chemical sense. (b) The receptors for taste sensations are found in taste buds located on the tongue, the soft palate, the pharynx, and the epiglottis. (c) Gustatory hairs are the sites of taste transduction. (d) The threshold for bitter substances is the highest. (e) Complete adaptation to taste can occur in 1 to 5 minutes.

8. When viewing an object close to your eyes, which of the following are required for proper image formation on the retina? (1) increased curvature of the lens, (2) contraction of the ciliary muscle, (3) divergence of the eyeballs, (4) refraction of light at the anterior and posterior surfaces of the cornea, (5) constriction of the pupil by contraction of the extrinsic eye muscles. (a) 1, 2, 3, 4, and 5; (b) 1, 2, and 4; (c) 1, 2, 3, and 4; (d) 2, 4, and 5; (e) 2, 3, and 4.

9. Which of the following are *mismatched*? (a) fungiform papillae: scattered over the entire tongue's surface, (b) filiform papillae: contain taste buds in early childhood, (c) vallate papillae: each houses 100–300 taste buds, (d) foliate papillae: located in trenches on the lateral margins of the tongue, (e) fungiform papillae: each houses about five taste buds.

10. Place in order the structures involved in the visual pathway. (a) optic tract, (b) ganglion cells, (c) cornea, (d) lens, (e) bipolar cells, (f) optic nerve, (g) visual cortex, (h) vitreous body, (i) optic chiasm, (j) aqueous humor, (k) pupil, (l) photoreceptors, (m) thalamus.

11. Which of the following statements is *incorrect*? (a) Retinal is the light-absorbing portion of all visual photopigments. (b) The only photopigment in rods is rhodopsin, but three different cone photopigments are present in the retina. (c) Retinal is a derivative of vitamin C. (d) Color vision results from different colors of light selectively activating different cone photopigments. (e) Bleaching and regeneration of the photopigments account for much but not all of the sensitivity changes during light and dark adaptation.

12. Which of the following is the *correct* sequence for the auditory pathway? (a) external auditory canal, tympanic membrane, auditory ossicles, oval window, cochlea and spiral organ; (b) tympanic membrane, external auditory canal, auditory ossicles, cochlea and spiral organ, round window; (c) auditory ossicles, tympanic membrane, cochlea and spiral organ, round window, oval window, external auditory canal; (d) auricle, tympanic membrane, round window, cochlea and spiral organ, oval window; (e) external auditory canal, tympanic membrane, auditory ossicles, internal auditory canal, spiral organ, oval window.

13. Match the following:

____ (a) upper and lower eyelids; shade the eyes during sleep, spread lubricating secretions over the eyeballs	(1) palpebrae
____ (b) produces and drains tears	(2) tarsal or Meibomian glands
____ (c) arch transversely above the eyeballs and help protect the eyeballs from foreign objects, perspiration, and the direct rays of the sun	(3) conjunctiva
____ (d) move the eyeball medially, laterally, superiorly, or inferiorly	(4) eyelashes
____ (e) a thick fold of connective tissue that gives form and support to the eyelids	(5) lacrimal apparatus
____ (f) modified sebaceous glands; secretion helps keep eyelids from adhering to one another	(6) extrinsic eye muscles
____ (g) project from the border of each eyelid; help protect the eyeballs from foreign objects, perspiration, and direct rays of the sun	(7) eyebrows
____ (h) a thin, protective mucous membrane that lines the inner aspect of the eyelids and passes from the eyelids onto the surface of the eyeball, where it covers the sclera	(8) tarsal plate

14. Match the following:

____(a) lines most of the internal surface of the sclera; provides nutrients to the posterior surface of the retina
____(b) colored portion of the eyeball; regulates the amount of light entering the posterior part of the eyeball
____(c) innermost layer of the eyeball; beginning of the visual pathway; contains rods and cones
____(d) biconvex transparent structure that fine tunes focusing of light rays for clear vision
____(e) transparent part of the eyeball that covers the iris; helps focus light
____(f) circular band of smooth muscle that alters the shape of the lens for near or far vision
____(g) site where the optic nerve exits the eyeball; the blind spot
____(h) watery fluid in the anterior cavity that helps nourish the lens and cornea; helps maintain shape of the eyeball
____(i) the hole in the center of the iris
____(j) jellylike substance in the vitreous chamber that helps prevent the eyeball from collapsing and holds the retina flush against the internal portions of the eyeball
____(k) white of the eye; gives shape to the eyeball, makes it more rigid, protects its inner parts
____(l) avascular superficial layer of the eyeball; includes cornea and sclera
____(m) small depression in the center of the macula lutea that contains only cone photoreceptors and is the area of highest visual acuity
____(n) contain blood capillaries that secrete aqueous humor; attach to suspensory ligaments of lens
____(o) middle, vascularized layer of the eyeball; includes choroid, ciliary body, and iris

(1) cornea
(2) sclera
(3) choroid
(4) ciliary processes
(5) ciliary muscle
(6) iris
(7) pupil
(8) uvea
(9) retina
(10) optic disc
(11) fibrous tunic
(12) central fovea
(13) aqueous humor
(14) lens
(15) vitreous body

15. Match the following:

____(a) partition between external auditory canal and middle ear; eardrum
____(b) oval central portion of the bony labyrinth; contains utricle and saccule
____(c) receptor for static equilibrium; also contributes to some aspects of dynamic equilibrium; consists of hair cells and supporting cells
____(d) spiral organ; organ for hearing
____(e) ear bones: malleus, incus, stapes
____(f) the pressure equalization tube that connects the middle ear to the nasopharynx
____(g) contains the spiral organ
____(h) fluid found within the membranous labyrinth; pressure waves in this fluid cause vibration of the basilar membrane
____(i) receptor organs for equilibrium; the saccule, utricle, and semicircular canals
____(j) swollen enlargement in semicircular canals; contains structures involved in dynamic equilibrium
____(k) opening between the middle ear and internal ear; is enclosed by a membrane called the secondary tympanic membrane
____(l) the flap of elastic cartilage covered by skin that captures sound waves; the pinna
____(m) fluid found inside bony labyrinth; bulging of the oval window causes pressure waves in this fluid
____(n) opening between the middle and inner ear; receives base of stapes

(1) auricle
(2) tympanic membrane
(3) auditory ossicles
(4) vestibular apparatus
(5) ampulla
(6) cochlea
(7) perilymph
(8) oval window
(9) round window
(10) auditory or eustachian tube
(11) vestibule
(12) endolymph
(13) spiral organ
(14) macula

CRITICAL THINKING QUESTIONS

1. Mario has experienced damage to his facial nerve. How would this affect his special senses?
2. The shift nurse brings ailing eighty-year-old Granny Gertrude her dinner. As Gertrude eats a small amount of her food, she comments that she isn't hungry and that "hospital food just doesn't taste good!" The nurse gives Gertrude a menu so she can choose her morning breakfast. Gertrude complains that she is having trouble reading the menu and asks the nurse to read it to her. As the nurse begins to read, Gertrude loudly asks her to "speak up and turn off the buzzing." What does the nurse know about aging and the special senses that allows her to maintain patience with this patient?
3. As you help your neighbor put drops in her six-year-old daughter's eyes, the daughter states, "That medicine tastes bad." How do you explain to the neighbor how her daughter can "taste" the eyedrops?

ANSWERS TO FIGURE QUESTIONS

17.1 The olfactory hairs detect odorant molecules.

17.2 Supporting cells develop into gustatory receptor cells.

17.3 The conjunctiva is continuous with the inner lining of the eyelids.

17.4 Lacrimal fluid, or tears, is a watery solution containing salts, some mucus, and lysozyme that protects, cleans, lubricates, and moistens the eyeball.

17.5 The fibrous tunic consists of the cornea and sclera; the vascular tunic consists of the choroid, ciliary body, and iris.

17.6 The parasympathetic division of the ANS causes pupillary constriction; the sympathetic division causes pupillary dilation.

17.7 An ophthalmoscopic examination can reveal evidence of hypertension, diabetes mellitus, cataract, and age-related macular disease.

17.8 The two types of photoreceptors are rods and cones. Rods provide black-and-white vision in dim light; cones provide high visual acuity and color vision in bright light.

17.9 After its secretion by the ciliary process, aqueous humor flows into the posterior chamber, around the iris, into the anterior chamber, and out of the eyeball through the scleral venous sinus.

17.10 During accommodation the ciliary muscle contracts, causing the zonular fibers to slacken. The lens then becomes more convex, increasing its focusing power.

17.11 Presbyopia is the loss of lens elasticity that occurs with aging.

17.12 Both rods and cones transduce light into receptor potentials, use a photopigment embedded in outer segment discs or folds, and release neurotransmitter at synapses with bipolar cells and horizontal cells.

17.13 The conversion of *cis*-retinal to *trans*-retinal is called isomerization.

17.14 Cyclic GMP is the ligand that opens Na^+ channels in photoreceptors, causing the dark current to flow.

17.15 Light rays from an object in the temporal half of the visual field fall on the nasal half of the retina.

17.16 The malleus of the middle ear is attached to the eardrum, which is part of the external ear.

17.17 The oval and round windows separate the middle ear from the internal ear.

17.18 The two sacs in the membranous labyrinth of the vestibule are the utricle and saccule.

17.19 The three subdivisions of the bony labyrinth are the semicircular canals, vestibule, and cochlea.

17.20 The region of the basilar membrane close to the oval and round windows vibrates most vigorously in response to high-frequency sounds.

17.21 The maculae are associated primarily with static equilibrium; they provide sensory information about the position of the head in space.

17.22 The semicircular ducts are associated with dynamic equilibrium.

17.23 The optic cup forms the neural and pigmented layers of the optic part of the retina.

17.24 The internal ear develops from surface ectoderm, the middle ear develops from pharyngeal pouches, and the external ear develops from a pharyngeal cleft.

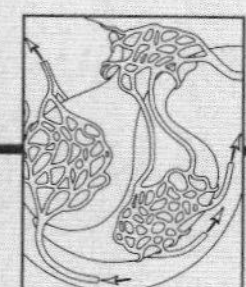

Chapter **18**

The Endocrine System

The Endocrine System and Homeostasis

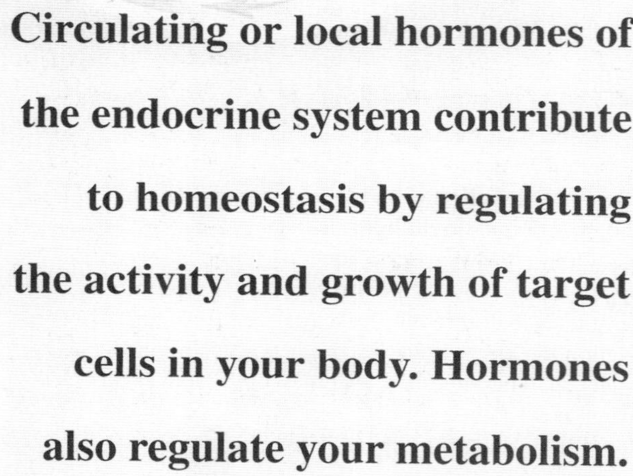

Circulating or local hormones of the endocrine system contribute to homeostasis by regulating the activity and growth of target cells in your body. Hormones also regulate your metabolism.

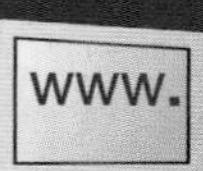

www. wiley.com/college/apcentral

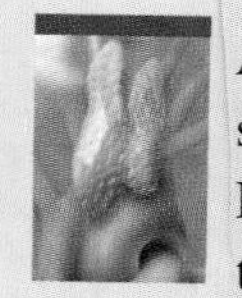

As girls and boys enter puberty, they start to develop striking differences in physical appearance and behavior. Perhaps no other period in life so dramatically shows the impact of the endocrine system in directing development and regulating body functions. In girls, estrogens promote accumulation of adipose tissue in the breasts and hips, sculpting a feminine shape. At the same time or a little later, increasing levels of testosterone in boys begin to help build muscle mass and enlarge the vocal cords, producing a lower-pitched voice. These changes are just a few examples of the powerful influence of endocrine secretions. Less dramatically, perhaps, multitudes of hormones help maintain homeostasis on a daily basis. They regulate the activity of smooth muscle, cardiac muscle, and some glands; alter metabolism; spur growth and development; influence reproductive processes; and participate in circadian (daily) rhythms established by the suprachiasmatic nucleus of the hypothalamus.

COMPARISON OF CONTROL BY THE NERVOUS AND ENDOCRINE SYSTEMS

▶ **OBJECTIVE**

Compare control of body functions by the nervous system and endocrine system.

The nervous and endocrine systems act together to coordinate functions of all body systems. Recall that the nervous system acts through nerve impulses conducted along axons of neurons. At synapses, nerve impulses trigger the release of mediator (messenger) molecules called *neurotransmitters* (shown in Figure 12.17 on page 425). The endocrine system also controls body activities by releasing mediators, called *hormones,* but the means of control of the two systems are very different.

A **hormone** (*hormon* = to excite or get moving) is a mediator molecule that is released in one part of the body but regulates the activity of cells in other parts of the body. Most hormones enter interstitial fluid and then the bloodstream. The circulating blood delivers hormones to cells throughout the body. Both neurotransmitters and hormones exert their effects by binding to receptors on or in their "target" cells. Several mediators act as both neurotransmitters and hormones. One familiar example is norepinephrine, which is released as a neurotransmitter by sympathetic postganglionic neurons and as a hormone by cells of the adrenal medullae.

Responses of the endocrine system often are slower than responses of the nervous system; although some hormones act within seconds, most take several minutes or more to cause a response. The effects of nervous system activation are generally briefer than those of the endocrine system. The nervous system acts on specific muscles and glands. The influence of the endocrine system is much broader; it helps regulate virtually all types of body cells.

We will also have several opportunities to see how the nervous and endocrine systems function together as an interlocking "supersystem." For example, certain parts of the nervous system stimulate or inhibit the release of hormones by the endocrine system.

Table 18.1 compares the characteristics of the nervous and endocrine systems. In this chapter, we focus on the major endocrine glands and hormone-producing tissues and examine how their hormones govern body activities.

▶ **CHECKPOINT**

1. List the similarities among and differences between the nervous and endocrine systems with regard to the control of homeostasis.

TABLE 18.1 Comparison of Control by the Nervous and Endocrine Systems

Characteristic	Nervous System	Endocrine System
Mediator molecules	Neurotransmitters released locally in response to nerve impulses.	Hormones delivered to tissues throughout the body by the blood.
Site of mediator action	Close to site of release, at a synapse; binds to receptors in postsynaptic membrane.	Far from site of release (usually); binds to receptors on or in target cells.
Types of target cells	Muscle (smooth, cardiac, and skeletal) cells, gland cells, other neurons.	Cells throughout the body.
Time to onset of action	Typically within milliseconds (thousandths of a second).	Seconds to hours or days.
Duration of action	Generally briefer (milliseconds).	Generally longer (seconds to days).

ENDOCRINE GLANDS

▶ **OBJECTIVE**

Distinguish between exocrine and endocrine glands.

Recall from Chapter 4 that the body contains two kinds of glands: exocrine glands and endocrine glands. **Exocrine glands** (*exo-* = outside) secrete their products into ducts that carry the secretions into body cavities, into the lumen of an organ, or to the outer surface of the body. Exocrine glands include sudoriferous (sweat), sebaceous (oil), mucous, and digestive glands. **Endocrine glands** (*endo-* = within) secrete their products (hormones) into the interstitial fluid surrounding the secretory cells rather than into ducts. From the interstitial fluid, hormones diffuse into capillaries and blood carries them to target cells throughout the body. Because most hormones are required in very small amounts, circulating levels typically are low.

The endocrine glands include the pituitary, thyroid, parathyroid, adrenal, and pineal glands (Figure 18.1). In addition, several organs and tissues are not exclusively classified as endocrine glands but contain cells that secrete hormones. These include the hypothalamus, thymus, pancreas, ovaries, testes, kidneys, stomach, liver, small intestine, skin, heart, adipose tissue, and placenta. Taken together, all endocrine glands and hormone-secreting cells constitute the **endocrine system.** The science of the structure and function of the endocrine glands and the diagnosis and treatment of disorders of the endocrine system is **endocrinology** (en′-dō-kri-NOL-ō-jē; *endo-* = within; *-crino* = to secrete; *-logy* = study of).

▶ **CHECKPOINT**

2. List 3 organs or tissues that are not exclusively classified as endocrine glands but contain cells that secrete hormones.

Figure 18.1 Location of many endocrine glands. Also shown are other organs that contain endocrine cells and associated structures.

Endocrine glands secrete hormones, which circulating blood delivers to target tissues.

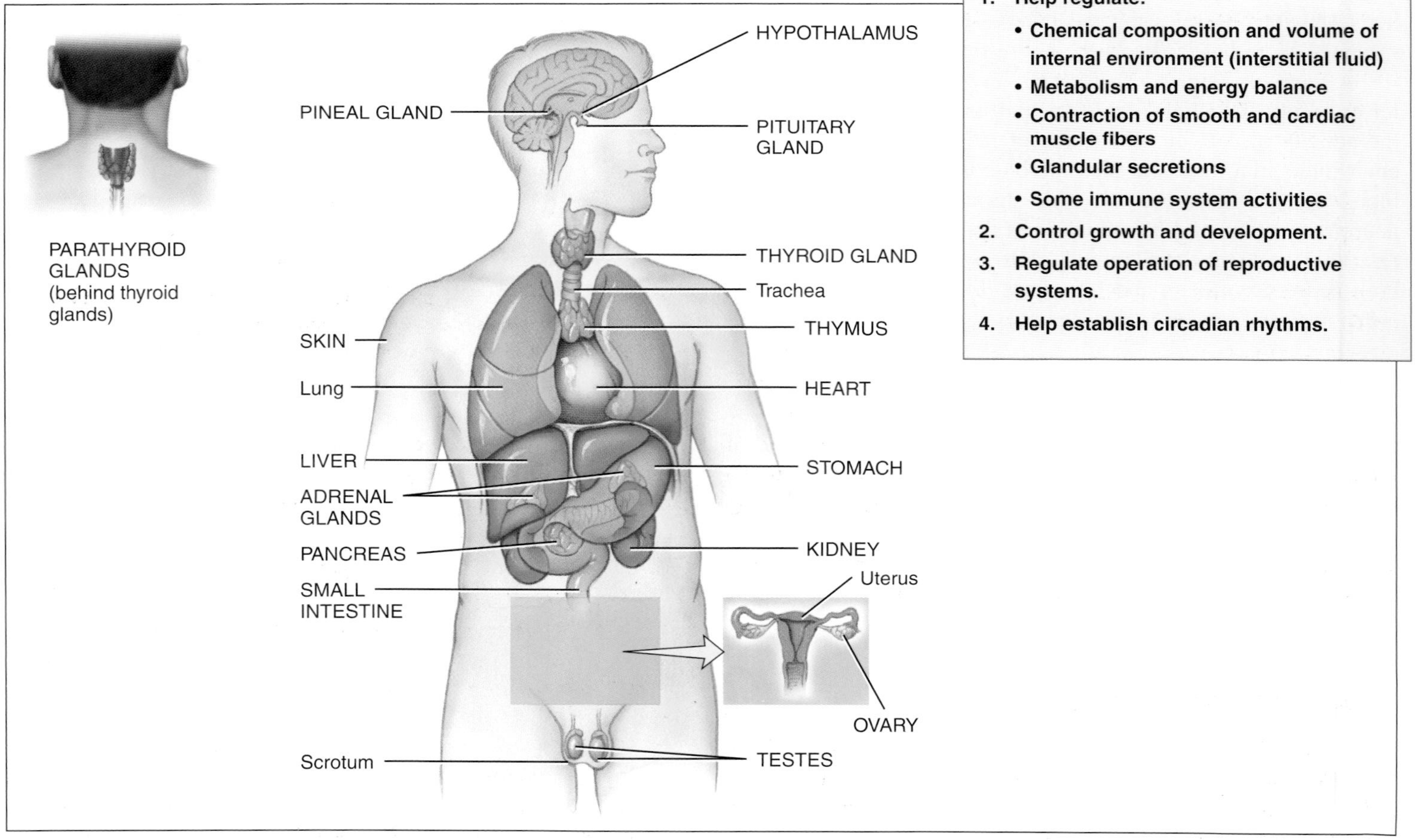

Functions of Hormones

1. **Help regulate:**
   - **Chemical composition and volume of internal environment (interstitial fluid)**
   - **Metabolism and energy balance**
   - **Contraction of smooth and cardiac muscle fibers**
   - **Glandular secretions**
   - **Some immune system activities**
2. **Control growth and development.**
3. **Regulate operation of reproductive systems.**
4. **Help establish circadian rhythms.**

What is the basic difference between endocrine glands and exocrine glands?

HORMONE ACTIVITY

▶ **OBJECTIVES**

Describe how hormones interact with target-cell receptors.

Compare the two chemical classes of hormones based on their solubility.

The Role of Hormone Receptors

Although a given hormone travels throughout the body in the blood, it affects only specific target cells. Hormones, like neurotransmitters, influence their target cells by chemically binding to specific protein or glycoprotein **receptors.** Only the target cells for a given hormone have receptors that bind and recognize that hormone. For example, thyroid-stimulating hormone (TSH) binds to receptors on cells of the thyroid gland, but it does not bind to cells of the ovaries because ovarian cells do not have TSH receptors.

Receptors, like other cellular proteins, are constantly being synthesized and broken down. Generally, a target cell has 2000 to 100,000 receptors for a particular hormone. If a hormone is present in excess, the number of target-cell receptors may decrease—an effect called **down-regulation.** For example, when certain cells of the testes are exposed to a high concentration of luteinizing hormone (LH), the number of LH receptors decreases. Down-regulation makes a target cell *less sensitive* to a hormone. In contrast, when a hormone is deficient, the number of receptors may increase. This phenomenon, known as **up-regulation,** makes a target cell *more sensitive* to a hormone.

Blocking Hormone Receptors

Synthetic hormones that *block the receptors* for some naturally occurring hormones are available as drugs. For example, RU486 (mifepristone), which is used to induce abortion, binds to the receptors for progesterone (a female sex hormone) and prevents progesterone from exerting its normal effect, in this case preparing the lining of the uterus for implantation. When RU486 is given to a pregnant woman, the uterine conditions needed for nurturing an embryo are not maintained, embryonic development stops, and the embryo is sloughed off along with the uterine lining. This example illustrates an important endocrine principle: If a hormone is prevented from interacting with its receptors, the hormone cannot perform its normal functions. ■

Circulating and Local Hormones

Most endocrine hormones are **circulating hormones**—they pass from the secretory cells that make them into interstitial fluid and then into the blood (Figure 18.2a). Other hormones, termed **local hormones,** act locally on neighboring cells or on the same cell that secreted them without first entering the bloodstream (Figure 18.2b). Local hormones that act on neighboring cells are called **paracrines** (*para-* = beside or near), and those that act on the same cell that secreted them are called **autocrines** (*auto-* = self). One example of a local hormone is interleukin 2 (IL-2), which is released by helper T cells (a type of white blood cell) during immune responses (see Chapter 22). IL-2 helps activate other nearby immune cells, a paracrine effect. But, it also acts as an autocrine by stimulating the same cell that released it to proliferate. This action generates more helper T cells that can secrete even more IL-2 and thus strengthen the immune response. Another example of a local hormone is the gas nitric oxide (NO), which is released by endothelial cells lining blood vessels. NO causes relaxation of nearby smooth muscle fibers in blood vessels, which in turn causes vasodilation (increase in blood vessel diameter). The effects of such vasodilation range

Figure 18.2 **Comparison between circulating hormones and local hormones (autocrines and paracrines).**

Circulating hormones are carried through the bloodstream to act on distant target cells. Paracrines act on neighboring cells and autocrines act on the same cell that produced them.

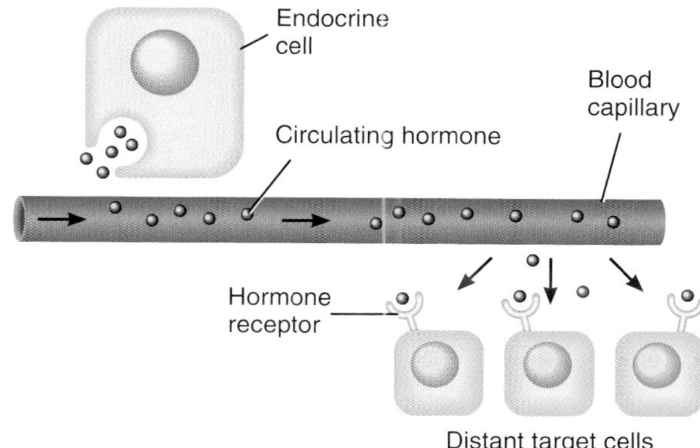

(a) Circulating hormones

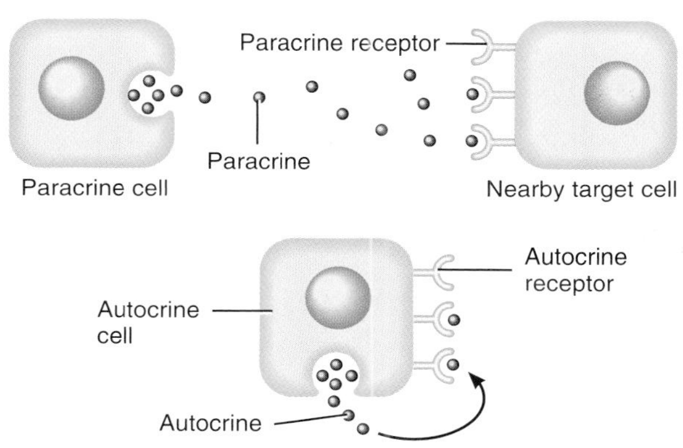

(b) Local hormones (paracrines and autocrines)

In the stomach, one stimulus for secretion of hydrochloric acid by parietal cells is the release of histamine by neighboring mast cells. Is histamine an autocrine or a paracrine in this situation?

from a lowering of blood pressure to erection of the penis in males. The drug *Viagra*® (sildenafil) enhances the effects stimulated by nitric oxide in the penis.

Local hormones usually are inactivated quickly; circulating hormones may linger in the blood and exert their effects for a few minutes or occasionally for a few hours. In time, circulating hormones are inactivated by the liver and excreted by the kidneys. In cases of kidney or liver failure, excessive levels of hormones may build up in the blood.

Chemical Classes of Hormones

Chemically, hormones can be divided into two broad classes: those that are soluble in lipids, and those that are soluble in water. This chemical classification is also useful functionally because the ways in which the two classes exert their effects are different.

Lipid-soluble Hormones

The lipid-soluble hormones include steroid hormones, thyroid hormones, and nitric oxide.

1. **Steroid hormones** are derived from cholesterol. Each steroid hormone is unique due to the presence of different chemical groups attached at various sites on the four rings at the core of its structure. These small differences allow for a large diversity of functions.

2. Two **thyroid hormones** (T_3 and T_4) are synthesized by attaching iodine to the amino acid tyrosine. The benzene ring of tyrosine plus the attached iodines make T_3 and T_4 very lipid soluble.

3. The gas **nitric oxide (NO)** is both a hormone and a neurotransmitter. Its synthesis is catalyzed by the enzyme nitric oxide synthase.

Water-soluble Hormones

The water-soluble hormones include amine hormones, peptide and protein hormones, and eicosanoid hormones.

1. **Amine hormones** are synthesized by decarboxylating (removing a molecule of CO_2) and otherwise modifying certain amino acids. They are called amines because they retain an amino group ($—NH_3^+$). The catecholamines—epinephrine, norepinephrine, and dopamine—are synthesized by modifying the amino acid tyrosine. Histamine is synthesized from the amino acid histidine by mast cells and platelets. Serotonin and melatonin are derived from tryptophan.

2. **Peptide hormones** and **protein hormones** are amino acid polymers. The smaller peptide hormones consist of chains of 3 to 49 amino acids; the larger protein hormones include 50 to 200 amino acids. Examples of peptide hormones are antidiuretic hormone and oxytocin; protein hormones include human growth hormone and insulin. Several of the protein hormones, such as thyroid-stimulating hormone, have attached carbohydrate groups and thus are **glycoprotein hormones.**

3. The **eicosanoid hormones** (ī-KŌ-sa-noid; *eicos-* = twenty forms; *-oid* = resembling) are derived from arachidonic acid, a 20-carbon fatty acid. The two major types of eicosanoids are **prostaglandins** and **leukotrienes.** The eicosanoids are important local hormones, and they may act as circulating hormones as well.

Table 18.2 summarizes the classes of lipid-soluble and water-soluble hormones and provides an overview of the major hormones and their sites of secretion.

Hormone Transport in the Blood

Most water-soluble hormone molecules circulate in the watery blood plasma in a "free" form (not attached to other molecules), but most lipid-soluble hormone molecules are bound to **transport proteins.** The transport proteins, which are synthesized by cells in the liver, have three functions:

1. They make lipid-soluble hormones temporarily water soluble, thus increasing their solubility in blood.

2. They retard passage of small hormone molecules through the filtering mechanism in the kidneys, thus slowing the rate of hormone loss in the urine.

3. They provide a ready reserve of hormone, already present in the bloodstream.

In general, 0.1–10% of the molecules of a lipid-soluble hormone are not bound to a transport protein. This **free fraction** diffuses out of capillaries, binds to receptors, and triggers responses. As free hormone molecules leave the blood and bind to their receptors, transport proteins release new ones to replenish the free fraction.

Administering Hormones

Both steroid hormones and thyroid hormones are effective when taken by mouth. They are not split apart during digestion and easily cross the intestinal lining because they are lipid-soluble. By contrast, peptide and protein hormones, such as insulin, are not effective oral medications because digestive enzymes destroy them by breaking their peptide bonds. This is why people who need insulin must take it by injection. ■

▶ CHECKPOINT

3. What is the difference between down-regulation and up-regulation?
4. Identify the chemical classes of hormones, and give an example of each.
5. How are hormones transported in the blood?

TABLE 18.2 Summary of Hormones by Chemical Class

Chemical Class	Hormones	Site of Secretion
Lipid-soluble		
Steroid hormones (Aldosterone)	Aldosterone, cortisol, and androgens.	Adrenal cortex.
	Calcitriol.	Kidneys.
	Testosterone.	Testes.
	Estrogens and progesterone.	Ovaries.
Thyroid hormones (Triiodothyronine (T_3))	T_3 (triiodothyronine) and T_4 (thyroxine).	Thyroid gland (follicular cells).
Gas	Nitric oxide (NO).	Endothelial cells lining blood vessels.
Water-soluble		
Amines (Norepinephrine)	Epinephrine and norepinephrine (catecholamines).	Adrenal medulla.
	Melatonin.	Pineal gland.
	Histamine.	Mast cells in connective tissues.
	Serotonin.	Platelets in blood.
Peptides and proteins (Oxytocin: Glutamine–Isoleucine–Tyrosine–Cysteine–S–S–Cysteine–Asparagine–Glutamine... Proline–Leucine–Glycine–NH_2)	All hypothalamic releasing and inhibiting hormones.	Hypothalamus.
	Oxytocin, antidiuretic hormone.	Posterior pituitary.
	Human growth hormone, thyroid-stimulating hormone, adrenocorticotropic hormone, follicle-stimulating hormone, luteinizing hormone, prolactin, melanocyte-stimulating hormone.	Anterior pituitary.
	Insulin, glucagon, somatostatin, pancreatic polypeptide.	Pancreas.
	Parathyroid hormone.	Parathyroid glands.
	Calcitonin.	Thyroid gland (parafollicular cells).
	Gastrin, secretin, cholecystokinin, GIP (glucose-dependent insulinotropic peptide).	Stomach and small intestine (enteroendocrine cells).
	Erythropoietin.	Kidneys.
	Leptin.	Adipose tissue.
Eicosanoids (A leukotriene (LTB_4))	Prostaglandins, leukotrienes.	All cells except red blood cells.

MECHANISMS OF HORMONE ACTION

▶ **OBJECTIVE**

Describe the two general mechanisms of hormone action.

The response to a hormone depends on both the hormone and the target cell. Various target cells respond differently to the same hormone. Insulin, for example, stimulates synthesis of glycogen in liver cells and synthesis of triglycerides in adipose cells.

The response to a hormone is not always the synthesis of new molecules, as is the case for insulin. Other hormonal effects include changing the permeability of the plasma membrane, stimulating transport of a substance into or out of the target cells, altering the rate of specific metabolic reactions, or causing contraction of smooth muscle or cardiac muscle. In part, these varied effects of hormones are possible because a single hormone can set in motion several different cellular responses. However, a hormone must first "announce its arrival" to a target cell by binding to its receptors. The receptors for lipid-soluble hormones are located inside target cells. The receptors for water-soluble hormones are part of the plasma membrane of target cells.

Action of Lipid-soluble Hormones

As you just learned, lipid-soluble hormones, including steroid hormones and thyroid hormones, bind to receptors within target cells. Their mechanism of action is as follows (Figure 18.3):

1. A free lipid-soluble hormone molecule diffuses from the blood, through interstitial fluid, and through the lipid bilayer of the plasma membrane into a cell.
2. If the cell is a target cell, the hormone binds to and activates receptors located within the cytosol or nucleus. The activated receptor-hormone complex then alters gene expression: It turns specific genes of the nuclear DNA on or off.
3. As the DNA is transcribed, new messenger RNA (mRNA) forms, leaves the nucleus, and enters the cytosol. There, it directs synthesis of a new protein, often an enzyme, on the ribosomes.
4. The new proteins alter the cell's activity and cause the responses typical of that hormone.

Action of Water-soluble Hormones

Because amine, peptide, protein, and eicosanoid hormones are not lipid-soluble, they cannot diffuse through the lipid bilayer of the plasma membrane and bind to receptors inside target cells. Instead, water-soluble hormones bind to receptors that protrude from the target cell surface. The receptors are integral transmembrane proteins in the plasma membrane. When a water-soluble hormone binds to its receptor at the outer surface of the plasma membrane, it acts as the **first messenger.** The first messenger (the hormone) then causes production of a **second messenger** inside the cell, where specific hormone-stimulated responses take place. One common second messenger is **cyclic AMP (cAMP).** Neurotransmitters, neuropeptides, and several sensory transduction mechanisms (for example, vision; see Figure 17.13 on page 591) also act via second-messenger systems.

The action of a typical water-soluble hormone occurs as follows (Figure 18.4):

1. A water-soluble hormone (the first messenger) diffuses from the blood through interstitial fluid and then binds to its receptor at the exterior surface of a target cell's plasma

Figure 18.3 Mechanism of action of the lipid-soluble steroid hormones and thyroid hormones.

Lipid-soluble hormones bind to receptors inside target cells.

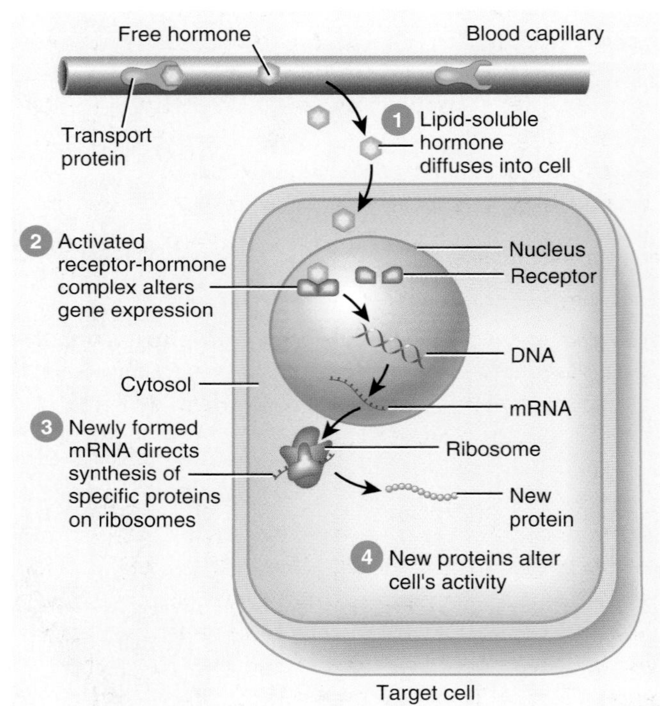

What is the action of the receptor-hormone complex?

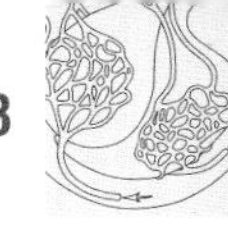

membrane. The hormone-receptor complex activates a membrane protein called a **G protein.** The activated G protein in turn activates **adenylate cyclase.**

2 Adenylate cyclase converts ATP into cyclic AMP (cAMP). Because the enzyme's active site is on the inner surface of the plasma membrane, this reaction occurs in the cytosol of the cell.

3 Cyclic AMP (the second messenger) activates one or more protein kinases, which may be free in the cytosol or bound to the plasma membrane. A **protein kinase** is an enzyme that phosphorylates (adds a phosphate group to) other cellular proteins (such as enzymes). The donor of the phosphate group is ATP, which is converted to ADP.

4 Activated protein kinases phosphorylate one or more cellular proteins. Phosphorylation activates some of these proteins and inactivates others, rather like turning a switch on or off.

5 Phosphorylated proteins, in turn, cause reactions that produce physiological responses. Different protein kinases exist within different target cells and within different organelles of the same target cell. Thus, one protein kinase might trigger glycogen synthesis, a second might cause the breakdown of triglyceride, a third may promote protein synthesis, and so forth. As noted in step 4, phosphorylation by a protein kinase can also inhibit certain proteins. For example, some of the kinases unleashed when epinephrine binds to liver cells inactivate an enzyme needed for glycogen synthesis.

6 After a brief period, an enzyme called **phosphodiesterase** inactivates cAMP. Thus, the cell's response is turned off unless new hormone molecules continue to bind to their receptors in the plasma membrane.

The binding of a hormone to its receptor activates many G-protein molecules, which in turn activate molecules of adenylate cyclase (as noted in step 1). Unless they are further stimulated by the binding of more hormone molecules to receptors, G proteins slowly inactivate, thus decreasing the activity of adenylate cyclase and helping to stop the hormone response. G proteins are a common feature of most second-messenger systems.

Cyclic AMP and other second messengers alter a cell's function in specific ways. For example, an increase in cAMP causes adipose cells to break down triglycerides and release fatty acids more rapidly, but this same molecule stimulates thyroid cells to secrete more thyroid hormone. Many hormones exert at least some of their physiological effects through the *increased* synthesis of cAMP. Examples include antidiuretic hormone (ADH), thyroid-stimulating hormone (TSH), adrenocorticotropic hormone (ACTH), glucagon, epinephrine, and hypothalamic-releasing hormones. In other cases, such as growth hormone-inhibiting hormone (GHIH), the level of cyclic AMP *decreases* in response to the binding of a hormone to its receptor.

Other second messengers include calcium ions (Ca^{2+}), cGMP (cyclic guanosine monophosphate, a cyclic nucleotide similar to cAMP), inositol trisphosphate (IP_3), and diacylglycerol (DAG). Nitric oxide, a lipid-soluble hormone, exerts its effect inside smooth muscle fibers by activating guanylyl cyclase. This enzyme, in turn, converts guanosine triphosphate (GTP) to cGMP, which causes calcium ions to enter storage areas of the smooth muscle fiber. The lowered Ca^{2+} level in the cytosol then causes muscle relaxation. A given hormone may use different second messengers in different target cells.

Figure 18.4 Mechanism of action of the water-soluble hormones (amines, peptides, proteins, and eicosanoids).

Water-soluble hormones bind to receptors embedded in the plasma membranes of target cells.

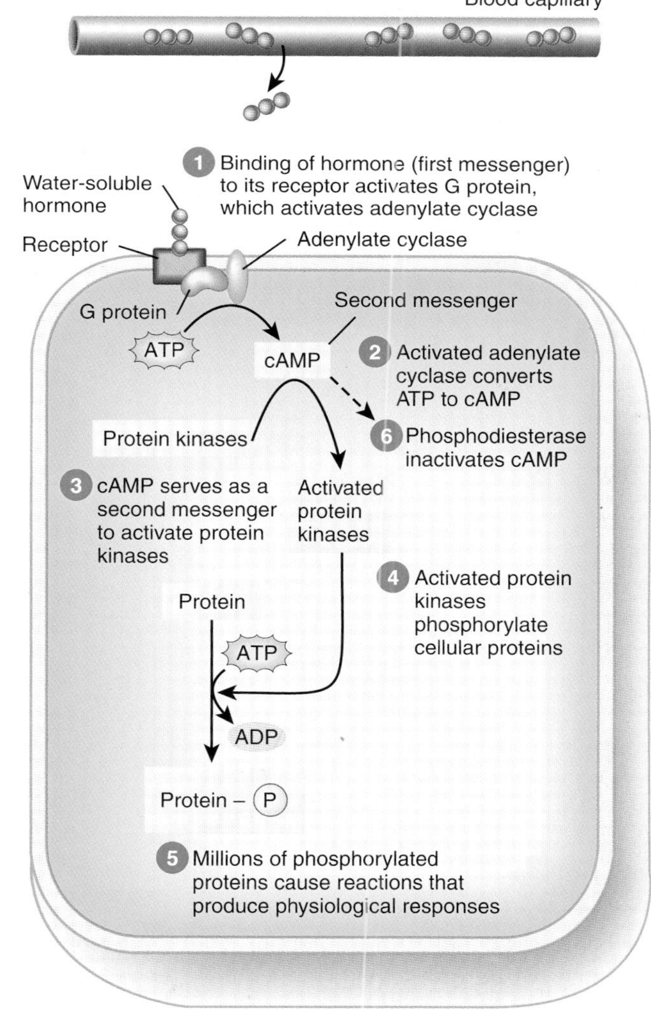

Why is cAMP called a "second messenger"?

Hormones that bind to plasma membrane receptors can induce their effects at very low concentrations because they initiate a cascade or chain reaction, each step of which multiplies or amplifies the initial effect. For example, the binding of a single molecule of epinephrine to its receptor on a liver cell may activate a hundred or so G proteins, each of which activates an adenylate cyclase molecule. If each adenylate cyclase produces even 1000 cAMP, then 100,000 of these second messengers will be liberated inside the cell. Each cAMP may activate a protein kinase, which in turn can act on hundreds or thousands of substrate molecules. Some of the kinases phosphorylate and activate a key enzyme needed for glycogen breakdown. The end result of the binding of a single molecule of epinephrine to its receptor is the breakdown of millions of glycogen molecules into glucose monomers.

Cholera Toxin and G Proteins

The toxin produced by cholera bacteria is deadly. It produces massive watery diarrhea, and an infected person can rapidly die from the resulting dehydration. The cholera toxin modifies the G proteins in intestinal epithelial cells so they become locked in an activated state. As a result, the intracellular cAMP concentration skyrockets. One of the effects of cAMP in these cells is to stimulate an active transport pump that ejects chloride ions (Cl^-) from the cells into the lumen of the intestines. Water follows the Cl^- by osmosis, and positively charged sodium ions follow the negatively charged Cl^-. Thus, cholera toxin causes a huge outflow of Na^+, Cl^-, and water into the feces. Treatment consists of replacement of lost fluids, either intravenously or by mouth (oral rehydration therapy), plus antibiotic therapy with tetracycline. ■

Hormone Interactions

The responsiveness of a target cell to a hormone depends on (1) the hormone's concentration, (2) the abundance of the target cell's hormone receptors, and (3) influences exerted by other hormones. A target cell responds more vigorously when the level of a hormone rises or when it has more receptors (up-regulation). In addition, the actions of some hormones on target cells require a simultaneous or recent exposure to a second hormone. In such cases, the second hormone is said to have a **permissive effect.** For example, epinephrine alone only weakly stimulates lipolysis (the breakdown of triglycerides), but when small amounts of thyroid hormones (T_3 and T_4) are present, the same amount of epinephrine stimulates lipolysis much more powerfully. Sometimes the permissive hormone increases the number of receptors for the other hormone, and sometimes it promotes the synthesis of an enzyme required for the expression of the other hormone's effects.

When the effect of two hormones acting together is greater or more extensive than the effect of each hormone acting alone, the two hormones are said to have a **synergistic effect.** For example, normal development of oocytes in the ovaries requires both follicle-stimulating hormone from the anterior pituitary and estrogens from the ovaries. Neither hormone alone is sufficient.

When one hormone opposes the actions of another hormone, the two hormones are said to have **antagonistic effects.** An example of an antagonistic pair of hormones is insulin, which promotes synthesis of glycogen by liver cells, and glucagon, which stimulates breakdown of glycogen in the liver.

► CHECKPOINT

6. What factors determine the responsiveness of a target cell to a hormone?
7. What are the differences among permissive effects, synergistic effects, and antagonistic effects of hormones?

CONTROL OF HORMONE SECRETION

► OBJECTIVE

Describe the mechanisms of control of hormone secretion.

The release of most hormones occurs in short bursts, with little or no secretion between bursts. When stimulated, an endocrine gland will release its hormone in more frequent bursts, increasing the concentration of the hormone in the blood. In the absence of stimulation, the blood level of the hormone decreases. Regulation of secretion normally prevents overproduction or underproduction of any given hormone.

Hormone secretion is regulated by (1) signals from the nervous system, (2) chemical changes in the blood, and (3) other hormones. For example, nerve impulses to the adrenal medullae regulate the release of epinephrine; blood Ca^{2+} level regulates the secretion of parathyroid hormone; and a hormone from the anterior pituitary (adrenocorticotropic hormone) stimulates the release of cortisol by the adrenal cortex. Most hormonal regulatory systems work via negative feedback (see Figure 1.3 on page 10), but a few operate via positive feedback (see Figure 1.4 on page 11). For example, during childbirth, the hormone oxytocin stimulates contractions of the uterus, and uterine contractions, in turn, stimulate more oxytocin release, a positive feedback effect.

Now that you have a general understanding of the roles of hormones in the endocrine system, we turn to discussions of the various endocrine glands and the hormones they secrete.

► CHECKPOINT

8. What three types of signals control hormone secretion?

HYPOTHALAMUS AND PITUITARY GLAND

▶ **OBJECTIVES**

Describe the locations of and relationships between the hypothalamus and pituitary gland.

Describe the location, histology, hormones, and functions of the anterior and posterior pituitary.

For many years, the **pituitary gland** (pi-TOO-i-tār-ē) or **hypophysis** (hī-POF-i-sis) was called the "master" endocrine gland because it secretes several hormones that control other endocrine glands. We now know that the pituitary gland itself has a master—the **hypothalamus.** This small region of the brain below the thalamus is the major link between the nervous and endocrine systems. It receives input from the limbic system, cerebral cortex, thalamus, and reticular activating system. It also receives sensory signals from internal organs and from the retina. Painful, stressful, and emotional experiences all cause changes in hypothalamic activity. In turn, the hypothalamus controls the autonomic nervous system and regulates body temperature, thirst, hunger, sexual behavior, and defensive reactions such as fear and rage.

The hypothalamus is an important regulatory center in the nervous system as well as a crucial endocrine gland. Cells in the hypothalamus synthesize at least nine different hormones, and the pituitary gland secretes seven. Together, these 16 hormones play important roles in the regulation of virtually all aspects of growth, development, metabolism, and homeostasis.

The pituitary gland is a pea-shaped structure that measures 1–1.5 cm (0.5 in.) in diameter and lies in the hypophyseal fossa of the sella turcica of the sphenoid bone. It attaches to the hypothalamus by a stalk, the **infundibulum** (= a funnel; Figure 18.5), and has two anatomically and functionally separate lobes. The **anterior pituitary (anterior lobe),** also called the **adenohypophysis,** accounts for about 75% of the total weight of the gland. The anterior pituitary consists of two parts in an adult: The **pars distalis** is the larger portion, and the **pars tuberalis** forms a sheath around the infundibulum. The **posterior pituitary (posterior lobe),** also called the **neurohypophysis,** also consists of two parts: the **pars nervosa,** the larger bulbar portion, and the infundibulum. The posterior pituitary contains axons and axon terminals of more than 10,000 neurons whose cell bodies are located in the supraoptic and paraventricular nuclei of the hypothalamus (see Figure 14.10 on page 490). The axon terminals in the posterior pituitary are associated with specialized neuroglia called **pituicytes** (pi-TOO-i-sītz). These cells have a supporting role similar to that of astrocytes (see Chapter 12).

A third region of the pituitary gland called the **pars intermedia** atrophies during human fetal development and ceases to exist as a separate lobe in adults (see Figure 18.21b). However, some of its cells migrate into adjacent parts of the anterior pituitary, where they persist.

Anterior Pituitary

The **anterior pituitary** or **adenohypophysis** (ad′-e-nō-hī-POF-i-sis; *adeno-* = gland; *-hypophysis* = undergrowth) secretes hormones that regulate a wide range of bodily activities, from growth to reproduction. Release of anterior pituitary hormones is stimulated by **releasing hormones** and suppressed by **inhibiting hormones** from the hypothalamus. Thus, the hypothalamic hormones are an important link between the nervous and endocrine systems.

Hypophyseal Portal System

Hypothalamic hormones reach the anterior pituitary through a portal system. Usually, blood passes from the heart through an artery to a capillary to a vein and back to the heart. In a *portal system,* blood flows from one capillary network into a portal vein, and then into a second capillary network without passing through the heart. The name of the portal system indicates the location of the second capillary network. In the **hypophyseal portal system** (hī′-pō-FIZ-ē-al), blood flows from capillaries in the hypothalamus into portal veins that carry blood to capillaries of the anterior pituitary.

The **superior hypophyseal arteries,** branches of the internal carotid arteries, bring blood into the hypothalamus (Figure 18.5). At the junction of the median eminence of the hypothalamus and the infundibulum, these arteries divide into a capillary network called the **primary plexus of the hypophyseal portal system.** From the primary plexus, blood drains into the **hypophyseal portal veins** that pass down the outside of the infundibulum. In the anterior pituitary, the hypophyseal portal veins divide again and form another capillary network called the **secondary plexus of the hypophyseal portal system.**

Near the median eminence and above the optic chiasm are clusters of specialized neurons, called **neurosecretory cells.** They synthesize the hypothalamic releasing and inhibiting hormones in their cell bodies and package the hormones inside vesicles, which reach the axon terminals by axonal transport. Nerve impulses stimulate the vesicles to undergo exocytosis. The hormones then diffuse into the primary plexus of the hypophyseal portal system. Quickly, the hypothalamic hormones flow with the blood through the portal veins and into the secondary plexus. This direct route permits hypothalamic hormones to act immediately on anterior pituitary cells, before the hormones are diluted or destroyed in the general circulation. Hormones secreted by anterior pituitary cells pass into the secondary plexus capillaries, which drain into the anterior hypophyseal veins and out into the general circulation. Anterior pituitary hormones then travel to target tissues throughout the body.

Types of Anterior Pituitary Cells

Five types of anterior pituitary cells—somatotrophs, thyrotrophs, gonadotrophs, lactotrophs, and corticotrophs—secrete seven hormones (Figure 18.5c and Table 18.3 on page 627):

Figure 18.5 Hypothalamus and pituitary gland, and their blood supply. Figure 18.5b indicates that releasing and inhibiting hormones synthesized by hypothalamic neurosecretory cells are transported within axons and released at the axon terminals. The hormones diffuse into capillaries of the primary plexus of the hypophyseal portal system and are carried by the hypophyseal portal veins to the secondary plexus of the hypophyseal portal system for distribution to target cells in the anterior pituitary.

Hypothalamic hormones are an important link between the nervous and endocrine systems.

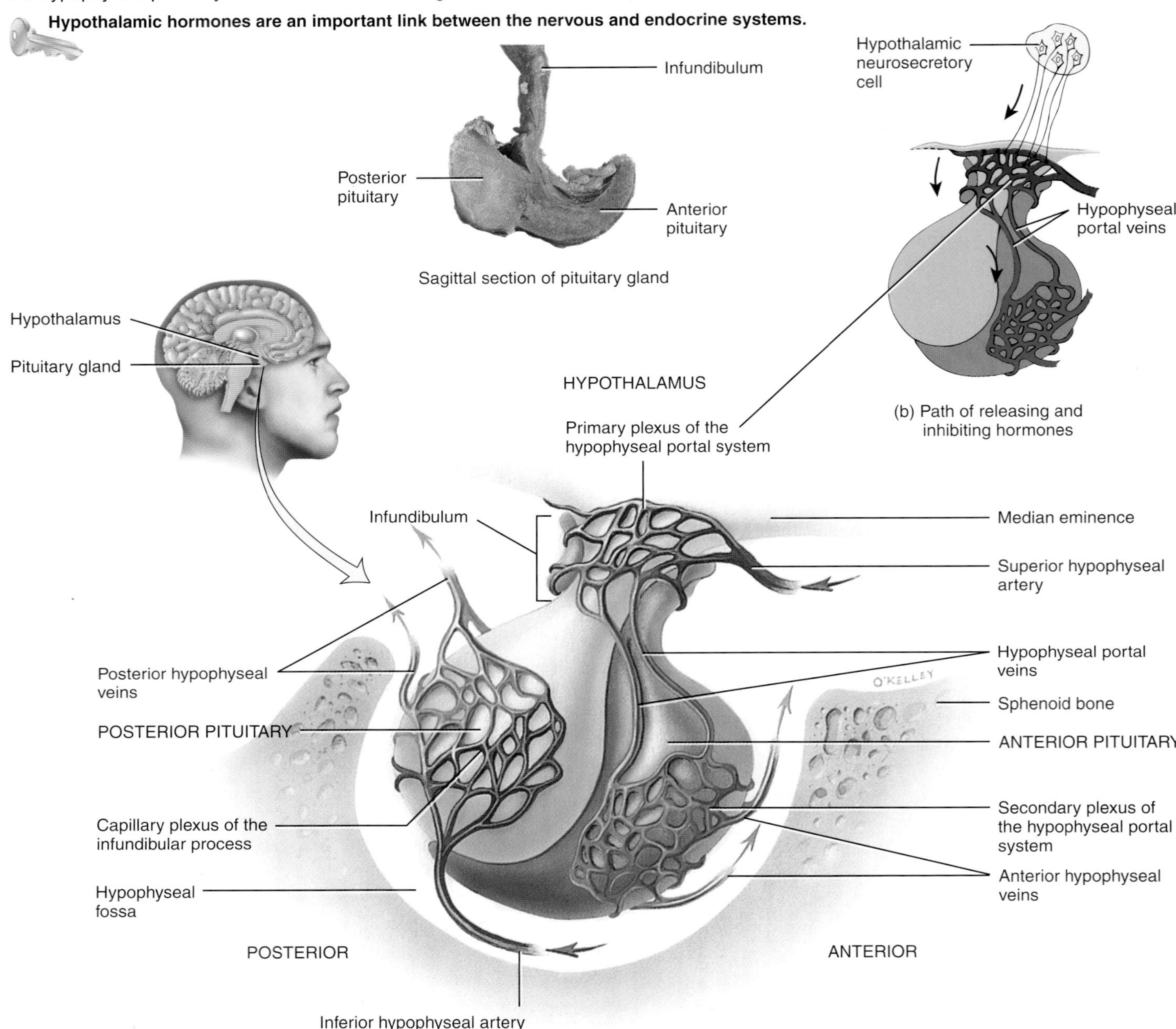

(a) Relationship of the hypothalamus to the pituitary gland

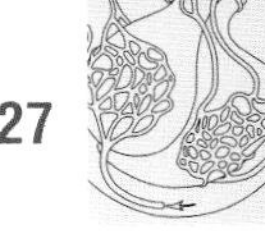

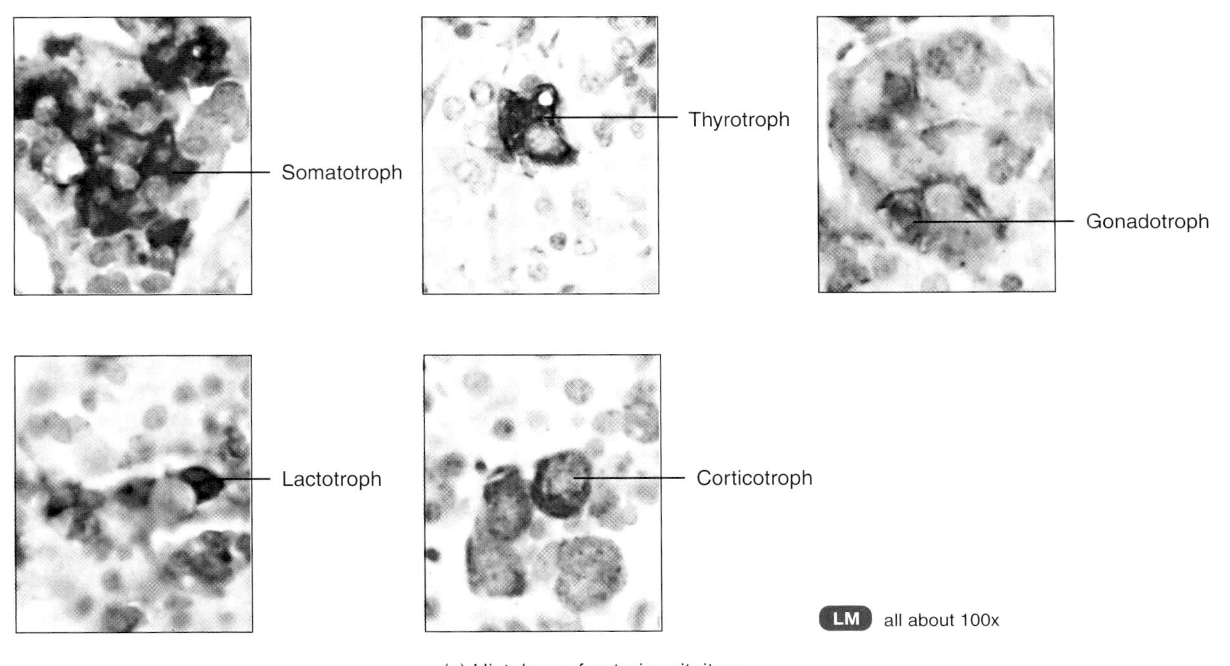

(c) Histology of anterior pituitary

What is the functional importance of the hypophyseal portal veins?

1. **Somatotrophs** secrete **human growth hormone (hGH)** or **somatotropin** (sō′-ma-tō-TRŌ-pin; *somato-* = body; *-tropin* = change). Human growth hormone in turn stimulates several tissues to secrete **insulinlike growth factors,** hormones that stimulate general body growth and regulate aspects of metabolism.
2. **Thyrotrophs** secrete **thyroid-stimulating hormone (TSH)** or **thyrotropin** (thī-rō-TRŌ-pin; *thyro-* = pertaining to the thyroid gland). TSH controls the secretions and other activities of the thyroid gland.
3. **Gonadotrophs** (*gonado-* = seed) secrete two hormones: **follicle-stimulating hormone (FSH)** and **luteinizing hormone (LH)** (LOO-tē-in′-īz-ing). FSH and LH both act on the gonads. They stimulate secretion of estrogens and progesterone and the

TABLE 18.3 Hormones of the Anterior Pituitary

Hormone	Secreted by	Releasing Hormone (Stimulates Secretion)	Inhibiting Hormone (Suppresses Secretion)
Human growth hormone (hGH) or **somatotropin**	Somatotrophs.	Growth hormone–releasing hormone (GHRH), also known as somatocrinin.	Growth hormone–inhibiting hormone (GHIH), also known as somatostatin.
Thyroid-stimulating hormone (TSH) or **thyrotropin**	Thyrotrophs.	Thyrotropin-releasing hormone (TRH).	Growth hormone–inhibiting hormone (GHIH).
Follicle-stimulating hormone (FSH)	Gonadotrophs.	Gonadotropin-releasing hormone (GnRH).	—
Luteinizing hormone (LH)	Gonadotrophs.	Gonadotropin-releasing hormone (GnRH).	—
Prolactin (PRL)	Lactotrophs.	Prolactin-releasing hormone (PRH); TRH.	Prolactin-inhibiting hormone (PIH), which is dopamine.
Adrenocorticotropic hormone (ACTH) or **corticotropin**	Corticotrophs.	Corticotropin-releasing hormone (CRH)	—
Melanocyte-stimulating hormone	Corticotrophs.	Corticotropin-releasing hormone (CRH).	Dopamine.

maturation of oocytes in the ovaries, and they stimulate sperm production and secretion of testosterone in the testes.

4. **Lactotrophs** (*lacto-* = milk) secrete **prolactin (PRL),** which initiates milk production in the mammary glands.

5. **Corticotrophs** secrete **adrenocorticotropic hormone (ACTH)** or **corticotropin** (kor′-ti-kō-TRŌ-pin; *cortico-* = rind or bark), which stimulates the adrenal cortex to secrete glucocorticoids such as cortisol. Some corticotrophs, remnants of the pars intermedia, also secrete **melanocyte-stimulating hormone (MSH).**

Hormones that influence another endocrine gland are called **tropic hormones** or **tropins.** Several of the anterior pituitary hormones are tropins. The two **gonadotropins,** follicle stimulating hormone and luteinizing hormone, specifically regulate the functions of the gonads (ovaries and testes). Thyrotropin stimulates the thyroid gland, and corticotropin acts on the cortex of the adrenal gland.

Control of Secretion by the Anterior Pituitary

Secretion of anterior pituitary hormones is regulated in two ways. First, neurosecretory cells in the hypothalamus secrete five releasing hormones, which stimulate secretion of anterior pituitary hormones, and two inhibiting hormones, which suppress secretion of anterior pituitary hormones (Table 18.3). Second, negative feedback in the form of hormones released by target glands decreases secretions of three types of anterior pituitary cells (Figure 18.6). In such negative feedback loops, the secretory activity of thyrotrophs, gonadotrophs, and corticotrophs decreases when blood levels of their target gland hormones rise. For example, adrenocorticotropic hormone (ACTH) stimulates the cortex of the adrenal gland to secrete glucocorticoids, mainly cortisol. In turn, an elevated blood level of cortisol decreases secretion of both corticotropin and corticotropin-releasing hormone (CRH) by suppressing the activity of the anterior pituitary corticotrophs and hypothalamic neurosecretory cells.

Figure 18.6 Negative feedback regulation of hypothalamic neurosecretory cells and anterior pituitary corticotrophs. Solid green arrows show stimulation of secretions; dashed red lines show inhibition of secretion via negative feedback.

Cortisol secreted by the adrenal cortex suppresses secretion of CRH and ACTH.

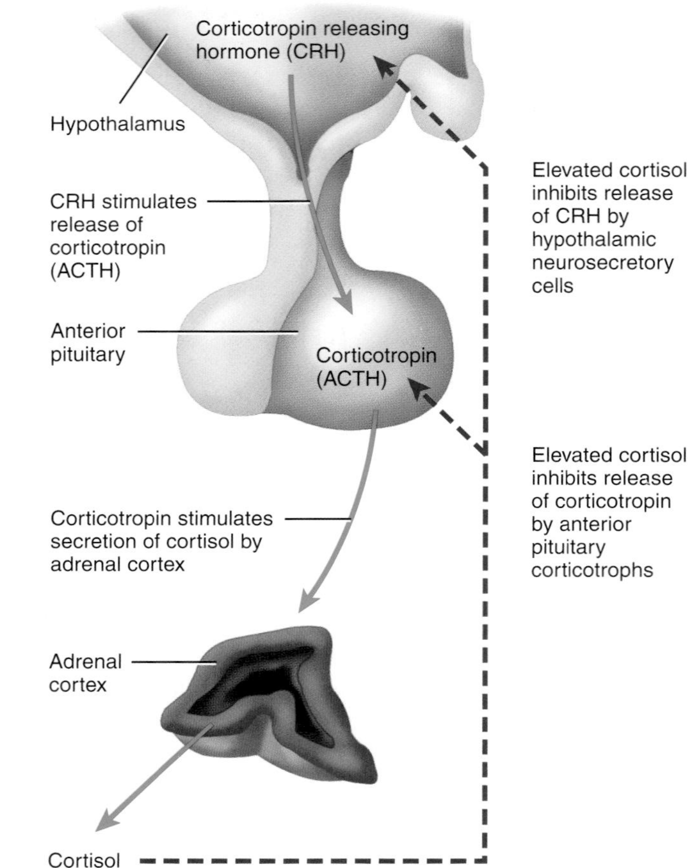
Corticotropin releasing hormone (CRH)
Hypothalamus
CRH stimulates release of corticotropin (ACTH)
Anterior pituitary
Corticotropin (ACTH)
Elevated cortisol inhibits release of CRH by hypothalamic neurosecretory cells
Elevated cortisol inhibits release of corticotropin by anterior pituitary corticotrophs
Corticotropin stimulates secretion of cortisol by adrenal cortex
Adrenal cortex
Cortisol

Which other target gland hormones suppress secretion of hypothalamic and anterior pituitary hormones by negative feedback?

Human Growth Hormone and Insulinlike Growth Factors

Somatotrophs are the most numerous cells in the anterior pituitary, and human growth hormone (hGH) is the most plentiful anterior pituitary hormone. The main function of hGH is to promote synthesis and secretion of small protein hormones called **insulinlike growth factors (IGFs)** or **somatomedins.** In response to human growth hormone, cells in the liver, skeletal muscles, cartilage, bones, and other tissues secrete IGFs, which may either enter the bloodstream from the liver or act locally in other tissues as autocrines or paracrines. IGFs cause cells to grow and multiply by increasing uptake of amino acids into cells and accelerating protein synthesis. IGFs also decrease the breakdown of proteins and the use of amino acids for ATP production. Due to these effects of the IGFs, human growth hormone increases the growth rate of the skeleton and skeletal muscles during childhood and the teenage years. In adults, human growth hormone and IGFs help maintain the mass of muscles and bones and promote healing of injuries and tissue repair.

Insulinlike growth factors also enhance lipolysis in adipose tissue, which results in increased use of the released fatty acids for ATP production by body cells. In addition to affecting protein and lipid metabolism, human growth hormone and IGFs

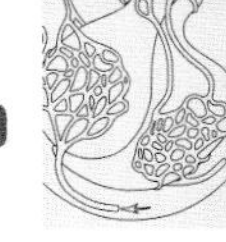

influence carbohydrate metabolism by decreasing glucose uptake, which decreases the use of glucose for ATP production by most body cells. This action spares glucose so that it is available to neurons for ATP production in times of glucose scarcity. IGFs and human growth hormone may also stimulate liver cells to release glucose into the blood.

Somatotrophs in the anterior pituitary release bursts of human growth hormone every few hours, especially during sleep. Their secretory activity is controlled mainly by two hypothalamic hormones: (1) growth hormone-releasing hormone (GHRH) promotes secretion of human growth hormone, and (2) growth hormone-inhibiting hormone (GHIH) suppresses it. A major regulator of GHRH and GHIH secretion is the blood glucose level (Figure 18.7):

1. **Hypoglycemia** (hī′-po-glī-SĒ-mē-a), an abnormally low blood glucose concentration, stimulates the hypothalamus to secrete GHRH, which flows toward the anterior pituitary in the hypophyseal portal veins.
2. Upon reaching the anterior pituitary, GHRH stimulates somatotrophs to release human growth hormone.
3. Human growth hormone stimulates secretion of insulinlike growth factors, which speed up breakdown of liver glycogen into glucose, causing glucose to enter the blood more rapidly.
4. As a result, blood glucose rises to the normal level (about 90 mg/100 mL of blood plasma).
5. An increase in blood glucose above the normal level inhibits release of GHRH.
6. **Hyperglycemia** (hī′-per-glī-SĒ-mē-a), an abnormally high blood glucose concentration, stimulates the hypothalamus to secrete GHIH (while inhibiting the secretion of GHRH).
7. Upon reaching the anterior pituitary in portal blood, GHIH inhibits secretion of human growth hormone by somatotrophs.
8. A low level of human growth hormone and IGFs slows breakdown of glycogen in the liver, and glucose is released into the blood more slowly.
9. Blood glucose falls to the normal level.
10. A decrease in blood glucose below the normal level (hypoglycemia) inhibits release of GHIH.

Other stimuli that promote secretion of human growth hormone include decreased fatty acids and increased amino acids in the blood; deep sleep (stages 3 and 4 of non-rapid eye movement sleep); increased activity of the sympathetic division of the autonomic nervous system, such as might occur with stress or vigorous physical exercise; and other hormones, including glucagon, estrogens, cortisol, and insulin. Factors that inhibit human growth hormone secretion are increased levels of fatty acids and decreased levels of amino acids in the blood; rapid eye movement sleep; emotional deprivation; obesity; low levels of thyroid hormones; and human growth hormone itself (through negative feedback).

Figure 18.7 **Effects of human growth hormone (hGH) and insulinlike growth factors (IGFs).** Dashed lines indicate inhibition.

Secretion of hGH is stimulated by growth hormone–releasing hormone (GHRH) and inhibited by growth hormone–inhibiting hormone (GHIH).

1 Low blood glucose (hypoglycemia) stimulates release of

6 High blood glucose (hyperglycemia) stimulates release of

GHRH GHIH

2 GHRH stimulates secretion of hGH by somatotrophs

7 GHIH inhibits secretion of hGH by somatotrophs

hGH

Anterior pituitary

3 hGH and IGFs speed up breakdown of liver glycogen into glucose, which enters the blood more rapidly

8 A low level of hGH and IGFs decreases the rate of glycogen breakdown in the liver and glucose enters the blood more slowly

4 Blood glucose level rises to normal (about 90 mg/100 mL)

9 Blood glucose level falls to normal (about 90 mg/100 mL)

5 If blood glucose continues to increase, hyperglycemia inhibits release of GHRH

10 If blood glucose continues to decrease, hypoglycemia inhibits release of GHIH

If a person has a pituitary tumor that secretes a large amount of hGH and the tumor cells are not responsive to regulation by GHRH and GHIH, will hyperglycemia or hypoglycemia be more likely?

Diabetogenic Effect of hGH

One symptom of excess human growth hormone (hGH) is hyperglycemia. Persistent hyperglycemia, in turn, stimulates the pancreas to secrete insulin continually. Such excessive stimulation, if it lasts for weeks or months, may cause "beta-cell burnout," a greatly decreased capacity of pancreatic beta cells to synthesize and secrete insulin. Thus, excess secretion of human growth hormone may have a **diabetogenic effect;** that is, it causes diabetes mellitus (lack of insulin activity). ■

Thyroid-stimulating Hormone

Thyroid-stimulating hormone (TSH) stimulates the synthesis and secretion of the two thyroid hormones, triiodothyronine (T_3) and thyroxine (T_4), both produced by the thyroid gland. Thyrotropin-releasing hormone (TRH) from the hypothalamus controls TSH secretion. Release of TRH, in turn, depends on blood levels of T_3 and T_4; high levels of T_3 and T_4 inhibit secretion of TRH via negative feedback. There is no thyrotropin-inhibiting hormone. The release of TRH is explained later in the chapter (see Figure 18.12).

Follicle-stimulating Hormone

In females, the ovaries are the targets for follicle-stimulating hormone (FSH). Each month FSH initiates the development of several ovarian follicles, saclike arrangements of secretory cells that surround a developing oocyte. FSH also stimulates follicular cells to secrete estrogens (female sex hormones). In males, FSH stimulates sperm production in the testes. Gonadotropin-releasing hormone (GnRH) from the hypothalamus stimulates FSH release. Release of GnRH and FSH is suppressed by estrogens in females and by testosterone (the principal male sex hormone) in males through negative feedback systems. There is no gonadotropin-inhibiting hormone.

Luteinizing Hormone

In females, luteinizing hormone (LH) triggers **ovulation,** the release of a secondary oocyte (future ovum) by an ovary. LH stimulates formation of the corpus luteum (structure formed after ovulation) in the ovary and the secretion of progesterone (another female sex hormone) by the corpus luteum. Together, FSH and LH also stimulate secretion of estrogens by ovarian cells. Estrogens and progesterone prepare the uterus for implantation of a fertilized ovum and help prepare the mammary glands for milk secretion. In males, LH stimulates cells in the testes to secrete testosterone. Secretion of LH, like that of FSH, is controlled by gonadotropin-releasing hormone (GnRH).

Prolactin

Prolactin (PRL), together with other hormones, initiates and maintains milk secretion by the mammary glands. By itself, prolactin has only a weak effect. Only after the mammary glands have been primed by estrogens, progesterone, glucocorticoids, human growth hormone, thyroxine, and insulin, which exert permissive effects, does PRL bring about milk secretion. Ejection of milk from the mammary glands depends on the hormone oxytocin, which is released from the posterior pituitary. Together, milk secretion and ejection constitute *lactation.*

The hypothalamus secretes both inhibitory and excitatory hormones that regulate prolactin secretion. Prolactin-inhibiting hormone (PIH), which is dopamine, inhibits the release of prolactin from the anterior pituitary. As the levels of estrogens and progesterone fall just before menstruation begins, the secretion of PIH diminishes and the blood level of prolactin rises. Breast tenderness just before menstruation may be caused by elevated prolactin. Because the prolactin level is high for only a short period, however, milk production does not start. As the menstrual cycle begins anew and the level of estrogens rises, PIH is again secreted and the prolactin level drops. Prolactin level rises during pregnancy, stimulated by prolactin-releasing hormone (PRH) from the hypothalamus. The sucking action of a nursing infant causes a reduction in hypothalamic secretion of PIH.

The function of prolactin is not known in males, but its hypersecretion causes erectile dysfunction (impotence, the inability to have an erection of the penis). In females, hypersecretion of prolactin causes galactorrhea (inappropriate lactation) and amenorrhea (absence of menstrual cycles).

Adrenocorticotropic Hormone

Corticotrophs secrete mainly adrenocorticotropic hormone (ACTH). ACTH controls the production and secretion of cortisol and other glucocorticoids by the cortex (outer portion) of the adrenal glands. Corticotropin-releasing hormone (CRH) from the hypothalamus stimulates secretion of ACTH by corticotrophs. Stress-related stimuli, such as low blood glucose or physical trauma, and interleukin-1, a substance produced by macrophages, also stimulate release of ACTH. Glucocorticoids inhibit CRH and ACTH release via negative feedback.

Melanocyte-stimulating Hormone

Melanocyte-stimulating hormone (MSH) increases skin pigmentation in amphibians by stimulating the dispersion of melanin granules in melanocytes. Its exact role in humans is unknown, but the presence of MSH receptors in the brain suggests it may influence brain activity. There is little circulating MSH in humans. However, continued administration of MSH for several days does produce a darkening of the skin. Excessive levels of corticotropin-releasing hormone (CRH) can stimulate MSH release; dopamine inhibits MSH release.

Table 18.4 summarizes the principal actions of the anterior pituitary hormones.

Posterior Pituitary

Although the **posterior pituitary** or **neurohypophysis** does not *synthesize* hormones, it does *store* and *release* two hormones. As noted earlier in the chapter, it consists of pituicytes and axon terminals of hypothalamic neurosecretory cells. The cell bodies of the neurosecretory cells are in the paraventricular and supraoptic nuclei of the hypothalamus; their axons form the **hypothalamohypophyseal tract** (hī′-pō-thal′-a-mō-hī-pō-FIZ-ē-al). This tract begins in the hypothalamus and ends near

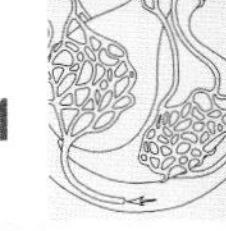

TABLE 18.4 Summary of the Principal Actions of Anterior Pituitary Hormones

Hormone and Target Tissues	Principal Actions
Human growth hormone (hGH) or **somatotropin** Liver	Stimulates liver, muscle, cartilage, bone, and other tissues to synthesize and secrete insulinlike growth factors (IGFs); IGFs promote growth of body cells, protein synthesis, tissue repair, lipolysis, and elevation of blood glucose concentration.
Thyroid-stimulating hormone (TSH) or **thyrotropin** Thyroid gland	Stimulates synthesis and secretion of thyroid hormones by thyroid gland.
Follicle-stimulating hormone (FSH) Ovaries Testes	In females, initiates development of oocytes and induces ovarian secretion of estrogens. In males, stimulates testes to produce sperm.
Luteinizing hormone (LH) Ovaries Testes	In females, stimulates secretion of estrogens and progesterone, ovulation, and formation of corpus luteum. In males, stimulates testes to produce testosterone.
Prolactin (PRL) Mammary glands	Together with other hormones, promotes milk secretion by the mammary glands.
Adrenocorticotropic hormone (ACTH) or **corticotropin** Adrenal cortex	Stimulates secretion of glucocorticoids (mainly cortisol) by adrenal cortex.
Melanocyte-stimulating hormone (MSH) Brain	Exact role in humans is unknown but may influence brain activity; when present in excess, can cause darkening of skin.

blood capillaries in the posterior pituitary (Figure 18.8). The paraventricular nucleus synthesizes the hormone **oxytocin** (**OT;** ok′-sē-TŌ-sin; *okytoc* = quick birth) and the supraoptic nucleus produces **antidiuretic hormone** (**ADH;** an-tī-dī-ū-RET-ik; *anti-* = against; *-diuretic* = increased urine production), also called **vasopressin** (vā-sō-PRES-in; *vaso-* = blood; *-pressus* = to press).

After their production in the cell bodies of neurosecretory cells, oxytocin and antidiuretic hormone are packaged into secretory vesicles, which move by fast axonal transport (described on page 407) to the axon terminals in the posterior pituitary, where they are stored until nerve impulses trigger exocytosis and release of the hormone.

Blood is supplied to the posterior pituitary by the **inferior hypophyseal arteries** (see Figure 18.5), which branch from the internal carotid arteries. In the posterior pituitary, the inferior hypophyseal arteries drain into the **capillary plexus of the infundibular process,** a capillary network that receives secreted oxytocin and antidiuretic hormone (see Figure 18.5). From this plexus, hormones pass into the **posterior hypophyseal veins** for distribution to target cells in other tissues.

Oxytocin

During and after delivery of a baby, oxytocin affects two target tissues: the mother's uterus and breasts. During delivery, oxytocin enhances contraction of smooth muscle cells in the wall of the uterus; after delivery, it stimulates milk ejection ("letdown") from the mammary glands in response to the mechanical stimulus provided by a suckling infant. The function of oxytocin in males and in nonpregnant females is not clear. Experiments with animals have suggested that it has actions within the brain that foster parental caretaking behavior toward young offspring. It may also be responsible, in part, for the feelings of sexual pleasure during and after intercourse.

Figure 18.8 Axons of hypothalamic neurosecretory cells form the hypothalamohypophyseal tract, which extends from the paraventricular and supraoptic nuclei to the posterior pituitary. Hormone molecules synthesized in the cell body of a neurosecretory cell are packaged into secretory vesicles that move down to the axon terminals. Nerve impulses trigger exocytosis of the vesicles, thereby releasing the hormone.

Oxytocin and antidiuretic hormone are synthesized in the hypothalamus and released into the capillary plexus of the infundibular process in the posterior pituitary.

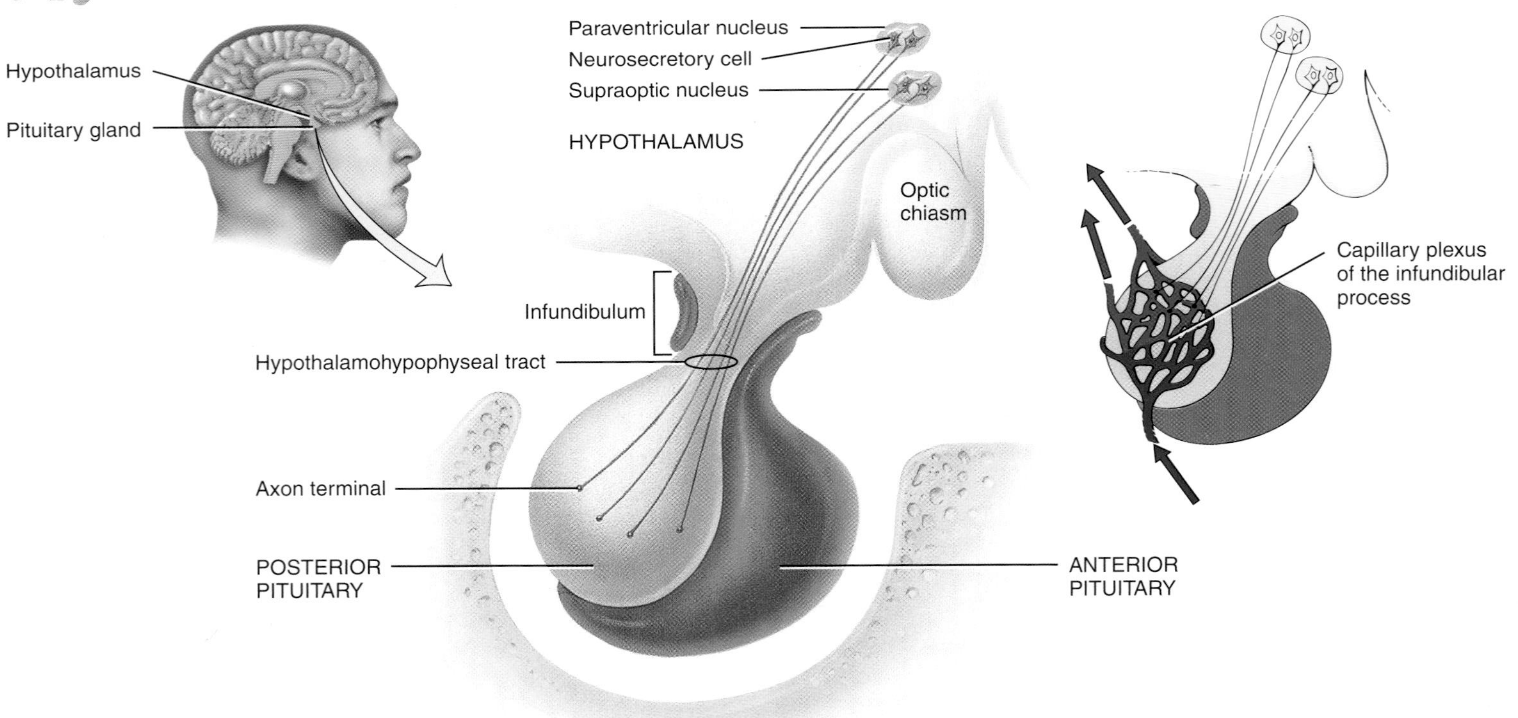

Functionally, how are the hypothalamohypophyseal tract and the hypophyseal portal veins similar? Structurally, how are they different?

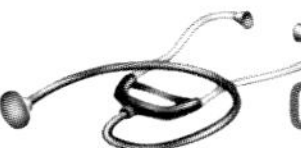

Oxytocin and Childbirth

Years before oxytocin was discovered, it was common practice in midwifery to let a first-born twin nurse at the mother's breast to speed the birth of the second child. Now we know why this practice is helpful—it stimulates the release of oxytocin. Even after a single birth, nursing promotes expulsion of the placenta (afterbirth) and helps the uterus regain its smaller size. Synthetic OT (Pitocin) often is given to induce labor or to increase uterine tone and control hemorrhage just after giving birth. ■

Antidiuretic Hormone

As its name implies, an **antidiuretic** (*anti-* = against; *dia* = throughout; *ouresis* = urination) is a substance that decreases urine production. ADH causes the kidneys to return more water to the blood, thus decreasing urine volume. In the absence of ADH, urine output increases more than tenfold, from the normal 1 to 2 liters to about 20 liters a day. Drinking alcohol often causes frequent and copious urination because alcohol inhibits secretion of ADH. ADH also decreases the water lost through sweating and causes constriction of arterioles, which increases blood pressure. This hormone's other name, vasopressin, reflects this effect on blood pressure.

The amount of ADH secreted varies with blood osmotic pressure and blood volume. Figure 18.9 shows regulation of ADH secretion and the actions of ADH.

1. High blood osmotic pressure—due to dehydration or a decline in blood volume because of hemorrhage, diarrhea, or excessive sweating—stimulates **osmoreceptors,** neurons in the hypothalamus that monitor blood osmotic pressure. Elevated blood osmotic pressure activates the osmoreceptors directly; they also receive excitatory input from other brain areas when blood volume decreases.
2. Osmoreceptors activate the hypothalamic neurosecretory cells that synthesize and release ADH.
3. When neurosecretory cells receive excitatory input from the osmoreceptors, they generate nerve impulses that cause exocytosis of ADH-containing vesicles from their axon terminals in the posterior pituitary. This liberates ADH, which diffuses into blood capillaries of the posterior pituitary.
4. The blood carries ADH to three target tissues: the kidneys, sudoriferous (sweat) glands, and smooth muscle in blood vessel walls. The kidneys respond by retaining more water, which decreases urine output. Secretory activity of sweat glands decreases, which lowers the rate of water loss by perspiration from the skin. Smooth muscle in the walls of arterioles (small arteries) contracts in response to high levels of ADH, which constricts (narrows) the lumen of these blood vessels and increases blood pressure.
5. Low osmotic pressure of blood (or increased blood volume) inhibits the osmoreceptors.

Figure 18.9 Regulation of secretion and actions of antidiuretic hormone (ADH).

ADH acts to retain body water and increase blood pressure.

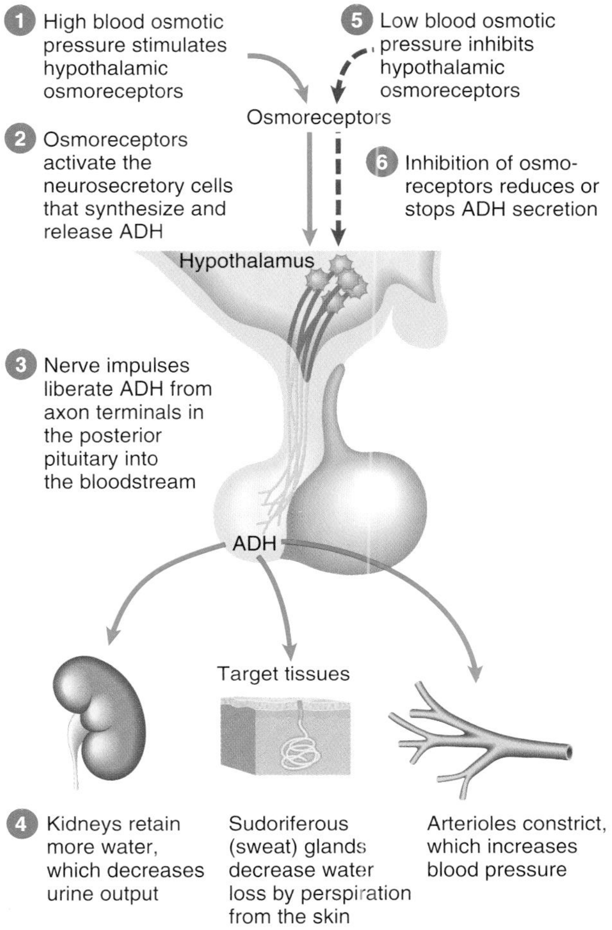

1 High blood osmotic pressure stimulates hypothalamic osmoreceptors
5 Low blood osmotic pressure inhibits hypothalamic osmoreceptors
Osmoreceptors
2 Osmoreceptors activate the neurosecretory cells that synthesize and release ADH
6 Inhibition of osmoreceptors reduces or stops ADH secretion
Hypothalamus
3 Nerve impulses liberate ADH from axon terminals in the posterior pituitary into the bloodstream
ADH
Target tissues
4 Kidneys retain more water, which decreases urine output
Sudoriferous (sweat) glands decrease water loss by perspiration from the skin
Arterioles constrict, which increases blood pressure

If you drank a liter of water, what effect would this have on the osmotic pressure of your blood, and how would the level of ADH change in your blood?

6. Inhibition of osmoreceptors reduces or stops ADH secretion. The kidneys then retain less water by forming a larger volume of urine, secretory activity of sweat glands increases, and arterioles dilate. The blood volume and osmotic pressure of body fluids return to normal.

Secretion of ADH can also be altered in other ways. Pain, stress, trauma, anxiety, acetylcholine, nicotine, and drugs such as morphine, tranquilizers, and some anesthetics stimulate ADH secretion. The dehydrating effect of alcohol, which has already been mentioned, may cause both the thirst and the headache typical of a hangover. Hyposecretion of ADH or nonfunctional ADH receptors causes diabetes insipidus (see page 659).

TABLE 18.5 Summary of Posterior Pituitary Hormones

Hormone and Target Tissues	Control of Secretion	Principal Actions
Oxytocin (OT) Uterus, Mammary glands	Neurosecretory cells of hypothalamus secrete OT in response to uterine distention and stimulation of nipples.	Stimulates contraction of smooth muscle cells of uterus during childbirth; stimulates contraction of myoepithelial cells in mammary glands to cause milk ejection.
Antidiuretic hormone (ADH) or **vasopressin** Kidneys, Sudoriferous (sweat) glands, Arterioles	Neurosecretory cells of hypothalamus secrete ADH in response to elevated blood osmotic pressure, dehydration, loss of blood volume, pain, or stress; low blood osmotic pressure, high blood volume, and alcohol inhibit ADH secretion.	Conserves body water by decreasing urine volume; decreases water loss through perspiration; raises blood pressure by constricting arterioles.

Table 18.5 lists the posterior pituitary hormones, control of their secretion, and their principal actions.

► CHECKPOINT

9. In what respect is the pituitary gland actually two glands?
10. How do hypothalamic releasing and inhibiting hormones influence secretions of the anterior pituitary?
11. Describe the structure and importance of the hypothalamo-hypophyseal tract.
12. Explain how blood levels of T_3/T_4, TSH, and TRH would change in a laboratory animal that has undergone a thyroidectomy (complete removal of its thyroid gland).

THYROID GLAND

► OBJECTIVE

Describe the location, histology, hormones, and functions of the thyroid gland.

The butterfly-shaped **thyroid gland** is located just inferior to the larynx (voice box). It is composed of right and left **lateral lobes,** one on either side of the trachea, that are connected by an **isthmus** (IS-mus = a narrow passage) anterior to the trachea (Figure 18.10a). A small, pyramidal-shaped lobe sometimes extends upward from the isthmus. The normal mass of the thyroid is about 30 g (1 oz). It is highly vascularized and receives 80–120 mL of blood per minute.

Microscopic spherical sacs called **thyroid follicles** (Figure 18.10b) make up most of the thyroid gland. The wall of each follicle consists primarily of cells called **follicular cells,** most of which extend to the lumen (internal space) of the follicle. A **basement membrane** surrounds each follicle. When the follicular cells are inactive, their shape is low cuboidal to squamous, but under the influence of TSH they become active in secretion and range from cuboidal to low columnar in shape. The follicular cells produce two hormones: **thyroxine** (thī-ROK-sēn), which is also called **tetraiodothyronine** (tet-ra-ī-ō-dō-THĪ-rō-nēn) or $\mathbf{T_4}$ because it contains four atoms of iodine, and **triiodothyronine** (trī-ī′-ō-dō-THĪ-rō-nēn) or $\mathbf{T_3}$, which contains three atoms of iodine. T_3 and T_4 are also known as **thyroid hormones.** A few cells called **parafollicular cells** or **C cells** lie between follicles. They produce the hormone **calcitonin** (kal-si-TŌ-nin), which helps regulate calcium homeostasis.

Formation, Storage, and Release of Thyroid Hormones

The thyroid gland is the only endocrine gland that stores its secretory product in large quantities—normally about a 100-day

Figure 18.10 Location, blood supply, and histology of the thyroid gland.

Thyroid hormones regulate (1) oxygen use and basal metabolic rate, (2) cellular metabolism, and (3) growth and development.

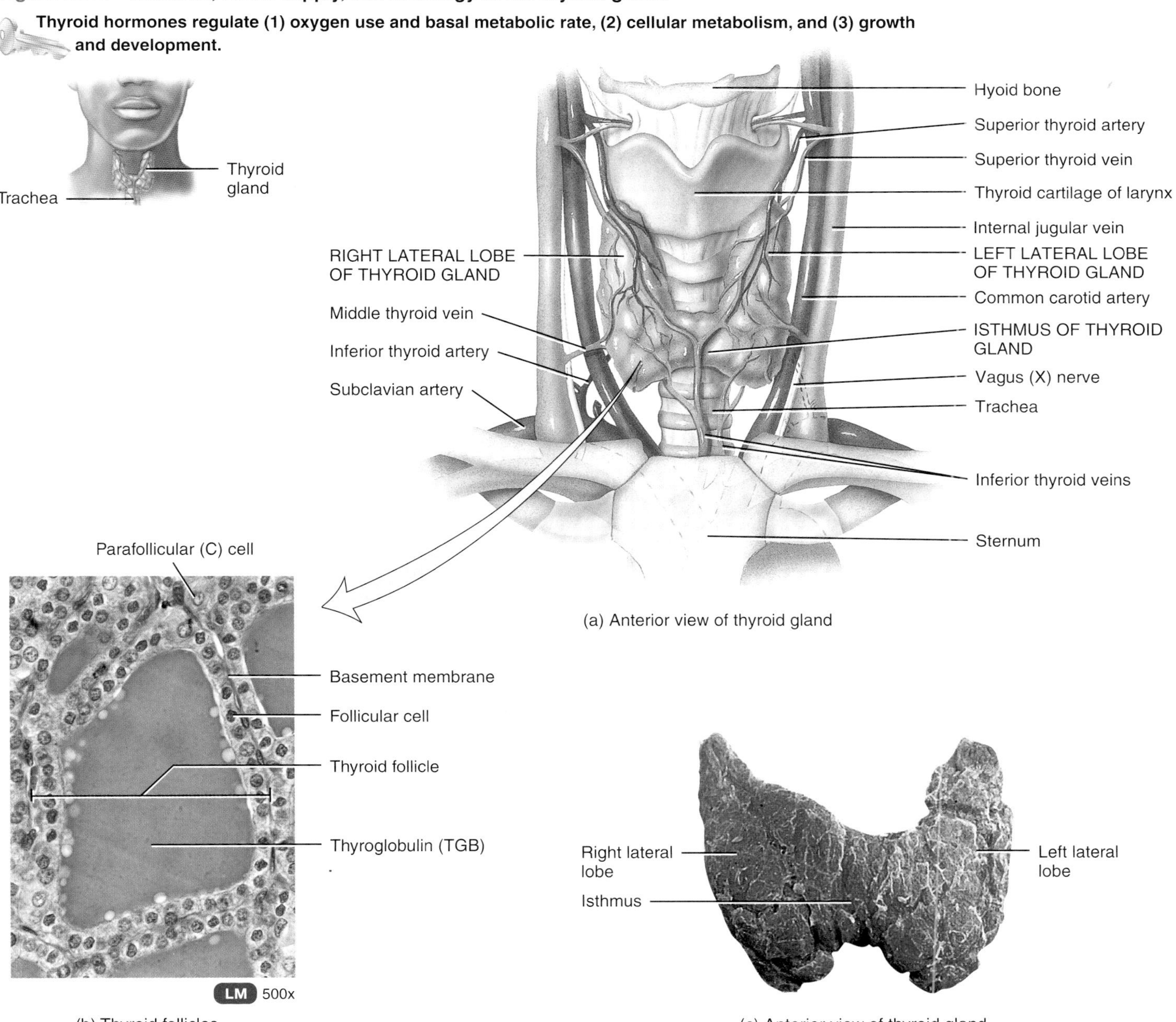

(a) Anterior view of thyroid gland

(b) Thyroid follicles

(c) Anterior view of thyroid gland

Which cells secrete T_3 and T_4? Which secrete calcitonin? Which of these hormones are also called thyroid hormones?

supply. Synthesis and secretion of T_3 and T_4 occurs as follows (Figure 18.11):

1. ***Iodide trapping.*** Thyroid follicular cells trap iodide ions (I^-) by actively transporting them from the blood into the cytosol. As a result, the thyroid gland normally contains most of the iodide in the body.
2. ***Synthesis of thyroglobulin.*** While the follicular cells are trapping I^-, they are also synthesizing **thyroglobulin (TGB),** a large glycoprotein that is produced in the rough endoplasmic reticulum, modified in the Golgi complex, and packaged into secretory vesicles. The vesicles then undergo exocytosis, which releases TGB into the lumen of the follicle.
3. ***Oxidation of iodide.*** Some of the amino acids in TGB are tyrosines that will become iodinated. However, negatively charged iodide ions cannot bind to tyrosine until they undergo oxidation (removal of electrons) to iodine: $2\ I^- \rightarrow I_2$. As the iodide ions are being oxidized, they pass through the membrane into the lumen of the follicle.
4. ***Iodination of tyrosine.*** As iodine molecules (I_2) form, they react with tyrosines that are part of thyroglobulin molecules. Binding of one iodine atom yields monoiodotyrosine (T_1), and a second iodination produces diiodotyrosine (T_2). The TGB with attached iodine atoms, a sticky material that accumulates and is stored in the lumen of the thyroid follicle, is termed **colloid.**
5. ***Coupling of T_1 and T_2.*** During the last step in the synthesis of thyroid hormone, two T_2 molecules join to form T_4 or one T_1 and one T_2 join to form T_3.
6. ***Pinocytosis and digestion of colloid.*** Droplets of colloid reenter follicular cells by pinocytosis and merge with lysosomes. Digestive enzymes in the lysosomes break down TGB, cleaving off molecules of T_3 and T_4.
7. ***Secretion of thyroid hormones.*** Because T_3 and T_4 are lipid soluble, they diffuse through the plasma membrane into interstitial fluid and then into the blood. T_4 normally is secreted in greater quantity than T_3, but T_3 is several times more potent. Moreover, after T_4 enters a body cell, most of it is converted to T_3 by removal of one iodine.
8. ***Transport in the blood.*** More than 99% of both the T_3 and the T_4 combine with transport proteins in the blood, mainly **thyroxine-binding globulin (TBG).**

Figure 18.11 Steps in the synthesis and secretion of thyroid hormones.

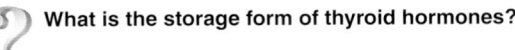

Thyroid hormones are synthesized by attaching iodine atoms to the amino acid tyrosine.

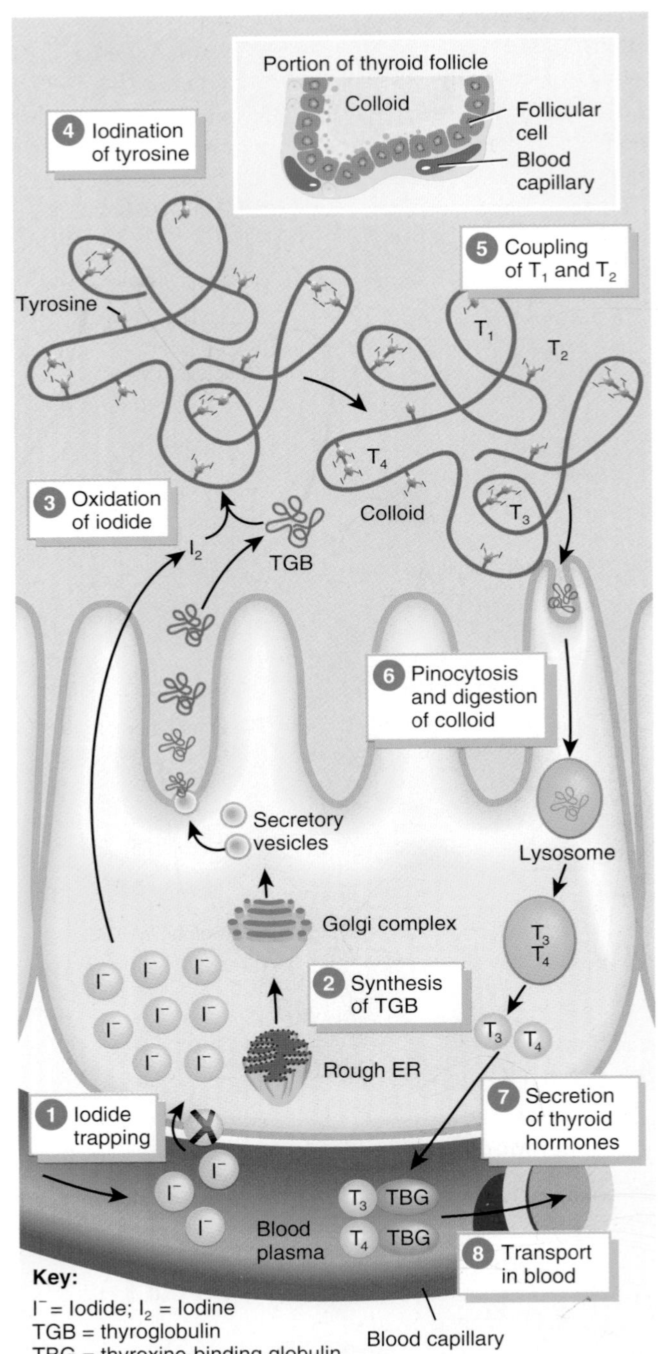

What is the storage form of thyroid hormones?

Actions of Thyroid Hormones

Because most body cells have receptors for thyroid hormones, T_3 and T_4 exert their effects throughout the body.

1. Thyroid hormones increase **basal metabolic rate (BMR),** the rate of oxygen consumption under standard or basal conditions (awake, at rest, and fasting), by stimulating the use of cellular oxygen to produce ATP. When the basal metabolic rate increases, cellular metabolism of carbohydrates, lipids, and proteins increases.

2. A second major effect of thyroid hormones is to stimulate synthesis of additional sodium-potassium pumps (Na^+/K^+ ATPase), which use large amounts of ATP to continually eject sodium ions (Na^+) from the cytosol into the extracellular fluid and potassium ions (K^+) from the extracellular fluid into the cytosol. As cells produce and use more ATP, more heat is given off, and body temperature rises. This phenomenon is called the **calorigenic effect.** In this way, thyroid hormones play an important role in the maintenance of normal body temperature. Normal mammals can survive in freezing temperatures, but those whose thyroid glands have been removed cannot.

3. In the regulation of metabolism, the thyroid hormones stimulate protein synthesis and increase the use of glucose and fatty acids for ATP production. They also increase lipolysis and enhance cholesterol excretion, thus reducing blood cholesterol level.

4. The thyroid hormones enhance some actions of the catecholamines (norepinephrine and epinephrine) because they up-regulate beta (β) receptors. For this reason, symptoms of hyperthyroidism include increased heart rate, more forceful heartbeats, and increased blood pressure.

5. Together with human growth hormone and insulin, thyroid hormones accelerate body growth, particularly the growth of the nervous and skeletal systems. Deficiency of thyroid hormones during fetal development, infancy, or childhood causes severe mental retardation and stunted bone growth.

Control of Thyroid Hormone Secretion

Thyrotropin-releasing hormone (TRH) from the hypothalamus and thyroid-stimulating hormone (TSH) from the anterior pituitary stimulate synthesis and release of thyroid hormones, as shown in Figure 18.12:

1. Low blood levels of T_3 and T_4 or low metabolic rate stimulate the hypothalamus to secrete TRH.
2. TRH enters the hypophyseal portal veins and flows to the anterior pituitary, where it stimulates thyrotrophs to secrete TSH.
3. TSH stimulates virtually all aspects of thyroid follicular cell activity, including iodide trapping (1 in Figure 18.11), hormone synthesis and secretion (2 and 7 in Figure 18.11), and growth of the follicular cells.
4. The thyroid follicular cells release T_3 and T_4 into the blood until the metabolic rate returns to normal.
5. An elevated level of T_3 inhibits release of TRH and TSH (negative feedback inhibition).

Conditions that increase ATP demand—a cold environment, hypoglycemia, high altitude, and pregnancy—also increase the secretion of the thyroid hormones.

Figure 18.12 Regulation of secretion and actions of thyroid hormones. TRH = thyrotropin-releasing hormone, TSH = thyroid-stimulating hormone, T_3 = triiodothyronine, and T_4 = thyroxine (tetraiodothyronine).

TSH promotes release of thyroid hormones (T_3 and T_4) by the thyroid gland.

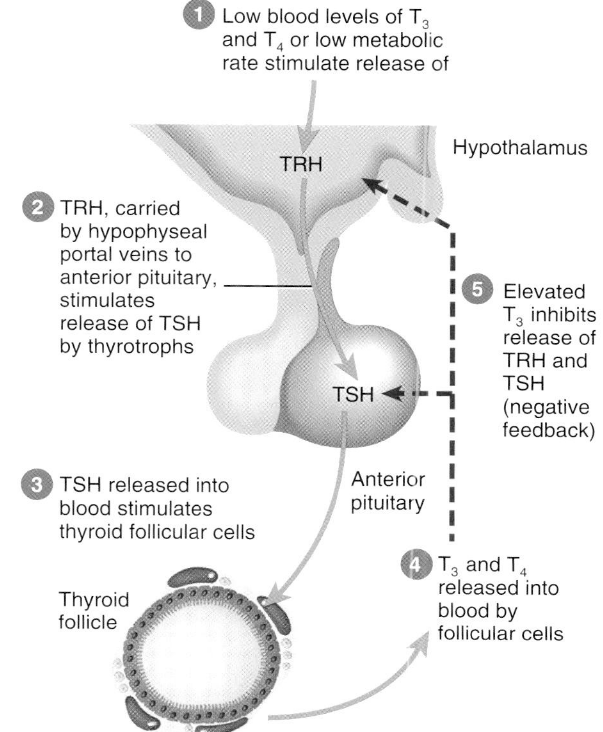
1 Low blood levels of T_3 and T_4 or low metabolic rate stimulate release of
Hypothalamus
TRH
2 TRH, carried by hypophyseal portal veins to anterior pituitary, stimulates release of TSH by thyrotrophs
5 Elevated T_3 inhibits release of TRH and TSH (negative feedback)
TSH
Anterior pituitary
3 TSH released into blood stimulates thyroid follicular cells
Thyroid follicle
4 T_3 and T_4 released into blood by follicular cells

Actions of Thyroid Hormones:

- Increase basal metabolic rate
- Stimulate synthesis of Na^+/K^+ ATPase
- Increase body temperature (calorigenic effect)
- Stimulate protein synthesis
- Increase the use of glucose and fatty acids for ATP production
- Stimulate lipolysis
- Enhance some actions of catecholamines
- Regulate development and growth of nervous tissue and bones

How could an iodine-deficient diet lead to goiter, which is an enlargement of the thyroid gland?

Calcitonin

The hormone produced by the **parafollicular cells** of the thyroid gland (see Figure 18.10b) is **calcitonin (CT)** (kal-si-TŌ-nin). CT can decrease the level of calcium in the blood by inhibiting the action of osteoclasts, the cells that break down bone extracel-

TABLE 18.6 Summary of Thyroid Gland Hormones

Hormone and Source	Control of Secretion	Principal Actions
T_3 (triiodothyronine) and **T_4 (thyroxine)** or **thyroid hormones** from follicular cells Thyroid follicle; Folliclular cells	Secretion is increased by thyrotropin-releasing hormone (TRH), which stimulates release of thyroid-stimulating hormone (TSH) in response to low thyroid hormone levels, low metabolic rate, cold, pregnancy, and high altitudes; TRH and TSH secretions are inhibited in response to high thyroid hormone levels; high iodine level suppresses T_3/T_4 secretion.	Increase basal metabolic rate, stimulate synthesis of proteins, increase use of glucose and fatty acids for ATP production, increase lipolysis, enhance cholesterol excretion, accelerate body growth, and contribute to development of the nervous system.
Calcitonin (CT) from parafollicular cells Thyroid follicle; Parafollicular cells	High blood Ca^{2+} levels stimulate secretion; low blood Ca^{2+} levels inhibit secretion.	Lowers blood levels of Ca^{2+} and HPO_4^{2-} by inhibiting bone resorption by osteoclasts and by accelerating uptake of calcium and phosphates into bone matrix.

lular matrix. The secretion of CT is controlled by a negative feedback system (see Figure 18.14).

When its blood level is high, calcitonin lowers the amount of blood calcium and phosphates by inhibiting bone resorption (breakdown of bone extracellular matrix) by osteoclasts and by accelerating uptake of calcium and phosphates into bone extracellular matrix. Miacalcin, a calcitonin extract derived from salmon that is ten times more potent than human calcitonin, is prescribed to treat osteoporosis.

Table 18.6 summarizes the hormones produced by the thyroid gland, control of their secretion, and their principal actions.

▶ CHECKPOINT

13. How are the thyroid hormones synthesized, stored, and secreted?
14. How is the secretion of T_3 and T_4 regulated?
15. What are the physiological effects of the thyroid hormones?

PARATHYROID GLANDS

▶ OBJECTIVE

Describe the location, histology, hormone, and functions of the parathyroid glands.

Partially embedded in the posterior surface of the lateral lobes of the thyroid gland are several small, round masses of tissue called the **parathyroid glands** (*para-* = beside). Each has a mass of about 40 mg (0.04 g). Usually, one superior and one inferior parathyroid gland are attached to each lateral thyroid lobe (Figure 18.13a), for a total of four.

Microscopically, the parathyroid glands contain two kinds of epithelial cells (Figure 18.13b, c). The more numerous cells, called **chief (principal) cells,** produce **parathyroid hormone (PTH),** also called **parathormone.** The function of the other kind of cell, called an *oxyphil cell,* is not known.

Parathyroid Hormone

Parathyroid hormone is the major regulator of the levels of calcium (Ca^{2+}), magnesium (Mg^{2+}), and phosphate (HPO_4^{2-}) ions in the blood. The specific action of PTH is to increase the number and activity of osteoclasts. The result is elevated bone *resorption,* which releases ionic calcium (Ca^{2+}) and phosphates (HPO_4^{2-}) into the blood. PTH also acts on the kidneys. First, it slows the rate at which Ca^{2+} and Mg^{2+} are lost from blood into the urine. Second, it increases loss of HPO_4^{2-} from blood into the urine. Because more HPO_4^{2-} is lost in the urine than is gained from the bones, PTH decreases blood HPO_4^{2-} level and increases blood Ca^{2+} and Mg^{2+} levels. A third effect of PTH on the kidneys is to promote formation of the hormone **calcitriol,** the active form of vitamin D. Calcitriol, also known as *1, 25-dihydroxy vitamin D_3,* increases the rate of Ca^{2+}, HPO_4^{2-}, and Mg^{2+} *absorption* from the gastrointestinal tract into the blood.

The blood calcium level directly controls the secretion of both calcitonin and parathyroid hormone via negative feedback loops that do not involve the pituitary gland (Figure 18.14 on page 640):

Figure 18.13 **Location, blood supply, and histology of the parathyroid glands.**

The parathyroid glands, normally four in number, are embedded in the posterior surface of the thyroid gland.

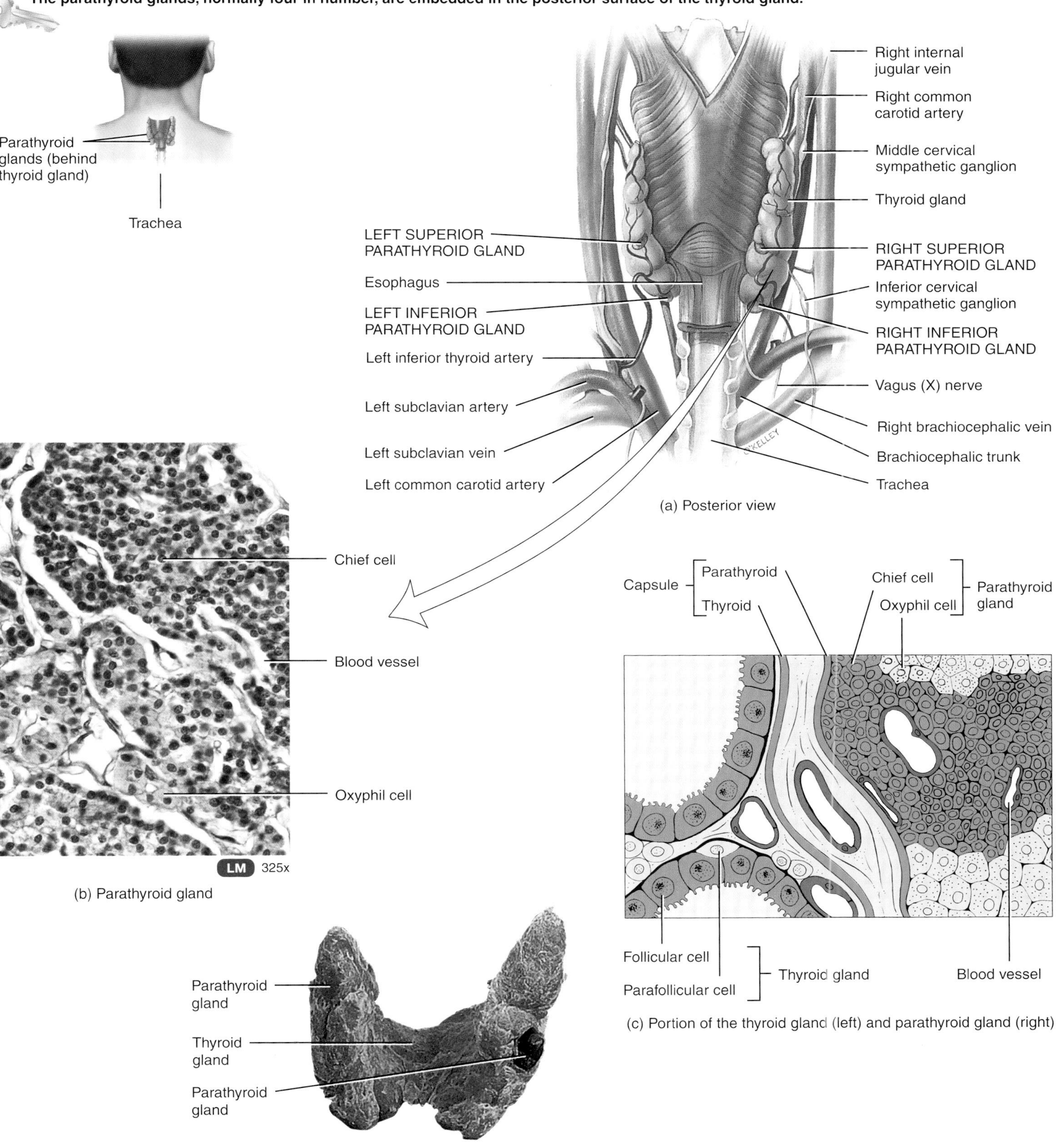

(a) Posterior view

(b) Parathyroid gland

(c) Portion of the thyroid gland (left) and parathyroid gland (right)

(d) Posterior view of parathyroid glands

What are the secretory products of (1) parafollicular cells of the thyroid gland and (2) chief (principal) cells of the parathyroid glands?

Figure 18.14 The roles of calcitonin (green arrows), parathyroid hormone (blue arrows), and calcitriol (orange arrows) in calcium homeostasis.

With respect to regulation of blood Ca^{2+} level, calcitonin and PTH are antagonists.

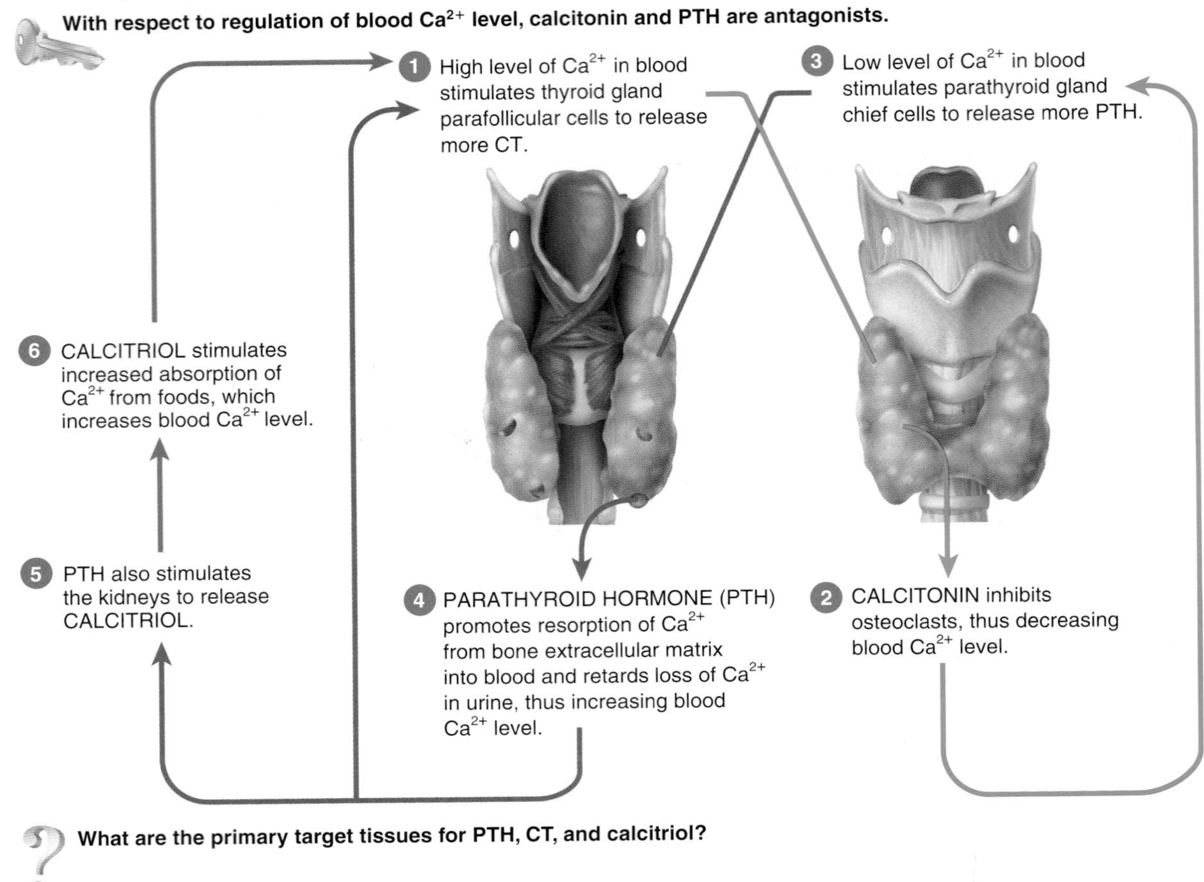

1 High level of Ca2+ in blood stimulates thyroid gland parafollicular cells to release more CT.
3 Low level of Ca2+ in blood stimulates parathyroid gland chief cells to release more PTH.
6 CALCITRIOL stimulates increased absorption of Ca2+ from foods, which increases blood Ca2+ level.
5 PTH also stimulates the kidneys to release CALCITRIOL.
4 PARATHYROID HORMONE (PTH) promotes resorption of Ca2+ from bone extracellular matrix into blood and retards loss of Ca2+ in urine, thus increasing blood Ca2+ level.
2 CALCITONIN inhibits osteoclasts, thus decreasing blood Ca2+ level.

What are the primary target tissues for PTH, CT, and calcitriol?

1. A higher-than-normal level of calcium ions (Ca^{2+}) in the blood stimulates parafollicular cells of the thyroid gland to release more calcitonin.
2. Calcitonin inhibits the activity of osteoclasts, thereby decreasing the blood Ca^{2+} level.
3. A lower-than-normal level of Ca^{2+} in the blood stimulates chief cells of the parathyroid gland to release more PTH.
4. PTH promotes resorption of bone extracellular matrix, which releases Ca^{2+} into the blood, and slows loss of Ca^{2+} in the urine, raising the blood level of Ca^{2+}.
5. PTH also stimulates the kidneys to synthesize calcitriol, the active form of vitamin D.
6. Calcitriol stimulates increased absorption of Ca^{2+} from foods in the gastrointestinal tract, which helps increase the blood level of Ca^{2+}.

TABLE 18.7 Summary of Parathyroid Gland Hormone

Hormone and Source	Control of Secretion	Principal Actions
Parathyroid hormone (PTH) from chief cells Chief cell	Low blood Ca^{2+} levels stimulate secretion. High blood Ca^{2+} levels inhibit secretion.	Increases blood Ca^{2+} and Mg^{2+} levels and decreases blood HPO_4^{2-} level; increases bone resorption by osteoclasts; increases Ca^{2+} reabsorption and HPO_4^{2-} excretion by kidneys; and promotes formation of calcitriol (active form of vitamin D), which increases rate of dietary Ca^{2+} and Mg^{2+} absorption.

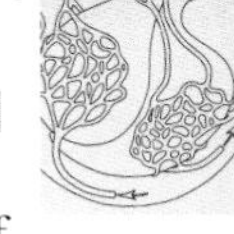

Table 18.7 on page 640 summarizes control of secretion and the principal actions of parathyroid hormone.

▶ CHECKPOINT

16. How is secretion of parathyroid hormone regulated?

17. In what ways are the actions of PTH and calcitriol similar? How are they different?

ADRENAL GLANDS

▶ OBJECTIVE

Describe the location, histology, hormones, and functions of the adrenal glands.

The paired **adrenal (suprarenal) glands,** one of which lies superior to each kidney (Figure 18.15a), have a flattened pyramidal shape. In an adult, each adrenal gland is 3–5 cm in height, 2–3 cm in width, and a little less than 1 cm thick, with a mass of 3.5–5 g, only half its size at birth. During embryonic development, the adrenal glands differentiate into two structurally and functionally distinct regions: A large, peripherally located **adrenal cortex,** comprising 80–90% of the gland and a small, centrally located **adrenal medulla** (Figure 18.15b). A connective tissue capsule covers the gland. The adrenal glands, like the thyroid gland, are highly vascularized.

The adrenal cortex produces steroid hormones that are essential for life. Complete loss of adrenocortical hormones leads to death due to dehydration and electrolyte imbalances in a few days to a week, unless hormone replacement therapy begins promptly. The adrenal medulla produces three catecholamine hormones—norepinephrine, epinephrine, and a small amount of dopamine.

Adrenal Cortex

The adrenal cortex is subdivided into three zones, each of which secretes different hormones (Figure 18.15d). The outer zone,

Figure 18.15 Location, blood supply, and histology of the adrenal (suprarenal) glands.

The adrenal cortex secretes steroid hormones that are essential for life; the adrenal medulla secretes norepinephrine and epinephrine.

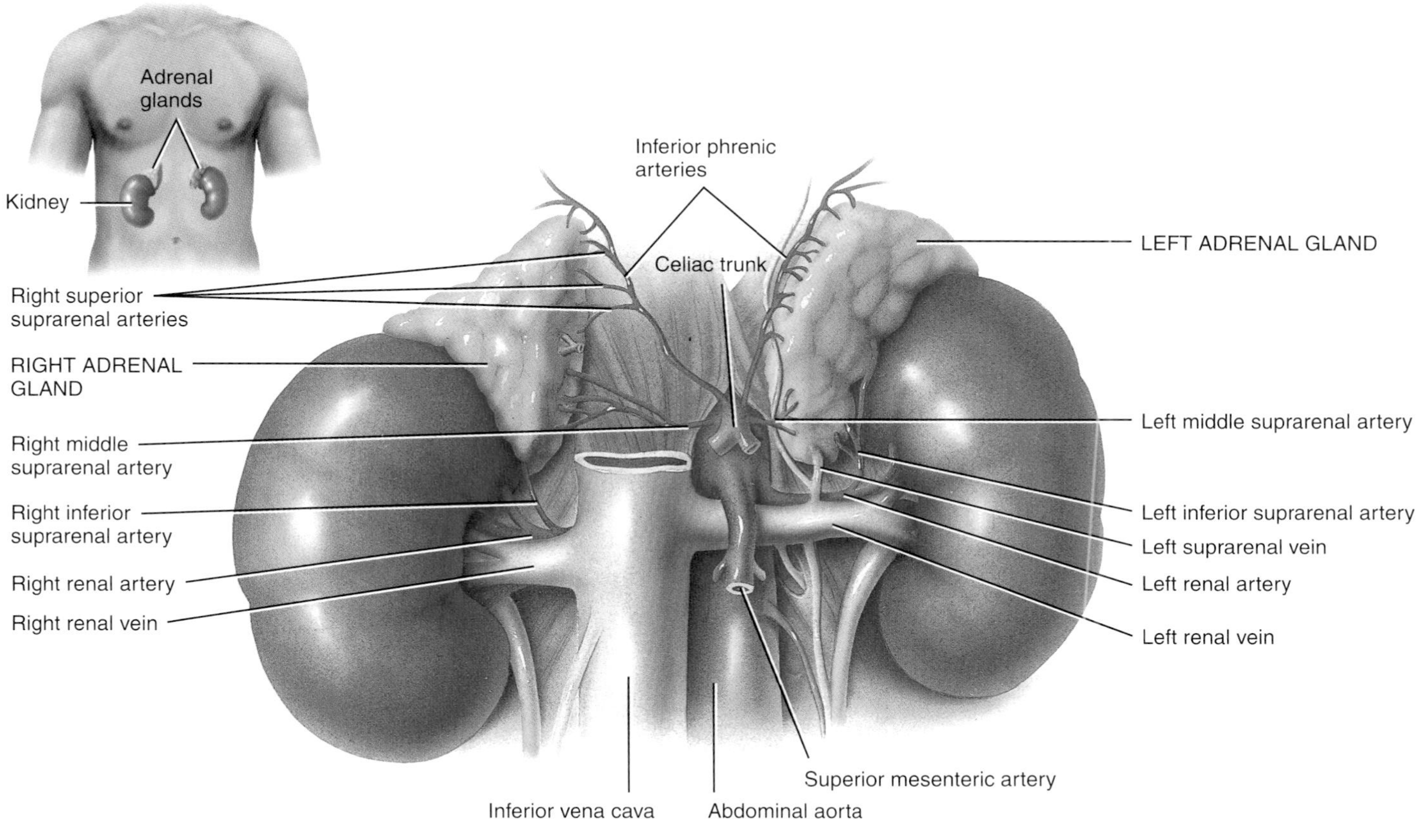

(a) Anterior view

continues

Figure 18.15 **(continued)**

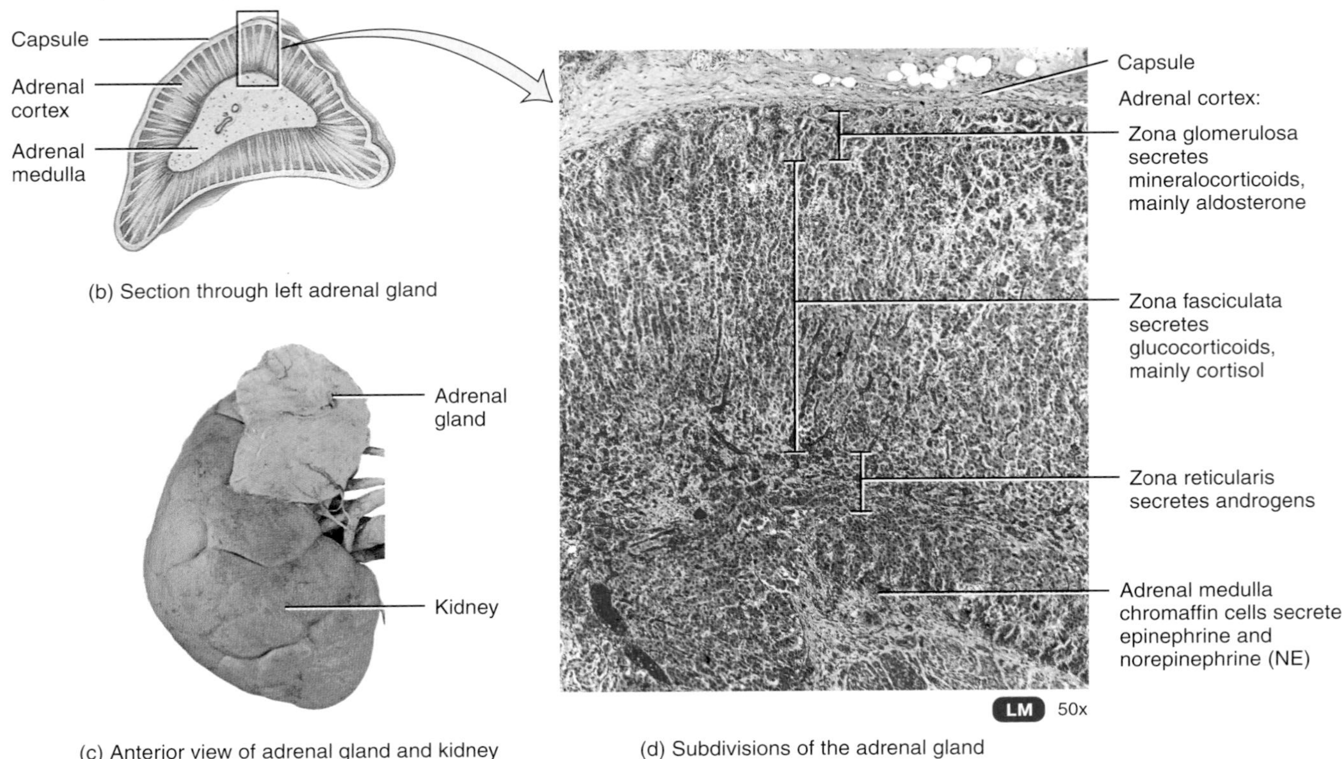

What is the position of the adrenal glands relative to the kidneys?

just deep to the connective tissue capsule, is the **zona glomerulosa** (*zona* = belt; *glomerul-* = little ball). Its cells, which are closely packed and arranged in spherical clusters and arched columns, secrete hormones called **mineralocorticoids** (min′-er-al-ō-KOR-ti-koyds) because they affect mineral homeostasis. The middle zone, or **zona fasciculata** (*fascicul-* = little bundle), is the widest of the three zones and consists of cells arranged in long, straight columns. The cells of the zona fasciculata secrete mainly **glucocorticoids** (gloo′-kō-KOR-ti-koyds), so named because they affect glucose homeostasis. The cells of the inner zone, the **zona reticularis** (*reticul-* = network), are arranged in branching cords. They synthesize small amounts of weak **androgens** (*andro-* = a man), steroid hormones that have masculinizing effects.

Mineralocorticoids

Aldosterone (al-DOS-ter-ōn) is the major mineralocorticoid. It regulates homeostasis of two mineral ions, namely sodium ions (Na^+) and potassium ions (K^+), and helps adjust blood pressure and blood volume. Aldosterone also promotes excretion of H^+ in the urine; this removal of acids from the body can help prevent acidosis (blood pH below 7.35), which is discussed in Chapter 27.

The **renin–angiotensin–aldosterone** or **RAA pathway** (RĒ-nin an′-jē-ō-TEN-sin) controls secretion of aldosterone (Figure 18.16):

1. Stimuli that initiate the renin–angiotensin–aldosterone pathway include dehydration, Na^+ deficiency, or hemorrhage.
2. These conditions cause a decrease in blood volume.
3. Decreased blood volume leads to decreased blood pressure.
4. Lowered blood pressure stimulates certain cells of the kidneys, called juxtaglomerular cells, to secrete the enzyme **renin.**
5. The level of renin in the blood increases.

Figure 18.16 Regulation of aldosterone secretion by the renin–angiotensin–aldosterone (RAA) pathway.

Aldosterone helps regulate blood volume, blood pressure, and levels of Na^+, K^+, and H^+ in the blood.

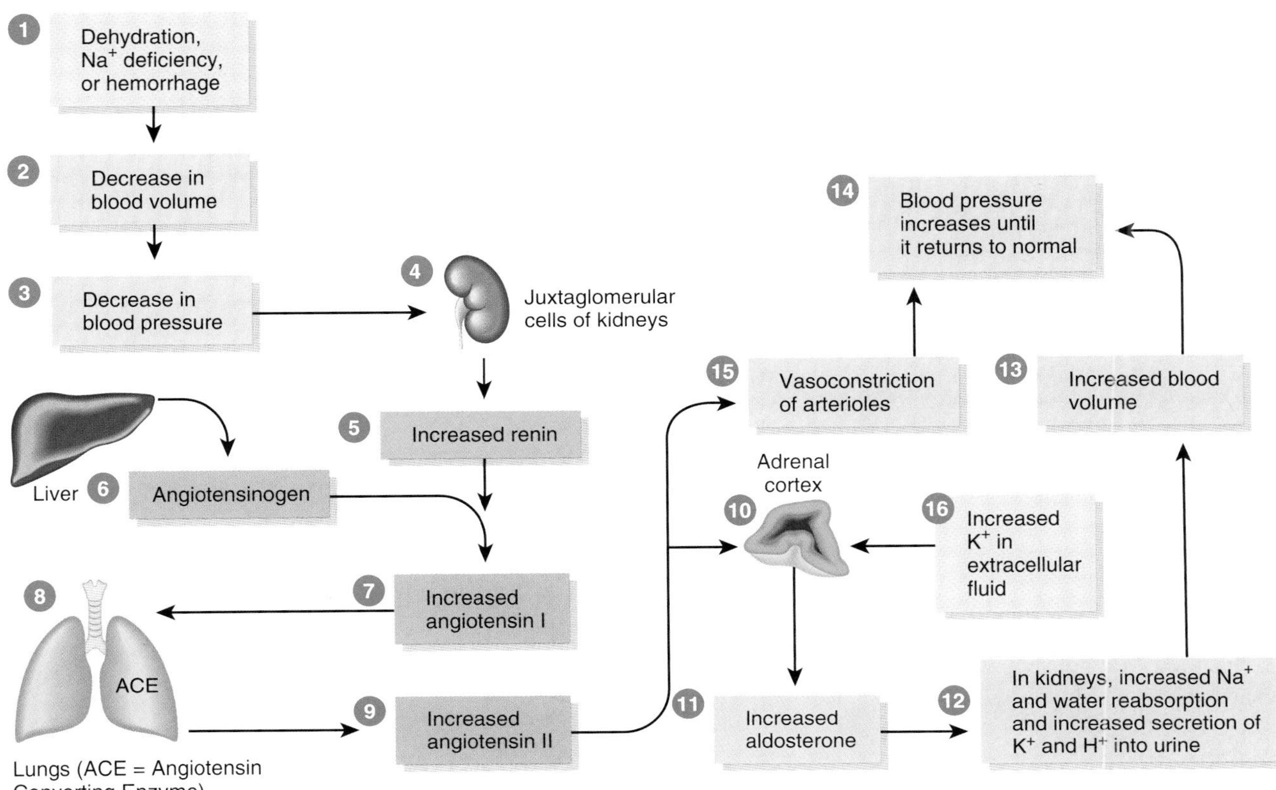

In what two ways can angiotensin II increase blood pressure, and what are its target tissues in each case?

6 Renin converts **angiotensinogen,** a plasma protein produced by the liver, into **angiotensin I.**

7 Blood containing increased levels of angiotensin I circulates in the body.

8 As blood flows through capillaries, particularly those of the lungs, the enzyme **angiotensin-converting enzyme (ACE)** converts angiotensin I into the hormone **angiotensin II.**

9 Blood level of angiotensin II increases.

10 Angiotensin II stimulates the adrenal cortex to secrete aldosterone.

11 Blood containing increased levels of aldosterone circulates to the kidneys.

12 In the kidneys, aldosterone increases reabsorption of Na^+ and water so that less is lost in the urine. Aldosterone also stimulates the kidneys to increase secretion of K^+ and H^+ into the urine.

13 With increased water reabsorption by the kidneys, blood volume increases.

14 As blood volume increases, blood pressure increases to normal.

15 Angiotensin II also stimulates contraction of smooth muscle in the walls of arterioles. The resulting vasoconstriction of the arterioles increases blood pressure and thus helps raise blood pressure to normal.

16 Besides angiotensin II, a second stimulator of aldosterone secretion is an increase in the K^+ concentration of blood (or interstitial fluid). A decrease in the blood K^+ level has the opposite effect.

Glucocorticoids

The glucocorticoids, which regulate metabolism and resistance to stress, include **cortisol (hydrocortisone), corticosterone,** and **cortisone.** Of these three hormones secreted by the zona fasciculata, cortisol is the most abundant, accounting for about 95% of glucocorticoid activity.

Control of glucocorticoid secretion occurs via a typical negative feedback system (Figure 18.17). Low blood levels of glucocorticoids, mainly cortisol, stimulate neurosecretory cells in the hypothalamus to secrete **corticotropin-releasing hormone (CRH).** CRH (together with a low level of cortisol) promotes the release of ACTH from the anterior pituitary. ACTH flows in the blood to the adrenal cortex, where it stimulates glucocorticoid secretion. (To a much smaller extent, ACTH also stimulates secretion of aldosterone.) The discussion of stress at the end of the chapter describes how the hypothalamus also increases CRH release in response to a variety of physical and emotional stresses.

Glucocorticoids have the following effects:

1. ***Protein breakdown.*** Glucocorticoids increase the rate of protein breakdown, mainly in muscle fibers, and thus increase the liberation of amino acids into the bloodstream. The amino acids may be used by body cells for synthesis of new proteins or for ATP production.

2. ***Glucose formation.*** Upon stimulation by glucocorticoids, liver cells may convert certain amino acids or lactic acid to glucose, which neurons and other cells can use for ATP production. Such conversion of a substance other than glycogen or another monosaccharide into glucose is called **gluconeogenesis** (gloo′-ko-nē′-ō-JEN-e-sis).

3. ***Lipolysis.*** Glucocorticoids stimulate **lipolysis,** the breakdown of triglycerides and release of fatty acids from adipose tissue into the blood.

4. ***Resistance to stress.*** Glucocorticoids work in many ways to provide resistance to stress. The additional glucose supplied by the liver cells provides tissues with a ready source of ATP to combat a range of stresses, including exercise, fasting, fright, temperature extremes, high altitude, bleeding, infection, surgery, trauma, and disease. Because glucocorticoids make blood vessels more sensitive to other hormones that cause vasoconstriction, they raise blood pressure. This effect would be an advantage in cases of severe blood loss, which causes blood pressure to drop.

5. ***Anti-inflammatory effects.*** Glucocorticoids inhibit white blood cells that participate in inflammatory responses. Unfortunately, glucocorticoids also retard tissue repair, and as a result, they slow wound healing. Although high doses can cause severe mental disturbances, glucocorticoids are very useful in the treatment of chronic inflammatory disorders such as rheumatoid arthritis.

6. ***Depression of immune responses.*** High doses of glucocorticoids depress immune responses. For this reason, glucocorticoids are prescribed for organ transplant recipients to retard tissue rejection by the immune system.

Figure 18.17 **Negative feedback regulation of glucocorticoid secretion.**

A high level of CRH and a low level of glucocorticoids promote the release of ACTH, which stimulates glucocorticoid secretion by the adrenal cortex.

Some stimulus disrupts homeostasis by
Decreasing
Glucocorticoid level in blood
Receptors
Neurosecretory cells in hypothalamus
Input
Increased CRH and decreased cortisol
Control center
Corticotrophs in anterior pituitary
Output
Increased ACTH
Effectors
Cells of zona fasciculata in adrenal cortex secrete glucocorticoids
Increased glucocorticoid level in blood
Return to homeostasis when response brings glucocorticoid level in blood back to normal

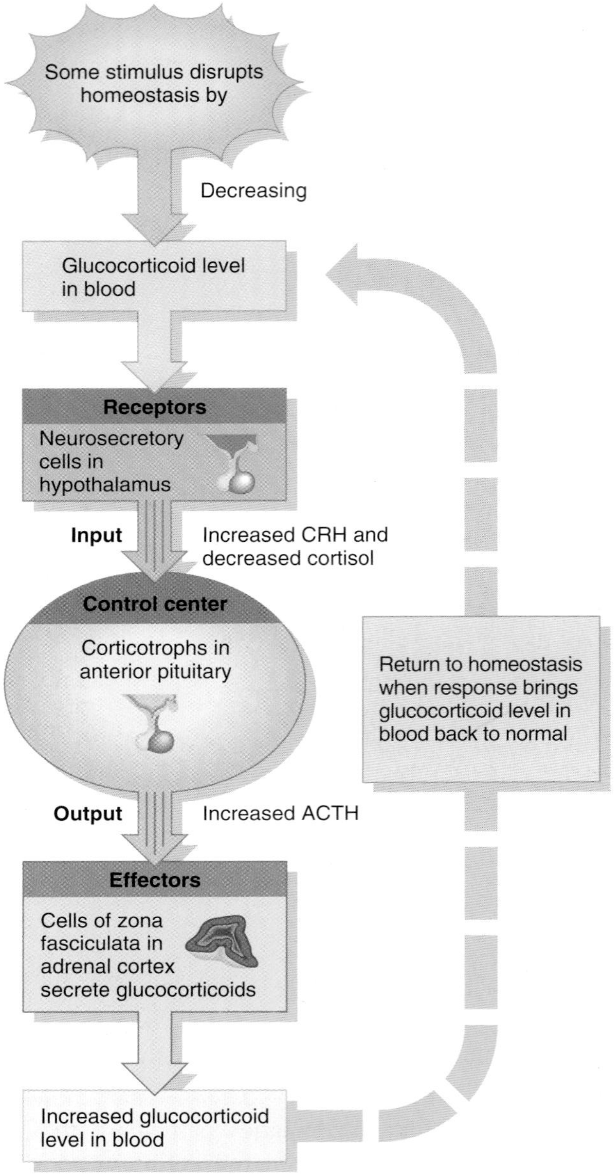

If a heart transplant patient receives prednisone (a glucocorticoid) to help prevent rejection of the transplanted tissue, will blood levels of ACTH and CRH be high or low? Explain.

Androgens

In both males and females, the adrenal cortex secretes small amounts of weak androgens. The major androgen secreted by the adrenal gland is **dehydroepiandrosterone (DHEA)** (dē-hī-drō-ep′-ē-an-DROS-ter-ōn). After puberty in males, the androgen testosterone is also released in much greater quantity by the testes. Thus, the amount of androgens secreted by the adrenal gland is usually so low that their effects are insignificant. In females, however, adrenal androgens play important roles. They promote libido (sex drive) and are converted into estrogens (feminizing sex steroids) by other body tissues. After menopause, when ovarian secretion of estrogens ceases, all female estrogens come from conversion of adrenal androgens. Adrenal androgens also stimulate growth of axillary and pubic hair in boys and girls and contribute to the prepubertal growth spurt. Although control of adrenal androgen secretion is not fully understood, the main hormone that stimulates its secretion is ACTH.

Congenital Adrenal Hyperplasia

Congenital adrenal hyperplasia (CAH) is a genetic disorder in which one or more enzymes needed for synthesis of cortisol are absent. Because the cortisol level is low, secretion of ACTH by the anterior pituitary is high due to lack of negative feedback inhibition. ACTH, in turn, stimulates growth and secretory activity of the adrenal cortex. As a result, both adrenal glands are enlarged. However, certain steps leading to synthesis of cortisol are blocked. Thus, precursor molecules accumulate, and some of these are weak androgens that can undergo conversion to testosterone. The result is **virilism,** or masculinization. In a female, virile characteristics include growth of a beard, development of a much deeper voice and a masculine distribution of body hair, growth of the clitoris so it may resemble a penis, atrophy of the breasts, and increased muscularity that produces a masculine physique. In prepubertal males, the syndrome causes the same characteristics as in females, plus rapid development of the male sexual organs and emergence of male sexual desires. In adult males, the virilizing effects of CAH are usually completely obscured by the normal virilizing effects of the testosterone secreted by the testes. As a result, CAH is often difficult to diagnose in adult males. Treatment involves cortisol therapy, which inhibits ACTH secretion and thus reduces production of adrenal androgens. ■

Adrenal Medulla

The inner region of the adrenal gland, the **adrenal medulla,** is a modified sympathetic ganglion of the autonomic nervous system (ANS). It develops from the same embryonic tissue as all other sympathetic ganglia, but its cells, which lack axons, form clusters around large blood vessels. Rather than releasing a neurotransmitter, the cells of the adrenal medulla secrete hormones. The hormone-producing cells, called **chromaffin cells** (KRŌ-maf-in; *chrom-* = color ; *-affin* = affinity for; see Figure 18.15d), are innervated by sympathetic preganglionic neurons in the splanchnic nerve. Because the ANS exerts direct control over the chromaffin cells, hormone release can occur very quickly.

The two major hormones synthesized by the adrenal medulla are **epinephrine** and **norepinephrine (NE),** also called adrenaline and noradrenaline, respectively. Cortisol secreted from the adrenal cortex induces synthesis of the enzyme needed to convert NE to epinephrine. Because the adrenal cortex surrounds the adrenal medulla, the cortisol level of blood in the medulla is normally quite high. Hence, about 80% of the medullary cells secrete epinephrine. Because they lack the converting enzyme, the remaining 20% secrete norepinephrine. Unlike the hormones of the adrenal cortex, the medullary hormones are not essential for life since they only intensify sympathetic responses in other parts of the body.

In stressful situations and during exercise, impulses from the hypothalamus stimulate sympathetic preganglionic neurons, which in turn stimulate the chromaffin cells to secrete epinephrine and norepinephrine. These two hormones greatly augment the fight-or-flight response that you learned about in Chapter 15. By increasing heart rate and force of contraction, epinephrine and norepinephrine increase the output of the heart, which increases blood pressure. They also increase blood flow to the heart, liver, skeletal muscles, and adipose tissue; dilate airways to the lungs; and increase blood levels of glucose and fatty acids.

Table 18.8 summarizes the hormones produced by the adrenal glands, control of their secretion, and their principal actions.

▶ CHECKPOINT

18. How do the adrenal cortex and adrenal medulla compare with regard to location and histology?
19. How is secretion of adrenal cortex hormones regulated?
20. How is the adrenal medulla related to the autonomic nervous system?

PANCREATIC ISLETS

▶ OBJECTIVE

Describe the location, histology, hormones, and functions of the pancreatic islets.

The **pancreas** (*pan-* = all; *-creas* = flesh) is both an endocrine gland and an exocrine gland. We discuss its endocrine functions

TABLE 18.8 Summary of Adrenal Gland Hormones

Hormones and Source	Control of Secretion	Principal Actions
Adrenal Cortex Hormones		
Mineralocorticoids (mainly aldosterone) from zona glomerulosa cells	Increased blood K^+ level and angiotensin II stimulate secretion.	Increase blood levels of Na^+ and water and decrease blood level of K^+.
Glucocorticoids (mainly cortisol) from zona fasciculata cells	ACTH stimulates release; corticotropin-releasing hormone (CRH) promotes ACTH secretion in response to stress and low blood levels of glucocorticoids.	Increase protein breakdown (except in liver), stimulate gluconeogenesis and lipolysis, provide resistance to stress, dampen inflammation, and depress immune responses.
Androgens (mainly dehydroepiandrosterone or **DHEA)** from zona reticularis cells	ACTH stimulates secretion.	Assist in early growth of axillary and pubic hair in both sexes; in females, contribute to libido and are source of estrogens after menopause.
Adrenal cortex		
Adrenal Medulla Hormones		
Epinephrine and **norepinephrine** from chromaffin cells	Sympathetic preganglionic neurons release acetylcholine, which stimulates secretion.	Produce effects that enhance those of the sympathetic division of the autonomic nervous system (ANS) during stress.
Adrenal medulla		

here and include its exocrine functions in Chapter 24 in the coverage of the digestive system. A flattened organ that measures about 12.5–15 cm (4.5–6 in.) in length, the pancreas is located in the curve of the duodenum, the first part of the small intestine, and consists of a head, a body, and a tail (Figure 18.18a). Roughly 99% of the cells of the pancreas are arranged in clusters called **acini.** The acini produce digestive enzymes, which flow into the gastrointestinal tract through a network of ducts. Scattered among the exocrine acini are 1–2 million tiny clusters of endocrine tissue called **pancreatic islets** or **islets of Langerhans** (LAHNG-er-hanz; Figure 18.18b, c). Abundant capillaries serve both the exocrine and endocrine portions of the pancreas.

Cell Types in the Pancreatic Islets

Each pancreatic islet includes four types of hormone-secreting cells:

1. **Alpha** or **A cells** constitute about 17% of pancreatic islet cells and secrete **glucagon** (GLOO-ka-gon).
2. **Beta** or **B cells** constitute about 70% of pancreatic islet cells and secrete **insulin** (IN-soo-lin).
3. **Delta** or **D cells** constitute about 7% of pancreatic islet cells and secrete **somatostatin** (identical to the growth hormone-inhibiting hormone secreted by the hypothalamus).
4. **F cells** constitute the remainder of pancreatic islet cells and secrete **pancreatic polypeptide.**

The interactions of the four pancreatic hormones are complex and not completely understood. We do know that glucagon raises blood glucose level, and insulin lowers it. Somatostatin acts in a paracrine manner to inhibit both insulin and glucagon release from neighboring beta and alpha cells. It may also act as a circulating hormone to slow absorption of nutrients from the gastrointestinal tract. Pancreatic polypeptide inhibits somatostatin secretion, gallbladder contraction, and secretion of digestive enzymes by the pancreas.

Regulation of Glucagon and Insulin Secretion

The principal action of glucagon is to increase blood glucose level when it falls below normal. Insulin, on the other hand, helps lower blood glucose level when it is too high. The level of blood glucose controls secretion of glucagon and insulin via negative feedback (Figure 18.19 on page 648):

Figure 18.18 Location, blood supply, and histology of the pancreas.

Pancreatic hormones regulate blood glucose level.

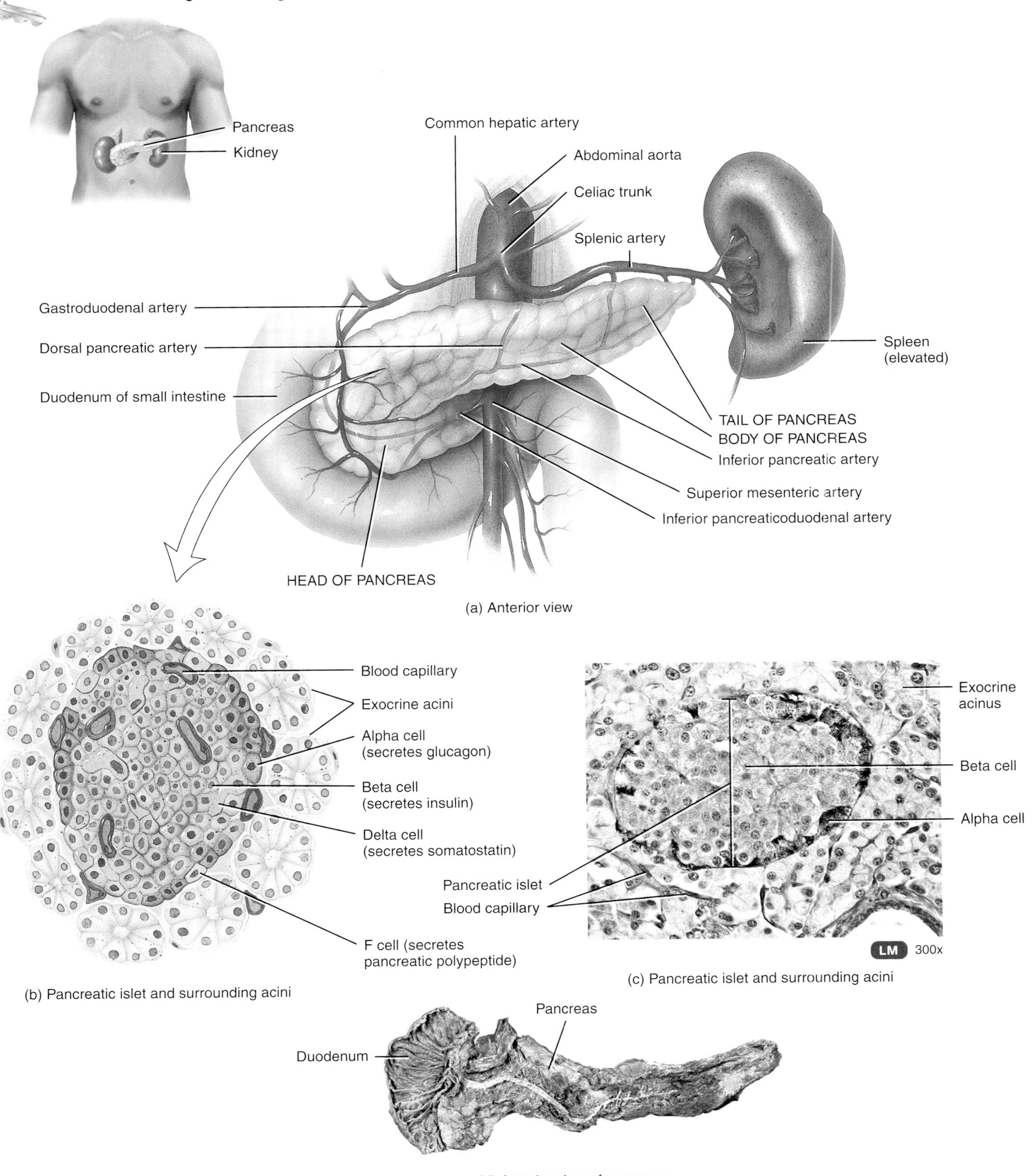

(a) Anterior view

(b) Pancreatic islet and surrounding acini

(c) Pancreatic islet and surrounding acini

(d) Anterior view of pancreas

Is the pancreas an exocrine gland or an endocrine gland?

1 Low blood glucose level (hypoglycemia) stimulates secretion of glucagon from alpha cells of the pancreatic islets.

2 Glucagon acts on hepatocytes (liver cells) to accelerate the conversion of glycogen into glucose (glycogenolysis) and to promote formation of glucose from lactic acid and certain amino acids (gluconeogenesis).

3 As a result, hepatocytes release glucose into the blood more rapidly, and blood glucose level rises.

4 If blood glucose continues to rise, high blood glucose level (hyperglycemia) inhibits release of glucagon (negative feedback).

5 High blood glucose (hyperglycemia) stimulates secretion of insulin by beta cells of the pancreatic islets.

6 Insulin acts on various cells in the body to accelerate facilitated diffusion of glucose into cells, especially skeletal muscle fibers; to speed conversion of glucose into glycogen (glycogenesis); to increase uptake of amino acids by cells and to increase protein synthesis; to speed synthesis of fatty acids (lipogenesis); to slow the conversion of glycogen to glucose (glycogenolysis); and to slow the formation of glucose from lactic acid and amino acids (gluconeogenesis).

7 As a result, blood glucose level falls.

8 If blood glucose level drops below normal, low blood glucose inhibits release of insulin (negative feedback) and stimulates release of glucagon.

Although blood glucose level is the most important regulator of insulin and glucagon, several hormones and neurotransmitters also regulate the release of these two hormones. In addition to the responses to blood glucose level just described, glucagon stimulates insulin release directly; insulin has the opposite effect, suppressing glucagon secretion. As blood glucose level declines and less insulin is secreted, the alpha cells of the pancreas are released from the inhibitory effect of insulin so they can secrete more glucagon. Indirectly, human growth hormone (hGH) and adrenocorticotropic hormone (ACTH) stimulate secretion of insulin because they act to elevate blood glucose.

Insulin secretion is also stimulated by:

- Acetylcholine, the neurotransmitter liberated from axon terminals of parasympathetic vagus nerve fibers that innervate the pancreatic islets,
- The amino acids arginine and leucine, which would be present in the blood at higher levels after a protein-containing meal, and
- Glucose-dependent insulinotropic peptide (GIP),* a hormone released by enteroendocrine cells of the small

*GIP—previously called gastric inhibitory peptide—was renamed because at physiological concentration its inhibitory effect on stomach function is negligible.

Figure 18.19 Negative feedback regulation of the secretion of glucagon (blue arrows) and insulin (orange arrows).

Low blood glucose stimulates release of glucagon; high blood glucose stimulates secretion of insulin.

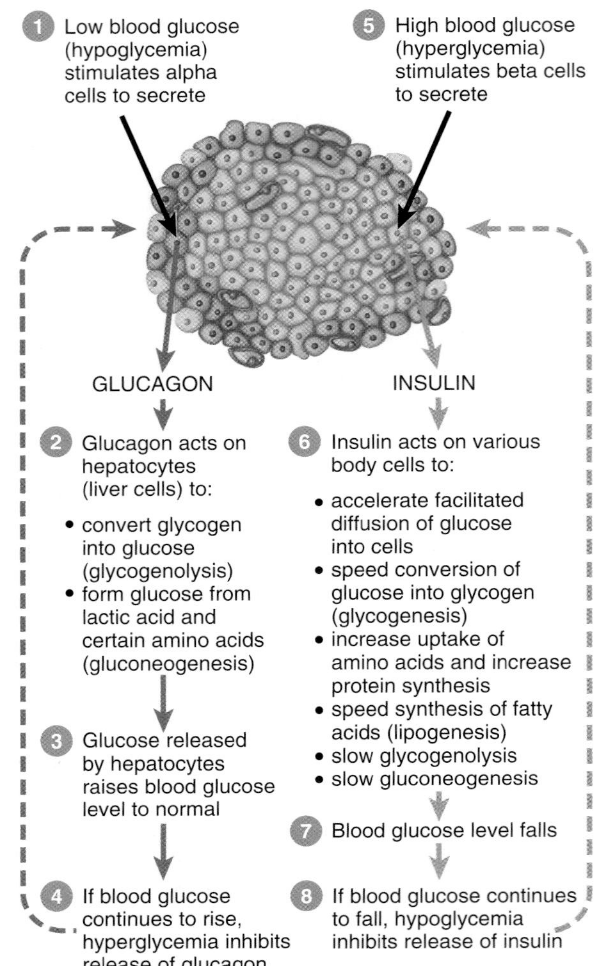
1 Low blood glucose (hypoglycemia) stimulates alpha cells to secrete
5 High blood glucose (hyperglycemia) stimulates beta cells to secrete
GLUCAGON
INSULIN
2 Glucagon acts on hepatocytes (liver cells) to:
• convert glycogen into glucose (glycogenolysis)
• form glucose from lactic acid and certain amino acids (gluconeogenesis)
3 Glucose released by hepatocytes raises blood glucose level to normal
4 If blood glucose continues to rise, hyperglycemia inhibits release of glucagon
6 Insulin acts on various body cells to:
• accelerate facilitated diffusion of glucose into cells
• speed conversion of glucose into glycogen (glycogenesis)
• increase uptake of amino acids and increase protein synthesis
• speed synthesis of fatty acids (lipogenesis)
• slow glycogenolysis
• slow gluconeogenesis
7 Blood glucose level falls
8 If blood glucose continues to fall, hypoglycemia inhibits release of insulin

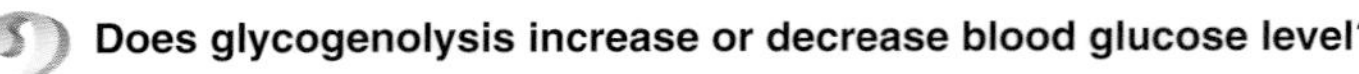
Does glycogenolysis increase or decrease blood glucose level?

intestine in response to the presence of glucose in the gastrointestinal tract.

Thus, digestion and absorption of food containing both carbohydrates and proteins provide strong stimulation for insulin release.

Glucagon secretion is stimulated by:

- Increased activity of the sympathetic division of the ANS, as occurs during exercise, and
- A rise in blood amino acids if blood glucose level is low, which could occur after a meal that contained mainly protein.

Table 18.9 summarizes the hormones produced by the pancreas, control of their secretion, and their principal actions.

▶ CHECKPOINT

21. How are blood levels of glucagon and insulin controlled?

22. What are the effects on secretion of insulin and glucagon of exercise versus eating a carbohydrate- and protein-rich meal?

TABLE 18.9 Summary of Pancreatic Islet Hormones

Hormone and Source	Control of Secretion	Principal Actions
Glucagon from alpha cells of pancreatic islets Alpha cell	Decreased blood level of glucose, exercise and mainly protein meals stimulate secretion; somatostatin and insulin inhibit secretion.	Raises blood glucose level by accelerating breakdown of glycogen into glucose in liver (glycogenolysis), converting other nutrients into glucose in liver (gluconeogenesis), and releasing glucose into the blood.
Insulin from beta cells of pancreatic islets Beta cell	Increased blood level of glucose, acetylcholine (released by parasympathetic vagus nerve fibers), arginine and leucine (two amino acids), glucagon, GIP, hGH, and ACTH stimulate secretion; somatostatin inhibits secretion.	Lowers blood glucose level by accelerating transport of glucose into cells, converting glucose into glycogen (glycogenesis), and decreasing glycogenolysis and gluconeogenesis; also increases lipogenesis and stimulates protein synthesis.
Somatostatin from delta cells of pancreatic islets Delta cell	Pancreatic polypeptide inhibits secretion.	Inhibits secretion of insulin and glucagon and slows absorption of nutrients from the gastrointestinal tract.
Pancreatic polypeptide from F cells of pancreatic islets F cell	Meals containing protein, fasting, exercise, and acute hypoglycemia stimulate secretion; somatostatin and elevated blood glucose level inhibit secretion.	Inhibits somatostatin secretion, gallbladder contraction, and secretion of pancreatic digestive enzymes.

OVARIES AND TESTES

► **OBJECTIVE**

Describe the location, hormones, and functions of the male and female gonads.

Gonads are the organs that produce gametes—sperm in males and oocytes in females. In addition to their reproductive function, the gonads secrete hormones. The **ovaries,** paired oval bodies located in the female pelvic cavity, produce several steroid hormones including two **estrogens** (estradiol and estrone) and **progesterone.** These female sex hormones, along with FSH and LH from the anterior pituitary, regulate the menstrual cycle, maintain pregnancy, and prepare the mammary glands for lactation. They also promote enlargement of the breasts and widening of the hips at puberty, and help maintain these female secondary sex characteristics. The ovaries also produce **inhibin,** a protein hormone that inhibits secretion of follicle-stimulating hormone (FSH). During pregnancy, the ovaries and placenta produce a peptide hormone called **relaxin,** which increases the flexibility of the pubic symphysis during pregnancy and helps dilate the uterine cervix during labor and delivery. These actions help ease the baby's passage by enlarging the birth canal.

The male gonads, the **testes,** are oval glands that lie in the scrotum. The main hormone produced and secreted by the testes is **testosterone,** an **androgen** or male sex hormone. Testosterone regulates production of sperm and stimulates the development and maintenance of male secondary sex characteristics, such as beard growth and deepening of the voice. The testes also produce inhibin, which inhibits secretion of FSH. The detailed structure of the ovaries and testes and the specific roles of sex hormones are discussed in Chapter 28.

Table 18.10 summarizes the hormones produced by the ovaries and testes and their principal actions.

TABLE 18.10 Summary of Hormones of the Ovaries and Testes

Hormone	Principal Actions
Ovarian Hormones	
Estrogens and **progesterone** Ovaries	Together with gonadotropic hormones of the anterior pituitary, regulate the female reproductive cycle, regulate oogenesis, maintain pregnancy, prepare the mammary glands for lactation, and promote development and maintenance of female secondary sex characteristics.
Relaxin	Increases flexibility of pubic symphysis during pregnancy and helps dilate uterine cervix during labor and delivery.
Inhibin	Inhibits secretion of FSH from anterior pituitary.
Testicular Hormones	
Testosterone 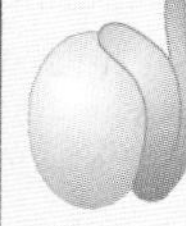 Testes	Stimulates descent of testes before birth, regulates spermatogenesis, and promotes development and maintenance of male secondary sex characteristics.
Inhibin	Inhibits secretion of FSH from anterior pituitary.

► **CHECKPOINT**

23. Why are the ovaries and testes classified as endocrine glands as well as reproductive organs?

PINEAL GLAND

► **OBJECTIVE**

Describe the location, histology, hormone, and functions of the pineal gland.

The **pineal gland** (PĪN-ē-al = pinecone shape) is a small endocrine gland attached to the roof of the third ventricle of the brain at the midline (see Figure 18.1). Part of the epithalamus, it is positioned between the two superior colliculi, has a mass of 0.1–0.2 g, and is covered by a capsule formed by the pia mater. The gland consists of masses of neuroglia and secretory cells called **pinealocytes** (pin-ē-AL̄-ō-sītz).

Although many anatomical features of the pineal gland have been known for years, its physiological roles are still unclear. We do know that the pineal gland secretes **melatonin,** an amine hormone derived from serotonin, and that more melatonin is released in darkness and less in strong sunlight. Sympathetic postganglionic axons from the superior cervical ganglion extend to the pineal gland and form synaptic contacts with pinealocytes. In darkness, norepinephrine released by the sympathetic fibers stimulates synthesis and secretion of melatonin, which may promote sleepiness.

Melatonin is thought to contribute to setting the body's biological clock, which is controlled from the suprachiasmatic nucleus of the hypothalamus. During sleep, plasma levels of melatonin increase tenfold and then decline to a low level again before awakening. Small doses of melatonin given orally can induce sleep and reset daily rhythms, which might benefit workers whose shifts alternate between daylight and nighttime hours. Melatonin also is a potent antioxidant that may provide some protection against damaging oxygen free radicals.

In animals that breed during specific seasons, melatonin inhibits reproductive functions, but it is unclear whether melatonin influences human reproductive function. Melatonin levels are higher in children and decline with age into adulthood, but there is no evidence that changes in melatonin secretion correlate with the onset of puberty and sexual maturation. Nevertheless, because melatonin causes atrophy of the gonads in several animal species, the possibility of adverse effects on human reproduction must be studied before its use to reset daily rhythms can be recommended.

Seasonal Affective Disorder and Jet Lag

Seasonal affective disorder (SAD) is a type of depression that afflicts some people during the winter months, when day length is short. It is thought to be due, in part, to overproduction of melatonin. Full-spectrum bright-light therapy—repeated doses of several hours of exposure to artificial light as bright as sunlight—provides relief for some people. Three to six hours of exposure to bright light also appears to speed recovery from jet lag, the fatigue suffered by travelers who quickly cross several time zones. ■

▶ CHECKPOINT

24. What is the relationship between melatonin and sleep?

THYMUS

The **thymus** is located behind the sternum between the lungs. Because of its role in immunity, the details of the structure and functions of the thymus are discussed in Chapter 22. The hormones produced by the thymus—**thymosin, thymic humoral factor (THF), thymic factor (TF),** and **thymopoietin**—promote the maturation of T cells (a type of white blood cell that destroys microbes and foreign substances) and may retard the aging process.

OTHER ENDOCRINE TISSUES AND ORGANS, EICOSANOIDS, AND GROWTH FACTORS

▶ OBJECTIVES

List the hormones secreted by cells in tissues and organs other than endocrine glands, and describe their functions.

Describe the actions of eicosanoids and growth factors.

Hormones from Other Endocrine Tissues and Organs

As you learned at the beginning of this chapter, cells in organs other than those usually classified as endocrine glands have an endocrine function and secrete hormones. You learned about several of these in this chapter: the hypothalamus, thymus, pancreas, ovaries, and testes. Table 18.11 provides an overview of these organs and tissues and their hormones and actions.

TABLE 18.11 Summary of Hormones Produced by Other Organs and Tissues that Contain Endocrine Cells

Hormone	Principal Actions
Gastrointestinal Tract	
Gastrin	Promotes secretion of gastric juice and increases movements of the stomach.
Glucose-dependent insulinotropic peptide (GIP)	Stimulates release of insulin by pancreatic beta cells.
Secretin	Stimulates secretion of pancreatic juice and bile.
Cholecystokinin (CCK)	Stimulates secretion of pancreatic juice, regulates release of bile from the gallbladder, and brings about a feeling of fullness after eating.
Placenta	
Human chorionic gonadotropin (hCG)	Stimulates the corpus luteum in the ovary to continue the production of estrogens and progesterone to maintain pregnancy.
Estrogens and **progesterone**	Maintain pregnancy and help prepare mammary glands to secrete milk.
Human chorionic somatomammotropin (hCS)	Stimulates the development of the mammary glands for lactation.
Kidneys	
Renin	Part of a sequence of reactions that raises blood pressure by bringing about vasoconstriction and secretion of aldosterone.
Erythropoietin (EPO)	Increases rate of red blood cell formation.
Calcitriol* (active form of vitamin D)	Aids in the absorption of dietary calcium and phosphorus.
Heart	
Atrial natriuretic peptide (ANP)	Decreases blood pressure.
Adipose Tissue	
Leptin	Suppresses appetite and may increase the activity of FSH and LH.

*Synthesis begins in the skin, continues in the liver, and ends in the kidneys.

Eicosanoids

Two families of eicosanoid molecules—the **prostaglandins** (pros′-ta-GLAN-dins), or **PGs,** and the **leukotrienes** (loo-kō-TRĪ-ēns), or **LTs**—are found in virtually all body cells except red blood cells, where they act as local hormones (paracrines or autocrines) in response to chemical or mechanical stimuli. They are synthesized by clipping a 20-carbon fatty acid called **arachidonic acid** from membrane phospholipid molecules. From arachidonic acid, different enzymatic reactions produce PGs or LTs. **Thromboxane (TX)** is a modified PG that constricts blood vessels and promotes platelet activation. Appearing in the blood in minute quantities, eicosanoids are present only briefly due to rapid inactivation.

To exert their effects, eicosanoids bind to receptors on target-cell plasma membranes and stimulate or inhibit the synthesis of second messengers such as cyclic AMP. Leukotrienes stimulate chemotaxis (attraction to a chemical stimulus) of white blood cells and mediate inflammation. The prostaglandins alter smooth muscle contraction, glandular secretions, blood flow, reproductive processes, platelet function, respiration, nerve impulse transmission, lipid metabolism, and immune responses.

They also have roles in promoting inflammation and fever, and in intensifying pain.

Nonsteroidal Anti-inflammatory Drugs

In 1971, scientists solved the long-standing puzzle of how aspirin works. Aspirin and related **nonsteroidal anti-inflammatory drugs (NSAIDs),** such as ibuprofen (Motrin®), inhibit a key enzyme in prostaglandin synthesis without affecting synthesis of leukotrienes. NSAIDs are used to treat a wide var-iety of inflammatory disorders, from rheumatoid arthritis to tennis elbow. The success of NSAIDs in reducing fever, pain, and inflammation shows how prostaglandins contribute to these woes. ■

Growth Factors

Several of the hormones we have described—insulinlike growth factor, thymosin, insulin, thyroid hormones, human growth hormone, and prolactin—stimulate cell growth and division. In addition, several more recently discovered hormones called **growth factors** play important roles in tissue development, growth, and repair. Growth factors are *mitogenic* substances—they cause growth by stimulating cell division. Many growth factors act locally, as autocrines or paracrines.

A summary of sources and actions of six important growth factors is presented in Table 18.12.

▶ CHECKPOINT

25. What hormones are secreted by the gastrointestinal tract, placenta, kidneys, skin, adipose tissue, and heart?

26. What are some functions of prostaglandins, leukotrienes, and growth factors?

TABLE 18.12 Summary of Selected Growth Factors

Growth Factor	Comment
Epidermal growth factor (EGF)	Produced in submaxillary (salivary) glands; stimulates proliferation of epithelial cells, fibroblasts, neurons, and astrocytes; suppresses some cancer cells and secretion of gastric juice by the stomach.
Platelet-derived growth factor (PDGF)	Produced in blood platelets; stimulates proliferation of neuroglia, smooth muscle fibers, and fibroblasts; appears to have a role in wound healing; may contribute to the development of atherosclerosis.
Fibroblast growth factor (FGF)	Found in pituitary gland and brain; stimulates proliferation of many cells derived from embryonic mesoderm (fibroblasts, adrenocortical cells, smooth muscle fibers, chondrocytes, and endothelial cells); also stimulates formation of new blood vessels (angiogenesis).
Nerve growth factor (NGF)	Produced in submaxillary (salivary) glands and hippocampus of brain; stimulates the growth of ganglia in embryonic life, maintains sympathetic nervous system; stimulates hypertrophy and differentiation of neurons.
Tumor angiogenesis factors (TAFs)	Produced by normal and tumor cells; stimulate growth of new capillaries, organ regeneration, and wound healing.
Transforming growth factors (TGFs)	Produced by various cells as separate molecules called TGF-alpha and TGF-beta. TGF-alpha has activities similar to epidermal growth factor, and TGF-beta inhibits proliferation of many cell types.

THE STRESS RESPONSE

▶ OBJECTIVE

Describe how the body responds to stress.

It is impossible to remove all of the stress from our everyday lives. Some stress, called **eustress,** prepares us to meet certain challenges and thus is helpful. Other stress, called **distress,** is harmful. Any stimulus that produces a stress response is called a **stressor.** A stressor may be almost any disturbance of the human body—heat or cold, environmental poisons, toxins given off by bacteria, heavy bleeding from a wound or surgery, or a strong emotional reaction. The responses to stressors may be pleasant or unpleasant, and they vary among people and even within the same person at different times.

Your body's homeostatic mechanisms attempt to counteract stress. When they are successful, the internal environment remains within normal physiological limits. If stress is extreme, unusual, or long lasting, the normal mechanisms may not be enough. In 1936, Hans Selye, a pioneer in stress research, showed that a variety of stressful conditions or noxious agents elicit a similar sequence of bodily changes. These changes, called the **stress response** or **general adaptation syndrome (GAS),** are controlled mainly by the hypothalamus. The stress response occurs in three stages: (1) an initial fight-or-flight response, (2) a slower resistance reaction, and eventually (3) exhaustion.

The Fight-or-Flight Response

The **fight-or-flight response,** initiated by nerve impulses from the hypothalamus to the sympathetic division of the autonomic nervous system (ANS), including the adrenal medulla, quickly mobilizes the body's resources for immediate physical activity (Figure 18.20a). It brings huge amounts of glucose and oxygen to the organs that are most active in warding off danger: the

Figure 18.20 Responses to stressors during the stress response. Red arrows (hormonal responses) and green arrows (neural responses) in (a) indicate immediate "fight-or- flight" reactions; black arrows in (b) indicate long-term resistance reactions.

Stressors stimulate the hypothalamus to initiate the stress response through the fight-or-flight response and the resistance reaction.

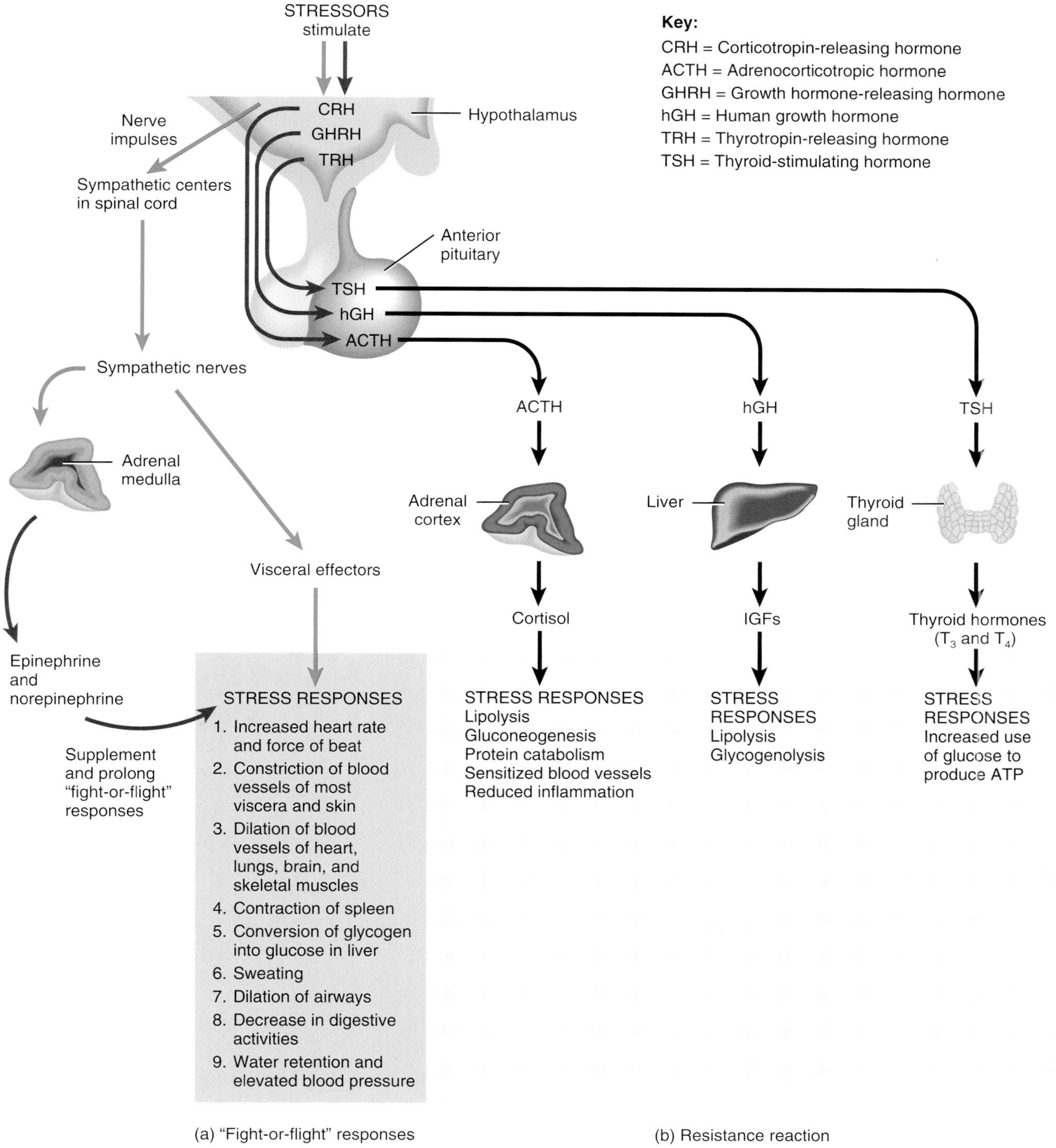

STRESSORS
stimulate

CRH
GHRH
TRH
Hypothalamus

Nerve impulses

Sympathetic centers in spinal cord

Anterior pituitary

TSH
hGH
ACTH

Sympathetic nerves

Adrenal medulla

Visceral effectors

Epinephrine and norepinephrine

Supplement and prolong "fight-or-flight" responses

STRESS RESPONSES
1. Increased heart rate and force of beat
2. Constriction of blood vessels of most viscera and skin
3. Dilation of blood vessels of heart, lungs, brain, and skeletal muscles
4. Contraction of spleen
5. Conversion of glycogen into glucose in liver
6. Sweating
7. Dilation of airways
8. Decrease in digestive activities
9. Water retention and elevated blood pressure

Key:
CRH = Corticotropin-releasing hormone
ACTH = Adrenocorticotropic hormone
GHRH = Growth hormone-releasing hormone
hGH = Human growth hormone
TRH = Thyrotropin-releasing hormone
TSH = Thyroid-stimulating hormone

ACTH
Adrenal cortex
Cortisol
STRESS RESPONSES
Lipolysis
Gluconeogenesis
Protein catabolism
Sensitized blood vessels
Reduced inflammation

hGH
Liver
IGFs
STRESS RESPONSES
Lipolysis
Glycogenolysis

TSH
Thyroid gland
Thyroid hormones (T_3 and T_4)
STRESS RESPONSES
Increased use of glucose to produce ATP

(a) "Fight-or-flight" responses

(b) Resistance reaction

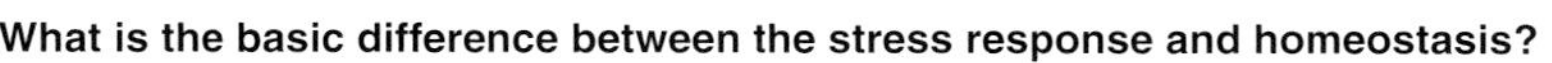

What is the basic difference between the stress response and homeostasis?

brain, which must become highly alert; the skeletal muscles, which may have to fight off an attacker or flee; and the heart, which must work vigorously to pump enough blood to the brain and muscles. During the fight-or-flight response, nonessential body functions such as digestive, urinary, and reproductive activities are inhibited. Reduction of blood flow to the kidneys promotes release of renin, which sets into the motion the renin–angiotensin–aldosterone pathway (see Figure 18.16). Aldosterone causes the kidneys to retain Na^+, which leads to water retention and elevated blood pressure. Water retention also helps preserve body fluid volume in the case of severe bleeding.

The Resistance Reaction

The second stage in the stress response is the **resistance reaction** (Figure 18.20b). Unlike the short-lived fight-or-flight response, which is initiated by nerve impulses from the hypothalamus, the resistance reaction is initiated in large part by hypothalamic releasing hormones and is a longer-lasting response. The hormones involved are corticotropin-releasing hormone (CRH), growth hormone–releasing hormone (GHRH), and thyrotropin-releasing hormone (TRH).

CRH stimulates the anterior pituitary to secrete ACTH, which in turn stimulates the adrenal cortex to increase release of cortisol. Cortisol then stimulates gluconeogenesis by liver cells, breakdown of triglycerides into fatty acids (lipolysis), and catabolism of proteins into amino acids. Tissues throughout the body can use the resulting glucose, fatty acids, and amino acids to produce ATP or to repair damaged cells. Cortisol also reduces inflammation.

A second hypothalamic releasing hormone, GHRH, causes the anterior pituitary to secrete human growth hormone (hGH). Acting via insulinlike growth factors, hGH stimulates lipolysis and glycogenolysis, the breakdown of glycogen to glucose, in the liver. A third hypothalamic releasing hormone, TRH, stimulates the anterior pituitary to secrete thyroid-stimulating hormone (TSH). TSH promotes secretion of thyroid hormones, which stimulate the increased use of glucose for ATP production. The combined actions of hGH and TSH supply additional ATP for metabolically active cells throughout the body.

The resistance stage helps the body continue fighting a stressor long after the fight-or-flight response dissipates. This is why your heart continues to pound for several minutes even after the stressor is removed. Generally, it is successful in seeing us through a stressful episode, and our bodies then return to normal. Occasionally, however, the resistance stage fails to combat the stressor, and the body moves into the state of exhaustion.

Exhaustion

The resources of the body may eventually become so depleted that they cannot sustain the resistance stage, and **exhaustion** ensues. Prolonged exposure to high levels of cortisol and other hormones involved in the resistance reaction causes wasting of muscle, suppression of the immune system, ulceration of the gastrointestinal tract, and failure of pancreatic beta cells. In addition, pathological changes may occur because resistance reactions persist after the stressor has been removed.

Stress and Disease

Although the exact role of stress in human diseases is not known, it is clear that stress can lead to particular diseases by temporarily inhibiting certain components of the immune system. Stress-related disorders include gastritis, ulcerative colitis, irritable bowel syndrome, hypertension, asthma, rheumatoid arthritis (RA), migraine headaches, anxiety, and depression. People under stress are at a greater risk of developing chronic disease or dying prematurely.

Interleukin-1, a substance secreted by macrophages of the immune system (see page 630), is an important link between stress and immunity. One action of interleukin-1 is to stimulate secretion of ACTH, which in turn stimulates the production of cortisol. Not only does cortisol provide resistance to stress and inflammation, but it also suppresses further production of interleukin-1. Thus, the immune system turns on the stress response, and the resulting cortisol then turns off one immune system mediator. This negative feedback system keeps the immune response in check once it has accomplished its goal. Because of this activity, cortisol and other glucocorticoids are used as immunosuppressive drugs for organ transplant recipients.

Posttraumatic Stress Disorder

Posttraumatic stress disorder (PTSD) is an anxiety disorder that may develop in an individual who has experienced, witnessed, or learned about a physically or psychologically distressing event. The immediate cause of PTSD appears to be the specific stressors associated with the events. Among the stressors are terrorism, hostage taking, imprisonment, military duty, serious accidents, torture, sexual or physical abuse, violent crimes, school shootings, massacres, and natural disasters. In the United States, PTSD affects 10% of females and 5% of males. Symptoms of PTSD include reexperiencing the event through nightmares or flashbacks; avoidance of any activity, person, place, or event associated with the stressors; loss of interest and lack of motivation; poor concentration; irritability; and insomnia. Treatment may include the use of antidepressants, mood stabilizers, and antianxiety and antipsychotic agents. ■

▶ CHECKPOINT

27. What is the central role of the hypothalamus during stress?
28. What body reactions occur during the fight-or-flight response, the resistance reaction, and exhaustion?
29. What is the relationship between stress and immunity?

DEVELOPMENT OF THE ENDOCRINE SYSTEM

▶ OBJECTIVE

Describe the development of endocrine glands.

The development of the endocrine system is not as localized as the development of other systems because, as you have already learned, endocrine organs are distributed throughout the body.

About three weeks after fertilization, the *pituitary gland (hypophysis)* begins to develop from two different regions of the **ectoderm**. The *posterior pituitary (neurohypophysis)* is derived from an outgrowth of ectoderm called the **neurohypophyseal bud,** located on the floor of the hypothalamus (Figure 18.21). The *infundibulum,* also an outgrowth of the neurohypophyseal bud, connects the posterior pituitary to the hypothalamus. The *anterior pituitary (adenohypophysis)* is derived from an outgrowth of ectoderm from the roof of the mouth called the **hypophyseal (Rathke's) pouch.** The pouch grows toward the neurohypophyseal bud and eventually loses its connection with the roof of the mouth.

The *thyroid gland* develops during the fourth week as a midventral outgrowth of **endoderm,** called the **thyroid diverticulum,** from the floor of the pharynx at the level of the second pair of pharyngeal pouches (Figure 18.21a). The outgrowth projects inferiorly and differentiates into the right and left lateral lobes and the isthmus of the gland.

The *parathyroid glands* develop during the fourth week from **endoderm** as outgrowths from the third and fourth **pharyngeal pouches,** which help to form structures of the head and neck.

Figure 18.21 Development of the endocrine system.

Glands of the endocrine system develop from all three primary germ layers: ectoderm, mesoderm, and endoderm.

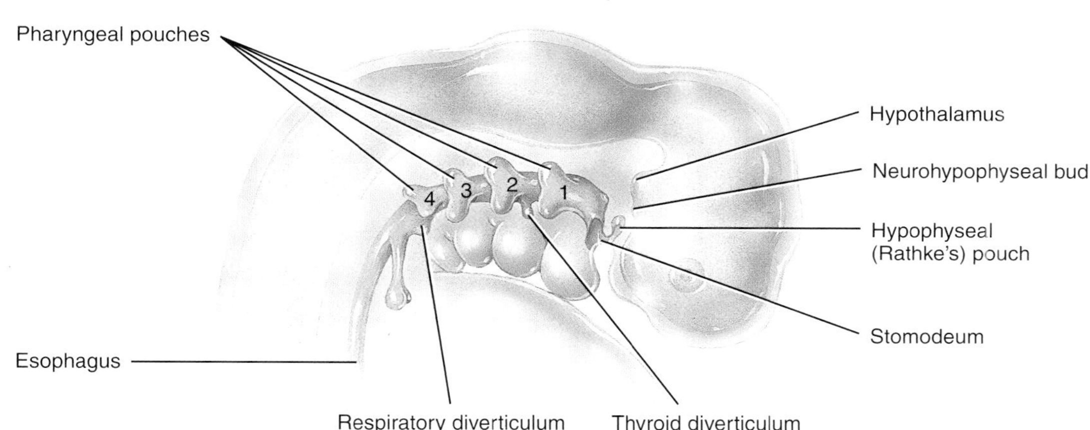

(a) Location of the neurohypophyseal bud, hypophyseal (Rathke's) pouch, thyroid diverticulum and pharyngeal pouches in a 28-day embryo

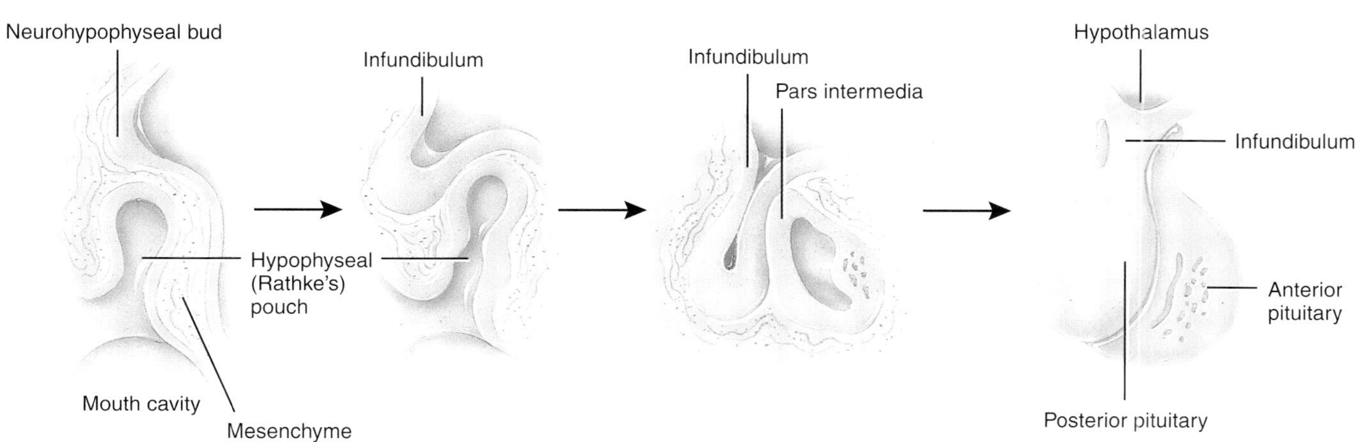

(b) Development of the pituitary gland between five and sixteen weeks

Which endocrine gland(s) develops from tissues with two different embryological origins?

The adrenal cortex and adrenal medulla develop during the fifth week and have completely different embryological origins. The *adrenal cortex* is derived from the same region of **mesoderm** that produces the gonads. Endocrine tissues that secrete steroid hormones all are derived from mesoderm. The *adrenal medulla* is derived from **ectoderm** from **neural crest** cells that migrate to the superior pole of the kidney. Recall that neural crest cells also give rise to sympathetic ganglia and other structures of the nervous system (see Figure 14.28b on page 515).

The *pancreas* develops during the fifth through seventh weeks from two outgrowths of **endoderm** from the part of the **foregut** that later becomes the duodenum (see Figure 29.12c on page 1119). The two outgrowths eventually fuse to form the pancreas. The origin of the ovaries and testes is discussed in the section on the reproductive system.

The *pineal gland* arises during the seventh week as an outgrowth between the thalamus and colliculi of the midbrain from **ectoderm** associated with the **diencephalon** (see Figure 14.29 on page 516).

The *thymus* arises during the fifth week from **endoderm** of the third pharyngeal pouches.

▶ CHECKPOINT

30. Compare the origins of the adrenal cortex and adrenal medulla.

AGING AND THE ENDOCRINE SYSTEM

▶ OBJECTIVE

Describe the effects of aging on the endocrine system.

Although some endocrine glands shrink as we get older, their performance may or may not be compromised. Production of human growth hormone by the anterior pituitary decreases, which is one cause of muscle atrophy as aging proceeds. The thyroid gland often decreases its output of thyroid hormones with age, causing a decrease in metabolic rate, an increase in body fat, and hypothyroidism, which is seen more often in older people. Because there is less negative feedback (lower levels of thyroid hormones), the level of thyroid-stimulating hormone increases with age (see Figure 18.12).

With aging, the blood level of PTH rises, perhaps due to inadequate dietary intake of calcium. In a study of older women who took 2,400 mg/day of supplemental calcium, blood levels of PTH were as low as those in younger women. Both calcitriol and calcitonin levels are lower in older persons. Together, the rise in PTH and the fall in calcitonin level heighten the age-related decrease in bone mass that leads to osteoporosis and increased risk of fractures (see Figure 18.14).

The adrenal glands contain increasingly more fibrous tissue and produce less cortisol and aldosterone with advancing age. However, production of epinephrine and norepinephrine remains normal. The pancreas releases insulin more slowly with age, and receptor sensitivity to glucose declines. As a result, blood glucose levels in older people increase faster and return to normal more slowly than in younger individuals.

The thymus is largest in infancy. After puberty, its size begins to decrease, and thymic tissue is replaced by adipose and areolar connective tissue. In older adults, the thymus has atrophied significantly. However, it still produces new T cells for immune responses.

The ovaries decrease in size with age, and they no longer respond to gonadotropins. The resultant decreased output of estrogens leads to conditions such as osteoporosis, high blood cholesterol, and atherosclerosis. FSH and LH levels are high due to less negative feedback inhibition of estrogens. Although testosterone production by the testes decreases with age, the effects are not usually apparent until very old age, and many elderly males can still produce active sperm in normal numbers.

▶ CHECKPOINT

31. Which hormone is related to the muscle atrophy that occurs with aging?

• • •

To appreciate the many ways the endocrine system contributes to homeostasis of other body systems, examine *Focus on Homeostasis: The Endocrine System* on page 657. Next, in Chapter 19, we will begin to explore the cardiovascular system, starting with a description of the composition and functions of blood.

FOCUS ON HOMEOSTASIS

The Endocrine System

BODY SYSTEM	CONTRIBUTION OF THE ENDOCRINE SYSTEM
For all body systems	Together with the nervous system, circulating and local hormones of the endocrine system regulate activity and growth of target cells throughout the body; several hormones regulate metabolism, uptake of glucose, and molecules used for ATP production by body cells.
Integumentary system	Androgens stimulate growth of axillary and pubic hair and activation of sebaceous glands; excess melanocyte-stimulating hormone (MSH) causes darkening of skin.
Skeletal system 	Human growth hormone (hGH) and insulinlike growth factors (IGFs) stimulate bone growth; estrogens cause closure of the epiphyseal plates at the end of puberty and help maintain bone mass in adults; parathyroid hormone (PTH) and calcitonin regulate levels of calcium and other minerals in bone matrix and blood; thyroid hormones are needed for normal development and growth of the skeleton.
Muscular system 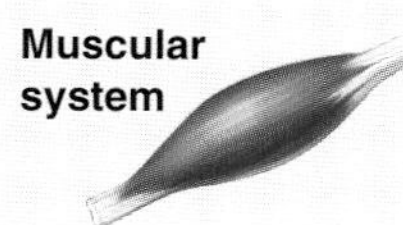	Epinephrine and norepinephrine help increase blood flow to exercising muscle; PTH maintains proper level of Ca^{2+}, needed for muscle contraction; glucagon, insulin, and other hormones regulate metabolism in muscle fibers; hGH, IGFs, and thyroid hormones help maintain muscle mass.
Nervous system 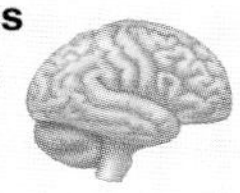	Several hormones, especially thyroid hormones, insulin, and growth hormone, influence growth and development of the nervous system; PTH maintains proper level of Ca^{2+}, needed for generation and conduction of nerve impulses.
Cardiovascular system 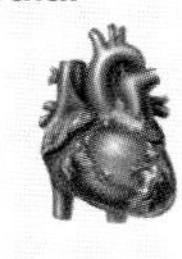	Erythropoietin (EPO) promotes formation of red blood cells; aldosterone and antidiuretic hormone (ADH) increase blood volume; epinephrine and norepinephrine increase heart rate and force of contraction; several hormones elevate blood pressure during exercise and other stresses.
Lymphatic system and immunity 	Glucocorticoids such as cortisol depress inflammation and immune responses; thymic hormones promote maturation of T cells (a type of white blood cell).
Respiratory system 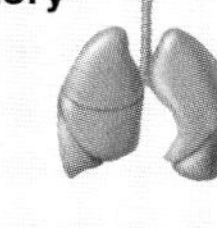	Epinephrine and norepinephrine dilate (widen) airways during exercise and other stresses; erythropoietin regulates amount of oxygen carried in blood by adjusting number of red blood cells.
Digestive system 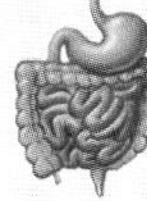	Epinephrine and norepinephrine depress activity of the digestive system; gastrin, cholecystokinin, secretin, and GIP help regulate digestion; calcitriol promotes absorption of dietary calcium; leptin suppresses appetite.
Urinary system 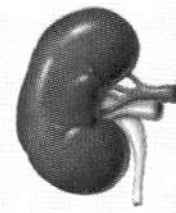	ADH, aldosterone, and atrial natriuretic peptide (ANP) adjust the rate of loss of water and ions in the urine, thereby regulating blood volume and ion content of the blood.
Reproductive systems 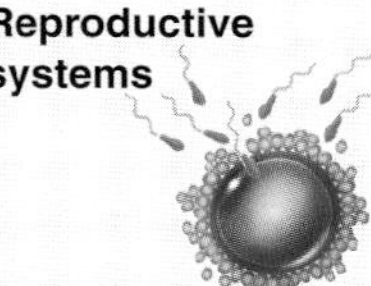	Hypothalamic releasing and inhibiting hormones, follicle-stimulating hormone (FSH), and luteinizing hormone (LH) regulate development, growth, and secretions of the gonads (ovaries and testes); estrogens and testosterone contribute to development of oocytes and sperm and stimulate development of secondary sex characteristics; prolactin promotes milk secretion in mammary glands; oxytocin causes contraction of the uterus and ejection of milk from the mammary glands.

DISORDERS: HOMEOSTATIC IMBALANCES

Disorders of the endocrine system often involve either **hyposecretion** (*hypo-* = too little or under), inadequate release of a hormone, or **hypersecretion** (*hyper-* = too much or above), excessive release of a hormone. In other cases, the problem is faulty hormone receptors, an inadequate number of receptors, or defects in second-messenger systems. Because hormones are distributed in the blood to target tissues throughout the body, problems associated with endocrine dysfunction may also be widespread.

Pituitary Gland Disorders

Pituitary Dwarfism, Giantism, and Acromegaly

Several disorders of the anterior pituitary involve human growth hormone (hGH). Hyposecretion of hGH during the growth years slows bone growth, and the epiphyseal plates close before normal height is reached. This condition is called **pituitary dwarfism.** Other organs of the body also fail to grow, and the body proportions are childlike. Treatment requires administration of hGH during childhood, before the epiphyseal plates close.

Hypersecretion of hGH during childhood causes **giantism,** an abnormal increase in the length of long bones. The person grows to be very tall, but body proportions are about normal. Figure 18.22a shows identical twins; one brother developed giantism due to a pituitary tumor. Hypersecretion of hGH during adulthood is called **acromegaly** (ak′-rō-MEG-a-lē). Although hGH cannot produce further lengthening of the long bones because the epiphyseal plates are already closed, the bones of the hands, feet, cheeks, and jaws thicken and other tissues enlarge. In addition, the eyelids, lips, tongue, and nose enlarge, and the skin thickens and develops furrows, especially on the forehead and soles (Figure 18.22b).

Figure 18.22 Various endocrine disorders.

Disorders of the endocrine system often involve hyposecretion or hypersecretion of hormones.

(a) A 22-year old man with pituitary giantism shown beside his identical twin

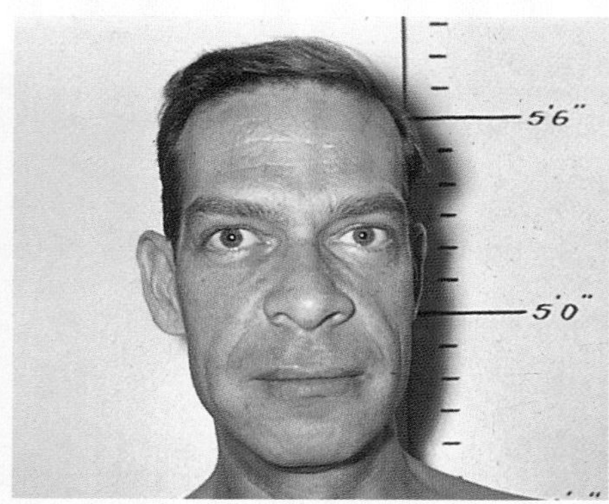

(b) Acromegaly (excess hGH during adulthood)

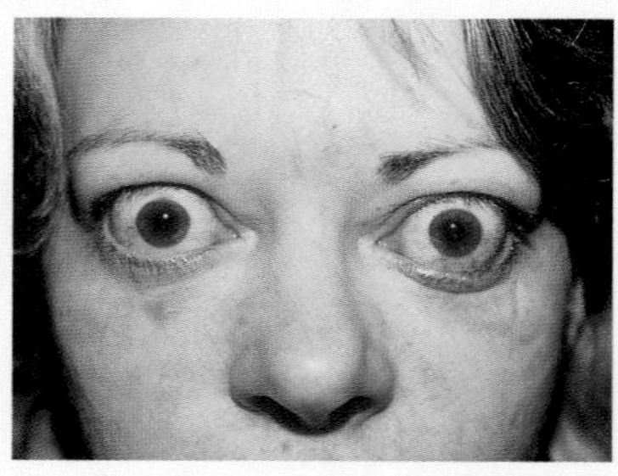

(c) Exophthalmos (excess thyroid hormones, as in Graves' disease)

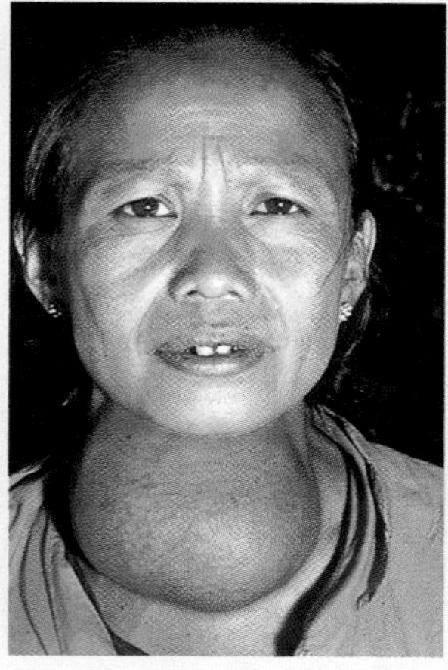

(d) Goiter (enlargement of thyroid gland)

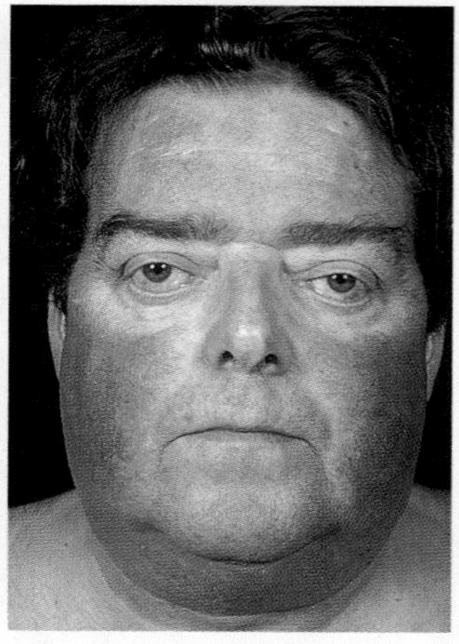

(e) Cushing's syndrome (excess glucocorticoids)

Which endocrine disorder is due to antibodies that mimic the action of TSH?

Diabetes Insipidus

The most common abnormality associated with dysfunction of the posterior pituitary is **diabetes insipidus** (dī-a-BĒ-tēs in-SIP-i-dus; *diabetes* = overflow; *insipidus* = tasteless) or **DI.** This disorder is due to defects in antidiuretic hormone (ADH) receptors or an inability to secrete ADH. *Neurogenic diabetes insipidus* results from hyposecretion of ADH, usually caused by a brain tumor, head trauma, or brain surgery that damages the posterior pituitary or the hypothalamus. In *nephrogenic diabetes insipidus,* the kidneys do not respond to ADH. The ADH receptors may be nonfunctional, or the kidneys may be damaged. A common symptom of both forms of DI is excretion of large volumes of urine, with resulting dehydration and thirst. Bed-wetting is common in afflicted children. Because so much water is lost in the urine, a person with DI may die of dehydration if deprived of water for only a day or so.

Treatment of neurogenic diabetes insipidus involves hormone replacement, usually for life. Either subcutaneous injection or nasal spray application of ADH analogs is effective. Treatment of nephrogenic diabetes insipidus is more complex and depends on the nature of the kidney dysfunction. Restriction of salt in the diet and, paradoxically, the use of certain diuretic drugs, are helpful.

Thyroid Gland Disorders

Thyroid gland disorders affect all major body systems and are among the most common endocrine disorders. **Congenital hypothyroidism,** hyposecretion of thyroid hormones that is present at birth, has devastating consequences if not treated promptly. Previously termed *cretinism,* this condition causes severe mental retardation and stunted bone growth. At birth, the baby typically is normal because lipid-soluble maternal thyroid hormones crossed the placenta during pregnancy and allowed normal development. Most states require testing of all newborns to ensure adequate thyroid function. If congenital hypothyroidism exists, oral thyroid hormone treatment must be started soon after birth and continued for life.

Hypothyroidism during the adult years produces **myxedema** (mix-e-DĒ-ma), which occurs about five times more often in females than in males. A hallmark of this disorder is edema (accumulation of interstitial fluid) that causes the facial tissues to swell and look puffy. A person with myxedema has a slow heart rate, low body temperature, sensitivity to cold, dry hair and skin, muscular weakness, general lethargy, and a tendency to gain weight easily. Because the brain has already reached maturity, mental retardation does not occur, but the person may be less alert. Oral thyroid hormones reduce the symptoms.

The most common form of hyperthyroidism is **Graves disease,** which also occurs seven to ten times more often in females than in males, usually before age 40. Graves disease is an autoimmune disorder in which the person produces antibodies that mimic the action of thyroid-stimulating hormone (TSH). The antibodies continually stimulate the thyroid gland to grow and produce thyroid hormones. A primary sign is an enlarged thyroid, which may be two to three times its normal size. Graves patients often have a peculiar edema behind the eyes, called **exophthalmos** (ek′-sof-THAL-mos), which causes the eyes to protrude (Figure 18.22c). Treatment may include surgical removal of part or all of the thyroid gland (thyroidectomy), the use of radioactive iodine (^{131}I) to selectively destroy thyroid tissue, and the use of antithyroid drugs to block synthesis of thyroid hormones.

A **goiter** (GOY-ter; *guttur* = throat) is simply an enlarged thyroid gland. It may be associated with hyperthyroidism, hypothyroidism, or **euthyroidism** (*eu* = good), which means normal secretion of thyroid hormone. In some places in the world, dietary iodine intake is inadequate; the resultant low level of thyroid hormone in the blood stimulates secretion of TSH, which causes thyroid gland enlargement (Figure 18.22d).

Parathyroid Gland Disorders

Hypoparathyroidism—too little parathyroid hormone—leads to a deficiency of blood Ca^{2+}, which causes neurons and muscle fibers to depolarize and produce action potentials spontaneously. This leads to twitches, spasms, and **tetany** (maintained contraction) of skeletal muscle. The leading cause of hypoparathyroidism is accidental damage to the parathyroid glands or to their blood supply during thyroidectomy surgery.

Hyperparathyroidism, an elevated level of parathyroid hormone, most often is due to a tumor of one of the parathyroid glands. An elevated level of PTH causes excessive resorption of bone matrix, raising the blood levels of calcium and phosphate ions and causing bones to become soft and easily fractured. High blood calcium level promotes formation of kidney stones. Fatigue, personality changes, and lethargy are also seen in patients with hyperparathyroidism.

Adrenal Gland Disorders

Cushing's Syndrome

Hypersecretion of cortisol by the adrenal cortex produces **Cushing's syndrome** (Figure 18.22e). Causes include a tumor of the adrenal gland that secretes cortisol, or a tumor elsewhere that secretes adrenocorticotropic hormone (ACTH), which in turn stimulates excessive secretion of cortisol. The condition is characterized by breakdown of muscle proteins and redistribution of body fat, resulting in spindly arms and legs accompanied by a rounded "moon face," "buffalo hump" on the back, and pendulous (hanging) abdomen. Facial skin is flushed, and the skin covering the abdomen develops stretch marks. The person also bruises easily, and wound healing is poor. The elevated level of cortisol causes hyperglycemia, osteoporosis, weakness, hypertension, increased susceptibility to infection, decreased resistance to stress, and mood swings. People who need long-term glucocorticoid therapy—for instance, to prevent rejection of a transplanted organ—may develop a cushinoid appearance.

Addison's Disease

Hyposecretion of glucocorticoids and aldosterone causes **Addison's disease (chronic adrenocortical insufficiency).** The majority of cases are autoimmune disorders in which antibodies cause adrenal cortex destruction or block binding of ACTH to its receptors. Pathogens, such as the bacterium that causes tuberculosis, also may trigger adrenal cortex destruction. Symptoms, which typically do not appear until 90% of the adrenal cortex has been destroyed, include mental lethargy, anorexia, nausea and vomiting, weight loss, hypoglycemia, and muscular weakness. Loss of aldosterone leads to elevated potassium and decreased sodium in the blood, low blood pressure, dehydration, decreased cardiac output, arrhythmias, and even cardiac arrest. The skin may have a "bronzed" appearance that often is mistaken for a suntan. Such was true in the case of President John F. Kennedy, whose Addison's disease was known to only a few while he was alive. Treatment consists of replacing glucocorticoids and mineralocorticoids and increasing sodium in the diet.

Pheochromocytomas

Usually benign tumors of the chromaffin cells of the adrenal medulla, called **pheochromocytomas** (fē-ō-krō′-mō-si-TŌ-mas; *pheo-* = dusky;

chromo- = color; *cyto-* = cell), cause hypersecretion of epinephrine and norepinephrine. The result is a prolonged version of the fight-or-flight response: rapid heart rate, high blood pressure, high levels of glucose in blood and urine, an elevated basal metabolic rate (BMR), flushed face, nervousness, sweating, and decreased gastrointestinal motility. Treatment is surgical removal of the tumor.

Pancreatic Islet Disorders

The most common endocrine disorder is **diabetes mellitus** (MEL-i-tus; *melli-* = honey sweetened), caused by an inability to produce or use insulin. Diabetes mellitus is the fourth leading cause of death by disease in the United States, primarily because of its damage to the cardiovascular system. Because insulin is unavailable to aid transport of glucose into body cells, blood glucose level is high and glucose "spills" into the urine (glucosuria). Hallmarks of diabetes mellitus are the three "polys": *polyuria,* excessive urine production due to an inability of the kidneys to reabsorb water; *polydipsia,* excessive thirst; and *polyphagia,* excessive eating.

Both genetic and environmental factors contribute to onset of the two types of diabetes mellitus—type 1 and type 2—but the exact mechanisms are still unknown. In **type 1 diabetes** insulin level is low because the person's immune system destroys the pancreatic beta cells. It is also called **insulin-dependent diabetes mellitus (IDDM)** because insulin injections are required to prevent death. Most commonly, IDDM develops in people younger than age 20, though it persists throughout life. By the time symptoms of IDDM arise, 80–90% of the islet beta cells have been destroyed. IDDM is most common in northern Europe, especially in Finland where nearly 1% of the population develops IDDM by 15 years of age. In the United States, IDDM is 1.5–2.0 times more common in whites than in African American or Asian populations.

The cellular metabolism of an untreated type 1 diabetic is similar to that of a starving person. Because insulin is not present to aid the entry of glucose into body cells, most cells use fatty acids to produce ATP. Stores of triglycerides in adipose tissue are catabolized to yield fatty acids and glycerol. The byproducts of fatty acid breakdown—organic acids called ketones or ketone bodies—accumulate. Buildup of ketones causes blood pH to fall, a condition known as **ketoacidosis.** Unless treated quickly, ketoacidosis can cause death.

The breakdown of stored triglycerides also causes weight loss. As lipids are transported by the blood from storage depots to cells, lipid particles are deposited on the walls of blood vessels, leading to atherosclerosis and a multitude of cardiovascular problems, including cerebrovascular insufficiency, ischemic heart disease, peripheral vascular disease, and gangrene. A major complication of diabetes is loss of vision due either to cataracts (excessive glucose attaches to lens proteins, causing cloudiness) or to damage to blood vessels of the retina. Severe kidney problems also may result from damage to renal blood vessels.

Type 1 diabetes is treated through self-monitoring of blood glucose level (up to 7 times daily), regular meals containing 45–50% carbohydrates and less than 30% fats, exercise, and periodic insulin injections (up to 3 times a day). Several implantable pumps are available to provide insulin without the need for repeated injections. Because they lack a reliable glucose sensor, however, the person must self-monitor blood glucose level to determine insulin doses. It is also possible to successfully transplant a pancreas, but immunosuppressive drugs must then be taken for life. Another promising approach under investigation is transplantation of isolated islets in semipermeable hollow tubes. The tubes allow glucose and insulin to enter and leave but prevent entry of immune system cells that might attack the islet cells.

Type 2 diabetes, also called **non-insulin-dependent diabetes mellitus (NIDDM),** is much more common than type 1, representing more than 90% of all cases. Type 2 diabetes most often occurs in obese people who are over age 35. However, the number of obese children and teenagers with type 2 diabetes is increasing. Clinical symptoms are mild, and the high glucose levels in the blood often can be controlled by diet, exercise, and weight loss. Sometimes, a drug such as *glyburide* (DiaBeta®) is used to stimulate secretion of insulin by pancreatic beta cells. Although some type 2 diabetics need insulin, many have a sufficient amount (or even a surplus) of insulin in the blood. For these people, diabetes arises not from a shortage of insulin but because target cells become less sensitive to it due to down-regulation of insulin receptors.

Hyperinsulinism most often results when a diabetic injects too much insulin. The main symptom is **hypoglycemia,** decreased blood glucose level, which occurs because the excess insulin stimulates too much uptake of glucose by body cells. The resulting hypoglycemia stimulates the secretion of epinephrine, glucagon, and human growth hormone. As a consequence, anxiety, sweating, tremor, increased heart rate, hunger, and weakness occur. When blood glucose falls, brain cells are deprived of the steady supply of glucose they need to function effectively. Severe hypoglycemia leads to mental disorientation, convulsions, unconsciousness, and shock. Shock due to an insulin overdose is termed **insulin shock.** Death can occur quickly unless blood glucose level is raised. From a clinical standpoint, a diabetic suffering from either a hyperglycemia or a hypoglycemia crisis can have very similar symptoms—mental changes, coma, seizures, and so on. It is important to quickly and correctly identify the cause of the underlying symptoms and treat them appropriately.

MEDICAL TERMINOLOGY

Gynecomastia (gī-ne′-kō-MAS-tē-a; *gyneco-* = woman; *mast-* = breast) Excessive development of mammary glands in a male. Sometimes a tumor of the adrenal gland may secrete sufficient amounts of estrogen to cause the condition.

Hirsutism (HER-soo-tizm; *hirsut-* = shaggy) Presence of excessive body and facial hair in a male pattern, especially in women; may be due to excess androgen production due to tumors or drugs.

Thyroid crisis (storm) A severe state of hyperthyroidism that can be life threatening. It is characterized by high body temperature, rapid heart rate, high blood pressure, gastrointestinal symptoms (abdominal pain, vomiting, diarrhea), agitation, tremors, confusion, seizures, and possibly coma.

Virilizing adenoma (*aden* = gland; *oma* = tumor) Tumor of the adrenal gland that liberates excessive androgens, causing virilism (masculinization) in females. Occasionally, adrenal tumor cells liberate estrogens to the extent that a male patient develops gynecomastia. Such a tumor is called a **feminizing adenoma.**

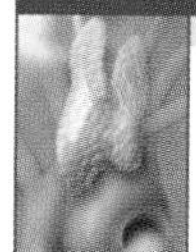

STUDY OUTLINE

INTRODUCTION (p. 617)

1. Hormones regulate the activity of smooth muscle, cardiac muscle, and some glands; alter metabolism; spur growth and development; influence reproductive processes; and participate in circadian (daily) rhythms.

COMPARISON OF CONTROL BY THE NERVOUS AND ENDOCRINE SYSTEMS (p. 617)

1. The nervous system controls homeostasis through nerve impulses and neurotransmitters, which act locally and quickly. The endocrine system uses hormones, which act more slowly in distant parts of the body. (See Table 18.1 on page 617)
2. The nervous system controls neurons, muscle cells, and glandular cells; the endocrine system regulates virtually all body cells.

ENDOCRINE GLANDS (p. 618)

1. Exocrine glands (sudoriferous, sebaceous, mucous, and digestive) secrete their products through ducts into body cavities or onto body surfaces. Endocrine glands secrete hormones into interstitial fluid. Then, the hormones diffuse into the blood.
2. The endocrine system consists of endocrine glands (pituitary, thyroid, parathyroid, adrenal, and pineal glands) and other hormone-secreting tissues (hypothalamus, thymus, pancreas, ovaries, testes, kidneys, stomach, liver, small intestine, skin, heart, adipose tissue, and placenta).

HORMONE ACTIVITY (p. 619)

1. Hormones affect only specific target cells that have receptors to recognize (bind) a given hormone. The number of hormone receptors may decrease (down-regulation) or increase (up-regulation).
2. Circulating hormones enter the bloodstream; local hormones (paracrines and autocrines) act locally on neighboring cells.
3. Chemically, hormones are either lipid-soluble (steroids, thyroid hormones, and nitric oxide) or water-soluble (amines; peptides, proteins, and glycoproteins; and eicosanoids). (See Table 18.2 on page 621.)
4. Water-soluble hormone molecules circulate in the watery blood plasma in a "free" form (not attached to plasma proteins); most lipid-soluble hormones are bound to transport proteins synthesized by the liver.

MECHANISMS OF HORMONE ACTION (p. 622)

1. Lipid-soluble steroid hormones and thyroid hormones affect cell function by altering gene expression.
2. Water-soluble hormones alter cell function by activating plasma membrane receptors, which elicit production of a second messenger that activates various enzymes inside the cell.
3. Hormonal interactions can have three types of effects: permissive, synergistic, or antagonistic.

CONTROL OF HORMONE SECRETION (p. 624)

1. Hormone secretion is controlled by signals from the nervous system, chemical changes in blood, and other hormones.
2. Negative feedback systems regulate the secretion of many hormones.

HYPOTHALAMUS AND PITUITARY GLAND (p. 625)

1. The hypothalamus is the major integrating link between the nervous and endocrine systems.
2. The hypothalamus and pituitary gland regulate virtually all aspects of growth, development, metabolism, and homeostasis.
3. The pituitary gland is located in the hypophyseal fossa and is divided into the anterior pituitary (glandular portion), the posterior pituitary (nervous portion), and the pars intermedia (avascular zone in between).
4. Secretion of anterior pituitary hormones is stimulated by releasing hormones and suppressed by inhibiting hormones from the hypothalamus.
5. The blood supply to the anterior pituitary is from the superior hypophyseal arteries. Hypothalamic releasing and inhibiting hormones enter the primary plexus and flow to the secondary plexus in the anterior pituitary by the hypophyseal portal veins.
6. The anterior pituitary consists of somatotrophs that produce human growth hormone (hGH); lactotrophs that produce prolactin (PRL); corticotrophs that secrete adrenocorticotropic hormone (ACTH) and melanocyte-stimulating hormone (MSH); thyrotrophs that secrete thyroid-stimulating hormone (TSH); and gonadotrophs that synthesize follicle-stimulating hormone (FSH) and luteinizing hormone (LH). (See Tables 18.3 and 18.4.)
7. Human growth hormone (hGH) stimulates body growth through insulinlike growth factors (IGFs). Secretion of hGH is inhibited by GHIH (growth hormone–inhibiting hormone, or somatostatin) and promoted by GHRH (growth hormone–releasing hormone).
8. TSH regulates thyroid gland activities. Its secretion is stimulated by TRH (thyrotropin-releasing hormone) and suppressed by GHIH.
9. FSH and LH regulate the activities of the gonads—ovaries and testes. Their secretion is controlled by GnRH (gonadotropin-releasing hormone).
10. Prolactin (PRL) helps initiate milk secretion. Prolactin-inhibiting hormone (PIH) suppresses secretion of PRL, whereas prolactin-releasing hormone (PRH) and TRH stimulate PRL secretion.
11. ACTH regulates the activities of the adrenal cortex and is controlled by CRH (corticotropin-releasing hormone).
12. Dopamine inhibits secretion of MSH.
13. The posterior pituitary contains axon terminals of neurosecretory cells whose cell bodies are in the hypothalamus.
14. Hormones made by the hypothalamus and stored in the posterior pituitary are oxytocin (OT), which stimulates contraction of the uterus and ejection of milk from the breasts, and antidiuretic hormone (ADH), which stimulates water reabsorption by the kidneys and constriction of arterioles. (See Table 18.5.)
15. Oxytocin secretion is stimulated by uterine stretching and suckling during nursing; ADH secretion is controlled by osmotic pressure of the blood and blood volume.

THYROID GLAND (p. 634)

1. The thyroid gland is located inferior to the larynx.
2. It consists of thyroid follicles composed of follicular cells, which secrete the thyroid hormones thyroxine (T_4) and triiodothyronine (T_3), and parafollicular cells, which secrete calcitonin (CT).
3. Thyroid hormones are synthesized from iodine and tyrosine within thyroglobulin (TGB). They are transported in the blood bound to plasma proteins, mostly thyroxine-binding globulin (TBG).
4. Secretion is controlled by TRH from the hypothalamus and thyroid-stimulating hormone (TSH) from the anterior pituitary.
5. Thyroid hormones regulate oxygen use and metabolic rate, cellular metabolism, and growth and development.
6. Calcitonin (CT) can lower the blood level of calcium ions (Ca^{2+}) and promote deposition of Ca^{2+}into bone matrix. Secretion of CT is controlled by the Ca^{2+}level in the blood. (See Table 18.6 on page 638.)

PARATHYROID GLANDS (p. 638)

1. The parathyroid glands are embedded in the posterior surfaces of the lateral lobes of the thyroid gland. They consist of chief cells and oxyphil cells.
2. Parathyroid hormone (PTH) regulates the homeostasis of calcium, magnesium, and phosphate ions by increasing blood calcium and magnesium levels and decreasing blood phosphate levels. PTH secretion is controlled by the level of calcium in the blood. (See Table 18.7 on page 640.)

ADRENAL GLANDS (p. 641)

1. The adrenal glands are located superior to the kidneys. They consist of an outer adrenal cortex and inner adrenal medulla.
2. The adrenal cortex is divided into a zona glomerulosa, a zona fasciculata, and a zona reticularis; the adrenal medulla consists of chromaffin cells and large blood vessels.
3. Cortical secretions include mineralocorticoids, glucocorticoids, and androgens.
4. Mineralocorticoids (mainly aldosterone) increase sodium and water reabsorption and decrease potassium reabsorption. Secretion is controlled by the renin–angiotensin–aldosterone (RAA) pathway and by K^+ level in the blood.
5. Glucocorticoids (mainly cortisol) promote protein breakdown, gluconeogenesis, and lipolysis; help resist stress; and serve as anti-inflammatory substances. Their secretion is controlled by ACTH.
6. Androgens secreted by the adrenal cortex stimulate growth of axillary and pubic hair, aid the prepubertal growth spurt, and contribute to libido.
7. The adrenal medulla secretes epinephrine and norepinephrine (NE), which are released during stress and produce effects similar to sympathetic responses. (See Table 18.8 on page 646.)

PANCREATIC ISLETS (p. 645)

1. The pancreas lies in the curve of the duodenum. It has both endocrine and exocrine functions.
2. The endocrine portion consists of pancreatic islets or islets of Langerhans, made up of four types of cells: alpha, beta, delta, and F cells.
3. Alpha cells secrete glucagon, beta cells secrete insulin, delta cells secrete somatostatin, and F cells secrete pancreatic polypeptide.
4. Glucagon increases blood glucose level; insulin decreases blood glucose level. Secretion of both hormones is controlled by the level of glucose in the blood. (See Table 18.9 on page 649.)

OVARIES AND TESTES (p. 650)

1. The ovaries are located in the pelvic cavity and produce estrogens, progesterone, and inhibin. These sex hormones govern the development and maintenance of female secondary sex characteristics, reproductive cycles, pregnancy, lactation, and normal female reproductive functions. (See Table 18.10 on page 650.)
2. The testes lie inside the scrotum and produce testosterone and inhibin. These sex hormones govern the development and maintenance of male secondary sex characteristics and normal male reproductive functions. (See Table 18.10 on page 650.)

PINEAL GLAND (p. 650)

1. The pineal gland is attached to the roof of the third ventricle of the brain. It consists of secretory cells called pinealocytes, neuroglia, and endings of sympathetic postganglionic axons.
2. The pineal gland secretes melatonin, which contributes to setting the body's biological clock (controlled in the suprachiasmatic nucleus). During sleep, plasma levels of melatonin increase.

THYMUS (p. 651)

1. The thymus secretes several hormones related to immunity.
2. Thymosin, thymic humoral factor (THF), thymic factor (TF), and thymopoietin promote the maturation of T cells.

OTHER ENDOCRINE TISSUES AND ORGANS, EICOSANOIDS, AND GROWTH FACTORS (p. 651)

1. Body tissues other than those normally classified as endocrine glands contain endocrine tissue and secrete hormones, including the gastrointestinal tract, placenta, kidneys, skin, and heart. (See Table 18.11 on page 651.)
2. Prostaglandins and leukotrienes are eicosanoids that act as local hormones in most body tissues.
3. Growth factors are local hormones that stimulate cell growth and division. (See Table 18.12 on page 652.)

THE STRESS RESPONSE (p. 652)

1. Productive stress is termed eustress, and harmful stress is termed distress.
2. If stress is extreme, it triggers the stress response (general adaptation syndrome), which occurs in three stages: the fight-or-flight response, resistance reaction, and exhaustion.
3. The stimuli that produce the stress response are called stressors. Stressors include surgery, poisons, infections, fever, and strong emotional responses.
4. The fight-or-flight response is initiated by nerve impulses from the hypothalamus to the sympathetic division of the autonomic nervous system and the adrenal medulla. This response rapidly increases circulation, promotes ATP production, and decreases nonessential activities.

5. The resistance reaction is initiated by releasing hormones secreted by the hypothalamus, most importantly CRH, TRH, and GHRH. Resistance reactions are longer lasting and accelerate breakdown reactions to provide ATP for counteracting stress.
6. Exhaustion results from depletion of body resources during the resistance stage.
7. Stress may trigger certain diseases by inhibiting the immune system. An important link between stress and immunity is interleukin-1, produced by macrophages; it stimulates secretion of ACTH.

DEVELOPMENT OF THE ENDOCRINE SYSTEM (p. 655)

1. The development of the endocrine system is not as localized as in other systems because endocrine organs develop in widely separated parts of the embryo.
2. The pituitary gland, adrenal medulla, and pineal gland develop from ectoderm; the adrenal cortex develops from mesoderm; and the thyroid gland, parathyroid glands, pancreas, and thymus develop from endoderm.

AGING AND THE ENDOCRINE SYSTEM (p. 656)

1. Although some endocrine glands shrink as we get older, their performance may or may not be compromised.
2. Production of human growth hormone, thyroid hormones, cortisol, aldosterone, and estrogens decrease with advancing age.
3. With aging, the blood levels of TSH, LH, FSH, and PTH rise.
4. The pancreas releases insulin more slowly with age, and receptor sensitivity to glucose declines.
5. After puberty, thymus size begins to decrease, and thymic tissue is replaced by adipose and areolar connective tissue.

SELF-QUIZ QUESTIONS

Fill in the blanks in the following statements.

1. The three stages of the stress response or general adaptation syndrome, in order of occurrence, are ____, ____, and ____.
2. The ____ is the major integrating link between the nervous and endocrine systems, acts as an endocrine gland itself, and helps control the stress response.
3. Down-regulation makes a target cell ____ sensitive to a hormone while up-regulation makes a target cell ____ sensitive to a hormone.

Indicate whether the following statements are true or false.

4. If the effect of two or more hormones acting together is greater than the sum of each acting alone, then the two hormones are said to have a permissive effect.
5. In the direct gene activation method of hormone action, the hormone enters the target cell and binds to an intracellular receptor. The activated receptor-hormone complex then alters gene expression to produce the protein that causes the physiological responses that are characteristic of the hormone.

Choose the one best answer to the following questions.

6. Which of the following comparisons are *true*? (1) Nerve impulses produce their effects quickly; hormonal responses generally are slower. (2) Nervous system effects are brief; endocrine system effects are longer lasting. (3) The nervous system controls homeostasis through mediator molecules called neurotransmitters; the endocrine system works through mediator molecules called hormones. (4) The nervous system can stimulate or inhibit the release of hormones; some hormones are released by neurons as neurotransmitters. (5) Unlike neurotransmitters, hormones must bind to receptors on or in target cells in order to exert their effects. (a) 1, 2, 3, 4, and 5; (b) 1, 2, 3, and 4; (c) 2, 3, 4, and 5; (d) 2, 4, and 5; (e) 1, 4, and 5.
7. Insulin and thyroxine arrive at an organ at the same time. Thyroxine causes an effect on the organ but insulin does not. Why? (a) Thyroxine is a lipid-soluble hormone and insulin is not. (b) The target cells in the organ have up-regulated for thyroxine. (c) Thyroxine is a local hormone and insulin is a circulating hormone. (d) Thyroxine inhibits the action of insulin. (e) The organ's cells have receptors for thyroxine but not for insulin.
8. Which of the following is *not* a category of water-soluble hormones? (a) peptides, (b) amines, (c) eicosanoids, (d) steroids, (e) proteins.
9. Place in correct order the action of a water-soluble hormone on its target cell. (1) Adenylate cyclase is activated, catalyzing the conversion of ATP to cAMP. (2) Enzymes catalyze reactions that produce a physiological response attributed to the hormone. (3) The hormone binds to a membrane receptor. (4) Activated protein kinases phosphorylate cellular proteins. (5) The hormone-receptor complex activates G proteins. (6) Cyclic AMP activates protein kinases. (a) 3, 5, 1, 6, 4, 2; (b) 3, 1, 5, 6, 4, 2; (c) 5, 1, 4, 2, 3, 6; (d) 3, 4, 5, 1, 6, 2; (e) 6, 3, 5, 1, 4, 2.
10. Hormones (1) generally utilize negative feedback mechanisms to regulate their secretion; (2) will only affect target cells far removed from the hormone-producing secretory cells; (3) must bind to transport proteins in order to circulate in the blood; (4) may be released in low concentrations but can produce large effects in the target cells because of amplification; (5) can regulate the responsiveness of the target tissue by controlling the number of receptor sites for the hormone. (a) 1, 2, and 3; (b) 1, 2, 4, and 5; (c) 2, 3, and 4; (d) 2, 3, 4, and 5; (e) 1, 4, and 5.
11. The pituitary gland (1) is located in the cribriform plate of the ethmoid bone, (2) is linked to the hypothalamus by the infundibulum, (3) has a posterior lobe that contains axon terminals from hypothalamic neurosecretory cells, (4) produces releasing and inhibiting hormones, (5) has a vascular connection with the hypothalamus known as the hypophyseal portal system. (a) 1, 2, and 4; (b) 2, 3, 4, and 5; (c) 2, 3, and 5; (d) 1, 2, 3, 4, and 5; (e) 2, 4, and 5.
12. The class of adrenal gland hormones that provide resistance to stress, produce anti-inflammatory effects, and promote normal metabolism to ensure adequate quantities of ATP is (a) glucocorticoids, (b) mineralocorticoids, (c) androgens, (d) catecholamines, (e) gonadocorticoids.

13. Match the following:

____(a) increases blood Ca^{2+} level
____(b) increases blood glucose level
____(c) decreases blood Ca^{2+} level
____(d) decreases blood glucose level
____(e) initiates and maintains milk secretion by the mammary glands
____(f) regulates the body's biological clock
____(g) stimulates sex hormone production; triggers ovulation
____(h) augments fight-or-flight responses
____(i) regulates metabolism and resistance to stress
____(j) helps control water and electrolyte homeostasis
____(k) suppresses release of FSH
____(l) stimulates growth of axillary and pubic hair
____(m) promotes T cell maturation
____(n) regulates oxygen use, basal metabolic rate, cellular metabolism, and growth and development
____(o) stimulates protein synthesis, inhibits protein breakdown, stimulates lipolysis, and retards use of glucose for ATP production
____(p) inhibits water loss through the kidneys
____(q) stimulates egg and sperm formation
____(r) enhances uterine contractions during labor; stimulates milk ejection
____(s) stimulates and inhibits secretion of anterior pituitary hormones
____(t) increases skin pigmentation when present in excess
____(u) stimulates synthesis and release of T_3 and T_4
____(v) local hormones involved in inflammation, smooth muscle contraction, blood flow, and inflammation

(1) insulin
(2) glucagon
(3) inhibin
(4) follicle-stimulating hormone
(5) luteinizing hormone
(6) thyroxine and triiodothyronine
(7) calcitonin
(8) parathormone
(9) melanocyte-stimulating hormone
(10) oxytocin
(11) antidiuretic hormone
(12) prolactin
(13) human growth hormone
(14) hypothalamic regulating hormones
(15) aldosterone
(16) thyroid stimulating hormone
(17) androgens
(18) epinephrine and norepinephrine
(19) prostaglandins
(20) melatonin
(21) thymosin
(22) cortisol

14. Match the following hormone-secreting cells to the hormone(s) they release:

____(a) ACTH and MSH
____(b) TSH
____(c) glucagon
____(d) PTH
____(e) glucocorticoids
____(f) calcitonin
____(g) insulin
____(h) androgens
____(i) progesterone
____(j) FSH and LH
____(k) epinephrine and norepinephrine
____(l) hGH
____(m) testosterone
____(n) mineralocorticoids
____(o) thyroxine and triiodothyronine
____(p) PRL

(1) beta cells of pancreatic islet
(2) alpha cells of pancreatic islet
(3) follicular cells of thyroid gland
(4) parafollicular cells of thyroid gland
(5) testes
(6) ovary
(7) somatotrophs
(8) thyrotrophs
(9) gonadotrophs
(10) corticotrophs
(11) lactotrophs
(12) chief cells
(13) chromaffin cells
(14) zona glomerulosa cells
(15) zona fasciculata cells
(16) zona reticularis cells

15. Match the endocrine disorder to the problem that produced the disorder:

____(a) hyposecretion of insulin or down-regulation of insulin receptors
____(b) hypersecretion of hGH before closure of epiphyseal plates
____(c) hyposecretion of thyroid hormone that is present at birth
____(d) hypersecretion of glucocorticoids
____(e) hyposecretion of hGH before closure of epiphyseal plates
____(f) hypersecretion of epinephrine and norepinephrine
____(g) hypersecretion of hGH after closure of epiphyseal plates
____(h) hyposecretion of glucocorticoids and aldosterone
____(i) hyposecretion of ADH
____(j) hypersecretion of melatonin
____(k) hyposecretion of thyroid hormone in adults
____(l) hyperthyroidism, an autoimmune disease

(1) giantism
(2) acromegaly
(3) pituitary dwarfism
(4) diabetes insipidus
(5) myxedema
(6) Graves' disease
(7) Cushing's syndrome
(8) seasonal affective disorder
(9) Addison's disease
(10) pheochromocytomas
(11) congenital hypothyroidism
(12) diabetes mellitus

CRITICAL THINKING QUESTIONS

1. Amanda hates her new student ID photo. Her hair looks dry, the extra weight that she's gained shows, and her neck looks fat. In fact, there's an odd butterfly shaped swelling across the front of her neck, under her chin. Amanda's also been feeling very tired and mentally "dull" lately, but she figures all new A&P students feel that way. Should she visit the clinic or just wear turtlenecks?
2. Amanda (from question 1 above) goes to the clinic and blood is drawn. The results show that her T_4 levels are low and her TSH levels are low. Later she is given a TSH stimulation test in which TSH is injected and the T_4 levels are monitored. After TSH injection, her T_4 levels rise. Does Amanda have problems with her pituitary or with her thyroid gland? How did you come to your conclusion?
3. Mr. Hernandez visited his doctor complaining that he is constantly thirsty and is "in the bathroom day and night" relieving his bladder. The doctor ordered blood and urine tests to check for glucose and ketones, which were all negative. What is the doctor's diagnosis of Mr. Hernandez and what gland(s) or organ(s) is(are) involved?

ANSWERS TO FIGURE QUESTIONS

18.1 Secretions of endocrine glands diffuse into interstitial fluid and then into the blood; exocrine secretions flow into ducts that lead into body cavities or to the body surface.

18.2 In the stomach, histamine is a paracrine because it acts on nearby parietal cells without entering the blood.

18.3 The receptor-hormone complex alters gene expression by turning specific genes of nuclear DNA on or off.

18.4 Cyclic AMP is termed a second messenger because it translates the presence of the first messenger, the water-soluble hormone, into a response inside the cell.

18.5 The hypophyseal portal veins carry blood from the median eminence of the hypothalamus, where hypothalamic releasing and inhibiting hormones are secreted, to the anterior pituitary, where these hormones act.

18.6 Thyroid hormones suppress secretion of TSH by thyrotrophs and of TRH by hypothalamic neurosecretory cells; gonadal hormones suppress secretion of FSH and LH by gonadotrophs and of GnRH by hypothalamic neurosecretory cells.

18.7 Excess levels of hGH would cause hyperglycemia.

18.8 Functionally, both the hypothalamohypophyseal tract and the hypophyseal portal veins carry hypothalamic hormones to the pituitary gland. Structurally, the tract is composed of axons of neurons that extend from the hypothalamus to the posterior pituitary; the portal veins are blood vessels that extend from the hypothalamus to the anterior pituitary.

18.9 Absorption of a liter of water in the intestines would decrease the osmotic pressure of your blood plasma, turning off secretion of ADH and decreasing the ADH level in your blood.

18.10 Follicular cells secrete T_3 and T_4, also known as thyroid hormones. Parafollicular cells secrete calcitonin.

18.11 The storage form of thyroid hormones is thyroglobulin.

18.12 Lack of iodine in the diet $\rightarrow$ diminished production of T_3 and T_4 $\rightarrow$ increased release of TSH $\rightarrow$ growth (enlargement) of the thyroid gland $\rightarrow$ goiter.

18.13 Parafollicular cells of the thyroid gland secrete calcitonin; chief (principal) cells of the parathyroid gland secrete PTH.

18.14 Target tissues for PTH are bones and the kidneys; target tissue for CT is bone; target tissue for calcitriol is the GI tract.

18.15 The adrenal glands are superior to the kidneys in the retroperitoneal space.

18.16 Angiotensin II acts to constrict blood vessels by causing contraction of vascular smooth muscle, and it stimulates secretion of aldosterone (by zona glomerulosa cells of the adrenal cortex), which in turn causes the kidneys to conserve water and thereby increase blood volume.

18.17 A transplant recipient who takes prednisone will have low blood levels of ACTH and CRH due to negative feedback suppression of the anterior pituitary and hypothalamus by the prednisone.

18.18 The pancreas is both an endocrine and an exocrine gland.

18.19 Glycogenolysis is the conversion of glycogen into glucose and therefore it increases blood glucose level.

18.20 Homeostasis maintains controlled conditions typical of a normal internal environment; the stress response resets controlled conditions at a different level to cope with various stressors.

18.21 The pituitary gland and the adrenal glands both include tissues having two different embryological origins.

18.22 Antibodies that mimic the action of TSH are produced in Graves disease.

Chapter 19

The Cardiovascular System: The Blood

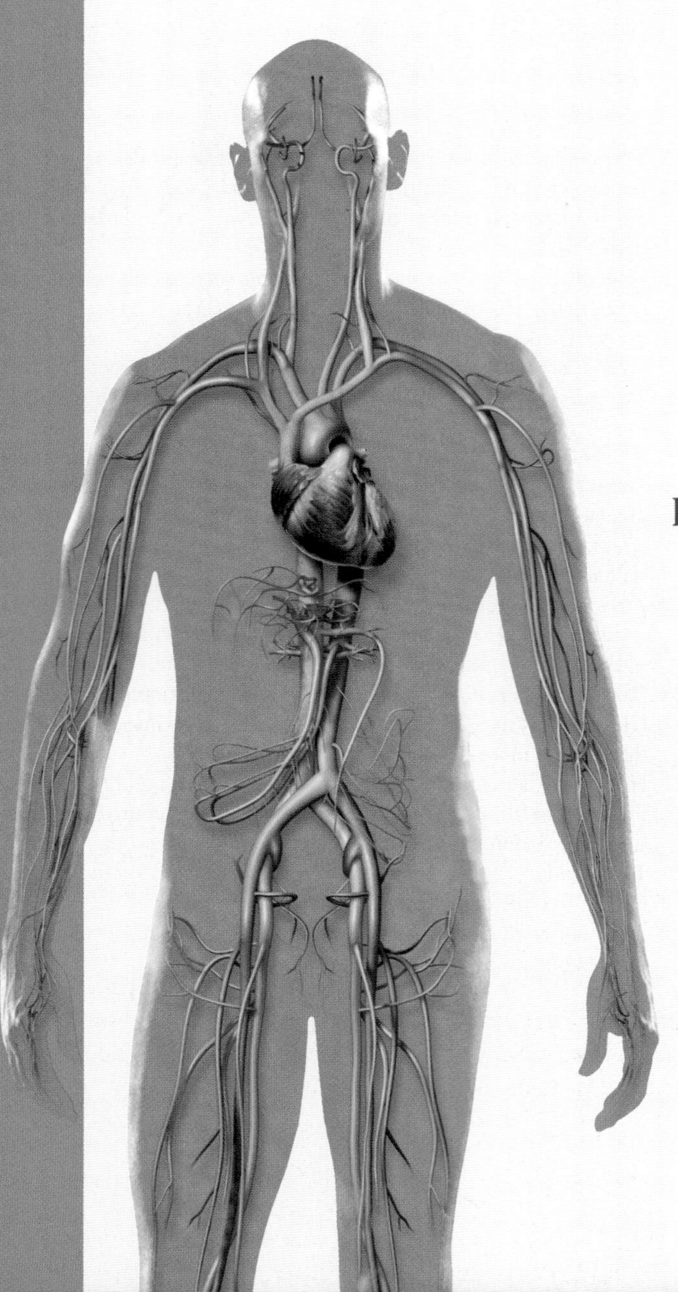

Blood and Homeostasis

Blood contributes to homeostasis by transporting oxygen, carbon dioxide, nutrients, and hormones to and from your body's cells. It helps regulate body pH and temperature, and provides protection against disease through phagocytosis and the production of antibodies.

wiley.com/college/apcentral

The **cardiovascular system** (*cardio-* = heart; *vascular* = blood vessels) consists of three interrelated components: blood, the heart, and blood vessels. The focus of this chapter is blood; the next two chapters will examine the heart and blood vessels, respectively. Blood transports various substances, helps regulate several life processes, and affords protection against disease. For all of its similarities in origin, composition, and functions, blood is as unique from one person to another as are skin, bone, and hair. Health-care professionals routinely examine and analyze its differences through various blood tests when trying to determine the cause of different diseases. The branch of science concerned with the study of blood, blood-forming tissues, and the disorders associated with them is **hematology** (hēm-a-TOL-ō-jē; *hema-* or *hemato-* = blood; *-logy* = study of).

FUNCTIONS AND PROPERTIES OF BLOOD

▶ **OBJECTIVES**

Describe the functions of blood.

Describe the physical characteristics and principal components of blood.

Most cells of a multicellular organism cannot move around to obtain oxygen and nutrients or eliminate carbon dioxide and other wastes. Instead, these needs are met by two fluids: blood and interstitial fluid. **Blood** is a connective tissue composed of a liquid extracellular matrix called blood plasma that dissolves and suspends various cells and cell fragments. **Interstitial fluid** is the fluid that bathes body cells (see Figure 27.1 on page 1037). Blood transports oxygen from the lungs and nutrients from the gastrointestinal tract. The oxygen and nutrients subsequently diffuse from the blood into the interstitial fluid and then into body cells. Carbon dioxide and other wastes move in the reverse direction, from body cells to interstitial fluid to blood. Blood then transports the wastes to various organs—the lungs, kidneys, and skin—for elimination from the body.

Functions of Blood

Blood, which is a liquid connective tissue, has three general functions:

1. ***Transportation.*** As you just learned, blood transports oxygen from the lungs to the cells of the body and carbon dioxide from the body cells to the lungs for exhalation. It carries nutrients from the gastrointestinal tract to body cells and hormones from endocrine glands to other body cells. Blood also transports heat and waste products to various organs for elimination from the body.
2. ***Regulation.*** Circulating blood helps maintain homeostasis of all body fluids. Blood helps regulate pH through the use of buffers. It also helps adjust body temperature through the heat-absorbing and coolant properties of the water (see page 40) in blood plasma and its variable rate of flow through the skin, where excess heat can be lost from the blood to the environment. In addition, blood osmotic pressure influences the water content of cells, mainly through interactions of dissolved ions and proteins.
3. ***Protection.*** Blood can clot, which protects against its excessive loss from the cardiovascular system after an injury. In addition, its white blood cells protect against disease by carrying on phagocytosis. Several types of blood proteins, including antibodies, interferons, and complement, help protect against disease in a variety of ways.

Physical Characteristics of Blood

Blood is denser and more viscous than water and feels slightly sticky. The temperature of blood is 38°C (100.4°F), about 1°C higher than oral or rectal body temperature, and it has a slightly alkaline pH ranging from 7.35 to 7.45. Blood constitutes about 20% of extracellular fluid, amounting to 8% of the total body mass. The blood volume is 5 to 6 liters (1.5 gal) in an average-sized adult male and 4 to 5 liters (1.2 gal) in an average-sized adult female. Several hormones, regulated by negative feedback, ensure that blood volume and osmotic pressure remain relatively constant. Especially important are the hormones aldosterone, antidiuretic hormone, and atrial natriuretic peptide, which regulate how much water is excreted in the urine (see pages 1139–1140).

Withdrawing Blood

Blood samples for laboratory testing may be obtained in several ways. The most common procedure is **venipuncture,** withdrawal of blood from a vein using a needle and collecting tube, which contains various additives. A tourniquet is wrapped around the arm above the venipuncture site, which causes blood to accumulate in the vein. This increased blood volume makes the vein stand out. Opening and closing the fist further causes it to stand out, making the venipuncture more successful. A common site for venipuncture is the median cubital vein anterior to the elbow (see Figure 21.25b on page 783). Another method of withdrawing blood is through a **finger** or **heel stick.** Diabetic patients who monitor their daily blood sugar typically perform a finger stick, and it is often used for drawing blood from infants and children. In an **arterial stick,** blood is withdrawn from an artery; this test is used to determine the level of oxygen in oxygenated blood. ■

Components of Blood

Blood has two components: (1) blood plasma, a watery liquid extracellular matrix that contains dissolved substances, and (2) formed elements, which are cells and cell fragments. If a sample of blood is centrifuged (spun) in a small glass tube, the cells sink to the bottom of the tube while the lighter-weight plasma forms a layer on top (Figure 19.1a). Blood is about 45% formed

Figure 19.1 Components of blood in a normal adult.

Blood is a connective tissue that consists of blood plasma (liquid) plus formed elements (red blood cells, white blood cells, and platelets).

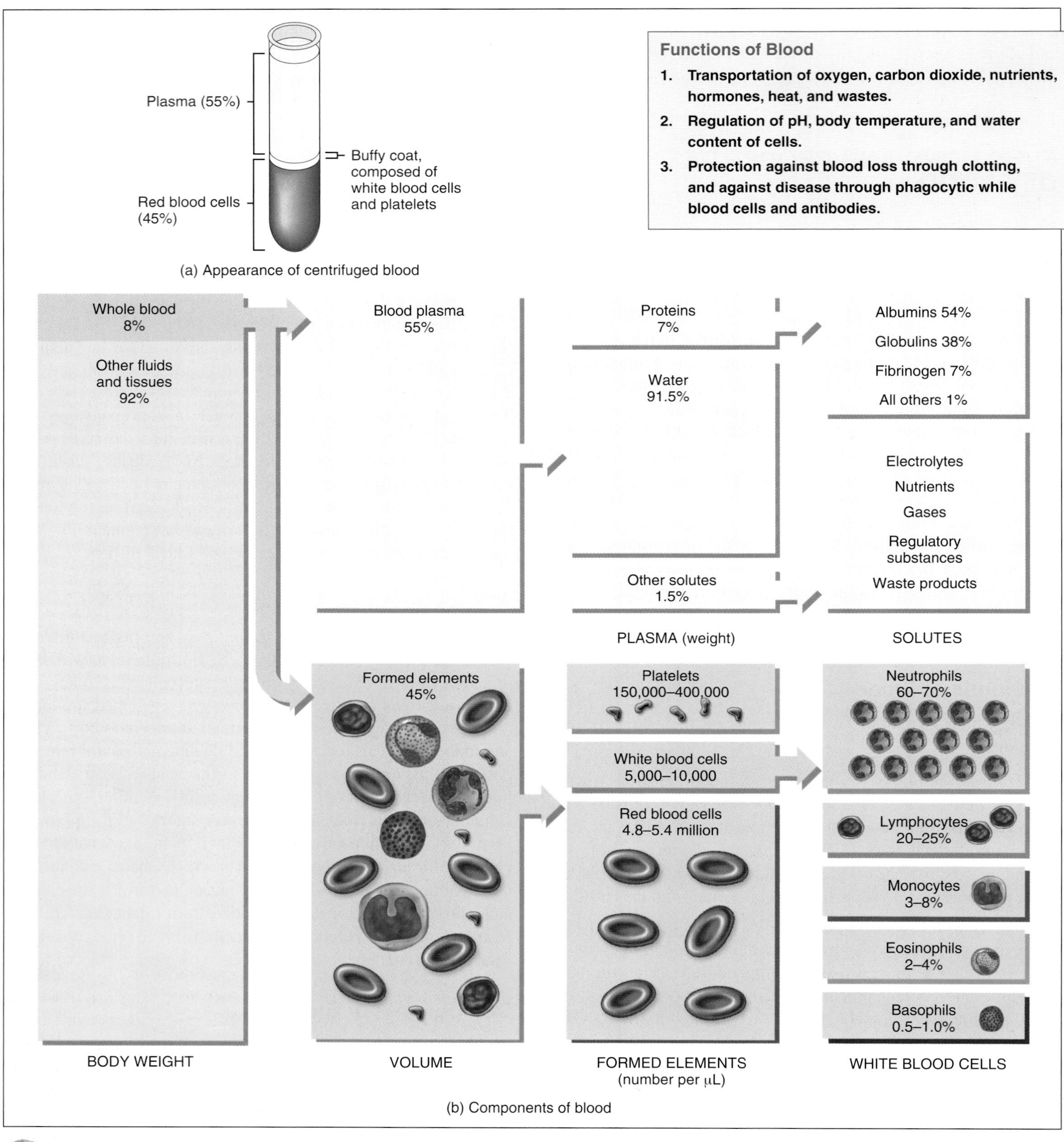

What is the approximate volume of blood in your body? (*Hint:* Each liter of blood has a mass of one kilogram.)

elements and 55% blood plasma. Normally, more than 99% of the formed elements are cells named for their red color—red blood cells (RBCs). Pale, colorless white blood cells (WBCs) and platelets occupy less than 1% of total blood volume. Because they are less dense than red blood cells but more dense than blood plasma, they form a very thin **buffy coat** layer between the packed RBCs and plasma in centrifuged blood. Figure 19.1b shows the composition of blood plasma and the numbers of the various types of formed elements in blood.

Blood Plasma

When the formed elements are removed from blood, a straw-colored liquid called **blood plasma** (or simply **plasma**) is left. Blood plasma is about 91.5% water and 8.5% solutes, most of which (7% by weight) are proteins. Some of the proteins in blood plasma are also found elsewhere in the body, but those confined to blood are called **plasma proteins.** Among other functions, these proteins play a role in maintaining proper blood osmotic pressure, which is an important factor in the exchange of fluids across capillary walls (discussed in Chapter 21).

Hepatocytes (liver cells) synthesize most of the plasma proteins, which include the **albumins** (54% of plasma proteins), **globulins** (38%), and **fibrinogen** (7%). Their functions are given in Table 19.1. Certain blood cells develop into cells that produce gamma globulins, an important type of globulin. These plasma proteins are also called **antibodies** or **immunoglobulins** because they are produced during certain immune responses. Foreign substances (antigens) such as bacteria and viruses stimulate production of millions of different antibodies. An antibody binds specifically to the antigen that stimulated its production and thus disables the invading antigen.

Besides proteins, other solutes in plasma include electrolytes, nutrients, regulatory substances such as enzymes and hormones, gases, and waste products such as urea, uric acid, creatinine, ammonia, and bilirubin.

Table 19.1 describes the chemical composition of blood plasma.

TABLE 19.1 Substances in Blood Plasma

Constituent	Description
Water (91.5%)	Liquid portion of blood. Acts as solvent and suspending medium for components of blood; absorbs, transports, and releases heat.
Plasma Proteins (7.0%)	Exert colloid osmotic pressure, which helps maintain water balance between blood and tissues and regulates blood volume.
Albumins	Smallest and most numerous blood plasma proteins; produced by liver. Function as transport proteins for several steroid hormones and for fatty acids.
Globulins	Produced by liver and by plasma cells, which develop from B lymphocytes. Antibodies (immunoglobulins) help attack viruses and bacteria. Alpha and beta globulins transport iron, lipids, and fat-soluble vitamins.
Fibrinogen	Produced by liver. Plays essential role in blood clotting.
Other Solutes (1.5%)	
Electrolytes	Inorganic salts. Positively charged ions (cations) include Na^+, K^+, Ca^{2+}, Mg^{2+}; negatively charged ions (anions) include Cl^-, HPO_4^{2-}, SO_4^{2-}, and HCO_3^-. Help maintain osmotic pressure and play essential roles in the function of cells.
Nutrients	Products of digestion pass into blood for distribution to all body cells. Include amino acids (from proteins), glucose (from carbohydrates), fatty acids and glycerol (from triglycerides), vitamins, and minerals.
Gases	Oxygen (O_2), carbon dioxide (CO_2), and nitrogen (N_2). More O_2 is associated with hemoglobin inside red blood cells; more CO_2 is dissolved in plasma. N_2 is present but has no known function in the body.
Regulatory substances	Enzymes, produced by body cells, catalyze chemical reactions. Hormones, produced by endocrine glands, regulate metabolism, growth, and development. Vitamins are cofactors for enzymatic reactions.
Waste products	Most are breakdown products of protein metabolism and are carried by blood to organs of excretion. Include urea, uric acid, creatine, creatinine, bilirubin, and ammonia.

Formed Elements

The **formed elements** of the blood include three principal components: **red blood cells (RBCs), white blood cells (WBCs),** and **platelets** (Figure 19.2). RBCs and WBCs are whole cells; platelets are cell fragments. RBCs and platelets have just a few roles, but WBCs have a number of specialized functions. Several distinct types of WBCs—neutrophils, lymphocytes, monocytes, eosinophils, and basophils—each with a unique microscopic appearance, carry out these functions, which are discussed later in this chapter.

The percentage of total blood volume occupied by RBCs is called the **hematocrit** (he-MAT-ō-krit); a hematocrit of 40 indicates that 40% of the volume of blood is composed of RBCs. The normal range of hematocrit for adult females is 38–46% (average = 42); for adult males, it is 40–54% (average = 47). The hormone testosterone, present in much higher concentration in males than in females, stimulates synthesis of erythropoietin (EPO), the hormone that in turn stimulates production of RBCs. Thus, testosterone contributes to higher hematocrits in males. Lower values in women during their reproductive years also may be due to excessive loss of blood during menstruation. A significant drop in hematocrit indicates *anemia,* a lower-than-normal number of RBCs. In *polycythemia* the percentage of RBCs is abnormally high, and the hematocrit may be 65% or higher. This raises the viscosity of blood, which increases the resistance to flow and makes the blood more difficult for the heart to pump.

Figure 19.2 Scanning electron micrograph of the formed elements of blood.

The formed elements of blood are red blood cells (RBCs), white blood cells (WBCs), and platelets.

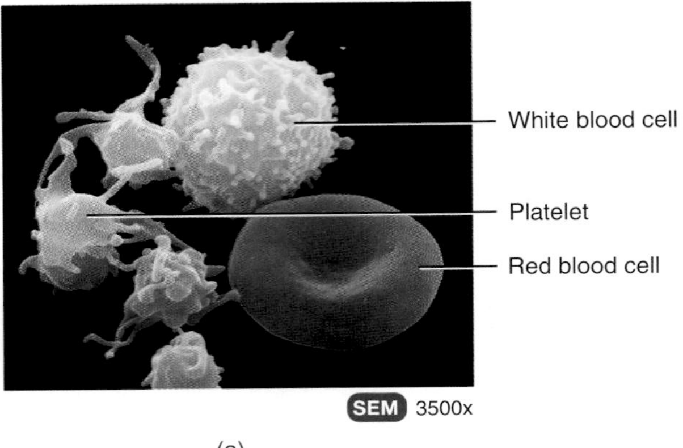

(a)

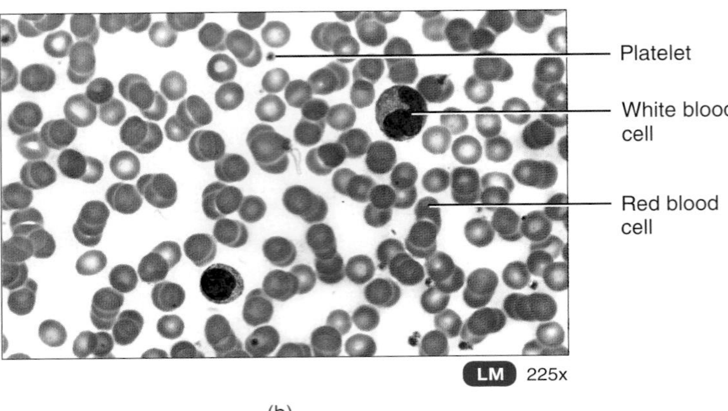

(b)

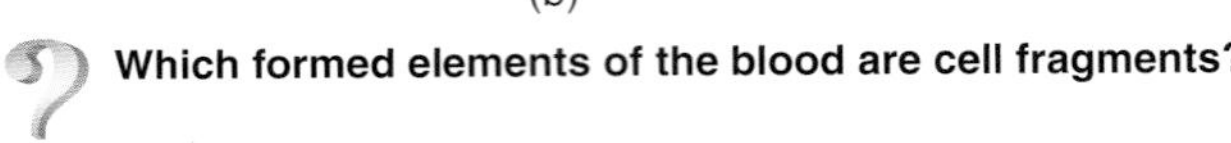

Which formed elements of the blood are cell fragments?

Increased viscosity also contributes to high blood pressure and increased risk of stroke. Causes of polycythemia include abnormal increases in RBC production, tissue hypoxia, dehydration, and blood doping or the use of EPO by athletes.

▶ CHECKPOINT

1. In what ways is blood plasma similar to interstitial fluid? How does it differ?
2. What substances does blood transport?
3. How many kilograms or pounds of blood are there in your body?
4. How does the volume of blood plasma in your body compare to the volume of fluid in a two-liter bottle of Coke?
5. List the formed elements in blood plasma and describe their functions.
6. What is the significance of lower-than-normal or higher-than-normal hematocrit?

FORMATION OF BLOOD CELLS

▶ OBJECTIVE

Explain the origin of blood cells.

Although some lymphocytes have a lifetime measured in years, most formed elements of the blood last only hours, days, or weeks, and must be replaced continually. Negative feedback systems regulate the total number of RBCs and platelets in circulation, and their numbers normally remain steady. The abundance of the different types of WBCs, however, varies in response to challenges by invading pathogens and other foreign antigens.

The process by which the formed elements of blood develop is called **hemopoiesis** (hē-mō-poy-Ē-sis; *-poiesis* = making) or *hematopoiesis.* Before birth, hemopoiesis first occurs in the yolk sac of an embryo and later in the liver, spleen, thymus, and lymph nodes of a fetus. Red bone marrow becomes the primary site of hemopoiesis in the last three months before birth, and continues as the main source of blood cells after birth and throughout life.

Red bone marrow is a highly vascularized connective tissue located in the microscopic spaces between trabeculae of spongy bone tissue. It is present chiefly in bones of the axial skeleton, pectoral and pelvic girdles, and the proximal epiphyses of the humerus and femur. About 0.05–0.1% of red bone marrow cells are derived from mesenchymal cells called **pluripotent stem cells** (ploo-RIP-ō-tent; *pluri-* = several) or *hemocytoblasts.* These cells have the capacity to develop into several different types of cells (Figure 19.3). In newborns all bone marrow is red and thus active in blood cell production. As an individual grows and in adulthood, the rate of blood cell formation decreases; the red bone marrow in the medullary (marrow) cavity of long bones becomes inactive and is replaced by yellow bone marrow, which is largely fat cells. Under certain conditions, such as severe bleeding, yellow bone marrow can revert to red bone marrow by extension of red bone marrow into yellow bone marrow and repopulation of yellow bone marrow by pluripotent stem cells.

Bone Marrow Examination

Sometimes a sample of red bone marrow must be obtained in order to diagnose certain blood disorders, such as leukemia and severe anemias. **Bone marrow examination** may involve *bone marrow aspiration* (withdrawal of a small amount of red bone marrow with a fine needle and syringe) or a *bone marrow biopsy* (removal of a core of red bone marrow with a larger needle).

Both types of samples are usually taken from the iliac crest of the hip bone, although samples are sometimes aspirated from the sternum. In young children, bone marrow samples are taken from a vertebra or tibia (shin bone). The tissue or cell sample is then sent to a pathology lab for analysis. Specifically, laboratory technicians look for signs of neoplastic (cancer) cells or other diseased cells to assist in diagnosis. ■

Figure 19.3 Origin, development, and structure of blood cells. Some of the generations of some cell lines have been omitted.

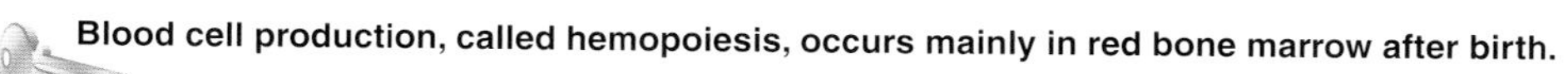

Blood cell production, called hemopoiesis, occurs mainly in red bone marrow after birth.

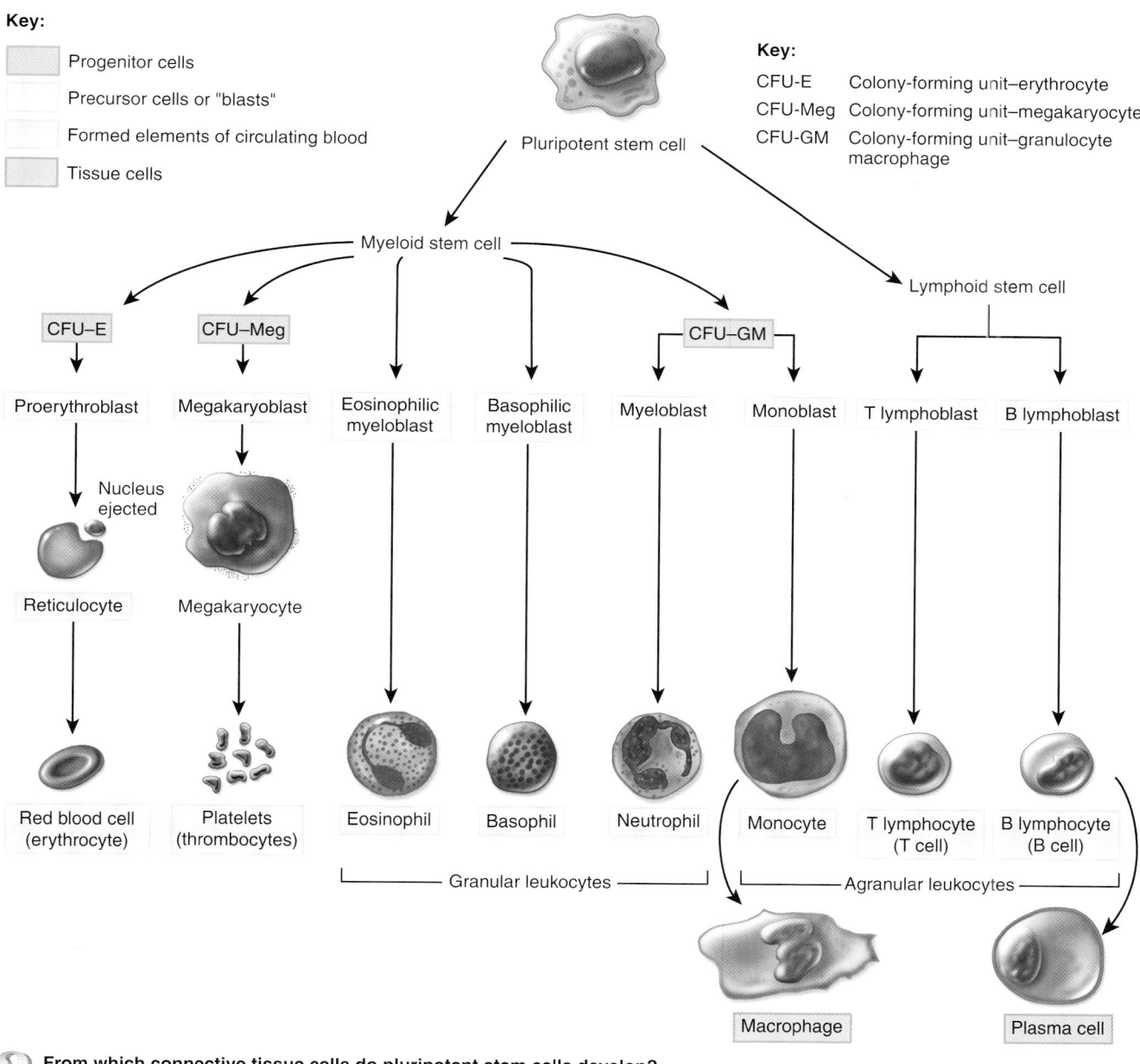

From which connective tissue cells do pluripotent stem cells develop?

Stem cells in red bone marrow reproduce themselves, proliferate, and differentiate into cells that give rise to blood cells, macrophages, reticular cells, mast cells, and adipocytes. Some of the stem cells can also form osteoblasts, chondroblasts, and muscle cells, and someday may be used as a source of bone, cartilage, and muscular tissue for tissue and organ replacement. The reticular cells produce reticular fibers, which form the stroma (framework) that supports red bone marrow cells. Once blood cells are produced in red bone marrow, they enter the bloodstream through *sinusoids* (also called *sinuses*), enlarged and leaky capillaries that surround red bone marrow cells and fibers. With the exception of lymphocytes, formed elements do not divide once they leave red bone marrow.

In order to form blood cells, pluripotent stem cells in red bone marrow produce two further types of stem cells, called *myeloid stem cells* and *lymphoid stem cells.* Myeloid stem cells begin their development in red bone marrow and give rise to red blood cells, platelets, monocytes, neutrophils, eosinophils, and basophils. Lymphoid stem cells begin their development in red bone marrow but complete it in lymphatic tissues; they give rise to lymphocytes. Although the various stem cells have distinctive cell identity markers in their plasma membranes, they cannot be distinguished histologically and resemble lymphocytes.

During hemopoiesis, some of the myeloid stem cells differentiate into **progenitor cells** (prō-JEN-i-tor). Other myeloid stem cells and the lymphoid stem cells develop directly into precursor cells (described shortly). Progenitor cells are no longer capable of reproducing themselves and are committed to giving rise to more specific elements of blood. Some progenitor cells are known as *colony-forming units (CFUs)*. Following the CFU designation is an abbreviation that indicates the mature elements in blood that they will produce: CFU–E ultimately produces erythrocytes (red blood cells), CFU–Meg produces megakaryocytes, the source of platelets, and CFU–GM ultimately produces granulocytes (specifically, neutrophils) and monocytes (see Figure 19.3). Progenitor cells, like stem cells, resemble lymphocytes and cannot be distinguished by their microscopic appearance alone.

In the next generation, the cells are called **precursor cells,** also known as **blasts.** Over several cell divisions they develop into the actual formed elements of blood. For example, monoblasts develop into monocytes, eosinophilic myeloblasts develop into eosinophils, and so on. Precursor cells have recognizable microscopic appearances.

Several hormones called **hemopoietic growth factors** regulate the differentiation and proliferation of particular progenitor cells. **Erythropoietin** (e-rith′-rō-POY-e-tin) or **EPO** increases the number of red blood cell precursors. EPO is produced primarily by cells in the kidneys that lie between the kidney tubules (peritubular interstitial cells). With renal failure, EPO release slows and RBC production is inadequate. **Thrombopoietin** (throm′-bō-POY-ē-tin) or **TPO** is a hormone produced by the liver that stimulates the formation of platelets (thrombocytes) from megakaryocytes. Several different cytokines regulate development of different blood cell types. **Cytokines** are small glycoproteins that are typically produced by cells such as red bone marrow cells, leukocytes, macrophages, fibroblasts, and endothelial cells. They generally act as local hormones (autocrines or paracrines; see Chapter 18). Cytokines stimulate proliferation of progenitor cells in red bone marrow and regulate the activities of cells involved in nonspecific defenses (such as phagocytes) and immune responses (such as B cells and T cells). Two important families of cytokines that stimulate white blood cell formation are **colony-stimulating factors (CSFs)** and **interleukins.**

Medical Uses of Hemopoietic Growth Factors

Hemopoietic growth factors made available through recombinant DNA technology hold tremendous potential for medical uses when a person's natural ability to form new blood cells is diminished or defective. The artificial form of erythropoietin (Epoetin alfa) is very effective in treating the diminished red blood cell production that accompanies end-stage kidney disease. Granulocyte-macrophage colony-stimulating factor and granulocyte CSF are given to stimulate white blood cell formation in cancer patients who are undergoing chemotherapy, which kills red bone marrow cells as well as cancer cells because both cell types are undergoing mitosis. (Recall that white blood cells help protect against disease.) Thrombopoietin shows great promise for preventing the depletion of platelets, which are needed to help blood clot, during chemotherapy. CSFs and thrombopoietin also improve the outcome of patients who receive bone marrow transplants. Hemopoietic growth factors are also used to treat thrombocytopenia in neonates, other clotting disorders, and various types of anemia. Research on these medications is ongoing and shows a great deal of promise. ■

▶ CHECKPOINT

7. Which hemopoietic growth factors regulate differentiation and proliferation of CFU–E and formation of platelets from megakaryocytes?
8. Describe the formation of platelets from pluripotent stem cells, including the influence of hormones.

RED BLOOD CELLS

▶ OBJECTIVE

Describe the structure, functions, life cycle, and production of red blood cells.

Red blood cells (RBCs) or **erythrocytes** (e-RITH-rō-sīts; *erythro-* = red; *-cyte* = cell) contain the oxygen-carrying protein **hemoglobin,** which is a pigment that gives whole blood its red color. A healthy adult male has about 5.4 million red blood cells per microliter (μL) of blood,* and a healthy adult female has about 4.8 million. (One drop of blood is about 50 μL.) To maintain normal numbers of RBCs, new mature cells must enter the circulation at the astonishing rate of at least 2 million per second, a pace that balances the equally high rate of RBC destruction.

RBC Anatomy

RBCs are biconcave discs with a diameter of 7–8 μm (Figure 19.4a). Mature red blood cells have a simple structure. Their plasma membrane is both strong and flexible, which allows them to deform without rupturing as they squeeze through narrow capillaries. As you will see later, certain glycolipids in the plasma membrane of RBCs are antigens that account for the various blood groups such as the ABO and Rh groups. RBCs lack

*1 $\mu L = 1\ mm^3 = 10^{-6}$ liter.

Figure 19.4 The shapes of a red blood cell (RBC) and a hemoglobin molecule, and the structure of a heme group. In (b), note that each of the four polypetide chains (blue) of a hemoglobin molecule has one heme group (gold), which contains an iron ion (Fe^{2+}), shown in red.

The iron portion of a heme group binds oxygen for transport by hemoglobin.

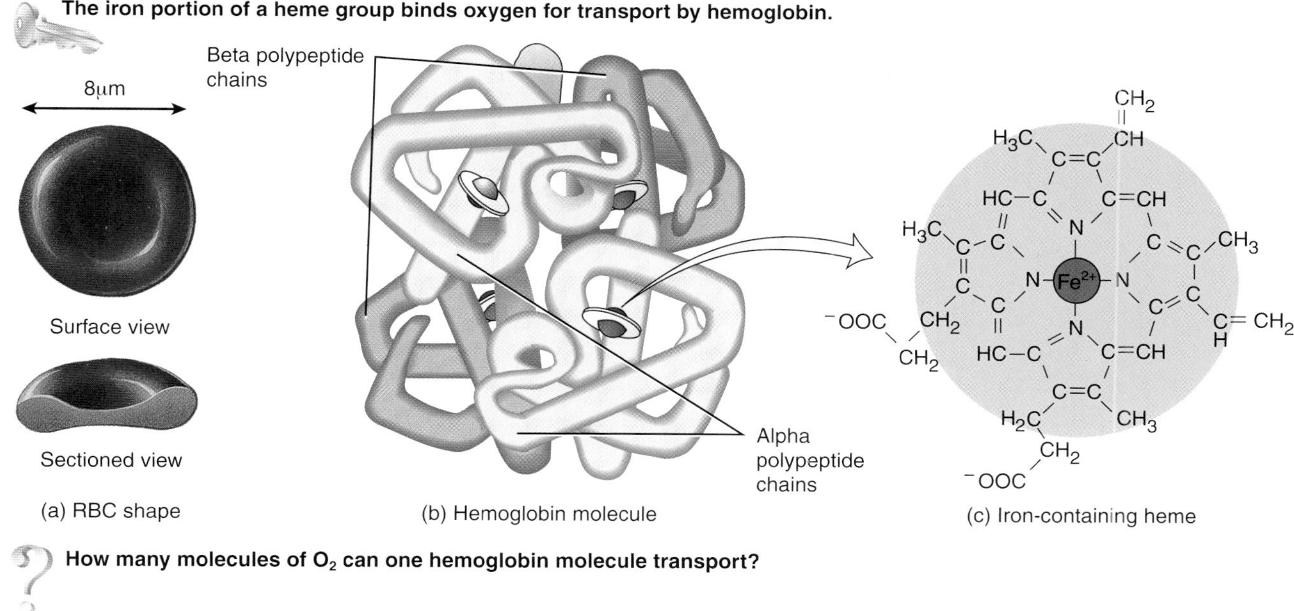

a nucleus and other organelles and can neither reproduce nor carry on extensive metabolic activities. The cytosol of RBCs contains hemoglobin molecules; these important molecules are synthesized before loss of the nucleus during RBC production and constitute about 33% of the cell's weight.

RBC Physiology

Red blood cells are highly specialized for their oxygen transport function. Because mature RBCs have no nucleus, all their internal space is available for oxygen transport. Because RBCs lack mitochondria and generate ATP anaerobically (without oxygen), they do not use up any of the oxygen they transport. Even the shape of an RBC facilitates its function. A biconcave disc has a much greater surface area for the diffusion of gas molecules into and out of the RBC than would, say, a sphere or a cube.

Each RBC contains about 280 million hemoglobin molecules. A hemoglobin molecule consists of a protein called **globin,** composed of four polypeptide chains (two alpha and two beta chains); a ringlike nonprotein pigment called a **heme** (Figure 19.4b) is bound to each of the four chains. At the center of the heme ring is an iron ion (Fe^{2+}) that can combine reversibly with one oxygen molecule (Figure 19.4c), allowing each hemoglobin molecule to bind four oxygen molecules. Each oxygen molecule picked up from the lungs is bound to an iron ion. As blood flows through tissue capillaries, the iron–oxygen reaction reverses. Hemoglobin releases oxygen, which diffuses first into the interstitial fluid and then into cells.

Hemoglobin also transports about 23% of the total carbon dioxide, a waste product of metabolism. Blood flowing through tissue capillaries picks up carbon dioxide, some of which combines with amino acids in the globin part of hemoglobin. As blood flows through the lungs, the carbon dioxide is released from hemoglobin and then exhaled.

In addition to its key role in transporting oxygen and carbon dioxide, hemoglobin also plays a role in the regulation of blood flow and blood pressure. The gaseous hormone **nitric oxide (NO),** produced by the endothelial cells that line blood vessels, binds to hemoglobin. Under some circumstances, hemoglobin releases NO. The released NO causes *vasodilation,* an increase in blood vessel diameter that occurs when the smooth muscle in the vessel wall relaxes. Vasodilation improves blood flow and enhances oxygen delivery to cells near the site of NO release.

RBC Life Cycle

Red blood cells live only about 120 days because of the wear and tear their plasma membranes undergo as they squeeze through blood capillaries. Without a nucleus and other organelles, RBCs cannot synthesize new components to replace damaged ones. The plasma membrane becomes more fragile with age, and the cells are more likely to burst, especially as they squeeze through narrow channels in the spleen. Ruptured red blood cells are removed from circulation and destroyed by fixed phagocytic macrophages in the spleen and liver, and the breakdown products are recycled, as follows (Figure 19.5):

Figure 19.5 Formation and destruction of red blood cells, and the recycling of hemoglobin components. RBCs circulate for about 120 days after leaving red bone marrow before they are phagocytized by macrophages.

The rate of RBC formation by red bone marrow equals the rate of RBC destruction by macrophages.

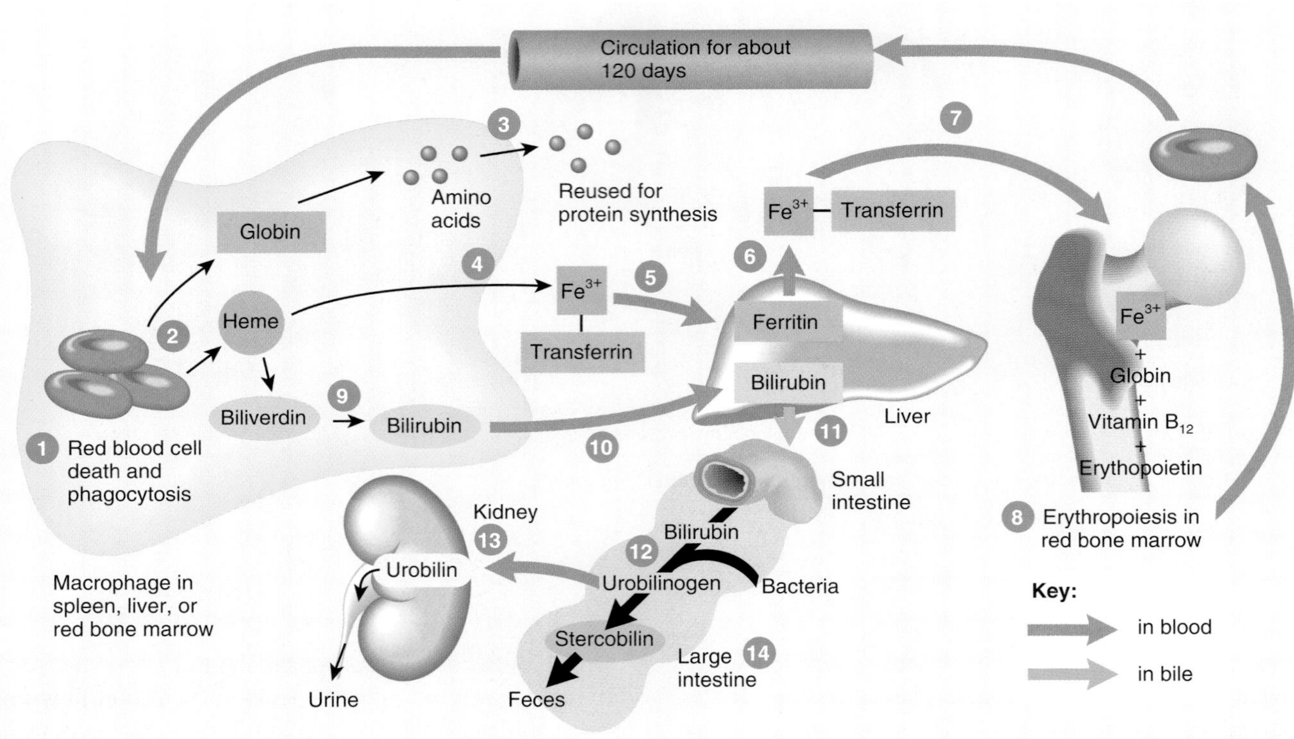

What is the function of transferrin?

1. Macrophages in the spleen, liver, or red bone marrow phagocytize ruptured and worn-out red blood cells.
2. The globin and heme portions of hemoglobin are split apart.
3. Globin is broken down into amino acids, which can be reused to synthesize other proteins.
4. Iron is removed from the heme portion in the form of Fe^{3+}, which associates with the plasma protein **transferrin** (trans-FER-in; *trans-* = across; *ferr-* = iron), a transporter for Fe^{3+} in the bloodstream.
5. In muscle fibers, liver cells, and macrophages of the spleen and liver, Fe^{3+} detaches from transferrin and attaches to an iron-storage protein called **ferritin.**
6. Upon release from a storage site or absorption from the gastrointestinal tract, Fe^{3+} reattaches to transferrin.
7. The Fe^{3+}-transferrin complex is then carried to red bone marrow, where RBC precursor cells take it up through receptor-mediated endocytosis (see Figure 3.10 on page 72) for use in hemoglobin synthesis. Iron is needed for the heme portion of the hemoglobin molecule, and amino acids are needed for the globin portion. Vitamin B_{12} is also needed for the synthesis of hemoglobin.
8. Erythropoiesis in red bone marrow results in the production of red blood cells, which enter the circulation.
9. When iron is removed from heme, the non-iron portion of heme is converted to **biliverdin** (bil′-i-VER-din), a green pigment, and then into **bilirubin** (bil′-i-ROO-bin), a yellow-orange pigment.
10. Bilirubin enters the blood and is transported to the liver.
11. Within the liver, bilirubin is released by liver cells into bile, which passes into the small intestine and then into the large intestine.
12. In the large intestine, bacteria convert bilirubin into **urobilinogen** (ūr-ō-bī-LIN-ō-jen).
13. Some urobilinogen is absorbed back into the blood, converted to a yellow pigment called **urobilin** (ūr-ō-BĪ-lin), and excreted in urine.

14 Most urobilinogen is eliminated in feces in the form of a brown pigment called **stercobilin** (ster′-kō-BĪ-lin), which gives feces its characteristic color.

Iron Overload and Tissue Damage

Because free iron ions (Fe^{2+} and Fe^{3+}) bind to and damage molecules in cells or in the blood, transferrin and ferritin act as protective "protein escorts" during transport and storage of iron ions. As a result, plasma contains virtually no free iron. Furthermore, only small amounts are available inside body cells for use in synthesis of iron-containing molecules such as the cytochrome pigments needed for ATP production in mitochondria (see Figure 25.9 on page 961). In cases of **iron overload,** the amount of iron present in the body builds up. Because we have no method for eliminating excess iron, any condition that increases dietary iron absorption can cause iron overload. At some point, the proteins transferrin and ferritin become saturated with iron ions, and free iron level rises. Common consequences of iron overload are diseases of the liver, heart, pancreatic islets, and gonads. Iron overload also allows certain iron-dependent microbes to flourish. Such microbes normally are not pathogenic, but they multiply rapidly and can cause lethal effects in a short time when free iron is present. ■

Erythropoiesis: Production of RBCs

Erythropoiesis (e-rith′-rō-poy-Ē-sis), the production of RBCs, starts in the red bone marrow with a precursor cell called a **proerythroblast** (see Figure 19.3). The proerythroblast divides several times, producing cells that begin to synthesize hemoglobin. Ultimately, a cell near the end of the development sequence ejects its nucleus and becomes a **reticulocyte** (re-TIK-ū-lō-sīt). Loss of the nucleus causes the center of the cell to indent, producing the red blood cell's distinctive biconcave shape. Reticulocytes retain some mitochondria, ribosomes, and endoplasmic reticulum. They pass from red bone marrow into the bloodstream by squeezing between the endothelial cells of blood capillaries. Reticulocytes develop into mature red blood cells within 1 to 2 days after their release from red bone marrow.

Normally, erythropoiesis and red blood cell destruction proceed at roughly the same pace. If the oxygen-carrying capacity of the blood falls because erythropoiesis is not keeping up with RBC destruction, a negative feedback system steps up RBC production (Figure 19.6). The controlled condition is the amount of oxygen delivered to body tissues. Cellular oxygen deficiency, called **hypoxia** (hī-POKS-ē-a), may occur if too little oxygen enters the blood. For example, the lower oxygen content of air at high altitudes reduces the amount of oxygen in the blood. Oxygen delivery may also fall due to anemia, which has many causes: Lack of iron, lack of certain amino acids, and lack of vitamin B_{12} are but a few (see page 689). Circulatory problems that reduce blood flow to tissues may also reduce oxygen delivery. Whatever the cause, hypoxia stimulates the kidneys to step up the release of erythropoietin, which speeds the development of proerythroblasts into reticulocytes in the red bone marrow. As the number of circulating RBCs increases, more oxygen can be delivered to body tissues.

Figure 19.6 **Negative feedback regulation of erythropoiesis (red blood cell formation).** Lower oxygen content of air at high altitudes, anemia, and circulatory problems may reduce oxygen delivery to body tissues.

The main stimulus for erythropoiesis is hypoxia, a decrease in the oxygen-carrying capacity of the blood.

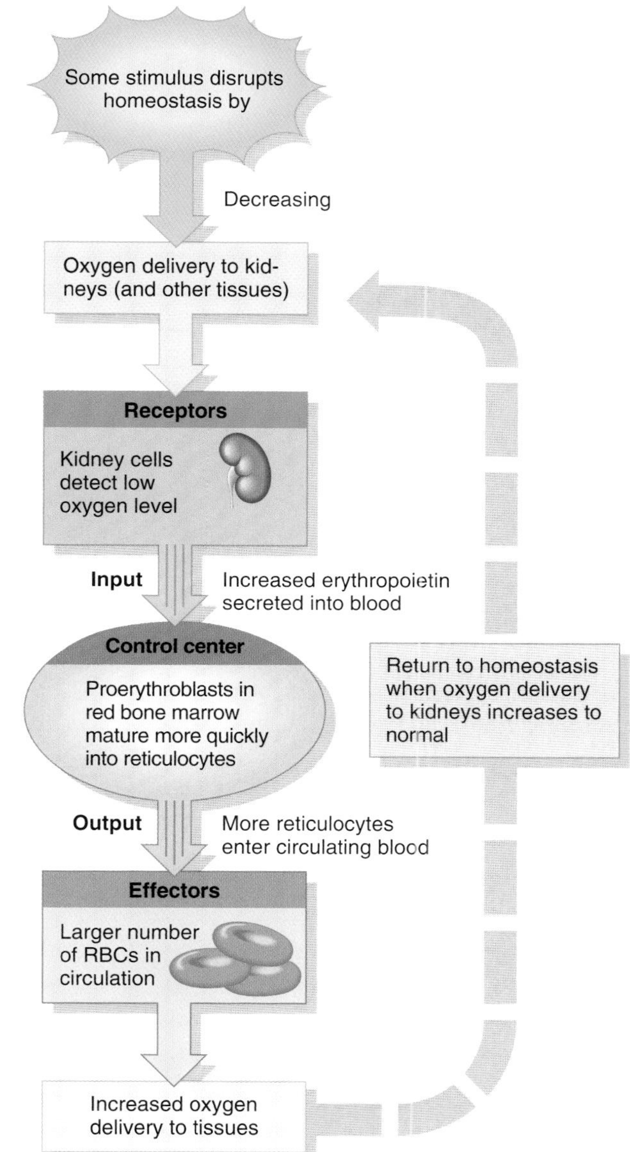

How might your hematocrit change if you moved from a town at sea level to a high mountain village?

Premature newborns often exhibit anemia, due in part to inadequate production of erythropoietin. During the first weeks after birth, the liver, not the kidneys, produces most EPO. Because the liver is less sensitive than the kidneys to hypoxia, newborns have a smaller EPO response to anemia than do adults.

Reticulocyte Count

The rate of erythropoiesis is measured by a **reticulocyte count.** Normally, a little less than 1% of the oldest RBCs are replaced by newcomer reticulocytes on any given day. It then takes 1 to 2 days for the reticulocytes to lose the last vestiges of endoplasmic reticulum and become mature RBCs. Thus, reticulocytes account for about 0.5–1.5% of all RBCs in a normal blood sample. A low "retic" count in a person who is anemic might indicate a shortage of erythropoietin or an inability of the red bone marrow to respond to EPO, perhaps because of a nutritional deficiency or leukemia. A high "retic" count might indicate a good red bone marrow response to previous blood loss or to iron therapy in someone who had been iron deficient. It could also point to illegal use of Epoetin alfa by an athlete. ■

▶ CHECKPOINT

9. Describe the size, microscopic appearance, and functions of RBCs.
10. How is hemoglobin recycled?
11. What is erythropoiesis? How does erythropoiesis affect hematocrit? What factors speed up and slow down erythropoiesis?

WHITE BLOOD CELLS

▶ OBJECTIVE

Describe the structure, functions, and production of white blood cells (WBCs).

Types of WBCs

Unlike red blood cells, white blood cells or **leukocytes** (LOO-kō-sīts; *leuko-* = white) have nuclei and do not contain hemoglobin (Figure 19.7). WBCs are classified as either granular or agranular, depending on whether they contain conspicuous chemical-filled cytoplasmic granules (vesicles) that are made visible by staining. *Granular leukocytes* include neutrophils, eosinophils, and basophils; *agranular leukocytes* include lymphocytes and monocytes. As shown in Figure 19.3, monocytes and granular leukocytes develop from a myeloid stem cell, and lymphocytes develop from a lymphoid stem cell.

Granular Leukocytes

After staining, each of the three types of granular leukocytes displays conspicuous granules with distinctive coloration that can be recognized under a light microscope. The large, uniform-sized granules within an **eosinophil** (ē-ō-SIN-ō-fil) are *eosinophilic* (= eosin-loving)—they stain red-orange with acidic dyes (Figure 19.7a). The granules usually do not cover or obscure the nucleus, which most often has two lobes connected by a thick strand of chromatin. The round, variable-sized granules of a **basophil** (BĀ-sō-fil) are *basophilic* (= basic loving)—they stain blue-purple with basic dyes (Figure 19.7b). The granules commonly obscure the nucleus, which has two lobes. The granules of a **neutrophil** (NOO-trō-fil) are smaller, evenly distributed, and pale lilac in color (Figure 19.7c); the nucleus has two to five lobes, connected by very thin strands of chromatin. As the cells age, the number of nuclear lobes increases. Because older neutrophils have several differently shaped nuclear lobes, they are often called *polymorphonuclear leukocytes (PMNs),* polymorphs, or "polys." Younger neutrophils are often called *bands* because their nucleus is more rod-shaped.

Agranular Leukocytes

Even though so-called agranular leukocytes possess cytoplasmic granules, the granules are not visible under a light microscope because of their small size and poor staining qualities.

The nucleus of a **lymphocyte** (LIM-fō-sīt) is round or slightly indented and stains darkly. The cytoplasm stains sky blue and forms a rim around the nucleus. The larger the cell, the more cytoplasm is visible. Lymphocytes are classified as small or large based on cell diameter: 6–9 μm in small lymphocytes and 10–14 μm in large lymphocytes (Figure 19.7d). (Although the functional significance of the size difference between small and large lymphocytes is unclear, the distinction is still clinically useful because an increase in the number of large lymphocytes has diagnostic significance in acute viral infections and in some immunodeficiency diseases.)

Monocytes (MON-ō-sīts) are 12–20 μm in diameter (Figure 19.7e). The nucleus of a monocyte is usually kidney shaped or horseshoe shaped, and the cytoplasm is blue-gray and has a foamy appearance. The color and appearance are due to very fine *azurophilic granules* (az′-ū-rō-FIL-ik; *azur-* = blue; *-philic* = loving), which are lysosomes. The blood transports monocytes from the blood into the tissues, where they enlarge and differentiate into **macrophages** (= large eaters). Some become **fixed macrophages,** which means they reside in a particular tissue; examples are alveolar macrophages in the lungs, macrophages in the spleen, and stellate reticuloendothelial (Kupffer) cells in the liver. Others become **wandering macrophages,** which roam the tissues and gather at sites of infection or inflammation.

White blood cells and all other nucleated cells in the body have proteins, called **major histocompatibility (MHC) antigens,** protruding from their plasma membrane into the extracellular fluid. These "cell identity markers" are unique for each person (except identical twins). Although RBCs possess blood group antigens, they lack the MHC antigens.

Figure 19.7 **Types of white blood cells.**

The shapes of their nuclei and the staining properties of their cytoplasmic granules distinguish white blood cells from one another.

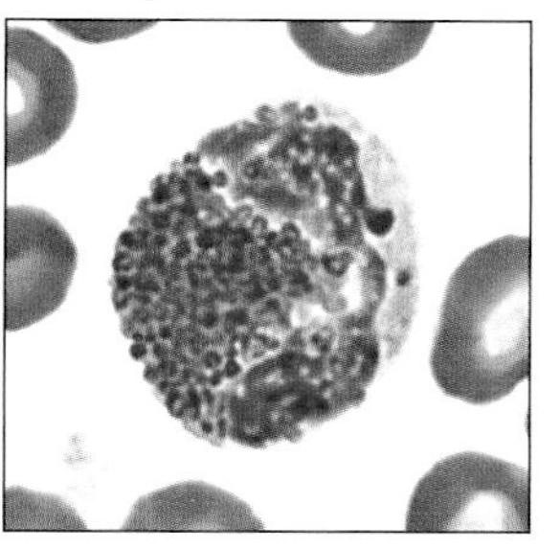
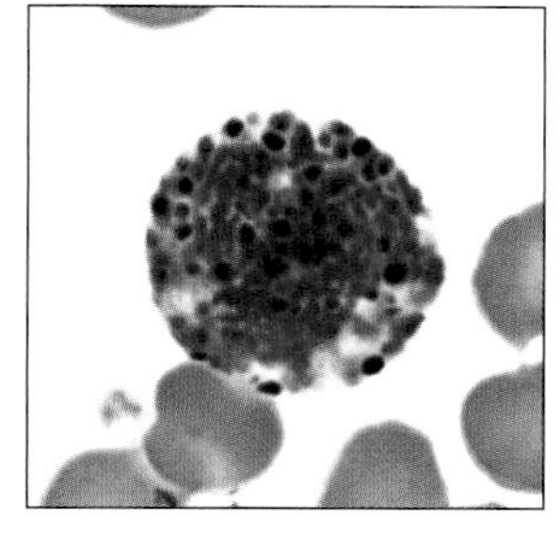
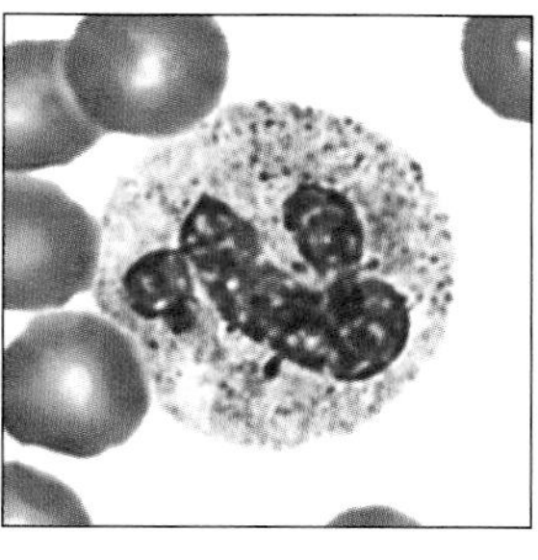
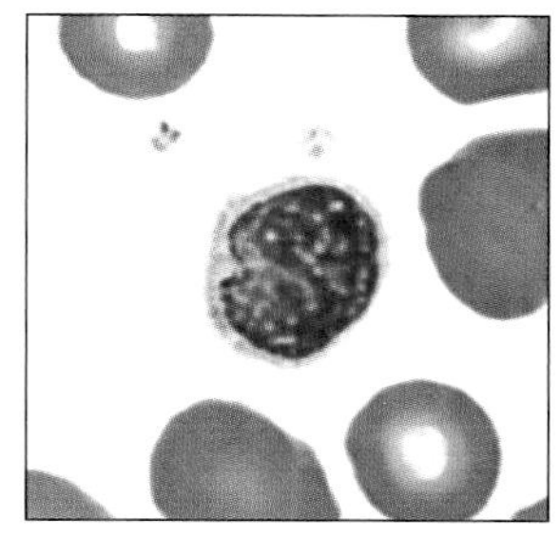
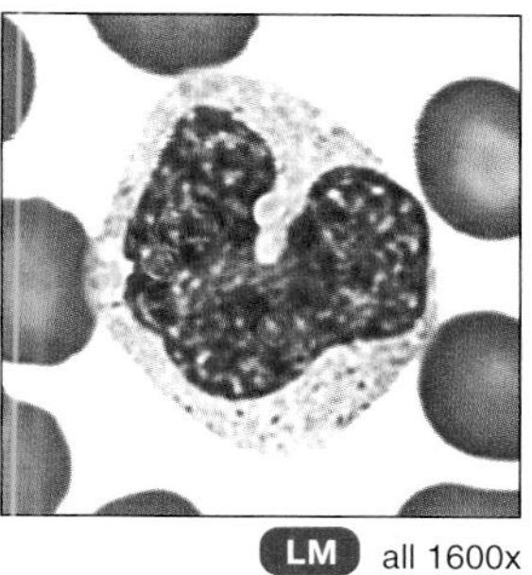

(a) Eosinophil (b) Basophil (c) Neutrophil (d) Lymphocyte (e) Monocyte

Which WBCs are called granular leukocytes? Why?

Functions of WBCs

In a healthy body, some WBCs, especially lymphocytes, can live for several months or years, but most live only a few days. During a period of infection, phagocytic WBCs may live only a few hours. WBCs are far less numerous than red blood cells; at about 5000–10,000 cells per μL of blood, they are outnumbered by RBCs by about 700:1. **Leukocytosis** (loo′-kō-sī-TŌ-sis), an increase in the number of WBCs above 10,000/μL, is a normal, protective response to stresses such as invading microbes, strenuous exercise, anesthesia, and surgery. An abnormally low level of white blood cells (below 5000/μL) is termed **leukopenia** (loo′-kō-PĒ-nē-a). It is never beneficial and may be caused by radiation, shock, and certain chemotherapeutic agents.

The skin and mucous membranes of the body are continuously exposed to microbes and their toxins. Some of these microbes can invade deeper tissues to cause disease. Once pathogens enter the body, the general function of white blood cells is to combat them by phagocytosis or immune responses. To accomplish these tasks, many WBCs leave the bloodstream and collect at sites of pathogen invasion or inflammation. Once granulocytes and monocytes leave the bloodstream to fight injury or infection, they never return to it. Lymphocytes, on the other hand, continually recirculate—from blood to interstitial spaces of tissues to lymphatic fluid and back to blood. Only 2% of the total lymphocyte population is circulating in the blood at any given time; the rest are in lymphatic fluid and organs such as the skin, lungs, lymph nodes, and spleen.

WBCs leave the bloodstream by a process termed **emigration** (em′-i-GRĀ-shun; *e-* = out; *migra-* = wander), formerly called *diapedesis* (dī-a-pe-DĒ-sis), in which they roll along the endothelium, stick to it, and then squeeze between endothelial cells (Figure 19.8). The precise signals that stimulate emigration through a particular blood vessel vary for the different types of WBCs. Molecules known as **adhesion molecules** help WBCs stick to the endothelium. For example, endothelial cells display adhesion molecules called *selectins* in response to nearby injury and inflammation. Selectins stick to carbohydrates on the surface of neutrophils, causing them to slow down and roll along the endothelial surface. On the neutrophil surface

Figure 19.8 **Emigration of white blood cells.**

Adhesion molecules (selectins and integrins) assist the emigration of WBCs from the bloodstream into interstitial fluid.

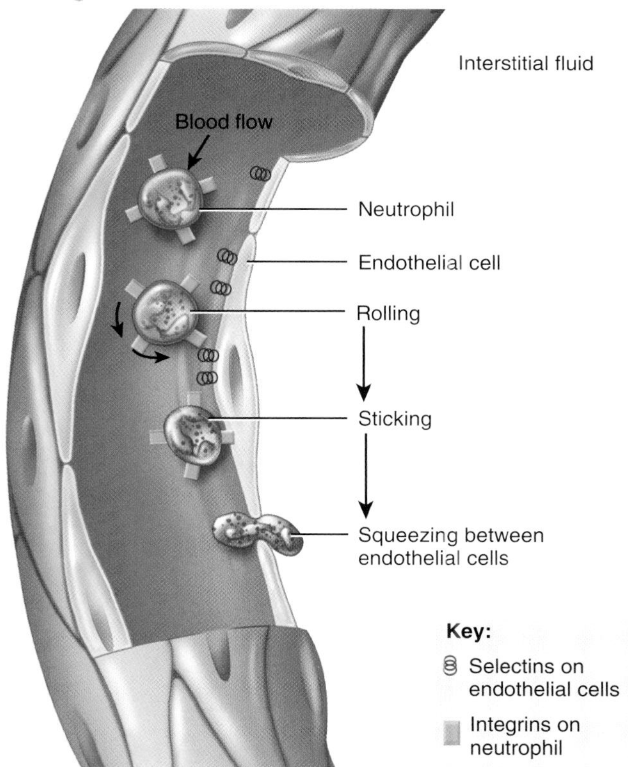

In what way is the "traffic pattern" of lymphocytes in the body different from that of other WBCs?

are other adhesion molecules called *integrins,* which tether neutrophils to the endothelium and assist their movement through the blood vessel wall and into the interstitial fluid of the injured tissue.

Neutrophils and macrophages are active in **phagocytosis;** they can ingest bacteria and dispose of dead matter (see Figure 3.11 on page 73). Several different chemicals released by microbes and inflamed tissues attract phagocytes, a phenomenon called **chemotaxis.** The substances that provide stimuli for chemotaxis include toxins produced by microbes; kinins, which are specialized products of damaged tissues; and some of the colony-stimulating factors (CSFs). The CSFs also enhance the phagocytic activity of neutrophils and macrophages.

Among WBCs, neutrophils respond most quickly to tissue destruction by bacteria. After engulfing a pathogen during phagocytosis, a neutrophil unleashes several chemicals to destroy the pathogen. These chemicals include the enzyme **lysozyme,** which destroys certain bacteria, and **strong oxidants,** such as the superoxide anion (O_2^-), hydrogen peroxide (H_2O_2), and the hypochlorite anion (OCl^-), which is similar to household bleach. Neutrophils also contain **defensins,** proteins that exhibit a broad range of antibiotic activity against bacteria and fungi. Within a neutrophil, vesicles containing defensins merge with phagosomes containing microbes. Defensins form peptide "spears" that poke holes in microbe membranes; the resulting loss of cellular contents kills the invader.

Monocytes take longer to reach a site of infection than do neutrophils, but they arrive in larger numbers and destroy more microbes. Upon arrival they enlarge and differentiate into wandering macrophages, which clean up cellular debris and microbes by phagocytosis after an infection.

At sites of inflammation, basophils leave capillaries, enter tissues, and release granules that contain heparin, histamine, and serotonin. These substances intensify the inflammatory reaction and are involved in hypersensitivity (allergic) reactions. Basophils are similar in function to mast cells, connective tissue cells that originate from pluripotent stem cells in red bone marrow. Like basophils, mast cells release substances involved in inflammation, including heparin, histamine, and proteases. Mast cells are widely dispersed in the body, particularly in connective tissues of the skin and mucous membranes of the respiratory and gastrointestinal tracts.

Eosinophils leave the capillaries and enter tissue fluid. They are believed to release enzymes, such as histaminase, that combat the effects of histamine and other substances involved in inflammation during allergic reactions. Eosinophils also phagocytize antigen–antibody complexes and are effective against certain parasitic worms. A high eosinophil count often indicates an allergic condition or a parasitic infection.

Lymphocytes are the major soldiers in immune system battles (described in detail in Chapter 22). Three main types of lymphocytes are B cells, T cells, and natural killer cells. B cells are particularly effective in destroying bacteria and inactivating their toxins. T cells attack viruses, fungi, transplanted cells, cancer cells, and some bacteria, and are responsible for transfusion reactions, allergies, and the rejection of transplanted organs. Immune responses carried out by both B cells and T cells help combat infection and provide protection against some diseases. Natural killer cells attack a wide variety of infectious microbes and certain spontaneously arising tumor cells.

As you have already learned, an increase in the number of circulating WBCs usually indicates inflammation or infection. A physician may order a **differential white blood cell count,** a count of each of the five types of white blood cells, to detect infection or inflammation, determine the effects of possible poisoning by chemicals or drugs, monitor blood disorders (for example, leukemia) and the effects of chemotherapy, or detect allergic reactions and parasitic infections. Because each type of white blood cell plays a different role, determining the *percentage* of each type in the blood assists in diagnosing the condition. Table 19.2 lists the significance of both elevated and depressed WBC counts.

▶ CHECKPOINT

12. What is the importance of emigration, chemotaxis, and phagocytosis in fighting bacterial invaders?

13. How are leukocytosis and leukopenia different?

14. What is a differential white blood cell count?

15. What functions do granular leukocytes, macrophages, B cells, T cells, and natural killer cells perform?

TABLE 19.2 Significance of High and Low White Blood Cell Counts

WBC Type	High Count May Indicate	Low Count May Indicate
Neutrophils	Bacterial infection, burns, stress, inflammation.	Radiation exposure, drug toxicity, vitamin B_{12} deficiency, or systemic lupus erythematosus (SLE).
Lymphocytes	Viral infections, some leukemias.	Prolonged illness, immunosuppression, or treatment with cortisol.
Monocytes	Viral or fungal infections, tuberculosis, some leukemias, other chronic diseases.	Bone marrow suppression, treatment with cortisol.
Eosinophils	Allergic reactions, parasitic infections, autoimmune diseases.	Drug toxicity, stress.
Basophils	Allergic reactions, leukemias, cancers, hypothyroidism.	Pregnancy, ovulation, stress, or hyperthyroidism.

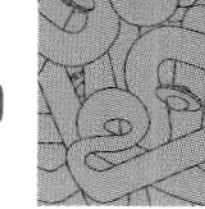

PLATELETS

► OBJECTIVE

Describe the structure, function, and origin of platelets.

Besides the immature cell types that develop into erythrocytes and leukocytes, hemopoietic stem cells also differentiate into cells that produce platelets. Under the influence of the hormone **thrombopoietin,** myeloid stem cells develop into megakaryocyte-colony-forming cells that, in turn, develop into precursor cells called megakaryoblasts (see Figure 19.3). Megakaryoblasts transform into megakaryocytes, huge cells that splinter into 2000 to 3000 fragments. Each fragment, enclosed by a piece of the plasma membrane, is a **platelet (thrombocyte).** Platelets break off from the megakaryocytes in red bone marrow and then enter the blood circulation. Between 150,000 and 400,000 platelets are present in each μL of blood. Each is disc-shaped, 2–4 μm in diameter, and has many vesicles but no nucleus.

Platelets help stop blood loss from damaged blood vessels by forming a platelet plug. Their granules also contain chemicals that, once released, promote blood clotting. Platelets have a short life span, normally just 5 to 9 days. Aged and dead platelets are removed by fixed macrophages in the spleen and liver.

Table 19.3 summarizes the formed elements in blood.

Complete Blood Count

A **complete blood count (CBC)** is a very valuable test that screens for anemia and various infections. Usually included are counts of RBCs, WBCs, and platelets per μL of whole blood; hematocrit; and differential white blood cell count. The amount of hemoglobin in grams per milliliter of blood also is determined. Normal hemoglobin ranges are: infants, 14–20 g/100 mL of blood; adult females, 12–16 g/100 mL of blood; and adult males, 13.5–18 g/100 mL of blood. ■

► CHECKPOINT

16. How do RBCs, WBCs, and platelets compare with respect to size, number per μL of blood, and life span?

STEM CELL TRANSPLANTS FROM BONE MARROW AND CORD-BLOOD

► OBJECTIVE

Explain the importance of bone marrow transplants and stem cell transplants.

A **bone marrow transplant** is the replacement of cancerous or abnormal red bone marrow with healthy red bone marrow in order to establish normal blood cell counts. In patients with cancer or certain genetic diseases, the defective red bone marrow is destroyed by high doses of chemotherapy and whole body radiation just before the transplant takes place. These treatments kill the cancer cells and destroy the patient's immune system in order to decrease the chance of transplant rejection.

Healthy red bone marrow for transplanting may be supplied by a donor or by the patient when the underlying disease is inactive, as when leukemia is in remission. The marrow from a donor is usually removed from the iliac crest of the hip bone under general anesthesia with a syringe and is then injected into the recipient's vein, much like a blood transfusion. The injected marrow migrates to the recipient's red bone marrow cavities and the stem cells in the marrow multiply. If all goes well, the recipient's red bone marrow is replaced entirely by healthy, noncancerous cells.

Bone marrow transplants have been used to treat aplastic anemia, certain types of leukemia, severe combined immunodeficiency disease (SCID), Hodgkin's disease, non-Hodgkin's lymphoma, multiple myeloma, thalassemia, sickle-cell disease, breast cancer, ovarian cancer, testicular cancer, and hemolytic anemia. However, there are some drawbacks. Since the recipients's white blood cells have been completely destroyed by chemotherapy and radiation, the patient is extremely vulnerable to infection. (It takes about 2–3 weeks for transplanted bone marrow to produce enough white blood cells to protect against infection.) In addition, transplanted red bone marrow may produce T cells that attack the recipient's tissues, a reaction called *graft-versus-host disease.* Similarly, any of the recipients's T cells that survived the chemotherapy and radiation can attack donor transplant cells. Another drawback is that patients must take immunosuppressive drugs for life. Because these drugs reduce the level of immune system activity, they increase the risk of infection. Immunosuppressive drugs also have side effects such as fever, muscle aches, headache, nausea, fatigue, depression, high blood pressure, and kidney and liver damage.

A more recent advance for obtaining stem cells involves a **cord-blood transplant.** The connection between the mother and embryo (and later the fetus) is the umbilical cord. Stem cells may be obtained from the umbilical cord shortly after birth. The stem cells are removed from the cord with a syringe and then frozen. Stem cells from the cord have several advantages over those obtained from red bone marrow:

1. They are easily collected following permission of the newborn's parents.

2. They are more abundant than stem cells in red bone marrow.

3. They are less likely to cause graft-versus-host disease, so the match between donor and recipient does not have to be as close as in a bone marrow transplant. This provides a larger number of potential donors.

4. They are less likely to transmit infections.

5. They can be stored indefinitely in cord-blood banks.

▶ CHECKPOINT

17. How are cord-blood transplants and bone marrow transplants similar? How do they differ?

TABLE 19.3 Summary of Formed Elements in Blood

Name and Appearance	Number	Charcteristics*	Functions
Red Blood Cells (RBC) or Erythrocytes	4.8 million/μL in females; 5.4 million/μL in males	7–8 μm diameter, biconcave discs, without nuclei; live for about 120 days.	Hemoglobin within RBCs transports most of the oxygen and part of the carbon dioxide in the blood.
White Blood Cells (WBCs) or Leukocytes	5000–10,000/μL	Most live for a few hours to a few days[†].	Combat pathogens and other foreign substances that enter the body.
Granular leukocytes			
Neutrophils	60–70% of all WBCs	10–12 μm diameter; nucleus has 2–5 lobes connected by thin strands of chromatin; cytoplasm has very fine, pale lilac granules.	Phagocytosis. Destruction of bacteria with lysozyme, defensins, and strong oxidants, such as superoxide anion, hydrogen peroxide, and hypochlorite anion.
Eosinophils	2–4% of all WBCs	10–12 μm diameter; nucleus usually has 2 lobes connected by a thick strand of chromatin; large, red-orange granules fill the cytoplasm.	Combat the effects of histamine in allergic reactions,phagocytize antigen–antibody complexes, and destroy certain parasitic worms.
Basophils	0.5–1% of all WBCs	8–10 μm diameter; nucleus has 2 lobes; large cytoplasmic granules appear deep blue-purple.	Liberate heparin, histamine,and serotonin in allergicreactions that intensify the overall inflammatory response.
Agranular leukocytes			
Lymphocytes (T cells, B cells, and **natural killer cells)**	20–25% of all WBCs	Small lymphocytes are 6–9 μm in diameter; large lymphocytes are 10–14 μm in diameter; nucleus is round or slightly indented; cytoplasm forms a rim around the nucleus that looks sky blue; the larger the cell, the more cytoplasm is visible.	Mediate immune responses, including antigen–antibody reactions. B cells develop into plasma cells, which secrete antibodies. T cells attack invading viruses, cancer cells, and transplanted tissue cells.Natural killer cells attack a wide variety of infectious microbes and certain spontaneously arising tumor cells.
Monocytes	3–8% of all WBCs	12–20 μm diameter; nucleus is kidney shaped or horseshoe shaped; cytoplasm is blue-gray and has foamy appearance.	Phagocytosis (after transforming into fixed or wandering macrophages).
Platelets (Thrombocytes)	150,000–400,000/μL	2–4 μm diameter cell fragments that live for 5–9 days; contain many vesicles but no nucleus.	Form platelet plug in hemostasis; release chemicals that promote vascular spasm and blood clotting.

*Colors are those seen when using Wright's stain.

[†]Some lymphocytes, called T and B memory cells, can live for many years once they are established.

HEMOSTASIS

► OBJECTIVES

Describe the three mechanisms that contribute to hemostasis.

Identify the stages of blood clotting and explain the various factors that promote and inhibit blood clotting.

Hemostasis (hē-mō-STĀ-sis), not to be confused with the very similar term homeostasis, is a sequence of responses that stops bleeding. When blood vessels are damaged or ruptured, the hemostatic response must be quick, localized to the region of damage, and carefully controlled in order to be effective. Three mechanisms reduce blood loss: (1) vascular spasm, (2) platelet plug formation, and (3) blood clotting (coagulation). When successful, hemostasis prevents **hemorrhage** (HEM-o-rij; *-rhage* = burst forth), the loss of a large amount of blood from the vessels. Hemostatic mechanisms can prevent hemorrhage from smaller blood vessels, but extensive hemorrhage from larger vessels usually requires medical intervention.

Vascular Spasm

When arteries or arterioles are damaged, the circularly arranged smooth muscle in their walls contracts immediately, a reaction called **vascular spasm.** This reduces blood loss for several minutes to several hours, during which time the other hemostatic mechanisms go into operation. The spasm is probably caused by damage to the smooth muscle, by substances released from activated platelets, and by reflexes initiated by pain receptors.

Platelet Plug Formation

Considering their small size, platelets store an impressive array of chemicals. Within many vesicles are clotting factors, ADP, ATP, Ca^{2+}, and serotonin. Also present are enzymes that produce thromboxane A2, a prostaglandin; *fibrin-stabilizing factor,* which helps to strengthen a blood clot; lysosomes; some mitochondria; membrane systems that take up and store calcium and provide channels for release of the contents of granules; and glycogen. Also within platelets is **platelet-derived growth factor (PDGF),** a hormone that can cause proliferation of vascular endothelial cells, vascular smooth muscle fibers, and fibroblasts to help repair damaged blood vessel walls.

Platelet plug formation occurs as follows (Figure 19.9):

1. Initially, platelets contact and stick to parts of a damaged blood vessel, such as collagen fibers of the connective tissue underlying the damaged endothelial cells. This process is called **platelet adhesion.**
2. Due to adhesion, the platelets become activated, and their characteristics change dramatically. They extend many projections that enable them to contact and interact with one another, and they begin to liberate the contents of their vesicles. This phase is called the **platelet release reaction.** Liberated ADP and thromboxane A2 play a major role by activating nearby platelets. Serotonin and thromboxane A2 function as vasoconstrictors, causing and sustaining contraction of vascular smooth muscle, which decreases blood flow through the injured vessel.
3. The release of ADP makes other platelets in the area sticky, and the stickiness of the newly recruited and activated

Figure 19.9 Platelet plug formation.

A platelet plug can stop blood loss completely if the hole in a blood vessel is small enough.

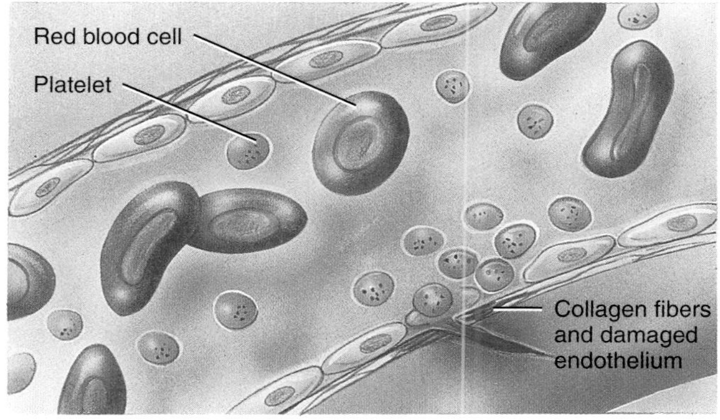

1 Platelet adhesion

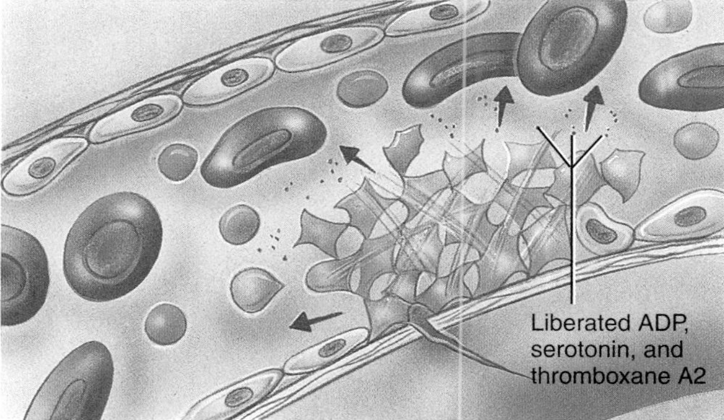

2 Platelet release reaction

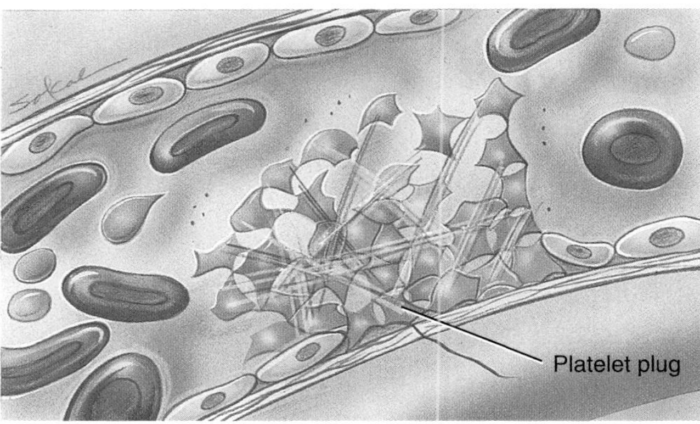

3 Platelet aggregation

Along with platelet plug formation, which two mechanisms contribute to hemostasis?

platelets causes them to adhere to the originally activated platelets. This gathering of platelets is called **platelet aggregation.** Eventually, the accumulation and attachment of large numbers of platelets form a mass called a **platelet plug.**

A platelet plug is very effective in preventing blood loss in a small vessel. Although initially the platelet plug is loose, it becomes quite tight when reinforced by fibrin threads formed during clotting (see Figure 19.10). A platelet plug can stop blood loss completely if the hole in a blood vessel is not too large.

Figure 19.10 Blood clot formation. Notice the platelet and red blood cells entrapped in fibrin threads.

A blood clot is a gel that contains formed elements of the blood entangled in fibrin threads.

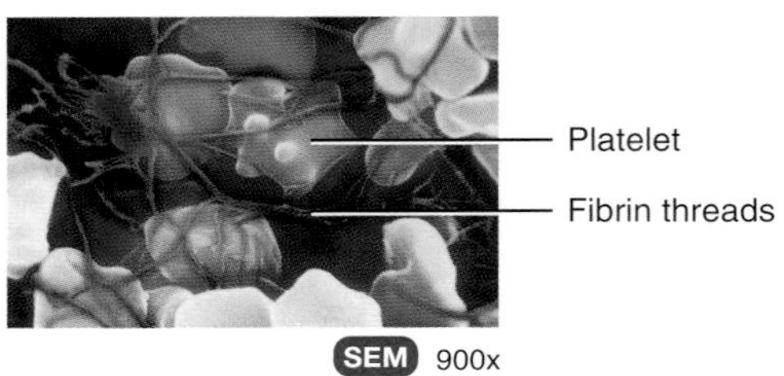

(a) Early stage

(b) Intermediate stage

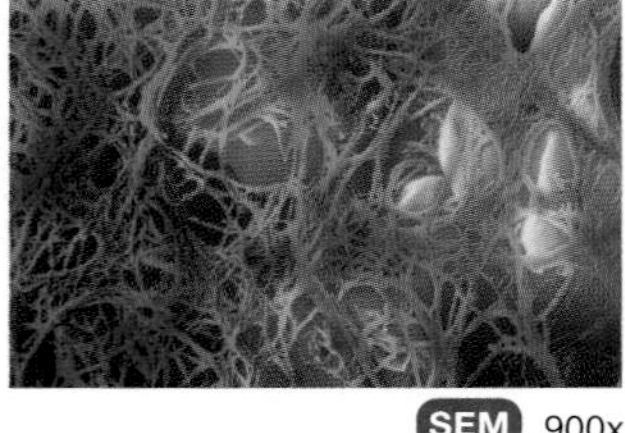

(c) Late stage

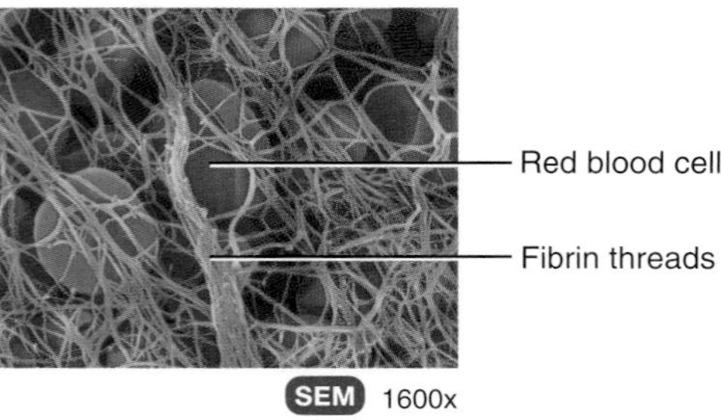

(d) Red blood cells trapped in fibrin threads

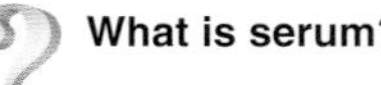

What is serum?

Blood Clotting

Normally, blood remains in its liquid form as long as it stays within its vessels. If it is drawn from the body, however, it thickens and forms a gel. Eventually, the gel separates from the liquid. The straw-colored liquid, called **serum,** is simply blood plasma minus the clotting proteins. The gel is called a **clot.** It consists of a network of insoluble protein fibers called fibrin in which the formed elements of blood are trapped (Figure 19.10).

The process of gel formation, called **clotting** or **coagulation,** is a series of chemical reactions that culminates in formation of fibrin threads. If blood clots too easily, the result can be **thrombosis**—clotting in an undamaged blood vessel. If the blood takes too long to clot, hemorrhage can occur.

Clotting involves several substances known as **clotting (coagulation) factors.** These factors include calcium ions (Ca^{2+}), several inactive enzymes that are synthesized by hepatocytes (liver cells) and released into the bloodstream, and various molecules associated with platelets or released by damaged tissues. Most clotting factors are identified by Roman numerals that indicate the order of their discovery (not necessarily the order of their participation in the clotting process).

Clotting is a complex cascade of enzymatic reactions in which each clotting factor activates many molecules of the next one in a fixed sequence. Finally, a large quantity of product (the insoluble protein fibrin) is formed. Clotting can be divided into three stages (Figure 19.11):

1. Two pathways, called the extrinsic pathway (Figure 19.11a) and the intrinsic pathway (Figure 19.11b), which will be described shortly, lead to the formation of prothrombinase. Once prothrombinase is formed, the steps involved in the next two stages of clotting are the same for both the extrinsic and intrinsic pathways, and together these two stages are referred to as the common pathway.
2. Prothrombinase converts prothrombin (a plasma protein formed by the liver) into the enzyme thrombin.
3. Thrombin converts soluble fibrinogen (another plasma protein formed by the liver) into insoluble fibrin. Fibrin forms the threads of the clot.

The Extrinsic Pathway

The **extrinsic pathway** of blood clotting has fewer steps than the intrinsic pathway and occurs rapidly—within a matter of seconds if trauma is severe. It is so named because a tissue protein called **tissue factor (TF),** also known as **thromboplastin,** leaks into the blood from cells *outside (extrinsic to)* blood vessels and initiates the formation of prothrombinase. TF is a complex mixture of lipoproteins and phospholipids released from the surfaces of damaged cells. In the presence of Ca^{2+}, TF begins a sequence of reactions that ultimately activates clotting

Figure 19.11 The blood-clotting cascade.

In blood clotting, coagulation factors are activated in sequence, resulting in a cascade of reactions that includes positive feedback cycles.

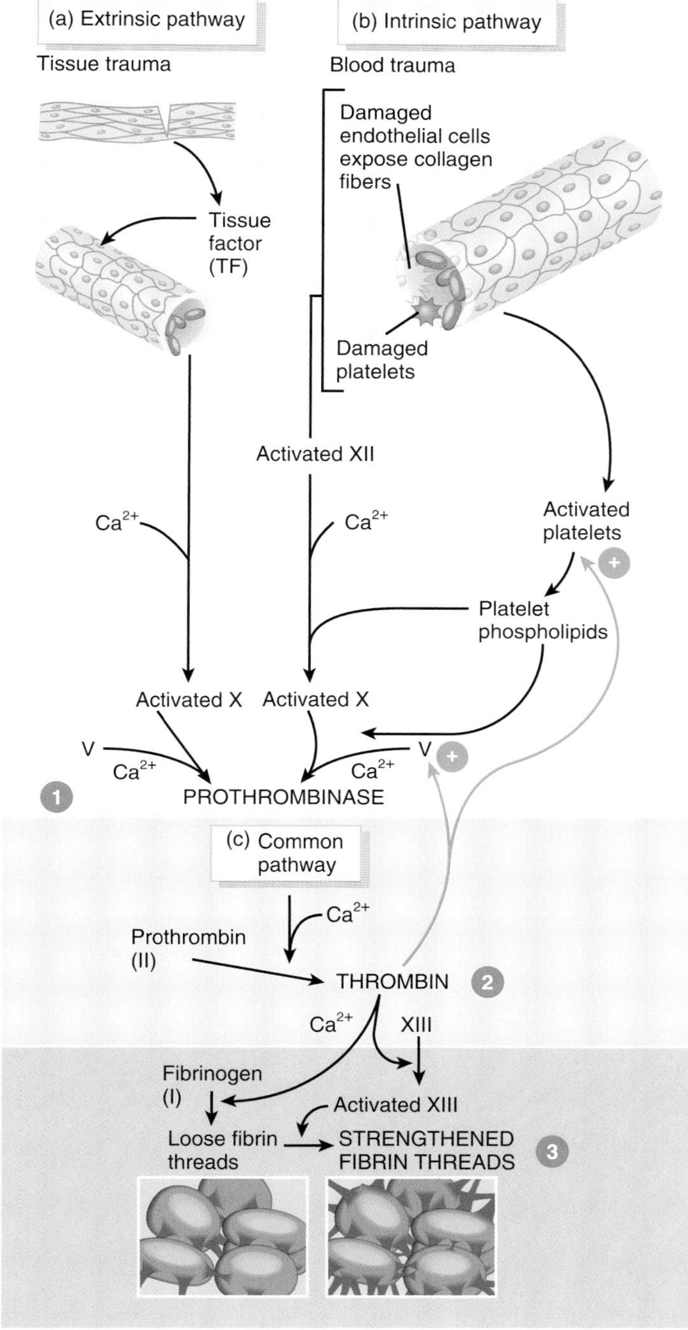

What is the outcome of the first stage of blood clotting?

factor X (Figure 19.11a). Once factor X is activated, it combines with factor V in the presence of Ca^{2+} to form the active enzyme prothrombinase, completing the extrinsic pathway.

The Intrinsic Pathway

The **intrinsic pathway** of blood clotting is more complex than the extrinsic pathway, and it occurs more slowly, usually requiring several minutes. The intrinsic pathway is so named because its activators are either in direct contact with blood or contained *within (intrinsic to)* the blood; outside tissue damage is not needed. If endothelial cells become roughened or damaged, blood can come in contact with collagen fibers in the connective tissue around the endothelium of the blood vessel. In addition, trauma to endothelial cells causes damage to platelets, resulting in the release of phospholipids by the platelets. Contact with collagen fibers (or with the glass sides of a blood collection tube) activates clotting factor XII (Figure 19.11b), which begins a sequence of reactions that eventually activates clotting factor X. Platelet phospholipids and Ca^{2+} can also participate in the activation of factor X. Once factor X is activated, it combines with factor V to form the active enzyme prothrombinase (just as occurs in the extrinsic pathway), completing the intrinsic pathway.

The Common Pathway

The formation of prothrombinase marks the beginning of the common pathway. In the second stage of blood clotting (Figure 19.11c), prothrombinase and Ca^{2+}catalyze the conversion of prothrombin to thrombin. In the third stage, thrombin, in the presence of Ca^{2+}, converts fibrinogen, which is soluble, to loose fibrin threads, which are insoluble. Thrombin also activates factor XIII (fibrin stabilizing factor), which strengthens and stabilizes the fibrin threads into a sturdy clot. Plasma contains some factor XIII, which is also released by platelets trapped in the clot.

Thrombin has two positive feedback effects. In the first positive feedback loop, which involves factor V, it accelerates the formation of prothrombinase. Prothrombinase, in turn, accelerates the production of more thrombin, and so on. In the second positive feedback loop, thrombin activates platelets, which reinforces their aggregation and the release of platelet phospholipids.

Clot Retraction

Once a clot is formed, it plugs the ruptured area of the blood vessel and thus stops blood loss. **Clot retraction** is the consolidation or tightening of the fibrin clot. The fibrin threads attached to the damaged surfaces of the blood vessel gradually contract as platelets pull on them. As the clot retracts, it pulls the edges of the damaged vessel closer together, decreasing the risk of further damage. During retraction, some serum can escape between the fibrin threads, but the formed elements in blood cannot. Normal retraction depends on an adequate number of platelets in the clot, which release factor

XIII and other factors, thereby strengthening and stabilizing the clot. Permanent repair of the blood vessel can then take place. In time, fibroblasts form connective tissue in the ruptured area, and new endothelial cells repair the vessel lining.

Role of Vitamin K in Clotting

Normal clotting depends on adequate levels of vitamin K in the body. Although vitamin K is not involved in actual clot formation, it is required for the synthesis of four clotting factors. Normally produced by bacteria that inhabit the large intestine, vitamin K is a fat-soluble vitamin that can be absorbed through the lining of the intestine and into the blood if absorption of lipids is normal. People suffering from disorders that slow absorption of lipids (for example, inadequate release of bile into the small intestine) often experience uncontrolled bleeding as a consequence of vitamin K deficiency.

The various clotting factors, their sources, and the pathways of activation are summarized in Table 19.4.

Hemostatic Control Mechanisms

Many times a day little clots start to form, often at a site of minor roughness or at a developing atherosclerotic plaque inside a blood vessel. Because blood clotting involves amplification and positive feedback cycles, a clot has a tendency to enlarge, creating the potential for impairment of blood flow through undamaged vessels. The **fibrinolytic system** (fī-bri-nō-LIT-ik) dissolves small, inappropriate clots; it also dissolves clots at a site of damage once the damage is repaired. Dissolution of a clot is called **fibrinolysis** (fī-bri-NOL-i-sis). When a clot is formed, an inactive plasma enzyme called **plasminogen** is incorporated into the clot. Both body tissues and blood contain substances that can activate plasminogen to **plasmin (fibrinolysin),** an active plasma enzyme. Among these substances are thrombin, activated factor XII, and tissue plasminogen activator (t-PA), which is synthesized in endothelial cells of most tissues and liberated into the blood. Once plasmin is formed, it can dissolve the clot by digesting fibrin threads and inactivating substances such as fibrinogen, prothrombin, and factors V and XII.

Even though thrombin has a positive feedback effect on blood clotting, clot formation normally remains localized at the site of damage. A clot does not extend beyond a wound site into the general circulation, in part because fibrin absorbs thrombin into the clot. Another reason for localized clot formation is that because of the dispersal of some of the clotting factors by the blood, their concentrations are not high enough to bring about widespread clotting.

Several other mechanisms also control blood clotting. For example, endothelial cells and white blood cells produce a prostaglandin called **prostacyclin** (pros-ta-SĪ-klin) that opposes the actions of thromboxane A2. Prostacyclin is a powerful inhibitor of platelet adhesion and release.

In addition, substances that delay, suppress, or prevent blood clotting, called **anticoagulants,** are present in blood. These include **antithrombin,** which blocks the action of several factors, including XII, X, and II (prothrombin). **Heparin,** an anticoagulant that is produced by mast cells and basophils, combines

TABLE 19.4 Clotting (Coagulation) Factors

Number*	Name(s)	Source	Pathway(s) of Activation
I	Fibrinogen.	Liver.	Common.
II	Prothrombin.	Liver.	Common.
III	Tissue factor (thromboplasm).	Damaged tissues and activated platelets.	Extrinsic.
IV	Calcium ions (Ca^{2+}).	Diet, bones, and platelets.	All.
V	Proaccelerin, labile factor, or accelerator globulin (AcG).	Liver and platelets.	Extrinsic and intrinsic.
VII	Serum prothrombin conversion accelerator (SPCA), stable factor, or proconvertin.	Liver.	Extrinsic.
VIII	Antihemophilic factor (AHF), antihemophilic factor A, or antihemophilic globulin (AHG).	Liver.	Intrinsic.
IX	Christmas factor, plasma thromboplastin component (PTC), or antihemophilic factor B.	Liver.	Intrinsic.
X	Stuart factor, Prower factor, or thrombokinase.	Liver.	Extrinsic and intrinsic.
XI	Plasma thromboplastin antecedent (PTA) or antihemophilic factor C.	Liver.	Intrinsic.
XII	Hageman factor, glass factor, contact factor, or antihemophilic factor D.	Liver.	Intrinsic.
XIII	Fibrin-stabilizing factor (FSF).	Liver and platelets.	Common.

*There is no factor VI. Prothrombinase (prothrombin activator) is a combination of activated factors V and X.

with antithrombin and increases its effectiveness in blocking thrombin. Another anticoagulant, **activated protein C (APC),** inactivates the two major clotting factors not blocked by antithrombin, and enhances activity of plasminogen activators. Babies that lack the ability to produce APC due to a genetic mutation usually die of blood clots in infancy.

Anticoagulants

Patients who are at increased risk of forming blood clots may receive anticoagulants. Examples are heparin or warfarin. Heparin is often administered during hemodialysis and open-heart surgery. **Warfarin (Coumadin®)** acts as an antagonist to vitamin K and thus blocks synthesis of four clotting factors. Warfarin is slower acting than heparin. To prevent clotting in donated blood, blood banks and laboratories often add substances that remove Ca^{2+}; examples are EDTA (ethylene diamine tetraacetic acid) and CPD (citrate phosphate dextrose). ■

Intravascular Clotting

Despite the anticoagulating and fibrinolytic mechanisms, blood clots sometimes form within the cardiovascular system. Such clots may be initiated by roughened endothelial surfaces of a blood vessel resulting from atherosclerosis, trauma, or infection. These conditions induce adhesion of platelets. Intravascular clots may also form when blood flows too slowly (stasis), allowing clotting factors to accumulate locally in high enough concentrations to initiate coagulation. Clotting in an unbroken blood vessel (usually a vein) is called **thrombosis** (throm-BŌ-sis; *thromb-* = clot; *-osis* = a condition of). The clot itself, called a **thrombus,** may dissolve spontaneously. If it remains intact, however, the thrombus may become dislodged and be swept away in the blood. A blood clot, bubble of air, fat from broken bones, or a piece of debris transported by the bloodstream is called an **embolus** (*em-* = in; *bolus* = a mass). An embolus that breaks away from an arterial wall may lodge in a smaller-diameter artery downstream and block blood flow to a vital organ. When an embolus lodges in the lungs, the condition is called **pulmonary embolism.**

Aspirin and Thrombolytic Agents

In patients with heart and blood vessel disease, the events of hemostasis may occur even without external injury to a blood vessel. At low doses, **aspirin** inhibits vasoconstriction and platelet aggregation by blocking synthesis of thromboxane A2. It also reduces the chance of thrombus formation. Due to these effects, aspirin reduces the risk of transient ischemic attacks (TIA), strokes, myocardial infarction, and blockage of peripheral arteries.

Thrombolytic agents are chemical substances that are injected into the body to dissolve blood clots that have already formed to restore circulation. They either directly or indirectly activate plasminogen. The first thrombolytic agent, approved in 1982 for dissolving clots in the coronary arteries of the heart, was **streptokinase,** which is produced by streptococcal bacteria. A genetically engineered version of human **tissue plasminogen activator (t-PA)** is now used to treat victims of both heart attacks and brain attacks (strokes) that are caused by blood clots. ■

▶ CHECKPOINT

18. What is hemostasis?
19. How do vascular spasm and platelet plug formation occur?
20. What is fibrinolysis? Why does blood rarely remain clotted inside blood vessels?
21. How do the extrinsic and intrinsic pathways of blood clotting differ?
22. Define each of the following terms: anticoagulant, thrombus, embolus, and thrombolytic agent.

BLOOD GROUPS AND BLOOD TYPES

▶ OBJECTIVES

Distinguish between the ABO and Rh blood groups.

Explain why it is so important to match donor and recipient blood types before administering a transfusion.

The surfaces of erythrocytes contain a genetically determined assortment of **antigens** composed of glycoproteins and glycolipids. These antigens, called **agglutinogens** (a-gloo-TIN-ō-jens), occur in characteristic combinations. Based on the presence or absence of various antigens, blood is categorized into different **blood groups.** Within a given blood group, there may be two or more different **blood types.** There are at least 24 blood groups and more than 100 antigens that can be detected on the surface of red blood cells. Here we discuss two major blood groups—ABO and Rh. Other blood groups include the Lewis, Kell, Kidd, and Duffy systems. The incidence of ABO and Rh blood types varies among different population groups, as indicated in Table 19.5.

TABLE 19.5 Blood Types in the United States

	Blood Type (percentage)				
Population Group	**O**	**A**	**B**	**AB**	**Rh^+**
European-American	45	40	11	4	85
African-American	49	27	20	4	95
Korean-American	32	28	30	10	100
Japanese-American	31	38	21	10	100
Chinese-American	42	27	25	6	100
Native American	79	16	4	1	100

ABO Blood Group

The **ABO blood group** is based on two glycolipid antigens called A and B (Figure 19.12). People whose RBCs display *only antigen A* have **type A** blood. Those who have *only antigen B* are **type B.** Individuals who have *both A and B antigens* are **type AB;** those who have *neither antigen A nor B* are **type O.**

Blood plasma usually contains **antibodies** called **agglutinogens** that react with the A or B antigens if the two are mixed. These are the **anti-A antibody,** which reacts with antigen A, and the **anti-B antibody,** which reacts with antigen B. The antibodies present in each of the four blood types are shown in Figure 19.12. You do not have antibodies that react with the antigens of your own RBCs, but you do have antibodies for any antigens that your RBCs lack. For example, if your blood type is B, you have B antigens on your red blood cells, and you have anti-A antibodies in your blood plasma. Although agglutinins start to appear in the blood within a few months after birth, the reason for their presence is not clear. Perhaps they are formed in response to bacteria that normally inhabit the gastrointestinal tract. Because the antibodies are large IgM-type antibodies (see Table 22.3 on page 831) that do not cross the placenta, ABO incompatibility between a mother and her fetus rarely causes problems.

Transfusions

Despite the differences in RBC antigens reflected in the blood group systems, blood is the most easily shared of human tissues, saving many thousands of lives every year through transfusions. A **transfusion** (trans-FŪ-shun) is the transfer of whole blood or blood components (red blood cells only or blood plasma only) into the bloodstream or directly into the red bone marrow. A transfusion is most often given to alleviate anemia, to increase blood volume (for example, after a severe hemorrhage), or to improve immunity. However, the normal components of one person's RBC plasma membrane can trigger damaging antigen–antibody responses in a transfusion recipient. In an incompatible blood transfusion, antibodies in the recipient's plasma bind to the antibodies on the donated RBCs, which causes **agglutination** (a-gloo-ti-NĀ-shun), or clumping, of the RBCs. Agglutination is an antigen–antibody response in which RBCs become cross-linked to one another. (Note that agglutination is not the same as blood clotting.) When these antigen–antibody complexes form, they activate plasma proteins of the complement family (described on page 831). In essence, complement molecules make the plasma membrane of the donated RBCs leaky, causing **hemolysis** (rupture) of the RBCs and the release of hemoglobin into the blood plasma. The liberated hemoglobin may cause kidney damage by clogging the filtration membranes. Although quite rare, it is possible for the viruses that cause AIDS and hepatitis B and C to be transmitted through transfusion of contaminated blood products.

Consider what happens if a person with type A blood receives a transfusion of type B blood. The recipient's blood (type A) contains A antigens on the red blood cells and anti-B antibodies in the plasma. The donor's blood (type B) contains B antigens and anti-A antibodies. In this situation, two things can happen. First, the anti-B antibodies in the recipient's plasma can bind to the B antigens on the donor's erythrocytes, causing agglutination and hemolysis of the red blood cells. Second, the

Figure 19.12 Antigens and antibodies of the ABO blood types.

The antibodies in your plasma do not react with the antigens on your red blood cells.

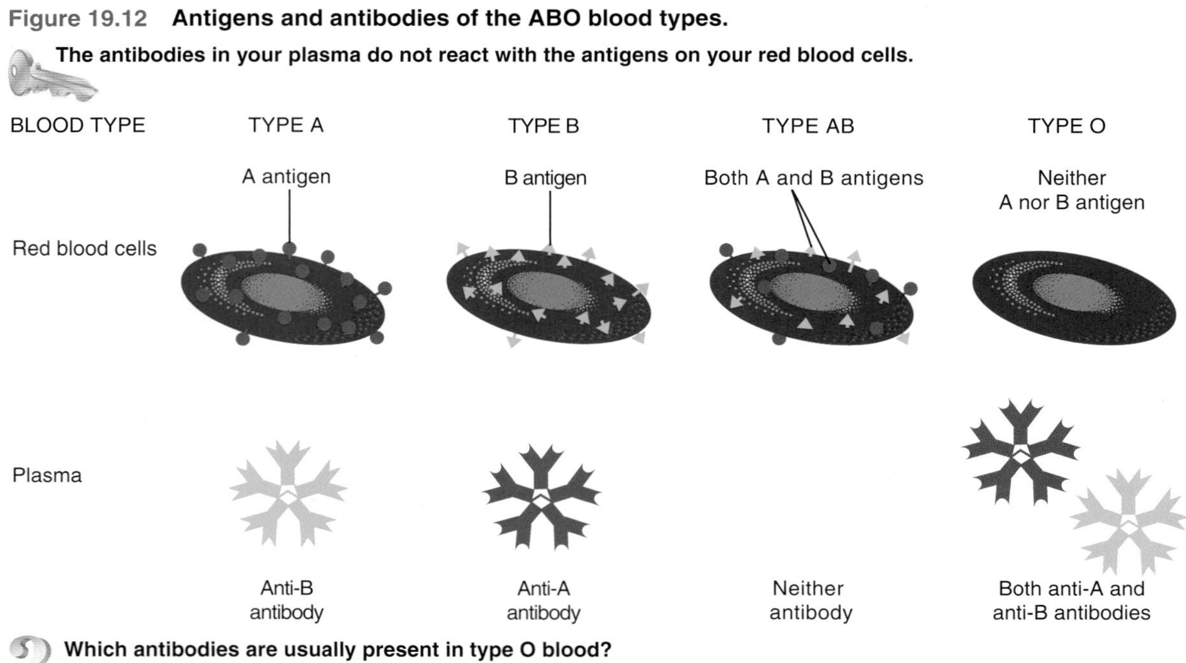

Which antibodies are usually present in type O blood?

TABLE 19.6 Summary of ABO Blood Group Interactions

Characteristic	Blood Type A	Blood Type B	Blood Type AB	Blood Type O
Agglutinogen (antigen) on RBCs	A	B	Both A and B	Neither A nor B
Agglutinin (antibody) in plasma	anti-B	anti-A	Neither anti-A nor anti-B	Both anti-A and anti-B
Compatible donor blood types (no hemolysis)	A, O	B, O	A, B, AB, O	O
Incompatible donor blood types (hemolysis)	B, AB	A, AB	—	A, B, AB

anti-A antibodies in the donor's plasma can bind to the A antigens on the recipient's red blood cells, a less serious reaction because the donor's anti-A antibodies become so diluted in the recipient's plasma that they do not cause significant agglutination and hemolysis of the recipient's RBCs.

Table 19.6 summarizes the interactions of the four blood types of the ABO system.

People with type AB blood do not have anti-A or anti-B antibodies in their blood plasma. They are sometimes called *universal recipients* because theoretically they can receive blood from donors of all four blood types. They have no antibodies to attack antigens on donated RBCs (Table 19.6). People with type O blood have neither A nor B antigens on their RBCs and are sometimes called *universal donors* because theoretically they can donate blood to all four ABO blood types. Type O persons requiring blood may receive only type O blood (Table 19.6). In practice, use of the terms universal recipient and universal donor is misleading and dangerous. Blood contains antigens and antibodies other than those associated with the ABO system that can cause transfusion problems. Thus, blood should be carefully cross-matched or screened before transfusion. In about 80% of the population, soluble antigens of the ABO type appear in saliva and other body fluids, in which case blood type can be identified from a sample of saliva.

Rh Blood Group

The **Rh blood group** is so named because the antigen was discovered in the blood of the *Rhesus* monkey. The alleles of three genes may code for the Rh antigen. People whose RBCs have Rh antigens are designated Rh^+ (Rh positive); those who lack Rh antigens are designated Rh^- (Rh negative). Table 19.5 shows the incidence of Rh^+ and Rh^- in various populations. Normally, blood plasma does not contain anti-Rh antibodies. If an Rh^- person receives an Rh^+ blood transfusion, however, the immune system starts to make anti-Rh antibodies that will remain in the blood. If a second transfusion of Rh^+ blood is given later, the previously formed anti-Rh antibodies will cause agglutination and hemolysis of the RBCs in the donated blood, and a severe reaction may occur.

Hemolytic Disease of the Newborn

The most common problem with Rh incompatibility, **hemolytic disease of the newborn (HDN),** may arise during pregnancy (Figure 19.13). Normally, no direct contact occurs between maternal and fetal blood while a woman is pregnant. However, if a small amount of Rh^+ blood leaks from the fetus through the placenta into the bloodstream of an Rh^- mother, the mother will start to make anti-Rh antibodies. Because the greatest possibility of fetal blood leakage into the maternal circulation occurs at delivery, the firstborn baby usually is not affected. If the mother becomes pregnant again, however, her anti-Rh antibodies can cross the placenta and enter the bloodstream of the fetus. If the fetus is Rh^-, there is no problem, because Rh^- blood does not have the Rh antigen. If the fetus is Rh^+, however, agglutination and hemolysis brought on by fetal–maternal incompatibility may occur in the fetal blood.

Figure 19.13 Development of hemolytic disease of the newborn (HDN). (a) At birth, a small quantity of fetal blood usually leaks across the placenta into the maternal bloodstream. A problem can arise when the mother is Rh^- and the baby is Rh^+, having inherited an allele for one of the Rh antigens from the father. (b) Upon exposure to Rh antigen, the mother's immune system responds by making anti-Rh antibodies. (c) During a subsequent pregnancy, the maternal antibodies cross the placenta into the fetal blood. If the second fetus is Rh^+, the ensuing antigen–antibody reaction causes agglutination and hemolysis of fetal RBCs. The result is HDN.

HDN occurs when maternal anti-Rh antibodies cross the placenta and cause hemolysis of fetal RBCs.

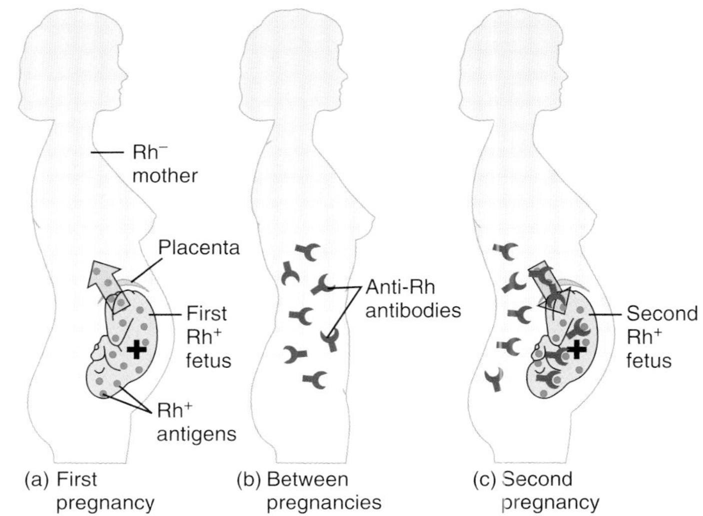

Why is the firstborn baby unlikely to have HDN?

An injection of anti-Rh antibodies called anti-Rh gamma globulin (RhoGAM®) can be given to prevent HDN. All Rh^- women should receive RhoGAM® soon after every delivery, miscarriage, or abortion. These antibodies bind to and inactivate the fetal Rh antigens before the mother's immune system can respond to the foreign antigens by producing her own anti-Rh antibodies. ■

Typing and Cross-Matching Blood for Transfusion

To avoid blood-type mismatches, laboratory technicians type the patient's blood and then either cross-match it to potential donor blood or screen it for the presence of antibodies. In the procedure for ABO blood typing, single drops of blood are mixed with different *antisera,* solutions that contain antibodies (Figure 19.14). One drop of blood is mixed with anti-A serum, which contains anti-A antibodies that will agglutinate red blood cells that possess A antigens. Another drop is mixed with anti-B serum, which contains anti-B antibodies that will agglutinate red blood cells that possess B antigens. If the red blood cells agglutinate only when mixed with anti-A serum, the blood is type A. If the red blood cells agglutinate only when mixed with anti-B serum, the blood is type B. The blood is type AB if both drops agglutinate; if neither drop agglutinates, the blood is type O.

In the procedure for determining Rh factor, a drop of blood is mixed with antiserum containing antibodies that will agglutinate RBCs displaying Rh antigens. If the blood agglutinates, it is Rh^+; no agglutination indicates Rh^-.

Once the patient's blood type is known, donor blood of the same ABO and Rh type is selected. In a **cross-match,** the possible donor RBCs are mixed with the recipient's serum. If agglutination does not occur, the recipient does not have antibodies that will attack the donor RBCs. Alternatively, the recipient's serum can be **screened** against a test panel of RBCs having antigens known to cause blood transfusion reactions to detect any antibodies that may be present.

► CHECKPOINT

23. What precautions must be taken before giving a blood transfusion?
24. What is hemolysis and how can it occur after a mismatched blood transfusion?
25. Explain the conditions that may cause hemolytic disease of the newborn.

Figure 19.14 **ABO blood typing.**

In the procedure for ABO blood typing, blood is mixed with anti-A serum and anti-B serum.

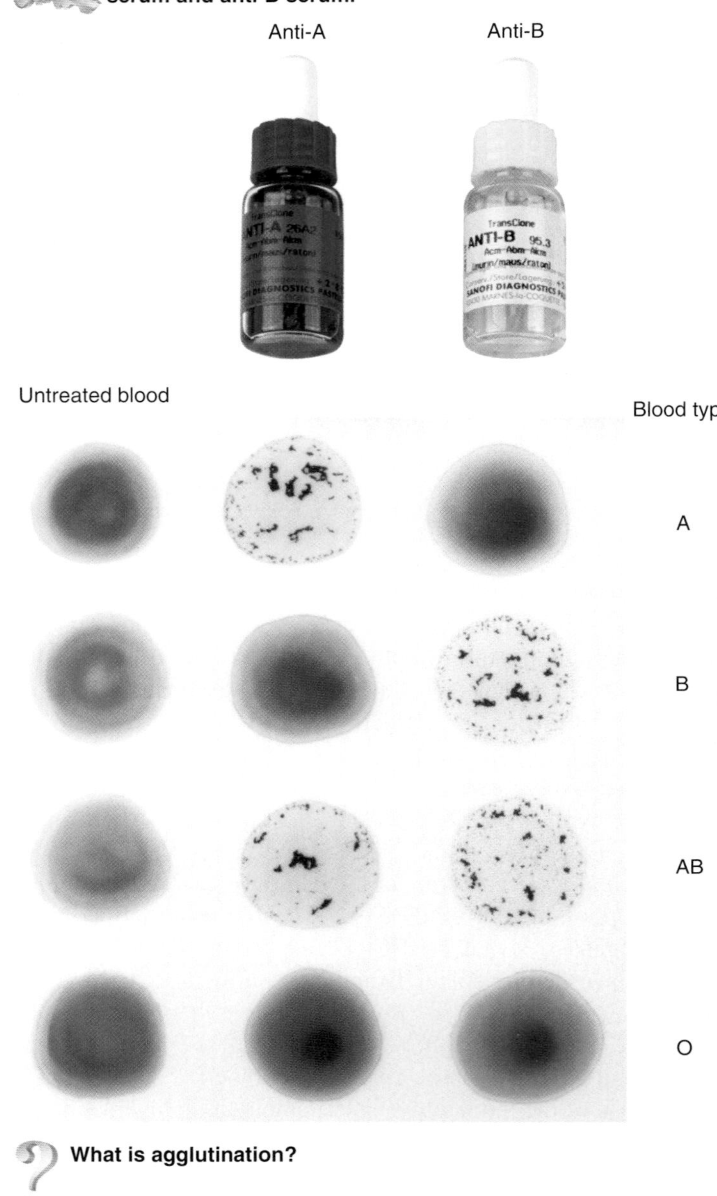

What is agglutination?

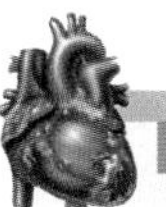

DISORDERS: HOMEOSTATIC IMBALANCES

Anemia

Anemia is a condition in which the oxygen-carrying capacity of blood is reduced. All of the many types of anemia are characterized by reduced numbers of RBCs or a decreased amount of hemoglobin in the blood. The person feels fatigued and is intolerant of cold, both of which are related to lack of oxygen needed for ATP and heat production. Also, the skin appears pale, due to the low content of red-colored hemoglobin circulating in skin blood vessels. Among the most important causes and types of anemia are the following:

- *Inadequate absorption of iron, excessive loss of iron, increased iron requirement, or insufficient intake of iron* causes **iron-deficiency anemia,** the most common type of anemia. Women are at greater risk for iron-deficiency anemia due to menstrual blood losses and increased iron demands of the growing fetus during pregnancy. Gastrointestinal losses, such as occurs with malignancy or ulceration, also contribute to this type of anemia.
- *Inadequate intake of vitamin B_{12} or folic acid* causes **megaloblastic anemia** in which red bone marrow produces large, abnormal red blood cells (megaloblasts). It may also be caused by drugs that alter gastric secretion or are used to treat cancer.
- *Insufficient hemopoiesis* resulting from an inability of the stomach to produce intrinsic factor, which is needed for absorption of vitamin B_{12} in the small intestine, causes **pernicious anemia.**
- *Excessive loss of RBCs* through bleeding resulting from large wounds, stomach ulcers, or especially heavy menstruation leads to **hemorrhagic anemia.**
- *RBC plasma membranes rupture prematurely* in **hemolytic anemia.** The released hemoglobin pours into the plasma and may damage the filtering units (glomeruli) in the kidneys. The condition may result from inherited defects such as abnormal red blood cell enzymes, or from outside agents such as parasites, toxins, or antibodies from incompatible transfused blood.
- *Deficient synthesis of hemoglobin* occurs in **thalassemia** (thal′-a-SĒ-mē-a), a group of hereditary hemolytic anemias. The RBCs are small (microcytic), pale (hypochromic), and short-lived. Thalassemia occurs primarily in populations from countries bordering the Mediterranean Sea.
- *Destruction of red bone marrow* results in **aplastic anemia.** It is caused by toxins, gamma radiation, and certain medications that inhibit enzymes needed for hemopoiesis.

Sickle-Cell Disease

The RBCs of a person with **sickle-cell disease (SCD)** contain Hb-S, an abnormal kind of hemoglobin. When Hb-S gives up oxygen to the interstitial fluid, it forms long, stiff, rodlike structures that bend the erythrocyte into a sickle shape (Figure 19.15). The sickled cells rupture easily. Even though erythropoiesis is stimulated by the loss of the cells, it cannot keep pace with hemolysis. People with sickle-cell disease always have some degree of anemia and mild jaundice and may experience joint or bone pain, breathlessness, rapid heart rate, abdominal pain, fever, and fatigue as a result of tissue damage caused by prolonged recovery oxygen uptake (oxygen debt). Any activity that reduces the amount of oxygen in the blood, such as vigorous exercise, may produce a **sickle-cell crisis** (worsening of the anemia, pain in the abdomen and long bones of the limbs, fever, and shortness of breath).

Sickle-cell disease is inherited. People with two sickle-cell genes have severe anemia; those with only one defective gene have minor problems. Sickle-cell genes are found primarily among populations, or descendants of populations, that live in the malaria belt around the world, including parts of Mediterranean Europe, sub-Saharan Africa, and tropical Asia. The gene responsible for the tendency of the RBCs to sickle also alters the permeability of the plasma membranes of sickled cells, causing potassium ions to leak out. Low levels of potassium kill the malaria parasites that may infect sickled cells. Because of this effect, a person with one normal gene and one sickle-cell gene has higher-than-average resistance to malaria. The possession of a single sickle-cell gene thus confers a survival advantage.

Treatment of SCD consists of administration of analgesics to relieve pain, fluid therapy to maintain hydration, oxygen to reduce the possibility of oxygen debt, antibiotics to counter infections, and blood transfusions. People who suffer from SCD have normal fetal hemoglobin (Hb-F), a slightly different form of hemoglobin that predominates at birth and is present in small amounts thereafter. In some patients with sickle-cell disease, a drug called hydroxyurea promotes transcription of the normal Hb-F gene, elevates the level of Hb-F, and reduces the chance that the RBCs will sickle. Unfortunately, this drug also has toxic effects on the bone marrow; thus, its safety for long-term use is questionable.

Hemophilia

Hemophilia (hē-mō-FIL-ē-a; *-philia* = loving) is an inherited deficiency of clotting in which bleeding may occur spontaneously or after

Figure 19.15 Red blood cells from a person with sickle-cell disease.

The red blood cells of a person with sickle-cell disease contain an abnormal type of hemoglobin called Hb-S.

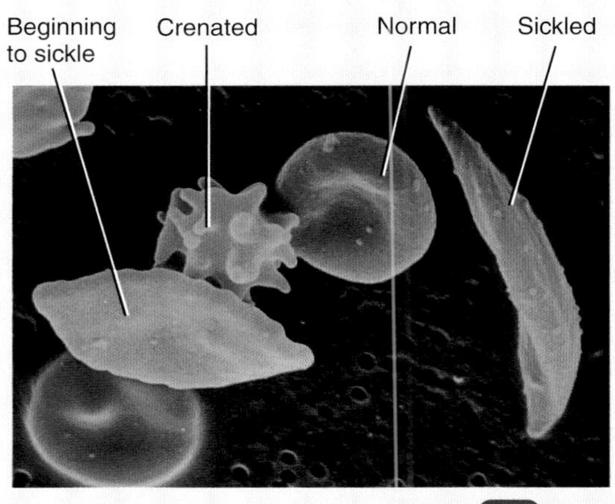

What are some symptoms of sickle-cell disease?

only minor trauma. It is the oldest known hereditary bleeding disorder; descriptions of the disease are found as early as the second century A.D. Hemophilia usually affects males and is sometimes referred to as "the royal disease" because many descendants of Queen Victoria, beginning with one of her sons, were affected by the disease. Different types of hemophilia are due to deficiencies of different blood clotting factors and exhibit varying degrees of severity, ranging from mild to severe bleeding tendencies. Hemophilia is characterized by spontaneous or traumatic subcutaneous and intramuscular hemorrhaging, nosebleeds, blood in the urine, and hemorrhages in joints that produce pain and tissue damage. Treatment involves transfusions of fresh blood plasma or concentrates of the deficient clotting factor to relieve the tendency to bleed. Another treatment is the drug desmopressin (DDAVP), which can boost the levels of the clotting factors.

Leukemia

The term **leukemia** (loo-KĒ-mē-a; *leuko-* = white) refers to a group of red bone marrow cancers in which abnormal white blood cells multiply uncontrollably. The accumulation of the cancerous white blood cells in red bone marrow interferes with the production of red blood cells, white blood cells, and platelets. As a result the oxygen-carrying capacity of the blood is reduced, an individual is more susceptible to infection, and blood clotting is abnormal. In most leukemias, the cancerous white blood cells spread to the lymph nodes, liver, and spleen, causing them to enlarge. All leukemias produce the usual symptoms of anemia (fatigue, intolerance to cold, and pale skin). In addition, weight loss, fever, night sweats, excessive bleeding, and recurrent infections may also occur.

In general, leukemias are classified as acute (symptoms develop rapidly) and chronic (symptoms may take years to develop). Adults may have either type, but children usually have the acute type.

The cause of most types of leukemia is unknown. However, certain risk factors have been implicated. These include exposure to radiation or chemotherapy for other cancers, genetics (some genetic disorders such as Down's syndrome), environmental factors (smoking and benzene), and microbes such as the human T cell leukemia-lymphoma virus-1 (HTLV-1) and the Epstein-Barr virus.

Treatment options include chemotherapy, radiation, stem cell transplantation, interferon, antibodies, and blood transfusion.

MEDICAL TERMINOLOGY

Acute normovolemic hemodilution (nor-mō-vō-LĒ-mik hē-mō-di-LOO-shun) Removal of blood immediately before surgery and its replacement with a cell-free solution to maintain sufficient blood volume for adequate circulation. At the end of surgery, once bleeding has been controlled, the collected blood is returned to the body.

Autologous preoperative transfusion (aw-TOL-o-gus trans-FŪ-zhun; *auto-* = self) Donating one's own blood; can be done up to 6 weeks before elective surgery. Also called **predonation.** This procedure eliminates the risk of incompatibility and blood-borne disease.

Blood bank A facility that collects and stores a supply of blood for future use by the donor or others. Because blood banks have additional and diverse functions (immunohematology reference work, continuing medical education, bone and tissue storage, and clinical consultation), they are more appropriately referred to as **centers of transfusion medicine.**

Cyanosis (sī-a-NŌ-sis; *cyano-* = blue) Slightly bluish/dark-purple skin discoloration, most easily seen in the nail beds and mucous membranes, due to an increased quantity of reduced hemoglobin (hemoglobin not combined with oxygen) in systemic blood.

Gamma globulin (GLOB-ū-lin) Solution of immunoglobulins from blood consisting of antibodies that react with specific pathogens, such as viruses. It is prepared by injecting the specific virus into animals, removing blood from the animals after antibodies have accumulated, isolating the antibodies, and injecting them into a human to provide short-term immunity.

Hemochromatosis (hē-mō-krō-ma-TŌ-sis; *chroma* = color) Disorder of iron metabolism characterized by excessive absorption of ingested iron and excess deposits of iron in tissues (especially the liver, heart, pituitary gland, gonads, and pancreas) that result in bronze discoloration of the skin, cirrhosis, diabetes mellitus, and bone and joint abnormalities.

Hemorrhage (HEM-or-ij; *rhegnynai* = bursting forth) Loss of a large amount of blood; can be either internal (from blood vessels into tissues) or external (from blood vessels directly to the surface of the body).

Jaundice (*jaund-* = yellow) An abnormal yellowish discoloration of the sclerae of the eyes, skin, and mucous membranes due to excess bilirubin (yellow-orange pigment) in the blood. The three main categories of jaundice are *prehepatic jaundice,* due to excess production of bilirubin; *hepatic jaundice,* abnormal bilirubin processing by the liver caused by congenital liver disease, cirrhosis (scar tissue formation) of the liver, or hepatitis (liver inflammation); and *extrahepatic jaundice,* due to blockage of bile drainage by gallstones or cancer of the bowel or pancreas.

Phlebotomist (fle-BOT-ō-mist; *phlebo-* = *vein; -tom* = cut) A technician who specializes in withdrawing blood.

Septicemia (sep′-ti-SĒ-mē-a; *septic-* = decay; *-emia* = condition of blood) Toxins or disease-causing bacteria in the blood. Also called "blood poisoning."

Thrombocytopenia (throm′-bō-sī-tō-PĒ-nē-a; *-penia* = poverty) Very low platelet count that results in a tendency to bleed from capillaries.

Venesection (vē′-ne-SEK-shun; *ven-* = vein) Opening of a vein for withdrawal of blood. Although **phlebotomy** (fle-BŌT-ō-mē) is a synonym for venesection, in clinical practice phlebotomy refers to therapeutic bloodletting, such as the removal of some blood to lower its viscosity in a patient with polycythemia.

Whole blood Blood containing all formed elements, plasma, and plasma solutes in natural concentrations.

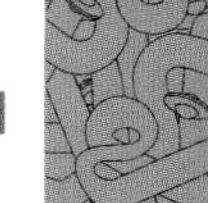

STUDY OUTLINE

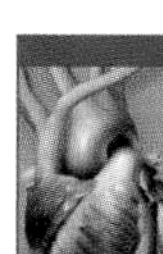

INTRODUCTION (p. 667)

1. The cardiovascular system consists of the blood, heart, and blood vessels.
2. Blood is a connective tissue composed of blood plasma (liquid portion) and formed elements (cells and cell fragments).

FUNCTIONS AND PROPERTIES OF BLOOD (p. 667)

1. Blood transports oxygen, carbon dioxide, nutrients, wastes, and hormones.
2. It helps regulate pH, body temperature, and water content of cells.
3. It provides protection through clotting and by combating toxins and microbes through certain phagocytic white blood cells or specialized blood plasma proteins.
4. Physical characteristics of blood include a viscosity greater than that of water; a temperature of 38°C (100.4°F); and a pH of 7.35–7.45.
5. Blood constitutes about 8% of body weight, and its volume is 4–6 liters in adults.
6. Blood is about 55% blood plasma and 45% formed elements.
7. The hematocrit is the percentage of total blood volume occupied by red blood cells.
8. Blood plasma consists of 91.5% water and 8.5% solutes. Principal solutes include proteins (albumins, globulins, fibrinogen), nutrients, vitamins, hormones, respiratory gases, electrolytes, and waste products.
9. The formed elements in blood include red blood cells (erythrocytes), white blood cells (leukocytes), and platelets.

FORMATION OF BLOOD CELLS (p. 670)

1. Hemopoiesis is the formation of blood cells from hemopoietic stem cells in red bone marrow.
2. Myeloid stem cells form RBCs, platelets, granulocytes, and monocytes. Lymphoid stem cells give rise to lymphocytes.
3. Several hemopoietic growth factors stimulate differentiation and proliferation of the various blood cells.

RED BLOOD CELLS (p. 672)

1. Mature RBCs are biconcave discs that lack nuclei and contain hemoglobin.
2. The function of the hemoglobin in red blood cells is to transport oxygen and some carbon dioxide.
3. RBCs live about 120 days. A healthy male has about 5.4 million RBCs/μL of blood; a healthy female has about 4.8 million/μL.
4. After phagocytosis of aged RBCs by macrophages, hemoglobin is recycled.
5. RBC formation, called erythropoiesis, occurs in adult red bone marrow of certain bones. It is stimulated by hypoxia, which stimulates the release of erythropoietin by the kidneys.
6. A reticulocyte count is a diagnostic test that indicates the rate of erythropoiesis.

WHITE BLOOD CELLS (p. 676)

1. WBCs are nucleated cells. The two principal types are granulocytes (neutrophils, eosinophils, and basophils) and agranulocytes (lymphocytes and monocytes).
2. The general function of WBCs is to combat inflammation and infection. Neutrophils and macrophages (which develop from monocytes) do so through phagocytosis.
3. Eosinophils combat the effects of histamine in allergic reactions, phagocytize antigen–antibody complexes, and combat parasitic worms. Basophils liberate heparin, histamine, and serotonin in allergic reactions that intensify the inflammatory response.
4. B lymphocytes, in response to the presence of foreign substances called antigens, differentiate into plasma cells that produce antibodies. Antibodies attach to the antigens and render them harmless. This antigen–antibody response combats infection and provides immunity. T lymphocytes destroy foreign invaders directly. Natural killer cells attack infectious microbes and tumor cells.
5. Except for lymphocytes, which may live for years, WBCs usually live for only a few hours or a few days. Normal blood contains 5000–10,000 WBCs/μL.

PLATELETS (p. 679)

1. Platelets (thrombocytes) are disc-shaped cell fragments that splinter from megakaryocytes. Normal blood contains 150,000–400,000 platelets/μL.
2. Platelets help stop blood loss from damaged blood vessels by forming a platelet plug.

STEM CELL TRANSPLANTS FROM BONE MARROW AND CORD-BLOOD (p. 679)

1. Bone marrow transplants involve removal of red bone marrow as a source of stem cells from the iliac crest.
2. In a cord-blood transplant, stem cells from the placenta are removed from the umbilical cord.
3. Cord-blood transplants have several advantages over bone marrow transplants.

HEMOSTASIS (p. 681)

1. Hemostasis refers to the stoppage of bleeding.
2. It involves vascular spasm, platelet plug formation, and blood clotting (coagulation).
3. In vascular spasm, the smooth muscle of a blood vessel wall contracts, which slows blood loss.
4. Platelet plug formation involves the aggregation of platelets to stop bleeding.
5. A clot is a network of insoluble protein fibers (fibrin) in which formed elements of blood are trapped.
6. The chemicals involved in clotting are known as clotting (coagulation) factors.
7. Blood clotting involves a cascade of reactions that may be divided into three stages: formation of prothrombinase, conversion of prothrombin into thrombin, and conversion of soluble fibrinogen into insoluble fibrin.

8. Clotting is initiated by the interplay of the extrinsic and intrinsic pathways of blood clotting.
9. Normal coagulation requires vitamin K and is followed by clot retraction (tightening of the clot) and ultimately fibrinolysis (dissolution of the clot).
10. Clotting in an unbroken blood vessel is called thrombosis. A thrombus that moves from its site of origin is called an embolus.

BLOOD GROUPS AND BLOOD TYPES (p. 685)

1. ABO and Rh blood groups are genetically determined and based on antigen–antibody responses.
2. In the ABO blood group, the presence or absence of A and B antigens on the surface of RBCs determines blood type.
3. In the Rh system, individuals whose RBCs have Rh antigens are classified as Rh^+; those who lack the antigen are Rh^-.
4. Hemolytic disease of the newborn (HDN) can occur when an Rh^- mother is pregnant with an Rh^+ fetus.
5. Before blood is transfused, a recipient's blood is typed and then either cross-matched to potential donor blood or screened for the presence of antibodies.

SELF-QUIZ QUESTIONS

Fill in the blanks in the following statements.

1. Plasma minus its clotting proteins is termed ____.
2. ____ is the consolidation or tightening of the fibrin clot that helps to bring the edges of a damaged vessel closer together.

Indicate whether the following statements are true or false.

3. Hemoglobin functions in transporting both oxygen and carbon dioxide and in regulating blood pressure.
4. The most numerous white blood cells in a differential white blood cell count of a healthy individual are the neutrophils.

Choose the one best answer to the following questions.

5. Which of the following are *not* required for clot formation? (1) vitamin K, (2) calcium, (3) prostacyclin, (4) plasmin, (5) fibrinogen. (a) 1, 2, and 5; (b) 3, 4, and 5; (c) 4 and 5; (d) 1, 2, and 3; (e) 3 and 4.
6. Place the steps involved in hemostasis in the correct order. (1) conversion of fibrinogen into fibrin, (2) conversion of prothrombin into thrombin, (3) adhesion and aggregation of platelets on damaged vessel, (4) prothrombinase formed by extrinsic or intrinsic pathway, (5) reduction of blood loss by initiation of a vascular spasm. (a) 5, 3, 4, 2, 1; (b) 5, 4, 3, 1, 2; (c) 3, 5, 4, 2, 1; (d) 5, 3, 2, 1, 4; (e) 5, 3, 2, 4, 1.
7. Which of the following statements explain why red blood cells (RBCs) are highly specialized for oxygen transport? (1) RBCs contain hemoglobin. (2) RBCs lack a nucleus. (3) RBCs have many mitochondria and thus generate ATP aerobically. (4) The biconcave shape of RBCs provides a large surface area for the inward and outward diffusion of gas molecules. (5) RBCs can carry up to four oxygen molecules for each hemoglobin molecule. (a) 1, 2, 3, and 5; (b) 1, 2, 4, and 5; (c) 2, 3, 4, and 5; (d) 1, 3, and 5; (e) 2, 4, and 5.
8. Which of the following are *true*? (1) White blood cells leave the bloodstream by emigration. (2) Adhesion molecules help white blood cells stick to the endothelium, which aids emigration. (3) Neutrophils and macrophages are active in phagocytosis. (4) The attraction of phagocytes to microbes and inflamed tissue is termed chemotaxis. (5) Leucopenia is an increase in white blood cell count that occurs during infection. (a) 1, 2, 4, and 5; (b) 2, 3, 4, and 5; (c) 1, 2, 3, and 4; (d) 1, 3, and 5; (e) 1, 2, and 4.
9. A person with type A Rh^- blood can receive a blood transfusion from which of the following types? (1) A Rh^+, (2) B Rh^-, (3) AB Rh^-, (4) O Rh^-, (5) A Rh^-. (a) 1 only; (b) 3 only; (c) 4 only; (d) 4 and 5; (e) 1 and 5.
10. A person with type B positive blood receives a transfusion of type AB positive blood. What will happen? (a) The recipient's antibodies will react with the donor's red blood cells. (b) The donor's antigens will destroy the recipient's antibodies. (c) The donor's antibodies will react with and destroy all of the recipient's red blood cells. (d) The recipient's blood type will change from Rh^+ to Rh^-. (e) These blood types are compatible and the transfusion will be accepted.
11. What happens to the iron (Fe^{3+}) that is released during the breakdown of damaged red blood cells? (a) It is used to synthesize proteins. (b) It is transported to the liver where it becomes part of bile. (c) It is converted into urobilin and excreted in urine. (d) It attaches to transferrin and is transported to bone marrow for use in hemoglobin synthesis. (e) It is utilized by intestinal bacteria to convert bilirubin into urobilinogen.
12. Which of the following would *not* cause an increase in erythropoietin? (a) anemia, (b) high altitude, (c) hemorrhage, (d) donating blood to a blood bank, (e) polycythemia.

13. Match the following:

____(a) contain hemoglobin and function in gas transport
____(b) cell fragments enclosed by a piece of the cell membrane of megakaryocytes; contain clotting factors
____(c) individual forms of progenitor cells; named on the basis of the mature elements in blood they will ultimately produce
____(d) white blood cell showing a kidney-shaped nucleus; capable of phagocytosis
____(e) monocytes that roam the tissues and gather at sites of infection or inflammation
____(f) occur as B cells, T cells, and natural killer cells
____(g) give rise to red blood cells, monocytes, neutrophils, eosinophils, basophils, and platelets
____(h) combat the effects of histamine and other mediators of inflammation in allergic reactions; also - phagocytize antigen-antibody complexes
____(i) respond to tissue destruction by bacteria; release lysozyme, strong oxidants, and defensins
____(j) older neutrophils with several differently shaped nuclear lobes
____(k) released from the red bone marrow, they develop into mature red blood cells
____(l) give rise to lymphocytes
____(m) cells no longer capable of replenishing themselves; can only give rise to more specific formed elements of blood
____(n) hormone that stimulates formation of platelets
____(o) monocytes that leave the blood and reside in a particular tissue such as alveolar macrophages in the lungs
____(p) involved in inflammatory and allergic reactions; are involved in hypersensitivity reactions
____(q) stimulate white blood cell formation
____(r) cells that give rise to all the formed elements of blood; derived from mesenchyme
____(s) hormone that increases the numbers of red blood cell precursors

(1) neutrophils
(2) lymphocytes
(3) monocytes
(4) eosinophils
(5) basophils
(6) pluripotent stem cells
(7) colony-forming units
(8) red blood cells
(9) reticulocytes
(10) polymorphs
(11) myeloid stem cells
(12) lymphoid stem cells
(13) progenitor cells
(14) platelets
(15) fixed macrophages
(16) wandering macrophages
(17) erythropoietin
(18) thrombopoietin
(19) cytokines

14. Match the following:

____(a) tissue protein that leaks into the blood from cells outside blood vessels and initiates the formation of prothrombinase
____(b) an anticoagulant
____(c) platelet hormone that stimulates repair of damaged vessel walls
____(d) its formation is initiated by either the extrinsic or intrinsic pathway or both; catalyzes the conversion of prothrombin to thrombin
____(e) glycoproteins and glycolipids on the surfaces of red blood cells that can act as antigens
____(f) forms the threads of a clot; produced from fibrinogen
____(g) can dissolve a clot by digesting fibrin threads
____(h) serves as the catalyst to form fibrin; formed from prothrombin

(1) prothrombinase
(2) thrombin
(3) fibrin
(4) thromboplastin
(5) plasmin
(6) heparin
(7) agglutinogens
(8) platelet-derived growth factor

15. Match the following:

____(a) the percentage of total blood volume occupied by red blood cells
____(b) the percentage of each type of white blood cell
____(c) measures numbers of RBCs, WBCs, platelets per μl of blood; hematocrit, and differential WBC count
____(d) measures the rate of erythropoiesis
____(e) withdrawal of blood from a vein using a needle and collecting tube
____(f) withdrawal of a small amount of red bone marrow with a fine needle and syringe
____(g) removal of a core of red bone marrow with a large needle

(1) reticulocyte count
(2) bone marrow biopsy
(3) venipuncture
(4) hematocrit
(5) bone marrow aspiration
(6) complete blood count
(7) differential white blood cell count

CRITICAL THINKING QUESTIONS

1. Shilpa has recently been on broad spectrum antibiotics for a recurrent urinary bladder infection. While slicing vegetables, she cut herself and had difficulty stopping the bleeding. How could the antibiotics have played a role in her bleeding?
2. Mrs. Brown is in renal failure. Her recent blood tests indicated a hematocrit of 22. Why is her hematocrit low? What can she be given to raise her hematocrit?
3. Thomas has hepatitis, which is disrupting his liver functions. What kinds of symptoms would he be experiencing based on the role(s) of the liver related to blood?

ANSWERS TO FIGURE QUESTIONS

19.1 Blood volume is about 8% of your body mass, roughly 5–6 liters in males and 4–5 liters in females. For instance, a 70-kg (150-lb.) person has a blood volume of 5.6 liters (70 kg $\times$ 8% $\times$ 1 liter/kg).

19.2 Platelets are cell fragments.

19.3 Pluripotent stem cells develop from mesenchyme.

19.4 One hemoglobin molecule can transport a maximum of four O_2 molecules, one O_2 bound to each heme group.

19.5 Transferrin is a plasma protein that transports iron in the blood.

19.6 Once you moved to high altitude, your hematocrit would increase due to increased secretion of erythropoietin.

19.7 Neutrophils, eosinophils, and basophils are called granular leukocytes because all have cytoplasmic granules that are visible through a light microscope when stained.

19.8 Lymphocytes recirculate from blood to tissues and back to blood. After leaving the blood, other WBCs remain in the tissues until they die.

19.9 Along with platelet plug formation, vascular spasm and blood clotting contribute to hemostasis.

19.10 Serum is blood plasma minus the clotting proteins.

19.11 The outcome of the first stage of clotting is the formation of prothrombinase.

19.12 Type O blood usually contains both anti-A and anti-B antibodies.

19.13 Because the mother is most likely to start making anti-Rh antibodies after the first baby is already born, that baby suffers no damage.

19.14 Agglutination refers to clumping of red blood cells.

19.15 Some symptoms of sickle-cell disease are anemia, mild jaundice, joint pain, shortness of breath, rapid heart rate, abdominal pain, fever, and fatigue.

Chapter 20

The Cardiovascular System: The Heart

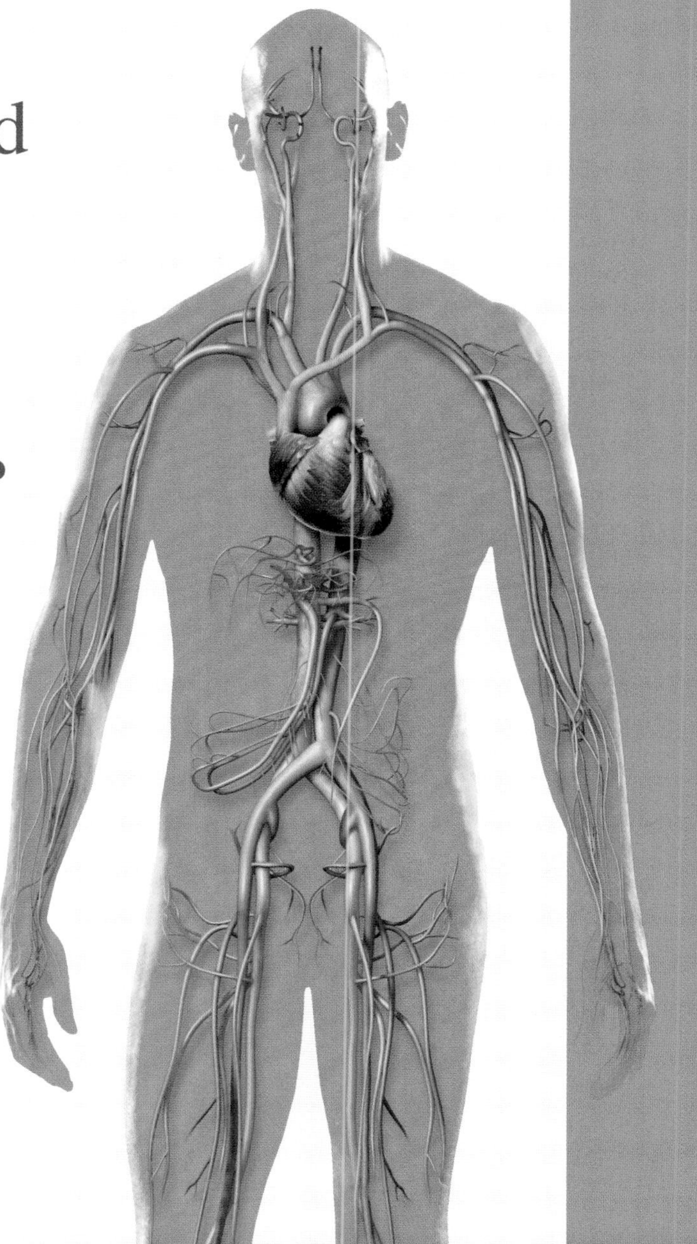

The Heart and Homeostasis

The heart pumps blood through blood vessels to all body tissues.

www. wiley.com/college/apcentral

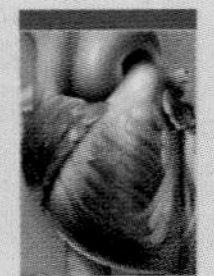

As you learned in the previous chapter, the **cardiovascular system** consists of the blood, the heart, and blood vessels. We also examined the composition and functions of blood, and in this chapter you will learn about the pump that circulates it throughout the body—the heart. Blood must be constantly pumped through the body's blood vessels so that it can reach body cells and exchange materials with them. To accomplish this, the heart beats about 100,000 times every day, which adds up to 35 million beats in a year. Even while you are sleeping, your heart pumps 30 times its own weight (5 L or 5.3 qt) each minute, which amounts to more than 14,000 liters (3600 gal) of blood in a day, and 10 million liters (2.6 million gal) in a year. Because you don't spend all your time sleeping, and your heart pumps more vigorously when you are active, the actual blood volume the heart pumps in a single day is even larger.

The scientific study of the normal heart and the diseases associated with it is known as **cardiology** (kar′-dē-OL-ō-jē; *cardio-* = heart; *-logy* = study of). This chapter explores the design of the heart and the unique properties that permit it to pump for a lifetime without rest.

ANATOMY OF THE HEART

▶ OBJECTIVES

Describe the location of the heart.

Describe the structure of the pericardium and the heart wall.

Discuss the external and internal anatomy of the chambers of the heart.

Location of the Heart

For all its might, the heart is relatively small, roughly the same size (but not the same shape) as your closed fist. It is about 12 cm (5 in.) long, 9 cm (3.5 in.) wide at its broadest point, and 6 cm (2.5 in.) thick, with an average mass of 250 g (8 oz) in adult females and 300 g (10 oz) in adult males. The heart rests on the diaphragm, near the midline of the thoracic cavity. It lies in the **mediastinum** (mē-dē-a-STĪ-num), a mass of tissue that extends from the sternum to the vertebral column between the lungs (Figure 20.1a). About two-thirds of the mass of the heart lies to the left of the body's midline (Figure 20.1b). You can visualize the heart as a cone lying on its side. The pointed **apex** is directed anteriorly, inferiorly, and to the left. The broad **base** is directed posteriorly, superiorly, and to the right.

In addition to the apex and base, the heart has several distinct surfaces and borders (margins). The **anterior surface** is deep to the sternum and ribs. The **inferior surface** is the part of the heart between the apex and right border and rests mostly on the diaphragm (Figure 20.1b). The **right border** faces the right lung and extends from the inferior surface to the base. The **left border,** also called the *pulmonary border,* faces the left lung and extends from the base to the apex.

Because the heart lies between two rigid structures—the vertebral column and the sternum (Figure 20.1a)—external pressure on the chest (compression) can be used to force blood out of the heart and into the circulation. In cases in which the heart suddenly stops beating, **cardiopulmonary resuscitation (CPR)**—properly applied cardiac compressions, performed with artificial ventilation of the lungs via mouth-to-mouth respiration—saves lives. CPR keeps oxygenated blood circulating until the heart can be restarted.

In a 2000 Seattle study, researchers found that chest compressions alone are equally as effective as, if not better than, traditional CPR with lung ventilation. This is good news because it is easier for an emergency dispatcher to give instructions limited to chest compressions to frightened, nonmedical bystanders. As public fear of contracting contagious diseases such as hepatitis, HIV, and tuberculosis continues to rise, bystanders are much more likely to perform chest compressions alone than treatment involving mouth-to-mouth rescue breathing. ■

Pericardium

The membrane that surrounds and protects the heart is the **pericardium** (*peri-* = around). It confines the heart to its position in the mediastinum, while allowing sufficient freedom of movement for vigorous and rapid contraction. The pericardium consists of two main parts: the fibrous pericardium and the serous pericardium (Figure 20.2a on page 698). The superficial **fibrous pericardium** is composed of tough, inelastic, dense irregular connective tissue. It resembles a bag that rests on and attaches to the diaphragm; its open end is fused to the connective tissues of the blood vessels entering and leaving the heart. The fibrous pericardium prevents overstretching of the heart, provides protection, and anchors the heart in the mediastinum.

The deeper **serous pericardium** is a thinner, more delicate membrane that forms a double layer around the heart (Figure 20.2a). The outer **parietal layer** of the serous pericardium is fused to the fibrous pericardium. The inner **visceral layer** of the serous pericardium, also called the **epicardium** (*epi-* = on top of), is one of the layers of the heart wall and adheres tightly to the surface of the heart. Between the parietal and visceral layers of the serous pericardium is a thin film of serous fluid. This slippery secretion of the pericardial cells, known as **pericardial fluid,** reduces friction between the layers of the serous pericardium as the heart moves. The space that contains the few milliliters of pericardial fluid is called the **pericardial cavity.**

Figure 20.1 Position of the heart and associated structures in the mediastinum (dashed outline). (See Tortora, *A Photographic Atlas of the Human Body, Second Edition,* Figures 6.5 and 6.6.)

The heart is located in the mediastinum, with two-thirds of its mass to the left of the midline.

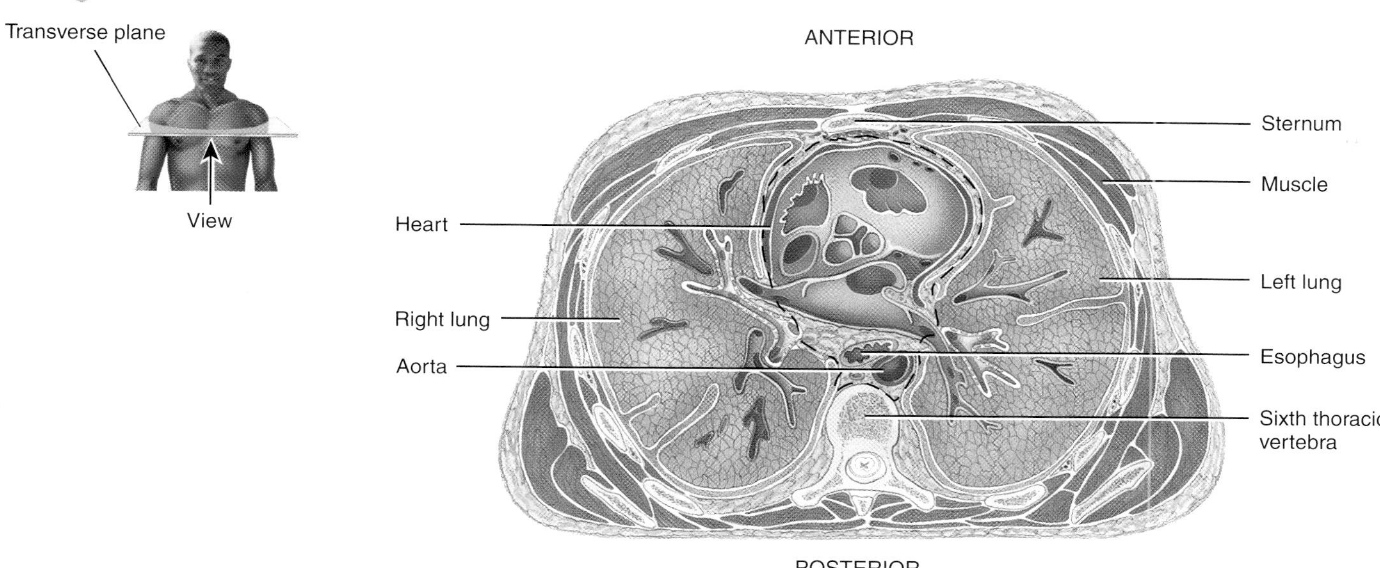

(a) Inferior view of transverse section of thoracic cavity showing the heart in the mediastinum

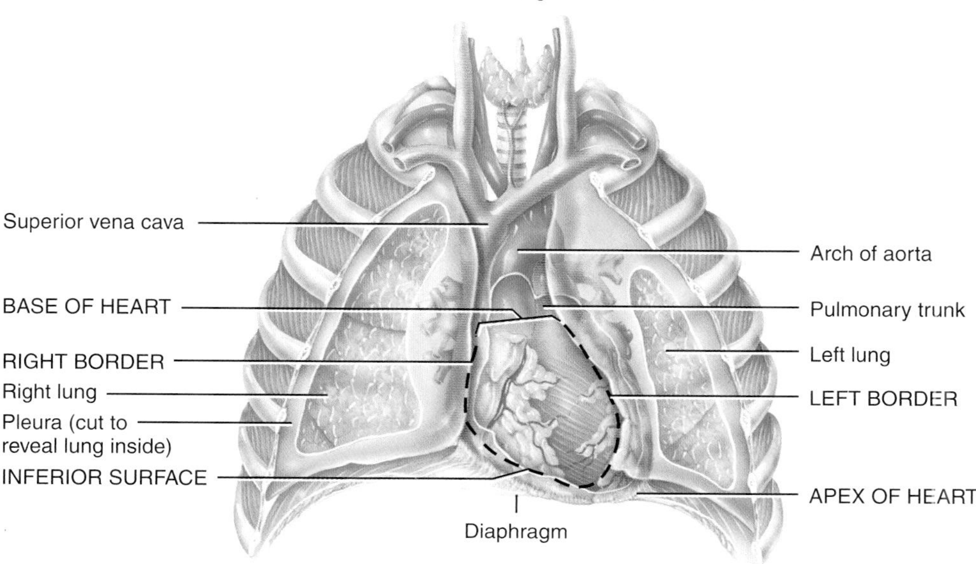

(b) Anterior view of the heart in the mediastinum

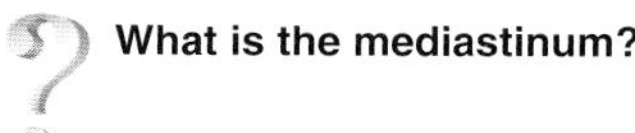

What is the mediastinum?

Pericarditis

Inflammation of the pericardium is called **pericarditis** (per′-i-kar-DĪ-tis). The most common type, *acute pericarditis,* begins suddenly and has no known cause in most cases but is sometimes linked to a viral infection. As a result of irritation to the pericardium, there is chest pain that may extend to the left shoulder and down the left arm (often mistaken for a heart attack) and pericardial friction rub (a scratchy or creaking sound heard through a stethoscope as the visceral layer of the serous pericardium rubs against the parietal layer of the serous pericardium). Acute pericarditis usually lasts for about one week and is treated with drugs that reduce inflammation and pain, such as ibuprofen or aspirin.

Figure 20.2 **Pericardium and heart wall.**

The pericardium is a triple-layered sac that surrounds and protects the heart.

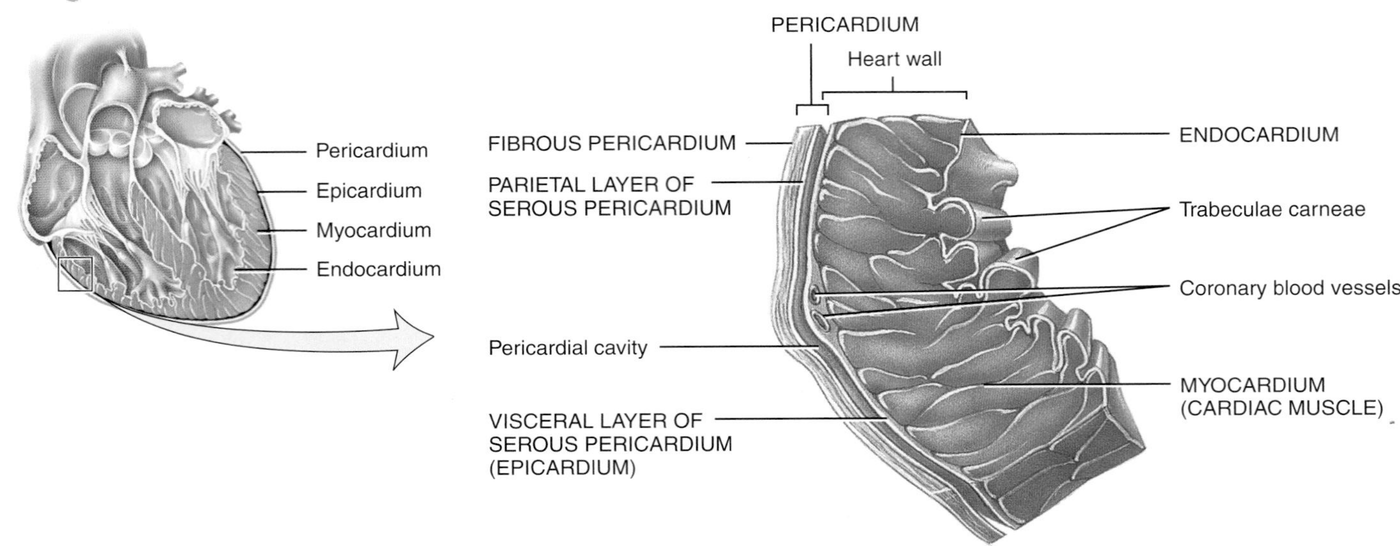

(a) Portion of pericardium and right ventricular heart wall showing the divisions of the pericardium and layers of the heart wall

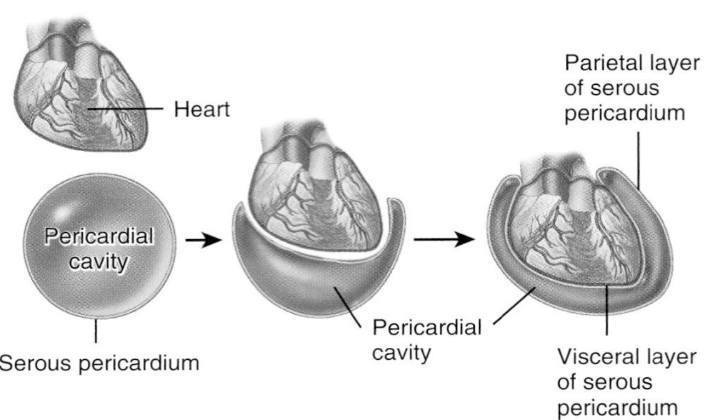

(b) Simplified relationship of the serous pericardium to the heart

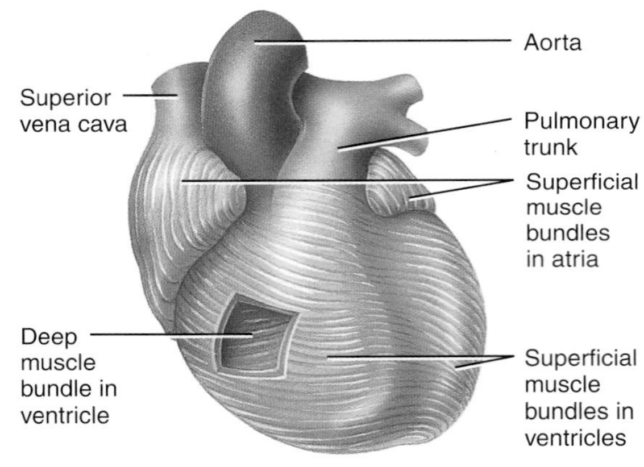

(c) Cardiac muscle bundles of the myocardium

Which layer is both a part of the pericardium and a part of the heart wall?

Chronic pericarditis begins gradually and is long-lasting. In one form of this condition, there is a buildup of pericardial fluid. If a great deal of fluid accumulates, this is a life-threatening condition because the fluid compresses the heart, a condition called *cardiac tamponade* (tam′-pon-ĀD). As a result of the compression, ventricular filling is decreased, cardiac output is reduced, venous return to the heart is diminished, blood pressure falls, and breathing is difficult. Most causes of chronic pericarditis involving cardiac tamponade are unknown, but it is sometimes caused by conditions such as cancer and tuberculosis. Treatment consists of draining the excess fluid through a needle passed into the pericardial cavity. ■

Layers of the Heart Wall

The wall of the heart consists of three layers (Figure 20.2a): the epicardium (external layer), the myocardium (middle layer), and the endocardium (inner layer). As noted earlier, the outermost **epicardium,** the thin, transparent outer layer of the heart wall, is also called the *visceral layer of the serous pericardium.* It is composed of mesothelium and delicate connective tissue that imparts a smooth, slippery texture to the outermost surface of the heart. The middle **myocardium** (*myo-* = muscle), which is cardiac muscle tissue, makes up the bulk of the heart and is responsible for its pumping action. Although it is striated like skeletal muscle, cardiac muscle is involuntary like smooth muscle. The cardiac muscle fibers swirl diagonally around the heart in bundles (Figure 20.2c). The innermost **endocardium** (*endo-* = within) is a thin layer of endothelium overlying a thin layer of connective tissue. It provides a smooth lining for the chambers of the heart and covers the valves of the heart. The endocardium is continuous with the endothelial lining of the large blood vessels attached to the heart.

Myocarditis and Endocarditis

Myocarditis (mī-ō-kar-DĪ-tis) is an inflammation of the myocardium that usually occurs as a complication of a viral infection, rheumatic fever, or exposure to radiation or certain chemicals or medications. Myocarditis often has no symptoms. However, if they do occur, they may include fever, fatigue, vague chest pain, irregular or rapid heartbeat, joint pain, and breathlessness. Myocarditis is usually mild and recovery occurs within two weeks. Severe cases can lead to cardiac failure and death. Treatment consists of avoiding vigorous exercise, a low-salt diet, electrocardiographic monitoring, and treatment of the cardiac failure. **Endocarditis** (en′-dō-kar-DĪ-tis) refers to an inflammation of the endocardium and typically involves the heart valves. Most cases are caused by bacteria (bacterial endocarditis). Signs and symptoms of endocarditis include fever, heart murmur, irregular or rapid heartbeat, fatigue, loss of appetitie, night sweats, and chills. Treatment is with intravenous antibiotics. ■

Chambers of the Heart

The heart has four chambers. The two superior chambers are the **atria** (= entry halls or chambers), and the two inferior chambers are the **ventricles** (= little bellies). On the anterior surface of each atrium is a wrinkled pouchlike structure called an **auricle** (OR-i-kul; *auri-* = ear), so named because of its resemblance to a dog's ear (Figure 20.3). Each auricle slightly increases the capacity of an atrium so that it can hold a greater volume of blood. Also on the surface of the heart are a series of grooves, called **sulci** (SUL-sē), that contain coronary blood vessels and a variable amount of fat. Each *sulcus* (SUL-kus) marks the external boundary between two chambers of the heart. The deep **coronary sulcus** (*coron-* = resembling a crown) encircles most of the heart and marks the boundary between the superior atria and

Figure 20.3 **Structure of the heart: surface features.** Throughout this book, blood vessels that carry oxygenated blood (which looks bright red) are colored red, whereas those that carry deoxygenated blood (which looks dark red) are colored blue.

Sulci are grooves that contain blood vessels and fat and mark the boundaries between the various chambers.

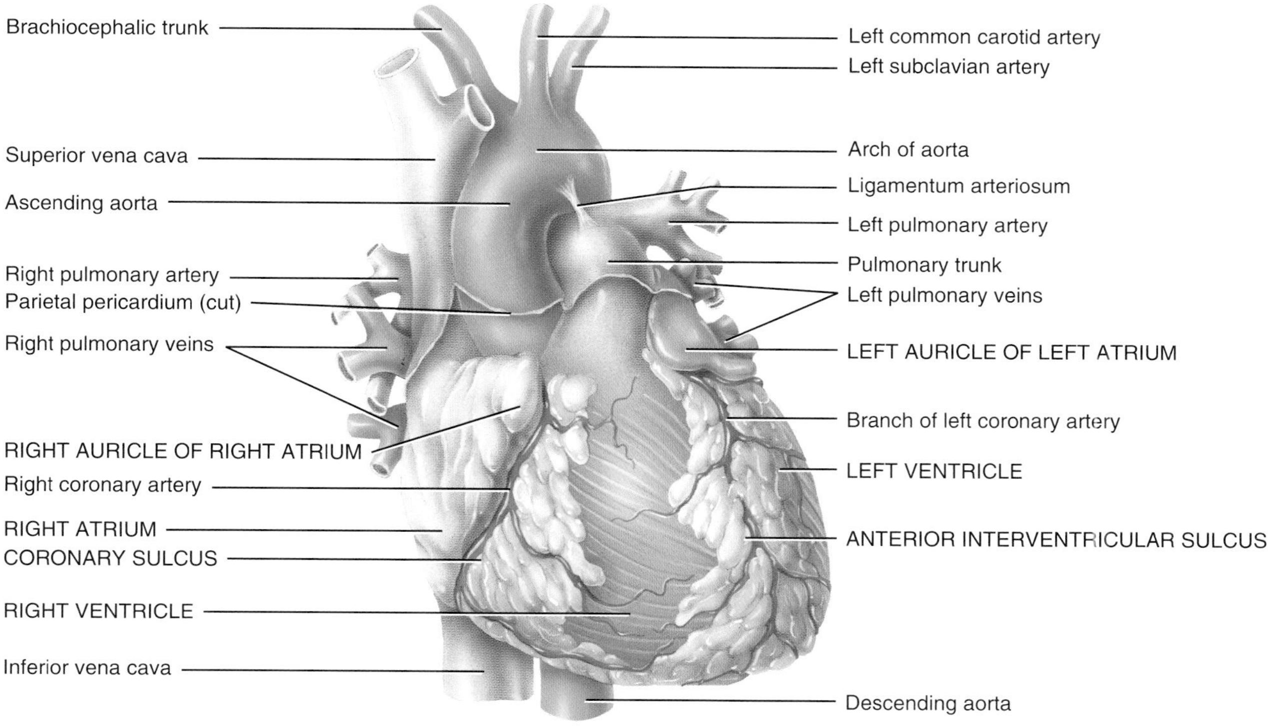

(a) Anterior external view showing surface features

continues

Figure 20.3 (continued)

Left subclavian artery
Left common carotid artery
Brachiocephalic trunk
Arch of aorta
Superior vena cava
Left pulmonary artery
Ligamentum arteriosum
Ascending aorta
Pulmonary trunk
RIGHT AURICLE OF RIGHT ATRIUM
LEFT AURICLE OF LEFT ATRIUM
ANTERIOR INTERVENTRICULAR SULCUS
RIGHT VENTRICLE
LEFT VENTRICLE

(b) Anterior external view showing surface features

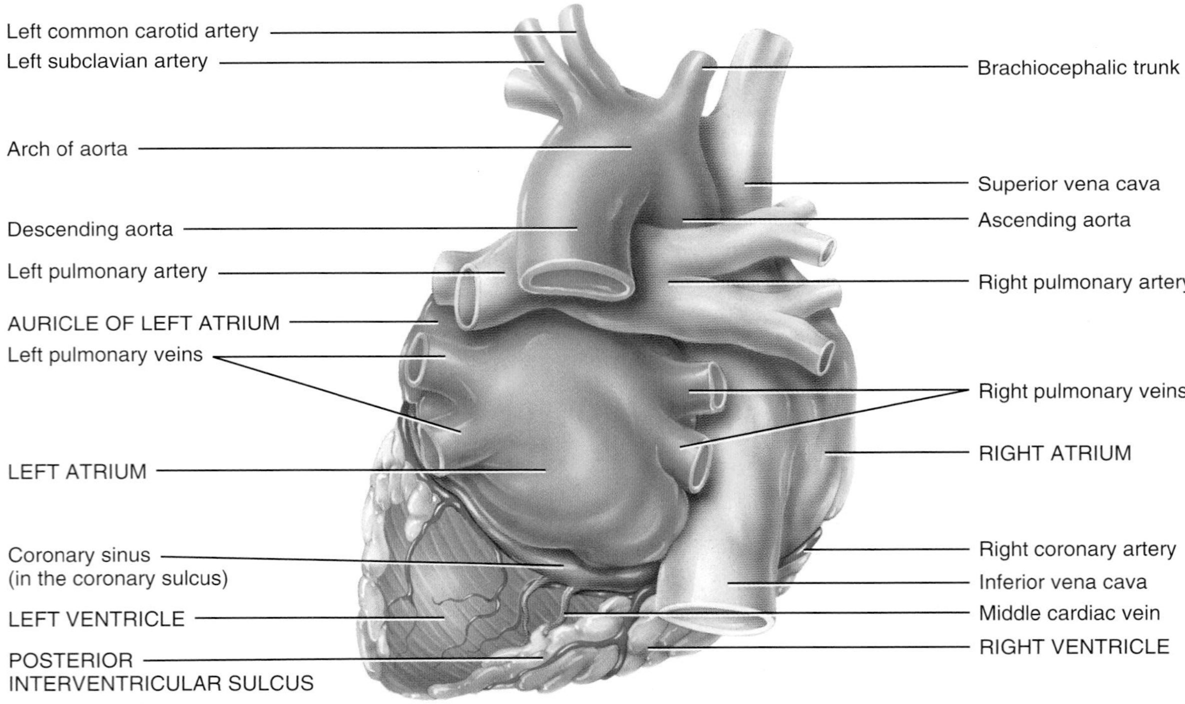

(c) Posterior external view showing surface features

The coronary sulcus forms a boundary between which chambers of the heart?

inferior ventricles. The **anterior interventricular sulcus** is a shallow groove on the anterior surface of the heart that marks the boundary between the right and left ventricles. This sulcus continues around to the posterior surface of the heart as the **posterior interventricular sulcus,** which marks the boundary between the ventricles on the posterior aspect of the heart (Figure 20.3c).

Right Atrium

The **right atrium** receives blood from three veins: the *superior vena cava, inferior vena cava,* and *coronary sinus* (Figure 20.4a). The anterior and posterior walls of the right atrium are very different. The posterior wall is smooth; the anterior wall is rough due to the presence of muscular ridges called **pectinate muscles** (PEK-tin-āt; *pectin* = comb), which also extend into the auricle (Figure 20.4b). Between the right atrium and left atrium is a thin partition called the **interatrial septum** (*inter-* = between; *septum* = a dividing wall or partition). A prominent feature of this septum is an oval depression called the **fossa ovalis,** the remnant of the *foramen ovale,* an opening in the intera-trial septum of the fetal heart that normally closes soon after birth (see Figure 21.30 on page 794). Blood passes from the right atrium into the right ventricle through a valve that is called the **tricuspid valve** (trī-KUS-pid; *tri-* = three; *cuspid* = point) because it consists of three leaflets or cusps (Figure 20.4a). It is also called the right **atrioventricular valve.** The valves of the heart are composed of dense connective tissue covered by endocardium.

Right Ventricle

The **right ventricle** forms most of the anterior surface of the heart. The inside of the right ventricle contains a series of ridges formed by raised bundles of cardiac muscle fibers called **trabeculae carneae** (tra-BEK-ū-lē KAR-nē-ē; trabeculae = little beams; carneae = fleshy; see Figure 20.2a). Some of the trabeculae carneae convey part of the conduction system of the heart, which you will learn about later in this chapter (see page 710). The cusps of the tricuspid valve are connected to tendon-like cords, the **chordae tendineae** (KOR-dē ten-DIN-ē-ē; *chord-* = cord; *tend-* = tendon), which in turn are connected to cone-shaped trabeculae carneae called **papillary muscles** (*papill-* = nipple). The right ventricle is separated from the left ventricle by a partition called the **interventricular septum.** Blood passes from the right ventricle through the **pulmonary valve** into a large artery called the *pulmonary trunk,* which divides into right and left *pulmonary arteries.*

Figure 20.4 Structure of the heart: internal anatomy.

Blood flows into the right atrium through the superior vena cava, inferior vena cava, and coronary sinus and into the left atrium through four pulmonary veins.

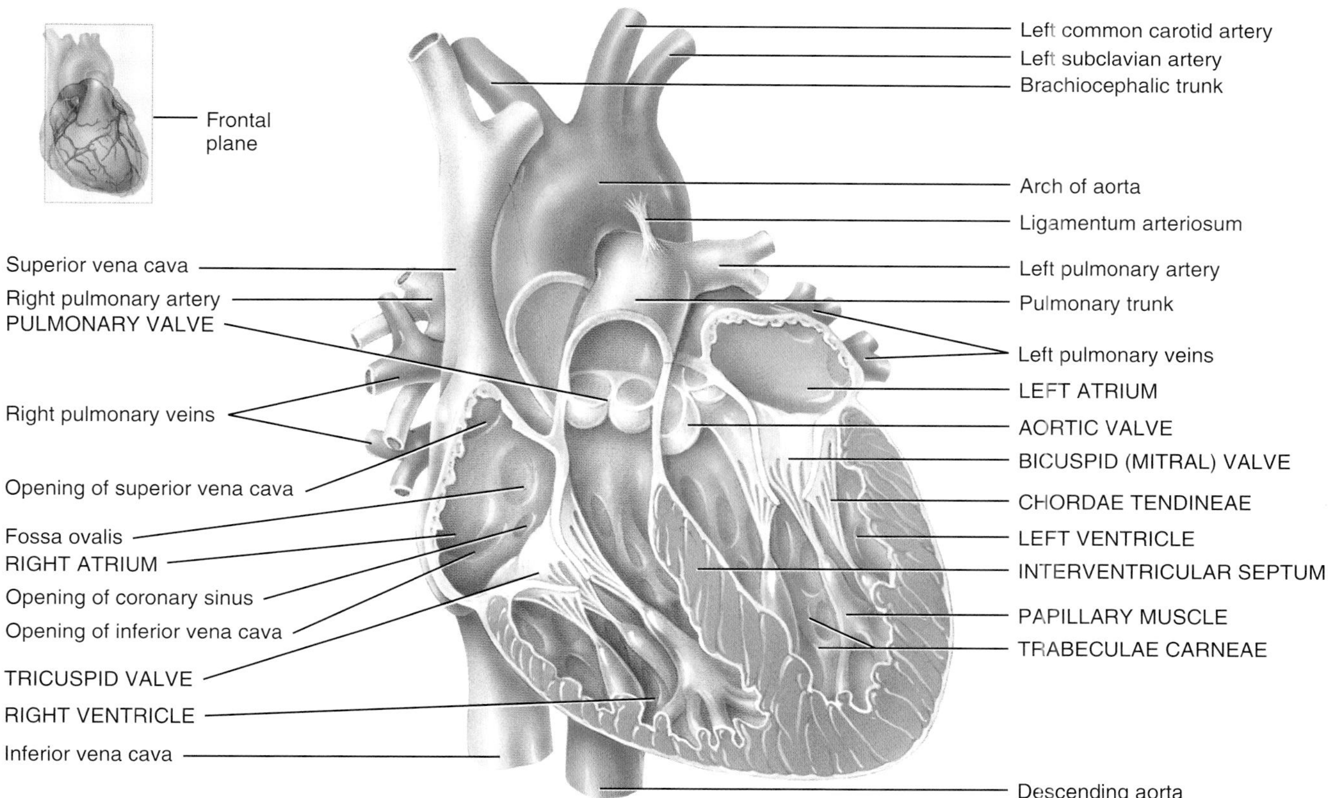

(a) Anterior view of frontal section showing internal anatomy

continues

Figure 20.4 **(continued)**

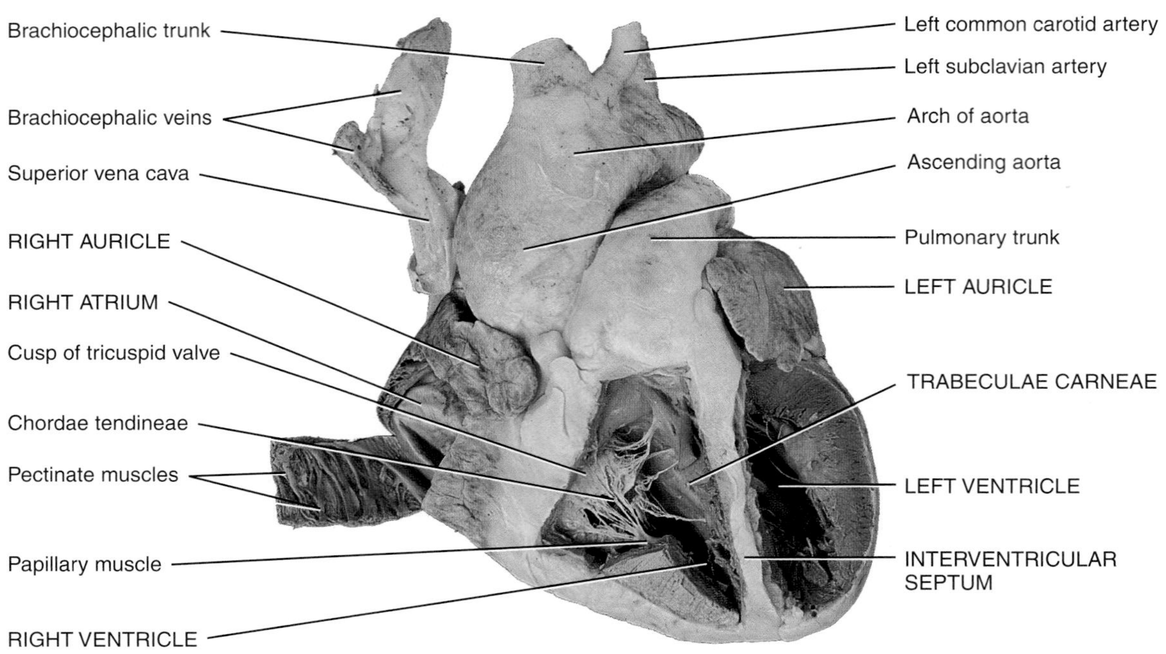

(b) Anterior view of partially sectioned heart showing internal anatomy

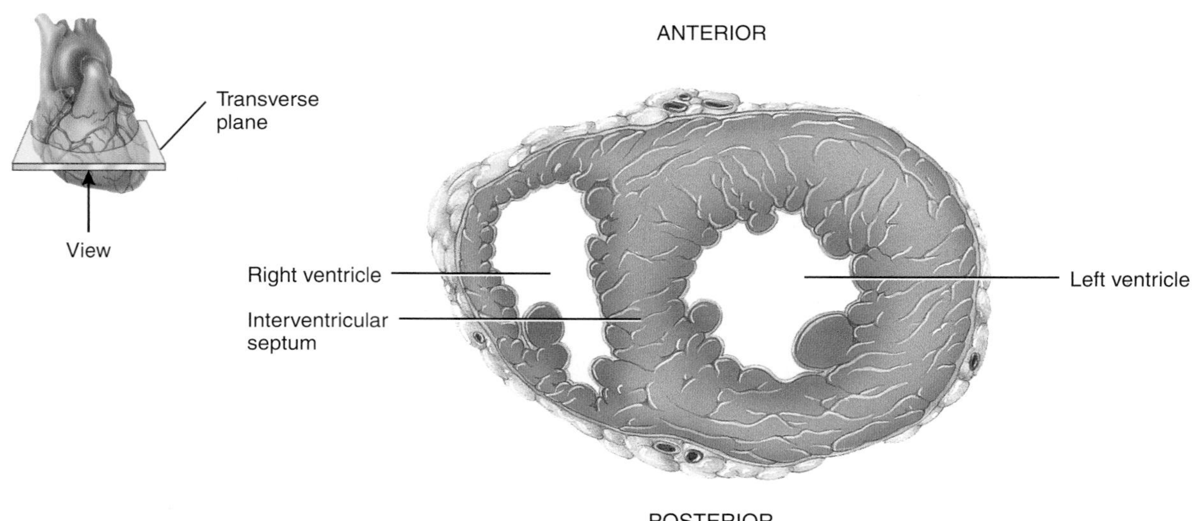

(c) Inferior view of transverse section showing differences in thickness of ventricular walls

How does thickness of the myocardium relate to the workload of a cardiac chamber?

Left Atrium

The **left atrium** forms most of the base of the heart (see Figure 20.1b). It receives blood from the lungs through four *pulmonary veins*. Like the right atrium, the inside of the left atrium has a smooth posterior wall. Because pectinate muscles are confined to the auricle of the left atrium, the anterior wall of the left atrium also is smooth. Blood passes from the left atrium into the left ventricle through the **bicuspid (mitral) valve** (bi- = two), which, as its name implies, has two cusps. The term *mitral* refers to the resemblance of the bicuspid valve to a bishop's miter (hat), which is two-sided. It is also called the left **atrioventricular valve.**

Left Ventricle

The **left ventricle** forms the apex of the heart (see Figure 20.1b). Like the right ventricle, the left ventricle contains trabeculae carneae and has chordae tendinae that anchor the cusps of the bicuspid valve to papillary muscles. Blood passes from the left ventricle through the **aortic valve** into the *ascending aorta* (*aorte* = to suspend, because the aorta once was believed to lift up the heart). Some of the blood in the aorta flows into the *coronary arteries*, which branch from the ascending aorta and carry blood to the heart wall. The remainder of the blood passes into the *arch of the aorta* and *descending aorta* (*thoracic aorta* and *abdominal aorta*). Branches of the arch of the aorta and descending aorta carry blood throughout the body.

During fetal life, a temporary blood vessel, called the *ductus arteriosus*, shunts blood from the pulmonary trunk into the aorta. Hence, only a small amount of blood enters the nonfunctioning fetal lungs (see Figure 21.30 on page 794). The ductus arteriosus normally closes shortly after birth, leaving a remnant known as the **ligamentum arteriosum,** which connects the arch of the aorta and pulmonary trunk (Figure 20.4a).

Myocardial Thickness and Function

The thickness of the myocardium of the four chambers varies according to each chamber's function. The thin-walled atria deliver blood into the adjacent ventricles. Because the ventricles pump blood greater distances, their walls are thicker (Figure 20.4a). Although the right and left ventricles act as two separate pumps that simultaneously eject equal volumes of blood, the right side has a much smaller workload. It pumps blood a short distance to the lungs at lower pressure, and the resist-ance to blood flow is small. The left ventricle pumps blood great distances to all other parts of the body at higher pressure, and the resistance to blood flow is larger. Therefore, the left ventricle works much harder than the right ventricle to maintain the same rate of blood flow. The anatomy of the two ventricles confirms this functional difference—the muscular wall of the left ventricle is considerably thicker than the wall of the right ventricle (Figure 20.4c). Note also that the perimeter of the lumen (space) of the left ventricle is roughly circular, whereas that of the right ventricle is somewhat crescent shaped.

Fibrous Skeleton of the Heart

In addition to cardiac muscle tissue, the heart wall also contains dense connective tissue that forms the **fibrous skeleton of the heart** (Figure 20.5). Essentially, the fibrous skeleton consists of four dense connective tissue rings that surround the valves of the heart, fuse with one another, and merge with the interventricular septum. In addition to forming a structural foundation for the heart valves, the fibrous skeleton prevents overstretching of the valves as blood passes through them. It also serves as a point of insertion for bundles of cardiac muscle fibers and acts as an electrical insulator between the atria and ventricles.

▶ CHECKPOINT

1. Define each of the following external features of the heart: auricle, coronary sulcus, anterior interventricular sulcus, and posterior interventricular sulcus.
2. Describe the structure of the pericardium and the layers of the wall of the heart.
3. What are the characteristic internal features of each chamber of the heart?
4. Which blood vessels deliver blood to the right and left atria?
5. What is the relationship between wall thickness and function among the various chambers of the heart?
6. What type of tissue composes the fibrous skeleton of the heart? What functions does this tissue perform?

Figure 20.5 Fibrous skeleton of the heart. Elements of the fibrous skeleton are shown in capital letters.

Fibrous rings support the four valves of the heart and are fused to one another.

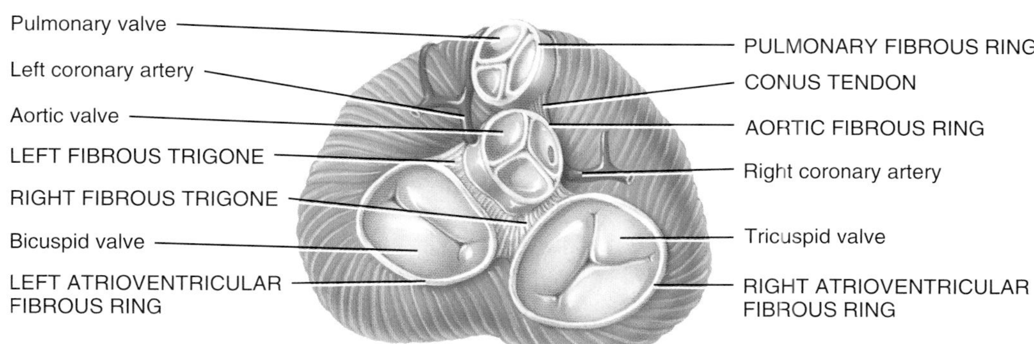

Superior view (the atria have been removed)

In what two ways does the fibrous skeleton contribute to the functioning of heart valves?

HEART VALVES AND CIRCULATION OF BLOOD

▶ OBJECTIVES

Describe the structure and function of the valves of the heart.

Outline the flow of blood through the chambers of the heart and through the systemic and pulmonary circulations.

Discuss the coronary circulation.

As each chamber of the heart contracts, it pushes a volume of blood into a ventricle or out of the heart into an artery. Valves open and close in response to *pressure changes* as the heart contracts and relaxes. Each of the four valves helps ensure the one-way flow of blood by opening to let blood through and then closing to prevent its backflow.

Operation of the Atrioventricular Valves

Because they are located between an atrium and a ventricle, the tricuspid and bicuspid valves are termed **atrioventricular (AV) valves.** When an AV valve is open, the pointed ends of the cusps project into the ventricle. When the ventricles are relaxed, the papillary muscles are relaxed, the chordae tendineae are slack, and blood moves from a higher pressure in the atria to a lower pressure in the ventricles through open AV valves (Figure 20.6a, c). When the ventricles contract, the pressure of the blood drives the cusps upward until their edges meet and close the opening (Figure 20.6b, d). At the same time, the papillary muscles contract, which pulls on and tightens the chordae tendineae. This prevents the valve cusps from everting (opening into the atria) in response to the high ventricular pressure. If the AV valves or chordae tendineae are damaged, blood may regurgitate (flow back) into the atria when the ventricles contract.

Operation of the Semilunar Valves

The aortic and pulmonary valves are known as the **semilunar (SL) valves** (*semi-* = half; *lunar* = moon-shaped) because they are made up of three crescent moon–shaped cusps (Figure 20.6c). Each cusp attaches to the arterial wall by its convex outer margin. The SL valves allow ejection of blood from the heart into arteries but prevent backflow of blood into the ventricles. The free borders of the cusps project into the lumen of the artery. When the ventricles contract, pressure builds up within the chambers. The semilunar valves open when pressure in the ventricles exceeds the pressure in the arteries, permitting ejection of blood from the ventricles into the pulmonary trunk and aorta (Figure 20.6d). As the ventricles relax, blood starts to flow back toward the heart. This backflowing blood fills the valve cusps, which causes the semilunar valves to close tightly (Figure 20.6c).

Surprisingly perhaps, there are no valves guarding the junctions between the venae cavae and the right atrium or the pulmonary veins and the left atrium. As the atria contract, a small amount of blood does flow backward from the atria into these vessels. However, backflow is minimized by a different mechanism; as the atrial muscle contracts, it compresses and nearly collapses the venous entry points.

Heart Valve Disorders

When heart valves operate normally, they open fully and close completely at the proper times. A narrowing of a heart valve opening that restricts blood flow is known as **stenosis** (ste-NŌ-sis = a narrowing); failure of a valve to close completely is termed **insufficiency** or **incompetence.** In **mitral stenosis,** scar formation or a congenital defect causes narrowing of the mitral valve. One cause of **mitral insufficiency,** in which there is backflow of blood from the left ventricle into the left atrium, is **mitral valve prolapse (MVP).** In MVP one or both cusps of the mitral valve protrude into the left atrium during ventricular contraction. Mitral valve prolapse is one of the most common valvular disorders, affecting as much as 30% of the population. It is more prevalent in women than in men, and does not always pose a serious threat. In **aortic stenosis** the aortic valve is narrowed, and in **aortic insufficiency** there is backflow of blood from the aorta into the left ventricle.

Certain infectious diseases can damage or destroy the heart valves. One example is **rheumatic fever,** an acute systemic inflammatory disease that usually occurs after a streptococcal infection of the throat. The bacteria trigger an immune response in which antibodies produced to destroy the bacteria instead attack and inflame the connective tissues in joints, heart valves, and other organs. Even though rheumatic fever may weaken the entire heart wall, most often it damages the mitral and aortic valves.

Systemic and Pulmonary Circulations

In postnatal (after birth) circulation, the heart pumps blood into two closed circuits—the **systemic circulation** and the **pulmonary circulation** (*pulmon-* = lung) with each beat. The two circuits are arranged in series: The output of one becomes the input of the other, as would happen if you attached two garden hoses (see Figure 21.17 on page 759). The left side of the heart is the pump for the systemic circulation; it receives bright red, oxygen-rich blood from the lungs. The left ventricle ejects blood into the *aorta* (Figure 20.7). From the aorta, the blood divides into separate streams, entering progressively smaller *systemic arteries* that carry it to all organs throughout the body—except for the air sacs (alveoli) of the lungs, which are supplied by the pulmonary circulation. In systemic tissues, arteries give rise to smaller-diameter *arterioles,* which finally lead into extensive beds of *systemic capillaries.* Exchange of nutrients and gases occurs across the thin capillary walls. Blood unloads O_2 (oxygen)

Figure 20.6 Responses of the valves to the pumping of the heart.

Heart valves prevent the backflow of blood.

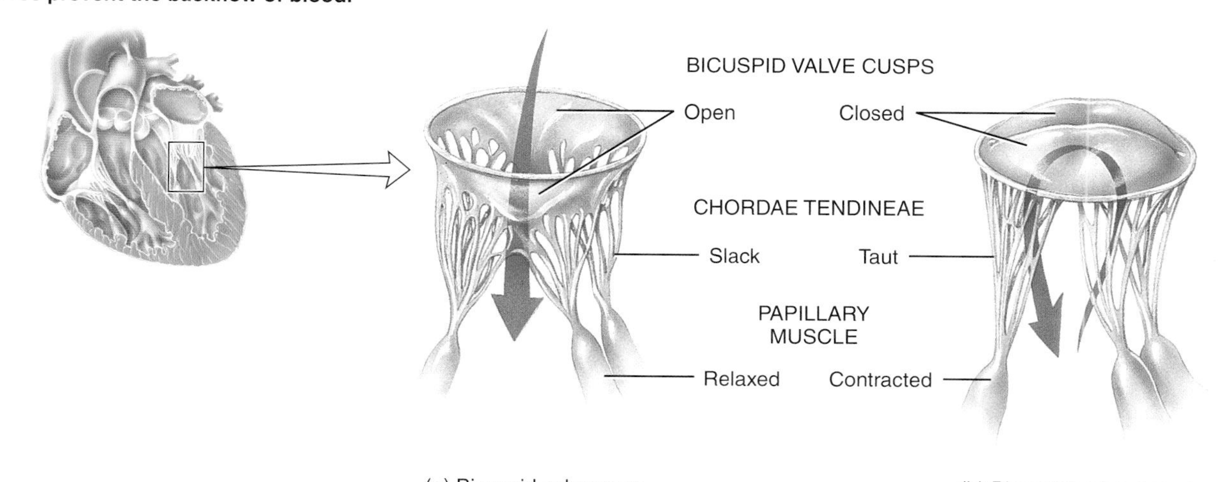

(a) Bicuspid valve open

(b) Bicuspid valve closed

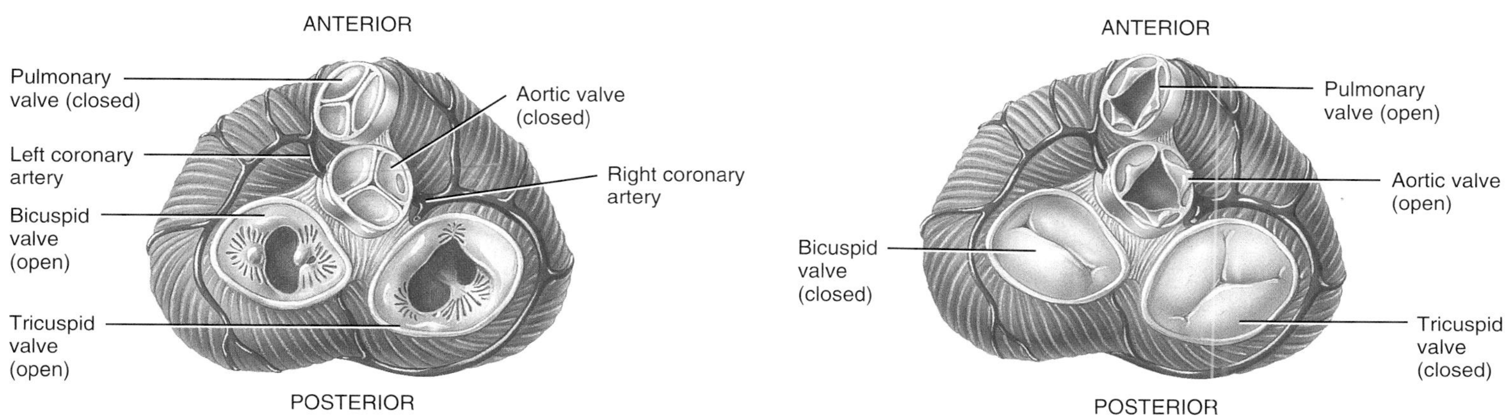

(c) Superior view with atria removed: pulmonary and aortic valves closed, bicuspid and tricuspid valves open.

(d) Superior view with atria removed: pulmonary and aortic valves open, bicuspid and tricuspid valves closed.

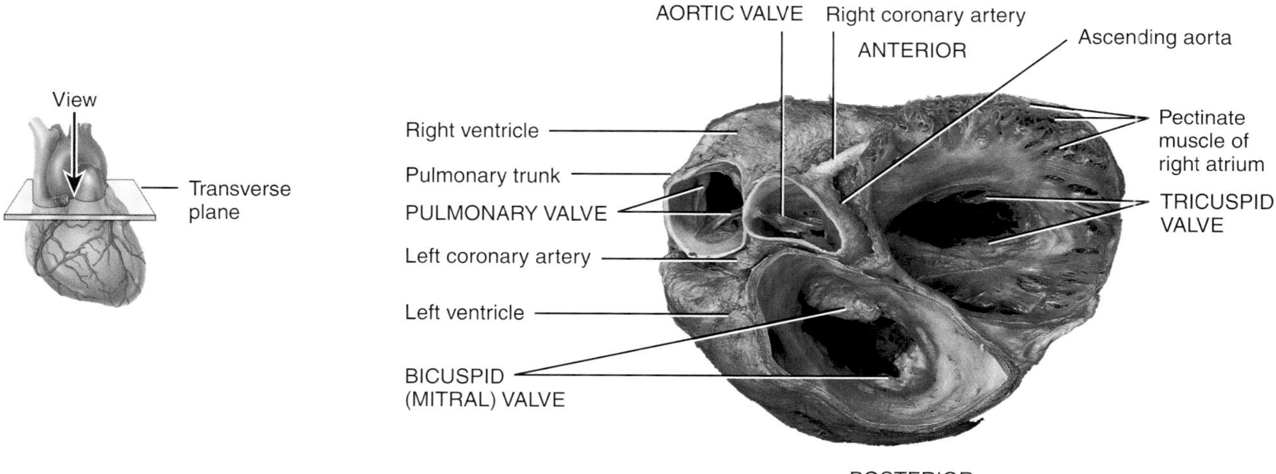

(e) Superior view of atrioventricular and semilunar valves

How do papillary muscles prevent AV valve cusps from everting (swinging upward) into the atria?

Figure 20.7 **Systemic and pulmonary circulations.**

The left side of the heart pumps oxygenated blood into the systemic circulation to all tissues of the body except the air sacs (alveoli) of the lungs. The right side of the heart pumps deoxygenated blood into the pulmonary circulation to the air sacs (alveoli) of the lungs.

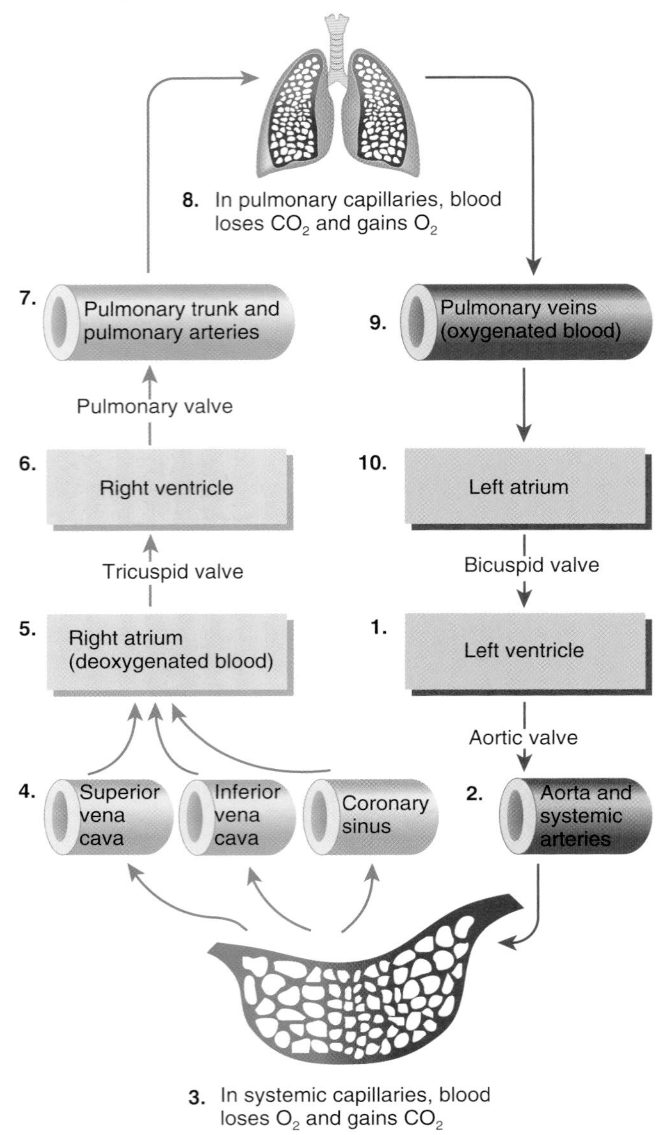

Diagram of blood flow

Which numbers constitute the pulmonary circulation? Which constitute the systemic circulation?

and picks up CO_2 (carbon dioxide). In most cases, blood flows through only one capillary and then enters a *systemic venule.* Venules carry deoxygenated (oxygen-poor) blood away from tissues and merge to form larger *systemic veins.* Ultimately the blood flows back to the right atrium.

The right side of the heart is the pump for the pulmonary circulation; it receives all the dark red, deoxygenated blood returning from the systemic circulation. Blood ejected from the right ventricle flows into the *pulmonary trunk,* which branches into *pulmonary arteries* that carry blood to the right and left lungs. In pulmonary capillaries, blood unloads CO_2, which is exhaled, and picks up inhaled O_2. The freshly oxygenated blood then flows into pulmonary veins and returns to the left atrium.

Coronary Circulation

Nutrients are not able to diffuse quickly enough from blood in the chambers of the heart to supply all the layers of cells that make up the heart wall. For this reason, the myocardium has its own network of blood vessels, the **coronary** or **cardiac circulation** (*coron-* = crown). The **coronary arteries** branch from the ascending aorta and encircle the heart like a crown encircles the head (Figure 20.8a). While the heart is contracting, little blood flows in the coronary arteries because they are squeezed shut. When the heart relaxes, however, the high pressure of blood in the aorta propels blood through the coronary arteries, into capillaries, and then into **coronary veins** (Figure 20.8b).

Coronary Arteries

Two coronary arteries, the right and left coronary arteries, branch from the ascending aorta and supply oxygenated blood to the myocardium (Figure 20.8a). The **left coronary artery** passes inferior to the left auricle and divides into the anterior interventricular and circumflex branches. The **anterior interventricular branch** or *left anterior descending (LAD) artery* is in the anterior interventricular sulcus and supplies oxygenated blood to the walls of both ventricles. The **circumflex branch** lies in the coronary sulcus and distributes oxygenated blood to the walls of the left ventricle and left atrium.

The **right coronary artery** supplies small branches *(atrial branches)* to the right atrium. It continues inferior to the right auricle and ultimately divides into the posterior interventricular and marginal branches. The **posterior interventricular branch** follows the posterior interventricular sulcus and supplies the walls of the two ventricles with oxygenated blood. The **marginal branch** in the coronary sulcus transports oxygenated blood to the myocardium of the right ventricle.

Most parts of the body receive blood from branches of more than one artery, and where two or more arteries supply the same region, they usually connect. These connections, called **anastomoses** (a-nas′-tō-MŌ-sēs), provide alternate routes for blood to reach a particular organ or tissue. The myocardium contains many anastomoses that connect branches of a given coronary artery or extend between branches of different coronary arteries.

Figure 20.8 The coronary circulation. The views of the heart from the anterior aspect in (a) and (b) are drawn as if the heart were transparent to reveal blood vessels on the posterior aspect. (See Tortora, *A Photographic Atlas of the Human Body, Second Edition,* Figures 6.8 and 6.9.)

The right and left coronary arteries deliver blood to the heart; the coronary veins drain blood from the heart into the coronary sinus.

Arch of aorta
Ascending aorta
Pulmonary trunk
RIGHT CORONARY
Right atrium
MARGINAL BRANCH
Right ventricle
LEFT CORONARY
Left auricle
CIRCUMFLEX BRANCH
ANTERIOR INTER-VENTRICULAR BRANCH
POSTERIOR INTER-VENTRICULAR BRANCH
Left ventricle

(a) Anterior view of coronary arteries

Superior vena cava
Right atrium
SMALL CARDIAC
ANTERIOR CARDIAC
MIDDLE CARDIAC
Right ventricle
Inferior vena cava
Pulmonary trunk
Left auricle
CORONARY SINUS
GREAT CARDIAC
Left ventricle

(b) Anterior view of coronary veins

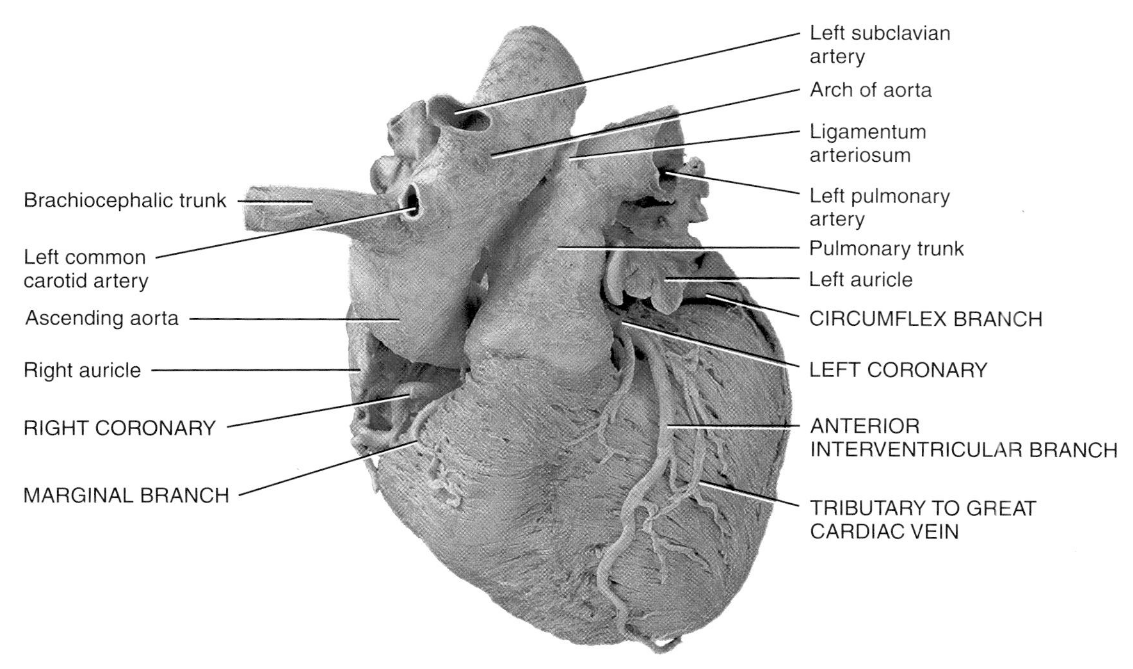

(c) Anterior view

Which coronary blood vessel delivers oxygenated blood to the left atrium and left ventricle?

They provide detours for arterial blood if a main route becomes obstructed. Thus, heart muscle may receive sufficient oxygen even if one of its coronary arteries is partially blocked.

Coronary Veins

After blood passes through the arteries of the coronary circulation, it flows into capillaries, where it delivers oxygen and nutrients to the heart muscle and collects carbon dioxide and waste, and then moves into coronary veins. Most of the deoxygenated blood from the myocardium drains into a large *vascular sinus* in the coronary sulcus on the posterior surface of the heart, called the **coronary sinus** (Figure 20.8b). (A *vascular sinus* is a thin-walled vein that has no smooth muscle to alter its diameter.) The deoxygenated blood in the coronary sinus empties into the right atrium. The principal tributaries carrying blood into the coronary sinus are the following:

- **Great cardiac vein** in the anterior interventricular sulcus, which drains the areas of the heart supplied by the left coronary artery (left and right ventricles and left atrium)
- **Middle cardiac vein** in the posterior interventricular sulcus, which drains the areas supplied by the posterior interventricular branch of the right coronary artery (left and right ventricles)
- **Small cardiac vein** in the coronary sulcus, which drains the right atrium and right ventricle
- **Anterior cardiac veins,** which drain the right ventricle and open directly into the right atrium

When blockage of a coronary artery deprives the heart muscle of oxygen, **reperfusion,** the reestablishment of blood flow, may damage the tissue further. This surprising effect is due to the formation of oxygen **free radicals** from the reintroduced oxygen. As you learned in Chapter 2, free radicals are electrically charged molecules that have an unpaired electron (see Figure 2.3b on page 32). These unstable, highly reactive molecules cause chain reactions that lead to cellular damage and death. To counter the effects of oxygen free radicals, body cells produce enzymes that convert free radicals to less reactive substances. Two such enzymes are *superoxide dismutase* and *catalase.* In addition, nutrients such as vitamin E, vitamin C, beta-carotene, zinc, and selenium serve as antioxidants, which remove oxygen free radicals. Drugs that lessen reperfusion damage after a heart attack or stroke are currently under development.

Myocardial Ischemia and Infarction

Partial obstruction of blood flow in the coronary arteries may cause **myocardial ischemia** (is-KĒ-mē-a; *ische-* = to obstruct; *-emia* = in the blood), a condition of reduced blood flow to the myocardium. Usually, ischemia causes **hypoxia** (reduced oxygen supply), which may weaken cells without killing them. **Angina pectoris** (an-JĪ-na, or AN-ji-na, PEK-to-ris), which literally means "strangled chest," is a severe pain that usually accompanies myocardial ischemia. Typically, sufferers describe it as a tightness or squeezing sensation, as though the chest were in a vise. The pain associated with angina pectoris is often referred to the neck, chin, or down the left arm to the elbow. **Silent myocardial ischemia,** ischemic episodes without pain, is particularly dangerous because the person has no forewarning of an impending heart attack.

A complete obstruction to blood flow in a coronary artery may result in a **myocardial infarction** (in-FARK-shun), or **MI,** commonly called a *heart attack. Infarction* means the death of an area of tissue because of interrupted blood supply. Because the heart tissue distal to the obstruction dies and is replaced by noncontractile scar tissue, the heart muscle loses some of its strength. Depending on the size and location of the infarcted (dead) area, an infarction may disrupt the conduction system of the heart and cause sudden death by triggering ventricular fibrillation. Treatment for a myocardial infarction may involve injection of a thrombolytic (clot-dissolving) agent such as streptokinase or t-PA, plus heparin (an anticoagulant), or performing coronary angioplasty or coronary artery bypass grafting. Fortunately, heart muscle can remain alive in a resting person if it receives as little as 10–15% of its normal blood supply. ■

▶ CHECKPOINT

7. What causes the heart valves to open and to close? What supporting structures ensure that the valves operate properly?
8. In correct sequence, which heart chambers, heart valves, and blood vessels would a drop of blood encounter as it flows from the right atrium to the aorta?
9. Which arteries deliver oxygenated blood to the myocardium of the left and right ventricles?

CARDIAC MUSCLE TISSUE AND THE CARDIAC CONDUCTION SYSTEM

▶ OBJECTIVES

- **Describe the structural and functional characteristics of cardiac muscle tissue and the conduction system of the heart.**
- **Describe how an action potential occurs in cardiac contractile fibers.**
- **Describe the electrical events of a normal electrocardiogram (ECG).**

Histology of Cardiac Muscle Tissue

Compared with skeletal muscle fibers, cardiac muscle fibers are shorter in length and less circular in transverse section (Figure 20.9). They also exhibit branching, which gives individual cardiac muscle fibers a "stair-step" appearance (see Table 4.5B on page 135). A typical cardiac muscle fiber is 50–100 μm long and has a diameter of about 14 μm. Usually one centrally located nucleus is present, although an occasional cell may have two nu-

Figure 20.9 Histology of cardiac muscle tissue. (See Table 4.5B on page 135 for a light microscopic view of cardiac muscle.)

Cardiac muscle fibers connect to neighboring fibers by intercalated discs, which contain desmosomes and gap junctions.

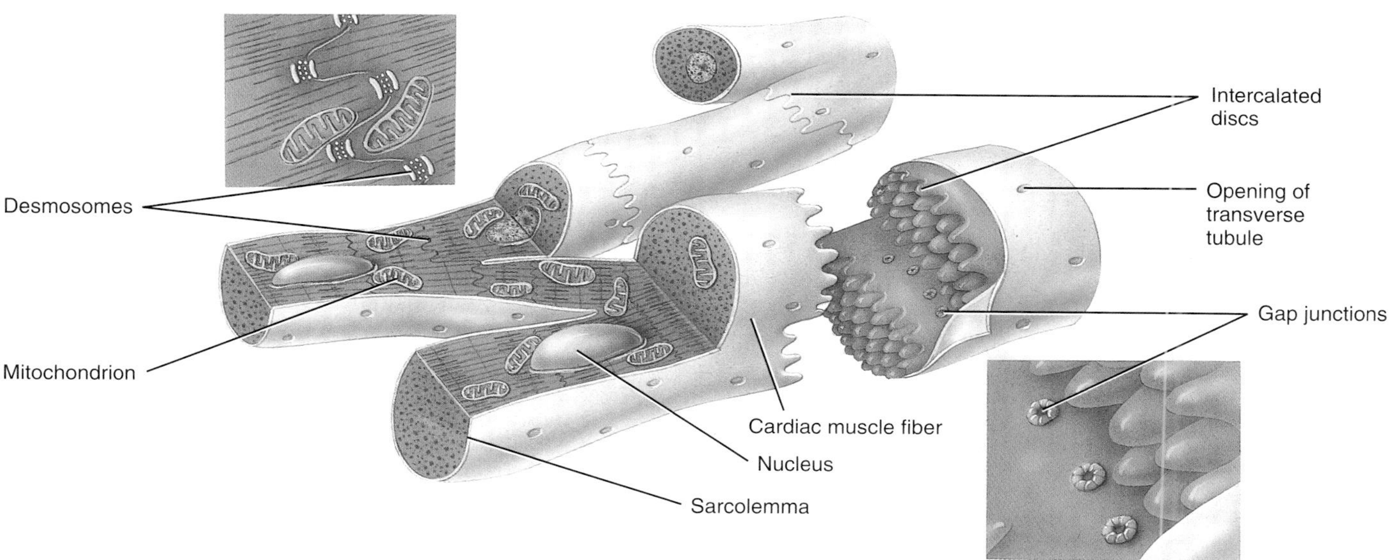

(a) Cardiac muscle fibers

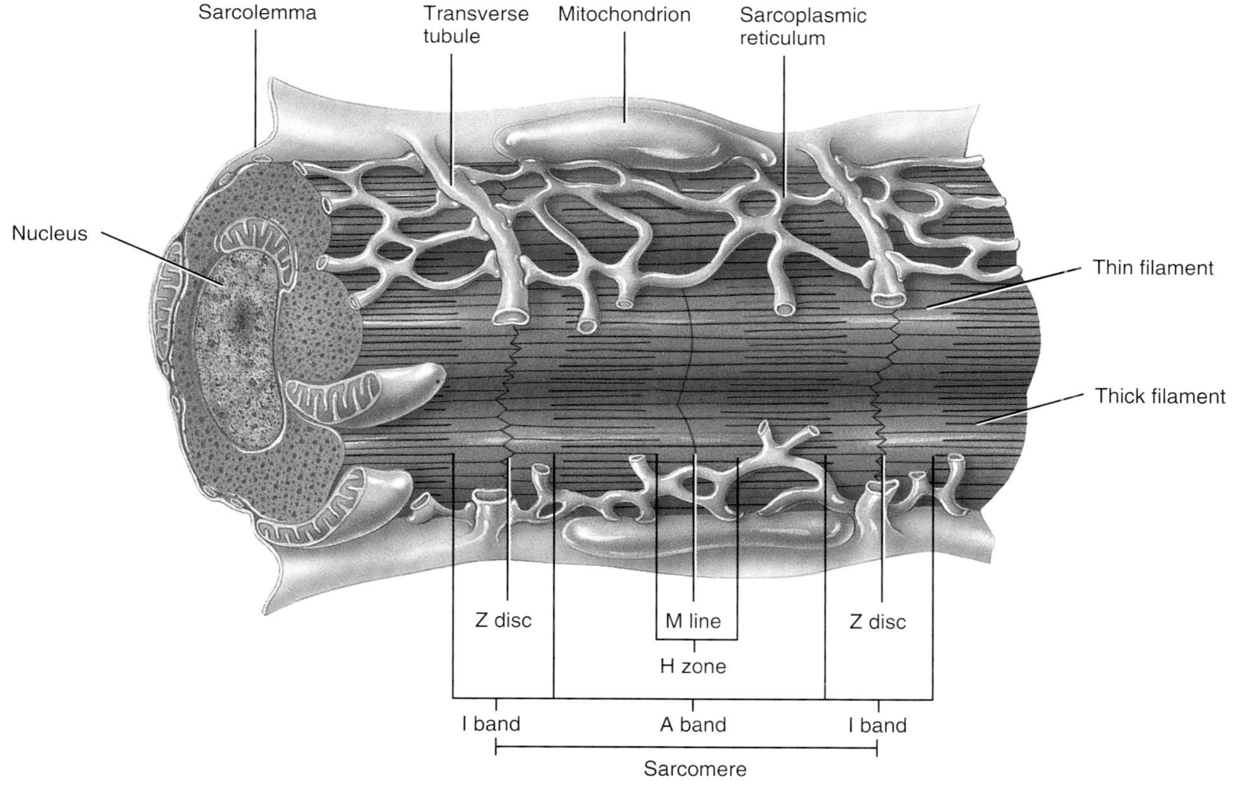

(b) Arrangement of components in a cardiac muscle fiber

What are the functions of intercalated discs in cardiac muscle fibers?

clei. The ends of cardiac muscle fibers connect to neighboring fibers by irregular transverse thickenings of the sarcolemma called **intercalated discs** (in-TER-kā-lāt-ed; *intercalat-* = to insert between). The discs contain **desmosomes,** which hold the fibers together, and **gap junctions,** which allow muscle action potentials to conduct from one muscle fiber to its neighbors.

Mitochondria are larger and more numerous in cardiac muscle fibers than in skeletal muscle fibers. In a cardiac muscle fiber, they take up 25% of the cytosolic space; in a skeletal muscle fiber only 2% of the cytosolic space is occupied by mitochondria. Cardiac muscle fibers have the same arrangement of actin and myosin, and the same bands, zones, and Z discs, as skeletal muscle fibers. The transverse tubules of cardiac muscle are wider but less abundant than those of skeletal muscle; the one transverse tubule per sarcomere is located at the Z disc. The sarcoplasmic reticulum of cardiac muscle fibers is somewhat smaller than the SR of skeletal muscle fibers. As a result, cardiac muscle has a smaller intracellular reserve of Ca^{2+}.

Regeneration of Heart Cells

As noted earlier in the chapter, the heart of a heart attack survivor often has regions of infarcted (dead) cardiac muscle tissue that typically are replaced with noncontractile fibrous scar tissue over time. Our inability to repair damage from a heart attack has been attributed to a lack of stem cells in cardiac muscle and to the absence of mitosis in mature cardiac muscle fibers. A recent study of heart transplant recipients by American and Italian scientists, however, provides evidence for significant replacement of heart cells. The researchers studied men who had received a heart from a female, and then looked for the presence of a Y chromosome in heart cells. (All female cells except gametes have two X chromosomes and lack the Y chromosome.) Several years after the transplant surgery, between 7% and 16% of the heart cells in the transplanted tissue, including cardiac muscle fibers and endothelial cells in coronary arterioles and capillaries, had been replaced by the recipient's own cells, as evidenced by the presence of a Y chromosome. The study also revealed cells with some of the characteristics of stem cells in both transplanted hearts and control hearts. Evidently, stem cells can migrate from the blood into the heart and differentiate into functional muscle and endothelial cells. The hope is that researchers can learn how to "turn on" such regeneration of heart cells to treat people with heart failure or cardiomyopathy (diseased heart). ■

Autorhythmic Fibers: The Conduction System

An inherent and rhythmical electrical activity is the reason for the heart's lifelong beat. The source of this electrical activity is a network of specialized cardiac muscle fibers called **autorhythmic fibers** (auto- = self) because they are self-excitable. Autorhythmic fibers repeatedly generate action potentials that trigger heart contractions. They continue to stimulate a heart to beat even after it is removed from the body—for example, to be transplanted into another person—and all of its nerves have been cut. (Note: Surgeons do not attempt to reattach heart nerves during heart transplant operations. For this reason, it has been said that heart surgeons are better "plumbers" than they are "electricians.")

During embryonic development, only about 1% of the cardiac muscle fibers become autorhythmic fibers; these relatively rare fibers have two important functions.

1. They act as a **pacemaker,** setting the rhythm of electrical excitation that causes contraction of the heart.
2. They form the **conduction system,** a network of specialized cardiac muscle fibers that provide a path for each cycle of cardiac excitation to progress through the heart. The conduction system ensures that cardiac chambers become stimulated to contract in a coordinated manner, which makes the heart an effective pump.

Cardiac action potentials propagate through the conduction system in the following sequence (Figure 20.10a):

1. Cardiac excitation normally begins in the **sinoatrial (SA) node,** located in the right atrial wall just inferior to the opening of the superior vena cava. SA node cells do not have a stable resting potential. Rather, they repeatedly depolarize to threshold spontaneously. The spontaneous depolarization is a **pacemaker potential.** When the pacemaker potential reaches threshold, it triggers an action potential (Figure 20.10b). Each action potential from the SA node propagates throughout both atria via gap junctions in the intercalated discs of atrial muscle fibers. Following the action potential, the atria contract.
2. By conducting along atrial muscle fibers, the action potential reaches the **atrioventricular (AV) node,** located in the septum between the two atria, just anterior to the opening of the coronary sinus (Figure 20.10a).
3. From the AV node, the action potential enters the **atrioventricular (AV) bundle** (also known as the **bundle of His**). This bundle is the only site where action potentials can conduct from the atria to the ventricles. (Elsewhere, the fibrous skeleton of the heart electrically insulates the atria from the ventricles.)
4. After propagating along the AV bundle, the action potential enters both the **right** and **left bundle branches.** The bundle branches extend through the interventricular septum toward the apex of the heart.
5. Finally, the large-diameter **Purkinje fibers** rapidly conduct the action potential from the apex of the heart upward to the remainder of the ventricular myocardium. Then the ventricles contract, pushing the blood upward toward the semilunar valves.

On their own, autorhythmic fibers in the SA node would initiate an action potential about every 0.6 second, or 100 times per minute. This rate is faster than that of any other autorhythmic fibers. Because action potentials from the SA node spread through the conduction system and stimulate other areas before

Figure 20.10 The conduction system of the heart. Autorhythmic fibers in the SA node, located in the right atrial wall (a), act as the heart's pacemaker, initiating cardiac action potentials (b) that cause contraction of the heart's chambers.

The conduction system ensures that the chambers of the heart contract in a coordinated manner.

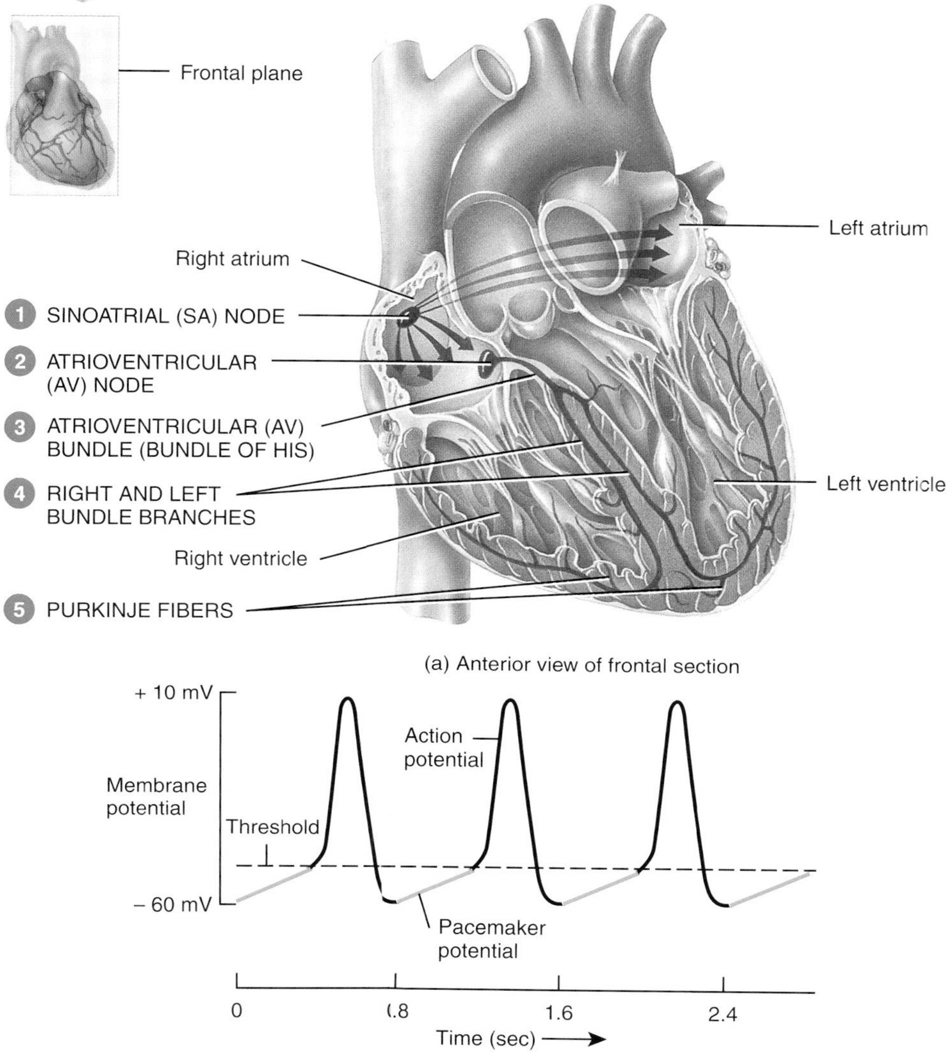

Frontal plane

Right atrium

1 SINOATRIAL (SA) NODE

2 ATRIOVENTRICULAR (AV) NODE

3 ATRIOVENTRICULAR (AV) BUNDLE (BUNDLE OF HIS)

4 RIGHT AND LEFT BUNDLE BRANCHES

Right ventricle

5 PURKINJE FIBERS

Left atrium

Left ventricle

(a) Anterior view of frontal section

+ 10 mV

Membrane potential

Threshold

– 60 mV

Action potential

Pacemaker potential

0 0.8 1.6 2.4

Time (sec) →

(b) Pacemaker potentials and action potentials in autorhythmic fibers of SA node

Which component of the conduction system provides the only electrical connection between the atria and the ventricles?

the other areas are able to generate an action potential at their own, slower rate, the SA node acts as the normal pacemaker of the heart. Nerve impulses from the autonomic nervous system (ANS) and blood-borne hormones (such as epinephrine) *modify the timing and strength* of each heartbeat, but they *do not establish the fundamental rhythm.* In a person at rest, for example, acetylcholine released by the parasympathetic division of the ANS slows SA node pacing to about 75 action potentials per minute, or one every 0.8 sec (Figure 20.10b).

Artificial Pacemakers

If the SA node becomes damaged or diseased, the slower AV node can pick up the pacemaking task. Its rate of spontaneous depolarization is 40 to 60 times per minute. If the activity of both nodes is suppressed, the heartbeat may still be maintained by autorhythmic fibers in the ventricles—the AV bundle, a bundle branch, or Purkinje fibers. However, the pacing rate is so slow (20–35 beats per minute) that blood flow to the brain is inadequate. When this condition occurs, normal heart rhythm can be restored and maintained by surgically implanting an **artificial pacemaker,** a device that sends out small electrical currents to stimulate the heart to contract. A pacemaker consists of a battery and impulse generator and is usually implanted beneath the skin just inferior to the clavicle. The pacemaker is connected to one or two flexible wires (leads) that are threaded through the superior vena cava and then passed into the right atrium and right ventricle. Many of the newer pacemakers, referred to as *activity-adjusted pacemakers,* automatically speed up the heartbeat during exercise. ■

Action Potential and Contraction of Contractile Fibers

The action potential initiated by the SA node travels along the conduction system and spreads out to excite the "working" atrial and ventricular muscle fibers, called **contractile fibers.** An action potential occurs in a contractile fiber as follows (Figure 20.11):

1. ***Depolarization.*** Unlike autorhythmic fibers, contractile fibers have a stable resting membrane potential that is close to −90 mV. When a contractile fiber is brought to threshold by an action potential from neighboring fibers, its **voltage-gated fast Na^+ channels** open. These sodium ion channels are referred to as "fast" because they open very rapidly in response to a threshold-level depolarization. Opening of these channels allows Na^+ inflow because the cytosol of contractile fibers is electrically more negative than interstitial fluid and Na^+ concentration is higher in interstitial fluid. Inflow of Na^+ down the electrochemical gradient produces a **rapid depolarization.** Within a few milliseconds, the fast Na^+ channels automatically inactivate and Na^+ inflow decreases.
2. ***Plateau.*** The next phase of an action potential in a contractile fiber is the **plateau,** a period of maintained depolarization. It is due in part to opening of **voltage-gated slow Ca^{2+} channels** in the sarcolemma. When these channels open, calcium ions move from the interstitial fluid (which has a higher Ca^{2+} concentration) into the cytosol. This inflow of Ca^{2+} causes even more Ca^{2+} to pour out of the sarcoplasmic reticulum into the cytosol through additional Ca^{2+} channels in the sarcoplasmic reticulum membrane. The increased Ca^{2+} concentration in the cytosol ultimately triggers contraction. Several different types of **voltage-gated K^+ channels** are also found in the sarcolemma of a contractile fiber. Just before the plateau phase begins, some of these K^+ channels open, allowing potassium ions to leave the contractile fiber. Therefore, depolarization is sustained during the plateau phase because Ca^{2+} inflow just balances K^+ outflow. The plateau phase lasts for about 0.25 sec, and the membrane potential of the contractile fiber is close to 0 mV. By comparison, depolarization in a neuron or skeletal muscle fiber is much briefer, about 1 msec (0.001 sec), because it lacks a plateau phase.
3. ***Repolarization.*** The recovery of the resting membrane potential during the **repolarization** phase of a cardiac action potential resembles that in other excitable cells. After a delay (which is particularly prolonged in cardiac muscle), voltage-gated K^+ channels open. Outflow of K^+ restores the negative resting membrane potential (−90 mV). At the same time, the calcium channels in the sarcolemma and the sarcoplasmic reticulum are closing, which also contributes to repolarization.

Figure 20.11 Action potential in a ventricular contractile fiber. The resting membrane potential is about −90 mV.

A long refractory period prevents tetanus in cardiac muscle fibers.

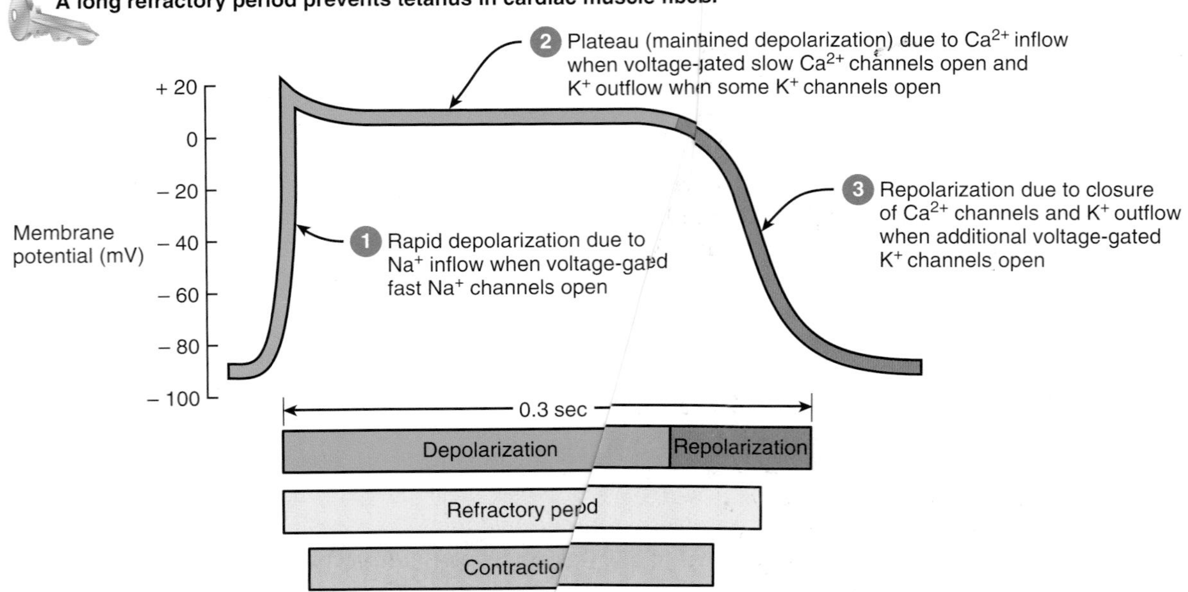

? **How does the duration of an action potential in ventricular contractile fiber compare with that in a skeletal muscle fiber?**

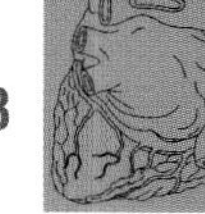

The mechanism of contraction is similar in cardiac and skeletal muscle: The electrical activity (action potential) leads to the mechanical response (contraction) after a short delay. As Ca^{2+} concentration rises inside a contractile fiber, Ca^{2+} binds to the regulatory protein troponin, which allows the actin and myosin filaments to begin sliding past one another, and tension starts to develop. Substances that alter the movement of Ca^{2+} through slow Ca^{2+} channels influence the strength of heart contractions. Epinephrine, for example, increases contraction force by enhancing Ca^{2+} flow into the cytosol.

In muscle, the **refractory period** is the time interval during which a second contraction cannot be triggered. The refractory period of a cardiac muscle fiber lasts longer than the contraction itself (Figure 20.11). As a result, another contraction cannot begin until relaxation is well underway. For this reason, tetanus (maintained contraction) cannot occur in cardiac muscle as it can in skeletal muscle. The advantage is apparent if you consider how the ventricles work. Their pumping function depends on alternating contraction (when they eject blood) and relaxation (when they refill). If heart muscle could undergo tetanus, blood flow would cease.

ATP Production in Cardiac Muscle

In contrast to skeletal muscle, cardiac muscle produces little of the ATP it needs by anaerobic cellular respiration (see Figure 10.12 on page 307). Instead, it relies almost exclusively on aerobic cellular respiration in its numerous mitochondria. The needed oxygen diffuses from blood in the coronary circulation and is released from myoglobin inside cardiac muscle fibers. Cardiac muscle fibers use several fuels to power mitochondrial ATP production. In a person at rest, the heart's ATP comes mainly from oxidation of fatty acids (60%) and glucose (35%), with smaller contributions from lactic acid, amino acids, and ketone bodies. During exercise, the heart's use of lactic acid, produced by actively contracting skeletal muscles, rises.

Like skeletal muscle, cardiac muscle also produces some ATP from creatine phosphate. One sign that a myocardial infarction (heart attack, see page 708) has occurred is the presence in blood of creatine kinase (CK), the enzyme that catalyzes transfer of a phosphate group from creatine phosphate to ADP to make ATP. Normally, CK and other enzymes are confined within cells. Injured or dying cardiac or skeletal muscle fibers release CK into the blood.

Electrocardiogram

As action potentials propagate through the heart, they generate electrical currents that can be detected at the surface of the body. An **electrocardiogram** (e-lek′-trō-KAR-dē-ō-gram), abbreviated either **ECG** or **EKG** (from the German word *Elektrokardiogram*) is a recording of these electrical signals. The ECG is a composite record of action potentials produced by all the heart muscle fibers during each heartbeat. The instrument used to record the changes is an **electrocardiograph.**

In clinical practice, electrodes are positioned on the arms and legs (limb leads) and at six positions on the chest (chest leads) to record the ECG. The electrocardiograph amplifies the heart's electrical signals and produces 12 different tracings from different combinations of limb and chest leads. Each limb and chest electrode records slightly different electrical activity because of the difference in its position relative to the heart. By comparing these records with one another and with normal records, it is possible to determine (1) if the conducting pathway is abnormal, (2) if the heart is enlarged, (3) if certain regions of the heart are damaged, and (4) the cause of chest pain.

In a typical record, three clearly recognizable waves appear with each heartbeat (Figure 20.12). The first, called the **P wave,** is a small upward deflection on the ECG. The P wave represents **atrial depolarization,** which spreads from the SA node through contractile fibers in both atria. The second wave, called the **QRS complex,** begins as a downward deflection, continues as a large, upright, triangular wave, and ends as a downward wave. The QRS complex represents **rapid ventricular depolarization,** as the action potential spreads through ventricular contractile fibers. The third wave is a dome-shaped upward deflection called the **T wave.** It indicates **ventricular**

Figure 20.12 Normal electrocardiogram or ECG (Lead II). P wave = atrial depolarization; QRS complex = onset of ventricular depolarization; T wave = ventricular repolarization.

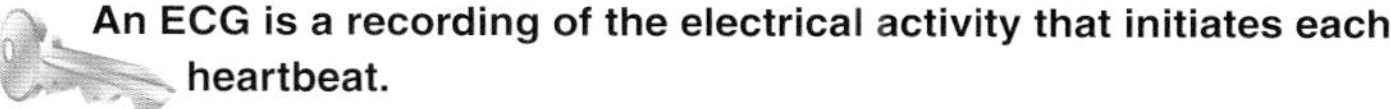

An ECG is a recording of the electrical activity that initiates each heartbeat.

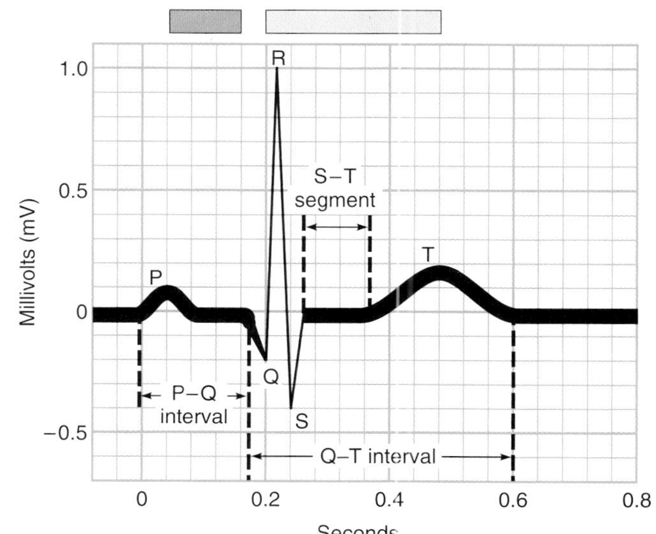

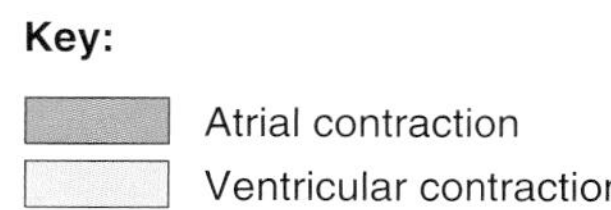

What is the significance of an enlarged Q wave?

repolarization and occurs just as the ventricles are starting to relax. The T wave is smaller and wider than the QRS complex because repolarization occurs more slowly than depolarization. During the plateau period of steady depolarization, the ECG tracing is flat.

In reading an ECG, the size of the waves can provide clues to abnormalities. Larger P waves indicate enlargement of an atrium; an enlarged Q wave may indicate a myocardial infarction; and an enlarged R wave generally indicates enlarged ventricles. The T wave is flatter than normal when the heart muscle is receiving insufficient oxygen—as, for example, in coronary artery disease. The T wave may be elevated in hyperkalemia (high blood K^+ level).

Analysis of an ECG also involves measuring the time spans between waves, which are called **intervals** or **segments.** For example, the **P-Q interval** is the time from the beginning of the P wave to the beginning of the QRS complex. It represents the conduction time from the beginning of atrial excitation to the beginning of ventricular excitation. Put another way, the P-Q interval is the time required for the action potential to travel through the atria, atrioventricular node, and the remaining fibers of the conduction system. As the action potential is forced to detour around scar tissue caused by disorders such as coronary artery disease and rheumatic fever, the P-Q interval lengthens.

The **S-T segment,** which begins at the end of the S wave and ends at the beginning of the T wave, represents the time when the ventricular contractile fibers are depolarized during the plateau phase of the action potential. The S-T segment is elevated (above the baseline) in acute myocardial infarction and depressed (below the baseline) when the heart muscle receives insufficient oxygen. The **Q-T interval** extends from the start of the QRS complex to the end of the T wave. It is the time from the beginning of ventricular depolarization to the end of ventricular repolarization. The Q-T interval may be lengthened by myocardial damage, myocardial ischemia (decreased blood flow), or conduction abnormalities.

Sometimes it is helpful to evaluate the heart's response to the stress of physical exercise (stress testing) (see page 727). Although narrowed coronary arteries may carry adequate oxygenated blood while a person is at rest, they will not be able to meet the heart's increased need for oxygen during strenuous exercise. This situation creates changes that can be seen on an electrocardiogram.

Abnormal heart rhythms and inadequate blood flow to the heart may occur only briefly or unpredictably. To detect these problems, **continuous ambulatory electrocardiographs** are used. With this procedure, a person wears a battery-operated monitor (Holter monitor) that records an ECG continuously for 24 hours. Electrodes attached to the chest are connected to the monitor and imformation on the heart's activity is stored in the monitor and retrieved later by medical personnel.

Correlation of ECG Waves with Atrial and Ventricular Systole

As we have seen, the atria and ventricles depolarize and then contract at different times because the conduction system routes cardiac action potentials along a specific pathway. The term **systole** (SIS-tō-lē = contraction) refers to the phase of contraction; the phase of relaxation is **diastole** (dī-AS-tō-lē = dilation or expansion). The ECG waves predict the timing of atrial and ventricular systole and diastole. At a heart rate of 75 beats per minute, the timing is as follows (Figure 20.13):

1. A cardiac action potential arises in the SA node. It propagates throughout the atrial muscle and down to the AV node in about 0.03 sec. As the atrial contractile fibers depolarize, the P wave appears in the ECG.
2. After the P wave begins, the atria contract (atrial systole). Conduction of the action potential slows at the AV node because the fibers there have much smaller diameters and fewer gap junctions. (Traffic slows in a similar way where a four-lane highway narrows to one lane in a construction zone!) The resulting 0.1-sec delay gives the atria time to contract, thus adding to the volume of blood in the ventricles, before ventricular diastole begins.
3. The action potential propagates rapidly again after entering the AV bundle. About 0.2 sec after onset of the P wave, it has propagated through the bundle branches, Purkinje fibers, and the entire ventricular myocardium. Depolarization progresses down the septum, upward from the apex, and outward from the endocardial surface, producing the QRS complex. At the same time, atrial repolarization is occurring, but it is not usually evident in an ECG because the larger QRS complex masks it.
4. Contraction of ventricular contractile fibers (ventricular systole) begins shortly after the QRS complex appears and continues during the S-T segment. As contraction proceeds from the apex toward the base of the heart, blood is squeezed upward toward the semilunar valves.
5. Repolarization of ventricular contractile fibers begins at the apex and spreads throughout the ventricular myocardium. This produces the T wave in the ECG about 0.4 sec after the onset of the P wave.
6. Shortly after the T wave begins, the ventricles start to relax (ventricular diastole). By 0.6 sec, ventricular repolarization is complete and ventricular contractile fibers are relaxed.

During the next 0.2 sec, contractile fibers in both the atria and ventricles are relaxed. At 0.8 sec, the P wave appears again in the ECG, the atria begin to contract, and the cycle repeats.

As you have just learned, events in the heart occur in cycles that repeat for as long as you live. Next, we will see how the pressure changes associated with relaxation and contraction of the heart chambers allow the heart to alternately fill with blood and then eject blood into the aorta and pulmonary trunk.

Figure 20.13 **Timing and route of action potential depolarization and repolarization through the conduction system and myocardium.** Green indicates depolarization and red indicates repolarization.

Depolarization causes contraction and repolarization causes relaxation of cardiac muscle fibers.

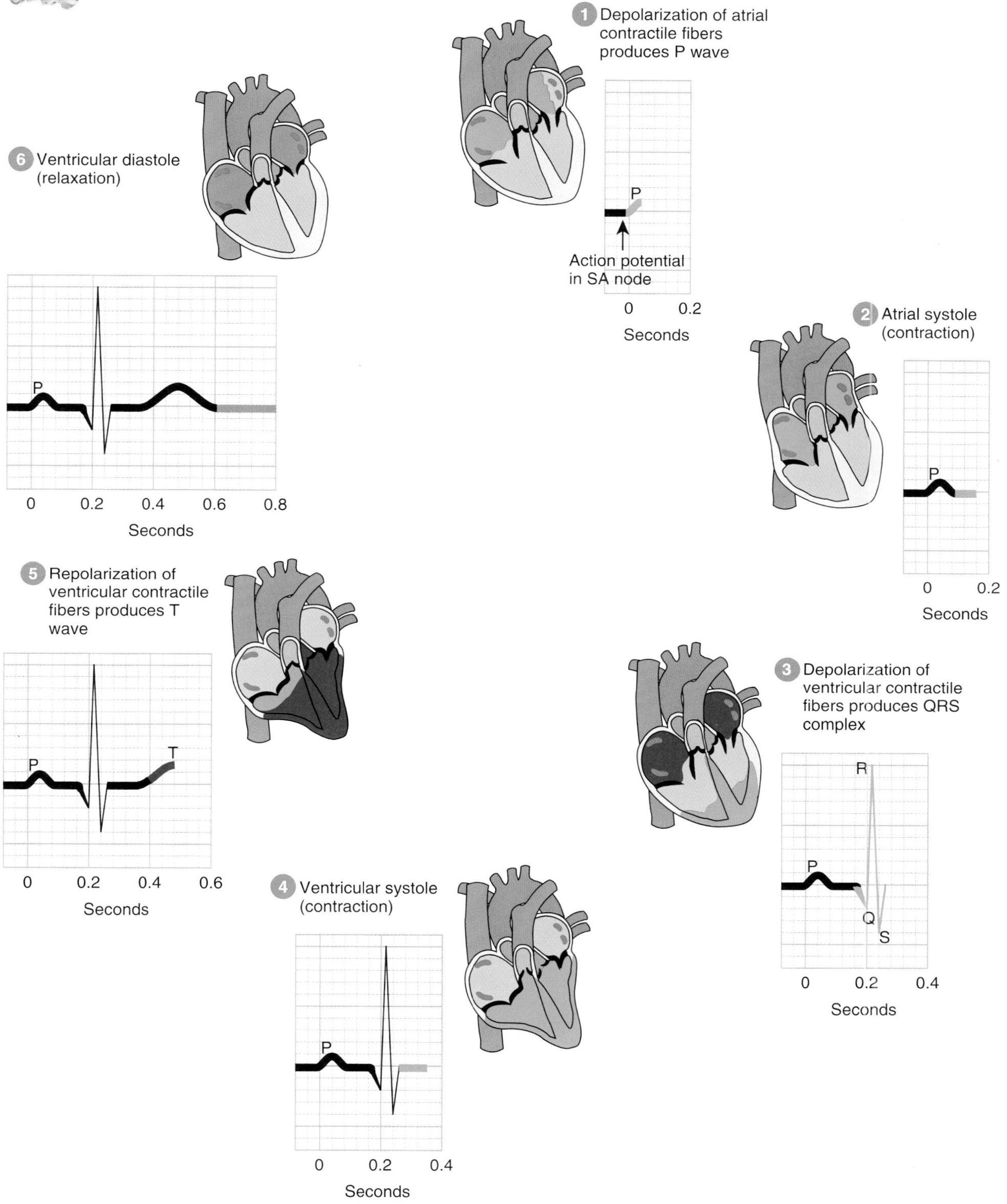

1 Depolarization of atrial contractile fibers produces P wave

P
Action potential in SA node
0 0.2
Seconds

2 Atrial systole (contraction)

P
0 0.2
Seconds

3 Depolarization of ventricular contractile fibers produces QRS complex

R
P
Q
S
0 0.2 0.4
Seconds

4 Ventricular systole (contraction)

P
0 0.2 0.4
Seconds

5 Repolarization of ventricular contractile fibers produces T wave

P
T
0 0.2 0.4 0.6
Seconds

6 Ventricular diastole (relaxation)

P
0 0.2 0.4 0.6 0.8
Seconds

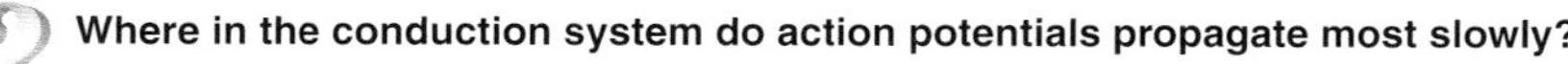

Where in the conduction system do action potentials propagate most slowly?

▶ CHECKPOINT

10. How do cardiac muscle fibers differ structurally and functionally from skeletal muscle fibers?

11. In what ways are autorhythmic fibers similar to and different from contractile fibers?

12. What happens during each of the three phases of an action potential in ventricular contractile fibers?

13. In what ways are ECGs helpful in diagnosing cardiac problems?

14. How does each ECG wave, interval, and segment relate to contraction (systole) and relaxation (diastole) of the atria and ventricles?

THE CARDIAC CYCLE

▶ OBJECTIVES

Describe the pressure and volume changes that occur during a cardiac cycle.

Relate the timing of heart sounds to the ECG waves and pressure changes during systole and diastole.

A single **cardiac cycle** includes all the events associated with one heartbeat. Thus, a cardiac cycle consists of systole and diastole of the atria plus systole and diastole of the ventricles.

Pressure and Volume Changes During the Cardiac Cycle

In each cardiac cycle, the atria and ventricles alternately contract and relax, forcing blood from areas of higher pressure to areas of lower pressure. As a chamber of the heart contracts, blood pressure within it increases. Figure 20.14 shows the relation between the heart's electrical signals (ECG) and changes in atrial pressure, ventricular pressure, aortic pressure, and ventricular volume during the cardiac cycle. The pressures given in Figure 20.14 apply to the left side of the heart; pressures on the right side are considerably lower. Each ventricle, however, expels the same volume of blood per beat, and the same pattern exists for both pumping chambers. When heart rate is 75 beats/min, a cardiac cycle lasts 0.8 sec. To examine and correlate the events taking place during a cardiac cycle, we will begin with atrial systole.

Atrial Systole

During **atrial systole,** which lasts about 0.1 sec, the atria are contracting. At the same time, the ventricles are relaxed.

1. Depolarization of the SA node causes atrial depolarization, marked by the P wave in the ECG.
2. Atrial depolarization causes atrial systole. As the atria contract, they exert pressure on the blood within, which forces blood through the open AV valves into the ventricles.
3. Atrial systole contributes a final 25 mL of blood to the volume already in each ventricle (about 105 mL). The end of atrial systole is also the end of ventricular diastole (relaxation). Thus, each ventricle contains about 130 mL at the end of its relaxation period (diastole). This blood volume is called the **end-diastolic volume (EDV).**
4. The QRS complex in the ECG marks the onset of ventricular depolarization.

Ventricular Systole

During **ventricular systole,** which lasts about 0.3 sec, the ventricles are contracting. At the same time, the atria are relaxed, in **atrial diastole.**

5. Ventricular depolarization causes ventricular systole. As ventricular systole begins, pressure rises inside the ventricles and pushes blood up against the atrioventricular (AV) valves, forcing them shut. For about 0.05 seconds, both the SL (semilunar) and AV valves are closed. This is the period of **isovolumetric contraction** (*iso-* = same). During this interval, cardiac muscle fibers are contracting and exerting force but are not yet shortening. Thus, the muscle contraction is isometric (same length). Moreover, because all four valves are closed, ventricular volume remains the same (isovolumic).
6. Continued contraction of the ventricles causes pressure inside the chambers to rise sharply. When left ventricular pressure surpasses aortic pressure at about 80 millimeters of mercury (mmHg) and right ventricular pressure rises above the pressure in the pulmonary trunk (about 20 mmHg), both SL valves open. At this point, ejection of blood from the heart begins. The period when the SL valves are open is **ventricular ejection** and lasts for about 0.25 sec. The pressure in the left ventricle continues to rise to about 120 mmHg, whereas the pressure in the right ventricle climbs to about 25–30 mmHg.
7. The left ventricle ejects about 70 mL of blood into the aorta and the right ventricle ejects the same volume of blood into the pulmonary trunk. The volume remaining in each ventricle at the end of systole, about 60 mL, is the **end-systolic volume (ESV). Stroke volume,** the volume ejected per beat from each ventricle, equals end-diastolic volume minus end-systolic volume: SV = EDV − ESV. At rest, the stroke volume is about 130 mL − 60 mL = 70 mL (a little more than 2 oz).
8. The T wave in the ECG marks the onset of ventricular repolarization.

Relaxation Period

During the **relaxation period,** which lasts about 0.4 sec, the atria and the ventricles are both relaxed. As the heart beats faster and faster, the relaxation period becomes shorter and shorter,

Figure 20.14 Cardiac cycle. (a) ECG. (b) Changes in left atrial pressure (green line), left ventricular pressure (blue line), and aortic pressure (red line) as they relate to the opening and closing of heart valves. (c) Heart sounds. (d) Changes in left ventricular volume. (e) Phases of the cardiac cycle.

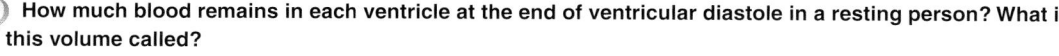

A cardiac cycle is composed of all the events associated with one heartbeat.

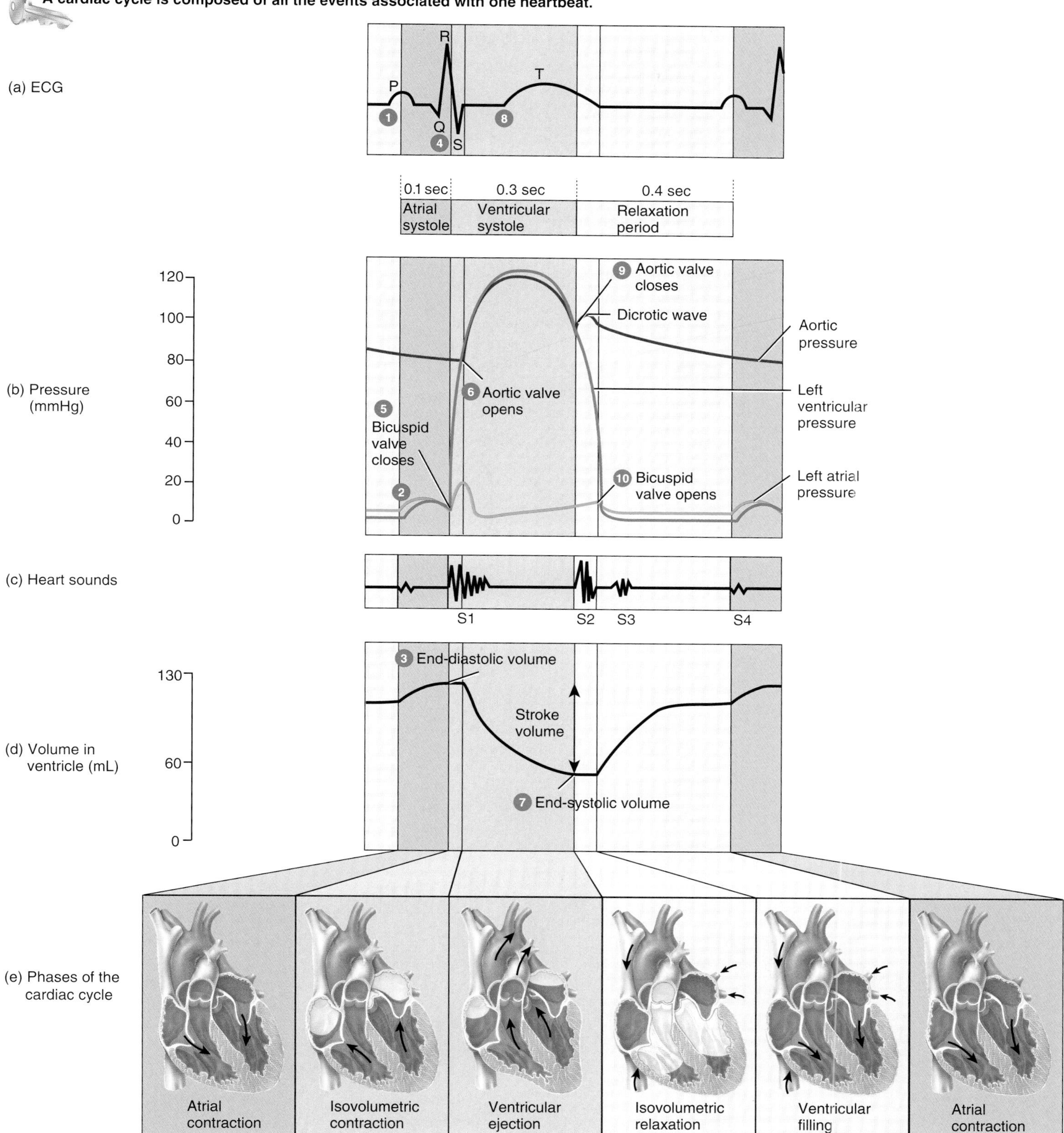

How much blood remains in each ventricle at the end of ventricular diastole in a resting person? What is this volume called?

whereas the durations of atrial systole and ventricular systole shorten only slightly.

9 Ventricular repolarization causes **ventricular diastole.** As the ventricles relax, pressure within the chambers falls, and blood in the aorta and pulmonary trunk begins to flow backward toward the regions of lower pressure in the ventricles. Backflowing blood catches in the valve cusps and closes the SL valves. The aortic valve closes at a pressure of about 100 mmHg. Rebound of blood off the closed cusps of the aortic valve produces the **dicrotic wave** on the aortic pressure curve. After the SL valves close, there is a brief interval when ventricular blood volume does not change because all four valves are closed. This is the period of **isovolumetric relaxation.**

10 As the ventricles continue to relax, the pressure falls quickly. When ventricular pressure drops below atrial pressure, the AV valves open, and **ventricular filling** begins. The major part of ventricular filling occurs just after the AV valves open. Blood that has been flowing into and building up in the atria during ventricular systole then rushes rapidly into the ventricles. At the end of the relaxation period, the ventricles are about three-quarters full. The P wave appears in the ECG, signaling the start of another cardiac cycle.

Heart Sounds

Auscultation (aws-kul-TĀ-shun; *ausculta-* = listening), the act of listening to sounds within the body, is usually done with a stethoscope. The sound of the heartbeat comes primarily from blood turbulence caused by the closing of the heart valves. Smoothly flowing blood is silent. Recall the sounds made by white-water rapids or a waterfall as compared with the silence of a smoothly flowing river. During each cardiac cycle, there are four **heart sounds,** but in a normal heart only the first and second heart sounds (S1 and S2) are loud enough to be heard by listening through a stethoscope. Figure 20.14c shows the timing of heart sounds relative to other events in the cardiac cycle.

The first sound (S1), which can be described as a **lubb** sound, is louder and a bit longer than the second sound. S1 is caused by blood turbulence associated with closure of the AV valves soon after ventricular systole begins. The second sound (S2), which is shorter and not as loud as the first, can be described as a **dupp** sound. S2 is caused by blood turbulence associated with closure of the SL valves at the beginning of ventricular diastole. Although S1 and S2 are due to blood turbulence associated with the closure of valves, they are best heard at the surface of the chest in locations that are slightly different from the locations of the valves (Figure 20.15). Normally not loud

Figure 20.15 Heart sounds. Location of valves (purple) and auscultation sites (red) for heart sounds.

Listening to sounds within the body is called auscultation; it is usually done with a stethoscope.

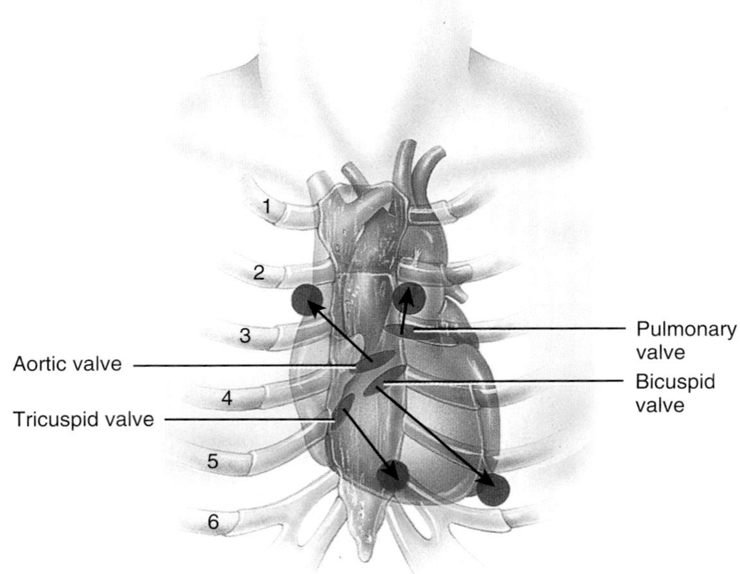

Anterior view of heart valve locations and auscultation sites

Which heart sound is related to blood turbulence associated with closure of the AV valves?

enough to be heard, S3 is due to blood turbulence during rapid ventricular filling, and S4 is due to blood turbulence during atrial systole.

Heart Murmurs

Heart sounds provide valuable information about the mechanical operation of the heart. A **heart murmur** is an abnormal sound consisting of a clicking, rushing, or gurgling noise that is heard before, between, or after the normal heart sounds, or that may mask the normal heart sounds. Heart murmurs in children are extremely common and usually do not represent a health condition. Murmurs are most frequently discovered in children between the ages of two and four. These types of heart murmurs are referred to as *innocent* or *functional heart murmurs;* they often subside or disappear with growth. Although some heart murmurs in adults are innocent, most often a murmur indicates a valve disorder. When a heart valve exhibits stenosis, the heart murmur is heard while the valve should be fully open but is not. For example, mitral stenosis (see page 704) produces a murmur during the relaxation period, between S2 and the next S1. An incompetent heart valve, by contrast, causes a murmur to appear when the valve should be fully closed but is not. So, a murmur due to mitral incompetence (see page 704) occurs during ventricular systole, between S1 and S2. ■

▶ CHECKPOINT

15. Why must left ventricular pressure be greater than aortic pressure during ventricular ejection?

16. Does more blood flow through the coronary arteries during ventricular diastole or ventricular systole? Explain why.

17. During which two periods of the cardiac cycle do the heart muscle fibers exhibit isometric contractions?

18. What events produce the four normal heart sounds? Which ones can usually be heard through a stethoscope?

CARDIAC OUTPUT

▶ OBJECTIVE

Define cardiac output.

Describe the factors that affect regulation of stroke volume.

Outline the factors that affect the regulation of heart rate.

Although the heart has autorhythmic fibers that enable it to beat independently, its operation is governed by events occurring throughout the body. All body cells must receive a certain amount of oxygenated blood each minute to maintain health and life. When cells are metabolically active, as during exercise, they take up even more oxygen from the blood. During rest periods, cellular metabolic need is reduced, and the workload of the heart decreases.

Cardiac output (CO) is the volume of blood ejected from the left ventricle (or the right ventricle) into the aorta (or pulmonary trunk) each minute. Cardiac output equals the **stroke volume (SV),** the volume of blood ejected by the ventricle during each contraction, multiplied by the **heart rate (HR),** the number of heartbeats per minute:

$$\underset{\text{(mL/min)}}{\text{CO}} = \underset{\text{mL/beat)}}{\text{SV}} \times \underset{\text{(beats/min)}}{\text{HR}}$$

In a typical resting adult male, stroke volume averages 70 mL/beat, and heart rate is about 75 beats/min. Thus, average cardiac output is

$$\begin{aligned} \text{CO} &= 70 \text{ mL/beat} \times 75 \text{ beats/min} \\ &= 5250 \text{ mL/min} \\ &= 5.25 \text{ L/min} \end{aligned}$$

This volume is close to the total blood volume, which is about 5 liters in a typical adult male. Thus, your entire blood volume flows through your pulmonary and systemic circulations each minute. Factors that increase stroke volume or heart rate normally increase CO. During mild exercise, for example, stroke volume may increase to 100 mL/beat, and heart rate to 100 beats/min. Cardiac output then would be 10 L/min. During intense (but still not maximal) exercise, the heart rate may accelerate to 150 beats/min, and stroke volume may rise to 130 mL/beat, resulting in a cardiac output of 19.5 L/min.

Cardiac reserve is the difference between a person's maximum cardiac output and cardiac output at rest. The average person has a cardiac reserve of four or five times the resting value. Top -endurance athletes may have a cardiac reserve seven or eight times their resting CO. People with severe heart disease may have little or no cardiac reserve, which limits their ability to carry out even the simple tasks of daily living.

Regulation of Stroke Volume

A healthy heart will pump out the blood that entered its chambers during the previous diastole. In other words, if more blood returns to the heart during diastole, then more blood is ejected during the next systole. At rest, the stroke volume is 50–60% of the end-diastolic volume because 40–50% of the blood remains in the ventricles after each contraction (end-systolic volume). Three factors regulate stroke volume and ensure that the left and right ventricles pump equal volumes of blood: (1) **preload,** the degree of stretch on the heart before it contracts; (2) **contractility,** the forcefulness of contraction of individual ventricular muscle fibers; and (3) **afterload,** the pressure that must be exceeded before ejection of blood from the ventricles can occur.

Preload: Effect of Stretching

A greater preload (stretch) on cardiac muscle fibers prior to contraction increases their force of contraction. Preload can be compared to the stretching of a rubber band. The more the rubber band is stretched, the more forcefully it will snap back. Within limits, the more the heart fills with blood during diastole, the greater the force of contraction during systole. This relationship is known as the **Frank–Starling law of the heart.** The preload is proportional to the end-diastolic volume (EDV) (the volume of blood that fills the ventricles at the end of diastole). Normally, the greater the EDV, the more forceful the next contraction.

Two key factors determine EDV: (1) the duration of ventricular diastole and (2) **venous return,** the volume of blood returning to the right ventricle. When heart rate increases, the duration of diastole is shorter. Less filling time means a smaller EDV, and the ventricles may contract before they are adequately filled. By contrast, when venous return increases, a greater volume of blood flows into the ventricles, and the EDV is increased.

When heart rate exceeds about 160 beats/min, stroke volume usually declines due to the short filling time. At such rapid heart rates, EDV is less, and the preload is lower. People who have slow resting heart rates usually have large resting stroke volumes because filling time is prolonged and preload is larger.

The Frank–Starling law of the heart equalizes the output of the right and left ventricles and keeps the same volume of blood flowing to both the systemic and pulmonary circulations. If the left side of the heart pumps a little more blood than the right side, the volume of blood returning to the right ventricle (venous return) increases. The increased EDV causes the right ventricle to contract more forcefully on the next beat, bringing the two sides back into balance.

Contractility

The second factor that influences stroke volume is myocardial **contractility,** the strength of contraction at any given preload. Substances that increase contractility are **positive inotropic agents;** those that decrease contractility are **negative inotropic agents.**Thus, for a constant preload, the stroke volume increases when a positive inotropic substance is present. Positive inotropic agents often promote Ca^{2+} inflow during cardiac action potentials, which strengthens the force of the next contraction. Stimulation of the sympathetic division of the autonomic nervous system (ANS), hormones such as epinephrine and norepinephrine, increased Ca^{2+} level in the interstitial fluid, and the drug digitalis all have positive inotropic effects. In contrast, inhibition of the sympathetic division of the ANS, anoxia, acidosis, some anesthetics, and increased K^{+} level in the interstitial fluid have negative inotropic effects. *Calcium channel blockers* are drugs that can have a negative inotropic effect by reducing Ca^{2+} inflow, thereby decreasing the strength of the heartbeat.

Afterload

Ejection of blood from the heart begins when pressure in the right ventricle exceeds the pressure in the pulmonary trunk (about 20 mmHg), and when the pressure in the left ventricle exceeds the pressure in the aorta (about 80 mmHg). At that point, the higher pressure in the ventricles causes blood to push the semilunar valves open. The pressure that must be overcome before a semilunar valve can open is termed the **afterload.** An increase in afterload causes stroke volume to decrease, so that more blood remains in the ventricles at the end of systole. Conditions that can increase afterload include hypertension (elevated blood pressure) and narrowing of arteries by atherosclerosis (see page 726).

In **congestive heart failure (CHF),** there is a loss of pumping efficiency by the heart. Causes of CHF include coronary artery disease (see page 726), congenital defects, long-term high blood pressure (which increases the afterload), myocardial infarctions (regions of dead heart tissue due to a previous heart attack), and valve disorders. As the pump becomes less effective, more blood remains in the ventricles at the end of each cycle, and gradually the end-diastolic volume (preload) increases. Initially, increased preload may promote increased force of contraction (the Frank–Starling law of the heart), but as the preload increases further, the heart is overstretched and contracts less forcefully. The result is a potentially lethal positive feedback loop: Less-effective pumping leads to even lower pumping capability.

Often, one side of the heart starts to fail before the other. If the left ventricle fails first, it can't pump out all the blood it receives. As a result, blood backs up in the lungs and causes *pulmonary edema,* fluid accumulation in the lungs that can cause suffocation if left untreated. If the right ventricle fails first, blood backs up in the systemic veins and over time, the kidneys cause an increase in blood volume. In this case, the resulting *peripheral edema* usually is most noticeable in the feet and ankles. ■

Regulation of Heart Rate

As you have just learned, cardiac output depends on both heart rate and stroke volume. Adjustments in heart rate are important in the short-term control of cardiac output and blood pressure. The sinoatrial (SA) node initiates contraction and, if left to itself, would set a constant heart rate of about 100 beats/min. However, tissues require different volumes of blood flow under different conditions. During exercise, for example, cardiac output rises to supply working tissues with increased amounts of oxygen and nutrients. Stroke volume may fall if the ventricular myocardium is damaged or if blood volume is reduced by bleeding. In these cases, homeostatic mechanisms maintain adequate cardiac output by increasing the heart rate and contractility. Among the several factors that contribute to regulation of heart rate, the most important are the autonomic nervous system and hormones released by the adrenal medullae (epinephrine and norepinephrine).

Autonomic Regulation of Heart Rate

Nervous system regulation of the heart originates in the **cardiovascular center** in the medulla oblongata. This region of the brain stem receives input from a variety of sensory receptors and from higher brain centers, such as the limbic system and cerebral cortex. The cardiovascular center then directs appropriate

Figure 20.16 **Nervous system control of the heart.**

The cardiovascular center in the medulla oblongata controls both sympathetic and parasympathetic nerves that innervate the heart.

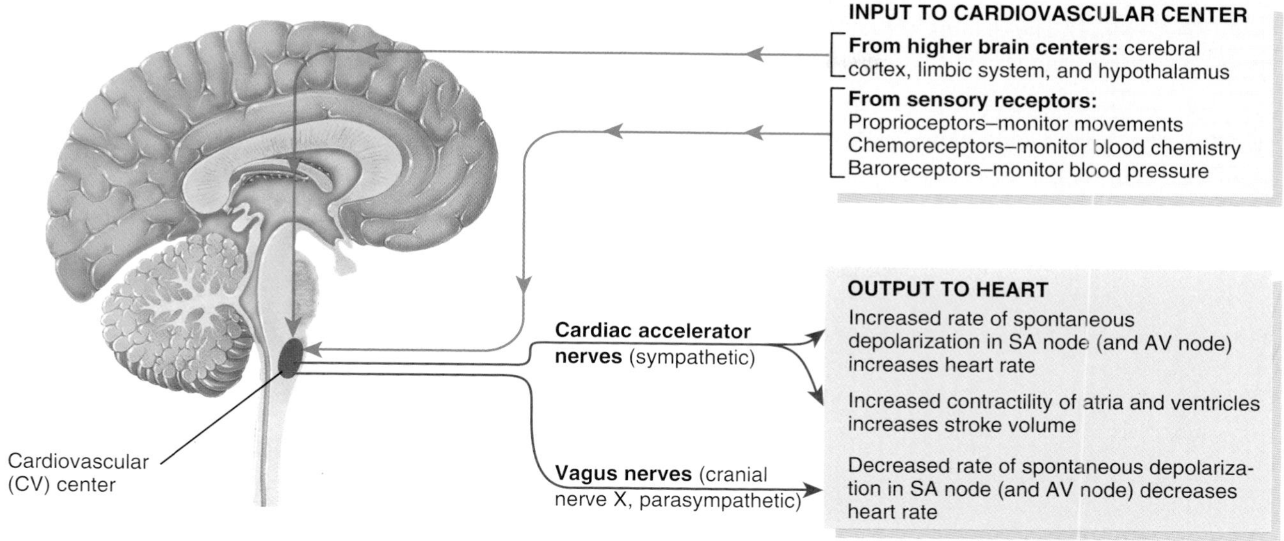
INPUT TO CARDIOVASCULAR CENTER
From higher brain centers: cerebral cortex, limbic system, and hypothalamus
From sensory receptors:
Proprioceptors–monitor movements
Chemoreceptors–monitor blood chemistry
Baroreceptors–monitor blood pressure
OUTPUT TO HEART
Increased rate of spontaneous depolarization in SA node (and AV node) increases heart rate
Increased contractility of atria and ventricles increases stroke volume
Decreased rate of spontaneous depolarization in SA node (and AV node) decreases heart rate
Cardiac accelerator nerves (sympathetic)
Vagus nerves (cranial nerve X, parasympathetic)
Cardiovascular (CV) center

What region of the heart is innervated by the sympathetic division but not by the parasympathetic division?

output by increasing or decreasing the frequency of nerve impulses in both the sympathetic and parasympathetic branches of the ANS (Figure 20.16).

Even before physical activity begins, especially in competitive situations, heart rate may climb. This anticipatory increase occurs because the limbic system sends nerve impulses to the cardiovascular center in the medulla. As physical activity begins, **proprioceptors** that are monitoring the position of limbs and muscles send nerve impulses at an increased frequency to the cardiovascular center. Proprioceptor input is a major stimulus for the quick rise in heart rate that occurs at the onset of physical activity. Other sensory receptors that provide input to the cardiovascular center include **chemoreceptors,** which monitor chemical changes in the blood, and **baroreceptors,** which monitor the stretching of major arteries and veins caused by the pressure of the blood flowing through them. Important baroreceptors located in the arch of the aorta and in the carotid arteries (see Figure 21.13 on page 751) detect changes in blood pressure and provide input to the cardiovascular center when it changes. The role of baroreceptors in the regulation of blood pressure is discussed in detail in Chapter 21. Here we focus on the innervation of the heart by the sympathetic and parasympathetic branches of the ANS.

Sympathetic neurons extend from the medulla oblongata into the spinal cord. From the thoracic region of the spinal cord, sympathetic **cardiac accelerator nerves** extend out to the SA node, AV node, and most portions of the myocardium. Impulses in the cardiac accelerator nerves trigger the release of norepinephrine, which binds to beta-1 receptors on cardiac muscle fibers. This interaction has two separate effects: (1) In SA (and AV) node fibers, norepinephrine speeds the rate of spontaneous depolarization so that these pacemakers fire impulses more rapidly and heart rate increases; (2) in contractile fibers throughout the atria and ventricles, norepinephrine enhances Ca^{2+} entry through the voltage-gated slow Ca^{2+} channels, thereby increasing contractility. As a result, a greater volume of blood is ejected during systole. With a moderate increase in heart rate, stroke volume does not decline because the increased contractility offsets the decreased preload. With maximal sympathetic stimulation, however, heart rate may reach 200 beats/min in a 20-year-old person. At such a high heart rate, stroke volume is lower than at rest due to the very short filling time. The maximal heart rate declines with age; as a rule, subtracting one's age from 220 provides a good estimate of maximal heart rate in beats per minute.

Parasympathetic nerve impulses reach the heart via the right and left **vagus (X) nerves.** Vagal axons terminate in the SA node, AV node, and atrial myocardium. They release acetylcholine, which decreases heart rate by slowing the rate of spontaneous depolarization in autorhythmic fibers. As only a few vagal fibers innervate ventricular muscle, changes in parasympathetic activity have little effect on contractility of the ventricles.

A continually shifting balance exists between sympathetic and parasympathetic stimulation of the heart. At rest, parasympathetic stimulation predominates. The resting heart rate—about 75 beats/min—is usually lower than the autorhythmic rate of the SA node (about 100 beats/min). With maximal stimulation by the parasympathetic division, the heart can slow to 20 or 30 beats/min, or can even stop momentarily.

Chemical Regulation of Heart Rate

Certain chemicals influence both the basic physiology of cardiac muscle and the heart rate. For example, hypoxia (lowered oxygen level), acidosis (low pH), and alkalosis (high pH) all depress cardiac activity. Several hormones and cations have major effects on the heart:

1. ***Hormones.*** Epinephrine and norepinephrine (from the adrenal medullae) enhance the heart's pumping effectiveness. These hormones affect cardiac muscle fibers in much the same way as does norepinephrine released by cardiac accelerator nerves—they increase both heart rate and contractility. Exercise, stress, and excitement cause the adrenal medullae to release more hormones. Thyroid hormones also enhance cardiac contractility and increase heart rate. One sign of hyperthyroidism (excessive thyroid hormone) is **tachycardia,** an elevated resting heart rate.

2. ***Cations.*** Given that differences between intracellular and extracellular concentrations of several cations (for example, Na^+ and K^+) are crucial for the production of action potentials in all nerve and muscle fibers, it is not surprising that ionic imbalances can quickly compromise the pumping effectiveness of the heart. In particular, the relative concentrations of three cations—K^+, Ca^{2+}, and Na^+—have a large effect on cardiac function. Elevated blood levels of K^+ or Na^+ decrease heart rate and contractility. Excess Na^+ blocks Ca^{2+} inflow during cardiac action potentials, thereby decreasing the force of contraction, whereas excess K^+ blocks generation of action potentials. A moderate increase in interstitial (and thus intracellular) Ca^{2+} level speeds heart rate and strengthens the heartbeat.

Other Factors in Heart Rate Regulation

Age, gender, physical fitness, and body temperature also influence resting heart rate. A newborn baby is likely to have a resting heart rate over 120 beats/min; the rate then gradually declines throughout life. Adult females often have slightly higher resting heart rates than adult males, although regular exercise tends to bring resting heart rate down in both sexes. A physically fit person may even exhibit **bradycardia** (brad-ē-KAR-dē-a; *bradys-* = slow), a resting heart rate under 50 beats/min. This is a beneficial effect of endurance-type training because a slowly beating heart is more energy efficient than one that beats more rapidly.

Increased body temperature, as occurs during a fever or strenuous exercise, causes the SA node to discharge impulses more quickly, thereby increasing heart rate. Decreased body temperature decreases heart rate and strength of contraction.

During surgical repair of certain heart abnormalities, it is helpful to slow a patient's heart rate by **hypothermia** (hī-pō-THER-mē-a), in which the person's body is deliberately cooled to a low core temperature. Hypothermia slows metabolism, which reduces the oxygen needs of the tissues, allowing the heart and brain to withstand short periods of interrupted or reduced blood flow during the procedure.

Figure 20.17 summarizes the factors that can increase stroke volume and heart rate to achieve an increase in cardiac output.

▶ CHECKPOINT

19. How is cardiac output calculated?
20. Define stroke volume (SV), and explain the factors that regulate it.
21. What is the Frank–Starling law of the heart? What is its significance?
22. Define cardiac reserve. How does it change with training or with heart failure?
23. How do the sympathetic and parasympathetic divisions of the autonomic nervous system adjust heart rate?

EXERCISE AND THE HEART

▶ OBJECTIVE

Explain the relationship between exercise and the heart.

Regardless of the current level, a person's cardiovascular fitness can be improved at any age with regular exercise. Some types of exercise are more effective than others for improving the health of the cardiovascular system. **Aerobics,** any activity that works large body muscles for at least 20 minutes, elevates cardiac output and accelerates metabolic rate. Three to five such sessions a week are usually recommended for improving the health of the cardiovascular system. Brisk walking, running, bicycling, cross-country skiing, and swimming are examples of aerobic activities.

Sustained exercise increases the oxygen demand of the muscles. Whether the demand is met depends mainly on the adequacy of cardiac output and proper functioning of the respiratory system. After several weeks of training, a healthy person increases maximal cardiac output, thereby increasing the maximal rate of oxygen delivery to the tissues. Oxygen delivery also rises because skeletal muscles develop more capillary networks in response to long-term training.

During strenuous activity, a well-trained athlete can achieve a cardiac output double that of a sedentary person, in part because training causes hypertrophy (enlargement) of the heart. Even though the heart of a well-trained athlete is larger, *resting* cardiac output is about the same as in a healthy untrained person, because stroke volume is increased while heart rate is decreased. The resting heart rate of a trained athlete often is only 40–60 beats per minute *(resting bradycardia)*. Regular exercise

Figure 20.17 Factors that increase cardiac output.

Cardiac output equals stroke volume multiplied by heart rate.

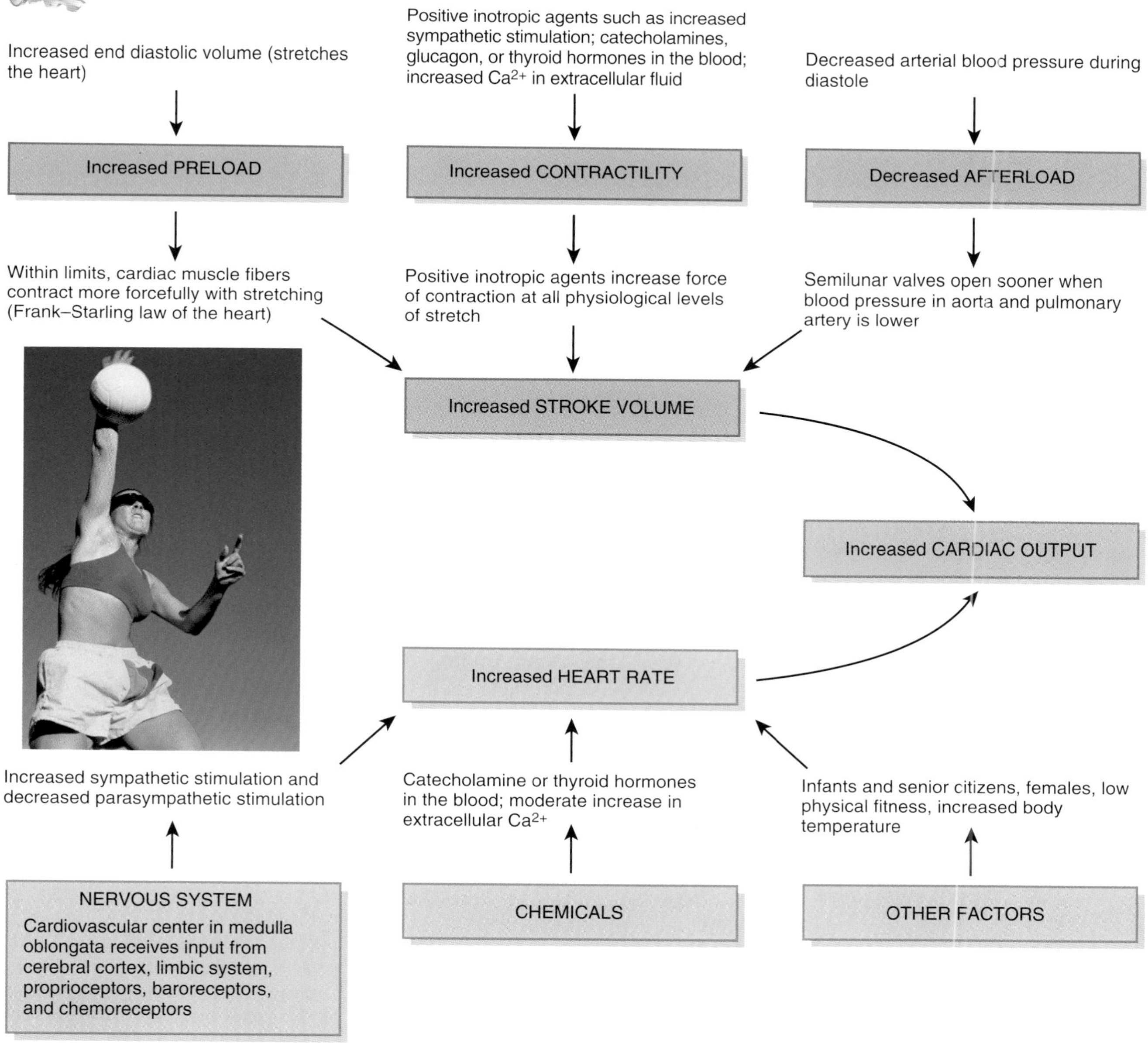

Increased end diastolic volume (stretches the heart)

Increased PRELOAD

Within limits, cardiac muscle fibers contract more forcefully with stretching (Frank–Starling law of the heart)

Positive inotropic agents such as increased sympathetic stimulation; catecholamines, glucagon, or thyroid hormones in the blood; increased Ca^{2+} in extracellular fluid

Increased CONTRACTILITY

Positive inotropic agents increase force of contraction at all physiological levels of stretch

Decreased arterial blood pressure during diastole

Decreased AFTERLOAD

Semilunar valves open sooner when blood pressure in aorta and pulmonary artery is lower

Increased STROKE VOLUME

Increased CARDIAC OUTPUT

Increased HEART RATE

Increased sympathetic stimulation and decreased parasympathetic stimulation

NERVOUS SYSTEM
Cardiovascular center in medulla oblongata receives input from cerebral cortex, limbic system, proprioceptors, baroreceptors, and chemoreceptors

Catecholamine or thyroid hormones in the blood; moderate increase in extracellular Ca^{2+}

CHEMICALS

Infants and senior citizens, females, low physical fitness, increased body temperature

OTHER FACTORS

When you are exercising, contraction of skeletal muscles helps return blood to the heart more rapidly. Would this tend to increase or decrease stroke volume?

also helps to reduce blood pressure, anxiety, and depression; control weight; and increase the body's ability to dissolve blood clots by increasing fibrinolytic activity.

Help for Failing Hearts

As the heart fails, a person has decreasing ability to exercise or even to move around. A variety of surgical techniques and medical devices exist to aid a failing heart. For some patients, even a 10% increase in the volume of blood ejected from the ventricles can mean the difference between being bedridden and having limited mobility. **Heart transplants** are common today and produce good results, but the availability of donor hearts is very limited. There are 50 potential candidates for each of the 2500 donor hearts available each year in the United States. Another approach is the use of **cardiac assist devices** and surgical procedures that augment heart function without removing the heart. Table 20.1 describes several of these.

Finally, scientists continue to develop and refine **artificial hearts,** mechanical devices that completely replace the functions of the natural heart. During the 1980s several patients received a Jarvik-7 artificial heart, which used an external power source to drive an internal pump by compressed air. In 1990, the U.S. Food and Drug Administration (FDA) banned use of the device because persistent problems with blood clotting caused strokes, and the chest tube led to infections. More than a decade later, in July 2001, the first person received a fully self-contained artificial heart called the AbioCor Implantable Replacement Heart. Made of titanium, plastic, and epoxy, the 2-pound AbioCor heart is powered by a battery pack worn outside but with no wires penetrating through the skin. It alternately pumps blood from the left and then the right side of the heart. Because a permanent opening through the chest is not needed, the risk of infection is much lower than with the Jarvik-7. The prognosis of the first recipient was little more than a month due to his congestive heart failure, kidney disease, and diabetes. After surgery, he lived for 151 days (almost 5 months), recovering enough to give several interviews and to enjoy a fishing trip. Internal bleeding and organ failure unrelated to the AbioCor heart caused his death. Since July 2001 several other patients have received the AbioCor device, and its use is still being monitored closely. ■

▶ **CHECKPOINT**

24. What are some of the cardiovascular benefits of regular exercise?

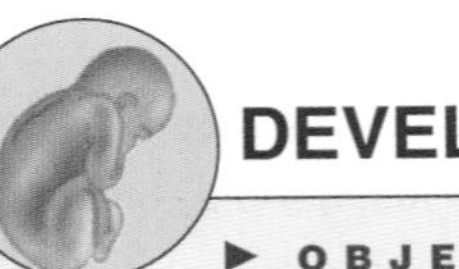

DEVELOPMENT OF THE HEART

▶ **OBJECTIVE**

Describe the development of the heart.

Listening to a fetal heartbeat for the first time is an exciting moment for prospective parents, but it is also an important diagnostic tool. The cardiovascular system is one of the first systems to form in an embryo, and the heart is the first functional organ. This order of development is essential because of the need of the rapidly growing embryo to obtain oxygen and nutrients and get rid of wastes. As you will learn shortly, the development of the heart is a complex process, and any disruptions along the way can result in congenital (present at birth) disorders of the heart. Such disorders, described on page 728, are responsible for almost half of all deaths from birth defects.

The *heart* begins its development from **mesoderm** on day 18 or 19 following fertilization. In the head end of the embryo, the heart develops from a group of mesodermal cells called the **cardiogenic area** (kar-dē-ō-JEN-ik; *cardio-* = heart; *-genic* = producing) (Figure 20.18a). In response to signals from the underlying endoderm, the mesoderm in the cardiogenic area forms a pair of elongated strands called **cardiogenic cords.** Shortly after, these cords develop a hollow center and then become known as **endocardial tubes** (Figure 20.18b). With lateral folding of the embryo, the paired endocardial tubes approach each other and fuse into a single tube called the **primitive heart tube** on day 21 following fertilization (Figure 20.18c).

TABLE 20.1 Cardiac Assist Devices and Procedures

Device	Description
Intra-aortic balloon pump (IABP)	A 40-mL polyurethane balloon mounted on a catheter is inserted into an artery in the groin and threaded into the thoracic aorta. An external pump inflates the balloon with gas at the beginning of ventricular diastole. As the balloon inflates, it pushes blood both backward toward the heart, which improves coronary blood flow, and forward toward peripheral tissues. The balloon then is rapidly deflated just before the next ventricular systole, making it easier for the left ventricle to eject blood. Because the balloon is inflated between heartbeats, this technique is called intra-aortic balloon counterpulsation.
Hemopump	This propeller-like pump is threaded through an artery in the groin and then into the left ventricle. There, the blades of the pump whirl at about 25,000 revolutions per minute, pulling blood out of the left ventricle and pushing it into the aorta.
Left ventricular assist device (LVAD)	The LVAD is a completely portable assist device. It is implanted within the abdomen and powered by a battery pack worn in a shoulder holster. The LVAD is connected to the patient's weakened left ventricle and pumps blood into the aorta. The pumping rate increases automatically during exercise.
Cardiomyoplasty	A large piece of the patient's own skeletal muscle (left latissimus dorsi) is partially freed from its connective tissue attachments and wrapped around the heart, leaving the blood and nerve supply intact. An implanted pacemaker stimulates the skeletal muscle's motor neurons to cause contraction 10–20 times per minute, in synchrony with some of the heartbeats.
Skeletal muscle assist device	A piece of the patient's own skeletal muscle is used to fashion a pouch that is inserted between the heart and the aorta, functioning as a booster heart. A pacemaker stimulates the muscle's motor neurons to elicit contraction.

Figure 20.18 **Development of the heart.** Arrows within the structures indicate the direction of blood flow.

The heart begins its development from a group of mesodermal cells called the cardiogenic area during the third week after fertilization.

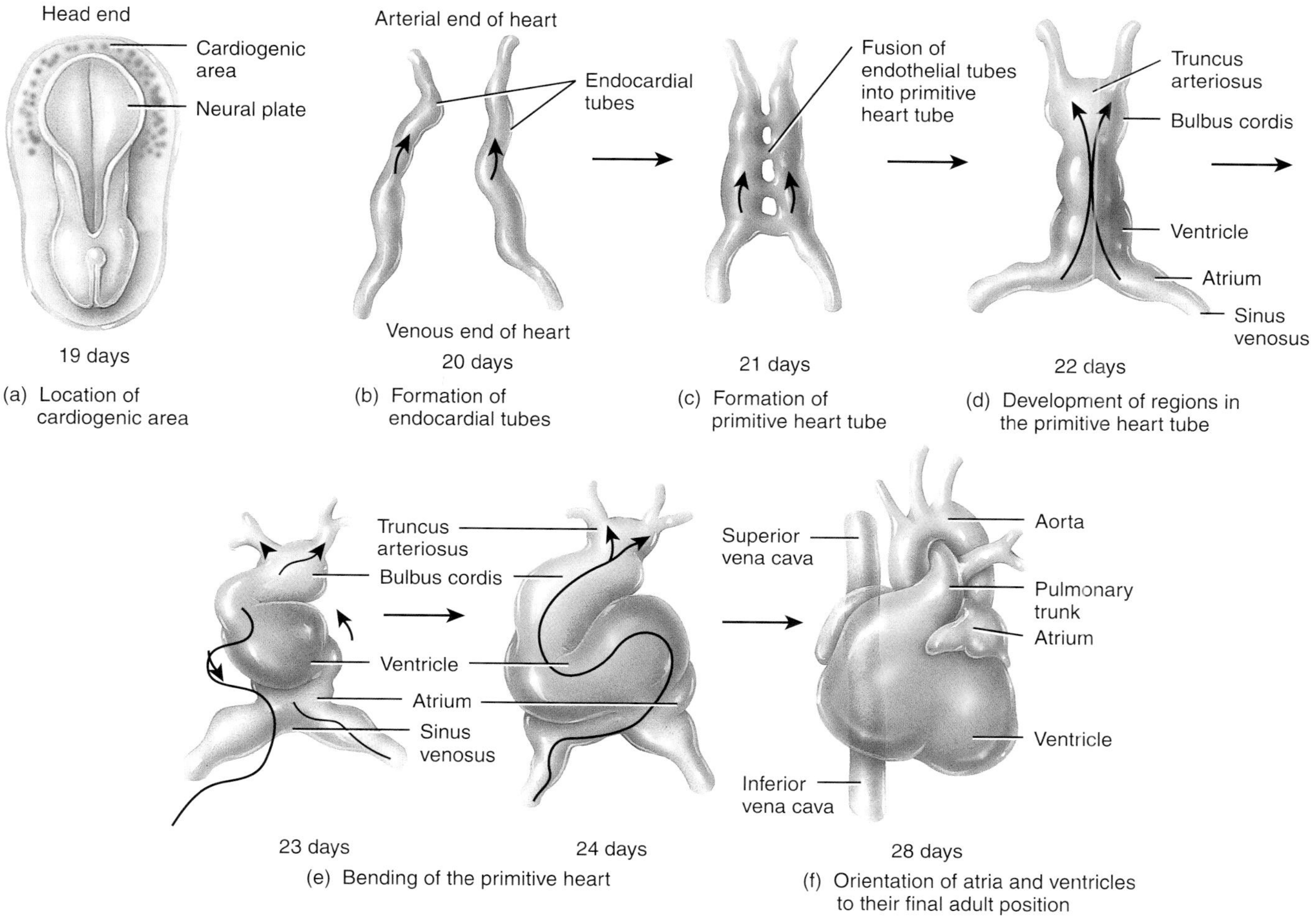

Head end
Cardiogenic area
Neural plate
19 days
(a) Location of cardiogenic area

Arterial end of heart
Endocardial tubes
Venous end of heart
20 days
(b) Formation of endocardial tubes

Fusion of endothelial tubes into primitive heart tube
21 days
(c) Formation of primitive heart tube

Truncus arteriosus
Bulbus cordis
Ventricle
Atrium
Sinus venosus
22 days
(d) Development of regions in the primitive heart tube

Truncus arteriosus
Bulbus cordis
Ventricle
Atrium
Sinus venosus
23 days
24 days
(e) Bending of the primitive heart

Superior vena cava
Aorta
Pulmonary trunk
Atrium
Ventricle
Inferior vena cava
28 days
(f) Orientation of atria and ventricles to their final adult position

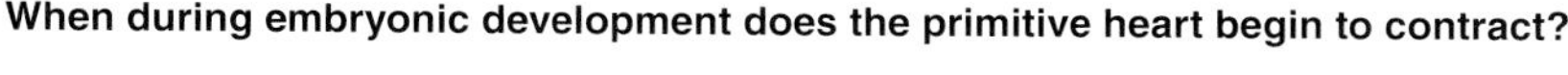

When during embryonic development does the primitive heart begin to contract?

On the twenty-second day, the primitive heart tube develops into five distinct regions and begins to pump blood. From tail end to head end (and the direction of blood flow) they are the (1) **sinus venosus,** (2) **atrium,** (3) **ventricle,** (4) **bulbus cordis,** and (5) **truncus arteriosus.** The sinus venosus initially receives blood from all the veins in the embryo; contractions of the heart begin in this region and follow sequentially in the other regions. Thus, at this stage, the heart consists of a series of unpaired regions. The fates of the regions are as follows:

1. The sinus venosus develops into part of the *right atrium, coronary sinus,* and *sinoatrial (SA) node.*
2. The atrium develops into part of the *right atrium* and the *left atrium.*
3. The ventricle gives rise to the *left ventricle.*
4. The bulbus cordis develops into the *right ventricle.*
5. The truncus arteriosus gives rise to the *ascending aorta* and *pulmonary trunk.*

On day 23, the primitive heart tube elongates. Because the bulbus cordis and ventricle grow more rapidly than other parts of the tube and because the atrial and venous ends of the tube are confined by the pericardium, the tube begins to loop and fold. At first, the primitive heart tube assumes a U-shape; later it becomes S-shaped (Figure 20.18e). As a result of these movements, which are completed by day 28, the atria and ventricles of the future heart are reoriented to assume their final adult positions. The remainder of heart development consists of reconstruction of the chambers and the formation of septa and valves to form a four-chambered heart.

Figure 20.19 Partitioning of the heart into four chambers.

Partitioning the heart begins on about the 28th day after fertilization.

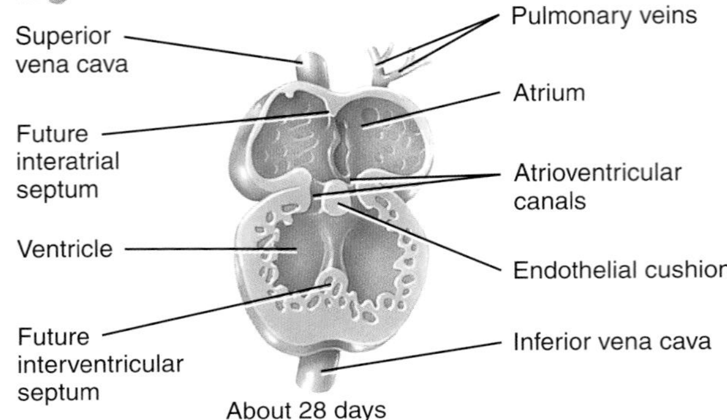

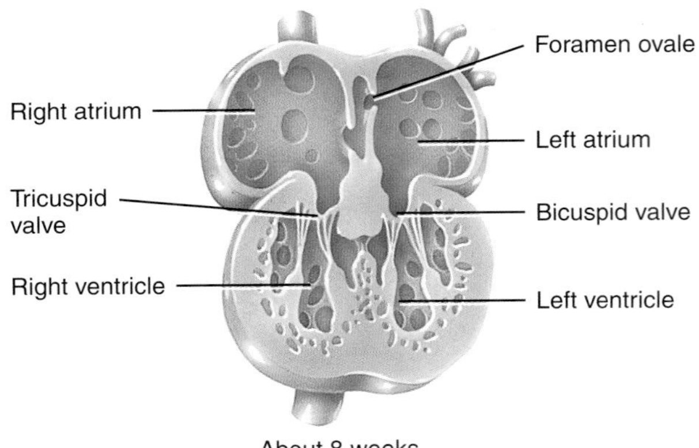

When is the partitioning of the heart complete?

On about day 28, thickenings of mesoderm of the inner lining of the heart wall, called **endocardial cushions,** appear (Figure 20.19). They grow toward each other, fuse, and divide the single **atrioventricular canal** (region between atria and ventricles) into smaller, separate left and right atrioventricular canals. Also, the *interatrial septum* begins its growth toward fused endocardial cushions. Ultimately, the interatrial septum and endocardial cushions unite and an opening in the septum, the **foramen ovale,** develops. The interatrial septum divides the atrial region into a *right atrium* and a *left atrium.* Before birth, the foramen ovale allows most blood entering the right atrium to pass into the left atrium. After birth, it normally closes so that the interatrial septum is a complete partition. The remnant of the foramen ovale is the fossa ovalis (Figure 20.19). Formation of the *interventricular septum* partitions the ventricular region into a *right ventricle* and a *left ventricle.* Partitioning of the atrioventricular canal, atrial region, and ventricular region is basically complete by the end of the fifth week. The *atrioventricular valves* form between the fifth and eighth weeks. The *semilunar valves* form between the fifth and ninth weeks.

▶ **CHECKPOINT**

25. Why is the cardiovascular system one of the first systems to develop?
26. From which tissue does the heart develop?

DISORDERS: HOMEOSTATIC IMBALANCES

Coronary Artery Disease

Coronary artery disease (CAD) is a serious medical problem that affects about 7 million people annually. Responsible for nearly three-quarters of a million deaths in the United States each year, it is the leading cause of death for both men and women. CAD results from the effects of the accumulation of atherosclerotic plaques (described shortly) in coronary arteries, which leads to a reduction in blood flow to the myocardiumSome individuals have no signs or symptoms; others experience angina pectoris (chest pain), and still others suffer heart attacks.

Risk Factors for CAD

People who possess combinations of certain risk factors are more likely to develop CAD. *Risk factors* (characteristics, symptoms, or signs present in a disease-free person that are statistically associated with a greater chance of developing a -disease) include smoking, high blood pressure, diabetes, high cholesterol levels, obesity, "type A" personality, sedentary lifestyle, and a family history of CAD. Most of these can be modified by changing diet and other habits or can be controlled by taking medications. However, other risk factors are unmodifiable—that is, beyond our control—including genetic predisposition (family history of CAD at an early age), age, and gender. For example, adult males are more likely than adult females to develop CAD; after age 70 the risks are roughly equal. Smoking is undoubtedly the number-one risk factor in all CAD-associated diseases, roughly doubling the risk of morbidity and mortality.

Development of Atherosclerotic Plaques

Although the following discussion applies to coronary arteries, the process can also occur in arteries out-side the heart. Thickening of the walls of arteries and loss of elasticity are the main characteristics of a group of diseases called **arteriosclerosis** (ar-tē-rē-ō-skle-RŌ-sis; *sclero-* = hardening). One form of arteriosclerosis is **atherosclerosis** (ath-er-ō-skle-RŌ-sis), a progressive disease characterized by the formation in the walls of large and medium-sized arteries of lesions called ***atherosclerotic plaques*** (Figure 20.20).

To understand how atherosclerotic plaques develop, you will need to learn about the role of molecules produced by the liver and small intestine called **lipoproteins.** These spherical particles consist of an inner core of triglycerides and other lipids and an outer shell of proteins,

Figure 20.20 **Photomicrographs of a transverse section of (a) a normal artery and (b) one partially obstructed by an atherosclerotic plaque.**

Inflammation plays a key role in the development of atherosclerotic plaques.

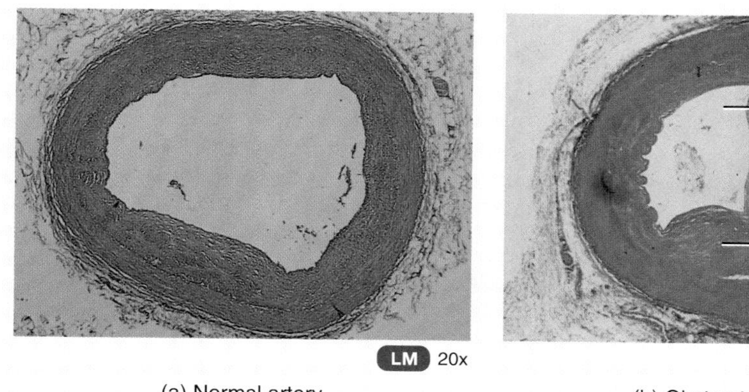

(a) Normal artery (b) Obstructed artery

What is the role of HDL?

phospholipids, and cholesterol. Like most lipids, cholesterol does not dissolve in water and must be made water-soluble in order to be transported in the blood. This is accomplished by combining it with lipoproteins. Two major lipoproteins are **low-density lipoproteins (LDLs)** and **high-density lipoproteins (HDLs).** LDLs transport cholesterol from the liver to body cells for use in cell membrane repair and the production of steroid hormones and bile salts. However, excessive amounts of LDLs promote atherosclerosis, so the cholesterol in these particles is commonly known as "bad cholesterol." HDLs, on the other hand, remove excess cholesterol from body cells and transport it to the liver for elimination. Because HDLs decrease blood cholesterol level, the cholesterol in HDLs is commonly referred to as "good cholesterol." Basically, you want your LDL concentration to be low and your HDL concentration to be high.

It has been learned recently that inflammation, a defensive response of the body to tissue damage, plays a key role in the development of atherosclerotic plaques. As a result of tissue damage, blood vessels dilate and increase their permeability, and phagocytes, including macrophages, appear in large numbers. The formation of atherosclerotic plaques begins when excess LDLs from the blood accumulate in the inner layer of an artery wall (layer closest to the bloodstream) and the lipids and proteins in the LDLs undergo oxidation and the proteins also bind to sugars. In response, endothelial and smooth muscle cells of the artery secrete substances that attract monocytes from the blood and convert them into macrophages. The macrophages then ingest and become so filled with oxidized LDL particles that they have a foamy appearance when viewed microscopically **(foam cells).** T cells (lymphocytes) follow monocytes into the inner lining of an artery where they release chemicals that intensify the inflammatory response. Together, the foam cells, macrophages, and T cells form a fatty streak, the beginning of an atherosclerotic plaque.

Macrophages secrete chemicals that cause smooth muscle cells of the middle layer of an artery to migrate to the top of the atherosclerotic plaque, forming a cap over it and thus walling it off from the blood.

Because most atherosclerotic plaques expand away from the bloodstream rather than into it, blood can flow through an artery with relative ease, often for decades. Relatively few heart attacks occur when plaque in a coronary artery expands into the bloodstream and restricts blood flow. Most heart attacks occur when the cap over the plaque breaks open in response to chemicals produced by foam cells. In addition, T cells induce foam cells to produce tissue factor (TF), a chemical that begins the cascade of reactions that result in blood clot formation. If the clot in a coronary artery is large enough, it can significantly decrease or stop the flow of blood and result in a heart attack.

In recent years, a number of new risk factors (all modifiable) have been identified as significant predictors of CAD. **C-reactive proteins (CRPs)** are proteins produced by the liver or present in blood in an inactive form that are converted to an active form during inflammation. CRPs may play a direct role in the development of atherosclerosis by promoting the uptake of LDLs by macrophages. **Lipoprotein (a)** is an LDL-like particle that binds to endothelial cells, macrophages, and blood platelets, may promote the proliferation of smooth muscle fibers, and inhibits the breakdown of blood clots. **Fibrinogen** is a glycoprotein involved in blood clotting that may help regulate cellular proliferation, vasoconstriction, and platelet aggregation. **Homocysteine** is an amino acid that may induce blood vessel damage by promoting platelet aggregation and smooth muscle fiber proliferation.

Diagnosis of CAD

Many procedures may be employed to diagnose CAD; the specific procedure used will depend on the signs and symptoms of the individual.

In addition to a resting electrocardiogram (see page 713), the standard test employed to diagnose CAD, **stress testing** can be performed. In an *exercise stress test,* the functioning of the heart is monitored when placed under physical stress by exercising using a treadmill, an exercise bicycle, or arm exercises. During the procedure, ECG recordings are monitored continuously and blood pressure is monitored at intervals. A *nonexercise (pharmacologic) stress test* is used for individuals who cannot exercise due to conditions such as arthritis. A medication is injected that stresses the heart mimicking the effects of exercise. During both exercise and nonexercise stress testing, **radionuclide**

imaging may be performed to evaluate blood flow through heart muscle (see page 23).

Diagnosis of CAD may also involve **echocardiography** (ek′-ō-kar-dē-OG-ra-fē), a technique that uses ultrasound waves to image the interior of the heart. Echocardiography allows the heart to be seen in motion and can be used to determine the size, shape, and functions of the chambers of the heart; the volume and velocity of blood pumped from the heart; the status of heart valves; the presence of birth defects; and abnormalities of the pericardium. A fairly recent technique for evaluating CAD is **electron beam computerized tomography (EBCT),** which detects calcium deposits in coronary arteries. These calcium deposits are indicators of atherosclerosis.

Cardiac catheterization (kath′-e-ter-i-ZĀ-shun) is an invasive procedure used to visualize the heart's chambers, valves, and great vessels in order to diagnose and treat disease not related to abnormalities of the coronary arteries. It may also be used to measure pressure in the heart and great vessels; to assess cardiac output; to measure the flow of blood through the heart and great vessels; to identify the location of septal and valvular defects; and to take tissue and blood samples. The basic procedure involves inserting a long, flexible, radiopaque **catheter** (plastic tube) into a peripheral vein (for *right heart catheterization*) or a peripheral artery (for *left heart catheterization*) and guiding it under fluoroscopy (x-ray observation).

Coronary angiography (an′-jē-OG-ra-fē; *angio-* = blood vessel; *-grapho* = to write) is another invasive procedure, but this one is used to obtain information about the coronary arteries. In the procedure, a catheter is inserted into an artery in the groin or wrist and threaded under fluroscopy toward the heart and then into the coronary arteries. After the tip of the catheter is in place, a radiopaque contrast medium is injected into the coronary arteries. The radiographs of the arteries, called *angiograms,* appear in motion on a monitor and the information is recorded on a videotape or computer disc. Coronary angiography may be used to visualize coronary arteries (see page 22) and to inject clot-dissolving drugs, such as streptokinase or tissue plasminogen activator (t-PA), into a coronary artery to dissolve an obstructing thrombus.

Treatment of CAD

Treatment options for CAD include **drugs** (antihypertensives, nitroglycerine, beta blockers, cholesterol-lowering drugs, and clot-dissolving agents) and various surgical and nonsurgical procedures designed to increase the blood supply to the heart.

Coronary artery bypass grafting (CABG) is a surgical procedure in which a blood vessel from another part of the body is attached ("grafted") to a coronary artery to bypass an area of blockage. A piece of the grafted blood vessel is sutured between the aorta and the unblocked portion of the coronary artery (Figure 20.21a).

A nonsurgical procedure used to treat CAD is **percutaneous transluminal coronary angioplasty (PTCA)** (percutaneous = through the skin; *trans-* = across; *lumen* = an opening or channel in a tube; *angio-* = blood vessel; *-plasty* = to mold or to shape). In this procedure, a balloon catheter is inserted into an artery of an arm or leg and gently guided into a coronary artery (Figure 20.21b). While dye is released, angiograms (x rays of blood vessels) are taken to locate the plaques. Next, the catheter is advanced to the point of obstruction, and a balloonlike device is inflated with air to squash the plaque against the blood vessel wall. Because 30–50% of PTCA-opened arteries fail due to restenosis (renarrowing) within six months after the procedure is done, a stent may be inserted via a catheter. A **stent** is a metallic, fine wire tube that is permanently placed in an artery to keep the artery patent (open), permitting blood to circulate (Figure 20.21c, d). Restenosis may be due to damage from the procedure itself, for PTCA may damage the arterial wall, leading to platelet activation, proliferation of smooth muscle fibers, and plaque formation. Recently, *drug-coated (drug-eluting) coronary stents* have been used to prevent restenosis. The stents are coated with one of several antiproliferative drugs (drugs that inhibit the proliferation of smooth muscle fibers of the middle layer of an artery) and anti-inflammatory drugs. It has been shown that drug-coated stents reduce the rate of restenosis when compared to bare-metal (non-coated) stents.

One area of current research involves cooling the body's core temperature during procedures such as coronary artery bypass grafting (CABG) and during other vascular disease processes or procedures. There have been some promising results from the application of cold therapy during a cerebral vascular accident (CVA or stroke). This research stemmed from observations of people who had suffered a hypothermic incident (such as cold water drowning) and recovered with relatively minimal neurologic deficits.

Congenital Heart Defects

A defect that is present at birth, and usually before, is called a **congenital defect.** Many such defects are not serious and may go unnoticed for a lifetime. Others are life threatening and must be surgically repaired. Among the several congenital defects that affect the heart are the following (Figure 20.22 on page 730):

- **Coarctation** (kō′-ark-TĀ-shun) **of the aorta.** In this condition, a segment of the aorta is too narrow, and thus the flow of oxygenated blood to the body is reduced, the left ventricle is forced to pump harder, and high blood pressure develops. Coarctation is usually repaired surgically by removing the area of obstruction. Surgical interventions that are done in childhood may require revisions in adulthood. Another surgical procedure is balloon dilation, insertion and inflation of a device in the aorta to stretch the vessel. A stent can be inserted and left in place to hold the vessel open.
- **Patent ductus arteriosus (PDA).** In some babies, the ductus arteriosus, a temporary blood vessel between the aorta and the pulmonary trunk, remains open rather than closing shortly after birth. As a result, aortic blood flows into the lower-pressure pulmonary trunk, thus increasing the pulmonary trunk blood pressure and overworking both ventricles. In uncomplicated PDA, medication can be used to facilitate the closure of the defect. In more severe cases, surgical intervention may be required.
- **Septal defect.** A septal defect is an opening in the septum that separates the interior of the heart into left and right sides. In an **atrial septal defect** the fetal foramen ovale between the two atria fails to close after birth. A **ventricular septal defect** is caused by incomplete development of the interventricular septum. In such cases, oxygenated blood flows directly from the left ventricle into the right ventricle, where it mixes with deoxygenated blood. The condition is treated surgically.
- **Tetralogy of Fallot** (tet-RAL-ō-jē of fal-Ō). This condition is a combination of four developmental defects: an interventricular septal defect, an aorta that emerges from both ventricles instead of from the left ventricle only, a stenosed pulmonary valve, and an enlarged right

Figure 20.21 Procedures for reestablishing blood flow in occluded coronary arteries.

Treatment options for CAD include drugs and various nonsurgical and surgical procedures.

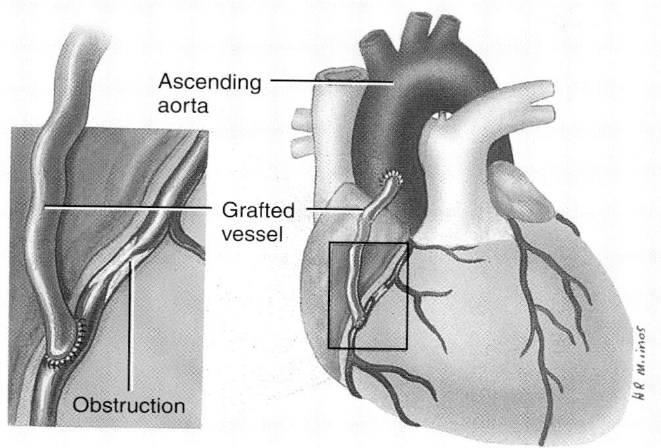

(a) Coronary artery bypass grafting (CABG)

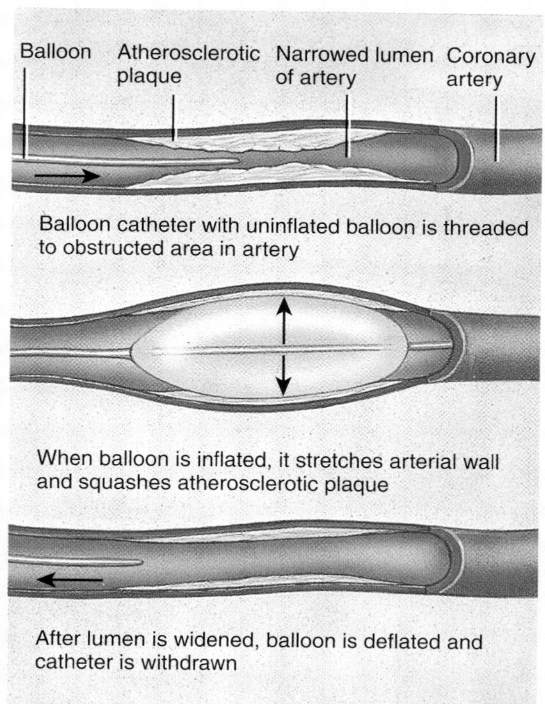

Balloon
Atherosclerotic plaque
Narrowed lumen of artery
Coronary artery
Balloon catheter with uninflated balloon is threaded to obstructed area in artery
When balloon is inflated, it stretches arterial wall and squashes atherosclerotic plaque
After lumen is widened, balloon is deflated and catheter is withdrawn

(b) Percutaneous transluminal coronary angioplasty (PTCA)

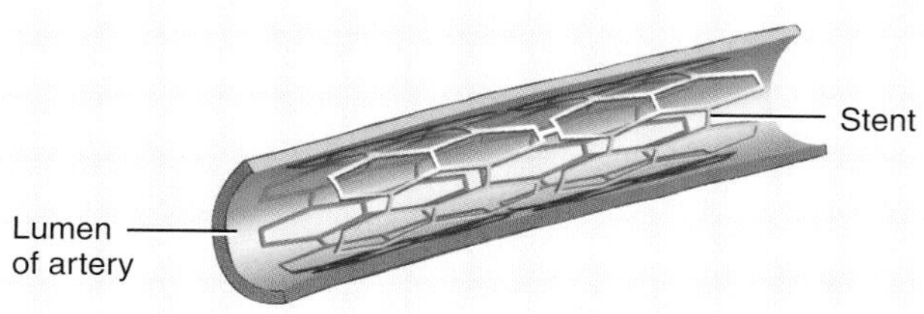

(c) Stent in an artery

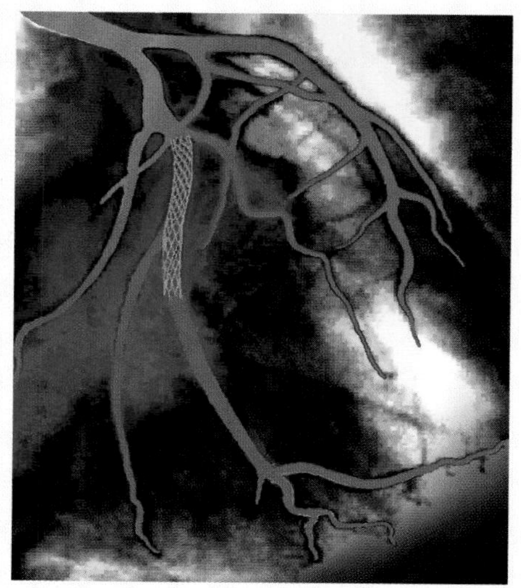

(d) Angiogram showing a stent in the circumflex artery

Which diagnostic procedure for CAD is used to visualize coronary blood vessels?

ventricle. There is a decreased flow of blood to the lungs and mixing of blood from both sides of the heart. This causes cyanosis, the bluish discoloration most easily seen in nail beds and mucous membranes when the level of deoxygenated hemoglobin is high; in infants, this condition is referred to as "blue baby." Despite the apparent complexity of this condition, surgical repair is usually successful.

Arrhythmias

The usual rhythm of heartbeats, established by the SA node, is called **normal sinus rhythm.** The term **arrhythmia** (a-RITH-mē-a) or **dysrhythmia** refers to an abnormal rhythm as a result of a defect in the conduction system of the heart. The heart may beat irregularly, too fast, or too slow. Symptoms include chest pain, shortness of breath, lightheadedness, dizziness, and fainting. Arrhythmias may be caused by factors that stimulate the heart such as stress, caffeine, alcohol, nicotine, cocaine, and certain drugs that contain caffeine or other stimulants. Arrhythmias may also be caused by a congenital defect, coronary artery disease, myocardial infarction, hypertension, defective heart valves, rheumatic heart disease, hyperthyroidism, and potassium deficiency.

Arrhythmias are categorized by their speed, rhythm, and origination of the problem. **Bradycardia** (brăd-e-KAR-dē-a; *brady-* = slow) refers to a slow heart rate (below 50 beats per minute); **tachycardia** (tak-i-KAR-dē-a; *tachy-* = quick) refers to a rapid heart rate (over 100 beats per minute); and **fibrillation** (fi-bri-LĀ-shun) refers to rapid, uncoordinated heart beats. Arrhythmias that begin in the atria are called **supraventricular** or **atrial arrhythmias;** those that originate in the ventricles are called **ventricular arrhythmias.**

- **Supraventricular tachycardia (SVT)** is a rapid but regular heart rate (160–200 beats) per minute that originates in the atria. The episodes begin and end suddenly and may last from a few minutes to many hours. SVTs can sometimes be stopped by maneuvers that stimulate the vagus (X) nerve and decrease heart rate. These include straining as if having a difficult bowel movement, rubbing the area over the carotid artery in the neck to stimulate the carotid sinus (not recommended for people over 50 since it may cause a stroke), and

Figure 20.22 Congenital heart defects.

A congenital defect is one that is present at birth, and usually before.

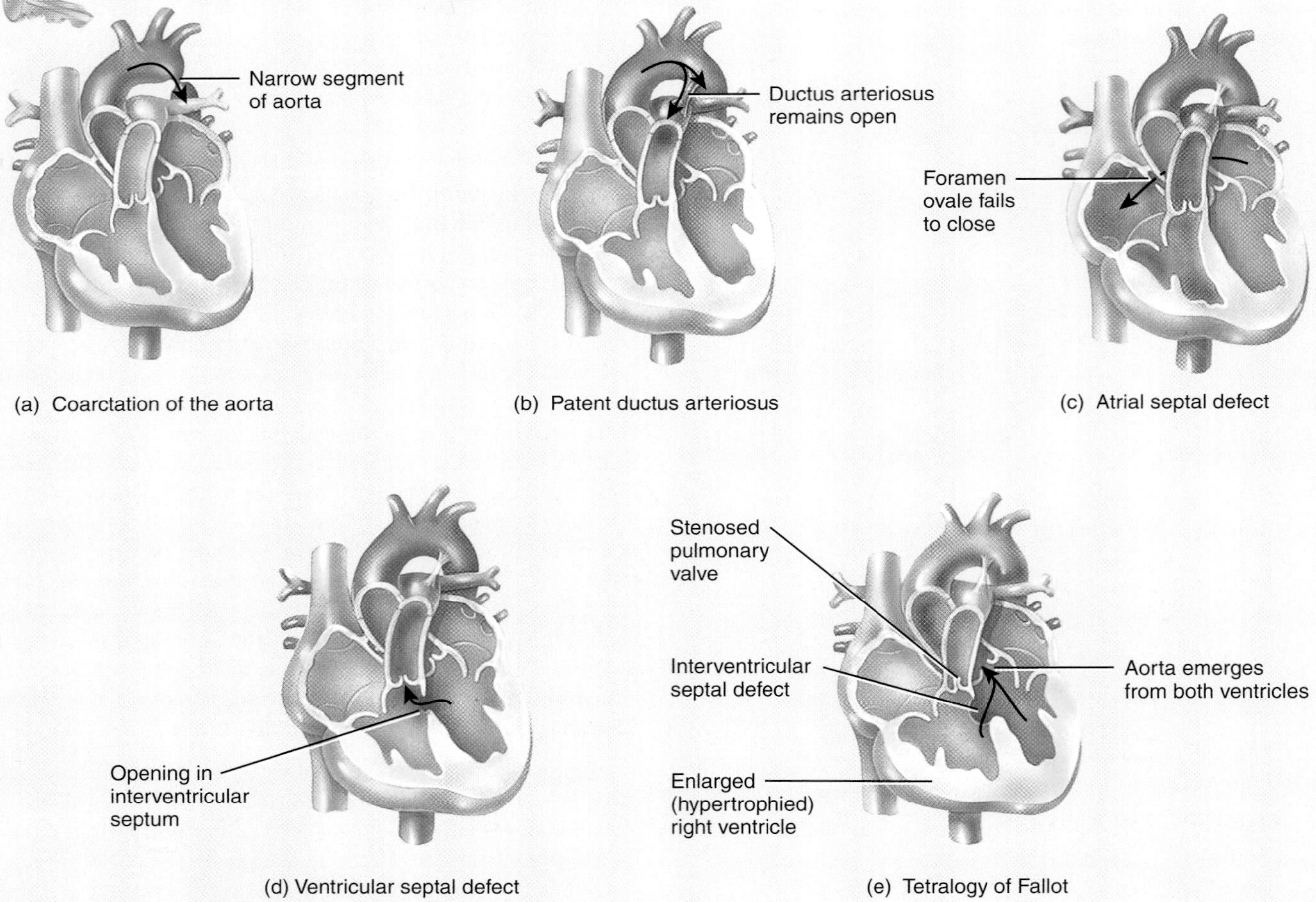

(a) Coarctation of the aorta

(b) Patent ductus arteriosus

(c) Atrial septal defect

(d) Ventricular septal defect

(e) Tetralogy of Fallot

Which four developmental defects occur in tetralogy of Fallot?

plunging the face into a bowl of ice-cold water. Treatment may also involve antiarrhythmic drugs and destruction of the abnormal pathway by radiofrequency ablation.

- **Heart block** is an arrhythmia that occurs when the electrical pathways between the atria and ventricles are blocked, slowing the transmission of nerve impulses. The most common site of blockage is the atrioventricular node, a condition called *atrioventricular (AV) block.* In *first-degree AV block,* the P-Q interval is prolonged, usually because conduction through the AV node is slower than normal. In *second-degree AV block,* some of the action potentials from the SA node are not conducted through the AV node. The result is "dropped" beats because excitation doesn't always reach the ventricles. Consequently, there are fewer QRS complexes than P waves on the ECG. In *third-degree (complete) AV block,* no SA node action potentials get through the AV node. Autorhythmic fibers in the atria and ventricles pace the upper and lower chambers separately. With complete AV block, the ventricular contraction rate is less than 40 beats/min.
- **Atrial flutter** consists of rapid, regular atrial contractions (240–360 beats/min) accompanied by an atrioventricular (AV) block in which some of the nerve impulses from the SA node are not conducted through the AV node.
- **Atrial fibrillation** is a common arrhythmia, affecting mostly older adults, in which contraction of the atrial fibers is asynchronous (not in unison) so that atrial pumping ceases altogether. The atria may beat 300–600 beats/min. The ventricles may also speed up, resulting in a rapid heartbeat (up to 160 beats/min). The ECG of an individual with atrial fibrillation typically has no clearly defined P waves and the QRS complexes are irregularly paced. Since the atria and ventricles do not beat in rhythm, the heart beat is irregular in timing and strength. In an otherwise strong heart, atrial fibrillation reduces the pumping effectiveness of the heart by 20–30%. The most dangerous complication of atrial fibrillation is stroke since blood may stagnate in the atria and form blood clots. A stroke occurs when part of a blood clot occludes an artery supplying the brain.
- **Ventricular tachycardia (VT)** is an arrhythmia that originates in the ventricles and causes the ventricles to beat too fast (at least 120 beats/min). VT is almost always associated with heart disease or a recent myocardial infarction and may develop in a very seri-

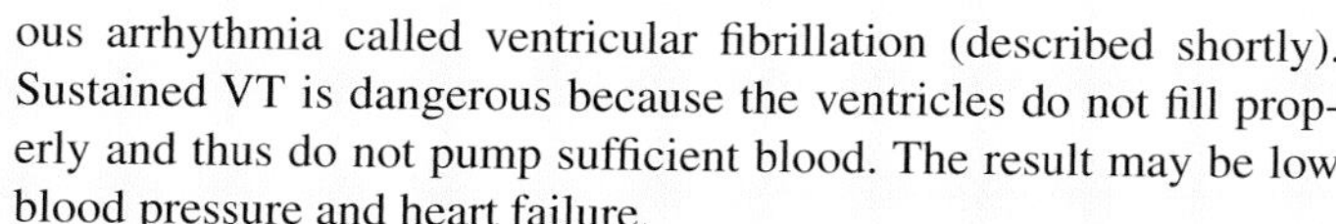

ous arrhythmia called ventricular fibrillation (described shortly). Sustained VT is dangerous because the ventricles do not fill properly and thus do not pump sufficient blood. The result may be low blood pressure and heart failure.

- **Ventricular fibrillation** is the most deadly arrhythmia, in which contractions of the ventricular fibers are completely asynchronous so that the ventricles quiver rather than contract in a coordinated way. As a result, ventricular pumping stops, blood ejection ceases, and circulatory failure and death occurs unless there is immediate medical intervention. During ventricular fibrillation, the ECG has no clearly defined P waves, QRS complexes, or T waves. The most common cause of ventricular fibrillation is inadequate blood flow to the heart due to coronary artery disease, as occurs during a myocardial infarction. Other causes are cardiovascular shock, electrical shock, drowning, and very low potassium levels. Ventricular fibrillation causes unconsciousness in seconds and, if untreated, seizures occur and irreversible brain damage may occur after five minutes. Death soon follows. Treatment involves cardiopulmonary resuscitation (CPR) and defibrillation. In **defibrillation** (dē-fib-re-LĀ-shun), also called **cardioversion** (kar′-dē-ō-VER-shun), a strong, brief electrical current is passed to the heart and often can stop the ventricular fibrillation. The electrical shock is generated by a device called a **defibrillator** (de-FIB-ri-lā-tor) and applied via two large paddle-shaped electrodes pressed against the skin of the chest. Patients who face a high risk of dying from heart rhythm disorders now can receive an **automatic implantable cardioverter defibrillator (AICD),** an implanted device that monitors their heart rhythm and delivers a small shock directly to the heart when a life-threatening rhythm disturbance occurs. Thousands of patients around the world have AICDs, including Dick Cheney, Vice President of the United States, who received a combination pacemaker-defibrillator in 2001. Also available are **automated external defibrillators (AEDs)** that function like AICDs, except that they are external devices. About the size of a laptop computer, AEDs are used by emergency response teams and are found increasingly in public places such as stadiums, casinos, airports, hotels, and shopping malls. Defibrillation may also be used as an emergency treatment for cardiac arrest.
- **Ventricular premature contraction.** Another form of arrhythmia arises when an *ectopic focus* (ek-TOP-ik), a region of the heart other than the conduction system, becomes more excitable than normal and causes an occasional abnormal action potential to occur. As a wave of depolarization spreads outward from the ectopic focus, it causes a **ventricular premature contraction (beat).** The contraction occurs early in diastole before the SA node is normally scheduled to discharge its action potential. Ventricular premature contractions may be relatively benign and may be caused by emotional stress, excessive intake of stimulants such as caffeine, alcohol, or nicotine, and lack of sleep. In other cases, the premature beats may reflect an underlying pathology.

MEDICAL TERMINOLOGY

Asystole (ā-SIS-tō-lē; *a-* = without) Failure of the myocardium to contract.

Cardiac arrest (KAR-dē-ak a-REST) A clinical term meaning cessation of an effective heartbeat. The heart may be completely stopped or in ventricular fibrillation.

Cardiomegaly (kar′-dē-ō-MEG-a-lē; *mega* = large) Heart enlargement.

Cardiac rehabilitation (rē-ha-bil-i-TĀ-shun) A supervised program of progressive exercise, psychological support, education, and training to enable a patient to resume normal activities following a myocardial infarction.

Cardiomyopathy (kar′-dē-ō-mī-OP-a-thē; *myo-* = muscle; *-pathos* = disease) A progressive disorder in which ventricular structure or function is impaired. In dilated cardiomyopathy, the ventricles enlarge (stretch) and become weaker and reduce the heart's pumping action. In hypertrophic cardiomyopathy, the ventricular walls thicken and the pumping efficiency of the ventricles is reduced.

Commotio cordis (kō-MŌ-shē-ō KOR-dis; *commotio-* = disturbance; *cordis* = heart) Damage to the heart, frequently fatal, as a result of a sharp, nonpenetrating blow to the chest while the ventricles are repolarizing.

Cor pulmonale (CP) (kor pul-mōn-ALE; *cor-* = heart; *pulmon-* = lung) A term referring to right ventricular hypertrophy from dis-orders that bring about hypertension (high blood pressure) in the pulmonary circulation.

Ejection fraction The fraction of the end-diastolic volume (EDV) that is ejected during an average heartbeat. Equal to stroke volume (SV) divided by EDV.

Electrophysiological testing (e-lek′-trō-fiz′-ē-OL-ō-ji-kal) A procedure in which a catheter with an electrode is passed through blood vessels and introduced into the heart. It is used to detect the exact locations of abnormal electrical conduction pathways. Once an abnormal pathway is located, it can be destroyed by sending a current through the electrode, a procedure called *radiofrequency ablation.*

Palpitation (pal′-pi-TĀ-shun) A fluttering of the heart or an abnormal rate or rhythm of the heart about which an individual is aware.

Paroxysmal tachycardia (par′-ok-SIZ-mal tak′-e-KAR-dē-a; *tachy-* = quick) A period of rapid heartbeats that begins and ends suddenly.

Sick sinus syndrome An abnormally functioning SA node that initiates heartbeats too slowly or rapidly, pauses too long between heartbeats, or stops producing heartbeats. Symptoms include lightheadedness, shortness of breath, loss of consciousness, and palpitations. It is caused by degeneration of cells in the SA node and is common in elderly persons. It is also related to coronary artery disease. Treatment consists of drugs to speed up or slow down the heart or implantation of an artificial pacemaker.

Sudden cardiac death The unexpected cessation of circulation and breathing due to an underlying heart disease such as ischemia, myocardial infarction, or a disturbance in cardiac rhythm.

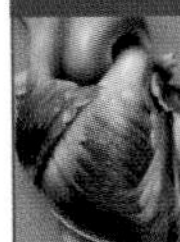

STUDY OUTLINE

ANATOMY OF THE HEART (p. 696)

1. The heart is located in the mediastinum; about two-thirds of its mass is to the left of the midline.
2. The heart is shaped like a cone lying on its side; its apex is the pointed, inferior part, whereas its base is the broad, superior part.
3. The pericardium is the membrane that surrounds and protects the heart; it consists of an outer fibrous layer and an inner serous pericardium, which is composed of a parietal and a visceral layer.
4. Between the parietal and visceral layers of the serous pericardium is the pericardial cavity, a potential space filled with a few milliliters of pericardial fluid that reduces friction between the two membranes.
5. Three layers make up the wall of the heart: epicardium (visceral layer of the serous pericardium), myocardium, and endocardium.
6. The epicardium consists of mesothelium and connective tissue, the myocardium is composed of cardiac muscle tissue, and the endocardium consists of endothelium and connective tissue.
7. The heart chambers include two superior chambers, the right and left atria, and two inferior chambers, the right and left ventricles.
8. External features of the heart include the auricles (flaps on each atrium that slightly increase their volume), the coronary sulcus between the atria and ventricles, and the anterior and posterior sulci between the ventricles on the anterior and posterior surfaces of the heart, respectively.
9. The right atrium receives blood from the superior vena cava, inferior vena cava, and coronary sinus. It is separated from the left atrium by the interatrial septum, which contains the fossa ovalis. Blood exits the right atrium through the tricuspid valve.
10. The right ventricle receives blood from the right atrium. It is separated from the left ventricle by the interventricular septum and pumps blood through the pulmonary valve into the pulmonary trunk.
11. Oxygenated blood enters the left atrium from the pulmonary veins and exits through the bicuspid (mitral) valve.
12. The left ventricle pumps oxygenated blood through the aortic valve into the aorta.
13. The thickness of the myocardium of the four chambers varies according to the chamber's function. The left ventricle, with the highest workload, has the thickest wall.
14. The fibrous skeleton of the heart is dense connective tissue that surrounds and supports the valves of the heart.

HEART VALVES AND CIRCULATION OF BLOOD (p. 704)

1. Heart valves prevent backflow of blood within the heart. The atrioventricular (AV) valves, which lie between atria and ventricles, are the tricuspid valve on the right side of the heart and the bicuspid (mitral) valve on the left. The semilunar (SL) valves are the aortic valve, at the entrance to the aorta, and the pulmonary valve, at the entrance to the pulmonary trunk.
2. The left side of the heart is the pump for the systemic circulation, the circulation of blood throughout the body except for the air sacs of the lungs. The left ventricle ejects blood into the aorta, and blood then flows into systemic arteries, arterioles, capillaries, venules, and veins, which carry it back to the right atrium.
3. The right side of the heart is the pump for pulmonary circulation, the circulation of blood through the lungs. The right ventricle ejects blood into the pulmonary trunk, and blood then flows into pulmonary arteries, pulmonary capillaries, and pulmonary veins, which carry it back to the left atrium.
4. The coronary circulation provides blood flow to the myocardium. The main arteries of the coronary circulation are left and right coronary arteries; the main veins are the cardiac vein and the coronary sinus.

CARDIAC MUSCLE TISSUE AND THE CARDIAC CONDUCTION SYSTEM (p. 708)

1. Cardiac muscle fibers usually contain a single centrally located nucleus. Compared to skeletal muscle fibers, cardiac muscle fibers have more and larger mitochondria, slightly smaller sarcoplasmic reticulum, and wider transverse tubules, which are located at Z discs.
2. Cardiac muscle fibers are connected via end-to-end intercalated discs. Desmosomes in the discs provide strength and gap junctions allow muscle action potentials to conduct from one muscle fiber to its neighbors.
3. Autorhythmic fibers form the conduction system, cardiac muscle fibers that spontaneously depolarize and generate action potentials.
4. Components of the conduction system are the sinoatrial (SA) node (pacemaker), atrioventricular (AV) node, atrioventricular (AV) bundle (bundle of His), bundle branches, and Purkinje fibers.
5. Phases of an action potential in a ventricular contractile fiber include rapid depolarization, a long plateau, and repolarization.
6. Cardiac muscle tissue has a long refractory period, which prevents tetanus.
7. The record of electrical changes during each cardiac cycle is called an electrocardiogram (ECG). A normal ECG consists of a P wave (atrial depolarization), a QRS complex (onset of ventricular depolarization), and a T wave (ventricular repolarization).
8. The P-Q interval represents the conduction time from the beginning of atrial excitation to the beginning of ventricular excitation. The S-T segment represents the time when ventricular contractile fibers are fully depolarized.

THE CARDIAC CYCLE (p. 716)

1. A cardiac cycle consists of the systole (contraction) and diastole (relaxation) of both atria, plus the systole and diastole of both ventricles. With an average heartbeat of 75 beats/min, a complete cardiac cycle requires 0.8 seconds.
2. The phases of the cardiac cycle are (a) atrial systole, (b) ventricular systole, and (c) relaxation period.
3. S1, the first heart sound (lubb), is caused by blood turbulence associated with the closing of the atrioventricular valves. S2, the second sound (dupp), is caused by blood turbulence associated with the closing of semilunar valves.

CARDIAC OUTPUT (p. 719)

1. Cardiac output (CO) is the amount of blood ejected per minute by the left ventricle into the aorta (or by the right ventricle into the pulmonary trunk). It is calculated as follows: CO (mL/min) = stroke volume (SV) in mL/beat × heart rate (HR) in beats per minute.

2. Stroke volume (SV) is the amount of blood ejected by a ventricle during each systole.
3. Cardiac reserve is the difference between a person's maximum cardiac output and his or her cardiac output at rest.
4. Stroke volume is related to preload (stretch on the heart before it contracts), contractility (forcefulness of contraction), and afterload (pressure that must be exceeded before ventricular ejection can begin).
5. According to the Frank–Starling law of the heart, a greater preload (end-diastolic volume) stretching cardiac muscle fibers just before they contract increases their force of contraction until the stretching becomes excessive.
6. Nervous control of the cardiovascular system originates in the cardiovascular center in the medulla oblongata.
7. Sympathetic impulses increase heart rate and force of contraction; parasympathetic impulses decrease heart rate.
8. Heart rate is affected by hormones (epinephrine, norepinephrine, thyroid hormones), ions (Na^+, K^+, Ca^{2+}), age, gender, physical fitness, and body temperature.

EXERCISE AND THE HEART (p. 722)

1. Sustained exercise increases oxygen demand on muscles.
2. Among the benefits of aerobic exercise are increased cardiac output, decreased blood pressure, weight control, and increased fibrinolytic activity.

DEVELOPMENT OF THE HEART (p. 724)

1. The heart develops from mesoderm.
2. The endocardial tubes develop into the four-chambered heart and great vessels of the heart.

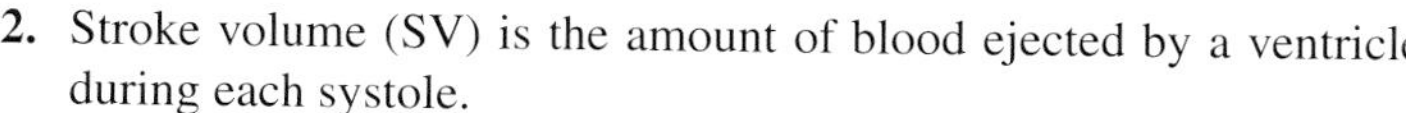

SELF-QUIZ QUESTIONS

Fill in the blanks in the following statements.

1. The chamber of the heart with the thickest myocardium is the ____.
2. The phase of heart contraction is called ____; the phase of relaxation is called ____.

Indicate whether the following statements are true or false.

3. In auscultation, the lubb represents closing of the semilunar valves and the dupp represents closing of the atrioventricular valves.
4. The Frank-Starling law of the heart equalizes the output of the right and left ventricles and keeps the same volume of blood flowing to both the systemic and pulmonary circulations.

Choose the one best answer to the following questions.

5. Which of the following is the correct route of blood through the heart from the systemic circulation to the pulmonary circulation and back to the systemic circulation? (a) right atrium, tricuspid valve, right ventricle, pulmonary semilunar valve, left atrium, mitral valve, left ventricle, aortic semilunar valve, (b) left atrium, tricuspid valve, left ventricle, pulmonary semilunar valve, right atrium, mitral valve, right ventricle, aortic semilunar valve, (c) left atrium, pulmonary semilunar valve, right atrium, tricuspid valve, left ventricle, aortic semilunar valve, right ventricle, mitral valve, (d) left ventricle, mitral valve, left atrium, pulmonary semilunar valve, right ventricle, tricuspid valve, right atrium, aortic semilunar valve, (e) right atrium, mitral valve, right ventricle, pulmonary semilunar valve, left atrium, tricuspid valve, left ventricle, aortic semilunar valve.
6. Which of the following represents the correct pathway for conduction of an action potential through the heart? (a) AV node, AV bundle, SA node, Purkinje fibers, bundle branches, (b) AV node, bundle branches, AV bundle, SA node, Purkinje fibers, (c) SA node, AV node, AV bundle, bundle branches, Purkinje fibers, (d) SA node, AV bundle, bundle branches, AV node, Purkinje fibers, (e) SA node, AV node, Purkinje fibers, bundle branches, AV bundle.
7. The external boundary between the atria and ventricles is the (a) anterior interventricular sulcus, (b) interventricular septum, (c) interatrial septum, (d) coronary sulcus, (e) posterior interventricular sulcus.
8. A softball player is found to have a resting cardiac output of 5.0 liters per minute and a heart rate of 50 beats per minute. What is her stroke volume? (a) 10 mL, (b) 100 mL, (c) 1000 mL, (d) 250 mL, (e) The information given is insufficient to calculate stroke volume.
9. Which of the following are true? (1) ANS regulation of heart rate originates in the cardiovascular center of the medulla oblongata. (2) Proprioceptor input is a major stimulus that accounts for the rapid rise in the heart rate at the onset of physical activity. (3) The vagus nerves release norepinephrine, causing the heart rate to increase. (4) Hormones from the adrenal medulla and the thyroid gland can increase the heart rate. (5) Hypothermia increases the heart rate. (a) 1, 2, 3, and 4, (b) 1, 2, and 4, (c) 2, 3, 4, and 5, (d) 3, 5, and 6, (e) 1, 2, 4, and 5.
10. Which of the following are true concerning action potentials and contraction in the myocardium? (1) The refractory period in a cardiac muscle fiber is very brief. (2) The binding of $Ca2^+$ to troponin allows the interaction of actin and myosin filaments, resulting in contraction. (3) Repolarization occurs when the voltage-gated K^+ channels open and calcium channels are closing. (4) Opening of voltage-gated fast Na^+ channels results in depolarization. (5) Opening of voltage-gated slow $Ca2^+$ channels results in a period of maintained depolarization, known as the plateau. (a) 1, 3, and 5, (b) 2, 3, and 4, (c) 2 and 5, (d) 3, 4, and 5, (e) 2, 3, 4, and 5.
11. Which of the following would not increase stroke volume? (a) increased $Ca2^+$ in the interstitial fluid, (b) epinephrine (c) increased K^+ in the interstitial fluid, (d) increase in venous return, (e) slow resting heart rate.

12. Match the following:

____(a) indicates ventricular repolarization
____(b) represents the time from the beginning of ventricular depolarization to the end of ventricular repolarization
____(c) represents atrial depolarization
____(d) represents the time when the ventricular contractile fibers are fully depolarized; occurs during the plateau phase of the action potential
____(e) represents the onset of ventricular depolarization
____(f) represents the conduction time from the beginning of atrial excitation to the beginning of ventricular excitation

(1) P wave
(2) QRS complex
(3) T wave
(4) P-Q interval
(5) S-T segment
(6) Q-T interval

13. Match the following:

____(a) major branch from the ascending aorta; passes inferior to the left auricle
____(b) lies in the posterior interventricular sulcus; supplies the walls of the ventricles with oxygenated blood
____(c) located in the coronary sulcus on the posterior surface of the heart; receives most of the deoxygenated blood from the myocardium
____(d) lies in the coronary sulcus; supplies oxygenated blood to the walls of the right ventricle
____(e) lies in the coronary sulcus; drains the right atrium and right ventricle
____(f) major branch from the ascending aorta; lies inferior to the right auricle
____(g) lies in the posterior interventricular sulcus; drains the right and left ventricles
____(h) lies in the anterior interventricular sulcus; supplies oxygenated blood to the walls of both ventricles
____(i) lies in the anterior interventricular sulcus; drains the walls of both ventricles and the left atrium
____(j) lies in the coronary sulcus; supplies oxygenated blood to the walls of the left ventricle and left atrium
____(k) drain the right ventricle and open directly into the right atrium

(1) small cardiac vein
(2) anterior interventricular branch (left anterior descending artery)
(3) anterior cardiac veins
(4) posterior interventricular branch
(5) marginal branch
(6) circumflex branch
(7) middle cardiac vein
(8) left coronary artery
(9) right coronary artery
(10) great cardiac vein
(11) coronary sinus

14. Match the following:

____(a) collects oxygenated blood from the pulmonary circulation
____(b) pumps deoxygenated blood to the lungs for oxygenation
____(c) their contraction pulls on and tightens the chordae tendineae, preventing the valve cusps from everting
____(d) cardiac muscle tissue
____(e) increase blood-holding capacity of the atria
____(f) tendonlike cords connected to the atrioventricular valve cusps which, along with the papillary muscles, prevent valve eversion
____(g) the superficial dense irregular connective tissue covering of the heart
____(h) outer layer of the serous pericardium; is fused to the fibrous pericardium
____(i) endothelial cells lining the interior of the heart; are continuous with the endothelium of the blood vessels
____(j) pumps oxygenated blood to all body cells, except the air sacs of the lungs
____(k) prevents backflow of blood from the right ventricle into the right atrium
____(l) collects deoxygenated blood from the systemic circulation
____(m) left atrioventricular valve
____(n) the remnant of the foramen ovale, an opening in the interatrial septum of the fetal heart
____(o) blood vessels that pierce the heart muscle and supply blood to the cardiac muscle fibers
____(p) grooves on the surface of the heart which delineate the external boundaries between the chambers
____(q) prevent backflow of blood from the arteries into the ventricles
____(r) the gap junction and desmosome connections between individual cardiac muscle fibers
____(s) internal wall dividing the chambers of the heart
____(t) separate the upper and lower heart chambers, preventing backflow of blood from the ventricles back into the atria
____(u) inner visceral layer of the pericardium; adheres tightly to the surface of the heart
____(v) ridges formed by raised bundles of cardiac muscle fibers

(1) right atrium
(2) right ventricle
(3) left atrium
(4) left ventricle
(5) tricuspid valve
(6) bicuspid (mitral) valve
(7) chordae tendineae
(8) auricles
(9) papillary muscles
(10) trabeculae carneae
(11) fibrous pericardium
(12) parietal pericardium
(13) epicardium
(14) myocardium
(15) endocardium
(16) atrioventricular valves
(17) semilunar valves
(18) intercalated discs
(19) sulci
(20) septum
(21) fossa ovalis
(22) coronary circulation

15. Match the following:

____(a) amount of blood contained in the ventricles at the end of ventricular relaxation
____(b) period of time when cardiac muscle fibers are contracting and exerting force but not shortening
____(c) amount of blood ejected per beat by each ventricle
____(d) amount of blood remaining in the ventricles following ventricular contraction
____(e) difference between a person's maximum cardiac output and cardiac output at rest
____(f) period of time when semilunar valves are open and blood flows out of the ventricles
____(g) period when all four valves are closed and ventricular blood volume does not change

(1) cardiac reserve
(2) stroke volume
(3) end-diastolic volume (EDV)
(4) isovolumetric relaxation
(5) end-systolic volume (ESV)
(6) ventricular ejection
(7) isovolumetric contraction

CRITICAL THINKING QUESTIONS

1. Gerald recently visited the dentist to have his teeth cleaned and checked. During the cleaning process, Gerald had some bleeding from his gums. A couple of days later, Gerald developed a fever, rapid heartbeat, sweating, and chills. He visited his family physician who detected a slight heart murmur. Gerald was given antibiotics and continued to have his heart monitored. How was Gerald's dental visit related to his illness?
2. Unathletic Sylvia makes a resolution to begin an exercise program. She tells you that she wants to make her heart "beat as fast as it can" during exercise. Explain to her why that may not be a good idea.
3. Mr. Perkins is a large, 62-year-old man with a weakness for sweets and fried foods. His idea of exercise is walking to the kitchen for more potato chips to eat while he is watching sports on television. Lately, he's been troubled by chest pains when he walks up stairs. His doctor told him to quit smoking and scheduled a cardiac angiography for next week. What is involved in performing this procedure? Why did the doctor order this test?

ANSWERS TO FIGURE QUESTIONS

20.1 The mediastinum is the mass of tissue that extends from the sternum to the vertebral column between the lungs.
20.2 The visceral layer of the serous pericardium (epicardium) is both a part of the pericardium and a part of the heart wall.
20.3 The coronary sulcus forms a boundary between the atria and ventricles.
20.4 The greater the workload of a heart chamber, the thicker its myocardium.
20.5 The fibrous skeleton attaches to the heart valves and prevents overstretching of the valves as blood passes through them.
20.6 The papillary muscles contract, which pulls on the chordae tendineae and prevents valve cusps from everting and letting blood flow back into the atria.
20.7 Numbers 6 (right ventricle) through 10 depict the pulmonary circulation, whereas numbers 1 (left ventricle) through 5 (right atrium) depict the systemic circulation.
20.8 The circumflex artery delivers oxygenated blood to the left atrium and left ventricle.
20.9 The intercalated discs hold the cardiac muscle fibers together and enable action potentials to propagate from one muscle fiber to another.
20.10 The only electrical connection between the atria and the ventricles is the atrioventricular bundle.
20.11 The duration of an action potential is much longer in a ventricular contractile fiber (0.3 sec = 300 msec) than in a skeletal muscle fiber (1–2 msec).
20.12 An enlarged Q wave may indicate a myocardial infarction (heart attack).
20.13 Action potentials propagate most slowly through the AV node.
20.14 The amount of blood in each ventricle at the end of ventricular diastole—called the end-diastolic volume—is about 130 mL in a resting person.
20.15 The first heart sound (S1), or lubb, is associated with closure of the AV valves.
20.16 The ventricular myocardium receives innervation from the sympathetic division only.
20.17 The skeletal muscle "pump" increases stroke volume by increasing preload (end-diastolic volume).
20.18 The heart begins to contract by the twenty-second day of gestation.
20.19 Partitioning of the heart is complete by the end of the fifth week.
20.20 HDL removes excess cholesterol from body cells and transports it to the liver for elimination.
20.21 Coronary angiography is used to visualize many blood vessels.
20.22 Tetralogy of Fallot involves an interventricular septal defect, an aorta that emerges from both ventricles, a stenosed pulmonary valve, and an enlarged right ventricle.

Chapter 21

The Cardiovascular System: Blood Vessels and Hemodynamics

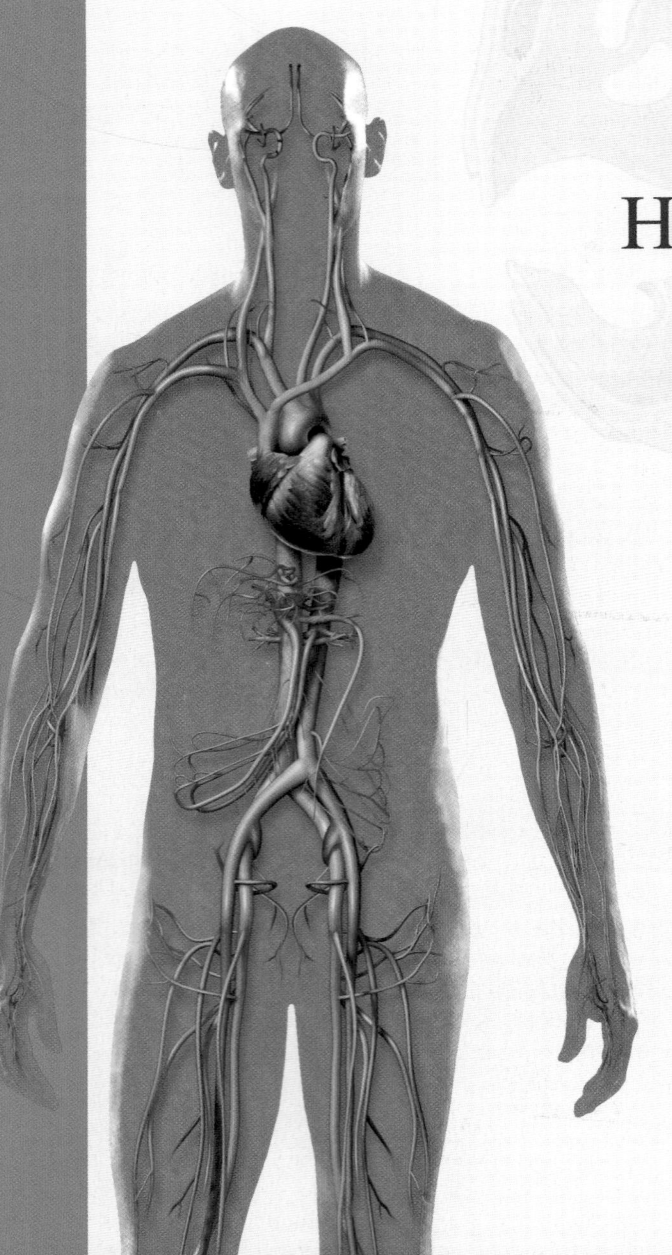

Blood Vessels, Hemodynamics, and Homeostasis

Blood vessels contribute to homeostasis by providing the structures for the flow of blood to and from the heart and the exchange of nutrients and wastes in tissues. They also play an important role in adjusting the velocity and volume of blood flow.

www. wiley.com/college/apcentral

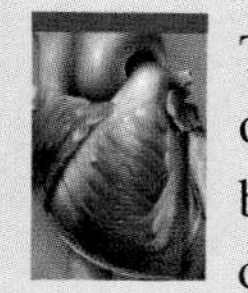

The cardiovascular system contributes to homeostasis of other body systems by transporting and distributing blood throughout the body to deliver materials (such as oxygen, nutrients, and hormones) and carry away wastes. The structures involved in these important tasks are the blood vessels, which form a closed system of tubes that carries blood away from the heart, transports it to the tissues of the body, and then returns it to the heart. The left side of the heart pumps blood through an estimated 100,000 km (60,000 mi) of blood vessels. The right side of the heart pumps blood through the lungs, enabling blood to pick up oxygen and unload carbon dioxide. Chapters 19 and 20 described the composition and functions of blood and the structure and function of the heart. In this chapter, we focus on the structure and functions of the various types of blood vessels; on **hemodynamics** (hē-mō-dī-NAM-iks; *hemo-* = blood; *dynamics* = power), the forces involved in circulating blood throughout the body; and on the blood vessels that constitute the major circulatory routes.

STRUCTURE AND FUNCTION OF BLOOD VESSELS

▶ **OBJECTIVES**

Contrast the structure and function of arteries, arterioles, capillaries, venules, and veins.

Outline the vessels through which the blood moves in its passage from the heart to the capillaries and back.

Distinguish between pressure reservoirs and blood reservoirs.

The five main types of blood vessels are arteries, arterioles, capillaries, venules, and veins. **Arteries** (AR-ter-ēz) carry blood *away from the heart* to other organs. Large, elastic arteries leave the heart and divide into medium-sized, muscular arteries that branch out into the various regions of the body. Medium-sized arteries then divide into small arteries, which in turn divide into still smaller arteries called **arterioles** (ar-TĒR-ē-ōls). As the arterioles enter a tissue, they branch into numerous tiny vessels called **capillaries** (KAP-i-lar′-ēz = hairlike). The thin walls of capillaries allow the exchange of substances between the blood and body tissues. Groups of capillaries within a tissue reunite to form small veins called **venules** (VEN-ūls). These, in turn, merge to form progressively larger blood vessels called veins. **Veins** (VĀNZ) are the blood vessels that convey blood from the tissues *back to the heart.* Because blood vessels require oxygen (O_2) and nutrients just like other tissues of the body, larger blood vessels are served by their own blood vessels, called **vasa vasorum** (literally, vasculature of vessels), located within their walls.

Angiogenesis and Disease

Angiogenesis (an′-jē-ō-JEN-e-sis; *angio-* = blood vessel; *-genesis* = production) refers to the growth of new blood vessels. It is an important process in embryonic and fetal development, and in postnatal life serves important functions such as wound healing, formation of a new uterine lining after menstruation, formation of the corpus luteum after ovulation, and development of blood vessels around obstructed arteries in the coronary circulation. Several proteins (peptides) are known to promote and inhibit angiogenesis.

Clinically angiogenesis is important because cells of a malignant tumor secrete proteins called *tumor angiogenesis factors (TAFs)* that stimulate blood vessel growth to provide nourishment for the tumor cells. Scientists are seeking chemicals that would inhibit angiogenesis and thus stop the growth of tumors. In diabetic retinopathy, angiogenesis may be important in the development of blood vessels that actually cause blindness, so finding inhibitors of angiogenesis may also prevent the blindness associated with diabetes. ■

Arteries

Because **arteries** (*ar-* = air; *ter-* = to carry) were found empty at death, in ancient times they were thought to contain only air. The wall of an artery has three coats or tunics: (1) tunica interna, (2) tunica media, and (3) tunica externa (Figure 21.1). The innermost coat, the **tunica interna** or **tunica intima,** contains a lining of simple squamous epithelium called *endothelium, a basement membrane,* and a layer of elastic tissue called the *internal elastic lamina.* The endothelium is a continuous layer of cells that line the inner surface of the entire cardiovascular system (the heart and all blood vessels). Normally, endothelium is the only tissue that makes contact with blood. The tunica interna is closest to the **lumen,** the hollow center through which blood flows. The middle coat, or **tunica media,** is usually the thickest layer. It consists of elastic fibers and smooth muscle fibers that extend circularly around the lumen, much like a ring encircles your finger. The tunica media also has an *external elastic lamina* composed of elastic tissue. Due to their plentiful elastic fibers, arteries normally have high *compliance,* which means that their walls stretch easily or expand without tearing in response to a small increase in pressure. The outer coat, the **tunica externa,** is composed mainly of elastic and collagen fibers.

Sympathetic neurons of the autonomic nervous system are distributed to the smooth muscle of the tunica media. An increase in sympathetic stimulation typically stimulates the smooth muscle to contract, squeezing the vessel wall and narrowing the lumen. Such a decrease in the diameter of the lumen of a blood vessel is called **vasoconstriction.** In contrast, smooth muscle fibers relax when sympathetic stimulation decreases or when certain chemicals, such as nitric oxide, H^+, and lactic acid, are present. The resulting increase in lumen diameter is called **vasodilation.** When an artery or arteriole is damaged, its smooth muscle contracts, producing vascular spasm (vasospasm) of the vessel, which limits blood flow through the damaged vessel and helps reduce blood loss if the vessel is small.

Figure 21.1 Comparative structure of blood vessels. The capillary in (c) is enlarged relative to the structures shown in parts (a) and (b).

Arteries carry blood from the heart to tissues; veins carry blood from tissues to the heart.

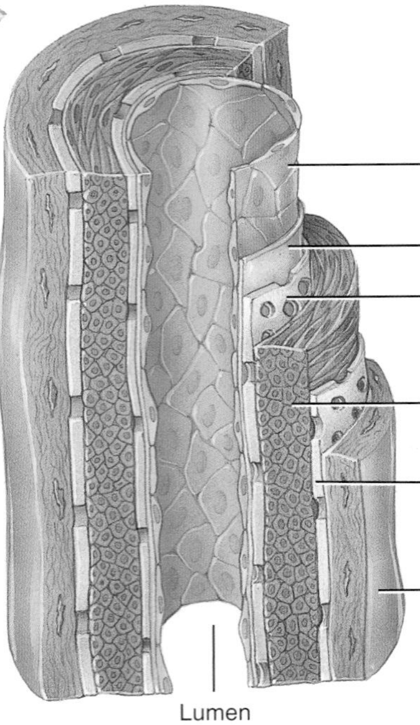

(a) Artery

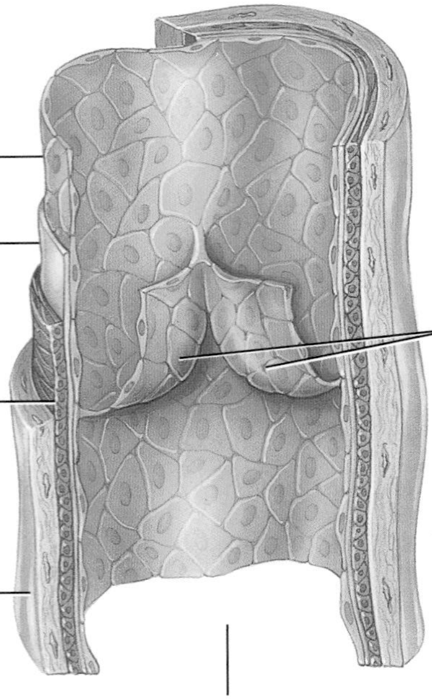

(b) Vein

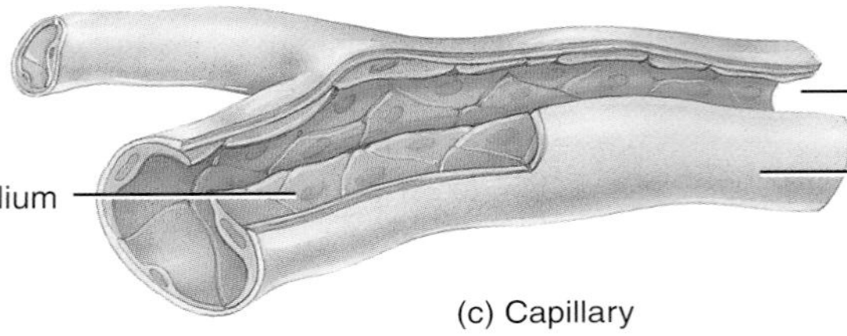

(c) Capillary

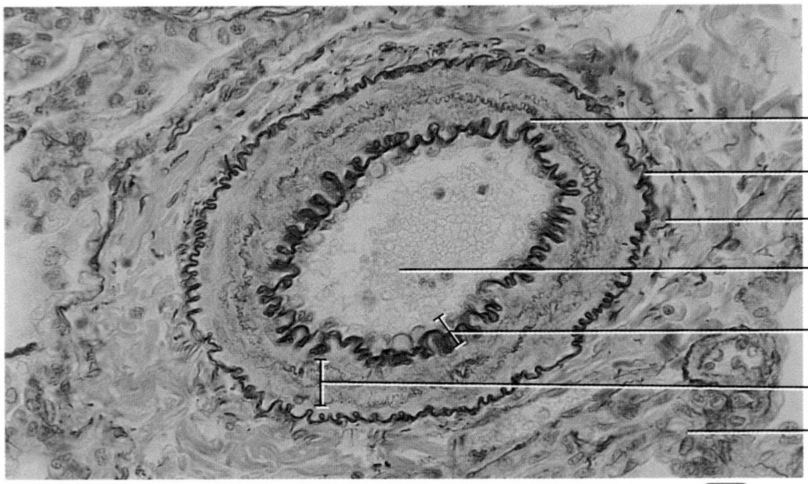

(d) Transverse section through a muscular artery

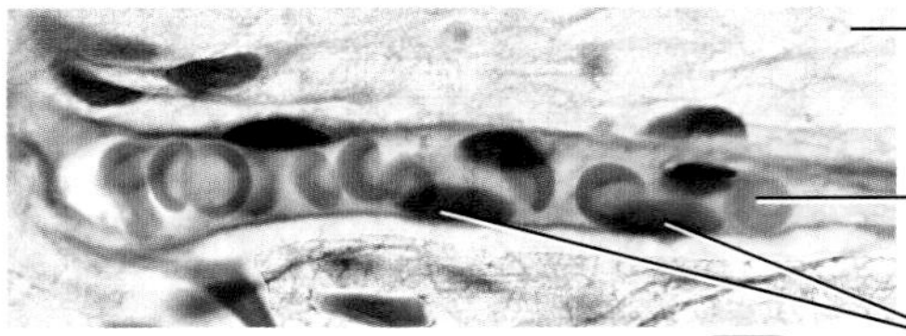

(e) Red blood cells passing through a capillary

Which vessel—the femoral artery or the femoral vein—has a thicker wall? Which has a wider lumen?

Elastic Arteries

The largest-diameter arteries (greater than 1 cm) are called **elastic arteries** because the tunica media contains a high proportion of elastic fibers. Elastic arteries have walls that are relatively thin in proportion to their overall diameter. Their internal elastic lamina is incomplete and their external elastic lamina is thin. Elastic arteries perform the important function of helping to propel blood onward while the ventricles are relaxing. As blood is ejected from the heart into elastic arteries, their walls stretch to accommodate the surge of blood, storing mechanical energy for a short time; the elastic fibers thus function as a **pressure reservoir** (Figure 21.2a). Then, the elastic fibers recoil and convert the stored (potential) energy into kinetic energy, causing the blood to flow. Thus, blood continues to move through the arteries even while the ventricles are relaxed (Figure 21.2b). Because they conduct blood from the heart to medium-sized, more-muscular arteries, elastic arteries also are called *conducting arteries.* The aorta and the brachiocephalic, common carotid, subclavian, vertebral, pulmonary, and common iliac arteries are elastic arteries (see Figure 21.18).

Figure 21.2 **Pressure reservoir function of elastic arteries.**

Recoil of elastic arteries keeps blood flowing during ventricular relaxation (diastole).

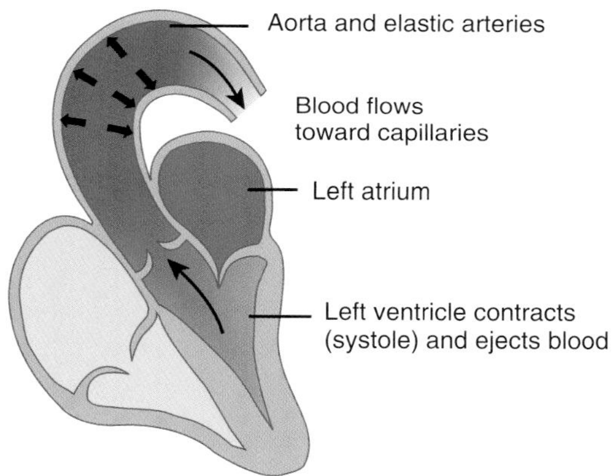

(a) Elastic aorta and arteries stretch during ventricular contraction

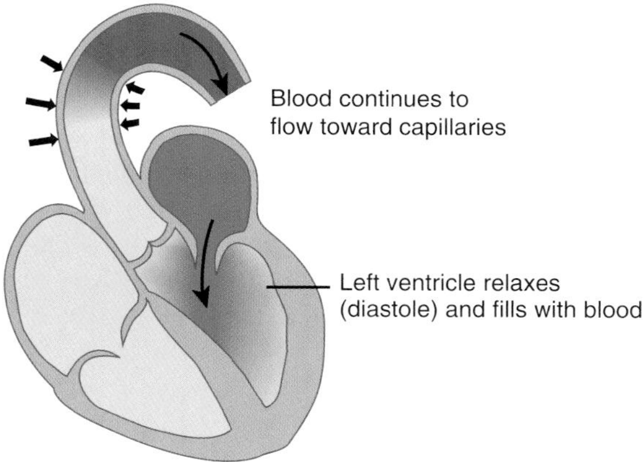

(b) Elastic aorta and arteries recoil during ventricular relaxation

In atherosclerosis, the walls of elastic arteries become less compliant (stiffer). What effect does reduced compliance have on the pressure reservoir function of arteries?

Muscular Arteries

Medium-sized arteries ranging in diameter from 0.1 to 10 mm are called **muscular arteries** because their tunica media contains more smooth muscle and fewer elastic fibers than elastic arteries. Thus, muscular arteries are capable of greater vasoconstriction and vasodilation to adjust the rate of blood flow. The large amount of smooth muscle makes the walls of muscular arteries relatively thick. They have a thin internal elastic lamina and a prominent external elastic lamina. Muscular arteries also are called *distributing arteries* because they distribute blood to various parts of the body. Examples include the brachial artery in the arm and radial artery in the forearm (see Figure 21.18).

Arterioles

An **arteriole** (= small artery) is a very small (almost microscopic) artery, ranging in diameter from 10 to 100 μm, that delivers blood to capillaries (Figure 21.3). Near the arteries from which they branch, arterioles have a tunica interna like that of arteries, a tunica media composed of smooth muscle and very few elastic fibers, and a tunica externa composed mostly of elastic and collagen fibers. In the smallest-diameter arterioles, which are closest to capillaries, the tunics consist of little more than a ring of endothelial cells surrounded by a few scattered smooth muscle fibers.

Arterioles play a key role in regulating blood flow from arteries into capillaries by regulating **resistance,** the opposition to blood flow. In a blood vessel, resistance is caused mainly by friction between blood and the inner walls of blood vessels. When the blood vessel diameter is smaller, the friction is greater. Because contraction and relaxation of the smooth muscle in arteriole walls can change their diameter, arterioles are known as *resistance vessels.* Contraction of arteriolar smooth muscle causes vasoconstriction, which increases vascular resistance and decreases blood flow into capillaries supplied by that arteriole. By contrast, relaxation of arteriolar smooth muscle causes vasodilation, which decreases vascular resistance and increases blood flow into capillaries. A change in arteriole diameter can also affect blood pressure: vasoconstriction of arterioles increases blood pressure, and vasodilation of arterioles decreases blood pressure.

Capillaries

Capillaries are microscopic vessels that connect arterioles to venules (Figure 21.3); they range in diameter from 4 to 10 μm. The flow of blood from arterioles to venules through capillaries is called the **microcirculation.** Capillaries are found near almost every cell in the body, but their number varies with the metabolic activity of the tissue they serve. Body tissues with high metabolic requirements, such as muscles, the liver, the kidneys, and the nervous system, use more O_2 and nutrients and thus have extensive

Figure 21.3 Arteriole, capillaries, and venule. Precapillary sphincters regulate the flow of blood through capillary beds.

In capillaries, nutrients, gases, and wastes are exchanged between the blood and interstitial fluid.

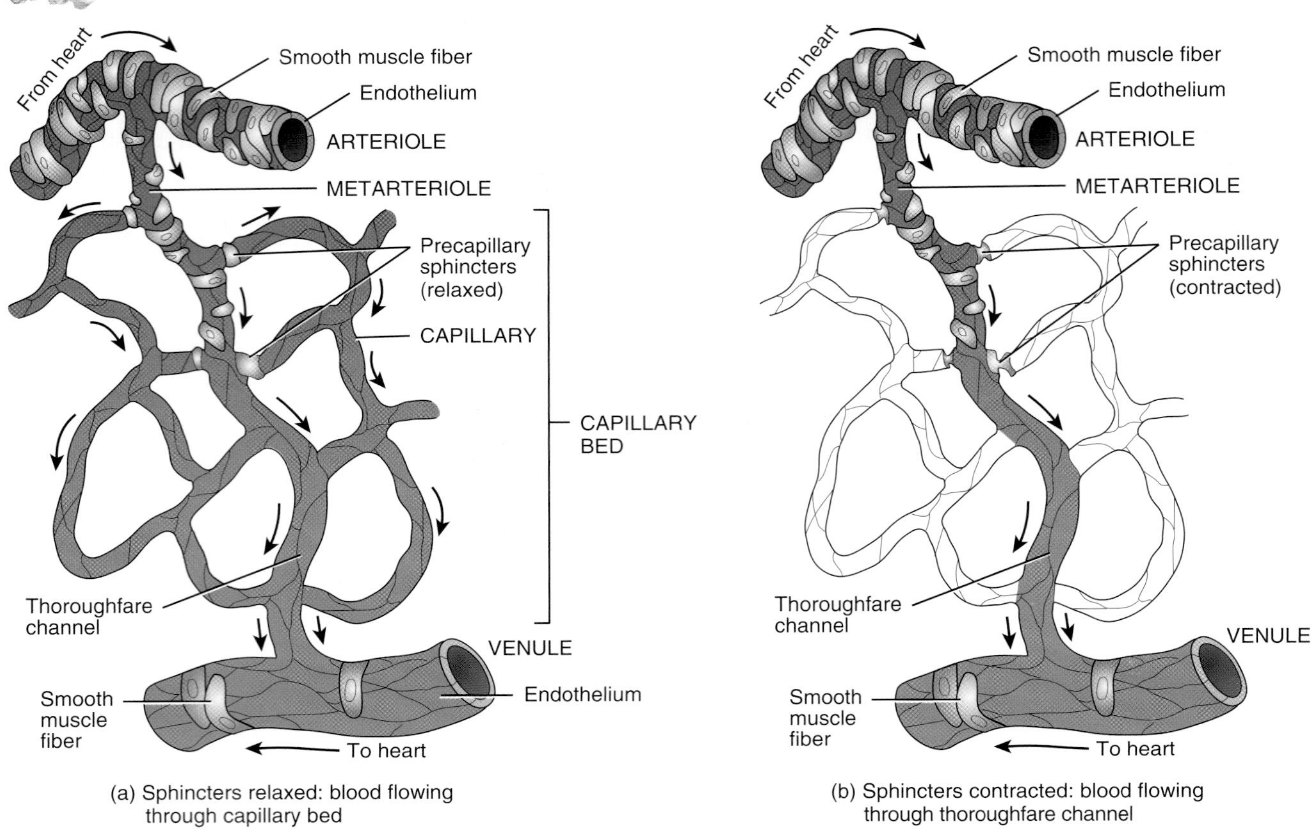

(a) Sphincters relaxed: blood flowing through capillary bed

(b) Sphincters contracted: blood flowing through thoroughfare channel

Why do metabolically active tissues have extensive capillary networks?

capillary networks. Tissues with lower metabolic requirements, such as tendons and ligaments, contain fewer capillaries. Capillaries are absent in a few tissues, such as all covering and lining epithelia, the cornea and lens of the eye, and cartilage.

Capillaries are known as *exchange vessels* because their prime function is the exchange of nutrients and wastes between the blood and tissue cells through the interstitial fluid. The structure of capillaries is well suited to this function. Capillary walls are composed of only a single layer of endothelial cells and a basement membrane (see Figure 21.1e). They have no tunica media or tunica externa. Thus, a substance in the blood must pass through just one cell layer to reach the interstitial fluid and tissue cells. Exchange of materials occurs only through the walls of capillaries and the beginning of venules; the walls of arteries, arterioles, most venules, and veins present too thick a barrier. Capillaries form extensive branching networks that increase the surface area available for rapid exchange of materials. In most tissues, blood flows through only a small part of the capillary network when metabolic needs are low. However, when a tissue such as contracting muscle is active, the entire capillary network fills with blood.

A **metarteriole** (*met-* = beyond) is a vessel that emerges from an arteriole and supplies a network of 10–100 capillaries called a **capillary bed** (Figure 21.3a). The proximal end of a metarteriole is surrounded by scattered smooth muscle fibers; the contraction and relaxation of the smooth muscle fibers help regulate blood flow through the capillary bed. The distal end of a metarteriole, which empties into a venule, has no smooth muscle fibers and is called a **thoroughfare channel.** Blood flowing through a thoroughfare channel bypasses the capillary bed.

At the junctions between the metarteriole and the capillaries of the capillary bed are rings of smooth muscle fibers called **precapillary sphincters** that control the flow of blood through the capillary bed. When the precapillary sphincters are relaxed (open), blood flows into the capillary bed (Figure 21.3a); when precapillary sphincters contract (close or partially close), blood flow through the capillary bed ceases or decreases (Figure 21.3b). Typically, blood flows intermittently through a capillary bed due to alternating contraction and relaxation of the smooth muscle of metarterioles and the precapillary sphincters. This intermittent contraction and relaxation, which may occur 5 to 10 times per minute, is called **vasomotion.** In part, vasomotion is due to chemicals released by the endothelial cells; nitric oxide is one example. At any given time, blood flows through only about 25% of a capillary bed.

The body contains three different types of capillaries: continuous capillaries, fenestrated capillaries, and sinusoids (Figure 21.4). Many capillaries are **continuous capillaries,** in which the plasma membranes of endothelial cells form a continuous tube that is interrupted only by **intercellular clefts,** which are gaps between neighboring endothelial cells (Figure 21.4a). Continuous capillaries are found in skeletal and smooth muscle, connective tissues, and the lungs.

Other capillaries of the body are **fenestrated capillaries** (*fenestr-* = window). The plasma membranes of the endothelial cells in these capillaries have many **fenestrations,** small pores (holes) ranging from 70 to 100 nm in diameter (Figure 21.4b) Fenestrated capillaries are found in the kidneys, villi of the small intestine, choroid plexuses of the ventricles in the brain, and some endocrine glands.

Sinusoids are wider and more winding than other capillaries. Their endothelial cells may have unusually large fenestrations. In addition to having an incomplete or absent basement membrane, sinusoids have very large intercellular clefts (Figure 21.4c). Sinusoids are found in the liver, red bone marrow, spleen, and some endocrine glands.

Usually blood passes from the heart and then in sequence through arteries, arterioles, capillaries, venules, and veins and then back to the heart. In some parts of the body, however, blood passes from one capillary network into another through a vein called a portal vein. Such a circulation of blood is called a **portal system.** The name of the portal system gives the name of the second capillary location. There are portal systems associated with the pituitary gland (hypophyseal portal system) and the liver (hepatic portal circulation).

Venules

When several capillaries unite, they form small veins called **venules** (= little veins). Venules, which range in diameter from 10 to 100 μm, collect blood from capillaries and deliver it to veins. The smallest venules, those closest to the capillaries, consist of a tunica interna of endothelium and a tunica media that has only a few scattered smooth muscle fibers. Like capillaries, the walls of the smallest venules are very porous; it is from the venules that many phagocytic white blood cells emigrate from the bloodstream into an inflamed or infected tissue. Larger venules that converge to form veins contain the tunica externa characteristic of veins (see Figure 21.1b).

Veins

The diameter of **veins** ranges from 0.1 mm to greater than 1 mm. Although veins are composed of essentially the same three coats (tunics) as arteries, the relative thicknesses of the layers are different. The tunica interna of veins is thinner than that of arteries; the tunica media of veins is much thinner than in arteries, with relatively little smooth muscle and elastic fibers. The tunica externa of veins is the thickest layer and consists of collagen

Figure 21.4 Types of capillaries.

Capillaries are microscopic blood vessels that connect arterioles and venules.

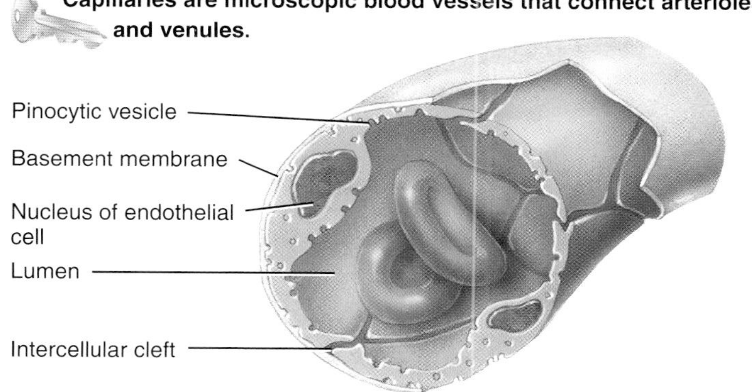

(a) Continuous capillary formed by endothelial cells

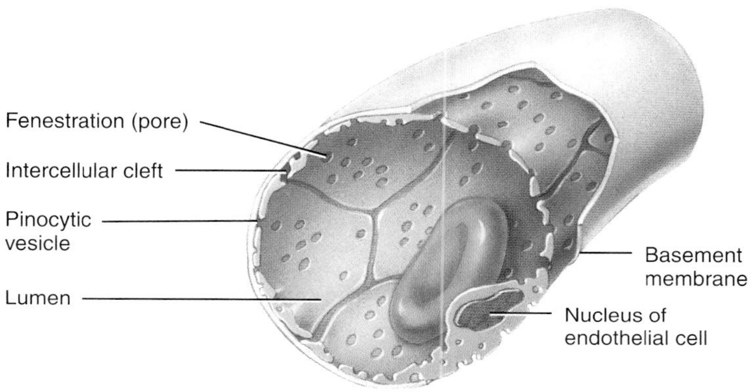

(b) Fenestrated capillary

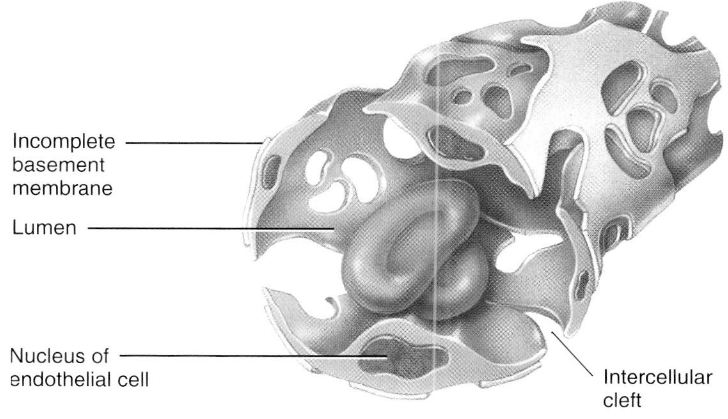

(c) Sinusoid

How do materials move through capillary walls?

and elastic fibers. Veins lack the external or internal elastic laminae found in arteries (see Figure 21.1b). They are distensible enough to adapt to variations in the volume and pressure of blood passing through them, but are not designed to withstand high pressure. The lumen of a vein is larger than that of a comparable artery, and veins often appear collapsed (flattened) when sectioned.

Many veins, especially those in the limbs, also feature **valves,** thin folds of tunica interna that form flaplike cusps. The valve cusps project into the lumen, pointing toward the heart (Figure 21.5). The low blood pressure in veins allows blood returning to the heart to slow and even back up; the valves aid in venous return by preventing the backflow of blood.

A **vascular (venous) sinus** is a vein with a thin endothelial wall that has no smooth muscle to alter its diameter. In a vascular sinus, the surrounding dense connective tissue replaces the tunica media and tunica externa in providing support. For example, dural venous sinuses, which are supported by the dura mater, convey deoxygenated blood from the brain to the heart. Another example of a vascular sinus is the coronary sinus of the heart (see Figure 20.3c page 700).

Figure 21.5 Venous valves.

Valves in veins allow blood to flow in one direction only—toward the heart.

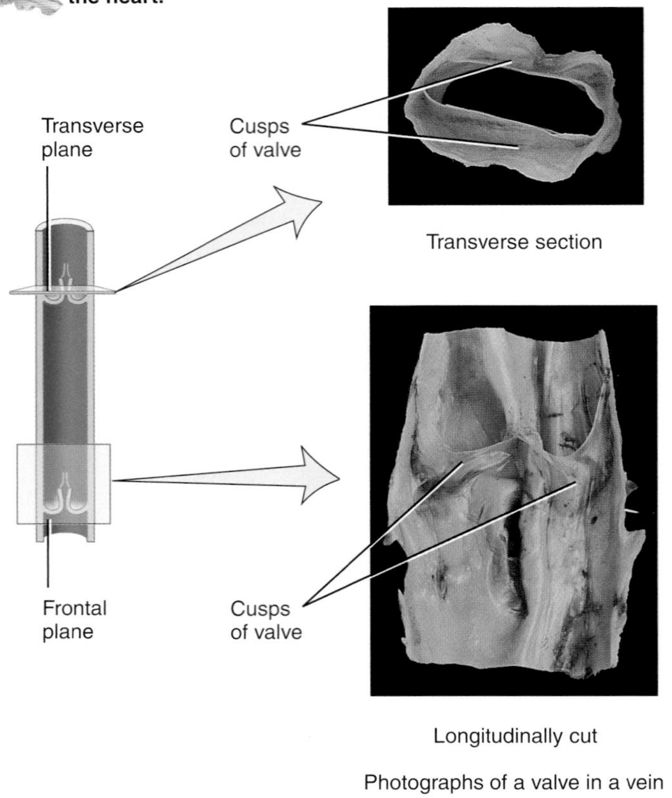

Transverse plane

Cusps of valve

Transverse section

Frontal plane

Cusps of valve

Longitudinally cut

Photographs of a valve in a vein

? Why are valves more important in arm veins and leg veins than in neck veins?

Varicose Veins

Leaky venous valves can cause veins to become dilated and twisted in appearance, a condition called **varicose veins** or **varices** (VAR-i-sēz; singular is **varix** (VAR-iks); *varic-* = a swollen vein). The condition may occur in the veins of almost any body part, but it is most common in the esophagus and in superficial veins of the lower limbs. Those in the lower limbs can range from cosmetic problems to serious medical conditions. The valvular defect may be congenital or may result from mechanical stress (prolonged standing or pregnancy) or aging. The leaking venous valves allow the backflow of blood, which causes pooling of blood. This, in turn, creates pressure that distends the vein and allows fluid to leak into the surrounding tissue. As a result, the affected vein and the tissue around it may become inflamed and painfully tender. Veins close to the surface, especially the saphenous vein, are highly susceptible to varicosities; deeper veins are not as vulnerable because surrounding skeletal muscles prevent their walls from stretching excessively. Varicosed veins in the anal canal are referred to as *hemorrhoids.* Esophageal varices result from dilated veins in the walls of the lower part of the esophagus and sometimes the upper part of the stomach. Bleeding esophageal varices, which are life-threatening, are usually a result of chronic liver disease.

Several treatment options are available for varicose veins in the lower limbs. *Elastic stockings* (support hose) may be used for individuals with mild symptoms or for whom other options are not recommended. *Sclerotherapy* involves injection of a solution into varicosed veins that damages the tunica interna by producing a harmless superficial thrombophlebitis (inflammation involving a blood clot). Healing of the damaged part leads to scar formation that occludes the vein. *Radiofrequency endovenous occlusion* involves the application of radiofrequency energy to heat up and close off varicosed veins. *Laser occlusion* uses laser therapy to shut down veins. In a surgical procedure called *stripping,* veins may be removed. In this procedure, a flexible wire is threaded through the vein and then pulled out to strip (remove) it from the body. ■

Anastomoses

Most tissues of the body receive blood from more than one artery. The union of the branches of two or more arteries supplying the same body region is called an **anastomosis** (a-nas-tō-MŌ-sis = connecting; plural is **anastomoses**). Anastomoses between arteries provide alternate routes for blood to reach a tissue or organ. If blood flow stops for a short time when normal movements compress a vessel, or if a vessel is blocked by disease, injury, or surgery, then circulation to a part of the body is not necessarily stopped. The alternate route of blood flow to a body part through an anastomosis is known as **collateral circulation.** Anastomoses may also occur between veins and between arterioles and venules. Arteries that do not anastomose are known as **end arteries.** Obstruction of an end artery interrupts the blood supply to a whole segment of an organ, produc-

TABLE 21.1 Distinguishing Features of Blood Vessels

	Diameter	Tunica Interna	Tunica Media	Tunica Externa	Function
Elastic arteries	Greater than 1 cm.	Endothelium, basement membrane, and incomplete internal elastic lamina.	Smooth muscle and higher proportion of elastic fibers and thin external elastic lamina.	Collagen and elastic fibers.	Conduct blood from the heart to muscular arteries.
Muscular arteries	0.1–10 mm	Endothelium, basement membrane, and thin internal elastic lamina.	Higher proportion of smooth muscle, fewer elastic fibers, and prominent external elastic lamina.	Collagen and elastic fibers.	Distribute blood to arterioles.
Anterioles (near arteries from which they branch)	10–100 μm.	Endothelium, basement membrane, and internal elastic lamina.	Smooth muscle and very few elastic fibers.	Collagen and elastic fibers.	Deliver blood to capillaries and help regulate blood flow.
Capillaries	4–10 μm.	Endothelium and basement membrane.	None.	None.	Permit exchange of nutrients and wastes between blood and interstitial fluid.
Venules (closer to convergence with veins)	10–100 μm.	Endothelium and basement membrane.	Smooth muscle.	Collagen and elastic fibers.	Collect blood from capillaries and pass it on to veins.
Veins	0.1mm–greater than 1 mm.	Endothelium and basement membrane; contains valves.	Smooth muscle and elastic fibers.	Collagen and elastic fibers.	Return blood to the heart, facilitated by valves in veins in limbs.

ing necrosis (death) of that segment. Alternate blood routes may also be provided by nonanastomosing vessels that supply the same region of the body.

A summary of the distinguishing features of blood vessels is presented in Table 21.1.

Blood Distribution

The largest portion of your blood volume at rest—about 64%—is in systemic veins and venules (Figure 21.6). Systemic arteries and arterioles hold about 13% of the blood volume, systemic capillaries hold about 7%, pulmonary blood vessels hold about 9%, and the heart holds about 7%. Because systemic veins and venules contain a large percentage of the blood volume, they function as **blood reservoirs** from which blood can be diverted quickly if the need arises. For example, during increased muscular activity, the cardiovascular center in the brain stem sends a larger number of sympathetic impulses to veins. The result is *venoconstriction,* constriction of veins, which reduces the volume of blood in reservoirs and allows a greater blood volume to flow to skeletal muscles, where it is needed most. A similar mechanism operates in cases of hemorrhage, when blood volume and pressure decrease; in this case, venoconstriction helps counteract the drop in blood pressure. Among the principal blood reservoirs are the veins of the abdominal organs (especially the liver and spleen) and the veins of the skin.

Figure 21.6 Blood distribution in the cardiovascular system at rest.

Because systemic veins and venules contain more than half the total blood volume, they are called blood reservoirs.

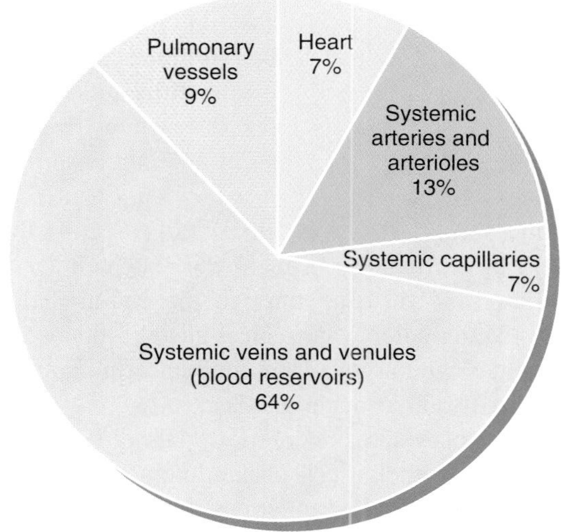

If your total blood volume is 5 liters, what volume is in your venules and veins right now? In your capillaries?

► CHECKPOINT

1. What is the function of elastic fibers and smooth muscle in the tunica media of arteries?
2. How are elastic arteries and muscular arteries different?
3. What structural features of capillaries allow the exchange of materials between blood and body cells?
4. What is the difference between pressure reservoirs and blood reservoirs? Why is each important?
5. What is the relationship between anastomoses and collateral circulation?

CAPILLARY EXCHANGE

► OBJECTIVE

Discuss the pressures that cause movement of fluids between capillaries and interstitial spaces.

The mission of the entire cardiovascular system is to keep blood flowing through capillaries to allow **capillary exchange,** the movement of substances between blood and interstitial fluid. The 7% of the blood in systemic capillaries at any given time is continually exchanging materials with interstitial fluid. Substances enter and leave capillaries by three basic mechanisms: diffusion, transcytosis, and bulk flow.

Diffusion

The most important method of capillary exchange is simple diffusion. Many substances, such as oxygen (O_2), carbon dioxide (CO_2), glucose, amino acids, and hormones, enter and leave capillaries by simple diffusion. Because O_2 and nutrients normally are present in higher concentrations in blood, they diffuse down their concentration gradients into interstitial fluid and then into body cells. CO_2 and other wastes released by body cells are present in higher concentrations in interstitial fluid, so they diffuse into blood.

Substances in blood or interstitial fluid can cross the walls of a capillary by diffusing through the intercellular clefts or fenestrations or by diffusing through the endothelial cells (see Figure 21.4). Water-soluble substances such as glucose and amino acids pass across capillary walls through intercellular clefts or fenestrations. Lipid-soluble materials, such as O_2, CO_2, and steroid hormones, may pass across capillary walls directly through the lipid bilayer of endothelial cell plasma membranes. Most plasma proteins and red blood cells cannot pass through capillary walls of continuous and fenestrated capillaries because they are too large to fit through the intercellular clefts and fenestrations.

In sinusoids, however, the intercellular clefts are so large that they allow even proteins and blood cells to pass through their walls. For example, hepatocytes (liver cells) synthesize and release many plasma proteins, such as fibrinogen (the main clotting protein) and albumin, which then diffuse into the bloodstream through sinusoids. In red bone marrow, blood cells are formed (hemopoiesis) and then enter the bloodstream through sinusoids.

In contrast to sinusoids, the capillaries of the brain allow only a few substances to move across their walls. Most areas of the brain contain continuous capillaries; however, these capillaries are very "tight." The endothelial cells of most brain capillaries are sealed together by tight junctions. The resulting blockade to movement of materials into and out of brain capillaries is known as the *blood–brain barrier* (see page 477). In brain areas that lack the blood–brain barrier, for example, the hypothalamus, pineal gland, and pituitary gland, materials undergo capillary exchange more freely.

Transcytosis

A small quantity of material crosses capillary walls by **transcytosis** (*trans-* = across; *cyt-* = cell; *-osis* = process). In this process, substances in blood plasma become enclosed within tiny pinocytic vesicles that first enter endothelial cells by endocytosis, then move across the cell and exit on the other side by exocytosis. This method of transport is important mainly for large, lipid-insoluble molecules that cannot cross capillary walls in any other way. For example, the hormone insulin (a small protein) enters the bloodstream by transcytosis, and certain antibodies (also proteins) pass from the maternal circulation into the fetal circulation by transcytosis.

Bulk Flow: Filtration and Reabsorption

Bulk flow is a passive process in which *large* numbers of ions, molecules, or particles in a fluid move together in the same direction. The substances move at rates far greater than can be accounted for by diffusion alone. Bulk flow occurs from an area of higher pressure to an area of lower pressure, and it continues as long as a pressure difference exists. Diffusion is more important for *solute exchange* between blood and interstitial fluid, but bulk flow is more important for regulation of the *relative volumes of blood and interstitial fluid.* Pressure-driven movement of fluid and solutes *from* blood capillaries *into* interstitial fluid is called **filtration.** Pressure-driven movement *from* interstitial fluid *into* blood capillaries is called **reabsorption.**

Two pressures promote filtration: **blood hydrostatic pressure (BHP),** the pressure generated by the pumping action of the heart, and **interstitial fluid osmotic pressure.** The main pressure promoting reabsorption of fluid is **blood colloid osmotic pressure.** The balance of these pressures, called **net filtration pressure (NFP),** determines whether the volumes of blood and interstitial fluid remain steady or change. Overall, the volume of fluid and solutes reabsorbed normally is almost as large as the volume filtered. This near equilibrium is known as **Starling's law of the capillaries.** Let's see how these hydrostatic and osmotic pressures balance.

Within vessels, the hydrostatic pressure is due to the pressure that water in blood plasma exerts against blood vessel walls. The **blood hydrostatic pressure (BHP)** is about 35 millimeters of mercury (mmHg) at the arterial end of a capillary, and about 16 mmHg at the capillary's venous end (Figure 21.7). BHP "pushes" fluid out of capillaries into interstitial fluid. The opposing pressure of the interstitial fluid, called **interstitial fluid hydrostatic pressure (IFHP),** "pushes" fluid from interstitial spaces back into capillaries. However, IFHP is close to zero. (IFHP is difficult to measure, and its reported values vary from small positive values to small negative values.) For our discussion we assume that IFHP equals 0 mmHg all along the capillaries.

The difference in osmotic pressure across a capillary wall is due almost entirely to the presence in blood of plasma proteins, which are too large to pass through either fenestrations or gaps between endothelial cells. **Blood colloid osmotic pressure (BCOP)** is a force caused by the colloidal suspension of these large proteins in plasma that averages 26 mmHg in most capillaries. The effect of BCOP is to "pull" fluid from interstitial spaces into capillaries. Opposing BCOP is **interstitial fluid osmotic pressure (IFOP),** which "pulls" fluid out of capillaries into interstitial fluid. Normally, IFOP is very small—0.1–5 mmHg—because only tiny amounts of protein are present in interstitial fluid. The small amount of protein that leaks from blood plasma into interstitial fluid does not accumulate there because it enters lymphatic fluid and is returned to the blood. For discussion, we can use a value of 1 mmHg for IFOP.

Whether fluids leave or enter capillaries depends on the balance of pressures. If the pressures that push fluid out of capil-

Figure 21.7 Dynamics of capillary exchange (Starling's law of the capillaries). Excess filtered fluid drains into lymphatic capillaries.

Blood hydrostatic pressure pushes fluid out of capillaries (filtration), and blood colloid osmotic pressure pulls fluid into capillaries (reabsorption).

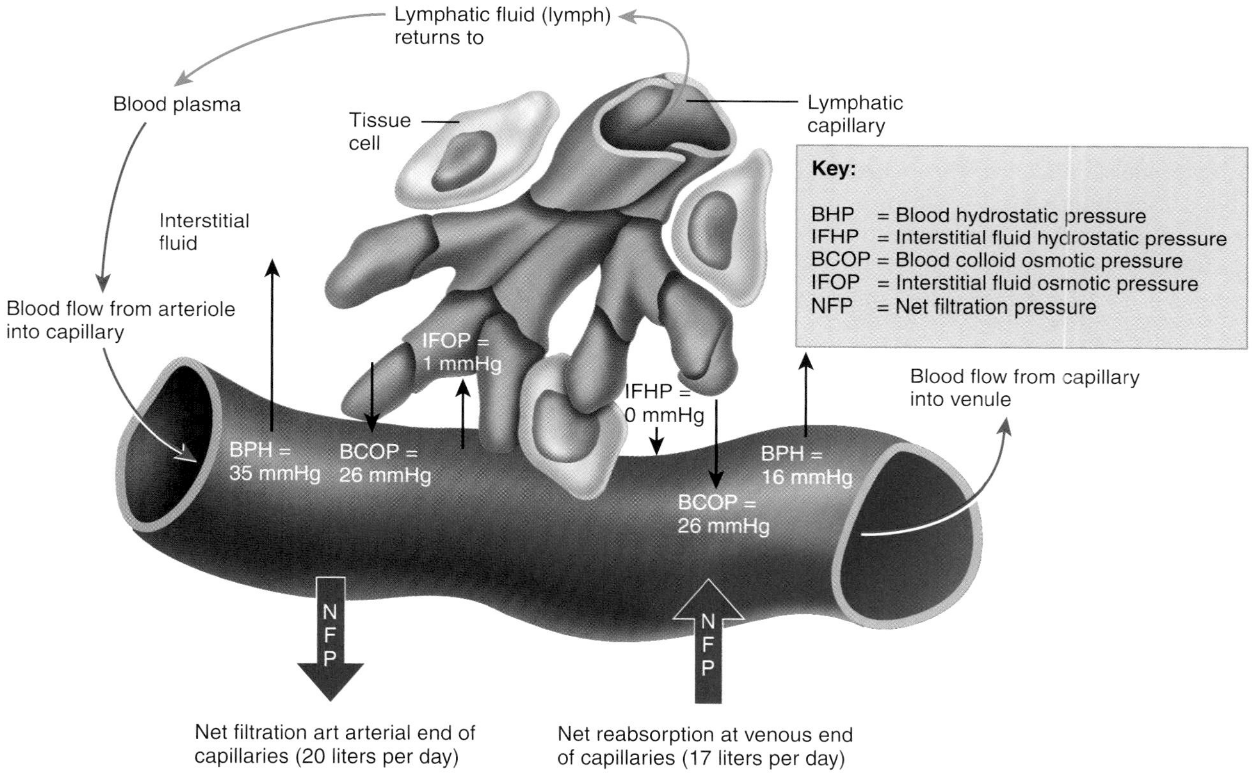

Lymphatic fluid (lymph) returns to
Blood plasma
Tissue cell
Lymphatic capillary
Interstitial fluid
Blood flow from arteriole into capillary
IFOP = 1 mmHg
IFHP = 0 mmHg
BPH = 35 mmHg
BCOP = 26 mmHg
BPH = 16 mmHg
BCOP = 26 mmHg
Blood flow from capillary into venule
NFP
NFP

Key:

BHP = Blood hydrostatic pressure
IFHP = Interstitial fluid hydrostatic pressure
BCOP = Blood colloid osmotic pressure
IFOP = Interstitial fluid osmotic pressure
NFP = Net filtration pressure

Net filtration art arterial end of capillaries (20 liters per day)

Net reabsorption at venous end of capillaries (17 liters per day)

Net filtration pressure (NFP) = (BHP + IFOP) – (BCOP + IFHP)

Pressures promoting filtration — Pressure promoting reabsorption

	Arterial end	Venous end
	NFP = (35 + 1) – (26 + 0) = 10 mmHg	NFP = (16 + 1) – (26 + 0) = – 9 mmHg
Result	Net filtration	Net reabsorption

A person who has liver failure cannot synthesize the normal amount of plasma proteins. How does a deficit of plasma proteins affect blood colloid osmotic pressure, and what is the effect on capillary filtration and reabsorption?

laries exceed the pressures that pull fluid into capillaries, fluid will move from capillaries into interstitial spaces (filtration). If, however, the pressures that push fluid out of interstitial spaces into capillaries exceed the pressures that pull fluid out of capillaries, then fluid will move from interstitial spaces into capillaries (reabsorption).

The net filtration pressure (NFP), which indicates the direction of fluid movement, is calculated as follows:

$$\text{NFP} = \underbrace{(\text{BHP} + \text{IFOP})}_{\text{Pressures that promote filtration}} - \underbrace{(\text{BCOP} + \text{IFHP})}_{\text{Pressures that promote reabsorption}}$$

At the arterial end of a capillary,

$$\begin{aligned}\text{NFP} &= (35 + 1)\ \text{mmHg} - (26 + 0)\ \text{mmHg}\\ &= 36 - 26\ \text{mmHg} = 10\ \text{mmHg}\end{aligned}$$

Thus, at the arterial end of a capillary, there is a *net outward pressure* of 10 mmHg, and fluid moves out of the capillary into interstitial spaces (filtration).

At the venous end of a capillary,

$$\begin{aligned}\text{NFP} &= (16 + 1)\ \text{mmHg} - (26 + 0)\ \text{mmHg}\\ &= 17 - 26\ \text{mmHg} = -9\ \text{mmHg}\end{aligned}$$

At the venous end of a capillary, the negative value (−9 mmHg) represents a *net inward pressure,* and fluid moves into the capillary from tissue spaces (reabsorption).

On average, about 85% of the fluid filtered out of capillaries is reabsorbed. The excess filtered fluid and the few plasma proteins that do escape from blood into interstitial fluid enter lymphatic capillaries (see Figure 22.2 on page 807). As lymph drains into the junction of the jugular and subclavian veins in the upper thorax (see Figure 22.3 on page 808), these materials return to the blood. Every day about 20 liters of fluid filter out of capillaries in tissues throughout the body. Of this fluid, 17 liters are reabsorbed and 3 liters enter lymphatic capillaries (excluding filtration during urine formation).

Edema

If filtration greatly exceeds reabsorption, the result is **edema** (= swelling), an abnormal increase in interstitial fluid volume. Edema is not usually detectable in tissues until interstitial fluid volume has risen to 30% above normal. Edema can result from either excess filtration or inadequate reabsorption.

Two situations may cause excess filtration:

- *Increased capillary blood pressure* causes more fluid to be filtered from capillaries.
- *Increased permeability of capillaries* raises interstitial fluid osmotic pressure by allowing some plasma proteins to escape. Such leakiness may be caused by the destructive effects of chemical, bacterial, thermal, or mechanical agents on capillary walls.

One situation commonly causes inadequate reabsorption:

- *Decreased concentration of plasma proteins* lowers the blood colloid osmotic pressure. Inadequate synthesis or loss of plasma proteins is associated with liver disease, burns, malnutrition, and kidney disease. ■

▶ CHECKPOINT

6. How can substances enter and leave blood plasma?
7. How do hydrostatic and osmotic pressures determine fluid movement across the walls of capillaries?
8. Define edema and describe how it develops.

HEMODYNAMICS: FACTORS AFFECTING BLOOD FLOW

▶ OBJECTIVES

Explain the factors that regulate the volume of blood flow.

Explain how blood pressure changes throughout the cardiovascular system.

Describe the factors that determine mean arterial pressure and systemic vascular resistance.

Describe the relationship between cross-sectional area and velocity of blood flow.

Blood flow is the volume of blood that flows through any tissue in a given time period (in mL/min). Total blood flow is cardiac output (CO), the volume of blood that circulates through systemic (or pulmonary) blood vessels each minute. In Chapter 20 we saw that cardiac output depends on heart rate and stroke volume: Cardiac output (CO) = heart rate (HR) × stroke volume (SV). How the cardiac output becomes distributed into circulatory routes that serve various body tissues depends on two more factors: (1) the *pressure difference* that drives the blood flow through a tissue and (2) the *resistance* to blood flow in specific blood vessels. Blood flows from regions of higher pressure to regions of lower pressure; the greater the pressure difference, the greater the blood flow. But the higher the resistance, the smaller the blood flow.

Blood Pressure

As you have just learned, blood flows from regions of higher pressure to regions of lower pressure; the greater the pressure difference, the greater the blood flow. Contraction of the ventricles generates **blood pressure (BP),** the hydrostatic pressure exerted by blood on the walls of a blood vessel. BP is highest in the aorta and large systemic arteries; in a resting, young adult, BP rises to about 110 mmHg during systole (ventricular contraction) and drops to about 70 mmHg during diastole (ventricular relaxation). **Systolic blood pressure** is the highest pressure attained in arteries during systole, and **diastolic blood pressure** is the lowest arterial pressure during diastole (Figure 21.8). As blood leaves the aorta and flows through the systemic circulation, its pressure falls progressively as the distance from the left ventricle increases. Blood pressure decreases to about 35 mmHg as blood passes from systemic arteries through systemic arterioles and into capillaries, where the pressure fluctuations disappear. At the venous end of capillaries, blood pressure has dropped to about 16 mmHg. Blood pressure continues to drop as blood enters systemic venules and then veins because these vessels are farthest from the left ventricle. Finally, blood pressure reaches 0 mmHg as blood flows into the right ventricle.

Mean arterial pressure (MAP), the average blood pressure in arteries, is roughly one-third of the way between the diastolic and systolic pressures. It can be estimated as follows:

$$\text{MAP} = \text{diastolic BP} + 1/3\ (\text{systolic BP} - \text{diastolic BP})$$

Thus, in a person whose BP is 110/70 mmHg, MAP is about 83 mmHg (70 + 1/3(110 − 70)).

We have already seen that cardiac output equals heart rate multiplied by stroke volume. Another way to calculate cardiac output is to divide mean arterial pressure (MAP) by resistance (R): $\text{CO} = \text{MAP} \div \text{R}$. By rearranging the terms of this equation, you can see that $\text{MAP} = \text{CO} \times \text{R}$. If cardiac output rises due to an increase in stroke volume or heart rate, then the mean arterial pressure rises as long as resistance remains steady. Likewise, a decrease in cardiac output causes a decrease in mean arterial pressure if resistance does not change.

Blood pressure also depends on the total volume of blood in the cardiovascular system. The normal volume of blood in an adult is about 5 liters (5.3 qt). Any decrease in this volume, as from hemorrhage, decreases the amount of blood that is circulated through the arteries each minute. A modest decrease can be

Figure 21.8 Blood pressures in various parts of the cardiovascular system. The dashed line is the mean (average) blood pressure in the aorta, arteries, and arterioles.

Blood pressure rises and falls with each heartbeat in blood vessels leading to capillaries.

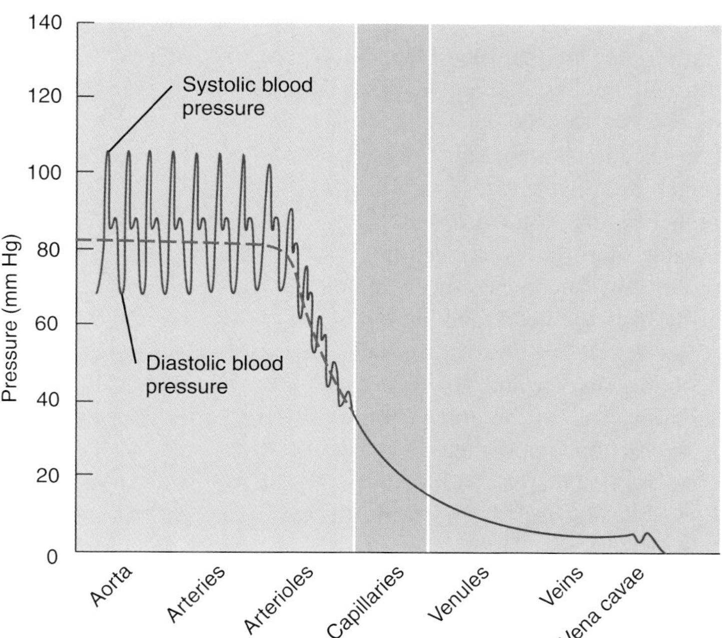

Is the mean blood pressure in the aorta closer to systolic or to diastolic pressure?

compensated for by homeostatic mechanisms that help maintain blood pressure (described on page 750), but if the decrease in blood volume is greater than 10% of the total, blood pressure drops. Conversely, anything that increases blood volume, such as water retention in the body, tends to increase blood pressure.

Resistance

As noted earlier, **vascular resistance** is the opposition to blood flow due to friction between blood and the walls of blood vessels. Vascular resistance depends on (1) size of the blood vessel lumen, (2) blood viscosity, and (3) total blood vessel length.

1. ***Size of the lumen.*** The smaller the lumen of a blood vessel, the greater its resistance to blood flow. Resistance is inversely proportional to the fourth power of the diameter (d) of the blood vessel's lumen ($R \propto 1/d^4$). The smaller the diameter of the blood vessel, the greater the resistance it offers to blood flow. For example, if the diameter of a blood vessel decreases by one-half, its resistance to blood flow increases 16 times. Vasoconstriction narrows the lumen, and vasodilation widens it. Normally, moment-to-moment fluctuations in blood flow through a given tissue are due to vasoconstriction and vasodilation of the tissue's arterioles. As arterioles dilate, resistance decreases, and blood pressure falls. As arterioles constrict, resistance increases, and blood pressure rises.
2. ***Blood viscosity.*** The viscosity (thickness) of blood depends mostly on the ratio of red blood cells to plasma (fluid) volume, and to a smaller extent on the concentration of proteins in plasma. The higher the blood's viscosity, the higher the resistance. Any condition that increases the viscosity of blood, such as dehydration or polycythemia (an unusually high number of red blood cells), thus increases blood pressure. A depletion of plasma proteins or red blood cells, due to anemia or hemorrhage, decreases viscosity and thus decreases blood pressure.
3. ***Total blood vessel length.*** Resistance to blood flow through a vessel is directly proportional to the length of the blood vessel. The longer a blood vessel, the greater the resistance. Obese people often have hypertension (elevated blood pressure) because the additional blood vessels in their adipose tissue increase their total blood vessel length. An estimated 650 km (about 400 miles) of additional blood vessels develop for each extra kilogram (2.2 lb) of fat.

Systemic vascular resistance (SVR), also known as *total peripheral resistance (TPR),* refers to all the vascular resistances offered by systemic blood vessels. The diameters of arteries and veins are large, so their resistance is very small because most of the blood does not come into physical contact with the walls of the blood vessel. The smallest vessels—arterioles, capillaries, and venules—contribute the most resistance. A major function of arterioles is to control SVR—and therefore blood pressure and blood flow to particular tissues—by changing their diameters. Arterioles need to vasodilate or vasoconstrict only slightly to have a large effect on SVR. The main center for regulation of SVR is the vasomotor center in the brain stem (described shortly).

Venous Return

Venous return, the volume of blood flowing back to the heart through the systemic veins, occurs due to the pressure generated by contractions of the heart's left ventricle. The pressure difference from venules (averaging about 16 mmHg) to the right ventricle (0 mmHg), although small, normally is sufficient to cause venous return to the heart. If pressure increases in the right atrium or ventricle, venous return will decrease. One cause of increased pressure in the right atrium is an incompetent (leaky) tricuspid valve, which lets blood regurgitate (flow backward) as the ventricles contract. The result is decreased venous return and buildup of blood on the venous side of the systemic circulation.

When you stand up, for example, at the end of an anatomy and physiology lecture, the pressure pushing blood up the veins in your lower limbs is barely enough to overcome the force of gravity pushing it back down. Besides the heart, two other mechanisms "pump" blood from the lower body back to the heart: (1) the skeletal muscle pump, and (2) the respiratory pump. Both pumps depend on the presence of valves in veins.

The **skeletal muscle pump** operates as follows (Figure 21.9):

Figure 21.9 Action of the skeletal muscle pump in returning blood to the heart. (1) At rest, both proximal and distal venous valves are open and blood flows toward the heart. (2) Contraction of leg muscles pushes blood through the proximal valve while closing the distal valve. (3) As the leg muscles relax, the proximal valve closes and the distal valve opens. When the vein fills with blood from the foot, the proximal valve reopens.

Milking refers to skeletal muscle contractions that drive venous blood toward the heart.

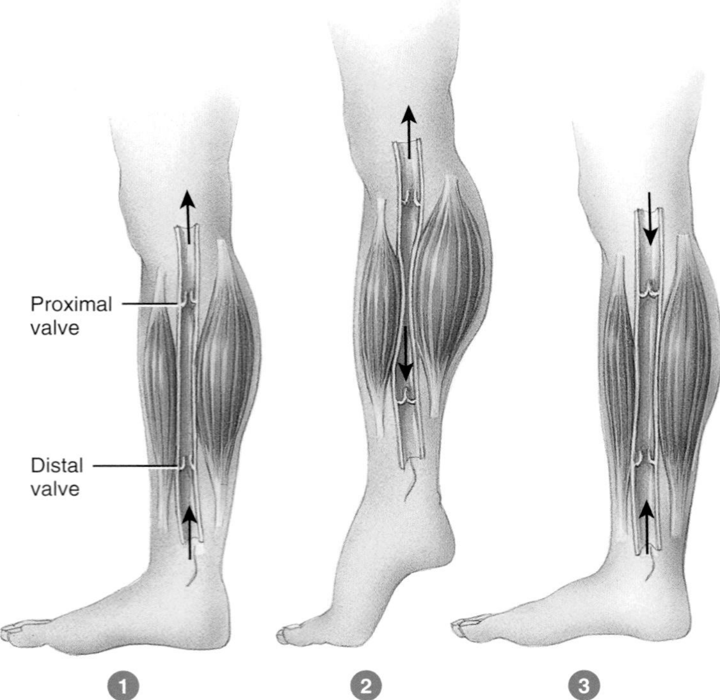

Aside from cardiac contractions, what mechanisms act as pumps to boost venous return?

1. While you are standing at rest, both the venous valve closer to the heart (proximal valve) and the one farther from the heart (distal valve) in this part of the leg are open, and blood flows upward toward the heart.
2. Contraction of leg muscles, such as when you stand on tiptoes or take a step, compresses the vein. The compression pushes blood through the proximal valve, an action called *milking*. At the same time, the distal valve in the uncompressed segment of the vein closes as some blood is pushed against it. People who are immobilized through injury or disease lack these contractions of leg muscles. As a result, their venous return is slower and they may develop circulation problems.
3. Just after muscle relaxation, pressure falls in the previously compressed section of vein, which causes the proximal valve to close. The distal valve now opens because blood pressure in the foot is higher than in the leg, and the vein fills with blood from the foot.

The **respiratory pump** is also based on alternating compression and decompression of veins. During inhalation, the diaphragm moves downward, which causes a decrease in pressure in the thoracic cavity and an increase in pressure in the abdominal cavity. As a result, abdominal veins are compressed, and a greater volume of blood moves from the compressed abdominal veins into the decompressed thoracic veins and then into the right atrium. When the pressures reverse during exhalation, the valves in the veins prevent backflow of blood from the thoracic veins to the abdominal veins.

Figure 21.10 summarizes the factors that increase blood pressure through increasing cardiac output or systemic vascular resistance.

Velocity of Blood Flow

Earlier we saw that blood flow is the *volume* of blood that flows through any tissue in a given time period (in mL/min). The speed or *velocity* of blood flow (in cm/sec) is inversely related to the cross-sectional area. Velocity is slowest where the total cross-sectional area is greatest (Figure 21.11). Each time an artery branches, the total cross-sectional area of all its branches is greater than the cross-sectional area of the original vessel, so blood flow becomes slower and slower as blood moves further away from the heart, and is slowest in the capillaries. Conversely, when venules unite to form veins, the total cross-sectional area becomes smaller and flow becomes faster. In an adult, the cross-sectional area of the aorta is only 3–5 cm^2, and the average velocity of the blood there is 40 cm/sec. In capillaries, the total cross-sectional area is 4500–6000 cm^2, and the velocity of blood flow is less than 0.1 cm/sec. In the two venae cavae combined, the cross-sectional area is about 14 cm^2, and the velocity is about 15 cm/sec. Thus, the velocity of blood flow decreases as blood flows from the aorta to arteries to arterioles to capillaries, and increases as it leaves capillaries and returns to the heart. The relatively slow rate of flow through capillaries aids the exchange of materials between blood and interstitial fluid.

Figure 21.10 Summary of factors that increase blood pressure. Changes noted within green boxes increase cardiac output; changes noted within blue boxes increase systemic vascular resistance.

Increases in cardiac output and increases in systemic vascular resistance will increase mean arterial pressure.

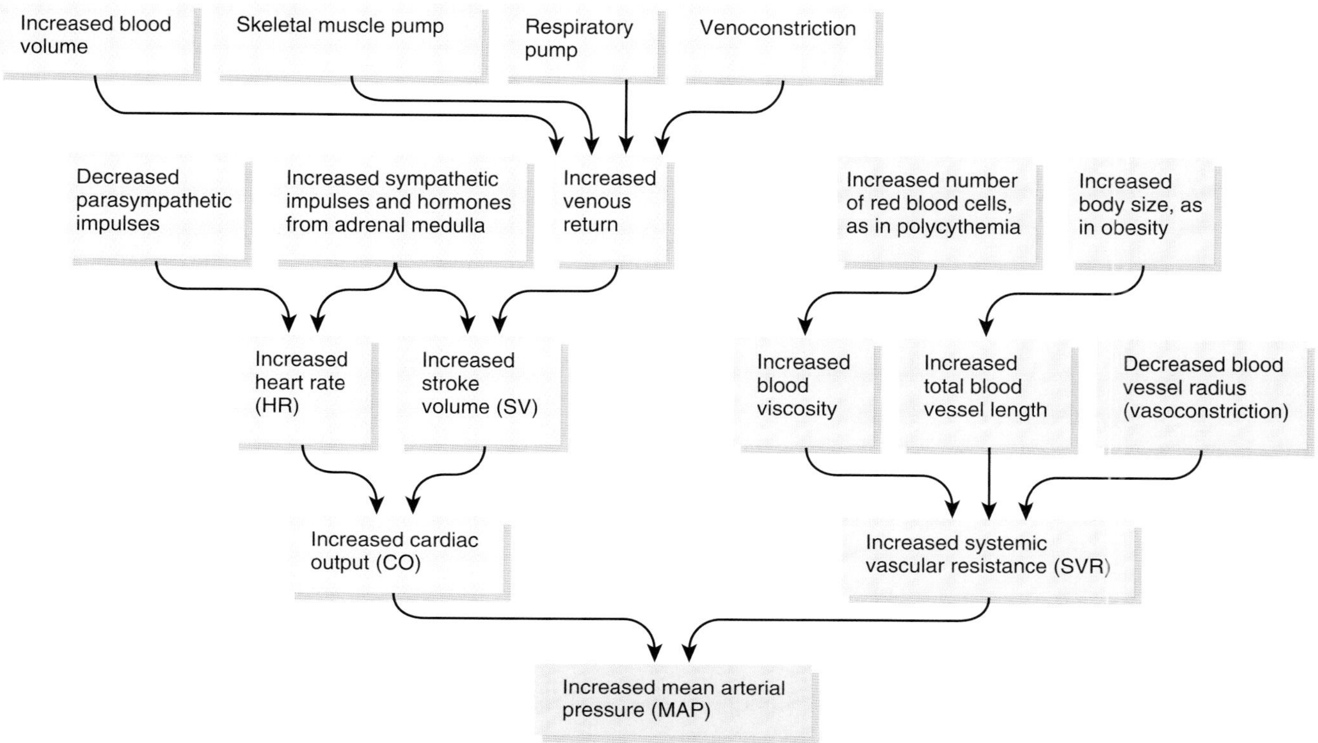

Which type of blood vessel exerts the major control of systemic vascular resistance, and how does it achieve this?

Circulation time is the time required for a drop of blood to pass from the right atrium, through the pulmonary circulation, back to the left atrium, through the systemic circulation down to the foot, and back again to the right atrium. In a resting person, circulation time normally is about 1 minute.

Syncope

Syncope (SIN-kō-pē), or fainting, is a sudden, temporary loss of consciousness that is not due to head trauma, followed by spontaneous recovery. It is most commonly due to cerebral ischemia, lack of sufficient blood flow to the brain. Syncope may occur for several reasons:

- *Vasodepressor syncope* is due to sudden emotional stress or real, threatened, or fantasized injury.
- *Situational syncope* is caused by pressure stress associated with urination, defecation, or severe coughing.
- *Drug-induced syncope* may be caused by drugs such as antihypertensives, diuretics, vasodilators, and tranquilizers.
- *Orthostatic hypotension,* an excessive decrease in blood pressure that occurs upon standing up, may cause fainting. ■

Figure 21.11 Relationship between velocity (speed) of blood flow and total cross-sectional area in different types of blood vessels.

Velocity of blood flow is slowest in the capillaries because they have the largest total cross-sectional area.

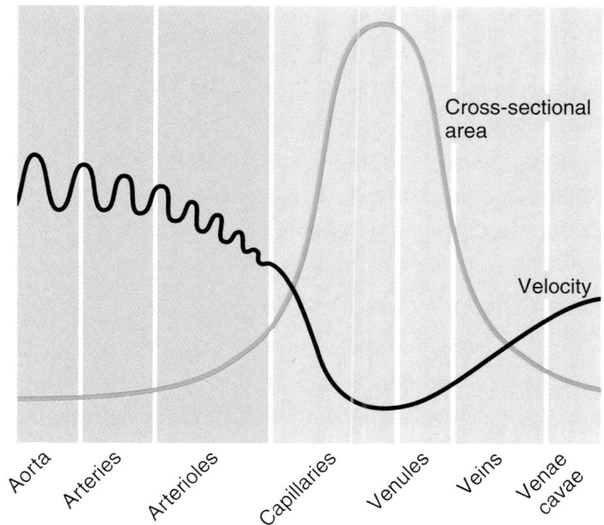

In which blood vessels is the velocity of flow fastest?

▶ CHECKPOINT

9. Explain how blood pressure and resistance determine volume of blood flow.
10. What is systemic vascular resistance and what factors contribute to it?
11. How is the return of venous blood to the heart accomplished?
12. Why is the velocity of blood flow faster in arteries and veins than in capillaries?

CONTROL OF BLOOD PRESSURE AND BLOOD FLOW

▶ OBJECTIVE

Describe how blood pressure is regulated.

Several interconnected negative feedback systems control blood pressure by adjusting heart rate, stroke volume, systemic vascular resistance, and blood volume. Some systems allow rapid adjustments to cope with sudden changes, such as the drop in blood pressure in the brain that occurs when you get out of bed; others act more slowly to provide long-term regulation of blood pressure. The body may also require adjustments to the distribution of blood flow. During exercise, for example, a greater percentage of the total blood flow is diverted to skeletal muscles.

Role of the Cardiovascular Center

In Chapter 20, we noted how the **cardiovascular (CV) center** in the medulla oblongata helps regulate heart rate and stroke volume. The CV center also controls neural, hormonal, and local negative feedback systems that regulate blood pressure and blood flow to specific tissues. Groups of neurons scattered within the CV center regulate heart rate, contractility (force of contraction) of the ventricles, and blood vessel diameter. Some neurons stimulate the heart (cardiostimulatory center); others inhibit the heart (cardioinhibitory center). Still others control blood vessel diameter by causing constriction (vasoconstrictor center) or dilation (vasodilator center); these neurons are referred to collectively as the vasomotor center. Because the CV center neurons communicate with one another, function together, and are not clearly separated anatomically, we discuss them here as a group.

The cardiovascular center receives input both from higher brain regions and from sensory receptors (Figure 21.12). Nerve impulses descend from the cerebral cortex, limbic system, and hypothalamus to affect the cardiovascular center. For example, even before you start to run a race, your heart rate may increase due to nerve impulses conveyed from the limbic system to the CV center. If your body temperature rises during a race, the hypothalamus sends nerve impulses to the CV center. The resulting vasodilation of skin blood vessels allows heat to dissipate more rapidly from the surface of the skin. The three main types of sensory receptors that provide input to the cardiovascular center are proprioceptors, baroreceptors, and chemoreceptors. *Proprioceptors* monitor movements of joints and muscles and provide input to the cardiovascular center during physical activity. Their activity accounts for the rapid increase in heart rate at the beginning of exercise. *Baroreceptors* monitor changes in pressure and stretch in the walls of blood vessels, and *chemoreceptors* monitor the concentration of various chemicals in the blood.

Output from the cardiovascular center flows along sympathetic and parasympathetic neurons of the ANS (Figure 21.12). Sympathetic impulses reach the heart via the **cardiac accelerator nerves.** An increase in sympathetic stimulation increases heart rate and contractility; a decrease in sympathetic stimulation decreases heart rate and contractility. Parasympathetic stimulation, conveyed along the **vagus (X) nerves,** decreases heart rate. Thus, opposing sympathetic (stimulatory) and parasympathetic (inhibitory) influences control the heart.

The cardiovascular center also continually sends impulses to smooth muscle in blood vessel walls via **vasomotor nerves.** These sympathetic neurons exit the spinal cord through all thoracic and the first one or two lumbar spinal nerves and then pass into the sympathetic trunk ganglia (see Figure 15.2 on page 528). From there, impulses propagate along sympathetic neurons that innervate blood vessels in viscera and peripheral areas. The vasomotor region of the cardiovascular center continually sends impulses over these routes to arterioles throughout the body, but especially to those in the skin and abdominal viscera. The result is a moderate state of tonic contraction or vasoconstriction, called **vasomotor tone,** that sets the resting level of systemic vascular resistance. Sympathetic stimulation of most veins causes constriction that moves blood out of venous blood reservoirs and increases blood pressure.

Neural Regulation of Blood Pressure

The nervous system regulates blood pressure via negative feedback loops that occur as two types of reflexes: baroreceptor reflexes and chemoreceptor reflexes.

Baroreceptor Reflexes

Baroreceptors, pressure-sensitive sensory receptors, are located in the aorta, internal carotid arteries (arteries in the neck that supply blood to the brain), and other large arteries in the neck and chest. They send impulses to the cardiovascular center to help regulate blood pressure. The two most important **baroreceptor reflexes** are the carotid sinus reflex and the aortic reflex.

Baroreceptors in the wall of the carotid sinuses initiate the **carotid sinus reflex,** which helps regulate blood pressure in the brain. The **carotid sinuses** are small widenings of the right and left internal carotid arteries just above the point where they branch from the common carotid arteries (Figure 21.13). Blood

Figure 21.12 Location and function of the cardiovascular (CV) center in the medulla oblongata. The CV center receives input from higher brain centers, proprioceptors, baroreceptors, and chemoreceptors. Then, it provides output to the sympathetic and parasympathetic divisions of the autonomic nervous system (ANS).

The cardiovascular center is the main region for nervous system regulation of the heart and blood vessels.

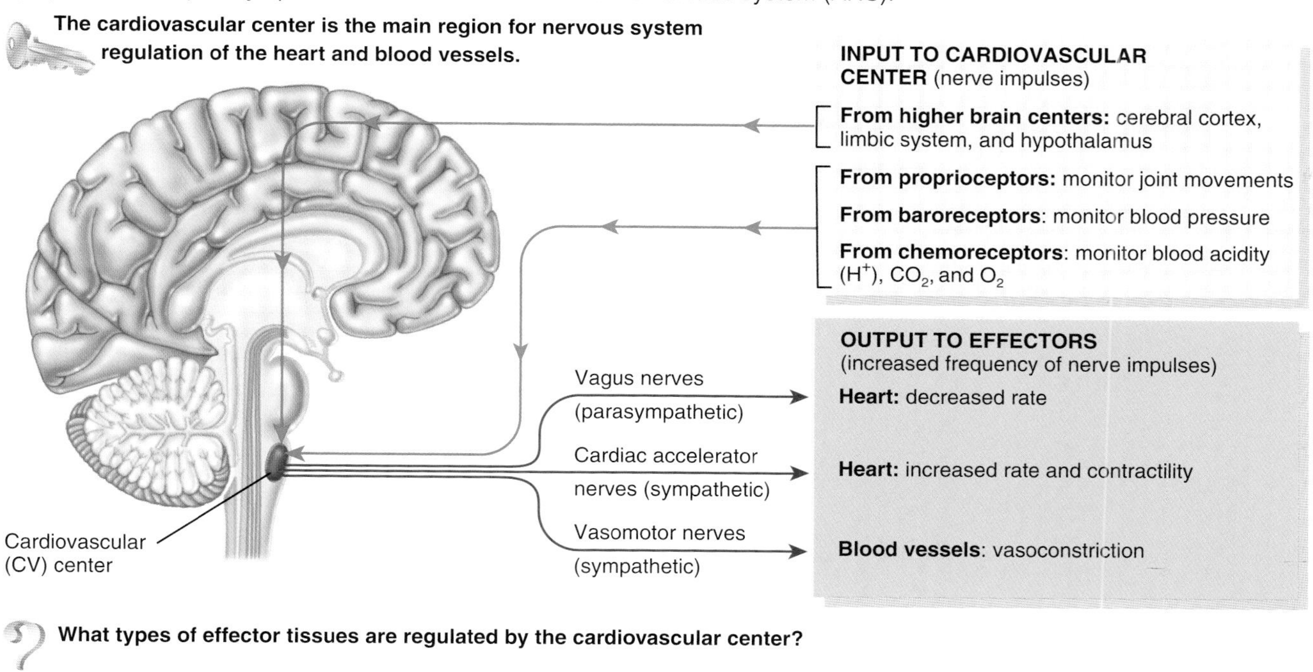

What types of effector tissues are regulated by the cardiovascular center?

Figure 21.13 ANS innervation of the heart and the baroreceptor reflexes that help regulate blood pressure.

Baroreceptors are pressure-sensitive neurons that monitor stretching.

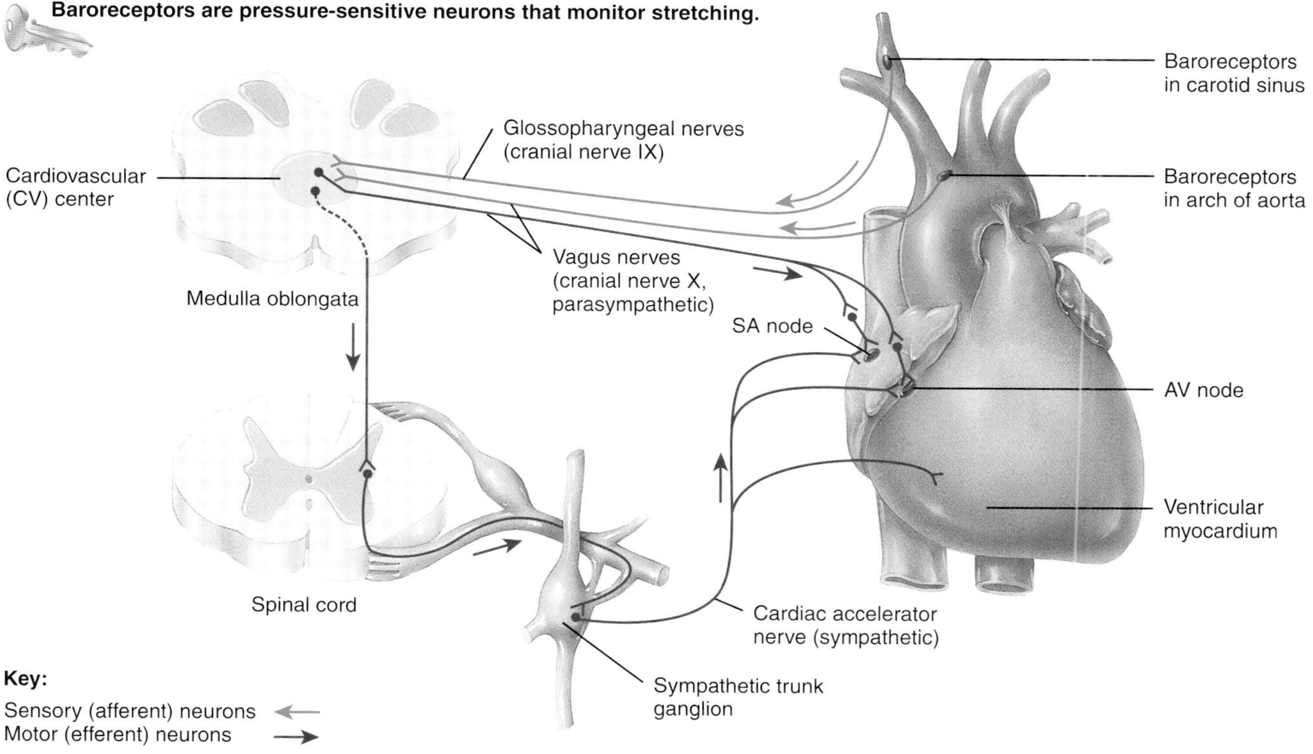

Which cranial nerves conduct impulses to the cardiovascular center from baroreceptors in the carotid sinuses and the arch of the aorta?

pressure stretches the wall of the carotid sinus, which stimulates the baroreceptors. Nerve impulses propagate from the carotid sinus baroreceptors over sensory axons in the **glossopharyngeal (IX) nerves** to the cardiovascular center in the medulla oblongata. Baroreceptors in the wall of the ascending aorta and arch of the aorta initiate the **aortic reflex,** which regulates systemic blood pressure. Nerve impulses from aortic baroreceptors reach the cardiovascular center via sensory axons of the **vagus (X) nerves.**

When blood pressure falls, the baroreceptors are stretched less, and they send nerve impulses at a slower rate to the cardiovascular center (Figure 21.14). In response, the CV center decreases parasympathetic stimulation of the heart by way of motor axons of the vagus nerves and increases sympathetic stimulation of the heart via cardiac accelerator nerves. Another consequence of increased sympathetic stimulation is increased secretion of epinephrine and norepinephrine by the adrenal medulla. As the heart beats faster and more forcefully, and as systemic vascular resistance increases, cardiac output and systemic vascular resistance rise, and blood pressure increases to the normal level.

Conversely, when an increase in pressure is detected, the baroreceptors send impulses at a faster rate. The CV center responds by increasing parasympathetic stimulation and decreasing sympathetic stimulation. The resulting decreases in heart rate and force of contraction reduce the cardiac output. The cardiovascular center also slows the rate at which it sends sympathetic impulses along vasomotor neurons that normally cause vasoconstriction. The resulting vasodilation lowers systemic vascular resistance. Decreased cardiac output and decreased systemic vascular resistance both lower systemic arterial blood pressure to the normal level.

Moving from a prone (lying down) to an erect position decreases blood pressure and blood flow in the head and upper part of the body. The baroreceptor reflexes, however, quickly counteract the drop in pressure. Sometimes these reflexes operate more slowly than normal, especially in the elderly, in which case a person can faint due to reduced brain blood flow upon standing up too quickly.

Carotid Sinus Massage and Carotid Sinus Syncope

Because the carotid sinus is close to the anterior surface of the neck, it is possible to stimulate the baroreceptors there by putting pressure on the neck. Physicians sometimes use **carotid sinus massage,** which involves carefully massaging the neck over the carotid sinus, to slow heart rate in a person who has paroxysmal superventricular tachycardia, a type of tachycardia that originates in the atria. Anything that stretches or puts pressure on the carotid sinus, such as hyperextension of the head, tight collars, or carrying heavy shoulder loads, may also slow heart rate and can cause **carotid sinus syncope,** fainting due to inappropriate stimulation of the carotid sinus baroreceptors. ■

Figure 21.14 Negative feedback regulation of blood pressure via baroreceptor reflexes.

When blood pressure decreases, heart rate increases.

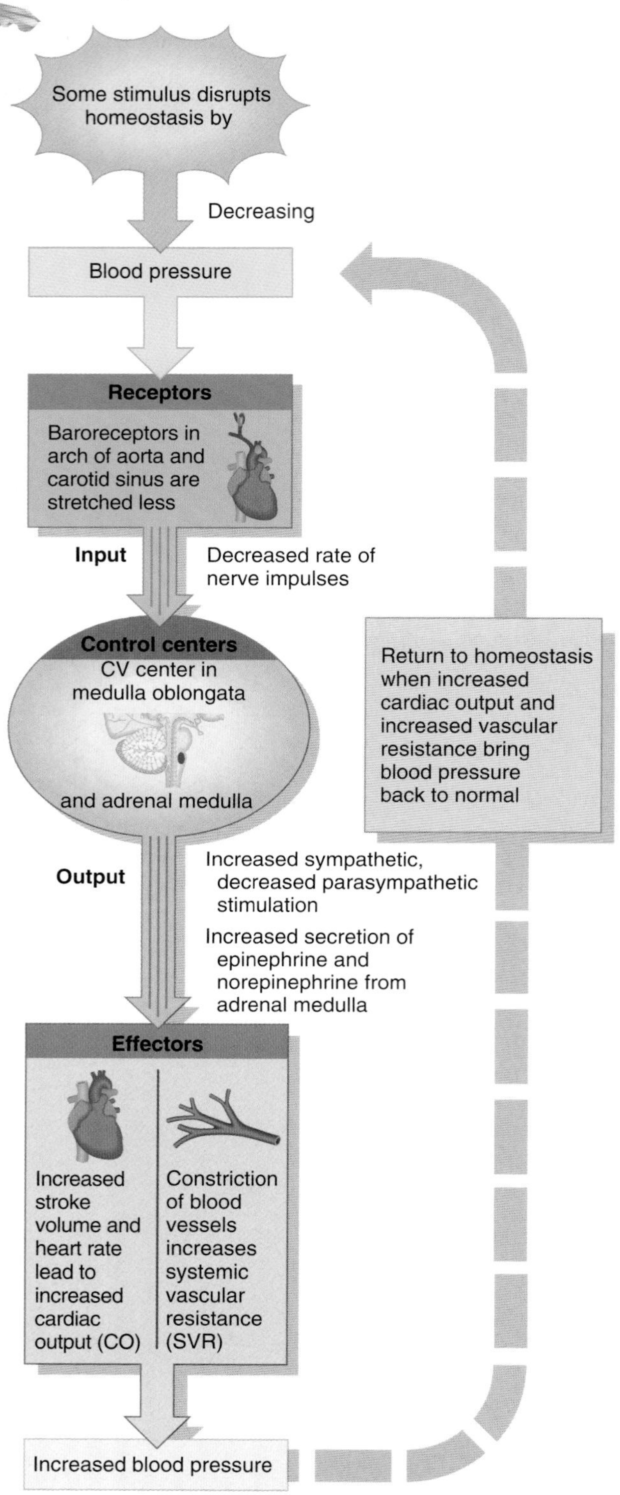

Does this negative feedback cycle represent the changes that occur when you lie down or when you stand up?

Chemoreceptor Reflexes

Chemoreceptors, sensory receptors that monitor the chemical composition of blood, are located close to the baroreceptors of the carotid sinus and arch of the aorta in small structures called **carotid bodies** and **aortic bodies,** respectively. These chemoreceptors detect changes in blood level of O_2, CO_2, and H^+. *Hypoxia* (lowered O_2 availability), *acidosis* (an increase in H^+ concentration), or *hypercapnia* (excess CO_2) stimulates the chemoreceptors to send impulses to the cardiovascular center. In response, the CV center increases sympathetic stimulation to arterioles and veins, producing vasoconstriction and an increase in blood pressure. These chemoreceptors also provide input to the respiratory center in the brain stem to adjust the rate of breathing.

Hormonal Regulation of Blood Pressure

As you learned in Chapter 18, several hormones help regulate blood pressure and blood flow by altering cardiac output, changing systemic vascular resistance, or adjusting the total blood volume:

1. ***Renin–angiotensin–aldosterone (RAA) system.*** When blood volume falls or blood flow to the kidneys decreases, juxtaglomerular cells in the kidneys secrete **renin** into the bloodstream. In sequence, renin and angiotensin converting enzyme (ACE) act on their substrates to produce the active hormone **angiotensin II,** which raises blood pressure in two ways. First, angiotensin II is a potent vasoconstrictor; it raises blood pressure by increasing systemic vascular resistance. Second, it stimulates secretion of **aldosterone,** which increases reabsorption of sodium ions (Na^+) and water by the kidneys. The water reabsorption increases total blood volume, which increases blood pressure. (See page 756.)
2. ***Epinephrine and norepinephrine.*** In response to sympathetic stimulation, the adrenal medulla releases epinephrine and norepinephrine. These hormones increase cardiac output by increasing the rate and force of heart contractions. They also cause vasoconstriction of arterioles and veins in the skin and abdominal organs and vasodilation of arterioles in cardiac and skeletal muscle, which helps increase blood flow to muscle during exercise (see Figure 18.20 on page 653).
3. ***Antidiuretic hormone (ADH).*** ADH is produced by the hypothalamus and released from the posterior pituitary in response to dehydration or decreased blood volume. Among other actions, ADH causes vasoconstriction, which increases blood pressure. For this reason ADH is also called **vasopressin.** (See Figure 18.9 on page 633)
4. ***Atrial natriuretic peptide (ANP).*** Released by cells in the atria of the heart, ANP lowers blood pressure by causing vasodilation and by promoting the loss of salt and water in the urine, which reduces blood volume.

Table 21.2 summarizes the regulation of blood pressure by hormones.

Autoregulation of Blood Pressure

In each capillary bed, local changes can regulate vasomotion. When vasodilators produce local dilation of arterioles and relaxation of precapillary sphincters, blood flow into capillary networks is increased, which increases O_2 level. Vasoconstrictors have the opposite effect. The ability of a tissue to automatically adjust its blood flow to match its metabolic demands is called **autoregulation.** In tissues such as the heart and skeletal muscle, where the demand for O_2 and nutrients and for the removal of wastes can increase as much as tenfold during physical activity, autoregulation is an important contributor to increased blood flow through the tissue. Autoregulation also controls regional blood flow in the brain; blood distribution to various parts of the brain changes dramatically for different mental and physical activities. During a conversation, for example, blood flow increases to your motor speech areas when you are talking and increases to the auditory areas when you are listening.

Two general types of stimuli cause autoregulatory changes in blood flow:

1. ***Physical changes.*** Warming promotes vasodilation, and cooling causes vasoconstriction. In addition, smooth muscle in arteriole walls exhibits a **myogenic response**—it contracts more forcefully when it is stretched and relaxes when stretching lessens. If, for

TABLE 21.2 Blood Pressure Regulation by Hormones

Factor Influencing Blood Pressure	Hormone	Effect on Blood Pressure
Cardiac Output		
Increased heart rate and contractility	Norepinephrine Epinephrine	Increase
Systemic Vascular Resistance		
Vasoconstriction	Angiotensin II Antidiuretic hormone (vasopressin) Norepinephrine* Epinephrine*	Increase
Vasodilation	Atrial natriuretic peptide Epinephrine† Nitric oxide	Decrease
Blood Volume		
Blood volume increase	Aldosterone Antidiuretic hormone	Increase
Blood volume decrease	Atrial natriuretic peptide	Decrease

*Acts at α_1 receptors in arterioles of abdomen and skin.
†Acts at β_2 receptors in arterioles of cardiac and skeletal muscle; norepinephrine has a much smaller vasodilating effect.

example, blood flow through an arteriole decreases, stretching of the arteriole walls decreases. As a result, the smooth muscle relaxes and produces vasodilation, which increases blood flow.

2. ***Vasodilating and vasoconstricting chemicals.*** Several types of cells—including white blood cells, platelets, smooth muscle fibers, macrophages, and endothelial cells—release a wide variety of chemicals that alter blood-vessel diameter. Vasodilating chemicals released by metabolically active tissue cells include K^+, H^+, lactic acid (lactate), and adenosine (from ATP). Another important vasodilator released by endothelial cells is nitric oxide (NO). Tissue trauma or inflammation causes release of vasodilating kinins and histamine. Vasoconstrictors include thromboxane A2, superoxide radicals, serotonin (from platelets), and endothelins (from endothelial cells).

An important difference between the pulmonary and systemic circulations is their autoregulatory response to changes in O_2 level. The walls of blood vessels in the systemic circulation *dilate* in response to low O_2. With vasodilation, O_2 delivery increases, which restores the normal O_2 level. By contrast, the walls of blood vessels in the pulmonary circulation *constrict* in response to low levels of O_2. This response ensures that blood mostly bypasses those alveoli (air sacs) in the lungs that are poorly ventilated by fresh air. Thus, most blood flows to better-ventilated areas of the lung.

► CHECKPOINT

13. What are the principal inputs to and outputs from the cardiovascular center?

14. Explain the operation of the carotid sinus reflex and the aortic reflex.

15. What is the role of chemoreceptors in the regulation of blood pressure?

16. How do hormones regulate blood pressure?

17. What is autoregulation and how does it differ in the systemic and pulmonary circulations?

CHECKING CIRCULATION

► OBJECTIVE

Define pulse, and define systolic, diastolic, and pulse pressures.

Pulse

The alternate expansion and recoil of elastic arteries after each systole of the left ventricle creates a traveling pressure wave that is called the **pulse.** The pulse is strongest in the arteries closest to the heart, becomes weaker in the arterioles, and disappears altogether in the capillaries. The pulse may be felt in any artery that lies near the surface of the body that can be compressed against a bone or other firm structure. Table 21.3 depicts some common pulse points.

The pulse rate normally is the same as the heart rate, about 70 to 80 beats per minute at rest. **Tachycardia** (tak′-i-KAR-dē-a; *tachy-* = fast) is a rapid resting heart or pulse rate over 100 beats/min. **Bradycardia** (brād′-i-KAR-dē-a; *brady-* = slow) is a slow resting heart or pulse rate under 50 beats/min. Endurance-trained athletes normally exhibit bradycardia.

Measuring Blood Pressure

In clinical use, the term **blood pressure** usually refers to the pressure in arteries generated by the left ventricle during systole and the pressure remaining in the arteries when the ventricle is in diastole. Blood pressure is usually measured in the brachial artery in the left arm (Table 21.3). The device used to measure blood pressure is a **sphygmomanometer** (sfig′-mō-ma-NOM-e-ter; *sphygmo-* = pulse; *manometer* = instrument used to measure pressure). It consists of a rubber cuff connected to a rubber bulb that is used to inflate the cuff and a meter that registers the pressure in the cuff. With the arm resting on a table so that it is about the same level as the heart, the cuff of the sphygmomanometer is wrapped around a bared arm. The cuff is inflated by squeezing the bulb until the brachial artery is compressed and blood flow stops, about 30 mmHg higher than the person's usual systolic pressure. The technician places a stethoscope below the cuff on the brachial artery, and slowly deflates the cuff. When the cuff is deflated enough to allow the artery to open, a spurt of blood passes through, resulting in the first sound heard through the stethoscope. This sound corresponds to **systolic blood pressure (SBP),** the force of blood pressure on arterial walls just after ventricular contraction (Figure 21.15). As the cuff is deflated further, the sounds suddenly become too faint to be heard through the stethoscope. This level, called the **diastolic blood pressure (DBP),** represents the force exerted by the blood remaining in arteries during ventricular relaxation. At pressures below diastolic blood pressure, sounds disappear altogether. The various sounds that are heard while taking blood pressure are called **Korotkoff sounds** (kō-ROT-kof).

The normal blood pressure of an adult male is less than 120 mmHg systolic and less than 80 mmHg diastolic. For example, "110 over 70" (written as 110/70) is a normal blood pressure. In young adult females, the pressures are 8 to 10 mmHg less. People who exercise regularly and are in good physical condition may have even lower blood pressures. Thus, blood pressure slightly lower than 120/80 may be a sign of good health and fitness.

The difference between systolic and diastolic pressure is called **pulse pressure.** This pressure, normally about 40 mmHg, provides information about the condition of the cardiovascular system. For example, conditions such as atherosclerosis and patent (open) ductus arteriosus greatly increase pulse pressure. The normal ratio of systolic pressure to diastolic pressure to pulse pressure is about 3 : 2 : 1.

TABLE 21.3 Pulse Points

Structure	Location
Superficial temporal artery	Lateral to orbit of eye.
Facial artery	Mandible (lower jawbone) on a line with the corners of the mouth.
Common carotid artery	Lateral to larynx (voice box).
Brachial artery	Medial side of biceps brachii muscle.
Femoral artery	Inferior to inguinal ligament.
Popliteal artery	Posterior to knee.
Radial artery	Distal aspect of wrist.
Dorsal artery of the foot (dorsalis pedis artery)	Superior to instep of foot.

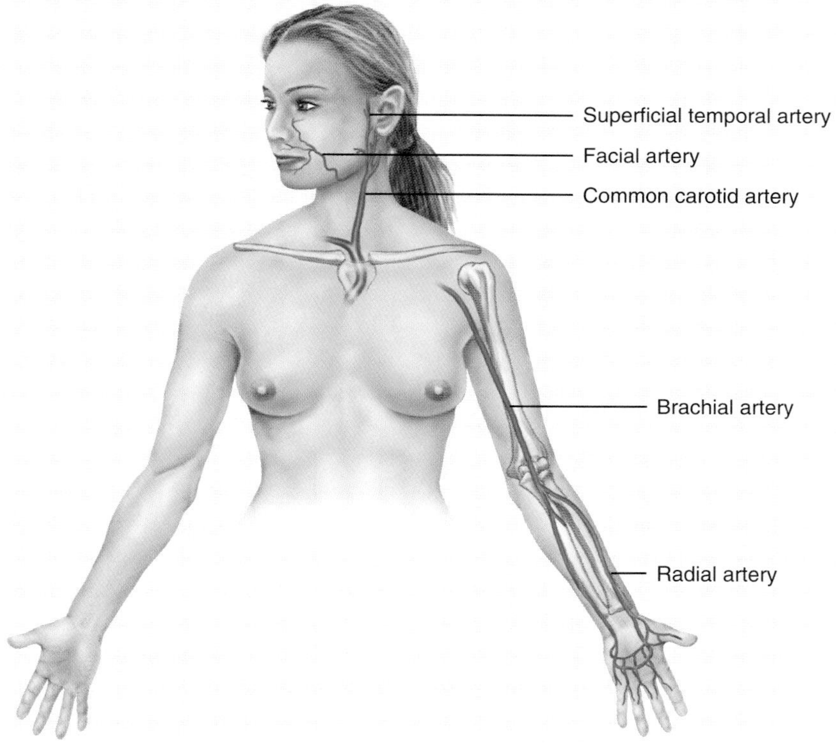

Superficial temporal artery
Facial artery
Common carotid artery
Brachial artery
Radial artery
Femoral artery 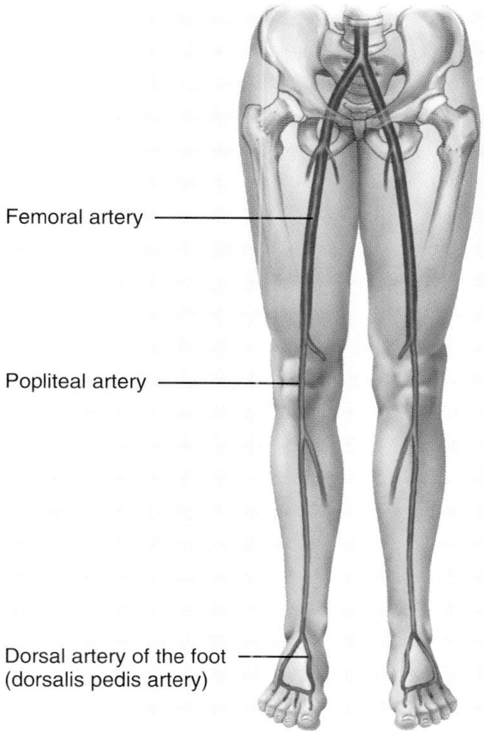
Popliteal artery
Dorsal artery of the foot (dorsalis pedis artery)

Figure 21.15 Relationship of blood pressure changes to cuff pressure.

As the cuff is deflated, sounds first occur at the systolic blood pressure; the sounds suddenly become faint at the diastolic blood pressure.

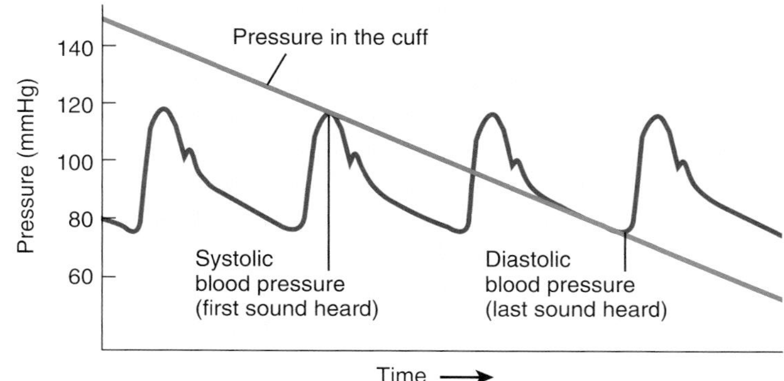

If a blood pressure is reported as "142 over 95," what are the diastolic, systolic, and pulse pressures? Does this person have hypertension as defined on page 798?

▶ CHECKPOINT

18. Where may the pulse be felt?

19. What do tachycardia and bradycardia mean?

20. How are systolic and diastolic blood pressures measured with a sphygmomanometer?

SHOCK AND HOMEOSTASIS

▶ OBJECTIVES

Define shock, and describe the four types of shock.

Explain how the body's response to shock is regulated by negative feedback.

Shock is a failure of the cardiovascular system to deliver enough O_2 and nutrients to meet cellular metabolic needs. The causes of shock are many and varied, but all are characterized by inadequate blood flow to body tissues. With inadequate oxygen delivery, cells switch from aerobic to anaerobic production of ATP, and lactic acid accumulates in body fluids. If shock persists, cells and organs become damaged, and cells may die unless proper treatment begins quickly.

Types of Shock

Shock can be of four different types: (1) **hypovolemic shock** (hī-pō-vō-LĒ-mik; *hypo-* = low; *-volemic* = volume) due to decreased blood volume, (2) **cardiogenic shock** due to poor heart function, (3) **vascular shock** due to inappropriate vasodilation, and (4) **obstructive shock** due to obstruction of blood flow.

A common cause of hypovolemic shock is acute (sudden) hemorrhage. The blood loss may be external, as occurs in trauma, or internal, as in rupture of an aortic aneurysm. Loss of body fluids through excessive sweating, diarrhea, or vomiting also can cause hypovolemic shock. Other conditions—for instance, diabetes mellitus—may cause excessive loss of fluid in the urine. Sometimes, hypovolemic shock is due to inadequate intake of fluid. Whatever the cause, when the volume of body fluids falls, venous return to the heart declines, filling of the heart lessens, stroke volume decreases, and cardiac output decreases.

In cardiogenic shock, the heart fails to pump adequately, most often because of a myocardial infarction (heart attack). Other causes of cardiogenic shock include poor perfusion of the heart (ischemia), heart valve problems, excessive preload or afterload, impaired contractility of heart muscle fibers, and arrhythmias.

Even with normal blood volume and cardiac output, shock may occur if blood pressure drops due to a decrease in systemic vascular resistance. A variety of conditions can cause inappropriate dilation of arterioles or venules. In *anaphylactic shock,* a severe allergic reaction—for example, to a bee sting—releases histamine and other mediators that cause vasodilation. In *neurogenic shock,* vasodilation may occur following trauma to the head that causes malfunction of the cardiovascular center in the medulla. Shock stemming from certain bacterial toxins that produce vasodilation is termed *septic shock.* In the United States, septic shock causes more than 100,000 deaths per year and is the most common cause of death in hospital critical care units.

Obstructive shock occurs when blood flow through a portion of the circulation is blocked. The most common cause is *pulmonary embolism,* a blood clot lodged in a blood vessel of the lungs.

Homeostatic Responses to Shock

The major mechanisms of compensation in shock are *negative feedback systems* that work to return cardiac output and arterial blood pressure to normal. When shock is mild, compensation by homeostatic mechanisms prevents serious damage. In an otherwise healthy person, compensatory mechanisms can maintain adequate blood flow and blood pressure despite an acute blood loss of as much as 10% of total volume. Figure 21.16 shows several of the negative feedback systems that respond to hypovolemic shock.

1. ***Activation of the renin–angiotensin–aldosterone system.*** Decreased blood flow to the kidneys causes the kidneys to secrete renin and initiates the renin–angiotensin–aldosterone system (see Figure 18.16 on page 643). Recall that angiotensin II causes vasoconstriction and stimulates the adrenal cortex to secrete aldosterone, a hormone that increases reabsorption of Na^+ and water by the kidneys. The increases in systemic vascular resistance and blood volume help raise blood pressure.

2. ***Secretion of antidiuretic hormone.*** In response to decreased blood pressure, the posterior pituitary releases more antidiuretic hormone (ADH). ADH enhances water reabsorption by the kidneys, which conserves remaining blood volume. It also causes vasoconstriction, which increases systemic vascular resistance. (See Figure 18.9 on page 633.)

3. ***Activation of the sympathetic division of the ANS.*** As blood pressure decreases, the aortic and carotid baroreceptors initiate powerful sympathetic responses throughout the body. One result is marked vasoconstriction of arterioles and veins of the skin, kidneys, and other abdominal viscera. (Vasoconstriction does not occur in the brain or heart.) The constriction of arterioles increases systemic vascular resistance and the constriction of veins increases venous return. Both effects help maintain an adequate blood pressure. Sympathetic stimulation also increases heart rate and contractility and increases secretion of epinephrine and norepinephrine by the adrenal medulla. These hormones intensify vasoconstriction and increase heart rate and contractility, all of which help raise blood pressure.

4. ***Release of local vasodilators.*** In response to *hypoxia,* cells liberate vasodilators—including K^+, H^+, lactic acid, adenosine, and nitric oxide—that dilate arterioles and relax precapillary

Figure 21.16 Negative feedback systems that can restore normal blood pressure during hypovolemic shock.

Homeostatic mechanisms can compensate for an acute blood loss of as much as 10% of total blood volume.

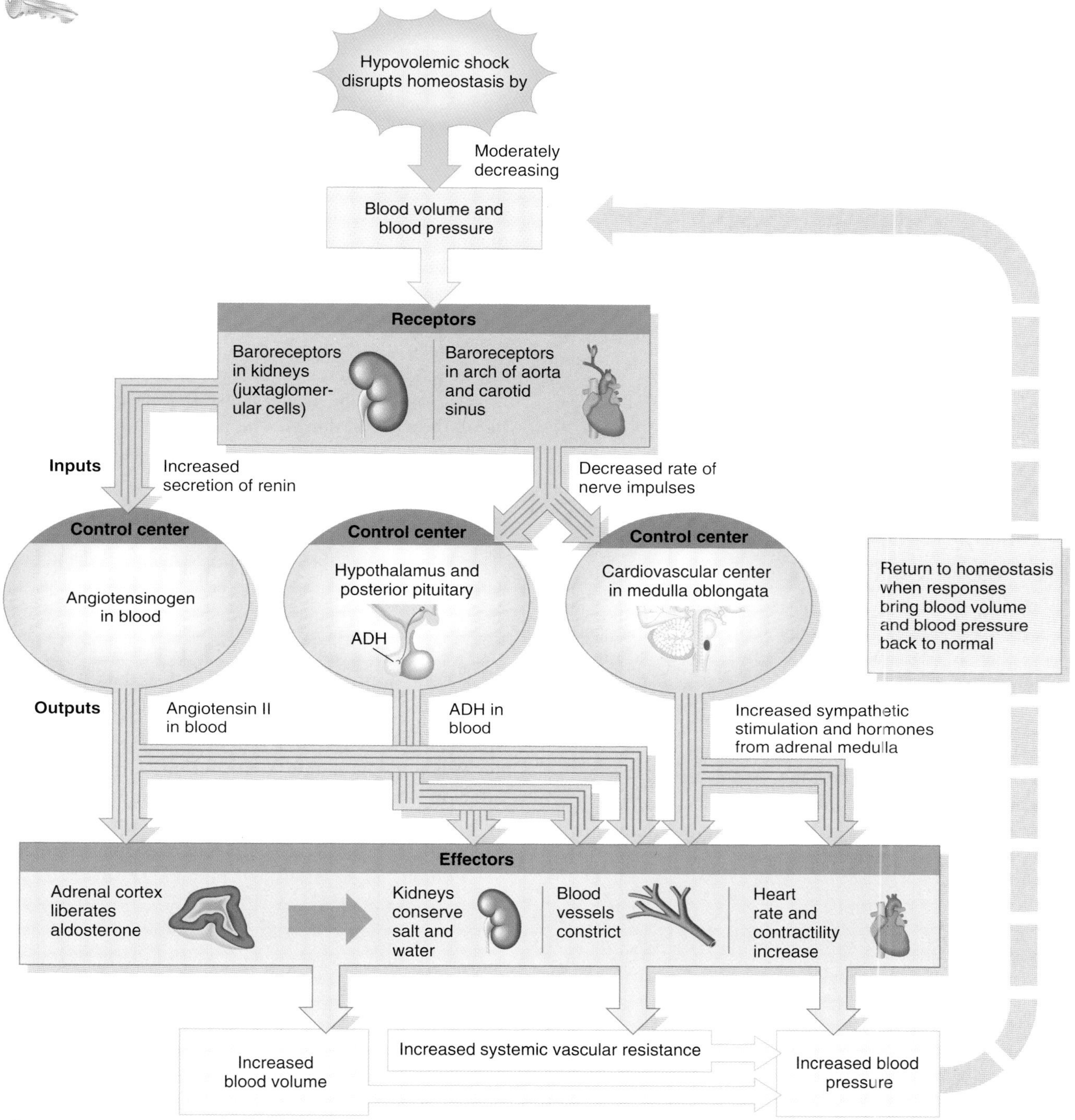

Hypovolemic shock disrupts homeostasis by

Moderately decreasing

Blood volume and blood pressure

Receptors

Baroreceptors in kidneys (juxtaglomerular cells)

Baroreceptors in arch of aorta and carotid sinus

Inputs

Increased secretion of renin

Decreased rate of nerve impulses

Control center

Angiotensinogen in blood

Control center

Hypothalamus and posterior pituitary

ADH

Control center

Cardiovascular center in medulla oblongata

Return to homeostasis when responses bring blood volume and blood pressure back to normal

Outputs

Angiotensin II in blood

ADH in blood

Increased sympathetic stimulation and hormones from adrenal medulla

Effectors

Adrenal cortex liberates aldosterone

Kidneys conserve salt and water

Blood vessels constrict

Heart rate and contractility increase

Increased blood volume

Increased systemic vascular resistance

Increased blood pressure

Does almost-normal blood pressure in a person who has lost blood indicate that the patient's tissues are receiving adequate perfusion (blood flow)?

sphincters. Such vasodilation increases local blood flow and may restore O_2 level to normal in part of the body. However, vasodilation also has the potentially harmful effect of decreasing systemic vascular resistance and thus lowering the blood pressure.

If blood volume drops more than 10–20%, or if the heart cannot bring blood pressure up sufficiently, compensatory mechanisms may fail to maintain adequate blood flow to tissues. At this point, shock becomes life-threatening as damaged cells start to die.

Signs and Symptoms of Shock

Even though the signs and symptoms of shock vary with the severity of the condition, most can be predicted in light of the responses generated by the negative feedback systems that attempt to correct the problem. Among the signs and symptoms of shock are the following:

- Systolic blood pressure is lower than 90 mmHg.
- Resting heart rate is rapid due to sympathetic stimulation and increased blood levels of epinephrine and norepinephrine.
- Pulse is weak and rapid due to reduced cardiac output and fast heart rate.
- Skin is cool, pale, and clammy due to sympathetic constriction of skin blood vessels and sympathetic stimulation of sweating.
- Mental state is altered due to reduced oxygen supply to the brain.
- Urine formation is reduced due to increased levels of aldosterone and antidiuretic hormone (ADH).
- The person is thirsty due to loss of extracellular fluid.
- The pH of blood is low (acidosis) due to buildup of lactic acid.
- The person may have nausea due to impaired blood flow to the digestive organs due to sympathetic vasoconstriction.

▶ CHECKPOINT

21. Which symptoms of hypovolemic shock relate to actual body fluid loss, and which relate to the negative feedback systems that attempt to maintain blood pressure and blood flow?

22. Describe the types of shock and their causes.

CIRCULATORY ROUTES

▶ OBJECTIVE

Describe and compare the major routes that blood takes through various regions of the body.

Blood vessels are organized into **circulatory routes** that carry blood to specific organs in the body (Figure 21.17). The routes are parallel—in most cases a portion of the cardiac output flows separately to each tissue of the body, so that each organ receives its own supply of freshly oxygenated blood. The two main circulatory routes, the systemic circulation and pulmonary circulation, differ in two important ways. First, blood in the pulmonary circulation need not be pumped as far as blood in the systemic circulation. Second, compared to systemic arteries, pulmonary arteries have larger diameters, thinner walls, and less elastic tissue. As a result, the resistance to pulmonary blood flow is very low, which means that less pressure is needed to move blood through the lungs. The peak systolic pressure in the right ventricle is only 20% of that in the left ventricle.

The Systemic Circulation

The **systemic circulation** includes the arteries and arterioles that carry oxygenated blood from the left ventricle to systemic capillaries, plus the veins and venules that return deoxygenated blood to the right atrium. Blood leaving the aorta and flowing through the systemic arteries is a bright red color. As blood flows through capillaries, it loses some of its oxygen and picks up carbon dioxide, becoming a dark red color. All systemic arteries branch from the **aorta.** Completing the circuit, all the veins of the systemic circulation drain into the **superior vena cava,** the **inferior vena cava,** or the **coronary sinus,** which in turn empty into the right atrium. The bronchial arteries, which carry nutrients to the lungs, also are part of the systemic circulation.

Exhibits 21.1–21.12 and Figures 21.18–21.27 describe the main arteries and veins of the systemic circulation. The blood-vessels are organized in the exhibits according to body regions. Figure 21.18a provides an overview of the major arteries, and Figure 21.23 provides an overview of the major veins. As you study the various blood vessels in the exhibits, refer to these two figures to see the relationships of the blood vessels under consideration to other regions of the body.

(text continues on page 791)

Figure 21.17 Circulatory routes. Large black arrows indicate the systemic circulation (detailed in Exhibits 21.3–21.12), small black arrows the pulmonary circulation (detailed in Figure 21.29), and red arrows the hepatic portal circulation (detailed in Figure 21.28). Refer to Figure 20.8 on page 707 for details of the coronary circulation, and to Figure 21.30 for details of the fetal circulation.

Blood vessels are organized into various routes that deliver blood to tissues of the body.

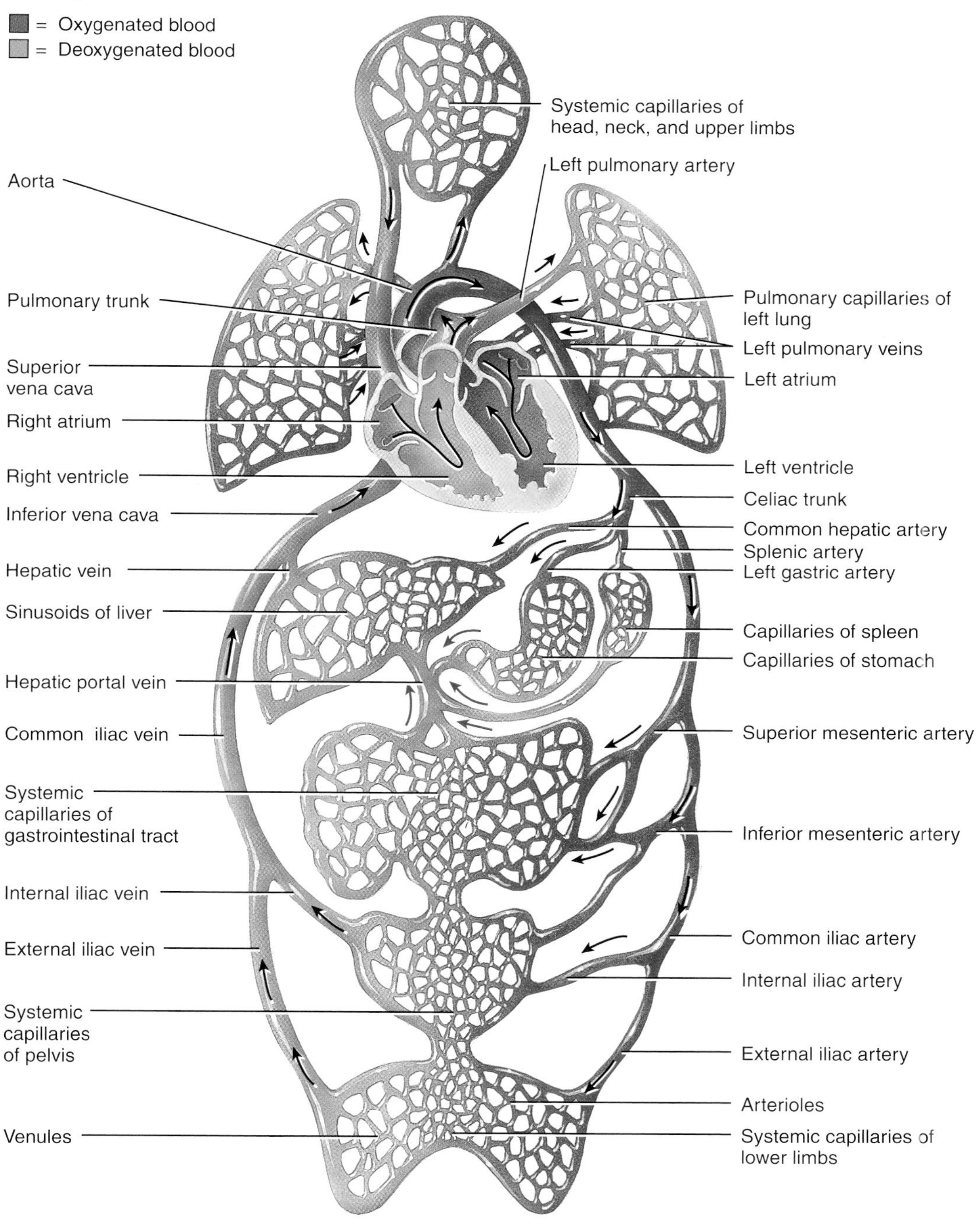

What are the two main circulatory routes?

▶ OBJECTIVES

Identify the four principal divisions of the aorta.

Locate the major arterial branches arising from each division.

The **aorta** (= to lift up) is the largest artery of the body, with a diameter of 2–3 cm (about 1 in.). Its four principal divisions are the ascending aorta, arch of the aorta, thoracic aorta, and abdominal aorta. The portion of the aorta that emerges from the left ventricle posterior to the pulmonary trunk is the **ascending aorta.** The beginning of the aorta contains the aortic valve (see Figure 20.4a on page 701). The ascending aorta gives off two coronary artery branches that supply the myocardium of the heart. Then the ascending aorta turns to the left, forming the **arch of the aorta,** which descends and ends at the level of the intervertebral disc between the fourth and fifth thoracic vertebrae. As the aorta continues to descend, it lies close to the vertebral bodies, passes through the aortic hiatus of the diaphragm, and divides at the level of the fourth lumbar vertebra into two **common iliac arteries,** which carry blood to the lower limbs. The section of the aorta between the arch of the aorta and the diaphragm is called the **thoracic aorta;** the section between the diaphragm and the common iliac arteries is the **abdominal aorta.** Each division of the aorta gives off arteries that branch into distributing arteries that lead to various organs. Within the organs, the arteries divide into arterioles and then into capillaries that service the systemic tissues (all tissues except the alveoli of the lungs).

▶ CHECKPOINT

What general regions do each of the four principal divisions of the aorta supply?

DIVISION AND BRANCHES	REGION SUPPLIED
Ascending Aorta	
Right and left coronary arteries	Heart.
Arch of the Aorta	
Brachiocephalic trunk (brā′-kē-ō-se-FAL-ik)	
Right common carotid artery (ka-ROT-id)	Right side of head and neck.
Right subclavian artery (sub-KLĀ-vē-an)	Right upper limb.
Left common carotid artery	Left side of head and neck.
Left subclavian artery	Left upper limb.
Thoracic Aorta (thorac- = chest)	
Pericardial arteries (per-i-KAR-dē-al)	Pericardium.
Bronchial arteries (BRONG-kē-al)	Bronchi of lungs.
Esophageal arteries (e-sof′-a-JĒ-al)	Esophagus.
Mediastinal arteries (mē′-dē-as-TĪ-nal)	Structures in mediastinum.
Posterior intercostal arteries (in′-ter-KOS-tal)	Intercostal and chest muscles.
Subcostal arteries (sub-KOS-tal)	Same as posterior intercostals.
Superior phrenic arteries (FREN-ik)	Superior and posterior surfaces of diaphragm.
Abdominal Aorta	
Inferior phrenic arteries (FREN-ik)	Inferior surface of diaphragm.
Celiac trunk (SĒ-lē-ak)	
Common hepatic artery (he-PAT-ik)	Liver.
Left gastric artery (GAS-trik)	Stomach and esophagus.
Splenic artery (SPLĒN-ik)	Spleen, pancreas, and stomach.
Superior mesenteric artery (MES-en-ter′-ik)	Small intestine, cecum, ascending and transverse colons, and pancreas.
Suprarenal arteries (soo-pra-RĒ-nal)	Adrenal (suprarenal) glands.
Renal arteries (RĒ-nal)	Kidneys.
Gonadal arteries (gō-NAD-al)	
Testicular arteries (tes-TIK-ū-lar)	Testes (male).
Ovarian arteries (ō-VAR-ē-an)	Ovaries (female).
Inferior mesenteric artery	Transverse, descending, and sigmoid colons; rectum.
Common iliac arteries (IL-ē-ak)	
External iliac arteries	Lower limbs.
Internal iliac arteries	Uterus (female), prostate (male), muscles of buttocks, and urinary bladder.

Figure 21.18 **Aorta and its principal branches.**

All systemic arteries branch from the aorta.

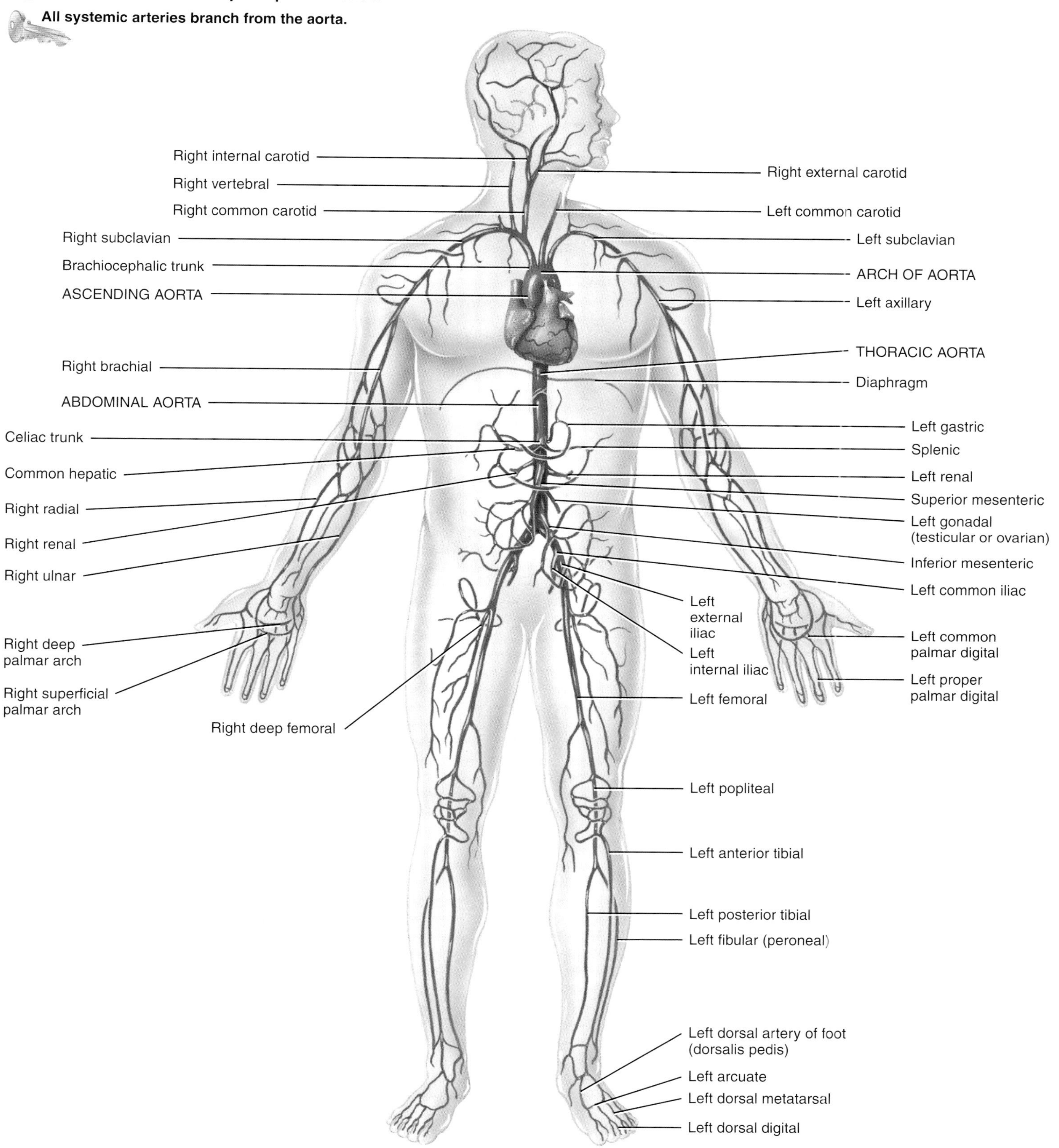

(a) Overall anterior view of the principal branches of the aorta

continues

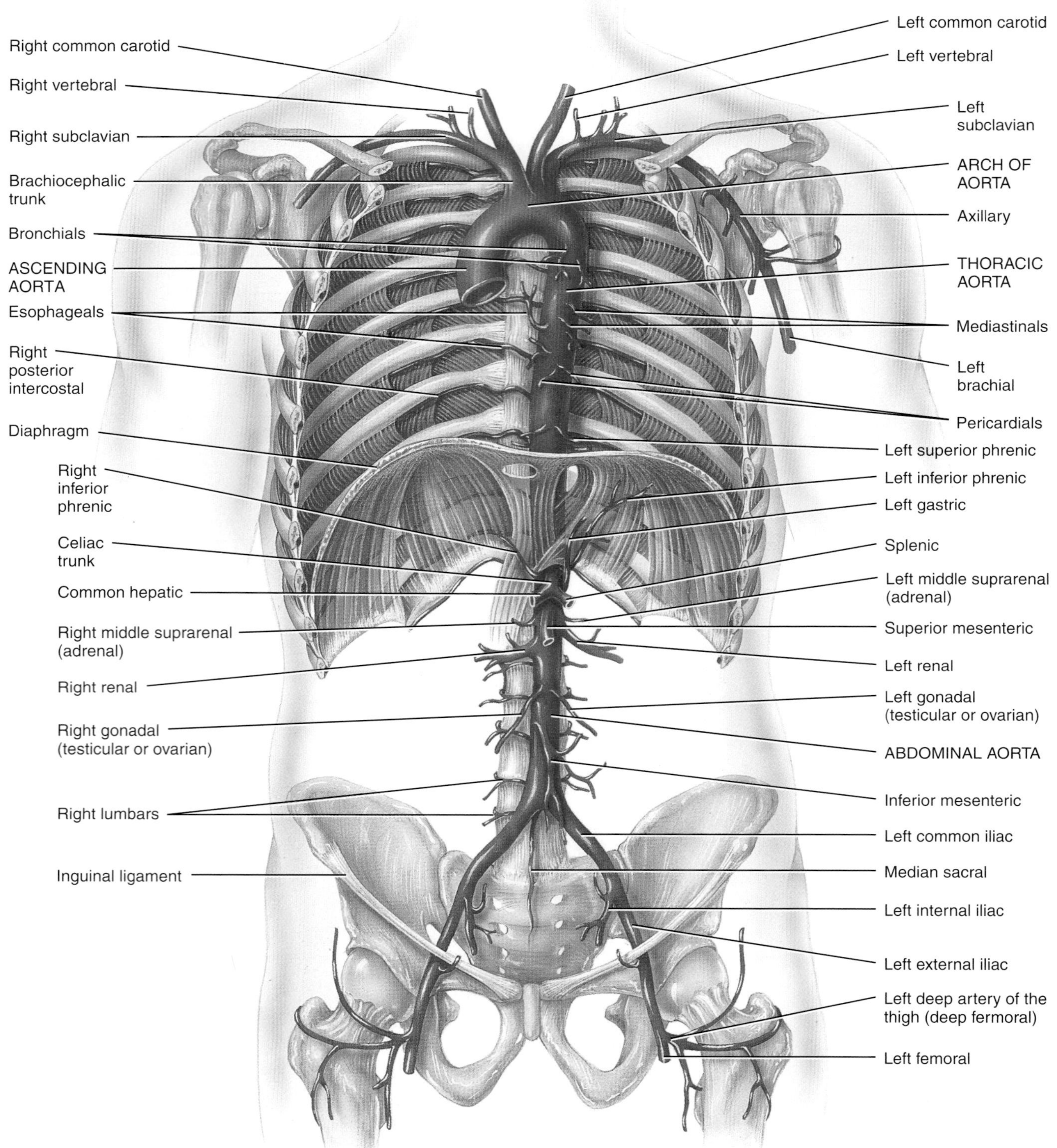

(b) Detailed anterior view of the principal branches of the aorta

What are the four subdivisions of the aorta?

ASCENDING AORTA

▶ **OBJECTIVE**

Identify the two primary arterial branches of the ascending aorta.

The **ascending aorta** is about 5 cm (2 in.) in length and begins at the aortic valve. It is directed superiorly, slightly anteriorly, and to the right. It ends at the level of the sternal angle, where it becomes the arch of the aorta. The beginning of the ascending aorta is posterior to the pulmonary trunk and right auricle; the right pulmonary artery is posterior to it. At its origin, the ascending aorta contains three dilations called *aortic sinuses.* Two of these, the right and left sinuses, give rise to the right and left coronary arteries, respectively.

The right and left **coronary arteries** (*coron-* = crown) arise from the ascending aorta just superior to the aortic valve (see Figure 20.8 on page 707). They form a crownlike ring around the heart, giving off branches to the atrial and ventricular myocardium. The **posterior interventricular branch** (in-ter-ven-TRIK-ū-lar; *inter-* = between) of the right coronary artery supplies both ventricles, and the **marginal branch** supplies the right ventricle. The **anterior interventricular branch,** also known as the **left anterior descending (LAD) branch,** of the left coronary artery supplies both ventricles, and the **circumflex branch** (SER-kum-flex; *circum-* = around; *-flex* = to bend) supplies the left atrium and left ventricle.

▶ **CHECKPOINT**

Which branches of the coronary arteries supply the left ventricle? Why does the left ventricle have such an extensive arterial blood supply?

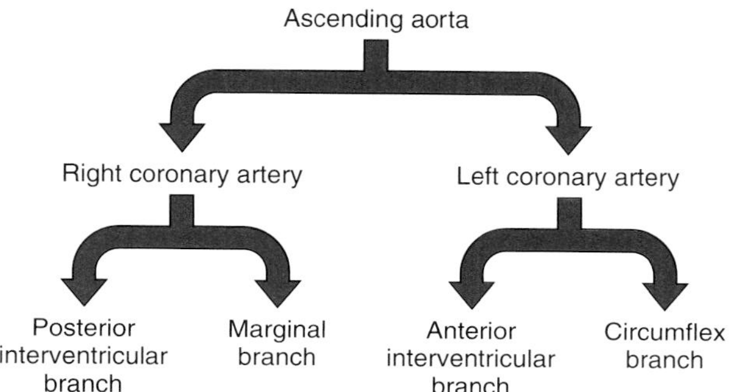

EXHIBIT 21.3 THE ARCH OF THE AORTA (FIGURE 21.19)

▶ **OBJECTIVE**

Identify the three principal arteries that branch from the arch of the aorta.

The **arch of the aorta** is 4–5 cm (almost 2 in.) in length and is the continuation of the ascending aorta. It emerges from the pericardium posterior to the sternum at the level of the sternal angle. The arch of the aorta is directed superiorly and posteriorly to the left and then inferiorly; it ends at the intervertebral disc between the fourth and fifth thoracic vertebrae, where it becomes the thoracic aorta. Three major arteries branch from the superior aspect of the arch of the aorta: the brachiocephalic trunk, the left common carotid, and the left subclavian. The first and largest branch from the arch of the aorta is the **brachiocephalic trunk** (brā′-kē-ō-se-FAL-ik; *brachio-* = arm; *-cephalic* = head). It extends superiorly, bending slightly to the right, and

BRANCH	DESCRIPTION AND REGION SUPPLIED
Brachiocephalic Trunk	The **brachiocephalic trunk** divides to form the right subclavian artery and right common carotid artery (Figure 21.19a).
Right subclavian artery (sub-KLĀ-vē-an)	The **right subclavian artery** extends from the brachiocephalic trunk to the first rib and then passes into the armpit (axilla). The general distribution of the artery is to the brain and spinal cord, neck, shoulder, thoracic viscera and wall, and scapular muscles.
Internal thoracic or **mammary artery** (thor-AS-ik; *thorac-* = chest)	The **internal thoracic artery** arises from the first part of the subclavian artery and descends posterior to the costal cartilages of the superior six ribs. It terminates at the sixth intercostal space. It supplies the anterior thoracic wall and structures in the mediastinum. In coronary artery bypass grafting, if only a single vessel is obstructed, the internal thoracic (usually the left) is used to create the bypass. The upper end of the artery is left attached to the subclavian artery and the cut end is connected to the coronary artery at a point distal to the blockage. The lower end of the internal thoracic artery is tied off. Artery grafts are preferred over vein grafts because arteries can withstand the greater pressure of blood flowing through coronary arteries and are less likely to become obstructed over time.
Vertebral artery (VER-te-bral)	Before passing into the axilla, the right subclavian artery gives off a major branch to the brain called the **right vertebral artery** (Figure 21.19b). The right vertebral artery passes through the foramina of the transverse processes of the sixth through first cervical vertebrae and enters the skull through the foramen magnum to reach the inferior surface of the brain. Here it unites with the left vertebral artery to form the **basilar** (BAS-i-lar) **artery.** The vertebral artery supplies the posterior portion of the brain with blood. The basilar artery passes along the midline of the anterior aspect of the brain stem. It gives off several branches (**posterior cerebral** and **cerebellar arteries**) that supply the cerebellum and pons of the brain and the inner ear.
Axillary artery (AK-sil-ār-ē = armpit)	The continuation of the right subclavian artery into the axilla is called the **axillary artery.** (Note that the right subclavian artery, which passes deep to the clavicle, is a good example of the practice of giving the same vessel different names as it passes through different regions.) Its general distribution is the shoulder, thoracic and scapular muscles, and humerus.
Brachial artery (BRĀ-kē-al = arm)	The **brachial artery** is the continuation of the axillary artery into the arm. The brachial artery provides the main blood supply to the arm and is superficial and palpable along its course. It begins at the tendon of the teres major muscle and ends just distal to the bend of the elbow. At first, the brachial artery is medial to the humerus, but as it descends it gradually curves laterally and passes through the cubital fossa, a triangular depression anterior to the elbow where you can easily detect the pulse of the brachial artery and listen to the various sounds when taking a person's blood pressure. Just distal to the bend in the elbow, the brachial artery divides into the radial artery and ulnar artery. Blood pressure is usually measured in the brachial artery. In order to control hemorrhage, the best place to compress the brachial artery is near the middle of the arm.
Radial artery (RĀ-dē-al = radius)	The **radial artery** is the smaller branch and is a direct continuation of the brachial artery. It passes along the lateral (radial) aspect of the forearm and then through the wrist and hand, supplying these structures with blood. At the wrist, the radial artery makes contact with the distal end of the radius, where it is covered only by fascia and skin. Because of its superficial location at this point, it is a common site for measuring the radial pulse.
Ulnar artery (UL-nar = ulna)	The **ulnar artery,** the larger branch of the brachial artery, passes along the medial (ulnar) aspect of the forearm and then into the wrist and hand, supplying these structures with blood. In the palm, branches of the radial and ulnar arteries anastomose to form the superficial palmar arch and the deep palmar arch.

divides at the right sternoclavicular joint to form the right subclavian artery and right common carotid artery. The second branch from the arch of the aorta is the **left common carotid artery** (ka-ROT-id), which divides into the same branches with the same names as the right common carotid artery. The third branch from the arch of the aorta is the **left subclavian artery** (sub-KLĀ-vē-an), which distributes blood to the left vertebral artery and vessels of the left upper limb. Arteries branching from the left subclavian artery are similar in distribution and name to those branching from the right subclavian artery. The following description focuses on the principal arteries originating from the brachiocephalic trunk.

▶ CHECKPOINT

What general regions do the arteries that arise from the arch of the aorta supply?

BRANCH	DESCRIPTION AND REGION SUPPLIED
Superficial palmar arch (*palma* = palm)	The **superficial palmar arch** is formed mainly by the ulnar artery, with a contribution from a branch of the radial artery. The arch is superficial to the long flexor tendons of the fingers and extends across the palm at the bases of the metacarpals. It gives rise to **common palmar digital arteries,** which supply the palm. Each divides into a pair of **proper palmar digital arteries,** which supply the fingers.
Deep palmar arch	Mainly the radial artery forms the **deep palmar arch,** with a contribution from a branch of the ulnar artery. The arch is deep to the long flexor tendons of the fingers and extends across the palm, just distal to the bases of the metacarpals. Arising from the deep palmar arch are **palmar metacarpal arteries,** which supply the palm and anastomose with the common palmar digital arteries of the superficial palmar arch.
Right common carotid artery	The **right common carotid artery** begins at the bifurcation (division into two branches) of the brachiocephalic trunk, posterior to the right sternoclavicular joint, and passes superiorly in the neck to supply structures in the head (Figure 21.19b). At the superior border of the larynx (voice box), it divides into the right external and right internal carotid arteries. Pulse may be detected in the common carotid artery, just lateral to the larynx. It is convenient to detect a carotid pulse when exercising or when administering cardiopulmonary resuscitation.
External carotid artery	The **external carotid artery** begins at the superior border of the larynx and terminates near the temporomandibular joint of the parotid gland, where it divides into two branches: the superficial temporal and maxillary arteries. The carotid pulse can be detected in the external carotid artery just anterior to the sternocleidomastoid muscle at the superior border of the larynx. The general distribution of the external carotid artery is to structures external to the skull.
Internal carotid artery	The **internal carotid artery** has no branches in the neck and supplies structures internal to the skull. It enters the cranial cavity through the carotid foramen in the temporal bone. The internal carotid artery supplies blood to the eyeball and other orbital structures, ear, most of the cerebrum of the brain, pituitary gland, and external nose. The terminal branches of the internal carotid artery are the **anterior cerebral artery,** which supplies most of the medial surface of the cerebrum and deep masses of gray matter within the cerebrum, and the **middle cerebral artery,** which supplies most of the lateral surface of the cerebrum (Figure 21.19c). Inside the cranium, anastomoses of the left and right internal carotid arteries along with the basilar artery form an arrangement of blood vessels at the base of the brain near the hypophyseal fossa called the **cerebral arterial circle (circle of Willis).** From this circle (Figure 21.19c) arise arteries supplying most of the brain. Essentially, the cerebral arterial circle is formed by the union of the **anterior cerebral arteries** (branches of internal carotids) and **posterior cerebral arteries** (branches of basilar artery). The posterior cerebral arteries supply the inferolateral surface of the temporal lobe and lateral and medial surfaces of the occipital lobe of the cerebrum, deep masses of gray matter within the cerebrum, and midbrain. The posterior cerebral arteries are connected with the internal carotid arteries by the **posterior communicating arteries.** The **anterior communicating arteries** connect the anterior cerebral arteries. The **internal carotid arteries** are also considered part of the cerebral arterial circle. The functions of the cerebral arterial circle are to equalize blood pressure to the brain and provide alternate routes for blood flow to the brain, should the arteries become damaged.
Left common carotid artery	See description in the introduction to this exhibit.
Left subclavian artery	See description in the introduction to this exhibit.

continues

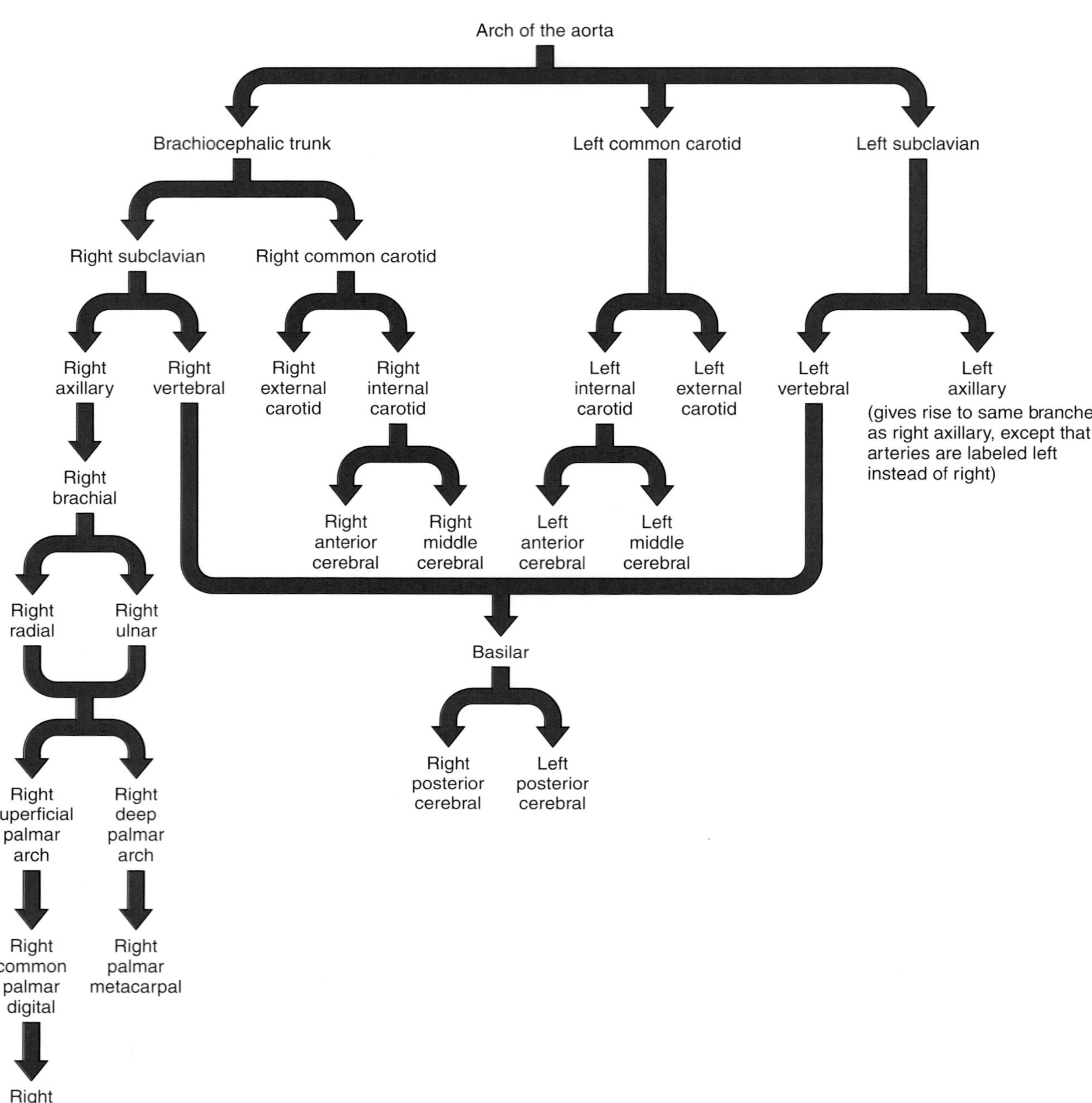
SCHEME OF DISTRIBUTION
Arch of the aorta
Brachiocephalic trunk
Left common carotid
Left subclavian
Right subclavian
Right common carotid
Right axillary
Right vertebral
Right external carotid
Right internal carotid
Left internal carotid
Left external carotid
Left vertebral
Left axillary
(gives rise to same branches as right axillary, except that arteries are labeled left instead of right)
Right brachial
Right anterior cerebral
Right middle cerebral
Left anterior cerebral
Left middle cerebral
Right radial
Right ulnar
Basilar
Right superficial palmar arch
Right deep palmar arch
Right posterior cerebral
Left posterior cerebral
Right common palmar digital
Right palmar metacarpal
Right proper palmar digital

Figure 21.19 Arch of the aorta and its branches. Note in (c) the arteries that constitute the cerebral arterial circle (circle of Willis).

The arch of the aorta ends at the level of the intervertebral disc between the fourth and fifth thoracic vertebrae.

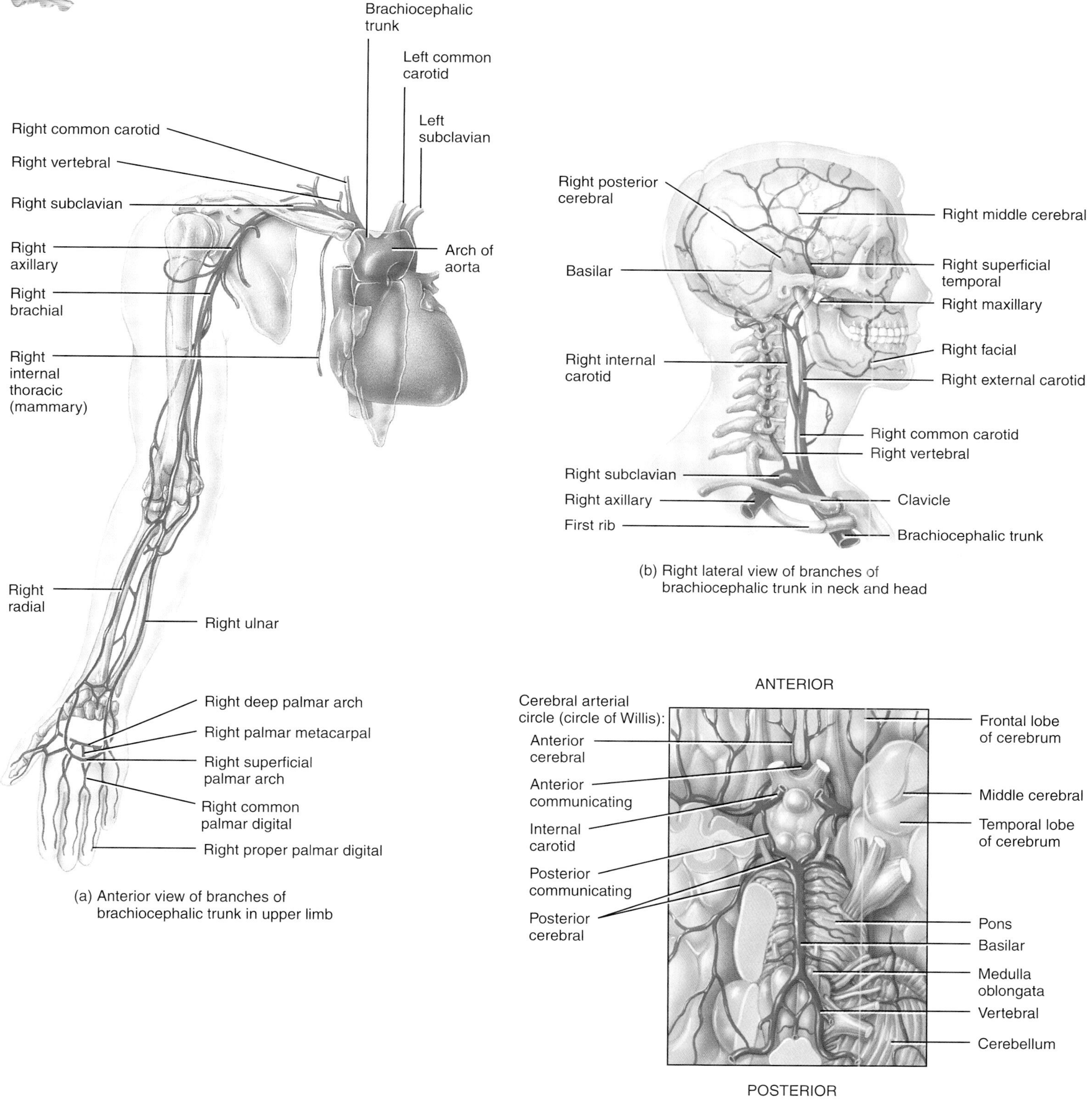

Brachiocephalic trunk
Left common carotid
Left subclavian
Right common carotid
Right vertebral
Right subclavian
Right axillary
Right brachial
Arch of aorta
Right internal thoracic (mammary)
Right radial
Right ulnar
Right deep palmar arch
Right palmar metacarpal
Right superficial palmar arch
Right common palmar digital
Right proper palmar digital

(a) Anterior view of branches of brachiocephalic trunk in upper limb

Right posterior cerebral
Right middle cerebral
Basilar
Right superficial temporal
Right maxillary
Right facial
Right internal carotid
Right external carotid
Right common carotid
Right vertebral
Right subclavian
Right axillary
Clavicle
First rib
Brachiocephalic trunk

(b) Right lateral view of branches of brachiocephalic trunk in neck and head

ANTERIOR
Cerebral arterial circle (circle of Willis):
Anterior cerebral
Anterior communicating
Internal carotid
Posterior communicating
Posterior cerebral
Frontal lobe of cerebrum
Middle cerebral
Temporal lobe of cerebrum
Pons
Basilar
Medulla oblongata
Vertebral
Cerebellum
POSTERIOR

(c) Inferior view of base of brain showing cerebral arterial circle

What are the three major branches of the arch of the aorta, in order of their origination?

▶ **OBJECTIVE**

Identify the visceral and parietal branches of the thoracic aorta.

The **thoracic aorta** is about 20 cm (8 in.) long and is a continuation of the arch of the aorta. It begins at the level of the intervertebral disc between the fourth and fifth thoracic vertebrae, where it lies to the left of the vertebral column. As it descends, it moves closer to the midline and extends through an opening in the diaphragm (aortic hiatus), which is located anterior to the vertebral column at the level of the intervertebral disc between the twelfth thoracic and first lumbar vertebrae.

Along its course, the thoracic aorta sends off numerous small arteries, **visceral branches** to visceral, and **parietal branches** to body wall structures.

▶ **CHECKPOINT**

What general regions do the visceral and parietal branches of the thoracic aorta supply?

BRANCH	DESCRIPTION AND REGION SUPPLIED
Visceral	
Pericardial arteries (per′-i-KAR-dē-al; *peri-* = around; *cardia-* = heart)	Two or three tiny **pericardial arteries** supply blood to the pericardium.
Bronchial arteries (BRONG-kē-al = windpipe)	One right and two left **bronchial arteries** supply the bronchial tubes, pleurae, bronchial lymph nodes, and esophagus. (The right bronchial artery arises from the third posterior intercostal artery; the two left bronchial arteries arise from the thoracic aorta.)
Esophageal arteries (e-sof′-a-JĒ-al; *eso-* = to carry; *phage-* = food)	Four or five **esophageal arteries** supply the esophagus.
Mediastinal arteries (mē′-dē-as-TĪ-nal)	Numerous small **mediastinal arteries** supply blood to structures in the mediastinum.
Parietal	
Posterior intercostal arteries (in′-ter-KOS-tal; *inter-* = between; *costa* = rib)	Nine pairs of **posterior intercostal arteries** supply the intercostal, pectoralis major and minor, and serratus anterior muscles; overlying subcutaneous tissue and skin; mammary glands; and vertebrae, meninges, and spinal cord.
Subcostal arteries (sub-KOS-tal; *sub-* = under)	The left and right **subcostal arteries** have a distribution similar to that of the posterior intercostals.
Superior phrenic arteries (FREN-ik = pertaining to the diaphragm)	Small **superior phrenic arteries** supply the superior and posterior surfaces of the diaphragm.

SCHEME OF DISTRIBUTION

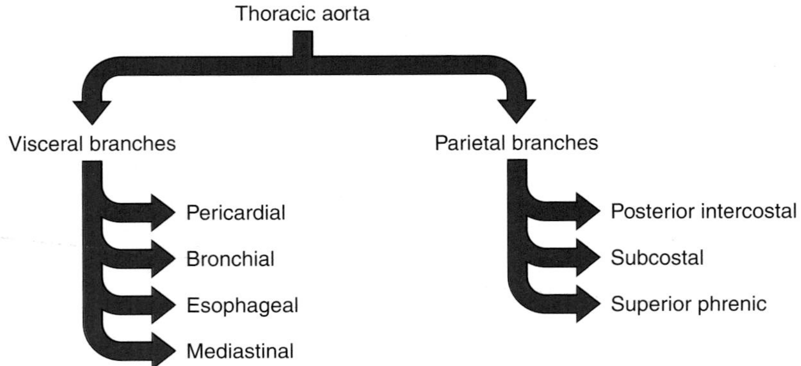

Figure 21.20 Thoracic aorta and abdominal aorta and their principal branches.

The thoracic aorta is the continuation of the ascending aorta.

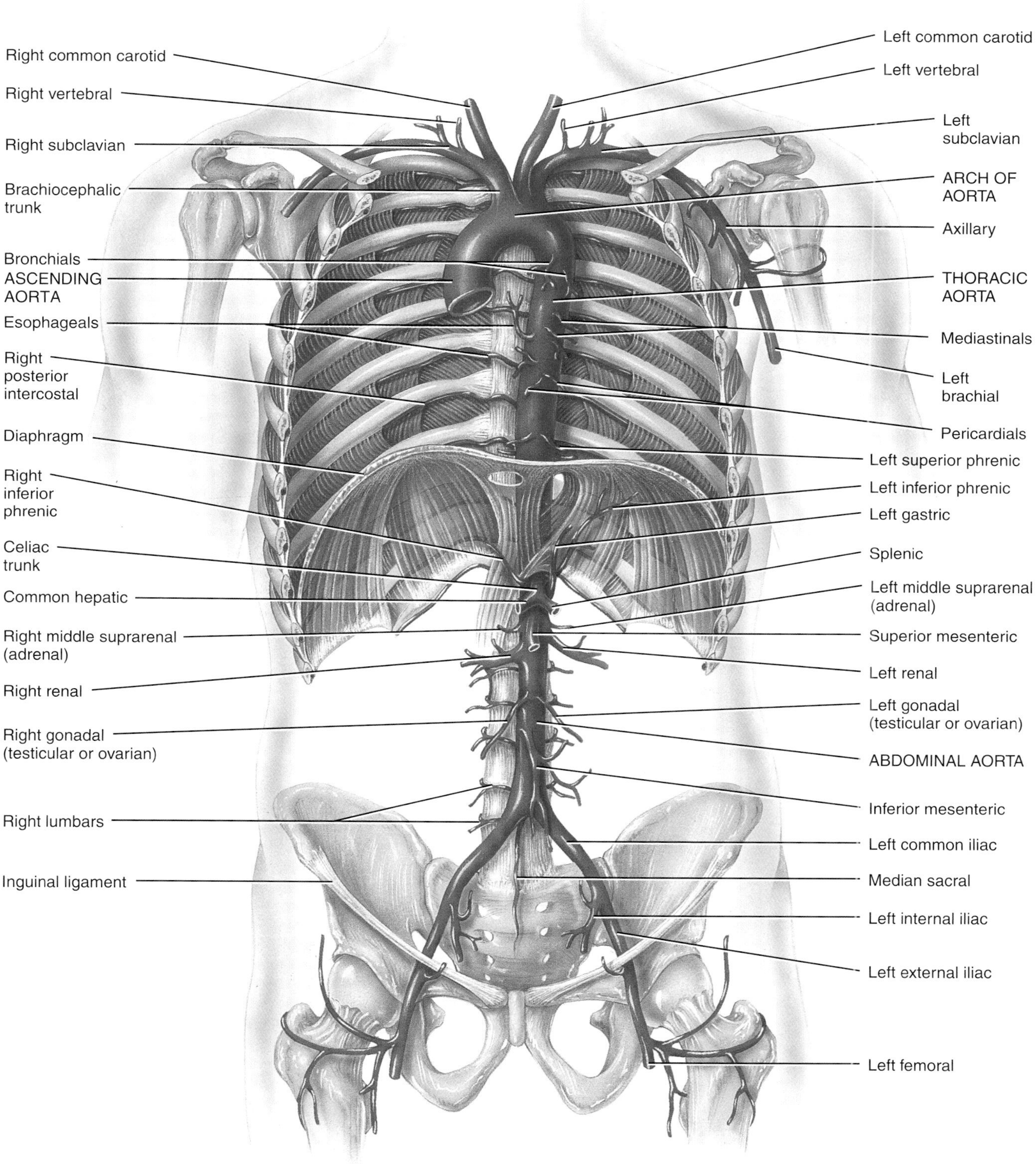

Right common carotid
Right vertebral
Right subclavian
Brachiocephalic trunk
Bronchials
ASCENDING AORTA
Esophageals
Right posterior intercostal
Diaphragm
Right inferior phrenic
Celiac trunk
Common hepatic
Right middle suprarenal (adrenal)
Right renal
Right gonadal (testicular or ovarian)
Right lumbars
Inguinal ligament

Left common carotid
Left vertebral
Left subclavian
ARCH OF AORTA
Axillary
THORACIC AORTA
Mediastinals
Left brachial
Pericardials
Left superior phrenic
Left inferior phrenic
Left gastric
Splenic
Left middle suprarenal (adrenal)
Superior mesenteric
Left renal
Left gonadal (testicular or ovarian)
ABDOMINAL AORTA
Inferior mesenteric
Left common iliac
Median sacral
Left internal iliac
Left external iliac
Left femoral

Anterior view of the principal branches of the aorta

Where does the thoracic aorta begin?

▶ **OBJECTIVE**

Identify the visceral and parietal branches of the abdominal aorta.

The **abdominal aorta** is the continuation of the thoracic aorta. It begins at the aortic hiatus in the diaphragm and ends at about the level of the fourth lumbar vertebra, where it divides into the right and left common iliac arteries. The abdominal aorta lies anterior to the vertebral column.

As with the thoracic aorta, the abdominal aorta gives off visceral and parietal branches. The unpaired visceral branches arise from the anterior surface of the aorta and include the **celiac trunk** and the **superior mesenteric** and **inferior mesenteric arteries** (see Figure 21.20). The paired visceral branches arise from the lateral surfaces of the aorta and include the **suprarenal, renal,** and **gonadal arteries.** The unpaired parietal branch is the **median sacral artery.** The paired parietal branches arise from the posterolateral surfaces of the aorta and include the **inferior phrenic** and **lumbar arteries.**

▶ **CHECKPOINT**

Name the paired visceral and parietal branches and the unpaired visceral and parietal branches of the abdominal aorta, and indicate the general regions they supply.

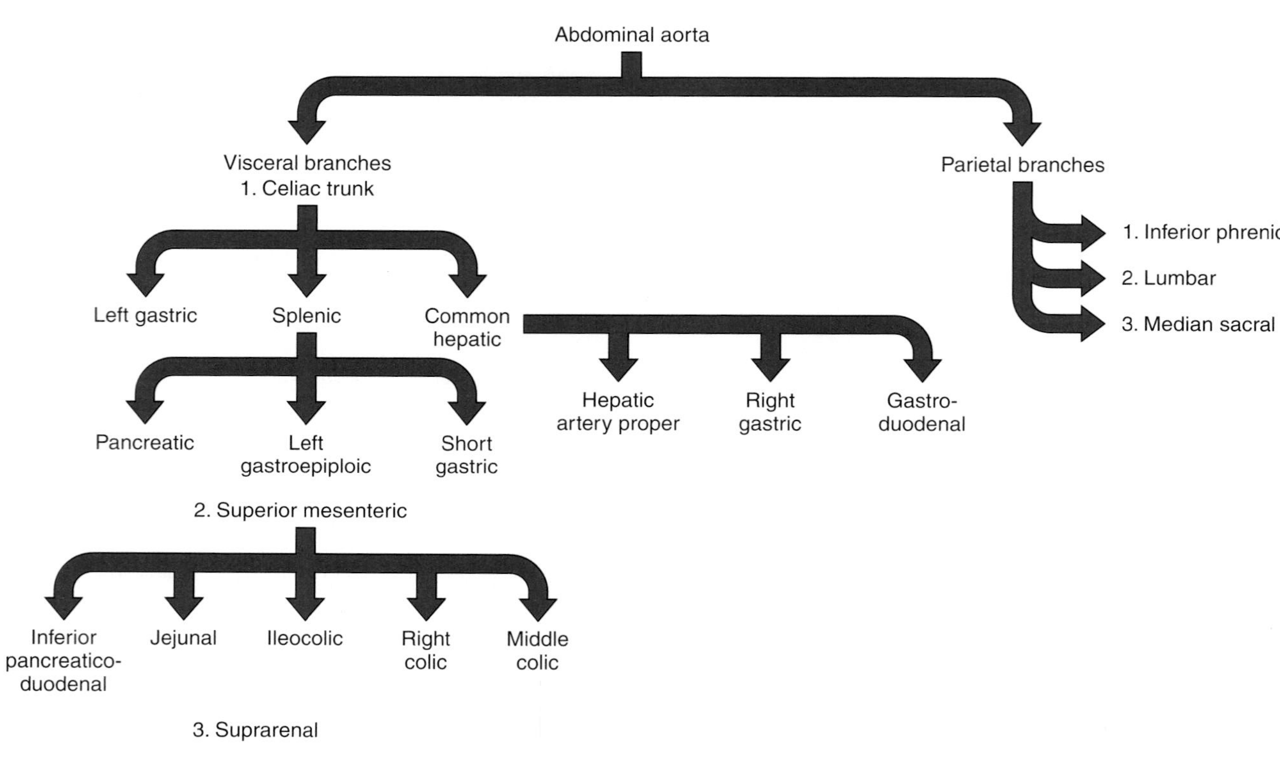

BRANCH	DESCRIPTION AND REGION SUPPLIED
Unpaired Visceral Branches	
Celiac trunk (SĒ-lē-ak)	The **celiac trunk (artery)** is the first visceral branch from the aorta inferior to the diaphragm, at about the level of the twelfth thoracic vertebra (Figure 21.21a). Almost immediately, the celiac trunk divides into three branches: the left gastric, splenic, and common hepatic arteries (Figure 21.21a). **1.** The **left gastric artery** (GAS-trik = stomach) is the smallest of the three branches. It passes superiorly to the left toward the esophagus and then turns to follow the lesser curvature of the stomach. It supplies the stomach and esophagus. **2.** The **splenic artery** (SPLĒN-ik = spleen) is the largest branch of the celiac trunk. It arises from the left side of the celiac trunk distal to the left gastric artery, and passes horizontally to the left along the pancreas. Before reaching the spleen, it gives rise to three arteries: • **Pancreatic artery** (pan-krē-AT-ik), which supplies the pancreas. • **Left gastroepiploic artery** (gas′-trō-ep′-i-PLŌ-ik; *epiplo-* = omentum), which supplies the stomach and greater omentum. • **Short gastric artery,** which supplies the stomach. **3.** The **common hepatic artery** (he-PAT-ik = liver) is intermediate in size between the left gastric and splenic arteries. Unlike the other two branches of the celiac trunk, the common hepatic artery arises from the right side. It gives rise to three arteries: • **Proper hepatic artery,** which supplies the liver, gallbladder, and stomach. • **Right gastric artery,** which supplies the stomach. • **Gastroduodenal artery** (gas′-trō-doo′-ō-DĒ-nal), which supplies the stomach, duodenum of the small intestine, pancreas, and greater omentum.
Superior mesenteric artery (MES-en-ter′-ik; *meso-* = middle; *enteric* = pertaining to the intestines)	The **superior mesenteric artery** (Figure 21.21b) arises from the anterior surface of the abdominal aorta about 1 cm inferior to the celiac trunk at the level of the first lumbar vertebra. It extends inferiorly and anteriorly and between the layers of mesentery, which is a portion of the peritoneum that attaches the small intestine to the posterior abdominal wall. It anastomoses extensively and has five branches: **1.** The **inferior pancreaticoduodenal artery** (pan′-krē-at′-i-kō-doo′-ō-DĒ-nal) supplies the pancreas and duodenum. **2.** The **jejunal** (je-JOO-nal) and **ileal arteries** (IL-ē-al) supply the jejunum and ileum of the small intestine, respectively. **3.** The **ileocolic artery** (il′ē-ō-KŌL-ik) supplies the ileum and ascending colon of the large intestine. **4.** The **right colic artery** (KŌL-ik) supplies the ascending colon. **5.** The **middle colic artery** supplies the transverse colon of the large intestine.
Inferior mesenteric artery	The **inferior mesenteric artery** (Figure 21.21c) arises from the anterior aspect of the abdominal aorta at the level of the third lumbar vertebra and then passes inferiorly to the left of the aorta. It anastomoses extensively and has three branches: **1.** The **left colic artery** supplies the transverse colon and descending colon of the large intestine. **2.** The **sigmoid arteries** (SIG-moyd) supply the descending colon and sigmoid colon of the large intestine. **3.** The **superior rectal artery** (REK-tal) supplies the rectum of the large intestine.
Suprarenal arteries (soo′-pra-RĒ-nal; *supra-* = above; *ren-* = kidney)	Although there are three pairs of **suprarenal (adrenal) arteries** that supply the adrenal (suprarenal) glands (superior, middle, and inferior), only the middle pair originates directly from the abdominal aorta (see Figure 21.20). The middle suprarenal arteries arise at the level of the first lumbar vertebra at or superior to the renal arteries. The superior suprarenal arteries arise from the inferior phrenic artery, and the inferior suprarenal arteries originate from the renal arteries.
Renal arteries (RĒ-nal = *ren-* = kidney)	The right and left **renal arteries** usually arise from the lateral aspects of the abdominal aorta at the superior border of the second lumbar vertebra, about 1 cm inferior to the superior mesenteric artery (see Figure 21.20). The right renal artery, which is longer than the left, arises slightly lower than the left and passes posterior to the right renal vein and inferior vena cava. The left renal artery is posterior to the left renal vein and is crossed by the inferior mesenteric vein. The renal arteries carry blood to the kidneys, adrenal (suprarenal) glands, and ureters. Their distribution within the kidneys is discussed in Chapter 26.
Gonadal (gō-NAD-al; *gon-* = seed) **[testicular** (test-TIK-ū-lar) or **ovarian arteries** (ō-VAR-ē-an)]	The **gonadal arteries** arise from the abdominal aorta at the level of the second lumbar vertebra just inferior to the renal arteries (see Figure 21.20). In males, the gonadal arteries are specifically referred to as the **testicular arteries.** They pass through the inguinal canal and supply the testes, epididymis, and ureters. In females, the gonadal arteries are called the **ovarian arteries.** They are much shorter than the testicular arteries and supply the ovaries, uterine (fallopian) tubes, and ureters.
Unpaired Parietal Branch	
Median sacral artery (SĀ-kral = pertaining to the sacrum)	The **median sacral artery** arises from the posterior surface of the abdominal aorta about 1 cm superior to the bifurcation (division into two branches) of the aorta into the right and left common iliac arteries (see Figure 21.20). The median sacral artery supplies the sacrum and coccyx.
Paired Parietal Branches	
Inferior phrenic arteries (FREN-ik = pertaining to the diaphragm)	The **inferior phrenic arteries** are the first paired branches of the abdominal aorta, immediately superior to the origin of the celiac trunk (see Figure 21.20). (They may also arise from the renal arteries.) The inferior phrenic arteries are distributed to the inferior surface of the diaphragm and adrenal (suprarenal) glands.
Lumbar arteries (LUM-bar = pertaining to the loin)	The four pairs of **lumbar arteries** arise from the posterolateral surface of the abdominal aorta (see Figure 21.20). They supply the lumbar vertebrae, spinal cord and its meninges, and the muscles and skin of the lumbar region of the back.

continues

Figure 21.21 Abdominal aorta and its principal branches.

The abdominal aorta is the continuation of the thoracic aorta.

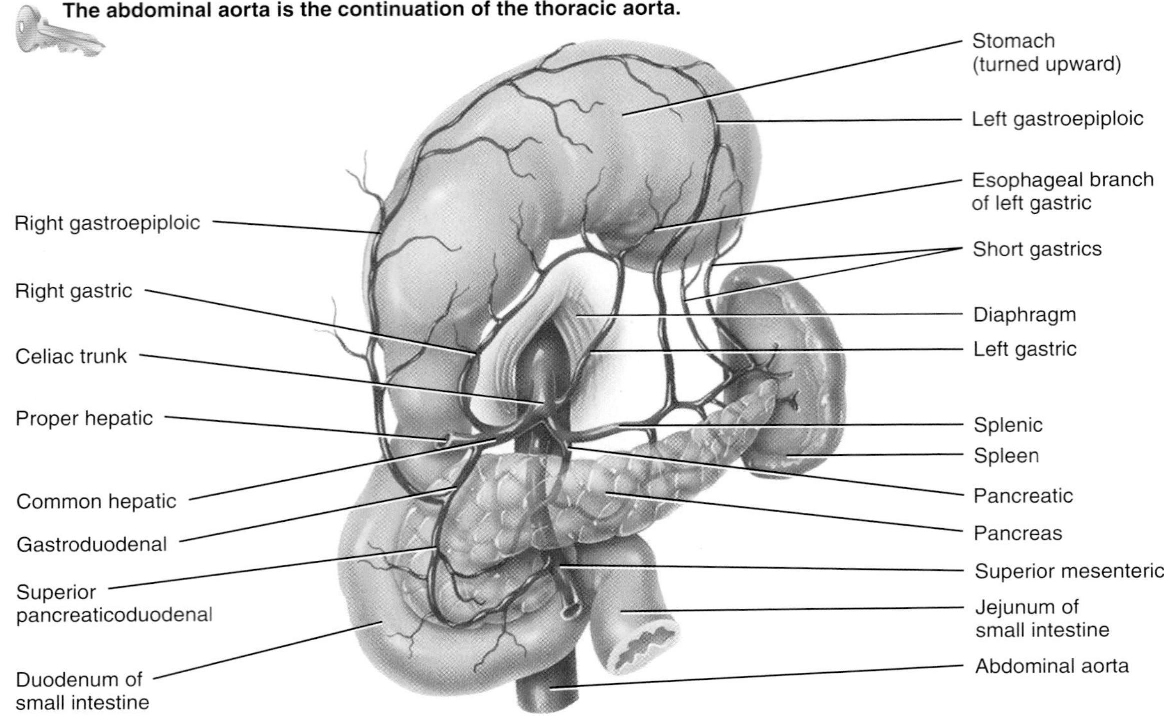

(a) Anterior view of celiac trunk and its branches

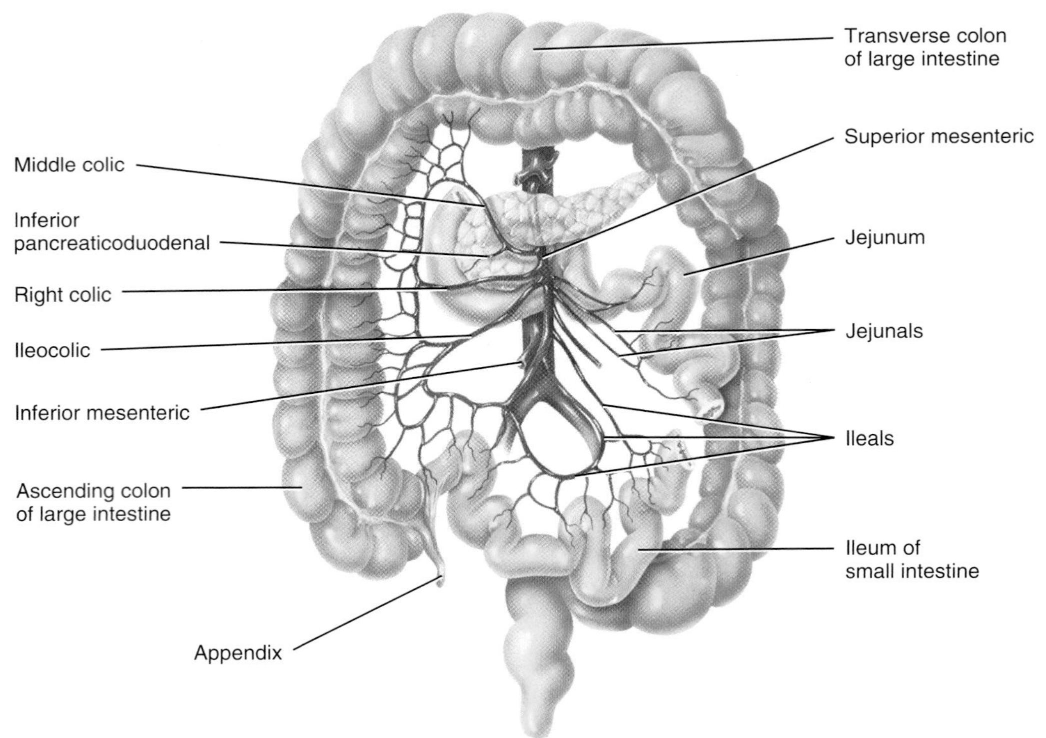

(b) Anterior view of superior mesenteric artery and its branches

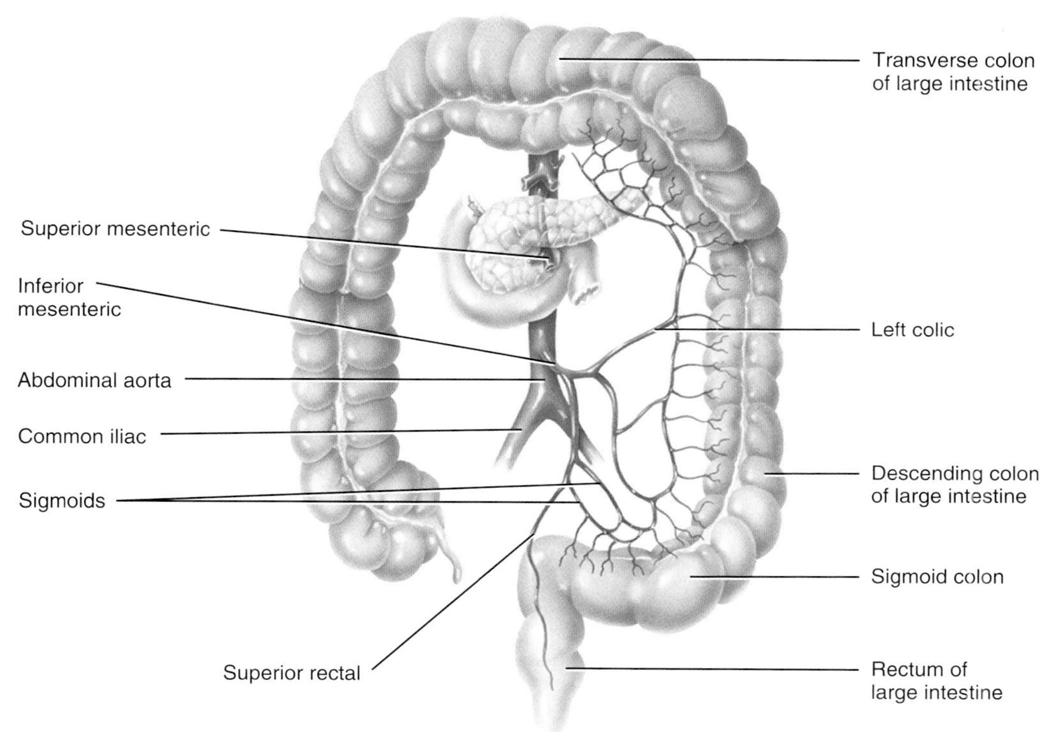

(c) Anterior view of inferior mesenteric artery and its branches

Where does the abdominal aorta begin?

EXHIBIT 21.6 ARTERIES OF THE PELVIS AND LOWER LIMBS (FIGURE 21.22)

▶ **OBJECTIVE**

Identify the two major branches of the common iliac arteries.

▶ **CHECKPOINT**

What general regions do the internal and external iliac arteries supply?

The abdominal aorta ends by dividing into the right and left **common iliac arteries.** These, in turn, divide into the **internal** and **external iliac arteries.** In sequence, the external iliacs become the **femoral arteries** in the thighs, the **popliteal arteries** posterior to the knee, and the **anterior** and **posterior tibial arteries** in the legs.

BRANCH	DESCRIPTION AND REGION SUPPLIED
Common iliac arteries (IL-ē-ak = pertaining to the ilium)	At about the level of the fourth lumbar vertebra, the abdominal aorta divides into the right and left **common iliac arteries,** the terminal branches of the abdominal aorta. Each passes inferiorly about 5 cm (2 in.) and gives rise to two branches: internal iliac and external iliac arteries. The general distribution of the common iliac arteries is to the pelvis, external genitals, and lower limbs.
Internal iliac arteries	The **internal iliac (hypogastric) arteries** are the primary arteries of the pelvis. They begin at the bifurcation (division into two branches) of the common iliac arteries anterior to the sacroiliac joint at the level of the lumbosacral intervertebral disc. They pass posteromedially as they descend in the pelvis and divide into anterior and posterior divisions. The general distribution of the internal iliac arteries is to the pelvis, buttocks,external genitals, and thigh.
External iliac arteries	The **external iliac arteries** are larger than the internal iliac arteries. Like the internal iliac arteries, they begin at the bifurcation of the common iliac arteries. They descend along the medial border of the psoas major muscles following the pelvic brim, pass posterior to the midportion of the inguinal ligaments, and become the femoral arteries. The general distribution of the external iliac arteries is to the lower limbs. Specifically, branches of the external iliac arteries supply the muscles of the anterior abdominal wall, the cremaster muscle in males and the round ligament of the uterus in females, and the lower limbs.
Femoral arteries (FEM-o-ral = pertaining to the thigh)	The **femoral arteries** descend along the anteromedial aspects of the thighs to the junction of the middle and lower third of the thighs. Here they pass through an opening in the tendon of the adductor magnus muscle, they emerge posterior to the femurs as the popliteal arteries. A pulse may be felt in the femoral artery where just inferior to the inguinal ligament. Recall from Chapter 11 that the femoral artery, along with the femoral vein and nerve and deep inguinal lymph nodes, are located in the *femoral triangle* (see Figure 11.20a on page 385). The general distribution of the femoral arteries is to the lower abdominal wall, groin, external genitals, and muscles of the thigh. A major branch of the femoral artery, the **deep artery of the thigh (deep femoral),** supplies most of the muscles of the thigh: quadriceps femoris, adductors, and hamstrings. Recall that in cardiac catheterization a catheter is inserted through a blood vessel and advanced into the major vessels and heart chamber. A catheter often contains a measuring instrument or other device at its tip. To reach the left side of the heart, the catheter is inserted into the femoral artery and passed into the aorta tothe coronary arteries or heart chamber.
Popliteal arteries (pop′-li-TĒ-al = posterior surface of the knee)	The **popliteal arteries** are the continuation of the femoral arteries through the popliteal fossa (space behind the knee). They descend to the inferior border of the popliteus muscles, where they divide into the anterior and posterior tibial arteries. A pulse may be detected in the popliteal arteries. In addition to supplying the adductor magnus and hamstring muscles and the skin on the posterior aspect of the legs, branches of the popliteal arteries also supply the gastrocnemius, soleus, and plantaris muscles of the calf, knee joint, femur, patella, and fibula.
Anterior tibial arteries (TIB-ē-al = pertaining to the shin bone)	The **anterior tibial arteries** descend from the bifurcation of the popliteal arteries. They are smaller than the posterior tibial arteries. The anterior tibial arteries descend through the anterior muscular compartment of the leg. They pass through the interosseous membrane that connects the tibia and fibula, lateral to the tibia. The anterior tibial arteries supply the knee joints, anterior compartment muscles of the legs, skin over the anterior aspects of the legs, and ankle joints. At the ankles, the anterior tibial arteries become the **dorsal arteries of the foot (dorsalis pedis arteries).** A pulse in this artery may be taken to evaluate the peripheral vascular system. The dorsal arteries of the foot supply the muscles, skin, and joints on the dorsal aspects of the feet. On the dorsum of the feet, the dorsal arteries of the foot give off a transverse branch at the first (medial) cuneiform bone called the **arcuate arteries** (*arcuat-* = bowed) that run laterally over the bases of the metatarsals. From the arcuate arteries branch the **dorsal metatarsal arteries,** which supply the feet. The dorsal metatarsal arteries terminate by dividing into the **dorsal digital arteries,** which supply the toes.
Posterior tibial arteries	The **posterior tibial arteries,** the direct continuations of the popliteal arteries, descend from the bifurcation of the popliteal arteries. They pass down the posterior muscular compartment of the legs posterior to the medial malleolus of the tibia. They terminate by dividing into the medial and lateral plantar arteries. Their general distribution is to the muscles, bones, and joints of the leg and foot. Major branches of the posterior tibial arteries are the **fibular (peroneal) arteries,** which supply the fibularis, soleus, tibialis posterior, and flexor hallucis muscles. They also supply the fibula, tarsus, and lateral aspect of the heel. The bifurcation of the posterior tibial arteries into the medial and lateral plantar arteries occurs deep to the flexor retinaculum on the medial side of the feet. The **medial plantar arteries** (PLAN-tar = sole) supply the abductor hallucis and flexor digitorum brevis muscles and the toes. The **lateral plantar arteries** unite with a branch of the dorsal arteries of the foot to form the **plantar arch.** The arch begins at the base of the fifth metatarsal and extends medially across the metacarpals. As the arch crosses the foot, it gives off **plantar metatarsal arteries,** which supply the feet. These terminate by dividing into **plantar digital arteries,** which supply the toes.

SCHEME OF DISTRIBUTION

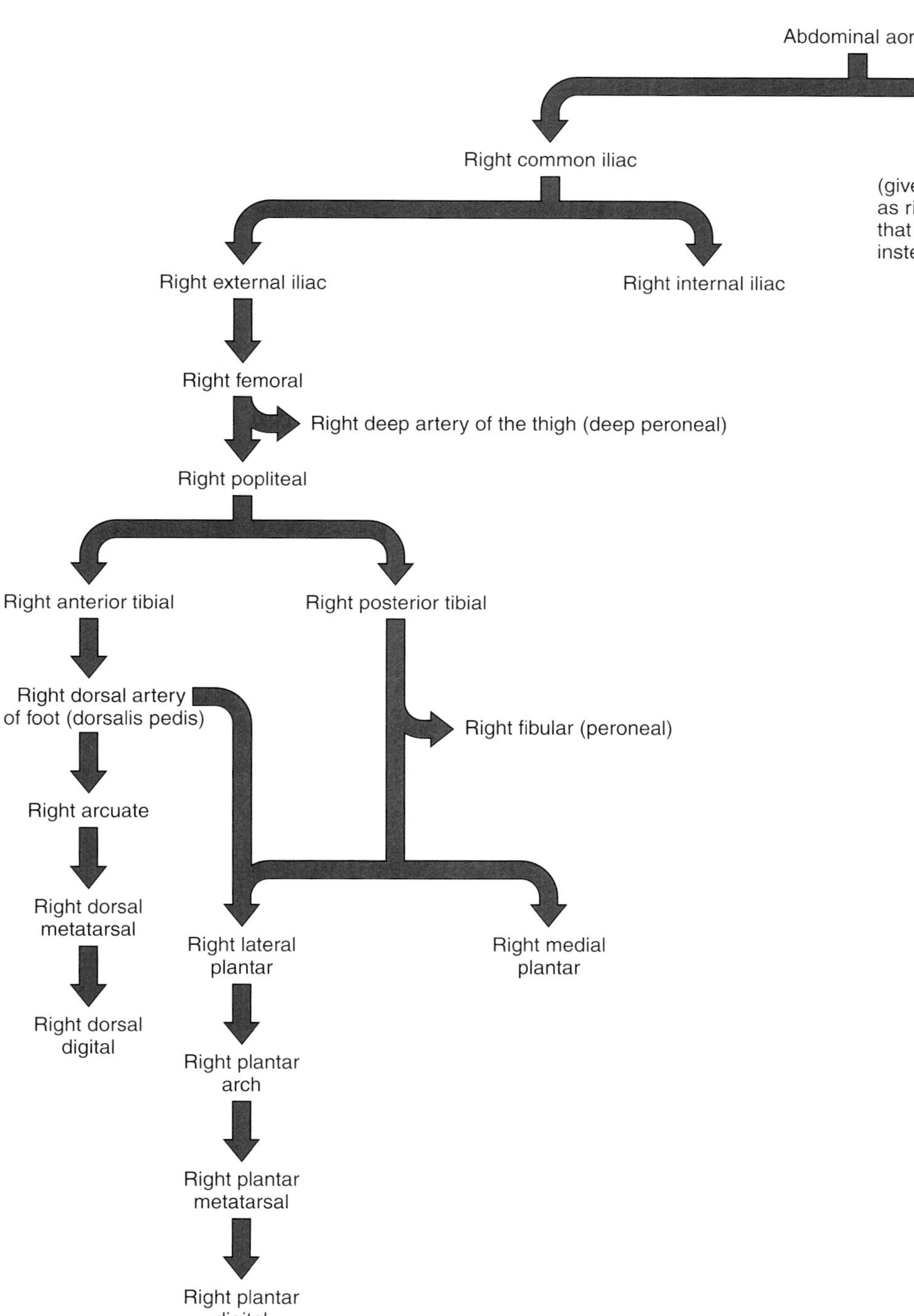
Abdominal aorta
Right common iliac
Left common iliac
(gives rise to same branches as right common iliac, except that arteries are labeled left instead of right)
Right external iliac
Right internal iliac
Right femoral
Right deep artery of the thigh (deep peroneal)
Right popliteal
Right anterior tibial
Right posterior tibial
Right dorsal artery of foot (dorsalis pedis)
Right fibular (peroneal)
Right arcuate
Right dorsal metatarsal
Right lateral plantar
Right medial plantar
Right dorsal digital
Right plantar arch
Right plantar metatarsal
Right plantar digital

continues

Figure 21.22 **Arteries of the pelvis and right lower limb.**

The internal iliac arteries carry most of the blood supply to the pelvic viscera and wall.

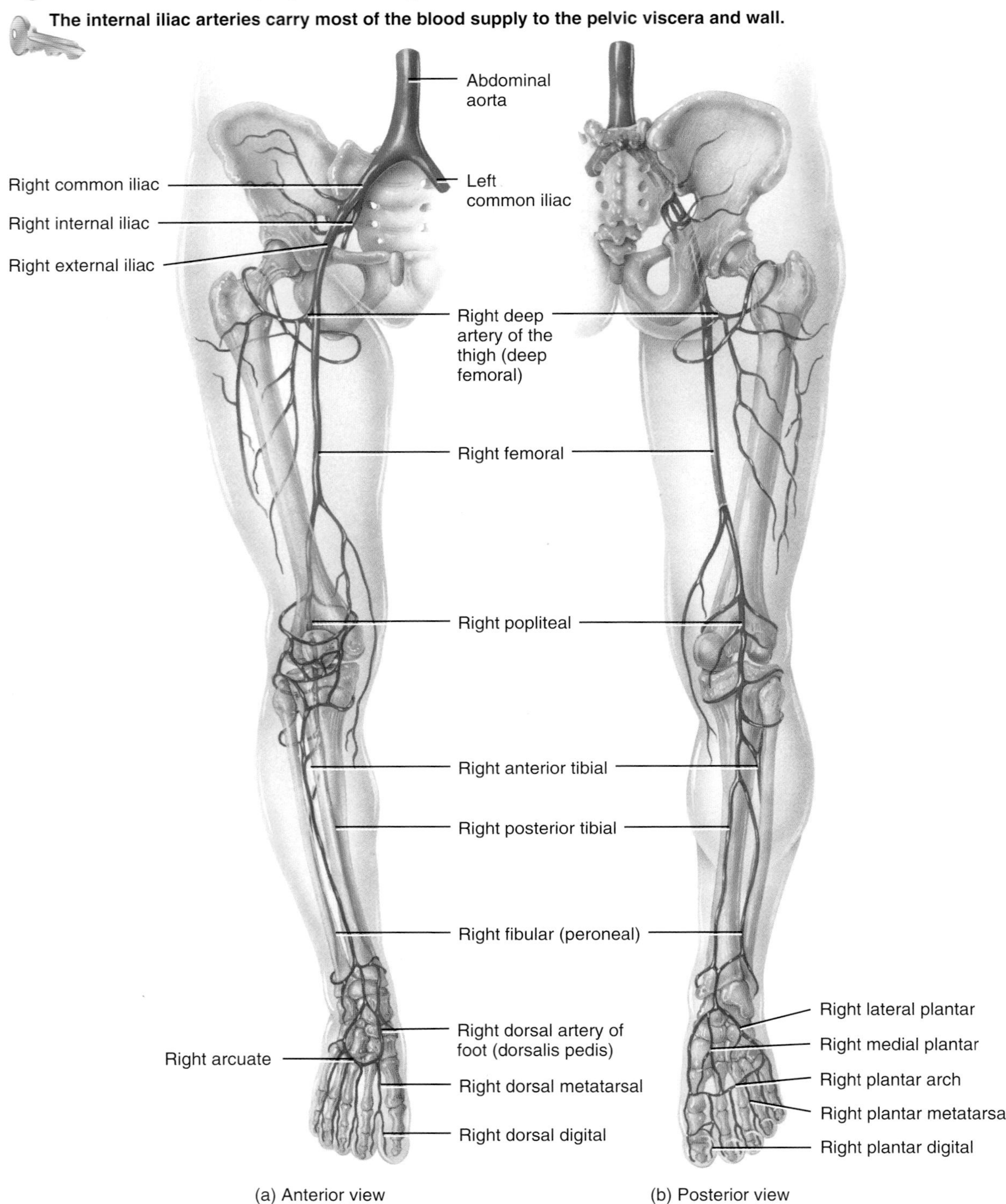

At what point does the abdominal aorta divide into the common iliac arteries?

EXHIBIT 21.7 VEINS OF THE SYSTEMIC CIRCULATION (FIGURE 21.23)

▶ **OBJECTIVE**

Identify the three systemic veins that return deoxygenated blood to the heart.

As you have already learned, arteries distribute blood to various parts of the body, and veins drain blood away from them. For the most part, arteries are deep, whereas veins may be superficial or deep. Superficial veins are located just beneath the skin and can be easily seen. Because there are no large superficial arteries, the names of superficial veins do not correspond to those of arteries. Superficial veins are clinically important as sites for withdrawing blood or giving injections. Deep veins generally travel alongside arteries and usually bear the same name. Arteries usually follow definite pathways; veins are more difficult to follow because they connect in irregular networks in which many tributaries merge to form a large vein. Although only one systemic artery, the aorta, takes oxygenated blood away from the heart (left ventricle), three systemic veins, the **coronary sinus, superior vena cava,** and **inferior vena cava,** return deoxygenated blood to the heart (right atrium). The coronary sinus receives blood from the cardiac veins; the superior vena cava receives blood from other veins superior to the diaphragm, except the air sacs (alveoli) of the lungs; the inferior vena cava receives blood from veins inferior to the diaphragm.

▶ **CHECKPOINT**

What are the three tributaries of the coronary sinus?

VEIN	DESCRIPTION AND REGION DRAINED
Coronary sinus (KOR-ō-nar-ē; *corona* = crown)	The **coronary sinus** is the main vein of the heart; it receives almost all venous blood from the myocardium. It is located in the coronary sulcus (see Figure 20.3c) and opens into the right atrium between the orifice of the inferior vena cava and the tricuspid valve. It is a wide venous channel into which three veins drain. It receives the **great cardiac vein** (in the anterior interventricular sulcus) into its left end, and the **middle cardiac vein** (in the posterior interventricular sulcus) and the **small cardiac vein** into its right end. Several **anterior cardiac veins** drain directly into the right atrium.
Superior vena cava (SVC) (VĒ-na CĀ-va; *vena* = vein; *cava* = cavelike)	The **superior vena cava** is about 7.5 cm (3 in.) long and 2 cm (1 in.) in diameter and empties its blood into the superior part of the right atrium. It begins posterior to the right first costal cartilage by the union of the right and left brachiocephalic veins and ends at the level of the right third costal cartilage, where it enters the right atrium. The SVC drains the head, neck, chest, and upper limbs.
Inferior vena cava (IVC)	The **inferior vena cava** is the largest vein in the body, about 3.5 cm (1.4 in.) in diameter. It begins anterior to the fifth lumbar vertebra by the union of the common iliac veins, ascends behind the peritoneum to the right of the midline, pierces the caval opening of the diaphragm at the level of the eighth thoracic vertebra, and enters the inferior part of the right atrium. The IVC drains the abdomen, pelvis, and lower limbs. The inferior vena cava is commonly compressed during the later stages of pregnancy by the enlarging uterus, producing edema of the ankles and feet and temporary varicose veins.

continues

Figure 21.23 Principal veins.

Deoxygenated blood returns to the heart via the superior vena cava, inferior vena cava, and the coronary sinus.

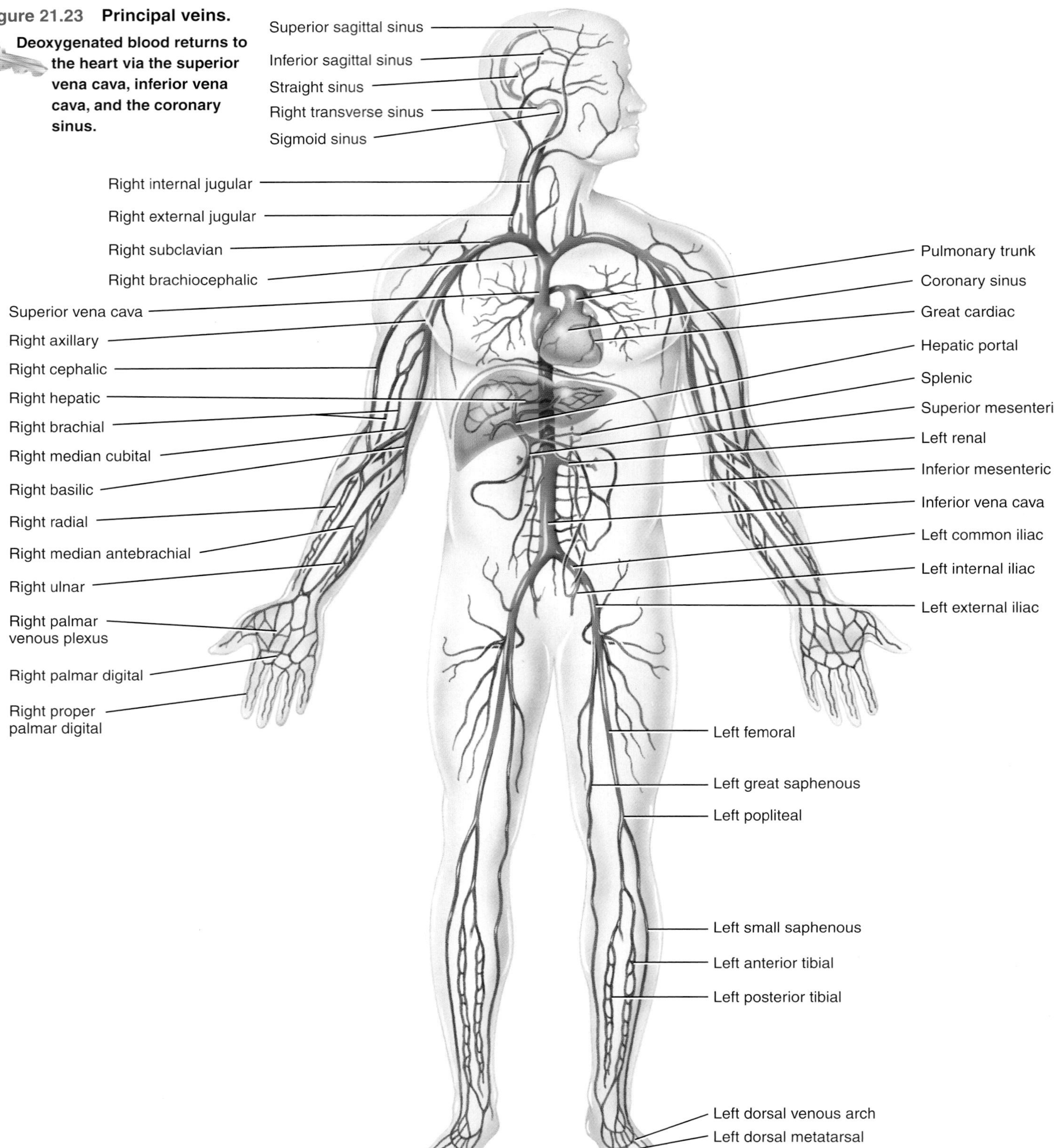

Overall anterior view of the principal veins

Which general regions of the body are drained by the superior vena cava and the inferior vena cava?

▶ **OBJECTIVE**

Identify the three major veins that drain blood from the head.

Most blood draining from the head passes into three pairs of veins: the **internal jugular, external jugular,** and **vertebral veins.** Within the brain, all veins drain into dural venous sinuses and then into the internal jugular veins. **Dural venous sinuses** are endothelial-lined venous channels between layers of the cranial dura mater.

▶ **CHECKPOINT**

Which general areas are drained by the internal jugular, external jugular, and vertebral veins?

VEIN	DESCRIPTION AND REGION DRAINED
Internal jugular veins (JUG-ū-lar; *jugular* = throat)	The flow of blood from the dural venous sinuses into the internal jugular veins is as follows (Figure 21.24): The **superior sagittal sinus** (SAJ-i-tal = straight) begins at the frontal bone, where it receives a vein from the nasal cavity, and passes posteriorly to the occipital bone. Along its course, it receives blood from the superior, medial, and lateral aspects of the cerebral hemispheres, meninges, and cranial bones. The superior sagittal sinus usually turns to the right and drains into the right transverse sinus. The **inferior sagittal sinus** is much smaller than the superior sagittal sinus; it begins posterior to the attachment of the falx cerebri and receives the great cerebral vein to become the straight sinus. The great cerebral vein drains the deeper parts of the brain. Along its course the inferior sagittal sinus also receives tributaries from the superior and medial aspects of the cerebral hemispheres. The **straight sinus** runs in the tentorium cerebelli and is formed by the union of the inferior sagittal sinus and the great cerebral vein. The straight sinus also receives blood from the cerebellum and usually drains into the left transverse sinus. The **transverse sinuses** begin near the occipital bone, pass laterally and anteriorly, and become the sigmoid sinuses near the temporal bone. The transverse sinuses receive blood from the cerebrum, cerebellum, and cranial bones. The **sigmoid sinuses** (SIG-moyd = S-shaped) are located along the temporal bone. They pass through the jugular foramina, where they terminate in the internal jugular veins. The sigmoid sinuses drain the transverse sinuses. The **cavernous sinuses** (KAV-er-nus = cavelike) are located on either side of the sphenoid bone. They receive blood from the ophthalmic veins from the orbits, and from the cerebral veins from the cerebral hemispheres. They ultimately empty into the transverse sinuses and internal jugular veins. The cavernous sinuses are unique because they have nerves and a major blood vessel passing through them on their way to the orbit and face. The oculomotor (III) nerve, trochlear (IV) nerve, and ophthalmic and maxillary branches of the trigeminal (V) nerve as well as the internal carotid arteries pass through the cavernous sinuses. The right and left **internal jugular veins** pass inferiorly on either side of the neck lateral to the internal carotid and common carotid arteries. They then unite with the subclavian veins posterior to the clavicles at the sternoclavicular joints to form the right and left brachiocephalic veins (brā′-kē-ō-se-FAL-ik; *brachio-* = arm; *cephal-* = head). From here blood flows into the superior vena cava. The general structures drained by the internal jugular veins are the brain (through the dural venous sinuses), face, and neck.
External jugular veins	The right and left **external jugular veins** begin in the parotid glands near the angle of the mandible. They are superficial veins that descend through the neck across the sternocleidomastoid muscles. They terminate at a point opposite the middle of the clavicles, where they empty into the subclavian veins. The general structures drained by the external jugular veins are external to the cranium, such as the scalp and superficial and deep regions of the face. When venous pressure rises, for example, during heavy coughing or straining or in cases of heart failure, the external jugular veins become very prominent along the side of the neck.
Vertebral veins (VER-te-bral; *vertebra* = vertebrae)	The right and left **vertebral veins** originate inferior to the occipital condyles. They descend through successive transverse foramina of the first six cervical vertebrae and emerge from the foramina of the sixth cervical vertebra to enter the brachiocephalic veins in the root of the neck. The vertebral veins drain deep structures in the neck such as the cervical vertebrae, cervical spinal cord, and some neck muscles.

continues

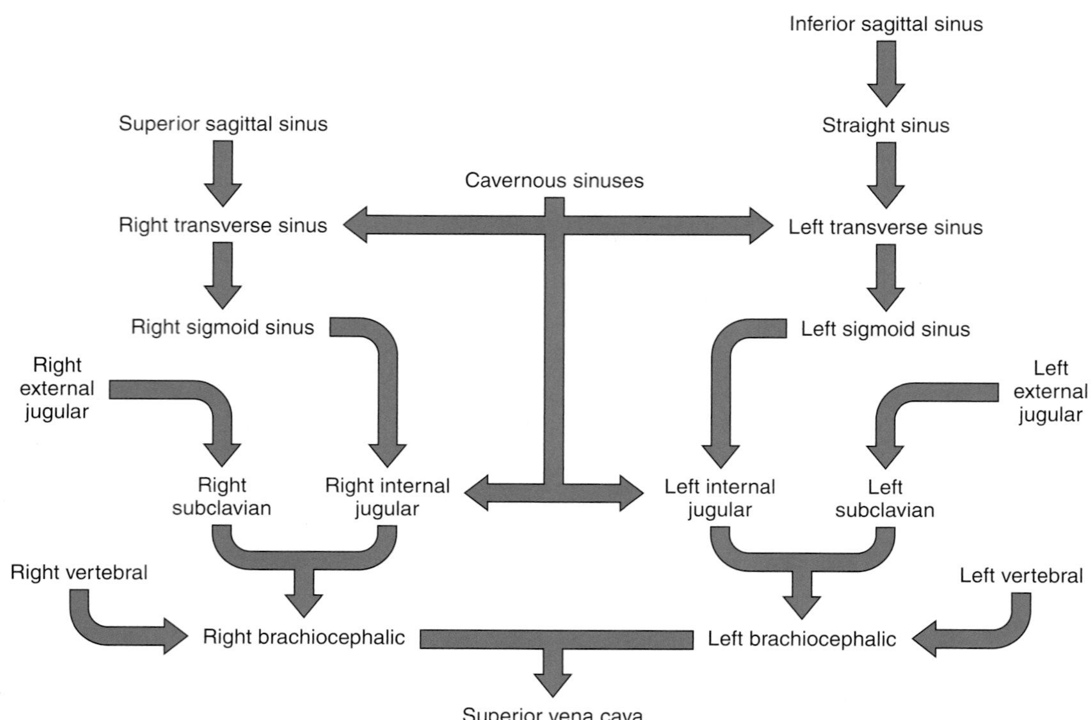

Figure 21.24 Principal veins of the head and neck.

Blood draining from the head passes into the internal jugular, external jugular, and vertebral veins.

Superior sagittal sinus
Inferior sagittal sinus
Great cerebral
Straight sinus
Right transverse sinus
Right sigmoid sinus
Right vertebral
Right internal jugular
Right external jugular
Right subclavian
Right axillary
Right cavernous sinus
Right ophthalmic
Right superficial temporal
Right facial
Right brachiocephalic

Right lateral view

Into which veins in the neck does all venous blood in the brain drain?

EXHIBIT 21.9 VEINS OF THE UPPER LIMBS (FIGURE 21.25)

▶ **OBJECTIVE**

Identify the principal veins that drain the upper limbs.

Both superficial and deep veins return blood from the upper limbs to the heart. **Superficial veins** are located just deep to the skin and are often visible. They anastomose extensively with one another and with deep veins, and they do not accompany arteries. Superficial veins are larger than deep veins and return most of the blood from the upper limbs. **Deep veins** are located deep in the body. They usually accompany arteries and have the same names as the corresponding arteries. Both superficial and deep veins have valves, but valves are more numerous in the deep veins.

▶ **CHECKPOINT**

Where do the cephalic, basilic, median antebrachial, radial, and ulnar veins originate?

VEIN	DESCRIPTION AND REGION DRAINED
Superficial	
Cephalic veins (se-FAL-ik = pertaining to the head)	The principal superficial veins that drain the upper limbs are the cephalic and basilic veins. They originate in the hand and convey blood from the smaller superficial veins into the axillary veins. The **cephalic veins** begin on the lateral aspect of the **dorsal venous networks of the hands (dorsal venous arches),** networks of veins on the dorsum of the hands formed by the **dorsal metacarpal veins** (Figure 21.25a). These veins, in turn, drain the **dorsal digital veins,** which pass along the sides of the fingers. Following their formation from the dorsal venous networks of the hands, the cephalic veins arch around the radial side of the forearms to the anterior surface and ascend through the entire limbs along the anterolateral surface. The cephalic veins end where they join the axillary veins, just inferior to the clavicles. **Accessory cephalic veins** originate either from a venous plexus on the dorsum of the forearms or from the medial aspects of the dorsal venous networks of the hands, and unite with the cephalic veins just inferior to the elbow. The cephalic veins drain blood from the lateral aspect of the upper limbs.
Basilic veins (ba-SIL-ik = royal, of prime importance)	The **basilic veins** begin on the medial aspects of the dorsal venous networks of the hands and ascend along the posteromedial surface of the forearm and anteromedial surface of the arm (Figure 21.25b). They drain blood from the medial aspects of the upper limbs. Anterior to the elbow, the basilic veins are connected to the cephalic veins by the **median cubital veins** (*cubital* = pertaining to the elbow), which drain the forearm. If veins must be punctured for an injection, transfusion, or removal of a blood sample, the medial cubital veins are preferred. After receiving the median cubital veins, the basilic veins continue ascending until they reach the middle of the arm. There they penetrate the tissues deeply and run alongside the brachial arteries until they join the brachial veins. As the basilic and brachial veins merge in the axillary area, they form the axillary veins.
Median antebrachial veins (an′-tē-BRĀ-kē-al; *ante-* = before, in front of; *brachi-* = arm)	The **median antebrachial veins (median veins of the forearm)** begin in the **palmar venous plexuses,** networks of veins on the palms. The plexuses drain the **palmar digital veins** in the fingers. The median ante-brachial veins ascend anteriorly in the forearms to join the basilic or median cubital veins, sometimes both. They drain the palms and forearms.
Deep	
Radial veins (RĀ-dē-al = pertaining to the radius)	The paired **radial veins** begin at the **deep palmar venous arches** (Figure 21.25c). These arches drain the **palmar metacarpal veins** in the palms. The radial veins drain the lateral aspects of the forearms and pass alongside the radial arteries. Just inferior to the elbow joint, the radial veins unite with the ulnar veins to form the brachial veins.
Ulnar veins (UL-nar = pertaining to the ulna)	The paired **ulnar veins,** which are larger than the radial veins, begin at the **superficial palmar venous arches.** These arches drain the **common palmar digital veins** and the **proper palmar digital veins** in the fingers. The ulnar veins drain the medial aspect of the forearms, pass alongside the ulnar arteries, and join with the radial veins to form the brachial veins.
Brachial veins (BRĀ-kē-al; *brachi-* = arm)	The paired **brachial veins** accompany the brachial arteries. They drain the forearms, elbow joints, arms, and humerus. They pass superiorly and join with the basilic veins to form the axillary veins.
Axillary veins (AK-sil-ār-ē; *axilla* = armpit)	The **axillary veins** ascend to the outer borders of the first ribs, where they become the subclavian veins. The axillary veins receive tributaries that correspond to the branches of the axillary arteries. The axillary veins drain the arms, axillas, and superolateral chest wall.
Subclavian veins (sub-KLĀ-vē-an; *sub-* = under; *clavian* = pertaining to the clavicle)	The **subclavian veins** are continuations of the axillary veins that terminate at the sternal end of the clavicles, where they unite with the internal jugular veins to form the brachiocephalic veins. The subclavian veins drain the arms, neck, and thoracic wall. The thoracic duct of the lymphatic system delivers lymph into the junction between the left subclavian and the left internal jugular veins. The right lymphatic duct delivers lymph into the junction between the right subclavian and right internal jugular veins (see Figure 22.3a). In a procedure called *central line placement,* the right subclavian vein is frequently used to administer nutrients and medication and measure venous pressure.

continues

SCHEME OF DRAINAGE

Superior vena cava
Right brachiocephalic
Right internal jugular
Right subclavian
Right axillary
Right brachial
Right cephalic
Right basilic
Right median cubital
Right radial
Right ulnar
Right accessory cephalic
Right median antebrachial
Right dorsal venous network of the hand (dorsal venous arch)
Right palmar venous plexus
Right deep palmar venous arch
Right superficial palmar venous arch
Right dorsal metacarpal
Right palmar digital
Right palmar metacarpal
Right common palmar digital
Right dorsal digital
Right proper palmar digital

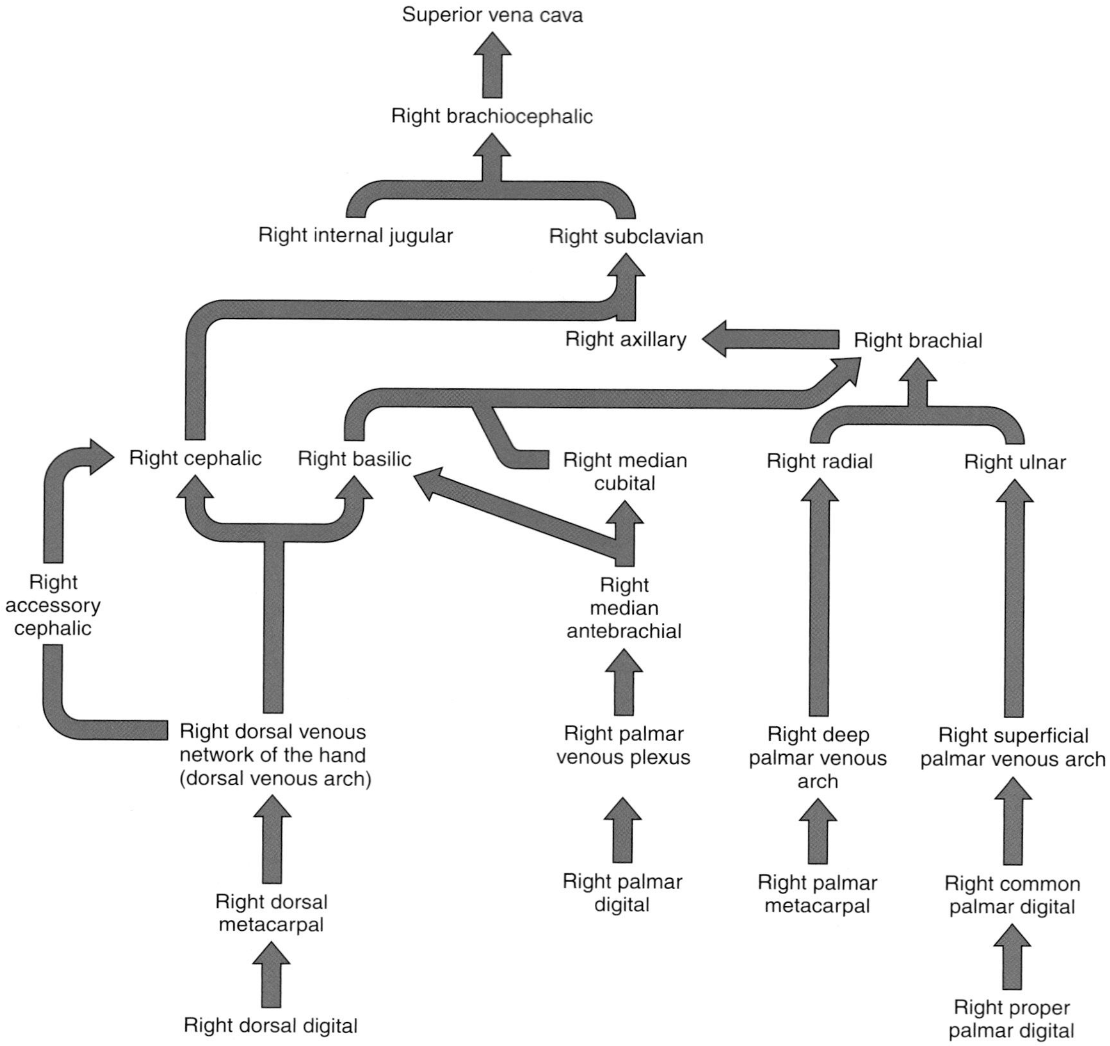

Figure 21.25 Principal veins of the right upper limb.

Deep veins usually accompany arteries that have similar names.

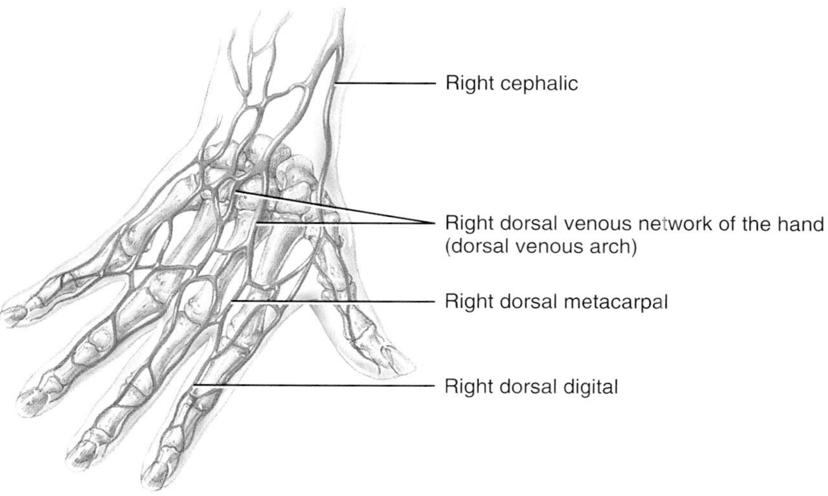

(a) Posterior view of superficial veins of the hand

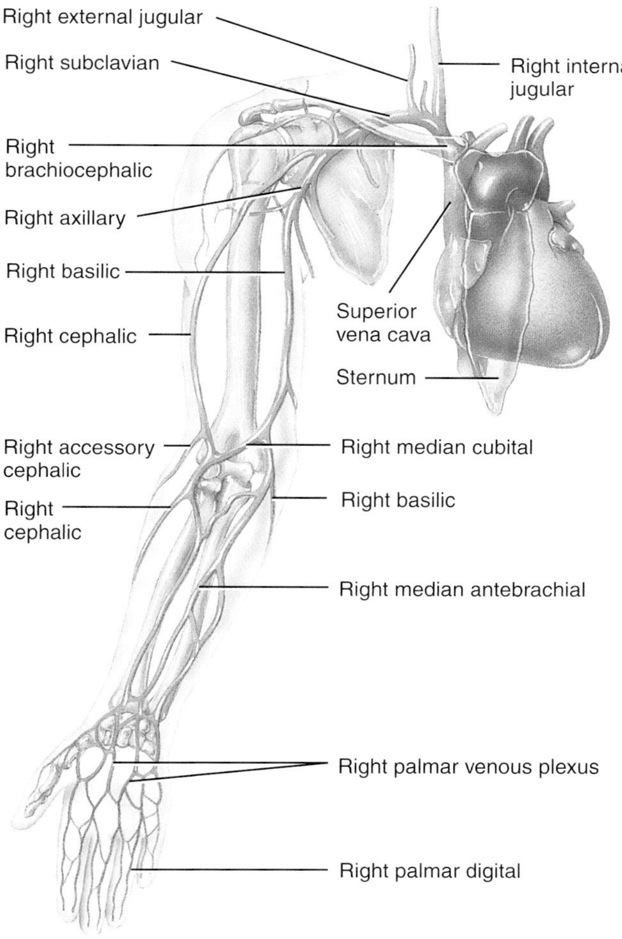

(b) Anterior view of superficial veins

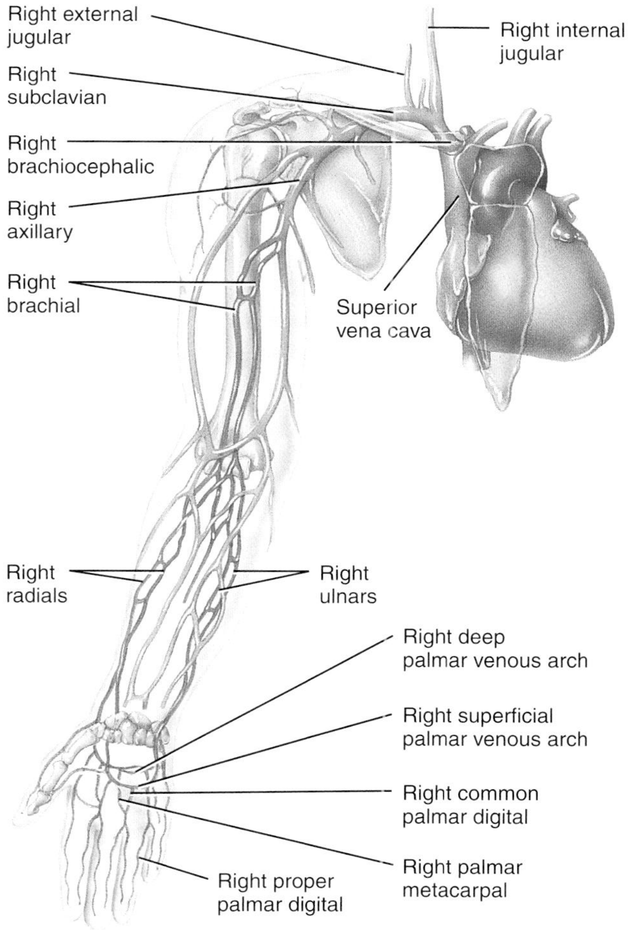

(c) Anterior view of deep veins

From which vein in the upper limb is a blood sample often taken?

VEINS OF THE THORAX (FIGURE 21.26)

▶ **OBJECTIVE**

Identify the components of the azygos system of veins.

Although the brachiocephalic veins drain some portions of the thorax, most thoracic structures are drained by a network of veins, called the **azygos system,** that runs on either side of the vertebral column. The system consists of three veins—the **azygos, hemiazygos,** and **accessory hemiazygos veins**—that show considerable variation in origin, course, tributaries, anastomoses, and termination. Ultimately they empty into the superior vena cava.

▶ **CHECKPOINT**

What is the importance of the azygos system relative to the inferior vena cava?

VEIN	DESCRIPTION AND REGION DRAINED
Brachiocephalic vein (brā′-kē-ō-se-FAL-ik; *brachio-* = arm; *cephalic* = pertaining to the head)	The right and left **brachiocephalic veins,** formed by the union of the subclavian and internal jugular veins, drain blood from the head, neck, upper limbs, mammary glands, and superior thorax. The brachiocephalic veins unite to form the superior vena cava. Because the superior vena cava is to the right of the body's midline, the left brachiocephalic vein is longer than the right. The right brachiocephalic vein is anterior and to the right of the brachiocephalic trunk. The left brachiocephalic vein is anterior to the brachiocephalic trunk, the left common carotid and left subclavian arteries, the trachea, the left vagus (X) nerve, and phrenic nerve.
Azygos system (az-Ī-gos = unpaired)	The **azygos system,** besides collecting blood from the thorax and abdominal wall, may serve as a bypass for the inferior vena cava that drains blood from the lower body. Several small veins directly link the azygos system with the inferior vena cava. Large veins that drain the lower limbs and abdomen conduct blood into the azygos system. If the inferior vena cava or hepatic portal vein becomes obstructed, the azygos system can return blood from the lower body to the superior vena cava.
Azygos vein	The **azygos vein** is anterior to the vertebral column, slightly to the right of the midline. It usually begins at the junction of the right ascending lumbar and right subcostal veins near the diaphragm. At the level of the fourth thoracic vertebra, it arches over the root of the right lung to end in the superior vena cava. Generally, the azygos vein drains the right side of the thoracic wall, thoracic viscera, and abdominal wall. Specifically, the azygos vein receives blood from most of the **right posterior intercostal, hemiazygos, accessory hemiazygos, esophageal, mediastinal, pericardial,** and **bronchial veins.**
Hemiazygos vein (HEM-ē-az-ī-gos; *hemi-* = half)	The **hemiazygos vein** is anterior to the vertebral column and slightly to the left of the midline. It often begins at the junction of the left ascending lumbar and left subcostal veins. It terminates by joining the azygos vein at about the level of the ninth thoracic vertebra. Generally, the hemiazygos vein drains the left side of the thoracic wall, thoracic viscera, and abdominal wall. Specifically, the hemiazygos vein receives blood from the ninth through eleventh **left posterior intercostal, esophageal, mediastinal,** and sometimes the **accessory hemiazygos veins.**
Accessory hemiazygos vein	The **accessory hemiazygos vein** is also anterior to the vertebral column and to the left of the midline. It begins at the fourth or fifth intercostal space and descends from the fifth to the eighth thoracic vertebra or ends in the hemiazygos vein. It terminates by joining the azygos vein at about the level of the eighth thoracic vertebra. The accessory hemiazygos vein drains the left side of the thoracic wall. It receives blood from the fourth through eighth **left posterior intercostal veins** (the first through third left posterior intercostal veins open into the left brachiocephalic vein), **left bronchial,** and **mediastinal veins.**

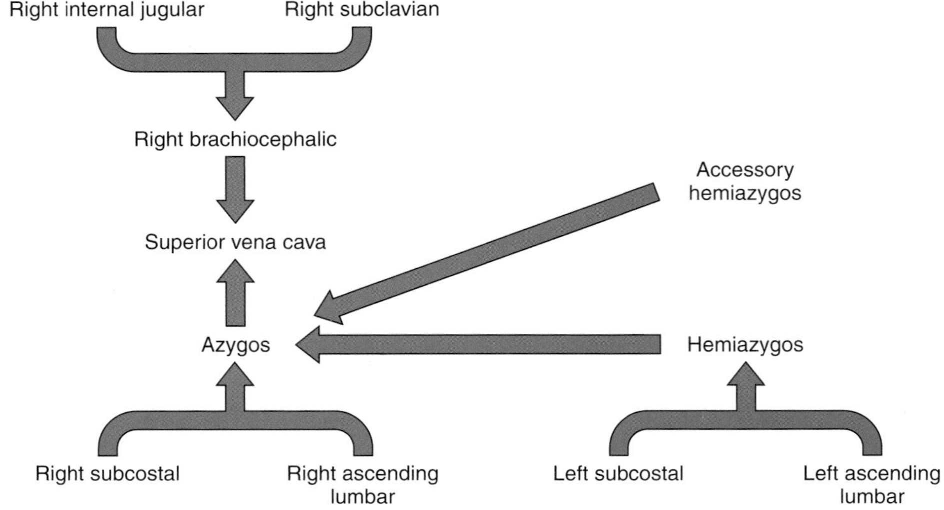

Figure 21.26 **Principal veins of the thorax, abdomen, and pelvis.**

Most thoracic structures are drained by the azygos system of veins.

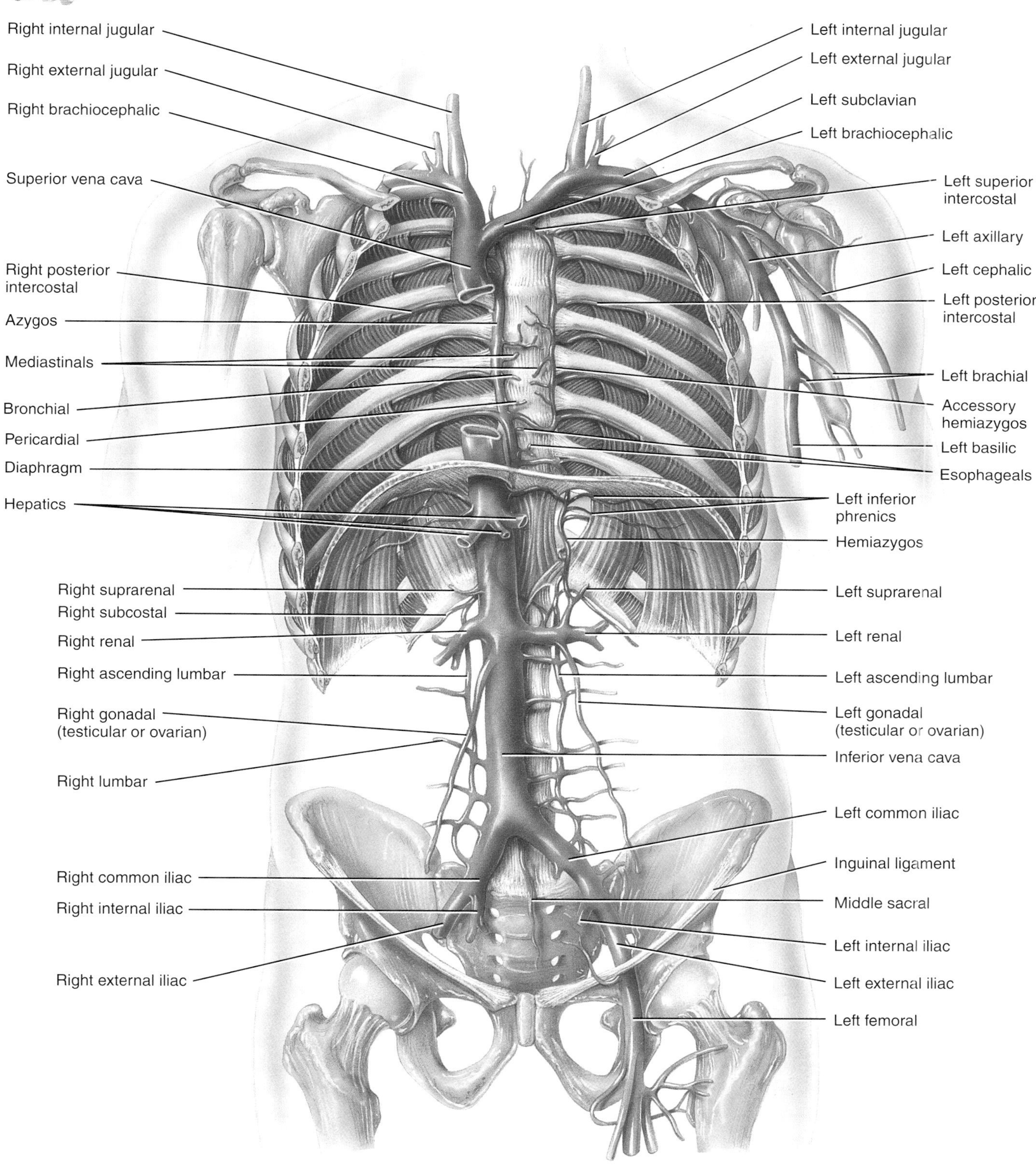

Anterior view

Which vein returns blood from the abdominopelvic viscera to the heart?

VEINS OF THE ABDOMEN AND PELVIS (SEE FIGURE 21.26)

▶ **OBJECTIVE**

Identify the principal veins that drain the abdomen and pelvis.

Blood from the abdominal and pelvic viscera and abdominal wall returns to the heart via the inferior vena cava. Many small veins enter the inferior vena cava. Most carry return flow from parietal branches of the abdominal aorta, and their names correspond to the names of the arteries.

The inferior vena cava does not receive veins directly from the gastrointestinal tract, spleen, pancreas, and gallbladder. These organs pass their blood into a common vein, the **hepatic portal vein,** which delivers the blood to the liver. The superior mesenteric and splenic veins unite to form the hepatic portal vein (see Figure 21.28). This special flow of venous blood, called the **hepatic portal circulation,** is described shortly. After passing through the liver for processing, blood drains into the hepatic veins, which empty into the inferior vena cava.

▶ **CHECKPOINT**

What structures do the lumbar, gonadal, renal, suprarenal, inferior phrenic, and hepatic veins drain?

VEIN	DESCRIPTION AND REGION DRAINED
Inferior vena cava (VĒ-na CĀ-va; *vena* = vein; *cava* = cavelike)	The two common iliac veins that drain the lower limbs, pelvis, and abdomen unite to form the **inferior vena cava.** The inferior vena cava extends superiorly through the abdomen and thorax to the right atrium.
Common iliac veins (IL-ē-ak = pertaining to the ilium)	The **common iliac veins** are formed by the union of the internal and external iliac veins anterior to the sacroiliac joint and represent the distal continuation of the inferior vena cava at their bifurcation. The right common iliac vein is much shorter than the left and is also more vertical. Generally, the common iliac veins drain the pelvis, external genitals, and lower limbs.
Internal iliac veins	The **internal iliac veins** begin near the superior portion of the greater sciatic notch and run medial to their corresponding arteries. Generally, the veins drain the thigh, buttocks, external genitals, and pelvis.
External iliac veins	The **external iliac veins** are companions of the internal iliac arteries and begin at the inguinal ligaments as continuations of the femoral veins. They end anterior to the sacroiliac joint where they join with the internal iliac veins to form the common iliac veins. The external iliac veins drain the lower limbs, cremaster muscle in males, and the abdominal wall.
Lumbar veins (LUM-bar = pertaining to the loin)	A series of parallel **lumbar veins,** usually four on each side, drain blood from both sides of the posterior abdominal wall, vertebral canal, spinal cord, and meninges. The lumbar veins run horizontally with the lumbar arteries. The lumbar veins connect at right angles with the right and left **ascending lumbar veins,** which form the origin of the corresponding azygos or hemiazygos vein. The lumbar veins drain blood into the ascending lumbars and then run to the inferior vena cava, where they release the remainder of the flow.
Gonadal veins (gō-NAD-al; *gono* = seed) [**testicular** (tes-TIK-ū-lar) or **ovarian** (ō-VAR-ē-an)]	The **gonadal veins** ascend with the gonadal arteries along the posterior abdominal wall. In the male, the gonadal veins are called the testicular veins. The **testicular veins** drain the testes (the left testicular vein empties into the left renal vein, and the right testicular vein drains into the inferior vena cava). In the female, the gonadal veins are called the ovarian veins. The **ovarian veins** drain the ovaries. The left ovarian vein empties into the left renal vein, and the right ovarian vein drains into the inferior vena cava.
Renal veins (RĒ-nal = *ren-* = kidney)	The large **renal veins** pass anterior to the renal arteries. The left renal vein is longer than the right renal vein and passes anterior to the abdominal aorta. It receives the left testicular (or ovarian), left inferior phrenic, and usually left suprarenal veins. The right renal vein empties into the inferior vena cava posterior to the duodenum. The renal veins drain the kidneys.
Suprarenal veins (soo′-pra-RĒ-nal; *supra-* = above)	The **suprarenal veins** drain the adrenal (suprarenal) glands (the left suprarenal vein empties into the left renal vein, and the right suprarenal vein empties into the inferior vena cava).
Inferior phrenic veins (FREN-ik = pertaining to the diaphragm)	The **inferior phrenic veins** drain the diaphragm (the left inferior phrenic vein usually sends one tributary to the left suprarenal vein, which empties into the left renal vein, and another tributary that empties into the inferior vena cava; the right inferior phrenic vein empties into the inferior vena cava).
Hepatic veins (he-PAT-ik = pertaining to the liver)	The **hepatic veins** drain the liver.

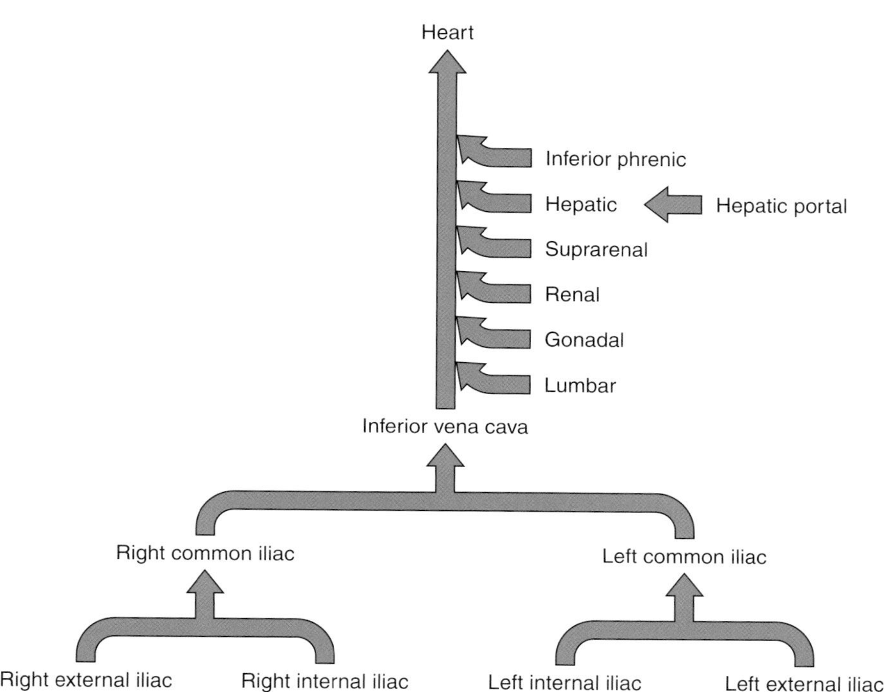
SCHEME OF DRAINAGE
Heart
Inferior phrenic
Hepatic
Hepatic portal
Suprarenal
Renal
Gonadal
Lumbar
Inferior vena cava
Right common iliac
Left common iliac
Right external iliac
Right internal iliac
Left internal iliac
Left external iliac

EXHIBIT 21.12 VEINS OF THE LOWER LIMBS (FIGURE 21.27)

▶ **OBJECTIVE**

Identify the principal superficial and deep veins that drain the lower limbs.

As with the upper limbs, blood from the lower limbs is drained by both **superficial** and **deep veins.** The superficial veins often anastomose with one another and with deep veins along their length. Deep veins, for the most part, have the same names as corresponding arteries. All veins of the lower limbs have valves, which are more numerous than in veins of the upper limbs.

▶ **CHECKPOINT**

What is the clinical importance of the great saphenous veins?

VEIN	DESCRIPTION AND REGION DRAINED
Superficial Veins	
Great saphenous veins (sa-FĒ-nus; *saphen-* = clearly visible)	The **great (long) saphenous veins,** the longest veins in the body, ascend from the foot to the groin in the subcutaneous layer. They begin at the medial end of the dorsal venous arches of the foot. The **dorsal venous arches** (VĒ-nus) are networks of veins on the dorsum of the foot formed by the **dorsal digital veins,** which collect blood from the toes, and then unite in pairs to form the **dorsal metatarsal veins,** which parallel the metatarsals. As the dorsal metatarsal veins approach the foot, they combine to form the dorsal venous arches. The great saphenous veins pass anterior to the medial malleolus of the tibia and then superiorly along the medial aspect of the leg and thigh just deep to the skin. They receive tributaries from superficial tissues and connect with the deep veins as well. They empty into the femoral veins at the groin. Generally, the great saphenous veins drain mainly the medial side of the leg and thigh, the groin, external genitals, and abdominal wall. Along their length, the great saphenous veins have from 10 to 20 valves, with more located in the leg than the thigh. These veins are more likely to be subject to varicosities than other veins in the lower limbs because they must support a long column of blood and are not well supported by skeletal muscles. The great saphenous veins are often used for prolonged administration of intravenous fluids. This is particularly important in very young children and in patients of any age who are in shock and whose veins are collapsed. In coronary artery bypass grafting, if multiple blood vessels need to be grafted, sections of the great saphenous vein are used along with at least one artery as a graft. After the great saphenous vein is removed and divided into sections, the sections are used to bypass the blockages. The vein grafts are reversed so that the valves do not obstruct the flow of blood.
Small saphenous veins	The **small (short) saphenous veins** begin at the lateral aspect of the dorsal venous arches of the foot. They pass posterior to the lateral malleolus of the fibula and ascend deep to the skin along the posterior aspect of the leg. They empty into the popliteal veins in the popliteal fossa, posterior to the knee. Along their length, the small saphenous veins have from 9 to 12 valves. The small saphenous veins drain the foot and posterior aspect of the leg. They may communicate with the great saphenous veins in the proximal thigh.
Deep Veins	
Posterior tibial veins (TIB-ē-al)	The **plantar digital veins** on the plantar surfaces of the toes unite to form the **plantar metatarsal veins,** which parallel the metatarsals. They in turn unite to form the **deep plantar venous arches.** From each arch emerges the **medial** and **lateral plantar veins.** The medial and lateral plantar veins, posterior to the medial malleolus of the tibia, form the paired **posterior tibial veins,** which sometimes merge into a single vessel. They accompany the posterior tibial artery through the leg. They ascend deep to the muscles in the posterior aspect of the leg and drain the foot and posterior compartment muscles. About two-thirds of the way up the leg, the posterior tibial veins drain blood from the **fibular (peroneal) veins,** which drain the lateral and posterior leg muscles. The posterior tibial veins unite with the anterior tibial veins just inferior to the popliteal fossa to form the popliteal veins.
Anterior tibial veins	The paired **anterior tibial veins** arise in the dorsal venous arch and accompany the anterior tibial artery. They ascend in the interosseous membrane between the tibia and fibula and unite with the posterior tibial veins to form the popliteal vein. The anterior tibial veins drain the ankle joint, knee joint, tibiofibular joint, and anterior portion of the leg.
Popliteal veins (pop′-li-TĒ-al = pertaining to the hollow behind knee)	The **popliteal veins,** formed by the union of the anterior and posterior tibial veins, also receive blood from the small saphenous veins and tributaries that correspond to branches of the popliteal artery. The popliteal veins drain the knee joint and the skin, muscles, and bones of portions of the calf and thigh around the knee joint.
Femoral veins (FEM-o-ral)	The **femoral veins** accompany the femoral arteries and are the continuations of the popliteal veins just superior to the knee. The femoral veins extend up the posterior surface of the thighs and drain the muscles of the thighs, femurs, external genitals, and superficial lymph nodes. The largest tributaries of the femoral veins are the **deep veins of the thigh (deep femoral veins).** Just before penetrating the abdominal wall, the femoral veins receive the deep femoral veins and the great saphenous veins. The veins formed from this union penetrate the body wall and enter the pelvic cavity. Here they are known as the **external iliac veins.** In order to take blood samples or pressure recordings from the right side of the heart, a catheter is inserted into the femoral vein as it passes through the femoral triangle. The catheter passes through the external and common iliac veins and inferior vena cava into the right atrium.

SCHEME OF DRAINAGE

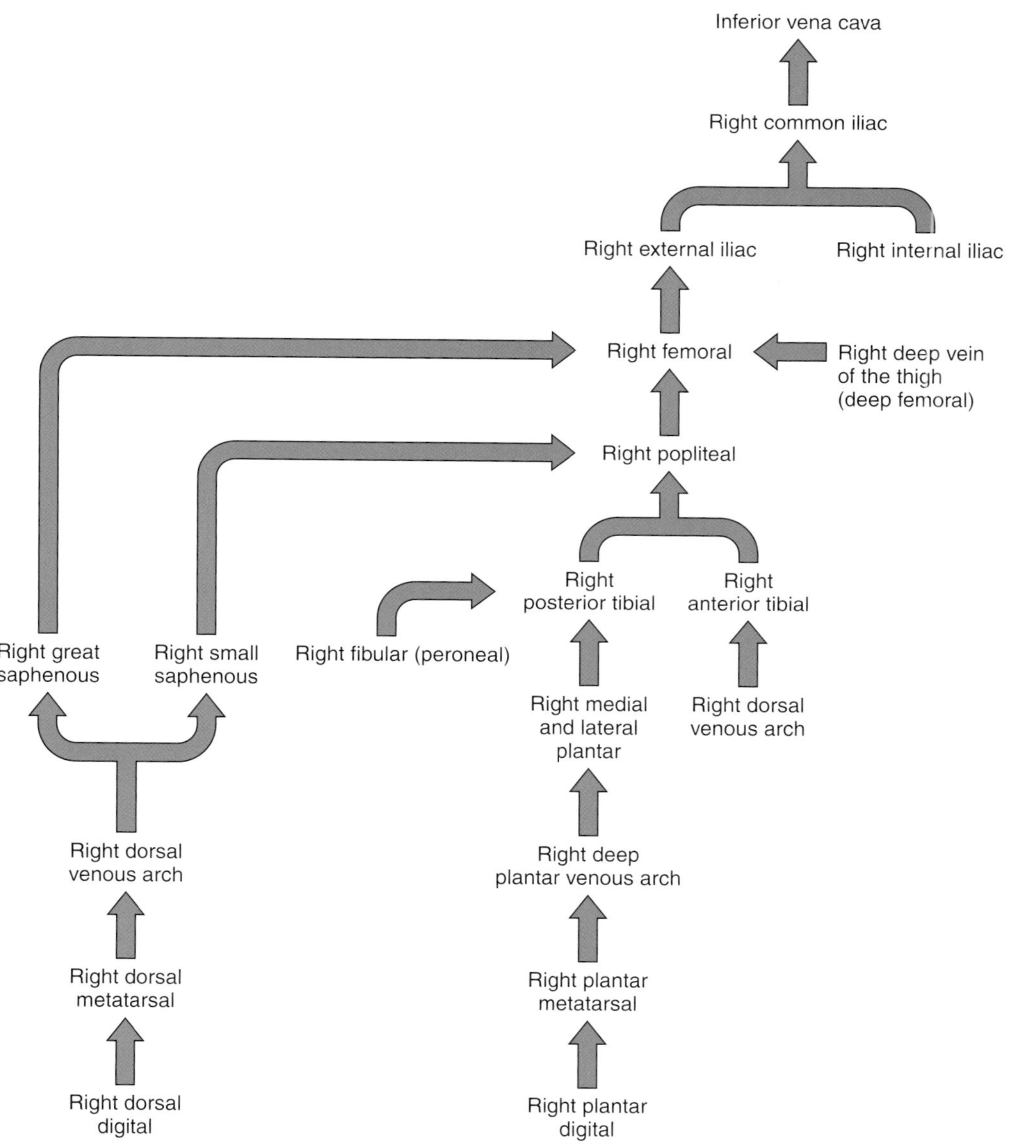

Inferior vena cava
Right common iliac
Right external iliac
Right internal iliac
Right femoral
Right deep vein
of the thigh
(deep femoral)
Right popliteal
Right
posterior tibial
Right
anterior tibial
Right great
saphenous
Right small
saphenous
Right fibular (peroneal)
Right medial
and lateral
plantar
Right dorsal
venous arch
Right dorsal
venous arch
Right deep
plantar venous arch
Right dorsal
metatarsal
Right plantar
metatarsal
Right dorsal
digital
Right plantar
digital

continues

Figure 21.27 Principal veins of the pelvis and lower limbs.

Deep veins usually bear the names of their companion arteries.

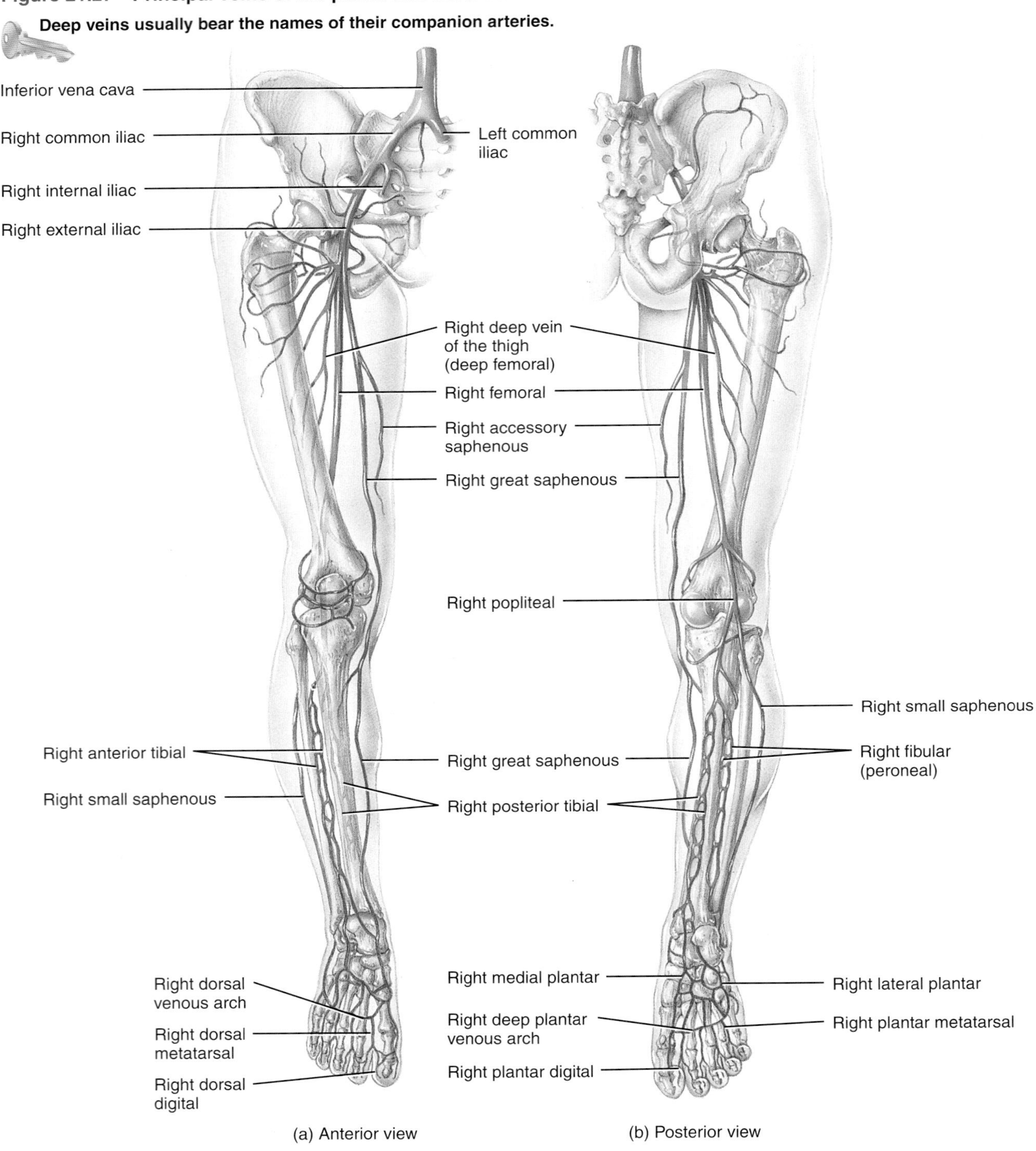

Which veins of the lower limb are superficial?

The Hepatic Portal Circulation

The **hepatic portal circulation** carries venous blood from the gastrointestinal organs and spleen to the liver. A vein that carries blood from one capillary network to another is called a **portal vein.** The **hepatic portal vein** (*hepat-* = liver) receives blood from capillaries of gastrointestinal organs and the spleen and delivers it to the sinusoids of the liver (Figure 21.28). After a meal, hepatic portal blood is rich in nutrients absorbed from the gastrointestinal tract. The liver stores some of them and modifies others before they pass into the general circulation. For example, the liver converts glucose into glycogen for storage, reducing blood glucose level shortly after a meal. The liver also detoxifies harmful substances, such as alcohol, that have been absorbed from the gastrointestinal tract and destroys bacteria by phagocytosis.

The superior mesenteric and splenic veins unite to form the hepatic portal vein. The **superior mesenteric vein** drains blood from the small intestine and portions of the large intestine, stomach, and pancreas through the *jejunal, ileal, ileocolic, right colic, middle colic, pancreaticoduodenal,* and *right gastroepiploic veins.* The

Figure 21.28 Hepatic portal circulation. A schematic diagram of blood flow through the liver, including arterial circulation, is shown in (b). As usual, deoxygenated blood is indicated in blue, oxygenated blood in red.

The hepatic portal circulation delivers venous blood from the organs of the gastrointestinal tract and spleen to the liver.

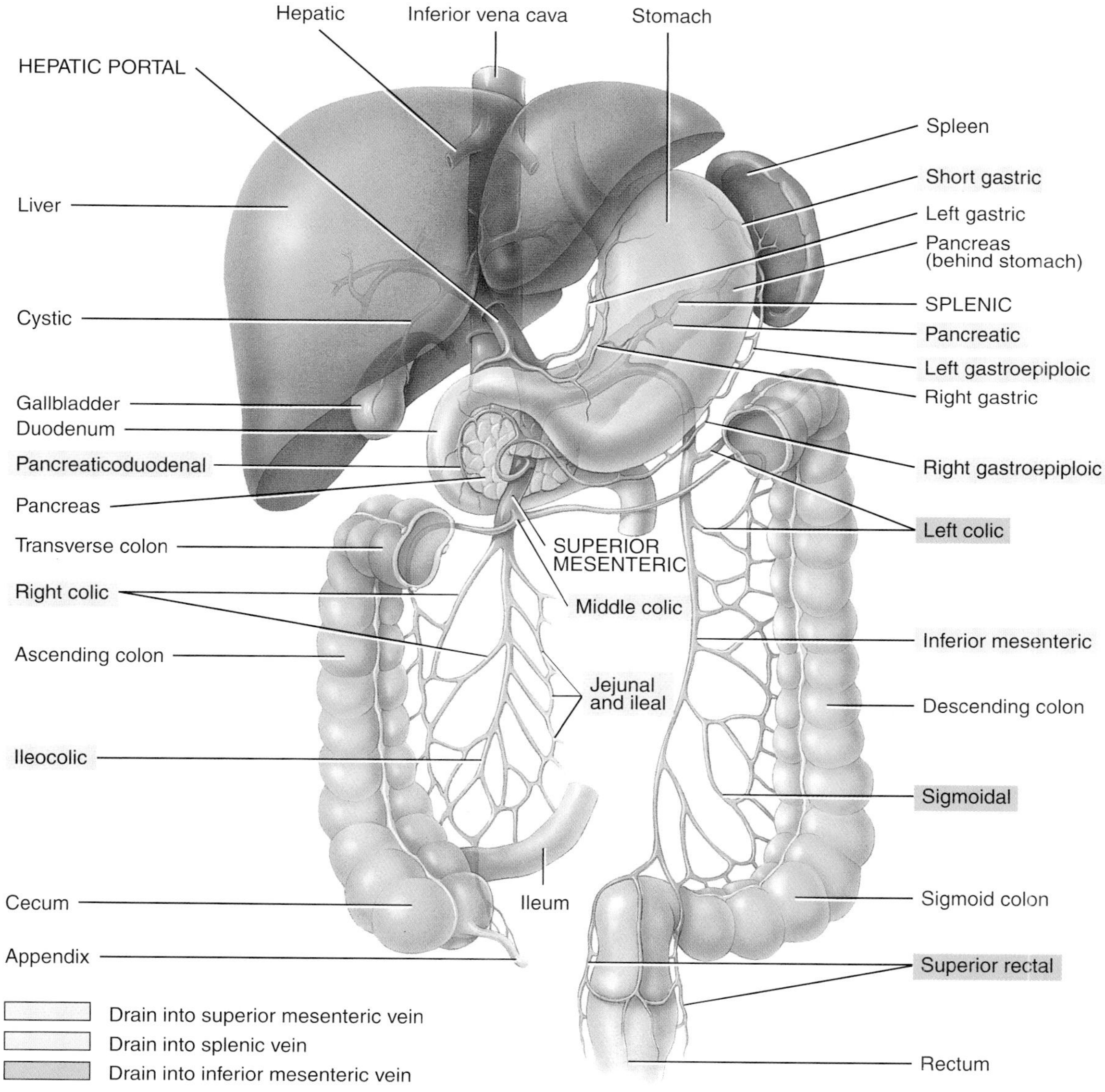

(a) Anterior view of veins draining into the hepatic portal vein

continues

Figure 21.28 **(Continued)**

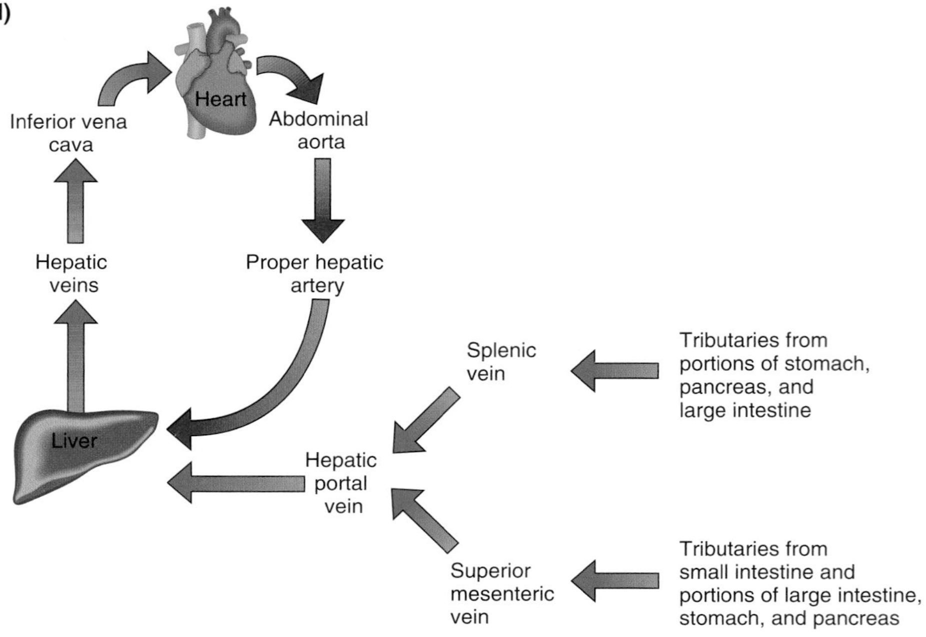

(b) Scheme of principal blood vessels of hepatic portal circulation and arterial supply and venous drainage of liver

Which veins carry blood away from the liver?

splenic vein drains blood from the stomach, pancreas, and portions of the large intestine through the *short gastric, left gastroepiploic, pancreatic,* and *inferior mesenteric veins.* The inferior mesenteric vein, which passes into the splenic vein, drains portions of the large intestine through the superior *rectal, sigmoidal,* and *left colic veins.* The *right* and *left gastric veins,* which open directly into the hepatic portal vein, drain the stomach. The *cystic vein,* which also opens into the hepatic portal vein, drains the gallbladder.

At the same time the liver is receiving nutrient-rich but deoxygenated blood via the hepatic portal vein, it also is receiving oxygenated blood via the hepatic artery, a branch of the celiac trunk. The oxygenated blood mixes with the deoxygenated blood in sinusoids. Eventually, blood leaves the sinusoids of the liver through the **hepatic veins,** which drain into the inferior vena cava.

The Pulmonary Circulation

The **pulmonary circulation** (*pulmo-* = lung) carries deoxygenated blood from the right ventricle to the air sacs (alveoli) within the lungs and returns oxygenated blood from the air sacs to the left atrium (Figure 21.29). The **pulmonary trunk** emerges from the right ventricle and passes superiorly, posteriorly, and to the left. It then divides into two branches: the **right pulmonary artery** to the right lung and the **left pulmonary artery** to the left lung. After birth, the pulmonary arteries are the only arteries that carry deoxygenated blood. On entering the lungs, the branches divide and subdivide until finally they form capillaries around the air sacs (alveoli) within the lungs. CO_2 passes from the blood into the air sacs and is exhaled. Inhaled O_2 passes from the air within the lungs into the blood. The pulmonary capillaries unite to form venules and eventually **pulmonary veins,** which exit the lungs and carry the oxygenated blood to the left atrium. Two left and two right pulmonary veins enter the left atrium. After birth, the pulmonary veins are the only veins that carry oxygenated blood. Contractions of the left ventricle then eject the oxygenated blood into the systemic circulation.

The Fetal Circulation

The circulatory system of a fetus, called the **fetal circulation,** exists only in the fetus and contains special structures that allow the developing fetus to exchange materials with its mother (Figure 21.30 on page 794). It differs from the postnatal (after birth) circulation because the lungs, kidneys, and gastrointestinal organs do not begin to function until birth. The fetus obtains O_2 and nutrients from and eliminates CO_2 and other wastes into the maternal blood.

The exchange of materials between fetal and maternal circulations occurs through the **placenta** (pla-SEN-ta), which forms inside the mother's uterus and attaches to the umbilicus (navel) of the fetus by the **umbilical cord** (um-BIL-i-kal). The placenta communicates with the mother's cardiovascular system through many small blood vessels that emerge from the uterine wall. The umbilical cord contains blood vessels that branch into capillaries in the placenta. Wastes from the fetal blood diffuse out of the capillaries, into spaces containing maternal blood (intervillous spaces) in the placenta, and finally into the mother's uterine veins.

Figure 21.29 **Pulmonary circulation.**

The pulmonary circulation brings deoxygenated blood from the right ventricle to the lungs and returns oxygenated blood from the lungs to the left atrium.

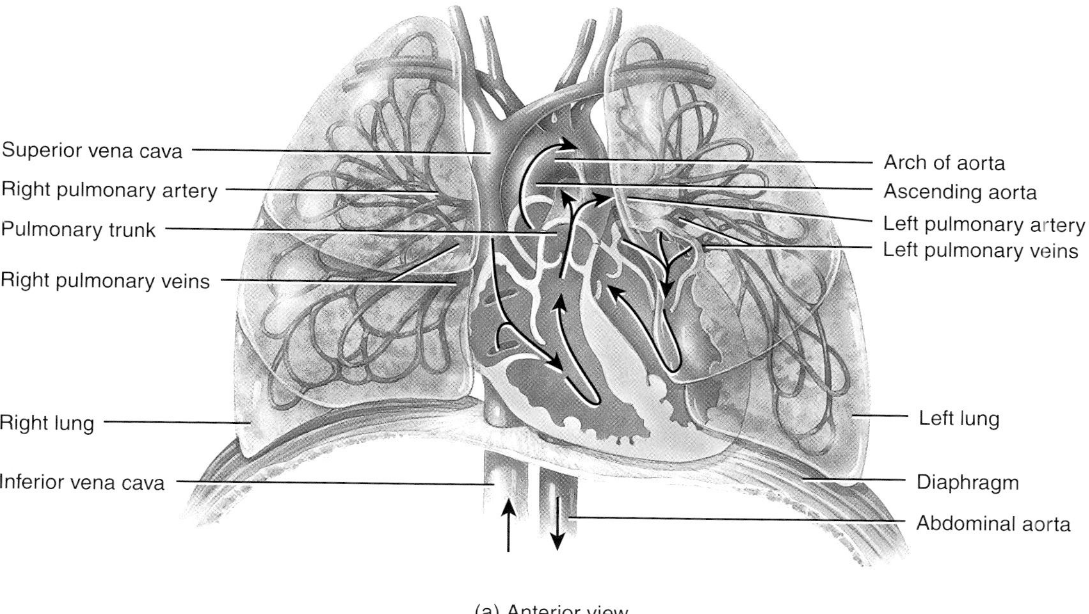

(a) Anterior view

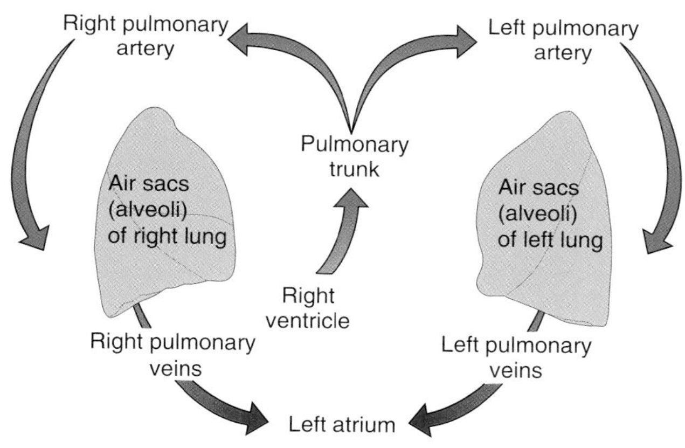

(b) Scheme of pulmonary circulation

After birth, which are the only arteries that carry deoxygenated blood?

Nutrients travel the opposite route—from the maternal blood vessels to the intervillous spaces to the fetal capillaries. Normally, there is no direct mixing of maternal and fetal blood because all exchanges occur by diffusion through capillary walls.

Blood passes from the fetus to the placenta via two **umbilical arteries** (Figure 21.30a, c). These branches of the internal iliac (hypogastric) arteries are within the umbilical cord. At the placenta, fetal blood picks up O_2 and nutrients and eliminates CO_2 and wastes. The oxygenated blood returns from the placenta via a single **umbilical vein.** This vein ascends to the liver of the fetus, where it divides into two branches. Some blood flows through the branch that joins the hepatic portal vein and enters the liver, but most of the blood flows into the second branch, the **ductus venosus** (DUK-tus ve-NŌ-sus), which drains into the inferior vena cava.

Deoxygenated blood returning from lower body regions of the fetus mingles with oxygenated blood from the ductus venosus in the inferior vena cava. This mixed blood then enters the right atrium. Deoxygenated blood returning from upper body regions of the fetus enters the superior vena cava and also passes into the right atrium.

Most of the fetal blood does not pass from the right ventricle to the lungs, as it does in postnatal circulation, because an opening called the **foramen ovale** (fō-RĀ-men ō-VAL-ē) exists in the septum between the right and left atria. Most of the blood

Figure 21.30 **Fetal circulation and changes at birth.** The gold boxes between parts (a) and (b) describe the fate of certain fetal structures once postnatal circulation is established.

The lungs and gastrointestinal organs do not begin to function until birth.

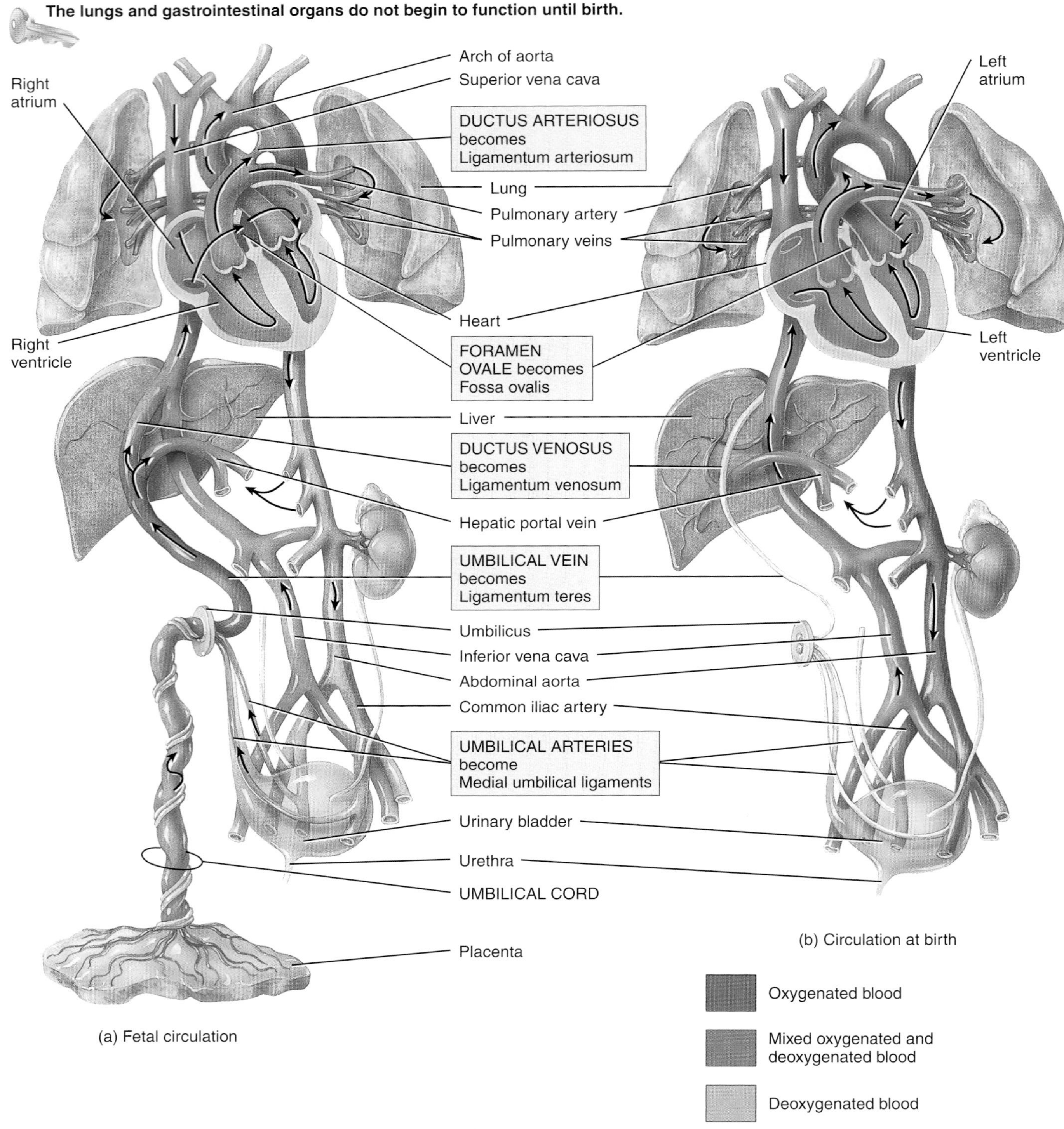

that enters the right atrium passes through the foramen ovale into the left atrium and joins the systemic circulation. The blood that does pass into the right ventricle is pumped into the pulmonary trunk, but little of this blood reaches the nonfunctioning fetal lungs. Instead, most is sent through the **ductus arteriosus** (ar-tē-rē-Ō-sus), a vessel that connects the pulmonary trunk with the aorta. The blood in the aorta is carried to all fetal tissues through the systemic circulation. When the common iliac arteries branch into the external and internal iliacs, part of the blood flows into the internal iliacs, into the umbilical arteries, and back to the placenta for another exchange of materials.

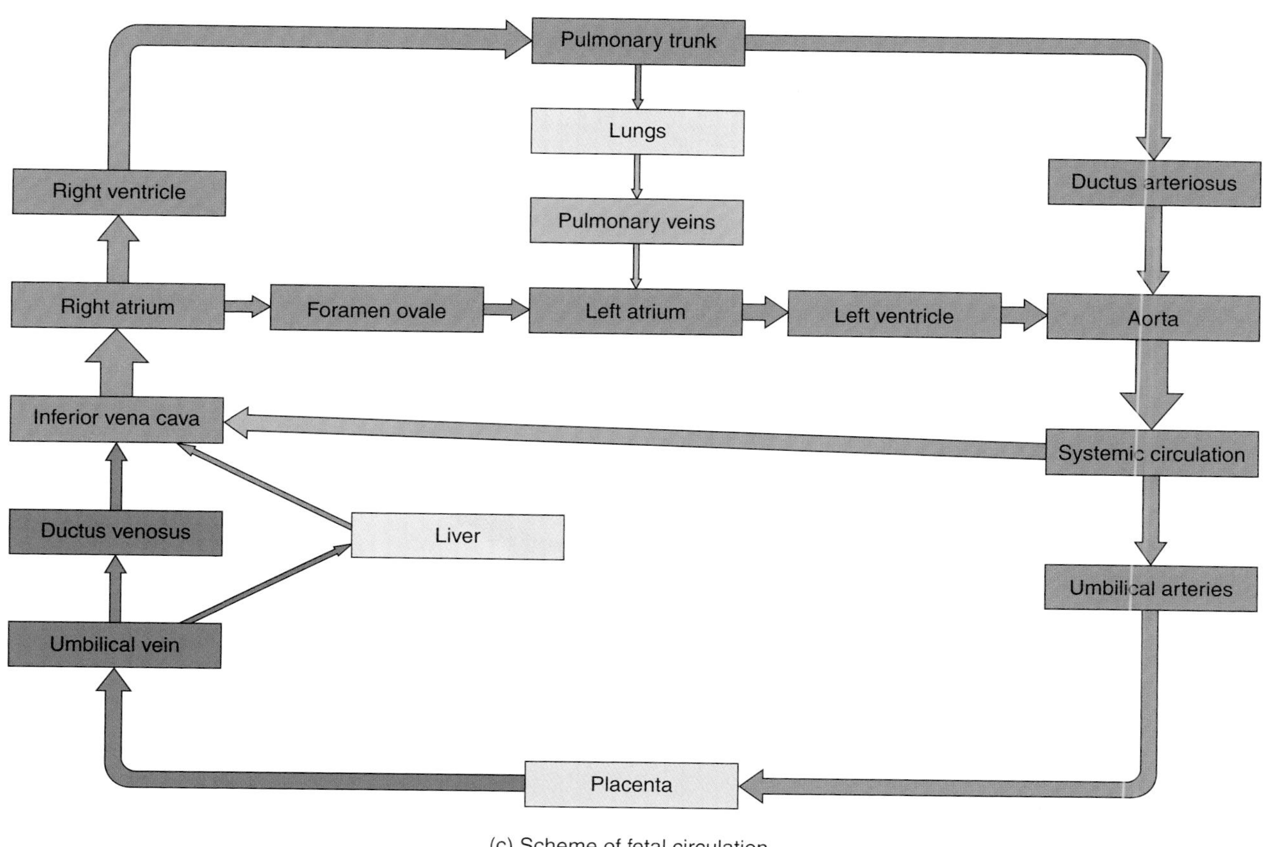

(c) Scheme of fetal circulation

Which structure provides for exchange of materials between mother and fetus?

After birth, when pulmonary (lung), renal (kidney), and digestive functions begin, the following vascular changes occur (Figure 21.30b):

1. When the umbilical cord is tied off, blood no longer flows through the umbilical arteries, they fill with connective tissue, and the distal portions of the umbilical arteries become fibrous cords called the **medial umbilical ligaments.** Although the arteries are closed functionally only a few minutes after birth, complete obliteration of the lumens may take 2 to 3 months.
2. The umbilical vein collapses but remains as the **ligamentum teres (round ligament),** a structure that attaches the umbilicus to the liver.
3. The ductus venosus collapses but remains as the **ligamentum venosum,** a fibrous cord on the inferior surface of the liver.
4. The placenta is expelled as the **"afterbirth."**
5. The foramen ovale normally closes shortly after birth to become the **fossa ovalis,** a depression in the interatrial septum. When an infant takes its first breath, the lungs expand and blood flow to the lungs increases. Blood returning from the lungs to the heart increases pressure in the left atrium. This closes the foramen ovale by pushing the valve that guards it against the interatrial septum. Permanent closure occurs in about a year.
6. The ductus arteriosus closes by vasoconstriction almost immediately after birth and becomes the **ligamentum arteriosum.** Complete anatomical obliteration of the lumen takes 1 to 3 months.

▶ CHECKPOINT

23. Diagram the hepatic portal circulation. Why is this route important?
24. Diagram the route of the pulmonary circulation.
25. Discuss the anatomy and physiology of the fetal circulation. Indicate the function of the umbilical arteries, umbilical vein, ductus venosus, foramen ovale, and ductus arteriosus.

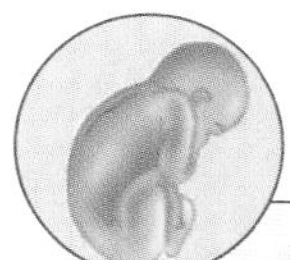

DEVELOPMENT OF BLOOD VESSELS AND BLOOD

▶ OBJECTIVE

Describe the development of blood vessels and blood.

The development of blood cells and the formation of blood vessels begins outside the embryo as early as 15 to 16 days in the **mesoderm** of the wall of the yolk sac, chorion, and connecting stalk. About two days later, blood vessels form within the embryo. The early formation of the cardiovascular system is linked to the small amount of yolk in the ovum and yolk sac. As the embryo develops rapidly during the third week, there is a greater need to develop a cardiovascular system to supply sufficient nutrients to the embryo and remove wastes from it.

Blood vessels and blood cells develop from the same precursor cell, called a **hemangioblast** (hē-MAN-jē-ō-blast; *hema-* = blood; *-blast* = immature stage). Once mesenchyme develops into hemangioblasts, they can give rise to cells that produce blood vessels (angioblasts) or cells that produce blood cells (pluripotent stem cells).

Blood vessels develop from **angioblasts,** which are derived from hemangioblasts (Figure 21.31). Angioblasts aggregate to form isolated masses and cords throughout the embryonic discs called **blood islands** (Figure 21.31). Spaces soon appear in the islands and become the lumens of the blood vessels. Some of the angioblasts immediately around the spaces give rise to the *endothelial lining of the blood vessels.* Angioblasts around the endothelium form the *tunics* (interna, media, and externa) of the larger blood vessels. Growth and fusion of blood islands form an extensive network of blood vessels throughout the embryo. By continuous branching, blood vessels outside the embryo connect with those inside the embryo, linking the embryo with the placenta.

Blood cells develop from **pluripotent stem cells** derived from hemangioblasts. This development occurs in the walls of blood vessels in the yolk sac, chorion, and allantois at about three weeks after fertilization. Blood formation in the embryo itself begins at about the fifth week in the liver and the twelfth week in the spleen, red bone marrow, and thymus.

▶ CHECKPOINT

26. What are the sites of blood cell production outside the embryo and within the embryo?

Figure 21.31 Development of blood vessels and blood cells from blood islands.

Blood vessel development begins in the embryo on about the 15th or 16th day.

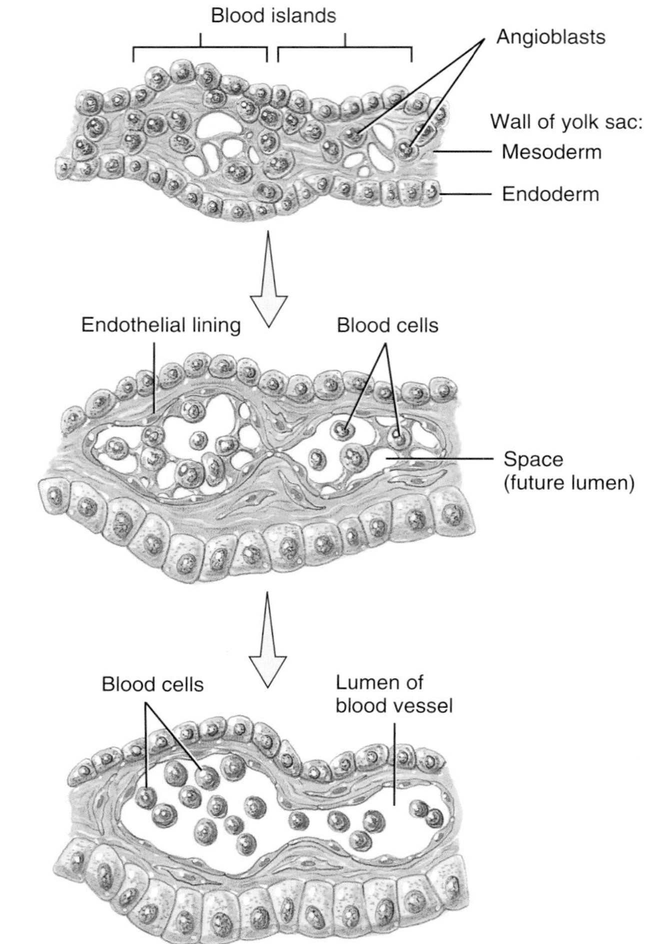

From which germ cell layer are blood vessels and blood derived?

AGING AND THE CARDIOVASCULAR SYSTEM

▶ OBJECTIVE

Explain the effects of aging on the cardiovascular system.

General changes in the cardiovascular system associated with aging include decreased compliance (distensibility) of the aorta, reduction in cardiac muscle fiber size, progressive loss of cardiac muscular strength, reduced cardiac output, a decline in maximum heart rate, and an increase in systolic blood pressure. Total blood cholesterol tends to increase with age, as does low-density lipoprotein (LDL); high-density lipoprotein (HDL) tends to decrease. There is an increase in the incidence of coronary artery disease (CAD), the major cause of heart disease and death in older Americans. Congestive heart failure, a set of symptoms associated with impaired pumping of the heart, is also prevalent in older individuals. Changes in blood vessels that serve brain tissue—for example, atherosclerosis—reduce nourishment to the brain and result in malfunction or death of brain cells. By age 80, cerebral blood flow is 20% less and renal blood flow is 50% less than in the same person at age 30.

▶ CHECKPOINT

27. How does aging affect the heart?

• • •

To appreciate the many ways the blood, heart, and blood vessels contribute to homeostasis of other body systems, examine *Focus on Homeostasis: The Cardiovascular System.*

FOCUS ON HOMEOSTASIS

The Cardiovascular System

BODY SYSTEM	CONTRIBUTION OF THE CARDIOVASCULAR SYSTEM
For all body systems 	The heart pumps blood through blood vessels to body tissues, delivering oxygen and nutrients and removing wastes by means of capillary exchange. Circulating blood keeps body tissues at a proper temperature.
Integumentary system 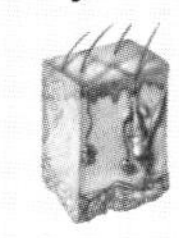	Blood delivers clotting factors and white blood cells that aid in hemostasis when skin is damaged and contribute to repair of injured skin. Changes in skin blood flow contribute to body temperature regulation by adjusting the amount of heat loss via the skin. Blood flowing in skin may give skin a pink hue.
Skeletal system 	Blood delivers calcium and phosphate ions that are needed for building bone extracellular matrix, hormones that govern building and breakdown of bone extracellular matrix, and erythropoietin that stimulates production of red blood cells by red bone marrow.
Muscular system 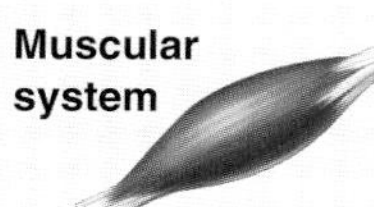	Blood circulating through exercising muscle removes heat and lactic acid.
Nervous system 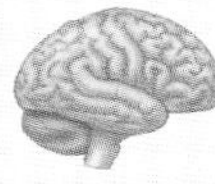	Endothelial cells lining choroid plexuses in brain ventricles help produce cerebrospinal fluid (CSF) and contribute to the blood–brain barrier.
Endocrine system 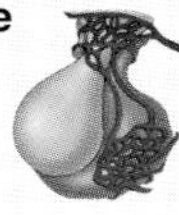	Circulating blood delivers most hormones to their target tissues. Atrial cells secrete atrial natriuretic peptide.
Lymphatic system and immunity 	Circulating blood distributes lymphocytes, antibodies, and macrophages that carry out immune functions. Lymph forms from excess interstitial fluid, which filters from blood plasma due to blood pressure generated by the heart.
Respiratory system 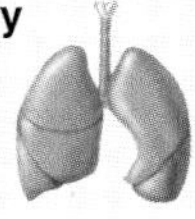	Circulating blood transports oxygen from the lungs to body tissues and carbon dioxide to the lungs for exhalation.
Digestive system 	Blood carries newly absorbed nutrients and water to liver. Blood distributes hormones that aid digestion.
Urinary system 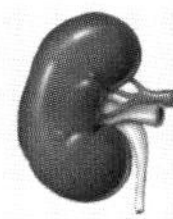	Heart and blood vessels deliver 20% of the resting cardiac output to the kidneys, where blood is filtered, needed substances are reabsorbed, and unneeded substances remain as part of urine, which is excreted.
Reproductive systems 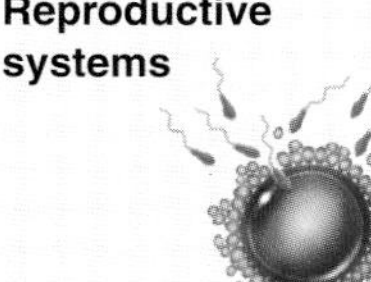	Vasodilation of arterioles in penis and clitoris cause erection during sexual intercourse. Blood distributes hormones that regulate reproductive functions.

DISORDERS: HOMEOSTATIC IMBALANCES

Hypertension

About 50 million Americans have **hypertension,** or persistently high blood pressure. It is the most common disorder affecting the heart and blood vessels and is the major cause of heart failure, kidney disease, and stroke. In May 2003, the Joint National Committee on Prevention, Detection, Evaluation, and Treatment of High Blood Pressure published new guidelines for hypertension because clinical studies have linked what were once considered fairly low blood pressure readings to an increased risk of cardiovascular disease. The new guidelines are as follows:

Category	*Systolic (mmHg)*	*Diastolic (mmHg)*
Normal	Less than 120 *and*	Less than 80
Prehypertension	120–139 *or*	80–89
Stage 1 hypertension	140–159 *or*	90–99
Stage 2 hypertension	Greater than 160 *or*	Greater than 100

Using the new guidelines, the normal classification was previously considered optimal; prehypertension now includes many more individuals previously classified as normal or high-normal; stage 1 hypertension is the same as in previous guidelines; and stage 2 hypertension now combines the previous stage 2 and stage 3 categories since treatment options are the same for the former stages 2 and 3.

Types and Causes of Hypertension

Between 90% and 95% of all cases of hypertension are **primary hypertension,** a persistently elevated blood pressure that cannot be attributed to any identifiable cause. The remaining 5–10% of cases are **secondary hypertension,** which has an identifiable underlying cause. Several disorders cause secondary hypertension:

- *Obstruction of renal blood flow* or disorders that damage renal tissue may cause the kidneys to release excessive amounts of renin into the blood. The resulting high level of angiotensin II causes vasoconstriction, thus increasing systemic vascular resistance.
- *Hypersecretion of aldosterone*—resulting, for instance, from a tumor of the adrenal cortex—stimulates excess reabsorption of salt and water by the kidneys, which increases the volume of body fluids.
- *Hypersecretion of epinephrine and norepinephrine* by a **pheochromocytoma** (fē-ō-krō′-mō-sī-TŌ-ma), a tumor of the adrenal medulla. Epinephrine and norepinephrine increase heart rate and contractility and increase systemic vascular resistance.

Damaging Effects of Untreated Hypertension

High blood pressure is known as the "silent killer" because it can cause considerable damage to the blood vessels, heart, brain, and kidneys before it causes pain or other noticeable symptoms. It is a major risk factor for the number-one (heart disease) and number-three (stroke) causes of death in the United States. In blood vessels, hypertension causes thickening of the tunica media, accelerates development of atherosclerosis and coronary artery disease, and increases systemic vascular resistance. In the heart, hypertension increases the afterload, which forces the ventricles to work harder to eject blood.

The normal response to an increased workload due to vigorous and regular exercise is hypertrophy of the myocardium, especially in the wall of the left ventricle. An increased afterload, however, leads to myocardial hypertrophy that is accompanied by muscle damage and fibrosis (a buildup of collagen fibers between the muscle fibers). As a result, the left ventricle enlarges, weakens, and dilates. Because arteries in the brain are usually less protected by surrounding tissues than are the major arteries in other parts of the body, prolonged hypertension can eventually cause them to rupture, resulting in a stroke. Hypertension also damages kidney arterioles, causing them to thicken, which narrows the lumen; because the blood supply to the kidneys is thereby reduced, the kidneys secrete more renin, which elevates the blood pressure even more.

Lifestyle Changes to Reduce Hypertension

Although several categories of drugs (described next) can reduce elevated blood pressure, the following lifestyle changes are also effective in managing hypertension:

- *Lose weight.* This is the best treatment for high blood pressure short of using drugs. Loss of even a few pounds helps reduce blood pressure in overweight hypertensive individuals.
- *Limit alcohol intake.* Drinking in moderation may lower the risk of coronary heart disease, mainly among males over 45 and females over 55. Moderation is defined as no more than one 12-oz beer per day for females and no more than two 12-oz beers per day for males.
- *Exercise.* Becoming more physically fit by engaging in moderate activity (such as brisk walking) several times a week for 30 to 45 minutes can lower systolic blood pressure by about 10 mmHg.
- *Reduce intake of sodium (salt).* Roughly half the people with hypertension are "salt sensitive." For them, a high-salt diet appears to promote hypertension, and a low-salt diet can lower their blood pressure.
- *Maintain recommended dietary intake of potassium, calcium, and magnesium.* Higher levels of potassium, calcium, and magnesium in the diet are associated with a lower risk of hypertension.
- *Don't smoke.* Smoking has devastating effects on the heart and can augment the damaging effects of high blood pressure by promoting vasoconstriction.
- *Manage stress.* Various meditation and biofeedback techniques help some people reduce high blood pressure. These methods may work by decreasing the daily release of epinephrine and norepinephrine by the adrenal medulla.

Drug Treatment of Hypertension

Drugs having several different mechanisms of action are effective in lowering blood pressure. Many people are successfully treated with *diuretics,* agents that decrease blood pressure by decreasing blood volume because they increase elimination of water and salt in the urine. *ACE (angiotensin converting enzyme) inhibitors* block formation of angiotensin II and thereby promote vasodilation and decrease the secretion of aldosterone. *Beta blockers* reduce blood pressure by inhibiting the secretion of renin and by decreasing heart rate and contractility. *Vasodilators* relax the smooth muscle in arterial walls, causing vasodilation and lowering blood pressure by lowering systemic vascular resistance. An important category of vasodilators are the *calcium channel blockers,* which slow the inflow of Ca^{2+} into vascular smooth muscle cells. They reduce the heart's workload by slowing Ca^{2+} entry into pacemaker cells and regular myocardial fibers, thereby decreasing heart rate and the force of myocardial contraction.

MEDICAL TERMINOLOGY

Aneurysm (AN-ū-rizm) A thin, weakened section of the wall of an artery or a vein that bulges outward, forming a balloonlike sac. Common causes are atherosclerosis, syphilis, congenital blood vessel defects, and trauma. If untreated, the aneurysm enlarges and the blood vessel wall becomes so thin that it bursts. The result is massive hemorrhage with shock, severe pain, stroke, or death. Treatment may involve surgery in which the weakened area of the blood vessel is removed and replaced with a graft of synthetic material.

Aortography (a′-or-TOG-ra-fē) X-ray examination of the aorta and its main branches after injection of a radiopaque dye.

Carotid endarterectomy (ka-ROT-id end′-ar-ter-EK-tō-mē) The removal of atherosclerotic plaque from the carotid artery to restore greater blood flow to the brain.

Claudication (klaw′-di-KĀ-shun) Pain and lameness or limping caused by defective circulation of the blood in the vessels of the limbs.

Deep vein thrombosis The presence of a thrombus (blood clot) in a deep vein of the lower limbs. It may lead to (1) pulmonary embolism, if the thrombus dislodges and then lodges within the pulmonary arterial blood flow, and (2) postphlebitic syndrome, which consists of edema, pain, and skin changes due to destruction of venous valves.

Doppler ultrasound scanning Imaging technique commonly used to measure blood flow. A transducer is placed on the skin and an image is displayed on a monitor that provides the exact position and severity of a blockage.

Femoral angiography An imaging technique in which a contrast medium is injected into the femoral artery and spreads to other arteries in the lower limb, and then a series of radiographs are taken of one or more sites. It is used to diagnose narrowing or blockage of arteries in the lower limbs.

Hypotension (hī-pō-TEN-shun) Low blood pressure; most commonly used to describe an acute drop in blood pressure, as occurs during excessive blood loss.

Normotensive (nor′-mō-TEN-siv) Characterized by normal blood pressure.

Occlusion (ō-KLOO-shun) The closure or obstruction of the lumen of a structure such as a blood vessel. An example is an atherosclerotic plaque in an artery.

Orthostatic hypotension (or′-thō-STAT-ik; *ortho-* = straight; *-static* = causing to stand) An excessive lowering of systemic blood pressure when a person assumes an erect or semi-erect posture; it is usually a sign of a disease. May be caused by excessive fluid loss, certain drugs, and cardiovascular or neurogenic factors. Also called **postural hypotension.**

Phlebitis (fle-BĪ-tis; *phleb-* = vein) Inflammation of a vein, often in a leg.

Thrombectomy (throm-BEK-tō-mē; *thrombo-* = clot) An operation to remove a blood clot from a blood vessel.

Thrombophlebitis (throm′-bō-fle-BĪ-tis) Inflammation of a vein involving clot formation. Superficial thrombophlebitis occurs in veins under the skin, especially in the calf.

Venipuncture (VEN-i-punk-chur; *vena-* = vein) The puncture of a vein, usually to withdraw blood for analysis or introduce a solution, for example, an antibiotic. The median cubital vein is frequently used.

White coat (office) hypertension A stress-induced syndrome found in patients who have elevated blood pressure when being examined by health-care personnel, but otherwise have normal blood pressure.

STUDY OUTLINE

STRUCTURE AND FUNCTION OF BLOOD VESSELS (p. 737)

1. Arteries carry blood away from the heart. The wall of an artery consists of a tunica interna, a tunica media (which maintains elasticity and contractility), and a tunica externa.
2. Large arteries are termed elastic (conducting) arteries, and medium-sized arteries are called muscular (distributing) arteries.
3. Many arteries anastomose: The distal ends of two or more vessels unite. An alternate blood route from an anastomosis is called collateral circulation. Arteries that do not anastomose are called end arteries.
4. Arterioles are small arteries that deliver blood to capillaries.
5. Through constriction and dilation, arterioles assume a key role in regulating blood flow from arteries into capillaries and in altering arterial blood pressure.
6. Capillaries are microscopic blood vessels through which materials are exchanged between blood and tissue cells; some capillaries are continuous, and others are fenestrated.
7. Capillaries branch to form an extensive network throughout a tissue. This network increases surface area, allowing a rapid exchange of large quantities of materials.
8. Precapillary sphincters regulate blood flow through capillaries.

9. Microscopic blood vessels in the liver are called sinusoids.
10. Venules are small vessels that continue from capillaries and merge to form veins.
11. Veins consist of the same three tunics as arteries but have a thinner tunica interna and a thinner tunica media. The lumen of a vein is also larger than that of a comparable artery.
12. Veins contain valves to prevent backflow of blood.
13. Weak valves can lead to varicose veins.
14. Vascular (venous) sinuses are veins with very thin walls.
15. Systemic veins are collectively called blood reservoirs because they hold a large volume of blood. If the need arises, this blood can be shifted into other blood vessels through vasoconstriction of veins.
16. The principal blood reservoirs are the veins of the abdominal organs (liver and spleen) and skin.

CAPILLARY EXCHANGE (p. 744)

1. Substances enter and leave capillaries by diffusion, transcytosis, or bulk flow.
2. The movement of water and solutes (except proteins) through capillary walls depends on hydrostatic and osmotic pressures.
3. The near equilibrium between filtration and reabsorption in capillaries is called Starling's law of the capillaries.
4. Edema is an abnormal increase in interstitial fluid.

HEMODYNAMICS: FACTORS AFFECTING BLOOD FLOW (p. 746)

1. The velocity of blood flow is inversely related to the cross-sectional area of blood vessels; blood flows slowest where cross-sectional area is greatest.
2. The velocity of blood flow decreases from the aorta to arteries to capillaries and increases in venules and veins.
3. Blood pressure and resistance determine blood flow.
4. Blood flows from regions of higher to lower pressure. The higher the resistance, however, the lower the blood flow.
5. Cardiac output equals the mean arterial pressure divided by total resistance ($CO = MAP \div R$).
6. Blood pressure is the pressure exerted on the walls of a blood vessel.
7. Factors that affect blood pressure are cardiac output, blood volume, viscosity, resistance, and the elasticity of arteries.
8. As blood leaves the aorta and flows through the systemic circulation, its pressure progressively falls to 0 mmHg by the time it reaches the right ventricle.
9. Resistance depends on blood vessel diameter, blood viscosity, and total blood vessel length.
10. Venous return depends on pressure differences between the venules and the right ventricle.
11. Blood return to the heart is maintained by several factors, including skeletal muscular contractions, valves in veins (especially in the limbs), and pressure changes associated with breathing.

CONTROL OF BLOOD PRESSURE AND BLOOD FLOW (p. 750)

1. The cardiovascular (CV) center is a group of neurons in the medulla oblongata that regulates heart rate, contractility, and blood vessel diameter.
2. The cardiovascular center receives input from higher brain regions and sensory receptors (baroreceptors and chemoreceptors).
3. Output from the cardiovascular center flows along sympathetic and parasympathetic axons. Sympathetic impulses propagated along cardioaccelerator nerves increase heart rate and contractility; parasympathetic impulses propagated along vagus nerves decrease heart rate.
4. Baroreceptors monitor blood pressure, and chemoreceptors monitor blood levels of O_2, CO_2, and hydrogen ions. The carotid sinus reflex helps regulate blood pressure in the brain. The aortic reflex regulates general systemic blood pressure.
5. Hormones that help regulate blood pressure are epinephrine, norepinephrine, ADH (vasopressin), angiotensin II, and ANP.
6. Autoregulation refers to local, automatic adjustments of blood flow in a given region to meet a particular tissue's need.
7. O_2 level is the principal stimulus for autoregulation.

CHECKING CIRCULATION (p. 754)

1. Pulse is the alternate expansion and elastic recoil of an artery wall with each heartbeat. It may be felt in any artery that lies near the surface or over a hard tissue.
2. A normal resting pulse (heart) rate is 70–80 beats/min.
3. Blood pressure is the pressure exerted by blood on the wall of an artery when the left ventricle undergoes systole and then diastole. It is measured by the use of a sphygmomanometer.
4. Systolic blood pressure (SBP) is the arterial blood pressure during ventricular contraction. Diastolic blood pressure (DBP) is the arterial blood pressure during ventricular relaxation. Normal blood pressure is less than 120/80.
5. Pulse pressure is the difference between systolic and diastolic blood pressure. It normally is about 40 mmHg.

SHOCK AND HOMEOSTASIS (p. 756)

1. Shock is a failure of the cardiovascular system to deliver enough O_2 and nutrients to meet the metabolic needs of cells.
2. Types of shock include hypovolemic, cardiogenic, vascular, and obstructive.
3. Signs and symptoms of shock include systolic blood pressure less than 90 mmHg; rapid resting heart rate; weak, rapid pulse; clammy, cool, pale skin; sweating; hypotension; altered mental state; decreased urinary output; thirst; and acidosis.

CIRCULATORY ROUTES (p. 758)

1. The two main circulatory routes are the systemic and pulmonary circulations.
2. Among the subdivisions of the systemic circulation are the coronary (cardiac) circulation and the hepatic portal circulation.
3. The systemic circulation carries oxygenated blood from the left ventricle through the aorta to all parts of the body, including some lung tissue, but *not* the air sacs (alveoli) of the lungs, and returns the deoxygenated blood to the right atrium.
4. The aorta is divided into the ascending aorta, the arch of the aorta, and the descending aorta. Each section gives off arteries that branch to supply the whole body.

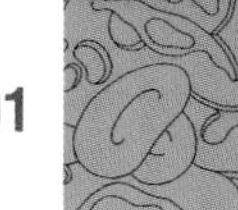

5. Blood returns to the heart through the systemic veins. All veins of the systemic circulation drain into the superior or inferior venae cavae or the coronary sinus, which, in turn, empty into the right atrium.
6. The major blood vessels of the systemic circulation may be reviewed in Exhibits 21.1–21.12.
7. The hepatic portal circulation directs venous blood from the gastrointestinal organs and spleen into the hepatic portal vein of the liver before it returns to the heart. It enables the liver to utilize nutrients and detoxify harmful substances in the blood.
8. The pulmonary circulation takes deoxygenated blood from the right ventricle to the alveoli within the lungs and returns oxygenated blood from the alveoli to the left atrium.
9. Fetal circulation exists only in the fetus. It involves the exchange of materials between fetus and mother via the placenta.
10. The fetus derives O_2 and nutrients from and eliminates CO_2 and wastes into maternal blood. At birth, when pulmonary (lung), digestive, and liver functions begin, the special structures of fetal circulation are no longer needed.

DEVELOPMENT OF BLOOD VESSELS AND BLOOD (p. 795)

1. Blood vessels develop from mesenchyme (hemangioblasts → angioblasts → blood islands) in mesoderm called blood islands.
2. Blood cells also develop from mesenchyme (hemangioblasts → pluripotent stem cells).
3. The development of blood cells from pluripotent stem cells derived from angioblasts occurs in the walls of blood vessels in the yolk sac, chorion, and allantois at about three weeks after fertilization. Within the embryo, blood is produced by the liver at about the fifth week and in the spleen, red bone marrow, and thymus at about the twelfth week.

AGING AND THE CARDIOVASCULAR SYSTEM (p. 796)

1. General changes associated with aging include reduced compliance (distensibility) of blood vessels, reduction in cardiac muscle size, reduced cardiac output, and increased systolic blood pressure.
2. The incidence of coronary artery disease (CAD), congestive heart failure (CHF), and atherosclerosis increases with age.

SELF-QUIZ QUESTIONS

Fill in the blanks in the following statements.

1. The ____ reflex helps maintain normal blood pressure in the brain; the ____ reflex governs general systemic blood pressure.
2. In addition to the pressure created by contraction of the left ventricle, venous return is aided by the ____ and the ____, both of which depend on the presence of valves in the veins.

Indicate whether the following statements are tru7e or false.

3. Baroreceptors and chemoreceptors are located in the aorta and carotid arteries.
4. The most important method of capillary exchange is simple diffusion.

Choose the one best answer to the following questions.

5. Which of the following are *not* true? (1) Muscular arteries are also known as conducting arteries. (2) Capillaries play a key role in regulating resistance. (3) The flow of blood through true capillaries is controlled by precapillary sphincters. (4) The lumen of an artery is larger than in a comparable vein. (5) Elastic arteries help propel blood. (6) The tunica media of arteries is thicker than the tunica media of veins. (a) 2, 3, and 6; (b) 1, 2, and 4; (c) 1, 2, 4, and 6; (d) 3, 4, and 5; (e) 1, 2, 3, and 4.
6. Which of the following are *true* concerning capillary exchange? (1) Large, lipid-insoluble molecules cross capillary walls by transcytosis. (2) The blood hydrostatic pressure promotes reabsorption of fluid into the capillaries. (3) If the pressures that promote filtration are greater than the pressures that promote reabsorption, fluid will move out of a capillary and into interstitial spaces. (4) A negative net filtration pressure results in reabsorption of fluid from interstitial spaces into a capillary. (5) The difference in osmotic pressure across a capillary wall is due primarily to red blood cells. (a) 1, 3, and 4; (b) 1, 2, 3, 4, and 5; (c) 1, 2, 3, and 4; (d) 3 and 4; (e) 2, 4, and 5.
7. Which of the following would *not* increase vascular resistance? (1) vasodilation, (2) polycythemia, (3) obesity, (4) dehydration, (5) anemia. (a) 1 and 2; (b) 1, 3, and 4; (c) 1 and 5; (d) 1, 4, and 5; (e) 1 only.
8. Capillary exchange is enhanced by (1) the slow rate of flow through the capillaries, (2) a small cross-sectional area, (3) the thinness of capillary walls, (4) the respiratory pump, (5) extensive branching, which increases the surface area. (a) 1, 2, 3, 4, and 5; (b) 1, 2, 3, and 5; (c) 1 and 3; (d) 3 and 5; (e) 1, 3, and 5.
9. Systemic vascular resistance depends on which of the following factors? (1) blood viscosity, (2) total blood vessel length, (3) size of the lumen, (4) type of blood vessel, (5) oxygen concentration of the blood. (a) 1, 2, and 3; (b) 2, 3, and 4; (c) 3, 4, and 5; (d) 1, 3, and 5; (e) 2, 4, and 5.
10. Which of the following help regulate blood pressure and help control regional blood flow? (1) baroreceptor and chemoreceptor reflexes, (2) hormones, (3) autoregulation, (4) H^+ concentration of blood, (5) oxygen concentration of the blood. (a) 1, 2, and 4; (b) 2, 4, and 5; (c) 1, 4, and 5; (d) 1, 2, 3, 4, and 5; (e) 3, 4, and 5.
11. For each of the following, indicate if it causes vasoconstriction or vasodilation. Use D for vasodilation and C for vasoconstriction. (a) atrial natriuretic peptide, (b) ADH, (c) decrease in body temperature, (d) lactic acid, (e) histamine, (f) hypoxia, (g) hypercapnia, (h) angiotensin II, (i) nitric oxide, (j) decreased sympathetic impulses, (k) acidosis.

12. Match the following:

____(a) pressure generated by the pumping of the heart; pushes fluids out of capillaries
____(b) pressure created by proteins present in the interstitial fluid; pulls fluid out of capillaries
____(c) balance of pressure; determines whether blood volume and interstitial fluid remain steady or change
____(d) force due to presence of plasma proteins; pulls fluid into capillaries from interstitial spaces
____(e) pressure due to fluid in interstitial spaces; pushes fluid back into capillaries

(1) net filtration pressure
(2) blood hydrostatic pressure
(3) interstitial fluid hydrostatic pressure
(4) blood colloid osmotic pressure
(5) interstitial fluid osmotic pressure

13. Match the following:

____(a) supplies blood to the kidney
____(b) drains blood from the small intestine, portions of the large intestine, stomach, and pancreas
____(c) main blood supply to arm; commonly used to measure blood pressure
____(d) supply blood to the lower limbs
____(e) drain oxygenated blood from the lungs and send it to the left atrium
____(f) supplies blood to the stomach, liver, and pancreas
____(g) supply blood to the brain
____(h) supplies blood to the large intestine
____(i) drain blood from the head
____(j) detours venous blood from the gastrointestinal organs and spleen through the liver before it returns to the heart
____(k) drains most of the thorax and abdominal wall; can serve as a bypass for the inferior vena cava
____(l) a part of the venous circulation of the leg; a vessel used in heart bypass surgery
____(m) carry deoxygenated blood from the right ventricle to the lungs

(1) superior mesenteric vein
(2) inferior mesenteric artery
(3) pulmonary veins
(4) brachial artery
(5) hepatic portal circulation
(6) carotid arteries
(7) jugular veins
(8) celiac trunk
(9) common iliac arteries
(10) azygos veins
(11) renal artery
(12) great saphenous vein
(13) pulmonary arteries

14. Match the following:

____(a) a traveling pressure wave created by the alternate expansion and recoil of elastic arteries after each systole of the left ventricle
____(b) the lowest blood pressure in arteries during ventricular relaxation
____(c) a slow resting heart rate or pulse rate
____(d) an inadequate cardiac output that results in a failure of the cardiovascular system to deliver enough oxygen and nutrients to meet the metabolic needs of body cells
____(e) a rapid resting heart rate or pulse rate
____(f) the highest force with which blood pushes against arterial walls as a result of ventricular contraction

(1) shock
(2) pulse
(3) tachycardia
(4) bradycardia
(5) systolic blood pressure
(6) diastolic blood pressure

15. Match the following (some answers will be used more than once):

____(a) returns oxygenated blood from the placenta to the fetal liver
____(b) an opening in the septum between the right and left atria
____(c) becomes the ligamentum venosum after birth
____(d) passes blood from the fetus to the placenta
____(e) bypasses the nonfunctioning lungs; becomes the ligamentum arteriosum at birth
____(f) become the medial umbilical ligaments at birth
____(g) transports oxygenated blood into the inferior vena cava
____(h) becomes the ligamentum teres at birth
____(i) becomes the fossa ovalis after birth

(1) ductus venosus
(2) ductus arteriosus
(3) foramen ovale
(4) umbilical arteries
(5) umbilical vein

CRITICAL THINKING QUESTIONS

1. Kim Sung was told that her baby was born with a hole in the upper chambers of his heart. Is this something Kim Sung should worry about?
2. Michael was brought into the emergency room suffering from a gunshot wound. He is bleeding profusely and exhibits the following: systolic blood pressure is 40 mmHg; weak pulse of 200 beats per minute; cool, pale, and clammy skin. Michael is not producing urine but is asking for water. He is confused and disoriented. What is his diagnosis and what, specifically, is causing these symptoms?
3. Maureen's job entails standing on a concrete floor for 10-hour days on an assembly line. Lately she has noticed swelling in her ankles at the end of the day and some tenderness in her calves. What do you suspect is Maureen's problem and how could she help counteract the problem?

ANSWERS TO FIGURE QUESTIONS

21.1 The femoral artery has the thicker wall; the femoral vein has the wider lumen.

21.2 Due to atherosclerosis, less energy is stored in the less-compliant elastic arteries during systole; thus, the heart must pump harder to maintain the same rate of blood flow.

21.3 Metabolically active tissues use O_2 and produce wastes more rapidly than inactive tissues, so they require more extensive capillary networks.

21.4 Materials cross capillary walls through intercellular clefts and fenestrations, via transcytosis in pinocytic vesicles, and through the plasma membranes of endothelial cells.

21.5 Valves are more important in arm veins and leg veins than in neck veins because, when you are standing, gravity causes pooling of blood in the veins of the limbs but aids the flow of blood in neck veins back toward the heart.

21.6 Blood volume in veins is about 64% of 5 liters, or 3.2 liters; blood volume in capillaries is about 7% of 5 liters, or 350 mL.

21.7 Blood colloid osmotic pressure is lower than normal in a person with a low level of plasma proteins, and therefore capillary reabsorption is low. The result is edema.

21.8 Mean blood pressure in the aorta is closer to diastolic than to systolic pressure.

21.9 The skeletal muscle pump and respiratory pump aid venous return.

21.10 Vasodilation and vasoconstriction of arterioles are the main regulators of systemic vascular resistance.

21.11 Velocity of blood flow is fastest in the aorta and arteries.

21.12 The effector tissues regulated by the cardiovascular center are cardiac muscle in the heart and smooth muscle in blood vessel walls.

21.13 Impulses to the cardiovascular center pass from baroreceptors in the carotid sinuses via the glossopharyngeal (IX) nerves and from baroreceptors in the arch of the aorta via the vagus (X) nerves.

21.14 It represents a change that occurs when you stand up because gravity causes pooling of blood in leg veins once you are upright, decreasing the blood pressure in your upper body.

21.15 Diastolic blood pressure = 95 mmHg; systolic blood pressure = 142 mmHg; pulse pressure = 47 mmHg. This person has stage I hypertension because the systolic blood pressure is greater than 140 mmHg and the diastolic blood pressure is greater than 90 mmHg.

21.16 Almost-normal blood pressure in a person who has lost blood does not necessarily indicate that the patient's tissues are receiving adequate blood flow; if systemic vascular resistance has increased greatly, tissue perfusion may be inadequate.

21.17 The two main circulatory routes are the systemic circulation and the pulmonary circulation.

21.18 The subdivisions of the aorta are the ascending aorta, arch of the aorta, thoracic aorta, and abdominal aorta.

21.19 Branches of the arch of aorta (in order of origination) are the brachiocephalic trunk, left common carotid artery, and left subclavian artery.

21.20 The thoracic aorta begins at the level of the intervertebral disc between T4 and T5.

21.21 The abdominal aorta begins at the aortic hiatus in the diaphragm.

21.22 The abdominal aorta divides into the common iliac arteries at about the level of L4.

21.23 The superior vena cava drains regions above the diaphragm, and the inferior vena cava drains regions below the diaphragm.

21.24 All venous blood in the brain drains into the internal jugular veins.

21.25 The median cubital vein is often used for withdrawing blood.

21.26 The inferior vena cava returns blood from abdominopelvic viscera to the heart.

21.27 Superficial veins of the lower limbs are the dorsal venous arch and the great saphenous and small saphenous veins.

21.28 The hepatic veins carry blood away from the liver.

21.29 After birth, the pulmonary arteries are the only arteries that carry deoxygenated blood.

21.30 Exchange of materials between mother and fetus occurs across the placenta.

21.31 Blood vessels and blood are derived from mesoderm.

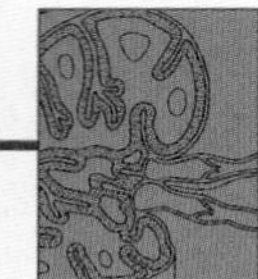

Chapter 22

The Lymphatic System and Immunity

The Lymphatic System, Disease Resistance, and Homeostasis

The lymphatic system contributes to homeostasis by draining interstitial fluid as well as providing the mechanisms for defense against disease.

wiley.com/college/apcentral

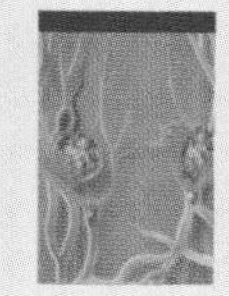

Maintaining homeostasis in the body requires continual combat against harmful agents in our internal and external environment. Despite constant exposure to a variety of **pathogens** (PATH-ō-jens), disease-producing microbes such as bacteria and viruses, most people remain healthy. The body surface also endures cuts and bumps, exposure to ultraviolet rays in sunlight, chemical toxins, and minor burns with an array of defensive ploys. **Resistance** is the ability to ward off damage or disease through our defenses. Vulnerability or lack of resistance is termed **susceptibility.**

The two general types of resistance are (1) nonspecific resistance or innate defenses and (2) specific resistance or immunity. **Nonspecific resistance (innate defenses)** are present at birth and include defense mechanisms that provide *immediate* but *general* protection against invasion by a wide range of pathogens. Mechanical and chemical barriers of the skin and mucous membranes provide the first line of defense in nonspecific resistance. The acidity of gastric juice in the stomach, for example, kills many bacteria in food. The second line of defense in nonspecific resistance consists of antimicrobial proteins (interferons, complement, and transferrins), phagocytes (mostly neutrophils and macrophages), natural killer cells, inflammation, and fever. **Specific resistance (immunity)** develops in response to contact with a *particular* invader. It occurs more *slowly* than nonspecific resistance mechanisms and involves activation of specific lymphocytes that can combat a specific invader.

The body system responsible for specific resistance (and some aspects of nonspecific resistance) is the lymphatic system. This system is closely allied with the cardiovascular system, and it also functions with the digestive system in the absorption of fatty foods. In this chapter, we will explore the mechanisms that provide defenses against intruders and promote the repair of damaged body tissues.

LYMPHATIC SYSTEM STRUCTURE AND FUNCTION

▶ **OBJECTIVES**

- **List the components and major functions of the lymphatic system.**
- **Describe the organization of lymphatic vessels.**
- **Explain the formation and flow of lymph.**
- **Compare the structure and functions of the primary and secondary lymphatic organs and tissues.**

The **lymphatic system** (lim-FAT-ik) consists of a fluid called lymph, vessels called lymphatic vessels that transport the lymph, a number of structures and organs containing lymphatic tissue, and red bone marrow, where stem cells develop into the various types of blood cells, including lymphocytes (Figure 22.1). It assists in circulating body fluids and helps defend the body against disease-causing agents. As you will see shortly, most components of blood plasma filter through blood capillary walls to form interstitial fluid. After interstitial fluid passes into lymphatic vessels, it is called **lymph** (LIMF = clear fluid). The major difference between interstitial fluid and lymph is location: Interstitial fluid is found between cells, and lymph is located within lymphatic vessels and lymphatic tissue.

Lymphatic tissue is a specialized form of reticular connective tissue (see Table 4.4C on page 127) that contains large numbers of lymphocytes. Recall from Chapter 19 that lymphocytes are agranular white blood cells (see page 676). Two types of lymphocytes participate in immune responses: B cells and T cells.

Functions of the Lymphatic System

The lymphatic system has three primary functions:

1. ***Draining excess interstitial fluid.*** Lymphatic vessels drain excess interstitial fluid from tissue spaces and return it to the blood.

2. ***Transporting dietary lipids.*** Lymphatic vessels transport lipids and lipid-soluble vitamins (A, D, E, and K) absorbed by the gastrointestinal tract to the blood.

3. ***Carrying out immune responses.*** Lymphatic tissue initiates highly specific responses directed against particular microbes or abnormal cells. T cells and B cells, aided by macrophages, recognize foreign cells, microbes, toxins, and cancer cells and respond to them in two basic ways: (1) In cell-mediated immune responses, T cells destroy the intruders by causing them to rupture or by releasing cytotoxic (cell-killing) substances. (2) In antibody-mediated immune responses, B cells differentiate into plasma cells that protect us against disease by producing antibodies, proteins that combine with and cause destruction of specific foreign substances.

Lymphatic Vessels and Lymph Circulation

Lymphatic vessels begin as lymphatic capillaries. These tiny vessels, which are located in the spaces between cells, are closed at one end (Figure 22.2 on page 807). Just as blood capillaries converge to form venules and then veins, lymphatic capillaries unite to form larger lymphatic vessels (see Figure 22.1), which resemble veins in structure but have thinner walls and more valves. At intervals along the lymphatic vessels, lymph flows through lymph nodes, encapsulated bean-shaped organs consisting of masses of B cells and T cells. In the skin, lymphatic vessels lie in the subcutaneous tissue and generally follow the same route as veins; lymphatic vessels of the viscera generally follow arteries, forming plexuses (networks) around them. Tissues that lack lymphatic capillaries include avascular tissues (such as cartilage, the epidermis, and the cornea of the eye), the central nervous system, portions of the spleen, and red bone marrow.

Figure 22.1 **Components of the lymphatic system.**

The lymphatic system consists of lymph, lymphatic vessels, lymphatic tissues, and red bone marrow.

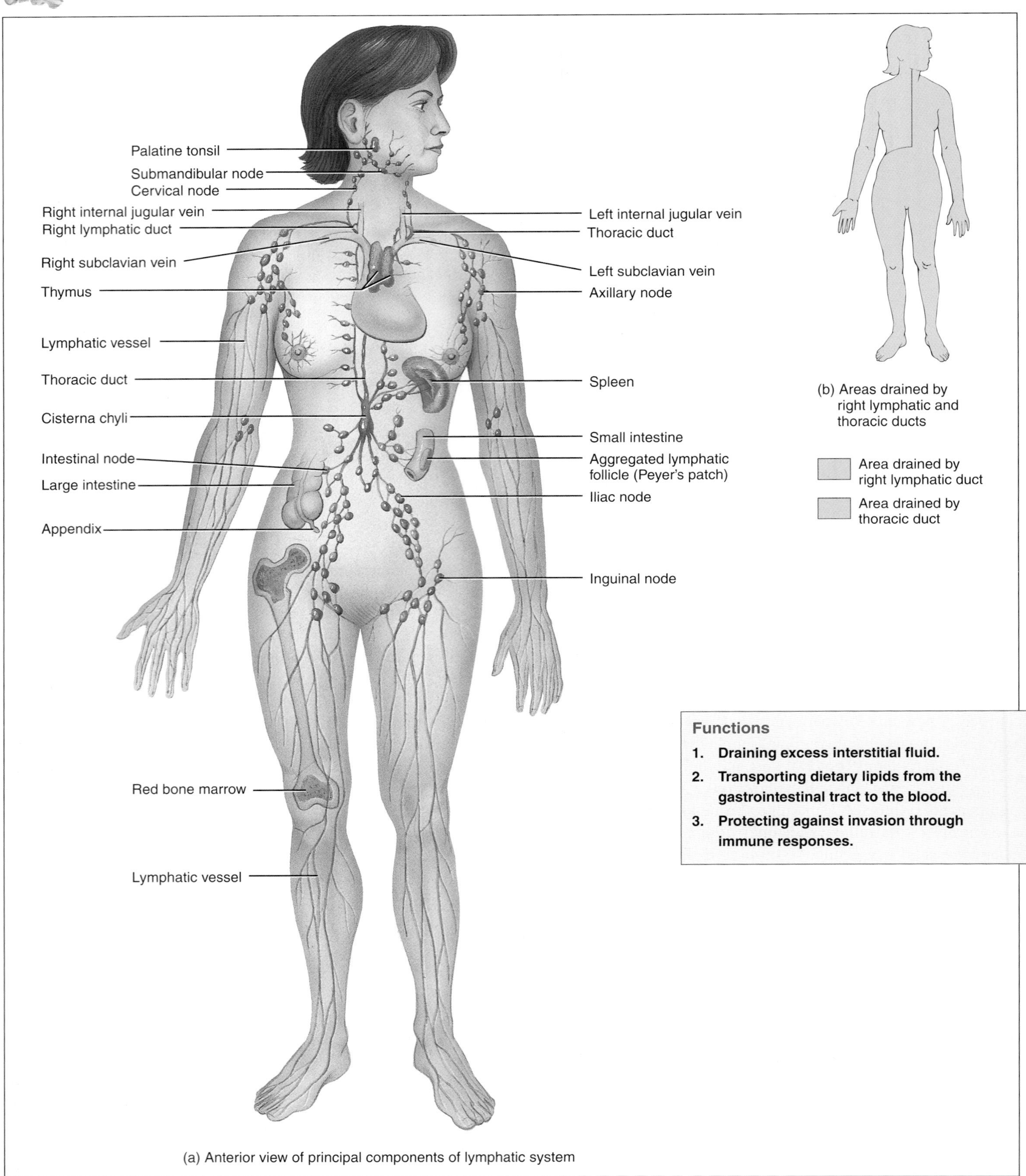

(a) Anterior view of principal components of lymphatic system

(b) Areas drained by right lymphatic and thoracic ducts

Functions

1. **Draining excess interstitial fluid.**
2. **Transporting dietary lipids from the gastrointestinal tract to the blood.**
3. **Protecting against invasion through immune responses.**

What tissue contains stem cells that develop into lymphocytes?

Figure 22.2 Lymphatic capillaries.

Lymphatic capillaries are found throughout the body except in avascular tissues, the central nervous system, portions of the spleen, and bone marrow.

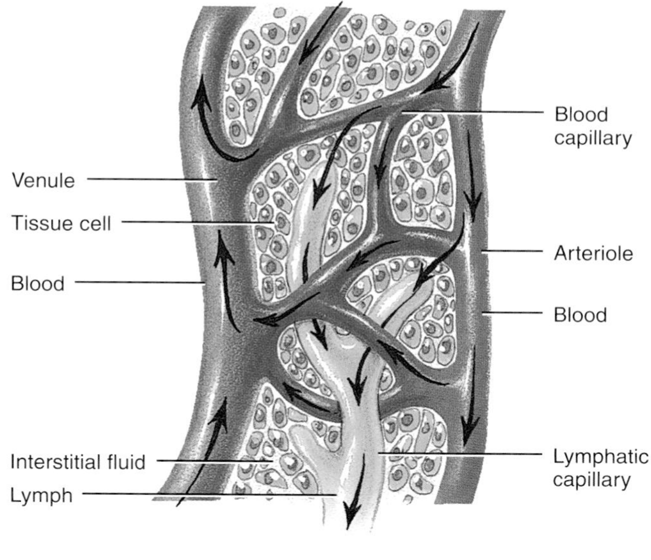

(a) Relationship of lymphatic capillaries to tissue cells and blood capillaries

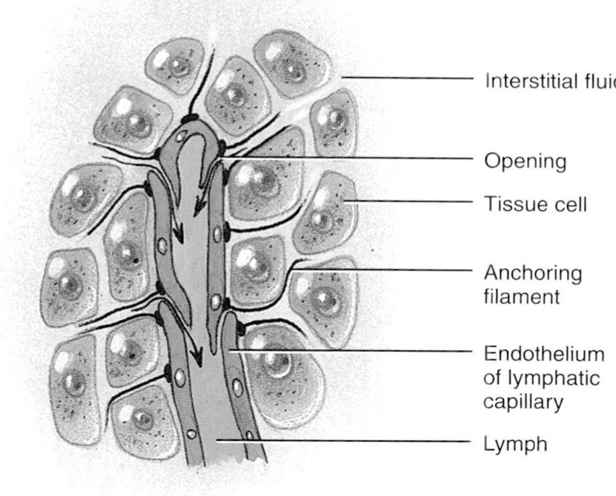

(b) Details of a lymphatic capillary

Is lymph more similar to blood plasma or to interstitial fluid? Why?

Lymphatic Capillaries

Lymphatic capillaries are slightly larger in diameter than blood capillaries and have a unique one-way structure that permits interstitial fluid to flow into them but not out. The ends of endothelial cells that make up the wall of a lymphatic capillary overlap (Figure 22.2b). When pressure is greater in the interstitial fluid than in lymph, the cells separate slightly, like the opening of a one-way swinging door, and interstitial fluid enters the lymphatic capillary. When pressure is greater inside the lymphatic capillary, the cells adhere more closely, and lymph cannot escape back into interstitial fluid. The pressure is relieved as lymph moves further down the lymphatic capillary. Attached to the lymphatic capillaries are *anchoring filaments,* which contain elastic fibers. They extend out from the lymphatic capillary, attaching lymphatic endothelial cells to surrounding tissues. When excess interstitial fluid accumulates and causes tissue swelling, the anchoring filaments are pulled, making the openings between cells even larger so that more fluid can flow into the lymphatic capillary.

In the small intestine, specialized lymphatic capillaries called **lacteals** (LAK-tē-als; *lact-* = milky) carry dietary lipids into lymphatic vessels and ultimately into the blood. The presence of these lipids causes the lymph draining from the small intestine to appear creamy white; such lymph is referred to as **chyle** (KĪL = juice). Elsewhere, lymph is a clear, pale-yellow fluid.

Lymph Trunks and Ducts

As you have already learned, lymph passes from lymphatic capillaries into lymphatic vessels and then through lymph nodes. As lymphatic vessels exit lymph nodes in a particular region of the body, they unite to form **lymph trunks.** The principal trunks are the lumbar, intestinal, bronchomediastinal, subclavian, and jugular trunks (Figure 22.3). The **lumbar trunks** drain lymph from the lower limbs, the wall and viscera of the pelvis, the kidneys, the adrenal glands, and the abdominal wall. The **intestinal trunk** drains lymph from the stomach, intestines, pancreas, spleen, and part of the liver. The **bronchomediastinal trunks** drain lymph from the thoracic wall, lung, and heart. The **subclavian trunks** drain the upper limbs. The **jugular trunks** drain the head and neck.

Lymph passes from lymph trunks into two main channels, the thoracic duct and the right lymphatic duct, and then drains into venous blood. The **thoracic (left lymphatic) duct** is about 38–45 cm (15–18 in.) long and begins as a dilation called the **cisterna chyli** (sis-TER-na KĪ-lē; *cisterna* = cavity or reservoir) anterior to the second lumbar vertebra. The thoracic duct is the main duct for the return of lymph to blood. The cisterna chyli receives lymph from the right and left lumbar trunks and from the intestinal trunk. In the neck, the thoracic duct also receives lymph from the left jugular, left subclavian, and left bronchomediastinal trunks. Therefore, the thoracic duct receives lymph from the left side of the head, neck, and chest, the left upper limb, and the

Figure 22.3 Routes for drainage of lymph from lymph trunks into the thoracic and right lymphatic ducts.

All lymph returns to the bloodstream through the thoracic (left) lymphatic duct and right lymphatic duct.

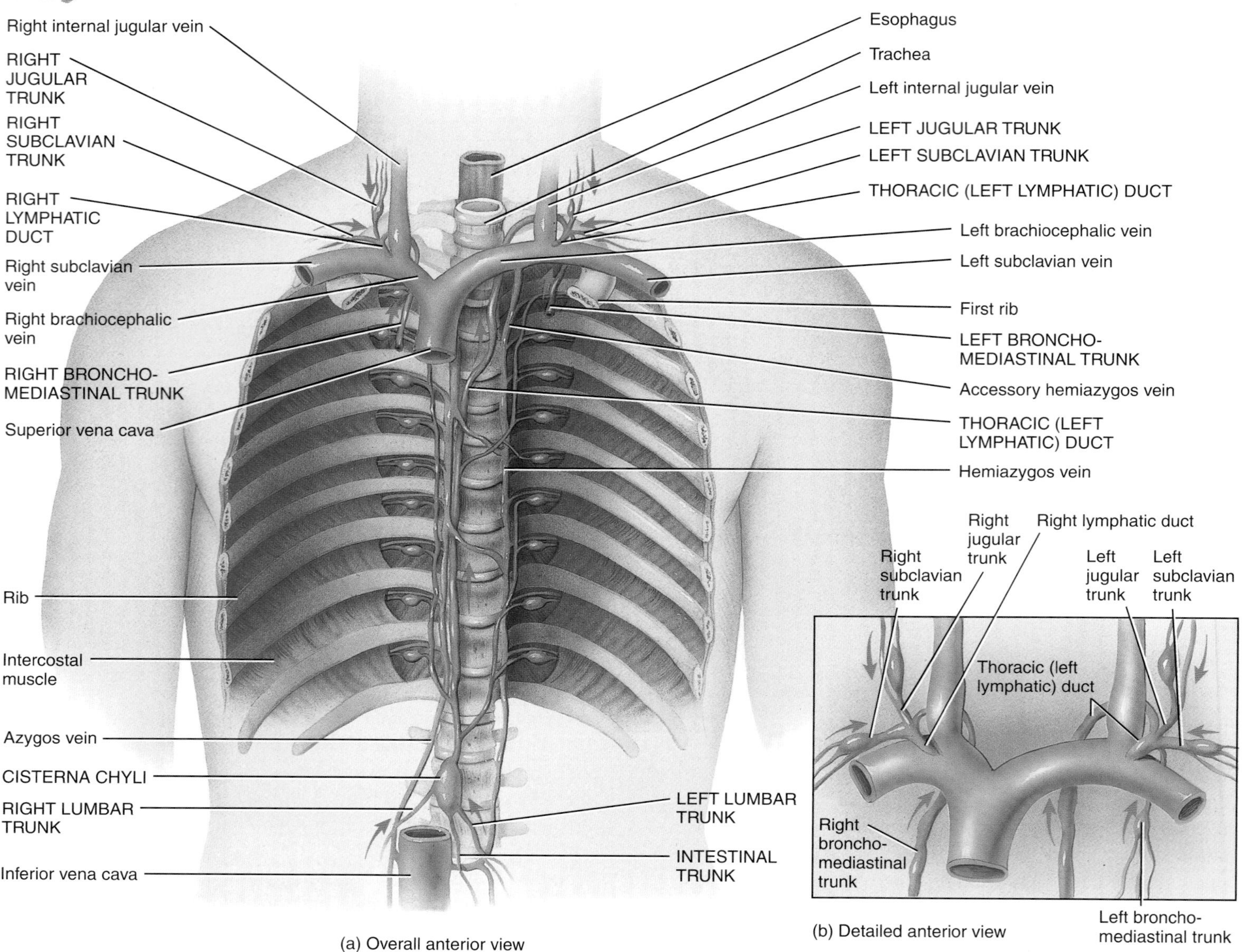

(a) Overall anterior view

(b) Detailed anterior view

Which lymphatic vessels empty into the cisterna chyli, and which duct receives lymph from the cisterna chyli?

entire body inferior to the ribs (see Figure 22.1b). The thoracic duct in turn drains lymph into venous blood at the junction of the left internal jugular and left subclavian veins.

The **right lymphatic duct** (Figure 22.3) is about 1.2 cm (0.5 in.) long and receives lymph from the right jugular, right subclavian, and right bronchomediastinal trunks. Thus, the right lymphatic duct receives lymph from the upper right side of the body (see Figure 22.1b). From the right lymphatic duct, lymph drains into venous blood at the junction of the right internal jugular and right subclavian veins.

Formation and Flow of Lymph

Most components of blood plasma filter freely through the capillary walls to form interstitial fluid, but more fluid filters out of blood capillaries than returns to them by reabsorption (see Figure 21.7 on page 745). The excess filtered fluid—about 3 liters per day—drains into lymphatic vessels and becomes lymph. Because most plasma proteins are too large to leave blood vessels, interstitial fluid contains only a small amount of protein. Proteins that do leave blood plasma cannot return to the blood by diffusion because the concentration gradient (high level of proteins inside blood capillaries, low level outside) opposes such movement. The proteins can, however, move readily through the more permeable lymphatic capillaries into lymph. Thus, an important function of lymphatic vessels is to return the lost plasma proteins to the bloodstream.

Like veins, lymphatic vessels contain valves, which ensure the one-way movement of lymph. As noted previously, lymph drains into venous blood through the right lymphatic duct and the thoracic duct at the junction of the internal jugular and subclavian veins (Figure 22.3). Thus, the sequence of fluid flow is blood capillaries (blood) → interstitial spaces (interstitial fluid) → lymphatic capillaries (lymph) → lymphatic vessels (lymph) → lymphatic ducts (lymph) → junction of the internal jugular and subclavian veins (blood). Figure 22.4 illustrates this sequence, along with the relationship of the lymphatic and cardiovascular systems.

The same two "pumps" that aid the return of venous blood to the heart maintain the flow of lymph.

1. ***Skeletal muscle pump.*** The "milking action" of skeletal muscle contractions (see Figure 21.9 on page 748) compresses lymphatic vessels (as well as veins) and forces lymph toward the junction of the internal jugular and subclavian veins.

2. ***Respiratory pump.*** Lymph flow is also maintained by pressure changes that occur during inhalation (breathing in). Lymph flows from the abdominal region, where the pressure is higher, toward the thoracic region, where it is lower. When the pressures reverse during exhalation (breathing out), the valves prevent backflow of lymph. In addition, when a lymphatic vessel distends, the smooth muscle in its wall contracts, which helps move lymph from one segment of the vessel to the next.

Lymphatic Organs and Tissues

The widely distributed lymphatic organs and tissues are classified into two groups based on their functions. **Primary lymphatic organs** are the sites where stem cells divide and become **immunocompetent,** that is, capable of mounting an immune response. The primary lymphatic organs are the red bone marrow (in flat bones and the epiphyses of long bones of adults) and the thymus. Pluripotent stem cells in red bone marrow give rise to mature, immunocompetent B cells and to pre-T cells,

Figure 22.4 **Schematic diagram showing the relationship of the lymphatic system to the cardiovascular system.** The arrows indicate the direction of flow of lymph and blood.

The sequence of fluid flow is blood capillaries (blood) → interstitial spaces (interstitial fluid) → lymphatic capillaries (lymph) → lymphatic vessels (lymph) → lymphatic ducts (lymph) → junction of the internal jugular and subclavian veins (blood).

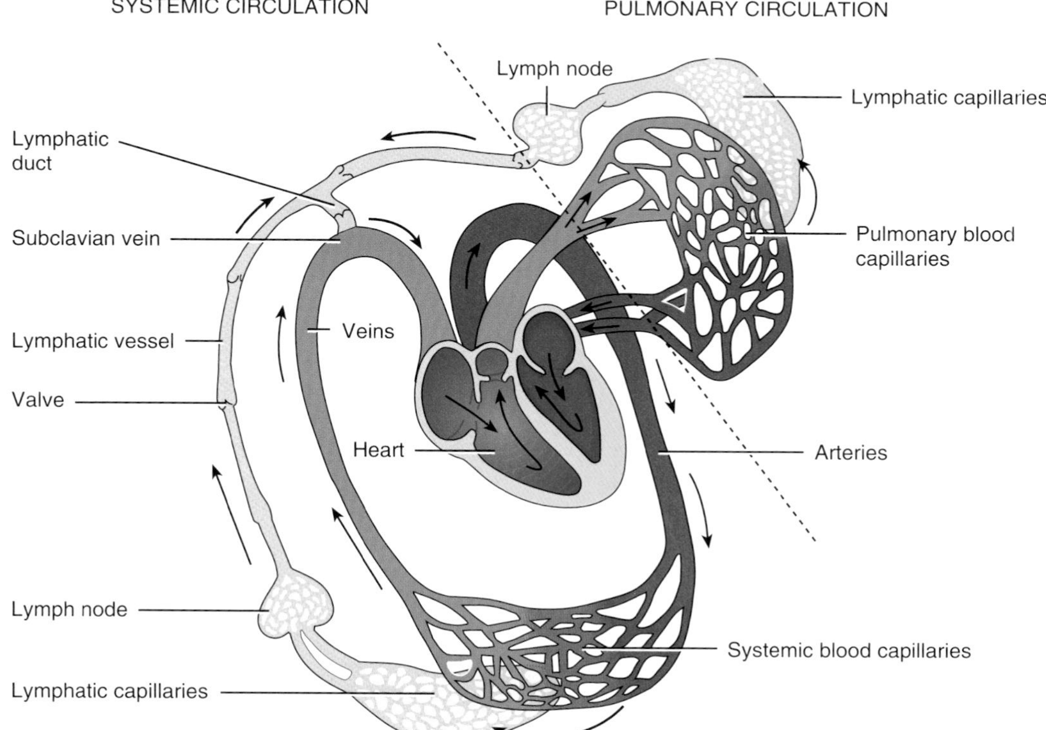

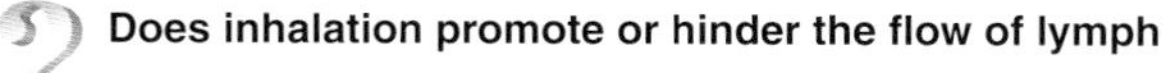

Does inhalation promote or hinder the flow of lymph?

which migrate to and become immunocompetent T cells in the thymus. The **secondary lymphatic organs** and **tissues** are the sites where most immune responses occur. They include lymph nodes, the spleen, and lymphatic nodules (follicles). The thymus, lymph nodes, and spleen are considered organs because each is surrounded by a connective tissue capsule; lymphatic nodules, in contrast, are not considered organs because they lack a capsule.

Thymus

The **thymus** is a bilobed organ located in the mediastinum between the sternum and the aorta (Figure 22.5a). An enveloping layer of connective tissue holds the two lobes closely together, but a connective tissue **capsule** separates the two. Extensions of the capsule, called **trabeculae** (tra-BEK-ū-lē = little beams), penetrate inward and divide each lobe into **lobules** (Figure 22.5b).

Each thymic lobule consists of a deeply staining outer cortex and a lighter-staining central medulla (Figure 22.5b). The **cortex** is composed of large numbers of T cells and scattered dendritic cells, epithelial cells, and macrophages. Immature T cells (pre-T cells) migrate from red bone marrow to the cortex of the thymus, where they proliferate and begin to mature. **Dendritic cells** (*dendr-* = a tree), so named because they have long, branched projections that resemble the dendrites of a neuron, assist the maturation process. As you will see shortly, dendritic cells in other parts of the body, such as lymph nodes, play another key role in immune responses. Each of the specialized **epithelial cells** in the cortex has several long processes that surround and serve as a framework for as many as 50 T cells. These epithelial cells help "educate" the pre-T cells in a process known as positive selection (see Figure 22.20). Additionally, they produce thymic hormones that are thought to aid in the maturation of T cells. Only about 2% of developing T cells survive in the cortex. The remaining cells die via apoptosis (programmed cell death). Thymic **macrophages** help clear out the debris of dead and dying cells. The surviving T cells enter the medulla.

The **medulla** consists of widely scattered, more mature T cells, epithelial cells, dendritic cells, and macrophages (Figure 22.5c). Some of the epithelial cells become arranged into concentric layers of flat cells that degenerate and become filled

Figure 22.5 **Thymus.** (See Tortora, *A Photographic Atlas of the Human Body, Second Edition,* Figure 7.2a.)

The bilobed thymus is largest at puberty and then atrophies with age.

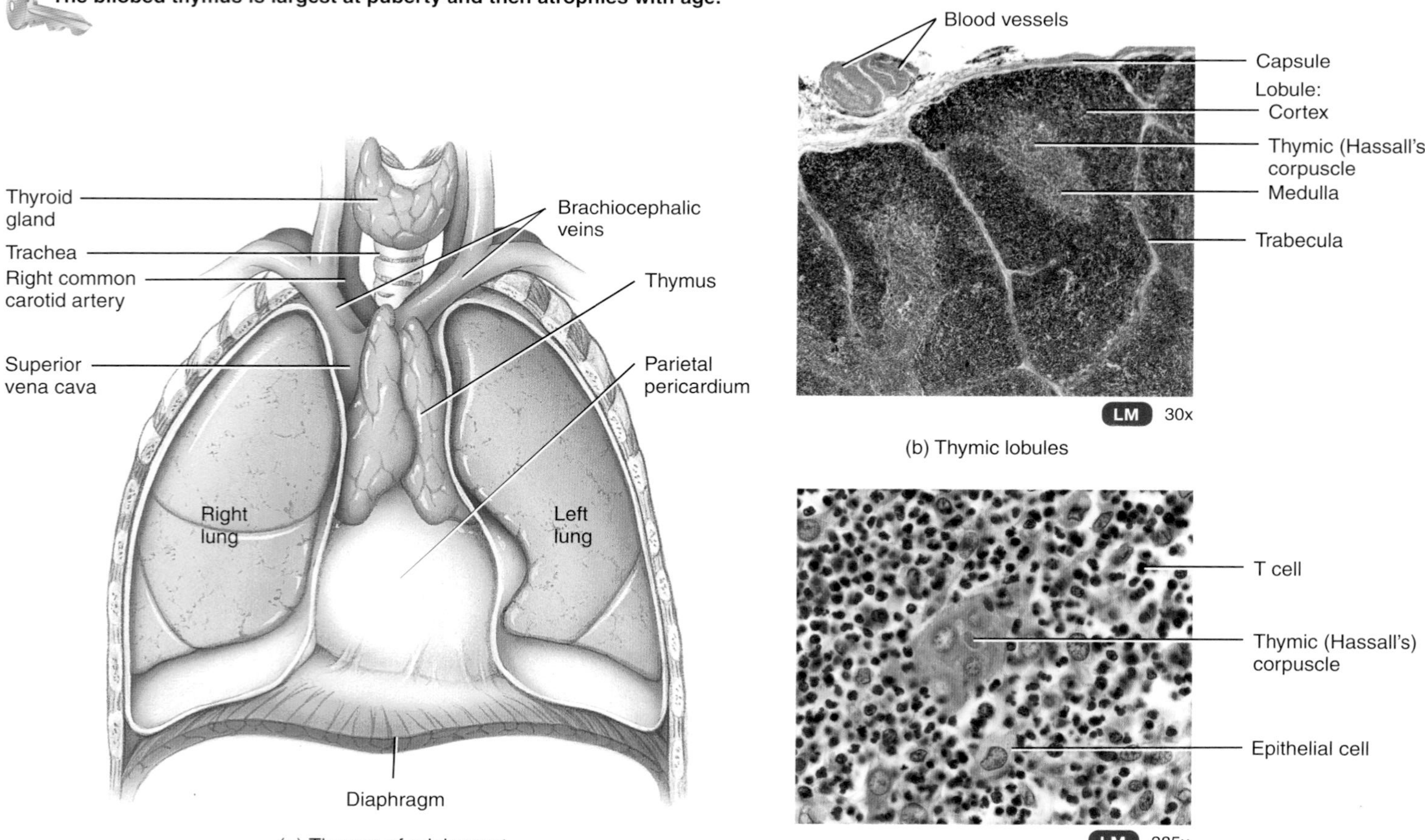

(a) Thymus of adolescent

(b) Thymic lobules

(c) Details of the thymic medulla

Which type of lymphocytes mature in the thymus?

with keratohyalin granules and keratin. These clusters are called **thymic (Hassall's) corpuscles.** Although their role is uncertain, they may serve as sites of T cell death in the medulla. T cells that leave the thymus via the blood migrate to lymph nodes, the spleen, and other lymphatic tissues where they colonize parts of these organs and tissues.

In infants, the thymus is large, with a mass of about 70 g (2.3 oz). After puberty, adipose and areolar connective tissue begin to replace the thymic tissue. By the time a person reaches maturity, the gland has atrophied considerably, and in old age it may weigh only 3 g (0.1 oz). Before the thymus atrophies, it populates the secondary lymphatic organs and tissues with T cells. However, some T cells continue to proliferate in the thymus throughout an individual's lifetime.

Lymph Nodes

Located along lymphatic vessels are about 600 bean-shaped **lymph nodes.** They are scattered throughout the body, both superficially and deep, and usually occur in groups (see Figure 22.1). Large groups of lymph nodes are present near the mammary glands and in the axillae and groin.

Lymph nodes are 1–25 mm (0.04–1 in.) long and, like the thymus, are covered by a **capsule** of dense connective tissue that extends into the node (Figure 22.6). The capsular extensions, called **trabeculae,** divide the node into compartments, provide support, and provide a route for blood vessels into the interior of a node. Internal to the capsule is a supporting network of reticular fibers and fibroblasts. The capsule, trabeculae, reticular fibers, and fibroblasts constitute the *stroma* (supporting connective tissue) of a lymph node.

The *parenchyma* (functioning part) of a lymph node is divided into a superficial cortex and a deep medulla. The cortex consists of an outer cortex and an inner cortex. Within the **outer cortex** are egg-shaped aggregates of B cells called **lymphatic nodules (follicles).** A lymphatic nodule consisting chiefly of B cells is called a *primary lymphatic nodule.* Most lymphatic nodules in the outer cortex are *secondary lymphatic nodules* (Figure 22.6), which form in response to an antigenic challenge and are sites of plasma cell and memory B cell formation. After B cells in a primary lymphatic nodule recognize an antigen, the primary lymphatic nodule develops into a secondary lymphatic nodule. The center of a secondary lymphatic nodule contains a region of light-staining cells called a *germinal center.* In the germinal center are B cells, follicular dendritic cells (a special type of dendritic cell), and macrophages. When follicular dendritic cells "present" an antigen (described later in the chapter), B cells proliferate and develop into antibody-producing plasma cells or develop into memory B cells. Memory B cells persist after an initial immune response and "remember" having encountered a specific antigen. B cells that do not develop properly undergo apoptosis (programmed cell death) and are destroyed by macrophages. The region of a secondary lymphatic nodule surrounding the germinal center is composed of dense accumulations of B cells that have migrated away from their site of origin within the nodule.

The **inner cortex** does not contain lymphatic nodules. It consists mainly of T cells and dendritic cells that enter a lymph node from other tissues. The dendritic cells present antigens to T cells, causing their proliferation. The newly formed T cells then migrate from the lymph node to areas of the body where there is antigenic activity.

The **medulla** of a lymph node contains B cells, antibody-producing plasma cells that have migrated out of the cortex into the medulla, and macrophages. The various cells are embedded in a network of reticular fibers and reticular cells.

As you have already learned, lymph flows through a node in one direction only (Figure 22.6a). It enters through **afferent lymphatic vessels** (*afferent* = to carry toward), which penetrate the convex surface of the node at several points. The afferent vessels contain valves that open toward the center of the node, directing the lymph *inward.* Within the node, lymph enters **sinuses,** series of irregular channels that contain branching reticular fibers, lymphocytes, and macrophages. From the afferent lymphatic vessels, lymph flows into the **subcapsular sinus,** immediately beneath the capsule. From here the lymph flows through **trabecular sinuses,** which extend through the cortex parallel to the trabeculae, and into **medullary sinuses,** which extend through the medulla. The medullary sinuses drain into one or two **efferent lymphatic vessels** (*efferent* = to carry away), which are wider and fewer in number than afferent vessels. They contain valves that open away from the center of the lymph node to convey lymph, antibodies secreted by plasma cells, and activated T cells *out* of the node. Efferent lymphatic vessels emerge from one side of the lymph node at a slight depression called a **hilum** (HĪ-lum). Blood vessels also enter and leave the node at the hilum.

Lymph nodes function as a type of filter. As lymph enters one end of a lymph node, foreign substances are trapped by the reticular fibers within the sinuses of the lymph node. Then macrophages destroy some foreign substances by phagocytosis while lymphocytes destroy others by immune responses. The filtered lymph then leaves the other end of the lymph node.

Metastasis Through Lymphatic Vessels

Metastasis (me-TAS-ta-sis; *meta-* = beyond; *stasis* = to stand), the spread of a disease from one part of the body to another, can occur via lymphatic vessels. All malignant tumors eventually metastasize. Cancer cells may travel in the blood or lymph and establish new tumors where they lodge. When metastasis occurs via lymphatic vessels, secondary tumor sites can be predicted according to the direction of lymph flow from the primary tumor site. Cancerous lymph nodes feel enlarged, firm, nontender, and fixed to underlying structures. By contrast, most lymph nodes that are enlarged due to an infection are softer, tender, and moveable. ■

Spleen

The oval **spleen** is the largest single mass of lymphatic tissue in the body, measuring about 12 cm (5 in.) in length (Figure 22.7a on page 813). It is located in the left hypochondriac region between

Figure 22.6 Structure of a lymph node. Arrows indicate the direction of lymph flow through a lymph node.

Lymph nodes are present throughout the body, usually clustered in groups.

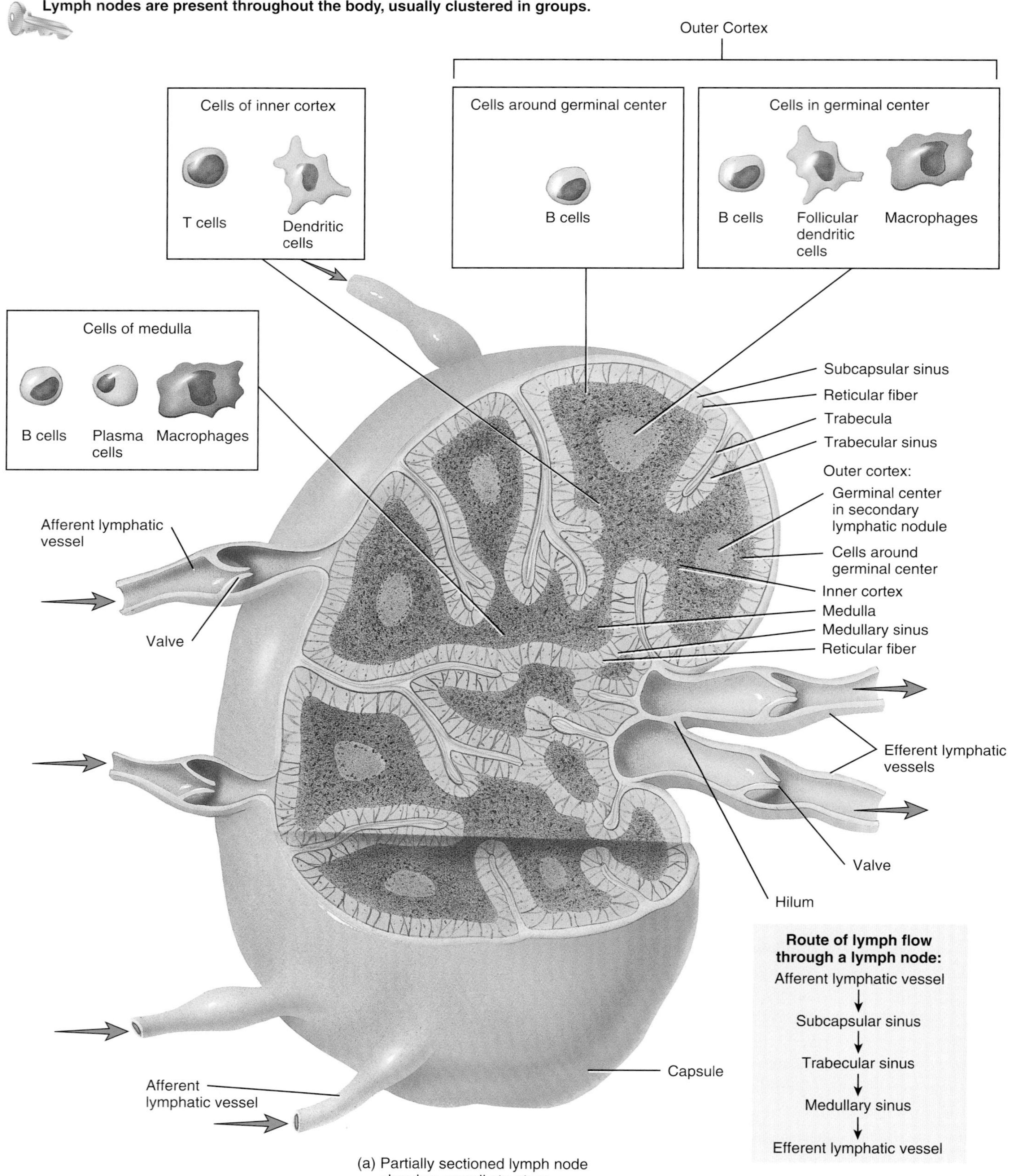

(a) Partially sectioned lymph node showing overall structure

Capsule
Subcapsular sinus
Trabecula
Trabecular sinus
Outer cortex
Germinal center in secondary lymphatic nodule
Inner cortex
Medullary sinus
Medulla
LM 55x

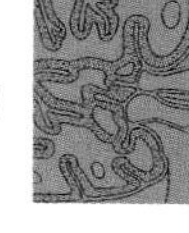

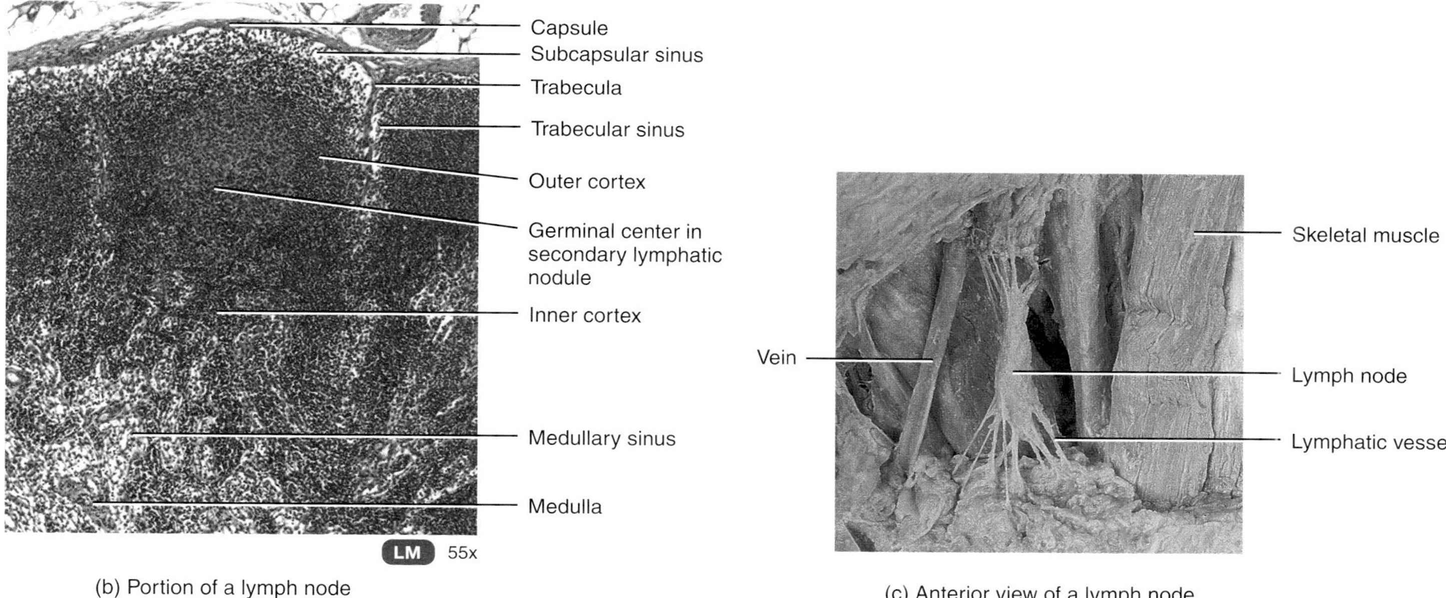

(b) Portion of a lymph node

Skeletal muscle
Vein
Lymph node
Lymphatic vessel

(c) Anterior view of a lymph node

What happens to foreign substances that enter a lymph node in lymph?

Figure 22.7 **Structure of the spleen.** (See Tortora, *A Photographic Atlas of the Human Body,* Second Edition, Figure 7.3a.)

The spleen is the largest single mass of lymphatic tissue in the body.

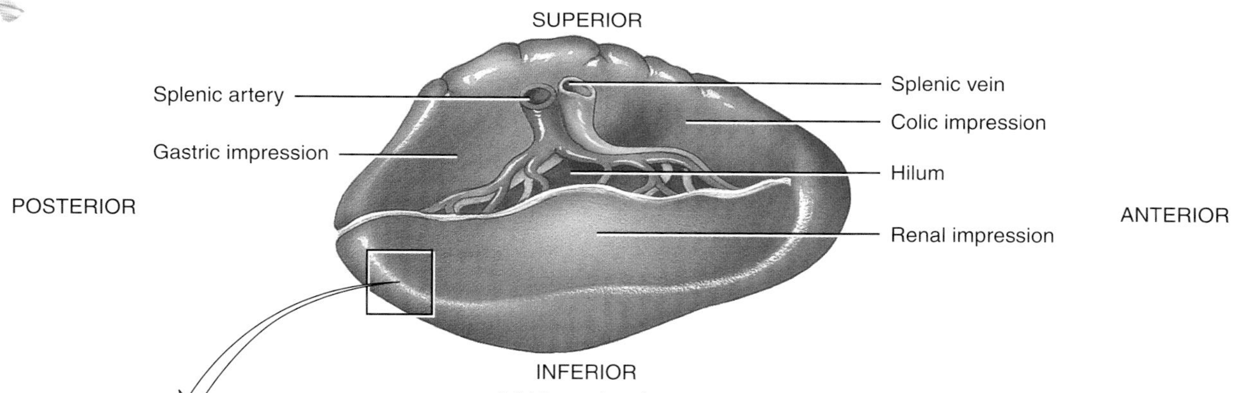

SUPERIOR
Splenic artery
Gastric impression
POSTERIOR
Splenic vein
Colic impression
Hilum
Renal impression
ANTERIOR
INFERIOR

(a) Visceral surface

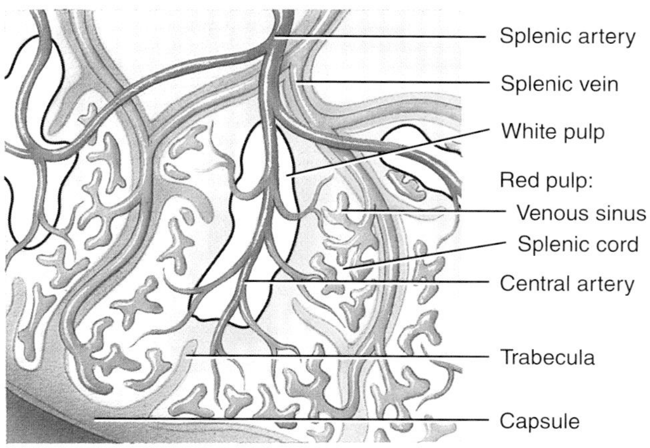

Splenic artery
Splenic vein
White pulp
Red pulp:
Venous sinus
Splenic cord
Central artery
Trabecula
Capsule

(b) Internal structure

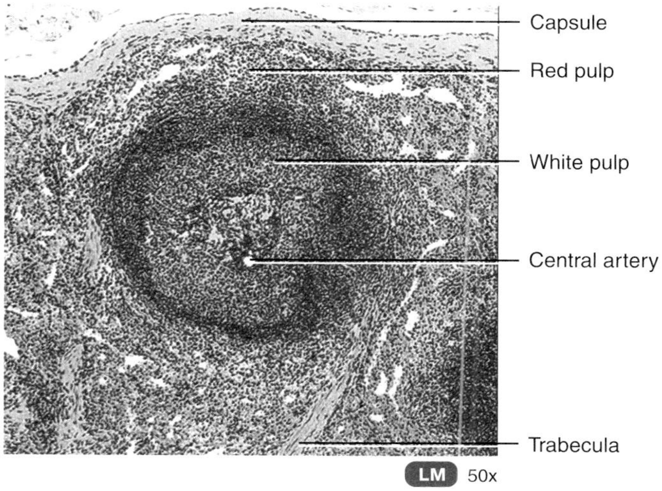

Capsule
Red pulp
White pulp
Central artery
Trabecula
LM 50x

(c) Portion of the spleen

After birth, what are the main functions of the spleen?

the stomach and diaphragm. The superior surface of the spleen is smooth and convex and conforms to the concave surface of the diaphragm. Neighboring organs make indentations in the visceral surface of the spleen—the *gastric impression* (stomach), the *renal impression* (left kidney), and the *colic impression* (left flexure of colon). Like lymph nodes, the spleen has a hilum. Through it pass the splenic artery, splenic vein, and efferent lymphatic vessels.

A capsule of dense connective tissue surrounds the spleen and is covered, in turn, by a serous membrane, the visceral peritoneum. Trabeculae extend inward from the capsule. The capsule plus trabeculae, reticular fibers, and fibroblasts constitute the stroma of the spleen; the parenchyma of the spleen consists of two different kinds of tissue called white pulp and red pulp (Figure 22.7b, c). **White pulp** is lymphatic tissue, consisting mostly of lymphocytes and macrophages arranged around branches of the splenic artery called **central arteries.** The **red pulp** consists of blood-filled **venous sinuses** and cords of splenic tissue called **splenic (Billroth's) cords.** Splenic cords consist of red blood cells, macrophages, lymphocytes, plasma cells, and granulocytes. Veins are closely associated with the red pulp.

Blood flowing into the spleen through the splenic artery enters the central arteries of the white pulp. Within the white pulp, B cells and T cells carry out immune functions, similar to lymph nodes, while spleen macrophages destroy blood-borne pathogens by phagocytosis. Within the red pulp, the spleen performs three functions related to blood cells: (1) removal by macrophages of ruptured, worn out, or defective blood cells and platelets; (2) storage of platelets, up to one-third of the body's supply; and (3) production of blood cells (hemopoiesis) during fetal life.

Ruptured Spleen

The spleen is the organ most often damaged in cases of abdominal trauma. Severe blows over the inferior left chest or superior abdomen can fracture the protecting ribs. Such crushing injury may result in a **ruptured spleen,** which causes significant hemorrhage and shock. Prompt removal of the spleen, called a **splenectomy,** is needed to prevent death due to bleeding. Other structures, particularly red bone marrow and the liver, can take over some functions normally carried out by the spleen. Immune functions, however, decrease in the absence of a spleen. The spleen's absence also places the patient at higher risk for **sepsis** (a blood infection) due to loss of the filtering and phagocytic functions of the spleen. To reduce the risk of sepsis, patients who have undergone a splenectomy take prophylactic (preventive) antibiotics before any invasive procedures. ■

Lymphatic Nodules

Lymphatic nodules (follicles) are egg-shaped masses of lymphatic tissue that are not surrounded by a capsule. Because they are scattered throughout the lamina propria (connective tissue) of mucous membranes lining the gastrointestinal, urinary, and reproductive tracts and the respiratory airways, lymphatic nodules in these areas are also referred to as **mucosa-associated lymphatic tissue (MALT).**

Although many lymphatic nodules are small and solitary, some occur in multiple large aggregations in specific parts of the body. Among these are the tonsils in the pharyngeal region and the aggregated lymphatic follicles (Peyer's patches) in the ileum of the small intestine. Aggregations of lymphatic nodules also occur in the appendix. Usually there are five **tonsils,** which form a ring at the junction of the oral cavity and oropharynx and at the junction of the nasal cavity and nasopharynx (see Figure 23.2b). The tonsils are strategically positioned to participate in immune responses against inhaled or ingested foreign substances. The single **pharyngeal tonsil** (fa-RIN-jē-al) or **adenoid** is embedded in the posterior wall of the nasopharynx. The two **palatine tonsils** (PAL-a-tīn) lie at the posterior region of the oral cavity, one on either side; these are the tonsils commonly removed in a tonsillectomy. The paired **lingual tonsils** (LIN-gwal), located at the base of the tongue, may also require removal during a tonsillectomy.

► CHECKPOINT

1. How are interstitial fluid and lymph similar, and how do they differ?
2. How do lymphatic vessels differ in structure from veins?
3. Diagram the route of lymph circulation.
4. What is the role of the thymus in immunity?
5. What functions do lymph nodes, the spleen, and the tonsils serve?

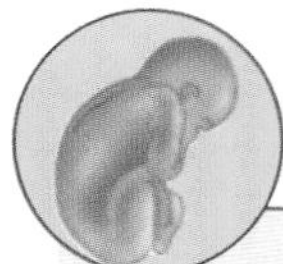

DEVELOPMENT OF LYMPHATIC TISSUES

► OBJECTIVE

Describe the development of lymphatic tissues.

Lymphatic tissues begin to develop by the end of the fifth week of embryonic life. *Lymphatic vessels* develop from **lymph sacs** that arise from developing veins, which are derived from **mesoderm.**

The first lymph sacs to appear are the paired **jugular lymph sacs** at the junction of the internal jugular and subclavian veins (Figure 22.8). From the jugular lymph sacs, lymphatic capillary plexuses spread to the thorax, upper limbs, neck, and head. Some of the plexuses enlarge and form lymphatic vessels in their respective regions. Each jugular lymph sac retains at least one connection with its jugular vein, the left one developing into the superior portion of the thoracic duct (left lymphatic duct).

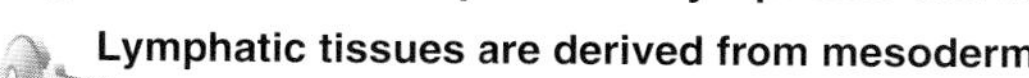

Figure 22.8 **Development of lymphatic tissues.**

Lymphatic tissues are derived from mesoderm.

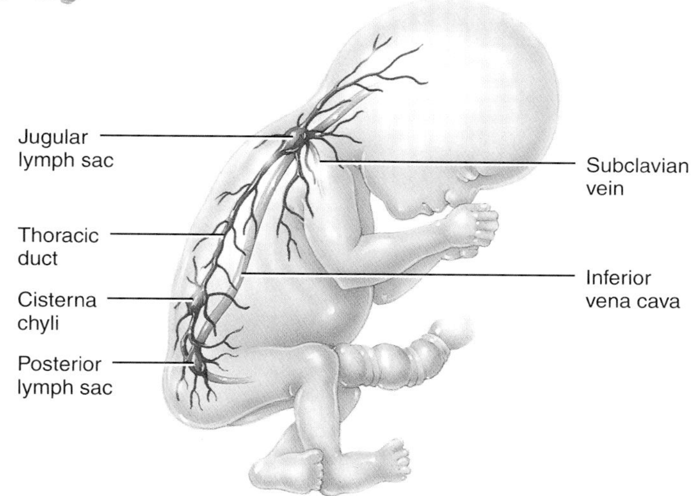

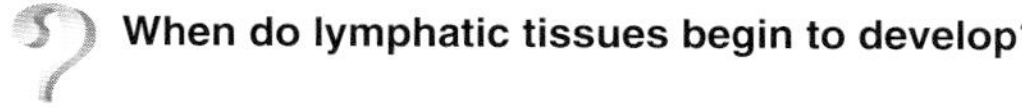

When do lymphatic tissues begin to develop?

The next lymph sac to appear is the unpaired **retroperitoneal lymph sac** at the root of the mesentery of the intestine. It develops from the primitive vena cava and mesonephric (primitive kidney) veins. Capillary plexuses and lymphatic vessels spread from the retroperitoneal lymph sac to the abdominal viscera and diaphragm. The sac establishes connections with the cisterna chyli but loses its connections with neighboring veins.

At about the time the retroperitoneal lymph sac is developing, another lymph sac, the **cisterna chyli,** develops inferior to the diaphragm on the posterior abdominal wall. It gives rise to the inferior portion of the *thoracic duct* and the *cisterna chyli* of the thoracic duct. Like the retroperitoneal lymph sac, the cisterna chyli also loses its connections with surrounding veins.

The last of the lymph sacs, the paired **posterior lymph sacs,** develop from the iliac veins. The posterior lymph sacs produce capillary plexuses and lymphatic vessels of the abdominal wall, pelvic region, and lower limbs. The posterior lymph sacs join the cisterna chyli and lose their connections with adjacent veins.

With the exception of the anterior part of the sac from which the cisterna chyli develops, all lymph sacs become invaded by **mesenchymal cells** and are converted into groups of *lymph nodes.*

The *spleen* develops from **mesenchymal cells** between layers of the dorsal mesentery of the stomach. The *thymus* arises as an outgrowth of the **third pharyngeal pouch** (see Figure 18.21a on page 655).

▶ CHECKPOINT

6. What are the names of the four lymph sacs from which lymphatic vessels develop?

NONSPECIFIC RESISTANCE: INNATE DEFENSES

▶ OBJECTIVE

Describe the mechanisms of nonspecific resistance to disease.

Although several mechanisms contribute to innate defenses, also called nonspecific resistance to disease, they all have two things in common. They are present at birth, and they offer immediate protection against a wide variety of pathogens and foreign substances. Nonspecific resistance, as its name suggests, lacks specific responses to specific invaders; instead, its protective mechanisms function the same way against any invader. Nonspecific resistance mechanisms include the external physical and chemical barriers provided by the skin and mucous membranes. They also include various internal nonspecific defenses, such as antimicrobial proteins, natural killer cells and phagocytes, inflammation, and fever.

First Line of Defense: Skin and Mucous Membranes

The skin and mucous membranes of the body are the first line of defense against pathogens. These structures provide both physical and chemical barriers that discourage pathogens and foreign substances from penetrating the body and causing disease.

With its many layers of closely packed, keratinized cells, the outer epithelial layer of the skin—the **epidermis**—provides a formidable physical barrier to the entrance of microbes (see Figure 5.1 on page 146). In addition, periodic shedding of epidermal cells helps remove microbes at the skin surface. Bacteria rarely penetrate the intact surface of healthy epidermis. If this surface is broken by cuts, burns, or punctures, however, pathogens can penetrate the epidermis and invade adjacent tissues or circulate in the blood to other parts of the body.

The epithelial layer of **mucous membranes,** which line body cavities, secretes a fluid called **mucus** that lubricates and moistens the cavity surface. Because mucus is slightly viscous, it traps many microbes and foreign substances. The mucous membrane of the nose has mucus-coated **hairs** that trap and filter microbes, dust, and pollutants from inhaled air. The mucous membrane of the upper respiratory tract contains **cilia,** microscopic hairlike projections on the surface of the epithelial cells. The waving action of cilia propels inhaled dust and microbes that have become trapped in mucus toward the throat. Coughing and sneezing accelerate movement of mucus and its entrapped pathogens out of the body. Swallowing mucus sends pathogens to the stomach where gastric juice destroys them.

Other fluids produced by various organs also help protect epithelial surfaces of the skin and mucous membranes. The **lacrimal apparatus** (LAK-ri-mal) of the eyes (see Figure 17.4 on page 581) manufactures and drains away tears in response to irritants. Blinking spreads tears over the surface of the eyeball, and the continual washing action of tears helps to dilute microbes and keep them from settling on the surface of the eye. Tears also contain **lysozyme,** an enzyme capable of breaking down the cell walls of certain bacteria. Besides tears, lysozyme is present in saliva, perspiration, nasal secretions, and tissue fluids. **Saliva,** produced by the salivary glands, washes microbes from the surfaces of the teeth and from the mucous membrane of the mouth, much as tears wash the eyes. The flow of saliva reduces colonization of the mouth by microbes.

The cleansing of the urethra by the **flow of urine** retards microbial colonization of the urinary system. **Vaginal secretions** likewise move microbes out of the body in females. **Defecation** and **vomiting** also expel microbes. For example, in response to some microbial toxins, the smooth muscle of the lower gastrointestinal tract contracts vigorously; the resulting diarrhea rapidly expels many of the microbes.

Certain chemicals also contribute to the high degree of resistance of the skin and mucous membranes to microbial invasion. Sebaceous (oil) glands of the skin secrete an oily substance called **sebum** that forms a protective film over the surface of the skin. The unsaturated fatty acids in sebum inhibit the growth of certain pathogenic bacteria and fungi. The acidity of the skin (pH 3–5) is caused in part by the secretion of fatty acids and lactic acid. **Perspiration** helps flush microbes from the surface of the skin. **Gastric juice,** produced by the glands of the stomach, is a mixture of hydrochloric acid, enzymes, and mucus. The strong acidity of gastric juice (pH 1.2–3.0) destroys many bacteria and most bacterial toxins. **Vaginal secretions** also are slightly acidic, which discourages bacterial growth.

Second Line of Defense: Internal Defenses

When pathogens penetrate the mechanical and chemical barriers of the skin and mucous membranes, they encounter a second line of defense: internal antimicrobial proteins, phagocytes, natural killer cells, inflammation, and fever.

Antimicrobial Proteins

Blood and interstitial fluids contain three main types of **antimicrobial proteins** that discourage microbial growth: interferons, complement, and transferrins.

1. Lymphocytes, macrophages, and fibroblasts infected with viruses produce proteins called **interferons** (in′-ter-FĒR-ons), or **IFNs.** Once released by virus-infected cells, IFNs diffuse to uninfected neighboring cells, where they induce synthesis of antiviral proteins that interfere with viral replication. Although IFNs do not prevent viruses from attaching to and penetrating host cells, they do stop replication. Viruses can cause disease only if they can replicate within body cells. IFNs are an important defense against infection by many different viruses. The three types of interferon are alpha-, beta-, and gamma-IFN.

2. A group of normally inactive proteins in blood plasma and on plasma membranes makes up the **complement system.** When activated, these proteins "complement" or enhance certain immune reactions (see page 831). The complement system causes cytolysis (bursting) of microbes, promotes phagocytosis, and contributes to inflammation.

3. Iron-binding proteins called **transferrins** inhibit the growth of certain bacteria by reducing the amount of available iron.

Natural Killer Cells and Phagocytes

When microbes penetrate the skin and mucous membranes or bypass the antimicrobial proteins in blood, the next nonspecific defense consists of natural killer cells and phagocytes. About 5–10% of lymphocytes in the blood are **natural killer (NK) cells.** They are also present in the spleen, lymph nodes, and red bone marrow. NK cells lack the membrane molecules that identify B and T cells, but they have the ability to kill a wide variety of infected body cells and certain tumor cells. NK cells attack any body cells that display abnormal or unusual plasma membrane proteins.

The binding of NK cells to a target cell, such as an infected human cell, causes the release of granules containing toxic substances from NK cells. Some granules contain a protein called **perforin** that inserts into the plasma membrane of the target cell and creates channels (perforations) in the membrane. As a result, extracellular fluid flows into the target cell and the cell bursts, a process called **cytolysis** (sī-TOL-i-sis; *cyto-* = cell; *-lysis* = loosening). Other granules of NK cells release **granzymes,** which are protein-digesting enzymes that induce the target cell to undergo apoptosis, or self-destruction. This type of attack kills infected cells, but not the microbes inside the cells; the released microbes, which may or may not be intact, can be destroyed by phagocytes.

Phagocytes (*phago-* = eat; *-cytes* = cells) are specialized cells that perform **phagocytosis** (*-osis* = process), the ingestion of microbes or other particles such as cellular debris (see Figure 3.11 on page 73). The two major types of phagocytes are **neutrophils** and **macrophages** (MAK-rō-fā-jez). When an infection occurs, neutrophils and monocytes migrate to the infected area. During this migration, the monocytes enlarge and develop into actively phagocytic macrophages called **wandering macrophages.** Other macrophages, called **fixed macrophages,** stand guard in specific tissues. Among the fixed macrophages are *histiocytes* (connective tissue macrophages), *stellate reticuloendothelial cells* (*Kupffer cells*) in the liver, *alveolar macrophages* in the lungs, *microglia* in the nervous system, and *tissue macrophages* in the spleen, lymph nodes, and red bone marrow. In addition to being an innate defense mechanism, phagocytosis plays a vital role in immunity (specific resistance), as discussed later in the chapter.

Phagocytosis occurs in five phases: chemotaxis, adherence, ingestion, digestion, and killing (Figure 22.9):

1. ***Chemotaxis.*** Phagocytosis begins with **chemotaxis,** a chemically stimulated movement of phagocytes to a site of damage. Chemicals that attract phagocytes might come from invading microbes, white blood cells, damaged tissue cells, or activated complement proteins.

2 ***Adherence.*** Attachment of the phagocyte to the microbe or other foreign material is termed **adherence.** The binding of complement proteins to the invading pathogen enhances adherence.

3 ***Ingestion.*** Following adherence, the plasma membrane of the phagocyte extends projections, called **pseudopods,** that engulf the microbe in a process called **ingestion.** When the pseudopods meet, they fuse, surrounding the microorganism with a sac called a **phagosome.**

4 ***Digestion.*** The phagosome enters the cytoplasm and merges with lysosomes to form a single, larger structure called a **phagolysosome.** The lysosome contributes lysozyme, which breaks down microbial cell walls, and other digestive enzymes that degrade carbohydrates, proteins, lipids, and nucleic acids. The phagocyte also forms lethal oxidants, such as superoxide anion (O_2^-), hypochlorite anion (OCl^-), and hydrogen peroxide (H_2O_2), in a process called an **oxidative burst.**

5 ***Killing.*** The chemical onslaught provided by lysozyme, digestive enzymes, and oxidants within a phagolysosome quickly kills many types of microbes. Any materials that cannot be degraded further remain in structures called **residual bodies.**

Microbial Evasion of Phagocytosis

Some microbes, such as the bacteria that cause pneumonia, have extracellular structures called capsules that prevent adherence. This makes it physically difficult for phagocytes to engulf the microbes. Other microbes, such as the toxin-producing bacteria that cause one kind of food poisoning, may be ingested but not killed; instead, the toxins they produce (leukocidins) may kill the phagocytes by causing the release of the phagocyte's own lysosomal enzymes into its cytoplasm. Still other microbes—such as the bacteria that cause tuberculosis—inhibit fusion of phagosomes and lysosomes and thus prevent exposure of the microbes to lysosomal enzymes. These bacteria apparently can also use chemicals in their cell walls to counter the effects of lethal oxidants produced by phagocytes. Subsequent multiplication of the microbes within phagosomes may eventually destroy the phagocyte. ■

Inflammation

Inflammation is a nonspecific, defensive response of the body to tissue damage. Among the conditions that may produce inflammation are pathogens, abrasions, chemical irritations, distortion or disturbances of cells, and extreme temperatures. The four characteristic signs and symptoms of inflammation are **redness, pain, heat,** and **swelling.** Inflammation can also cause the **loss of function** in the injured area (for example, the inability to detect sensations), depending on the site and extent of the injury. Inflammation is an attempt to dispose of microbes, toxins, or foreign material at the site of injury, to prevent their spread to other tissues, and to prepare the site for tissue repair in an attempt to restore tissue homeostasis.

Because inflammation is one of the body's nonspecific resistance mechanisms, the response of a tissue to a cut is similar to the response to damage caused by burns, radiation, or bacterial or viral invasion. In each case, the inflammatory response has

Figure 22.9 Phagocytosis of a microbe.

The major types of phagocytes are neutrophils and macrophages.

1 CHEMOTAXIS
Microbe
Phagocyte
2 ADHERENCE
3 INGESTION
Pseudopod
Lysosome
4 DIGESTION
Digested microbe in phagolysosome
Plasma membrane
Residual body (indigestible material)
5 KILLING
Digestive enzymes

(a) Phases of phagocytosis

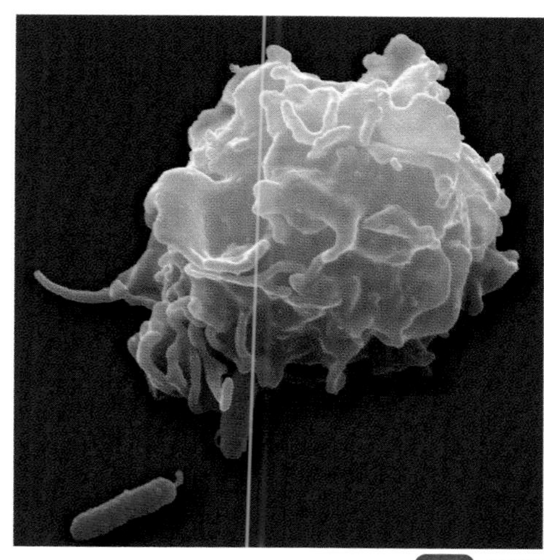

(b) Phagocyte (white blood cell) engulfing a microbe.

What chemicals are responsible for killing ingested microbes?

three basic stages: (1) vasodilation and increased permeability of blood vessels, (2) emigration (movement) of phagocytes from the blood into interstitial fluid, and, ultimately, (3) tissue repair.

VASODILATION AND INCREASED PERMEABILITY OF BLOOD VESSELS Two immediate changes occur in the blood vessels in a region of tissue injury: **vasodilation** (increase in the diameter) of arterioles and **increased permeability** of capillaries (Figure 22.10). Increased permeability means that substances normally retained in blood are permitted to pass from the blood vessels. Vasodilation allows more blood to flow through the damaged area, and increased permeability permits defensive proteins such as antibodies and clotting factors to enter the injured area from the blood. The increased blood flow also helps remove microbial toxins and dead cells.

Among the substances that contribute to vasodilation, increased permeability, and other aspects of the inflammatory response are the following:

Figure 22.10 **Inflammation.**

The three stages of inflammation are as follows: (1) vasodilation and increased permeability of blood vessels, (2) phagocyte emigration, and (3) tissue repair.

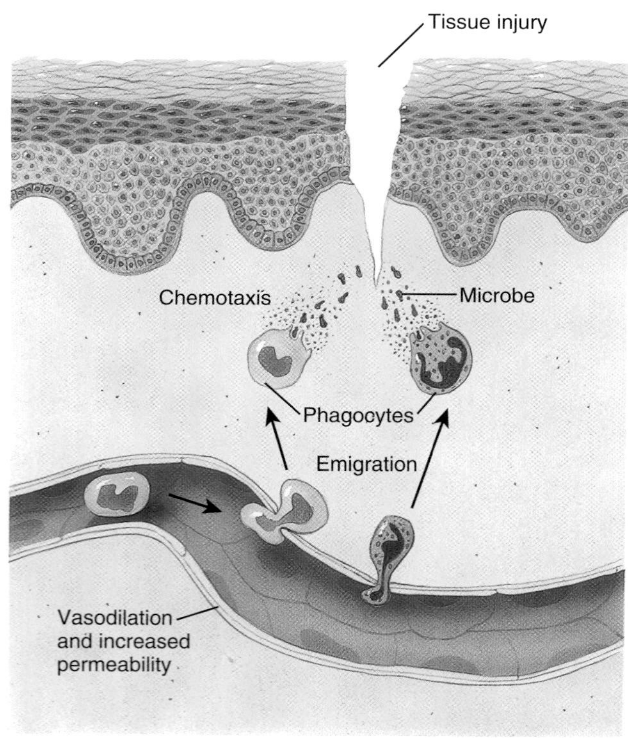

Phagocytes migrate from blood to site of tissue injury

What causes each of the following signs and symptoms of inflammation: redness, pain, heat, and swelling?

- ***Histamine.*** In response to injury, mast cells in connective tissue and basophils and platelets in blood release histamine. Neutrophils and macrophages attracted to the site of injury also stimulate the release of histamine, which causes vasodilation and increased permeability of blood vessels.
- ***Kinins.*** These polypeptides, formed in blood from inactive precursors called kininogens, induce vasodilation and increased permeability and serve as chemotactic agents for phagocytes. An example of a kinin is bradykinin.
- ***Prostaglandins (PGs).*** These lipids, especially those of the E series, are released by damaged cells and intensify the effects of histamine and kinins. PGs also may stimulate the emigration of phagocytes through capillary walls.
- ***Leukotrienes (LTs).*** Produced by basophils and mast cells, LTs cause increased permeability; they also function in adherence of phagocytes to pathogens and as chemotactic agents that attract phagocytes.
- ***Complement.*** Different components of the complement system stimulate histamine release, attract neutrophils by chemotaxis, and promote phagocytosis; some components can also destroy bacteria.

Dilation of arterioles and increased permeability of capillaries produce three of the symptoms of inflammation: heat, redness (erythema), and swelling (edema). Heat and redness result from the large amount of blood that accumulates in the damaged area. As the local temperature rises slightly, metabolic reactions proceed more rapidly and release additional heat. Edema results from increased permeability of blood vessels, which permits more fluid to move from blood plasma into tissue spaces.

Pain is a prime symptom of inflammation. It results from injury to neurons and from toxic chemicals released by microbes. Kinins affect some nerve endings, causing much of the pain associated with inflammation. Prostaglandins intensify and prolong the pain associated with inflammation. Pain may also be due to increased pressure from edema.

The increased permeability of capillaries allows leakage of blood-clotting factors into tissues. The clotting sequence is set into motion, and fibrinogen is ultimately converted to an insoluble, thick mesh of fibrin threads that localizes and traps invading microbes and blocks their spread.

EMIGRATION OF PHAGOCYTES Within an hour after the inflammatory process starts, phagocytes appear on the scene. As large amounts of blood accumulate, neutrophils begin to stick to the inner surface of the endothelium (lining) of blood vessels (Figure 22.10). Then the neutrophils begin to squeeze through the wall of the blood vessel to reach the damaged area. This process, called **emigration,** depends on chemotaxis. Neutrophils attempt to destroy the invading microbes by phagocytosis. A steady stream of neutrophils is ensured by the production and release of additional cells from red bone marrow. Such an increase in white blood cells in the blood is termed **leukocytosis.**

Although neutrophils predominate in the early stages of infection, they die off rapidly. As the inflammatory response continues, monocytes follow the neutrophils into the infected area. Once in the tissue, monocytes transform into wandering macrophages that add to the phagocytic activity of the fixed macrophages already present. True to their name, macrophages are much more potent phagocytes than neutrophils. They are large enough to engulf damaged tissue, worn-out neutrophils, and invading microbes.

Eventually, macrophages also die. Within a few days, a pocket of dead phagocytes and damaged tissue forms; this collection of dead cells and fluid is called **pus.** Pus formation occurs in most inflammatory responses and usually continues until the infection subsides. At times, pus reaches the surface of the body or drains into an internal cavity and is dispersed; on other occasions the pus remains even after the infection is terminated. In this case, the pus is gradually destroyed over a period of days and is absorbed.

Abscesses and Ulcers

If pus cannot drain out of an inflamed region, the result is an **abscess**—an excessive accumulation of pus in a confined space. Common examples are pimples and boils. When superficial inflamed tissue sloughs off the surface of an organ or tissue, the resulting open sore is called an **ulcer.** People with poor circulation—for instance, diabetics with advanced atherosclerosis—are susceptible to ulcers in the tissues of their legs. These ulcers, which are called stasis ulcers, develop because of poor oxygen and nutrient supply to tissues that then become very susceptible to even a very mild injury or an infection. ■

Fever

Fever is an abnormally high body temperature that occurs because the hypothalamic thermostat is reset. It commonly occurs during infection and inflammation. Many bacterial toxins elevate body temperature, sometimes by triggering release of fever-causing cytokines such as interleukin-1 from macrophages. Elevated body temperature intensifies the effects of interferons, inhibits the growth of some microbes, and speeds up body reactions that aid repair.

Table 22.1 summarizes the components of nonspecific resistance.

▶ CHECKPOINT

7. What physical and chemical factors provide protection from disease in the skin and mucous membranes?
8. What internal defenses provide protection against microbes that penetrate the skin and mucous membranes?
9. How are the activities of natural killer cells and phagocytes similar and different?
10. What are the main signs, symptoms, and stages of inflammation?

TABLE 22.1 Summary of Nonspecific Resistance (Innate Defenses)

Component	Functions
First Line of Defense: Skin and Mucous Membranes	
Physical Factors	
Epidermis of skin	Forms a physical barrier to the entrance of microbes.
Mucous membranes	Inhibit the entrance of many microbes, but not as effective as intact skin.
Mucus	Traps microbes in respiratory and gastrointestinal tracts.
Hairs	Filter out microbes and dust in nose.
Cilia	Together with mucus, trap and remove microbes and dust from upper respiratory tract.
Lacrimal apparatus	Tears dilute and wash away irritating substances and microbes.
Saliva	Washes microbes from surfaces of teeth and mucous membranes of mouth.
Urine	Washes microbes from urethra.
Defecation and vomiting	Expel microbes from body.
Chemical Factors	
Sebum	Forms a protective acidic film over the skin surface that inhibits growth of many microbes.
Lysozyme	Antimicrobial substance in perspiration, tears, saliva, nasal secretions, and tissue fluids.
Gastric juice	Destroys bacteria and most toxins in stomach.
Vaginal secretions	Slight acidity discourages bacterial growth; flush microbes out of vagina.
Second Line of Defense: Internal Defenses	
Antimicrobial Proteins	
Interferons (IFNs)	Protect uninfected host cells from viral infection.
Complement system	Causes cytolysis of microbes, promotes phagocytosis, and contributes to inflammation.
Transferrins	Inhibit growth of certain bacteria by reducing the amount of available iron.
Natural killer (NK) cells	Kill infected target cells by releasing granules that contain perforin and granzymes. Phagocytes then kill the released microbes.
Phagocytes	Ingest foreign particulate matter.
Inflammation	Confines and destroys microbes and initiates tissue repair.
Fever	Intensifies the effects of interferons, inhibits growth of some microbes, and speeds up body reactions that aid repair.

SPECIFIC RESISTANCE: IMMUNITY

▶ **OBJECTIVES**

Define immunity, and describe how T cells and B cells arise.

Explain the relationship between an antigen and an antibody.

Compare the functions of cell-mediated immunity and antibody-mediated immunity.

The ability of the body to defend itself against specific invading agents such as bacteria, toxins, viruses, and foreign tissues is called **specific resistance** or **immunity.** Substances that are recognized as foreign and provoke immune responses are called **antigens (Ags).** Two properties distinguish immunity from nonspecific defenses: (1) *specificity* for particular foreign molecules (antigens), which also involves distinguishing self from nonself molecules, and (2) *memory* for most previously encountered antigens so that a second encounter prompts an even more rapid and vigorous response. The branch of science that deals with the responses of the body when challenged by antigens is called **immunology** (im′-ū-NOL-ō-jē; *immuno-* = free from service or exempt; *-logy* = study of). The **immune system** includes the cells and tissues that carry out immune responses.

Maturation of T Cells and B Cells

The cells that develop **immunocompetence,** the ability to carry out immune responses when properly stimulated, are lymphocytes called B cells and T cells. Both develop in primary lymphatic organs (red bone marrow and the thymus) from pluripotent stem cells that originate in red bone marrow (see Figure 19.3 on page 671). B cells complete their development in red bone marrow, a process that continues throughout life. T cells develop from pre-T cells that migrate from red bone marrow into the thymus, where they mature (Figure 22.11). Most T cells arise before puberty, but they continue to mature and leave the thymus throughout life.

Before T cells leave the thymus or B cells leave red bone marrow, they begin to make several distinctive proteins that are inserted into their plasma membranes. Some of these proteins function as **antigen receptors**—molecules capable of recognizing specific antigens (Figure 22.11). T cells exit the thymus as either CD4 or CD8 cells, which means that, in addition to the antigen receptors, their plasma membrane includes a protein called either CD4 or CD8. As we will see later in this chapter, these two types of T cells have very different functions.

Types of Immune Responses

Immunity consists of two kinds of closely allied responses, both triggered by antigens. In **cell-mediated immune responses,** CD8 T cells proliferate into cytotoxic T cells that directly attack the invading antigen. In **antibody-mediated immune responses,** B cells transform into plasma cells, which synthesize and secrete specific proteins called **antibodies (Abs)** or **immunoglobulins** (im′-ū-nō-GLOB-ū-lins). A given antibody can bind to and inactivate a specific antigen. Most CD4 T cells become helper T cells that aid both cell-mediated and antibody-mediated immune responses.

Cell-mediated immunity is particularly effective against (1) intracellular pathogens, which include any viruses, bacteria, or fungi that are inside cells; (2) some cancer cells; and (3) foreign tissue transplants. Thus, cell-mediated immunity always involves cells attacking cells. Antibody-mediated immunity works mainly against (1) antigens present in body fluids and (2) extracellular pathogens, which include any viruses, bacteria, or fungi that are outside cells. Depending on its location, a given pathogen can provoke both types of immune responses.

Antigens and Antigen Receptors

Antigens have two important characteristics: immunogenicity and reactivity. **Immunogenicity** (im-ū-nō-je-NIS-i-tē; *-genic* = producing) is the ability to provoke an immune response by stimulating the production of specific antibodies, the proliferation of specific T cells, or both. The term *antigen* derives from its function as an *anti*body *gen*erator. **Reactivity** is the ability of the antigen to react specifically with the antibodies or cells it provoked. Strictly speaking, immunologists define antigens as substances that have reactivity; substances with both immunogenicity and reactivity are considered **complete antigens.** Commonly, however, the term *antigen* implies both immunogenicity and reactivity, and we use the word in this way.

Entire microbes or parts of microbes may act as antigens. Chemical components of bacterial structures such as flagella, capsules, and cell walls are antigenic, as are bacterial toxins. Nonmicrobial examples of antigens include chemical components of pollen, egg white, incompatible blood cells, and transplanted tissues and organs. The huge variety of antigens in the environment provides myriad opportunities for provoking immune responses. Typically, just certain small parts of a large antigen molecule act as the triggers for immune responses. These small parts are called **epitopes** (EP-i-tōps), or *antigenic determinants* (Figure 22.12). Most antigens have many epitopes, each of which induces production of a specific antibody or activates a specific T cell.

Antigens that get past the nonspecific defenses generally follow one of three routes into lymphatic tissue: (1) Most antigens that enter the bloodstream (for example, through an injured blood vessel) are trapped as they flow through the spleen. (2) Antigens that penetrate the skin enter lymphatic vessels and lodge in lymph nodes. (3) Antigens that penetrate mucous membranes are entrapped by mucosa-associated lymphatic tissue (MALT).

Chemical Nature of Antigens

Antigens are large, complex molecules. Most often, they are proteins. However, nucleic acids, lipoproteins, glycoproteins, and certain large polysaccharides may also act as antigens. T cells

Figure 22.11 B cells and pre-T cells arise from pluripotent stem cells in red bone marrow. B cells and T cells develop in primary lymphatic tissues (red bone marrow and the thymus) and are activated in secondary lymphatic tissues (lymph nodes, spleen, and lymphatic nodules).

The two types of immune responses are cell-mediated immune responses and antibody-mediated immune responses.

CELL-MEDIATED IMMUNE RESPONSES
Directed against intracellular pathogens, some cancer cells, and tissue transplants

ANTIBODY-MEDIATED IMMUNE RESPONSES
Directed against extracellular pathogens

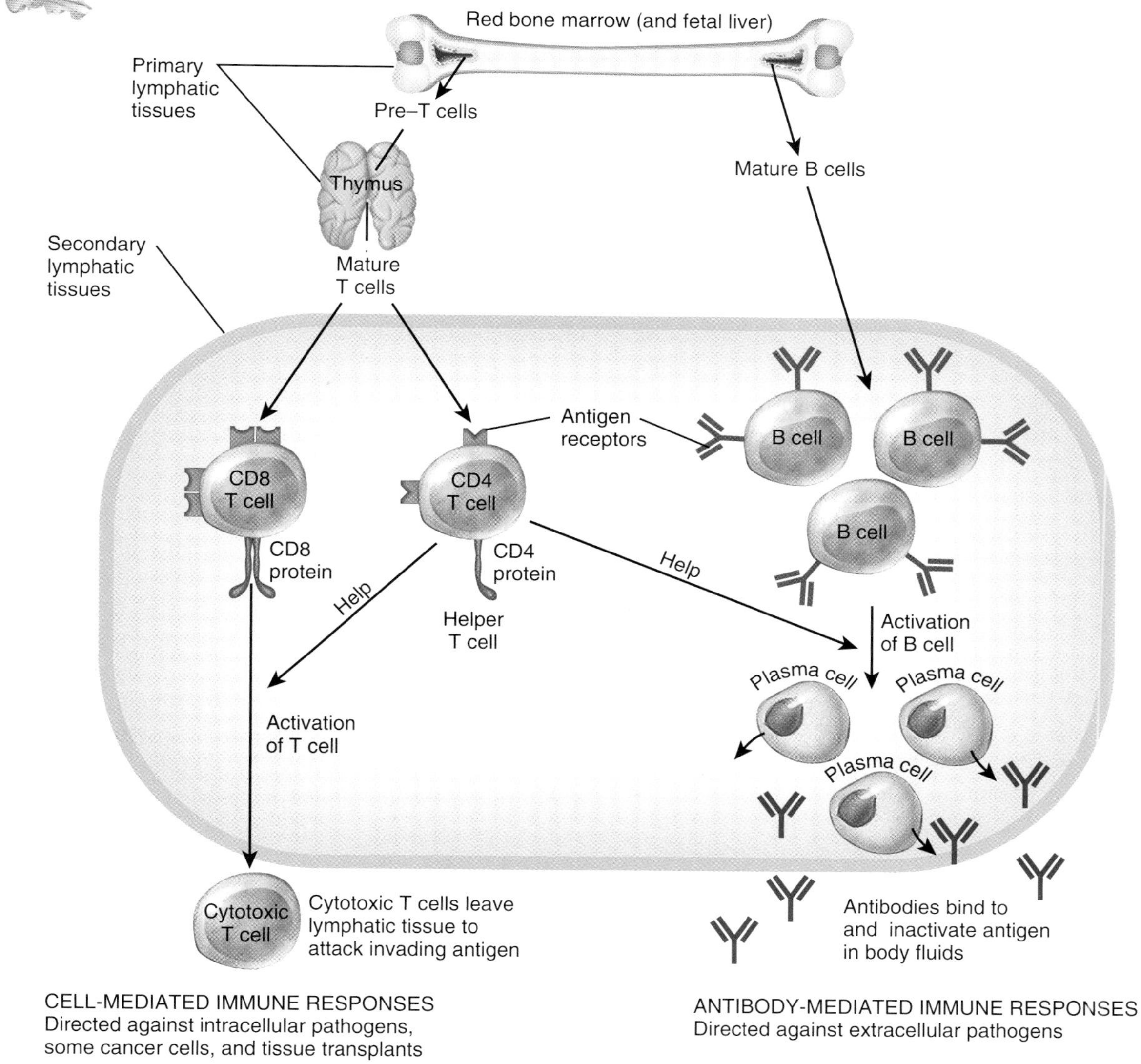

Which type of T cell participates in both cell-mediated and antibody-mediated immune responses?

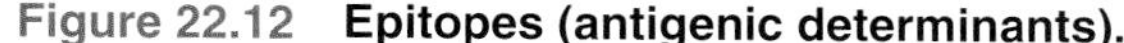

Figure 22.12 Epitopes (antigenic determinants).

Most antigens have several epitopes that induce the production of different antibodies or activate different T cells.

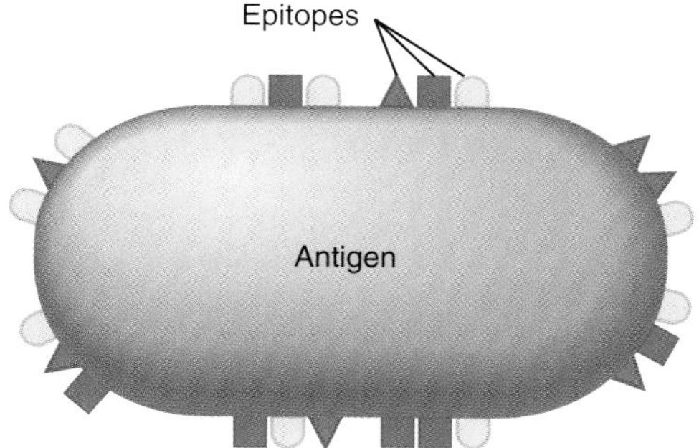

What is the difference between an epitope and a hapten?

respond only to antigens made up of proteins; B cells respond to antigens made of proteins, certain lipids, carbohydrates, and nucleic acids. Complete antigens usually have large molecular weights of 10,000 daltons or more, but large molecules that have simple, repeating subunits—for example, cellulose and most plastics—are not usually antigenic. This is why plastic materials can be used in artificial heart valves or joints.

A smaller substance that has reactivity but lacks immunogenicity is called a **hapten** (HAP-ten = to grasp). A hapten can stimulate an immune response only if it is attached to a larger carrier molecule. An example is the small lipid toxin in poison ivy, which triggers an immune response after combining with a body protein. Likewise, some drugs, such as penicillin, may combine with proteins in the body to form immunogenic complexes. Such hapten-stimulated immune responses are responsible for some allergic reactions to drugs and other substances in the environment (see page 839).

As a rule, antigens are foreign substances; they are not usually part of body tissues. However, sometimes the immune system fails to distinguish "friend" (self) from "foe" (nonself). The result is an autoimmune disorder (see page 839) in which self-molecules or cells are attacked as though they were foreign.

Diversity of Antigen Receptors

An amazing feature of the human immune system is its ability to recognize and bind to at least a billion (10^9) different epitopes. Before a particular antigen ever enters the body, T cells and B cells that can recognize and respond to that intruder are ready and waiting. Cells of the immune system can even recognize artificially made molecules that do not exist in nature. The basis for the ability to recognize so many epitopes is an equally large diversity of antigen receptors. Given that human cells contain only about 35,000 genes, how could a billion or more different antigen receptors possibly be generated?

The answer to this puzzle turned out to be simple in concept. The diversity of antigen receptors in both B cells and T cells is the result of shuffling and rearranging a few hundred versions of several small gene segments. This process is called **genetic recombination.** The gene segments are put together in different combinations as the lymphocytes are developing from stem cells in red bone marrow and the thymus. The situation is similar to shuffling a deck of 52 cards and then dealing out three cards. If you did this over and over, you could generate many more than 52 different sets of three cards. Because of genetic recombination, each B cell or T cell has a unique set of gene segments that codes for its unique antigen receptor. After transcription and translation, the receptor molecules are inserted into the plasma membrane.

Major Histocompatibility Complex Antigens

Located in the plasma membrane of body cells are "self-antigens," the **major histocompatibility complex (MHC)** antigens. These transmembrane glycoproteins are also called *human leukocyte antigens (HLA)* because they were first identified on white blood cells. Unless you have an identical twin, your MHC antigens are unique. Thousands to several hundred thousand MHC molecules mark the surface of each of your body cells except red blood cells. Although MHC antigens are the reason that tissues may be rejected when they are transplanted from one person to another, their normal function is to help T cells recognize that an antigen is foreign, not self. Such recognition is an important first step in any immune response.

The two types of major histocompatibility complex antigens are class I and class II. Class I MHC (MHC-I) molecules are built into the plasma membranes of all body cells except red blood cells. Class II MHC (MHC-II) molecules appear on the surface of antigen-presenting cells (described in the next section).

Pathways of Antigen Processing

For an immune response to occur, B cells and T cells must recognize that a foreign antigen is present. B cells can recognize and bind to antigens in lymph, interstitial fluid, or blood plasma. T cells only recognize fragments of antigenic proteins that are processed and presented in a certain way. In **antigen processing,** antigenic proteins are broken down into peptide fragments that then associate with MHC molecules. Next the antigen-MHC complex is inserted into the plasma membrane of a body cell. The insertion of the complex into the plasma membrane is called **antigen presentation.** When a peptide fragment comes from a *self-protein,* T cells ignore the antigen-MHC complex. However, if the peptide fragment comes from a *foreign protein,* T cells recognize the antigen-MHC complex as an intruder, and an immune response takes place. Antigen processing and presentation occurs in two ways, depending on whether the antigen is located outside or inside body cells.

Processing of Exogenous Antigens

Foreign antigens that are present in fluids *outside* body cells are termed *exogenous antigens.* They include intruders such as bacteria and bacterial toxins, parasitic worms, inhaled pollen and dust, and viruses that have not yet infected a body cell. A special class of cells called **antigen-presenting cells (APCs)** process and present exogenous antigens. APCs include dendritic cells, macrophages, and B cells. They are strategically located in places where antigens are likely to penetrate nonspecific defenses and enter the body, such as the epidermis and dermis of the skin (Langerhans cells are a type of dendritic cell); mucous membranes that line the respiratory, gastrointestinal, urinary, and reproductive tracts; and lymph nodes. After processing and presenting an antigen, APCs migrate from tissues via lymphatic vessels to lymph nodes.

The steps in the processing and presenting of an exogenous antigen by an antigen-presenting cell occur as follows (Figure 22.13):

1. ***Ingestion of the antigen.*** Antigen-presenting cells ingest antigens by phagocytosis or endocytosis. Ingestion could occur almost anywhere in the body that invaders, such as microbes, have penetrated the nonspecific defenses.
2. ***Digestion of antigen into peptide fragments.*** Within the phagosome or endosome protein-digesting enzymes split large antigens into short peptide fragments.
3. ***Synthesis of MHC-II molecules.*** At the same time, the APC synthesizes MHC-II molecules and packages them into vesicles.
4. ***Fusion of vesicles.*** The vesicles containing antigen peptide fragments and MHC-II molecules merge and fuse.
5. ***Binding of peptide fragments to MHC-II molecules.*** After fusion of the two types of vesicles, antigen peptide fragments bind to MHC-II molecules.
6. ***Insertion of antigen–MHC-II complex into the plasma membrane.*** The combined vesicle that contains antigen–MHC-II complexes undergoes exocytosis. As a result, the antigen–MHC-II complexes are inserted into the plasma membrane.

After processing an antigen, the antigen-presenting cell migrates to lymphatic tissue to present the antigen to T cells. Within lymphatic tissue, a small number of T cells that have compatibly shaped receptors recognize and bind to the antigen fragment–MHC-II complex, triggering either a cell-mediated or an antibody-mediated immune response. The presentation of exogenous antigen together with MHC-II molecules by antigen-presenting cells informs T cells that intruders are present in the body and that combative action should begin.

Processing of Endogenous Antigens

Foreign antigens that are synthesized inside body cells are termed *endogenous antigens*. Such antigens may be viral proteins produced after a virus infects the cell and takes over the cell's metabolic machinery, or abnormal proteins synthesized by a cancerous cell. Fragments of endogenous antigens associate with major histocompatibility complex-I molecules inside infected cells. The resulting endogenous antigen fragment–MHC-I complex then moves to the plasma membrane, where it is presented at the surface of the cell. Most cells of the body can process and present endogenous antigens. The display of an endogenous antigen bound to an MHC-I molecule signals that a cell has been infected and needs help.

Cytokines

Cytokines are small protein hormones that stimulate or inhibit many normal cell functions, such as cell growth and differentiation. Lymphocytes and antigen-presenting cells secrete

Figure 22.13 Processing and presenting of exogenous antigen by an antigen-presenting cell (APC).

Except for identical twins, major histocompatibility complex (MHC) molecules are unique in each person. They help T cells recognize foreign invaders.

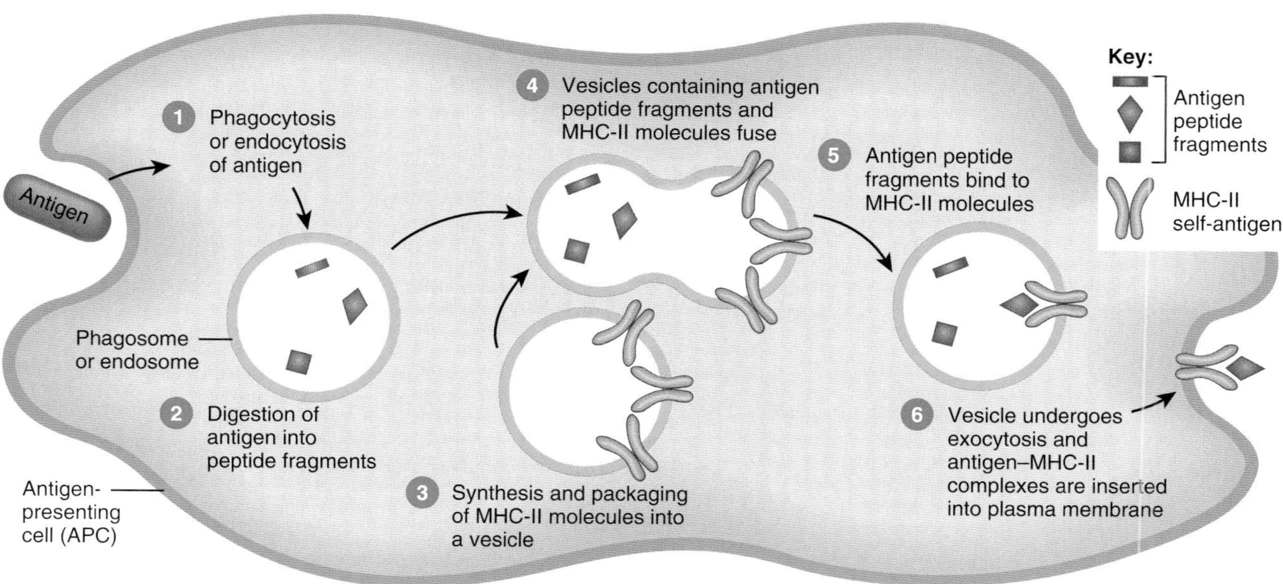

APCs present exogenous antigens in association with MHC-II molecules

What types of cells are APCs, and where in the body are they found?

cytokines, as do fibroblasts, endothelial cells, monocytes, hepatocytes, and kidney cells. Some cytokines stimulate proliferation of progenitor blood cells in red bone marrow. Others regulate activities of cells involved in nonspecific defenses or immune responses, as described in Table 22.2.

Cytokine Therapy

Cytokine therapy is the use of cytokines to treat medical conditions. Interferons were the first cytokines shown to have limited effects against some human cancers. Alpha-interferon (Intron A®) is approved in the United States for treating Kaposi (KAP-ō-sē) sarcoma, a cancer that often occurs in patients infected with HIV, the virus that causes AIDS. Other approved uses for alpha-interferon include treating genital herpes caused by the herpes virus; treating hepatitis B and C, caused by the hepatitis B and C viruses; and treating hairy cell leukemia. A form of beta-interferon (Betaseron®) slows the progression of multiple sclerosis (MS) and lessens the frequency and severity of MS attacks. Of the interleukins, the one most widely used to fight cancer is interleukin-2. Although this treatment is effective in causing tumor regression in some patients, it also can be very toxic. Among the adverse effects are high fever, severe weakness, difficulty breathing due to pulmonary edema, and hypotension leading to shock. ■

▶ CHECKPOINT

11. What is immunocompetence, and which body cells display it?

12. How do the major histocompatibility complex class I and class II self-antigens function?

13. How do antigens arrive at lymphatic tissues?

14. How do antigen-presenting cells process exogenous antigens?

15. What are cytokines, where do they arise, and how do they function?

TABLE 22.2 Summary of Cytokines Participating in Immune Responses

Cytokine	Origins and Functions
Interleukin-1 (IL-1)	Produced by monocytes and macrophages; promotes proliferation of helper T cells; acts on hypothalamus to cause fever.
Interleukin-2 (IL-2) (T cell growth factor)	Secreted by helper T cells; costimulates the proliferation of helper T cells, cytotoxic T cells, and B cells; activates NK cells.
Interleukin-4 (IL-4) (B cell stimulating factor)	Produced by activated helper T cells; costimulator for B cells; causes plasma cells to secrete IgE antibodies (see Table 22.3); promotes growth of T cells.
Interleukin-5 (IL-5)	Produced by certain activated CD4 T cells and activated mast cells; costimulator for B cells; causes plasma cells to secrete IgA antibodies.
Tumor necrosis factor (TNF)	Produced mainly by macrophages; stimulates accumulation of neutrophils and macrophages at sites of inflammation and stimulates their killing of microbes; stimulates macrophages to produce IL-1; induces synthesis of colony-stimulating factors by endothelial cells and fibroblasts; exerts an interferon-like protective effect against viruses; and functions as an endogenous pyrogen to induce fever.
Transforming growth factor beta (TGF-β)	Secreted by T cells and macrophages; has some positive effects but is thought to be important for turning off immune responses; inhibits proliferation of T cells and activation of macrophages.
Gamma-interferon (γ-IFN)	Secreted by helper and cytotoxic T cells and NK cells; strongly stimulates phagocytosis by neutrophils and macrophages; activates NK cells; enhances both cell-mediated and antibody-mediated immune responses.
Alpha- and beta-interferons (α-IFN and β-IFN)	Produced by virus-infected cells to inhibit viral replication in uninfected cells; produced by antigen-stimulated macrophages to stimulate T cell growth; activate NK cells, inhibit cell growth, and suppress formation of some tumors.
Lymphotoxin (LT)	Secreted by cytotoxic T cells; kills infected target cells by activating enzymes that cause fragmentation of DNA.
Perforin	Secreted by cytotoxic T cells and by NK cells; perforates cell membranes of infected target cells, which causes cytolysis.
Macrophage migration inhibiting factor	Produced by cytotoxic T cells; prevents macrophages from leaving site of infection.
Granzymes	Secreted by cytotoxic T cells and NK cells; cause infected target cells to undergo apoptosis.
Granulysin	Secreted by cytotoxic T cells; creates holes in microbial plasma membranes that kill the microbe.

CELL-MEDIATED IMMUNITY

▶ **OBJECTIVES**

Outline the steps in a cell-mediated immune response.

Distinguish between the action of natural killer cells and cytotoxic T cells.

Define immunological surveillance.

A cell-mediated immune response begins with *activation* of a small number of T cells by a specific antigen. Once a T cell has been activated, it undergoes *proliferation* and *differentiation* into a clone of **effector cells,** a population of identical cells that can recognize the same antigen and carry out some aspect of the immune attack. Finally, the immune response results in *elimination* of the intruder.

Activation, Proliferation, and Differentiation of T Cells

At any given time, most T cells are inactive. As you learned in the last section, antigen receptors on the surface of T cells, called **T-cell receptors (TCRs),** recognize and bind to specific foreign antigen fragments that are presented in antigen–MHC complexes. There are millions of different T cells; each has its own unique TCRs that can recognize a specific antigen–MHC complex. When an antigen enters the body, only a few T cells have TCRs that can recognize and bind to the antigen. Antigen recognition also involves other surface proteins on T cells, the CD4 or CD8 proteins. These proteins interact with the MHC antigens and help maintain the TCR-MHC coupling. For this reason, they are referred to as *coreceptors.* Antigen recognition by a TCR with CD4 or CD8 proteins is the *first signal* in activation of a T cell.

A T cell becomes activated only if it binds to the foreign antigen and at the same time receives a *second signal,* a process known as **costimulation.** Of the more than 20 known costimulators, some are cytokines, such as **interleukin-2.** Other costimulators include pairs of plasma membrane molecules, one on the surface of the T cell and a second on the surface of an antigen-presenting cell, that enable the two cells to adhere to one another for a period of time.

The need for two signals is a little like starting and driving a car: When you insert the correct key (antigen) in the ignition (TCR) and turn it, the car starts (recognition of specific antigen), but it cannot move forward until you move the gear shift into drive (costimulation). The need for costimulation may prevent immune responses from occurring accidentally. Different costimulators affect the activated T cell in different ways, just as shifting a car into reverse has a different effect than shifting it into drive. Moreover, recognition (antigen binding to a receptor) without costimulation leads to a prolonged *state of inactivity* called **anergy** in both T cells and B cells. Anergy is rather like leaving a car in neutral gear with its engine running until it's out of gas!

Once a T cell has received these two signals (antigen recognition and costimulation), it is **activated.** An activated T cell enlarges and begins to **proliferate** (divide several times) and to **differentiate** (form more highly specialized cells). The result is a population of identical cells, called a **clone,** that can recognize the same specific antigen. Before the first exposure to a given antigen, only a few T cells might be able to recognize it, but once an immune response has begun, there are thousands. Activation, proliferation, and differentiation of T cells occur in the secondary lymphatic organs and tissues. The swollen tonsils or lymph nodes in your neck you experienced the last time you were sick were probably caused by the proliferation of lymphocytes participating in an immune response.

Types of T Cells

The three main types of differentiated T cells are helper T cells, cytotoxic T cells, and memory T cells.

Helper T Cells

Most T cells that display CD4 develop into **helper T cells,** also known as **CD4 T cells.** Inactive (resting) helper T cells recognize exogenous antigen fragments associated with major histocompatibility complex class II (MHC-II) molecules at the surface of an APC (Figure 22.14a). With the aid of the CD4 protein, the helper T cell and APC interact with each other (antigenic recognition), costimulation occurs, and the helper T cell becomes activated.

Within hours after costimulation, activated helper T cells start secreting a variety of cytokines (see Table 22.2). Different subsets of helper T cells specialize in the production of particular cytokines. One very important cytokine produced by helper T cells is interleukin-2 (IL-2), which is needed for virtually all immune responses and is the prime trigger of T cell proliferation. IL-2 can act as a costimulator for resting helper T cells or cytotoxic T cells, and it enhances activation and proliferation of T cells, B cells, and natural killer cells.

Some actions of interleukin-2 provide a good example of a beneficial positive feedback system. As noted earlier, activation of a helper T cell stimulates it to start secreting IL-2, which then acts in an autocrine manner by binding to IL-2 receptors on the plasma membrane of the cell that secreted it. One effect is stimulation of cell division. As the helper T cells proliferate, a positive feedback effect occurs because they secrete more IL-2, which causes further cell division. IL-2 may also act in a paracrine manner by binding to IL-2 receptors on neighboring helper T cells, cytotoxic T cells, or B cells. If any of these neighboring cells have already bound an antigen, IL-2 serves as a costimulator.

Cytotoxic T Cells

T cells that display CD8 develop into **cytotoxic T cells,** also termed **CD8 cells.** Cytotoxic T cells recognize foreign antigens combined with major histocompatibility complex class I (MHC-I) molecules on the surfaces of (1) body cells infected by microbes, (2) some tumor cells, and (3) cells of a tissue transplant (Figure 22.14b). Recognition requires the TCR and CD8 protein to maintain the coupling with MHC-I. Following antigenic recognition, costimulation occurs. In order to become activated, cytotoxic T cells require costimulation by interleukin-2 or other cytokines produced by helper T cells. (Recall that helper T cells are activated by antigen associated with MHC-II molecules.) Thus, *maximal activation* of cytotoxic T cells requires presentation of antigen associated with both MHC-I and MHC-II molecules.

Memory T Cells

T cells that remain from a proliferated clone after a cell-mediated immune response are termed **memory T cells.** Should a pathogen bearing the same foreign antigen invade the body later, thousands of memory cells are available to initiate a far swifter reaction than occurred during the first invasion. The

Figure 22.14 Activation, proliferation, and differentiation of T cells.

The binding of CD4 to MHC-II and CD8 to MHC-I helps anchor the T-cell receptor (TCR)–antigen interaction so that antigen recognition can occur.

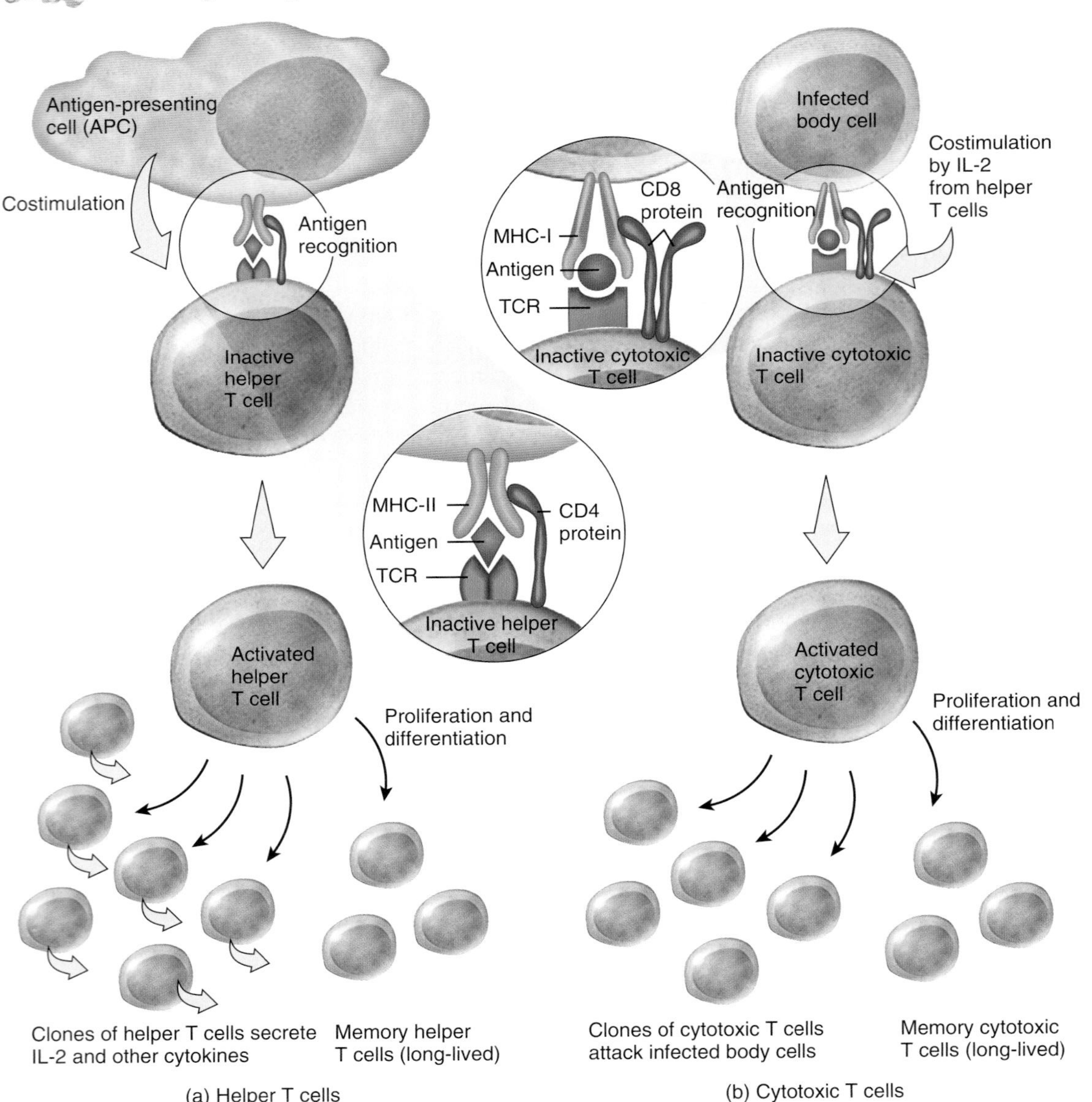

What are the first and second signals in activation of a T cell?

second response usually is so fast and so vigorous that the pathogens are destroyed before any signs or symptoms of disease can occur.

Elimination of Invaders

Cytotoxic T cells are the soldiers that march forth to do battle with foreign invaders in cell-mediated immune responses. They leave secondary lymphatic organs and tissues and migrate to seek out and destroy infected target cells, cancer cells, and transplanted cells (Figure 22.15). Cytotoxic T cells recognize and attach to target cells. Then, the cytotoxic T cells deliver a "lethal hit" that kills the target cells.

Cytotoxic T cells kill infected target body cells much like natural killer cells do. The major difference is that cytotoxic T cells have receptors specific for a particular microbe and thus kill only target body cells infected with *one* particular type of microbe; natural killer cells can destroy a wide variety of microbe-infected body cells. Cytotoxic T cells have two principal mechanisms for killing infected target cells.

1. Cytotoxic T cells, using receptors on their surfaces, recognize and bind to infected target cells that have microbial antigens

Figure 22.15 Activity of cytotoxic T cells. After delivering a "lethal hit," a cytotoxic T cell can detach and attack another infected target cell displaying the same antigen.

Cytotoxic T cells release granzymes that trigger apoptosis and perforin that triggers cytolysis of infected target cells.

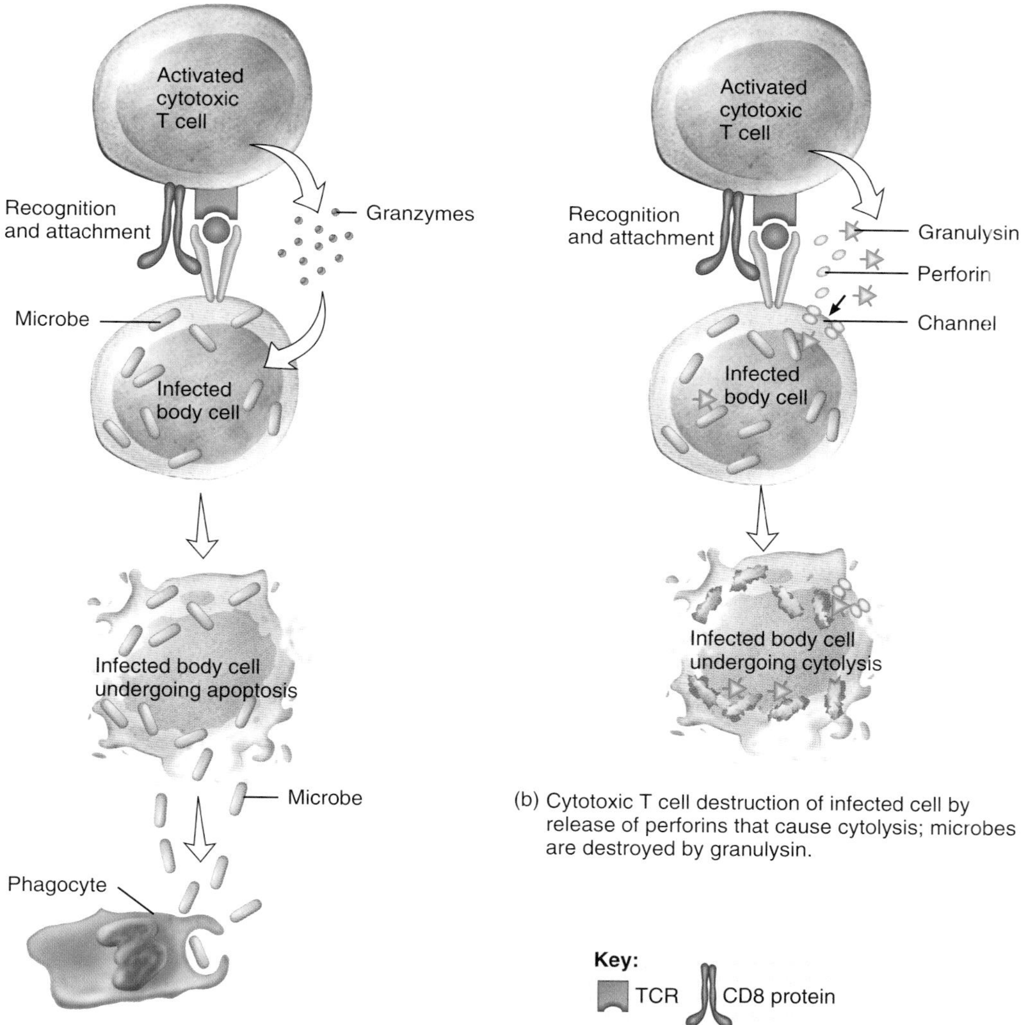

Activated cytotoxic T cell
Recognition and attachment
Granzymes
Microbe
Infected body cell
Infected body cell undergoing apoptosis
Microbe
Phagocyte

(a) Cytotoxic T cell destruction of infected cell by release of granzymes that cause apoptosis; released microbes are destroyed by phagocyte.

Activated cytotoxic T cell
Recognition and attachment
Granulysin
Perforin
Channel
Infected body cell
Infected body cell undergoing cytolysis

(b) Cytotoxic T cell destruction of infected cell by release of perforins that cause cytolysis; microbes are destroyed by granulysin.

Key:
TCR
CD8 protein
Antigen–MHC-I complex

In addition to cells infected by microbes, what other types of target cells are attacked by cytotoxic T cells?

displayed on their surface. The cytotoxic T cell then releases **granzymes,** protein-digesting enzymes that trigger apoptosis (Figure 22.15a). Once the infected cell is destroyed, the released microbes are killed by phagocytes.

2. Alternatively, cytotoxic T cells bind to infected body cells and release two proteins from their granules: perforin and granulysin. **Perforin** inserts into the plasma membrane of the target cell and creates channels in the membrane (Figure 22.15b). As a result, extracellular fluid flows into the target cell and cytolysis (cell bursting) occurs. Other granules in cytotoxic T cells release **granulysin,** which enters through the channels and destroys the microbes by creating holes in their plasma membranes. Cytotoxic T cells may also destroy target cells by releasing a toxic molecule called **lymphotoxin,** which activates enzymes in the target cell. These enzymes cause the target cell's DNA to fragment and the cell dies. In addition, cytotoxic T cells secrete gamma-interferon, which attracts and activates phagocytic cells, and macrophage migration inhibition factor, which prevents migration of phagocytes from the infection site. After detaching from a target cell, a cytotoxic T cell can seek out and destroy another target cell.

Immunological Surveillance

When a normal cell transforms into a cancerous cell, it often displays novel cell surface components called **tumor antigens.** These molecules are rarely, if ever, displayed on the surface of normal cells. If the immune system recognizes a tumor antigen as nonself, it can destroy any cancer cells carrying that antigen. Such immune responses, called **immunological surveillance,** are carried out by cytotoxic T cells, macrophages, and natural killer cells. Immunological surveillance is most effective in eliminating tumor cells due to cancer-causing viruses. For this reason, transplant recipients who are taking immunosuppressive drugs to prevent transplant rejection have an increased incidence of virus-associated cancers. Their risk for other types of cancer is not increased.

Graft Rejection and Tissue Typing

Organ transplantation involves the replacement of an injured or diseased organ, such as the heart, liver, kidney, lungs, or pancreas, with an organ donated by another individual. Usually, the immune system recognizes the proteins in the transplanted organ as foreign and mounts both cell-mediated and antibody-mediated immune responses against them. This phenomenon is known as **graft rejection.**

The success of an organ or tissue transplant depends on **histocompatibility** (his′-tō-kom-pat-i-BIL-i-tē)—that is, the tissue compatibility between the donor and the recipient. The more similar the MHC antigens, the greater the histocompatibility, and thus the greater the probability that the transplant will not be rejected. **Tissue typing (histocompatibility testing)** is done before any organ transplant. In the United States, a nationwide computerized registry helps physicians select the most histocompatible and needy organ transplant recipients whenever donor organs become available. The closer the match between the major histocompatibility complex proteins of the donor and recipient, the weaker is the graft rejection response.

To reduce the risk of graft rejection, organ transplant recipients receive immunosuppressive drugs. One such drug is *cyclosporine,* derived from a fungus, which inhibits secretion of interleukin-2 by helper T cells but has only a minimal effect on B cells. Thus, the risk of rejection is diminished while resistance to some diseases is maintained. ■

CHECKPOINT

16. What are the functions of helper, cytotoxic, and memory T cells?

17. How do cytotoxic T cells kill infected target cells?

18. How is immunological surveillance useful?

ANTIBODY-MEDIATED IMMUNITY

OBJECTIVES

- **Describe the steps in an antibody-mediated immune response.**
- **List the chemical characteristics and actions of antibodies.**
- **Explain how the complement system operates.**
- **Distinguish between a primary response and a secondary response to infection.**

The body contains not only millions of different T cells but also millions of different B cells, each capable of responding to a specific antigen. Cytotoxic T cells leave lymphatic tissues to seek out and destroy a foreign antigen, but B cells stay put. In the presence of a foreign antigen, specific B cells in lymph nodes, the spleen, or mucosa-associated lymphatic tissue become activated. They then differentiate into plasma cells that secrete specific antibodies, which in turn circulate in the lymph and blood to reach the site of invasion.

Activation, Proliferation, and Differentiation of B Cells

During activation of a B cell, an antigen binds to **B-cell receptors (BCRs)** (Figure 22.16). These integral transmembrane proteins are chemically similar to the antibodies that eventually are secreted by plasma cells. Although B cells can respond to an unprocessed antigen present in lymph or interstitial fluid, their response is much more intense when they process the antigen. Antigen processing in a B cell occurs in the following way: the

Figure 22.16 Activation, proliferation, and differentiation of B cells into plasma cells and memory cells. Plasma cells are actually much larger than B cells.

Plasma cells secrete antibodies.

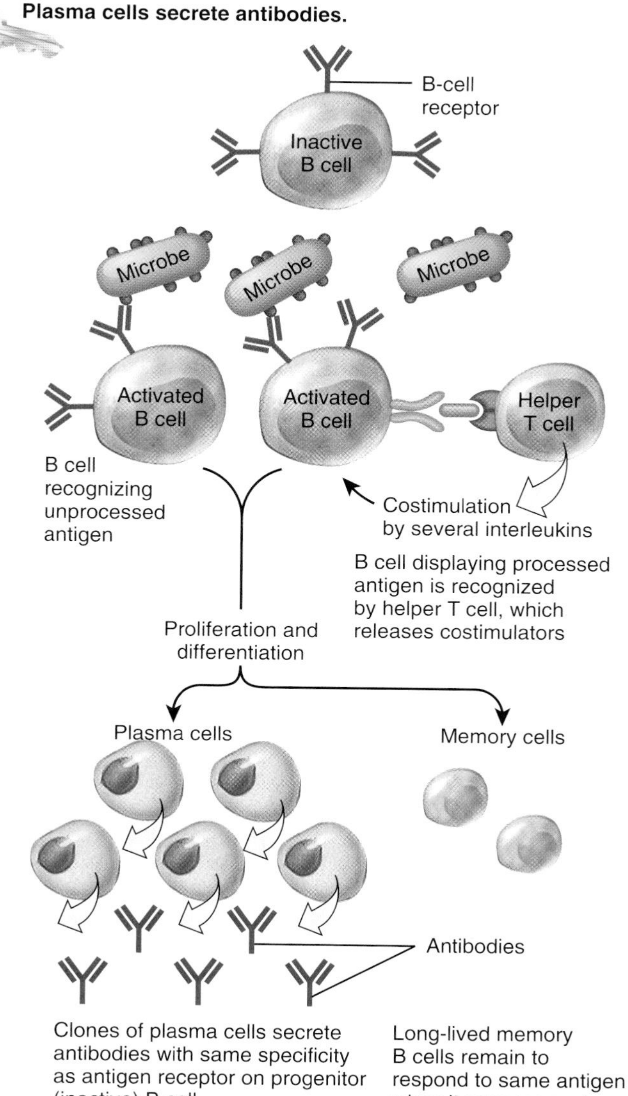

B-cell receptor
Inactive B cell
Microbe
Microbe
Microbe
Activated B cell
Activated B cell
Helper T cell
B cell recognizing unprocessed antigen
Costimulation by several interleukins
B cell displaying processed antigen is recognized by helper T cell, which releases costimulators
Proliferation and differentiation
Plasma cells
Memory cells
Antibodies
Clones of plasma cells secrete antibodies with same specificity as antigen receptor on progenitor (inactive) B cell.
Long-lived memory B cells remain to respond to same antigen when it appears again.

How many different kinds of antibodies will be secreted by the plasma cells in the clone shown here?

antigen is taken into the B cell, broken down into peptide fragments and combined with MHC-II self-antigens, and moved to the B cell plasma membrane. Helper T cells recognize the antigen–MHC-II complex and deliver the costimulation needed for B cell proliferation and differentiation. The helper T cell produces interleukin-2 and other cytokines that function as costimulators to activate B cells. Interleukin-4 and interleukin-6, also produced by helper T cells, enhance B cell proliferation, B cell differentiation into plasma cells, and secretion of antibodies by plasma cells.

Some of the activated B cells enlarge, divide, and differentiate into a clone of antibody-secreting **plasma cells.** A few days after exposure to an antigen, a plasma cell secretes hundreds of millions of antibodies each day for about 4 or 5 days, until the plasma cell dies. Most antibodies travel in lymph and blood to the invasion site. Some activated B cells do not differentiate into plasma cells but remain as **memory B cells** that are ready to respond more rapidly and forcefully should the same antigen reappear at a future time.

Different antigens stimulate different B cells to develop into plasma cells and their accompanying memory B cells. All of the B cells of a particular clone are capable of secreting only one type of antibody, which is identical to the antigen receptor displayed by the B cell that first responded. Each specific antigen activates only those B cells that are predestined (by the combination of gene segments they carry) to secrete antibody specific to that antigen. Antibodies produced by a clone of plasma cells enter the circulation and form antigen–antibody complexes with the antigen that initiated their production.

Antibodies

An **antibody (Ab)** can combine specifically with the epitope on the antigen that triggered its production. The antibody's structure matches its antigen much as a lock accepts a specific key. In theory, plasma cells could secrete as many different antibodies as there are different B-cell receptors because the same recombined gene segments code for both the BCR and the antibodies eventually secreted by plasma cells.

Antibody Structure

Antibodies belong to a group of glycoproteins called globulins, and for this reason they are also known as **immunoglobulins (Igs).** Most antibodies contain four polypeptide chains (Figure 22.17). Two of the chains are identical to each other and are called **heavy (H) chains;** each consists of about 450 amino acids. Short carbohydrate chains are attached to each heavy polypeptide chain. The two other polypeptide chains, also identical to each other, are called **light (L) chains,** and each consists of about 220 amino acids. A disulfide bond (S—S) holds each light chain to a heavy chain. Two disulfide bonds also link the midregion of the two heavy chains; this part of the

Figure 22.17 Chemical structure of the immunoglobulin G (IgG) class of antibody. Each molecule is composed of four polypeptide chains (two heavy and two light) plus a short carbohydrate chain attached to each heavy chain. In (a), each circle represents one amino acid. In (b), V_L = variable regions of light chain, C_L = constant region of light chain, V_H = variable region of heavy chain, and C_H = constant region of heavy chain.

An antibody combines only with the epitope on the antigen that triggered its production.

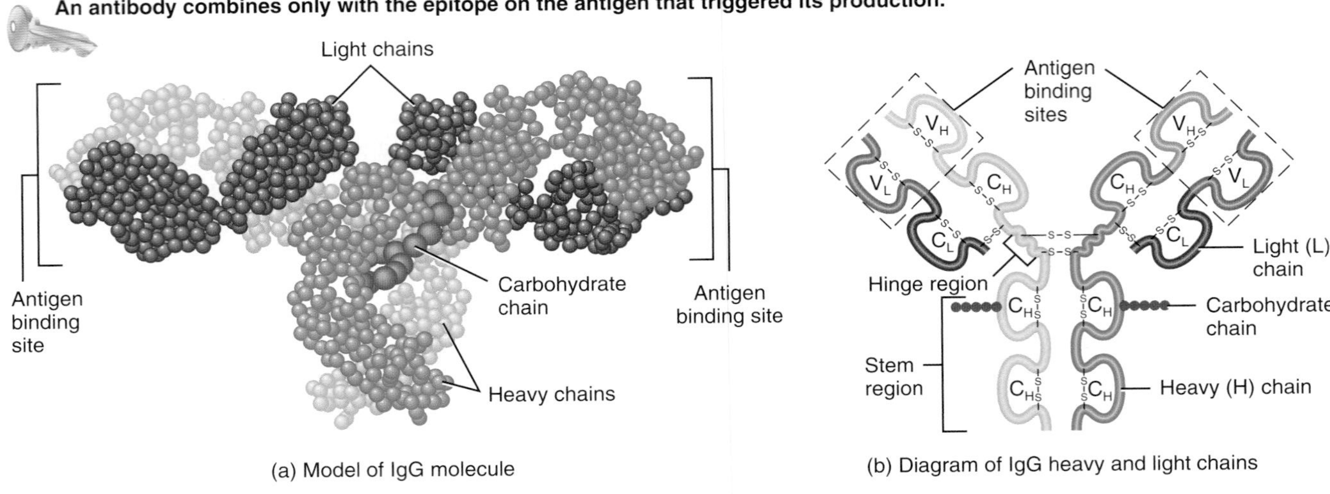

(a) Model of IgG molecule

(b) Diagram of IgG heavy and light chains

What is the function of the variable regions in an antibody molecule?

antibody displays considerable flexibility and is called the **hinge region.** Because the antibody "arms" can move somewhat as the hinge region bends, an antibody can assume either a T shape (Figure 22.17a) or a Y shape (Figure 22.17b). Beyond the hinge region, parts of the two heavy chains form the **stem region.**

Within each H and L chain are two distinct regions. The tips of the H and L chains, called the **variable (V) regions,** constitute the **antigen-binding site.** The variable region, which is different for each kind of antibody, is the part of the antibody that recognizes and attaches specifically to a particular antigen. Because most antibodies have two antigen binding sites, they are said to be bivalent. Flexibility at the hinge allows the antibody to simultaneously bind to two epitopes that are some distance apart—for example, on the surface of a microbe.

The remainder of each H and L chain, called the **constant (C) region,** is nearly the same in all antibodies of the same class and is responsible for the type of antigen–antibody reaction that occurs. However, the constant region of the H chain differs from one class of antibody to another, and its structure serves as a basis for distinguishing five different classes, designated IgG, IgA, IgM, IgD, and IgE. Each class has a distinct chemical structure and a specific biological role. Because they appear first and are relatively short-lived, IgM antibodies indicate a recent invasion. In a sick patient, the responsible pathogen may be suggested by the presence of high levels of IgM specific to a particular organism. Resistance of the fetus and newborn baby to infection stems mainly from maternal IgG antibodies that cross the placenta before birth and IgA antibodies in breast milk after birth. Table 22.3 summarizes the structures and functions of the five classes of antibodies.

Antibody Actions

The actions of the five classes of immunoglobulins differ somewhat, but all of them act to disable antigens in some way. Actions of antibodies include:

- ***Neutralizing antigen.*** The reaction of antibody with antigen blocks or neutralizes some bacterial toxins and prevents attachment of some viruses to body cells.
- ***Immobilizing bacteria.*** If antibodies form against antigens on the cilia or flagella of motile bacteria, the antigen–antibody reaction may cause the bacteria to lose their motility, which limits their spread into nearby tissues.
- ***Agglutinating and precipitating antigen.*** Because antibodies have two or more sites for binding to antigen, the antigen–antibody reaction may cross-link pathogens to one another, causing agglutination (clumping together). Phagocytic cells ingest agglutinated microbes more readily. Likewise, soluble antigens may come out of solution and form a more-easily phagocytized precipitate when cross-linked by antibodies.
- ***Activating complement.*** Antigen–antibody complexes initiate the classical pathway of the complement system (discussed shortly).
- ***Enhancing phagocytosis.*** The stem region of an antibody acts as a "flag" that attracts phagocytes once antigens have bound to the antibody's variable region. Antibodies enhance the activity of phagocytes by causing agglutination and precipitation, by activating complement, and by coating microbes so that they are more susceptible to phagocytosis.

TABLE 22.3 Classes of Immunoglobulins (Igs)

Name and Structure	Characteristics and Functions
IgG	Most abundant, about 80% of all antibodies in the blood; found in blood, lymph, and the intestines; monomer (one-unit) structure. Protects against bacteria and viruses by enhancing phagocytosis, neutralizing toxins, and triggering the complement system. It is the only class of antibody to cross the placenta from mother to fetus, conferring considerable immune protection in newborns.
IgA	Found mainly in sweat, tears, saliva, mucus, breast milk and gastrointestinal secretions. Smaller quantities are present in blood and lymph. Makes up 10–15% of all antibodies in the blood; occurs as monomers and dimers (two units). Levels decrease during stress, lowering resistance to infection. Provides localized protection on mucous membranes against bacteria and viruses.
IgM	About 5–10% of all antibodies in the blood; also found in lymph. Occurs as pentamers (five units); first antibody class to be secreted by plasma cells after an initial exposure to any antigen. Activates complement and causes agglutination and lysis of microbes. Also present as monomers on the surfaces of B cells, where they serve as antigen receptors. In blood plasma, the anti-A and anti-B antibodies of the ABO blood group, which bind to A and B antigens during incompatible blood transfusions, are also IgM antibodies (see Figure 19.12 on page 686).
IgD	Mainly found on the surfaces of B cells as antigen receptors, where it occurs as monomers; involved in activation of B cells. About 0.2% of all antibodies in the blood.
IgE	Less than 0.1% of all antibodies in the blood; occurs as monomers; located on mast cells and basophils. Involved in allergic and hypersensitivity reactions; provides protection against parasitic worms.

Monoclonal Antibodies

The antibodies produced against a given antigen by plasma cells can be harvested from an individual's blood. However, because an antigen typically has many epitopes, several different clones of plasma cells produce different antibodies against the antigen. If a single plasma cell could be isolated and induced to proliferate into a clone of identical plasma cells, then a large quantity of identical antibodies could be produced. Unfortunately, lymphocytes and plasma cells are difficult to grow in culture, so scientists sidestepped this difficulty by fusing B cells with tumor cells that grow easily and proliferate endlessly. The resulting hybrid cell is called a **hybridoma** (hī-bri-DŌ-ma). Hybridomas are long-term sources of large quantities of pure, identical antibodies, called **monoclonal antibodies (MAbs)** because they come from a single clone of identical cells. One clinical use of monoclonal antibodies is for measuring levels of a drug in a patient's blood. Other uses include the diagnosis of strep throat, pregnancy, allergies, and diseases such as hepatitis, rabies, and some sexually transmitted diseases. MAbs have also been used to detect cancer at an early stage and to ascertain the extent of metastasis. They may also be useful in preparing vaccines to counteract the rejection associated with transplants, to treat autoimmune diseases, and perhaps to treat AIDS. ■

Role of the Complement System in Immunity

The **complement system** is a defensive system made up of over 30 proteins produced by the liver and found circulating in blood plasma and within tissues thoughout the body. Collectively, the complement proteins destroy microbes by causing phagocytosis, cytolysis, and inflammation; they also prevent excessive damage to body tissues.

Most complement proteins are designated by an uppercase letter C, numbered C1 through C9, named for the order in which they were discovered. The C1–C9 complement proteins are inactive and only become activated when split by enzymes into active fragments, which are indicated by lowercase letters *a* and *b*. For example, inactive complement protein C3 is split into the activated fragments, C3a and C3b. The active fragments carry out the destructive actions of the C1–C9 complement proteins. Other complement proteins are referred to as factors B, D, and P (properdin).

Complement proteins act in a *cascade*—one reaction triggers another reaction, which, in turn, triggers another reaction, and so on. With each succeeding reaction, more and more product is formed so that the net effect is amplified many times.

Complement activation may begin by three different pathways (described shortly), all of which activate C3. Once activated, C3 begins a cascade of reactions that brings about phagocytosis, cytolysis, and inflammation as follows (Figure 22.18):

1. Inactivated C3 splits into activated C3a and C3b.
2. C3b binds to the surface of a microbe and receptors on phagocytes attach to the C3b. Thus C3b enhances **phagocytosis** by coating a microbe, a process called **opsonization** (op-sō-ni-ZĀ-shun). Opsonization promotes attachment of a phagocyte to a microbe.
3. C3b also initiates a series of reactions that bring about cytolysis. First, C3b splits C5. The C5b fragment then binds to C6 and C7, which attach to the plasma membrane of an invading microbe. Then C8 and several C9 molecules join the other complement proteins and together form a cylinder-shaped **membrane attack complex,** which inserts into the plasma membrane.
4. The membrane attack complex creates channels in the plasma membrane that result in **cytolysis,** the bursting of the microbial cells due to the inflow of extracellular fluid through the channels.
5. C3a and C5a bind to mast cells and cause them to release histamine that increases blood vessel permeability during **inflammation.** C5a also attracts phagocytes to the site of inflammation (chemotaxis).

C3 can be activated in three ways: (1) The **classical pathway** starts when antibodies bind to antigens (microbes). The antigen–antibody complex binds and activates C1. Eventually,

Figure 22.18 Complement activation and results of activation. (Adapted from Tortora, Funke, and Case, *Microbiology: An Introduction, Eighth Edition,* Figure 16.10, Pearson Benjamin-Cummings, 2004.)

When activated, complement proteins enhance phagocytosis, cytolysis, and inflammation.

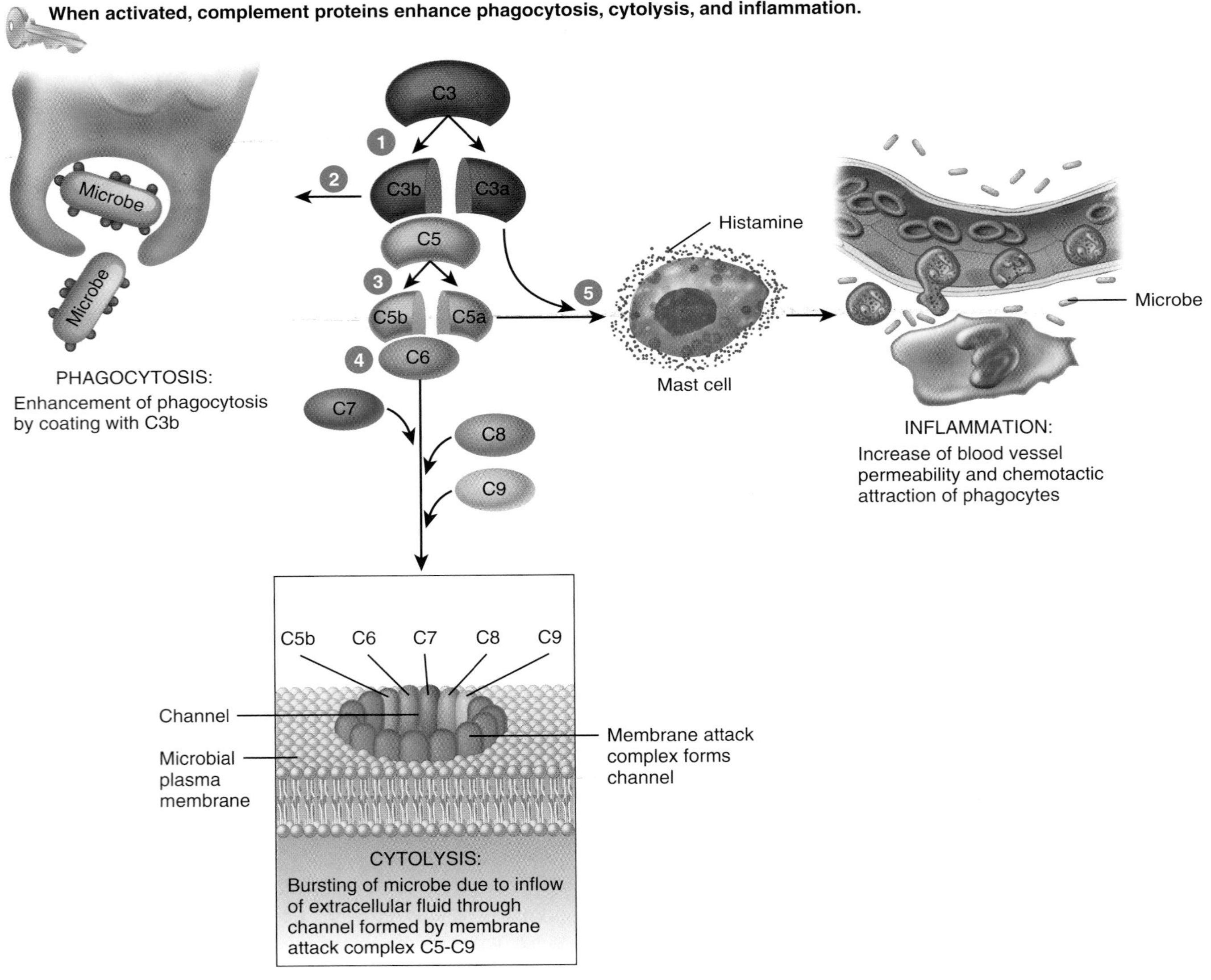

Which pathway for activation of complement involves antibodies? Explain why.

C3 is activated and the C3 fragments initiate phagocytosis, cytolysis, and inflammation. (2) The **alternative pathway** does not involve antibodies. It is initiated by an interaction between lipid–carbohydrate complexes on the surface of microbes and complement protein factors B, D, and P. This interaction activates C3. (3) In the **lectin pathway,** macrophages that digest microbes release chemicals that cause the liver to produce proteins called **lectins.** Lectins bind to the carbohydrates on the surface of microbes, ultimately causing the activation of C3.

Once complement is activated, proteins in blood and on body cells such as blood cells break down activated C3. In this way, its destructive capabilities cease very quickly so that damage to body cells is minimized.

Immunological Memory

A hallmark of immune responses is memory for specific antigens that have triggered immune responses in the past. Immunological memory is due to the presence of long-lasting antibodies and very long-lived lymphocytes that arise during proliferation and differentiation of antigen-stimulated B cells and T cells.

Immune responses, whether cell-mediated or antibody-mediated, are much quicker and more intense after a second or subsequent exposure to an antigen than after the first exposure. Initially, only a few cells have the correct specificity to respond, and the immune response may take several days to build to maximum intensity. Because thousands of memory cells exist after an initial encounter with an antigen, the next time the same antigen appears they can proliferate and differentiate into plasma cells or cytotoxic T cells within hours.

One measure of immunological memory is *antibody titer* (TĪ-ter), the amount of antibody in serum. After an initial contact with an antigen, no antibodies are present for a period of several days. Then, a slow rise in the antibody titer occurs, first IgM and then IgG, followed by a gradual decline in antibody titer (Figure 22.19). This is the **primary response.**

Memory cells may remain for decades. Every new encounter with the same antigen results in a rapid proliferation of memory cells. After subsequent encounters, the antibody titer is far greater than during a primary response and consists mainly of IgG antibodies. This accelerated, more intense response is called the **secondary response.** Antibodies produced during a secondary response have an even higher affinity for the antigen than those produced during a primary response, and thus they are more successful in disposing of it.

Primary and secondary responses occur during microbial infection. When you recover from an infection without taking antimicrobial drugs, it is usually because of the primary response. If the same microbe infects you later, the secondary response could be so swift that the microbes are destroyed before you exhibit any signs or symptoms of infection.

Immunological memory provides the basis for immunization by vaccination against certain diseases (for example, polio).

Figure 22.19 Production of antibodies in the primary (after first exposure) and secondary (after second exposure) responses to a given antigen.

Immunological memory is the basis for successful immunization by vaccination.

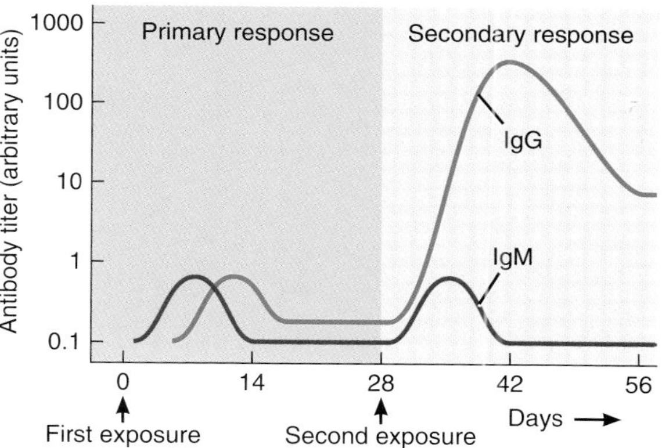

According to this graph, how much more IgG is circulating in the blood in the secondary response than in the primary response? (*Hint:* Notice that each mark on the antibody titer axis represents a 10-fold increase.)

When you receive the vaccine, which may contain **attenuated** (weakened) or killed whole microbes or portions of microbes, your B cells and T cells are activated. Should you subsequently encounter the living pathogen as an infecting microbe, your body initiates a secondary response. Table 22.4 summarizes the various types of antigen encounters that provide naturally and artificially acquired immunity.

TABLE 22.4 Types of Immunity

Type of Immunity	How Acquired
Naturally acquired active immunity	Following exposure to a microbe, antigen recognition by B cells and T cells and costimulation lead to antibody-secreting plasma cells, cytotoxic T cells, and B and T memory cells.
Naturally acquired passive immunity	Transfer of IgG antibodies from mother to fetus across placenta, or of IgA antibodies from mother to baby in milk during breast-feeding.
Artificially acquired active immunity	Antigens introduced during a vaccination stimulate cell-mediated and antibody-mediated immune responses, leading to production of memory cells. The antigens are pretreated to be immunogenic but not pathogenic; that is, they will trigger an immune response but not cause significant illness.
Artificially acquired passive immunity	Intravenous injection of immunoglobulins (antibodies).

▶ CHECKPOINT

19. How do the five classes of antibodies differ in structure and function?
20. How are cell-mediated and antibody-mediated immune responses similar and different?
21. In what ways does the complement system augment antibody-mediated immune responses?
22. How is the secondary response to an antigen different from the primary response?

SELF-RECOGNITION AND SELF-TOLERANCE

▶ OBJECTIVE

Describe how self-recognition and self-tolerance develop.

To function properly, your T cells must have two traits: (1) They must be able to *recognize* your own major histocompatibility complex (MHC) proteins, a process known as **self-recognition,** and (2) they must *lack reactivity* to peptide fragments from your own proteins, a condition known as **self-tolerance** (Figure 22.20). B

Figure 22.20 **Development of self-recognition and self-tolerance.** MHC = major histocompatibility complex. TCR = T-cell receptor.

Positive selection allows recognition of self-MHC proteins; negative selection provides self-tolerance of your own peptides and other self-antigens.

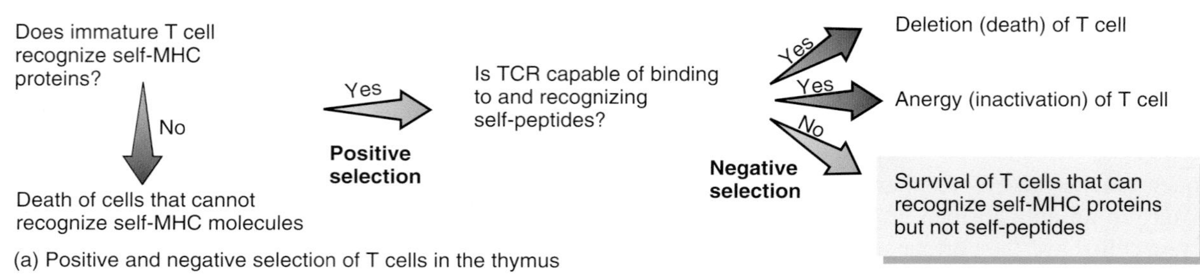

(a) Positive and negative selection of T cells in the thymus

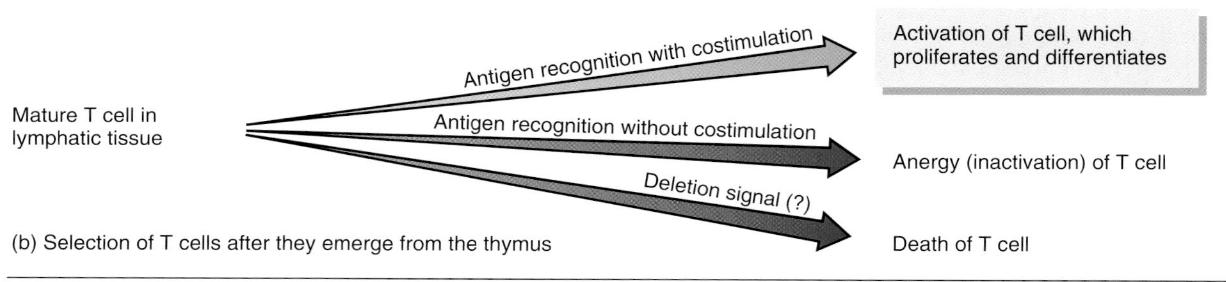

(b) Selection of T cells after they emerge from the thymus

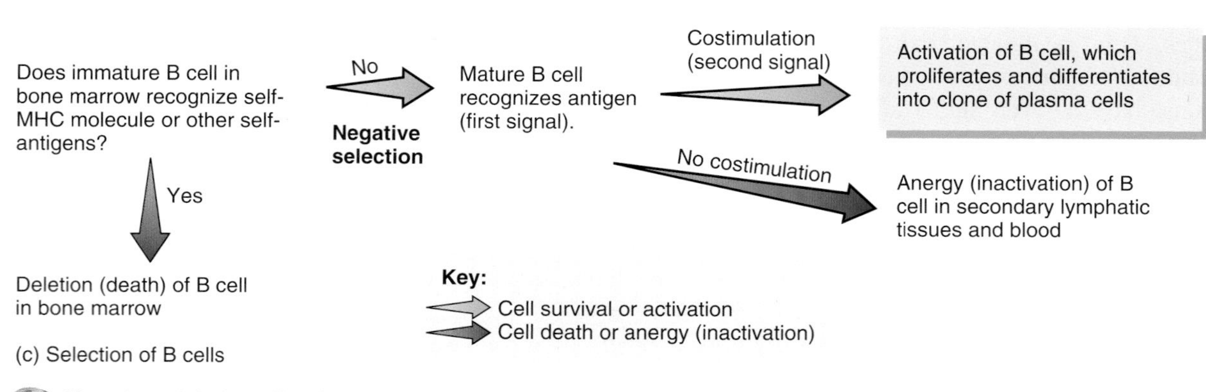

(c) Selection of B cells

How does deletion differ from anergy?

cells also display self-tolerance. Loss of self-tolerance leads to the development of autoimmune diseases (see page 839).

Pre-T cells in the thymus develop the capability for self-recognition via **positive selection** (Figure 22.20a). In this process, some pre-T cells express T-cell receptors (TCRs) that interact with self-MHC proteins on epithelial cells in the thymic cortex. Because of this interaction, the T cells can recognize the MHC part of an antigen–MHC complex. These T cells survive. Other immature T cells that fail to interact with thymic epithelial cells are not able to recognize self-MHC proteins. These cells undergo apoptosis.

The development of self-tolerance occurs by a weeding-out process called **negative selection** in which the T cells interact with dendritic cells located at the junction of the cortex and medulla in the thymus. In this process, T cells with receptors that recognize self-peptide fragments or other self-antigens are eliminated or inactivated (Figure 22.20a). The T cells selected to survive do not respond to self-antigens, the fragments of molecules that are normally present in the body. Negative selection occurs via both deletion and anergy. In **deletion,** self-reactive T cells undergo apoptosis and die; in anergy they remain alive but are unresponsive to antigenic stimulation. Only 1–5% of the immature T cells in the thymus receive the proper signals to survive apoptosis during both positive and negative selection and emerge as mature, immunocompetent T cells.

Once T cells have emerged from the thymus, they may still encounter an unfamiliar self-protein; in such cases they may also become anergic if there is no costimulator (Figure 22.20b). Deletion of self-reactive T cells may also occur after they leave the thymus.

B cells also develop tolerance through deletion and anergy (Figure 22.20c). While B cells are developing in bone marrow, those cells exhibiting antigen receptors that recognize common self-antigens (such as MHC proteins or blood group antigens) are deleted. Once B cells are released into the blood, however, anergy appears to be the main mechanism for preventing responses to self-proteins. When B cells encounter an antigen not associated with an antigen-presenting cell, the necessary costimulation signal often is missing. In this case, the B cell is likely to become anergic (inactivated) rather than activated.

Table 22.5 summarizes the activities of cells involved in immune responses.

Cancer Immunology

Although the immune system responds to cancerous cells, often immunity provides inadequate protection, as evidenced by the number of people dying each year from cancer. Over the past 25 years, considerable research has focused on *cancer immunology,* the study of ways to use immune responses for detecting, monitoring, and treating cancer. For example, some tumors of the colon release *carcinoembryonic antigen (CEA)* into the blood and prostate cancer cells release *prostate-specific antigen (PSA).* Detecting these antigens in blood does not provide definitive diagnosis of cancer, because both antigens are also released in certain noncancerous conditions. However, high levels of cancer-related antigens in the blood often do indicate the presence of a malignant tumor.

Finding ways to induce our immune system to mount vigorous attacks against cancerous cells has been an elusive goal. Many different techniques have been tried, with only modest success. In one method, inactive lymphocytes are removed in a blood sample and cultured with interleukin-2. The resulting

TABLE 22.5 Summary of Functions of Cells Participating in Immune Responses

Cell	Functions
Antigen-Presenting Cells (APCs)	
Macrophage	Phagocytosis; processing and presentation of foreign antigens to T cells; secretion of interleukin-1, which stimulates secretion of interleukin-2 by helper T cells and induces proliferation of B cells; secretion of interferons that stimulate T cell growth.
Dendritic cell	Processes and presents antigen to T cells and B cells; found in mucous membranes, skin, and lymph nodes.
B cell	Processes and presents antigen to helper T cells.
Lymphocytes	
Cytotoxic T cell	Kills host target cells by releasing granzymes that induce apoptosis, perforin that forms channels to cause cytolysis, granulysin that destroys microbes, lymphotoxin that destroys target cell DNA, gamma-interferon that attracts macrophages and increases their phagocytic activity, and macrophage migration inhibition factor that prevents macrophage migration from site of infection.
Helper T cell	Cooperates with B cells to amplify antibody production by plasma cells and secretes interleukin-2, which stimulates proliferation of T cells and B cells. May secrete gamma-IFN and tumor necrosis factor (TNF), which stimulate inflammatory response.
Memory T cell	Remains in lymphatic tissue and recognizes original invading antigens, even years after the first encounter.
B cell	Differentiates into antibody-producing plasma cell.
Plasma cell	Descendant of B cell that produces and secretes antibodies.
Memory B cell	Descendant of B cell that remains after an immune response and is ready to respond rapidly and forcefully should the same antigen enter the body in the future.

lymphokine-activated killer (LAK) cells are then transfused back into the patient's blood. Although LAK cells have produced dramatic improvement in a few cases, severe complications affect most patients. In another method, lymphocytes procured from a small biopsy sample of a tumor are cultured with interleukin-2. After their proliferation in culture, such *tumor-infiltrating lymphocytes (TILs)* are reinjected. About a quarter of patients with malignant melanoma and renal-cell carcinoma who received TIL therapy showed significant improvement. The many studies currently underway provide reason to hope that immune-based methods will eventually lead to cures for cancer. ■

▶ CHECKPOINT

23. What do positive selection, negative selection, and anergy accomplish?

STRESS AND IMMUNITY

▶ OBJECTIVE

Describe the effects of stress on immunity.

The field of **psychoneuroimmunology (PNI)** deals with communication pathways that link the nervous, endocrine, and immune systems. PNI research appears to justify what people have long observed: Your thoughts, feelings, moods, and beliefs influence your level of health and the course of disease. For example, cortisol, a hormone secreted by the adrenal cortex in association with the stress response, inhibits immune system activity.

If you want to observe the relationship between lifestyle and immune function, visit a college campus. As the semester progresses and the workload accumulates, an increasing number of students can be found in the waiting rooms of student health services. When work and stress pile up, health habits can change. Many people smoke or consume more alcohol when stressed, two habits detrimental to optimal immune function. Under stress, people are less likely to eat well or exercise regularly, two habits that enhance immunity.

People resistant to the negative health effects of stress are more likely to experience a sense of control over the future, a commitment to their work, expectations of generally positive outcomes for themselves, and feelings of social support. To increase your stress resistance, cultivate an optimistic outlook, get involved in your work, and build good relationships with others.

Adequate sleep and relaxation are especially important for a healthy immune system. But when there aren't enough hours in the day, you may be tempted to steal some from the night. While skipping sleep may give you a few more hours of productive time in the short run, in the long run you end up even farther behind, especially if getting sick keeps you out of commission for several days, blurs your concentration, and blocks your creativity.

Even if you make time to get eight hours of sleep, stress can cause insomnia. If you find yourself tossing and turning at night, it's time to improve your stress management and relaxation skills! Be sure to unwind from the day before going to bed.

▶ CHECKPOINT

24. Have you ever observed a connection between stress and illness in your own life?

AGING AND THE IMMUNE SYSTEM

▶ OBJECTIVE

Describe the effects of aging on the immune system.

With advancing age, most people become more susceptible to all types of infections and malignancies. Their response to vaccines is decreased, and they tend to produce more autoantibodies (antibodies against their body's own molecules). In addition, the immune system exhibits lowered levels of function. For example, T cells become less responsive to antigens, and fewer T cells respond to infections. This may result from age-related atrophy of the thymus or decreased production of thymic hormones. Because the T cell population decreases with age, B cells are also less responsive. Consequently, antibody levels do not increase as rapidly in response to a challenge by an antigen, resulting in increased susceptibility to various infections. It is for this key reason that elderly individuals are encouraged to get influenza (flu) vaccinations each year.

▶ CHECKPOINT

25. How are T cells affected by aging?

• • •

To appreciate the many ways that the lymphatic system contributes to homeostasis of other body systems, examine *Focus on Homeostasis: The Lymphatic System and Immunity* on page 837.

Next, in Chapter 23, we will explore the structure and function of the respiratory system and see how its operation is regulated by the nervous system. Most importantly, the respiratory system provides for gas exchange—taking in oxygen and blowing off carbon dioxide. The cardiovascular system aids gas exchange by transporting blood containing these gases between the lungs and tissue cells.

FOCUS ON HOMEOSTASIS

BODY SYSTEM	CONTRIBUTION OF THE LYMPHATIC SYSTEM AND IMMUNITY
For all body systems 	B cells, T cells, and antibodies protect all body systems from attack by harmful foreign invaders (pathogens), foreign cells, and cancer cells.
Integumentary system 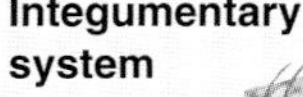 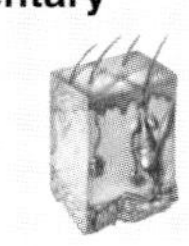	Lymphatic vessels drain excess interstitial fluid and leaked plasma proteins from dermis of skin. Immune system cells (Langerhans cells) in skin help protect skin. Lymphatic tissue provides IgA antibodies in sweat.
Skeletal system 	Lymphatic vessels drain excess interstitial fluid and leaked plasma proteins from connective tissue around bones.
Muscular system 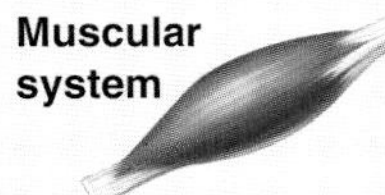	Lymphatic vessels drain excess interstitial fluid and leaked plasma proteins from muscles.
Endocrine system 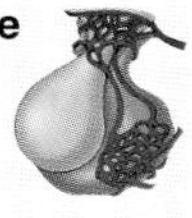	Flow of lymph helps distribute some hormones and cytokines. Lymphatic vessels drain excess interstitial fluid and leaked plasma proteins from endocrine glands.
Cardiovascular system 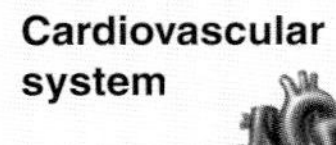 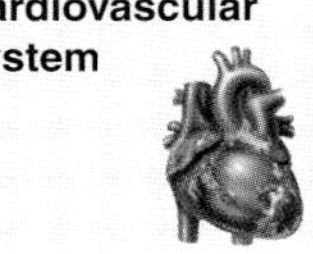	Lymph returns excess fluid filtered from blood capillaries and leaked plasma proteins to venous blood. Macrophages in spleen destroy aged red blood cells and remove debris in blood.
Respiratory system 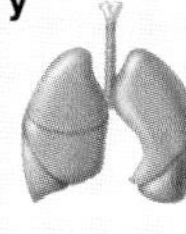	Tonsils, alveolar macrophages, and MALT (mucosa-associated lymphatic tissue) help protect lungs from pathogens. Lymphatic vessels drain excess interstitial fluid from lungs.
Digestive system 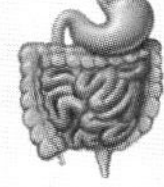	Tonsils and MALT help defend against toxins and pathogens that penetrate the body from the gastrointestinal tract. Digestive system provides IgA antibodies in saliva and gastrointestinal secretions. Lymphatic vessels pick up absorbed dietary lipids and fat-soluble vitamins from the small intestine and transport them to the blood. Lymphatic vessels drain excess interstitial fluid and leaked plasma proteins from organs of the digestive system.
Urinary system 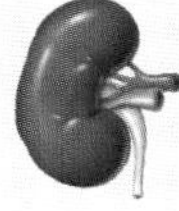	Lymphatic vessels drain excess interstitial fluid and leaked plasma proteins from organs of the urinary system. MALT helps defend against toxins and pathogens that penetrate the body via the urethra.
Reproductive systems 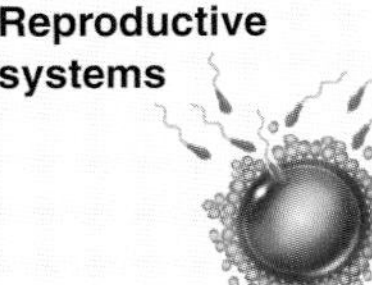	Lymphatic vessels drain excess interstitial fluid and leaked plasma proteins from organs of the reproductive system. MALT helps defend against toxins and pathogens that penetrate the body via the vagina and penis. In females, sperm deposited in the vagina are not attacked as foreign invaders due to inhibition of immune responses. IgG antibodies can cross the placenta to provide protection to a developing fetus. Lymphatic tissue provides IgA antibodies in the milk of a nursing mother.

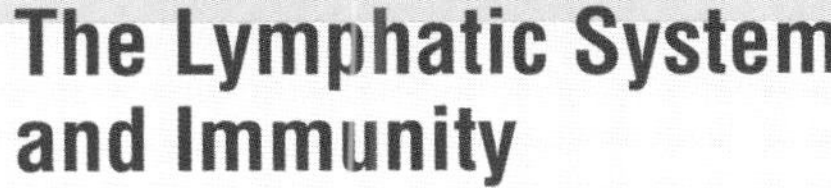

The Lymphatic System and Immunity

DISORDERS: HOMEOSTATIC IMBALANCES

AIDS: Acquired Immunodeficiency Syndrome

Acquired immunodeficiency syndrome (AIDS) is a condition in which a person experiences a telltale assortment of infections due to the progressive destruction of immune system cells by the **human immunodeficiency virus (HIV).** AIDS represents the end stage of infection by HIV. A person who is infected with HIV may be symptom-free for many years, even while the virus is actively attacking the immune system. In the two decades after the first five cases were reported in 1981, 22 million people died of AIDS. Worldwide, 35 to 40 million people are currently infected with HIV.

HIV Transmission

Because HIV is present in the blood and some body fluids, it is most effectively transmitted (spread from one person to another) by actions or practices that involve the exchange of blood or body fluids between people. HIV is transmitted in semen or vaginal fluid during unprotected (without a condom) anal, vaginal, or oral sex. HIV also is transmitted by direct blood-to-blood contact, such as occurs among intravenous drug users who share hypodermic needles or health-care professionals who may be accidentally stuck by HIV-contaminated hypodermic needles. In addition, HIV can be transmitted from an HIV-infected mother to her baby at birth or during breast-feeding.

The chance of transmitting or of being infected by HIV during vaginal or anal intercourse can be greatly reduced—although not entirely eliminated—by the use of latex condoms. Public health programs aimed at encouraging drug users not to share needles have proven effective at checking the increase in new HIV infections in this population. Also, giving certain drugs to pregnant HIV-infected women greatly reduces the risk of transmission of the virus to their babies.

HIV is a very fragile virus; it cannot survive for long outside the human body. The virus is not transmitted by insect bites. One cannot become infected by casual physical contact with an HIV-infected person, such as by hugging or sharing household items. The virus can be eliminated from personal care items and medical equipment by exposing them to heat (135°F for 10 minutes) or by cleaning them with common disinfectants such as hydrogen peroxide, rubbing alcohol, household bleach, or germicidal cleansers such as Betadine or Hibiclens. Standard dishwashing and clothes washing also kills HIV.

HIV: Structure and Infection

HIV consists of an inner core of ribonucleic acid (RNA) covered by a protein coat (capsid). HIV is classified as a **retrovirus** since its genetic information is carried in RNA instead of DNA. Surrounding the HIV capsid is an envelope composed of a lipid bilayer that is penetrated by glycoproteins (Figure 22.21).

Figure 22.21 Human immunodeficiency virus (HIV), the causative agent of AIDS.

HIV is most effectively transmitted by practices that involve the exchange of body fluids.

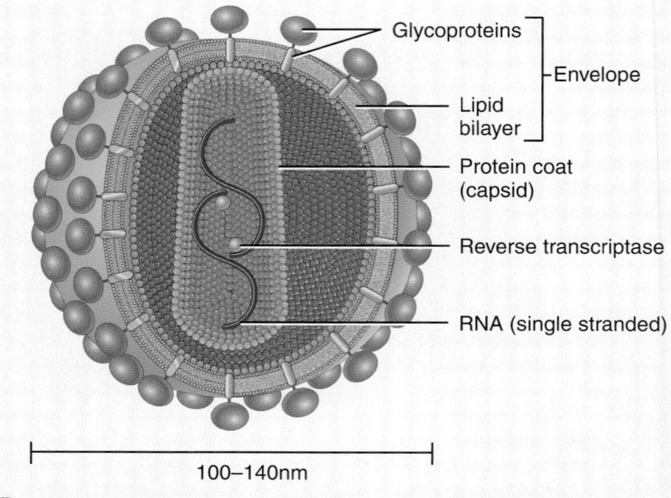

Which cells of the immune system are attacked by HIV?

Outside a living host cell, a virus is unable to replicate. However, when a virus infects and enters a host cell, it uses the host cell's enzymes and ribosomes to make thousands of copies of the virus. New viruses eventually leave and then infect other cells. HIV infection of a host cell begins with the binding of HIV glycoproteins to receptors in the host cell's plasma membrane. This causes the cell to transport the virus into its cytoplasm via receptor-mediated endocytosis. Once inside the host cell, HIV sheds its protein coat and a viral enzyme called **reverse transcriptase** reads the viral RNA strand and makes a DNA copy. The viral DNA copy then becomes integrated into the host cell's DNA. Thus, the viral DNA is duplicated along with the host cell's DNA during normal cell division. In addition, the viral DNA can cause the infected cell to begin producing millions of copies of viral RNA and to assemble new protein coats for each copy. The new HIV copies bud off from the cell's plasma membrane and circulate in the blood to infect other cells.

HIV mainly damages helper T cells, and it does so in various ways. Over 10 billion viral copies may be produced each day. The viruses bud so rapidly from an infected cell's plasma membrane that cell lysis eventually occurs. In addition, the body's defenses attack the infected cells, killing them but not all the viruses they harbor. In most HIV-infected individuals, helper T cells are initially replaced as fast as they are destroyed. After several years, however, the body's ability to replace helper T cells is slowly exhausted, and the number of helper T cells in circulation progressively declines.

Signs, Symptoms, and Diagnosis of HIV Infection

Soon after being infected with HIV, most people experience a brief flu-like illness. Common signs and symptoms are fever, fatigue, rash, headache, joint pain, sore throat, and swollen lymph nodes. About 50% of infected people also experience night sweats. As early as three to four weeks after HIV infection, plasma cells begin secreting antibodies against HIV. These antibodies are detectable in blood plasma and form the basis for some of the screening tests for HIV. When people test "HIV-positive," it usually means they have antibodies to HIV antigens in their bloodstream.

Progression to AIDS

After a period of 2 to 10 years, HIV destroys enough helper T cells that most infected people begin to experience symptoms of immunodeficiency. HIV-infected people commonly have enlarged lymph nodes and experience persistent fatigue, involuntary weight loss, night sweats, skin rashes, diarrhea, and various lesions of the mouth and gums. In addition, the virus may begin to infect neurons in the brain, affecting the person's memory and producing visual disturbances.

As the immune system slowly collapses, an HIV-infected person becomes susceptible to a host of *opportunistic infections.* These are diseases caused by microorganisms that are normally held in check but now proliferate because of the defective immune system. AIDS is diagnosed when the helper T cell count drops below 200 cells per microliter (= cubic millimeter) of blood or when opportunistic infections arise, whichever occurs first. In time, opportunistic infections usually are the cause of death.

Treatment of HIV Infection

At present, infection with HIV cannot be cured. Vaccines designed to block new HIV infections and to reduce the viral load (the number of copies of HIV RNA in a microliter of blood plasma) in those who are already infected are in clinical trials. Meanwhile, two categories of drugs have proved successful in extending the life of many HIV-infected people:

1. **Reverse transcriptase inhibitors** interfere with the action of the reverse transcriptase enzyme, the enzyme that the virus uses to convert its RNA into a DNA copy. Among the drugs in this category are zidovudine (ZDV, previously called AZT), didanosine (ddI), and stavudine (d4T®). Trizivir, approved in 2000 for treatment of HIV infection, combines three reverse transcriptase inhibitors in one pill.
2. **Protease inhibitors** interfere with the action of protease, a viral enzyme that cuts proteins into pieces to assemble the protein coat of newly produced HIV particles. Drugs in this category include nelfinavir, saquinavir, ritonavir, and indinavir.

In 1996, physicians treating HIV-infected patients widely adopted *highly active antiretroviral therapy (HAART)*—a combination of two differently acting reverse transcriptase inhibitors and one protease inhibitor. Most HIV-infected individuals receiving HAART experience a drastic reduction in viral load and an increase in the number of helper T cells in their blood. Not only does HAART delay the progression of HIV infection to AIDS, but many individuals with AIDS have seen the remission or disappearance of opportunistic infections and an apparent return to health. Unfortunately, HAART is very costly (exceeding $10,000 per year), the dosing schedule is grueling, and not all people can tolerate the toxic side effects of these drugs. Although HIV may virtually disappear from the blood with drug treatment (and thus a blood test may be "negative" for HIV), the virus typically still lurks in various lymphatic tissues. In such cases, the infected person can still transmit the virus to another person.

Allergic Reactions

A person who is overly reactive to a substance that is tolerated by most other people is said to be **allergic** or **hypersensitive.** Whenever an allergic reaction takes place, some tissue injury occurs. The antigens that induce an allergic reaction are called **allergens.** Common allergens include certain foods (milk, peanuts, shellfish, eggs), antibiotics (penicillin, tetracycline), vaccines (pertussis, typhoid), venoms (honeybee, wasp, snake), cosmetics, chemicals in plants such as poison ivy, pollens, dust, molds, iodine-containing dyes used in certain x-ray procedures, and even microbes.

There are four basic types of hypersensitivity reactions: type I (anaphylactic), type II (cytotoxic), type III (immune-complex), and type IV (cell-mediated). The first three are antibody-mediated immune responses; the last is a cell-mediated immune response.

Type I (anaphylactic) reactions are the most common and occur within a few minutes after a person sensitized to an allergen is reexposed to it. In response to the first exposure to certain allergens, some people produce IgE antibodies that bind to the surface of mast cells and basophils. The next time the same allergen enters the body, it attaches to the IgE antibodies already present. In response, the mast cells and basophils release histamine, prostaglandins, leukotrienes, and kinins. Collectively, these mediators cause vasodilation, increased blood capillary permeability, increased smooth muscle contraction in the airways of the lungs, and increased mucus secretion. As a result, a person may experience inflammatory responses, difficulty in breathing through the constricted airways, and a runny nose from excess mucus secretion. In **anaphylactic shock,** which may occur in a susceptible individual who has just received a triggering drug or been stung by a wasp, wheezing and shortness of breath as airways constrict are usually accompanied by shock due to vasodilation and fluid loss from blood. This life-threatening emergency is usually treated by injecting epinephrine to dilate the airways and strengthen the heartbeat.

Type II (cytotoxic) reactions are caused by antibodies (IgG or IgM) directed against antigens on a person's blood cells (red blood cells, lymphocytes, or platelets) or tissue cells. The reaction of antibodies and antigens usually leads to activation of complement. Type II reactions, which may occur in incompatible blood transfusion reactions, damage cells by causing lysis.

Type III (immune-complex) reactions involve antigens, antibodies (IgA or IgM), and complement. When certain ratios of antigen to antibody occur, the immune complexes are small enough to escape phagocytosis, but they become trapped in the basement membrane under the endothelium of blood vessels, where they activate complement and cause inflammation. Glomerulonephritis and rheumatoid arthritis (RA) arise in this way.

Type IV (cell-mediated) reactions or **delayed hypersensitivity reactions** usually appear 12–72 hours after exposure to an allergen. Type IV reactions occur when allergens are taken up by antigen-presenting cells (such as Langerhans cells in the skin) that migrate to lymph nodes and present the allergen to T cells, which then proliferate. Some of the new T cells return to the site of allergen entry into the body, where they produce gamma-interferon, which activates macrophages, and tumor necrosis factor, which stimulates an inflammatory response. Intracellular bacteria such as *Mycobacterium tuberculosis* trigger this type of cell-mediated immune response, as do certain haptens, such as poison ivy toxin. The skin test for tuberculosis also is a delayed hypersensitivity reaction.

Autoimmune Diseases

In an **autoimmune disease** (aw-tō-i-MŪN) or **autoimmunity,** the immune system fails to display self-tolerance and attacks the person's own tissues. Autoimmune diseases usually arise in early adulthood and are common, afflicting an estimated 5% of adults in North America and Europe. Females suffer autoimmune diseases twice as often as males. Recall that self-reactive B cells and T cells normally are deleted or undergo anergy during negative selection (see Figure 22.20). Apparently, this process is not 100% effective. Under the influence of unknown environmental triggers and certain genes that make some people more susceptible, self-tolerance breaks down, leading to activation of self-reactive clones of T cells and B cells. These cells then generate cell-mediated or antibody-mediated immune responses against self-antigens.

A variety of mechanisms produce different autoimmune diseases. Some involve production of **autoantibodies,** antibodies that bind to and stimulate or block self-antigens. For example, autoantibodies that mimic TSH (thyroid-stimulating hormone) are present in Graves disease and stimulate secretion of thyroid hormones (thus producing hyperthyroidism); autoantibodies that bind to and block acetylcholine receptors cause the muscle weakness characteristic of myasthenia gravis. Other autoimmune diseases involve activation of cytotoxic T cells that destroy certain body cells. Examples include type 1 diabetes mellitus, in which T cells attack the insulin-producing pancreatic beta cells, and multiple sclerosis (MS), in which T cells attack myelin sheaths around axons of neurons. Inappropriate activation of helper T cells or excessive production of gamma-interferon also occur in certain autoimmune diseases. Other autoimmune disorders include rheumatoid arthritis (RA), systemic lupus erythematosus (SLE), rheumatic fever, hemolytic and pernicious anemias, Addison's disease, Hashimoto's thyroiditis, and ulcerative colitis.

Therapies for various autoimmune diseases include removal of the thymus gland (thymectomy), injections of beta interferon, immunosuppressive drugs, and plasmapheresis, in which the person's blood plasma is filtered to remove antibodies and antigen–antibody complexes.

Infectious Mononucleosis

Infectious mononucleosis or "mono" is a contagious disease caused by the *Epstein–Barr virus (EBV).* It occurs mainly in children and young adults, and more often in females than in males. The virus most commonly enters the body through intimate oral contact such as kissing, which accounts for its common name, the "kissing disease." EBV then multiplies in lymphatic tissues and filters into the blood, where it infects and multiplies in B cells, the primary host cells. Because of this infection, the B cells become so enlarged and abnormal in appearance that they resemble monocytes, the primary reason for the term **mononucleosis.** In addition to an elevated white blood cell count with an abnormally high percentage of lymphocytes, signs and symptoms include fatigue, headache, dizziness, sore throat, enlarged and tender lymph nodes, and fever. There is no cure for infectious mononucleosis, but the disease usually runs its course in a few weeks.

Lymphomas

Lymphomas (lim-FŌ-mas; *lymph-* = clear water; *-oma* = tumor) are cancers of the lymphatic organs, especially the lymph nodes. Most have no known cause. The two main types of lymphomas are Hodgkin disease and non-Hodgkin lymphoma.

Hodgkin disease (HD) is characterized by a painless, nontender enlargement of one or more lymph nodes, most commonly in the neck, chest, and axilla. If the disease has metastasized from these sites, fever, night sweats, weight loss, and bone pain also occur. HD primarily affects individuals between ages 15 and 35 and those over 60, and it is more common in males. If diagnosed early, HD has a 90–95% cure rate.

Non-Hodgkin lymphoma (NHL), which is more common than HD, occurs in all age groups, the incidence increasing with age to a maximum between ages 45 and 70. NHL may start the same way as HD but may also include an enlarged spleen, anemia, and general malaise. Up to half of all individuals with NHL are cured or survive for a lengthy period. Treatment options for both HD and NHL include radiation therapy, chemotherapy, and bone marrow transplantation.

Systemic Lupus Erythematosus

Systemic lupus erythematosus (er-e′-thēm-a-TŌ-sus), **SLE,** or simply **lupus** (*lupus* = wolf) is a chronic autoimmune, inflammatory disease that affects multiple body systems. Lupus is characterized by periods of active disease and remission; symptoms range from mild to life-threatening. Lupus most often develops between ages 15 and 44 and is 10–15 times more common in females than males. It is also 2–3 times more common in African-Americans, Hispanics, Asian-Americans, and Native Americans than in European-Americans. Although the cause of SLE is not known, both a genetic predisposition to the disease and environmental factors (infections, antibiotics, ultraviolet light, stress, and hormones) may trigger it. Sex hormones appear to influence the development of SLE. The disorder often occurs in females who exhibit extremely low levels of androgens.

Signs and symptoms of SLE include joint pain, muscle pain, chest pain with deep breaths, headaches, pale or purple fingers or toes, kidney dysfunction, low blood cell count, nerve or brain dysfunction, slight fever, fatigue, oral ulcers, weight loss, swelling in the legs or around the eyes, enlarged lymph nodes and spleen, photosensitivity, rapid loss of large amounts of scalp hair, and sometimes an eruption across the bridge of the nose and cheeks called a "butterfly rash." The erosive nature of some of the SLE skin lesions was thought to resemble the damage inflicted by the bite of a wolf—thus, the term lupus.

Two immunological features of SLE are excessive activation of B cells and inappropriate production of autoantibodies against DNA (anti-DNA antibodies) and other components of cellular nuclei such as histone proteins. Triggers of B cell activation are thought to include various chemicals and drugs, viral and bacterial antigens, and exposure to sunlight. Circulating complexes of abnormal autoantibodies and their "antigens" cause damage in tissues throughout the body. Kidney damage occurs as the complexes become trapped in the basement membrane of kidney capillaries, obstructing blood filtering. Renal failure is the most common cause of death.

There is no cure for lupus, but drug therapy can minimize symptoms, reduce inflammation, and forestall flare-ups. The most commonly used lupus medications are pain relievers (nonsteroidal anti-inflammatory drugs such as aspirin and ibuprofen), antimalarial drugs (hydroxychloroquine), and corticosteroids (prednisone and hydrocortisone).

MEDICAL TERMINOLOGY

Adenitis (ad′-e-NĪ-tis; *aden-* = gland; *-itis* = inflammation of) Enlarged, tender, and inflamed lymph nodes resulting from an infection.

Allograft (AL-ō-graft; *allo-* = other) A transplant between genetically distinct individuals of the same species. Skin transplants from other people and blood transfusions are allografts.

Autograft (AW-tō-graft; *auto-* = self) A transplant in which one's own tissue is grafted to another part of the body (such as skin grafts for burn treatment or plastic surgery).

Chronic fatigue syndrome (CFS) A disorder, usually occurring in young adults and primarily in females, characterized by

(1) extreme fatigue that impairs normal activities for at least 6 months and (2) the absence of other known diseases (cancer, infections, drug abuse, toxicity, or psychiatric disorders) that might produce similar symptoms.

Gamma globulin (GLOB-ū-lin) Suspension of immunoglobulins from blood consisting of antibodies that react with a specific pathogen. It is prepared by injecting the pathogen into animals, removing blood from the animals after antibodies have been produced, isolating the antibodies, and injecting them into a human to provide short-term immunity.

Hypersplenism (hī-per-SPLĒN-izm; *hyper-* = over) Abnormal splenic activity due to splenic enlargement and associated with an increased rate of destruction of normal blood cells.

Lymphadenopathy (lim-fad′-e-NOP-a-thē; *lymph-* = clear fluid; *-pathy* = disease) Enlarged, sometimes tender lymph glands as a response to infection; also called ***swollen glands.***

Lymphangitis (lim-fan-JĪ-tis; *-itis* = inflammation of) Inflammation of lymphatic vessels.

Lymphedema (lim′-fe-DĒ-ma; *edema* = swelling) Accumulation of lymph in lymphatic vessels, causing painless swelling of a limb.

Severe combined immunodeficiency disease (SCID) A rare inherited disorder in which both B cells and T cells are missing or inactive. Scientists have now identified mutations in several genes that are responsible for some types of SCID. In some cases, an infusion of red bone marrow cells from a sibling having very similar MHC (HLA) antigens can provide normal stem cells that give rise to normal B and T cells. The result can be a complete cure. Less than 30% of afflicted patients, however, have a compatible sibling who could serve as a donor.

Splenomegaly (splē′-nō-MEG-a-lē; *mega-* = large) Enlarged spleen.

Tonsillectomy (ton′-si-LEK-tō-mē; *-ectomy* = excision) Removal of a tonsil.

Xenograft (ZEN-ō-graft; *xeno-* = strange or foreign) A transplant between animals of different species. Xenografts from porcine (pig) or bovine (cow) tissue may be used in humans as a physiological dressing for severe burns. Other xenografts include pig heart valves and baboon hearts.

STUDY OUTLINE

INTRODUCTION (p. 805)

1. The ability to ward off disease is called resistance. Lack of resistance is called susceptibility.
2. Nonspecific resistance refers to a wide variety of body responses to a wide range of pathogens; specific resistance or immunity involves activation of specific lymphocytes to combat a particular foreign substance.

LYMPHATIC SYSTEM STRUCTURE AND FUNCTION (p. 805)

1. The lymphatic system carries out immune responses and consists of lymph, lymphatic vessels, and structures and organs that contain lymphatic tissue (specialized reticular tissue containing many lymphocytes).
2. The lymphatic system drains interstitial fluid, transports dietary lipids, and protects against invasion through immune responses.
3. Lymphatic vessels begin as closed-ended lymphatic capillaries in tissue spaces between cells.
4. Interstitial fluid drains into lymphatic capillaries, thus forming lymph.
5. Lymph capillaries merge to form larger vessels, called lymphatic vessels, which convey lymph into and out of lymph nodes.
6. The route of lymph flow is from lymph capillaries to lymphatic vessels to lymph trunks to the thoracic duct (or right lymphatic duct) to the subclavian veins.
7. Lymph flows because of skeletal muscle contractions and respiratory movements. Valves in lymphatic vessels also aid flow of lymph.
8. The primary lymphatic organs are red bone marrow and the thymus. Secondary lymphatic organs are lymph nodes, the spleen, and lymphatic nodules.
9. The thymus lies between the sternum and the large blood vessels above the heart. It is the site of T cell maturation.
10. Lymph nodes are encapsulated, egg-shaped structures located along lymphatic vessels.
11. Lymph enters nodes through afferent lymphatic vessels, is filtered, and exits through efferent lymphatic vessels.
12. Lymph nodes are the site of proliferation of plasma cells and T cells.
13. The spleen is the largest single mass of lymphatic tissue in the body. It is a site of B cell proliferation into plasma cells and phagocytosis of bacteria and worn-out red blood cells.
14. Lymphatic nodules are scattered throughout the mucosa of the gastrointestinal, respiratory, urinary, and reproductive tracts. This lymphatic tissue is termed mucosa-associated lymphatic tissue (MALT).

DEVELOPMENT OF LYMPHATIC TISSUES (p. 814)

1. Lymphatic vessels develop from lymph sacs, which arise from developing veins. Thus, they are derived from mesoderm.
2. Lymph nodes develop from lymph sacs that become invaded by mesenchymal cells.

NONSPECIFIC RESISTANCE: INNATE DEFENSES (p. 815)

1. Mechanisms of nonspecific resistance include mechanical factors, chemical factors, antimicrobial proteins, natural killer cells, phagocytes, inflammation, and fever.

2. The skin and mucous membranes are the first line of defense against entry of pathogens.
3. Antimicrobial proteins include interferons, the complement system, and transferrins.
4. Natural killer cells and phagocytes attack and kill pathogens and defective cells in the body.
5. Inflammation aids disposal of microbes, toxins, or foreign material at the site of an injury, and prepares the site for tissue repair.
6. Fever intensifies the antiviral effects of interferons, inhibits growth of some microbes, and speeds up body reactions that aid repair.
7. Table 22.1 on page 819 summarizes the components of nonspecific resistance.

SPECIFIC RESISTANCE: IMMUNITY (p. 820)

1. Specific resistance to disease involves the production of a specific lymphocyte or antibody against a specific antigen and is also called immunity.
2. B cells and T cells arise from stem cells in red bone marrow.
3. T cells complete their maturation and develop immunocompetence in the thymus.
4. In cell-mediated immune responses, cytotoxic T cells directly attack the invading antigen; in antibody-mediated immune responses, plasma cells secrete antibodies.
5. Antigens (Ags) are chemical substances that are recognized as foreign by the immune system.
6. Antigen receptors exhibit great diversity due to genetic recombination.
7. "Self-antigens" called major histocompatibility complex (MHC) antigens are unique to each person's body cells. All cells except red blood cells display MHC-I molecules; some cells also display MHC-II molecules.
8. Cells called antigen-presenting cells (APCs), which include macrophages, B cells, and dendritic cells, process antigens.
9. Exogenous antigens (formed outside the body) are presented with MHC-II molecules to T cells; endogenous (formed inside a body cell) antigens are presented with MHC-I molecules.
10. Cytokines are small protein hormones that may stimulate or inhibit many normal cell functions such as growth and differentiation. Other cytokines regulate immune responses (see Table 22.2 on page 824).

CELL-MEDIATED IMMUNITY (p. 825)

1. In a cell-mediated immune response, an antigen is recognized, specific T cells proliferate and differentiate into effector cells, and the antigen is eliminated.
2. T-cell receptors (TCRs) recognize antigen fragments associated with MHC molecules on the surface of a body cell.
3. Proliferation of T cells requires costimulation, either by cytokines such as interleukin-2 or by pairs of plasma membrane molecules.
4. T cells consist of several subpopulations. Helper T cells display CD4 protein, recognize antigen fragments associated with MHC-II molecules, and secrete several cytokines, most importantly interleukin-2, which acts as a costimulator for other helper T cells, cytotoxic T cells, and B cells. Cytotoxic T cells display CD8 protein and recognize antigen fragments associated with MHC-I molecules. Memory T cells remain after a cell-mediated immune response and initiate a faster response when a pathogen bearing the same foreign antigen invades the body again.
5. Cytotoxic T cells eliminate invaders by (1) releasing granzymes that cause target cell apoptosis (phagocytes then kill the microbes) and (2) releasing perforin, which causes cytolysis, and granulysin that destroys the microbes.
6. Cytotoxic T cells, macrophages, and natural killer cells carry out immunological surveillance, recognizing and destroying cancerous cells that display tumor antigens.

ANTIBODY-MEDIATED IMMUNITY (p. 828)

1. B cells can respond to unprocessed antigens, but their response is more intense when they process the antigen. Interleukin-2 and other cytokines secreted by helper T cells provide costimulation for the proliferation of B cells.
2. An activated B cell develops into a clone of antibody-producing plasma cells.
3. An antibody (Ab) is a protein that combines specifically with the antigen that triggered its production.
4. Antibodies consist of heavy and light chains and variable and constant regions.
5. Based on chemistry and structure, antibodies are grouped into five principal classes, (IgG, IgA, IgM, IgD, and IgE), each with specific biological roles.
6. Actions of antibodies include neutralization of antigen, immobilization of bacteria, agglutination and precipitation of antigen, activation of complement, and enhancement of phagocytosis.
7. Complement is a group of proteins that complement immune responses and help clear antigens from the body.
8. Immunization against certain microbes is possible because memory B cells and memory T cells remain after a primary response to an antigen. The secondary response provides protection should the same microbe enter the body again.

SELF-RECOGNITION AND SELF-TOLERANCE (p. 834)

1. T cells undergo positive selection to ensure that they can recognize self-MHC proteins (self-recognition), and negative selection to ensure that they do not react to other self-proteins (self-tolerance). Negative selection involves both deletion and anergy.
2. B cells develop tolerance through deletion and anergy.

STRESS AND IMMUNITY (p. 836)

1. Psychoneuroimmunology (PNI) deals with communication pathways that link the nervous, endocrine, and immune systems. Thoughts, feelings, moods, and beliefs influence health and the course of disease.
2. Under stress, people are less likely to eat well or exercise regularly, two habits that enhance immunity.

AGING AND THE IMMUNE SYSTEM (p. 836)

1. With advancing age, individuals become more susceptible to infections and malignancies, respond less well to vaccines, and produce more autoantibodies.
2. Immune responses also diminish with age.

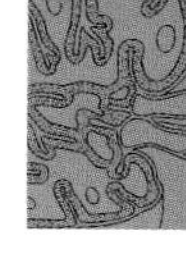

SELF-QUIZ QUESTIONS

Fill in the blanks in the following statements.

1. The first line of nonspecific defense against pathogens are the ____ and ____; the second line of nonspecific defense are the ____, ____, and ____.
2. Substances that are recognized as foreign and provoke immune responses are known as ____.

Indicate whether the following statements are true or false.

3. The body's ability to ward off damage or disease through our defenses is known as resistance; vulnerability to disease is susceptibility.
4. A person's T cells must be able to recognize the person's own MHC molecules, a process known as self-recognition, and lack reactivity to peptide fragments from the person's own proteins, a condition known as self-tolerance.

Choose the one best answer to the following questions.

5. Trace the sequence of fluid from blood vessel to blood vessel by way of the lymphatic system. (1) lymphatic vessels, (2) blood capillaries, (3) subclavian veins, (4) lymphatic capillaries, (5) interstitial spaces, (6) arteries, (7) lymphatic ducts. (a) 2, 5, 4, 1,7, 6, 3; (b) 3, 6, 2, 4, 5, 1, 7; (c) 6, 2, 5, 4, 1, 7, 3; (d) 6, 2, 5, 4, 7, 1, 3; (e) 2, 5, 4, 7, 1, 3, 6.
6. Which of the following describe lymph nodes? (1) Lymph enters the nodes through efferent lymphatic vessels and leaves through afferent lymphatic vessels. (2) The outer cortex consists of lymphatic nodules that contain B cells and are the sites of plasma cell and memory B cell formation. (3) The inner cortex contains lymphatic nodules with mature T cells. (4) The reticular fibers within the sinuses of the lymph nodes trap foreign substances in the lymph. (5) The sinuses of lymph nodes are known as red pulp. (a) 1, 2, 3, and 4; (b) 2, 4, and 5; (c) 1, 2, 3, 4, and 5; (d) 2 and 4; (e) 1, 2, and 4.
7. Which of the following statements are *correct*? (1) Lymphatic vessels are found throughout the body, except in avascular tissues, the CNS, portions of the spleen, and red bone marrow. (2) Lymphatic capillaries allow interstitial fluid to flow into them but not out of them. (3) Anchoring filaments attach lymphatic endothelial cells to surrounding tissues. (4) Lymphatic vessels freely receive all the components of blood, including the formed elements. (5) Lymph ducts directly connect to blood vessels by way of the subclavian veins. (a) 1, 3, 4, and 5; (b) 2, 3, 4, and 5; (c) 1, 2, 3, and 4; (d) 1, 2, 4, and 5; (e) 1, 2, 3, and 5.
8. Which of the following are mechanical factors that help fight pathogens and disease? (1) tight junctions of epidermal cells, (2) mucus of mucous membranes, (3) saliva, (4) interferons, (5) complement. (a) 1, 3, and 4; (b) 2, 4, and 5; (c) 1, 4, and 5; (d) 1, 2, and 3; (e) 1, 2, and 4.
9. Which of the following are functions of antibodies? (1) neutralization of antigens, (2) immobilization of bacteria, (3) agglutination and precipitation of antigens, (4) activation of complement, (5) enhancement of phagocytosis. (a) 1, 3, and 4; (b) 2, 4, and 5; (c) 1, 2, 3, and 4; (d) 1, 2, 3, and 5; (e) 1, 2, 3, 4, and 5.
10. Which of the following are *true*? (1) Lymphatic vessels resemble arteries. (2) Lymph is very similar to interstitial fluid. (3) Lacteals are specialized lymphatic capillaries responsible for transporting dietary lipids. (4) Lymph is normally a cloudy, pale yellow fluid. (5) The thoracic duct drains lymph from the upper right side of the body. (6) Lymph flow is maintained by skeletal muscle contractions, one-way valves, and breathing movements. (a) 1, 2, 5, and 6; (b) 2, 3, and 6; (c) 2, 3, 4, and 6; (d) 2, 4, and 6; (e) 3, 5, and 6.
11. Place the stages of phagocytosis in the correct order of occurrence. (1) formation of phagolysosome, (2) adherence to microbe, (3) destruction of microbe, (4) ingestion to form a phagosome, (5) chemotactic attraction of phagocyte. (a) 2, 5, 4, 1, 3; (b) 4, 5, 2, 1, 3; (c) 5, 2, 4, 1, 3; (d) 5, 4, 2, 3, 1; (e) 2, 5, 1, 4, 3.
12. Place in order the steps involved in cell-mediated immune response to an exogenous antigen. (a) costimulation and activation of T cells; (b) presentation of antigen to T cells; (c) elimination of invaders through the release of granzymes, perforin, granulysin, or lymphotoxin or by attraction and activation of phagocytes; (d) proliferation and differentiation of T cells to produce T cell clones; (e) antigen processing by dendritic cells, macrophages, or B cells; (f) recognition of antigen fragments associated with MHC-II molecules by T-cell receptors; (g) secretion of cytokines such as interleukin-2 by activated helper T cells; (h) migration of antigen-presenting cells to lymphatic tissue; (i) activation of cytotoxic T cells.
13. Match the following:

____(a) encapsulated bean-shaped structures located along the length of lymphatic vessels; contain T and B cells, macrophages, and follicular dendritic cells; filter lymph	(1) red bone marrow
____(b) produces pre-T cells and B cells; found in flat bones and epiphyses of long bones	(2) thymus
____(c) clusters of lymphatic nodules involved in immune responses against inhaled or ingested foreign substances	(3) lymph nodes
____(d) the single largest mass of lymphatic tissue in the body; consists of red and white pulp	(4) spleen
____(e) responsible for the maturation of T cells	(5) mucosa-associated lymphatic tissue
____(f) lymphatic nodules associated with mucous membranes of the digestive, urinary, reproductive, and respiratory systems	(6) lymphatic nodules
____(g) nonencapsulated clusters of lymphocytes	(7) tonsils

14. Match the following:

____ (a) recognize foreign antigens combined with MHC-1 molecules on the surface of body cells infected by microbes, some tumor cells, and cells of a tissue transplant; display CD8 proteins

____ (b) are programmed to recognize the reappearance of a previously encountered antigen

____ (c) differentiate into plasma cells that secrete specific antibodies

____ (d) process and present exogenous antigens; include macrophages, B cells, and dendritic cells

____ (e) secrete cytokines as costimulators; display CD4 proteins

____ (f) ingest microbes or any foreign particulate matter; include neutrophils and macrophages

____ (g) lymphocytes that have the ability to kill a wide variety of infectious microbes plus certain spontaneously arising tumor cells; lack antigen receptors

(1) helper T cells
(2) cytotoxic T cells
(3) memory T cells
(4) B cells
(5) NK cells
(6) phagocytes
(7) antigen-presenting cells

15. Match the following (answers may be used more than once):

____ (a) participate in inflammation, opsonization, and cytolysis

____ (b) stimulate histamine release, attract neutrophils by chemotaxis, promote phagocytosis, and destroy bacteria

____ (c) glycoproteins that mark the surface of all body cells except for RBCs; distinguish self from nonself

____ (d) foreign antigens present in fluids outside body cells

____ (e) foreign antigens synthesized within a body cell

____ (f) small protein hormones that stimulate or inhibit many normal cell functions; serve as costimulators for B cell and T cell activity

____ (g) a substance that has reactivity but lacks immunogenicity

____ (h) causes vasodilation and increased permeability of blood vessels; is found in mast cells in connective tissue and in basophils and platelets in blood

____ (i) polypeptides formed in blood; induce vasodilation and increased permeability of blood vessels; serve as chemotactic agents for phagocytes

____ (j) released by damaged cells; intensify the effects of histamine and kinins

____ (k) small parts of antigens that initiate immune responses

____ (l) produced by virus-infected cells, they interfere with viral replication in host cells

____ (m) chemicals released by NK and cytotoxic T cells that can cause apoptosis in target cells

____ (n) glycoproteins that contain four polypeptide chains, two of which are identical to each other and two of which are variable and contain the antigen-binding site

____ (o) produced by basophils and mast cells; involved in chemotaxis, adherence, and increased permeability

(1) exogenous antigens
(2) endogenous antigens
(3) interferons
(4) hapten
(5) cytokines
(6) granzymes
(7) histamine
(8) major histocompatibility complex (MHC) antigens
(9) kinins
(10) leukotrienes
(11) antibodies
(12) complement proteins
(13) epitopes
(14) prostaglandins

CRITICAL THINKING QUESTIONS

1. Esperanza watched as her mother got her "flu shot." "Why do you need a shot if you're not sick?" she asked. "So I won't get sick," answered her mom. Explain how the influenza vaccination prevents illness.
2. Due to the presence of breast cancer, Mrs. Franco had a right radical mastectomy in which her right breast and underlying muscle, right axillary lymph nodes and vessels were removed. Now she is experiencing severe swelling in her right arm. Why did the surgeon remove lymph tissue as well as the breast? Why is Mrs. Franco's right arm swollen?
3. Tariq's little sister has the mumps. Tariq can't remember if he has had mumps or not, but he is feeling slightly feverish. How could Tariq's doctor determine if he is getting sick with mumps or if he has previously had mumps?

ANSWERS TO FIGURE QUESTIONS

22.1 Red bone marrow contains stem cells that develop into lymphocytes.

22.2 Lymph is more similar to interstitial fluid than to blood plasma because the protein content of lymph is low.

22.3 The left and right lumbar trunks and the intestinal trunk empty into the cisterna chyli, which then drains into the thoracic duct.

22.4 Inhalation promotes the movement of lymph from abdominal lymphatic vessels toward the thoracic region because the pressure in the vessels of the thoracic region is lower than the pressure in the abdominal region when you inhale.

22.5 T cells mature in the thymus.

22.6 Foreign substances that enter a lymph node in lymph may be phagocytized by macrophages or attacked by lymphocytes that mount immune responses.

22.7 White pulp of the spleen functions in immunity; red pulp of the spleen performs functions related to blood cells.

22.8 Lymphatic tissues begin to develop by the end of the fifth week of gestation.

22.9 Lysozyme, digestive enzymes, and oxidants can kill microbes ingested during phagocytosis.

22.10 Redness results from increased blood flow due to vasodilation; pain, from injury of nerve fibers, irritation by microbial toxins, kinins, and prostaglandins, and pressure due to edema; heat, from increased blood flow and heat released by locally increased metabolic reactions; swelling, from leakage of fluid from capillaries due to increased permeability.

22.11 Helper T cells participate in both cell-mediated and antibody-mediated immune responses.

22.12 Epitopes are small immunogenic parts of a larger antigen; haptens are small molecules that become immunogenic only when they attach to a body protein.

22.13 APCs include macrophages in tissues throughout the body, B cells in blood and lymphatic tissue, and dendritic cells in mucous membranes and the skin.

22.14 The first signal in T cell activation is antigen binding to a TCR; the second signal is a costimulator, such as a cytokine or another pair of plasma membrane molecules.

22.15 Cytotoxic T cells attack some tumor cells and transplanted tissue cells, as well as cells infected by microbes.

22.16 A clone of plasma cells secretes just one kind of antibody.

22.17 The variable regions recognize and bind to a specific antigen.

22.18 The classical pathway for the activation of complement is linked to antibody-mediated immunity because Ag–Ab complexes activate C1.

22.19 At peak secretion, approximately 1000 times more IgG is produced in the secondary response than in the primary response.

22.20 In deletion, self-reactive T cells or B cells die; in anergy, T cells or B cells are alive but are unresponsive to antigenic stimulation.

22.21 HIV attacks helper T cells.

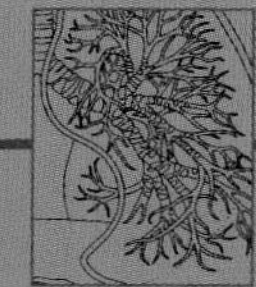

Chapter 23

The Respiratory System

The Respiratory System and Homeostasis

The respiratory system contributes to homeostasis by providing for the exchange of gases – oxygen and carbon dioxide – between the atmospheric air, blood, and tissue cells. It also helps adjust the pH of body fluids.

www. wiley.com/college/apcentral

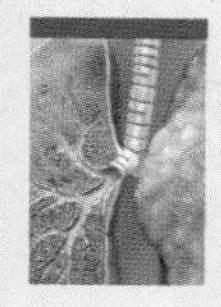

Your body's cells continually use oxygen (O_2) for the metabolic reactions that release energy from nutrient molecules and produce ATP. At the same time, these reactions release carbon dioxide (CO_2). Because an excessive amount of CO_2 produces acidity that can be toxic to cells, excess CO_2 must be eliminated quickly and efficiently. The cardiovascular and respiratory systems cooperate to supply O_2 and eliminate CO_2. The respiratory system provides for gas exchange—intake of O_2 and elimination of CO_2—and the cardiovascular system transports blood containing the gases between the lungs and body cells. Failure of either system disrupts homeostasis by causing rapid death of cells from oxygen starvation and buildup of waste products. In addition to functioning in gas exchange, the respiratory system also participates in regulating blood pH, contains receptors for the sense of smell, filters inspired air, produces sounds, and rids the body of some water and heat in exhaled air.

RESPIRATORY SYSTEM ANATOMY

▶ **OBJECTIVES**

Describe the anatomy and histology of the nose, pharynx, larynx, trachea, bronchi, and lungs.

Identify the functions of each respiratory system structure.

The **respiratory system** consists of the nose, pharynx (throat), larynx (voice box), trachea (windpipe), bronchi, and lungs (Figure 23.1). Its parts can be classified according to either structure or function. *Structurally,* the respiratory system consists of two parts: (1) The **upper respiratory system** includes the nose, pharynx, and associated structures. (2) The **lower respiratory system** includes the larynx, trachea, bronchi, and lungs. *Functionally,* the respiratory system also consists of two parts: (1) The **conducting zone** consists of a series of interconnecting cavities and tubes both outside and within the lungs—the nose, pharynx, larynx, trachea, bronchi, bronchioles, and terminal bronchioles—that filter, warm, and moisten air and conduct it into the lungs. (2) The **respiratory zone** consists of tissues within the lungs where gas exchange occurs—the respiratory bronchioles, alveolar ducts, alveolar sacs, and alveoli, the main sites of gas exchange between air and blood.

The branch of medicine that deals with the diagnosis and treatment of diseases of the ears, nose, and throat (ENT) is called **otorhinolaryngology** (ō′-tō-rī′-nō-lar′-in-GOL-ō-jē; *oto-* = ear; *rhino-* = nose; *laryngo-* = voice box; *-logy* = study of). A **pulmonologist** is a specialist in the diagnosis and treatment of diseases of the lungs.

Nose

The **nose** can be divided into external and internal portions. The **external nose** consists of a supporting framework of bone and hyaline cartilage covered with muscle and skin and lined by a mucous membrane. The frontal bone, nasal bones, and maxillae form the bony framework of the external nose (Figure 23.2a on page 849). The cartilaginous framework of the external nose consists of the **septal cartilage,** which forms the anterior portion of the nasal septum; the **lateral nasal cartilages** inferior to the nasal bones; and the **alar cartilages,** which form a portion of the walls of the nostrils. Because it consists of pliable hyaline cartilage, the cartilaginous framework of the external nose is somewhat flexible. On the undersurface of the external nose are two openings called the **external nares** (NĀ-rez; singular is **naris**) or **nostrils.** Figure 23.3 on page 850 shows the surface anatomy of the nose.

The interior structures of the external nose have three functions: (1) warming, moistening, and filtering incoming air; (2) detecting olfactory stimuli; and (3) modifying speech vibrations as they pass through the large, hollow resonating chambers. *Resonance* refers to prolonging, amplifying, or modifying a sound by vibration.

Rhinoplasty

Rhinoplasty (RĪ-nō-plas′-tē; *-plasty* = to mold or to shape), commonly called a "nose job," is a surgical procedure in which the structure of the external nose is altered. Although rhinoplasty is often done for cosmetic reasons, it is sometimes performed to repair a fractured nose or a deviated nasal septum. In the procedure, both local and general anesthetics are given. Instruments are then inserted through the nostrils, the nasal cartilage is reshaped, and the nasal bones are fractured and repositioned to achieve the desired shape. An internal packing and splint are inserted to keep the nose in the desired position as it heals. ■

The **internal nose** is a large cavity in the anterior aspect of the skull that lies inferior to the nasal bone and superior to the mouth; it is lined with muscle and mucous membrane. Anteriorly, the internal nose merges with the external nose, and posteriorly it communicates with the pharynx through two openings called the **internal nares** or **choanae** (kō-Ā-nē) (see Figure 23.2b). Ducts from the paranasal sinuses and the nasolacrimal ducts also open into the internal nose. Recall from Chapter 7 that the paranasal sinuses are cavities in certain cranial and facial bones lined with mucous membranes that are continuous with the lining of the nasal cavity. Skull bones containing the paranasal sinuses are the frontal, sphenoid, ethmoid, and maxillae. Besides producing mucus, the paranasal sinuses serve as resonating chambers for sound as we speak or sing. The lateral walls of the internal nose are formed by the ethmoid, maxillae, lacrimal, palatine, and inferior nasal conchae bones (see Figure 7.9 on page 206); the ethmoid bone also forms the roof. The palatine bones and palatine processes of the maxillae, which together constitute the hard palate, form the floor of the internal nose.

Figure 23.1 **Structures of the respiratory system.**

The upper respiratory system includes the nose, pharynx, and associated structures; the lower respiratory system includes the larynx, trachea, bronchi, and lungs.

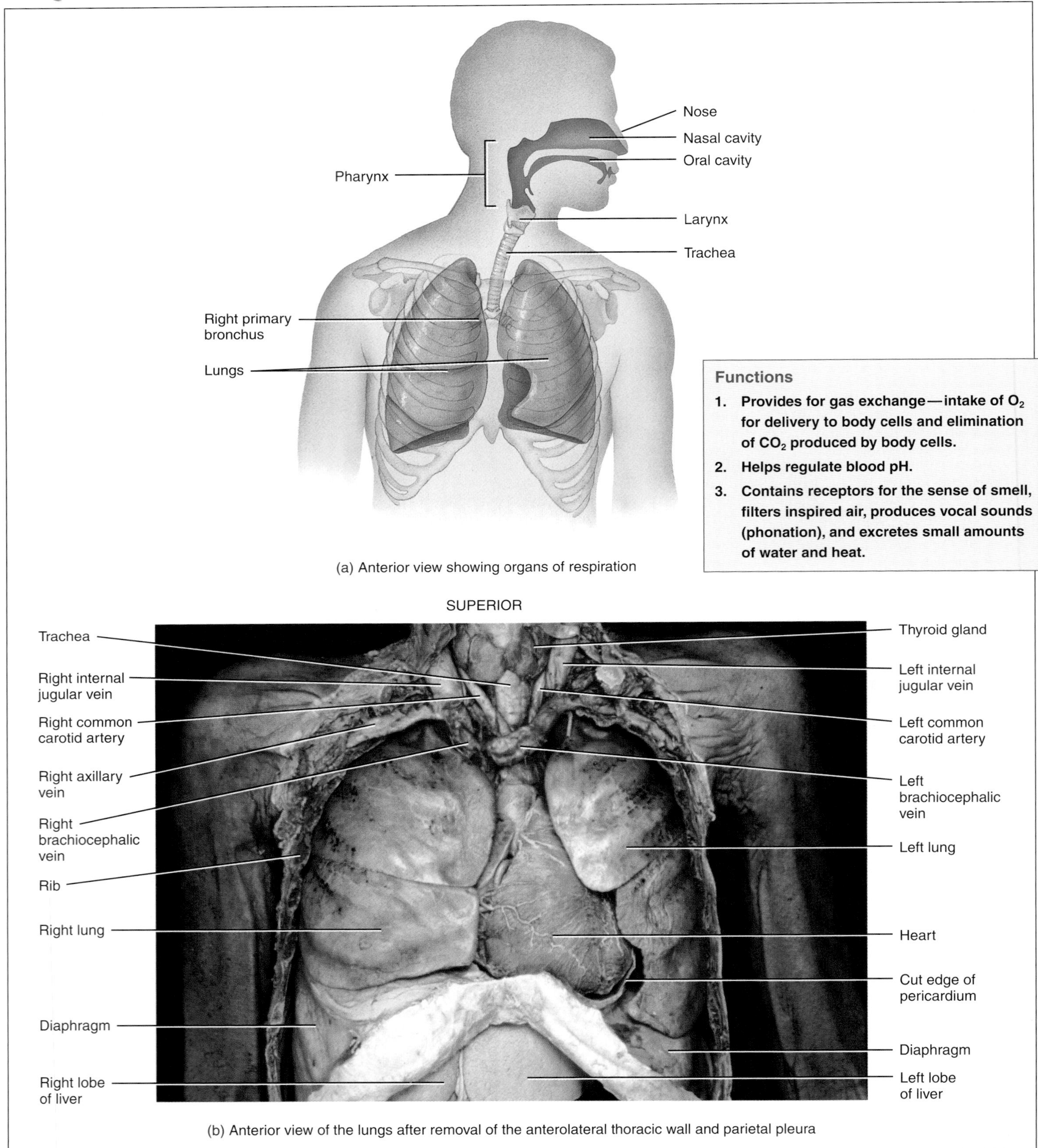

(a) Anterior view showing organs of respiration

(b) Anterior view of the lungs after removal of the anterolateral thoracic wall and parietal pleura

Which structures are part of the conducting zone of the respiratory system?

Figure 23.2 Respiratory structures in the head and neck. (See Tortora, *A Photographic Atlas of the Human Body, Second Edition,* Figures 11.2 and 11.3.)

As air passes through the nose, it is warmed, filtered, moistened, and olfaction occurs.

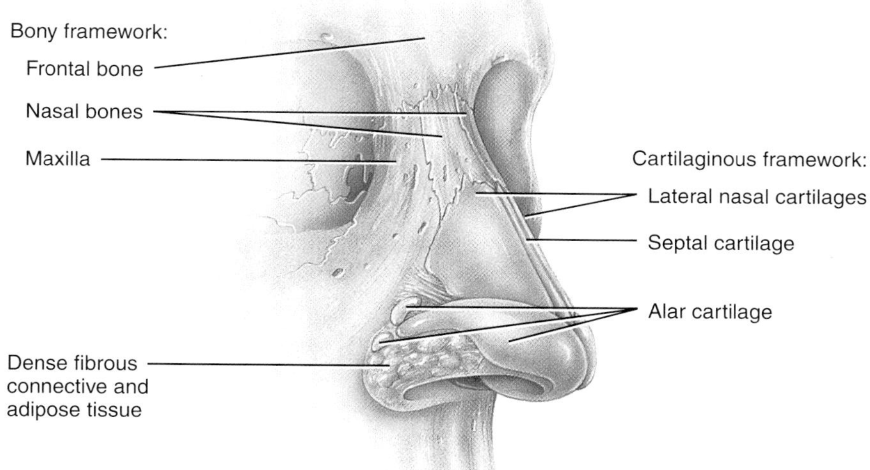

(a) Anterolateral view of external portion of nose showing cartilaginous and bony framework

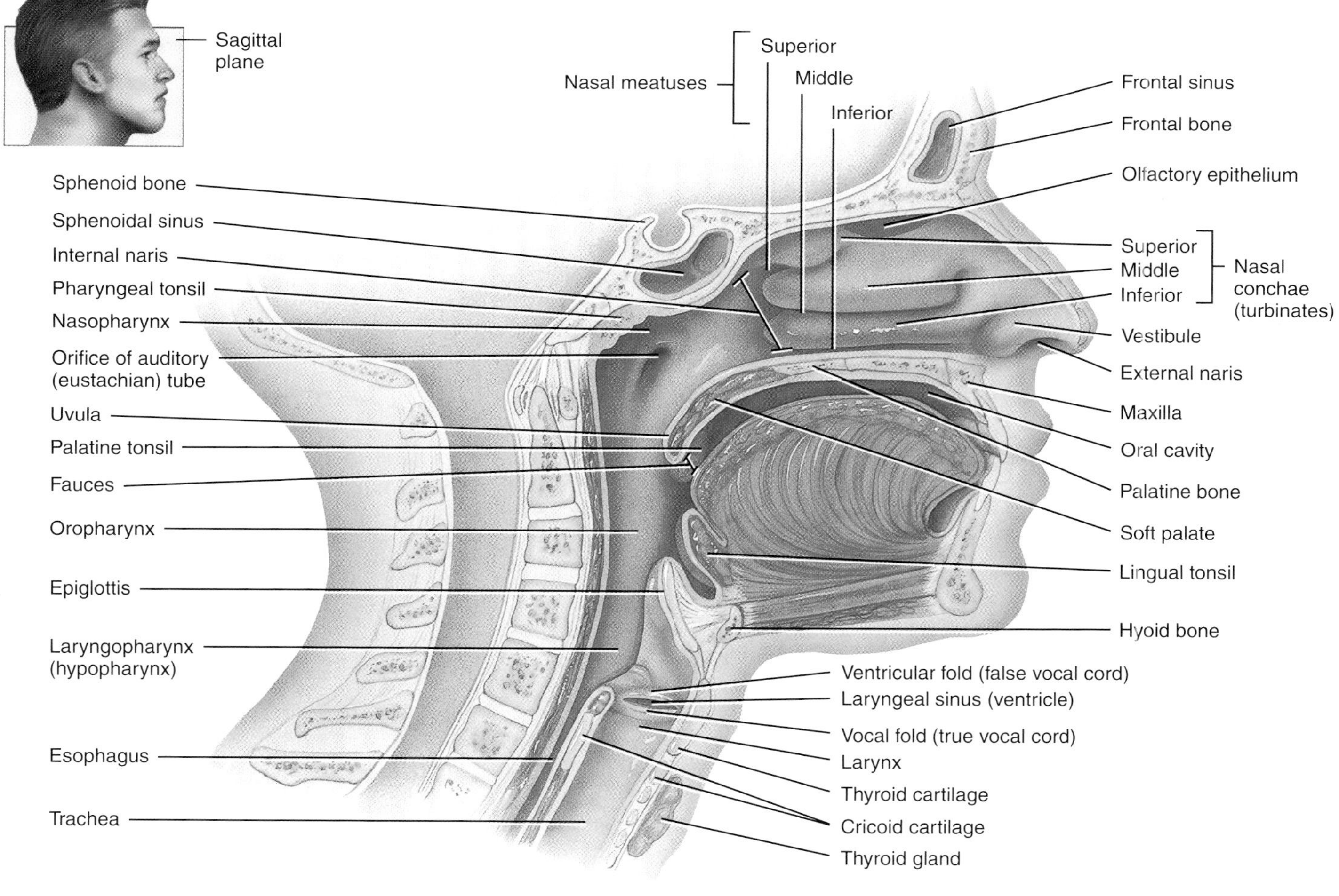

Sagittal plane

Nasal meatuses: Superior, Middle, Inferior

Frontal sinus
Frontal bone
Olfactory epithelium
Nasal conchae (turbinates): Superior, Middle, Inferior
Vestibule
External naris
Maxilla
Oral cavity
Palatine bone
Soft palate
Lingual tonsil
Hyoid bone

Sphenoid bone
Sphenoidal sinus
Internal naris
Pharyngeal tonsil
Nasopharynx
Orifice of auditory (eustachian) tube
Uvula
Palatine tonsil
Fauces
Oropharynx
Epiglottis
Laryngopharynx (hypopharynx)
Esophagus
Trachea

Ventricular fold (false vocal cord)
Laryngeal sinus (ventricle)
Vocal fold (true vocal cord)
Larynx
Thyroid cartilage
Cricoid cartilage
Thyroid gland

(b) Sagittal section of the left side of the head and neck showing the location of respiratory structures

What is the path taken by air molecules into and through the nose?

Figure 23.3 Surface anatomy of the nose.

The external nose has a cartilaginous framework and a bony framework.

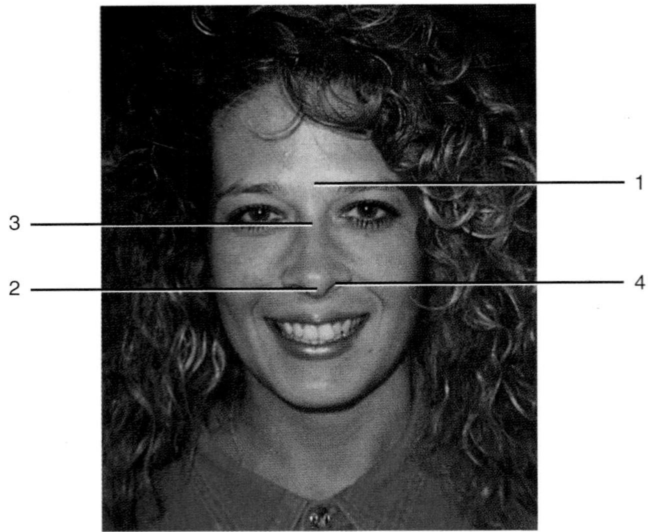

Anterior view

1. **Root**: Superior attachment of the nose to the frontal bone
2. **Apex**: Tip of nose
3. **Bridge**: Bony framework of nose formed by nasal bones
4. **External naris**: Nostril; external opening into nasal cavity

Which part of the nose is attached to the frontal bone?

The space within the internal nose is called the **nasal cavity.** The anterior portion of the nasal cavity just inside the nostrils, called the **vestibule,** is surrounded by cartilage; the superior part of the nasal cavity is surrounded by bone. A vertical partition, the **nasal septum,** divides the nasal cavity into right and left sides. The anterior portion of the septum consists primarily of hyaline cartilage; the remainder is formed by the vomer, perpendicular plate of the ethmoid, maxillae, and palatine bones (see Figure 7.11 on page 208).

When air enters the nostrils, it passes first through the vestibule, which is lined by skin containing coarse hairs that filter out large dust particles. Three shelves formed by projections of the superior, middle, and inferior nasal conchae extend out of each lateral wall of the cavity. The conchae, almost reaching the septum, subdivide each side of the nasal cavity into a series of groovelike passageways—the **superior, middle,** and **inferior meatuses** (mē-Ā-tus-ēz = openings or passages; singular is **meatus**). Mucous membrane lines the cavity and its shelves. The arrangement of conchae and meatuses increases surface area in the internal nose and prevents dehydration by trapping water droplets during exhalation.

The olfactory receptors lie in a region of the membrane lining the superior nasal conchae and adjacent septum called the **olfactory epithelium.** Inferior to the olfactory epithelium, the mucous membrane contains capillaries and pseudostratified ciliated columnar epithelium with many goblet cells. As inhaled air whirls around the conchae and meatuses, it is warmed by blood in the capillaries. Mucus secreted by the goblet cells moistens the air and traps dust particles. Drainage from the nasolacrimal ducts also helps moisten the air, and is sometimes assisted by secretions from the paranasal sinuses. The cilia move the mucus and trapped dust particles toward the pharynx, at which point they can be swallowed or spit out, thus removing the particles from the respiratory tract.

▶ **CHECKPOINT**

1. What functions do the respiratory and cardiovascular systems have in common?
2. What structural and functional features are different in the upper and lower respiratory systems? Which are the same?
3. Compare the structure and functions of the external nose and the internal nose.

Pharynx

The **pharynx** (FAIR-inks), or throat, is a funnel-shaped tube about 13 cm (5 in.) long that starts at the internal nares and extends to the level of the cricoid cartilage, the most inferior cartilage of the larynx (voice box) (Figure 23.4). The pharynx lies just posterior to the nasal and oral cavities, superior to the larynx, and just anterior to the cervical vertebrae. Its wall is composed of skeletal muscles and is lined with a mucous membrane. The pharynx functions as a passageway for air and food, provides a resonating chamber for speech sounds, and houses the tonsils, which participate in immunological reactions against foreign invaders.

The pharynx can be divided into three anatomical regions: (1) nasopharynx, (2) oropharynx, and (3) laryngopharynx. (See the lower orientation diagram in Figure 23.4.) The muscles of the entire pharynx are arranged in two layers, an outer circular layer and an inner longitudinal layer.

The superior portion of the pharynx, called the **nasopharynx,** lies posterior to the nasal cavity and extends to the soft palate. There are five openings in its wall: two internal nares, two openings that lead into the auditory (pharyngotympanic) tubes (commonly known as the eustachian tubes), and the opening into the oropharynx. The posterior wall also contains the **pharyngeal tonsil.** Through the internal nares, the nasopharynx receives air from the nasal cavity along with packages of dust-laden mucus. The nasopharynx is lined with pseudostratified ciliated columnar epithelium, and the cilia move the mucus down toward the most inferior part of the pharynx. The nasopharynx also exchanges small amounts of air with the auditory tubes to equalize air pressure between the pharynx and the middle ear.

The intermediate portion of the pharynx, the **oropharynx,** lies posterior to the oral cavity and extends from the soft palate

Figure 23.4 Pharynx. (See Tortora, *A Photographic Atlas of the Human Body, Second Edition,* Figure 11.4.)

The three subdivisions of the pharynx are the (1) nasopharynx, (2) oropharynx, and (3) laryngopharynx.

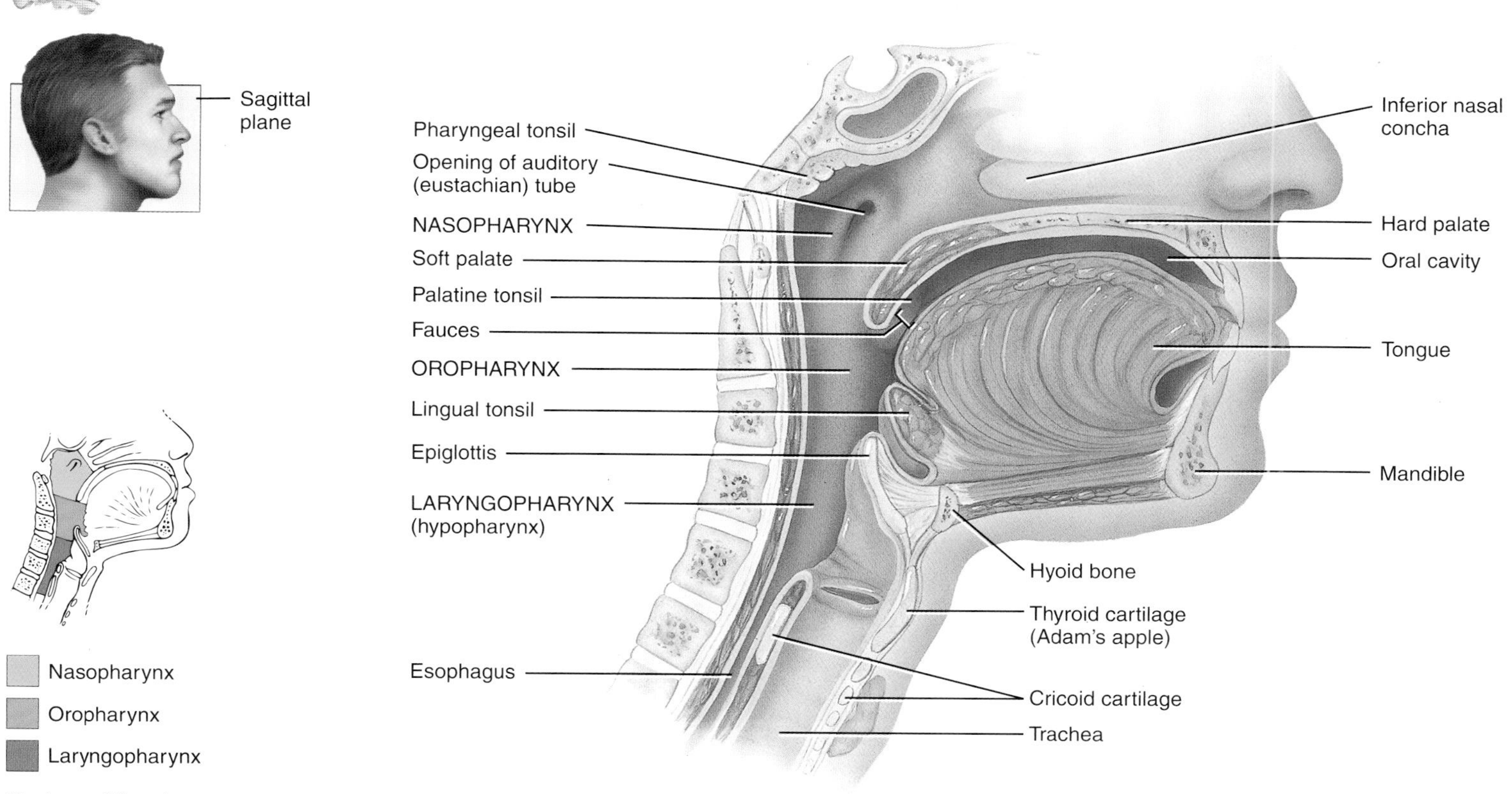

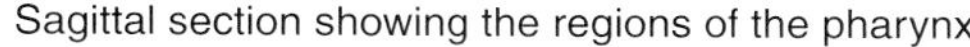

Sagittal section showing the regions of the pharynx

What are the superior and inferior borders of the pharynx?

inferiorly to the level of the hyoid bone. It has only one opening into it, the **fauces** (FAW-sēz = throat), the opening from the mouth. This portion of the pharynx has both respiratory and digestive functions, serving as a common passageway for air, food, and drink. Because the oropharynx is subject to abrasion by food particles, it is lined with nonkeratinized stratified squamous epithelium. Two pairs of tonsils, the **palatine** and **lingual tonsils,** are found in the oropharynx.

The inferior portion of the pharynx, the **laryngopharynx** (la-rin′-gō-FAIR-inks), or **hypopharynx,** begins at the level of the hyoid bone. It opens into the esophagus (food tube) posteriorly and the larynx (voice box) anteriorly. Like the oropharynx, the laryngopharynx is both a respiratory and a digestive pathway and is lined by nonkeratinized stratified squamous epithelium.

Larynx

The **larynx** (LAIR-inks), or voice box, is a short passageway that connects the laryngopharynx with the trachea. It lies in the midline of the neck anterior to the esophagus and the fourth through sixth cervical vertebrae (C4–C6).

The wall of the larynx is composed of nine pieces of cartilage (Figure 23.5). Three occur singly (thyroid cartilage, epiglottis, and cricoid cartilage), and three occur in pairs (arytenoid, cuneiform, and corniculate cartilages). Of the paired cartilages, the arytenoid cartilages are the most important because they influence changes in position and tension of the vocal folds (true vocal cords for speech). The extrinsic muscles of the larynx connect the cartilages to other structures in the throat; the intrinsic muscles connect the cartilages to one another.

The **thyroid cartilage (Adam's apple)** consists of two fused plates of hyaline cartilage that form the anterior wall of the larynx and give it a triangular shape. It is present in both males and females but is usually larger in males due to the influence of male sex hormones on its growth during puberty. The ligament that connects the thyroid cartilage to the hyoid bone is called the **thyrohyoid membrane.**

Figure 23.5 Larynx. (See Tortora, *A Photographic Atlas of the Human Body, Second Edition,* Figures 11.5 and 11.6.)

The larynx is composed of nine pieces of cartilage.

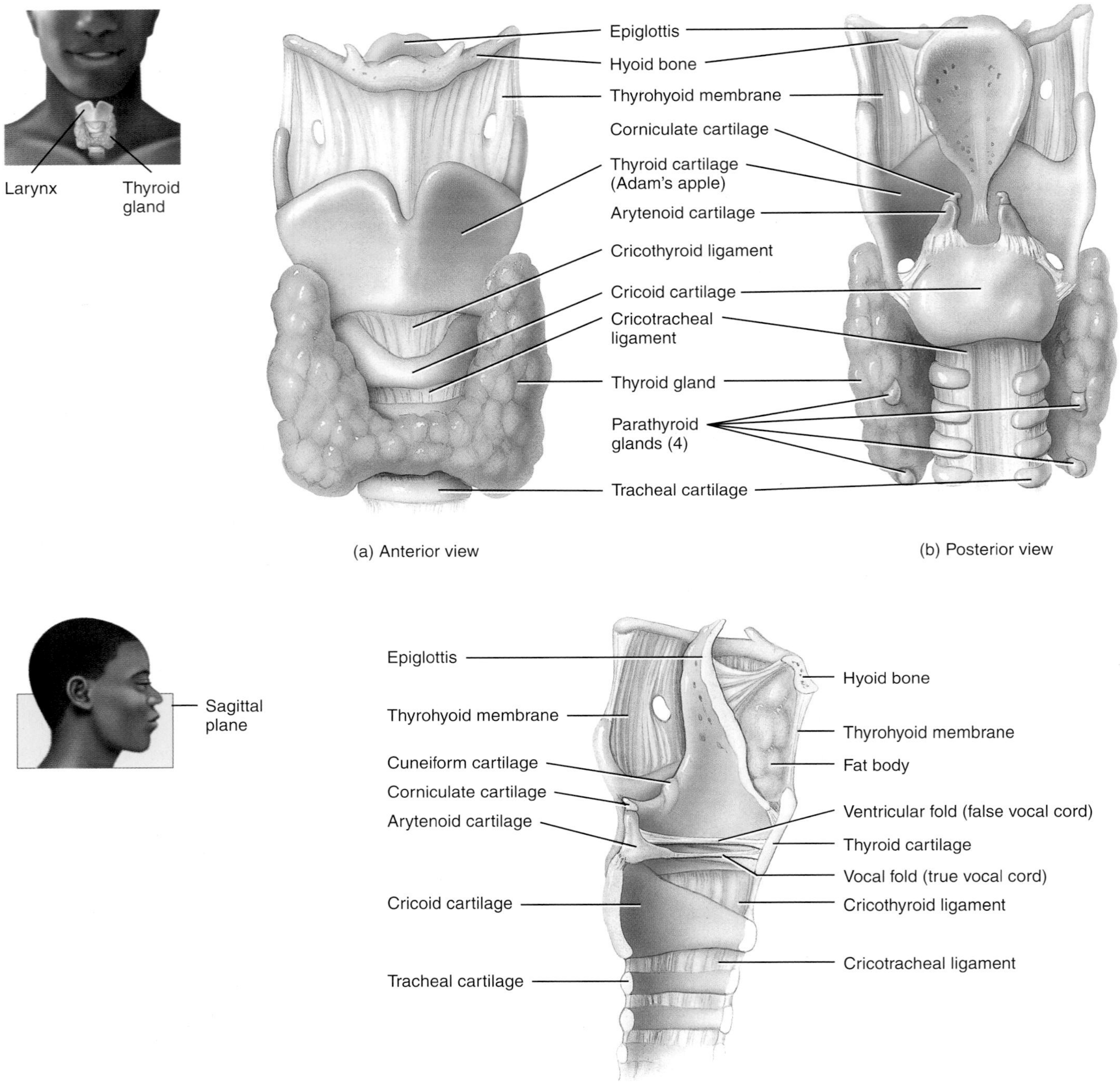

(c) Sagittal section

How does the epiglottis prevent aspiration of foods and liquids?

The **epiglottis** (*epi-* = over; *glottis* = tongue) is a large, leaf-shaped piece of elastic cartilage that is covered with epithelium (see also Figure 23.4). The "stem" of the epiglottis is the tapered inferior portion that is attached to the anterior rim of the thyroid cartilage and hyoid bone. The broad superior "leaf" portion of the epiglottis is unattached and is free to move up and down like a trap door. During swallowing, the pharynx and larynx rise. Elevation of the pharynx widens it to receive food or drink; elevation of the larynx causes the epiglottis to move down and form a lid over the glottis, closing it off. The **glottis** consists of a pair of folds of mucous membrane, the vocal folds (true vocal cords) in the larynx, and the space between them called the **rima glottidis** (RĪ-ma GLOT-ti-dis). The closing of the larynx in this way during swallowing routes liquids and foods into the esophagus and keeps them out of the larynx and airways. When small particles of dust, smoke, food, or liquids pass into the larynx, a cough reflex occurs, usually expelling the material.

The **cricoid cartilage** (KRĪ-koyd = ringlike) is a ring of hyaline cartilage that forms the inferior wall of the larynx. It is attached to the first ring of cartilage of the trachea by the **cricotracheal ligament.** The thyroid cartilage is connected to the cricoid cartilage by the **cricothyroid ligament.** The cricoid cartilage is the landmark for making an emergency airway (a tracheotomy; see page 856).

The paired **arytenoid cartilages** (ar′-i-TĒ-noyd = ladle-like) are triangular pieces of mostly hyaline cartilage located at the posterior, superior border of the cricoid cartilage. They attach to the vocal folds and intrinsic pharyngeal muscles. Supported by the arytenoid cartilages, the intrinsic pharyngeal muscles contract and move the vocal folds to produce sounds.

The paired **corniculate cartilages** (kor-NIK-ū-lāt = shaped like a small horn), which are horn-shaped pieces of elastic cartilage, are located at the apex of each arytenoid cartilage. They are supporting structures for the epiglottis.

The paired **cuneiform cartilages** (KŪ-nē-i-form = wedge-shaped), club-shaped elastic cartilages anterior to the corniculate cartilages, support the vocal folds and lateral aspects of the epiglottis.

The lining of the larynx superior to the vocal folds is nonkeratinized stratified squamous epithelium. The lining of the larynx inferior to the vocal folds is pseudostratified ciliated columnar epithelium consisting of ciliated columnar cells, goblet cells, and basal cells. The mucus produced by the goblet cells helps trap dust not removed in the upper passages. The cilia in the upper respiratory tract move mucus and trapped particles *down* toward the pharynx; the cilia in the lower respiratory tract move them *up* toward the pharynx.

The Structures of Voice Production

The mucous membrane of the larynx forms two pairs of folds (Figure 23.5c): a superior pair called the **ventricular folds (false vocal cords)** and an inferior pair called the **vocal folds (true vocal cords).** The space between the ventricular folds is known as the **rima vestibuli.** The **laryngeal sinus (ventricle)** is a lateral expansion of the middle portion of the laryngeal cavity inferior to the ventricular folds and superior to the vocal folds (see Figure 23.2b).

When the ventricular folds are brought together, they function in holding the breath against pressure in the thoracic cavity, such as might occur when you strain to lift a heavy object. Deep to the mucous membrane of the vocal folds, which is lined by nonkeratinized stratified squamous epithelium, bands of elastic ligaments are stretched between pieces of rigid cartilage like the strings on a guitar. Intrinsic laryngeal muscles attach to both the rigid cartilage and the vocal folds. When the muscles contract, they pull the elastic ligaments tight and stretch the vocal folds out into the airways so that the rima glottidis is narrowed. If air is directed against the vocal folds, they vibrate and produce sounds (phonation) and set up sound waves in the column of air in the pharynx, nose, and mouth. The greater the pressure of air, the louder the sound.

When the intrinsic muscles of the larynx contract, they pull on the arytenoid cartilages, which causes them to pivot. Contraction of the posterior cricoarytenoid muscles, for example, moves the vocal folds apart (abduction), thereby opening the rima glottidis (Figure 23.6a). By contrast, contraction of the lateral cricoarytenoid muscles moves the vocal folds together (adduction), thereby closing the rima glottidis (Figure 23.6b). Other intrinsic muscles can elongate (and place tension on) or shorten (and relax) the vocal folds.

Pitch is controlled by the tension on the vocal folds. If they are pulled taut by the muscles, they vibrate more rapidly, and a higher-pitch results. Decreasing the muscular tension on the vocal folds causes them to vibrate more slowly and produce lower-pitch sounds. Due to the influence of androgens (male sex hormones), vocal folds are usually thicker and longer in males than in females, and therefore they vibrate more slowly. This is why a man's voice generally has a lower range of pitch than that of a woman.

Sound originates from the vibration of the vocal folds, but other structures are necessary for converting the sound into recognizable speech. The pharynx, mouth, nasal cavity, and paranasal sinuses all act as resonating chambers that give the voice its human and individual quality. We produce the vowel sounds by constricting and relaxing the muscles in the wall of the pharynx. Muscles of the face, tongue, and lips help us enunciate words.

Whispering is accomplished by closing all but the posterior portion of the rima glottidis. Because the vocal folds do not vibrate during whispering, there is no pitch to this form of speech. However, we can still produce intelligible speech while whispering by changing the shape of the oral cavity as we enunciate. As the size of the oral cavity changes, its resonance qualities change, which imparts a vowel-like pitch to the air as it rushes toward the lips.

Figure 23.6 Movement of the vocal folds.

The glottis consists of a pair of folds of mucous membrane in the larynx (the vocal folds) and the space between them (the rima glottidis).

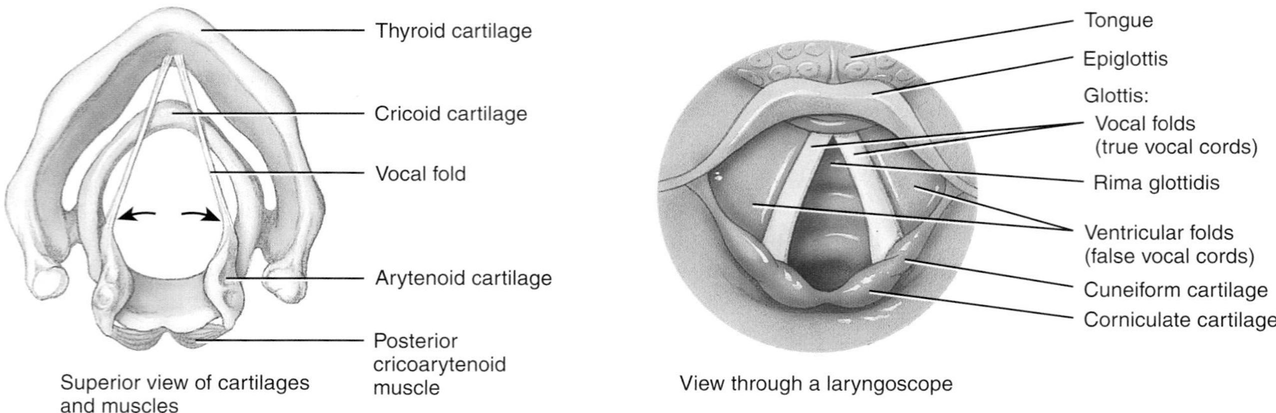

(a) Movement of vocal folds apart (abduction)

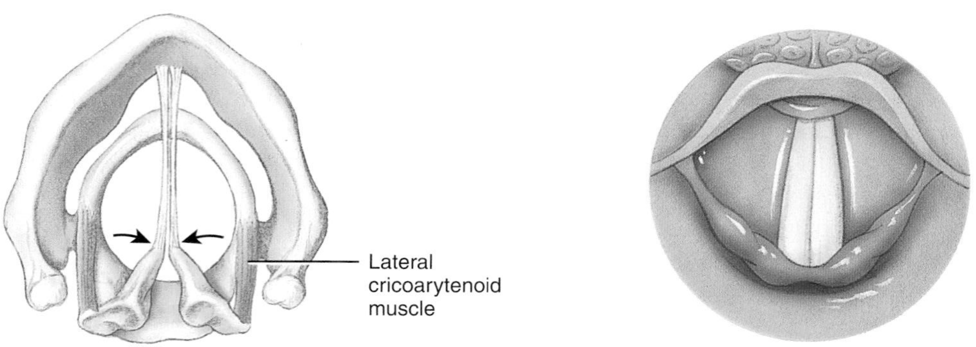

(b) Movement of vocal folds together (adduction)

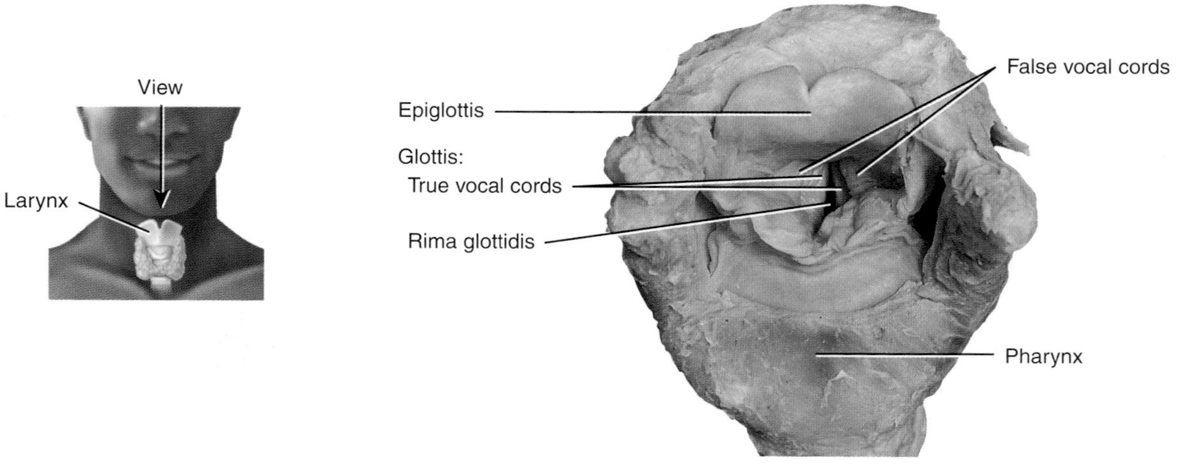

(c) Superior view

What is the main function of the vocal folds?

Laryngitis and Cancer of the Larynx

Laryngitis is an inflammation of the larynx that is most often caused by a respiratory infection or irritants such as cigarette smoke. Inflammation of the vocal folds causes hoarseness or loss of voice by interfering with the contraction of the folds or by causing them to swell to the point where they cannot vibrate freely. Many long-term smokers acquire a permanent hoarseness from the damage done by chronic inflammation. **Cancer of the larynx** is found almost exclusively in individuals who smoke. The condition is characterized by hoarseness, pain on swallowing, or pain radiating to an ear. Treatment consists of radiation therapy and/or surgery. ■

Trachea

The **trachea** (TRĀ-kē-a = sturdy), or windpipe, is a tubular passageway for air that is about 12 cm (5 in.) long and 2.5 cm (1 in.) in diameter. It is located anterior to the esophagus (Figure 23.7) and extends from the larynx to the superior border of the fifth thoracic vertebra (T5), where it divides into right and left primary bronchi (see Figure 23.8).

The layers of the tracheal wall, from deep to superficial, are the (1) mucosa, (2) submucosa, (3) hyaline cartilage, and (4) adventitia (composed of areolar connective tissue). The mucosa of the trachea consists of an epithelial layer of pseudostratified ciliated columnar epithelium and an underlying layer of lamina propria that contains elastic and reticular fibers. Pseudostratified ciliated columnar epithelium consists of ciliated columnar cells and goblet cells that reach the luminal surface, plus basal cells that do not (see Table 4.1E on page 115); it provides the same protection against dust as the membrane lining the nasal cavity and larynx. The submucosa consists of areolar connective tissue that contains seromucous glands and their ducts.

The 16–20 incomplete, horizontal rings of hyaline cartilage resemble the letter C and are stacked one on top of another. They may be felt through the skin inferior to the larynx. The open part of each C-shaped cartilage ring faces the esophagus (Figure 23.7), an arrangement that accommodates slight expansion of the esophagus into the trachea during swallowing. Transverse smooth muscle fibers, called the **trachealis muscle,** and elastic connective tissue stabilize the open ends of the cartilage rings. The solid C-shaped cartilage rings provide a semirigid support so that the tracheal wall does not collapse inward (especially during inhalation) and obstruct the air passageway. The adventitia of the trachea consists of areolar connective tissue that joins the trachea to surrounding tissues.

Figure 23.7 Location of the trachea in relation to the esophagus.

The trachea is anterior to the esophagus and extends from the larynx to the superior border of the fifth thoracic vertebra.

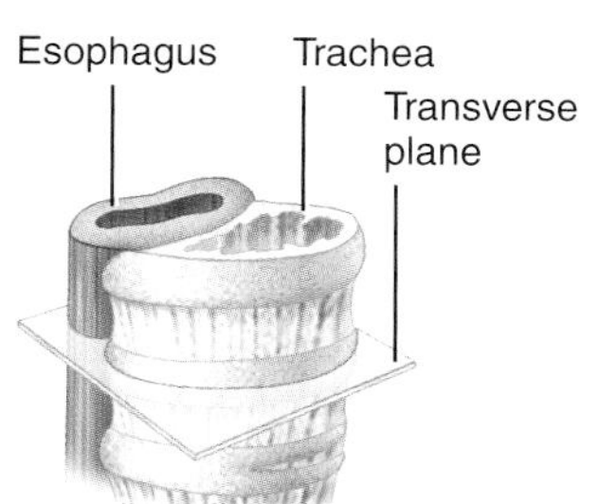

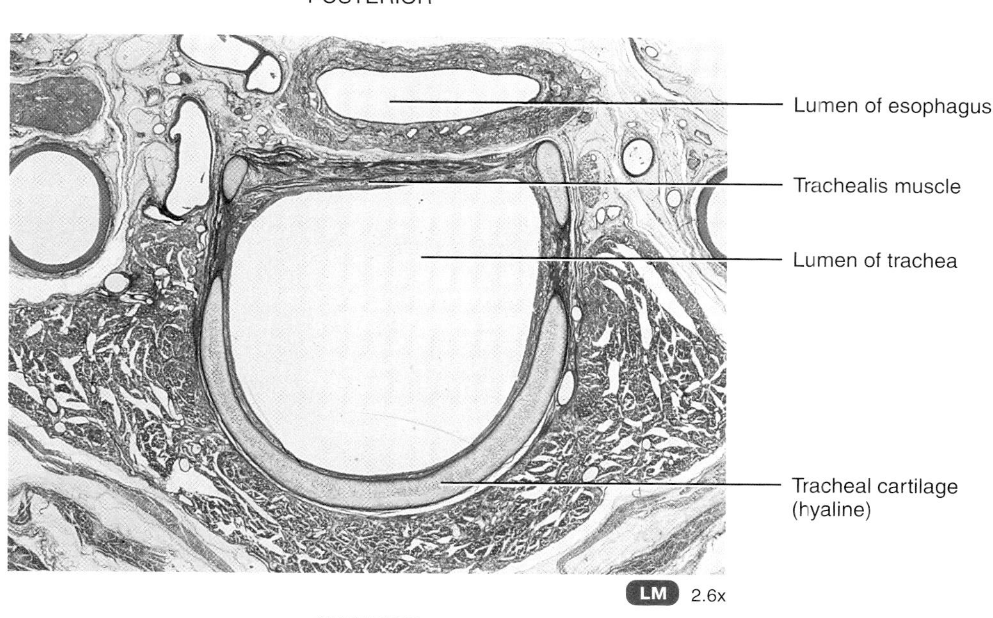

Transverse section of the trachea and esophagus

What is the benefit of not having cartilage between the trachea and the esophagus?

Tracheotomy and Intubation

Several conditions may block airflow by obstructing the trachea. For example, the rings of cartilage that support the trachea may collapse due to a crushing injury to the chest, inflammation of the mucous membrane may cause it to swell so much that the airway closes, vomit or a foreign object may be aspirated into it, or a cancerous tumor may protrude into the airway. Two methods are used to reestablish airflow past a tracheal obstruction. If the obstruction is superior to the level of the larynx, a **tracheotomy** (trā-kē-O-tō-mē; *-tome* = cutting), an operation to make an opening into the trachea, may be performed. In this procedure, also called a *tracheostomy,* a skin incision is followed by a short longitudinal incision into the trachea inferior to the cricoid cartilage. The patient can then breathe through a metal or plastic tracheal tube inserted through the incision. The second method is **intubation,** in which a tube is inserted into the mouth or nose and passed inferiorly through the larynx and trachea. The firm wall of the tube pushes aside any flexible obstruction, and the lumen of the tube provides a passageway for air; any mucus clogging the trachea can be suctioned out through the tube. ■

Bronchi

At the superior border of the fifth thoracic vertebra, the trachea divides into a **right primary bronchus** (BRON-kus = windpipe), which goes into the right lung, and a **left primary bronchus,** which goes into the left lung (Figure 23.8). The right

Figure 23.8 **Branching of airways from the trachea: the bronchial tree.** (See Tortora, *A Photographic Atlas of the Human Body, Second Edition,* Figure 11.8.)

The bronchial tree begins at the trachea and ends at the terminal bronchioles.

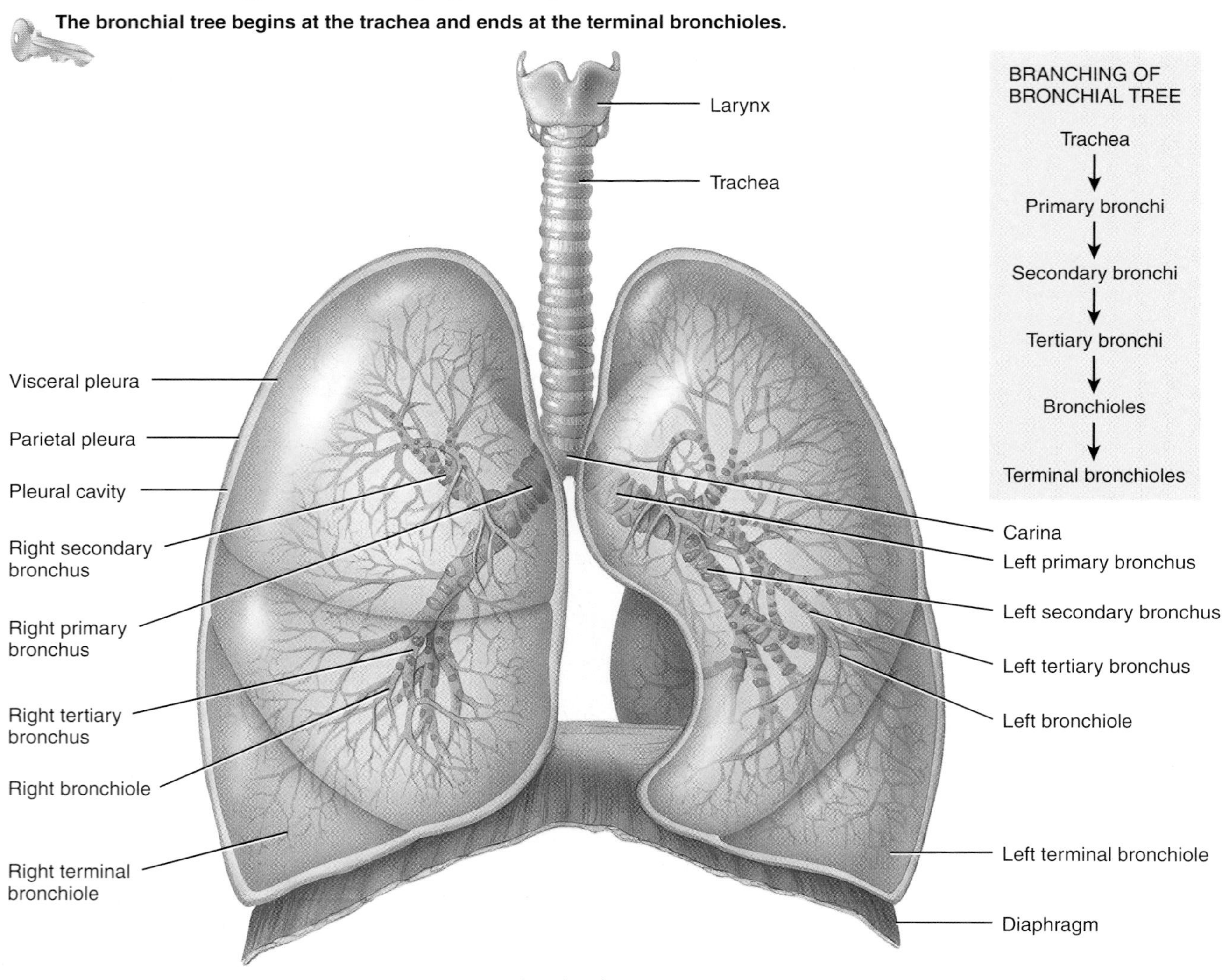

How many lobes and secondary bronchi are present in each lung?

primary bronchus is more vertical, shorter, and wider than the left. As a result, an aspirated object is more likely to enter and lodge in the right primary bronchus than the left. Like the trachea, the primary bronchi (BRON-kē) contain incomplete rings of cartilage and are lined by pseudostratified ciliated columnar epithelium.

At the point where the trachea divides into right and left primary bronchi an internal ridge called the **carina** (ka-RĪ-na = keel of a boat) is formed by a posterior and somewhat inferior projection of the last tracheal cartilage. The mucous membrane of the carina is one of the most sensitive areas of the entire larynx and trachea for triggering a cough reflex. Widening and distortion of the carina is a serious sign because it usually indicates a carcinoma of the lymph nodes around the region where the trachea divides.

On entering the lungs, the primary bronchi divide to form smaller bronchi—the **secondary (lobar) bronchi,** one for each lobe of the lung. (The right lung has three lobes; the left lung has two.) The secondary bronchi continue to branch, forming still smaller bronchi, called **tertiary (segmental) bronchi,** that divide into **bronchioles.** Bronchioles, in turn, branch repeatedly, and the smallest ones branch into even smaller tubes called **terminal bronchioles.** This extensive branching from the trachea resembles an inverted tree and is commonly referred to as the **bronchial tree.**

As the branching becomes more extensive in the bronchial tree, several structural changes may be noted.

1. The mucous membrane in the bronchial tree changes from pseudostratified ciliated columnar epithelium in the primary bronchi, secondary bronchi, and tertiary bronchi to ciliated simple columnar epithelium with some goblet cells in larger bronchioles, to mostly ciliated simple cuboidal epithelium with no goblet cells in smaller bronchioles, to mostly nonciliated simple cuboidal epithelium in terminal bronchioles. (In regions where simple nonciliated cuboidal epithelium is present, inhaled particles are removed by macrophages.)
2. Plates of cartilage gradually replace the incomplete rings of cartilage in primary bronchi and finally disappear in the distal bronchioles.
3. As the amount of cartilage decreases, the amount of smooth muscle increases. Smooth muscle encircles the lumen in spiral bands. Because there is no supporting cartilage, however, muscle spasms can close off the airways. This is what happens during an asthma attack, which can be a life-threatening situation.

During exercise, activity in the sympathetic division of the autonomic nervous system (ANS) increases and the adrenal medulla releases the hormones epinephrine and norepinephrine; both of these events cause relaxation of smooth muscle in the bronchioles, which dilates the airways. Because air reaches the alveoli more quickly, lung ventilation improves. The parasympathetic division of the ANS and mediators of allergic reactions such as histamine have the opposite effect, causing contraction of bronchiolar smooth muscle, which results in constriction of distal bronchioles.

▶ CHECKPOINT

4. List the roles of each of the three anatomical regions of the pharynx in respiration.
5. How does the larynx function in respiration and voice production?
6. Describe the location, structure, and function of the trachea.
7. Describe the structure of the bronchial tree.

Lungs

The **lungs** (= lightweights, because they float) are paired cone-shaped organs in the thoracic cavity. They are separated from each other by the heart and other structures in the mediastinum, which divides the thoracic cavity into two anatomically distinct chambers. As a result, if trauma causes one lung to collapse, the other may remain expanded. Two layers of serous membrane, collectively called the **pleural membrane** (PLOOR-al; *pleur-* = side), enclose and protect each lung. The superficial layer, called the **parietal pleura,** lines the wall of the thoracic cavity; the deep layer, the **visceral pleura,** covers the lungs themselves (Figure 23.9). Between the visceral and parietal pleurae is a small space, the **pleural cavity,** which contains a small amount of lubricating fluid secreted by the membranes. This pleural fluid reduces friction between the membranes, allowing them to slide easily over one another during breathing. Pleural fluid also causes the two membranes to adhere to one another just as a film of water causes two glass slides to stick together, a phenomenon called surface tension. Separate pleural cavities surround the left and right lungs. Inflammation of the pleural membrane, called **pleurisy** or **pleuritis,** may in its early stages cause pain due to friction between the parietal and visceral layers of the pleura. If the inflammation persists, excess fluid accumulates in the pleural space, a condition known as **pleural effusion.**

Pneumothorax and Hemothorax

In certain conditions, the pleural cavities may fill with air (**pneumothorax;** *pneumo-* = air or breath), blood (**hemothorax**), or pus. Air in the pleural cavities, most commonly introduced in a surgical opening of the chest or as a result of a stab or gunshot wound, may cause the lungs to collapse. This collapse of a part of a lung, or rarely an entire lung, is called **atelectasis** (at′-e-LEK-ta-sis; *ateles-* = incomplete; *-ectasis-* = expansion). The goal of treatment is the evacuation of air (or blood) from the pleural space, which allows the lung to reinflate. A small pneumothorax may resolve on its own, but it is often necessary to insert a chest tube to assist in evacuation. ■

Figure 23.9 **Relationship of the pleural membranes to the lungs.** The arrow in the inset indicates the direction from which the lungs are viewed (inferior).

The parietal pleura lines the thoracic cavity, whereas the visceral pleura covers the lungs.

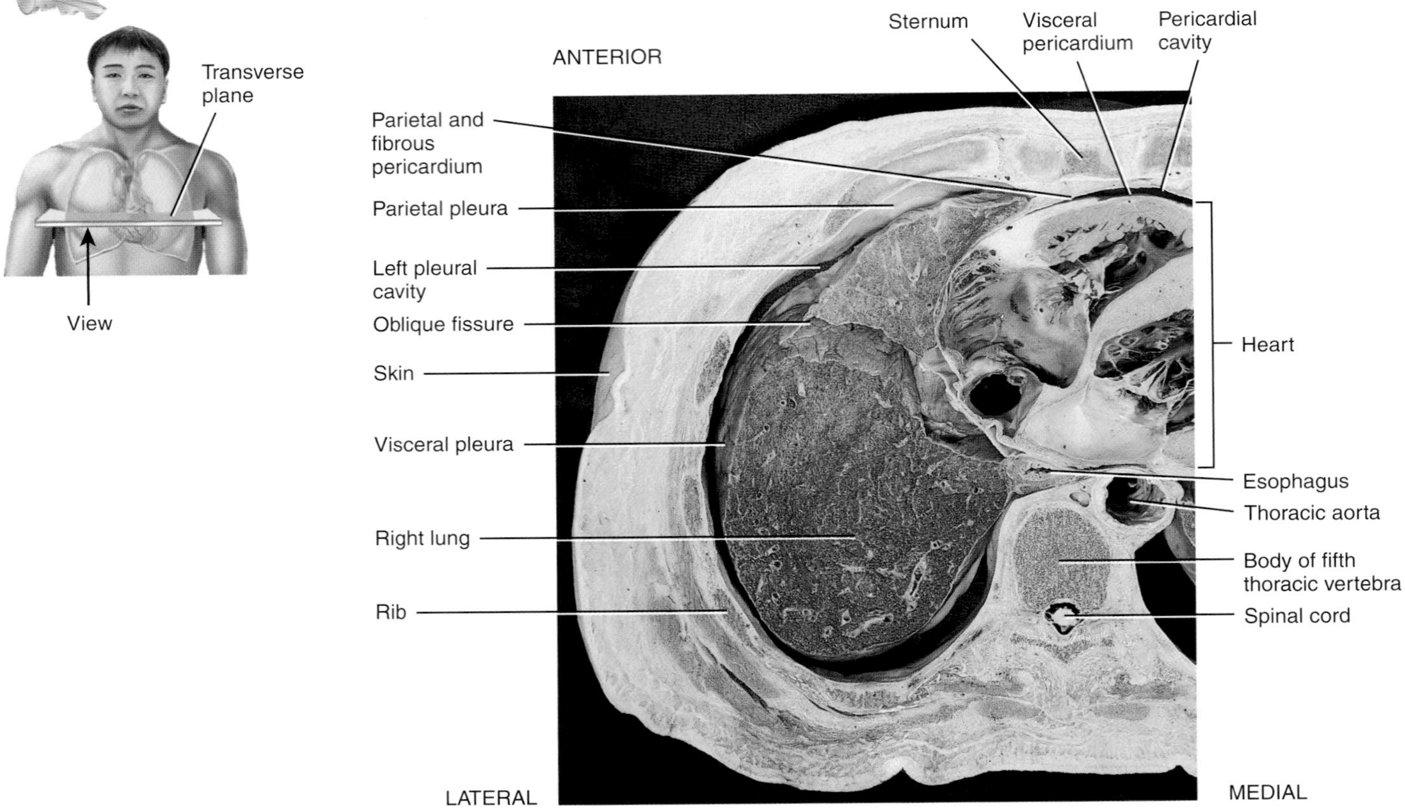

Transverse plane

View

ANTERIOR

Sternum

Visceral pericardium

Pericardial cavity

Parietal and fibrous pericardium

Parietal pleura

Left pleural cavity

Oblique fissure

Skin

Visceral pleura

Right lung

Rib

Heart

Esophagus

Thoracic aorta

Body of fifth thoracic vertebra

Spinal cord

LATERAL

MEDIAL

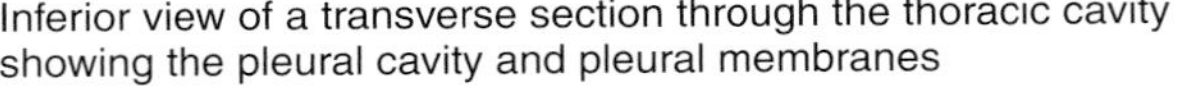

Inferior view of a transverse section through the thoracic cavity showing the pleural cavity and pleural membranes

What type of membrane is the pleural membrane?

The lungs extend from the diaphragm to just slightly superior to the clavicles and lie against the ribs anteriorly and posteriorly (Figure 23.10a). The broad inferior portion of the lung, the **base,** is concave and fits over the convex area of the diaphragm. The narrow superior portion of the lung is the **apex.** The surface of the lung lying against the ribs, the **costal surface,** matches the rounded curvature of the ribs. The **mediastinal (medial) surface** of each lung contains a region, the **hilum,** through which bronchi, pulmonary blood vessels, lymphatic vessels, and nerves enter and exit (Figure 23.10e). These structures are held together by the pleura and connective tissue and constitute the **root** of the lung. Medially, the left lung also contains a concavity, the **cardiac notch,** in which the heart lies. Due to the space occupied by the heart, the left lung is about 10% smaller than the right lung. Although the right lung is thicker and broader, it is also somewhat shorter than the left lung because the diaphragm is higher on the right side, accommodating the liver that lies inferior to it.

The lungs almost fill the thorax (Figure 23.10a). The apex of the lungs lies superior to the medial third of the clavicles and is the only area that can be palpated. The anterior, lateral, and posterior surfaces of the lungs lie against the ribs. The base of the lungs extends from the sixth costal cartilage anteriorly to the spinous process of the tenth thoracic vertebra posteriorly. The pleura extends about 5 cm (2 in) below the base from the sixth costal cartilage anteriorly to the twelfth rib posteriorly. Thus, the lungs do not completely fill the pleural cavity in this area. Removal of excessive fluid in the pleural cavity can be accomplished without injuring lung tissue by inserting a needle posteriorly through the seventh intercostal space, a procedure called **thoracentesis** (thor′-a-sen-TĒ-sis; *-centesis* = puncture). The needle is passed along the superior border of the lower rib to avoid damage to the intercostal nerves and blood vessels. Inferior to the seventh intercostal space there is danger of penetrating the diaphragm.

Lobes, Fissures, and Lobules

One or two fissures divide each lung into lobes (Figure 23.10b–e). Both lungs have an **oblique fissure,** which extends

Figure 23.10 **Surface anatomy of the lungs.** (See Tortora, *A Photographic Atlas of the Human Body, Second Edition,* Figures 11.12 and 11.14.)

The oblique fissure divides the left lung into two lobes. The oblique and horizontal fissures divide the right lung into three lobes.

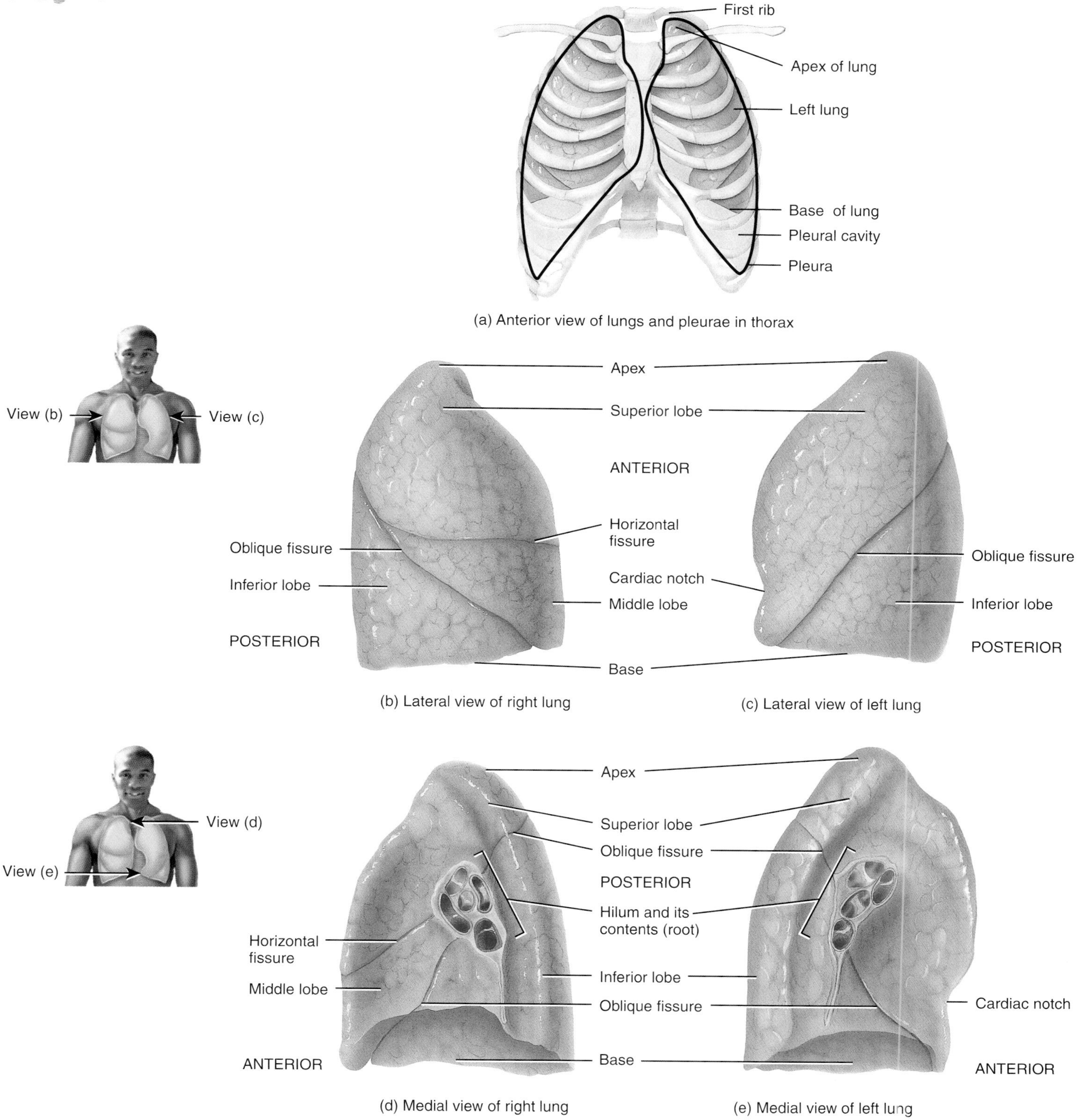

First rib
Apex of lung
Left lung
Base of lung
Pleural cavity
Pleura

(a) Anterior view of lungs and pleurae in thorax

View (b)
View (c)

Apex
Superior lobe
ANTERIOR
Horizontal fissure
Oblique fissure
Inferior lobe
Cardiac notch
Middle lobe
POSTERIOR
Oblique fissure
Inferior lobe
POSTERIOR
Base

(b) Lateral view of right lung

(c) Lateral view of left lung

View (d)
View (e)

Apex
Superior lobe
Oblique fissure
POSTERIOR
Hilum and its contents (root)
Horizontal fissure
Middle lobe
Inferior lobe
Oblique fissure
Cardiac notch
ANTERIOR
Base
ANTERIOR

(d) Medial view of right lung

(e) Medial view of left lung

Why are the right and left lungs slightly different in size and shape?

inferiorly and anteriorly; the right lung also has a **horizontal fissure.** The oblique fissure in the left lung separates the **superior lobe** from the **inferior lobe.** In the right lung, the superior part of the oblique fissure separates the superior lobe from the inferior lobe; the inferior part of the oblique fissure separates the inferior lobe from the **middle lobe,** which is bordered superiorly by the horizontal fissure.

Each lobe receives its own secondary (lobar) bronchus. Thus, the right primary bronchus gives rise to three secondary (lobar) bronchi called the **superior, middle,** and **inferior secondary (lobar) bronchi,** and the left primary bronchus gives rise to **superior** and **inferior secondary (lobar) bronchi.** Within the lung, the secondary bronchi give rise to the **tertiary (segmental) bronchi,** which are constant in both origin and distribution—there are 10 tertiary bronchi in each lung. The segment of lung tissue that each tertiary bronchus supplies is called a **bronchopulmonary segment.** Bronchial and pulmonary disorders (such as tumors or abscesses) that are localized in a bronchopulmonary segment may be surgically removed without seriously disrupting the surrounding lung tissue.

Each bronchopulmonary segment of the lungs has many small compartments called **lobules;** each lobule is wrapped in elastic connective tissue and contains a lymphatic vessel, an arteriole, a venule, and a branch from a terminal bronchiole (Figure 23.11a). Terminal bronchioles subdivide into microscopic branches called **respiratory bronchioles** (Figure 23.11b). As the respiratory bronchioles penetrate more deeply into the lungs, the epithelial lining changes from simple cuboidal to simple squamous. Respiratory bronchioles, in turn, subdivide into several (2–11) **alveolar ducts.** The respiratory passages from the trachea to the alveolar ducts contain about 25 orders of branching; branching from the trachea into primary bronchi is called first-order branching, from primary bronchi into secondary bronchi is called second-order branching, and so on down to the alveolar ducts.

Alveoli

Around the circumference of the alveolar ducts are numerous alveoli and alveolar sacs. An **alveolus** (al-VĒ-ō-lus) is a cup-

Figure 23.11 Microscopic anatomy of a lobule of the lungs.

Alveolar sacs consist of two or more alveoli that share a common opening.

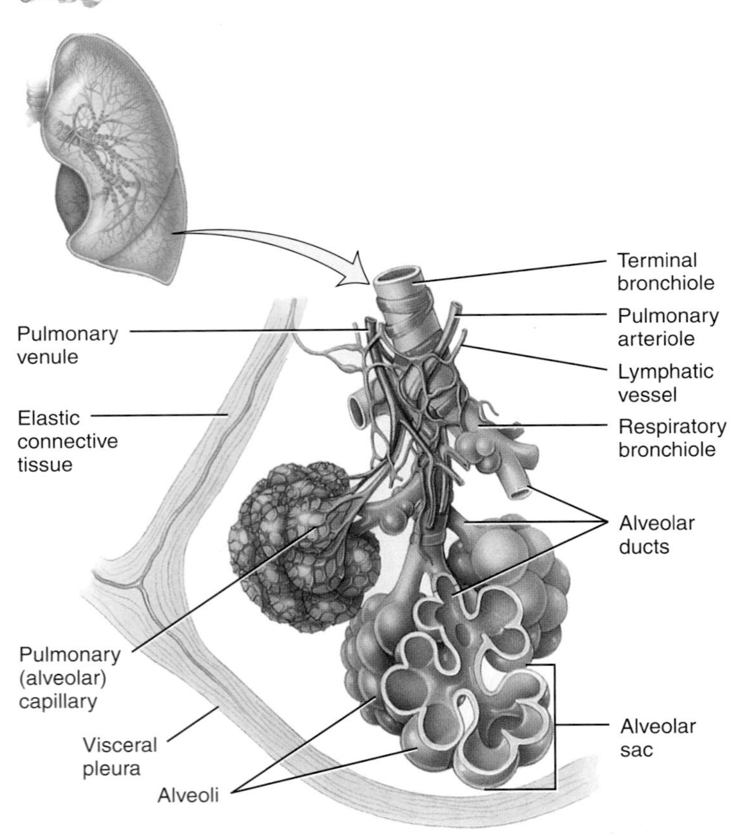

(a) Diagram of a portion of a lobule of the lung

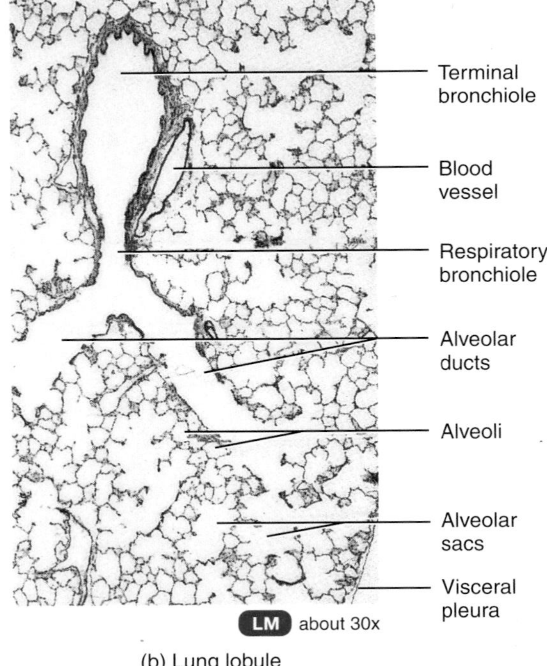

(b) Lung lobule

What types of cells make up the wall of an alveolus?

shaped outpouching lined by simple squamous epithelium and supported by a thin elastic basement membrane; an **alveolar sac** consists of two or more alveoli that share a common opening (Figure 23.11a, b). The walls of alveoli consist of two types of alveolar epithelial cells (Figure 23.12). The more numerous **type I alveolar cells** are simple squamous epithelial cells that form a nearly continuous lining of the alveolar wall. **Type II alveolar cells,** also called **septal cells,** are fewer in number and are found between type I alveolar cells. The thin type I alveolar cells are the main sites of gas exchange. Type II alveolar cells, rounded or cuboidal epithelial cells with free surfaces containing microvilli, secrete alveolar fluid, which keeps the surface between the cells and the air moist. Included in the alveolar fluid is **surfactant** (sur-FAK-tant), a complex mixture of phospholipids and lipoproteins. Surfactant lowers the surface tension of alveolar fluid, which reduces the tendency of alveoli to collapse (described later).

Associated with the alveolar wall are **alveolar macrophages (dust cells),** phagocytes that remove fine dust particles and other debris from the alveolar spaces. Also present are fibroblasts that produce reticular and elastic fibers. Underlying the layer of type I alveolar cells is an elastic basement membrane. On the outer surface of the alveoli, the lobule's arteriole and venule disperse into a network of blood capillaries (see Figure 23.11a) that consist of a single layer of endothelial cells and basement membrane.

Figure 23.12 Structural components of an alveolus. The respiratory membrane consists of a layer of type I and type II alveolar cells, an epithelial basement membrane, a capillary basement membrane, and the capillary endothelium.

The exchange of respiratory gases occurs by diffusion across the respiratory membrane.

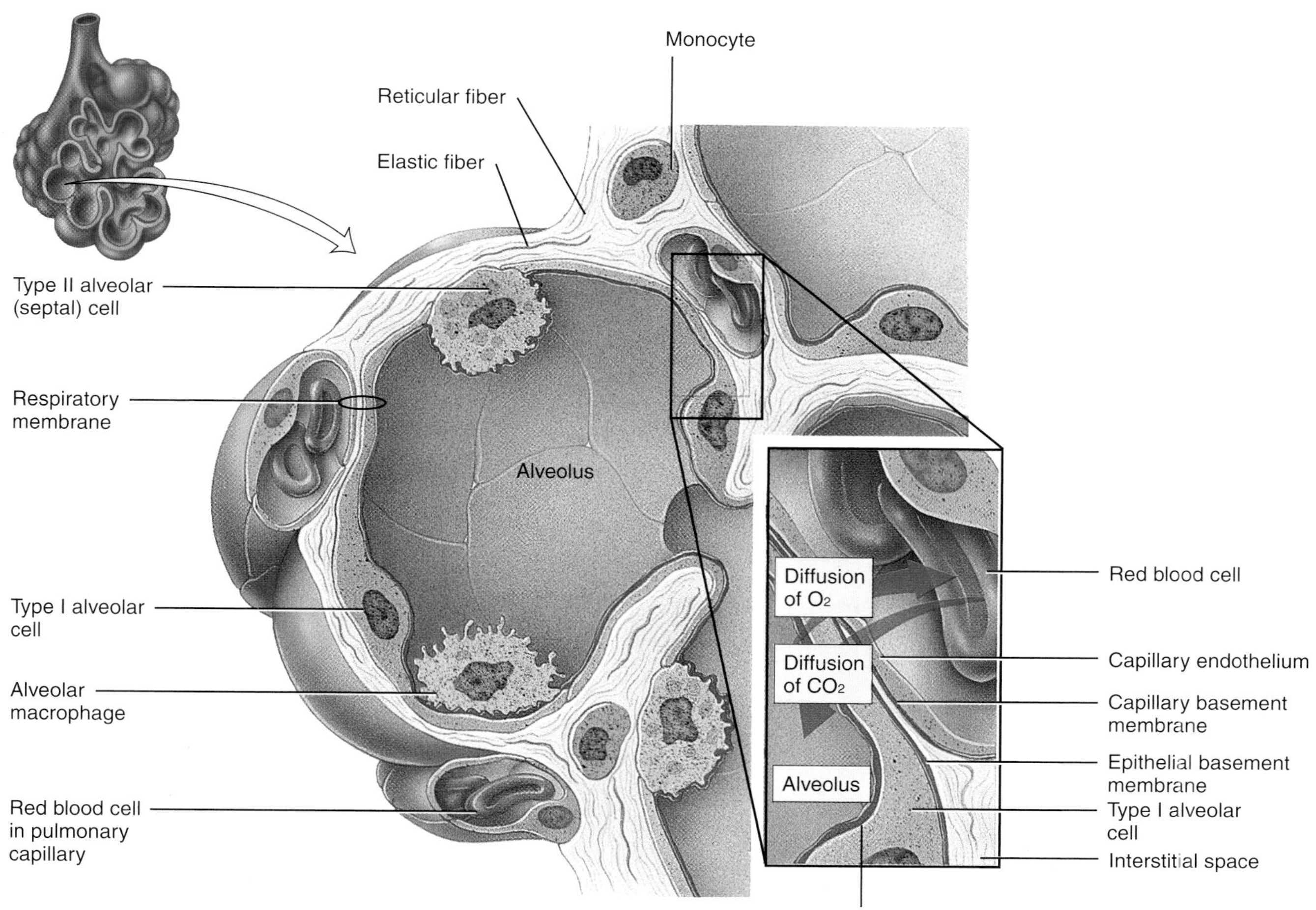

(a) Section through an alveolus showing its cellular components

(b) Details of respiratory membrane

continues

Figure 23.12 **(Continued)**

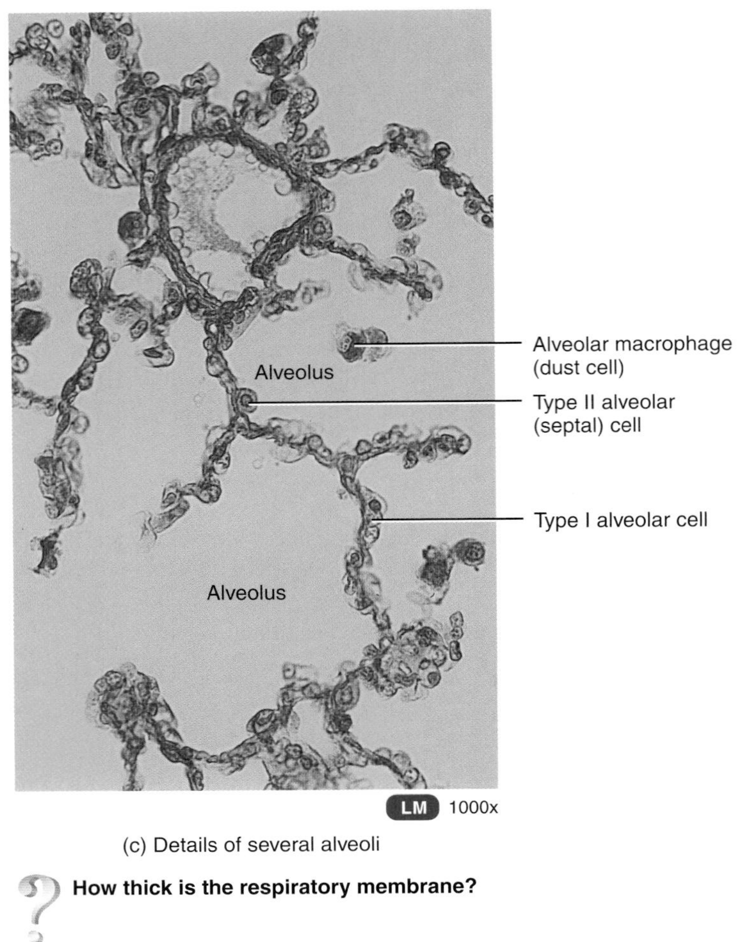

(c) Details of several alveoli

How thick is the respiratory membrane?

The exchange of O_2 and CO_2 between the air spaces in the lungs and the blood takes place by diffusion across the alveolar and capillary walls, which together form the **respiratory membrane.** Extending from the alveolar air space to blood plasma, the respiratory membrane consists of four layers (Figure 23.12b):

1. A layer of type I and type II alveolar cells and associated alveolar macrophages that constitutes the **alveolar wall**
2. An **epithelial basement membrane** underlying the alveolar wall
3. A **capillary basement membrane** that is often fused to the epithelial basement membrane
4. The **capillary endothelium**

Despite having several layers, the respiratory membrane is very thin—only 0.5 μm thick, about one-sixteenth the diameter of a red blood cell—to allow rapid diffusion of gases. It has been estimated that the lungs contain 300 million alveoli, providing an immense surface area of 70 m^2 (750 ft^2)—about the size of a racquetball court—for gas exchange.

Blood Supply to the Lungs

The lungs receive blood via two sets of arteries: pulmonary arteries and bronchial arteries. Deoxygenated blood passes through the pulmonary trunk, which divides into a left pulmonary artery that enters the left lung and a right pulmonary artery that enters the right lung. (The pulmonary arteries are the only arteries in the body that carry deoxygenated blood.) Return of the oxygenated blood to the heart occurs by way of the four pulmonary veins, which drain into the left atrium (see Figure 21.29 on page 793). A unique feature of pulmonary blood vessels is their constriction in response to localized hypoxia (low O_2 level). In all other body tissues, hypoxia causes dilation of blood vessels to increase blood flow. In the lungs, however, vasoconstriction in response to hypoxia diverts pulmonary blood from poorly ventilated areas of the lungs to well-ventilated regions. This phenomenon is known as **ventilation-perfusion coupling** because the perfusion (blood flow) to each area of the lungs matches the extent of ventilation (airflow) to alveoli in that area.

Bronchial arteries, which branch from the aorta, deliver oxygenated blood to the lungs. This blood mainly perfuses the walls of the bronchi and bronchioles. Connections exist between branches of the bronchial arteries and branches of the pulmonary arteries, however, and most blood returns to the heart via pulmonary veins. Some blood, however, drains into bronchial veins, branches of the azygos system, and returns to the heart via the superior vena cava.

▶ CHECKPOINT

8. Where are the lungs located? Distinguish the parietal pleura from the visceral pleura.
9. Define each of the following parts of a lung: base, apex, costal surface, medial surface, hilum, root, cardiac notch, lobe, and lobule.
10. What is a bronchopulmonary segment?
11. Describe the histology and function of the respiratory membrane.

PULMONARY VENTILATION

▶ OBJECTIVE

Describe the events that cause inhalation and exhalation.

The process of gas exchange in the body, called **respiration,** has three basic steps:

1. **Pulmonary ventilation** (*pulmon-* = lung), or **breathing,** is the inhalation (inflow) and exhalation (outflow) of air between the atmosphere and the alveoli of the lungs.
2. **External (pulmonary) respiration** is the exchange of gases between the alveoli of the lungs and the blood in pulmonary capillaries across the respiratory membrane. In this process, pulmonary capillary blood gains O_2 and loses CO_2.
3. **Internal (tissue) respiration** is the exchange of gases between blood in systemic capillaries and tissue cells. In this step the blood loses O_2 and gains CO_2. Within cells, the metabolic reactions that consume O_2 and give off CO_2 during the production of ATP are termed *cellular respiration* (discussed in Chapter 25).

In pulmonary ventilation, air flows between the atmosphere and the alveoli of the lungs because of alternating pressure differences created by contraction and relaxation of respiratory muscles. The rate of airflow and the amount of effort needed for breathing is also influenced by alveolar surface tension, compliance of the lungs, and airway resistance.

Pressure Changes During Pulmonary Ventilation

Air moves into the lungs when the air pressure inside the lungs is less than the air pressure in the atmosphere. Air moves out of the lungs when the air pressure inside the lungs is greater than the air pressure in the atmosphere.

Inhalation

Breathing in is called **inhalation (inspiration).** Just before each inhalation, the air pressure inside the lungs is equal to the air pressure of the atmosphere, which at sea level is about 760 millimeters of mercury (mmHg), or 1 atmosphere (atm). For air to flow into the lungs, the pressure inside the alveoli must become lower than the atmospheric pressure. This condition is achieved by increasing the volume of the lungs.

The pressure of a gas in a closed container is inversely proportional to the volume of the container. This means that if the size of a closed container is increased, the pressure of the gas inside the container decreases, and that if the size of the container is decreased, then the pressure inside it increases. This inverse relationship between volume and pressure, called **Boyle's law,** may be demonstrated as follows (Figure 23.13): Suppose we place a gas in a cylinder that has a movable piston and a pressure gauge, and that the initial pressure created by the gas molecules striking the wall of the container is 1 atm. If the piston is pushed down, the gas is compressed into a smaller volume, so that the same number of gas molecules strike less wall area. The gauge shows that the pressure doubles as the gas is compressed to half its original volume. In other words, the same number of molecules in half the volume produces twice the pressure. Conversely, if the piston is raised to increase the volume, the pressure decreases. Thus, the pressure of a gas varies inversely with volume.

Differences in pressure caused by changes in lung volume force air into our lungs when we inhale and out when we exhale. For inhalation to occur, the lungs must expand, which increases lung volume and thus decreases the pressure in the lungs to below atmospheric pressure. The first step in expanding the lungs during normal quiet inhalation involves contraction of the

Figure 23.13 **Boyle's law.**

The volume of a gas varies inversely with its pressure.

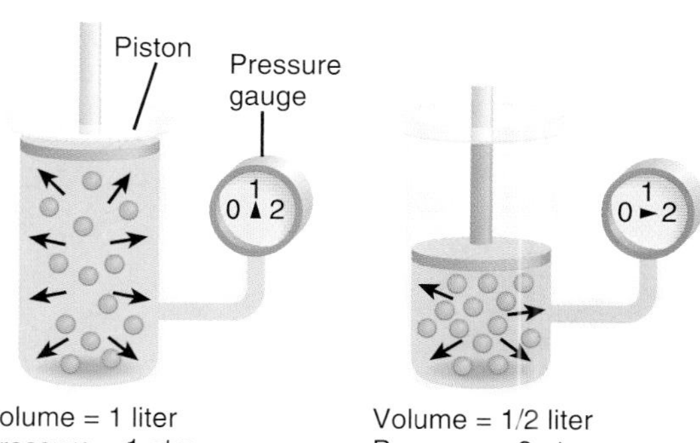

If the volume is decreased from 1 liter to 1/4 liter, how would the pressure change?

main muscles of inhalation, the diaphragm and external intercostals (Figure 23.14).

The most important muscle of inhalation is the diaphragm, the dome-shaped skeletal muscle that forms the floor of the thoracic cavity. It is innervated by fibers of the phrenic nerves, which emerge from the spinal cord at cervical levels 3, 4, and 5. Contraction of the diaphragm causes it to flatten, lowering its dome. This increases the vertical diameter of the thoracic cavity. During normal quiet inhalation, the diaphragm descends about 1 cm (0.4 in.), producing a pressure difference of 1–3 mmHg and the inhalation of about 500 mL of air. In strenuous breathing, the diaphragm may descend 10 cm (4 in.), which produces a pressure difference of 100 mmHg and the inhalation of 2–3 liters of air. Contraction of the diaphragm is responsible for about 75% of the air that enters the lungs during quiet breathing. Advanced pregnancy, excessive obesity, or confining abdominal clothing can prevent complete descent of the diaphragm.

The next most important muscles of inhalation are the external intercostals. When these muscles contract, they elevate the ribs. As a result, there is an increase in the anteroposterior and

Figure 23.14 Muscles of inhalation and exhalation and their actions. The pectoralis minor muscle (not shown here) is illustrated in Figure 11.14a on page 361.

During deep, labored breathing, accessory muscles of inhalation (sternocleidomastoids, scalenes, and pectoralis minors) participate.

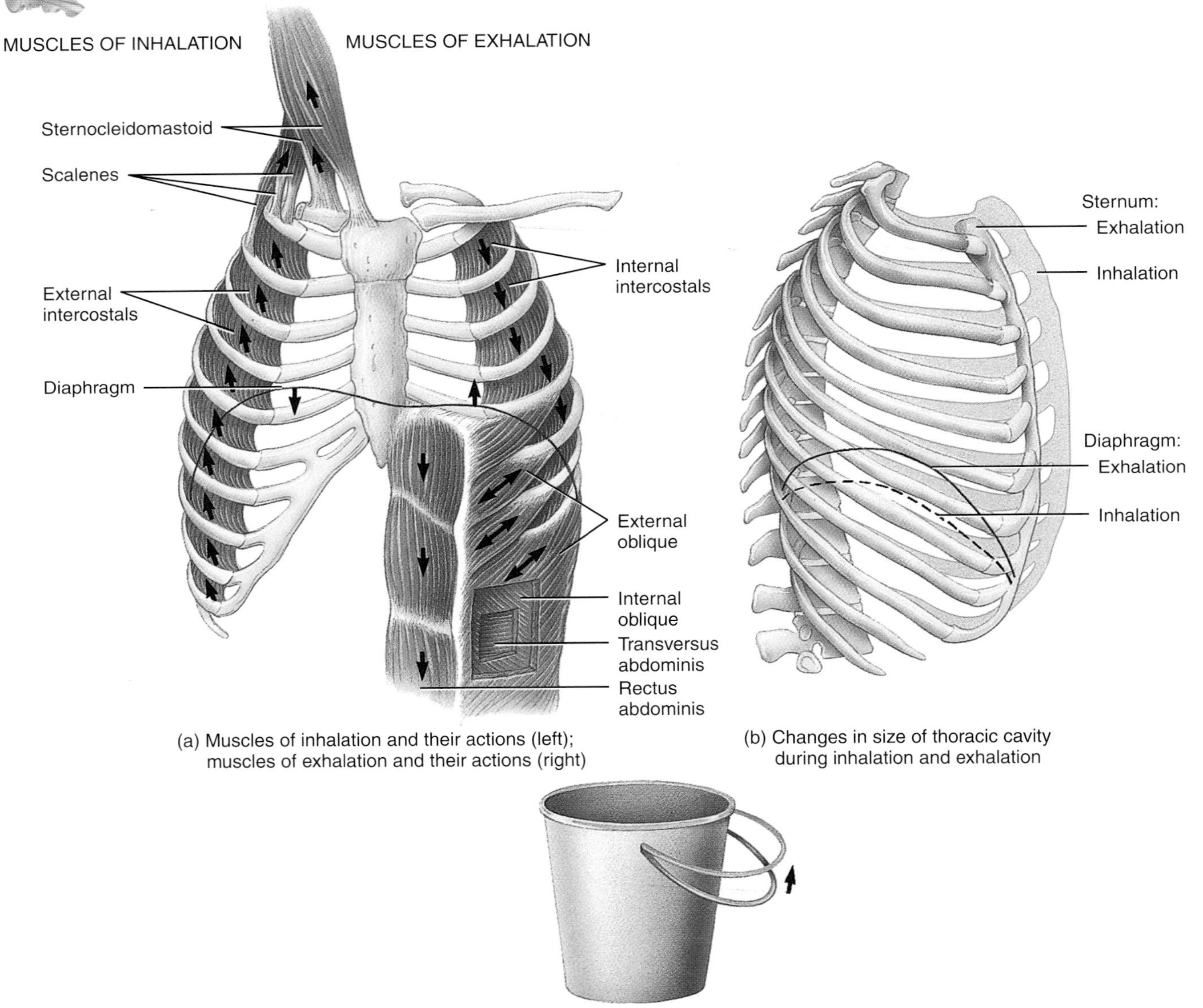

(a) Muscles of inhalation and their actions (left); muscles of exhalation and their actions (right)

(b) Changes in size of thoracic cavity during inhalation and exhalation

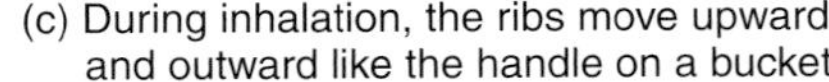

(c) During inhalation, the ribs move upward and outward like the handle on a bucket

Right now, what is the main muscle that powers your breathing?

lateral diameters of the chest cavity. Contraction of the external intercostals is responsible for about 25% of the air that enters the lungs during normal quiet breathing.

During quiet inhalations, the pressure between the two pleural layers in the pleural cavity, called **intrapleural (intrathoracic) pressure,** is always subatmospheric (lower than atmospheric pressure). Just before inhalation, it is about 4 mmHg less than the atmospheric pressure, or about 756 mmHg at an atmospheric pressure of 760 mmHg (Figure 23.15). As the diaphragm and external intercostals contract and the overall size of the thoracic cavity increases, the volume of the pleural cavity also increases, which causes intrapleural pressure to decrease to about 754 mmHg. During expansion of the thorax, the parietal and visceral pleurae normally adhere tightly because of the subatmospheric pressure between them and because of the surface tension created by their moist adjoining surfaces. As the thoracic cavity expands, the parietal pleura lining the cavity is pulled outward in all directions, and the visceral pleura and lungs are pulled along with it.

As the volume of the lungs increases in this way, the pressure inside the lungs, called the **alveolar (intrapulmonic) pressure,** drops from 760 to 758 mmHg. A pressure difference is thus established between the atmosphere and the alveoli. Because air always flows from a region of higher pressure to a region of lower pressure, inhalation takes place. Air continues to flow into the lungs as long as a pressure difference exists. During deep, forceful inhalations, accessory muscles of inspiration also participate in increasing the size of the thoracic cavity (see Figure 23.14a). The muscles are so named because they make little, if any, contribution during normal quiet inhalation, but during exercise or forced ventilation they may contract vigorously. The accessory muscles of inhalation include the sternocleidomastoid muscles, which elevate the sternum; the

Figure 23.15 Pressure changes in pulmonary ventilation. During inhalation, the diaphragm contracts, the chest expands, the lungs are pulled outward, and alveolar pressure decreases. During exhalation, the diaphragm relaxes, the lungs recoil inward, and alveolar pressure increases, forcing air out of the lungs.

Air moves into the lungs when alveolar pressure is less than atmospheric pressure, and out of the lungs when alveolar pressure is greater than atmospheric pressure.

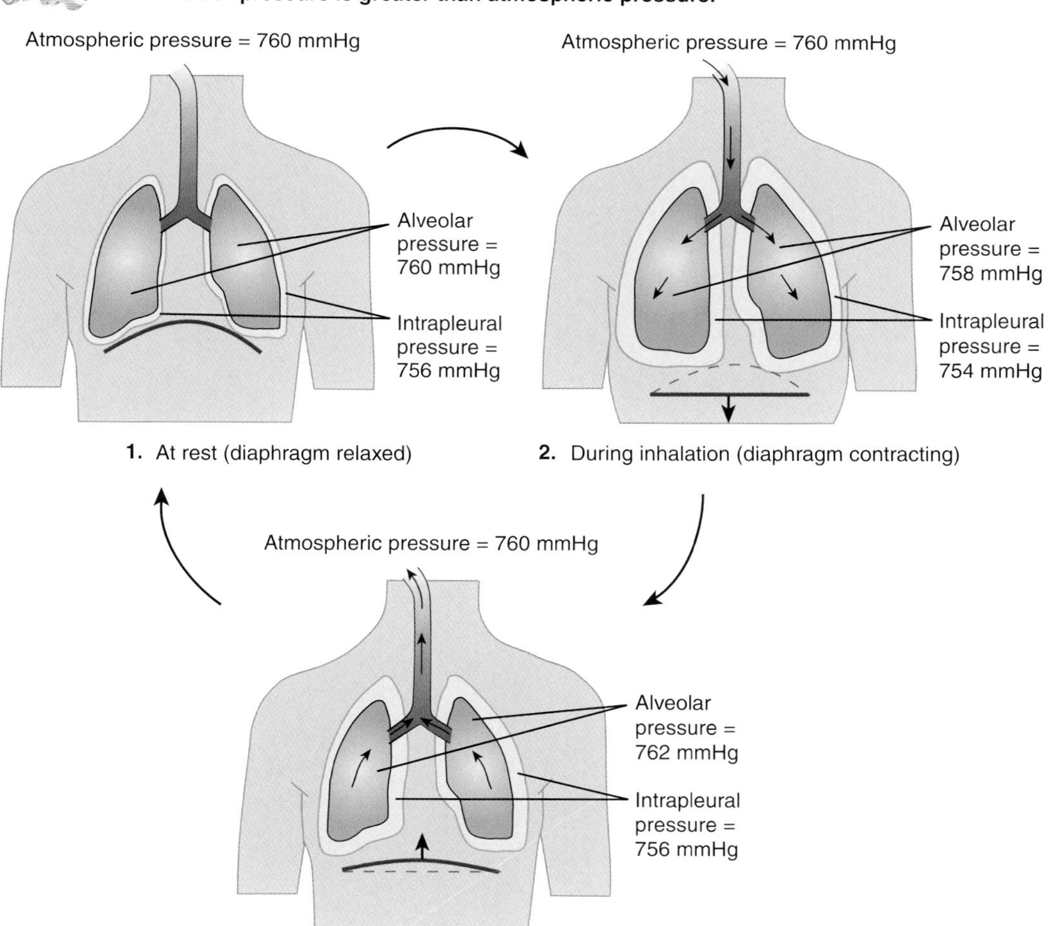

How does the intrapleural pressure change during a normal, quiet breath?

scalene muscles, which elevate the first two ribs; and the pectoralis minor muscles, which elevate the third through fifth ribs. Because both normal quiet inhalation and inhalation during exercise or forced ventilation involve muscular contraction, the process of inhalation is said to be *active.*

Figure 23.16a summarizes the events of inhalation.

Exhalation

Breathing out, called **exhalation (expiration),** is also due to a pressure gradient, but in this case the gradient is in the opposite direction: The pressure in the lungs is greater than the pressure of the atmosphere. Normal exhalation during quiet breathing, unlike inhalation, is a *passive process* because no muscular contractions are involved. Instead, exhalation results from **elastic recoil** of the chest wall and lungs, both of which have a natural tendency to spring back after they have been stretched. Two inwardly directed forces contribute to elastic recoil: (1) the recoil of elastic fibers that were stretched during inhalation and (2) the inward pull of surface tension due to the film of alveolar fluid.

Exhalation starts when the inspiratory muscles relax. As the diaphragm relaxes, its dome moves superiorly owing to its elasticity. As the external intercostals relax, the ribs are depressed. These movements decrease the vertical, lateral, and anteroposterior diameters of the thoracic cavity, which decreases lung volume. In turn, the alveolar pressure increases, to about 762 mmHg. Air then flows from the area of higher pressure in the alveoli to the area of lower pressure in the atmosphere (see Figure 23.15).

Exhalation becomes active only during forceful breathing, as occurs while playing a wind instrument or during exercise. During these times, muscles of exhalation—the abdominals and internal intercostals (see Figure 23.14a)—contract, which increases pressure in the abdominal region and thorax. Contraction of the abdominal muscles moves the inferior ribs downward and compresses the abdominal viscera, thereby forcing the diaphragm superiorly. Contraction of the internal intercostals, which extend inferiorly and posteriorly between adjacent ribs, pulls the ribs inferiorly. Although intrapleural pressure is always less than alveolar pressure, it may briefly exceed atmospheric pressure during a forceful exhalation, such as during a cough.

Figure 23.16b summarizes the events of exhalation.

Other Factors Affecting Pulmonary Ventilation

As you have just learned, air pressure differences drive airflow during inhalation and exhalation. However, three other factors affect the rate of airflow and the ease of pulmonary ventilation: surface tension of the alveolar fluid, compliance of the lungs, and airway resistance.

Figure 23.16 Summary of events of inhalation and exhalation.

Inhalation and exhalation are caused by changes in alveolar pressure.

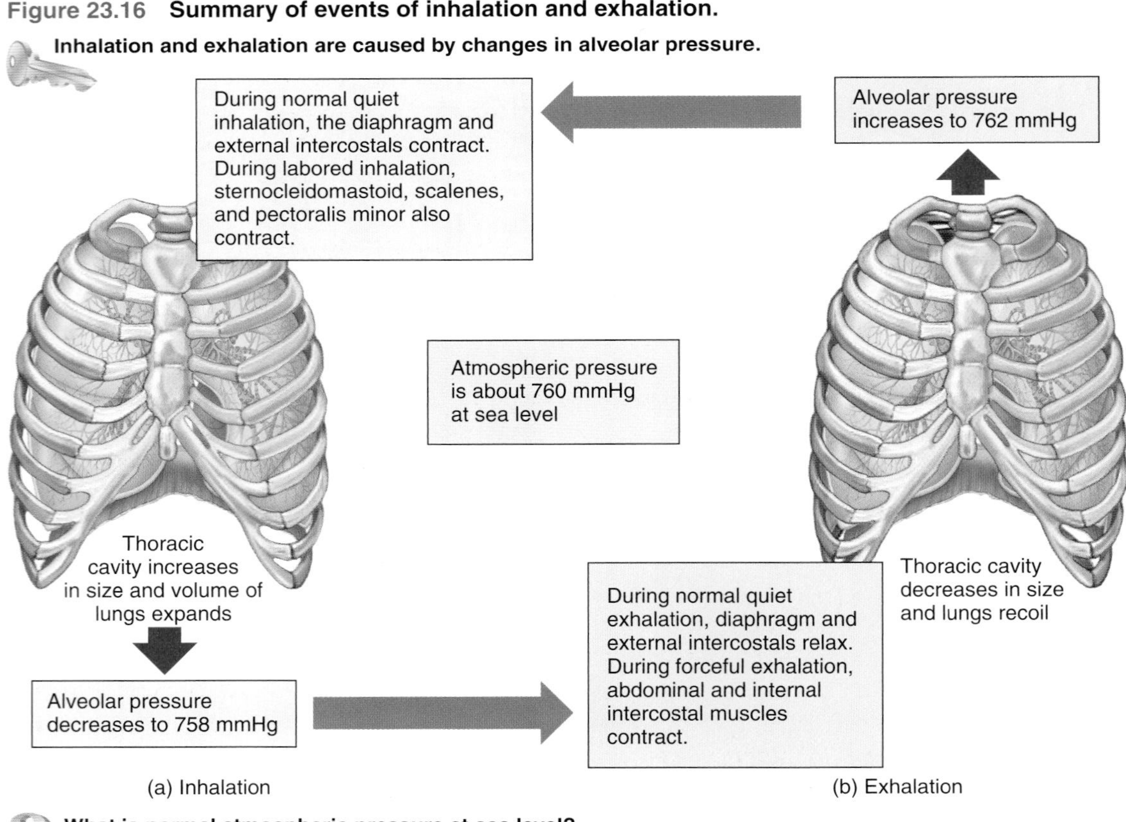

During normal quiet inhalation, the diaphragm and external intercostals contract. During labored inhalation, sternocleidomastoid, scalenes, and pectoralis minor also contract.
Alveolar pressure increases to 762 mmHg
Atmospheric pressure is about 760 mmHg at sea level
Thoracic cavity increases in size and volume of lungs expands
Alveolar pressure decreases to 758 mmHg
During normal quiet exhalation, diaphragm and external intercostals relax. During forceful exhalation, abdominal and internal intercostal muscles contract.
Thoracic cavity decreases in size and lungs recoil
(a) Inhalation
(b) Exhalation

What is normal atmospheric pressure at sea level?

Surface Tension of Alveolar Fluid

As noted earlier, a thin layer of alveolar fluid coats the luminal surface of alveoli and exerts a force known as **surface tension.** Surface tension arises at all air–water interfaces because the polar water molecules are more strongly attracted to each other than they are to gas molecules in the air. When liquid surrounds a sphere of air, as in an alveolus or a soap bubble, surface tension produces an inwardly directed force. Soap bubbles "burst" because they collapse inward due to surface tension. In the lungs, surface tension causes the alveoli to assume the smallest possible diameter. During breathing, surface tension must be overcome to expand the lungs during each inhalation. Surface tension also accounts for two-thirds of lung elastic recoil, which decreases the size of alveoli during exhalation.

The surfactant (a mixture of phospholipids and lipoproteins) present in alveolar fluid reduces its surface tension below the surface tension of pure water. A deficiency of surfactant in premature infants causes *respiratory distress syndrome,* in which the surface tension of alveolar fluid is greatly increased, so that many alveoli collapse at the end of each exhalation. Great effort is then needed at the next inhalation to reopen the collapsed alveoli.

Respiratory Distress Syndrome

Respiratory distress syndrome (RDS) is a breathing disorder of premature newborns in which the alveoli do not remain open due to a lack of surfactant. Recall that surfactant reduces surface tension and is necessary to prevent the collapse of alveoli during exhalation. The more premature the newborn, the greater the chance that RDS will develop. The condition is also more common in infants whose mothers have diabetes, in males, and occurs more often in European Americans than African Americans. Symptoms of RDS include labored and irregular breathing, flaring of the nostrils during inhalation, grunting during exhalation, and perhaps a blue skin color. Besides the symptoms, RDS is diagnosed on the basis of chest radiographs and a blood test. A newborn with mild RDS may require only supplemental oxygen administered through an oxygen hood or through a tube placed in the nose. In severe cases oxygen may be delivered by continuous positive airway pressure (CPAP) through tubes in the nostrils or a mask on the face. In such cases surfactant may be administered directly into the lungs. ■

Compliance of the Lungs

Compliance refers to how much effort is required to stretch the lungs and chest wall. High compliance means that the lungs and chest wall expand easily; low compliance means that they resist expansion. By analogy, a thin balloon that is easy to inflate has high compliance, whereas a heavy and stiff balloon that takes a lot of effort to inflate has low compliance. In the lungs, compliance is related to two principal factors: elasticity and surface tension. The lungs normally have high compliance and expand easily because elastic fibers in lung tissue are easily stretched and surfactant in alveolar fluid reduces surface tension. Decreased compliance is a common feature in pulmonary conditions that (1) scar lung tissue (for example, tuberculosis), (2) cause lung tissue to become filled with fluid (pulmonary edema), (3) produce a deficiency in surfactant, or (4) impede lung expansion in any way (for example, paralysis of the intercostal muscles). Decreased lung compliance occurs in emphysema (see page 887) due to destruction of elastic fibers in alveolar walls.

Airway Resistance

Like the flow of blood through blood vessels, the rate of airflow through the airways depends on both the pressure difference and the resistance: Airflow equals the pressure difference between the alveoli and the atmosphere divided by the resistance. The walls of the airways, especially the bronchioles, offer some resistance to the normal flow of air into and out of the lungs. As the lungs expand during inhalation, the bronchioles enlarge because their walls are pulled outward in all directions. Larger-diameter airways have decreased resistance. Airway resistance then increases during exhalation as the diameter of bronchioles decreases. Airway diameter is also regulated by the degree of contraction or relaxation of smooth muscle in the walls of the airways. Signals from the sympathetic division of the autonomic nervous system cause relaxation of this smooth muscle, which results in bronchodilation and decreased resistance.

Any condition that narrows or obstructs the airways increases resistance, so that more pressure is required to maintain the same airflow. The hallmark of asthma or chronic obstructive pulmonary disease (COPD)—emphysema or chronic bronchitis—is increased airway resistance due to obstruction or collapse of airways.

Breathing Patterns and Modified Respiratory Movements

The term for the normal pattern of quiet breathing is **eupnea** (ūp-NĒ-a; *eu-* = good, easy, or normal; *-pnea* = breath). Eupnea can consist of shallow, deep, or combined shallow and deep breathing. A pattern of shallow (chest) breathing, called **costal breathing,** consists of an upward and outward movement of the chest due to contraction of the external intercostal muscles. A pattern of deep (abdominal) breathing, called **diaphragmatic breathing,** consists of the outward movement of the abdomen due to the contraction and descent of the diaphragm.

Respirations also provide humans with methods for expressing emotions such as laughing, sighing, and sobbing. Moreover, respiratory air can be used to expel foreign matter from the lower air passages through actions such as sneezing and coughing. Respiratory movements are also modified and controlled during talking and singing. Some of the modified respiratory movements that express emotion or clear the airways are listed in Table 23.1. All these movements are reflexes, but some of them also can be initiated voluntarily.

▶ CHECKPOINT

12. What are the basic differences among pulmonary ventilation, external respiration, and internal respiration?
13. Compare what happens during quiet versus forceful pulmonary ventilation.
14. Describe how alveolar surface tension, compliance, and airway resistance affect pulmonary ventilation.
15. Demonstrate the various types of modified respiratory movements.

TABLE 23.1 Modified Respiratory Movements

Movement	Description
Coughing	A long-drawn and deep inhalation followed by a complete closure of the rima glottidis, which results in a strong exhalation that suddenly pushes the rima glottidis open and sends a blast of air through the upper respiratory passages. Stimulus for this reflex act may be a foreign body lodged in the larynx, trachea, or epiglottis.
Sneezing	Spasmodic contraction of muscles of exhalation that forcefully expels air through the nose and mouth. Stimulus may be an irritation of the nasal mucosa.
Sighing	A long-drawn and deep inhalation immediately followed by a shorter but forceful exhalation.
Yawning	A deep inhalation through the widely opened mouth producing an exaggerated depression of the mandible. It may be stimulated by drowsiness, or someone else's yawning, but the precise cause is unknown.
Sobbing	A series of convulsive inhalations followed by a single prolonged exhalation. The rima glottidis closes earlier than normal after each inhalation so only a little air enters the lungs with each inhalation.
Crying	An inhalation followed by many short convulsive exhalations, during which the rima glottidis remains open and the vocal folds vibrate; accompanied by characteristic facial expressions and tears.
Laughing	The same basic movements as crying, but the rhythm of the movements and the facial expressions usually differ from those of crying. Laughing and crying are sometimes indistinguishable.
Hiccupping	Spasmodic contraction of the diaphragm followed by a spasmodic closure of the rima glottidis, which produces a sharp sound on inhalation. Stimulus is usually irritation of the sensory nerve endings of the gastrointestinal tract.
Valsalva (val-SAL-va) **maneuver**	Forced exhalation against a closed rima glottidis as may occur during periods of straining while defecating.

LUNG VOLUMES AND CAPACITIES

▶ OBJECTIVES

Explain the difference between tidal volume, inspiratory reserve volume, expiratory reserve volume, and residual volume.

Differentiate between inspiratory capacity, functional residual capacity, vital capacity, and total lung capacity.

While at rest, a healthy adult averages 12 breaths a minute, with each inhalation and exhalation moving about 500 mL of air into and out of the lungs. The volume of one breath is called the **tidal volume (V_T).** The **minute ventilation (MV)**—the total volume of air inhaled and exhaled each minute—is respiratory rate multiplied by tidal volume:

$$\begin{aligned} MV &= 12 \text{ breaths/min} \times 500 \text{ mL/breath} \\ &= 6 \text{ liters/min} \end{aligned}$$

A lower-than-normal minute ventilation usually is a sign of pulmonary malfunction. The apparatus commonly used to measure the volume of air exchanged during breathing and the respiratory rate is a **spirometer** (*spiro-* = breathe; *meter* = measuring device) or **respirometer.** The record is called a **spirogram.** Inhalation is recorded as an upward deflection, and exhalation is recorded as a downward deflection (Figure 23.17).

Tidal volume varies considerably from one person to another and in the same person at different times. In a typical adult, about 70% of the tidal volume (350 mL) actually reaches the respiratory zone of the respiratory system—the respiratory bronchioles, alveolar ducts, alveolar sacs, and alveoli—and participates in external respiration. The other 30% (150 mL) remains in the conducting airways of the nose, pharynx, larynx, trachea, bronchi, bronchioles, and terminal bronchioles. Collectively, the conducting airways with air that does not undergo respiratory exchange are known as the **anatomic (respiratory) dead space.** (An easy rule of thumb for determining the volume of your anatomic dead space is that it is about the same in milliliters as your ideal weight in pounds.) Not all of the minute ventilation can be used in gas exchange because some of it remains in the anatomic dead space. The **alveolar ventilation rate** is the volume of air per minute that actually reaches the respiratory zone. In the example just given, alveolar ventilation rate would be 350 mL/breath × 12 breaths/min = 4200 mL/min.

Several other lung volumes are defined relative to forceful breathing. In general, these volumes are larger in males, taller individuals, and younger adults, and smaller in females, shorter individuals, and the elderly. Various disorders also may be diagnosed by comparison of actual and predicted normal values for a patient's gender, height, and age. The values given here are averages for young adults.

By taking a very deep breath, you can inhale a good deal more than 500 mL. This additional inhaled air, called the **inspiratory reserve volume,** is about 3100 mL in an average adult male and 1900 mL in an average adult female

(Figure 23.17). Even more air can be inhaled if inhalation follows forced exhalation. If you inhale normally and then exhale as forcibly as possible, you should be able to push out considerably more air in addition to the 500 mL of tidal volume. The extra 1200 mL in males and 700 mL in females is called the **expiratory reserve volume.** The **$FEV_{1.0}$** is the **forced expiratory volume in 1 second,** the volume of air that can be exhaled from the lungs in 1 second with maximal effort following a maximal inhalation. Typically, chronic obstructive pulmonary disease (COPD) greatly reduces $FEV_{1.0}$ because COPD increases airway resistance.

Even after the expiratory reserve volume is exhaled, considerable air remains in the lungs because the subatmospheric intrapleural pressure keeps the alveoli slightly inflated, and some air also remains in the noncollapsible airways. This volume, which cannot be measured by spirometry, is called the **residual volume** and amounts to about 1200 mL in males and 1100 mL in females.

If the thoracic cavity is opened, the intrapleural pressure rises to equal the atmospheric pressure and forces out some of the residual volume. The air remaining is called the **minimal volume.** Minimal volume provides a medical and legal tool for determining whether a baby is born dead (stillborn) or died after birth. The presence of minimal volume can be demonstrated by placing a piece of lung in water and observing if it floats. Fetal lungs contain no air, so the lung of a stillborn baby will not float in water.

Lung capacities are combinations of specific lung volumes (Figure 23.17). **Inspiratory capacity** is the sum of tidal volume and inspiratory reserve volume (500 mL + 3100 mL = 3600 mL in males and 500 mL + 1900 mL = 2400 mL in females). **Functional residual capacity** is the sum of residual volume and expiratory reserve volume (1200 mL + 1200 mL = 2400 mL in males and 1100 mL + 700 mL = 1800 mL in females). **Vital capacity** is the sum of inspiratory reserve volume, tidal volume, and expiratory reserve volume (4800 mL in males and 3100 mL in females). Finally, **total lung capacity** is the sum of vital capacity and residual volume (4800 mL + 1200 mL = 6000 mL in males and 3100 mL + 1100 mL = 4200 mL in females).

▶ **CHECKPOINT**

16. What is a spirometer?

17. What is the difference between a lung volume and a lung capacity?

18. How is minute ventilation calculated?

19. Define alveolar ventilation rate and $FEV_{1.0}$.

Figure 23.17 Spirogram of lung volumes and capacities. The average values for a healthy adult male and female are indicated, with the values for a female in parentheses. Note that the spirogram is read from right (start of record) to left (end of record).

Lung capacities are combinations of various lung volumes.

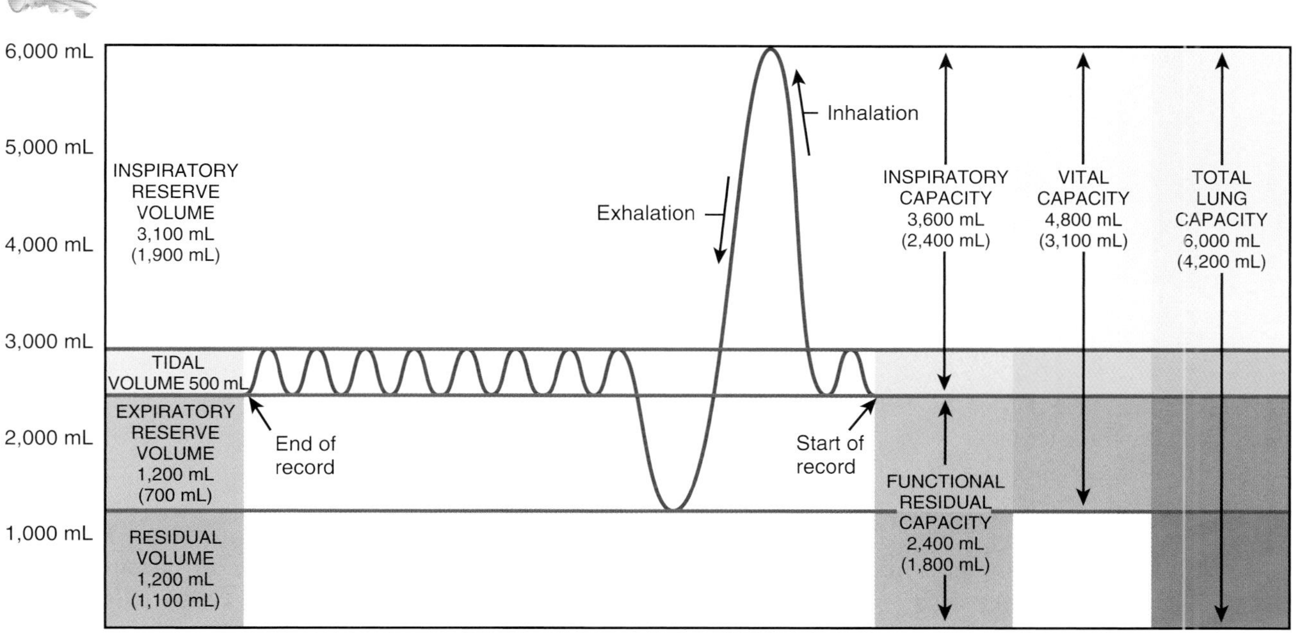

If you breathe in as deeply as possible and then exhale as much air as you can, which lung capacity have you demonstrated?

EXCHANGE OF OXYGEN AND CARBON DIOXIDE

▶ OBJECTIVES

Explain Dalton's law and Henry's law.

Describe the exchange of oxygen and carbon dioxide in external and internal respiration.

The exchange of oxygen and carbon dioxide between alveolar air and pulmonary blood occurs via passive diffusion, which is governed by the behavior of gases as described by two gas laws, Dalton's law and Henry's law. Dalton's law is important for understanding how gases move down their pressure differences by diffusion, and Henry's law helps explain how the solubility of a gas relates to its diffusion.

Gas Laws: Dalton's Law and Henry's Law

According to **Dalton's law,** each gas in a mixture of gases exerts its own pressure as if no other gases were present. The pressure of a specific gas in a mixture is called its *partial pressure* (P_x); the subscript is the chemical formula of the gas. The total pressure of the mixture is calculated simply by adding all the partial pressures. Atmospheric air is a mixture of gases—nitrogen (N_2), oxygen (O_2), water vapor (H_2O), and carbon dioxide (CO_2), plus other gases present in small quantities. Atmospheric pressure is the sum of the pressures of all these gases:

$$\text{Atmospheric pressure (760 mmHg)} = P_{N_2} + P_{O_2} + P_{H_2O} + P_{CO_2} + P_{\text{other gases}}$$

We can determine the partial pressure exerted by each component in the mixture by multiplying the percentage of the gas in the mixture by the total pressure of the mixture. Atmospheric air is 78.6% nitrogen, 20.9% oxygen, 0.04% carbon dioxide, and 0.06% other gases; a variable amount of water vapor is also present, about 0.4% on a cool, dry day. Thus, the partial pressures of the gases in inhaled air are as follows:

$$\begin{aligned} P_{N_2} &= 0.786 \times 760 \text{ mmHg} = 597.4 \text{ mmHg} \\ P_{O_2} &= 0.209 \times 760 \text{ mmHg} = 158.8 \text{ mmHg} \\ P_{H_2O} &= 0.004 \times 760 \text{ mmHg} = 3.0 \text{ mmHg} \\ P_{CO_2} &= 0.0004 \times 760 \text{ mmHg} = 0.3 \text{ mmHg} \\ P_{\text{other gases}} &= 0.0006 \times 760 \text{ mmHg} = 0.5 \text{ mmHg} \\ \text{Total} &= 760.0 \text{ mmHg} \end{aligned}$$

These partial pressures determine the movement of O_2 and CO_2 between the atmosphere and lungs, between the lungs and blood, and between the blood and body cells. Each gas diffuses across a permeable membrane from the area where its partial pressure is greater to the area where its partial pressure is less. The greater the difference in partial pressure, the faster the rate of diffusion.

Compared with inhaled air, alveolar air has less O_2 (13.6% versus 20.9%) and more CO_2 (5.2% versus 0.04%) for two reasons. First, gas exchange in the alveoli increases the CO_2 content and decreases the O_2 content of alveolar air. Second, when air is inhaled it becomes humidified as it passes along the moist mucosal linings. As water vapor content of the air increases, the relative percentage that is O_2 decreases. In contrast, exhaled air contains more O_2 than alveolar air (16% versus 13.6%) and less CO_2 (4.5% versus 5.2%) because some of the exhaled air was in the anatomic dead space and did not participate in gas exchange. Exhaled air is a mixture of alveolar air and inhaled air that was in the anatomic dead space.

Henry's law states that the quantity of a gas that will dissolve in a liquid is proportional to the partial pressure of the gas and its solubility. In body fluids, the ability of a gas to stay in solution is greater when its partial pressure is higher and when it has a high solubility in water. The higher the partial pressure of a gas over a liquid and the higher the solubility, the more gas will stay in solution. In comparison to oxygen, much more CO_2 is dissolved in blood plasma because the solubility of CO_2 is 24 times greater than that of O_2. Even though the air we breathe contains mostly N_2, this gas has no known effect on bodily functions, and at sea level pressure very little of it dissolves in blood plasma because its solubility is very low.

An everyday experience gives a demonstration of Henry's law. You have probably noticed that a soft drink makes a hissing sound when the top of the container is removed, and bubbles rise to the surface for some time afterward. The gas dissolved in carbonated beverages is CO_2. Because the soft drink is bottled or canned under high pressure and capped, the CO_2 remains dissolved as long as the container is unopened. Once you remove the cap, the pressure decreases and the gas begins to bubble out of solution.

Henry's law explains two conditions resulting from changes in the solubility of nitrogen in body fluids. Even though the air we breathe contains about 79% nitrogen, this gas has no known effect on bodily functions, and very little of it dissolves in blood plasma because of its low solubility at sea level pressure. As the total air pressure increases, the partial pressures of all its gases increase. When a scuba diver breathes air under high pressure, the nitrogen in the mixture can have serious negative effects. Because the partial pressure of nitrogen is higher in a mixture of compressed air than in air at sea level pressure, a considerable amount of nitrogen dissolves in plasma and interstitial fluid. Excessive amounts of dissolved nitrogen may produce giddiness and other symptoms similar to alcohol intoxication. The condition is called **nitrogen narcosis** or "rapture of the deep."

If a diver comes to the surface slowly, the dissolved nitrogen can be eliminated by exhaling it. However, if the ascent is too rapid, nitrogen comes out of solution too quickly and forms gas bubbles in the tissues, resulting in **decompression sickness** (the **bends**). The effects of decompression sickness typically result from bubbles in nervous tissue and can be mild or severe, depending on the number of bubbles formed. Symptoms include joint pain, especially in the arms and legs, dizziness, shortness of breath, extreme fatigue, paralysis, and unconsciousness.

Hyperbaric Oxygenation

A major clinical application of Henry's law is **hyperbaric oxygenation** (*hyper* = over; *baros* = pressure), the use of pressure to cause more O_2 to dissolve in the blood. It is an effective technique in treating patients infected by anaerobic bacteria, such as those that cause tetanus and gangrene. (Anaerobic bacteria cannot live in the presence of free O_2.) A person undergoing hyperbaric oxygenation is placed in a hyperbaric chamber, which contains O_2 at a pressure greater than one atmosphere (760 mmHg). As body tissues pick up the O_2, the bacteria are killed. Hyperbaric chambers may also be used for treating certain heart disorders, carbon monoxide poisoning, gas embolisms, crush injuries, cerebral edema, certain hard-to-treat bone infections caused by anaerobic bacteria, smoke inhalation, near-drowning, asphyxia, vascular insufficiencies, and burns. ■

External and Internal Respiration

External respiration or **pulmonary gas exchange** is the diffusion of O_2 from air in the alveoli of the lungs to blood in pulmonary capillaries and the diffusion of CO_2 in the opposite direction (Figure 23.18a). External respiration in the lungs converts **deoxygenated blood** (depleted of some O_2) coming from the right side of the heart into **oxygenated blood** (saturated with O_2) that returns to the left side of the heart (see Figure 21.29 on page 793). As blood flows through the pulmonary capillaries, it picks up O_2 from alveolar air and unloads CO_2 into alveolar air. Although this process is commonly called an "exchange" of gases, each gas diffuses independently from the area where its partial pressure is higher to the area where its partial pressure is lower.

As Figure 23.18a shows, O_2 diffuses from alveolar air, where its partial pressure is 105 mmHg, into the blood in pulmonary capillaries, where P_{O_2} is only 40 mmHg in a resting person. If you have been exercising, the P_{O_2} will be even lower because contracting muscle fibers are using more O_2. Diffusion continues until the P_{O_2} of pulmonary capillary blood increases to match the P_{O_2} of alveolar air, 105 mmHg. Because blood leaving pulmonary capillaries near alveolar air spaces mixes with a small volume of blood that has flowed through conducting portions of the respiratory system, where gas exchange does not occur, the P_{O_2} of blood in the pulmonary veins is slightly less than the P_{O_2} in pulmonary capillaries, about 100 mmHg.

While O_2 is diffusing from alveolar air into deoxygenated blood, CO_2 is diffusing in the opposite direction. The P_{CO_2} of deoxygenated blood is 45 mmHg in a resting person, whereas P_{CO_2} of alveolar air is 40 mmHg. Because of this difference in P_{CO_2}, carbon dioxide diffuses from deoxygenated blood into the alveoli until the P_{CO_2} of the blood decreases to 40 mmHg. Exhalation keeps alveolar P_{CO_2} at 40 mmHg. Oxygenated blood returning to the left side of the heart in the pulmonary veins thus has a P_{CO_2} of 40 mmHg.

The number of capillaries near alveoli in the lungs is very large, and blood flows slowly enough through these capillaries that it picks up a maximal amount of O_2. During vigorous exercise, when cardiac output is increased, blood flows more rapidly through both the systemic and pulmonary circulations. As a result, blood's transit time in the pulmonary capillaries is shorter. Still, the P_{O_2} of blood in the pulmonary veins normally reaches 100 mmHg. In diseases that decrease the rate of gas diffusion, however, the blood may not come into full equilibrium with alveolar air, especially during exercise. When this happens, the P_{O_2} declines and P_{CO_2} rises in systemic arterial blood.

The left ventricle pumps oxygenated blood into the aorta and through the systemic arteries to systemic capillaries. The exchange of O_2 and CO_2 between systemic capillaries and tissue cells is called **internal respiration** or **systemic gas exchange** (Figure 23.18b). As O_2 leaves the bloodstream, oxygenated blood is converted into deoxygenated blood. Unlike external respiration, which occurs only in the lungs, internal respiration occurs in tissues throughout the body.

The P_{O_2} of blood pumped into systemic capillaries is higher (100 mmHg) than the P_{O_2} in tissue cells (40 mmHg at rest) because the cells constantly use O_2 to produce ATP. Due to this pressure difference, oxygen diffuses out of the capillaries into tissue cells and blood P_{O_2} drops to 40 mmHg by the time the blood exits systemic capillaries.

While O_2 diffuses from the systemic capillaries into tissue cells, CO_2 diffuses in the opposite direction. Because tissue cells are constantly producing CO_2, the P_{CO_2} of cells (45 mmHg at rest) is higher than that of systemic capillary blood (40 mmHg). As a result, CO_2 diffuses from tissue cells through interstitial fluid into systemic capillaries until the P_{CO_2} in the blood increases to 45 mmHg. The deoxygenated blood then returns to the heart and is pumped to the lungs for another cycle of external respiration.

In a person at rest, tissue cells, on average, need only 25% of the available O_2 in oxygenated blood; despite its name, deoxygenated blood retains 75% of its O_2 content. During exercise, more O_2 diffuses from the blood into metabolically active cells, such as contracting skeletal muscle fibers. Active cells use more O_2 for ATP production, causing the O_2 content of deoxygenated blood to drop below 75%.

The *rate* of pulmonary and systemic gas exchange depends on several factors.

- ***Partial pressure difference of the gases.*** Alveolar P_{O_2} must be higher than blood P_{O_2} for oxygen to diffuse from alveolar air into the blood. The rate of diffusion is faster when the difference between P_{O_2} in alveolar air and pulmonary capillary blood is larger; diffusion is slower when the difference is smaller. The differences between P_{O_2} and P_{CO_2} in alveolar air versus pulmonary blood increase during exercise. The larger partial pressure differences accelerate the rates of gas diffusion. The partial pressures of O_2 and CO_2 in alveolar air also depend on the rate of airflow into and out of the lungs. Certain drugs (such as morphine) slow ventilation, thereby decreasing the amount of O_2 and CO_2 that can be exchanged between alveolar air and blood. With increasing altitude,

Figure 23.18 Changes in partial pressures of oxygen and carbon dioxide (in mmHg) during external and internal respiration.

Gases diffuse from areas of higher partial pressure to areas of lower partial pressure.

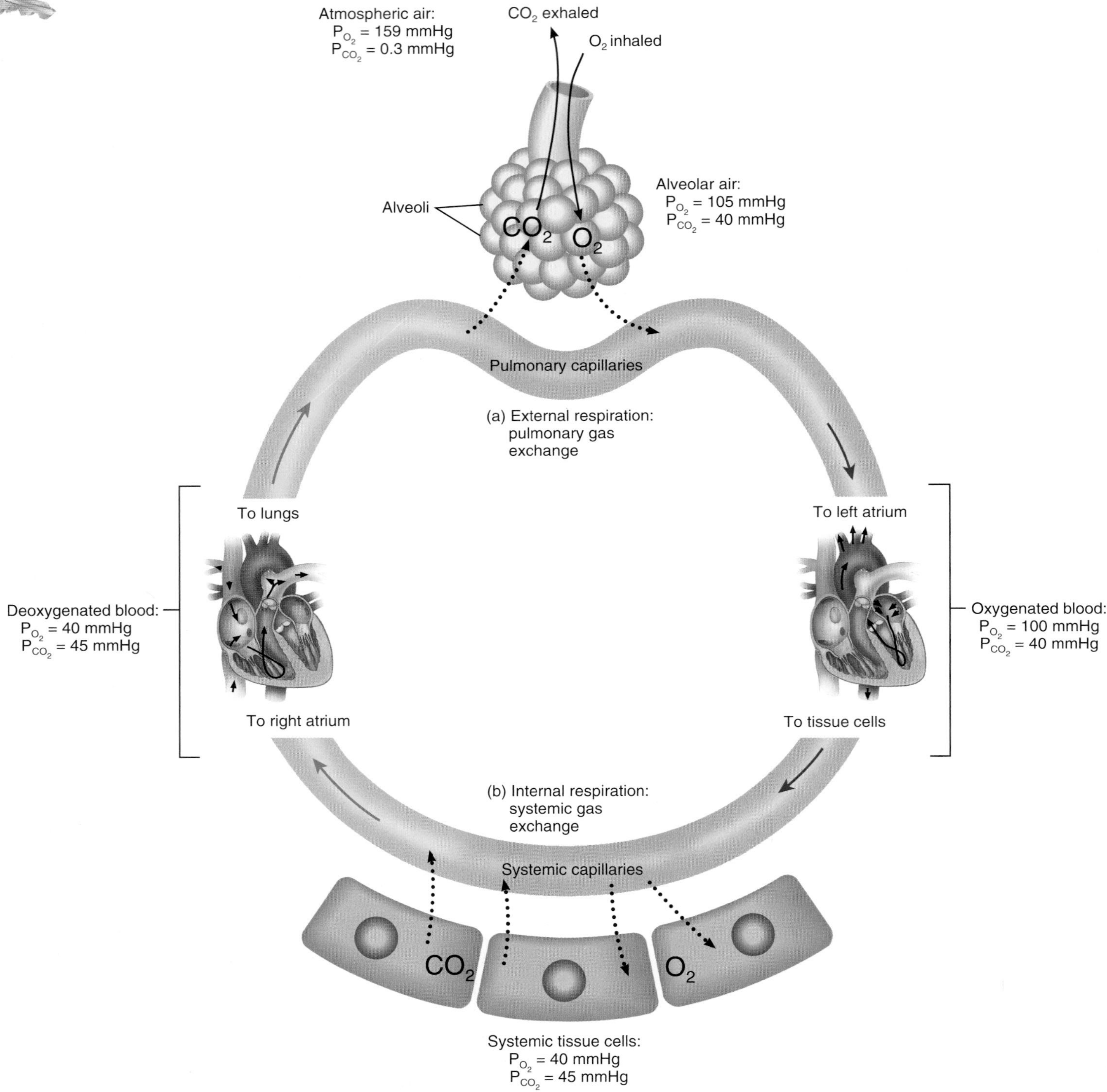

What causes oxygen to enter pulmonary capillaries from alveoli and to enter tissue cells from systemic capillaries?

the total atmospheric pressure decreases, as does the partial pressure of O_2—from 159 mmHg at sea level, to 110 mmHg at 10,000 ft, to 73 mmHg at 20,000 ft. Although O_2 still is 20.9% of the total, the P_{O_2} of inhaled air decreases with increasing altitude. Alveolar P_{O_2} decreases correspondingly, and O_2 diffuses into the blood more slowly. The common signs and symptoms of **high altitude sickness**—shortness of breath, headache, fatigue, insomnia, nausea, and dizziness—are due to a lower level of oxygen in the blood.

- ***Surface area available for gas exchange.*** As you learned earlier in the chapter, the surface area of the alveoli is huge (about 70 m^2 or 750 ft^2). In addition, many capillaries surround each alveolus, so many that as much as 900 mL of blood is able to participate in gas exchange at any instant. Any pulmonary disorder that decreases the functional surface area of the respiratory membranes decreases the rate of external respiration. In emphysema (page 887), for example, alveolar walls disintegrate, so surface area is smaller than normal and pulmonary gas exchange is slowed.
- ***Diffusion distance.*** The respiratory membrane is very thin, so diffusion occurs quickly. Also, the capillaries are so narrow that the red blood cells must pass through them in single file, which minimizes the diffusion distance from an alveolar air space to hemoglobin inside red blood cells. Buildup of interstitial fluid between alveoli, as occurs in pulmonary edema (page 888), slows the rate of gas exchange because it increases diffusion distance.
- ***Molecular weight and solubility of the gases.*** Because O_2 has a lower molecular weight than CO_2, it could be expected to diffuse across the respiratory membrane about 1.2 times faster. However, the solubility of CO_2 in the fluid portions of the respiratory membrane is about 24 times greater than that of O_2. Taking both of these factors into account, net outward CO_2 diffusion occurs 20 times more rapidly than net inward O_2 diffusion. Consequently, when diffusion is slower than normal, for example, in emphysema or pulmonary edema, O_2 insufficiency (hypoxia) typically occurs before there is significant retention of CO_2 (hypercapnia).

▶ CHECKPOINT

20. Distinguish between Dalton's law and Henry's law and give a practical application of each.

21. How does the partial pressure of oxygen change as altitude changes?

22. What are the diffusion paths of oxygen and carbon dioxide during external and internal respiration?

23. What factors affect the rate of diffusion of oxygen and carbon dixode?

TRANSPORT OF OXYGEN AND CARBON DIOXIDE

▶ OBJECTIVE

Describe how the blood transports oxygen and carbon dioxide.

As you have already learned, the blood transports gases between the lungs and body tissues. When O_2 and CO_2 enter the blood, certain chemical reactions occur that aid in gas transport and gas exchange.

Oxygen Transport

Oxygen does not dissolve easily in water, so only about 1.5% of inhaled O_2 is dissolved in blood plasma, which is mostly water. About 98.5% of blood O_2 is bound to hemoglobin in red blood cells (Figure 23.19). Each 100 mL of oxygenated blood contains the equivalent of 20 mL of gaseous O_2. Using the percentages just given, the amount dissolved in the plasma is 0.3 mL and the amount bound to hemoglobin is 19.7 mL.

The heme portion of hemoglobin contains four atoms of iron, each capable of binding to a molecule of O_2 (see Figure 19.4b, c on page 673). Oxygen and hemoglobin bind in an easily reversible reaction to form **oxyhemoglobin:**

$$\underset{\substack{\text{Reduced hemoglobin}\\\text{(deoxyhemoglobin)}}}{\text{Hb}} + \underset{\text{Oxygen}}{O_2} \underset{\substack{\text{Dissociation}\\\text{of } O_2}}{\overset{\text{Binding of } O_2}{\rightleftharpoons}} \underset{\text{Oxyhemoglobin}}{\text{Hb}-O_2}$$

The 98.5% of the O_2 that is bound to hemoglobin is trapped inside RBCs, so only the dissolved O_2 (1.5%) can diffuse out of tissue capillaries into tissue cells. Thus, it is important to understand the factors that promote O_2 binding to and dissociation (separation) from hemoglobin.

The Relation Between Hemoglobin and Oxygen Partial Pressure

The most important factor that determines how much O_2 binds to hemoglobin is the P_{O_2}; the higher the P_{O_2}, the more O_2 combines with Hb. When reduced hemoglobin (Hb) is completely converted to oxyhemoglobin (Hb–O_2), the hemoglobin is said to be **fully saturated;** when hemoglobin consists of a mixture of Hb and Hb–O_2, it is **partially saturated.** The **percent saturation of hemoglobin** expresses the average saturation of hemoglobin with oxygen. For instance, if each hemoglobin molecule has bound two O_2 molecules, then the hemoglobin is 50% saturated because each Hb can bind a maximum of four O_2.

The relation between the percent saturation of hemoglobin and P_{O_2} is illustrated in the oxygen–hemoglobin dissociation curve in Figure 23.20 on page 875. Note that when the P_{O_2} is high, hemoglobin binds with large amounts of O_2 and is almost 100% saturated. When P_{O_2} is low, hemoglobin is only partially saturated. In other words, the greater the P_{O_2}, the more

Figure 23.19 **Transport of oxygen (O_2) and carbon dioxide (CO_2) in the blood.**

Most O_2 is transported by hemoglobin as oxyhemoglobin ($Hb-O_2$) within red blood cells; most CO_2 is transported in blood plasma as bicarbonate ions (HCO_3^-).

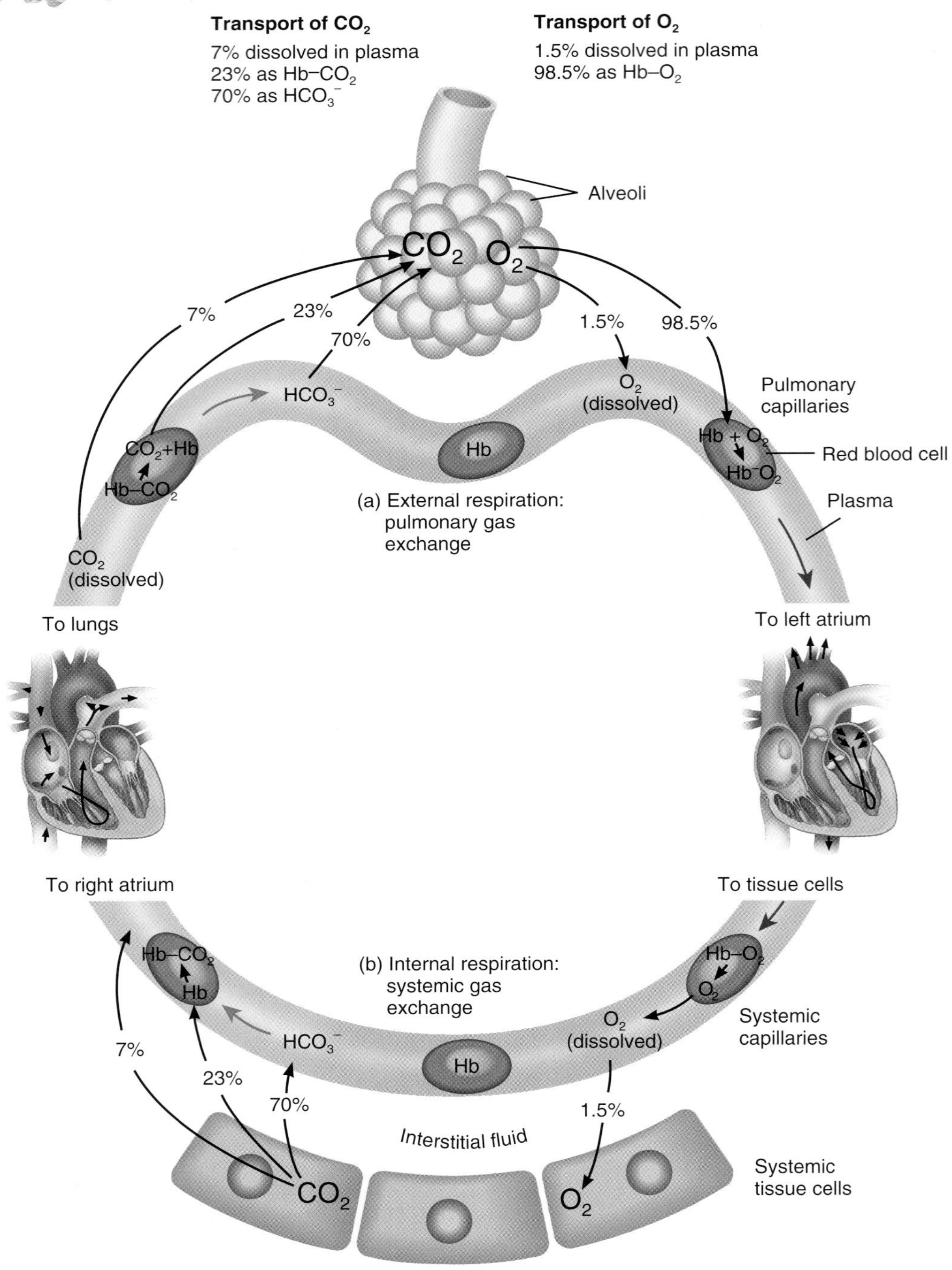

What is the most important factor that determines how much O_2 binds to hemoglobin?

O_2 will bind to hemoglobin, until all the available hemoglobin molecules are saturated. Therefore, in pulmonary capillaries, where P_{O_2} is high, a lot of O_2 binds to hemoglobin. In tissue capillaries, where the P_{O_2} is lower, hemoglobin does not hold as much O_2, and the dissolved O_2 is unloaded via diffusion into tissue cells (see Figure 23.19b). Note that hemoglobin is still 75% saturated with O_2 at a P_{O_2} of 40 mmHg, the average P_{O_2} of tissue cells in a person at rest. This is the basis for the earlier statement that only 25% of the available O_2 unloads from hemoglobin and is used by tissue cells under resting conditions.

When the P_{O_2} is between 60 and 100 mmHg, hemoglobin is 90% or more saturated with O_2 (Figure 23.20). Thus, blood picks up a nearly full load of O_2 from the lungs even when the P_{O_2} of alveolar air is as low as 60 mmHg. The $Hb-P_{O_2}$ curve

Figure 23.20 Oxygen–hemoglobin dissociation curve showing the relationship between hemoglobin saturation and P_{O_2} at normal body temperature.

As P_{O_2} increases, more O_2 combines with hemoglobin.

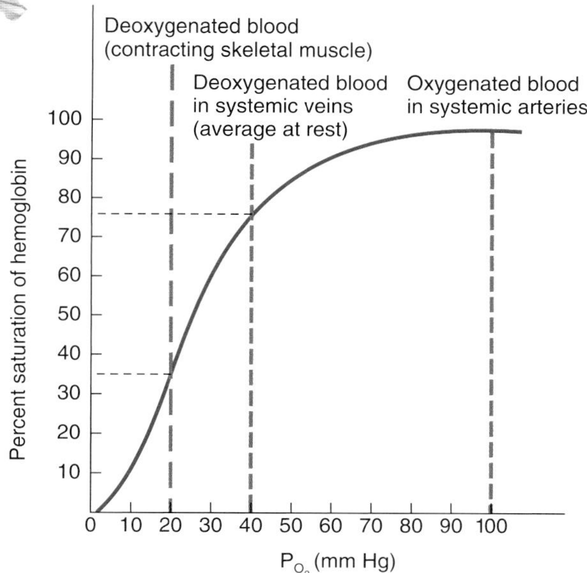

What point on the curve represents blood in your pulmonary veins right now? In your pulmonary veins if you were jogging?

explains why people can still perform well at high altitudes or when they have certain cardiac and pulmonary diseases, even though P_{O_2} may drop as low as 60 mmHg. Note also in the curve that at a considerably lower P_{O_2} of 40 mmHg, hemoglobin is still 75% saturated with O_2. However, oxygen saturation of Hb drops to 35% at 20 mmHg. Between 40 and 20 mmHg, large amounts of O_2 are released from hemoglobin in response to only small decreases in P_{O_2}. In active tissues such as contracting muscles, P_{O_2} may drop well below 40 mmHg. Then, a large percentage of the O_2 is released from hemoglobin, providing more O_2 to metabolically active tissues.

Other Factors Affecting Hemoglobin's Affinity for Oxygen

Although P_{O_2} is the most important factor that determines the percent O_2 saturation of hemoglobin, several other factors influence the tightness or **affinity** with which hemoglobin binds O_2. In effect, these factors shift the entire curve either to the left (higher affinity) or to the right (lower affinity). The changing affinity of hemoglobin for O_2 is another example of how homeostatic mechanisms adjust body activities to cellular needs. Each one makes sense if you keep in mind that metabolically active tissue cells need O_2 and produce acids, CO_2, and heat as wastes. The following four factors affect hemoglobin's affinity for O_2:

1. ***Acidity (pH).*** As acidity increases (pH decreases), the affinity of hemoglobin for O_2 decreases, and O_2 dissociates more readily from hemoglobin (Figure 23.21a). In other words, increasing acidity enhances the unloading of oxygen from hemoglobin. The main acids produced by metabolically active tissues are lactic acid and carbonic acid. When pH decreases, the entire oxygen–hemoglobin dissociation curve shifts to the right; at any given P_{O_2}, Hb is less saturated with O_2, a change termed the **Bohr effect.** The Bohr effect works both ways: An increase in H^+ in blood causes O_2 to unload from hemoglobin, and the

Figure 23.21 Oxygen–hemoglobin dissociation curves showing the relationship of (a) pH and (b) P_{CO_2} to hemoglobin saturation at normal body temperature. As pH increases or P_{CO_2} decreases, O_2 combines more tightly with hemoglobin, so that less is available to tissues. The broken lines emphasize these relationships.

As pH decreases or P_{CO_2} increases, the affinity of hemoglobin for O_2 declines, so less O_2 combines with hemoglobin and more is available to tissues.

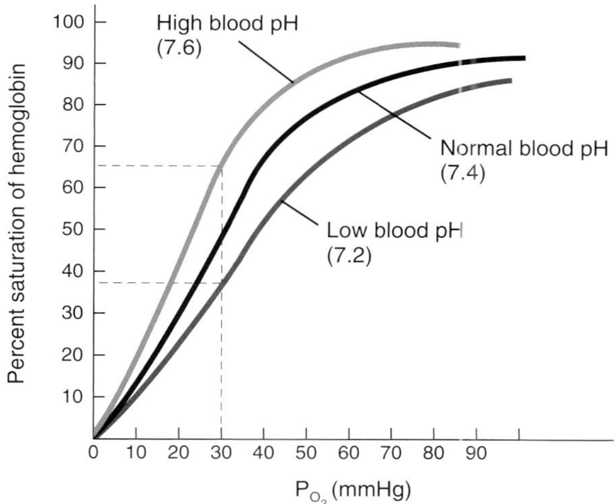

(a) Effect of pH on affinity of hemoglobin for oxygen

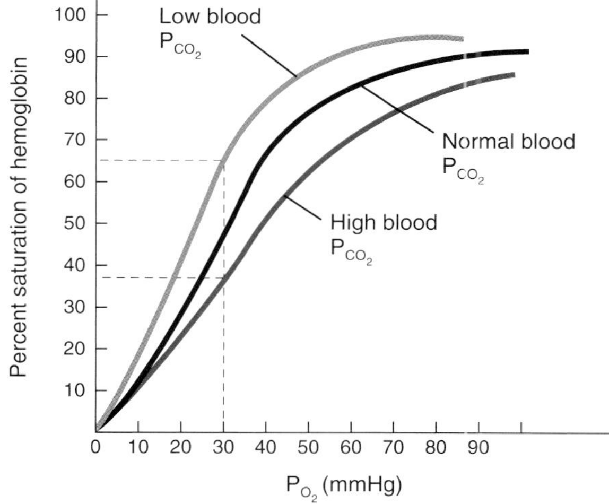

(b) Effect of P_{CO_2} on affinity of hemoglobin for oxygen

In comparison to the value when you are sitting, is the affinity of your hemoglobin for O_2 higher or lower when you are exercising? How does this benefit you?

binding of O_2 to hemoglobin causes unloading of H^+ from hemoglobin. The explanation for the Bohr effect is that hemoglobin can act as a buffer for hydrogen ions (H^+). But when H^+ ions bind to amino acids in hemoglobin, they alter its structure slightly, decreasing its oxygen-carrying capacity. Thus, lowered pH drives O_2 off hemoglobin, making more O_2 available for tissue cells. By contrast, elevated pH increases the affinity of hemoglobin for O_2 and shifts the oxygen–hemoglobin dissociation curve to the left.

2. ***Partial pressure of carbon dioxide.*** CO_2 also can bind to hemoglobin, and the effect is similar to that of H^+ (shifting the curve to the right). As P_{CO_2} rises, hemoglobin releases O_2 more readily (Figure 23.21b). P_{CO_2} and pH are related factors because low blood pH (acidity) results from high P_{CO_2}. As CO_2 enters the blood, much of it is temporarily converted to carbonic acid (H_2CO_3), a reaction catalyzed by an enzyme in red blood cells called *carbonic anhydrase (CA):*

$$\underset{\text{Carbon dioxide}}{CO_2} + \underset{\text{Water}}{H_2O} \overset{CA}{\rightleftharpoons} \underset{\text{Carbonic acid}}{H_2CO_3} \rightleftharpoons \underset{\text{Hydrogen ion}}{H^+} + \underset{\text{Bicarbonate ion}}{HCO_3^-}$$

The carbonic acid thus formed in red blood cells dissociates into hydrogen ions and bicarbonate ions. As the H^+ concentration increases, pH decreases. Thus, an increased P_{CO_2} produces a more acidic environment, which helps release O_2 from hemoglobin. During exercise, lactic acid—a byproduct of anaerobic metabolism within muscles—also decreases blood pH. Decreased P_{CO_2} (and elevated pH) shifts the saturation curve to the left.

3. ***Temperature.*** Within limits, as temperature increases, so does the amount of O_2 released from hemoglobin (Figure 23.22). Heat is a byproduct of the metabolic reactions of all cells, and the heat released by contracting muscle fibers tends to raise body temperature. Metabolically active cells require more O_2 and liberate more acids and heat. The acids and heat, in turn, promote release of O_2 from oxyhemoglobin. Fever produces a similar result. In contrast, during hypothermia (lowered body temperature) cellular metabolism slows, the need for O_2 is reduced, and more O_2 remains bound to hemoglobin (a shift to the left in the saturation curve).

4. ***BPG.*** A substance found in red blood cells called **2,3-bisphosphoglycerate (BPG),** previously called diphosphoglycerate (DPG), decreases the affinity of hemoglobin for O_2 and thus helps unload O_2 from hemoglobin. BPG is formed in red blood cells when they break down glucose to produce ATP in a process called glycolysis (described on page 955). When BPG combines with hemoglobin by binding to the terminal amino groups of the two beta globin chains, the hemoglobin binds O_2 less tightly at the heme group sites. The greater the level of BPG, the more O_2 is unloaded from hemoglobin. Certain hormones, such as thyroxine, human growth hormone, epinephrine, norepinephrine, and testosterone, increase the formation of BPG. The level of BPG also is higher in people living at higher altitudes.

Figure 23.22 Oxygen–hemoglobin dissociation curves showing the effect of temperature changes.

As temperature increases, the affinity of hemoglobin for O_2 decreases.

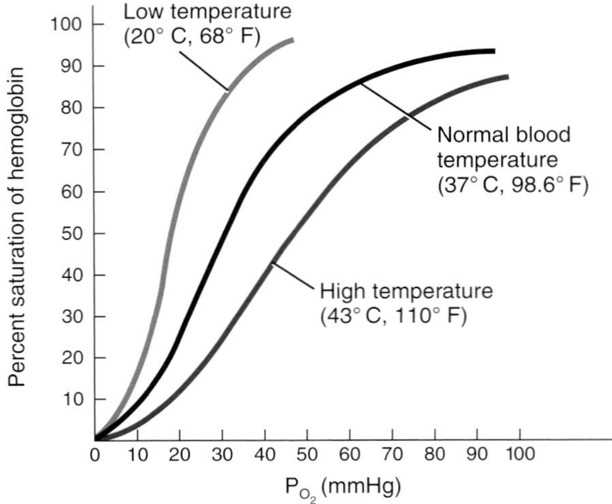

Is O_2 more available or less available to tissue cells when you have a fever? Why?

Oxygen Affinity of Fetal and Adult Hemoglobin

Fetal hemoglobin (Hb-F) differs from **adult hemoglobin (Hb-A)** in structure and in its affinity for O_2. Hb-F has a higher affinity for O_2 because it binds BPG less strongly. Thus, when P_{O_2} is low, Hb-F can carry up to 30% more O_2 than maternal Hb-A (Figure 23.23). As the maternal blood enters the placenta, O_2 is readily transferred to fetal blood. This is very important because the O_2 saturation in maternal blood in the placenta is quite low, and the fetus might suffer hypoxia were it not for the greater affinity of fetal hemoglobin for O_2.

Carbon Monoxide Poisoning

Carbon monoxide (CO) is a colorless and odorless gas found in exhaust fumes from automobiles, gas furnaces, and space heaters and in tobacco smoke. It is a byproduct of the combustion of carbon-containing materials such as coal, gas, and wood. CO binds to the heme group of hemoglobin, just as O_2 does, except that the binding of carbon monoxide to hemoglobin is over 200 times as strong as the binding of O_2 to hemoglobin. Thus, at a concentration as small as 0.1% (P_{CO} = 0.5 mmHg), CO will combine with half the available hemoglobin molecules and reduce the oxygen-carrying capacity of the blood by 50%. Elevated blood levels of CO cause **carbon**

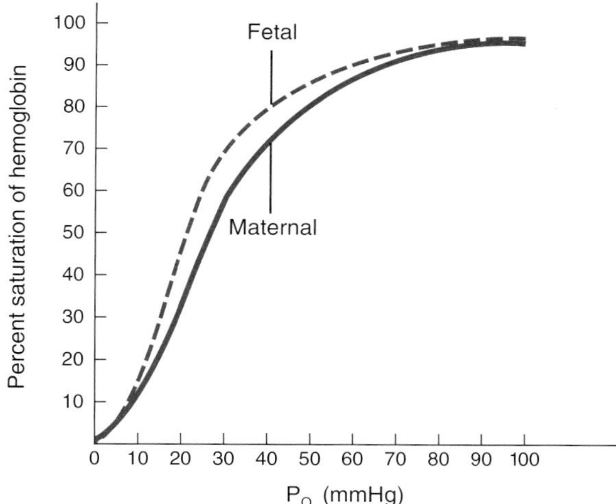

Figure 23.23 **Oxygen–hemoglobin dissociation curves comparing fetal and maternal hemoglobin.**

Fetal hemoglobin has a higher affinity for O_2 than does adult hemoglobin.

The P_{O_2} of placental blood is about 40 mmHg. What are the O_2 saturations of maternal and fetal hemoglobin at this P_{O_2}?

monoxide poisoning, which can cause the lips and oral mucosa to appear bright, cherry-red (the color of hemoglobin with carbon monoxide bound to it). Without prompt treatment, carbon monoxide poisoning is fatal. It is possible to rescue a victim of CO poisoning by administering pure oxygen, which speeds up the separation of carbon monoxide from hemoglobin. ■

Carbon Dioxide Transport

Under normal resting conditions, each 100 mL of deoxygenated blood contains the equivalent of 53 mL of gaseous CO_2, which is transported in the blood in three main forms (see Figure 23.19):

1. ***Dissolved CO_2.*** The smallest percentage—about 7%—is dissolved in blood plasma. Upon reaching the lungs, it diffuses into alveolar air and is exhaled.

2. ***Carbamino compounds.*** A somewhat higher percentage, about 23%, combines with the amino groups of amino acids and proteins in blood to form **carbamino compounds.** Because the most prevalent protein in blood is hemoglobin (inside red blood cells), most of the CO_2 transported in this manner is bound to hemoglobin. The main CO_2 binding sites are the terminal amino acids in the two alpha and two beta globin chains. Hemoglobin that has bound CO_2 is termed **carbaminohemoglobin ($Hb-CO_2$):**

$$\underset{\text{Hemoglobin}}{Hb} + \underset{\text{Carbon dioxide}}{CO_2} \rightleftharpoons \underset{\text{Carbaminohemoglobin}}{Hb-CO_2}$$

The formation of carbaminohemoglobin is greatly influenced by P_{CO_2}. For example, in tissue capillaries P_{CO_2} is relatively high, which promotes formation of carbaminohemoglobin. But in pulmonary capillaries, P_{CO_2} is relatively low, and the CO_2 readily splits apart from globin and enters the alveoli by diffusion.

3. ***Bicarbonate ions.*** The greatest percentage of CO_2—about 70%—is transported in blood plasma as **bicarbonate ions (HCO_3^-).** As CO_2 diffuses into systemic capillaries and enters red blood cells, it reacts with water in the presence of the enzyme carbonic anhydrase (CA) to form carbonic acid, which dissociates into H^+ and HCO_3^-:

$$\underset{\substack{\text{Carbon}\\\text{dioxide}}}{CO_2} + \underset{\text{Water}}{H_2O} \overset{CA}{\rightleftharpoons} \underset{\substack{\text{Carbonic}\\\text{acid}}}{H_2CO_3} \rightleftharpoons \underset{\substack{\text{Hydrogen}\\\text{ion}}}{H^+} + \underset{\substack{\text{Bicarbonate}\\\text{ion}}}{HCO_3^-}$$

Thus, as blood picks up CO_2, HCO_3^- accumulates inside RBCs. Some HCO_3^- moves out into the blood plasma, down its concentration gradient. In exchange, chloride ions (Cl^-) move from plasma into the RBCs. This exchange of negative ions, which maintains the electrical balance between blood plasma and RBC cytosol, is known as the **chloride shift** (see Figure 23.24b). The net effect of these reactions is that CO_2 is removed from tissue cells and transported in blood plasma as HCO_3^-. As blood passes through pulmonary capillaries in the lungs, all these reactions reverse and CO_2 is exhaled.

The amount of CO_2 that can be transported in the blood is influenced by the percent saturation of hemoglobin with oxygen. The lower the amount of oxyhemoglobin ($Hb-O_2$), the higher the CO_2 carrying capacity of the blood, a relationship known as the **Haldane effect.** Two characteristics of deoxyhemoglobin give rise to the Haldane effect: (1) Deoxyhemoglobin binds to and thus transports more CO_2 than does $Hb-O_2$. (2) Deoxyhemoglobin also buffers more H^+ than does $Hb-O_2$, thereby removing H^+ from solution and promoting conversion of CO_2 to HCO_3^- via the reaction catalyzed by carbonic anhydrase.

Summary of Gas Exchange and Transport in Lungs and Tissues

Deoxygenated blood returning to the pulmonary capillaries in the lungs (Figure 23.24a) contains CO_2 dissolved in blood plasma, CO_2 combined with globin as carbaminohemoglobin ($Hb-CO_2$), and CO_2 incorporated into HCO_3^- within RBCs. The RBCs have also picked up H^+, some of which binds to and therefore is buffered by hemoglobin (Hb–H). As blood passes through the pulmonary capillaries, molecules of CO_2 dissolved in blood plasma and CO_2 that dissociates from the globin portion of hemoglobin diffuse into alveolar air and are exhaled. At the same time, inhaled O_2 is diffusing from alveolar air into RBCs and is binding to hemoglobin to form oxyhemoglobin (Hb–O_2). Carbon dioxide also is released from HCO_3^- when H^+ combines with HCO_3^- inside RBCs. The H_2CO_3 formed from this reaction then splits into CO_2, which is exhaled, and H_2O. As the concentration of HCO_3^- declines inside RBCs in pulmonary capillaries, HCO_3^- diffuses in from the blood plasma, in exchange for Cl^-. In sum, oxygenated blood leaving the lungs has increased O_2 content and decreased amounts of CO_2 and H^+. In systemic capillaries, as cells use O_2 and produce CO_2, the chemical reactions reverse (Figure 23.24b).

Figure 23.24 Summary of chemical reactions that occur during gas exchange. (a) As carbon dioxide (CO_2) is exhaled, hemoglobin (Hb) inside red blood cells in pulmonary capillaries unloads CO_2 and picks up O_2 from alveolar air. Binding of O_2 to Hb—H releases hydrogen ions (H^+). Bicarbonate ions (HCO_3^-) pass into the RBC and bind to released H^+, forming carbonic acid (H_2CO_3). The H_2CO_3 dissociates into water (H_2O) and CO_2, and the CO_2 diffuses from blood into alveolar air. To maintain electrical balance, a chloride ion (Cl^-) exits the RBC for each HCO_3^- that enters (reverse chloride shift). (b) CO_2 diffuses out of tissue cells that produce it and enters red blood cells, where some of it binds to hemoglobin, forming carbaminohemoglobin (Hb–CO_2). This reaction causes O_2 to dissociate from oxyhemoglobin (Hb–O_2). Other molecules of CO_2 combine with water to produce bicarbonate ions (HCO_3^-) and hydrogen ions (H^+). As Hb buffers H^+, the Hb releases O_2 (Bohr effect). To maintain electrical balance, a chloride ion (Cl^-) enters the RBC for each HCO_3^- that exits (chloride shift).

Hemoglobin inside red blood cells transports O_2, CO_2, and H^+.

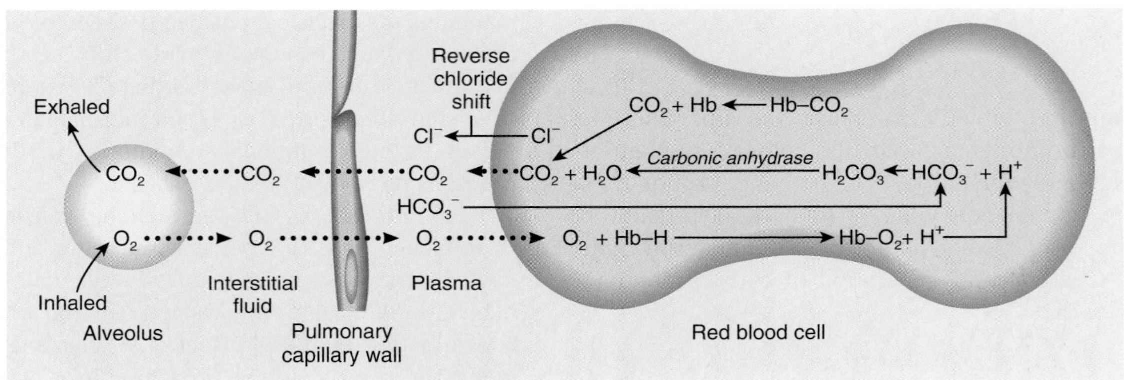

(a) Exchange of O_2 and CO_2 in pulmonary capillaries (external respiration)

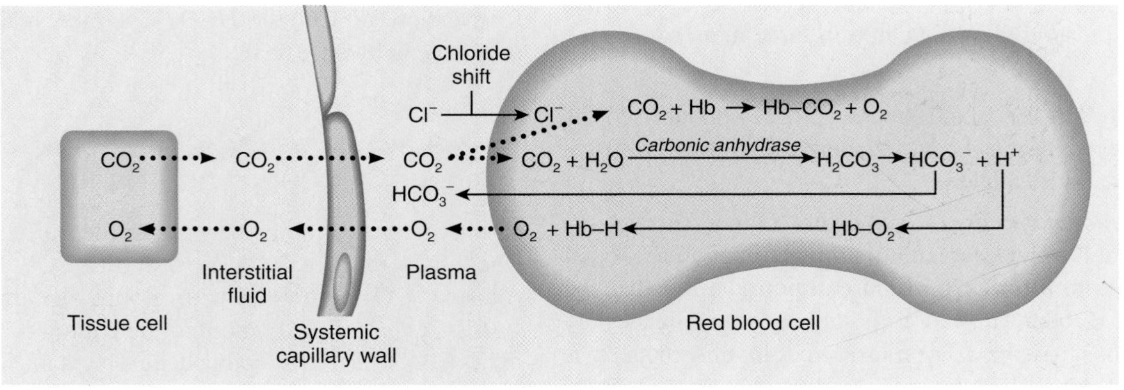

(b) Exchange of O_2 and CO_2 in systemic capillaries (internal respiration)

Would you expect the concentration of HCO_3^- to be higher in blood plasma taken from a systemic artery or a systemic vein?

▶ CHECKPOINT

24. In a resting person, how many O_2 molecules are attached to each hemoglobin molecule, on average, in blood in the pulmonary arteries? In blood in the pulmonary veins?

25. What is the relationship between hemoglobin and P_{O_2}? How do temperature, H^+, P_{CO_2}, and BPG influence the affinity of Hb for O_2?

26. Why can hemoglobin unload more oxygen as blood flows through capillaries of metabolically active tissues, such as skeletal muscle during exercise, than is unloaded at rest?

CONTROL OF RESPIRATION

▶ OBJECTIVES

Explain how the nervous system controls breathing.

List the factors that can alter the rate and depth of breathing.

At rest, about 200 mL of O_2 are used each minute by body cells. During strenuous exercise, however, O_2 use typically increases 15- to 20-fold in normal healthy adults, and as much as 30-fold in elite endurance-trained athletes. Several mechanisms help match respiratory effort to metabolic demand.

Figure 23.25 Locations of areas of the respiratory center.

The respiratory center is composed of neurons in the medullary rhythmicity area in the medulla oblongata plus the pneumotaxic and apneustic areas in the pons.

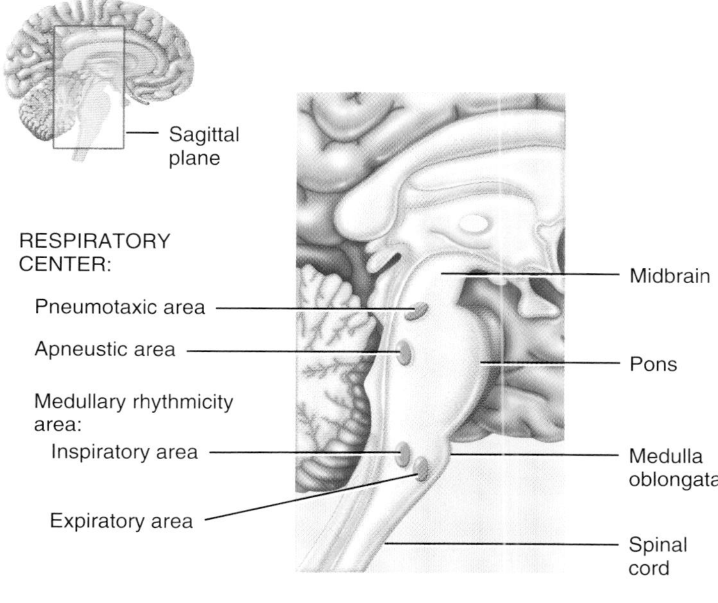

Which area contains neurons that are active and then inactive in a repeating cycle?

Respiratory Center

The size of the thorax is altered by the action of the respiratory muscles, which contract as a result of nerve impulses transmitted to them from centers in the brain and relax in the absence of nerve impules. These nerve impulses are sent from clusters of neurons located bilaterally in the medulla oblongata and pons of the brain stem. This widely dispersed group of neurons, collectively called the **respiratory center,** can be divided into three areas on the basis of their functions: (1) the medullary rhythmicity area in the medulla oblongata; (2) the pneumotaxic area in the pons; and (3) the apneustic area, also in the pons (Figure 23.25).

Medullary Rhythmicity Area

The function of the **medullary rhythmicity area** (rith-MIS-i-tē) is to control the basic rhythm of respiration. There are inspiratory and expiratory areas within the medullary rhythmicity area. Figure 23.26 shows the relationships of the inspiratory and expiratory areas during normal quiet breathing and forceful breathing.

During quiet breathing, inhalation lasts for about 2 seconds and exhalation lasts for about 3 seconds. Nerve impulses generated in the **inspiratory area** establish the basic rhythm of breathing. While the inspiratory area is active, it generates nerve impulses for about 2 seconds (Figure 23.26a). The impulses propagate to the external intercostal muscles via intercostal nerves and to the diaphragm via the phrenic nerves. When the nerve impulses reach the diaphragm and external intercostal muscles, the muscles contract and inhalation occurs. Even when all incoming nerve connections to the inspiratory area are cut or blocked, neurons in this area still rhythmically discharge impulses that cause inhalation. At the end of 2 seconds, the inspiratory area becomes inactive and nerve impulses cease. With no impulses arriving, the diaphragm and external intercostal muscles relax for about 3 seconds, allowing passive elastic recoil of the lungs and thoracic wall. Then, the cycle repeats.

The neurons of the **expiratory area** remain inactive during quiet breathing. However, during forceful breathing nerve impulses from the inspiratory area activate the expiratory area (Figure 23.26b). Impulses from the expiratory area cause contraction of the internal intercostal and abdominal muscles, which decreases the size of the thoracic cavity and causes forceful exhalation.

Pneumotaxic Area

Although the medullary rhythmicity area controls the basic rhythm of respiration, other sites in the brain stem help coordinate the transition between inhalation and exhalation. One of these sites is the **pneumotaxic area** (noo-mō-TAK-sik;

Figure 23.26 Roles of the medullary rhythmicity area in controlling (a) the basic rhythm of respiration and (b) forceful breathing.

During normal, quiet breathing, the expiratory area is inactive; during forceful breathing, the inspiratory area activates the expiratory area.

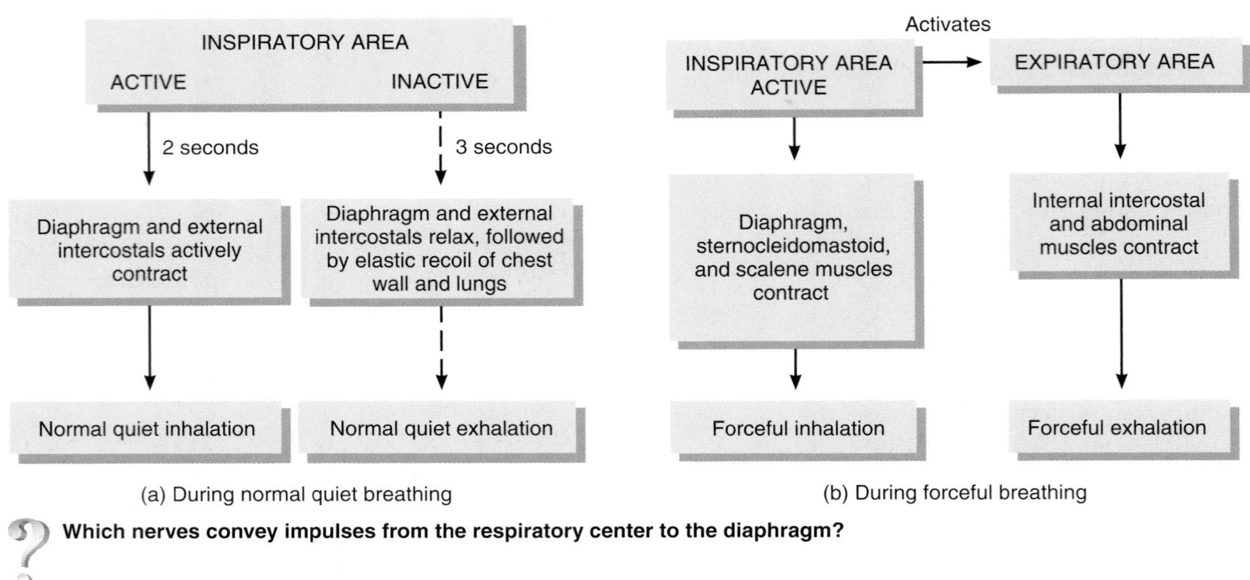

(a) During normal quiet breathing (b) During forceful breathing

Which nerves convey impulses from the respiratory center to the diaphragm?

pneumo- = air or breath; *-taxic* = arrangement) in the upper pons (see Figure 23.25), which transmits inhibitory impulses to the inspiratory area. The major effect of these nerve impulses is to help turn off the inspiratory area before the lungs become too full of air. In other words, the impulses shorten the duration of inhalation. When the pneumotaxic area is more active, breathing rate is more rapid.

Apneustic Area

Another part of the brain stem that coordinates the transition between inhalation and exhalation is the **apneustic area** (ap-NOO-stik) in the lower pons (see Figure 23.25). This area sends stimulatory impulses to the inspiratory area that activate it and prolong inhalation. The result is a long, deep inhalation. When the pneumotaxic area is active, it overrides signals from the apneustic area.

Regulation of the Respiratory Center

The basic rhythm of respiration set and coordinated by the inspiratory area can be modified in response to inputs from other brain regions, receptors in the peripheral nervous system, and other factors.

Cortical Influences on Respiration

Because the cerebral cortex has connections with the respiratory center, we can voluntarily alter our pattern of breathing. We can even refuse to breathe at all for a short time. Voluntary control is protective because it enables us to prevent water or irritating gases from entering the lungs. The ability to not breathe, however, is limited by the buildup of CO_2 and H^+ in the body. When P_{CO_2} and H^+ concentrations increase to a certain level, the inspiratory area is strongly stimulated, nerve impulses are sent along the phrenic and intercostal nerves to inspiratory muscles, and breathing resumes, whether the person wants it to or not. It is impossible for small children to kill themselves by voluntarily holding their breath, even though many have tried in order to get their way. If breath is held long enough to cause fainting, breathing resumes when consciousness is lost. Nerve impulses from the hypothalamus and limbic system also stimulate the respiratory center, allowing emotional stimuli to alter respirations as, for example, in laughing and crying.

Chemoreceptor Regulation of Respiration

Certain chemical stimuli modulate how quickly and how deeply we breathe. The respiratory system functions to maintain proper levels of CO_2 and O_2 and is very responsive to changes in the levels of these gases in body fluids. We introduced sensory neurons that are responsive to chemicals, called **chemoreceptors,** in Chapter 21. Chemoreceptors in two locations monitor levels of CO_2, H^+, and O_2 and provide input to the respiratory center (Figure 23.27). **Central chemoreceptors** are located in or near the medulla oblongata in the *central* nervous system. They respond to changes in H^+ concentration or P_{CO_2}, or both, in cerebrospinal fluid. **Peripheral chemoreceptors** are located in

Figure 23.27 Locations of peripheral chemoreceptors.

Chemoreceptors are sensory neurons that respond to changes in the levels of certain chemicals in the body.

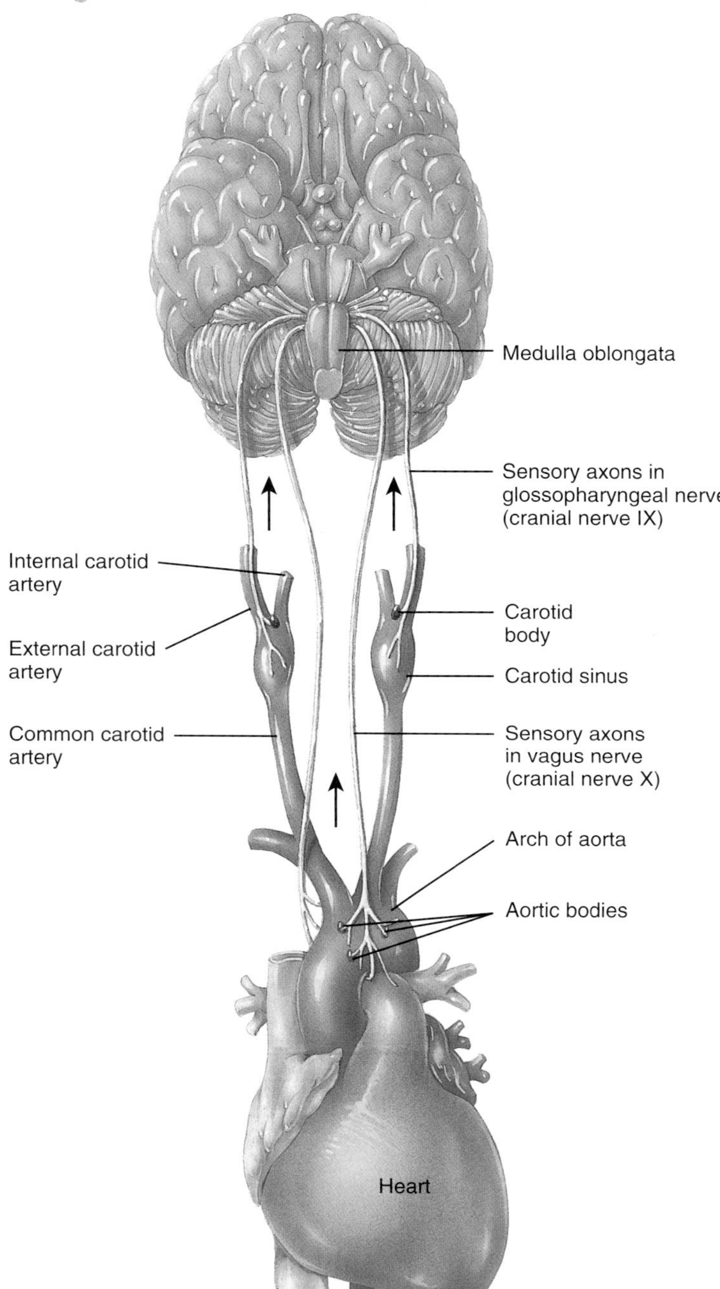

Which chemicals stimulate peripheral chemoreceptors?

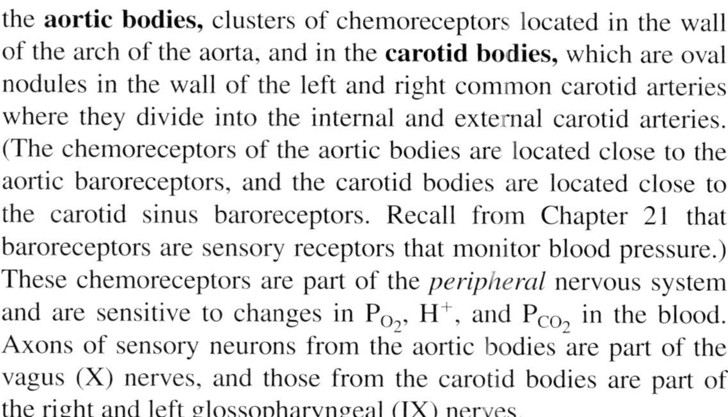

the **aortic bodies,** clusters of chemoreceptors located in the wall of the arch of the aorta, and in the **carotid bodies,** which are oval nodules in the wall of the left and right common carotid arteries where they divide into the internal and external carotid arteries. (The chemoreceptors of the aortic bodies are located close to the aortic baroreceptors, and the carotid bodies are located close to the carotid sinus baroreceptors. Recall from Chapter 21 that baroreceptors are sensory receptors that monitor blood pressure.) These chemoreceptors are part of the *peripheral* nervous system and are sensitive to changes in P_{O_2}, H^+, and P_{CO_2} in the blood. Axons of sensory neurons from the aortic bodies are part of the vagus (X) nerves, and those from the carotid bodies are part of the right and left glossopharyngeal (IX) nerves.

Because CO_2 is lipid-soluble, it easily diffuses into cells, where in the presence of carbonic anhydrase it combines with water (H_2O) to form carbonic acid (H_2CO_3). Carbonic acid quickly breaks down into H^+ and HCO_3^-. Thus, an increase in CO_2 in the blood causes an increase in H^+ inside cells, and a decrease in CO_2 causes a decrease in H^+.

Normally, the P_{CO_2} in arterial blood is 40 mmHg. If even a slight increase in P_{CO_2} occurs—a condition called **hypercapnia** or **hypercarbia**—the central chemoreceptors are stimulated and respond vigorously to the resulting increase in H^+ level. The peripheral chemoreceptors also are stimulated by both the high P_{CO_2} and the rise in H^+. In addition, the peripheral chemoreceptors (but not the central chemoreceptors) respond to a deficiency of O_2. When P_{O_2} in arterial blood falls from a normal level of 100 mmHg but is still above 50 mmHg, the peripheral chemoreceptors are stimulated. Severe deficiency of O_2 depresses activity of the central chemoreceptors and inspiratory area, which then do not respond well to any inputs and send fewer impulses to the muscles of inhalation. As the breathing rate decreases or breathing ceases altogether, P_{O_2} falls lower and lower, establishing a positive feedback cycle with a possibly fatal result.

The chemoreceptors participate in a negative feedback system that regulates the levels of CO_2, O_2, and H^+ in the blood (Figure 23.28). As a result of increased P_{CO_2}, decreased pH (increased H^+), or decreased P_{O_2}, input from the central and peripheral chemoreceptors causes the inspiratory area to become highly active, and the rate and depth of breathing increase. Rapid and deep breathing, called **hyperventilation,** allows the inhalation of more O_2 and exhalation of more CO_2 until P_{CO_2} and H^+ are lowered to normal.

If arterial P_{CO_2} is lower than 40 mmHg—a condition called **hypocapnia** or **hypocarbia**—the central and peripheral chemoreceptors are not stimulated, and stimulatory impulses are not sent to the inspiratory area. As a result, the area sets its own moderate pace until CO_2 accumulates and the P_{CO_2} rises to 40 mmHg. The inspiratory center is more strongly stimulated when P_{CO_2} is rising above normal than when P_{O_2} is falling below normal. As a result, people who hyperventilate voluntarily and cause hypocapnia can hold their breath for an unusually long period. Swimmers were once encouraged to hyperventilate just before diving in to compete. However, this practice is risky because the O_2 level may fall dangerously low and cause fainting before the P_{CO_2} rises high enough to stimulate inhalation. If you faint on land you may suffer bumps and bruises, but if you faint in the water you could drown.

Hypoxia

Hypoxia (hī-POK-sē-a; *hypo-* = under) is a deficiency of O_2 at the tissue level. Based on the cause, we can classify hypoxia into four types, as follows:

1. **Hypoxic hypoxia** is caused by a low P_{O_2} in arterial blood as a result of high altitude, airway obstruction, or fluid in the lungs.
2. In **anemic hypoxia,** too little functioning hemoglobin is present in the blood, which reduces O_2 transport to tissue cells. Among the causes are hemorrhage, anemia, and failure of hemoglobin to carry its normal complement of O_2, as in carbon monoxide poisoning.
3. In **ischemic hypoxia,** blood flow to a tissue is so reduced that too little O_2 is delivered to it, even though P_{O_2} and oxyhemoglobin levels are normal.
4. In **histotoxic hypoxia,** the blood delivers adequate O_2 to tissues, but the tissues are unable to use it properly because of the action of some toxic agent. One cause is cyanide poisoning, in which cyanide blocks an enzyme required for the use of O_2 during ATP synthesis. ■

Proprioceptor Stimulation of Respiration

As soon as you start exercising, your rate and depth of breathing increase, even before changes in P_{O_2}, P_{CO_2}, or H^+ level occur. The main stimulus for these quick changes in respiratory effort is input from proprioceptors, which monitor movement of joints and muscles. Nerve impulses from the proprioceptors stimulate the inspiratory area of the medulla oblongata. At the same time, axon collaterals (branches) of upper motor neurons that originate in the primary motor cortex (precentral gyrus) also feed excitatory impulses into the inspiratory area.

The Inflation Reflex

Similar to those in the blood vessels, stretch-sensitive receptors called **baroreceptors** or **stretch receptors** are located in the walls of bronchi and bronchioles. When these receptors become stretched during overinflation of the lungs, nerve impulses are sent along the vagus (X) nerves to the inspiratory and apneustic areas. In response, the inspiratory area is inhibited directly, and the apneustic area is inhibited from activating the inspiratory area. As a result, exhalation begins. As air leaves the lungs during exhalation, the lungs deflate and the stretch receptors are no longer stimulated. Thus, the inspiratory and apneustic areas are no longer inhibited, and a new inhalation begins. This reflex, referred to as the **inflation (Hering–Breuer) reflex,** is mainly a

Figure 23.28 **Regulation of breathing in response to changes in blood P_{CO_2}, P_{O_2}, and pH (H^+ concentration) via negative feedback control.**

An increase in arterial blood P_{CO_2} stimulates the inspiratory center.

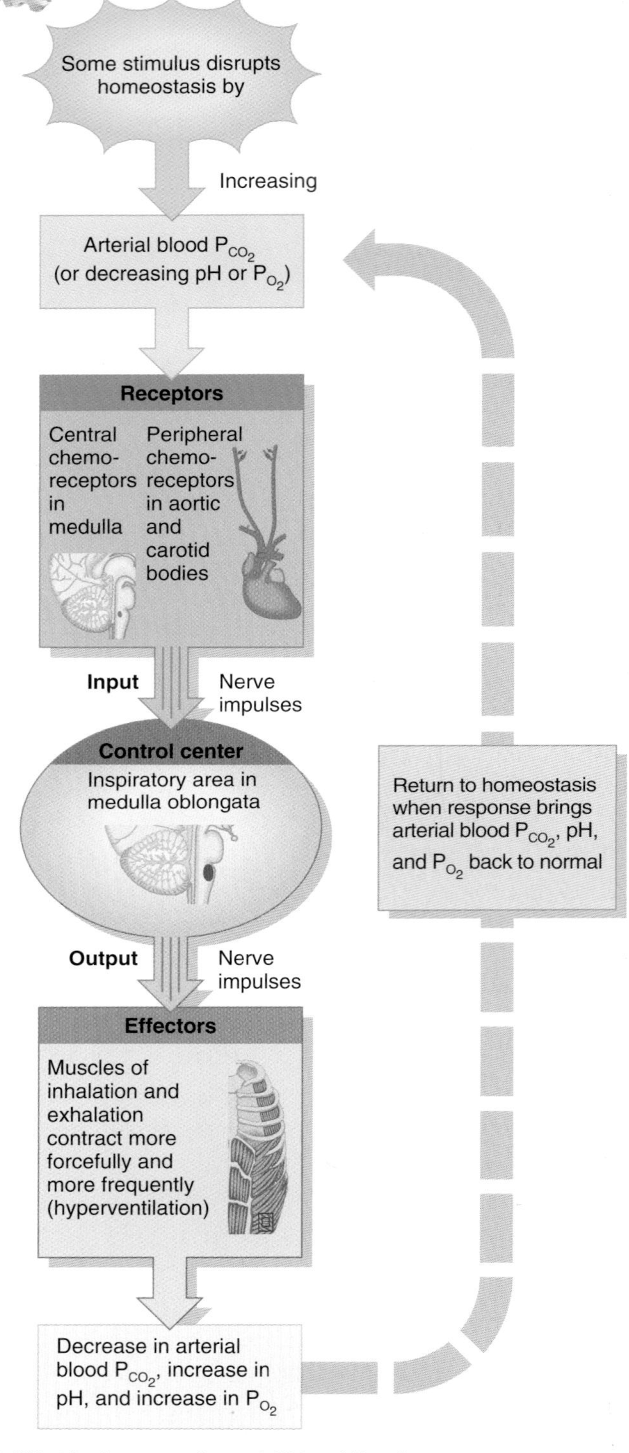
Some stimulus disrupts homeostasis by
Increasing
Arterial blood P_{CO_2} (or decreasing pH or P_{O_2})
Receptors
Central chemo- receptors in medulla
Peripheral chemo- receptors in aortic and carotid bodies
Input
Nerve impulses
Control center
Inspiratory area in medulla oblongata
Return to homeostasis when response brings arterial blood P_{CO_2}, pH, and P_{O_2} back to normal
Output
Nerve impulses
Effectors
Muscles of inhalation and exhalation contract more forcefully and more frequently (hyperventilation)
Decrease in arterial blood P_{CO_2}, increase in pH, and increase in P_{O_2}

What is the normal arterial blood P_{CO_2}?

TABLE 23.2 Summary of Stimuli that Affect Ventilation Rate and Depth

Stimuli that Increase Ventilation Rate and Depth	Stimuli that Decrease Ventilation Rate and Depth
Voluntary hyperventilation controlled by the cerebral cortex and anticipation of activity bystimulation of the limbic system.	Voluntary hypoventilation controlled by the cerebral cortex.
Increase in arterial blood P_{CO_2} above 40 mmHg (causes an increase in H^+) detected by peripheral and central chemoreceptors.	Decrease in arterial blood P_{CO_2} below 40 mmHg (causes a decrease in H^+) detected by peripheral and central chemoreceptors.
Decrease in arterial blood P_{O_2} from 105 mmHg to 50 mmHg.	Decrease in arterial blood P_{O_2} below 50 mmHg.
Increased activity of proprioceptors.	Decreased activity of proprioceptors.
Increase in body temperature.	Decrease in body temperature decreases the rate of respiration, and a sudden cold stimulus causes apnea.
Prolonged pain.	Severe pain causes apnea.
Decrease in blood pressure.	Increase in blood pressure.
Stretching the anal sphincter.	Irritation of pharynx or larynx by touch or chemicals causes brief apnea followed by coughing or sneezing.

protective mechanism for preventing excessive inflation of the lungs rather than a key component in the normal regulation of respiration.

Other Influences on Respiration

Other factors that contribute to regulation of respiration include the following:

- ***Limbic system stimulation.*** Anticipation of activity or emotional anxiety may stimulate the limbic system, which then sends excitatory input to the inspiratory area, increasing the rate and depth of ventilation.
- ***Temperature.*** An increase in body temperature, as occurs during a fever or vigorous muscular exercise, increases the rate of respiration. A decrease in body temperature decreases respiratory rate. A sudden cold stimulus (such as plunging into cold water) causes temporary **apnea** (AP-nē-a; *a-* = without; *-pnea* = breath), an absence of breathing.
- ***Pain.*** A sudden, severe pain brings about brief apnea, but a prolonged somatic pain increases respiratory rate. Visceral pain may slow the rate of respiration.
- ***Stretching the anal sphincter muscle.*** This action increases the respiratory rate and is sometimes used to stimulate respiration in a newborn baby or a person who has stopped breathing.
- ***Irritation of airways.*** Physical or chemical irritation of the pharynx or larynx brings about an immediate cessation of breathing followed by coughing or sneezing.
- ***Blood pressure.*** The carotid and aortic baroreceptors that detect changes in blood pressure have a small effect on respiration. A sudden rise in blood pressure decreases the rate of respiration, and a drop in blood pressure increases the respiratory rate.

Table 23.2 summarizes the stimuli that affect the rate and depth of ventilation.

▶ **CHECKPOINT**

27. How does the medullary rhythmicity area regulate respiration?

28. How are the apneustic and pneumotaxic areas related to the control of respiration?

29. How do the cerebral cortex, levels of CO_2 and O_2, proprioceptors, inflation reflex, temperature changes, pain, and irritation of the airways modify respiration?

EXERCISE AND THE RESPIRATORY SYSTEM

▶ **OBJECTIVE**

Describe the effects of exercise on the respiratory system.

The respiratory and cardiovascular systems make adjustments in response to both the intensity and duration of exercise. The effects of exercise on the heart are discussed in Chapter 20. Here we focus on how exercise affects the respiratory system.

Recall that the heart pumps the same amount of blood to the lungs as to all the rest of the body. Thus, as cardiac output rises, the blood flow to the lungs, termed **pulmonary perfusion,** increases as well. In addition, the **O_2 diffusing capacity,** a measure of the rate at which O_2 can diffuse from alveolar air into the blood, may increase threefold during maximal exercise because more pulmonary capillaries become maximally perfused. As a result, there is a greater surface area available for diffusion of O_2 into pulmonary blood capillaries.

When muscles contract during exercise, they consume large amounts of O_2 and produce large amounts of CO_2. During vigorous exercise, O_2 consumption and pulmonary ventilation both increase dramatically. At the onset of exercise, an abrupt increase in pulmonary ventilation is followed by a more gradual increase. With moderate exercise, the increase is due mostly to an increase in the depth of ventilation rather than to increased breathing rate. When exercise is more strenuous, the frequency of breathing also increases.

The abrupt increase in ventilation at the start of exercise is due to *neural* changes that send excitatory impulses to the inspiratory area in the medulla oblongata. These changes include (1) anticipation of the activity, which stimulates the limbic system; (2) sensory impulses from proprioceptors in muscles, tendons, and joints; and (3) motor impulses from the primary motor cortex (precentral gyrus). The more gradual increase in ventilation during moderate exercise is due to *chemical* and *physical* changes in the bloodstream, including (1) slightly decreased P_{O_2}, due to increased O_2 consumption; (2) slightly increased P_{CO_2}, due to increased CO_2 production by contracting muscle fibers; and (3) increased temperature, due to liberation of more heat as more O_2 is utilized. During strenuous exercise, HCO_3^- buffers H^+ released by lactic acid in a reaction that liberates CO_2, which further increases P_{CO_2}.

At the end of an exercise session, an abrupt decrease in pulmonary ventilation is followed by a more gradual decline to the resting level. The initial decrease is due mainly to changes in neural factors when movement stops or slows; the more gradual phase reflects the slower return of blood chemistry levels and temperature to the resting state.

The Effect of Smoking on Respiratory Efficiency

Smoking may cause a person to become easily "winded" during even moderate exercise because several factors decrease respiratory efficiency in smokers: (1) Nicotine constricts terminal bronchioles, which decreases airflow into and out of the lungs. (2) Carbon monoxide in smoke binds to hemoglobin and reduces its oxygen-carrying capability. (3) Irritants in smoke cause increased mucus secretion by the mucosa of the bronchial tree and swelling of the mucosal lining, both of which impede airflow into and out of the lungs. (4) Irritants in smoke also inhibit the movement of cilia and destroy cilia in the lining of the respiratory system. Thus, excess mucus and foreign debris are not easily removed, which further adds to the difficulty in breathing. (5) With time, smoking leads to destruction of elastic fibers in the lungs and is the prime cause of emphysema (described on page 887). These changes cause collapse of small bronchioles and trapping of air in alveoli at the end of exhalation. The result is less efficient gas exchange. ■

▶ CHECKPOINT

30. How does exercise affect the inspiratory area?

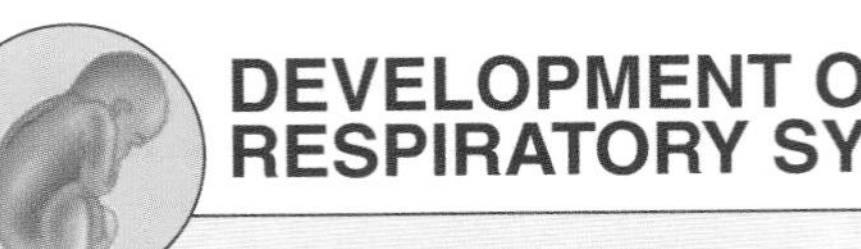

DEVELOPMENT OF THE RESPIRATORY SYSTEM

▶ OBJECTIVE

Describe the development of the respiratory system.

The development of the mouth and pharynx are discussed in Chapter 24. Here we consider the development of the other structures of the respiratory system that you learned about in this chapter.

At about four weeks of development, the respiratory system begins as an outgrowth of the foregut (precursor of some digestive organs) just anterior to the pharynx. This outgrowth is called the **respiratory diverticulum** (Figure 23.29). The **endoderm** lining the respiratory diverticulum gives rise to the epithelium and glands of the trachea, bronchi, and alveoli. **Mesoderm** surrounding the respiratory diverticulum gives rise to the connective tissue, cartilage, and smooth muscle of these structures.

The epithelial lining of the *larynx* develops from the **endoderm** of the respiratory diverticulum; the cartilages and muscles originate from the **fourth** and **sixth pharyngeal arches,** swellings on the surface of the embryo.

As the respiratory diverticulum elongates, its distal end enlarges to form a globular **tracheal bud,** which gives rise to the *trachea.* Soon after, the tracheal bud divides into **bronchial buds,** which branch repeatedly and develop with the *bronchi.* By 24 weeks, 17 orders of branches have formed and *respiratory bronchioles* have developed.

During weeks 6 to 16, all major elements of the *lungs* have formed, except for those involved in gaseous exchange (respiratory bronchioles, alveolar ducts, and alveoli). Since respiration is not possible at this stage, fetuses born during this time cannot survive.

During weeks 16 to 26, lung tissue becomes highly vascular and respiratory bronchioles, alveolar ducts, and some primitive alveoli develop. Although it is possible for a fetus born near the end of this period to survive if given intensive care, death frequently occurs due to the immaturity of the respiratory and other systems.

From 26 weeks to birth, many more primitive alveoli develop; they consist of type I alveolar cells (main sites of gaseous exchange) and type II surfactant-producing cells. Blood capillaries also establish close contact with the primitive alveoli. Recall that surfactant is necessary to lower surface tension of alveolar fluid and thus reduce the tendency of alveoli to collapse on exhalation. Although surfactant production begins by 20 weeks, it is present in only small quantities. Amounts sufficient to permit survival of a premature (preterm) infant are not produced until 26 to 28 weeks gestation. Infants born before 26 to 28 weeks are at high risk of respiratory distress syndrome (RDS), in which the alveoli collapse during exhalation and must be reinflated during inhalation (see page 867).

At about 30 weeks, mature alveoli develop. However, it is estimated that only about one-sixth of the full complement of

Figure 23.29 **Development of the bronchial tubes and lungs.**

The respiratory system develops from endoderm and mesoderm.

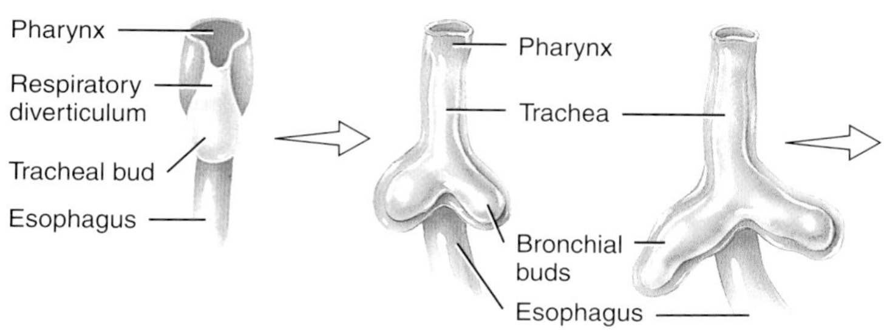

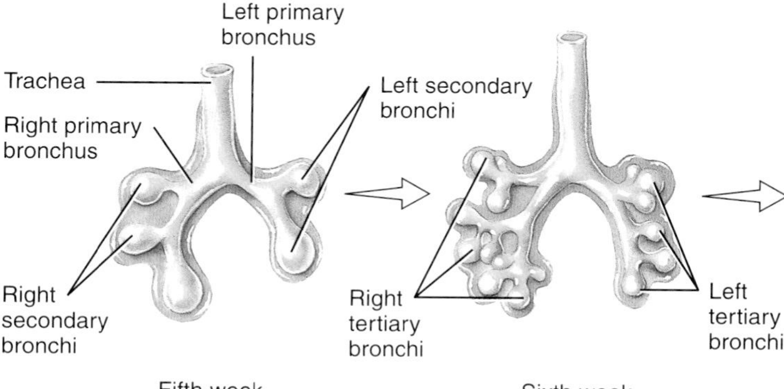

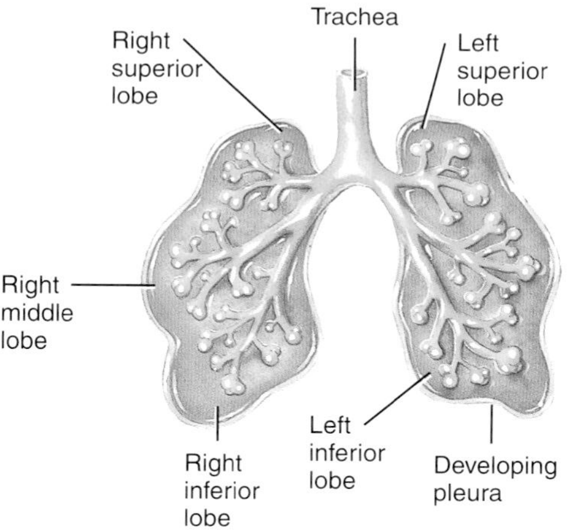

When does the respiratory system begin to develop in an embryo?

alveoli develop before birth; the remainder develop after birth during the first eight years.

As the lungs develop, they acquire their *pleural sacs.* The *visceral pleura* and the *parietal pleura* develop from **mesoderm.** The space between the pleural layers is the *pleural cavity.*

During development, breathing movements of the fetus cause the aspiration of fluid into the lungs. This fluid is a mixture of amniotic fluid, mucus from bronchial glands, and surfactant. At birth, the lungs are about half-filled with fluid. When breathing begins at birth, most of the fluid is rapidly reabsorbed by blood and lymph capillaries and a small amount is expelled through the nose and mouth during delivery.

▶ CHECKPOINT

31. What structures develop from the laryngotracheal bud?

AGING AND THE RESPIRATORY SYSTEM

▶ OBJECTIVE

Describe the effects of aging on the respiratory system.

With advancing age, the airways and tissues of the respiratory tract, including the alveoli, become less elastic and more rigid; the chest wall becomes more rigid as well. The result is a decrease in lung capacity. In fact, vital capacity (the maximum amount of air that can be expired after maximal inhalation) can decrease as much as 35% by age 70. A decrease in blood level of O_2, decreased activity of alveolar macrophages, and diminished ciliary action of the epithelium lining the respiratory tract occur. Owing to all these age-related factors, elderly people are more susceptible to pneumonia, bronchitis, emphysema, and other pulmonary disorders. Age-related changes in the structure and functions of the lung can also contribute to an older person's reduced ability to perform vigorous exercises, such as running.

▶ CHECKPOINT

32. What accounts for the decrease in lung capacity with aging?

• • •

To appreciate the many ways that the respiratory system contributes to homeostasis of other body systems, examine *Focus on Homeostasis: The Respiratory System.* Next, in Chapter 24, we will see how the digestive system makes nutrients available to body cells so that oxygen provided by the respiratory system can be used for ATP production.

FOCUS ON HOMEOSTASIS

The Respiratory System

BODY SYSTEM	CONTRIBUTION OF THE RESPIRATORY SYSTEM
For all body systems	Provides oxygen and removes carbon dioxide. Helps adjust pH of body fluids through exhalation of carbon dioxide.
Muscular system 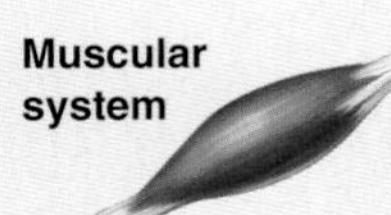	Increased rate and depth of breathing support increased activity of skeletal muscles during exercise.
Nervous system 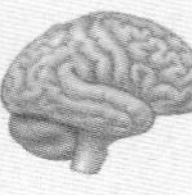	Nose contains receptors for sense of smell (olfaction). Vibrations of air flowing across vocal folds produce sounds for speech.
Endocrine system 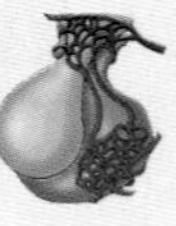	Angiotensin converting enzyme (ACE) in lungs catalyzes formation of the hormone angiotensin II from angiotensin I.
Cardiovascular system	During inhalations, respiratory pump aids return of venous blood to the heart.
Lymphatic system and immunity 	Hairs in nose, cilia and mucus in trachea, bronchi, and smaller airways, and alveolar macrophages contribute to nonspecific resistance o disease. Pharynx (throat) contains lymphatic ttissue (tonsils). Respiratory pump (during inhalation) promotes flow of lymph.
Digestive system 	Forceful contraction of respiratory muscles can assist in defecation.
Urinary system 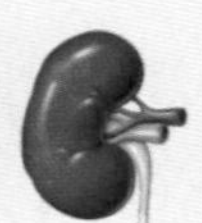	Together, respiratory and urinary systems regulate pH of body fluids.
Reproductive systems	Increased rate and depth of breathing support activity during sexual intercourse. Internal respiration provides oxygen to developing fetus.

DISORDERS: HOMEOSTATIC IMBALANCES

Asthma

Asthma (AZ-ma = panting) is a disorder characterized by chronic airway inflammation, airway hypersensitivity to a variety of stimuli, and airway obstruction. It is at least partially reversible, either spontaneously or with treatment. Asthma affects 3–5% of the U.S. population and is more common in children than in adults. Airway obstruction may be due to smooth muscle spasms in the walls of smaller bronchi and bronchioles, edema of the mucosa of the airways, increased mucus secretion, and/or damage to the epithelium of the airway.

Individuals with asthma typically react to concentrations of agents too low to cause symptoms in people without asthma. Sometimes the trigger is an allergen such as pollen, house dust mites, molds, or a particular food. Other common triggers of asthma attacks are emotional upset, aspirin, sulfiting agents (used in wine and beer and to keep greens fresh in salad bars), exercise, and breathing cold air or cigarette smoke. In the early phase (acute) response, smooth muscle spasm is accompanied by excessive secretion of mucus that may clog the bronchi and bronchioles and worsen the attack. The late phase (chronic) response is characterized by inflammation, fibrosis, edema, and necrosis (death) of bronchial epithelial cells. A host of mediator chemicals, including leukotrienes, prostaglandins, thromboxane, platelet-activating factor, and histamine, take part.

Symptoms include difficult breathing, coughing, wheezing, chest tightness, tachycardia, fatigue, moist skin, and anxiety. An acute attack is treated by giving an inhaled $beta_2$-adrenergic agonist (albuterol) to help relax smooth muscle in the bronchioles and open up the airways. However, long-term therapy of asthma strives to suppress the underlying inflammation. The anti-inflammatory drugs that are used most often are inhaled corticosteroids (glucocorticoids), cromolyn sodium (Intal®), and leukotriene blockers (Accolate®).

Chronic Obstructive Pulmonary Disease

Chronic obstructive pulmonary disease (COPD) is a type of respiratory disorder characterized by chronic and recurrent obstruction of airflow, which increases airway resistance. COPD affects about 30 million Americans and is the fourth leading cause of death behind heart disease, cancer, and cerebrovascular disease. The principal types of COPD are emphysema and chronic bronchitis. In most cases, COPD is preventable because its most common cause is cigarette smoking or breathing secondhand smoke. Other causes include air pollution, pulmonary infection, occupational exposure to dusts and gases, and genetic factors. Because men, on average, have more years of exposure to cigarette smoke than women, they are twice as likely as women to suffer from COPD; still, the incidence of COPD in women has risen sixfold in the past 50 years, a reflection of increased smoking among women.

Emphysema

Emphysema (em-fi-SĒ-ma = blown up or full of air) is a disorder characterized by destruction of the walls of the alveoli, producing abnormally large air spaces that remain filled with air during exhalation. With less surface area for gas exchange, O_2 diffusion across the damaged respiratory membrane is reduced. Blood O_2 level is somewhat lowered, and any mild exercise that raises the O_2 requirements of the cells leaves the patient breathless. As increasing numbers of alveolar walls are damaged, lung elastic recoil decreases due to loss of elastic fibers, and an increasing amount of air becomes trapped in the lungs at the end of exhalation. Over several years, added exertion during inhalation increases the size of the chest cage, resulting in a "barrel chest."

Emphysema is generally caused by a long-term irritation; cigarette smoke, air pollution, and occupational exposure to industrial dust are the most common irritants. Some destruction of alveolar sacs may be caused by an enzyme imbalance. Treatment consists of cessation of smoking, removal of other environmental irritants, exercise training under careful medical supervision, breathing exercises, use of bronchodilators, and oxygen therapy.

Chronic Bronchitis

Chronic bronchitis is a disorder characterized by excessive secretion of bronchial mucus accompanied by a productive cough (sputum is raised) that lasts for at least three months of the year for two successive years. Cigarette smoking is the leading cause of chronic bronchitis. Inhaled irritants lead to chronic inflammation with an increase in the size and number of mucous glands and goblet cells in the airway epithelium. The thickened and excessive mucus produced narrows the airway and impairs ciliary function. Thus, inhaled pathogens become embedded in airway secretions and multiply rapidly. Besides a productive cough, symptoms of chronic bronchitis are shortness of breath, wheezing, cyanosis, and pulmonary hypertension. Treatment for chronic bronchitis is similar to that for emphysema.

Lung Cancer

In the United States **lung cancer** is the leading cause of cancer death in both males and females, accounting for 160,000 deaths annually. At the time of diagnosis, lung cancer is usually well advanced, with distant metastases present in about 55% of patients, and regional lymph node involvement in an additional 25%. Most people with lung cancer die within a year of the initial diagnosis; the overall survival rate is only 10–15%. Cigarette smoke is the most common cause of lung cancer. Roughly 85% of lung cancer cases are related to smoking, and the disease is 10 to 30 times more common in smokers than nonsmokers. Exposure to secondhand smoke is also associated with lung cancer and heart disease. In the United States, secondhand smoke causes an estimated 4000 deaths a year from lung cancer, and nearly 40,000 deaths a year from heart disease. Other causes of lung cancer are ionizing radiation and inhaled irritants, such as asbestos and radon gas. Emphysema is a common precursor to the development of lung cancer.

The most common type of lung cancer, **bronchogenic carcinoma,** starts in the epithelium of the bronchial tubes. Bronchogenic tumors are named based on where they arise. For example, *adenocarcinomas* develop in peripheral areas of the lungs from bronchial glands and alveolar cells, *squamous cell carcinomas* develop from the epithelium of larger bronchial tubes, and *small (oat) cell carcinomas* develop from epithelial cells in primary bronchi near the hilum of the lungs and tend to involve the mediastinum early on. Depending on the type of bronchogenic tumors, they may be aggressive, locally invasive, and undergo widespread metastasis. The tumors begin as epithelial lesions that grow to form masses that obstruct the bronchial tubes or invade adjacent lung tissue. Bronchogenic carcinomas metastasize to lymph nodes, the brain, bones, liver, and other organs.

Symptoms of lung cancer are related to the location of the tumor. These may include a chronic cough, spitting blood from the respiratory tract, wheezing, shortness of breath, chest pain, hoarseness, difficulty swallowing, weight loss, anorexia, fatigue, bone pain, confusion, problems with balance, headache, anemia, thrombocytopenia, and jaundice.

Treatment consists of partial or complete surgical removal of a diseased lung (pulmonectomy), radiation therapy, and chemotherapy.

Pneumonia

Pneumonia is an acute infection or inflammation of the alveoli. It is the most common infectious cause of death in the United States, where an estimated 4 million cases occur annually. When certain microbes enter the lungs of susceptible individuals, they release damaging toxins, stimulating inflammation and immune responses that have damaging side effects. The toxins and immune response damage alveoli and bronchial mucous membranes; inflammation and edema cause the alveoli to fill with fluid, interfering with ventilation and gas exchange.

The most common cause of pneumonia is the pneumococcal bacterium *Streptococcus pneumoniae,* but other microbes may also cause pneumonia. Those who are most susceptible to pneumonia are the elderly, infants, immunocompromised individuals (AIDS or cancer patients, or those taking immunosuppressive drugs), cigarette smokers, and individuals with an obstructive lung disease. Most cases of pneumonia are preceded by an upper respiratory infection that often is viral. Individuals then develop fever, chills, productive or dry cough, malaise, chest pain, and sometimes dyspnea (difficult breathing) and hemoptysis (spitting blood).

Treatment may involve antibiotics, bronchodilators, oxygen therapy, increased fluid intake, and chest physiotherapy (percussion, vibration, and postural drainage).

Tuberculosis

The bacterium *Mycobacterium tuberculosis* produces an infectious, communicable disease called **tuberculosis (TB)** that most often affects the lungs and the pleurae but may involve other parts of the body. Once the bacteria are inside the lungs, they multiply and cause inflammation, which stimulates neutrophils and macrophages to migrate to the area and engulf the bacteria to prevent their spread. If the immune system is not impaired, the bacteria remain dormant for life, but impaired immunity may enable the bacteria to escape into blood and lymph to infect other organs. In many people, symptoms—fatigue, weight loss, lethargy, anorexia, a low-grade fever, night sweats, cough, dyspnea, chest pain, and hemoptysis—do not develop until the disease is advanced.

During the past several years, the incidence of TB in the United States has risen dramatically. Perhaps the single most important factor related to this increase is the spread of the human immunodeficiency virus (HIV). People infected with HIV are much more likely to develop tuberculosis because their immune systems are impaired. Among the other factors that have contributed to the increased number of cases are homelessness, increased drug abuse, increased immigration from countries with a high prevalence of tuberculosis, increased crowding in housing among the poor, and airborne transmission of tuberculosis in prisons and shelters. In addition, recent outbreaks of tuberculosis involving multi-drug-resistant strains of *Mycobacterium tuberculosis* have occurred because patients fail to complete their antibiotic and other treatment regimens. TB is treated with the medication isoniazid.

Coryza and Influenza

Hundreds of viruses can cause **coryza** (ko-RĪ-za) or the **common cold,** but a group of viruses called *rhinoviruses* is responsible for about 40% of all colds in adults. Typical symptoms include sneezing, excessive nasal secretion, dry cough, and congestion. The uncomplicated common cold is not usually accompanied by a fever. Complications include sinusitis, asthma, bronchitis, ear infections, and laryngitis. Recent investigations suggest an association between emotional stress and the common cold. The higher the stress level, the greater the frequency and duration of colds.

Influenza (flu) is also caused by a virus. Its symptoms include chills, fever (usually higher than 101°F = 39°C), headache, and muscular aches. Influenza can become life-threatening and may develop into pneumonia. It is important to recognize that influenza is a respiratory disease, not a gastrointestinal (GI) disease. Many people mistakenly report having "the flu" when they are suffering from a GI illness.

Pulmonary Edema

Pulmonary edema is an abnormal accumulation of fluid in the interstitial spaces and alveoli of the lungs. The edema may arise from increased permeability of the pulmonary capillaries (pulmonary origin) or increased pressure in the pulmonary capillaries (cardiac origin); the latter cause may coincide with congestive heart failure. The most common symptom is dyspnea. Others include wheezing, tachypnea (rapid breathing rate), restlessness, a feeling of suffocation, cyanosis, pallor (paleness), diaphoresis (excessive perspiration), and pulmonary hypertension. Treatment consists of administering oxygen, drugs that dilate the bronchioles and lower blood pressure, diuretics to rid the body of excess fluid, and drugs that correct acid–base imbalance; suctioning of airways; and mechanical ventilation. One of the recent culprits for causing pulmonary edema was the "phen-fen" diet pills.

Cystic Fibrosis

Cystic fibrosis (CF) is an inherited disease of secretory epithelia that affects the airways, liver, pancreas, small intestine, and sweat glands. It is the most common lethal genetic disease in whites: 5% of the population are thought to be genetic carriers. The cause of cystic fibrosis is a genetic mutation affecting a transporter protein that carries chloride ions across the plasma membranes of many epithelial cells. Because dysfunction of sweat glands causes perspiration to contain excessive sodium chloride (salt), measurement of the excess chloride is one index for diagnosing CF. The mutation also disrupts the normal functioning of several organs by causing ducts within them to become obstructed by thick mucus secretions that do not drain easily from the passageways. Buildup of these secretions leads to inflammation and replacement of injured cells with connective tissue that further blocks the ducts. Clogging and infection of the airways leads to difficulty in breathing and eventual destruction of lung tissue. Lung disease accounts for most deaths from CF. Obstruction of small bile ducts in the liver interferes with digestion and disrupts liver function; clogging of pancreatic ducts prevents digestive enzymes from reaching the small intestine. Because pancreatic juice contains the main fat-digesting enzyme, the person fails to absorb fats or fat-soluble vitamins and thus suffers from vitamin A, D, and K deficiency diseases. With respect to the reproductive systems, blockage of the ductus (vas) deferens leads to infertility in males; the formation of dense mucus plugs in the vagina restricts the entry of sperm into the uterus and can lead to infertility in females.

A child suffering from cystic fibrosis is given pancreatic extract and large doses of vitamins A, D, and K. The recommended diet is high in calories, fats, and proteins, with vitamin supplementation and liberal use of salt. One of the newest treatments for CF is heart-lung transplants.

Asbestos-related Diseases

Asbestos-related diseases are serious lung disorders that develop as a result of inhaling asbestos particles decades earlier. When asbestos particles are inhaled, they penetrate lung tissue. In response, white blood cells attempt to destroy them by phagocytosis. However, the fibers usually destroy the white blood cells and scarring of lung tissue may follow. Asbestos-related diseases include **asbestosis** (widespread scarring of lung tissue), **diffuse pleural thickening** (thickening of the pleurae), and **mesothelioma** (cancer of the pleurae or, less commonly, the peritoneum).

Sudden Infant Death Syndrome

Sudden infant death syndrome (SIDS) is the sudden, unexpected death of an apparently healthy infant during sleep. It rarely occurs before 2 weeks or after 6 months of age, with the peak incidence between the second and fourth months. SIDS is more common in premature infants, male babies, low-birth-weight babies, babies of drug users or smokers, babies who have stopped breathing and have had to be resuscitated, babies with upper respiratory tract infections, and babies who have had a sibling die of SIDS. African American and Native American babies are also at high risk. The exact cause of SIDS is unknown. However, it may be due to an abnormality in the mechanisms that control respiration or low levels of oxygen in the blood. SIDS may also be linked to hypoxia while sleeping in a prone position (on the stomach) and the rebreathing of exhaled air trapped in a depression of a mattress. It is recommended that for the first six months infants be placed on their backs for sleeping ("back to sleep").

Severe Acute Respiratory Syndrome

Severe acute respiratory syndrome (SARS) is an example of an *emerging infectious disease,* that is, a disease that is new or changing. Other examples of emerging infectious diseases are West Nile encephalitis, mad cow disease, and AIDS. SARS first appeared in Southern China in late 2002 and has subsequently spread worldwide. It is a respiratory illness caused by a new variety of coronavirus. Symptoms of SARS include fever, malaise, muscle aches, nonproductive (dry) cough, difficulty in breathing, chills, headache, and diarrhea. About 10–20% of patients require mechanical ventilation and in some cases death may result. The disease is primarily spread through person-to-person contact. There is no effective treatment for SARS and the death rate is 5–10%, usually among the elderly and in persons with other medical problems.

MEDICAL TERMINOLOGY

Abdominal thrust (Heimlich) maneuver (HĪM-lik ma-NOO-ver) First-aid procedure designed to clear the airways of obstructing objects. It is performed by applying a quick upward thrust between the navel and costal margin that causes sudden elevation of the diaphragm and forceful, rapid expulsion of air in the lungs; this action forces air out the trachea to eject the obstructing object. The Heimlich maneuver is also used to expel water from the lungs of near-drowning victims before resuscitation is begun.

Asphyxia (as-FIK-sē-a; *sphyxia* = pulse) Oxygen starvation due to low atmospheric oxygen or interference with ventilation, external respiration, or internal respiration.

Aspiration (as′-pi-RĀ-shun) Inhalation of a foreign substance such as water, food, or a foreign body into the bronchial tree; also, the drawing of a substance in or out by suction.

Bronchiectasis (bron′-kē-EK-ta-sis; *-ektasis* = stretching) A chronic dilation of the bronchi or bronchioles resulting from damage to the bronchial wall, for example, from respiratory infections.

Bronchography (bron-KOG-ra-fē) An imaging technique used to visualize the bronchial tree using x rays. After an opaque contrast medium is inhaled through an intratracheal catheter, radiographs of the chest in various positions are taken, and the developed film, a **bronchogram** (BRON-kō-gram), provides a picture of the bronchial tree.

Bronchoscopy (bron-KOS-kō-pē) Visual examination of the bronchi through a **bronchoscope,** an illuminated, flexible tubular instrument that is passed through the mouth (or nose), larynx, and trachea into the bronchi. The examiner can view the interior of the trachea and bronchi to biopsy a tumor, clear an obstructing object or secretions from an airway, take cultures or smears for microscopic examination, stop bleeding, or deliver drugs.

Cheyne–Stokes respiration (CHĀN STŌKS res′-pi-RĀ-shun) A repeated cycle of irregular breathing that begins with shallow breaths that increase in depth and rapidity and then decrease and cease altogether for 15 to 20 seconds. Cheyne–Stokes is normal in infants; it is also often seen just before death from pulmonary, cerebral, cardiac, and kidney disease.

Dyspnea (DISP-nē-a; *dys-* = painful, difficult) Painful or labored breathing.

Epistaxis (ep′-i-STAK-sis) Loss of blood from the nose due to trauma, infection, allergy, malignant growths, or bleeding disorders. It can be arrested by cautery with silver nitrate, electrocautery, or firm packing. Also called **nosebleed.**

Hypoventilation (*hypo-* = below) Slow and shallow breathing.

Mechanical ventilation The use of an automatically cycling device (ventilator or respirator) to assist breathing. A plastic tube is inserted into the nose or mouth and the tube is attached to a device that forces air into the lungs. Exhalation occurs passively due to the elastic recoil of the lungs.

Rales (RĀLS) Sounds sometimes heard in the lungs that resemble bubbling or rattling. Rales are to the lungs what murmurs are to the heart. Different types are due to the presence of an abnormal type or amount of fluid or mucus within the bronchi or alveoli, or to bronchoconstriction that causes turbulent airflow.

Respirator (RES-pi-rā′-tor) An apparatus fitted to a mask over the nose and mouth, or hooked directly to an endotracheal or tracheotomy tube, that is used to assist or support ventilation or to provide nebulized medication to the air passages.

Respiratory failure A condition in which the respiratory system either cannot supply sufficient O_2 to maintain metabolism or cannot eliminate enough CO_2 to prevent respiratory acidosis (a lower-than-normal pH in interstitial fluid).

Rhinitis (rī-NĪ-tis; *rhin-* = nose) Chronic or acute inflammation of the mucous membrane of the nose due to viruses, bacteria, or irritants. Excessive mucus production produces a runny nose, nasal congestion, and postnasal drip.

Sleep apnea (AP-nē-a *a-* = without; *-pnea* = breath) A disorder in which a person repeatedly stops breathing for 10 or more seconds while sleeping. Most often, it occurs because loss of muscle tone in pharyngeal muscles allows the airway to collapse.

Sputum (SPŪ-tum = to spit) Mucus and other fluids from the air passages that is expectorated (expelled by coughing).

Strep throat Inflammation of the pharynx caused by the bacterium *Streptococcus pyogenes.* It may also involve the tonsils and middle ear.

Tachypnea (tak′-ip-NĒ-a; *tachy-* = rapid; *-pnea* = breath) Rapid breathing rate.

Wheeze (HWĒZ) A whistling, squeaking, or musical high-pitched sound during breathing resulting from a partially obstructed airway.

STUDY OUTLINE

RESPIRATORY SYSTEM ANATOMY (p. 847)

1. The respiratory system consists of the nose, pharynx, larynx, trachea, bronchi, and lungs. They act with the cardiovascular system to supply oxygen (O_2) and remove carbon dioxide (CO_2) from the blood.
2. The external portion of the nose is made of cartilage and skin and is lined with a mucous membrane. Openings to the exterior are the external nares.
3. The internal portion of the nose communicates with the paranasal sinuses and nasopharynx through the internal nares.
4. The nasal cavity is divided by a septum. The anterior portion of the cavity is called the vestibule. The nose warms, moistens, and filters air and functions in olfaction and speech.
5. The pharynx (throat) is a muscular tube lined by a mucous membrane. The anatomic regions are the nasopharynx, oropharynx, and laryngopharynx.
6. The nasopharynx functions in respiration. The oropharynx and laryngopharynx function both in digestion and in respiration.
7. The larynx (voice box) is a passageway that connects the pharynx with the trachea. It contains the thyroid cartilage (Adam's apple); the epiglottis, which prevents food from entering the larynx; the cricoid cartilage, which connects the larynx and trachea; and the paired arytenoid, corniculate, and cuneiform cartilages.
8. The larynx contains vocal folds, which produce sound as they vibrate. Taut folds produce high pitches, and relaxed ones produce low pitches.
9. The trachea (windpipe) extends from the larynx to the primary bronchi. It is composed of C-shaped rings of cartilage and smooth muscle and is lined with pseudostratified ciliated columnar epithelium.
10. The bronchial tree consists of the trachea, primary bronchi, secondary bronchi, tertiary bronchi, bronchioles, and terminal bronchioles. Walls of bronchi contain rings of cartilage; walls of bronchioles contain increasingly smaller plates of cartilage and increasing amounts of smooth muscle.
11. Lungs are paired organs in the thoracic cavity enclosed by the pleural membrane. The parietal pleura is the superficial layer that lines the thoracic cavity; the visceral pleura is the deep layer that covers the lungs.
12. The right lung has three lobes separated by two fissures; the left lung has two lobes separated by one fissure and a depression, the cardiac notch.
13. Secondary bronchi give rise to branches called segmental bronchi, which supply segments of lung tissue called bronchopulmonary segments.
14. Each bronchopulmonary segment consists of lobules, which contain lymphatics, arterioles, venules, terminal bronchioles, respiratory bronchioles, alveolar ducts, alveolar sacs, and alveoli.
15. Alveolar walls consist of type I alveolar cells, type II alveolar cells, and associated alveolar macrophages.
16. Gas exchange occurs across the respiratory membranes.

PULMONARY VENTILATION (p. 863)

1. Pulmonary ventilation, or breathing, consists of inhalation and exhalation.
2. The movement of air into and out of the lungs depends on pressure changes governed in part by Boyle's law, which states that the volume of a gas varies inversely with pressure, assuming that temperature remains constant.
3. Inhalation occurs when alveolar pressure falls below atmospheric pressure. Contraction of the diaphragm and external intercostals increases the size of the thorax, thereby decreasing the intrapleural pressure so that the lungs expand. Expansion of the lungs decreases alveolar pressure so that air moves down a pressure gradient from the atmosphere into the lungs.
4. During forceful inhalation, accessory muscles of inhalation (sternocleidomastoids, scalenes, and pectoralis minors) are also used.
5. Exhalation occurs when alveolar pressure is higher than atmospheric pressure. Relaxation of the diaphragm and external intercostals results in elastic recoil of the chest wall and lungs, which increases intrapleural pressure; lung volume decreases and alveolar pressure increases, so air moves from the lungs to the atmosphere.

6. Forceful exhalation involves contraction of the internal intercostal and abdominal muscles.
7. The surface tension exerted by alveolar fluid is decreased by the presence of surfactant.
8. Compliance is the ease with which the lungs and thoracic wall can expand.
9. The walls of the airways offer some resistance to breathing.
10. Normal quiet breathing is termed eupnea; other patterns are costal breathing and diaphragmatic breathing. Modified respiratory movements, such as coughing, sneezing, sighing, yawning, sobbing, crying, laughing, and hiccupping, are used to express emotions and to clear the airways. (See Table 23.1 on page 868.)

LUNG VOLUMES AND CAPACITIES (p. 868)

1. Lung volumes exchanged during breathing and the rate of respiration are measured with a spirometer.
2. Lung volumes measured by spirometry include tidal volume, minute ventilation, alveolar ventilation rate, inspiratory reserve volume, expiratory reserve volume, and $FEV_{1.0}$. Other lung volumes include anatomic dead space, residual volume, and minimal volume.
3. Lung capacities, the sum of two or more lung volumes, include inspiratory, functional, residual, vital, and total lung capacities.

EXCHANGE OF OXYGEN AND CARBON DIOXIDE (p. 870)

1. The partial pressure of a gas is the pressure exerted by that gas in a mixture of gases. It is symbolized by P_x, where the subscript is the chemical formula of the gas.
2. According to Dalton's law, each gas in a mixture of gases exerts its own pressure as if all the other gases were not present.
3. Henry's law states that the quantity of a gas that will dissolve in a liquid is proportional to the partial pressure of the gas and its solubility (given that the temperature remains constant).
4. In internal and external respiration, O_2 and CO_2 diffuse from areas of higher partial pressures to areas of lower partial pressures.
5. External respiration or pulmonary gas exchange is the exchange of gases between alveoli and pulmonary blood capillaries. It depends on partial pressure differences, a large surface area for gas exchange, a small diffusion distance across the respiratory membrane, and the rate of airflow into and out of the lungs.
6. Internal respiration or systemic gas exchange is the exchange of gases between systemic blood capillaries and tissue cells.

TRANSPORT OF OXYGEN AND CARBON DIOXIDE (p. 873)

1. In each 100 mL of oxygenated blood, 1.5% of the O_2 is dissolved in blood plasma and 98.5% is bound to hemoglobin as oxyhemoglobin ($Hb-O_2$).
2. The binding of O_2 to hemoglobin is affected by P_{O_2}, acidity (pH), P_{CO_2}, temperature, and 2,3-bisphosphoglycerate (BPG).
3. Fetal hemoglobin differs from adult hemoglobin in structure and has a higher affinity for O_2.
4. In each 100 mL of deoxygenated blood, 7% of CO_2 is dissolved in blood plasma, 23% combines with hemoglobin as carbaminohemoglobin ($Hb-CO_2$), and 70% is converted to bicarbonate ions (HCO_3^-).
5. In an acidic environment, hemoglobin's affinity for O_2 is lower, and O_2 dissociates more readily from it (Bohr effect).
6. In the presence of O_2, less CO_2 binds to hemoglobin (Haldane effect).

CONTROL OF RESPIRATION (p. 879)

1. The respiratory center consists of a medullary rhythmicity area in the medulla oblongata and a pneumotaxic area and an apneustic area in the pons.
2. The inspiratory area sets the basic rhythm of respiration.
3. The pneumotaxic and apneustic areas coordinate the transition between inhalation and exhalation.
4. Respirations may be modified by a number of factors, including cortical influences; the inflation reflex; chemical stimuli, such as O_2 and CO_2 and H levels; proprioceptor input; blood pressure changes; limbic system stimulation; temperature; pain; and irritation to the airways. (See Table 23.2 on page 882.)

EXERCISE AND THE RESPIRATORY SYSTEM (p. 883)

1. The rate and depth of ventilation change in response to both the intensity and duration of exercise.
2. An increase in pulmonary perfusion and O_2-diffusing capacity occurs during exercise.
3. The abrupt increase in ventilation at the start of exercise is due to neural changes that send excitatory impulses to the inspiratory area in the medulla oblongata. The more gradual increase in ventilation during moderate exercise is due to chemical and physical changes in the bloodstream.

DEVELOPMENT OF THE RESPIRATORY SYSTEM (p. 884)

1. The respiratory system begins as an outgrowth of endoderm called the respiratory diverticulum.
2. Smooth muscle, cartilage, and connective tissue of the bronchial tubes and pleural sacs develop from mesoderm.

AGING AND THE RESPIRATORY SYSTEM (p. 885)

1. Aging results in decreased vital capacity, decreased blood level of O_2, and diminished alveolar macrophage activity.
2. Elderly people are more susceptible to pneumonia, emphysema, bronchitis, and other pulmonary disorders.

SELF-QUIZ QUESTIONS

Fill in the blanks in the following statements.

1. Oxygen in blood is carried primarily in the form of ____; carbon dioxide is carried as ____, ____, and ____.
2. Write the equation for the chemical reaction that occurs for the transport of carbon dioxide as bicarbonate ions in blood: ____.

Indicate whether the following statements are true or false.

3. The three basic steps of respiration are pulmonary ventilation, external respiration, and cellular respiration.
4. For inhalation to occur, air pressure in the alveoli must be less than atmospheric pressure; for exhalation to occur, air pressure in the alveoli must be greater than atmospheric pressure.

Choose the one best answer to the following questions.

5. What structural changes occur from primary bronchi to terminal bronchioles? (1) The mucous membrane changes from pseudostratified ciliated columnar epithelium to nonciliated simple cuboidal epithelium. (2) The number of goblet cells increases. (3) The amount of smooth muscle increases. (4) Incomplete rings of cartilage disappear. (5) The amount of branching decreases. (a) 1, 2, 3, 4, and 5; (b) 2, 3, and 4; (c) 1, 3, and 4; (d) 1, 3, 4, and 5; (e) 1, 2, 3, and 4.
6. Which of the following would cause oxygen to dissociate more readily from hemoglobin? (1) low pO_2, (2) an increase in H^+ in blood, (3) hypercapnia, (4) hypothermia, (5) low levels of BPG (2,3-bisphosphoglycerate). (a) 1 and 2; (b) 2, 3, and 4; (c) 1, 2, 3, and 5; (d) 1, 3, and 5; (e) 1, 2, and 3.
7. Which of the following statements are *correct*? (1) Normal exhalation during quiet breathing is an active process involving intensive muscle contraction. (2) Passive exhalation results from elastic recoil of the chest wall and lungs. (3) Air flow during breathing is due to a pressure gradient between the lungs and the atmospheric air. (4) During normal breathing, the pressure between the two pleural layers (intrapleural pressure) is always subatmospheric. (5) Surface tension of alveolar fluid facilitates inhalation. (a) 1, 2, and 3; (b) 2, 3, and 4; (c) 3, 4, and 5; (d) 1, 3, and 5; (e) 2, 3, and 5.
8. Which of the following factors affect the rate of external respiration? (1) partial pressure differences of the gases, (2) surface area for gas exchange, (3) diffusion distance, (4) solubility and molecular weight of the gases, (5) presence of bisphosphoglycerate (BPG). (a) 1, 2, and 3; (b) 2, 4, and 5; (c) 1, 2, 4, and 5; (d) 1, 2, 3, and 4; (e) 2, 3, 4, and 5.
9. The most important factor in determining the percent oxygen saturation of hemoglobin is (a) the partial pressure of oxygen, (b) acidity, (c) the partial pressure of carbon dioxide, (d) temperature, (e) BPG.
10. Which of the following statements are *true*? (1) Peripheral and central chemoreceptors are stimulated by an increase in pCO_2 and H^+ and a decrease in O_2. (2) Respiratory rate increases during the initial onset of exercise due to input to the inspiratory area from proprioceptors. (3) When baroreceptors in the lungs are stimulated, the expiratory area is activated. (4) Stimulation of the limbic system can result in excitation of the inspiratory area. (5) Sudden severe pain causes brief apnea, while prolonged somatic pain causes an increase in respiratory rate. (6) The respiratory rate increases during fever. (a) 1, 2, 3, and 6; (b) 1, 4, and 5; (c) 1, 2, 4, 5, and 6; (d) 2, 3, 4, 5, and 6; (e) 2, 4, 5, and 6.
11. Place the steps for normal inhalation in order. (a) decrease in intrapleural pressure to 754 mmHg, (b) increase in the size of the thoracic cavity, (c) flow of air from higher to lower pressure, (d) outward pull of pleurae, resulting in lung expansion, (e) stimulation of primary breathing muscles by phrenic and intercostal nerves, (f) decrease in alveolar pressure to 758 mmHg, (g) contraction of the diaphragm and external intercostals, (h) increase in the volume of the pleural cavity.
12. Match the following:

____(a) functions as a passageway for air and food, provides a resonating chamber for speech sounds, and houses the tonsils
____(b) site of external respiration
____(c) connects the laryngopharynx with the trachea; houses the vocal cords
____(d) serous membrane that surrounds the lungs
____(e) functions in warming, moistening, and filtering air; receives olfactory stimuli; is a resonating chamber for sound
____(f) simple squamous epithelial cells that form a continuous lining of the alveolar wall; sites of gas exchange
____(g) forms anterior wall of the larynx
____(h) a tubular passageway for air connecting the larynx to the bronchi
____(i) secrete alveolar fluid and surfactant
____(j) forms inferior wall of larynx; landmark for tracheotomy
____(k) prevents food or fluid from entering the airways
____(l) air passageways entering the lungs
____(m) ridge covered by a sensitive mucous membrane; irritation triggers cough reflex

(1) nose
(2) pharynx
(3) larynx
(4) epiglottis
(5) trachea
(6) bronchi
(7) carina
(8) cricoid cartilage
(9) pleura
(10) thyroid cartilage
(11) alveoli
(12) type I alveolar cells
(13) type II alveolar cells

13. Match the following:
- ____ (a) a deficiency of oxygen at the tissue level
- ____ (b) above-normal partial pressure of carbon dioxide
- ____ (c) normal quiet breathing
- ____ (d) deep, abdominal breathing
- ____ (e) the ease with which the lungs and thoracic wall can be expanded
- ____ (f) hypoxia-induced vasoconstriction to divert pulmonary blood from poorly ventilated to well-ventilated regions of the lungs
- ____ (g) absence of breathing
- ____ (h) rapid and deep breathing
- ____ (i) shallow, chest breathing

(1) eupnea
(2) apnea
(3) hyperventilation
(4) costal breathing
(5) diaphragmatic breathing
(6) compliance
(7) hypoxia
(8) hypercapnia
(9) ventilation-perfusion coupling

14. Match the following:
- ____ (a) total volume of air inhaled and exhaled each minute
- ____ (b) tidal volume + inspiratory reserve volume + expiratory reserve volume
- ____ (c) additional amount of air inhaled beyond tidal volume when taking a very deep breath
- ____ (d) residual volume + expiratory reserve volume
- ____ (e) amount of air remaining in lungs after expiratory reserve volume is expelled
- ____ (f) tidal volume + inspiratory reserve volume
- ____ (g) vital capacity + residual volume
- ____ (h) volume of air in one breath
- ____ (i) amount of air exhaled in forced exhalation
- ____ (j) provides a medical and legal tool for determining if a baby was born dead or died after birth

(1) tidal volume
(2) residual volume
(3) minute ventilation
(4) expiratory reserve volume
(5) inspiratory reserve volume
(6) minimal volume
(7) inspiratory capacity
(8) vital capacity
(9) functional residual volume
(10) total lung capacity

15. Match the following:
- ____ (a) prevents excessive inflation of the lungs
- ____ (b) the lower the amount of oxyhemoglobin, the higher the carbon dioxide carrying capacity of the blood
- ____ (c) controls the basic rhythm of respiration
- ____ (d) active during normal inhalation; sends nerve impulses to external intercostals and diaphragm
- ____ (e) sends stimulatory impulses to the inspiratory area that activate it and prolong inhalation
- ____ (f) as acidity increases, the affinity of hemoglobin for oxygen decreases and oxygen dissociates more readily from hemoglobin; shifts oxygen-dissociation curve to the right
- ____ (g) active during forceful exhalation
- ____ (h) pressure of a gas in a closed container is inversely proportional to the volume of the container
- ____ (i) transmits inhibitory impulses to turn off the inspiratory area before the lungs become too full of air
- ____ (j) the quantity of a gas that dissolves in a liquid is proportional to the partial pressure of the gas and its solubility
- ____ (k) relates to the partial pressure of a gas in a mixture of gases whereby each gas in a mixture exerts its own pressure as if all the other gases were not present

(1) Bohr effect
(2) Dalton's law
(3) medullary rhythmicity area
(4) inspiratory area
(5) expiratory area
(6) apneustic area
(7) pneumotaxic area
(8) Henry's law
(9) inflation (Hering-Breuer) reflex
(10) Boyle's law
(11) Haldane effect

CRITICAL THINKING QUESTIONS

1. Aretha loves to sing. Right now she has a cold, a severely runny nose, and a "sore throat" that is affecting her ability to sing and talk. What structures are involved and how are they affected by her cold?
2. Ms. Brown has smoked cigarettes for years and is having breathing difficulties. She has been diagnosed with emphysema. Describe specific kinds of structural changes you would expect to observe in Mrs. Brown's respiratory system. How is air flow and gas exchange affected by these structural changes?
3. The Robinson family went to bed one frigid winter night and were found deceased the next day. A squirrel's nest was found in their chimney. What happened to the Robinsons?

ANSWERS TO FIGURE QUESTIONS

23.1 The conducting zone of the respiratory system includes the nose, pharynx, larynx, trachea, bronchi, and bronchioles (except the respiratory bronchioles).

23.2 The path of air is external nares → vestibule → nasal cavity → internal nares.

23.3 The root of the nose attaches it to the frontal bone.

23.4 The superior border of the pharynx is the internal nares; the inferior border of the pharynx is the cricoid cartilage, the most inferior cartilage of the larynx (voice box).

23.5 During swallowing, the epiglottis closes over the rima glottidis, the entrance to the trachea, to prevent aspiration of food and liquids into the lungs.

23.6 The main function of the vocal folds is voice production.

23.7 Because the tissues between the esophagus and trachea are soft, the esophagus can bulge and press against the trachea during swallowing.

23.8 The left lung has two lobes and two secondary bronchi; the right lung has three of each.

23.9 The pleural membrane is a serous membrane.

23.10 Because two-thirds of the heart lies to the left of the midline, the left lung contains a cardiac notch to accommodate the presence of the heart. The right lung is shorter than the left because the diaphragm is higher on the right side to accommodate the liver.

23.11 The wall of an alveolus is made up of type I alveolar cells, type II alveolar cells, and associated alveolar macrophages.

23.12 The respiratory membrane averages 0.5 μm in thickness.

23.13 The pressure would increase to 4 atm.

23.14 If you are at rest while reading, your diaphragm is responsible for about 75% of each inhalation.

23.15 At the start of inhalation, intrapleural pressure is about 756 mmHg. With contraction of the diaphragm, it decreases to about 754 mmHg as the volume of the space between the two pleural layers expands. With relaxation of the diaphragm, it increases back to 756 mmHg.

23.16 Normal atmospheric pressure at sea level is 760 mmHg.

23.17 Breathing in and then exhaling as much air as possible demonstrates vital capacity.

23.18 A difference in P_{O_2} promotes oxygen diffusion into pulmonary capillaries from alveoli and into tissue cells from systemic capillaries.

23.19 The most important factor that determines how much O_2 binds to hemoglobin is the P_{O_2}.

23.20 Both during exercise and at rest, hemoglobin in your pulmonary veins would be fully saturated with O_2, a point which is at the upper right of the curve.

23.21 Because lactic acid (lactate) and CO_2 are produced by active skeletal muscles, blood pH decreases slightly and P_{CO_2} increases when you are actively exercising. The result is lowered affinity of hemoglobin for O_2, so more O_2 is available to the working muscles.

23.22 O_2 is more available to your tissue cells when you have a fever because the affinity of hemoglobin for O_2 decreases with increasing temperature.

23.23 At a P_{O_2} of 40 mmHg, fetal Hb is 80% saturated with O_2 and maternal Hb is about 75% saturated.

23.24 Blood in a systemic vein would have a higher concentration of HCO_3^-.

23.25 The medullary inspiratory area contains autorhythmic neurons that are active and then inactive in a repeating cycle.

23.26 The phrenic nerves innervate the diaphragm.

23.27 Peripheral chemoreceptors are responsive to changes in blood levels of oxygen, carbon dioxide, and H^+.

23.28 Normal arterial P_{CO_2} is 40 mmHg.

23.29 The respiratory system begins to develop about 4 weeks after fertilization.

Chapter 24

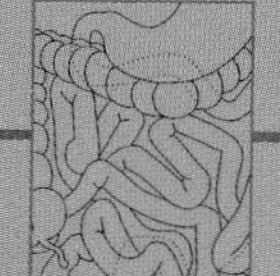

The Digestive System

The Digestive System and Homeostasis

The digestive system contributes to homeostasis by breaking down food into forms that can be absorbed and used by body cells. It also absorbs water, vitamins, and minerals, and eliminates wastes from the body.

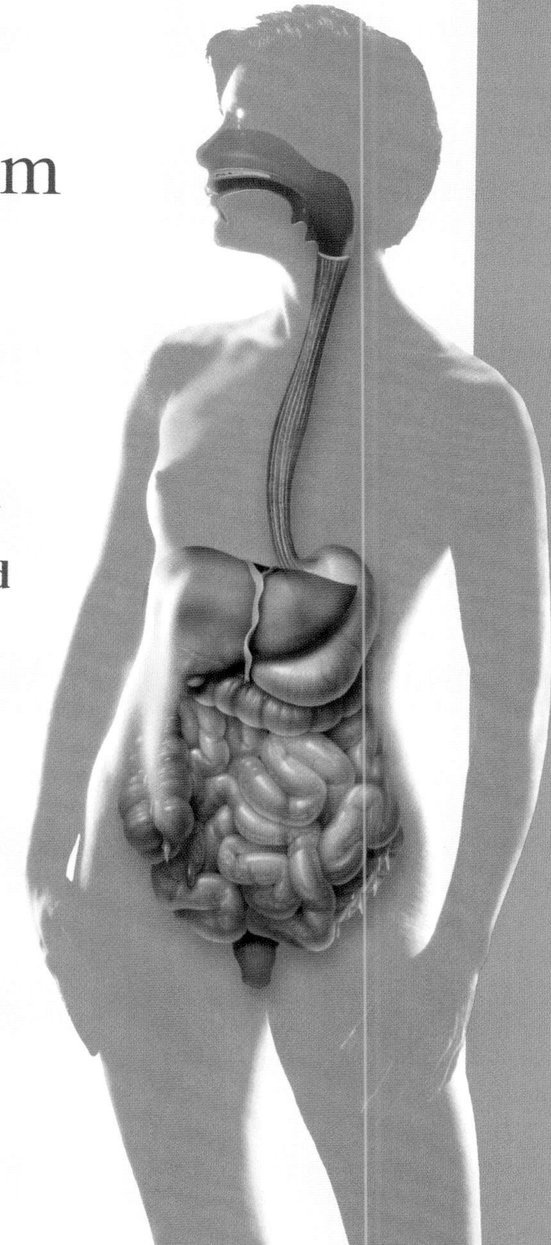

www. wiley.com/college/apcentral

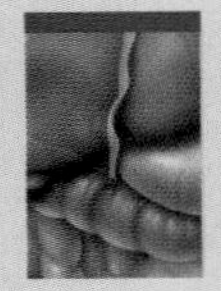

The food we eat contains a variety of nutrients, which are used for building new body tissues and repairing damaged tissues. Food is also vital for life because it is our only source of chemical energy. However, most of the food we eat consists of molecules that are too large to be used by body cells. Therefore, foods must be broken down into molecules that are small enough to enter body cells, a process known as **digestion.** The organs involved in the breakdown of food are collectively known as the **digestive system.**

The medical specialty that deals with the structure, function, diagnosis, and treatment of diseases of the stomach and intestines is called **gastroenterology** (gas′-trō-en′-ter-OL-ō-jē; *gastro-* = stomach; *entero-* = intestines; *-logy* = study of). The medical specialty that deals with the diagnosis and treatment of disorders of the rectum and anus is called **proctology** (prok-TOL-ō-jē; *proct-* = rectum).

OVERVIEW OF THE DIGESTIVE SYSTEM

▶ OBJECTIVES

Identify the organs of the digestive system.

Describe the basic processes performed by the digestive system.

Two groups of organs compose the digestive system (Figure 24.1): the gastrointestinal (GI) tract and the accessory digestive organs. The **gastrointestinal (GI) tract,** or **alimentary canal** (*alimentary* = nourishment), is a continuous tube that extends from the mouth to the anus. Organs of the gastrointestinal tract include the mouth, most of the pharynx, esophagus, stomach, small intestine, and large intestine. The length of the GI tract taken from a cadaver is about 9 m (30 ft). In a living person, it is much shorter because the muscles along the walls of GI tract organs are in a state of tonus (sustained contraction). The **accessory digestive organs** include the teeth, tongue, salivary glands, liver, gallbladder, and pancreas. Teeth aid in the physical breakdown of food, and the tongue assists in chewing and swallowing. The other accessory digestive organs, however, never come into direct contact with food. They produce or store secretions that flow into the GI tract through ducts; the secretions aid in the chemical breakdown of food.

Overall, the digestive system performs six basic processes:

1. ***Ingestion.*** This process involves taking foods and liquids into the mouth (eating).
2. ***Secretion.*** Each day, cells within the walls of the GI tract and accessory digestive organs secrete a total of about 7 liters of water, acid, buffers, and enzymes into the lumen (interior space) of the tract.
3. ***Mixing and propulsion.*** Alternating contractions and relaxations of smooth muscle in the walls of the GI tract mix food and secretions and propel them toward the anus. This capability of the GI tract to mix and move material along its length is called **motility.**
4. ***Digestion.*** Mechanical and chemical processes break down ingested food into small molecules. In **mechanical digestion** the teeth cut and grind food before it is swallowed, and then smooth muscles of the stomach and small intestine churn the food. As a result, food molecules become dissolved and thoroughly mixed with digestive enzymes. In **chemical digestion** the large carbohydrate, lipid, protein, and nucleic acid molecules in food are split into smaller molecules by hydrolysis (see Figure 2.15 on page 45). Digestive enzymes produced by the salivary glands, tongue, stomach, pancreas, and small intestine catalyze these catabolic reactions. A few substances in food can be absorbed without chemical digestion. These include vitamins, ions, cholesterol, and water.
5. ***Absorption.*** The entrance of ingested and secreted fluids, ions, and the products of digestion into the epithelial cells lining the lumen of the GI tract is called **absorption.** The absorbed substances pass into blood or lymph and circulate to cells throughout the body.
6. ***Defecation.*** Wastes, indigestible substances, bacteria, cells sloughed from the lining of the GI tract, and digested materials that were not absorbed in their journey through the digestive tract leave the body through the anus in a process called **defecation.** The eliminated material is termed **feces.**

▶ CHECKPOINT

1. Which components of the digestive system are GI tract organs and which are accessory digestive organs?
2. Which organs of the digestive system come in contact with food, and what are some of their digestive functions?
3. Which kinds of food molecules undergo chemical digestion, and which do not?

LAYERS OF THE GI TRACT

▶ OBJECTIVE

Describe the structure and function of the layers that form the wall of the gastrointestinal tract.

The wall of the GI tract from the lower esophagus to the anal canal has the same basic, four-layered arrangement of tissues. The four layers of the tract, from deep to superficial, are the mucosa, submucosa, muscularis, and serosa (Figure 24.2 on page 898).

Mucosa

The **mucosa,** or inner lining of the GI tract, is a mucous membrane. It is composed of (1) a layer of epithelium in direct contact with the contents of the GI tract, (2) a layer of connective tissue called the lamina propria, and (3) a thin layer of smooth muscle (muscularis mucosae).

Figure 24.1 Organs of the digestive system.

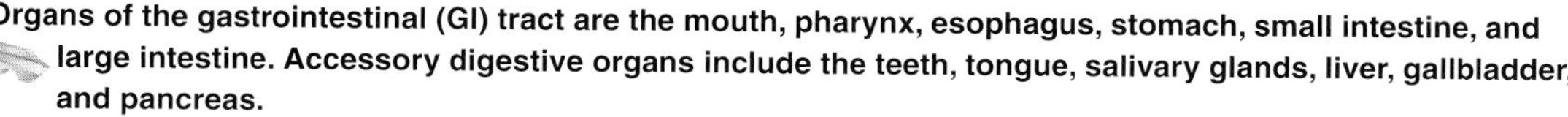

Organs of the gastrointestinal (GI) tract are the mouth, pharynx, esophagus, stomach, small intestine, and large intestine. Accessory digestive organs include the teeth, tongue, salivary glands, liver, gallbladder, and pancreas.

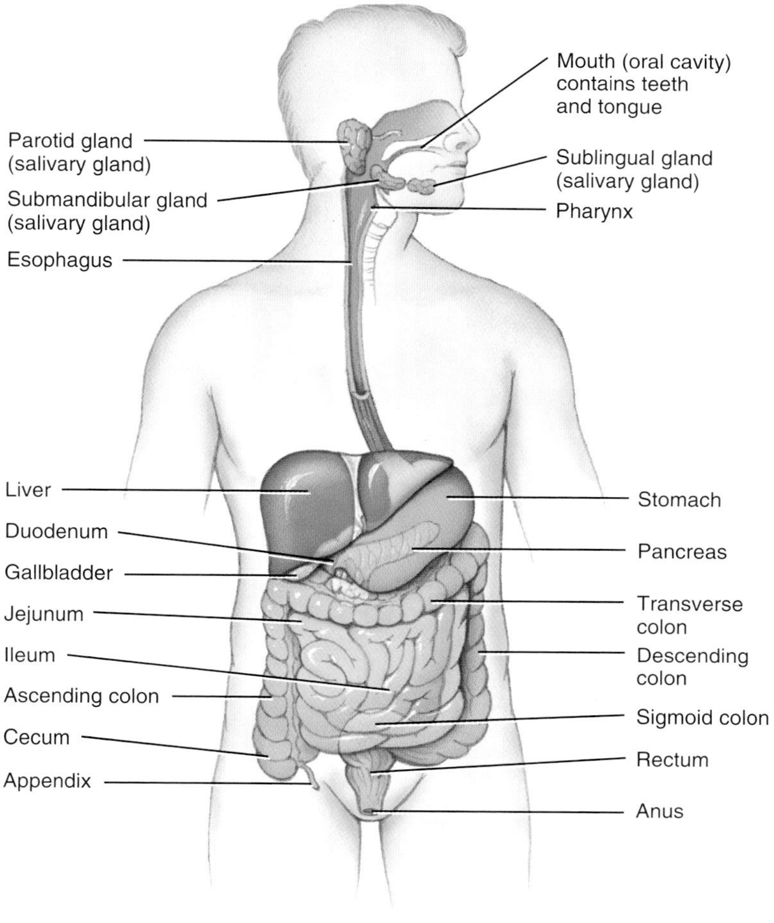

Parotid gland (salivary gland)
Submandibular gland (salivary gland)
Esophagus
Mouth (oral cavity) contains teeth and tongue
Sublingual gland (salivary gland)
Pharynx
Liver
Duodenum
Gallbladder
Jejunum
Ileum
Ascending colon
Cecum
Appendix
Stomach
Pancreas
Transverse colon
Descending colon
Sigmoid colon
Rectum
Anus

Right lateral view of head and neck and anterior view of trunk

Functions

1. **Ingestion: taking food into the mouth.**
2. **Secretion: release of water, acid, buffers, and enzymes into the lumen of the GI tract.**
3. **Mixing and propulsion: churning and propulsion of food through the GI tract.**
4. **Digestion: mechanical and chemical breakdown of food.**
5. **Absorption: passage of digested products from the GI tract into the blood and lymph.**
6. **Defecation: the elimination of feces from the GI tract.**

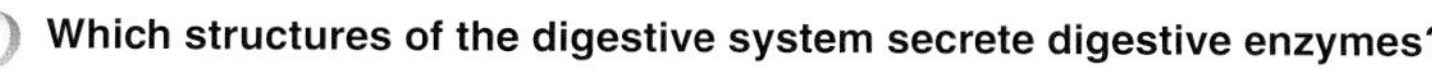

Which structures of the digestive system secrete digestive enzymes?

1. The **epithelium** in the mouth, pharynx, esophagus, and anal canal is mainly nonkeratinized stratified squamous epithelium that serves a protective function. Simple columnar epithelium, which functions in secretion and absorption, lines the stomach and intestines. The tight junctions that firmly seal neighboring simple columnar epithelial cells to one another restrict leakage between the cells. The rate of renewal of GI tract epithelial cells is rapid: Every 5 to 7 days they slough off and are replaced by new cells. Located among the epithelial cells are exocrine cells that secrete mucus and fluid into the lumen of the tract, and several types of endocrine cells, collectively called **enteroendocrine cells,** that secrete hormones into the bloodstream.

2. The **lamina propria** (*lamina* = thin, flat plate; *propria* = one's own) is areolar connective tissue containing many blood and lymphatic vessels, which are the routes by which nutrients absorbed into the GI tract reach the other tissues of the body. This layer supports the epithelium and binds it to the muscularis mucosae (discussed next). The lamina propria also contains the majority of the cells of the **mucosa-associated lymphatic tissue (MALT).** These prominent lymphatic nodules contain immune system cells that protect against disease (see Chapter 22). MALT is present all along the GI tract, especially in the tonsils, small intestine, appendix, and large intestine.

3. A thin layer of smooth muscle fibers called the **muscularis mucosae** throws the mucous membrane of the stomach and small intestine into many small folds, which increase the surface area for digestion and absorption. Movements of the muscularis

Figure 24.2 Layers of the gastrointestinal tract. Variations in this basic plan may be seen in the esophagus (Figure 24.9), stomach (Figure 24.12), small intestine (Figure 24.18), and large intestine (Figure 24.23).

The four layers of the GI tract, from deep to superficial, are the mucosa, submucosa, muscularis, and serosa.

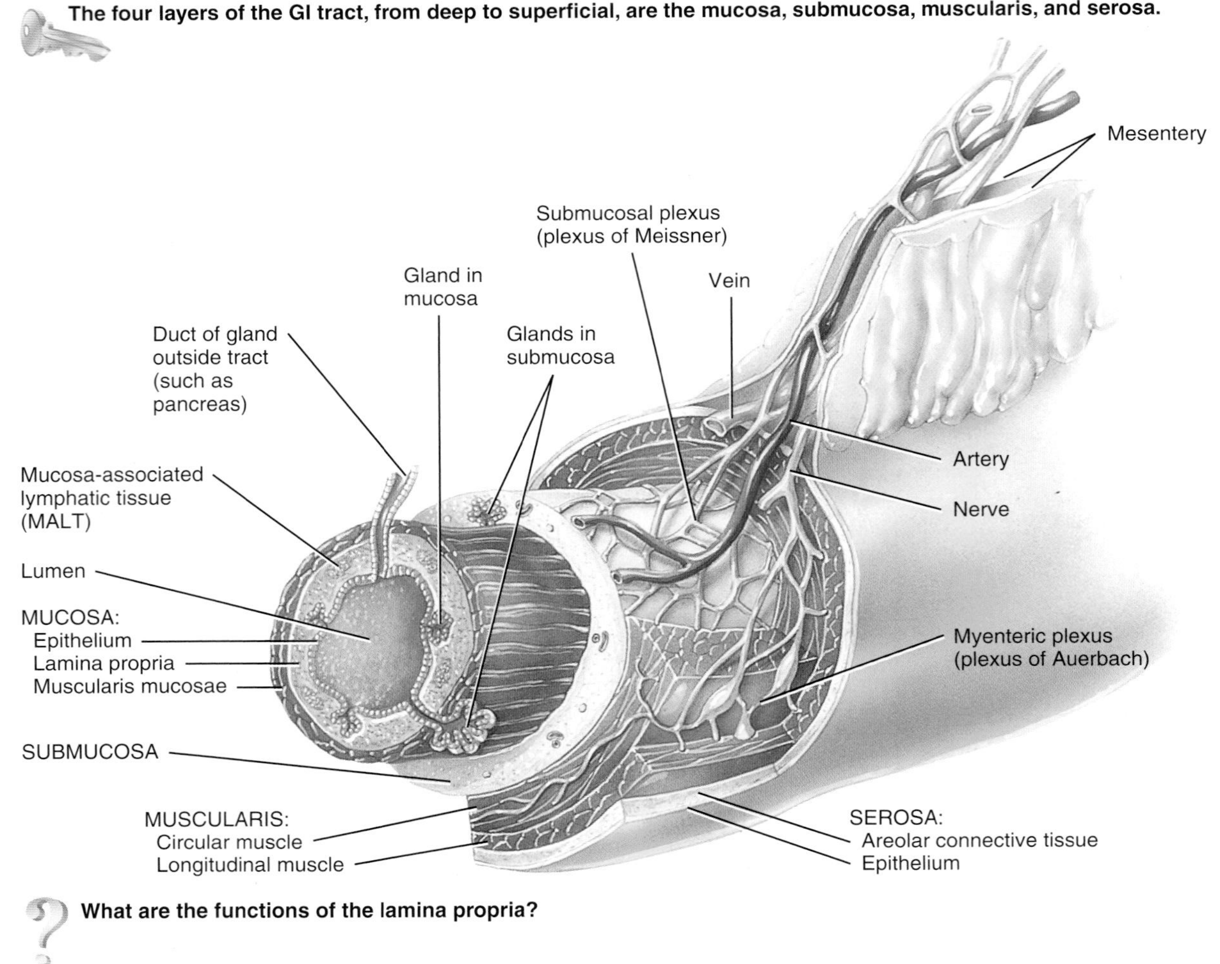

What are the functions of the lamina propria?

mucosae ensure that all absorptive cells are fully exposed to the contents of the GI tract.

Submucosa

The **submucosa** consists of areolar connective tissue that binds the mucosa to the muscularis. It contains many blood and lymphatic vessels that receive absorbed food molecules. Also located in the submucosa is an extensive network of neurons known as the submucosal plexus (to be described shortly). The submucosa may also contain glands and lymphatic tissue.

Muscularis

The **muscularis** of the mouth, pharynx, and superior and middle parts of the esophagus contains *skeletal muscle* that produces voluntary swallowing. Skeletal muscle also forms the external anal sphincter, which permits voluntary control of defecation. Throughout the rest of the tract, the muscularis consists of *smooth muscle* that is generally found in two sheets: an inner sheet of circular fibers and an outer sheet of longitudinal fibers. Involuntary contractions of the smooth muscle help break down food, mix it with digestive secretions, and propel it along the tract. Between the layers of the muscularis is a second plexus of neurons—the myenteric plexus (to be described shortly).

Serosa

Those portions of the GI tract that are suspended in the abdominopelvic cavity have a superficial layer called the **serosa.** As its name implies, the serosa is a serous membrane composed of areolar connective tissue and simple squamous epithelium (mesothelium). The serosa is also called the *visceral peritoneum* because it forms a portion of the peritoneum, which we examine in detail shortly. The esophagus lacks a serosa; instead only a single layer of areolar connective tissue called the *adventitia* forms the superficial layer of this organ.

▶ CHECKPOINT

4. Where along the GI tract is the muscularis composed of skeletal muscle? Is control of this skeletal muscle voluntary or involuntary?
5. Name the four layers of the gastrointestinal tract, and describe their functions.

NEURAL INNERVATION OF THE GI TRACT

▶ OBJECTIVE

Describe the nerve supply of the GI tract.

The gastrointestinal tract is regulated by an intrinsic set of nerves known as the enteric nervous system and by an extrinsic set of nerves that are part of the autonomic nervous system.

Enteric Nervous System

We first introduced you to the **enteric nervous system (ENS),** the "brain of the gut," in Chapter 12. It consists of about 100 million neurons that extend from the esophagus to the anus. The neurons of ENS are arranged into two plexuses: the myenteric plexus and submucosal plexus (see Figure 24.2). The **myenteric plexus** (*myo-* = muscle), or *plexus of Auerbach,* is located between the longitudinal and circular smooth muscle layers of the muscularis. The **submucosal plexus,** or *plexus of Meissner,* is found within the submucosa. The plexuses of the ENS consist of motor neurons, interneurons, and sensory neurons (Figure 24.3). Because the motor neurons of the myenteric plexus supply the longitudinal and circular smooth muscle layers of the muscularis, this plexus mostly controls GI tract motility (movement), particularly the frequency and strength of contraction of the muscularis. The motor neurons of the submucosal plexus supply the secretory cells of the mucosal epithelium, controlling the secretions of the organs of the GI tract. The interneurons of the ENS interconnect the neurons of the myenteric and submucosal plexuses. The sensory neurons of the ENS supply the mucosal epithelium. Some of these sensory neurons function as *chemoreceptors,* receptors that are activated by the presence of certain chemicals in food located in the lumen of a GI organ. Other sensory neurons function as *stretch receptors,* receptors that are activated when food distends (stretches) the wall of a GI organ.

Autonomic Nervous System

Although the neurons of the ENS can function independently, they are subject to regulation by the neurons of the autonomic nervous system. The vagus (X) nerves supply parasympathetic fibers to most parts of the GI tract, with the exception of the last half of the large intestine, which is supplied with parasympathetic fibers from the sacral spinal cord. The parasympathetic nerves that supply the GI tract form neural connections with the ENS. Parasympathetic preganglionic neurons of the vagus or pelvic splanchnic nerves synapse with parasympathetic postganglionic neurons located in the myenteric and submucosal plexuses. Some of the parasympathetic postganglionic neurons in turn synapse with neurons in the ENS; others directly innervate smooth muscle and glands within the wall of the GI tract. In general, stimulation of the parasympathetic nerves that innervate the GI tract causes an increase in GI secretion and motility by increasing the activity of ENS neurons.

Sympathetic nerves that supply the GI tract arise from the thoracic and upper lumbar regions of the spinal cord. Like the parasympathetic nerves, these sympathetic nerves form neural connections with the ENS. Sympathetic postganglionic neurons synapse with neurons located in the myenteric plexus and the submucosal plexus. In general, the sympathetic nerves that supply the GI tract cause a decrease in GI secretion and motility by inhibiting the neurons of the ENS. Emotions such as anger, fear, and anxiety may slow digestion because they stimulate the sympathetic nerves that supply the GI tract.

Figure 24.3 **Organization of the enteric nervous system.**

The enteric nervous system consists of neurons arranged into the myenteric and submucosal plexuses.

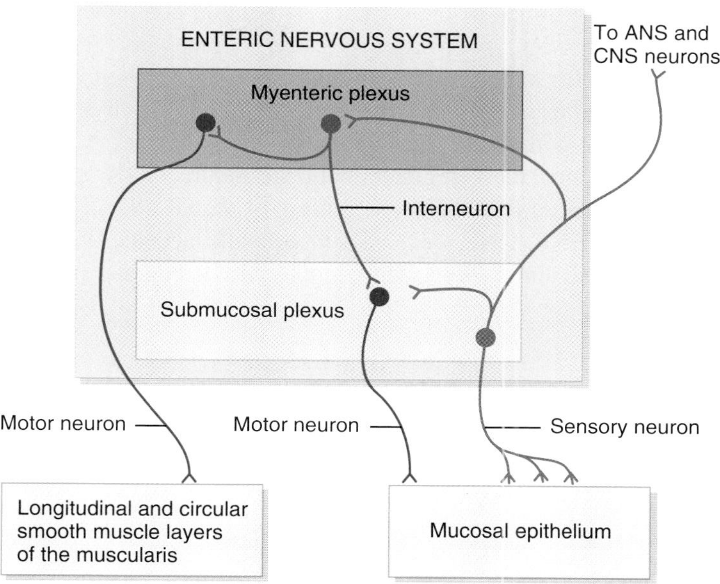

What are the functions of the myenteric and submucosal plexuses of the enteric nervous system?

Gastrointestinal Reflex Pathways

Many neurons of the ENS are components of *GI reflex pathways* that regulate GI secretion and motility in response to stimuli present in the lumen of the GI tract. The initial components of a typical GI reflex pathway are sensory receptors (such as chemoreceptors and stretch receptors) that are associated with the sensory neurons of the ENS. The axons of these sensory neurons can synapse with other neurons located in the ENS, CNS, or ANS, informing these regions about the nature of the contents and the degree of distension (stretching) of the GI tract. The neurons of the ENS, CNS, or ANS subsequently activate or inhibit GI glands and smooth muscle, altering GI secretion and motility.

▶ CHECKPOINT

6. How is the enteric nervous system regulated by the autonomic nervous system?

7. What is a gastrointestinal reflex pathway?

PERITONEUM

▶ OBJECTIVE

Describe the peritoneum and its folds.

The **peritoneum** (per′-i-tō-NĒ-um; *peri-* = around) is the largest serous membrane of the body; it consists of a layer of simple squamous epithelium (mesothelium) with an underlying supporting layer of areolar connective tissue. The peritoneum is divided into the **parietal peritoneum,** which lines the wall of the abdominopelvic cavity, and the **visceral peritoneum,** which covers some of the organs in the cavity and is their serosa (Figure 24.4a). The slim space containing serous fluid that is between the parietal and visceral portions of the peritoneum is called the **peritoneal cavity.** In certain diseases, the peritoneal cavity may become distended by the accumulation of several liters of fluid, a condition called **ascites** (a-SĪ-tēz).

As you will see shortly, some organs lie on the posterior abdominal wall and are covered by peritoneum only on their

Figure 24.4 **Relationship of the peritoneal folds to one another and to organs of the digestive system.** The size of the peritoneal cavity in (a) is exaggerated for emphasis.

The peritoneum is the largest serous membrane in the body.

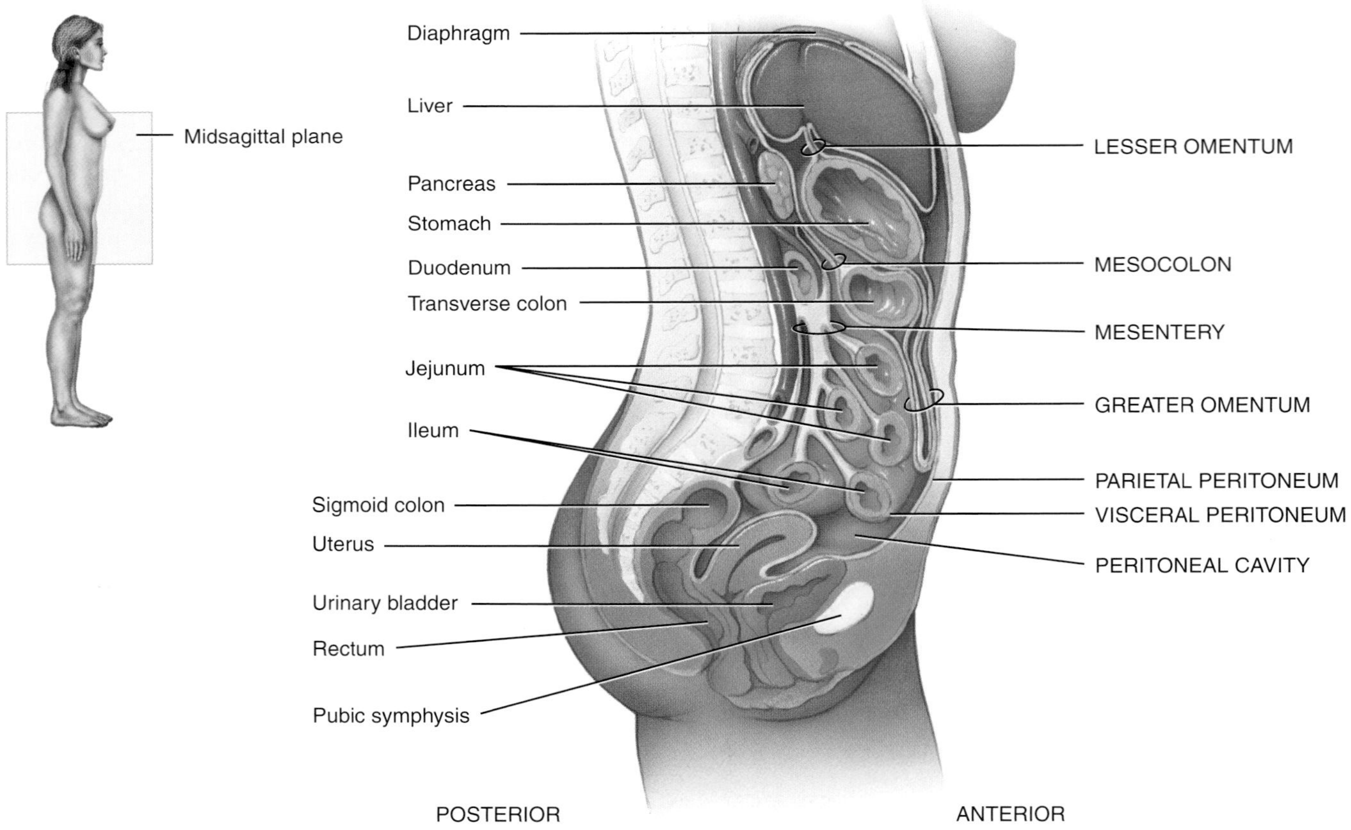

Midsagittal plane

Diaphragm
Liver
Pancreas
Stomach
Duodenum
Transverse colon
Jejunum
Ileum
Sigmoid colon
Uterus
Urinary bladder
Rectum
Pubic symphysis
LESSER OMENTUM
MESOCOLON
MESENTERY
GREATER OMENTUM
PARIETAL PERITONEUM
VISCERAL PERITONEUM
PERITONEAL CAVITY
POSTERIOR
ANTERIOR

(a) Midsagittal section showing the peritoneal folds

anterior surfaces. Such organs, including the kidneys and pancreas, are said to be **retroperitoneal** (*retro-* = behind).

Unlike the pericardium and pleurae, which smoothly cover the heart and lungs, the peritoneum contains large folds that weave between the viscera. The folds bind the organs to one another and to the walls of the abdominal cavity. They also contain blood vessels, lymphatic vessels, and nerves that supply the abdominal organs. There are five major peritoneal folds: the greater omentum, falciform ligament, lesser omentum, mesentery, and mesocolon.

1. The **greater omentum** (ō-MEN-tum = fat skin), the largest peritoneal fold, drapes over the transverse colon and coils of the small intestine like a "fatty apron" (Figure 24.4a, d). The greater omentum is a double sheet that folds back on itself, giving it a total of four layers. From attachments along the stomach and duodenum, the greater omentum extends downward anterior to the small intestine, then turns and extends upward and attaches to the transverse colon. The greater omentum normally contains a considerable amount of adipose

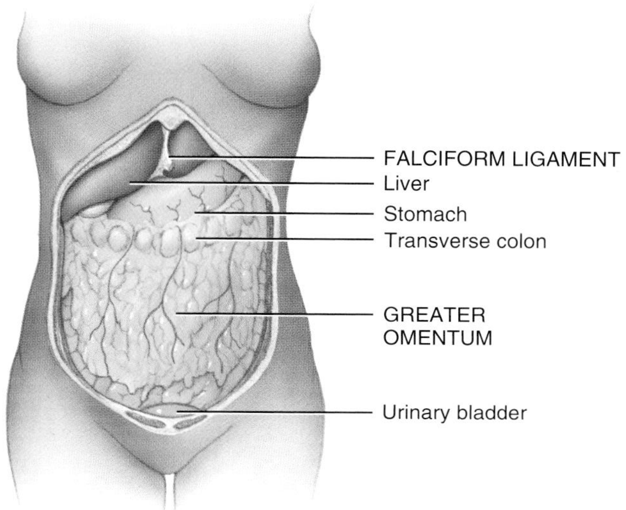

(b) Anterior view

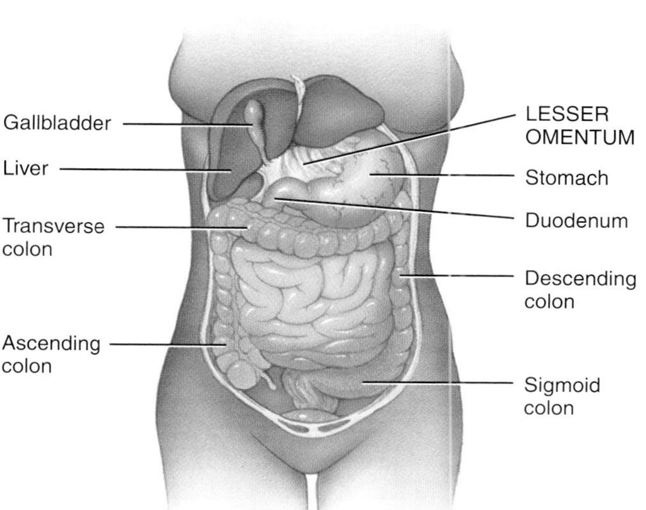

(c) Lesser omentum, anterior view (liver and gallbladder lifted)

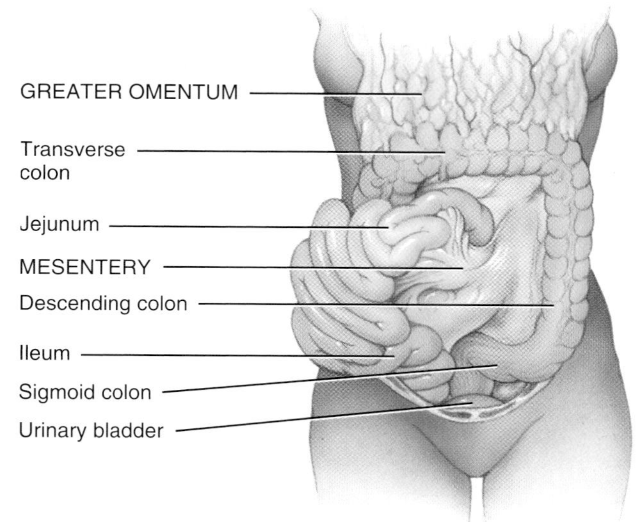

(d) Anterior view (greater omentum lifted and small intestine reflected to right side)

Which peritoneal fold binds the small intestine to the posterior abdominal wall?

tissue. Its adipose tissue content can greatly expand with weight gain, giving rise to the characteristic "beer belly" seen in some overweight individuals. The many lymph nodes of the greater omentum contribute macrophages and antibody-producing plasma cells that help combat and contain infections of the GI tract.

2. The **falciform ligament** (FAL-si-form; *falc-* = sickle-shaped) attaches the liver to the anterior abdominal wall and diaphragm (Figure 24.4b). The liver is the only digestive organ that is attached to the anterior abdominal wall.

3. The **lesser omentum** arises as two folds in the serosa of the stomach and duodenum, and it suspends the stomach and duodenum from the liver (Figure 24.4a, c). It contains some lymph nodes.

4. A fan-shaped fold of the peritoneum, called the **mesentery** (MEZ-en-ter′-ē; *mes-* = middle), binds the small intestine to the posterior abdominal wall (Figure 24.4a, d). It extends from the posterior abdominal wall to wrap around the small intestine and then returns to its origin, forming a double-layered structure. Between the two layers are blood and lymphatic vessels and lymph nodes.

5. A fold of peritoneum, the **mesocolon** (mez′-ō-KŌ-lon), binds the large intestine to the posterior abdominal wall (Figure 24.4a). It also carries blood and lymphatic vessels to the intestines. Together, the mesentery and mesocolon hold the intestines loosely in place, allowing movement as muscular contractions mix and move the luminal contents along the GI tract.

Peritonitis

A common cause of **peritonitis,** an acute inflammation of the peritoneum, is contamination of the peritoneum by infectious microbes, which can result from accidental or surgical wounds in the abdominal wall, or from perforation or rupture of abdominal organs. If, for example, bacteria gain access to the peritoneal cavity through an intestinal perforation or rupture of the appendix, they can produce an acute, life-threatening form of peritonitis. A less serious (but still painful) form of peritonitis can result from the rubbing together of inflamed peritoneal surfaces. Peritonitis is of particularly grave concern to those who rely on peritoneal dialysis, a procedure in which the peritoneum is used to filter the blood when the kidneys do not function properly (see page 1022). ■

▶ CHECKPOINT

8. Where are the visceral peritoneum and parietal peritoneum located?
9. Describe the attachment sites and functions of the mesentery, mesocolon, falciform ligament, lesser omentum, and greater omentum.

MOUTH

▶ OBJECTIVES

Identify the locations of the salivary glands, and describe the functions of their secretions.

Describe the structure and functions of the tongue.

Identify the parts of a typical tooth, and compare deciduous and permanent dentitions.

The **mouth,** also referred to as the **oral** or **buccal cavity** (BUK-al; *bucca* = cheeks), is formed by the cheeks, hard and soft palates, and tongue (Figure 24.5). The **cheeks** form the lateral walls of the oral cavity. They are covered externally by skin and internally by a mucous membrane, which consists of nonkeratinized stratified squamous epithelium. Buccinator muscles and connective tissue lie between the skin and mucous membranes of the cheeks. The anterior portions of the cheeks end at the lips.

The **lips** or **labia** (= fleshy borders) are fleshy folds surrounding the opening of the mouth. They contain the orbicularis oris muscle and are covered externally by skin and internally by a mucous membrane. The inner surface of each lip is attached to its corresponding gum by a midline fold of mucous membrane called the **labial frenulum** (LĀ-bē-al FREN-ū-lum; *frenulum* = small bridle). During chewing, contraction of the buccinator muscles in the cheeks and orbicularis oris muscle in the lips helps keep food between the upper and lower teeth. These muscles also assist in speech.

The **vestibule** (= entrance to a canal) of the oral cavity is a space bounded externally by the cheeks and lips and internally by the gums and teeth. The **oral cavity proper** is a space that extends from the gums and teeth to the **fauces** (FAW-sēs = passages), the opening between the oral cavity and the pharynx (throat).

The **hard palate**—the anterior portion of the roof of the mouth—is formed by the maxillae and palatine bones and is covered by a mucous membrane; it forms a bony partition between the oral and nasal cavities. The **soft palate,** which forms the posterior portion of the roof of the mouth, is an arch-shaped muscular partition between the oropharynx and nasopharynx that is lined with mucous membrane.

Hanging from the free border of the soft palate is a conical muscular process called the **uvula** (Ū-vū-la = little grape). During swallowing, the soft palate and uvula are drawn superiorly, closing off the nasopharynx and preventing swallowed foods and liquids from entering the nasal cavity. Lateral to the base of the uvula are two muscular folds that run down the lateral sides of the soft palate: Anteriorly, the **palatoglossal arch** extends to the side of the base of the tongue; posteriorly, the **palatopharyngeal arch** (PAL-a-tō-fa-rin′-jē-al) extends to the side of the pharynx. The palatine tonsils are situated between the arches, and the lingual tonsils are situated at the base of the tongue. At the posterior border

Figure 24.5 **Structures of the mouth (oral cavity).**

The mouth is formed by the cheeks, hard and soft palates, and tongue.

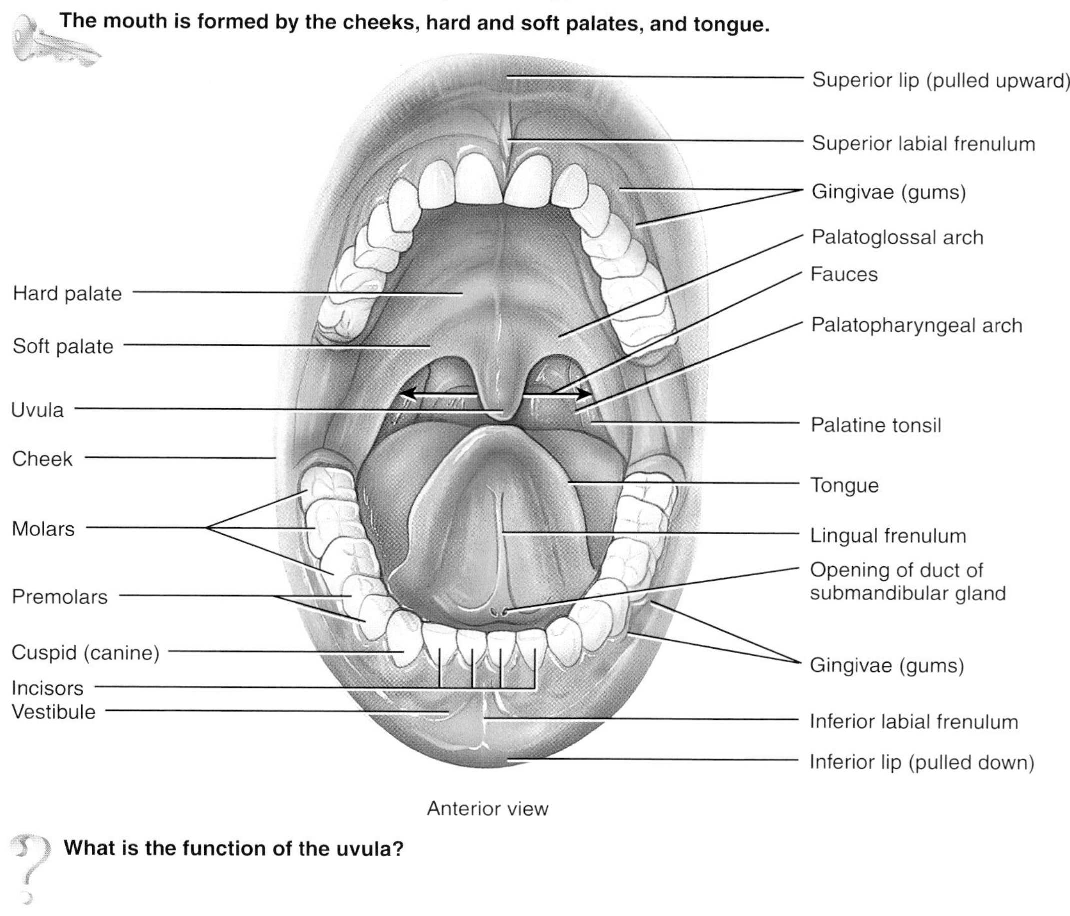

What is the function of the uvula?

of the soft palate, the mouth opens into the oropharynx through the fauces (Figure 24.5).

Salivary Glands

A **salivary gland** is a gland that releases a secretion called saliva into the oral cavity. Ordinarily, just enough saliva is secreted to keep the mucous membranes of the mouth and pharynx moist and to cleanse the mouth and teeth. When food enters the mouth, however, secretion of saliva increases, and it lubricates, dissolves, and begins the chemical breakdown of the food.

The mucous membrane of the mouth and tongue contains many small salivary glands that open directly, or indirectly via short ducts, to the oral cavity. These glands include *labial, buccal,* and *palatal glands* in the lips, cheeks, and palate, respectively, and *lingual glands* in the tongue, all of which make a small contribution to saliva.

However, most saliva is secreted by the **major salivary glands,** which lie beyond the oral mucosa, into ducts that lead to the oral cavity. There are three pairs of major salivary glands: the parotid, submandibular, and sublingual glands (Figure 24.6a). The **parotid glands** (*par-* = near; *to-* = ear) are located inferior and anterior to the ears, between the skin and the masseter muscle. Each secretes saliva into the oral cavity via a **parotid duct** that pierces the buccinator muscle to open into the vestibule opposite the second maxillary (upper) molar tooth. The **submandibular glands** are found in the floor of the mouth; they are medial and partly inferior to the body of the mandible. Their ducts, the **submandibular ducts,** run under the mucosa on either side of the midline of the floor of the mouth and enter the oral cavity proper lateral to the lingual frenulum. The **sublingual glands** are beneath the tongue and superior to the submandibular glands. Their ducts, the **lesser sublingual ducts,** open into the floor of the mouth in the oral cavity proper.

Composition and Functions of Saliva

Chemically, **saliva** is 99.5% water and 0.5% solutes. Among the solutes are ions, including sodium, potassium, chloride, bicarbonate, and phosphate. Also present are some dissolved gases and various organic substances, including urea and uric acid, mucus, immunoglobulin A, the bacteriolytic enzyme lysozyme, and salivary amylase, a digestive enzyme that acts on starch.

Not all salivary glands supply the same ingredients. The parotid glands secrete a watery (serous) liquid containing salivary amylase. Because the submandibular glands contain cells similar to those found in the parotid glands, plus some mucous cells, they secrete a fluid that contains amylase but is thickened with mucus. The sublingual glands contain mostly mucous cells, so they secrete a much thicker fluid that contributes only a small amount of salivary amylase.

The water in saliva provides a medium for dissolving foods so that they can be tasted by gustatory receptors and so that digestive reactions can begin. Chloride ions in the saliva activate salivary amylase, an enzyme that starts the breakdown of starch. Bicarbonate and phosphate ions buffer acidic foods that enter the mouth, so saliva is only slightly acidic (pH 6.35–6.85). Salivary glands (like the sweat glands of the skin) help remove waste molecules from the body, which accounts for the presence of urea and uric acid in saliva. Mucus lubricates food so it can be moved around easily in the mouth, formed into a ball, and swallowed. Immunoglobulin A (IgA) prevents attachment of microbes so they cannot penetrate the epithelium, and the enzyme lysozyme kills bacteria; however, these substances are not present in large enough quantities to eliminate all oral bacteria.

Salivation

Secretion of saliva, or **salivation** (sal-i-VĀ-shun), is controlled by the autonomic nervous system. Amounts of saliva secreted daily vary considerably but average 1000–1500 mL (1–1.6 qt). Normally, parasympathetic stimulation promotes continuous secretion of a moderate amount of saliva, which keeps the mucous membranes moist and lubricates the movements of the tongue and lips during speech. The saliva is then swallowed and helps moisten the esophagus. Eventually, most components of saliva are reabsorbed, which prevents fluid loss. Sympathetic stimulation dominates during stress, resulting in dryness of the mouth. If the body becomes dehydrated, the salivary glands stop secreting saliva to conserve water; the resulting dryness in the mouth contributes to the sensation of thirst. Drinking not only restores the homeostasis of body water but also moistens the mouth.

Figure 24.6 **The three major salivary glands—parotid, sublingual, and submandibular.** The submandibular glands, shown in the light micrograph in (b), consist mostly of serous acini (serous-fluid-secreting portions of gland) and a few mucous acini (mucus-secreting portions of gland); the parotid glands consist of serous acini only; and the sublingual glands consist of mostly mucous acini and a few serous acini. (See Tortora, *A Photographic Atlas of the Human Body, Second Edition*, Figure 12.6a.)

Saliva lubricates and dissolves foods and begins the chemical breakdown of carbohydrates and lipids.

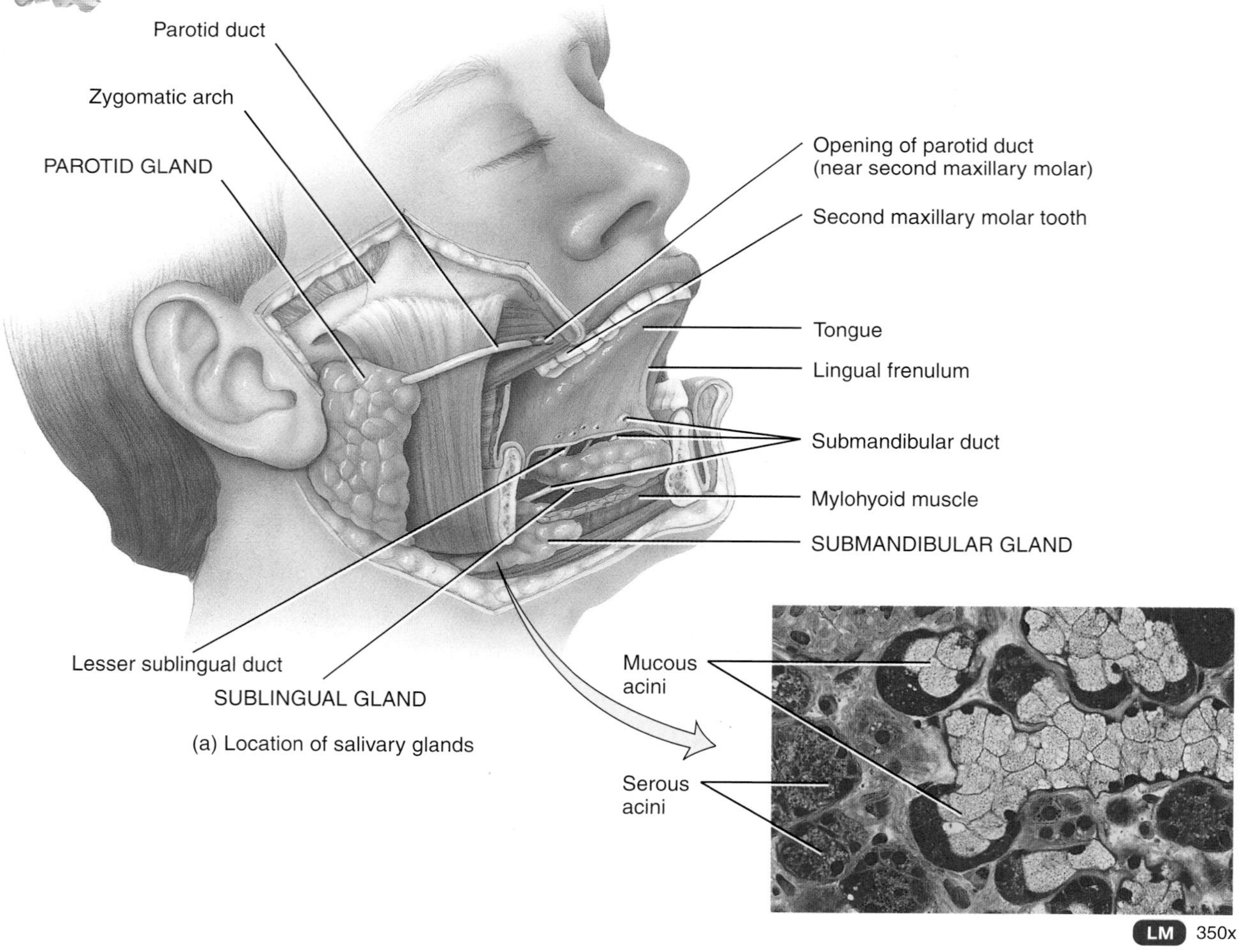

Parotid duct
Zygomatic arch
PAROTID GLAND
Opening of parotid duct (near second maxillary molar)
Second maxillary molar tooth
Tongue
Lingual frenulum
Submandibular duct
Mylohyoid muscle
SUBMANDIBULAR GLAND
Lesser sublingual duct
SUBLINGUAL GLAND

(a) Location of salivary glands

Mucous acini
Serous acini
LM 350x

(b) Submandibular gland

What is the function of the chloride ions in saliva?

The feel and taste of food also are potent stimulators of salivary gland secretions. Chemicals in the food stimulate receptors in taste buds on the tongue, and impulses are conveyed from the taste buds to two salivary nuclei in the brain stem (**superior** and **inferior salivatory nuclei**). Returning parasympathetic impulses in fibers of the facial (VII) and glossopharyngeal (IX) nerves stimulate the secretion of saliva. Saliva continues to be secreted heavily for some time after food is swallowed; this flow of saliva washes out the mouth and dilutes and buffers the remnants of irritating chemicals such as that tasty (but hot!) salsa. The smell, sight, sound, or thought of food may also stimulate secretion of saliva.

Mumps

Although any of the salivary glands may be the target of a nasopharyngeal infection, the mumps virus *(paramyxovirus)* typically attacks the parotid glands. **Mumps** is an inflammation and enlargement of the parotid glands accompanied by moderate fever, malaise (general discomfort), and extreme pain in the throat, especially when swallowing sour foods or acidic juices. Swelling occurs on one or both sides of the face, just anterior to the ramus of the mandible. In about 30% of males past puberty, the testes may also become inflamed; sterility rarely occurs because testicular involvement is usually unilateral (one testis only). Since a vaccine became available for mumps in 1967, the incidence of the disease has declined dramatically. ■

Tongue

The **tongue** is an accessory digestive organ composed of skeletal muscle covered with mucous membrane. Together with its associated muscles, it forms the floor of the oral cavity. The tongue is divided into symmetrical lateral halves by a median septum that extends its entire length, and it is attached inferiorly to the hyoid bone, styloid process of the temporal bone, and mandible. Each half of the tongue consists of an identical complement of extrinsic and intrinsic muscles.

The **extrinsic muscles** of the tongue, which originate outside the tongue (attach to bones in the area) and insert into connective tissues in the tongue, include the hyoglossus, genioglossus, and styloglossus muscles (see Figure 11.7 on page 345). The extrinsic muscles move the tongue from side to side and in and out to maneuver food for chewing, shape the food into a rounded mass, and force the food to the back of the mouth for swallowing. They also form the floor of the mouth and hold the tongue in position. The **intrinsic muscles** originate in and insert into connective tissue within the tongue. They alter the shape and size of the tongue for speech and swallowing. The intrinsic muscles include the longitudinalis superior, longitudinalis inferior, transversus linguae, and verticalis linguae muscles. The **lingual frenulum** (*lingua* = the tongue), a fold of mucous membrane in the midline of the undersurface of the tongue, is attached to the floor of the mouth and aids in limiting the movement of the tongue posteriorly (see Figures 24.5 and 24.6). If a person's lingual frenulum is abnormally short or rigid—a condition called **ankyloglossia** (ang′-kē-lō-GLOSS-ēa)—the person is said to be "tongue-tied" because of the resulting impairment to speech.

The dorsum (upper surface) and lateral surfaces of the tongue are covered with **papillae** (pa-PIL-ē = nipple-shaped projections), projections of the lamina propria covered with keratinized epithelium (see Figure 17.2 on page 578). Many papillae contain taste buds, the receptors for gustation (taste). Some papillae lack taste buds, but they contain receptors for touch and increase friction between the tongue and food, making it easier for the tongue to move food in the oral cavity. The different types of taste buds are described in detail in Chapter 17. **Lingual glands** in the lamina propria of the tongue secrete both mucus and a watery serous fluid that contains the enzyme **lingual lipase,** which acts on triglycerides.

Teeth

The **teeth,** or **dentes** (Figure 24.7), are accessory digestive organs located in sockets of the alveolar processes of the

Figure 24.7 A typical tooth and surrounding structures.

Teeth are anchored in sockets of the alveolar processes of the mandible and maxillae.

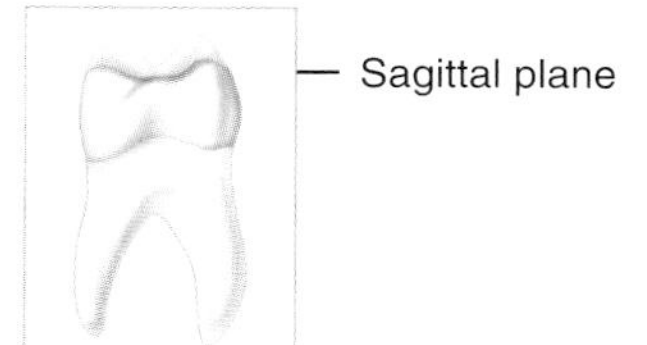

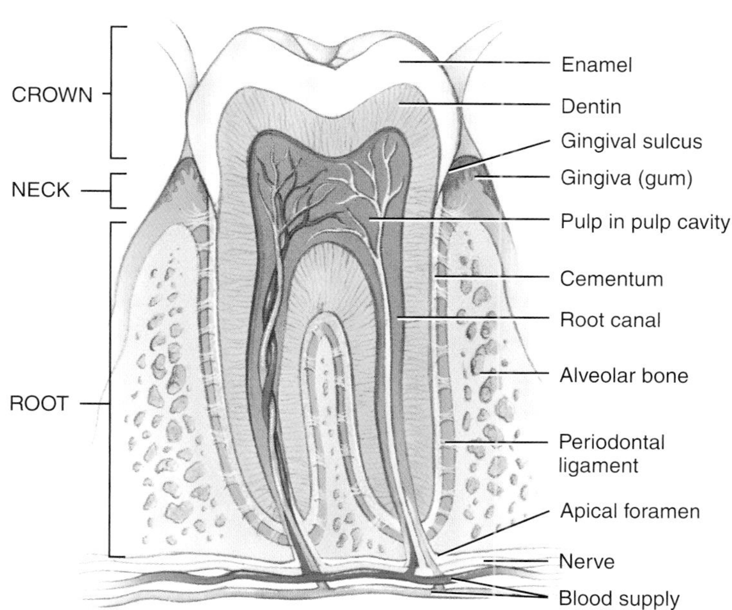

Sagittal section of a mandibular (lower) molar

What type of tissue is the main component of teeth?

mandible and maxillae. The alveolar processes are covered by the **gingivae** (JIN-ji-vē), or gums, which extend slightly into each socket. The sockets are lined by the **periodontal ligament** or **membrane** (*odont-* = tooth), which consists of dense fibrous connective tissue that anchors the teeth to the socket walls.

A typical tooth has three major external regions: the crown, root, and neck. The **crown** is the visible portion above the level of the gums. Embedded in the socket are one to three **roots.** The **neck** is the constricted junction of the crown and root near the gum line.

Internally, **dentin** forms the majority of the tooth. Dentin consists of a calcified connective tissue that gives the tooth its basic shape and rigidity. It is harder than bone because of its higher content of calcium salts (70% of dry weight).

The dentin of the crown is covered by **enamel,** which consists primarily of calcium phosphate and calcium carbonate. Enamel is also harder than bone because of its higher content of calcium salts (about 95% of dry weight). In fact, enamel is the hardest substance in the body. It serves to protect the tooth from the wear and tear of chewing. It also protects against acids that can easily dissolve dentin. The dentin of the root is covered by **cementum,** another bonelike substance, which attaches the root to the periodontal ligament.

The dentin of a tooth encloses a space. The enlarged part of the space, the **pulp cavity,** lies within the crown and is filled with **pulp,** a connective tissue containing blood vessels, nerves, and lymphatic vessels. Narrow extensions of the pulp cavity, called **root canals,** run through the root of the tooth. Each root canal has an opening at its base, the **apical foramen,** through which blood vessels, lymphatic vessels, and nerves extend.

Root Canal Therapy

Root canal therapy is a multistep procedure in which all traces of pulp tissue are removed from the pulp cavity and root canals of a badly diseased tooth. After a hole is made in the tooth, the root canals are filed out and irrigated to remove bacteria. Then, the canals are treated with medication and sealed tightly. The damaged crown is then repaired. ■

The branch of dentistry that is concerned with the prevention, diagnosis, and treatment of diseases that affect the pulp, root, periodontal ligament, and alveolar bone is known as **endodontics** (en′-dō-DON-tiks; *endo-* = within). **Orthodontics** (or′-thō-DON-tiks; *ortho-* = straight) is a branch of dentistry that is concerned with the prevention and correction of abnormally aligned teeth; **periodontics** (per′-ē-ō-DON-tiks) is a branch of dentistry concerned with the treatment of abnormal conditions of the tissues immediately surrounding the teeth, such as gingivitis (gum disease).

Humans have two **dentitions,** or sets of teeth: deciduous and permanent. The first of these—the **deciduous teeth** (*decidu-* = falling out), also called **primary teeth, milk teeth,** or **baby teeth**—begin to erupt at about 6 months of age, and approximately two teeth appear each month thereafter, until all 20 are present (Figure 24.8a). The incisors, which are closest to the midline, are chisel-shaped and adapted for cutting into food. They are referred to as either **central** or **lateral incisors** based on their position. Next to the incisors, moving posteriorly, are the **cuspids (canines),** which have a pointed surface called a *cusp.* Cuspids are used to tear and shred food. Incisors and cuspids have only one root apiece. Posterior to the cuspids lie the **first** and **second molars,** which have four cusps. Maxillary (upper) molars have three roots; mandibular (lower) molars have two roots. The molars crush and grind food to prepare it for swallowing.

All the deciduous teeth are lost—generally between ages 6 and 12 years—and are replaced by the **permanent (secondary) teeth** (Figure 24.8b). The permanent dentition contains 32 teeth that erupt between age 6 and adulthood. The pattern resembles the deciduous dentition, with the following exceptions. The deciduous molars are replaced by the **first** and **second premolars (bicuspids),** which have two cusps and one root (upper first bicuspids have two roots) and are used for crushing and grinding. The permanent molars, which erupt into the mouth posterior to the bicuspids, do not replace any deciduous teeth and erupt as the jaw grows to accommodate them—the **first molars** at age 6 (six-year molars), the **second molars** at age 12 (twelve-year molars), and the **third molars (wisdom teeth)** after age 17.

Often the human jaw does not have enough room posterior to the second molars to accommodate the eruption of the third molars. In this case, the third molars remain embedded in the alveolar bone and are said to be "impacted." They often cause pressure and pain and must be removed surgically. In some people, third molars may be dwarfed in size or may not develop at all.

Mechanical and Chemical Digestion in the Mouth

Mechanical digestion in the mouth results from chewing, or **mastication** (mas′-ti-KĀ-shun = to chew), in which food is manipulated by the tongue, ground by the teeth, and mixed with saliva. As a result, the food is reduced to a soft, flexible, easily swallowed mass called a **bolus** (= lump). Food molecules begin to dissolve in the water in saliva, an important activity because enzymes can react with food molecules in a liquid medium only.

Two enzymes, salivary amylase and lingual lipase, contribute to chemical digestion in the mouth. **Salivary amylase,** which is secreted by the salivary glands, initiates the breakdown of starch. Dietary carbohydrates are either monosaccharide and disaccharide sugars or complex polysaccharides such as starches. Most of the carbohydrates we eat are starches, but only monosaccharides can be absorbed into the bloodstream. Thus, ingested disaccharides and starches must be broken down into monosaccharides. The function of salivary amylase is to begin starch digestion by breaking down starch into smaller molecules such as the disaccharide maltose, the trisaccharide maltotriose, and short-chain glucose polymers called α-dextrins. Even though food is usually swallowed too quickly for all the starches to be broken down in the mouth, salivary amylase in the swallowed food continues to act on the starches for about another hour, at which time stomach acids inactivate it. Saliva also contains **lingual lipase,** which is secreted by lingual glands in the tongue. This enzyme becomes activated in the acidic environment of the stomach and thus starts to work after food is swallowed. It breaks down dietary triglycerides into fatty acids and diglycerides. A diglyceride consists of a glycerol molecule that is attached to two fatty acids.

Table 24.1 summarizes the digestive activities in the mouth.

▶ **CHECKPOINT**

10. What structures form the mouth (oral cavity)?

11. How are the major salivary glands distinguished on the basis of location?

12. How is the secretion of saliva regulated?

13. What functions do incisors, cuspids, premolars, and molars perform?

Figure 24.8 **Dentitions and times of eruptions (indicated in parentheses).** A designated letter (deciduous teeth) or number (permanent teeth) uniquely identifies each tooth. Deciduous teeth begin to erupt at 6 months of age, and approximately two teeth appear each month thereafter, until all 20 are present. (See Tortora, *A Photographic Atlas of the Human Body, Second Edition*, Figure 12.7.)

There are 20 teeth in a complete deciduous set and 32 teeth in a complete permanent set.

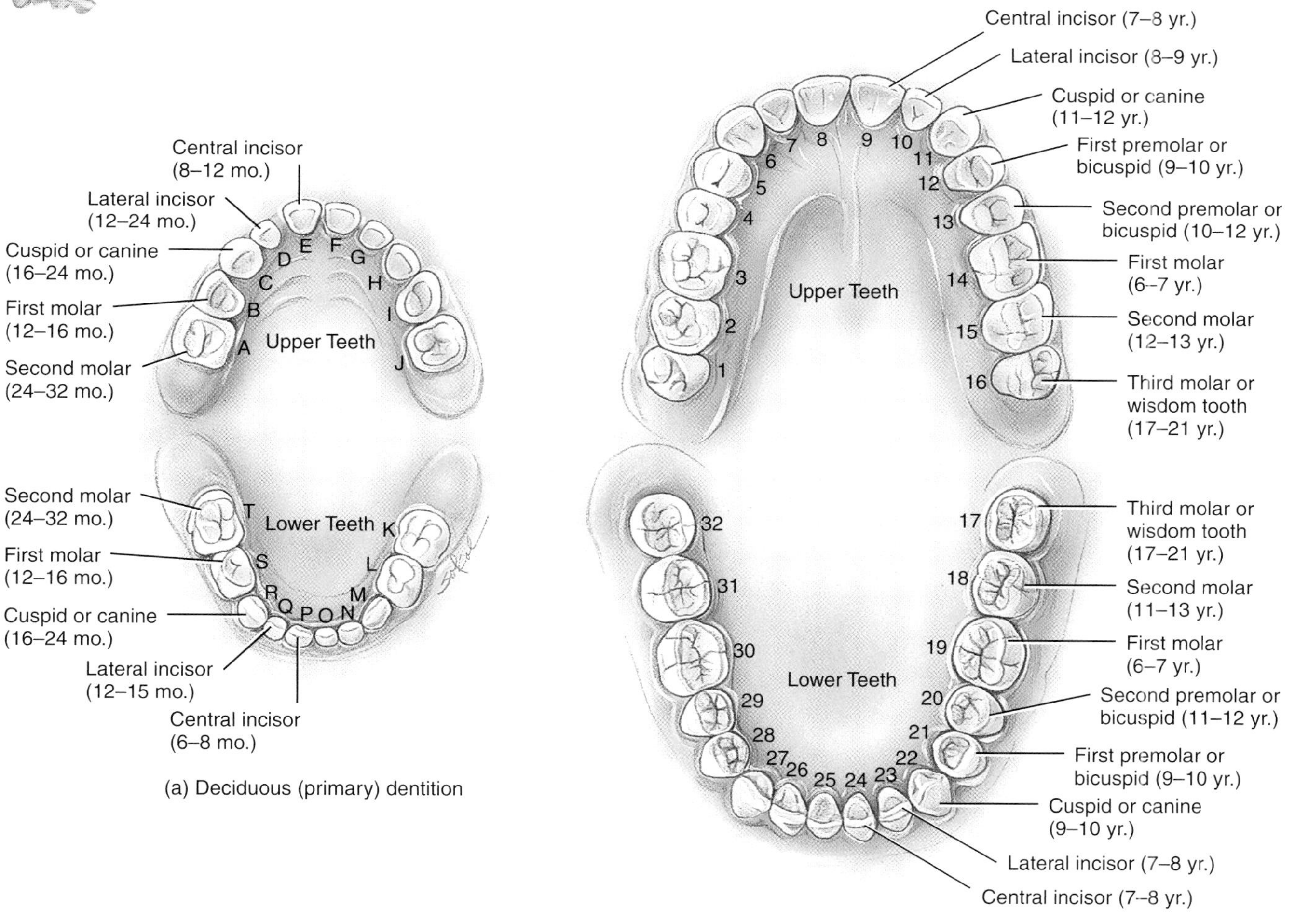

(a) Deciduous (primary) dentition

(b) Permanent (secondary) dentition

Which permanent teeth do not replace any deciduous teeth?

TABLE 24.1 Summary of Digestive Activities in the Mouth

Structure	Activity	Result
Cheeks and **Lips**	Keep food between teeth.	Foods uniformly chewed during mastication.
Salivary glands	Secrete saliva.	Lining of mouth and pharynx moistened and lubricated. Saliva softens, moistens, and dissolves food and cleanses mouth and teeth. Salivary amylase splits starch into smaller fragments.
Tongue		
Extrinsic tongue muscles	Move tongue from side-to-side and in and out.	Food maneuvered for mastication, shaped into bolus, and maneuvered for swallowing.
Intrinsic tongue muscles	Alter shape of tongue	Swallowing and speech.
Taste buds	Serve as receptors for gustation (taste) and presence of food in mouth.	Secretion of saliva stimulated by nerve impulses from taste buds to salivary nuclei in brain stem to salivary glands.
Lingual glands	Secrete lingual lipase.	Triglycerides broken down into fatty acids and diglycerides.
Teeth	Cut, tear, and pulverize food.	Solid foods reduced to smaller particles for swallowing.

PHARYNX

▶ **OBJECTIVE**

Describe the location and function of the pharynx.

When food is first swallowed, it passes from the mouth into the **pharynx** (= throat), a funnel-shaped tube that extends from the internal nares to the esophagus posteriorly and to the larynx anteriorly (see Figure 23.4 on page 851). The pharynx is composed of skeletal muscle and lined by mucous membrane, and is divided into three parts: the nasopharynx, the oropharynx, and the laryngopharynx. The nasopharynx functions only in respiration, but both the oropharynx and laryngopharynx have digestive as well as respiratory functions. Swallowed food passes from the mouth into the oropharynx and laryngopharynx; the muscular contractions of these areas help propel food into the esophagus and then into the stomach.

▶ **CHECKPOINT**

14. To which two organ systems does the pharynx belong?

ESOPHAGUS

▶ **OBJECTIVE**

Describe the location, anatomy, histology, and functions of the esophagus.

The **esophagus** (e-SOF-a-gus = eating gullet) is a collapsible muscular tube, about 25 cm (10 in.) long, that lies posterior to the trachea. The esophagus begins at the inferior end of the laryngopharynx and passes through the mediastinum anterior to the vertebral column. Then it pierces the diaphragm through an opening called the **esophageal hiatus,** and ends in the superior portion of the stomach (see Figure 24.1). Sometimes, part of the stomach protrudes above the diaphragm through the esophageal hiatus. This condition, termed a **hiatus hernia** (HER-nē-a), is described on page 944.

Histology of the Esophagus

The **mucosa** of the esophagus consists of nonkeratinized stratified squamous epithelium, lamina propria (areolar connective tissue), and a muscularis muscosae (smooth muscle) (Figure 24.9). Near the stomach, the mucosa of the esophagus also contains mucous glands. The stratified squamous epithelium associated with the lips, mouth, tongue, oropharynx, laryngopharynx, and esophagus affords considerable protection against abrasion and wear-and-tear from food particles that are chewed, mixed with secretions, and swallowed. The **submucosa** contains areolar connective tissue, blood vessels, and mucous glands. The **muscularis** of the superior third of the esophagus is skeletal muscle, the intermediate third is skeletal and smooth muscle, and the inferior third is smooth muscle. At each end of the esophagus, the muscularis becomes slightly more prominent and forms two sphincters—the **upper esophageal sphincter (UES)** (e-sof′-a-JĒ-al), which consists of skeletal muscle, and the **lower esophageal sphincter (LES),** which consists of smooth muscle. The upper esophageal sphincter regulates the movement of food from the pharynx into the esophagus; the lower esophageal sphincter regulates the movement of food from the esophagus into the stomach. The superficial layer of the esophagus is known as the **adventitia** (ad-ven-TISH-a), rather than the serosa, because the areolar connective tissue of this layer is not covered by mesothelium and because the connective tissue merges with the connective tissue of surrounding structures of the mediastinum, through which it passes. The adventitia attaches the esophagus to surrounding structures.

Physiology of the Esophagus

The esophagus secretes mucus and transports food into the stomach. It does not produce digestive enzymes, and it does not carry on absorption.

Figure 24.9 Histology of the esophagus. A higher magnification view of nonkeratinized stratified squamous epithelium is shown in Table 4.1F on page 116. (See Tortora, *A Photographic Atlas of the Human Body, Second Edition,* Figure 12.8a.)

The esophagus secretes mucus and transports food to the stomach.

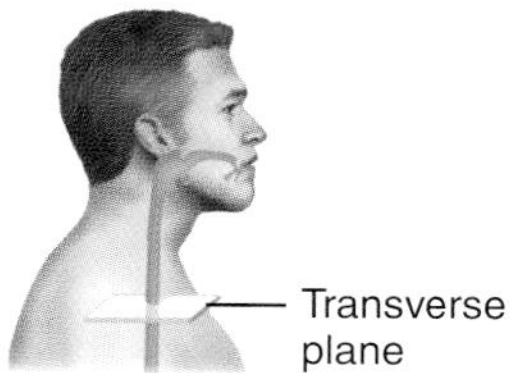

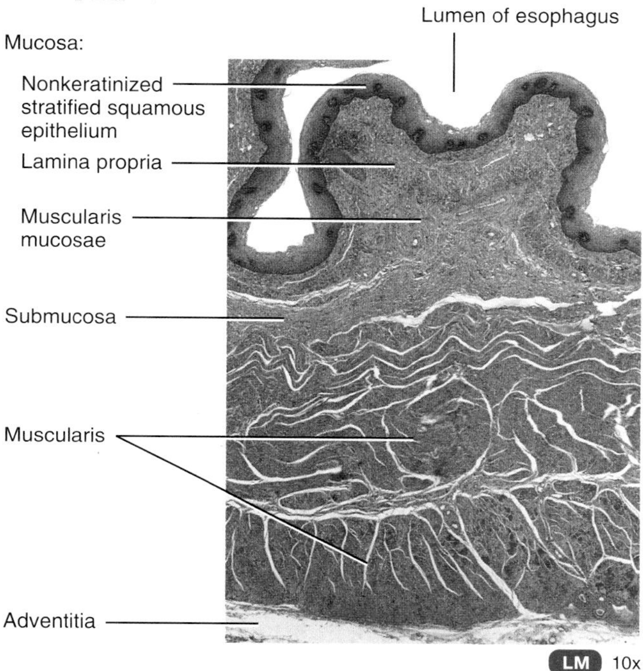

Transverse section through the wall of the esophagus

In which layers of the esophagus are the glands that secrete lubricating mucus located?

▶ **CHECKPOINT**

15. Describe the location and histology of the esophagus. What is its role in digestion?

16. What are the functions of the upper and lower esophageal sphincters?

DEGLUTITION

▶ **OBJECTIVE**

Describe the three phases of deglutition.

The movement of food from the mouth into the stomach is achieved by the act of swallowing, or **deglutition** (dē-gloo-TISH-un) (Figure 24.10). Deglutition is facilitated by the secretion of saliva and mucus and involves the mouth, pharynx, and esophagus. Swallowing occurs in three stages: (1) the voluntary stage, in which the bolus is passed into the oropharynx; (2) the pharyngeal stage, the involuntary passage of the bolus through the pharynx into the esophagus; and (3) the esophageal stage, the involuntary passage of the bolus through the esophagus into the stomach.

Swallowing starts when the bolus is forced to the back of the oral cavity and into the oropharynx by the movement of the tongue upward and backward against the palate; these actions constitute the **voluntary stage** of swallowing. With the passage of the bolus into the oropharynx, the involuntary **pharyngeal stage** of swallowing begins (Figure 24.10b). The bolus stimulates receptors in the oropharynx, which send impulses to the **deglutition center** in the medulla oblongata and lower pons of the brain stem. The returning impulses cause the soft palate and uvula to move upward to close off the nasopharynx, which prevents swallowed foods and liquids from entering the nasal cavity. In addition, the epiglottis closes off the opening to the larynx, which prevents the bolus from entering the rest of the respiratory tract. The bolus moves through the oropharynx and the laryngopharynx. Once the upper esophageal sphincter relaxes, the bolus moves into the esophagus.

The **esophageal stage** of swallowing begins once the bolus enters the esophagus. During this phase, **peristalsis** (per′-i-STAL-sis; *stalsis* = constriction), a progression of coordinated contractions and relaxations of the circular and longitudinal layers of the muscularis, pushes the bolus onward (Figure 24.10c). (Peristalsis occurs in other tubular structures, including other parts of the GI tract and the ureters, bile ducts, and uterine tubes; in the esophagus it is controlled by the medulla oblongata.) In the section of the esophagus just superior to the bolus, the circular muscle fibers contract, constricting the esophageal wall and squeezing the bolus toward the stomach. Meanwhile, longitudinal fibers inferior to the bolus also contract, which shortens this inferior section and pushes its walls outward so it can receive the bolus. The contractions are repeated in waves that push the food toward the stomach. As the bolus approaches the end of the esophagus, the lower esophageal sphincter relaxes and the bolus moves into the stomach. Mucus secreted by esophageal glands lubricates the bolus and reduces friction. The passage of solid or semisolid food from the mouth to the stomach takes 4 to 8 seconds; very soft foods and liquids pass through in about 1 second.

Figure 24.10 Deglutition (swallowing). During the pharyngeal stage of deglutition (b) the tongue rises against the palate, the nasopharynx is closed off, the larynx rises, the epiglottis seals off the larynx, and the bolus is passed into the esophagus. During the esophageal stage of deglutition (c), food moves through the esophagus into the stomach via peristalsis.

Deglutition is a mechanism that moves food from the mouth into the stomach.

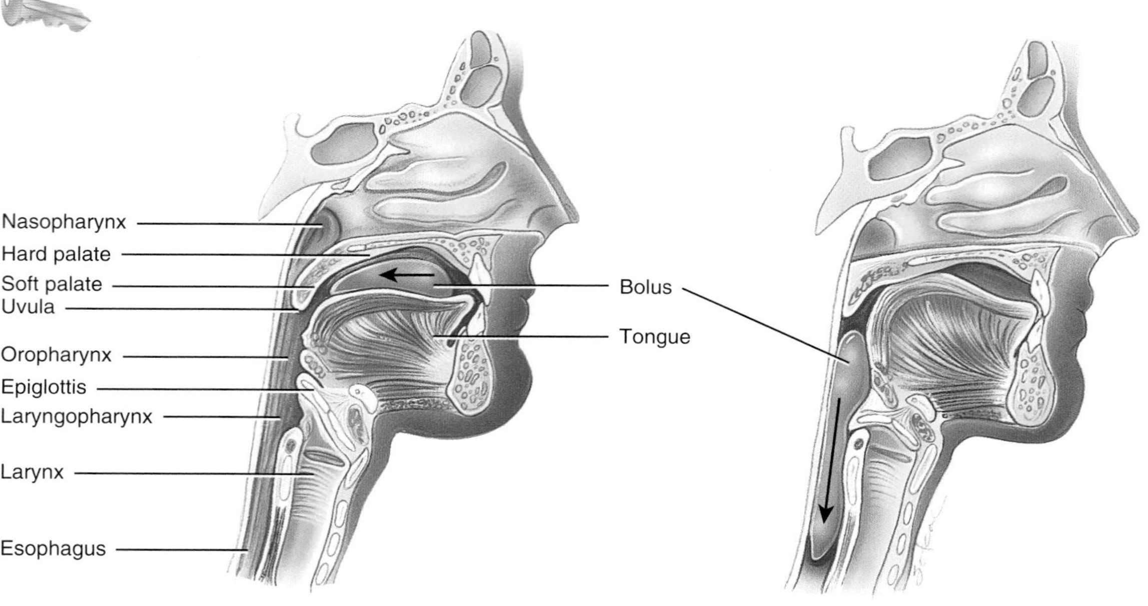

(a) Position of structures before swallowing

(b) During the pharyngeal stage of swallowing

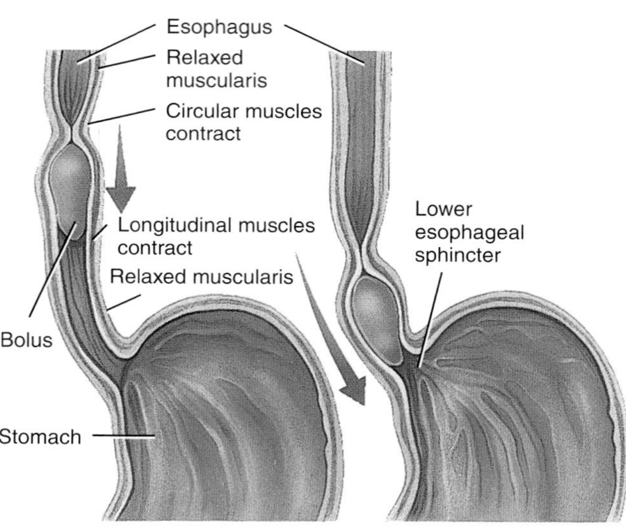

(c) Anterior view of frontal sections of peristalsis in esophagus

Is swallowing a voluntary action or an involuntary action?

Table 24.2 summarizes the digestive activities of the pharynx and esophagus.

Gastroesophageal Reflux Disease

If the lower esophageal sphincter fails to close adequately after food has entered the stomach, the stomach contents can reflux (back up) into the inferior portion of the esophagus. This condition is known as **gastroesophageal reflux disease (GERD).** Hydrochloric acid (HCl) from the stomach contents can irritate the esophageal wall, resulting in a burning sensation that is called **heartburn** because it is experienced in a region very near the heart; it is unrelated to any cardiac problem. Drinking alcohol and smoking can cause the sphincter to relax, worsening the problem. The symptoms of GERD often can be controlled by avoiding foods that strongly stimulate stomach acid secretion

TABLE 24.2 Summary of Digestive Activities in the Pharynx and Esophagus

Structure	Activity	Result
Pharynx	Pharyngeal stage of deglutition.	Moves bolus from oropharynx to laryngopharynx and into esophagus; closes air passageways.
Esophagus	Relaxation of upper esophageal sphincter.	Permits entry of bolus from laryngopharynx into esophagus.
	Esophageal stage of deglutition (peristalsis).	Pushes bolus down esophagus.
	Relaxation of lower esophageal sphincter.	Permits entry of bolus into stomach.
	Secretion of mucus.	Lubricates esophagus for smooth passage of bolus.

(coffee, chocolate, tomatoes, fatty foods, orange juice, peppermint, spearmint, and onions). Other acid-reducing strategies include taking over-the-counter histamine-2 (H_2) blockers such as Tagamet HB® or Pepcid AC® 30 to 60 minutes before eating to block acid secretion, and neutralizing acid that has already been secreted with antacids such as Tums® or Maalox®. Symptoms are less likely to occur if food is eaten in smaller amounts and if the person does not lie down immediately after a meal. GERD may be associated with cancer of the esophagus. ■

▶ **CHECKPOINT**

17. What does deglutition mean?

18. What occurs during the voluntary and pharyngeal phases of swallowing?

19. Does peristalsis "push" or "pull" food along the gastrointestinal tract?

STOMACH

▶ **OBJECTIVE**

Describe the location, anatomy, histology, and functions of the stomach.

The **stomach** is a J-shaped enlargement of the GI tract directly inferior to the diaphragm in the epigastric, umbilical, and left hypochondriac regions of the abdomen (see Figure 1.12a on page 20). The stomach connects the esophagus to the duodenum, the first part of the small intestine (Figure 24.11). Because a meal can be eaten much more quickly than the intestines can digest and absorb it, one of the functions of the stomach is to serve as a mixing chamber and holding reservoir. At appropriate intervals after food is ingested, the stomach forces a small quantity of material into the first portion of the small intestine. The position and size of the stomach vary continually; the diaphragm pushes it inferiorly with each inhalation and pulls it superiorly with each exhalation. Empty, it is about the size of a large sausage, but it is the most distensible part of the GI tract and can accommodate a large quantity of food. In the stomach, digestion of starch continues, digestion of proteins and triglycerides begins, the semisolid bolus is converted to a liquid, and certain substances are absorbed.

Anatomy of the Stomach

The stomach has four main regions: the cardia, fundus, body, and pylorus (Figure 24.11). The **cardia** (CAR-dē-a) surrounds the superior opening of the stomach. The rounded portion superior to and to the left of the cardia is the **fundus** (FUN-dus). Inferior to the fundus is the large central portion of the stomach, called the **body.** The region of the stomach that connects to the duodenum is the **pylorus** (pī-LOR-us; *pyl-* = gate; *-orus* = guard); it has two parts, the **pyloric antrum** (AN-trum = cave), which connects to the body of the stomach, and the **pyloric canal,** which leads into the duodenum. When the stomach is empty, the mucosa lies in large folds, called **rugae** (ROO-gē = wrinkles), that can be seen with the unaided eye. The pylorus communicates with the duodenum of the small intestine via a sphincter called the **pyloric sphincter.** The concave medial border of the stomach is called the **lesser curvature,** and the convex lateral border is called the **greater curvature.**

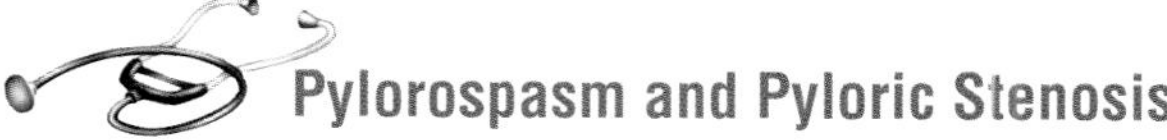

Two abnormalities of the pyloric sphincter can occur in infants. In **pylorospasm** (pī-LOR-ō-spazm), the muscle fibers of the sphincter fail to relax normally, so food does not pass easily from the stomach to the small intestine, the stomach becomes overly full, and the infant vomits often to relieve the pressure. Pylorospasm is treated by drugs that relax the muscle fibers of the pyloric sphincter. **Pyloric stenosis** (ste-NŌ-sis) is a narrowing of the pyloric sphincter that must be corrected surgically. The hallmark symptom is *projectile vomiting*—the spraying of liquid vomitus some distance from the infant. ■

Histology of the Stomach

The stomach wall is composed of the same four basic layers as the rest of the GI tract, with certain modifications. The surface of the **mucosa** is a layer of simple columnar epithelial cells called **surface mucous cells** (Figure 24.12b on page 913). The mucosa contains a **lamina propria** (areolar connective tissue) and a **muscularis mucosae** (smooth muscle) (Figure 24.12b). Epithelial cells extend down into the lamina propria, where they form columns of secretory cells called **gastric glands** that line many narrow channels called **gastric pits.** Secretions from several gastric glands flow into each gastric pit and then into the lumen of the stomach.

Figure 24.11 External and internal anatomy of the stomach. (See Tortora, *A Photographic Atlas of the Human Body, Second Edition,* Figure 12.9.)

The four regions of the stomach are the cardia, fundus, body, and pylorus.

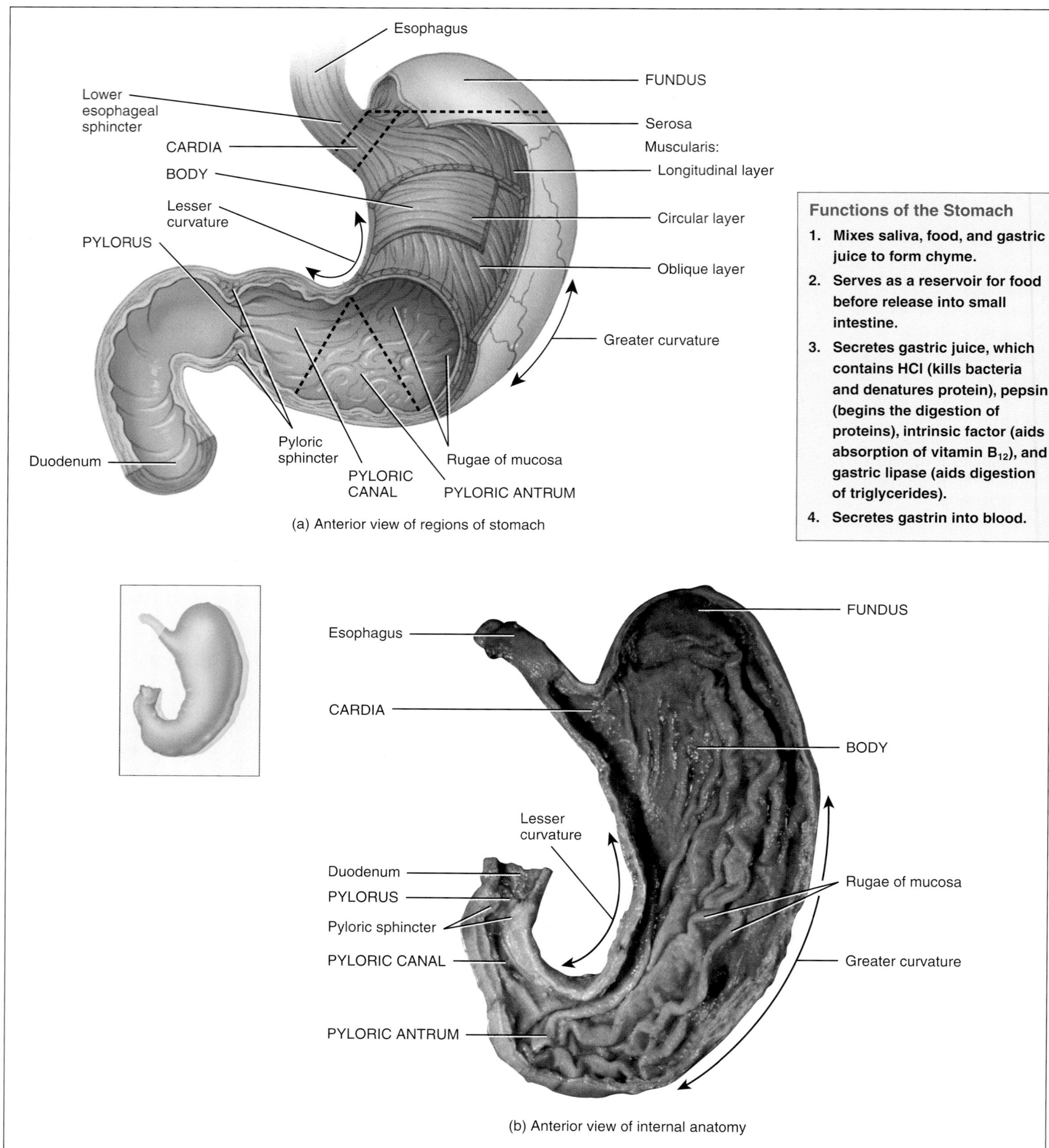

After a very large meal, does your stomach still have rugae?

The gastric glands contain three types of *exocrine gland cells* that secrete their products into the stomach lumen: mucous neck cells, chief cells, and parietal cells. Both surface mucous cells and **mucous neck cells** secrete mucus (Figure 24.12b). **Parietal cells** produce intrinsic factor (needed for absorption of vitamin B_{12}) and hydrochloric acid. The **chief cells** secrete pepsinogen and gastric lipase. The secretions of the mucous, parietal, and chief cells form **gastric juice,** which totals 2000–3000 mL (roughly 2–3 qt.) per day. In addition, gastric glands include a type of enteroendocrine cell, the **G cell,** which is located mainly in the pyloric antrum and secretes the hormone gastrin into the bloodstream. As we will see shortly, this hormone stimulates several aspects of gastric activity.

Three additional layers lie deep to the mucosa. The **submucosa** of the stomach is composed of areolar connective tissue. The **muscularis** has three layers of smooth muscle (rather than the two found in the small and large intestines): an outer longitudinal layer, a middle circular layer, and an inner oblique layer. The oblique layer is limited primarily to the body of the stomach. The **serosa** is composed of simple squamous epithelium (mesothelium) and areolar connective tissue; the portion of the serosa covering the stomach is part of the visceral peritoneum. At the lesser curvature of the stomach, the visceral peritoneum extends upward to the liver as the lesser omentum. At the greater curvature of the stomach, the visceral peritoneum continues downward as the greater omentum and drapes over the intestines.

Mechanical and Chemical Digestion in the Stomach

Several minutes after food enters the stomach, gentle, rippling, peristaltic movements called **mixing waves** pass over the stomach every 15 to 25 seconds. These waves macerate food, mix it with secretions of the gastric glands, and reduce it to a soupy liquid called **chyme** (KĪM = juice). Few mixing waves are observed in the fundus, which primarily has a storage function. As digestion proceeds in the stomach, more vigorous mixing waves begin at the body of the stomach and intensify as they reach the pylorus. The pyloric sphincter normally remains almost, but not completely, closed. As food reaches the pylorus, each mixing wave periodically forces about 3 mL of chyme into the duodenum through the pyloric sphincter, a phenomenon known as **gastric emptying.** Most of the chyme is forced back into the body of the stomach, where mixing continues. The next wave pushes the chyme forward again and forces a little more into the duodenum. These

Figure 24.12 Histology of the stomach.

Gastric juice is the combined secretions of mucous cells, parietal cells, and chief cells.

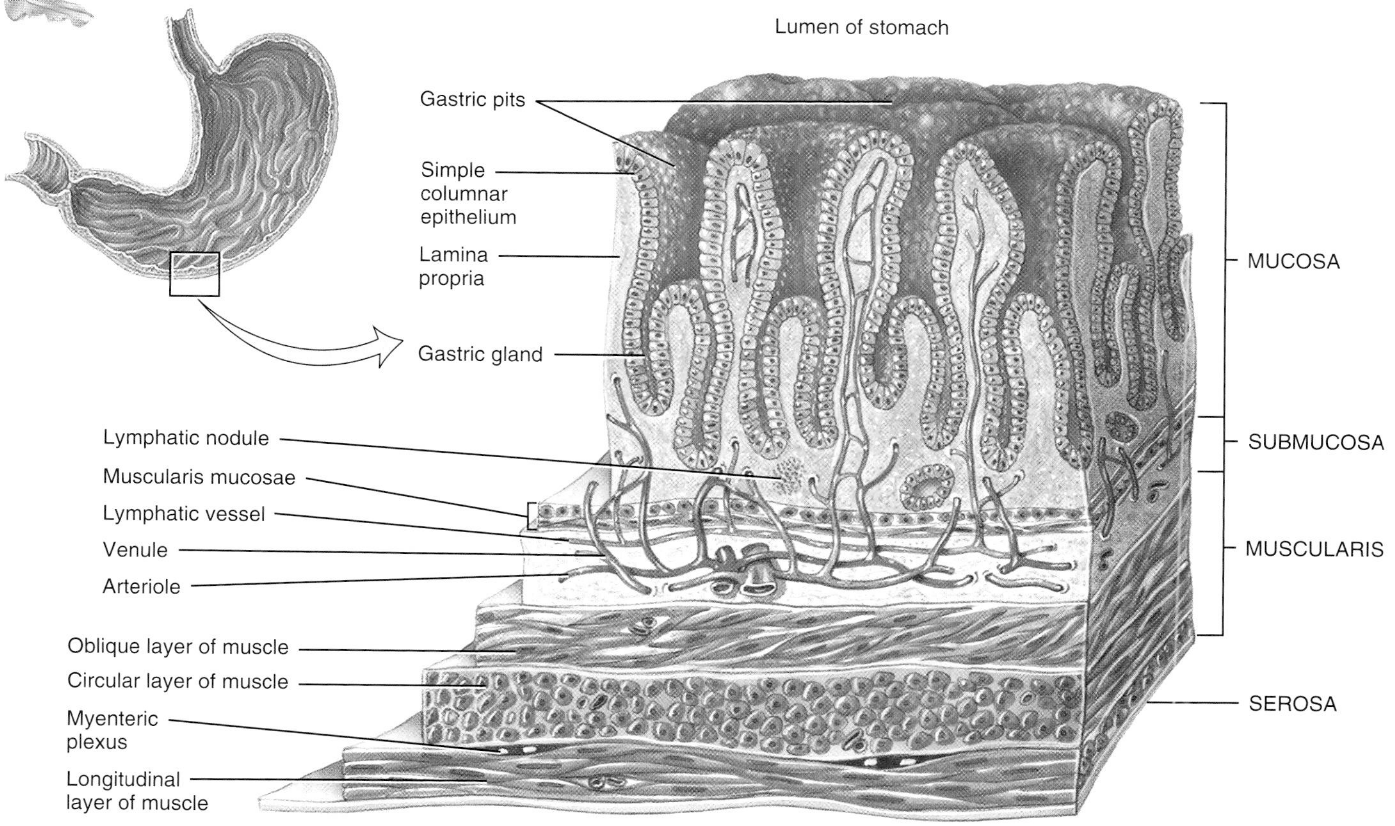

(a) Three-dimensional view of layers of the stomach

continues

Figure 24.12 **(Continued)**

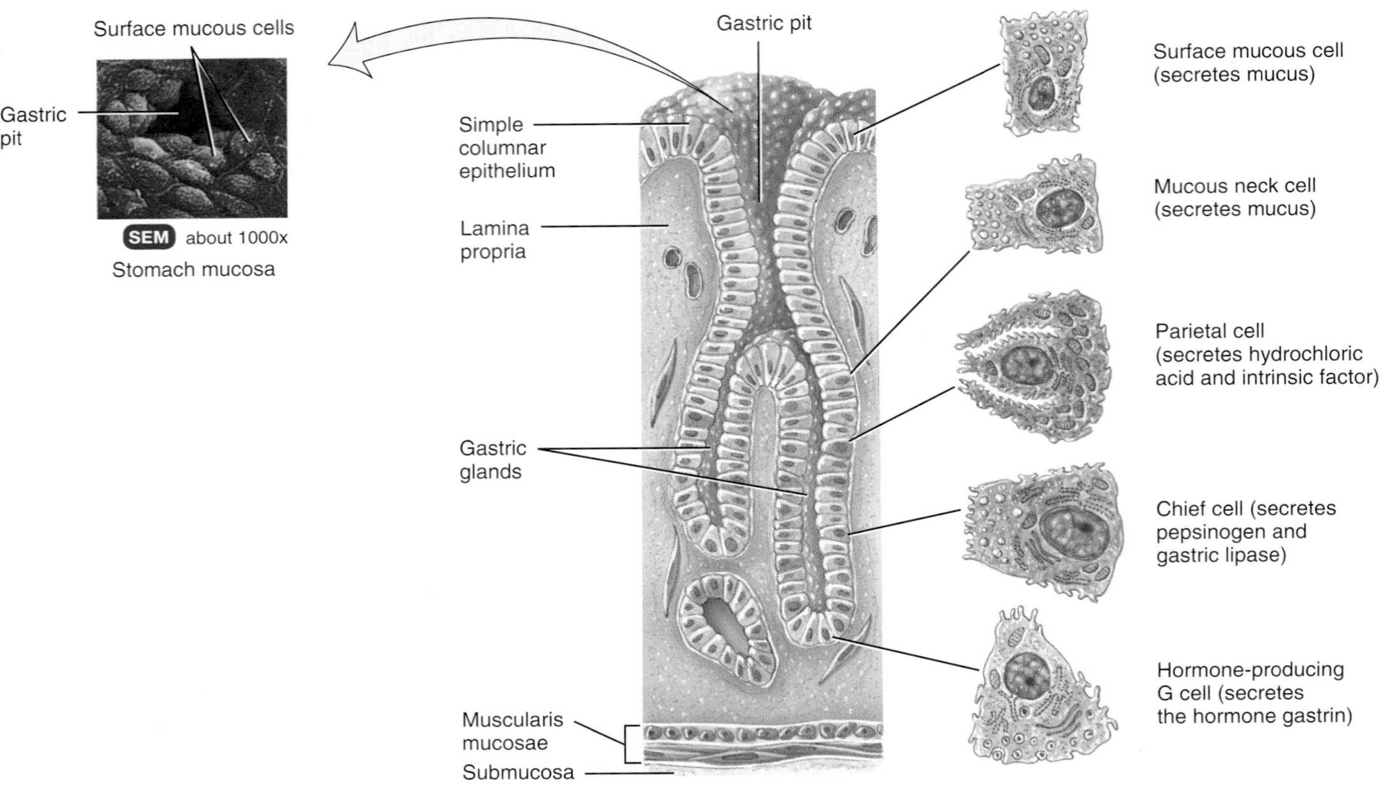

(b) Sectional view of the stomach mucosa showing gastric glands and cell types

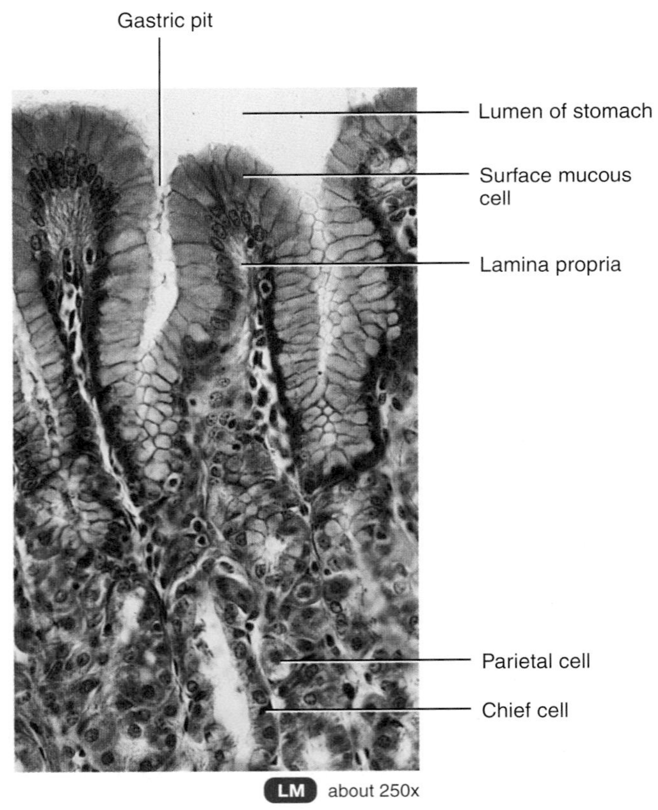

(c) Fundic mucosa

Where is HCl secreted, and what are its functions?

forward and backward movements of the gastric contents are responsible for most mixing in the stomach.

Foods may remain in the fundus for about an hour without becoming mixed with gastric juice. During this time, digestion by salivary amylase continues. Soon, however, the churning action mixes chyme with acidic gastric juice, inactivating salivary amylase and activating lingual lipase, which starts to digest triglycerides into fatty acids and diglycerides.

Although parietal cells secrete hydrogen ions (H^+) and chloride ions (Cl^-) separately into the stomach lumen, the net effect is secretion of hydrochloric acid (HCl). **Proton pumps** powered by H^+/K^+ ATPases actively transport H^+ into the lumen while bringing potassium ions (K^+) into the cell (Figure 24.13). At the same time, Cl^- and K^+ diffuse out into the lumen through Cl^- and K^+ channels in the apical membrane. The enzyme *carbonic anhydrase,* which is especially plentiful in parietal cells, catalyzes the formation of carbonic acid (H_2CO_3) from water (H_2O) and carbon dioxide (CO_2). As carbonic acid dissociates, it provides a ready source of H^+ for the proton pumps but also generates bicarbonate ions (HCO_3^-). As HCO_3^- builds up in the cytosol, it exits the parietal cell in exchange for Cl^- via Cl^-/HCO_3^- antiporters in the basolateral membrane (next to the lamina propria). HCO_3^- diffuses into nearby blood capillaries. This "alkaline tide" of bicarbonate ions entering the bloodstream after a meal may be large enough to elevate blood pH slightly and make urine more alkaline.

Figure 24.13 Secretion of HCl (hydrochloric acid) by parietal cells in the stomach.

Proton pumps, powered by ATP, secrete the H^+; Cl^- diffuses into the stomach lumen through Cl^- channels.

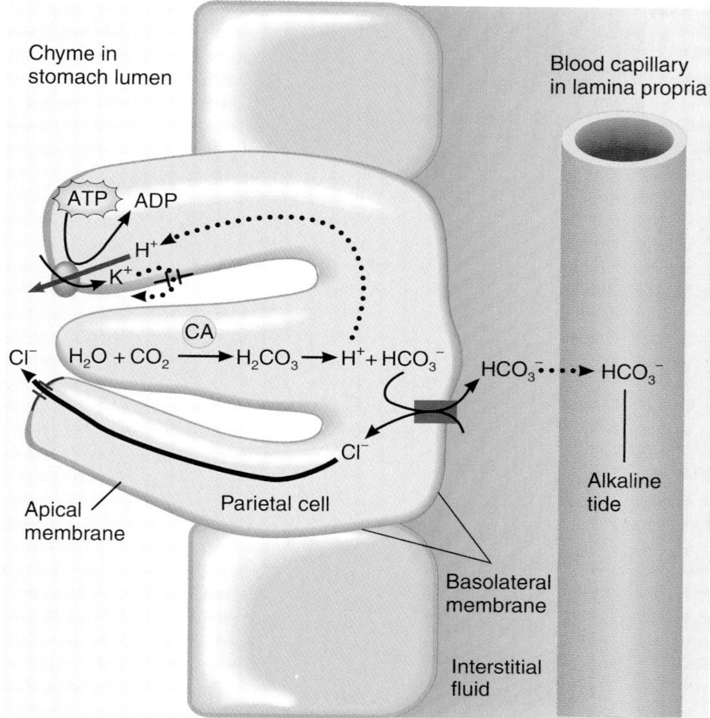

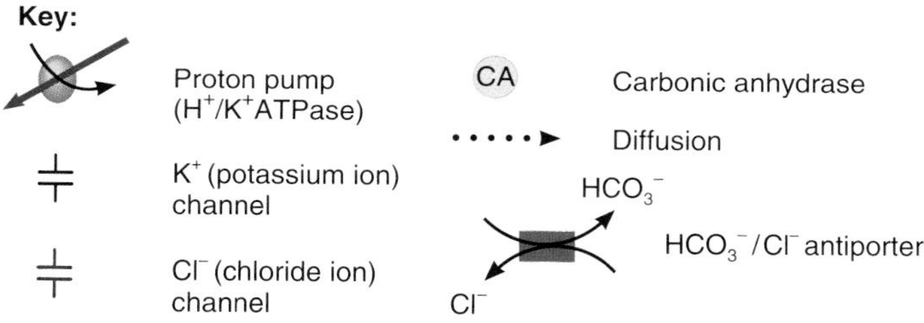

What molecule is the source of the hydrogen ions that are secreted into gastric juice?

HCl secretion by parietal cells can be stimulated by several sources: acetylcholine (ACh) released by parasympathetic neurons, gastrin secreted by G cells, and histamine, which is a paracrine substance released by mast cells in the nearby lamina propria. Acetylcholine and gastrin stimulate parietal cells to secrete more HCl in the presence of histamine. In other words, histamine acts synergistically, enhancing the effects of acetylcholine and gastrin. Receptors for all three substances are present in the plasma membrane of parietal cells. The histamine receptors on parietal cells are called H_2 receptors; they mediate different responses than do the H_1 receptors involved in allergic responses.

The strongly acidic fluid of the stomach kills many microbes in food. HCl partially denatures (unfolds) proteins in food and stimulates the secretion of hormones that promote the flow of bile and pancreatic juice. Enzymatic digestion of proteins also begins in the stomach. The only proteolytic (protein-digesting) enzyme in the stomach is **pepsin,** which is secreted by chief cells. Pepsin severs certain peptide bonds between amino acids, breaking down a protein chain of many amino acids into smaller peptide fragments. Pepsin is most effective in the very acidic environment of the stomach (pH 2); it becomes inactive at a higher pH.

What keeps pepsin from digesting the protein in stomach cells along with the food? First, pepsin is secreted in an inactive form called *pepsinogen;* in this form, it cannot digest the proteins in the chief cells that produce it. Pepsinogen is not converted into active pepsin until it comes in contact with hydrochloric acid secreted by parietal cells or active pepsin molecules. Second, the stomach epithelial cells are protected from gastric juices by a 1–3 mm thick layer of alkaline mucus secreted by surface mucous cells and mucous neck cells.

Another enzyme of the stomach is **gastric lipase,** which splits the short-chain triglycerides in fat molecules (such as those found in milk) into fatty acids and monoglycerides. A monoglyceride consists of a glycerol molecule that is attached to one fatty acid molecule. This enzyme, which has a limited role in the adult stomach, operates best at a pH of 5–6. More important than either lingual lipase or gastric lipase is pancreatic lipase, an enzyme secreted by the pancreas into the small intestine.

Only a small amount of nutrients are absorbed in the stomach because its epithelial cells are impermeable to most materials. However, mucous cells of the stomach absorb some water, ions, and short-chain fatty acids, as well as certain drugs (especially aspirin) and alcohol.

Within 2 to 4 hours after eating a meal, the stomach has emptied its contents into the duodenum. Foods rich in carbohydrate spend the least time in the stomach; high-protein foods remain somewhat longer, and emptying is slowest after a fat-laden meal containing large amounts of triglycerides.

Table 24.3 summarizes the digestive activities of the stomach.

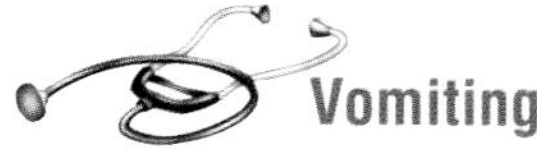

Vomiting

Vomiting or *emesis* is the forcible expulsion of the contents of the upper GI tract (stomach and sometimes duodenum) through the mouth. The strongest stimuli for vomiting are irritation and distension of the stomach; other stimuli include unpleasant sights, general anesthesia, dizziness, and certain drugs such as morphine and derivatives of digitalis. Nerve impulses are transmitted to the vomiting center in the medulla oblongata, and returning impulses propagate to the upper GI tract organs, diaphragm, and abdominal muscles. Vomiting involves squeezing the stomach between the diaphragm and abdominal muscles and expelling the contents through open esophageal sphincters. Prolonged vomiting, especially in infants and elderly people, can be serious because the loss of acidic gastric juice can lead to alkalosis (higher than normal blood pH), dehydration, and damage to the esophagus and teeth. ■

▶ **CHECKPOINT**

20. Compare the epithelium of the esophagus with that of the stomach. How is each adapted to the function of the organ?

21. What is the importance of rugae, surface mucous cells, mucous neck cells, chief cells, parietal cells, and G cells in the stomach?

22. What is the role of pepsin? Why is it secreted in an inactive form?

23. What are the functions of gastric lipase and lingual lipase in the stomach?

TABLE 24.3 Summary of Digestive Activities in the Stomach

Structure	Activity	Result
Mucosa		
Chief cells	Secrete pepsinogen.	Pepsin, the activated form, breaks down proteins into peptides.
	Secrete gastric lipase.	Split triglycerides into fatty acids and monoglycerides.
Parietal cells	Secrete hydrochloric acid.	Kills microbes in food; denature proteins; convert pepsinogen into pepsin.
	Secrete intrinsic factor.	Needed for absorption of vitamin B_{12}, which is used in red blood cell formation (erythropoiesis).
Surface mucous cells and **mucous neck cells**	Secrete mucus.	Form a protective barrier that prevents digestion of stomach wall.
	Absorption.	Small quantity of water, ions, short-chain fatty acids, and some drugs enter the bloodstream.
G cells	Secrete gastrin.	Stimulate parietal cells to secrete HCl and chief cells to secrete pepsinogen; contracts lower esophageal sphincter, increases motility of the stomach, and relaxes pyloric sphincter.
Muscularis	Mixing waves.	Macerate food and mix it with gastric juice, forming chyme.
	Peristalsis.	Forces chyme through pyloric sphincter.
Pyloric sphincter	Opens to permit passage of chyme into duodenum.	Regulates passage of chyme from stomach to duodenum; prevents backflow of chyme from duodenum to stomach.

PANCREAS

▶ **OBJECTIVE**

Describe the location, anatomy, histology, and function of the pancreas.

From the stomach, chyme passes into the small intestine. Because chemical digestion in the small intestine depends on activities of the pancreas, liver, and gallbladder, we first consider the activities of these accessory digestive organs and their contributions to digestion in the small intestine.

Anatomy of the Pancreas

The **pancreas** (*pan-* = all; *-creas* = flesh), a retroperitoneal gland that is about 12–15 cm (5–6 in.) long and 2.5 cm (1 in.) thick, lies posterior to the greater curvature of the stomach. The pancreas consists of a head, a body, and a tail and is usually connected to the duodenum by two ducts (Figure 24.14a). The **head** is the expanded portion of the organ near the curve of the duodenum; superior to and to the left of the head are the central **body** and the tapering **tail.**

Pancreatic juices are secreted by exocrine cells into small ducts that ultimately unite to form two larger ducts, the pancreatic duct and the accessory duct. These in turn convey the secretions into the small intestine. The **pancreatic duct (duct of Wirsung)** is the larger of the two ducts. In most people, the pancreatic duct joins the common bile duct from the liver and gallbladder and enters the duodenum as a common duct called the **hepatopancreatic ampulla (ampulla of Vater).** The ampulla opens on an elevation of the duodenal mucosa known as the **major duodenal papilla,** which lies about 10 cm (4 in.) inferior to the pyloric sphincter of the stomach. The passage of pancreatic juice and bile through the hepatopancreatic ampulla into the small intestine is regulated by a mass of smooth muscle known as the **sphincter of the hepatopancreatic ampulla (sphincter of Oddi).** The other major duct of the pancreas, the **accessory duct (duct of Santorini),** leads from the pancreas and empties into the duodenum about 2.5 cm (1 in.) superior to the hepatopancreatic ampulla.

Histology of the Pancreas

The pancreas is made up of small clusters of glandular epithelial cells. About 99% of the clusters, called **acini** (AS-i-nē), constitute the *exocrine* portion of the organ (see Figure 18.18b, c on page 647). The cells within acini secrete a mixture of fluid and digestive enzymes called **pancreatic juice.** The remaining 1% of the clusters, called **pancreatic islets (islets of Langerhans),** form the *endocrine* portion of the pancreas. These cells secrete the hormones glucagon, insulin, somatostatin, and pancreatic polypeptide. The functions of these hormones are discussed in Chapter 18.

Figure 24.14 Relation of the pancreas to the liver, gallbladder, and duodenum. The inset shows details of the common bile duct and pancreatic duct forming the hepatopancreatic ampulla (ampulla of Vater) and emptying into the duodenum. (See Tortora, *A Photographic Atlas of the Human Body, Second Edition,* Figures 12.10, 12.11.)

Pancreatic enzymes digest starches (polysaccharides), proteins, triglycerides, and nucleic acids.

(a) Anterior view

(b) Details of hepatopancreatic ampulla

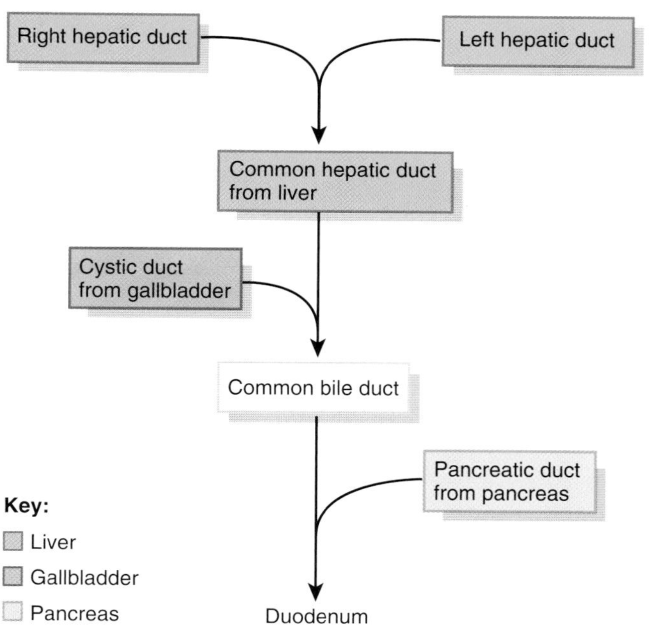

(c) Ducts carrying bile from liver and gallbladder and pancreatic juice from pancreas to the duodenum

What type of fluid is found in the pancreatic duct? The common bile duct? The hepatopancreatic ampulla?

Composition and Functions of Pancreatic Juice

Each day the pancreas produces 1200–1500 mL (about 1.2–1.5 qt) of **pancreatic juice,** a clear, colorless liquid consisting mostly of water, some salts, sodium bicarbonate, and several enzymes. The sodium bicarbonate gives pancreatic juice a slightly alkaline pH (7.1–8.2) that buffers acidic gastric juice in chyme, stops the action of pepsin from the stomach, and creates the proper pH for the action of digestive enzymes in the small intestine. The enzymes in pancreatic juice include a starch-digesting enzyme called **pancreatic amylase;** several protein-digesting enzymes called **trypsin** (TRIP-sin), **chymotrypsin** (kī′-mō-TRIP-sin), **carboxypeptidase** (kar-bok′-sē-PEP-ti-dās), and **elastase** (ē-LAS-tās); the principal triglyceride-digesting enzyme in adults, called **pancreatic lipase;** and nucleic acid–digesting enzymes called **ribonuclease** and **deoxyribonuclease.**

The protein-digesting enzymes of the pancreas are produced in an inactive form just as pepsin is produced in the stomach as pepsinogen. Because they are inactive, the enzymes do not digest cells of the pancreas itself. Trypsin is secreted in an inactive form called **trypsinogen** (trip-SIN-ō-jen). Pancreatic acinar cells also secrete a protein called **trypsin inhibitor** that combines with any trypsin formed accidentally in the pancreas or in pancreatic juice and blocks its enzymatic activity. When trypsinogen reaches the lumen of the small intestine, it encounters an activating brush-border enzyme called **enterokinase** (en′-ter-ō-KĪ-nās), which splits off part of the trypsinogen molecule to form trypsin. In turn, trypsin acts on the inactive precursors (called **chymotrypsinogen, procarboxypeptidase,** and **proelastase**) to produce chymotrypsin, carboxypeptidase, and elastase, respectively.

Pancreatitis and Pancreatic Cancer

Inflammation of the pancreas, as may occur in association with alcohol abuse or chronic gallstones, is called **pancreatitis** (pan′-krē-a-TĪ-tis). In a more severe condition known as **acute pancreatitis,** which is associated with heavy alcohol intake or biliary tract obstruction, the pancreatic cells may release either trypsin instead of trypsinogen or insufficient amounts of trypsin inhibitor, and the trypsin begins to digest the pancreatic cells. Patients with acute pancreatitis usually respond to treatment, but recurrent attacks are the rule. In some people pancreatitis is idiopathic, meaning that the cause is unknown. Other causes of pancreatitis include cystic fibrosis, high levels of calcium in the blood (hypercalcemia), high levels of blood fats (hyperlipidemia or hypertriglyceridemia), some drugs, and certain autoimmune conditions. However, in roughly 70 percent of adults with pancreatitis, the cause is alcoholism. Often the first episode happens between ages 30 and 40.

Pancreatic cancer usually affects people over 50 years of age and occurs more frequently in males. Typically, there are few symptoms until the disorder reaches an advanced stage and often not until it has metastasized to other parts of the body such as the lymph nodes, liver, or lungs. The disease is nearly always fatal and is the fourth most common cause of death from cancer in the United States. Pancreatic cancer has been linked to fatty foods, high alcohol consumption, genetic factors, smoking, and chronic pancreatitis. ■

► **CHECKPOINT**

24. Describe the duct system connecting the pancreas to the duodenum.

25. What are pancreatic acini? How do their functions differ from those of the pancreatic islets (islets of Langerhans)?

26. What are the digestive functions of the components of pancreatic juice?

LIVER AND GALLBLADDER

► **OBJECTIVE**

Describe the location, anatomy, histology, and functions of the liver and gallbladder.

The **liver** is the heaviest gland of the body, weighing about 1.4 kg (about 3 lb) in an average adult. Of all of the organs of the body, it is second only to the skin in size. The liver is inferior to the diaphragm and occupies most of the right hypochondriac and part of the epigastric regions of the abdominopelvic cavity (see Figure 1.12a on page 20).

The **gallbladder** (*gall-* = bile) is a pear-shaped sac that is located in a depression of the posterior surface of the liver. It is 7–10 cm (3–4 in.) long and typically hangs from the anterior inferior margin of the liver (Figure 24.14a).

Anatomy of the Liver and Gallbladder

The liver is almost completely covered by visceral peritoneum and is completely covered by a dense irregular connective tissue layer that lies deep to the peritoneum. The liver is divided into two principal lobes—a large **right lobe** and a smaller **left lobe**—by the **falciform ligament,** a fold of the peritoneum (Figure 24.14a). Although the right lobe is considered by many anatomists to include an inferior **quadrate lobe** and a posterior **caudate lobe,** based on internal morphology (primarily the distribution of blood vessels), the quadrate and caudate lobes more appropriately belong to the left lobe. The falciform ligament extends from the undersurface of the diaphragm between the two principal lobes of the liver to the superior surface of the liver, helping to suspend the liver in the abdominal cavity. In the free border of the falciform ligament is the **ligamentum teres (round ligament),** a remnant of the umbilical vein of the fetus (see Figure 21.30a, b on page 794); this fibrous cord extends from the liver to the umbilicus. The right and left **coronary ligaments** are narrow extensions of the parietal peritoneum that suspend the liver from the diaphragm.

The parts of the gallbladder include the broad **fundus,** which projects inferiorly beyond the inferior border of the liver; the **body,** the central portion; and the **neck,** the tapered portion. The body and neck project superiorly.

Histology of the Liver and Gallbladder

The lobes of the liver are made up of many functional units called **lobules** (Figure 24.15). A lobule is typically a six-sided structure (hexagon) that consists of specialized epithelial cells, called **hepatocytes** (*hepat-* = liver; *-cytes* = cells), arranged in irregular, branching, interconnected plates around a **central vein.** In addition, the liver lobule contains highly-permeable capillaries called **sinusoids,** through which blood passes. Also present in the sinusoids are fixed phagocytes called **stellate reticuloendothelial (Kupffer) cells,** which destroy worn-out white blood cells and red blood cells, bacteria, and other foreign matter in the venous blood draining from the gastrointestinal tract.

Bile, which is secreted by hepatocytes, enters **bile canaliculi** (kan′-a-LIK-ū-lī = small canals), narrow intercellular canals that empty into small **bile ductules** (Figure 24.15a). The ductules pass bile into **bile ducts** at the periphery of the lobules. The bile ducts merge and eventually form the larger **right** and **left hepatic ducts,** which unite and exit the liver as the **common hepatic duct** (see Figure 24.14). The common hepatic duct joins

Figure 24.15 Histology of a lobule, the functional unit of the liver.

A lobule consists of hepatocytes arranged around a central vein.

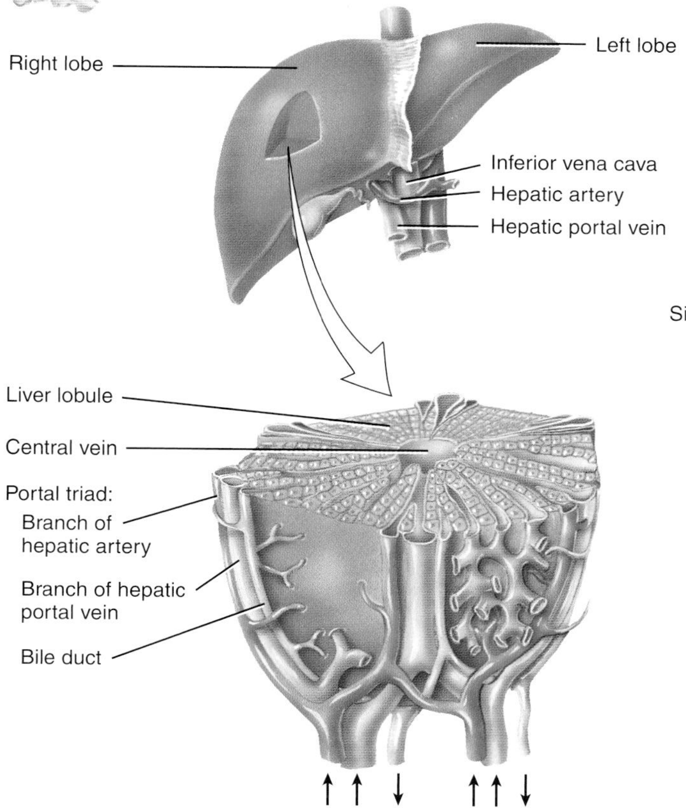

Right lobe
Left lobe
Inferior vena cava
Hepatic artery
Hepatic portal vein
Liver lobule
Central vein
Portal triad:
Branch of hepatic artery
Branch of hepatic portal vein
Bile duct

(a) Overview of a single lobule

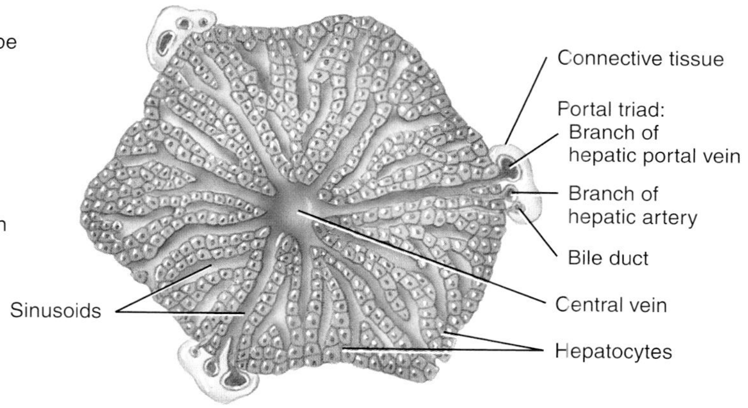

Connective tissue
Portal triad:
Branch of hepatic portal vein
Branch of hepatic artery
Bile duct
Central vein
Hepatocytes
Sinusoids

(b) Details of a single liver lobule

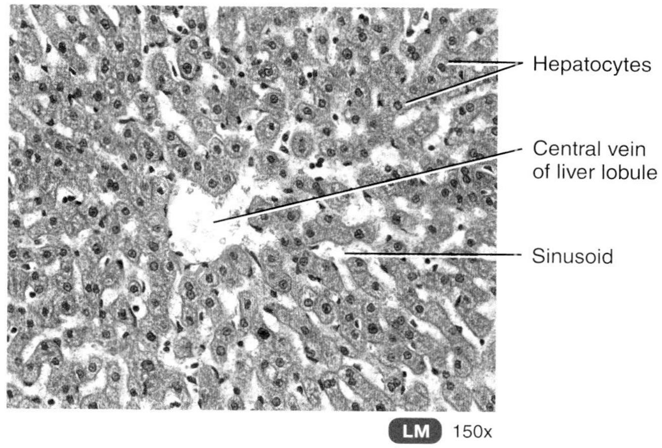

Hepatocytes
Central vein of liver lobule
Sinusoid
LM 150x

(c) Portion of a liver lobule

Sinusoid
To hepatic vein
Bile canaliculi
Portal triad:
Bile duct
Branch of hepatic portal vein
Branch of hepatic artery
Central vein
Hepatocyte
Reticuloendothelial (Kupffer) cell
Sinusoids
Connective tissue

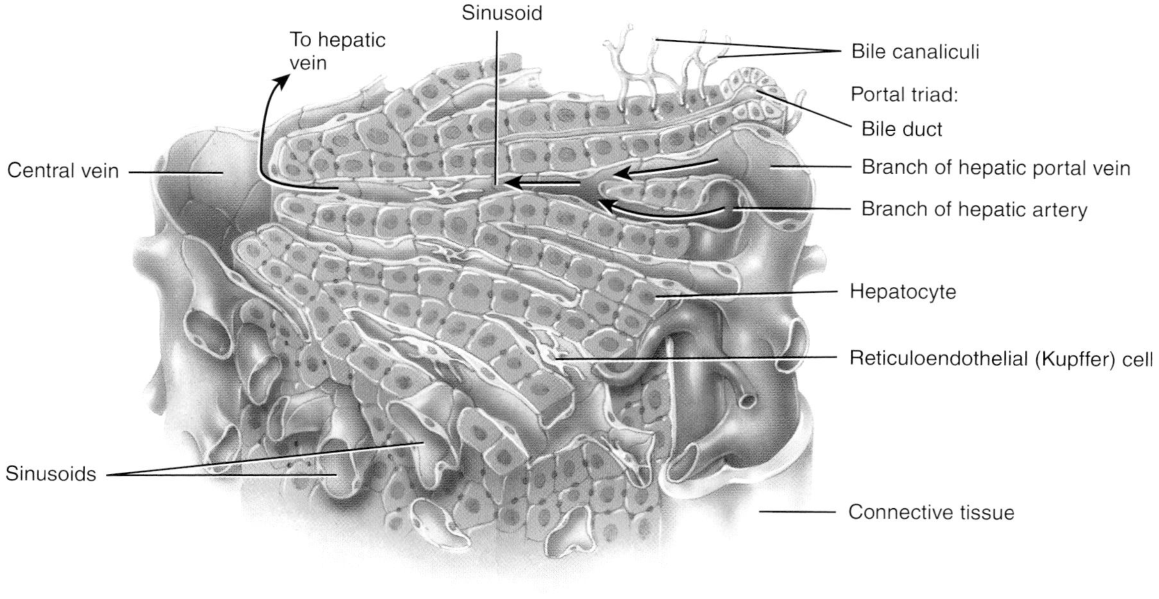

(d) Details of a portion of a liver lobule

Which type of cell in the liver is phagocytic?

the **cystic duct** (*cystic* = bladder) from the gallbladder to form the **common bile duct.**

The mucosa of the gallbladder consists of simple columnar epithelium arranged in rugae resembling those of the stomach. The wall of the gallbladder lacks a submucosa. The middle, muscular coat of the wall consists of smooth muscle fibers. Contraction of the smooth muscle fibers ejects the contents of the gallbladder into the **cystic duct.** The gallbladder's outer coat is the visceral peritoneum. The functions of the gallbladder are to store and concentrate the bile produced by the liver (up to tenfold) until it is needed in the small intestine. In the concentration process, water and ions are absorbed by the gallbladder mucosa.

Jaundice

Jaundice (JAWN-dis = yellowed) is a yellowish coloration of the sclerae (whites of the eyes), skin, and mucous membranes due to a buildup of a yellow compound called bilirubin. After bilirubin is formed from the breakdown of the heme pigment in aged red blood cells, it is transported to the liver, where it is processed and eventually excreted into bile. The three main categories of jaundice are (1) *prehepatic jaundice,* due to excess production of bilirubin; (2) *hepatic jaundice,* due to congenital liver disease, cirrhosis of the liver, or hepatitis; and (3) *extrahepatic jaundice,* due to blockage of bile drainage by gallstones or cancer of the bowel or the pancreas.

Because the liver of a newborn functions poorly for the first week or so, many babies experience a mild form of jaundice called *neonatal (physiological) jaundice* that disappears as the liver matures. Usually, it is treated by exposing the infant to blue light, which converts bilirubin into substances the kidneys can excrete. ■

Blood Supply of the Liver

The liver receives blood from two sources (Figure 24.16). From the hepatic artery it obtains oxygenated blood, and from the hepatic portal vein it receives deoxygenated blood containing newly absorbed nutrients, drugs, and possibly microbes and toxins from the gastrointestinal tract (see Figure 21.28 on page 791). Branches of both the hepatic artery and the hepatic portal vein carry blood into liver sinusoids, where oxygen, most of the nutrients, and certain toxic substances are taken up by the hepatocytes. Products manufactured by the hepatocytes and nutrients needed by other cells are secreted back into the blood, which then drains into the central vein and eventually passes into a hepatic vein. Because blood from the gastrointestinal tract passes through the liver as part of the hepatic portal circulation, the liver is often a site for metastasis of cancer that originates in the GI tract. Branches of the hepatic portal vein, hepatic artery, and bile duct typically accompany each other in their distribution through the liver. Collectively, these three structures are called a **portal triad** (see Figure 24.15). Portal triads are located at the corners of the liver lobules.

Role and Composition of Bile

Each day, hepatocytes secrete 800–1000 mL (about 1 qt) of **bile,** a yellow, brownish, or olive-green liquid. It has a pH of 7.6–8.6 and consists mostly of water, bile salts, cholesterol, a phospholipid called lecithin, bile pigments, and several ions.

The principal bile pigment is **bilirubin.** The phagocytosis of aged red blood cells liberates iron, globin, and bilirubin (derived from heme) (see Figure 19.5 on page 674). The iron and globin are recycled; the bilirubin is secreted into the bile and is eventually broken down in the intestine. One of its breakdown products—**stercobilin**—gives feces their normal brown color.

Bile is partially an excretory product and partially a digestive secretion. Bile salts, which are sodium salts and potassium salts of bile acids (mostly chenodeoxycholic acid and cholic acid), play a role in **emulsification,** the breakdown of large lipid globules into a suspension of small lipid globules. The small lipid globules present a very large surface area that allows pancreatic lipase to more rapidly accomplish digestion of triglycerides. Bile salts also aid in the absorption of lipids following their digestion.

Figure 24.16 Hepatic blood flow: sources, path through the liver, and return to the heart.

The liver receives oxygenated blood via the hepatic artery and nutrient-rich deoxygenated blood via the hepatic portal vein.

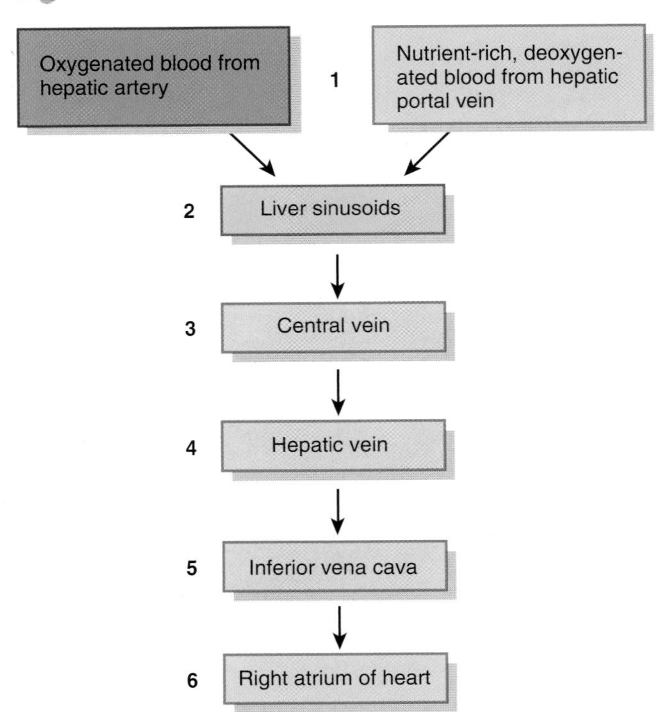

During the first few hours after a meal, how does the chemical composition of blood change as it flows through the liver sinusoids?

Although hepatocytes continually release bile, they increase production and secretion when the portal blood contains more bile acids; thus, as digestion and absorption continue in the small intestine, bile release increases. Between meals, after most absorption has occurred, bile flows into the gallbladder for storage because the sphincter of the hepatopancreatic ampulla (sphincter of Oddi; see Figure 24.14) closes off the entrance to the duodenum.

Gallstones

If bile contains either insufficient bile salts or lecithin or excessive cholesterol, the cholesterol may crystallize to form **gallstones.** As they grow in size and number, gallstones may cause minimal, intermittent, or complete obstruction to the flow of bile from the gallbladder into the duodenum. Treatment consists of using gallstone-dissolving drugs, lithotripsy (shock-wave therapy), or surgery. For people with recurrent gallstones or for whom drugs or lithotripsy is not indicated, *cholecystectomy*—the removal of the gallbladder and its contents—is necessary. More than half a million cholecystectomies are performed each year in the United States. ■

Functions of the Liver

In addition to secreting bile, which is needed for absorption of dietary fats, the liver performs many other vital functions:

- ***Carbohydrate metabolism.*** The liver is especially important in maintaining a normal blood glucose level. When blood glucose is low, the liver can break down glycogen to glucose and release the glucose into the bloodstream. The liver can also convert certain amino acids and lactic acid to glucose, and it can convert other sugars, such as fructose and galactose, into glucose. When blood glucose is high, as occurs just after eating a meal, the liver converts glucose to glycogen and triglycerides for storage.
- ***Lipid metabolism.*** Hepatocytes store some triglycerides; break down fatty acids to generate ATP; synthesize lipoproteins, which transport fatty acids, triglycerides, and cholesterol to and from body cells; synthesize cholesterol; and use cholesterol to make bile salts.
- ***Protein metabolism.*** Hepatocytes *deaminate* (remove the amino group, NH_2, from) amino acids so that the amino acids can be used for ATP production or converted to carbohydrates or fats. The resulting toxic ammonia (NH_3) is then converted into the much less toxic urea, which is excreted in urine. Hepatocytes also synthesize most plasma proteins, such as alpha and beta globulins, albumin, prothrombin, and fibrinogen.
- ***Processing of drugs and hormones.*** The liver can detoxify substances such as alcohol and excrete drugs such as penicillin, erythromycin, and sulfonamides into bile. It can also chemically alter or excrete thyroid hormones and steroid hormones such as estrogens and aldosterone.
- ***Excretion of bilirubin.*** As previously noted, bilirubin, derived from the heme of aged red blood cells, is absorbed by the liver from the blood and secreted into bile. Most of the bilirubin in bile is metabolized in the small intestine by bacteria and eliminated in feces.
- ***Synthesis of bile salts.*** Bile salts are used in the small intestine for the emulsification and absorption of lipids.
- ***Storage.*** In addition to glycogen, the liver is a prime storage site for certain vitamins (A, B_{12}, D, E, and K) and minerals (iron and copper), which are released from the liver when needed elsewhere in the body.
- ***Phagocytosis.*** The stellate reticuloendothelial (Kupffer) cells of the liver phagocytize aged red blood cells, white blood cells, and some bacteria.
- ***Activation of vitamin D.*** The skin, liver, and kidneys participate in synthesizing the active form of vitamin D.

The liver functions related to metabolism are discussed more fully in Chapter 25.

► CHECKPOINT

27. Draw and label a diagram of a liver lobule.
28. Describe the pathways of blood flow into, through, and out of the liver.
29. How are the liver and gallbladder connected to the duodenum?
30. Once bile has been formed by the liver, how is it collected and transported to the gallbladder for storage?
31. What is the function of bile?

SMALL INTESTINE

► OBJECTIVE

Describe the location, anatomy, histology, and functions of the small intestine.

The major events of digestion and absorption occur in a long tube called the **small intestine.** Because most digestion and absorption of nutrients occur in the small intestine, its structure is specially adapted for these functions. Its length alone provides a large surface area for digestion and absorption, and that area is further increased by circular folds, villi, and microvilli. The small intestine begins at the pyloric sphincter of the stomach, coils through the central and inferior part of the abdominal cavity, and eventually opens into the large intestine. It averages 2.5 cm (1 in.) in diameter; its length is about 3 m (10 ft) in a living person and about 6.5 m (21 ft) in a cadaver due to the loss of smooth muscle tone after death.

Anatomy of the Small Intestine

The small intestine is divided into three regions (Figure 24.17). The **duodenum** (doo-ō-DĒ-num), the shortest region, is retroperitoneal. It starts at the pyloric sphincter of the stomach and extends about 25 cm (10 in.) until it merges with the jejunum. *Duodenum* means "12"; it is so named because it is about as long as the width of 12 fingers. The **jejunum** (je-JOO-num) is about 1 m (3 ft) long and extends to the ileum. *Jejunum* means "empty," which is how it is found at death. The final and longest region of the small intestine, the **ileum** (IL-ē-um = twisted), measures about 2 m (6 ft) and joins the large intestine at the **ileocecal sphincter** (il′-ē-ō-SĒ-kal).

Histology of the Small Intestine

The wall of the small intestine is composed of the same four layers that make up most of the GI tract: mucosa, submucosa, muscularis, and serosa (Figure 24.18a). The **mucosa** is composed of a layer of epithelium, lamina propria, and muscularis mucosae. The epithelial layer of the small intestinal mucosa consists of simple columnar epithelium that contains many types of cells (Figure 24.18b). **Absorptive cells** of the epithelium digest and absorb nutrients in small intestinal chyme. Also present in the epithelium are **goblet cells,** which secrete mucus. The small intestinal mucosa contains many deep crevices lined with glandular epithelium. Cells lining the crevices form the **intestinal glands (crypts of Lieberkühn)** and secrete intestinal juice (to be discussed shortly). Besides absorptive cells and goblet cells, the intestinal glands also contain paneth cells and enteroendocrine cells. **Paneth cells** secrete lysozyme, a bactericidal enzyme, and are capable of phagocytosis. Paneth cells may have a role in regulating the microbial population in the small intestine. Three types of enteroendocrine cells are found in the intestinal glands of the small intestine: **S cells, CCK cells,** and **K cells,** which secrete the hormones **secretin** (se-KRĒ-tin), **cholecystokinin** (kō-lē-sis′-tō-KĪN-in) or **CCK,** and **glucose-dependent insulinotropic peptide** or **GIP** respectively.

The lamina propria of the small intestinal mucosa contains areolar connective tissue and has an abundance of mucosa-associated lymphoid tissue (MALT). **Solitary lymphatic nodules** are most numerous in the distal part of the ileum (Figure 24.19c on page 925). Groups of lymphatic nodules referred to as **aggregated lymphatic follicles (Peyer's patches)** are also present in the ileum. The muscularis mucosae of the small intestinal mucosa consists of smooth muscle.

Figure 24.17 Anatomy of the small intestine. (a) Regions of the small intestine are the duodenum, jejunum, and ileum. (See Tortora, *A Photographic Atlas of the Human Body, Second Edition*, Figure 12.12a.) (b) Circular folds increase the surface area for digestion and absorption in the small intestine.

Most digestion and absorption occur in the small intestine.

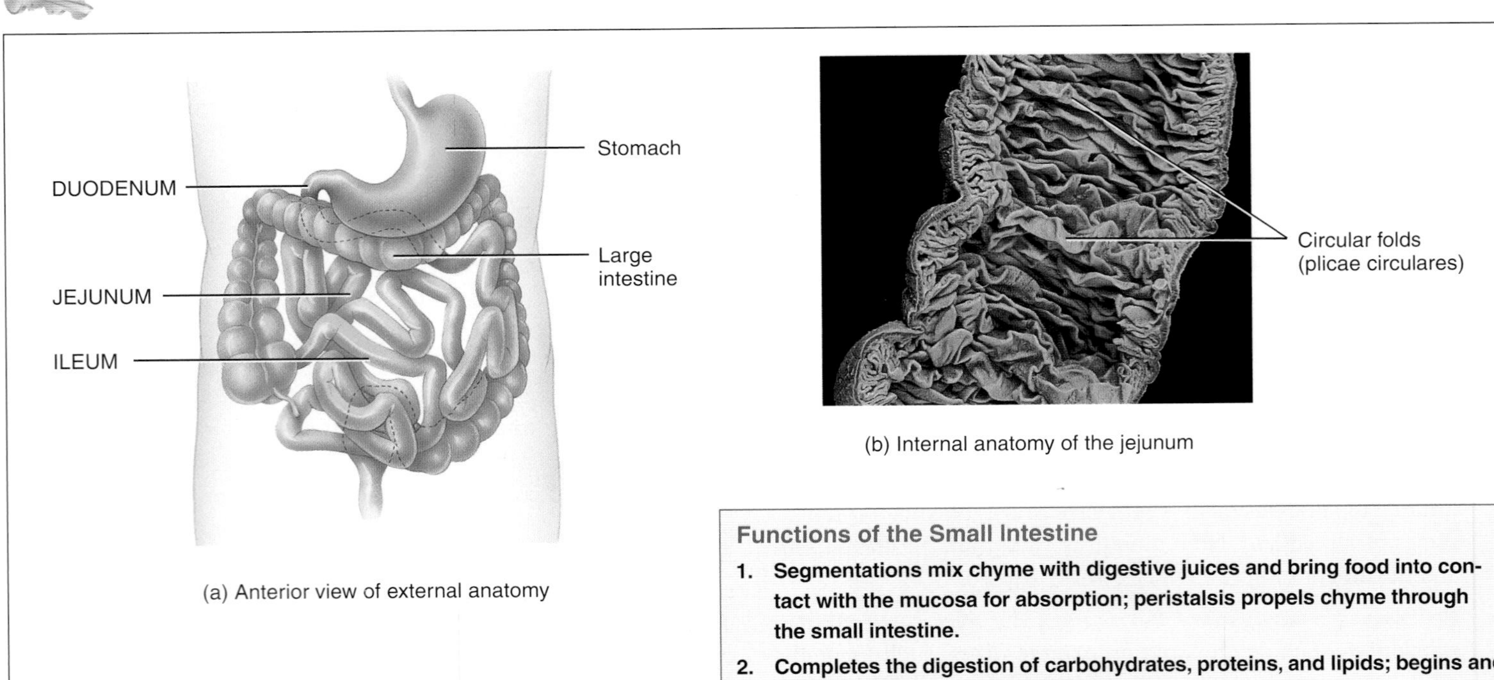

Functions of the Small Intestine

1. **Segmentations mix chyme with digestive juices and bring food into contact with the mucosa for absorption; peristalsis propels chyme through the small intestine.**
2. **Completes the digestion of carbohydrates, proteins, and lipids; begins and completes the digestion of nucleic acids.**
3. **Absorbs about 90% of nutrients and water that pass through the digestive system.**

Which portion of the small intestine is the longest?

Figure 24.18 **Histology of the small intestine.**

Circular folds, villi and microvilli increase the surface area of the small intestine for digestion and absorption.

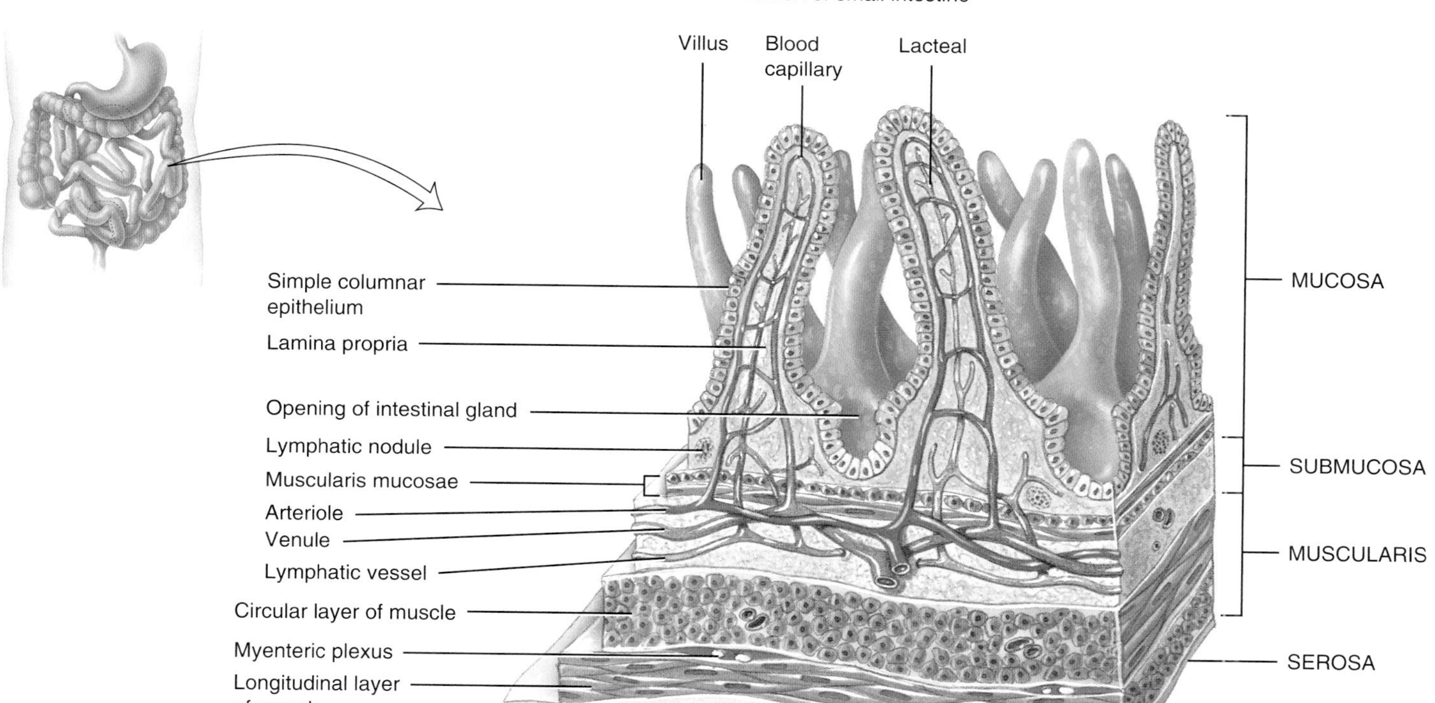

(a) Three-dimensional view of layers of the small intestine showing villi

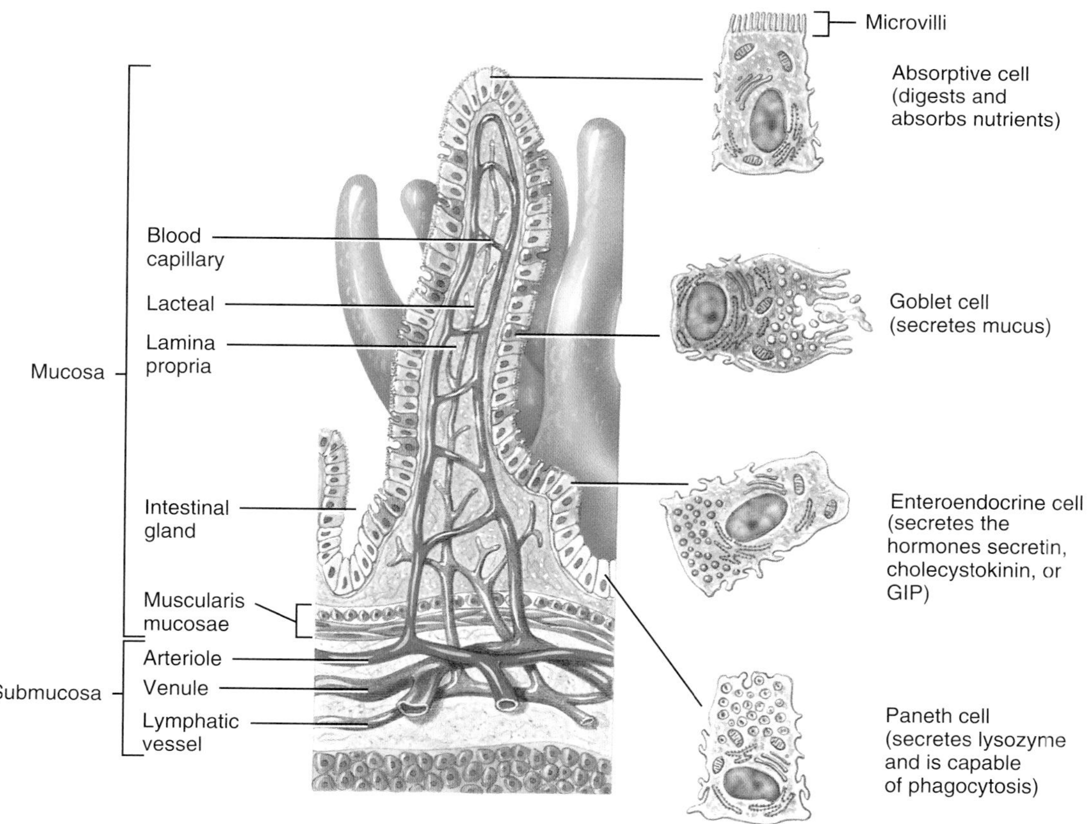

(b) Enlarged villus showing lacteal, capillaries, intestinal glands, and cell types

What is the functional significance of the blood capillary network and lacteal in the center of each villus?

The **submucosa** of the duodenum contains **duodenal (Brunner's) glands** (see Figure 24.19a), which secrete an alkaline mucus that helps neutralize gastric acid in the chyme. Sometimes the lymphatic tissue of the lamina propria extends through the muscularis mucosae into the submucosa. The **muscularis** of the small intestine consists of two layers of smooth muscle. The outer, thinner layer contains longitudinal fibers; the inner, thicker layer contains circular fibers. Except for a major portion of the duodenum, the **serosa** (or visceral peritoneum) completely surrounds the small intestine.

Even though the wall of the small intestine is composed of the same four basic layers as the rest of the GI tract, special structural features of the small intestine facilitate the process of digestion and absorption. These structural features include circular folds, villi, and microvilli. **Circular folds** or *plicae circulares* are folds of the mucosa and submucosa (see Figure 24.17b). These permanent ridges, which are about 10 mm (0.4 in.) long, begin near the proximal portion of the duodenum and end at about the midportion of the ileum. Some extend all the way around the circumference of the intestine; others extend only part of the way around. Circular folds enhance absorption by increasing surface area and causing the chyme to spiral, rather than move in a straight line, as it passes through the small intestine.

Also present in the small intestine are **villi** (= tufts of hair), which are fingerlike projections of the mucosa that are 0.5–1 mm long (see Figure 24.18a). The large number of villi (20–40 per square millimeter) vastly increases the surface area of the epithelium available for absorption and digestion and gives the intestinal mucosa a velvety appearance. Each villus (singular form) is covered by epithelium and has a core of lamina propria; embedded in the connective tissue of the lamina propria are an arteriole, a venule, a blood capillary network, and a **lacteal** (LAK-tē-al = milky), which is a lymphatic capillary. Nutrients absorbed by the epithelial cells covering the villus pass through the wall of a capillary or a lacteal to enter blood or lymph, respectively.

Besides circular folds and villi, the small intestine also has **microvilli** (mī-krō-VIL-ī; *micro-* = small), which are projections of the apical (free) membrane of the absorptive cells. Each microvillus is a 1 μm-long cylindrical, membrane-covered projection that contains a bundle of 20–30 actin filaments. When viewed through a light microscope, the microvilli are too small to be seen individually; instead they form a fuzzy line, called the **brush border,** extending into the lumen of the small intestine (Figure 24.19d). There are an estimated 200 million microvilli per square millimeter of small intestine. Because the microvilli greatly increase the surface area of the plasma membrane, larger amounts of digested nutrients can diffuse into absorptive cells in a given period. The brush border also contains several brush-border enzymes that have digestive functions (discussed shortly).

Figure 24.19 Histology of the duodenum and ileum.

Microvilli in the small intestine contain several brush-border enzymes that help digest nutrients.

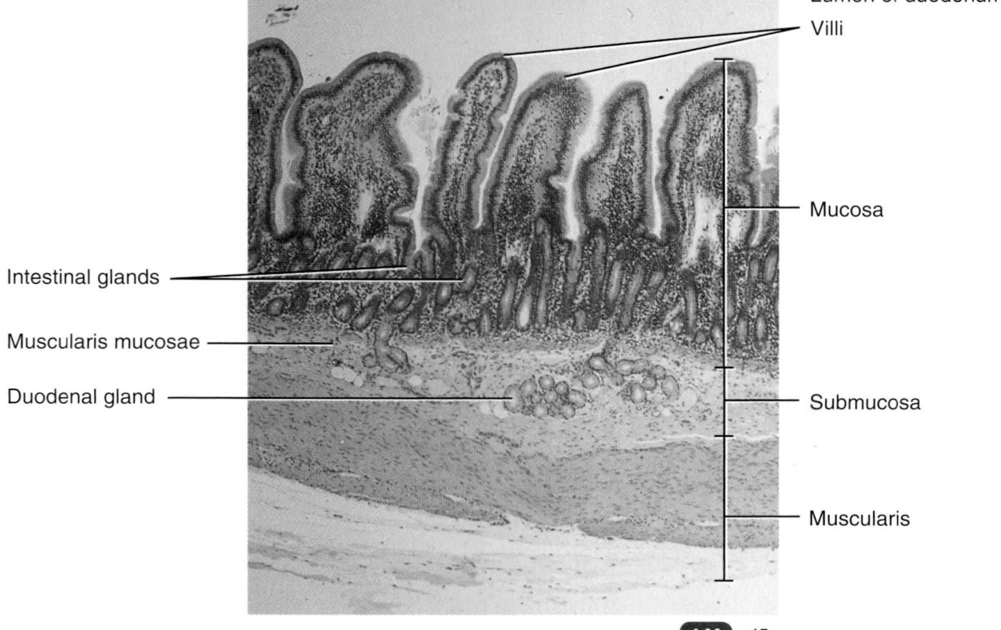

(a) Wall of the duodenum

Role of Intestinal Juice and Brush-Border Enzymes

About 1–2 liters (1–2 qt) of **intestinal juice,** a clear yellow fluid, are secreted each day. Intestinal juice contains water and mucus and is slightly alkaline (pH 7.6). Together, pancreatic and intestinal juices provide a liquid medium that aids the absorption of substances from chyme in the small intestine. The absorptive cells of the small intestine synthesize several digestive enzymes, called **brush-border enzymes,** and insert them in the plasma membrane of the microvilli. Thus, some enzymatic digestion occurs at the surface of the absorptive cells that line the villi, rather than in the lumen exclusively, as occurs in other parts of the GI tract. Among the brush-border enzymes are four carbohydrate-digesting enzymes called α-dextrinase, maltase, sucrase, and lactase; protein-digesting enzymes called peptidases (aminopeptidase and dipeptidase); and two types of nucleotide-digesting enzymes, nucleosidases and phosphatases. Also, as absorptive cells slough off into the lumen of the small intestine, they break apart and release enzymes that help digest nutrients in the chyme.

Mechanical Digestion in the Small Intestine

The two types of movements of the small intestine—segmentations and a type of peristalsis called migrating motility complexes—are governed mainly by the myenteric plexus. **Segmentations** are localized, mixing contractions that occur in portions of intestine distended by a large volume of chyme. Segmentations mix chyme with the digestive juices and bring the particles of food into contact with the mucosa for absorption; they do not push the intestinal contents along the tract. A segmentation starts with the contractions of circular muscle fibers in a portion of the small intestine, an action that constricts the intestine into segments. Next, muscle fibers that encircle the middle of each segment also contract, dividing each segment again. Finally, the fibers that first contracted relax, and each small segment unites with an adjoining small segment so that large segments are formed again. As this sequence of events repeats, the chyme sloshes back and forth. Segmentations occur most rapidly in the duodenum, about 12 times per minute, and progressively slow to about 8 times per minute in the ileum. This movement is similar to alternately squeezing the middle and then the ends of a capped tube of toothpaste.

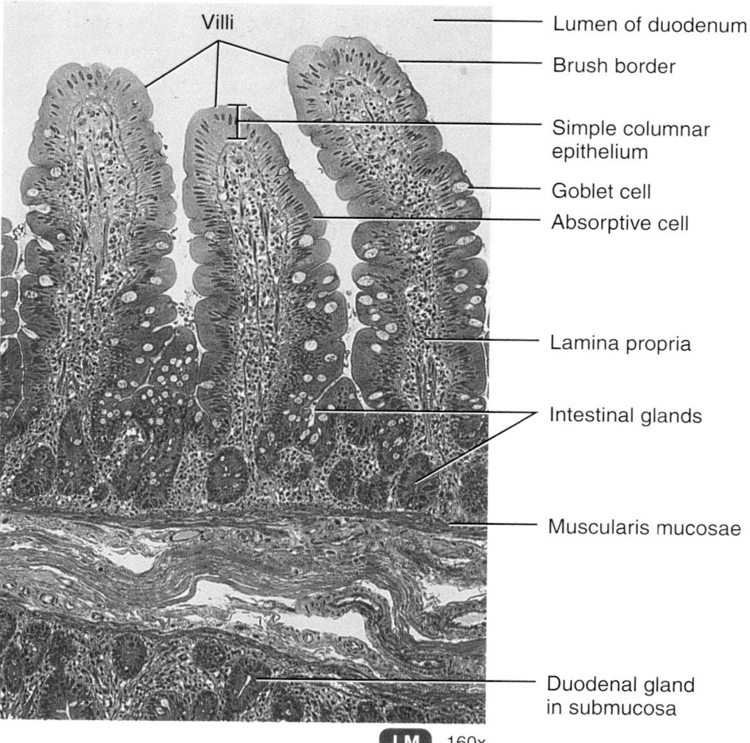

(b) Three villi from the duodenum of the small intestine

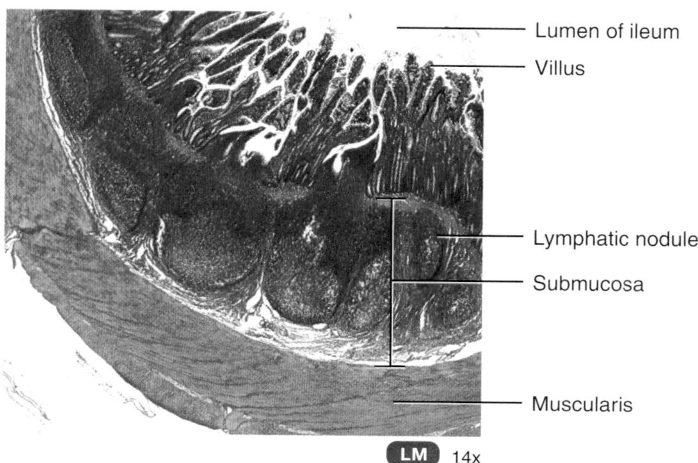

(c) Lymphatic nodules in the ileum

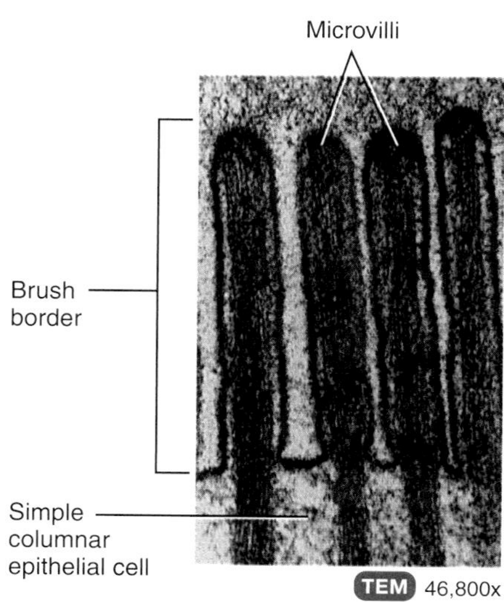

(d) Several microvilli from the duodenum

What is the function of the fluid secreted by duodenal (Brunner's) glands?

After most of a meal has been absorbed, which lessens distension of the wall of the small intestine, segmentation stops and peristalsis begins. The type of peristalsis that occurs in the small intestine, termed a **migrating motility complex (MMC),** begins in the lower portion of the stomach and pushes chyme forward along a short stretch of small intestine before dying out. The MMC slowly migrates down the small intestine, reaching the end of the ileum in 90–120 minutes. Then another MMC begins in the stomach. Altogether, chyme remains in the small intestine for 3–5 hours.

Chemical Digestion in the Small Intestine

In the mouth, salivary amylase converts starch (a polysaccharide) to maltose (a disaccharide), maltotriose (a trisaccharide), and α-dextrins (short-chain, branched fragments of starch with 5 to 10 glucose units). In the stomach, pepsin converts proteins to peptides (small fragments of proteins), and lingual and gastric lipases convert some triglycerides into fatty acids, diglycerides, and monoglycerides. Thus, chyme entering the small intestine contains partially digested carbohydrates, proteins, and lipids. The completion of the digestion of carbohydrates, proteins, and lipids is a collective effort of pancreatic juice, bile, and intestinal juice in the small intestine.

Digestion of Carbohydrates

Even though the action of **salivary amylase** may continue in the stomach for a while, the acidic pH of the stomach destroys salivary amylase and ends its activity. Thus, only a few starches are broken down by the time chyme leaves the stomach. Those starches not already broken down into maltose, maltotriose, and α-dextrins are cleaved by **pancreatic amylase,** an enzyme in pancreatic juice that acts in the small intestine. Although pancreatic amylase acts on both glycogen and starches, it has no effect on another polysaccharide called cellulose, an indigestible plant fiber that is commonly referred to as "roughage" as it moves through the digestive system. After amylase (either salivary or pancreatic) has split starch into smaller fragments, a brush-border enzyme called **α-dextrinase** acts on the resulting α-dextrins, clipping off one glucose unit at a time.

Ingested molecules of sucrose, lactose, and maltose—three disaccharides—are not acted on until they reach the small intestine. Three brush-border enzymes digest the disaccharides into monosaccharides. **Sucrase** breaks sucrose into a molecule of glucose and a molecule of fructose; **lactase** digests lactose into a molecule of glucose and a molecule of galactose; and **maltase** splits maltose and maltotriose into two or three molecules of glucose, respectively. Digestion of carbohydrates ends with the production of monosaccharides, which the digestive system is able to absorb.

Lactose Intolerance

In some people the mucosal cells of the small intestine fail to produce enough lactase, which, as you just learned, is essential for the digestion of lactose. This results in a condition called **lactose intolerance,** in which undigested lactose in chyme causes fluid to be retained in the feces; bacterial fermentation of the undigested lactose results in the production of gases. Symptoms of lactose intolerance include diarrhea, gas, bloating, and abdominal cramps after consumption of milk and other dairy products. The symptoms can be relatively minor or serious enough to require medical attention. The *hydrogen breath test* is often used to aid in diagnosis of lactose intolerance. Very little hydrogen can be detected in the breath of a normal person, but hydrogen is among the gases produced when undigested lactose in the colon is fermented by bacteria. The hydrogen is absorbed from the intestines and carried through the bloodstream to the lungs, where it is exhaled. Persons with lactose intolerance can take dietary supplements to aid in the digestion of lactose. ■

Digestion of Proteins

Recall that protein digestion starts in the stomach, where proteins are fragmented into peptides by the action of **pepsin.** Enzymes in pancreatic juice—**trypsin, chymotrypsin, carboxypeptidase,** and **elastase**—continue to break down proteins into peptides. Although all these enzymes convert whole proteins into peptides, their actions differ somewhat because each splits peptide bonds between different amino acids. Trypsin, chymotrypsin, and elastase all cleave the peptide bond between a specific amino acid and its neighbor; carboxypeptidase splits off the amino acid at the carboxyl end of a peptide. Protein digestion is completed by two **peptidases** in the brush border: aminopeptidase and dipeptidase. **Aminopeptidase** cleaves off the amino acid at the amino end of a peptide. **Dipeptidase** splits dipeptides (two amino acids joined by a peptide bond) into single amino acids.

Digestion of Lipids

The most abundant lipids in the diet are triglycerides, which consist of a molecule of glycerol bonded to three fatty acid molecules (see Figure 2.17 on page 47. Enzymes that split triglycerides and phospholipids are called **lipases.** Recall that there are three types of lipases that can participate in lipid digestion: **lingual lipase, gastric lipase,** and **pancreatic lipase.** Although some lipid digestion occurs in the stomach through the action of lingual and gastric lipases, most occurs in the small intestine through the action of pancreatic lipase. Triglycerides are broken down by pancreatic lipase into fatty acids and monoglycerides. The liberated fatty acids can be either short-chain fatty acids (with fewer than 10–12 carbons) or long-chain fatty acids.

Before a large lipid globule containing triglycerides can be digested in the small intestine, it must first undergo **emulsification**—a process in which the large lipid globule is broken down into several small lipid globules. Recall that bile contains bile salts, the sodium salts and potassium salts of bile acids (mainly chenodeoxycholic acid and cholic acid). Bile salts are **amphipathic,** which means that each bile salt has a hydrophobic (nonpolar) region and a hydrophilic (polar) region. The amphipathic nature of bile salts allows them to emulsify a large lipid globule: The hydrophobic regions of bile salts interact with the large lipid globule, while the hydrophilic regions of bile salts interact with the watery intestinal chyme. Consequently, the large lipid globule is broken apart into several small lipid globules, each about 1 μm in diameter. The small lipid globules formed from emulsification provide a large surface area that allows pancreatic lipase to function more effectively.

Digestion of Nucleic Acids

Pancreatic juice contains two nucleases: **ribonuclease,** which digests RNA, and **deoxyribonuclease,** which digests DNA. The nucleotides that result from the action of the two nucleases are further digested by brush-border enzymes called **nucleosidases** and **phosphatases** into pentoses, phosphates, and nitrogenous bases. These products are absorbed via active transport.

Table 24.4 summarizes the sources, substrates, and products of the digestive enzymes.

TABLE 24.4 Summary of Digestive Enzymes

Enzyme	Source	Substrates	Products
Saliva			
Salivary amylase	Salivary glands.	Starches (polysaccharides).	Maltose (disaccharide), maltotriose (trisaccharide), and α-dextrins.
Lingual lipase	Lingual glands in the tongue.	Triglycerides (fats and oils) and other lipids.	Fatty acids and diglycerides.
Gastric Juice			
Pepsin (activated from pepsinogen by pepsin and hydrochloric acid)	Stomach chief cells.	Proteins.	Peptides.
Gastric lipase	Stomach chief cells.	Triglycerides (fats and oils).	Fatty acids and monoglycerides.
Pancreatic Juice			
Pancreatic amylase	Pancreatic acinar cells.	Starches (polysaccharides).	Maltose (disaccharide), maltotriose (trisaccharide), and α-dextrins.
Trypsin (activated from trypsinogen by enterokinase)	Pancreatic acinar cells.	Proteins.	Peptides.
Chymotrypsin (activated from chymotrypsinogen by trypsin)	Pancreatic acinar cells.	Proteins.	Peptides.
Elastase (activated from proelastase by trypsin)	Pancreatic acinar cells.	Proteins.	Peptides.
Carboxypeptidase (activated from procarboxypeptidase by trypsin)	Pancreatic acinar cells.	Amino acid at carboxyl end of peptides.	Amino acids and peptides.
Pancreatic lipase	Pancreatic acinar cells.	Triglycerides (fats and oils) that have been emulsified by bile salts.	Fatty acids and monoglycerides.
Nucleases			
Ribonuclease	Pancreatic acinar cells.	Ribonucleic acid.	Nucleotides.
Deoxyribonuclease	Pancreatic acinar cells.	Deoxyribonucleic acid.	Nucleotides.
Brush Border			
α-Dextrinase	Small intestine.	α-Dextrins.	Glucose.
Maltase	Small intestine.	Maltose.	Glucose.
Sucrase	Small intestine.	Sucrose.	Glucose and fructose.
Lactase	Small intestine.	Lactose.	Glucose and galactose.
Enterokinase	Small intestine.	Trypsinogen.	Trypsin.
Peptidases			
Aminopeptidase	Small intestine.	Amino acid at amino end of peptides.	Amino acids and peptides.
Dipeptidase	Small intestine.	Dipeptides.	Amino acids.
Nucleosidases and **phosphatases**	Small intestine.	Nucleotides.	Nitrogenous bases, pentoses, and phosphates.

Absorption in the Small Intestine

All the chemical and mechanical phases of digestion from the mouth through the small intestine are directed toward changing food into forms that can pass through the absorptive epithelial cells lining the mucosa and into the underlying blood and lymphatic vessels. These forms are monosaccharides (glucose, fructose, and galactose) from carbohydrates; single amino acids, dipeptides, and tripeptides from proteins; and fatty acids, glycerol, and monoglycerides from triglycerides. Passage of these digested nutrients from the gastrointestinal tract into the blood or lymph is called **absorption.**

Absorption of materials occurs via diffusion, facilitated diffusion, osmosis, and active transport. About 90% of all absorption of nutrients occurs in the small intestine; the other 10% occurs in the stomach and large intestine. Any undigested or unabsorbed material left in the small intestine passes on to the large intestine.

Absorption of Monosaccharides

All carbohydrates are absorbed as monosaccharides. The capacity of the small intestine to absorb monosaccharides is huge—an estimated 120 grams per hour. As a result, all dietary carbohydrates that are digested normally are absorbed, leaving only indigestible cellulose and fibers in the feces. Monosaccharides pass from the lumen through the apical membrane via *facilitated diffusion* or *active transport.* Fructose, a monosaccharide found in fruits, is transported via *facilitated diffusion;* glucose and galactose are transported into absorptive cells of the villi via *secondary active transport* that is coupled to the active transport of Na^+ (Figure 24.20a). The transporter has binding sites for one glucose molecule and two sodium ions; unless all three sites are filled, neither substance is transported. Galactose competes with glucose to ride the same transporter. (Because both Na^+ and glucose or galactose move in the same direction, this is a *symporter.* Monosaccharides then move out of the absorptive cells through their basolateral surfaces via *facilitated diffusion* and enter the capillaries of the villi (see Figure 24.20b).

Absorption of Amino Acids, Dipeptides, and Tripeptides

Most proteins are absorbed as amino acids via *active transport* processes that occur mainly in the duodenum and jejunum. About half of the absorbed amino acids are present in food; the other half come from the body itself as proteins in digestive juices and dead cells that slough off the mucosal surface! Normally, 95–98% of the protein present in the small intestine is digested and absorbed. Different transporters carry different types of amino acids. Some amino acids enter absorptive cells of the villi via Na^+-dependent secondary active transport processes that are similar to the glucose transporter; other amino acids are actively transported by themselves. At least one symporter brings in dipeptides and tripeptides together with H^+; the peptides then are hydrolyzed to single amino acids inside the absorptive cells. Amino acids move out of the absorptive cells via diffusion and enter capillaries of the villus (Figure 24.20a, b). Both monosaccharides and amino acids are transported in the blood to the liver by way of the hepatic portal system. If not removed by hepatocytes, they enter the general circulation.

Absorption of Lipids

All dietary lipids are absorbed via *simple diffusion.* Adults absorb about 95% of the lipids present in the small intestine; due to their lower production of bile, newborn infants absorb only about 85% of lipids. As a result of their emulsification and digestion, triglycerides are mainly broken down into monoglycerides and fatty acids, which can be either short-chain fatty acids or long-chain fatty acids. Although short-chain fatty acids are hydrophobic, they are very small in size. Because of their size, they can dissolve in the watery intestinal chyme, pass through the absorptive cells via simple diffusion, and follow the same route taken by monosaccharides and amino acids into a blood capillary of a villus (Figure 24.20a). Long-chain fatty acids and monoglycerides are large and hydrophobic and have difficulty being suspended in the watery environment of the intestinal chyme. Besides their role in emulsification, bile salts also help to make these long-chain fatty acids and monoglycerides more soluble. The bile salts in intestinal chyme surround the long-chain fatty acids and monoglycerides, forming tiny spheres called **micelles** (mī-SELZ = small morsels), each of which is 2–10 nm in diameter and includes 20–50 bile salt molecules (Figure 24.20a). Micelles are formed due to the amphipathic nature of bile salts: The hydrophobic regions of bile salts interact with the long-chain fatty acids and monoglycerides, and the hydrophilic regions of bile salts interact with the watery intestinal chyme. Once formed, the micelles move from the interior of the small intestinal lumen to the brush border of the absorptive cells. At that point, the long-chain fatty acids and monoglycerides diffuse out of the micelles into the absorptive cells, leaving the micelles behind in the chyme. The micelles continually repeat this ferrying function as they move from the brush border back through the chyme to the interior of the small intestinal lumen to pick up more long-chain fatty acids and monoglycerides. Micelles also solubilize other large hydrophobic molecules such as fat-soluble vitamins (A, D, E, and K) and cholesterol that may be present in intestinal chyme and aid in their absorption. These fat-soluble vitamins and cholesterol molecules are packed in the micelles along with the long-chain fatty acids and monoglycerides.

Once inside the absorptive cells, long-chain fatty acids and monoglycerides are recombined to form triglycerides, which aggregate into globules along with phospholipids and choles-terol and become coated with proteins. These large spherical masses, about 80 nm in diameter, are called **chylomicrons.** Chylomicrons leave the absorptive cell via exocytosis. Because they are so large and bulky, chylomicrons cannot enter blood capillaries—the pores in the walls of blood

Figure 24.20 Absorption of digested nutrients in the small intestine. For simplicity, all digested foods are shown in the lumen of the small intestine, even though some nutrients are digested by brush-border enzymes.

Long-chain fatty acids and monoglycerides are absorbed into lacteals; other products of digestion enter blood capillaries.

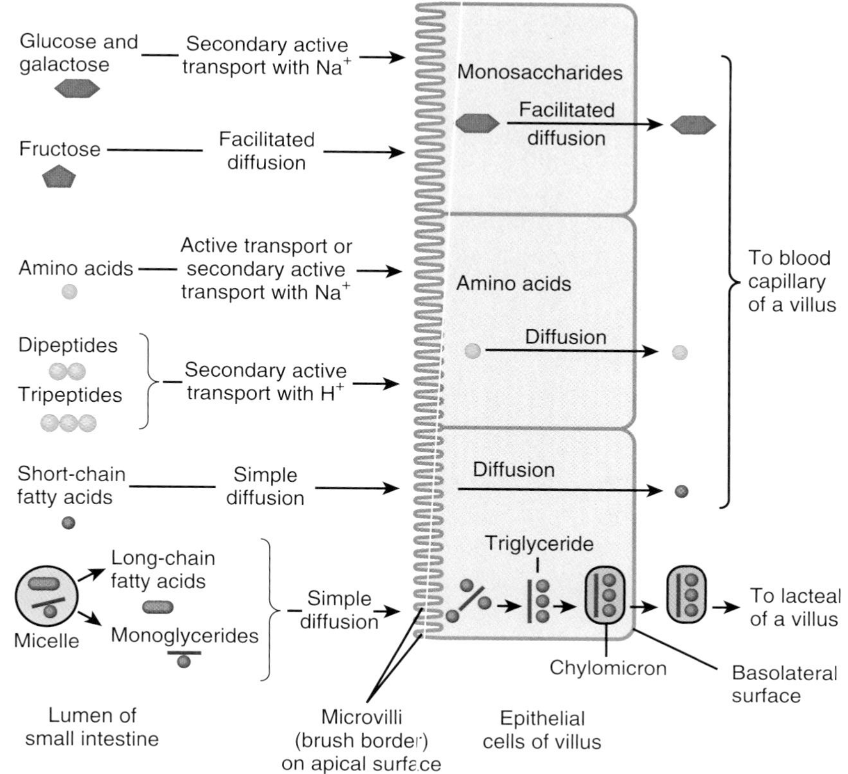

(a) Mechanisms for movement of nutrients through absorptive epithelial cells of the villi

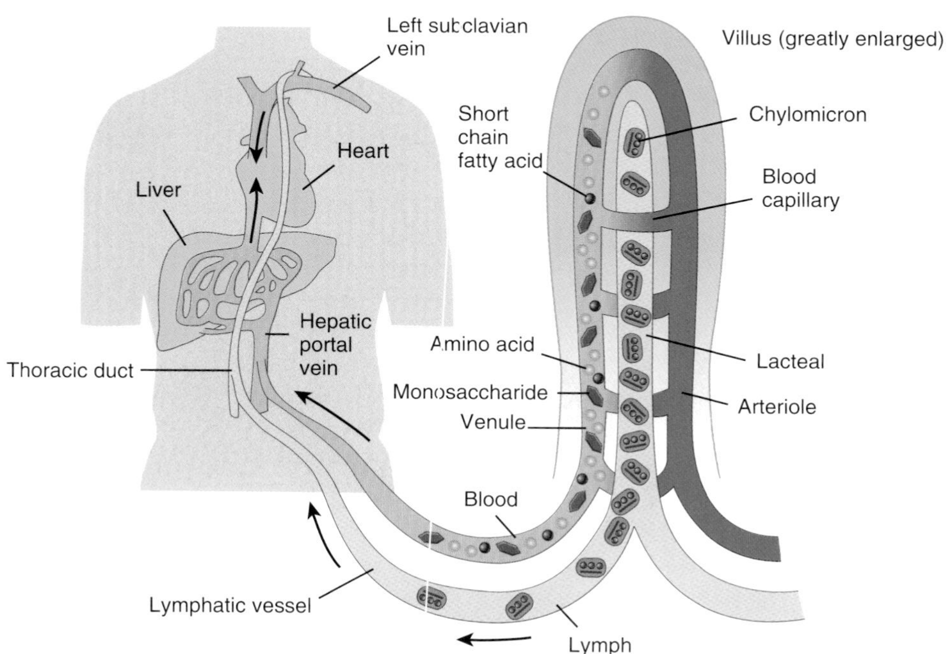

(b) Movement of absorbed nutrients into the blood and lymph

A monoglyceride may be larger than an amino acid. Why can monoglycerides be absorbed by simple diffusion, whereas amino acids cannot?

capillaries are too small. Instead, chylomicrons enter lacteals, which have much larger pores than blood capillaries. From lacteals, chylomicrons are transported by way of lymphatic vessels to the thoracic duct and enter the blood at the left subclavian vein (Figure 24.20b). The hydrophilic protein coat that surrounds each chylomicron keeps the chylomicrons suspended in blood and prevents them from sticking to each other.

Within 10 minutes after absorption, about half of the chylomicrons have already been removed from the blood as they pass through blood capillaries in the liver and adipose tissue. This removal is accomplished by an enzyme attached to the apical surface of capillary endothelial cells, called **lipoprotein lipase,** that breaks down triglycerides in chylomicrons and other lipoproteins into fatty acids and glycerol. The fatty acids diffuse into hepatocytes and adipose cells and combine with glycerol during resynthesis of triglycerides. Two or three hours after a meal, few chylomicrons remain in the blood.

After participating in the emulsification and absorption of lipids, 90–95% of the bile salts are reabsorbed by active transport in the final segment of the small intestine (ileum) and returned by the blood to the liver through the hepatic portal system for recycling. This cycle of bile salt secretion by hepatocytes into bile, reabsorption by the ileum, and resecretion into bile is called the **enterohepatic circulation.** Insufficient bile salts, due either to obstruction of the bile ducts or removal of the gallbladder, can result in the loss of up to 40% of dietary lipids in feces due to diminished lipid absorption. When lipids are not absorbed properly, the fat-soluble vitamins are not adequately absorbed.

Absorption of Electrolytes

Many of the electrolytes absorbed by the small intestine come from gastrointestinal secretions, and some are part of ingested foods and liquids. Recall that electrolytes are compounds that separate into ions in water and conduct electricity. Sodium ions are actively transported out of absorptive cells by basolateral sodium-potassium pumps (Na^+/K^+ ATPase) after they have moved into absorptive cells via diffusion and secondary active transport. Thus, most of the sodium ions (Na^+) in gastrointestinal secretions are reclaimed and not lost in the feces. Negatively charged bicarbonate, chloride, iodide, and nitrate ions can passively follow Na^+ or be actively transported. Calcium ions also are absorbed actively in a process stimulated by calcitriol. Other electrolytes such as iron, potassium, magnesium, and phosphate ions also are absorbed via active transport mechanisms.

Absorption of Vitamins

As you have just learned, the fat-soluble vitamins A, D, E, and K are included with ingested dietary lipids in micelles and are absorbed via simple diffusion. Most water-soluble vitamins, such as most B vitamins and vitamin C, also are absorbed via simple diffusion. Vitamin B_{12}, however, combines with intrinsic factor produced by the stomach, and the combination is absorbed in the ileum via an active transport mechanism.

Absorption of Water

The total volume of fluid that enters the small intestine each day—about 9.3 liters (9.8 qt)—comes from ingestion of liquids (about 2.3 liters) and from various gastrointestinal secretions (about 7.0 liters). Figure 24.21 depicts the amounts of fluid ingested, secreted, absorbed, and excreted by the GI tract. The small intestine absorbs about 8.3 liters of the fluid; the remainder passes into the large intestine, where most of the rest of it—about 0.9 liter—is also absorbed. Only 0.1 liter (100 mL) of water is excreted in the feces each day. Most is excreted via the urinary system.

All water absorption in the GI tract occurs via *osmosis* from the lumen of the intestines through absorptive cells and into

Figure 24.21 Daily volumes of fluid ingested, secreted, absorbed, and excreted from the GI tract.

All water absorption in the GI tract occurs via osmosis.

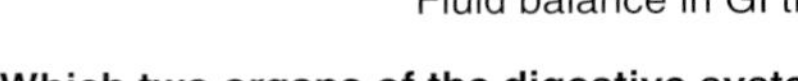

Fluid balance in GI tract

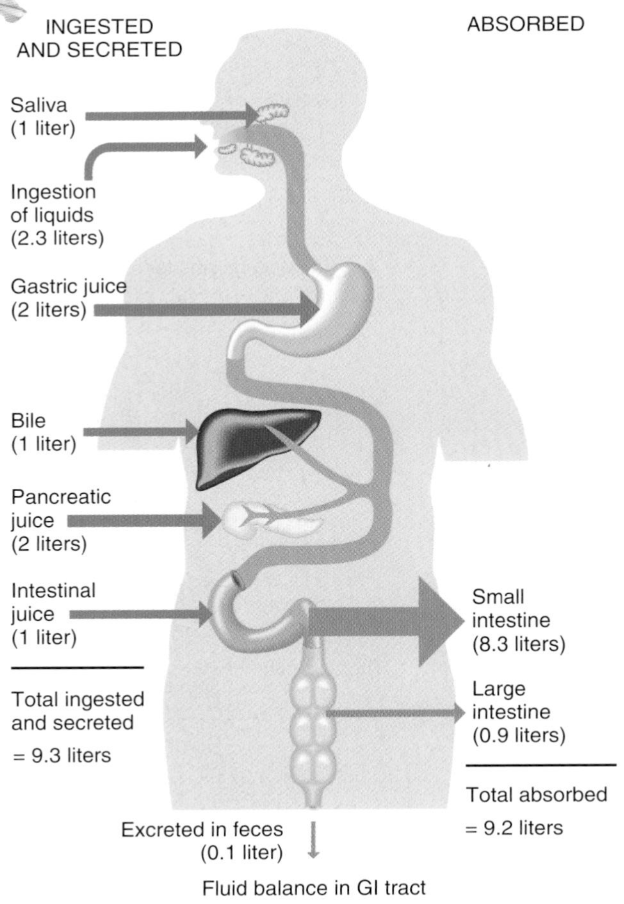

Which two organs of the digestive system secrete the most fluid?

blood capillaries. Because water can move across the intestinal mucosa in both directions, the absorption of water from the small intestine depends on the absorption of electrolytes and nutrients to maintain an osmotic balance with the blood. The absorbed electrolytes, monosaccharides, and amino acids establish a concentration gradient for water that promotes water absorption via osmosis.

Table 24.5 summarizes the digestive activities of the pancreas, liver, gallbladder, and small intestine.

Absorption of Alcohol

The intoxicating and incapacitating effects of alcohol depend on the blood alcohol level. Because it is lipid-soluble, alcohol begins to be absorbed in the stomach. However, the surface area available for absorption is much greater in the small intestine than in the stomach, so when alcohol passes into the duodenum, it is absorbed more rapidly. Thus, the longer the alcohol remains in the stomach, the more slowly blood alcohol level rises. Because fatty acids in chyme slow gastric emptying, blood alcohol level will rise more slowly when fat-rich foods, such as pizza, hamburgers, or nachos, are consumed with alcoholic beverages. Also, the enzyme alcohol dehydrogenase, which is present in gastric mucosa cells, breaks down some of the alcohol to acetaldehyde, which is not intoxicating. When the rate of gastric emptying is slower, proportionally more alcohol will be absorbed and converted to acetaldehyde in the stomach, and thus less alcohol will reach the bloodstream. Given identical consumption of alcohol, females often develop higher blood alcohol levels (and therefore experience greater intoxication) than males of comparable size because the activity of gastric alcohol dehydrogenase is up to 60% lower in females than in males. Asian males may also have lower levels of this gastric enzyme. ■

TABLE 24.5 Summary of Digestive Activities in the Pancreas, Liver, Gallbladder, and Small Intestine

Structure	Activity
Pancreas	Delivers pancreatic juice into the duodenum via the pancreatic duct (see Table 24.4 for pancreatic enzymes and their functions).
Liver	Produces bile (bile salts) necessary for emulsification and absorption of lipids.
Gallbladder	Stores, concentrates, and delivers bile into the duodenum via the common bile duct.
Small Intestine	Major site of digestion and absorption of nutrients and water in the gastrointestinal tract.
Mucosa/submucosa	
Intestinal glands	Secrete intestinal juice.
Duodenal (Brunner's) glands	Secrete alkaline fluid to buffer stomach acids, and mucus for protection and lubrication.
Microvilli	Microscopic, membrane-covered projections of absorptive epithelial cells that contain brush-border enzymes (listed in Table 24.4) and that increase the surface area for digestion and absorption.
Villi	Fingerlike projections of mucosa that are the sites of absorption of digested food and increase the surface area for digestion and absorption.
Circular folds	Folds of mucosa and submucosa that increase the surface area for digestion and absorption.
Muscularis	
Segmentation	Consists of alternating contractions of circular smooth muscle fibers that produce segmentation and resegmentation of sections of the small intestine; mixes chyme with digestive juices and brings food into contact with the mucosa for absorption.
Migrating motility complex (MMC)	A type of peristalsis consisting of waves of contraction and relaxation of circular and longitudinal smooth muscle fibers passing the length of the small intestine; moves chyme toward ileocecal sphincter.

▶ **CHECKPOINT**

32. List the regions of the small intestine and describe their functions.
33. In what ways are the mucosa and submucosa of the small intestine adapted for digestion and absorption?
34. Describe the types of movement that occur in the small intestine.
35. Explain the functions of pancreatic amylase, aminopeptidase, gastric lipase, and deoxyribonuclease.
36. What is the difference between digestion and absorption? How are the end products of carbohydrate, protein, and lipid digestion absorbed?
37. By what routes do absorbed nutrients reach the liver?
38. Describe the absorption of electrolytes, vitamins, and water by the small intestine.

LARGE INTESTINE

▶ **OBJECTIVE**

Describe the anatomy, histology, and functions of the large intestine.

The large intestine is the terminal portion of the GI tract. The overall functions of the large intestine are the completion of absorption, the production of certain vitamins, the formation of feces, and the expulsion of feces from the body.

Anatomy of the Large Intestine

The **large intestine,** which is about 1.5 m (5 ft) long and 6.5 cm (2.5 in.) in diameter, extends from the ileum to the anus. It is attached to the posterior abdominal wall by its **mesocolon,** which is a double layer of peritoneum. Structurally, the four major regions of the large intestine are the cecum, colon, rectum, and anal canal (Figure 24.22a).

The opening from the ileum into the large intestine is guarded by a fold of mucous membrane called the **ileocecal sphincter (valve),** which allows materials from the small intestine to pass into the large intestine. Hanging inferior to the ileocecal valve is the **cecum,** a small pouch about 6 cm (2.4 in.) long. Attached to the cecum is a twisted, coiled tube, measuring about 8 cm (3 in.) in length, called the **appendix** or **vermiform appendix** (*vermiform* = worm-shaped; *appendix* = appendage). The mesentery of the appendix, called the **mesoappendix,** attaches the appendix to the inferior part of the mesentery of the ileum.

The open end of the cecum merges with a long tube called the **colon** (= food passage), which is divided into ascending, transverse, descending, and sigmoid portions. Both the ascending and descending colon are retroperitoneal; the transverse and sigmoid colon are not. True to its name, the **ascending colon** ascends on the right side of the abdomen, reaches the inferior surface of the liver, and turns abruptly to the left to form the **right colic (hepatic) flexure.** The colon continues across the abdomen to the left side as the **transverse colon.** It curves beneath the inferior end of the spleen on the left side as the **left colic (splenic) flexure** and passes inferiorly to the level of the iliac crest as the **descending colon.** The **sigmoid colon** (*sigm-* = S-shaped) begins near the left iliac crest, projects medially to the midline, and terminates as the rectum at about the level of the third sacral vertebra.

The **rectum,** the last 20 cm (8 in.) of the GI tract, lies anterior to the sacrum and coccyx. The terminal 2–3 cm (1 in.) of the rectum is called the **anal canal** (Figure 24.22b). The mucous membrane of the anal canal is arranged in longitudinal folds called **anal columns** that contain a network of arteries and veins. The opening of the anal canal to the

Functions of the Large Intestine

1. **Haustral churning, peristalsis, and mass peristalsis drive the contents of the colon into the rectum.**
2. **Bacteria in the large intestine convert proteins to amino acids, break down amino acids, and produce some B vitamins and vitamin K.**
3. **Absorbing some water, ions, and vitamins.**
4. **Forming feces.**
5. **Defecating (emptying the rectum).**

Figure 24.22 Anatomy of the large intestine. (See Tortora, *A Photographic Atlas of the Human Body, Second Edition,* Figure 12.13.)

The regions of the large intestine are the cecum, colon, rectum, and anal canal.

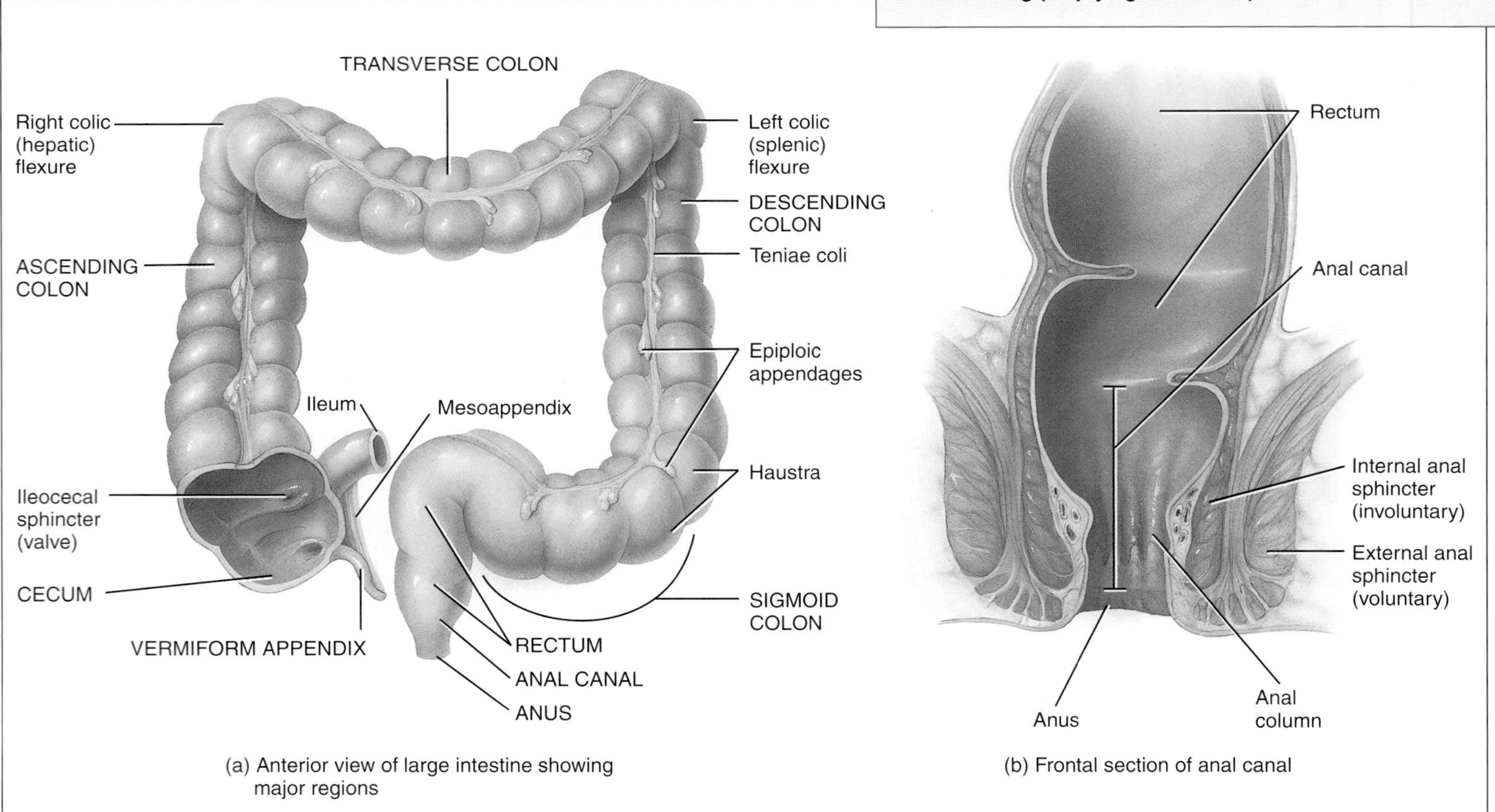

(a) Anterior view of large intestine showing major regions

(b) Frontal section of anal canal

Which portions of the colon are retroperitoneal?

exterior, called the **anus,** is guarded by an **internal anal sphincter** of smooth muscle (involuntary) and an **external anal sphincter** of skeletal muscle (voluntary). Normally these sphincters keep the anus closed except during the elimination of feces.

Appendicitis

Inflammation of the appendix, termed **appendicitis,** is preceded by obstruction of the lumen of the appendix by chyme, inflammation, a foreign body, a carcinoma of the cecum, stenosis, or kinking of the organ. It is characterized by high fever, elevated white blood cell count, and a neutrophil count higher than 75%. The infection that follows may result in edema and ischemia and may progress to gangrene and perforation within 24 hours. Typically, appendicitis begins with referred pain in the umbilical region of the abdomen, followed by anorexia (loss of appetite for food), nausea, and vomiting. After several hours the pain localizes in the right lower quadrant (RLQ) and is continuous, dull or severe, and intensified by coughing, sneezing, or body movements. Early appendectomy (removal of the appendix) is recommended because it is safer to operate than to risk rupture, peritonitis, and gangrene. Although it required major abdominal surgery in the past, today appendectomies are usually performed laparoscopically. ■

Histology of the Large Intestine

The wall of the large intestine contains the typical four layers found in the rest of the GI tract: mucosa, submucosa, muscularis, and serosa. The **mucosa** consists of simple columnar epithelium, lamina propria (areolar connective tissue), and muscularis mucosae (smooth muscle) (Figure 24.23a). The epithelium contains mostly absorptive and goblet cells (Figure 24.23b and c). The absorptive cells function primarily in water absorption; the goblet cells secrete mucus that lubricates the passage of the colonic contents. Both absorptive and goblet cells are located in long, straight, tubular intestinal glands (crypts of Lieberkühn) that extend the full thickness of the mucosa. Solitary lymphatic nodules are also found in the lamina propria of the mucosa and may extend through the muscularis mucosae into the submucosa. Compared to the small intestine, the mucosa of the large intestine does not have as many structural adaptations that increase surface area. There are no circular folds or villi; however, micro-villi of the absorptive cells are present. Consequently, much more absorption occurs in the small intestine than in the large intestine.

The **submucosa** of the large intestine consists of areolar connective tissue. The **muscularis** consists of an external layer of longitudinal smooth muscle and an internal layer of circular

Figure 24.23 Histology of the large intestine.

Intestinal glands formed by simple columnar epithelial cells and goblet cells extend the full thickness of the mucosa.

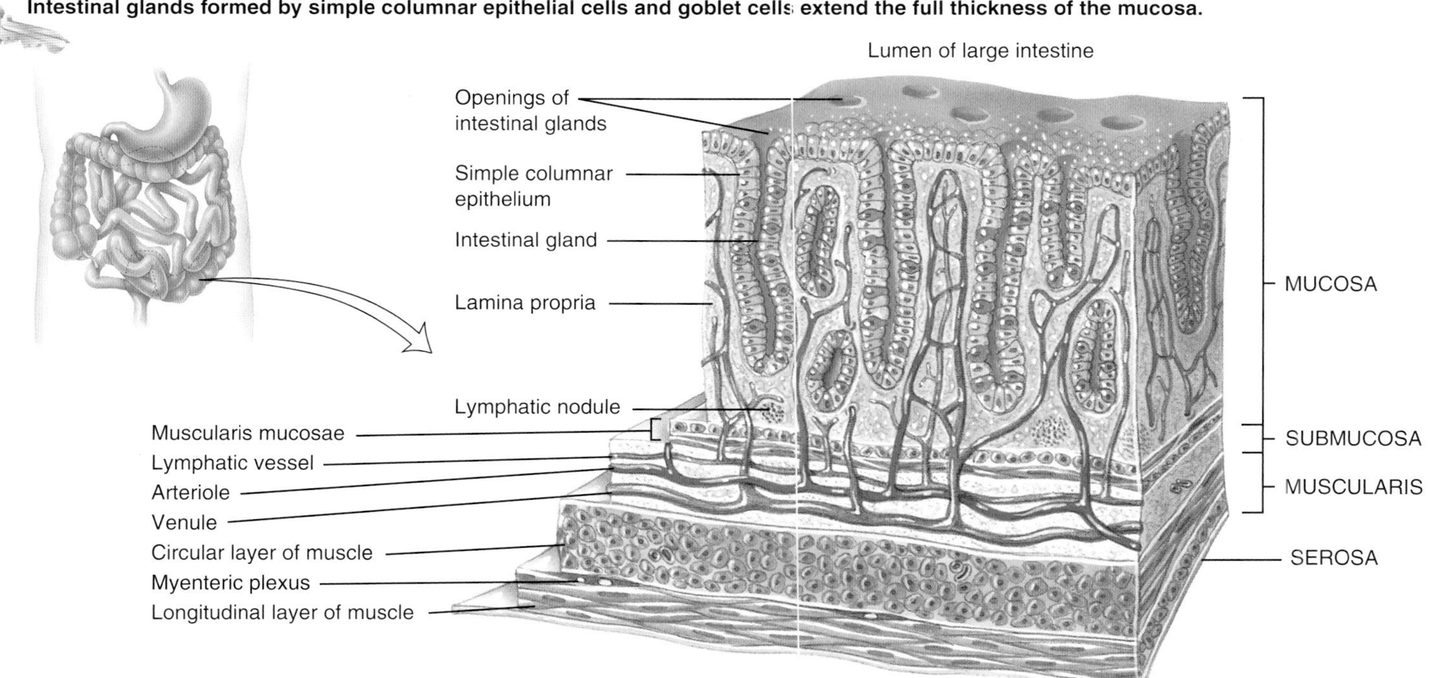

(a) Three-dimensional view of layers of the large intestine

continues

Figure 24.23 **(continued)**

Openings of intestinal glands
Lamina propria
Microvilli
Absorptive cell (absorbs water)
Intestinal gland
Goblet cell (secretes mucus)
Lymphatic nodule
Muscularis mucosae
Submucosa

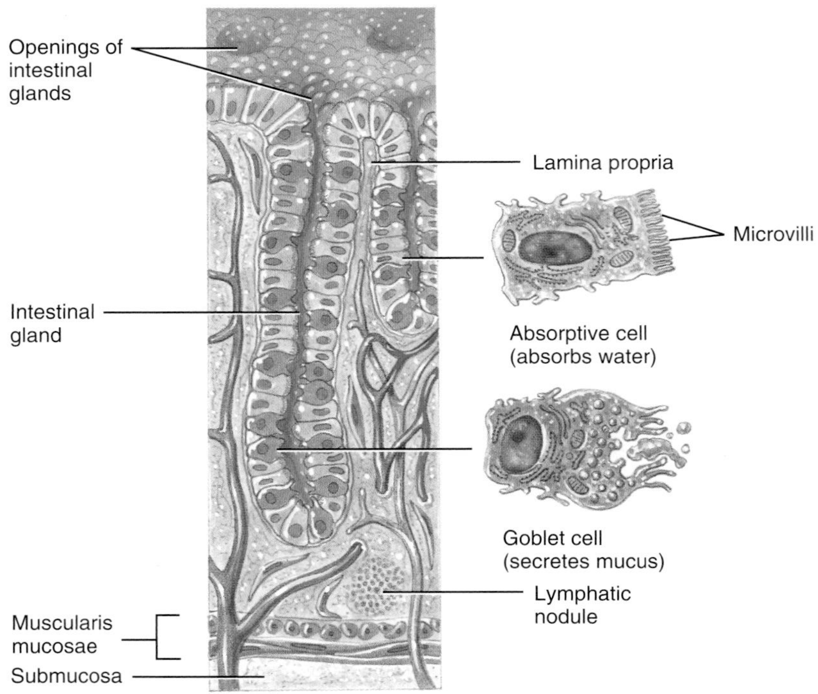

(b) Sectional view of intestinal glands and cell types

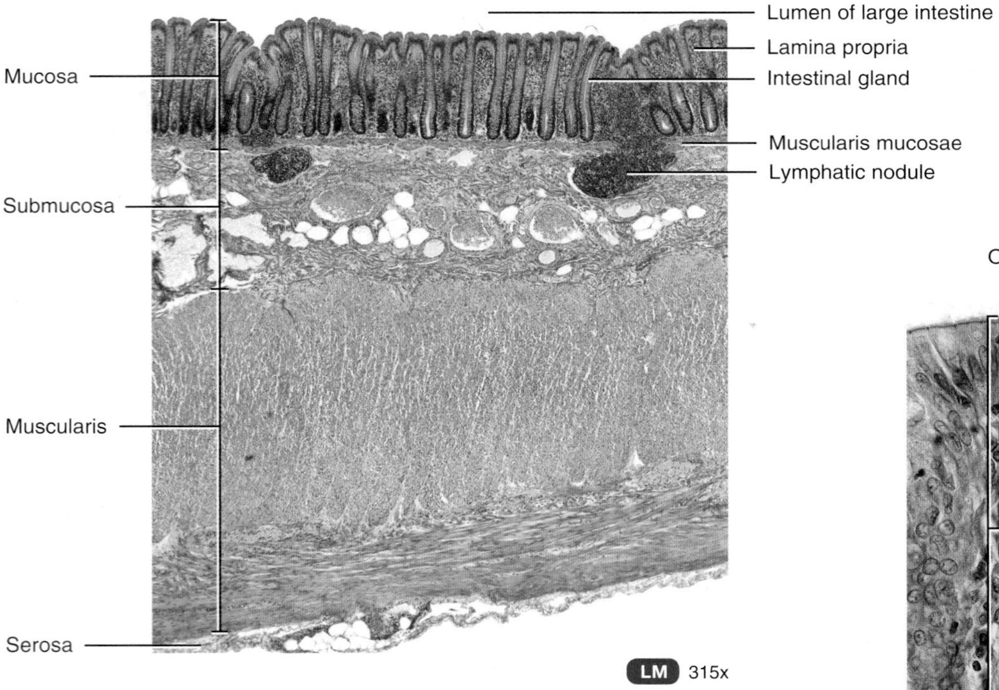

Lumen of large intestine
Lamina propria
Mucosa
Intestinal gland
Muscularis mucosae
Lymphatic nodule
Submucosa
Muscularis
Serosa
LM 315x

(c) Portion of the wall of the large intestine

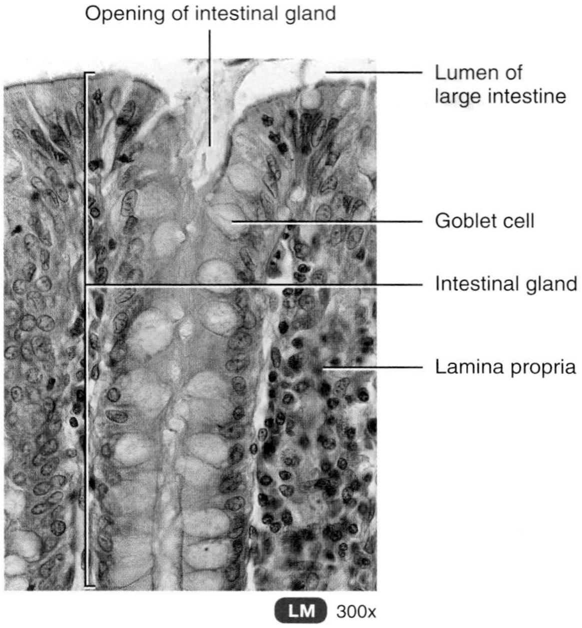

Opening of intestinal gland
Lumen of large intestine
Goblet cell
Intestinal gland
Lamina propria
LM 300x

(d) Details of mucosa of large intestine

What is the function of the goblet cells in the large intestine?

smooth muscle. Unlike other parts of the GI tract, portions of the longitudinal muscles are thickened, forming three conspicuous longitudinal bands called the **teniae coli** (TĒ-nē-ē KŌ-lī; *teniae* = flat bands), that run most of the length of the large intestine (see Figure 24.22a). The teniae coli are separated by portions of the wall with less or no longitudinal muscle. Tonic contractions of the bands gather the colon into a series of pouches called **haustra** (HAWS-tra = shaped like pouches; singular is **haustrum**), which give the colon a puckered appearance. A single layer of circular smooth muscle lies between teniae coli. The **serosa** of the large intestine is part of the visceral peritoneum. Small pouches of visceral peri-toneum filled with fat are attached to teniae coli and are called **epiploic appendages.**

Polyps in the Colon

Polyps in the colon are generally slow-developing benign growths that arise from the mucosa of the large intestine. Often, they do not cause symptoms. If symptoms do occur, they include diarrhea, blood in the feces, and mucus discharged from the anus. The polyps are removed by colonoscopy or surgery because some of them may become cancerous. ■

Mechanical Digestion in the Large Intestine

The passage of chyme from the ileum into the cecum is regulated by the action of the ileocecal sphincter. Normally, the valve remains partially closed so that the passage of chyme into the cecum usually occurs slowly. Immediately after a meal, a **gastroileal reflex** intensifies peristalsis in the ileum and forces any chyme into the cecum. The hormone gastrin also relaxes the sphincter. Whenever the cecum is distended, the degree of contraction of the ileocecal sphincter intensifies.

Movements of the colon begin when substances pass the ileocecal sphincter. Because chyme moves through the small intestine at a fairly constant rate, the time required for a meal to pass into the colon is determined by gastric emptying time. As food passes through the ileocecal sphincter, it fills the cecum and accumulates in the ascending colon.

One movement characteristic of the large intestine is **haustral churning.** In this process, the haustra remain relaxed and become distended while they fill up. When the distension reaches a certain point, the walls contract and squeeze the contents into the next haustrum. **Peristalsis** also occurs, although at a slower rate (3–12 contractions per minute) than in more proximal portions of the tract. A final type of movement is **mass peristalsis,** a strong peristaltic wave that begins at about the middle of the transverse colon and quickly drives the contents of the colon into the rectum. Because food in the stomach initiates this **gastrocolic reflex** in the colon, mass peristalsis usually takes place three or four times a day, during or immediately after a meal.

Chemical Digestion in the Large Intestine

The final stage of digestion occurs in the colon through the activity of bacteria that inhabit the lumen. Mucus is secreted by the glands of the large intestine, but no enzymes are secreted. Chyme is prepared for elimination by the action of bacteria, which ferment any remaining carbohydrates and release hydrogen, carbon dioxide, and methane gases. These gases contribute to flatus (gas) in the colon, termed *flatulence* when it is excessive. Bacteria also convert any remaining proteins to amino acids and break down the amino acids into simpler substances: indole, skatole, hydrogen sulfide, and fatty acids. Some of the indole and skatole is eliminated in the feces and contributes to their odor; the rest is absorbed and transported to the liver, where these compounds are converted to less toxic compounds and excreted in the urine. Bacteria also decompose bilirubin to simpler pigments, including stercobilin, which give feces their brown color. Bacterial products that are absorbed in the colon include several vitamins needed for normal metabolism, among them some B vitamins and vitamin K.

Absorption and Feces Formation in the Large Intestine

By the time chyme has remained in the large intestine 3–10 hours, it has become solid or semisolid because of water absorption and is now called **feces.** Chemically, feces consist of water, inorganic salts, sloughed-off epithelial cells from the mucosa of the gastrointestinal tract, bacteria, products of bacterial decomposition, unabsorbed digested materials, and indigestible parts of food.

Although 90% of all water absorption occurs in the small intestine, the large intestine absorbs enough to make it an important organ in maintaining the body's water balance. Of the 0.5–1.0 liter of water that enters the large intestine, all but about 100–200 mL is normally absorbed via osmosis. The large intestine also absorbs ions, including sodium and chloride, and some vitamins.

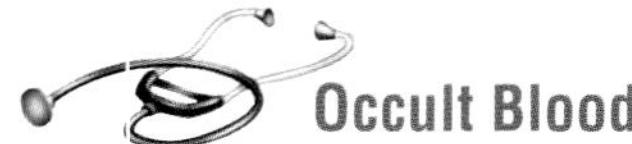

Occult Blood

The term **occult blood** refers to blood that is hidden; it is not detectable by the human eye. The main diagnostic value of occult blood testing is to screen for colorectal cancer. Two substances often examined for occult blood are feces and urine. Several types of products are available for at-home testing for hidden blood in feces. The tests are based on color changes when reagents are added to feces. The presence of occult blood in urine may be detected at home by using dip-and-read reagent strips. ■

The Defecation Reflex

Mass peristaltic movements push fecal material from the sigmoid colon into the rectum. The resulting distension of the rectal wall stimulates stretch receptors, which initiates a **defecation reflex** that empties the rectum. The defecation reflex occurs as follows: In response to distension of the rectal wall, the receptors send sensory nerve impulses to the sacral spinal cord. Motor impulses from the cord travel along parasympathetic nerves back to the descending colon, sigmoid colon, rectum, and anus. The resulting contraction of the longitudinal rectal muscles shortens the rectum, thereby increasing the pressure within it. This pressure, along with voluntary contractions of the diaphragm and abdominal muscles, plus parasympathetic stimulation, opens the internal anal sphincter.

The external anal sphincter is voluntarily controlled. If it is voluntarily relaxed, defecation occurs and the feces are expelled through the anus; if it is voluntarily constricted, defecation can be postponed. Voluntary contractions of the diaphragm and abdominal muscles aid defecation by increasing the pressure within the abdomen, which pushes the walls of the sigmoid colon and rectum inward. If defecation does not occur, the feces back up into the sigmoid colon until the next wave of mass peristalsis stimulates the stretch receptors, again creating the urge to defecate. In infants, the defecation reflex causes automatic emptying of the rectum because voluntary control of the external anal sphincter has not yet developed.

The amount of bowel movements that a person has over a given period of time depends on various factors such as diet, health, and stress. The normal range of bowel activity varies from two or three bowel movements per day to three or four bowel movements per week.

Diarrhea (dī-a-RĒ-a; *dia-* = through; *rrhea* = flow) is an increase in the frequency, volume, and fluid content of the feces caused by increased motility of and decreased absorption by the intestines. When chyme passes too quickly through the small intestine and feces pass too quickly through the large intestine, there is not enough time for absorption. Frequent diarrhea can result in dehydration and electrolyte imbalances. Excessive motility may be caused by lactose intolerance, stress, and microbes that irritate the gastrointestinal mucosa.

Constipation (kon-sti-PĀ-shun; *con-* = together; *stip-* = to press) refers to infrequent or difficult defecation caused by decreased motility of the intestines. Because the feces remain in the colon for prolonged periods, excessive water absorption occurs, and the feces become dry and hard. Constipation may be caused by poor habits (delaying defecation), spasms of the colon, insufficient fiber in the diet, inadequate fluid intake, lack of exercise, emotional stress, and certain drugs. A common treatment is a mild laxative, such as milk of magnesia, which induces defecation. However, many physicians maintain that laxatives are habit-forming, and that adding fiber to the diet, increasing the amount of exercise, and increasing fluid intake are safer ways of controlling this common problem.

TABLE 24.6 Summary of Digestive Activities in the Large Intestine

Structure	Activity	Function(s)
Lumen	Bacterial activity.	Breaks down undigested carbohydrates, proteins, and amino acids into products that can be expelled in feces or absorbed and detoxified by liver; synthesizes certain B vitamins and vitamin K.
Mucosa	Secretes mucus.	Lubricates colon and protects mucosa.
	Absorption.	Water absorption solidifies feces and contributes to the body's water balance; solutes absorbed include ions and some vitamins.
Muscularis	Haustral churning.	Moves contents from haustrum to haustrum by muscular contractions.
	Peristalsis.	Moves contents along length of colon by contractions of circular and longitudinal muscles.
	Mass peristalsis.	Forces contents into sigmoid colon and rectum.
	Defecation reflex.	Eliminates feces by contractions in sigmoid colon and rectum.

Table 24.6 summarizes the digestive activities in the large intestine, and Table 24.7 summarizes the functions of all digestive system organs.

Dietary fiber consists of indigestible plant carbohydrates—such as cellulose, lignin, and pectin—found in fruits, vegetables, grains, and beans. **Insoluble fiber,** which does not dissolve in water, includes the woody or structural parts of plants such as the skins of fruits and vegetables and the bran coating around wheat and corn kernels. Insoluble fiber passes through the GI tract largely unchanged but speeds up the passage of material through the tract. **Soluble fiber,** which does dissolve in water, forms a gel that slows the passage of material through the tract. It is found in abun-dance in beans, oats, barley, broccoli, prunes, apples, and citrus fruits.

People who choose a fiber-rich diet may reduce their risk of developing obesity, diabetes, atherosclerosis, gallstones, hemorrhoids, diverticulitis, appendicitis, and colorectal cancer. Soluble fiber also may help lower blood cholesterol. The liver normally converts cholesterol to bile salts, which are released into the small intestine to help fat digestion. Having accomplished their task, the bile salts are reabsorbed by the small intestine and recycled back to the liver. Since soluble fiber

TABLE 24.7 Summary of Organs of the Digestive System and Their Functions

Organ	Functions
Mouth	See other listings in this table for the functions of the tongue, salivary glands, and teeth, all of which are in the mouth. Additionally, the lips and cheeks keep food between the teeth during mastication, and buccal glands lining the mouth produce saliva.
Tongue	Maneuvers food for mastication, shapes food into a bolus, maneuvers food for deglutition, detects taste and touch sensations, and initiates digestion of triglycerides.
Salivary glands	Produce saliva, which softens, moistens, and dissolves foods; cleanses mouth and teeth; and initiates the digestion of starch.
Teeth	Cut, tear, and pulverize food to reduce solids to smaller particles for swallowing.
Pharynx	Receives a bolus from the oral cavity and passes it into the esophagus.
Esophagus	Receives a bolus from the pharynx and moves it into the stomach. This requires relaxation of the upper esophageal sphincter and secretion of mucus.
Stomach	Mixing waves macerate food, mix it with secretions of gastric glands (gastric juice), and reduce food to chyme. Gastric juice activates pepsin and kills many microbes in food. intrinsic factor aids absorption of vitamin B_{12}. The stomach serves as a reservoir for food before releasing it into the small intestine.
Pancreas	Pancreatic juice buffers acidic gastric juice in chyme (creating the proper pH for digestion in the small intestine), stops the action of pepsin from the stomach, and contains enzymes that digest carbohydrates, proteins, triglycerides, and nucleic acids.
Liver	Produces bile, which is needed for the emulsification and absorption of lipids in the small intestine.
Gallbladder	Stores and concentrates bile and releases it into the small intestine.
Small intestine	Segmentations mix chyme with digestive juices; migrating motility complexes propel chyme toward the ileocecal sphincter; digestive secretions from the small intestine, pancreas, and liver complete the digestion of carbohydrates, proteins, lipids, and nucleic acids; circular folds, villi, and microvilli increase surface area for absorption; site where about 90% of nutrients and water are absorbed.
Large intestine	Haustral churning, peristalsis, and mass peristalsis drive the contents of the colon into the rectum; bacteria produce some B vitamins and vitamin K; absorption of some water, ions, and vitamins; defecation.

binds to bile salts to prevent their reabsorption, the liver makes more bile salts to replace those lost in feces. Thus, the liver uses more cholesterol to make more bile salts and blood cholesterol level is lowered. ■

▶ CHECKPOINT

39. What are the major regions of the large intestine?
40. How does the muscularis of the large intestine differ from that of the rest of the gastrointestinal tract? What are haustra?
41. Describe the mechanical movements that occur in the large intestine.
42. What is defecation and how does it occur?
43. What activities occur in the large intestine to change its contents into feces?

PHASES OF DIGESTION

▶ OBJECTIVES

Describe the three phases of digestion.

Describe the major hormones that regulate digestive activities.

Digestive activities occur in three overlapping phases: the cephalic phase, the gastric phase, and the intestinal phase.

Cephalic Phase

During the **cephalic phase** of digestion, the smell, sight, thought, or initial taste of food activates neural centers in the cerebral cortex, hypothalamus, and brain stem. The brain stem then activates the facial (VII), glossopharyngeal (IX), and vagus (X) nerves. The facial and glossopharyngeal nerves stimulate the salivary glands to secrete saliva, while the vagus nerves stimulate the gastric glands to secrete gastric juice. The purpose of the cephalic phase of digestion is to prepare the mouth and stomach for food that is about to be eaten.

Gastric Phase

Once food reaches the stomach, the **gastric phase** of digestion begins. Neural and hormonal mechanisms regulate the gastric phase of digestion to promote gastric secretion and gastric motility.

- ***Neural Regulation*** Food of any kind distends the stomach and stimulates stretch receptors in its walls. Chemoreceptors in the stomach monitor the pH of the stomach chyme. When the stomach walls are distended or pH increases because proteins have entered the stomach and buffered some of the stomach acid, the stretch receptors and chemoreceptors are activated, and a neural negative feedback loop is set in motion (Figure 24.24). From the stretch receptors and chemoreceptors, nerve impulses propagate to the submu-

Figure 24.24 Neural negative feedback regulation of the pH of gastric juice and gastric motility during the gastric phase of digestion.

Food entering the stomach stimulates secretion of gastric juice and causes vigorous waves of peristalsis.

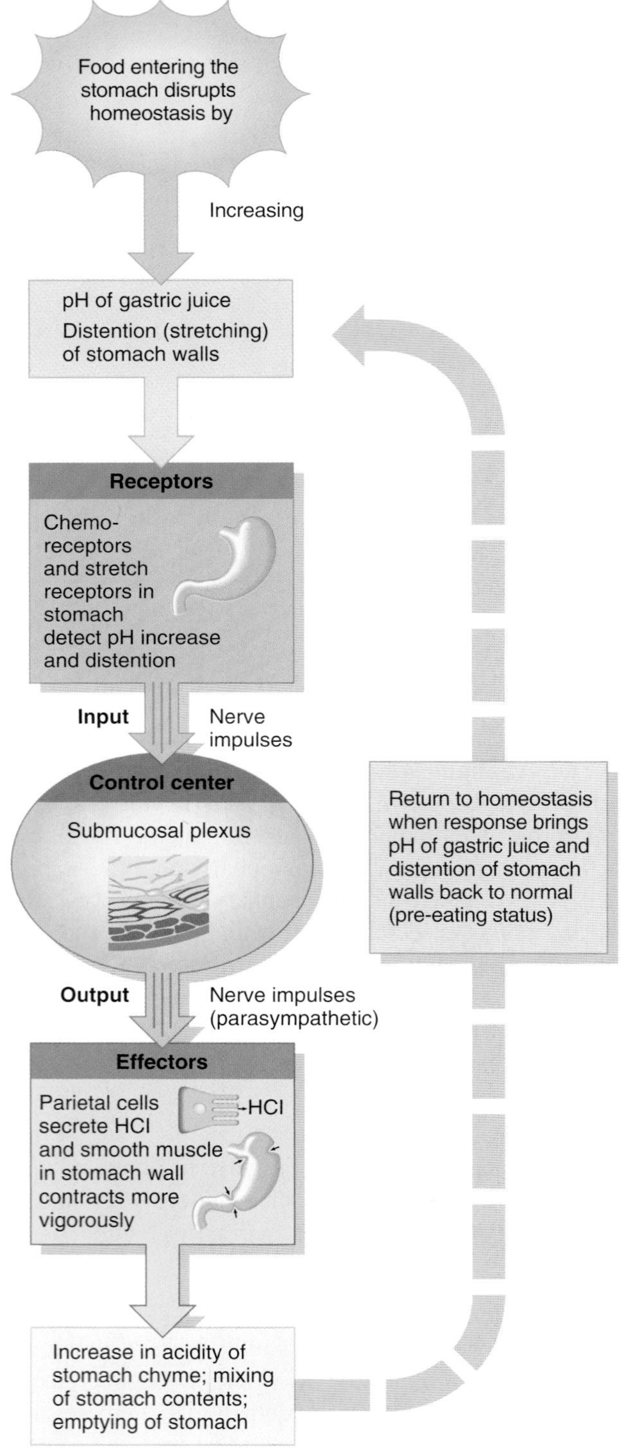

Why does food initially cause the pH of the gastric juice to rise?

cosal plexus, where they activate parasympathetic and enteric neurons. The resulting nerve impulses cause waves of peristalsis and continue to stimulate the flow of gastric juice from gastric glands. The peristaltic waves mix the food with gastric juice; when the waves become strong enough, a small quantity of chyme undergoes gastric emptying into the duodenum. The pH of the stomach chyme decreases (becomes more acidic) and the distension of the stomach walls lessens because chyme has passed into the small intestine, suppressing secretion of gastric juice.

- ***Hormonal Regulation*** Gastric secretion during the gastric phase is also regulated by the hormone **gastrin.** Gastrin is released from the **G cells** of the gastric glands in response to several stimuli: distension of the stomach by chyme, partially digested proteins in chyme, the high pH of chyme due to the presence of food in the stomach, caffeine in gastric chyme, and acetycholine released from parasympathetic neurons. Once it is released, gastrin enters the bloodstream, makes a round-trip through the body, and finally reaches its target organs in the digestive system. Gastrin stimulates gastric glands to secrete large amounts of gastric juice. It also strengthens the contraction of the lower esophageal sphincter to prevent reflux of acid chyme into the esophagus, increases motility of the stomach, and relaxes the pyloric sphincter, which promotes gastric emptying. Gastrin secretion is inhibited when the pH of gastric juice drops below 2.0 and is stimulated when the pH rises. This negative feedback mechanism helps provide an optimal low pH for the functioning of pepsin, the killing of microbes, and the denaturing of proteins in the stomach.

Intestinal Phase

The **intestinal phase** of digestion begins once food enters into the small intestine. In contrast to reflexes initiated during the cephalic and gastric phases, which stimulate stomach secretory activity and motility, those occurring during the intestinal phase have inhibitory effects that slow the exit of chyme from the stomach. This prevents the duodenum from being overloaded with more chyme than it can handle. In addition, responses occurring during the intestinal phase promote the continued digestion of foods that have reached the small intestine. These activities of the intestinal phase of digestion are regulated by neural and hormonal mechanisms.

- ***Neural Regulation*** Distension of the duodenum by the presence of chyme causes the **enterogastric reflex.** Stretch receptors in the duodenal wall send nerve impulses to the medulla oblongata, where they inhibit parasympathetic stimulation and stimulate the sympathetic nerves to the stomach. As a result, gastric motility is inhibited and there is an increase in the contraction of the pyloric sphincter, which decreases gastric emptying.
- ***Hormonal Regulation***The intestinal phase of digestion is mediated by two major hormones secreted by the small intes-

tine: cholecystokinin and secretin. **Cholecystokinin (CCK)** is secreted by the **CCK cells** of the small intestinal crypts of Lieberkühn in response to chyme containing amino acids from partially digested proteins and fatty acids from partially digested triglycerides. CCK stimulates secretion of pancreatic juice that is rich in digestive enzymes. It also causes contraction of the wall of the gallbladder, which squeezes stored bile out of the gallbladder into the cystic duct and through the common bile duct. In addition, CCK causes relaxation of the sphincter of the hepatopancreatic ampulla (sphincter of Oddi), which allows pancreatic juice and bile to flow into the duodenum. CCK also slows gastric emptying by promoting contraction of the pyloric sphincter, produces satiety (a feeling of fullness) by acting on the hypothalamus in the brain, promotes normal growth and maintenance of the pancreas, and enhances the effects of secretin. Acidic chyme entering the duodenum stimulates the release of **secretin** from the **S cells** of the small intestinal crypts of Lieberkühn. In turn, secretin stimulates the flow of pancreatic juice that is rich in bicarbonate (HCO_3^-) ions to buffer the acidic chyme that enters the duodenum from the small intestine. Besides this major effect, secretin inhibits secretion of gastric juice, promotes normal growth and maintenance of the pancreas, and enhances the effects of CCK. Overall, secretin causes buffering of acid in chyme that reaches the duodenum and slows production of acid in the stomach.

Other Hormones of the Digestive System

Besides gastrin, CCK, and secretin, at least 10 other so-called "gut hormones" are secreted by and have effects on the GI tract. They include *motilin, substance P,* and *bombesin,* which stimulate motility of the intestines; *vasoactive intestinal polypeptide (VIP),* which stimulates secretion of ions and water by the intestines and inhibits gastric acid secretion; *gastrin-releasing peptide,* which stimulates release of gastrin; and *somatostatin,* which inhibits gastrin release. Some of these hormones are thought to act as local hormones (paracrines), whereas others are secreted into the blood or even into the lumen of the GI tract. The physiological roles of these and other gut hormones are still under investigation.

Table 24.8 summarizes the major hormones that control digestion.

▶ CHECKPOINT

44. What is the purpose of the cephalic phase of digestion?

45. Describe the role of gastrin in the gastric phase of digestion.

46. Outline the steps of the enterogastric reflex.

47. Explain the roles of CCK and secretin in the intestinal phase of digestion.

DEVELOPMENT OF THE DIGESTIVE SYSTEM

▶ OBJECTIVE

Describe the development of the digestive system.

During the fourth week of development, the cells of the **endoderm** form a cavity called the **primitive gut,** the forerunner of the gastrointestinal tract (see Figure 29.12b on page 1119). Soon afterwards the mesoderm forms and splits into two layers

TABLE 24.8 Major Hormones that Control Digestion

Hormone	Stimulus and Site of Secretion	Actions
Gastrin	Distension of stomach, partially digested proteins and caffeine in stomach, and high pH of stomach chyme stimulate gastrin secretion by enteroendocrine G cells, located mainly in the mucosa of pyloric antrum of stomach.	*Major effects:* Promotes secretion of gastric juice, increases gastric motility, and promotes growth of gastric mucosa. *Minor effects:* Constricts lower esophageal sphincter, relaxes pyloric sphincter.
Secretin	Acidic (high H^+ level) chyme that enters the small intestine stimulates secretion of secretin by enteroendocrine S cells in the mucosa of the duodenum.	*Major effects:* Stimulates secretion of pancreatic juice and bile that are rich in HCO_3^- (bicarbonate ions). *Minor effects:* Inhibits secretion of gastric juice, promotes normal growth and maintenance of the pancreas, and enhances effects of CCK.
Cholecystokinin (CCK)	Partially digested proteins (amino acids),triglycerides, and fatty acids that enter the small intestine stimulate secretion of CCK by enteroendocrine CCK cells in the mucosa of the small intestine; CCK is also released in the brain.	*Major effects:* Stimulates secretion of pancreatic juice rich in digestive enzymes, causes ejection of bile from the gallbladder and opening of the sphincter of the hepatopancreatic ampulla (sphincter of Oddi), and induces satiety (feeling full to satisfaction). *Minor effects:* Inhibits gastric emptying, promotes normal growth and maintenance of the pancreas, and enhances effects of secretin.

(somatic and splanchnic), as shown in Figure 29.9d on page 1115. The splanchnic mesoderm associates with the endoderm of the primitive gut; as a result, the primitive gut has a double-layered wall. The **endodermal layer** gives rise to the *epithelial lining* and *glands* of most of the gastrointestinal tract; the **mesodermal layer** produces the *smooth muscle* and *connective tissue* of the tract.

The primitive gut elongates and differentiates into an anterior **foregut,** an intermediate **midgut,** and a posterior **hindgut** (see Figure 29.12c). Until the fifth week of development, the midgut opens into the yolk sac; after that time, the yolk sac constricts and detaches from the midgut, and the midgut seals. In the region of the foregut, a depression consisting of ectoderm, the **stomodeum** (stō-mō-DĒ-um), appears (see Figure 29.12d), which develops into the *oral cavity.* The **oropharyngeal membrane** is a depression of fused ectoderm and endoderm on the surface of the embryo that separates the foregut from the stomodeum. The membrane ruptures during the fourth week of development, so that the foregut is continuous with the outside of the embryo through the oral cavity. Another depression consisting of ectoderm, the **proctodeum** (prok-tō-DĒ-um), forms in the hindgut and goes on to develop into the *anus* (see Figure 29.12d). The **cloacal membrane** (klō-Ā-kul) is a fused membrane of ectoderm and endoderm that separates the hindgut from the proctodeum. After it ruptures during the seventh week, the hindgut is continuous with the outside of the embryo through the anus. Thus, the gastrointestinal tract forms a continuous tube from mouth to anus.

The foregut develops into the *pharynx, esophagus, stomach,* and *part of the duodenum.* The midgut is transformed into the *remainder of the duodenum,* the *jejunum,* the *ileum,* and *portions of the large intestine* (cecum, appendix, ascending colon, and most of the transverse colon). The hindgut develops into the *remainder of the large intestine,* except for a portion of the anal canal that is derived from the proctodeum.

As development progresses, the endoderm at various places along the foregut develops into hollow buds that grow into the mesoderm. These buds will develop into the *salivary glands, liver, gallbladder,* and *pancreas.* Each of these organs retains a connection with the gastrointestinal tract through ducts.

▶ CHECKPOINT

48. What structures develop from the foregut, midgut, and hindgut?

AGING AND THE DIGESTIVE SYSTEM

▶ OBJECTIVE

Describe the effects of aging on the digestive system.

Overall changes of the digestive system associated with aging include decreased secretory mechanisms, decreased motility of the digestive organs, loss of strength and tone of the muscular tissue and its supporting structures, changes in neurosensory feedback regarding enzyme and hormone release, and diminished response to pain and internal sensations. In the upper portion of the GI tract, common changes include reduced sensitivity to mouth irritations and sores, loss of taste, periodontal disease, difficulty in swallowing, hiatal hernia, gastritis, and peptic ulcer disease. Changes that may appear in the small intestine include duodenal ulcers, malabsorption, and maldigestion. Other pathologies that increase in incidence with age are appendicitis, gallbladder problems, jaundice, cirrhosis, and acute pancreatitis. Large intestinal changes such as constipation, hemorrhoids, and diverticular disease may also occur. Cancer of the colon or rectum is quite common.

▶ CHECKPOINT

49. What are the general effects of aging on the digestive system?

• • •

Now that our exploration of the digestive system is complete, you can appreciate the many ways that this system contributes to homeostasis of other body systems by examining *Focus on Homeostasis: The Digestive System.* Next, in Chapter 25, you will discover how the nutrients absorbed by the GI tract enter into metabolic reactions in the body tissues.

FOCUS ON HOMEOSTASIS

The Digestive System

BODY SYSTEM	CONTRIBUTION OF THE DIGESTIVE SYSTEM
For all body systems 	The digestive system breaks down dietary nutrients into forms that can be absorbed and used by body cells for producing ATP and building body tissues. Absorbs water, minerals, and vitamins needed for growth and function of body tissues; and eliminates wastes from body tissues in feces.
Integumentary system	Small intestine absorbs vitamin D, which skin and kidneys modify to produce the hormone calcitriol. Excess dietary calories are stored as triglycerides in adipose cells in dermis and subcutaneous layer.
Skeletal system 	Small intestine absorbs dietary calcium and phosphorus salts needed to build bone extracellular matrix.
Muscular system 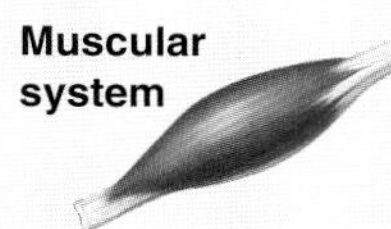	Liver can convert lactic acid (produced by muscles during exercise) to glucose.
Nervous system 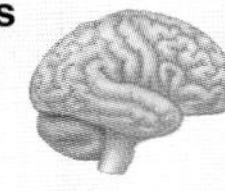	Gluconeogenesis (synthesis of new glucose molecules) in liver plus digestion and absorption of dietary carbohydrates provide glucose, needed for ATP production by neurons.
Endocrine system 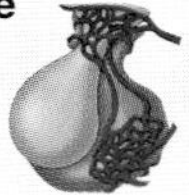	Liver inactivates some hormones, ending their activity. Pancreatic islets release insulin and glucagon. Cells in mucosa of stomach and small intestine release hormones that regulate digestive activities. Liver produces angiotensinogen.
Cardiovascular system 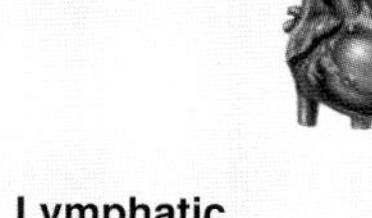	GI tract absorbs water that helps maintain blood volume and iron that is needed for synthesis of hemoglobin in red blood cells. Bilirubin from hemoglobin breakdown is partially excreted in feces. Liver synthesizes most plasma proteins.
Lymphatic system and immunity 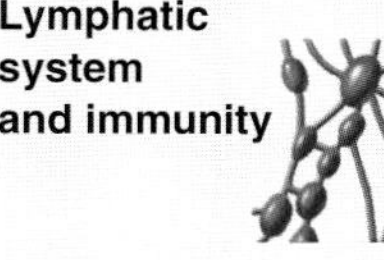	Acidity of gastric juice destroys bacteria and most toxins in stomach.
Respiratory system 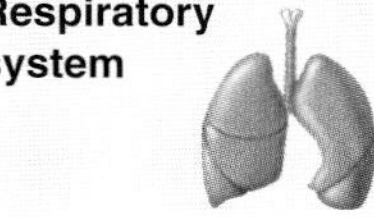	Pressure of abdominal organs against the diaphragm helps expel air quickly during a forced exhalation.
Urinary system 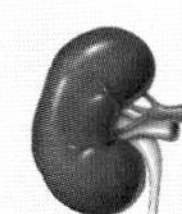	Absorption of water by GI tract provides water needed to excrete waste products in urine.
Reproductive system 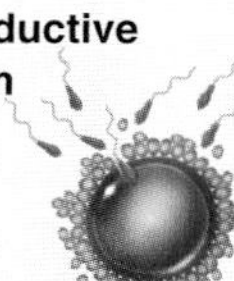	Digestion and absorption provides adequate nutrients, including fats, for normal development of reproductive structures, for production of gametes (oocytes and sperm), and for fetal growth and development during pregnancy.

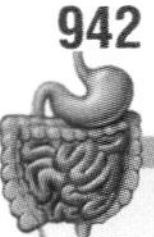

DISORDERS: HOMEOSTATIC IMBALANCES

Dental Caries

Dental caries, or tooth decay, involves a gradual demineralization (softening) of the enamel and dentin. If untreated, microorganisms may invade the pulp, causing inflammation and infection, with subsequent death of the pulp and abscess of the alveolar bone surrounding the root's apex, requiring root canal therapy (see page 906).

Dental caries begin when bacteria, acting on sugars, produce acids that demineralize the enamel. **Dextran,** a sticky polysaccharide produced from sucrose, causes the bacteria to stick to the teeth. Masses of bacterial cells, dextran, and other debris adhering to teeth constitute **dental plaque.** Saliva cannot reach the tooth surface to buffer the acid because the plaque covers the teeth. Brushing the teeth after eating removes the plaque from flat surfaces before the bacteria can produce acids. Dentists also recommend that the plaque between the teeth be removed every 24 hours with dental floss.

Periodontal Disease

Periodontal disease is a collective term for a variety of conditions characterized by inflammation and degeneration of the gingivae, alveolar bone, periodontal ligament, and cementum. In one such condition, called **pyorrhea,** initial symptoms include enlargement and inflammation of the soft tissue and bleeding of the gums. Without treatment, the soft tissue may deteriorate and the alveolar bone may be resorbed, causing loosening of the teeth and recession of the gums. Periodontal diseases are often caused by poor oral hygiene; by local irritants, such as bacteria, impacted food, and cigarette smoke; or by a poor "bite."

Peptic Ulcer Disease

In the United States, 5–10% of the population develops **peptic ulcer disease (PUD).** An **ulcer** is a craterlike lesion in a membrane; ulcers that develop in areas of the GI tract exposed to acidic gastric juice are called **peptic ulcers.** The most common complication of peptic ulcers is bleeding, which can lead to anemia if enough blood is lost. In acute cases, peptic ulcers can lead to shock and death. Three distinct causes of PUD are recognized: (1) the bacterium *Helicobacter pylori;* (2) nonsteroidal anti-inflammatory drugs (NSAIDs) such as aspirin; and (3) hypersecretion of HCl, as occurs in Zollinger–Ellison syndrome, a gastrin-producing tumor, usually of the pancreas.

Helicobacter pylori (previously named *Campylobacter pylori*) is the most frequent cause of PUD. The bacterium produces an enzyme called urease, which splits urea into ammonia and carbon dioxide. While shielding the bacterium from the acidity of the stomach, the ammonia also damages the protective mucous layer of the stomach and the underlying gastric cells. *H. pylori* also produces catalase, an enzyme that may protect the microbe from phagocytosis by neutrophils, plus several adhesion proteins that allow the bacterium to attach itself to gastric cells.

Several therapeutic approaches are helpful in the treatment of PUD. Cigarette smoke, alcohol, caffeine, and NSAIDs should be avoided because they can impair mucosal defensive mechanisms, which increases mucosal susceptibility to the damaging effects of HCl. In cases associated with *H. pylori,* treatment with an antibiotic drug often resolves the problem. Oral antacids such as Tums® or Maalox® can help temporarily by buffering gastric acid. When hypersecretion of HCl is the cause of PUD, H_2 blockers (such as Tagamet®) or proton pump inhibitors such as omeprazole (Prilosec®), which block secretion of H^+ from parietal cells, may be used.

Diverticular Disease

In **diverticular disease,** saclike outpouchings of the wall of the colon, termed **diverticula,** occur in places where the muscularis has weakened and may become inflamed. Development of diverticula is known as **diverticulosis.** Many people who develop diverticulosis have no symptoms and experience no complications. Of those people known to have diverticulosis, 10–25% eventually develop an inflammation known as **diverticulitis.** This condition may be characterized by pain, either constipation or increased frequency of defecation, nausea, vomiting, and low-grade fever. Because diets low in fiber contribute to development of diverticulitis, patients who change to high-fiber diets show marked relief of symptoms. In severe cases, affected portions of the colon may require surgical removal. If diverticula rupture, the release of bacteria into the abdominal cavity can cause peritonitis.

Colorectal Cancer

Colorectal cancer is among the deadliest of malignancies, ranking second to lung cancer in males and third after lung cancer and breast cancer in females. Genetics plays a very important role; an inherited predisposition contributes to more than half of all cases of colorectal cancer. Intake of alcohol and diets high in animal fat and protein are associated with increased risk of colorectal cancer, whereas dietary fiber, retinoids, calcium, and selenium may be protective. Signs and symptoms of colorectal cancer include diarrhea, constipation, cramping, abdominal pain, and rectal bleeding, either visible or occult (hidden in feces). Precancerous growths on the mucosal surface, called **polyps,** also increase the risk of developing colorectal cancer. Screening for colorectal cancer includes testing for blood in the feces, digital rectal examination, sigmoidoscopy, colonoscopy, and barium enema. Tumors may be removed endoscopically or surgically.

Hepatitis

Hepatitis is an inflammation of the liver that can be caused by viruses, drugs, and chemicals, including alcohol. Clinically, several types of viral hepatitis are recognized. **Hepatitis A (infectious hepatitis)** is caused by the hepatitis A virus and is spread via fecal contamination of objects such as food, clothing, toys, and eating utensils (fecal–oral route). It is generally a mild disease of children and young adults characterized by loss of appetite, malaise, nausea, diarrhea, fever, and chills. Eventually, jaundice appears. This type of hepatitis does not cause lasting liver damage. Most people recover in 4 to 6 weeks.

Hepatitis B is caused by the hepatitis B virus and is spread primarily by sexual contact and contaminated syringes and transfusion equipment. It can also be spread via saliva and tears. Hepatitis B virus can be present for years or even a lifetime, and it can produce cirrhosis and possibly cancer of the liver. Individuals who harbor the active hepatitis B virus also become carriers. Vaccines produced through recombinant DNA technology (for example, Recombivax HB®) are available to prevent hepatitis B infection.

Hepatitis C, caused by the hepatitis C virus, is clinically similar to hepatitis B. Hepatitis C can cause cirrhosis and possibly liver cancer. In developed nations, donated blood is screened for the presence of hepatitis B and C.

Hepatitis D is caused by the hepatitis D virus. It is transmitted like hepatitis B and, in fact, a person must have been co-infected with hepatitis B before contracting hepatitis D. Hepatitis D results in severe liver damage and has a higher fatality rate than infection with hepatitis B virus alone.

Hepatitis E is caused by the hepatitis E virus and is spread like hepatitis A. Although it does not cause chronic liver disease, hepatitis E virus has a very high mortality rate among pregnant women.

Anorexia Nervosa

Anorexia nervosa is a chronic disorder characterized by self-induced weight loss, negative perception of body image, and physiological changes that result from nutritional depletion. Patients with anorexia nervosa have a fixation on weight control and often insist on having a bowel movement every day despite inadequate food intake. They often abuse laxatives, which worsens the fluid and electrolyte imbalances and nutrient deficiencies. The disorder is found predominantly in young, single females, and it may be inherited. Abnormal patterns of menstruation, amenorrhea (absence of menstruation), and a lowered basal metabolic rate reflect the depressant effects of starvation. Individuals may become emaciated and may ultimately die of starvation or one of its complications. Also associated with the disorder are osteoporosis, depression, and brain abnormalities coupled with impaired mental performance. Treatment consists of psychotherapy and dietary regulation.

MEDICAL TERMINOLOGY

Achalasia (ak′-a-LĀ-zē-a; *a-* = without; *chalasis* = relaxation) A condition caused by malfunction of the myenteric plexus in which the lower esophageal sphincter fails to relax normally as food approaches. A whole meal may become lodged in the esophagus and enter the stomach very slowly. Distension of the esophagus results in chest pain that is often confused with pain originating from the heart.

Borborygmus (bor′-bō-RIG-mus) A rumbling noise caused by the propulsion of gas through the intestines.

Bulimia (*bu-* = ox; *limia* = hunger) or ***binge–purge syndrome*** A disorder that typically affects young, single, middle-class, white females, characterized by overeating at least twice a week followed by purging by self-induced vomiting, strict dieting or fasting, vigorous exercise, or use of laxatives or diuretics; it occurs in response to fears of being overweight or to stress, depression, and physiological disorders such as hypothalamic tumors.

Canker sore (KANG-ker) Painful ulcer on the mucous membrane of the mouth that affects females more often than males, usually between ages 10 and 40; may be an autoimmune reaction or a food allergy.

Cirrhosis Distorted or scarred liver as a result of chronic inflammation due to hepatitis, chemicals that destroy hepatocytes, parasites that infect the liver, or alcoholism; the hepatocytes are replaced by fibrous or adipose connective tissue. Symptoms include jaundice, edema in the legs, uncontrolled bleeding, and increased sensitivity to drugs.

Colitis (ko-LĪ-tis) Inflammation of the mucosa of the colon and rectum in which absorption of water and salts is reduced, producing watery, bloody feces and, in severe cases, dehydration and salt depletion. Spasms of the irritated muscularis produce cramps. It is thought to be an autoimmune condition.

Colonoscopy (kō-lon-OS-kō-pē; *-skopes* = to view) The visual examination of the lining of the colon using an elongated, flexible, fiberoptic endoscope called a *colonoscope*. It is used to detect disorders such as polyps, cancer, and diverticulosis, to take tissue samples, and to remove small polyps. Most tumors of the large intestine occur in the rectum.

Colostomy (kō-LOS-tō-mē; *-stomy* = provide an opening) The diversion of feces through an opening in the colon, creating a surgical "stoma" (artificial opening) that is made in the exterior of the abdominal wall. This opening serves as a substitute anus through which feces are eliminated into a bag worn on the abdomen.

Dysphagia (dis-FĀ-jē-a; *dys-* = abnormal; *phagia* = to eat) Difficulty in swallowing that may be caused by inflammation, paralysis, obstruction, or trauma.

Flatus (FLĀ-tus) Air (gas) in the stomach or intestine, usually expelled through the anus. If the gas is expelled through the mouth, it is called **eructation** or **belching** (burping). Flatus may result from gas released during the breakdown of foods in the stomach or from swallowing air or gas-containing substances such as carbonated drinks.

Food poisoning A sudden illness caused by ingesting food or drink contaminated by an infectious microbe (bacterium, virus, or protozoan) or a toxin (poison). The most common cause of food poisoning is the toxin produced by the bacterium *Staphylococcus aureus*. Most types of food poisoning cause diarrhea and/or vomiting, often associated with abdominal pain.

Gastroenteritis (gas′-trō-en-ter-Ī-tis; *gastro-* = stomach; *enteron* = intestine; *-itis* = inflammation) Inflammation of the lining of the stomach and intestine (especially the small intestine). It is usually caused by a viral or bacterial infection that may be acquired by contaminated food or water or by people in close contact. Symptoms include diarrhea, vomiting, fever, loss of appetite, cramps, and abdominal discomfort.

Gastroscopy (gas-TROS-kō-pē; *-scopy* = to view with a lighted instrument) Endoscopic examination of the stomach in which the examiner can view the interior of the stomach directly to evaluate an ulcer, tumor, inflammation, or source of bleeding.

Halitosis (hal′-ī-TŌ-sis; *halitus-* = breath; *-osis* = condition) A foul odor from the mouth; also called **bad breath.**

Heartburn A burning sensation in a region near the heart due to irritation of the mucosa of the esophagus from hydrochloric acid in stomach contents. It is caused by failure of the lower esophageal sphincter to close properly, so that the stomach contents enter the inferior esophagus. It is not related to any cardiac problem.

Hemorrhoids (HEM-ō-royds; *hemi* = blood; *rhoia* = flow) Varicosed (enlarged and inflamed) superior rectal veins. Hemorrhoids develop when the veins are put under pressure and become engorged with blood. If the pressure continues, the wall of the vein stretches. Such a distended vessel oozes blood; bleeding or itching is usually the first sign that a hemorrhoid has developed. Stretching of a vein also favors clot formation, further aggravating swelling and pain. Hemorrhoids may be caused by constipation, which may be brought on by low-fiber diets. Also, repeated straining during defecation forces blood down into the rectal veins, increasing pressure in those veins and possibly causing hemorrhoids. Also called **piles.**

Hernia (HER-nē-a) Protrusion of all or part of an organ through a membrane or cavity wall, usually the abdominal cavity. *Hiatus (diaphragmatic) hernia* is the protrusion of a part of the stomach into the thoracic cavity through the esophageal hiatus of the diaphragm. *Inguinal hernia* is the protrusion of the hernial sac into the inguinal opening; it may contain a portion of the bowel in an advanced stage and may extend into the scrotal compartment in males, causing strangulation of the herniated part.

Inflammatory bowel disease (in-FLAM-a-tō′-rē BOW-el) Inflammation of the gastrointestinal tract that exists in two forms. (1) ***Crohn's disease*** is an inflammation of any part of the gastrointestinal tract in which the inflammation extends from the mucosa through the submucosa, muscularis, and serosa. (2) ***Ulcerative colitis*** is an inflammation of the mucosa of the colon and rectum, usually accompanied by rectal bleeding. Curiously, cigarette smoking increases the risk of Crohn's disease but decreases the risk of ulcerative colitis.

Irritable bowel syndrome (IBS) Disease of the entire gastrointestinal tract in which a person reacts to stress by developing symptoms (such as cramping and abdominal pain) associated with alternating patterns of diarrhea and constipation. Excessive amounts of mucus may appear in feces; other symptoms include flatulence, nausea, and loss of appetite. The condition is also known as **irritable colon** or **spastic colitis.**

Malabsorption (mal-ab-SORP-shun; *mal-* = bad) A number of disorders in which nutrients from food are not absorbed properly. It may be due to disorders that result in the inadequate breakdown of food during digestion (due to inadequate digestive enzymes or juices), damage to the lining of the small intestine (from surgery, infections, and drugs like neomycin and alcohol), and impairment of motility. Symptoms may include diarrhea, weight loss, weakness, vitamin deficiencies, and bone demineralization.

Malocclusion (mal′-ō-KLOO-zhun; *mal-* = bad; *occlusion* = to fit together) Condition in which the surfaces of the maxillary (upper) and mandibular (lower) teeth fit together poorly.

Nausea (NAW-sē-a; *nausia* = seasickness) Discomfort characterized by a loss of appetite and the sensation of impending vomiting. Its causes include local irritation of the gastrointestinal tract, a systemic disease, brain disease or injury, overexertion, or the effects of medication or drug overdosage.

Traveler's diarrhea Infectious disease of the gastrointestinal tract that results in loose, urgent bowel movements, cramping, abdominal pain, malaise, nausea, and occasionally fever and dehydration. It is acquired through ingestion of food or water contaminated with fecal material typically containing bacteria (especially *Escherichia coli*); viruses or protozoan parasites are less common causes.

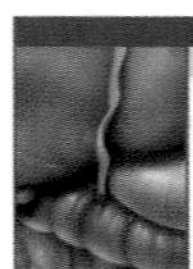

STUDY OUTLINE

INTRODUCTION (p. 896)

1. The breaking down of larger food molecules into smaller molecules is called digestion.
2. The organs involved in the breakdown of food are collectively known as the *digestive system* and are composed of two main groups: the gastrointestinal (GI) tract and accessory digestive organs.
3. The GI tract is a continuous tube extending from the mouth to the anus.
4. The accessory digestive organs include the teeth, tongue, salivary glands, liver, gallbladder, and pancreas.

OVERVIEW OF THE DIGESTIVE SYSTEM (p. 896)

1. Digestion includes six basic processes: ingestion, secretion, mixing and propulsion, mechanical and chemical digestion, absorption, and defecation.
2. Mechanical digestion consists of mastication and movements of the gastrointestinal tract that aid chemical digestion.
3. Chemical digestion is a series of hydrolysis reactions that break down large carbohydrates, lipids, proteins, and nucleic acids in foods into smaller molecules that are usable by body cells.

LAYERS OF THE GI TRACT (p. 896)

1. The basic arrangement of layers in most of the gastrointestinal tract, from deep to superficial, is the mucosa, submucosa, muscularis, and serosa.
2. Associated with the lamina propria of the mucosa are extensive patches of lymphatic tissue called mucosa-associated lymphoid tissue (MALT).

NEURAL INNERVATION OF THE GI TRACT (p. 899)

1. The gastrointestinal tract is regulated by an intrinsic set of nerves known as the enteric nervous system (ENS) and by an extrinsic set of nerves that are part of the autonomic nervous system (ANS).
2. The ENS consists of neurons arranged into two plexuses: the myenteric plexus and the submucosal plexus.
3. The myenteric plexus, which is located between the longitudinal and circular smooth muscle layers of the muscularis, regulates GI tract motility.
4. The submucosal plexus, which is located in the submucosa, regulates GI secretion.
5. Although the neurons of the ENS can function independently, they are subject to regulation by the neurons of the ANS.

6. Parasympathetic fibers of the vagus (X) nerves and pelvic splanchnic nerves increase GI tract secretion and motility by increasing the activity of ENS neurons.
7. Sympathetic fibers from the thoracic and upper lumbar regions of the spinal cord decrease GI tract secretion and motility by inhibiting ENS neurons.

PERITONEUM (p. 900)

1. The peritoneum is the largest serous membrane of the body; it lines the wall of the abdominal cavity and covers some abdominal organs.
2. Folds of the peritoneum include the mesentery, mesocolon, falciform ligament, lesser omentum, and greater omentum.

MOUTH (p. 902)

1. The mouth is formed by the cheeks, hard and soft palates, lips, and tongue.
2. The vestibule is the space bounded externally by the cheeks and lips and internally by the teeth and gums.
3. The oral cavity proper extends from the vestibule to the fauces.
4. The tongue, together with its associated muscles, forms the floor of the oral cavity. It is composed of skeletal muscle covered with mucous membrane.
5. The upper surface and sides of the tongue are covered with papillae, some of which contain taste buds.
6. The major portion of saliva is secreted by the major salivary glands, which lie outside the mouth and pour their contents into ducts that empty into the oral cavity.
7. There are three pairs of major salivary glands: parotid, submandibular, and sublingual glands.
8. Saliva lubricates food and starts the chemical digestion of carbohydrates.
9. Salivation is controlled by the nervous system.
10. The teeth (dentes) project into the mouth and are adapted for mechanical digestion.
11. A typical tooth consists of three principal regions: crown, root, and neck.
12. Teeth are composed primarily of dentin and are covered by enamel, the hardest substance in the body.
13. There are two dentitions: deciduous and permanent.
14. Through mastication, food is mixed with saliva and shaped into a soft, flexible mass called a bolus.
15. Salivary amylase begins the digestion of starches, and lingual lipase acts on triglycerides.

PHARYNX (p. 908)

1. The pharynx is a funnel-shaped tube that extends from the internal nares to the esophagus posteriorly and to the larynx anteriorly.
2. The pharynx has both respiratory and digestive functions.

ESOPHAGUS (p. 908)

1. The esophagus is a collapsible, muscular tube that connects the pharynx to the stomach.
2. It contains an upper and a lower esophageal sphincter.

DEGLUTITION (p. 909)

1. Deglutition, or swallowing, moves a bolus from the mouth to the stomach.
2. Swallowing consists of a voluntary stage, a pharyngeal stage (involuntary), and an esophageal stage (involuntary).

STOMACH (p. 911)

1. The stomach connects the esophagus to the duodenum.
2. The principal anatomic regions of the stomach are the cardia, fundus, body, and pylorus.
3. Adaptations of the stomach for digestion include rugae; glands that produce mucus, hydrochloric acid, pepsin, gastric lipase, and intrinsic factor; and a three-layered muscularis.
4. Mechanical digestion consists of mixing waves.
5. Chemical digestion consists mostly of the conversion of proteins into peptides by pepsin.
6. The stomach wall is impermeable to most substances.
7. Among the substances the stomach can absorb are water, certain ions, drugs, and alcohol.

PANCREAS (p. 916)

1. The pancreas consists of a head, a body, and a tail and is connected to the duodenum via the pancreatic duct and accessory duct.
2. Endocrine pancreatic islets (islets of Langerhans) secrete hormones, and exocrine acini secrete pancreatic juice.
3. Pancreatic juice contains enzymes that digest starch (pancreatic amylase), proteins (trypsin, chymotrypsin, carboxypeptidase, and elastase), triglycerides (pancreatic lipase), and nucleic acids (ribonuclease and deoxyribonuclease).

LIVER AND GALLBLADDER (p. 918)

1. The liver has left and right lobes; the right lobe includes a quadrate lobe and a caudate lobe. The gallbladder is a sac located in a depression on the posterior surface of the liver that stores and concentrates bile.
2. The lobes of the liver are made up of lobules that contain hepatocytes (liver cells), sinusoids, stellate reticuloendothelial (Kupffer) cells, and a central vein.
3. Hepatocytes produce bile that is carried by a duct system to the gallbladder for concentration and temporary storage.
4. Bile's contribution to digestion is the emulsification of dietary lipids.
5. The liver also functions in carbohydrate, lipid, and protein metabolism; processing of drugs and hormones; excretion of bilirubin; synthesis of bile salts; storage of vitamins and minerals; phagocytosis; and activation of vitamin D.

SMALL INTESTINE (p. 921)

1. The small intestine extends from the pyloric sphincter to the ileocecal sphincter.
2. It is divided into duodenum, jejunum, and ileum.
3. Its glands secrete fluid and mucus, and the circular folds, villi, and microvilli of its wall provide a large surface area for digestion and absorption.
4. Brush-border enzymes digest α-dextrins, maltose, sucrose, lactose, peptides, and nucleotides at the surface of mucosal epithelial cells.

5. Pancreatic and intestinal brush-border enzymes break down starches into maltose, maltotriose, and α-dextrins (pancreatic amylase), α-dextrins into glucose (α-dextrinase), maltose to glucose (maltase), sucrose to glucose and fructose (sucrase), lactose to glucose and galactose (lactase), and proteins into peptides (trypsin, chymotrypsin, and elastase). Also, enzymes break off amino acids at the carboxyl ends of peptides (carboxypeptidases) and break off amino acids at the amino ends of peptides (aminopeptidases). Finally, enzymes split dipeptides into amino acids (dipeptidases), triglycerides to fatty acids and monoglycerides (lipases), and nucleotides to pentoses and nitrogenous bases (nucleosidases and phosphatases).
6. Mechanical digestion in the small intestine involves segmentation and migrating motility complexes.
7. Absorption occurs via diffusion, facilitated diffusion, osmosis, and active transport; most absorption occurs in the small intestine.
8. Monosaccharides, amino acids, and short-chain fatty acids pass into the blood capillaries.
9. Long-chain fatty acids and monoglycerides are absorbed from micelles, resynthesized to triglycerides, and formed into chylomicrons.
10. Chylomicrons move into lymph in the lacteal of a villus.
11. The small intestine also absorbs electrolytes, vitamins, and water.

LARGE INTESTINE (p. 931)

1. The large intestine extends from the ileocecal sphincter to the anus.
2. Its regions include the cecum, colon, rectum, and anal canal.
3. The mucosa contains many goblet cells, and the muscularis consists of teniae coli and haustra.
4. Mechanical movements of the large intestine include haustral churning, peristalsis, and mass peristalsis.
5. The last stages of chemical digestion occur in the large intestine through bacterial action. Substances are further broken down, and some vitamins are synthesized.
6. The large intestine absorbs water, ions, and vitamins.
7. Feces consist of water, inorganic salts, epithelial cells, bacteria, and undigested foods.
8. The elimination of feces from the rectum is called defecation.
9. Defecation is a reflex action aided by voluntary contractions of the diaphragm and abdominal muscles and relaxation of the external anal sphincter.

PHASES OF DIGESTION (p. 937)

1. Digestive activities occur in three overlapping phases: cephalic phase, gastric phase, and intestinal phase.
2. During the cephalic phase of digestion, salivary glands secrete saliva and gastric glands secrete gastric juice in order to prepare the mouth and stomach for food that is about to be eaten.
3. The presence of food in the stomach causes the gastric phase of digestion, which promotes gastric juice secretion and gastric motility.
4. During the intestinal phase of digestion, food is digested in the small intestine. In addition, gastric motility and gastric secretion decrease in order to slow the exit of chyme from the stomach, which prevents the small intestine from being overloaded with more chyme that it can handle.
5. The activities that occur during the various phases of digestion are coordinated by neural pathways and by hormones. Table 24.8 summarizes the major hormones that control digestion.

DEVELOPMENT OF THE DIGESTIVE SYSTEM (p. 939)

1. The endoderm of the primitive gut forms the epithelium and glands of most of the gastrointestinal tract.
2. The mesoderm of the primitive gut forms the smooth muscle and connective tissue of the gastrointestinal tract.

AGING AND THE DIGESTIVE SYSTEM (p. 940)

1. General changes include decreased secretory mechanisms, decreased motility, and loss of tone.
2. Specific changes may include loss of taste, pyorrhea, hernias, peptic ulcer disease, constipation, hemorrhoids, and diverticular diseases.

Q SELF-QUIZ QUESTIONS

Fill in the blanks in the following statements.

1. The end-products of chemical digestion of carbohydrates are ____, of proteins are ____, of lipids are ____ and ____, and of nucleic acids are ____, ____, and ____.
2. List the mechanisms of absorption of materials in the small intestine: ____, ____, ____, and ____.

Indicate whether the following statements are true or false.

3. The soft palate, uvula, and epiglottis prevent swallowed foods and liquids from entering the respiratory passages.
4. The coordinated contractions and relaxations of the muscularis which propels materials through the GI tract is known as peristalsis.

Choose the one best answer to the following questions.

5. Which of the following are mismatched? (a) chemical digestion: splitting food molecules into simple substances by hydrolysis and aided by digestive enzymes, (b) motility: mechanical processes that break apart ingested food into small molecules, (c) ingestion: taking foods and liquids into the mouth, (d) propulsion: movement of food through GI tract due to smooth muscle contraction, (e) absorption: passage into blood or lymph of ions, fluids and small molecules into the epithelial lining of the GI tract lumen.
6. Which of the following are *true* concerning the peritoneum? (1) The kidneys and pancreas are retroperitoneal. (2) The greater omentum is the largest of the peritoneal folds. (3) The lesser omentum binds the large intestine to the posterior abdominal wall.

(4) The falciform ligament attaches the liver to the anterior abdominal wall and diaphragm. (5) The mesentery is associated with the small intestine. (a) 1, 2, 3, and 5; (b) 1, 2, and 5; (c) 2 and 5; (d) 1, 2, 4, and 5; (e) 3, 4, and 5.

7. When a surgeon makes an incision in the small intestine, in what order would the physician encounter these structures? (1) epithelium, (2) submucosa, (3) serosa, (4) muscularis, (5) lamina propria, (6) muscularis mucosae. (a) 3, 4, 5, 6, 2, 1; (b) 1, 2, 3, 4, 6, 5; (c) 1, 5, 6, 2, 4, 3; (d) 5, 1, 2, 6, 4, 3; (e) 3, 4, 2, 6, 5, 1.

8. Which of the following are functions of the liver? (1) carbohydrate, lipid, and protein metabolism, (2) nucleic acid metabolism, (3) excretion of bilirubin, (4) synthesis of bile salts, (5) activation of vitamin D. (a) 1, 2, 3, and 5; (b) 1, 2, 3, and 4; (c) 1, 3, 4, and 5; (d) 2, 3, 4, and 5; (e) 1, 2, 4, and 5.

9. Which of the following statements regarding the regulation of gastric secretion and motility are *true*? (1) The sight, smell, taste, or thought of food can initiate the cephalic phase of gastric activity. (2) The gastric phase begins when food enters the small intestine. (3) Once activated, stretch receptors and chemoreceptors in the stomach trigger the flow of gastric juice and peristalsis. (4) The intestinal phase reflexes inhibit gastric activity. (5) The enterogastric reflex stimulates gastric emptying. (a) 1, 3, and 4; (b) 2, 4, and 5; (c) 1, 3, 4, and 5; (d) 1, 2, and 5; (e) 1, 2, 3, and 4.

10. Which of the following are *true*? (1) Segmentations in the small intestine help propel chyme through the intestinal tract. (2) The migrating motility complex is a type of peristalsis in the small intestine. (3) The large surface area for absorption in the small intestine is due to the presence of circular folds, villi, and microvilli. (4) The mucus-producing cells of the small intestine are Paneth cells. (5) Most long-chain fatty acid and monoglyceride absorption in the small intestine requires the presence of bile salts. (a) 1, 2, and 3; (b) 2, 3, and 5; (c) 1, 2, 3, 4, and 5; (d) 1, 3, and 5; (e) 1, 2, 3, and 5.

11. The release of feces from the large intestine is dependent on (1) stretching of the rectal walls, (2) voluntary relaxation of the external anal sphincter, (3) involuntary contraction of the diaphragm and abdominal muscles, (4) activity of the intestinal bacteria, (5) sympathetic stimulation of the internal sphincter. (a) 2, 4, and 5; (b) 1, 2, and 5; (c) 1, 2, 3, and 5; (d) 1 and 2; (e) 3, 4, and 5.

12. Which of the following is *not* true concerning the liver? (a) The left hepatic duct joins the cystic duct from the gallbladder. (b) As blood passes through the sinusoids, it is processed by hepatocytes and phagocytes. (c) Processed blood returns from the liver to systemic circulation through the hepatic vein. (d) The liver receives oxygenated blood through the hepatic artery. (e) The hepatic portal vein delivers deoxygenated blood from the GI tract to the liver.

13. Match the following:

____(a) collapsed, muscular tube involved in deglutition and peristalsis
____(b) coiled tube attached to the cecum
____(c) contains duodenal glands in the submucosa
____(d) produces and secretes bile
____(e) contains aggregated lymphatic follicles in the submucosa
____(f) responsible for ingestion, mastication, and deglutition
____(g) responsible for churning, peristalsis, storage, and chemical digestion with the enzyme pepsin
____(h) storage area for bile
____(i) contain acini that release juices containing several digestive enzymes for protein, carbohydrate, lipid, and nucleic acid digestion and sodium bicarbonate to buffer stomach acid
____(j) composed of enamel, dentin, and pulp cavity; used in mastication
____(k) passageway for food, fluid, and air; involved in deglutition
____(l) forms a semisolid waste material through haustral churning and peristalsis
____(m) forces the food to the back of the mouth for swallowing; places food in contact with the teeth
____(n) produce a fluid in the mouth that helps cleanse the mouth and teeth and that lubricates, dissolves, and begins the chemical breakdown of food

(1) mouth
(2) teeth
(3) salivary glands
(4) pharynx
(5) esophagus
(6) tongue
(7) stomach
(8) duodenum
(9) ileum
(10) colon
(11) liver
(12) gallbladder
(13) appendix
(14) pancreas

14. Match the following:

____(a) an activating brush-border enzyme that splits off part of the trypsinogen molecule to form trypsin, a protease
____(b) an enzyme that initiates carbohydrate digestion in the mouth
____(c) the principal triglyceride-digesting enzyme in adults
____(d) stimulates secretion of gastric juices and promotes gastric emptying
____(e) secreted by chief cells in the stomach; a proteolytic enzyme
____(f) stimulates the flow of pancreatic juice rich in bicarbonates; decreases gastric secretions
____(g) a nonenzymatic fat-emulsifying agent
____(h) causes contraction of the gallbladder and stimulates the production of pancreatic juice rich in digestive enzymes
____(i) inhibits gastrin release
____(j) stimulates secretion of ions and water by the intestines and inhibits gastric acid secretion
____(k) secreted by glands in the tongue; begins breakdown of triglycerides in the stomach

(1) gastrin
(2) cholecystokinin
(3) secretin
(4) enterokinase
(5) pepsin
(6) salivary amylase
(7) pancreatic lipase
(8) lingual lipase
(9) bile
(10) vasoactive intestinal polypeptide
(11) somatostatin

15. Match the following:

____(a) microvilli of the small intestine that increase surface area for absorption; also contain some digestive enzymes
____(b) finger-like projections of the mucosa of the small intestine that increase surface area for digestion and absorption
____(c) produce hydrochloric acid and intrinsic factor in the stomach
____(d) secrete lysozyme; help regulate microbial population in the intestines
____(e) stomach enteroendocrine cells that secrete gastrin
____(f) longitudinal muscular bands in the large intestine; tonic contractions produce haustra
____(g) lymphatic capillary used for chylomicron absorption in the small intestine
____(h) groups of lymphatic nodules in the small intestine
____(i) controls the GI tract motility and secretions of GI tract organs
____(j) large mucosal folds in the stomach
____(k) secrete pepsinogen and gastric lipase in the stomach
____(l) permanent ridges in the mucosa of the small intestine; enhance absorption by increasing surface area and causing chyme to spiral rather than move in a straight line
____(m) phagocytic cells of the liver; destroy worn-out white blood cells and red blood cells, bacteria, and other foreign matter in the blood draining the GI tract

(1) lacteal
(2) parietal cells
(3) chief cells
(4) brush border
(5) stellate reticuloendothelial cells
(6) rugae
(7) teniae coli
(8) villi
(9) circular folds
(10) Paneth cells
(11) G cells
(12) enteric nervous system
(13) Peyer's patches

CRITICAL THINKING QUESTIONS

1. Why would you *not* want to completely suppress HCl secretion in the stomach?
2. Trey has cystic fibrosis, a genetic disorder that is characterized by the production of excessive mucus, affecting several body systems (e.g., respiratory, digestive, reproductive). In the digestive system, the excess mucus blocks bile ducts in the liver and pancreatic ducts. How would this affect Trey's digestive processes?
3. Antonio had dinner at his favorite Italian restaurant. His menu consisted of a salad, large plate of spaghetti, garlic bread and wine. For dessert, he consumed "death by chocolate" cake and a cup of coffee. He topped off his evening with a cigarette and brandy. He returned home and, while lying on his couch watching television, he experienced a pain in his chest. He called 911 because he was certain he was having a heart attack. Antonio was told his heart was fine, but he needed to watch his diet. What happened to Antonio?

ANSWERS TO FIGURE QUESTIONS

24.1 Digestive enzymes are produced by the salivary glands, tongue, stomach, pancreas, and small intestine.

24.2 The lamina propria has the following functions: (1) It contains blood vessels and lymphatic vessels, which are the routes by which nutrients are absorbed from the GI tract; (2) it supports the mucosal epithelium and binds it to the muscularis mucosae; and (3) it contains mucosa-associated lymphatic tissue (MALT), which helps protect against disease.

24.3 The neurons of the myenteric plexus regulate GI tract motility, while the neurons of the submucosal plexus regulate GI secretion.

24.4 Mesentery binds the small intestine to the posterior abdominal wall.

24.5 The uvula helps prevent foods and liquids from entering the nasal cavity during swallowing.

24.6 Chloride ions in saliva activate salivary amylase.

24.7 The main component of teeth is connective tissue, specifically dentin.

24.8 The first, second, and third molars do not replace any deciduous teeth.

24.9 The esophageal mucosa and submucosa contain mucus-secreting glands.

24.10 Both. Initiation of swallowing is voluntary and the action is carried out by skeletal muscles. Completion of swallowing—moving a bolus along the esophagus and into the stomach—is involuntary and involves peristalsis by smooth muscle.

24.11 After a large meal, the rugae stretch and disappear as the stomach fills.

24.12 Parietal cells secrete HCl, which is a component of gastric juice. HCl kills microbes in food, denatures proteins, and converts pepsinogen into pepsin.

24.13 Hydrogen ions secreted into gastric juice are derived from carbonic acid (H_2CO_3).

24.14 The pancreatic duct contains pancreatic juice (fluid and digestive enzymes); the common bile duct contains bile; the hepatopancreatic ampulla contains pancreatic juice and bile.

24.15 The phagocytic cell in the liver is the stellate reticuloendothelial (Kupffer) cell.

24.16 While a meal is being absorbed, nutrients, O_2, and certain toxic substances are removed by hepatocytes from blood flowing through liver sinusoids.

24.17 The ileum is the longest part of the small intestine.

24.18 Nutrients being absorbed in the small intestine enter the blood via capillaries or the lymph via lacteals.

24.19 The fluid secreted by duodenal (Brunner's glands)—alkaline mucus—neutralizes gastric acid and protects the mucosal lining of the duodenum.

24.20 Because monoglycerides are hydrophobic (nonpolar) molecules, they can dissolve in and diffuse through the lipid bilayer of the plasma membrane.

24.21 The stomach and pancreas are the two digestive system organs that secrete the largest volumes of fluid.

24.22 The ascending and descending portions of the colon are retroperitoneal.

24.23 Goblet cells in the large intestine secrete mucus to lubricate colonic contents.

24.24 The pH of gastric juice rises due to the buffering action of some amino acids in food proteins.

Chapter 25

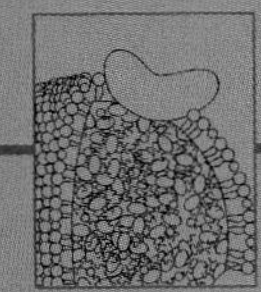

Metabolism and Nutrition

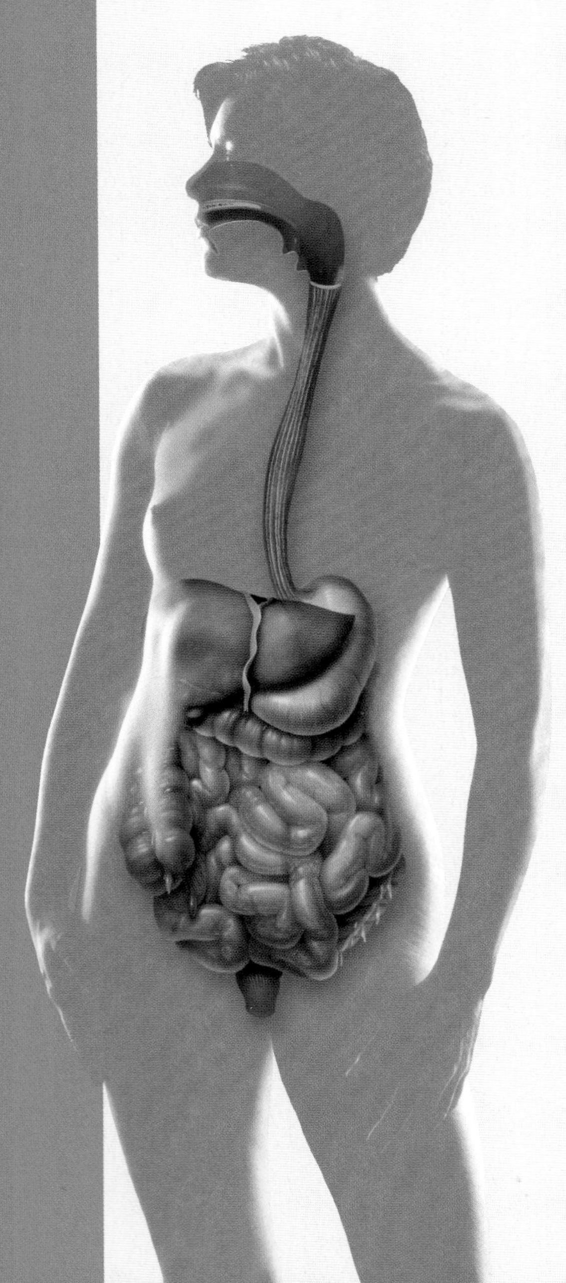

Metabolism and Nutrition and Homeostasis

Metabolic reactions contribute to homeostasis by harvesting chemical energy from consumed nutrients to contribute to the body's growth, repair, and normal functioning.

www. wiley.com/college/apcentral

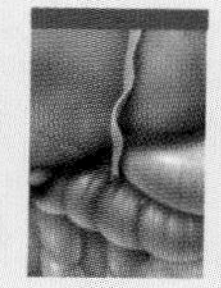

Plants use the green pigment chlorophyll to trap energy in sunlight. We don't have a similarly functioning pigment in our skin, so the food we eat is our only source of energy for running, walking, and even breathing. Many molecules needed to maintain cells and tissues can be made from simpler precursors by the body's metabolic reactions; others—the essential amino acids, essential fatty acids, vitamins, and minerals—must be obtained from the food we eat. As you learned in Chapter 24, carbohydrates, lipids, and proteins in food are digested by enzymes and absorbed in the gastrointestinal tract. The products of digestion that reach body cells are monosaccharides, fatty acids, glycerol, monoglycerides, and amino acids. Some minerals and many vitamins are part of enzyme systems that catalyze the breakdown and synthesis of carbohydrates, lipids, and proteins. Food molecules absorbed by the gastrointestinal (GI) tract have three main fates:

1. Most food molecules are used to *supply energy* for sustaining life processes, such as active transport, DNA replication, protein synthesis, muscle contraction, maintenance of body temperature, and mitosis.
2. Some food molecules *serve as building blocks* for the synthesis of more complex structural or functional molecules, such as muscle proteins, hormones, and enzymes.
3. Other food molecules are *stored for future use.* For example, glycogen is stored in liver cells, and triglycerides are stored in adipose cells.

In this chapter we discuss how metabolic reactions harvest the chemical energy stored in foods, how each group of food molecules contributes to the body's growth, repair, and energy needs, and how heat and energy balance is maintained in the body. Finally, we explore some aspects of nutrition to discover why you should opt for fish instead of a burger the next time you eat out.

METABOLIC REACTIONS

► **OBJECTIVES**

Define metabolism.

Explain the role of ATP in anabolism and catabolism.

Metabolism (me-TAB-ō-lizm; *metabol-* = change) refers to all of the chemical reactions that occur in the body. There are two types of metabolism: catabolism and anabolism. Those chemical reactions that break down complex organic molecules into simpler ones are collectively known as **catabolism** (ka-TAB-ō-lizm; *cata-* = downward). Overall, catabolic (decomposition) reactions are *exergonic;* they produce more energy than they consume, releasing the chemical energy stored in organic molecules. Important sets of catabolic reactions occur in glycolysis, the Krebs cycle, and the electron transport chain, each of which will be discussed later in the chapter.

Chemical reactions that combine simple molecules and monomers to form the body's complex structural and functional components are collectively known as **anabolism** (a-NAB-ō-lizm; *ana-* = upward). Examples of anabolic reactions are the formation of peptide bonds between amino acids during protein synthesis, the building of fatty acids into phospholipids that form the plasma membrane bilayer, and the linkage of glucose monomers to form glycogen. Anabolic reactions are *endergonic;* they consume more energy than they produce.

Metabolism is an energy-balancing act between catabolic (decomposition) reactions, and anabolic (synthesis) reactions. The molecule that participates most often in energy exchanges in living cells is **ATP (adenosine triphosphate),** which couples energy-releasing catabolic reactions to energy-requiring anabolic reactions.

The metabolic reactions that occur depend on which enzymes are active in a particular cell at a particular time, or even in a particular part of the cell. Catabolic reactions can be occurring in the mitochondria of a cell at the same time as anabolic reactions are taking place in the endoplasmic reticulum.

A molecule synthesized in an anabolic reaction has a limited lifetime. With few exceptions, it will eventually be broken down and its component atoms recycled into other molecules or excreted from the body. Recycling of biological molecules occurs continuously in living tissues, more rapidly in some than in others. Individual cells may be refurbished molecule by molecule, or a whole tissue may be rebuilt cell by cell.

Coupling of Catabolism and Anabolism by ATP

The chemical reactions of living systems depend on the efficient transfer of manageable amounts of energy from one molecule to another. The molecule that most often performs this task is ATP, the "energy currency" of a living cell. Like money, it is readily available to "buy" cellular activities; it is spent and earned over and over. A typical cell has about a billion molecules of ATP, each of which typically lasts for less than a minute before being used. Thus, ATP is not a long-term-storage form of currency, like gold in a vault, but rather convenient cash for moment-to-moment transactions.

Recall from Chapter 2 that a molecule of ATP consists of an adenine molecule, a ribose molecule, and three phosphate groups bonded to one another (see Figure 2.25 on page 55). Figure 25.1 shows how ATP links anabolic and catabolic reactions. When the terminal phosphate group is split off ATP, adenosine diphosphate (ADP) and a phosphate group (symbolized as Ⓟ) are formed. Some of the energy released is used to drive anabolic reactions such as the formation of glycogen from glucose. In addition, energy from complex molecules is used in catabolic reactions to combine ADP and a phosphate group to resynthesize ATP:

$$\text{ADP} + \text{Ⓟ} + \text{energy} \longrightarrow \text{ATP}$$

Figure 25.1 Role of ATP in linking anabolic and catabolic reactions. When complex molecules and polymers are split apart (catabolism, at left), some of the energy is transferred to form ATP and the rest is given off as heat. When simple molecules and monomers are combined to form complex molecules (anabolism, at right), ATP provides the energy for synthesis, and again some energy is given off as heat.

The coupling of energy-releasing and energy-requiring reactions is achieved through ATP.

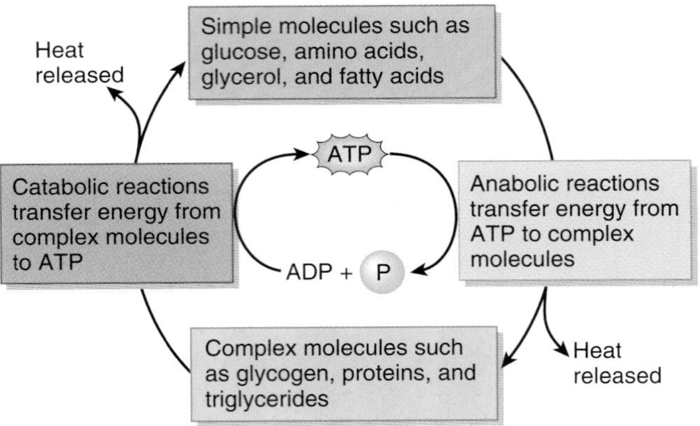

In a pancreatic cell that produces digestive enzymes, does anabolism or catabolism predominate?

About 40% of the energy released in catabolism is used for cellular functions; the rest is converted to heat, some of which helps maintain normal body temperature. Excess heat is lost to the environment. Compared with machines, which typically convert only 10–20% of energy into work, the 40% efficiency of the body's metabolism is impressive. Still, the body has a continuous need to take in and process external sources of energy so that cells can synthesize enough ATP to sustain life.

▶ CHECKPOINT

1. What is metabolism? Distinguish between anabolism and catabolism, and give examples of each.
2. How does ATP link anabolism and catabolism?

ENERGY TRANSFER

▶ OBJECTIVES

Describe oxidation–reduction reactions.

Explain the role of ATP in metabolism.

Various catabolic reactions transfer energy into the "high-energy" phosphate bonds of ATP. Although the amount of energy in these bonds is not exceptionally large, it can be released quickly and easily. Before discussing metabolic pathways, it is important to understand how this transfer of energy occurs. Two important aspects of energy transfer are oxidation–reduction reactions and mechanisms of ATP generation.

Oxidation–Reduction Reactions

Oxidation is the *removal of electrons* from an atom or molecule; the result is a *decrease* in the potential energy of the atom or molecule. Because most biological oxidation reactions involve the loss of hydrogen atoms, they are called *dehydrogenation reactions*. An example of an oxidation reaction is the conversion of lactic acid into pyruvic acid:

$$\underset{\text{Lactic acid}}{\text{H—C(COOH)(CH}_3\text{)—OH}} \xrightarrow[\text{Remove 2H (H}^+ + \text{H}^-)]{\text{Oxidation}} \underset{\text{Pyruvic acid}}{\text{(COOH)C=O(CH}_3)}$$

In the preceding reaction, 2H ($H^+ + H^-$) means that two neutral hydrogen atoms (2H) are removed as one hydrogen ion (H^+) plus one hydride ion (H^-).

Reduction is the opposite of oxidation; it is the *addition of electrons* to a molecule. Reduction results in an *increase* in the potential energy of the molecule. An example of a reduction reaction is the conversion of pyruvic acid into lactic acid:

$$\underset{\text{Pyruvic acid}}{\text{(COOH)C=O(CH}_3)} \xrightarrow[\text{Add 2H (H}^+ + \text{H}^-)]{\text{Reduction}} \underset{\text{Lactic acid}}{\text{H—C(COOH)(CH}_3\text{)—OH}}$$

When a substance is oxidized, the liberated hydrogen atoms do not remain free in the cell but are transferred immediately by coenzymes to another compound. Two coenzymes are commonly used by animal cells to carry hydrogen atoms: **nicotinamide adenine dinucleotide (NAD),** a derivative of the B vitamin niacin, and **flavin adenine dinucleotide (FAD),** a derivative of vitamin B_2 (riboflavin). The oxidation and reduction states of NAD^+ and FAD can be represented as follows:

$$\underset{\text{Oxidized}}{NAD^+} \underset{-\,2H\,(H^+ + H^-)}{\overset{+\,2H\,(H^+ + H^-)}{\rightleftharpoons}} \underset{\text{Reduced}}{NADH + H^+}$$

$$\underset{\text{Oxidized}}{FAD} \underset{-\,2H\,(H^+ + H^-)}{\overset{+\,2H\,(H^+ + H^-)}{\rightleftharpoons}} \underset{\text{Reduced}}{FADH_2}$$

When NAD^+ is reduced to $NADH + H^+$, the NAD^+ gains a hydride ion (H^-), neutralizing its charge, and the H^+ is released into the surrounding solution. When NADH is oxidized to NAD^+, the loss of the hydride ion results in one less hydrogen atom and an additional positive charge. FAD is reduced to $FADH_2$ when it gains a hydrogen ion and a hydride ion, and $FADH_2$ is oxidized to FAD when it loses the same two ions.

Oxidation and reduction reactions are always coupled; each time one substance is oxidized, another is simultaneously reduced. Such paired reactions are called **oxidation–reduction** or **redox reactions.** For example, when lactic acid is *oxidized* to form pyruvic acid, the two hydrogen atoms removed in the reaction are used to *reduce* NAD^+. This coupled redox reaction may be written as follows:

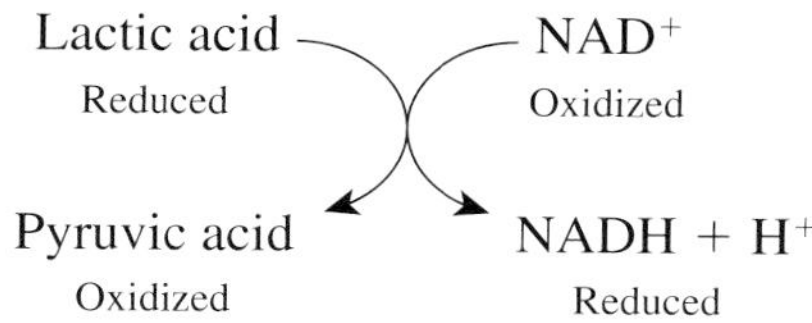

An important point to remember about oxidation–reduction reactions is that oxidation is usually an exergonic (energy-releasing) reaction. Cells use multistep biochemical reactions to release energy from energy-rich, highly reduced compounds (with many hydrogen atoms) to lower-energy, highly oxidized compounds (with many oxygen atoms or multiple bonds). For example, when a cell oxidizes a molecule of glucose ($C_6H_{12}O_6$), the energy in the glucose molecule is removed in a stepwise manner. Ultimately, some of the energy is captured by transferring it to ATP, which then serves as an energy source for energy-requiring reactions within the cell. Compounds with many hydrogen atoms such as glucose contain more chemical potential energy than oxidized compounds. For this reason, glucose is a valuable nutrient.

Mechanisms of ATP Generation

Some of the energy released during oxidation reactions is captured within a cell when ATP is formed. Briefly, a phosphate group (Ⓟ) is added to ADP, with an input of energy, to form ATP. The two high-energy phosphate bonds that can be used to transfer energy are indicated by "squiggles" (~):

$$\underset{\text{ADP}}{\text{Adenosine} - \text{Ⓟ} \sim \text{Ⓟ}} + \text{Ⓟ} + \text{energy} \longrightarrow \underset{\text{ATP}}{\text{Adenosine} - \text{Ⓟ} \sim \text{Ⓟ} \sim \text{Ⓟ}}$$

The high-energy phosphate bond that attaches the third phosphate group contains the energy stored in this reaction. The addition of a phosphate group to a molecule, called **phosphorylation** (fos′-for-i-LĀ-shun), increases its potential energy. Organisms use three mechanisms of phosphorylation to generate ATP:

1. **Substrate-level phosphorylation** generates ATP by transferring a high-energy phosphate group from an intermediate phosphorylated metabolic compound—a substrate—directly to ADP. In human cells, this process occurs in the cytosol.
2. **Oxidative phosphorylation** removes electrons from organic compounds and passes them through a series of electron acceptors, called the **electron transport chain,** to molecules of oxygen (O_2). This process occurs in the inner mitochondrial membrane of cells.
3. **Photophosphorylation** occurs only in chlorophyll-containing plant cells or in certain bacteria that contain other light-absorbing pigments.

▶ CHECKPOINT

3. How is a hydride ion different from a hydrogen ion? What is the involvement of both ions in redox reactions?
4. What are three ways that ATP can be generated?

CARBOHYDRATE METABOLISM

▶ OBJECTIVE

Describe the fate, metabolism, and functions of carbohydrates.

As you learned in Chapter 24, both polysaccharides and disaccharides are hydrolyzed into the monosaccharides glucose (about 80%), fructose, and galactose during the digestion of carbohydrates. (Some fructose is converted into glucose as it is absorbed through the intestinal epithelial cells.) Hepatocytes (liver cells) convert most of the remaining fructose and practically all the galactose to glucose. So the story of carbohydrate metabolism is really the story of glucose metabolism. Because negative feedback systems maintain blood glucose at about 90 mg/100 mL of plasma (5 mmol/liter), a total of 2–3 g of glucose normally circulates in the blood.

The Fate of Glucose

Because glucose is the body's preferred source for synthesizing ATP, its use depends on the needs of body cells, which include the following:

- ***ATP production.*** In body cells that require immediate energy, glucose is oxidized to produce ATP. Glucose not needed for immediate ATP production can enter one of several other metabolic pathways.
- ***Amino acid synthesis.*** Cells throughout the body can use glucose to form several amino acids, which then can be incorporated into proteins.
- ***Glycogen synthesis.*** Hepatocytes and muscle fibers can perform **glycogenesis** (glī′-kō-JEN-e-sis; *glyco-* = sugar or sweet; *-genesis* = to generate), in which hundreds of glucose monomers are combined to form the polysaccharide glycogen. Total storage capacity of glycogen is about 125 g in the liver and 375 g in skeletal muscles.
- ***Triglyceride synthesis.*** When the glycogen storage areas are filled up, hepatocytes can transform the glucose to glycerol and fatty acids that can be used for **lipogenesis** (lip′-ō-JEN-ē-sis), the synthesis of triglycerides. Triglycerides then are deposited in adipose tissue, which has virtually unlimited storage capacity.

Glucose Movement into Cells

Before glucose can be used by body cells, it must first pass through the plasma membrane and enter the cytosol. Glucose absorption in the gastrointestinal tract (and kidney tubules) is accomplished via secondary active transport (Na^+-glucose symporters). Glucose entry into most other body cells occurs via GluT molecules, a family of transporters that bring glucose into cells via facilitated diffusion (see page 69). A high level of insulin increases the insertion of one type of GluT, called GluT4, into the plasma membranes of most body cells, thereby increasing the rate of facilitated diffusion of glucose into cells. In neurons and hepatocytes, however, another type of GluT is always present in the plasma membrane, so glucose entry is always "turned on." Upon entering a cell, glucose becomes phosphorylated. Because GluT cannot transport phosphorylated glucose, this reaction traps glucose within the cell.

Glucose Catabolism

The oxidation of glucose to produce ATP is also known as **cellular respiration,** and it involves four sets of reactions: glycolysis, the formation of acetyl coenzyme A, the Krebs cycle, and the electron transport chain (Figure 25.2).

1. ***Glycolysis*** is a set of reactions in which one glucose molecule is oxidized and two molecules of pyruvic acid are produced. The reactions also produce two molecules of ATP and two energy-containing NADH + H^+. Because glycolysis does not require oxygen, it is a way to produce ATP anaerobically (without oxygen) and is known as **anaerobic cellular respiration** (an-ar-Ō-bik; *an-* = not; *aer-* = air; *-bios* = life).
2. ***Formation of acetyl coenzyme A*** is a transition step that prepares pyruvic acid for entrance into the Krebs cycle. This

Figure 25.2 Overview of cellular respiration (oxidation of glucose). A modified version of this figure appears in several places in this chapter to indicate the relationships of particular reactions to the overall process of cellular respiration.

The oxidation of glucose involves glycolysis, the formation of acetyl coenzyme A, the Krebs cycle, and the electron transport chain.

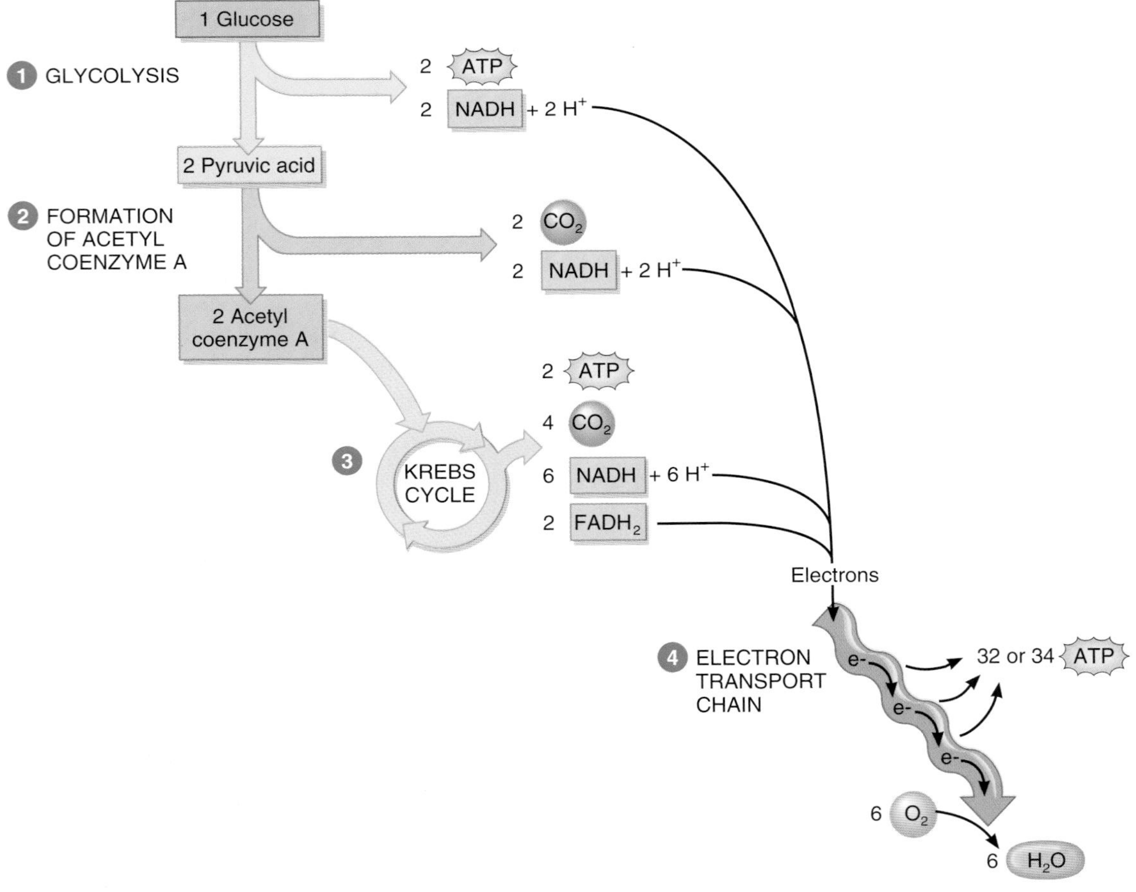

Which of the four processes shown here is also called anaerobic cellular respiration?

step also produces energy-containing NADH + H^+ plus carbon dioxide (CO_2).

3 ***Krebs cycle reactions*** oxidize acetyl coenzyme A and produce CO_2, ATP, energy-containing NADH + H^+, and $FADH_2$.

4 ***Electron transport chain reactions*** oxidize NADH + H^+ and $FADH_2$ and transfer their electrons through a series of electron carriers. The Krebs cycle and the electron transport chain both require oxygen to produce ATP and are collectively known as **aerobic cellular respiration.**

Glycolysis

During **glycolysis** (glī-KOL-i-sis; *-lysis* = breakdown), chemical reactions split a 6-carbon molecule of glucose into two 3-carbon molecules of pyruvic acid (Figure 25.3). Even though glycolysis consumes two ATP molecules, it produces four ATP molecules, for a net gain of two ATP molecules for each glucose molecule that is oxidized.

Figure 25.4 shows the 10 reactions that comprise glycolysis. In the first half of the sequence (reactions 1 through 5), energy in the form of ATP is "invested" and the 6-carbon glucose is split into two 3-carbon molecules of glyceraldehyde 3-phosphate. *Phosphofructokinase,* the enzyme that catalyzes step 3, is the key regulator of the rate of glycolysis. The activity of this enzyme is high when ADP concentration is high, in which case ATP is produced rapidly. When the activity of phosphofructokinase is low, most glucose does not enter the reactions of glycolysis but instead undergoes conversion to glycogen for storage. In the second half of the sequence (reactions 6 through 10), the two glyceraldehyde 3-phosphate molecules are converted to two pyruvic acid molecules and ATP is generated.

Figure 25.3 Cellular respiration begins with glycolysis.

During glycolysis, each molecule of glucose is converted to two molecules of pyruvic acid.

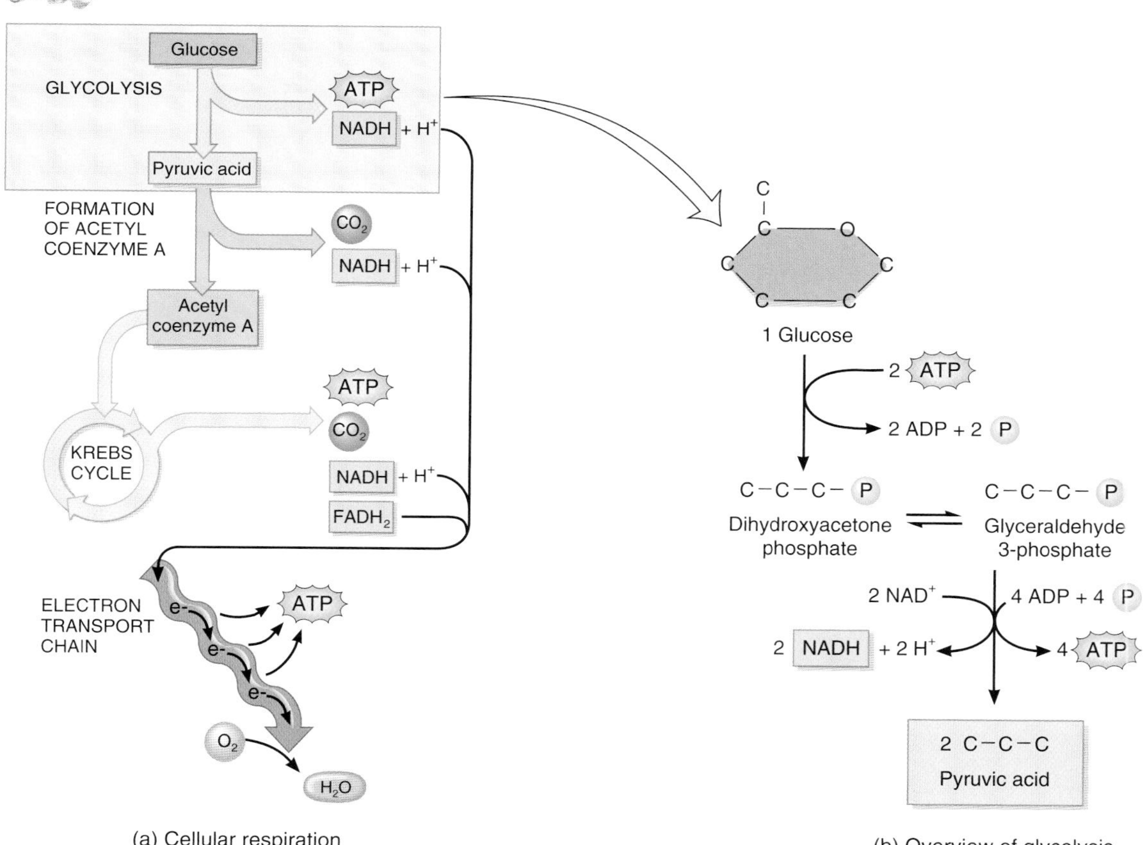

(a) Cellular respiration

(b) Overview of glycolysis

For each glucose molecule that undergoes glycolysis, how many ATP molecules are generated?

Figure 25.4 The 10 reactions of glycolysis. (1) Glucose is phosphorylated, using a phosphate group from an ATP molecule to form glucose 6-phosphate. (2) Glucose 6-phosphate is converted to fructose 6-phosphate. (3) A second ATP is used to add a second phosphate group to fructose 6-phosphate to form fructose 1,6-bisphosphate. (4) and (5) Fructose splits into two three-carbon molecules, glyceraldehyde 3-phosphate (G 3-P) and dihydroxyacetone phosphate, each having one phosphate group. (6) Oxidation occurs as two molecules of NAD^+ accept two pairs of electrons and hydrogen ions from two molecules of G 3-P; each molecule of G 3-P forms two molecules of NADH. Many body cells use the two NADH produced in this step to generate four ATPs in the electron transport chain. Hepatocytes, kidney cells, and cardiac muscle fibers can generate six ATPs from the two NADH. A second phosphate group attaches to G 3-P, forming 1,3-bisphosphoglyceric acid (BPG). (7) through (10). These reactions generate four molecules of ATP and produce two molecules of pyruvic acid (pyruvate*).

Glycolysis results in a net gain of two ATP, two NADH, and two H^+.

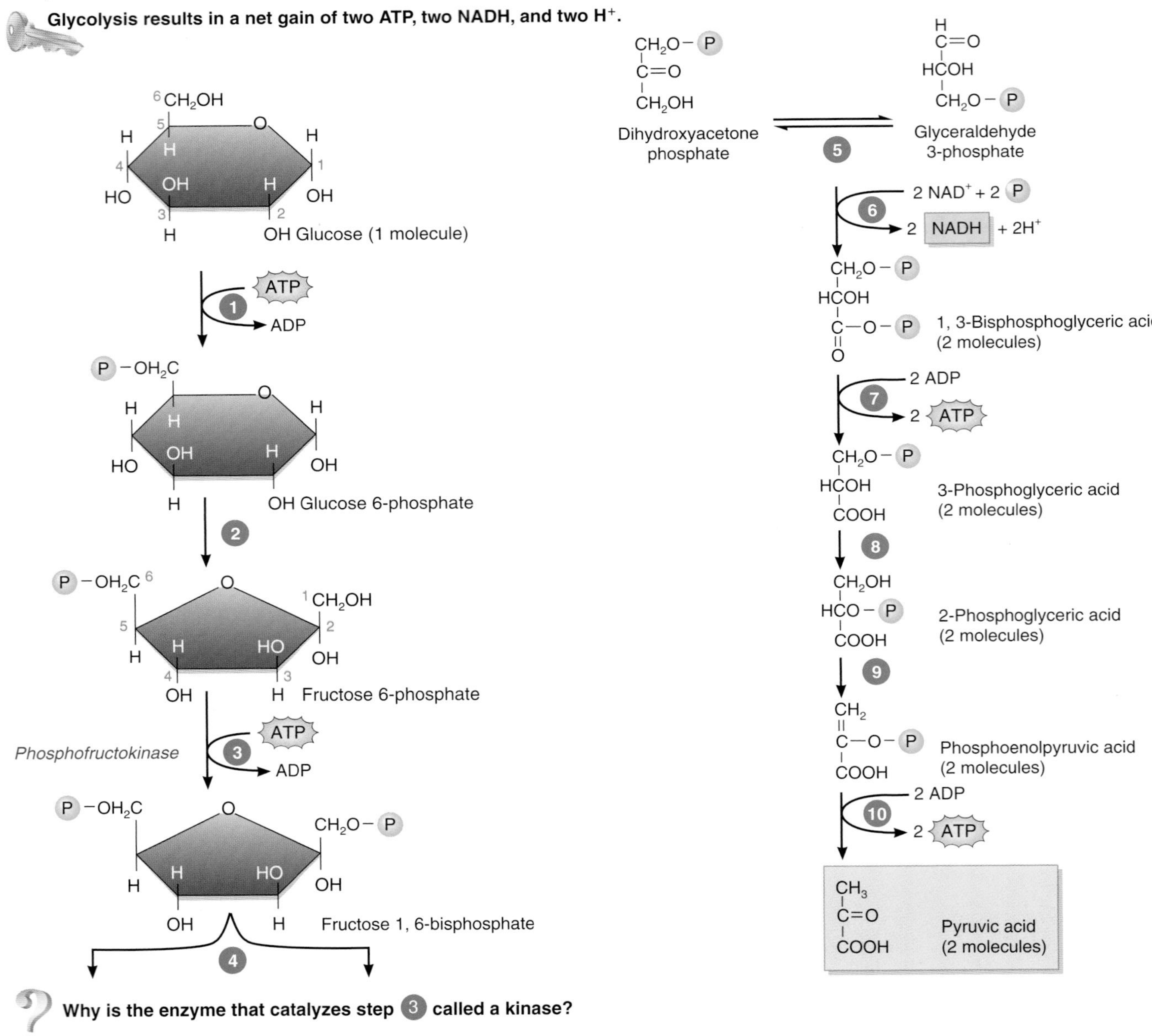

Why is the enzyme that catalyzes step (3) called a kinase?

*The carboxyl groups (—COOH) of intermediates in glycolysis and in the citric acid cycle are mostly ionized at the pH of body fluids to $—COO^-$. The suffix "-ic acid" indicates the non-ionized form, whereas the ending "-ate" indicates the ionized form. Although the "-ate" names are more correct, we will use the "acid" names because these terms are more familiar.

The Fate of Pyruvic Acid

The fate of pyruvic acid produced during glycolysis depends on the availability of oxygen (Figure 25.5). If oxygen is scarce (anaerobic conditions)—for example, in skeletal muscle fibers during strenuous exercise—then pyruvic acid is reduced via an anaerobic pathway by the addition of two hydrogen atoms to form lactic acid (lactate):

$$\underset{\text{Oxidized}}{2\ \text{Pyruvic acid} + 2\ \text{NADH}} + 2\ \text{H}^+ \longrightarrow \underset{\text{Reduced}}{2\ \text{Lactic acid}} + 2\text{NAD}^+$$

This reaction regenerates the NAD^+ that was used in the oxidation of glyceraldehyde 3-phosphate (see step 6 in Figure 25.4) and thus allows glycolysis to continue. As lactic acid is produced, it rapidly diffuses out of the cell and enters the blood. Hepatocytes remove lactic acid from the blood and convert it back to pyruvic acid. Recall that a buildup of lactic acid is one factor that contributes to muscle fatigue.

When oxygen is plentiful (aerobic conditions), most cells convert pyruvic acid to acetyl coenzyme A. This molecule links glycolysis, which occurs in the cytosol, with the Krebs cycle, which occurs in the matrix of mitochondria. Pyruvic acid enters the mitochondrial matrix with the help of a special transporter protein. Because they lack mitochondria, red blood cells can only produce ATP through glycolysis.

Formation of Acetyl Coenzyme A

Each step in the oxidation of glucose requires a different enzyme, and often a coenzyme as well. The coenzyme used at this point in cellular respiration is **coenzyme A (CoA),** which is derived from pantothenic acid, a B vitamin. During the transitional step between glycolysis and the Krebs cycle, pyruvic acid is prepared for entrance into the cycle. The enzyme *pyruvate dehydrogenase,* which is located exclusively in the mitochondrial matrix, converts pyruvic acid to a two-carbon fragment called an **acetyl group** by removing a molecule of carbon dioxide (Figure 25.5). The loss of a molecule of CO_2 by a substance is called **decarboxylation** (dē-kar-bok′-si-LĀ-shun). This is the first reaction in cellular respiration that releases CO_2. During this reaction, pyruvic acid is also oxidized. Each pyruvic acid loses two hydrogen atoms in the form of one hydride ion (H^-) plus one hydrogen ion (H^+). The coenzyme NAD^+ is reduced as it picks up the H^- from pyruvic acid; the H^+ is released into the mitochondrial matrix. The reduction of NAD^+ to NADH + H^+ is indicated in Figure 25.5 by the curved arrow entering and then leaving the reaction. Recall that the oxidation of one glucose molecule produces two molecules of pyruvic acid, so for each molecule of glucose, two molecules of carbon dioxide are lost and two NADH + H^+ are produced. The acetyl group attaches to coenzyme A, producing a molecule called **acetyl coenzyme A (acetyl CoA).**

The Krebs Cycle

Once the pyruvic acid has undergone decarboxylation and the remaining acetyl group has attached to CoA, the resulting com-

Figure 25.5 Fate of pyruvic acid.

When oxygen is plentiful, pyruvic acid enters mitochondria, is converted to acetyl coenzyme A, and enters the Krebs cycle (aerobic pathway). When oxygen is scarce, most pyruvic acid is converted to lactic acid via an anaerobic pathway.

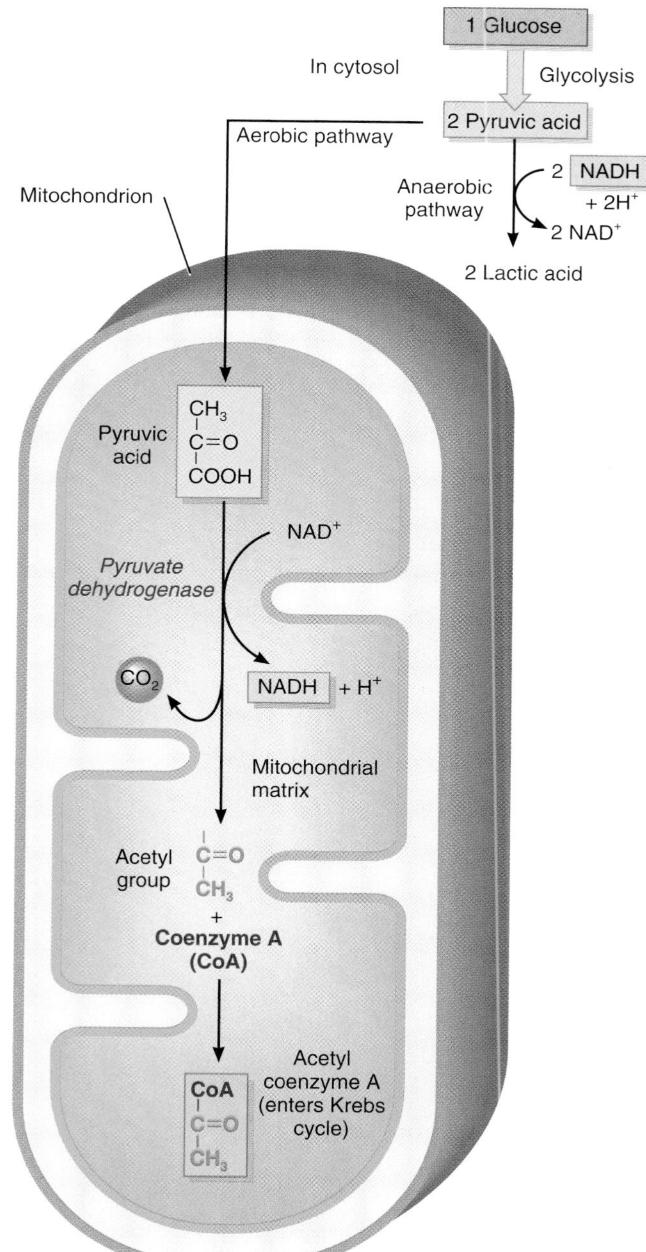

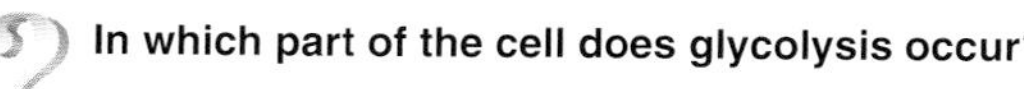
In which part of the cell does glycolysis occur?

pound (acetyl CoA) is ready to enter the Krebs cycle (Figure 25.6). The **Krebs cycle**—named for the biochemist Hans Krebs, who described these reactions in the 1930s—is also known as the **citric acid cycle,** for the first molecule formed when an acetyl group joins the cycle. The reactions occur in the matrix of mitochondria and consist of a series of oxidation–reduction reactions and decarboxylation reactions that release CO_2. In the Krebs cycle, the oxidation–reduction reactions transfer chemical energy, in the form of electrons, to two coenzymes—NAD^+ and FAD. The pyruvic acid derivatives are oxidized, and the coenzymes are reduced. In addition, one step generates ATP. Figure 25.7 shows the reactions of the Krebs cycle in more detail.

The reduced coenzymes (NADH and $FADH_2$) are the most important outcome of the Krebs cycle because they contain the energy originally stored in glucose and then in pyruvic acid. Overall, for every acetyl CoA that enters the Krebs cycle, three NADH, three H^+, and one $FADH_2$ are produced by oxidation–reduction reactions, and one molecule of ATP is generated by substrate-level phosphorylation (see Figure 25.6). In the electron transport chain, the three NADH + three H^+ will later yield nine ATP molecules, and the $FADH_2$, will later yield two ATP molecules. Thus, each "turn" of the Krebs cycle eventually generates 12 molecules of ATP. Because each glucose molecule provides two acetyl CoA molecule, glucose catabolism via the Krebs cycle and the electron transport chain yields 24 molecules of ATP per glucose molecule.

Liberation of CO_2 occurs as pyruvic acid is converted to acetyl CoA and during the two decarboxylation reactions of the Krebs cycle (see Figure 25.6). Because each molecule of glucose generates two molecules of pyruvic acid, six molecules of CO_2 are liberated from each original glucose molecule catabolized along this pathway. The molecules of CO_2 diffuse out of the mitochondria, through the cytosol and plasma membrane, and then into the blood. Blood transports the CO_2 to the lungs, where it eventually is exhaled.

Figure 25.6 After formation of acetyl coenzyme A, the next stage of cellular respiration is the Krebs cycle.

Reactions of the Krebs cycle occur in the matrix of mitochondria.

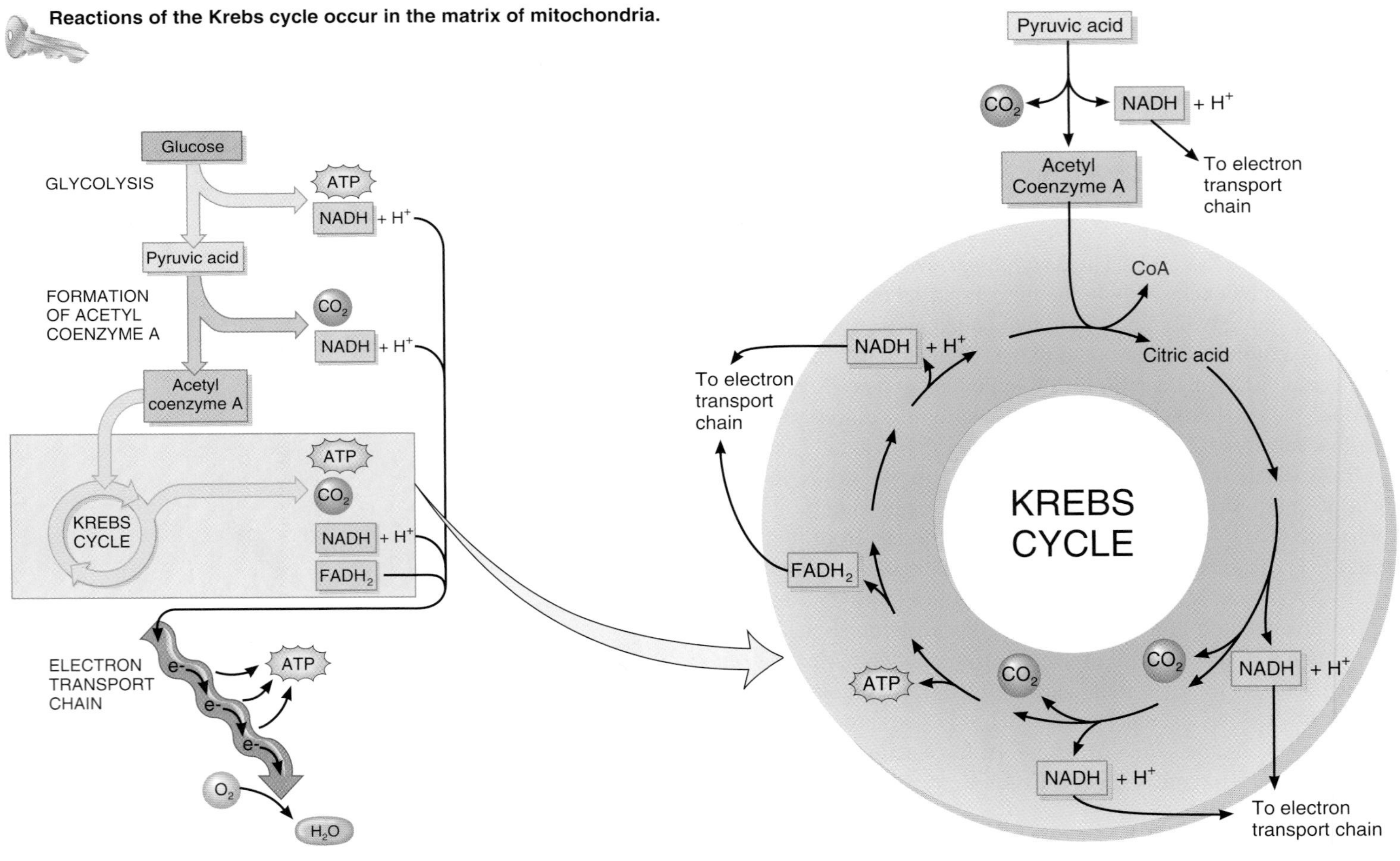

When in cellular respiration is carbon dioxide given off? What happens to this gas?

Figure 25.7 The eight reactions of the Krebs cycle. ① ***Entry of the acetyl group.*** The chemical bond that attaches the acetyl group to coenzyme A (CoA) breaks, and the two-carbon acetyl group attaches to a four-carbon molecule of oxaloacetic acid to form a six-carbon molecule called citric acid. CoA is free to combine with another acetyl group from pyruvic acid and repeat the process. ② ***Isomerization.*** Citric acid undergoes isomerization to isocitric acid, which has the same molecular formula as citrate. Notice, however, that the hydroxyl group (—OH) is attached to a different carbon. ③ ***Oxidative decarboxylation.*** Isocitric acid is oxidized and loses a molecule of CO_2, forming alpha-ketoglutaric acid. The H^- from the oxidation is passed on to NAD^+, which is reduced to NADH + H^+. ④ ***Oxidative decarboxylation.*** Alpha-ketoglutaric acid is oxidized, loses a molecule of CO_2, and picks up CoA to form succinyl CoA. ⑤ ***Substrate-level phosphorylation.*** CoA is displaced by a phosphate group, which is then transferred to guanosine diphosphate (GDP) to form guanosine triphosphate (GTP). GTP can donate a phosphate group to ADP to form ATP. ⑥ ***Dehydrogenation.*** Succinic acid is oxidized to fumaric acid as two of its hydrogen atoms are transferred to the coenzyme flavin adenine dinucleotide (FAD), which is reduced to $FADH_2$. ⑦ ***Hydration.*** Fumaric acid is converted to malic acid by the addition of a molecule of water. ⑧ ***Dehydrogenation.*** In the final step in the cycle, malic acid is oxidized to re-form oxaloacetic acid. Two hydrogen atoms are removed and one is transferred to NAD^+, which is reduced to NADH + H^+. The regenerated oxaloacetic acid can combine with another molecule of acetyl CoA, beginning a new cycle.

The three main results of the Krebs cycle are the production of reduced coenzymes (NADH + H^+ and $FADH_2$), which contain stored energy; the generation of GTP, a high-energy compound that is used to produce ATP; and the formation of CO_2, which is transported to the lungs and exhaled.

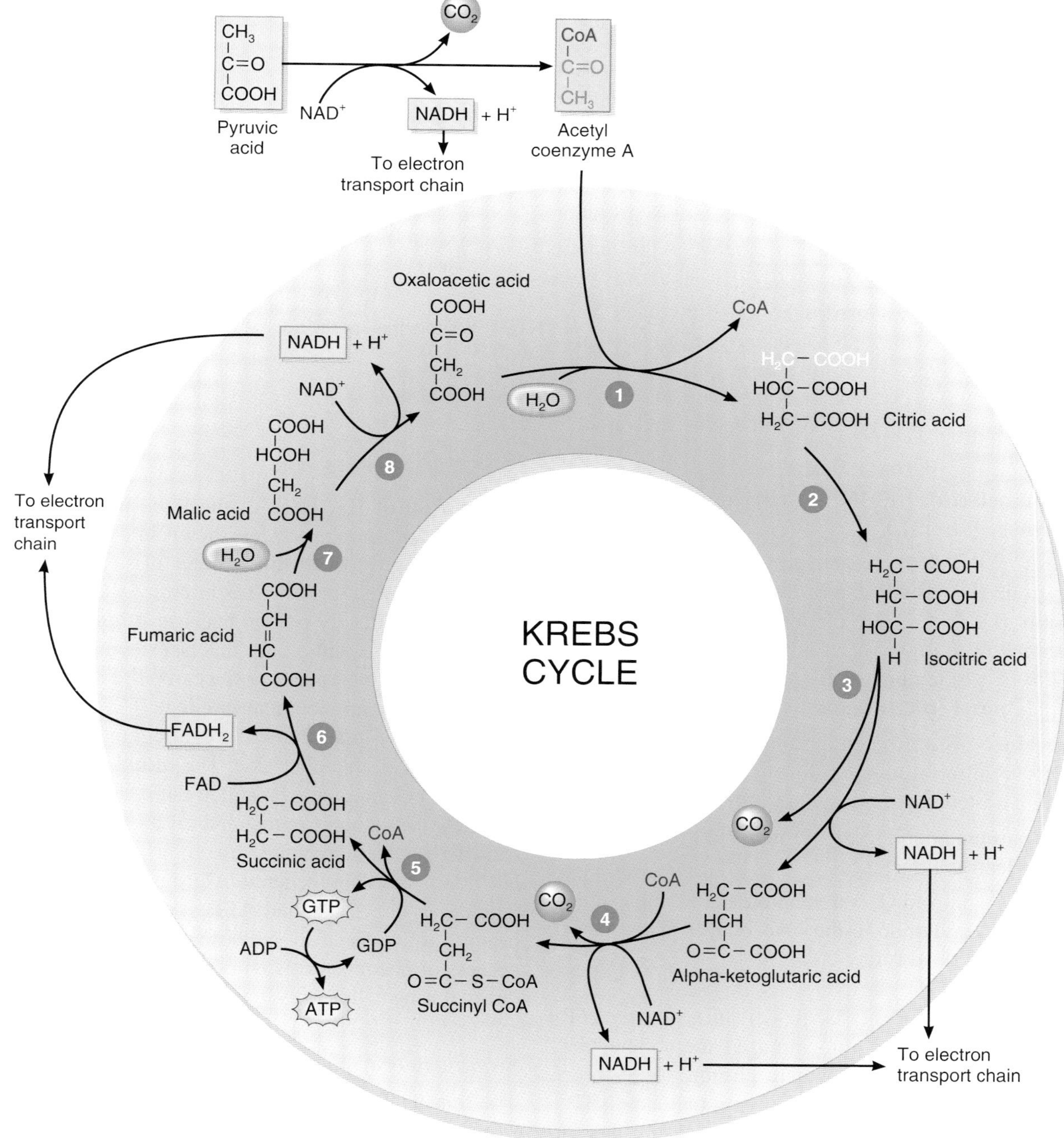

Why is the production of reduced coenzymes important in the Krebs cycle?

The Electron Transport Chain

The **electron transport chain** is a series of **electron carriers,** integral membrane proteins in the inner mitochondrial membrane. This membrane is folded into cristae that increase its surface area, accommodating thousands of copies of the transport chain in each mitochondrion. Each carrier in the chain is reduced as it picks up electrons and oxidized as it gives up electrons. As electrons pass through the chain, a series of exergonic reactions release small amounts of energy; this energy is used to form ATP. In aerobic cellular respiration, the final electron acceptor of the chain is oxygen. Because this mechanism of ATP generation links chemical reactions (the passage of electrons along the transport chain) with the pumping of hydrogen ions, it is called **chemiosmosis** (kem′-ē-oz-MŌ-sis; *chemi-* = chemical; *-osmosis* = pushing). Briefly, chemiosmosis works as follows (Figure 25.8):

1. Energy from NADH + H^+ passes along the electron transport chain and is used to pump H^+ from the matrix of the mitochondrion into the space between the inner and outer mitochondrial membranes. This mechanism is called a **proton pump** because H^+ ions consist of a single proton.
2. A high concentration of H^+ accumulates between the inner and outer mitochondrial membranes.
3. ATP synthesis then occurs as hydrogen ions flow back into the mitochondrial matrix through a special type of H^+ channel in the inner membrane.

Figure 25.8 **Chemiosmosis.**

In chemiosmosis, ATP is produced when hydrogen ions diffuse back into the mitochondrial matrix.

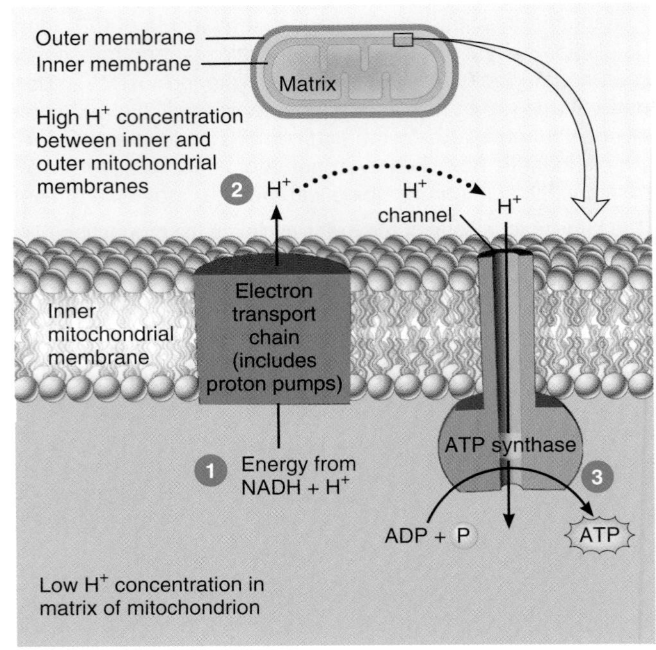

What is the energy source that powers the proton pumps?

Electron Carriers Several types of molecules and atoms serve as electron carriers:

- **Flavin mononucleotide (FMN)** is a flavoprotein derived from riboflavin (vitamin B_2).
- **Cytochromes** (SĪ-tō-krōmz) are proteins with an iron-containing group (heme) capable of existing alternately in a reduced form (Fe^{2+}) and an oxidized form (Fe^{3+}). The cytochromes involved in the electron transport chain include cytochrome *b* (cyt *b*), cytochrome c_1 (cyt c_1), cytochrome *c* (cyt *c*), cytochrome *a* (cyt *a*), and cytochrome a_3 (cyt a_3).
- **Iron-sulfur (Fe-S) centers** contain either two or four iron atoms bound to sulfur atoms that form an electron transfer center within a protein.
- **Copper (Cu) atoms** bound to two proteins in the chain also participate in electron transfer.
- **Coenzyme Q,** symbolized **Q,** is a nonprotein, low-molecular-weight carrier that is mobile in the lipid bilayer of the inner membrane.

Steps in Electron Transport and Chemiosmotic ATP Generation Within the inner mitochondrial membrane, the carriers of the electron transport chain are clustered into three complexes, each of which acts as a proton pump that expels H^+ from the mitochondrial matrix and helps create an electrochemical gradient of H^+. Each of the three proton pumps transports electrons and pumps H^+, as shown in Figure 25.9. Notice that oxygen is used to help form water in step 3. This is the only point in aerobic cellular respiration where O_2 is consumed. **Cyanide** is a deadly poison because it binds to the cytochrome oxidase complex and blocks this last step in electron transport.

The pumping of H^+ produces both a concentration gradient of protons and an electrical gradient. The buildup of H^+ makes one side of the inner mitochondrial membrane positively charged compared with the other side. The resulting electrochemical gradient has potential energy, called the *proton motive force.* Proton channels in the inner mitochondrial membrane allow H^+ to flow back across the membrane, driven by the proton motive force. As H^+ flow back, they generate ATP because the H^+ channels also include an enzyme called **ATP synthase.** The enzyme uses the proton motive force to synthesize ATP from ADP and Ⓟ. The process of chemiosmosis is responsible for most of the ATP produced during cellular respiration.

Summary of Cellular Respiration

The various electron transfers in the electron transport chain generate either 32 or 34 ATP molecules from each molecule of

Figure 25.9 The actions of the three proton pumps and ATP synthase in the inner membrane of mitochondria. Each pump is a complex of three or more electron carriers. (1) The first proton pump is the *NADH dehydrogenase complex,* which contains flavin mononucleotide (FMN) and five or more Fe-S centers. NADH + H^+ is oxidized to NAD^+, and FMN is reduced to $FMNH_2$, which in turn is oxidized as it passes electrons to the iron-sulfur centers. Q, which is mobile in the membrane, shuttles electrons to the second pump complex. (2) The second proton pump is the *cytochrome b-c_1 complex,* which contains cytochromes and an iron-sulfur center. Electrons are passed successively from Q to cyt *b,* to Fe-S, to cyt c_1. The mobile shuttle that passes electrons from the second pump complex to the third is cytochrome *c* (cyt *c*). (3) The third proton pump is the *cytochrome oxidase complex,* which contains cytochromes *a* and a_3 and two copper atoms. Electrons pass from cyt *c,* to Cu, to cyt *a,* and finally to cyt a_3. Cyt a_3 passes its electrons to one-half of a molecule of oxygen (O_2), which becomes negatively charged and then picks up two H^+ from the surrounding medium to form H_2O.

As the three proton pumps pass electrons from one carrier to the next, they also move protons (H^+) from the matrix into the space between the inner and outer mitochondrial membranes. As protons flow back into the mitochondrial matrix through the H^+ channel in ATP synthase, ATP is synthesized.

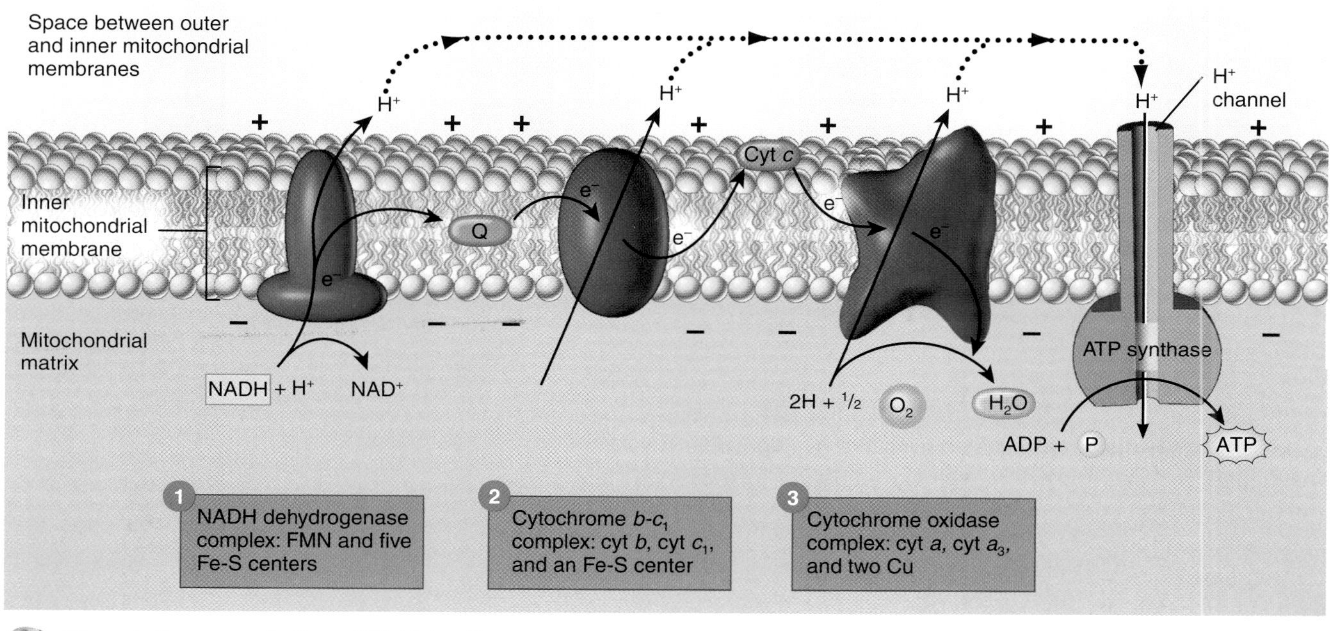

Where is the concentration of H^+ highest?

glucose that is oxidized: either 28 or 30* from the 10 molecules of NADH + H^+ and two from each of the two molecules of $FADH_2$ (four total). Thus, during cellular respiration, 36 or 38 ATPs can be generated from one molecule of glucose. Note that two of those ATPs come from substrate-level phosphorylation in glycolysis, and two come from substrate-level phosphorylation in the Krebs cycle. The overall reaction is:

$$\underset{\text{Glucose}}{C_6H_{12}O_6} + \underset{\text{Oxygen}}{6\ O_2} + 36 \text{ or } 38 \text{ ADPs} + 36 \text{ or } 38 \text{ (P)} \longrightarrow \underset{\text{Carbon dioxide}}{6\ CO_2} + \underset{\text{Water}}{6\ H_2O} + 36 \text{ or } 38 \text{ ATPs}$$

Table 25.1 summarizes the ATP yield during cellular respiration. A schematic depiction of the principal reactions of cellular respiration is presented in Figure 25.10. The actual ATP yield may be lower than 36 or 38 ATPs per glucose. One uncertainty is the exact number of H^+ that must be pumped out to generate one ATP during chemiosmosis. In addition, the ATP

*The two NADH produced in the cytosol during glycolysis cannot enter mitochondria. Instead, they donate their electrons to one of two transfer molecules, known as the malate shuttle and the glycerol phosphate shuttle. In cells of the liver, kidneys, and heart, use of the malate shuttle results in three ATPs synthesized for each NADH. In other body cells, such as skeletal muscle fibers and neurons, use of the glycerol phosphate shuttle results in two ATPs synthesized for each NADH.

TABLE 25.1 Summary of ATP Produced in Cellular Respiration

Source	ATP Yield Per Glucose Molecule (Process)
Glycolysis	
Oxidation of one glucose molecule to two pyruvic acid molecules	2 ATPs (substrate-level phosphorylation)
Production of 2 NADH + H^+	4 or 6 ATPs (oxidative phosphorylation in electron transport chain)
Formation of Two Molecules of Acetyl Coenzyme A	
2 NADH + 2 H^+	6 ATPs (oxidative phosphorylation in electron transport chain)
Krebs Cycle and Electron Transport Chain	
Oxidation of succinyl CoA to succinic acid	2 GTPs that are converted to 2 ATPs (substrate-level phosphorylation)
Production of 6 NADH + 6 H^+	18 ATPs (oxidative phosphorylation in electron transport chain)
Production of 2 $FADH_2$	4 ATPs (oxidative phosphorylation in electron transport chain)
Total:	36 or 38 ATPs per glucose molecule (theoretical maximum)

generated in mitochondria must be transported out of these organelles into the cytosol for use elsewhere in a cell. Exporting ATP in exchange for the inward movement of ADP formed from metabolic reactions in the cytosol uses up part of the proton motive force.

Glycolysis, the Krebs cycle, and especially the electron transport chain provide all the ATP for cellular activities. Because the Krebs cycle and electron transport chain are aerobic processes, cells cannot carry on their activities for long if oxygen is lacking.

Glucose Anabolism

Even though most of the glucose in the body is catabolized to generate ATP, glucose may take part in or be formed via several anabolic reactions. One is the synthesis of glycogen; another is the synthesis of new glucose molecules from some of the products of protein and lipid breakdown.

Glucose Storage: Glycogenesis

If glucose is not needed immediately for ATP production, it combines with many other molecules of glucose to form **glycogen,** a polysaccharide that is the only stored form of carbohydrate in our bodies. The hormone insulin, from pancreatic beta cells, stimulates hepatocytes and skeletal muscle cells to carry out **glycogenesis,** the synthesis of glycogen (Figure 25.11). The body can store about 500 g (about 1.1 lb) of glycogen, roughly 75% in skeletal muscle fibers and the rest in liver cells. During glycogenesis, glucose is first phosphorylated to glucose 6-phosphate by hexokinase. Glucose 6-phosphate is converted to glucose 1-phosphate, then to uridine diphosphate glucose, and finally to glycogen.

Figure 25.10 **Summary of the principal reactions of cellular respiration.** ETC = electron transport chain and chemiosmosis.

Except for glycolysis, which occurs in the cytosol, all other reactions of cellular respiration occur within mitochondria.

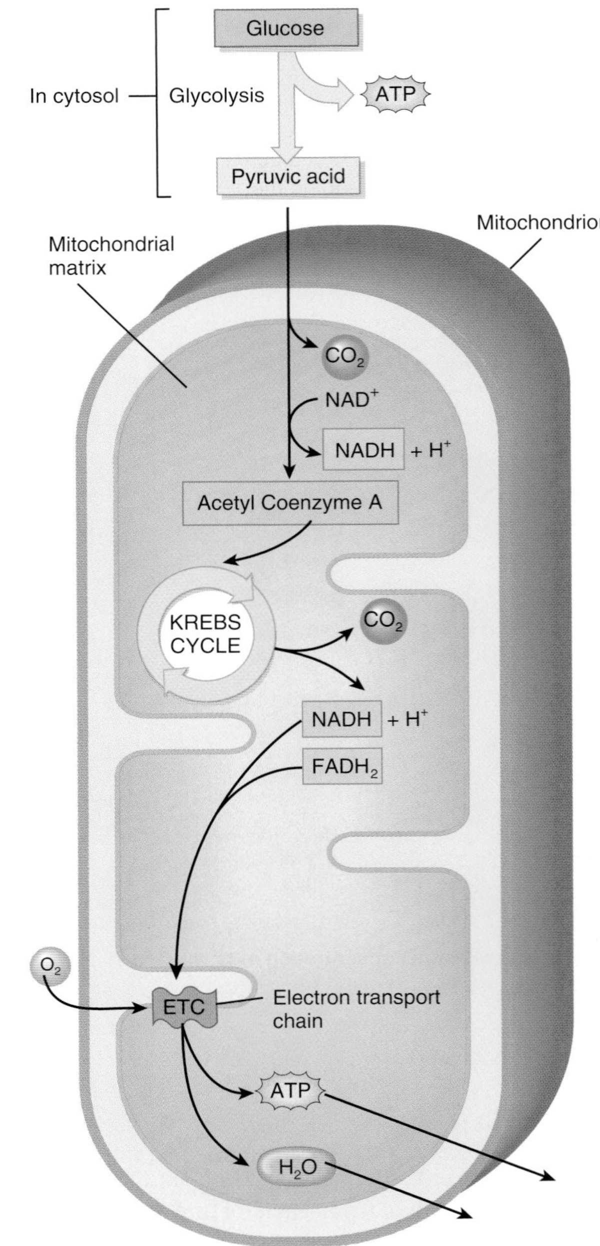

How many molecules of O_2 are used and how many molecules of CO_2 are produced during the complete oxidation of one glucose molecule?

Figure 25.11 Glycogenesis and glycogenolysis.

The glycogenesis pathway converts glucose into glycogen; the glycogenolysis pathway breaks down glycogen into glucose.

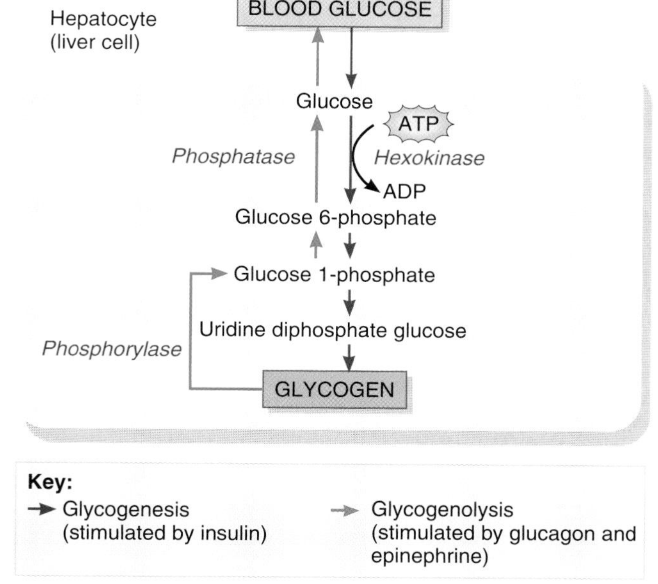

Besides hepatocytes, which body cells can synthesize glycogen? Why can't they release glucose into the blood?

Glucose Release: Glycogenolysis

When body activities require ATP, glycogen stored in hepatocytes is broken down into glucose and released into the blood to be transported to cells, where it will be catabolized by the processes of cellular respiration already described. The process of splitting glycogen into its glucose subunits is called **glycogenolysis** (glī′-kō-je-NOL-e-sis). (Note: Do not confuse *glycogenolysis,* which is the breakdown of glycogen to glucose, with *glycolysis,* the 10 reactions that convert glucose to pyruvic acid.)

Glycogenolysis is not a simple reversal of the steps of glycogenesis (Figure 25.11). It begins by splitting glucose molecules off the branched glycogen molecule via phosphorylation to form glucose 1-phosphate. Phosphorylase, the enzyme that catalyzes this reaction, is activated by glucagon from pancreatic alpha cells and epinephrine from the adrenal medulla. Glucose 1-phosphate is then converted to glucose 6-phosphate and finally to glucose, which leaves hepatocytes via glucose transporters (GluT) in the plasma membrane. Phosphorylated glucose molecules cannot ride aboard the GluT transporters, however, and *phosphatase,* the enzyme that converts glucose 6-phosphate into glucose, is absent in skeletal muscle cells. Thus, hepatocytes, which have phosphatase, can release glucose derived from glycogen to the bloodstream, but skeletal muscle cells cannot. In skeletal muscle cells, glycogen is broken down into glucose 1-phosphate, which is then catabolized for ATP production via glycolysis and the Krebs cycle. However, the lactic acid produced by glycolysis in muscle cells can be converted to glucose in the liver. In this way, muscle glycogen can be an indirect source of blood glucose.

Carbohydrate Loading

The amount of glycogen stored in the liver and skeletal muscles varies and can be completely exhausted during long-term athletic endeavors. Thus, many marathon runners and other endurance athletes follow a precise exercise and dietary regimen that includes eating large amounts of complex carbohydrates, such as pasta and potatoes, in the three days before an event. This practice, called **carbohydrate loading,** helps maximize the amount of glycogen available for ATP production in muscles. For athletic events lasting more than an hour, carbohydrate loading has been shown to increase an athlete's endurance. The increased endurance is due to increased glycogenolysis, which results in more glucose that can be catabolized for energy. ■

Formation of Glucose from Proteins and Fats: Gluconeogenesis

When your liver runs low on glycogen, it is time to eat. If you don't, your body starts catabolizing triglycerides (fats) and proteins. Actually, the body normally catabolizes some of its triglycerides and proteins, but large-scale triglyceride and protein catabolism does not happen unless you are starving, eating very few carbohydrates, or suffering from an endocrine disorder.

The glycerol part of triglycerides, lactic acid, and certain amino acids can be converted in the liver to glucose (Figure 25.12). The process by which glucose is formed from these non-

Figure 25.12 Gluconeogenesis, the conversion of noncarbohydrate molecules (amino acids, lactic acid, and glycerol) into glucose.

About 60% of the amino acids in the body can be used for gluconeogenesis.

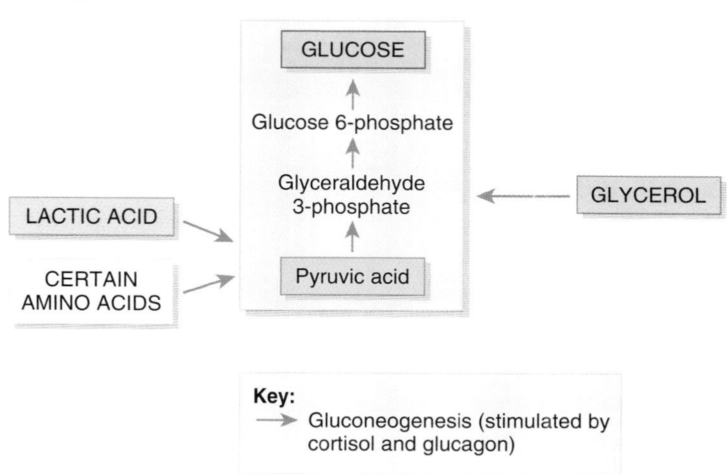

What cells can carry out gluconeogenesis and glycogenesis?

carbohydrate sources is called **gluconeogenesis** (gloo′-kō-nē′-ō-JEN-e-sis; *neo* = new). An easy way to distinguish this term from glycogenesis or glycogenolysis is to remember that in this case glucose is not converted back from glycogen, but is instead *newly formed.* About 60% of the amino acids in the body can be used for gluconeogenesis. Amino acids such as alanine, cysteine, glycine, serine, and threonine, and lactic acid are converted to pyruvic acid, which then may be synthesized into glucose or enter the Krebs cycle. Glycerol may be converted into glyceraldehyde 3-phosphate, which may form pyruvic acid or be used to synthesize glucose.

Gluconeogenesis is stimulated by cortisol, the main glucocorticoid hormone of the adrenal cortex, and by glucagon from the pancreas. In addition, cortisol stimulates the breakdown of proteins into amino acids, thus expanding the pool of amino acids available for gluconeogenesis. Thyroid hormones (thyroxine and triiodothyronine) also mobilize proteins and may mobilize triglycerides from adipose tissue, thereby making glycerol available for gluconeogenesis.

▶ CHECKPOINT

5. How does glucose move into or out of body cells?
6. What happens during glycolysis?
7. How is acetyl coenzyme A formed?
8. Outline the principal events and outcomes of the Krebs cycle.
9. What happens in the electron transport chain and why is this process called chemiosomosis?
10. Which reactions produce ATP during the complete oxidation of a molecule of glucose?
11. Under what circumstances do glycogenesis and glycogenolysis occur?
12. What is gluconeogenesis, and why is it important?

LIPID METABOLISM

▶ OBJECTIVES

Describe the lipoproteins that transport lipids in the blood.

Describe the fate, metabolism, and functions of lipids.

Transport of Lipids by Lipoproteins

Most lipids, such as triglycerides, are nonpolar and therefore very hydrophobic molecules. They do not dissolve in water. To be transported in watery blood, such molecules first must be made more water-soluble by combining them with proteins produced by the liver and intestine. The lipid and protein combinations thus formed are *lipoproteins,* spherical particles with an outer shell of proteins, phospholipids, and cholesterol molecules surrounding an inner core of triglycerides and other lipids (Figure 25.13). The proteins in the outer shell are called **apoproteins (apo)** and are designated by the letters A, B, C, D, and E plus a number. In addition to helping solubilize the lipoprotein in body fluids, each apoprotein has specific functions.

Each of the several types of lipoproteins has different functions, but all essentially are transport vehicles. They provide delivery and pickup services so that lipids can be available when cells need them or removed from circulation when they are not needed. Lipoproteins are categorized and named mainly according to their density, which varies with the ratio of lipids (which have a low density) to proteins (which have a high density). From largest and lightest to smallest and heaviest, the four major classes of lipoproteins are chylomicrons, very low-density lipoproteins (VLDLs), low-density lipoproteins (LDLs), and high-density lipoproteins (HDLs).

Chylomicrons, which form in mucosal epithelial cells of the small intestine, transport *dietary* (ingested) lipids to adipose tissue for storage. They contain about 1–2% proteins, 85% triglycerides, 7% phospholipids, and 6–7% cholesterol, plus a small amount of fat-soluble vitamins. Chylomicrons enter lacteals of intestinal villi and are carried by lymph into venous blood and then into the systemic circulation. Their presence gives blood plasma a milky appearance, but they remain in the blood for only a few minutes. As chylomicrons circulate through the capillaries of adipose tissue, one of their apoproteins, **apo C-2,** activates *endothelial lipoprotein lipase,* an enzyme that removes fatty acids from chylomicron triglycerides. The free fatty acids are then taken up by adipocytes for synthesis and storage as triglycerides and by muscle cells for ATP production. Hepatocytes remove chylomicron remnants from the blood via receptor-mediated endocytosis, in which another chylomicron apoprotein, **apo E,** is the docking protein.

Figure 25.13 A lipoprotein. Shown here is a VLDL.

A single layer of amphipathic phospholipids, cholesterol, and proteins surrounds a core of nonpolar lipids.

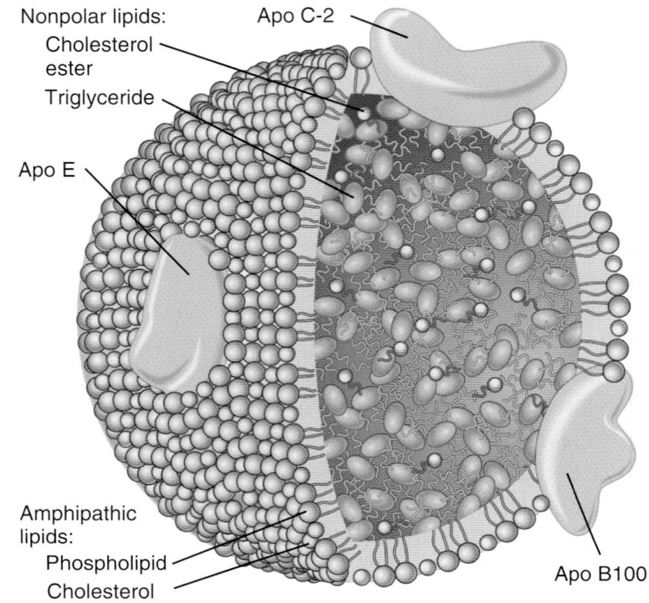

Which type of lipoprotein delivers cholesterol to body cells?

Very low-density lipoproteins (VLDLs), which form in hepatocytes, contain mainly *endogenous* (made in the body) lipids. VLDLs contain about 10% proteins, 50% triglycerides, 20% phospholipids, and 20% cholesterol. VLDLs transport triglycerides synthesized in hepatocytes to adipocytes for storage. Like chylomicrons, they lose triglycerides as their apo C-2 activates endothelial lipoprotein lipase, and the resulting fatty acids are taken up by adipocytes for storage and by muscle cells for ATP production. As they deposit some of their triglycerides in adipose cells, VLDLs are converted to LDLs.

Low-density lipoproteins (LDLs) contain 25% proteins, 5% triglycerides, 20% phospholipids, and 50% cholesterol. They carry about 75% of the total cholesterol in blood and deliver it to cells throughout the body for use in repair of cell membranes and synthesis of steroid hormones and bile salts. LDLs contain a single apoprotein, **apo B100,** which is the docking protein that binds to LDL receptors on the plasma membrane of body cells so that LDL can enter the cell via receptor-mediated endocytosis. Within the cell, the LDL is broken down, and the cholesterol is released to serve the cell's needs. Once a cell has sufficient cholesterol for its activities, a negative feedback system inhibits the cell's synthesis of new LDL receptors.

When present in excessive numbers, LDLs also deposit cholesterol in and around smooth muscle fibers in arteries, forming fatty plaques that increase the risk of coronary artery disease (see page 726). For this reason, the cholesterol in LDLs, called LDL-cholesterol, is known as "bad" cholesterol. Because some people have too few LDL receptors, their body cells remove LDL from the blood less efficiently; as a result, their plasma LDL level is abnormally high, and they are more likely to develop fatty plaques. Eating a high-fat diet increases the production of VLDLs, which elevates the LDL level and increases the formation of fatty plaques.

High-density lipoproteins (HDLs), which contain 40–45% proteins, 5–10% triglycerides, 30% phospholipids, and 20% cholesterol, remove excess cholesterol from body cells and the blood and transport it to the liver for elimination. Because HDLs prevent accumulation of cholesterol in the blood, a high HDL level is associated with decreased risk of coronary artery disease. For this reason, HDL-cholesterol is known as "good" cholesterol.

Sources and Significance of Blood Cholesterol

There are two sources of cholesterol in the body. Some is present in foods (eggs, dairy products, organ meats, beef, pork, and processed luncheon meats), but most is synthesized by hepatocytes. Fatty foods that don't contain any cholesterol at all can still dramatically increase blood cholesterol level in two ways. First, a high intake of dietary fats stimulates reabsorption of cholesterol-containing bile back into the blood, so less cholesterol is lost in the feces. Second, when saturated fats are broken down in the body, hepatocytes use some of the breakdown products to make cholesterol.

A lipid profile test usually measures total cholesterol (TC), HDL-cholesterol, and triglycerides (VLDLs). LDL-cholesterol then is calculated by using the following formula: LDL-cholesterol = TC − HDL-cholesterol − (triglycerides/5). In the United States, blood cholesterol is usually measured in milligrams per deciliter (mg/dL); a deciliter is 0.1 liter or 100 mL. For adults, desirable levels of blood cholesterol are total cholesterol under 200 mg/dL, LDL-cholesterol under 130 mg/dL, and HDL-cholesterol over 40 mg/dL. Normally, triglycerides are in the range of 10–190 mg/dL.

As total cholesterol level increases, the risk of coronary artery disease begins to rise. When total cholesterol is above 200 mg/dL (5.2 mmol/liter), the risk of a heart attack doubles with every 50 mg/dL (1.3 mmol/liter) increase in total cholesterol. Total cholesterol of 200–239 mg/dL and LDL of 130–159 mg/dL are borderline-high; total cholesterol above 239 mg/dL and LDL above 159 mg/dL are classified as high blood cholesterol. The ratio of total cholesterol to HDL-cholesterol predicts the risk of developing coronary artery disease. For example, a person with a total cholesterol of 180 mg/dL and HDL of 60 mg/dL has a risk ratio of 3. Ratios above 4 are considered undesirable; the higher the ratio, the greater the risk of developing coronary artery disease.

Among the therapies used to reduce blood cholesterol level are exercise, diet, and drugs. Regular physical activity at aerobic and nearly aerobic levels raises HDL level. Dietary changes are aimed at reducing the intake of total fat, saturated fats, and cholesterol. Drugs used to treat high blood cholesterol levels include cholestyramine (Questran) and colestipol (Colestid), which promote excretion of bile in the feces; nicotinic acid (Liponicin); and the "statin" drugs—atorvastatin (Lipitor), lovastatin (Mevacor), and simvastatin (Zocor), which block the key enzyme (HMG-CoA reductase) needed for cholesterol synthesis.

The Fate of Lipids

Lipids, like carbohydrates, may be oxidized to produce ATP. If the body has no immediate need to use lipids in this way, they are stored in adipose tissue (fat depots) throughout the body and in the liver. A few lipids are used as structural molecules or to synthesize other essential substances. Some examples include phospholipids, which are constituents of plasma membranes; lipoproteins, which are used to transport cholesterol throughout the body; thromboplastin, which is needed for blood clotting; and myelin sheaths, which speed up nerve impulse conduction. Two **essential fatty acids** that the body cannot synthesize are linoleic acid and linolenic acid. Dietary sources include vegetable oils and leafy vegetables. The various functions of lipids in the body may be reviewed in Table 2.7 on page 46.

Triglyceride Storage

A major function of adipose tissue is to remove triglycerides from chylomicrons and VLDLs and store them until they are needed for ATP production in other parts of the body. Triglycerides stored in adipose tissue constitute 98% of all body energy reserves. They are stored more readily than glycogen, in part because triglycerides are hydrophobic and do not exert osmotic pressure on cell membranes. Adipose tissue also insulates and protects various parts of the body. Adipocytes in the subcutaneous layer contain about 50% of the stored triglycerides. Other adipose tissues account for the other half: about 12% around the kidneys, 10–15% in the omenta, 15% in genital areas, 5–8% between muscles, and 5% behind the eyes, in the sulci of the heart, and attached to the outside of the large intestine. Triglycerides in adipose tissue are continually broken down and resynthesized. Thus, the triglycerides stored in adipose tissue today are not the same molecules that were present last month because they are continually released from storage, transported in the blood, and redeposited in other adipose tissue cells.

Lipid Catabolism: Lipolysis

In order for muscle, liver, and adipose tissue to oxidize the fatty acids derived from triglycerides to produce ATP, the triglycerides must first be split into glycerol and fatty acids, a process called **lipolysis** (li-POL-i-sis). Lipolysis is catalyzed by enzymes called *lipases.* Epinephrine and norepinephrine enhance triglyceride breakdown into fatty acids and glycerol. These hormones are released when sympathetic tone increases, as occurs, for example, during exercise. Other lipolytic hormones include cortisol, thyroid hormones, and insulinlike growth factors. By contrast, insulin inhibits lipolysis.

The glycerol and fatty acids that result from lipolysis are catabolized via different pathways (Figure 25.14). Glycerol is converted by many cells of the body to glyceraldehyde 3-phosphate, one of the compounds also formed during the catabolism of glucose. If ATP supply in a cell is high, glyceraldehyde 3-phosphate is converted into glucose, an example of gluconeogenesis. If ATP supply in a cell is low, glyceraldehyde 3-phosphate enters the catabolic pathway to pyruvic acid.

Fatty acids are catabolized differently than glycerol and yield more ATP. The first stage in fatty acid catabolism is a series of reactions, collectively called **beta oxidation,** that occurs in the matrix of mitochondria. Enzymes remove two carbon atoms at a time from the long chain of carbon atoms composing a fatty acid and attach the resulting two-carbon fragment to coenzyme A, forming acetyl CoA. Then, acetyl CoA enters the Krebs cycle (Figure 25.14). A 16-carbon fatty acid such as palmitic acid can yield as many as 129 ATPs upon its complete oxidation via beta oxidation, the Krebs cycle, and the electron transport chain.

As part of normal fatty acid catabolism, hepatocytes can take two acetyl CoA molecules at a time and condense them to form **acetoacetic acid.** This reaction liberates the bulky CoA

Figure 25.14 Pathways of lipid metabolism. Glycerol may be converted to glyceraldehyde 3-phosphate, which can then be converted to glucose or enter the Krebs cycle for oxidation. Fatty acids undergo beta oxidation and enter the Krebs cycle via acetyl coenzyme A. The synthesis of lipids from glucose or amino acids is called lipogenesis.

Glycerol and fatty acids are catabolized in separate pathways.

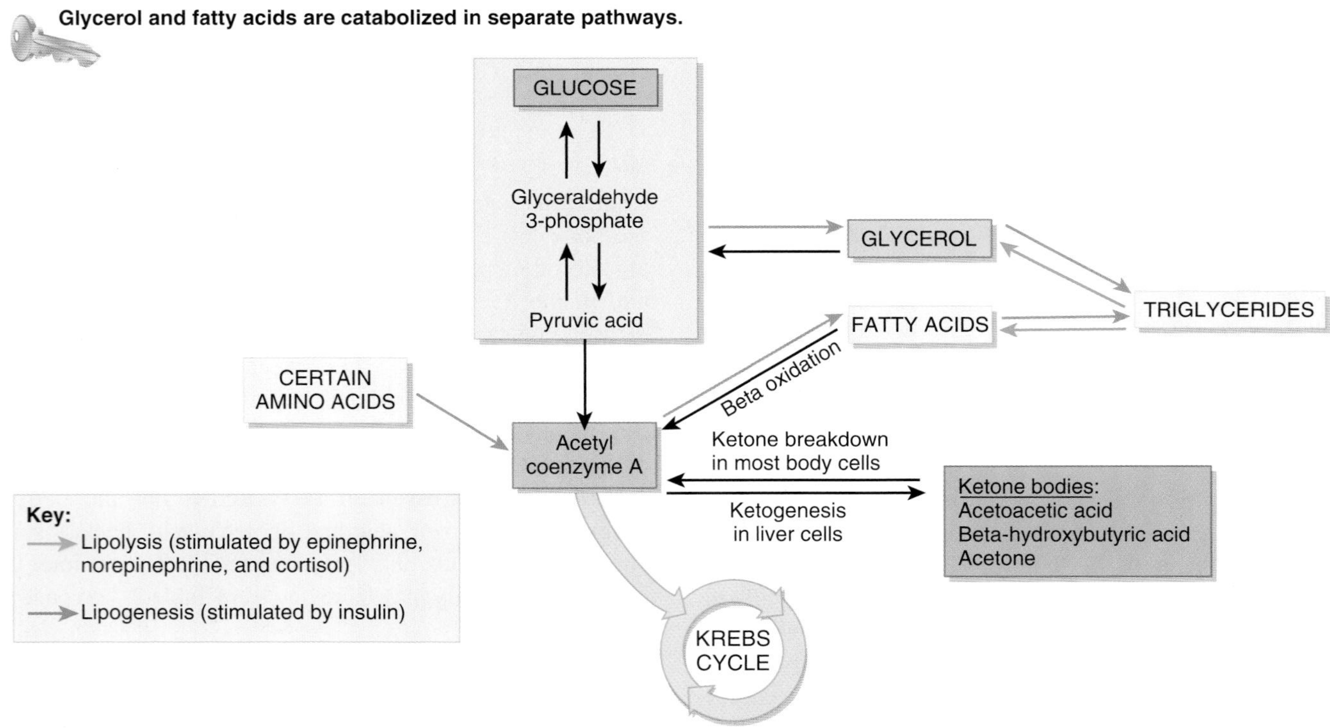

What types of cells can carry out lipogenesis, beta oxidation, and lipolysis? What type of cell can carry out ketogenesis?

portion, which cannot diffuse out of cells. Some acetoacetic acid is converted into **beta-hydroxybutyric acid** and **acetone.** The formation of these three substances, collectively known as **ketone bodies,** is called **ketogenesis** (Figure 25.14). Because ketone bodies freely diffuse through plasma membranes, they leave hepatocytes and enter the bloodstream.

Other cells take up acetoacetic acid and attach its four carbons to two coenzyme A molecules to form two acetyl CoA molecules, which can then enter the Krebs cycle for oxidation. Heart muscle and the cortex (outer part) of the kidneys use acetoacetic acid in preference to glucose for generating ATP. Hepatocytes, which make acetoacetic acid, cannot use it for ATP production because they lack the enzyme that transfers acetoacetic acid back to coenzyme A.

Lipid Anabolism: Lipogenesis

Liver cells and adipose cells can synthesize lipids from glucose or amino acids through **lipogenesis** (Figure 25.14), which is stimulated by insulin. Lipogenesis occurs when individuals consume more calories than are needed to satisfy their ATP needs. Excess dietary carbohydrates, proteins, and fats all have the same fate—they are converted into triglycerides. Certain amino acids can undergo the following reactions: amino acids → acetyl CoA → fatty acids → triglycerides. The use of glucose to form lipids takes place via two pathways: (1) glucose → glyceraldehyde 3-phosphate → glycerol and (2) glucose → glyceraldehyde 3-phosphate → acetyl CoA → fatty acids. The resulting glycerol and fatty acids can undergo anabolic reactions to become stored triglycerides, or they can go through a series of anabolic reactions to produce other lipids such as lipoproteins, phospholipids, and cholesterol.

Ketosis

The level of ketone bodies in the blood normally is very low because other tissues use them for ATP production as fast as they are generated from the breakdown of fatty acids in the liver. During periods of excessive beta oxidation, however, the production of ketone bodies exceeds their uptake and use by body cells. This might occur after a meal rich in triglycerides, or during fasting or starvation, because few carbohydrates are available for catabolism. Excessive beta oxidation may also occur in poorly controlled or untreated diabetes mellitus for two reasons: (1) Because adequate glucose cannot get into cells, triglycerides are used for ATP production, and (2) because insulin normally inhibits lipolysis, a lack of insulin accelerates the pace of lipolysis. When the concentration of ketone bodies in the blood rises above normal—a condition called **ketosis**—the ketone bodies, most of which are acids, must be buffered. If too many accumulate, they decrease the concentration of buffers such as bicarbonate ions, and blood pH falls. Extreme or prolonged ketosis can lead to **acidosis (ketoacidosis),** an abnormally low blood pH. The decreased blood pH in turn causes depression of the central nervous system, which can result in disorientation, coma, and even death if the condition is not treated. When a diabetic becomes seriously insulin deficient, one of the telltale signs is the sweet smell on the breath from the ketone body acetone. ■

▶ CHECKPOINT

13. What are the functions of the apoproteins in lipoproteins?

14. Which lipoprotein particles contain "good" and "bad" cholesterol, and why are these terms used?

15. Where are triglycerides stored in the body?

16. Explain the principal events of the catabolism of glycerol and fatty acids.

17. What are ketone bodies? What is ketosis?

18. Define lipogenesis and explain its importance.

PROTEIN METABOLISM

▶ OBJECTIVE

Describe the fate, metabolism, and functions of proteins.

During digestion, proteins are broken down into amino acids. Unlike carbohydrates and triglycerides, which are stored, proteins are not warehoused for future use. Instead, amino acids are either oxidized to produce ATP or used to synthesize new proteins for body growth and repair. Excess dietary amino acids are not excreted in the urine or feces but instead are converted into glucose (gluconeogenesis) or triglycerides (lipogenesis).

The Fate of Proteins

The active transport of amino acids into body cells is stimulated by insulinlike growth factors (IGFs) and insulin. Almost immediately after digestion, amino acids are reassembled into proteins. Many proteins function as enzymes; others are involved in transportation (hemoglobin) or serve as antibodies, clotting chemicals (fibrinogen), hormones (insulin), or contractile elements in muscle fibers (actin and myosin). Several proteins serve as structural components of the body (collagen, elastin, and keratin). The various functions of proteins in the body may be reviewed in Table 2.8 on page 50.

Protein Catabolism

A certain amount of protein catabolism occurs in the body each day, stimulated mainly by cortisol from the adrenal cortex. Proteins from worn-out cells (such as red blood cells) are broken down into amino acids. Some amino acids are converted into other amino acids, peptide bonds are re-formed, and new proteins are synthesized as part of the recycling process. Hepatocytes convert some amino acids to fatty acids, ketone bodies, or glucose. Cells throughout the body oxidize a small amount of amino acids to generate ATP via the Krebs cycle and the electron transport chain. However, before amino acids can be oxidized, they must first be converted to molecules that are part of the Krebs cycle or can enter the Krebs cycle, such as acetyl CoA (Figure 25.15). Before amino acids can enter the Krebs cycle, their amino group (NH_2) must first be removed—a process called **deamination** (dē-am′-i-NĀ-shun). Deamination occurs in hepatocytes and produces ammonia (NH_3). The liver cells then convert the highly toxic ammonia to urea, a relatively harmless substance that is excreted in the urine. The conversion of amino acids into glucose (gluconeogenesis) may be reviewed in Figure 25.12; the conversion of amino acids into fatty acids (lipogenesis) or ketone bodies (ketogenesis) is shown in Figure 25.14.

Protein Anabolism

Protein anabolism, the formation of peptide bonds between amino acids to produce new proteins, is carried out on the ribosomes of

Figure 25.15 **Various points at which amino acids (shown in yellow boxes) enter the Krebs cycle for oxidation.**

Before amino acids can be catabolized, they must first be converted to various substances that can enter the Krebs cycle.

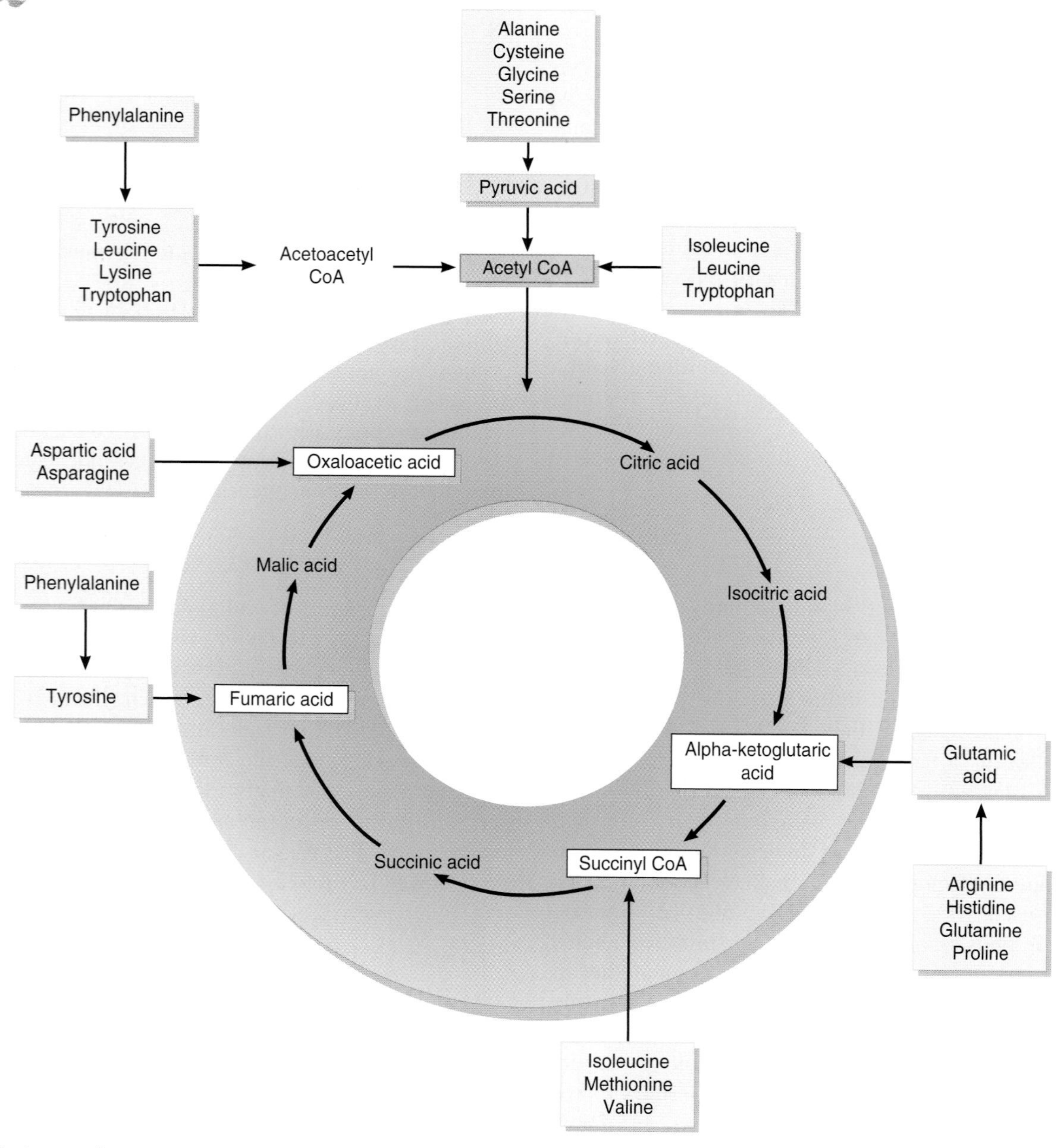

What group is removed from an amino acid before it can enter the Krebs cycle, and what is this process called?

almost every cell in the body, directed by the cells' DNA and RNA (see Figure 3.27 on page 90). Insulinlike growth factors, thyroid hormones (T3 and T4), insulin, estrogen, and testosterone stimulate protein synthesis. Because proteins are a main component of most cell structures, adequate dietary protein is especially essential during the growth years, during pregnancy, and when tissue has been damaged by disease or injury. Once dietary intake of protein is adequate, eating more protein will not increase bone or muscle mass; only a regular program of forceful, weight-bearing muscular activity accomplishes that goal.

Of the 20 amino acids in the human body, 10 are **essential amino acids:** They must be present in the diet because they cannot be synthesized in the body in adequate amounts. It is *essential* to include them in your diet. Humans are unable to synthesize eight amino acids (isoleucine, leucine, lysine, methionine, phenylalanine, threonine, tryptophan, and valine) and synthesize two others (arginine and histidine) in inadequate amounts, especially in childhood. A **complete protein** contains sufficient amounts of all essential amino acids. Beef, fish, poultry, eggs, and milk are examples of foods that contain complete proteins. An **incomplete protein** does not contain all essential amino acids. Examples of incomplete proteins are leafy green vegetables, legumes (beans and peas), and grains. **Nonessential amino acids** can be synthesized by body cells. They are formed by **transamination,** the transfer of an amino group from an amino acid to pyruvic acid or to an acid in the Krebs cycle. Once the appropriate essential and nonessential amino acids are present in cells, protein synthesis occurs rapidly.

Phenylketonuria

Phenylketonuria (fen′-il-kē′-tō-NOO-rē-a) or **PKU** is a genetic error of protein metabolism characterized by elevated blood levels of the amino acid phenylalanine. Most children with phenylketonuria have a mutation in the gene that codes for the enzyme phenylalanine hydroxylase, the enzyme needed to convert phenylalanine into the amino acid tyrosine, which can enter the Krebs cycle (Figure 25.15). Because the enzyme is deficient, phenylalanine cannot be metabolized, and what is not used in protein synthesis builds up in the blood. If untreated, the disorder causes vomiting, rashes, seizures, growth deficiency, and severe mental retardation. Newborns are screened for PKU, and mental retardation can be prevented by restricting the affected child to a diet that supplies only the amount of phenylalanine needed for growth, although learning disabilities may still ensue. Because the artificial sweetener aspartame (NutraSweet) contains phenylalanine, its consumption must be restricted in children with PKU. ■

▶ CHECKPOINT

19. What is deamination and why does it occur?

20. What are the possible fates of the amino acids from protein catabolism?

21. How are essential and nonessential amino acids different?

KEY MOLECULES AT METABOLIC CROSSROADS

▶ OBJECTIVE

Identify the key molecules in metabolism, and describe the reactions and the products they may form.

Although there are thousands of different chemicals in cells, three molecules—glucose 6-phosphate, pyruvic acid, and acetyl coenzyme A—play pivotal roles in metabolism (Figure 25.16). These molecules stand at "metabolic crossroads"; as you will learn shortly, the reactions that occur (or do not occur) depend on the nutritional or activity status of the individual. Reactions 1 through 7 in Figure 25.16 occur in the cytosol, reactions 8 and 9 occur inside mitochondria, and reactions indicated by 10 occur on smooth endoplasmic reticulum.

The Role of Glucose 6-Phosphate

Shortly after glucose enters a body cell, a kinase converts it to **glucose 6-phosphate.** Four possible fates await glucose 6-phosphate (see Figure 25.16):

1. ***Synthesis of glycogen.*** When glucose is abundant in the bloodstream, as it is just after a meal, a large amount of glucose 6-phosphate is used to synthesize glycogen, the storage form of carbohydrate in animals. Subsequent breakdown of glycogen into glucose 6-phosphate occurs through a slightly different series of reactions. Synthesis and breakdown of glycogen occur mainly in skeletal muscle fibers and hepatocytes.
2. ***Release of glucose into the bloodstream.*** If the enzyme glucose 6-phosphatase is present and active, glucose 6-phosphate can be dephosphorylated to glucose. Once glucose is released from the phosphate group, it can leave the cell and enter the bloodstream. Hepatocytes are the main cells that can provide glucose to the bloodstream in this way.
3. ***Synthesis of nucleic acids.*** Glucose 6-phosphate is the precursor used by cells throughout the body to make ribose 5-phosphate, a 5-carbon sugar that is needed for synthesis of RNA (ribonucleic acid) and DNA (deoxyribonucleic acid). The same sequence of reactions also produces NADPH. This molecule is a hydrogen and electron donor in certain reduction reactions, such as synthesis of fatty acids and steroid hormones.
4. ***Glycolysis.*** Some ATP is produced anaerobically via glycolysis, in which glucose 6-phosphate is converted to pyruvic acid, another key molecule in metabolism. Most body cells carry out glycolysis.

The Role of Pyruvic Acid

Each 6-carbon molecule of glucose that undergoes glycolysis yields two 3-carbon molecules of **pyruvic acid.** This molecule, like glucose 6-phosphate, stands at a metabolic crossroads: Given

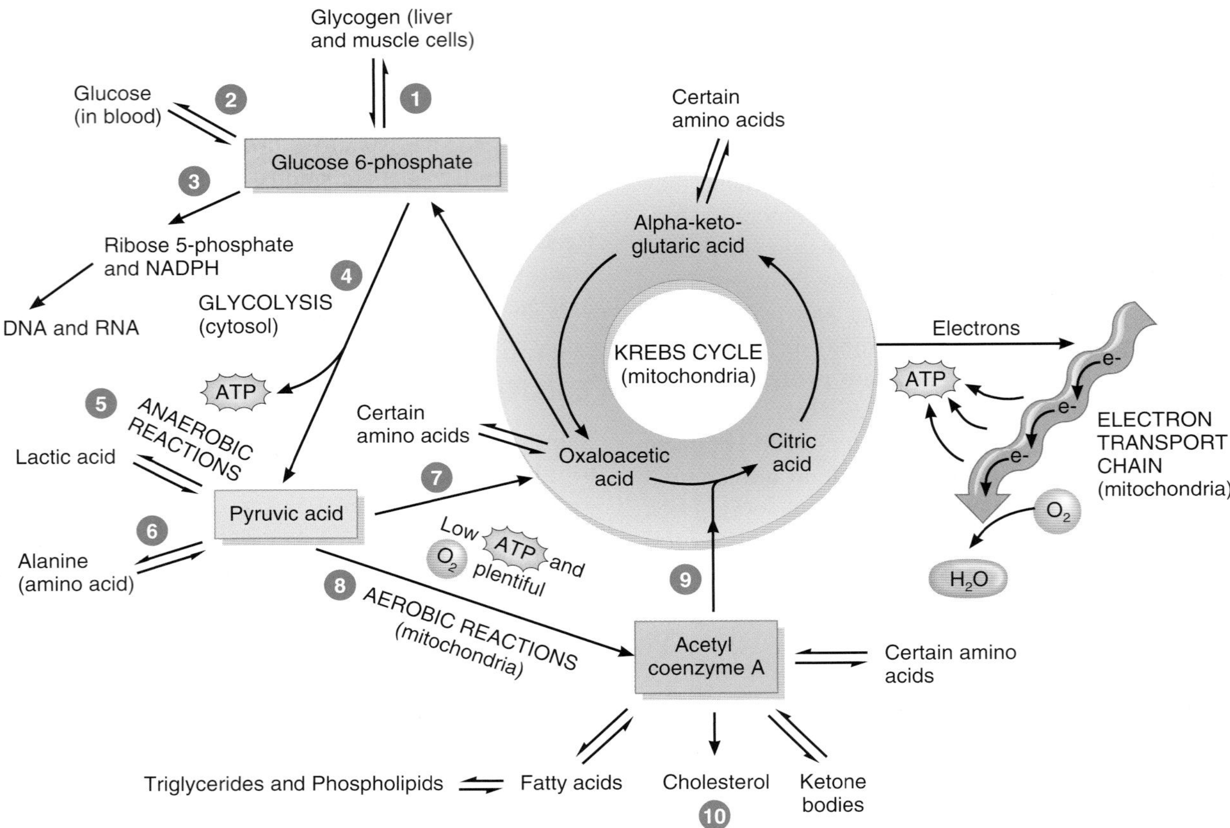

Figure 25.16 **Summary of the roles of the key molecules in metabolic pathways.** Double-headed arrows indicate that reactions between two molecules may proceed in either direction, if the appropriate enzymes are present and the conditions are favorable; single-headed arrows signify the presence of an irreversible step.

Three molecules—glucose 6-phosphate, pyruvic acid, and acetyl coenzyme A—stand at "metabolic crossroads." They can undergo different reactions depending on your nutritional or activity status.

Which substance is the gateway into the Krebs cycle for molecules that are being oxidized to generate ATP?

enough oxygen, the aerobic (oxygen-consuming) reactions of cellular respiration can proceed; if oxygen is in short supply, anaerobic reactions can occur (Figure 25.16):

5 ***Production of lactic acid.*** When oxygen is in short supply in a tissue, as in actively contracting skeletal or cardiac muscle, some pyruvic acid is changed to lactic acid. The lactic acid then diffuses into the bloodstream and is taken up by hepatocytes, which eventually convert it back to pyruvic acid.

6 ***Production of alanine.*** Carbohydrate and protein metabolism are linked by pyruvic acid. Through transamination, an amino group (NH_3) can either be added to pyruvic acid (a carbohydrate) to produce the amino acid alanine, or be removed from alanine to generate pyruvic acid.

7 ***Gluconeogenesis.*** Pyruvic acid and certain amino acids also can be converted to oxaloacetic acid, one of the Krebs cycle intermediates, which in turn can be used to form glucose 6-phosphate. This sequence of gluconeogenesis reactions bypasses certain one-way reactions of glycolysis.

The Role of Acetyl Coenzyme A

8 When the ATP level in a cell is low but oxygen is plentiful, most pyruvic acid streams toward ATP-producing reactions—the Krebs cycle and electron transport chain—via conversion to **acetyl coenzyme A.**

9 ***Entry into the Krebs cycle.*** Acetyl CoA is the vehicle for 2-carbon acetyl groups to enter the Krebs cycle. Oxidative Krebs cycle reactions convert acetyl CoA to CO_2 and produce reduced coenzymes (NADH and $FADH_2$) that transfer electrons into the electron transport chain. Oxidative reactions in the electron transport chain, in turn, generate ATP. Most fuel molecules that will be oxidized to generate ATP—glucose, fatty acids, and ketone bodies—are first converted to acetyl CoA.

10 ***Synthesis of lipids.*** Acetyl CoA also can be used for synthesis of certain lipids, including fatty acids, ketone bodies, and cholesterol. Because pyruvic acid can be converted to acetyl CoA, carbohydrates can be turned into

triglycerides; this metabolic pathway stores some excess carbohydrate calories as fat. Mammals, including humans, cannot reconvert acetyl CoA to pyruvic acid, however, so fatty acids cannot be used to generate glucose or other carbohydrate molecules.

Table 25.2 summarizes carbohydrate, lipid, and protein metabolism.

▶ CHECKPOINT

22. What are the possible fates of glucose 6-phosphate, pyruvic acid, and acetyl coenzyme A in a cell?

METABOLIC ADAPTATIONS

▶ OBJECTIVE

Compare metabolism during the absorptive and postabsorptive states.

Regulation of metabolic reactions depends both on the chemical environment within body cells, such as the levels of ATP and oxygen, and on signals from the nervous and endocrine systems. Some aspects of metabolism depend on how much time has passed since the last meal. During the **absorptive state,** ingested nutrients are entering the bloodstream, and glucose is readily available for ATP production. During the **postabsorptive state,** absorption of nutrients from the GI tract is complete, and energy needs must be met by fuels already in the body. A typical meal requires about 4 hours for complete absorption; given three meals a day, the absorptive state exists for about 12 hours each day. Assuming no between-meal snacks, the other 12 hours—typically late morning, late afternoon, and most of the night—are spent in the postabsorptive state.

Because the nervous system and red blood cells continue to depend on glucose for ATP production during the postabsorptive state, maintaining a steady blood glucose level is critical during this period. Hormones are the major regulators of metabolism in each state. The effects of insulin dominate in the absorptive state; several other hormones regulate metabolism in the postabsorptive state. During fasting and starvation, many body cells turn to ketone bodies for ATP production, as noted in the Clinical Application on page 967.

Metabolism During the Absorptive State

Soon after a meal, nutrients start to enter the blood. Recall that ingested food reaches the bloodstream mainly as glucose, amino acids, and triglycerides (in chylomicrons). Two metabolic hallmarks of the absorptive state are the oxidation of glucose

TABLE 25.2 Summary of Metabolism

Process	Comments
Carbohydrates	
Glucose catabolism	Complete oxidation of glucose (cellular respiration) is the chief source of ATP in cells and consists of glycolysis, the Krebs cycle, and the electron transport chain. Complete oxidation of 1 molecule of glucose yields a maximum of 36 or 38 molecules of ATP.
Glycolysis	Conversion of glucose into pyruvic acid results in the production of some ATP. Reactions do not require oxygen (anaerobic cellular respiration).
Krebs cycle	Cycle includes a series of oxidation–reduction reactions in which coenzymes (NAD^+ and FAD) pick up hydrogen ions and hydride ions from oxidized organic acids, and some ATP is produced. CO_2 and H_2O are byproducts. Reactions are aerobic.
Electron transport chain	Third set of reactions in glucose catabolism is another series of oxidation–reduction reactions, in which electrons are passed from one carrier to the next, and most of the ATP is produced. Reactions require oxygen (aerobic cellular respiration).
Glucose anabolism	Some glucose is converted into glycogen (glycogenesis) for storage if not needed immediately for ATP production. Glycogen can be reconverted to glucose (glycogenolysis). The conversion of amino acids, glycerol, and lactic acid into glucose is called gluconeogenesis.
Lipids	
Triglyceride catabolism	Triglycerides are broken down into glycerol and fatty acids. Glycerol may be converted into glucose (gluconeogenesis) or catabolized via glycolysis. Fatty acids are catabolized via beta oxidation into acetyl coenzyme A that can enter the Krebs cycle for ATP production or be converted into ketone bodies (ketogenesis).
Triglyceride anabolism	The synthesis of triglycerides from glucose and fatty acids is called lipogenesis. Triglycerides are stored in adipose tissue.
Proteins	
Protein catabolism	Amino acids are oxidized via the Krebs cycle after deamination. Ammonia resulting from deamination is converted into urea in the liver, passed into blood, and excreted in urine. Amino acids may be converted into glucose (gluconeogenesis), fatty acids, or ketone bodies.
Protein anabolism	Protein synthesis is directed by DNA and utilizes cells' RNA and ribosomes.

for ATP production, which occurs in most body cells, and the storage of excess fuel molecules for future between-meal use, which occurs mainly in hepatocytes, adipocytes, and skeletal muscle fibers.

Absorptive State Reactions

The following reactions dominate during the absorptive state (Figure 25.17):

1. About 50% of the glucose absorbed from a typical meal is oxidized by cells throughout the body to produce ATP via glycolysis, the Krebs cycle, and the electron transport chain.
2. Most glucose that enters hepatocytes is converted to glycogen. Small amounts may be used for synthesis of fatty acids and glyceraldehyde 3-phosphate.
3. Some fatty acids and triglycerides synthesized in the liver remain there, but hepatocytes package most into VLDLs, which carry lipids to adipose tissue for storage.
4. Adipocytes also take up glucose not picked up by the liver and convert it into triglycerides for storage. Overall, about 40% of the glucose absorbed from a meal is converted to triglycerides, and about 10% is stored as glycogen in skeletal muscles and hepatocytes.

Figure 25.17 Principal metabolic pathways during the absorptive state.

During the absorptive state, most body cells produce ATP by oxidizing glucose to CO_2 and H_2O.

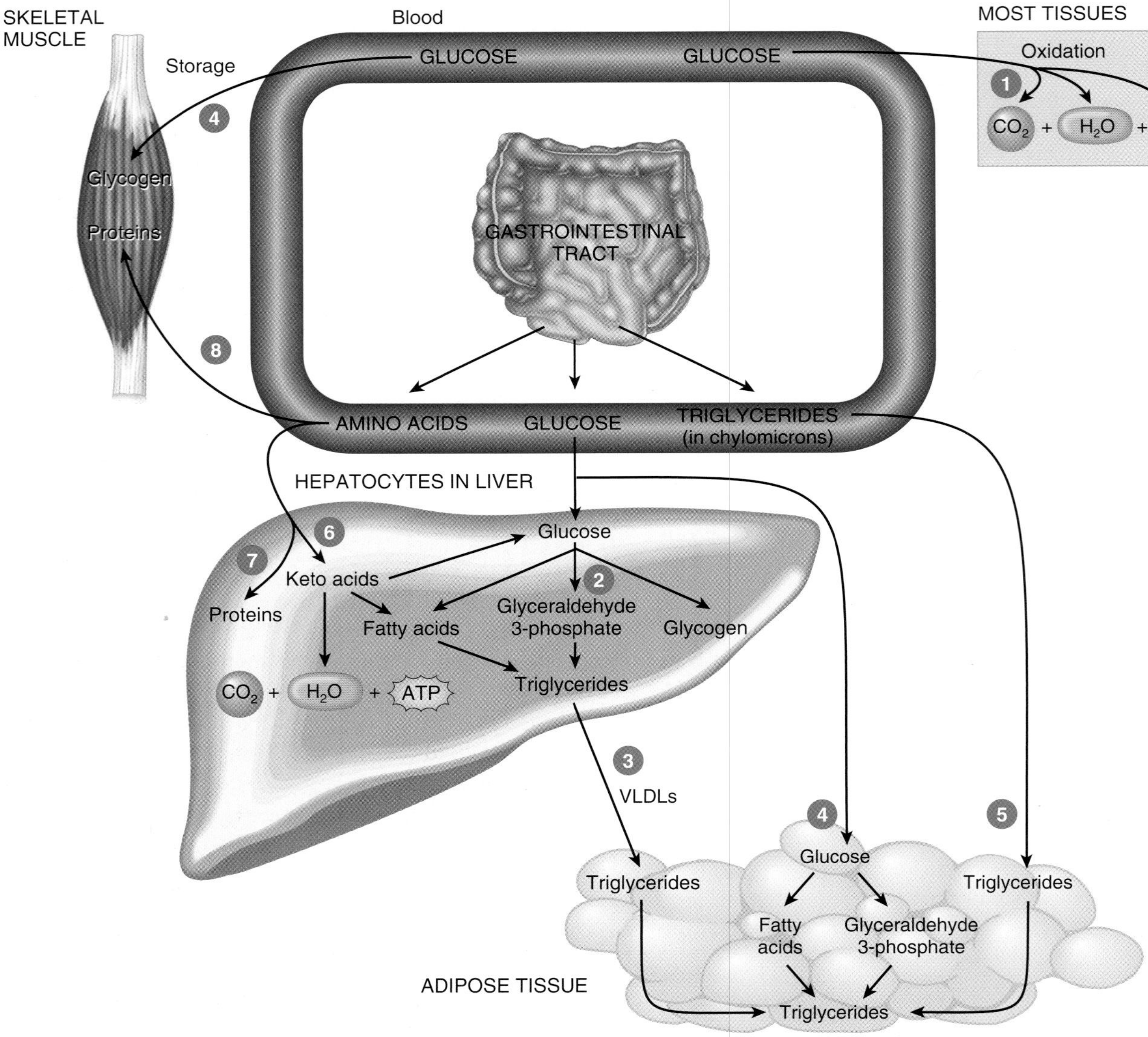

Are the reactions shown in this figure mainly anabolic or catabolic?

1. Most dietary lipids (mainly triglycerides and fatty acids) are stored in adipose tissue; only a small portion is used for synthesis reactions. Adipocytes obtain the lipids from chylomi-crons, from VLDLs, and from their own synthesis reactions.
2. Many absorbed amino acids that enter hepatocytes are deaminated to keto acids, which can either enter the Krebs cycle for ATP production or be used to synthesize glucose or fatty acids.
3. Some amino acids that enter hepatocytes are used to synthesize proteins (for example, plasma proteins).
4. Amino acids not taken up by hepatocytes are used in other body cells (such as muscle cells) for synthesis of proteins or regulatory chemicals such as hormones or enzymes.

Regulation of Metabolism During the Absorptive State

Soon after a meal, glucose-dependent insulinotropic peptide (GIP), plus the rising blood levels of glucose and certain amino acids, stimulate pancreatic beta cells to release insulin. In general, insulin increases the activity of enzymes needed for anabolism and the synthesis of storage molecules; at the same time it decreases the activity of enzymes needed for catabolic or breakdown reactions. Insulin promotes the entry of glucose and amino acids into cells of many tissues, and it stimulates the phosphorylation of glucose in hepatocytes and the conversion of glucose 6-phosphate to glycogen in both liver and muscle cells. In liver and adipose tissue, insulin enhances the synthesis of triglycerides, and in cells throughout the body insulin stimulates protein synthesis. (See page 645 to review the effects of insulin.) Insulinlike growth factors and the thyroid hormones (T3 and T4) also stimulate protein synthesis. Table 25.3 summarizes the hormonal regulation of metabolism in the absorptive state.

TABLE 25.3 Hormonal Regulation of Metabolism in the Absorptive State

Process	Location(s)	Main Stimulating Hormone(s)
Facilitated diffusion of glucose into cells	Most cells.	Insulin.*
Active transport of amino acids into cells	Most cells.	Insulin.
Glycogenesis (glycogen synthesis)	Hepatocytes and muscle fibers.	Insulin.
Protein synthesis	All body cells.	Insulin, thyroid hormones, and insulinlike growth factors.
Lipogenesis (triglyceride synthesis)	Adipose cells and hepatocytes.	Insulin.

*Facilitated diffusion of glucose into hepatocytes (liver cells) and neurons is always "turned on" and does not require insulin.

Metabolism During the Postabsorptive State

About 4 hours after the last meal, absorption of nutrients from the small intestine is nearly complete, and blood glucose level starts to fall because glucose continues to leave the bloodstream and enter body cells while none is being absorbed from the GI tract. Thus, the main metabolic challenge during the postabsorptive state is to maintain the normal blood glucose level of 70–110 mg/100 mL (3.9–6.1 mmol/liter). Homeostasis of blood glucose concentration is especially important for the nervous system and for red blood cells for the following reasons:

- The dominant fuel molecule for ATP production in the nervous system is glucose, because fatty acids are unable to pass the blood–brain barrier.
- Red blood cells derive all their ATP from glycolysis of glucose because they have no mitochondria, so the Krebs cycle and the electron transport chain are not available to them.

Postabsorptive State Reactions

During the postabsorptive state, both *glucose production* and *glucose conservation* help maintain blood glucose level: Hepatocytes produce glucose molecules and export them into the blood, and other body cells switch from glucose to alternative fuels for ATP production to conserve scarce glucose. The major reactions of the postabsorptive state that produce glucose are the following (Figure 25.18):

1. ***Breakdown of liver glycogen.*** During fasting, a major source of blood glucose is liver glycogen, which can provide about a 4-hour supply of glucose. Liver glycogen is continually being formed and broken down as needed.
2. ***Lipolysis.*** Glycerol, produced by breakdown of triglycerides in adipose tissue, is also used to form glucose.
3. ***Gluconeogenesis using lactic acid.*** During exercise, skeletal muscle tissue breaks down stored glycogen (see step 9) and produces some ATP anaerobically via glycolysis. Some of the pyruvic acid that results is converted to acetyl CoA, and some is converted to lactic acid, which diffuses into the blood. In the liver, lactic acid can be used for gluconeogenesis, and the resulting glucose is released into the blood.
4. ***Gluconeogenesis using amino acids.*** Modest breakdown of proteins in skeletal muscle and other tissues releases large amounts of amino acids, which then can be converted to glucose by gluconeogenesis in the liver.

Despite all of these ways the body produces glucose, blood glucose level cannot be maintained for very long without further metabolic changes. Thus, a major adjustment must be made during

Figure 25.18 **Principal metabolic pathways during the postabsorptive state.**

The principal function of postabsorptive state reactions is to maintain a normal blood glucose level.

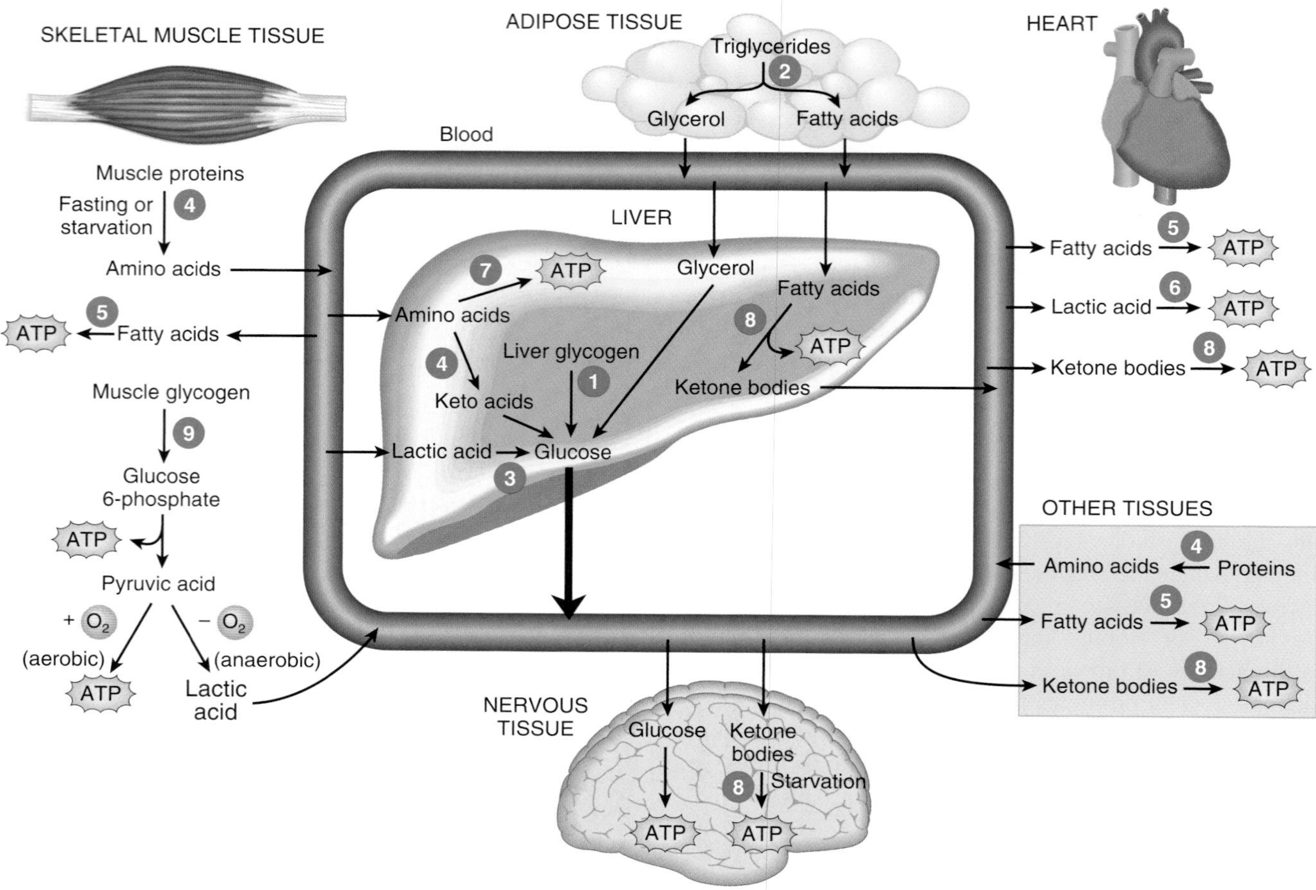

What processes directly elevate blood glucose during the postabsorptive state, and where does each occur?

the postabsorptive state to produce ATP while conserving glucose. The following reactions produce ATP without using glucose:

5. ***Oxidation of fatty acids.*** The fatty acids released by lipolysis of triglycerides cannot be used for glucose production because acetyl CoA cannot be readily converted to pyruvic acid. But most cells can oxidize the fatty acids directly, feed them into the Krebs cycle as acetyl CoA, and produce ATP through the electron transport chain.
6. ***Oxidation of lactic acid.*** Cardiac muscle can produce ATP aerobically from lactic acid.
7. ***Oxidation of amino acids.*** In hepatocytes, amino acids may be oxidized directly to produce ATP.
8. ***Oxidation of ketone bodies.*** Hepatocytes also convert fatty acids to ketone bodies, which can be used by the heart, kidneys, and other tissues for ATP production.
9. ***Breakdown of muscle glycogen.*** Skeletal muscle cells break down glycogen to glucose 6-phosphate, which undergoes glycolysis and provides ATP for muscle contraction.

Regulation of Metabolism During the Postabsorptive State

Both hormones and the sympathetic division of the autonomic nervous system (ANS) regulate metabolism during the postabsorptive state. The hormones that regulate postabsorptive state metabolism sometimes are called anti-insulin hormones because they counter the effects of insulin during the absorptive state. As blood glucose level declines, the secretion of insulin falls and the release of anti-insulin hormones rises.

When blood glucose concentration starts to drop, the pancreatic alpha cells release glucagon at a faster rate, and the beta cells secrete insulin more slowly. The primary target tissue of glucagon is the liver; the major effect is increased release of glucose into the bloodstream due to gluconeogenesis and glycogenolysis.

Low blood glucose also activates the sympathetic branch of the ANS. Glucose-sensitive neurons in the hypothalamus detect low blood glucose and increase sympathetic output. As a result, sympathetic nerve endings release the neurotransmitter norepinephrine, and the adrenal medulla releases two catecholamine hormones—epinephrine and norepinephrine—into the bloodstream. Like glucagon, epinephrine stimulates glycogen breakdown. Epinephrine and norepinephrine are both potent stimulators of lipolysis. These actions of the catecholamines help to increase glucose and free fatty acid levels in the blood. As a result, muscle uses more fatty acids for ATP production, and more glucose is available to the nervous system. Table 25.4 summarizes the hormonal regulation of metabolism in the postabsorptive state.

Metabolism During Fasting and Starvation

The term **fasting** means going without food for many hours or a few days; **starvation** implies weeks or months of food deprivation or inadequate food intake. People can survive without food for two months or more if they drink enough water to prevent dehydration. Although glycogen stores are depleted within a few hours of beginning a fast, catabolism of stored triglycerides and structural proteins can provide energy for several weeks. The amount of adipose tissue the body contains determines the lifespan possible without food.

During fasting and starvation, nervous tissue and RBCs continue to use glucose for ATP production. There is a ready supply of amino acids for gluconeogenesis because lowered insulin and increased cortisol levels slow the pace of protein synthesis and promote protein catabolism. Most cells in the body, especially skeletal muscle cells because of their high protein content, can spare a fair amount of protein before their performance is adversely affected. During the first few days of fasting, protein catabolism outpaces protein synthesis by about 75 grams daily as some of the "old" amino acids are being deaminated and used for gluconeogenesis and "new" (that is, dietary) amino acids are lacking.

By the second day of a fast, blood glucose level has stabilized at about 65 mg/100 mL (3.6 mmol/liter); at the same time the level of fatty acids in plasma has risen fourfold. Lipolysis of triglycerides in adipose tissue releases glycerol, which is used for gluconeogenesis, and fatty acids. The fatty acids diffuse into muscle fibers and other body cells, where they are used to produce acetyl-CoA, which enters the Krebs cycle. ATP then is synthesized as oxidation proceeds via the Krebs cycle and the electron transport chain.

The most dramatic metabolic change that occurs with fasting and starvation is the increase in the formation of ketone bodies by hepatocytes. During fasting, only small amounts of glucose undergo glycolysis to pyruvic acid, which in turn can be converted to oxaloacetic acid. Acetyl-CoA enters the Krebs cycle by combining with oxaloacetic acid (see Figure 25.16); when oxaloacetic acid is scarce due to fasting, only some of the available acetyl-CoA can enter the Krebs cycle. Surplus acetyl-CoA is used for ketogenesis, mainly in hepatocytes. Ketone body production thus increases as catabolism of fatty acids rises. Lipid-soluble ketone bodies can diffuse through plasma membranes and across the blood–brain barrier and be used as an alternate fuel for ATP production, especially by cardiac and skeletal muscle fibers and neurons. Normally, only a trace of ketone bodies (0.01 mmol/liter) are present in the blood, so they are a negligible fuel source. After two days of fasting, however, the level of ketones is 100–300 times higher and supplies roughly a third of the brain's fuel for ATP production. By 40 days of starvation, ketones provide up to two-thirds of the brain's energy needs. The presence of ketones actually reduces the use of glucose for ATP production, which in turn decreases the demand for gluconeogenesis and slows the catabolism of muscle proteins later in starvation to about 20 grams daily.

TABLE 25.4 Hormonal Regulation of Metabolism in the Postabsorptive State

Process	Location(s)	Main Stimulating Hormone(s)
Glycogenolysis (glycogen breakdown)	Hepatocytes and skeletal muscle fibers.	Glucagon and epinephrine.
Lipolysis (triglyceride breakdown)	Adipocytes.	Epinephrine, norepinephrine, cortisol, insulinlike growth factors, thyroid hormones, and others.
Protein breakdown	Most body cells, but especially skeletal muscle fibers.	Cortisol.
Gluconeogenesis (synthesis of glucose from noncarbohydrates)	Hepatocytes and kidney cortex cells.	Glucagon and cortisol.

▶ CHECKPOINT

23. What are the roles of insulin, glucagon, epinephrine, insulinlike growth factors, thyroxine, cortisol, estrogen, and testosterone in regulation of metabolism?

24. Why is ketogenesis more significant during fasting or starvation than during normal absorptive and postabsorptive states?

HEAT AND ENERGY BALANCE

▶ OBJECTIVES

Define basal metabolic rate (BMR), and explain several factors that affect it.

Describe the factors that influence body heat production.

Explain how normal body temperature is maintained by negative feedback loops involving the hypothalamic thermostat.

Your body produces more or less heat depending on the rates of its metabolic reactions. Because homeostasis of body temperature can be maintained only if the rate of heat loss from the body equals the rate of heat production by metabolism, it is important to understand the ways in which heat can be lost, gained, or conserved. **Heat** is a form of energy that can be measured as **temperature** and expressed in units called calories. A **calorie (cal)** is defined as the amount of heat required to raise the temperature of 1 gram of water 1°C. Because the calorie is a relatively small unit, the **kilocalorie (kcal)** or **Calorie (Cal)** (always spelled with an uppercase C) is often used to measure the body's metabolic rate and to express the energy content of foods. A kilocalorie equals 1000 calories. Thus, when we say that a particular food item contains 500 Calories, we are actually referring to kilocalories.

Metabolic Rate

The overall rate at which metabolic reactions use energy is termed the **metabolic rate.** As you have already learned, some of the energy is used to produce ATP, and some is released as heat. Because many factors affect metabolic rate, it is measured under standard conditions, with the body in a quiet, resting, and fasting condition called the **basal state.** The measurement obtained under these conditions is the **basal metabolic rate (BMR).** The most common way to determine BMR is by measuring the amount of oxygen used per kilocalorie of food metabolized. When the body uses 1 liter of oxygen to oxidize a typical dietary mixture of triglycerides, carbohydrates, and proteins, about 4.8 Cal of energy is released. BMR is 1200–1800 Cal/day in adults, or about 24 Cal/kg of body mass in adult males and 22 Cal/kg in adult females. The added calories needed to support daily activities, such as digestion and walking, range from 500 Cal for a small, relatively sedentary person to over 3000 Cal for a person in training for Olympic-level competitions or mountain climbing.

Body Temperature Homeostasis

Despite wide fluctuations in environmental temperature, homeostatic mechanisms can maintain a normal range for internal body temperature. If the rate of body heat production equals the rate of heat loss, the body maintains a constant core temperature near 37°C (98.6°F). **Core temperature** is the temperature in body structures deep to the skin and subcutaneous layer. **Shell temperature** is the temperature near the body surface—in the skin and subcutaneous layer. Depending on the environmental temperature, shell temperature is 1–6°C lower than core temperature. A core temperature that is too high kills by denaturing body proteins; a core temperature that is too low causes cardiac arrhythmias that result in death.

Heat Production

The production of body heat is proportional to metabolic rate. Several factors affect the metabolic rate and thus the rate of heat production:

- ***Exercise.*** During strenuous exercise, the metabolic rate may increase to as much as 15 times the basal rate. In well-trained athletes, the rate may increase up to 20 times.
- ***Hormones.*** Thyroid hormones (thyroxine and triiodothyronine) are the main regulators of BMR; BMR increases as the blood levels of thyroid hormones rise. The response to changing levels of thyroid hormones is slow, however, taking several days to appear. Thyroid hormones increase BMR in part by stimulating aerobic cellular respiration. As cells use more oxygen to produce ATP, more heat is given off, and body temperature rises. Other hormones have minor effects on BMR. Testosterone, insulin, and human growth hormone can increase the metabolic rate by 5–15%.
- ***Nervous system.*** During exercise or in a stressful situation, the sympathetic division of the autonomic nervous system is stimulated. Its postganglionic neurons release norepinephrine (NE), and it also stimulates release of the hormones epinephrine and norepinephrine by the adrenal medulla. Both epinephrine and norepinephrine increase the metabolic rate of body cells.
- ***Body temperature.*** The higher the body temperature, the higher the metabolic rate. Each 1°C rise in core temperature increases the rate of biochemical reactions by about 10%. As a result, metabolic rate may be increased substantially during a fever.
- ***Ingestion of food.*** The ingestion of food raises the metabolic rate 10–20% due to the energy "costs" of digesting, absorbing, and storing nutrients. This effect, *food-induced thermogenesis,* is greatest after eating a high-protein meal and is less after eating carbohydrates and lipids.
- ***Age.*** The metabolic rate of a child, in relation to its size, is about double that of an elderly person due to the high rates of reactions related to growth.
- ***Other factors.*** Other factors that affect metabolic rate are gender (lower in females, except during pregnancy and lactation), climate (lower in tropical regions), sleep (lower), and malnutrition (lower).

Mechanisms of Heat Transfer

Maintaining normal body temperature depends on the ability to lose heat to the environment at the same rate as it is produced by metabolic reactions. Heat can be transferred from the body to its surroundings in four ways: via conduction, convection, radiation, and evaporation.

1. Conduction is the heat exchange that occurs between molecules of two materials that are in direct contact with each other. At rest, about 3% of body heat is lost via conduction to solid materials in contact with the body, such as a chair, clothing, and jewelry. Heat can also be gained via conduction—for example, while soaking in a hot tub. Because water conducts heat 20 times more effectively than air, heat loss or heat gain via conduction is much greater when the body is submerged in cold or hot water.

2. Convection is the transfer of heat by the movement of a fluid (a gas or a liquid) between areas of different temperature. The contact of air or water with your body results in heat transfer by both conduction and convection. When cool air makes contact with the body, it becomes warmed and therefore less dense and is carried away by convection currents created as the less dense air rises. The faster the air moves—for example, by a breeze or a fan—the faster the rate of convection. At rest, about 15% of body heat is lost to the air via conduction and convection.

3. **Radiation** is the transfer of heat in the form of infrared rays between a warmer object and a cooler one without physical contact. Your body loses heat by radiating more infrared waves than it absorbs from cooler objects. If surrounding objects are warmer than you are, you absorb more heat than you lose by radiation. In a room at 21°C (70°F), about 60% of heat loss occurs via radiation in a resting person.

4. **Evaporation** is the conversion of a liquid to a vapor. Every milliliter of evaporating water takes with it a great deal of heat—about 0.58 Cal/mL. Under typical resting conditions, about 22% of heat loss occurs through evaporation of about 700 mL of water per day—300 mL in exhaled air and 400 mL from the skin surface. Because we are not normally aware of this water loss through the skin and mucous membranes of the mouth and respiratory system, it is termed **insensible water loss.** The rate of evaporation is inversely related to relative humidity, the ratio of the actual amount of moisture in the air to the maximum amount it can hold at a given temperature. The higher the relative humidity, the lower the rate of evaporation. At 100% humidity, heat is gained via condensation of water on the skin surface as fast as heat is lost via evaporation. Evaporation provides the main defense against overheating during exercise. Under extreme conditions, a maximum of about 3 liters of sweat can be produced each hour, removing more than 1700 Calories of heat if all of it evaporates. (Note: Sweat that drips off the body rather than evaporating removes very little heat.)

Hypothalamic Thermostat

The control center that functions as the body's thermostat is a group of neurons in the anterior part of the hypothalamus, the **preoptic area.** This area receives impulses from thermoreceptors in the skin and mucous membranes and in the hypothalamus. Neurons of the preoptic area generate nerve impulses at a higher frequency when blood temperature increases, and at a lower frequency when blood temperature decreases.

Nerve impulses from the preoptic area propagate to two other parts of the hypothalamus known as the **heat-losing center** and the **heat-promoting center,** which, when stimulated by the preoptic area, set into operation a series of responses that lower body temperature and raise body temperature, respectively.

Thermoregulation

If core temperature declines, mechanisms that help conserve heat and increase heat production act via several negative feedback loops to raise the body temperature to normal (Figure 25.19). Thermoreceptors in the skin and hypothalamus send nerve impulses to the preoptic area and the heat-promoting center in the hypothalamus, and to hypothalamic neurosecretory cells that produce thyrotropin-releasing hormone (TRH). In response, the hypothalamus discharges nerve impulses and secretes TRH, which in turn stimulates thyrotrophs in the anterior pituitary gland to release thyroid-stimulating hormone (TSH). Nerve impulses from the hypothalamus and TSH then activate several effectors.

Each effector responds in a way that helps increase core temperature to the normal value:

- Nerve impulses from the heat-promoting center stimulate sympathetic nerves that cause blood vessels of the skin to constrict. Vasoconstriction decreases the flow of warm blood, and thus the transfer of heat, from the internal organs to the skin. Slowing the rate of heat loss allows the internal body temperature to increase as metabolic reactions continue to produce heat.
- Nerve impulses in sympathetic nerves leading to the adrenal medulla stimulate the release of epinephrine and norepinephrine into the blood. The hormones, in turn, bring about an increase in cellular metabolism, which increases heat production.
- The heat-promoting center stimulates parts of the brain that increase muscle tone and hence heat production. As muscle tone increases in one muscle (the agonist), the small contractions stretch muscle spindles in its antagonist, initiating a stretch reflex. The resulting contraction in the antagonist stretches muscle spindles in the agonist, and it too develops a stretch reflex. This repetitive cycle—called **shivering**—greatly increases the rate of heat production. During maximal shivering, body heat production can rise to about four times the basal rate in just a few minutes.
- The thyroid gland responds to TSH by releasing more thyroid hormones into the blood. As increased levels of thyroid hormones slowly increase the metabolic rate, body temperature rises.

If core body temperature rises above normal, a negative feedback loop opposite to the one depicted in Figure 25.19 goes into action. The higher temperature of the blood stimulates thermoreceptors that send nerve impulses to the preoptic area, which in turn stimulate the heat-losing center and inhibit the heat-promoting center. Nerve impulses from the heat-losing center cause dilation of blood vessels in the skin. The skin becomes warm, and the excess heat is lost to the environment via radiation and conduction as an increased volume of blood flows from the warmer core of the body into the cooler skin. At the same time, metabolic rate decreases, and shivering does not occur. The high temperature of the blood stimulates sweat glands of the skin via hypothalamic activation of sympathetic nerves. As the water in perspiration evaporates from the surface of the skin, the skin is cooled. All these responses counteract heat-promoting effects and help return body temperature to normal.

Figure 25.19 Negative feedback mechanisms that conserve heat and increase heat production.

Core temperature is the temperature in body structures deep to the skin and subcutaneous layer; shell temperature is the temperature near the body surface.

Some stimulus disrupts homeostasis by
Decreasing
Body temperature
Receptors
Thermoreceptors in skin and hypothalamus
Input
Nerve impulses
Control centers
Preoptic area, heat promoting center, and neurosecretory cells in hypothalamus and thyrotropes in anterior pituitary gland
Output
Nerve impulses and TSH
Effectors
Vasoconstriction decreases heat loss through the skin
Adrenal medulla releases hormones that increase cellular metabolism
Skeletal muscles contract in a repetitive cycle called shivering
Thyroid gland releases thyroid hormones, which increase metabolic rate
Increase in body temperature
Return to homeostasis when response brings body temperature back to normal

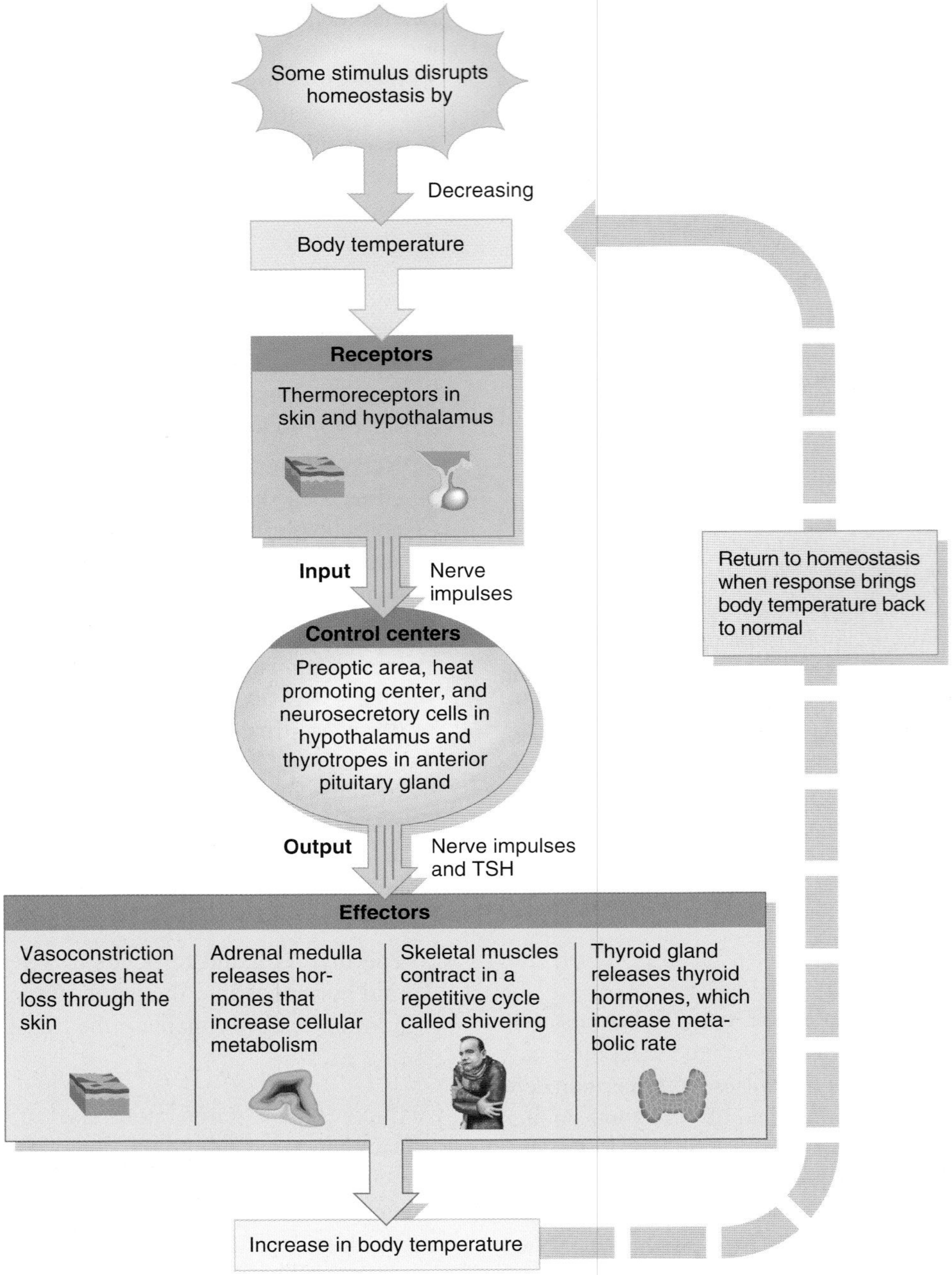

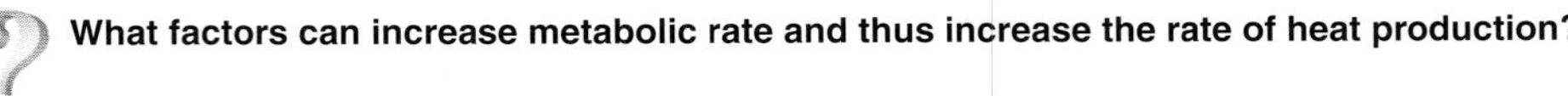

What factors can increase metabolic rate and thus increase the rate of heat production?

Hypothermia

Hypothermia is a lowering of core body temperature to 35°C (95°F) or below. Causes of hypothermia include an overwhelming cold stress (immersion in icy water), metabolic diseases (hypoglycemia, adrenal insufficiency, or hypothyroidism), drugs (alcohol, antidepressants, sedatives, or tranquilizers), burns, and malnutrition. Hypothermia is characterized by the following as core body temperature falls: sensation of cold, shivering, confusion, vasoconstriction, muscle rigidity, bradycardia, acidosis, hypoventilation, hypotension, loss of spontaneous movement, coma, and death (usually caused by cardiac arrhythmias). Because the elderly have reduced metabolic protection against a cold environment coupled with a reduced perception of cold, they are at greater risk for developing hypothermia. ■

Energy Homeostasis and Regulation of Food Intake

Most mature animals and many men and women maintain **energy homeostasis,** the precise matching of energy intake (in food) to energy expenditure over time. When the energy content of food balances the energy used by all the cells of the body, body weight remains constant (unless there is a gain or loss of water). In many people, weight stability persists despite large day-to-day variations in activity and food intake. In the more affluent nations, however, a large fraction of the population is overweight. Easy access to tasty, high-calorie foods and a "couch-potato" lifestyle promote weight gain. Being overweight increases the risk of dying from a variety of cardiovascular and metabolic disorders, including hypertension, varicose veins, diabetes mellitus, arthritis, certain cancers, and gallbladder disease.

Energy intake depends only on the amount of food consumed (and absorbed), but three components contribute to total energy expenditure.

1. The basal metabolic rate accounts for about 60% of energy expenditure.
2. Physical activity typically adds 30–35% but can be lower in sedentary people. The energy expenditure is partly from voluntary exercise, such as walking, and partly from **nonexercise activity thermogenesis (NEAT),** the energy costs for maintaining muscle tone, posture while sitting or standing, and involuntary fidgeting movements.
3. **Food-induced thermogenesis,** the heat produced while food is being digested, absorbed, and stored, represents 5–10% of total energy expenditure.

The major site of stored chemical energy in the body is adipose tissue. When energy use exceeds energy input, triglycerides in adipose tissue are catabolized to provide the extra energy, and when energy input exceeds energy expenditure, triglycerides are stored. Over time, the amount of stored triglycerides indicates the excess of energy intake over energy expenditure. Even small differences add up over time. A gain of 20 lb (9 kg) between ages 25 and 55 represents only a tiny imbalance, about 0.3% more energy intake in food than energy expenditure.

Clearly, negative feedback mechanisms are regulating both our energy intake and our energy expenditure. But no sensory receptors exist to monitor our weight or size. How, then, is food intake regulated? The answer to this question is incomplete, but important advances in understanding regulation of food intake have occurred in the past decade. It depends on many factors, including neural and endocrine signals, levels of certain nutrients in the blood, psychological elements such as stress or depression, signals from the GI tract and the special senses, and neural connections between the hypothalamus and other parts of the brain.

Within the hypothalamus are clusters of neurons that play key roles in regulating food intake. Although these neurons receive signals that indicate hunger or satiety, they are not precisely organized into "feeding" and "satiety" centers, as was once thought. **Satiety** (sa-TĪ-i-tē) is a feeling of fullness accompanied by lack of desire to eat. Two hypothalamic areas involved in regulation of food intake are the *arcuate nucleus* and the *paraventricular nucleus* (see Figure 14.10 on page 490). In 1994, the first experiments were reported on a mouse gene, named *obese,* that causes overeating and severe obesity in its mutated form. The product of this gene is the hormone **leptin.** In both mice and humans, leptin helps decrease **adiposity,** total body-fat mass. Leptin is synthesized and secreted by adipocytes in proportion to adiposity; as more triglycerides are stored, more leptin is secreted into the bloodstream. Leptin acts on the hypothalamus to inhibit circuits that stimulate eating while also activating circuits that increase energy expenditure. The hormone insulin has a similar, but smaller, effect. Both leptin and insulin are able to pass through the blood–brain barrier.

When leptin and insulin levels are *low,* neurons that extend from the arcuate nucleus to the paraventricular nucleus release a neurotransmitter called **neuropeptide Y** that stimulates food intake. Other neurons that extend between the arcuate and paraventricular nuclei release a neurotransmitter called **melanocortin,** which is similar to melanocyte-stimulating hormone (MSH). Leptin stimulates release of melanocortin, which acts to inhibit food intake. Although leptin, neuropeptide Y, and melanocortin are key signaling molecules for maintaining energy homeostasis, several other hormones and neurotransmitters also contribute. An understanding of the brain circuits involved is still far from complete. Other areas of the hypothalamus plus nuclei in the brainstem, limbic system, and cerebral cortex take part.

Achieving energy homeostasis requires regulation of energy intake. Most increases and decreases in food intake are due to changes in meal size rather than changes in number of meals. Many experiments have demonstrated the presence of satiety signals, chemical or neural changes that help terminate eating when "fullness" is attained. For example, an increase in blood glucose level, as occurs after a meal, decreases appetite. Several hormones, such as glucagon, cholecystokinin, estrogens, and epinephrine (acting via beta receptors) act to signal satiety and to increase energy expenditure. Distension of the GI tract, particularly the stomach and duodenum, also contributes to termination of food intake. Other hormones increase appetite and decrease energy expenditure. These include growth hormone-releasing hormone (GHRH), androgens, glucocorticoids, epinephrine (acting via alpha receptors), and progesterone.

Emotional Eating

In addition to keeping us alive, eating serves countless psychological, social, and cultural purposes. We eat to celebrate, punish, comfort, defy, and deny. Eating in response to emotional drives, such as feeling stressed, bored, or tired, is called **emotional eating.** Emotional eating is so common that, within limits, it is considered well within the range of normal behavior. Who hasn't at one time or another headed for the refrigerator after a bad day? Problems arise when emotional eating becomes so excessive that it interferes with health. Physical health problems include obesity and associated disorders such as hypertension and heart disease. Psychological health problems include poor self-esteem, an inability to cope effectively with feelings of stress, and in extreme cases, eating disorders, such as anorexia nervosa, bulimia, and obesity.

Eating provides comfort and solace, numbing pain and "feeding the hungry heart." Eating may provide a biochemical "fix" as well. Emotional eaters typically overeat carbohydrate foods (sweets and starches), which may raise brain serotonin levels and lead to feelings of relaxation. Food becomes a way to self-medicate when negative emotions arise. ■

▶ CHECKPOINT

25. Define a kilocalorie (kcal). How is the unit used? How does it relate to a calorie?
26. Distinguish between core temperature and shell temperature.
27. In what ways can a person lose heat to or gain heat from the surroundings? How is it possible for a person to lose heat on a sunny beach when the temperature is 40°C (104°F) and the humidity is 85%?
28. What does the term energy homeostasis mean?
29. How is food intake regulated?

NUTRITION

▶ OBJECTIVES

- **Describe how to select foods to maintain a healthy diet.**
- **Compare the sources, functions, and importance of minerals and vitamins in metabolism.**

Nutrients are chemical substances in food that body cells use for growth, maintenance, and repair. The six main types of nutrients are water, carbohydrates, lipids, proteins, minerals, and vitamins. Water is the nutrient needed in the largest amount—about 2–3 liters per day. As the most abundant compound in the body, water provides the medium in which most metabolic reactions occur, and it also participates in some reactions (for example, hydrolysis reactions). The important roles of water in the body can be reviewed on pages 39–40. Three organic nutrients—carbohydrates, lipids, and proteins—provide the energy needed for metabolic reactions and serve as building blocks to make body structures. Some minerals and many vitamins are components of the enzyme systems that catalyze metabolic reactions. *Essential nutrients* are specific nutrient molecules that the body cannot make in sufficient quantity to meet its needs and thus must be obtained from the diet. Some amino acids, fatty acids, vitamins, and minerals are essential nutrients.

Next, we discuss some guidelines for healthy eating and the roles of minerals and vitamins in metabolism.

Guidelines for Healthy Eating

Each gram of protein or carbohydrate in food provides about 4 Calories; a gram of fat (lipids) provides about 9 Calories. On a daily basis, many women and older people need about 1600 Calories; children, teenage girls, active women, and most men need about 2200 Calories; and teenage boys and active men need about 2800 Calories.

We do not know with certainty what levels and types of carbohydrate, fat, and protein are optimal in the diet. Different populations around the world eat radically different diets that are adapted to their particular lifestyles. However, many experts recommend the following distribution of calories: 50–60% from carbohydrates, with less than 15% from simple sugars; less than 30% from fats (triglycerides are the main type of dietary fat), with no more than 10% as saturated fats; and about 12–15% from proteins.

The guidelines for healthy eating are to:

- Eat a variety of foods.
- Maintain a healthy weight.
- Choose foods low in fat, saturated fat, and cholesterol.
- Eat plenty of vegetables, fruits, and grain products.
- Use sugars in moderation only.
- Use salt and sodium in moderation only (less than 6 grams daily).
- If you drink alcoholic beverages, do so in moderation (less than 1 ounce of the equivalent of pure alcohol per day).

To help people achieve a good balance of vitamins, minerals, carbohydrates, fats, and proteins in their food, the U.S. Department of Agriculture developed the food guide pyramid (Figure 25.20). The sections of the pyramid indicate how many servings of each of the five major food groups to eat each day. The smallest number of servings corresponds to a 1600 Cal/day diet; the largest number of servings corresponds to a 2800 Cal/day diet. Because they should be consumed in largest quantity, foods rich in complex carbohydrates—the bread, cereal, rice, and pasta group—form the base of the pyramid. Vegetables and fruits form the next level. The health benefits of eating generous amounts of these foods are well documented. Foods on the next level up—the milk, yogurt, and cheese group and the meat, poultry, fish, dry beans, eggs, and nuts group—should be eaten in smaller quantities. These two food groups have higher fat and protein content than the food groups below them. To lower your daily intake of

Figure 25.20 The food guide pyramid. The smallest number of servings corresponds to 1600 Calories per day; the largest number of servings corresponds to 2800 Calories per day. Each example given equals one serving.

The sections of the pyramid show how many servings of five major food groups to eat each day.

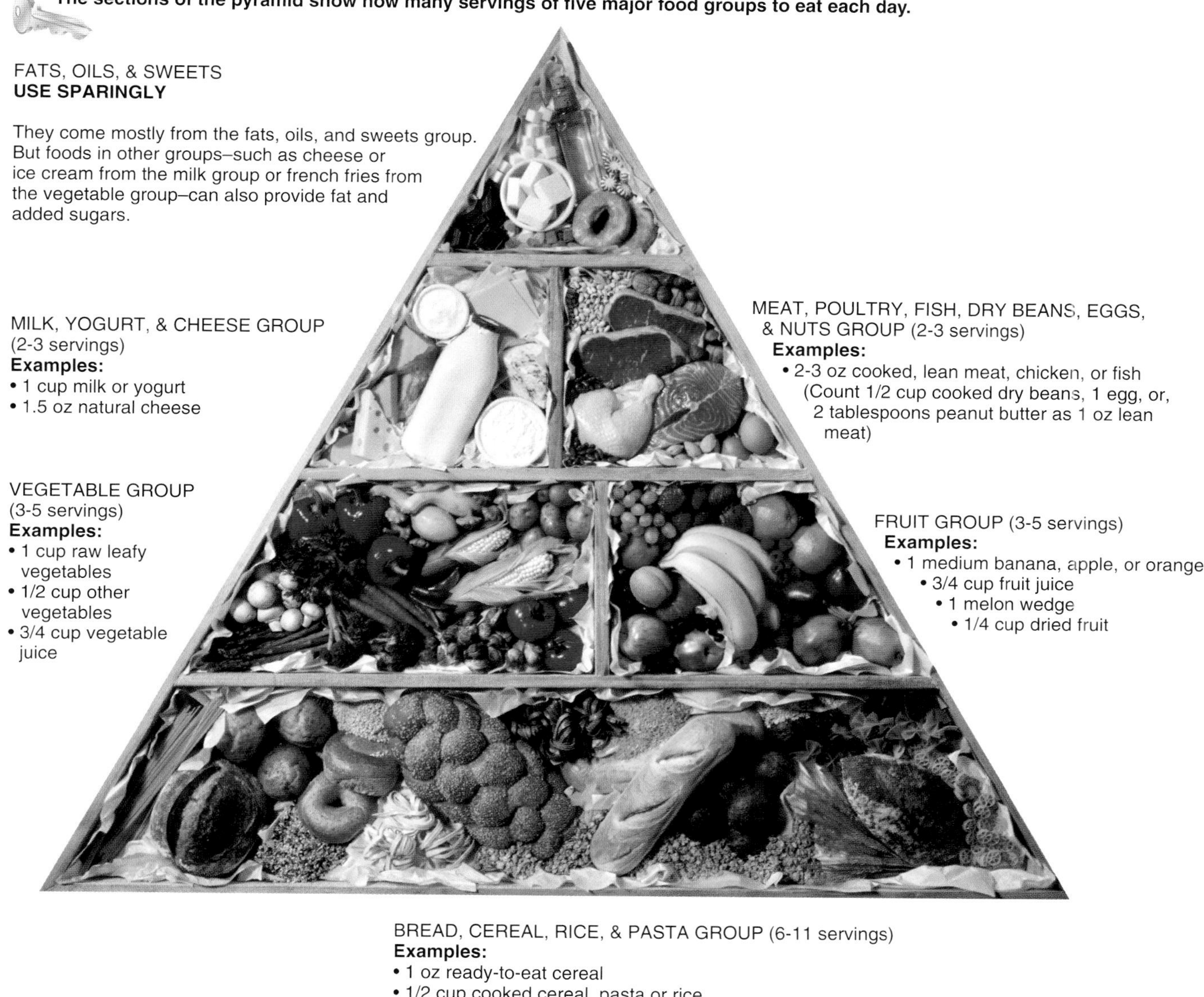

Which foods shown contain cholesterol and most of the saturated fatty acids in the diet?

fats, choose low-fat foods from these groups—nonfat milk and yogurt, low-fat cheese, fish, and poultry with the skin removed.

The apex of the pyramid is not a food group but rather a caution to use fats, oils, and sweets sparingly. The food guide pyramid does not distinguish among the different types of fatty acids—saturated, polyunsaturated, and monounsaturated—in dietary fats. However, atherosclerosis and coronary artery disease are prevalent in populations that consume large amounts of saturated fats and cholesterol. Populations living around the Mediterranean Sea, in contrast, have low rates of coronary artery disease despite eating a diet in which up to 40% of the calories are fats. Most of their fat comes from olive oil, which is rich in monounsaturated fatty acids and has no cholesterol. Canola oil, avocados, nuts, and peanut oil are also rich in monounsaturated fatty acids.

Minerals

Minerals are inorganic elements that occur naturally in the Earth's crust. In the body they appear in combination with one another, in combination with organic compounds, or as ions in solution. Minerals constitute about 4% of total body mass and are concentrated most heavily in the skeleton. Minerals with known functions in the body include calcium, phosphorus, potassium, sulfur, sodium, chloride, magnesium, iron, iodide, manganese, copper, cobalt, zinc, fluoride, selenium, and chromium. Table 25.5 describes the vital functions of these minerals. Note that the body generally uses the ions of the minerals rather than the non-ionized form. Some minerals, such as chlorine, are toxic or even fatal if ingested in the non-ionized form. Other minerals—aluminum, boron, silicon, and molybdenum—are present but their functions are unclear. Typical diets supply adequate amounts of potassium, sodium, chloride, and magnesium. Some attention must be paid to eating foods that provide enough calcium, phosphorus, iron, and iodide. Excess amounts of most minerals are excreted in the urine and feces.

Calcium and phosphorus form part of the matrix of bone. Because minerals do not form long-chain compounds, they are otherwise poor building materials. A major role of minerals is to help regulate enzymatic reactions. Calcium, iron, magnesium, and manganese are constituents of some coenzymes. Magnesium also serves as a catalyst for the conversion of ADP to ATP. Minerals such as sodium and phosphorus work in buffer systems, which help control the pH of body fluids. Sodium also helps regulate the osmosis of water and, along with other ions, is involved in the generation of nerve impulses.

Vitamins

Organic nutrients required in small amounts to maintain growth and normal metabolism are called **vitamins.** Unlike carbohydrates, lipids, or proteins, vitamins do not provide energy or serve as the body's building materials. Most vitamins with known functions are coenzymes.

Most vitamins cannot be synthesized by the body and must be ingested in food. Other vitamins, such as vitamin K, are produced by bacteria in the GI tract and then absorbed. The body can assemble some vitamins if the raw materials, called **provitamins,** are provided. For example, vitamin A is produced by the body from the provitamin beta-carotene, a chemical present in yellow vegetables such as carrots and in dark green vegetables such as spinach. No single food contains all the required vitamins—one of the best reasons to eat a varied diet.

Vitamins are divided into two main groups: fat-soluble and water-soluble. The **fat-soluble vitamins,** vitamins A, D, E, and K, are absorbed along with other dietary lipids in the small intestine and packaged into chylomicrons. They cannot be absorbed in adequate quantity unless they are ingested with other lipids. Fat-soluble vitamins may be stored in cells, particularly hepatocytes. The **water-soluble vitamins,** including several B vitamins and vitamin C, are dissolved in body fluids. Excess quantities of these vitamins are not stored but instead are excreted in the urine.

Besides their other functions, three vitamins—C, E, and beta-carotene (a provitamin)—are termed **antioxidant vitamins** because they inactivate oxygen free radicals. Recall that free radicals are highly reactive ions or molecules that carry an unpaired electron in their outermost electron shell (see Figure 2.3 on page 32). Free radicals damage cell membranes, DNA, and other cellular structures and contribute to the formation of artery-narrowing atherosclerotic plaques. Some free radicals arise naturally in the body, and others come from environmental hazards such as tobacco smoke and radiation. Antioxidant vitamins are thought to play a role in protecting against some kinds of cancer, reducing the buildup of atherosclerotic plaque, delaying some effects of aging, and decreasing the chance of cataract formation in the lens of the eyes. Table 25.6 on pages 984–985 lists the major vitamins, their sources, their functions, and related deficiency disorders.

Vitamin and Mineral Supplements

Most nutritionists recommend eating a balanced diet that includes a variety of foods rather than taking vitamin or mineral supplements, except in special circumstances. Common examples of necessary supplementations include iron for women who have excessive menstrual bleeding; iron and calcium for women who are pregnant or breast-feeding; folic acid (folate) for all women who may become pregnant, to reduce the risk of fetal neural tube defects; calcium for most adults, because they do not receive the recommended amount in their diets; and vitamin B_{12} for strict vegetarians, who eat no meat. Because most North Americans do not ingest in their food the high levels of antioxidant vitamins thought to have beneficial effects, some experts recommend supplementing vitamins C and E. More is not always better, however; larger doses of vitamins or minerals can be very harmful.

Hypervitaminosis (hī-per-vī-ta-mi-NŌ-sis; *hyper-* = too much or above) refers to dietary intake of a vitamin that exceeds that ability of the body to utilize, store, or excrete the vitamin. Since water-soluble vitamins are not stored in the body, few of them cause any problems related to hypervitaminosis. However, because lipid-soluble vitamins are stored in the body, excessive consumption may cause problems. For example, excess intake of vitamin A can cause drowsiness, general weakness, irritability, headache, vomiting, dry and peeling skin, partial hair loss, joint pain, liver and spleen enlargement, coma, and even death. Excessive intake of vitamin D may result in loss of appetite, nausea, vomiting, excessive thirst, general weakness, irritability, hypertension, and kidney damage and malfunction. **Hypovitaminosis** (*hypo-* = too little or below), or vitamin deficiency, is discussed in Table 25.6 for the various vitamins. ■

TABLE 25.5 Minerals Vital to the Body

Mineral	Comments	Importance
Calcium	Most abundant mineral in body. Appears in combination with phosphates. About 99% is stored in bone and teeth. Blood Ca^{2+} level is controlled by parathyroid hormone (PTH). Calcitriol promotes absorption of dietary calcium. Excess is excreted in feces and urine. Sources are milk, egg yolk, shellfish, and leafy green vegetables.	Formation of bones and teeth, blood clotting, normal muscle and nerve activity, endocytosis and exocytosis, cellular motility, chromosome movement during cell division, glycogen metabolism, and release of neurotransmitters and hormones.
Phosphorus	About 80% is found in bones and teeth as phosphate salts. Blood phosphate level is controlled by parathyroid hormone (PTH). Excess is excreted in urine; small amount is eliminated in feces. Sources are dairy products, meat, fish, poultry, and nuts.	Formation of bones and teeth. Phosphates ($H_2PO_4^-$, HPO_4^-, and PO_4^{3-}) constitute a major buffer system of blood. Plays important role in muscle contraction and nerve activity. Component of many enzymes. Involved in energy transfer (ATP). Component of DNA and RNA.
Potassium	Major cation (K^+) in intracellular fluid. Excess excreted in urine. Present in most foods (meats, fish, poultry, fruits, and nuts).	Needed for generation and conduction of action potentials in neurons and muscle fibers.
Sulfur	Component of many proteins (such as insulin and chrondroitin sulfate), electron carriers in electron transport chain, and some vitamins (thiamine and biotin). Excreted in urine. Sources include beef, liver, lamb, fish, poultry, eggs, cheese, and beans.	As component of hormones and vitamins, regulates various body activities. Needed for ATP production by electron transport chain.
Sodium	Most abundant cation (Na^+) in extracellular fluids; some found in bones. Excreted in urine and perspiration. Normal intake of NaCl (table salt) supplies more than the required amounts.	Strongly affects distribution of water through osmosis. Part of bicarbonate buffer system. Functions in nerve and muscle action potential conduction.
Chloride	Major anion (Cl^-) in extracellular fluid. Excess excreted in urine. Sources include table salt (NaCl), soy sauce, and processed foods.	Plays role in acid–base balance of blood, water balance, and formation of HCl in stomach.
Magnesium	Important cation (Mg^{2+}) in intracellular fluid. Excreted in urine and feces. Widespread in various foods, such as green leafy vegetables, seafood, and whole-grain cereals.	Required for normal functioning of muscle and nervous tissue. Participates in bone formation. Constituent of many coenzymes.
Iron	About 66% found in hemoglobin of blood. Normal losses of iron occur by shedding of hair, epithelial cells, and mucosal cells, and in sweat, urine, feces, bile, and blood lost during menstruation. Sources are meat, liver, shellfish, egg yolk, beans, legumes, dried fruits, nuts, and cereals.	As component of hemoglobin, reversibly binds O_2. Component of cytochromes involved in electron transport chain.
Iodide	Essential component of thyroid hormones. Excreted in urine. Sources are seafood, iodized salt, and vegetables grown in iodine-rich soils.	Required by thyroid gland to synthesize thyroid hormones, which regulate metabolic rate.
Manganese	Some stored in liver and spleen. Most excreted in feces.	Activates several enzymes. Needed for hemoglobin synthesis, urea formation, growth, reproduction, lactation, bone formation, and possibly production and release of insulin, and inhibition of cell damage.
Copper	Some stored in liver and spleen. Most excreted in feces. Sources include eggs, whole-wheat flour, beans, beets, liver, fish, spinach, and asparagus.	Required with iron for synthesis of hemoglobin. Component of coenzymes in electron transport chain and enzyme necessary for melanin formation.
Cobalt	Constituent of vitamin B_{12}.	As part of vitamin B_{12}, required for erythropoiesis.
Zinc	Important component of certain enzymes. Widespread in many foods, especially meats.	As a component of carbonic anhydrase, important in carbon dioxide metabolism. Necessary for normal growth and wound healing, normal taste sensations and appetite, and normal sperm counts in males. As a component of peptidases, it is involved in protein digestion.
Fluoride	Components of bones, teeth, other tissues.	Appears to improve tooth structure and inhibit tooth decay.
Selenium	Important component of certain enzymes. Found in seafood, meat, chicken, tomatoes, egg yolk, milk, mushrooms, and garlic, and cereal grains grown in selenium-rich soil.	Needed for synthesis of thyroid hormones, sperm motility, and proper functioning of the immune system. Also functions as an antioxidant. Prevents chromosome breakage and may play a role in preventing certain birth defects, miscarriage, prostate cancer, and coronary artery disease.
Chromium	Found in high concentrations in brewer's yeast. Also found in wine and some brands of beer.	Needed for normal activity of insulin in carbohydrate and lipid metabolism.

TABLE 25.6 The Principal Vitamins

Vitamin	Comment and Source	Functions	Deficiency Symptoms and Disorders
Fat-soluble	All require bile salts and some dietary lipids for adequate absorption.		
A	Formed from provitamin beta-carotene (and other provitamins) in GI tract. Stored in liver. Sources of carotene and other provitamins include orange, yellow, and green vegetables; sources of vitamin A include liver and milk.	Maintains general health and vigor of epithelial cells. Beta-carotene acts as an antioxidant to inactivate free radicals. Essential for formation of light-sensitive pigments in photoreceptors of retina. Aids in growth of bones and teeth by helping to regulate activity of osteoblasts and osteoclasts.	Deficiency results in atrophy and keratinization of epithelium, leading to dry skin and hair; increased incidence of ear, sinus, respiratory, urinary, and digestive system infections; inability to gain weight; drying of cornea; and skin sores. **Night blindness** or decreased ability for dark adaptation. Slow and faulty development of bones and teeth.
D	Sunlight converts 7-dehydrocholesterol in the skin to cholecalciferol (vitamin D_3). A liver enzyme then converts cholecalciferol to 25-hydroxycholecalciferol. A second enzyme in the kidneys converts 25-hydroxycholecalciferol to calcitriol (1,25-dihydroxycalciferol), which is the active form of vitamin D. Most is excreted in bile. Dietary sources include fish-liver oils, egg yolk, and fortified milk.	Essential for absorption of calcium and phosphorus from GI tract. Works with parathyroid hormone (PTH) to maintain Ca^{2+} homeostasis.	Defective utilization of calcium by bones leads to **rickets** in children and **osteomalacia** in adults. Possible loss of muscle tone.
E (tocopherols)	Stored in liver, adipose tissue, and muscles. Sources include fresh nuts and wheat germ, seed oils, and green leafy vegetables.	Inhibits catabolism of certain fatty acids that help form cell structures, especially membranes. Involved in formation of DNA, RNA, and red blood cells. May promote wound healing, contribute to the normal structure and functioning of the nervous system, and prevent scarring. May help protect liver from toxic chemicals such as carbon tetrachloride. Acts as an antioxidant to inactivate free radicals.	May cause oxidation of monounsaturated fats, resulting in abnormal structure and function of mitochondria, lysosomes, and plasma membranes. A possible consequence is hemolytic anemia.
K	Produced by intestinal bacteria. Stored in liver and spleen. Dietary sources include spinach, cauliflower, cabbage, and liver.	Coenzyme essential for synthesis of several clotting factors by liver, including prothrombin.	Delayed clotting time results in excessive bleeding.
Water-soluble	Dissolved in body fluids. Most are not stored in body. Excess intake is eliminated in urine.		
B_1 (thiamine)	Rapidly destroyed by heat. Sources include whole-grain products, eggs, pork, nuts, liver, and yeast.	Acts as coenzyme for many different enzymes that break carbon-to-carbon bonds and are involved in carbohydrate metabolism of pyruvic acid to CO_2 and H_2O. Essential for synthesis of the neurotransmitter acetylcholine.	Improper carbohydrate metabolism leads to buildup of pyruvic and lactic acids and insufficient production of ATP for muscle and nerve cells. Deficiency leads to: (1) **beriberi,** partial paralysis of smooth muscle of GI tract, causing digestive disturbances; skeletal muscle paralysis; and atrophy of limbs; (2) **polyneuritis,** due to degeneration of myelin sheaths; impaired reflexes, impaired sense of touch, stunted growth in children, and poor appetite.

Vitamin	Comment and Source	Functions	Deficiency Symptoms and Disorders
Water-soluble (continued)			
B_2 (riboflavin)	Small amounts supplied by bacteria of GI tract. Dietary sources include yeast, liver, beef, veal, lamb, eggs, whole-grain products, asparagus, peas, beets, and peanuts.	Component of certain coenzymes (for example, FAD and FMN) in carbohydrate and protein metabolism, especially in cells of eye, integument, mucosa of intestine, and blood.	Deficiency may lead to improper utilization of oxygen resulting in blurred vision, cataracts, and corneal ulcerations. Also dermatitis and cracking of skin, lesions of intestinal mucosa, and one type of anemia.
Niacin (nicotinamide)	Derived from amino acid tryptophan. Sources include yeast, meats, liver, fish, whole-grain products, peas, beans, and nuts.	Essential component of NAD and NADP, coenzymes in oxidation–reduction reactions. In lipid metabolism, inhibits production of cholesterol and assists in triglyceride breakdown.	Principal deficiency is **pellagra,** characterized by dermatitis, diarrhea, and psychological disturbances.
B_6 (pyridoxine)	Synthesized by bacteria of GI tract. Stored in liver, muscle, and brain. Other sources include salmon, yeast, tomatoes, yellow corn, spinach, whole grain products, liver, and yogurt.	Essential coenzyme for normal amino acid metabolism. Assists production of circulating antibodies. May function as coenzyme in triglyceride metabolism.	Most common deficiency symptom is dermatitis of eyes, nose, and mouth. Other symptoms are retarded growth and nausea.
B_{12} (cyanocobalamin)	Only B vitamin not found in vegetables; only vitamin containing cobalt. Absorption from GI tract depends on intrinsic factor secreted by gastric mucosa. Sources include liver, kidney, milk, eggs, cheese, and meat.	Coenzyme necessary for red blood cell formation, formation of the amino acid methionine, entrance of some amino acids into Krebs cycle, and manufacture of choline (used to synthesize acetylcholine).	Pernicious anemia, neuropsychiatric abnormalities (ataxia, memory loss, weakness, personality and mood changes, and abnormal sensations), and impaired activity of osteoblasts.
Pantothenic acid	Some produced by bacteria of GI tract. Stored primarily in liver and kidneys. Other sources include kidney, liver, yeast, green vegetables, and cereal.	Constituent of coenzyme A, which is essential for transfer of acetyl group from pyruvic acid into the Krebs cycle, conversion of lipids and amino acids into glucose, and synthesis of cholesterol and steroid hormones.	Fatigue, muscle spasms, insufficient production of adrenal steroid hormones, vomiting, and insomnia.
Folic acid (folate, folacin)	Synthesized by bacteria of GI tract. Dietary sources include green leafy vegetables, broccoli, asparagus, breads, dried beans, and citrus fruits.	Component of enzyme systems synthesizing nitrogenous bases of DNA and RNA. Essential for normal production of red and white blood cells.	Production of abnormally large red blood cells (macrocytic anemia). Higher risk of neural tube defects in babies born to folate-deficient mothers.
Biotin	Synthesized by bacteria of GI tract. Dietary sources include yeast, liver, egg yolk, and kidneys.	Essential coenzyme for conversion of pyruvic acid to oxaloacetic acid and synthesis of fatty acids and purines.	Mental depression, muscular pain, dermatitis, fatigue, and nausea.
C (ascorbic acid)	Rapidly destroyed by heat. Some stored in glandular tissue and plasma. Sources include citrus fruits, tomatoes, and green vegetables.	Promotes protein synthesis including laying down of collagen in formation of connective tissue. As coenzyme, may combine with poisons, rendering them harmless until excreted. Works with antibodies, promotes wound healing, and functions as an antioxidant.	Scurvy; anemia; many symptoms related to poor collagen formation, including tender swollen gums, loosening of teeth (alveolar processes also deteriorate), poor wound healing, bleeding (vessel walls are fragile because of connective tissue degeneration), and retardation of growth.

▶ CHECKPOINT

30. What is a nutrient?

31. Describe the food guide pyramid and give examples of foods from each food group.

32. What is a mineral? Briefly describe the functions of the following minerals: calcium, phosphorus, potassium, sulfur, sodium, chloride, magnesium, iron, iodine, copper, zinc, fluoride, manganese, cobalt, chromium, and selenium.

33. Define a vitamin. Explain how we obtain vitamins. Distinguish between a fat-soluble vitamin and a water-soluble vitamin.

34. For each of the following vitamins, indicate its principal function and the effect(s) of deficiency: A, D, E, K, B_1, B_2, niacin, B_6, B_{12}, pantothenic acid, folic acid, biotin, and C.

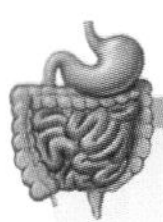

DISORDERS: HOMEOSTATIC IMBALANCES

Fever

A **fever** is an elevation of core temperature caused by a resetting of the hypothalamic thermostat. The most common causes of fever are viral or bacterial infections and bacterial toxins; other causes are ovulation, excessive secretion of thyroid hormones, tumors, and reactions to vaccines. When phagocytes ingest certain bacteria, they are stimulated to secrete a **pyrogen** (PĪ-rō-gen; *pyro-* = fire; *-gen* = produce), a fever-producing substance. One pyrogen is interleukin-1. It circulates to the hypothalamus and induces neurons of the preoptic area to secrete prostaglandins. Some prostaglandins can reset the hypothalamic thermostat at a higher temperature, and temperature-regulating reflex mechanisms then act to bring the core body temperature up to this new setting. *Antipyretics* are agents that relieve or reduce fever. Examples include aspirin, acetaminophen (Tylenol), and ibuprofen (Advil), all of which reduce fever by inhibiting synthesis of certain prostaglandins.

Suppose that due to production of pyrogens the thermostat is reset at 39°C (103°F). Now the heat-promoting mechanisms (vasoconstriction, increased metabolism, shivering) are operating at full force. Thus, even though core temperature is climbing higher than normal—say, 38°C (101°F)—the skin remains cold, and shivering occurs. This condition, called a **chill,** is a definite sign that core temperature is rising. After several hours, core temperature reaches the setting of the thermostat, and the chills disappear. But now the body will continue to regulate temperature at 39°C (103°F). When the pyrogens disappear, the thermostat is reset at normal—37.0°C (98.6°F). Because core temperature is high in the beginning, the heat-losing mechanisms (vasodilation and sweating) go into operation to decrease core temperature. The skin becomes warm, and the person begins to sweat. This phase of the fever is called the **crisis,** and it indicates that core temperature is falling.

Although death results if core temperature rises above 44–46°C (112–114°F), up to a point, fever is beneficial. For example, a higher temperature intensifies the effects of interferons and the phagocytic activities of macrophages while hindering replication of some pathogens. Because fever increases heart rate, infection-fighting white blood cells are delivered to sites of infection more rapidly. In addition, antibody production and T cell proliferation increase. Moreover, heat speeds up the rate of chemical reactions, which may help body cells repair themselves more quickly.

Obesity

Obesity is body weight more than 20% above a desirable standard due to an excessive accumulation of adipose tissue. About one-third of the adult population in the United States is obese. (An athlete may be *overweight* due to higher-than-normal amounts of muscle tissue without being obese.) Even moderate obesity is hazardous to health; it is a risk factor in cardiovascular disease, hypertension, pulmonary disease, non-insulin-dependent diabetes mellitus, arthritis, certain cancers (breast, uterus, and colon), varicose veins, and gallbladder disease.

In a few cases, obesity may result from trauma of or tumors in the food-regulating centers in the hypothalamus. In most cases of obesity, no specific cause can be identified. Contributing factors include genetic factors, eating habits taught early in life, overeating to relieve tension, and social customs. Studies indicate that some obese people burn fewer calories during digestion and absorption of a meal, a smaller food-induced thermogenesis effect. Additionally, obese people who lose weight require about 15% fewer calories to maintain normal body weight than do people who have never been obese. Interestingly, people who gain weight easily when deliberately fed excess calories exhibit less NEAT (nonexercise activity thermogenesis, such as occurs with fidgeting) than people who resist weight gains in the face of excess calories. Although leptin suppresses appetite and produces satiety in experimental animals, it is not deficient in most obese people.

Most surplus calories in the diet are converted to triglycerides and stored in adipose cells. Initially, the adipocytes increase in size, but at a maximal size, they divide. As a result, proliferation of adipocytes occurs in extreme obesity. The enzyme endothelial lipoprotein lipase regulates triglyceride storage. The enzyme is very active in abdominal fat but less active in hip fat. Accumulation of fat in the abdomen is associated with higher blood cholesterol level and other cardiac risk factors because adipose cells in this area appear to be more metabolically active.

Treatment of obesity is difficult because most people who are successful at losing weight gain it back within two years. Yet, even modest weight loss is associated with health benefits. Treatments for obesity include behavior modification programs, very-low-calorie diets, drugs, and surgery. Behavior modification programs, offered at many hospitals, strive to alter eating behaviors and increase exercise activity. The nutrition program includes a “heart-healthy” diet that includes abundant vegetables but is low in fats, especially saturated fats. A typical exercise program suggests walking for 30 minutes a day, five to seven times a week. Regular exercise enhances both weight loss and weight-loss maintenance. Very-low-calorie (VLC) diets include 400 to 800 kcal/day in a commercially made liquid mixture. The VLC diet is usually prescribed for 12 weeks, under close medical supervision. Two drugs are available to treat obesity. Sibutramine is an appetite suppressant that works by inhibiting reuptake of serotonin and norepinephrine in brain areas that govern eating behavior. Orlistat works by inhibiting the lipases released into the lumen of the GI tract. With less lipase activity, fewer dietary triglycerides are absorbed. For those with extreme obesity who have not responded to other treatments, a surgical procedure may be considered. The two operations most commonly performed—gastric bypass and gastroplasty—both greatly reduce the stomach size so that it can hold just a tiny quantity of food.

MEDICAL TERMINOLOGY

Heat cramps Cramps that result from profuse sweating. The salt lost in sweat causes painful contractions of muscles; such cramps tend to occur in muscles used while working but do not appear until the person relaxes once the work is done. Drinking salted liquids usually leads to rapid improvement.

Heat exhaustion (heat prostration) A condition in which the core temperature is generally normal, or a little below, and the skin is cool and moist due to profuse perspiration. Heat exhaustion is usually characterized by loss of fluid and electrolytes, especially salt (NaCl). The salt loss results in muscle cramps, dizziness, vomiting, and fainting; fluid loss may cause low blood pressure. Complete rest, rehydration, and electrolyte replacement are recommended.

Heatstroke (sunstroke) A severe and often fatal disorder caused by exposure to high temperatures, especially when the relative humidity is high, which makes it difficult for the body to lose heat. Blood flow to the skin is decreased, perspiration is greatly reduced, and body temperature rises sharply because of failure of the hypothalamic thermostat. Body temperature may reach 43°C (110°F). Treatment, which must be undertaken immediately, consists of cooling the body by immersing the victim in cool water and by administering fluids and electrolytes.

Kwashiorkor (kwash-ē-OR-kor) A disorder in which protein intake is deficient despite normal or nearly normal caloric intake, characterized by edema of the abdomen, enlarged liver, decreased blood pressure, low pulse rate, lower-than-normal body temperature, and sometimes mental retardation. Because the main protein in corn (zein) lacks two essential amino acids, which are needed for growth and tissue repair, many African children whose diet consists largely of cornmeal develop kwashiorkor.

Malnutrition (*mal-* = bad) An imbalance of total caloric intake or intake of specific nutrients, which can be either inadequate or excessive.

Marasmus (mar-AZ-mus) A type of protein–calorie undernutrition that results from inadequate intake of both protein and calories. Its characteristics include retarded growth, low weight, muscle wasting, emaciation, dry skin, and thin, dry, dull hair.

STUDY OUTLINE

INTRODUCTION (p. 951)

1. Our only source of energy for performing biological work is the food we eat. Food also provides essential substances that we cannot synthesize.
2. Most food molecules absorbed by the gastrointestinal tract are used to supply energy for life processes, serve as building blocks during synthesis of complex molecules, or are stored for future use.

METABOLIC REACTIONS (p. 951)

1. Metabolism refers to all chemical reactions of the body and is of two types: catabolism and anabolism.
2. Catabolism is the term for reactions that break down complex organic compounds into simple ones. Overall, catabolic reactions are exergonic; they produce more energy than they consume.
3. Chemical reactions that combine simple molecules into more complex ones that form the body's structural and functional components are collectively known as anabolism. Overall, anabolic reactions are endergonic; they consume more energy than they produce.
4. The coupling of anabolism and catabolism occurs via ATP.

ENERGY TRANSFER (p. 952)

1. Oxidation is the removal of electrons from a substance; reduction is the addition of electrons to a substance.
2. Two coenzymes that carry hydrogen atoms during coupled oxidation–reduction reactions are nicotinamide adenine dinucleotide (NAD^+) and flavin adenine dinucleotide (FAD).
3. ATP can be generated via substrate-level phosphorylation, oxidative phosphorylation, and photophosphorylation.

CARBOHYDRATE METABOLISM (p. 953)

1. During digestion, polysaccharides and disaccharides are hydrolyzed into the monosaccharides glucose (about 80%), fructose, and galactose; the latter two are then converted to glucose.
2. Some glucose is oxidized by cells to provide ATP. Glucose also can be used to synthesize amino acids, glycogen, and triglycerides.
3. Glucose moves into most body cells via facilitated diffusion through glucose transporters (GluT) and becomes phosphorylated to glucose 6-phosphate. In muscle cells, this process is stimulated by insulin. Glucose entry into neurons and hepatocytes is always "turned on."
4. Cellular respiration, the complete oxidation of glucose to CO_2 and H_2O, involves glycolysis, the Krebs cycle, and the electron transport chain.

5. Glycolysis is the breakdown of glucose into two molecules of pyruvic acid; there is a net production of two molecules of ATP.
6. When oxygen is in short supply, pyruvic acid is reduced to lactic acid; under aerobic conditions, pyruvic acid enters the Krebs cycle.
7. Pyruvic acid is prepared for entrance into the Krebs cycle by conversion to a two-carbon acetyl group followed by the addition of coenzyme A to form acetyl coenzyme A.
8. The Krebs cycle involves decarboxylations, oxidations, and reductions of various organic acids.
9. Each molecule of pyruvic acid that is converted to acetyl coenzyme A and then enters the Krebs cycle produces three molecules of CO_2, four molecules of NADH and four H^+, one molecule of $FADH_2$, and one molecule of ATP.
10. The energy originally stored in glucose and then in pyruvic acid is transferred primarily to the reduced coenzymes NADH and $FADH_2$.
11. The electron transport chain involves a series of oxidation–reduction reactions in which the energy in NADH and $FADH_2$ is liberated and transferred to ATP.
12. The electron carriers include FMN, cytochromes, iron–sulfur centers, copper atoms, and coenzyme Q.
13. The electron transport chain yields a maximum of 32 or 34 molecules of ATP and six molecules of H_2O.
14. Table 25.1 on page 962 summarizes the ATP yield during cellular respiration. The complete oxidation of glucose can be represented as follows:

$$C_6H_{12}O_6 + 6\,O_2 + 36 \text{ or } 38 \text{ ADPs} + 36 \text{ or } 38 \text{ Ⓟ} \longrightarrow 6\,CO_2 + 6\,H_2O + 36 \text{ or } 38 \text{ ATPs}$$

15. The conversion of glucose to glycogen for storage in the liver and skeletal muscle is called glycogenesis. It is stimulated by insulin.
16. The conversion of glycogen to glucose is called glycogenolysis. It occurs between meals and is stimulated by glucagon and epinephrine.
17. Gluconeogenesis is the conversion of noncarbohydrate molecules into glucose. It is stimulated by cortisol and glucagon.

LIPID METABOLISM (p. 964)

1. Lipoproteins transport lipids in the bloodstream. Types of lipoproteins include chylomicrons, which carry dietary lipids to adipose tissue; very low-density lipoproteins (VLDLs), which carry triglycerides from the liver to adipose tissue; low-density lipoproteins (LDLs), which deliver cholesterol to body cells; and high-density lipoproteins (HDLs), which remove excess cholesterol from body cells and transport it to the liver for elimination.
2. Cholesterol in the blood comes from two sources: from food and from synthesis by the liver.
3. Lipids may be oxidized to produce ATP or stored as triglycerides in adipose tissue, mostly in the subcutaneous layer.
4. A few lipids are used as structural molecules or to synthesize essential molecules.
5. Adipose tissue contains lipases that catalyze the deposition of triglycerides from chylomicrons and hydrolyze triglycerides into fatty acids and glycerol.
6. In lipolysis, triglycerides are split into fatty acids and glycerol and released from adipose tissue under the influence of epinephrine, norepinephrine, cortisol, thyroid hormones, and insulinlike growth factors.
7. Glycerol can be converted into glucose by conversion into glyceraldehyde 3-phosphate.
8. In beta-oxidation of fatty acids, carbon atoms are removed in pairs from fatty acid chains; the resulting molecules of acetyl coenzyme A enter the Krebs cycle.
9. The conversion of glucose or amino acids into lipids is called lipogenesis; it is stimulated by insulin.

PROTEIN METABOLISM (p. 967)

1. During digestion, proteins are hydrolyzed into amino acids, which enter the liver via the hepatic portal vein.
2. Amino acids, under the influence of insulinlike growth factors and insulin, enter body cells via active transport.
3. Inside cells, amino acids are synthesized into proteins that function as enzymes, hormones, structural elements, and so forth; stored as fat or glycogen; or used for energy.
4. Before amino acids can be catabolized, they must be deaminated and converted to substances that can enter the Krebs cycle.
5. Amino acids may also be converted into glucose, fatty acids, and ketone bodies.
6. Protein synthesis is stimulated by insulinlike growth factors, thyroid hormones, insulin, estrogen, and testosterone.
7. Table 25.2 on page 971 summarizes carbohydrate, lipid, and protein metabolism.

KEY MOLECULES AT METABOLIC CROSSROADS (p. 969)

1. Three molecules play a key role in metabolism: glucose 6-phosphate, pyruvic acid, and acetyl coenzyme A.
2. Glucose 6-phosphate may be converted to glucose, glycogen, ribose 5-phosphate, and pyruvic acid.
3. When ATP is low and oxygen is plentiful, pyruvic acid is converted to acetyl coenzyme A; when oxygen supply is low, pyruvic acid is converted to lactic acid. Carbohydrate and protein metabolism are linked by pyruvic acid.
4. Acetyl coenzyme A is the molecule that enters the Krebs cycle; it is also used to synthesize fatty acids, ketone bodies, and cholesterol.

METABOLIC ADAPTATIONS (p. 971)

1. During the absorptive state, ingested nutrients enter the blood and lymph from the GI tract.
2. During the absorptive state, blood glucose is oxidized to form ATP, and glucose transported to the liver is converted to glycogen or triglycerides. Most triglycerides are stored in adipose tissue. Amino acids in hepatocytes are converted to carbohydrates, fats, and proteins. Table 25.3 on page 973 summarizes the hormonal regulation of metabolism during the absorptive state.
3. During the postabsorptive state, absorption is complete and the ATP needs of the body are satisfied by nutrients already present in the body. The major task is to maintain normal blood glucose level by converting glycogen in the liver and skeletal muscle into glucose, converting glycerol into glucose, and converting amino acids into glucose. Fatty acids, ketone bodies, and amino acids are oxidized to supply ATP. Table 25.4 on page 975 summarizes the hormonal regulation of metabolism during the postabsorptive state.

4. Fasting is going without food for a few days; starvation implies weeks or months of inadequate food intake. During fasting and starvation, fatty acids and ketone bodies are increasingly utilized for ATP production.

HEAT AND ENERGY BALANCE (p. 975)

1. Measurement of the metabolic rate under basal conditions is called the basal metabolic rate (BMR).
2. A kilocalorie (kcal) or Calorie is the amount of energy required to raise the temperature of 1000 g of water 1°C.
3. Normal core temperature is maintained by a delicate balance between heat-producing and heat-losing mechanisms.
4. Exercise, hormones, the nervous system, body temperature, ingestion of food, age, gender, climate, sleep, and malnutrition affect metabolic rate.
5. Mechanisms of heat transfer include conduction, convection, radiation, and evaporation. Conduction is the transfer of heat between two substances or objects in contact with each other. Convection is the transfer of heat by a liquid or gas between areas of different temperatures. Radiation is the transfer of heat from a warmer object to a cooler object without physical contact. Evaporation is the conversion of a liquid to a vapor; in the process, heat is lost.
6. The hypothalamic thermostat is in the preoptic area.
7. Responses that produce, conserve, or retain heat when core temperature falls are vasoconstriction; release of epinephrine, norepinephrine, and thyroid hormones; and shivering.
8. Responses that increase heat loss when core temperature increases include vasodilation, decreased metabolic rate, and evaporation of perspiration.
9. Two nuclei in the hypothalamus that help regulate food intake are the arcuate and paraventricular nuclei. The hormone leptin, released by adipocytes, inhibits release of neuropeptide Y from the arcuate nucleus and thereby decreases food intake. Melanocortin also decreases food intake.

NUTRITION (p. 980)

1. Nutrients include water, carbohydrates, lipids, proteins, minerals, and vitamins.
2. Most teens and adults need between 1600 and 2800 Calories each day.
3. Nutrition experts suggest dietary calories be 50–60% from carbohydrates, 30% or less from fats, and 12–15% from proteins, although the optimal levels of these nutrients may vary.
4. The food guide pyramid indicates how many servings of five food groups are recommended each day to attain the number of calories and variety of nutrients needed for wellness.
5. Minerals known to perform essential functions include calcium, phosphorus, potassium, sulfur, sodium, chloride, magnesium, iron, iodide, manganese, copper, cobalt, zinc, fluoride, selenium, and chromium. Their functions are summarized in Table 25.5.
6. Vitamins are organic nutrients that maintain growth and normal metabolism. Many function in enzyme systems.
7. Fat-soluble vitamins are absorbed with fats and include vitamins A, D, E, and K; water-soluble vitamins include the B vitamins and vitamin C.
8. The functions and deficiency disorders of the principal vitamins are summarized in Table 25.6.

SELF-QUIZ QUESTIONS

Fill in the blanks in the following statements.

1. The thermostat and food intake regulating center of the body is in the ______ of the brain.
2. The three key molecules of metabolism are ____, ____, and ____.

Indicate whether the following statements are true or false.

3. Foods that we eat are used to supply energy for life processes, serve as building blocks for synthesis reactions, or are stored for future use.
4. Vitamins A, B, D, and K are fat-soluble vitamins.

Choose the one best answer to the following questions.

5. NAD^+ and FAD (1) are both derivatives of B vitamins, (2) are used to carry hydrogen atoms released during oxidation reactions, (3) become NADH and $FADH_2$ in their reduced forms, (4) act as coenzymes in the Krebs cycle, (5) are the final electron acceptors in the electron transport chain. (a) 1, 2, 3, 4, and 5; (b) 2, 3, and 4; (c) 2 and 4; (d) 1, 2, and 3; (e) 1, 2, 3, and 4.
6. During glycolysis, (1) a six-carbon glucose is split into two three-carbon pyruvic acids, (2) there is a net gain of two ATP molecules, (3) two NADH molecules are oxidized, (4) moderately high levels of oxygen are needed, (5) the activity of phosphofructokinase determines the rate of the chemical reactions. (a) 1, 2, and 3; (b) 1 and 2; (c) 1, 2, and 5; (d) 2, 3, 4, and 5; (e) 1, 2, 3, 4, and 5.
7. If glucose is not needed for immediate ATP production, it can be used for (1) vitamin synthesis, (2) amino acid synthesis, (3) gluconeogenesis, (4) glycogenesis, (5) lipogenesis. (a) 1, 3, and 5; (b) 2, 4, and 5; (c) 2, 3, 4, and 5; (d) 1, 2, and 3; (e) 2 and 5.
8. Which of the following is the *correct* sequence for the oxidation of glucose to produce ATP? (a) electron transport chain, Krebs cycle, glycolysis, formation of acetyl CoA; (b) Krebs cycle, formation of acetyl CoA, electron transport chain, glycolysis; (c) glycolysis, electron transport chain, Krebs cycle, formation of acetyl CoA; (d) glycolysis, formation of acetyl CoA, Krebs cycle, electron transport chain; (e) formation of acetyl CoA, Krebs cycle, glycolysis, electron transport chain.
9. Which of the following would you *not* expect to experience during fasting or starvation? (a) decrease in plasma fatty acid levels, (b) increase in ketone body formation, (c) lipolysis, (d) increased use of ketones for ATP production in the brain, (e) depletion of glycogen.
10. If core body temperature rises above normal, which of the following would occur to cool the body? (1) dilation of vessels in the skin, (2) increased radiation and conduction of heat to the environment, (3) increased metabolic rate, (4) evaporation of perspiration, (5) increased secretion of thyroid hormones. (a) 3, 4, and 5; (b) 1, 2, and 4; (c) 1, 2, and 5; (d) 1, 2, 3, 4, and 5; (e) 1, 2, 4, and 5.

11. In which of the following situations would the metabolic rate increase? (1) sleep, (2) after ingesting food, (3) increased secretion of thyroid hormones, (4) parasympathetic nervous system stimulation, (5) fever. (a) 3 and 4; (b) 1, 3, and 5; (c) 2 and 3; (d) 2, 3, and 4; (e) 2, 3, and 5.
12. Which of the following are absorptive state reactions? (1) aerobic cellular respiration, (2) glycogenesis, (3) glycogenolysis, (4) gluconeogenesis using lactic acid, (5) lipolysis. (a) 1 and 2; (b) 2 and 3; (c) 3 and 4; (d) 4 and 5; (e) 1 and 5.
13. Match the hormones with the reactions they regulate (answers may be used more than once; some reactions have more than one answer):

____(a) gluconeogenesis	(1) insulin
____(b) glycogenesis	(2) cortisol
____(c) glycogenolysis	(3) glucagon
____(d) lipolysis	(4) thyroid hormones
____(e) lipogenesis	(5) epinephrine
____(f) protein catabolism	(6) insulinlike growth factors
____(g) protein anabolism	

14. Match the following:

____(a) deliver cholesterol to body cells for use in repair of membranes and synthesis of steroid hormones and bile salts
____(b) remove excess cholesterol from body cells and transport it to the liver for elimination
____(c) organic nutrients required in small amounts for growth and normal metabolism
____(d) the energy-transferring molecule of the body
____(e) nutrient molecules that can be oxidized to produce ATP or stored in adipose tissue
____(f) transport endogenous lipids to adipocytes for storage
____(g) the body's preferred source for synthesizing ATP
____(h) composed of amino acids and are the primary regulatory molecules in the body
____(i) acetoacetic acid, beta-hydroxybutyric acid, and acetone
____(j) hormone secreted by adipocytes that acts to decrease total body-fat mass
____(k) neurotransmitter that stimulates food intake
____(l) inorganic substances that perform many vital functions in the body
____(m) carriers of electrons in the electron transport chain

(1) leptin
(2) minerals
(3) glucose
(4) lipids
(5) proteins
(6) neuropeptide Y
(7) cytochromes
(8) ketone bodies
(9) low-density lipoproteins
(10) ATP
(11) vitamins
(12) high-density lipoproteins
(13) very low-density lipoproteins

15. Match the following:

____(a) the mechanism of ATP generation that links chemical reactions with pumping of hydrogen ions
____(b) the removal of electrons from an atom or molecule resulting in a decrease in potential energy
____(c) the transfer of an amino group from an amino acid to a substance such as pyruvic acid
____(d) the formation of glucose from noncarbohydrate sources
____(e) refers to all the chemical reactions in the body
____(f) the oxidation of glucose to produce ATP
____(g) the splitting of a triglyceride into glycerol and fatty acids
____(h) the synthesis of lipids
____(i) the addition of electrons to a molecule resulting in an increase in potential energy content of the molecule
____(j) the formation of ketone bodies
____(k) the breakdown of glycogen back to glucose
____(l) exergonic chemical reactions that break down complex organic molecules into simpler ones
____(m) overall rate at which metabolic reactions use energy
____(n) the breakdown of glucose into two molecules of pyruvic acid
____(o) removal of CO_2 from a molecule
____(p) endergonic chemical reactions that combine simple molecules and monomers to make more complex ones
____(q) the addition of a phosphate group to a molecule
____(r) the removal of the amino group from an amino acid
____(s) the cleavage of one pair of carbon atoms at a time from a fatty acid
____(t) the conversion of glucose into glycogen

(1) metabolism
(2) catabolism
(3) beta oxidation
(4) lipolysis
(5) phosphorylation
(6) glycolysis
(7) cellular respiration
(8) transamination
(9) anabolism
(10) lipogenesis
(11) glycogenolysis
(12) glycogenesis
(13) metabolic rate
(14) ketogenesis
(15) oxidation
(16) reduction
(17) chemiosmosis
(18) deamination
(19) gluconeogenesis
(20) decarboxylation

CRITICAL THINKING QUESTIONS

1. Jane Doe's deceased body was found at her dining room table. Her death was considered suspicious. Lab results from the medical investigation revealed cyanide in her blood. How did the cyanide cause her death?
2. During a recent physical, 55-year-old Glenn's blood serum lab results showed the following: total cholesterol = 300 mg/dL; LDL = 175 mg/dL; HDL = 20 mg/dL. Interpret these results for Glenn and indicate to him what changes, if any, he needs to make in his lifestyle. Why are these changes important?
3. Marissa has joined a weight loss program. As part of the program, she regularly submits a urine sample which is tested for ketones. She went to the clinic today, had her urine checked, and was confronted by the nurse who accused Marissa of "cheating" on her diet. How did the nurse know Marissa was not following her diet?

ANSWERS TO FIGURE QUESTIONS

25.1 In pancreatic acinar cells, anabolism predominates because the primary activity is synthesis of complex molecules (digestive enzymes).

25.2 Glycolysis is also called anaerobic cellular respiration.

25.3 The reactions of glycolysis consume two molecules of ATP but generate four molecules of ATP, for a net gain of two.

25.4 Kinases are enzymes that phosphorylate (add phosphate to) their substrate.

25.5 Glycolysis occurs in the cytosol.

25.6 CO_2 is given off during the production of acetyl coenzyme A and during the Krebs cycle. It diffuses into the blood, is transported by the blood to the lungs, and is exhaled.

25.7 The production of reduced coenzymes is important in the Krebs cycle because they will subsequently yield ATP in the electron transport chain.

25.8 The energy source that powers the proton pumps is electrons provided by NADH + H^+.

25.9 The concentration of H^+ is highest in the space between the inner and outer mitochondrial membranes.

25.10 During the complete oxidation of one glucose molecule, six molecules of O_2 are used and six molecules of CO_2 are produced.

25.11 Skeletal muscle fibers can synthesize glycogen, but they cannot release glucose into the blood because they lack the enzyme phosphatase required to remove the phosphate group from glucose.

25.12 Hepatocytes can carry out gluconeogenesis and glycogenesis.

25.13 LDLs deliver cholesterol to body cells.

25.14 Hepatocytes and adipose cells carry out lipogenesis, beta oxidation, and lipolysis; hepatocytes carry out ketogenesis.

25.15 Before an amino acid can enter the Krebs cycle. an amino group must be removed via deamination.

25.16 Acetyl coenzyme A is the gateway into the Krebs cycle for molecules being oxidized to generate ATP.

25.17 Reactions of the absorptive state are mainly anabolic.

25.18 Processes that directly elevate blood glucose during the postabsorptive state include lipolysis (in adipocytes and hepatocytes), gluconeogenesis (in hepatocytes), and glycogenolysis (in hepatocytes).

25.19 Exercise, the sympathetic nervous system, hormones (epinephrine, norepinephrine, thyroxine, testosterone, human growth hormone), elevated body temperature, and ingestion of food increase metabolic rate, which results in an increase in body temperature.

25.20 Foods that contain cholesterol and most of the saturated fatty acids in the diet are milk, yogurt, cheeses, and meats.

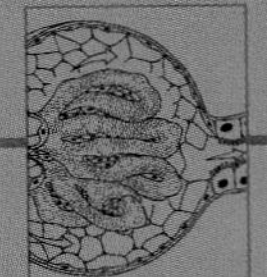

Chapter 26

The Urinary System

The Urinary System and Homeostasis

The urinary system contributes to homeostasis by altering blood composition, pH, volume, and pressure; maintaining blood osmolarity; excreting wastes and foreign substances; and producing hormones.

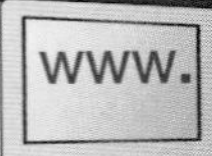

wiley.com/college/apcentral

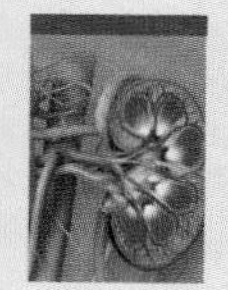

The **urinary system** consists of two kidneys, two ureters, one urinary bladder, and one urethra (Figure 26.1). After the kidneys filter blood plasma, they return most of the water and solutes to the bloodstream. The remaining water and solutes constitute **urine,** which passes through the ureters and is stored in the urinary bladder until it is excreted from the body through the urethra. **Nephrology** (nef-ROL-ō-jē; *nephr-* = kidney; *-ology* = study of) is the scientific study of the anatomy, physiology, and pathology of the kidneys. The branch of medicine that deals with the male and female urinary systems and the male reproductive system is **urology** (ū-ROL-ō-jē; *uro-* = urine). A physician who specializes in this branch of medicine is called a **urologist** (ū-ROL-ō-jist).

OVERVIEW OF KIDNEY FUNCTIONS

► OBJECTIVE

List the functions of the kidneys.

The kidneys do the major work of the urinary system. The other parts of the system are mainly passageways and storage areas. Functions of the kidneys include the following:

- ***Regulation of blood ionic composition.*** The kidneys help regulate the blood levels of several ions, most importantly sodium ions (Na^+), potassium ions (K^+), calcium ions (Ca^{2+}), chloride ions (Cl^-), and phosphate ions (HPO_4^{2-}).
- ***Regulation of blood pH.*** The kidneys excrete a variable amount of hydrogen ions (H^+) into the urine and conserve bicarbonate ions (HCO_3^-), which are an important buffer of H^+ in the blood. Both of these activities help regulate blood pH.

Figure 26.1 Organs of the urinary system in a female. (See Tortora, *A Photographic Atlas of the Human Body, Second Edition,* Figure 13.2.)

Urine formed by the kidneys passes first into the ureters, then to the urinary bladder for storage, and finally through the urethra for elimination from the body.

RIGHT KIDNEY
Right renal artery
RIGHT URETER
URINARY BLADDER
URETHRA
Diaphragm
Esophagus
Left adrenal (suprarenal) gland
Left renal vein
LEFT KIDNEY
Abdominal aorta
Inferior vena cava
LEFT URETER
Rectum
Left ovary
Uterus
Anterior view

Functions of the Urinary System

1. **The kidneys regulate blood volume and composition, help regulate blood pressure, synthesize glucose, release erythropoietin, participate in vitamin D synthesis, and excrete wastes in the urine.**
2. **The ureters transport urine from the kidneys to the urinary bladder.**
3. **The urinary bladder stores urine.**
4. **The urethra discharges urine from the body.**

Which organs constitute the urinary system?

- ***Regulation of blood volume.*** The kidneys adjust blood volume by conserving or eliminating water in the urine. An increase in blood volume increases blood pressure; a decrease in blood volume decreases blood pressure.
- ***Regulation of blood pressure.*** The kidneys also help regulate blood pressure by secreting the enzyme renin, which activates the renin–angiotensin–aldosterone pathway (see Figure 18.16 on page 643). Increased renin causes an increase in blood pressure.
- ***Maintenance of blood osmolarity.*** By separately regulating loss of water and loss of solutes in the urine, the kidneys maintain a relatively constant blood osmolarity close to 300 milliosmoles per liter (mOsm/liter).*
- ***Production of hormones.*** The kidneys produce two hormones. *Calcitriol,* the active form of vitamin D, helps regulate calcium homeostasis (see Figure 18.14 on page 640), and *erythropoietin* stimulates production of red blood cells (see Figure 19.5 on page 674).
- ***Regulation of blood glucose level.*** Like the liver, the kidneys can use the amino acid glutamine in *gluconeogenesis,* the synthesis of new glucose molecules. They can then release glucose into the blood to help maintain a normal blood glucose level.
- ***Excretion of wastes and foreign substances.*** By forming urine, the kidneys help excrete **wastes**—substances that have no useful function in the body. Some wastes excreted in urine result from metabolic reactions in the body. These include ammonia and urea from the deamination of amino acids; bilirubin from the catabolism of hemoglobin; creatinine from the breakdown of creatine phosphate in muscle fibers; and uric acid from the catabolism of nucleic acids. Other wastes excreted in urine are foreign substances from the diet, such as drugs and environmental toxins.

▶ CHECKPOINT

1. What are wastes, and how do the kidneys participate in their removal from the body?

*The **osmolarity** of a solution is a measure of the total number of dissolved particles per liter of solution. The particles may be molecules, ions, or a mixture of both. To calculate osmolarity, multiply molarity (see page 41) by the number of particles per molecule, once the molecule dissolves. A similar term, *osmolality,* is the number of particles of solute per *kilogram* of water. Because it is easier to measure volumes of solutions than to determine the mass of water they contain, osmolarity is used more commonly than osmolality. Most body fluids and solutions used clinically are dilute, in which case there is less than a 1% difference between the two measures.

ANATOMY AND HISTOLOGY OF THE KIDNEYS

▶ OBJECTIVES

Describe the external and internal gross anatomical features of the kidneys.

Trace the path of blood flow through the kidneys.

Describe the structure of renal corpuscles and renal tubules.

The paired **kidneys** are reddish, kidney-bean-shaped organs located just above the waist between the peritoneum and the posterior wall of the abdomen. Because their position is posterior to the peritoneum of the abdominal cavity, they are said to be **retroperitoneal** (re′-trō-per-i-tō-NĒ-al; *retro-* = behind) organs (Figure 26.2). The kidneys are located between the levels of the last thoracic and third lumbar vertebrae, a position where they are partially protected by the eleventh and twelfth pairs of ribs. The right kidney is slightly lower than the left (see Figure 26.1) because the liver occupies considerable space on the right side superior to the kidney.

External Anatomy of the Kidneys

A typical adult kidney is 10–12 cm (4–5 in.) long, 5–7 cm (2–3 in.) wide, and 3 cm (1 in.) thick—about the size of a bar of bath soap—and has a mass of 135–150 g (4.5–5 oz). The concave medial border of each kidney faces the vertebral column (see Figure 26.1). Near the center of the concave border is a deep vertical fissure called the **renal hilum** (see Figure 26.3), through which the ureter emerges from the kidney along with blood vessels, lymphatic vessels, and nerves.

Three layers of tissue surround each kidney (Figure 26.2). The deep layer, the **renal capsule** (*ren-* = kidney), is a smooth, transparent sheet of dense irregular connective tissue that is continuous with the outer coat of the ureter. It serves as a barrier against trauma and helps maintain the shape of the kidney. The middle layer, the **adipose capsule,** is a mass of fatty tissue surrounding the renal capsule. It also protects the kidney from trauma and holds it firmly in place within the abdominal cavity. The superficial layer, the **renal fascia,** is another thin layer of dense irregular connective tissue that anchors the kidney to the surrounding structures and to the abdominal wall. On the anterior surface of the kidneys, the renal fascia is deep to the peritoneum.

Nephroptosis (Floating Kidney)

Nephroptosis (nef′-rōp-TŌ-sis; *ptosis* = falling), or **floating kidney,** is an inferior displacement or dropping of the kidney. It occurs when the kidney slips from its normal position because it is not securely held in place by adjacent organs or its covering of fat. Nephroptosis develops most often in very thin people whose adipose capsule or renal fascia is deficient. It is dangerous because the ureter may kink and block urine flow. The resulting

Figure 26.2 Position and coverings of the kidneys. (See Tortora, *A Photographic Atlas of the Human Body, Second Edition,* Figure 13.3.)

The kidneys are surrounded by a renal capsule, adipose capsule, and renal fascia.

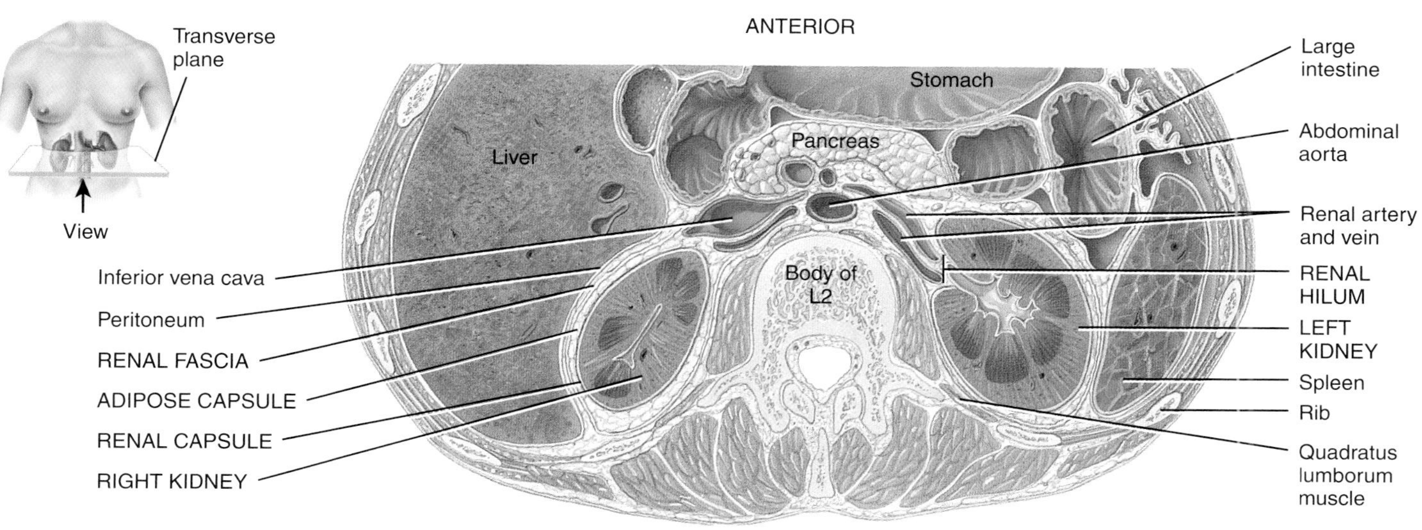

(a) Inferior view of transverse section of abdomen (L2)

SUPERIOR

Sagittal plane

Lung

Liver

Diaphragm

Adrenal (suprarenal) gland

Twelfth rib

Peritoneum

RENAL FASCIA

Right kidney

ADIPOSE CAPSULE

RENAL CAPSULE

Quadratus lumborum muscle

Large intestine

Hip bone

POSTERIOR

ANTERIOR

(b) Sagittal section through the right kidney

Why are the kidneys said to be retroperitoneal?

Figure 26.3 **Internal anatomy of the kidneys.**

The two main regions of the kidney parenchyma are the renal cortex and the renal pyramids in the renal medulla.

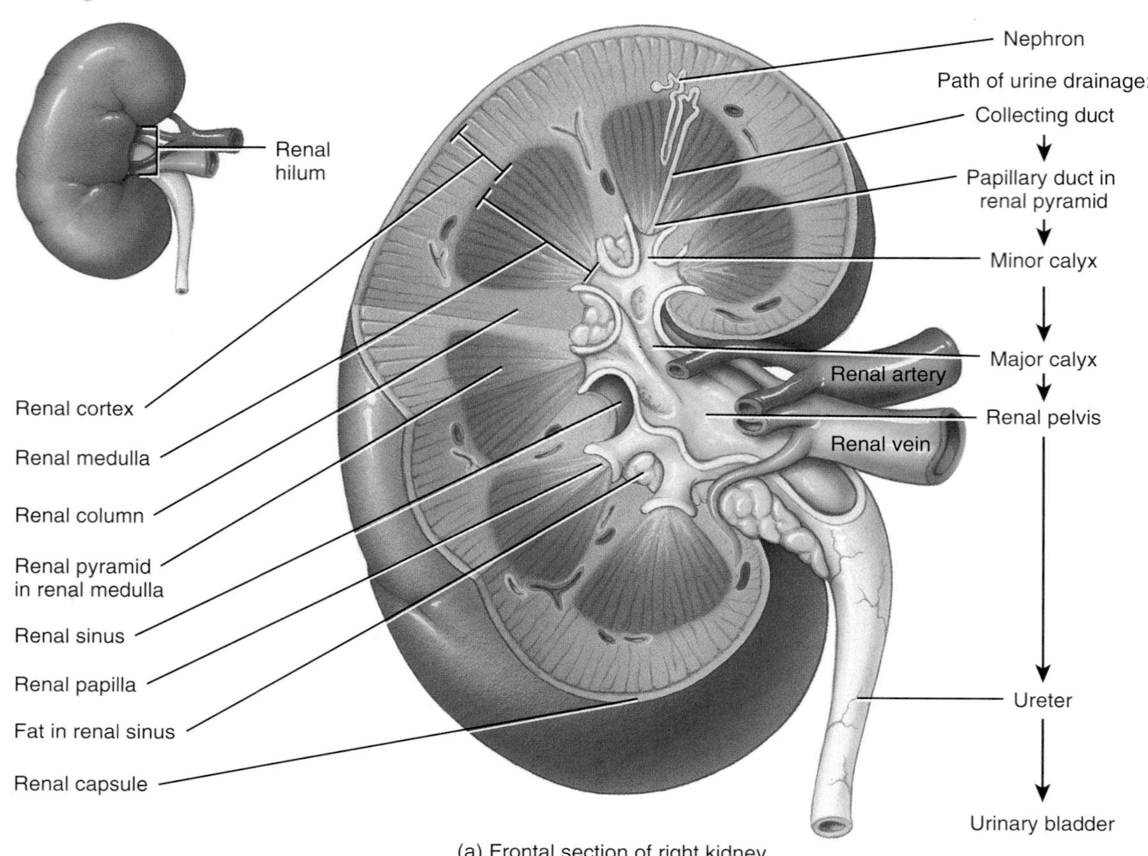

(a) Frontal section of right kidney

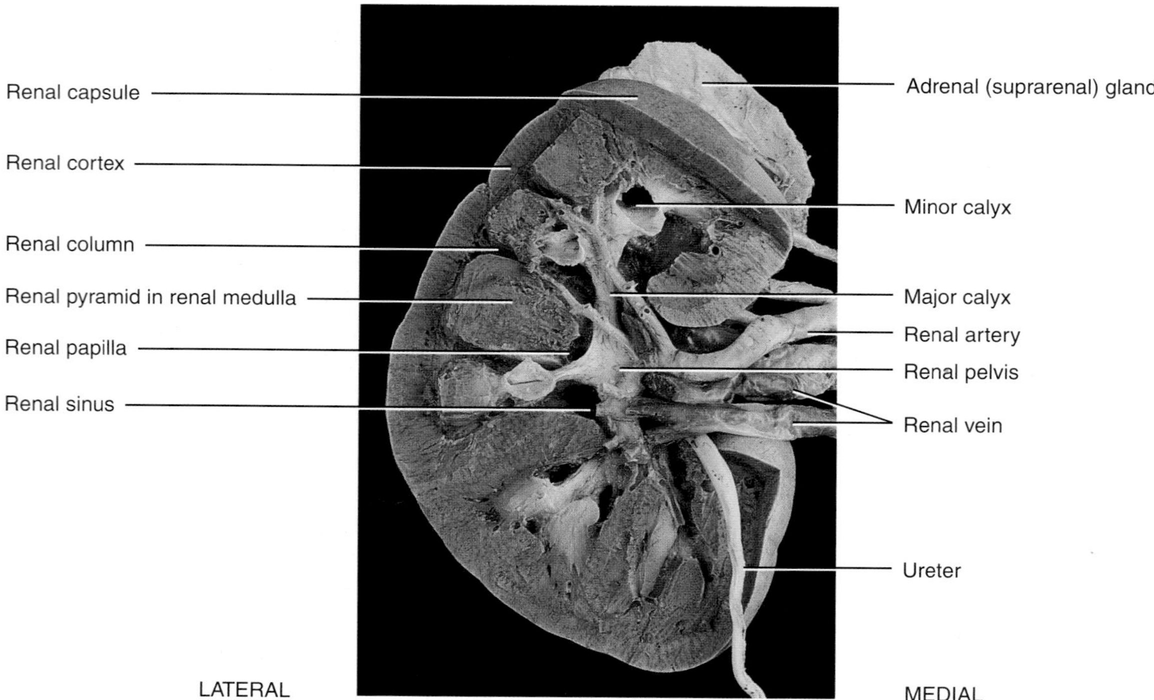

(b) Frontal section of right kidney

What structures pass through the renal hilum?

backup of urine puts pressure on the kidney, which damages the tissue. Twisting of the ureter also causes pain. Nephroptosis is very common; about one in four people has some degree of weakening of the fibrous bands that hold the kidney in place. It is 10 times more common in females than males. Because it happens during life it is very easy to distinguish from congenital anomalies. ■

Internal Anatomy of the Kidneys

A frontal section through the kidney reveals two distinct regions: a superficial, smooth-textured reddish area called the **renal cortex** (*cortex* = rind or bark) and a deep, reddish-brown inner region called the **renal medulla** (*medulla* = inner portion) (Figure 26.3). The renal medulla consists of 8 to 18 cone-shaped **renal pyramids.** The base (wider end) of each pyramid faces the renal cortex, and its apex (narrower end), called a **renal papilla,** points toward the renal hilum. The renal cortex is the smooth-textured area extending from the renal capsule to the bases of the renal pyramids and into the spaces between them. It is divided into an outer *cortical zone* and an inner *juxtamedullary zone.* Those portions of the renal cortex that extend between renal pyramids are called **renal columns.** A **renal lobe** consists of a renal pyramid, its overlying area of renal cortex, and one-half of each adjacent renal column.

Together, the renal cortex and renal pyramids of the renal medulla constitute the **parenchyma** (functional portion) of the kidney. Within the parenchyma are the functional units of the kidney—about 1 million microscopic structures called **nephrons** (NEF-rons). Urine formed by the nephrons drains into large **papillary ducts,** which extend through the renal papillae of the pyramids. The papillary ducts drain into cuplike structures called **minor** and **major calyces** (KĀ-li-sēz = cups; singular is *calyx*). Each kidney has 8 to 18 minor calyces and 2 to 3 major calyces. A minor calyx receives urine from the papillary ducts of one renal papilla and delivers it to a major calyx. From the major calyces, urine drains into a single large cavity called the **renal pelvis** (*pelv-* = basin) and then out through the ureter to the urinary bladder.

The hilum expands into a cavity within the kidney called the **renal sinus,** which contains part of the renal pelvis, the calyces, and branches of the renal blood vessels and nerves. Adipose tissue helps stabilize the position of these structures in the renal sinus.

Blood and Nerve Supply of the Kidneys

Because the kidneys remove wastes from the blood and regulate its volume and ionic composition, it is not surprising that they are abundantly supplied with blood vessels. Although the kidneys constitute less than 0.5% of total body mass, they receive 20–25% of the resting cardiac output via the right and left **renal arteries** (Figure 26.4). In adults, **renal blood flow,** the blood flow through both kidneys, is about 1200 mL per minute.

Within the kidney, the renal artery divides into several **segmental arteries,** which supply different segments (areas) of the kidney. Each segmental artery gives off several branches that enter the parenchyma and pass through the renal columns between the renal pyramids as the **interlobar arteries.** At the bases of the renal pyramids, the interlobar arteries arch between the renal medulla and cortex; here they are known as the **arcuate arteries** (AR-kū-āt = shaped like a bow). Divisions of the arcuate arteries produce a series of **interlobular arteries.** These arteries are so named because they pass between renal lobules. Interlobular arteries enter the renal cortex and give off branches called **afferent arterioles** (*af-* = toward; *-ferrent* = to carry).

Each nephron receives one afferent arteriole, which divides into a tangled, ball-shaped capillary network called the **glomerulus** (glō-MER-ū-lus = little ball; plural is *glomeruli*). The glomerular capillaries then reunite to form an **efferent arteriole** (*ef-* = out) that carries blood out of the glomerulus. Glomerular capillaries are unique among capillaries in the body because they are positioned between two arterioles, rather than between an arteriole and a venule. Because they are capillary networks and they also play an important role in urine formation, the glomeruli are considered part of both the cardiovascular and the urinary systems.

The efferent arterioles divide to form the **peritubular capillaries** (*peri-* = around), which surround tubular parts of the nephron in the renal cortex. Extending from some efferent arterioles are long loop-shaped capillaries called **vasa recta** (VĀ-sa REK-ta; *vasa* = vessels; *recta* = straight) that supply tubular portions of the nephron in the renal medulla (see Figure 26.5b).

The peritubular capillaries eventually reunite to form **peritubular venules** and then **interlobular veins,** which also receive blood from the vasa recta. Then the blood drains through the **arcuate veins** to the **interlobar veins** running between the renal pyramids. Blood leaves the kidney through a single **renal vein** that exits at the renal hilum and carries venous blood to the inferior vena cava.

Most renal nerves originate in the *celiac ganglion* and pass through the *renal plexus* into the kidneys along with the renal arteries. Renal nerves are part of the sympathetic division of the autonomic nervous system. Most are vasomotor nerves that regulate the flow of blood through the kidney by causing vasodilation or vasoconstriction of renal arterioles.

Kidney Transplant

A **kidney transplant** is the transfer of a kidney from a donor to a recipient whose kidneys no longer function. In the procedure, the donor kidney is placed in the pelvis of the recipient through an abdominal incision. The renal artery and vein of the transplanted kidney are attached to the renal artery and vein of the recipient. The ureter of the transplanted kidney is then attached to the urinary bladder. During a kidney transplant, the patient receives only one donor kidney, since only one kidney is needed to maintain sufficient renal function. The nonfunctioning

Figure 26.4 Blood supply of the kidneys. (See Tortora, *A Photographic Atlas of the Human Body, Second Edition,* Figure 13.6.)

The renal arteries deliver 20–25% of the resting cardiac output to the kidneys.

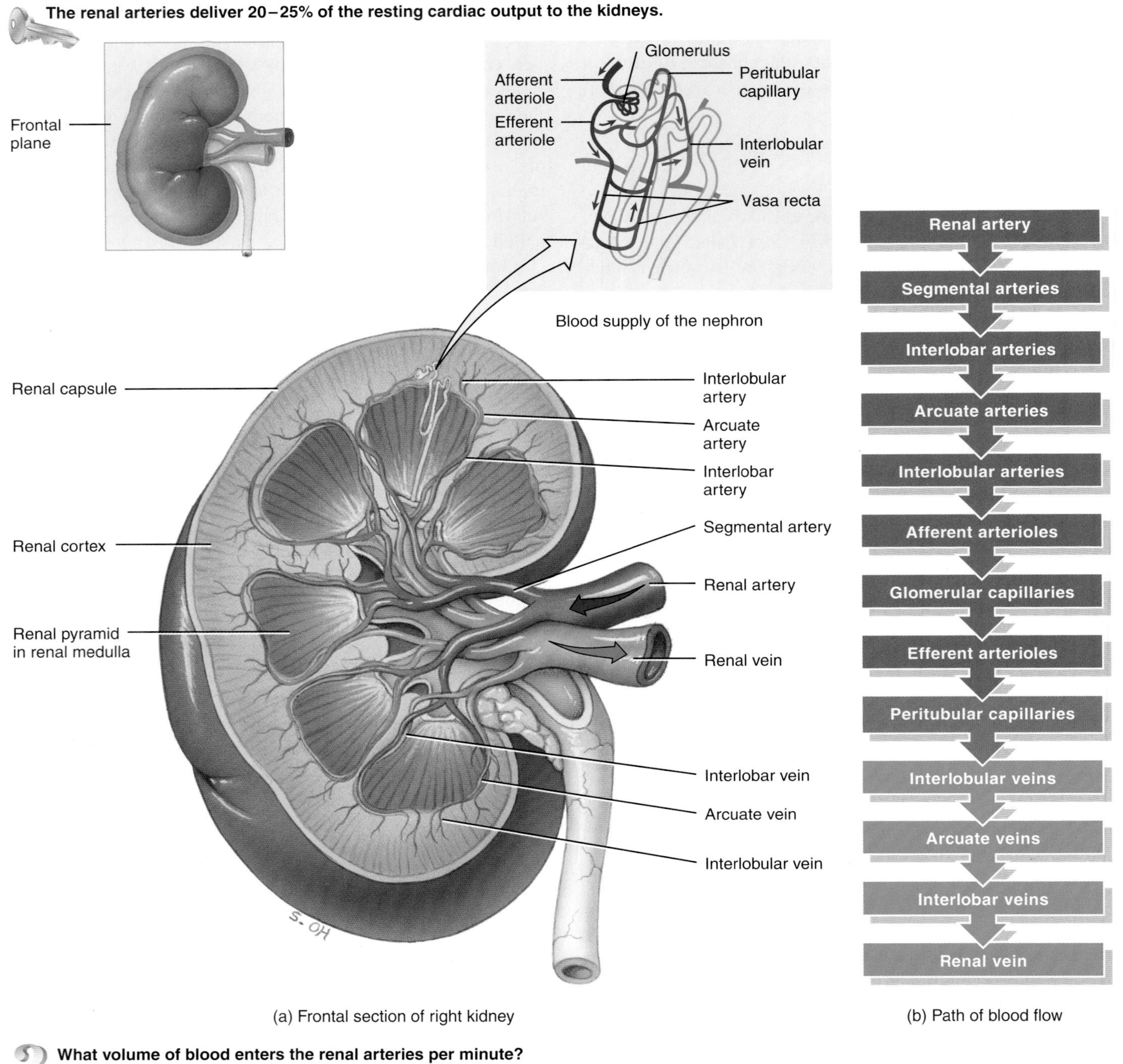

(a) Frontal section of right kidney

(b) Path of blood flow

What volume of blood enters the renal arteries per minute?

diseased kidneys are usually left in place. As with all organ transplants, kidney transplant recipients must be ever vigilant for signs of infection or organ rejection. The transplant recipient will take immunosuppressive drugs for the rest of his or her life to avoid rejection of the "foreign" organ. ■

The Nephron

Parts of a Nephron

Nephrons are the functional units of the kidneys. Each nephron (Figure 26.5) consists of two parts: a **renal corpuscle** (KOR-pus-sul = tiny body), where blood plasma is filtered, and a **renal tubule** into which the filtered fluid passes. The two components of a renal corpuscle are the **glomerulus** (capillary network) and the **glomerular (Bowman's) capsule,** a double-walled epithelial cup that surrounds the glomerular capillaries. Blood plasma is filtered in the glomerular capsule, and then the filtered fluid passes into the renal tubule, which has three main sections. In the order that fluid passes through them, the renal tubule consists of a (1) **proximal convoluted tubule,** (2) **loop of Henle (nephron loop),** and (3) **distal convoluted tubule.** *Proximal* denotes the part of the tubule attached to the glomerular capsule, and *distal* denotes the part that is further away. *Convoluted* means the tubule is tightly coiled rather than straight. The renal corpuscle and both convoluted tubules lie within the renal cortex; the loop of Henle extends into the renal medulla, makes a hairpin turn, and then returns to the renal cortex.

The distal convoluted tubules of several nephrons empty into a single **collecting duct.** Collecting ducts then unite and converge into several hundred large **papillary ducts,** which drain into the minor calyces. The collecting ducts and papillary ducts extend from the renal cortex through the renal medulla to the renal pelvis. So one kidney has about 1 million nephrons, but a much smaller number of collecting ducts and even fewer papillary ducts.

Figure 26.5 The structure of nephrons (colored gold) and associated blood vessels. (a) A cortical nephron. (b) A juxtamedullary nephron.

Nephrons are the functional units of the kidneys.

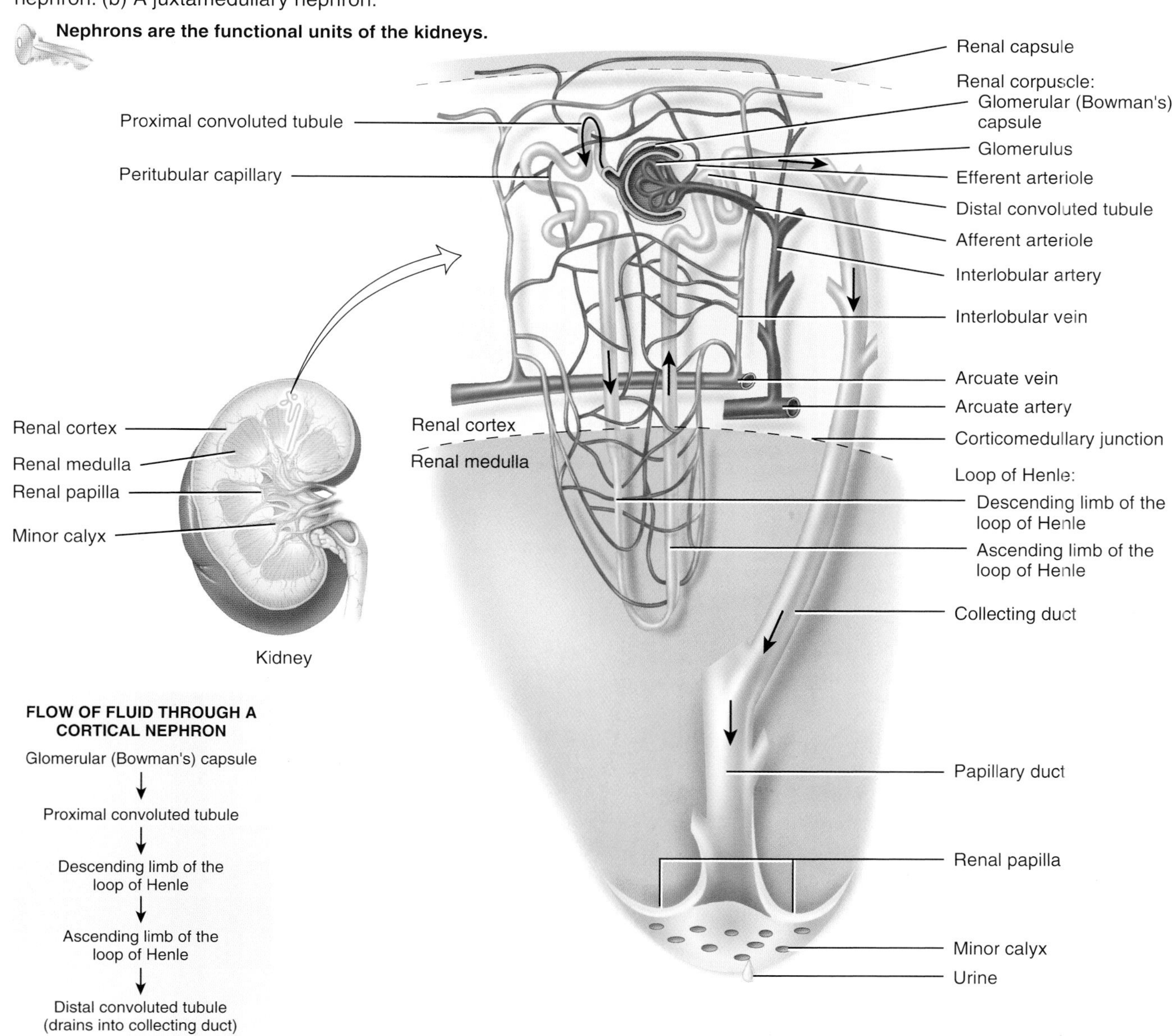

(a) Cortical nephron and vascular supply

continued

Figure 26.5 **(Continued)**

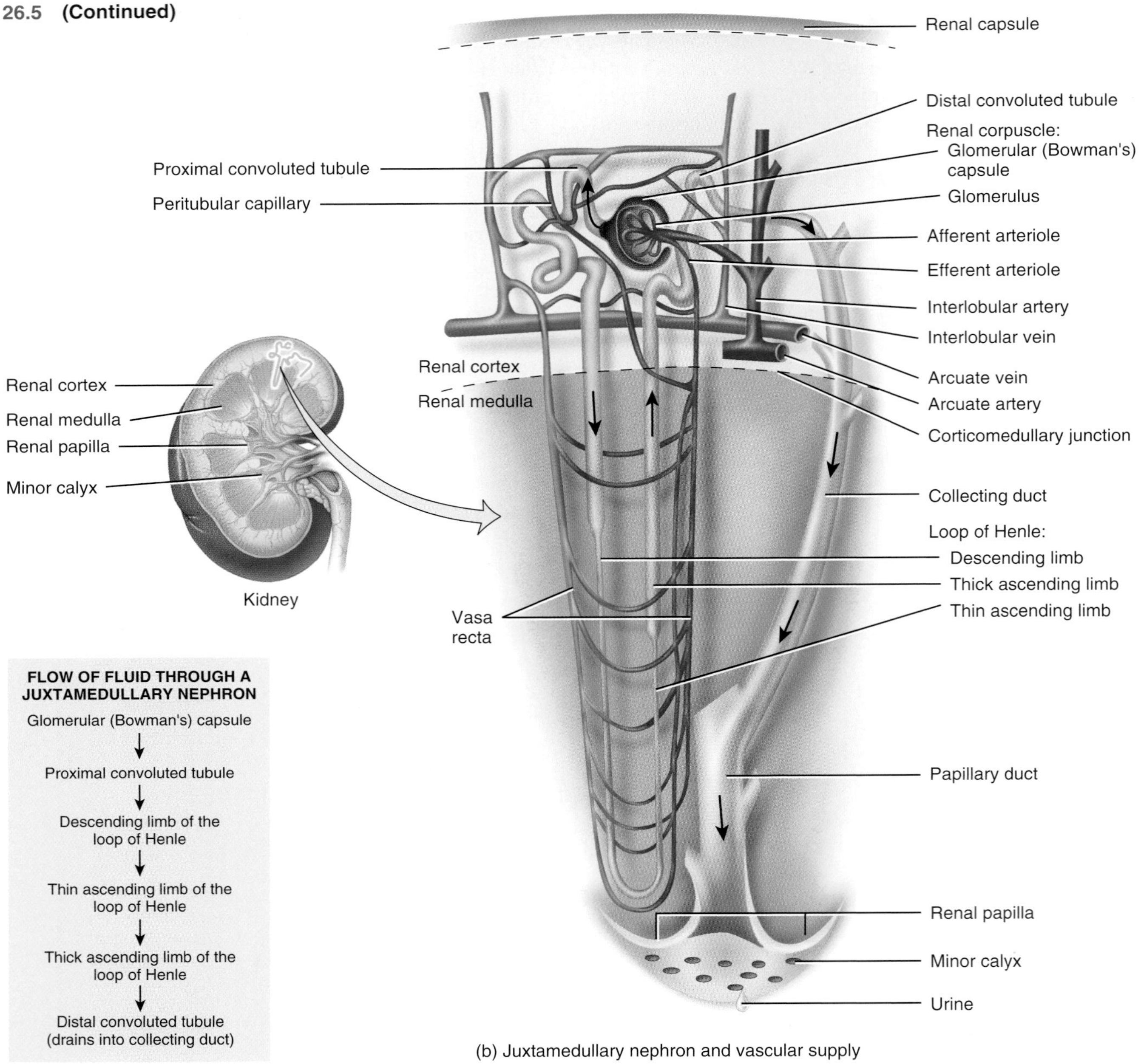

FLOW OF FLUID THROUGH A JUXTAMEDULLARY NEPHRON

Glomerular (Bowman's) capsule
↓
Proximal convoluted tubule
↓
Descending limb of the loop of Henle
↓
Thin ascending limb of the loop of Henle
↓
Thick ascending limb of the loop of Henle
↓
Distal convoluted tubule (drains into collecting duct)

(b) Juxtamedullary nephron and vascular supply

What are the basic differences between cortical and juxtamedullary nephrons?

In a nephron, the loop of Henle connects the proximal and distal convoluted tubules. The first part of the loop of Henle dips into the renal medulla, where it is called the **descending limb of the loop of Henle** (Figure 26.5). It then makes that hairpin turn and returns to the renal cortex as the **ascending limb of the loop of Henle.** About 80–85% of the nephrons are **cortical nephrons.** Their renal corpuscles lie in the outer portion of the renal cortex, and they have *short* loops of Henle that lie mainly in the cortex and penetrate only into the outer region of the renal medulla (Figure 26.5a). The short loops of Henle receive their blood supply from peritubular capillaries that arise from efferent arterioles. The other 15–20% of the nephrons are **juxtamedullary nephrons** (*juxta-* = near to). Their renal corpuscles lie deep in the cortex, close to the medulla, and they have a *long* loop of Henle that extends into the deepest region of the medulla (Figure 26.5b). Long loops of Henle receive their blood supply from peritubular capillaries and from the vasa recta that arise from efferent arterioles. In addition, the ascending limb of the loop of Henle of juxtamedullary nephrons consists of two portions: a **thin ascending limb** followed by a **thick ascending limb** (Figure 26.5b). The lumen of the thin ascending limb is the same as in other areas of the renal tubule; it is only the epithelium that is thinner. Nephrons with long loops of Henle enable the kidneys to excrete very dilute or very concentrated urine (described on pages 1016–1019).

Histology of the Nephron and Collecting Duct

A single layer of epithelial cells forms the entire wall of the glomerular capsule, renal tubule, and ducts. However, each part has distinctive histological features that reflect its particular functions. We will discuss them in the order that fluid flows through them: glomerular capsule, renal tubule, and collecting duct.

GLOMERULAR CAPSULE The glomerular (Bowman's) capsule consists of visceral and parietal layers (Figure 26.6a). The visceral layer consists of modified simple squamous epithelial cells called

Figure 26.6 Histology of a renal corpuscle.

A renal corpuscle consists of a glomerular (Bowman's) capsule and a glomerulus.

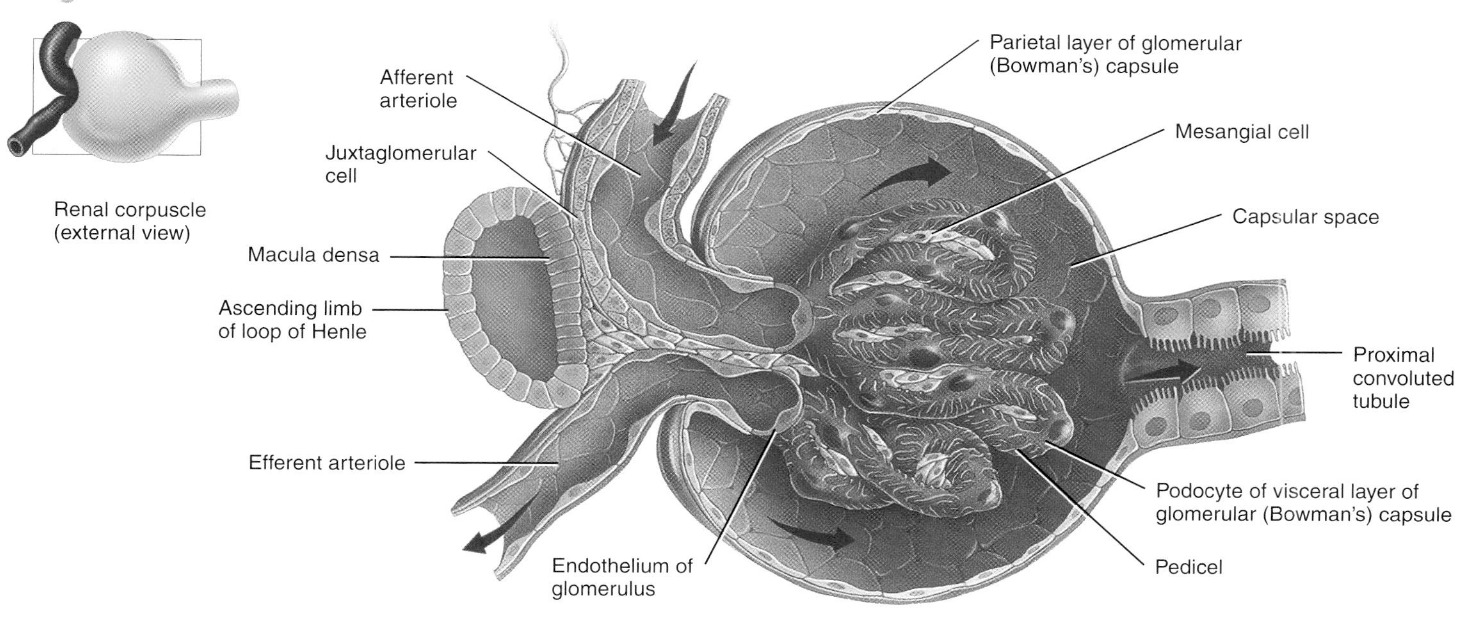

(a) Renal corpuscle (internal view)

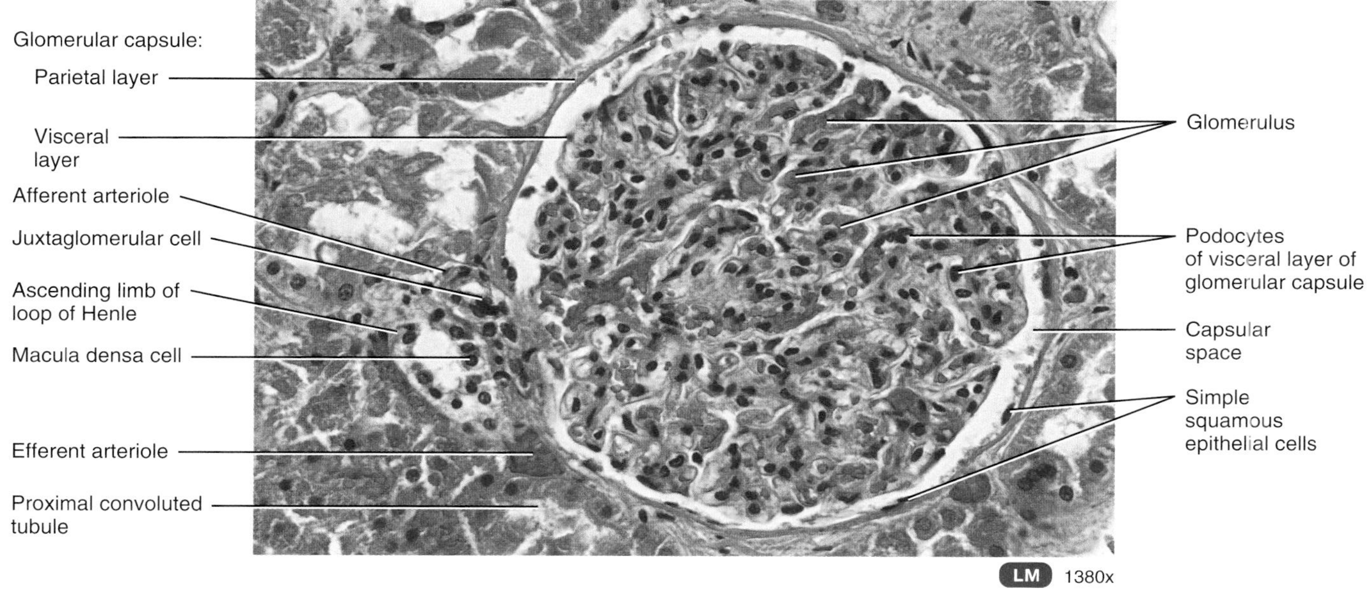

(b) Renal corpuscle

Is the photomicrograph in (b) from a section through the renal cortex or renal medulla? How can you tell?

podocytes (PŌ-dō-cīts; *podo-* = foot; *-cytes* = cells). The many footlike projections of these cells (pedicels) wrap around the single layer of endothelial cells of the glomerular capillaries and form the inner wall of the capsule. The parietal layer of the glomerular capsule consists of simple squamous epithelium and forms the outer wall of the capsule. Fluid filtered from the glomerular capillaries enters the **capsular (Bowman's) space,** the space between the two layers of the glomerular capsule. Think of the glomerulus as a fist punched into a limp balloon (the glomerular capsule) until the fist is covered by two layers of balloon (visceral and parietal layers) with a space in between (the capsular space).

Renal Tubule and Collecting Duct Table 26.1 illustrates the histology of the cells that form the renal tubule and collecting duct. In the proximal convoluted tubule, the cells are simple cuboidal epithelial cells with a prominent brush border of microvilli on their apical surface (surface facing the lumen). These microvilli, like those of the small intestine, increase the surface area for reabsorption and secretion. The descending limb of the loop of Henle and the first part of the ascending limb of the loop of Henle (the thin ascending limb) are composed of simple squamous epithelium. (Recall that cortical or short-loop nephrons lack the thin ascending limb.) The thick ascending limb of the loop of Henle is composed of simple cuboidal to low columnar epithelium.

In each nephron, the final part of the ascending limb of the loop of Henle makes contact with the afferent arteriole serving that renal corpuscle (Figure 26.6a). Because the columnar tubule cells in this region are crowded together, they are known as the **macula densa** (*macula* = spot; *densa* = dense). Alongside the macula densa, the wall of the afferent arteriole (and sometimes the efferent arteriole) contains modified smooth muscle fibers called **juxtaglomerular (JG) cells.** Together with the macula densa, they constitute the **juxtaglomerular apparatus (JGA).** As you will see later, the JGA helps regulate blood pres-

TABLE 26.1 Histological Features of the Renal Tubule and Collecting Duct

Region and Histology		Description
Proximal convoluted tubule (PCT)	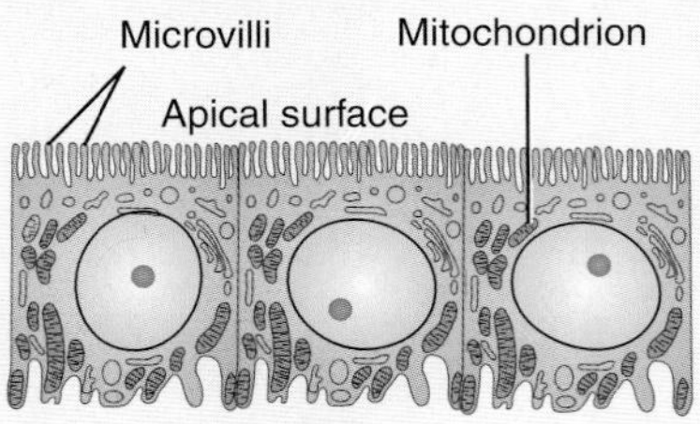 Microvilli Mitochondrion Apical surface	Simple cuboidal epithelial cells with prominent brush borders of microvilli.
Loop of Henle: descending limb and **thin ascending limb**		Simple squamous epithelial cells.
Loop of Henle: thick ascending limb	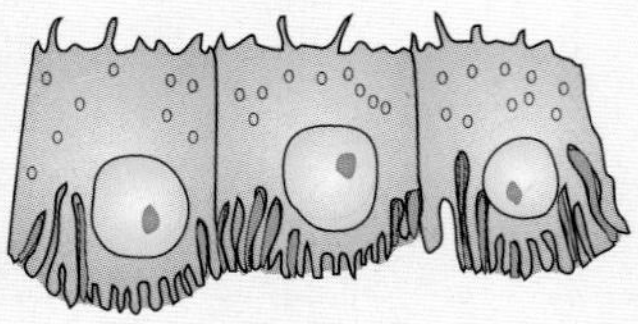	Simple cuboidal to low columnar epithelial cells.
Most of distal convoluted tubule (DCT)	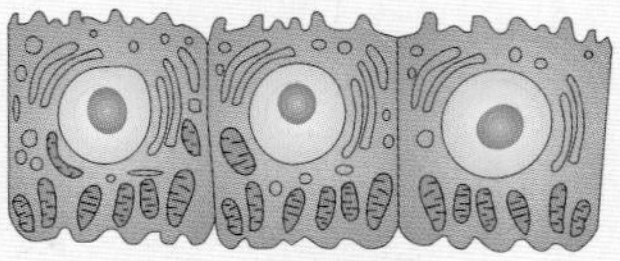	Simple cuboidal epithelial cells.
Last part of DCT and **all of collecting duct (CD)**	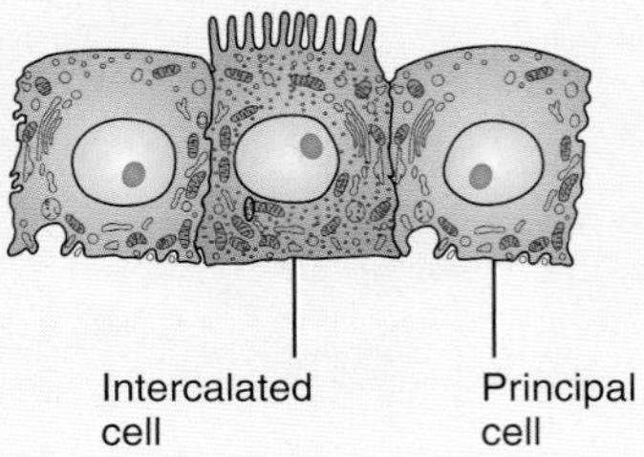 Intercalated cell Principal cell	Simple cuboidal epithelium consisting of principal cells and intercalated cells.

sure within the kidneys. The distal convoluted tubule (DCT) begins a short distance past the macula densa. In the last part of the DCT and continuing into the collecting ducts, two different types of cells are present. Most are **principal cells,** which have receptors for both antidiuretic hormone (ADH) and aldosterone, two hormones that regulate their functions. A smaller number are **intercalated cells,** which play a role in the homeostasis of blood pH. The collecting ducts drain into large papillary ducts, which are lined by simple columnar epithelium.

The number of nephrons is constant from birth. Any increase in kidney size is due solely to the growth of individual nephrons. If nephrons are injured or become diseased, new ones do not form. Signs of kidney dysfunction usually do not become apparent until function declines to less than 25% of normal because the remaining functional nephrons adapt to handle a larger-than-normal load. Surgical removal of one kidney, for example, stimulates hypertrophy (enlargement) of the remaining kidney, which eventually is able to filter blood at 80% of the rate of two normal kidneys.

▶ CHECKPOINT

2. Why are the kidneys said to be retroperitoneal?
3. What are the major parts of a nephron?
4. How do cortical nephrons and juxtamedullary nephrons differ structurally?
5. Where is the juxtaglomerular apparatus (JGA) located and what is its structure?

OVERVIEW OF RENAL PHYSIOLOGY

▶ OBJECTIVE

Identify the three basic functions performed by nephrons and collecting ducts, and indicate where each occurs.

To produce urine, nephrons and collecting ducts perform three basic processes—glomerular filtration, tubular reabsorption, and tubular secretion (Figure 26.7):

1. ***Glomerular filtration.*** In the first step of urine production, water and most solutes in blood plasma move across the wall of glomerular capillaries into the glomerular capsule and then into the renal tubule.
2. ***Tubular reabsorption.*** As filtered fluid flows along the renal tubule and through the collecting duct, tubule cells reabsorb about 99% of the filtered water and many useful solutes. The water and solutes return to the blood as it flows through the peritubular capillaries and vasa recta. Note that the term *reabsorption* refers to the return of substances to the bloodstream. The term *absorption,* by contrast, means entry of new substances into the body, as occurs in the gastrointestinal tract.
3. ***Tubular secretion.*** As fluid flows along the renal tubule and through the collecting duct, the tubule and duct cells secrete other materials, such as wastes, drugs, and excess ions, into the fluid. Notice that tubular secretion *removes* a substance from the blood. In other instances of secretion—for instance, secretion of hormones—cells release substances into interstitial fluid and blood.

Figure 26.7 **Relation of a nephron's structure to its three basic functions: glomerular filtration, tubular reabsorption, and tubular secretion.** Excreted substances remain in the urine and subsequently leave the body. For any substance S, excretion rate of S = filtration rate of S − reabsorption rate of S + secretion rate of S.

Glomerular filtration occurs in the renal corpuscle, whereas tubular reabsorption and tubular secretion occur all along the renal tubule and collecting duct.

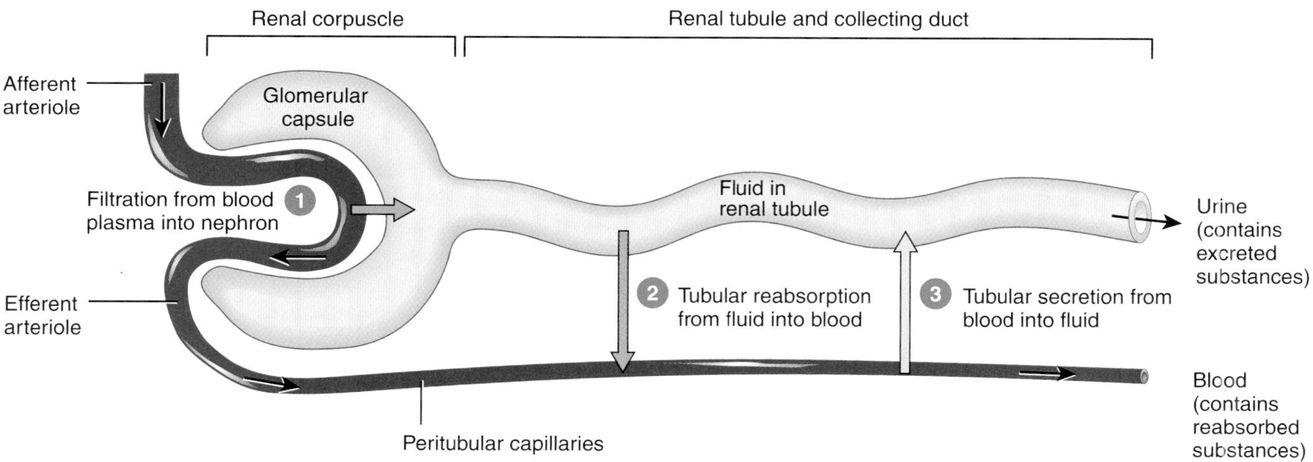

When cells of the renal tubules secrete the drug penicillin, is the drug being added to or removed from the bloodstream?

Solutes in the fluid that drains into the renal pelvis remain in the urine and are excreted. The rate of urinary excretion of any solute is equal to its rate of glomerular filtration, plus its rate of secretion, minus its rate of reabsorption.

By filtering, reabsorbing, and secreting, nephrons help maintain homeostasis of the blood's volume and composition. The situation is somewhat analogous to a recycling center: Garbage trucks dump refuse into an input hopper, where the smaller refuse passes onto a conveyor belt (glomerular filtration of plasma). As the conveyor belt carries the garbage along, workers remove useful items, such as aluminum cans, plastics, and glass containers (reabsorption). Other workers place additional garbage left at the center and larger items onto the conveyor belt (secretion). At the end of the belt, all remaining garbage falls into a truck for transport to the landfill (excretion of wastes in urine).

GLOMERULAR FILTRATION

▶ **OBJECTIVES**

Describe the filtration membrane.

Discuss the pressures that promote and oppose glomerular filtration.

The fluid that enters the capsular space is called the **glomerular filtrate.** The fraction of blood plasma in the afferent arterioles of the kidneys that becomes glomerular filtrate is the **filtration fraction.** Although a filtration fraction of 0.16–0.20 (16–20%) is typical, the value varies considerably in both health and disease. On average, the daily volume of glomerular filtrate in adults is 150 liters in females and 180 liters in males. More than 99% of the glomerular filtrate returns to the bloodstream via tubular reabsorption, so only 1–2 liters (about 1–2 qt) are excreted as urine.

The Filtration Membrane

Together, the endothelial cells of glomerular capillaries and the podocytes, which completely encircle the capillaries, form a leaky barrier known as the **filtration membrane.** This sandwichlike assembly permits filtration of water and small solutes but prevents filtration of most plasma proteins, blood cells, and platelets. Substances filtered from the blood cross three barriers—a glomerular endothelial cell, the basal lamina, and a filtration slit formed by a podocyte (Figure 26.8):

1. Glomerular endothelial cells are quite leaky because they have large **fenestrations** (pores) that measure 70–100 nm (0.07–0.1 μm) in diameter. This size permits all solutes in blood plasma to exit glomerular capillaries but prevents filtration of blood cells and platelets. Located among the glomerular capillaries and in the cleft between afferent and efferent arterioles are **mesangial cells** (*mes-* = in the middle; *-angi* = blood vessel) (see Figure 26.6a). These contractile cells help regulate glomerular filtration.
2. The **basal lamina,** a layer of acellular material between the endothelium and the podocytes, consists of minute collagen fibers and proteoglycans in a glycoprotein matrix; it prevents filtration of larger plasma proteins.
3. Extending from each podocyte are thousands of footlike processes termed **pedicels** (PED-i-sels = little feet) that wrap around glomerular capillaries. The spaces between pedicels are the **filtration slits.** A thin membrane, the **slit membrane,** extends across each filtration slit; it permits the passage of molecules having a diameter smaller than 6–7 nm (0.006–0.007 μm), including water, glucose, vitamins, amino acids, very small plasma proteins, ammonia, urea, and ions. Less than 1% of albumin, the most plentiful plasma protein, passes the slit membrane because, with a diameter of 7.1 nm, it is slightly too big to get through.

The principle of *filtration*—the use of pressure to force fluids and solutes through a membrane—is the same in glomerular capillaries as in capillaries elsewhere in the body (see Starling's law of the capillaries, page 744). However, the volume of fluid filtered by the renal corpuscle is much larger than in other capillaries of the body for three reasons:

1. Glomerular capillaries present a large surface area for filtration because they are long and extensive. The mesangial cells regulate how much of this surface area is available for filtration. When mesangial cells are relaxed, surface area is maximal, and glomerular filtration is very high. Contraction of mesangial cells reduces the available surface area, and glomerular filtration decreases.

2. The filtration membrane is thin and porous. Despite having several layers, the thickness of the filtration membrane is only 0.1 μm. Glomerular capillaries also are about 50 times leakier than capillaries in most other tissues, mainly because of their large fenestrations.

3. Glomerular capillary blood pressure is high. Because the efferent arteriole is smaller in diameter than the afferent arteriole, resistance to the outflow of blood from the glomerulus is high. As a result, blood pressure in glomerular capillaries is considerably higher than in capillaries elsewhere in the body.

Net Filtration Pressure

Glomerular filtration depends on three main pressures. One pressure *promotes* filtration and two pressures *oppose* filtration (Figure 26.9 on page 1006).

1. **Glomerular blood hydrostatic pressure (GBHP)** is the blood pressure in glomerular capillaries. Generally, GBHP is about 55 mmHg. It promotes filtration by forcing water and solutes in blood plasma through the filtration membrane.

Figure 26.8 The filtration membrane. The size of the endothelial fenestrations and filtration slits in (a) have been exaggerated for emphasis.

During glomerular filtration, water and solutes pass from blood plasma into the capsular space.

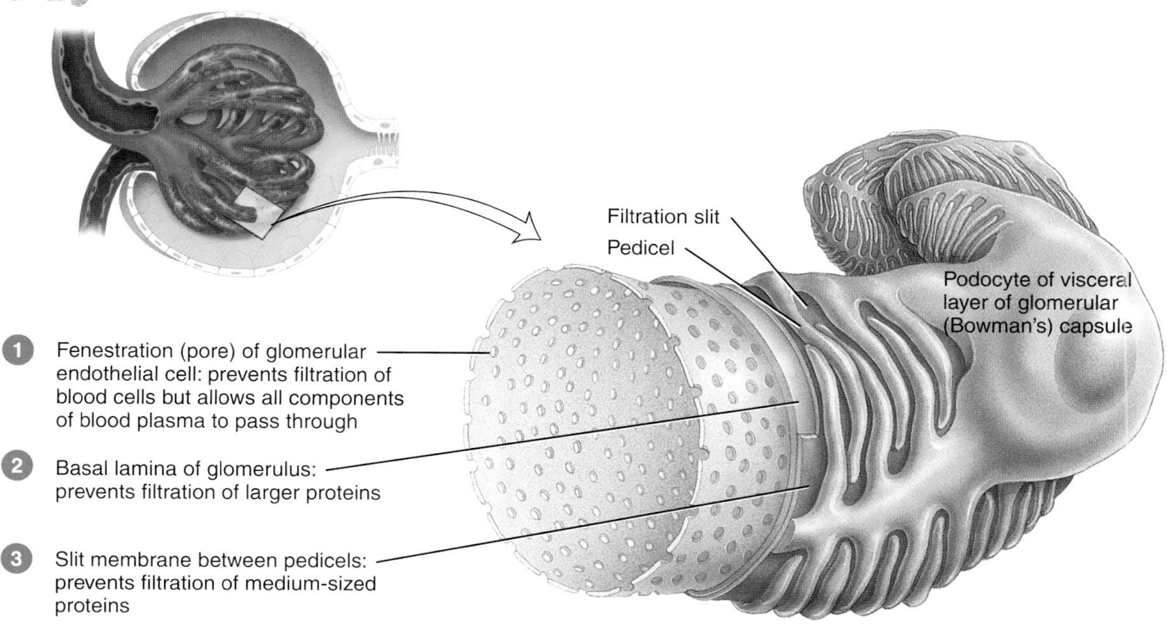

(a) Details of filtration membrane

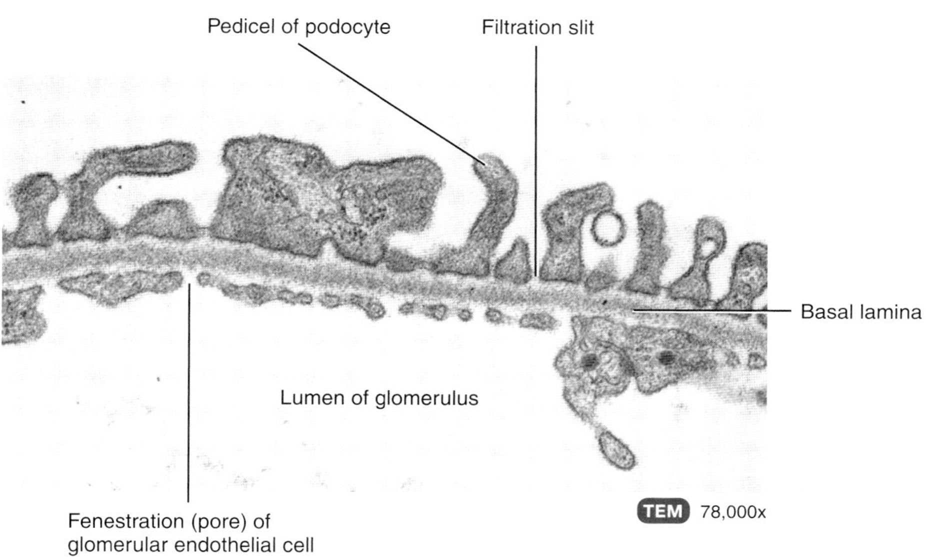

(b) Filtration membrane

Which part of the filtration membrane prevents red blood cells from entering the capsular space?

Figure 26.9 **The pressures that drive glomerular filtration.** Taken together, these pressures determine net filtration pressure (NFP).

Glomerular blood hydrostatic pressure promotes filtration, whereas capsular hydrostatic pressure and blood colloid osmotic pressure oppose filtration.

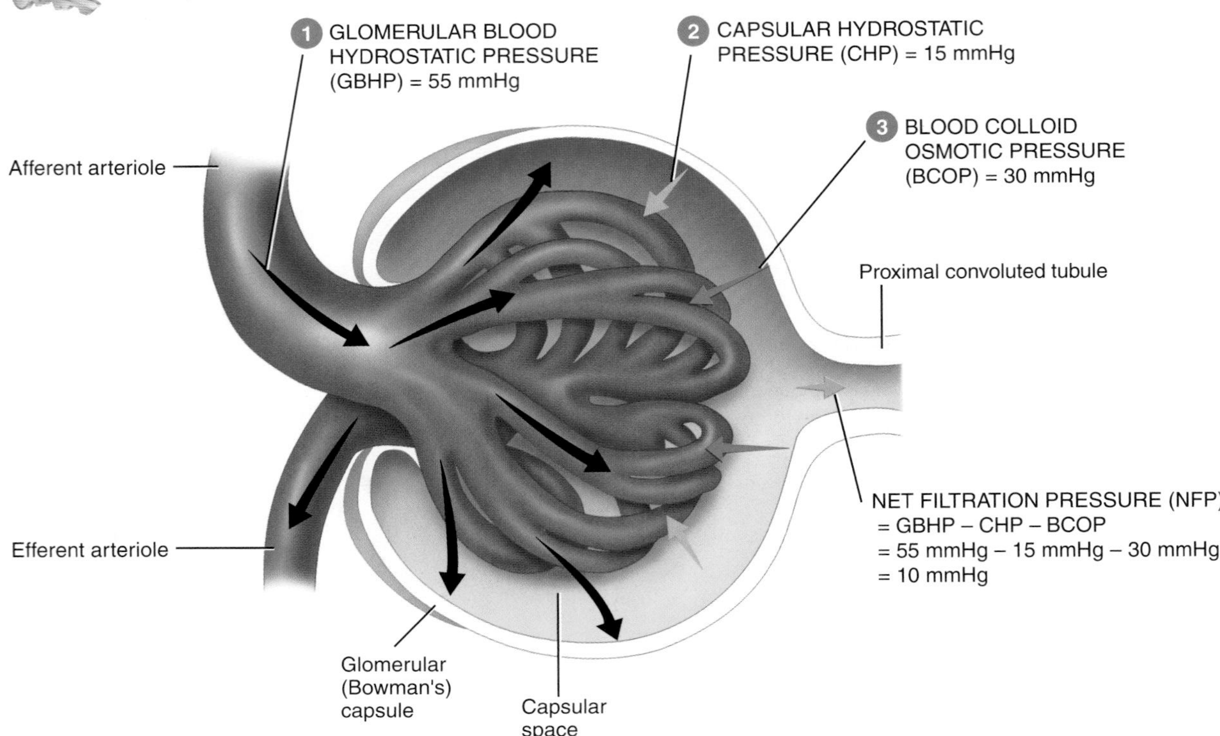

Suppose a tumor is pressing on and obstructing the right ureter. What effect might this have on CHP and thus on NFP in the right kidney? Would the left kidney also be affected?

2. **Capsular hydrostatic pressure (CHP)** is the hydrostatic pressure exerted against the filtration membrane by fluid already in the capsular space and renal tubule. CHP opposes filtration and represents a "back pressure" of about 15 mmHg.
3. **Blood colloid osmotic pressure (BCOP),** which is due to the presence of proteins such as albumin, globulins, and fibrinogen in blood plasma, also opposes filtration. The average BCOP in glomerular capillaries is 30 mmHg.

Net filtration pressure (NFP), the total pressure that promotes filtration, is determined as follows:

$$\text{Net Filtration Pressure (NFP)} = \text{GBHP} - \text{CHP} - \text{BCOP}$$

By substituting the values just given, normal NFP may be calculated:

$$\begin{aligned}\text{NFP} &= 55 \text{ mmHg} - 15 \text{ mmHg} - 30 \text{ mmHg} \\ &= 10 \text{ mmHg}\end{aligned}$$

Thus, a pressure of only 10 mmHg causes a normal amount of blood plasma (minus plasma proteins) to filter from the glomerulus into the capsular space.

Loss of Plasma Proteins in Urine Causes Edema

In some kidney diseases, glomerular capillaries are damaged and become so permeable that plasma proteins enter glomerular filtrate. As a result, the filtrate exerts a colloid osmotic pressure that draws water out of the blood. In this situation, the NFP increases, which means more fluid is filtered. At the same time, blood colloid osmotic pressure decreases because plasma proteins are being lost in the urine. Because more fluid filters out of blood capillaries into tissues throughout the body than returns via reabsorption, blood volume decreases and interstitial fluid volume increases. Thus, loss of plasma proteins in urine causes *edema,* an abnormally high volume of interstitial fluid. ■

Glomerular Filtration Rate

The amount of filtrate formed in all the renal corpuscles of both kidneys each minute is the **glomerular filtration rate (GFR).** In adults, the GFR averages 125 mL/min in males and 105 mL/min in females. Homeostasis of body fluids requires that the kidneys maintain a relatively constant GFR. If the GFR is too high, needed substances may pass so quickly through the renal tubules that some are not reabsorbed and are lost in the urine. If the GFR is too low, nearly all the filtrate may be reabsorbed and certain waste products may not be adequately excreted.

GFR is directly related to the pressures that determine net filtration pressure; any change in net filtration pressure will affect GFR. Severe blood loss, for example, reduces mean arterial blood pressure and decreases the glomerular blood hydrostatic pressure. Filtration ceases if glomerular blood hydrostatic pressure drops to 45 mmHg because the opposing pressures add up to 45 mmHg. Amazingly, when systemic blood pressure rises above normal, net filtration pressure and GFR increase very little. GFR is nearly constant when the mean arterial blood pressure is anywhere between 80 and 180 mmHg.

The mechanisms that regulate glomerular filtration rate operate in two main ways: (1) by adjusting blood flow into and out of the glomerulus and (2) by altering the glomerular capillary surface area available for filtration. GFR increases when blood flow into the glomerular capillaries increases. Coordinated control of the diameter of both afferent and efferent arterioles regulates glomerular blood flow. Constriction of the afferent arteriole decreases blood flow into the glomerulus; dilation of the afferent arteriole increases it. Three mechanisms control GFR: renal autoregulation, neural regulation, and hormonal regulation.

Renal Autoregulation of GFR

The kidneys themselves help maintain a constant renal blood flow and GFR despite normal, everyday changes in blood pressure, like those that occur during exercise. This capability is called **renal autoregulation** and consists of two mechanisms—the myogenic mechanism and tubuloglomerular feedback. Working together, they can maintain nearly constant GFR over a wide range of systemic blood pressures.

The **myogenic mechanism** (*myo-* = muscle; *-genic* = producing) occurs when stretching triggers contraction of smooth muscle cells in the walls of afferent arterioles. As blood pressure rises, GFR also rises because renal blood flow increases. However, the elevated blood pressure stretches the walls of the afferent arterioles. In response, smooth muscle fibers in the wall of the afferent arteriole contract, which narrows the arteriole's lumen. As a result, renal blood flow decreases, thus reducing GFR to its previous level. Conversely, when arterial blood pressure drops, the smooth muscle cells are stretched less and thus relax. The afferent arterioles dilate, renal blood flow increases, and GFR increases. The myogenic mechanism normalizes renal blood flow and GFR within seconds after a change in blood pressure.

The second contributor to renal autoregulation, **tubuloglomerular feedback,** is so named because part of the renal tubules—the macula densa—provides feedback to the glomerulus (Figure 26.10). When GFR is above normal due to elevated systemic blood pressure, filtered fluid flows more rapidly along the renal tubules. As a result, the proximal convoluted tubule and loop of Henle have less time to reabsorb Na^+, Cl^-, and water. Macula densa cells are thought to detect the increased delivery of Na^+, Cl^-, and water and to inhibit release of nitric oxide (NO) from cells in the juxtaglomerular apparatus (JGA). Because NO causes vasodilation, afferent arterioles constrict when the level of NO declines. As a result, less blood flows into the glomerular capillaries, and GFR decreases. When blood pressure falls, causing GFR to be lower than normal, the opposite sequence of events occurs, although to a lesser degree. Tubuloglomerular feedback operates more slowly than the myogenic mechanism.

Figure 26.10 Tubuloglomerular feedback.

Macula densa cells of the juxtaglomerular apparatus provide negative feedback regulation of glomerular filtration rate.

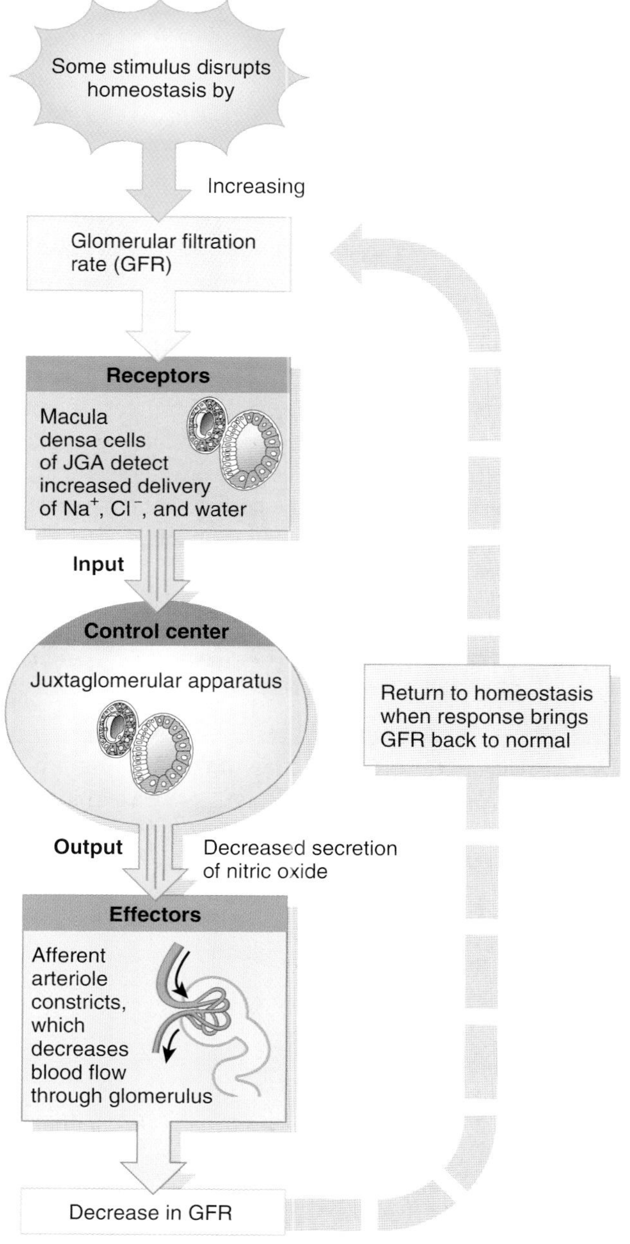

Why is this process termed autoregulation?

Neural Regulation of GFR

Like most blood vessels of the body, those of the kidneys are supplied by sympathetic ANS fibers that release norepinephrine. Norepinephrine causes vasoconstriction through the activation of α_1 receptors, which are particularly plentiful in the smooth muscle fibers of afferent arterioles. At rest, sympathetic stimulation is moderately low, the afferent and efferent arterioles are dilated, and renal autoregulation of GFR prevails. With moderate sympathetic stimulation, both afferent and efferent arterioles constrict to the same degree. Blood flow into and out of the glomerulus is restricted to the same extent, which decreases GFR only slightly. With greater sympathetic stimulation, however, as occurs during exercise or hemorrhage, vasoconstriction of the afferent arterioles predominates. As a result, blood flow into glomerular capillaries is greatly decreased, and GFR drops. This lowering of renal blood flow has two consequences: (1) It reduces urine output, which helps conserve blood volume. (2) It permits greater blood flow to other body tissues.

Hormonal Regulation of GFR

Two hormones contribute to regulation of GFR. Angiotensin II reduces GFR; atrial natriuretic peptide (ANP) increases GFR. **Angiotensin II** is a very potent vasoconstrictor that narrows both afferent and efferent arterioles and reduces renal blood flow, thereby decreasing GFR. Cells in the atria of the heart secrete **atrial natriuretic peptide (ANP).** Stretching of the atria, as occurs when blood volume increases, stimulates secretion of ANP. By causing relaxation of the glomerular mesangial cells, ANP increases the capillary surface area available for filtration. Glomerular filtration rate rises as the surface area increases.

Table 26.2 summarizes the regulation of glomerular filtration rate.

▶ CHECKPOINT

6. If the urinary excretion rate of a drug such as penicillin is greater than the rate at which it is filtered at the glomerulus, how else is it getting into the urine?
7. What is the major chemical difference between blood plasma and glomerular filtrate?
8. Why is there much greater filtration through glomerular capillaries than through capillaries elsewhere in the body?
9. Write the equation for the calculation of net filtration pressure (NFP) and explain the meaning of each term.
10. How is glomerular filtration rate regulated?

TUBULAR REABSORPTION AND TUBULAR SECRETION

▶ OBJECTIVES

Describe the routes and mechanisms of tubular reabsorption and secretion.

Describe how specific segments of the renal tubule and collecting duct reabsorb water and solutes.

Describe how specific segments of the renal tubule and collecting duct secrete solutes into the urine.

Principles of Tubular Reabsorption and Secretion

The volume of fluid entering the proximal convoluted tubules in just half an hour is greater than the total blood plasma volume because the normal rate of glomerular filtration is so high. Obviously some of this fluid must be returned somehow to the

TABLE 26.2 Regulation of Glomerular Filteration Rate (GFR)

Type of Regulation	Major Stimulus	Mechanism and Site of Action	Effect on GFR
Renal Autoregulation			
Myogenic mechanism	Increased stretching of smooth muscle fibers in afferent arteriole walls due to increased blood pressure.	Stretched smooth muscle fibers contract, thereby narrowing the lumen of the afferent arterioles.	Decrease.
Tubuloglomerular feedback	Rapid delivery of Na^+ and Cl^- to the macula densa due to high systemic blood pressure.	Decreased release of nitric oxide (NO) by the juxtaglomerular apparatus causes constriction of afferent arterioles.	Decrease.
Neural Regulation	Increase in level of activity of renal sympathetic nerves releases norepinephrine.	Constriction of afferent arterioles through activation of α_1 receptors and increased release of renin.	Decrease.
Hormone Regulation			
Angiotensin II	Decreased blood volume or blood pressure stimulates production of angiotensin II.	Constriction of both afferent and efferent arterioles.	Decrease.
Atrial natriuretic peptide (ANP)	Stretching of the atria of the heart stimulates secretion of ANP.	Relaxation of mesangial cells in glomerulus increases capillary surface area available for filtration.	Increase.

TABLE 26.3 Substances Filtered, Reabsorbed, and Excreted in Urine

Substance	Filtered* (Enters Glomerular Capsule per Day)	Reabsorbed (Returned to Blood per Day)	Urine (Excreted per Day)
Water	180 liters	178–179 liters	1–2 liters
Proteins	2.0 g	1.9 g	0.1 g
Sodium ions (Na^+)	579 g	575 g	4 g
Chloride ions (Cl^-)	640 g	633.7 g	6.3 g
Bicarbonate ions (HCO_3^-)	275 g	274.97 g	0.03 g
Glucose	162 g	162 g	0 g
Urea	54 g	24 g	30 g†
Potassium ions (K^+)	29.6 g	29.6 g	2.0 g‡
Uric acid	8.5 g	7.7 g	0.8 g
Creatinine	1.6 g	0 g	1.6 g

*Assuming GFR is 180 liters per day.
†In addition to being filtered and reabsorbed, urea is secreted.
‡After virtually all filtered K^+ is reabsorbed in the convoluted tubules and loop of Henle, a variable amount of K^+ is secreted by principal cells in the collecting duct.

bloodstream. Reabsorption—the return of most of the filtered water and many of the filtered solutes to the bloodstream—is the second basic function of the nephron and collecting duct. Normally, about 99% of the filtered water is reabsorbed. Epithelial cells all along the renal tubule and duct carry out reabsorption, but proximal convoluted tubule cells make the largest contribution. Solutes that are reabsorbed by both active and passive processes include glucose, amino acids, urea, and ions such as Na^+ (sodium), K^+ (potassium), Ca^{2+} (calcium), Cl^- (chloride), HCO_3^- (bicarbonate), and HPO_4^{2-} (phosphate). Once fluid passes through the proximal convoluted tubule, cells located more distally fine-tune the reabsorption processes to maintain homeostatic balances of water and selected ions. Most small proteins and peptides that pass through the filter also are reabsorbed, usually via pinocytosis. To appreciate the magnitude of tubular reabsorption, look at Table 26.3 and compare the amounts of substances that are filtered, reabsorbed, and excreted in urine.

The third function of nephrons and collecting ducts is tubular secretion, the transfer of materials from the blood and tubule cells into tubular fluid. Secreted substances include hydrogen ions (H^+), K^+, ammonium ions (NH_4^+), creatinine, and certain drugs such as penicillin. Tubular secretion has two important outcomes: (1) The secretion of H^+ helps control blood pH. (2) The secretion of other substances helps eliminate them from the body.

Reabsorption Routes

A substance being reabsorbed from the fluid in the tubule lumen can take one of two routes before entering a peritubular capillary: It can move *between* adjacent tubule cells or *through* an individual tubule cell (Figure 26.11). Along the renal tubule,

Figure 26.11 Reabsorption routes: paracellular reabsorption and transcellular reabsorption.

In paracellular reabsorption, water and solutes in tubular fluid return to the bloodstream by moving between tubule cells; in transcellular reabsorption, solutes and water in tubular fluid return to the bloodstream by passing through a tubule cell.

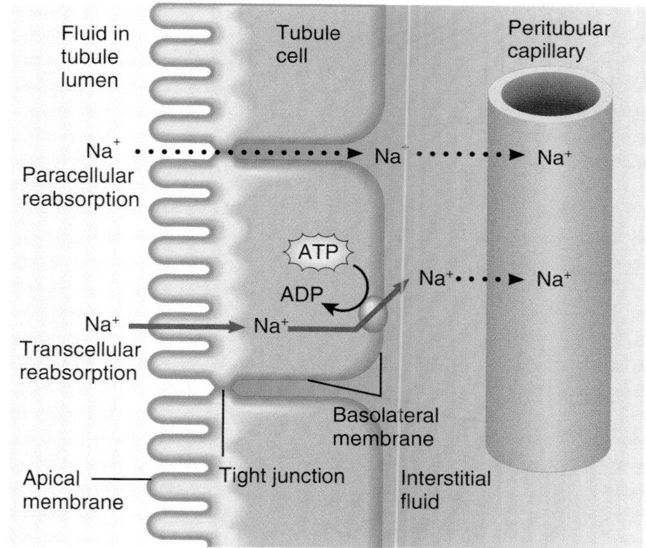

Key:

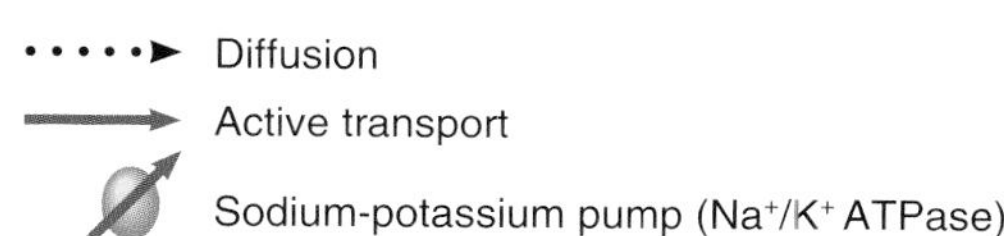

What is the main function of the tight junctions between tubule cells?

tight junctions surround and join neighboring cells to one another, much like the plastic rings that hold a six-pack of soda cans together. The **apical membrane** (the tops of the soda cans) contacts the tubular fluid, and the **basolateral membrane** (the bottoms and sides of the soda cans) contacts interstitial fluid at the base and sides of the cell.

The tight junctions do not completely seal off the interstitial fluid from the fluid in the tubule lumen. Fluid can leak *between* the cells in a passive process known as **paracellular reabsorption** (*para-* = beside). In some parts of the renal tubule, the paracellular route is thought to account for up to 50% of the reabsorption of certain ions and the water that accompanies them via osmosis. In **transcellular reabsorption** (*trans-* = across), a substance passes from the fluid in the tubular lumen through the apical membrane of a tubule cell, across the cytosol, and out into interstitial fluid through the basolateral membrane.

Transport Mechanisms

When renal cells transport solutes out of or into tubular fluid, they move specific substances in one direction only. Not surprisingly, different types of transport proteins are present in the apical and basolateral membranes. The tight junctions form a barrier that prevents mixing of proteins in the apical and basolateral membrane compartments. Reabsorption of Na^+ by the renal tubules is especially important because of the large number of sodium ions that pass through the glomerular filters.

Cells lining the renal tubules, like other cells throughout the body, have a low concentration of Na^+ in their cytosol due to the activity of sodium-potassium pumps (Na^+/K^+ ATPases). These pumps are located in the basolateral membranes and eject Na^+ from the renal tubule cells (Figure 26.11). The absence of sodium-potassium pumps in the apical membrane ensures that reabsorption of Na^+ is a one-way process. Most sodium ions that cross the apical membrane will be pumped into interstitial fluid at the base and sides of the cell. The amount of ATP used by sodium-potassium pumps in the renal tubules is about 6% of the total ATP consumption of the body at rest. This may not sound like much, but it is about the same amount of energy used by the diaphragm as it contracts during quiet breathing.

As we noted in Chapter 3, transport of materials across membranes may be either active or passive. Recall that in **primary active transport** the energy derived from hydrolysis of ATP is used to "pump" a substance across a membrane; the sodium-potassium pump is one such pump. In **secondary active transport** the energy stored in an ion's electrochemical gradient, rather than hydrolysis of ATP, drives another substance across a membrane. Secondary active transport couples the movement of an ion down its electrochemical gradient to the "uphill" movement of a second substance against its electrochemical gradient. *Symporters* are membrane proteins that move two or more substances in the same direction across a membrane. *Antiporters* move two or more substances in opposite directions across a membrane. Each type of transporter has an upper limit on how fast it can work, just as an escalator has a limit on how many people it can carry from one level to another in a given period. This limit, called the **transport maximum (T_m)**, is measured in mg/min.

Solute reabsorption drives water reabsorption because all water reabsorption occurs via osmosis. About 90% of the reabsorption of water filtered by the kidneys occurs along with the reabsorption of solutes such as Na^+, Cl^-, and glucose. Water reabsorbed with solutes in tubular fluid is termed **obligatory water reabsorption** (ob-LIG-a-tor′-ē) because the water is "obliged" to follow the solutes when they are reabsorbed. This type of water reabsorption occurs in the proximal convoluted tubule and the descending limb of the loop of Henle because these segments of the nephron are always permeable to water. Reabsorption of the final 10% of the water, a total of 10–20 liters per day, is termed **facultative water reabsorption** (FAK-ul-tā′-tiv). The word *facultative* means "capable of adapting to a need." Facultative water reabsorption is regulated by antidiuretic hormone and occurs mainly in the collecting ducts.

Glucosuria

When the blood concentration of glucose is above 200 mg/mL, the renal symporters cannot work fast enough to reabsorb all the glucose that enters the glomerular filtrate. As a result, some glucose remains in the urine, a condition called **glucosuria** (gloo′-kō-SOO-rē-a). The most common cause of glucosuria is diabetes mellitus, in which the blood glucose level may rise far above normal because insulin activity is deficient. Rare genetic mutations in the renal Na^+-glucose symporter greatly reduce its T_m and cause glucosuria. In these cases, glucose appears in the urine even though the blood glucose level is normal. Excessive glucose in the glomerular filtrate inhibits water reabsorption by kidney tubules. This leads to increased urinary output (polyuria), decreased blood voume, and dehydration. ■

Now that we have discussed the principles of renal transport, we will follow the filtered fluid from the proximal convoluted tubule, into the loop of Henle, on to the distal convoluted tubule, and through the collecting ducts. In each segment, we will examine where and how specific substances are reabsorbed and secreted. The filtered fluid becomes *tubular fluid* once it enters the proximal convoluted tubule. The composition of tubular fluid changes as it flows along the nephron tubule and through the collecting duct due to reabsorption and secretion. The fluid that drains from papillary ducts into the renal pelvis is *urine.*

Reabsorption and Secretion in the Proximal Convoluted Tubule

The largest amount of solute and water reabsorption from filtered fluid occurs in the proximal convoluted tubules, and most absorptive processes involve Na^+. Na^+ transport occurs via symport and antiport mechanisms in the proximal convoluted tubule. Normally, filtered glucose, amino acids, lactic acid, water-soluble vitamins, and other nutrients are not lost in the

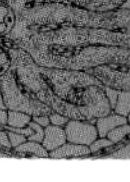

Figure 26.12 Reabsorption of glucose by Na^+-glucose symporters in cells of the proximal convoluted tubule (PCT).

Normally, all filtered glucose is reabsorbed in the PCT.

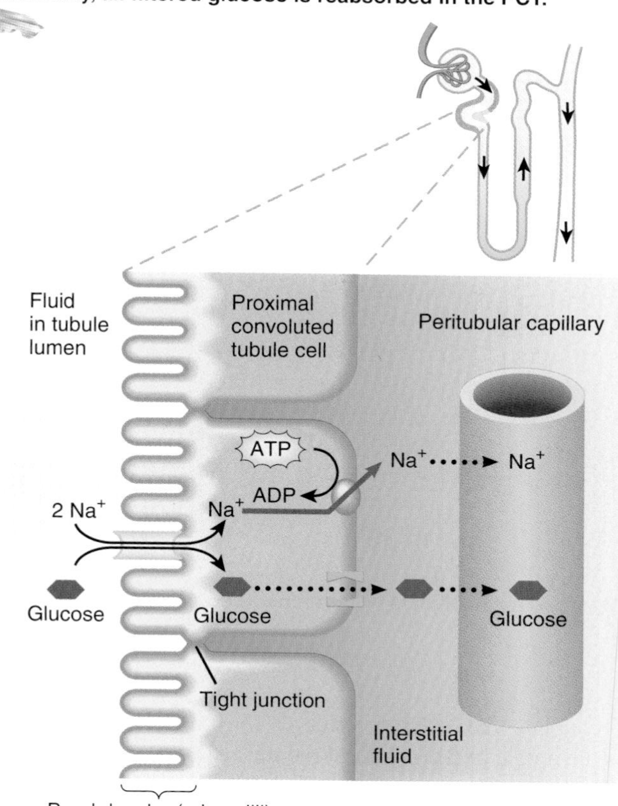

Key:

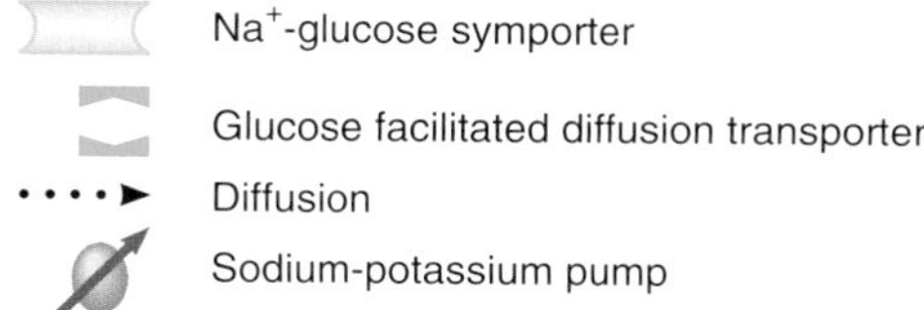

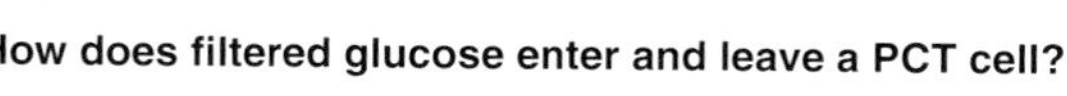

How does filtered glucose enter and leave a PCT cell?

urine. Rather, they are completely reabsorbed in the first half of the proximal convoluted tubule (PCT) by several types of **Na^+ symporters** located in the apical membrane. Figure 26.12 depicts the operation of one such symporter, the Na^+-glucose symporter in the apical membrane of a cell in the PCT. Two Na^+ and a molecule of glucose attach to the symporter protein, which carries them from the tubular fluid into the tubule cell. The glucose molecules then exit the basolateral membrane via facilitated diffusion and they diffuse into peritubular capillaries. Other Na^+ symporters in the PCT reclaim filtered HPO_4^{2-} (phosphate) and SO_4^{2-} (sulfate) ions, all amino acids, and lactic acid in a similar way.

In another secondary active transport process, the **Na^+/H^+ antiporters** carry filtered Na^+ down its concentration gradient into a PCT cell as H^+ is moved from the cytosol into the lumen (Figure 26.13a), causing Na^+ to be reabsorbed into

Figure 26.13 Actions of Na^+/H^+ antiporters in proximal convoluted tubule cells. (a) Reabsorption of sodium ions (Na^+) and secretion of hydrogen ions (H^+) via secondary active transport through the apical membrane; (b) reabsorption of bicarbonate ions (HCO_3^-) via facilitated diffusion through the basolateral membrane. CO_2 = carbon dioxide; H_2CO_3 = carbonic acid; CA = carbonic anhydrase.

Na^+/H^+ antiporters promote transcellular reabsorption of Na^+, HCO_3^-, and water in the proximal convoluted tubule.

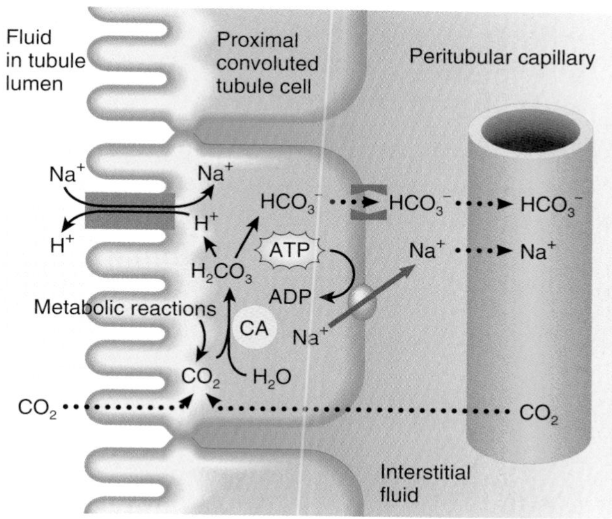

(a) Na^+ reabsorption and H^+ secretion

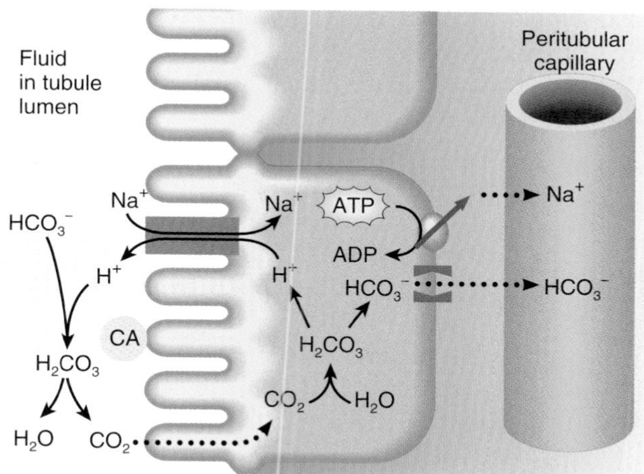

(b) HCO_3^- reabsorption

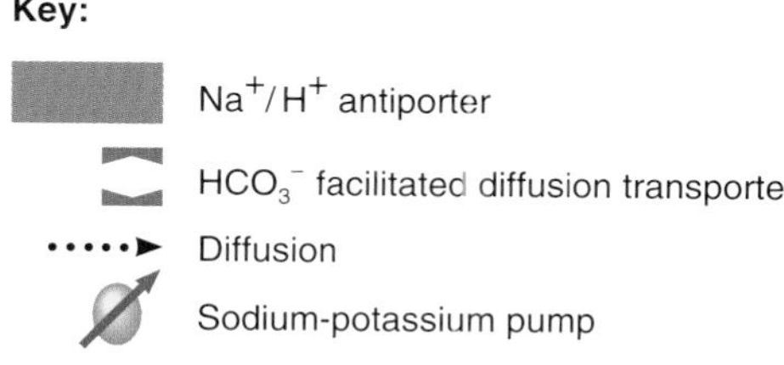

Which step in Na^+ movement in part (a) is promoted by the electrochemical gradient?

blood and H^+ to be secreted into tubular fluid. PCT cells produce the H^+ needed to keep the antiporters running in the following way. Carbon dioxide (CO_2) diffuses from peritubular blood or tubular fluid or is produced by metabolic reactions within the cells. As also occurs in red blood cells (see Figure 23.24 on page 878), the enzyme *carbonic anhydrase (CA)* catalyzes the reaction of CO_2 with water (H_2O) to form carbonic acid (H_2CO_3), which then dissociates into H^+ and HCO_3^-:

$$CO_2 + H_2O \xrightarrow{\text{Carbonic anhydrase}} H_2CO_3 \longrightarrow H^+ + HCO_3^-$$

This mechanism achieves reabsorption of 80–90% of the filtered bicarbonate ions, thereby safeguarding the body's supply of an important buffer. Figure 26.13b depicts HCO_3^- reabsorption in the proximal convoluted tubule. After H^+ is secreted into the fluid within the lumen of the proximal convoluted tubule, it reacts with filtered HCO_3^- to form H_2CO_3, which readily dissociates into CO_2 and H_2O. Carbon dioxide then diffuses into the tubule cells and joins with H_2O to form H_2CO_3, which dissociates into H^+ and HCO_3^-. As the level of HCO_3^- rises in the cytosol, it exits via facilitated diffusion transporters in the basolateral membrane and diffuses into the blood with Na^+. Thus, for every H^+ secreted into the tubular fluid of the proximal convoluted tubule, one HCO_3^- and one Na^+ are reabsorbed.

Besides achieving reabsorption of sodium ions, the Na^+ symporters and Na^+/H^+ antiporters promote osmosis of water and passive reabsorption of other solutes (Figure 26.14). They normally achieve reabsorption of 100% of most organic solutes, such as glucose and amino acids, from the filtrate, and reclaim 80–90% of the HCO_3^-; 65% of the water, Na^+, and K^+; 50% of the Cl^-; and a variable amount of Ca^{2+}, Mg^{2+}, and HPO_4^{2-} from the filtered fluid.

As water leaves the tubular fluid, the concentrations of the remaining filtered solutes increase. In the second half of the PCT, electrochemical gradients for Cl^-, K^+, Ca^{2+}, Mg^{2+}, and urea promote their passive diffusion into peritubular capillaries via both paracellular and transcellular routes. Among these ions, Cl^- is present in the highest concentration. Diffusion of negatively charged Cl^- into interstitial fluid via the paracellular route makes the interstitial fluid electrically more negative than the tubular fluid. This negativity promotes passive paracellular reabsorption of cations, such as K^+, Ca^{2+}, and Mg^{2+}.

Each reabsorbed solute increases the osmolarity, first inside the tubule cell, then in interstitial fluid, and finally in the blood. Water thus moves rapidly from the tubular fluid, via both the paracellular and transcellular routes, into the peritubular capillaries and restores osmotic balance (Figure 26.14). In other words, reabsorption of the solutes creates an osmotic gradient that promotes the reabsorption of water via osmosis. Cells lining the proximal convoluted tubule and the descending limb of the loop of Henle are especially permeable to water because they have many molecules of *aquaporin-1*. This integral protein in the plasma membrane is a

Figure 26.14 Passive reabsorption of Cl^-, K^+, Ca^{2+}, Mg^{2+}, urea, and water in the second half of the proximal convoluted tubule.

Electrochemical gradients promote passive reabsorption of solutes via both paracellular and transcellular routes.

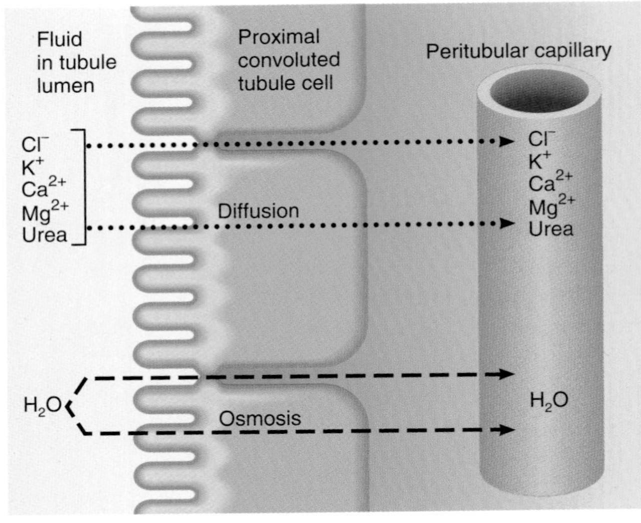

By what mechanism is water reabsorbed from tubular fluid?

water channel that greatly increases the rate of water movement across the apical and basolateral membranes.

Ammonia (NH_3) is a poisonous waste product derived from the deamination (removal of an amino group) of various amino acids, a reaction that occurs mainly in hepatocytes (liver cells). Hepatocytes convert most of this ammonia to urea, a less-toxic compound. Although tiny amounts of urea and ammonia are present in sweat, most excretion of these nitrogen-containing waste products occurs via the urine. Urea and ammonia in blood are both filtered at the glomerulus and secreted by proximal convoluted tubule cells into the tubular fluid.

Proximal convoluted tubule cells can produce additional NH_3 by deaminating the amino acid glutamine in a reaction that also generates HCO_3^-. The NH_3 quickly binds H^+ to become an ammonium ion (NH_4^+), which can substitute for H^+ aboard Na^+/H^+ antiporters in the apical membrane and be secreted into the tubular fluid. The HCO_3^- generated in this reaction moves through the basolateral membrane and then diffuses into the bloodstream, providing additional buffers in blood plasma.

Reabsorption in the Loop of Henle

Because all of the proximal convoluted tubules reabsorb about 65% of the filtered water (about 80 mL/min), fluid enters the next part of the nephron, the loop of Henle, at a rate of 40–45 mL/min. The chemical composition of the tubular fluid now is quite different from that of glomerular filtrate because glucose, amino acids, and other nutrients are no longer present. The osmolarity of the tubular fluid is still close to the osmolarity of blood, however, because reabsorption of water by osmosis keeps pace with reabsorption of solutes all along the proximal convoluted tubule.

The loop of Henle reabsorbs about 20–30% of the filtered Na^+, K^+, Ca^{2+}; 10–20% of the filtered HCO_3^-; 35% of the filtered Cl^-; and 15% of the filtered water. Here, for the first time, reabsorption of water via osmosis is *not* automatically coupled to reabsorption of filtered solutes because part of the loop of Henle is relatively impermeable to water. The loop of Henle thus sets the stage for *independent* regulation of both the *volume* and *osmolarity* of body fluids.

The apical membranes of cells in the thick ascending limb of the loop of Henle have **Na^+-K^+-$2Cl^-$ symporters** that simultaneously reclaim one Na^+, one K^+, and two Cl^- from the fluid in the tubular lumen (Figure 26.15). Na^+ that is actively transported into interstitial fluid at the base and sides of the cell diffuses into the vasa recta. Cl^- moves through leakage channels in the basolateral membrane. Because many K^+ leakage channels are present in the apical membrane, most K^+ brought in by the symporters moves down its concentration gradient back into the tubular fluid. Thus, the main effect of the Na^+-K^+-$2Cl^-$ symporters is reabsorption of Na^+ and Cl^-.

The movement of positively charged K^+ into the tubular fluid through the apical membrane channels leaves the interstitial fluid and blood with a negative charge relative to fluid in the ascending limb of the loop of Henle. This relative negativity promotes reabsorption of cations—Na^+, K^+, Ca^{2+}, and Mg^{2+}—via the paracellular route.

Although about 15% of the filtered water is reabsorbed in the *descending* limb of the loop of Henle, little or no water is reabsorbed in the *ascending* limb. In this segment of the tubule, the apical membranes are virtually impermeable to water. Because ions but not water molecules are reabsorbed, the osmolarity of the tubular fluid progressively decreases as fluid flows toward the end of the ascending limb.

Reabsorption in the Distal Convoluted Tubule

Fluid enters the distal convoluted tubules (DCT) at a rate of about 25 mL/min because 80% of the filtered water has now been reabsorbed. As fluid flows along the DCT, reabsorption of Na^+ and Cl^- continues by means of **Na^+-Cl^- symporters** in the apical membranes. Sodium-potassium pumps and Cl^- leakage channels in the basolateral membranes then permit reabsorption of Na^+ and Cl^- into the peritubular capillaries. The DCT also is the major site where parathyroid hormone (PTH) stimulates reabsorption of Ca^{2+}. Overall, cells of the DCT reabsorb 10–15% of the filtered water.

Reabsorption and Secretion in the Collecting Duct

By the time fluid reaches the end of the distal convoluted tubule, 90–95% of the filtered solutes and water have returned to the bloodstream. Recall that two different types of cells—principal

Figure 26.15 Na^+-K^+-$2Cl^-$ symporter in the thick ascending limb of the loop of Henle.

Cells in the thick ascending limb have symporters that simultaneously reabsorb one Na^+, one K^+, and two Cl^-.

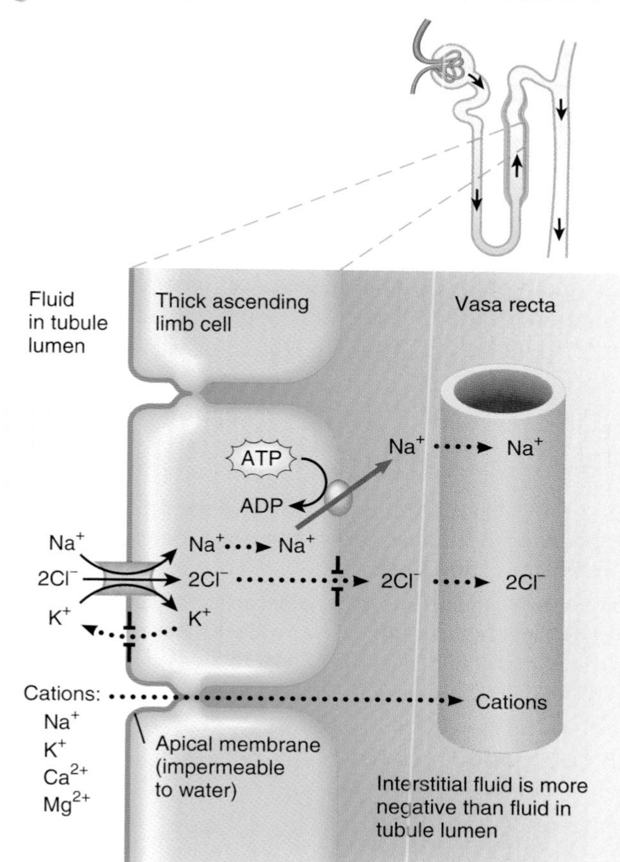

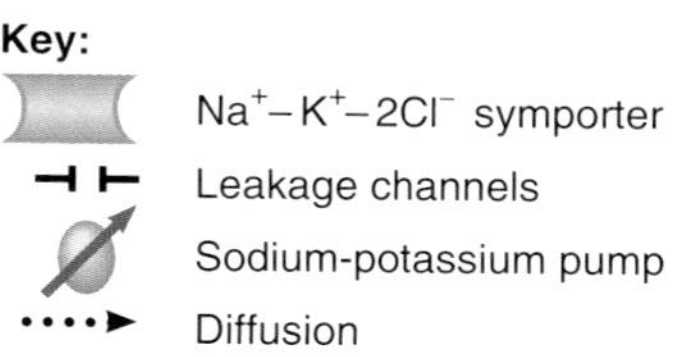

Why is this process considered secondary active transport? Does water reabsorption accompany ion reabsorption in this region of the nephron?

cells and intercalated cells—are present at the end of the distal convoluted tubule and throughout the collecting duct. The principal cells reabsorb Na^+ and secrete K^+; the intercalated cells reabsorb K^+ and HCO_3^- and secrete H^+.

In contrast to earlier segments of the nephron, Na^+ passes through the apical membrane of principal cells via Na^+ leakage channels rather than by means of symporters or antiporters (Figure 26.16). The concentration of Na^+ in the cytosol remains low, as usual, because the sodium-potassium pumps actively transport Na^+ across the basolateral membranes. Then Na^+ passively diffuses into the peritubular capillaries from the interstitial spaces around the tubule cells.

Normally, transcellular and paracellular reabsorption in the proximal convoluted tubule and loop of Henle return most filtered K^+ to the bloodstream. To adjust for varying dietary intake of potassium and to maintain a stable level of K^+ in body fluids, principal cells secrete a variable amount of K^+ (Figure 26.16).

Figure 26.16 Reabsorption of Na^+ and secretion of K^+ by principal cells in the last part of the distal convoluted tubule and in the collecting duct.

In the apical membrane of principal cells, Na^+ leakage channels allow entry of Na^+ while K^+ leakage channels allow exit of K^+ into the tubular fluid.

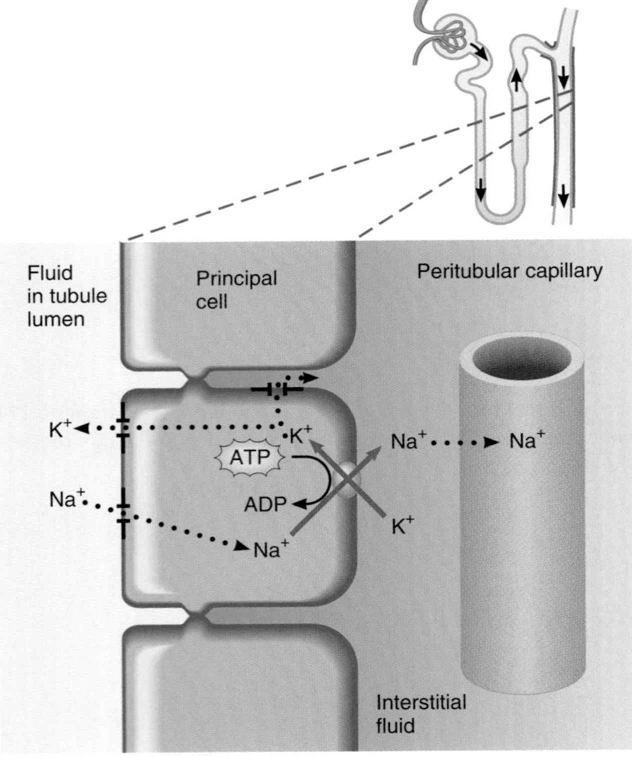

Which hormone stimulates reabsorption and secretion by principal cells, and how does this hormone exert its effect?

Because the basolateral sodium-potassium pumps continually bring K^+ into principal cells, the intracellular concentration of K^+ remains high. K^+ leakage channels are present in both the apical and basolateral membranes. Thus, some K^+ diffuses down its concentration gradient into the tubular fluid, where the K^+ concentration is very low. This secretion mechanism is the main source of K^+ excreted in the urine.

Hormonal Regulation of Tubular Reabsorption and Tubular Secretion

Four hormones affect the extent of Na^+, Cl^-, and water reabsorption as well as K^+ secretion by the renal tubules. The most important hormonal regulators of electrolyte reabsorption and secretion are angiotensin II and aldosterone. The major hormone that regulates water reabsorption is antidiuretic hormone. Atrial natriuretic peptide plays a minor role in inhibiting both electrolyte and water reabsorption.

Renin–Angiotensin–Aldosterone System

When blood volume and blood pressure decrease, the walls of the afferent arterioles are stretched less, and the juxtaglomerular cells secrete the enzyme **renin** into the blood. Sympathetic stimulation also directly stimulates release of renin from juxtaglomerular cells. Renin clips off a 10-amino-acid peptide called angiotensin I from angiotensinogen, which is synthesized by hepatocytes. By clipping off two more amino acids, *angiotensin converting enzyme (ACE)* converts angiotensin I to **angiotensin II,** which is the active form of the hormone.

Angiotensin II affects renal physiology in three main ways:

1. It decreases the glomerular filtration rate by causing vasoconstriction of the afferent arterioles.
2. It enhances reabsorption of Na^+, Cl^-, and water in the proximal convoluted tubule by stimulating the activity of Na^+/H^+ antiporters.
3. It stimulates the adrenal cortex to release **aldosterone,** a hormone that in turn stimulates the principal cells in the collecting ducts to reabsorb more Na^+ and Cl^- and secrete more K^+. The osmotic consequence of reabsorbing more Na^+ and Cl^- is excreting less water, which increases blood volume.

Antidiuretic Hormone

Antidiuretic hormone (ADH or **vasopressin)** is released by the posterior pituitary. It regulates facultative water reabsorption by increasing the water permeability of principal cells in the last part of the distal convoluted tubule and throughout the collecting duct. In the absence of ADH, the apical membranes of principal cells have a very low permeability to water. Within principal cells are tiny vesicles containing many copies of a water channel protein known as **aquaporin-2.*** ADH stimulates insertion of

*ADH does not govern the previously mentioned water channel (aquaporin-1).

the aquaporin-2–containing vesicles into the apical membranes via exocytosis. As a result, the water permeability of the principal cell's apical membrane increases, and water molecules move more rapidly from the tubular fluid into the cells. Because the basolateral membranes are always relatively permeable to water, water molecules then move rapidly into the blood. The kidneys can produce as little as 400–500 mL of very concentrated urine each day when ADH concentration is maximal, for instance during severe dehydration. When ADH level declines, the aquaporin-2 channels are removed from the apical membrane via endocytosis. The kidneys produce a large volume of dilute urine when ADH level is low.

A negative feedback system involving ADH regulates facultative water reabsorption (Figure 26.17). When the osmolarity or osmotic pressure of plasma and interstitial fluid increases—that is, when water concentration decreases—by as little as 1%, osmoreceptors in the hypothalamus detect the change. Their nerve impulses stimulate secretion of more ADH into the blood, and the principal cells become more permeable to water. As facultative water reabsorption increases, plasma osmolarity decreases to normal. A second powerful stimulus for ADH secretion is a decrease in blood volume, as occurs in hemorrhaging or severe dehydration. In the pathological absence of ADH activity, a condition known as *diabetes insipidus*, a person may excrete up to 20 liters of very dilute urine daily.

Atrial Natriuretic Peptide

A large increase in blood volume promotes release of atrial natriuretic peptide (ANP) from the heart. Although the importance of ANP in normal regulation of tubular function is unclear, it can inhibit reabsorption of Na^+ and water in the proximal convoluted tubule and collecting duct. ANP also suppresses the secretion of aldosterone and ADH. These effects increase the excretion of Na^+ in urine (natriuresis) and increase urine output (diuresis), which decreases blood volume and blood pressure.

Table 26.4 summarizes hormonal regulation of tubular reabsorption and tubular secretion.

CHECKPOINT

11. Diagram the reabsorption of substances via the transcellular and paracellular routes. Label the apical membrane and the basolateral membrane. Where are the sodium-potassium pumps located?
12. Describe two mechanisms in the PCT, one in the LOH, one in the DCT, and one in the collecting duct for reabsorption of Na^+. What other solutes are reabsorbed or secreted with Na^+ in each mechanism?
13. How do intercalated cells secrete hydrogen ions?
14. Graph the percentages of filtered water and filtered Na^+ that are reabsorbed in the PCT, LOH, DCT, and collecting duct. Indicate which hormones, if any, regulate reabsorption in each segment.

Figure 26.17 Negative feedback regulation of facultative water reabsorption by ADH.

Most water reabsorption (90%) is obligatory; 10% is facultative.

Some stimulus disrupts homeostasis by
Increasing
Osmolarity of plasma and interstitial fluid
Receptors
Osmoreceptors in hypothalamus
Input
Nerve impulses
Control center
Hypothalamus and posterior pituitary
ADH
Output
Increased release of ADH
Effectors
Principal cells become more permeable to water, which increases facultative water reabsorption
H_2O
Decrease in plasma osmolarity
Return to homeostasis when response brings plasma osmolarity back to normal

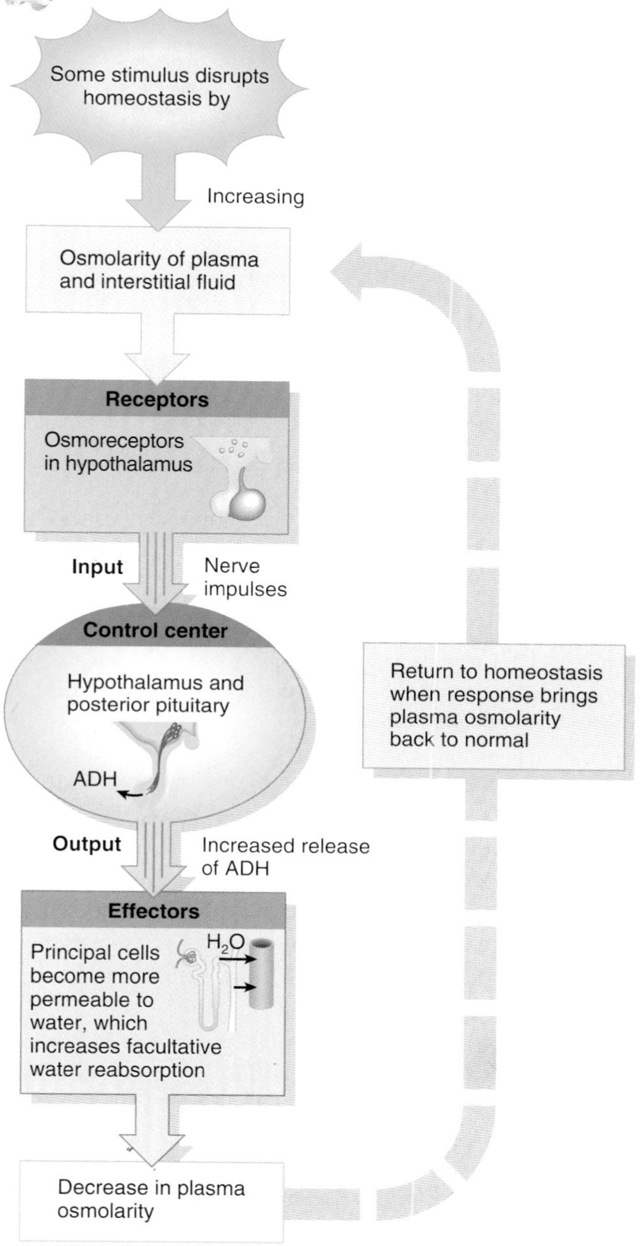

In addition to ADH, which other hormones contribute to the regulation of water reabsorption?

TABLE 26.4 Hormonal Regulation of Tubular Reabsorption and Tubular Secretion

Hormone	Major Stimuli that Trigger Release	Mechanism and Site of Action	Effects
Angiotensin II	Low blood volume or low blood pressure stimulates renin-induced production of angiotensin II.	Stimulates activity of Na^+/H^+ antiporters in proximal tubule cells.	Increases reabsorption of Na^+, other solutes, and water, which increases blood volume.
Aldosterone	Increased angiotensin II level and increased level of plasma K^+ promote release of aldosterone by adrenal cortex.	Enhances activity of sodium-potassium pumps in basolateral membrane and Na^+ channels in apical membrane of principal cells in collecting duct.	Increases secretion of K^+ and reabsorption of Na^+, Cl^-; increases reabsorption of water, which increases blood volume.
Antidiuretic hormone (ADH) or **vasopressin**	Increased osmolarity of extracellular fluid or decreased blood volume promote release of ADH from the posterior pituitary gland.	Stimulates insertion of water-channel proteins (aquaporin-2) into the apical membranes of principal cells.	Increases facultative reabsorption of water, which decreases osmolarity of body fluids.
Atrial natriuretic peptide (ANP)	Stretching of atria of heart stimulates secretion of ANP.	Suppresses reabsorption of Na^+ and water in proximal tubule and collecting duct; also inhibits secretion of aldosterone and ADH.	Increases excretion of Na^+ in urine (natriuresis); increases urine output (diuresis) and thus decreases blood volume.

PRODUCTION OF DILUTE AND CONCENTRATED URINE

▶ **OBJECTIVE**

Describe how the renal tubule and collecting ducts produce dilute and concentrated urine.

Even though your fluid intake can be highly variable, the total volume of fluid in your body normally remains stable. Homeostasis of body fluid volume depends in large part on the ability of the kidneys to regulate the rate of water loss in urine. Normally functioning kidneys produce a large volume of dilute urine when fluid intake is high, and a small volume of concentrated urine when fluid intake is low or fluid loss is large. ADH controls whether dilute urine or concentrated urine is formed. In the absence of ADH, urine is very dilute. However, a high level of ADH stimulates reabsorption of more water into blood, producing a concentrated urine.

Formation of Dilute Urine

Glomerular filtrate has the same ratio of water and solute particles as blood; its osmolarity is about 300 mOsm/liter. As previously noted, fluid leaving the proximal convoluted tubule is still isotonic to plasma. When *dilute* urine is being formed (Figure 26.18), the osmolarity of the fluid in the tubular lumen *increases* as it flows down the descending limb of the loop of Henle, *decreases* as it flows up the ascending limb, and *decreases* still more as it flows through the rest of the nephron and collecting duct. These changes in osmolarity result from the following conditions along the path of tubular fluid:

1. Because the osmolarity of the interstitial fluid of the renal medulla becomes progressively greater, more and more water is reabsorbed by osmosis as tubular fluid flows along the descending limb toward the tip of the loop. (The source of this medullary osmotic gradient is explained shortly.) As a result, the fluid remaining in the lumen becomes progressively more concentrated.

2. Cells lining the thick ascending limb of the loop have symporters that actively reabsorb Na^+, K^+, and Cl^- from the tubular fluid (see Figure 26.15). The ions pass from the tubular fluid into thick ascending limb cells, then into interstitial fluid, and finally some diffuse into the blood inside the vasa recta.

3. Although solutes are being reabsorbed in the thick ascending limb, the water permeability of this portion of the nephron is always quite low, so water cannot follow by osmosis. As solutes—but not water molecules—are leaving the tubular fluid, its osmolarity drops to about 150 mOsm/liter. The fluid entering the distal convoluted tubule is thus more dilute than plasma.

4. While the fluid continues flowing along the distal convoluted tubule, additional solutes but only a few water molecules are reabsorbed. The distal convoluted tubule cells are not very permeable to water and are not regulated by ADH.

5. Finally, the principal cells of the collecting ducts are impermeable to water when the ADH level is very low. Thus, tubular fluid becomes progressively more dilute as it flows onward. By the time the tubular fluid drains into the renal pelvis, its concentration can be as low as 65–70 mOsm/liter. This is four times more dilute than blood plasma or glomerular filtrate.

Formation of Concentrated Urine

When water intake is low or water loss is high (such as during heavy sweating), the kidneys must conserve water while still eliminating wastes and excess ions. Under the influence of ADH, the kidneys produce a small volume of highly concentrated urine. Urine can be four times more concentrated (up to 1200 mOsm/liter) than blood plasma or glomerular filtrate (300 mOsm/liter).

Figure 26.18 Formation of dilute urine. Numbers indicate osmolarity in milliosmoles per liter (mOsm/liter). Heavy brown lines in the ascending limb of the loop of Henle and in the distal convoluted tubule indicate impermeability to water; heavy blue lines indicate the last part of the distal convoluted tubule and the collecting duct, which are impermeable to water in the absence of ADH; light blue areas around the nephron represent interstitial fluid. When ADH is absent, the osmolarity of urine can be as low as 65 mOsm/liter.

When ADH level is low, urine is dilute and has an osmolarity less than the osmolarity of blood.

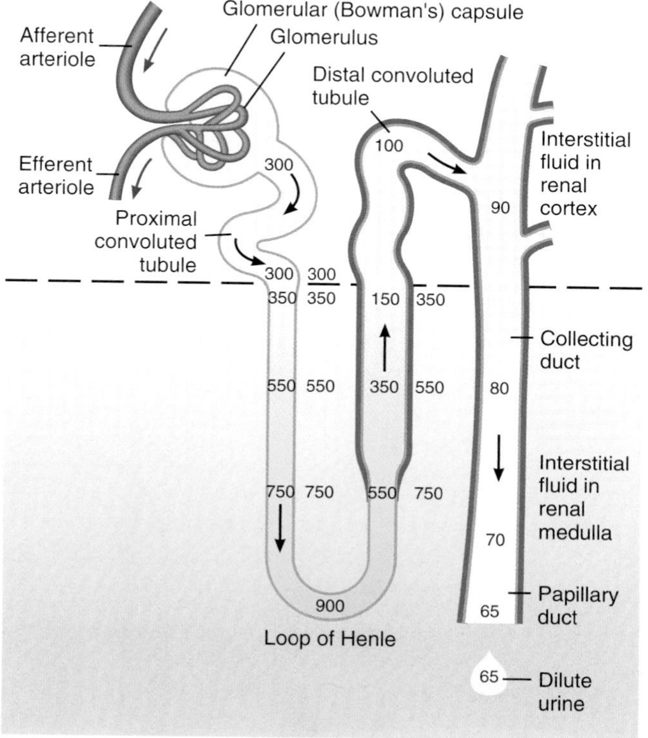

Which portions of the renal tubule and collecting duct reabsorb more solutes than water to produce dilute urine?

The ability of ADH to cause excretion of concentrated urine depends on the presence of an **osmotic gradient** of solutes in the interstitial fluid of the renal medulla. Notice in Figure 26.19 that the solute concentration of the interstitial fluid in the kidney increases from about 300 mOsm/liter in the renal cortex to about 1200 mOsm/liter deep in the renal medulla. The three major solutes that contribute to this high osmolarity are Na^+, Cl^-, and urea. Two main factors contribute to building and maintaining this osmotic gradient: (1) differences in solute and water permeability and reabsorption in different sections of the long loops of Henle and the collecting duct, and (2) the countercurrent flow (flow in opposite directions) of fluid in neighboring descending and ascending limbs of the loop of Henle.

Production of concentrated urine occurs as follows (Figure 26.19):

1. ***In long-loop nephrons, symporters in thick ascending limb cells of the loop of Henle establish the osmotic gradient in the renal medulla.*** In the thick ascending limb of the LOH, the Na^+-K^+-$2Cl^-$ symporters reabsorb Na^+ and Cl^- from the tubular fluid (Figure 26.19a). Water is not reabsorbed in this segment, however, because the cells are impermeable to water. As a result, the reabsorbed ions become increasingly concentrated in the interstitial fluid of the outer medulla. Cells in the thin ascending limb of the LOH also appear to contribute to building the osmotic gradient in the inner medulla. Those ions that diffuse into the vasa recta are carried deep into the inner medulla by the blood flow (Figure 26.19b). Because blood flow is sluggish in the vasa recta, however, there is time for diffusion of solutes to occur among tubular fluid, interstitial fluid, and blood at each level in the medulla. Thus, fluids in the descending limb, interstitial fluid, and plasma attain the same osmolarity.
2. ***Cells in the collecting ducts reabsorb more water and urea.*** When ADH increases the water permeability of the principal cells, water quickly moves via osmosis out of the collecting duct tubular fluid, into the interstitial fluid of the inner medulla, and then into the vasa recta. With loss of water, the urea left behind in the tubular fluid of the collecting duct becomes increasingly concentrated. Because duct cells deep in the medulla are permeable to it, urea diffuses from the fluid in the duct into the interstitial fluid of the medulla.
3. ***Urea recycling causes a buildup of urea in the renal medulla.*** As urea accumulates in the interstitial fluid, some of it diffuses into the tubular fluid in the descending and thin ascending limbs of the long loops of Henle, which also are permeable to urea (Figure 26.19a). However, while the fluid flows through the thick ascending limb, DCT, connecting tubule, and cortical portion of the collecting duct, urea remains in the lumen because cells in these segments are impermeable to it. As fluid flows along the collecting ducts, water reabsorption continues via osmosis because ADH is present. This water reabsorption *further increases* the concentration of urea in the tubular fluid, more urea diffuses

Figure 26.19 **Mechanism of urine concentration in long-loop juxtamedullary nephrons.** The green line indicates the presence of Na^+-K^+-$2Cl^-$ symporters that simultaneously reabsorb these ions into the interstitial fluid of the renal medulla; this portion of the nephron is also relatively impermeable to water and urea. All concentrations are in milliosmoles per liter (mOsm/liter).

The formation of concentrated urine depends on high concentrations of solutes in interstitial fluid in the renal medulla.

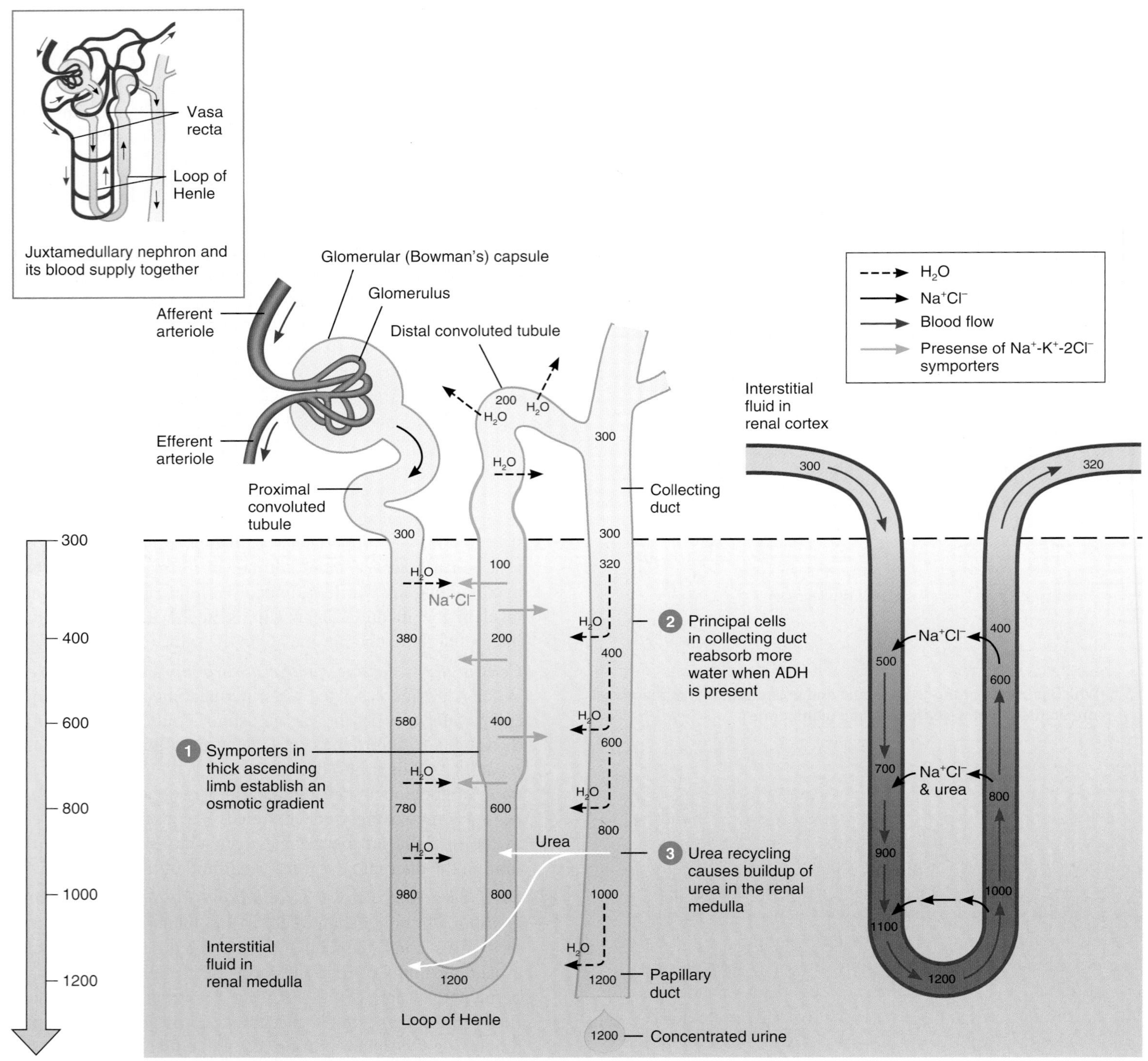

(a) Reabsorption of Na^+, Cl^- and water in a long-loop juxtamedullary nephron

(b) Recycling of salts and urea in the vasa recta

Which solutes are the main contributors to the high osmolarity of interstitial fluid in the renal medulla?

into the interstitial fluid of the inner renal medulla, and the cycle repeats. The constant transfer of urea between segments of the renal tubule and the interstitial fluid of the medulla is termed *urea recycling*. In this way, reabsorption of water from the tubular fluid of the ducts promotes the buildup of urea in the interstitial fluid of the renal medulla, which in turn promotes water reabsorption. The solutes left behind in the lumen thus become very concentrated, and a small volume of concentrated urine is excreted.

The second contributor to the osmotic gradient in the renal medulla is the **countercurrent mechanism,** which has its basis in the hairpin shape of the long loops of Henle of juxtamedullary nephrons. Note in Figure 26.19a that the descending limb of the loop of Henle carries tubular fluid from the renal cortex deep into the medulla, and the ascending limb carries it in the opposite direction. Thus, fluid flowing in one tube runs counter (opposite) to fluid flowing in a nearby parallel tube, an arrangement called *countercurrent flow.*

The descending limb of the loop of Henle is very permeable to water but impermeable to solutes except urea. Because the osmolarity of the interstitial fluid outside the descending limb is higher than the tubular fluid within it, water moves out of the descending limb via osmosis. This causes the osmolarity of the tubular fluid to increase. As the fluid continues along the descending limb, its osmolarity increases even more: At the hairpin turn of the loop, the osmolarity can be as high as 1200 mOsm/liter in juxtamedullary nephrons.

As previously noted, the ascending limb of the loop is impermeable to water, but its symporters reabsorb Na^+ and Cl^- from the tubular fluid into the interstitial fluid of the renal medulla, so the osmolarity of the tubular fluid progressively decreases as it flows through the ascending limb. At the junction of the medulla and cortex, the osmolarity of the tubular fluid has fallen to about 100 mOsm/liter. Overall, tubular fluid becomes progressively more concentrated as it flows along the descending limb and progressively more dilute as it moves along the ascending limb.

Note in Figure 26.19b that the vasa recta also consists of descending and ascending limbs that are parallel to each other and to the loop of Henle. Just as tubular fluid flows in opposite directions in the loop of Henle, blood flows in opposite directions in the ascending and descending parts of the vasa recta. Blood entering the vasa recta has an osmolarity of about 300 mOsm/liter. As it flows along the descending part into the renal medulla, where the interstitial fluid becomes increasingly concentrated, Na^+, Cl^-, and urea diffuse from interstitial fluid into the blood. But after its osmolarity increases, the blood flows into the ascending part of the vasa recta. Here blood flows through a region where the interstitial fluid becomes increasingly less concentrated. As a result, ions and urea diffuse from the blood into interstitial fluid, and reabsorbed water diffuses from interstitial fluid into the vasa recta. The osmolarity of blood leaving the vasa recta is only slightly higher than the osmolarity of blood entering the vasa recta. Thus, the vasa recta provides oxygen and nutrients to the renal medulla without washing out or diminishing the osmotic gradient.

Figure 26.20 summarizes the processes of filtration, reabsorption, and secretion in each segment of the nephron and collecting duct.

Diuretics

Diuretics are substances that slow renal reabsorption of water and thereby cause *diuresis,* an elevated urine flow rate, which in turn reduces blood volume. Diuretic drugs often are prescribed to treat *hypertension* (high blood pressure) because lowering blood volume usually reduces blood pressure. Naturally occurring diuretics include *caffeine* in coffee, tea, and sodas, which inhibits Na^+ reabsorption, and *alcohol* in beer, wine, and mixed drinks, which inhibits secretion of ADH. Most diuretic drugs act by interfering with a mechanism for reabsorption of filtered Na^+. For example, loop diuretics, such as furosemide (Lasix®), selectively inhibit the Na^+-K^+-$2Cl^-$ symporters in the thick ascending limb of the loop of Henle (see Figure 26.15). The thiazide diuretics, such as chlorthiazide (Diuril®), act in the distal convoluted tubule, where they promote loss of Na^+ and Cl^- in the urine by inhibiting Na^+-Cl^- symporters. ■

▶ CHECKPOINT

15. How do symporters in the ascending limb of the LOH and principal cells in the collecting duct contribute to the formation of concentrated urine?
16. How does ADH regulate facultative water reabsorption?
17. What is the countercurrent mechanism? Why is it important?

EVALUATION OF KIDNEY FUNCTION

▶ OBJECTIVES

- **Define urinalysis and describe its importance.**
- **Define renal plasma clearance and describe its importance.**

Routine assessment of kidney function involves evaluating both the quantity and quality of urine and the levels of wastes in the blood.

Urinalysis

An analysis of the volume and physical, chemical, and microscopic properties of urine, called a **urinalysis,** reveals much about the state of the body. Table 26.5 on page 1021 summarizes the major characteristics of normal urine. The volume of urine eliminated per day in a normal adult is 1–2 liters (about

Figure 26.20 Summary of filtration, reabsorption, and secretion in the nephron and collecting duct.

Filtration occurs in the renal corpuscle; reabsorption occurs all along the renal tubule and collecting ducts.

PROXIMAL CONVOLUTED TUBULE

Reabsorption (into blood) of filtered:

Water	65% (osmosis)
Na^+	65% (sodium-potassium pumps, symporters, antiporters)
K^+	65% (diffusion)
Glucose	100% (symporters and facilitated diffusion)
Amino acids	100% (symporters and facilitated diffusion)
Cl^-	50% (diffusion)
HCO_3^-	80–90% (facilitated diffusion)
Urea	50% (diffusion)
Ca^{2+}, Mg^{2+}	variable (diffusion)

Secretion (into urine) of:

H^+	variable (antiporters)
NH_4^+	variable, increases in acidosis (antiporters)
Urea	variable (diffusion)
Creatinine	small amount

At end of PCT, tubular fluid is still isotonic to blood (300 mOsm/liter).

RENAL CORPUSCLE

Glomerular filtration rate: 105–125 mL/min of fluid that is isotonic to blood

Filtered substances: water and all solutes present in blood (except proteins) including ions, glucose, amino acids, creatinine, uric acid

DISTAL CONVOLUTED TUBULE

Reabsorption (into blood) of:

Water	10–15% (osmosis)
Na^+	5% (symporters)
Cl^-	5% (symporters)
Ca^{2+}	variable (stimulated by parathyroid hormone)

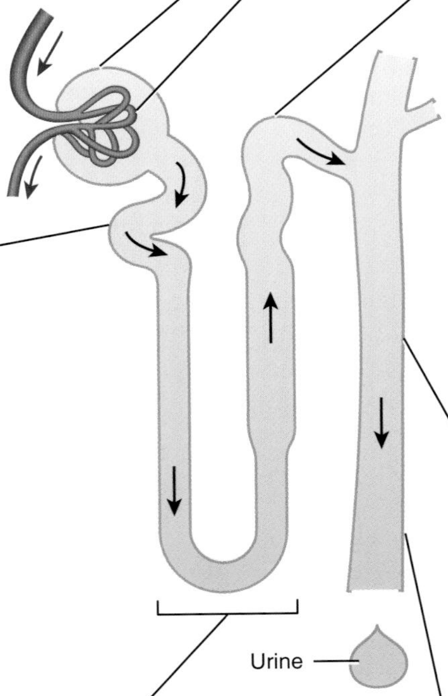

PRINCIPAL CELLS IN LATE DISTAL TUBULE AND COLLECTING DUCT

Reabsorption (into blood) of:

Water	5–9% (insertion of water channels stimulated by ADH)
Na^+	1–4% (sodium-potassium pumps)
Urea	variable (recycling to loop of Henle)

Secretion (into urine) of:

K^+	variable amount to adjust for dietary intake (leakage channels)

Tubular fluid leaving the collecting duct is dilute when ADH level is low and concentrated when ADH level is high.

Urine

LOOP OF HENLE

Reabsorption (into blood) of:

Water	15% (osmosis in descending limb)
Na^+	20–30% (symporters in ascending limb)
K^+	20–30% (symporters in ascending limb)
Cl^-	35% (symporters in ascending limb)
HCO_3^-	10–20% (facilitated diffusion)
Ca^{2+}, Mg^{2+}	variable (diffusion)

Secretion (into urine) of:

Urea	variable (recycling from collecting duct)

At end of loop of Henle, tubular fluid is hypotonic (100–150 mOsm/liter).

INTERCALATED CELLS IN LATE DISTAL TUBULE AND COLLECTING DUCT

Reabsorption (into blood) of:

HCO_3^- (new)	varible amount, depends on H^+ secretion (antiporters)
Urea	variable (recycling to loop of Henle)

Secretion (into urine) of:

H^+	variable amounts to maintain acid-base homeostasis (H^+ pumps)

In which segments of the nephron and collecting duct does secretion occur?

TABLE 26.5 Characteristics of Normal Urine

Characteristic	Description
Volume	One to two liters in 24 hours but varies considerably.
Color	Yellow or amber but varies with urine concentration and diet. Color is due to urochrome (pigment produced from breakdown of bile) and urobilin (from breakdown of hemoglobin). Concentrated urine is darker in color. Diet (reddish-colored urine from beets), medications, and certain diseases affect color. Kidney stones may produce blood in urine.
Turbidity	Transparent when freshly voided but becomes turbid (cloudy) upon standing.
Odor	Mildly aromatic but becomes ammonia-like upon standing. Some people inherit the ability to form methylmercaptan from digested asparagus that gives urine a characteristic odor. Urine of diabetics has a fruity odor due to presence of ketone bodies.
pH	Ranges between 4.6 and 8.0; average 6.0; varies considerably with diet. High-protein diets increase acidity; vegetarian diets increase alkalinity.
Specific gravity	Specific gravity (density) is the ratio of the weight of a volume of a substance to the weight of an equal volume of distilled water. In urine, it ranges from 1.001 to 1.035. The higher the concentration of solutes, the higher the specific gravity.

1–2 qt). Fluid intake, blood pressure, blood osmolarity, diet, body temperature, diuretics, mental state, and general health influence urine volume. For example, low blood pressure triggers the renin–angiotensin–aldosterone pathway. Aldosterone increases reabsorption of water and salts in the renal tubules and decreases urine volume. By contrast, when blood osmolarity decreases—for example, after drinking a large volume of water—secretion of ADH is inhibited and a larger volume of urine is excreted.

Water accounts for about 95% of the total volume of urine. The remaining 5% consists of electrolytes, solutes derived from cellular metabolism, and exogenous substances such as drugs. Normal urine is virtually protein free. Typical solutes normally present in urine include filtered and secreted electrolytes that are not reabsorbed, urea (from breakdown of proteins), creatinine (from breakdown of creatine phosphate in muscle fibers), uric acid (from breakdown of nucleic acids), urobilinogen (from breakdown of hemoglobin), and small quantities of other substances, such as fatty acids, pigments, enzymes, and hormones.

If disease alters body metabolism or kidney function, traces of substances not normally present may appear in the urine, or normal constituents may appear in abnormal amounts. Table 26.6 lists several abnormal constituents in urine that may be detected as part of a urinalysis. Normal values of urine components and the clinical implications of deviations from normal are listed in Appendix D.

TABLE 26.6 Summary of Abnormal Constituents in Urine

Abnormal Constituent	Comments
Albumin	A normal constituent of plasma, it usually appears in only very small amounts in urine because it is too large to pass through capillary fenestrations. The presence of excessive albumin in the urine—**albuminuria** (al′-bū-mi-NOO-rē-a)—indicates an increase in the permeability of filtration membranes due to injury or disease, increased blood pressure, or irritation of kidney cells by substances such as bacterial toxins, ether, or heavy metals.
Glucose	The presence of glucose in the urine is called **glucosuria** (gloo-kō-SOO-rē-a) and usually indicates diabetes mellitus. Occasionally it may be caused by stress, which can cause excessive amounts of epinephrine to be secreted. Epinephrine stimulates the breakdown of glycogen and liberation of glucose from the liver.
Red blood cells (erythrocytes)	The presence of red blood cells in the urine is called **hematuria** (hēm-a-TOO-rē-a) and generally indicates a pathological condition. One cause is acute inflammation of the urinary organs as a result of disease or irritation from kidney stones. Other causes include tumors, trauma, and kidney disease, or possible contamination of the sample by menstrual blood.
Ketone bodies	High levels of ketone bodies in the urine, called **ketonuria** (kē-tō-NOO-rē-a), may indicate diabetes mellitus, anorexia, starvation, or simply too little carbohydrate in the diet.
Bilirubin	When red blood cells are destroyed by macrophages, the globin portion of hemoglobin is split off and the heme is converted to biliverdin. Most of the biliverdin is converted to bilirubin, which gives bile its major pigmentation. An above-normal level of bilirubin in urine is called **bilirubinuria** (bil′-ē-roo-bi-NOO-rē-a).
Urobilinogen	The presence of urobilinogen (breakdown product of hemoglobin) in urine is called **urobilinogenuria** (ū′-rō-bi-lin′-ō-je-NOO-rē-a). Trace amounts are normal, but elevated urobilinogen may be due to hemolytic or pernicious anemia, infectious hepatitis, biliary obstruction, jaundice, cirrhosis, congestive heart failure, or infectious mononucleosis.
Casts	**Casts** are tiny masses of material that have hardened and assumed the shape of the lumen of the tubule in which they formed. They are then flushed out of the tubule when filtrate builds up behind them. Casts are named after the cells or substances that compose them or based on their appearance. For example, there are white blood cell casts, red blood cell casts, and epithelial cell casts that contain cells from the walls of the tubules.
Microbes	The number and type of bacteria vary with specific infections in the urinary tract. One of the most common is *E. coli.* The most common fungus to appear in urine is the yeast *Candida albicans,* a cause of vaginitis. The most frequent protozoan seen is *Trichomonas vaginalis,* a cause of vaginitis in females and urethritis in males.

Blood Tests

Two blood-screening tests can provide information about kidney function. One is the **blood urea nitrogen (BUN)** test, which measures the blood nitrogen that is part of the urea resulting from catabolism and deamination of amino acids. When glomerular filtration rate decreases severely, as may occur with renal disease or obstruction of the urinary tract, BUN rises steeply. One strategy in treating such patients is to minimize their protein intake, thereby reducing the rate of urea production.

Another test often used to evaluate kidney function is measurement of **plasma creatinine,** which results from catabolism of creatine phosphate in skeletal muscle. Normally, the blood creatinine level remains steady because the rate of creatinine excretion in the urine equals its discharge from muscle. A creatinine level above 1.5 mg/dL (135 mmol/liter) usually is an indication of poor renal function. Normal values for selected blood tests are listed in Appendix C along with situations that may cause the values to increase or decrease.

Renal Plasma Clearance

Even more useful than BUN and blood creatinine values in the diagnosis of kidney problems is an evaluation of how effectively the kidneys are removing a given substance from blood plasma. **Renal plasma clearance** is the volume of blood that is "cleaned" or cleared of a substance per unit of time, usually expressed in units of *milliliters per minute.* High renal plasma clearance indicates efficient excretion of a substance in the urine; low clearance indicates inefficient excretion. For example, the clearance of glucose normally is zero because it is not excreted at all. Instead, 100% of the filtered glucose is returned to the blood via tubular reabsorption (see Table 26.3). Knowing a drug's clearance is essential for determining the correct dosage. If clearance is high (one example is penicillin), then the dosage must also be high, and the drug must be given several times a day to maintain an adequate therapeutic level in the blood.

The following equation is used to calculate clearance:

$$\text{Renal plasma clearance of substance S} = \left(\frac{\text{U} \times \text{V}}{\text{P}}\right)$$

where U and P are the concentrations of the substance in urine and plasma, respectively (both expressed in the same units, such as mg/mL), and V is the urine flow rate in mL/min.

The clearance of a solute depends on the three basic processes of a nephron: glomerular filtration, tubular reabsorption, and tubular secretion. Consider a substance that is filtered but neither reabsorbed nor secreted. Its clearance equals the glomerular filtration rate because all the molecules that pass the filtration membrane appear in the urine. This is very nearly the situation for creatinine; it easily passes the filter, it is not reabsorbed, and it is secreted only to a very small extent. Measuring the creatinine clearance, which normally is 120– 140 mL/min, is the easiest way to assess glomerular filtration rate. The waste product urea is filtered, reabsorbed, and secreted to varying extents. Its clearance typically is less than the GFR, about 70 mL/min.

Dialysis

If a person's kidneys are so impaired by disease or injury that he or she is unable to function adequately, then blood must be cleansed artificially by **dialysis** (dī-AL-i-sis; *dialyo* = to separate), the separation of large solutes from smaller ones by diffusion through a selectively permeable membrane. One method of dialysis is **hemodialysis** (hē-mō-dī-AL-i-sis; *hemo-* = blood), which directly filters the patient's blood by removing wastes and excess electrolytes and fluid and then returning the cleansed blood to the patient. Blood removed from the body is delivered to a *hemodialyzer* (artificial kidney). Inside the hemodialyzer, blood flows through a *dialysis membrane,* which contains pores large enough to permit the diffusion of small solutes. A special solution, called the *dialysate* (dī-AL-i-sāt) is pumped into the hemodialyzer so that it surrounds the dialysis membrane. The dialysate is specially formulated to maintain diffusion gradients that remove wastes from the blood (for example, urea, creatinine, uric acid, excess phosphate, potassium, and sulfate ions) and add needed substances (for example, glucose and bicarbonate ions) to it. The cleansed blood is passed through an air embolus detector to remove air and then returned to the body. An anticoagulant (heparin) is added to prevent blood from clotting in the hemodialyzer. As a rule, most people on hemodialysis require about 6–12 hours a week, typically divided into three sessions.

Another method of dialysis, called **peritoneal dialysis,** uses the peritoneum of the abdominal cavity as the dialysis membrane to filter the blood. The peritoneum has a large surface area and numerous blood vessels, and is a very effective filter. A catheter is inserted into the peritoneal cavity and connected to a bag of dialysate. The fluid flows into the peritoneal cavity by gravity and is left there for sufficient time to permit wastes and excess electrolytes and fluids to diffuse into the dialysate. Then the dialysate is drained out into a bag, discarded, and replaced with fresh dialysate.

Each cycle is called an *exchange.* One variation of peritoneal dialysis, called **continuous ambulatory peritoneal dialysis (CAPD),** can be performed at home. Usually, the dialysate is drained and replenished four times a day and once at night during sleep. Between exchanges the person can move about freely with the dialysate in the peritoneal cavity. ■

▶ CHECKPOINT

18. What are the characteristics of normal urine?

19. What chemical substances normally are present in urine?

20. How may kidney function be evaluated?

21. Why are the renal plasma clearances of glucose, urea, and creatinine different? How does each clearance compare to glomerular filtration rate?

URINE TRANSPORTATION, STORAGE, AND ELIMINATION

▶ **OBJECTIVE**

Describe the anatomy, histology, and physiology of the ureters, urinary bladder, and urethra.

From collecting ducts, urine drains through papillary ducts into the minor calyces, which join to become major calyces that unite to form the renal pelvis (see Figure 26.3). From the renal pelvis, urine first drains into the ureters and then into the urinary bladder. Urine is then discharged from the body through the single urethra (see Figure 26.1).

Ureters

Each of the two **ureters** (Ū-rē-ters) transports urine from the renal pelvis of one kidney to the urinary bladder. Peristaltic contractions of the muscular walls of the ureters push urine toward the urinary bladder, but hydrostatic pressure and gravity also contribute. Peristaltic waves that pass from the renal pelvis to the urinary bladder vary in frequency from one to five per minute, depending on how fast urine is being formed.

The ureters are 25–30 cm (10–12 in.) long and are thick-walled, narrow tubes that vary in diameter from 1 mm to 10 mm along their course between the renal pelvis and the urinary bladder. Like the kidneys, the ureters are retroperitoneal. At the base of the urinary bladder, the ureters curve medially and pass obliquely through the wall of the posterior aspect of the urinary bladder (Figure 26.21).

Even though there is no anatomical valve at the opening of each ureter into the urinary bladder, a physiological one is quite effective. As the urinary bladder fills with urine, pressure within it compresses the oblique openings into the ureters and prevents the backflow of urine. When this physiological valve is not operating properly, it is possible for microbes to travel up the ureters from the urinary bladder to infect one or both kidneys.

Three layers of tissue form the wall of the ureters. The deepest coat, the **mucosa,** is a mucous membrane with **transitional epithelium** (see Table 4.1I on page 117) and an underlying **lamina propria** of areolar connective tissue with considerable collagen, elastic fibers, and lymphatic tissue. Transitional epithelium is able to stretch—a marked advantage for any organ that must accommodate a variable volume of fluid. Mucus secreted by the goblet cells of the mucosa prevents the cells from coming in contact with urine, the solute concentration and pH of which may differ drastically from the cytosol of cells that form the wall of the ureters.

Throughout most of the length of the ureters, the intermediate coat, the **musculariss,** is composed of inner longitudinal and outer circular layers of smooth muscle fibers. This arrangement is opposite to that of the gastrointestinal tract, which contains inner circular and outer longitudinal layers. The muscularis of

Figure 26.21 **Ureters, urinary bladder, and urethra in a female.** (See Tortora, *A Photographic Atlas of the Human Body, Second Edition,* Figures 13.8, 13.9.)

Urine is stored in the urinary bladder before being expelled by micturition.

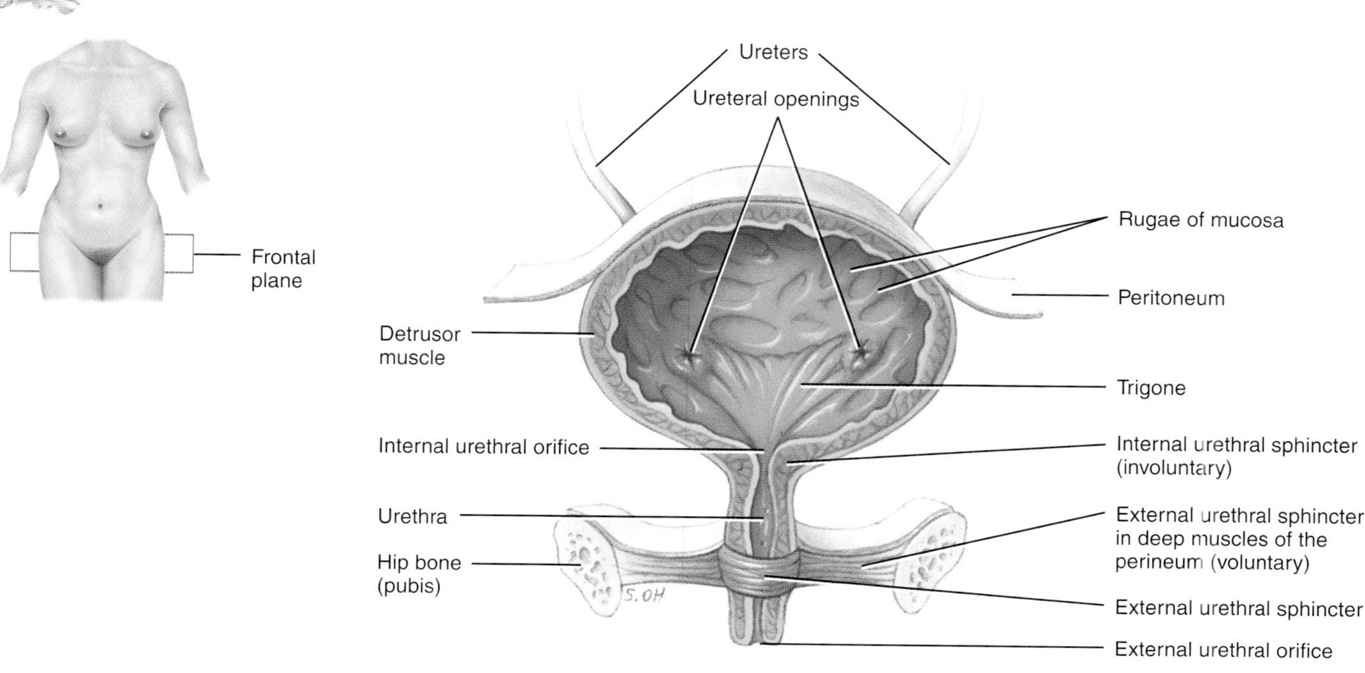

What is a lack of voluntary control over micturition called?

the distal third of the ureters also contains an outer layer of longitudinal muscle fibers. Thus, the muscularis in the distal third of the ureter is inner longitudinal, middle circular, and outer longitudinal. Peristalsis is the major function of the muscularis.

The superficial coat of the ureters is the **adventitia,** a layer of areolar connective tissue containing blood vessels, lymphatic vessels, and nerves that serve the muscularis and mucosa. The adventitia blends in with surrounding connective tissue and anchors the ureters in place.

Urinary Bladder

The **urinary bladder** is a hollow, distensible muscular organ situated in the pelvic cavity posterior to the pubic symphysis. In males, it is directly anterior to the rectum; in females, it is anterior to the vagina and inferior to the uterus (see Figure 26.22). Folds of the peritoneum hold the urinary bladder in position. When slightly distended due to the accumulation of urine, the urinary bladder is spherical. When it is empty, it collapses. As urine volume increases, it becomes pear-shaped and rises into the abdominal cavity. Urinary bladder capacity averages 700–800 mL. It is smaller in females because the uterus occupies the space just superior to the urinary bladder.

Anatomy and Histology of the Urinary Bladder

In the floor of the urinary bladder is a small triangular area called the **trigone** (TRĪ-gōn = triangle). The two posterior corners of the trigone contain the two ureteral openings; the opening into the urethra, the **internal urethral orifice,** lies in the anterior corner (Figure 26.21). Because its mucosa is firmly bound to the muscularis, the trigone has a smooth appearance.

Three coats make up the wall of the urinary bladder. The deepest is the **mucosa,** a mucous membrane composed of **transitional epithelium** and an underlying **lamina propria** similar to that of the ureters. Rugae (the folds in the mucosa) are also present to permit expansion of the urinary bladder. Surrounding the mucosa is the intermediate **muscularis,** also called the **detrusor muscle** (de-TROO-ser = to push down), which consists of three layers of smooth muscle fibers: the inner longitudinal, middle circular, and outer longitudinal layers. Around the opening to the urethra the circular fibers form an **internal urethral sphincter;** inferior to it is the **external urethral sphincter,** which is composed of skeletal muscle and is a modification of the deep muscles of the perineum (see Figure 11.12 on page 357). The most superficial coat of the urinary bladder on the posterior and inferior surfaces is the **adventitia,** a layer of areolar connective tissue that is continuous with that of the ureters. Over the superior surface of the urinary bladder is the **serosa,** a layer of visceral peritoneum.

The Micturition Reflex

Discharge of urine from the urinary bladder, called **micturition** (mik′-too-RISH-un; *mictur-* = urinate), is also known as *urination* or *voiding.* Micturition occurs via a combination of involuntary and voluntary muscle contractions. When the volume of urine in the urinary bladder exceeds 200–400 mL, pressure within the bladder increases considerably, and stretch receptors in its wall transmit nerve impulses into the spinal cord. These impulses propagate to the **micturition center** in sacral spinal cord segments S2 and S3 and trigger a spinal reflex called the **micturition reflex.** In this reflex arc, parasympathetic impulses from the micturition center propagate to the urinary bladder wall and internal urethral sphincter. The nerve impulses cause *contraction* of the detrusor muscle and *relaxation* of the internal urethral sphincter muscle. Simultaneously, the micturition center inhibits somatic motor neurons that innervate skeletal muscle in the external urethral sphincter. Upon contraction of the urinary bladder wall and relaxation of the sphincters, urination takes place. Urinary bladder filling causes a sensation of fullness that initiates a conscious desire to urinate before the micturition reflex actually occurs. Although emptying of the urinary bladder is a reflex, in early childhood we learn to initiate it and stop it voluntarily. Through learned control of the external urethral sphincter muscle and certain muscles of the pelvic floor, the cerebral cortex can initiate micturition or delay its occurrence for a limited period.

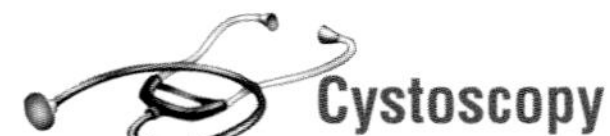

Cystoscopy

Cystoscopy (sis-TOS-kō-pē; *cysto-* = bladder; *-skopy* = to examine) is a very important procedure for direct examination of the mucosa of the urethra and urinary bladder and prostate in males. In the procedure, a *cystoscope* (a flexible narrow tube with a light) is inserted into the urethra to examine the structures through which it passes. With special attachments, tissue samples can be removed for examination (biopsy) and small stones can be removed. Cystoscopy is useful for evaluating urinary bladder problems such as cancer and infections. It can also evaluate the degree of obstruction resulting from an enlarged prostate. ■

Urethra

The **urethra** (ū-RĒ-thra) is a small tube leading from the internal urethral orifice in the floor of the urinary bladder to the exterior of the body. In both males and females, the urethra is the terminal portion of the urinary system and the passageway for discharging urine from the body. In males, it discharges semen (fluid that contains sperm) as well.

In females, the urethra lies directly posterior to the pubic symphysis, is directed obliquely inferiorly and anteriorly, and has a length of 4 cm (1.5 in.) (Figure 26.22a). The opening of the urethra to the exterior, the **external urethral orifice,** is located between the clitoris and the vaginal opening (see Figure 28.11a on page 1070). The wall of the female urethra consists of a deep **mucosa** and a superficial **muscularis.** The mucosa is a mucous membrane composed of **epithelium** and **lamina**

Figure 26.22 Comparison between female and male urethras.

The female urethra is about 4 cm (1.5 in.) in length, while the male urethra is about 20 cm (8 in.) in length.

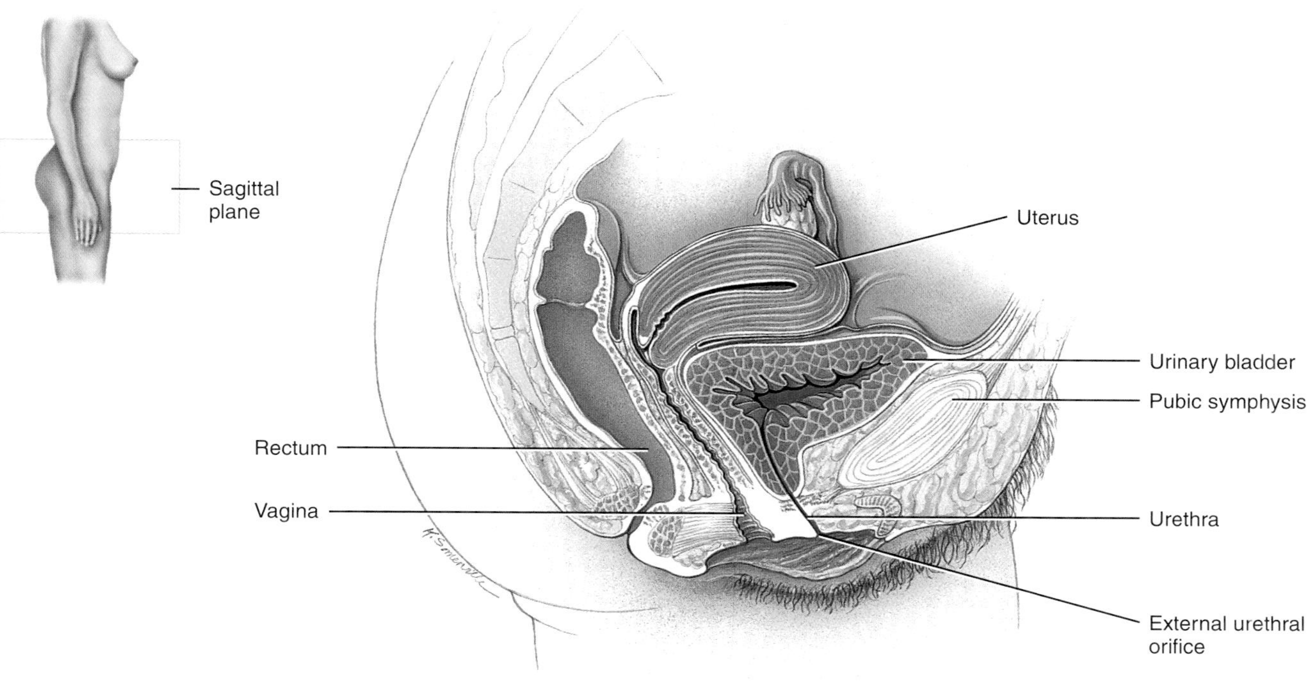

(a) Sagittal section

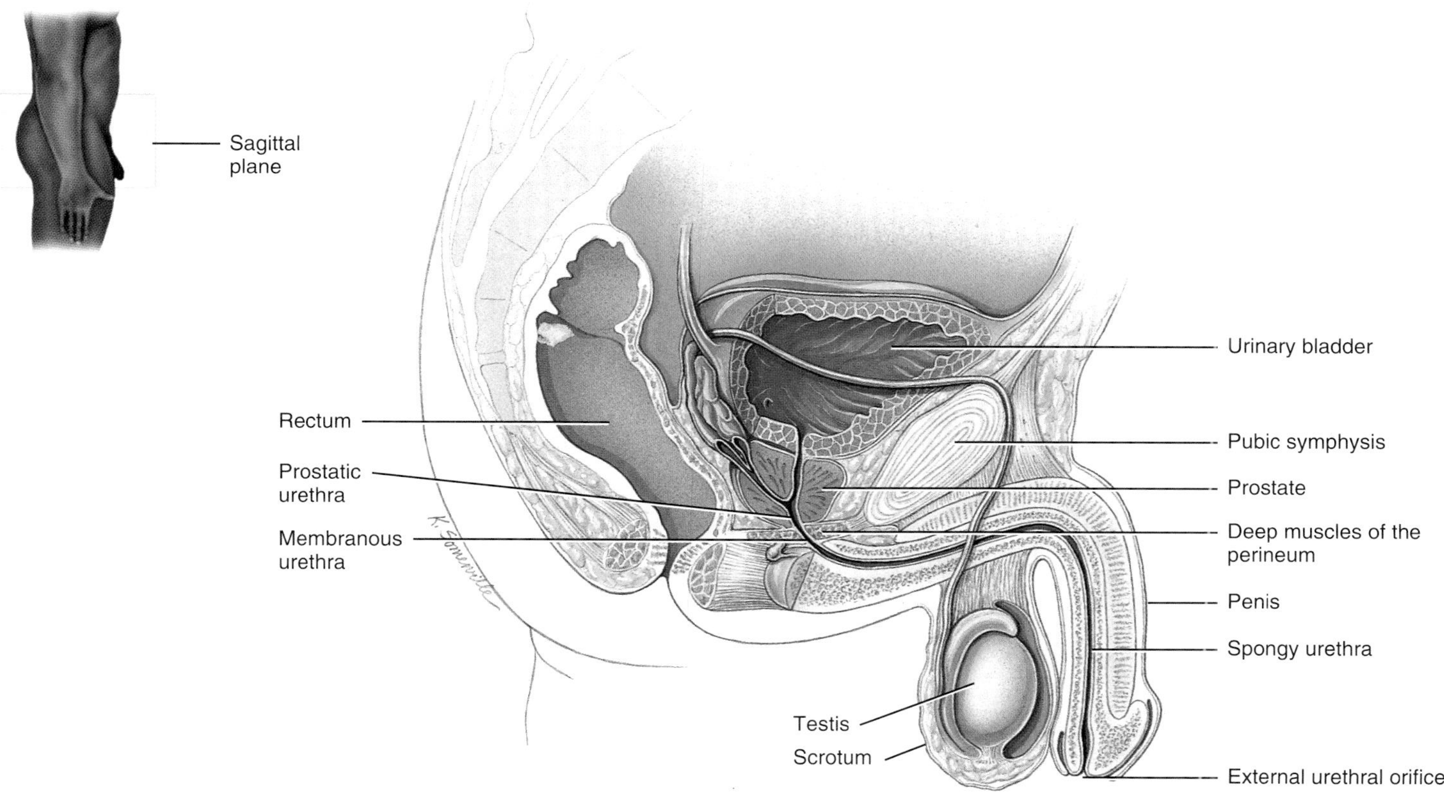

(b) Sagittal section

What are the three subdivisions of the male urethra?

propria (areolar connective tissue with elastic fibers and a plexus of veins). The muscularis consists of circularly arranged smooth muscle fibers and is continuous with that of the urinary bladder. Near the urinary bladder, the mucosa contains transitional epithelium that is continuous with that of the urinary bladder; near the external urethral orifice, the epithelium is nonkeratinized stratified squamous epithelium. Between these areas, the mucosa contains stratified columnar or pseudostratified columnar epithelium.

In males, the urethra also extends from the internal urethral orifice to the exterior, but its length and passage through the body are considerably different than in females (Figure 26.22b). The male urethra first passes through the prostate, then through the deep muscles of the perineum, and finally through the penis, a distance of about 20 cm (8 in.).

The male urethra, which also consists of a deep **mucosa** and a superficial **muscularis,** is subdivided into three anatomical regions: (1) The **prostatic urethra** passes through the prostate. (2) The **membranous urethra,** the shortest portion, passes through the deep muscles of the perineum. (3) The **spongy urethra,** the longest portion, passes through the penis. The epithelium of the prostatic urethra is continuous with that of the urinary bladder and consists of transitional epithelium that becomes stratified columnar or pseudostratified columnar epithelium more distally. The mucosa of the membranous urethra contains stratified columnar or pseudostratified columnar epithelium. The epithelium of the spongy urethra is stratified columnar or pseudostratified columnar epithelium, except near the external urethral orifice. There it is nonkeratinized stratified squamous epithelium. The **lamina propria** of the male urethra is areolar connective tissue with elastic fibers and a plexus of veins.

The muscularis of the prostatic urethra is composed of mostly circular smooth muscle fibers superficial to the lamina propria; these circular fibers help form the internal urethral sphincter of the urinary bladder. The muscularis of the membranous urethra consists of circularly arranged skeletal muscle fibers of the deep muscles of the perineum that help form the external urethral sphincter of the urinary bladder.

Several glands and other structures associated with reproduction (described in detail in Chapter 28) deliver their contents into the male urethra. The prostatic urethra receives secretions that contain sperm, neutralize the acidity of the female reproductive tract, and contribute to sperm motility and viability. The spongy urethra receives an alkaline substance before ejaculation that neutralizes the acidity of the urethra and mucus, which lubricates the end of the penis during sexual arousal. The entire urethra, but especially the spongy urethra, receives mucus during sexual arousal or ejaculation.

Urinary Incontinence

A lack of voluntary control over micturition is called **urinary incontinence.** In infants and children under 2–3 years old, incontinence is normal because neurons to the external urethral sphincter muscle are not completely developed; voiding occurs whenever the urinary bladder is sufficiently distended to stimulate the micturition reflex. Urinary incontinence also occurs in adults. There are four types of urinary incontinence—stress, urge, overflow, and functional. **Stress incontinence** is the most common type of incontinence in young and middle-aged females, and results from weakness of the deep muscles of the pelvic floor. As a result, any physical stress that increases abdominal pressure, such as coughing, sneezing, laughing, exercising, straining, lifting heavy objects, and pregnancy, causes leakage of urine from the urinary bladder. **Urge incontinence** is most common in older people and is characterized by an abrupt and intense urge to urinate followed by an involuntary loss of urine. It may be caused by irritation of the urinary bladder wall by infection or stones, stroke, multiple sclerosis, spinal cord injury, or anxiety. **Overflow incontinence** refers to the involuntary leakage of small amounts of urine caused by some type of blockage or weak contractions of the musculature of the urinary bladder. When urine flow is blocked (for example, from an enlarged prostate or stones) or the urinary bladder muscles can no longer contract, the urinary bladder becomes overfilled and the pressure inside increases until small amounts of urine dribble out. **Functional incontinence** is urine loss resulting from the inability to get to a toilet facility in time as a result of conditions such as stroke, severe arthritis, and Alzheimer disease. Choosing the right treatment option depends on correct diagnosis of the type of incontinence. Treatments include Kegel exercises (see page 364), urinary bladder training, medication, and possibly even surgery. ■

► CHECKPOINT

22. What forces help propel urine from the renal pelvis to the urinary bladder?

23. What is micturition? How does the micturition reflex occur?

24. How do the location, length, and histology of the urethra compare in males and females?

WASTE MANAGEMENT IN OTHER BODY SYSTEMS

► OBJECTIVE

Describe the ways that body wastes are handled.

As we have seen, just one of the many functions of the urinary system is to help rid the body of some kinds of waste materials. Besides the kidneys, several other tissues, organs, and processes contribute to the temporary confinement of wastes, the transport of waste materials for disposal, the recycling of materials, and the excretion of excess or toxic substances in the body. These waste management systems include the following:

- ***Body buffers.*** Buffers in body fluids bind excess hydrogen ions (H^+), thereby preventing an increase in the acidity of body fluids. Buffers, like wastebaskets, have a limited capacity; eventually the H^+, like the paper in a wastebasket, must be eliminated from the body by excretion.

- ***Blood.*** The bloodstream provides pickup and delivery services for the transport of wastes, in much the same way that garbage trucks and sewer lines serve a community.
- ***Liver.*** The liver is the primary site for metabolic recycling, as occurs, for example, in the conversion of amino acids into glucose or of glucose into fatty acids. The liver also converts toxic substances into less toxic ones, such as ammonia into urea. These functions of the liver are described in Chapters 24 and 25.
- ***Lungs.*** With each exhalation, the lungs excrete CO_2, and expel heat and a little water vapor.
- ***Sweat (sudoriferous) glands.*** Especially during exercise, sweat glands in the skin help eliminate excess heat, water, and CO_2, plus small quantities of salts and urea as well.
- ***Gastrointestinal tract.*** Through defecation, the gastrointestinal tract excretes solid, undigested foods; wastes; some CO_2; water; salts; and heat.

▶ CHECKPOINT

25. What roles do the liver and lungs play in the elimination of wastes?

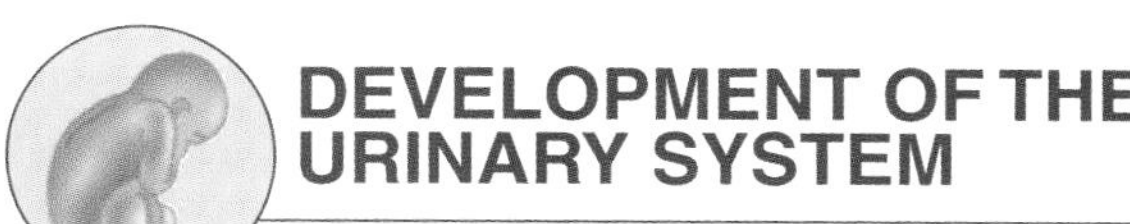

DEVELOPMENT OF THE URINARY SYSTEM

▶ OBJECTIVE

Describe the development of the urinary system.

Starting in the third week of fetal development, a portion of the mesoderm along the posterior aspect of the embryo, the **intermediate mesoderm,** differentiates into the kidneys. The intermediate mesoderm is located in paired elevations called **urogenital ridges.** Three pairs of kidneys form within the intermediate mesoderm in succession: the pronephros, the mesonephros, and the metanephros (Figure 26.23). Only the last pair remains as the functional kidneys of the newborn.

The first kidney to form, the **pronephros** (prō-NEF-rōs; *pro-* = before; *-nephros* = kidney), is the most superior of the three and has an associated **pronephric duct.** This duct empties into the **cloaca,** the expanded terminal part of the hindgut, which functions as a common outlet for the urinary, digestive, and reproductive ducts. The pronephros begins to degenerate during the fourth week and is completely gone by the sixth week.

The second kidney, the **mesonephros** (mez′-ō-NEF-rōs; *meso-* = middle), replaces the pronephros. The retained portion of the pronephric duct, which connects to the mesonephros, develops into the **mesonephric duct.** The mesonephros begins to degenerate by the sixth week and is almost gone by the eighth week.

At about the fifth week, a mesodermal outgrowth, called a **ureteric bud** (ū-rē-TER-ik), develops from the distal portion of the mesonephric duct near the cloaca. The **metanephros** (met-a-NEF-rōs; *meta-* = after), or ultimate kidney, develops from the ureteric bud and metanephric mesoderm. The ureteric bud forms the *collecting ducts, calyces, renal pelvis,* and *ureter.* The **metanephric mesoderm** forms the *nephrons* of the kidneys. By the third month, the fetal kidneys begin excreting urine into the surrounding amniotic fluid; indeed, fetal urine makes up most of the amniotic fluid.

During development, the cloaca divides into a **urogenital sinus,** into which urinary and genital ducts empty, and a *rectum* that discharges into the anal canal. The *urinary bladder* develops from the urogenital sinus. In females, the *urethra* develops as a result of lengthening of the short duct that extends from the urinary bladder to the urogenital sinus. In males, the urethra is considerably longer and more complicated, but it is also derived from the urogenital sinus.

Although the metanephric kidneys form in the pelvis, they ascend to their ultimate destination in the abdomen. As they do so, they receive renal blood vessels. Although the inferior blood vessels usually degenerate as superior ones appear, sometimes the inferior vessels do not degenerate. Consequently, some individuals (about 30%) develop multiple renal vessels.

▶ CHECKPOINT

26. Which type of embryonic tissue develops into nephrons?

27. Which tissue gives rise to collecting ducts, calyces, renal pelves, and ureters?

AGING AND THE URINARY SYSTEM

▶ OBJECTIVE

Describe the effects of aging on the urinary system.

With aging, the kidneys shrink in size, have a decreased blood flow, and filter less blood. These age-related changes in kidney size and function seem to be linked to a progressive reduction in blood supply to the kidneys as an individual gets older; for example, blood vessels such as the glomeruli become damaged or decrease in number. The mass of the two kidneys decreases from an average of nearly 300 g in 20-year-olds to less than 200 g by age 80, a decrease of about one-third. Likewise, renal blood flow and filtration rate decline by 50% between ages 40 and 70. By age 80, about 40% of glomeruli are not functioning and thus filtration, reabsorption, and secretion decrease. Kidney diseases that become more common with age include acute and chronic kidney inflammations and renal calculi (kidney stones). Because the sensation of thirst diminishes with age, older individuals also are susceptible to dehydration. Urinary bladder changes that occur with aging include a reduction in size and capacity and weakening of the muscles. Urinary tract infections are more common among the elderly, as are

Figure 26.23 Development of the urinary system.

Three pairs of kidneys form within intermediate mesoderm in succession: pronephros, mesonephros, and metanephros.

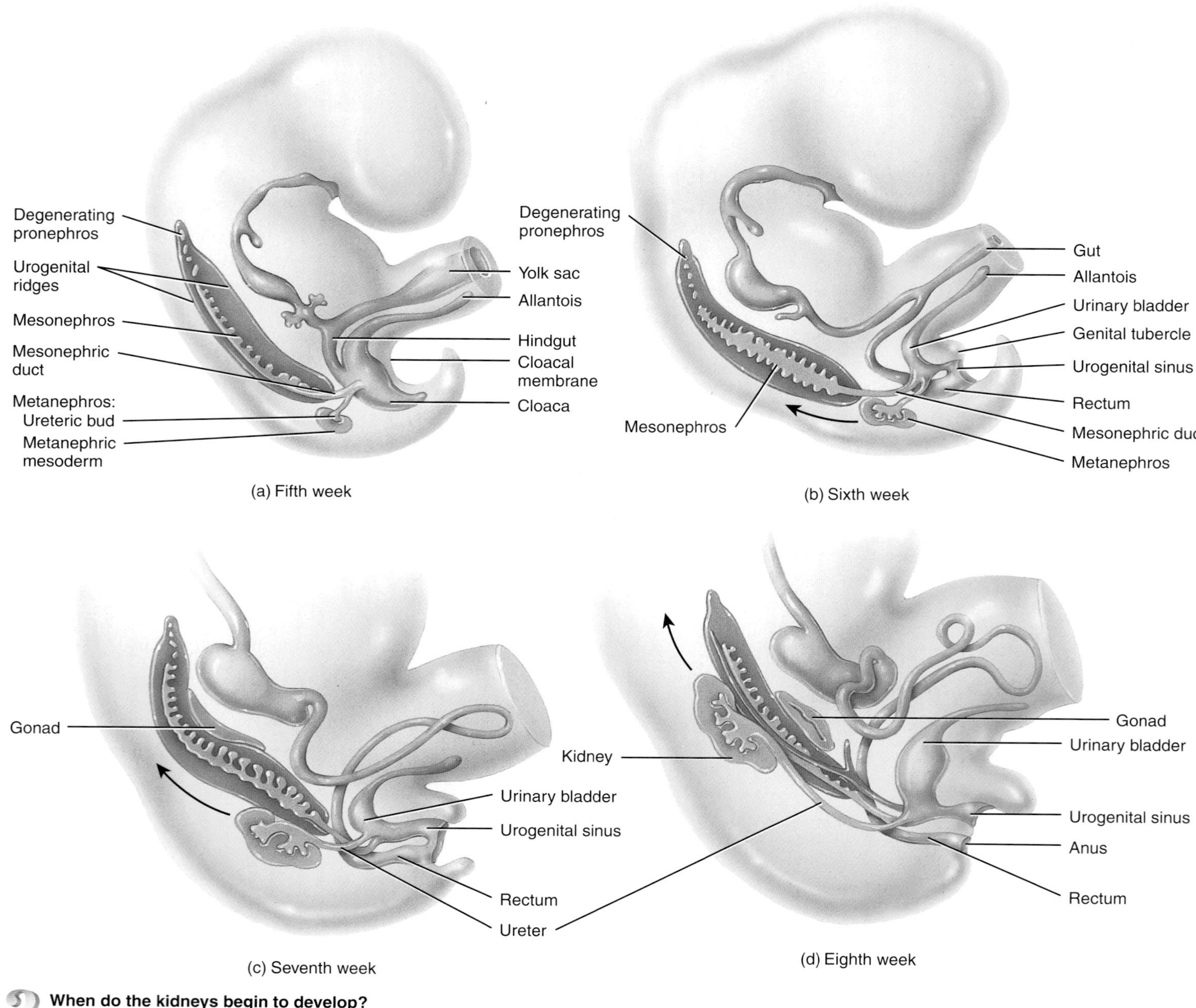

When do the kidneys begin to develop?

polyuria (excessive urine production), nocturia (excessive urination at night), increased frequency of urination, dysuria (painful urination), urinary retention or incontinence, and hematuria (blood in the urine).

▶ CHECKPOINT

28. To what extent do kidney mass and filtration rate decrease with age?

• • •

To appreciate the many ways that the urinary system contributes to homeostasis of other body systems, examine *Focus on Homeostasis: The Urinary System* on page 1029. Next, in Chapter 27, we will see how the kidneys and lungs contribute to maintenance of homeostasis of body fluid volume, electrolyte levels in body fluids, and acid–base balance.

FOCUS ON HOMEOSTASIS

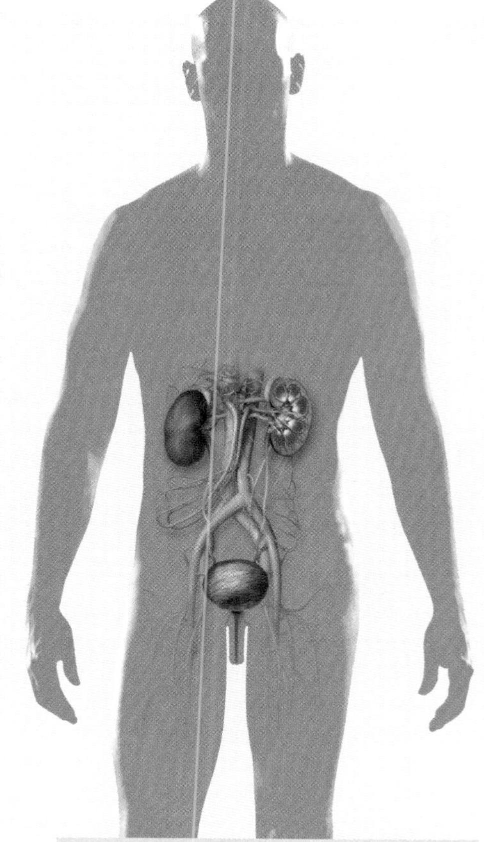

The Urinary System

BODY SYSTEM	CONTRIBUTION OF THE URINARY SYSTEM
For all body systems 	Kidneys regulate the volume, composition, and pH of body fluids by removing wastes and excess substances from blood and excreting them in urine; the ureters transport urine from the kidneys to the urinary bladder, which stores urine until it is eliminated through the urethra.
Integumentary system 	Kidneys and skin both contribute to synthesis of calcitriol, the active form of vitamin D.
Skeletal system	Kidneys help adjust levels of blood calcium and phosphates, needed for building extracellular bone matrix.
Muscular system 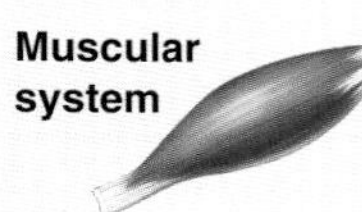	Kidneys help adjust level of blood calcium, needed for contraction of muscle.
Nervous system 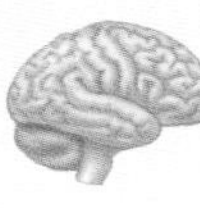	Kidneys perform gluconeogenesis, which provides glucose for ATP production in neurons, especially during fasting or starvation.
Endocrine system 	Kidneys participate in synthesis of calcitriol, the active form of vitamin D, and release erythropoietin, the hormone that stimulates production of red blood cells.
Cardiovascular system 	By increasing or decreasing their reabsorption of water filtered from blood, the kidneys help adjust blood volume and blood pressure; renin released by juxtaglomerular cells in the kidneys raises blood pressure; some bilirubin from hemoglobin breakdownis converted to a yellow pigment (urobilin), which is excreted in urine.
Lymphatic system and immunity 	By increasing or decreasing their reabsorption of water filtered from blood, the kidneys help adjust the volume of interstitial fluid and lymph; urine flushes microbes out of the urethra.
Respiratory system 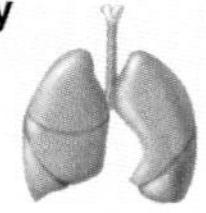	Kidneys and lungs cooperate in adjusting pH of body fluids.
Digestive system 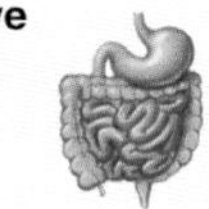	Kidneys help synthesize calcitriol, the active form of vitamin D, which is needed for absorption of dietary calcium.
Reproductive systems 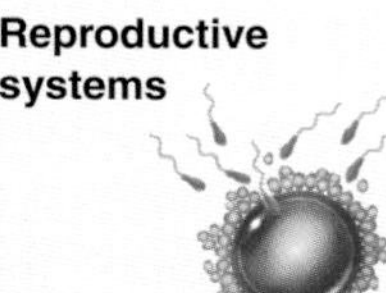	In males, the portion of the urethra that extends through the prostate and penis is a passageway for semen as well as urine.

DISORDERS: HOMEOSTATIC IMBALANCES

Renal Calculi

The crystals of salts present in urine occasionally precipitate and solidify into insoluble stones called **renal calculi** (*calculi* = pebbles) or **kidney stones.** They commonly contain crystals of calcium oxalate, uric acid, or calcium phosphate. Conditions leading to calculus formation include the ingestion of excessive calcium, low water intake, abnormally alkaline or acidic urine, and overactivity of the parathyroid glands. When a stone lodges in a narrow passage, such as a ureter, the pain can be intense. **Shock-wave lithotripsy** (LITH-ō-trip′-sē; *litho* = stone) is a procedure that uses high-energy shock waves to disintegrate kidney stones and offers an alternative to surgical removal. Once the kidney stone is located using x rays, a device called a *lithotripter* delivers brief, high-intensity sound waves through a water- or gel-filled cushion placed under the back. Over a period of 30 to 60 minutes, 1000 or more shock waves pulverize the stone, creating fragments that are small enough to wash out in the urine.

Urinary Tract Infections

The term **urinary tract infection (UTI)** is used to describe either an infection of a part of the urinary system or the presence of large numbers of microbes in urine. UTIs are more common in females due to the shorter length of the urethra. Symptoms include painful or burning urination, urgent and frequent urination, low back pain, and bedwetting. UTIs include *urethritis* (inflammation of the urethra), *cystitis* (inflammation of the urinary bladder), and *pyelonephritis* (inflammation of the kidneys). If pyelonephritis becomes chronic, scar tissue can form in the kidneys and severely impair their function. Drinking cranberry juice can prevent the attachment of *E. coli* bacteria to the lining of the urinary bladder so that they are more readily flushed away during urination.

Glomerular Diseases

A variety of conditions may damage the kidney glomeruli, either directly or indirectly because of disease elsewhere in the body. Typically, the filtration membrane sustains damage, and its permeability increases.

Glomerulonephritis is an inflammation of the kidney that involves the glomeruli. One of the most common causes is an allergic reaction to the toxins produced by streptococcal bacteria that have recently infected another part of the body, especially the throat. The glomeruli become so inflamed, swollen, and engorged with blood that the filtration membranes allow blood cells and plasma proteins to enter the filtrate. As a result, the urine contains many erythrocytes (hematuria) and a lot of protein. The glomeruli may be permanently damaged, leading to chronic renal failure.

Nephrotic syndrome is a condition characterized by *proteinuria* (protein in the urine) and *hyperlipidemia* (high blood levels of cholesterol, phospholipids, and triglycerides). The proteinuria is due to an increased permeability of the filtration membrane, which permits proteins, especially albumin, to escape from blood into urine. Loss of albumin results in *hypoalbuminemia* (low blood albumin level) once liver production of albumin fails to meet increased urinary losses. Edema, usually seen around the eyes, ankles, feet, and abdomen, occurs in nephrotic syndrome because loss of albumin from the blood decreases blood colloid osmotic pressure. Nephrotic syndrome is associated with several glomerular diseases of unknown cause, as well as with systemic disorders such as diabetes mellitus, systemic lupus erythematosus (SLE), a variety of cancers, and AIDS.

Renal Failure

Renal failure is a decrease or cessation of glomerular filtration. In **acute renal failure (ARF),** the kidneys abruptly stop working entirely (or almost entirely). The main feature of ARF is the suppression of urine flow, usually characterized either by *oliguria* (daily urine output between 50 mL and 250 mL), or by *anuria* (daily urine output less than 50 mL). Causes include low blood volume (for example, due to hemorrhage), decreased cardiac output, damaged renal tubules, kidney stones, the dyes used to visualize blood vessels in angiograms, nonsteroidal anti-inflammatory drugs, and some antibiotic drugs. It is also common in people who suffer a devastating illness or overwhelming traumatic injury; in such cases it may be related to a more general organ failure known as *multiple organ dysfunction syndrome (MODS).*

Renal failure causes a multitude of problems. There is edema due to salt and water retention and acidosis due to an inability of the kidneys to excrete acidic substances. In the blood, urea builds up due to impaired renal excretion of metabolic waste products and potassium level rises, which can lead to cardiac arrest. Often, there is anemia because the kidneys no longer produce enough erythropoietin for adequate red blood cell production. Because the kidneys are no longer able to convert vitamin D to calcitriol, which is needed for adequate calcium absorption from the small intestine, osteomalacia also may occur.

Chronic renal failure (CRF) refers to a progressive and usually irreversible decline in glomerular filtration rate (GFR). CRF may result from chronic glomerulonephritis, pyelonephritis, polycystic kidney disease, or traumatic loss of kidney tissue. CRF develops in three stages. In the first stage, *diminished renal reserve,* nephrons are destroyed until about 75% of the functioning nephrons are lost. At this stage, a person may have no signs or symptoms because the remaining nephrons enlarge and take over the function of those that have been lost. Once 75% of the nephrons are lost, the person enters the second stage, called *renal insufficiency,* characterized by a decrease in GFR and increased blood levels of nitrogen-containing wastes and creatinine. Also, the kidneys cannot effectively concentrate or dilute the urine. The final stage, called *end-stage renal failure,* occurs when about 90% of the nephrons have been lost. At this stage, GFR diminishes to 10–15% of normal, oliguria is present, and blood levels of nitrogen-containing wastes and creatinine increase further. People with end-stage renal failure need dialysis therapy and are possible candidates for a kidney transplant operation.

Polycystic Kidney Disease

Polycystic kidney disease (PKD) is one of the most common inherited disorders. In PKD, the kidney tubules become riddled with hundreds or thousands of cysts (fluid-filled cavities). In addition, inappropriate apoptosis (programmed cell death) of cells in noncystic tubules leads to progressive impairment of renal function and eventually to end-stage renal failure.

People with PKD also may have cysts and apoptosis in the liver, pancreas, spleen, and gonads; increased risk of cerebral aneurysms; heart valve defects; and diverticuli in the colon. Typically, symptoms are not noticed until adulthood, when patients may have back pain,

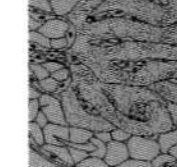

urinary tract infections, blood in the urine, hypertension, and large abdominal masses. Using drugs to restore normal blood pressure, restricting protein and salt in the diet, and controlling urinary tract infections may slow progression to renal failure.

Urinary Bladder Cancer

Each year, nearly 12,000 Americans die from **urinary bladder cancer.** It generally strikes people over 50 years of age and is three times more likely to develop in males than females. The disease is typically painless as it develops, but in most cases blood in the urine is a primary sign of the disease. Less often, people experience painful and/or frequent urination.

As long as the disease is identified early and treated promptly, the prognosis is favorable. Fortunately, about 75% of the urinary bladder cancers are confined to the epithelium of the urinary bladder and are easily removed by surgery. The lesions tend to be low-grade, meaning that they have only a small potential for metastasis.

Urinary bladder cancer is frequently the result of a carcinogen. About half of all cases occur in people who smoke or have at some time smoked cigarettes. The cancer also tends to develop in people who are exposed to chemicals called aromatic amines. Workers in the leather, dye, rubber, and aluminum industries, as well as painters, are often exposed to these chemicals.

MEDICAL TERMINOLOGY

Azotemia (az-ō-TĒ-mē-a; *azot-* = nitrogen; *-emia* = condition of blood) Presence of urea or other nitrogen-containing substances in the blood.

Cystocele (SIS-tō-sēl; *cysto-* = bladder; *-cele* = hernia or rupture) Hernia of the urinary bladder.

Diabetic kidney disease A disorder caused by diabetes mellitus in which glomeruli are damaged. The result is the leakage of proteins into the urine and a reduction in the ability of the kidney to remove water and waste.

Dysuria (dis-Ū-rē-a; *dys-* = painful; *uria* = urine) Painful urination.

Enuresis (en′-ū-RĒ-sis = to void urine) Involuntary voiding of urine after the age at which voluntary control has typically been attained.

Hydronephrosis (hi′-drō-ne-FRŌ-sis; *hydro-* = water; *nephros* = kidney; *-osis* = condition) Swelling of the kidney due to dilation of the renal pelvis and calyces as a result of an obstruction to the flow of urine. It may be due to a congenital abnormality, a narrowing of the ureter, a kidney stone, or an enlarged prostate.

Intravenous pyelogram (in′-tra-VĒ-nus PĪ-e-lō-gram′; *intra-* = within; *veno-* = vein; *pyelo-* = pelvis of kidney; *-gram* = record) (or **IVP**) Radiograph (x ray) of the kidneys, ureters, and urinary bladder after venous injection of a radiopaque contrast medium.

Nephropathy (ne-FROP-a-thē; *neph-* = kidney; *-pathos* = suffering) Any disease of the kidneys. Types include analgesic (from long-term and excessive use of drugs such as ibuprofen), lead (from ingestion of lead-based paint), and solvent (from carbon tetrachloride and other solvents).

Nocturnal enuresis (nok-TUR-nal en′-ū-RĒ-sis) Discharge of urine during sleep, resulting in bed-wetting; occurs in about 15% of 5-year-old children and generally resolves spontaneously, afflicting only about 1% of adults. It may have a genetic basis, as bed-wetting occurs more often in identical twins than in fraternal twins and more often in children whose parents or siblings were bed-wetters. Possible causes include smaller-than-normal bladder capacity, failure to awaken in response to a full bladder, and above-normal production of urine at night. Also referred to as **nocturia.**

Polyuria (pol′-ē-Ū-rē-a; *poly-* = too much) Excessive urine formation. It may occur in conditions such as diabetes mellitus and glomerulonephritis.

Stricture (STRIK-chur) Narrowing of the lumen of a canal or hollow organ, as may occur in the ureter, urethra, or any other tubular structure in the body.

Uremia (ū-RĒ-mē-a; *emia* = condition of blood) Toxic levels of urea in the blood resulting from severe malfunction of the kidneys.

Urinary retention A failure to completely or normally void urine; may be due to an obstruction in the urethra or neck of the urinary bladder, to nervous contraction of the urethra, or to lack of urge to urinate. In men, an enlarged prostate may constrict the urethra and cause urinary retention. If urinary retention is prolonged, a catheter (slender rubber drainage tube) must be placed into the urethra to drain the urine.

STUDY OUTLINE

INTRODUCTION (p. 993)

1. The organs of the urinary system are the kidneys, ureters, urinary bladder, and urethra.
2. After the kidneys filter blood and return most water and many solutes to the bloodstream, the remaining water and solutes constitute urine.

OVERVIEW OF KIDNEY FUNCTIONS (p. 993)

1. The kidneys regulate blood ionic composition, blood osmolarity, blood volume, blood pressure, and blood pH.
2. The kidneys also perform gluconeogenesis, release calcitriol and erythropoietin, and excrete wastes and foreign substances.

ANATOMY AND HISTOLOGY OF THE KIDNEYS (p. 994)

1. The kidneys are retroperitoneal organs attached to the posterior abdominal wall.
2. Three layers of tissue surround the kidneys: renal capsule, adipose capsule, and renal fascia.
3. Internally, the kidneys consist of a renal cortex, a renal medulla, renal pyramids, renal papillae, renal columns, calyces, and a renal pelvis.
4. Blood flows into the kidney through the renal artery and successively into segmental, interlobar, arcuate, and interlobular arteries; afferent arterioles; glomerular capillaries; efferent arterioles; per-itubular capillaries and vasa recta; and interlobular, arcuate, and interlobar veins before flowing out of the kidney through the renal vein.
5. Vasomotor nerves from the sympathetic division of the autonomic nervous system supply kidney blood vessels; they help regulate the flow of blood through the kidney.
6. The nephron is the functional unit of the kidneys. A nephron consists of a renal corpuscle (glomerulus and glomerular or Bowman's capsule) and a renal tubule.
7. A renal tubule consists of a proximal convoluted tubule, a loop of Henle, and a distal convoluted tubule, which drains into a collecting duct (shared by several nephrons). The loop of Henle consists of a descending limb and an ascending limb.
8. A cortical nephron has a short loop that dips only into the superficial region of the renal medulla; a juxtamedullary nephron has a long loop of Henle that stretches through the renal medulla almost to the renal papilla.
9. The wall of the entire glomerular capsule, renal tubule, and ducts consists of a single layer of epithelial cells. The epithelium has distinctive histological features in different parts of the tubule. Table 26.1 on page 1002 summarizes the histological features of the renal tubule and collecting duct.
10. The juxtaglomerular apparatus (JGA) consists of the juxtaglomerular cells of an afferent arteriole and the macula densa of the final portion of the ascending limb of the loop of Henle.

OVERVIEW OF RENAL PHYSIOLOGY (p. 1003)

1. Nephrons perform three basic tasks: glomerular filtration, tubular secretion, and tubular reabsorption.

GLOMERULAR FILTRATION (p. 1004)

1. Fluid that enters the capsular space is glomerular filtrate.
2. The filtration membrane consists of the glomerular endothelium, basal lamina, and filtration slits between pedicels of podocytes.
3. Most substances in blood plasma easily pass through the glomerular filter. However, blood cells and most proteins normally are not filtered.
4. Glomerular filtrate amounts to up to 180 liters of fluid per day. This large amount of fluid is filtered because the filter is porous and thin, the glomerular capillaries are long, and the capillary blood pressure is high.
5. Glomerular blood hydrostatic pressure (GBHP) promotes filtration; capsular hydrostatic pressure (CHP) and blood colloid osmotic pressure (BCOP) oppose filtration. Net filtration pressure (NFP) = GBHP − CHP − BCOP. NFP is about 10 mmHg.
6. Glomerular filtration rate (GFR) is the amount of filtrate formed in both kidneys per minute; it is normally 105–125 mL/min.
7. Glomerular filtration rate depends on renal autoregulation, neural regulation, and hormonal regulation. Table 26.2 on page 1008 summarizes regulation of GFR.

TUBULAR REABSORPTION AND TUBULAR SECRETION (p. 1008)

1. Tubular reabsorption is a selective process that reclaims materials from tubular fluid and returns them to the bloodstream. Reabsorbed substances include water, glucose, amino acids, urea, and ions, such as sodium, chloride, potassium, bicarbonate, and phosphate (Table 26.3 on page 1009).
2. Some substances not needed by the body are removed from the blood and discharged into the urine via tubular secretion. Included are ions (K^+, H^+, and NH_4^+), urea, creatinine, and certain drugs.
3. Reabsorption routes include both paracellular (between tubule cells) and transcellular (across tubule cells) routes.
4. The maximum amount of a substance that can be reabsorbed per unit time is called the transport maximum (T_m).
5. About 90% of water reabsorption is obligatory; it occurs via osmosis, together with reabsorption of solutes, and is not hormonally regulated. The remaining 10% is facultative water reabsorption, which varies according to body needs and is regulated by ADH.
6. Na^+ are reabsorbed throughout the basolateral membrane via primary active transport.
7. In the proximal convoluted tubule, sodium ions are reabsorbed through the apical membranes via Na^+-glucose symporters and Na^+/H^+ antiporters; water is reabsorbed via osmosis; Cl^-, K^+, Ca^{2+}, Mg^{2+}, and urea are reabsorbed via passive diffusion; and NH_3 and NH_4^+ are secreted.
8. The loop of Henle reabsorbs 20–30% of the filtered Na^+, K^+, Ca^{2+}, and HCO_3^-; 35% of the filtered Cl^-; and 15% of the filtered water.
9. The distal convoluted tubule reabsorbs sodium and chloride ions via Na^+-Cl^- symporters.
10. In the collecting duct, principal cells reabsorb Na^+ and secrete K^+; intercalated cells reabsorb K^+ and HCO_3^- and secrete H^+.
11. Angiotensin II, aldosterone, antidiuretic hormone, and atrial natriuretic peptide regulate solute and water reabsorption, as summarized in Table 26.4 on page 1016.

PRODUCTION OF DILUTE AND CONCENTRATED URINE (p. 1016)

1. In the absence of ADH, the kidneys produce dilute urine; renal tubules absorb more solutes than water.
2. In the presence of ADH, the kidneys produce concentrated urine; large amounts of water are reabsorbed from the tubular fluid into interstitial fluid, increasing solute concentration of the urine.
3. The countercurrent mechanism establishes an osmotic gradient in the interstitial fluid of the renal medulla that enables production of concentrated urine when ADH is present.

EVALUATION OF KIDNEY FUNCTION (p. 1019)

1. A urinalysis is an analysis of the volume and physical, chemical, and microscopic properties of a urine sample. Table 26.5 on page 1021 summarizes the principal physical characteristics of normal urine.
2. Chemically, normal urine contains about 95% water and 5% solutes. The solutes normally include urea, creatinine, uric acid, urobilinogen, and various ions.

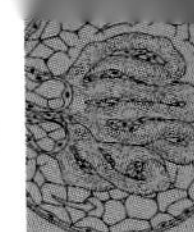

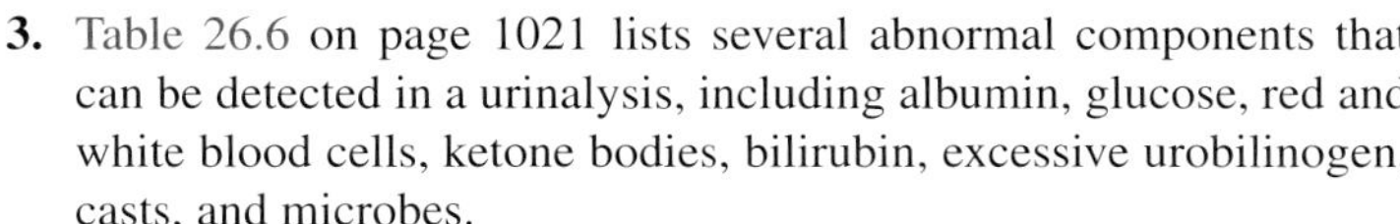

3. Table 26.6 on page 1021 lists several abnormal components that can be detected in a urinalysis, including albumin, glucose, red and white blood cells, ketone bodies, bilirubin, excessive urobilinogen, casts, and microbes.
4. Renal clearance refers to the ability of the kidneys to clear (remove) a specific substance from blood.

URINE TRANSPORTATION, STORAGE, AND ELIMINATION (p. 1023)

1. The ureters are retroperitoneal and consist of a mucosa, muscularis, and adventitia. They transport urine from the renal pelvis to the urinary bladder, primarily via peristalsis.
2. The urinary bladder is located in the pelvic cavity posterior to the pubic symphysis; its function is to store urine before micturition.
3. The urinary bladder consists of a mucosa with rugae, a muscularis (detrusor muscle), and an adventitia (serosa over the superior surface).
4. The micturition reflex discharges urine from the urinary bladder via parasympathetic impulses that cause contraction of the detrusor muscle and relaxation of the internal urethral sphincter muscle and via inhibition of impulses in somatic motor neurons to the external urethral sphincter.
5. The urethra is a tube leading from the floor of the urinary bladder to the exterior. Its anatomy and histology differ in females and males. In both sexes, the urethra functions to discharge urine from the body; in males, it discharges semen as well.

WASTE MANAGEMENT IN OTHER BODY SYSTEMS (p. 1026)

1. Besides the kidneys, several other tissues, organs, and processes temporarily confine wastes, transport waste materials for disposal, recycle materials, and excrete excess or toxic substances.
2. Buffers bind excess H^+, the blood transports wastes, the liver converts toxic substances into less toxic ones, the lungs exhale CO_2, sweat glands help eliminate excess heat, and the gastrointestinal tract eliminates solid wastes.

DEVELOPMENT OF THE URINARY SYSTEM (p. 1027)

1. The kidneys develop from intermediate mesoderm.
2. The kidneys develop in the following sequence: pronephros, mesonephros, and metanephros. Only the metanephros remains and develops into a functional kidney.

AGING AND THE URINARY SYSTEM (p. 1027)

1. With aging, the kidneys shrink in size, have a decreased blood flow, and filter less blood.
2. Common problems related to aging include urinary tract infections, increased frequency of urination, urinary retention or incontinence, and renal calculi.

SELF-QUIZ QUESTIONS

Fill in the blanks in the following statements.

1. The renal corpuscle consists of the ____ and ____.
2. Discharge of urine from the urinary bladder is called ____.

Indicate whether the following statements are true or false.

3. The most superficial region of the internal kidney is the renal medulla.
4. When dilute urine is being formed, the osmolarity of the fluid in the tubular lumen increases as it flows down the descending limb of the loop of Henle, decreases as it flows up the ascending limb, and continues to decrease as it flows through the rest of the nephron and collecting duct.

Choose the one best answer to the following questions.

5. Which of the following statements are *correct*? (1) Glomerular filtration rate (GFR) is directly related to the pressures that determine net filtration pressure. (2) Angiotensin II and atrial natriuretic peptide help regulate GFR. (3) Mechanisms that regulate GFR work by adjusting blood flow into and out of the glomerulus and by altering the glomerular capillary surface area available for filtration. (4) GFR increases when blood flow into glomerular capillaries decreases. (5) Normally, GFR increases very little when systemic blood pressure rises. (a) 1, 2, and 3; (b) 2, 3, and 4; (c) 3, 4, and 5; (d) 1, 2, 3, and 5; (e) 2, 3, 4, and 5.
6. Which of the following hormones affect Na^+, Cl^-, and water reabsorption and K^+ secretion by the renal tubules? (1) angiotensin II, (2) aldosterone, (3) ADH, (4) atrial natriuretic peptide, (5) thyroid hormone. (a) 1, 3, and 5; (b) 2, 3, and 4; (c) 2, 4, and 5; (d) 1, 2, 4, and 5; (e) 1, 2, 3, and 4.
7. Which of the following are features of the renal corpuscle that enhance its filtering capacity? (1) large glomerular capillary surface area, (2) thick, selectively permeable filtration membrane, (3) high capsular hydrostatic pressure, (4) high glomerular capillary pressure, (5) mesangial cells regulating the filtering surface area. (a) 1, 2, and 3; (b) 2, 4, and 5; (c) 1, 4, and 5; (d) 2, 3, and 4; (e) 2, 3, and 5.
8. Given the following values, calculate the net filtration pressure: (1) glomerular blood hydrostatic pressure = 40 mmHg, (2) capsular hydrostatic pressure = 10 mmHg, (3) blood colloid osmotic pressure = 30 mmHg. (a) −20 mmHg, (b) 0 mmHg, (c) 20 mmHg, (d) 60 mmHg, (e) 80 mmHg.
9. The micturition reflex (1) is initiated by stretch receptors in the ureters, (2) relies on parasympathetic impulses from the micturition center in S2 and S3, (3) results in contraction of the detrusor muscle, (4) results in contraction of the internal urethral sphincter muscle, (5) inhibits motor neurons in the external urethral sphincter. (a) 1, 2, 3, 4, and 5; (b) 1, 3, and 4; (c) 2, 3, 4, and 5; (d) 2 and 5; (e) 2, 3, and 5.
10. Which of the following are mechanisms that control GFR? (1) renal autoregulation, (2) neural regulation, (3) hormonal regulation, (4) chemical regulation of ions, (5) presence or absence of a transporter. (a) 1, 2, and 3; (b) 2, 3, and 4; (c) 3, 4, and 5; (d) 1, 3, and 5; (e) 1, 3, and 4.

11. Place the route of blood flow through the kidney in the correct order: (a) segmental arteries, (b) vasa recta, (c) arcuate arteries, (d) peritubular venules, (e) interlobular veins, (f) renal vein, (g) renal artery, (h) interlobar arteries, (i) peritubular capillaries, (j) efferent arterioles, (k) interlobar veins, (l) glomeruli, (m) arcuate veins, (n) afferent arterioles, (o) interlobular arteries.

12. Place the route of filtrate flow in the correct order from its origin to the ureter: (a) minor calyx, (b) ascending limb of loop of Henle, (c) papillary duct, (d) distal convoluted tubule, (e) major calyx, (f) descending limb of loop of Henle, (g) proximal convoluted tubule, (h) collecting duct, (i) renal pelvis.

13. Match the following:

____(a) cells in the last portion of the distal convoluted tubule and in the collecting ducts; regulated by ADH and aldosterone	(1) podocytes
____(b) a capillary network lying in the glomerular capsule and functioning in filtration	(2) glomerulus
____(c) the functional unit of the kidney	(3) renal corpuscle
____(d) drains into a collecting duct	(4) proximal convoluted tubule
____(e) combined glomerulus and glomerular capsule; where plasma is filtered	(5) distal convoluted tubule
____(f) the visceral layer of the glomerular capsule consisting of modified simple squamous epithelial cells	(6) juxtaglomerular cells
____(g) cells of the final portion of the ascending limb of the loop of Henle that make contact with the afferent arteriole	(7) macula densa
____(h) site of obligatory water reabsorption	(8) principal cells
____(i) pores in the glomerular endothelial cells that allow filtration of blood solutes but not blood cells and platelets	(9) intercalated cells
____(j) can secrete H^+ against a concentration gradient	(10) nephron
____(k) modified smooth muscle cells in the wall of the afferent arteriole	(11) fenestrations

14. Match the following:

____(a) measure of blood nitrogen resulting from the catabolism and deamination of amino acids	(1) incontinence
____(b) produced from the catabolism of creatine phosphate in skeletal muscle	(2) renal calculi
____(c) volume of blood that is cleared of a substance per unit of time	(3) plasma creatinine
____(d) can result from diabetes mellitus	(4) BUN test
____(e) insoluble stones of crystallized salts	(5) albuminuria
____(f) usually indicates a pathological condition	(6) glucosuria
____(g) lack of voluntary control of micturition	(7) renal plasma clearance
____(h) can be caused by damage to the filtration membranes	(8) hematuria

15. Match the following:

____(a) membrane proteins that function as water channels	(1) angiotensin II
____(b) a secondary active transport process that achieves Na^+ reabsorption, returns filtered HCO_3^- and water to the peritubular capillaries, and secretes H^+	(2) atrial natriuretic peptide
____(c) stimulates principal cells to secrete more K^+ into tubular fluid and absorb more Na^+ and Cl^- into tubular fluid	(3) Na^+ symporters
____(d) enzyme secreted by juxtaglomerular cells	(4) Na^+/H^+ antiporters
____(e) reduces glomerular filtration rate; increases blood volume and pressure	(5) aquaporins
____(f) inhibits Na^+ and H_2O reabsorption in the proximal convoluted tubules and collecting ducts	(6) aldosterone
____(g) regulates facultative water reabsorption by increasing the water permeability of principal cells in the distal convoluted tubules and collecting ducts	(7) ADH
____(h) reabsorb Na^+ together with a variety of other solutes	(8) renin

CRITICAL THINKING QUESTIONS

1. Imagine the discovery of a new toxin that blocks renal tubule reabsorption but does not affect filtration. Predict the short-term effects of this toxin.
2. For each of the following urinalysis results, indicate whether you should be concerned or not and why: (a) dark yellow urine that is turbid; (b) ammonia-like odor of the urine; (c) presence of excessive albumin; (d) presence of epithelial cell casts; (e) pH of 5.5; (f) hematuria.
3. Bruce is experiencing sudden, rhythmic waves of pain in his groin area. He has noticed that, although he is consuming fluids, his urine output has decreased. From what condition is Bruce suffering? How is it treated? How can he prevent future episodes?

ANSWERS TO FIGURE QUESTIONS

26.1 The kidneys, ureters, urinary bladder, and urethra are the components of the urinary system.

26.2 The kidneys are retroperitoneal because they are posterior to the peritoneum.

26.3 Blood vessels, lymphatic vessels, nerves, and a ureter pass through the renal hilum.

26.4 About 1200 mL of blood enters the renal arteries each minute.

26.5 Cortical nephrons have glomeruli in the superficial renal cortex, and their short loops of Henle penetrate only into the superficial renal medulla. Juxtamedullary nephrons have glomeruli deep in the renal cortex, and their long loops of Henle extend through the renal medulla nearly to the renal papilla.

26.6 This section must pass through the renal cortex because there are no renal corpuscles in the renal medulla.

26.7 Secreted penicillin is being removed from the bloodstream.

26.8 Endothelial fenestrations (pores) in glomerular capillaries are too small for red blood cells to pass through.

26.9 Obstruction of the right ureter would increase CHP and thus decrease NFP in the right kidney; the obstruction would have no effect on the left kidney.

26.10 *Auto* means self; tubuloglomerular feedback is an example of autoregulation because it takes place entirely within the kidneys.

26.11 The tight junctions between tubule cells form a barrier that prevents diffusion of transporter, channel, and pump proteins between the apical and basolateral membranes.

26.12 Glucose enters a PCT cell via a Na^+-glucose symporter in the apical membrane and leaves via facilitated diffusion through the basolateral membrane.

26.13 The electrochemical gradient promotes movement of Na^+ into the tubule cell through the apical membrane antiporters.

26.14 Reabsorption of the solutes creates an osmotic gradient that promotes the reabsorption of water via osmosis.

26.15 This is considered secondary active transport because the symporter uses the energy stored in the concentration gradient of Na^+ between extracellular fluid and the cytosol. No water is reabsorbed here because the thick ascending limb of the loop of Henle is virtually impermeable to water.

26.16 In principal cells, aldosterone stimulates secretion of K^+ and reabsorption of Na^+ by increasing the activity of sodium-potassium pumps and number of leakage channels for Na^+ and K^+.

26.17 Aldosterone and atrial natriuretic peptide influence renal water reabsorption along with ADH.

26.18 Dilute urine is produced when the thick ascending limb of the loop of Henle, the distal convoluted tubule, and the collecting duct reabsorb more solutes than water.

26.19 The high osmolarity of interstitial fluid in the renal medulla is due mainly to Na^+, Cl^-, and urea.

26.20 Secretion occurs in the proximal convoluted tubule, the loop of Henle, and the collecting duct.

26.21 Lack of voluntary control over micturition is termed urinary incontinence.

26.22 The three subdivisions of the male urethra are the prostatic urethra, membranous urethra, and spongy urethra.

26.23 The kidneys start to form during the third week of development.

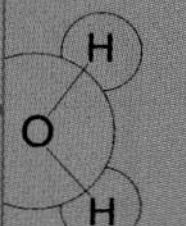

Chapter 27

Fluid, Electrolyte, and Acid–Base Homeostasis

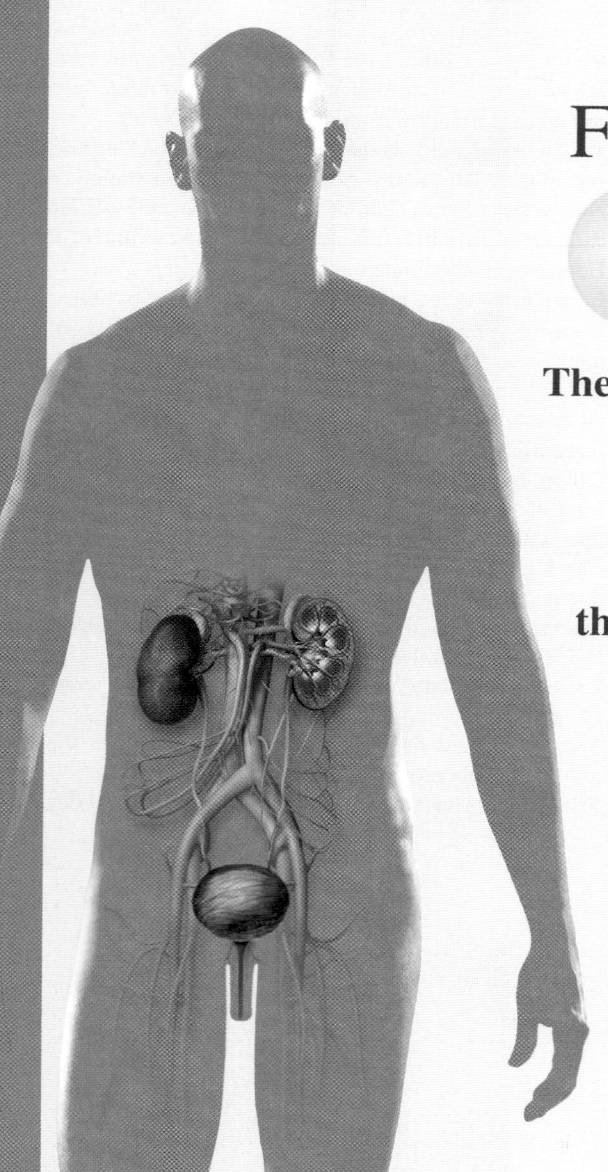

Fluid, Electrolyte, and Acid–Base Homeostasis

The regulation of the volume and composition of body fluids, their distribution throughout the body, and balancing the pH of body fluids is crucial to maintaining overall homeostasis and health.

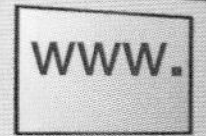

wiley.com/college/apcentral

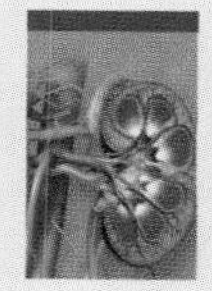

In Chapter 26 you learned how the kidneys form urine. One important function of the kidneys is to help maintain fluid balance in the body. The water and dissolved solutes throughout the body constitute the **body fluids.** Regulatory mechanisms involving the kidneys and other organs normally maintain homeostasis of the body fluids. Malfunction in any or all of them may seriously endanger the functioning of organs throughout the body. In this chapter, we will explore the mechanisms that regulate the volume and distribution of body fluids and examine the factors that determine the concentrations of solutes and the pH of body fluids.

FLUID COMPARTMENTS AND FLUID BALANCE

▶ **OBJECTIVES**

- **Compare the locations of intracellular fluid (ICF) and extracellular fluid (ECF), and describe the various fluid compartments of the body.**
- **Describe the sources of water and solute gain and loss, and explain how each is regulated.**
- **Explain how fluids move between compartments.**

In lean adults, body fluids constitute between 55% and 60% of total body mass in females and males, respectively (Figure 27.1). Body fluids are present in two main "compartments"—inside cells and outside cells. About two-thirds of body fluid is **intracellular fluid (ICF)** (*intra-* = within) or **cytosol,** the fluid within cells. The other third, called **extracellular fluid (ECF)** (*extra-* = outside) is outside cells and includes all other body fluids. About 80% of the ECF is **interstitial fluid** (*inter-* = between), which occupies the microscopic spaces between tissue cells, and 20% of the ECF is **plasma,** the liquid portion of the blood. Other extracellular fluids that are grouped with interstitial fluid include lymph in lymphatic vessels; cerebrospinal fluid in the nervous system; synovial fluid in joints; aqueous humor and vitreous body in the eyes; endolymph and perilymph in the ears; and pleural, pericardial, and peritoneal fluids between serous membranes.

Figure 27.1 **Body fluid compartments.**

The term body fluid refers to body water and its dissolved substances.

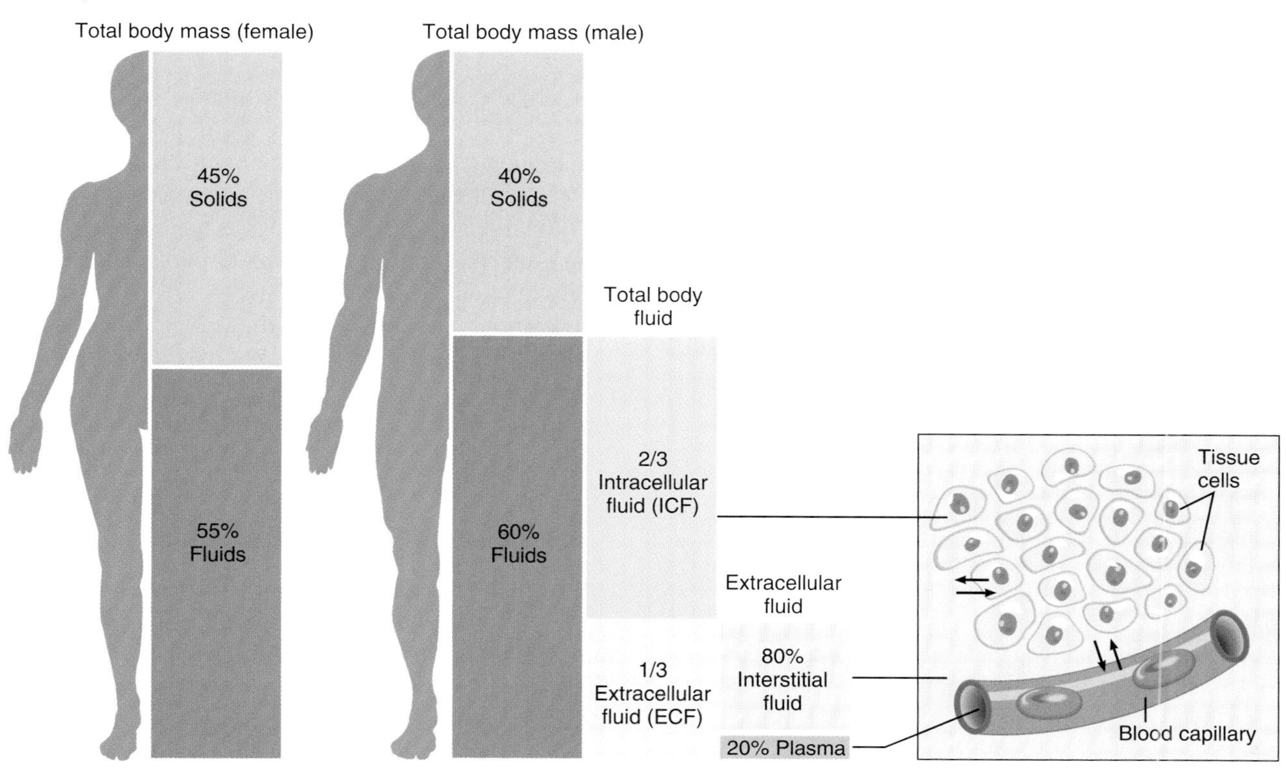

(a) Distribution of body solids and fluids in an average lean, adult female and male

(b) Exchange of water among body fluid compartments

What is the approximate volume of blood plasma in a lean 60-kg male? In a lean 60-kg female? *(Note: One liter of body fluid has a mass of 1 kilogram.)*

Two general "barriers" separate intracellular fluid, interstitial fluid, and blood plasma.

1. The *plasma membrane* of individual cells separates intracellular fluid from the surrounding interstitial fluid. You learned in Chapter 3 that the plasma membrane is a selectively permeable barrier: It allows some substances to cross but blocks the movement of other substances. In addition, active transport pumps work continuously to maintain different concentrations of certain ions in the cytosol and interstitial fluid.

2. *Blood vessel walls* divide the interstitial fluid from blood plasma. Only in capillaries, the smallest blood vessels, are the walls thin enough and leaky enough to permit the exchange of water and solutes between blood plasma and interstitial fluid.

The body is in **fluid balance** when the required amounts of water and solutes are present and are correctly proportioned among the various compartments. **Water** is by far the largest single component of the body, making up 45–75% of total body mass, depending on age and gender.

The processes of filtration, reabsorption, diffusion, and osmosis allow continual exchange of water and solutes among body fluid compartments (Figure 27.1b). Yet, the volume of fluid in each compartment remains remarkably stable. The pressures that promote filtration of fluid from blood capillaries and reabsorption of fluid back into capillaries can be reviewed in Figure 21.7 on page 745. Because osmosis is the primary means of water movement between intracellular fluid and interstitial fluid, the concentration of solutes in these fluids determines the *direction* of water movement. Because most solutes in body fluids are **electrolytes,** inorganic compounds that dissociate into ions, fluid balance is closely related to electrolyte balance. Because intake of water and electrolytes rarely occurs in exactly the same proportions as their presence in body fluids, the ability of the kidneys to excrete excess water by producing dilute urine, or to excrete excess electrolytes by producing concentrated urine, is of utmost importance in the maintenance of homeostasis.

Figure 27.2 Sources of daily water gain and loss under normal conditions. Numbers are average volumes for adults.

Normally, daily water loss equals daily water gain.

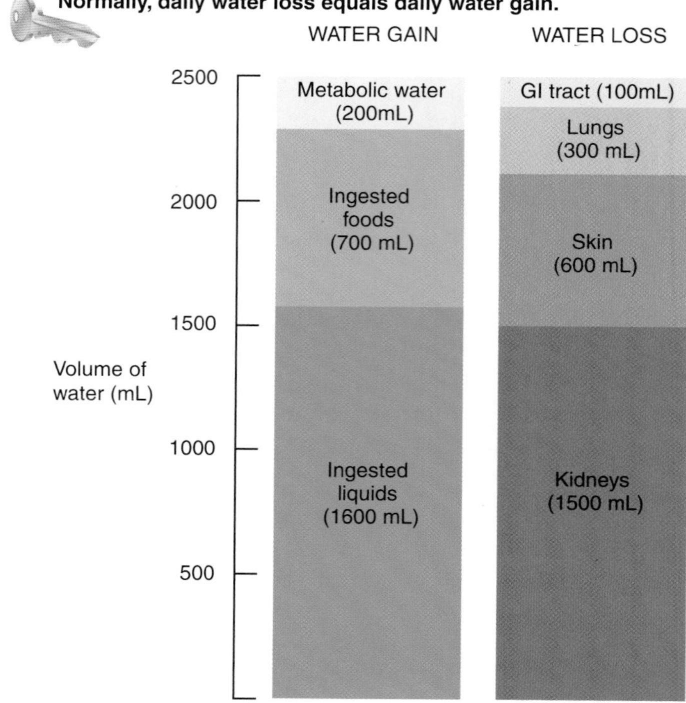

How does each of the following affect fluid balance: Hyperventilation? Vomiting? Fever? Diuretics?

Sources of Body Water Gain and Loss

The body can gain water by ingestion and by metabolic synthesis (Figure 27.2). The main sources of body water are ingested liquids (about 1600 mL) and moist foods (about 700 mL) absorbed from the gastrointestinal (GI) tract, which total about 2300 mL/day. The other source of water is **metabolic water** that is produced in the body mainly when electrons are accepted by oxygen during aerobic cellular respiration (see Figure 25.2 on page 954) and to a smaller extent during dehydration synthesis reactions (see Figure 2.15 on page 45). Metabolic water gain accounts for only 200 mL/day. Daily water gain from these two sources totals about 2500 mL.

Normally, body fluid volume remains constant because water loss equals water gain. Water loss occurs in four ways (Figure 27.2). Each day the kidneys excrete about 1500 mL in urine, the skin evaporates about 600 mL (400 mL through insensible perspiration, sweat that evaporates before it is perceived as moisture, and 200 mL as sweat), the lungs exhale about 300 mL as water vapor, and the gastrointestinal tract eliminates about 100 mL in feces. In women of reproductive age, additional water is lost in menstrual flow. On average, daily water loss totals about 2500 mL. The amount of water lost by a given route can vary considerably over time. For example, water may literally pour from the skin in the form of sweat during strenuous exertion. In other cases, water may be lost in diarrhea during a GI tract infection.

Regulation of Body Water Gain

The volume of metabolic water formed in the body depends entirely on the level of aerobic cellular respiration, which reflects the demand for ATP in body cells. When more ATP is produced, more water is formed. Body water gain is regulated mainly by the volume of water intake, or how much fluid you drink. An area in the hypothalamus known as the **thirst center** governs the urge to drink.

When water loss is greater than water gain, **dehydration**—a decrease in volume and an increase in osmolarity of body fluids—stimulates thirst (Figure 27.3). When body mass decreases by 2% due to fluid loss, mild dehydration exists.

Figure 27.3 Pathways through which dehydration stimulates thirst.

Dehydration occurs when water loss is greater than water gain.

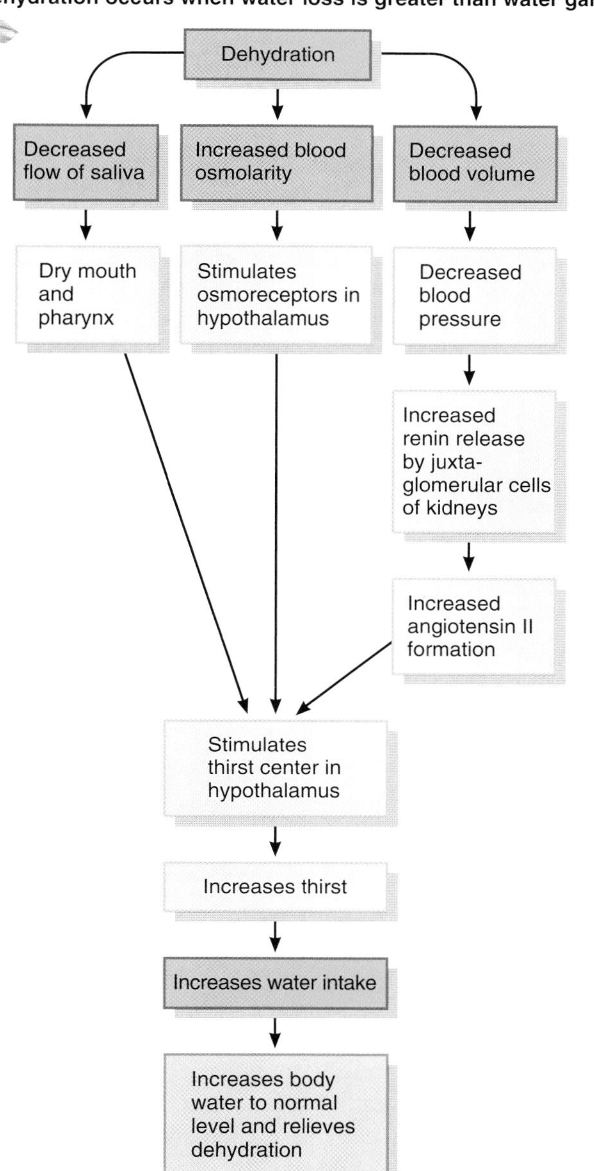

Does regulation of these pathways occur via negative or positive feedback? Why?

A decrease in blood volume causes blood pressure to fall. This change stimulates the kidneys to release renin, which promotes the formation of angiotensin II. Increased nerve impulses from osmoreceptors in the hypothalamus, triggered by increased blood osmolarity, and increased angiotensin II in the blood both stimulate the thirst center in the hypothalamus. Other signals that stimulate thirst come from (1) neurons in the mouth that detect dryness due to a decreased flow of saliva and (2) baroreceptors that detect lowered blood pressure in the heart and blood vessels. As a result, the sensation of thirst increases, which usually leads to increased fluid intake (if fluids are available) and restoration of normal fluid volume. Overall, fluid gain balances fluid loss. Sometimes, however, the sensation of thirst does not occur quickly enough or access to fluids is restricted, and significant dehydration ensues. This happens most often in elderly people, in infants, and in those who are in a confused mental state. When heavy sweating or fluid loss from diarrhea or vomiting occurs, it is wise to start replacing body fluids by drinking fluids even before the sensation of thirst occurs.

Regulation of Water and Solute Loss

Even though the loss of water and solutes through sweating and exhalation increases during exercise, elimination of *excess* body water or solutes occurs mainly by control of their loss in urine. The extent of *urinary salt (NaCl) loss* is the main factor that determines body fluid *volume*. The reason for this is that "water follows solutes" in osmosis, and the two main solutes in extracellular fluid (and in urine) are sodium ions (Na^+) and chloride ions (Cl^-). In a similar way, the main factor that determines body fluid *osmolarity* is the extent of *urinary water loss*.

Because our daily diet contains a highly variable amount of NaCl, urinary excretion of Na^+ and Cl^- must also vary to maintain homeostasis. Hormonal changes regulate the urinary loss of these ions, which in turn affects blood volume. Figure 27.4 depicts the sequence of changes that occur after a salty meal. The increased intake of NaCl produces an increase in plasma levels of Na^+ and Cl^- (the major contributors to osmolarity of extracellular fluid). As a result, the osmolarity of interstitial fluid increases, which causes movement of water from intracellular fluid into interstitial fluid and then into plasma. Such water movement increases blood volume.

The three most important hormones that regulate the extent of renal Na^+ and Cl^- reabsorption (and thus how much is lost in the urine) are **angiotensin II, aldosterone,** and **atrial natriuretic peptide (ANP).** When your body is dehydrated, angiotensin II and aldosterone promote urinary reabsorption of Na^+ and Cl^- (and water by osmosis with the electrolytes), conserving the volume of body fluids by reducing urinary loss. An increase in blood volume, as might occur after you finish one or more supersized drinks, stretches the atria of the heart and promotes release of atrial natriuretic peptide. ANP promotes **natriuresis,** elevated urinary excretion of Na^+ (and Cl^-) followed by water excretion, which decreases blood volume. An increase in blood volume also slows release of renin from juxtaglomerular cells of the kidneys. When renin level declines, less angiotensin II is formed. Decline in angiotensin II from a moderate level to a low level increases glomerular filtration rate and reduces Na^+, Cl^-, and water reabsorption in the kidney tubules. In addition, less angiotensin II leads to lower levels of aldosterone, which causes reabsorption of filtered Na^+ and Cl^- to slow in the renal collecting ducts. More filtered Na^+ and Cl^- thus remain in the tubular fluid to be excreted in the urine. The

Figure 27.4 Hormonal regulation of renal Na^+ and Cl^- reabsorption.

The three main hormones that regulate renal Na^+ and Cl^- reabsorption (and thus the amount lost in the urine) are angiotensin II, aldosterone, and atrial natriuretic peptide.

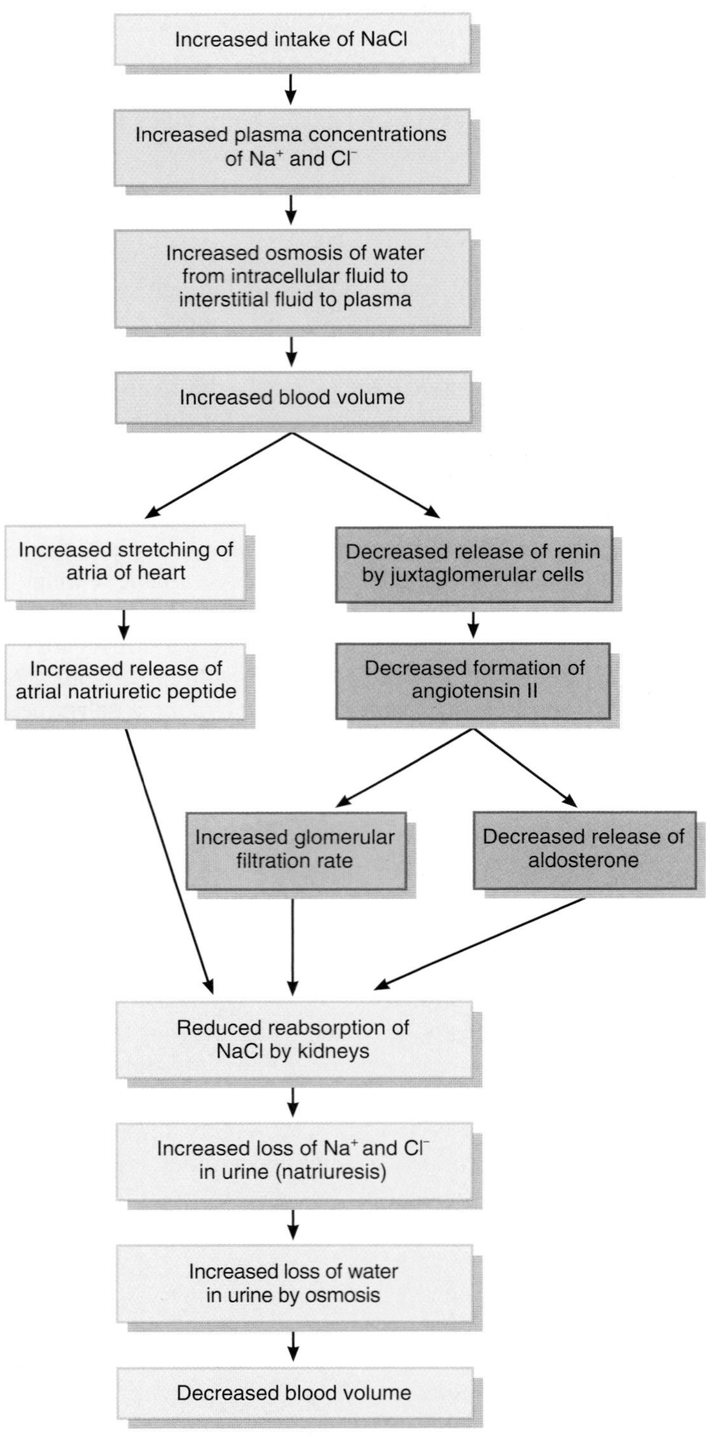

How does hyperaldosteronism (excessive aldosterone secretion) cause edema?

osmotic consequence of excreting more Na^+ and Cl^- is loss of more water in urine, which decreases blood volume and blood pressure.

The major hormone that regulates water loss is **antidiuretic hormone (ADH).** This hormone, also known as **vasopressin,** is produced by neurosecretory cells that extend from the hypothalamus to the posterior pituitary. In addition to stimulating the thirst mechanism, an increase in the osmolarity of body fluids stimulates release of ADH (see Figure 26.17 on page 1015). ADH promotes the insertion of water-channel proteins (aquaporin-2) into the apical membranes of principal cells in the collecting ducts of the kidneys. As a result, the permeability of these cells to water increases. Water molecules move by osmosis from the renal tubular fluid into the cells and then from the cells into the bloodstream. The result is production of a small volume of very concentrated urine (see page 1016). Intake of water in response to the thirst mechanism decreases the osmolarity of blood and interstitial fluid. Within minutes, ADH secretion shuts down, and soon its blood level is close to zero. When the principal cells are not stimulated by ADH, aquaporin-2 molecules are removed from the apical membrane by endocytosis. As the number of water channels decreases, the water permeability of the principal cells' apical membrane falls, and more water is lost in the urine.

Under some conditions, factors other than blood osmolarity influence ADH secretion. A large decrease in blood volume, which is detected by baroreceptors (sensory neurons that respond to stretching) in the left atrium and in blood vessel walls, also stimulates ADH release. In severe dehydration, glomerular filtration rate decreases because blood pressure falls, so that less water is lost in the urine. Conversely, the intake of too much water increases blood pressure, causing the rate of glomerular filtration to rise, and more water to be lost in the urine. Hyperventilation (abnormally fast and deep breathing) can increase fluid loss through the exhala-tion of more water vapor. Vomiting and diarrhea result in fluid loss from the GI tract. Finally, fever, heavy sweating, and destruction of extensive areas of the skin from burns can cause excessive water loss through the skin. In all of these conditions, an increase in ADH secretion will help conserve body fluids.

Table 27.1 summarizes the factors that maintain body water balance.

Movement of Water Between Body Fluid Compartments

Normally, cells neither shrink nor swell because intracellular and interstitial fluids have the same osmolarity. Changes in the osmolarity of interstitial fluid, however, causes fluid imbalances. An increase in the osmolarity of interstitial fluid draws water out of cells, and they shrink slightly. A decrease in the osmolarity of interstitial fluid, by contrast, causes cells to swell. Changes in osmolarity most often result from changes in the concentration of Na^+.

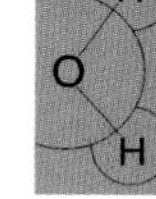

TABLE 27.1 Summary of Factors that Maintain Body Water Balance

Factor	Mechanism	Effect
Thirst center in hypothalamus	Stimulates desire to drink fluids.	Water gain if thirst is quenched.
Angiotensin II	Stimulates secretion of aldosterone.	Reduces loss of water in urine.
Aldosterone	By promoting urinary reabsorption of Na^+ and Cl^-, increases water reabsorption via osmosis.	Reduces loss of water in urine.
Atrial natriuretic peptide (ANP)	Promotes natriuresis, elevated urinary excretion of Na^+ (and Cl^-), accompanied by water.	Increases loss of water in urine.
Antidiuretic hormone (ADH), also known as vasopressin	Promotes insertion of water-channel proteins (aquaporin-2) into the apical membranes of principal cells in the collecting ducts of the kidneys. As a result, the water permeability of these cells increases and more water is reabsorbed.	Reduces loss of water in urine.

Figure 27.5 Series of events in water intoxication.

Water intoxication is a state in which excessive body water causes cells to swell.

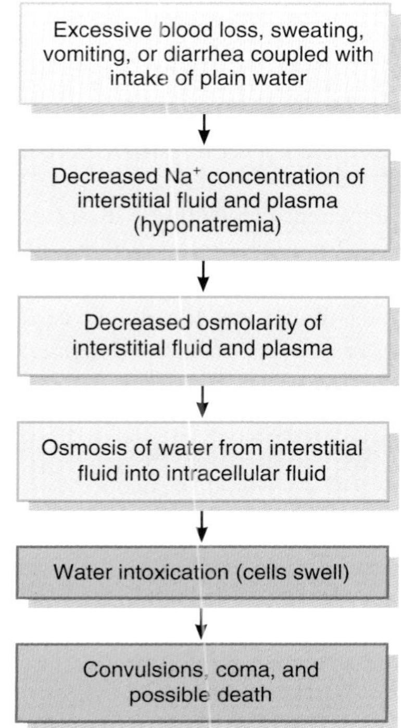

Why do solutions used for oral rehydration therapy contain a small amount of table salt (NaCl)?

A decrease in the osmolarity of interstitial fluid, as may occur after drinking a large volume of water, inhibits secretion of ADH. Normally, the kidneys then excrete a large volume of dilute urine, which restores the osmotic pressure of body fluids to normal. As a result, body cells swell only slightly, and only for a brief period. But when a person steadily consumes water faster than the kidneys can excrete it (the maximum urine flow rate is about 15 mL/min) or when renal function is poor, the result may be **water intoxication,** a state in which excessive body water causes cells to swell dangerously (Figure 27.5). If the body water and Na^+ lost during blood loss or excessive sweating, vomiting, or diarrhea is replaced by drinking plain water, then body fluids become more dilute. This dilution can cause the Na^+ concentration of plasma and then of interstitial fluid to fall below the normal range. When the Na^+ concentration of interstitial fluid decreases, its osmolarity also falls. The net result is osmosis of water from interstitial fluid into the cytosol. Water entering the cells causes them to swell, producing convulsions, coma, and possibly death. To prevent this dire sequence of events in cases of severe electrolyte and water loss, solutions given for intravenous or oral rehydration therapy (ORT) include a small amount of table salt (NaCl).

Enemas and Fluid Balance

An **enema** (EN-e-ma) is the introduction of a solution into the rectum to draw water (and electrolytes) into the colon osmotically. The increased volume increases peristalsis, which evacuates feces. Enemas are used to treat constipation. Repeated enemas, especially in young children, increase the risk of fluid and electrolyte imbalances. ■

▶ CHECKPOINT

1. What is the approximate volume of each of your body fluid compartments?
2. How are the routes of water gain and loss from the body regulated?
3. By what mechanism does thirst help regulate water intake?
4. How do angiotensin II, aldosterone, atrial natriuretic peptide, and antidiuretic hormone regulate the volume and osmolarity of body fluids?
5. What factors control the movement of water between interstitial fluid and intracellular fluid?

ELECTROLYTES IN BODY FLUIDS

OBJECTIVES

Compare the electrolyte composition of the three major fluid compartments: plasma, interstitial fluid, and intracellular fluid.

Discuss the functions of sodium, chloride, potassium, bicarbonate, calcium, phosphate, and magnesium ions, and explain how their concentrations are regulated.

The ions formed when electrolytes dissolve and dissociate serve four general functions in the body. (1) Because they are largely confined to particular fluid compartments and are more numerous than nonelectrolytes, certain ions *control the osmosis of water between fluid compartments.* (2) Ions *help maintain the acid–base balance* required for normal cellular activities. (3) Ions *carry electrical current,* which allows production of action potentials and graded potentials. (4) Several ions *serve as cofactors* needed for optimal activity of enzymes.

Concentrations of Electrolytes in Body Fluids

To compare the charge carried by ions in different solutions, the concentration of ions is typically expressed in units of **milliequivalents per liter (mEq/liter).** These units give the concentration of cations or anions in a given volume of solution. One equivalent is the positive or negative charge equal to the amount of charge in one mole of H^+; a milliequivalent is one-thousandth of an equivalent. Recall that a mole of a substance is its molecular weight expressed in grams. For ions such as sodium (Na^+), potassium (K^+), and bicarbonate (HCO_3^-), which have a single positive or negative charge, the number of mEq/liter is equal to the number of mmol/liter. For ions such as calcium (Ca^{2+}) or phosphate (HPO_4^{2-}), which have two positive or negative charges, the number of mEq/liter is twice the number of mmol/liter.

Figure 27.6 compares the concentrations of the main electrolytes and protein anions in blood plasma, interstitial fluid, and intracellular fluid. The chief difference between the

Figure 27.6 Electrolyte and protein anion concentrations in plasma, interstitial fluid, and intracellular fluid. The height of each column represents the milliequivalents per liter (mEq/liter).

The electrolytes present in extracellular fluids are different from those present in intracellular fluid.

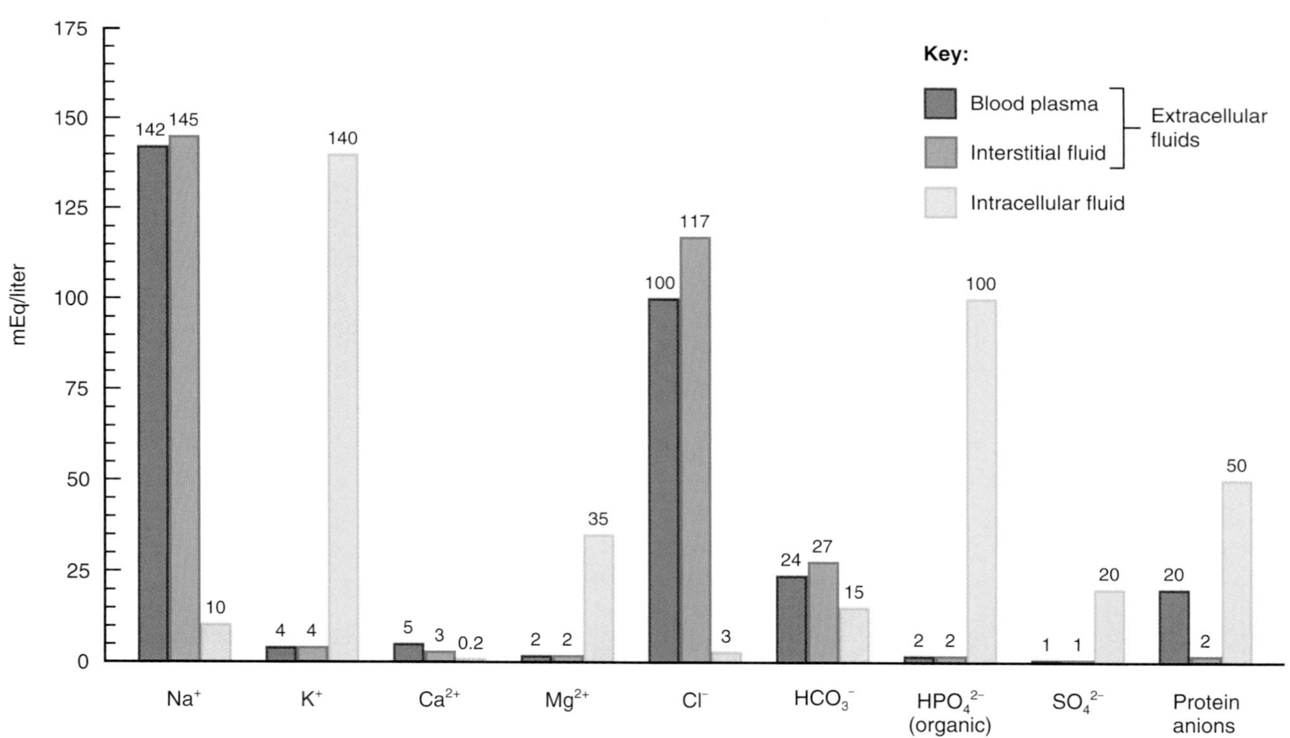

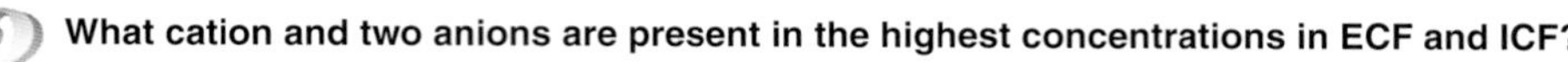
What cation and two anions are present in the highest concentrations in ECF and ICF?

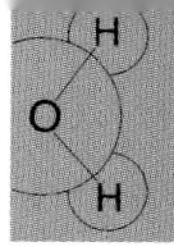

two extracellular fluids—blood plasma and interstitial fluid—is that blood plasma contains many protein anions, whereas interstitial fluid has very few. Because normal capillary membranes are virtually impermeable to proteins, only a few plasma proteins leak out of blood vessels into the interstitial fluid. This difference in protein concentration is largely responsible for the blood colloid osmotic pressure exerted by blood plasma. In other respects, the two fluids are similar.

The electrolyte content of intracellular fluid differs considerably from that of extracellular fluid. In extracellular fluid, the most abundant cation is Na^+, and the most abundant anion is Cl^-. In intracellular fluid, the most abundant cation is K^+, and the most abundant anions are proteins and phosphates (HPO_4^{2-}). By actively transporting Na^+ out of cells and K^+ into cells, sodium-potassium pumps (Na^+/K^+ ATPase) play a major role in maintaining the high intracellular concentration of K^+ and high extracellular concentration of Na^+.

Sodium

Sodium ions (Na^+) are the most abundant ions in extracellular fluid, accounting for 90% of the extracellular cations. The normal blood plasma Na^+ concentration is 136–148 mEq/liter. As we have already seen, Na^+ plays a pivotal role in fluid and electrolyte balance because it accounts for almost half of the osmolarity of extracellular fluid (142 of about 300 mOsm/liter). The flow of Na^+ through voltage-gated channels in the plasma membrane also is necessary for the generation and conduction of action potentials in neurons and muscle fibers. The typical daily intake of Na^+ in North America often far exceeds the body's normal daily requirements, due largely to excess dietary salt. The kidneys excrete excess Na^+, but they also can conserve it during periods of shortage.

The Na^+ level in the blood is controlled by aldosterone, antidiuretic hormone (ADH), and atrial natriuretic peptide (ANP). Aldosterone increases renal reabsorption of Na^+. When the blood plasma concentration of Na^+ drops below 135 mEq/liter, a condition called *hyponatremia,* ADH release ceases. The lack of ADH, in turn, permits greater excretion of water in urine and restoration of the normal Na^+ level in ECF. Atrial natriuretic peptide (ANP) increases Na^+ excretion by the kidneys when Na^+ level is above normal, a condition called *hypernatremia.*

Indicators of Na^+ Imbalance

If excess sodium ions remain in the body because the kidneys fail to excrete enough of them, water is also osmotically retained. The result is increased blood volume, increased blood pressure, and **edema,** an abnormal accumulation of interstitial fluid. Renal failure and hyperaldosteronism (excessive aldosterone secretion) are two causes of Na^+ retention. Excessive urinary loss of Na^+, by contrast, causes excessive water loss, which results in **hypovolemia,** an abnormally low blood volume. Hypovolemia related to Na^+ loss is most frequently due to the inadequate secretion of aldosterone associated with adrenal insufficiency or overly vigorous therapy with diuretic drugs. ■

Chloride

Chloride ions (Cl^-) are the most prevalent anions in extracellular fluid. The normal blood plasma Cl^- concentration is 95–105 mEq/liter. Cl^- moves relatively easily between the extracellular and intracellular compartments because most plasma membranes contain many Cl^- leakage channels and antiporters. For this reason, Cl^- can help balance the level of anions in different fluid compartments. One example is the chloride shift that occurs between red blood cells and blood plasma as the blood level of carbon dioxide either increases or decreases (see Figure 23.24 on page 878). In this case, the antiporter exchange of Cl^- for HCO_3^- maintains the correct balance of anions between ECF and ICF. Chloride ions also are part of the hydrochloric acid secreted into gastric juice. ADH helps regulate Cl^- balance in body fluids because it governs the extent of water loss in urine. Processes that increase or decrease renal reabsorption of sodium ions also affect reabsorption of chloride ions. (Recall that reabsorption of Na^+ and Cl^- occurs by means of Na^+-Cl^- symporters.)

Potassium

Potassium ions (K^+) are the most abundant cations in intracellular fluid (140 mEq/liter). K^+ plays a key role in establishing the resting membrane potential and in the repolarization phase of action potentials in neurons and muscle fibers; K^+ also helps maintain normal intracellular fluid volume. When K^+ moves into or out of cells, it often is exchanged for H^+ and thereby helps regulate the pH of body fluids.

The normal blood plasma K^+ concentration is 3.5–5.0 mEq/liter and is controlled mainly by aldosterone. When blood plasma K^+ concentration is high, more aldosterone is secreted into the blood. Aldosterone then stimulates principal cells of the renal collecting ducts to secrete more K^+ so excess K^+ is lost in the urine. Conversely, when blood plasma K^+ concentration is low, aldosterone secretion decreases and less K^+ is excreted in urine. Because K^+ is needed during the repolarization phase of action potentials, abnormal K^+ levels can be lethal. For instance, *hyperkalemia* (above-normal concentration of K^+ in blood) can cause death due to ventricular fibrillation.

Bicarbonate

Bicarbonate ions (HCO_3^-) are the second most prevalent extracellular anions. Normal blood plasma HCO_3^- concentration is 22–26 mEq/liter in systemic arterial blood and 23–

27 mEq/liter in systemic venous blood. HCO_3^- concentration increases as blood flows through systemic capillaries because the carbon dioxide released by metabolically active cells combines with water to form carbonic acid; the carbonic acid then dissociates into H^+ and HCO_3^-. As blood flows through pulmonary capillaries, however, the concentration of HCO_3^- decreases again as carbon dioxide is exhaled. (Figure 23.24 on page 878 shows these reactions.) Intracellular fluid also contains a small amount of HCO_3^-. As previously noted, the exchange of Cl^- for HCO_3^- helps maintain the correct balance of anions in extracellular fluid and intracellular fluid.

The kidneys are the main regulators of blood HCO_3^- concentration. The intercalated cells of the renal tubule can either form HCO_3^- and release it into the blood when the blood level is low (see Figure 27.8) or excrete excess HCO_3^- in the urine when the level in blood is too high. Changes in the blood level of HCO_3^- are considered later in this chapter in the section on acid–base balance.

Calcium

Because such a large amount of calcium is stored in bone, it is the most abundant mineral in the body. About 98% of the calcium in adults is located in the skeleton and teeth, where it is combined with phosphates to form a crystal lattice of mineral salts. In body fluids, calcium is mainly an extracellular cation (Ca^{2+}). The normal concentration of free or unattached Ca^{2+} in blood plasma is 4.5–5.5 mEq/liter. About the same amount of Ca^{2+} is attached to various plasma proteins. Besides contributing to the hardness of bones and teeth, Ca^{2+} plays important roles in blood clotting, neurotransmitter release, maintenance of muscle tone, and excitability of nervous and muscle tissue.

The two main regulators of Ca^{2+} concentration in blood plasma are parathyroid hormone (PTH) and calcitriol (1,25-dihydroxy vitamin D_3), the form of vitamin D that acts as a hormone (see Figure 18.14 on page 640). A low level of Ca^{2+} in blood plasma promotes release of more PTH, which stimulates osteoclasts in bone tissue to release calcium (and phosphate) from bone matrix. Thus, PTH increases bone *resorption*. Parathyroid hormone also enhances *reabsorption* of Ca^{2+} from glomerular filtrate through renal tubule cells and back into blood, and increases production of calcitriol, which in turn increases Ca^{2+} *absorption* from food in the gastrointestinal tract.

Phosphate

About 85% of the phosphate in adults is present as calcium phosphate salts, which are structural components of bone and teeth. The remaining 15% is ionized. Three phosphate ions ($H_2PO_4^-$, HPO_4^{2-}, and PO_4^{3-}) are important intracellular anions. At the normal pH of body fluids, HPO_4^{2-} is the most prevalent form. Phosphates contribute about 100 mEq/liter of anions to intracellular fluid. HPO_4^{2-} is an important buffer of H^+, both in body fluids and in the urine. Although some are "free," most phosphate ions are covalently bound to organic molecules such as lipids (phospholipids), proteins, carbohydrates, nucleic acids (DNA and RNA), and adenosine triphosphate (ATP).

The normal blood plasma concentration of ionized phosphate is only 1.7–2.6 mEq/liter. The same two hormones that govern calcium homeostasis—parathyroid hormone (PTH) and calcitriol—also regulate the level of HPO_4^{2-} in blood plasma. PTH stimulates resorption of bone matrix by osteoclasts, which releases both phosphate and calcium ions into the bloodstream. In the kidneys, however, PTH inhibits reabsorption of phosphate ions while stimulating reabsorption of calcium ions by renal tubular cells. Thus, PTH increases urinary excretion of phosphate and lowers blood phosphate level. Calcitriol promotes absorption of both phosphates and calcium from the digestive tract.

Magnesium

In adults, about 54% of the total body magnesium is part of bone matrix as magnesium salts. The remaining 46% occurs as magnesium ions (Mg^{2+}) in intracellular fluid (45%) and extracellular fluid (1%). Mg^{2+} is the second most common intracellular cation (35 mEq/liter). Functionally, Mg^{2+} is a cofactor for certain enzymes needed for the metabolism of carbohydrates and proteins and for the sodium-potassium pump. Mg^{2+} is essential for normal neuromuscular activity, synaptic transmission, and myocardial functioning. In addition, secretion of parathyroid hormone (PTH) depends on Mg^{2+}.

Normal blood plasma Mg^{2+} concentration is low, only 1.3–2.1 mEq/liter. Several factors regulate the blood plasma level of Mg^{2+} by varying the rate at which it is excreted in the urine. The kidneys increase urinary excretion of Mg^{2+} in response to hypercalcemia, hypermagnesemia, increases in extracellular fluid volume, decreases in parathyroid hormone, and acidosis. The opposite conditions decrease renal excretion of Mg^{2+}.

Table 27.2 describes the imbalances that result from the deficiency or excess of several electrolytes.

People at risk for fluid and electrolyte imbalances include those who depend on others for fluid and food, such as infants, the elderly, and the hospitalized; individuals undergoing medical treatment that involves intravenous infusions, drainages or suctions, and urinary catheters; and people who receive diuretics, experience excessive fluid losses and require increased fluid intake, or experience fluid retention and have fluid restrictions. Finally, athletes and military personnel in extremely hot environments, postoperative individuals, severe burn or trauma cases, individuals with chronic diseases (congestive heart failure, diabetes, chronic obstructive lung disease, and cancer), people in confinement, and individuals with altered

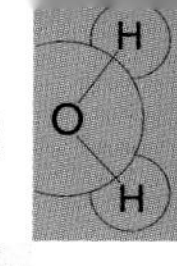

TABLE 27.2 Blood Electrolyte Imbalances

	Deficiency		Excess	
Electrolyte*	**Name and Causes**	**Signs and Symptoms**	**Name and Causes**	**Signs and Symptoms**
Sodium (Na^+) 136–148 mEq/liter	**Hyponatremia** (hī-pō-na-TRĒ-mē-a) may be due to decreased sodium intake; increased sodium loss through vomiting, diarrhea, aldosterone deficiency, or taking certain diuretics; and excessive water intake.	Muscular weakness; dizziness, headache, and hypotension; tachycardia and shock; mental confusion, stupor, and coma.	**Hypernatremia** may occur with dehydration, water deprivation, or excessive sodium in diet or intravenous fluids; causes hypertonicity of ECF, which pulls water out of body cells into ECF, causing cellular dehydration.	Intense thirst, hypertension, edema, agitation, and convulsions.
Chloride (Cl^-) 95–105 mEq/liter	**Hypochloremia** (hī-pō-klō-RĒ-mē-a) may be due to excessive vomiting, overhydration, aldosterone deficiency, congestive heart failure, and therapy with certain diuretics such as furosemide (Lasix®).	Muscle spasms, metabolic alkalosis, shallow respirations, hypotension, and tetany.	**Hyperchloremia** may result from dehydration due to water loss or water deprivation; excessive chloride intake; or severe renal failure, hyperaldosteronism, certain types of acidosis, and some drugs.	Lethargy, weakness, metabolic acidosis, and rapid, deep breathing.
Potassium (K^+) 3.5–5.0 mEq/liter	**Hypokalemia** (hī-pō-ka-LĒ-mē-a) may result from excessive loss due to vomiting or diarrhea, decreased potassium intake, hyperaldosteronism, kidney disease, and therapy with some diuretics.	Muscle fatigue, flaccid paralysis, mental confusion, increased urine output, shallow respirations, and changes in the electrocardiogram, including flattening of the T wave.	**Hyperkalemia** may be due to excessive intake, renal failure, aldosterone deficiency, crushing injuries to body tissues, or transfusion of hemolyzed blood.	Irritability, nausea, vomiting, diarrhea, muscular weakness; can cause death by inducing ventricular fibrillation.
Calcium (Ca^{2+}) Total 5 9–10.5 mg/dL; ionized = 4.5–5.5 mEq/liter	**Hypocalcemia** (hī-pō-kal-SĒ-mē-a) may be due to increased calcium loss, reduced calcium intake, elevated levels of phosphate, or hypoparathyroidism.	Numbness and tingling of the fingers; hyperactive reflexes, muscle cramps, tetany, and convulsions; bone fractures; spasms of laryngeal muscles that can cause death by asphyxiation.	**Hypercalcemia** may result from hyperparathyroidism, some cancers, excessive intake of vitamin D, and Paget's disease of bone.	Lethargy, weakness, anorexia, nausea, vomiting, polyuria, itching, bone pain, depression, confusion, paresthesia, stupor, and coma.
Phosphate (HPO_4^{2-}) 1.7–2.6 mEq/liter	**Hypophosphatemia** (hī-pō-fos′-fa-TĒ-mē-a) may occur through increased urinary losses, decreased intestinal absorption, or increased utilization.	Confusion, seizures, coma, chest and muscle pain, numbness and tingling of the fingers, decreased coordination, memory loss, and lethargy.	**Hyperphosphatemia** occurs when the kidneys fail to excrete excess phosphate, as happens in renal failure; can also result from increased intake of phosphates or destruction of body cells, which releases phosphates into the blood.	Anorexia, nausea, vomiting, muscular weakness, hyperactive reflexes, tetany, and tachycardia.
Magnesium (Mg^{2+}) 1.3–2.1 mEq/liter	**Hypomagnesemia** (hī′-pō-mag′-ne-SĒ-mē-a) may be due to inadequate intake or excessive loss in urine or feces; also occurs in alcoholism, malnutrition, diabetes mellitus, and diuretic therapy.	Weakness, irritability, tetany, delirium, convulsions, confusion, anorexia, nausea, vomiting, paresthesia, and cardiac arrhythmias.	**Hypermagnesemia** occurs in renal failure or due to increased intake of Mg^{2+}, such as Mg^{2+}-containing antacids; also occurs in aldosterone deficiency and hypothyroidism.	Hypotension, muscular weakness or paralysis, nausea, vomiting, and altered mental functioning.

*Values are normal ranges of blood plasma levels in adults.

levels of consciousness who may be unable to communicate needs or respond to thirst are also subject to fluid and electrolyte imbalances.

▶ CHECKPOINT

6. What are the functions of electrolytes in the body?

7. Name three important extracellular electrolytes and three important intracellular electrolytes and indicate how each is regulated.

ACID–BASE BALANCE

▶ OBJECTIVES

Compare the roles of buffers, exhalation of carbon dioxide, and kidney excretion of H^+ in maintaining pH of body fluids.

Define acid–base imbalances, describe their effects on the body, and explain how they are treated.

From our discussion thus far, it should be clear that various ions play different roles that help maintain homeostasis. A major homeostatic challenge is keeping the H^+ concentration (pH) of body fluids at an appropriate level. This task—the maintenance of acid–base balance—is of critical importance to normal cellular function. For example, the three-dimensional shape of all body proteins, which enables them to perform specific functions, is very sensitive to pH changes. When the diet contains a large amount of protein, as is typical in North America, cellular metabolism produces more acids than bases, which tends to acidify the blood. Before proceeding with this section of the chapter, you may wish to review the discussion of acids, bases, and pH on pages 42–43.

In a healthy person, several mechanisms help maintain the pH of systemic arterial blood between 7.35 and 7.45. (A pH of 7.4 corresponds to a H^+ concentration of 0.00004 mEq/liter = 40 nEq /liter.) Because metabolic reactions often produce a huge excess of H^+, the lack of any mechanism for the disposal of H^+ would cause H^+ level in body fluids to rise quickly to a lethal level. Homeostasis of H^+ concentration within a narrow range is thus essential to survival. The removal of H^+ from body fluids and its subsequent elimination from the body depend on the following three major mechanisms:

1. ***Buffer systems.*** Buffers act quickly to temporarily bind H^+, removing the highly reactive, excess H^+ from solution. Buffers thus raise pH of body fluids but do not remove H^+ from the body.
2. ***Exhalation of carbon dioxide.*** By increasing the rate and depth of breathing, more carbon dioxide can be exhaled. Within minutes this reduces the level of carbonic acid in blood, which raises the blood pH (reduces blood H^+ level).
3. ***Kidney excretion of H^+.*** The slowest mechanism, but the only way to eliminate acids other than carbonic acid, is through their excretion in urine.

We will examine each of these mechanisms in more detail in the following sections.

The Actions of Buffer Systems

Most buffer systems in the body consist of a weak acid and the salt of that acid, which functions as a weak base. Buffers prevent rapid, drastic changes in the pH of body fluids by converting strong acids and bases into weak acids and weak bases within fractions of a second. Strong acids lower pH more than weak acids because strong acids release H^+ more readily and thus contribute more free hydrogen ions. Similarly, strong bases raise pH more than weak ones. The principal buffer systems of the body fluids are the protein buffer system, the carbonic acid–bicarbonate buffer system, and the phosphate buffer system.

Protein Buffer System

The **protein buffer system** is the most abundant buffer in intracellular fluid and blood plasma. For example, the protein hemoglobin is an especially good buffer within red blood cells, and albumin is the main protein buffer in blood plasma. Proteins are composed of amino acids, organic molecules that contain at least one carboxyl group (—COOH) and at least one amino group (—NH_2); these groups are the functional components of the protein buffer system. The free carboxyl group at one end of a protein acts like an acid by releasing H^+ when pH rises; it dissociates as follows:

$$NH_2-\underset{\displaystyle H}{\overset{\displaystyle R}{\underset{|}{\overset{|}{C}}}}-COOH \longrightarrow NH_2-\underset{\displaystyle H}{\overset{\displaystyle R}{\underset{|}{\overset{|}{C}}}}-COO^- + H^+$$

The H^+ is then able to react with any excess OH^- in the solution to form water. The free amino group at the other end of a protein can act as a base by combining with H^+ when pH falls, as follows:

$$NH_2-\underset{\displaystyle H}{\overset{\displaystyle R}{\underset{|}{\overset{|}{C}}}}-COOH + H^+ \longrightarrow {}^+NH_3-\underset{\displaystyle H}{\overset{\displaystyle R}{\underset{|}{\overset{|}{C}}}}-COOH$$

So proteins can buffer both acids and bases. In addition to the terminal carboxyl and amino groups, side chains that can buffer H^+ are present on seven of the 20 amino acids.

As we have already noted, the protein hemoglobin is an important buffer of H^+ in red blood cells (see Figure 23.24 on page 878). As blood flows through the systemic capillaries, carbon dioxide (CO_2) passes from tissue cells into red blood

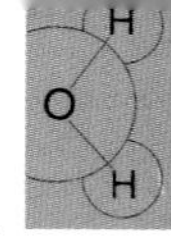

cells, where it combines with water (H_2O) to form carbonic acid (H_2CO_3). Once formed, H_2CO_3 dissociates into H^+ and HCO_3^-. At the same time that CO_2 is entering red blood cells, oxyhemoglobin (Hb—O_2) is giving up its oxygen to tissue cells. Reduced hemoglobin (deoxyhemoglobin) picks up most of the H^+. For this reason, reduced hemoglobin usually is written as Hb—H. The following reactions summarize these relations:

$$H_2O + CO_2 \longrightarrow H_2CO_3$$

H_2O	CO_2	H_2CO_3
Water	Carbon dioxide (entering RBCs)	Carbonic acid

$$H_2CO_3 \longrightarrow H^+ + HCO_3^-$$

H_2CO_3	H^+	HCO_3^-
Carbonic acid	Hydrogen ion	Bicarbonate ion

$$Hb{-}O_2 + H^+ \longrightarrow Hb{-}H + O_2$$

$Hb{-}O_2$	H^+	$Hb{-}H$	O_2
Oxyhemoglobin (in RBCs)	Hydrogen ion (from carbonic acid)	Reduced hemoglobin	Oxygen (released to tissue cells)

Carbonic Acid–Bicarbonate Buffer System

The **carbonic acid–bicarbonate buffer system** is based on the *bicarbonate ion* (HCO_3^-), which can act as a weak base, and *carbonic acid* (H_2CO_3), which can act as a weak acid. As you have already learned, HCO_3^- is a significant anion in both intracellular and extracellular fluids (Figure 27.6). Because the kidneys also synthesize new HCO_3^- and reabsorb filtered HCO_3^-, this important buffer is not lost in the urine. If there is an excess of H^+, the HCO_3^- can function as a weak base and remove the excess H^+ as follows:

$$H^+ + HCO_3^- \longrightarrow H_2CO_3$$

H^+	HCO_3^-	H_2CO_3
Hydrogen ion	Bicarbonate ion (weak base)	Carbonic acid

Then, H_2CO_3 dissociates into water and carbon dioxide, and the CO_2 is exhaled from the lungs.

Conversely, if there is a shortage of H^+, the H_2CO_3 can function as a weak acid and provide H^+ as follows:

$$H_2CO_3 \longrightarrow H^+ + HCO_3^-$$

H_2CO_3	H^+	HCO_3^-
Carbonic acid (weak acid)	Hydrogen ion	Bicarbonate ion

At a pH of 7.4, HCO_3^- concentration is about 24 mEq/liter and H_2CO_3 concentration is about 1.2 mmol/liter, so bicarbonate ions outnumber carbonic acid molecules by 20 to 1. Because CO_2 and H_2O combine to form H_2CO_3, this buffer system cannot protect against pH changes due to respiratory problems in which there is an excess or shortage of CO_2.

Phosphate Buffer System

The **phosphate buffer system** acts via a mechanism similar to the one for the carbonic acid–bicarbonate buffer system. The components of the phosphate buffer system are the ions *dihydrogen phosphate* ($H_2PO_4^-$) and *monohydrogen phosphate* (HPO_4^{2-}). Recall that phosphates are major anions in intracellular fluid and minor ones in extracellular fluids (Figure 27.6). The dihydrogen phosphate ion acts as a weak acid and is capable of buffering strong bases such as OH^-, as follows:

$$OH^- + H_2PO_4^- \longrightarrow H_2O + HPO_4^{2-}$$

OH^-	$H_2PO_4^-$	H_2O	HPO_4^{2-}
Hydroxide ion (strong base)	Dihydrogen phosphate (weak acid)	Water	Monohydrogen phosphate (weak base)

The monohydrogen phosphate ion is capable of buffering the H^+ released by a strong acid such as hydrochloric acid (HCl) by acting as a weak base:

$$H^+ + HPO_4^{2-} \longrightarrow H_2PO_4^-$$

H^+	HPO_4^{2-}	$H_2PO_4^-$
Hydrogen ion (strong acid)	Monohydrogen phosphate (weak base)	Dihydrogen phosphate (weak acid)

Because the concentration of phosphates is highest in intracellular fluid, the phosphate buffer system is an important regulator of pH in the cytosol. It also acts to a smaller degree in extracellular fluids, and buffers acids in urine. $H_2PO_4^{2-}$ is formed when excess H^+ in the kidney tubule fluid combines with HPO_4^{2-} (see Figure 27.8). The H^+ that becomes part of the $H_2PO_4^-$ passes into the urine. This reaction is one way the kidneys help maintain blood pH by excreting H^+ in the urine.

Exhalation of Carbon Dioxide

The simple act of breathing also plays an important role in maintaining the pH of body fluids. An increase in the carbon dioxide (CO_2) concentration in body fluids increases H^+ concentration and thus lowers the pH (makes body fluids more acidic). Because H_2CO_3 can be eliminated by exhaling CO_2, it is called a **volatile acid.** Conversely, a decrease in the CO_2 concentration of body fluids raises the pH (makes body fluids more alkaline). This chemical interaction is illustrated by the following reversible reactions:

$$CO_2 + H_2O \rightleftharpoons H_2CO_3 \rightleftharpoons H^+ + HCO_3^-$$

CO_2	H_2O	H_2CO_3	H^+	HCO_3^-
Carbon dioxide	Water	Carbonic acid	Hydrogen ion	Bicarbonate ion

Changes in the rate and depth of breathing can alter the pH of body fluids within a couple of minutes. With increased ventilation, more CO_2 is exhaled; the reaction is driven to the left, H^+ concentration falls, and blood pH rises. Doubling the ventilation increases pH by about 0.23 units, from 7.4 to 7.63. If ventilation is slower than normal, less carbon dioxide is exhaled, and the blood pH falls. Reducing ventilation to one-quarter of normal lowers the pH by 0.4 units, from 7.4 to 7.0. These examples show the powerful effect of alterations in breathing on the pH of body fluids.

The pH of body fluids and the rate and depth of breathing interact via a negative feedback loop (Figure 27.7). When the blood acidity increases, the decrease in pH (increase in concentration of H^+) is detected by central chemoreceptors in the medulla oblongata and peripheral chemoreceptors in the aortic and carotid bodies, both of which stimulate the inspiratory area in the medulla oblongata. As a result, the diaphragm and other respiratory muscles contract more forcefully and frequently, so

Figure 27.7 Negative feedback regulation of blood pH by the respiratory system.

Exhalation of carbon dioxide lowers the H^+ concentration of blood.

Some stimulus disrupts homeostasis by
Decreasing
Blood pH (increase in H^+ concentration)
Receptors
Central chemo-receptors in medulla oblongata
Peripheral chemo-receptors in aortic and carotid bodies
Input
Nerve impulses
Control center
Inspiratory area in medulla oblongata
Output
Nerve impulses
Effectors
Diaphragm contracts more force-fully and frequently so more CO_2 is exhaled
As less H_2CO_3 forms and fewer H^+ are present, blood pH increases (H^+ concentration decreases)
Return to homeostasis when response brings blood pH or H^+ con-centration back to normal

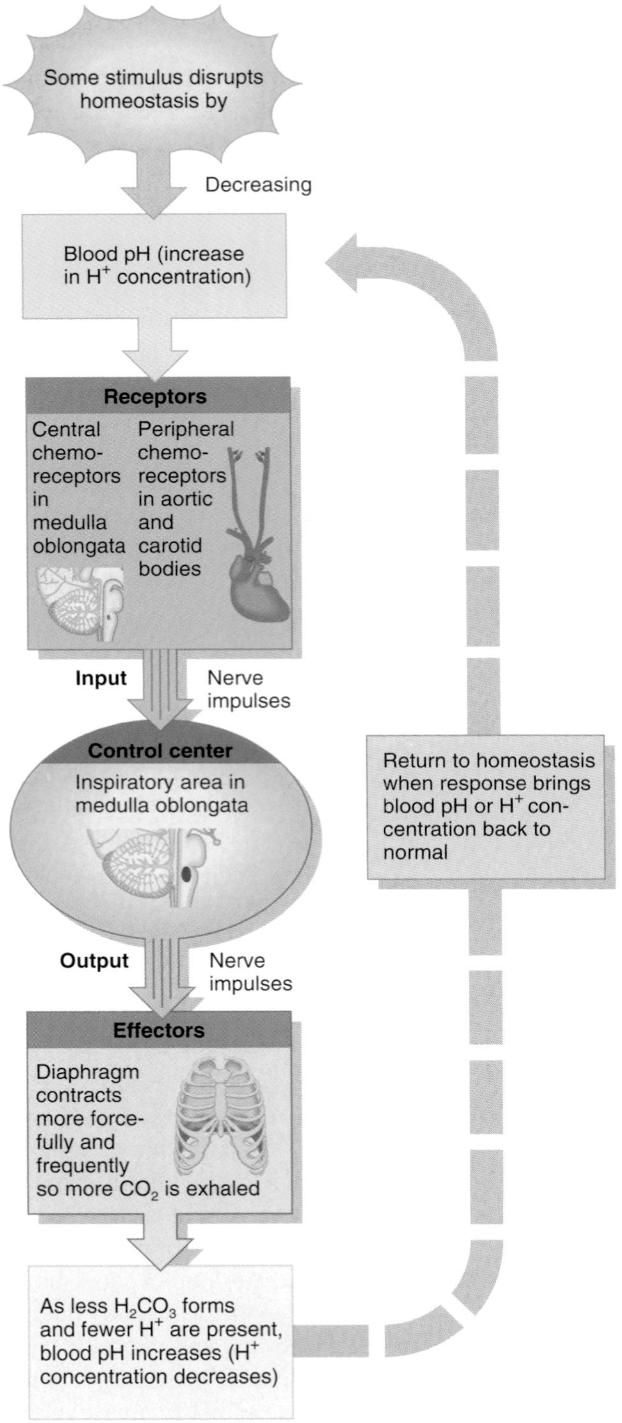

If you hold your breath for 30 seconds, what is likely to happen to your blood pH?

more CO_2 is exhaled. As less H_2CO_3 forms and fewer H^+ are present, blood pH increases. When the response brings blood pH (H^+ concentration) back to normal, there is a return to acid–base homeostasis. The same negative feedback loop operates if the blood level of CO_2 increases. Ventilation increases, which removes more CO_2, reducing the H^+ concentration and increasing the blood's pH.

By contrast, if the pH of the blood increases, the respiratory center is inhibited and the rate and depth of breathing decreases. A decrease in the CO_2 concentration of the blood has the same effect. When breathing decreases, CO_2 accumulates in the blood so its H^+ concentration increases.

Kidney Excretion of H^+

Metabolic reactions produce **nonvolatile acids** such as sulfuric acid at a rate of about 1 mEq of H^+ per day for every kilogram of body mass. The only way to eliminate this huge acid load is to excrete H^+ in the urine. Given the magnitude of these contributions to acid–base balance, it's not surprising that renal failure can quickly cause death.

As you learned in Chapter 26, cells in both the proximal convoluted tubules (PCT) and the collecting ducts of the kidneys secrete hydrogen ions into the tubular fluid. In the PCT, Na^+/H^+ antiporters secrete H^+ as they reabsorb Na^+ (see Figure 26.13 on page 1011). Even more important for regulation of pH of body fluids, however, are the intercalated cells of the collecting duct. The *apical* membranes of some intercalated cells include **proton pumps (H^+ ATPases)** that secrete H^+ into the tubular fluid (Figure 27.8). Intercalated cells can secrete H^+ against a concentration gradient so effectively that urine can be up to 1000 times (3 pH units) more acidic than blood. HCO_3^- produced by dissociation of H_2CO_3 inside intercalated cells crosses the basolateral membrane by means of **Cl^-/HCO_3^- antiporters** and then diffuses into peritubular capillaries (Figure 27.8a). The HCO_3^- that enters the blood in this way is *new* (not filtered). For this reason, blood leaving the kidney in the renal vein may have a higher HCO_3^- concentration than blood entering the kidney in the renal artery.

Interestingly, a second type of intercalated cell has proton pumps in its *basolateral* membrane and Cl^-/HCO_3^- antiporters in its apical membrane. These intercalated cells secrete HCO_3^- and reabsorb H^+. Thus, the two types of intercalated cells help maintain the pH of body fluids in two ways—by excreting excess H^+ when pH of body fluids is too low and by excreting excess HCO_3^- when pH is too high.

Some H^+ secreted into the tubular fluid of the collecting duct are buffered, but not by HCO_3^-, most of which has been filtered and reabsorbed. Two other buffers combine with H^+ in the collecting duct (Figure 27.8b). The most plentiful buffer in the tubular fluid of the collecting duct is HPO_4^{2-} (monohydrogen phosphate ion). In addition, a small amount of NH_3 (ammonia) also is present. H^+ combines with

Figure 27.8 Secretion of H^+ by intercalated cells in the collecting duct. HCO_3^- = bicarbonate ion; CO_2 = carbon dioxide; H_2O = water; H_2CO_3 = carbonic acid; Cl^- = chloride ion; NH_3 = ammonia; NH_4^+ = ammonium ion; HPO_4^{2-} = monohydrogen phosphate ion; $H_2PO_4^-$ = dihydrogen phosphate ion.

Urine can be up to 1000 times more acidic than blood due to the operation of the proton pumps in the collecting ducts of the kidneys.

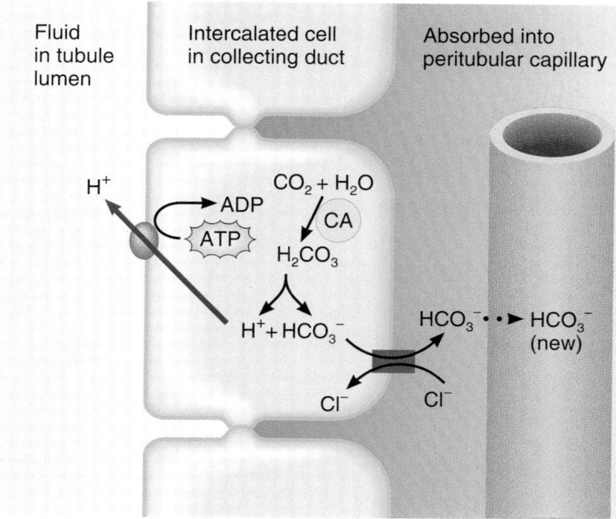

(a) Secretion of H^+

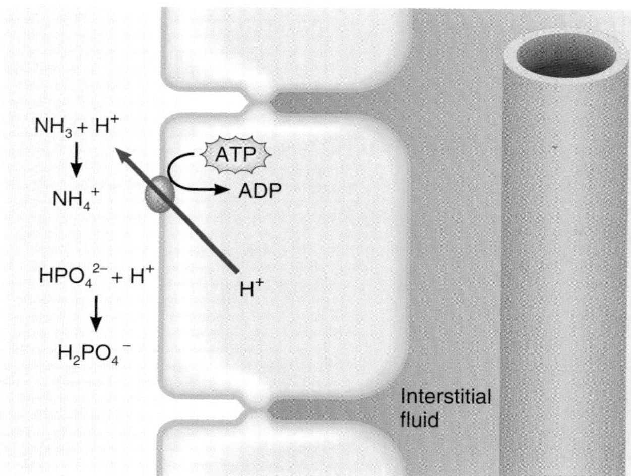

(b) Buffering of H^+ in urine

Key:

 Proton pump (H^+ ATPase) in apical membrane

 HCO_3^-/Cl^- antiporter in basolateral membrane

Diffusion

What would be the effects of a drug that blocks the activity of carbonic anhydrase?

HPO_4^{2-} to form $H_2PO_4^-$ (dihydrogen phosphate ion) and with NH_3 to form NH_4^+ (ammonium ion). Because these ions cannot diffuse back into tubule cells, they are excreted in the urine.

Table 27.3 summarizes the mechanisms that maintain pH of body fluids.

Acid–Base Imbalances

The normal pH range of systemic arterial blood is between 7.35 (= 45 nEq of H^+/liter) and 7.45 (= 35 nEq of H^+/liter). **Acidosis** (or **acidemia**) is a condition in which blood pH is below 7.35; **alkalosis** (or **alkalemia**) is a condition in which blood pH is higher than 7.45.

The major physiological effect of acidosis is depression of the central nervous system through depression of synaptic transmission. If the systemic arterial blood pH falls below 7, depression of the nervous system is so severe that the individual becomes disoriented, then comatose, and may die. Patients with severe acidosis usually die while in a coma. A major physiological effect of alkalosis, by contrast, is overexcitability in both the central nervous system and peripheral nerves. Neurons conduct impulses repetitively, even when not stimulated by normal stimuli; the results are nervousness, muscle spasms, and even convulsions and death.

A change in blood pH that leads to acidosis or alkalosis may be countered by **compensation**, the physiological response to an acid–base imbalance that acts to normalize arterial blood pH. Compensation may be either *complete*, if pH indeed is brought within the normal range, or *partial*, if systemic arterial blood pH

TABLE 27.3 Mechanisms that Maintain pH of Body Fluids

Mechanism	Comments
Buffer Systems	Most consist of a weak acid and the salt of that acid, which functions as a weak base. They prevent drastic changes in body fluid pH.
Proteins	The most abundant buffers in body cells and blood. Hemoglobin inside red blood cells is a good buffer.
Carbonic acid–bicarbonate	Important regulator of blood pH. The most abundant buffers in extracellular fluid (ECF).
Phosphates	Important buffers in intracellular fluid and in urine.
Exhalation of CO_2	With increased exhalation of CO_2, pH rises (fewer H^+). With decreased exhalation of CO_2, pH falls (more H^+).
Kidneys	Renal tubules secrete H^+ into the urine and reabsorb HCO_3^- so it is not lost in the urine.

is still lower than 7.35 or higher than 7.45. If a person has altered blood pH due to metabolic causes, hyperventilation or hypoventilation can help bring blood pH back toward the normal range; this form of compensation, termed **respiratory compensation,** occurs within minutes and reaches its maximum within hours. If, however, a person has altered blood pH due to respiratory causes, then **renal compensation**—changes in secretion of H^+ and reabsorption of HCO_3^- by the kidney tubules—can help reverse the change. Renal compensation may begin in minutes, but it takes days to reach maximum effectiveness.

In the discussion that follows, note that both respiratory acidosis and respiratory alkalosis are disorders resulting from changes in the partial pressure of CO_2 (P_{CO_2}) in systemic arterial blood (normal range is 35–45 mmHg). By contrast, both metabolic acidosis and metabolic alkalosis are disorders resulting from changes in HCO_3^- concentration (normal range is 22– 26 mEq/liter in systemic arterial blood).

Respiratory Acidosis

The hallmark of **respiratory acidosis** is an abnormally high P_{CO_2} in systemic arterial blood—above 45 mmHg. Inadequate exhalation of CO_2 causes the blood pH to drop. Any condition that decreases the movement of CO_2 from the blood to the alveoli of the lungs to the atmosphere causes a buildup of CO_2, H_2CO_3, and H^+. Such conditions include emphysema, pulmonary edema, injury to the respiratory center of the medulla oblongata, airway obstruction, or disorders of the muscles involved in breathing. If the respiratory problem is not too severe, the kidneys can help raise the blood pH into the normal range by increasing excretion of H^+ and reabsorption of HCO_3^- (renal compensation). The goal in treatment of respiratory acidosis is to increase the exhalation of CO_2, as, for instance, by providing ventilation therapy. In addition, intravenous administration of HCO_3^- may be helpful.

Respiratory Alkalosis

In **respiratory alkalosis,** systemic arterial blood P_{CO_2} falls below 35 mmHg. The cause of the drop in P_{CO_2} and the resulting increase in pH is hyperventilation, which occurs in conditions that stimulate the inspiratory area in the brain stem. Such conditions include oxygen deficiency due to high altitude or pulmonary disease, cerebrovascular accident (stroke), or severe anxiety. Again, renal compensation may bring blood pH into the normal range if the kidneys are able to decrease excretion of H^+ and reabsorption of HCO_3^-. Treatment of respiratory alkalosis is aimed at increasing the level of CO_2 in the body. One simple treatment is to have the person inhale and exhale into a paper bag for a short period; as a result, the person inhales air containing a higher-than-normal concentration of CO_2.

Metabolic Acidosis

In **metabolic acidosis,** the systemic arterial blood HCO_3^- level drops below 22 mEq/liter. Such a decline in this important buffer causes the blood pH to decrease. Three situations may lower the blood level of HCO_3^-: (1) actual loss of HCO_3^-, such as may occur with severe diarrhea or renal dysfunction; (2) accumulation of an acid other than carbonic acid, as may occur in ketosis (described on page 967); or (3) failure of the kidneys to excrete H^+ from metabolism of dietary proteins. If the problem is not too severe, hyperventilation can help bring blood pH into the normal range (respiratory compensation). Treatment of metabolic acidosis consists of administering intravenous solutions of sodium bicarbonate and correcting the cause of the acidosis.

Metabolic Alkalosis

In **metabolic alkalosis,** the systemic arterial blood HCO_3^- concentration is above 26 mEq/liter. A nonrespiratory loss of acid or excessive intake of alkaline drugs causes the blood pH to increase above 7.45. Excessive vomiting of gastric contents, which results in a substantial loss of hydrochloric acid, is probably the most frequent cause of metabolic alkalosis. Other causes include gastric suctioning, use of certain diuretics, endocrine disorders, excessive intake of alkaline drugs (antacids), and severe dehydration. Respiratory compensation through hypoventilation may bring blood pH into the normal range. Treatment of metabolic alkalosis consists of giving fluid solutions to correct Cl^-, K^+, and other electrolyte deficiencies plus correcting the cause of alkalosis.

Table 27.4 summarizes respiratory and metabolic acidosis and alkalosis.

Diagnosis of Acid–Base Imbalances

One can often pinpoint the cause of an acid–base imbalance by careful evaluation of three factors in a sample of systemic arterial blood: pH, concentration of HCO_3^-, and P_{CO_2}. These three blood chemistry values are examined in the following four-step sequence:

1. Note whether the pH is high (alkalosis) or low (acidosis).

2. Then decide which value—P_{CO_2} or HCO_3^-—is out of the normal range and could be the *cause* of the pH change. For example, *elevated pH* could be caused by *low* P_{CO_2} or *high* HCO_3^-.

3. If the cause is a *change in* P_{CO_2}, the problem is *respiratory;* if the cause is a *change in* HCO_3^-, the problem is *metabolic.*

4. Now look at the value that doesn't correspond with the observed pH change. If it is within its normal range, there is no compensation. If it is outside the normal range, compensation is occurring and partially correcting the pH imbalance. ■

► CHECKPOINT

8. Explain how each of the following buffer systems helps to maintain the pH of body fluids: proteins, carbonic acid–bicarbonate buffers, and phosphates.

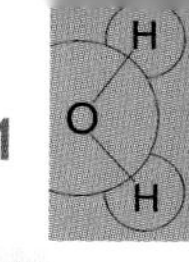

TABLE 27.4 Summary of Acidosis and Alkalosis

Condition	Definition	Common Causes	Compensatory Mechanism
Respiratory acidosis	Increased P_{CO_2} (above 45 mmHg) and decreased pH (below 7.35) if there is no compensation.	Hypoventilation due to emphysema, pulmonary edema, trauma to respiratory center, airway obstructions, or dysfunction of muscles of respiration.	*Renal:* increased excretion of H^+; increased reabsorption of HCO_3^-. If compensation is complete, pH will be within the normal range but P_{CO_2} will be high.
Respiratory alkalosis	Decreased P_{CO_2} (below 35 mmHg) and increased pH (above 7.45) if there is no compensation.	Hyperventilation due to oxygen deficiency, pulmonary disease, cerebrovascular accident (CVA), or severe anxiety.	*Renal:* decreased excretion of H^+; decreased reabsorption of HCO_3^-. If compensation is complete, pH will be within the normal range but P_{CO_2} will be low.
Metabolic acidosis	Decreased HCO_3^- (below 22 mEq/liter) and decreased pH (below 7.35) if there is no compensation.	Loss of bicarbonate ions due to diarrhea, accumulation of acid (ketosis), renal dysfunction.	*Respiratory:* hyperventilation, which increases loss of CO_2. If compensation is complete, pH will be within the normal range but HCO_3^- will be low.
Metabolic alkalosis	Increased HCO_3^- (above 26 mEq/liter) and increased pH (above 7.45) if there is no compensation.	Loss of acid due to vomiting, gastric suctioning, or use of certain diuretics; excessive intake of alkaline drugs.	*Respiratory:* hypoventilation, which slows loss of CO_2. If compensation is complete, pH will be within the normal range but HCO_3^- will be high.

*Values are normal ranges of blood plasma levels in adults.

9. Define acidosis and alkalosis. Distinguish among respiratory and metabolic acidosis and alkalosis.

10. What are the principal physiological effects of acidosis and alkalosis?

AGING AND FLUID, ELECTROLYTE, AND ACID–BASE BALANCE

▶ **OBJECTIVE**

Describe the changes in fluid, electrolyte, and acid–base balance that may occur with aging.

There are significant differences between adults and infants, especially premature infants, with respect to fluid distribution, regulation of fluid and electrolyte balance, and acid–base homeostasis. Accordingly, infants experience more problems than adults in these areas. The differences are related to the following conditions:

- ***Proportion and distribution of water.*** A newborn's total body mass is about 75% water (and can be as high as 90% in a premature infant); an adult's total body mass is about 55–60% water. (The "adult" percentage is achieved at about 2 years of age.) Adults have twice as much water in ICF as ECF, but the opposite is true in premature infants. Because ECF is subject to more changes than ICF, rapid losses or gains of body water are much more critical in infants. Given that the rate of fluid intake and output is about seven times higher in infants than in adults, the slightest changes in fluid balance can result in severe abnormalities.
- ***Metabolic rate.*** The metabolic rate of infants is about double that of adults. This results in the production of more metabolic wastes and acids, which can lead to the development of acidosis in infants.
- ***Functional development of the kidneys.*** Infant kidneys are only about half as efficient in concentrating urine as those of adults. (Functional development is not complete until close to the end of the first month after birth.) As a result, the kidneys of newborns can neither concentrate urine nor rid the body of excess acids as effectively as those of adults.
- ***Body surface area.*** The ratio of body surface area to body volume of infants is about three times greater than that of adults. Water loss through the skin is significantly higher in infants than in adults.
- ***Breathing rate.*** The higher breathing rate of infants (about 30 to 80 times a minute) causes greater water loss from the lungs. Respiratory alkalosis may occur because greater ventilation eliminates more CO_2 and lowers the P_{CO_2}.
- ***Ion concentrations.*** Newborns have higher K^+ and Cl^- concentrations than adults. This creates a tendency toward metabolic acidosis.

By comparison with children and younger adults, older adults often have an impaired ability to maintain fluid, electrolyte, and acid–base balance. With increasing age, many people have a decreased volume of intracellular fluid and decreased total body K^+ due to declining skeletal muscle mass and increasing mass of adipose tissue (which contains very little water). Age-related decreases in respiratory and renal functioning may compromise acid–base balance by slowing the exhalation of CO_2 and the excretion of excess acids in urine. Other kidney changes, such as decreased blood flow, decreased glomerular filtration rate, and reduced sensitivity to antidiuretic hormone, have an adverse effect on the ability to maintain fluid and electrolyte balance. Due to a decrease in the number and efficiency of sweat glands, water loss from the skin declines with age. Because of these age-related changes, older adults are susceptible to several fluid and electrolyte disorders:

- *Dehydration* and *hypernatremia* often occur due to inadequate fluid intake or loss of more water than Na^+ in vomit, feces, or urine.
- *Hyponatremia* may occur due to inadequate intake of Na^+; elevated loss of Na^+ in urine, vomit, or diarrhea; or impaired ability of the kidneys to produce dilute urine.
- *Hypokalemia* often occurs in older adults who chronically use laxatives to relieve constipation or who take K^+-depleting diuretic drugs for treatment of hypertension or heart disease.
- *Acidosis* may occur due to impaired ability of the lungs and kidneys to compensate for acid–base imbalances. One cause of acidosis is decreased production of ammonia (NH_3) by renal tubule cells, which then is not available to combine with H^+ and be excreted in urine as $NH_4{}^+$; another cause is reduced exhalation of CO_2.

CHECKPOINT

11. Why do infants experience greater problems with fluid, electrolyte, and acid–base balance than adults?

STUDY OUTLINE

FLUID COMPARTMENTS AND FLUID BALANCE (p. 1037)

1. Body fluid includes water and dissolved solutes.
2. About two-thirds of the body's fluid is located within cells and is called intracellular fluid (ICF). The other one-third, called extracellular fluid (ECF), includes interstitial fluid; blood plasma and lymph; cerebrospinal fluid; gastrointestinal tract fluids; synovial fluid; fluids of the eyes and ears; pleural, pericardial, and peritoneal fluids; and glomerular filtrate.
3. Fluid balance means that the required amounts of water and solutes are present and are correctly proportioned among the various compartments.
4. An inorganic substance that dissociates into ions in solution is called an electrolyte.
5. Water is the largest single constituent in the body. It makes up 45–75% of total body mass, depending on age, gender, and the amount of adipose tissue present.
6. Daily water gain and loss are each about 2500 mL. Sources of water gain are ingested liquids and foods, and water produced by cellular respiration and dehydration synthesis reactions (metabolic water). Water is lost from the body via urination, evaporation from the skin surface, exhalation of water vapor, and defecation. In women, menstrual flow is an additional route for loss of body water.
7. Body water gain is regulated by adjusting the volume of water intake, mainly by drinking more or less fluid. The thirst center in the hypothalamus governs the urge to drink.
8. Although increased amounts of water and solutes are lost through sweating and exhalation during exercise, loss of excess body water or excess solutes depends mainly on regulating excretion in the urine. The extent of urinary NaCl loss is the main determinant of body fluid volume, whereas the extent of urinary water loss is the main determinant of body fluid osmolarity.
9. Table 27.1 on page 1041 summarizes the factors that regulate water gain and water loss in the body.
10. Angiotensin II and aldosterone reduce urinary loss of Na^+ and Cl^- and thereby increase the volume of body fluids. ANP promotes natriuresis, elevated excretion of Na^+ (and Cl^-), which decreases blood volume.
11. The major hormone that regulates water loss and thus body fluid osmolarity is antidiuretic hormone (ADH).
12. An increase in the osmolarity of interstitial fluid draws water out of cells, and they shrink slightly. A decrease in the osmolarity of interstitial fluid causes cells to swell. Most often a change in osmolarity is due to a change in the concentration of Na^+, the dominant solute in interstitial fluid.
13. When a person consumes water faster than the kidneys can excrete it or when renal function is poor, the result may be water intoxication, in which cells swell dangerously.

ELECTROLYTES IN BODY FLUIDS (p. 1042)

1. Ions formed when electrolytes dissolve in body fluids control the osmosis of water between fluid compartments, help maintain acid–base balance, and carry electrical current.
2. The concentrations of cations and anions is expressed in units of milliequivalents/liter (mEq/liter).
3. Blood plasma, interstitial fluid, and intracellular fluid contain varying types and amounts of ions.
4. Sodium ions (Na^+) are the most abundant extracellular ions. They are involved in impulse transmission, muscle contraction, and fluid and electrolyte balance. Na^+ level is controlled by aldosterone, antidiuretic hormone, and atrial natriuretic peptide.
5. Chloride ions (Cl^-) are the major extracellular anions. They play a role in regulating osmotic pressure and forming HCl in gastric juice. Cl^- level is controlled indirectly by antidiuretic hormone and by processes that increase or decrease renal reabsorption of Na^+.
6. Potassium ions (K^+) are the most abundant cations in intracellular fluid. They play a key role in the resting membrane potential and action potential of neurons and muscle fibers; help maintain intracellular fluid volume; and contribute to regulation of pH. K^+ level is controlled by aldosterone.

7. Bicarbonate ions (HCO_3^-) are the second most abundant anions in extracellular fluid. They are the most important buffer in blood plasma.
8. Calcium is the most abundant mineral in the body. Calcium salts are structural components of bones and teeth. Ca^{2+}, which are principally extracellular cations, function in blood clotting, neurotransmitter release, and contraction of muscle. Ca^{2+} level is controlled mainly by parathyroid hormone and calcitriol.
9. Phosphate ions ($H_2PO_4^-$, HPO_4^{2-}, and PO_4^{3-}) are principally intracellular anions, and their salts are structural components of bones and teeth. They are also required for the synthesis of nucleic acids and ATP and participate in buffer reactions. Their level is controlled by parathyroid hormone and calcitriol.
10. Magnesium ions (Mg^{2+}) are primarily intracellular cations. They act as cofactors in several enzyme systems.
11. Table 27.2 on page 1045 describes the imbalances that result from deficiency or excess of important body electrolytes.

ACID–BASE BALANCE (p. 1046)

1. The overall acid–base balance of the body is maintained by controlling the H^+ concentration of body fluids, especially extracellular fluid.
2. The normal pH of systemic arterial blood is 7.35–7.45.
3. Homeostasis of pH is maintained by buffer systems, via exhalation of carbon dioxide, and via kidney excretion of H^+ and reabsorption of HCO_3^-.
4. The important buffer systems include proteins, carbonic acid–bicarbonate buffers, and phosphates.
5. An increase in exhalation of carbon dioxide increases blood pH; a decrease in exhalation of CO_2 decreases blood pH.
6. In the proximal convoluted tubules of the kidneys, Na^+/H^+ antiporters secrete H^+ as they reabsorb Na^+. In the collecting ducts of the kidneys, some intercalated cells reabsorb K^+ and HCO_3^- and secrete H^+; other intercalated cells secrete HCO_3^-. In these ways, the kidneys can increase or decrease the pH of body fluids.
7. Table 27.3 on page 1049 summarizes the mechanisms that maintain pH of body fluids.
8. Acidosis is a systemic arterial blood pH below 7.35; its principal effect is depression of the central nervous system (CNS). Alkalosis is a systemic arterial blood pH above 7.45; its principal effect is overexcitability of the CNS.
9. Respiratory acidosis and alkalosis are disorders due to changes in blood P_{CO_2}; metabolic acidosis and alkalosis are disorders associated with changes in blood HCO_3^- concentration.
10. Metabolic acidosis or alkalosis can be compensated by respiratory mechanisms (respiratory compensation); respiratory acidosis or alkalosis can be compensated by renal mechanisms (renal compensation).
11. Table 27.4 on page 1051 summarizes the effects of respiratory and metabolic acidosis and alkalosis.
12. By examining systemic arterial blood pH, HCO_3^-, and P_{CO_2} values, it is possible to pinpoint the cause of an acid–base imbalance.

AGING AND FLUID, ELECTROLYTE, AND ACID–BASE BALANCE (p. 1051)

1. With increasing age, there is decreased intracellular fluid volume and decreased K^+ due to declining skeletal muscle mass.
2. Decreased kidney function with aging adversely affects fluid and electrolyte balance.

SELF-QUIZ QUESTIONS

Fill in the blanks in the following statements.

1. The source of water that is derived from aerobic cellular respiration and dehydration synthesis reactions is ____ water.
2. In the carbonic acid–bicarbonate buffer system, the ____ acts as a weak base, and ____ acts as a weak acid.

Indicate whether the following statements are true or false.

3. The phosphate buffer system is an important regulator of pH in the cytosol.
4. The two compartments in which water can be found are plasma and cytosol.

Choose the one best answer to the following questions.

5. The primary means of regulating body water gain is adjusting (a) the volume of water intake, (b) the rate of cellular respiration, (c) the formation of metabolic water, (d) the volume of metabolic water, (e) the metabolic use of water.
6. Which of the following stimulate thirst? (1) a decreased production of saliva, (2) a decrease in nerve impulses from hypothalamic osmoreceptors, (3) an increase in osmolarity of body fluids, (4) angiotensin II release, (5) release of atrial natriuretic peptide, (6) an increase in blood volume. (a) 1, 2, 4, and 6; (b) 1, 3, 5, and 6; (c) 1, 3, and 4; (d) 2, 4, and 6; (e) 1, 4, 5, and 6.
7. Which of the following is *not* true concerning the protein buffer system? (a) Albumin is considered the main protein buffer in blood plasma. (b) Albumin is the most abundant buffer in blood plasma and intracellular fluid. (c) The functional components of a protein buffer system are the carboxyl group and the amino group. (d) Protein buffers are primary buffers of acids in urine. (e) Proteins can buffer both acids and bases.
8. Which of the following statements are *true*? (1) Buffers prevent rapid, drastic changes in pH of a body fluid. (2) Buffers work slowly. (3) Strong acids lower pH more than weak acids because strong acids contribute fewer H^+. (4) Most buffers consist of a weak acid and the salt of that acid, which acts as weak base. (5) Hemoglobin is an important buffer. (a) 1, 2, 3, and 5; (b) 1, 3, 4, and 5; (c) 1, 3, and 5; (d) 1, 4, and 5; (e) 2, 3, and 5.
9. Which of the following hormones regulate fluid loss? (1) antidiuretic hormone, (2) aldosterone, (3) atrial natriuretic peptide, (4) thyroxine, (5) cortisol. (a) 1, 3, and 5; (b) 1, 2, and 3; (c) 2, 4, and 5; (d) 2, 3, and 4; (e) 1, 3, and 4.
10. Which of the following are *true* concerning ions in the body? (1) They control osmosis of water between fluid compartments. (2) They help maintain acid–base balance. (3) They carry electrical current. (4) They serve as cofactors for enzyme activity. (5) They serve as neurotransmitters under special circumstances. (a) 1, 3, and 5; (b) 2, 4, and 5; (c) 1, 4, and 5; (d) 1, 2, and 4; (e) 1, 2, 3, and 4.

11. Which of the following statements are *true*? (1) An increase in the carbon dioxide concentration in body fluids increases H^+ concentration and thus lowers pH. (2) Breath holding results in a decline in blood pH. (3) The respiratory buffer mechanism can eliminate a single volatile acid: carbonic acid. (4) The only way to eliminate nonvolatile acids is to excrete H^+ in the urine. (5) When the diet contains a large amount of protein, normal metabolism produces more acids than bases. (a) 1, 2, 3, 4, and 5; (b) 1, 3, 4, and 5; (c) 1, 2, 3, and 4; (d) 1, 2, 4, and 5; (e) 1, 3, and 4.

12. Concerning acid–base imbalances: (1) Acidosis can cause depression of the central nervous system through depression of synaptic transmission. (2) Renal compensation can resolve respiratory alkalosis or acidosis. (3) A major physiological effect of alkalosis is lack of excitability in the central nervous system and peripheral nerves. (4) Resolution of metabolic acidosis and alkalosis occurs through renal compensation. (5) In adjusting blood pH, renal compensation occurs quickly whereas respiratory compensation takes days. (a) 1, 2, and 5; (b) 1 and 2; (c) 2, 3, and 4; (d) 2, 3, and 5; (e) 1, 2, 3, and 5.

13. Which of the following are *mismatched*? (a) Hypoventilation: respiratory alkalosis, (b) severe diarrhea: metabolic acidosis, (c) excessive vomiting: metabolic alkalosis, (d) airway obstruction: respiratory acidosis, (e) inability of kidneys to excrete H^+ from dietary protein metabolism: metabolic acidosis.

14. Match the following:

____(a) the most abundant cation in intracellular fluid; plays a key role in establishing the resting membrane potential
____(b) the most abundant mineral in the body; plays important roles in blood clotting, neurotransmitter release, maintenance of muscle tone, and excitability of nervous and muscle tissue
____(c) second most common intracellular cation; is a cofactor for enzymes involved in carbohydrate, protein, and Na^+/K^+ ATPase metabolism
____(d) the most abundant extracellular cation; essential in fluid and electrolyte balance
____(e) ions that are mostly combined with lipids, proteins, carbohydrates, nucleic acids, and ATP inside cells
____(f) most prevalent extracellular anion; can help balance the level of anions in different fluid compartments
____(g) second most prevalent extracellular anion; mainly regulated by the kidneys; important for acid–base balance
____(h) substances that act to prevent rapid, drastic changes in the pH of a body fluid
____(i) inorganic substances that dissociate into ions when in solution

(1) sodium
(2) chloride
(3) electrolytes
(4) bicarbonate
(5) buffers
(6) phosphate
(7) magnesium
(8) potassium
(9) calcium

15. Match the following:

____(a) an abnormal increase in the volume of interstitial fluid
____(b) can occur during renal failure or destruction of body cells, which releases phosphates into the blood
____(c) the swelling of cells due to water moving from plasma into interstitial fluid and then into cells
____(d) occurs when water loss is greater than water gain
____(e) can be caused by excessive sodium in diet or with dehydration
____(f) condition that can occur as water moves out of plasma into interstitial fluid and blood volume decreases
____(g) can be caused by decreased potassium intake or kidney disease; results in muscle fatigue, increased urine output, changes in electrocardiogram
____(h) can occur from hypoparathyroidism
____(i) can be caused by emphysema, pulmonary edema, injury to the respiratory center of the medulla oblongata, airway destruction, or disorders of the muscles involved in breathing
____(j) can be caused by excessive water intake, excessive vomiting, or aldosterone deficiency
____(k) can be caused by actual loss of bicarbonate ions, ketosis, or failure of kidneys to excrete H^+
____(l) can be caused by excessive vomiting of gastric contents, gastric suctioning, use of certain diuretics, severe dehydration, or excessive intake of alkaline drugs
____(m) can be caused by oxygen deficiency at high altitude, stroke, or severe anxiety

(1) respiratory acidosis
(2) respiratory alkalosis
(3) metabolic acidosis
(4) metabolic alkalosis
(5) dehydration
(6) hypovolemia
(7) water intoxication
(8) edema
(9) hypokalemia
(10) hypernatremia
(11) hyponatremia
(12) hyperphosphatemia
(13) hypocalcemia

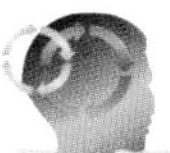

CRITICAL THINKING QUESTIONS

1. Robin is in the early stages of pregnancy and has been vomiting excessively for several days. She became weak, was confused and was taken to the emergency room. What do you suspect has happened to Robin's acid–base balance? How would her body attempt to compensate? What electrolytes would be affected by her vomiting, and how do her symptoms reflect those imbalances?
2. Henry is in the intensive care unit because he suffered a severe myocardial infarction three days ago. The lab reports the following values from an arterial blood sample: pH 7.30, $HCO_3^- =$ 20 mEq/liter, $P_{CO_2} = 32$ mmHg. Diagnose Henry's acid–base status and decide whether compensation is occurring.
3. This summer, Sam is training for a marathon by running 10 miles a day. Describe changes in his fluid balance as he trains.

ANSWERS TO FIGURE QUESTIONS

27.1 Plasma volume equals body mass × percent of body mass that is body fluid × proportion of body fluid that is ECF × proportion of ECF that is plasma × a conversion factor (1 liter/kg).

For males: blood plasma volume = 60 kg × 0.60 × 1/3 × 0.20 × 1 liter/kg = 2.4 liters. Using similar calculations, female blood plasma volume is 2.2 liters.

27.2 Hyperventilation, vomiting, fever, and diuretics all increase fluid loss.

27.3 Negative feedback is in operation because the result (an increase in fluid intake) is opposite to the initiating stimulus (dehydration).

27.4 An elevated aldosterone level promotes abnormally high renal reabsorption of NaCl and water, which expands blood volume and increases blood pressure. Because of the increased blood pressure, more fluid filters out of capillaries and accumulates in the interstitial fluid, causing edema.

27.5 If a solution used for oral rehydration therapy contains a small amount of salt, both the salt and water are absorbed in the gastrointestinal tract, blood volume increases without a decrease in osmolarity, and water intoxication does not occur.

27.6 In ECF, the major cation is Na^+, and the major anions are Cl^- and HCO_3^-. In ICF, the major cation is K^+, and the major anions are proteins and organic phosphates (for example, ATP).

27.7 Holding your breath causes blood pH to decrease slightly as CO_2 and H^+ accumulate in the blood.

27.8 A carbonic anhydrase inhibitor reduces secretion of H^+ into the urine and reduces reabsorption of Na^+ and HCO_3^- into the blood. It has a diuretic effect and can cause acidosis (lowered pH of the blood) due to loss of HCO_3^- in the urine.

Chapter 28

The Reproductive Systems

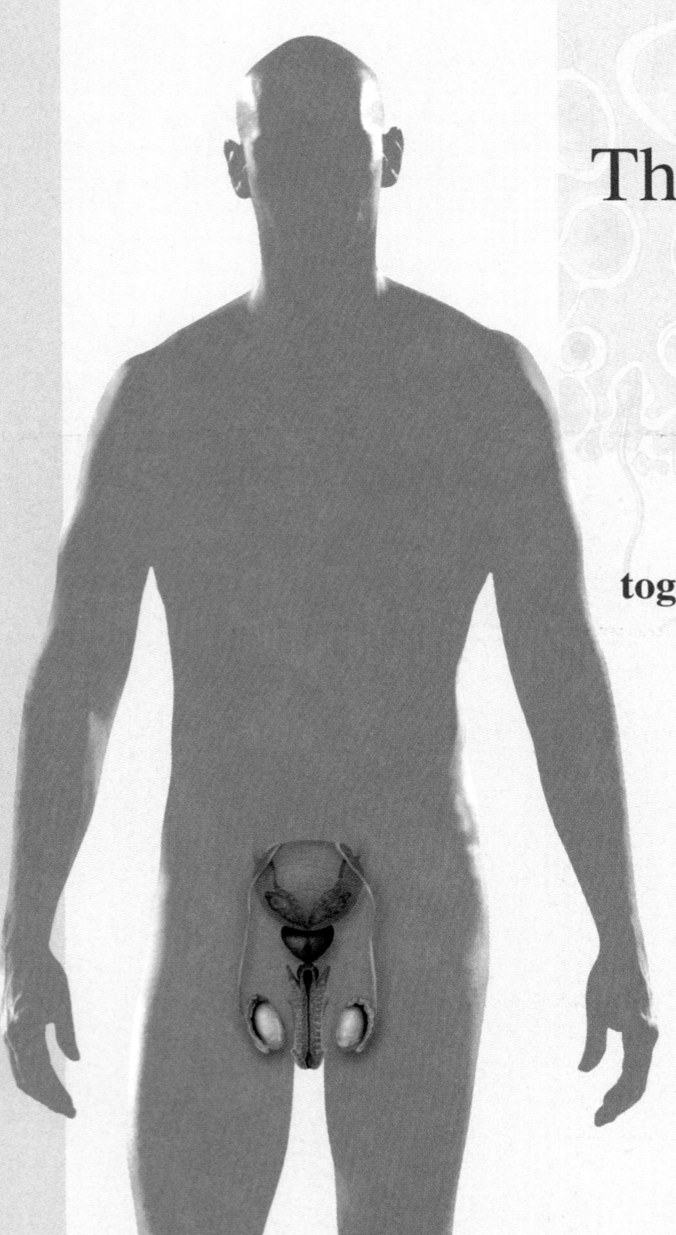

The Reproductive Systems and Homeostasis

The male and female reproductive organs work together to produce offspring. In addition, the female reproductive organs contribute to sustaining the growth of embryos and fetuses.

www. wiley.com/college/apcentral

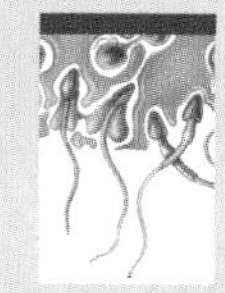

Sexual reproduction is the process by which organisms produce offspring by making germ cells called **gametes** (GAM-ēts = spouses). After the male gamete (sperm cell) unites with the female gamete (secondary oocyte)—an event called **fertilization**—the resulting cell contains one set of chromosomes from each parent. Males and females have anatomically distinct reproductive organs that are adapted for producing gametes, facilitating fertilization, and, in females, sustaining the growth of the embryo and fetus.

The male and female reproductive organs can be grouped by function. The **gonads**—testes in males and ovaries in females—produce gametes and secrete sex hormones. Various **ducts** then store and transport the gametes, and **accessory sex glands** produce substances that protect the gametes and facilitate their movement. Finally, **supporting structures,** such as the penis and the uterus, assist the delivery and joining of gametes and, in females, the growth of the embryo and fetus during pregnancy.

Gynecology (gī-ne-KOL-ō-jē; *gyneco-* = woman; *-logy* = study of) is the specialized branch of medicine concerned with the diagnosis and treatment of diseases of the female reproductive system. As noted in Chapter 26, **urology** (ū-ROL-ō-jē) is the study of the urinary system. Urologists also diagnose and treat diseases and disorders of the male reproductive system. The branch of medicine that deals with male disorders, especially infertility and sexual dysfunction, is called **andrology** (an-DROL-ō-jē; *andro-* = masculine).

MALE REPRODUCTIVE SYSTEM

▶ **OBJECTIVES**

Describe the location, structure, and functions of the organs of the male reproductive system.

Discuss the process of spermatogenesis in the testes.

The organs of the male reproductive system include the testes, a system of ducts (including the epididymis, ductus deferens, ejaculatory ducts, and urethra), accessory sex glands (seminal vesicles, prostate, and bulbourethral glands), and several supporting structures, including the scrotum and the penis (Figure 28.1). The testes (male gonads) produce sperm and secrete hormones. The duct system transports and stores sperm, assists in their maturation, and conveys them to the exterior. Semen contains sperm plus the secretions provided by the accessory sex glands. The supporting structures have various functions. The penis delivers sperm into the female reproductive tract and the scrotum supports the testes.

Scrotum

The **scrotum** (SKRŌ-tum = bag), the supporting structure for the testes, consists of loose skin and superficial fascia that hangs from the root (attached portion) of the penis (Figure 28.1a). Externally, the scrotum looks like a single pouch of skin separated into lateral portions by a median ridge called the **raphe** (RĀ-fē = seam). Internally, the **scrotal septum** divides the scrotum into two sacs, each containing a single testis (Figure 28.2 on page 1059). The septum is made up of superficial fascia and muscle tissue called the **dartos muscle** (DAR-tōs = skinned), which is composed of bundles of smooth muscle fibers. The dartos muscle is also found in the subcutaneous tissue of the scrotum. Associated with each testis in the scrotum is the **cremaster muscle** (krē-MAS-ter = suspender), a small band of skeletal muscle that is a continuation of the internal oblique muscle.

The location of the scrotum and the contraction of its muscle fibers regulate the temperature of the testes. Normal sperm production requires a temperature about 2–3°C below core body temperature. This lowered temperature is maintained within the scrotum because it is outside the pelvic cavity. In response to cold temperatures, the cremaster and dartos muscles contract. Contraction of the cremaster muscles moves the testes closer to the body, where they can absorb body heat. Contraction of the dartos muscle causes the scrotum to become tight (wrinkled in appearance), which reduces heat loss. Exposure to warmth reverses these actions.

Testes

The **testes** (TES-tēz), or **testicles,** are paired oval glands in the scrotum measuring about 5 cm (2 in.) long and 2.5 cm (1 in.) in diameter (Figure 28.3 on page 1060). Each testis (singular) has a mass of 10–15 grams. The testes develop near the kidneys, in the posterior portion of the abdomen, and they usually begin their descent into the scrotum through the inguinal canals (passageways in the anterior abdominal wall; see Figure 28.2) during the latter half of the seventh month of fetal development.

A serous membrane called the **tunica vaginalis** (*tunica* = sheath), which is derived from the peritoneum and forms during the descent of the testes, partially covers the testes. A collection of serous fluid in the tunica vaginalis is called a **hydrocele** (HĪ-drō-sēl; *hydro-* = water; *-kele* = hernia). It may be caused by injury to the testes or inflammation of the epididymis. Usually, no treatment is required. Internal to the tunica vaginalis is a dense white fibrous capsule composed of dense irregular connective tissue, the **tunica albuginea** (al′-bū-JIN-ē-a; *albu-* = white); it extends inward, forming septa that divide the testis into a series of internal compartments called **lobules.** Each of the 200–300 lobules contains one to three tightly coiled tubules, the **seminiferous tubules** (*semin-* = seed; *fer-* = to carry), where sperm are produced. The process by which the seminiferous tubules of the testes produce sperm is called **spermatogenesis** (sper′-ma-tō-JEN-e-sis; *genesis* = beginning process or production).

The seminiferous tubules contain two types of cells: **spermatogenic cells,** the sperm-forming cells, and **Sertoli cells,** which have several functions in supporting spermatogenesis (Figure 28.4 on page 1061). Stem cells called **spermatogonia** (sper′-ma-tō-GŌ-nē-a; *-gonia* = offspring; singular is *sper-*

Figure 28.1 Male organs of reproduction and surrounding structures.

Reproductive organs are adapted for producing new individuals and passing on genetic material from one generation to the next.

Functions of the Male Reproductive System

1. **The testes produce sperm and the male sex hormone testosterone.**
2. **The ducts transport, store, and assist in maturation of sperm.**
3. **The accessory sex glands secrete most of the liquid portion of semen.**
4. **The penis contains the urethra, a passageway for ejaculation of semen and excretion of urine.**

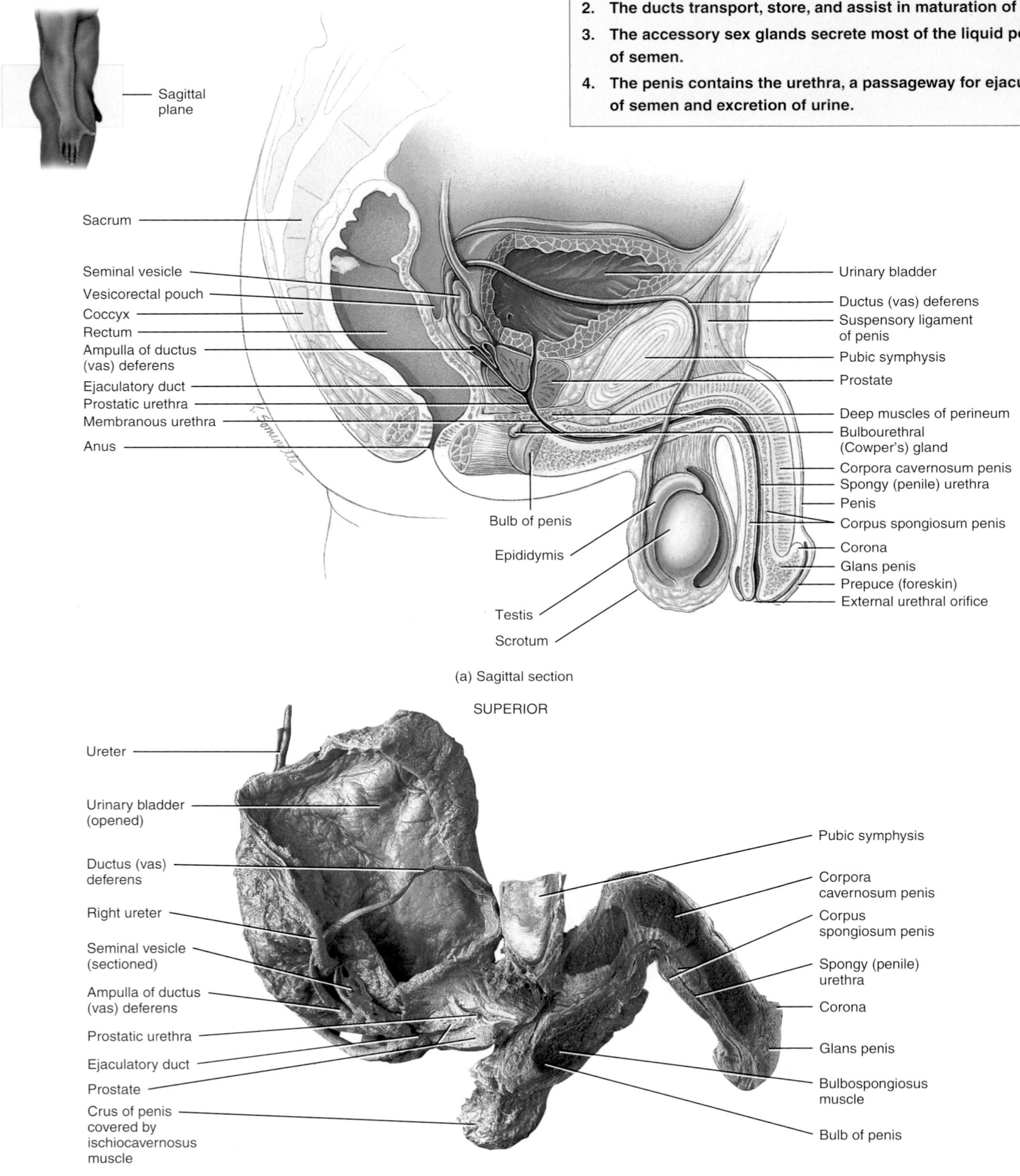

(a) Sagittal section

(b) Sagittal dissection

What are the groups of reproductive organs in males, and what are the functions of each group?

matogonium) develop from **primordial germ cells** (*primordi-* = primitive or early form) that arise from the yolk sac and enter the testes during the fifth week of development. In the embryonic testes, the primordial germ cells differentiate into spermatogonia, which remain dormant during childhood and actively begin producing sperm at puberty. Toward the lumen of the seminiferous tubule are layers of progressively more mature cells. In order of advancing maturity, these are primary spermatocytes, secondary spermatocytes, spermatids, and sperm cells. After a **sperm cell,** or **spermatozoon** (sper′-ma-tō-ZŌ-on; *-zoon* = life) has formed, it is released into the lumen of the seminiferous tubule. (The plural terms are *sperm* and *spermatozoa.*)

Embedded among the spermatogenic cells in the seminiferous tubules are large **Sertoli cells** or *sustentacular cells* (sus′-ten-TAK-ū-lar), which extend from the basement membrane to the lumen of the tubule. Internal to the basement membrane and spermatogonia, tight junctions join neighboring Sertoli cells to one another. These junctions form an obstruction known as the **blood–testis barrier** because substances must first pass through the Sertoli cells before they can reach the developing sperm. By isolating the developing gametes from the blood, the blood–testis barrier prevents an immune response against the spermatogenic cell's surface antigens, which are recognized as "foreign" by the immune system. The blood–testis barrier does not include spermatogonia.

Sertoli cells support and protect developing spermatogenic cells in several ways. They nourish spermatocytes, spermatids, and sperm; phagocytize excess spermatid cytoplasm as development proceeds; and control movements of spermatogenic cells and the release of sperm into the lumen of the seminiferous tubule. They also produce fluid for sperm transport, secrete the hormone inhibin, and mediate the effects of testosterone and FSH (follicle-stimulating hormone).

In the spaces between adjacent seminiferous tubules are clusters of cells called **Leydig (interstitial) cells** (Figure 28.4). These cells secrete testosterone, the most prevalent androgen. An **androgen** is a hormone that promotes the development of mas-

Figure 28.2 The scrotum, the supporting structure for the testes.

The scrotum consists of loose skin and superficial fascia and supports the testes.

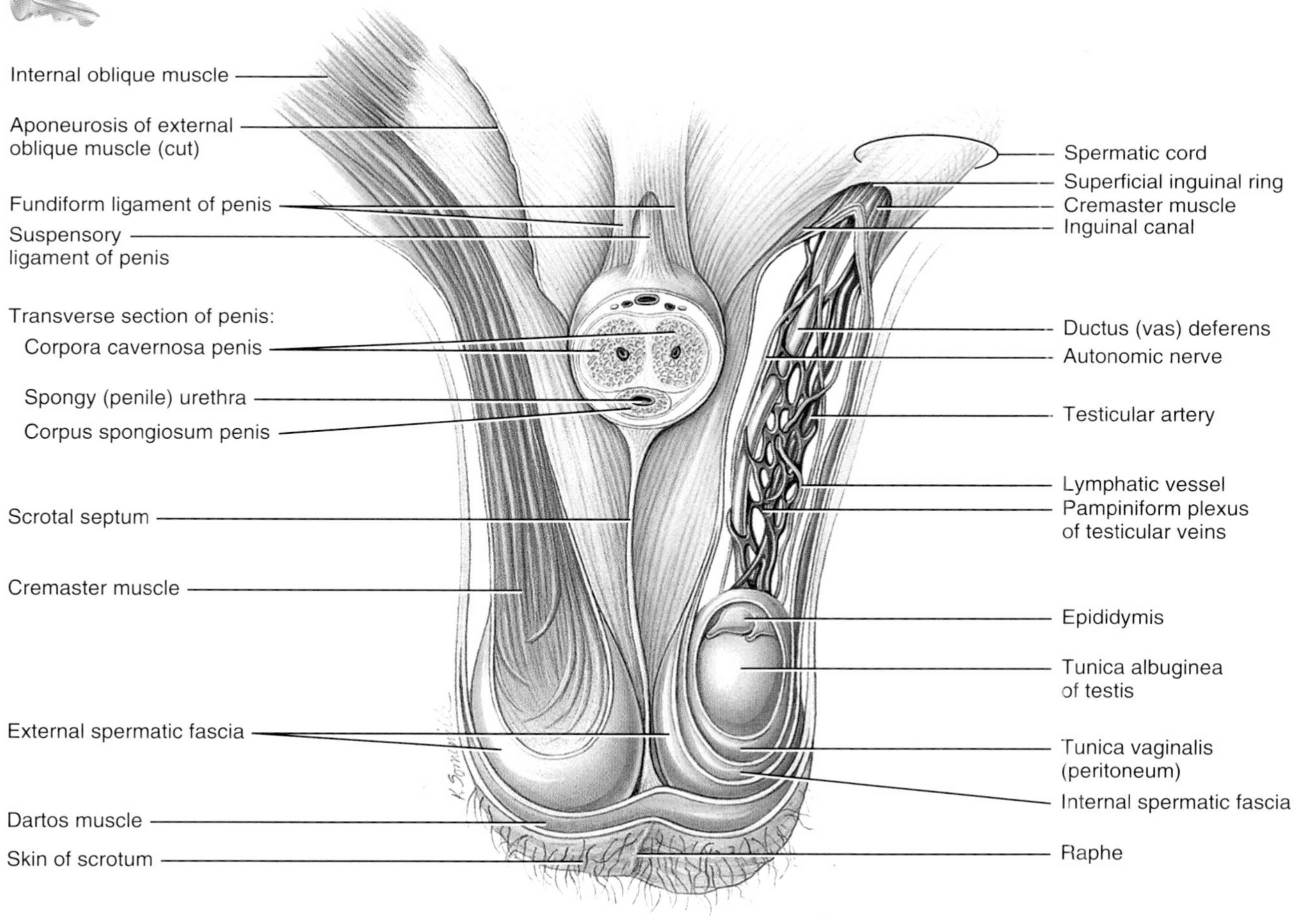

Anterior view of scrotum and testes and transverse section of penis

Which muscles help regulate the temperature of the testes?

Figure 28.3 Internal and external anatomy of a testis.

The testes are the male gonads, which produce haploid sperm.

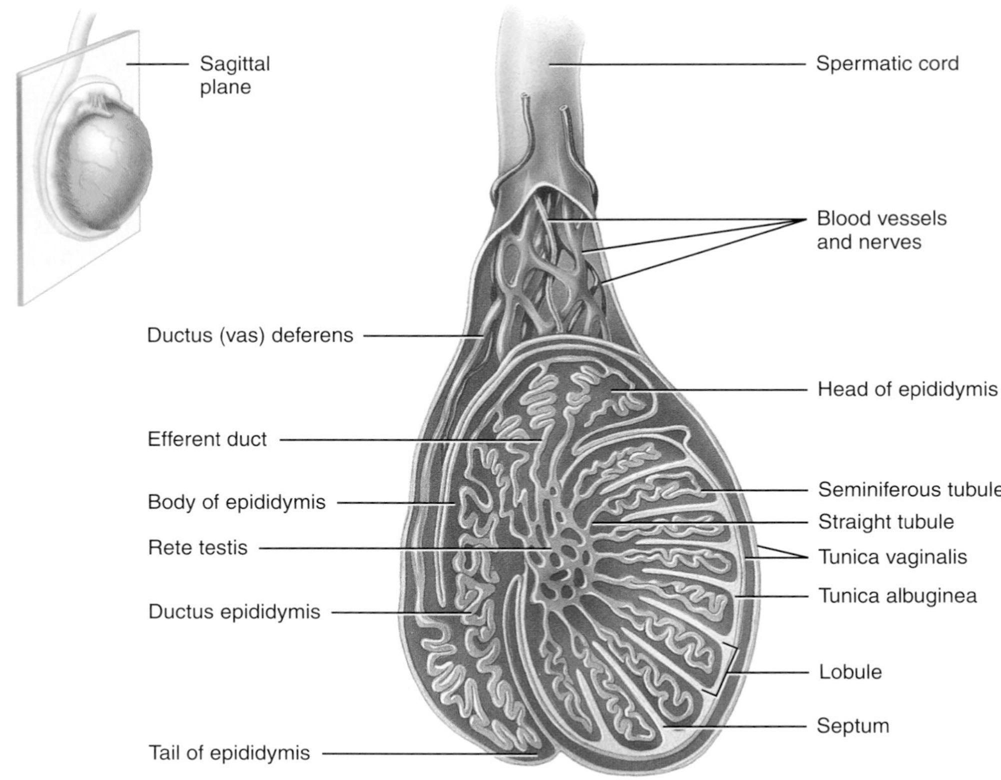

(a) Sagittal section of a testis showing seminiferous tubules

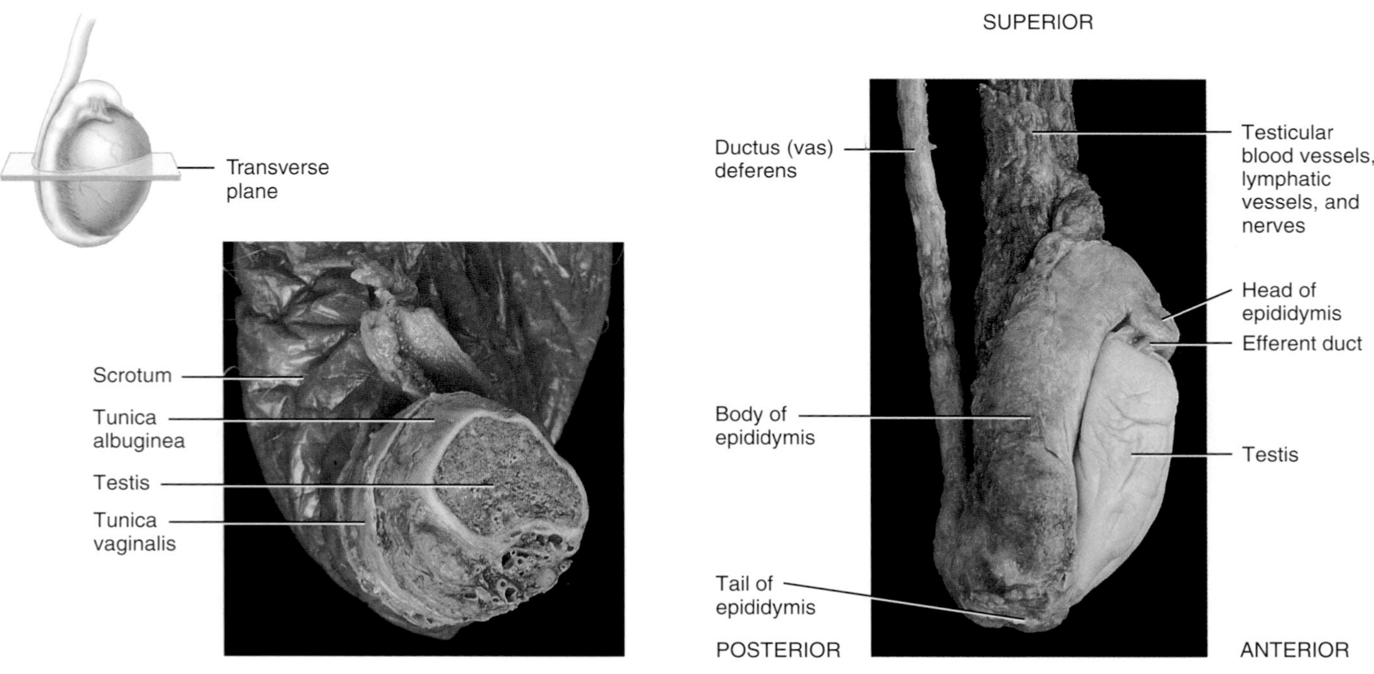

(b) Transverse section

(c) Testis and associated structures (lateral view)

What tissue layers cover and protect the testes?

Figure 28.4 Microscopic anatomy of the seminiferous tubules and stages of sperm production (spermatogenesis). Arrows in (b) indicate the progression of spermatogenic cells from least mature to most mature. The (*n*) and (2*n*) refer to haploid and diploid numbers of chromosomes, respectively.

Spermatogenesis occurs in the seminiferous tubules of the testes.

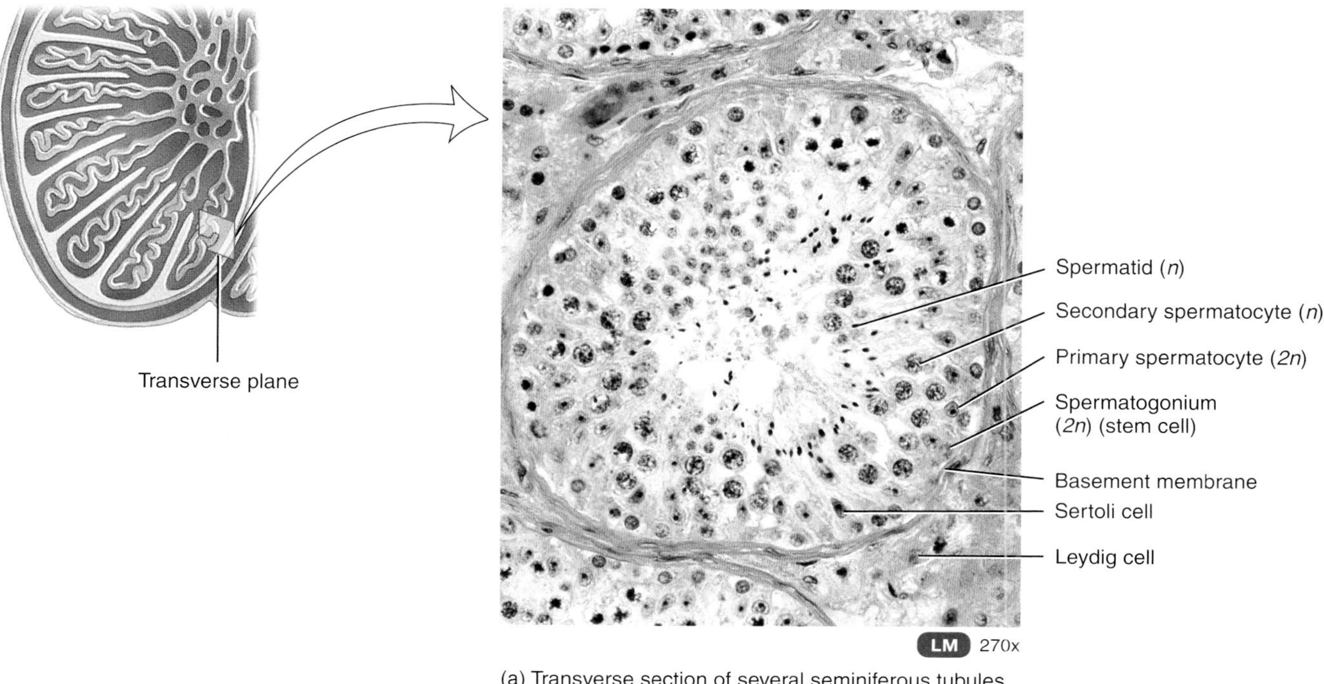

(a) Transverse section of several seminiferous tubules

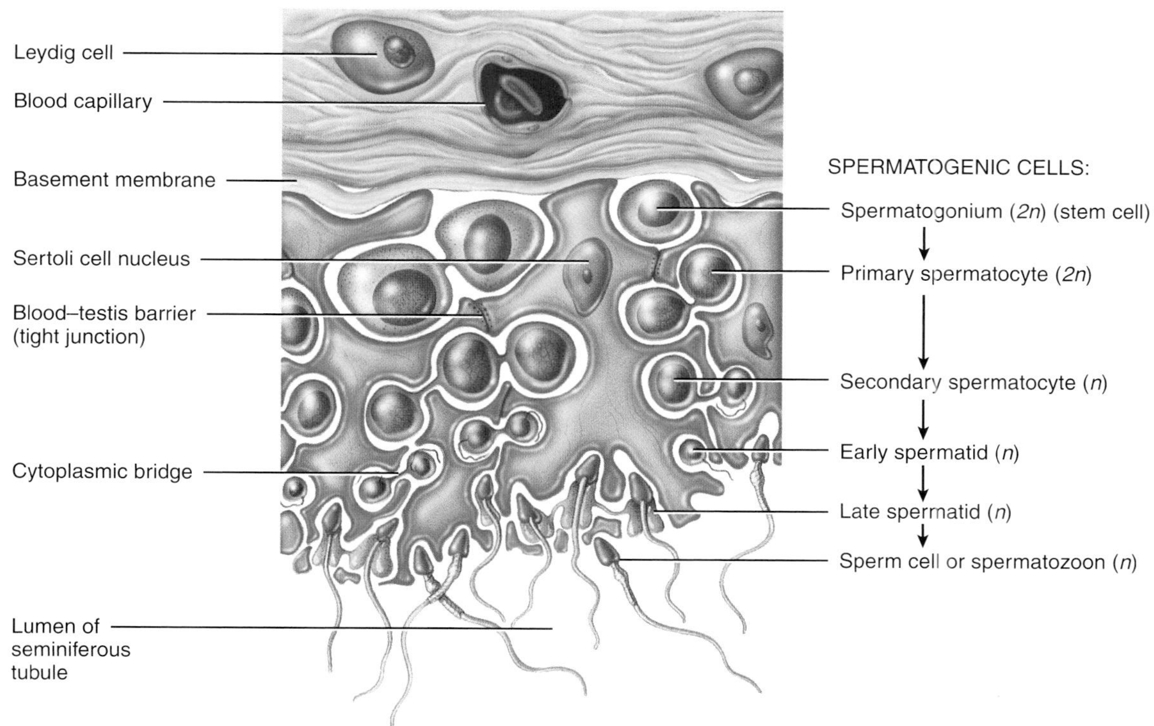

(b) Transverse section of a portion of a seminiferous tubule

Which cells secrete testosterone?

culine characteristics. Testosterone also promotes a man's libido (sexual drive).

Cryptorchidism

The condition in which the testes do not descend into the scrotum is called **cryptorchidism** (krip-TOR-ki-dizm; *crypt-* = hidden; *orchid* = testis); it occurs in about 3% of full-term infants and about 30% of premature infants. Untreated bilateral cryptorchidism results in sterility because the cells involved in the initial stages of spermatogenesis are destroyed by the higher temperature of the pelvic cavity. The chance of testicular cancer is 30–50 times greater in cryptorchid testes. The testes of about 80% of boys with cryptorchidism will descend spontaneously during the first year of life. When the testes remain undescended, the condition can be corrected surgically, ideally before 18 months of age. ■

Spermatogenesis

Before you read this section, please review the topic of reproductive cell division in Chapter 3 on pages 95–97. Pay particular attention to Figures 3.31 and 3.32, which appear on pages 96 and 97, respectively.

In humans, spermatogenesis takes 65–75 days. It begins with the spermatogonia, which contain the diploid ($2n$) number of chromosomes (Figure 28.5). Spermatogonia are types of *stem cells;* when they undergo mitosis, some spermatogonia remain near the basement membrane of the seminiferous tubule in an undifferentiated state to serve as a reservoir of cells for future mitosis and subsequent sperm production. The rest of the spermatogonia lose contact with the basement membrane, squeeze through the tight junctions of the blood–testis barrier, undergo developmental changes, and differentiate into **primary spermatocytes** (SPER-ma-tō-sītz′). Primary spermatocytes, like spermatogonia, are diploid ($2n$); that is, they have 46 chromosomes.

Shortly after it forms, each primary spermatocyte replicates its DNA and then meiosis begins (Figure 28.5). In meiosis I, homologous pairs of chromosomes line up at the metaphase plate, and crossing-over occurs. Then, the meiotic spindle pulls one (duplicated) chromosome of each pair to an opposite pole of the dividing cell. The two cells formed by meiosis I are called **secondary spermatocytes.** Each secondary spermatocyte has 23 chromosomes, the haploid number. Each chromosome within a secondary spermatocyte, however, is made up of two chromatids (two copies of the DNA) still attached by a centromere. No further replication of DNA occurs in the secondary spermatocytes.

In meiosis II, the chromosomes line up in single file along the metaphase plate, and the two chromatids of each chromosome separate. The four haploid cells resulting from meiosis II are called **spermatids.** A single primary spermatocyte therefore produces four spermatids via two rounds of cell division (meiosis I and meiosis II).

A unique process occurs during spermatogenesis. As spermatogenic cells proliferate, they fail to complete cytoplasmic separation (cytokinesis). The cells remain in contact via cytoplasmic bridges through their entire development (see Figures 28.4b and 28.5). This pattern of development most likely accounts for the synchronized production of sperm in any given area of seminiferous tubule. It may also have survival value in that half of the sperm contain an X chromosome and half contain a Y chromosome. The larger X chromosome may carry genes needed for spermatogenesis that are lacking on the smaller Y chromosome.

Figure 28.5 Events in spermatogenesis. Diploid cells ($2n$) have 46 chromosomes; haploid cells (n) have 23 chromosomes.

Spermiogenesis involves the maturation of spermatids into sperm.

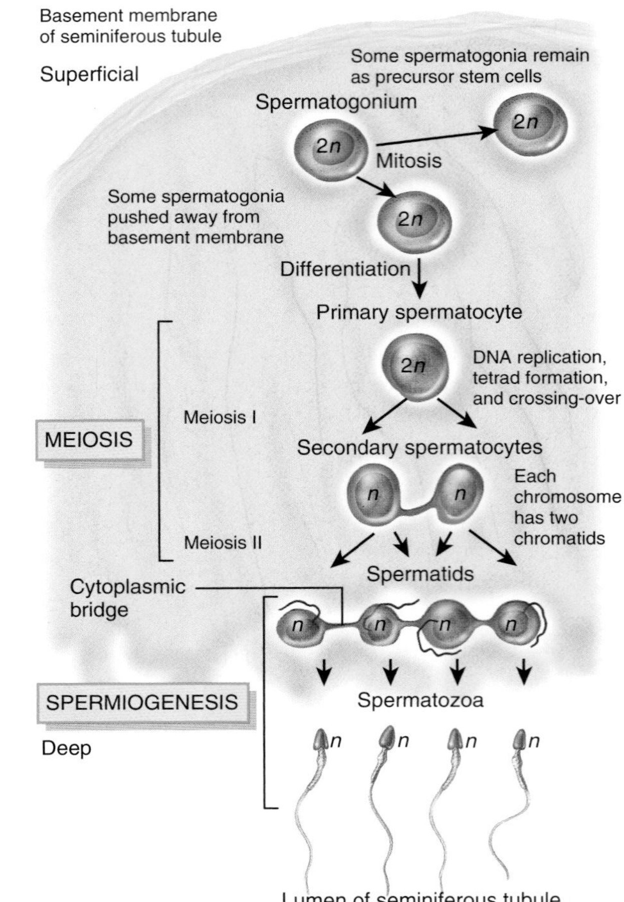

What is "reduced" during meiosis I?

The final stage of spermatogenesis, **spermiogenesis** (sper′-mē-ō-JEN-e-sis), is the development of haploid spermatids into sperm. No cell division occurs in spermiogenesis; each spermatid becomes a single **sperm cell.** During this process, spherical spermatids transform into elongated, slender sperm. An acrosome (described shortly) forms atop the nucleus, which condenses and elongates, a flagellum develops, and mitochondria multiply. Sertoli cells dispose of the excess cytoplasm that sloughs off. Finally, sperm are released from their connections to Sertoli cells, an event known as **spermiation.** Sperm then enter the lumen of the seminiferous tubule. Fluid secreted by Sertoli cells pushes sperm along their way, toward the ducts of the testes.

Sperm

Each day about 300 million sperm complete the process of spermatogenesis. A sperm is about 60 μm long and contains several structures that are highly adapted for reaching and penetrating a secondary oocyte (Figure 28.6). The major parts of a sperm are the head and the tail. The flattened, pointed **head** of the sperm is about 4–5 μm long. It contains a **nucleus** with 23 highly condensed chromosomes. Covering the anterior two-thirds of the nucleus is the **acrosome** (*acro-* = atop; *-some* = body), a caplike vesicle filled with enzymes that help a sperm to penetrate a secondary oocyte to bring about fertilization. Among the enzymes are hyaluronidase and proteases. The **tail** of a sperm is subdivided into four parts: neck, middle piece, principal piece, and end piece. The **neck** is the constricted region just behind the head that contains centrioles. The centrioles form the microtubules that comprise the remainder of the tail. The **middle piece** contains mitochondria arranged in a spiral, which provide the energy (ATP) for locomotion of sperm to the site of fertilization and for sperm metabolism. The **principal piece** is the longest portion of the tail and the **end piece** is the terminal, tapering portion of the tail. Once ejaculated, most sperm do not survive more than 48 hours within the female reproductive tract.

Figure 28.6 Parts of a sperm cell.

About 300 million sperm mature each day.

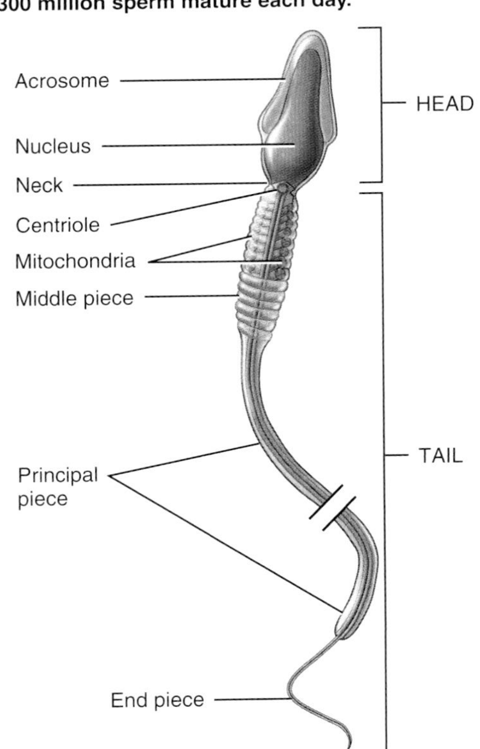

What are the functions of each part of a sperm cell?

Hormonal Control of the Testes

Although the initiating factors are unknown, at puberty certain hypothalamic neurosecretory cells increase their secretion of **gonadotropin-releasing hormone (GnRH).** This hormone, in turn, stimulates gonadotrophs in the anterior pituitary to increase their secretion of the two gonadotropins, **luteinizing hormone (LH)** and **follicle-stimulating hormone (FSH).** Figure 28.7

Figure 28.7 Hormonal control of spermatogenesis and actions of testosterone and dihydrotestosterone (DHT). In response to stimulation by FSH and testosterone, Sertoli cells secrete androgen-binding protein (ABP). Dashed red lines indicate negative feedback inhibition.

Release of FSH is stimulated by GnRH and inhibited by inhibin; release of LH is stimulated by GnRH and inhibited by testosterone.

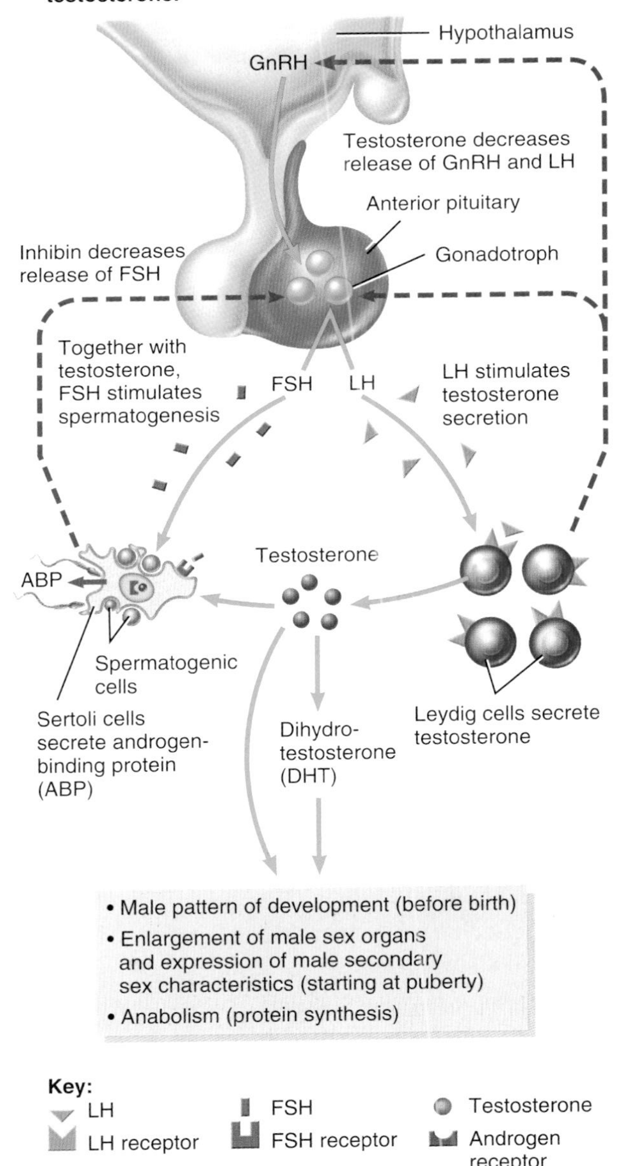

Which cells secrete inhibin?

shows the hormones and negative feedback loops that control secretion of testosterone and spermatogenesis.

LH stimulates Leydig cells, which are located between sem-iniferous tubules, to secrete the hormone **testosterone** (tes-TOS-te-rōn). This steroid hormone is synthesized from cholesterol in the testes and is the principal androgen. It is lipid-soluble and readily diffuses out of Leydig cells into the interstitial fluid and then into blood. Via negative feedback, testosterone suppresses secretion of LH by anterior pituitary gonadotrophs and suppresses secretion of GnRH by hypothalamic neurosecretory cells. In some target cells, such as those in the external genitals and prostate, the enzyme 5 alpha-reductase converts testosterone to another androgen called **dihydrotestosterone (DHT).**

FSH acts indirectly to stimulate spermatogenesis (Figure 28.7). FSH and testosterone act synergistically on the Sertoli cells to stimulate secretion of **androgen-binding protein (ABP)** into the lumen of the seminiferous tubules and into the interstitial fluid around the spermatogenic cells. ABP binds to testosterone, keeping its concentration high. Testosterone stimulates the final steps of spermatogenesis in the seminiferous tubules. Once the degree of spermatogenesis required for male reproductive functions has been achieved, Sertoli cells release **inhibin,** a protein hormone named for its role in inhibiting FSH secretion by the anterior pituitary (Figure 28.7). If spermatogenesis is proceeding too slowly, less inhibin is released, which permits more FSH secretion and an increased rate of spermatogenesis.

Testosterone and dihydrotestosterone both bind to the same androgen receptors, which are found within the nuclei of target cells. The hormone–receptor complex regulates gene expression, turning some genes on and others off. Because of these changes, the androgens produce several effects:

- ***Prenatal development.*** Before birth, testosterone stimulates the male pattern of development of reproductive system ducts and the descent of the testes. Dihydrotestosterone stimulates development of the external genitals (described on page 1093). Testosterone also is converted in the brain to estrogens (feminizing hormones), which may play a role in the development of certain regions of the brain in males.
- ***Development of male sexual characteristics.*** At puberty, testosterone and dihydrotestosterone bring about development and enlargement of the male sex organs and the development of masculine secondary sexual characteristics. These include muscular and skeletal growth that results in wide shoulders and narrow hips; pubic, axillary, facial, and chest hair (within hereditary limits); thickening of the skin; increased sebaceous (oil) gland secretion; and enlargement of the larynx and consequent deepening of the voice.
- ***Development of sexual function.*** Androgens contribute to male sexual behavior and spermatogenesis and to sex drive (libido) in both males and females. Recall that the adrenal cortex is the main source of androgens in females.
- ***Stimulation of anabolism.*** Androgens are anabolic hormones; that is, they stimulate protein synthesis. This effect is obvious in the heavier muscle and bone mass of most men as compared to women.

A negative feedback system regulates testosterone production (Figure 28.8). When testosterone concentration in the blood increases to a certain level, it inhibits the release of GnRH by cells in the hypothalamus. As a result, there is less GnRH in the portal blood that flows from the hypothalamus to the anterior

Figure 28.8 Negative feedback control of blood level of testosterone.

Gonadotrophs of the anterior pituitary produce luteinizing hormone (LH).

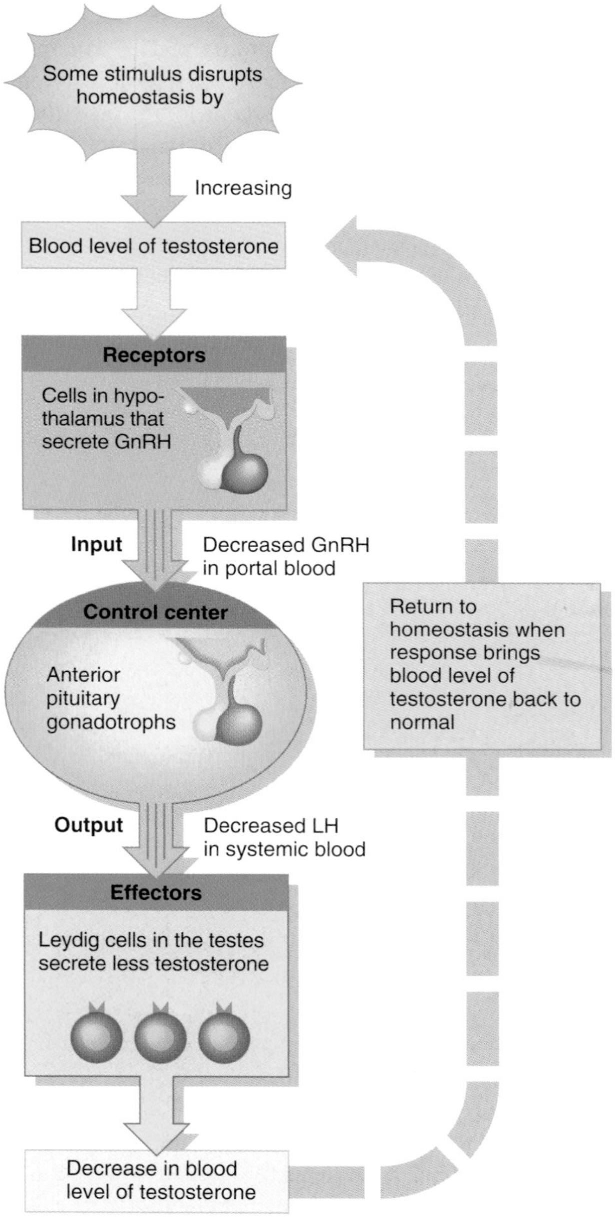

Some stimulus disrupts homeostasis by
Increasing
Blood level of testosterone
Receptors
Cells in hypothalamus that secrete GnRH
Input Decreased GnRH in portal blood
Control center
Anterior pituitary gonadotrophs
Output Decreased LH in systemic blood
Effectors
Leydig cells in the testes secrete less testosterone
Decrease in blood level of testosterone
Return to homeostasis when response brings blood level of testosterone back to normal

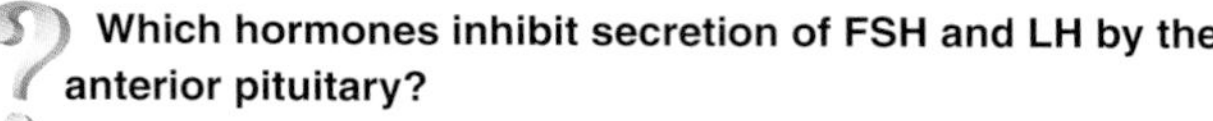

Which hormones inhibit secretion of FSH and LH by the anterior pituitary?

pituitary. Gonadotrophs in the anterior pituitary then release less LH, so the concentration of LH in systemic blood falls. With less stimulation by LH, the Leydig cells in the testes secrete less testosterone, and there is a return to homeostasis. If the testosterone concentration in the blood falls too low, however, GnRH is again released by the hypothalamus and stimulates secretion of LH by the anterior pituitary. LH, in turn, stimulates testosterone production by the testes.

▶ CHECKPOINT

1. Describe the function of the scrotum in protecting the testes from temperature fluctuations.
2. Describe the internal structure of a testis. Where are sperm cells produced? What are the functions of Sertoli cells and Leydig cells?
3. Describe the principal events of spermatogenesis.
4. Identify the parts of a sperm cell, and list the functions of each.
5. What are the roles of FSH, LH, testosterone and inhibin in the male reproductive system? How is secretion of these hormones controlled?

Reproductive System Ducts in Males

Ducts of the Testis

Pressure generated by the fluid secreted by Sertoli cells pushes sperm and fluid along the lumen of seminiferous tubules and then into a series of very short ducts called **straight tubules** (see Figure 28.3a). The straight tubules lead to a network of ducts in the testis called the **rete testis** (RĒ-tē = network). From the rete testis, sperm move into a series of coiled **efferent ducts** in the epididymis that empty into a single tube called the **ductus epididymis.**

Epididymis

The **epididymis** (ep′-i-DID-i-mis; *epi-* = above or over; *-didymis* = testis) is a comma-shaped organ about 4 cm (1.5 in.) long that lies along the posterior border of each testis (see Figure 28.3a). The plural is **epididymides** (ep′-i-did-ĪM-i-dēs). Each epididymis consists mostly of the tightly coiled **ductus epididymis.** The efferent ducts from the testis join the ductus epididymis at the larger, superior portion of the epididymis called the **head.** The **body** is the narrow midportion of the epididymis, and the **tail** is the smaller, inferior portion. At its distal end, the tail of the epididymis continues as the ductus (vas) deferens (discussed shortly).

The ductus epididymis would measure about 6 m (20 ft) in length if it were uncoiled. It is lined with pseudostratified columnar epithelium and encircled by layers of smooth muscle. The free surfaces of the columnar cells contain **stereocilia,** which despite their name are long, branching microvilli (not cilia) that increase the surface area for the reabsorption of degenerated sperm. Connective tissue around the muscle layer attaches the loops of the ductus epididymis and carries blood vessels and nerves.

Functionally, the epididymis is the site of **sperm maturation,** the process by which sperm acquire motility and the ability to fertilize an ovum. This occurs over a period of about 14 days. The epididymis also helps propel sperm into the ductus (vas) deferens during sexual arousal by peristaltic contraction of its smooth muscle. In addition, the epididymis stores sperm, which remain viable here for up to several months. Any stored sperm that are not ejaculated by that time are eventually reabsorbed.

Ductus Deferens

Within the tail of the epididymis, the ductus epididymis becomes less convoluted, and its diameter increases. Beyond this point, the duct is known as the **ductus deferens** or **vas deferens** (see Figure 28.3a). The ductus deferens, which is about 45 cm (18 in.) long, ascends along the posterior border of the epididymis, passes through the inguinal canal (see Figure 28.2), and enters the pelvic cavity. There it loops over the ureter and passes over the side and down the posterior surface of the urinary bladder (see Figure 28.1a). The dilated terminal portion of the ductus deferens is the **ampulla** (am-PUL-la = little jar; Figure 28.9). The mucosa of the ductus deferens consists of pseudostratified columnar epithelium and lamina propria (areolar connective tissue). The muscularis is composed of three layers of smooth muscle; the inner and outer layers are longitudinal, and the middle layer is circular.

Functionally, the ductus deferens conveys sperm during sexual arousal from the epididymis toward the urethra by peristaltic contractions of the muscular coat. Like the epididymis, the ductus deferens also can store sperm for several months. Any stored sperm that are not ejaculated by that time are eventually reabsorbed.

Spermatic Cord

The **spermatic cord** is a supporting structure of the male reproductive system that ascends out of the scrotum (see Figure 28.2). It consists of the ductus (vas) deferens as it ascends through the scrotum, the testicular artery, veins that drain the testes and carry testosterone into circulation (the pampiniform plexus), autonomic nerves, lymphatic vessels, and the cremaster muscle. The term **varicocele** (VAR-i-kō-sēl; *varico-* = varicose; *-kele* = hernia) refers to a swelling in the scrotum due to a dilation of the veins that drain the testes. It is usually more apparent when the person is standing and typically does not require treatment. The spermatic cord and ilioinguinal nerve pass through the **inguinal canal** (IN-gwin-al = groin), an oblique passageway in the anterior abdominal wall just superior and parallel to the medial half of the inguinal ligament. The canal, which is about 4–5 cm (about 2 in.) long, originates at the **deep (abdominal) inguinal ring,** a slitlike opening in the aponeurosis of the transversus abdominis muscle; the canal ends at the **superficial (subcutaneous) inguinal ring** (see Figure 28.2), a somewhat triangular opening in the aponeurosis of the external oblique muscle. In females, the round ligament of the uterus and ilioinguinal nerve pass through the inguinal canal.

Figure 28.9 Locations of several accessory reproductive organs in males. The prostate, urethra, and penis have been sectioned to show internal details.

The male urethra has three subdivisions: the prostatic, membranous, and spongy (penile) urethra.

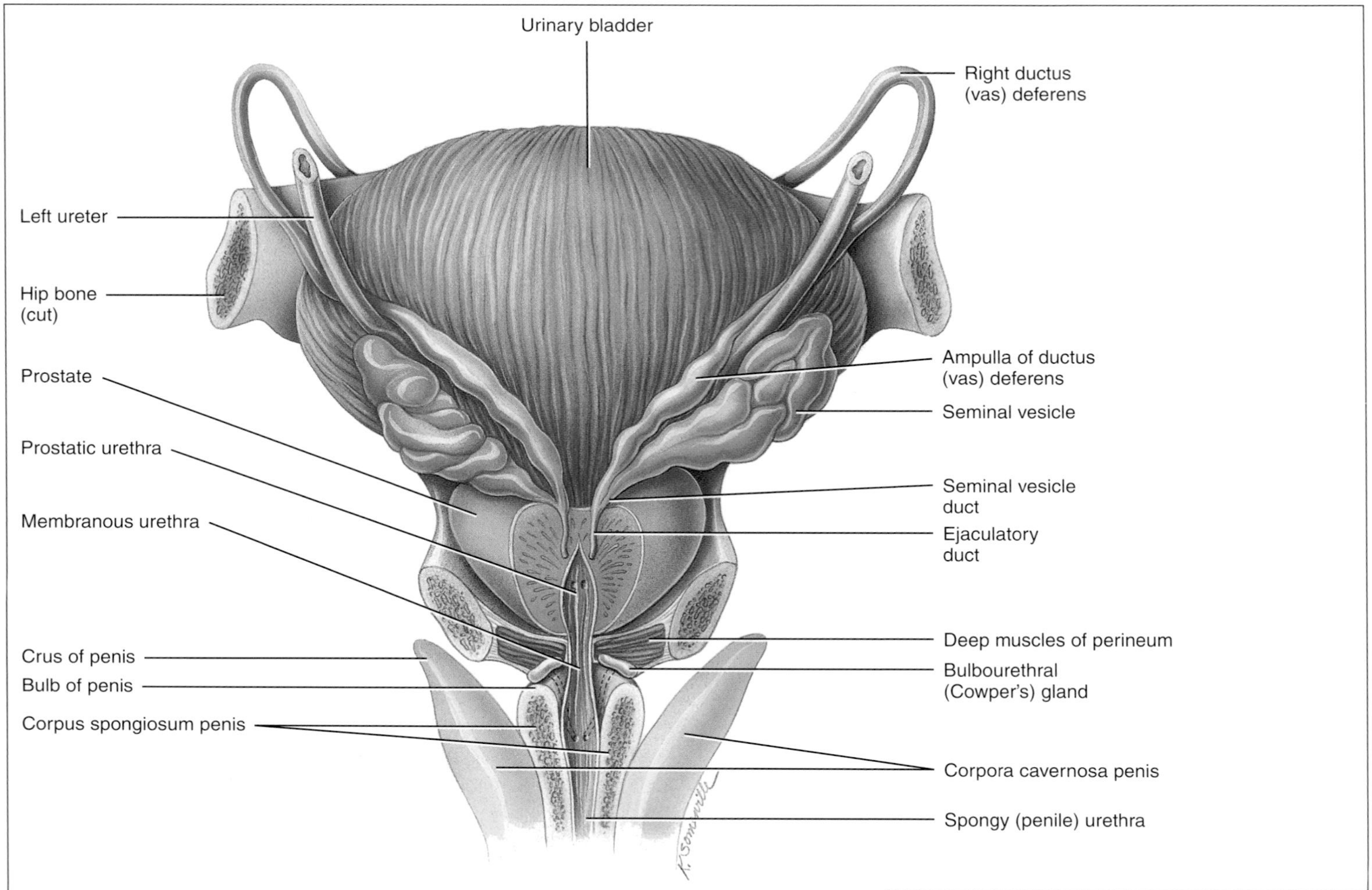

Urinary bladder
Right ductus (vas) deferens
Left ureter
Hip bone (cut)
Prostate
Prostatic urethra
Membranous urethra
Ampulla of ductus (vas) deferens
Seminal vesicle
Seminal vesicle duct
Ejaculatory duct
Deep muscles of perineum
Crus of penis
Bulb of penis
Bulbourethral (Cowper's) gland
Corpus spongiosum penis
Corpora cavernosa penis
Spongy (penile) urethra

Posterior view of male accessory organs of reproduction

Functions of Accessory Sex Gland Secretions

1. **The seminal vesicles secrete alkaline, viscous fluid that helps neutralize acid in the female reproductive tract, provides fructose for ATP production by sperm, contributes to sperm motility and viability, and helps semen coagulate after ejaculation.**
2. **The prostate secretes a milky, slightly acidic fluid that helps semen coagulate after ejaculation and subsequently breaks down the clot.**
3. **The bulbourethral (Cowper's) glands secrete alkaline fluid that neutralizes the acidic environment of the urethra and mucus that lubricates the lining of the urethra and the tip of the penis during sexual intercourse.**

What accessory sex gland contributes the majority of the seminal fluid?

Vasectomy

The principal method for sterilization of males is a **vasectomy** (vas-EK-tō-mē; *-ectomy* = cut out), in which a portion of each ductus deferens is removed. An incision is made on either side of the scrotum, the ducts are located and cut, each is tied (ligated) in two places with stitches, and the portion between the ties is removed. Although sperm production continues in the testes, sperm can no longer reach the exterior. The sperm degenerate and are destroyed by phagocytosis. Because the blood vessels are not cut, testosterone levels in the blood remain normal, so vasectomy has no effect on sexual desire and performance. If done correctly, it is close to 100% effective. The procedure can be reversed, but the chance of regaining fertility is only 30–40%. ■

Ejaculatory Ducts

Each **ejaculatory duct** (e-JAK-ū-la-tō′-rē; *ejacul-* = to expel) is about 2 cm (1 in.) long and is formed by the union of the duct from the seminal vesicle and the ampulla of the ductus (vas) deferens (Figure 28.9). The ejaculatory ducts form just superior to the base (superior portion) of the prostate and pass inferiorly and anteriorly through the prostate. They terminate in the prostatic urethra, where they eject sperm and seminal vesicle secretions just before the release of semen from the urethra to the exterior.

Urethra

In males, the **urethra** is the shared terminal duct of the reproductive and urinary systems; it serves as a passageway for both semen and urine. About 20 cm (8 in.) long, it passes through the prostate, the deep muscles of the perineum, and the penis, and is subdivided into three parts (see Figures 28.1 and 26.22). The **prostatic urethra** is 2–3 cm (1 in.) long and passes through the prostate. As this duct continues inferiorly, it passes through the deep muscles of the perineum, where it is known as the **membranous urethra.** The membranous urethra is about 1 cm (0.5 in.) in length. As this duct passes through the corpus spongiosum of the penis, it is known as the **spongy (penile) urethra,** which is about 15–20 cm (6–8 in.) long. The spongy urethra ends at the **external urethral orifice.** The histology of the male urethra may be reviewed on page 1026 of Chapter 26.

▶ CHECKPOINT

6. Which ducts transport sperm within the testes?
7. Describe the location, structure, and functions of the ductus epididymis, ductus (vas) deferens, and ejaculatory duct.
8. Give the locations of the three subdivisions of the male urethra.
9. Trace the course of sperm through the system of ducts from the seminiferous tubules through the urethra.
10. List the structures within the spermatic cord.

Accessory Sex Glands

The ducts of the male reproductive system store and transport sperm cells, but the **accessory sex glands** secrete most of the liquid portion of semen. The accessory sex glands include the seminal vesicles, the prostate, and the bulbourethral glands.

Seminal Vesicles

The paired **seminal vesicles** (VES-i-kuls) or **seminal glands** are convoluted pouchlike structures, about 5 cm (2 in.) in length, lying posterior to the base of the urinary bladder and anterior to the rectum (Figure 28.9). They secrete an alkaline, viscous fluid that contains fructose (a monosaccharide sugar), prostaglandins, and clotting proteins that are different from those in blood. The alkaline nature of the seminal fluid helps to neutralize the acidic environment of the male urethra and female reproductive tract that otherwise would inactivate and kill sperm. The fructose is used for ATP production by sperm. Prostaglandins contribute to sperm motility and viability and may stimulate smooth muscle contractions within the female reproductive tract. The clotting proteins help semen coagulate after ejaculation. Fluid secreted by the seminal vesicles normally constitutes about 60% of the volume of semen.

Prostate

The **prostate** (PROS-tāt) is a single, doughnut-shaped gland about the size of a golf ball. It measures about 4 cm (1.6 in.) from side to side, about 3 cm (1.2 in.) from top to bottom, and about 2 cm (0.8 in.) from front to back. It is inferior to the urinary bladder and surrounds the prostatic urethra (Figure 28.9). The prostate slowly increases in size from birth to puberty. It then expands rapidly until about age 30, after which time its size typically remains stable until about age 45, when further enlargement may occur.

The prostate secretes a milky, slightly acidic fluid (pH about 6.5) that contains several substances. (1) *Citric acid* in prostatic fluid is used by sperm for ATP production via the Krebs cycle. (2) Several *proteolytic enzymes,* such as *prostate-specific antigen (PSA),* pepsinogen, lysozyme, amylase, and hyaluronidase, eventually break down the clotting proteins from the seminal vesicles. (3) The function of the *acid phosphatase* secreted by the prostate is unknown. (4) *Seminalplasmin* in prostatic fluid is an antibiotic that can destroy bacteria. Seminalplasmin may help decrease the number of naturally occurring bacteria in semen and in the lower female reproductive tract. Secretions of the prostate enter the prostatic urethra through many prostatic ducts. Prostatic secretions make up about 25% of the volume of semen and contribute to sperm motility and viability.

Bulbourethral Glands

The paired **bulbourethral glands** (bul′-bō-ū-RĒ-thral), or **Cowper's glands,** are about the size of peas. They are located inferior to the prostate on either side of the membranous urethra within the deep muscles of the perineum, and their ducts open into the spongy urethra (Figure 28.9). During sexual arousal, the bulbourethral glands secrete an alkaline fluid into the urethra that protects the passing sperm by neutralizing acids from urine in the urethra. They also secrete mucus that lubricates the end of the penis and the lining of the urethra, decreasing the number of sperm damaged during ejaculation.

Semen

Semen (= seed) is a mixture of sperm and **seminal fluid,** a liquid that consists of the secretions of the seminiferous tubules, seminal vesicles, prostate, and bulbourethral glands. The volume of semen in a typical ejaculation is 2.5–5 milliliter (mL), with 50–150 million sperm per mL. When the number falls below 20 million/mL, the male is likely to be infertile. A very large number of sperm is required for successful fertilization because only a tiny fraction ever reaches the secondary oocyte.

Despite the slight acidity of prostatic fluid, semen has a slightly alkaline pH of 7.2–7.7 due to the higher pH and larger volume of fluid from the seminal vesicles. The prostatic secretion gives semen a milky appearance, and fluids from the seminal vesicles and bulbourethral glands give it a sticky consistency. Seminal fluid provides sperm with a transportation medium, nutrients, and protection from the hostile acidic environment of the male's urethra and the female's vagina.

Once ejaculated, liquid semen coagulates within 5 minutes due to the presence of clotting proteins from the seminal vesicles. The functional role of semen coagulation is not known, but the proteins involved are different from those that cause blood coagulation. After about 10 to 20 minutes, semen reliquefies because prostate-specific antigen (PSA) and other proteolytic enzymes produced by the prostate break down the clot. Abnormal or delayed liquefaction of clotted semen may cause complete or partial immobilization of sperm, thereby inhibiting their movement through the cervix of the uterus. The presence of blood in semen is called **hemospermia** (hē-mō-SPER-mē-a; *hemo-* = blood; *-sperma* = seed). In most cases, it is caused by inflammation of the blood vessels lining the seminal vesicles; it is usually treated with antibiotics.

Penis

The **penis** contains the urethra and is a passageway for the ejaculation of semen and the excretion of urine (Figure 28.10). It is cylindrical in shape and consists of a body, glans penis, and a root. The **body of the penis** is composed of three cylindrical masses of tissue, each surrounded by fibrous tissue called the **tunica albuginea** (Figure 28.10). The two dorsolateral masses are called the **corpora cavernosa penis** (*corpora* = main bodies; *cavernosa* = hollow). The smaller midventral mass, the **corpus spongiosum penis,** contains the spongy urethra and keeps it open during ejaculation. Fascia and skin enclose all three masses, which consist of erectile tissue. *Erectile tissue* is composed of numerous blood sinuses (vascular spaces) lined by endothelial cells and surrounded by smooth muscle and elastic connective tissue.

The distal end of the corpus spongiosum penis is a slightly enlarged, acorn-shaped region called the **glans penis;** its margin is the **corona.** The distal urethra enlarges within the glans penis and forms a terminal slitlike opening, the **external urethral orifice.** Covering the glans in an uncircumcised penis is the loosely fitting **prepuce** (PRĒ-poos), or **foreskin.**

The **root of the penis** is the attached portion (proximal portion). It consists of the **bulb of the penis,** the expanded portion of the base of the corpus spongiosum penis, and the **crura of the penis** (singular is *crus* = resembling a leg), the two separated and tapered portions of the corpora cavernosa penis. The bulb of the penis is attached to the inferior surface of the deep muscles of the perineum and is enclosed by the bulbospongiosus muscle. Each crus of the penis is attached to the ischial and inferior pubic rami and is surrounded by the ischiocavernosus muscle (see Figure 11.13 on page 359). Contraction of these skeletal muscles aids ejaculation. The weight of the penis is supported by two ligaments that are continuous with the fascia of the penis. (1) The **fundiform ligament** arises from the inferior part of the linea alba. (2) The **suspensory ligament of the penis** arises from the pubic symphysis.

Circumcision

Circumcision (= to cut around) is a surgical procedure in which part of or the entire prepuce is removed. It is usually performed just after delivery, 3 to 4 days after birth, or on the eighth day as part of a Jewish religious rite. Although most health-care professionals find no medical justification for circumcision, some feel that it has benefits, such as a lower risk of urinary tract infections, protection against penile cancer, and possibly a lower risk for sexually transmitted diseases. Indeed, studies in several African villages have found lower rates of HIV infection among circumcised men. ■

Upon sexual stimulation (visual, tactile, auditory, olfactory, or imagined), parasympathetic fibers from the sacral portion of the spinal cord initiate and maintain an **erection,** the enlargement and stiffening of the penis. The parasympathetic fibers release and cause local production of nitric oxide (NO). The NO causes smooth muscle in the walls of arterioles supplying erectile tissue to relax, which allows these blood vessels to dilate. This, in turn, causes large amounts of blood to enter the erectile tissue of the penis. NO also causes the smooth muscle within the erectile tissue to relax, resulting in widening of the blood sinuses. The combination of increased blood flow and widening of the blood sinuses results in an erection. Expansion of the blood sinuses also compresses the veins that drain the penis; the slowing of blood outflow helps to maintain the erection.

The term **priapism** (PRĪ-a-pizm) refers to a persistent and usually painful erection of the corpora cavernosa of the penis that does not involve sexual desire or excitement. The condition may last up to several hours and is accompanied by pain and tenderness. It results from abnormalities of blood vessels and nerves, usually in response to medication used to produce erections in males who otherwise cannot attain them. Other causes include a spinal cord disorder, leukemia, sickle-cell disease, or a pelvic tumor.

Ejaculation (ē-jak-ū-LĀ-shun; *ejectus-* = to throw out), the powerful release of semen from the urethra to the exterior, is a sympathetic reflex coordinated by the lumbar portion of the spinal cord. As part of the reflex, the smooth muscle

Figure 28.10 Internal structure of the penis. The inset in (b) shows details of the skin and fasciae. (See Tortora, *A Photographic Atlas of the Human Body, Second Edition,* Figure 14.6.)

The penis contains the urethra, a pathway for the ejaculation of semen and the excretion of urine.

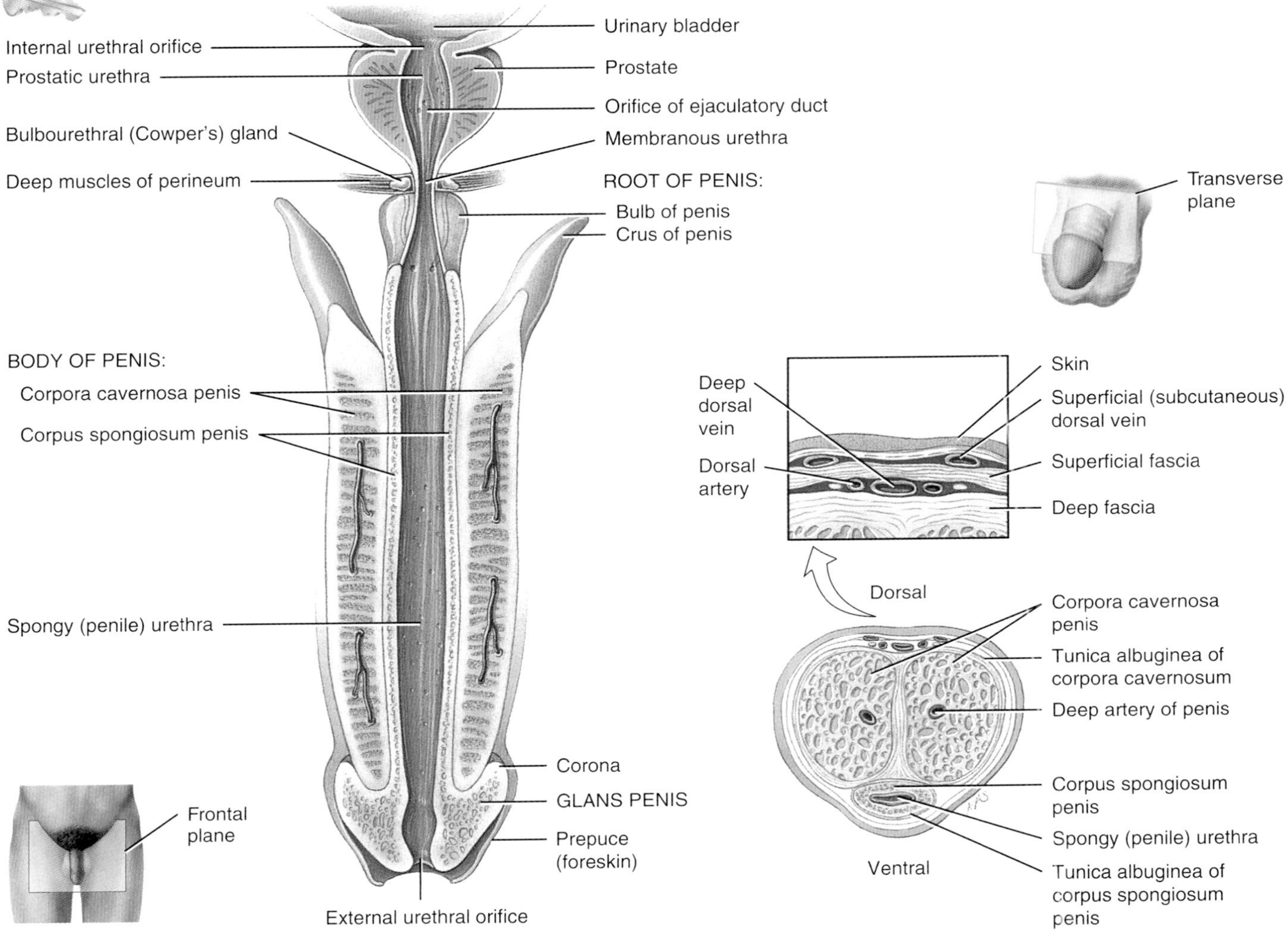

Urinary bladder
Internal urethral orifice
Prostate
Prostatic urethra
Orifice of ejaculatory duct
Bulbourethral (Cowper's) gland
Membranous urethra
Deep muscles of perineum
ROOT OF PENIS:
Bulb of penis
Crus of penis
BODY OF PENIS:
Corpora cavernosa penis
Corpus spongiosum penis
Spongy (penile) urethra
Corona
GLANS PENIS
Prepuce (foreskin)
External urethral orifice
Frontal plane

(a) Frontal section

Transverse plane
Deep dorsal vein
Dorsal artery
Skin
Superficial (subcutaneous) dorsal vein
Superficial fascia
Deep fascia
Dorsal
Corpora cavernosa penis
Tunica albuginea of corpora cavernosum
Deep artery of penis
Corpus spongiosum penis
Spongy (penile) urethra
Tunica albuginea of corpus spongiosum penis
Ventral

(b) Transverse section

Which tissue masses form the erectile tissue in the penis, and why do they become rigid during sexual arousal?

sphincter at the base of the urinary bladder closes, preventing urine from being expelled during ejaculation, and semen from entering the urinary bladder. Even before ejaculation occurs, peristaltic contractions in the epididymis, ductus (vas) deferens, seminal vesicles, ejaculatory ducts, and prostate propel semen into the penile portion of the urethra (spongy urethra). Typically, this leads to **emission** (ē-MISH-un), the discharge of a small volume of semen before ejaculation. Emission may also occur during sleep (nocturnal emission). The musculature of the penis (bulbospongiosus, ischiocavernosus, and superficial transverse perineus muscles), which is supplied by the pudendal nerve, also contracts at ejaculation (see Figure 11.13 on page 359).

Once sexual stimulation of the penis has ended, the arterioles supplying the erectile tissue of the penis constrict and the smooth muscle within erectile tissue contracts, making the blood sinuses smaller. This relieves pressure on the veins supplying the penis and allows the blood to drain through them. Consequently, the penis returns to its flaccid (relaxed) state.

Premature Ejaculation

A **premature ejaculation** is ejaculation that occurs too early, for example, during foreplay or upon or shortly after penetration. It is usually caused by anxiety, other psychological causes, or an unusually sensitive foreskin or glans penis. For most males, premature ejaculation can be overcome by various techniques (such as squeezing the penis between the glans penis and shaft as ejaculation approaches), behavioral therapy, or medication. ■

▶ **CHECKPOINT**

11. Briefly explain the locations and functions of the seminal vesicles, the prostate, and the bulbourethral (Cowper's) glands.
12. What is semen? What is its function?
13. Explain the physiological processes involved in erection and ejaculation.

FEMALE REPRODUCTIVE SYSTEM

▶ **OBJECTIVES**

Describe the location, structure, and functions of the organs of the female reproductive system.

Discuss the process of oogenesis in the ovaries.

The organs of the female reproductive system (Figure 28.11) include the ovaries (female gonads); the uterine (fallopian) tubes, or oviducts; the uterus; the vagina; and external organs, which are collectively called the vulva, or pudendum. The mammary glands are considered part of both the integumentary system and the female reproductive system.

Functions of the Female Reproductive System

1. **The ovaries produce secondary oocytes and hormones, including progesterone and estrogens (female sex hormones), inhibin, and relaxin.**
2. **The uterine tubes transport a secondary oocyte to the uterus and normally are the sites where fertilization occurs.**
3. **The uterus is the site of implantation of a fertilized ovum, development of the fetus during pregnancy, and labor.**
4. **The vagina receives the penis during sexual intercourse and is a passageway for childbirth.**
5. **The mammary glands synthesize, secrete, and eject milk for nourishment of the newborn.**

Figure 28.11 **Organs of reproduction and surrounding structures in females.**

The organs of reproduction in females include the ovaries, uterine (fallopian) tubes, uterus, vagina, vulva, and mammary glands.

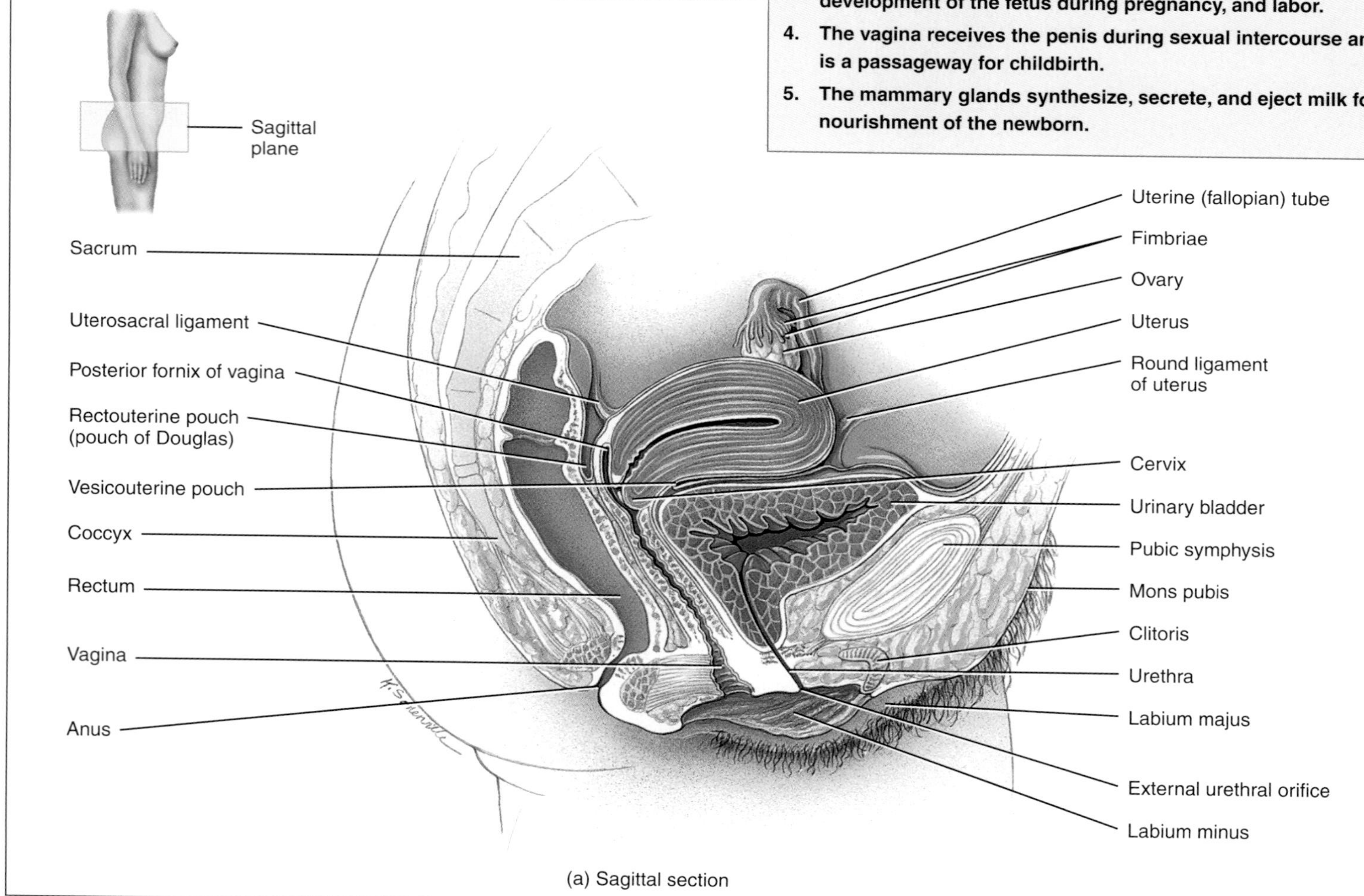

(a) Sagittal section

Ovaries

The **ovaries** (= egg receptacles), which are the female gonads, are paired glands that resemble unshelled almonds in size and shape; they are homologous to the testes. (Here *homologous* means that two organs have the same embryonic origin.) The ovaries produce (1) gametes, secondary oocytes that develop into mature ova (eggs) after fertilization, and (2) hormones, including progesterone and estrogens (the female sex hormones), inhibin, and relaxin.

The ovaries, one on either side of the uterus, descend to the brim of the superior portion of the pelvic cavity during the third month of development. A series of ligaments holds them in position (Figure 28.12). The **broad ligament** of the uterus (see also Figure 28.11b), which is itself part of the parietal peritoneum, attaches to the ovaries by a double-layered fold of peritoneum called the **mesovarium.** The **ovarian ligament** anchors the ovaries to the uterus, and the **suspensory ligament** attaches them to the pelvic wall. Each ovary contains a **hilum,** the point of entrance and exit for blood vessels and nerves along which the mesovarium is attached.

Histology of the Ovary

Each ovary consists of the following parts (Figure 28.13 on page 1073):

- The **germinal epithelium** (*germen* = sprout or bud) is a layer of simple epithelium (low cuboidal or squamous) that covers the surface of the ovary. The term germinal epithelium is a misnomer because it does not give rise to ova; the name came about because, at one time, people believed that it did. Now we know that the progenitors of ova arise from the yolk sac and migrate to the ovaries during embryonic development.
- The **tunica albuginea** is a whitish capsule of dense irregular connective tissue located immediately deep to the germinal epithelium.
- The **ovarian cortex** is a region just deep to the tunica albuginea. It consists of ovarian follicles (described shortly) surrounded by dense irregular connective tissue that contains scattered smooth muscle cells.
- The **ovarian medulla** is deep to the ovarian cortex. The border between the cortex and medulla is indistinct, but

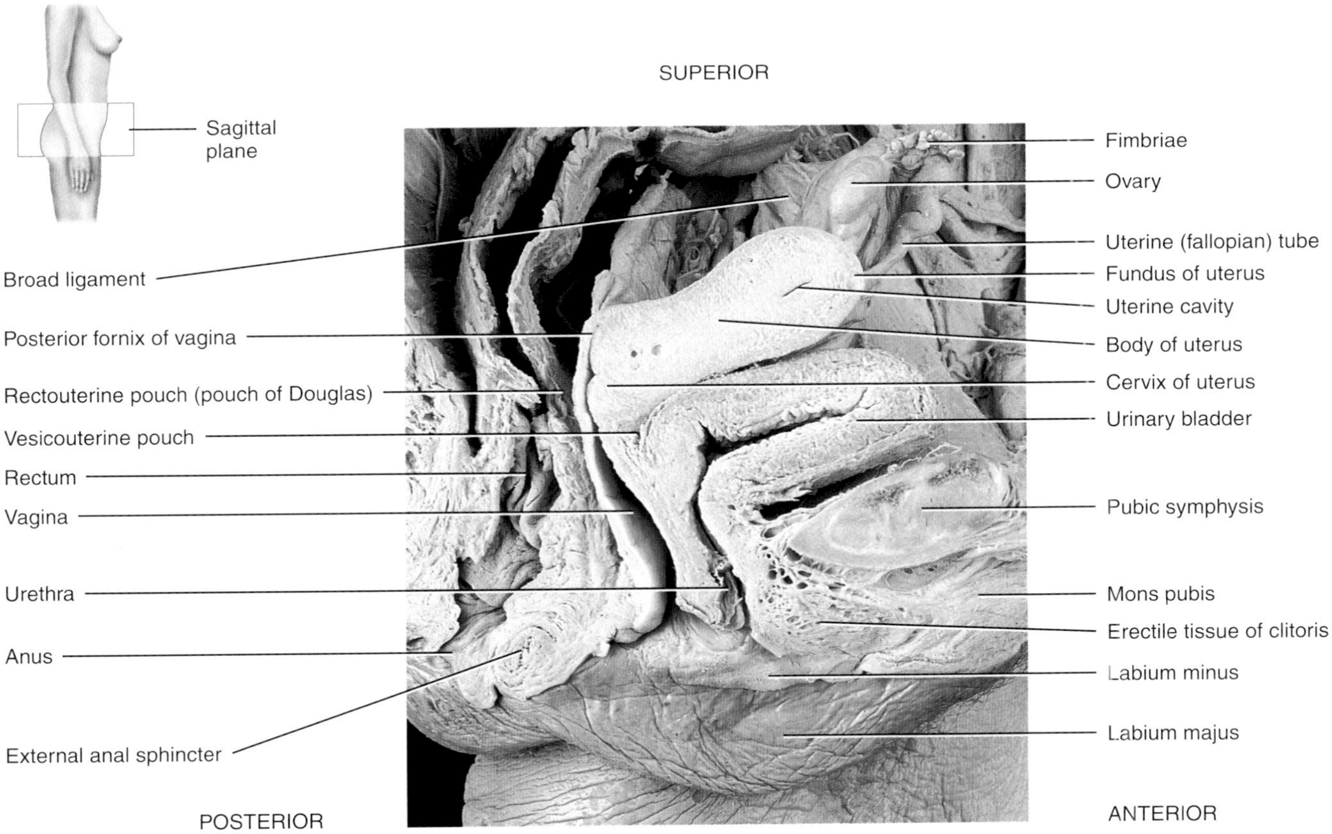

SUPERIOR

Sagittal plane

Broad ligament
Posterior fornix of vagina
Rectouterine pouch (pouch of Douglas)
Vesicouterine pouch
Rectum
Vagina
Urethra
Anus
External anal sphincter

Fimbriae
Ovary
Uterine (fallopian) tube
Fundus of uterus
Uterine cavity
Body of uterus
Cervix of uterus
Urinary bladder
Pubic symphysis
Mons pubis
Erectile tissue of clitoris
Labium minus
Labium majus

POSTERIOR

ANTERIOR

(b) Sagittal section

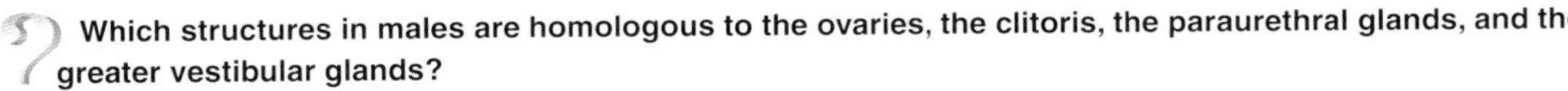

Which structures in males are homologous to the ovaries, the clitoris, the paraurethral glands, and the greater vestibular glands?

Figure 28.12 Relative positions of the ovaries, the uterus, and the ligaments that support them.

Ligaments holding the ovaries in position are the mesovarium, the ovarian ligament, and the suspensory ligament.

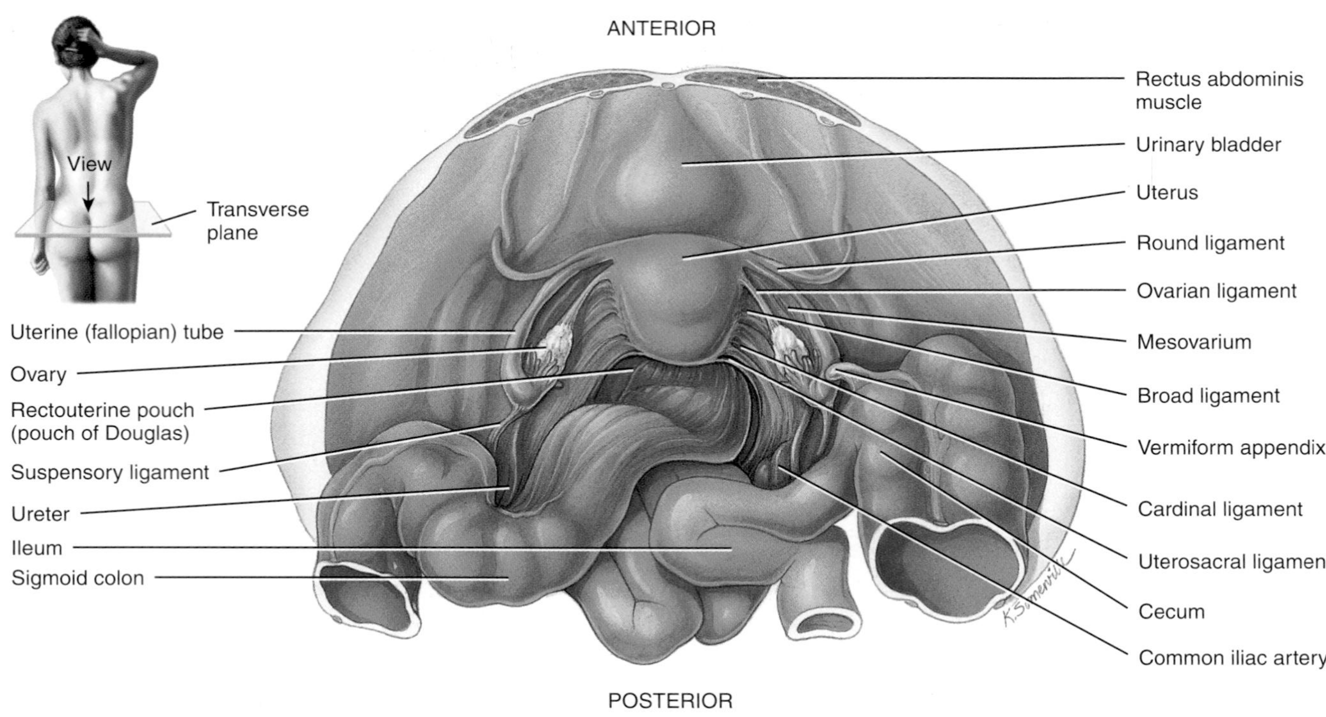

Superior view of transverse section

To which structures do the mesovarium, ovarian ligament, and suspensory ligament anchor the ovary?

the medulla consists of more loosely arranged connective tissue and contains blood vessels, lymphatic vessels, and nerves.

- **Ovarian follicles** (*folliculus* = little bag) are in the cortex and consist of **oocytes** in various stages of development, plus the cells surrounding them. When the surrounding cells form a single layer, they are called **follicular cells;** later in development, when they form several layers, they are referred to as **granulosa cells.** The surrounding cells nourish the developing oocyte and begin to secrete estrogens as the follicle grows larger.
- A **mature (graafian) follicle** is a large, fluid-filled follicle that is ready to rupture and expel its secondary oocyte, a process known as **ovulation.**
- A **corpus luteum** (= yellow body) contains the remnants of a mature follicle after ovulation. The corpus luteum produces progesterone, estrogens, relaxin, and inhibin until it degenerates into fibrous scar tissue called the **corpus albicans** (= white body).

Oogenesis and Follicular Development

The formation of gametes in the ovaries is termed **oogenesis** (ō-ō-JEN-e-sis; *oo-* = egg). In contrast to spermatogenesis, which begins in males at puberty, oogenesis begins in females before they are even born. Oogenesis occurs in essentially the same manner as spermatogenesis; meiosis (see Chapter 3) takes place and the resulting germ cells undergo maturation.

During early fetal development, primordial (primitive) germ cells migrate from the yolk sac to the ovaries. There, germ cells differentiate within the ovaries into **oogonia** (ō′-ō-GŌ-nē-a; singular is *oogonium*). Oogonia are diploid ($2n$) stem cells that divide mitotically to produce millions of germ cells. Even before birth, most of these germ cells degenerate in a process known as **atresia** (a-TRĒ-zē-a). A few, however, develop into larger cells called **primary oocytes** (Ō-ō-sītz) that enter prophase of meiosis I during fetal development but do not complete that phase until after puberty. During this arrested stage of development, each primary oocyte is surrounded by a single layer of follicular cells, and the entire structure is called a **primordial follicle** (Figure 28.14a on page 1074). At birth, approximately 200,000 to 2,000,000 pri-

Figure 28.13 Histology of the ovary. The arrows in (a) indicate the sequence of developmental stages that occur as part of the maturation of an ovum during the ovarian cycle.

The ovaries are the female gonads; they produce haploid oocytes.

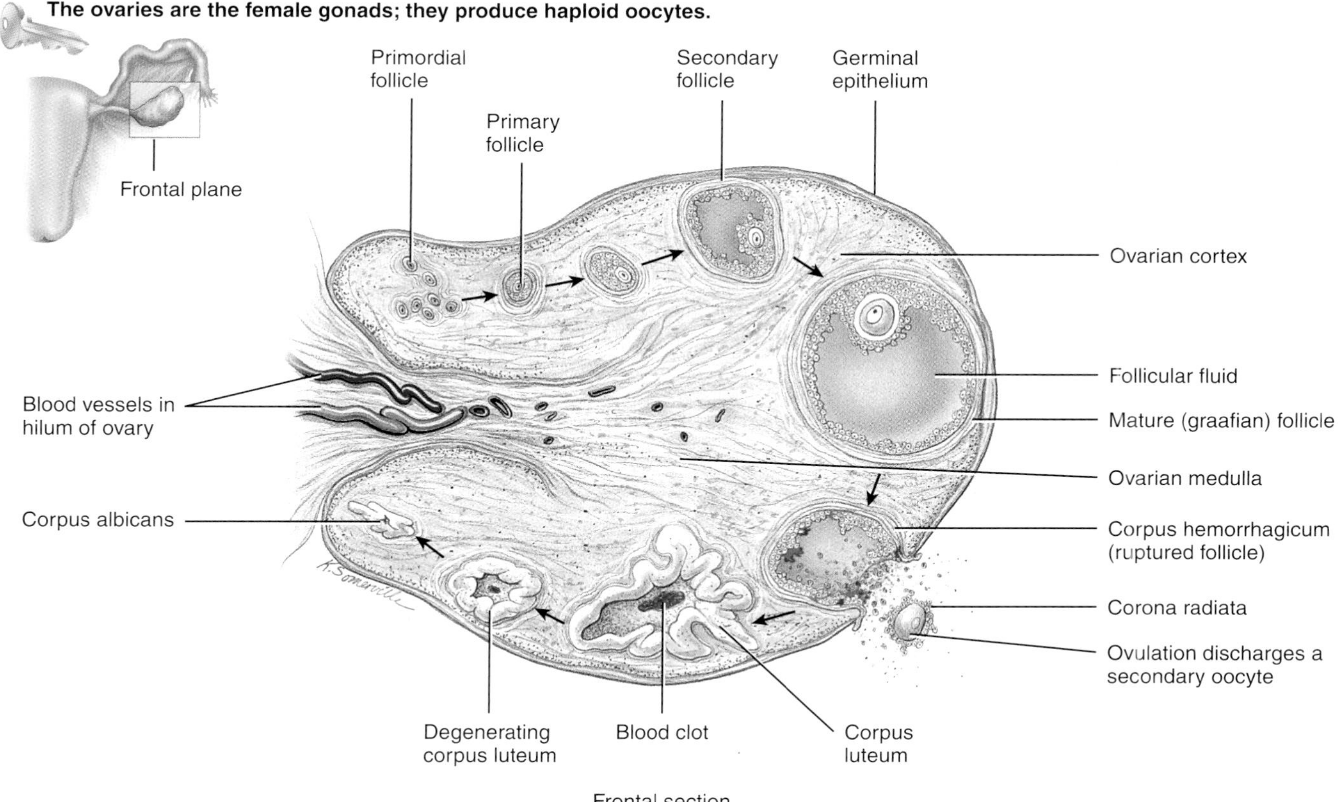

Frontal section

What structures in the ovary contain endocrine tissue, and what hormones do they secrete?

mary oocytes remain in each ovary. Of these, about 40,000 are still present at puberty, and around 400 will mature and ovulate during a woman's reproductive lifetime. The remainder of the primary oocytes undergo atresia.

Each month after puberty until menopause, gonadotropins (FSH and LH) secreted by the anterior pituitary further stimulate the development of several primordial follicles, although only one will typically reach the maturity needed for ovulation. A few primordial follicles start to grow, developing into **primary follicles** (Figure 28.14a). Each primary follicle consists of a primary oocyte that is surrounded by several layers of cuboidal and low-columnar cells called **granulosa cells.** As a primary follicle grows, it forms a clear glycoprotein layer, called the **zona pellucida** (pe-LOO-si-da), between the primary oocyte and the granulosa cells.

The outermost layer of granulosa cells rest on a basement membrane. Encircling the basement membrane is a region called the **theca folliculi.** As a primary follicle develops into a **secondary follicle,** the theca differentiates into two layers of cells: (1) the **theca interna,** a highly vascularized internal layer of cuboidal secretory cells and (2) the **theca externa,** an outer layer of connective tissue cells and collagen fibers. In addition, the granulosa cells begin to secrete follicular fluid, which builds up in a cavity called the **antrum** in the center of the secondary follicle. Furthermore, the innermost layer of granulosa cells becomes firmly attached to the zona pellucida and is now called the **corona radiata** (*corona* = crown; *radiata* = radiation) (Figure 28.14b).

The secondary follicle eventually becomes larger, turning into a **mature (graafian) follicle.** While in this follicle, the diploid primary oocyte completes meiosis I, producing two haploid cells of unequal size—each with 23 chromosomes (Figure 28.15). The smaller cell produced by meiosis I, called the **first polar body,** is essentially a packet of discarded nuclear material. The larger cell, known as the **secondary oocyte,** receives most of the cytoplasm. Once a secondary oocyte is formed, it begins meiosis II but then stops in metaphase. The mature (graafian) follicle soon ruptures and releases its secondary oocyte, a process known as **ovulation.**

At ovulation, the secondary oocyte is expelled into the pelvic cavity together with the first polar body and corona radiata. Normally these cells are swept into the uterine tube. If fertil-

Figure 28.14 Ovarian follicles. (a) Primordial and primary follicles in the ovarian cortex. (b) A secondary follicle.

As an ovarian follicle enlarges, follicular fluid accumulates in a cavity called the antrum.

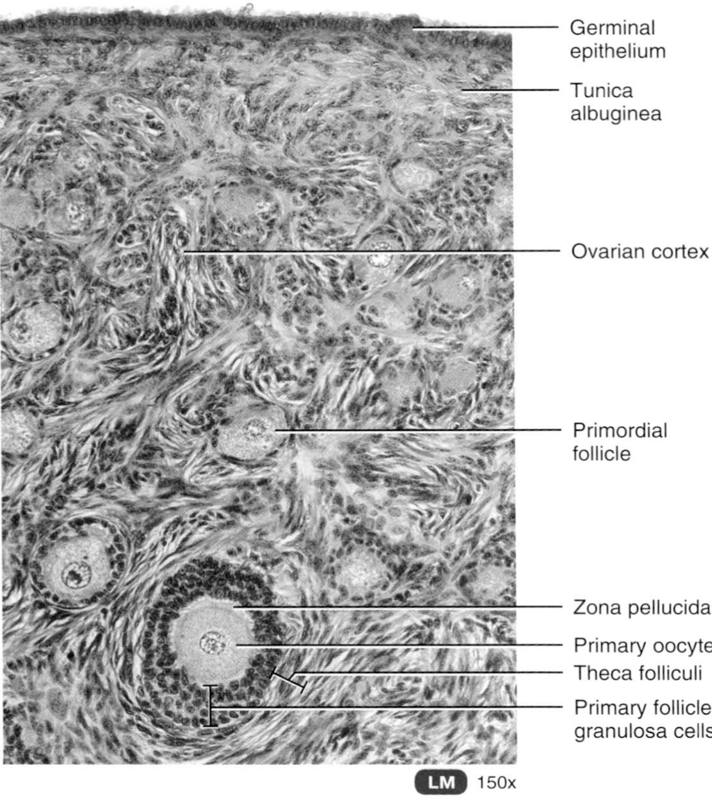

(a) Ovarian cortex

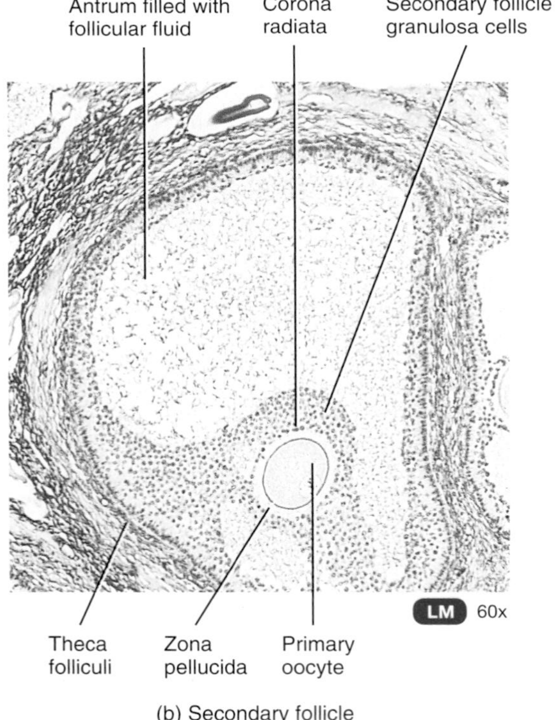

(b) Secondary follicle

What happens to most ovarian follicles?

Figure 28.15 Oogenesis. Diploid cells ($2n$) have 46 chromosomes; haploid cells (n) have 23 chromosomes.

In a secondary oocyte, meiosis II is completed only if fertilization occurs.

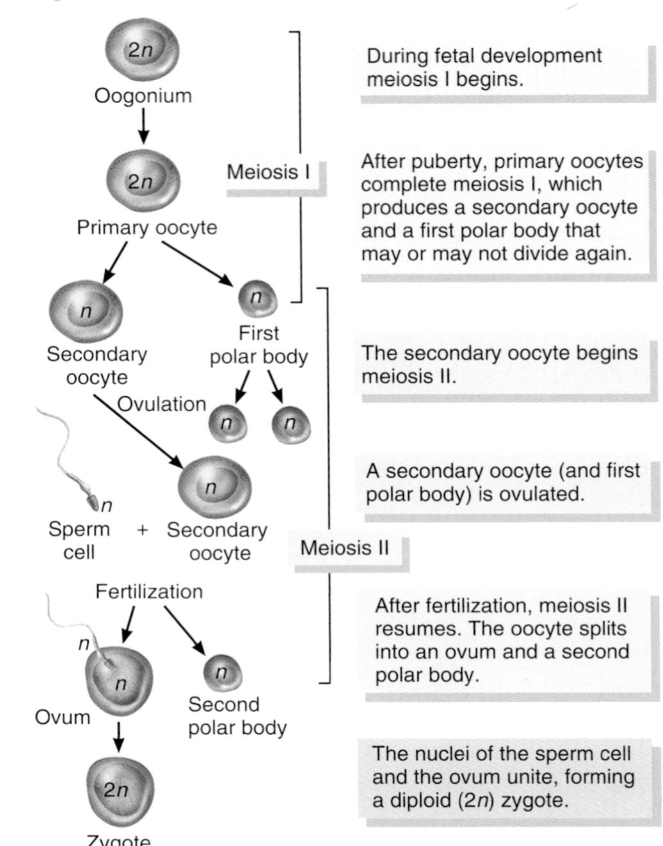

How does the age of a primary oocyte in a female compare with the age of a primary spermatocyte in a male?

ization does not occur, the cells degenerate. If sperm are present in the uterine tube and one penetrates the secondary oocyte, however, meiosis II resumes. The secondary oocyte splits into two haploid (n) cells, again of unequal size. The larger cell is the **ovum,** or mature egg; the smaller one is the **second polar body.** The nuclei of the sperm cell and the ovum then unite, forming a diploid ($2n$) **zygote.** If the first polar body undergoes another division to produce two polar bodies, then the primary oocyte ultimately gives rise to three haploid (n) polar bodies, which all degenerate, and a single haploid (n) ovum. Thus, one primary oocyte gives rise to a single gamete (an ovum). By contrast, recall that in males one primary spermatocyte produces four gametes (sperm).

Table 28.1 on page 1075 summarizes the events of oogenesis and follicular development.

TABLE 28.1 Summary of Oogenesis and Follicular Development

Age	Oogenesis	Follicular development
Fetal period	Oogonium ($2n$) Mitosis ↓ Primary oocyte ($2n$) Meiosis in progress ↓ Primary oocyte (in prophase I) ($2n$)	Primordial follicle
Childhood (no development of follicles)	↓	↓
Puberty to menopause each month	Primary oocyte (still in prophase I) ($2n$) ↓ Primary oocyte ($2n$) Meiosis I completed by one primary oocyte each month First polar body (n) Secondary oocyte (in metaphase II) (n) Meiosis II of first polar body may or may not occur Sperm cell (n) Meiosis II completed if fertilization occurs n n Second polar body (n) Ovum (n n) All polar bodies degenerate	Primary follicle ↓ Secondary follicle ↓ Mature (graafian) follicle Ovulation ↓ Ovulated secondary oocyte

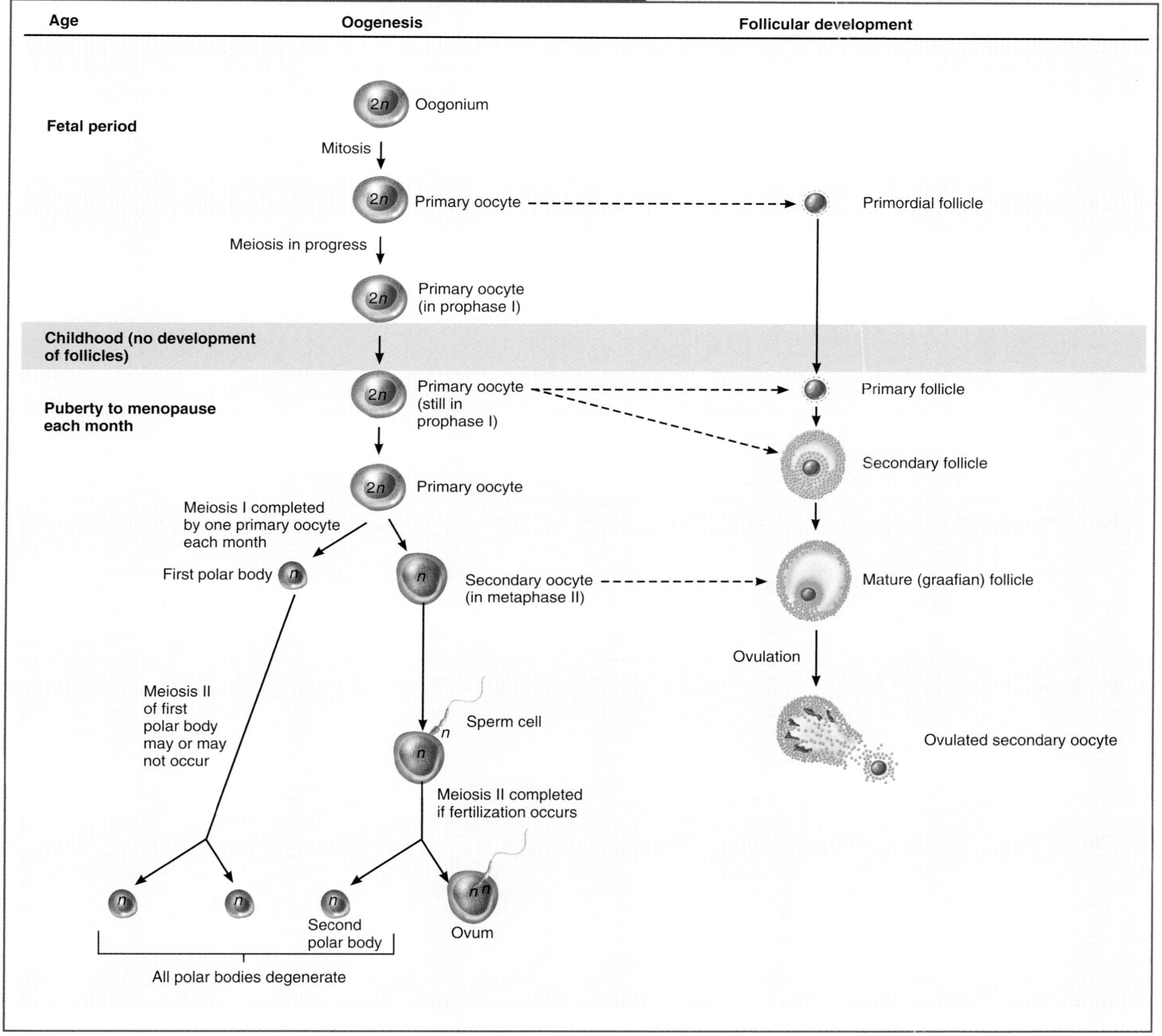

Ovarian Cysts

An **ovarian cyst** is a fluid-filled sac in or on an ovary. Such cysts are relatively common, are usually noncancerous, and frequently disappear on their own. Cancerous cysts are more likely to occur in women over 40. Ovarian cysts may cause pain, pressure, a dull ache, or fullness in the abdomen; pain during sexual intercourse; delayed, painful, or irregular menstrual periods; abrupt onset of sharp pain in the lower abdomen; and/or vaginal bleeding. Most ovarian cysts require no treatment, but larger ones (more than 5 cm or 2 in.) may be removed surgically. ■

▶ CHECKPOINT

14. How are the ovaries held in position in the pelvic cavity?
15. Describe the microscopic structure and functions of an ovary.
16. Describe the principal events of oogenesis.

Uterine Tubes

Females have two **uterine (fallopian) tubes,** or **oviducts,** that extend laterally from the uterus (Figure 28.16). The tubes, which measure about 10 cm (4 in.) long, lie between the folds of the broad ligaments of the uterus. They provide a route for sperm to reach an ovum and transport secondary oocytes and fertilized ova from the ovaries to the uterus. The funnel-shaped portion of each tube, called the **infundibulum,** is close to the ovary but is open to the pelvic cavity. It ends in a fringe of fingerlike projections called **fimbriae** (FIM-brē-ē = fringe), one of which is attached to the lateral end of the ovary. From the infundibulum, the uterine tube extends medially and eventually inferiorly and attaches to the superior lateral angle of the uterus. The **ampulla** of the uterine tube is the widest, longest portion, making up about the lateral two-thirds of its length. The **isthmus** of the uterine tube is the more medial, short, narrow, thick-walled portion that joins the uterus.

Histologically, the uterine tubes are composed of three layers: mucosa, muscularis, and serosa. The mucosa consists of epithelium and lamina propria (areolar connective tissue). The epithelium contains ciliated simple columnar cells, which function as a "ciliary conveyor belt" to help move a fertilized ovum

Figure 28.16 Relationship of the uterine (fallopian) tubes to the ovaries, uterus, and associated structures. In the left side of the drawing the uterine tube and uterus have been sectioned to show internal structures. (See Tortora, *A Photographic Atlas of the Human Body, Second Edition,* Figure 14.9a.)

After ovulation, a secondary oocyte and its corona radiata move from the pelvic cavity into the infundibulum of the uterine tube. The uterus is the site of menstruation, implantation of a fertilized ovum, development of the fetus, and labor.

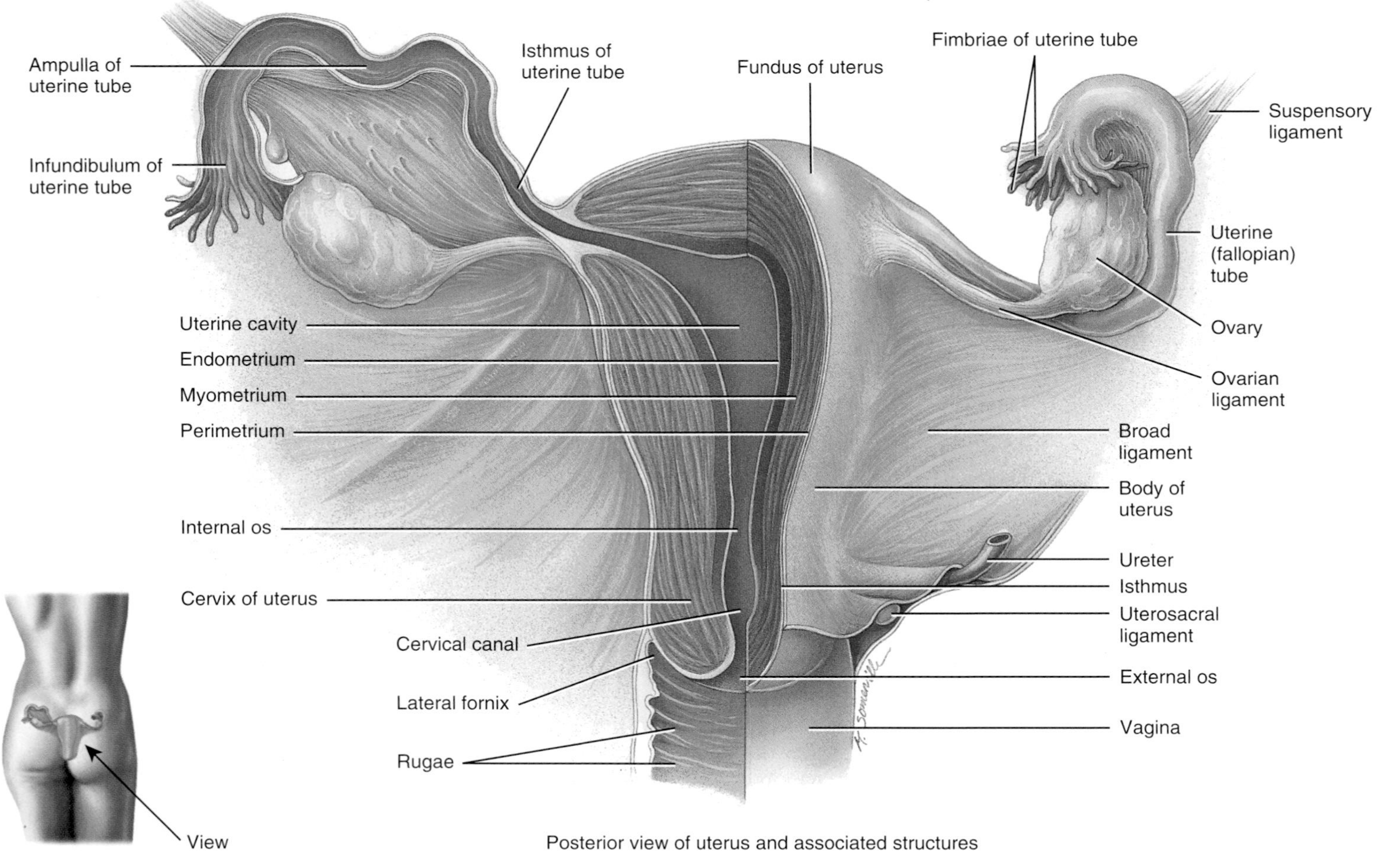

Posterior view of uterus and associated structures

Where does fertilization usually occur?

(or secondary oocyte) along the uterine tube toward the uterus, and nonciliated (peg) cells that have microvilli and secrete a fluid that provides nutrition for the ovum (Figure 28.17). The middle layer, the muscularis, is composed of an inner, thick, circular ring of smooth muscle and an outer, thin region of longitudinal smooth muscle. Peristaltic contractions of the muscularis and the ciliary action of the mucosa help move the oocyte or fertilized ovum toward the uterus. The outer layer of the uterine tubes is a serous membrane, the serosa.

Local currents produced by movements of the fimbriae, which surround the ovary during ovulation, sweep the ovulated secondary oocyte from the pelvic cavity into the uterine tube. A sperm cell usually encounters and fertilizes a secondary oocyte in the ampulla of the uterine tube, although fertilization in the pelvic cavity is not uncommon. Fertilization can occur at any time up to about 24 hours after ovulation. Some hours after fertilization, the nuclear materials of the haploid ovum and sperm unite. The diploid fertilized ovum is now called a **zygote** and begins to undergo cell divisions while moving toward the uterus. It arrives at the uterus 6 to 7 days after ovulation.

Uterus

The **uterus** (womb) serves as part of the pathway for sperm deposited in the vagina to reach the uterine tubes. It is also the site of implantation of a fertilized ovum, development of the fetus during pregnancy, and labor. During reproductive cycles when implantation does not occur, the uterus is the source of menstrual flow.

Anatomy of the Uterus

Situated between the urinary bladder and the rectum, the uterus is the size and shape of an inverted pear (see Figure 28.16). In females who have never been pregnant, it is about 7.5 cm (3 in.) long, 5 cm (2 in.) wide, and 2.5 cm (1 in.) thick. The uterus is larger in females who have recently been pregnant, and smaller (atrophied) when sex hormone levels are low, as occurs after menopause.

Anatomical subdivisions of the uterus include: (1) a dome-shaped portion superior to the uterine tubes called the **fundus,** (2) a tapering central portion called the **body,** and (3) an inferior narrow portion called the **cervix** that opens into the vagina. Between the body of the uterus and the cervix is the **isthmus** (IS-mus), a constricted region about 1 cm (0.5 in.) long. The interior of the body of the uterus is called the **uterine cavity,** and the interior of the cervix is called the **cervical canal.** The cervical canal opens into the uterine cavity at the **internal os** (*os* = mouthlike opening) and into the vagina at the **external os.**

Normally, the body of the uterus projects anteriorly and superiorly over the urinary bladder in a position called **anteflexion.** The cervix projects inferiorly and posteriorly and enters the anterior wall of the vagina at nearly a right angle (see

Figure 28.17 **Histology of the uterine (fallopian) tube.** (See Tortora, *A Photographic Atlas of the Human Body, Second Edition,* Figure 14.11a.)

Peristaltic contractions of the muscularis and ciliary action of the mucosa of the uterine tube help move the oocyte or fertilized ovum toward the uterus.

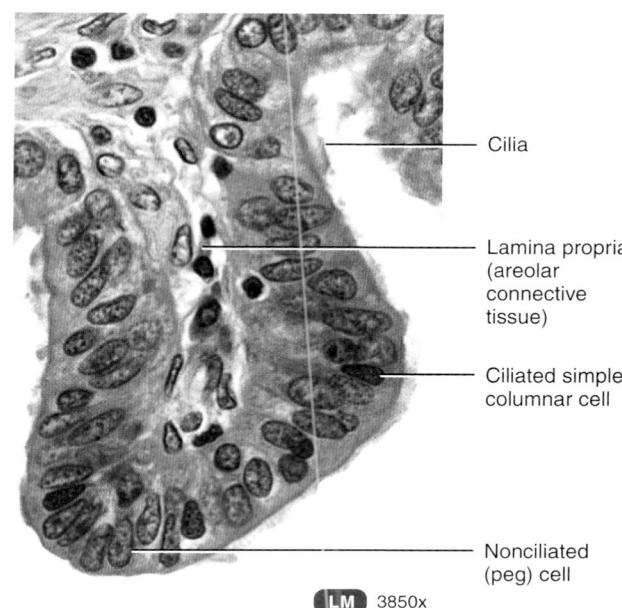

Cilia
Lamina propria (areolar connective tissue)
Ciliated simple columnar cell
Noncilated (peg) cell
LM 3850x

(a) Details of epithelium in sectional view

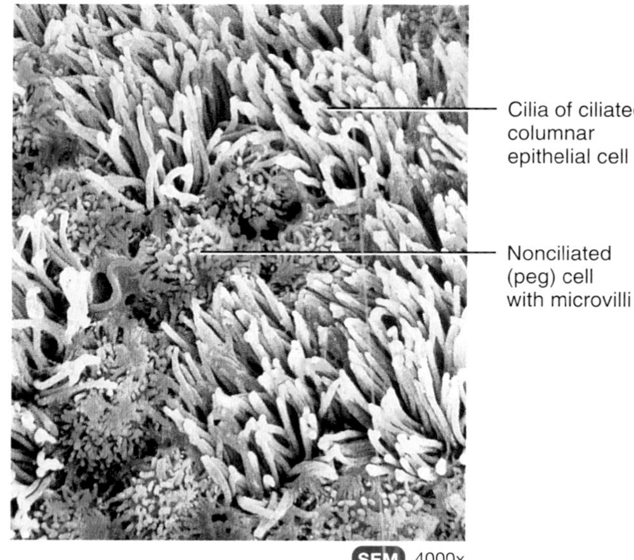

(b) Details of epithelium in surface view

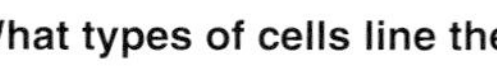

What types of cells line the uterine tubes?

Figure 28.11). Several ligaments that are either extensions of the parietal peritoneum or fibromuscular cords maintain the position of the uterus (see Figure 28.12). The paired **broad ligaments** are double folds of peritoneum attaching the uterus to either side of the pelvic cavity. The paired **uterosacral ligaments,** also peritoneal extensions, lie on either side of the rectum and connect the uterus to the sacrum. The **cardinal (lateral cervical) ligaments** are located inferior to the bases of the broad ligaments and extend from the pelvic wall to the cervix and vagina. The **round ligaments** are bands of fibrous connective tissue between the layers of the broad ligament; they extend from a point on the uterus just inferior to the uterine tubes to a portion of the labia majora of the external genitalia. Although the ligaments normally maintain the anteflexed position of the uterus, they also afford the uterine body enough movement such that the uterus may become malpositioned. A posterior tilting of the uterus is called **retroflexion** (*retro-* = backward or behind). It is a harmless variation of the normal position of the uterus. There is often no cause for the condition, but it may occur after childbirth or because of an ovarian cyst.

Uterine Prolapse

A condition called **uterine prolapse** (*prolapse* = falling down or downward displacement) may result from weakening of supporting ligaments and pelvic musculature associated with age or disease, traumatic vaginal delivery, chronic straining from coughing or difficult bowel movements, or pelvic tumors. The prolapse may be characterized as *first degree (mild),* in which the cervix remains within the vagina; *second degree (marked),* in which the cervix protrudes through the vagina to the exterior; and *third degree (complete),* in which the entire uterus is outside the vagina. Depending on the degree of prolapse, treatment may involve pelvic exercises, dieting if a patient is overweight, a stool softener to minimize straining during defecation, pessary therapy (placement of a rubber device around the uterine cervix that helps prop up the uterus), or surgery. ■

Histology of the Uterus

Histologically, the uterus consists of three layers of tissue: perimetrium, myometrium, and endometrium (Figure 28.18).

Figure 28.18 Histology of the uterus. (See Tortora, *A Photographic Atlas of the Human Body, Second Edition,* Figure 14.9a.)

The three layers of the uterus from superficial to deep are the perimetrium (serosa), the myometrium, and the endometrium.

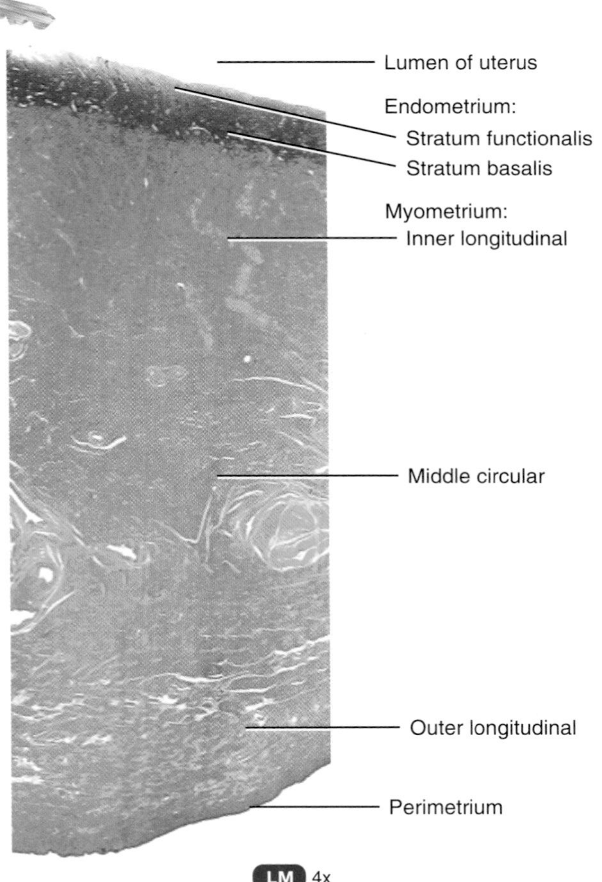

(a) Transverse section through the uterus

Lumen of uterus

Simple columnar epithelium

Stratum functionalis

Endometrial gland

Stratum basalis

LM 115x

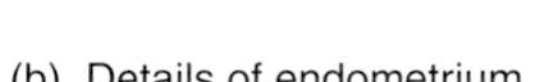

(b) Details of endometrium

What structural features of the endometrium and myometrium contribute to their functions?

The outer layer—the **perimetrium** (*peri-* = around; *-metrium* = uterus) or serosa—is part of the visceral peritoneum; it is composed of simple squamous epithelium and areolar connective tissue. Laterally, it becomes the broad ligament. Anteriorly, it covers the urinary bladder and forms a shallow pouch, the **vesicouterine pouch** (ves′-i-kō-Ū-ter-in; *vesico-* = bladder; see Figure 28.11). Posteriorly, it covers the rectum and forms a deep pouch, the **rectouterine pouch** (rek-tō-Ū-ter-in; *recto-* = rectum) or *pouch of Douglas*—the most inferior point in the pelvic cavity.

The middle layer of the uterus, the **myometrium** (*myo-* = muscle), consists of three layers of smooth muscle fibers that are thickest in the fundus and thinnest in the cervix. The thicker middle layer is circular; the inner and outer layers are longitudinal or oblique. During labor and childbirth, coordinated contractions of the myometrium in response to oxytocin from the posterior pituitary help expel the fetus from the uterus.

The inner layer of the uterus, the **endometrium** (*endo-* = within), is highly vascularized and has three components: (1) An innermost layer of simple columnar epithelium (ciliated and secretory cells) lines the lumen. (2) An underlying endometrial stroma is a very thick region of lamina propria (areolar connective tissue). (3) Endometrial (uterine) glands develop as invaginations of the luminal epithelium and extend almost to the myometrium. The endometrium is divided into two layers. The **stratum functionalis** *(functional layer)* lines the uterine cavity and sloughs off during menstruation. The deeper layer, the **stratum basalis** *(basal layer)*, is permanent and gives rise to a new stratum functionalis after each menstruation.

Branches of the internal iliac artery called **uterine arteries** (Figure 28.19) supply blood to the uterus. Uterine arteries give off branches called **arcuate arteries** (= shaped like a bow) that are arranged in a circular fashion in the myometrium. These arteries branch into **radial arteries** that penetrate deeply into the myometrium. Just before the branches enter the endometrium, they divide into two kinds of arterioles: **Straight arterioles** supply the stratum basalis with the materials needed to regenerate the stratum functionalis; **spiral arterioles** supply the stratum functionalis and change markedly during the menstrual cycle. Blood leaving the uterus is drained by the **uterine veins** into the internal iliac veins. The extensive blood supply of the uterus is essential to support regrowth of a new stratum functionalis after menstruation, implantation of a fertilized ovum, and development of the placenta.

Figure 28.19 **Blood supply of the uterus.** The inset shows histological details of the blood vessels of the endometrium.

Straight arterioles supply the materials needed for regeneration of the stratum functionalis.

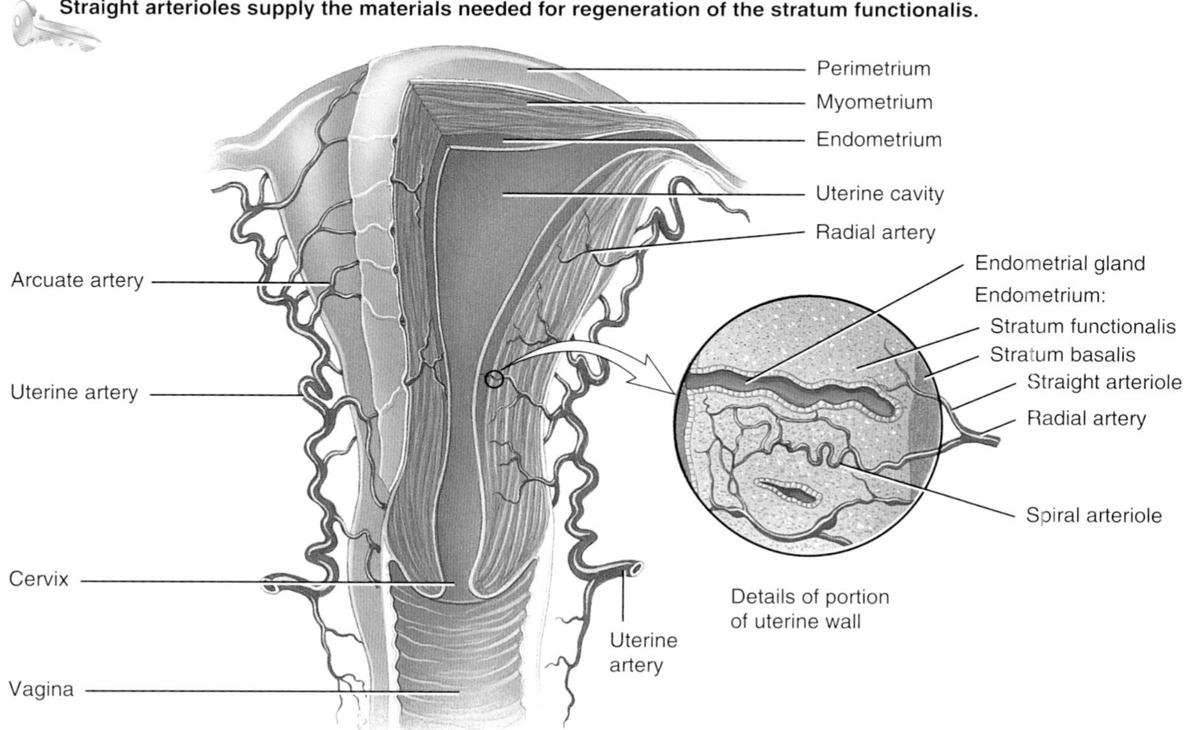

What is the functional significance of the stratum basalis of the endometrium?

Cervical Mucus

The secretory cells of the mucosa of the cervix produce a secretion called **cervical mucus,** a mixture of water, glycoproteins, lipids, enzymes, and inorganic salts. During their reproductive years, females secrete 20–60 mL of cervical mucus per day. Cervical mucus is more hospitable to sperm at or near the time of ovulation because it is then less viscous and more alkaline (pH 8.5). At other times, viscous mucus forms a cervical plug that physically impedes sperm penetration. Cervical mucus supplements the energy needs of sperm, and both the cervix and cervical mucus protect sperm from phagocytes and the hostile environment of the vagina and uterus. Cervical mucus may also play a role in *capacitation*—a series of functional changes that sperm undergo in the female reproductive tract before they are able to fertilize a secondary oocyte. Capacitation causes a sperm cell's tail to beat even more vigorously and it prepares the sperm cell's plasma membrane to fuse with the oocyte's plasma membrane.

Hysterectomy

Hysterectomy (hiss-ter-EK-tō-mē; *hyster-* = uterus), the surgical removal of the uterus, is the most common gynecological operation. It may be indicated in conditions such as fibroids, which are noncancerous tumors composed of muscular and fibrous tissue, endometriosis, pelvic inflammatory disease, recurrent ovarian cysts, excessive uterine bleeding, and cancer of the cervix, uterus, or ovaries. In a *partial (subtotal) hysterectomy,* the body of the uterus is removed but the cervix is left in place. A *complete hysterectomy* is the removal of both the body and cervix of the uterus. A *radical hysterectomy* includes removal of the body and cervix of the uterus, uterine tubes, possibly the ovaries, the superior portion of the vagina, pelvic lymph nodes, and supporting structures, such as ligaments. A hysterectomy can be performed either through an incision in the abdominal wall, or through the vagina. ■

▶ CHECKPOINT

17. Where are the uterine tubes located, and what is their function?
18. What are the principal parts of the uterus? Where are they located in relation to one another?
19. Describe the arrangement of ligaments that hold the uterus in its normal position.
20. Describe the histology of the uterus.
21. Why is an abundant blood supply important to the uterus?

Vagina

The **vagina** (= sheath) is a tubular, 10-cm (4-in.) long fibromuscular canal lined with mucous membrane that extends from the exterior of the body to the uterine cervix (see Figures 28.11 and 28.16). It is the receptacle for the penis during sexual intercourse, the outlet for menstrual flow, and the passageway for childbirth. Situated between the urinary bladder and the rectum, the vagina is directed superiorly and posteriorly, where it attaches to the uterus. A recess called the **fornix** (= arch or vault) surrounds the vaginal attachment to the cervix. When properly inserted, a contraceptive diaphragm rests on the fornix, covering the cervix.

The **mucosa** of the vagina is continuous with that of the uterus. Histologically, it consists of nonkeratinized stratified squamous epithelium and areolar connective tissue that lies in a series of transverse folds called **rugae** (ROO-gē). Dendritic cells in the mucosa are antigen-presenting cells (described on page 822). Unfortunately, they also participate in the transmission of viruses—for example, HIV (the virus that causes AIDS)—to a female during intercourse with an infected male. The mucosa of the vagina contains large stores of glycogen, the decomposition of which produces organic acids. The resulting acidic environment retards microbial growth, but it also is harmful to sperm. Alkaline components of semen, mainly from the seminal vesicles, raise the pH of fluid in the vagina and increase viability of the sperm.

The **muscularis** is composed of an outer circular layer and an inner longitudinal layer of smooth muscle that can stretch considerably to accommodate the penis during sexual intercourse and a child during birth.

The **adventitia,** the superficial layer of the vagina, consists of areolar connective tissue. It anchors the vagina to adjacent organs such as the urethra and urinary bladder anteriorly and the rectum and anal canal posteriorly.

A thin fold of vascularized mucous membrane, called the **hymen** (= membrane), forms a border around and partially closes the inferior end of the vaginal opening to the exterior, the **vaginal orifice** (see Figure 28.20). Sometimes the hymen completely covers the orifice, a condition called **imperforate hymen** (im-PER-fō-rāt). Surgery may be needed to open the orifice and permit the discharge of menstrual flow.

Vulva

The term **vulva** (VUL-va = to wrap around), or **pudendum** (pū-DEN-dum), refers to the external genitals of the female (Figure 28.20). The following components comprise the vulva:

- Anterior to the vaginal and urethral openings is the **mons pubis** (MONZ PŪ-bis; *mons* = mountain), an elevation of adipose tissue covered by skin and coarse pubic hair that cushions the pubic symphysis.
- From the mons pubis, two longitudinal folds of skin, the **labia majora** (LĀ-bē-a ma-JŌ-ra; *labia* = lips; *majora* = larger), extend inferiorly and posteriorly. The singular term is *labium majus.* The labia majora are covered by pubic hair and contain an abundance of adipose tissue, sebaceous (oil) glands, and apocrine sudoriferous (sweat) glands. They are homologous to the scrotum.

Figure 28.20 Components of the vulva (pudendum). (See Tortora, *A Photographic Atlas of the Human Body, Second Edition,* Figure 14.7.)

The vulva refers to the external genitals of the female.

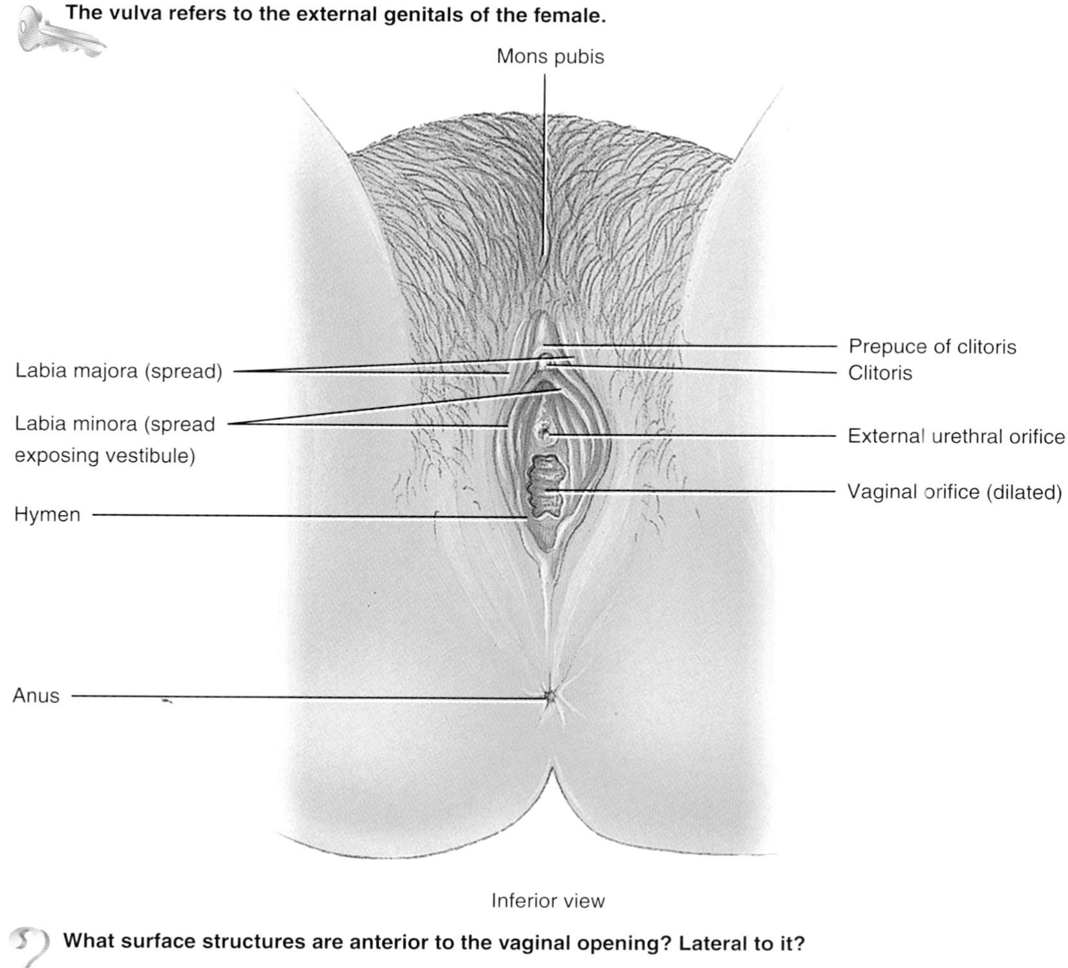

What surface structures are anterior to the vaginal opening? Lateral to it?

- Medial to the labia majora are two smaller folds of skin called the **labia minora** (mī-NŌ-ra; *minora* = smaller). The singular term is *labium minus.* Unlike the labia majora, the labia minora are devoid of pubic hair and fat and have few sudoriferous glands, but they do contain many sebaceous glands. The labia minora are homologous to the spongy (penile) urethra.
- The **clitoris** (KLI-to-ris) is a small cylindrical mass of erectile tissue and nerves located at the anterior junction of the labia minora. A layer of skin called the **prepuce of the clitoris** is formed at the point where the labia minora unite and covers the body of the clitoris. The exposed portion of the clitoris is the **glans.** The clitoris is homologous to the glans penis in males. Like the male structure, it is capable of enlargement upon tactile stimulation and has a role in sexual excitement in the female.
- The region between the labia minora is the **vestibule.** Within the vestibule are the hymen (if still present), the vaginal orifice, the external urethral orifice, and the openings of the ducts of several glands. The vestibule is homologous to the membranous urethra of males. The **vaginal orifice,** the opening of the vagina to the exterior, occupies the greater portion of the vestibule and is bordered by the hymen. Anterior to the vaginal orifice and posterior to the clitoris is the **external urethral orifice,** the opening of the urethra to the exterior. On either side of the external urethral orifice are the openings of the ducts of the **paraurethral (Skene's) glands.** These mucus-secreting glands are embedded in the wall of the urethra. The paraurethral glands are homologous to the prostate. On either side of the vaginal orifice itself are the **greater vestibular (Bartholin's) glands** (see Figure 28.21), which open by ducts into a groove between the hymen and labia minora. They produce a small quantity of mucus during sexual arousal and intercourse that adds to cervical mucus and provides lubrication. The greater vestibular

Figure 28.21 Perineum of a female. (Figure 11.13 on page 359 shows the perineum of a male.)

The perineum is a diamond-shaped area that includes the urogenital triangle and the anal triangle.

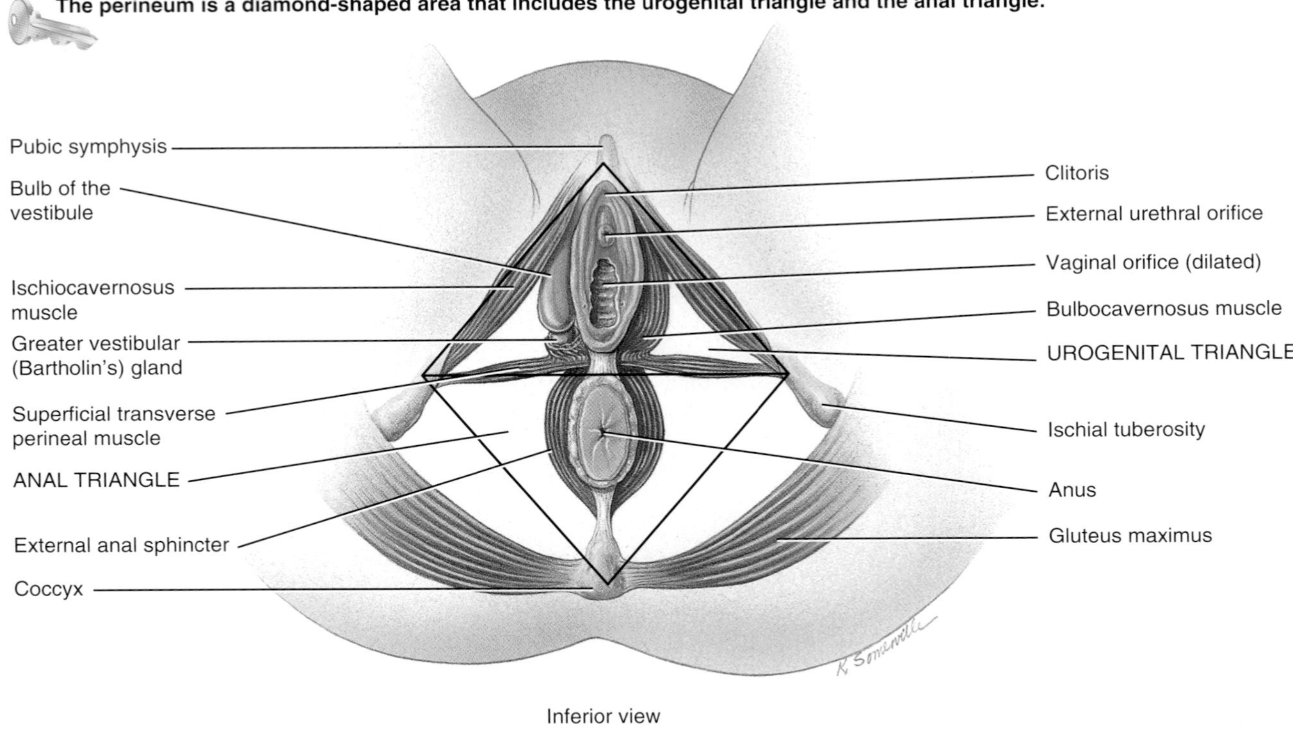

Inferior view

Why is the anterior portion of the perineum called the urogenital triangle?

glands are homologous to the bulbourethral glands in males. Several **lesser vestibular glands** also open into the vestibule.

- The **bulb of the vestibule** (see Figure 28.21) consists of two elongated masses of erectile tissue just deep to the labia on either side of the vaginal orifice. The bulb of the vestibule becomes engorged with blood during sexual arousal, narrowing the vaginal orifice and placing pressure on the penis during intercourse. The bulb of the vestibule is homologous to the corpus spongiosum and bulb of the penis in males.

Table 28.2 summarizes the homologous structures of the female and male reproductive systems.

Perineum

The **perineum** (per′-i-NĒ-um) is the diamond-shaped area medial to the thighs and buttocks of both males and females (Figure 28.21). It contains the external genitals and anus. The perineum is bounded anteriorly by the pubic symphysis, laterally by the ischial tuberosities, and posteriorly by the coccyx. A transverse line drawn between the ischial tuberosities divides the perineum into an anterior **urogenital triangle** (ū′-rō-JEN-i-tal) that contains the external genitals and a posterior **anal triangle** that contains the anus.

Episiotomy

During childbirth, the emerging fetus stretches the perineal region. To prevent excessive stretching and even tearing of this region, a physician sometimes performs an **episiotomy** (e-piz-ē-OT-ō-mē; *episi-* = vulva or pubic region; *-otomy* = incision), a perineal cut made with surgical scissors. The cut may be made along the midline or at an angle of approximately 45 degrees to

TABLE 28.2 Summary of Homologous Structures of the Female and Male Reproductive Systems

Female Structures	Male Structures
Ovaries	Testes
Ovum	Sperm cell
Labia majora	Scrotum
Labia minora	Spongy (penile) urethra
Vestibule	Membranous urethra
Bulb of vestibule	Corpus spongiosum penis and bulb of penis
Clitoris	Glans penis
Paraurethral glands	Prostate
Greater vestibular glands	Bulbourethral (Cowper's) glands

the midline. In effect, a straight, more easily sutured cut is substituted for the jagged tear that would otherwise be caused by passage of the fetus. The incision is closed in layers with sutures that are absorbed within a few weeks, so that the busy new mom does not have to worry about making time to have them removed. ■

Mammary Glands

Each **breast** is a hemispheric projection of variable size anterior to the pectoralis major and serratus anterior muscles and attached to them by a layer of deep fascia composed of dense irregular connective tissue.

Each breast has one pigmented projection, the **nipple,** that has a series of closely spaced openings of ducts called **lactiferous ducts,** where milk emerges. The circular pigmented area of skin surrounding the nipple is called the **areola** (a-RĒ-ō-la = small space); it appears rough because it contains modified sebaceous (oil) glands. Strands of connective tissue called the **suspensory ligaments of the breast (Cooper's ligaments)** run between the skin and deep fascia and support the breast. These ligaments become looser with age or with the excessive strain that can occur in long-term jogging or high-impact aerobics. Wearing a supportive bra slows the appearance of "Cooper's droop."

Within each breast is a **mammary gland,** a modified sudoriferous (sweat) gland that produces milk (Figure 28.22). A mammary gland consists of 15 to 20 lobes, or compartments, separated by a variable amount of adipose tissue. In each lobe are several smaller compartments called **lobules,** composed of grapelike clusters of milk-secreting glands termed **alveoli** (= small cavities) embedded in connective tissue. Contraction of **myoepithelial cells** surrounding the alveoli helps propel milk toward the nipples. When milk is being produced, it passes from the alveoli into a series of **secondary tubules** and then into the **mammary ducts.** Near the nipple, the mammary ducts expand to form sinuses called **lactiferous sinuses** (*lact-* = milk), where some milk may be stored before draining into a lactiferous duct. Each lactiferous duct typically carries milk from one of the lobes to the exterior.

The functions of the mammary glands are the synthesis, secretion, and ejection of milk; these functions, called **lactation,** are associated with pregnancy and childbirth. Milk production is stimulated largely by the hormone prolactin from the anterior pituitary, with contributions from progesterone and estrogens.

Figure 28.22 **Mammary glands within the breasts.** (See Tortora, *A Photographic Atlas of the Human Body, Second Edition,* Figure 14.14.)

The mammary glands function in the synthesis, secretion, and ejection of milk (lactation).

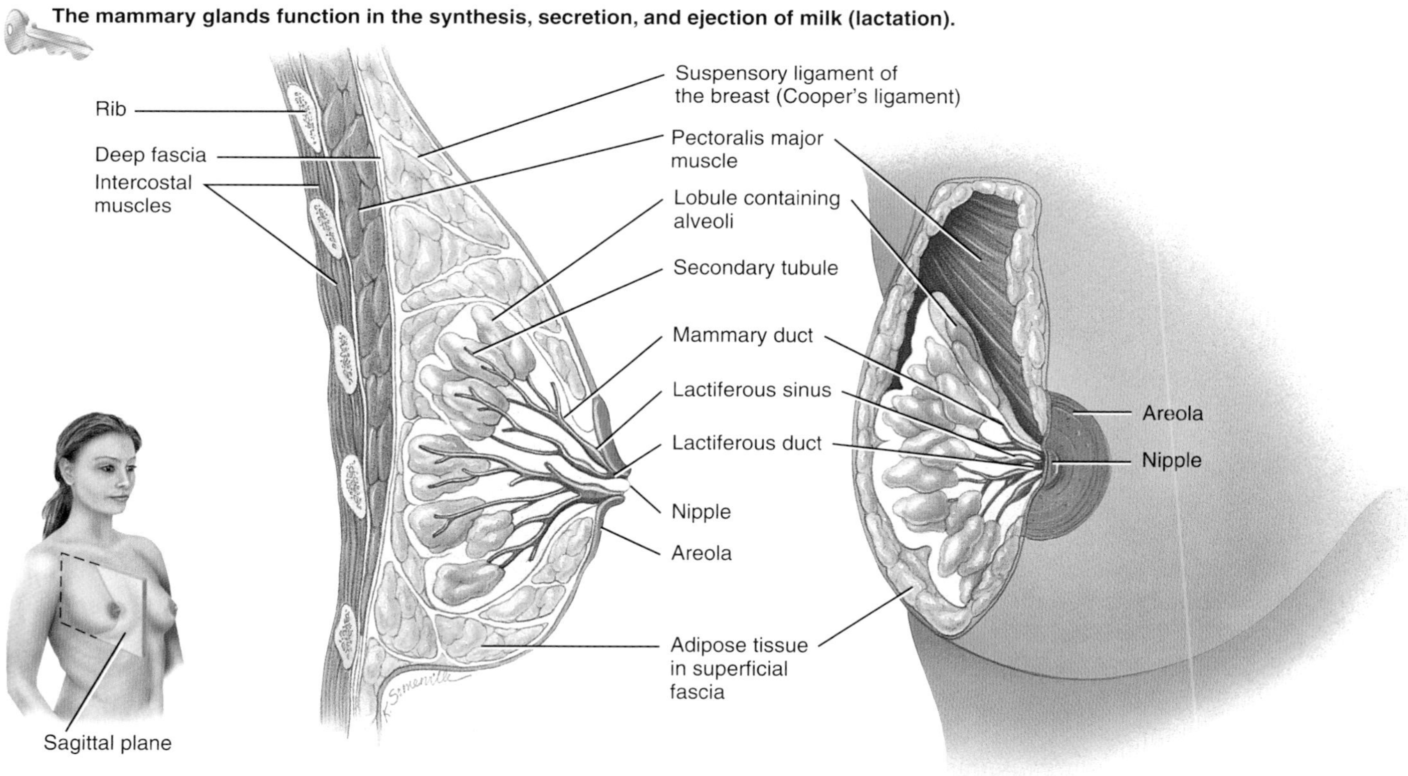

What hormones regulate the synthesis and ejection of milk?

The ejection of milk is stimulated by oxytocin, which is released from the posterior pituitary in response to the sucking of an infant on the mother's nipple (suckling).

Fibrocystic Disease of the Breasts

The breasts of females are highly susceptible to cysts and tumors. In **fibrocystic disease,** the most common cause of breast lumps in females, one or more cysts (fluid-filled sacs) and thickenings of alveoli develop. The condition, which occurs mainly in females between the ages of 30 and 50, is probably due to a relative excess of estrogens or a deficiency of progesterone in the postovulatory (luteal) phase of the reproductive cycle (discussed shortly). Fibrocystic disease usually causes one or both breasts to become lumpy, swollen, and tender a week or so before menstruation begins. ■

▶ CHECKPOINT

22. How does the histology of the vagina contribute to its function?

23. What are the structures and functions of each part of the vulva?

24. Describe the components of the mammary glands and the structures that support them.

25. Outline the route milk takes from the alveoli of the mammary gland to the nipple.

THE FEMALE REPRODUCTIVE CYCLE

▶ OBJECTIVE

Compare the major events of the ovarian and uterine cycles.

During their reproductive years, nonpregnant females normally exhibit cyclical changes in the ovaries and uterus. Each cycle takes about a month and involves both oogenesis and preparation of the uterus to receive a fertilized ovum. Hormones secreted by the hypothalamus, anterior pituitary, and ovaries control the main events. The **ovarian cycle** is a series of events in the ovaries that occur during and after the maturation of an oocyte. The **uterine (menstrual) cycle** is a concurrent series of changes in the endometrium of the uterus to prepare it for the arrival of a fertilized ovum that will develop there until birth. If fertilization does not occur, ovarian hormones wane, which causes the stratum functionalis of the endometrium to slough off. The general term **female reproductive cycle** encompasses the ovarian and uterine cycles, the hormonal changes that regulate them, and the related cyclical changes in the breasts and cervix.

Hormonal Regulation of the Female Reproductive Cycle

Gonadotropin-releasing hormone (GnRH) secreted by the hypothalamus controls the ovarian and uterine cycles (Figure 28.23). GnRH stimulates the release of **follicle-stimulating hormone (FSH)** and **luteinizing hormone (LH)** from the anterior pituitary. FSH initiates follicular growth, while LH stimulates further development of the ovarian follicles. In addition, both FSH and LH stimulate the ovarian follicles to secrete estrogens. LH stimulates the theca cells of a developing follicle to produce androgens. Under the influence of FSH, the androgens are taken up by the granulosa cells of the follicle and then converted into estrogens. At midcycle, LH triggers ovulation and then promotes formation of the corpus luteum, the reason for the name luteinizing hormone. Stimulated by LH, the corpus luteum produces and secretes estrogens, progesterone, relaxin, and inhibin.

At least six different estrogens have been isolated from the plasma of human females, but only three are present in significant quantities: *beta (β)-estradiol, estrone,* and *estriol.* In a nonpregnant woman, the most abundant estrogen is β-estradiol, which is synthesized from cholesterol in the ovaries.

Estrogens secreted by ovarian follicles have several important functions (Figure 28.23).

- Estrogens promote the development and maintenance of female reproductive structures, secondary sex characteristics, and the breasts. The secondary sex characteristics include distribution of adipose tissue in the breasts, abdomen, mons pubis, and hips; voice pitch; a broad pelvis; and pattern of hair growth on the head and body.
- Estrogens increase protein anabolism, including the building of strong bones. In this regard, estrogens are synergistic with human growth hormone (hGH).
- Estrogens lower blood cholesterol level, which is probably the reason that women under age 50 have a much lower risk of coronary artery disease than do men of comparable age.
- Moderate levels of estrogens in the blood inhibit both the release of GnRH by the hypothalamus and secretion of LH and FSH by the anterior pituitary.

Progesterone, secreted mainly by cells of the corpus luteum, cooperates with estrogens to prepare and maintain the endometrium for implantation of a fertilized ovum and to prepare the mammary glands for milk secretion. High levels of progesterone also inhibit secretion of GnRH and LH.

The small quantity of **relaxin** produced by the corpus luteum during each monthly cycle relaxes the uterus by inhibiting contractions of the myometrium. Presumably, implantation of a fertilized ovum occurs more readily in a "quiet" uterus. During pregnancy, the placenta produces much more relaxin, and it continues to relax uterine smooth muscle. At the end of pregnancy, relaxin also increases the flexibility of the pubic symphysis and may help dilate the uterine cervix, both of which ease delivery of the baby.

Inhibin is secreted by granulosa cells of growing follicles and by the corpus luteum after ovulation. It inhibits secretion of FSH and, to a lesser extent, LH.

Figure 28.23 **Secretion and physiological effects of estrogens, progesterone, relaxin, and inhibin in the female reproductive cycle.** Dashed red lines indicate negative feedback inhibition.

The uterine and ovarian cycles are controlled by gonadotropin-releasing hormone (GnRH) and ovarian hormones (estrogens and progesterone).

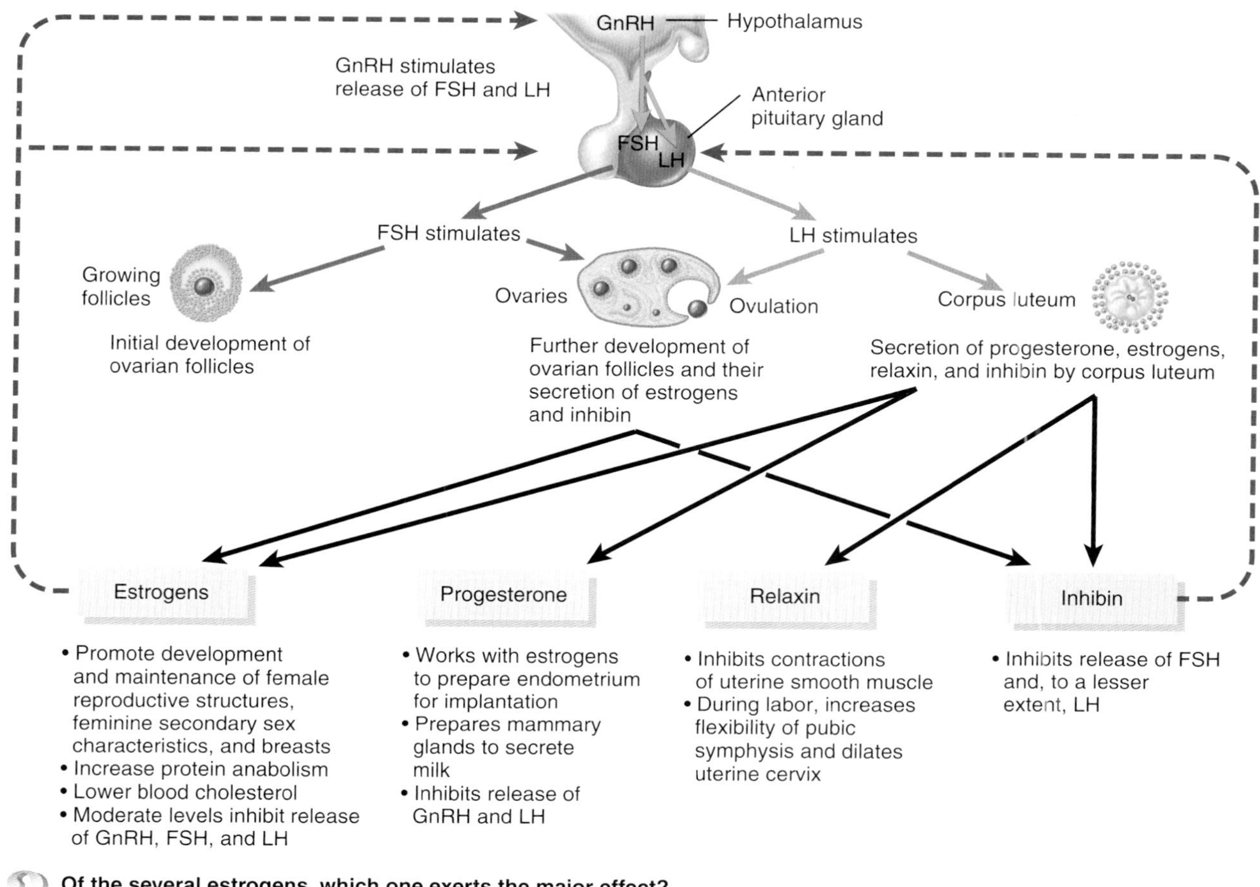

Of the several estrogens, which one exerts the major effect?

Phases of the Female Reproductive Cycle

The duration of the female reproductive cycle typically ranges from 24 to 35 days. For this discussion, we assume a duration of 28 days and divide it into four phases: the menstrual phase, the preovulatory phase, ovulation, and the postovulatory phase (Figure 28.24).

Menstrual Phase

The **menstrual phase** (MEN-stroo-al), also called **menstruation** (men′-stroo-Ā-shun) or **menses** (= month), lasts for roughly the first 5 days of the cycle. (By convention, the first day of menstruation is day one of a new cycle.)

Events in the Ovaries Under the influence of FSH, several primordial follicles develop into primary follicles and then into secondary follicles. This developmental process may take several months to occur. Therefore, a follicle that begins to develop at the beginning of a particular menstrual cycle may not reach maturity and ovulate until several menstrual cycles later.

Events in the Uterus Menstrual flow from the uterus consists of 50–150 mL of blood, tissue fluid, mucus, and epithelial cells shed from the endometrium. This discharge occurs because the declining levels of progesterone and estrogens stimulate release of prostaglandins that cause the uterine spiral arterioles to constrict. As a result, the cells they supply become oxygen-deprived and start to die. Eventually, the entire stratum functionalis sloughs off. At this time the endometrium is very thin, about 2–5 mm, because only the stratum basalis remains. The menstrual flow passes from the uterine cavity through the cervix and vagina to the exterior.

Preovulatory Phase

The **preovulatory phase** is the time between the end of menstruation and ovulation. The preovulatory phase of the cycle is more variable in length than the other phases and accounts for

Figure 28.24 The female reproductive cycle. The length of the female reproductive cycle typically is 24 to 36 days; the preovulatory phase is more variable in length than the other phases. (a) Events in the ovarian and uterine cycles and the release of anterior pituitary hormones are correlated with the sequence of the cycle's four phases. In the cycle shown, fertilization and implantation have not occurred. (b) Relative concentrations of anterior pituitary hormones (FSH and LH) and ovarian hormones (estrogens and progesterone) during the phases of a normal female reproductive cycle.

Estrogens are the primary ovarian hormones before ovulation; after ovulation, both progesterone and estrogens are secreted by the corpus luteum.

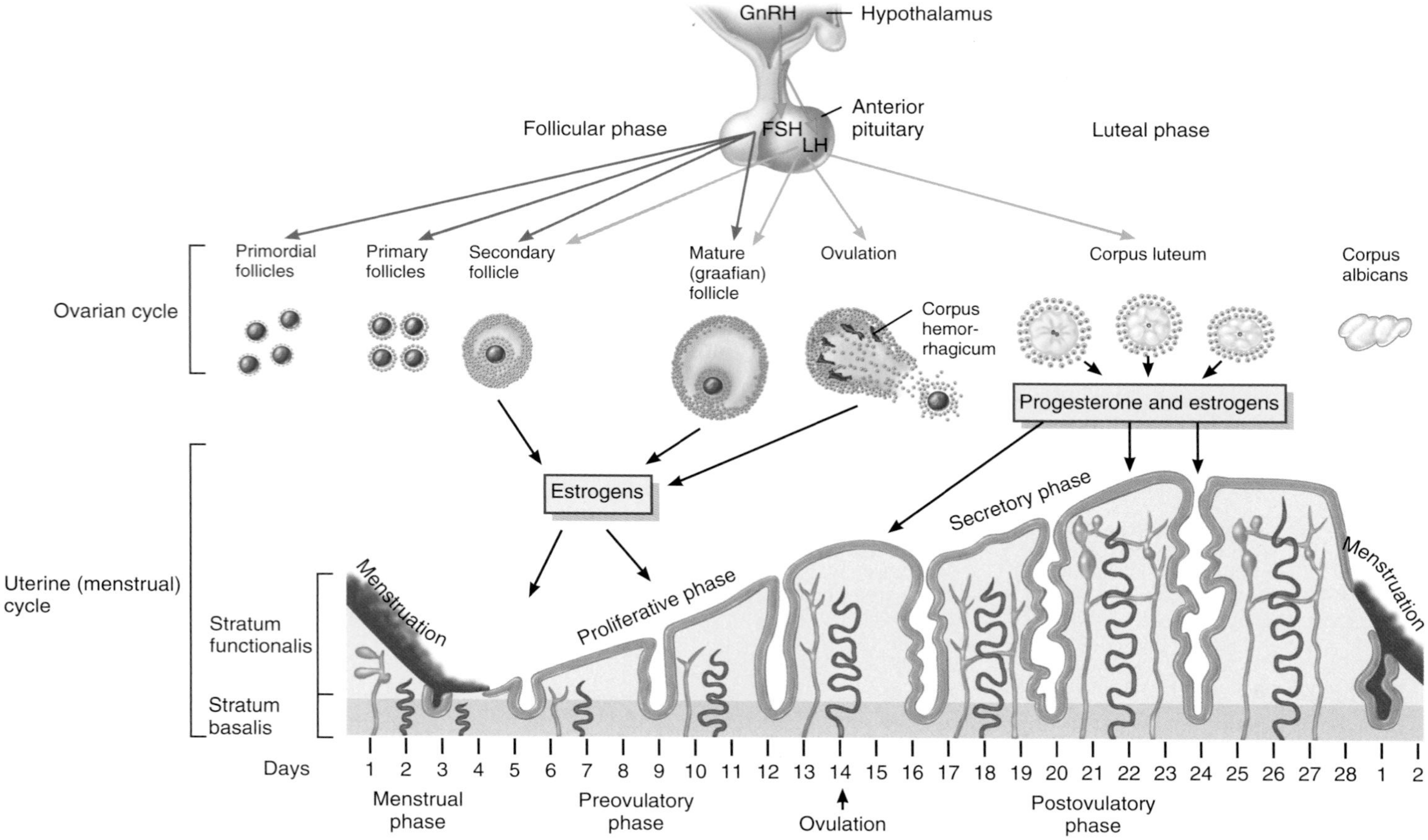

(a) Hormonal regulation of changes in the ovary and uterus

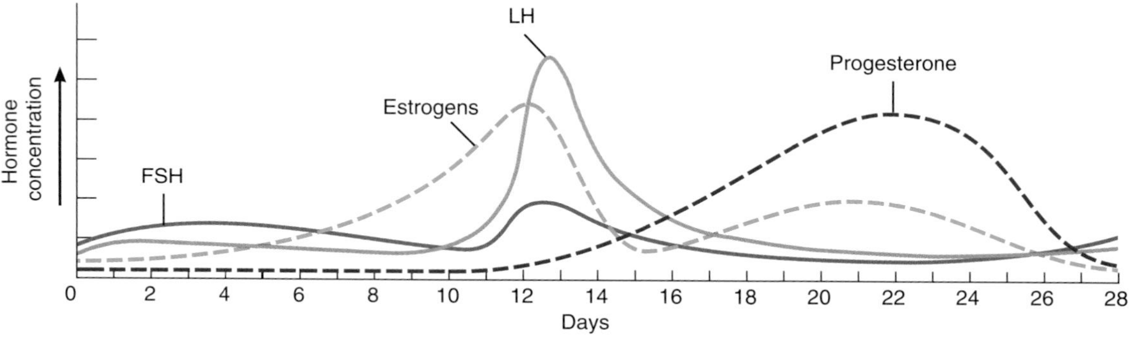

(b) Changes in concentration of anterior pituitary and ovarian hormones

Which hormones are responsible for the proliferative phase of endometrial growth, for ovulation, for growth of the corpus luteum, and for the surge of LH at midcycle?

most of the differences in length of the cycle. It lasts from days 6 to 13 in a 28-day cycle.

EVENTS IN THE OVARIES Some of the secondary follicles in the ovaries begin to secrete estrogens and inhibin. By about day 6, a single secondary follicle in one of the two ovaries has outgrown all the others to become the **dominant follicle.** Estrogens and inhibin secreted by the dominant follicle decrease the secretion of FSH, which causes other, less well-developed follicles to stop growing and undergo atresia. Fraternal (nonidentical) twins or triplets result when two or three secondary follicles become codominant and later are ovulated and fertilized at about the same time.

Normally, the one dominant secondary follicle becomes the **mature (graafian) follicle,** which continues to enlarge until it is more than 20 mm in diameter and ready for ovulation (see Figure 28.13). This follicle forms a blisterlike bulge due to the swelling antrum on the surface of the ovary. During the final maturation process, the mature follicle continues to increase its estrogen production (Figure 28.24).

With reference to the ovarian cycle, the menstrual and preovulatory phases together are termed the **follicular phase** (fō-LIK-ū-lar) because ovarian follicles are growing and developing.

EVENTS IN THE UTERUS Estrogens liberated into the blood by growing ovarian follicles stimulate the repair of the endometrium; cells of the stratum basalis undergo mitosis and produce a new stratum functionalis. As the endometrium thickens, the short, straight endometrial glands develop, and the arterioles coil and lengthen as they penetrate the stratum functionalis. The thickness of the endometrium approximately doubles, to about 4–10 mm. With reference to the uterine cycle, the preovulatory phase is also termed the **proliferative phase** because the endometrium is proliferating.

Ovulation

Ovulation, the rupture of the mature (graafian) follicle and the release of the secondary oocyte into the pelvic cavity, usually occurs on day 14 in a 28-day cycle. During ovulation, the secondary oocyte remains surrounded by its zona pellucida and corona radiata.

The *high levels of estrogens* during the last part of the preovulatory phase exert a *positive* feedback effect on the cells that secrete LH and gonadotropin-releasing hormone (GnRH) and cause ovulation, as follows (Figure 28.25):

1. A high concentration of estrogens stimulates more frequent release of GnRH from the hypothalamus. It also directly stimulates gonadotrophs in the anterior pituitary to secrete LH.
2. GnRH promotes the release of FSH and additional LH by the anterior pituitary.
3. LH causes rupture of the mature (graafian) follicle and expulsion of a secondary oocyte about 9 hours after the peak of the LH surge. The ovulated oocyte and its corona radiata cells are usually swept into the uterine tube.

From time to time, an oocyte is lost into the pelvic cavity, where it later disintegrates. The small amount of blood that sometimes leaks into the pelvic cavity from the ruptured follicle can cause pain, known as **mittelschmerz** (MIT-el-shmārts = pain in the middle), at the time of ovulation.

An over-the-counter home test that detects a rising level of LH can be used to predict ovulation a day in advance.

Postovulatory Phase

The **postovulatory phase** of the female reproductive cycle is the time between ovulation and onset of the next menses. In duration, it is the most constant part of the female reproductive cycle. It lasts for 14 days in a 28-day cycle, from day 15 to day 28 (see Figure 28.24).

EVENTS IN ONE OVARY After ovulation, the mature follicle collapses, and the basement membrane between the granulosa cells and theca interna breaks down. Once a blood clot forms from minor bleeding of the ruptured follicle, the follicle becomes the **corpus hemorrhagicum** (*hemo-* = blood; *rrhagic-* = bursting forth) (see Figure 28.13). Theca interna cells mix with the granulosa cells as they all become transformed into **corpus luteum cells** under the influence of LH. Stimulated by LH, the corpus luteum secretes progesterone, estrogen, relaxin, and inhibin. The luteal cells also absorb

Figure 28.25 High levels of estrogens exert a positive feedback effect (green arrows) on the hypothalamus and anterior pituitary, thereby increasing secretion of GnRH and LH.

At midcycle, a surge of LH triggers ovulation.

GnRH
Hypothalamus
1 High levels of estrogens from almost mature follicle stimulate release of more GnRH and LH
2 GnRH promotes release of FSH and more LH
LH
Anterior pituitary
3 LH surge brings about ovulation
Ovulated secondary oocyte
Ovary
Almost mature (graafian) follicle
Corpus hemorrhagicum (ruptured follicle)

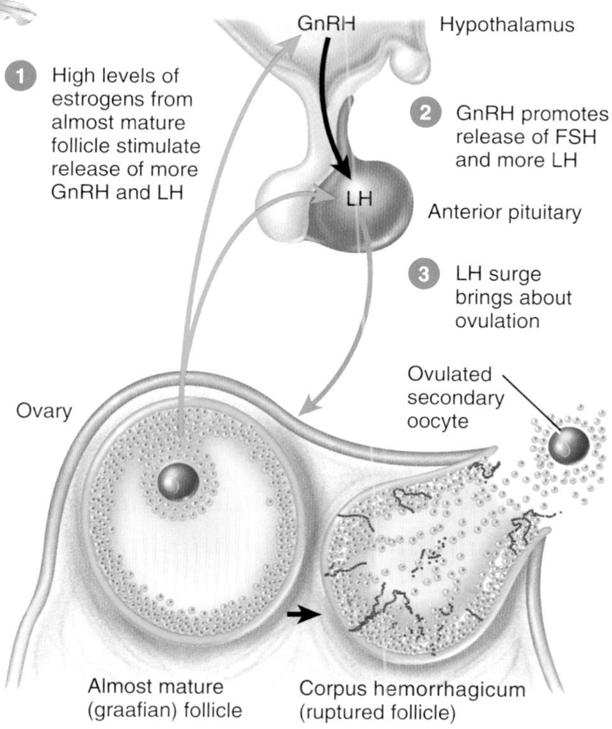

What is the effect of rising but still moderate levels of estrogens on the secretion of GnRH, LH, and FSH?

the blood clot. With reference to the ovarian cycle, this phase is also called the **luteal phase.**

Later events in an ovary that has ovulated an oocyte depend on whether the oocyte is fertilized. If the oocyte *is not fertilized,* the corpus luteum has a lifespan of only 2 weeks. Then, its secretory activity declines, and it degenerates into a corpus albicans (see Figure 28.13). As the levels of progesterone, estrogens, and inhibin decrease, release of GnRH, FSH, and LH rise due to loss of negative feedback suppression by the ovarian hormones. Follicular growth resumes and a new ovarian cycle begins.

If the secondary oocyte *is fertilized* and begins to divide, the corpus luteum persists past its normal 2-week lifespan. It is "rescued" from degeneration by **human chorionic gonadotropin** (kō-rē-ON-ik) **(hCG).** This hormone is produced by the chorion of the embryo beginning about 8 days after fertilization. Like LH, hCG stimulates the secretory activity of the corpus luteum. The presence of hCG in maternal blood or urine is an indicator of pregnancy and is the hormone detected by home pregnancy tests.

Events in the Uterus Progesterone and estrogens produced by the corpus luteum promote growth and coiling of the endometrial glands, vascularization of the superficial endometrium, and thickening of the endometrium to 12–18 mm (0.48–0.72 in.). Because of the secretory activity of the endometrial glands, which begin to secrete glycogen, this period is called the **secretory phase** of the uterine cycle. These preparatory changes peak about one week after ovulation, at the time a fertilized ovum might arrive in the uterus. If fertilization does not occur, the levels of progesterone and estrogens decline due to degeneration of the corpus luteum. Withdrawal of progesterone and estrogens causes menstruation.

Figure 28.26 summarizes the hormonal interactions and cyclical changes in the ovaries and uterus during the ovarian and uterine cycles.

Female Athlete Triad: Disordered Eating, Amenorrhea, and Premature Osteoporosis

The female reproductive cycle can be disrupted by many factors, including weight loss, low body weight, disordered eating, and vigorous physical activity. The observation that three conditions—disordered eating, amenorrhea, and osteoporosis—often occur together in female athletes led researchers to coin the term **female athlete triad.**

Many athletes experience intense pressure from coaches, parents, peers, and themselves to lose weight to improve performance. Hence, they may develop disordered eating behaviors and engage in other harmful weight-loss practices in a struggle to maintain a very low body weight. **Amenorrhea** (ā-men′-ō-RĒ-a; *a-* = without; *men-* = month; *-rrhea* = a flow) is the absence of menstruation. The most common causes of amenorrhea are pregnancy and menopause. In female athletes, amenorrhea results from reduced secretion of gonadotropin-releasing hormone, which decreases the release of LH and FSH. As a result, ovarian follicles fail to develop, ovulation does not occur, synthesis of estrogens and progesterone wanes, and monthly menstrual bleeding ceases. Most cases of the female athlete triad occur in young women with very low amounts of body fat. Low levels of the hormone leptin, secreted by adipose cells, may be a contributing factor.

Because estrogens help bones retain calcium and other minerals, chronically low levels of estrogens are associated with loss of bone mineral density. The female athlete triad causes "old bones in young women." In one study, amenorrheic runners in their twenties had low bone mineral densities, similar to those of postmenopausal women 50 to 70 years old! Short periods of amenorrhea in young athletes may cause no lasting harm. However, long-term cessation of the reproductive cycle may be accompanied by a loss of bone mass, and adolescent athletes may fail to achieve an adequate bone mass; both of these situations can lead to premature osteoporosis and irreversible bone damage. ■

▶ **CHECKPOINT**

26. Describe the function of each of the following hormones in the uterine and ovarian cycles: GnRH, FSH, LH, estrogens, progesterone, and inhibin.

27. Briefly outline the major events of each phase of the uterine cycle, and correlate them with the events of the ovarian cycle.

28. Prepare a labeled diagram of the major hormonal changes that occur during the uterine and ovarian cycles.

BIRTH CONTROL METHODS

▶ **OBJECTIVE**

Explain the differences among the various types of birth control methods and compare their effectiveness.

Birth control refers to restricting the number of children by various methods designed to control fertility and prevent conception. No single, ideal method of birth control exists. The only method of preventing pregnancy that is 100% reliable is total **abstinence,** the avoidance of sexual intercourse. Several other methods are available; each has its advantages and disadvantages. These include surgical sterilization, hormonal methods, intrauterine devices, spermicides, barrier methods, and periodic abstinence. Table 28.3 on page 1090 provides the failure rates for various methods of birth control. Although it is not a form of birth control, in this section we will also discuss abortion, the premature expulsion of the products of conception from the uterus.

Surgical Sterilization

Sterilization is a procedure that renders an individual incapable of reproduction. The principal method for sterilization of males is a vasectomy (described in the Clinical Application on page 1066). Sterilization in females most often is achieved by performing a **tubal ligation** (lī-GĀ-shun), in which both uterine tubes are tied closed and then cut. This can be achieved in a few different ways. "Clips" or "clamps" can be placed on the uterine tubes, the tubes can be tied and/or cut, and sometimes they are cauterized. In any case the result is that the secondary oocyte cannot pass through the uterine tubes, and sperm cannot reach the oocyte. Tubal ligation reduces the risk of pelvic inflammatory disease in women who are exposed to sexually transmitted infections; it may also reduce the risk of ovarian cancer.

Hormonal Methods

Aside from total abstinence or surgical sterilization, hormonal methods are the most effective means of birth control. Used by 50 million women worldwide, **oral contraceptives** ("the pill") contain various mixtures of synthetic estrogens and progestins (chemicals with actions similar to those of progesterone). They prevent pregnancy mainly by negative feedback inhibition of secretion of the gonadotropins FSH and LH from the anterior pituitary. The low levels of FSH and LH usually prevent development of a dominant follicle. As a result, estrogen level does not rise, the midcycle LH surge does not occur, and ovulation is not triggered. Thus, there is no secondary oocyte available for fertilization. Even if ovulation does occur, as it does in some cases, oral contraceptives also alter cervical mucus so that it is more hostile to sperm and blocks implantation in the uterus. If taken properly, the pill is close to 100% effective.

Figure 28.26 Summary of hormonal interactions in the ovarian and uterine cycles.

Hormones from the anterior pituitary regulate ovarian function, and hormones from the ovaries regulate the changes in the endometrial lining of the uterus.

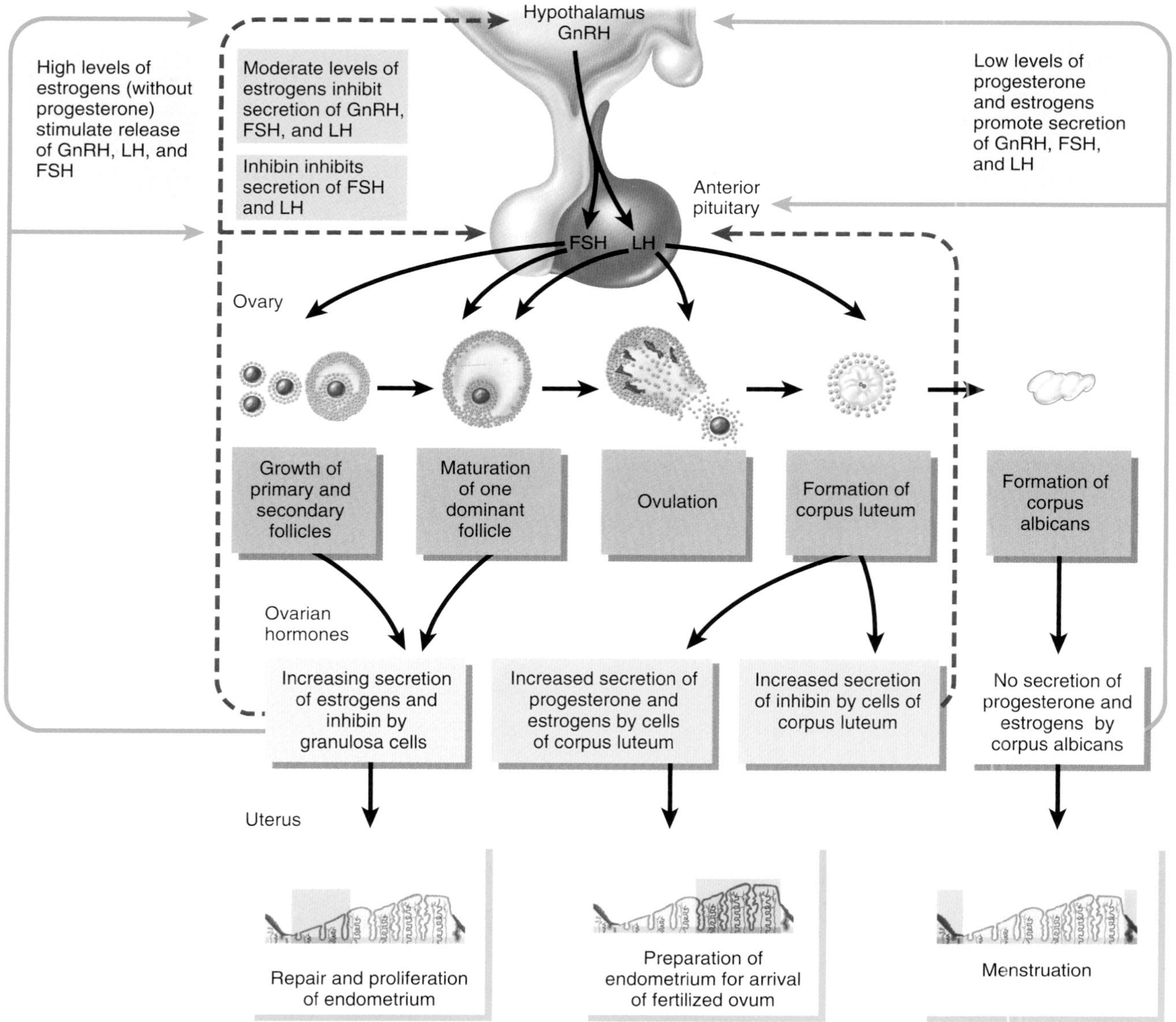

Hypothalamus
GnRH
High levels of estrogens (without progesterone) stimulate release of GnRH, LH, and FSH
Moderate levels of estrogens inhibit secretion of GnRH, FSH, and LH
Inhibin inhibits secretion of FSH and LH
Low levels of progesterone and estrogens promote secretion of GnRH, FSH, and LH
Anterior pituitary
FSH
LH
Ovary
Growth of primary and secondary follicles
Maturation of one dominant follicle
Ovulation
Formation of corpus luteum
Formation of corpus albicans
Ovarian hormones
Increasing secretion of estrogens and inhibin by granulosa cells
Increased secretion of progesterone and estrogens by cells of corpus luteum
Increased secretion of inhibin by cells of corpus luteum
No secretion of progesterone and estrogens by corpus albicans
Uterus
Repair and proliferation of endometrium
Preparation of endometrium for arrival of fertilized ovum
Menstruation

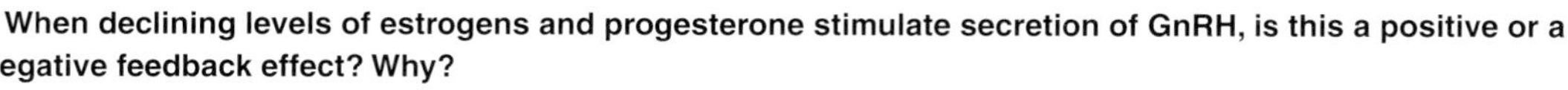

When declining levels of estrogens and progesterone stimulate secretion of GnRH, is this a positive or a negative feedback effect? Why?

TABLE 28.3 Failure Rates of Several Birth Control Methods

	Failure Rates*	
Method	**Perfect Use†**	**Typical Use**
None	85%	85%
Complete abstinence	0%	0%
Surgical sterilization		
Vasectomy	0.10%	0.15%
Tubal ligation	0.5%	0.5%
Hormonal methods		
Oral contraceptives	0.1%	3%‡
Norplant	0.3%	0.3%
Depo-provera	0.05%	0.05%
Intrauterine device		
Copper T 380A	0.6%	0.8%
Spermicides	6%	26%‡
Barrier methods		
Male condom	3%	14%‡
Vaginal pouch	5%	21%‡
Diaphragm	6%	20%‡
Periodic abstinence		
Rhythm	9%	25%‡
Sympto-thermal	2%	20%‡

*Defined as percentage of women having an unintended pregnancy during the first year of use.

†Failure rate when the method is used correctly and consistently.

‡Includes couples who forgot to use the method.

Among the noncontraceptive benefits of oral contraceptives are regulation of the length of menstrual cycles and decreased menstrual flow (and therefore decreased risk of anemia). The pill also provides protection against endometrial and ovarian cancers and reduces the risk of endometriosis. However, oral contraceptives may not be advised for women with a history of blood clotting disorders, cerebral blood vessel damage, migraine headaches, hypertension, liver malfunction, or heart disease. Women who take the pill and smoke face far higher odds of having a heart attack or stroke than do nonsmoking pill users. Smokers should quit smoking or use an alternative method of birth control.

Oral contraceptives also may be used for **emergency contraception (EC),** the so-called "morning-after pill." The relatively high levels of estrogens and progestin in EC pills provide negative feedback inhibition of FSH and LH secretion. Loss of the stimulating effects of these gonadotropic hormones causes the ovaries to cease secretion of their own estrogens and progesterone. In turn, declining levels of estrogens and progesterone induce shedding of the uterine lining, thereby blocking implantation. In consultation with a physician, when two pills are taken within 72 hours after unprotected intercourse, and another two pills are taken 12 hours later, the chance of pregnancy is reduced by 75%.

Other hormonal methods of contraception are also available:

- **Norplant®** consists of six slender hormone-containing capsules that are surgically implanted under the skin of the arm using local anesthesia. They slowly and continually release a progestin, which inhibits ovulation and thickens the cervical mucus. The effects last for 5 years, and Norplant is about as reliable as sterilization. Removing the Norplant capsules restores fertility.
- **Depo-provera®,** which is given as an intramuscular injection once every 3 months, contains progestin that prevents maturation of the ovum and causes changes in the uterine lining that make pregnancy less likely.
- **Lunelle®** is a once-a-month intramuscular injection. It contains estrogen and progestin and acts like an oral contraceptive.
- **Birth control skin patches** contain estrogens and progestin and are placed on the skin once a week for three weeks. Each week the patch is removed and a new one is placed on a different area of the skin. During the fourth week no patch is used so that menstruation can occur.
- The **vaginal ring** is a doughnut-shaped ring that fits in the vagina and releases either a progestin alone or a progestin and an estrogen. It is worn for 3 weeks and removed for 1 week to allow menstruation to occur.

Intrauterine Devices

An **intrauterine device (IUD)** is a small object made of plastic, copper, or stainless steel that is inserted into the cavity of the uterus. IUDs cause changes in the uterine lining that prevent implantation of a fertilized ovum. The IUD most commonly used in the United States today is the Copper T 380A, which is approved for up to 10 years of use and has long-term effectiveness comparable to that of tubal ligation. Some women cannot use IUDs because of expulsion, bleeding, or discomfort.

Spermicides

Various foams, creams, jellies, suppositories, and douches that contain sperm-killing agents, or **spermicides,** make the vagina and cervix unfavorable for sperm survival and are available without prescription. The most widely used spermicide is nonoxynol-9, which kills sperm by disrupting their plasma membrane. A spermicide is more effective when used with a barrier method such as a diaphragm or a condom.

Barrier Methods

Barrier methods are designed to prevent sperm from gaining access to the uterine cavity and uterine tubes. In addition to preventing pregnancy, certain barrier methods (condom and vaginal pouch) may also provide some protection against sexually transmitted diseases (STDs) such as AIDS. In contrast, oral contraceptives and IUDs confer no such protection. Among the barrier methods are the condom, the vaginal pouch, and the diaphragm.

A **condom** is a nonporous, latex covering placed over the penis that prevents deposition of sperm in the female reproductive tract. A **vaginal pouch,** sometimes called a female condom, is made of two flexible rings connected by a polyurethane sheath. One ring lies inside the sheath and is inserted to fit over the cervix; the other ring remains outside the vagina and covers the female external genitals. A **diaphragm** is a rubber, dome-shaped structure that fits over the cervix and is used in conjunction with a spermicide. It can be inserted up to 6 hours before intercourse. The diaphragm stops most sperm from passing into the cervix and the spermicide kills most sperm that do get by. Although diaphragm use does decrease the risk of some STDs, it does not fully protect against HIV infection.

Periodic Abstinence

A couple can use their knowledge of the physiological changes that occur during the female reproductive cycle to decide either to abstain from intercourse on those days when pregnancy is a likely result, or to plan intercourse on those days if they wish to conceive a child. In females with normal and regular menstrual cycles, these physiological events help to predict the day on which ovulation is likely to occur.

The first physiologically based method, developed in the 1930s, is known as the **rhythm method.** It involves abstaining from sexual activity on the days that ovulation is likely to occur in each reproductive cycle. During this time (3 days before ovulation, the day of ovulation, and 3 days after ovulation) the couple abstains from intercourse. The effectiveness of the rhythm method for birth control is poor in many women due to the irregularity of the female reproductive cycle.

Another system is the **sympto-thermal method,** in which couples must learn and understand certain signs of fertility. The signs of ovulation include increased basal body temperature; the production of abundant clear, stretchy cervical mucus; and pain associated with ovulation (mittelschmerz). If a couple abstains from sexual intercourse when the signs of ovulation are present and for 3 days afterward, the chance of pregnancy is decreased. A big problem with this method is that fertilization is very likely if intercourse occurs one or two days *before* ovulation.

Abortion

Abortion refers to the premature expulsion of the products of conception from the uterus, usually before the twentieth week of pregnancy. An abortion may be spontaneous (naturally occurring; also called a miscarriage) or induced (intentionally performed). Induced abortions may be performed by vacuum aspiration (suction), infusion of a saline solution, or surgical evacuation (scraping).

Certain drugs, most notably RU 486, can induce a so-called nonsurgical abortion. **RU 486 (mifepristone)** is an antiprogestin; it blocks the action of progesterone by binding to and blocking progesterone receptors. Recall that progesterone prepares the uterine endometrium for implantation and then maintains the uterine lining after implantation. If the level of progesterone falls during pregnancy or if the action of the hormone is blocked, menstruation occurs, and the embryo sloughs off along with the uterine lining. Within 12 hours after taking RU 486, the endometrium starts to degenerate, and within 72 hours, it begins to slough off. A form of prostaglandin E (misoprostol), which stimulates uterine contractions, is given after RU 486 to aid in expulsion of the endometrium. RU 486 can be taken up to 5 weeks after conception. One side effect of the drug is uterine bleeding.

▶ CHECKPOINT

29. How do oral contraceptives reduce the likelihood of pregnancy?

30. How do some methods of birth control protect against sexually transmitted diseases?

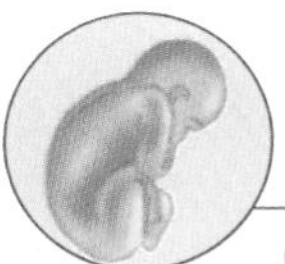

DEVELOPMENT OF THE REPRODUCTIVE SYSTEMS

▶ OBJECTIVE

Describe the development of the male and female reproductive systems.

The *gonads* develop from the **intermediate mesoderm** during the fifth week of development and they appear as bulges (Figure 28.27). Adjacent to the gonads are the **mesonephric (Wolffian) ducts,** which eventually develop into structures of the reproductive system in males. A second pair of ducts, the **paramesonephric (Müllerian) ducts,** develop lateral to the mesonephric ducts and eventually form structures of the reproductive system in females. Both sets of ducts empty into the urogenital sinus. An early embryo has the potential to follow either the male or the female pattern of development because it contains both sets of ducts and primitive gonads that can differentiate into either testes or ovaries.

Cells of a male embryo have one X chromosome and one Y chromosome. The male pattern of development is initiated by a "master switch" gene on the Y chromosome named ***SRY,*** which stands for *S*ex-determining *R*egion of the *Y* chromosome. When the *SRY* gene is expressed during development, its protein product causes the primitive Sertoli cells to begin to differentiate in the gonadal tissues during the seventh week. The developing Sertoli cells secrete a hormone called **Müllerian-inhibiting substance (MIS),** which causes apoptosis of cells within the paramesonephric (Müllerian) ducts. As a result, those cells do not contribute any functional structures to the male reproductive system. Stimulated by human chorionic gonadotropin (hCG),

Figure 28.27 **Development of the internal reproductive systems.**

The gonads develop from intermediate mesoderm.

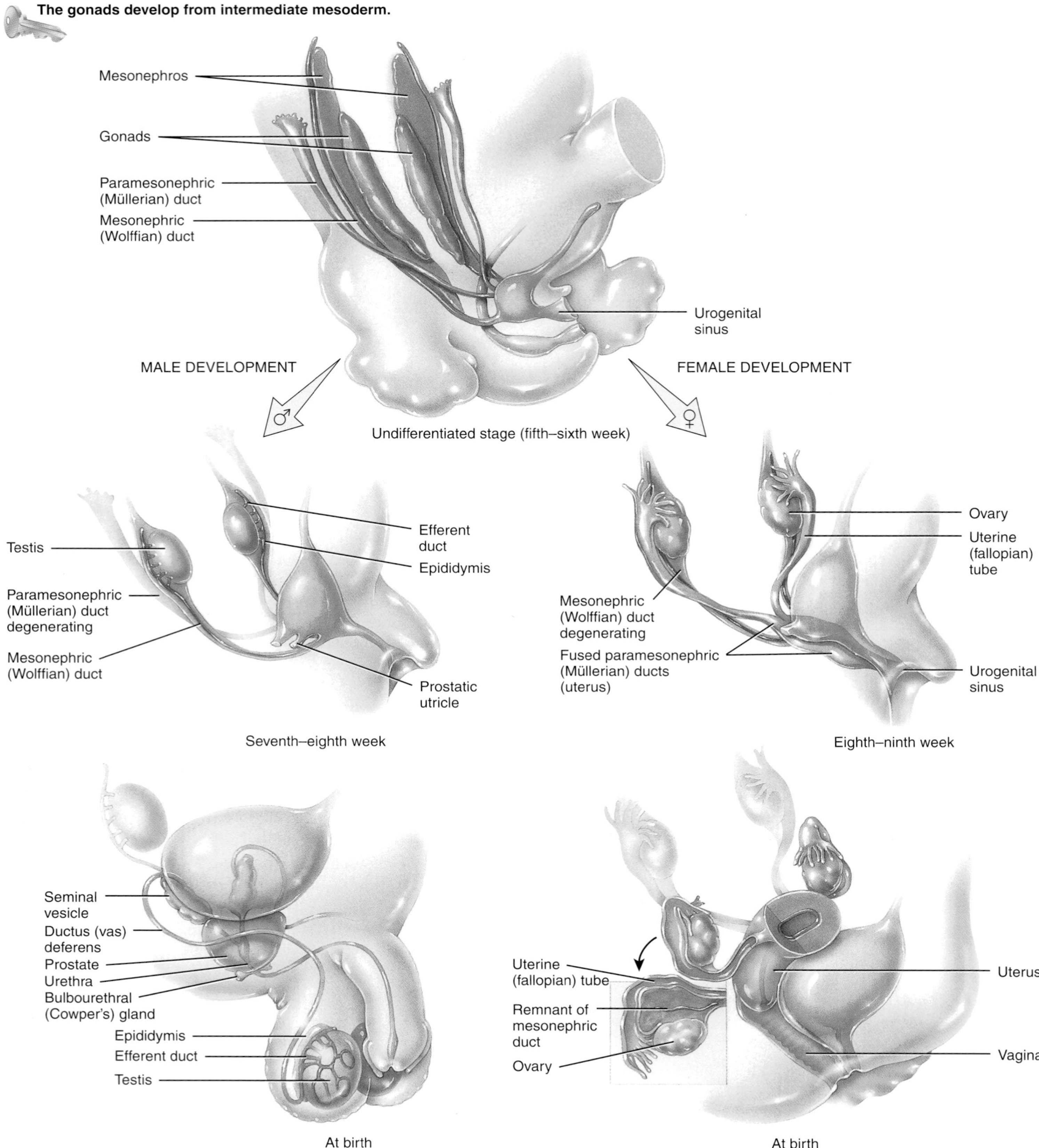

Which gene is responsible for the development of the gonads into testes?

primitive Leydig cells in the gonadal tissue begin to secrete the androgen **testosterone** during the eighth week. Testosterone then stimulates development of the mesonephric duct on each side into the *epididymis, ductus (vas) deferens, ejaculatory duct,* and *seminal vesicle.* The *testes* connect to the mesonephric duct through a series of tubules that eventually become the *seminiferous tubules.* The *prostate* and *bulbourethral glands* are **endodermal** outgrowths of the urethra.

Cells of a female embryo have two X chromosomes and no Y chromosome. Because *SRY* is absent, the gonads develop into *ovaries,* and because MIS is not produced, the paramesonephric ducts flourish. The distal ends of the paramesonephric ducts fuse to form the *uterus* and *vagina;* the unfused proximal portions of the ducts become the *uterine (fallopian) tubes.* The mesonephric ducts degenerate without contributing any functional structures to the female reproductive system because of the absence of testosterone. The *greater* and *lesser vestibular glands* develop from **endodermal** outgrowths of the vestibule.

The *external genitals* of both male and female embryos (penis and scrotum in males and clitoris, labia, and vaginal orifice in females) also remain undifferentiated until about the eighth week. Before differentiation, all embryos have an elevated midline swelling called the **genital tubercle** (Figure 28.28). The tubercle consists of the **urethral groove** (opening into the urogenital sinus), paired **urethral folds,** and paired **labioscrotal swellings.**

In male embryos, some testosterone is converted to a second androgen called **dihydrotestosterone (DHT).** DHT stimulates development of the urethra, prostate, and external genitals (scrotum and penis). Part of the genital tubercle elongates and develops into a penis. Fusion of the urethral folds forms the *spongy (penile) urethra* and leaves an opening to the exterior only at the distal end of the penis, the *external urethral orifice.* The labioscrotal swellings develop into the *scrotum.* In the absence of DHT, the genital tubercle gives rise to the *clitoris* in female embryos. The urethral folds remain open as the *labia minora,* and the labioscrotal swellings become the *labia majora.* The urethral groove becomes the *vestibule.* After birth, androgen levels decline because hCG is no longer present to stimulate secretion of testosterone.

▶ **CHECKPOINT**

31. Describe the role of hormones in differentiation of the gonads, the mesonephric ducts, the paramesonephric ducts, and the external genitals.

Figure 28.28 Development of the external genitals.

The external genitals of male and female embryos remain undifferentiated until about the eighth week.

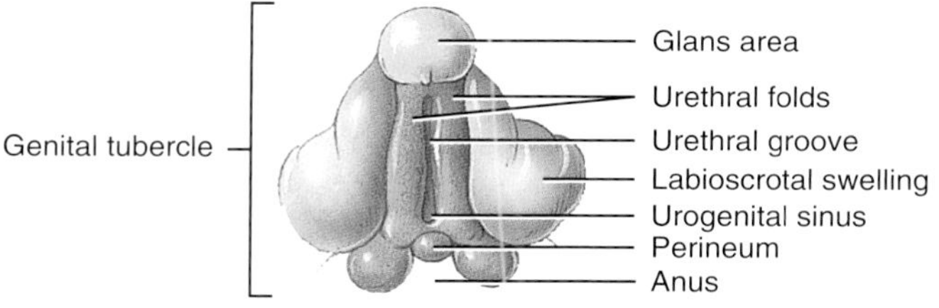

Undifferentiated stage (about five-week embryo)

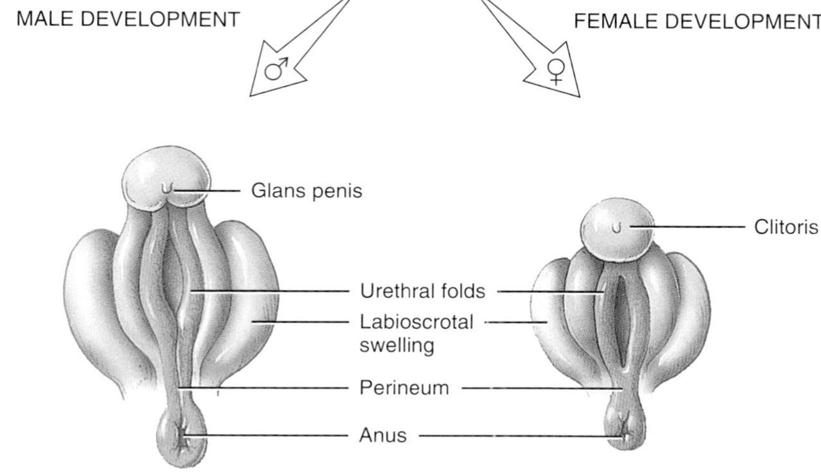

Ten-week embryo

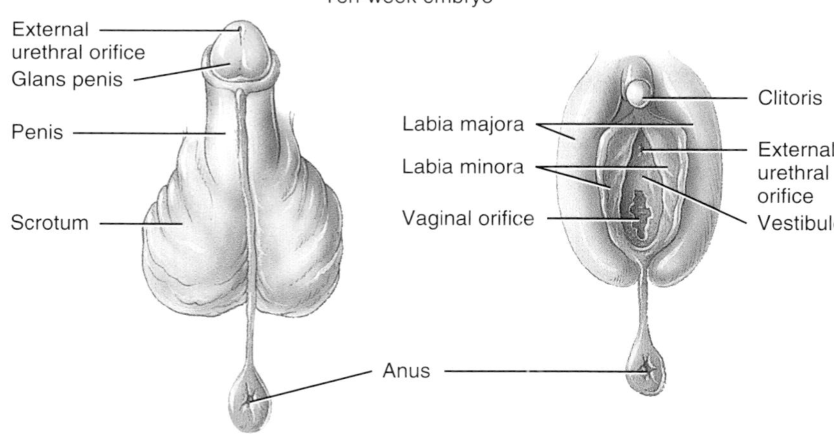

Near birth

Which hormone is responsible for the differentiation of the external genitals?

AGING AND THE REPRODUCTIVE SYSTEMS

► OBJECTIVE

Describe the effects of aging on the reproductive systems.

During the first decade of life, the reproductive system is in a juvenile state. At about age 10, hormone-directed changes start to occur in both sexes. **Puberty** (PŪ-ber-tē = a ripe age) is the period when secondary sexual characteristics begin to develop and the potential for sexual reproduction is reached. The onset of puberty is marked by pulses or bursts of LH and FSH secretion, each triggered by a pulse of GnRH. Most pulses occur during sleep. As puberty advances, the hormone pulses occur during the day as well as at night. The pulses increase in frequency during a three- to four-year period until the adult pattern is established. The stimuli that cause the GnRH pulses are still unclear, but a role for the hormone leptin is starting to unfold. Just before puberty, leptin levels rise in proportion to adipose tissue mass. Interestingly, leptin receptors are present in both the hypothalamus and anterior pituitary. Mice that lack a functional leptin gene from birth are sterile and remain in a prepubertal state. Giving leptin to such mice elicits secretion of gonadotropins, and they become fertile. Leptin may signal the hypothalamus that long-term energy stores (triglycerides in adipose tissue) are adequate for reproductive functions to begin.

In females, the reproductive cycle normally occurs once each month from **menarche** (me-NAR-kē), the first menses, to **menopause,** the permanent cessation of menses. Thus, the female reproductive system has a time-limited span of fertility between menarche and menopause. For the first 1 to 2 years after menarche, ovulation only occurs in about 10% of the cycles and the luteal phase is short. Gradually, the percentage of ovulatory cycles increases, and the luteal phase reaches its normal duration of 14 days. With age, fertility declines. Between the ages of 40 and 50 the pool of remaining ovarian follicles becomes exhausted. As a result, the ovaries become less responsive to hormonal stimulation. The production of estrogens declines, despite copious secretion of FSH and LH by the anterior pituitary. Many women experience hot flashes and heavy sweating, which coincide with bursts of GnRH release. Other symptoms of menopause are headache, hair loss, muscular pains, vaginal dryness, insomnia, depression, weight gain, and mood swings. Some atrophy of the ovaries, uterine tubes, uterus, vagina, external genitalia, and breasts occurs in postmenopausal women. Due to loss of estrogens, most women experience a decline in bone mineral density after menopause. Sexual desire (libido) does not show a parallel decline; it may be maintained by adrenal sex steroids. The risk of having uterine cancer peaks at about 65 years of age, but cervical cancer is more common in younger women.

In males, declining reproductive function is much more subtle than in females. Healthy men often retain reproductive capacity into their eighties or nineties. At about age 55 a decline in testosterone synthesis leads to reduced muscle strength, fewer viable sperm, and decreased sexual desire. Although sperm production decreases 50–70% between ages 60 and 80, abundant sperm may still be present even in old age.

Enlargement of the prostate to two to four times its normal size occurs in most males over age 60. This condition, called **benign prostatic hyperplasia (BPH),** decreases the size of the prostatic urethra and is characterized by frequent urination, nocturia (bed-wetting), hesitancy in urination, decreased force of urinary stream, postvoiding dribbling, and a sensation of incomplete emptying.

► CHECKPOINT

32. What changes occur in males and females at puberty?

33. What do the terms menarche and menopause mean?

DISORDERS: HOMEOSTATIC IMBALANCES

Reproductive System Disorders in Males

Testicular Cancer

Testicular cancer is the most common cancer in males between the ages of 20 and 35. More than 95% of testicular cancers arise from spermatogenic cells within the seminiferous tubules. An early sign of testicular cancer is a mass in the testis, often associated with a sensation of testicular heaviness or a dull ache in the lower abdomen; pain usually does not occur. To increase the chance for early detection of a testicular cancer, all males should perform regular self-examinations of the testes. The examination should be done starting in the teen years and once each month thereafter. After a warm bath or shower (when the scrotal skin is loose and relaxed) each testicle should be examined as follows. The testicle is grasped and gently rolled between the index finger and thumb, feeling for lumps, swellings, hardness, or other changes. If a lump or other change is detected, a physician should be consulted as soon as possible.

Prostate Disorders

Because the prostate surrounds part of the urethra, any infection, enlargement, or tumor can obstruct the flow of urine. Acute and chronic infections of the prostate are common in postpubescent males, often in association with inflammation of the urethra. Symptoms may include fever, chills, urinary frequency, frequent urination at night, difficulty in urinating, burning or painful urination, low back pain, joint and muscle pain, blood in the urine, or painful ejaculation. However, often there are no symptoms. Antibiotics are used to treat most cases that result from a bacterial infection. In **acute prostatitis,** the prostate becomes swollen and tender. **Chronic prostatitis** is one of the most common chronic infections in men of the middle and later years. On examination, the prostate feels enlarged, soft, and very tender, and its surface outline is irregular.

Prostate cancer is the leading cause of death from cancer in men in the United States, having surpassed lung cancer in 1991. Each year it is diagnosed in almost 200,000 U.S. men and causes nearly 40,000 deaths. The amount of PSA (prostate-specific antigen), which is produced only by prostate epithelial cells, increases with enlargement of the prostate and may indicate infection, benign enlargement, or prostate cancer. A blood test can measure the level of PSA in the blood. Males over the age of 40 should have an annual examination of the prostate gland. In a **digital rectal exam,** a physician palpates the gland through the rectum with the fingers (digits). Many physicians also recommend an annual PSA test for males over age 50. Treatment for prostate cancer may involve surgery, cryotherapy, radiation, hormonal therapy, and chemotherapy. Because many prostate cancers grow very slowly, some urologists recommend "watchful waiting" before treating small tumors in men over age 70.

Erectile Dysfunction

Erectile dysfunction (ED), previously termed *impotence,* is the consistent inability of an adult male to ejaculate or to attain or hold an erection long enough for sexual intercourse. Many cases of impotence are caused by insufficient release of nitric oxide (NO), which relaxes the smooth muscle of the penile arterioles and erectile tissue. The drug *Viagra®* (sildenafil) enhances smooth muscle relaxation by nitric oxide in the penis. Other causes of erectile dysfunction include diabetes mellitus, physical abnormalities of the penis, systemic disorders such as syphilis, vascular disturbances (arterial or venous obstructions), neurological disorders, surgery, testosterone deficiency, and drugs (alcohol, antidepressants, antihistamines, antihypertensives, narcotics, nicotine, and tranquilizers). Psychological factors such as anxiety or depression, fear of causing pregnancy, fear of sexually transmitted diseases, religious inhibitions, and emotional immaturity may also cause ED.

Reproductive System Disorders in Females

Premenstrual Syndrome and Premenstrual Dysphoric Disorder

Premenstrual syndrome (PMS) is a cyclical disorder of severe physical and emotional distress. It appears during the postovulatory (luteal) phase of the female reproductive cycle and dramatically disappears when menstruation begins. The signs and symptoms are highly variable from one woman to another. They may include edema, weight gain, breast swelling and tenderness, abdominal distension, backache, joint pain, constipation, skin eruptions, fatigue and lethargy, greater need for sleep, depression or anxiety, irritability, mood swings, headache, poor coordination and clumsiness, and cravings for sweet or salty foods. The cause of PMS is unknown. For some women, getting regular exercise; avoiding caffeine, salt, and alcohol; and eating a diet that is high in complex carbohydrates and lean proteins can bring considerable relief.

Premenstrual dysphoric disorder (PMDD) is a more severe syndrome in which PMS-like signs and symptoms do not resolve after the onset of menstruation. Clinical research studies have found that suppression of the reproductive cycle by a drug that interferes with GnRH (leuprolide) decreases symptoms significantly. Because symptoms reappear when estradiol or progesterone is given together with leuprolide, researchers propose that PMDD is caused by abnormal responses to normal levels of these ovarian hormones. *SSRIs* (selective serotonin receptor inhibitors) have shown promise in treating both PMS and PMDD.

Endometriosis

Endometriosis (en-dō-mē-trē-Ō-sis; *endo-* = within; *metri-* = uterus; *osis* = condition) is characterized by the growth of endometrial tissue outside the uterus. The tissue enters the pelvic cavity via the open uterine tubes and may be found in any of several sites—on the ovaries, the rectouterine pouch, the outer surface of the uterus, the sigmoid colon, pelvic and abdominal lymph nodes, the cervix, the abdominal wall, the kidneys, and the urinary bladder. Endometrial tissue responds to hormonal fluctuations, whether it is inside or outside the uterus. With each reproductive cycle, the tissue proliferates and then breaks down and bleeds. When this occurs outside the uterus, it can cause inflammation, pain, scarring, and infertility. Symptoms include premenstrual pain or unusually severe menstrual pain.

Breast Cancer

One in eight women in the United States faces the prospect of **breast cancer.** After lung cancer, it is the second-leading cause of death from cancer in U.S. women. Breast cancer can occur in males but is rare. In females, breast cancer is seldom seen before age 30; its incidence rises rapidly after menopause. An estimated 5% of the 180,000 cases diagnosed each year in the United States, particularly those that arise in younger women, stem from inherited genetic mutations (changes in the DNA). Researchers have now identified two genes that increase susceptibility to breast cancer: *BRCA1* (*br*east *ca*ncer *1*) and *BRCA2.* Mutation of *BRCA1* also confers a high risk for ovarian cancer. In addition, mutations of the *p53* gene increase the risk of breast cancer in both males and females, and mutations of the androgen receptor gene are associated with the occurrence of breast cancer in some males. Because breast cancer generally is not painful until it becomes quite advanced, any lump, no matter how small, should be reported to a physician at once. Early detection—by breast self-examination and mammograms—is the best way to increase the chance of survival.

The most effective technique for detecting tumors less than 1 cm (0.4 in.) in diameter is **mammography** (mam-OG-ra-fē; *-graphy* = to record), a type of radiography using very sensitive x-ray film. The image of the breast, called a **mammogram** (see Table 1.3 on page 21), is best obtained by compressing the breasts, one at a time, using flat plates. A supplementary procedure for evaluating breast abnormalities is **ultrasound.** Although ultrasound cannot detect tumors smaller than 1 cm in diameter (which mammography can detect), it can be used to determine whether a lump is a benign, fluid-filled cyst or a solid (and therefore possibly malignant) tumor.

Among the factors that increase the risk of developing breast cancer are (1) a family history of breast cancer, especially in a mother or sister; (2) nulliparity (never having borne a child) or having a first child after age 35; (3) previous cancer in one breast; (4) exposure to ionizing radiation, such as x rays; (5) excessive alcohol intake; and (6) cigarette smoking.

The American Cancer Society recommends the following steps to help diagnose breast cancer as early as possible:

- All women over 20 should develop the habit of monthly breast self-examination.
- A physician should examine the breasts every 3 years when a woman is between the ages of 20 and 40, and every year after age 40.
- A mammogram should be taken in women between the ages of 35 and 39, to be used later for comparison (baseline mammogram).
- Women with no symptoms should have a mammogram every year or two between ages 40 and 49, and every year after age 50.
- Women of any age with a history of breast cancer, a strong family history of the disease, or other risk factors should consult a physician to determine a schedule for mammography.

Treatment for breast cancer may involve hormone therapy, chemotherapy, radiation therapy, **lumpectomy** (removal of the tumor and the immediate surrounding tissue), a modified or radical mastectomy, or a combination of these approaches. A **radical mastectomy** (*mast-* = breast) involves removal of the affected breast along with the underlying pectoral muscles and the axillary lymph nodes. (Lymph nodes are removed because metastasis of cancerous cells usually occurs through lymphatic or blood vessels.) Radiation treatment and chemotherapy may follow the surgery to ensure the destruction of any stray cancer cells. Several types of chemotherapeutic drugs are used to decrease the risk of relapse or disease progression. *Nolvadex® (tamoxifen)* is an estrogen antagonist that binds to and blocks estrogen receptors, thus decreasing the stimulating effect of estrogen on breast cancer cells. Tamoxifen has been used for 20 years and greatly reduces the risk of cancer recurrence. *Herceptin®*, a monoclonal antibody drug, targets an antigen on the surface of breast cancer cells. It is effective in causing regression of tumors and retarding progression of the disease. The early data from clinical trials of two new drugs, *Femara®* and *Amimidex®*, show relapse rates that are lower than those for tamoxifen. These drugs are inhibitors of aromatase, the enzyme needed for the final step in synthesis of estrogens. Finally, two drugs—tamoxifen and *Evista® (raloxifene)*—are being marketed for breast cancer *prevention.* Interestingly, raloxifene blocks estrogen receptors in the breasts and uterus but activates estrogen receptors in bone, providing effective treatment for osteoporosis while possibly decreasing the risk of breast or endometrial (uterine) cancer.

Ovarian Cancer

Even though **ovarian cancer** is the sixth most common form of cancer in females, it is the leading cause of death from all gynecological malignancies (excluding breast cancer) because it is difficult to detect before it metastasizes (spreads) beyond the ovaries. Risk factors associated with ovarian cancer include age (usually over age 50); race (whites are at highest risk); family history of ovarian cancer; more than 40 years of active ovulation; nulliparity or first pregnancy after age 30; a high-fat, low-fiber, vitamin A–deficient diet; and prolonged exposure to asbestos or talc. Early ovarian cancer has no symptoms or only mild ones associated with other common problems, such as abdominal discomfort, heartburn, nausea, loss of appetite, bloating, and flatulence. Later-stage signs and symptoms include an enlarged abdomen, abdominal and/or pelvic pain, persistent gastrointestinal disturbances, urinary complications, menstrual irregularities, and heavy menstrual bleeding.

Cervical Cancer

Cervical cancer, carcinoma of the cervix of the uterus, starts with **cervical dysplasia** (dis-PLĀ-sē-a), a change in the shape, growth, and number of cervical cells. The cells may either return to normal or progress to cancer. In most cases, cervical cancer may be detected in its earliest stages by a Pap test (see page 118). Some evidence links cervical cancer to the virus that causes genital warts, human papillomavirus (HPV). Increased risk is associated with having a large number of sexual partners, having first intercourse at a young age, and smoking cigarettes.

Vulvovaginal Candidiasis

Candida albicans is a yeastlike fungus that commonly grows on mucous membranes of the gastrointestinal and genitourinary tracts. The organism is responsible for **vulvovaginal candidiasis** (vul-vō-VAJ-i-nal can-di-DĪ-a-sis), the most common form of **vaginitis** (vaj-i-NĪ-tis), inflammation of the vagina. Candidiasis is characterized by severe itching; a thick, yellow, cheesy discharge; a yeasty odor; and pain. The disorder, experienced at least once by about 75% of females, is usually a result of proliferation of the fungus following antibiotic therapy for another condition. Predisposing conditions include the use of oral contraceptives or cortisone-like medications, pregnancy, and diabetes.

Sexually Transmitted Diseases

A **sexually transmitted disease (STD)** is one that is spread by sexual contact. In most developed countries of the world, such as those of Western Europe, Japan, Australia, and New Zealand, the incidence of STDs has declined markedly during the past 25 years. In the United States, by contrast, STDs have been rising to near-epidemic proportions; they currently affect more than 65 million people. AIDS and hepatitis B, which are sexually transmitted diseases that also may be contracted in other ways, are discussed in Chapters 22 and 24, respectively.

Chlamydia

Chlamydia (kla-MID-ē-a) is a sexually transmitted disease caused by the bacterium *Chlamydia trachomatis* (*chlamy-* = cloak). This unusual bacterium cannot reproduce outside body cells; it "cloaks" itself inside cells, where it divides. At present, chlamydia is the most prevalent sexually transmitted disease in the United States. In most cases, the initial infection is asymptomatic and thus difficult to recognize clinically. In males, urethritis is the principal result, causing a clear discharge, burning on urination, frequent urination, and painful urination. Without treatment, the epididymides may also become inflamed, leading to sterility. In 70% of females with chlamydia, symptoms are absent, but chlamydia is the leading cause of pelvic inflammatory disease. The uterine tubes may also become inflamed, which increases the risk of ectopic pregnancy (implantation of a fertilized ovum outside the uterus) and infertility due to the formation of scar tissue in the tubes.

Gonorrhea

Gonorrhea (gon-ō-RĒ-a) or **"the clap"** is caused by the bacterium *Neisseria gonorrhoeae.* In the United States, 1–2 million new cases of gonorrhea appear each year, most among individuals aged 15–29

years. Discharges from infected mucus membranes are the source of transmission of the bacteria either during sexual contact or during the passage of a newborn through the birth canal. The infection site can be in the mouth and throat after oral-genital contact, in the vagina and penis after genital intercourse, or in the rectum after recto-genital contact.

Males usually experience urethritis with profuse pus drainage and painful urination. The prostate and epididymis may also become infected. In females, infection typically occurs in the vagina, often with a discharge of pus. Both infected males and females may harbor the disease without any symptoms, however, until it has progressed to a more advanced stage; about 5–10% of males and 50% of females are asymptomatic. In females, the infection and consequent inflammation can proceed from the vagina into the uterus, uterine tubes, and pelvic cavity. An estimated 50,000 to 80,000 women in the United States are made infertile by gonorrhea every year as a result of scar tissue formation that closes the uterine tubes. If bacteria in the birth canal are transmitted to the eyes of a newborn, blindness can result. Administration of a 1% silver nitrate solution in the infant's eyes prevents infection.

Syphilis

Syphilis, caused by the bacterium *Treponema pallidum,* is transmitted through sexual contact or exchange of blood, or through the placenta to a fetus. The disease progresses through several stages. During the *primary stage,* the chief sign is a painless open sore, called a **chancre** (SHANG-ker), at the point of contact. The chancre heals within 1 to 5 weeks. From 6 to 24 weeks later, signs and symptoms such as a skin rash, fever, and aches in the joints and muscles usher in the *secondary stage,* which is systemic—the infection spreads to all major body systems. When signs of organ degeneration appear, the disease is said to be in the *tertiary stage.* If the nervous system is involved, the tertiary stage is called **neurosyphilis.** As motor areas become damaged extensively, victims may be unable to control urine and bowel movements. Eventually they may become bedridden and unable even to feed themselves. In addition, damage to the cerebral cortex produces memory loss and personality changes that range from irritability to hallucinations.

Genital Herpes

Genital herpes is an incurable STD. Type II herpes simplex virus (HSV-2) causes genital infections, producing painful blisters on the prepuce, glans penis, and penile shaft in males and on the vulva or sometimes high up in the vagina in females. The blisters disappear and reappear in most patients, but the virus itself remains in the body. A related virus, type I herpes simplex virus (HSV-1), causes cold sores on the mouth and lips. Infected individuals typically experience recurrences of symptoms several times a year.

Genital Warts

Warts are an infectious disease caused by viruses. *Human papillomavirus (HPV)* causes **genital warts,** which is commonly transmitted sexually. Nearly one million people a year develop genital warts in the United States. Patients with a history of genital warts may be at increased risk for cancers of the cervix, vagina, anus, vulva, and penis. There is no cure for genital warts.

MEDICAL TERMINOLOGY

Castration (kas-TRĀ-shun = to prune) Removal, inactivation, or destruction of the gonads; commonly used in reference to removal of the testes only.

Colposcopy (kol-POS-kō-pē; *colpo-* = vagina; *-scopy* = to view) Visual inspection of the vagina and cervix of the uterus using a culposcope, an instrument that has a magnifying lens (between 5 and 50×) and a light. The procedure generally takes place after an unusual Pap smear.

Culdoscopy (kul-DOS-kō-pē; *-cul-* = cul-de-sac; *-scopy* = to examine) A procedure in which a culdoscope (endoscope) is inserted through the posterior wall of the vagina to view the rectouterine pouch in the pelvic cavity.

Dysmenorrhea (dis′-men-or-Ē-a; *dys-* = difficult or painful) Pain associated with menstruation; the term is usually reserved to describe menstrual symptoms that are severe enough to prevent a woman from functioning normally for one or more days each month. Some cases are caused by uterine tumors, ovarian cysts, pelvic inflammatory disease, or intrauterine devices.

Dyspareunia (dis-pa-ROO-nē-a; *dys-* = difficult; *para* = beside; *-enue* = bed) Pain during sexual intercourse. It may occur in the genital area or in the pelvic cavity, and may be due to inadequate lubrication, inflammation, infection, an improperly fitting diaphragm or cervical cap, endometriosis, pelvic inflammatory disease, pelvic tumors, or weakened uterine ligaments.

Endocervical curettage (kū-re-TAHZH; *curette* = scraper) A procedure in which the cervix is dilated and the endometrium of the uterus is scraped with a spoon-shaped instrument called a curette; commonly called a D and C (dilation and curettage).

Fibroids (FĪ-broyds; *fibro-* = fiber; *-eidos* = resemblance) Noncancerous tumors in the myometrium of the uterus composed of muscular and fibrous tissue. Their growth appears to be related to high levels of estrogens. They do not occur before puberty and usually stop growing after menopause. Symptoms include abnormal menstrual bleeding, and pain or pressure in the pelvic area.

Hermaphroditism (her-MAF-rō-dīt-izm) The presence of both ovarian and testicular tissue in one individual.

Hypospadias (hī′-pō-SPĀ-dē-as; *hypo-* = below) A common congenital abnormality in which the urethral opening is displaced. In males, the displaced opening may be on the underside of the penis, at the penoscrotal junction, between the scrotal folds, or in the perineum; in females, the urethra opens into the vagina. The problem can be corrected surgically.

Leukorrhea (loo′-kō-RĒ-a; *leuko-* = white) A whitish (nonbloody) vaginal discharge containing mucus and pus cells that may occur at any age and affects most women at some time.

Menorrhagia (men-ō-RA-jē-a; *meno-* = menstruation; *-rhage* = to burst forth) Excessively prolonged or profuse menstrual period. May be due to a disturbance in hormonal regulation of the menstrual cycle, pelvic infection, medications (anticoagulants), fibroids (noncancerous uterine tumors composed of muscle and fibrous tissue), endometriosis, or intrauterine devices.

Oophorectomy (ō′-of-ō-REK-tō-mē; *oophor-* = bearing eggs) Removal of the ovaries.

Orchitis (or-KĪ-tis; *orchi-* = testes; *-itis* = inflammation) Inflammation of the testes, for example, as a result of the mumps virus or a bacterial infection.

Ovarian cyst The most common form of ovarian tumor, in which a fluid-filled follicle or corpus luteum persists and continues growing.

Pelvic inflammatory disease (PID) A collective term for any extensive bacterial infection of the pelvic organs, especially the uterus, uterine tubes, or ovaries, which is characterized by pelvic soreness, lower back pain, abdominal pain, and urethritis. Often the early symptoms of PID occur just after menstruation. As infection spreads, fever may develop, along with painful abscesses of the reproductive organs.

Salpingectomy (sal′-pin-JEK-tō-mē; *salpingo* = tube) Removal of a uterine (fallopian) tube.

Smegma (SMEG-ma) The secretion, consisting principally of desquamated epithelial cells, found chiefly around the external genitalia and especially under the foreskin of the male.

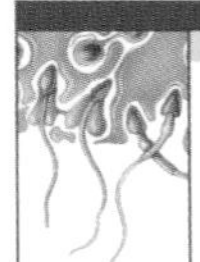

STUDY OUTLINE

MALE REPRODUCTIVE SYSTEM (p. 1057)

1. Reproduction is the process by which new individuals of a species are produced and the genetic material is passed from generation to generation.
2. The organs of reproduction are grouped as gonads (produce gametes), ducts (transport and store gametes), accessory sex glands (produce materials that support gametes), and supporting structures (have various roles in reproduction).
3. The male structures of reproduction include the testes, ductus epididymis, ductus (vas) deferens, ejaculatory duct, urethra, seminal vesicles, prostate, bulbourethral (Cowper's) glands, and penis.
4. The scrotum is a sac that hangs from the root of the penis and consists of loose skin and superficial fascia; it supports the testes.
5. The temperature of the testes is regulated by contraction of the cremaster muscle and dartos muscle, which either elevates them and brings them closer to the pelvic cavity or relaxes and moves them farther from the pelvic cavity.
6. The testes are paired oval glands (gonads) in the scrotum containing seminiferous tubules, in which sperm cells are made; Sertoli cells (sustentacular cells), which nourish sperm cells and secrete inhibin; and Leydig (interstitial) cells, which produce the male sex hormone testosterone.
7. The testes descend into the scrotum through the inguinal canals during the seventh month of fetal development. Failure of the testes to descend is called cryptorchidism.
8. Secondary oocytes and sperm, both of which are called gametes, are produced in the gonads.
9. Spermatogenesis, which occurs in the testes, is the process whereby immature spermatogonia develop into sperm. The spermatogenesis sequence, which includes meiosis I, meiosis II, and spermiogenesis, results in the formation of four haploid sperm (spermatozoa) from each primary spermatocyte.
10. Mature sperm consist of a head and a tail. Their function is to fertilize a secondary oocyte.
11. At puberty, gonadotropin-releasing hormone (GnRH) stimulates anterior pituitary secretion of FSH and LH. LH stimulates production of testosterone; FSH and testosterone stimulate spermatogenesis. Sertoli cells secrete androgen-binding protein (ABP), which binds to testosterone and keeps its concentration high in the seminiferous tubule.
12. Testosterone controls the growth, development, and maintenance of sex organs; stimulates bone growth, protein anabolism, and sperm maturation; and stimulates development of masculine secondary sex characteristics.
13. Inhibin is produced by Sertoli cells; its inhibition of FSH helps regulate the rate of spermatogenesis.
14. The duct system of the testes includes the seminiferous tubules, straight tubules, and rete testis. Sperm flow out of the testes through the efferent ducts.
15. The ductus epididymis is the site of sperm maturation and storage.
16. The ductus (vas) deferens stores sperm and propels them toward the urethra during ejaculation.
17. Each ejaculatory duct, formed by the union of the duct from the seminal vesicle and ampulla of the ductus (vas) deferens, is the passageway for ejection of sperm and secretions of the seminal vesicles into the first portion of the urethra, the prostatic urethra.
18. The urethra in males is subdivided into three portions: the prostatic, membranous, and spongy (penile) urethra.
19. The seminal vesicles secrete an alkaline, viscous fluid that contains fructose (used by sperm for ATP production). Seminal fluid constitutes about 60% of the volume of semen and contributes to sperm viability.
20. The prostate secretes a slightly acidic fluid that constitutes about 25% of the volume of semen and contributes to sperm motility.
21. The bulbourethral (Cowper's) glands secrete mucus for lubrication and an alkaline substance that neutralizes acid.
22. Semen is a mixture of sperm and seminal fluid; it provides the fluid in which sperm are transported, supplies nutrients, and neutralizes the acidity of the male urethra and the vagina.
23. The penis consists of a root, a body, and a glans penis.
24. Engorgement of the penile blood sinuses under the influence of sexual excitation is called erection.

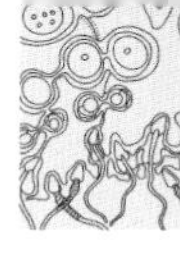

FEMALE REPRODUCTIVE SYSTEM (p. 1070)

1. The female organs of reproduction include the ovaries (gonads), uterine (fallopian) tubes or oviducts, uterus, vagina, and vulva.
2. The mammary glands are part of the integumentary system and also are considered part of the reproductive system in females.
3. The ovaries, the female gonads, are located in the superior portion of the pelvic cavity, lateral to the uterus.
4. Ovaries produce secondary oocytes, discharge secondary oocytes (the process of ovulation), and secrete estrogens, progesterone, relaxin, and inhibin.
5. Oogenesis (the production of haploid secondary oocytes) begins in the ovaries. The oogenesis sequence includes meiosis I and meiosis II, which goes to completion only after an ovulated secondary oocyte is fertilized by a sperm cell.
6. The uterine (fallopian) tubes transport secondary oocytes from the ovaries to the uterus and are the normal sites of fertilization. Ciliated cells and peristaltic contractions help move a secondary oocyte or fertilized ovum toward the uterus.
7. The uterus is an organ the size and shape of an inverted pear that functions in menstruation, implantation of a fertilized ovum, development of a fetus during pregnancy, and labor. It also is part of the pathway for sperm to reach the uterine tubes to fertilize a secondary oocyte. Normally, the uterus is held in position by a series of ligaments.
8. Histologically, the layers of the uterus are an outer perimetrium (serosa), a middle myometrium, and an inner endometrium.
9. The vagina is a passageway for sperm and the menstrual flow, the receptacle of the penis during sexual intercourse, and the inferior portion of the birth canal. It is capable of considerable stretching.
10. The vulva, a collective term for the external genitals of the female, consists of the mons pubis, labia majora, labia minora, clitoris, vestibule, vaginal and urethral orifices, hymen, bulb of the vestibule, and three sets of glands: the paraurethral (Skene's), greater vestibular (Bartholin's), and lesser vestibular glands.
11. The perineum is a diamond-shaped area at the inferior end of the trunk medial to the thighs and buttocks.
12. The mammary glands are modified sweat glands lying superficial to the pectoralis major muscles. Their function is to synthesize, secrete, and eject milk (lactation).
13. Mammary gland development depends on estrogens and progesterone.
14. Milk production is stimulated by prolactin, estrogens, and progesterone; milk ejection is stimulated by oxytocin.

THE FEMALE REPRODUCTIVE CYCLE (p. 1084)

1. The function of the ovarian cycle is to develop a secondary oocyte; the function of the uterine (menstrual) cycle is to prepare the endometrium each month to receive a fertilized egg. The female reproductive cycle includes both the ovarian and uterine cycles.
2. The uterine and ovarian cycles are controlled by GnRH from the hypothalamus, which stimulates the release of FSH and LH by the anterior pituitary.
3. FSH and LH stimulate development of follicles and secretion of estrogens by the follicles. LH also stimulates ovulation, formation of the corpus luteum, and the secretion of progesterone and estrogens by the corpus luteum.
4. Estrogens stimulate the growth, development, and maintenance of female reproductive structures; stimulate the development of secondary sex characteristics; and stimulate protein synthesis.
5. Progesterone works with estrogens to prepare the endometrium for implantation and the mammary glands for milk synthesis.
6. Relaxin relaxes the myometrium at the time of possible implantation. At the end of a pregnancy, relaxin increases the flexibility of the pubic symphysis and helps dilate the uterine cervix to facilitate delivery.
7. During the menstrual phase, the stratum functionalis of the endometrium is shed, discharging blood, tissue fluid, mucus, and epithelial cells.
8. During the preovulatory phase, a group of follicles in the ovaries begins to undergo final maturation. One follicle outgrows the others and becomes dominant while the others degenerate. At the same time, endometrial repair occurs in the uterus. Estrogens are the dominant ovarian hormones during the preovulatory phase.
9. Ovulation is the rupture of the mature (graafian) follicle and the release of a secondary oocyte into the pelvic cavity. It is brought about by a surge of LH. Signs and symptoms of ovulation include increased basal body temperature; clear, stretchy cervical mucus; changes in the uterine cervix; and abdominal pain.
10. During the postovulatory phase, both progesterone and estrogens are secreted in large quantity by the corpus luteum of the ovary, and the uterine endometrium thickens in readiness for implantation.
11. If fertilization and implantation do not occur, the corpus luteum degenerates, and the resulting low levels of progesterone and estrogens allow discharge of the endometrium followed by the initiation of another reproductive cycle.
12. If fertilization and implantation do occur, the corpus luteum is maintained by hCG. The corpus luteum and later the placenta secrete progesterone and estrogens to support pregnancy and breast development for lactation.

BIRTH CONTROL METHODS (p. 1088)

1. Birth control refers to restricting the number of children by various methods designed to control fertility and prevent conception.
2. Birth control methods include surgical sterilization (vasectomy, tubal ligation), hormonal methods, intrauterine devices, spermicides, barrier methods (condom, vaginal pouch, diaphragm), periodic abstinence (rhythm and sympto-thermal methods), and induced abortion. See Table 28.3 on page 1090 for failure rates for these methods.
3. Contraceptive pills of the combination type contain estrogens and progestins in concentrations that decrease the secretion of FSH and LH, inhibiting development of ovarian follicles and ovulation.
4. An abortion is a spontaneous or induced premature expulsion of the products of conception from the uterus. RU 486 can induce abortion by blocking the action of progesterone.

DEVELOPMENT OF THE REPRODUCTIVE SYSTEMS (p. 1091)

1. The gonads develop from intermediate mesoderm. In the presence of the *SRY* gene, the gonads begin to differentiate into testes during the seventh week. The gonads differentiate into ovaries when the *SRY* gene is absent.

2. In males, testosterone stimulates development of each mesonephric duct into an epididymis, ductus (vas) deferens, ejaculatory duct, and seminal vesicle, and Müllerian-inhibiting substance (MIS) causes the paramesonephric duct cells to die. In females, testosterone and MIS are absent; the paramesonephric ducts develop into the uterine tubes, uterus, and vagina and the mesonephric ducts degenerate.
3. The external genitals develop from the genital tubercle and are stimulated to develop into typical male structures by the hormone dihydrotestosterone (DHT). The external genitals develop into female structures when DHT is not produced, the normal situation in female embryos.

AGING AND THE REPRODUCTIVE SYSTEMS (p. 1094)

1. Puberty is the period when secondary sex characteristics begin to develop and the potential for sexual reproduction is reached.
2. The onset of puberty is marked by pulses or bursts of LH and FSH secretion, each triggered by a pulse of GnRH. The hormone leptin, released by adipose tissue, may signal the hypothalamus that long-term energy stores (triglycerides in adipose tissue) are adequate for reproductive functions to begin.
3. In females, the reproductive cycle normally occurs once each month from menarche, the first menses, to menopause, the permanent cessation of menses.
4. Between the ages of 40 and 50, the pool of remaining ovarian follicles becomes exhausted and levels of progesterone and estrogens decline. Most women experience a decline in bone mineral density after menopause, together with some atrophy of the ovaries, uterine tubes, uterus, vagina, external genitalia, and breasts. Uterine and breast cancer increase in incidence with age.
5. In older males, decreased levels of testosterone are associated with decreased muscle strength, waning sexual desire, and fewer viable sperm; prostate disorders are common.

SELF-QUIZ QUESTIONS

Fill in the blanks in the following statement.

1. The period of time when secondary sexual characteristics begin to develop and the potential for sexual reproduction is reached is called ____. The first menses is called ____, and the permanent cessation of menses is called ____.

Indicate whether the following statements are true or false.

2. Spermatogenesis does not occur at normal core body temperature.
3. The route of sperm from the production in the testes to the exterior of the body is: seminiferous tubules, straight tubules, rete testes, epididymis, ductus (vas) deferens, ejaculatory duct, prostatic urethra, membranous urethra, spongy urethra, external urethral orifice.

Choose the one best answer to the following questions.

4. Which of the following are functions of Sertoli cells? (1) protection of developing spermatogenic cells, (2) nourishment of spermatocytes, spermatids, and sperm, (3) phagocytosis of excess sperm cytoplasm as development proceeds, (4) media-tion of the effects of testosterone and FSH, (5) control of movements of spermatogenic cells and release of sperm into the lumen of seminiferous tubules. (a) 1, 2, 4, and 5; (b) 1, 2, 3, and 5; (c) 2, 3, 4, and 5; (d) 1, 2, 3, and 4; (e) 1, 2, 3, 4, and 5.
5. Which of the following are *true*? (1) An erection is a sympathetic response initiated by sexual stimulation. (2) Dilation of blood vessels supplying erectile tissue results in erection. (3) Nitric oxide causes smooth muscle within erectile tissue to relax, which results in widening of blood sinuses. (4) Ejaculation is a sympathetic reflex coordinated by the sacral region of the spinal cord. (5) The purpose of the corpus cavernosa penis is to keep the spongy urethra open during ejaculation. (a) 1, 2, and 3; (b) 1, 2, 3, 4, and 5; (c) 2 and 3; (d) 2, 4, and 5; (e) 1, 2, 3, and 4.
6. Which of the following are *true* concerning estrogens? (1) They promote development and maintenance of female reproductive structures and secondary sex characteristics. (2) They help control fluid and electrolyte balance. (3) They increase protein catabolism. (4) They lower blood cholesterol. (5) In moderate levels, they inhibit the release of GnRH and the secretion of LH and FSH. (a) 1, 4, and 5; (b) 1, 3, 4, and 5; (c) 1, 2, 3, and 5; (d) 1, 2, 3, and 4; (e) 1, 2, 3, 4, and 5.
7. Which of the following statements are *correct*? (1) A sperm head contains DNA and an acrosome. (2) An acrosome is a specialized lysosome that contains enzymes that enable sperm to produce the ATP needed to propel themselves out of the male reproductive tract. (3) Mitochondria in the midpiece of a sperm produce ATP for sperm motility. (4) A sperm's tail, a flagellum, propels it along its way. (5) Once ejaculated, sperm are viable and normally are able to fertilize a secondary oocyte for 5 days. (a) 1, 2, 3, and 4; (b) 2, 3, 4, and 5; (c) 1, 3, and 4; (d) 2, 4, and 5; (e) 2, 3, and 4.
8. Which of the following statements are *correct*? (1) Spermatogonia are stem cells because when they undergo mitosis, some of the daughter cells remain to serve as a reservoir of cells for future mitosis. (2) Meiosis I is a division of pairs of chromosomes resulting in daughter cells with only one member of each chromosome pair. (3) Meiosis II separates the chromatids of each chromosome. (4) Spermiogenesis involves the maturation of spermatids into sperm. (5) The process by which the seminiferous tubules produce haploid sperm is called spermatogenesis. (a) 1, 2, 3, and 5; (b) 1, 2, 3, 4, and 5; (c) 1, 3, 4, and 5; (d) 1, 2, 3, and 4; (e) 1, 3, and 5.
9. Which of the following statements are *correct*? (1) Cells from the yolk sac give rise to oogonia. (2) Ova arise from the germinal epithelium of the ovary. (3) Primary oocytes enter prophase of meiosis I during fetal development but do not complete it until after puberty. (4) Once a secondary oocyte is formed, it proceeds

to metaphase of meiosis II and stops at this stage. (5) The secondary oocyte resumes meiosis II and forms the ovum and a polar body only if fertilization occurs. (6) A primary oocyte gives rise to an ovum and four polar bodies. (a) 1, 3, 4, and 5; (b) 1, 3, 4, and 6; (c) 1, 2, 4, and 6; (d) 1, 2, 4, and 5; (e) 1, 2, 5, and 6.

10. Which of the following statements are *correct*? (1) The female reproductive cycle consists of a menstrual phase, a preovulatory phase, ovulation, and a postovulatory phase. (2) During the menstrual phase, small secondary follicles in the ovary begin to enlarge while the uterus is shedding its lining. (3) During the preovulatory phase, a dominant follicle continues to grow and begins to secrete estrogens and inhibin while the uterine lining begins to rebuild. (4) Ovulation results in the release of an ovum and the shedding of the uterine lining to nourish and support the released ovum. (5) After ovulation, a corpus luteum forms from the ruptured follicle and begins to secrete progesterone and estrogens, which it will continue to do throughout pregnancy if the egg is fertilized. (6) If pregnancy does not occur, then the corpus luteum degenerates into a scar called the corpus albicans, and the uterine lining is prepared to be shed again. (a) 1, 2, 4, and 5; (b) 2, 4, 5, and 6; (c) 1, 4, 5, and 6; (d) 1, 3, 4, and 6; (e) 1, 2, 3, and 6.

11. Oral contraceptives work by (1) causing a thickening of the cervical mucus, (2) blocking the uterine tubes, (3) inhibiting the release of FSH and LH, (4) preventing ovulation, (5) disrupting the plasma membranes of sperm, (6) irritating the endometrial lining so that it is inhospitable for fetal development. (a) 3 only; (b) 3 and 4; (c) 1, 2, and 5; (d) 1, 3, and 4; (e) 1, 2, 3, 4, and 5.

12. Match the following:

____(a) modified sudoriferous glands involved in lactation
____(b) a small, cylindrical mass of erectile tissue and nerves in the female; homologue of the male glans penis
____(c) produce mucus in the female during sexual arousal and intercourse; homologous to the male bulbourethral glands
____(d) the group of cells that nourish the developing oocyte and begin to secrete estrogens
____(e) a pathway for sperm to reach the uterine tubes; the site of menstruation; the site of implantation of a fertilized ovum; the womb
____(f) produces progesterone, estrogens, relaxin, and inhibin
____(g) draw the ovum into the uterine tube
____(h) the opening between the uterus and vagina
____(i) muscular layer of uterus; responsible for expulsion of fetus from uterus
____(j) mucus-secreting glands in the female that are homologous to the prostate gland
____(k) the female copulatory organ; the birth canal
____(l) passageway for the ovum to the uterus; usual site of fertilization; site of tubal ligation
____(m) refers to the external genitals of the female
____(n) the layer of the uterine lining that is partially shed during each monthly cycle

(1) follicle
(2) corpus luteum
(3) uterine tube
(4) fimbriae
(5) uterus
(6) cervix
(7) endometrium
(8) vagina
(9) vulva
(10) clitoris
(11) paraurethral glands
(12) greater vestibular glands
(13) mammary glands
(14) myometrium

13. Match the following:

____(a) site of sperm maturation
____(b) the male copulatory organ; a passageway for ejaculation of sperm and excretion of urine
____(c) sperm-forming cells
____(d) produce an alkaline substance that protects sperm by neutralizing acids in the urethra
____(e) ejects sperm into the urethra just before ejaculation
____(f) the supporting structure for the testes
____(g) carries the sperm from the scrotum into the abdominopelvic cavity for release by ejaculation; is cut and tied as a means of sterilization
____(h) the shared terminal duct of the reproductive and urinary systems in the male
____(i) surrounds the urethra at the base of the bladder; produces secretions that contribute to sperm motility and viability
____(j) produce testosterone
____(k) supporting structure that consists of the ductus deferens, testicular artery, autonomic nerves, veins that drain the testes, lymphatic vessels, and cremaster muscle
____(l) support and protect developing spermatogenic cells; secrete inhibin; form the blood–testis barrier
____(m) secrete an alkaline fluid to help neutralize acids in the female reproductive tract; secrete fructose for use in ATP production by sperm
____(n) contraction and relaxation moves testes near to or away from pelvic cavity
____(o) site of spermatogenesis

(1) spermatogenic cells
(2) Sertoli cells
(3) Leydig cells
(4) penis
(5) scrotum
(6) epididymis
(7) ductus deferens
(8) ejaculatory duct
(9) seminiferous tubules
(10) seminal vesicles
(11) prostate gland
(12) bulbourethral glands
(13) urethra
(14) spermatic cord
(15) cremaster muscle

14. Match the following:

____(a) relaxes the uterus by inhibiting myometrial contractions during monthly cycles; increases flexibility of the pubic symphysis during childbirth
____(b) stimulates Leydig cells to secrete testosterone in males and triggers ovulation in females
____(c) inhibits production of FSH by the anterior pituitary gland
____(d) posterior pituitary hormone responsible for uterine contraction and release of milk from mammary glands
____(e) stimulates male pattern of development; stimulates protein synthesis; contributes to sex drive
____(f) stimulates male external genital development
____(g) maintains the corpus luteum during the first trimester of pregnancy
____(h) contribute to male sexual behavior, spermatogenesis, and libido
____(i) promotes development of female reproductive structures; lowers blood cholesterol
____(j) stimulates the initial secretion of estrogens by growing follicles; promotes follicle growth
____(k) is secreted by the corpus luteum to maintain the uterine lining during the first trimester of pregnancy
____(l) anterior pituitary hormone that stimulates milk production

(1) inhibin
(2) LH
(3) FSH
(4) testosterone
(5) estrogens
(6) progesterone
(7) relaxin
(8) human chorionic gonadotropin
(9) prolactin
(10) oxytocin
(11) androgens
(12) dihydrotestosterone

15. Match the following:

____(a) the process during meiosis when portions of homologous chromosomes may be exchanged with each other
____(b) refers to cells containing one-half the chromosome number
____(c) the cell produced by the union of an egg and a sperm
____(d) the degeneration of oogonia before and after birth
____(e) a packet of discarded nuclear material from the first or second meiotic division of the egg
____(f) refers to cells containing the full chromosome number

(1) zygote
(2) haploid
(3) diploid
(4) crossing-over
(5) polar body
(6) atresia

CRITICAL THINKING QUESTIONS

1. Twenty-three year old Monica and her husband Bill are ready to start a family. They are both avid bicyclists and weight-lifters who carefully watch what they eat and pride themselves on their "buff" bodies. However, Monica is having difficulty becoming pregnant. She thinks it is Bill's fault. Monica hasn't had a menstrual period for some time but informs the doctor that is normal for her. After consulting with her physician, the doctor tells Monica that she needs to cut back on her exercise routine and "put on some weight" in order to get pregnant. Monica is outraged because she figures she will gain enough weight when she is pregnant! Explain to Monica what has happened to her and why weight gain could help her achieve her goal of pregnancy.
2. The term "progesterone" means "for gestation (or pregnancy)." Describe how progesterone helps prepare the female body for pregnancy and helps maintain pregnancy.
3. After having borne five children, Mark's wife Isabella insists that he have a vasectomy. Mark is afraid that he will "dry up" and won't be able to perform sexually. How can you reassure him that his reproductive organs will function fine? Will Mark and Isabella be guaranteed that she will not become pregnant immediately after the vasectomy?

ANSWERS TO FIGURE QUESTIONS

28.1 The gonads (testes) produce gametes (sperm) and hormones; the ducts transport, store, and receive gametes; the accessory sex glands secrete materials that support gametes; and the penis assists in the delivery and joining of gametes.

28.2 The cremaster and dartos muscles help regulate the temperature of the testes.

28.3 The tunica vaginalis and tunica albuginea are tissue layers that cover and protect the testes.

28.4 The Leydig cells of the testes secrete testosterone.

28.5 During meiosis I, the number of chromosomes in each cell is reduced by half.

28.6 The sperm head contains the nucleus with 23 highly condensed chromosomes and an acrosome that contains enzymes for penetration of a secondary oocyte; the neck contains centrioles that produce microtubules for the rest of the tail; the midpiece contains mitochondria for ATP production for locomotion and metabolism; the principal and end pieces of the tail provide motility.

28.7 Sertoli cells secrete inhibin.

28.8 Testosterone inhibits secretion of LH, and inhibin inhibits secretion of FSH.

28.9 The seminal vesicles are the accessory sex glands that contribute the largest volume to seminal fluid.

28.10 Two corpora cavernosa penis and one corpus spongiosum penis contain blood sinuses that fill with blood that cannot flow out of the penis as quickly as it flows in. The trapped blood engorges and stiffens the tissue, producing an erection. The corpus spongiosum penis keeps the spongy urethra open so that ejaculation can occur.

28.11 The testes are homologous to the ovaries; the glans penis is homologous to the clitoris; the prostate is homologous to the paraurethral glands; and the bulbourethral glands are homologous to the greater vestibular glands.

28.12 The mesovarium anchors the ovary to the broad ligament of the uterus and the uterine tube; the ovarian ligament anchors it to the uterus; the suspensory ligament anchors it to the pelvic wall.

28.13 Ovarian follicles secrete estrogens; the corpus luteum secretes progesterone, estrogens, relaxin, and inhibin.

28.14 Most ovarian follicles undergo atresia (degeneration).

28.15 Primary oocytes are present in the ovary at birth, so they are as old as the woman is. In males, primary spermatocytes are continually being formed from stem cells (spermatogonia) and thus are only a few days old.

28.16 Fertilization most often occurs in the ampulla of the uterine tube.

28.17 Ciliated columnar epithelial cells and nonciliated (peg) cells with microvilli line the uterine tubes.

28.18 The endometrium is a highly vascularized, secretory epithelium that provides the oxygen and nutrients needed to sustain a fertilized egg; the myometrium is a thick smooth muscle layer that supports the uterine wall during pregnancy and contracts to expel the fetus at birth.

28.19 The stratum basalis of the endometrium provides cells to replace those that are shed (the stratum functionalis) during each menstruation.

28.20 Anterior to the vaginal opening are the mons pubis, clitoris, and prepuce. Lateral to the vaginal opening are the labia minora and labia majora.

28.21 The anterior portion of the perineum is called the urogenital triangle because its borders form a triangle that encloses the urethral (uro-) and vaginal (-genital) orifices.

28.22 Prolactin, estrogens, and progesterone regulate the synthesis of milk. Oxytocin regulates the ejection of milk.

28.23 The principal estrogen is β-estradiol.

28.24 The hormones responsible for the proliferative phase of endometrial growth are estrogens; for ovulation, LH; for growth of the corpus luteum, LH; and for the midcycle surge of LH, estrogens.

28.25 The effect of rising but moderate levels of estrogens is negative feedback inhibition of the secretion of GnRH, LH, and FSH.

28.26 This is negative feedback, because the response is opposite to the stimulus. A reduced amount of negative feedback due to declining levels of estrogens and progesterone stimulates release of GnRH, which in turn increases the production and release of FSH and LH, ultimately stimulating the secretion of estrogens.

28.27 The *SRY* gene on the Y chromosome is responsible for the development of the gonads into testes.

28.28 The presence of dihydrotestosterone (DHT) stimulates differentiation of the external genitals in males; its absence allows differentiation of the external genitals in females.

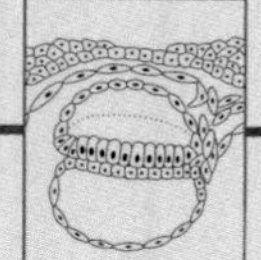

Chapter 29

Development and Inheritance

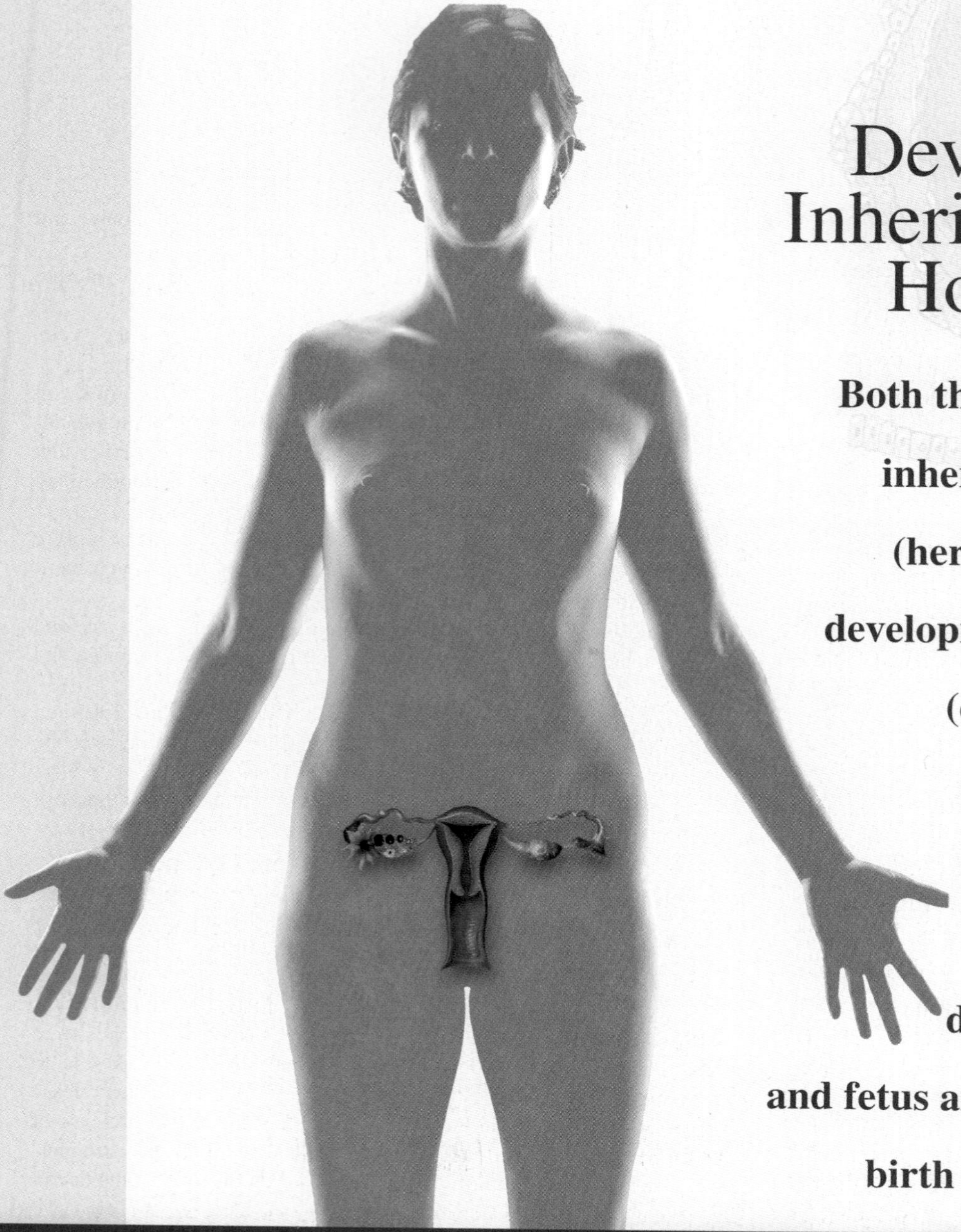

Development, Inheritance, and Homeostasis

Both the genetic material inherited from parents (heredity) and normal development in the uterus (environment) play important roles in determining the homeostasis of a developing embryo and fetus and the subsequent birth of a healthy child.

www. wiley.com/college/apcentral

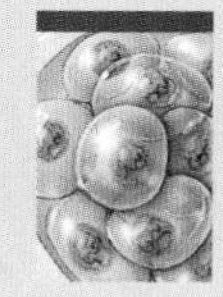

Developmental biology is the study of the sequence of events from the fertilization of a secondary oocyte by a sperm cell to the formation of an adult organism. From fertilization through the eighth week of development, a stage called the **embryonic period,** the developing human is called an **embryo** (*em-* = into; *-bryo* = grow). **Embryology** (em-brē-OL-ō-jē) is the study of development from the fertilized egg through the eighth week. The **fetal period** begins at week nine and continues until birth. During this time, the developing human is called a **fetus** (FĒ-tus = offspring).

Once sperm and a secondary oocyte have developed through meiosis and maturation, and the sperm have been deposited in the vagina, pregnancy can occur. **Pregnancy** is a sequence of events that begins with fertilization, proceeds to implantation, embryonic development, and fetal development, and normally ends with birth about 38 weeks later, or 40 weeks after the last menstrual period.

Obstetrics (ob-STET-riks; *obstetrix* = midwife) is the branch of medicine that deals with the management of pregnancy, labor, and the **neonatal period,** the first 28 days after birth. **Prenatal development** (prē-NĀ-tal; *pre-* = before; *natal* = birth) is the time from fertilization to birth and includes both embryological and fetal development. It is divided into three periods of three calendar months each, called **trimesters.**

1. The **first trimester** is the most critical stage of development during which the rudiments of all the major organs systems appear, and also during which the developing organism is the most vulnerable to the effects of drugs, radiation, and microbes.
2. The **second trimester** is characterized by the nearly complete development of organ systems. By the end of this stage, the fetus assumes distinctively human features.
3. The **third trimester** represents a period of rapid fetal growth. During the early stages of this period, most of the organ systems are becoming fully functional.

In this chapter, we focus on the developmental sequence from fertilization through implantation, embryonic and fetal development, labor, birth, and the principles of inheritance, the passage of hereditary traits from one generation to another.

EMBRYONIC PERIOD

► **OBJECTIVE**

Explain the major developmental events that occur during the embryonic period.

First Week of Development

The first week of development is characterized by several significant events including fertilization, cleavage of the zygote, blastocyst formation, and implantation.

Fertilization

During **fertilization** (fer-til-i-ZĀ-shun; *fertil-* = fruitful), the genetic material from a haploid sperm cell (spermatozoon) and a haploid secondary oocyte merges into a single diploid nucleus. Of the about 300 million sperm introduced into the vagina, fewer than 2 million (1%) reach the cervix of the uterus and only about 200 reach the secondary oocyte. Fertilization normally occurs in the uterine (fallopian) tube within 12 to 24 hours after ovulation. Sperm can remain viable for about 48 hours after deposition in the vagina, although a secondary oocyte is viable for only about 24 hours after ovulation. Thus, pregnancy is *most likely* to occur if intercourse takes place during a 3-day "window"—from 2 days before ovulation to 1 day after ovulation.

Sperm are propelled from the vagina into the cervical canal by the whiplike movements of their tails (flagella). The passage of sperm through the rest of the uterus and then into the uterine tube results mainly from contractions of the walls of these organs. Prostaglandins in semen are believed to stimulate uterine motility at the time of intercourse and to aid in the movement of sperm through the uterus and into the uterine tube. Sperm that reach the vicinity of the oocyte within minutes after ejaculation *are not capable* of fertilizing it until about seven hours later. During this time in the female reproductive tract, mostly in the uterine tube, sperm undergo **capacitation** (ka-pas′-i-TĀ-shun; *capacit-* = capable of), a series of functional changes that cause the sperm's tail to beat even more vigorously and prepare its plasma membrane to fuse with the oocyte's plasma membrane. During capacitation, sperm are acted upon by secretions in the female reproductive tract that result in the removal of cholesterol, glycoproteins, and proteins from the plasma membrane around the head of the sperm.

For fertilization to occur, a sperm cell first must penetrate two layers: the **corona radiata** (kō-RŌ-na = crown; rā-dē-A-ta = to shine), the granulosa cells that surround the secondary oocyte, and the **zona pellucida** (ZŌ-na = zone; pe-LOO-si-da = allowing passage of light), the clear glycoprotein layer between the corona radiata and the oocyte's plasma membrane (Figure 29.1a). One of the glycoproteins in the zona pellucida called ZP3 acts as a sperm receptor. Its binding to specific membrane proteins in the sperm head triggers the **acrosomal reaction,** the release of the contents of the acrosome. The acrosome is a helmetlike structure that covers the head of the sperm and contains several enzymes. The acrosomal enzymes digest a path through the zona pellucida as the lashing sperm tail pushes the sperm cell onward. Although many sperm bind to ZP3 molecules and undergo acrosomal reactions, only the first sperm cell to penetrate the entire zona pellucida and reach the oocyte's plasma membrane fuses with the oocyte.

The fusion of a sperm cell with a secondary oocyte, called **syngamy** (*syn-* = coming together; *-gamy* = marriage), sets in motion events that block **polyspermy,** fertilization by more

Figure 29.1 **Selected structures and events in fertilization.** (a) A sperm cell penetrating the corona radiata and zona pellucida around a secondary oocyte. (b) A sperm cell in contact with a secondary oocyte. (c) Male and female pronuclei.

During fertilization, genetic material from a sperm cell and a secondary oocyte merge to form a single diploid nucleus.

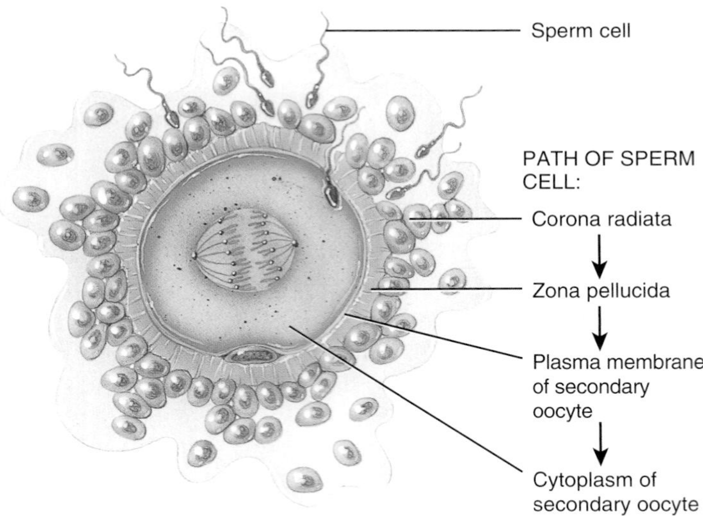

(a) Sperm cell penetrating a secondary oocyte

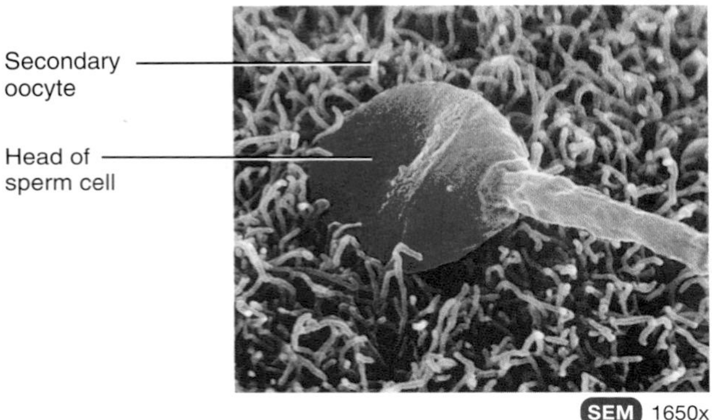

(b) Sperm cell in contact with a secondary oocyte

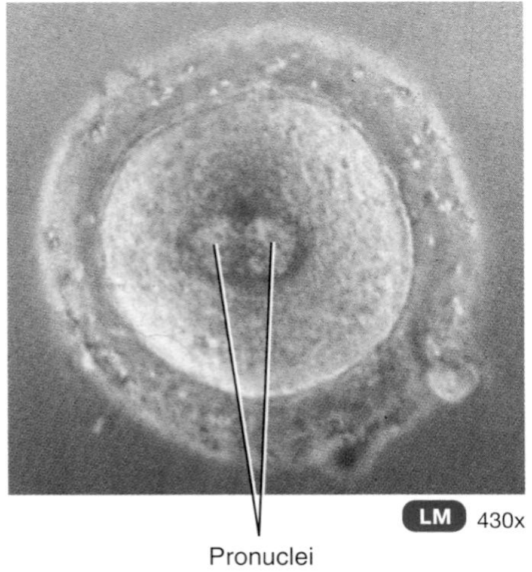

(c) Male and female pronuclei

What is capacitation?

than one sperm cell. Within a few seconds, the cell membrane of the oocyte depolarizes, which acts as a *fast block to polyspermy*—a depolarized oocyte cannot fuse with another sperm. Depolarization also triggers the intracellular release of calcium ions, which stimulate exocytosis of secretory vesicles from the oocyte. The molecules released by exocytosis inactivate ZP3 and harden the entire zona pellucida, events called the *slow block to polyspermy.*

Once a sperm cell enters a secondary oocyte, the oocyte first must complete meiosis II. It divides into a larger ovum (mature egg) and a smaller second polar body that fragments and disintegrates (see Figure 28.15 on page 1074). The nucleus in the head of the sperm develops into the **male pronucleus,** and the nucleus of the fertilized ovum develops into the **female pronucleus** (Figure 29.1c). After the male and female pronuclei form, they fuse, producing a single diploid nucleus that contains 23 chromosomes from each pronucleus. Thus, the fusion of the haploid (n) pronuclei restores the diploid number ($2n$) of 46 chromosomes. The fertilized ovum now is called a **zygote** (*zygon* = yolk).

Dizygotic (fraternal) twins are produced from the independent release of two secondary oocytes and the subsequent fertilization of each by different sperm. They are the same age and in the uterus at the same time, but genetically they are as dissimilar as any other siblings. Dizygotic twins may or may not be the same sex. Because **monozygotic (identical) twins** develop from a single fertilized ovum, they contain exactly the same genetic material and are always the same sex. Monozygotic twins arise from separation of the developing cells into two embryos, which in 99% of the cases occurs before 8 days have passed. Separations that occur later than 8 days are likely to produce **conjoined twins,** a situation in which the twins are joined together and share some body structures.

Cleavage of the Zygote

After fertilization, rapid mitotic cell divisions of the zygote called **cleavage** (KLĒV-ij) take place (Figure 29.2). The first division of the zygote begins about 24 hours after fertilization and is completed about 6 hours later. Each succeeding division takes slightly less time. By the second day after fertilization, the second cleavage is completed and there are four cells (Figure 29.2b). By the end of the third day, there are 16 cells. The progressively smaller cells produced by cleavage are called **blastomeres** (BLAS-tō-mērz; *blasto-* = germ or sprout; *-meres* = parts). Successive cleavages eventually produce a solid sphere of cells called the **morula** (MOR-ū-la; *morula* = mulberry). The morula is still surrounded by the zona pellucida and is about the same size as the original zygote (Figure 29.2c).

Blastocyst Formation

By the end of the fourth day, the number of cells in the morula increases as it continues to move through the uterine tube toward the uterine cavity. When the morula enters the uterine cavity on day 4 or 5, a glycogen-rich secretion from the glands of the

Figure 29.2 Cleavage and the formation of the morula and blastocyst.

Cleavage refers to the early, rapid mitotic divisions of a zygote.

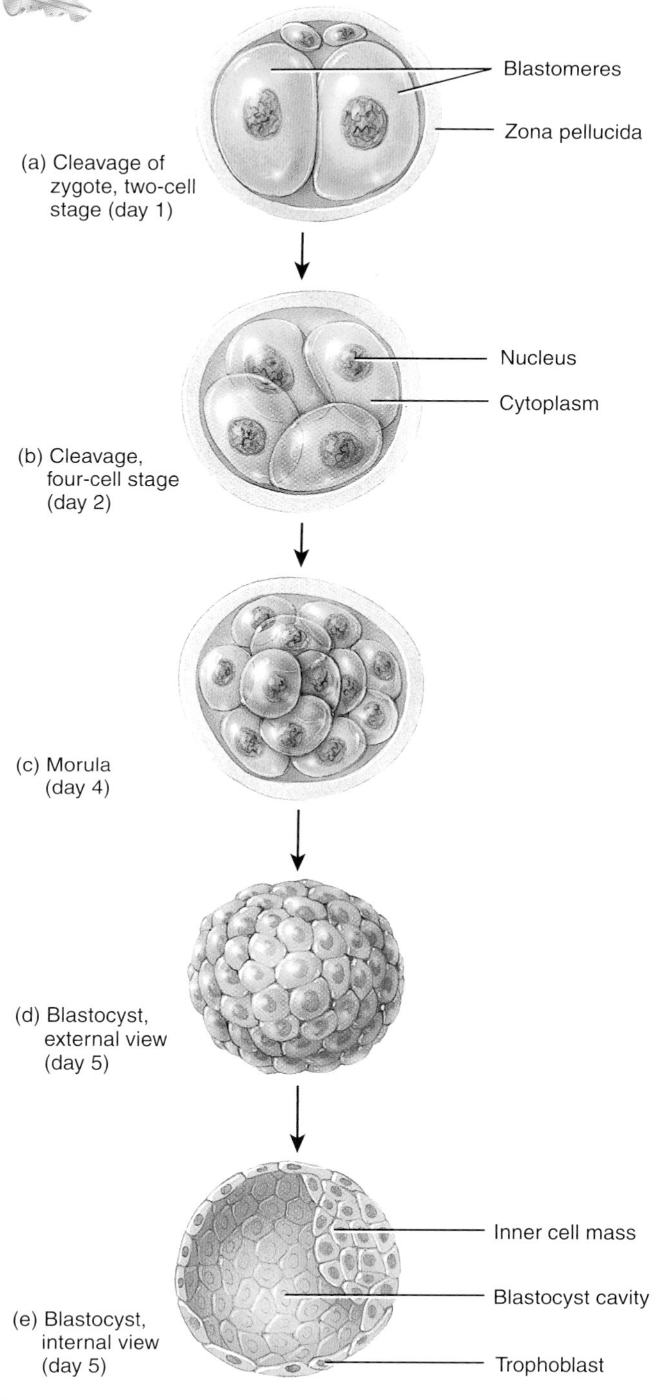

What is the histological difference between a morula and a blastocyst?

endometrium of the uterus passes into the uterine cavity and enters the morula through the zona pellucida. This fluid, called **uterine milk,** along with nutrients stored in the cytoplasm of the blastomeres of the morula, provides nourishment for the developing morula. At the 32-cell stage, the fluid enters the morula, collects between the blastomeres, and reorganizes them around a large fluid-filled cavity called the **blastocyst cavity** (BLAS-tō-sist; *blasto-* = germ or sprout; *-cyst* = bag) (Figure 29.2e). Once the blastocyst cavity is formed, the developing mass is called the **blastocyst.** Though it now has hundreds of cells, the blastocyst is still about the same size as the original zygote.

Further rearrangement of the blastomeres results in the formation of two distinct structures: the inner cell mass and trophoblast (Figure 29.2e). The **inner cell mass** is located inside the blastocyst and eventually develops into the embryo. The **trophoblast** (TRŌF-ō-blast; *tropho-* = develop or nourish) is an outer superficial layer of cells that forms the wall of the blastocyst. It will ultimately develop into the fetal portion of the placenta, the site of exchange of nutrients and wastes between the mother and fetus. On about the fifth day after fertilization, the blastocyst digests a hole in the zona pellucida with an enzyme, and then squeezes through it. Shedding of the zona pellucida is necessary to permit the next step, implantation.

Stem Cell Research and Therapeutic Cloning

Stem cells are unspecialized cells that have the ability to divide for indefinite periods and give rise to specialized cells. In the context of human development, a zygote (fertilized ovum) is a stem cell. Because it has the potential to form an entire organism, a zygote is known as a *totipotent stem cell* (tō-TIP-ō-tent; *totus-* = whole; *-potentia* = power). Inner cell mass cells, called *pluripotent stem cells* (ploo-RIP-ō-tent; *plur-* = several), can give rise to many (but not all) different types of cells. Later, pluripotent stem cells can undergo further specialization into *multipotent stem cells* (mul-TIP-ō-tent), stem cells with a specific function. Examples include keratinocytes that produce new skin cells, myeloid and lymphoid stem cells that develop into blood cells, and spermatogonia that give rise to sperm. Pluripotent stem cells currently used in research are derived from (1) the inner cell mass of embryos in the blastocyst stage that were destined to be used for infertility treatments but were not needed and from (2) nonliving fetuses terminated during the first three months of pregnancy.

On October 13, 2001, researchers reported cloning of the first human embryo to grow cells to treat human diseases. **Therapeutic cloning** is envisioned as a procedure in which the genetic material of a patient with a particular disease is used to create pluripotent stem cells to treat the disease. Using the principles of therapeutic cloning, scientists hope to make an embryo clone of a patient, remove the pluripotent stem cells from the embryo, and then use them to grow tissues to treat particular diseases and disorders, such as cancer, Parkinson disease, Alzheimer disease, spinal cord injury, diabetes, heart diease, stroke, burns, birth defects, osteoarthritis, and rheumatoid arthritis. Presumably, the tissues would not be rejected since they would contain the patient's own genetic material.

Scientists are also investigating the potential clinical applications of *adult stem cells*—stem cells that remain in the body throughout adulthood. Recent experiments suggest that the ovaries of adult mice contain stem cells that can develop into new ova (eggs). If these same types of stem cells are found in the ovaries of adult women, scientists could potentially harvest some of them from a woman about to undergo a sterilizing medical treatment (such as chemotherapy), store them, and then return the stem cells to her ovaries after the medical treatment is completed in order to restore fertility. Studies have also suggested that stem cells in human adult red bone marrow have the ability to differentiate into cells of the liver, kidney, heart, lung, skeletal muscle, skin, and organs of the gastrointestinal tract. In theory, adult stem cells from red bone marrow could be harvested from a patient and then used to repair other tissues and organs in that patient's body without having to use stem cells from embryos. ■

Implantation

The blastocyst remains free within the uterine cavity for about 2 days before it attaches to the uterine wall. At this time the endometrium is in its secretory phase. About 6 days after fertilization, the blastocyst loosely attaches to the endometrium in a process called **implantation** (Figure 29.3). As the blastocyst implants, usually in either the posterior portion of the fundus or the body of the uterus, it orients with the inner cell mass toward the endometrium (Figure 29.3b). About seven days after fertilization, the blastocyst attaches to the endometrium more firmly, endometrial glands in the vicinity enlarge, and the endometrium becomes more vascularized (forms new blood vessels).

Following implantation, the endometrium is known as the **decidua** (dē-SID-ū-a = falling off). The decidua separates from the endometrium after the fetus is delivered much as it does in normal menstruation. Different regions of the decidua are named based on their positions relative to the site of the implanted blastocyst (Figure 29.4). The **decidua basalis** is the portion of the endometrium between the embryo and the stratum basalis of the uterus; it provides large amounts of glycogen and lipids for the developing embryo and fetus and later becomes the maternal part of the placenta. The **decidua capsularis** is the portion of the endometrium located between the embryo and the uterine cavity. The **decidua parietalis** (par-rī-e-TAL-is) is the remaining modified endometrium that lines the noninvolved areas of the rest of the uterus. As the embryo and later the fetus enlarges, the decidua capsularis bulges into the uterine cavity and fuses with the decidua parietalis, thereby obliterating the uterine cavity. By about 27 weeks, the decidua capsularis degenerates and disappears.

The major events associated with the first week of development are summarized in Figure 29.5.

Figure 29.3 Relation of a blastocyst to the endometrium of the uterus at the time of implantation.

Implantation, the attachment of a blastocyst to the endometrium, occurs about 6 days after fertilization.

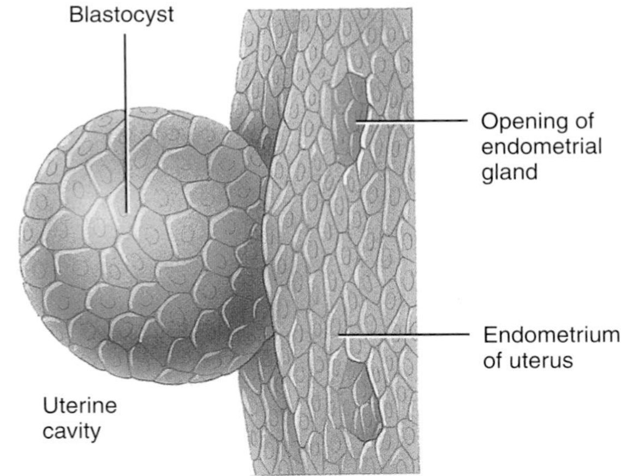

(a) External view of blastocyst, about 6 days after fertilization

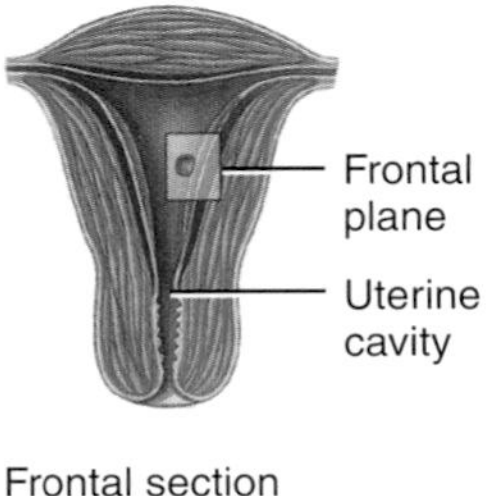

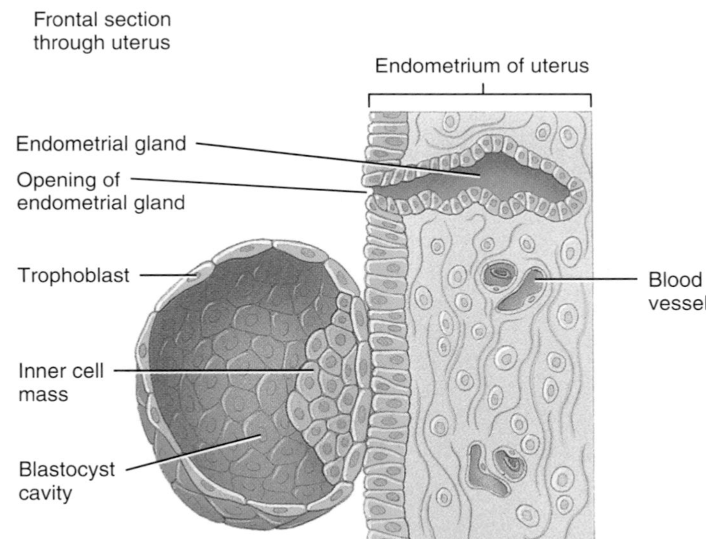

(b) Frontal section through endometrium of uterus and blastocyst, about 6 days after fertilization

How does the blastocyst merge with and burrow into the endometrium?

Figure 29.4 **Regions of the decidua.**

The decidua is a modified portion of the endometrium that develops after implantation.

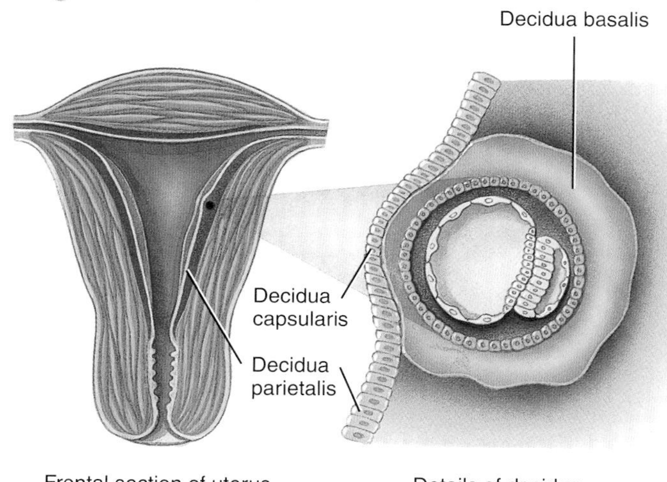

Which part of the decidua helps form the maternal part of the placenta?

Ectopic Pregnancy

Ectopic pregnancy (ek-TOP-ik; *ec-* = out of; *-topic* = place) is the development of an embryo or fetus outside the uterine cavity. An ectopic pregnancy usually occurs when movement of the fertilized ovum through the uterine tube is impaired by scarring due to a prior tubal infection, decreased movement of the uterine tube smooth muscle, or abnormal tubal anatomy. Although the most common site of ectopic pregnancy is the uterine tube, ectopic pregnancies may also occur in the ovary, abdominal cavity, or uterine cervix. Women who smoke are twice as likely to have an ectopic pregnancy because nicotine in cigarette smoke paralyzes the cilia in the lining of the uterine tube (as it does those in the respiratory airways). Scars from pelvic inflammatory disease, previous uterine tube surgery, and previous ectopic pregnancy may also hinder movement of the fertilized ovum.

The signs and symptoms of ectopic pregnancy include one or two missed menstrual cycles followed by bleeding and acute abdominal and pelvic pain. Unless removed, the developing embryo can rupture the uterine tube, often resulting in death of the mother. Treatment options include surgery or the use of a cancer drug called methotrexate, which causes embryonic cells to stop dividing and eventually disappear. ■

▸ CHECKPOINT

1. Where does fertilization normally occur?
2. How is polyspermy prevented?
3. What is a morula, and how is it formed?
4. Describe the layers of a blastocyst and their eventual fates.
5. When, where, and how does implantation occur?

Figure 29.5 **Summary of events associated with the first week of development.**

Fertilization usually occurs in the uterine tube.

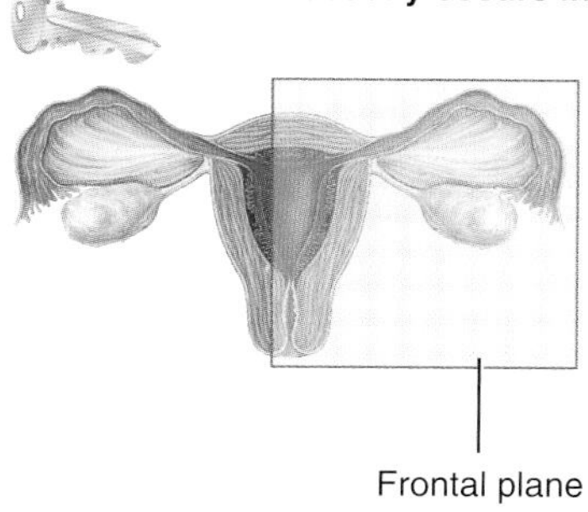

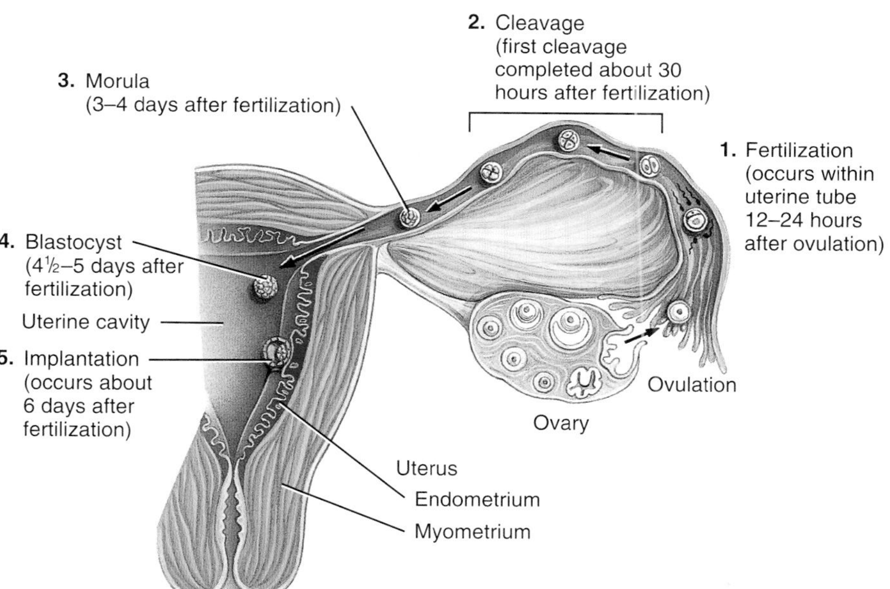

Frontal section through uterus, uterine tube, and ovary

In which phase of the uterine cycle does implantation occur?

Second Week of Development

Development of the Trophoblast

About 8 days after fertilization, the trophoblast develops into two layers in the region of contact between the blastocyst and endometrium. These are a **syncytiotrophoblast** (sin-sīt′-ē-ō-TRŌF-ō-blast) that contains no distinct cell boundaries, and a **cytotrophoblast** (sī-tō-TRŌF-ō-blast) between the inner cell mass and syncytiotrophoblast that is composed of distinct cells (Figure 29.6a). The two layers of trophoblast become part of the chorion (one of the fetal membranes) as they undergo further growth (see Figure 29.11a inset). During implantation, the syncytiotrophoblast secretes enzymes that enable the blastocyst to penetrate the uterine lining by digesting and liquefying the endometrial cells. Eventually, the blastocyst becomes buried in the endometrium and inner one-third of the myometrium. Another secretion of the trophoblast is human chorionic gonadotropin (hCG), which has actions similar to LH. Human chorionic gonadotropin rescues the corpus luteum from degeneration and sustains its secretion of progesterone and estrogens. These hormones maintain the uterine lining in a secretory state, preventing menstruation. Peak secretion of hCG occurs about the ninth week of pregnancy at which time the placenta is fully developed and produces the progesterone and estrogens that continue to sustain the pregnancy. The presence of hCG in maternal blood or urine is an indicator of pregnancy and is the hormone detected by home pregnancy tests.

Development of the Bilaminar Embryonic Disc

Like those of the trophoblast, cells of the inner cell mass also differentiate into two layers around 8 days after fertilization: a **hypoblast (primitive endoderm)** and **epiblast (primitive ectoderm)** (Figure 29.6a). Cells of the hypoblast and epiblast together form a flat disc referred to as the **bilaminar embryonic disc** (bī-LAM-in-ar = two-layered). In addition, a small cavity appears within the epiblast and eventually enlarges to form the **amniotic cavity** (am-nē-OT-ik; *amnio-* = lamb).

Development of the Amnion

As the amniotic cavity enlarges, a thin protective membrane called the **amnion** (AM-nē-on) develops from the epiblast (Figure 29.6a). The amnion forms the roof of the amniotic cavity, and the epiblast forms the floor. Initially, the amnion overlies only the bilaminar embryonic disc. However, as the embryo grows, the amnion eventually surrounds the entire embryo (see Figure 29.11a inset), creating the amniotic cavity that becomes filled with **amniotic fluid.** Most amniotic fluid is initially derived from a filtrate of maternal blood. Later, the fetus contributes to the fluid by excreting urine into the amniotic cavity. Amniotic fluid serves as a shock absorber for the fetus, helps regulate fetal body temperature, helps prevent desiccation, and prevents adhesions between the skin of the fetus and surrounding tissues. The amnion usually ruptures just before birth; it and its fluid constitute the "bag of waters." Embryonic cells that are sloughed off into amniotic fluid can be examined in a procedure called **amniocentesis** (am′-nē-ō-sen-TĒ-sis; *amnio-* = amnion; *-centesis* = puncture to remove fluid), in which some of the amniotic fluid is withdrawn and analyzed (see page 1126).

Development of the Yolk Sac

Also on the eighth day after fertilization, cells at the edge of the hypoblast migrate and cover the inner surface of the blastocyst wall (Figure 29.6a). The migrating columnar cells become squamous (flat) and then form a thin membrane referred to as the **exocoelomic membrane** (ek′-sō-sē-LŌ-mik; *exo-* = outside; *-koilos* = space). Together with the hypoblast, the exocoelomic membrane forms the wall of the **yolk sac,** formerly called the blastocyst cavity (Figure 29.6b). As a result, the bilaminar embryonic disc is now positioned between the amniotic cavity and yolk sac.

Since human embryos receive their nutrients from the endometrium, the yolk sac is relatively empty, small, and decreases in size as development progresses (see Figure 29.11a inset). Nevertheless, the yolk sac has several important functions in humans: supplies nutrients to the embryo during the second and third weeks of development; is the source of blood cells from the third through sixth weeks; contains the first cells (primordial germ cells) that will eventually migrate into the developing gonads, differentiate into the primitive germ cells, and form gametes; forms part of the gut (gastrointestinal tract); functions as a shock absorber; and helps prevent drying out of the embryo.

Development of Sinusoids

On the ninth day after fertilization, the blastocyst becomes completely embedded in the endometrium. As the syncytiotrophoblast expands, small spaces called **lacunae** (la-KOO-nē = little lakes) develop within it (Figure 29.6b).

By the twelfth day of development, the lacunae fuse to form larger, interconnecting spaces called **lacunar networks** (Figure 29.6c). Endometrial capillaries around the developing embryo become dilated and are referred to as **sinusoids.** As the syncytiotrophoblast erodes some of the sinusoids and endometrial glands, maternal blood and secretions from the glands enter the lacunar networks and flow through them. Maternal blood is both a rich source of materials for embryonic nutrition and a disposal site for the embryo's wastes.

Development of the Extraembryonic Coelom

About the twelfth day after fertilization, the **extraembryonic mesoderm** develops. These mesodermal cells are derived from the yolk sac and form a connective tissue layer (mesenchyme) around the amnion and yolk sac (Figure 29.6c). Soon a number of large cavities develop in the extraembryonic mesoderm, which then fuse to form a single, larger cavity called the **extraembryonic coelom** (SĒ-lōm).

Development of the Chorion

The extraembryonic mesoderm, together with the two layers of the trophoblast (the cytotrophoblast and syncytiotrophoblast),

Figure 29.6 Principal events of the second week of development.

About 8 days after fertilization, the trophoblast develops into a syncytiotrophoblast and a cytotrophoblast; the inner cell mass develops into a hypoblast and epiblast (bilaminar embryonic disc).

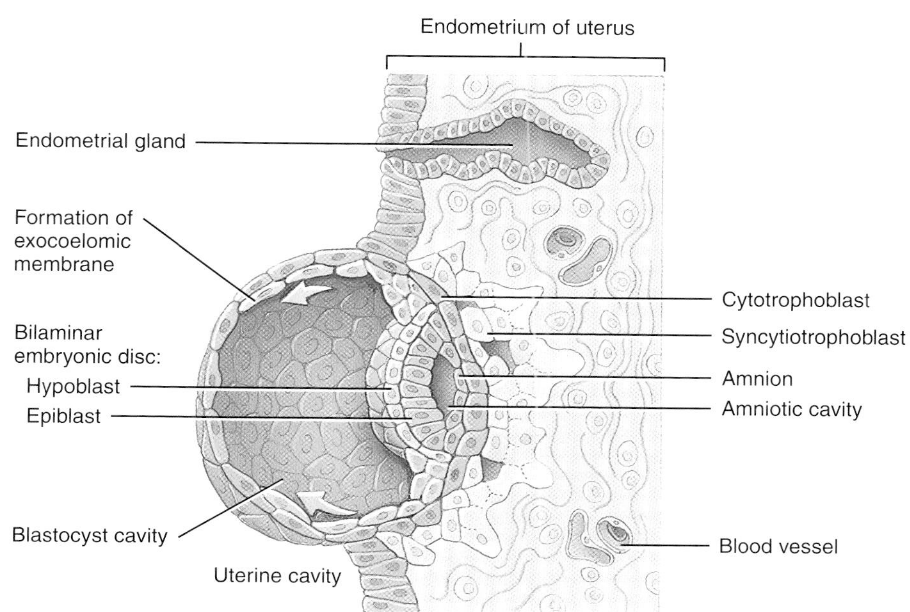

(a) Frontal section through endometrium of uterus showing blastocyst, about 8 days after fertilization

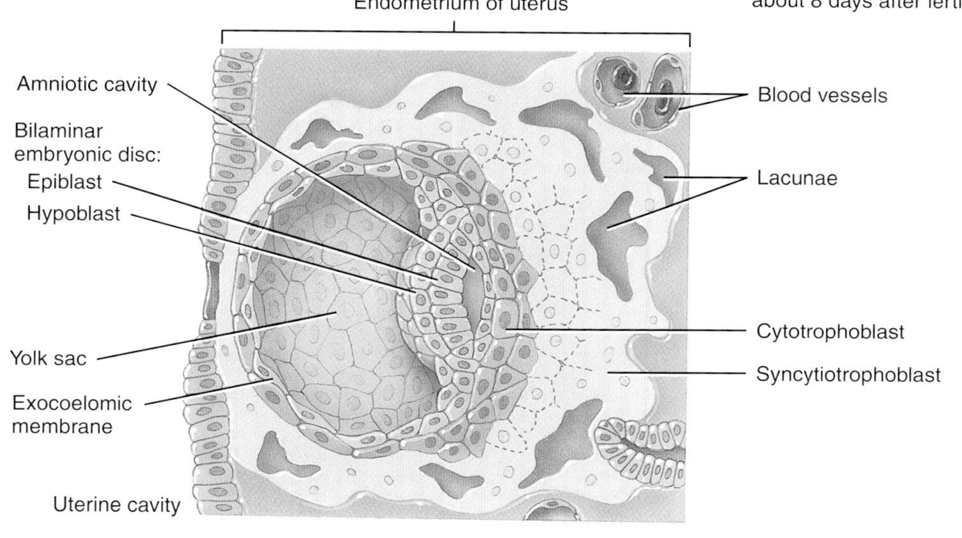

(b) Frontal section through endometrium of uterus showing blastocyst, about 9 days after fertilization

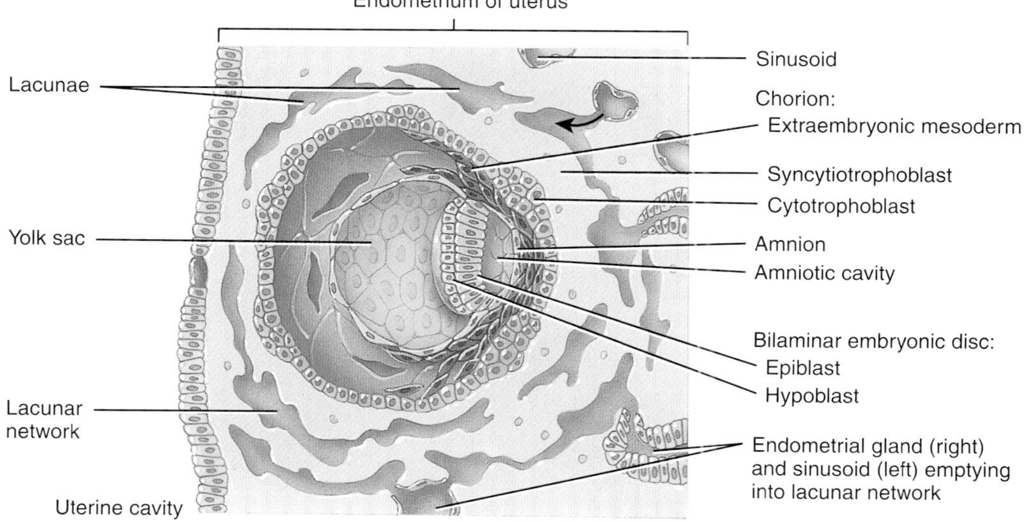

(c) Frontal section through endometrium of uterus showing blastocyst, about 12 days after fertilization

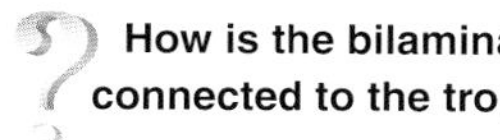

How is the bilaminar embryonic disc connected to the trophoblast?

together form the **chorion** (KOR-ē-on = membrane) (Figure 29.6c). The chorion surrounds the embryo and, later, the fetus (see Figure 29.11a inset). Eventually it becomes the principal embryonic part of the placenta, the structure for exchange of materials between mother and fetus. The chorion also protects the embryo and fetus from the immune responses of the mother in two ways: (1) It secretes proteins that block antibody production by the mother. (2) It promotes the production of T lymphocytes that suppress the normal immune response in the uterus. Finally, the chorion produces human chorionic gonadotropin (hCG), an important hormone of pregnancy (see Figure 29.16).

The inner layer of the chorion eventually fuses with the amnion. With the development of the chorion, the extraembryonic coelom is now referred to as the **chorionic cavity.** By the end of the second week of development, the bilaminar embryonic disc becomes connected to the trophoblast by a band of extraembryonic mesoderm called the **connecting (body) stalk** (see Figure 29.7, inset). The connecting stalk is the future umbilical cord.

▶ CHECKPOINT

6. What are the functions of the trophoblast?
7. How is the bilaminar embryonic disc formed?
8. Describe the formation of the amnion, yolk sac, and chorion and explain their functions.
9. Why are sinusoids important during embryonic development?

Third Week of Development

The third week of development begins a six-week period of very rapid embryonic development and differentiation. During the third week, the three primary germ layers are established and lay the groundwork for organ development in weeks four through eight.

Gastrulation

The first major event of the third week of development, **gastrulation** (gas′-troo-LĀ-shun), occurs about 15 days after fertilization. In this process, the bilaminar (two-layered) embryonic disc, consisting of epiblast and hypoblast, transforms into a **trilaminar** (three-layered) **embryonic disc** consisting of three primary germ layers: the ectoderm, mesoderm, and endoderm. The primary germ layers are the major embryonic tissues from which the various tissues and organs of the body develop.

Gastrulation involves the rearrangement and migration of cells from the epiblast. The first evidence of gastrulation is the formation of the **primitive streak,** a faint groove on the dorsal surface of the epiblast that elongates from the posterior to the anterior part of the embryo (Figure 29.7a). The primitive streak clearly establishes the head and tail ends of the embryo, as well as its right and left sides. At the head end of the primitive streak a small group of epiblastic cells forms a rounded structure called the **primitive node.**

Following formation of the primitive streak, cells of the epiblast move inward below the primitive streak and detach from the epiblast (Figure 29.7b) in a process called **invagination** (in-vaj-i-NĀ-shun). Once the cells have invaginated, some of them displace the hypoblast, forming the **endoderm** (*endo-* = inside; *-derm* = skin). Other cells remain between the epiblast and newly formed endoderm to form the **mesoderm** (*meso-* = middle). Cells remaining in the epiblast then form the **ectoderm** (*ecto-* = outside). The ectoderm and endoderm are epithelia composed of tightly packed cells; the mesoderm is a loosely organized connective tissue (mesenchyme). As the embryo develops, the endoderm ultimately becomes the epithelial lining of the gastrointestinal tract, respiratory tract, and several other organs. The mesoderm gives rise to muscles, bones, and other connective tissues, and the peritoneum. The ectoderm develops into the epidermis of the skin and the nervous system. Table 29.1 provides more details about the fates of these primary germ layers.

TABLE 29.1 Structures Produced by the Three Primary Germ Layers

Endoderm	Mesoderm	Ectoderm
Epithelial lining of gastrointestinal tract (except the oral cavity and anal canal) and the epithelium of its glands. Epithelial lining of urinary bladder, gallbladder, and liver. Epithelial lining of pharynx, auditory (eustachian) tubes, tonsils, larynx, trachea, bronchi, and lungs. Epithelium of thyroid gland, parathyroid glands, pancreas, and thymus. Epithelial lining of prostate and bulbourethral (Cowper's) glands, vagina, vestibule, urethra, and associated glands such as the greater (Bartholin's) vestibular and lesser vestibular glands.	All skeletal and cardiac muscle tissue and most smooth muscle tissue. Cartilage, bone, and other connective tissues. Blood, red bone marrow, and lymphatic tissue. Endothelium of blood vessels and lymphatic vessels. Dermis of skin. Fibrous tunic and vascular tunic of eye. Middle ear. Mesothelium of thoracic, abdominal, and pelvic cavities. Epithelium of kidneys and ureters. Epithelium of adrenal cortex. Epithelium of gonads and genital ducts.	All nervous tissue. Epidermis of skin. Hair follicles, arrector pili muscles, nails, epithelium of skin glands (sebaceous and sudoriferous), and mammary glands. Lens, cornea, and internal eye muscles. Internal and external ear. Neuroepithelium of sense organs. Epithelium of oral cavity, nasal cavity, paranasal sinuses, salivary glands, and anal canal. Epithelium of pineal gland, pituitary gland, and adrenal medullae.

Figure 29.7 Gastrulation.

Gastrulation involves the rearrangement and migration of cells from the epiblast.

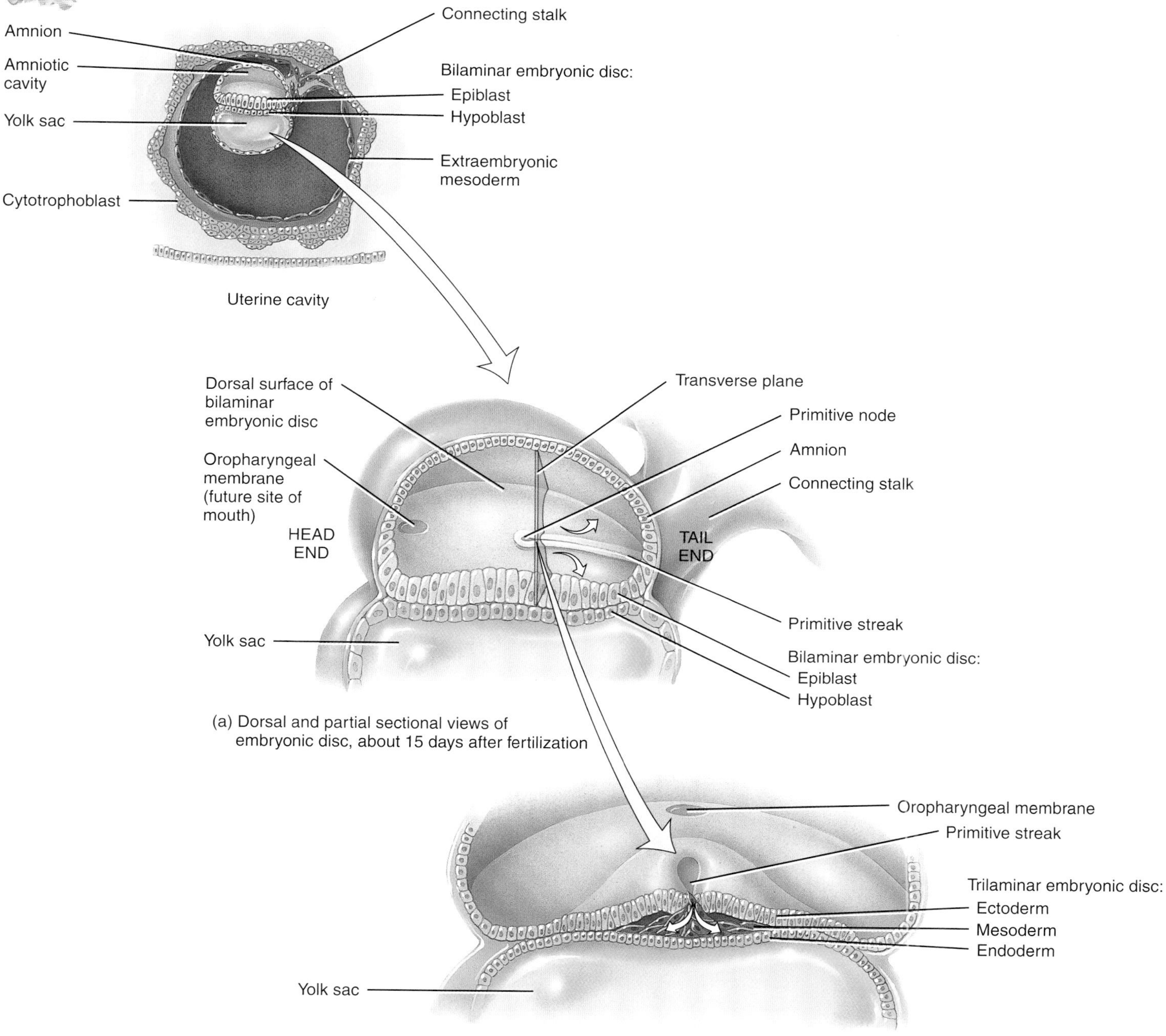

Amnion
Amniotic cavity
Yolk sac
Cytotrophoblast
Connecting stalk
Bilaminar embryonic disc:
Epiblast
Hypoblast
Extraembryonic mesoderm
Uterine cavity

Dorsal surface of bilaminar embryonic disc
Oropharyngeal membrane (future site of mouth)
HEAD END
Yolk sac
Transverse plane
Primitive node
Amnion
Connecting stalk
TAIL END
Primitive streak
Bilaminar embryonic disc:
Epiblast
Hypoblast

(a) Dorsal and partial sectional views of embryonic disc, about 15 days after fertilization

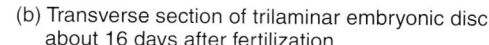

Oropharyngeal membrane
Primitive streak
Trilaminar embryonic disc:
Ectoderm
Mesoderm
Endoderm
Yolk sac

(b) Transverse section of trilaminar embryonic disc, about 16 days after fertilization

What is the significance of gastrulation?

About 16 days after fertilization, mesodermal cells from the primitive node migrate toward the head end of the embryo and form a hollow tube of cells in the midline called the **notochordal process** (nō-tō-KOR-dal) (Figure 29.8). By days 22–24, the notochordal process becomes a solid cylinder of cells called the **notochord** (nō-tō-KORD; *noto-* = back; *-chord* = cord). This structure plays an extremely important role in **induction** (in-DUK-shun), the process by which one tissue *(inducing tissue)* stimulates the development of an adjacent unspecialized tissue *(responding tissue)* into a specialized one. An inducing tissue usually produces a chemical substance that influences the responding tissue. The notochord induces certain mesodermal cells to develop into the vertebral bodies. It also forms the nucleus pulposus of the intervertebral discs (see Figure 7.16d on page 213).

During the third week of development, two faint depressions appear on the dorsal surface of the embryo. The structure closer to the head end is called the **oropharyngeal membrane** (or-ō-fa-RIN-jē-al; *oro-* = mouth; *-pharyngeal* = pertaining to the pharynx) (Figure 29.8a, b). It breaks down during the fourth week to connect the mouth cavity to the pharynx and the remainder of the gastrointestinal tract. The structure closer to the tail end is called the **cloacal membrane** (klō-Ā-kul = sewer), which degenerates in the seventh week to form the openings of the anus and urinary and reproductive tracts.

When the cloacal membrane appears, the wall of the yolk sac forms a small vascularized outpouching called the **allantois** (a-LAN-tō-is; *allant-* = sausage) that extends into the connecting stalk (Figure 29.8b). In most other mammals, the allantois is used for gas exchange and waste removal. Because of the role of the human placenta in these activities, the allantois is not a prominent structure in humans (see Figure 29.11a, inset). Nevertheless, it does function in the early formation of blood and blood vessels and it is associated with the development of the urinary bladder.

Neurulation

In addition to inducing mesodermal cells to develop into vertebral bodies, the notochord also induces ectodermal cells over it to form the **neural plate** (Figure 29.9a). (Also see Figure 14.28 on page 515). By the end of the third week, the lateral edges of the neural plate become more elevated and form the **neural fold** (Figure 29.9b). The depressed midregion is called the **neural groove** (Figure 29.9c). Generally, the neural folds approach each other and fuse, thus converting the neural plate into a **neural tube** (Figure 29.9d). This occurs first near the middle of the embryo and then progresses toward the head and tail ends. Neural tube cells then develop into the brain and spinal cord. The process by which the neural plate, neural folds, and neural tube form is called **neurulation** (noor-oo-LĀ-shun).

As the neural tube forms, some of the ectodermal cells from the tube migrate to form several layers of cells called the **neural crest** (see Figure 14.28b on page 515). Neural crest cells give rise to spinal and cranial nerves and their ganglia, autonomic nervous system ganglia, the meninges of the brain and spinal cord, the adrenal medullae, and several skeletal and muscular components of the head.

Figure 29.8 Development of the notochordal process.

The notochordal process develops from the primitive node and later becomes the notochord.

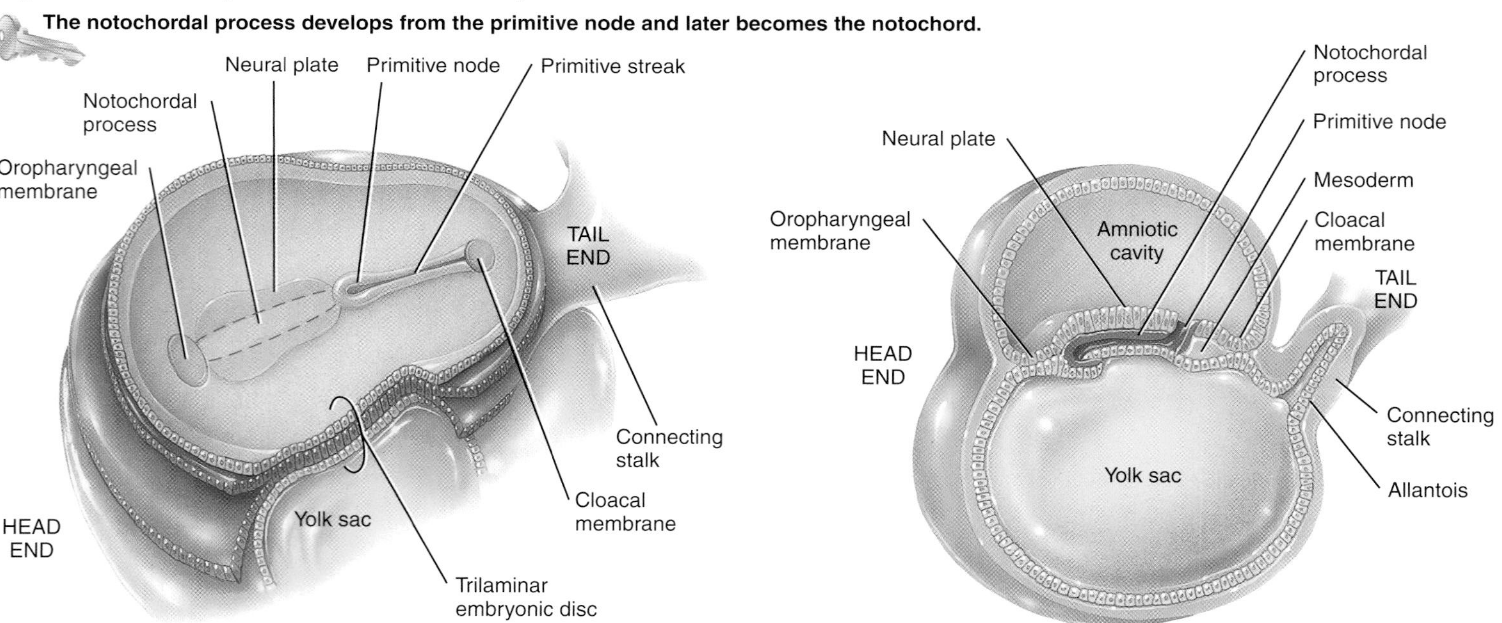

(a) Dorsal and partial sectional views of trilaminar embryonic disc, about 16 days after fertilization

(b) Sagittal section of trilaminar embryonic disc, about 16 days after fertilization

What is the significance of the notochord?

At about four weeks after fertilization, the head end of the neural tube develops into three enlarged areas called **primary brain vesicles** (see Figure 14.29 on pages 516–517): the **prosencephalon (forebrain), mesencephalon (midbrain),** and **rhombencephalon (hindbrain).** At about five weeks, the prosencephalon develops into **secondary brain vesicles** called the **telencephalon** and **diencephalon** and the rhombencephalon develops into secondary brain vesicles called the **metencephalon** and **myelencephalon.** The areas of the neural tube adjacent to the myelencephalon develop into the spinal cord. The parts of the brain that develop from the various brain vesicles are described on page 516–517 in Chapter 14.

Figure 29.9 Neurulation and the development of somites.

Neurulation is the process by which the neural plate, neural folds, and neural tube form.

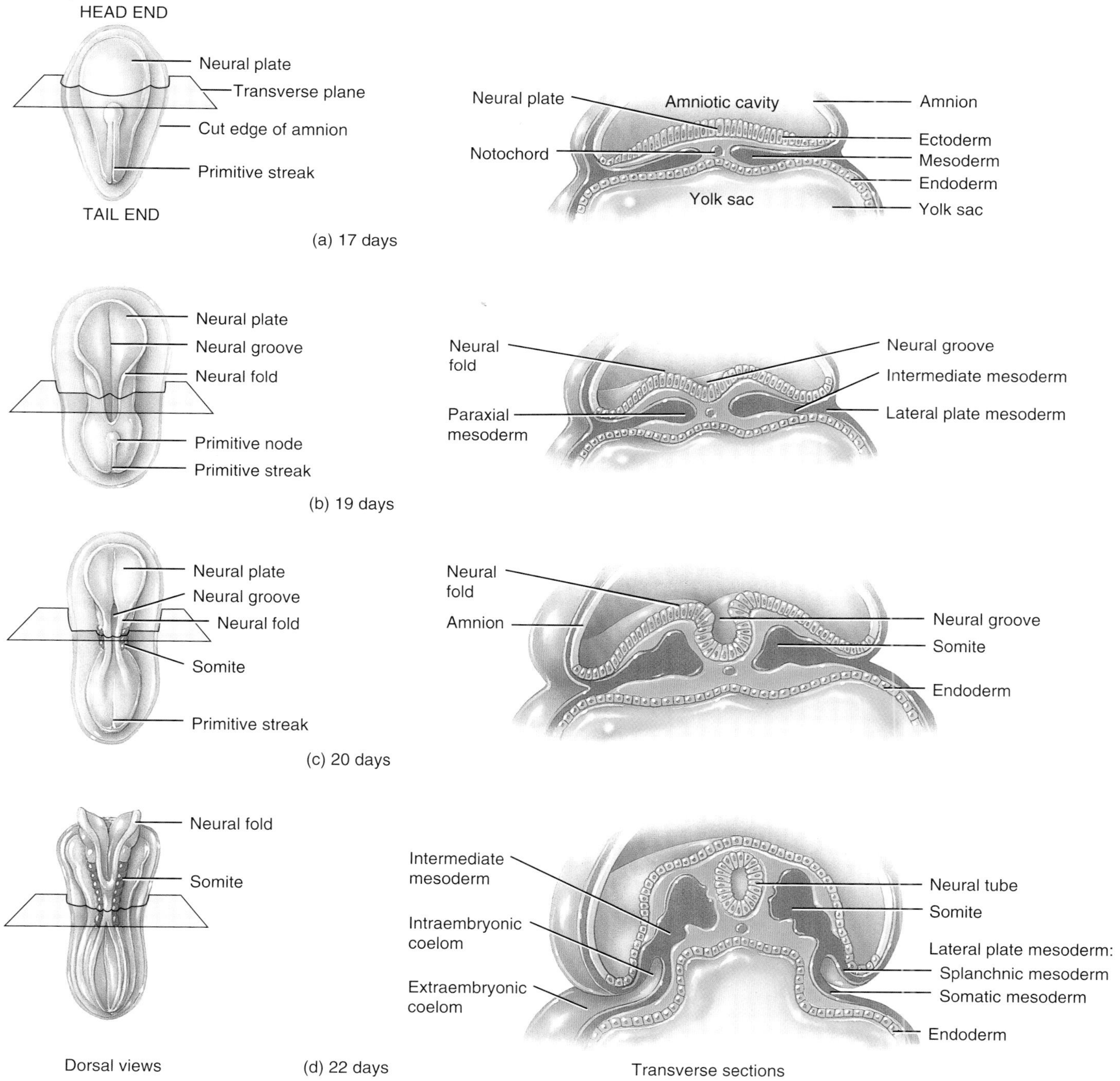

HEAD END
Neural plate
Transverse plane
Cut edge of amnion
Primitive streak
TAIL END
Neural plate
Amniotic cavity
Amnion
Ectoderm
Notochord
Mesoderm
Endoderm
Yolk sac
Yolk sac

(a) 17 days

Neural plate
Neural groove
Neural fold
Primitive node
Primitive streak
Neural fold
Paraxial mesoderm
Neural groove
Intermediate mesoderm
Lateral plate mesoderm

(b) 19 days

Neural plate
Neural groove
Neural fold
Somite
Primitive streak
Neural fold
Amnion
Neural groove
Somite
Endoderm

(c) 20 days

Neural fold
Somite
Intermediate mesoderm
Intraembryonic coelom
Extraembryonic coelom
Neural tube
Somite
Lateral plate mesoderm:
Splanchnic mesoderm
Somatic mesoderm
Endoderm

Dorsal views
(d) 22 days
Transverse sections

Which structures develop from the neural tube and somites?

Neural tube defects (NTDs) are caused by arrest of the normal development and closure of the neural tube. These include spina bifida (discussed on page 225) and **anencephaly** (an′-en-SEPH-a-lē; *an-* = without; *encephal* = brain). In anencephaly, the cranial bones fail to develop and certain parts of the brain remain in contact with amniotic fluid and degenerate. Usually, a part of the brain that controls vital functions such as breathing and regulation of the heart is also affected. Infants with anencephaly are stillborn or die within a few days after birth. The condition occurs about once in every 1000 births and is 2 to 4 times more common in female infants than males. ■

Development of Somites

By about the 17th day after fertilization, the mesoderm adjacent to the notochord and neural tube forms paired longitudinal columns of **paraxial mesoderm** (par-AK-sē-al; *para-* = near) (Figure 29.9b). The mesoderm lateral to the paraxial mesoderm forms paired cylindrical masses called **intermediate mesoderm.** The mesoderm lateral to the intermediate mesoderm consists of a pair of flattened sheets called **lateral plate mesoderm.** The paraxial mesoderm soon segments into a series of paired, cube-shaped structures called **somites** (SŌ-mīts = little bodies). By the end of the fifth week, 42–44 pairs of somites are present. The number of somites that develop over a given period can be correlated to the approximate age of the embryo.

Each somite differentiates into three regions: a **myotome,** a **dermatome,** and a **sclerotome** (see Figure 10.19b on page 318). The myotomes develop into the skeletal muscles of the neck, trunk, and limbs; the dermatomes form connective tissue, including the dermis of the skin; and the sclerotomes give rise to the vertebrae and ribs.

Development of the Intraembryonic Coelom

In the third week of development, small spaces appear in the lateral plate mesoderm. These spaces soon merge to form a larger cavity called the **intraembryonic coelom** (SĒ-lom = cavity). This cavity splits the lateral plate mesoderm into two parts called the splanchnic mesoderm and somatic mesoderm (Figure 29.9d). **Splanchnic mesoderm** (SPLANGK-nik = visceral) forms the heart and the visceral layer of the serous pericardium, blood vessels, the smooth muscle and connective tissues of the respiratory and digestive organs, and the visceral layer of the serous membrane of the pleurae and peritoneum. **Somatic mesoderm** (sō-MAT-ik; *soma-* = body) gives rise to the bones, ligaments, and dermis of the limbs and the parietal layer of the serous membrane of the pericardium, pleurae, and peritoneum.

Development of the Cardiovascular System

At the beginning of the third week, **angiogenesis** (an-jē-ō-JEN-e-sis; *angio-* = vessel; *-genesis* = production), the formation of blood vessels, begins in the extraembryonic mesoderm in the yolk sac, connecting stalk, and chorion. This early development is necessary because there is insufficient yolk in the yolk sac and ovum to provide adequate nutrition for the rapidly developing embryo. Angiogenesis is initiated when mesodermal cells differentiate into **hemoangioblasts.** These then develop into cells called **angioblasts,** which aggregate to form isolated masses of cells referred to as **blood islands** (see Figure 21.31 on page 796). Spaces soon develop in the blood islands and form the lumens of blood vessels. Some angioblasts arrange themselves around each space to form the endothelium and the tunics (layers) of the developing blood vessels. As the blood islands grow and fuse, they soon form an extensive system of blood vessels throughout the embryo.

About 3 weeks after fertilization, blood cells and blood plasma begin to develop *outside* the embryo from hemoangioblasts in the blood vessels in the walls of the yolk sac, allantois, and chorion. These then develop into pluripotent stem cells that form blood cells. Blood formation begins *within* the embryo at about the fifth week in the liver and the twelfth week in the spleen, red bone marrow, and thymus.

The heart forms from splanchnic mesoderm in the head end of the embryo on days 18 and 19. This region of mesodermal cells is called the **cardiogenic area** (kar-dē-ō-JEN-ik; *cardio-* = heart; *-genic* = producing). In response to induction signals from the underlying endoderm, these mesodermal cells form a pair of **endocardial tubes** (see Figure 20.18 on page 725). The tubes then fuse to form a single **primitive heart tube.** By the end of the third week, the primitive heart tube bends on itself, becomes S-shaped, and begins to beat. It then joins blood vessels in other parts of the embryo, connecting stalk, chorion, and yolk sac to form a primitive cardiovascular system.

Development of the Chorionic Villi and Placenta

By the end of the second week of development, **chorionic villi** (ko-rē-ON-ik VIL-ī) begin to develop. These fingerlike projections consist of chorion (syncytiotrophoblast surrounded by cytotrophoblast) (Figure 29.10a). By the end of the third week, blood capillaries develop in the chorionic villi (Figure 29.10b). Blood vessels in the chorionic villi connect to the embryonic heart by way of the umbilical arteries and umbilical vein (Figure 29.10c). As a result, maternal and fetal blood vessels are in close proximity. Note, however, that maternal and fetal blood vessels do not join, and the blood they carry does not normally mix. Instead, oxygen and nutrients in the blood of the mother's **intervillous spaces,** the spaces between chorionic villi, diffuse across the cell membranes into the capillaries of the villi. Waste products such as carbon dioxide diffuse in the opposite direction.

Placentation (plas′-en-TĀ-shun) is the process of forming the **placenta** (pla-SEN-ta = flat cake), the site of exchange of nutrients and wastes between the mother and fetus. The placenta also produces hormones needed to sustain the pregnancy (see Figure 29.16). The placenta is unique because it develops from two separate individuals, the mother and the fetus.

By the beginning of the twelfth week, the placenta has two distinct parts: (1) the fetal portion formed by the chorionic villi of the chorion and (2) the maternal portion formed by the decidua basalis of the endometrium (Figure 29.11a on page 1118). When

Figure 29.10 Development of chorionic villi.

Blood vessels in chorionic villi connect to the embryonic heart via the umbilical arteries and umbilical vein.

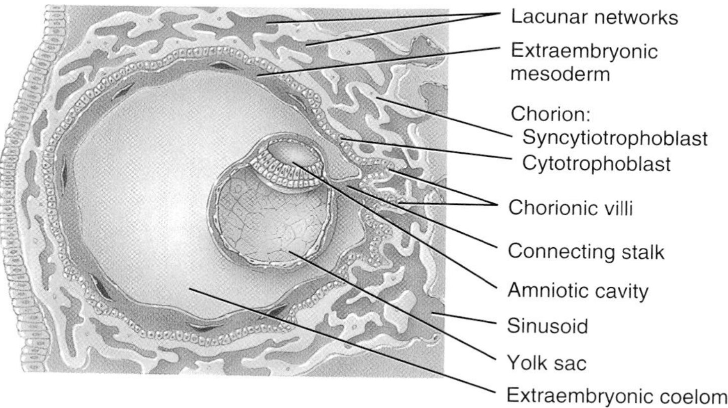

(a) Frontal section through uterus showing blastocyst, about 13 days after fertilization

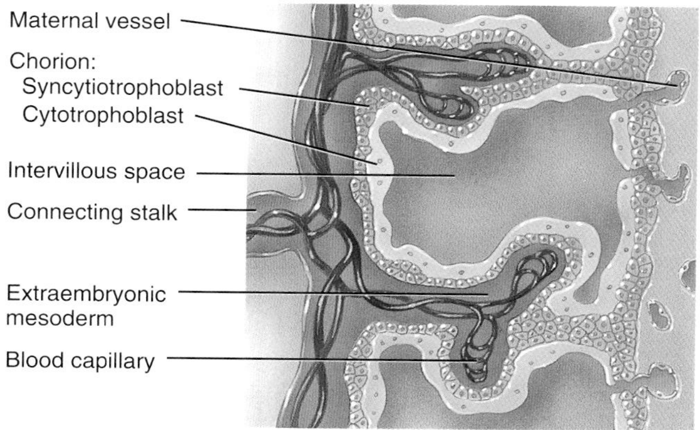

(b) Details of two chorionic villi, about 21 days after fertilization

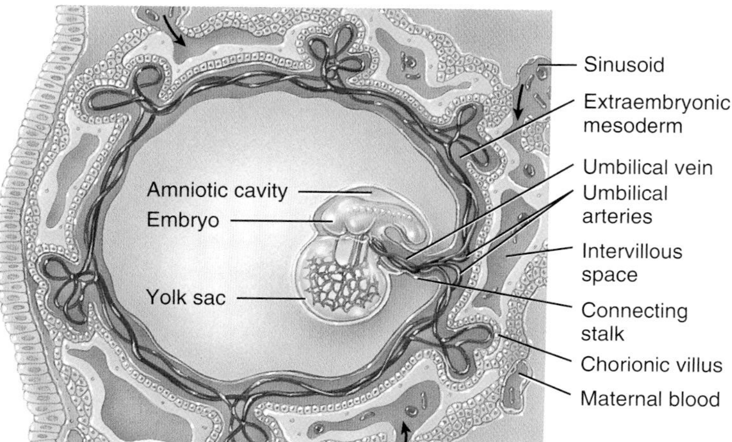

(c) Frontal section through uterus showing an embryo and its vascular supply, about 21 days after fertilization

Why is development of chorionic villi important?

fully developed, the placenta is shaped like a pancake (Figure 29.11b). Functionally, the placenta allows oxygen and nutrients to diffuse from maternal blood into fetal blood while carbon dioxide and wastes diffuse from fetal blood into maternal blood. The placenta also is a protective barrier because most microorganisms cannot pass through it. However, certain viruses, such as those that cause AIDS, German measles, chickenpox, measles, encephalitis, and poliomyelitis, can cross the placenta. Many drugs, alcohol, and some substances that can cause birth defects also pass freely. The placenta stores nutrients such as carbohydrates, proteins, calcium, and iron, which are released into fetal circulation as required.

The actual connection between the placenta and embryo, and later the fetus, is through the **umbilical cord** (um-BIL-i-kul = navel), which develops from the connecting stalk and is usually about 2 cm (1 in.) wide and about 50–60 cm (20–24 in.) in length. The umbilical cord consists of two umbilical arteries that carry deoxygenated fetal blood to the placenta, one umbilical vein that carries oxygen and nutrients acquired from the mother's intervillous spaces into the fetus, and supporting mucous connective tissue called **Wharton's jelly** derived from the allantois. A layer of amnion surrounds the entire umbilical cord and gives it a shiny appearance (Figure 29.11a). In some cases, the umbilical vein is used to transfuse blood into a fetus or to introduce drugs for various medical treatments.

In about 1 in 200 newborns, only one of the two umbilical arteries is present in the umbilical cord. It may be due to failure of the artery to develop or degeneration of the vessel early in development. Nearly 20% of infants with this condition develop cardiovascular defects.

After the birth of the baby, the placenta detaches from the uterus and is therefore termed the **afterbirth.** At this time, the umbilical cord is tied off and then severed. The small portion (about an inch) of the cord that remains attached to the infant begins to wither and falls off, usually within 12 to 15 days after birth. The area where the cord was attached becomes covered by a thin layer of skin, and scar tissue forms. The scar is the **umbilicus** (navel).

Pharmaceutical companies use human placentas as a source of hormones, drugs, and blood; portions of placentas are also used for burn coverage. The placental and umbilical cord veins can also be used in blood vessel grafts, and cord blood can be frozen to provide a future source of pluripotent stem cells, for example, to repopulate red bone marrow following radiotherapy for cancer.

Placenta Previa

In some cases, the entire placenta or part of it may become implanted in the inferior portion of the uterus, near or covering the internal os of the cervix. This condition is called **placenta previa** (PRĒ-vē-a = before or in front of). Although placenta previa may lead to spontaneous abortion, it also occurs in approximately 1 in 250 live births. It is dangerous to the fetus because it may cause premature birth and intrauterine hypoxia due to maternal bleeding. Maternal mortality is increased due to hemorrhage and infection. The most important symptom is sudden, painless, bright-red vaginal bleeding in the third trimester. Cesarean section is the preferred method of delivery in placenta previa. ■

Figure 29.11 **Placenta and umbilical cord.**

The placenta is formed by the chorionic villi of the embryo and the decidua basalis of the endometrium of the mother.

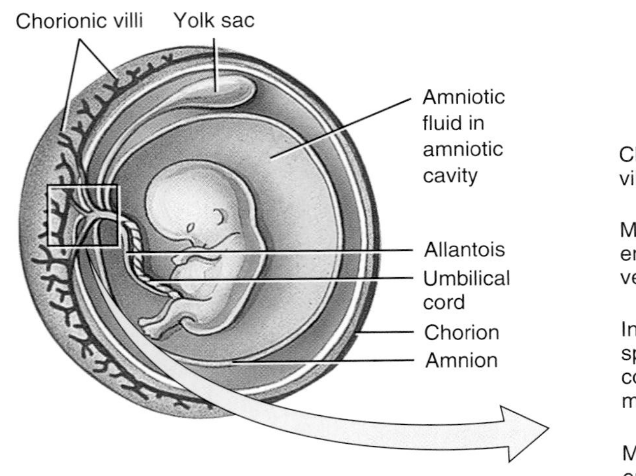

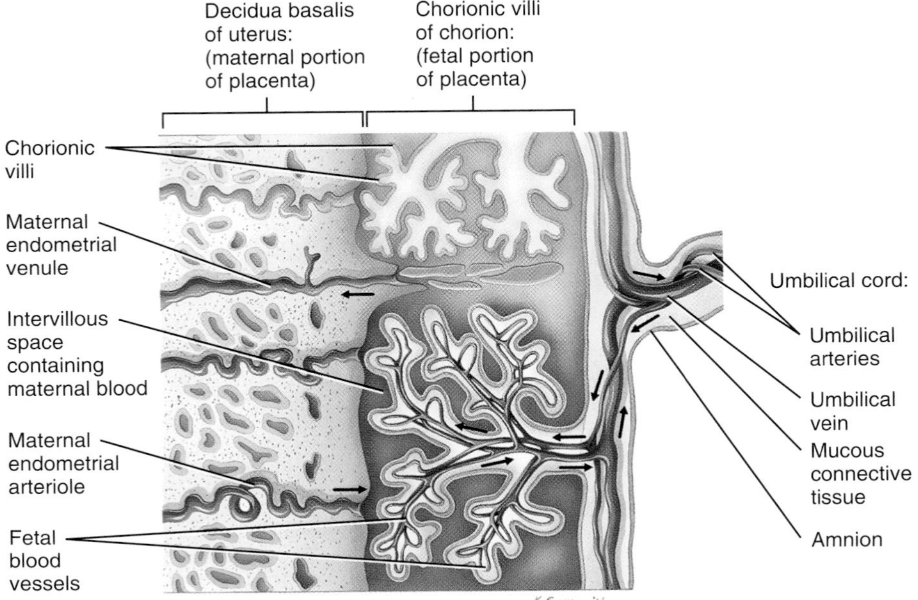

(a) Details of placenta and umbilical cord

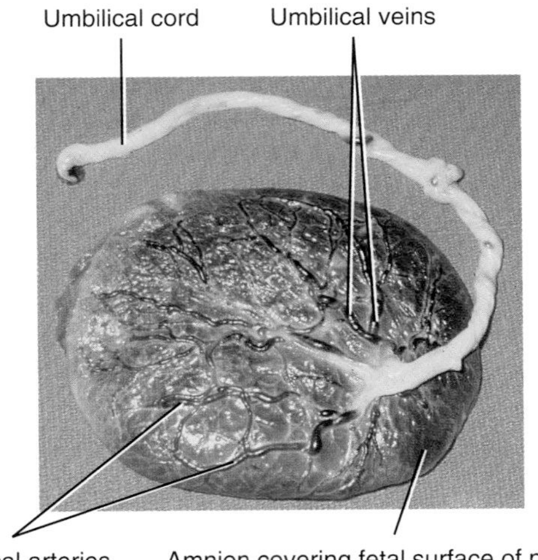

(b) Fetal aspect of placenta

What is the function of the placenta?

▶ CHECKPOINT

10. What is the significance of gastrulation?

11. How do the three primary germ layers form? Why are they important?

12. What is the function of the notochord?

13. Describe how neurulation occurs. Why is it important?

14. What are the functions of somites?

15. How does the cardiovascular system develop?

16. How does the placenta form and what is its importance?

Fourth Week of Development

The fourth through eighth weeks of development are very significant in embryonic development because all major organs appear during this time. The term **organogenesis** (or′-ga-nō-JEN-e-sis) refers to the formation of body organs and systems. By the end of the eighth week, all the major body systems have begun to develop, although their functions for the most part are minimal. Organogenesis requires the presence of blood vessels to supply developing organs with oxygen and other nutrients. However, recent studies suggest that blood vessels play a significant role in organogenesis even before blood begins to flow within them. The endothelial cells of blood vessels apparently provide some type of developmental signal, either a secreted substance or a direct cell-to-cell interaction, that is necessary for organogenesis.

During the fourth week after fertilization, the embryo undergoes very dramatic changes in shape and size, nearly tripling its size. It is essentially converted from a flat, two-dimensional trilaminar embryonic disc to a three-dimensional cylinder, a process called **embryonic folding** (Figure 29.12a–d). The cylinder consists of endoderm in the center (gut), ectoderm on the outside (epidermis), and mesoderm in between. The main force responsible for embryonic folding is the different rates of growth of various parts of the embryo, especially the rapid longitudinal growth of the nervous system (neural tube). Folding in the median plane produces a **head fold** and a **tail fold;** folding in the horizontal plane results in

Figure 29.12 Embryonic folding.

Embryonic folding converts the two-dimensional trilaminar embryonic disc into a three-dimensional cylinder.

Neural tube
Amniotic cavity
Notochord
Connecting stalk
HEAD END
TAIL END
Oropharyngeal membrane
Yolk sac
Cloacal membrane

Amniotic cavity
Neural tube
Notochord
Intraembryonic coelom
Yolk sac

(a) 22 days

Heart
Primitive gut
Yolk sac

(b) 24 days

Foregut
Head fold
Midgut
Tail fold
Hindgut
Stomodeum
Proctodeum
Yolk sac

Midgut
Intraembryonic coelom
Yolk sac

(c) 26 days

Future pharynx
Oropharyngeal membrane
Amniotic cavity
Cloacal membrane
Proctodeum
Stomodeum
Future umbilical cord

Aorta
Dorsal mesentery
Midgut
Ventral mesentery
Abdominal wall

Sagittal sections (d) 28 days Transverse sections

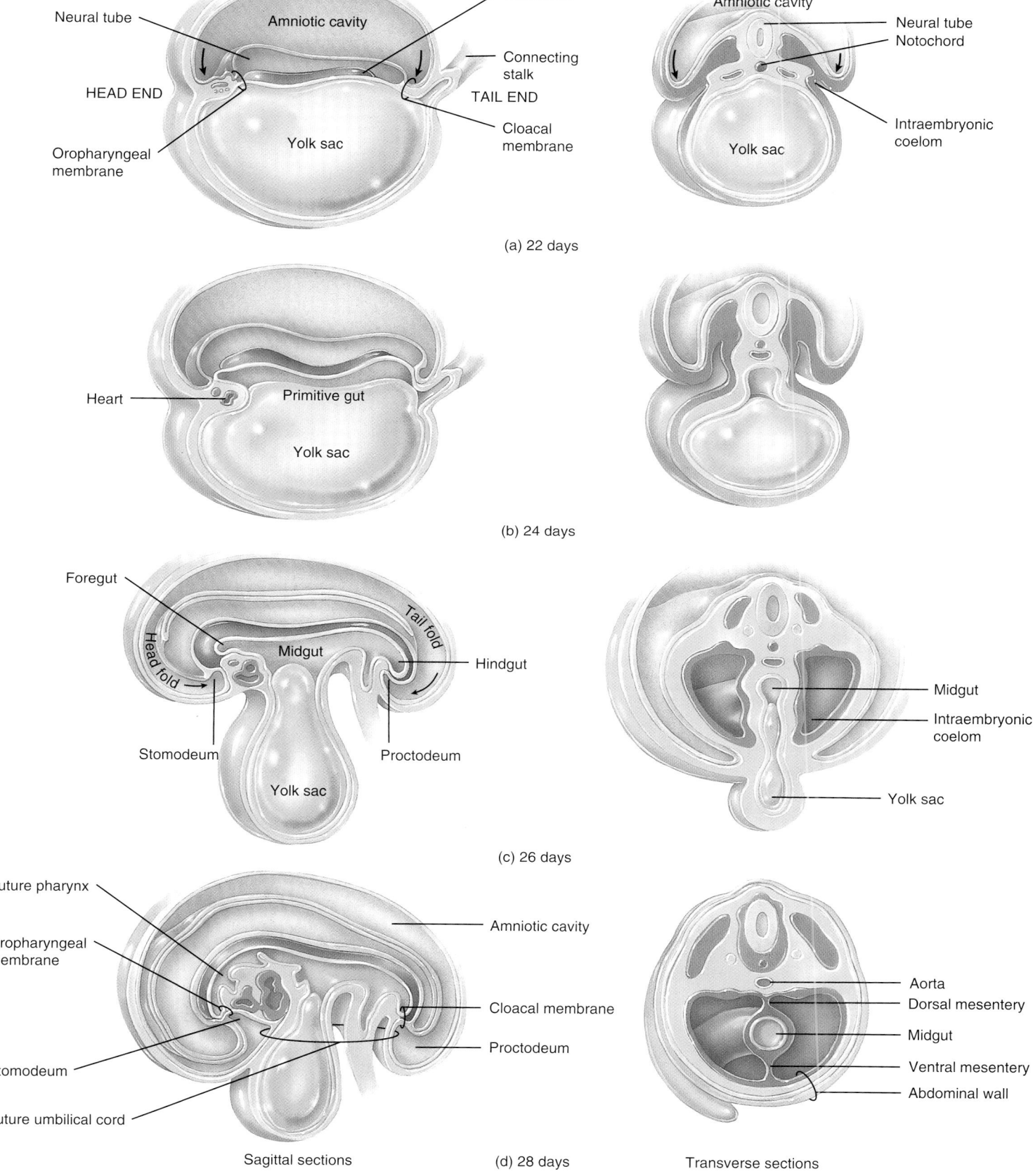

What are the results of embryonic folding?

the two **lateral folds.** Overall, due to the foldings, the embryo curves into a C-shape.

The head fold brings the developing heart and mouth into their eventual adult positions. The tail fold brings the developing anus into its eventual adult position. The lateral folds form as the lateral margins of the trilaminar embryonic disc bend ventrally. As they move toward the midline, the lateral folds incorporate the dorsal part of the yolk sac into the embryo as the **primitive gut,** the forerunner of the gastrointestinal tract (Figure 29.12b). The primitive gut differentiates into an anterior **foregut,** an intermediate **midgut,** and a posterior **hindgut** (Figure 29.12c). The fates of the foregut, midgut, and hindgut are described on page 940. Recall that the oropharyngeal membrane is located in the head end of the embryo (see Figure 29.8). It separates the future pharyngeal (throat) region of the foregut from the **stomodeum** (stō-mō-DĒ-um; *stomo-* = mouth), the future oral cavity. Because of head folding, the oropharyngeal membrane moves downward and the foregut and stomodeum move closer to their final positions. When the oropharyngeal membrane ruptures during the fourth week, the pharyngeal region of the pharynx is brought into contact with the stomodeum.

In a developing embryo, the last part of the hindgut expands into a cavity called the **cloaca** (see Figure 26.23 on page 1028). On the outside of the embryo is a small cavity in the tail region called the **proctodeum** (prok-tō-DĒ-um; *procto-* = anus) (Figure 29.12c). Separating the cloaca from the proctodeum is the **cloacal membrane** (see Figure 29.8). During embryonic development, the cloaca divides into a ventral urogenital sinus and a dorsal anorectal canal. As a result of tail folding, the cloacal membrane moves downward and the urogenital sinus, anorectal canal, and proctodeum move closer to their final positions. When the cloacal membrane ruptures during the seventh week of development, the urogenital and anal openings are created.

Along with embryonic folding, development of somites and the neural tube (previously described) occurs during the fourth week of development. In addition, several **pharyngeal (branchial) arches** develop on each side of the future head and neck regions (Figure 29.13). These five paired structures form swellings on the surface of the embryo beginning on the 22nd day after fertilization. Each pharyngeal arch consists of an outer covering of ectoderm, an inner covering of endoderm, and mesoderm in between. The pharyngeal arches also contain an artery, a cranial nerve, cartilage, and muscular tissue. Also on the outside of the embryo are a series of grooves between the pharyngeal arches called **pharyngeal clefts,** which separate the pharyngeal arches (Figure 29.13a). Simultaneous with the development of the pharyngeal arches and cleft\s, four distinct pairs of **pharyngeal (branchial) pouches** develop inside the embryo (Figure 29.13b). The pharyngeal pouches are endoderm-lined, balloon-like outgrowths of the primitive pharynx, the most cranial part of the foregut.

Together, the pharyngeal arches, clefts, and pouches give rise to the structures of the head and neck. The first sign of a developing ear is a thickened area of ectoderm, the **otic placode** (future internal ear), which can be distinguished about 22 days after fertilization (Figure 29.13a). The eyes also begin their development about 22 days after fertilization, evidenced by a thickened area of ectoderm refered to as the **lens placode** (Figure 29.13a).

By the middle of the fourth week, the upper limbs begin their development as outgrowths of mesoderm covered by ectoderm called **upper limb buds** (see Figure 8.18b on page 253). By the end of the fourth week, the **lower limb buds** develop. The heart also forms a distinct projection on the ventral surface of the embryo called the **heart prominence** (see Figure 8.18a). At the end of the fourth week the embryo has a distinctive **tail** (see Figure 8.18b).

Figure 29.13 Development of pharyngeal arches, pharyngeal clefts, and pharyngeal pouches.

The five pairs of pharyngeal pouches consist of ectoderm, mesoderm, and endoderm and contain blood vessels, cranial nerves, cartilage, and muscle tissue.

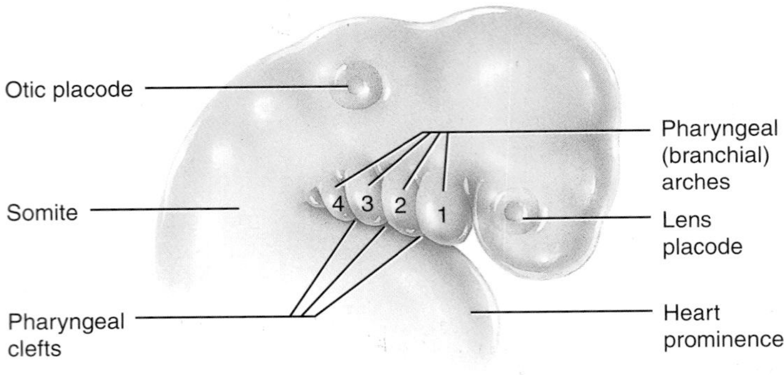

(a) External view, about 28-day embryo

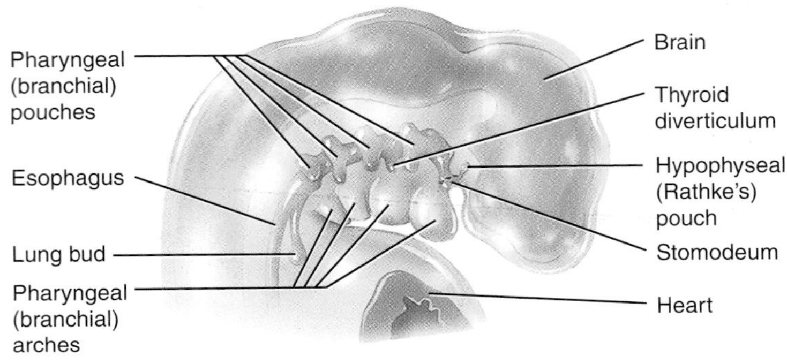

(b) Sagittal section, about 28-day embryo

Why are pharyngeal arches, clefts, and pouches important?

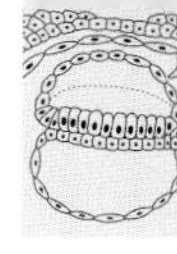

Fifth Through Eighth Weeks of Development

During the fifth week of development, there is a very rapid development of the brain, so growth of the head is considerable. By the end of the sixth week, the head grows even larger relative to the trunk, and the limbs show substantial development (see Figure 8.18c on page 253). In addition, the neck and trunk begin to straighten, and the heart is now four-chambered. By the seventh week, the various regions of the limbs become distinct and the beginnings of digits appear (see Figure 8.18d on page 253). At the start of the eighth week (the final week of the embryonic period), the digits of the hands are short and webbed, the tail is shorter but still visible, the eyes are open, and the auricles of the ears are visible (see Figure 8.18e on page 253). By the end of the eighth week, all regions of limbs are apparent; the digits are distinct and no longer webbed due to removal of cells via apoptosis. Also, the eyelids come together and may fuse, the tail disappears, and the external genitals begin to differentiate. The embryo now has clearly human characteristics.

▶ CHECKPOINT

17. How does embryonic folding occur?

18. How does the primitive gut form and what is its significance?

19. What is the origin of the structures of the head and neck?

20. What are limb buds?

21. What changes occur in the limbs during the second half of the embryonic period?

FETAL PERIOD

▶ OBJECTIVE

Describe the major events of the fetal period.

During the fetal period, tissues and organs that developed during the embryonic period grow and differentiate. Very few new structures appear during the fetal period, but the rate of body growth is remarkable, especially during the second half of intrauterine life. For example, during the last two and one half months of intrauterine life, half of the full-term weight is added. At the beginning of the fetal period, the head is half the length of the body. By the end of the fetal period, the head size is only one-quarter the length of the body. During the same period, the limbs also increase in size from one-eighth to one-half the fetal length. The fetus is also less vulnerable to the damaging effects of drugs, radiation, and microbes than it was as an embryo.

A summary of the major developmental events of the embryonic and fetal period is presented in Table 29.2 and illustrated in Figure 29.14 on page 1124.

Throughout the text we have discussed the developmental anatomy of the various body systems in their respective chapters. The following list of these sections is presented here for your review.

▶ CHECKPOINT

22. What are the general developmental trends during the fetal period?

23. Using Table 29.2 as a guide, select any one body structure in weeks 9 through 12 and trace its development through the remainder of the fetal period.

TERATOGENS

▶ OBJECTIVE

Define a teratogen and list several examples of teratogens.

Exposure of a developing embryo or fetus to certain environmental factors can damage the developing organism or even cause death. A **teratogen** (TER-a-tō-jen; *terato-* = monster; *-gen* = creating) is any agent or influence that causes developmental defects in the embryo. In the following sections we briefly discuss several examples.

Chemicals and Drugs

Because the placenta is not an absolute barrier between the maternal and fetal circulations, any drug or chemical that is dangerous to an infant should be considered potentially dangerous to the fetus when given to the mother. Alcohol is by far the number-one fetal teratogen. Intrauterine exposure to even a small amount of alcohol may result in **fetal alcohol syndrome (FAS),** one of the most common causes of mental retardation and the most common preventable cause of birth defects in the United States. The symptoms of FAS may include slow growth before and after birth, characteristic facial features (short palpebral fissures, a thin upper lip, and sunken nasal bridge), defective heart and other organs, malformed limbs, genital abnormalities, and central nervous system damage. Behavioral problems, such as hyperactivity, ex-

TABLE 29.2 Summary of Changes During Embryonic and Fetal Development

Time	Approximate Size and Weight	Representative Changes
Embryonic Period		
1–4 weeks	0.6 cm (3/16 in.)	Primary germ layers and notochord develop. Neurulation occurs. Primary brain vesicles, somites, and intraembryonic coelom develop. Blood vessel formation begins and blood forms in yolk sac, allantois, and chorion. Heart forms and begins to beat. Chorionic villi develop and placental formation begins. The embryo folds. The primitive gut, pharyngeal arches, and limb buds develop. Eyes and ears begin to develop, tail forms, and body systems begin to form.
5–8 weeks	3 cm (1.25 in.) 1 g (1/30 oz)	Primary brain vesicles develop into secondary brain vesicles. Limbs become distinct and digits appear. Heart becomes four-chambered. Eyes are far apart and eyelids are fused. Nose develops is and flat. Face is more human-like. Ossification begins. Blood cells start to form in liver. External genitals begin to differentiate. Tail disappears. Major blood vessels form. Many internal organs continue to develop.
Fetal period		
9–12 weeks	7.5 cm (3 in.) 30 g (1 oz)	Head constitutes about half the length of the fetal body, and fetal length nearly doubles. Brain continues to enlarge. Face is broad, with eyes fully developed, closed, and widely separated. Nose develops a bridge. External ears develop and are low set. Ossification continues. Upper limbs almost reach final relative length but lower limbs are not quite as well developed. Heartbeat can be detected. Gender is distinguishable from external genitals. Urine secreted by fetus is added to amniotic fluid. Red bone marrow, thymus, and spleen participate in blood cell formation. Fetus begins to move, but its movements cannot yet be felt by the mother. Body systems continue to develop.
13–16 weeks	18 cm (6.5–7 in.) 100 g (4 oz)	Head is relatively smaller than rest of body. Eyes move medially to their final positions, and ears move to their final positions on the sides of the head. Lower limbs lengthen. Fetus appears even more human-like. Rapid development of body systems occurs.
17–20 weeks	25–30 cm (10–12 in.) 200–450 g (0.5–1 lb)	Head is more proportionate to rest of body. Eyebrows and head hair are visible. Growth slows but lower limbs continue to lengthen. Vernix caseosa (fatty secretions of sebaceous glands and dead epithelial cells) and lanugo (delicate fetal hair) cover fetus. Brown fat forms and is the site of heat production. Fetal movements are commonly felt by mother (quickening).
21–25 weeks	27–35 cm (11–14 in.) 550–800 g (1.25–1.5 lb)	Head becomes even more proportionate to rest of body. Weight gain is substantial, and skin is pink and wrinkled. By 24 weeks, Type II alveolar cells begin to produce surfactant.
26–29 weeks	32–42 cm (13–17 in.) 1110–1350 g (2.5–3 lb)	Head and body are more proportionate and eyes are open. Toenails are visible. Body fat is 3.5% of total body mass and additional subcutaneous fat smoothes out some wrinkles. Testes begin to descend toward scrotum at 28 to 32 weeks. Red bone marrow is major site of blood cell production. Many fetuses born prematurely during this period survive if given intensive care because lungs can provide adequate ventilation and central nervous system is developed enough to control breathing and body temperature.
30–34 weeks	41–45 cm (16.5–18 in.) 2000–2300 g (4.5–5 lb)	Skin is pink and smooth. Fetus assumes upside down position. Pupillary reflex is present by 30 weeks. Body fat is 8% of total body mass. Fetuses 33 weeks and older usually survive if born prematurely.
35–38 weeks	50 cm (20 in.) 3200–3400 g (7–7.5 lb)	By 38 weeks circumference of fetal abdomen is greater than that of head. Skin is usually bluish-pink, and growth slows as birth approaches. Body fat is 16% of total body mass. Testes are usually in scrotum in full-term male infants. Even after birth, an infant is not completely developed; an additional year is required, especially for complete development of the nervous system.

treme nervousness, reduced ability to concentrate, and an inability to appreciate cause-and-effect relationships, are common.

Other teratogens include certain viruses (hepatitis B and C and certain papilloma viruses that cause sexually transmitted diseases); pesticides; defoliants (chemicals that cause plants to shed their leaves prematurely); industrial chemicals; some hormones; antibiotics; oral anticoagulants, anticonvulsants, antitumor agents, thyroid drugs, thalidomide, diethylstilbestrol (DES), and numerous other prescription drugs; LSD; and cocaine. A pregnant woman who uses cocaine, for example, subjects the fetus to higher risk of retarded growth, attention and orientation problems, hyperirritability, a tendency to stop breathing, malformed or missing organs, strokes, and seizures. The risks of spontaneous abortion, premature birth, and stillbirth also increase with fetal exposure to cocaine.

Cigarette Smoking

Strong evidence implicates cigarette smoking during pregnancy as a cause of low infant birth weight; there is also a strong asso-

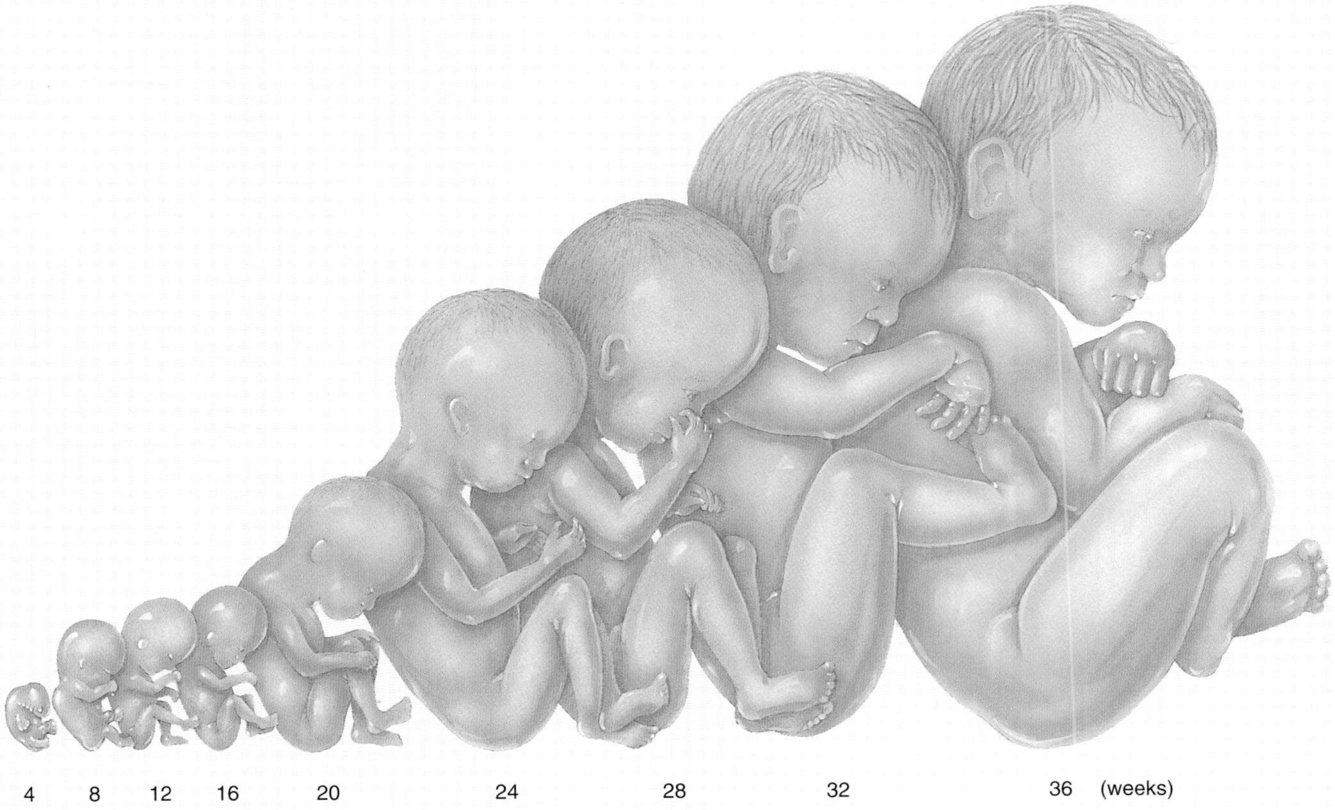

4 8 12 16 20 24 28 32 36 (weeks)

ciation between smoking and a higher fetal and infant mortality rate. Women who smoke have a much higher risk of an ectopic pregnancy. Cigarette smoke may be teratogenic and may cause cardiac abnormalities as well as and anencephaly (see page 1116). Maternal smoking also is a significant factor in the development of cleft lip and palate and has been linked with sudden infant death syndrome (SIDS). Infants nursing from smoking mothers have also been found to have an increased incidence of gastrointestinal disturbances. Even a mother's exposure to secondhand cigarette smoke (breathing air containing tobacco smoke) during pregnancy or while nursing predisposes her baby to increased incidence of respiratory problems, including bronchitis and pneumonia, during the first year of life.

Irradiation

Ionizing radiation of various kinds is a potent teratogen. Exposure of pregnant mothers to x rays or radioactive isotopes during the embryo's susceptible period of development may cause microcephaly (small head size relative to the rest of the

Figure 29.14 **Summary of representative developmental events of the embryonic and fetal periods.** The embryos and fetuses are not shown at their actual sizes.

Development during the fetal period is mostly concerned with the growth and differentiation of tissues and organs formed during the embryonic period.

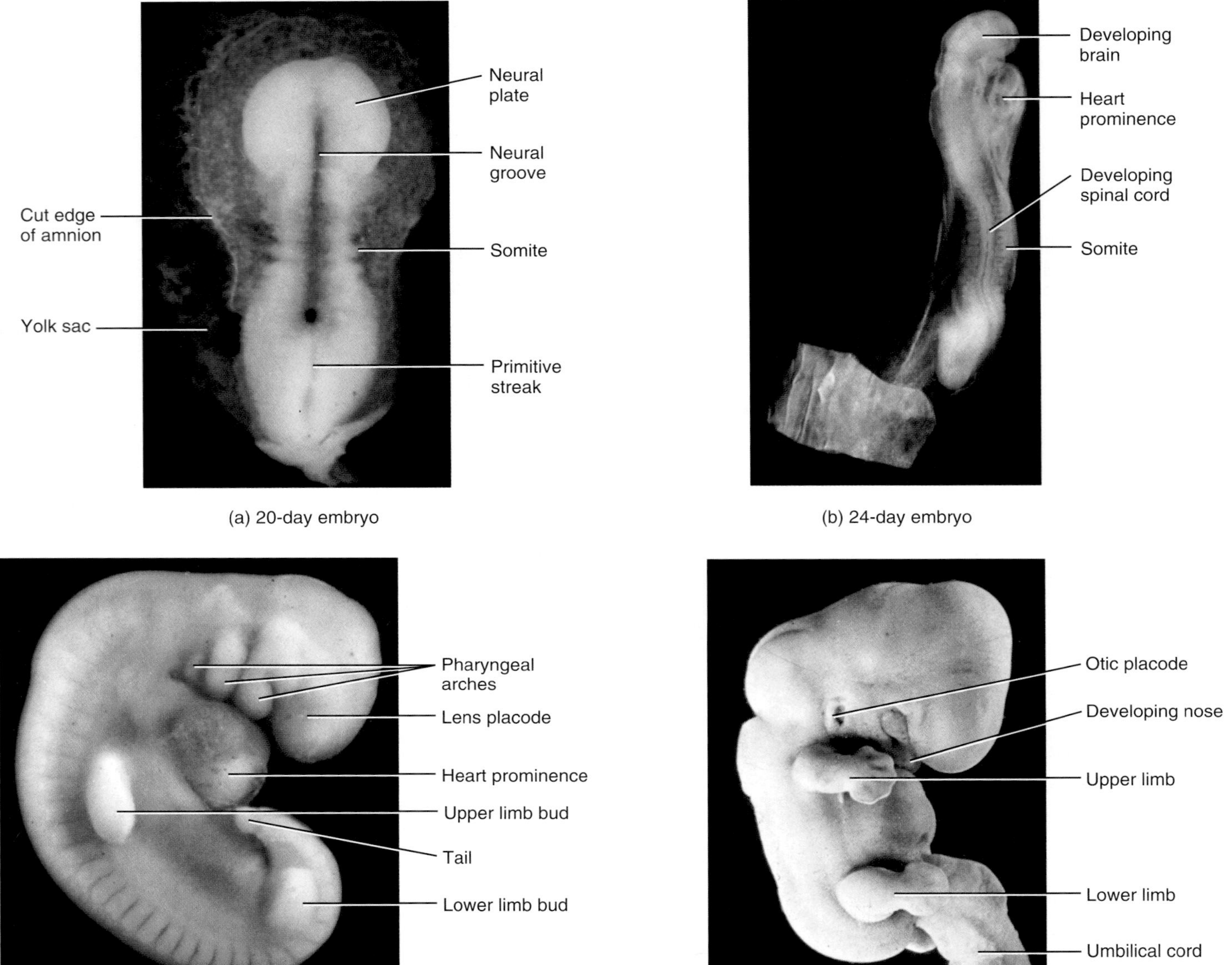

(a) 20-day embryo

(b) 24-day embryo

(c) 32-day embryo

(d) 44-day embryo

body), mental retardation, and skeletal malformations. Caution is advised, especially during the first trimester of pregnancy.

► CHECKPOINT

24. What are some of the symptoms of fetal alcohol syndrome?
25. How does cigarette smoking affect embryonic and fetal development?

PRENATAL DIAGNOSTIC TESTS

► OBJECTIVE

Describe the procedures for fetal ultrasonography, amniocentesis, and chorionic villi sampling.

Several tests are available to detect genetic disorders and assess fetal well-being. Here we describe fetal ultrasonography, amniocentesis, and chorionic villi sampling (CVS).

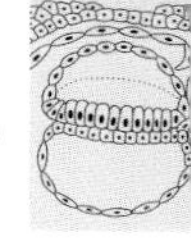

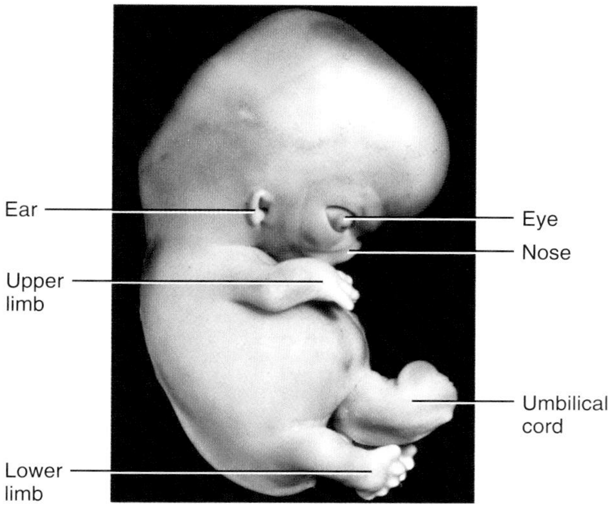

Ear
Eye
Nose
Upper limb
Umbilical cord
Lower limb

(e) 52-day embryo

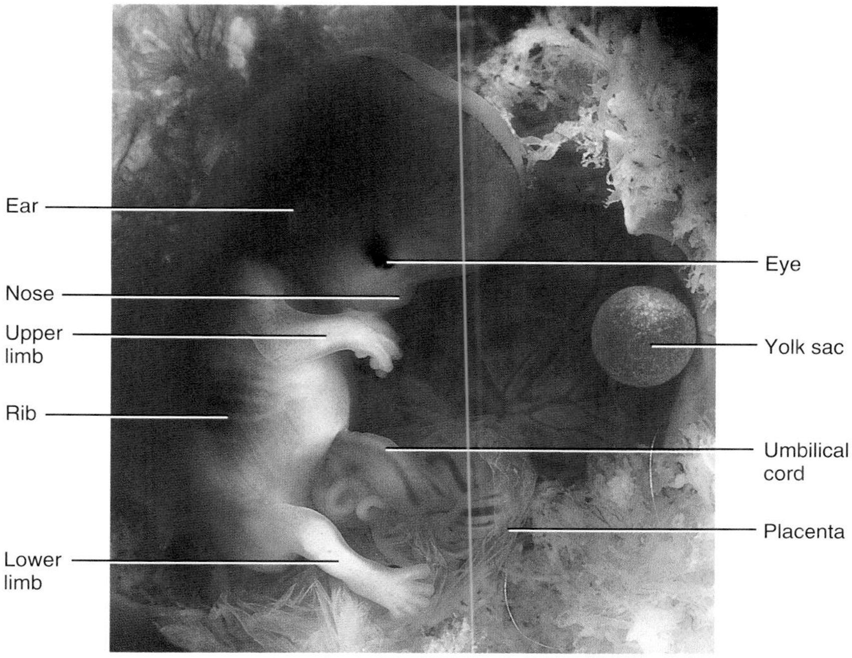

Ear
Eye
Nose
Yolk sac
Upper limb
Rib
Umbilical cord
Placenta
Lower limb

(f) Ten-week fetus

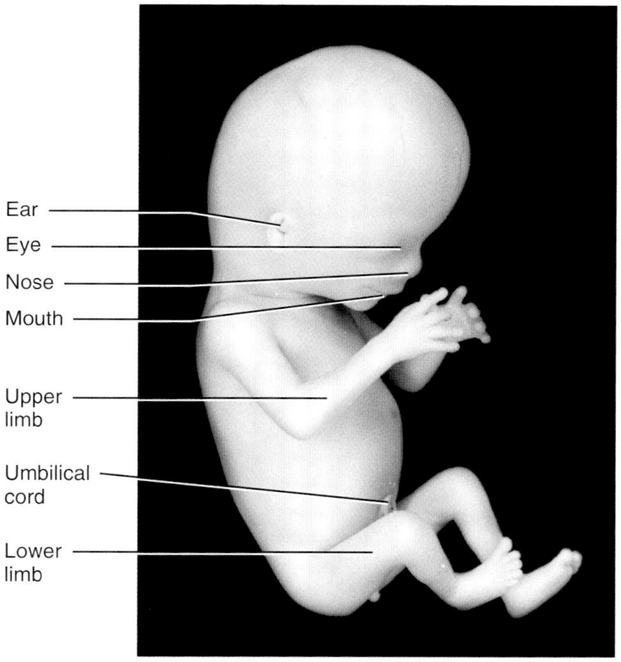

Ear
Eye
Nose
Mouth
Upper limb
Umbilical cord
Lower limb

(g) Thirteen-week fetus

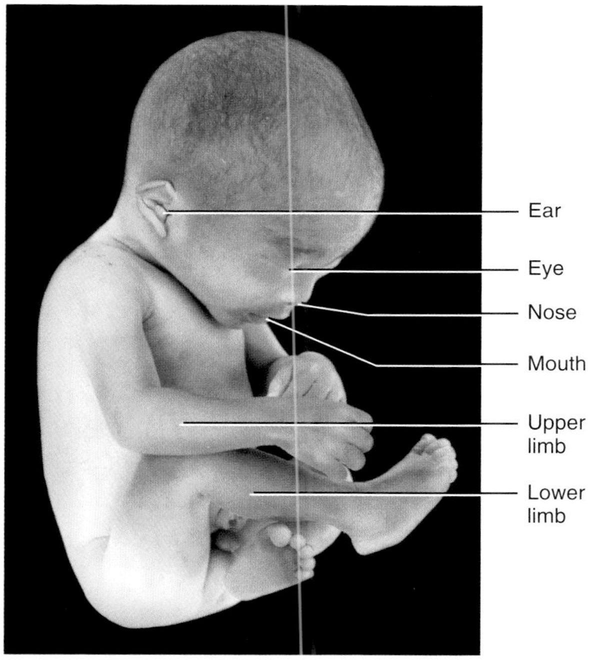

Ear
Eye
Nose
Mouth
Upper limb
Lower limb

(h) Twenty-six-week fetus

How does mid-fetal weight compare to end-fetal weight?

Fetal Ultrasonography

If there is a question about the normal progress of a pregnancy, **fetal ultrasonography** (ul′-tra-son-OG-ra-fē) may be performed. By far the most common use of diagnostic ultrasound is to determine a more accurate fetal age when the date of conception is unclear. It is also used to confirm pregnancy, evaluate fetal viability and growth, determine fetal position, identify multiple pregnancies, identify fetal–maternal abnormalities, and serve as an adjunct to special procedures such as amniocentesis.

During fetal ultrasonography, a transducer, an instrument that emits high-frequency sound waves, is passed back and forth over the abdomen. The reflected sound waves from the developing fetus are picked up by the transducer and converted to an onscreen image called a **sonogram** (see Table 1.3 on page 23). Because the urinary bladder serves as a landmark during the procedure, the patient needs to drink liquids before the procedure and not void urine to maintain a full bladder.

Amniocentesis

Amniocentesis (am′-nē-ō-sen-TĒ-sis; *amnio-* = amnion; *-centesis* = puncture to remove fluid) involves withdrawing some of the amniotic fluid that bathes the developing fetus and analyzing the fetal cells and dissolved substances. It is used to test for the presence of certain genetic disorders, such as Down syndrome (DS), hemophilia, Tay-Sachs disease, sickle-cell disease, and certain muscular dystrophies. It is also used to help determine survivability of the fetus. The test is usually done at 14–18 weeks of gestation. All gross chromosomal abnormalities and over 50 biochemical defects can be detected through amniocentesis. It can also reveal the baby's gender, which is important information for the diagnosis of sex-linked disorders, in which an abnormal gene carried by the mother affects her male offspring only (described on page 1139).

During amniocentesis, the position of the fetus and placenta is first identified using ultrasound and palpation. After the skin is prepared with an antiseptic and a local anesthetic is given, a hypodermic needle is inserted through the mother's abdominal wall and into the amniotic cavity within the uterus. Then, 10 to 30 mL of fluid and suspended cells are aspirated (Figure 29.15a) for microscopic examination and biochemical testing. Elevated levels of alphafetoprotein (AFP) and acetylcholinesterase may indicate failure of the nervous system to develop properly, as occurs in spina bifida or anencephaly (absence of the cerebrum) or may be due to other developmental or chromosomal problems. Chromosome studies, which require growing the cells for 2–4 weeks in a culture medium, may reveal rearranged, missing, or extra chromosomes. Amniocentesis is performed only when a risk for genetic defects is suspected, because there is about a 0.5% chance of spontaneous abortion after the procedure.

Chorionic Villi Sampling

In **chorionic villi sampling** (ko-rē-ON-ik VIL-ī) or **CVS,** a catheter is guided through the vagina and cervix of the uterus and then advanced to the chorionic villi under ultrasound guidance (Figure 29.15b). About 30 milligrams of tissue are suctioned out and prepared for chromosomal analysis. Alternatively, the chorionic villi can be sampled by inserting a needle through the abdominal cavity, as performed in amniocentesis.

CVS can identify the same defects as amniocentesis because chorion cells and fetal cells contain the same genome. CVS offers several advantages over amniocentesis: It can be performed as

Figure 29.15 Amniocentesis and chorionic villi sampling.

To detect genetic abnormalities, amniocentesis is performed at 14–16 weeks of gestation; chorionic villi sampling may be performed as early as 8 weeks of gestation.

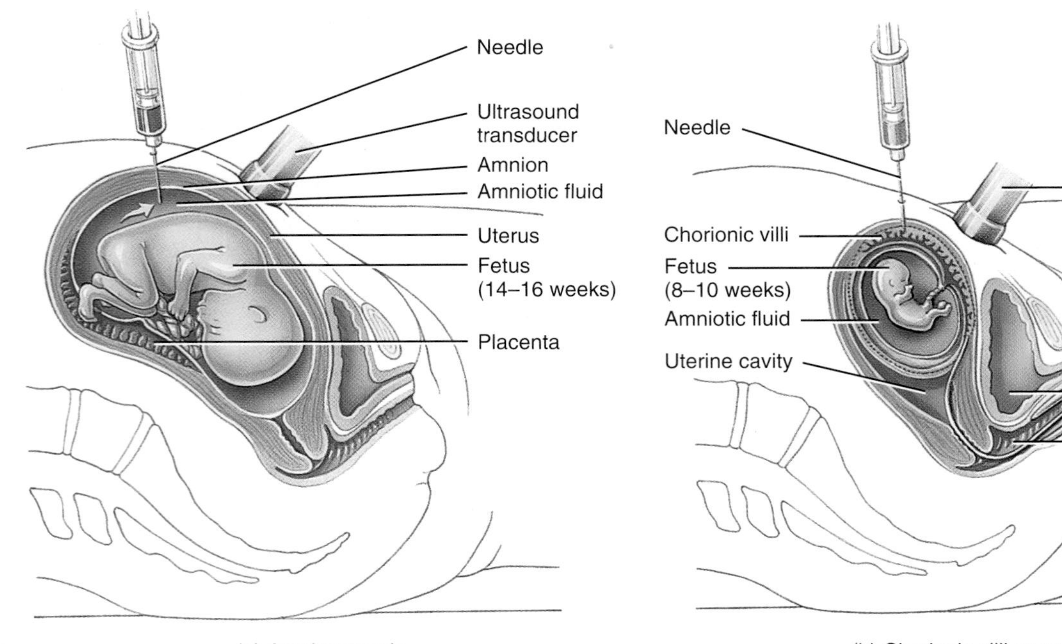

What information can be provided by amniocentesis?

early as eight weeks of gestation, and test results are available in only a few days, permitting an earlier decision on whether to continue the pregnancy. In addition, the procedure does not require penetration of the abdomen, uterus, or amniotic cavity by a needle. However, CVS is slightly riskier than amniocentesis; after the procedure there is a 1–2% chance of spontaneous abortion.

Noninvasive Prenatal Tests

Currently, chorionic villi testing and amniocentesis are the only useful ways to obtain fetal tissue for prenatal testing of gene defects. While these invasive procedures pose relatively little risk when performed by experts, much work has been done to develop **noninvasive prenatal tests,** which do not require the penetration of any embryonic structure. The goal is to develop accurate, safe, more efficient, and less expensive tests for screening a large population.

The first such test developed was the **maternal alpha-fetoprotein (AFP) test.** In this test, the mother's blood is analyzed for the presence of AFP, a protein synthesized in the fetus that passes into the maternal circulation. The highest levels of AFP normally occur during weeks 12 through 15 of pregnancy. Later, AFP is not produced, and its concentration decreases to a very low level both in the fetus and in maternal blood. A high level of AFP after week 16 usually indicates that the fetus has a neural tube defect, such as spina bifida or anencephaly. Because the test is 95% accurate, it is now recommended that all pregnant women be tested for AFP. A newer test (Quad AFP Plus) probes maternal blood for AFP and three other molecules. The test permits prenatal screening for Down syndrome, trisomy 18, and neural tube defects; it also helps predict the delivery date and may reveal the presence of twins.

▶ CHECKPOINT

26. What conditions can be detected using fetal ultrasonography, amniocentesis, and chorionic villi sampling? What are the advantages of noninvasive prenatal tests?

MATERNAL CHANGES DURING PREGNANCY

▶ OBJECTIVES

Describe the sources and functions of the hormones secreted during pregnancy.

Describe the hormonal, anatomical, and physiological changes in the mother during pregnancy.

Hormones of Pregnancy

During the first 3 to 4 months of pregnancy, the corpus luteum in the ovary continues to secrete **progesterone** and **estrogens,** which maintain the lining of the uterus during pregnancy and prepare the mammary glands to secrete milk. The amounts secreted by the corpus luteum, however, are only slightly more than those produced after ovulation in a normal menstrual cycle. From the third month through the remainder of the pregnancy, the placenta itself provides the high levels of progesterone and estrogens required. As noted previously, the chorion secretes **human chorionic gonadotropin (hCG)** into the blood. In turn, hCG stimulates the corpus luteum to continue production of progesterone and estrogens—an activity required to prevent menstruation and for the continued attachment of the embryo and fetus to the lining of the uterus (Figure 29.16a). By the eighth day after fertilization, hCG can be detected in the blood and urine of a pregnant woman. Peak secretion of hCG occurs at about the ninth week of pregnancy (Figure 29.16b). During the fourth and fifth months the hCG level decreases sharply and then levels off until childbirth.

The chorion begins to secrete estrogens after the first 3 or 4 weeks of pregnancy and progesterone by the sixth week. These hormones are secreted in increasing quantities until the time of birth (Figure 29.16b). By the fourth month, when the placenta is fully established, the secretion of hCG is greatly reduced, and the secretions of the corpus luteum are no longer essential. A high level of progesterone ensures that the uterine myometrium is relaxed and that the cervix is tightly closed. After delivery, estrogens and progesterone in the blood decrease to normal levels.

Relaxin, a hormone produced first by the corpus luteum of the ovary and later by the placenta, increases the flexibility of the pubic symphysis and ligaments of the sacroiliac and sacrococcygeal joints and helps dilate the uterine cervix during labor. Both of these actions ease delivery of the baby.

A third hormone produced by the chorion of the placenta is **human chorionic somatomammotropin (hCS),** also known as **human placental lactogen (hPL).** The rate of secretion of hCS increases in proportion to placental mass, reaching maximum levels after 32 weeks and remaining relatively constant after that. It is thought to help prepare the mammary glands for lactation, enhance maternal growth by increasing protein synthesis, and regulate certain aspects of metabolism in both mother and fetus. For example, hCS decreases the use of glucose by the mother and promotes the release of fatty acids from her adipose tissue, making more glucose available to the fetus.

The hormone most recently found to be produced by the placenta is **corticotropin-releasing hormone (CRH),** which in nonpregnant people is secreted only by neurosecretory cells in the hypothalamus. CRH is now thought to be part of the "clock" that establishes the timing of birth. Secretion of CRH by the placenta begins at about 12 weeks and increases enormously toward the end of pregnancy. Women who have higher levels of CRH earlier in pregnancy are more likely to deliver prematurely; those who have low levels are more likely to deliver after their due date. CRH from the placenta has a second important effect: It increases secretion of cortisol, which is needed for maturation of the fetal lungs and the production of surfactant (see page 861).

Figure 29.16 Hormones during pregnancy.

The corpus luteum produces progesterone and estrogens during the first 3–4 months of pregnancy, after which time the placenta assumes this function.

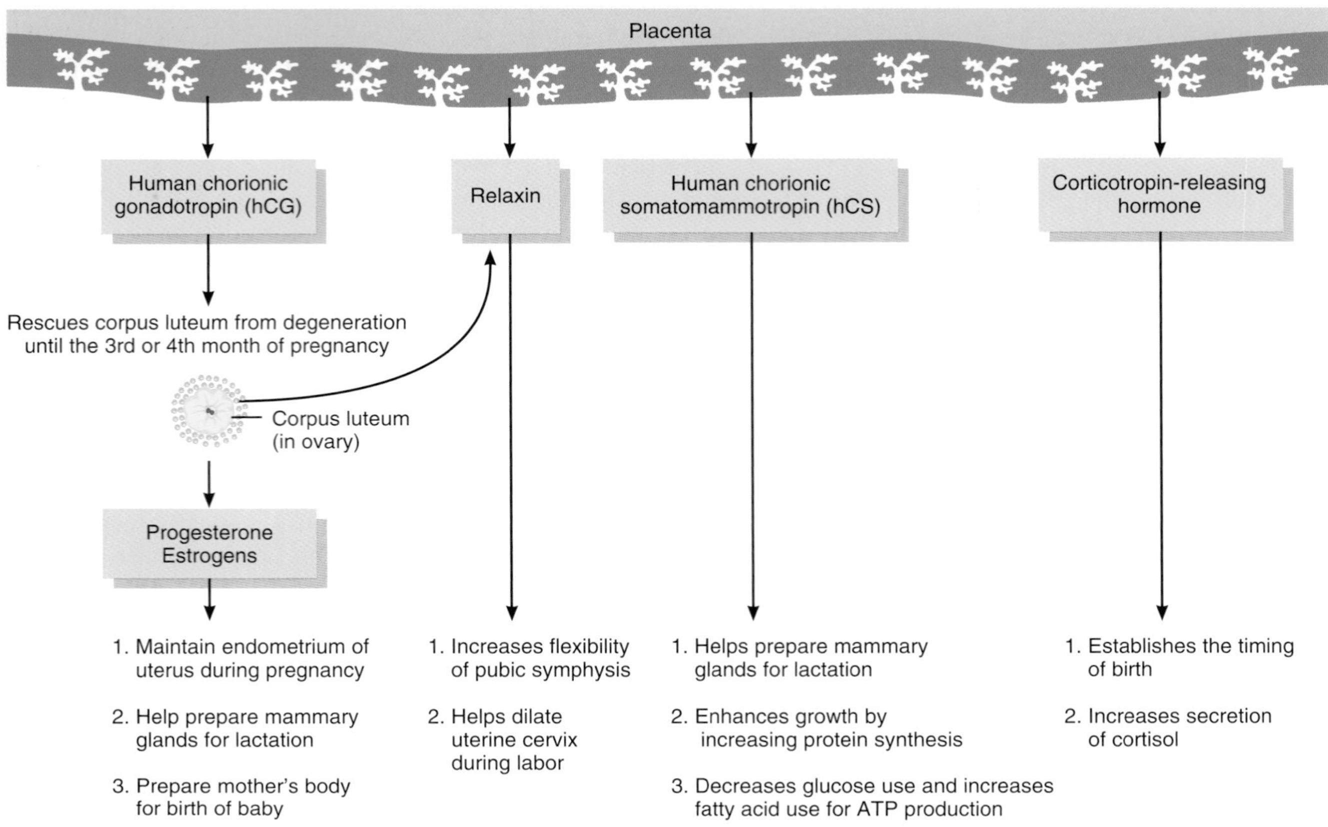

(a) Sources and functions of hormones

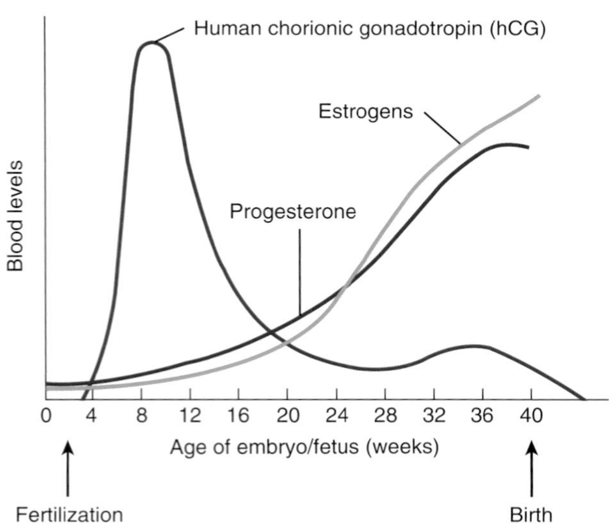

(b) Blood levels of hormones during pregnancy

Which hormone is detected by early pregnancy tests?

Early pregnancy tests detect the tiny amounts of human chorionic gonadotropin (hCG) in the urine that begin to be excreted about 8 days after fertilization. The test kits can detect pregnancy as early as the first day of a missed menstrual period—that is, at about 14 days after fertilization. Chemicals in the kits produce a color change if a reaction occurs between hCG in the urine and hCG antibodies included in the kit.

Several of the test kits available at pharmacies are as sensitive and accurate as test methods used in many hospitals. Still, false-negative and false-positive results can occur. A false-negative result (the test is negative, but the woman is pregnant) may be due to testing too soon or to an ectopic pregnancy. A false-positive result (the test is positive, but the woman is not pregnant) may be due to excess protein or blood in the urine or to hCG production due to a rare type of uterine cancer. Thiazide diuretics, hormones, steroids, and thyroid drugs may also affect the outcome of an early pregnancy test. ■

Changes During Pregnancy

By about the end of the third month of pregnancy, the uterus occupies most of the pelvic cavity. As the fetus continues to grow, the uterus extends higher and higher into the abdominal cavity. Toward the end of a full-term pregnancy, the uterus fills nearly the entire abdominal cavity, reaching above the costal margin nearly to the xiphoid process of the sternum (Figure 29.17). It pushes the maternal intestines, liver, and stomach superiorly, elevates the diaphragm, and widens the thoracic cavity. Pressure on the stomach may force the stomach contents superiorly into the esophagus, resulting in heartburn. In the pelvic cavity, compression of the ureters and urinary bladder occurs.

Pregnancy-induced physiological changes also occur, including weight gain due to the fetus, amniotic fluid, the placenta, uterine enlargement, and increased total body water; increased storage of proteins, triglycerides, and minerals; marked breast enlargement in preparation for lactation; and lower back pain due to lordosis (hollow back).

Several changes occur in the maternal cardiovascular system. Stroke volume increases by about 30% and cardiac output rises by 20–30% due to increased maternal blood flow to the placenta and increased metabolism. Heart rate increases 10–15% and blood volume increases 30–50%, mostly during the second half of pregnancy. These increases are necessary to meet the additional demands of the fetus for nutrients and oxygen. When a pregnant woman is lying on her back, the enlarged uterus may compress the aorta, resulting in diminished blood flow to the uterus. Compression of the inferior vena cava also decreases venous return, which leads to edema in the lower limbs and may produce varicose veins. Compression of the renal artery can lead to renal hypertension.

Figure 29.17 Normal fetal location and position at the end of a full-term pregnancy.

The gestation period is the time interval (about 38 weeks) from fertilization to birth.

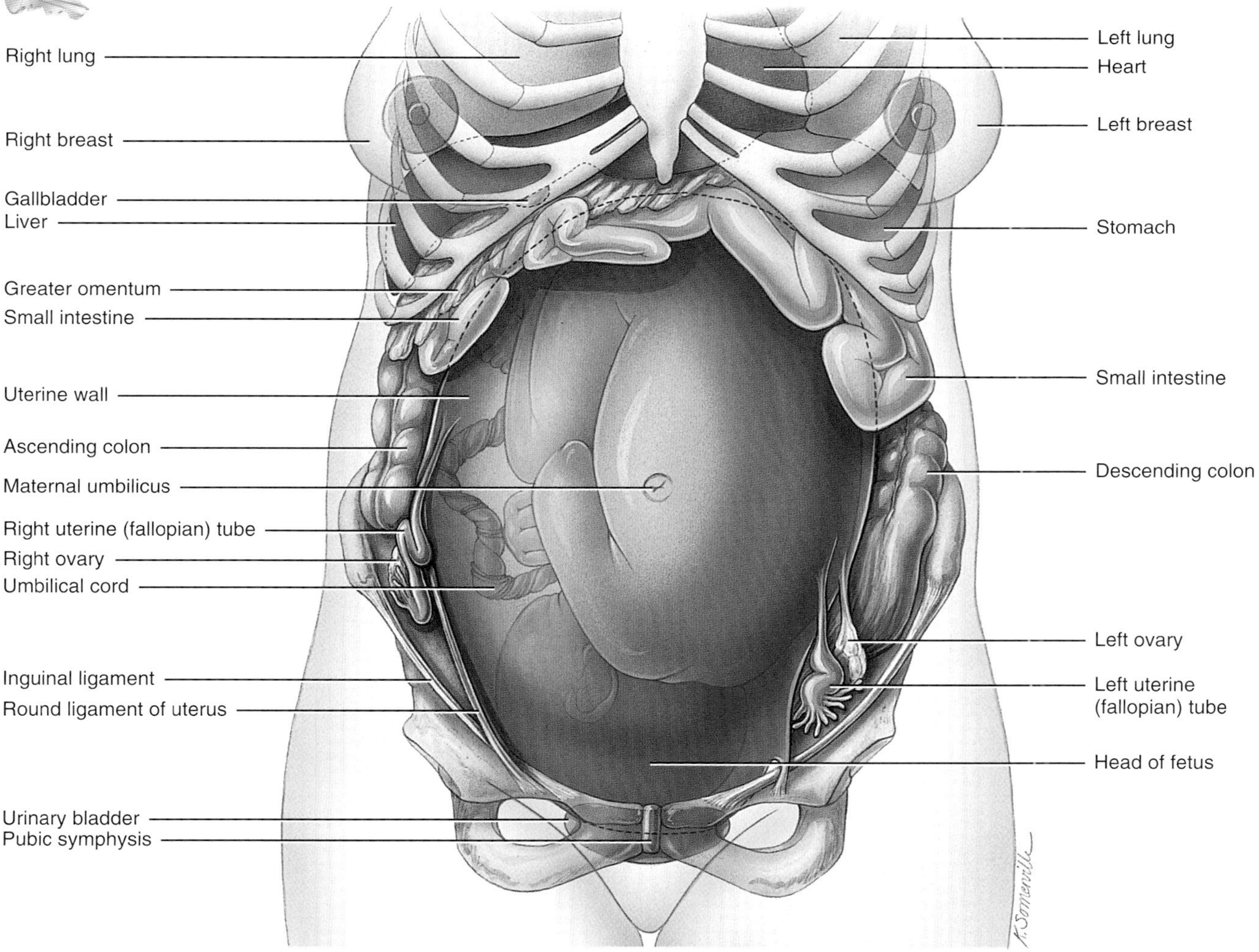

Anterior view of position of organs at end of full-term pregnancy

What hormone increases the flexibility of the pubic symphysis and helps dilate the cervix of the uterus to ease delivery of the baby?

Respiratory function is also altered during pregnancy to meet the added oxygen demands of the fetus. Tidal volume can increase by 30–40%, expiratory reserve volume can be reduced by up to 40%, functional residual capacity can decline by up to 25%, minute ventilation (the total volume of air inhaled and exhaled each minute) can increase by up to 40%, airway resistance in the bronchial tree can decline by 30–40%, and total body oxygen consumption can increase by about 10–20%. Dyspnea (difficult breathing) also occurs.

The digestive system also undergoes changes. Pregnant women experience an increase in appetite due to the added nutritional demands of the fetus. A general decrease in GI tract motility can cause constipation, delay gastric emptying time, and produce nausea, vomiting, and heartburn.

Pressure on the urinary bladder by the enlarging uterus can produce urinary symptoms, such as increased frequency and urgency of urination, and stress incontinence. An increase in renal plasma flow up to 35% and an increase in glomerular filtration rate up to 40% increase the renal filtering capacity, which allows faster elimination of the extra wastes produced by the fetus.

Changes in the skin during pregnancy are more apparent in some women than in others. Some women experience increased pigmentation around the eyes and cheekbones in a masklike pattern (chloasma), in the areolae of the breasts, and in the linea alba of the lower abdomen (linea nigra). Striae (stretch marks) over the abdomen can occur as the uterus enlarges, and hair loss increases.

Changes in the reproductive system include edema and increased vascularity of the vulva and increased pliability and vascularity of the vagina. The uterus increases from its nonpregnant mass of 60–80 g to 900–1200 g at term because of hyperplasia of muscle fibers in the myometrium in early pregnancy and hypertrophy of muscle fibers during the second and third trimesters.

Pregnancy-Induced Hypertension

About 10–15% of all pregnant women in the United States experience **pregnancy-induced hypertension (PIH),** an elevated blood pressure that is associated with pregnancy. The major cause is **preeclampsia** (prē-ē-KLAMP-sē-a), an abnormal condition of pregnancy characterized by sudden hypertension, large amounts of protein in the urine, and generalized edema that typically appears after the 20th week of pregnancy. Other signs and symptoms are generalized edema, blurred vision, and headaches. Preeclampsia might be related to an autoimmune or allergic reaction resulting from the presence of a fetus. Treatment involves bed rest and various drugs. When the condition is also associated with convulsions and coma, it is termed **eclampsia.** ■

▶ CHECKPOINT

27. List the hormones involved in pregnancy, and describe the functions of each.

28. What structural and functional changes occur in the mother during pregnancy?

EXERCISE AND PREGNANCY

▶ OBJECTIVE

Explain the effects of pregnancy on exercise and of exercise on pregnancy.

Only a few changes in early pregnancy affect the ability to exercise. A pregnant woman may tire more easily than usual, or morning sickness may interfere with regular exercise. As the pregnancy progresses, weight is gained and posture changes, so more energy is needed to perform activities, and certain maneuvers (sudden stopping, changes in direction, rapid movements) are more difficult to execute. In addition, certain joints, especially the pubic symphysis, become less stable in response to the increased level of the hormone relaxin. As compensation, many mothers-to-be walk with widely spread legs and a shuffling motion.

Although blood shifts from viscera (including the uterus) to the muscles and skin during exercise, there is no evidence of inadequate blood flow to the placenta. The heat generated during exercise may cause dehydration and further increase body temperature. Especially during early pregnancy, excessive exercise and heat buildup should be avoided because elevated body temperature has been implicated in neural tube defects. Exercise has no known effect on lactation, provided a woman remains hydrated and wears a bra that provides good support. Overall, moderate physical activity does not endanger the fetus of a healthy woman who has a normal pregnancy. However, any physical activity that might endanger the fetus should be avoided.

Among the benefits of exercise to the mother during pregnancy are a greater sense of well-being and fewer physical complaints.

▶ CHECKPOINT

29. Which changes in pregnancy have an effect on the ability to exercise?

LABOR

▶ OBJECTIVE

Explain the events associated with the three stages of labor.

Labor is the process by which the fetus is expelled from the uterus through the vagina, also referred to as giving birth. A synonym for labor is **parturition** (par′-toor-ISH-un; *parturit-* = childbirth).

The onset of labor is determined by complex interactions of several placental and fetal hormones. Because progesterone inhibits uterine contractions, labor cannot take place until its effects are diminished. Toward the end of gestation, the levels of estrogens in the mother's blood rise sharply, producing changes that overcome the inhibiting effects of progesterone. The rise in estrogens results from increasing secretion by the placenta of corticotropin-releasing hormone, which stimulates the anterior pituitary gland of the fetus to secrete ACTH (adrenocorticotropic hormone). In turn, ACTH stimulates the fetal adrenal gland to secrete cortisol and dehydroepiandrosterone (DHEA), the major adrenal androgen. The placenta then converts DHEA into an estrogen. High levels of estrogens cause the number of receptors for oxytocin on uterine muscle fibers to increase, and cause uterine muscle fibers to form gap junctions with one another. Oxytocin released by the posterior pituitary stimulates uterine contractions, and relaxin from the placenta assists by increasing the flexibility of the pubic symphysis and helping dilate the uterine cervix. Estrogen also stimulates the placenta to release prostaglandins, which induce production of enzymes that digest collagen fibers in the cervix, causing it to soften.

Control of labor contractions during parturition occurs via a positive feedback cycle (see Figure 1.4 on page 11). Contractions of the uterine myometrium force the baby's head or body into the cervix, distending (stretching) the cervix. Stretch receptors in the cervix send nerve impulses to neurosecretory cells in the hypothalamus, causing them to release oxytocin into blood capillaries of the posterior pituitary gland. Oxytocin then is carried by the blood to the uterus, where it stimulates the myometrium to contract more forcefully. As the contractions intensify, the baby's body stretches the cervix still more, and the resulting nerve impulses stimulate the secretion of yet more oxytocin. With birth of the infant, the positive feedback cycle is broken because cervical distension suddenly lessens.

Uterine contractions occur in waves (quite similar to the peristaltic waves of the gastrointestinal tract) that start at the top of the uterus and move downward, eventually expelling the fetus. **True labor** begins when uterine contractions occur at regular intervals, usually producing pain. As the interval between contractions shortens, the contractions intensify. Another symptom of true labor in some women is localization of pain in the back that is intensified by walking. The most reliable indicator of true labor is dilation of the cervix and the "show," a discharge of a blood-containing mucus into the cervical canal. In **false labor,** pain is felt in the abdomen at irregular intervals, but it does not intensify and walking does not alter it significantly. There is no "show" and no cervical dilation.

True labor can be divided into three stages (Figure 29.18):

1. ***Stage of dilation.*** The time from the onset of labor to the complete dilation of the cervix is the **stage of dilation.** This stage, which typically lasts 6–12 hours, features regular contractions of the uterus, usually a rupturing of the amniotic sac, and complete dilation (to 10 cm) of the cervix. If the amniotic sac does not rupture spontaneously, it is ruptured intentionally.
2. ***Stage of expulsion.*** The time (10 minutes to several hours) from complete cervical dilation to delivery of the baby is the **stage of expulsion.**
3. ***Placental stage.*** The time (5–30 minutes or more) after delivery until the placenta or "afterbirth" is expelled by

Figure 29.18 Stages of true labor.

The term *parturition* refers to birth.

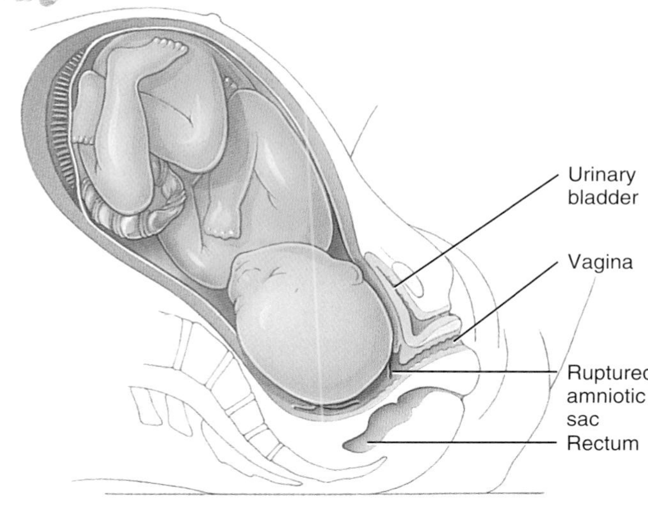

1 Stage of dilation

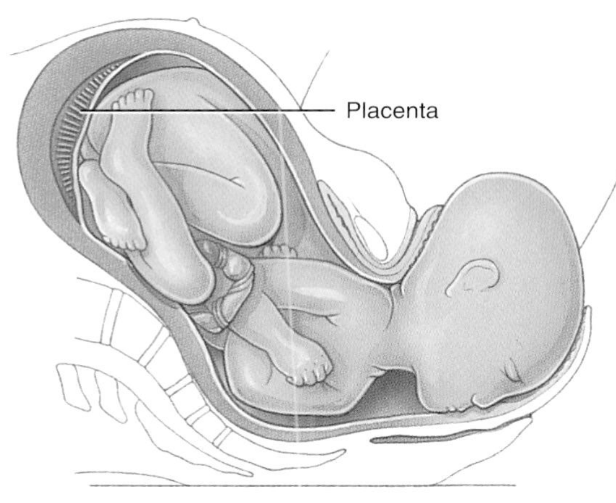

2 Stage of expulsion

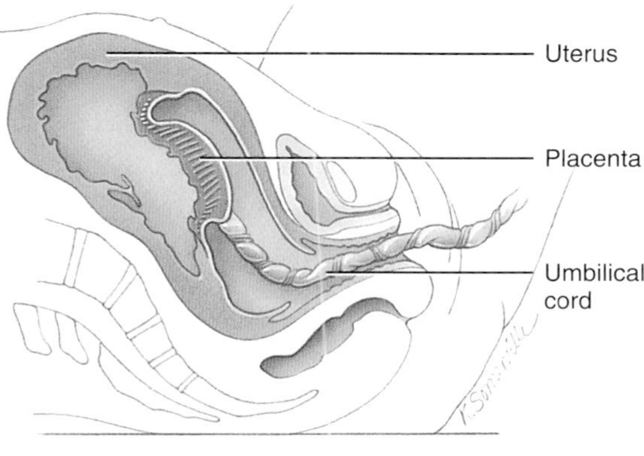

3 Placental stage

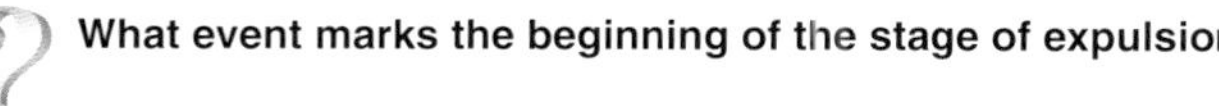

What event marks the beginning of the stage of expulsion?

powerful uterine contractions is the **placental stage.** These contractions also constrict blood vessels that were torn during delivery, reducing the likelihood of hemorrhage.

As a rule, labor lasts longer with first babies, typically about 14 hours. For women who have previously given birth, the average duration of labor is about 8 hours—although the time varies enormously among births. Because the fetus may be squeezed through the birth canal (cervix and vagina) for up to several hours, the fetus is stressed during childbirth: The fetal head is compressed, and the fetus undergoes some degree of intermittent hypoxia due to compression of the umbilical cord and the placenta during uterine contractions. In response to this stress, the fetal adrenal medullae secrete very high levels of epinephrine and norepinephrine, the "fight-or-flight" hormones. Much of the protection against the stresses of parturition, as well as preparation of the infant for surviving extrauterine life, is provided by these hormones. Among other functions, epinephrine and norepinephrine clear the lungs and alter their physiology in readiness for breathing air, mobilize readily usable nutrients for cellular metabolism, and promote an increased flow of blood to the brain and heart.

About 7% of pregnant women do not deliver by 2 weeks after their due date. Such cases carry an increased risk of brain damage to the fetus, and even fetal death, due to inadequate supplies of oxygen and nutrients from an aging placenta. Post-term deliveries may be facilitated by inducing labor, initiated by administration of oxytocin (Pitocin®), or by surgical delivery (cesarean section).

Following the delivery of the baby and placenta is a 6-week period during which the maternal reproductive organs and physiology return to the prepregnancy state. This period is called the **puerperium** (pū′-er-PER-ē-um). Through a process of tissue catabolism, the uterus undergoes a remarkable reduction in size, called **involution** (in′-vō-LOO-shun), especially in lactating women. The cervix loses its elasticity and regains its prepregnancy firmness. For 2–4 weeks after delivery, women have a uterine discharge called **lochia** (LŌ-kē-a), which consists initially of blood and later of serous fluid derived from the former site of the placenta.

Dystocia and Cesarean Section

Dystocia (dis-TŌ-sē-a; *dys-* = painful or difficult; *toc-* = birth), or difficult labor, may result either from an abnormal position (presentation) of the fetus or a birth canal of inadequate size to permit vaginal delivery. In a **breech presentation,** for example, the fetal buttocks or lower limbs, rather than the head, enter the birth canal first; this occurs most often in premature births. If fetal or maternal distress prevents a vaginal birth, the baby may be delivered surgically through an abdominal incision. A low, horizontal cut is made through the abdominal wall and lower portion of the uterus, through which the baby and placenta are removed. Even though it is popularly associated with the birth of Julius Caesar, the true reason this procedure is termed a **cesarean section (C-section)** is because it was described in Roman Law, *lex cesarea,* about 600 years before Julius Caesar was born. Even a history of multiple C-sections need not exclude a pregnant woman from attempting a vaginal delivery. ■

► CHECKPOINT

30. What hormonal changes induce labor?

31. What is the difference between false labor and true labor?

32. What happens during the stage of dilation, the stage of expulsion, and the placental stage of true labor?

ADJUSTMENTS OF THE INFANT AT BIRTH

► OBJECTIVE

Explain the respiratory and cardiovascular adjustments that occur in an infant at birth.

During pregnancy, the embryo (and later the fetus) is totally dependent on the mother for its existence. The mother supplies the fetus with oxygen and nutrients, eliminates its carbon dioxide and other wastes, protects it against shocks and temperature changes, and provides antibodies that confer protection against certain harmful microbes. At birth, a physiologically mature baby becomes much more self-supporting, and the newborn's body systems must make various adjustments. The most dramatic changes occur in the respiratory and cardiovascular systems.

Respiratory Adjustments

The reason that the fetus depends entirely on the mother for obtaining oxygen and eliminating carbon dioxide is that the fetal lungs are either collapsed or partially filled with amniotic fluid. The production of surfactant begins by the end of the sixth month of development. Because the respiratory system is fairly well developed at least 2 months before birth, premature babies delivered at 7 months are able to breathe and cry. After delivery, the baby's supply of oxygen from the mother ceases, and any amniotic fluid in the fetal lungs is absorbed. Because carbon dioxide is no longer being removed, it builds up in the blood. A rising CO_2 level stimulates the respiratory center in the medulla oblongata, causing the respiratory muscles to contract, and the baby to draw his or her first breath. Because the first inspiration is unusually deep, as the lungs contain no air, the baby also exhales vigorously and naturally cries. A full-term baby may breathe 45 times a minute for the first 2 weeks after birth. Breathing rate gradually declines until it approaches a normal rate of 12 breaths per minute.

Cardiovascular Adjustments

After the baby's first inspiration, the cardiovascular system must make several adjustments (see Figure 21.30 on page 794). Closure of the foramen ovale between the atria of the fetal heart, which occurs at the moment of birth, diverts deoxygenated blood to the lungs for the first time. The foramen ovale is closed by two flaps of septal heart tissue that fold together and permanently fuse. The remnant of the foramen ovale is the fossa ovalis.

Once the lungs begin to function, the ductus arteriosus shuts off due to contractions of smooth muscle in its wall, and it becomes the ligamentum arteriosum. The muscle contraction is probably mediated by the polypeptide bradykinin, released from the lungs during their initial inflation. The ductus arteriosus generally does not close completely until about 3 months after birth. Prolonged incomplete closure results in a condition called **patent ductus arteriosus** (see Figure 20.22b on page 730.

After the umbilical cord is tied off and severed and blood no longer flows through the umbilical arteries, they fill with connective tissue, and their distal portions become the medial umbilical ligaments. The umbilical vein then becomes the ligamentum teres (round ligament) of the liver.

In the fetus, the ductus venosus connects the umbilical vein directly with the inferior vena cava, allowing blood from the placenta to bypass the fetal liver. When the umbilical cord is severed, the ductus venosus collapses, and venous blood from the viscera of the fetus flows into the hepatic portal vein to the liver and then via the hepatic vein to the inferior vena cava. The remnant of the ductus venosus becomes the ligamentum venosum.

At birth, an infant's pulse may range from 120 to 160 beats per minute and may go as high as 180 upon excitation. After birth, oxygen use increases, which stimulates an increase in the rate of red blood cell and hemoglobin production. The white blood cell count at birth is very high—sometimes as much as 45,000 cells per microliter—but the count decreases rapidly by the seventh day. Recall that the white blood cell count of an adult is 5000–10,000 cells per microliter.

Premature Infants

Delivery of a physiologically immature baby carries certain risks. A **premature infant** or "preemie" is generally considered a baby who weighs less than 2500 g (5.5 lb) at birth. Poor prenatal care, drug abuse, history of a previous premature delivery, and mother's age below 16 or above 35 increase the chance of premature delivery. The body of a premature infant is not yet ready to sustain some critical functions, and thus its survival is uncertain without medical intervention. The major problem after delivery of an infant under 36 weeks of gestation is respiratory distress syndrome (RDS) of the newborn due to insufficient surfactant. RDS can be eased by use of artificial surfactant and a ventilator that delivers oxygen until the lungs can operate on their own. ■

▶ CHECKPOINT

33. Why are respiratory and cardiovascular adjustments so important at birth?

THE PHYSIOLOGY OF LACTATION

▶ OBJECTIVE

Discuss the physiology and hormonal control of lactation.

Lactation (lak′-TĀ-shun) is the secretion and ejection of milk from the mammary glands. A principal hormone in promoting milk synthesis and secretion is **prolactin (PRL),** which is secreted from the anterior pituitary gland. Even though prolactin levels increase as the pregnancy progresses, no milk secretion occurs because progesterone inhibits the effects of prolactin. After delivery, the levels of estrogens and progesterone in the mother's blood decrease, and the inhibition is removed. The principal stimulus in maintaining prolactin secretion during lactation is the sucking action of the infant. Suckling initiates nerve impulses from stretch receptors in the nipples to the hypothalamus; the impulses decrease hypothalamic release of prolactin-inhibiting hormone (PIH) and increase release of prolactin-releasing hormone (PRH), so more prolactin is released by the anterior pituitary.

Oxytocin causes release of milk into the mammary ducts via the **milk ejection reflex** (Figure 29.19). Milk formed by the glandular cells of the breasts is stored until the baby begins active suckling. Stimulation of touch receptors in the nipple initiates sensory nerve impulses that are relayed to the hypothalamus. In response, secretion of oxytocin from the posterior pituitary increases. Carried by the bloodstream to the mammary glands, oxytocin stimulates contraction of myoepithelial (smooth-muscle-like) cells surrounding the glandular cells and ducts. The resulting compression moves the milk from the alveoli of the mammary glands into the mammary ducts, where it can be suckled. This process is termed **milk ejection (let-down).** Even though the actual ejection of milk does not occur until 30–60 seconds after nursing begins (the latent period), some milk stored in lactiferous sinuses near the nipple is available during the latent period. Stimuli other than suckling, such as hearing a baby's cry or touching the mother's genitals, also can trigger oxytocin release and milk ejection. The suckling stimulation that produces the release of oxytocin also inhibits the release of PIH; this results in increased secretion of prolactin, which maintains lactation.

During late pregnancy and the first few days after birth, the mammary glands secrete a cloudy fluid called **colostrum.** Although it is not as nutritious as milk—it contains less lactose and virtually no fat—colostrum serves adequately until the appearance of true milk on about the fourth day. Colostrum and maternal milk contain important antibodies that protect the infant during the first few months of life.

Figure 29.19 **The milk ejection reflex, a positive feedback cycle.**

Oxytocin stimulates contraction of myoepithelial cells in the breasts, which squeezes the glandular and duct cells and causes milk ejection.

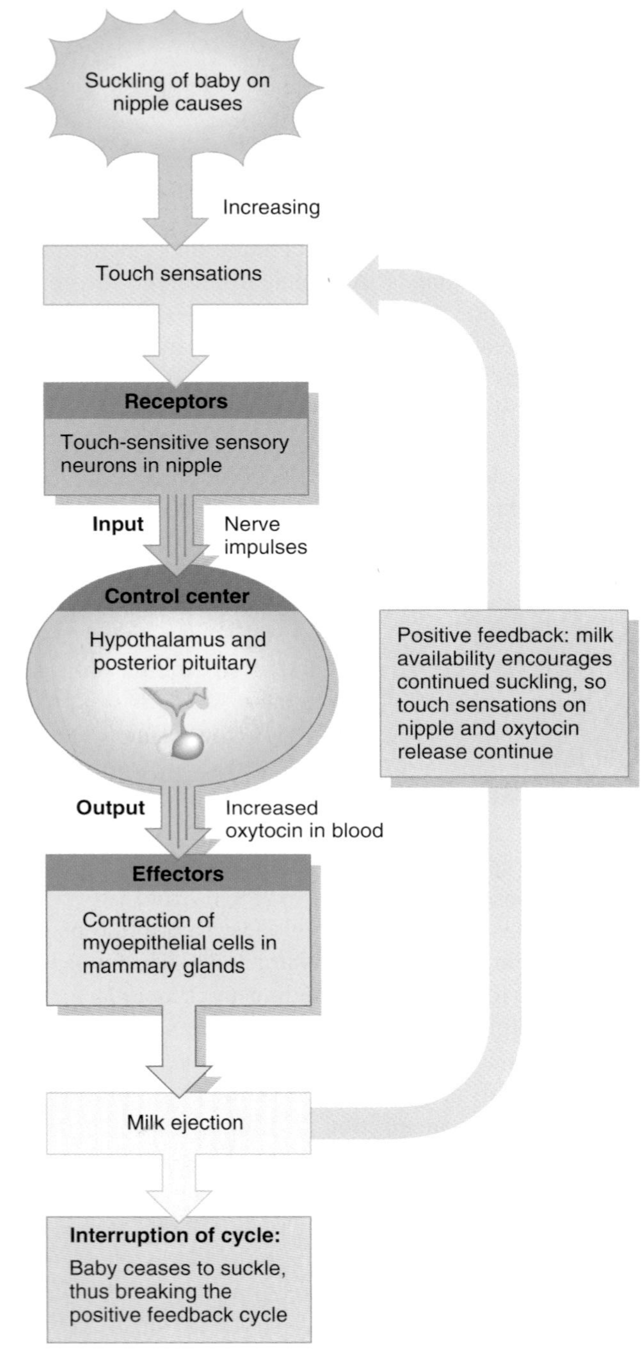

What is another function of oxytocin?

Following birth of the infant, the prolactin level starts to return to the nonpregnant level. However, each time the mother nurses the infant, nerve impulses from the nipples to the hypothalamus increase the release of PRH (and decrease the release of PIH), resulting in a tenfold increase in prolactin secretion by the anterior pituitary that lasts about an hour. Prolactin acts on the mammary glands to provide milk for the next nursing period. If this surge of prolactin is blocked by injury or disease, or if nursing is discontinued, the mammary glands lose their ability to secrete milk in only a few days. Even though milk secretion normally decreases considerably within 7–9 months after birth, it can continue for several years if nursing continues.

Lactation often blocks ovarian cycles for the first few months following delivery, if the frequency of sucking is about 8–10 times a day. This effect is inconsistent, however, and ovulation commonly precedes the first menstrual period after delivery of a baby. As a result, the mother can never be certain she is not fertile. Breast-feeding is therefore an unreliable birth control measure. The suppression of ovulation during lactation is believed to occur as follows: During breast-feeding, neural input from the nipple reaches the hypothalamus and causes it to produce neurotransmitters that suppress the release of gonadotropin-releasing hormone (GnRH). As a result, production of LH and FSH decreases, and ovulation is inhibited.

A primary benefit of **breast-feeding** is nutritional: Human milk is a sterile solution that contains amounts of fatty acids, lactose, amino acids, minerals, vitamins, and water that are ideal for the baby's digestion, brain development, and growth. Breast-feeding also benefits infants by providing the following:

- ***Beneficial cells.*** Several types of white blood cells are present in breast milk. Neutrophils and macrophages serve as phagocytes, ingesting microbes in the baby's gastrointestinal tract. Macrophages also produce lysozyme and other immune system components. Plasma cells, which develop from B lymphocytes, produce antibodies against specific microbes, and T lymphocytes kill microbes directly or help mobilize other defenses.
- ***Beneficial molecules.*** Breast milk also contains an abundance of beneficial molecules. Maternal IgA antibodies in breast milk bind to microbes in the baby's gastrointestinal tract and prevent their migration into other body tissues. Because a mother produces antibodies to whatever disease-causing microbes are present in her environment, her breast milk affords protection against the specific infectious agents to which her baby is also exposed. Additionally, two milk proteins bind to nutrients that many bacteria need to grow and survive: B_{12}-binding protein ties up vitamin B_{12}, and lactoferrin ties up iron. Some fatty acids can kill certain viruses by disrupting their membranes, and lysozyme kills bacteria by disrupting their cell walls. Finally, interferons enhance the antimicrobial activity of immune cells.
- ***Decreased incidence of diseases later in life.*** Breast-feeding provides children with a slight reduction in risk of lymphoma, heart disease, allergies, respiratory and gastrointestinal infections, ear infections, diarrhea, diabetes mellitus, and meningitis. Breast-feeding also protects the mother against osteoporosis and breast cancer.
- ***Miscellaneous benefits.*** Breast-feeding supports optimal infant growth, enhances intellectual and neurological devel-

opment, and fosters mother–infant relations by establishing early and prolonged contact between them. Compared to cow's milk, the fats and iron in breast milk are more easily absorbed, the proteins in breast milk are more readily metabolized, and the lower sodium content of breast milk is more suited to an infant's needs. Premature infants benefit even more from breast-feeding because the milk produced by mothers of premature infants seems to be specially adapted to the infant's needs; it has a higher protein content than the milk of mothers of full-term infants. Finally, a baby is less likely to have an allergic reaction to its mother's milk than to milk from another source.

Years before oxytocin was discovered, it was common practice in midwifery to let a first-born twin nurse at the mother's breast to speed the birth of the second child. Now we know why this practice is helpful—it stimulates the release of oxytocin. Even after a single birth, nursing promotes expulsion of the placenta (afterbirth) and helps the uterus return to its normal size. Synthetic oxytocin (Pitocin) is often given to induce labor or to increase uterine tone and control hemorrhage just after parturition.

▶ CHECKPOINT

34. Which hormones contribute to lactation? What is the function of each?

35. What are the benefits of breast-feeding over bottle-feeding?

INHERITANCE

▶ OBJECTIVE

Define inheritance, and explain the inheritance of dominant, recessive, complex, and sex-linked traits.

As previously indicated, the genetic material of a father and a mother unite when a sperm cell fuses with a secondary oocyte to form a zygote. Children resemble their parents because they inherit traits passed down from both parents. We now examine some of the principles involved in that process, called inheritance.

Inheritance is the passage of hereditary traits from one generation to the next. It is the process by which you acquired your characteristics from your parents and may transmit some of your traits to your children. The branch of biology that deals with inheritance is called **genetics** (je-NET-iks). The area of health care that offers advice on genetic problems (or potential problems) is called **genetic counseling.**

Genotype and Phenotype

As you have already learned, the nuclei of all human cells except gametes contain 23 pairs of chromosomes—the diploid number ($2n$). One chromosome in each pair came from the mother, and the other came from the father. Each of these two homologues contains genes that control the same traits. If one chromosome of the pair contains a gene for body hair, for example, its homologue will contain a gene for body hair in the same position. Alternative forms of a gene that code for the same trait and are at the same location on homologous chromosomes are called **alleles** (ah-LĒLZ). One allele of the previously mentioned body hair gene might code for coarse hair, and another might code for fine hair. A **mutation** (mū-TĀ-shun; *muta-* = change) is a permanent heritable change in an allele that produces a different variant of the same trait.

The relationship of genes to heredity is illustrated by examining the alleles involved in a disorder called **phenylketonuria (PKU).** People with PKU (see page 969) are unable to manufacture the enzyme phenylalanine hydroxylase. The allele that codes for phenylalanine hydroxylase is symbolized as *P*; the mutated allele that fails to produce a functional enzyme is represented by *p*. The chart in Figure 29.20, which shows the possible combinations of gametes from two parents who each have one *P* and one *p* allele, is called a **Punnett square.** In constructing a Punnett square, the possible paternal alleles in sperm are written at the left side and the possible maternal alleles in ova (or secondary oocytes) are written at the top. The four spaces on the

Figure 29.20 **Inheritance of phenylketonuria (PKU).**

Genotype refers to genetic makeup; phenotype refers to the physical or outward expression of a gene.

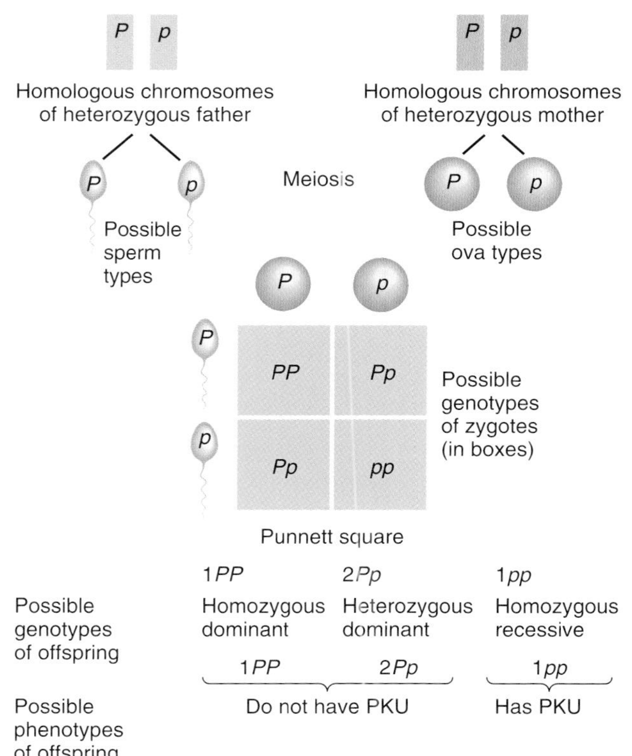

If parents have the genotypes shown here, what is the chance that their first child will have PKU? What is the chance of PKU occurring in their second child?

chart show how the alleles can combine in zygotes formed by the union of these sperm and ova to produce the three different combinations of genes, or **genotypes** (JĒ-nō-tīps): *PP, Pp,* or *pp*. Notice from the Punnett square that 25% of the offspring will have the *PP* genotype, 50% will have the *Pp* genotype, and 25% will have the *pp* genotype. (These percentages are probabilities only; parents who have four children won't necessarily end up with one with PKU.) People who inherit *PP* or *Pp* genotypes do not have PKU; those with a *pp* genotype suffer from the disorder. Although people with a *Pp* genotype have one PKU allele (*p*), the allele that codes for the normal trait (*P*), masks the presence of the PKU allele. An allele that dominates or masks the presence of another allele and is fully expressed (*P* in this example) is said to be a **dominant allele,** and the trait expressed is called a dominant trait. The allele whose presence is completely masked (*p* in this example) is said to be a **recessive allele,** and the trait it controls is called a recessive trait.

By tradition, the symbols for genes are written in italics, with dominant alleles written in capital letters and recessive alleles in lowercase letters. A person with the same alleles on homologous chromosomes (for example, *PP* or *pp*) is said to be **homozygous** for the trait. *PP* is homozygous dominant, and *pp* is homozygous recessive. An individual with different alleles on homologous chromosomes (for example, *Pp*) is said to be **heterozygous** for the trait.

Phenotype (FĒ-nō-tīp; *pheno-* = showing) refers to how the genetic makeup is expressed in the body; it is the physical or outward expression of a gene. A person with *Pp* (a heterozygote) has a different *genotype* from a person with *PP* (a homozygote), but both have the same *phenotype*—normal production of phenylalanine hydroxylase. Heterozygous individuals who carry a recessive gene but do not express it (*Pp*) can pass the gene on to their offspring. Such individuals are called **carriers** of the recessive gene.

Most genes give rise to the same phenotype whether they are inherited from the mother or the father. In a few cases, however, the phenotype is dramatically different, depending on the parental origin. This surprising phenomenon, first appreciated in the 1980s, is called **genomic imprinting.** In humans, the abnormalities most clearly associated with mutation of an imprinted gene are *Angelman syndrome* (mental retardation, ataxia, seizures, and minimal speech), which results when the gene for a particular abnormal trait is inherited from the mother, and *Prader-Willi syndrome* (short stature, mental retardation, obesity, poor responsiveness to external stimuli, and sexual immaturity), which results when it is inherited from the father.

Alleles that code for normal traits do not always dominate over those that code for abnormal ones, but dominant alleles for severe disorders usually are lethal and cause death of the embryo or fetus. One exception is Huntington disease (HD) (see page 564), which is caused by a dominant allele whose effects do not become manifest until adulthood. Both homozygous dominant and heterozygous people exhibit the disease; homozygous recessive people are normal. HD causes progressive degeneration of the nervous system and eventual death, but because symptoms typically do not appear until after age 30 or 40, many afflicted individuals will already have passed the allele for the condition on to their children by the time they discover they have the disease.

Occasionally an error in cell division, called **nondisjunction,** results in an abnormal number of chromosomes. In this situation, homologous chromosomes (during meiosis I) or sister chromatids (during anaphase of mitosis or meiosis II) fail to separate properly. See Figure 3.32 on page 97. A cell from which one or more chromosomes has been added or deleted is called an **aneuploid** (AN-ū-ployd). A monosomic cell ($2n - 1$) is missing a chromosome; a trisomic cell ($2n + 1$) has an extra chromosome. Most cases of Down syndrome (see page 1141) are aneuploid disorders in which there is trisomy of chromosome 21. Nondisjunction usually occurs during gametogenesis (meiosis), but about 2% of Down syndrome cases result from nondisjunction during mitotic divisions in early embryonic development.

Another error in meiosis is a **translocation.** In this case, two chromosomes that are *not* homologous break and interchange portions of their chromosomes. The individual who has a translocation may be perfectly normal if no loss of genetic material took place when the rearrangement occurred. However, some of the person's gametes may not contain the correct amount and type of genetic material. About 3% of Down syndrome cases result from a translocation of part of chromosome 21 to another chromosome, usually chromosome 14 or 15. The individual who has this translocation is normal and does not even know that he or she is a "carrier." When such a carrier produces gametes, however, some gametes end up with a whole chromosome 21 plus another chromosome with the translocated fragment of chromosome 21. Upon fertilization, the zygote then has three, rather than two, copies of that part of chromosome 21.

Table 29.3 lists some dominant and recessive inherited structural and functional traits in humans.

Variations on Dominant–Recessive Inheritance

Most patterns of inheritance do not conform to the simple **dominant–recessive inheritance** we have just described, in which only dominant and recessive alleles interact. The phenotypic expression of a particular gene may be influenced not only by which alleles are present, but also by other genes and by the environment. Most inherited traits are influenced by more than one gene, and, to complicate matters, most genes can influence more than one trait. Variations on dominant–recessive inheritance include incomplete dominance, multiple-allele inheritance, and complex inheritance.

Incomplete Dominance

In **incomplete dominance,** neither member of a pair of alleles is dominant over the other, and the heterozygote has a phenotype intermediate between the homozygous dominant and the homozygous recessive phenotypes. An example of incomplete dominance in humans is the inheritance of **sickle-cell disease**

TABLE 29.3 Selected Hereditary Traits in Humans

Dominant	Recessive
Normal skin pigmentation	Albinism
Near- or farsightedness	Normal vision
PTC taster*	PTC nontaster
Polydactyly (extra digits)	Normal digits
Brachydactyly (short digits)	Normal digits
Syndactylism (webbed digits)	Normal digits
Diabetes insipidus	Normal urine excretion
Huntington disease	Normal nervous system
Widow's peak	Straight hairline
Curved (hyperextended) thumb	Straight thumb
Normal Cl^- transport	Cystic fibrosis
Hypercholesterolemia (familial)	Normal cholesterol level

*Ability to taste a chemical compound called phenylthiocarbamide (PTC).

(SCD) (Figure 29.21). People with the homozygous dominant genotype Hb^AHb^A form normal hemoglobin; those with the homozygous recessive genotype Hb^SHb^S have sickle-cell disease and severe anemia. Although they are usually healthy, those with the heterozygous genotype Hb^AHb^S have minor problems with anemia because half their hemoglobin is normal and half is not. Heterozygotes are carriers, and they are said to have *sickle-cell trait.*

Figure 29.21 Inheritance of sickle-cell disease.

Sickle-cell disease is an example of incomplete dominance.

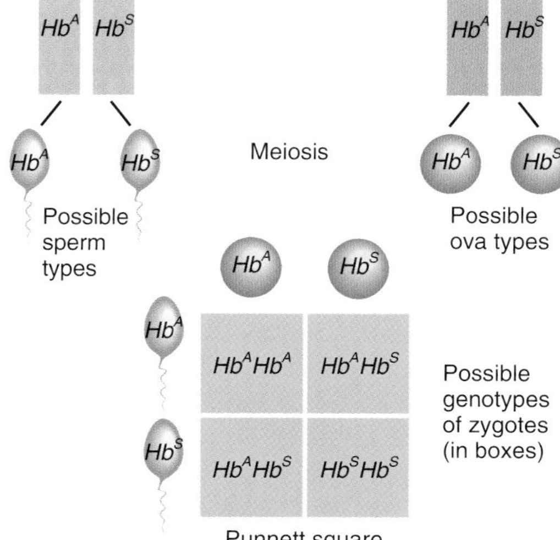

Hb^AHb^A = normal

Hb^AHb^S = carrier of sickle-cell disease

Hb^SHb^S = has sickle-cell disease

What are the distinguishing features of incomplete dominance?

Multiple-Allele Inheritance

Although a single individual inherits only two alleles for each gene, some genes may have more than two alternative forms; this is the basis for **multiple-allele inheritance.** One example of multiple-allele inheritance is the inheritance of the ABO blood group. The four blood types (phenotypes) of the ABO group—A, B, AB, and O—result from the inheritance of six combinations of three different alleles of a single gene called the *I* gene: (1) allele I^A produces the A antigen, (2) allele I^B produces the B antigen, and (3) allele *i* produces neither A nor B antigen. Each person inherits two *I*-gene alleles, one from each parent, that give rise to the various phenotypes. The six possible genotypes produce four blood types, as follows:

Genotype	*Blood type (phenotype)*
I^AI^A or I^Ai	A
I^BI^B or I^Bi	B
I^AI^B	AB
ii	O

Notice that both I^A and I^B are inherited as dominant traits, and *i* is inherited as a recessive trait. Because an individual with type AB blood has characteristics of both type A and type B red blood cells expressed in the phenotype, alleles I^A and I^B are said to be **codominant.** In other words, both genes are expressed equally in the heterozygote. Depending on the parental blood types, different offspring may have blood types different from each other. Figure 29.22 shows the blood types offspring could inherit, given the blood types of their parents.

Figure 29.22 The ten possible combinations of parental ABO blood types and the blood types their offspring could inherit. For each possible set of parents, the blue letters represent the blood types their offspring could inherit.

Inheritance of ABO blood types is an example of multiple-allele inheritance.

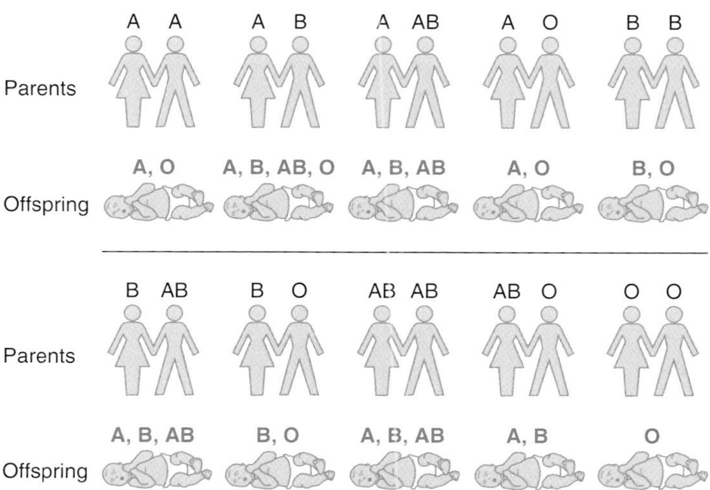

How is it possible for a baby to have type O blood if neither parent is type O?

Complex Inheritance

Most inherited traits are not controlled by one gene, but instead by the combined effects of two or more genes, a situation referred to as **polygenic inheritance** (*poly-* = many), or the combined effects of many genes and environmental factors, a situation referred to as **complex inheritance.** Examples of complex traits include skin color, hair color, eye color, height, metabolism rate, and body build. In complex inheritance, one genotype can have many possible phenotypes, depending on the environment, or one phenotype can include many possible genotypes. For example, even though a person inherits several genes for tallness, full height potential may not be reached due to environmental factors, such as disease or malnutrition during the growth years. You have already learned that the risk of having a child with a neural tube defect is greater in pregnant women who lack adequate folic acid in their diet; this is also considered an environmental factor. Because neural tube defects are more prevalent in some families than in others, however, one or more genes may also contribute.

Often, a complex trait shows a continuous gradation of small differences between extremes among individuals. It is relatively easy to predict the risk of passing on an undesirable trait that is due to a single dominant or recessive gene, but it is very difficult to make this prediction when the trait is complex. Such traits are difficult to follow in a family because the range of variation is large, the number of different genes involved usually is not known, and the impact of environmental factors may be incompletely understood.

Skin color is a good example of a complex trait. It depends on environmental factors such as sun exposure and nutrition, as well as on several genes. Suppose that skin color is controlled by three separate genes, each having two alleles: *A, a*; *B, b*; and *C, c* (Figure 29.23). A person with the genotype *AABBCC* is very dark skinned, an individual with the genotype *aabbcc* is very light skinned, and a person with the genotype *AaBbCc* has an intermediate skin color. Parents having an intermediate skin color may have children with very light, very dark, or intermediate skin color. Note that the **P generation** (parental generation) is the starting generation, the $\mathbf{F_1}$ **generation** (first filial generation) is produced from the P generation, and the $\mathbf{F_2}$ **generation** (second filial generation) is produced from the F_1 generation.

Figure 29.23 Complex inheritance of skin color.

In complex inheritance, a trait is controlled by the combined effects of many genes and environmental factors.

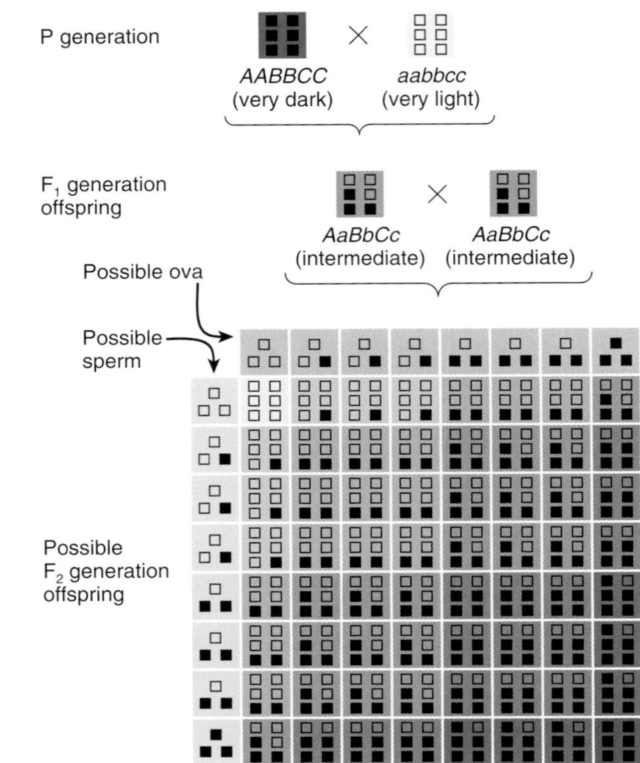

What other traits are transmitted by complex inheritance?

Autosomes, Sex Chromosomes, and Sex Determination

When viewed under a microscope, the 46 human chromosomes in a normal somatic cell can be identified by their size, shape, and staining pattern to be members of 23 different pairs. In 22 of the pairs, the homologous chromosomes look alike and have the same appearance in both males and females; these 22 pairs are called **autosomes.** The two members of the 23rd pair are termed the **sex chromosomes;** they look different in males and females (Figure 29.24). In females, the pair consists of two chromosomes called X chromosomes. One X chromosome is also present in males, but its mate is a much smaller chromosome called a Y chromosome. The Y chromosome has only 231 genes, less than 10% of the 2968 genes present on chromosome 1, the largest autosome.

When a spermatocyte undergoes meiosis to reduce its chromosome number, it gives rise to two sperm that contain an X chromosome and two sperm that contain a Y chromosome. Oocytes have no Y chromosomes and produce only X-containing gametes. If the secondary oocyte is fertilized by an X-bearing sperm, the offspring normally is female (XX). Fertilization by a Y-bearing sperm produces a male (XY). Thus, an individual's sex is determined by the father's chromosomes (Figure 29.25).

Both female and male embryos develop identically until about 7 weeks after fertilization. At that point, one or more genes set into motion a cascade of events that leads to the development of a male; in the absence of normal expression of the gene or genes, the female pattern of development occurs. It has been known since 1959 that the Y chromosome is needed to initiate male development. Experiments published in 1991 established that the prime male-determining gene is one called ***SRY* (sex-determining region of the Y chromosome).** When a small DNA fragment containing this gene was inserted into 11 female mouse embryos, three of them developed as males. (The researchers suspected that the gene failed to be integrated into the genetic material in the other eight.) *SRY* acts as a molecular switch to turn on the male pattern of development. Only if the *SRY* gene is present

Figure 29.24 **Autosomes and sex chromosomes.**

Human somatic cells contain 23 different pairs of chromosomes.

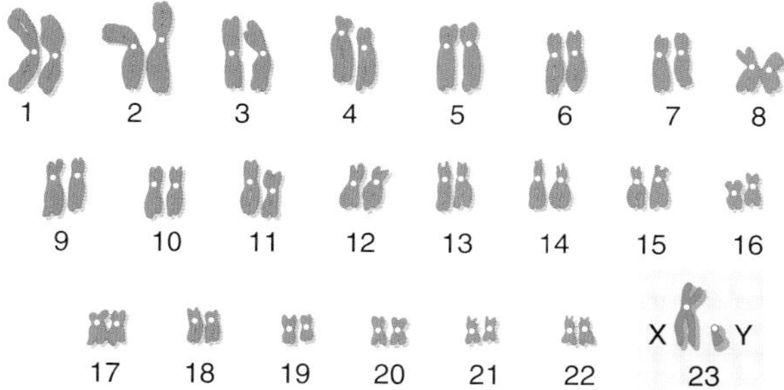

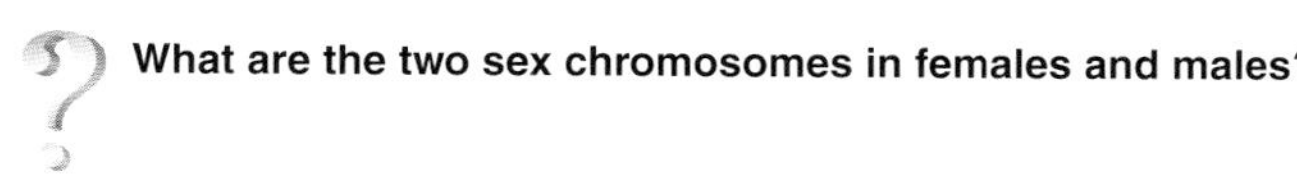

What are the two sex chromosomes in females and males?

and functional in a fertilized ovum will the fetus develop testes and differentiate into a male; in the absence of *SRY*, the fetus will develop ovaries and differentiate into a female.

Case studies have confirmed the key role of *SRY* in directing the male pattern of development in humans. In some cases, phenotypic females with an XY genotype were found to have mutated *SRY* genes. These individuals failed to develop normally as males because their *SRY* gene was defective. In other cases, phenotypic males with an XX genotype were found to have a small piece of the Y chromosome, including the *SRY* gene, inserted into one of their X chromosomes.

Figure 29.25 **Sex determination.**

Sex is determined at the time of fertilization by the presence or absence of a Y chromosome in the sperm.

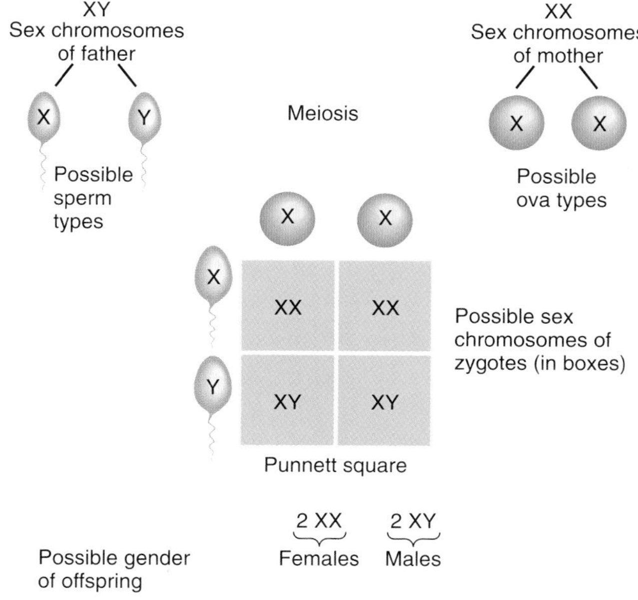

What are all chromosomes other than the sex chromosomes called?

Sex-Linked Inheritance

In addition to determining the sex of the offspring, the sex chromosomes are responsible for the transmission of several nonsexual traits. Many of the genes for these traits are present on X chromosomes but are absent from Y chromosomes. This feature produces a pattern of heredity called **sex-linked inheritance** that is different from the patterns already described.

Red–Green Color Blindness

One example of sex-linked inheritance is **red–green color blindness,** the most common type of color blindness. This condition is characterized by a deficiency in either red- or green-sensitive cones, so red and green are seen as the same color (either red or green, depending on which cone is present). The gene for red–green color blindness is a recessive one designated *c*. Normal color vision, designated *C*, dominates. The *C/c* genes are located only on the X chromosome, so the ability to see colors depends entirely on the X chromosomes. The possible combinations are as follows:

Genotype	*Phenotype*
X^CX^C	Normal female
X^CX^c	Normal female (but a carrier of the recessive gene)
X^cX^c	Red–green color-blind female
X^CY	Normal male
X^cY	Red–green color-blind male

Only females who have two X^c genes are red–green color blind. This rare situation can result only from the mating of a color-blind male and a color-blind or carrier female. Because males do not have a second X chromosome that could mask the trait, all males with an X^c gene will be red–green color blind. Figure 29.26 illustrates the inheritance of red–green color blindness in the offspring of a normal male and a carrier female.

Traits inherited in the manner just described are called **sex-linked traits.** The most common type of **hemophilia**—a condition in which the blood fails to clot or clots very slowly after an injury—is also a sex-linked trait. Like the trait for red–green color blindness, hemophilia is caused by a recessive gene. Other sex-linked traits in humans are fragile X syndrome nonfunctional sweat glands, certain forms of diabetes, some types of deafness, uncontrollable rolling of the eyeballs, absence of central incisors, night blindness, one form of cataract, juvenile glaucoma, and juvenile muscular dystrophy.

X-Chromosome Inactivation

Because they have two X chromosomes in every cell (except developing oocytes), females have a double set of all genes on the X chromosome. A mechanism termed **X-chromosome inactivation (lyonization)** in effect reduces the X-chromosome genes to a single set in females. In each cell of a female's body,

Figure 29.26 An example of the inheritance of red–green color blindness.

Red–green color blindness and hemophilia are examples of sex-linked traits.

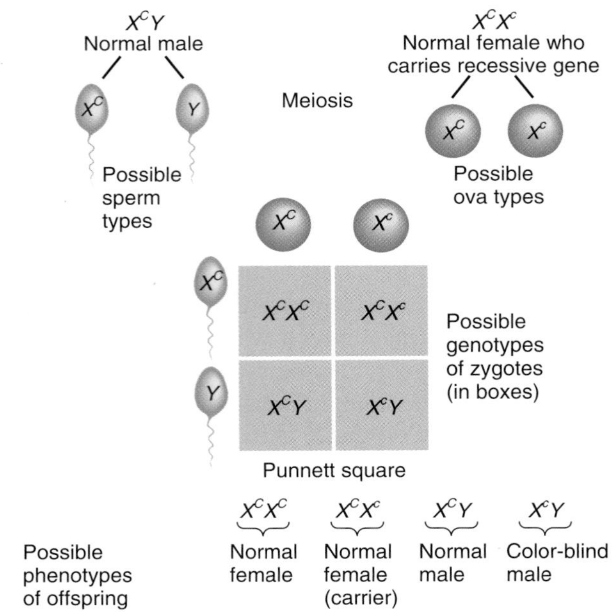

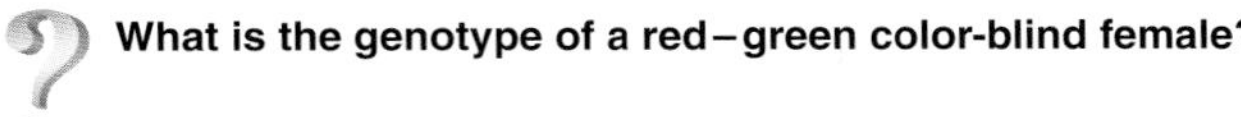

What is the genotype of a red–green color-blind female?

one X chromosome is randomly and permanently inactivated early in development, and most of the genes of the inactivated X chromosome are not expressed (transcribed and translated). The nuclei of cells in female mammals contain a dark-staining body, called a **Barr body,** that is not present in the nuclei of cells in males. Geneticist Mary Lyon correctly predicted in 1961 that the Barr body is the inactivated X chromosome. During inactivation, chemical groups that prevent transcription into RNA are added to the X chromosome's DNA. As a result, an inactivated X chromosome reacts differently to histological stains and has a different appearance than the rest of the DNA. In nondividing (interphase) cells, it remains tightly coiled and can be seen as a dark-staining body within the nucleus. In a blood smear, the Barr body of neutrophils is termed a "drumstick" because it looks like a tiny drumstick-shaped projection of the nucleus.

▶ CHECKPOINT

36. What do the terms genotype, phenotype, dominant, recessive, homozygous, and heterozygous mean?

37. What are genomic imprinting and nondisjunction?

38. Define incomplete dominance. Give an example.

39. What is multiple-allele inheritance? Give an example.

40. Define complex inheritance and give an example.

41. Why does X-chromosome inactivation occur?

DISORDERS: HOMEOSTATIC IMBALANCES

Infertility

Female infertility, or the inability to conceive, occurs in about 10% of all women of reproductive age in the United States. Female infertility may be caused by ovarian disease, obstruction of the uterine tubes, or conditions in which the uterus is not adequately prepared to receive a fertilized ovum. **Male infertility (sterility)** is an inability to fertilize a secondary oocyte; it does not imply erectile dysfunction (impotence). Male fertility requires production of adequate quantities of viable, normal sperm by the testes, unobstructed transport of sperm though the ducts, and satisfactory deposition in the vagina. The seminiferous tubules of the testes are sensitive to many factors—x rays, infections, toxins, malnutrition, and higher-than-normal scrotal temperatures—that may cause degenerative changes and produce male sterility.

One cause of infertility in females is inadequate body fat. To begin and maintain a normal reproductive cycle, a female must have a minimum amount of body fat. Even a moderate deficiency of fat—10% to 15% below normal weight for height—may delay the onset of menstruation (menarche), inhibit ovulation during the reproductive cycle, or cause amenorrhea (cessation of menstruation). Both dieting and intensive exercise may reduce body fat below the minimum amount and lead to infertility that is reversible, if weight gain or reduction of intensive exercise or both occurs. Studies of very obese women indicate that they, like very lean ones, experience problems with amenorrhea and infertility. Males also experience reproductive problems in response to undernutrition and weight loss. For example, they produce less prostatic fluid and reduced numbers of sperm having decreased motility.

Many fertility-expanding techniques now exist for assisting infertile couples to have a baby. The birth of Louise Joy Brown on July 12, 1978, near Manchester, England, was the first recorded case of **in vitro fertilization (IVF)**—fertilization in a laboratory dish. In the IVF procedure, the mother-to-be is given follicle-stimulating hormone (FSH) soon after menstruation, so that several secondary oocytes, rather than the typical single oocyte, will be produced (superovulation). When several follicles have reached the appropriate size, a small incision is made near the umbilicus, and the secondary oocytes are aspirated from the stimulated follicles and transferred to a solution containing sperm, where the oocytes undergo fertilization. Alternatively, an oocyte may be fertilized in vitro by suctioning a sperm or even a spermatid obtained from the testis into a tiny pipette and then injecting it into the oocyte's cytoplasm. This procedure, termed **intracytoplasmic sperm injection (ICSI),** has been used when infertility is due to impairments in sperm motility or to the failure of spermatids to develop into spermatozoa. When the zygote achieved

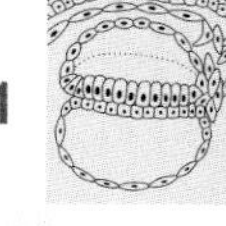

by IVF reaches the 8-cell or 16-cell stage, it is introduced into the uterus for implantation and subsequent growth.

In **embryo transfer,** a man's semen is used to artificially inseminate a fertile secondary oocyte donor. After fertilization in the donor's uterine tube, the morula or blastocyst is transferred from the donor to the infertile woman, who then carries it (and subsequently the fetus) to term. Embryo transfer is indicated for women who are infertile or who do not want to pass on their own genes because they are carriers of a serious genetic disorder.

In **gamete intrafallopian transfer (GIFT)** the goal is to mimic the normal process of conception by uniting sperm and secondary oocyte in the prospective mother's uterine tubes. It is an attempt to bypass conditions in the female reproductive tract that might prevent fertilization, such as high acidity or inappropriate mucus. In this procedure, a woman is given FSH and LH to stimulate the production of several secondary oocytes, which are aspirated from the mature follicles, mixed outside the body with a solution containing sperm, and then immediately inserted into the uterine tubes.

Congenital Defects

An abnormality that is present at birth, and usually before, is called a **congenital defect.** Such defects occur during the formation of structures that develop during the period of organogenesis, the fourth through eighth weeks of development, when all major organs appear. During organogenesis stem cells are establishing the basic patterns of organ development and it is during this time that developing structures are very susceptible to genetic and environmental influences.

Major structural defects occur in 2–3% of liveborn infants and they are the leading cause of infant mortality, accounting for about 21% of infant deaths. Many congenital defects can be prevented by supplementation or avoidance of certain substances. For example, neural tube defects, such as spina bifida and anencephaly, can be prevented by having a pregnant female take folic acid. Iodine supplementation can prevent the mental retardation and bone deformation associated with cretinism. Avoidance of teratogens is also very important in preventing congenital defects.

Down Syndrome

Down syndrome (DS) is a disorder characterized by three, rather than two, copies of at least part of chromosome 21. Overall, one infant in 900 is born with Down syndrome. However, older women are more likely to have a DS baby. The chance of having a baby with this syndrome, which is less than 1 in 3000 for women under age 30, increases to 1 in 300 in the 35–39 age group and to 1 in 9 at age 48.

Down syndrome is characterized by mental retardation, retarded physical development (short stature and stubby fingers), distinctive facial structures (large tongue, flat profile, broad skull, slanting eyes, and round head), kidney defects, suppressed immune system, and malformations of the heart, ears, hands, and feet. Sexual maturity is rarely attained, and life expectancy is shorter.

MEDICAL TERMINOLOGY

Breech presentation A malpresentation in which the fetal buttocks or lower limbs present into the maternal pelvis; the most common cause is prematurity.

Conceptus (kon-SEP-tus) Includes all structures that develop from a zygote and includes an embryo plus the embryonic part of the placenta and associated membranes (chorion, amnion, yolk sac, and allantois).

Cryopreserved embryo (krī-ō-PRĒ-servd; *cryo-* = cold) An early embryo produced by in vitro fertilization (fertilization of a secondary oocyte in a laboratory dish) that is preserved for a long period by freezing it. After thawing, the early embryo is implanted into the uterine cavity. Also called a **frozen embryo.**

Deformation (dē-for-MĀ-shun; *de-* = without; *-forma* = form) A developmental abnormality due to mechanical forces that mold a part of the fetus over a prolonged period of time. Deformations usually involve the skeletal and/or muscular system and may be corrected after birth. An example is clubfeet.

Emesis gravidarum (EM-e-sis gra-VID-ar-um; *emeo* = to vomit; *gravida* = a pregnant woman) Episodes of nausea and possibly vomiting that are most likely to occur in the morning during the early weeks of pregnancy; also called **morning sickness.** Its cause is unknown, but the high levels of human chorionic gonadotropin (hCG) secreted by the placenta, and of progesterone secreted by the ovaries, have been implicated. If the severity of these symptoms requires hospitalization for intravenous feeding, the condition is known as **hyperemesis gravidarum.**

Epigenesis (ep-i-GEN-e-sis; *epi-* = upon; *-genesis* = creation) The development of an organism from an undifferentiated cell.

Fertilization age Two weeks less than the gestational age since a secondary oocyte is not fertilized until about two weeks after the last normal menstrual period (LNMP).

Fetal surgery A surgical procedure performed on a fetus; in some cases the uterus is opened and the fetus is operated on directly. Fetal surgery has been used to repair diaphragmatic hernias and remove lesions in the lungs.

Gestational age (jes-TĀ-shun-al; *gestatus* = to bear) The age of an embryo or fetus calculated from the presumed first day of the last normal menstrual period (LNMP).

Karyotype (KAR-ē-ō-tīp; *karyo-* = nucleus) The chromosomal characteristics of an individual presented as a systematic arrangement of pairs of metaphase chromosomes arrayed in descending order of size and according to the position of the centromere (see Figure 29.24); useful in judging whether chromosomes are normal in number and structure.

Klinefelter's syndrome A sex chromosome aneuploidy, usually due to trisomy XXY, that occurs once in every 500 births. Such individuals are somewhat mentally disadvantaged, sterile males with undeveloped testes, scant body hair, and enlarged breasts.

Lethal gene (LĒ-thal jēn; *lethum* = death) A gene that, when expressed, results in death either in the embryonic state or shortly after birth.

Metafemale syndrome A sex chromosome aneuploidy characterized by at least three X chromosomes (XXX) that occurs about once in every 700 births. These females have underdeveloped genital organs and limited fertility, and most are mentally retarded.

Primordium (prī-MOR-dē-um; *primus-* = first; *-ordior* = to begin) The beginning or first discernible indication of the development of an organ or structure.

Puerperal fever (pū-ER-per-al; *puer* = child) An infectious disease of childbirth, also called puerperal sepsis and childbed fever. The disease, which results from an infection originating in the birth canal, affects the mother's endometrium. It may spread to other pelvic structures and lead to septicemia.

Turner's syndrome A sex chromosome aneuploidy caused by the presence of a single X chromosome (designated XO); occurring about once in every 5000 births, it produces a sterile female with virtually no ovaries and limited development of secondary sex characteristics. Other features include short stature, webbed neck, underdeveloped breasts, and widely spaced nipples. Intelligence usually is normal.

STUDY OUTLINE

EMBRYONIC PERIOD (p. 1105)

1. Pregnancy is a sequence of events that begins with fertilization, and proceeds to implantation, embryonic development, and fetal development. It normally ends in birth.
2. During fertilization a sperm cell penetrates a secondary oocyte and their pronuclei unite. Penetration of the zona pellucida is facilitated by enzymes in the sperm's acrosome. The resulting cell is a zygote.
3. Normally, only one sperm cell fertilizes a secondary oocyte because of the fast and slow blocks to polyspermy.
4. Early rapid cell division of a zygote is called cleavage, and the cells produced by cleavage are called blastomeres. The solid sphere of cells produced by cleavage is a morula.
5. The morula develops into a blastocyst, a hollow ball of cells differentiated into a trophoblast and an inner cell mass.
6. The attachment of a blastocyst to the endometrium is termed implantation; it occurs as a result of enzymatic degradation of the endometrium.
7. After implantation, the endometrium becomes modified and is known as the decidua.
8. The trophoblast develops into the syncytiotrophoblast and cytotrophoblast, both of which become part of the chorion.
9. The inner cell mass differentiates into hypoblast and epiblast, the bilaminar (two-layered) embryonic disc.
10. The amnion is a thin protective membrane that develops from the cytotrophoblast.
11. The exocoelomic membrane and hypoblast form the yolk sac, which transfers nutrients to the embryo, forms blood cells, produces primordial germ cells, and forms part of the gut.
12. Erosion of sinusoids and endometrial glands provides blood and secretions, which enter lacunar networks to supply nutrition to and remove wastes from the embryo.
13. The extraembryonic coelom forms within extraembryonic mesoderm.
14. The extraembryonic mesoderm and trophoblast form the chorion, the principal embryonic part of the placenta.
15. The third week of development is characterized by gastrulation, the conversion of the bilaminar disc into a trilaminar (three-layered) embryo consisting of ectoderm, mesoderm, and endoderm.
16. The first evidence of gastrulation is formation of the primitive streak, after which the primitive node, notochordal process, and notochord develop.
17. The three primary germ layers form all tissues and organs of the developing organism. Table 29.1 summarizes the structures that develop from the primary germ layers.
18. Also during the third week, the oropharyngeal and cloacal membranes form. The wall of the yolk sac forms a small vascularized outpouching called the allantois, which functions in blood formation and development of the urinary bladder.
19. The process by which the neural plate, neural folds, and neural tube form is called neurulation. The brain and spinal cord develop from the neural tube.
20. Paraxial mesoderm segments to form somites from which skeletal muscles of the neck, trunk, and limbs develop. Somites also form connective tissues and vertebrae.
21. Blood vessel formation, called angiogenesis, begins in mesodermal cells called angioblasts.
22. The heart forms from mesodermal cells called the cardiogenic area. By the end of the third week, the primitive heart beats and circulates blood.
23. Chorionic villi, projections of the chorion, connect to the embryonic heart so that maternal and fetal blood vessels are brought into close proximity, allowing the exchange of nutrients and wastes between maternal and fetal blood.
24. Placentation refers to formation of the placenta, the site of exchange of nutrients and wastes between the mother and fetus. The placenta also functions as a protective barrier, stores nutrients, and produces several hormones to maintain pregnancy.
25. The actual connection between the placenta and embryo (and later the fetus) is the umbilical cord.
26. Organogenesis refers to the formation of body organs and systems and occurs during the fourth week of development.
27. Conversion of the flat, two-dimensional trilaminar embryonic disc to a three-dimensional cylinder occurs by a process called embryonic folding.
28. Embryonic folding brings various organs into their final adult positions and helps form the gastrointestinal tract.
29. Pharyngeal arches, clefts, and pouches give rise to the structures of the head and neck.
30. By the end of the fourth week, upper and lower limb buds develop, and by the end of the eighth week the embryo has clearly human features.

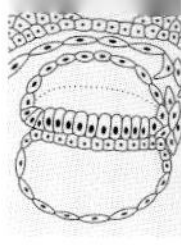

FETAL PERIOD (p. 1121)

1. The fetal period is primarily concerned with the growth and differentiation of tissues and organs that developed during the embryonic period.
2. The rate of body growth is remarkable, especially during the ninth and sixteenth weeks.
3. The principal changes associated with embryonic and fetal growth are summarized in Table 29.2.

TERATOGENS (p. 1121)

1. Teratogens are agents that cause physical defects in developing embryos.
2. Among the more important teratogens are alcohol, pesticides, industrial chemicals, some prescription drugs, cocaine, LSD, nicotine, and ionizing radiation.

PRENATAL DIAGNOSTIC TESTS (p. 1124)

1. Several prenatal diagnostic tests are used to detect genetic disorders and to assess fetal well-being. These include fetal ultrasonography, in which an image of a fetus is displayed on a screen; amniocentesis, the withdrawal and analysis of amniotic fluid and the fetal cells within it; and chorionic villi sampling (CVS), which involves withdrawal of chorionic villi tissue for chromosomal analysis.
2. CVS can be done earlier than amniocentesis, and the results are available more quickly, but it is also slightly riskier than amniocentesis.
3. Noninvasive prenatal tests include the maternal alpha-fetoprotein (AFP) test to detect neural tube defects and the Quad AFP Plus test to detect Down syndrome, trisomy 18, and neural tube defects.

MATERNAL CHANGES DURING PREGNANCY (p. 1127)

1. Pregnancy is maintained by human chorionic gonadotropin (hCG), estrogens, and progesterone.
2. Human chorionic somatomammotropin (hCS) contributes to breast development, protein anabolism, and catabolism of glucose and fatty acids.
3. Relaxin increases flexibility of the pubic symphysis and helps dilate the uterine cervix near the end of pregnancy.
4. Corticotropin-releasing hormone, produced by the placenta, is thought to establish the timing of birth, and stimulates the secretion of cortisol by the fetal adrenal gland.
5. During pregnancy, several anatomical and physiological changes occur in the mother.

EXERCISE AND PREGNANCY (p. 1130)

1. During pregnancy, some joints become less stable, and certain physical activities are more difficult to execute.
2. Moderate physical activity does not endanger the fetus in a normal pregnancy.

LABOR (p. 1130)

1. Labor is the process by which the fetus is expelled from the uterus through the vagina to the outside. True labor involves dilation of the cervix, expulsion of the fetus, and delivery of the placenta.
2. Oxytocin stimulates uterine contractions via a positive feedback cycle.

ADJUSTMENTS OF THE INFANT AT BIRTH (p. 1132)

1. The fetus depends on the mother for oxygen and nutrients, the removal of wastes, and protection.
2. Following birth, an infant's respiratory and cardiovascular systems undergo changes to enable them to become self-supporting during postnatal life.

THE PHYSIOLOGY OF LACTATION (p. 1133)

1. Lactation refers to the production and ejection of milk by the mammary glands.
2. Milk production is influenced by prolactin (PRL), estrogens, and progesterone.
3. Milk ejection is stimulated by oxytocin.
4. A few of the many benefits of breast-feeding include ideal nutrition for the infant, protection from disease, and decreased likelihood of developing allergies.

INHERITANCE (p. 1135)

1. Inheritance is the passage of hereditary traits from one generation to the next.
2. The genetic makeup of an organism is called its genotype; the traits expressed are called its phenotype.
3. Dominant genes control a particular trait; expression of recessive genes is masked by dominant genes.
4. Many patterns of inheritance do not conform to the simple dominant–recessive patterns.
5. In incomplete dominance, neither member of an allelic pair dominates; phenotypically, the heterozygote is intermediate between the homozygous dominant and the homozygous recessive.
6. In multiple-allele inheritance, genes have more than two alternative forms. An example is the inheritance of ABO blood groups.
7. In complex inheritance, a trait such as skin or eye color is controlled by the combined effects of two or more genes and may be influenced by environmental factors.
8. Each somatic cell has 46 chromosomes—22 pairs of autosomes and 1 pair of sex chromosomes.
9. In females, the sex chromosomes are two X chromosomes; in males, they are one X chromosome and a much smaller Y chromosome, which normally includes the prime male-determining gene, called *SRY*.
10. If the *SRY* gene is present and functional in a fertilized ovum, the fetus will develop testes and differentiate into a male. In the absence of *SRY*, the fetus will develop ovaries and differentiate into a female.
11. Red–green color blindness and hemophilia result from recessive genes located on the X chromosome. These sex-linked traits occur primarily in males because of the absence of any counterbalancing dominant genes on the Y chromosome.
12. A mechanism termed X-chromosome inactivation (lyonization) balances the difference in number of X chromosomes between males (one X) and females (two Xs). In each cell of a female's body, one X chromosome is randomly and permanently inactivated early in development and becomes a Barr body.
13. A given phenotype is the result of the interactions of genotype and the environment.

SELF-QUIZ QUESTIONS

Fill in the blanks in the following statements.

1. The three stages of true labor, in order of occurrence, are ____, ____, and ____.
2. Hormones produced by the ____ are responsible for maintaining the pregnancy during the first 3–4 months. The secretion responsible for preventing degeneration of the corpus luteum is ____ produced by the trophoblast.
3. Indicate the germ layers responsible for development of the following structures: (a) muscle, bone, and peritoneum: ____; (b) nervous system and epidermis: ____; (c) epithelial linings of respiratory and gastrointestinal tracts: ____.

Indicate whether the following statement is true or false.

4. Labor is an example of a negative feedback cycle that ends with the birth of the infant.

Choose the one best answer to the following questions.

5. Which of the following are *true*? (1) During implantation the outer cell mass of the blastocyst orients toward the endometrium. (2) The decidua basalis provides glycogen and lipids for the developing fetus. (3) The decidua parietalis becomes the maternal part of the placenta. (4) During implantation, the syncytiotrophoblast secretes enzymes that allow the blastocyst to penetrate the uterine lining. (5) After fetal delivery, the decidua separates from the endometrium and is released from the uterus. (a) 2, 4, and 5; (b) 1, 2, and 3; (c) 2, 3, 4, and 5; (d) 1, 2, 3, 4, and 5; (e) 1, 3, and 5.
6. Which of the following are maternal changes that occur during pregnancy? (1) altered pulmonary function; (2) increased stroke volume, cardiac output, and heart rate, and decreased blood volume; (3) weight gain; (4) increased gastric motility, causing a delay in gastric emptying time; (5) edema and possible varicose veins. (a) 1, 2, 3, and 4; (b) 2, 3, 4, and 5; (c) 1, 3, 4, and 5; (d) 1, 3, and 5; (e) 2, 4, and 5.
7. Which of the following statements is *correct*? (a) Normal traits always dominate over abnormal traits. (b) Occasionally an error in meiosis called nondisjunction results in an abnormal number of chromosomes. (c) The mother always determines the sex of the child because she has either an X or Y gene in her oocytes. (d) Most patterns of inheritance are simple dominant-recessive inheritances. (e) Genes are expressed normally regardless of any outside influence such as chemicals or radiation.
8. Which of the following are *true* concerning fertilization? (1) The sperm first penetrate the zona pellucida and then the corona radiata. (2) The binding of specific membrane proteins in the sperm head to ZP3 causes the release of acrosomal contents. (3) Sperm are able to fertilize the oocyte within minutes after ejaculation. (4) Depolarization of the cell membrane of the secondary oocyte inhibits fertilization by more than one sperm. (5) The oocyte completes meiosis II after fertilization. (a) 1, 2, 4, and 5; (b) 1, 3, and 5; (c) 1, 2, 3, and 4; (d) 1, 4, and 5; (e) 2, 4, and 5.
9. Amniotic fluid (1) is derived entirely from a filtrate of maternal blood, (2) acts as a fetal shock absorber, (3) provides nutrients to the fetus, (4) helps regulate fetal body temperature, (5) prevents adhesions between the skin of the fetus and surrounding tissues. (a) 1, 2, 3, 4, and 5; (b) 2, 4, and 5; (c) 2, 3, 4, and 5; (d) 1, 4, and 5; (e) 1, 2, 4, and 5.
10. Which of the following structures develop during the fourth week after fertilization? (1) embryonic folding, (2) the neural tube, (3) otic placode (beginning of the ear), (4) beginning of the eyes, (5) upper and lower limb buds. (a) 1 and 2; (b) 1, 2, and 5; (c) 1, 2, 3, 4, and 5; (d) 2, 3, and 5; (e) 1, 3, 4, and 5.
11. Match the following:

____(a) a fluid-filled sphere of cells that enters the uterine cavity	(1) cleavage
____(b) cells produced by cleavage	(2) blastomeres
____(c) the developing individual from week nine of pregnancy until birth	(3) morula
____(d) the outer covering of cells of the blastocyst	(4) angiogenesis
____(e) membrane derived from trophoblast	(5) trophoblast
____(f) early divisions of the zygote	(6) blastocyst
____(g) a solid sphere of cells still surrounded by the zona pellicuda	(7) zygote
____(h) event in which differentiation into the three primary germ layers occurs	(8) gastrulation
____(i) embryonic development of structures that will become the nervous system	(9) neurulation
____(j) the formation of blood vessels to support the developing embryo	(10) chorion
____(k) result of the fusion of female and male pronuclei	(11) fetus

12. Match the following:

____(a) stimulates the corpus luteum to continue production of progesterone and estrogens	(1) oxytocin
____(b) increases the flexibility of the pubic symphysis and helps dilate the uterine cervix during labor	(2) human chorionic somatomammotropin
____(c) secreted by the placenta; helps establish the timing of birth and increases the secretion of cortisol for fetal lung maturation	(3) human chorionic gonadotropin
____(d) helps prepare mammary glands for lactation; regulates certain aspects of maternal and fetal metabolism	(4) prolactin
____(e) stimulates uterine contractions; responsible for the milk ejection reflex	(5) corticotropin-releasing hormone
____(f) promotes milk synthesis and secretion; inhibited by progesterone during pregnancy	(6) relaxin

13. Match the following:

____(a) the penetration of a secondary oocyte by a single sperm cell
____(b) fertilization of a secondary oocyte by more than one sperm
____(c) the attachment of a blastocyst to the endometrium
____(d) the fusion of the genetic material from a haploid sperm and a haploid secondary oocyte into a single diploid nucleus
____(e) the induction by the female reproductive tract of functional changes in sperm that allow them to fertilize a secondary oocyte
____(f) the examination of embryonic or fetal cells sloughed off into the amniotic fluid
____(g) an abnormal condition of pregnancy characterized by sudden hypertension, large amounts of protein in urine, and generalized edema
____(h) noninvasive test that can detect fetal neural tube defects
____(i) the process of giving birth
____(j) the period of time (about 6 weeks) during which the maternal reproductive organs and physiology return to the prepregnancy state

(1) fertilization
(2) capacitation
(3) syngamy
(4) polyspermy
(5) implantation
(6) amniocentesis
(7) preeclampsia
(8) parturition
(9) puerperium
(10) maternal AFP test

14. Match the following:

____(a) the control of inherited traits by the combined effects of many genes
____(b) the two alternative forms of a gene that code for the same trait and are at the same location on homologous chromosomes
____(c) abnormal number of chromosomes due to failure of homologous chromosomes or chromatids to separate
____(d) inheritance based on genes that have more than two alternative forms; an example is the inheritance of blood type
____(e) a cell in which one or more chromosomes of a set is added or deleted
____(f) refers to an individual with different alleles on homologous chromosomes
____(g) traits linked to the X chromosome
____(h) permanent inheritable change in an allele that produces a different variant of the same trait
____(i) neither member of the allelic pair is dominant over the other, and the heterozygote has a phenotype intermediate between the homozygous dominant and the homozygous recessive
____(j) refers to how the genetic makeup is expressed in the body; the physical or outward expression of a gene
____(k) a homozygous dominant, homozygous recessive, or heterozygous genetic makeup; the actual gene arrangement
____(l) refers to a person with the same alleles on homologous chromosomes
____(m) inactivated X chromosome in females
____(n) heterozygous individuals who possess a recessive gene (but do not express it) and can pass the gene on to their offspring
____(o) interchange of portions of nonhomologous chromosomes
____(p) an allele that masks the presence of another allele and is fully expressed

(1) genotype
(2) phenotype
(3) alleles
(4) aneuploid
(5) incomplete dominance
(6) multiple-allele inheritance
(7) polygenic inheritance
(8) sex-linked inheritance
(9) homozygous
(10) heterozygous
(11) carriers
(12) dominant trait
(13) mutation
(14) nondisjunction
(15) translocation
(16) Barr body

15. Match the following:

____(a) the embryonic membrane that entirely surrounds the embryo
____(b) functions as an early site of blood formation; contains cells that migrate into the gonads and differentiate into the primitive germ cells
____(c) becomes the principal part of the embryonic placenta; produces human chorionic gonadotropin
____(d) modified endometrium after implantation has occurred; separates from the endometrium after the fetus is delivered
____(e) contains the vascular connections between mother and fetus
____(f) the fetal portion is formed by the chorionic villi and the maternal portion is formed by the decidua basalis of the endometrium; allows oxygen and nutrients to diffuse from maternal blood into fetal blood
____(g) serves as an early site of blood vessel formation
____(h) finger-like projections of the chorion that bring maternal and fetal blood vessels into close proximity
____(i) plays an important role in induction whereby an inducing tissue stimulates the development of an unspecialized responding tissue into a specialized tissue

(1) decidua
(2) placenta
(3) amnion
(4) chorion
(5) allantois
(6) yolk sac
(7) notochord
(8) chorionic villi
(9) umbilical cord

CRITICAL THINKING QUESTIONS

1. Kathy is breast-feeding her infant and is experiencing what feels like early labor pains. What is causing these painful feelings? Is there a benefit to them?
2. Jack has hemophilia, which is a sex-linked blood clotting disorder. He blames his father for passing on the gene for hemophilia. Explain to Jack why his reasoning is wrong. How can Jack have hemophilia if his parents do not?
3. Alisa has asked her obstetrician to save and freeze her baby's cord blood after delivery in case the child needs a future bone marrow transplant. What is in the baby's cord blood that could be used to treat future disorders in the child?

ANSWERS TO FIGURE QUESTIONS

29.1 Capacitation is the group of functional changes in sperm that enable them to fertilize a secondary oocyte, which occur after they have been deposited in the female reproductive tract.

29.2 A morula is a solid ball of cells; a blastocyst consists of a rim of cells (trophoblast) surrounding a cavity (blastocyst cavity) and an inner cell mass.

29.3 The blastocyst secretes digestive enzymes that eat away the endometrial lining at the site of implantation.

29.4 The decidua basalis helps form the maternal part of the placenta.

29.5 Implantation occurs during the secretory phase of the uterine cycle.

29.6 The bilaminar embryonic disc is attached to the trophoblast by the connecting stalk.

29.7 Gastrulation converts a bilaminar embryonic disc into a trilaminar embryonic disc.

29.8 The notochord induces mesodermal cells to develop into vertebral bodies and forms the nucleus pulposus of intervertebral discs.

29.9 The neural tube forms the brain and spinal cord; somites develop into skeletal muscles, connective tissue, and the vertebrae.

29.10 Chorionic villi help to bring the fetal and maternal blood vessels into close proximity.

29.11 The placenta participates in the exchange of materials between fetus and mother, serves as a protective barrier against many microbes, and stores nutrients.

29.12 As a result of embryonic folding, the embryo curves into a C-shape, various organs are brought into their eventual adult positions, and the primitive gut is formed.

29.13 Pharyngeal arches, clefts, and pouches give rise to structures of the head and neck.

29.14 Fetal weight doubles between the mid-fetal period and birth.

29.15 Amniocentesis is used primarily to detect genetic disorders, but it also provides information concerning the maturity (and survivability) of the fetus.

29.16 Early pregnancy tests detect elevated levels of human chorionic gonadotropin (hCG).

29.17 Relaxin increases the flexibility of the pubic symphysis and helps dilate the cervix of the uterus to ease delivery.

29.18 Complete dilation of the cervix marks the onset of the stage of expulsion.

29.19 Oxytocin also stimulates contraction of the uterus during delivery of a baby.

29.20 The odds that a child will have PKU are the same for each child—25%.

29.21 In incomplete dominance, neither member of an allelic pair is dominant; the heterozygote has a phenotype intermediate between the homozygous dominant and the homozygous recessive phenotypes.

29.22 A baby can have blood type O if each parent is heterozygous and has one *i* allele.

29.23 Hair color, height, and body build, among others, are traits passed on by complex inheritance.

29.24 The female sex chromosomes are XX, and the male sex chromosomes are XY.

29.25 The chromosomes that are not sex chromosomes are called autosomes.

29.26 A red–green color-blind female has an X^cX^c genotype.

Appendix A

Measurements

U.S. Customary System

Parameter	Unit	Relation to Other U.S. Units	SI (Metric) Equivalent
Length	inch	1/12 foot	2.54 centimeters
foot	12 inches	0.305 meter	
yard	36 inches	9.144 meters	
mile	5,280 feet	1.609 kilometers	
Mass	grain	1/1000 pound	64.799 milligrams
dram	1/16 ounce	1.772 grams	
ounce	16 drams	28.350 grams	
pound	16 ounces	453.6 grams	
ton	2,000 pounds	907.18 kilograms	
Volume (Liquid)	ounce	1/16 pint	29.574 milliliters
pint	16 ounces	0.473 liter	
quart	2 pints	0.946 liter	
gallon	4 quarts	3.785 liters	
Volume (Dry)	pint	1/2 quart	0.551 liter
quart	2 pints	1.101 liters	
peck	8 quarts	8.810 liters	
bushel	4 pecks	35.239 liters	

International System (SI)

Base Units

Unit	Quantity	Symbol
meter	length	M
kilogram	mass	Kg
second	time	S
liter	volume	L
mole	amount of matter	Mol

Prefixes

Prefix	Multiplier	Symbol
tera-	10^{12} = 1,000,000,000,000	T
giga-	10^{9} = 1,000,000,000	G
mega-	10^{6} = 1,000,000	M
kilo-	10^{3} = 1,000	k
hecto-	10^{2} = 100	h
deca-	10^{1} = 10	da
deci-	10^{-1} = 0.1	d
centi-	10^{-2} = 0.01	c
milli-	10^{-3} = 0.001	m
micro-	10^{-6} = 0.000,001	μ
nano-	10^{-9} = 0.000,000,001	n
pico-	10^{-12} = 0.000,000,000,001	p

Temperature Conversion

Fahrenheit (F) To Celsius (C)

$$°C = (°F - 32) \div 1.8$$

Celsius (C) To Fahrenheit (F)

$$°F = (°C \times 1.8) + 32$$

U.S To SI (Metric) Conversion

When you know	Multiply by	To find
inches	2.54	centimeters
feet	30.48	centimeters
yards	0.91	meters
miles	1.61	kilometers
ounces	28.35	grams
pounds	0.45	kilograms
tons	0.91	metric tons
fluid ounces	29.57	milliliters
pints	0.47	liters
quarts	0.95	liters
gallons	3.79	liters

SI (Metric) To U.S. Conversion

When you know	Multiply by	To find
millimeters	0.04	inches
centimeters	0.39	inches
meters	3.28	feet
kilometers	0.62	miles
liters	1.06	quarts
cubic meters	35.32	cubic feet
grams	0.035	ounces
kilograms	2.21	pounds

Appendix B

Periodic Table

The periodic table lists the known **chemical elements,** the basic units of matter. The elements in the table are arranged left-to-right in rows in order of their **atomic number,** the number of protons in the nucleus. Each horizontal row, numbered from 1 to 7, is a **period.** All elements in a given period have the same number of electron shells as their period number. For example, an atom of hydrogen or helium each has one electron shell, while an atom of potassium or calcium each has four electron shells. The elements in each column, or **group,** share chemical properties. For example, the elements in column IA are very chemically reactive, whereas the elements in column VIIIA have full electron shells and thus are chemically inert.

Scientists now recognize 113 different elements; 92 occur naturally on Earth, and the rest are produced from the natural elements using particle accelerators or nuclear reactors. Elements are designated by **chemical symbols,** which are the first one or two letters of the element's name in English, Latin, or another language.

Twenty-six of the 92 naturally occurring elements normally are present in your body. Of these, just four elements—oxygen (O), carbon (C), hydrogen (H), and nitrogen (N) (coded blue)—constitute about 96% of the body's mass. Eight others—calcium (Ca), phosphorus (P), potassium (K), sulfur (S), sodium (Na), chlorine (Cl), magnesium (Mg), and iron (Fe) (coded pink)—contribute 3.8% of the body's mass. An additional 14 elements, called **trace elements** because they are present in tiny amounts, account for the remaining 0.2% of the body's mass. The trace elements are aluminum, boron, chromium, cobalt, copper, fluorine, iodine, manganese, molybdenum, selenium, silicon, tin, vanadium, and zinc (coded yellow). Table 2.1 on page 29 provides information about the main chemical elements in the body.

23 — Atomic number
V — Chemical symbol
50.942 — Atomic mass (weight)

Percentage of body mass
- 96% (4 elements)
- 3.8% (8 elements)
- 0.2% (14 elements)

	IA	IIA	IIIB	IVB	VB	VIB	VIIB	VIIIB	VIIIB	VIIIB	IB	IIB	IIIA	IVA	VA	VIA	VIIA	VIIIA
1	1 Hydrogen **H** 1.0079																	2 Helium **He** 4.003
2	3 Lithium **Li** 6.941	4 Beryllium **Be** 9.012											5 Boron **B** 10.811	6 Carbon **C** 12.011	7 Nitrogen **N** 14.007	8 Oxygen **O** 15.999	9 Fluorine **F** 18.998	10 Neon **Ne** 20.180
3	11 Sodium **Na** 22.989	12 Magnesium **Mg** 24.305											13 Aluminum **Al** 26.9815	14 Silicon **Si** 28.086	15 Phosphorus **P** 30.974	16 Sulfur **S** 32.066	17 Chlorine **Cl** 35.453	18 Argon **Ar** 39.948
4	19 Potassium **K** 39.098	20 Calcium **Ca** 40.08	21 Scandium **Sc** 44.956	22 Titanium **Ti** 47.87	23 Vanadium **V** 50.942	24 Chromium **Cr** 51.996	25 Manganese **Mn** 54.938	26 Iron **Fe** 55.845	27 Cobalt **Co** 58.933	28 Nickel **Ni** 58.69	29 Copper **Cu** 63.546	30 Zinc **Zn** 65.38	31 Gallium **Ga** 69.723	32 Germanium **Ge** 72.59	33 Arsenic **As** 74.992	34 Selenium **Se** 78.96	35 Bromine **Br** 79.904	36 Krypton **Kr** 83.80
5	37 Rubidium **Rb** 85.468	38 Strontium **Sr** 87.62	39 Yttrium **Y** 88.905	40 Zirconium **Zr** 91.22	41 Niobium **Nb** 92.906	42 Molybdenum **Mo** 95.94	43 Technetium **Tc** (99)	44 Ruthenium **Ru** 101.07	45 Rhodium **Rh** 102.905	46 Palladium **Pd** 106.42	47 Silver **Ag** 107.868	48 Cadmium **Cd** 112.40	49 Indium **In** 114.82	50 Tin **Sn** 118.69	51 Antimony **Sb** 121.75	52 Tellurium **Te** 127.60	53 Iodine **I** 126.904	54 Xenon **Xe** 131.30
6	55 Cesium **Cs** 132.905	56 Barium **Ba** 137.33		72 Hafnium **Hf** 178.49	73 Tantalum **Ta** 180.948	74 Tungsten **W** 183.85	75 Rhenium **Re** 186.2	76 Osmium **Os** 190.2	77 Iridium **Ir** 192.22	78 Platinum **Pt** 195.08	79 Gold **Au** 196.967	80 Mercury **Hg** 200.59	81 Thallium **Tl** 204.38	82 Lead **Pb** 207.19	83 Bismuth **Bi** 208.980	84 Polonium **Po** (209)	85 Astatine **At** (210)	86 Radon **Rn** (222)
7	87 Francium **Fr** (223)	88 Radium **Ra** (226)		104 Rutherfordium **Rf** (261)	105 Dubnium **Db** (262)	106 Seaborgium **Sg** (263)	107 Bohrium **Bh** (264)	108 Hassium **Hs** (269)	109 Meitnerium **Mt** (268)	110 Unnamed (271)	111 Unnamed (272)	112 Unnamed (277)		114 Unnamed (289)				

57–71, Lanthanides

57 Lanthanum **La** 138.91	58 Cerium **Ce** 140.12	59 Praseodymium **Pr** 140.907	60 Neodymium **Nd** 144.24	61 Promethium **Pm** 144.913	62 Samarium **Sm** 150.35	63 Europium **Eu** 151.96	64 Gadolinium **Gd** 157.25	65 Terbium **Tb** 158.925	66 Dysprosium **Dy** 162.50	67 Holmium **Ho** 164.930	68 Erbium **Er** 167.26	69 Thulium **Tm** 168.934	70 Ytterbium **Yb** 173.04	71 Lutetium **Lu** 174.97

89–103, Actinides

89 Actinium **Ac** (227)	90 Thorium **Th** 232.038	91 Protactinium **Pa** (231)	92 Uranium **U** 238.03	93 Neptunium **Np** (237)	94 Plutonium **Pu** 244.064	95 Americium **Am** (243)	96 Curium **Cm** (247)	97 Berkelium **Bk** (247)	98 Californium **Cf** 242.058	99 Einsteinium **Es** (254)	100 Fermium **Fm** 257.095	101 Mendelevium **Md** 258.10	102 Nobelium **No** 259.10	103 Lawrencium **Lr** 260.105

C Appendix

Normal Values for Selected Blood Tests

The system of international (SI) units (Système Internationale d'Unités) is used in most countries and in many medical and scientific journals. Clinical laboratories in the United States, by contrast, usually report values for blood and urine tests in conventional units. The laboratory values in this Appendix give conventional units first, followed by SI equivalents in parentheses. Values listed for various blood tests should be viewed as reference values rather than absolute "normal" values for all well people. Values may vary due to age, gender, diet, and environment of the subject or the equipment, methods, and standards of the lab performing the measurement.

Key To Symbols

g = gram	mL = milliliter
mg = milligram = 10^{-3} gram	μL = microliter
μg = microgram = 10^{-6} gram	mEq/L = milliequivalents per liter
U = units	mmol/L = millimoles per liter
L = liter	μmol/L = micromoles per liter
dL = deciliter	$>$ = greater than; $<$ = less than

Blood Tests

Test (Specimen)	U.S. Reference Values (SI Units)	Values Increase In	Values Decrease In
Aminotransferases (serum)			
Alanine aminotransferase (ALT)	0–35 U/L (same)	Liver disease or liver damage due to toxic drugs.	
Aspartate aminotransferase (AST)	0–35 U/L (same)	Myocardial infarction, liver disease, trauma to skeletal muscles, severe burns.	Beriberi, uncontrolled diabetes mellitus with acidosis, pregnancy.
Ammonia (plasma)	20–120 μg/dL (12–55 μmol/L)	Liver disease, heart failure, emphysema, pneumonia, hemolytic disease of newborn.	Hypertension.
Bilirubin (serum)	Conjugated: $<$0.5 mg/dL ($<$5.0 μmol/L)	Conjugated bilirubin: liver dysfunction or gallstones.	
	Unconjugated: 0.2–1.0 mg/dL (18–20 μmol/L) Newborn: 1.0–12.0 mg/dL ($<$200 μmol/L)	Unconjugated bilirubin: excessive hemolysis of red blood cells.	
Blood urea nitrogen (BUN) (serum)	8–26 mg/dL (2.9–9.3 mmol/L)	Kidney disease, urinary tract obstruction, shock, diabetes, burns, dehydration, myocardial infarction.	Liver failure, malnutrition, overhydration, pregnancy.
Carbon dioxide content (bicarbonate + dissolved CO_2) (whole blood)	Arterial: 19–24 mEq/L (19–24 mmol/L) Venous: 22–26 mEq/L (22–26 mmol/L)	Severe diarrhea, severe vomiting, starvation, emphysema, aldosteronism.	Renal failure, diabetic ketoacidosis, shock.

Blood Tests (continued)

Test (Specimen)	U.S. Reference Values (SI Units)	Values Increase In	Values Decrease In
Cholesterol, total (plasma) **HDL cholesterol** (plasma) **LDL cholesterol** (plasma)	<200 mg/dL (<5.2 mmol/L) is desirable >40 mg/dL (>1.0 mmol/L) is desirable <130 mg/dL (<3.2 mmol/L) is desirable	Hypercholesterolemia, uncontrolled diabetes mellitus, hypothyroidism, hypertension, atherosclerosis, nephrosis.	Liver disease, hyperthyroidism, fat malabsorption, pernicious or hemolytic anemia, severe infections.
Creatine (serum)	Males: 0.15–0.5 mg/dL (10–40 μmol/L) Females: 0.35–0.9 mg/dL (30–70 μmol/L)	Muscular dystrophy, damage to muscle tissue, electric shock, chronic alcoholism.	
Creatine Kinase (CK), also known as Creatine phosphokinase (CPK) (serum)	0–130 U/L (same)	Myocardial infarction, progressive muscular dystrophy, hypothyroidism, pulmonary edema.	
Creatinine (serum)	0.5–1.2 mg/dL (45–105 μmol/L)	Impaired renal function, urinary tract obstruction, giantism, acromegaly.	Decreased muscle mass, as occurs in muscular dystrophy or myasthenia gravis.
Electrolytes (plasma)	See Table 27.2 on page 1045		
Gamma-glutamyl transferase (GGT) (serum)	0–30 U/L (same)	Bile duct obstruction, cirrhosis, alcoholism, metastatic liver cancer, congestive heart failure.	
Glucose (plasma)	70–110 mg/dL (3.9–6.1 mmol/L)	Diabetes mellitus, acute stress, hyperthyroidism, chronic liver disease,Cushing's disease.	Addison's disease, hypothyroidism, hyperinsulinism.
Hemoglobin (whole blood)	Males: 14–18 g/100 mL (140–180 g/L) Females: 12–16 g/100 mL (120–160 g/L) Newborns: 14–20 g/100 mL (140–200 g/L)	Polycythemia, congestive heart failure, chronic obstructive pulmonary disease, living at high altitude.	Anemia, severe hemorrhage, cancer, hemolysis, Hodgkin's disease, nutritional deficiency of vitamin B_{12}, systemic lupus erythematosus, kidney disease.
Iron, total (serum)	Males: 80–180 μg/dL (14–32 μmol/L) Females: 60–160 μg/dL (11–29 μmol/L)	Liver disease, hemolytic anemia, iron poisoning.	Iron-deficiency anemia, chronic blood loss, pregnancy (late), chronic heavy menstruation.
Lactic dehydrogenase (LDH) (serum)	71–207 U/L (same)	Myocardial infarction, liver disease, skeletal muscle necrosis, extensive cancer.	
Lipids (serum) **Total** **Triglycerides**	 400–850 mg/dL (4.0–8.5 g/L) 10–190 mg/dL (0.1–1.9 g/L)	Hyperlipidemia, diabetes mellitus.	Fat malabsorption, hypothyroidism.
Platelet (thrombocyte) count (whole blood)	150,000–400,000/μL	Cancer, trauma, leukemia, cirrhosis.	Anemias, allergic conditions, hemorrhage.
Protein (serum) **Total** **Albumin** **Globulin**	 6–8 g/dL (60–80 g/L) 4–6 g/dL (40–60 g/L) 2.3–3.5 g/dL (23–35 g/L)	Dehydration, shock, chronic infections.	Liver disease, poor protein intake, hemorrhage, diarrhea, malabsorption, chronic renal failure, severe burns.
Red blood cell (erythrocyte) count (whole blood)	Males: 4.5–6.5 million/μL Females: 3.9–5.6 million/μL	Polycythemia, dehydration, living at high altitude.	Hemorrhage, hemolysis, anemias, cancer, overhydration.
Uric acid (urate) (serum)	2.0–7.0 mg/dL (120–420 μmol/L	Impaired renal function, gout, metastatic cancer, shock, starvation.	
White blood cell (leukocyte) count, total (whole blood)	5,000–10,000/μL (See Table 19.3 on page 680 for relative percentages of different types of WBCs.)	Acute infections, trauma, malignant diseases, cardiovascular diseases. (See also Table 19.2 on page 678.)	Diabetes mellitus, anemia. (See also Table 19.2 on page 678.)

D Appendix

Normal Values for Selected Urine Tests

Urine Tests

Test (Specimen)	U.S. Reference Values (SI Units)	Clinical Implications
Amylase (2 hour)	35–260 Somogyi units/hr (6.5–48.1 units/hr)	Values increase in inflammation of the pancreas (pancreatitis) or salivary glands, obstruction of the pancreatic duct, and perforated peptic ulcer.
Bilirubin* (random)	Negative	Values increase in liver disease and obstructive biliary disease.
Blood* (random)	Negative	Values increase in renal disease, extensive burns, transfusion reactions, and hemolytic anemia.
Calcium (Ca^{21}) (random)	10 mg/dL (2.5 mmol/liter); up to 300 mg/24 hr (7.5 mmol/24 hr)	Amount depends on dietary intake; values increase in hyperparathyroidism, metastatic malignancies, and primary cancer of breasts and lungs; values decrease in hypoparathyroidism and vitamin D deficiency.
Casts (24 hour)		
Epithelial	Occasional	Values increase in nephrosis and heavy metal poisoning.
Granular	Occasional	Values increase in nephritis and pyelonephritis.
Hyaline	Occasional	Values increase in kidney infections.
Red blood cell	Occasional	Values increase in glomerular membrane damage and fever.
White blood cell	Occasional	Values increase in pyelonephritis, kidney stones, and cystitis.
Chloride (Cl^-) (24 hour)	140–250 mEq/24 hr (140–250 mmol/24 hr)	Amount depends on dietary salt intake; values increase in Addison's disease, dehydration, and starvation; values decrease in pyloric obstruction, diarrhea, and emphysema.
Color (random)	Yellow, straw, amber	Varies with many disease states, hydration, and diet.
Creatinine (24 hour)	Males: 1.0–2.0 g/24 hr (9–18 mmol/24 hr) Females: 0.8–1.8 g/24 hr (7–16 mmol/24 hr)	Values increase in infections; values decrease in muscular atrophy, anemia, and kidney diseases.
Glucose*	Negative	Values increase in diabetes mellitus, brain injury, and myocardial infarction.
Hydroxycorticosteroids (17-hydroxysteroids) (24 hour)	Males: 5–15 mg/24 hr (13–41 μmol/24 hr) Females: 2–13 mg/24 hr (5–36 μmol/24 hr)	Values increase in Cushing's syndrome, burns, and infections; values decrease in Addison's disease.
Ketone bodies* (random)	Negative	Values increase in diabetic acidosis, fever, anorexia, fasting, and starvation.
17-ketosteroids (24 hour)	Males: 8–25 mg/24 hr (28–87 μmol/24 hr) Females: 5–15 mg/24 hr (17–53 μmol/24 hr)	Values decrease in surgery, burns, infections, adrenogenital syndrome, and Cushing's syndrome.

Test (Specimen)	U.S. Reference Values (SI Units)	Clinical Implications
Odor (random)	Aromatic	Becomes acetonelike in diabetic ketosis.
Osmolality (24 hour)	500–1400 mOsm/kg water (500–1400 mmol/kg water)	Values increase in cirrhosis, congestive heart failure (CHF), and high-protein diets; values decrease in aldosteronism, diabetes insipidus, and hypokalemia.
pH* (random)	4.6–8.0	Values increase in urinary tract infections and severe alkalosis; values decrease in acidosis, emphysema, starvation, and dehydration.
Phenylpyruvic acid (random)	Negative	Values increase in phenylketonuria (PKU).
Potassium (K^+) (24 hour)	40–80 mEq/24 hr (40–80 mmol/24 hr)	Values increase in chronic renal failure, dehydration, starvation, and Cushing's syndrome; values decrease in diarrhea, malabsorption syndrome, and adrenal cortical insufficiency
Protein* (albumin) (random)	Negative	Values increase in nephritis, fever, severe anemias, trauma, and hyperthyroidism.
Sodium (Na^+) (24 hour)	75–200 mg/24 hr (75–200 mmol/24 hr)	Amount depends on dietary salt intake; values increase in dehydration, starvation, and diabetic acidosis; values decrease in diarrhea, acute renal failure, emphysema, and Cushing's syndrome.
Specific gravity* (random)	1.001–1.035 (same)	Values increase in diabetes mellitus and excessive water loss; values decrease in absence of antidiuretic hormone (ADH) and severe renal damage.
Urea (random)	25–35 g/24 hr (420–580 mmol/24 hr)	Values increase in response to increased protein intake; values decrease in impaired renal function.
Uric acid (24 hour)	0.4–1.0 g/24 hr (1.5–4.0 mmol/24 hr)	Values increase in gout, leukemia, and liver disease; values decrease in kidney disease.
Urobilinogen* (2 hour)	0.3–1.0 Ehrlich units (1.7–6.0 μmol/24 hr)	Values increase in anemias, hepatitis A (infectious), biliary disease, and cirrhosis; values decrease in cholelithiasis and renal insufficiency.
Volume, total (24 hour)	1000–2000 mL/24 hr (1.0–2.0 liters/24 hr)	Varies with many factors.

* Test often performed using a **dipstick,** a plastic strip impregnated with chemicals that is dipped into a urine specimen to detect particular substances. Certain colors indicate the presence or absence of a substance and sometimes give a rough estimate of the amount(s) present.

E Appendix

Answers

Answers to Self-Quiz Questions

Chapter 1

1. tissue **2.** metabolism, anabolism, catabolism **3.** intracellular fluid (ICF), extracellular fluid (ECF) **4.** true **5.** true **6.** false **7.** e **8.** d **9.** a **10.** c **11.** c **12.** (a) 1, (b) 12, (c) 1, 6 , (d) 6, (e) 4, (f) 8, (g) 7, (h) 3, (i) 2, (j) 10 **13.** (a) 4, (b) 1, (c) 3, (d) 6, (e) 5, (f) 7, (g) 2 **14.** (a) 6, (b) 1, (c) 11, (d) 5, (e) 10, (f) 8, (g) 7, (h) 9, (i) 4, (j) 3, (k) 2 **15.** (a) 4, (b) 6, (c) 8, (d) 1, (e) 9, (f) 5, (g) 2, (h) 7, (i), 3, (j) 10

Chapter 2

1. 8 **2.** solid, liquid, gas **3.** monosaccharides, amino acids **4.** true **5.** false **6.** true **7.** c **8.** a **9.** d **10.** b **11.** e **12.** a **13.** e **14.** (a) 1, (b), 2, (c) 1, (d) 4, (e) 3 **15.** (a) 11, (b) 1, (c) 8, (d) 3, (e) 7, (f) 4, (g) 5, (h) 9, (i) 10, (j) 12, (k) 6, (l) 2

Chapter 3

1. plasma membrane, cytoplasm, nucleus **2.** apoptosis, necrosis **3.** Telomeres **4.** UAG **5.** false **6.** true **7.** true **8.** e **9.** c **10.** c, g, i, b, d, k, f, j, a, e, h **11.** a **12.** c **13.** (a) 2, (b) 3, (c) 5, (d) 7, (e) 6, (f) 8, (g) 1, (h) 4 **14.** (a) 2, (b) 9, (c) 3, (d) 5, (e) 11, (f) 8, (g) 1, (h) 6, (i) 10, (j) 7, (k) 13, (l) 4, (m) 12 **15.** (a) 3, (b) 9, (c) 1, (d) 5, (e) 11, (f) 4, (g) 8, (h) 7, (i) 2, (j) 10, (k) 6

Chapter 4

1. epithelial, connective, muscle, nervous **2.** arrangement of cells in layers, cell shape **3.** true **4.** true **5.** e **6.** b **7.** a **8.** c **9.** e **10.** b **11.** d **12.** c **13.** (a) C, (b) M, (c) N, (d) E, (e) C, (f) E, (g) M, (h) E, (i) C, (j) M, (k) N, (l) E, (m) C, (n) E, (o) M and N **14.** (a) 4, (b) 8, (c) 5, (d) 2, (e) 6, (f) 3, (g) 1, (h) 7 **15.** (a) 3, (b) 5, (c) 8, (d) 13, (e) 9, (f) 7, (g) 11, (h) 6, (i) 2, (j) 4, (k) 10, (l) 12, (m) 1

Chapter 5

1. stratum lucidum **2.** eccrine, ceruminous, apocrine **3.** false **4.** true **5.** c **6.** e **7.** a **8.** c **9.** b **10.** e **11.** a **12.** c **13.** (a) 3, (b) 5, (c) 4, (d) 1, (e) 6, (f) 11, (g) 2, (h) 8, (i) 9, (j) 10, (k) 7 **14.** (a) 3, (b) 4, (c) 1, (d) 2 **15.** (a) 4, (b) 3, (c) 2, (d) 1, inflammatory, migratory, proliferative, maturation

Chapter 6

1. interstitial, appositional **2.** hardness, tensile strength **3.** true **4.** true **5.** true **6.** d **7.** a **8.** e **9.** c **10.** a **11.** (a) 3, (b) 9, (c) 8, (d) 1, (e) 5, (f) 4, (g) 6, (h) 7, (i) 12, (j) 2, (k) 11 (l) 10 **12.** (a) 12, (b) 4, (c) 8, (d) 6, (e) 3, (f) 9, (g) 13, (h) 10, (i) 7, (j) 5, (k) 2, (l) 11, (m) 1 **13.** (a) 2, (b) 6, (c) 4, (d) 5, (e) 7, (f) 3, (g) 1 **14.** (a) 1, (b) 4, (c) 3, (d) 2 **15.** (a) 3, (b) 7, (c) 6, (d) 1, (e) 4, (f) 2, (g) 5, (h) 9, (i) 8, (j) 10

Chapter 7

1. fontanels **2.** pituitary gland **3.** sacrum, coccyx **4.** false **5.** false **6.** b **7.** c **8.** a **9.** e **10.** d **11.** e **12.** (a) 4, (b) 9, (c) 7, (d) 5, (e) 3, (f) 1, (g) 2, (h) 8, (i) 6 **13.** (a) 7, (b) 5, (c) 1, (d) 6, (e) 2, (f) 4, (g) 8, (h) 9, (i) 3, (j) 10, (k) 11, (l) 13, (m) 12 **14.** (a) 2, (b) 3, (c) 5, (d) 6, (e) 4, (f) 1, (g) 5, (h) 4, (i) 2, (j) 4, (k) 3 **15.** (a) 3, (b) 1, (c) 6, (d) 9, (e) 13, (f) 12, (g) 2, (h) 4, (i) 5, (j) 7, (k) 10, (l) 15, (m) 8, (n) 11, (o) 14

Chapter 8

1. metacarpals **2.** ilium, ischium, pubis **3.** true (lesser), false (greater) **4.** false **5.** true **6.** b **7.** c **8.** e **9.** c **10.** a **11.** d **12.** a **13.** (a) 2, (b) 6, (c) 9, (d) 7, (e) 4, (f) 5, (g) 8, (h) 10, (i) 1, (j) 3 **14.** (a) 3, (b) 8, (c) 4, (d) 11, (e) 9, (f) 13, (g) 5, (h) 6, (i) 10, (j) 14, (k) 2, (l) 1, (m) 7, (n) 12 **15.** (a) 4, (b) 3, (c) 3, (d) 6, (e) 7, (f) 1, (g) 3, (h) 2, (i) 5, (j) 9, (k) 8, (l) 2, (m) 4, (n) 6, (o) 7, (p) 9, (q) 6, (r) 3, (s) 4, (t) 4 and 5

Chapter 9

1. joint, articulation or arthrosis **2.** arthroplasty **3.** false **4.** false **5.** false **6.** e **7.** d **8.** b **9.** c **10.** a **11.** c **12.** e **13.** (a) 5, (b) 3, (c) 7, (d) 2, (e) 6, (f) 4, (g) 1 **14.** (a) 6, (b) 4, (c) 5, (d) 1, (e) 3, (f) 2 **15.** (a) 8, (b) 11, (c) 10, (d) 13, (e) 15, (f) 9, (g) 6, (h) 12, (i) 3, (j) 4, (k) 16, (l) 2, (m) 18, (n) 1, (o) 7, (p) 14, (q) 17, (r) 5

Chapter 10

1. motor unit **2.** muscular atrophy, fibrosis **3.** acetylcholine **4.** true **5.** true **6.** e **7.** a **8.** c **9.** e **10.** d **11.** b **12.** (a) 5, (b) 6, (c) 9, (d) 7, (e) 2, (f) 4, (g) 10, (h) 3, (i) 1, (j) 8 **13.** (a) 7, (b) 10, (c) 9, (d) 12, (e) 8, (f) 11, (g) 6, (h) 1, (i) 2, (j) 3, (k) 4, (l) 13, (m) 5 **14.** (a) 10, (b) 2, (c) 4, (d) 3, (e) 6, (f) 5, (g) 1, (h) 12, (i) 7, (j) 9, (k) 11, (l) 8 **15.** (a) 2, (b) 3, (c) 1, (d) 1 and 2, (e) 3, (f) 2, (g) 1, (h) 3, (i) 1 and 2, (j) 3, (k) 2 and 3, (l) 3

Chapter 11

1. buccinator **2.** gastrocnemius, soleus, plantaris **3.** true **4.** true **5.** b **6.** c **7.** d **8.** a **9.** e **10.** e **11.** (a) 6, (b) 2, (c) 8, (d) 5, (e) 3, (f) 1, (g) 7, (h) 4 **12.** (a) 13, (b) 9, (c) 8, (d) 6, (e) 3, (f) 11, (g) 10, (h) 1, (i) 2, (j) 7, (k) 12, (l) 4, (m) 5 **13.** (a) 6, (b) 3, (c) 7, (d) 4, (e) 2, (f) 9, (g) 5, (h) 1, (i) 8 **14.** (a) 10, (b) 1, (c) 9, (d) 8, (e) 12, (f) 17, (g) 2, (h) 6, (i) 8, (j) 14, (k) 5, (l) 4, (m) 2, (n) 15, (o) 1, (p) 11, (q) 13, (r) 12, (s) 7, (t) 16, (u) 11, (v) 17, (w) 16, (x) 15, (y) 3, (z) 10 **15.** (a) 3, (b) 1, (c) 2, (d) 1, (e) 2, (f) 3, (g) 3

Chapter 12

1. somatic, autonomic, enteric **2.** sympathetic, parasympathetic **3.** false **4.** false **5.** c **6.** d **7.** c **8.** e **9.** e **10.** d **11.** e **12.** b **13.** (a) 6, (b) 12, (c) 1, (d) 2, (e) 9, (f) 14, (g) 4, (h) 8, (i) 7, (j) 13, (k) 5, (l) 3, (m) 10, (n) 15, (o) 11 **14.** (a) 2, (b) 1, (c) 10, (d) 9, (e) 6, (f) 3, (g) 4, (h) 5, (i) 12, (j) 8,

(k) 7, (l) 13, (m) 11 **15.** (a) 4, (b) 5, (c) 16, (d) 8, (e) 7, (f) 1, (g) 2, (h) 10, (i) 15, (j) 6, (k) 3, (l) 13, (m) 9, (n) 11, (o) 14, (p) 12

Chapter 13

1. mixed **2.** sensory receptor, sensory neuron, integrating center, motor neuron, effector **3.** true **4.** false **5.** c **6.** c **7.** a **8.** c **9.** d **10.** e **11.** a **12.** d **13.** (a) 1, (b) 8, (c) 4, (d) 2, (e) 11, (f) 1, (g) 6, (h) 5, (i) 3, (j) 9, (k) 1, (l) 12, (m) 7, (n) 2, (o) 10 **14.** (a) 14, (b) 12, (c) 13, (d) 1, (e) 2, (f) 5, (g) 11, (h) 8, (i) 10, (j) 9, (k) 15, (l) 4, (m) 7, (n) 3, (o) 6 **15.** (a) 2, (b) 1, (c) 3, (d) 4, (e) 1, (f) 5, (g) 3, (h) 2, (i) 4, (j) 1, (k) 2, (l) 4, (m) 3, (n) 5, (o) 1

Chapter 14

1. corpus callosum **2.** frontal, temporal, parietal, occipital, insula **3.** longitudinal fissure **4.** false **5.** true **6.** d **7.** c **8.** d **9.** e **10.** d **11.** e **12.** (a) 3, (b) 5, (c) 6, (d) 8, (e) 11, (f) 10, (g) 7, (h) 9, (i) 1, (j) 4, (k) 2, (l) 12, (m) 1, (n) 8, (o) 5, (p) 7, (q) 12, (r) 10, (s) 9, (t) 1 and 2, (u) 3, 4, and 6, (v) 11 **13.** (a) 9, (b) 2, (c) 6, (d) 10, (e) 4, (f) 11, (g) 1, (h) 2, (i) 5, (j) 8, (k) 12, (l) 7, (m) 3, (n) 6 and 8, (o) 13, (p) 7, (q) 1 **14.** (a) 5, (b) 9, (c) 11, (d) 6, (e) 3, (f) 1, (g) 10, (h) 8, (i) 2, (j) 4, (k) 7 **15.** (a) 10, (b) 2, (c) 6, (d) 8, (e) 7, (f) 5, (g) 3, (h) 11, (i) 14, (j) 13, (k) 4, (l) 1, (m) 12, (n) 9

Chapter 15

1. acetylcholine, epinephrine or norepinephrine **2.** thoracolumbar, craniosacral **3.** true **4.** true **5.** d **6.** d **7.** b **8.** c **9.** e **10.** a **11.** a **12.** c **13.** d, b, f, e, d, a, c **14.** (a) 3, (b) 2, (c) 1, (d) 1, (e) 2, (f) 3, (g) 3, (h) 1, (i) 4, (j) 2, (k) 5 **15.** (a) 2, (b) 1, (c) 1, (d) 2, (e) 1, (f) 1, (g) 2, (h) 2

Chapter 16

1. sensation, perception **2.** decussation **3.** false **4.** true **5.** c **6.** a **7.** d **8.** b **9.** d **10.** e **11.** e **12.** d **13.** (a) 9, (b) 8, (c) 4, (d) 7, (e) 10, (f) 2, (g) 3, (h) 1, (i) 5, (j) 6, (k) 11 **14.** (a) 3, (b) 2, (c) 5, (d) 7, (e) 1, (f) 9, (g) 3, (h) 11, (i) 8, (j) 4, (k) 6, (l) 10 **15.** (a) 10, (b) 8, (c) 7, (d) 1, (e) 4, (f) 3, (g) 5, (h) 6, (i) 9, (j) 2

Chapter 17

1. sweet, sour, salty, bitter, umami **2.** static, dynamic **3.** true **4.** false **5.** d **6.** a **7.** d **8.** b **9.** b **10.** c, j, k, d, h, l, e, b, f, i, a, m, g **11.** c **12.** a **13.** (a) 1, (b) 5, (c) 7, (d) 6, (e) 8, (f) 2, (g) 4, (h) 3 **14.** (a) 3, (b) 6, (c) 9, (d) 14, (e) 1, (f) 5, (g) 10, (h) 13, (i) 7, (j) 15, (k) 2, (l) 11, (m) 12, (n) 4, (o) 8 **15.** (a) 2, (b) 11, (c) 14, (d) 13, (e) 3, (f) 10, (g) 6, (h) 12, (i) 4, (j) 5, (k) 9, (l) 1, (m) 7, (n) 8

Chapter 18

1. fight-or-flight response, resistance reaction, exhaustion **2.** hypothalamus **3.** less, more **4.** false **5.** true **6.** b **7.** e **8.** d **9.** a **10.** e **11.** c **12.** a **13.** (a) 8, (b) 2, (c) 7, (d) 1, (e) 12, (f) 20, (g) 5, (h) 18, (i) 22, (j) 15, (k) 3, (l) 17, (m) 21, (n) 6, (o) 13, (p) 11, (q) 4, (r) 10, (s) 14, (t) 9, (u) 16, (v) 19 **14.** (a) 10, (b) 8, (c) 2, (d) 12, (e) 15, (f) 4, (g) 1, (h) 16, (i) 6, (j) 9, (k) 13, (l) 7, (m) 5, (n) 14, (o) 3, (p) 11 **15.** (a) 12, (b) 1, (c) 11, (d) 7, (e) 3, (f) 10, (g) 2, (h) 9, (i) 4, (j) 8, (k) 5, (l) 6

Chapter 19

1. serum **2.** clot retraction **3.** true **4.** true **5.** e **6.** a **7.** b **8.** c **9.** d **10.** a **11.** d **12.** e **13.** (a) 8, (b) 14, (c) 7, (d) 3, (e) 16, (f) 2, (g) 11, (h) 4, (i) 1, (j) 10, (k) 9, (l) 12, (m) 13, (n) 18, (o) 15, (p) 5, (q) 19, (r) 6, (s) 17 **14.** (a) 4, (b) 6, (c) 8, (d) 1, (e) 7, (f) 3, (g) 5, (h) 2 **15.** (a) 4, (b) 7, (c) 6, (d) 1, (e) 3, (f) 5, (g) 2

Chapter 20

1. left ventricle **2.** systole, diastole **3.** false **4.** true **5.** a **6.** c **7.** d **8.** b **9.** b **10.** e **11.** c **12.** (a) 3, (b) 6, (c) 1, (d) 5, (e) 2, (f) 4 **13.** (a) 8, (b) 4, (c) 11, (d) 5, (e) 1, (f) 9, (g) 7, (h) 2, (i) 10, (j) 6, (k) 3 **14.** (a) 3, (b) 2, (c) 9, (d) 14, (e) 8, (f) 7, (g) 11, (h) 12, (i) 15, (j) 4, (k) 5, (l) 1, (m) 6, (n) 21, (o) 22, (p) 19, (q) 17, (r) 18, (s) 20, (t) 16, (u) 13, (v) 10 **15.** (a) 3, (b) 7, (c) 2, (d) 5, (e) 1, (f) 6, (g) 4 and 7

Chapter 21

1. carotid sinus, aortic **2.** skeletal muscle pump, respiratory pump **3.** true **4.** true **5.** b **6.** a **7.** c **8.** e **9.** a **10.** d **11.** (a) D, (b) C, (c) C, (d) D, (e) D, (f) C, (g) C, (h) C, (i) D, (j) D, (k) C **12.** (a) 2, (b) 5, (c) 1, (d) 4, (e) 3 **13.** (a) 11, (b) 1, (c) 4, (d) 9, (e) 3, (f) 8, (g) 6, (h) 2, (i) 7, (j) 5, (k) 10, (l) 12, (m) 13 **14.** (a) 2, (b) 6, (c) 4, (d) 1, (e) 3, (f) 5 **15.** (a) 5, (b) 3, (c) 1, (d) 4, (e) 2, (f) 4, (g) 1, (h) 5, (i) 3

Chapter 22

1. skin, mucous membranes, antimicrobial proteins, natural killer cells, phagocytes **2.** antigens **3.** true **4.** true **5.** c **6.** d **7.** e **8.** d **9.** e **10.** b **11.** c **12.** e, h, b, f, a, d, g, i, c **13.** (a) 3, (b) 1, (c) 7, (d) 4, (e) 2, (f) 5, (g) 6 **14.** (a) 2, (b) 3, (c) 4, (d) 7, (e) 1, (f) 6, (g) 5 **15.** (a) 12, (b) 12, (c) 8, (d) 1, (e) 2, (f) 5, (g) 4, (h) 7, (i) 9, (j) 14, (k) 13, (l) 3, (m) 6, (n) 11, (o) 10

Chapter 23

1. oxyhemoglobin; dissolved CO_2, carbamino compounds (primarily carbaminohemoglobin), and bicarbonate ion **2.** $CO_2 + H_2O \rightarrow H_2CO_3 \rightarrow H^+ + HCO_3^-$ **3.** false **4.** true **5.** c **6.** e **7.** b **8.** d **9.** a **10.** e **11.** e, g, b, h, a, d, f, c **12.** (a) 2, (b) 11, (c) 3, (d) 9, (e) 1, (f) 12, (g) 10, (h) 5, (i) 13, (j) 8, (k) 4, (l) 6, (m) 7 **13.** (a) 7, (b) 8, (c) 1, (d) 5, (e) 6, (f) 9, (g) 2, (h) 3, (i) 4 **14.** (a) 3, (b) 8, (c) 5, (d) 9, (e) 2, (f) 7, (g) 10, (h) 1, (i) 4, (j) 6 **15.** (a) 9, (b) 11, (c) 3, (d) 4, (e) 6, (f) 1, (g) 5, (h) 10, (i) 7, (j) 8, (k) 2

Chapter 24

1. monosaccharides; amino acids; monoglycerides, fatty acids; pentoses, phosphates, nitrogenous bases **2.** diffusion, facilitated diffusion, osmosis, active transport **3.** true **4.** true **5.** b **6.** d **7.** e **8.** c **9.** a **10.** b **11.** d **12.** a **13.** (a) 5, (b) 13, (c) 8, (d) 11, (e) 9, (f) 1, (g) 7, (h) 12, (i) 14, (j) 2, (k) 4, (l) 10, (m) 6, (n) 3 **14.** (a) 4, (b) 6, (c) 7, (d) 1, (e) 5, (f) 3, (g) 9, (h) 2, (i) 11, (j) 10, (k) 8 **15.** (a) 4, (b) 8, (c) 2, (d) 10, (e) 11, (f) 7, (g) 1, (h) 13, (i) 12, (j) 6, (k) 3, (l) 9, (m) 5

Chapter 25

1. hypothalamus **2.** glucose 6-phosphate, pyruvic acid, acetyl coenzyme A **3.** true **4.** false **5.** e **6.** c **7.** b **8.** d **9.** a **10.** b **11.** e **12.** a **13.** (a) 2 and 3, (b) 1, (c) 3 and 5, (d) 2, 4, 5, and 6, (e) 1, (f) 2, (g) 1, 4, and 6 **14.** (a) 9, (b) 12, (c) 11, (d) 10, (e) 4, (f) 13, (g) 3, (h) 5, (i) 8, (j) 1, (k) 6, (l) 2, (m) 7 **15.** (a) 17, (b) 15, (c) 8, (d) 19, (e) 1, (f) 7, (g) 4, (h) 10, (i) 16, (j) 14, (k) 11, (l) 2, (m) 13, (n) 6, (o) 20, (p) 9, (q) 5, (r) 18, (s) 3, (t) 12

Chapter 26

1. glomerulus, glomerular (Bowman's) capsule **2.** micturition **3.** false **4.** true **5.** d **6.** e **7.** c **8.** b **9.** e **10.** a **11.** g, a, h, c, o, n, l, j, i, b, d, e, m, k, f **12.** g, f, b, d, h, c, a, e, i **13.** (a) 8, (b) 2, (c) 10, (d) 5, (e) 3, (f) 1, (g) 7, (h) 4, (i) 11, (j) 9, (k) 6 **14.** (a) 4, (b) 3, (c) 7, (d) 6, (e) 2, (f) 8, (g) 1, (h) 5 **15.** (a) 5, (b) 4, (c) 6, (d) 8, (e) 1, (f) 2, (g) 7, (h) 3

Chapter 27

1. metabolic **2.** bicarbonate ion, carbonic acid **3.** true **4.** false **5.** a **6.** c **7.** d **8.** d **9.** b **10.** e **11.** a **12.** b **13.** a **14.** (a) 8, (b) 9, (c) 7, (d) 1, (e) 6, (f) 2, (g) 4, (h) 5, (i) 3 **15.** (a) 8, (b) 12, (c) 7, (d) 5, (e) 10, (f) 6, (g) 9, (h) 13, (i) 1, (j) 11, (k) 3, (l) 4, (m) 2

Chapter 28

1. puberty, menarche, menopause **2.** true **3.** true **4.** e **5.** c **6.** a **7.** c **8.** b **9.** a **10.** e **11.** d **12.** (a) 13, (b) 10, (c) 12, (d) 1, (e) 5, (f) 2, (g) 4, (h) 6, (i) 14, (j) 11, (k) 8, (l) 3, (m) 9, (n) 7 **13.** (a) 6, (b) 4, (c) 1, (d) 12, (e) 8, (f) 5, (g) 7, (h) 13, (i) 11, (j) 3, (k) 14, (l) 2, (m) 10, (n) 15, (o) 9

14. (a) 7, (b) 2, (c) 1, (d) 10, (e) 4, (f) 12, (g) 8, (h) 11, (i) 5, (j) 3, (k) 6, (l) 9 **15.** (a) 4, (b) 2, (c) 1, (d) 6, (e) 5, (f) 3

Chapter 29

1. dilation, expulsion, placental **2.** corpus luteum, human chorionic gonadotropin **3.** mesoderm, ectoderm, endoderm **4.** false **5.** a **6.** d **7.** b **8.** e **9.** b **10.** c **11.** (a) 6, (b) 2, (c) 11, (d) 5, (e) 10, (f) 1, (g) 3, (h) 8, (i) 9, (j) 4, (k) 7 **12.** (a) 3, (b) 6, (c) 5, (d) 2, (e) 1, (f) 4 **13.** (a) 3, (b) 4, (c) 5, (d) 1, (e) 2, (f) 6, (g) 7, (h) 10, (i) 8, (j) 9 **14.** (a) 7, (b) 3, (c) 14, (d) 6, (e) 4, (f) 10, (g) 8, (h) 13, (i) 5, (j) 2, (k) 1, (l) 9, (m) 16, (n) 11, (o) 15, (p) 12 **15.** (a) 3, (b) 6, (c) 4, (d) 1, (e) 9, (f) 2, (g) 5, (h) 8, (i) 7

Answers to Critical Thinking Questions

Chapter 1

1. No. Computed tomography is used to look at differences in tissue density. To assess activity in an organ such as the brain, a positron emission tomography (PET) scan or a single-photo-emission computerized tomography (SPECT) scan would both provide a colorized visual assessment of brain activity.

2. Stem cells are undifferentiated cells. Research using stem cells has shown that these undifferentiated may be prompted to differentiate into the specific cells needed to replace those which are damaged or malfunctioning.

3. Homeostasis is the relative constancy of the body's internal environment. Homeostasis is maintained as the body changes in response to shifting external and internal conditions. Body temperature should vary within a narrow range around normal body temperature (38 °C or 98.6 °F), which is above normal room temperature (usually around 72 °F (25 °C).

Chapter 2

1. Neither butter nor margarine are particularly good choices for frying eggs. Butter contains saturated fats that are associated with heart disease. However, many margarines contain hydrogenated or partially hydrogenated trans-fatty acids that also increase the risk of heart disease. An alternative would be frying the eggs in any of the mono- or polyunsaturated fats such as olive oil, peanut oil, or corn oil.

2. High body temperatures can be life-threatening, especially in infants. The increased temperature can cause denaturing of structural proteins and vital enzymes. When this happens, the proteins become nonfunctional. If the denatured enzymes are required for reactions that are necessary for life, then the infant could die.

3. Simply adding water to the table sugar does not cause it to break apart into monosaccharides. The water acts as a solvent, dissolving the sucrose, and forming a sugar-water solution. To complete the breakdown of table sugar to glucose and fructose would require the presence of the enzyme sucrase.

Chapter 3

1. Synthesis of mucin by ribosomes on rough endoplasmic reticulum, to transport vesicle, to entry face of Golgi complex, to transfer vesicle, to medial cisternae where protein is modified, to transfer vesicle, to exit face, to secretory vesicle, to plasma membrane where it undergoes exocytosis.

2. The cells that have been pierced are surrounded by membranes that have some fluidity. This fluidity allows the lipid bilayer to seal itself after being punctured.

3. Removing part of the small intestine will drastically reduce the cellular membrane surface area for absorption of digested nutrients. Although it can cause weight loss, it also can result in a lack of absorption of essential vitamins and minerals; dietary supplements are generally required.

4. In order to restore water balance to the cells, the runners need to consume hypotonic solutions. The water in the hypotonic solution will move from the blood, into the interstitial fluid, and then into the cells. Plain water works well; sports drinks contain water and some electrolytes (which may have been lost due to sweating) but will still be hypotonic in relation to the body cells.

Chapter 4

1. There are many possible adaptations, including: more adipose tissue for insulation; thicker bones for support; more red blood cells for oxygen transport; increased thickness of skin to prevent water loss; etc.

2. Infants tend to have a high proportion of brown, fat which contains many mitochondria and is highly vascularized. When broken down, brown fat produces heat that helps to maintain infants' body temperatures. This heat can also warm the blood, which then distributes the heat throughout the body.

3. Your bread-and-water diet is not providing you with the necessary nutrients to encourage tissue repair. You need proper amounts of many essential vitamins, especially vitamin C, which is required for repair of the matrix and blood vessels. Vitamin A is needed to help properly maintain epithelial tissue. Adequate protein is also needed in order to synthesis the structural proteins of the damaged tissue.

Chapter 5

1. The dust particles are primarily keratinocytes that are shed from the stratum corneum of the skin.

2. Cutting hair does not make the hair thicker, although certain hairstyles and application of hair products can give the appearance of thicker hair. Janet was born with a certain number of hair follicles on her head in which hair grows; cutting her hair does not produce additional hair follicles.

3. Chef Eduardo has damaged the nail matrix—the part of the nail that produces growth. Because the damaged area has not regrown properly, the nail matrix may be permanently damaged.

Chapter 6

1. Due to the strenuous, repetitive activity, Taryn has probably developed a stress fracture of her right tibia (lower leg bone). Stress fractures are due to repeated stress on a bone that causes microscopic breaks in the bone without any evidence of injury to other tissue. An x-ray would not reveal the stress fracture, but a bone scan would. Thus the bone scan would either confirm or negate the physician's diagnosis.

2. When Marcus broke his arm as a child, he injured his epiphyseal plate. Damage to the cartilage in the epiphyseal plate, resulted in premature closure of the plate, which interfered with the lengthwise growth of the arm bone.

3. Exercise causes mechanical stress on bones, but because there is effectively zero gravity in space, the pull of gravity on bones is missing. The lack of stress from gravity results in bone demineralization and weakness.

Chapter 7

1. Inability to open mouth—damage to the mandible, probably at temporal-mandibular joint; black eye—trauma to the ridge over the supraorbital margin; broken nose—probably damage to the nasal septum (includes the vomer, septal cartilage, and perpendicular plate of the ethmoid) and possibly the nasal bones; broken cheek—fracture of zygomatic bone; broken upper jaw—fracture of maxilla; damaged eye socket—fracture of parts of the sphenoid, frontal, ethmoid, palatine, zygomatic, lacrimal and maxilla (all compose the eye socket); punctured lung—damage to the thoracic vertebrae, which have punctured the lung.

2. Due to the repeated and extensive tension on his bone surfaces, Bubba would experience deposition of new bone tissue. His arm bones would be thicker and with increased raised areas (projections) where the tendons attach his muscles to bone.

3. The "soft area" being referred to is the anterior fontanel, located between the parietal and frontal bones. This is one of several areas of fibrous connective tissue in the skull that has not ossified; it should complete its ossification at 18–24 months after birth. Fontanels allow flexibility of the skull for childbirth and for brain growth after birth. The connective tissue will not allow passage of water, thus no brain damage will occur through simply washing the baby's hair.

Chapter 8

1. There are several characteristics of the bony pelves that can be used to differentiate male from female: (1) The pelvis in the female is wider and more shallow that the male's; (2) the pelvic brim of the female is larger and more oval; (3) the pubic arch has an angle greater than 90°; (4) the pelvic outlet is wider than in a male's; (5) the female's iliac crest is less curved and the ilium less vertical. Table 8.1 provides additional differences between female and male pelves. Age of the skeleton can be determined by the size of the bones, the presence or absence of epiphyseal plates, the degree of demineralization of the bones, and the general appearance of the "bumps" and ridges of bones.

2. Infants do have "flat feet" because their arches have not yet developed. As they begin to stand and walk, the arches should begin to develop in order to accommodate and support their body weight. The arches are usually fully developed by age 12 or 13, so Dad doesn't need to worry yet!

3. There are 14 phalanges in each hand: two bones in the thumb and three in each of the other fingers. Farmer White has lost five phalanges on his left hand (two in his thumb and three in his index finger), so he has nine remaining on his left and 14 remaining on his right for a total of 23.

Chapter 9

1. Katie's vertebral column, head, thighs, lower legs, lower arms, and fingers are flexed. Her lower arms and shoulders are medially rotated.

2. The knee joint is commonly injured, especially amongst athletes. The twisting of Jeremiah's leg could have resulted in a multitude of internal injuries to the knee joint but often football players suffer tearing of the anterior cruciate ligament and medial meniscus. The immediate swelling is due to blood from damaged blood vessels, damaged synovial membranes, and the torn meniscus. Continued swelling is a result of a buildup synovial fluid, which can result in pain and decreased mobility. Jeremiah's doctor may aspirate some of the fluid ("draining the water off his knee") and might want to perform arthroscopy to check for the extent of the knee damage.

3. Aunt Agnes had a hip replacement. The elderly are prone to hip degeneration caused by arthritis. Replacement of the damaged acetabulum and head of the femur with artificial devices can often restore movement at the hip joint, which is the most movable joint in the body. Although Agnes may not be "swinging" her legs behind her head, she probably will enjoy increased mobility!

Chapter 10

1. Muscle cells lose their ability to undergo mitosis after birth. Therefore, the increase in size is not due to an increase in the number of muscle cells but rather is due to enlargement of the existing muscle fibers (hypertrophy). This enlargement can occur from forceful, repetitive muscular activity. It will cause the muscle fibers to increase their production of internal structures such as mitochondria and myofibrils and produce an increase in the muscle fiber diameter.

2. The "dark meat" of both chickens and ducks is composed primarily of slow oxidative (SO) muscle fibers. These fibers contain large amounts of myoglobin and capillaries, which accounts for their dark color. In addition, these fibers contain large numbers of mitochondria and generate ATP by aerobic respiration. SO fibers are resistant to fatigue and can produce sustained contractions for many hours. The legs of chickens and ducks are used for support, walking, and swimming (in ducks), all activities in which endurance is needed. In addition, migrating ducks require SO fibers in their breasts to enable them to have enough energy to fly for extremely long distances while migrating. There may be some fast oxidative-glycolytic (FOG) fibers in the dark meat. FOG fibers also contain large amounts of myoglobin and capillaries, contributing to the dark color. They can use aerobic or anaerobic cellular respiration to generate ATP and have high-to-moderate resistance to fatigue. These fibers would be good for the occasional "sprint" that ducks and chickens undergo to escape dangerous situations. In contrast, the white meat of a chicken breast is composed primarily of fast glycolytic (FG) fibers. FG fibers have lower amounts of myoglobin and capillaries that give the meat its white color. There are also few mitochondria in FG fibers so these fibers generate ATP mainly by glycolysis. These fibers contract strongly and quickly and are adapted for intense anaerobic movements of short duration. Chickens occasionally use their breasts for flying extremely short distances, usually to escape prey or perceived danger, so FG fibers are appropriate for their breast muscle.

3. Destruction of the somatic motor neurons to skeletal muscle fibers will result in a loss of stimulation to the skeletal muscles. When not stimulated on a regular basis, a muscle begins to lose its muscle tone. Through lack of use, the muscle fibers will weaken, begin to decrease in size, and can be replaced by fibrous connective tissue, resulting in a type of denervation atrophy. A lack of stimulation of the breathing muscles (especially the diaphragm) from motor neurons can result in inability of the breathing muscles to contract, thus causing respiratory paralysis and possibly death of the individual from respiratory failure.

Chapter 11

1. All of the following could occur on the affected (right) side of the face: (1) drooping of eyelid—levator palpebrae superioris; (2) drooping of the mouth, drooling, keeping food in mouth—orbicularis oris, buccinator; (3) uneven smile—zgyomaticus major, levator labii superioris, risorius; (4) unable to wrinkle forehead—occipitofrontalis; (5) trouble sucking through a straw—buccinator.

2. Bulbospongiosus, external urethral sphincter, and deep transverse perineal.

3. The rotator cuff is formed by a combination of the tendons of four deep muscles of the shoulder—supscapularis, supraspinatus, infraspinatus, and teres minor. These muscles add strength and stability to the shoulder joint. Although any of the muscles' tendons can be injured, the subscapularis is most often damaged. Dependent upon the injured muscle, Jose may have trouble medially rotating his arm (subscapularis), abducting his arm (supraspinatus), laterally rotating his arm (infraspinatus, teres minor), adducting his arm (infraspinatus, teres minor), or extending his arm (teres minor).

Chapter 12

1. Smelling the coffee and hearing the alarm are somatic sensory, stretching and yawning are somatic motor, salivating is autonomic (parasympathetic) motor, stomach rumble is enteric motor.
2. Demyelinaton or destruction of the myelin sheath can lead to multiple problems, especially in infants and children whose myelin sheaths are still in the process of developing. The affected axons deteriorate, which will interfere with function in both the CNS and PNS. There will be lack of sensation and loss of motor control with less rapid and less coordinated body responses. Damage to the axons in the CNS can be permanent and Ming's brain development may be irreversibly affected.
3. Dr. Moro could develop a drug that: (1) is an agonist of substance P; (2) blocks the breakdown of substance P; (3) blocks the reuptake of substance P; (4) promotes the release of substance P; (5) suppresses the release of enkephalins.

Chapter 13

1. The needles will pierce the epidermis, dermis, and subcutaneous layer and then go between the vertebrae through the epidural space, the dura mater, the subdural space, the arachnoid mater, and into the CSF in the subarchnoid space. CSF is produced in the brain, and the spinal meninges are continuous with the cranial meninges.
2. The anterior gray horns contain cell bodies of somatic motor neurons and motor nuclei that are responsible for the nerve impulses for skeletal muscle contraction. Because the lower cervical region is affected (brachial plexus, C5–C8), you would expect that Sunil may have trouble with movement in his shoulder, arm, and hand on the affected side.
3. Allyson has damaged her posterior columns in the lower (lumbar) region of the spinal cord. The posterior columns are responsible for transmitting nerve impulses responsible for awareness of muscle position (proprioception) and discriminative touch—which are affected in Allyson—as well as other functions such as two-point discrimination, light pressure sensations, and vibration sensations.

Chapter 14

1. Movement of the right arm is controlled by the left hemisphere's primary motor area, located in the precentral gyrus. Speech is controlled by Broca's area in the left hemisphere's frontal lobe just superior to the lateral cerebral sulcus.
2. Nicky's right facial nerve has been affected; she is suffering from Bell's palsy due to the viral infection. The facial nerve controls contraction of skeletal muscles of the face, tear gland and salivary gland secretion, as well as conveying sensory impulses from many of the taste buds on the tongue.
3. You will need to design a drug that can get through the brain's blood–brain barrier (BBB). The drug should be lipid- or water-soluble. If the drug can open a gap between the tight junctions of the endothelial cells of the brain capillaries, it would be more likely to pass through the BBB. Targeting the drug to enter the brain in certain areas near the third and fourth ventricles (the circumventricular organs) might be an option as the BBB is entirely absent in those areas and the capillary endothelium is more permeable, allowing the blood-borne drug to more readily enter the brain tissue.

Chapter 15

1. Digestion and relaxation are controlled by increased stimulation of the parasympathetic division of the ANS. The salivary glands, pancreas, and liver will show increased secretion; the stomach and intestines will have increased activity; the gallbladder will have increased contractions; heart contractions will have decreased force and rate. Following is the nerve supply to each listed organ: salivary glands—facial nerves (cranial nerve VII) and glossopharyngeal nerves (cranial nerve IX); pancreas, liver, stomach, gallbladder, intestines and heart—vagus nerves (cranial nerve X).
2. Ciara experienced one of the "E situations" (emergency in her case), which has activated the fight-or-flight response. Some noticeable effects of increased sympathetic activity include an increase in heart rate, sweating on the palms, and contraction of the arrector pili muscles, which causes the goose flesh. Secretion of epinephrine and norepinphrine from the adrenal medullae will intensify and prolong the responses.
3. Mrs. Young needs to slow down the activity of her digestive system, which seems to be experiencing increased parasympathetic response. A parasympathetic blocking agent is needed. Because the stomach and intestines have muscarinic receptors, she needs to be provided with a muscarinic blocking agent (such as atropine), which will result in decreased motility in the stomach and intestines.

Chapter 16

1. Chemoreceptors in the nose detect odors. Proprioceptors detect body position and are involved in equilibrium. The chemoreceptors in the nose are rapidly adapting, whereas proprioceptors are slowly adapting. Thus the smell faded while the sensation of motion remained.
2. Thermal (heat) receptors in her left hand detect the stimulus. A nerve impulse is transmitted to the spinal cord through first-order neurons with cell bodies in posterior root ganglia. The impulses travel into the spinal cord where the first-order neurons synapse with second-order neurons, whose cell bodies are located in the posterior gray horn of the spinal cord. The axons of the second-order neurons decussate to the right side in the spinal cord and then the impulses ascend through the lateral spinothalamic tract. The axons of the second-order neurons end in the ventral posterior nucleus of the thalamus where they synapse with the third-order neurons. Axons of the third-order neurons transmit impulses to the specific primary somatosensory areas in the postcentral gyrus of the right parietal lobe.
3. When Marvin settled down for the night, he passed through Stages 1–3 of NREM sleep. Sleepwalking occurred when he was in Stage 4 (slow-wave sleep). Because this is the deepest stage of sleep, his mother was able to return him to his bed without awakening him. Marvin then cycled through REM and NREM sleep. His dreaming occurred during the REM phases of sleep. The noise of the alarm clock provided the sensory stimulus that stimulated the reticular activating system. Activation of this system sends numerous nerve impulses to widespread areas of the cerebral cortex, both directly and via the thalamus. The result is the state of wakefulness.

Chapter 17

1. Damage to the facial nerve would affect smell, taste, and hearing. Within the nasal epithelium and connective tissue, both the supporting cells and olfactory glands are innervated by branches of the facial nerve. Without input from the facial nerve, there will be a lack of mucus production required to dissolve odorants. The facial nerve also serves taste bids in the anterior two-thirds of the tongue, so damage can affect taste sensations. Hearing will be affected by

a damaged facial nerve because the stapedius muscle, which is attached to the stapes, is innervated by the facial nerve. Contraction of the stapedius muscle helps to protect the inner from prolonged loud noises. Damage to the facial nerve will result in sounds that are excessively loud, resulting in more susceptibility to damage by prolonged loud noises.

2. With age, Gertrude has lost much of her sense of smell and taste due to a decline in olfactory and gustatory receptors. Since smell and taste are intimately linked, food no longer smells nor tastes as good to Gertrude. Gertrude has presbyopia, a loss of lens elasticity, which makes it difficult to read. She may also be experiencing age-related loss of sharpness of vision and depth perception. Gertrude's hearing difficulties could be a result of damage to hair cells in the organ of Corti or degeneration of the nerve pathway for hearing. The "buzzing" Gertrude hears may be tinnitus, which also occurs more frequently in the elderly.
3. Some of the eyedrops placed in the eye may pass through the nasolacrimal duct into the nasal cavity where olfactory receptors are stimulated. Because most "tastes" are actually smells, the child will "taste" the medicine from her eye.

Chapter 18

1. Amanda has an enlarged thyroid gland, or goiter. The goiter is probably due to hypothyroidism, which is causing the weight gain, fatigue, mental dullness, and other symptoms.
2. Amanda's problem is her pituitary gland, which is not secreting normal levels of TSH. Rising thyroxine (T_4) levels after the TSH injection indicates that her thyroid is functioning normally and able to respond to the increased TSH levels. If the thyroxine levels had not risen, then the problem would have been her thyroid gland.
3. Mr. Hernandez has diabetes insipidus caused by either insufficient production or release of ADH due to hypothalamus or posterior pituitary damage. He also could have defective ADH receptors in the kidneys. Diabetes insipidus is characterized by production of large volumes of urine, dehydration, and increased thirst, but with no glucose or ketones present in the urine (which would be indicative of diabetes mellitus rather than diabetes insipidus).

Chapter 19

1. The broad spectrum antibiotics may have destroyed the bacteria that caused Shilpa's bladder infection but also destroyed the naturally occurring large intestine bacteria that produce vitamin K. Vitamin K is required for the synthesis of four clotting factors (II, VII, IX and X). Without these clotting factors present in normal amounts, Shilpa will experience clotting problems until the intestinal bacteria reach normal levels and produce additional vitamin K.
2. Mrs. Brown's kidney failure is interfering with her ability to produce erythropoietin (EPO). Her physician can prescribe Epoetin alfa, a recombinant EPO, which is very effective in treating the decline in RBC production with renal failure.
3. A primary problem Thomas may experience is with clotting. Clotting time becomes longer because the liver is responsible for producing many of the clotting factors and clotting proteins such as fibrinogen. Thrombopoietin, which stimulates the formation of platelets, is also produced in the liver. In addition, the liver is responsible for eliminating bilirubin, produced from the breakdown of RBCs. With a malfunctioning liver, the bilirubin will accumulate, resulting in jaundice. In addition, there can be decreased concentrations of the plasma protein albumin, which can affect blood pressure.

Chapter 20

1. The dental procedures introduced bacteria into Gerald's blood. The bacteria colonized his endocardium and heart valves, resulting in bacterial endocarditis. Gerald may have had a previously undetected heart murmur, or the heart murmur may have resulted from his endocarditis. His physician will want to monitor his heart to watch for further damage to the valve.
2. Extremely rapid heart rates can result in a decreased stroke volume due to insufficient ventricular filling. As a result, the cardiac output will decline to the point where there may not be enough blood reaching the central nervous system. She initially may experience lightheadedness but could lose consciousness if the cardiac output declines dramatically.
3. Mr. Perkins is suffering from angina pectoris and has several risk factors for coronary artery disease such as smoking, obesity, sedentary lifestyle, and male gender. Cardiac angiography involves the use of a cardiac catheter to inject a radiopaque medium into the heart and its vessels. The angiogram may reveal blockages such as atherosclerotic plaques in his coronary arteries.

Chapter 21

1. The hole in the heart was the foramen ovale, which is an opening between the right and left atria. In fetal circulation it allows blood to bypass the right ventricle, enter the left atrium and join systemic circulation. The "hole" should close shortly after birth to become the fossa ovalis. Closure of the foramen ovale after birth will allow the deoxgenated blood from the right atrium to enter pulmonary circulation so that the blood can become oxygenated prior to entering systemic circulation. If closure doesn't occur, surgery may be required.
2. Michael is suffering from hypovolemic shock due to the loss of blood. The low blood pressure is a result of low blood volume and a subsequent decrease in cardiac output. His rapid, weak pulse is an attempt of the heart to compensate for the decrease in cardiac output through sympathetic stimulation of the heart and increased blood levels of epinephrine and norepinephrine. His pale, cool, and clammy skin is a result of sympathetic constriction of the blood vessels of the skin and sympathetic stimulation of sweat glands. The lack of urine production is due to increased secretion of aldosterone and ADH, both of which are produced to increase blood volume in order to compensate for Michael's hypotension. The fluid loss from his bleeding results in activation of the thirst center in the hypothalamus. His confusion and disorientation is caused by a reduced oxygen supply to the brain from the decreased cardiac output.
3. Maureen has varicose veins, a condition in which the venous valves become leaky. The leaking valves allow the backflow of blood and an increased pressure that distends the veins and allows fluid to leak into the surrounding tissue. Standing on hard surfaces for long periods of time can cause varicosities to develop. Maureen needs to elevate her legs when possible to counteract the effects of gravity on the blood flow in the lower legs. She could also utilize support hose, which adds external support for the superficial veins, much like skeletal muscle does for deeper veins. If the varices become severe, Maureen may require more extensive treatment such as sclerotherapy, radiofrequency endovenous occlusion, laser occlusion, or stripping.

Chapter 22

1. The influenza vaccination introduces a weakened or killed virus (which will not cause the disease) to the body. The immune system recognizes the antigen and mounts a primary immune response. Upon exposure to the same flu virus that was in the vaccine, the body

will produce a secondary response, which will usually prevent a case of the flu. This is artificially acquired active immunity.

2. Mrs. Franco's lymph nodes were removed because metastasis of cancerous cells can occur through the lymph nodes and lymphatic vessels. Mrs. Franco's swelling is a lymphedema that is occurring due to the buildup of interstitial fluid from interference with drainage in the lymph vessels.
3. Tariq's physician would need to perform an antibody titer, which is a measure of the amount of antibody in the serum. If Tariq has previously been exposed to mumps (or been vaccinated for mumps), his blood should have elevated levels of IgG antibodies after this exposure from his sister. His immune system would be experiencing a secondary response. If he has not previously had mumps and has contracted mumps from his sister, his immune system would initiate a primary response. In that case, his blood would show an elevated titer of IgM antibodies, which are secreted by plasma cells after an initial exposure to the mumps antigen.

Chapter 23

1. Aretha's excess mucus production is causing blockage of the paranasal sinuses, which are used as hollow resonating chambers for singing and speech. In addition, her sore throat could be due to inflammation of the pharynx and larynx, which will affect their normal functions. Normally, the pharynx also acts as a resonating chamber and the true vocal cords, located in the larynx, vibrate for speech and singing. Inflammation of the true vocal cords (laryngitis) interferes with their ability to freely vibrate, which will affect both singing and speech.
2. In emphysema, there is destruction of the alveolar walls, producing abnormally large air spaces that remain filled with air during exhalation. The destruction of alveoli decreases the surface area for gas exchange across the respiratory membrane, resulting in a decreased blood O_2 level. Damage to the alveolar walls also causes a loss of elasticity, making exhalation more difficult. This can result in a buildup of CO_2. Emphysema causes constriction of bronchioles. This narrowing of the breathing airways increases resistance and results in more pressure required to maintain airflow. Cigarette smoke contains nicotine, carbon monoxide, and a variety of irritants, all of which affect the lungs. Nicotine constricts terminal bronchioles, decreasing the air flow into and out of the lungs; carbon monoxide binds to hemoglobin, reducing its ability to carry oxygen; irritants such as tar and fine particulate matter destroy cilia and increase mucus secretion, interfering with the ability of the respiratory passages to cleanse themselves.
3. The squirrel's nest blocked the passage of exhaust gas from the furnace, causing an accumulation of carbon monoxide (CO), a colorless, odorless gas, in the home. As they were sleeping, their blood was saturated with CO, which has a stronger affinity for hemoglobin than oxygen. As a result, the Robinsons became oxygen deficient. Without adequate oxygenation of the brain, the Robinsons died during their sleep.

Chapter 24

1. HCl has several important roles in digestion. HCl stimulates the secretion of hormones that promote the flow of bile and pancreatic juice. The presence of HCl destroys certain microbes that may have been ingested with food. HCl begins denaturing proteins in food, and provides the proper chemical environment for activating pepsinogen into pepsin, which breaks apart certain peptide bonds in proteins. It also helps in the action of gastric lipase, which splits triglycerides in fat molecules found in milk into fatty acids and monoglycerides.
2. Blockage of the pancreatic and bile ducts prevent pancreatic digestive enzymes and bile from reaching the duodenum. As a consequence, there will be problems digesting carbohydrates, proteins, nucleic acids, and lipids. Of particular concern is lipid digestion since the pancreatic juices contain the primary lipid-digesting enzyme. Fats will not be adequately digested, and Trey's feces will contain larger than normal amounts of lipids. In addition, the lack of bile salts will affect the body's ability to emulsify lipids and to form micelles required for absorption of fatty acids and monoglycerides (from lipid breakdown). When lipids are not absorbed properly, then there will be malabsorption of the lipid-soluble vitamins (A, D, E, and K).
3. Antonio experienced gastroesophageal reflux. The stomach contents backed up (refluxed) into Antonio's esophagus due to a failure of the lower esophageal sphincter to fully close. The HCl from the stomach irritated the esophageal wall, which resulted in the burning sensation he felt; this is commonly known as "heartburn," even though it is not related to the heart. Antonio's recent meal worsened the problem. Alcohol and smoking both can cause the sphincter to relax, while certain foods such tomatoes, chocolate, and coffee can stimulate stomach acid secretion. In addition, lying down immediately after a meal can exacerbate the problem.

Chapter 25

1. Ingestion of cyanide affects cellular respiration. The cyanide binds to the cytochrome oxidase complex in the inner membrane of mitochondria. Blocking this complex interferes with the last step in electron transport in aerobic ATP production. Jane Doe's body would quickly run out of energy to perform vital functions, resulting in her death.
2. Glenn's total cholesterol and LDL levels are very high, while his HDL levels are low. Total cholesterol above 239 mg/dL and LDL above 159 mg/dL are considered high. The ratio of total cholesterol (TC) to HDL-cholesterol is a predictor of the risk of developing coronary artery disease. Glenn's TC to HDL is 15; a ratio above 4 is undesirable. His ratio places him at high risk of developing coronary artery disease. In addition, for every 50 mg/dL TC over 200 mg/dL, the risk of a heart attack doubles. Glenn needs to reduce his TC and LDL-cholesterol while raising his HDL-cholesterol levels. LDLs contribute to fatty plaque formation on coronary artery walls. On the other hand, HDLs help remove excess cholesterol from the blood, which helps decrease the risk of coronary artery disease. Glenn will need to reduce his dietary intake of total fat, saturated fats, and cholesterol, all of which contribute to raising LDL levels. Exercise will raise HDL levels. If those changes are not successful, drug therapy may be required.
3. The goal of weight loss programs is to reduce caloric intake so that the body utilizes stored lipids as an energy source. As part of that desired lipid metabolism, ketone bodies are produced. Some of these ketone bodies will be excreted in the urine. If no ketones are present, then Marissa's body is not breaking down lipids. Only through using fewer calories than needed will her body break down the stored fat and release ketones. Thus, she must be eating more calories than needed to support her daily activities—she is "cheating."

Chapter 26

1. Without reabsorption, initially 105–125 mL of filtrate would be lost per minute, assuming normal glomerular filtration rate. Fluid loss from the blood would cause a decrease in blood pressure, and therefore a decrease in GBHP. When GBHP dropped below 45 mmHg, filtration would stop (assuming normal CHP and BCOP) because NFP would be zero.

2. a. Although normally pale yellow, urine color can vary based upon concentration, diet, drugs, and disease. A dark yellow color would not necessarily indicate a problem, but further investigation may be needed. Turbidity or cloudiness can be caused by urine that has been standing for a period of time, from certain foods, or from bacterial infections. Further investigation is needed. b. Ammonia-like odor occurs when the urine sample is allowed to stand. c. Albumin should not be present in urine (or present in very small amounts) because it is too large to pass through the filtration membranes. The presence of high levels of albumin is cause for concern as it indicates damage to the filtration membranes. d. Casts are hardened masses of material that are flushed out in the urine. The presence of casts is not normal and indicates a pathology. e. The pH of normal urine ranges from 4.8 to 8.0. A pH of 5.5 is in normal range. f. Hematuria is the presence of red blood cells in the urine. It can occur with certain pathological conditions or from kidney trauma. Hematuria may occur if the urine sample was contaminated with menstrual blood.
3. Bruce has developed renal calculi (kidney stones), which are blocking his ureters and interfering with the flow of urine from the kidneys to the bladder. The rhythmic pains are a result of the peristaltic contractions of the ureters as they attempt to move the stones toward the bladder. Bruce can wait for the stones to pass, can have them surgically removed, or can use shock-wave lithotripsy to break apart the stones into smaller fragments that can be eliminated with urine. To prevent future episodes, Bruce needs to watch his diet (limit calcium), and drink fluids, and may need drug intervention.

Chapter 27

1. The loss of stomach acids can result in metabolic alkalosis. Robin's HCO_3^- levels would be higher than normal. She would be hypoventilating in order to decrease her pH by slowing the loss of CO_2. Excessive vomiting can result in hyponatremia, hypokalemia, and hypochloremia. Both hyponatremia and hypokalemia can cause mental confusion.
2. (Step 1) pH = 7.30 indicates slight acidosis, which could be caused by elevated PCO_2 of lowered HCO_3^-. (Step 2) The HCO_3^- is lower than normal (20 mEq/liter), so (Step 3) the cause is metabolic. (Step 4) The PCO_2 is lower than normal (32 mmHg), so hyperventilation is providing some compensation. Diagnosis: Henry has partially compensated metabolic acidosis. A possible cause is kidney damage that resulted from interruption of blood flow during the heart attack.
3. Sam will experience increased fluid loss through increased evaporation from the skin and water vapor from the respiratory system through his increased respiratory rate. His insensible water loss will also increase (loss of water from mucous membranes of the mouth and respiratory system). Sam will have a decrease in urine formation.

Chapter 28

1. Monica's excessive training has resulted in an abnormally low amount of body fat. A certain amount of body fat is needed in order to produce the hormones required for the ovarian cycle. Several hormones are involved. Her ammenorrhea is due to a lack of gonadotropin-releasing hormone, which in turn reduces the release of LH and FSH. Her follicles with their enclosed ova fail to develop and ovulation will not occur. In addition, synthesis of estrogens and progesterone declines from the lack of hormonal feedback. Usually a gain of weight will allow normal hormonal feedback mechanisms to return.
2. Along with estrogens, progesterone helps to prepare the endometrium for possible implantation of a zygote by promoting growth of the endometrium. The endometrial glands secrete glycogen, which will help sustain an embryo if one should implant. If implantation occurs, progesterone helps maintain the endometrium for the developing fetus. In addition, it helps prepare mammary glands to secrete milk. It inhibits the release of GnRH and LH, which stops a new ovarian cycle from occurring.
3. The ductus deferens is cut and tied in a vasectomy. This stops the release of sperm into the ejaculatory duct and urethra. Mark will still produce the secretions from his accessory glands (prostate, seminal vesicles, bulbourethral glands) in his ejaculate. In addition, a vasectomy does not affect sexual performance; he will be able to achieve erection and ejaculation as those events are nervous system responses. Mark and Isabella will need to use additional protection for a period of time after the vasectomy as sperm can be stored in the ductus deferens and remain viable for several months. The physician will want to monitor his ejaculate for sperm levels for a short period of time.

Chapter 29

1. As part of the feedback mechanism for lactation, oxytocin is released from the posterior pituitary. It is carried to the mammary glands where it causes milk to be released into the mammary ducts (milk ejection). The oxytocin is also transported in the blood to the uterus, which contains oxytocin receptors on the myometrium. The oxytocin causes contraction of the myometrium, resulting in the painful sensations that Kathy is experiencing. The uterine contractions can help return the uterus back to its prepregnancy size.
2. Sex-linked genetic traits, such as hemophilia, are present on the X chromosomes but not on the Y chromosomes. In males, the X chromosome is always inherited from the mother, and the Y chromosome from the father. Thus, Jack's hemophilia gene was inherited from his mother on his X chromosome. The gene for hemophilia is a recessive gene. His mother would need two recessive genes, one on each of her X chromosomes, to be hemophiliac. His father must carry the dominant (nonhemophiliac) gene on his X chromosome, so he also would not have hemophilia.
3. The cord blood is a source of pluripotent stem cells, which are unspecialized cells that have the potential to specialize into cells with specific functions. The hope is that stem cells can be used to generate cells and tissues to treat a variety of disorders. It is assumed that the tissues would not be rejected since they would contain the same genetic material as the patient—in this case Alisa's baby.

G

Glossary

Pronunciation Key

1. The most strongly accented syllable appears in capital letters, for example, bilateral (bī-LAT-er-al) and diagnosis (dī-ag-NŌ-sis).
2. If there is a secondary accent, it is noted by a prime (′), for example, constitution (kon′-sti-TOO-shun) and physiology (fiz′-ē-OL-ō-jē). Any additional secondary accents are also noted by a prime, for example, decarboxylation (dē′-kar-bok′-si-LĀ-shun).
3. Vowels marked by a line above the letter are pronounced with the long sound, as in the following common words:
   ā as in *māke* — *ō* as in *pōle*
   ē as in *bē* — *ū* as in *cūte*
   ī as in *īvy*
4. Vowels not marked by a line above the letter are pronounced with the short sound, as in the following words:
   a as in *above* or *at* — *o* as in *not*
   e as in *bet* — *u* as in *bud*
   i as in *sip*
5. Other vowel sounds are indicated as follows:
   oy as in *oil*
   oo as in *root*
6. Consonant sounds are pronounced as in the following words:
   b as in *bat* — *m* as in *mother*
   ch as in *chair* — *n* as in *no*
   d as in *dog* — *p* as in *pick*
   f as in *father* — *r* as in *rib*
   g as in *get* — *s* as in *so*
   h as in *hat* — *t* as in *tea*
   j as in *jump* — *v* as in *very*
   k as in *can* — *w* as in *welcome*
   ks as in *tax* — *z* as in *zero*
   kw as in *quit* — *zh* as in *lesion*
   l as in *let*

A

Abdomen (ab-DŌ-men *or* AB-dō-men) The area between the diaphragm and pelvis.

Abdominal (ab-DŌM-i-nal) **cavity** Superior portion of the abdominopelvic cavity that contains the stomach, spleen, liver, gallbladder, most of the small intestine, and part of the large intestine.

Abdominal thrust maneuver A first-aid procedure for choking. Employs a quick, upward thrust against the diaphragm that forces air out of the lungs with sufficient force to eject any lodged material. Also called the **Heimlich** (HĪM-lik) **maneuver.**

Abdominopelvic (ab-dom′-i-nō-PEL-vik) **cavity** Inferior to the diaphragm that is subdivided into a superior abdominal cavity and an inferior pelvic cavity.

Abduction (ab-DUK-shun) Movement away from the midline of the body.

Abortion (a-BOR-shun) The premature loss (spontaneous) or removal (induced) of the embryo or nonviable fetus; miscarriage due to a failure in the normal process of developing or maturing.

Abscess (AB-ses) A localized collection of pus and liquefied tissue in a cavity.

Absorption (ab-SORP-shun) Intake of fluids or other substances by cells of the skin or mucous membranes; the passage of digested foods from the gastrointestinal tract into blood or lymph.

Accessory duct A duct of the pancreas that empties into the duodenum about 2.5 cm (1 in.) superior to the ampulla of Vater (hepatopancreatic ampulla). Also called the **duct of Santorini** (san′-tō-RĒ-nē).

Acetabulum (as′-e-TAB-ū-lum) The rounded cavity on the external surface of the hip bone that receives the head of the femur.

Acetylcholine (as′-ē-til-KŌ-lēn) **(ACh)** A neurotransmitter liberated by many peripheral nervous system neurons and some central nervous system neurons. It is excitatory at neuromuscular junctions but inhibitory at some other synapses (for example, it slows heart rate).

Achalasia (ak′-a-LĀ-zē-a) A condition, caused by malfunction of the myenteric plexus, in which the lower esophageal sphincter fails to relax normally as food approaches. A whole meal may become lodged in the esophagus and enter the stomach very slowly. Distension of the esophagus results in chest pain that is often confused with pain originating from the heart.

Achilles tendon *See* **Calcaneal tendon.**

Acini (AS-i-nē) Groups of cells in the pancreas that secrete digestive enzymes.

Acoustic (a-KOOS-tik) Pertaining to sound or the sense of hearing.

Acquired immunodeficiency syndrome (AIDS) A fatal disease caused by the human immunodeficiency virus (HIV). Characterized by a positive HIV-antibody test, low helper T cell count, and certain indicator diseases (for example Kaposi's sarcoma, pneumocystis carinii pneumonia, tuberculosis, fungal diseases). Other symptoms include fever or night sweats, coughing, sore throat, fatigue, body aches, weight loss, and enlarged lymph nodes.

Acrosome (AK-rō-sōm) A lysosomelike organelle in the head of a sperm cell containing enzymes that facilitate the penetration of a sperm cell into a secondary oocyte.

Actin (AK-tin) A contractile protein that is part of thin filaments in muscle fibers.

Action potential (AP) An electrical signal that propagates along the membrane of a neuron or muscle fiber (cell); a rapid change in membrane potential that involves a depolarization followed by a repolarization. Also called a **nerve action potential** or **nerve impulse** as it relates to a neuron, and a **muscle action potential** as it relates to a muscle fiber.

Activation (ak′-ti-VĀ-shun) **energy** The minimum amount of energy required for a chemical reaction to occur.

Active transport The movement of substances across cell membranes against a concentration gradient, requiring the expenditure of cellular energy (ATP).

Acute (a-KŪT) Having rapid onset, severe symptoms, and a short course; not chronic.

Adaptation (ad′-ap-TĀ-shun) The adjustment of the pupil of the eye to changes in light intensity. The property by which a sensory neuron relays a decreased frequency of action potentials from a receptor, even though the strength of the stimulus remains constant; the decrease in perception of a sensation over time while the stimulus is still present.

Adduction (ad-DUK-shun) Movement toward the midline of the body.

Adenoids (AD-e-noyds) The pharyngeal tonsils.

Adenosine triphosphate (a-DEN-ō-sēn trī-FOS-fāt) **(ATP)** The main energy currency in living cells; used to transfer the chemical energy needed for metabolic reactions. ATP consists of the purine base *adenine* and the five-carbon sugar *ribose,* to which are added, in linear array, three *phosphate* groups.

Adhesion (ad-HĒ-zhun) Abnormal joining of parts to each other.

Adipocyte (AD-i-pō-sīt) Fat cell, derived from a fibroblast.

Adipose (AD-i-pōz) **tissue** Tissue composed of adipocytes specialized for triglyceride storage and present in the form of soft pads between various organs for support, protection, and insulation.

Adrenal cortex (a-DRĒ-nal KOR-teks) The outer portion of an adrenal gland, divided into three zones; the zona glomerulosa secretes mineralocorticoids, the zona fasciculata secretes glucocorticoids, and the zona reticularis secretes androgens.

Adrenal glands Two glands located superior to each kidney. Also called the **suprarenal** (soo′-pra-RĒ-nal) **glands.**

Adrenal medulla (me-DUL-a) The inner part of an adrenal gland, consisting of cells that secrete epinephrine, norepinephrine, and a small amount of dopamine in response to stimulation by sympathetic preganglionic neurons.

Adrenergic (ad′-ren-ER-jik) **neuron** A neuron that releases epinephrine (adrenaline) or norepinephrine (noradrenaline) as its neurotransmitter.

Adrenocorticotropic (ad-rē′-nō-kor-ti-kō-TRŌP-ik) **hormone (ACTH)** A hormone produced by the anterior pituitary that influences the production and secretion of certain hormones of the adrenal cortex.

Adventitia (ad-ven-TISH-a) The outermost covering of a structure or organ.

Aerobic (air-Ō-bik) Requiring molecular oxygen.

Afferent arteriole (AF-er-ent ar-TĒ-rē-ōl) A blood vessel of a kidney that divides into the capillary network called a glomerulus; there is one afferent arteriole for each glomerulus.

Agglutination (a-gloo-ti-NĀ-shun) Clumping of microorganisms or blood cells, typically due to an antigen–antibody reaction.

Aggregated lymphatic follicles Clusters of lymph nodules that are most numerous in the ileum. Also called **Peyer's** (PĪ-erz) **patches.**

Albinism (AL-bin-izm) Abnormal, nonpathological, partial, or total absence of pigment in skin, hair, and eyes.

Aldosterone (al-DOS-ter-ōn) A mineralocorticoid produced by the adrenal cortex that promotes sodium and water reabsorption by the kidneys and potassium excretion in urine.

Allantois (a-LAN-tō-is) A small, vascularized outpouching of the yolk sac that serves as an early site for blood formation and development of the urinary bladder.

Alleles (a-LĒLZ) Alternate forms of a single gene that control the same inherited trait (such as type A blood) and are located at the same position on homologous chromosomes.

Allergen (AL-er-jen) An antigen that evokes a hypersensitivity reaction.

Alopecia (al′-ō-PĒ-shē-a) The partial or complete lack of hair as a result of factors such as genetics, aging, endocrine disorders, chemotherapy, and skin diseases.

Alpha (AL-fa) **cell** A type of cell in the pancreatic islets (islets of Langerhans) in the pancreas that secretes the hormone glucagon. Also termed an **A cell.**

Alpha receptor A type of receptor for norepinephrine and epinephrine; present on visceral effectors innervated by sympathetic postganglionic neurons.

Alveolar duct Branch of a respiratory bronchiole around which alveoli and alveolar sacs are arranged.

Alveolar macrophage (MAK-rō-fāj) Highly phagocytic cell found in the alveolar walls of the lungs. Also called a **dust cell.**

Alveolar sac A cluster of alveoli that share a common opening.

Alveolus (al-VĒ-ō-lus) A small hollow or cavity; an air sac in the lungs; milk-secreting portion of a mammary gland. *Plural* is **alveoli** (al-VĒ-ol-ī).

Alzheimer (ALTZ-hī-mer) **disease (AD)** Disabling neurological disorder characterized by dysfunction and death of specific cerebral neurons, resulting in widespread intellectual impairment, personality changes, and fluctuations in alertness.

Amenorrhea (ā-men-ō-RĒ-a) Absence of menstruation.

Amnesia (am-NĒ-zē-a) A lack or loss of memory.

Amnion (AM-nē-on) A thin, protective fetal membrane that develops from the epiblast; holds the fetus suspended in amniotic fluid. Also called the **"bag of waters."**

Amniotic (am′-nē-OT-ik) **fluid** Fluid in the amniotic cavity, the space between the developing embryo (or fetus) and amnion; the fluid is initially produced as a filtrate from maternal blood and later includes fetal urine. It functions as a shock absorber, helps regulate fetal body temperature, and helps prevent desiccation.

Amphiarthrosis (am′-fē-ar-THRŌ-sis) A slightly movable joint, in which the articulating bony surfaces are separated by fibrous connective tissue or fibrocartilage to which both are attached; types are syndesmosis and symphysis.

Ampulla (am-PUL-la) A saclike dilation of a canal or duct.

Ampulla of Vater *See* **Hepatopancreatic ampulla.**

Anabolism (a-NAB-ō-lizm) Synthetic, energy-requiring reactions whereby small molecules are built up into larger ones.

Anaerobic (an-ar-Ō-bik) Not requiring oxygen.

Anal (Ā-nal) **canal** The last 2 or 3 cm (1 in.) of the rectum; opens to the exterior through the anus.

Anal column A longitudinal fold in the mucous membrane of the anal canal that contains a network of arteries and veins.

Anal triangle The subdivision of the female or male perineum that contains the anus.

Analgesia (an-al-JĒ-zē-a) Pain relief; absence of the sensation of pain.

Anaphase (AN-a-fā z) The third stage of mitosis in which the chromatids that have separated at the centromeres move to opposite poles of the cell.

Anaphylaxis (an′-a-fi-LAK-sis) A hypersensitivity (allergic) reaction in which IgE antibodies attach to mast cells and basophils, causing them to produce mediators of anaphylaxis (histamine, leukotrienes, kinins, and prostaglandins) that bring about increased blood permeability, increased smooth muscle contraction, and increased mucus production. Examples are hay fever, hives, and anaphylactic shock.

Anastomosis (a-nas-tō-MŌ-sis) An end-to-end union or joining of blood vessels, lymphatic vessels, or nerves.

Anatomic dead space Spaces of the nose, pharynx, larynx, trachea, bronchi, and bronchioles totaling about 150 mL of the 500 mL in a quiet breath (tidal volume); air in the anatomic dead space does not reach the alveoli to participate in gas exchange.

Anatomical (an′-a-TOM-i-kal) **position** A position of the body universally used in anatomical descriptions in which the body is erect, the head is level, the eyes face forward, the upper limbs are at the sides, the palms face forward, and the feet are flat on the floor.

Anatomy (a-NAT-ō-mē) The structure or study of the structure of the body and the relation of its parts to each other.

Androgens (AN-drō-jenz) Masculinizing sex hormones produced by the testes in males and the adrenal cortex in both sexes; responsible for libido (sexual desire); the two main androgens are testosterone and dihydrotestosterone.

Anemia (a-NĒ-mē-a) Condition of the blood in which the number of functional red blood cells or their hemoglobin content is below normal.

Anesthesia (an′-es-THĒ-zē-a) A total or partial loss of feeling or sensation; may be general or local.

Aneurysm (AN-ū-rizm) A saclike enlargement of a blood vessel caused by a weakening of its wall.

Angina pectoris (an-JI-na *or* AN-ji-na PEK-tō-ris) A pain in the chest related to reduced coronary circulation due to coronary artery disease (CAD) or spasms of vascular smooth muscle in coronary arteries.

Angiogenesis (an′-jē-ō-JEN-e-sis) The formation of blood vessels in the extraembryonic mesoderm of the yolk sac, connecting stalk, and chorion at the beginning of the third week of development.

Ankylosis (ang′-ki-LŌ-sis) Severe or complete loss of movement at a joint as the result of a disease process.

Antagonist (an-TAG-ō-nist) A muscle that has an action opposite that of the prime mover (agonist) and yields to the movement of the prime mover.

Antagonistic (an-tag-ō-NIST-ik) **effect** A hormonal interaction in which the effect of one hormone on a target cell is opposed by another hormone. For example, calcitonin (CT) lowers blood calcium level, whereas parathyroid hormone (PTH) raises it.

Anterior (an-TĒR-ē-or) Nearer to or at the front of the body. Equivalent to **ventral** in bipeds.

Anterior pituitary Anterior lobe of the pituitary gland. Also called the **adenohypophysis** (ad′-e-nō-hī-POF-i-sis).

Anterior root The structure composed of axons of motor (efferent) neurons that emerges from the anterior aspect of the spinal cord and extends laterally to join a posterior root, forming a spinal nerve. Also called a **ventral root.**

Anterolateral (an′-ter-ō-LAT-er-al) **pathway** Sensory pathway that conveys information related to pain, temperature, crude touch, pressure, tickle, and itch.

Antibody (AN-ti-bod′-ē) A protein produced by plasma cells in response to a specific antigen; the antibody combines with that antigen to neutralize, inhibit, or destroy it. Also called an **immunoglobulin** (im-ū-nō-GLOB-ū-lin) or **Ig.**

Anticoagulant (an-tī-cō-AG-ū-lant) A substance that can delay, suppress, or prevent the clotting of blood.

Antidiuretic (an′-ti-dī-ū-RET-ik) Substance that inhibits urine formation.

Antidiuretic hormone (ADH) Hormone produced by neurosecretory cells in the paraventricular and supraoptic nuclei of the hypothalamus that stimulates water reabsorption from kidney tubule cells into the blood and vasoconstriction of arterioles. Also called **vasopressin** (vāz-ō-PRES-in).

Antigen (AN-ti-jen) A substance that has immunogenicity (the ability to provoke an immune response) and reactivity (the ability to react with the antibodies or cells that result from the immune response); contraction of *anti*body *gen*erator. Also termed a **complete antigen.**

Antigen-presenting cell (APC) Special class of migratory cell that processes and presents antigens to T cells during an immune response; APCs include macrophages, B cells, and dendritic cells, which are present in the skin, mucous membranes, and lymph nodes.

Antrum (AN-trum) Any nearly closed cavity or chamber, especially one within a bone, such as a sinus.

Anuria (an-Ū-rē-a) Absence of urine formation or daily urine output of less than 50 mL.

Anus (Ā-nus) The distal end and outlet of the rectum.

Aorta (ā-OR-ta) The main systemic trunk of the arterial system of the body that emerges from the left ventricle.

Aortic (ā-OR-tik) **body** Cluster of chemoreceptors on or near the arch of the aorta that respond to changes in blood levels of oxygen, carbon dioxide, and hydrogen ions (H^+).

Aortic reflex A reflex that helps maintain normal systemic blood pressure; initiated by baroreceptors in the wall of the ascending aorta and arch of the aorta. Nerve impulses from aortic baroreceptors reach the cardiovascular center via sensory axons of the vagus (X) nerves.

Apex (Ā-peks) The pointed end of a conical structure, such as the apex of the heart.

Aphasia (a-FĀ-zē-a) Loss of ability to express oneself properly through speech or loss of verbal comprehension.

Apnea (AP-nē-a) Temporary cessation of breathing.

Apneustic (ap-NOO-stik) **area** A part of the respiratory center in the pons that sends stimulatory nerve impulses to the inspiratory area that activate and prolong inhalation and inhibit exhalation.

Apocrine (AP-ō-krin) **gland** A type of gland in which the secretory products gather at the free end of the secreting cell and are pinched off, along with some of the cytoplasm, to become the secretion, as in mammary glands.

Aponeurosis (ap′-ō-noo-RŌ-sis) A sheetlike tendon joining one muscle with another or with bone.

Apoptosis (ap-ō-TŌ-sis *or* ap′-ōp-TŌ-sis) Programmed cell death; a normal type of cell death that removes unneeded cells during embryological development, regulates the number of cells in tissues, and eliminates many potentially dangerous cells such as cancer cells. During apoptosis, the DNA fragments, the nucleus condenses, mitochondria cease to function, and the cytoplasm shrinks, but the plasma membrane remains intact. Phagocytes engulf and digest the apoptotic cells, and an inflammatory response does not occur.

Appositional (a-pō-ZISH-o-nal) **growth** Growth due to surface deposition of material, as in the growth in diameter of cartilage and bone. Also called **exogenous** (eks-OJ-e-nus) **growth.**

Aqueous humor (AK-wē-us HŪ-mer) The watery fluid, similar in composition to cerebrospinal fluid, that fills the anterior cavity of the eye.

Arachnoid (a-RAK-noyd) **mater** The middle of the three meninges (coverings) of the brain and spinal cord. Also termed the **arachnoid.**

Arachnoid villus (VIL-us) Berrylike tuft of the arachnoid mater that protrudes into the superior sagittal sinus and through which cerebrospinal fluid is reabsorbed into the bloodstream.

Arbor vitae (AR-bor VĪ-tē) The white matter tracts of the cerebellum, which have a treelike appearance when seen in midsagittal section.

Arch of the aorta The most superior portion of the aorta, lying between the ascending and descending segments of the aorta.

Areola (a-RĒ-ō-la) Any tiny space in a tissue. The pigmented ring around the nipple of the breast.

Arm The part of the upper limb from the shoulder to the elbow.

Arousal (a-ROW-zal) Awakening from sleep, a response due to stimulation of the reticular activating system (RAS).

Arrector pili (a-REK-tor PI-lē) Smooth muscles attached to hairs; contraction pulls the hairs into a vertical position, resulting in "goose bumps."

Arrhythmia (a-RITH-mē-a) An irregular heart rhythm. Also called a **dysrhythmia.**

Arteriole (ar-TĒ-rē-ōl) A small, almost microscopic, artery that delivers blood to a capillary.

Arteriosclerosis (ar-tē-rē-ō-skle-RŌ-sis) Group of diseases characterized by thickening of the walls of arteries and loss of elasticity.

Artery (AR-ter-ē) A blood vessel that carries blood away from the heart.

Arthritis (ar-THRI-tis) Inflammation of a joint.

Arthrology (ar-THROL-ō-jē) The study or description of joints.

Arthroplasty (AR-thrō-plas′-tē) Surgical replacement of joints, for example, the hip and knee joints.

Arthroscopy (ar-THROS-kō-pē) A procedure for examining the interior of a joint, usually the knee, by inserting an arthroscope into a small incision; used to determine extent of damage, remove torn cartilage, repair cruciate ligaments, and obtain samples for analysis.

Arthrosis (ar-THRŌ-sis) A joint or articulation.

Articular (ar-TIK-ū-lar) **capsule** Sleevelike structure around a synovial joint composed of a fibrous capsule and a synovial membrane.

Articular cartilage (KAR-ti-lij) Hyaline cartilage attached to articular bone surfaces.

Articular disc Fibrocartilage pad between articular surfaces of bones of some synovial joints. Also called a **meniscus** (men-IS-kus).

Articulation (ar-tik-ū-LĀ-shun) A joint; a point of contact between bones, cartilage and bones, or teeth and bones.

Arytenoid (ar′-i-TĒ-noyd) **cartilages** A pair of small, pyramidal cartilages of the larynx that attach to the vocal folds and intrinsic pharyngeal muscles and can move the vocal folds.

Ascending colon (KŌ-lon) The part of the large intestine that passes superiorly from the cecum to the inferior border of the liver, where it bends at the right colic (hepatic) flexure to become the transverse colon.

Ascites (as-SĪ-tēz) Abnormal accumulation of serous fluid in the peritoneal cavity.

Association areas Large cortical regions on the lateral surfaces of the occipital, parietal, and temporal lobes and on the frontal lobes anterior to the motor areas connected by many motor and sensory axons to other parts of the cortex. The association areas are concerned with motor patterns, memory, concepts of word-hearing and word-seeing, reasoning, will, judgment, and personality traits.

Asthma (AZ-ma) Usually allergic reaction characterized by smooth muscle spasms in bronchi resulting in wheezing and difficult breathing. Also called **bronchial asthma.**

Astigmatism (a-STIG-ma-tizm) An irregularity of the lens or cornea of the eye causing the image to be out of focus and producing faulty vision.

Astrocyte (AS-trō-sīt) A neuroglial cell having a star shape that participates in brain development and the metabolism of neurotransmitters, helps form the blood–brain barrier, helps maintain the proper balance of K^+ for generation of nerve impulses, and provides a link between neurons and blood vessels.

Ataxia (a-TAK-sē-a) A lack of muscular coordination, lack of precision.

Atherosclerotic (ath′-er-ō-skle-RO-tic) **plaque** (PLAK) A lesion that results from accumulated cholesterol and smooth muscle fibers (cells) of the tunica media of an artery; may become obstructive.

Atom Unit of matter that makes up a chemical element; consists of a nucleus (containing positively charged protons and uncharged neutrons) and negatively charged electrons that orbit the nucleus.

Atresia (a-TRĒ-zē-a) Degeneration and reabsorption of an ovarian follicle before it fully matures and ruptures; abnormal closure of a passage, or absence of a normal body opening.

Atrial fibrillation (Ā-trē-al fib-ri-LĀ-shun) Asynchronous contraction of cardiac muscle fibers in the atria that results in the cessation of atrial pumping.

Atrial natriuretic (na′-trē-ū-RET-ik) **peptide (ANP)** Peptide hormone, produced by the atria of the heart in response to stretching, that inhibits aldosterone production and thus lowers blood pressure; causes natriuresis, increased urinary excretion of sodium.

Atrioventricular (AV) (ā′-trē-ō-ven-TRIK-ū-lar) **bundle** The part of the conduction system of the heart that begins at the atrioventricular (AV) node, passes through the cardiac skeleton separating the atria and the ventricles, then extends a short distance down the interventricular septum before splitting into right and left bundle branches. Also called the **bundle of His** (HISS).

Atrioventricular (AV) node The part of the conduction system of the heart made up of a compact mass of conducting cells located in the septum between the two atria.

Atrioventricular (AV) valve A heart valve made up of membranous flaps or cusps that allows blood to flow in one direction only, from an atrium into a ventricle.

Atrium (Ā-trē-um) A superior chamber of the heart.

Atrophy (AT-rō-fē) Wasting away or decrease in size of a part, due to a failure, abnormality of nutrition, or lack of use.

Auditory ossicle (AW-di-tō-rē OS-si-kul) One of the three small bones of the middle ear called the **malleus, incus,** and **stapes.**

Auditory tube The tube that connects the middle ear with the nose and nasopharynx region of the throat. Also called the **eustachian** (ū-STĀ-shun *or* ū-STĀ-kē-an) **tube** or **pharyngotympanic tube.**

Auscultation (aws-kul-TĀ-shun) Examination by listening to sounds in the body.

Autoimmunity An immunological response against a person's own tissues.

Autolysis (aw-TOL-i-sis) Self-destruction of cells by their own lysosomal digestive enzymes after death or in a pathological process.

Autonomic ganglion (aw′-tō-NOM-ik GANG-lē-on) A cluster of cell bodies of sympathetic or parasympathetic neurons located outside the central nervous system.

Autonomic nervous system (ANS) Visceral sensory (afferent) and visceral motor (efferent) neurons. Autonomic motor neurons, both sympathetic and parasympathetic, conduct nerve impulses from the central nervous system to smooth muscle, cardiac muscle, and glands. So named because this part of the nervous system was thought to be self-governing or spontaneous.

Autonomic plexus (PLEK-sus) A network of sympathetic and parasympathetic axons; examples are the cardiac, celiac, and pelvic plexuses, which are located in the thorax, abdomen, and pelvis, respectively.

Autophagy (aw-TOF-a-jē) Process by which worn-out organelles are digested within lysosomes.

Autopsy (AW-top-sē) The examination of the body after death.

Autorhythmic cells Cardiac or smooth muscle fibers that are self-excitable (generate impulses without an external stimulus); act as the heart's pacemaker and conduct the pacing impulse through the conduction system of the heart; self-excitable neurons in the central nervous system, as in the inspiratory area of the brain stem.

Autosome (AW-tō-sōm) Any chromosome other than the X and Y chromosomes (sex chromosomes).

Axilla (ak-SIL-a) The small hollow beneath the arm where it joins the body at the shoulders. Also called the **armpit.**

Axon (AK-son) The usually single, long process of a nerve cell that propagates a nerve impulse toward the axon terminals.

Axon terminal Terminal branch of an axon where synaptic vesicles undergo exocytosis to release neurotransmitter molecules.

B

B cell A lymphocyte that can develop into a clone of antibody-producing plasma cells or memory cells when properly stimulated by a specific antigen.

Babinski (ba-BIN-skē) **sign** Extension of the great toe, with or without fanning of the other toes, in response to stimulation of the outer margin of the sole; normal up to 18 months of age and indicative of damage to descending motor pathways such as the corticospinal tracts after that.

Back The posterior part of the body; the dorsum.

Ball-and-socket joint A synovial joint in which the rounded surface of one bone moves within a cup-shaped depression or socket of another bone, as in the shoulder or hip joint. Also called a **spheroid** (SFĒ-royd) **joint.**

Baroreceptor (bar′-ō-re-SEP-tor) Neuron capable of responding to changes in blood, air, or fluid pressure. Also called a **pressoreceptor.**

Basal ganglia (GANG-glē-a) Paired clusters of gray matter deep in each cerebral hemisphere including the globus pallidus, putamen, and caudate nucleus. Together, the caudate nucleus and putamen are known as the **corpus striatum.** Nearby structures that are functionally linked to the basal ganglia are the substantia nigra of the midbrain and the subthalamic nuclei of the diencephalon.

Basement membrane Thin, extracellular layer between epithelium and connective tissue consisting of a basal lamina and a reticular lamina.

Basilar (BĀS-i-lar) **membrane** A membrane in the cochlea of the internal ear that separates the cochlear duct from the scala tympani and on which the spiral organ (organ of Corti) rests.

Basophil (BĀ-sō-fil) A type of white blood cell characterized by a pale nucleus and large granules that stain blue-purple with basic dyes.

Belly The abdomen. The gaster or prominent, fleshy part of a skeletal muscle.

Beta (BĀ-ta) **cell** A type of cell in the pancreatic islets (islets of Langerhans) in the pancreas that secretes the hormone insulin.

Beta receptor A type of adrenergic receptor for epinephrine and norepinephrine; found on visceral effectors innervated by sympathetic postganglionic neurons.

Bicuspid (bī-KUS-pid) **valve** Atrioventricular (AV) valve on the left side of the heart. Also called the **mitral valve.**

Bilateral (bī-LAT-er-al) Pertaining to two sides of the body.

Bile (BĪL) A secretion of the liver consisting of water, bile salts, bile pigments, cholesterol, lecithin, and several ions; it emulsifies lipids prior to their digestion.

Bilirubin (bil-ē-ROO-bin) An orange pigment that is one of the end products of hemoglobin breakdown in the hepatocytes and is excreted as a waste material in bile.

Blastocele (BLAS-tō-sēl) The fluid-filled cavity within the blastocyst.

Blastocyst (BLAS-tō-sist) In the development of an embryo, a hollow ball of cells that consists of a blastocele (the internal cavity), trophoblast (outer cells), and inner cell mass.

Blastomere (BLAS-tō-mēr) One of the cells resulting from the cleavage of a fertilized ovum.

Blastula (BLAS-tyū-la) An early stage in the development of a zygote.

Blind spot Area in the retina at the end of the optic (II) nerve in which there are no photoreceptors.

Blood The fluid that circulates through the heart, arteries, capillaries, and veins and that constitutes the chief means of transport within the body.

Blood–brain barrier (BBB) A barrier consisting of specialized brain capillaries and astrocytes that prevents the passage of materials from the blood to the cerebrospinal fluid and brain.

Blood island Isolated mass of mesoderm derived from angioblasts and from which blood vessels develop.

Blood pressure (BP) Force exerted by blood against the walls of blood vessels due to contraction of the heart and influenced by the elasticity of the vessel walls; clinically, a measure of the pressure in arteries during ventricular systole and ventricular diastole.

Blood reservoir (REZ-er-vwar) Systemic veins and venules that contain large amounts of blood that can be moved quickly to parts of the body requiring the blood.

Blood–testis barrier (BTB) A barrier formed by Sertoli cells that prevents an immune response against antigens produced by spermatogenic cells by isolating the cells from the blood.

Body cavity A space within the body that contains various internal organs.

Bolus (BŌ-lus) A soft, rounded mass, usually food, that is swallowed.

Bony labyrinth (LAB-i-rinth) A series of cavities within the petrous portion of the temporal bone forming the vestibule, cochlea, and semicircular canals of the inner ear.

Bowman's capsule *See* **Glomerular capsule.**

Brachial plexus (BRĀ-kē-al PLEK-sus) A network of nerve axons of the ventral rami of spinal nerves C5, C6, C7, C8, and T1. The nerves that emerge from the brachial plexus supply the upper limb.

Bradycardia (brād′-i-KAR-dē-a) A slow resting heart or pulse rate (under 50 beats per minute).

Brain The part of the central nervous system contained within the cranial cavity.

Brain stem The portion of the brain immediately superior to the spinal cord, made up of the medulla oblongata, pons, and midbrain.

Brain waves Electrical signals that can be recorded from the skin of the head due to electrical activity of brain neurons.

Broad ligament A double fold of parietal peritoneum attaching the uterus to the side of the pelvic cavity.

Broca's (BRŌ-kaz) **area** Motor area of the brain in the frontal lobe that translates thoughts into speech. Also called the **motor speech area.**

Bronchi (BRON-kē) Branches of the respiratory passageway including primary bronchi (the two divisions of the trachea), secondary or lobar bronchi (divisions of the primary bronchi that are distributed to the lobes of the lung), and tertiary or segmental bronchi (divisions of the secondary bronchi that are distributed to bronchopulmonary segments of the lung). *Singular* is **bronchus.**

Bronchial tree The trachea, bronchi, and their branching structures up to and including the terminal bronchioles.

Bronchiole (BRONG-kē-ōl) Branch of a tertiary bronchus further dividing into terminal bronchioles (distributed to lobules of the lung), which divide into respiratory bronchioles (distributed to alveolar sacs).

Bronchitis (brong-KĪ-tis) Inflammation of the mucous membrane of the bronchial tree; characterized by hypertrophy and hyperplasia of seromucous glands and goblet cells that line the bronchi which results in a productive cough.

Bronchopulmonary (brong′-kō-PUL-mō-ner-ē) **segment** One of the smaller divisions of a lobe of a lung supplied by its own branches of a bronchus.

Brunner's gland *See* **Duodenal gland.**

Buccal (BUK-al) Pertaining to the cheek or mouth.

Bulb of penis Expanded portion of the base of the corpus spongiosum penis.

Bulbourethral (bul′-bō-ū-RĒ-thral) **gland** One of a pair of glands located inferior to the prostate on either side of the urethra that secretes an alkaline fluid into the cavernous urethra. Also called a **Cowper's** (KOW-perz) **gland.**

Bulimia (boo-LIM-ē-a *or* boo-LĒ-mē-a) A disorder characterized by overeating at least twice a week followed by purging by self-induced vomiting, strict dieting or fasting, vigorous exercise, or use of laxatives or diuretics. Also called **binge–purge syndrome.**

Bulk-phase endocytosis A process by which most body cells can ingest membrane-surrounded droplets of interstitial fluid.

Bundle branch One of the two branches of the atrioventricular (AV) bundle made up of specialized muscle fibers (cells) that transmit electrical impulses to the ventricles.

Bundle of His *See* **Atrioventricular (AV) bundle.**

Bursa (BUR-sa) A sac or pouch of synovial fluid located at friction points, especially about joints.

Bursitis (bur-SĪ-tis) Inflammation of a bursa.

Buttocks (BUT-oks) The two fleshy masses on the posterior aspect of the inferior trunk, formed by the gluteal muscles.

C

Calcaneal (kal-KĀ-nē-al) **tendon** The tendon of the soleus, gastrocnemius, and plantaris muscles at the back of the heel. Also called the **Achilles** (a-KIL-ēz) **tendon.**

Calcification (kal′-si-fi-KĀ-shun) Deposition of mineral salts, primarily hydroxyapatite, in a framework formed by collagen fibers in which the tissue hardens. Also called **mineralization** (min′-e-ral-i-ZĀ-shun).

Calcitonin (kal-si-TŌ-nin) **(CT)** A hormone produced by the parafollicular cells of the thyroid gland that can lower the amount of blood calcium and phosphates by inhibiting bone resorption (breakdown of bone extracellular matrix) and by accelerating uptake of calcium and phosphates into bone matrix.

Calculus (KAL-kū-lus) A stone, or insoluble mass of crystallized salts or other material, formed within the body, as in the gallbladder, kidney, or urinary bladder.

Callus (KAL-lus) A growth of new bone tissue in and around a fractured area, ultimately replaced by mature bone. An acquired, localized thickening.

Calyx (KĀL-iks) Any cuplike division of the kidney pelvis. *Plural* is **calyces** (KĀ-li-sēz).

Canal (ka-NAL) A narrow tube, channel, or passageway.

Canaliculus (kan′-a-LIK-ū-lus) A small channel or canal, as in bones, where they connect lacunae. *Plural* is **canaliculi** (kan′-a-LIK-ū-lī).

Canal of Schlemm *See* **Scleral venous sinus.**

Capacitation (ka′-pas-i-TĀ-shun) The functional changes that sperm undergo in the female reproductive tract that allow them to fertilize a secondary oocyte.

Capillary (KAP-i-lar′-ē) A microscopic blood vessel located between an arteriole and venule through which materials are exchanged between blood and interstitial fluid.

Carcinogen (car-SIN-ō-jen) A chemical substance or radiation that causes cancer.

Cardiac (KAR-dē-ak) **arrest** Cessation of an effective heartbeat in which the heart is completely stopped or in ventricular fibrillation.

Cardiac cycle A complete heartbeat consisting of systole (contraction) and diastole (relaxation) of both atria plus systole and diastole of both ventricles.

Cardiac muscle Striated muscle fibers (cells) that form the wall of the heart; stimulated by an intrinsic conduction system and autonomic motor neurons.

Cardiac notch An angular notch in the anterior border of the left lung into which part of the heart fits.

Cardinal ligament A ligament of the uterus, extending laterally from the cervix and vagina as a continuation of the broad ligament.

Cardiogenic area (kar-dē-ō-JEN-ik) A group of mesodermal cells in the head end of an embryo that gives rise to the heart.

Cardiology (kar-dē-OL-ō-jē) The study of the heart and diseases associated with it.

Cardiovascular (kar-dē-ō-VAS-kū-lar) **center** Groups of neurons scattered within the medulla oblongata that regulate heart rate, force of contraction, and blood vessel diameter.

Carotene (KAR-ō-tēn) Antioxidant precursor of vitamin A, which is needed for synthesis of photopigments; yellow-orange pigment present in the stratum corneum of the epidermis. Accounts for the yellowish coloration of skin. Also termed **beta-carotene.**

Carotid (ka-ROT-id) **body** Cluster of chemoreceptors on or near the carotid sinus that respond to changes in blood levels of oxygen, carbon dioxide, and hydrogen ions.

Carotid sinus A dilated region of the internal carotid artery just superior to where it branches from the common carotid artery; it contains baroreceptors that monitor blood pressure.

Carpal bones The eight bones of the wrist. Also called **carpals.**

Carpus (KAR-pus) A collective term for the eight bones of the wrist.

Cartilage (KAR-ti-lij) A type of connective tissue consisting of chondrocytes in lacunae embedded in a dense network of collagen and elastic fibers and an extracellular matrix of chondroitin sulfate.

Cartilaginous (kar-ti-LAJ-i-nus) **joint** A joint without a synovial (joint) cavity where the articulating bones are held tightly together by cartilage, allowing little or no movement.

Catabolism (ka-TAB-ō-lizm) Chemical reactions that break down complex organic compounds into simple ones, with the net release of energy.

Cataract (KAT-a-rakt) Loss of transparency of the lens of the eye or its capsule or both.

Cauda equina (KAW-da ē-KWĪ-na) A tail-like array of roots of spinal nerves at the inferior end of the spinal cord.

Caudal (KAW-dal) Pertaining to any tail-like structure; inferior in position.

Cecum (SĒ-kum) A blind pouch at the proximal end of the large intestine that attaches to the ileum.

Celiac plexus (PLEK-sus) A large mass of autonomic ganglia and axons located at the level of the superior part of the first lumbar vertebra. Also called the **solar plexus.**

Cell The basic structural and functional unit of all organisms; the smallest structure capable of performing all the activities vital to life.

Cell cycle Growth and division of a single cell into two identical cells; consists of interphase and cell division.

Cell division Process by which a cell reproduces itself that consists of a nuclear division (mitosis) and a cytoplasmic division (cytokinesis); types include somatic and reproductive cell division.

Cell junction Point of contact between plasma membranes of tissue cells.

Cementum (se-MEN-tum) Calcified tissue covering the root of a tooth.

Central canal A microscopic tube running the length of the spinal cord in the gray commissure. A circular channel running longitudinally in the center of an osteon (haversian system) of mature compact bone, containing blood and lymphatic vessels and nerves. Also called an **haversian** (ha-VER-shun) **canal.**

Central fovea (FŌ-vē-a) A depression in the center of the macula lutea of the retina, containing cones only and lacking blood vessels; the area of highest visual acuity (sharpness of vision).

Central nervous system (CNS) That portion of the nervous system that consists of the brain and spinal cord.

Centrioles (SEN-trē-ōlz) Paired, cylindrical structures of a centrosome, each consisting of a ring of microtubules and arranged at right angles to each other.

Centromere (SEN-trō-mēr) The constricted portion of a chromosome where the two chromatids are joined; serves as the point of attachment for the microtubules that pull chromatids during anaphase of cell division.

Centrosome (SEN-trō-sōm) A dense network of small protein fibers near the nucleus of a cell, containing a pair of centrioles and pericentriolar material.

Cephalic (se-FAL-ik) Pertaining to the head; superior in position.

Cerebellar peduncle (ser-e-BEL-ar pe-DUNG-kul) A bundle of nerve axons connecting the cerebellum with the brain stem.

Cerebellum (ser′-e-BEL-um) The part of the brain lying posterior to the medulla oblongata and pons; governs balance and coordinates skilled movements.

Cerebral aqueduct (SER-ē-bral AK-we-dukt) A channel through the midbrain connecting the third and fourth ventricles and containing cerebrospinal fluid. Also termed the **aqueduct of Sylvius.**

Cerebral arterial circle A ring of arteries forming an anastomosis at the base of the brain between the internal carotid and basilar arteries and arteries supplying the cerebral cortex. Also called the **circle of Willis.**

Cerebral cortex The surface of the cerebral hemispheres, 2–4 mm thick, consisting of gray matter; arranged in six layers of neuronal cell bodies in most areas.

Cerebral peduncle One of a pair of nerve axon bundles located on the anterior surface of the midbrain, conducting nerve impulses between the pons and the cerebral hemispheres.

Cerebrospinal (se-rē′-brō-SPĪ-nal) **fluid (CSF)** A fluid produced by ependymal cells that cover choroid plexuses in the ventricles of the brain; the fluid circulates in the ventricles, the central canal, and the subarachnoid space around the brain and spinal cord.

Cerebrovascular (se rē′-brō-VAS-kū-lar) **accident (CVA)** Destruction of brain tissue (infarction) resulting from obstruction or rupture of blood vessels that supply the brain. Also called a **stroke** or **brain attack.**

Cerebrum (SER-e-brum *or* se-RĒ-brum) The two hemispheres of the forebrain (derived from the telencephalon), making up the largest part of the brain.

Cerumen (se-ROO-men) Waxlike secretion produced by ceruminous glands in the external auditory meatus (ear canal). Also termed **ear wax.**

Ceruminous (se-RŪ-mi-nus) **gland** A modified sudoriferous (sweat) gland in the external auditory meatus that secretes cerumen (ear wax).

Cervical ganglion (SER-vi-kul GANG-glē-on) A cluster of cell bodies of postganglionic sympathetic neurons located in the neck, near the vertebral column.

Cervical plexus (PLEK-sus) A network formed by nerve axons from the ventral rami of the first four cervical nerves and receiving gray rami communicantes from the superior cervical ganglion.

Cervix (SER-viks) Neck; any constricted portion of an organ, such as the inferior cylindrical part of the uterus.

Chemoreceptor (kē′-mō-rē-SEP-tor) Sensory receptor that detects the presence of a specific chemical.

Chiasm (KĪ-azm) A crossing; especially the crossing of axons in the optic (II) nerve.

Chief cell The secreting cell of a gastric gland that produces pepsinogen, the precursor of the enzyme pepsin, and the enzyme gastric lipase. Also called a **zymogenic** (zī′-mō-JEN-ik) **cell.** Cell in the parathyroid glands that secretes parathyroid hormone (PTH). Also called a **principal cell.**

Cholecystectomy (kō′-lē-sis-TEK-tō-mē) Surgical removal of the gallbladder.

Cholecystitis (kō′-lē-sis-TĪ-tis) Inflammation of the gallbladder.

Cholesterol (kō-LES-te-rol) Classified as a lipid, the most abundant steroid in animal tissues; located in cell membranes and used for the synthesis of steroid hormones and bile salts.

Cholinergic (kō′-lin-ER-jik) **neuron** A neuron that liberates acetylcholine as its neurotransmitter.

Chondrocyte (KON-drō-sīt) Cell of mature cartilage.

Chondroitin (kon-DROY-tin) **sulfate** An amorphous extracellular matrix material found outside connective tissue cells.

Chordae tendineae (KOR-dē TEN-di-nē-ē) Tendonlike, fibrous cords that connect atrioventricular valves of the heart with papillary muscles.

Chorion (KŌrē-on) The most superficial fetal membrane that becomes the principal embryonic portion of the placenta; serves a protective and nutritive function.

Chorionic villi (kō-rē-ON-ik VIL-lī) Fingerlike projections of the chorion that grow into the decidua basalis of the endometrium and contain fetal blood vessels.

Chorionic villi sampling (CVS) The removal of a sample of chorionic villus tissue by means of a catheter to analyze the tissue for prenatal genetic defects.

Choroid (KŌ-royd) One of the vascular coats of the eyeball.

Choroid plexus (PLEK-sus) A network of capillaries located in the roof of each of the four ventricles of the brain; ependymal cells around choroid plexuses produce cerebrospinal fluid.

Chromaffin (KRŌ-maf-in) **cell** Cell that has an affinity for chrome salts, due in part to the presence of the precursors of the neurotransmitter epinephrine; found, among other places, in the adrenal medulla.

Chromatid (KRŌ-ma-tid) One of a pair of identical connected nucleoprotein strands that are joined at the centromere and separate during cell division, each becoming a chromosome of one of the two daughter cells.

Chromatin (KRŌ-ma-tin) The threadlike mass of genetic material, consisting of DNA and histone proteins, that is present in the nucleus of a nondividing or interphase cell.

Chromatolysis (krō′-ma-TOL-i-sis) The breakdown of Nissl bodies into finely granular masses in the cell body of a neuron whose axon has been damaged.

Chromosome (KRŌ-mō-sōm) One of the small, threadlike structures in the nucleus of a cell, normally 46 in a human diploid cell, that bears the genetic material; composed of DNA and proteins (histones) that form a delicate chromatin thread during interphase; becomes packaged into compact rodlike structures that are visible under the light microscope during cell division.

Chronic (KRON-ik) Long term or frequently recurring; applied to a disease that is not acute.

Chronic obstructive pulmonary disease (COPD) A disease, such as bronchitis or emphysema, in which there is some degree of obstruction of airways and consequent increase in airway resistance.

Chyle (KĪL) The milky-appearing fluid found in the lacteals of the small intestine after absorption of lipids in food.

Chyme (KĪM) The semifluid mixture of partly digested food and digestive secretions found in the stomach and small intestine during digestion of a meal.

Ciliary (SIL-ē-ar′-ē) **body** One of the three parts of the vascular tunic of the eyeball, the others being the choroid and the iris; includes the ciliary muscle and the ciliary processes.

Ciliary ganglion (GANG-glē-on) A very small parasympathetic ganglion whose preganglionic axons come from the oculomotor (III) nerve and whose postganglionic axons carry nerve impulses to the ciliary muscle and the sphincter muscle of the iris.

Cilium (SIL-ē-um) A hair or hairlike process projecting from a cell that may be used to move the entire cell or to move substances along the surface of the cell. *Plural* is **cilia.**

Circle of Willis *See* **Cerebral arterial circle.**

Circular folds Permanent, deep, transverse folds in the mucosa and submucosa of the small intestine that increase the surface area for absorption. Also called **plicae circulares** (PLĪ-kē SER-kū-lar-ēs).

Circumduction (ser-kum-DUK-shun) A movement at a synovial joint in which the distal end of a bone moves in a circle while the proximal end remains relatively stable.

Cirrhosis (si-RŌ-sis) A liver disorder in which the parenchymal cells are destroyed and replaced by connective tissue.

Cisterna chyli (sis-TER-na KĪ-lē) The origin of the thoracic duct.

Cleavage The rapid mitotic divisions following the fertilization of a secondary oocyte, resulting in an increased number of progressively smaller cells, called blastomeres.

Clitoris (KLI-to-ris) An erectile organ of the female, located at the anterior junction of the labia minora, that is homologous to the male penis.

Clone (KLŌN) A population of identical cells.

Coarctation (kō′-ark-TĀ-shun) **of the aorta** A congenital heart defect in which a segment of the aorta is too narrow. As a result, the flow of oxygenated blood to the body is reduced, the left ventricle is forced to pump harder, and high blood pressure develops.

Coccyx (KOK-siks) The fused bones at the inferior end of the vertebral column.

Cochlea (KOK-lē-a) A winding, cone-shaped tube forming a portion of the inner ear and containing the spiral organ (organ of Corti).

Cochlear duct The membranous cochlea consisting of a spirally arranged tube enclosed in the bony cochlea and lying along its outer wall. Also called the **scala media** (SCA-la MĒ-dē-a).

Collagen (KOL-a-jen) A protein that is the main organic constituent of connective tissue.

Collateral circulation The alternate route taken by blood through an anastomosis.

Colliculus (ko-LIK-ū-lus) A small elevation.

Colon The portion of the large intestine consisting of ascending, transverse, descending, and sigmoid portions.

Colony-stimulating factor (CSF) One of a group of molecules that stimulates development of white blood cells. Examples are macrophage CSF and granulocyte CSF.

Colostrum (kō-LOS-trum) A thin, cloudy fluid secreted by the mammary glands a few days prior to or after delivery before true milk is produced.

Column (KOL-um) Group of white matter tracts in the spinal cord.

Common bile duct A tube formed by the union of the common hepatic duct and the cystic duct that empties bile into the duodenum at the hepatopancreatic ampulla (ampulla of Vater).

Compact (dense) bone tissue Bone tissue that contains few spaces between osteons (haversian systems); forms the external portion of all bones and the bulk of the diaphysis (shaft) of long bones; is found immediately deep to the periosteum and external to spongy bone.

Concha (KONG-ka) A scroll-like bone found in the skull. *Plural is* **conchae** (KONG-kē).

Concussion (kon-KUSH-un) Traumatic injury to the brain that produces no visible bruising but may result in abrupt, temporary loss of consciousness.

Conduction system A group of autorhythmic cardiac muscle fibers that generates and distributes electrical impulses to stimulate coordinated contraction of the heart chambers; includes the sinoatrial (SA) node, the atrioventricular (AV) node, the atrioventricular (AV) bundle, the right and left bundle branches, and the Purkinje fibers.

Condyloid (KON-di-loyd) **joint** A synovial joint structured so that an oval-shaped condyle of one bone fits into an elliptical cavity of another bone, permitting side-to-side and back-and-forth movements, such as the joint at the wrist between the radius and carpals. Also called an **ellipsoidal** (ē-lip-SOYD-al) **joint.**

Cone (KŌN) The type of photoreceptor in the retina that is specialized for highly acute color vision in bright light.

Congenital (kon-JEN-i-tal) Present at the time of birth.

Conjunctiva (kon′-junk-TĪ-va) The delicate membrane covering the eyeball and lining the eyes.

Connective tissue One of the most abundant of the four basic tissue types in the body, performing the functions of binding and supporting; consists of relatively few cells in a generous matrix (the ground substance and fibers between the cells).

Consciousness (KON-shus-nes) A state of wakefulness in which an individual is fully alert, aware, and oriented, partly as a result of feedback between the cerebral cortex and reticular activating system.

Continuous conduction (kon-DUK-shun) Propagation of an action potential (nerve impulse) in a step-by-step depolarization of each adjacent area of an axon membrane.

Contraception (kon′-tra-SEP-shun) The prevention of fertilization or impregnation without destroying fertility.

Contractility (kon′-trak-TIL-i-tē) The ability of cells or parts of cells to actively generate force to undergo shortening for movements. Muscle fibers (cells) exhibit a high degree of contractility.

Contralateral (CON-tra-lat-er-al) On the opposite side; affecting the opposite side of the body.

Conus medullaris (KŌ-nus med-ū-LAR-is) The tapered portion of the spinal cord inferior to the lumbar enlargement.

Convergence (con-VER-jens) A synaptic arrangement in which the synaptic end bulbs of several presynaptic neurons terminate on one postsynaptic neuron. The medial movement of the two eyeballs so that both are directed toward a near object being viewed in order to produce a single image.

Cornea (KOR-nē-a) The nonvascular, transparent fibrous coat through which the iris of the eye can be seen.

Corona (kō-RŌ-na) Margin of the glans penis.

Corona radiata The innermost layer of granulosa cells that is firmly attached to the zona pellucida around a secondary oocyte.

Coronary artery disease (CAD) A condition such as atherosclerosis that causes narrowing of coronary arteries so that blood flow to the heart is reduced. The result is **coronary heart disease (CHD),** in which the heart muscle receives inadequate blood flow due to an interruption of its blood supply.

Coronary circulation The pathway followed by the blood from the ascending aorta through the blood vessels supplying the heart and returning to the right atrium. Also called **cardiac circulation.**

Coronary sinus (SĪ-nus) A wide venous channel on the posterior surface of the heart that collects the blood from the coronary circulation and returns it to the right atrium.

Corpus albicans (KOR-pus AL-bi-kanz) A white fibrous patch in the ovary that forms after the corpus luteum regresses.

Corpus callosum (kal-LŌ-sum) The great commissure of the brain between the cerebral hemispheres.

Corpuscle of touch The sensory receptor for the sensation of touch; found in the dermal papillae, especially in palms and soles. Also called a **Meissner** (MĪS-ner) **corpuscle.**

Corpus luteum (LOO-tē-um) A yellowish body in the ovary formed when a follicle has discharged its secondary oocyte; secretes estrogens, progesterone, relaxin, and inhibin.

Corpus striatum (strī-Ā-tum) An area in the interior of each cerebral hemisphere composed of the caudate and putamen of the basal ganglia and white matter of the internal capsule, arranged in a striated manner.

Cortex (KOR-teks) An outer layer of an organ. The convoluted layer of gray matter covering each cerebral hemisphere.

Costal (KOS-tal) Pertaining to a rib.

Cramp A spasmodic, usually painful contraction of a muscle.

Cranial (KRĀ-ne-al) **cavity** A subdivision of the dorsal body cavity formed by the cranial bones and containing the brain.

Cranial nerve One of 12 pairs of nerves that leave the brain; pass through foramina in the skull; and supply sensory and motor neurons to the head, neck, part of the trunk, and viscera of the thorax and abdomen. Each is designated by a Roman numeral and a name.

Craniosacral (krā-nē-ō-SĀK-ral) **outflow** The axons of parasympathetic preganglionic neurons, which have their cell bodies located in nuclei in the brain stem and in the lateral gray matter of the sacral portion of the spinal cord.

Cranium (KRĀ-nē-um) The skeleton of the skull that protects the brain and the organs of sight, hearing, and balance; includes the frontal, parietal, temporal, occipital, sphenoid, and ethmoid bones.

Crista (KRIS-ta) A crest or ridged structure. A small elevation in the ampulla of each semicircular duct that contains receptors for dynamic equilibrium. *Plural* is **cristae.**

Crossing-over The exchange of a portion of one chromatid with another during meiosis. It permits an exchange of genes among chromatids and is one factor that results in genetic variation of progeny.

Crus (KRUS) **of penis** Separated, tapered portion of the corpora cavernosa penis. *Plural* is **crura** (KROO-ra).

Crypt of Lieberkühn *See* **Intestinal gland.**

Cryptorchidism (krip-TOR-ki-dizm) The condition of undescended testes.

Cuneate (KŪ-nē-āt) **nucleus** A group of neurons in the inferior part of the medulla oblongata in which axons of the cuneate fasciculus terminate.

Cupula (KU-pū-la) A mass of gelatinous material covering the hair cells of a crista; a sensory receptor in the ampulla of a semicircular canal stimulated when the head moves.

Cushing's syndrome Condition caused by a hypersecretion of glucocorticoids characterized by spindly legs, "moon face," "buffalo hump," pendulous abdomen, flushed facial skin, poor wound healing, hyperglycemia, osteoporosis, hypertension, and increased susceptibility to disease.

Cutaneous (kū-TĀ-nē-us) Pertaining to the skin.

Cyanosis (sī-a-NŌ-sis) A blue or dark purple discoloration, most easily seen in nail beds and mucous membranes, that results from an increased concentration of deoxygenated (reduced) hemoglobin (more than 5 gm/dL).

Cyst (SIST) A sac with a distinct connective tissue wall, containing a fluid or other material.

Cystic (SIS-tik) **duct** The duct that carries bile from the gallbladder to the common bile duct.

Cystitis (sis-TĪ-tis) Inflammation of the urinary bladder.

Cytokinesis (sī′-tō-ki-NĒ-sis) Distribution of the cytoplasm into two separate cells during cell division; coordinated with nuclear division (mitosis).

Cytolysis (sī-TOL-i-sis) The rupture of living cells in which the contents leak out.

Cytoplasm (SĪ-tō-plasm) Cytosol plus all organelles except the nucleus.

Cytoskeleton Complex internal structure of cytoplasm consisting of microfilaments, microtubules, and intermediate filaments.

Cytosol (SĪ-tō-sol) Semifluid portion of cytoplasm in which organelles and inclusions are suspended and solutes are dissolved. Also called **intracellular fluid.**

D

Dartos (DAR-tōs) The contractile tissue deep to the skin of the scrotum.

Decidua (dē-SID-ū-a) That portion of the endometrium of the uterus (all but the deepest layer) that is modified during pregnancy and shed after childbirth.

Deciduous (dē-SID-ū-us) Falling off or being shed seasonally or at a particular stage of development. In the body, referring to the first set of teeth.

Decussation (dē′-ku-SĀ-shun) A crossing-over to the opposite (contralateral) side; an example is the crossing of 90% of the axons in the large motor tracts to opposite sides in the medullary pyramids.

Deep Away from the surface of the body or an organ.

Deep fascia (FASH-ē-a) A sheet of connective tissue wrapped around a muscle to hold it in place.

Deep inguinal (IN-gwi-nal) **ring** A slitlike opening in the aponeurosis of the transversus abdominis muscle that represents the origin of the inguinal canal.

Deep-venous thrombosis (DVT) The presence of a thrombus in a vein, usually a deep vein of the lower limbs.

Defecation (def-e-KĀ-shun) The discharge of feces from the rectum.

Deglutition (dē-gloo-TISH-un) The act of swallowing.

Dehydration (dē-hī-DRĀ-shun) Excessive loss of water from the body or its parts.

Delta cell A cell in the pancreatic islets (islets of Langerhans) in the pancreas that secretes somatostatin. Also termed a **D cell.**

Demineralization (de-min′-er-al-i-ZĀ-shun) Loss of calcium and phosphorus from bones.

Dendrite (DEN-drīt) A neuronal process that carries electrical signals, usually graded potentials, toward the cell body.

Dendritic (den-DRIT-ik) **cell** One type of antigen-presenting cell with long branchlike projections that commonly is present in mucosal linings such as the vagina, in the skin (Langerhans cells in the epidermis), and in lymph nodes (follicular dendritic cells).

Dental caries (KA-rēz) Gradual demineralization of the enamel and dentin of a tooth that may invade the pulp and alveolar bone. Also called **tooth decay.**

Denticulate (den-TIK-ū-lāt) Finely toothed or serrated; characterized by a series of small, pointed projections.

Dentin (DEN-tin) The bony tissues of a tooth enclosing the pulp cavity.

Dentition (den-TI-shun) The eruption of teeth. The number, shape, and arrangement of teeth.

Deoxyribonucleic (dē-ok′-sē-rī-bō-nū-KLĒ-ik) **acid (DNA)** A nucleic acid constructed of nucleotides consisting of one of four bases (adenine, cytosine, guanine, or thymine), deoxyribose, and a phosphate group; encoded in the nucleotides is genetic information.

Depression (de-PRESH-un) Movement in which a part of the body moves inferiorly.

Dermal papilla (pa-PILL-a) Fingerlike projection of the papillary region of the dermis that may contain blood capillaries or corpuscles of touch (Meissner corpuscles).

Dermatology (der′-ma-TOL-ō-jē) The medical specialty dealing with diseases of the skin.

Dermatome (DER-ma-tōm) The cutaneous area developed from one embryonic spinal cord segment and receiving most of its sensory innervation from one spinal nerve. An instrument for incising the skin or cutting thin transplants of skin.

Dermis (DER-mis) A layer of dense irregular connective tissue lying deep to the epidermis.

Descending colon (KŌ-lon) The part of the large intestine descending from the left colic (splenic) flexure to the level of the left iliac crest.

Detrusor (de-TROO-ser) **muscle** Smooth muscle that forms the wall of the urinary bladder.

Developmental biology The study of development from the fertilized egg to the adult form.

Diagnosis (dī′-ag-NŌ-sis) Distinguishing one disease from another or determining the nature of a disease from signs and symptoms by inspection, palpation, laboratory tests, and other means.

Dialysis (dī-AL-i-sis) The removal of waste products from blood by diffusion through a selectively permeable membrane.

Diaphragm (DĪ-a-fram) Any partition that separates one area from another, especially the dome-shaped skeletal muscle between the thoracic and abdominal cavities. Also a dome-shaped device that is placed over the cervix, usually with a spermicide, to prevent conception.

Diaphysis (dī-AF-i-sis) The shaft of a long bone.

Diarrhea (dī-a-RE-a) Frequent defecation of liquid feces caused by increased motility of the intestines.

Diarthrosis (dī-ar-THRŌ-sis) A freely movable joint; types are gliding, hinge, pivot, condyloid, saddle, and ball-and-socket.

Diastole (dī-AS-tō-lē) In the cardiac cycle, the phase of relaxation or dilation of the heart muscle, especially of the ventricles.

Diastolic (dī-as-TOL-ik) **blood pressure** The force exerted by blood on arterial walls during ventricular relaxation; the lowest blood pressure measured in the large arteries, normally about 80 mmHg in a young adult.

Diencephalon (dī′-en-SEF-a-lon) A part of the brain consisting of the thalamus, hypothalamus, and epithalamus.

Diffusion (di-FŪ-zhun) A passive process in which there is a net or greater movement of molecules or ions from a region of high concentration to a region of low concentration until equilibrium is reached.

Digestion (dī-JES-chun) The mechanical and chemical breakdown of food to simple molecules that can be absorbed and used by body cells.

Dilate (DĪ-lāt) To expand or swell.

Diploid (DIP-loid) Having the number of chromosomes characteristically found in the somatic cells of an organism; having two haploid sets of chromosomes, one each from the mother and father. Symbolized $2n$.

Direct motor pathways Collections of upper motor neurons with cell bodies in the motor cortex that project axons into the spinal cord, where they synapse with lower motor neurons or interneurons in the anterior horns. Also called the **pyramidal pathways.**

Disease Any change from a state of health.

Dislocation (dis′-lō-KĀ-shun) Displacement of a bone from a joint with tearing of ligaments, tendons, and articular capsules. Also called **luxation** (luks-Ā-shun).

Dissect (di-SEKT) To separate tissues and parts of a cadaver or an organ for anatomical study.

Distal (DIS-tal) Farther from the attachment of a limb to the trunk; farther from the point of origin or attachment.

Diuretic (dī-ū-RET-ik) A chemical that increases urine volume by decreasing reabsorption of water, usually by inhibiting sodium reabsorption.

Divergence (dī-VER-jens) A synaptic arrangement in which the synaptic end bulbs of one presynaptic neuron terminate on several postsynaptic neurons.

Diverticulum (dī-ver-TIK-ū-lum) A sac or pouch in the wall of a canal or organ, especially in the colon.

Dorsal ramus (RĀ-mus) A branch of a spinal nerve containing motor and sensory axons supplying the muscles, skin, and bones of the posterior part of the head, neck, and trunk.

Dorsiflexion (dor-si-FLEK-shun) Bending the foot in the direction of the dorsum (upper surface).

Down-regulation Phenomenon in which there is a decrease in the number of receptors in response to an excess of a hormone or neurotransmitter.

Duct of Santorini *See* **Accessory duct.**

Duct of Wirsung *See* **Pancreatic duct.**

Ductus arteriosus (DUK-tus ar-tē-rē-O-sus) A small vessel connecting the pulmonary trunk with the aorta; found only in the fetus.

Ductus (vas) deferens (DEF-er-ens) The duct that carries sperm from the epididymis to the ejaculatory duct. Also called the **seminal duct.**

Ductus epididymis (ep′-i-DID-i-mis) A tightly coiled tube inside the epididymis, distinguished into a head, body, and tail, in which sperm undergo maturation.

Ductus venosus (ve-NŌ-sus) A small vessel in the fetus that helps the circulation bypass the liver.

Duodenal (doo-ō-DĒ-nal) **gland** Gland in the submucosa of the duodenum that secretes an alkaline mucus to protect the lining of the small intestine from the action of enzymes and to help neutralize the acid in chyme. Also called **Brunner's** (BRUN-erz) **gland.**

Duodenal papilla (pa-PILL-a) An elevation on the duodenal mucosa that receives the hepatopancreatic ampulla (ampulla of Vater).

Duodenum (doo′-ō-DĒ-num *or* doo-OD-e-num) The first 25cm (10 in.) of the small intestine, which connects the stomach and the ileum.

Dura mater (DOO-ra MĀ-ter) The outermost of the three meninges (coverings) of the brain and spinal cord.

Dynamic equilibrium (ē-kwi-LIB-rē-um) The maintenance of body position, mainly the head, in response to sudden movements such as rotation.

Dysmenorrhea (dis′-men-ō-RĒ-a) Painful menstruation.

Dysplasia (dis-PLĀ-zē-a) Change in the size, shape, and organization of cells due to chronic irritation or inflammation; may either revert to normal if stress is removed or progress to neoplasia.

Dyspnea (DISP-nē-a) Shortness of breath; painful or labored breathing.

E

Eardrum A thin, semitransparent partition of fibrous connective tissue between the external auditory meatus and the middle ear. Also called the **tympanic membrane.**

Ectoderm The primary germ layer that gives rise to the nervous system and the epidermis of skin and its derivatives.

Ectopic (ek-TOP-ik) Out of the normal location, as in ectopic pregnancy.

Edema (e-DĒ-ma) An abnormal accumulation of interstitial fluid.

Effector (e-FEK-tor) An organ of the body, either a muscle or a gland, that is innervated by somatic or autonomic motor neurons.

Efferent arteriole (EF-er-ent ar-TĒ-rē-ōl) A vessel of the renal vascular system that carries blood from a glomerulus to a peritubular capillary.

Efferent (EF-er-ent) **ducts** A series of coiled tubes that transport sperm from the rete testis to the epididymis.

Ejaculation (e-jak-ū-LĀ-shun) The reflex ejection or expulsion of semen from the penis.

Ejaculatory (e-JAK-ū-la-tō-rē) **duct** A tube that transports sperm from the ductus (vas) deferens to the prostatic urethra.

Elasticity (e-las-TIS-i-tē) The ability of tissue to return to its original shape after contraction or extension.

Electrocardiogram (e-lek′-trō-KAR-dē-ō-gram) (**ECG** or **EKG**) A recording of the electrical changes that accompany the cardiac cycle that can be detected at the surface of the body; may be resting, stress, or ambulatory.

Elevation (el-e-VĀ-shun) Movement in which a part of the body moves superiorly.

Embolus (EM-bō-lus) A blood clot, bubble of air or fat from broken bones, mass of bacteria, or other debris or foreign material transported by the blood.

Embryo (EM-brē-ō) The young of any organism in an early stage of development; in humans, the developing organism from fertilization to the end of the eighth week of development.

Embryology (em′-brē-OL-ō-jē) The study of development from the fertilized egg to the end of the eighth week of development.

Emesis (EM-e-sis) Vomiting.

Emigration (em′-i-GRĀ-shun) Process whereby white blood cells (WBCs) leave the bloodstream by rolling along the endothelium, sticking to it, and squeezing between the endothelial cells. Adhesion molecules help WBCs stick to the endothelium. Also known as **migration** or **extravasation.**

Emission (ē-MISH-un) Propulsion of sperm into the urethra due to peristaltic contractions of the ducts of the testes, epididymides, and ductus (vas) deferens as a result of sympathetic stimulation.

Emphysema (em-fi-SĒ-ma) A lung disorder in which alveolar walls disintegrate, producing abnormally large air spaces and loss of elasticity in the lungs; typically caused by exposure to cigarette smoke.

Emulsification (ē-mul′-si-fi-KĀ-shun) The dispersion of large lipid globules into smaller, uniformly distributed particles in the presence of bile.

Enamel (e-NAM-el) The hard, white substance covering the crown of a tooth.

Endocardium (en-dō-KAR-dē-um) The layer of the heart wall, composed of endothelium and smooth muscle, that lines the inside of the heart and covers the valves and tendons that hold the valves open.

Endochondral ossification (en′-dō-KON-dral os′-i-fi-KĀ-shun) The replacement of cartilage by bone. Also called **intracartilaginous** (in′-tra-kar′-ti-LAJ-i-nus) **ossification.**

Endocrine (EN-dō-krin) **gland** A gland that secretes hormones into interstitial fluid and then the blood; a ductless gland.

Endocrinology (en′-dō-kri-NOL-ō-jē) The science concerned with the structure and functions of endocrine glands and the diagnosis and treatment of disorders of the endocrine system.

Endocytosis (en′-dō-sī-TŌ-sis) The uptake into a cell of large molecules and particles in which a segment of plasma membrane surrounds the substance, encloses it, and brings it in; includes phagocytosis, pinocytosis, and receptormediated endocytosis.

Endoderm (EN-dō-derm) A primary germ layer of the developing embryo; gives rise to the gastrointestinal tract, urinary bladder, urethra, and respiratory tract.

Endodontics (en′-dō-DON-tiks) The branch of dentistry concerned with the prevention, diagnosis, and treatment of diseases that affect the pulp, root, periodontal ligament, and alveolar bone.

Endolymph (EN-dō-limf′) The fluid within the membranous labyrinth of the internal ear.

Endometriosis (en′-dō-MĒ-trē-ō′-sis) The growth of endometrial tissue outside the uterus.

Endometrium (en′-dō-MĒ-trē-um) The mucous membrane lining the uterus.

Endomysium (en′-dō-MĪZ-ē-um) Invagination of the perimysium separating each individual muscle fiber (cell).

Endoneurium (en′-dō-NOO-rē-um) Connective tissue wrapping around individual nerve axons.

Endoplasmic reticulum (en′-dō-PLAS-mik re-TIK-ū-lum) **(ER)** A network of channels running through the cytoplasm of a cell that serves in intracellular transportation, support, storage, synthesis, and packaging of molecules. Portions of ER where ribosomes are attached to the outer surface are called **rough ER;** portions that have no ribosomes are called **smooth ER.**

End organ of Ruffini *See* **Type II cutaneous mechanoreceptor.**

Endosteum (end-OS-tē-um) The membrane that lines the medullary (marrow) cavity of bones, consisting of osteogenic cells and scattered osteoclasts.

Endothelium (en′-dō-THĒ-lē-um) The layer of simple squamous epithelium that lines the cavities of the heart, blood vessels, and lymphatic vessels.

Enteric (EN-ter-ik) **nervous system** The part of the nervous system that is embedded in the submucosa and muscularis of the gastrointestinal (GI) tract; governs motility and secretions of the GI tract.

Enteroendocrine (en-ter-ō-EN-dō-krin) **cell** A cell of the mucosa of the gastrointestinal tract that secretes a hormone that governs function of the GI tract; hormones secreted include gastrin, cholecystokinin, glucose-dependent insulinotropic peptide (GIP), and secretin.

Enzyme (EN-zīm) A substance that accelerates chemical reactions; an organic catalyst, usually a protein.

Eosinophil (ē-ō-SIN-ō-fil) A type of white blood cell characterized by granules that stain red or pink with acid dyes.

Ependymal (ep-EN-de-mal) **cells** Neuroglial cells that cover choroid plexuses and produce cerebrospinal fluid (CSF); they also line the ventricles of the brain and probably assist in the circulation of CSF.

Epicardium (ep′-i-KAR-dē-um) The thin outer layer of the heart wall, composed of serous tissue and mesothelium. Also called the **visceral pericardium.**

Epidemiology (ep′-i-dē-mē-OL-ō-jē) Study of the occurrence and transmission of diseases and disorders in human populations.

Epidermis (ep′-i-DERM-is) The superficial, thinner layer of skin, composed of keratinized stratified squamous epithelium.

Epididymis (ep′-i-DID-i-mis) A comma-shaped organ that lies along the posterior border of the testis and contains the ductus epididymis, in which sperm undergo maturation. *Plural* is **epididymides** (ep′-i-di-DIM-i-dēz).

Epidural (ep′-i-DOO-ral) **space** A space between the spinal dura mater and the vertebral canal, containing areolar connective tissue and a plexus of veins.

Epiglottis (ep′-i-GLOT-is) A large, leaf-shaped piece of cartilage lying on top of the larynx, attached to the thyroid cartilage; its unattached portion is free to move up and down to cover the glottis (vocal folds and rima glottidis) during swallowing.

Epimysium (ep-i-MĪZ-ē-um) Fibrous connective tissue around muscles.

Epinephrine (ep-ē-NEF-rin) Hormone secreted by the adrenal medulla that produces actions similar to those that result from sympathetic stimulation. Also called **adrenaline** (a-DREN-a-lin).

Epineurium (ep′-i-NOO-rē-um) The superficial connective tissue covering around an entire nerve.

Epiphyseal (ep′-i-FIZ-ē-al) **line** The remnant of the epiphyseal plate in the metaphysis of a long bone.

Epiphyseal plate The hyaline cartilage plate in the metaphysis of a long bone; site of lengthwise growth of long bones.

Epiphysis (e-PIF-i-sis) The end of a long bone, usually larger in diameter than the shaft (diaphysis).

Epiphysis cerebri (se-RĒ-brē) Pineal gland.

Episiotomy (e-piz′-ē-OT-ō-mē) A cut made with surgical scissors to avoid tearing of the perineum at the end of the second stage of labor.

Epistaxis (ep′-i-STAK-sis) Loss of blood from the nose due to trauma, infection, allergy, neoplasm, and bleeding disorders. Also called **nosebleed.**

Epithalamus (ep′-i-THAL-a-mus) Part of the diencephalon superior and posterior to the thalamus, comprising the pineal gland and associated structures.

Epithelial (ep-i-THĒ-lē-al) **tissue** The tissue that forms the innermost and outermost surfaces of body structures and forms glands.

Eponychium (ep′-o-NIK-ē-um) Narrow band of stratum corneum at the proximal border of a nail that extends from the margin of the nail wall. Also called the **cuticle.**

Erectile dysfunction Failure to maintain an erection long enough for sexual intercourse. Also known as **impotence** (IM-pō-tens).

Erection (ē-REK-shun) The enlarged and stiff state of the penis or clitoris resulting from the engorgement of the spongy erectile tissue with blood.

Eructation (e-ruk′-TĀ-shun) The forceful expulsion of gas from the stomach. Also called **belching.**

Erythema (er-e-THĒ-ma) Skin redness usually caused by dilation of the capillaries.

Erythrocyte (e-RITH-rō-sīt) A mature red blood cell.

Erythropoietin (e-rith′-rō-POY-e-tin) A hormone released by the juxtaglomerular cells of the kidneys that stimulates red blood cell production.

Esophagus (e-SOF-a-gus) The hollow muscular tube that connects the pharynx and the stomach.

Estrogens (ES-tro-jenz) Feminizing sex hormones produced by the ovaries; govern development of oocytes, maintenance of female reproductive structures, and appearance of secondary sex characteristics; also affect fluid and electrolyte balance, and protein anabolism. Examples are β-estradiol, estrone, and estriol.

Eupnea (ŪP-nē-a) Normal quiet breathing.

Eustachian tube *See* **Auditory tube.**

Eversion (ē-VER-zhun) The movement of the sole laterally at the ankle joint or of an atrioventricular valve into an atrium during ventricular contraction.

Excitability (ek-sīt′-a-BIL-i-tē) The ability of muscle fibers to receive and respond to stimuli; the ability of neurons to respond to stimuli and generate nerve impulses.

Excretion (eks-KRĒ-shun) The process of eliminating waste products from the body; also the products excreted.

Exocrine (EK-sō-krin) **gland** A gland that secretes its products into ducts that carry the secretions into body cavities, into the lumen of an organ, or to the outer surface of the body.

Exocytosis (ex′-ō-sī-TŌ-sis) A process in which membrane-enclosed secretory vesicles form inside the cell, fuse with the plasma membrane, and release their contents into the interstitial fluid; achieves secretion of materials from a cell.

Exhalation (eks-ha-LĀ-shun) Breathing out; expelling air from the lungs into the atmosphere. Also called **expiration.**

Extensibility (ek-sten′-si-BIL-i-tē) The ability of muscle tissue to stretch when it is pulled.

Extension (eks-TEN-shun) An increase in the angle between two bones; restoring a body part to its anatomical position after flexion.

External Located on or near the surface.

External auditory (AW-di-tōr-ē) **canal** or **meatus** (mē-Ā-tus) A curved tube in the temporal bone that leads to the middle ear.

External ear The outer ear, consisting of the pinna, external auditory canal, and tympanic membrane (eardrum).

External nares (NĀ-rez) The openings into the nasal cavity on the exterior of the body. Also called the **nostrils.**

External respiration The exchange of respiratory gases between the lungs and blood. Also called **pulmonary respiration.**

Exteroceptor (EKS-ter-ō-sep′-tor) A sensory receptor adapted for the reception of stimuli from outside the body.

Extracellular fluid (ECF) Fluid outside body cells, such as interstitial fluid and plasma.

Extracellular matrix (MĀ-triks) The ground substance and fibers between cells in a connective tissue.

Eyebrow The hairy ridge superior to the eye.

F

F cell A cell in the pancreatic islets (islets of Langerhans) that secretes pancreatic polypeptide.

Face The anterior aspect of the head.

Falciform ligament (FAL-si-form LIG-a-ment) A sheet of parietal peritoneum between the two principal lobes of the liver. The ligamentum teres, or remnant of the umbilical vein, lies within its fold.

Falx cerebelli (FALKS cer-e-BEL-li) A small triangular process of the dura mater attached to the occipital bone in the posterior cranial fossa and projecting inward between the two cerebellar hemispheres.

Falx cerebri (FALKS CER-e-brē) A fold of the dura mater extending deep into the longitudinal fissure between the two cerebral hemispheres.

Fascia (FASH-ē-a) A fibrous membrane covering, supporting, and separating muscles.

Fascicle (FAS-i-kul) A small bundle or cluster, especially of nerve or muscle fibers (cells). Also called a **fasciculus** (fa-SIK-ū-lus). *Plural* is **fasciculi** (fa-SIK-yoo-lī).

Fasciculation (fa-sik-ū-LĀ-shun) Abnormal, spontaneous twitch of all skeletal muscle fibers in one motor unit that is visible at the skin surface; not associated with movement of the affected muscle; present in progressive diseases of motor neurons, for example, poliomyelitis.

Fauces (FAW-sēs) The opening from the mouth into the pharynx.

Feces (FĒ-sēz) Material discharged from the rectum and made up of bacteria, excretions, and food residue. Also called **stool.**

Female reproductive cycle General term for the ovarian and uterine cycles, the hormonal changes that accompany them, and cyclic changes in the breasts and cervix; includes changes in the endometrium of a nonpregnant female that prepares the lining of the uterus to receive a fertilized ovum. Less correctly termed the **menstrual cycle.**

Fertilization (fer′-ti-li-ZĀ-shun) Penetration of a secondary oocyte by a sperm cell, meiotic division of secondary oocyte to form an ovum, and subsequent union of the nuclei of the gametes.

Fetal circulation The cardiovascular system of the fetus, including the placenta and special blood vessels involved in the exchange of materials between fetus and mother.

Fetus (FE-tus) In humans, the developing organism *in utero* from the beginning of the third month to birth.

Fever An elevation in body temperature above the normal temperature of 37 °C (98.6 °F) due to a resetting of the hypothalamic thermostat.

Fibroblast (FĪ-brō-blast) A large, flat cell that secretes most of the extracellular matrix of areolar and dense connective tissues.

Fibrous (FĪ-brus) **joint** A joint that allows little or no movement, such as a suture or a syndesmosis.

Fibrous tunic (TOO-nik) The superficial coat of the eyeball, made up of the posterior sclera and the anterior cornea.

Fight-or-flight response The effects produced upon stimulation of the sympathetic division of the autonomic nervous system.

Filiform papilla (FIL-i-form pa-PIL-a) One of the conical projections that are distributed in parallel rows over the anterior two-thirds of the tongue and lack taste buds.

Filtration (fil-TRĀ-shun) The flow of a liquid through a filter (or membrane that acts like a filter) due to a hydrostatic pressure; occurs in capillaries due to blood pressure.

Filum terminale (FĪ-lum ter-mi-NAL-ē) Non-nervous fibrous tissue of the spinal cord that extends inferiorly from the conus medullaris to the coccyx.

Fimbriae (FIM-brē-ē) Fingerlike structures, especially the lateral ends of the uterine (Fallopian) tubes.

Fissure (FISH-ur) A groove, fold, or slit that may be normal or abnormal.

Fixator A muscle that stabilizes the origin of the prime mover so that the prime mover can act more efficiently.

Fixed macrophage (MAK-rō-fāj) Stationary phagocytic cell found in the liver, lungs, brain, spleen, lymph nodes, subcutaneous tissue, and red bone marrow. Also called a **histiocyte** (HIS-tē-ō-sīt).

Flaccid (FLAS-sid) Relaxed, flabby, or soft; lacking muscle tone.

Flagellum (fla-JEL-um) A hairlike, motile process on the extremity of a bacterium, protozoan, or sperm cell. *Plural* is **flagella** (fla-JEL-a).

Flatus (FLĀ-tus) Gas in the stomach or intestines; commonly used to denote expulsion of gas through the anus.

Flexion (FLEK-shun) Movement in which there is a decrease in the angle between two bones.

Follicle (FOL-i-kul) A small secretory sac or cavity; the group of cells that contains a developing oocyte in the ovaries.

Follicle-stimulating hormone (FSH) Hormone secreted by the anterior pituitary; it initiates development of ova and stimulates the ovaries to secrete estrogens in females, and initiates sperm production in males.

Fontanel (fon-ta-NEL) A mesenchyme-filled space where bone formation is not yet complete, especially between the cranial bones of an infant's skull.

Foot The terminal part of the lower limb, from the ankle to the toes.

Foramen (fō-RĀ-men) A passage or opening; a communication between two cavities of an organ, or a hole in a bone for passage of vessels or nerves. *Plural* is **foramina** (fō-RAM-i-na).

Foramen ovale (fō-RĀ-men ō-VAL-ē) An opening in the fetal heart in the septum between the right and left atria. A hole in the greater wing of the sphenoid bone that transmits the mandibular branch of the trigeminal (V) nerve.

Forearm (FOR-arm) The part of the upper limb between the elbow and the wrist.

Fornix (FOR-niks) An arch or fold; a tract in the brain made up of association fibers, connecting the hippocampus with the mammillary bodies; a recess around the cervix of the uterus where it protrudes into the vagina.

Fossa (FOS-a) A furrow or shallow depression.

Fourth ventricle (VEN-tri-kul) A cavity filled with cerebrospinal fluid within the brain lying between the cerebellum and the medulla oblongata and pons.

Fracture (FRAK-choor) Any break in a bone.

Frontal plane A plane at a right angle to a midsagittal plane that divides the body or organs into anterior and posterior portions. Also called a **coronal** (kō-RŌ-nal) **plane.**

Fundus (FUN-dus) The part of a hollow organ farthest from the opening.

Fungiform papilla (FUN-ji-form pa-PIL-a) A mushroomlike elevation on the upper surface of the tongue appearing as a red dot; most contain taste buds.

Furuncle (FŪ-rung-kul) A boil; painful nodule caused by bacterial infection and inflammation of a hair follicle or sebaceous (oil) gland.

G

Gallbladder A small pouch, located inferior to the liver, that stores bile and empties by means of the cystic duct.

Gallstone A solid mass, usually containing cholesterol, in the gallbladder or a bile-containing duct; formed anywhere between bile canaliculi in the liver and the hepatopancreatic ampulla (ampulla of Vater), where bile enters the duodenum. Also called a **biliary calculus.**

Gamete (GAM-ēt) A male or female reproductive cell; a sperm cell or secondary oocyte.

Ganglion (GANG-glē-on) Usually, a group of neuronal cell bodies lying outside the central nervous system (CNS). *Plural* is **ganglia** (GANG-glē-a).

Gastric (GAS-trik) **glands** Glands in the mucosa of the stomach composed of cells that empty their secretions into narrow channels called gastric pits. Types of cells are chief cells (secrete pepsinogen), parietal cells (secrete hydrochloric acid and intrinsic factor), surface mucous and mucous neck cells (secrete mucus), and G cells (secrete gastrin).

Gastroenterology (gas′-trō-en′-ter-OL-ō-jē) The medical specialty that deals with the structure, function, diagnosis, and treatment of diseases of the stomach and intestines.

Gastrointestinal (gas-trō-in-TES-ti-nal) **(GI) tract** A continuous tube running through the ventral body cavity extending from the mouth to the anus. Also called the **alimentary** (al′-i-MEN-tar-ē) **canal.**

Gastrulation (gas′-troo-LĀ-shun) The migration of groups of cells from the epiblast that transform a bilaminar embryonic disc into a trilaminar embryonic disc with three primary germ layers; transformation of the blastula into the gastrula.

Gene (JĒN) Biological unit of heredity; a segment of DNA located in a definite position on a particular chromosome; a sequence of DNA that codes for a particular mRNA, rRNA, or tRNA.

Genetic engineering The manufacture and manipulation of genetic material.

Genetics The study of genes and heredity.

Genome (JĒ-nōm) The complete set of genes of an organism.

Genotype (JĒ-nō-tīp) The genetic makeup of an individual; the combination of alleles present at one or more chromosomal locations, as distinguished from the appearance, or phenotype, that results from those alleles.

Geriatrics (jer′-ē-AT-riks) The branch of medicine devoted to the medical problems and care of elderly persons.

Gestation (jes-TĀ-shun) The period of development from fertilization to birth.

Gingivae (jin-JI-vē) Gums. They cover the alveolar processes of the mandible and maxilla and extend slightly into each socket.

Gland Specialized epithelial cell or cells that secrete substances; may be exocrine or endocrine.

Glans penis (glanz PĒ-nis) The slightly enlarged region at the distal end of the penis.

Glaucoma (glaw-KŌ-ma) An eye disorder in which there is increased intraocular pressure due to an excess of aqueous humor.

Gliding joint A synovial joint having articulating surfaces that are usually flat, permitting only side-to-side and back-and-forth movements, as between carpal bones, tarsal bones, and the scapula and clavicle. Also called an **arthrodial** (ar-THRŌ-dē-al) **joint.**

Glomerular (glō-MER-ū-lar) **capsule** A double-walled globe at the proximal end of a nephron that encloses the glomerular capillaries. Also called **Bowman's** (BŌ-manz) **capsule.**

Glomerular filtrate (glō-MER-ū-lar FIL-trāt) The fluid produced when blood is filtered by the filtration membrane in the glomeruli of the kidneys.

Glomerular filtration The first step in urine formation in which substances in blood pass through the filtration membrane and the filtrate enters the proximal convoluted tubule of a nephron.

Glomerulus (glō-MER-ū-lus) A rounded mass of nerves or blood vessels, especially the microscopic tuft of capillaries that is surrounded by the glomerular (Bowman's) capsule of each kidney tubule. *Plural* is **glomeruli.**

Glottis (GLOT-is) The vocal folds (true vocal cords) in the larynx plus the space between them (rima glottidis).

Glucagon (GLOO-ka-gon) A hormone produced by the alpha cells of the pancreatic islets (islets of Langerhans) that increases blood glucose level.

Glucocorticoids (gloo′-kō-KOR-ti-koyds) Hormones secreted by the cortex of the adrenal gland, especially cortisol, that influence glucose metabolism.

Glucose (GLOO-kōs) A hexose (six-carbon sugar), $C_6H_{12}O_6$, that is a major energy source for the production of ATP by body cells.

Glucosuria (gloo′-kō-SOO-rē-a) The presence of glucose in the urine; may be temporary or pathological. Also called **glycosuria.**

Glycogen (GLĪ-kō-jen) A highly branched polymer of glucose containing thousands of subunits; functions as a compact store of glucose molecules in liver and muscle fibers (cells).

Goblet cell A goblet-shaped unicellular gland that secretes mucus; present in epithelium of the airways and intestines.

Goiter (GOY-ter) An enlarged thyroid gland.

Golgi (GOL-jē) **complex** An organelle in the cytoplasm of cells consisting of four to six flattened sacs (cisternae), stacked on one another, with expanded areas at their ends; functions in processing, sorting, packaging, and delivering proteins and lipids to the plasma membrane, lysosomes, and secretory vesicles.

Golgi tendon organ *See* **Tendon organ.**

Gomphosis (gom-FŌ-sis) A fibrous joint in which a cone-shaped peg fits into a socket.

Gonad (GŌ-nad) A gland that produces gametes and hormones; the ovary in the female and the testis in the male.

Gonadotropic hormone Anterior pituitary hormone that affects the gonads.

Gout (GOWT) Hereditary condition associated with excessive uric acid in the blood; the acid crystallizes and deposits in joints, kidneys, and soft tissue.

Graafian follicle *See* **Vesicular ovarian follicle.**

Gracile (GRAS-il) **nucleus** A group of nerve cells in the inferior part of the medulla oblongata in which axons of the gracile fasciculus terminate.

Gray commissure (KOM-mi-shur) A narrow strip of gray matter connecting the two lateral gray masses within the spinal cord.

Gray matter Areas in the central nervous system and ganglia containing neuronal cell bodies, dendrites, unmyelinated axons, axon terminals, and neuroglia; Nissl bodies impart a gray color and there is little or no myelin in gray matter.

Gray ramus communicans (RĀ-mus kō-MŪ-ni-kans) A short nerve containing axons of sympathetic postganglionic neurons; the cell bodies of the neurons are in a sympathetic chain ganglion, and the unmyelinated axons extend via the gray ramus to a spinal nerve and then to the periphery to supply smooth muscle in blood vessels, arrector pili muscles, and sweat glands. *Plural* is **rami communicantes** (RĀ-mē kō-mū-ni-KAN-tēz).

Greater omentum (ō-MEN-tum) A large fold in the serosa of the stomach that hangs down like an apron anterior to the intestines.

Greater vestibular (ves-TIB-ū-lar) **glands** A pair of glands on either side of the vaginal orifice that open by a duct into the space between the hymen and the labia minora. Also called **Bartholin's** (BAR-to-linz) **glands.**

Groin (GROYN) The depression between the thigh and the trunk; the inguinal region.

Gross anatomy The branch of anatomy that deals with structures that can be studied without using a microscope. Also called **macroscopic anatomy.**

Growth An increase in size due to an increase in (1) the number of cells, (2) the size of existing cells as internal components increase in size, or (3) the size of intercellular substances.

Gustatory (GUS-ta-tō′-rē) Pertaining to taste.

Gynecology (gī′-ne-KOL-ō-jē) The branch of medicine dealing with the study and treatment of disorders of the female reproductive system.

Gynecomastia (gīn′e-kō-MAS-tē-a) Excessive growth (benign) of the male mammary glands due to secretion of estrogens by an adrenal gland tumor (feminizing adenoma).

Gyrus (JĪ-rus) One of the folds of the cerebral cortex of the brain. *Plural* is **gyri** (JĪ-rī). Also called a **convolution.**

H

Hair A threadlike structure produced by hair follicles that develops in the dermis. Also called a **pilus** (PĪ-lus).

Hair follicle (FOL-li-kul) Structure composed of epithelium and surrounding the root of a hair from which hair develops.

Hair root plexus (PLEK-sus) A network of dendrites arranged around the root of a hair as free or naked nerve endings that are stimulated when a hair shaft is moved.

Hand The terminal portion of an upper limb, including the carpus, metacarpus, and phalanges.

Haploid (HAP-loyd) **cell** Having half the number of chromosomes characteristically found in the somatic cells of an organism; characteristic of mature gametes. Symbolized *n*.

Hard palate (PAL-at) The anterior portion of the roof of the mouth, formed by the maxillae and palatine bones and lined by mucous membrane.

Haustra (HAWS-tra) A series of pouches that characterize the colon; caused by tonic contractions of the teniae coli. *Singular is* **haustrum.**

Haversian canal *See* **Central canal.**

Haversian system *See* **Osteon.**

Head The superior part of a human, cephalic to the neck. The superior or proximal part of a structure.

Heart A hollow muscular organ lying slightly to the left of the midline of the chest that pumps the blood through the cardiovascular system.

Heart block An arrhythmia (dysrhythmia) of the heart in which the atria and ventricles contract independently because of a blocking of electrical impulses through the heart at some point in the conduction system.

Heart murmur (MER-mer) An abnormal sound that consists of a flow noise that is heard before, between, or after the normal heart sounds, or that may mask normal heart sounds.

Hemangioblast (hē-MAN-jē-ō-blast) A precursor mesodermal cell that develops into blood and blood vessels.

Hematocrit (hē-MAT-ō-krit) **(Hct)** The percentage of blood made up of red blood cells. Usually measured by centrifuging a blood sample in a graduated tube and then reading the volume of red blood cells and dividing it by the total volume of blood in the sample.

Hematology (hēm-a-TOL-ō-jē) The study of blood.

Hematoma (hē′-ma-TŌ-ma) A tumor or swelling filled with blood.

Hemiplegia (hem-i-PLĒ-jē-a) Paralysis of the upper limb, trunk, and lower limb on one side of the body.

Hemoglobin (hē′-mō-GLŌ-bin) **(Hb)** A substance in red blood cells consisting of the protein globin and the iron-containing red pigment heme that transports most of the oxygen and some carbon dioxide in blood.

Hemolysis (hē-MOL-i-sis) The escape of hemoglobin from the interior of a red blood cell into the surrounding medium; results from disruption of the cell membrane by toxins or drugs, freezing or thawing, or hypotonic solutions.

Hemolytic disease of the newborn A hemolytic anemia of a newborn child that results from the destruction of the infant's erythrocytes (red blood cells) by antibodies produced by the mother; usually the antibodies are due to an Rh blood type incompatibility. Also called **erythroblastosis fetalis** (e-rith′-rō-blas-TŌ-sis fe-TAL-is).

Hemophilia (hē-mō-FIL-ē-a) A hereditary blood disorder where there is a deficient production of certain factors involved in blood clotting, resulting in excessive bleeding into joints, deep tissues, and elsewhere.

Hemopoiesis (hēm-ō-poy-Ē-sis) Blood cell production, which occurs in red bone marrow after birth. Also called **hematopoiesis** (hem′-a-tō-poy-Ē-sis).

Hemorrhage (HEM-o-rij) Bleeding; the escape of blood from blood vessels, especially when the loss is profuse.

Hemorrhoids (HEM-ō-royds) Dilated or varicosed blood vessels (usually veins) in the anal region. Also called **piles.**

Hepatic (he-PAT-ik) Refers to the liver.

Hepatic duct A duct that receives bile from the bile capillaries. Small hepatic ducts merge to form the larger right and left hepatic ducts that unite to leave the liver as the common hepatic duct.

Hepatic portal circulation The flow of blood from the gastrointestinal organs to the liver before returning to the heart.

Hepatocyte (he-PAT-ō-cyte) A liver cell.

Hepatopancreatic (hep′-a-tō-pan′-krē-A-tik) **ampulla** A small, raised area in the duodenum where the combined common bile duct and main pancreatic duct empty into the duodenum. Also called the **ampulla of Vater** (VA-ter).

Hernia (HER-nē-a) The protrusion or projection of an organ or part of an organ through a membrane or cavity wall, usually the abdominal cavity.

Herniated (HER-nē-ā′-ted) **disc** A rupture of an intervertebral disc so that the nucleus pulposus protrudes into the vertebral cavity. Also called a **slipped disc.**

Hiatus (hī-Ā-tus) An opening; a foramen.

Hilum (HĪ-lus) An area, depression, or pit where blood vessels and nerves enter or leave an organ. Also called a **hilus.**

Hinge joint A synovial joint in which a convex surface of one bone fits into a concave surface of another bone, such as the elbow, knee, ankle, and interphalangeal joints. Also called a **ginglymus** (JIN-gli-mus) **joint.**

Hirsutism (HER-soo-tizm) An excessive growth of hair in females and children, with a distribution similar to that in adult males, due to the conversion of vellus hairs into large terminal hairs in response to higher-than-normal levels of androgens.

Histamine (HISS-ta-mēn) Substance found in many cells, especially mast cells, basophils, and platelets, that is released when the cells are injured; results in vasodilation, increased permeability of blood vessels, and constriction of bronchioles.

Histology (hiss′-TOL-ō-jē) Microscopic study of the structure of tissues.

Holocrine (HŌ-lō-krin) **gland** A type of gland in which entire secretory cells, along with their accumulated secretions, make up the secretory product of the gland, as in the sebaceous (oil) glands.

Homeostasis (hō′-mē-ō-STĀ-sis) The condition in which the body's internal environment remains relatively constant within physiological limits.

Homologous chromosomes Two chromosomes that belong to a pair. Also called **homologs.**

Hormone (HOR-mōn) A secretion of endocrine cells that alters the physiological activity of target cells of the body.

Horn An area of gray matter (anterior, lateral, or posterior) in the spinal cord.

Human chorionic gonadotropin (kō-rē-ON-ik gō-nad-ō-TRŌ-pin) **(hCG)** A hormone produced by the developing placenta that maintains the corpus luteum.

Human chorionic somatomammotropin (sō-mat-ō-mam-ō-TRŌ-pin) **(hCS)** Hormone produced by the chorion of the placenta that stimulates breast tissue for lactation, enhances body growth, and regulates metabolism. Also called **human placental lactogen (hPL).**

Human growth hormone (hGH) Hormone secreted by the anterior pituitary that stimulates growth of body tissues, especially skeletal and muscular tissues. Also known as **somatotropin** and **somatotropic hormone (STH).**

Hyaluronic (hī′-a-loo-RON-ik) **acid** A viscous, amorphous extracellular material that binds cells together, lubricates joints, and maintains the shape of the eyeballs.

Hymen (HĪ-men) A thin fold of vascularized mucous membrane at the vaginal orifice.

Hyperextension (hī′-per-ek-STEN-shun) Continuation of extension beyond the anatomical position, as in bending the head backward.

Hyperplasia (hī-per-PLĀ-zē-a) An abnormal increase in the number of normal cells in a tissue or organ, increasing its size.

Hypersecretion (hī′-per-se-KRĒ-shun) Overactivity of glands resulting in excessive secretion.

Hypersensitivity (hī′-per-sen-si-TI-vi-tē) Overreaction to an allergen that results in pathological changes in tissues. Also called **allergy.**

Hypertension (hī′-per-TEN-shun) High blood pressure.

Hyperthermia (hī′-per-THERM-ē-a) An elevated body temperature.

Hypertonia (hī′-per-TŌ-nē-a) Increased muscle tone that is expressed as spasticity or rigidity.

Hypertonic (hī′-per-TON-ik) Solution that causes cells to shrink due to loss of water by osmosis.

Hypertrophy (hī-PER-trō-fē) An excessive enlargement or overgrowth of tissue without cell division.

Hyperventilation (hī′-per-ven-ti-LĀ-shun) A rate of inhalation and exhalation higher than that required to maintain a normal partial pressure of carbon dioxide in the blood.

Hyponychium (hī′-pō-NIK-ē-um) Free edge of the fingernail.

Hypophyseal fossa (hī′-pō-FIZ-ē-al FOS-a) A depression on the superior surface of the sphenoid bone that houses the pituitary gland.

Hypophyseal (hī′-pō-FIZ-ē-al) **pouch** An outgrowth of ectoderm from the roof of the mouth from which the anterior pituitary develops.

Hypophysis (hī-POF-i-sis) Pituitary gland.

Hyposecretion (hī′-pō-se-KRĒ-shun) Underactivity of glands resulting in diminished secretion.

Hypothalamohypophyseal (hī′-pō-thal′-a-mō-hī-pō-FIZ-ē-al) **tract** A bundle of axons containing secretory vesicles filled with oxytocin or antidiuretic hormone that extend from the hypothalamus to the posterior pituitary.

Hypothalamus (hī′-pō-THAL-a-mus) A portion of the diencephalon, lying beneath the thalamus and forming the floor and part of the wall of the third ventricle.

Hypothermia (hī′-pō-THER-mē-a) Lowering of body temperature below 35 °C (95 °F); in surgical procedures, it refers to deliberate cooling of the body to slow down metabolism and reduce oxygen needs of tissues.

Hypotonia (hī′-pō-TŌ-nē-a) Decreased or lost muscle tone in which muscles appear flaccid.

Hypotonic (hī′-pō-TON-ik) Solution that causes cells to swell and perhaps rupture due to gain of water by osmosis.

Hypoventilation (hī-pō-ven-ti-LĀ-shun) A rate of inhalation and exhalation lower than that required to maintain a normal partial pressure of carbon dioxide in plasma.

Hypoxia (hī-POKS-ē-a) Lack of adequate oxygen at the tissue level.

Hysterectomy (hiss-te-REK-tō-mē) The surgical removal of the uterus.

I

Ileocecal (il-ē-ō-SĒ-kal) **sphincter** A fold of mucous membrane that guards the opening from the ileum into the large intestine. Also called the **ileocecal valve.**

Ileum (IL-ē-um) The terminal part of the small intestine.

Immunity (im-Ū-ni-tē) The state of being resistant to injury, particularly by poisons, foreign proteins, and invading pathogens.

Immunoglobulin (im-ū-nō-GLOB-ū-lin) **(Ig)** An antibody synthesized by plasma cells derived from B lymphocytes in response to the introduction of an antigen. Immunoglobulins are divided into five kinds (IgG, IgM, IgA, IgD, IgE).

Immunology (im′-ū-NOL-ō-jē) The study of the responses of the body when challenged by antigens.

Imperforate (im-PER-fō-rāt) Abnormally closed.

Implantation (im-plan-TĀ-shun) The insertion of a tissue or a part into the body. The attachment of the blastocyst to the stratum basalis of the endometrium about 6 days after fertilization.

Incontinence (in-KON-ti-nens) Inability to retain urine, semen, or feces through loss of sphincter control.

Indirect motor pathways Motor tracts that convey information from the brain down the spinal cord for automatic movements, coordination of body movements with visual stimuli, skeletal muscle tone and posture, and balance. Also known as **extrapyramidal pathways.**

Induction (in-DUK-shun) The process by which one tissue (inducting tissue) stimulates the development of an adjacent unspecialized tissue (responding tissue) into a specialized one.

Infarction (in-FARK-shun) A localized area of necrotic tissue, produced by inadequate oxygenation of the tissue.

Infection (in-FEK-shun) Invasion and multiplication of microorganisms in body tissues, which may be inapparent or characterized by cellular injury.

Inferior (in-FĒR-ē-or) Away from the head or toward the lower part of a structure. Also called **caudad** (KAW-dad).

Inferior vena cava (VĒ-na CĀ-va) **(IVC)** Large vein that collects blood from parts of the body inferior to the heart and returns it to the right atrium.

Infertility Inability to conceive or to cause conception. Also called **sterility.**

Inflammation (in′-fla-MĀ-shun) Localized, protective response to tissue injury designed to destroy, dilute, or wall off the infecting agent or injured tissue; characterized by redness, pain, heat, swelling, and sometimes loss of function.

Infundibulum (in-fun-DIB-ū-lum) The stalklike structure that attaches the pituitary gland to the hypothalamus of the brain. The funnel-shaped, open, distal end of the uterine (Fallopian) tube.

Ingestion (in-JES-chun) The taking in of food, liquids, or drugs, by mouth.

Inguinal (IN-gwi-nal) Pertaining to the groin.

Inguinal canal An oblique passageway in the anterior abdominal wall just superior and parallel to the medial half of the inguinal ligament that transmits the spermatic cord and ilioinguinal nerve in the male and round ligament of the uterus and ilioinguinal nerve in the female.

Inhalation (in-ha-LĀ-shun) The act of drawing air into the lungs. Also termed **inspiration.**

Inheritance The acquisition of body traits by transmission of genetic information from parents to offspring.

Inhibin A hormone secreted by the gonads that inhibits release of follicle-stimulating hormone (FSH) by the anterior pituitary.

Inhibiting hormone Hormone secreted by the hypothalamus that can suppress secretion of hormones by the anterior pituitary.

Inner cell mass A region of cells of a blastocyst that differentiates into the three primary germ layers—ectoderm, mesoderm, and endoderm—from which all tissues and organs develop; also called an **embryoblast.**

Insertion (in-SER-shun) The attachment of a muscle tendon to a movable bone or the end opposite the origin.

Insula (IN-soo-la) A triangular area of the cerebral cortex that lies deep within the lateral cerebral fissue, under the parietal, frontal, and temporal lobes.

Insulin (IN-soo-lin) A hormone produced by the beta cells of a pancreatic islet (islet of Langerhans) that decreases the blood glucose level.

Integrins (IN-te-grinz) A family of transmembrane glycoproteins in plasma membranes that function in cell adhesion; they are present in hemidesmosomes, which anchor cells to a basement membrane, and they mediate adhesion of neutrophils to endothelial cells during emigration.

Integumentary (in-teg-ū-MEN-tar-ē) Relating to the skin.

Intercalated (in-TER-ka-lā t-ed) **disc** An irregular transverse thickening of sarcolemma that contains desmosomes, which hold cardiac muscle fibers (cells) together, and gap junctions, which aid in conduction of muscle action potentials from one fiber to the next.

Intercostal (in′-ter-KOS-tal) **nerve** A nerve supplying a muscle located between the ribs.

Intermediate Between two structures, one of which is medial and one of which is lateral.

Intermediate filament Protein filament, ranging from 8 to 12 nm in diameter, that may provide structural reinforcement, hold organelles in place, and give shape to a cell.

Internal Away from the surface of the body.

Internal capsule A large tract of projection fibers lateral to the thalamus that is the major connection between the cerebral cortex and the brain stem and spinal cord; contains axons of sensory neurons carrying auditory, visual, and somatic sensory signals to the cerebral cortex plus axons of motor neurons descending from the cerebral cortex to the thalamus, subthalamus, brain stem, and spinal cord.

Internal ear The inner ear or labyrinth, lying inside the temporal bone, containing the organs of hearing and balance.

Internal nares (NĀ-rez) The two openings posterior to the nasal cavities opening into the nasopharynx. Also called the **choanae** (kō-Ā-nē).

Internal respiration The exchange of respiratory gases between blood and body cells. Also called **tissue respiration.**

Interneurons (in′-ter-NOO-ronz) Neurons whose axons extend only for a short distance and contact nearby neurons in the brain, spinal cord, or a ganglion; they comprise the vast majority of neurons in the body.

Interoceptor (IN-ter-ō-sep′-tor) Sensory receptor located in blood vessels and viscera that provides information about the body's internal environment.

Interphase (IN-ter-fāz) The period of the cell cycle between cell divisions, consisting of the G_1-(gap or growth) phase, when the cell is engaged in growth, metabolism, and production of substances required for division; S-(synthesis) phase, during which chromosomes are replicated; and G_2-phase.

Interstitial cell of Leydig *See* **Interstitial endocrinocyte.**

Interstitial (in′-ter-STISH-al) **endocrinocyte** A cell that is located in the connective tissue between seminiferous tubules in a mature testis that secretes testosterone. Also called an **interstitial cell of Leydig** (LĪ-dig).

Interstitial (in′-ter-STISH-al) **fluid** The portion of extracellular fluid that fills the microscopic spaces between the cells of tissues; the internal environment of the body. Also called **intercellular** or **tissue fluid.**

Interstitial growth Growth from within, as in the growth of cartilage. Also called **endogenous** (en-DOJ-e-nus) **growth.**

Interventricular (in′-ter-ven-TRIK-ū-lar) **foramen** A narrow, oval opening through which the lateral ventricles of the brain communicate with the third ventricle. Also called the **foramen of Monro.**

Intervertebral (in′-ter-VER-te-bral) **disc** A pad of fibrocartilage located between the bodies of two vertebrae.

Intestinal gland A gland that opens onto the surface of the intestinal mucosa and secretes digestive enzymes. Also called a **crypt of Lieberkühn** (LĒ-ber-kūn).

Intracellular (in′-tra-SEL-yū-lar) **fluid** (**ICF**) Fluid located within cells.

Intrafusal (in′-tra-FŪ-sal) **fibers** Three to ten specialized muscle fibers (cells), partially enclosed in a spindle-shaped connective tissue capsule, that make up a muscle spindle.

Intramembranous ossification (in′-tra-MEM-bra-nus os′-i-fi-KĀ-shun) The method of bone formation in which the bone is formed directly in mesenchyme arranged sheet like layers that resemble membranes.

Intraocular (in′-tra-OK-ū-lar) **pressure (IOP)** Pressure in the eyeball, produced mainly by aqueous humor.

Intrinsic factor (IF) A glycoprotein, synthesized and secreted by the parietal cells of the gastric mucosa, that facilitates vitamin B_{12} absorption in the small intestine.

Invagination (in-vaj′-i-NĀ-shun) The pushing of the wall of a cavity into the cavity itself.

Inversion (in-VER-zhun) The movement of the sole medially at the ankle joint.

In vitro (VĒ-trō) Literally, in glass; outside the living body and in an artificial environment such as a laboratory test tube.

Ipsilateral (ip-si-LAT-er-al) On the same side, affecting the same side of the body.

Iris The colored portion of the vascular tunic of the eyeball seen through the cornea that contains circular and radial smooth muscle; the hole in the center of the iris is the pupil.

Irritable bowel syndrome (IBS) Disease of the entire gastrointestinal tract in which a person reacts to stress by developing symptoms (such as cramping and abdominal pain) associated with alternating patterns of diarrhea and constipation. Excessive amounts of mucus may appear in feces, and other symptoms include flatulence, nausea, and loss of appetite. Also known as **irritable colon** or **spastic colitis.**

Ischemia (is-KĒ-mē-a) A lack of sufficient blood to a body part due to obstruction or constriction of a blood vessel.

Islet of Langerhans *See* **Pancreatic islet.**

Isotonic (ī′-sō-TON-ik) Having equal tension or tone. A solution having the same concentration of impermeable solutes as cytosol.

Isthmus (IS-mus) A narrow strip of tissue or narrow passage connecting two larger parts.

J

Jaundice (JON-dis) A condition characterized by yellowness of the skin, the white of the eyes, mucous membranes, and body fluids because of a buildup of bilirubin.

Jejunum (je-JOO-num) The middle part of the small intestine.

Joint kinesthetic (kin′-es-THET-ik) **receptor** A proprioceptive receptor located in a joint, stimulated by joint movement.

Juxtaglomerular (juks-ta-glō-MER-ū-lar) **apparatus (JGA)** Consists of the macula densa (cells of the distal convoluted tubule adjacent to the afferent and efferent arteriole) and juxtaglomerular cells (modified cells of the afferent and sometimes efferent arteriole); secretes renin when blood pressure starts to fall.

K

Keratin (KER-a-tin) An insoluble protein found in the hair, nails, and other keratinized tissues of the epidermis.

Keratinocyte (ker-a-TIN-ō-sīt) The most numerous of the epidermal cells; produces keratin.

Kidney (KID-nē) One of the paired reddish organs located in the lumbar region that regulates the composition, volume, and pressure of blood and produces urine.

Kidney stone A solid mass, usually consisting of calcium oxalate, uric acid, or calcium phosphate crystals, that may form in any portion of the urinary tract. Also called a **renal calculus.**

Kinesiology (ki-nē-sē′-OL-ō-jē) The study of the movement of body parts.

Kinesthesia (kin′-es-THĒ-zē-a) The perception of the extent and direction of movement of body parts; this sense is possible due to nerve impulses generated by proprioceptors.

Kinetochore (ki-NET-ō-kor) Protein complex attached to the outside of a centromere to which kinetochore microtubules attach.

Kupffer's cell *See* **Stellate reticuloendothelial cell.**

Kyphosis (kī-FŌ-sis) An exaggeration of the thoracic curve of the vertebral column, resulting in a "round-shouldered" appearance. Also called **hunchback.**

L

Labial frenulum (LĀ-bē-al FREN-ū-lum) A medial fold of mucous membrane between the inner surface of the lip and the gums.

Labia majora (LĀ-bē-a ma-JŌ-ra) Two longitudinal folds of skin extending downward and backward from the mons pubis of the female.

Labia minora (min-OR-a) Two small folds of mucous membrane lying medial to the labia majora of the female.

Labium (LĀ-bē-um) A lip. A liplike structure. *Plural* is **labia** (LA-bē-a).

Labor The process of giving birth in which a fetus is expelled from the uterus through the vagina.

Labyrinth (LAB-i-rinth) Intricate communicating passageway, especially in the internal ear.

Lacrimal canal A duct, one on each eyelid, beginning at the punctum at the medial margin of an eyelid and conveying tears medially into the nasolacrimal sac.

Lacrimal gland Secretory cells, located at the superior anterolateral portion of each orbit, that secrete tears into excretory ducts that open onto the surface of the conjunctiva.

Lacrimal sac The superior expanded portion of the nasolacrimal duct that receives the tears from a lacrimal canal.

Lactation (lak-TĀ-shun) The secretion and ejection of milk by the mammary glands.

Lacteal (LAK-tē-al) One of many lymphatic vessels in villi of the intestines that absorb triglycerides and other lipids from digested food.

Lacuna (la-KOO-na) A small, hollow space, such as that found in bones in which the osteocytes lie. *Plural* is **lacunae** (la-KOO-nē).

Lambdoid (LAM-doyd) **suture** The joint in the skull between the parietal bones and the occipital bone; sometimes contains sutural (Wormian) bones.

Lamellae (la-MEL-ē) Concentric rings of hard, calcified extracellular matrix found in compact bone.

Lamellated corpuscle Oval-shaped pressure receptor located in the dermis or subcutaneous tissue and consisting of concentric layers of connective tissue wrapped around the dendrites of a sensory neuron. Also called a **pacinian** (pa-SIN-ē-an) **corpuscle.**

Lamina (LAM-i-na) A thin, flat layer or membrane, as the flattened part of either side of the arch of a vertebra. *Plural* is **laminae** (LAM-i-nē).

Lamina propria (PRŌ-prē-a) The connective tissue layer of a mucosa.

Langerhans (LANG-er-hans) **cell** Epidermal dendritic cell that functions as an antigen-presenting cell (APC) during an immune response.

Lanugo (la-NOO-gō) Fine downy hairs that cover the fetus.

Large intestine The portion of the gastrointestinal tract extending from the ileum of the small intestine to the anus, divided structurally into the cecum, colon, rectum, and anal canal.

Laryngopharynx (la-rin′-gō-FAR-inks) The inferior portion of the pharynx, extending downward from the level of the hyoid bone that divides posteriorly into the esophagus and anteriorly into the larynx. Also called the **hypopharynx.**

Larynx (LAR-inks) The voice box, a short passageway that connects the pharynx with the trachea.

Lateral (LAT-er-al) Farther from the midline of the body or a structure.

Lateral ventricle (VEN-tri-kul) A cavity within a cerebral hemisphere that communicates with the lateral ventricle in the other cerebral hemisphere and with the third ventricle by way of the interventricular foramen.

Leg The part of the lower limb between the knee and the ankle.

Lens A transparent organ constructed of proteins (crystallins) lying posterior to the pupil and iris of the eyeball and anterior to the vitreous body.

Lesion (LĒ-zhun) Any localized, abnormal change in a body tissue.

Lesser omentum (ō-MEN-tum) A fold of the peritoneum that extends from the liver to the lesser curvature of the stomach and the first part of the duodenum.

Lesser vestibular (ves-TIB-ū-lar) **gland** One of the paired mucus-secreting glands with ducts that open on either side of the urethral orifice in the vestibule of the female.

Leukemia (loo-KĒ-mē-a) A malignant disease of the blood-forming tissues characterized by either uncontrolled production and accumulation of immature leukocytes in which many cells fail to reach maturity (acute) or an accumulation of mature leukocytes in the blood because they do not die at the end of their normal life span (chronic).

Leukocyte (LOO-kō-sīt) A white blood cell.

Leydig (LĪ-dig) **cell** A type of cell that secretes testosterone; located in the connective tissue between seminiferous tubules in a mature testis. Also known as **interstitial cell of Leydig** or **interstitial endocrinocyte.**

Ligament (LIG-a-ment) Dense regular connective tissue that attaches bone to bone.

Ligand (LĪ-gand) A chemical substance that binds to a specific receptor.

Limbic system A part of the forebrain, sometimes termed the visceral brain, concerned with various aspects of emotion and behavior; includes the limbic lobe, dentate gyrus, amygdala, septal nuclei, mammillary bodies, anterior thalamic nucleus, olfactory bulbs, and bundles of myelinated axons.

Lingual frenulum (LIN-gwal FREN-ū-lum) A fold of mucous membrane that connects the tongue to the floor of the mouth.

Lipase An enzyme that splits fatty acids from triglycerides and phospholipids.

Lipid (LIP-id) An organic compound composed of carbon, hydrogen, and oxygen that is usually insoluble in water, but soluble in alcohol, ether, and chloroform; examples include triglycerides (fats and oils), phospholipids, steroids, and eicosanoids.

Lipid bilayer Arrangement of phospholipid, glycolipid, and cholesterol molecules in two parallel sheets in which the hydrophilic "heads" face outward and the hydrophobic "tails" face inward; found in cellular membranes.

Lipoprotein (lip′-ō-PRŌ-tēn) One of several types of particles containing lipids (cholesterol and triglycerides) and proteins that make it water soluble for transport in the blood; high levels of **low-density lipoproteins (LDLs)** are associated with increased risk of atherosclerosis, whereas high levels of **high-density lipoproteins (HDLs)** are associated with decreased risk of atherosclerosis.

Liver Large organ under the diaphragm that occupies most of the right hypochondriac region and part of the epigastric region. Functionally, it produces bile and synthesizes most plasma proteins; interconverts nutrients; detoxifies substances; stores glycogen, iron, and vitamins; carries on phagocytosis of worn-out blood cells and bacteria; and helps synthesize the active form of vitamin D.

Long-term potentiation (LTP) Prolonged, enhanced synaptic transmission that occurs at certain synapses within the hippocampus of the brain; believed to underlie some aspects of memory.

Lordosis (lor-DŌ-sis) An exaggeration of the lumbar curve of the vertebral column. Also called **hollow back.**

Lower limb The appendage attached at the pelvic (hip) girdle, consisting of the thigh, knee, leg, ankle, foot, and toes. Also called the **lower extremity.**

Lumbar (LUM-bar) Region of the back and side between the ribs and pelvis; loin.

Lumbar plexus (PLEK-sus) A network formed by the anterior (ventral) branches of spinal nerves L1 through L4.

Lumen (LOO-men) The space within an artery, vein, intestine, renal tubule, or other tubular structure.

Lungs Main organs of respiration that lie on either side of the heart in the thoracic cavity.

Lunula (LOO-noo-la) The moon-shaped white area at the base of a nail.

Luteinizing (LOO-tē-in′-īz-ing) **hormone (LH)** A hormone secreted by the anterior pituitary that stimulates ovulation, stimulates progesterone secretion by the corpus luteum, and readies the mammary glands for milk secretion in females; stimulates testosterone secretion by the testes in males.

Lymph (LIMF) Fluid confined in lymphatic vessels and flowing through the lymphatic system until it is returned to the blood.

Lymph node An oval or bean-shaped structure located along lymphatic vessels.

Lymphatic (lim-FAT-ik) **capillary** Closed-ended microscopic lymphatic vessel that begins in spaces between cells and converges with other lymphatic capillaries to form lymphatic vessels.

Lymphatic tissue A specialized form of reticular tissue that contains large numbers of lymphocytes.

Lymphatic vessel A large vessel that collects lymph from lymphatic capillaries and converges with other lymphatic vessels to form the thoracic and right lymphatic ducts.

Lymphocyte (LIM-fō-sīt) A type of white blood cell that helps carry out cell-mediated and antibody-mediated immune responses; found in blood and in lymphatic tissues.

Lysosome (LĪ-sō-sōm) An organelle in the cytoplasm of a cell, enclosed by a single membrane and containing powerful digestive enzymes.

Lysozyme (LĪ-sō-zīm) A bactericidal enzyme found in tears, saliva, and perspiration.

M

Macrophage (MAK-rō-fāj) Phagocytic cell derived from a monocyte; may be fixed or wandering.

Macula (MAK-ū-la) A discolored spot or a colored area. A small, thickened region on the wall of the utricle and saccule that contains receptors for static equilibrium.

Macula lutea (LOO-tē-a) The yellow spot in the center of the retina.

Major histocompatibility (MHC) antigens Surface proteins on white blood cells and other nucleated cells that are unique for each person (except for identical siblings); used to type tissues and help prevent rejection of transplanted tissues. Also known as **human leukocyte antigens (HLA).**

Malignant (ma-LIG-nant) Referring to diseases that tend to become worse and cause death, especially the invasion and spreading of cancer.

Mammary (MAM-ar-ē) **gland** Modified sudoriferous (sweat) gland of the female that produces milk for the nourishment of the young.

Mammillary (MAM-i-ler-ē) **bodies** Two small rounded bodies on the inferior aspect of the hypothalamus that are involved in reflexes related to the sense of smell.

Marrow (MAR-ō) Soft, spongelike material in the cavities of bone. Red bone marrow produces blood cells; yellow bone marrow contains adipose tissue that stores triglycerides.

Mast cell A cell found in areolar connective tissue that releases histamine, a dilator of small blood vessels, during inflammation.

Mastication (mas′-ti-KĀ-shun) Chewing.

Mature follicle A large, fluid-filled follicle containing a secondary oocyte and surrounding granulosa cells that secrete estrogens. Also called a **graafian** (GRAF-ē-an) **follicle.**

Meatus (mē-Ā-tus) A passage or opening, especially the external portion of a canal.

Mechanoreceptor (me-KAN-ō-rē-sep-tor) Sensory receptor that detects mechanical deformation of the receptor itself or adjacent cells; stimuli so detected include those related to touch, pressure, vibration, proprioception, hearing, equilibrium, and blood pressure.

Medial (MĒ-dē-al) Nearer the midline of the body or a structure.

Medial lemniscus (lem-NIS-kus) A white matter tract that originates in the gracile and cuneate nuclei of the medulla oblongata and extends to the thalamus on the same side; sensory axons in this tract conduct nerve impulses for the sensations of proprioception, fine touch, vibration, hearing, and equilibrium.

Median aperture (AP-er-choor) One of the three openings in the roof of the fourth ventricle through which cerebrospinal fluid enters the subarachnoid space of the brain and cord. Also called the **foramen of Magendie.**

Median plane A vertical plane dividing the body into right and left halves. Situated in the middle.

Mediastinum (mē′-dē-as-TĪ-num) The broad, median partition between the pleurae of the lungs that extends from the sternum to the vertebral column in the thoracic cavity.

Medulla (me-DOOL-la) An inner layer of an organ, such as the medulla of the kidneys.

Medulla oblongata (me-DOOL-la ob′-long-GA-ta) The most inferior part of the brain stem. Also termed the **medulla.**

Medullary (MED-ū-lar′-ē) **cavity** The space within the diaphysis of a bone that contains yellow bone marrow. Also called the **marrow cavity.**

Medullary rhythmicity (rith-MIS-i-tē) **area** The neurons of the respiratory center in the medulla oblongata that control the basic rhythm of respiration.

Meibomian gland *See* **Tarsal gland.**

Meiosis (mī-Ō-sis) A type of cell division that occurs during production of gametes, involving two successive nuclear divisions that result in cells with the haploid (n) number of chromosomes.

Meissner corpuscle *See* **Corpuscle of touch.**

Melanin (MEL-a-nin) A dark black, brown, or yellow pigment found in some parts of the body such as the skin, hair, and pigmented layer of the retina.

Melanocyte (MEL-a-nō-sīt′) A pigmented cell, located between or beneath cells of the deepest layer of the epidermis, that synthesizes melanin.

Melanocyte-stimulating hormone (MSH) A hormone secreted by the anterior pituitary that stimulates the dispersion of melanin granules in melanocytes in amphibians; continued administration produces darkening of skin in humans.

Melatonin (mel-a-TŌN-in) A hormone secreted by the pineal gland that helps set the timing of the body's biological clock.

Membrane A thin, flexible sheet of tissue composed of an epithelial layer and an underlying connective tissue layer, as in an epithelial membrane, or of areolar connective tissue only, as in a synovial membrane.

Membranous labyrinth (mem-BRA-nus LAB-i-rinth) The part of the labyrinth of the internal ear that is located inside the bony labyrinth and separated from it by the perilymph; made up of the semicircular ducts, the saccule and utricle, and the cochlear duct.

Memory The ability to recall thoughts; commonly classifed as short-term (activated) and long-term.

Menarche (me-NAR-kē) The first menses (menstrual flow) and beginning of ovarian and uterine cycles.

Meninges (me-NIN-jēz) Three membranes covering the brain and spinal cord, called the dura mater, arachnoid mater, and pia mater. *Singular* is **meninx** (MEN-inks).

Menopause (MEN-ō-pawz) The termination of the menstrual cycles.

Menstrual (MEN-strū-al) **cycle** A series of changes in the endometrium of a nonpregnant female that prepares the lining of the uterus to receive a fertilized ovum.

Menstruation (men′-stroo-Ā-shun) Periodic discharge of blood, tissue fluid, mucus, and epithelial cells that usually lasts for 5 days; caused by a sudden reduction in estrogens and progesterone. Also called the **menstrual phase** or **menses.**

Merkel (MER-kel) **cell** Type of cell in the epidermis of hairless skin that makes contact with a tactile (Merkel) disc, which functions in touch.

Merocrine (MER-ō-krin) **gland** Gland made up of secretory cells that remain intact throughout the process of formation and discharge of the secretory product, as in the salivary and pancreatic glands.

Mesenchyme (MEZ-en-kīm) An embryonic connective tissue from which all other connective tissues arise.

Mesentery (MEZ-en-ter′-ē) A fold of peritoneum attaching the small intestine to the posterior abdominal wall.

Mesocolon (mez′-ō-KŌ-lon) A fold of peritoneum attaching the colon to the posterior abdominal wall.

Mesoderm The middle primary germ layer that gives rise to connective tissues, blood and blood vessels, and muscles.

Mesothelium (mez′-ō-THĒ-lē-um) The layer of simple squamous epithelium that lines serous membranes.

Mesovarium (mez′-ō-VAR-ē-um) A short fold of peritoneum that attaches an ovary to the broad ligament of the uterus.

Metabolism (me-TAB-ō-lizm) All the biochemical reactions that occur within an organism, including the synthetic (anabolic) reactions and decomposition (catabolic) reactions.

Metacarpus (met′-a-KAR-pus) A collective term for the five bones that make up the palm.

Metaphase (MET-a-phāz) The second stage of mitosis, in which chromatid pairs line up on the metaphase plate of the cell.

Metaphysis (me-TAF-i-sis) Region of a long bone between the diaphysis and epiphysis that contains the epiphyseal plate in a growing bone.

Metarteriole (met′-ar-TĒ-rē-ōl) A blood vessel that emerges from an arteriole, traverses a capillary network, and empties into a venule.

Metastasis (me-TAS-ta-sis) The spread of cancer to surrounding tissues (local) or to other body sites (distant).

Metatarsus (met′-a-TAR-sus) A collective term for the five bones located in the foot between the tarsals and the phalanges.

Microfilament (mī-krō-FIL-a-ment) Rodlike protein filament about 6 nm in diameter; constitutes contractile units in muscle fibers (cells) and provides support, shape, and movement in nonmuscle cells.

Microglia (mī-KROG-lē-a) Neuroglial cells that carry on phagocytosis.

Microtubule (mī-krō-TOO-būl′) Cylindrical protein filament, from 18 to 30 nm in diameter, consisting of the protein tubulin; provides support, structure, and transportation.

Microvilli (mī′-krō-VIL-ē) Microscopic, fingerlike projections of the plasma membranes of cells that increase surface area for absorption, especially in the small intestine and proximal convoluted tubules of the kidneys.

Micturition (mik′-choo-RISH-un) The act of expelling urine from the urinary bladder. Also called **urination** (ū-ri-NĀ-shun).

Midbrain The part of the brain between the pons and the diencephalon. Also called the **mesencephalon** (mes′-en-SEF-a-lon).

Middle ear A small, epithelial-lined cavity hollowed out of the temporal bone, separated from the external ear by the eardrum and from the internal ear by a thin bony partition containing the oval and round windows; extending across the middle ear are the three auditory ossicles. Also called the **tympanic** (tim-PAN-ik) **cavity.**

Midline An imaginary vertical line that divides the body into equal left and right sides.

Midsagittal plane A vertical plane through the midline of the body that divides the body or organs into *equal* right and left sides. Also called a **median plane.**

Mineralocorticoids (min′-er-al-ō-KOR-ti-koyds) A group of hormones of the adrenal cortex that help regulate sodium and potassium balance.

Mitochondrion (mī-tō-KON-drē-on) A double-membraned organelle that plays a central role in the production of ATP; known as the "powerhouse" of the cell. *Plural* is **mitochondria.**

Mitosis (mī-TŌ-sis) The orderly division of the nucleus of a cell that ensures that each new nucleus has the same number and kind of chromosomes as the original nucleus. The process includes the replication of chromosomes and the distribution of the two sets of chromosomes into two separate and equal nuclei.

Mitotic spindle Collective term for a football-shaped assembly of microtubules (nonkinetochore, kinetochore, and aster) that is responsible for the movement of chromosomes during cell division.

Modality (mō-DAL-i-tē) Any of the specific sensory entities, such as vision, smell, taste, or touch.

Modiolus (mō-DĪ-ō′-lus) The central pillar or column of the cochlea.

Monocyte (MON-ō-sit′) The largest type of white blood cell, characterized by agranular cytoplasm.

Monounsaturated fat A fatty acid that contains one double covalent bond between its carbon atoms; it is not completely saturated with hydrogen atoms. Plentiful in triglycerides of olive and peanut oils.

Mons pubis (MONZ PŪ-bis) The rounded, fatty prominence over the pubic symphysis, covered by coarse pubic hair.

Morula (MOR-ū-la) A solid sphere of cells produced by successive cleavages of a fertilized ovum about four days after fertilization.

Motor area The region of the cerebral cortex that governs muscular movement, particularly the precentral gyrus of the frontal lobe.

Motor end plate Region of the sarcolemma of a muscle fiber (cell) that includes acetylcholine (ACh) receptors, which bind ACh released by synaptic end bulbs of somatic motor neurons.

Motor neurons (NOO-ronz) Neurons that conduct impulses from the brain toward the spinal cord or out of the brain and spinal cord into cranial or spinal nerves to effectors that may be either muscles or glands. Also called **efferent neurons.**

Motor unit A motor neuron together with the muscle fibers (cells) it stimulates.

Mucosa-associated lymphatic tissue (MALT) Lymphatic nodules scattered throughout the lamina propria (connective tissue) of mucous membranes lining the gastrointestinal tract, respiratory airways, urinary tract, and reproductive tract.

Mucous (MŪ-kus) **cell** A unicellular gland that secretes mucus. Two types are mucous neck cells and surface mucous cells in the stomach.

Mucous membrane A membrane that lines a body cavity that opens to the exterior. Also called the **mucosa** (mū-KŌ-sa).

Mucus The thick fluid secretion of goblet cells, mucous cells, mucous glands, and mucous membranes.

Muscarinic (mus′-ka-RIN-ik) **receptor** Receptor for the neurotransmitter acetylcholine found on all effectors innervated by parasympathetic postganglionic axons and on sweat glands innervated by cholinergic sympathetic postganglionic axons; so named because muscarine activates these receptors but does not activate nicotinic receptors for acetylcholine.

Muscle An organ composed of one of three types of muscle tissue (skeletal, cardiac, or smooth), specialized for contraction to produce voluntary or involuntary movement of parts of the body.

Muscle action potential A stimulating impulse that propagates along the sarcolemma and transverse tubules; in skeletal muscle, it is generated by acetylcholine, which increases the permeability of the sarcolemma to cations, especially sodium ions (Na^+).

Muscle fatigue (fa-TĒG) Inability of a muscle to maintain its strength of contraction or tension; may be related to insufficient oxygen, depletion of glycogen, and/or lactic acid buildup.

Muscle spindle An encapsulated proprioceptor in a skeletal muscle, consisting of specialized intrafusal muscle fibers and nerve endings; stimulated by changes in length or tension of muscle fibers.

Muscle tone A sustained, partial contraction of portions of a skeletal or smooth muscle in response to activation of stretch receptors or a baseline level of action potentials in the innervating motor neurons.

Muscular dystrophies (DIS-trō-fēz′) Inherited muscle-destroying diseases, characterized by degeneration of muscle fibers (cells), which causes progressive atrophy of the skeletal muscle.

Muscularis (MUS-kū-la′-ris) A muscular layer (coat or tunic) of an organ.

Muscularis mucosae (mū-KŌ-sē) A thin layer of smooth muscle fibers that underlie the lamina propria of the mucosa of the gastrointestinal tract.

Muscular tissue A tissue specialized to produce motion in response to muscle action potentials by its qualities of contractility, extensibility, elasticity, and excitability; types include skeletal, cardiac, and smooth.

Mutation (mū-TĀ-shun) Any change in the sequence of bases in a DNA molecule resulting in a permanent alteration in some inheritable trait.

Myasthenia (mī-as-THĒ-nē-a) **gravis** Weakness and fatigue of skeletal muscles caused by antibodies directed against acetylcholine receptors.

Myelin (MĪ-e-lin) **sheath** Multilayered lipid and protein covering, formed by Schwann cells and oligodendrocytes, around axons of many peripheral and central nervous system neurons.

Myenteric plexus A network of autonomic axons and postganglionic cell bodies located in the muscularis of the gastrointestinal tract. Also called the **plexus of Auerbach** (OW-er-bak).

Myocardial infarction (mī′-ō-KAR-dē-al in-FARK-shun) **(MI)** Gross necrosis of myocardial tissue due to interrupted blood supply. Also called a **heart attack.**

Myocardium (mī′-ō-KAR-dē-um) The middle layer of the heart wall, made up of cardiac muscle tissue, lying between the epicardium and the endocardium and constituting the bulk of the heart.

Myofibril (mī-ō-FĪ-bril) A threadlike structure, extending longitudinally through a muscle fiber (cell) consisting mainly of thick filaments (myosin) and thin filaments (actin, troponin, and tropomyosin).

Myoglobin (mī-ō-GLŌB-in) The oxygen-binding, iron-containing protein present in the sarcoplasm of muscle fibers (cells); contributes the red color to muscle.

Myogram (MĪ-ō-gram) The record or tracing produced by a myograph, an apparatus that measures and records the force of muscular contractions.

Myology (mī-OL-ō-jē) The study of muscles.

Myometrium (mī′-ō-MĒ-trē-um) The smooth muscle layer of the uterus.

Myopathy (mī-OP-a-thē) Any abnormal condition or disease of muscle tissue.

Myopia (mī-Ō-pē-a) Defect in vision in which objects can be seen distinctly only when very close to the eyes; nearsightedness.

Myosin (MĪ-ō-sin) The contractile protein that makes up the thick filaments of muscle fibers.

Myotome (MĪ-ō-tōm) A group of muscles innervated by the motor neurons of a single spinal segment. In an embryo, the portion of a somite that develops into some skeletal muscles.

N

Nail A hard plate, composed largely of keratin, that develops from the epidermis of the skin to form a protective covering on the dorsal surface of the distal phalanges of the fingers and toes.

Nail matrix (MĀ-triks) The part of the nail beneath the body and root from which the nail is produced.

Nasal (NĀ-zal) **cavity** A mucosa-lined cavity on either side of the nasal septum that opens onto the face at the external nares and into the nasopharynx at the internal nares.

Nasal septum (SEP-tum) A vertical partition composed of bone (perpendicular plate of ethmoid and vomer) and cartilage, covered with a mucous membrane, separating the nasal cavity into left and right sides.

Nasolacrimal (nā′-zō-LAK-ri-mal) **duct** A canal that transports the lacrimal secretion (tears) from the nasolacrimal sac into the nose.

Nasopharynx (nā′-zō-FAR-inks) The superior portion of the pharynx, lying posterior to the nose and extending inferiorly to the soft palate.

Neck The part of the body connecting the head and the trunk. A constricted portion of an organ, such as the neck of the femur or uterus.

Necrosis (ne-KRŌ-sis) A pathological type of cell death that results from disease, injury, or lack of blood supply in which many adjacent cells swell, burst, and spill their contents into the interstitial fluid, triggering an inflammatory response.

Neonatal (nē-ō-NĀ-tal) Pertaining to the first four weeks after birth.

Neoplasm (NĒ-ō-plazm) A new growth that may be benign or malignant.

Nephron (NEF-ron) The functional unit of the kidney.

Nerve A cordlike bundle of neuronal axons and/or dendrites and associated connective tissue coursing together outside the central nervous system.

Nerve fiber General term for any process (axon or dendrite) projecting from the cell body of a neuron.

Nerve impulse A wave of depolarization and repolarization that self-propagates along the plasma membrane of a neuron; also called a **nerve action potential.**

Nervous tissue Tissue containing neurons that initiate and conduct nerve impulses to coordinate homeostasis, and neuroglia that provide support and nourishment to neurons.

Neuralgia (noo-RŌG-lē-a) Attacks of pain along the entire course or branch of a peripheral sensory nerve.

Neural plate A thickening of ectoderm, induced by the notochord, that forms early in the third week of development and represents the beginning of the development of the nervous system.

Neural tube defect (NTD) A developmental abnormality in which the neural tube does not close properly. Examples are spina bifida and anencephaly.

Neuritis (noo-RĪ-tis) Inflammation of one or more nerves.

Neurofibral node *See* **Node of Ranvier.**

Neuroglia (noo-RŌG-lē-a) Cells of the nervous system that perform various supportive functions. The neuroglia of the central nervous system are the astrocytes, oligodendrocytes, microglia, and ependymal cells; neuroglia of the peripheral nervous system include Schwann cells and satellite cells. Also called **glial** (GLĒ-al) **cells.**

Neurohypophyseal (noo′-rō-hī′-pō-FIZ-ē-al) **bud** An outgrowth of ectoderm located on the floor of the hypothalamus that gives rise to the posterior pituitary.

Neurolemma (noo-rō-LEM-ma) The peripheral, nucleated cytoplasmic layer of the Schwann cell. Also called **sheath of Schwann** (SCHWON).

Neurology (noo-ROL-ō-jē) The study of the normal functioning and disorders of the nervous system.

Neuromuscular (noo-rō-MUS-kū-lar) **junction (NMJ)** A synapse between the axon terminals of a motor neuron and the sarcolemma of a muscle fiber (cell).

Neuron (NOO-ron) A nerve cell, consisting of a cell body, dendrites, and an axon.

Neurosecretory (noo-rō-SĒC-re-tō-rē) **cell** A neuron that secretes a hypothalamic releasing hormone or inhibiting hormone into blood capillaries of the hypothalmus; a neuron that secretes oxytocin or antidiuretic hormone into blood capillaries of the posterior pituitary.

Neurotransmitter One of a variety of molecules within axon terminals that are released into the synaptic cleft in response to a nerve impulse and that change the membrane potential of the postsynaptic neuron.

Neurulation (noor-oo-LĀ-shun) The process by which the neural plate, neural folds, and neural tube develop.

Neutrophil (NOO-trō-fil) A type of white blood cell characterized by granules that stain pale lilac with a combination of acidic and basic dyes.

Nicotinic (nik′-ō-TIN-ik) **receptor** Receptor for the neurotransmitter acetylcholine found on both sympathetic and parasympathetic postganglionic neurons and on skeletal muscle in the motor end plate; so named because nicotine activates these receptors but does not activate muscarinic receptors for acetylcholine.

Nipple A pigmented, wrinkled projection on the surface of the breast that is the location of the openings of the lactiferous ducts for milk release.

Nociceptor (nō′-sē-SEP-tor) A free (naked) nerve ending that detects painful stimuli.

Node of Ranvier (RON-vē-ā) A space along a myelinated axon between the individual Schwann cells that form the myelin sheath and the neurolemma. Also called a **neurofibral node.**

Norepinephrine (nor′-ep-ē-NEF-rin) **(NE)** A hormone secreted by the adrenal medulla that produces actions similar to those that result from sympathetic stimulation. Also called **noradrenaline** (nor-a-DREN-a-lin).

Notochord (NŌ-tō-cord) A flexible rod of mesodermal tissue that lies where the future vertebral column will develop and plays a role in induction.

Nucleic (noo-KLĒ-ic) **acid** An organic compound that is a long polymer of nucleotides, with each nucleotide containing a pentose sugar, a phosphate group, and one of four possible nitrogenous bases (adenine, cytosine, guanine, and thymine or uracil).

Nucleolus (noo′-KLĒ-ō-lus) Spherical body within a cell nucleus composed of protein, DNA, and RNA that is the site of the assembly of small and large ribosomal subunits. *Plural* is **nucleoli.**

Nucleosome (NOO-klē-ō-sōm) Structural subunit of a chromosome consisting of histones and DNA.

Nucleus (NOO-klē-us) A spherical or oval organelle of a cell that contains the hereditary factors of the cell, called genes. A cluster of unmyelinated nerve cell bodies in the central nervous system. The central part of an atom made up of protons and neutrons.

Nucleus pulposus (pul-PŌ-sus) A soft, pulpy, highly elastic substance in the center of an intervertebral disc; a remnant of the notochord.

Nutrient A chemical substance in food that provides energy, forms new body components, or assists in various body functions.

O

Obesity (ō-BĒS-i-tē) Body weight more than 20% above a desirable standard due to excessive accumulation of fat.

Oblique (ō-BLĒK) **plane** A plane that passes through the body or an organ at an angle between the transverse plane and either the midsagittal, parasagittal, or frontal plane.

Obstetrics (ob-STET-riks) The specialized branch of medicine that deals with pregnancy, labor, and the period of time immediately after delivery (about 6 weeks).

Olfactory (ōl-FAK-tō-rē) Pertaining to smell.

Olfactory bulb A mass of gray matter containing cell bodies of neurons that form synapses with neurons of the olfactory (I) nerve, lying inferior to the frontal lobe of the cerebrum on either side of the crista galli of the ethmoid bone.

Olfactory receptor A bipolar neuron with its cell body lying between supporting cells located in the mucous membrane lining the superior portion of each nasal cavity; transduces odors into neural signals.

Olfactory tract A bundle of axons that extends from the olfactory bulb posteriorly to olfactory regions of the cerebral cortex.

Oligodendrocyte (OL-i-gō-den′-drō-sīt) A neuroglial cell that supports neurons and produces a myelin sheath around axons of neurons of the central nervous system.

Oliguria (ol′-i-GŪ-rē-a) Daily urinary output usually less than 250 ml.

Olive A prominent oval mass on each lateral surface of the superior part of the medulla oblongata.

Oncogenes (ON-kō-jēnz) Cancer-causing genes; they derive from normal genes, termed proto-oncogenes, that encode proteins involved in cell growth or cell regulation but have the ability to transform a normal cell into a cancerous cell when they are mutated or inappropriately activated. One example is *p53.*

Oncology (on-KOL-ō-jē) The study of tumors.

Oogenesis (ō′-ō-JEN-e-sis) Formation and development of female gametes (oocytes).

Oophorectomy (ō′-of-ō-REK-tō-me) Surgical removal of the ovaries.

Ophthalmic (of-THAL-mik) Pertaining to the eye.

Ophthalmologist (of′-thal-MOL-ō-jist) A physician who specializes in the diagnosis and treatment of eye disorders using drugs, surgery, and corrective lenses.

Ophthalmology (of-thal-MOL-ō-jē) The study of the structure, function, and diseases of the eye.

Optic (OP-tik) Refers to the eye, vision, or properties of light.

Optic chiasm (kī-AZ-m) A crossing point of the two branches of the optic (II) nerve, anterior to the pituitary gland. Also called **optic chiasma.**

Optic disc A small area of the retina containing openings through which the axons of the ganglion cells emerge as the optic (II) nerve. Also called the **blind spot.**

Optic tract A bundle of axons that carry nerve impulses from the retina of the eye between the optic chiasm and the thalamus.

Ora serrata (Ō-ra ser-RĀ-ta) The irregular margin of the retina lying internal and slightly posterior to the junction of the choroid and ciliary body.

Orbit (OR-bit) The bony, pyramidal-shaped cavity of the skull that holds the eyeball.

Organ A structure composed of two or more different kinds of tissues with a specific function and usually a recognizable shape.

Organelle (or-gan-EL) A permanent structure within a cell with characteristic morphology that is specialized to serve a specific function in cellular activities.

Organism (OR-ga-nizm) A total living form; one individual.

Organogenesis (or′-ga-nō-JEN-e-sis) The formation of body organs and systems. By the end of the eighth week of development, all major body systems have begun to develop.

Orifice (OR-i-fis) Any aperture or opening.

Origin (OR-i-jin) The attachment of a muscle tendon to a stationary bone or the end opposite the insertion.

Oropharynx (or′-ō-FAR-inks) The intermediate portion of the pharynx, lying posterior to the mouth and extending from the soft palate to the hyoid bone.

Orthopedics (or′-thō-PĒ-diks) The branch of medicine that deals with the preservation and restoration of the skeletal system, articulations, and associated structures.

Osmoreceptor (oz′-mō-re-CEP-tor) Receptor in the hypothalamus that is sensitive to changes in blood osmolarity and, in response to high osmolarity (low water concentration), stimulates synthesis and release of antidiuretic hormone (ADH).

Osmosis (oz-MŌ-sis) The net movement of water molecules through a selectively permeable membrane from an area of higher water concentration to an area of lower water concentration until equilibrium is reached.

Osseous (OS-ē-us) Bony.

Ossicle (OS-si-kul) One of the small bones of the middle ear (malleus, incus, stapes).

Ossification (os′-i-fi-KĀ-shun) Formation of bone. Also called **osteogenesis.**

Ossification (os′-i-fi-KĀ-shun) **center** An area in the cartilage model of a future bone where the cartilage cells hypertrophy, secrete enzymes that calcify their extracellular matrix, and die, and the area they occupied is invaded by osteoblasts that then lay down bone.

Osteoblast (OS-tē-ō-blast′) Cell formed from an osteogenic cell that participates in bone formation by secreting some organic components and inorganic salts.

Osteoclast (OS-tē-ō-clast′) A large, multinuclear cell that resorbs (destroys) bone matrix.

Osteocyte (OS-tē-ō-sīt′) A mature bone cell that maintains the daily activities of bone tissue.

Osteogenic (os′-tē-ō-JEN-ik) **cell** Stem cell derived from mesenchyme that has mitotic potential and the ability to differentiate into an osteoblast.

Osteogenic layer The inner layer of the periosteum that contains cells responsible for forming new bone during growth and repair.

Osteology (os-tē-OL-ō-jē) The study of bones.

Osteon (OS-tē-on) The basic unit of structure in adult compact bone, consisting of a central (haversian) canal with its concentrically arranged lamellae, lacunae, osteocytes, and canaliculi. Also called a **haversian** (ha-VER-shan) **system.**

Osteoporosis (os′-tē-ō-pō-RŌ-sis) Age-related disorder characterized by decreased bone mass and increased susceptibility to fractures, often as a result of decreased levels of estrogens.

Otic (Ō-tik) Pertaining to the ear.

Otolith (Ō-tō-lith) A particle of calcium carbonate embedded in the otolithic membrane that functions in maintaining static equilibrium.

Otolithic (ō-tō-LITH-ik) **membrane** Thick, gelatinous, glycoprotein layer located directly over hair cells of the macula in the saccule and utricle of the internal ear.

Otorhinolaryngology (ō-tō-rī′-nō-lar-in-GOL-ō-jē) The branch of medicine that deals with the diagnosis and treatment of diseases of the ears, nose, and throat.

Oval window A small, membrane-covered opening between the middle ear and inner ear into which the footplate of the stapes fits.

Ovarian (ō-VAR-ē-an) **cycle** A monthly series of events in the ovary associated with the maturation of a secondary oocyte.

Ovarian follicle (FOL-i-kul) A general name for oocytes (immature ova) in any stage of development, along with their surrounding epithelial cells.

Ovarian ligament (LIG-a-ment) A rounded cord of connective tissue that attaches the ovary to the uterus.

Ovary (Ō-var-ē) Female gonad that produces oocytes and the estrogens, progesterone, inhibin, and relaxin hormones.

Ovulation (ov-ū-LĀ-shun) The rupture of a mature ovarian (Graafian) follicle with discharge of a secondary oocyte into the pelvic cavity.

Ovum (Ō-vum) The female reproductive or germ cell; an egg cell; arises through completion of meiosis in a secondary oocyte after penetration by a sperm.

Oxyhemoglobin (ok′-sē-HĒ-mō-glō-bin) (**Hb–O_2**) Hemoglobin combined with oxygen.

Oxytocin (ok′-sē-TŌ-sin) **(OT)** A hormone secreted by neurosecretory cells in the paraventricular and supraoptic nuclei of the hypothalamus that stimulates contraction of smooth muscle in the pregnant uterus and myoepithelial cells around the ducts of mammary glands.

P

P wave The deflection wave of an electrocardiogram that signifies atrial depolarization.

Pacinian corpuscle *See* **Lamellated corpuscle.**

Palate (PAL-at) The horizontal structure separating the oral and the nasal cavities; the roof of the mouth.

Palpate (PAL-pāt) To examine by touch; to feel.

Pancreas (PAN-krē-as) A soft, oblong organ lying along the greater curvature of the stomach and connected by a duct to the duodenum. It is both an exocrine gland (secreting pancreatic juice) and an endocrine gland (secreting insulin, glucagon, somatostatin, and pancreatic polypeptide).

Pancreatic (pan′-krē-AT-ik) **duct** A single large tube that unites with the common bile duct from the liver and gallbladder and drains pancreatic juice into the duodenum at the hepatopancreatic ampulla (ampulla of Vater). Also called the **duct of Wirsung.**

Pancreatic islet A cluster of endocrine gland cells in the pancreas that secretes insulin, glucagon, somatostatin, and pancreatic polypeptide. Also called an **islet of Langerhans** (LANG-er-hanz).

Papanicolaou (pa-pa-NI-kō-lō) **test** A cytological staining test for the detection and diagnosis of premalignant and malignant conditions of the female genital tract. Cells scraped from the epithelium of the cervix of the uterus are examined microscopically. Also called a **Pap test** or **Pap smear.**

Papilla (pa-PIL-a) A small nipple-shaped projection or elevation.

Paralysis (pa-RAL-a-sis) Loss or impairment of motor function due to a lesion of nervous or muscular origin.

Paranasal sinus (par′-a-NĀ-zal SĪ-nus) A mucus-lined air cavity in a skull bone that communicates with the nasal cavity. Paranasal sinuses are located in the frontal, maxillary, ethmoid, and sphenoid bones.

Paraplegia (par-a-PLĒ-jē-a) Paralysis of both lower limbs.

Parasagittal plane (par-a-SAJ-i-tal) A vertical plane that does not pass through the midline and that divides the body or organs into *unequal* left and right portions.

Parasympathetic (par′-a-sim-pa-THET-ik) **division** One of the two subdivisions of the autonomic nervous system, having cell bodies of preganglionic neurons in nuclei in the brain stem and in the lateral gray horn of the sacral portion of the spinal cord; primarily concerned with activities that conserve and restore body energy.

Parathyroid (par′-a-THĪ-royd) **gland** One of usually four small endocrine glands embedded in the posterior surfaces of the lateral lobes of the thyroid gland.

Parathyroid hormone (PTH) A hormone secreted by the chief (principal) cells of the parathyroid glands that increases blood calcium level and decreases blood phosphate level.

Paraurethral (par′-a-ū-RĒ-thral) **gland** Gland embedded in the wall of the urethra whose duct opens on either side of the urethral orifice and secretes mucus. Also called **Skene's** (SKĒNZ) **gland.**

Parenchyma (par-EN-ki-ma) The functional parts of any organ, as opposed to tissue that forms its stroma or framework.

Parietal (pa-RĪ-e-tal) Pertaining to or forming the outer wall of a body cavity.

Parietal cell A type of secretory cell in gastric glands that produces hydrochloric acid and intrinsic factor. Also called an **oxyntic cell.**

Parietal pleura (PLOO-ra) The outer layer of the serous pleural membrane that encloses and protects the lungs; the layer that is attached to the wall of the pleural cavity.

Parkinson disease (PD) Progressive degeneration of the basal ganglia and substantia nigra of the cerebrum resulting in decreased production of dopamine (DA) that leads to tremor, slowing of voluntary movements, and muscle weakness.

Parotid (pa-ROT-id) **gland** One of the paired salivary glands located inferior and anterior to the ears and connected to the oral cavity via a duct (Stensen's) that opens into the inside of the cheek opposite the maxillary (upper) second molar tooth.

Pars intermedia A small avascular zone between the anterior and posterior pituitary glands.

Parturition (par′-too-RISH-un) Act of giving birth to young; childbirth, delivery.

Patent ductus arteriosus A congenital heart defect in which the ductus arteriosus remains open. As a result, aortic blood flows into the lower-pressure pulmonary trunk, increasing pulmonary trunk pressure and overworking both ventricles.

Pathogen (PATH-ō-jen) A disease-producing microbe.

Pathological (path′-ō-LOJ-i-kal) **anatomy** The study of structural changes caused by disease.

Pectinate (PEK-ti-nāt) **muscles** Projecting muscle bundles of the anterior atrial walls and the lining of the auricles.

Pectoral (PEK-tō-ral) Pertaining to the chest or breast.

Pedicel (PED-i-sel) Footlike structure, as on podocytes of a glomerulus.

Pelvic (PEL-vik) **cavity** Inferior portion of the abdominopelvic cavity that contains the urinary bladder, sigmoid colon, rectum, and internal female and male reproductive structures.

Pelvic splanchnic (PEL-vik SPLANGK-nik) **nerves** Consist of preganglionic parasympathetic axons from the levels of S2, S3, and S4 that supply the urinary bladder, reproductive organs, and the descending and sigmoid colon and rectum.

Pelvis The basinlike structure formed by the two hip bones, the sacrum, and the coccyx. The expanded, proximal portion of the ureter, lying within the kidney and into which the major calyces open.

Penis (PĒ-nis) The organ of urination and copulation in males; used to deposit semen into the female vagina.

Pepsin Protein-digesting enzyme secreted by chief cells of the stomach in the inactive form pepsinogen, which is converted to active pepsin by hydrochloric acid.

Peptic ulcer An ulcer that develops in areas of the gastrointestinal tract exposed to hydrochloric acid; classified as a gastric ulcer if in the lesser curvature of the stomach and as a duodenal ulcer if in the first part of the duodenum.

Percussion (pur-KUSH-un) The act of striking (percussing) an underlying part of the body with short, sharp taps as an aid in diagnosing the part by the quality of the sound produced.

Perforating canal A minute passageway by means of which blood vessels and nerves from the periosteum penetrate into compact bone. Also called **Volkmann's** (FŌLK-mans) **canal.**

Pericardial (per′-i-KAR-dē-al) **cavity** Small potential space between the visceral and parietal layers of the serous pericardium that contains pericardial fluid.

Pericardium (per-i-KAR-dē-um) A loose-fitting membrane that encloses the heart, consisting of a superficial fibrous layer and a deep serous layer.

Perichondrium (per′-i-KON-drē-um) The membrane that covers cartilage.

Perilymph (PER-i-limf) The fluid contained between the bony and membranous labyrinths of the inner ear.

Perimetrium (per′-i-MĒ-trē-um) The serosa of the uterus.

Perimysium (per-i-MĪZ-ē-um) Invagination of the epimysium that divides muscles into bundles.

Perineum (per′-i-NĒ-um) The pelvic floor; the space between the anus and the scrotum in the male and between the anus and the vulva in the female.

Perineurium (per′-i-NOO-rē-um) Connective tissue wrapping around fascicles in a nerve.

Periodontal (per-ē-ō-DON-tal) **disease** A collective term for conditions characterized by degeneration of gingivae, alveolar bone, periodontal ligament, and cementum.

Periodontal ligament The periosteum lining the alveoli (sockets) for the teeth in the alveolar processes of the mandible and maxillae.

Periosteum (per′-ē-OS-tē-um) The membrane that covers bone and consists of connective tissue, osteogenic cells, and osteoblasts; is essential for bone growth, repair, and nutrition.

Peripheral (pe-RIF-er-al) Located on the outer part or a surface of the body.

Peripheral nervous system (PNS) The part of the nervous system that lies outside the central nervous system, consisting of nerves and ganglia.

Peristalsis (per′-i-STAL-sis) Successive muscular contractions along the wall of a hollow muscular structure.

Peritoneum (per-i-tō-NĒ-um) The largest serous membrane of the body that lines the abdominal cavity and covers the viscera within it.

Peritonitis (per′-i-tō-NĪ-tis) Inflammation of the peritoneum.

Peroxisome (pe-ROKS-i-sōm) Organelle similar in structure to a lysosome that contains enzymes that use molecular oxygen to oxidize various organic compounds; such reactions produce hydrogen peroxide; abundant in liver cells.

Perspiration Sweat; produced by sudoriferous (sweat) glands and containing water, salts, urea, uric acid, amino acids, ammonia, sugar, lactic acid, and ascorbic acid. Helps maintain body temperature and eliminate wastes.

Peyer's patches *See* **Aggregated lymphatic follicles.**

pH A measure of the concentration of hydrogen ions (H^+) in a solution. The pH scale extends from 0 to 14, with a value of 7 expressing neutrality, values lower than 7 expressing increasing acidity, and values higher than 7 expressing increasing alkalinity.

Phagocytosis (fag′-ō-sī-TŌ-sis) The process by which phagocytes ingest and destroy microbes, cell debris, and other foreign matter.

Phalanx (FĀ-lanks) The bone of a finger or toe. *Plural* is **phalanges** (fa-LAN-jēz).

Pharmacology (far′-ma-KOL-ō-jē) The science of the effects and uses of drugs in the treatment of disease.

Pharynx (FAR-inks) The throat; a tube that starts at the internal nares and runs partway down the neck, where it opens into the esophagus posteriorly and the larynx anteriorly.

Phenotype (FĒ-nō-tīp) The observable expression of genotype; physical characteristics of an organism determined by genetic makeup and influenced by interaction between genes and internal and external environmental factors.

Phlebitis (fle-BĪ-tis) Inflammation of a vein, usually in a lower limb.

Photopigment A substance that can absorb light and undergo structural changes that can lead to the development of a receptor potential. An example is rhodopsin. In the eye, also called **visual pigment.**

Photoreceptor Receptor that detects light shining on the retina of the eye.

Physiology (fiz′-ē-OL-o-jē) Science that deals with the functions of an organism or its parts.

Pia mater (PĪ-a MĀ-ter *or* PĒ-a MA-ter) The innermost of the three meninges (coverings) of the brain and spinal cord.

Pineal (PĪN-ē-al) **gland** A cone-shaped gland located in the roof of the third ventricle that secretes melatonin. Also called the **epiphysis cerebri** (ē-PIF-i-sis se-RĒ-brē).

Pinealocyte (pin-ē-AL-ō-sīt) Secretory cell of the pineal gland that releases melatonin.

Pinna (PIN-na) The projecting part of the external ear composed of elastic cartilage and covered by skin and shaped like the flared end of a trumpet. Also called the **auricle** (OR-i-kul).

Pituicyte (pi-TOO-i-sīt) Supporting cell of the posterior pituitary.

Pituitary (pi-TOO-i-tā r-ē) **gland** A small endocrine gland occupying the hypophyseal fossa of the sphenoid bone and attached to the hypothalamus by the infundibulum. Also called the **hypophysis** (hī-POF-i-sis).

Pivot joint A synovial joint in which a rounded, pointed, or conical surface of one bone articulates with a ring formed partly by another bone and partly by a ligament, as in the joint between the atlas and axis and between the proximal ends of the radius and ulna. Also called a **trochoid** (TRŌ-koyd) **joint.**

Placenta (pla-SEN-ta) The special structure through which the exchange of materials between fetal and maternal circulations occurs. Also called the **afterbirth.**

Plantar flexion (PLAN-tar FLEK-shun) Bending the foot in the direction of the plantar surface (sole).

Plaque (PLAK) A layer of dense proteins on the inside of a plasma membrane in adherens junctions and desmosomes. A mass of bacterial cells, dextran (polysaccharide), and other debris that adheres to teeth (dental plaque). *See* also **Atherosclerotic plaque.**

Plasma (PLAZ-ma) The extracellular fluid found in blood vessels; blood minus the formed elements.

Plasma cell Cell that develops from a B cell (lymphocyte) and produces antibodies.

Plasma (cell) membrane Outer, limiting membrane that separates the cell's internal parts from extracellular fluid or the external environment.

Platelet (PLĀT-let) A fragment of cytoplasm enclosed in a cell membrane and lacking a nucleus; found in the circulating blood; plays a role in hemostasis. Also called a **thrombocyte** (THROM-bō-sīt).

Platelet plug Aggregation of platelets (thrombocytes) at a site where a blood vessel is damaged that helps stop or slow blood loss.

Pleura (PLOO-ra) The serous membrane that covers the lungs and lines the walls of the chest and the diaphragm.

Pleural cavity Small potential space between the visceral and parietal pleurae.

Plexus (PLEK-sus) A network of nerves, veins, or lymphatic vessels.

Plexus of Auerbach *See* **Myenteric plexus.**

Plexus of Meissner *See* **Submucosal plexus.**

Pluripotent stem cell Immature stem cell in red bone marrow that gives rise to precursors of all the different mature blood cells.

Pneumotaxic (noo-mō-TAK-sik) **area** A part of the respiratory center in the pons that continually sends inhibitory nerve impulses to the inspiratory area, limiting inhalation and facilitating exhalation.

Polycythemia (pol′-ē-sī-THĒ-mē-a) Disorder characterized by an above-normal hematocrit (above 55%) in which hypertension, thrombosis, and hemorrhage can occur.

Polyunsaturated fat A fatty acid that contains more than one double covalent bond between its carbon atoms; abundant in triglycerides of corn oil, safflower oil, and cottonseed oil.

Polyuria (pol′-ē-Ū-rē-a) An excessive production of urine.

Pons (PONZ) The part of the brain stem that forms a "bridge" between the medulla oblongata and the midbrain, anterior to the cerebellum.

Portal system The circulation of blood from one capillary network into another through a vein.

Postcentral gyrus Gyrus of cerebral cortex located immediately posterior to the central sulcus; contains the primary somatosensory area.

Posterior (pos-TĒR-ē-or) Nearer to or at the back of the body. Equivalent to **dorsal** in bipeds.

Posterior column–medial lemniscus pathways Sensory pathways that carry information related to proprioception, fine touch, two-point discrimination, pressure, and vibration. First-order neurons project from the spinal cord to the ipsilateral medulla in the posterior columns (gracile fasciculus and cuneate fasciculus). Second-order neurons project from the medulla to the contralateral thalamus in the medial lemniscus. Third-order neurons project from the thalamus to the somatosensory cortex (postcentral gyrus) on the same side.

Posterior pituitary Posterior lobe of the pituitary gland. Also called the **neurohypophysis** (noo-rō-hī-POF-i-sis).

Posterior root The structure composed of sensory axons lying between a spinal nerve and the dorsolateral aspect of the spinal cord. Also called the **dorsal (sensory) root.**

Posterior root ganglion (GANG-glē-on) A group of cell bodies of sensory neurons and their supporting cells located along the posterior root of a spinal nerve. Also called a **dorsal (sensory) root ganglion.**

Postganglionic neuron (pōst′-gang-lē-ON-ik NOO-ron) The second autonomic motor neuron in an autonomic pathway, having its cell body and dendrites located in an autonomic ganglion and its unmyelinated axon ending at cardiac muscle, smooth muscle, or a gland.

Postsynaptic (pōst-sin-AP-tik) **neuron** The nerve cell that is activated by the release of a neurotransmitter from another neuron and carries nerve impulses away from the synapse.

Pouch of Douglas *See* **Rectouterine pouch.**

Precapillary sphincter (SFINGK-ter) A ring of smooth muscle fibers (cells) at the site of origin of true capillaries that regulate blood flow into true capillaries.

Precentral gyrus Gyrus of cerebral cortex located immediately anterior to the central sulcus; contains the primary motor area.

Preganglionic (pre′-gang-lē-ON-ik) **neuron** The first autonomic motor neuron in an autonomic pathway, with its cell body and dendrites in the brain or spinal cord and its myelinated axon ending at an autonomic ganglion, where it synapses with a postganglionic neuron.

Pregnancy Sequence of events that normally includes fertilization, implantation, embryonic growth, and fetal growth and terminates in birth.

Premenstrual syndrome (PMS) Severe physical and emotional stress ocurring late in the postovulatory phase of the menstrual cycle and sometimes overlapping with menstruation.

Prepuce (PRĒ-poos) The loose-fitting skin covering the glans of the penis and clitoris. Also called the **foreskin.**

Presbyopia (prez-bē-Ō-pē-a) A loss of elasticity of the lens of the eye due to advancing age with resulting inability to focus clearly on near objects.

Presynaptic (prē-sin-AP-tik) **neuron** A neuron that propagates nerve impulses toward a synapse.

Prevertebral ganglion (prē-VER-te-bral GANG-glē-on) A cluster of cell bodies of postganglionic sympathetic neurons anterior to the spinal column and close to large abdominal arteries. Also called a **collateral ganglion.**

Primary germ layer One of three layers of embryonic tissue, called ectoderm, mesoderm, and endoderm, that give rise to all tissues and organs of the body.

Primary motor area A region of the cerebral cortex in the precentral gyrus of the frontal lobe of the cerebrum that controls specific muscles or groups of muscles.

Primary somatosensory area A region of the cerebral cortex posterior to the central sulcus in the postcentral gyrus of the parietal lobe of the cerebrum that localizes exactly the points of the body where somatic sensations originate.

Prime mover The muscle directly responsible for producing a desired motion. Also called an **agonist** (AG-ō-nist).

Primitive gut Embryonic structure formed from the dorsal part of the yolk sac that gives rise to most of the gastrointestinal tract.

Primordial (prī-MŌR-dē-al) Existing first; especially primordial egg cells in the ovary.

Principal cell Cell type in the distal convoluted tubules and collecting ducts of the kidneys that is stimulated by aldosterone and antidiuretic hormone.

Proctology (prok-TOL-ō-jē) The branch of medicine concerned with the rectum and its disorders.

Progeny (PROJ-e-nē) Offspring or descendants.

Progesterone (prō-JES-te-rōn) A female sex hormone produced by the ovaries that helps prepare the endometrium of the uterus for implantation of a fertilized ovum and the mammary glands for milk secretion.

Prognosis (prog-NŌ-sis) A forecast of the probable results of a disorder; the outlook for recovery.

Prolactin (prō-LAK-tin) **(PRL)** A hormone secreted by the anterior pituitary that initiates and maintains milk secretion by the mammary glands.

Prolapse (PRŌ-laps) A dropping or falling down of an organ, especially the uterus or rectum.

Proliferation (prō-lif′-er-Ā-shun) Rapid and repeated reproduction of new parts, especially cells.

Pronation (prō-NĀ-shun) A movement of the forearm in which the palm is turned posteriorly.

Prophase (PRŌ-fāz) The first stage of mitosis during which chromatid pairs are formed and aggregate around the metaphase plate of the cell.

Proprioception (prō-prē-ō-SEP-shun) The perception of the position of body parts, especially the limbs, independent of vision; this sense is possible due to nerve impulses generated by proprioceptors.

Proprioceptor (PRŌ-prē-ō-sep′-tor) A receptor located in muscles, tendons, joints, or the internal ear (muscle spindles, tendon organs, joint kinesthetic receptors, and hair cells of the vestibular apparatus) that provides information about body position and movements.

Prostaglandin (pros′-ta-GLAN-din) **(PG)** A membrane-associated lipid; released in small quantities and acts as a local hormone.

Prostate (PROS-tā t) A doughnut-shaped gland inferior to the urinary bladder that surrounds the superior portion of the male urethra and secretes a slightly acidic solution that contributes to sperm motility and viability.

Proteasome (PRŌ-tē-a-sōm) Tiny cellular organelle in cytosol and nucleus containing proteases that destroy unneeded, damaged, or faulty proteins.

Protein An organic compound consisting of carbon, hydrogen, oxygen, nitrogen, and sometimes sulfur and phosphorus; synthesized on ribosomes and made up of amino acids linked by peptide bonds.

Prothrombin (prō-THROM-bin) An inactive blood-clotting factor synthesized by the liver, released into the blood, and converted to active thrombin in the process of blood clotting by the activated enzyme prothrombinase.

Proto-oncogene (prō′-tō-ON-kō-jēn) Gene responsible for some aspect of normal growth and development; it may transform into an oncogene, a gene capable of causing cancer.

Protraction (prō-TRAK-shun) The movement of the mandible or shoulder girdle forward on a plane parallel with the ground.

Proximal (PROK-si-mal) Nearer the attachment of a limb to the trunk; nearer to the point of origin or attachment.

Pseudopods (SOO-dō-pods) Temporary protrusions of the leading edge of a migrating cell; cellular projections that surround a particle undergoing phagocytosis.

Pterygopalatine ganglion (ter′-i-gō-PAL-a-tīn GANG-glē-on) A cluster of cell bodies of parasympathetic postganglionic neurons ending at the lacrimal and nasal glands.

Ptosis (TŌ-sis) Drooping, as of the eyelid or the kidney.

Puberty (PŪ-ber-tē) The time of life during which the secondary sex characteristics begin to appear and the capability for sexual reproduction is possible; usually occurs between the ages of 10 and 17.

Pubic symphysis A slightly movable cartilaginous joint between the anterior surfaces of the hip bones.

Puerperium (pū′-er-PER-ē-um) The period immediately after childbirth, usually 4–6 weeks.

Pulmonary (PUL-mo-ner′-ē) Concerning or affected by the lungs.

Pulmonary circulation The flow of deoxygenated blood from the right ventricle to the lungs and the return of oxygenated blood from the lungs to the left atrium.

Pulmonary edema (e-DĒ-ma) An abnormal accumulation of interstitial fluid in the tissue spaces and alveoli of the lungs due to increased pulmonary capillary permeability or increased pulmonary capillary pressure.

Pulmonary embolism (EM-bō-lizm) **(PE)** The presence of a blood clot or a foreign substance in a pulmonary arterial blood vessel that obstructs circulation to lung tissue.

Pulmonary ventilation The inflow (inhalation) and outflow (exhalation) of air between the atmosphere and the lungs. Also called **breathing.**

Pulp cavity A cavity within the crown and neck of a tooth, which is filled with pulp, a connective tissue containing blood vessels, nerves, and lymphatic vessels.

Pulse (PULS) The rhythmic expansion and elastic recoil of a systemic artery after each contraction of the left ventricle.

Pupil The hole in the center of the iris, the area through which light enters the posterior cavity of the eyeball.

Purkinje (pur-KIN-jē) **fiber** Muscle fiber (cell) in the ventricular tissue of the heart specialized for conducting an action potential to the myocardium; part of the conduction system of the heart.

Pus The liquid product of inflammation containing leukocytes or their remains and debris of dead cells.

Pyloric (pī-LOR-ik) **sphincter** A thickened ring of smooth muscle through which the pylorus of the stomach communicates with the duodenum. Also called the **pyloric valve.**

Pyorrhea (pī-ō-RĒ-a) A discharge or flow of pus, especially in the alveoli (sockets) and the tissues of the gums.

Pyramid (PIR-a-mid) A pointed or cone-shaped structure. One of two roughly triangular structures on the anterior aspect of the medulla oblongata composed of the largest motor tracts that run from the cerebral cortex to the spinal cord. A triangular structure in the renal medulla.

Pyramidal (pi-RAM-i-dal) **tracts (pathways).** *See* **Direct motor pathways.**

Q

QRS wave The deflection waves of an electrocardiogram that represent onset of ventricular depolarization.

Quadrant (KWOD-rant) One of four parts.

Quadriplegia (kwod′-ri-PLĒ-jē-a) Paralysis of four limbs: two upper and two lower.

R

Radiographic (rā′-dē-ō-GRAF-ic) **anatomy** Diagnostic branch of anatomy that includes the use of x rays.

Rami communicantes (RĀ-mē kō-mū-ni-KAN-tēz) Branches of a spinal nerve. *Singular* is **ramus communicans** (RĀ-mus kō-MŪ-ni-kans).

Rathke's pouch *See* **Hypophyseal pouch.**

Receptor A specialized cell or a distal portion of a neuron that responds to a specific sensory modality, such as touch, pressure, cold, light, or sound, and converts it to an electrical signal (generator or receptor potential). A specific molecule or cluster of molecules that recognizes and binds a particular ligand.

Receptor-mediated endocytosis A highly selective process whereby cells take up specific ligands, which usually are large molecules or particles, by enveloping them within a sac of plasma membrane. Ligands are eventually broken down by enzymes in lysosomes.

Recombinant DNA Synthetic DNA, formed by joining a fragment of DNA from one source to a portion of DNA from another.

Rectouterine pouch A pocket formed by the parietal peritoneum as it moves posteriorly from the surface of the uterus and is reflected onto the rectum; the most inferior point in the pelvic cavity. Also called the **pouch** or **cul de sac of Douglas.**

Rectum (REK-tum) The last 20 cm (8 in.) of the gastrointestinal tract, from the sigmoid colon to the anus.

Recumbent (re-KUM-bent) Lying down.

Red bone marrow A highly vascularized connective tissue located in microscopic spaces between trabeculae of spongy bone tissue.

Red nucleus A cluster of cell bodies in the midbrain, occupying a large part of the tectum from which axons extend into the rubroreticular and rubrospinal tracts.

Red pulp That portion of the spleen that consists of venous sinuses filled with blood and thin plates of splenic tissue called splenic (Billroth's) cords.

Referred pain Pain that is felt at a site remote from the place of origin.

Reflex Fast response to a change (stimulus) in the internal or external environment that attempts to restore homeostasis.

Reflex arc The most basic conduction pathway through the nervous system, connecting a receptor and an effector and consisting of a receptor, a sensory neuron, an integrating center in the central nervous system, a motor neuron, and an effector.

Regional anatomy The division of anatomy dealing with a specific region of the body, such as the head, neck, chest, or abdomen.

Regurgitation (rē-gur′-ji-TĀ-shun) Return of solids or fluids to the mouth from the stomach; backward flow of blood through incompletely closed heart valves.

Relaxin (RLX) A female hormone produced by the ovaries and placenta that increases flexibility of the pubic symphysis and helps dilate the uterine cervix to ease delivery of a baby.

Releasing hormone Hormone secreted by the hypothalamus that can stimulate secretion of hormones of the anterior pituitary.

Remodeling Replacement of old bone by new bone tissue.

Renal (RĒ-nal) Pertaining to the kidneys.

Renal corpuscle (KOR-pus-l) A glomerular (Bowman's) capsule and its enclosed glomerulus.

Renal pelvis A cavity in the center of the kidney formed by the expanded, proximal portion of the ureter, lying within the kidney, and into which the major calyces open.

Renal pyramid A triangular structure in the renal medulla containing the straight segments of renal tubules and the vasa recta.

Reproduction (rē-prō-DUK-shun) The formation of new cells for growth, repair, or replacement; the production of a new individual.

Reproductive cell division Type of cell division in which gametes (sperm and oocytes) are produced; consists of meiosis and cytokinesis.

Respiration (res-pi-RĀ-shun) Overall exchange of gases between the atmosphere, blood, and body cells consisting of pulmonary ventilation, external respiration, and internal respiration.

Respiratory center Neurons in the pons and medulla oblongata of the brain stem that regulate the rate and depth of pulmonary ventilation.

Retention (rē-TEN-shun) A failure to void urine due to obstruction, nervous contraction of the urethra, or absence of sensation of desire to urinate.

Rete (RĒ-tē) **testis** The network of ducts in the testes.

Reticular (re-TIK-ū-lar) **activating system (RAS)** A portion of the reticular formation that has many ascending connections with the cerebral cortex; when this area of the brain stem is active, nerve impulses pass to the thalamus and widespread areas of the cerebral cortex, resulting in generalized alertness or arousal from sleep.

Reticular formation A network of small groups of neuronal cell bodies scattered among bundles of axons (mixed gray and white matter) beginning in the medulla oblongata and extending superiorly through the central part of the brain stem.

Reticulocyte (re-TIK-ū-lō-sīt) An immature red blood cell.

Reticulum (re-TIK-ū-lum) A network.

Retina (RET-i-na) The deep coat of the posterior portion of the eyeball consisting of nervous tissue (where the process of vision begins) and a pigmented layer of epithelial cells that contact the choroid.

Retinaculum (ret-i-NAK-ū-lum) A thickening of deep fascia that holds structures in place, for example, the superior and inferior retinacula of the ankle.

Retraction (rē-TRAK-shun) The movement of a protracted part of the body posteriorly on a plane parallel to the ground, as in pulling the lower jaw back in line with the upper jaw.

Retroperitoneal (re′-trō-per-i-tō-NĒ-al) External to the peritoneal lining of the abdominal cavity.

Rh factor An inherited antigen on the surface of red blood cells in Rh^+ individuals; not present in Rh^- individuals.

Rhinology (rī-NOL-ō-jē) The study of the nose and its disorders.

Ribonucleic (rī-bō-noo-KLĒ-ik) **acid (RNA)** A single-stranded nucleic acid made up of nucleotides, each consisting of a nitrogenous base (adenine, cytosine, guanine, or uracil), ribose, and a phosphate group; three types are messenger RNA (mRNA), transfer RNA (tRNA), and ribosomal RNA (rRNA), each of which has a specific role during protein synthesis.

Ribosome (RĪ-bō-sōm) A cellular structure in the cytoplasm of cells, composed of a small subunit and a large subunit that contain ribosomal RNA and ribosomal proteins; the site of protein synthesis.

Right lymphatic (lim-FAT-ik) **duct** A vessel of the lymphatic system that drains lymph from the upper right side of the body and empties it into the right subclavian vein.

Rigidity (ri-JID-i-tē) Hypertonia characterized by increased muscle tone, but reflexes are not affected.

Rigor mortis State of partial contraction of muscles after death due to lack of ATP; myosin heads (crossbridges) remain attached to actin, thus preventing relaxation.

Rod One of two types of photoreceptor in the retina of the eye; specialized for vision in dim light.

Root canal A narrow extension of the pulp cavity lying within the root of a tooth.

Root of penis Attached portion of penis that consists of the bulb and crura.

Rotation (rō-TĀ-shun) Moving a bone around its own axis, with no other movement.

Round ligament (LIG-a-ment) A band of fibrous connective tissue enclosed between the folds of the broad ligament of the uterus, emerging from the uterus just inferior to the uterine tube, extending laterally along the pelvic wall and through the deep inguinal ring to end in the labia majora.

Round window A small opening between the middle and internal ear, directly inferior to the oval window, covered by the secondary tympanic membrane.

Rugae (ROO-gē) Large folds in the mucosa of an empty hollow organ, such as the stomach and vagina.

S

Saccule (SAK-ūl) The inferior and smaller of the two chambers in the membranous labyrinth inside the vestibule of the internal ear containing a receptor organ for static equilibrium.

Sacral plexus (SĀ-kral PLEK-sus) A network formed by the ventral branches of spinal nerves L4 through S3.

Sacral promontory (PROM-on-tor′-ē) The superior surface of the body of the first sacral vertebra that projects anteriorly into the pelvic cavity; a line from the sacral promontory to the superior border of the pubic symphysis divides the abdominal and pelvic cavities.

Saddle joint A synovial joint in which the articular surface of one bone is saddle-shaped and the articular surface of the other bone is shaped like the legs of the rider sitting in the saddle, as in the joint between the trapezium and the metacarpal of the thumb.

Sagittal (SAJ-i-tal) **plane** A plane that divides the body or organs into left and right portions. Such a plane may be **midsagittal (median),** in which the divisions are equal, or **parasagittal,** in which the divisions are unequal.

Saliva (sa-LĪ-va) A clear, alkaline, somewhat viscous secretion produced mostly by the three pairs of salivary glands; contains various salts, mucin, lysozyme, salivary amylase, and lingual lipase (produced by glands in the tongue).

Salivary amylase (SAL-i-ver-ē AM-i-lās) An enzyme in saliva that initiates the chemical breakdown of starch.

Salivary gland One of three pairs of glands that lie external to the mouth and pour their secretory product (saliva) into ducts that empty into the oral cavity; the parotid, submandibular, and sublingual glands.

Sarcolemma (sar′-kō-LEM-ma) The cell membrane of a muscle fiber (cell), especially of a skeletal muscle fiber.

Sarcomere (SAR-kō-mēr) A contractile unit in a striated muscle fiber (cell) extending from one Z disc to the next Z disc.

Sarcoplasm (SAR-kō-plazm) The cytoplasm of a muscle fiber (cell).

Sarcoplasmic reticulum (sar′-kō-PLAZ-mik re-TIK-ū-lum) **(SR)** A network of saccules and tubes surrounding myofibrils of a muscle fiber (cell), comparable to endoplasmic reticulum; functions to reabsorb calcium ions during relaxation and to release them to cause contraction.

Satellite cell (SAT-i-līt) Flat neuroglial cells that surround cell bodies of peripheral nervous system ganglia to provide structural support and regulate the exchange of material between a neuronal cell body and interstitial fluid.

Saturated fat A fatty acid that contains only single bonds (no double bonds) between its carbon atoms; all carbon atoms are bonded to the maximum number of hydrogen atoms; prevalent in triglycerides of animal products such as meat, milk, milk products, and eggs.

Scala tympani (SKA-la TIM-pan-ē) The inferior spiral-shaped channel of the bony cochlea, filled with perilymph.

Scala vestibuli (ves-TIB-ū-lē) The superior spiral-shaped channel of the bony cochlea, filled with perilymph.

Schwann (SCHWON) **cell** A neuroglial cell of the peripheral nervous system that forms the myelin sheath and neurolemma around a nerve axon by wrapping around the axon in a jelly-roll fashion.

Sciatica (sī-AT-i-ka) Inflammation and pain along the sciatic nerve; felt along the posterior aspect of the thigh extending down the inside of the leg.

Sclera (SKLE-ra) The white coat of fibrous tissue that forms the superficial protective covering over the eyeball except in the most anterior portion; the posterior portion of the fibrous tunic.

Scleral venous sinus A circular venous sinus located at the junction of the sclera and the cornea through which aqueous humor drains from the anterior chamber of the eyeball into the blood. Also called the **canal of Schlemm** (SHLEM).

Sclerosis (skle-RŌ-sis) A hardening with loss of elasticity of tissues.

Scoliosis (skō-lē-Ō-sis) An abnormal lateral curvature from the normal vertical line of the backbone.

Scrotum (SKRŌ-tum) A skin-covered pouch that contains the testes and their accessory structures.

Sebaceous (se-BĀ-shus) **gland** An exocrine gland in the dermis of the skin, almost always associated with a hair follicle, that secretes sebum. Also called an **oil gland.**

Sebum (SĒ-bum) Secretion of sebaceous (oil) glands.

Secondary sex characteristic A characteristic of the male or female body that develops at puberty under the influence of sex hormones but is not directly involved in sexual reproduction; examples are distribution of body hair, voice pitch, body shape, and muscle development.

Secretion (se-KRĒ-shun) Production and release from a cell or a gland of a physiologically active substance.

Selective permeability (per′-mē-a-BIL-i-tē) The property of a membrane by which it permits the passage of certain substances but restricts the passage of others.

Semen (SĒ-men) A fluid discharged at ejaculation by a male that consists of a mixture of sperm and the secretions of the seminiferous tubules, seminal vesicles, prostate, and bulbourethral (Cowper's) glands.

Semicircular canals Three bony channels (anterior, posterior, lateral), filled with perilymph, in which lie the membranous semicircular canals filled with endolymph. They contain receptors for equilibrium.

Semicircular ducts The membranous semicircular canals filled with endolymph and floating in the perilymph of the bony semicircular canals; they contain cristae that are concerned with dynamic equilibrium.

Semilunar (sem′-ē-LOO-nar) **valve** A valve between the aorta or the pulmonary trunk and a ventricle of the heart.

Seminal vesicle (SEM-i-nal VES-i-kul) One of a pair of convoluted, pouchlike structures, lying posterior and inferior to the urinary bladder and anterior to the rectum, that secrete a component of semen into the ejaculatory ducts. Also termed **seminal gland.**

Seminiferous tubule (sem′-i-NI-fer-us TOO-būl) A tightly coiled duct, located in the testis, where sperm are produced.

Sensation A state of awareness of external or internal conditions of the body.

Sensory area A region of the cerebral cortex concerned with the interpretation of sensory impulses.

Sensory neurons (NOO-ronz) Neurons that carry sensory information from cranial and spinal nerves into the brain and spinal cord or from a lower to a higher level in the spinal cord and brain. Also called **afferent neurons.**

Septal defect An opening in the atrial septum (atrial septal defect) because the foramen ovale fails to close, or the ventricular septum (ventricular septal defect) due to incomplete development of the ventricular septum.

Septum (SEP-tum) A wall dividing two cavities.

Serous (SĒR-us) **membrane** A membrane that lines a body cavity that does not open to the exterior. The external layer of an organ formed by a serous membrane. The membrane that lines the pleural, pericardial, and peritoneal cavities. Also called a **serosa** (se-RŌ-sa).

Sertoli (ser-TŌ-lē) **cell** A supporting cell in the seminiferous tubules that secretes fluid for supplying nutrients to sperm and the hormone inhibin, removes excess cytoplasm from spermatogenic cells, and mediates the effects of FSH and testosterone on spermatogenesis. Also called a **sustentacular** (sus′-ten-TAK-ū-lar) **cell.**

Serum Blood plasma minus its clotting proteins.

Sesamoid (SES-a-moyd) **bones** Small bones usually found in tendons.

Sex chromosomes The twenty-third pair of chromosomes, designated X and Y, which determine the genetic sex of an individual; in males, the pair is XY; in females, XX.

Sexual intercourse The insertion of the erect penis of a male into the vagina of a female. Also called **coitus** (KŌ-i-tus).

Sheath of Schwann *See* **Neurolemma.**

Shock Failure of the cardiovascular system to deliver adequate amounts of oxygen and nutrients to meet the metabolic needs of the body due to inadequate cardiac output. It is characterized by hypotension; clammy, cool, and pale skin; sweating; reduced urine formation; altered mental state; acidosis; tachycardia; weak, rapid pulse; and thirst. Types include hypovolemic, cardiogenic, vascular, and obstructive.

Shoulder joint A synovial joint where the humerus articulates with the scapula.

Sigmoid colon (SIG-moyd KŌ-lon) The S-shaped part of the large intestine that begins at the level of the left iliac crest, projects medially, and terminates at the rectum at about the level of the third sacral vertebra.

Sign Any objective evidence of disease that can be observed or measured, such as a lesion, swelling, or fever.

Sinoatrial (si-nō-Ā-trē-al) **(SA) node** A small mass of cardiac muscle fibers (cells) located in the right atrium inferior to the opening of the superior vena cava that spontaneously depolarize and generate a cardiac action potential about 100 times per minute. Also called the **pacemaker.**

Sinus (SĪ-nus) A hollow in a bone (paranasal sinus) or other tissue; a channel for blood (vascular sinus); any cavity having a narrow opening.

Sinusoid (SĪ-nū-soyd) A large, thin-walled, and leaky type of capillary, having large intercellular clefts that may allow proteins and blood cells to pass from a tissue into the bloodstream; present in the liver, spleen, anterior pituitary, parathyroid glands, and red bone marrow.

Skeletal muscle An organ specialized for contraction, composed of striated muscle fibers (cells), supported by connective tissue, attached to a bone by a tendon or an aponeurosis, and stimulated by somatic motor neurons.

Skene's gland *See* **Paraurethral gland.**

Skin The external covering of the body that consists of a superficial, thinner epidermis (epithelial tissue) and a deep, thicker dermis (connective tissue) that is anchored to the subcutaneous layer.

Skull The skeleton of the head consisting of the cranial and facial bones.

Sleep A state of partial unconsciousness from which a person can be aroused; associated with a low level of activity in the reticular activating system.

Small intestine A long tube of the gastrointestinal tract that begins at the pyloric sphincter of the stomach, coils through the central and inferior part of the abdominal cavity, and ends at the large intestine; divided into three segments: duodenum, jejunum, and ileum.

Smooth muscle A tissue specialized for contraction, composed of smooth muscle fibers (cells), located in the walls of hollow internal organs, and innervated by autonomic motor neurons.

Sodium-potassium ATPase An active transport pump located in the plasma membrane that transports sodium ions out of the cell and potassium ions into the cell at the expense of cellular ATP. It functions to keep the ionic concentrations of these ions at physiological levels. Also called the **sodium-potassium pump.**

Soft palate (PAL-at) The posterior portion of the roof of the mouth, extending from the palatine bones to the uvula. It is a muscular partition lined with mucous membrane.

Somatic (sō-MAT-ik) **cell division** Type of cell division in which a single starting cell duplicates itself to produce two identical cells; consists of mitosis and cytokinesis.

Somatic nervous system (SNS) The portion of the peripheral nervous system consisting of somatic sensory (afferent) neurons and somatic motor (efferent) neurons.

Somite (SŌ-mīt) Block of mesodermal cells in a developing embryo that is distinguished into a myotome (which forms most of the skeletal muscles), dermatome (which forms connective tissues), and sclerotome (which forms the vertebrae).

Spasm (SPAZM) A sudden, involuntary contraction of large groups of muscles.

Spasticity (spas-TIS-i-tē) Hypertonia characterized by increased muscle tone, increased tendon reflexes, and pathological reflexes (Babinski sign).

Spermatic (sper-MAT-ik) **cord** A supporting structure of the male reproductive system, extending from a testis to the deep inguinal ring, that includes the ductus (vas) deferens, arteries, veins, lymphatic vessels, nerves, cremaster muscle, and connective tissue.

Spermatogenesis (sper′-ma-tō-JEN-e-sis) The formation and development of sperm in the seminiferous tubules of the testes.

Sperm cell A mature male gamete. Also termed **spermatozoon** (sper′-ma-tō-ZŌ-on).

Spermiogenesis (sper′-mē-ō-JEN-e-sis) The maturation of spermatids into sperm.

Sphincter (SFINGK-ter) A circular muscle that constricts an opening.

Sphincter of Oddi *See* **Sphincter of the hepatopancreatic ampulla.**

Sphincter of the hepatopancreatic ampulla A circular muscle at the opening of the common bile and main pancreatic ducts in the duodenum. Also called the **sphincter of Oddi** (OD-ē).

Spinal (SPĪ-nal) **cord** A mass of nerve tissue located in the vertebral canal from which 31 pairs of spinal nerves originate.

Spinal nerve One of the 31 pairs of nerves that originate on the spinal cord from posterior and anterior roots.

Spinal shock A period from several days to several weeks following transection of the spinal cord that is characterized by the abolition of all reflex activity.

Spinothalamic (spī-nō-tha-LAM-ik) **tracts** Sensory (ascending) tracts that convey information up the spinal cord to the thalamus for sensations of pain, temperature, crude touch, and deep pressure.

Spinous (SPĪ-nus) **process** A sharp or thornlike process or projection. Also called a **spine.** A sharp ridge running diagonally across the posterior surface of the scapula.

Spiral organ The organ of hearing, consisting of supporting cells and hair cells that rest on the basilar membrane and extend into the endolymph of the cochlear duct. Also called the **organ of Corti** (KOR-tē).

Splanchnic (SPLANK-nik) Pertaining to the viscera.

Spleen (SPLĒN) Large mass of lymphatic tissue between the fundus of the stomach and the diaphragm that functions in formation of blood cells during early fetal development, phagocytosis of ruptured blood cells, and proliferation of B cells during immune responses.

Spongy (cancellous) bone tissue Bone tissue that consists of an irregular latticework of thin plates of bone called trabeculae; spaces between trabeculae of some bones are filled with red bone marrow; found inside short, flat, and irregular bones and in the epiphyses (ends) of long bones.

Sprain Forcible wrenching or twisting of a joint with partial rupture or other injury to its attachments without dislocation.

Squamous (SKWĀ-mus) Flat or scalelike.

Starvation (star-VĀ-shun) The loss of energy stores in the form of glycogen, triglycerides, and proteins due to inadequate intake of nutrients or inability to digest, absorb, or metabolize ingested nutrients.

Static equilibrium (ē-kwi-LIB-rē-um) The maintenance of posture in response to changes in the orientation of the body, mainly the head, relative to the ground.

Stellate reticuloendothelial (STEL-āt re-tik′-ū-lō-en′-dō-THĒ-lē-al) **cell** Phagocytic cell bordering a sinusoid of the liver. Also called a **Kupffer** (KOOP-fer) **cell.**

Stem cell An unspecialized cell that has the ability to divide for indefinite periods and give rise to a specialized cell.

Stenosis (sten-Ō-sis) An abnormal narrowing or constriction of a duct or opening.

Stereocilia (ste′-rē-ō-SIL-ē-a) Groups of extremely long, slender, nonmotile microvilli projecting from epithelial cells lining the epididymis.

Sterile (STE-ril) Free from any living microorganisms. Unable to conceive or produce offspring.

Sterilization (ster′-i-li-ZĀ-shun) Elimination of all living microorganisms. Any procedure that renders an individual incapable of reproduction (for example, castration, vasectomy, hysterectomy, or oophorectomy).

Stimulus Any stress that changes a controlled condition; any change in the internal or external environment that excites a sensory receptor, a neuron, or a muscle fiber.

Stomach The J-shaped enlargement of the gastrointestinal tract directly inferior to the diaphragm in the epigastric, umbilical, and left hypochondriac regions of the abdomen, between the esophagus and small intestine.

Straight tubule (TOO-būl) A duct in a testis leading from a convoluted seminiferous tubule to the rete testis.

Stratum (STRĀ-tum) A layer.

Stratum basalis (ba-SAL-is) The layer of the endometrium next to the myometrium that is maintained during menstruation and gestation and produces a new stratum functionalis following menstruation or parturition.

Stratum functionalis (funk′-shun-AL-is) The layer of the endometrium next to the uterine cavity that is shed during menstruation and that forms the maternal portion of the placenta during gestation.

Stretch receptor Receptor in the walls of blood vessels, airways, or organs that monitors the amount of stretching. Also termed **baroreceptor.**

Stroma (STRŌ-ma) The tissue that forms the ground substance, foundation, or framework of an organ, as opposed to its functional parts (parenchyma).

Subarachnoid (sub′-a-RAK-noyd) **space** A space between the arachnoid mater and the pia mater that surrounds the brain and spinal cord and through which cerebrospinal fluid circulates.

Subcutaneous (sub′-kū-TĀ-nē-us) Beneath the skin. Also called **hypodermic** (hi-pō-DER-mik).

Subcutaneous layer A continuous sheet of areolar connective tissue and adipose tissue between the dermis of the skin and the deep fascia of the muscles. Also called the **superficial fascia** (FASH-ē-a).

Subdural (sub-DOO-ral) **space** A space between the dura mater and the arachnoid mater of the brain and spinal cord that contains a small amount of fluid.

Sublingual (sub-LING-gwal) **gland** One of a pair of salivary glands situated in the floor of the mouth deep to the mucous membrane and to the side of the lingual frenulum, with a duct (Rivinus') that opens into the floor of the mouth.

Submandibular (sub′-man-DIB-ū-lar) **gland** One of a pair of salivary glands found inferior to the base of the tongue deep to the mucous membrane in the posterior part of the floor of the mouth, posterior to the sublingual glands, with a duct (Wharton's) situated to the side of the lingual frenulum. Also called the **submaxillary** (sub′-MAK-si-ler-ē) **gland.**

Submucosa (sub-mū-KŌ-sa) A layer of connective tissue located deep to a mucous membrane, as in the gastrointestinal tract or the urinary bladder; the submucosa connects the mucosa to the muscularis layer.

Submucosal plexus A network of autonomic nerve fibers located in the superficial part of the submucous layer of the small intestine. Also called the **plexus of Meissner** (MĪZ-ner).

Substrate A molecule upon which an enzyme acts.

Subthalamus (sub-THAL-a-mus) Part of the diencephalon inferior to the thalamus; the substantia nigra and red nucleus extend from the midbrain into the subthalamus.

Sudoriferous (soo′-dor-IF-er-us) **gland** An apocrine or eccrine exocrine gland in the dermis or subcutaneous layer that produces perspiration. Also called a **sweat gland.**

Sulcus (SUL-kus) A groove or depression between parts, especially between the convolutions of the brain. *Plural* is **sulci** (SUL-sī).

Superficial (soo′-per-FISH-al) Located on or near the surface of the body or an organ.

Superficial fascia (FASH-ē-a) A continuous sheet of fibrous connective tissue between the dermis of the skin and the deep fascia of the muscles. Also called **subcutaneous** (sub′-kū-TĀ-nē-us) **layer** or **hypodermis.**

Superficial inguinal (IN-gwi-nal) **ring** A triangular opening in the aponeurosis of the external oblique muscle that represents the termination of the inguinal canal.

Superior (soo-PĒR-ē-or) Toward the head or upper part of a structure.

Superior vena cava (VĒ-na CĀ-va) **(SVC)** Large vein that collects blood from parts of the body superior to the heart and returns it to the right atrium.

Supination (soo-pi-NĀ-shun) A movement of the forearm in which the palm is turned anteriorly.

Surface anatomy The study of the structures that can be identified from the outside of the body.

Surfactant (sur-FAK-tant) Complex mixture of phospholipids and lipoproteins, produced by type II alveolar (septal) cells in the lungs, that decreases surface tension.

Suspensory ligament (sus-PEN-so-rē LIG-a-ment) A fold of peritoneum extending laterally from the surface of the ovary to the pelvic wall.

Sutural (SOO-chur-al) **bone** A small bone located within a suture between certain cranial bones. Also called **Wormian** (WER-mē-an) **bone.**

Suture (SOO-chur) An immovable fibrous joint that joins skull bones.

Sympathetic (sim′-pa-THET-ik) **division** One of the two subdivisions of the autonomic nervous system, having cell bodies of preganglionic neurons in the lateral gray columns of the thoracic segment and the first two or three lumbar segments of the spinal cord; primarily concerned with processes involving the expenditure of energy.

Sympathetic trunk ganglion (GANG-glē-on) A cluster of cell bodies of sympathetic postganglionic neurons lateral to the vertebral column, close to the body of a vertebra. These ganglia extend inferiorly through the neck, thorax, and abdomen to the coccyx on both sides of the vertebral column and are connected to one another to form a chain on each side of the vertebral column. Also called **sympathetic chain** or **vertebral chain ganglia.**

Symphysis (SIM-fi-sis) A line of union. A slightly movable cartilaginous joint such as the pubic symphysis.

Symptom (SIMP-tum) A subjective change in body function not apparent to an observer, such as pain or nausea, that indicates the presence of a disease or disorder of the body.

Synapse (SIN-aps) The functional junction between two neurons or between a neuron and an effector, such as a muscle or gland; may be electrical or chemical.

Synapsis (sin-AP-sis) The pairing of homologous chromosomes during prophase I of meiosis.

Synaptic (sin-AP-tik) **cleft** The narrow gap at a chemical synapse that separates the axon terminal of one neuron from another neuron or muscle fiber (cell) and across which a neurotransmitter diffuses to affect the postsynaptic cell.

Synaptic end bulb Expanded distal end of an axon terminal that contains synaptic vesicles. Also called a **synaptic knob.**

Synaptic vesicle Membrane-enclosed sac in a synaptic end bulb that stores neurotransmitters.

Synarthrosis (sin′-ar-THRŌ-sis) An immovable joint such as a suture, gomphosis, or synchondrosis.

Synchondrosis (sin′-kon-DRŌ-sis) A cartilaginous joint in which the connecting material is hyaline cartilage.

Syndesmosis (sin′-dez-MŌ-sis) A slightly movable joint in which articulating bones are united by fibrous connective tissue.

Synergist (SIN-er-gist) A muscle that assists the prime mover by reducing undesired action or unnecessary movement.

Synergistic (syn-er-JIS-tik) **effect** A hormonal interaction in which the effects of two or more hormones acting together is greater or more extensive than the sum of each hormone acting alone.

Synostosis (sin′-os-TŌ-sis) A joint in which the dense fibrous connective tissue that unites bones at a suture has been replaced by bone, resulting in a complete fusion across the suture line.

Synovial (sī-NŌ-vē-al) **cavity** The space between the articulating bones of a synovial joint, filled with synovial fluid. Also called a **joint cavity.**

Synovial fluid Secretion of synovial membranes that lubricates joints and nourishes articular cartilage.

Synovial joint A fully movable or diarthrotic joint in which a synovial (joint) cavity is present between the two articulating bones.

Synovial membrane The deeper of the two layers of the articular capsule of a synovial joint, composed of areolar connective tissue that secretes synovial fluid into the synovial (joint) cavity.

System An association of organs that have a common function.

Systemic (sis-TEM-ik) Affecting the whole body; generalized.

Systemic anatomy The anatomic study of particular systems of the body, such as the skeletal, muscular, nervous, cardiovascular, or urinary systems.

Systemic circulation The routes through which oxygenated blood flows from the left ventricle through the aorta to all the organs of the body and deoxygenated blood returns to the right atrium.

Systole (SIS-tō-lē) In the cardiac cycle, the phase of contraction of the heart muscle, especially of the ventricles.

Systolic (sis-TOL-ik) **blood pressure** The force exerted by blood on arterial walls during ventricular contraction; the highest pressure measured in the large arteries, about 120 mmHg under normal conditions for a young adult.

T

T cell A lymphocyte that becomes immunocompetent in the thymus and can differentiate into a helper T cell or a cytotoxic T cell, both of which function in cell-mediated immunity.

T wave The deflection wave of an electrocardiogram that represents ventricular repolarization.

Tachycardia (tak′-i-KAR-dē-a) An abnormally rapid resting heartbeat or pulse rate (over 100 beats per minute).

Tactile (TAK-tīl) Pertaining to the sense of touch.

Tactile disc Modified epidermal cell in the stratum basale of hairless skin that functions as a cutaneous receptor for discriminative touch. Also called a **Merkel** (MER-kel) **disc.**

Target cell A cell whose activity is affected by a particular hormone.

Tarsal bones The seven bones of the ankle. Also called **tarsals.**

Tarsal gland Sebaceous (oil) gland that opens on the edge of each eyelid. Also called a **Meibomian** (mī-BŌ-mē-an) **gland.**

Tarsal plate A thin, elongated sheet of connective tissue, one in each eyelid, giving the eyelid form and support. The aponeurosis of the levator palpebrae superioris is attached to the tarsal plate of the superior eyelid.

Tarsus (TAR-sus) A collective term for the seven bones of the ankle.

Tectorial (tek-TŌ-rē-al) **membrane** A gelatinous membrane projecting over and in contact with the hair cells of the spiral organ (organ of Corti) in the cochlear duct.

Teeth (TĒTH) Accessory structures of digestion, composed of calcified connective tissue and embedded in bony sockets of the mandible and maxilla, that cut, shred, crush, and grind food. Also called **dentes** (DEN-tēz).

Telophase (TEL-ō-fāz) The final stage of mitosis.

Tendon (TEN-don) A white fibrous cord of dense regular connective tissue that attaches muscle to bone.

Tendon organ A proprioceptive receptor, sensitive to changes in muscle tension and force of contraction, found chiefly near the junctions of tendons and muscles. Also called a **Golgi** (GOL-jē) **tendon organ.**

Tendon reflex A polysynaptic, ipsilateral reflex that protects tendons and their associated muscles from damage that might be brought about by excessive tension. The receptors involved are called tendon organs (Golgi tendon organs).

Teniae coli (TĒ-nē-ē KŌ-lī) The three flat bands of thickened, longitudinal smooth muscle running the length of the large intestine, except in the rectum. *Singular* is **tenia coli.**

Tentorium cerebelli (ten-TŌ-rē-um ser′-e-BEL-ē) A transverse shelf of dura mater that forms a partition between the occipital lobe of the cerebral hemispheres and the cerebellum and that covers the cerebellum.

Teratogen (TER-a-tō-jen) Any agent or factor that causes physical defects in a developing embryo.

Terminal ganglion (TER-min-al GANG-glē-on) A cluster of cell bodies of parasympathetic postganglionic neurons either lying very close to the visceral effectors or located within the walls of the visceral effectors supplied by the postganglionic neurons.

Testis (TES-tis) Male gonad that produces sperm and the hormones testosterone and inhibin. Also called a **testicle.**

Testosterone (tes-TOS-te-rōn) A male sex hormone (androgen) secreted by interstitial endocrinocytes (Leydig cells) of a mature testis; needed for development of sperm; together with a second androgen termed **dihydrotestosterone (DHT),** controls the growth and development of male reproductive organs, secondary sex characteristics, and body growth.

Tetralogy of Fallot (tet-RAL-ō-jē of fal-Ō) A combination of four congenital heart defects: (1) constricted pulmonary semilunar valve, (2) interventricular septal opening, (3) emergence of the aorta from both ventricles instead of from the left only, and (4) enlarged right ventricle.

Thalamus (THAL-a-mus) A large, oval structure located bilaterally on either side of the third ventricle, consisting of two masses of gray matter organized into nuclei; main relay center for sensory impulses ascending to the cerebral cortex.

Thermoreceptor (THER-mō-rē-sep-tor) Sensory receptor that detects changes in temperature.

Thigh The portion of the lower limb between the hip and the knee.

Third ventricle (VEN-tri-kul) A slitlike cavity between the right and left halves of the thalamus and between the lateral ventricles of the brain.

Thoracic (thor-AS-ik) **cavity** Cavity superior to the diaphragm that contains two pleural cavities, the mediastinum, and the pericardial cavity.

Thoracic duct A lymphatic vessel that begins as a dilation called the cisterna chyli, receives lymph from the left side of the head, neck, and chest, left arm, and the entire body below the ribs, and empties into the junction between the internal jugular and left subclavian veins. Also called the **left lymphatic** (lim-FAT-ik) **duct.**

Thoracolumbar (thōr′-a-kō-LUM-bar) **outflow** The axons of sympathetic preganglionic neurons, which have their cell bodies in the lateral gray columns of the thoracic segments and first two or three lumbar segments of the spinal cord.

Thorax (THŌ-raks) The chest.

Thrombosis (throm-BŌ-sis) The formation of a clot in an unbroken blood vessel, usually a vein.

Thrombus A stationary clot formed in an unbroken blood vessel, usually a vein.

Thymus (THĪ-mus) A bilobed organ, located in the superior mediastinum posterior to the sternum and between the lungs, in which T cells develop immunocompetence.

Thyroid cartilage (THĪ-royd KAR-ti-lij) The largest single cartilage of the larynx, consisting of two fused plates that form the anterior wall of the larynx.

Thyroid follicle (FOL-i-kul) Spherical sac that forms the parenchyma of the thyroid gland and consists of follicular cells that produce thyroxine (T_4) and triiodothyronine (T_3).

Thyroid gland An endocrine gland with right and left lateral lobes on either side of the trachea connected by an isthmus; located anterior to the trachea just inferior to the cricoid cartilage; secretes thyroxine (T_4), triiodothyronine (T_3), and calcitonin.

Thyroid-stimulating hormone (TSH) A hormone secreted by the anterior pituitary that stimulates the synthesis and secretion of thyroxine (T_4) and triiodothyronine (T_3).

Thyroxine (thī-ROK-sēn) **(T_4)** A hormone secreted by the thyroid gland that regulates metabolism, growth and development, and the activity of the nervous system.

Tic Spasmodic, involuntary twitching of muscles that are normally under voluntary control.

Tissue A group of similar cells and their intercellular substance joined together to perform a specific function.

Tissue rejection Phenomenon by which the body recognizes the protein (HLA antigens) in transplanted tissues or organs as foreign and produces antibodies against them.

Tongue A large skeletal muscle covered by a mucous membrane located on the floor of the oral cavity.

Tonsil (TON-sil) An aggregation of large lymphatic nodules embedded in the mucous membrane of the throat.

Topical (TOP-i-kal) Applied to the surface rather than ingested or injected.

Torn cartilage A tearing of an articular disc (meniscus) in the knee.

Trabecula (tra-BEK-ū-la) Irregular latticework of thin plates of spongy bone tissue. Fibrous cord of connective tissue serving as supporting fiber by forming a septum extending into an organ from its wall or capsule. *Plural* is **trabeculae** (tra-BEK-ū-lē).

Trabeculae carneae (KAR-nē-ē) Ridges and folds of the myocardium in the ventricles.

Trachea (TRĀ-kē-a) Tubular air passageway extending from the larynx to the fifth thoracic vertebra. Also called the **windpipe.**

Tract A bundle of nerve axons in the central nervous system.

Transplantation (tranz-plan-TĀ-shun) The transfer of living cells, tissues, or organs from a donor to a recipient or from one part of the body to another in order to restore a lost function.

Transverse colon (trans-VERS KŌ-lon) The portion of the large intestine extending across the abdomen from the right colic (hepatic) flexure to the left colic (splenic) flexure.

Transverse fissure (FISH-er) The deep cleft that separates the cerebrum from the cerebellum.

Transverse plane A plane that divides the body or organs into superior and inferior portions. Also called a **cross-sectional** or **horizontal plane.**

Transverse tubules (TOO-būls) **(T tubules)** Small, cylindrical invaginations of the sarcolemma of striated muscle fibers (cells) that conduct muscle action potentials toward the center of the muscle fiber.

Tremor (TREM-or) Rhythmic, involuntary, purposeless contraction of opposing muscle groups.

Triad (TRĪ-ad) A complex of three units in a muscle fiber composed of a transverse tubule and the sarcoplasmic reticulum terminal cisterns on both sides of it.

Tricuspid (trī-KUS-pid) **valve** Atrioventricular (AV) valve on the right side of the heart.

Triglyceride (trī-GLI-cer-īd) A lipid formed from one molecule of glycerol and three molecules of fatty acids that may be either solid (fats) or liquid (oils) at room temperature; the body's most highly concentrated source of chemical potential energy. Found mainly within adipocytes. Also called a **neutral fat** or a **triacylglycerol.**

Trigone (TRĪ-gon) A triangular region at the base of the urinary bladder.

Triiodothyronine (trī-ī-ō-dō-THĪ-rō-nēn) **(T_3)** A hormone produced by the thyroid gland that regulates metabolism, growth and development, and the activity of the nervous system.

Trophoblast (TRŌF-ō-blast) The superficial covering of cells of the blastocyst.

Tropic (TRŌ-pik) **hormone** A hormone whose target is another endocrine gland.

Trunk The part of the body to which the upper and lower limbs are attached.

Tubal ligation (lī-GĀ-shun) A sterilization procedure in which the uterine (fallopian) tubes are tied and cut.

Tubular reabsorption The movement of filtrate from renal tubules back into blood in response to the body's specific needs.

Tubular secretion The movement of substances in blood into renal tubular fluid in response to the body's specific needs.

Tumor suppressor gene A gene coding for a protein that normally inhibits cell division; loss or alteration of a tumor suppressor gene called *p53* is the most common genetic change in a wide variety of cancer cells.

Tunica albuginea (TOO-ni-ka al′-bū-JIN-ē-a) A dense white fibrous capsule covering a testis or deep to the surface of an ovary.

Tunica externa (eks-TER-na) The superficial coat of an artery or vein, composed mostly of elastic and collagen fibers. Also called the **adventitia.**

Tunica interna (in-TER-na) The deep coat of an artery or vein, consisting of a lining of endothelium, basement membrane, and internal elastic lamina. Also called the **tunica intima** (IN-ti-ma).

Tunica media (MĒ-dē-a) The intermediate coat of an artery or vein, composed of smooth muscle and elastic fibers.

Tympanic antrum (tim-PAN-ik AN-trum) An air space in the middle ear that leads into the mastoid air cells or sinus.

Tympanic (tim-PAN-ik) **membrane** A thin, semitransparent partition of fibrous connective tissue between the external auditory meatus and the middle ear. Also called the **eardrum.**

Type II cutaneous mechanoreceptor A sensory receptor embedded deeply in the dermis and deeper tissues that detects stretching of skin. Also called a **Ruffini corpuscle.**

U

Umbilical cord The long, ropelike structure containing the umbilical arteries and vein that connect the fetus to the placenta.

Umbilicus (um-bi-LĪ-kus *or* um-bil-Ī-kus) A small scar on the abdomen that marks the former attachment of the umbilical cord to the fetus. Also called the **navel.**

Upper limb The appendage attached at the shoulder girdle, consisting of the arm, forearm, wrist, hand, and fingers. Also called **upper extremity.**

Uremia (ū-RĒ-mē-a) Accumulation of toxic levels of urea and other nitrogenous waste products in the blood, usually resulting from severe kidney malfunction.

Ureter (Ū-rē-ter) One of two tubes that connect the kidney with the urinary bladder.

Urethra (ū-RĒ-thra) The duct from the urinary bladder to the exterior of the body that conveys urine in females and urine and semen in males.

Urinalysis (ū-ri-NAL-i-sis) An analysis of the volume and physical, chemical, and microscopic properties of urine.

Urinary (Ū-ri-ner-ē) **bladder** A hollow, muscular organ situated in the pelvic cavity posterior to the pubic symphysis; receives urine via two ureters and stores urine until it is excreted through the urethra.

Urine The fluid produced by the kidneys that contains wastes and excess materials; excreted from the body through the urethra.

Urogenital (ū′-rō-JEN-i-tal) **triangle** The region of the pelvic floor inferior to the pubic symphysis, bounded by the pubic symphysis and the ischial tuberosities, and containing the external genitalia.

Urology (ū-ROL-ō-jē) The specialized branch of medicine that deals with the structure, function, and diseases of the male and female urinary systems and the male reproductive system.

Uterine (Ū-ter-in) **tube** Duct that transports ova from the ovary to the uterus. Also called the **fallopian** (fal-LŌ-pē-an) **tube** or **oviduct.**

Uterosacral ligament (ū′-ter-ō-SĀ-kral LIG-a-ment) A fibrous band of tissue extending from the cervix of the uterus laterally to the sacrum.

Uterovesical (ū′-ter-ō-VES-i-kal) **pouch** A shallow pouch formed by the reflection of the peritoneum from the anterior surface of the uterus, at the junction of the cervix and the body, to the posterior surface of the urinary bladder.

Uterus (Ū-te-rus) The hollow, muscular organ in females that is the site of menstruation, implantation, development of the fetus, and labor. Also called the **womb.**

Utricle (Ū-tri-kul) The larger of the two divisions of the membranous labyrinth located inside the vestibule of the inner ear, containing a receptor organ for static equilibrium.

Uvea (Ū-vē-a) The three structures that together make up the vascular tunic of the eye.

Uvula (Ū-vū-la) A soft, fleshy mass, especially the V-shaped pendant part, descending from the soft palate.

V

Vagina (va-JĪ-na) A muscular, tubular organ that leads from the uterus to the vestibule, situated between the urinary bladder and the rectum of the female.

Vallate papilla (VAL-at pa-PIL-a) One of the circular projections that is arranged in an inverted V-shaped row at the back of the tongue; the largest of the elevations on the upper surface of the tongue containing taste buds. Also called **circumvallate papilla.**

Varicocele (VAR-i-kō-sēl) A twisted vein; especially, the accumulation of blood in the veins of the spermatic cord.

Varicose (VAR-i-kōs) Pertaining to an unnatural swelling, as in the case of a varicose vein.

Vas A vessel or duct.

Vasa recta (VĀ-sa REK-ta) Extensions of the efferent arteriole of a juxtamedullary nephron that run alongside the loop of the nephron (Henle) in the medullary region of the kidney.

Vasa vasorum (va-SŌ-rum) Blood vessels that supply nutrients to the larger arteries and veins.

Vascular (VAS-kū-lar) Pertaining to or containing many blood vessels.

Vascular (venous) sinus A vein with a thin endothelial wall that lacks a tunica media and externa and is supported by surrounding tissue.

Vascular spasm Contraction of the smooth muscle in the wall of a damaged blood vessel to prevent blood loss.

Vascular tunic (TOO-nik) The middle layer of the eyeball, composed of the choroid, ciliary body, and iris. Also called the **uvea** (Ū-ve-a).

Vasectomy (va-SEK-tō-mē) A means of sterilization of males in which a portion of each ductus (vas) deferens is removed.

Vasoconstriction (vāz-ō-kon-STRIK-shun) A decrease in the size of the lumen of a blood vessel caused by contraction of the smooth muscle in the wall of the vessel.

Vasodilation (vā z′-ō-DĪ-lā-shun) An increase in the size of the lumen of a blood vessel caused by relaxation of the smooth muscle in the wall of the vessel.

Vein A blood vessel that conveys blood from tissues back to the heart.

Vena cava (VĒ-na KĀ-va) One of two large veins that open into the right atrium, returning to the heart all of the deoxygenated blood from the systemic circulation except from the coronary circulation.

Ventral (VEN-tral) Pertaining to the anterior or front side of the body; opposite of dorsal.

Ventral ramus (RĀ-mus) The anterior branch of a spinal nerve, containing sensory and motor fibers to the muscles and skin of the anterior surface of the head, neck, trunk, and the limbs.

Ventricle (VEN-tri-kul) A cavity in the brain filled with cerebrospinal fluid. An inferior chamber of the heart.

Ventricular fibrillation (ven-TRIK-ū-lar fib-ri-LĀ-shun) Asynchronous ventricular contractions; unless reversed by defibrillation, results in heart failure.

Venule (VEN-ūl) A small vein that collects blood from capillaries and delivers it to a vein.

Vermiform appendix (VER-mi-form a-PEN-diks) A twisted, coiled tube attached to the cecum.

Vermis (VER-mis) The central constricted area of the cerebellum that separates the two cerebellar hemispheres.

Vertebral (VER-te-bral) **canal** A cavity within the vertebral column formed by the vertebral foramina of all the vertebrae and containing the spinal cord. Also called the **spinal canal.**

Vertebral column The 26 vertebrae of an adult and 33 vertebrae of a child; encloses and protects the spinal cord and serves as a point of attachment for the ribs and back muscles. Also called the **backbone, spine,** or **spinal column.**

Vesicle (VES-i-kul) A small bladder or sac containing liquid.

Vesicouterine (ves′-ik-ō-Ū-ter-in) **pouch** A shallow pouch formed by the reflection of the peritoneum from the anterior surface of the uterus, at the junction of the cervix and the body, to the posterior surface of the urinary bladder.

Vestibular (ves-TIB-ū-lar) **apparatus** Collective term for the organs of equilibrium, which includes the saccule, utricle, and semicircular ducts.

Vestibular membrane The membrane that separates the cochlear duct from the scala vestibuli.

Vestibule (VES-ti-būl) A small space or cavity at the beginning of a canal, especially the inner ear, larynx, mouth, nose, and vagina.

Villus (VIL-lus) A projection of the intestinal mucosal cells containing connective tissue, blood vessels, and a lymphatic vessel; functions in the absorption of the end products of digestion. *Plural* is **villi** (VIL-ī).

Viscera (VIS-er-a) The organs inside the ventral body cavity. *Singular* is **viscus** (VIS-kus).

Visceral (VIS-er-al) Pertaining to the organs or to the covering of an organ.

Visceral effectors (e-FEK-torz) Organs of the ventral body cavity that respond to neural stimulation, including cardiac muscle, smooth muscle, and glands.

Vitamin An organic molecule necessary in trace amounts that acts as a catalyst in normal metabolic processes in the body.

Vitreous (VIT-rē-us) **body** A soft, jellylike substance that fills the vitreous chamber of the eyeball, lying between the lens and the retina.

Vocal folds Pair of mucous membrane folds below the ventricular folds that function in voice production. Also called **true vocal cords.**

Volkmann's canal *See* **Perforating canal.**

Vulva (VUL-va) Collective designation for the external genitalia of the female. Also called the **pudendum** (poo-DEN-dum).

W

Wallerian (wal-LE-rē-an) **degeneration** Degeneration of the portion of the axon and myelin sheath of a neuron distal to the site of injury.

Wandering macrophage (MAK-rō-fāj) Phagocytic cell that develops from a monocyte, leaves the blood, and migrates to infected tissues.

White matter Aggregations or bundles of myelinated and unmyelinated axons located in the brain and spinal cord.

White pulp The regions of the spleen composed of lymphatic tissue, mostly B lymphocytes.

White ramus communicans (RĀ-mus kō-MŪ-ni-kans) The portion of a preganglionic sympathetic axon that branches from the anterior ramus of a spinal nerve to enter the nearest sympathetic trunk ganglion.

X

Xiphoid (ZĪ-foyd) Sword-shaped. The inferior portion of the sternum is the **xiphoid process.**

Y

Yolk sac An extraembryonic membrane composed of the exocoelomic membrane and hypoblast. It transfers nutrients to the embryo, is a source of blood cells, contains primordial germ cells that migrate into the gonads to form primitive germ cells, forms part of the gut, and helps prevent desiccation of the embryo.

Z

Zona fasciculata (ZŌ-na fa-sik′-ū-LA-ta) The middle zone of the adrenal cortex consisting of cells arranged in long, straight cords that secrete glucocorticoid hormones, mainly cortisol.

Zona glomerulosa (glo-mer′-ū-LŌ-sa) The outer zone of the adrenal cortex, directly under the connective tissue covering, consisting of cells arranged in arched loops or round balls that secrete mineralocorticoid hormones, mainly aldosterone.

Zona pellucida (pe-LOO-si-da) Clear glycoprotein layer between a secondary oocyte and the surrounding granulosa cells of the corona radiata.

Zona reticularis (ret-ik′-ū-LAR-is) The inner zone of the adrenal cortex, consisting of cords of branching cells that secrete sex hormones, chiefly androgens.

Zygote (ZĪ-got) The single cell resulting from the union of male and female gametes; the fertilized ovum.

C

Credits

Illustration Credits

Chapter 1 Table 1.2: Keith Kasnot. 1.1: Kevin Somerville. 1.2–1.4: Jared Schneidman Design. 1.5: Molly Borman. 1.6: Kevin Somerville. 1.7: Molly Borman. 1.8, 1.9, 1.10a, 1.10b: Imagineering. 1.10c, 1.11, 1.12a-c: Kevin Somerville.

Chapter 2 2.1–2.25: Imagineering.

Chapter 3 3.1, 3.2: Tomo Narashima. 3.3, 3.5, 3.6: Imagineering. 3.7: Jared Schneidman Design. 3.8–3.13: Imagineering. 3.14–3.18: Tomo Narashima. 3.19: Imagineering. 3.20–3.22: Tomo Narashima. 3.23–3.32: Imagineering. 3.33: Hilda Muinos.

Chapter 4 Table 4.1–Table 4.6, 4.1–4.7: Imagineering.

Chapter 5 5.1–5.7: Kevin Somerville. 5.10: Imagineering

Chapter 6 6.1: Leonard Dank/Imagineering. 6.2: Lauren Keswick. 6.3: Kevin Somerville/Imagineering. 6.4–6.8: Kevin Somerville. 6.9: Leonard Dank. 6.10: Kevin Somerville. 6.11: Jared Schneidman Design.

Chapter 7 Tables 7.1, 7.2: Imagineering. 7.1–7.12: Leonard Dank/Imagineering. 7.13: Kevin Somerville. 7.14–7.24: Leonard Dank/Imagineering. 7.25: Imagineering. 7.26:

Chapter 8 Table 8.1: Leonard Dank. 8.1–8.17: Leonard Dank/Imagineering. 8.18a: Kevin Somerville. 8.18b-e: Leonard Dank

Chapter 9 9.1–9.3, 9.12–9.16: Leonard Dank/Imagineering.

Chapter 10 Table 10.2: Imagineering. 10.1, 10.2: Kevin Somerville. 10.3: Imagineering. 10.5–10.9: Imagineering. 10.10a-c: Kevin Somerville. 10.11: Imagineering. 10.12–10.14: Jared Schneidman Design. 10.15: Imagineering. 10.17, 10.18: Imagineering. 10.19: Kevin Somerville.

Chapter 11 Table 11.1: Kevin Somerville. 11.1–11.9: Leonard Dank. 11.10a: Leonard Dank/Imagineering. 11.10bc, 11.11–11.16a-b: Leonard Dank. 11.16c: Leonard Dank/Imagineering. 11.17–11.18a-b: Leonard Dank. 11.18c: Leonard Dank/Imagineering. 11.18d: Kevin Somerville. 11.19–11.20: Leonard Dank. 11.21: Leonard Dank/ Imagineering. 11.22, 11.23: Leonard Dank.

Chapter 12 12.1: Kevin Somerville/Imagineering. 12.2: Jared Schneidman Design. 12.3a-b: Kevin Somerville. 12.4, 12.5: Imagineering. 12.6–12.8: Kevin Somerville. 12.9–12.20: Imagineering.

Chapter 13 Table 13.1: Jared Schneidman Design. 13.1a: Kevin Somerville. 13.1c: Imagineering. 13.2, 13.3a: Kevin Somerville. 13.4a: Kevin Somerville. 13.5: Kevin Somerville. 13.6, 13.7a: Steve Oh. 13.7b: Kevin Somerville. 13.8: Imagineering. 13.9, 13.10: Steve Oh/Imagineering. 13.11: Imagineering. 13.12, 13.13: Kevin Somerville. 13.14: Leonard Dank. 13.15–13.17: Leonard Dank

Chapter 14 Tables 14.1, 14.2, 14.4: Imagineering. 14.1a: Kevin Somerville/Imagineering. 14.2, 14.3: Kevin Somerville. 14.4a, b: Kevin Somerville/Imagineering. 14.4c: Imagineering. 14.5–14.8: Kevin Somerville/Imagineering. 14.9: Kevin Somerville. 14.10: Kevin Somerville/Imagineering. 14.11: Kevin Somerville. 14.12: Imagineering. 14.13, 14.14: Kevin Somerville/Imagineering. 14.15: Kevin Somerville. 14.17: Imagineering. 14.18: Kevin Somerville/Tomo Narashima. 14.19–14.27: Kevin Somerville/Sharon Ellis. 14.28, 14.29 Kevin Somerville.

Chapter 15 15.1–15.3: Imagineering. 15.4, 15.5: Kevin Somerville. 15.6: Imagineering.

Chapter 16 Tables 16.3, 16.4: Imagineering. 16.1: Imagineering. 16.2, 16.3: Kevin Somerville. 16.4: Leonard Dank. 16.5: Kevin Somerville. 16.6a-b: Imagineering. 16.7: Jared Schneidman Design. 16.8: Kevin Somerville. 16.9, 16.10: Sharon Ellis.

Chapter 17 Tables 17.1, 17.2: Imagineering. 17.1: Tomo Narashima. 17.2: Molly Borman. 17.3: 17.4: Sharon Ellis/Imagineering. 17.5: Tomo Narashima/Imagineering. 17.8: Lynn O'Kelley/Imagineering. 17.9: Tomo Narashima/Imagineering. 17.10, 17.11: Jared Schneidman Design. 17.12: Lynn O'Kelley. 17.13: Jared Schneidman Design. 17.15: Imagineering. 17.16–17.19: Tomo Narashima. 17.20: Tomo Narashima. 17.21, 17.22: Tomo Narashima/Imagineering/Sharon Ellis. 17.23, 17.24: Kevin Somerville.

Chapter 18 Table 18.2: Jared Schneidman Design. Tables 18.4, 18.5, 18.8, 18.10: Nadine Sokol. Tables 18.6, 18.7, 18.9: Imagineering. 18.1: Steve Oh/Imagineering. 18.2: Jared Schneidman Design. 18.3, 18.4: Imagineering. 18.5ab: Lynn O'Kelley/Imagineering. 18.6, 18.7: Jared Schneidman Design. 18.8: Lynn O'Kelley/Imagineering. 18.9: Jared Schneidman Design. 18.10: Molly Borman/Imagineering. 18.11, 18.12: Jared Schneidman Design. 18.13: Molly Borman/ Imagineering. 18.14: Imagineering. 18.15: Molly Borman/ Imagineering. 18.16, 18.17: Jared Schneidman Design. 18.18: Molly Borman/Imagineering. 18.19, 18.20: Jared Schneidman Design. 18.21: Kevin Somerville.

Chapter 19 Table 19.3: Imagineering. 19.1, 19.3: Imagineering. 19.4: Nadine Sokol. 19.5, 19.6: Jared Schneidman Design. 19.8: Imagineering. 19.9: Nadine Sokol. 19.11: Imagineering. 19.12: Jean Jackson. 19.13: Nadine Sokol.

Chapter 20 20.1–20.6: Kevin Somerville/Imagineering. 20.7: Imagineering. 20.8–20.16: Kevin Somerville/Imagineering. 20.18, 20.19: Kevin Somerville. 20.21a-c: Hilda Muinos/Imagineering. 20.22: Kevin Somerville.

Chapter 21 Table 21.3: Imagineering. 21.1: Kevin Somerville. 21.2: Hilda Muinos. 21.3: Nadine Sokol/Imagineering. 21.4: Kevin Somerville. 21.5: Imagineering. 21.6: Jared Schneidman Design. 21.7, 21.8: Imagineering. 21.9: Kevin Somerville. 21.10, 21.11: Jared Schneidman Design. 21.12: Imagineering. 21.13: Kevin Somerville. 21.14–21.16: Jared Schneidman Design. 21.17–21.30: Kevin Somerville.

Chapter 22 22.1: Molly Borman. 22.2: Sharon Ellis. 22.3: Molly Borman. 22.5: Steve Oh. 22.6a: Molly Borman. 22.7: Steve Oh. 22.8: Kevin Somerville. 22.9, 22.10: Molly Borman. 22.11–22.16: Imagineering. 22.17: Jared Schneidman Design. 22.18: Imagineering. 22.19: Jared Schneidman Design. 22.20: Jared Schneidman Design. 22.21: Nadine Sokol/Imagineering.

Chapter 23 23.1a: Molly Borman. 23.2: Kevin Somerville/Imagineering. 23.4–23.5: Molly Borman/Imagineering. 23.6: Steve Oh/Imagineering. 23.7: Imagineering. 23.8: Molly Borman. 23.9: Imagineering. 23.10a-e: Molly Borman/Imagineering. 23.11, 23.12: Kevin Somerville/ Imagineering. 23.13: Jared Schneidman Design. 23.14: Kevin Somerville. 23.15–23.24: Jared Schneidman Design. 23.25: Imagineering. 23.26: Jared Schneidman Design. 23.27: Kevin Somerville. 23.28: Jared Schneidman Design. 23.29: Kevin Somerville.

Chapter 24 24.1: Steve Oh. 24.2: Kevin Somerville. 24.3: Imagineering. 24.4: Steve Oh/Imagineering. 24.5: Nadine Sokol. 24.6: Molly Borman. 24.7: Steve Oh/Imagineering. 24.8: Nadine Sokol. 24.9: Imagineering. 24.10: Nadine Sokol. 24.11a: Steve Oh. 24.11b: Imagineering. 24.12: Kevin Somerville. 24.13: Imagineering. 24.14ab: Steve Oh. 24.14c: Jared Schneidman Design. 24.15: Kevin Somerville. 24.16: Jared Schneidman Design. 24.17a: Kevin Somerville. 24.18: Kevin Somerville. 24.20, 24.21: Jared Schneidman Design. 24.22: Molly Borman. 24.23: Kevin Somerville. 24.24: Jared Schneidman Design.

Chapter 25 25.1–25.20: Imagineering

Chapter 26 Table 26.1: Nadine Sokol. 26.1: Kevin Somerville. 26.2: Kevin Somerville/Imagineering. 26.3: Steve Oh. 26.4: Steve Oh/ Imagineeering. 26.5: Imagineering. 26.6a: Kevin Somerville/ Imagineering. 26.8: Kevin Somerville. 26.9: Imagineering. 26.10–26.18: Jared Schneidman Design. 26.19: Imagineering. 26.20: Jared Schneidman Design. 26.21: Steve Oh/Imagineering. 26.22: Kevin Somerville/Imagineering. 26.23: Kevin Somerville.

Chapter 27 27.1–27.8: Jared Schneidman Design.

Chapter 28 Table 28.1: Imagineering. 28.1: Kevin Somerville/ Imagineering. 28.2: Kevin Somerville. 28.3a: Kevin Somerville/ Imagineering. 28.4–28.7: Imagineering. 28.8: Jared Schneidman Design. 28.9: Kevin Somerville. 28.10–28.13: Kevin Somerville/ Imagineering. 28.15: Imagineering. 28.16: Kevin Somerville/ Imagineering. 28.19–28.21: Kevin Somerville. 28.22: Kevin Somerville/Imagineering. 28.23–28.26: Imagineering. 28.27, 28.28: Kevin Somerville.

Chapter 29 Table 29.2: Kevin Somerville. 29.1–29.4: Kevin Somerville. 29.5: Kevin Somerville/Imagineering. 29.6–29.15: Kevin Somerville. 29.16: Jared Schneidman Design. 29.17, 29.18: Kevin Somerville. 29.18: Kevin Somerville. 29.19–29.26: Jared Schneidmen Design.

Focus on Homeostasis icons: Imagineering.

Photo Credits

Chapter 1 Figure 1-1 (top and bottom): Rubberball Productions/Getty Images. Figure 1-8a: Stephen A. Kieffer and E. Robert Heitzman, *An Atlas of Cross-Sectional Anatomy*. Harper & Row, New York, 1979. Figure 1-8b: Lester V. Bergman/Project Masters, Inc. Figure 1-8c: Martin Rotker. Figure 1-10c: Mark Nielsen. Figure 1-12a: Andy Washnik. Table 1-4a: Biophoto Associates/Photo Researchers. Table 1-4b: Breast Cancer Unit, Kings College Hospital, London/Photo Researchers. Table 1-4c: Zephyr/Photo Researchers. Figure 1-4d: Cardio-Thoracic Centre, Freeman Hospital, Newcastle-Upon-Tyne/ Photo Researchers. Table 1-4e: CNRI/Science Photo Library/Photo Researchers. Table 1-4f: Science Photo Library/Photo Researchers. Table 1-4g: Scott Camazine/Photo Researchers. Table 1-4h: Simon Fraser/Photo Researchers. Table 1-4i: Courtesy Andrew Joseph Tortora and Damaris Soler. Table 1-4j: Howard Sochurek/ Medical Images, Inc. Table 1-4k: SIU/Visuals Unlimited. Table 1-4l: Dept. of Nuclear Medicine, Charing Cross Hospital/Photo Researchers. Table 1-4m: ©Camal/Phototake

Chapter 3 Figure 3-4: Andy Washnik. Figure 3-7a-c: David Phillips/ Photo Researchers. Figure 3-11b,c: Courtesy Abbott Laboratories. Figure 3-14c: Donald Fawcett/Visuals Unlimited. Figure 3-15d: P. Motta/ Photo Researchers. Figure 3-15e: David M. Phillips/Visuals Unlimited. Figure 3-17: D. W. Fawcett/Photo Researchers. Figure 3-18: Biophoto Associates/Photo Researchers. Figure 3-20b: Dr. Gopal Murti/Visuals Unlimited. Figures 3–21 and 3–22: D.W. Fawcett/ Photo Researchers. Figure 3-30: Courtesy Michael Ross, University of Florida.

Chapter 4 Table 4-1a (left), f: Biophoto Associates/Photo Researchers. Table 4-1a (right), b-e, g-i: Courtesy Michael Ross, University of Florida. Table 4-2a: Lester V. Bergman/The Bergman Collection. Table 4-2b: Courtesy Michael Ross, University of Florida. Table 4-3a,b: Courtesy Michael Ross, University of Florida. Table 4-4a-c, f-I, k: Courtesy Michael Ross, University of Florida. Table 4-4d: Courtesy Andrew J. Kuntzman. Table 4-4e: Ed Reschke. Table 4-4j: John Burbidge/Photo Researchers. Table 4-5a-c: Courtesy Michael Ross, University of Florida. Table 4-6: Science VU/Visuals Unlimited.

Chapter 5 Figures 5-1b and 5-3b: Courtesy Michael Ross, University of Florida. Figures 5-4b: VVG/Science Photo Library/Photo Researchers. Figures 5-8a: Alain Dex/Photo Researchers. Figures 5-8b: Biophoto Associates/Photo Researchers. Figures 5-9a: Sheila Terry/ Science Photo Library/Photo Researchers. Figures 5-9b,c: St. Stephen's Hospital/Science Photo Library/Photo Researchers. Figures 5-11: Dr. P. Marazzi/Science Photo Library/Photo Researchers.

Chapter 6 Figure 6-1b: Mark Nielsen. Figure 6-2b: CNRI/Photo Researchers. Figure 6-2c, d: Dr. Richard Kessel and Randy Kardon/ Tissues & Organs/Visuals Unlimited. Figure 6-7a: The Bergman Collection. Figure 6-7b: Biophoto Associates/Photo Researchers. Figure 6-9a-c, e, f: Courtesy Department of Medical Illustration, University of Wisconsin Medical School. Figure 6-9d: Dr. Andrew Schmidt, Hennepin County Medical Center/The Bergman Collection/Project Masters, Inc. Figure 6-12a, b: P. Motta/Photo Researchers.

Chapter 7 Figure 7-25a: Princess Margaret Rose Orthopaedic Hospital/Photo Researchers. Figure 7-25b: Dr. P. Marazzi/Photo Researchers. Figure 7-25c: Custom Medical Stock Photo, Inc. Figure 7-26: Center for Disease Control/Project Masters, Inc.

Chapter 9 Figures 9-4 to 9-9: John Wilson White.

Chapter 10 Figure 10-4: Courtesy D.E. Kelley. Figure 10-6: Courtesy Hiroyouki Sasaki, Yale E. Goldman and Clara Franzini-Armstrong. Figure 10-10d: Don Fawcett/Photo Researchers. Figure 10-16: ©John Wiley & Sons.

Chapter 12 Figure 12-3b: Science VU/Visuals Unlimited. Figure 12-8c: Dennis Kunkel/Phototake. Figure 12-8d: Martin Rotker/Phototake.

Chapter 13 Figure 13-1b, c: Mark Nielsen. Figure 13-3b: Courtesy Michael Ross, University of Florida. Figure 13-4b: ©Dr.Richard Kessel and Dr. Randy Kardon/Visuals Unlimited.

Chapter 14 Figure 14-1b: Mark Nielsen. Figure 14-9e: From Stephen A. Kieffer and E. Robert Heitzman, *An Atlas of Cross-Sectional Anatomy*, Harper and Row, Publishers, 1979. Reproduced with permission. Figure 14-12: N. Gluhbegovic and T.H. Williams, *The Human Brain: A Photographic Guide*, Harper and Row, Publishers, Inc. Hagerstown, MD, 1980. Figure 14-16: From *Nature*, Vol. 360, November 26, 1992, page 340. Reproduced with permission from Nature and Robert Zatorre, Department of Neuropsychology, McGill University.

Chapter 17 Figure 17-3: John Moore. Figure 17-7: Courtesy Michael Ross, University of Florida. Figure 17-15a: N. Gluhbegovic and T. H. Williams, *The Human Brain: A Photographic Guide*, Harper and Row, Publishers, Inc., Hagerstown, MD, 1980.

Chapter 18 Figures 18-5a, 18–10c, d and 18–15c: Mark Nielsen. Figure 18–5c: Courtesy James Lowe, University of Nottingham, Nottingham, United Kingdom. Figures 18-10b, 18-13b, 18-15d and 18-18c: Courtesy Michael Ross, University of Florida. Figure 18-18d: Courtesy James Sheetz, Department of Cell Biology, University of Alabama, Birmingham. Figure 18-22a: From *New England Journal of Medicine*, February 18, 1999, vol. 340, No. 7, page 524. Photo provided courtesy of Robert Gagel, Department of Internal Medicine, University of Texas M.D. Anderson Cancer Center, Houston, Texas. Figure 18-22b-d: ©The Bergman Collection/Project Masters, Inc. Figure 18-22e: Biophoto Associates/Photo Researchers.

Chapter 19 Figure 19-2a: Juergen Berger/Photo Researchers. Figures 19-2b and 19-7: Courtesy Michael Ross, University of Florida. Figure 19-10a-c: David M. Phillips/Photo Researchers. Figure 19-10d: Dennis Kunkel/Phototake. Figure 19-15: Lewin/Royal Free Hospital/Photo Researchers. Figure 19-14: Jean Claude Revy/Phototake.

Chapter 20 Figures 20-3b, 20-4b, 20-6e and 20-8c: Mark Nielsen. Figure 20-17: Gregg Adams/Stone/Getty Images. Figure 20-20a: ©Vu/Cabisco/Visuals Unlimited. Figure 20-20b: W. Ober/Visuals Unlimited. Figure 20-21d: ©ISM/Phototake.

Chapter 21 Figure 21-1d: Dennis Strete. Figure 21-1e: Courtesy Michael Ross, University of Florida. Figure 21-5: Mark Nielsen.

Chapter 22 Figures 22-5b, c and 22-7c: Courtesy Michael Ross, University of Florida. Figure 22-6b: Leroy, Biocosmos/Photo Researchers. Figure 22-6c: Mark Nielsen. Figure 22-9: Science Photo Library/Photo Researchers.

Chapter 23 Figure 23-1b: From J. W. Rohen, Ch. Yokochi, E. Lüetjen, Drecoll, *Color Atlas of Anatomy*, 5e., Schattauer Publishing, Stuttgart, Germany. Reproduced with permission. Figure 23-3: Courtesy Lynne Marie Barghesi. Figures 23-6 and 23-9: Mark Nielsen. Figure 23-7: John Cunningham/Visuals Unlimited. Figure 23-11: Biophoto Associates/Photo Researchers. Figure 23-12c: Biophoto Associates/Photo Researchers.

Chapter 24 Figures 24-6b, 24-9, 24-15c, 24-19 c, d and 24-23c, d : Courtesy Michael Ross, University of Florida. Figure 24-11b: From Johannes W. Rohen, Chihiro Yokochi and Elke Lütjen-Drecoll, *Color Atlas of Anatomy*, Schattauer Publishing, Stuttgart, Germany. Reproduced with permission. Figure 24-12 (inset): Hessler/Vu/Visuals Unlimited. Figure 24-12c: Ed Reschke. Figure 24-17b: Mark Nielsen. Figure 24-19a: Fred E. Hossler/Visuals Unlimited. Figure 24-19b: G. W. Willis/Visuals Unlimited.

Chapter 25 Figure 25-20: John E. Kelly/Stone/Getty Images.

Chapter 26 Figure 26-3b: Mark Nielsen. Figure 26-6b: Dennis Strete. Figure 26-8: Courtesy Michael Ross, University of Florida.

Chapter 28 Figures 28-1b, 28-3b-c, 28-11b: Mark Nielsen. Figures 28-4a, 28-10c, 28-14a, 28-17a and 28-18a, b: Courtesy Michael Ross, University of Florida. Figure 28-14b: Biophoto Associates/Photo Researchers. Figure 28-17b: P. Motta/Photo Researchers.

Chapter 29 Figure 29-1b: David Phillips/Photo Researchers. Figure 29-1c: Myriam Wharman/Phototake. Figure 29-11b: Siu, Biomedical Comm./ Custom Medical Stock Photo, Inc. Figures 29-14a, g and h: Photo provided courtesy of Kohei Shiota, Congenital Anomaly Research Center, Kyoto University, Graduate School of Medicine. Figures 29-14b-e: Courtesy National Museum of Health and Medicine, Armed Forces Institute of Pathology. Figure 29-14f: Photo by Lennart Nilsson/Albert Bonniers Förlag AB, *A Child is Born*, Dell Publishing Company. Reproduced with permission.

I

Index

Note Page numbers in boldface type indicate a major discussion. A t following a page number indicates a table, an f following a page number indicates a figure, and an e following a page number indicates an exhibit.

EPONYMS USED IN THIS TEXT

In the life sciences, an eponym is the name of a structure, drug, or disease that is based on the name of a person. For example, you may be more familiar with the Achilles tendon than you are with its more anatomically descriptive term, the calcaneal tendon. Because eponyms remain in frequent use, this listing correlates common eponyms with their anatomical terms.

EPONYM	ANATOMICAL TERM
Achilles tendon	calcaneal tendon
Adam's apple	thyroid cartilage
ampulla of Vater (VA-ter)	hepatopancreatic ampulla
Bartholin's (BAR-tō-linz) gland	greater vestibular gland
Billroth's (BIL-rōtz) cord	splenic cord
Bowman's (BŌ-manz) capsule	glomerular capsule
Bowman's (BŌ-manz) gland	olfactory gland
Broca's (BRŌ-kaz) area	motor speech area
Brunner's (BRUN-erz) gland	duodenal gland
bundle of His (HISS)	artrioventricular (AV) bundle
canal of Schlemm (SHLEM)	scleral venous sinus
circle of Willis (WIL-is)	cerebral arterial circle
Cooper's (KOO-perz) ligament	suspensory ligament of the breast
Cowper's (KOW-perz) gland	bulbourethral gland
crypt of Lieberkühn (LE-ber-kyūn)	intestinal gland
duct of Santorini (san′-tō-RĒ-nē)	accessory duct
duct of Wirsung (VĒR-sung)	pancreatic duct
Eustachian (yoo-STĀ-kē-an)	auditory tube
Fallopian (fal-LŌ-pē-an) tube	uterine tube
gland of Littré (LĒ-tra)	urethral gland
Golgi (GOL-jē) tendon organ	tendon organ
Graafian (GRAF-ē-an) follicle	mature ovarian follicle
Hassall's (HAS-alz) corpuscle	thymic corpuscle
Haversian (ha-VĒR-shun) canal	central canal
Haversian (ha-VĒR-shun) system	osteon
Heimlich (HĪM-lik) maneuver	abdomial thrust maneuver
islet of Langerhans (LANG-er-hanz)	pancreatic islet

EPONYM	ANATOMICAL TERM
Kupffer (KOOP-fer) cell	stellate reticuloendothelial cell
Leydig (LĪ-dig) cell	interstitial endocrinocyte
loop of Henle (HEN-lē)	loop of the nephron
Luschka's (LUSH-kaz) aperture	lateral aperture
Magendie's (ma-JEN-dēz) aperture	median aperture
Meibomian (mi-BŌ-mē-an) gland	tarsal gland
Meissner (MĪS-ner) corpuscle	corpuscle of touch
Merkel (MER-kel) disc	tactile disc
Müllerian (mil-E rē-an) duct	paramesonephric duct
organ of Corti (KOR-tē)	spiral organ
Pacinian (pa-SIN-ē-an) corpuscle	lamellated corpuscle
Peyer's (PĪ-erz) patch	aggregated lymphatic follicle
plexus of Auerbach (OW-er-bak)	myenteric plexus
plexus of Meissner (MĪS-ner)	submucosal plexus
pouch of Douglas	rectouterine pouch
Purkinje (pur-KIN-jē) fiber	conduction myofiber
Rathke's (rath-KĒZ) pouch	hypophyseal pouch
Ruffini (roo-FĒ-nē) corpuscle	type II cutaneous mechanoreceptor
Sertoli (ser-TŌ-lē) cell	sustentacular cell
Skene's (SKĒNZ) gland	paraurethral gland
sphincter of Oddi (OD-dē)	sphincter of the hepatopancreatic ampulla
Volkmann's (FŌLK-manz) canal	perforating canal
Wernicke's (VER-ni-kēz) area	auditory association area
Wharton's (HWAR-tunz) jelly	mucous connective tissue
Wolffian duct	mesonephric duct
Wormian (WER-mē-an) bone	sutural bone

COMBINING FORMS, WORD ROOTS, PREFIXES, AND SUFFIXES

Many of the terms used in anatomy and phsiology are compound words; that is, they are made up of word roots and one or more prefixes or suffixes. For example, *leukocyte* is formed from the word roots *leuk-* meaning "white", a connecting vowel (o), and *cyte* meaning "cell." Thus, a leukocyte is a white blood cell. The following list includes some of the most commonly used combining forms, word roots, prefixes, ad suffixes used in the study of anatomy and physiology. Each entry includes a usage example. Learning the meanings of these fundamental word parts will help you remember terms that, at first glance, may seem long or complicated.

COMBINING FORMS AND WORD ROOTS

Acous-, Acu- hearing Acoustics.
Acr- extremity Acromegaly.
Aden- gland Adenoma.
Alg-, Algia- pain Neuralgia.
Angi- vessel Angiocardiography.
Anthr- joint Arthropathy.
Aut-, Auto- self Autolysis.
Audit- hearing Auditory canal.

Bio- life, living Biopsy.
Blast- germ, bud Blastula.
Blephar- eyelid Blepharitis.
Brachi- arm Brachial plexus.
Bronch- trachea, windpipe Bronchoscopy.
Bucc- cheek Buccal.

Capit- head Decapitate.
Carcin- cancer Carcinogenic.
Cardi-, Cardia-, Cardio- heart Cardiogram.
Cephal- head Hydrocephalus.
Cerebro- brain Cerebrospinal fluid.
Chole- bile, gall Cholecystogram.
Chondr-, cartilage Chondrocyte.
Cor-, Coron- heart Coronary.
Cost- rib Costal.
Crani- skull Craniotomy.
Cut- skin Subcutaneous.
Cyst- sac, bladder Cystoscope.

Derma-, Dermato- skin Dermatosis.
Dura- hard Dura mater.

Enter- intestine Enteritis.
Erythr- red Erythrocyte.

Gastr- stomach Gastrointestinal.
Gloss- tongue Hypoglossal.
Glyco- sugar Glycogen.
Gyn-, Gynec- female, woman Gynecology.

Hem-, Hemat- blood Hematoma.
Hepar-, Hepat- liver Hepatitis.
Hist-, Histio- tissue Histology.
Hydr- water Dehydration.
Hyster- uterus Hysterectomy.

Ischi- hip, hip joint Ischium.

Kines- motion Kinesiology.

Labi- lip Labial.
Lacri- tears Lacrimal glands.
Laparo- loin, flank, abdomen Laparoscopy.
Leuko- white Leukocyte.
Lingu- tongue Sublingual glands.
Lip- fat Lipid.
Lumb- lower back, loin Lumbar.

Macul- spot, blotch Macula.
Malign- bad, harmful Malignant.
Mamm-, Mast- breast Mammography, Mastitis.
Meningo- membrane Meningitis.
Myel- marrow, spinal cord Myeloblast.
My-, Myo- muscle Myocardium.

Necro- corpse, dead Necrosis.
Nephro- kidney Nephron.
Neuro- nerve Neurotransmitter.

Ocul- eye Binocular.
Odont- tooth Orthodontic
Onco- mass, tumor Oncology.
Oo- egg Oocyte.
Opthalm- eye Ophthalmology.
Or- mouth Oral.
Osm- odor, sense of small Anosmia.
Os-, Osseo-, Osteo- bone Osteocyte.
Ot- ear Otitus media.

Palpebr- eyelid Palpebra.
Patho- disease Pathogen.
Pelv- basin Renal pelvis.
Phag- to eat Phagocytosis.
Phleb- vein Phlebitis.
Phren- diaphragm Phrenic.
Pilo- hair Depilatory.
Pneumo- lung, air Pneumothorax.
Pod- foot Podocyte.
Procto- anus, rectum Proctology.
Pulmon- lung Pulmonary.

Ren- kidneys Renal artery.
Rhin- nose Rhinitis.

Scler-, Sclero- hard Atherosclerosis.
Sep-, Spetic- toxic condition due to micoorganisms Septicemia.
Soma-, Somato- body Somatotropin.
Sten- narrow Stenosis.
Stasis-, Stat- stand still Homeostasis.

Tegument- skin, covering Integumentary.
Therm- heat Thermogenesis.
Thromb- clot, lump Thrombus.

Vas- vessel, duct Vasoconstriction.

Zyg- joined Zygote.

PREFIXES

A-, An- without, lack of, deficient Anesthesia.
Ab- away from, from Abnormal.
Ad-, Af- to, toward Adduction, Afferent neuron.
Alb- white Albino.
Alveol- cavity, socket Alveolus.
Andro- male, masculine Androgen.
Ante- before Antebrachial vein.
Anti- against Anticoagulant.

Bas- base, foundation Basal ganglia.
Bi- two, double Biceps.
Brady- slow Bradycardia.

Cata- down, lower, under Catabolism.
Circum- around Circumduction.
Cirrh- yellow Cirrhosis of the liver.
Co-, Con-, Com with, together Congenital.
Contra- against, opposite Contraception.
Crypt- hidden, concealed Cryptorchidism.
Cyano- blue Cyanosis.

De- down, from Deciduous.
Demi-, hemi- half Hemiplegia.
Di-, Diplo- two Diploid.
Dis- separation, apart, away from Dissection.
Dys- painful, difficult Dyspnea.

E-, Ec-, Ef- out from, out of Efferent neuron.
Ecto-, Exo- outside Ectopic pregnancy.
Em-, En- in, on Emmetropia.
End-, Endo- within, inside Endocardium.
Epi- upon, on, above Epidermis.
Eu- good, easy, normal Eupnea.
Ex-, Exo- outside, beyond Exocrine gland.
Extra- outside, beyond, in addition to Extracellular fluid.

Fore- before, in front of Forehead.

Gen- originate, produce, form Genitalia.
Gingiv- gum Gingivitis.

Hemi- half Hemiplegia.
Heter-, Hetero- other, different Heterozygous.
Homeo-, Homo- unchanging, the same, steady Homeostasis.
Hyper- over, above, excessive Hyperglycemia.
Hypo- under, beneath, deficient Hypothalamus.

In-, Im- in, inside, not Incontinent.
Infra- beneath Infraorbital.
Inter- among, between Intercostal.
Intra- within, inside Intracellular fluid.
Ipsi- same Ipsilateral.
Iso- equal, like Isotonic.

Juxta- near to Juxtaglomerular apparatus.

Later- side Lateral.

Macro- large, great Macrophage.
Mal- bad, abnormal Malnutrition.
Medi-, Meso- middle Medial.
Mega-, Megalo- great, large Magakaryocyte.
Melan- black Melanin.
Meta- after, beyond Metacarpus.
Micro- small Microfilament.
Mono- one Monounsaturated fat.

Neo- new Neonatal.

P9-ELH-988

VEGAN *holiday* COOKBOOK

VEGAN *holiday* COOKBOOK

Festive Plant-Based Meals and Desserts
for the Thanksgiving and Christmas Table

KATIE CULPIN

Skyhorse Publishing

Skyhorse Publishing books may be purchased in bulk at special discounts for sales promotion, corporate gifts, fund-raising, or educational purposes. Special editions can also be created to specifications. For details, contact the Special Sales Department, Skyhorse Publishing, 307 West 36th Street, 11th Floor, New York, NY 10018 or info@skyhorsepublishing.com.

Visit our website at www.skyhorsepublishing.com.

10 9 8 7 6 5 4 3 2 1

Library of Congress Cataloging-in-Publication Data is available on file.

Cover design by Erin Seaward-Hiatt
Cover photo credit by Josh Bailey
Photography by Josh Bailey

Print ISBN: 978-1-5107-5631-1
Ebook ISBN: 978-1-5107-6167-4

Printed in China

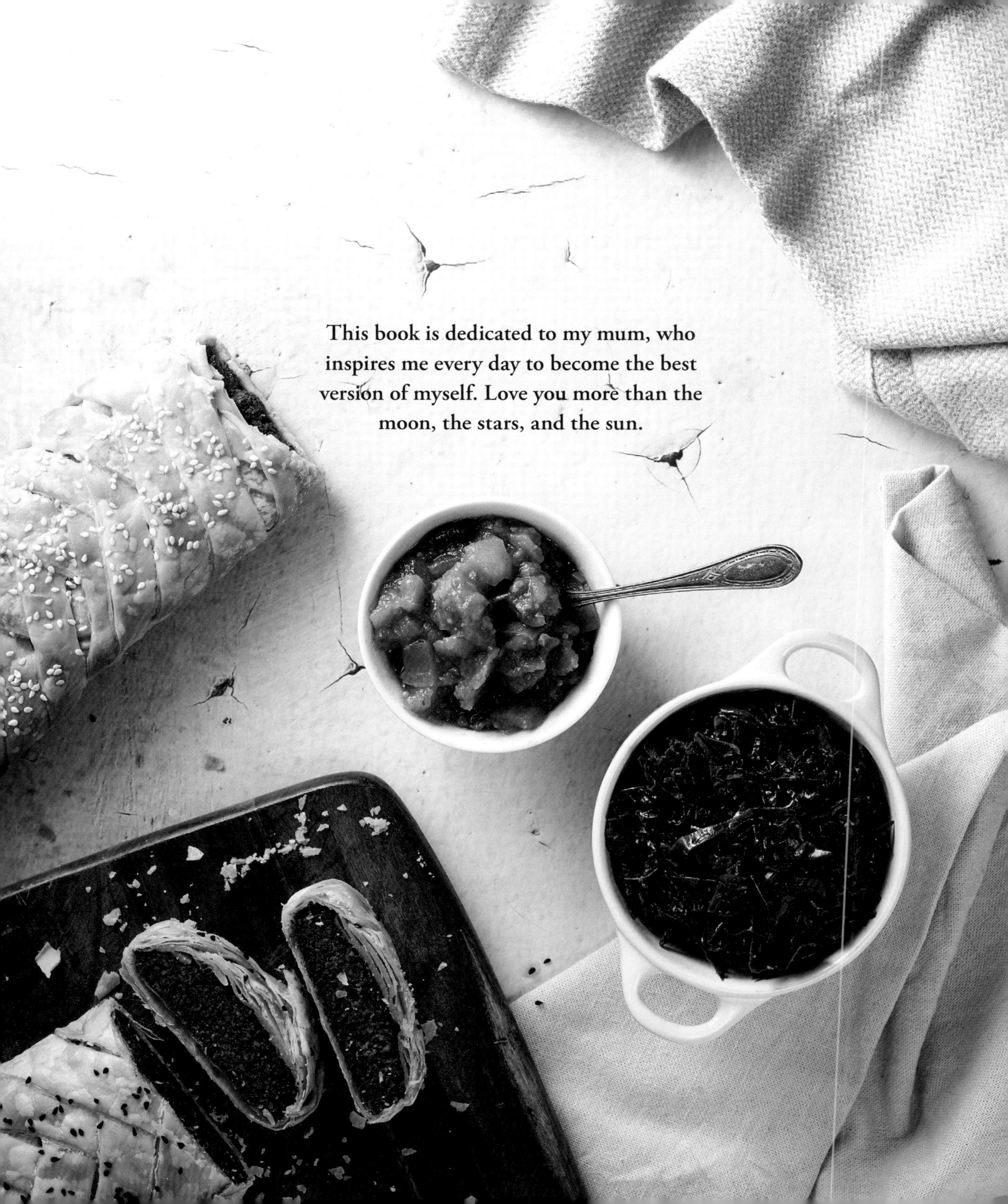

This book is dedicated to my mum, who inspires me every day to become the best version of myself. Love you more than the moon, the stars, and the sun.

CONTENTS

INTRODUCTION

Hello, there! My name's Katie, and welcome to my first published cookbook. It's been a dream of mine to create and have a book published for many years, and I'm finally pleased to be able to share it with you. Hopefully, it'll be the first of many.

I run a food and travel blog called *Delightful Vegans* (delightfulvegans.com) with my partner, Josh (the man behind the photography in this book!), which we've been posting on for over four years.

First, I have a confession to make: I'm a massive foodie! Haha, well, I guess that may seem obvious owing to the fact that you are reading my cookbook. But really, here's the thing—I'm not a qualified chef. There, I spilled the beans. I just love cooking and creating new things in the kitchen, and I love experimenting with new ingredients, a few of which you may find in this book (don't let that put you off though: most of them are easy to find, see page xv). My culinary journey started many years ago, influenced by my mum, who is a wonderful cook. I remember her having guests over for dinner and making delicious food, but we kids always had to go to bed early. Luckily, we still got to taste-test her food beforehand. My mum and I often share great recipes, and she even saves recipes she finds in magazines or the newspaper to make when I visit. My grandmother was also an incredible cook: whenever we'd visit, the house would always smell like freshly baked cookies, and she always had jars of homemade cookies waiting for us to devour—the best cookies ever!

I've worked in many vegan food places over the years, from wholesale stores selling vegan goods to restaurants. I even worked in my best friend's vegan ice cream factory for many years—yes, I ate *a lot* of ice cream; and yes, it was great! (Believe it or not, I didn't get sick of it and ice cream is still one of my favorite foods.) Over the past few years, we've seen so many more vegan-friendly products on the shelves of grocery stores, and so many more people are either vegan or are interested in veganism. I believe that our growing awareness of the health consequences of eating animal products, along with the environmental impact of animal agriculture, are forcing many people to look carefully at their food and lifestyle choices.

At university, I studied ethics and environmental philosophy, and many of my subjects highlighted the environmental impact of eating animals. When I was in my twenties, I moved in with a friend, Luke, the director of the Sustainability Festival in Melbourne, who was already vegan at the time. Up until that point, I hadn't thought too much about where my food came from or how it ended up on my plate. Living with him really opened my eyes to veganism and sustainability. We'd have big Sunday-night gatherings and invite over a bunch of people, and everyone would bring a vegan dish to share. I remember looking up recipes and making delicious food I'd never made before. What I loved most about this experience was the sense of community it created. I learned that vegan food is inclusive and delicious, and that preparing and sharing vegan meals with others is as satisfying as sharing *any* type of food! To this day, one of my main motivations to cook food is to share it. Through my actions in the kitchen, I make connections with others, gain inspiration, and also inspire others on their cooking journeys.

Thank you so much for picking up this book. Whether you are vegan, vegan-curious, or not vegan at all, this book is for everyone and anyone who wishes to make delicious food over the holidays for friends and family. I want people to get into the kitchen, be excited about cooking, and prepare delicious food that will change people's minds and bring us together. Just have fun and enjoy!

Katie Culpin

WHY VEGAN HOLIDAYS?

The holiday season is always a time to cherish and remember. Holiday gatherings are all about spending time with loved ones, celebrating, and . . . eating!

However, since becoming vegan, I've personally found festive occasions a little difficult. As much as I love spending time with my family, I don't like being surrounded by meat-based food, which can often be the case during these times of the year.

Luckily, it doesn't have to be this way. I've found that making and sharing vegan food is just as inclusive as sharing any type of food—and just as tasty and creative. In this book, I will show you how to prepare delicious, compassionate food that tastes just as good—if not better—than meaty dishes. I find that vegan meals also leave you feeling full and satisfied, rather than heavy and lethargic.

It's hard not to notice this snowball movement of people moving toward a plant-based diet. Perhaps it's for health reasons, or perhaps they're athletes trying to maximize their potential, or maybe they have just made a more heartfelt connection to the world and want to minimize the impacts of destruction to our planet and promote the well-being of its inhabitants. Even if you're not vegan yourself, there's a good chance that a vegan will be invited to your Christmas or Thanksgiving lunch or dinner. Wondering what to cook? This book is filled with plenty of recipes and ideas for you.

And if you are already vegan (whether new or a veteran), you'll also find some great inspiration here!

When preparing a plant-based meal, the question that can stump many people trying it out for the first time is "how do I start?" Don't worry—this book is designed for the average person who doesn't like to spend too long in the kitchen, all the way to the delicate operator who likes to style every leaf on their dish.

Something that I find helps at this time of the year is simply being prepared. Christmas and Thanksgiving can be a stressful time, and we want to be able to be relaxed and spend quality time with family and friends. Some of the recipes in this book can be made a few days before or the day before, which is a handy time-saver, as well as a stress saver! Also, check out my meal plans on page xix for some great lunch, dinner, and breakfast ideas.

So, get ready to celebrate your plant-based adventure with this festive vegan cookbook as the sharpest tool in your kitchen!

INGREDIENTS AND WHERE TO GET THEM

ABC Nut Butter

This is a nut butter made from almonds, Brazil nuts, and cashews. It's a tasty alternative to almond butter. It can be found at health-food shops.

Almond Flour

A gluten-free low-carb substitute for flour. You can get this at most health-food shops.

Agar powder

Found at specialty shops or health-food shops, this is a great ingredient for binding things together.

Aquafaba

This is the brine out of a can of chickpeas! Don't discard this precious liquid—you can make many wonderful things out of it, including meringue.

Black Salt

Also called *kala namak*, this sulfurous salt adds an eggy flavor to dishes. It's usually found at Indian grocers or specialty shops.

Buckwheat Flour

High in fiber, buckwheat flour is a great gluten-free alternative to use in baking—and pancakes! Available at most grocery stores.

Chickpea Flour

A great binding ingredient, chickpea flour is usually found at larger supermarkets or health-food shops.

Chinese Five-Spice Powder

A fragrant spice powder, this is an excellent ingredient to keep stocked in your pantry. It's found at most supermarkets.

Chocolate Vincotto

A dark, sweet, and musty syrup, we use this to give a full-bodied flavor in our Irish Dream Liqueur recipe (page 161). You can find it in specialty shops or buy it online.

Cream of Tartar

Commonly used as a leavener or a stabilizer, cream of tartar is a handy ingredient to have on hand, especially for baking. Find it at your local supermarket.

Condensed Coconut Milk

A fantastic alternative to condensed milk, condensed coconut milk is a delicious, sweet ingredient perfect for sweet treats. Find it in specialty shops or online.

Coconut Flour

Another great gluten-free option for baking and cooking, coconut flour is found at most health-food shops.

Coconut Sugar

This sugar is made from coconut palm sap, and its sweetness is almost caramel in flavor. It is found at health-food shops and specialized grocers.

Corn Flour

Fantastic when used as a thickener in recipes, this ingredient is easy to find and a convenient addition to your pantry.

Elderflower Cordial

A beautiful, sweet, and fragrant cordial, this is a great addition to drinks. Look out for it at specialty shops or supermarkets.

Liquid Smoke

This ingredient gives a wonderful smoky flavor to dishes. It is found at most health-food shops or specialty shops.

Mirin

Mirin is a sweet rice wine, used in Japanese cuisine. You can find it at an Asian grocer or your local supermarket.

Molasses

A thick, dark, and sticky syrup, molasses adds a distinct flavor to baking. It's particularly nice in gingerbread and cookies. Search for it at your local supermarket.

Nutritional Yeast Flakes

Sometimes called *savory yeast*, or also known as *nooch*, this ingredient adds a great cheesy flavor to dishes. High in vitamin B and protein, it's also a great addition to your pantry!

Porcini Powder

This is a handy pantry staple that gives a great depth of flavor to stews, soups, and rice or pasta dishes.

Rice Flour

Made from rice, this flour is great for gluten-free baking and for thickening, and it's perfect for making shortbread.

Rice Malt Syrup

A sweet syrup found at health-food stores and supermarkets, this is great for adding sweetness to desserts without using refined sugar.

Spirulina Powder

This is a nutrient-dense blue-green algae that is great for adding to smoothies or protein balls. It is generally found in health-food stores or online.

Sumac Powder

A Middle Eastern spice, this adds a nice tang to savory dishes. It is generally found at your local supermarket.

Tamari

A gluten-free substitute for soy sauce, tamari is a fermented sauce made out of soybeans. It is available at most Asian grocers or health-food shops.

Tapioca Flour

Tapioca is a starch derived from the cassava root. It is great for thickening dishes and is often used in gluten-free cooking.

Vegan Stock

Found in cubes, liquid, or powder form, vegan stock is easy to find at your local supermarket or health-food shop.

Young Jackfruit

Generally found at an Asian grocer, this is a wonderful ingredient that takes on the flavors you pair with it.

MENU PLANS

Thanksgiving Lunch

Roast Sweet Potato Salad, 19
Cheesy Broccoli Bake, 37
Mashed Potato and Homemade Gravy, 61
Jackfruit "Chicken" Salad, 77
Mushroom Steaks, 33
Mini Raw or Cooked Apple Pies, 141

Thanksgiving Dinner

Baked Cashew Cheese, 85
Green Bean and Olive Dish, 75
Parsnip and Mushroom Potpie, 39
Braised Red Cabbage, 67
Mashed Potato and Homemade Gravy, 61
Roast Vegetable Stuffing, 65
Pumpkin Pie Meringue Cheesecake, 139

Christmas Lunch

Warm Lemon Artichoke Dip, 83
Colorful Garden Salad, 21
Mushroom, Corn, and Zucchini Bake, 43
Spicy Orange Ginger Baked Cauliflower Wings, 73
Colorful Trifles, 149

Christmas Dinner

Cashew Dip, 97
Pecan and Mushroom Wellington, 47
Lemon and Thyme Roast Potatoes, 59
Whole Roasted Cauliflower, 35
Spiced Carrots, 63
Pecan Coconut Caramel Pie, 137

Festive Breakfast/Brunch

Orange and Vanilla Chia Puddings, 9
Vegan Smoked Salmon (Carrot Lox), 11
Macadamia Dill Cheese, 93
Silky Scrambled Tofu, 7
Cherry Orange Punch, 163

Happy Hour

Irish Dream Liqueur, 161
Lychee Mint Mocktails (for a nonalcoholic version), 157
Chai Spiced Nuts, 103
Pesto Pinwheels, 81
Spiced Pumpkin Cupcakes, 145

After-Dinner Treats

Golden Turmeric Milk, 159
White Rum Balls, 105
Nutmeg Cookies, 101

Gluten-Free Lunch or Dinner

Macadamia Dill Cheese, 93
Jackfruit Bourguignon, 49
Gluten-Free Festive Roast, 53
Cranberry Sauce, 133
Fried Asian Greens, 69
Gluten-Free Funfetti Cake, 143

BREAKFAST

ROSEMARY AND RED ONION POTATO ROSTI

SERVES 2

Makes one large rosti, perfect for sharing. The rosemary gives it a unique flavor, and paired with the onion and potato, it's a match made in heaven. Try it with some vegan sour cream!

2 medium potatoes, peeled and grated

½ red onion

1 teaspoon dried rosemary

1 tablespoon olive oil

Squeeze as much water as you can from the potatoes. Mix the potatoes, onion, and rosemary in a bowl.

Heat the olive oil in an 8-inch nonstick frying pan. Turn down to medium heat and place the potato mixture into the pan until the whole pan is covered and all the potato is used. Cook for 10 minutes and then carefully flip the rosti and cook for a further 10 minutes.

BLUEBERRY AND VANILLA BUCKWHEAT PANCAKES

SERVES 2 (DOUBLE THE AMOUNT TO SERVE 4–5)

Perfect for breakfast or dessert, these delicious pancakes are also gluten-free. Excellent for sharing on Christmas morning!

½ cup buckwheat flour
1 teaspoon baking powder
¼ cup coconut sugar
¼ teaspoon cinnamon
½–¾ cup soy milk
1 teaspoon vanilla extract
½ cup fresh or frozen blueberries

Mix the buckwheat flour, baking powder, coconut sugar, and cinnamon in a bowl. Add the milk, vanilla extract, and blueberries.

Cook in a nonstick frying pan until bubbles appear, and then flip and cook until golden brown.

Serve with coconut yogurt and a drizzle of maple syrup.

SILKY SCRAMBLED TOFU

SERVES 2

This delicious, silky scrambled tofu is the perfect way to indulge and relax before kicking off any festivities for the day.

⅓ cup aquafaba
⅓ cup soy milk
⅛ teaspoon black salt
Pinch turmeric
1 tablespoon chickpea flour
1 tablespoon nutritional yeast
¾ cup silken tofu
1 tablespoon vegan butter

Mix the aquafaba and soy milk in a bowl and add the black salt and turmeric. Whisk in the chickpea flour and nutritional yeast. Crumble the silken tofu and add this to the mix.

Melt the butter in a nonstick frying pan. Add the mixture and cook, stirring frequently for 5 to 7 minutes.

Serve on warm toast with cracked black pepper.

ORANGE AND VANILLA CHIA PUDDINGS

MAKES 2 PUDDINGS

A lighter start to the day, but still filling and absolutely delicious. The vanilla and orange are a winning combination!

⅓ cup white chia seeds
¼ teaspoon orange zest
¼ teaspoon cinnamon
¼ teaspoon vanilla extract
1½ cups soy milk
1 orange, peeled and diced
2 tablespoons vegan coconut yogurt

Mix the chia, orange zest, cinnamon, and vanilla in a bowl. Cover with the soy milk and stir well, so there are no lumps. Let sit for 20 to 30 minutes.

Add some orange to the bottom of 2 glasses—reserve some for the top. Spoon the chia mixture evenly into the glasses. Top with the coconut yogurt and more orange.

TIP

Try this with any other fruit—blueberries, raspberries, or stewed apples!

VEGAN SMOKED SALMON (CARROT LOX)

SERVES 4–6

One of the most popular recipes on my blog, Delightful Vegans, *this is truly a winning dish—and it's easy and cheap to make! Try it on a festive platter with some vegan cheese.*

2–3 carrots, shaved
3 tablespoons olive oil
Sprinkle black pepper
¼ teaspoon salt
1 tablespoon lemon juice
¼ teaspoon liquid smoke
Small handful fresh dill

Preheat the oven to 350°F. Blanch the carrots for 5 minutes in a double boiler. Add remaining ingredients, toss well, and bake for 20 minutes.

Eat warm or let cool and store in the fridge.

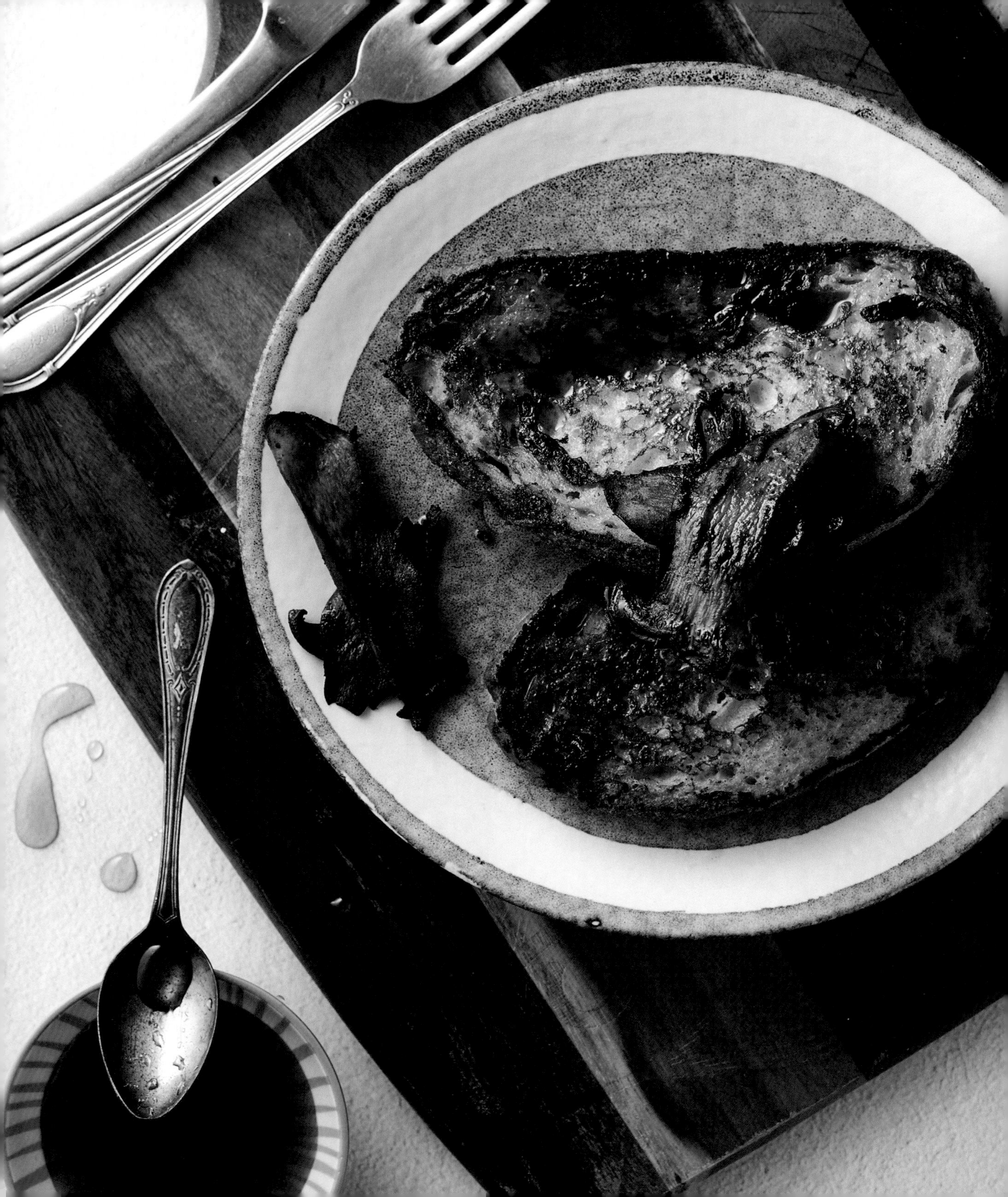

FRENCH TOAST WITH KING OYSTER BACON AND MAPLE SYRUP

SERVES 2

Next-level French toast! Try this savory-sweet version with king oysters and maple syrup.

KING OYSTER BACON

2 king oyster mushrooms

1 tablespoon oil

¼ teaspoon liquid smoke

½ tablespoon nutritional yeast flakes

¼ teaspoon paprika

⅛ teaspoon chipotle powder

⅛ teaspoon white pepper

½ tablespoon tamari

1 tablespoon maple syrup

FRENCH TOAST

½ cup silken tofu

¼ teaspoon cinnamon

¼ teaspoon vanilla

½ cup soy milk

2 tablespoons olive oil

4 slices nice bread, such as Vienna

1–2 tablespoons vegan butter

Maple syrup

Slice the oyster mushrooms lengthways. Mix the remaining ingredients for the king oyster bacon in a bowl and coat the mushrooms in it. Marinate for 20 minutes or longer.

Fry in a nonstick frying pan for a few minutes each side or bake in a moderate oven for 5 to 7 minutes each side, being careful not to burn. Set aside.

To make the French toast, blend the silken tofu, cinnamon, vanilla, milk, and olive oil in a blender until smooth. Pour into a bowl. Dip the bread in the mixture until it coats both sides.

Heat some of the butter in a nonstick frying pan. Fry the bread on medium heat for 2 to 3 minutes each side, until cooked. Add more butter if required. Repeat with all the slices of bread.

Serve topped with the king oyster bacon and maple syrup.

SALADS & SOUPS

PECAN MANGO SALAD

SERVES 5

Fresh and delicious is what this salad is all about. This is a variation on a salad my mum makes, which is absolutely scrumptious! If mangoes are hard to find, use any type of stone fruit, or sweet oranges.

2–3 cups of lettuce (oak leaf or butter lettuce work well)

½ cup pecans, toasted

½ avocado

½ cup fresh mango pieces

DRESSING

½ cup fresh mango

½ teaspoon fresh ginger, grated

1 tablespoon white wine vinegar

¼ teaspoon seeded mustard

Pinch salt and pepper

Pinch red pepper flakes

1 tablespoon water

Wash and spin the lettuce. Add the lettuce to a salad bowl and then add the pecans, avocado, and mango.

For the dressing, blend all ingredients in a blender. Pour over the salad and mix well.

ROAST SWEET POTATO SALAD

SERVES 6

A simple, yet tasty salad, this is a great winter warmer, with roasted sweet potato, arugula, and the crunch of toasted pepitas.

1 large sweet potato
1½ cups arugula
1 cup cherry tomatoes, halved
1 tablespoon capers
⅓ cup pepitas, toasted
1–2 tablespoons olive oil
Juice of ½ lime
Salt and pepper to taste

Preheat the oven to 350°F. Roast sweet potato for 45 minutes, or until cooked through. Mix through arugula when sweet potatoes are still warm. Then add the cherry tomatoes, capers, and toasted pepitas.

Drizzle generously with olive oil and lime juice. Add salt and pepper to taste.

STRAWBERRY FENNEL SALAD

SERVES 5

A light salad with delicious flavors, the shaved fennel gives this salad a unique taste. If strawberries are too hard to find, try pitted cherries or sliced pear.

2 cups arugula, washed

5 ounces strawberries, washed, ends chopped and quartered

½ fennel, shaved or very finely sliced

½ avocado, sliced

1 tablespoon fresh mint

1 tablespoon hemp seeds

DRESSING

2 tablespoons olive oil

1 tablespoon balsamic vinegar

1 tablespoon maple syrup

Pinch salt

1 teaspoon fennel fronds

Combine the arugula, strawberries, shaved fennel, and avocado in a bowl. Top with fresh mint and hemp seeds.

Whisk the dressing ingredients in a small bowl and pour over the salad.

APPLE KALE SALAD

SERVES 6

Packed full of goodness, this is a wonderful recipe that my good friend Malinda made for us. She prepared it with watermelon, which is a great alternative to apples when it's in season!

2 cups kale, stalks removed and washed

Pinch salt

2 tablespoons olive oil

1 tablespoon fresh lime juice

2 medium apples

½ cup vegan feta cheese

Add the kale, salt, olive oil, and lime juice to a bowl; massage the kale for about 10 minutes, until quite soft. Core and slice apples and add to the salad, along with the vegan feta cheese.

TIP

Replace the feta in this recipe with Macadamia Dill Cheese (page 93), which works wonderfully in this salad.

SPICED LENTIL SOUP

SERVES 4

This is such an easy soup to make—it's full of goodness and tasty, too. Throw everything in a saucepan, cook, and serve!

1 tablespoon olive oil
1 red onion, diced
1 clove garlic, minced
2 medium potatoes, peeled and diced
2 medium carrots, peeled and diced
1 teaspoon curry powder
1 teaspoon cumin powder
1 teaspoon ground coriander
1¾ cups red lentils
1 vegan beef stock cube
4 cups water
1 (14-oz) can tomatoes

Add olive oil to a saucepan, and cook onion, garlic, potatoes, and carrots for around 5 minutes. Add the curry powder, cumin powder, coriander, and lentils and mix through.

Mix the stock cube in 1 cup of water and add to the saucepan; add the remainder of the water and the tomatoes. Cook on medium heat for 30 minutes, stirring occasionally.

Serve with a natural vegan coconut yogurt if desired.

MUSHROOM AND PARSNIP SOUP

SERVES 4–6

A rich and creamy soup, this serves as a great starter.

4 parsnips, peeled and chopped

1 clove garlic, minced

1 onion, diced

1 tablespoons fresh sage leaves, chopped

1½ cups mushrooms, sliced

3 cups vegan stock

1 cup coconut cream

In a large saucepan, cook the parsnips, garlic, onion, and sage leaves for 5 to 10 minutes. Add mushrooms and cook for a further 5 minutes, until mushrooms are soft.

Add the stock and coconut cream. Simmer on medium-low heat for 15 to 20 minutes, until parsnips are soft.

Let cool slightly and blend until smooth. If too thick for your liking, add some water, stock, or vegan milk to thin it out.

THE MAINS

MUSHROOM STEAKS

SERVES 2–4

These mushroom steaks really do stand on their own in terms of flavor and satisfaction. They have a real "meaty" texture and, combined with these spices, create an amazing dish.

3 large portobello mushrooms
2 tablespoons tamari
1 tablespoon olive oil
Dash sesame oil
2 tablespoons maple syrup
1 tablespoon nutritional yeast
¼ teaspoon garlic powder
Pinch salt
⅛ teaspoon chipotle powder
¼ teaspoon liquid smoke
¼ teaspoon Chinese five-spice powder
Sprinkle black pepper

Preheat the oven to 375°F.

Slice the mushrooms into 1-inch-thick slices. Mix all of the remaining ingredients in a bowl and marinate the mushrooms in the mixture for 20 minutes or longer.

Line a baking tray with parchment paper and bake the mushrooms for 25 minutes.

WHOLE ROASTED CAULIFLOWER

SERVES 8–10

This is the perfect showstopping centerpiece for your Thanksgiving or Christmas table. It has a lovely crispy outer crunch and a delicious flavor.

¾ cup macadamia nuts
1 small lemon, zest and juice
1 tablespoon fresh thyme, chopped
1 tablespoon fresh rosemary
1 tablespoon grated ginger
1 tablespoon sweet paprika
Pinch salt
¼ cup olive oil
1 large cauliflower
½ tablespoon sliced almonds, toasted

Preheat the oven to 350°F.

Blend the macadamia nuts, lemon zest and juice, thyme, rosemary, ginger, paprika, salt, and olive oil in a food processor until it resembles a paste. Get messy and use your hands to cover the whole cauliflower.

Bake uncovered for 30 minutes. Then cover with aluminum foil and bake for 1 hour. Uncover again and let it crisp up by baking for a further 10 minutes or so.

Decorate the top with the toasted sliced almonds.

CHEESY BROCCOLI BAKE

SERVES 4–6

Creamy and yummy! A baked cheesy vegetable dish is a must for any holiday gathering, and this one is perfect. Also, try it out with cauliflower as an alternative to broccoli—or a mixture of both!

2 heads broccoli, cut into florets
2 cups vegan milk
1 cup liquid stock
¼ cup nutritional yeast
¼ teaspoon white pepper
¼ teaspoon salt
½ teaspoon turmeric
⅓ cup vegan cheese, grated

Steam the broccoli in a double boiler for about 5 minutes.

Blend the milk, stock, nutritional yeast, white pepper, salt, and turmeric in a blender and pour into a saucepan. Bring to a boil and add the vegan cheese. Cook for a further few minutes until cheese has melted.

Put the broccoli in a baking dish and pour the cheesy mixture on top. Bake at 390°F for 25 to 30 minutes.

PARSNIP AND MUSHROOM POTPIE

SERVES 6–8

Parsnips are such a tasty seasonal vegetable, and I love using them in this potpie. Paired with mushrooms, this pie is a hearty treat to share with friends and family.

1 tablespoon olive oil

1 onion, diced

1 large carrot, peeled and finely chopped

3 parsnips, peeled and finely chopped

2 cups mushrooms, chopped (use a mixture)

1 cup peas

2 teaspoons curry powder

Pinch salt

⅔ cup soy milk

2 tablespoons corn flour

⅓ cup water

1 sheet puff pastry

Heat olive oil in a frying pan on medium heat. Add the onion, carrot, and parsnips and cook for 5 minutes. Add the mushrooms (we use a mixture of enoki, portobello, and king oyster) and peas and cook for a further 5 minutes. Add the curry powder, salt, and soy milk and stir.

In a separate bowl, whisk the corn flour with the water until smooth. Add this to the mixture and stir through well.

Preheat the oven to 350°F. Place the mixture in a pie dish and cover with puff pastry. Bake for 45 minutes until cooked through.

CABBAGE STEAKS WITH MUSTARD DRESSING

SERVES 4–6

If you haven't tried cabbage steaks yet, now is the time! For the mustard dressing, I modified a recipe that my mum makes, and it pairs perfectly.

1 tablespoon olive oil

1 sugarloaf cabbage, washed and ends trimmed, cut into ½-inch slices

Pinch salt

MUSTARD SAUCE

½ cup soy milk

1 teaspoon vegan chicken stock powder

1 tablespoon lemon juice

1 teaspoon seeded mustard

Sprinkle white pepper

2 tablespoons coconut cream

1 tablespoon corn flour

Preheat the oven to 375°F. Line a baking tray with parchment paper. Brush or spray parchment paper with olive oil. Add cabbage rounds, then brush or spray with more olive oil. Sprinkle with salt and bake for 30 to 40 minutes, being careful not to burn.

For the mustard sauce, blend all ingredients, heat in a saucepan until thickened, and simmer for 5 minutes.

Once the cabbage is cooked, serve with the mustard sauce.

MUSHROOM, CORN, AND ZUCCHINI BAKE

SERVES 6–8

This creamy dish is comfort food at its finest. I used king oyster mushrooms to give this dish more texture, but play around with the ingredients and add whatever you have on hand.

1 leek, sliced
2 cloves garlic, minced
1 tablespoon plant-based butter
2 zucchini, diced
1½ cups mushrooms, chopped
1½ cups frozen corn
10.5 ounces silken tofu
1 cup plant-based milk
½ cup liquid stock
2 tablespoons nutritional yeast
Dash tamari
¼ teaspoon turmeric
Salt and pepper to taste
¾ cup bread crumbs
2 tablespoons fresh parsley, finely chopped

Gently fry the leek and garlic on medium heat in butter for 5 minutes or until softened. Add the zucchini, mushrooms, and corn and cook for a further 5 to 10 minutes.

Preheat the oven to 350°F.

Blend the remaining ingredients except for the bread crumbs and parsley in a blender, add the mixture to the pan with the vegetables, and combine well. Pour into a baking dish, cover with bread crumbs and parsley, and bake for 45 minutes until golden brown.

FENNEL RATATOUILLE

SERVES 5–7

A flavorful spin on a classic ratatouille, this is quick to throw together; plus, the fennel gives a unique and delicious flavor! The leftovers are perfect on toast, too.

1–2 eggplant, diced
Salt, for the eggplant
1 medium fennel, diced
3 zucchini, diced
2 bell peppers (1 green, 1 red)
Olive oil, to drizzle
1 teaspoon Italian seasoning
1 teaspoon dried rosemary
1 teaspoon dried thyme
¼ teaspoon salt
1 (14-oz) can chopped tomatoes
1 tablespoon tomato paste

To salt the eggplant, place chopped eggplant in a bowl and sprinkle generously with salt. Let sit for 30 minutes. Rinse and pat dry.

Preheat the oven to 300°F. Place all the vegetables in two baking trays lined with parchment paper. Drizzle generously with olive oil, add Italian seasoning, rosemary, thyme, and salt. Bake for 1 hour, 20 minutes.

Once cooked, add the vegetables to a large saucepan or frying pan. Add the tomatoes and tomato paste and cook through.

PECAN AND MUSHROOM WELLINGTON

SERVES 6–8

One of the most popular recipes on my blog, Delightful Vegans, *this dish has been tried and tested by many readers who make it year after year. The beauty of this recipe is that it's not complicated to make, but it tastes amazing.*

1 onion, finely diced
2 cloves garlic, crushed
1 tablespoon olive oil
2 cups mushrooms, chopped (I use mostly cremini)
1 tablespoon fresh thyme, chopped
Pinch red pepper flakes
1 tablespoon tamari
1½ cups pecans, toasted
1 cup bread crumbs
2 sheets vegan puff pastry
Sesame or poppy seeds (optional)

TIP

Pair this with the Green Tomato and Apple Relish (page 123).

Fry the onion and garlic in olive oil for 5 minutes, or until soft. Add the mushrooms, thyme, red pepper flakes, and tamari and cook for 5 to 10 minutes, until the mushrooms soften. Then let the mixture cool for 10 to 15 minutes.

In a food processor, blend the pecans finely. Add the mushroom mixture and blend again. Add the bread crumbs and blend until all the ingredients are combined.

Preheat the oven to 350°F. Let the pastry thaw a little. Spread half the mixture in the middle of one of the pastry sheets. Cut the pastry in ½-inch diagonal strips on either side of the mixture. Cross them over each other until they are covering the mixture. Alternatively, you can put half the mixture on one edge of the sheet and roll the pastry over it to form a loaf. Pierce the top with a fork or score with a knife. Top with sesame seeds or poppy seeds if desired.

Bake for 50 minutes. If the pastry browns too much on the top while cooking, cover with aluminum foil until ready to come out of the oven.

JACKFRUIT BOURGUIGNON

SERVES 6–8

Using young jackfruit, which will take on the flavors you pair with it, this dish is a rich and tasty meal that will leave you feeling full and content.

JACKFRUIT

2 cans young jackfruit (net weight 20 oz, drained weight 10 oz)

¼ teaspoon garlic powder

½ teaspoon salt

1 teaspoon sage powder

1 teaspoon dried thyme

SAUCE

3 carrots, cut into rounds

7 new potatoes, cut into quarters or smaller

1 onion, sliced

1 tablespoon olive oil

1 teaspoon vegan beef stock powder

Few sprigs fresh thyme

3 cups water, divided

1 cup vegan white wine

1 (14-oz) can chopped tomatoes

4 portobello mushrooms, halved and sliced

2 tablespoons tomato paste

Preheat the oven to 350°F. Drain the jackfruit and pat dry. Combine with the garlic powder, salt, sage powder, and thyme and bake on a lined baking tray for 30 minutes on each side.

To make the sauce, in a large saucepan, cook the carrots, potatoes, and onions in a drizzle of olive oil for 5 to 10 minutes. Add the stock, thyme sprigs, 2 cups of water, and wine. Cover and cook on medium heat for 15 minutes.

Add the tomatoes, jackfruit, mushrooms, and the remaining 1 cup of water. Cover and cook for a further 15 minutes.

Add the tomato paste and stir through. Cook for another 5 minutes to thicken slightly.

STUFFED BUTTERNUT SQUASH

SERVES 6–8

This is a great Thanksgiving recipe to share during the fall. I love the flavor of butternuts; they are delicious roasted to perfection and stuffed with a tasty filling.

- 2 butternut squash
- 2 tablespoons olive oil, divided
- ½ onion, finely chopped
- ½ tablespoon fresh sage leaves, chopped
- 1–1½ cups cooked brown rice
- 2 tablespoons cranberries
- 1 tablespoon slivered almonds, toasted

Preheat the oven to 350°F. Scoop out the seeds from each squash, spray or coat squash with olive oil, and bake in the oven for 1½ hours until you can push a knife or skewer in easily.

Cook the onion and sage in a frying pan with 1 tablespoon olive oil for 5 minutes. Add the cooked brown rice, cranberries, and toasted almonds and toss through until warm.

Once the squash are done, take out of the oven and fill the middle with the rice mixture. Serve immediately, or keep warm in the oven until ready to serve.

GLUTEN-FREE FESTIVE ROAST

SERVES 4–6

A gluten-free version of a roast, this is a great dish for those who have celiac disease or are intolerant to gluten. It's more flavorful the next day, so you can make it ahead and reheat. It's fantastic as leftovers, too—especially on toast with cranberry sauce.

ROAST

1½ tablespoons dried rosemary
1½ tablespoons dried thyme
1 tablespoon dried oregano
1 teaspoon salt
1 teaspoon vegan chicken stock powder
1 teaspoon onion powder
1 teaspoon garlic powder
1 (14-oz) can chickpeas, drained
9 ounces firm tofu
1½ teaspoons agar powder
2½ tablespoons corn flour
¼ cup rice flour
1 tablespoon olive oil

GLAZE

1 tablespoon olive oil
1 teaspoon tamari
1 tablespoon maple syrup
¼ teaspoon dried thyme
¼ teaspoon dried rosemary

Blend the rosemary, thyme, oregano, salt, stock powder, onion powder, and garlic powder into a fine powder.

Preheat the oven to 350°F. In a food processor, blend the chickpeas and tofu until smooth. Add the blended herb mixture, agar powder, and corn flour. Transfer to a bowl and mix through with the rice flour. It should be a sticky consistency.

Place a large piece of aluminum foil on a baking tray and a piece of parchment paper on top of that the foil. Brush or spray with olive oil. Place the mixture on top and form into a loaf shape. Brush or spray with olive oil. Wrap in the parchment paper and aluminum foil and bake in the oven for 45 minutes.

Meanwhile, make the glaze by mixing all the glaze ingredients in a bowl. Uncover the roast and cover with the glaze. Bake uncovered for a further 5 to 10 minutes.

ZUCCHINI SLICE

SERVES 6–8

Great warm or cold, this slice is good to have on hand for snacks or as an addition to any main meal. It's best to let it cool before slicing and reheat if desired.

1 cup silken tofu
¼ cup nutritional yeast flakes
½ teaspoon black salt
¼ cup sunflower oil
¼ cup soy milk
2 medium zucchini, grated
3 scallions, finely chopped
Pinch nutmeg
½ teaspoon chipotle powder
1 cup self-rising flour
½ cup vegan cheese, grated
5 cherry tomatoes, halved

Preheat the oven to 350°F.

Blend the tofu, nutritional yeast flakes, black salt, oil, and milk until smooth.

In a bowl, place the grated zucchini, scallions (reserve some for decoration), nutmeg, chipotle powder, and self-rising flour. Mix well. Pour over the blended tofu mix and stir well. Stir in the grated vegan cheese.

Pour into a greased container lined with parchment paper, decorate with the remaining scallions and cherry tomatoes, and bake for 45 minutes. Let cool before slicing.

TIP

If you don't have self-rising flour, replace with 1 cup of all-purpose flour and 1 teaspoon of baking powder. Try adding asparagus or other in-season vegetables to this recipe.

SIDES

LEMON AND THYME ROAST POTATOES

SERVES 4–6

A twist on a classic roasted potato recipe, this zingy dish will complement any festive meal!

1.75 pounds potatoes, peeled and cut
1 small lemon, zest and juice
2 tablespoons extra virgin olive oil
Salt and cracked black pepper to taste
1 tablespoon fresh thyme

Preheat the oven to 350°F.

Boil the potatoes in salted water for 10 minutes. Drain and let cool for 5 minutes. Add remaining ingredients and bake in the oven for 1 hour to 1 hour, 15 minutes

TIP

After you have juiced the lemon, cut it into quarters and add the quarters to the pan to bake for an extra lemony taste.

MASHED POTATO AND HOMEMADE GRAVY

SERVES 3–4

It's so easy to make your own gravy; try it on the Gluten-Free Festive Roast (page 53) or Pecan and Mushroom Wellington (page 47). It's the ultimate comfort festive food!

3–4 potatoes
1 tablespoon vegan butter
1 tablespoon soy milk

GRAVY

1 cup liquid vegan chicken stock
1 bay leaf
1 sprig rosemary
1 teaspoon porcini powder
1 teaspoon curry powder
1½ tablespoons corn flour
¼ cup water

Boil the potatoes in salted hot water for 15 minutes. When straining, reserve ¼ cup of the water. Mash with the reserved water, butter, and milk.

To make the gravy, pour the stock into a saucepan and add the bay leaf, rosemary, porcini powder, and curry powder. Bring to a boil. In a separate bowl add the corn flour and water and whisk well. Pour this into the saucepan and whisk frequently for 4 to 5 minutes until thickened.

Pour the gravy over the mashed potatoes and serve.

SPICED CARROTS

SERVES 4

A colorful dish that is full of flavor—just mix the spices together, coat, and bake. A great addition to any festive feast!

1 tablespoon olive oil
Pinch salt
1 teaspoon tamari
⅛ teaspoon turmeric
⅛ teaspoon mustard powder
¼ teaspoon coriander powder
¼ teaspoon cumin powder
A bunch of carrots (yellow, purple and orange)

Preheat the oven to 375°F.

Mix the olive oil, salt, tamari, and powdered spices in a bowl. Peel and wash the carrots and coat them with the mixture. Bake in the oven for 60 minutes, turning halfway through.

ROAST VEGETABLE STUFFING

SERVES 6–8

Try this for a different twist on stuffing—full of flavor, it will go well with any main dish.

- 2 carrots, peeled and finely chopped
- 1½ cups pumpkin, peeled and finely chopped
- 1 medium sweet potato, peeled and finely chopped
- 2 tablespoons olive oil, divided
- 1 medium onion, peeled and diced
- ¾ cup celery, finely chopped
- 2 cloves garlic
- 1 bay leaf
- 4 thick slices sourdough bread, cut into cubes
- ¼ teaspoon salt
- 1 teaspoon dried thyme
- 1 teaspoon dried rosemary
- 14 ounces liquid vegan chicken stock

Preheat the oven to 390°F. Place the carrots, pumpkin, and sweet potato on a baking tray and coat or spray with 1 tablespoon olive oil. Bake for 40 minutes.

Heat 1 tablespoon of olive oil in a frying pan. Fry the onion, celery, garlic, and bay leaf for 5 minutes, or until onion has softened. Add the cooked vegetables and bread to the pan and stir through. Add the salt, thyme, and rosemary and mix through.

Add the stock a little at a time until it is absorbed. Once you have added all the stock, the stuffing is ready.

BRAISED RED CABBAGE

SERVES 10+

My friends Jess and Sean make this dish every year for Thanksgiving, a recipe passed down from Jess's grandmother. I've modified their recipe slightly. The longer you cook this, the better it tastes.

- 2 tablespoons vegan butter
- 1 teaspoon vegan stock powder
- 1 large red cabbage or 2 small cabbages (2 lb total), shredded
- 1 onion, sliced
- 3 small green apples, peeled, cored, and sliced
- 2 teaspoons caraway seeds
- ⅓ cup raw sugar
- ½ cup white wine vinegar
- 1 cup water

Melt butter and stock in a large saucepan. Add cabbage and stir through.

Add remaining ingredients and cook on medium heat for 1½ hours, stirring occasionally.

FRIED ASIAN GREENS

SERVES 2

A quick and tasty side, these fried Asian greens will be a welcome addition to your table.

2 bunches bok choy
1 tablespoon sesame oil
½ tablespoon mirin
1 tablespoon tamari
1½ tablespoons sesame seeds, toasted

Slice the bok choy into 1-inch-thick pieces. Heat the sesame oil in a frying pan and cook the stalks of the bok choy for 5 to 7 minutes. Add the mirin and tamari, stir through, and add the remaining bok choy leaves. Cook for another minute and turn off the heat.

Add the toasted sesame seeds, toss through, and serve.

SPICY RAMEN NOODLES

SERVES 4

For a quick and easy side dish, try whipping up these noodles! It takes less than 10 minutes to throw this one together.

SAUCE

2 tablespoons tamari

⅛ teaspoon chili powder

1 tablespoon maple syrup

Sprinkle white pepper

1 tablespoon sesame oil

1 tablespoon tomato paste

NOODLES

3 single-serving packets instant noodles

1 small onion, sliced

½ tablespoon ginger, grated

1 tablespoon olive oil

Mix all the sauce ingredients together in a bowl.

Start cooking the noodles according to the packet instructions. Cook the onion and ginger in the oil for a few minutes. Once the noodles are done, strain them and add to the frying pan. Add the sauce and mix through well.

SPICY ORANGE GINGER BAKED CAULIFLOWER WINGS

SERVES 8–10

A little bit of time, a whole lot of flavor! I've mainly seen BBQ cauli-wing recipes, so I thought I'd mix it up a bit by making a spicy orange chili ginger sauce.

CAULIFLOWER

1 whole cauliflower

BATTER

1½ cups plain flour

½ teaspoon chili powder

½ teaspoon ginger powder

¼ teaspoon salt

½ teaspoon white pepper

1 cup water, plus extra if needed

ORANGE CHILI GINGER SAUCE

Juice of 2 oranges

¼ teaspoon chili powder

1 teaspoon orange zest

1 tablespoon sugar

1 tablespoon ginger

1 cup plus 2 tablespoons of water, divided

2 tablespoons corn flour

2 tablespoons sesame seeds

DIPPING SAUCE

1 cup cashews

½ cup fresh orange juice

1 clove garlic

1 tablespoon nutritional yeast

Pinch salt

¼ teaspoon mustard powder

Black pepper

½ cup water

Preheat the oven to 350°F. Cut the cauliflower into florets and wash well.

Mix the batter ingredients. Dip each floret of cauliflower into the batter and place on a lined baking tray. Bake for 30 to 40 minutes.

While the cauliflower is cooking, make the orange chili ginger sauce. In a saucepan, add the orange juice, chili powder, orange zest, sugar, ginger, and 1 cup of water, and bring to a boil. Reduce to a simmer. In a small bowl, add the corn flour and 2 tablespoons of water and whisk. Add this to the saucepan and whisk well. The mixture will thicken slightly. Add the sesame seeds and set the mixture aside.

Once the cauliflower has finished cooking, use a fork to pick up each floret and dip it in the orange chili ginger sauce. Return to the oven and bake for a further 35 minutes.

To make the dipping sauce, blend all of the ingredients until smooth. Serve with the cauliflower wings.

GREEN BEAN AND OLIVE DISH

SERVES 6

A green bean dish is always a favorite, especially at Thanksgiving. Serve this one hot or cold.

- 1 pound green beans, washed and ends trimmed, cut into 1-inch pieces
- 1 tablespoon vegan butter
- 1 clove garlic, crushed
- ½ cup pitted green olives, finely chopped
- 1 tablespoon fried shallots

Cook the green beans in boiling water for 3 to 4 minutes. Drain and transfer to a bowl of cold water with ice until cool. Drain and pat dry.

Heat butter in a saucepan until sizzling, add the garlic and beans, and cook for a few minutes, stirring frequently. Add the green olives and cook for a further few minutes.

Serve topped with fried shallots.

JACKFRUIT "CHICKEN" SALAD

SERVES 4–6

This is something that I created a while ago, and it's really an amazing dish—fantastic as a side to a main meal, on cold sandwiches or a toasted sandwich, or even on crackers as a snack. You'll love it.

2 cans jackfruit (net weight 20 oz, drained weight 10 oz)
1 teaspoon dried oregano
1 teaspoon dried thyme
1 teaspoon dried parsley
½ teaspoon onion powder
½ teaspoon garlic powder
1 teaspoon sumac powder
¼ teaspoon black salt
⅛ teaspoon white pepper
½ cup vegan mayonnaise
1 scallion, finely chopped
1 tablespoon fresh lemon juice
Generous sprinkle black pepper
¼ teaspoon Dijon mustard

Preheat the oven to 350°F. Drain the jackfruit and pat dry. Combine with the oregano, thyme, parsley, onion powder, garlic powder, sumac powder, salt, and pepper and bake on a lined baking tray for 30 minutes on each side. Allow to cool.

In a small bowl, add the vegan mayonnaise, scallion, lemon juice, black pepper, and Dijon mustard. Mix well. When the cooked jackfruit is cool, dice into ½-inch pieces. Stir the jackfruit into the mayonnaise mixture. Store in the fridge until ready to use.

SAVORY SNACKS

PESTO PINWHEELS

SERVES 6–8

These festive wheels are great for kids and adults alike! They're super easy to make, and you can get creative with different ingredients.

KALE PESTO

¾ cup pine nuts
2 cups packed kale leaves (2.85 oz)
1 clove garlic
½ lemon, juice and zest
¼ teaspoon salt
¼ teaspoon red pepper flakes
¼ cup olive oil

PINWHEELS

1 to 2 pastry sheets
Plant-based milk or oil, for brushing

Blitz all of the ingredients for the kale pesto except the oil in a food processor. Gradually add in the oil while blending. Scrape down the sides as necessary.

Preheat the oven to 350°F. To assemble the pinwheels, lightly defrost the pastry sheets. Spread half of the pesto mixture onto the pastry sheet, leaving about ½ inch around the side. Baste the edge with some plant-based milk or oil to make it stick. Roll up the pastry sheet and cut into ⅓-inch rounds with a sharp knife.

Place pinwheels on a tray lined with parchment paper. Bake in the oven for 30 to 35 minutes until puffy and golden brown.

TIP

Use any of the pesto recipes on page 118 to make these pinwheels. Pesto can be refrigerated for up to five days or frozen.

WARM LEMON ARTICHOKE DIP

SERVES 6

A lovely addition to a festive platter, this warm artichoke dip goes well with crudités and crackers or on sandwiches.

2 (14-oz) cans whole artichokes
½ teaspoon lemon zest
1 tablespoon lemon juice
2 tablespoons olive oil
Pinch salt
Pinch black pepper
¼ teaspoon paprika
⅓ cup soy milk
2 tablespoons nutritional yeast

Preheat the oven to 350°F. Blend all ingredients in a food processor to a chunky consistency. Pour into a greased ovenproof 5" x 7" container and bake for 25 to 30 minutes until golden brown on top.

Serve warm with crudités or crackers.

BAKED CASHEW CHEESE

SERVES 6

This is one of my favorite recipes. It's hard not to eat it straight out of the oven, but it's best left overnight and eaten the next day. Perfect on a cheese and fruit platter, or just with crackers, this is a crowd-pleaser.

½ cup cashews, soaked
2.5 ounces firm tofu
2 tablespoons fresh lemon juice
¼ teaspoon salt
Sprinkle white pepper
¼ teaspoon garlic powder
1 tablespoon nutritional yeast powder
¼ cup oil
2 teaspoons soy lecithin
⅓ cup water
½ teaspoon agar
Oil, for brushing
Generous sprinkle cracked black pepper

TIP

Try adding 1 teaspoon of dried dill or Italian seasoning into the mix for a more flavorful baked cheese.

Soak the cashews for 6 hours or overnight. If you don't have time to soak them, boil them for 10 minutes and then strain and rinse. Add cashews to a blender with the tofu, lemon juice, salt, white pepper, garlic powder, nutritional yeast, oil, and soy lecithin.

In a small saucepan, mix the water and agar. Bring to a boil and pour into the remaining ingredients. Blend until smooth.

Preheat the oven to 350°F. Brush a 4-inch springform pan with oil to keep the mixture from sticking. Pour the mixture into the pan and smooth the top. Generously sprinkle with cracked black pepper and bake for 30 minutes until golden on top.

Let the cheese cool for 30 minutes before you release the spring. You can eat it warm, although it will harden more as it cools, so it's best to leave it in the fridge overnight and eat the next day. Store in an airtight container.

BUBBLE AND SQUEAK BITES

SERVES 4

This is a fantastic way to use up leftovers after Christmas or Thanksgiving.

- 1–2 tablespoons vegan butter
- 4–5 cups leftover vegetables or steamed vegetables (potatoes, pumpkin, carrots, onion and peas are all great)
- Generous pinch salt
- 1 cup panko bread crumbs

Melt butter in a frying pan until sizzling. Add vegetables and salt and cook well for 5 to 10 minutes.

Let mixture cool and form into little patties. Coat with panko bread crumbs and bake in the oven for 10 minutes each side at 350°F or in an air fryer for 10 minutes.

LEFTOVER TOASTIES

MAKES 2 SANDWICHES

Nothing better than a quick, easy dinner or lunch thrown together with leftovers! In this recipe, I've used Jackfruit "Chicken" Salad (page 77), but you could use the Gluten-Free Festive Roast (page 53), which would go beautifully with some cranberry sauce, or Baked Cashew Cheese (page 85) paired with Roast Vegetable Stuffing (page 65)—the possibilities are endless!

1 tablespoon pesto (page 118)

½ cup Jackfruit "Chicken" Salad (page 77)

2 slices vegan cheese

4 slices bread

Plant-based butter, for spreading (optional)

Assemble each sandwich with half the pesto, half the Jackfruit "Chicken" Salad, and 1 slice of vegan cheese. Butter the outside of the sandwich (optional).

Toast in an air fryer or sandwich press.

COB LOAF SPINACH DIP

SERVES 8–10

This loaf dip is perfect for sharing. Take it to a party or gathering and wow your guests with how great it looks and tastes.

1 small brown onion, diced
2 cloves garlic, minced
1 tablespoon olive oil
1 bunch fresh spinach, washed
1 block (10-oz) silken tofu
¼ teaspoon black salt or Himalayan sea salt
¼ teaspoon white pepper
2 tablespoons nutritional yeast flakes
Pinch nutmeg
1 wedge preserved lemon, chopped
1 medium cob loaf

Cook onion and garlic in a frying pan in olive oil for 5 minutes. Add the spinach and cook until wilted.

Add this mixture to a food processor with the silken tofu, salt, white pepper, nutritional yeast, nutmeg, and preserved lemon. Blend until well incorporated.

Cut the top out of a cob loaf with a bread knife and pull the insides out, creating a hole for the dip. Fill the hole with the spinach dip. Cut the remaining bread into bite-size pieces to use for dipping (toast if desired).

MACADAMIA DILL CHEESE

SERVES 4–6

Make your own plant-based cheese with only five ingredients! If you can't get hold of macadamia nuts, try this with cashews; it works just as well. Store in the fridge and use when needed. Perfect in salads or as a spread. Try it on bagels—it's delicious!

1 cup macadamia nuts
¼ cup fresh lemon juice
¼ cup coconut oil, melted
½ teaspoon Himalayan salt
1 tablespoon fresh dill

Soak macadamia nuts for at least 2 hours and strain.

Blend with the remaining ingredients. Add water 1 tablespoon at a time as needed until the mixture is thick and smooth.

Line a 5" x 7" container with parchment paper, scoop the mixture into it, and flatten out the top. Set in the refrigerator for a few hours.

PUMPKIN AND DATE BISCUITS

MAKES 8 BISCUITS

Mashed pumpkin gives these biscuits their great color and a hint of flavor. I added dates, but you could also add cinnamon or other sweet spices. Or, try a savory version with chives or scallions.

1 cup self-rising flour
Pinch salt
2 tablespoons sugar
½ cup mashed pumpkin
2 tablespoons coconut cream
¼ cup soda water
2 tablespoons dates, chopped

Preheat the oven to 350°F.

Put self-rising flour, salt, and sugar in a mixing bowl and mix well. In a separate bowl, stir pumpkin and coconut cream until well mixed, then add the soda water.

Add this wet mixture to the dry ingredients until it forms a dough (add more flour if too sticky). Add dates and mix through. On a floured surface, press out the dough to 1-inch thickness and cut into rounds with a cookie cutter.

Place on a tray lined with parchment paper and bake for 20 minutes.

Serve with vegan cream or butter.

CASHEW DIP

SERVES 4–6

This cashew dip is simple and tasty just as it is—or you could use it as a base and add other ingredients to make it more flavorful, such as fresh herbs: try dill, coriander, or oregano. You can even try blending some pesto through for a flavor hit.

1 cup cashews
¼ teaspoon salt
1 scallion
1 clove garlic
2 tablespoons fresh lemon juice
¼ cup water

Add all ingredients into a blender and blend until smooth. Add more water if it is too thick. Pour into a ramekin and serve with crackers or vegetables.

SWEET SNACKS

NUTMEG COOKIES

MAKES 12 COOKIES

It's always handy to have a stash of cookies at the ready for any drop-in guests or late-night snack attacks! These have a shortbread-like consistency, are super tasty, and are perfect to leave out for Santa—with a glass of rice milk!

½ cup vegan butter
¼ cup brown sugar
¼ cup white sugar
¼ teaspoon vanilla extract
½ tablespoon molasses
1⅓ cups plain flour
⅓ cup rice flour
½ cup almond flour
⅛ teaspoon nutmeg
⅛ teaspoon ginger
¼ teaspoon cinnamon
Pinch salt
1 teaspoon orange zest

Preheat the oven to 350°F.

Beat the butter and sugars in a standing mixer or with a hand mixer until mixed well. Add the vanilla extract and molasses and beat that in.

In a separate bowl, add the flours, spices, and salt. Mix well. Scoop spoonfuls of this mixture into the butter mixture and beat it in gradually. Mix in the orange zest.

Line a baking tray with parchment paper. Roll out dough and cut with a cookie cutter. Place cookies on baking tray and bake for 12 minutes until golden brown. Let cool before eating.

CHAI SPICED NUTS

MAKES 4 CUPS

These nuts are totally addictive and so incredibly tasty that you'll want them on hand all the time for a festive snack! Try making a batch of these and gifting them to your loved ones.

Mixture of nuts (approx. 14 oz nuts total)
- 1 cup pecans
- 1 cup macadamia nuts
- 1 cup cashews
- 1 cup walnuts

3 tablespoons maple syrup
1 teaspoon cinnamon powder
1 teaspoon cardamom powder
¼ teaspoon nutmeg powder
½ teaspoon ginger powder
¼ teaspoon clove powder
1 teaspoon orange zest

Preheat the oven to 350°F.

Mix all of the nuts in a bowl. Drizzle over the maple syrup and mix through. Mix the cinnamon, cardamom, nutmeg, ginger, and clove powders in a small bowl. Add these to the nuts and mix through. Add the orange zest and stir.

Spread the nuts on a baking tray lined with parchment paper. Make sure they are spread out well and not clumping. Bake for 10 to 12 minutes, making sure not to burn them. Let the nuts cool completely before eating them.

BOW STREET DISTILLERY, DUBLIN, IRELAND
JQ-058548
TRIPLE DISTILLED BY JOHN JAMESON & SON
JOHN JAMESON & SON LIMITED
JJ&S

WHITE RUM BALLS

MAKES 16

What's a Christmas without you-can't-have-just-one rum balls? This is a take on the traditional chocolate rum balls, but they're just as tasty.

- 1 (7-oz) packet vegan arrowroot cookies
- 1½ cups desiccated coconut
- ½ cup vegan condensed coconut milk
- ½ teaspoon salt
- ½ teaspoon vanilla extract
- 1–2 tablespoons rum or whiskey

Process the cookies to make a fine crumb, then add coconut and process some more. Add remaining ingredients and process until mixture is combined.

Form into balls and roll in desiccated coconut. Store in the refrigerator. They taste better the next day!

THE MOST EPIC VEGAN ROCKY ROAD

SERVES 4–6

Take ordinary rocky road to the next level with this recipe—it's got everything: honeycomb, marshmallows, nuts, and more. Definitely a hit with everyone!

½ cup macadamia nuts

½ cup hazelnuts

¾ cup vegan honeycomb, roughly chopped

½ cup vegan white chocolate, roughly chopped

1 cup vegan mini marshmallows

½ cup desiccated coconut

8 ounces vegan dark chocolate, melted

Preheat the oven to 350°F. Place the macadamia nuts and hazelnuts on a baking tray and bake for about 10 minutes, until slightly brown. Be careful not to burn them. Let cool.

Put the honeycomb, white chocolate, marshmallows, and cooled macadamia nuts and hazelnuts into a bowl and mix well.

After melting the dark chocolate, let it cool slightly so it doesn't melt the vegan white chocolate when you add them together. Pour into the mixture and mix well.

Line a 7" x 5" baking pan or container with parchment paper. Pour in the rocky road mixture and even it out. Put in the refrigerator for at least 1 hour.

ALMOND CARDAMOM FUDGE

SERVES 4–6

This fudge is the perfect decadent sweet treat to take to a Christmas or Thanksgiving party! Keep this one in the freezer until the last minute, as it has a low melting point.

½ cup blanched almonds, toasted
1 cup ABC nut butter
½ cup coconut oil, melted
¼ cup maple syrup
½–1 tablespoon cardamom powder
Pinch Himalayan sea salt
¼ teaspoon vanilla extract

Preheat the oven to 350°F. Toast the almonds for 7 to 10 minutes, being careful not to burn them. Let cool.

Mix all of the ingredients in a bowl. Line a 5" x 8" container with parchment paper. Pour in the mixture and let set in the freezer for 30 minutes to 1 hour. This fudge is best kept in the freezer until you want to eat it.

TIP

Try this with other nuts, such as walnuts, cashews, or pecans! Note: ABC nut butter is made from almonds, Brazil nuts, and cashew nuts. This recipe would also work just as well with almond butter.

PEPITA, DATE, AND SPIRULINA BALLS

MAKES 12 BALLS

With only four ingredients, these are a good healthy treat for your holiday table. Plus, the green is the perfect color for Christmas!

1 cup pepitas, soaked for 6 hours or overnight
½ cup Medjool dates, pitted
½ cup shredded coconut
½ tablespoon spirulina powder

Drain the pepitas well, then blend them in a food processor for a few minutes until crumbly. Add the dates and process until well incorporated. Next add the coconut and do the same. Add the spirulina powder and blend until mixed in well.

Form into balls and store in the refrigerator.

CHOCOLATE CRACKLES

MAKES 12

Quick and easy to throw together, these crispy, chocolatey treats are one of my childhood favorites. They're perfect for kids, who would enjoy helping to make and eating them.

2 cups Rice Krispies/puffed rice
⅔ cup shredded coconut
¾ cup vegan candied cherries, chopped
7 ounces vegan dark chocolate, melted
⅓ cup vegan white chocolate, melted

Place muffin liners in a muffin tin. Combine the Rice Krispies, shredded coconut, and candied cherries (reserve 2 tablespoons for garnish) in a large bowl. Pour the melted dark chocolate into the bowl, and mix well. Scoop the mixture evenly into the muffin liners and let set in the refrigerator for 1 hour or more.

Once set, remove from the muffin liners. Drizzle the melted white chocolate over the top. Decorate with the reserved candied cherries.

OIL-FREE, SUGAR-FREE BLUEBERRY AND LEMON MUFFINS

MAKES 12 MUFFINS

It's great to have some sugar-free options at the holiday table, too . . . for when you're all sugared out!

- 2 cups self-rising flour
- 1 teaspoon cinnamon
- Pinch salt
- 1 cup dates, soaked in water for 1–2 hours
- 1 teaspoon vanilla extract
- 1 tablespoon applesauce
- 1 teaspoon lemon zest
- ¾ cup sparkling mineral water
- 1 cup frozen blueberries

Preheat the oven to 390°F.

Combine the self-rising flour, cinnamon, and salt in a large mixing bowl. Strain the dates, reserving ½ cup of date water. Blend the dates with ½ cup of date water and add to the mixing bowl. Add the vanilla extract, applesauce, and lemon zest. Add in the sparkling water and mix well. Add the blueberries and mix gently.

Fill a muffin pan with muffin liners. Spoon the mixture evenly between the pans. Bake for 20 to 25 minutes.

COMPANION DISHES

PESTO THREE WAYS

Pesto is a great staple to have on hand. Here are three amazing pesto recipes that will rock your world. I'm especially fond of the mushroom one—toss through some pasta with coconut milk and you have a beautiful, creamy mushroom dish. Pestos are so versatile—use in Pesto Pinwheels (page 81), in a soup, on Leftover Toasties (page 89), or in pasta.

MUSHROOM HAZELNUT PESTO

4 cups cremini mushrooms, chopped
1 tablespoon fresh thyme
Small handful parsley
½ cup hazelnuts, toasted and skin removed as much as possible
1 cup cashews, raw
2 tablespoons olive oil
¼ teaspoon salt

In a frying pan, cook mushrooms, thyme, and parsley in a splash of water for 10 minutes. Let cool.

Blend in a food processor with the remaining ingredients, scraping down the sides as necessary.

Store in an airtight container in the fridge.

TIP
Use a lemon or truffle oil instead of olive oil for extra flavor.

KALE PESTO

¾ cup pine nuts, toasted
2 cups packed kale leaves
1 clove garlic
½ lemon, juice and zest
¼ teaspoon salt
¼ teaspoon red pepper flakes
¼ cup olive oil

Blend all the ingredients except for the oil in a food processor. Add the oil gradually and process until well combined. You may need to scrape down the sides a few times.

TIP
Use basil in place of kale to make a more traditional pesto.

SUN-DRIED TOMATO AND WALNUT PESTO

1 cup walnuts, soaked for 2–4 hours
½ cup sun-dried tomatoes
¼ cup olives
3–4 dates, chopped
1 tablespoon nutritional yeast
1 clove garlic
2 tablespoons olive oil
Salt and pepper to taste

Add ingredients to the food processor and blend until well combined, scraping down the sides as required.

MARSHMALLOW FLUFF

SERVES 8–10

Aquafaba (the brine from chickpeas) is a fantastic ingredient to work with. You'll be amazed at this marshmallow fluff that can be used as a meringue topping in hot chocolate or in trifles, or to jazz up a fruit salad.

½ cup aquafaba
1¼ cups powdered sugar
½ teaspoon vanilla extract
¼ teaspoon cream of tartar

Mix the aquafaba in a standing mixer for 10 minutes. Add the powdered sugar (1 tablespoon at a time), the vanilla, and the cream of tartar, until it is thick and stiff peaks form.

TIP

Use this marshmallow fluff for the Pumpkin Pie Meringue Cheesecake (page 139) or the Hot Chocolate with Marshmallow Fluff (page 155).

GREEN TOMATO AND APPLE RELISH

MAKES APPROXIMATELY 3 JARS

Great as a gift, and wonderful paired with the Pecan and Mushroom Wellington (page 47) or Gluten-Free Festive Roast (page 53), this chutney is easy to make—and tasty, too.

2.2 pounds green apples, peeled, cored, and diced

2.2 pounds green tomatoes, washed and diced

1 large brown onion

1 cinnamon stick

1 star anise pod

2 bay leaves

1 teaspoon yellow mustard seeds

½ cup sultanas (raisins)

1 teaspoon salt

1 teaspoon grated ginger

½ lemon

⅓ cup white wine vinegar

½ cup brown sugar

¼ cup raw sugar

½ cup water

Put all the ingredients into a large saucepan. Cook on medium heat for 1 hour and 40 minutes, stirring occasionally.

Let cool and then store in sterilized jars.

ROAST EGGPLANT

SERVES 4–6

Sometimes you need a few quick and easy recipes to have on hand, and this is one of those. Great as a side or in a toasted sandwich, you'll want this eggplant around anytime.

2 large eggplants (17 oz total)
Salt, for sprinkling
1 tablespoon olive oil
¼ teaspoon sweet paprika
½ teaspoon dried oregano

Cut eggplant into rounds and put in a large bowl. Sprinkle salt over the eggplant and leave for 30 minutes. Rinse the salted eggplant and pat dry.

Preheat the oven to 350°F. Line a tray with parchment paper and brush with olive oil. Lay the eggplant on the paper and brush again with olive oil. Then sprinkle paprika, dried oregano, and more salt on top.

Place eggplant on the middle tray in the oven. Bake for 5 to 7 minutes on each side, being careful not to burn.

ROAST TOMATOES

SERVES 8–10

Perfect for risottos, pasta dishes, on toast, or just as a side, these are a delightful burst of flavor.

2.5 pounds tomatoes, halved
Generous pinch salt
1 teaspoon Italian seasoning
Sprigs rosemary or thyme, or both
Generous drizzle olive oil

Preheat the oven to 350°F. Place the tomatoes faceup on a lined baking tray. Sprinkle over the salt and Italian seasoning, and place the herb sprigs on top.

Bake in the oven for 2 hours, rotating the tray every 30 minutes.

TIP

Use a flavored salt, such as garlic or rosemary salt, for more flavor on these tomatoes.

HOMEMADE CREAM

SERVES 8

Sometimes vegan cream can be hard to come by, so why not make your own with these simple ingredients? Useful to have on hand for all your festive needs!

½ cup cashews
1 (14-oz) can coconut cream
2 tablespoons corn flour
2 tablespoons maple syrup
½ tablespoon vanilla

Blend all ingredients for 1 minute. Pour mixture into a saucepan and cook on medium heat, whisking continuously, for 2 to 3 minutes until thick.

Pour mixture into a separate bowl, let cool, and store in the refrigerator.

HOMEMADE CUSTARD

SERVES 6

Serve this custard cold or warm with your favorite dessert. Easy to make, it can be stored in the refrigerator for a few days.

½ cup cashews
½ cup coconut cream
⅛ teaspoon turmeric
1 tablespoon corn flour
½ tablespoon vanilla extract
1 tablespoon maple syrup
½ cup water

Blend all ingredients in a high-powered blender for 1 minute until smooth. Pour into a saucepan and cook on medium heat for 3 to 4 minutes until thickened.

Store in the refrigerator until ready to use.

CRANBERRY SAUCE

SERVES 8–10

Another must for the holiday table! Cranberry sauce with a zesty orange flavor—a great accompaniment to many main dishes.

2 cups dried cranberries
1 orange, zest and juice
1 cinnamon stick
2 pods star anise
1 cup water

Put all the ingredients in a saucepan and boil for 20 minutes on medium heat. For smoother consistency, blend in a food processor.

Store in the refrigerator until ready to eat. To serve warm, add water while heating in a saucepan over low heat.

DESSERTS

PECAN COCONUT CARAMEL PIE

SERVES 8–12

This is one of the tastiest things I've ever made! Don't be daunted at the list of ingredients; it isn't too difficult to throw together, and your guests will love it. Many people have tried this recipe—and they all love it. Someone even requested me to make it for them last Christmas!

CRUST

1¼ cups flour

½ cup coconut sugar

½ cup vegan butter

Pinch salt

2 tablespoons rice malt syrup

½ cup almond meal

⅓ cup coconut flour

1 tablespoon water

Vegan butter, for the baking tin

FILLING

2 cups desiccated coconut

1 (11.25-oz) can condensed coconut milk

1 cup almond butter

⅓ cup coconut oil

1 tablespoon vanilla extract

Pinch salt

½ cup coconut sugar

TOPPING

1¼ cup pecans

¾ cup desiccated coconut

¼ cup coconut sugar

2 tablespoons rice malt syrup

Preheat the oven to 350°F. Add all ingredients for the crust except the water and butter into a food processor and process. You may need to scrape down the sides a few times. Add the water and it should come together into a dough.

Grease a 10-inch flan or tart pan with butter. Roll out the dough and press into the baking pan. Pierce with a fork a few times and bake for 12 minutes. Let cool.

To make the filling, blend the coconut in the food processor until fine. Place the remaining ingredients in a saucepan and mix well on medium heat until well incorporated. Add the coconut, mix well, and pour this mixture into the baked piecrust.

Mix all the ingredients for the topping in a bowl and place on top of the pie. Bake for 25 minutes.

PUMPKIN PIE MERINGUE CHEESECAKE

SERVES 8–12

This cheesecake takes the traditional pumpkin pie to the next level by adding a marshmallow meringue layer on top.

CRUST

1¼ cups plain flour

1¾ cups almond flour

Pinch salt

¼ cup vegan butter

¼ cup rice malt syrup

1 tablespoon ice cold water

FILLING

1 cup mashed pumpkin

14 ounces vegan cream cheese

7 ounces silken tofu

¾ cup sugar

½ teaspoon cinnamon

⅓ cup corn flour

1 teaspoon pumpkin pie spice

Marshmallow Fluff (page 121)

Preheat the oven to 350°F. Process the ingredients for the crust in a food processor until it comes together. Press into a 10-inch pie pan. Pierce with a fork a few times and bake for 10 minutes. Let cool.

Blend the ingredients for the filling in a food processor until smooth. Scoop into the baked crust. Bake for 40 minutes. Let the pie cool down completely before topping with marshmallow fluff.

For the marshmallow topping, mix the aquafaba in a standing mixer for 10 minutes. Add the powdered sugar 1 tablespoon at a time, the vanilla extract, and the cream of tartar, until it is thick and stiff peaks form. Pipe the marshmallow fluff onto the pie with a wide nozzle piping tip. Brown the edges with a cooking torch, or alternatively broil in the oven for up to 1 minute, being careful not to burn it.

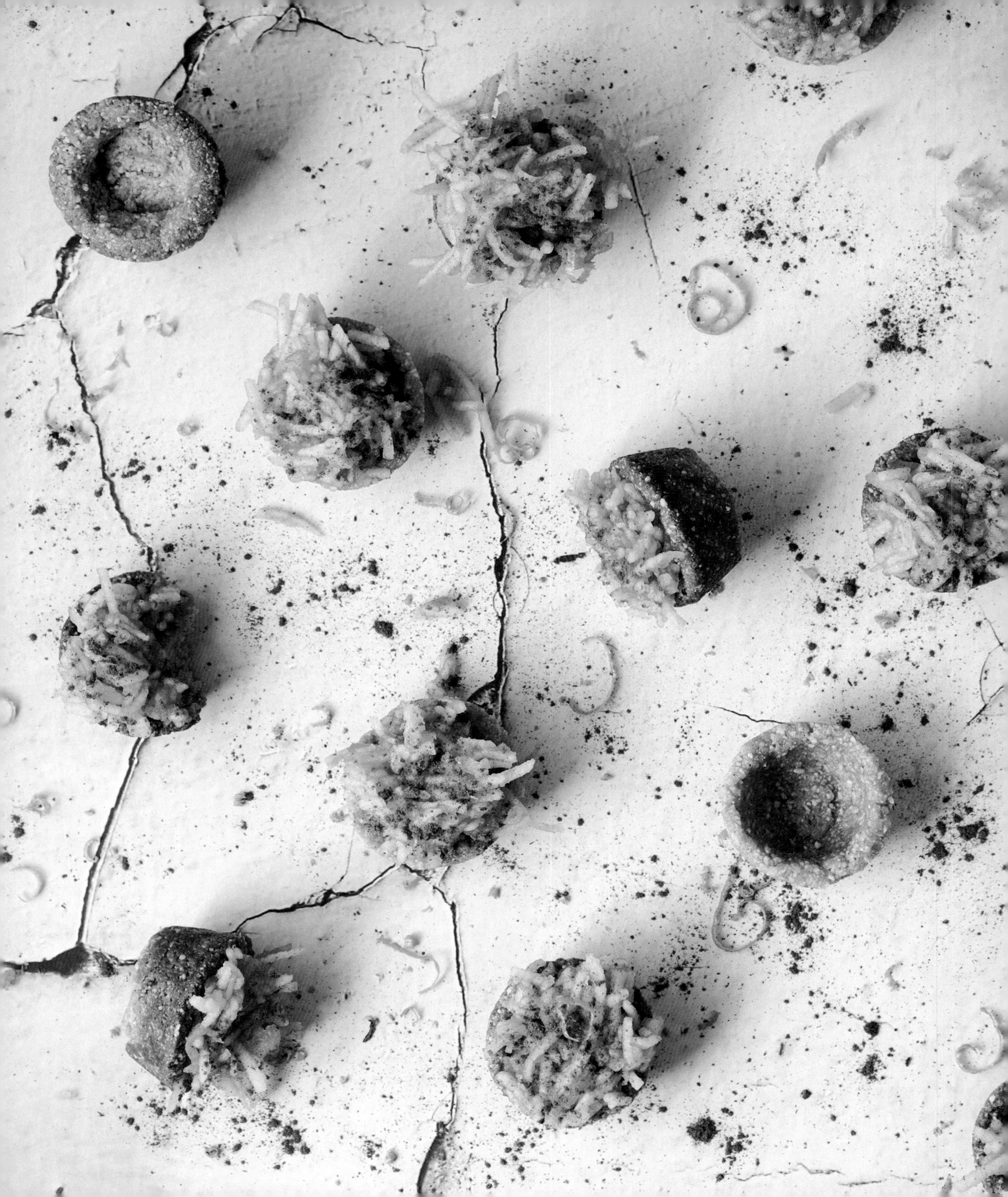

MINI RAW OR COOKED APPLE PIES

MAKES 12–15 MINI PIES

For a lighter dessert, give these mini apple pies a go. You can cook the base or leave it raw, but either way, they're a cute little treat—and gluten-free!

BASE

1½ cups raw cashews
½ cup almond meal
¼ teaspoon cinnamon
Pinch salt
2 tablespoons maple syrup
1 tablespoon coconut oil

FILLING

3 green apples, cored, peeled and grated
Zest of 1 lemon
2 tablespoons fresh lemon juice
½ teaspoon vanilla extract
⅛ teaspoon cinnamon

To make the base, blitz the cashews until fine. Add the remaining ingredients until the mixture forms a dough. Roll a small amount of mixture into a ball in your palm, flatten out, and press into the base of a well-greased mini muffin pan. Repeat with the remainder of the mixture.

At this stage you can bake the shells for 7 to 8 minutes at 350°F, or leave raw. If you leave them raw, put them in the refrigerator to harden up. If you bake them, let them cool. Remove the bases from the pan using a knife if necessary.

For the filling, add the grated apples in a bowl with the remaining ingredients and mix well. Scoop this filling into the shells. Store in the refrigerator until ready to eat.

GLUTEN-FREE FUNFETTI CAKE

SERVES 10–14

This is a fabulous, fun cake that has a distinct vanilla taste—and your guests won't even know it's gluten-free!

¾ cup plant-based butter

¾ cup sugar

½ cup coconut flour

⅓ cup rice flour

½ cup buckwheat flour

1 cup almond flour

2 tablespoons tapioca flour

½ tablespoon baking powder

Pinch salt

¾ cup soy milk

1 teaspoon apple cider vinegar

1 teaspoon vanilla extract

1 tablespoon applesauce

4 tablespoons vegan sprinkles, *divided*

Preheat the oven to 350°F.

Mix the butter and sugar in a standing mixer or hand mixer until well combined. Mix the rest of the dry ingredients—the flours, baking powder, and salt—in a bowl.

Mix the wet ingredients—the soy milk, apple cider vinegar, vanilla extract, and applesauce—together in a separate bowl. Gradually add a bit of each dry and wet mixture to the butter mixture while mixing, until it forms a batter.

Add 2 tablespoons of the sprinkles into the batter and gently stir through. Pour batter into a lined 8" x 8" baking dish and bake for 25 to 30 minutes. Once baked, add the remaining sprinkles over the top. Let cool.

SPICED PUMPKIN CUPCAKES

MAKES 12

Cupcakes! The feeling of biting into the delicate flavorsome delights is like no other. These cupcakes are sure to disappear quickly.

2 cups plain flour
1 teaspoon baking soda
Pinch salt
1 teaspoon cinnamon
¼ teaspoon nutmeg
⅛ teaspoon clove powder
½ cup white sugar
½ cup mashed pumpkin
½ cup olive oil
½ teaspoon vanilla extract
1 cup soda water

ICING

1¼ cup vegan butter
1 tablespoon coconut cream
¼ cup mashed pumpkin
1½ cups powdered sugar, sifted

Mix all the dry ingredients—flour, baking soda, salt, cinnamon, nutmeg, clove powder, and sugar—in a bowl.

Mix mashed pumpkin, olive oil, and vanilla together. Add this to the dry ingredients, then slowly add the soda water until well incorporated.

Divide the mixture into 12 muffin liners in a muffin pan and bake at 350°F for 25 minutes or until a skewer comes out clean.

For the icing, in a standing mixer or hand mixer, mix the butter first and add the coconut cream, then the mashed pumpkin. Add the powdered sugar, 1 tablespoon at a time, until well incorporated.

Pipe this icing onto the cooled cupcakes and store in the refrigerator until ready to serve.

RHUBARB AND CUSTARD ROLLS

MAKES 10 ROLLS

A little bit more time goes into making these rolls, but with a lot of reward! Homemade rolls are some of the tastiest treats you can make.

RHUBARB

2½–3 cups rhubarb, washed and cut into 1-inch pieces

4 tablespoons water

½ cup sugar

1 teaspoon vanilla extract

CUSTARD

½ cup cashews

½ cup coconut cream

⅛ teaspoon turmeric

1 tablespoon corn flour

½ tablespoon vanilla extract

1 tablespoon maple syrup

½ cup water

ROLLS

1 cup soy milk, warmed

2 tablespoons coconut oil, plus extra for brushing

2 teaspoons yeast

3 cups plain flour

Pinch salt

1 tablespoon sugar

ICING

1¼ cup powdered sugar

4 teaspoons soy milk

Drop food coloring

Add all the ingredients for the rhubarb into a small saucepan and cook on medium-low heat for about 10 to 12 minutes or until rhubarb is soft enough to put a fork through.

Blend all the ingredients for the custard in a high-powered blender for 1 minute until smooth. Pour into a saucepan and cook on medium heat for 3 to 4 minutes until thickened. Pour into a separate bowl and let cool.

To make the rolls, put the warmed milk in a bowl and add the coconut oil. Add the yeast, stir, and let sit for 10 minutes. Add the flour, salt, and sugar gradually, forming into a dough.

Place the dough in a bowl brushed with extra coconut oil and let sit with a cover over it for 1 or 2 hours, until it has doubled in size. Once the dough has risen, knead a few more times. Sprinkle a working surface with flour and roll out the dough into a large rectangular shape.

Spread the custard on the dough in a thin layer, and then dollop the rhubarb mixture onto that and spread out evenly. Starting from the long edge, roll the dough into a cylindrical shape, then cut into rounds a couple of inches long. Place the rounds faceup in an 8" x 8" ovenproof dish. Let rest for about 15 minutes. Preheat the oven to 350°F. Brush the rolls with coconut oil and bake for 25 minutes. Let cool before adding the icing.

To make the icing, sift the powdered sugar into a bowl. Add the soy milk gradually until it becomes a smooth consistency. Add a couple of drops of food coloring of your choice. A pink color goes nicely with rhubarb. Drizzle over the rolls.

COLORFUL TRIFLES

SERVES 2–4

This recipe is so easy once you have everything prepared—just layer it up! I used individual glasses for this, but you could use a big glass dish if you want to make a big one; just increase the amount of ingredients to suit.

½ cup cookies, processed into crumbs

½ cup canned peaches

1 cup Simply Delish Jel Dessert

1 cup vegan custard

¾ cup vegan cream

Using large glasses, start by spooning 2 to 3 tablespoons of cookie crumbs in the bottom of each glass. Add 2 to 3 slices of peach in each glass. On top of that, layer a few spoons of Jel Dessert, a couple of spoons of custard, and some cream.

Repeat this layer again and, voilà, you have a trifle!

TIP

Use the custard recipe (page 131) and the cream recipe (page 129). You can use chocolate or ginger cookies to mix up the flavor, and you can use any type of canned fruit.

DRINKS

MULLED APPLE CIDER

SERVES 6–8

This flavorsome cider will be sure to please all your guests with its warm spices and fruity overtones.

3 (12-oz) bottles hard apple cider
1 orange, cut into rounds
2 cinnamon sticks
½ lemon, cut into rounds
2 cardamom pods, smashed
2 cloves
½ apple, cored and sliced
1 mandarin, peeled and segmented
2 pods star anise

Combine all the ingredients in a large saucepan and slowly heat on medium heat for 15 minutes.

Serve warm.

TIP
Use blood oranges for a great taste and color.

HOT CHOCOLATE WITH MARSHMALLOW FLUFF

SERVES 2

A decadent drink sure to warm your bodies and hearts, this is perfect for relaxing before or after all the festivities.

Marshmallow Fluff (page 121)

HOT CHOCOLATE

⅓ cup dark vegan chocolate, broken into pieces

½ teaspoon vanilla extract

¼ teaspoon cinnamon powder

2 tablespoons coconut cream

1½ cups soy milk

To make the hot chocolate, add chocolate, vanilla, cinnamon, and coconut cream to a pan and whisk until chocolate has melted. Add milk gradually and whisk until all the milk is added and the consistency is smooth.

Top with Marshmallow Fluff.

LYCHEE MINT MOCKTAILS

SERVES 2–4

I first made this delicious mocktail with my friend Doris at a dumpling party. It's such a refreshing and tasty drink, perfect to kick off any celebration.

3–4 cups ice
1 (4-oz) can lychees in juice
Handful fresh mint
1–2 tablespoons fresh lime juice
1 tablespoon elderflower cordial
Lemonade, to top up

Add the ice, lychees (reserve some for decoration), lychee juice, mint (reserve some for decoration), lime juice, and elderflower cordial into a blender and blend until it achieves a slushy consistency.

In each cocktail glass, place one of the reserved lychees. Fill three-quarters of each glass with the lychee drink. Top up with lemonade. Decorate with mint leaves.

TIP

For a less sweet version, try replacing the lemonade with soda water.

GOLDEN TURMERIC MILK

SERVES 2

This deliciously creamy golden turmeric milk will warm you inside and out. Not only that, it's jam-packed with nutrients to boost your immune system.

1 cup water
1 tablespoon fresh turmeric, grated
1 teaspoon fresh ginger, grated
1½ cups rice milk
2 pods star anise
½ teaspoon cardamom powder
1 teaspoon cinnamon powder
1 teaspoon ground turmeric powder
¼ teaspoon grated nutmeg
2 grinds black pepper
½ cup coconut milk
1 teaspoon coconut oil
1 tablespoon coconut sugar

Add water to a medium saucepan on medium heat. Add the fresh turmeric and ginger. Add the rice milk. Add the star anise, cardamom, cinnamon, turmeric powder, nutmeg, and ground black pepper. Add the coconut milk, coconut oil, and coconut sugar.

Bring to a slow boil, pour into mugs, and enjoy.

ml 500
400
300
200
100

IRISH DREAM LIQUEUR

SERVES 6–8

Try making your own creamy and delicious liqueur at home! It's an easy recipe, and great to share. It is also a perfect homemade gift to give to loved ones.

1 cup Jameson Irish whiskey

½ cup coconut sugar

1 teaspoon instant coffee (optional)

14 ounces boiling water

28 ounces coconut cream

1 tablespoon chocolate vincotto (or rich chocolate sauce)

Pour the whiskey into a large bowl (preferably one with a pouring lip). Put the coconut sugar and instant coffee, if using, in a measuring cup; pour in the boiling water and stir to dissolve. Pour this in with the whiskey.

Add the coconut cream to the bowl. Whisk out any lumps that may be present from the coconut cream. Add the chocolate vincotto and whisk that in.

Pour the mixture into a clean and sterilized glass bottle (using a funnel may be best!) and store in the refigerator, where it will last for about a week.

CHERRY ORANGE PUNCH

SERVES 6–8

You can double or triple this recipe, which is perfect if you're wanting to make a drink in large quantities. This definitely has a festive vibe going on with its flavors!

½ cup canned cherries

¼ cup cherry syrup from the canned cherries

1 cup orange juice

2 cups lemonade

Pour the cherries and cherry syrup into a 16-ounce carafe. Gently pour the orange juice in. You should see a really cool layer effect! Top off with lemonade.

Double or triple this recipe if using a larger carafe or a punch bowl.

ACKNOWLEDGMENTS

Firstly, I'd like to thank the readers of my blog, *Delightful Vegans*, for making our recipes, sharing them with others, and supporting me on this journey.

Huge thanks to my mum for always listening, for inspiration and motivation, and for helping out in so many ways.

Thanks to my friend Matt—even though he lives so far away, he is a great help with inspiration, ideas, and suggestions.

Thanks to Lou for some excellent and wise advice at a time when I was under pressure working ten-hour days and writing this book on the side.

Thanks to Sean and Jess for sharing their cabbage recipe, passed down from generations, and for sharing valuable knowledge with me.

Thanks to the *Vegan Food & Living* magazine, which features our recipes regularly. Thanks to the Australian Vegan Foodies Facebook page for inspiration and for the amazing fellow foodies! And thanks to the many food blogs I follow for the great community, creativity, and imaginativeness on which I thrive.

Thanks to Swami, Alex, Marg, and Ken for taste-testing some of our recipes and giving valuable feedback.

Massive thank you to my bestie, Swami, for being an inspiration in so many ways—I have tremendous gratitude for her words of encouragement and her confidence in me.

Thanks to many other friends who were understanding while my social life took a bit of a step back while I wrote this book.

Last but not least, an extra special thanks to my partner, Josh, for helping me on this journey, for the epic photography and technical wizardry, for being the chief dishwasher, and for his amazing knowledge and support.

ABOUT THE AUTHOR

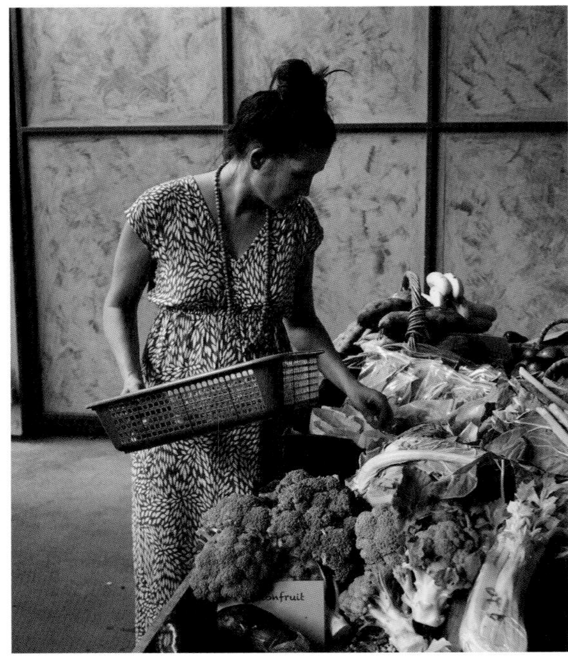

Katie Culpin has had a passion for plant-based food for over ten years. She is the co-founder and chief recipe developer for the successful blog *Delightful Vegans*. Katie and her partner, Josh, created *Delightful Vegans* nearly five years ago and it now has many mouthwatering plant-based recipes as well as the odd travel adventure thrown in! As well as working in many vegan restaurants and helping with menu plans and development, Katie's recipes have also been featured in *Vegan Food and Living Magazine* and on numerous other blogger websites. She has performed cooking demonstrations at the World Vegan Day and the Buddha's Day & Multicultural Festival in Melbourne. Katie and Josh live on the Gold Coast in Australia with their three cats.

CONVERSION CHARTS

Metric and Imperial Conversions

(These conversions are rounded for convenience)

Ingredient	Cups/Tablespoons/ Teaspoons	Ounces	Grams/Milliliters
Butter	1 cup/ 16 tablespoons/ 2 sticks	8 ounces	230 grams
Cheese, shredded	1 cup	4 ounces	110 grams
Cornstarch	1 tablespoon	0.3 ounce	8 grams
Cream cheese	1 tablespoon	0.5 ounce	14.5 grams
Flour, all-purpose	1 cup/1 tablespoon	4.5 ounces/0.3 ounce	125 grams/8 grams
Flour, whole wheat	1 cup	4 ounces	120 grams
Fruit, dried	1 cup	4 ounces	120 grams
Fruits or veggies, chopped	1 cup	5 to 7 ounces	145 to 200 grams
Fruits or veggies, pureed	1 cup	8.5 ounces	245 grams
Liquids: cream, milk, water, or juice	1 cup	8 fluid ounces	240 milliliters
Maple syrup or corn syrup	1 tablespoon	0.75 ounce	20 grams
Oats	1 cup	5.5 ounces	150 grams
Salt	1 teaspoon	0.2 ounce	6 grams
Spices: cinnamon, cloves, ginger, or nutmeg (ground)	1 teaspoon	0.2 ounce	5 milliliters
Sugar, brown, firmly packed	1 cup	7 ounces	200 grams
Sugar, white	1 cup/1 tablespoon	7 ounces/0.5 ounce	200 grams/12.5 grams
Vanilla extract	1 teaspoon	0.2 ounce	4 grams

Oven Temperatures

Fahrenheit	Celsius	Gas Mark
225°	110°	¼
250°	120°	½
275°	140°	1
300°	150°	2
325°	160°	3
350°	180°	4
375°	190°	5
400°	200°	6
425°	220°	7
450°	230°	8

INDEX